Clinical Engineering Handbook

This is a volume in the
ACADEMIC PRESS SERIES IN BIOMEDICAL ENGINEERING

JOSEPH BRONZINO, SERIES EDITOR
Trinity College—Hartford, Connecticut

Clinical Engineering Handbook

Edited by

Joseph F Dyro, PhD, FACCE, FAIMBE, CCE
Biomedical Research Group
Setauket, NY

ELSEVIER
ACADEMIC
PRESS

Amsterdam • Boston • Heidelberg • London
New York • Oxford • Paris • San Diego
San Francisco • Singapore • Sydney • Tokyo

Elsevier Academic Press
200 Wheeler Road, 6th Floor, Burlington, MA 01803, USA
525 B Street, Suite 1900, San Diego, California 92101-4495, USA
84 Theobald's Road, London WC1X 8RR, UK

This book is printed on acid-free paper. ∞

Copyright © 2004, Elsevier Inc. All rights reserved.

Except chapters 50, 51, 55, 74, 80, 86, 118, 135 and Section VI Introduction

Chapters 50, 51, 55, 74 and Section VI Introduction
Copyright © U.S. Government

Chapters 80, 86 and 118
Copyright © 2004 ECRI

Chapter 135
Copyright © IEEE EMBS

No part of this publication may be reproduced or transmitted in any form or by any means, electronic or mechanical, including photocopy, recording, or any information storage and retrieval system, without permission in writing from the publisher.

Permissions may be sought directly from Elsevier's Science & Technology Rights Department in Oxford, UK: phone: (+44) 1865 843830, fax: (+44) 1865 853333, e-mail: permissions@elsevier.com.uk. You may also complete your request on-line via the Elsevier homepage (http://elsevier.com), by selecting "Customer Support" and then "Obtaining Permissions."

Library of Congress Cataloging-in-Publication Data
Application Submitted

British Library Cataloguing in Publication Data
A catalogue record for this book is available from the British Library

ISBN: 0-12-226570-X

For all information on all Academic Press publications
visit our Web site at www.academicpress.com

Printed in the United States of America
04 05 06 07 08 09 9 8 7 6 5 4 3 2 1

Contents

Contributors xiii

Introduction xvii

Section I   Clinical Engineering   1
Introduction   Raymond P. Zambuto

1. **Clinical Engineering: Evolution of a Discipline   3**
   Joseph D. Bronzino

2. **History of Engineering and Technology in Health Care   7**
   Malcolm G. Ridgway, George I. Johnston, and Joseph P. McClain

3. **The Health Care Environment   11**
   Lee O. Welter

4. **Enhancing Patient Safety: The Role of Clinical Engineering   14**
   American College of Clinical Engineering

5. **A Model Clinical Engineering Department   16**
   Caroline A. Campbell

6. **Clinical Engineering in an Academic Medical Center   18**
   Ira Soller

7. **Regional Clinical Engineering Shared Services and Cooperatives   26**
   J. Tobey Clark

8. **Nationwide Clinical Engineering System   28**
   Diego Bravar and Teresa dell'Aquila

9. **Clinical Engineering and Biomedical Maintenance in the United States Military   34**
   Joseph P. McClain

10. **Careers, Roles, and Responsibilities   36**
    Nicholas Cram

11. **Clinical Engineering at the Bedside   39**
    Saul Miodownik

12. **The Clinical Engineer as Consultant   41**
    J. Sam Miller

13. **The Clinical Engineer as Investigator and Expert Witness   43**
    Jerome T. Anderson

14. **Careers in Facilities   49**
    Bruce Hyndman

Section II   Worldwide Clinical Engineering Practice   51
Introduction   Enrico Nunziata

15. **World Clinical Engineering Survey   53**
    Mariana Glouhova and Nicolas Pallikarakis

16. **Clinical Engineering in the United Kingdom   58**
    Stuart J. Meldrum

17. **Clinical Engineering in Canada   62**
    William M. Gentles

18. **Clinical Engineering in Estonia   65**
    Siim Aid and Ole Golubjatnikov

19. **Clinical Engineering in Germany   67**
    Vera Dammann

20. **Clinical Engineering in Brazil   69**
    Lúcio Flávio de Magalhães Brito

21. **Clinical Engineering in Colombia   72**
    Jorge Enrique Villamil Gutiérrez

22. **Clinical Engineering in Ecuador   78**
    Juan Gomez

23. **Clinical Engineering in Mexico   80**
    Adriana Velásquez

24. **Clinical Engineering in Paraguay   84**
    Pedro Galvan

25. **Clinical Engineering in Peru   87**
    Luis Vilcahuaman and Javier Brandán

26. **Clinical Engineering in Venezuela   89**
    Ricardo Silva and Luis Lara-Estrella

27. **Clinical Engineering in Japan   91**
    Hiroshi Kanai

28. **Clinical Engineering in Mozambique** 93
   Enrico Nunziata and Momade Sumalgy

29. **Clinical Engineering in the Middle East** 97
   Hashem O. Al-Fadel

Section III Health Technology Management 99
   Introduction   Thomas M. Judd

30. **Introduction to Medical Technology Management Practices** 101
   Yadin David, Thomas M. Judd, and Raymond P. Zambuto

31. **Good Management Practice for Medical Equipment** 108
   Michael Cheng and Joseph F. Dyro

32. **Health Care Strategic Planning Utilizing Technology Assessment** 110
   Nicholas Cram

33. **Technology Evaluation** 114
   Gary H. Harding and Alice L. Epstein

34. **Technology Procurement** 118
   Gary H. Harding and Alice L. Epstein

35. **Equipment Control and Asset Management** 122
   Matthew F. Baretich

36. **Computerized Maintenance Management Systems** 124
   Ted Cohen and Nicholas Cram

37. **Maintenance and Repair of Medical Devices** 130
   James McCauley

38. **A Strategy to Maintain Essential Medical Equipment in Developing Countries** 133
   Michael Cheng

39. **Outsourcing Clinical Engineering Service** 135
   Peter Smithson and David Dickey

40. **New Strategic Directions in Acquiring and Outsourcing High-Tech Services by Hospitals and Implications for Clinical Engineering Organizations and ISOs** 137
   Donald F. Blumberg

41. **Vendor and Service Management** 147
   Joseph F. Dyro

42. **Health Care Technology Replacement Planning** 153
   J. Tobey Clark

43. **Donation of Medical Device Technologies** 155
   Joseph F. Dyro

44. **National Health Technology Policy** 159
   Thomas M. Judd

45. **The Essential Health Care Technology Package** 163
   Peter Heimann, Andrei Issakov, and S. Yunkap Kwankam

46. **Impact Analysis** 171
   Thomas M. Judd

Section IV Management 179
   Introduction   Alfred M. Dolan

47. **Industrial/Management Engineering in Healthcare** 181
   George Seaman

48. **Financial Management of Clinical Engineering Services** 188
   Binseng Wang

49. **Cost-Effectiveness and Productivity** 199
   Larry Fennigkoh

50. **Clinical Engineering Program Indicators** 202
   Dennis Autio

51. **Personnel Management** 206
   James O. Wear

52. **Skills Identification** 212
   Nicholas Cram

53. **Management Styles and Human Resource Development** 213
   Alice L. Epstein and Gary H. Harding

54. **Quality** 219
   Thomas M. Judd

Section V Safety 225
   Introduction   Marvin Shepherd

55. **Patient Safety and the Clinical Engineer** 227
   Bryanne Patail

56. **Risk Management** 235
   Alice L. Epstein and Gary H. Harding

57. **Patient Safety Best Practices Model** 241
   Paul Vegoda and Carl Abramson

58. **Hospital Safety Programs** 243
   Matthew F. Baretich

59. **Systems Approach to Medical Device Safety** 246
   Marvin Shepherd

Contents vii

60. **Interactions Between Medical Devices** 249
    Saul Miodownik

61. **Single Use Injection Devices** 251
    Michael Cheng

62. **Electromagnetic Interference with Medical Devices:** *In Vitro* **Laboratory Studies and Electromagnetic Compatibility Standards** 254
    Kok-Swang Tan and Irwin Hinberg

63. **Electromagnetic Interference in the Hospital** 263
    W. David Paperman, Yadin David, and James Hibbetts

64. **Accident Investigation** 269
    Joseph F. Dyro

65. **The Great Debate on Electrical Safety—In Retrospect** 281
    Malcolm G. Ridgway

Section VI   Education and Training 285
Introduction   James O. Wear

66. **Academic Programs in North America** 287
    Tim Baker

67. **Clinical Engineering Education in Germany** 294
    Vera Dammann

68. **Clinical Engineering Internship** 297
    Izabella A. Gieras and Frank R. Painter

69. **Biomedical Engineering Technology Program** 299
    Bruce J. Morgan

70. **Advanced Clinical Engineering Workshops** 301
    Joseph F. Dyro, Thomas M. Judd, and James O. Wear

71. **Advanced Health Technology Management Workshop** 305
    Thomas M. Judd, Joseph F. Dyro, and James O. Wear

72. **Distance Education** 309
    James O. Wear and Alan Levenson

73. **Emerging Technologies: Internet and Interactive Video conferencing** 312
    Albert Lozano-Nieto

74. **In-Service Education** 315
    James O. Wear

75. **Technical Service Schools** 317
    Manny Roman

76. **Clinical Engineering and Nursing** 321
    Thomas J. Bauld III, Joseph F. Dyro, and Stephen L. Grimes

77. **Retraining Programs** 328
    James Gilchriest

78. **Techno-Bio-Psycho-Socio-Medical Approach to Health Care** 332
    T. G. Krishnamurthy

Section VII   Medical Devices: Design, Manufacturing, Evaluation, and Control 337
Introduction   Gary H. Harding and Alice L. Epstein

79. **Evolution of Medical Device Technology** 339
    Nandor Richter

80. **Technology in Health Care** 342
    Jonathan A. Gaev

81. **Medical Device Design and Control in the Hospital** 346
    Joel R. Canlas, Jay W. Hall, and Pam Shuck-Holmes

82. **Medical Device Research and Design** 350
    P. Åke Öberg

83. **Human Factors: Environment** 353
    William A. Hyman and Valory Wangler

84. **Medical Devices: Failure Modes, Accidents, and Liability** 355
    Leslie A. Geddes

85. **Medical Device Software Development** 359
    Richard C. Fries and Andre E. Bloesch

86. **Comparative Evaluations of Medical Devices** 366
    James P. Keller

87. **Evaluating Investigational Devices for Institutional Review Boards** 369
    Salil D. Balar

Section VIII   Medical Devices: Utilization and Service 371
Introduction   Joseph F. Dyro

88. **Intensive Care** 373
    Saul Miodownik

89. **Operating Room** 376
    Chad J. Smith, Raj Rane, and Luis Melendez

90. **Anesthesiology** 384
    Luis Melendez and Raj Rane

91. **Imaging Devices** 392
   David Harrington

92. **Machine Vision** 401
   Eric Rosow and Melissa Burns

93. **Perinatology** 410
   Vinnie DeFrancesco

94. **Cardiovascular Techniques and Technology** 417
   Gerald Goodman

95. **General Hospital Devices: Beds, Stretchers, and Wheelchairs** 421
   Joseph F. Dyro

96. **Medical Device Troubleshooting** 436
   Joseph F. Dyro and Robert L. Morris

Section IX   Information   449
Introduction   Elliot B. Sloane

97. **Information Systems Management** 451
   Elliot B. Sloane

98. **Physiologic Monitoring and Clinical Information Systems** 456
   Sunder Subramanian

99. **Advanced Diagnostics and Artificial Intelligence** 464
   Donald F. Blumberg

100. **Real-Time Executive Dashboards and Virtual Instrumentation: Solutions for Health Care Systems** 476
   Eric Rosow and Joseph Adam

101. **Telemedicine: Clinical and Operational Issues** 484
   Yadin David

102. **Picture Archiving and Communication Systems (PACS)** 487
   Ted Cohen

103. **Wireless Medical Telemetry: Addressing the Interference Issue and the New Wireless Medical Telemetry Service (WMTS)** 492
   Donald Witters and Caroline A. Campbell

104. **Health Insurance Portability and Accountability Act (HIPAA) and its Implications for Clinical Enginerring** 498
   Stephen L. Grimes

105. **Y2K and Clinical Engineering** 506
   Stephen L. Grimes

106. **The Integration and Convergence of Medical and Information Technologies** 509
   Ted Cohen and Colleen Ward

Section X   Engineering the Clinical Environment   513
Introduction   Matthew F. Baretich

107. **Physical Plant** 515
   Bruce Hyndman

108. **Heating, Ventilation, and Air Conditioning** 517
   Bruce Hyndman

109. **Electrical Power** 520
   Matthew F. Baretich

110. **Medical Gas Systems** 522
   William Frank

111. **Support Services** 525
   Nicholas Cram

112. **Construction and Renovation** 527
   Matthew F. Baretich

113. **Radiation Safety** 529
   Jadwiga (Jodi) Strzelczyk

114. **Sanitation** 532
   Lúcio Flávio de Magalhães Brito and Douglas Magagna

115. **Water Systems in Health Care Facilities** 546
   Diógenes Hernández

116. **Disaster Planning** 549
   Gary H. Harding and Alice L. Epstein

Section XI   Medical Device Standards, Regulations, and the Law   555
Introduction   David A. Simmons

117. **Primer on Standards and Regulations** 557
   Michael Cheng

118. **Medical Device Regulatory and Technology Assessment Agencies** 559
   Mark E. Bruley and Vivian H. Coates

119. **Health Care Quality and ISO 9001:2000** 565
   David A. Simmons

120. **Hospital Facilities Safety Standards** 568
   Gerald Goodman

121. **JCAHO Accreditation** 570
   Britt Berek

122. **Medical Equipment Management Program and ANSI/AAMI EQ56** 573
   Ethan Hertz

123. **Clinical Engineering Standards of Practice for Canada** 576
   Tony Easty and William M. Gentles

124. **Regulations and the Law** 578
   Michael Cheng

125. **European Union Medical Device Directives and Vigilance System** 582
   Nicolas Pallikarakis

126. **United States Food & Drug Administration** 586
   F. Blix Winston

127. **Tort Liability for Clinical Engineers and Device Manufacturers** 590
   Edward P. Richards, III and Charles Walter

Section XII   Professionalism and Ethics   593
   Introduction   Gerald Goodman

128. **Professionalism** 595
   Gerald Goodman

129. **Clinical Engineering Advocacy** 598
   Thomas J. O'Dea

130. **American College of Clinical Engineering** 600
   Jennifer C. Ott and Joseph F. Dyro

131. **The New England Society of Clinical Engineering** 610
   Nicholas T. Noyes

132. **New York City Metropolitan Area Clinical Engineering Directors Group** 613
   Ira Soller and Michael B. Mirsky

133. **Clinical Engineering Certification in the United States** 617
   Thomas Nicoud and Eben Kermit

134. **Clinical Engineering Certification in Germany** 619
   Vera Dammann

Section XIII   The Future   621
   Introduction   Jennifer C. Ott

135. **The Future of Clinical Engineering: The Challenge of Change** 623
   Stephen L. Grimes

136. **Virtual Instrumentation—Applications to Health Care** 627
   Eric Rosow

137. **Clinical Engineers in Non-Traditional Roles** 644
   Eben Kermit

138. **Clinical Support: The Forgotten Function** 646
   Stan Scahill

139. **Postmarket Surveillance and Vigilance on Medical Devices** 647
   Michael Cheng

140. **Small Business Development: Business Plan Development Fundamentals for the Entrepreneur** 649
   Peter W. Dyro

141. **Engineering Primary Health Care: The Sickle Cell Business Case** 652
   Yancy Y. Phillips and Thomas M. Judd

142. **Global Hospital in 2050—A Vision** 655
   Yasushi Nagasawa, Edward Sivak, and Errki Vauramo

Appendices   661

Index   665

This Handbook is dedicated to the memory of my father, Sigmund Stanislaus Dyro, a model for intellectual curiosity, artistry, diligence, and compassion.

Contributors

Carl Abramson
Malvern Group, Malvern, PA

Joseph Adam
President, Premise Development Corporation, Hartford, CT

Siim Aid
Department of Medical Devices, State Agency of Medicines, Tartu, Estonia

Hashem O. Al-Fadel
Managing Director, ISNAD Medical Systems, Amman, Jordan

Jerome T. Anderson
Principal, Biomedical Consulting Services, San Clemente, CA

Dennis Autio
Sr. Clinical Engineer, Department of Veterans Affairs, Portland VA Medical Center, Portland, OR

Tim Baker
Managing Editor, *The Journal of Clinical Engineering*, University Heights, OH

Salil Balar
Clinical Engineering and Technology Management Department, Beaumont Services Company. LLC, Royal Oak, MI

Mathew F. Baretich
President, Baretich Engineering, Inc., Fort Collins, CO

Thomas J. Bauld, III
Biomedical Engineering Manager, Riverside Health System, ARAMARK/Clinical Technology Services, Newport News, VA

Britt Berek
JCAHO, Oakbrook Terrace, IL

Andre E. Bloesch
Datex-Ohmeda, Inc., Madison, WI

Donald F. Blumberg
President, D. F. Blumberg & Associates, Inc., Fort Washington, PA

Javier Brandán
Jefe de la Unidad de Bioingenieria, Hospital Nacional Guillermo Almenara Irigoyen, La Victoria-Lima, Peru

Diego Bravar
Research Area of Trieste, CIVAB, ITAL TBS SPA, Trieste, Italy

Lúcio Flávio de Magalhães Brito
Certified Clinical Engineer, Engenharia Clínica Limitada, São Paulo, Brazil

Joseph D. Bronzino
Vernon Roosa Professor of Applied Science, Department of Engineering, Trinity College, Hartford, CT

Mark E. Bruley
Vice President, Accident and Forensic Investigation ECRI, Plymouth Meeting, PA

Melissa Burns
Holyoke, MA

Caroline A. Campbell
ARAMARK Clarian Health Partners, Indianapolis, IN

Joel R. Canlas
Clinical Engineering and Technology Management Department, Beaumont Services Company, LLC, Royal Oak, MI

Michael Cheng
Health Technology Management, Ottawa, Ontario, Canada

J. Tobey Clark
Director, Technical Services Program, University of Vermont, Burlington, VT

Vivian H. Coates
Vice President, Technology Assessment and Information Services, ECRI, Plymouth Meeting, PA

Ted Cohen
Manager, Clinical Engineering Department, Sacramento Medical Center, University of California, Sacramento, CA

Nicholas Cram
Texas A&M University, College Station, TX

Vera Dammann
Department of Hospital and Clinical Engineering, Environmental, and BioTechnology, University of Applied Sciences Giessen-Friedberg, Giessen, Germany

Yadin David
Director, Center for TeleHealth and Biomedical Engineering Department, Texas Children's Hospital, Houston, TX

Vinnie DeFrancesco
Clinical Engineer, Biomedical Engineering Department, Hartford Hospital, Hartford, CT

Teresa dell'Aquila
Research Area of Trieste, CIVAB, ITAL TBS SPA, Trieste, Italy

David Dickey
President & CEO, Medical Technology Management, Inc., Clarkston, MI

Alfred M. Dolan
Samuel Lunenfeld Associate Professor in Clinical Engineering, Associate Director, Institute of Biomaterials and Biomedical Engineering, University of Toronto, Toronto, Ontario, Canada

Joseph F. Dyro
President, Biomedical Resource Group, Setauket, NY

Peter W. Dyro
Principal, Re-Source Builders, Seattle, WA

Tony Easty
Head, Medical Engineering Department, The Toronto Hospital, Toronto, Ontario, Canada

Alice L. Epstein
CNA HealthPro, Durango, CO

Larry Fennigkoh
Associate Professor, Electrical Engineering and Computer Science Department, Milwaukee School of Engineering, Milwaukee, WI

William Frank
Medical Gas Services, Inc., Webster, NH

Richard C. Fries
Manager, Support Engineering, Datex-Ohmeda, Inc., Madison, WI

Jonathan A. Gaev
International Programs, ECRI, Plymouth Meeting, PA

Pedro Galvan
Departmento de Mantenimiento, Instituto de Investigaciones en Ciencias de la Salud, Ascuncion, Paraguay

Leslie A. Geddes
Showalter Distinguished Professor Emeritus of Bioengineering, Purdue University, West Lafayette, IN

William M. Gentles
BT Medical Technology Consulting, Toronto, Ontario, Canada

Izabella A. Gieras
Clinical Engineer, Clinical Engineering and Technology Management Department, Beaumont Services Company, LLC, Royal Oak, MI

James Gilchriest
Touro College School of Health Sciences, Bay Shore, NY

Mariana Glouhova
Clinical Engineering Consultant, Sofia, Bulgaria

Ole Golubjatnikov
Estonian American Fund, Syracuse, NY

Juan Gomez
Director, Hospital Maintenance Division, Ministry of Health, Quito, Ecuador

Gerald Goodman
Assistant Professor, Health Care Administration, Texas Women's University, Houston, TX

Stephen L. Grimes
Senior Consultant and Analyst, GENTECH, Saratoga Springs, NY

Jorge Enrique Villamil Gutiérrez
Independent Technology Consultant for Health Projects, Bogota, Colombia

Jay W. Hall
Manager, Clinical Engineering, St. John Health, Detroit, MI

Gary H. Harding
Greener Pastures, Inc., Durango, CO

David Harrington
President, SBT Technology, Inc., Medway, MA

Peter Heimann
Medical Research Council, Cape Town, South Africa

Diógenes Hernández
PAHO/WHO, Panama City, Panama

Ethan Hertz
Duke University Health System, Durham, NC

James Hibbetts
Texas Children's Hospital, Houston, TX

Irwin Hinberg
Medical Devices Bureau, Therapeutic Products Directorate, Health Canada, Ottawa, Ontario, Canada

William A. Hyman
Biomedical Engineering Program, Texas A&M University, College Station, TX

Bruce Hyndman
Director of Engineering Services, Community Hospital of the Monterey Peninsula, Monterey, CA

Andrei Issakov
Program Manager (Physical Resources), WHO, Health Systems Development, Geneva, Switzerland

George I. Johnston
President, Dybonics, Inc., Portland, OR

Thomas M. Judd
Director, Quality Assessment, Improvement and Reporting, Kaiser Permanente Georgia Region, Atlanta, GA

Hiroshi Kanai
Institute of Applied Superconductivity, Tokyo Denki University, Tokyo, Japan

James P. Keller
Director, Health Devices Group, ECRI, Plymouth Meeting, PA

Eben Kermit
Biomechaical Engineer and Consultant, Palo Alto, CA

T. G. Krishnamurthy
Chairman, Paramedical Discipline Board, CEP-AICTE, UVCE Campus, Bangalore, India

Yunkap Kwankam
Health Systems Development, WHO, Geneva, Switzerland

Luis Lara-Estrella
Simon Bolivar University, Caracas, Venezuela

Alan Levenson
Mediq PRN, Morristown, NJ

Albert Lozano-Nieto
Assistant Professor of Engineering, The Pennsylvania State University, Wilkes-Barre Campus, Wilkes-Barre, PA

Joseph P. McClain
Gilbert, AZ

James McCauley
Director, Biomedical Engineering, Central Coast Hospital, Gosford, New South Wales, Australia

Douglas Magagna
Clinical Engineer, Enginaria Clínica Limitada, São Paulo, Brazil

Stuart J. Meldrum
Director of Medical Physics and Bioengineering, Norfolk and Norwich University Hospital, NHS Trust, Norwich, England

Luis Melendez
Massachusetts General Hospital, Boston, MA

J. Sam Miller
Clinical Engineering Consultant, Buffalo, NY

Saul Miodownik
Director, Clinical Engineering, Memorial Sloan-Kettering Cancer Center, New York, NY

Michael B. Mirsky
Clinical Engineering Solutions, Yorktown Heights, NY

Bruce J. Morgan
Clinical Engineering Consultant, Riverside, CA

Robert L. Morris
Oregon Health and Science University, Portland, OR

Yasushi Nagasawa
Department of Architecture, The University of Tokyo, Tokyo, Japan

Thomas Nicoud
Design Planning Information, Tempe, Arizona

Nicholas T. Noyes
Director, Department of Clinical Engineering, University of Connecticut Health Center, Farmington, CT

Enrico Nunziata
Health Care Technology Management Consultant, Torino, Italy

Åke Öberg
Department of Biomedical Engineering, Linköping University, Linköping, Sweden

Thomas J. O'Dea
Hemoxy, LLC, Shoreview, MN

Jennifer C. Ott
Director, Clinical Engineering Department, St. Louis University Hospital, St Louis, MO

Frank R. Painter
President, Technology Management Solutions LLC, Trumbull, CT

Nicolas Pallikarakis
Department of Medical Physics, University of Patras, Patras, Greece

W. David Paperman
Clinical Engineering Consultant, Cut 'N Shoot, TX

Bryanne Patail
Biomedical Engineer, National Center for Patient Safety, Department of Veterans Affairs, Ann Arbor, MI

Yancy Y. Phillips
The Southeast Permanente Medical Group, Atlanta, GA

Raj Rane
Massachusetts General Hospital, Boston, MA

Edward P. Richards, III
Director, Program in Law, Science, and Public Health, Paul M. Hebert Law Center, Louisiana State University, Baton Rouge, LA

Nandor Richter
ORKI, Budapest, Hungary

Malcolm G. Ridgway
Sr. Vice President, Technology Management and Chief Technology Officer, MasterPlan Inc., Chatsworth, CA

Manny Roman
President, DITEC, Inc., Solon, OH

Eric Rosow
Director of Biomedical Engineering, Hartford Hospital, Hartford, CT

Stan Scahill
Director of Biomedical Engineering, Concord Hospital, Concord, New South Wales, Australia

Pam Schuck-Holmes
Product Development Engineer, Detroit, MI

George Seaman
President, Seaman Associates, Northport, NY

Marvin Shepherd
DEVTEQ Consulting, Walnut Creek, CA

Ricardo Silva
The Pennsylvania State University, State College, PA

David A. Simmons
Health Care Engineering, Inc., Glen Allen, VA.

Edward Sivak
Upstate Medical University, Syracuse, NY

Elliot B. Sloane
Assistant Professor, Department of Decision and Information Technology, Villanova University, Villanova, PA

Chad J. Smith
Product Development Engineer, Kensey Nash Corporation, Exton, PA

Peter Smithson
QA & Organizational Development Manager, Principal Clinical Scientist, Medical Equipment Management Organization, Bristol, England, UK

Ira Soller
Director, Scientific and Medical SUNY Health Science Center at Brooklyn, Instrumentation, Brooklyn, NY

Jadwiga (Jodi) Strzelczyk
Assistant Professor of Radiology, Radiological Sciences Division, University of Colorado, Health Sciences Center, Denver, CO

Sunder Subramanian
Director, Clinical Information Systems, Information Technology Division, University Hospital and Medical Center, SUNY at Stony Brook, East Setauket, NY

Momade Sumalgy
Director, Department of Maintenance, Ministry of Health, Mozambique

Kok-Swang Tan
Medical Devices Bureau, Therapeutic Products Directorate, Health Canada, Ottawa, Ontario, Canada

Errki Vauramo
Care Facilities SOTERA, Research Institute for Health, Hut, Finland

Paul Vegoda
Manhasset, NY

Adriana Velásquez
Clinical Engineering Consultant, Mexico City, Mexico

Luis Vilcahuaman
Coordinado Bioingeniería, Pontificia Universidad Católica del Perú, Lima, Peru

Charles Walter
College of Engineering, University of Houston, Houston, TX

Binseng Wang
Senior Director, CE & QA, MEDIQ/PRN Life Support Services Inc., Pennsauken, NJ

Valory Wangler
Biomedical Engineering Program, Texas A&M University, College Station, TX

Colleen Ward
Clinical Engineering Department, Sacramento Medical Center, University of California, Sacramento, CA

James O. Wear
Veterans Administration, North Little Rock, AR

Lee O. Welter
Health Care Consultant, Sacramento, CA

F. Blix Winston
Mount Airy, MD

Donald Witters
Physicist, FDA, Center for Devices and Radiological Health, Rockville, MD

Raymond P. Zambuto
CEO, Technology in Medicine, Inc., Holliston, MA

Introduction

Joseph F. Dyro
Editor-in-Chief
Setauket, NY

Clinical Engineering: A Definition

The American College of Clinical Engineering defines a clinical engineer (CE) as: *"a professional who supports and advances patient care by applying engineering and managerial skills to health care technology."* Clinical engineering is a subset of biomedical engineering. Whereas biomedical engineering is practiced primarily in academic institutions, the research laboratory, and manufacturing, clinical engineering is practiced in hospitals and other environments where medical device technologies are utilized. The purpose of the *Clinical Engineering Handbook* is to provide a central core of knowledge and essential information the clinical engineer needs to practice the profession.

Health Care and Clinical Engineering

Clinical engineering emerged as a discipline in the latter half of the twentieth century as increasing numbers of complex electronic and mechanical medical devices entered the health care environment for preventive, therapeutic, diagnostic and restorative applications. Within the complex environment of the modern hospital, clinical engineering is concerned primarily with devices but recognizes that interactions between drugs, procedures, and devices commonly occur and must be understood and managed to ensure safe and effective patient care. Clinical engineering, first practiced in the hospital, has extended its sphere of influence as medical devices are utilized increasingly in extended care facilities and in the home. Career mobility and diversity has increased as manufacturers and regulatory agencies have recognized the important role clinical engineers play in the research, development, manufacturing and regulation of medical devices. Clinical engineers apply their talents in all areas of health care including the following:

- Medical device industry
- Government regulatory agencies
- Standards organizations
- Hospitals, clinics, extended care and long-term care facilities
- Regional and national service organizations
- Consultancies for litigation support, regulatory compliance, program evaluation, and research
- Academic institutions
- Technical training schools

Engineers who are active in the health care sector have great opportunities to contribute to long-term quality enhancement by developing new techniques or improving existing ones. By working daily with commercially available medical devices and observing how products are used in practice, the CE gains valuable knowledge that, when coupled with keen insight and creativity, can lead to ways to improve existing techniques or to solve long-standing problems. The CE has a perspective on developmental requirements, an extremely valuable attribute in order to function as an innovator. Furthermore, the CE's proximity to the point of health care delivery provides ample opportunities for the testing and trial of new products in the end-user environment. In the pursuit of improvements in health care, many with expertise in the medical sciences, doctors, nurses, therapists, collaborate well with engineers; for they understand the engineer's gifted ability to analyze problems and synthesize solutions. The beneficial result of such analysis and synthesis, the warp and weft of the engineer's mantel, is invention, the creation of something useful, an object, a machine, or a technique, that did not exist before.

Clinical engineering was born from the concept elaborated by Cesar Caceres over two decades ago, that engineering attributes, i.e., analysis and synthesis, are relevant and necessary to improve health care. Over the last two decades, however, the engineering aspects of the profession were eroded as hospital administrators and engineers themselves viewed the primary clinical engineering activity as medical device inspection, maintenance and repair. Today, the original concept of clinical engineering has reemerged, largely in response to the driving forces of cost-control, utilization optimization, regulatory requirements, patient safety and human error awareness, and increasing complexity of the technological environment. For example, among the forces of change were the revelations of inadequacies in the so-called *health care delivery system* contained in the Institute of Medicine's year 2000 report entitled *To Err is Human*. Systemic ailments in this *system* cry out for reengineering. No longer can clinical engineering be content with following routine maintenance and repair procedures when the need is so great for making engineering changes in the healthcare arena that will support and advance patient care. The Handbook contains many illustrations of the benefits of practicing the engineering aspects of the profession.

The enormous potential of highly competent, well-educated, talented, and skilled clinical engineers is now tapped for inventions, technical development, and systems analysis. Through the advocacy and public policy efforts of the American College of Clinical Engineering (ACCE) and the ACCE Healthcare Technology Foundation, the value of clinical engineering to the health care community is clearly recognized and enthusiastically embraced. Clinical support, engineering at the bedside, health care delivery process improvement, enhanced technology utilization and patient care, and device design are at the core of the clinical engineering discipline. The listing of the many facets of the discipline of clinical engineering is shown in the accompanying figure. Clinical engineering will continue to adapt as technology advances. Engineering in primary health care is but one example of how clinical engineering can proactively influence the quality of life by disease prevention and health improvement at a reasonable cost.

This Handbook will serve as an instructional tool in clinical engineering academic programs and a desk reference for all clinical engineers and others involved with health care technologies. The author, with experience as a faculty member of many ACCE Advanced Clinical Engineering Workshops, a practicing clinical engineer, and a university educator, is keenly aware of the educational requirements of the clinical engineer. This Handbook incorporates all the essential elements of clinical engineering and will serve as a solid platform upon which to build the management and technical skills of the profession. It includes all the subjects that are necessary for a comprehensive educational curriculum for clinical engineers and others who manage medical device technologies

Within each Handbook section can be found an elaboration upon each of the facets of the discipline of clinical engineering, which are summarized in the accompanying figure.

Safety and Efficacy of Medical Device Technology

The methodology for making the patient environment safer is described in the section on Safety. Clinical engineering response to the patient safety movement involves active participation in risk management, hospital safety programs, and accident investigation. The complex interactions involving patient, devices, users, facilities, and the environment require a systems approach to medical device safety. For example, the emergence of a wide variety of wireless technologies and the proliferation of devices that produce and are receptive to electromagnetic energy has drawn attention to deleterious effects of interference caused by this energy. Means for assessing and managing the risks posed by EMI are presented. A historical account of the emergence of, reaction to, and effect of the electrical safety scare of the 1970s puts electrical safety in clear perspective *vis-à-vis* other more significant safety issues and demonstrates the folly of unreasonable assumptions and wild extrapolations that are not well-grounded in scientific fact.

Medical Devices

The different phases in the medical device technology life cycle, such as research, design, manufacture, evaluation, control, utilization, service, standards, regulations, and law are addressed in several sections: Medical Devices: Design, Manufacture, Evaluation and Control; Medical Devices: Utilization and Service; and Standards, Regulations, and the Law. The evolution of medical technology from the earliest recorded history to the present day clearly demonstrates the contributions of engineering to the advancement of health care. Development of medical devices must involve research, design, clinical testing, and good manufacturing practices. Human factors, failure mode analysis, and software development are discussed as they affect device performance and safety. Medical device utilization is described in the hospital. In addition to general hospital devices, specialty devices are described that are utilized in the following selected areas and services of the hospital: intensive care, operating room, anesthesiology, imaging, perinatology, and cardiovascular services. Standards and regulations applicable to medical devices and clinical engineering are presented. International and national perspectives on standards are addressed including those promulgated by the Joint Commission on Accreditation of Healthcare Organizations, United States Food and Drug Administration, International Standards Organization, the American National Standards Institute, and the Association for the Advancement of Medical Instrumentation.

Health Care Technology Management

Health care technology management is a core clinical engineering discipline entailing assessment, strategic planning, evaluation, acquisition, utilization, maintenance, asset control, replacement planning, and quality assurance. The first chapter, Introduction to Medical Technology Management Practices, is followed by chapters on strategic planning, evaluation, procurement, asset management, management systems, maintenance, sources of service and service management, and replacement planning. The chapters on national health technology policy, impact analysis, donations, good management practice for medical equipment, strategies to maintain essential medical equipment and essential healthcare technology package are of particular significance to countries with developing and emerging economies.

Management

Several chapters describe the management skills required for effective clinical engineering practice including utilization analysis, financial management, cost-effectiveness and productivity measurement, personnel management, and quality assessment and improvement.

Professionalism and Ethics

Advocacy efforts, certification programs, and active participation in professional societies are discussed in the section on Professionalism and Ethics. The American College of Clinical Engineering, the first organization in the world founded to represent solely the interest of clinical engineers, is described in detail. In addition, examples of local and regional clinical engineering organizations are presented in order that they may serve as models of organizations for countries and regions where clinical engineering is developing.

Information

The section on Information reveals the effect that the explosion of information technology (i.e., computers and methods for utilizing information to advance patient care) has had upon clinical engineering. Following the chapter on Information Systems Management, chapters address Physiological Monitoring and Clinical Information Systems, Advanced Diagnostics and Artificial Intelligence, Real-Time Executive Dashboards and Virtual Instrumentation, Telemedicine, Picture Archiving and Communication Systems, Wireless Medical Telemetry, Health Insurance Portability and Accountability Act, Y2K and Clinical Engineering, and The Integration and Convergence of Medical and Information Technologies.

Facilities and the Clinical Environment

The safe and effective operation of medical devices and systems depends to a great extent on the physical plant and its utilities and support systems such as heating, ventilation, and air conditioning, electrical distribution systems, medical gas systems, sanitation systems, and water supply. Radiation safety, facilities design and construction, and disaster preparedness are all issues with which the clinical engineer must knowledgeable as they directly relate to the medical devices utilized within a health care institution.

Education

Many journal articles and several textbooks have been written over the years on the subject of clinical engineering. These have served well to educate generations of clinical engineers. The author owes a debt of gratitude to those who have contributed to the literature with books on clinical engineering including Barry Feinberg, Cesar Caceres, Joseph Bronzino, John Webster, and Albert Cook. In the course of discussions with colleagues at national and international clinical engineering conferences, colleagues recognized the need for a comprehensive text that encompasses all aspects of clinical engineering. This Handbook is built on the solid foundation of information presented over the years by experts in the field. The section on Education includes chapters on Academic Programs in North America, Advanced Clinical Engineering Workshops, Distance Learning, In-Service Education, Technical Training Schools, Retraining Programs, and Clinical Engineering Internship. Educational programs in Germany and India are described.

Worldwide Clinical Engineering Practice

Clinical engineering is practiced around the globe in countries at all stages of economic development. The lead chapter of the section on Worldwide Clinical Engineering Practice is World Clinical Engineering Survey. This is followed by descriptions of the practice of clinical engineering in fifteen countries throughout different regions of the world. Developing countries and emerging economies, whose resources tend to be scarce and where lack of effective technology management is keenly felt, have relatively more to gain from the implementation of clinical engineering than economically advanced nations.

The Future

As systems and methods of addressing the well-being of the inhabitants of this planet change, so too will the discipline of clinical engineering. The section on The Future leads the reader one step in that direction by suggesting ways by which clinical engineering can contribute to improvement in health care. It is this editor's hope that those entering the field of clinical engineering today will one day be among those that write the next edition of this Handbook.

Acknowledgments

I thank my friend and colleague, Joseph D. Bronzino, for recommending to the publisher that I undertake the task of editing the *Clinical Engineering Handbook*. He told me at a meeting several years ago that when we leave this earth the knowledge in our brain is lost forever. So, write! I heeded that advice and urge all others to do likewise. I thank all authors from across the globe who contributed their time and energy to write section introductions and chapters. I thank the following members of the Handbook Advisory Board for their review of and comment on the initial table of contents: Matthew Baretich, Joseph D. Bronzino, Andrei Issakov, Bruce J. Morgan, Robert Morris, Åke Öberg, Jennifer C. Ott, Frank R. Painter, Nicolas Pallikarakis, Nandor Richter, Malcolm G. Ridgway, Binseng Wang, and Raymond P. Zambuto. Their comments and suggestions helped refine the Handbook's scope and content.

I give special thanks to my kind and understanding wife, Betsy, for her patience, forbearance, encouragement, and support throughout my long hours at the keyboard.

The Discipline of Clinical Engineering

Health Technology Management

Technology assessment, evaluation, strategic planning, acquisition, life cycle cost analysis, upgrades & replacement planning, utilization analysis, resources optimization, regional & national health technology policy, program & personnel administration

Safety

Systems analysis, hospital safety programs, incident investigation, root cause analysis, user error, risk analysis & management, hazard & recall reporting systems, post-market device surveillance, device-device interactions, electromagnetic compatibility, disaster preparedness

Medical Device Service

Equipment control, computerized assets & maintenance management systems, inspection, maintenance, repair, in-house and outsourced programs, independent service organizations, vendor and service management, spare parts management

Technology Application

Engineering at the bedside, specialization in clinical areas, quality assurance & improvement, clinical applications support, home care support, help desk, installation & integration

Information Technology

Information systems integration and management, patient data management, artificial intelligence, telemedicine, picture archiving and communication systems, wireless networks (telemetry), Health Insurance Portability and Accountability Act

Education & Training

Credentialing, health care provider technology training, distance education, in-service education, training schools, academic programs, international training, professional development, volunteer work

Research & Development

Medical device design & manufacturing, evaluations, modeling & simulation, human factors, failure mode and effect analysis, clinical trials & institutional review board support

Clinical Facilities

Clinical space design, electrical power, medical gases, water, HVAC (heating, ventilation, and air conditioning), sanitation, construction, renovation, communications infrastructure

Standards & Regulations

Compliance assurance, medical device and facilities standards, quality standards, regulations, consensus standards and guidelines, accreditation, expert witness, certification

Section I: Clinical Engineering

Raymond P. Zambuto
CEO, Technology in Medicine, Inc.
Holliston, MA

The place is Olduvai Gorge. The time is 2 million BC. Homo habilis, "handy man," uses a sharpened stone to remove a splinter. Fast forward to the twenty-first century. The place is Boston, Massachusetts. A cardiologist evaluates a child in distress in São Paolo, Brazil using real-time video teleconferencing with the child's physician.

Human beings have used tools in their medical care from the beginning. Like people, those tools have evolved over the ages. The sharpened stones have become complex instrumentation- and technology-based clinical systems of care. As medical "tools" became more sophisticated, the technologies that they employed grew apart from the core competencies of the users. Today, medical technology is rooted in science, designed using engineering, and applied interactively in a demanding environment. Furthermore, clinical systems go beyond the technology to include policies, procedures, and protocol that ensure safety, security, and privacy as well as efficacy. This is the realm of clinical engineering.

Clinical engineering began with the application of engineering principles to the economic solution of clinical problems. It relied heavily on a body of knowledge that drew on electrical and mechanical engineering, physiology, human factors, and chemistry.

The clinical engineer was envisioned to be the professional who would bridge the communications gaps among medical, administrative, and engineering personnel in hospitals, thus playing a direct role in the improvement of care by leveraging technological solutions to patient diagnosis and therapy. Employment opportunities were seen in research and design as well as in the clinical environment. As such, the clinical engineer was required to possess specialized knowledge, such as medical terminology and sterile technique, that was not provided in the traditional engineering curriculum.

Today, as in the past, in typical practice the engineering foundation gives the clinical engineer analytical tools to solve problems from the perspective of an "outsider," while the specialized knowledge makes the clinical engineer a member of the health care team. For example, the engineering education tells the clinical engineer that an intensive care room is properly sized to hold the patient, equipment, and personnel. Human factors training might reveal that the room does not give adequate space for the staff to work in an emergency.

As an interdisciplinary practitioner, the clinical engineer is something of a chameleon. To one person, the clinical engineer is an advisor on technology selection; to another, an incident investigator or patient-safety expert; to a third person, a partner in clinical studies. Clinical engineers are everywhere on the health scene, often in jobs that parallel the work of other professionals, including medical physicists, biomedical equipment managers, and information-technology specialists.

This combination of diverse training with varied job experiences can confuse the casual observer who wishes to "define" a clinical engineer in the same way that one would identify a doctor, nurse, or accountant. As clinical engineering has evolved and matured as a unique field, its members have applied their professional skills to an ever-changing panorama of applications, driven by the changing landscape of the health care industry.

When the first formally trained clinical engineers came into the work force, the industrialized world was in the midst of a post-war health care boom. That boom eventually ended as the cost of health care reached the point where businesses and governments sought to control it. The result has been a plethora of systems including state-sponsored health care, single-payer systems, Health Maintenance Organization (HMO)-based care, capitated payment schemes, systems of hospitals, both for- and nonprofit, and in some cases, rationed care.

While all of these changes were occurring, the developing world was struggling against poverty, basic diseases, and AIDS, all the while hindered by insufficient resources for care. While one part of the world began to shut down excess capacity by closing and merging hospitals, another part was still building its health care infrastructure. Clinical engineers have been called upon to apply their skills in each of these situations, always looking for economic solutions to technology-based problems. The technology may be high or low, but the quality-of-care-vs.-cost tightrope is always being walked.

While economic forces continue to shape health care, another external factor that is rising in influence is the growth of electronic data processing and communications. Health care systems are establishing enterprise-wide data processing, storage, and transfer, and are bringing complex clinical systems on line. In tomorrow's environment, information management will continue to play a wider role in the management of health care and its cost and therefore will shape many aspects of equipment design and technology management. The implementation of smarter technology in clinical equipment, as well as distributed technologies such as telemedicine and remote diagnosis, will create new applications for clinical engineers in terms of deployment, economics, efficacy, and security.

This wide range and rapid change in clinical engineering practice is atypical for a relatively young profession. While the more mature engineering disciplines saw major changes in their practice as their "body of knowledge" expanded, those changes occurred over many decades or even centuries. Clinical engineering, however, being born in the second half of the twentieth century, has been impacted from the start by the most prolific growth in medical discovery and technology advance in history. And yet, the basic elements of the profession have remained the same through all of the diverse ways that those elements are applied.

Clinical engineers are, particularly in the industrialized nations, commonly perceived primarily as managers of the biomedical equipment maintenance function. While this job is often filled by a clinical engineer, the stereotype does a disservice to both the clinical engineering and the biomedical equipment technician fields. Management of the maintenance shop can be a career path for the management-oriented technician as well as for the engineer.

While these perceptions reflect the most visible application of clinical engineering's skill set, they exclude much of what a clinical engineer brings to the workplace. For example, in the mid-1970s, clinical engineers were intimately involved in health care industry issues unrelated to biomedical department management, including: electrical safety of invasive medical equipment, incident investigations, design of medical equipment, isolated power in the operating room, fire hazards in oxygen-rich atmospheres, patient-connection standards and interlocks, infection control, in-service education, electromagnetic interference with instrumentation, codes and standards writing-and-interpretation, and assessment of new technologies. At the same time, clinical engineers also participated in the development of equipment-related systems for assuring the safety and efficacy through planned maintenance.

More recently, these same clinical engineers are confronting issues that range from acquisition and installation of Picture Archiving and Communications Systems (PACS) to standards for the refurbishing of medical equipment. Taken in the aggregate, these applications cover a broad, ever-changing body of knowledge.

Every profession possesses a unique body of knowledge consisting of a basic skill set plus derivative knowledge based on current practices and applications. The derivative knowledge is often the more visible trait, but it is also the more volatile. The underpinnings of the profession are the basic skills, and they remain constant.

For clinical engineering, the basic skill set is the combination of classical engineering education supplemented with life-science and management course work such as physiology, human factors, psychology, and economics. With those basic tools, the clinical engineer is prepared to address a variety of issues.

Despite this broad education, in the real world clinical engineering work often entails decisions that must be made without knowing all of the facts, and with no crystal ball regarding the outcome or consequences. In some situations, such as the investigation of an incident, the need to use one's judgment is apparent. In other situations, such as the balancing of equipment maintenance schedules, the clinical engineer must consider a variety of sometimes conflicting factors such as accreditation requirements, real need, budget, and potential harm to the patient. The use of professional judgment is more transparent in this type of situation. Similarly, the "repair or replace" decision can be complicated by staff capability, the hospital's census, and the strategic medical plan of the institution, not just factors of cost and budget.

The common denominator for clinical engineers is the patient, regardless of job description or employer. The fundamental drive to put their talents to a better purpose is what usually drives engineers to choose health care over a more lucrative career in the defense or computer industries. That fundamental drive to improve patient care through the use of engineering can be exercised in many venues. It might be by finding economic solutions to problems, managing medical equipment to ensure its safe operation and continuous availability, or designing the next generation of technology into that equipment. In every situation, however, the patient is always seen as the ultimate customer.

This focus on the patient is the centerpiece of the clinical engineer's professional ethics. In addition to ensuring the individual patient's right to safe and efficacious equipment, clinical engineers also function in applications that more broadly impact patient welfare and the public good. These applications include: incident investigations, design of medical equipment, infection control, in-service education, electromagnetic interference with equipment function, codes and standards interpretation, and assessment of new technologies. In each of these areas, clinical engineers are working to improve safety, to reduce costs, and to continuously improve the health care system.

Additionally, non–hospital-based clinical engineers, such as those in clinical research, education, forensics, consulting, and manufacturing, often play a direct role in situations that directly impact the public good. These clinical engineers work different applications, but their expertise is brought to bear directly on patient outcomes, clinician performance, and hospital management.

The result of this ethical commitment by clinical engineers is the engendering of a public trust. That public trust in turn imputes a degree of personal responsibility for the actions or omissions of the clinical engineer, thus completing the picture of the professional. This mindset, shaped in the educational process and reinforced by professional groups such as the American College of Clinical Engineering, gives the clinical engineer a special identification with the good of the patient. As a result, clinical engineers, while generally protected by the corporate shield of the hospital or another employer, will make every effort to ensure that the decisions they make, and the actions they take, are thoroughly investigated and based on firm foundations.

Today, clinical engineering continues to mature as a profession. Its earlier over-identification with equipment maintenance is gradually giving way to an understanding of the broader tools that the clinical engineer brings to the table. One of the primary drivers behind this effect is the financial pressure on hospitals and the need to improve utilization. As a result, equipment and system design is being driven to more complexity and automation. This increased complexity ties the technology to the financial performance of the institution, as well as to patient diagnosis and treatment. The management of health care continues to become intertwined with higher levels of technology and more complex systems, placing increased demands for the skills of the clinical engineer, while the recognition that properly engineered technology solutions are more cost-effective has improved the visibility of clinical engineering to the management structure.

As technology continues to assume a greater role in care, its management will continue to be elevated in stature. The cross-linking of medical technology with communications and data processing technologies, will foster the development of a central management function to span the "technology bundle" as the medical equipment of today becomes the "front end" of tomorrow's clinical information system. This "chief technology officer" will need to interface with the medical, administrative, and engineering disciplines of the future. Eventually, another final shift will occur from a focus on the technology to a focus on the information itself. This will create a new set of challenges and opportunities for clinical engineering.

In the nine chapters that comprise this first section of the *Handbook of Clinical Engineering*, a team of authors unparalleled in the industry critically and expertly defines the profession. Covering a range of topics, they examine the origins of the profession, career paths, work environments, and topics concerning clinical engineers' relation to health care and to technology in health care.

The section also includes practical accounts of the daily activities of clinical engineers and a detailed exploration of the expanding role of the clinical engineer in an important activity that benefits the public good, that of process improvement for the reduction of medical errors.

Information on the need for and practice of clinical engineering around the world will be emphasized here, including a chapter that exclusively concerns World Health Organization (WHO) technology guidelines. These chapters demonstrate the differences and the wide range of opportunities that exist for the future of a profession that looks very bright, indeed.

1
Clinical Engineering: Evolution of a Discipline

Joseph D. Bronzino
Vernon Roosa Professor of Applied Sciene,
Department of Engineering
Trinity College
Hartford, CT

Technological innovation has continually reshaped the field of medicine and the delivery of health care services. Throughout history, advances in medical technology have provided a wide range of positive diagnostic, therapeutic, and rehabilitative tools. With the dramatic role that technology has played in shaping medical care during the latter part of the 20th century, engineering professionals have become intimately involved in many medical ventures. As a result, the discipline of biomedical engineering has emerged as an integrating medium for two dynamic professions: medicine and engineering. Today, biomedical engineers assist in the struggle against illness and disease by providing materials, tools, and techniques (such as medical imaging and artificial intelligence) that can be utilized for research, diagnosis, and treatment by health care professionals. In addition, one subset of the biomedical engineering community, namely clinical engineers, has become an integral part of the health care delivery team by managing the use of medical equipment within the hospital environment. The purpose of this chapter is to discuss the evolution of clinical engineering, to define the role played by clinical engineers, and to present the status of the professionalization of the discipline and to reflect upon its future.

What Is Clinical Engineering?

Many of the problems confronting health care professionals today are of extreme interest to engineers because they involve the design and practical application of medical devices and systems—processes that are fundamental to engineering practice. These medically related problems can range from very complex, large-scale constructs, such as the design and implementation of automated clinical laboratories, multiphasic screening facilities, and hospital information systems, to the creation of relatively small and "simple" devices, such as recording electrodes and biosensors that are used to monitor the activity of specific physiological processes in a clinical setting. Furthermore, these problems often involve addressing the many complexities found in specific clinical areas, such as emergency vehicles, operating rooms, and intensive care units.

The field of biomedical engineering, as it has evolved, now involves applying the concepts, knowledge, and approaches of virtually all engineering disciplines (e.g., electrical, mechanical, and chemical engineering) to solve specific health care-related problems (Bronzino, 1995, 2000). When biomedical engineers work within a hospital or clinical environment, they are more properly called clinical engineers.

But what exactly is the definition of a "clinical engineer"? Over the years, a number of organizations have attempted to provide an appropriate definition (Schaffer and Schaffer, 1992). For example, the AHA defines a clinical engineer as:

- *"a person who adapts, maintains, and improves the safe use of equipment and instruments in the hospital,"* (AHA, 1986).

The American College of Clinical Engineering defines a clinical engineer as:

- *"a professional who supports and advances patient care by applying engineering and managerial skills to health care technology,"* (Bauld, 1991).

The definition that the AAMI originally applied to board certified practitioners describes a clinical engineer as:

- *"a professional who brings to health care facilities a level of education, experience, and accomplishment which will enable him to responsibly, effectively, and safety manage and interface with medical devices, instruments, and systems and the user thereof during patient care...,"* (Goodman, 1989).

For the purpose of certification, the Board of Examiners for Clinical Engineering Certification considers a clinical engineer to be:

- *"an engineer whose professional focus is on patient-device interfacing; one who applies engineering principles in managing medical systems and devices in the patient setting,* (1CC, 1991).

The *Journal of Clinical Engineering* has defined the distinction between a biomedical engineer and a clinical engineer by suggesting that the biomedical engineer:

- *"applies a wide spectrum of engineering level knowledge and principles to the understanding, modification or control of human or animal biological systems* (Pacela, 1991).

Finally, in the book "Management of Medical Technology," a clinical engineer was defined as:

- *"an engineer who has graduated from an accredited academic program in engineering or who is licensed as a professional engineer or engineer-in-training and is engaged in the application of scientific and technological knowledge developed through engineering education and subsequent professional experience within the health care environment in support of clinical activities* (Bronzino, 1992).

It is important to emphasize that one of the major features of these definitions is the clinical environment (i.e., that portion of the health care system in which patient care is delivered). Clinical activities include direct patient care, research, teaching, and management activities that are intended to enhance patient care.

Engineers were first encouraged to enter the clinical scene during the late 1960s in response to concerns about patient safety as well as the rapid proliferation of clinical equipment, especially in academic medical centers. In the process, a new engineering discipline—clinical engineering—evolved to provide the technological support that was necessary to meet these new needs. During the 1970s, a major expansion of clinical engineering occurred, primarily due to the following events (Bronzino and Hayes, 1988; Bronzino, 1992):

- The Veterans' Administration (VA), convinced that clinical engineers were vital to the overall operation of the VA hospital system, divided the country into biomedical engineering districts, with a chief biomedical engineer overseeing all engineering activities in the hospitals in a district.
- Throughout the United States, clinical engineering departments were established in most large medical centers and hospitals and in some smaller clinical facilities with at least 300 beds.
- Clinical engineers were hired in increasing numbers to help these facilities to use existing technology and to incorporate new technology.

Having entered the hospital environment, routine electrical safety inspections exposed the clinical engineer to all types of patient equipment that were not being properly maintained. It soon became obvious that electrical safety failures represented only a small part of the overall problem posed by the presence of medical equipment in the clinical environment. This equipment was neither totally understood nor properly maintained. Simple visual inspections often revealed broken knobs, frayed wires, and even evidence of liquid spills. Investigating further, it was found that many devices did not perform in accordance with manufacturers' specifications and were not maintained in accordance with manufacturers' recommendations. In short, electrical safety problems were only the tip of the iceberg. By the mid-1970s, complete performance inspections before and after use became the norm, and sensible inspection procedures were developed (Newhouse et al., 1989). Clinical engineering departments became the logical support center for all medical technologies. As a result, clinical engineers assumed additional responsibilities, including the management of high-technology instruments and systems used in hospitals, the training of medical personnel in equipment use and safety, and the design, selection, and use of technology to deliver safe and effective health care.

In the process, hospitals and major medical centers formally established clinical engineering departments to address these new technical responsibilities and to train and supervise biomedical engineering technicians to carry out these tasks. Hospitals that established centralized clinical engineering departments to meet these responsibilities used clinical engineers to provide the hospital administration with an objective opinion of equipment function, purchase, application, overall system analysis, and preventive maintenance policies. With the in-house availability of such talent and expertise, the hospital was in a far better position to make more effective use of its technological resources (Jacobs, 1975; Bronzino, 1977; 1986; 1992). It is also important to note that competent clinical engineers, as part of the health care system, also created a more unified and predictable market for biomedical equipment. By providing health care professionals with needed assurance of safety, reliability, and efficiency in using new and innovative equipment, clinical engineers identified poor quality and ineffective equipment much more readily. These activities, in turn, led to a faster, more appropriate utilization of new medical equipment and provided a natural incentive for greater industrial involvement—a step that is an essential prerequisite to widespread use of any technology (Newhouse et al., 1989, Bronzino, 1992). Thus, the presence of clinical engineers not only ensured the establishment of a safer environment, but also facilitated the use of modern medical technology to make patient care more efficient and effective.

Today, clinical engineers are an integral part of the health care delivery team. In fact, their role is multifaceted. Figure 1-1 illustrates the multifaceted role played by clinical engineers. They must successfully interface with clinical staff, hospital administrators and regulatory agencies to ensure that the medical equipment within the hospital is safely and effectively used.

To further illustrate the diversity of their tasks, some typical pursuits of clinical engineers are provided below:

- Supervision of a hospital clinical engineering department that includes clinical engineers and biomedical equipment technicians (BMETs)
- Pre-purchase evaluation and planning for new medical technology
- Design, modification, or repair of sophisticated medical instruments or systems
- Cost-effective management of a medical equipment calibration and repair service
- Safety and performance testing of medical equipment by BMETs
- Inspection of all incoming equipment (new and returning repairs)
- Establishment of performance benchmarks for all equipment
- Medical equipment inventory control
- Coordination of outside services and vendors
- Training of medical personnel in the safe and effective use of medical devices and systems
- Clinical applications engineering, such as custom modification of medical devices for clinical research or evaluation of new noninvasive monitoring systems
- Biomedical computer support
- Input to the design of clinical facilities where medical technology is used (e.g., operating rooms (ORs) or intensive-care units).
- Development and implementation of documentation protocols required by external accreditation and licensing agencies

Clinical engineers thereby provide extensive engineering services for the clinical staff and, in recent years, physicians, nurses, and other clinical professionals have increasingly accepted them as valuable team members. The acceptance of clinical engineers in the hospital setting has led to different types of engineering-medicine interactions, which in turn have improved health care delivery. Furthermore, clinical engineers serve as a significant resource for the entire hospital. Since they possess in-depth knowledge regarding available in-house technological capabilities and the technical resources available from outside firms, the hospital is able to make more effective and efficient use of all of its technological resources.

The Role of Clinical Engineering Within the Hospital Organization

Over the years, management organization within hospitals has evolved into a diffuse authority structure that is commonly referred to as the "triad model." The three primary components are the governing board (trustees), hospital administration (CEO and administrative staff), and the medical staff organization.

Clinical Engineering Program

In many hospitals, administrators have established clinical engineering departments to manage effectively all the technological resources, especially those relating to medical equipment, that are necessary for providing patient care. The primary objective of these departments is to provide a broad-based engineering program that addresses all aspects of medical instrumentation and systems support.

Figure 1-2 illustrates the organizational chart of the medical support services division of a typical major medical facility. Note that within this organizational structure, the director of clinical engineering reports directly to the vice president of medical support services. This administrative relationship is extremely important because it recognizes the important role that clinical engineering departments play in delivering quality care. It should be noted, however, that in other organizational structures, clinical engineering services fall under the category of "facilities," "materials management," or even simply "support services."

In practice, there is an alternative capacity in which clinical engineers can function. They can work directly with clinical departments, thereby bypassing much of the hospital hierarchy. In this situation, clinical departments can offer the clinical engineer both the chance for intense specialization and, at the same time, the opportunity to develop a personal relationship with specific clinicians based on mutual concerns and interests (Wald, 1989). What is important today is the presence of clinical engineering at the appropriate point in the organizational structure for it to have a maximum impact on the proper use and management of modern medical technology (Bronzino, 1992; 1995; 2000).

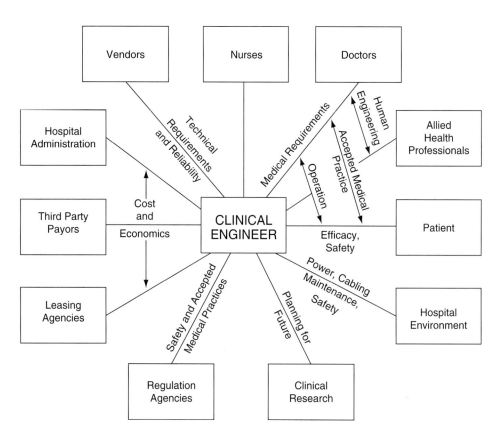

Figure 1-1 Diagram illustrating the range of interactions in which a clinical engineer might be required to engage in a hospital setting.

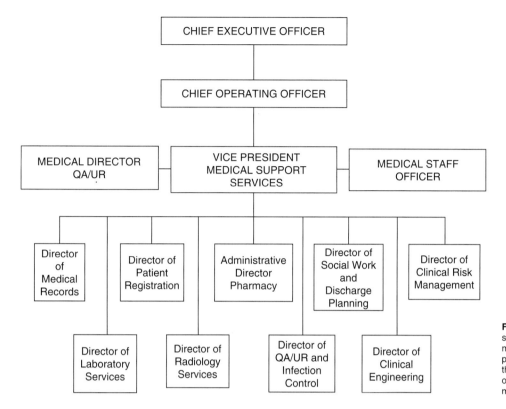

Figure 1-2 Organizational chart of medical support services division for a typical, major medical facility. This organizational structure points out the critical interrelationship between the clinical engineering department and the other primary services provided by the medical facility.

Major Functions of a Clinical Engineering Department

The role of the clinical engineer in today's hospital can be both challenging and gratifying because the care of patients requires a greater partnership between medical staff and modern technology. As previously discussed, this interchange has led to a close working relationship between the clinical engineer and many members of the medical and hospital staff. The team approach is key to the successful operation of any clinical engineering program. Figure 1-3 illustrates the degree of teamwork and interdependence that is required in order to maintain constructive interrelationships. In this matrix presentation, it is important to note that the health care team approach to the delivery of patient care creates both vertical and lateral reporting relationships. Although clinical engineers report hierarchically to their hospital administrator, they also interact with hospital staff to meet patients' requirements.

As a result of the wide-ranging scope of interrelationships within the medical setting, the duties and responsibilities of clinical engineering directors continue to be extremely diversified. Yet, a common thread is provided by the very nature of the technology that they manage. Directors of clinical engineering departments are usually involved in the following areas:

- Developing, implementing, and directing equipment management programs. Specific tasks include evaluating and selecting new technology, accepting and installing new equipment, and managing the inventory of medical instrumentation, all in keeping with the responsibilities and duties defined by the hospital administration. The clinical engineering director advises the administrator of the budgetary, personnel, space, and test equipment requirements that are necessary to support this equipment management program.
- Advising administration and medical and nursing staffs in areas such as safety, the purchase of new medical instrumentation and equipment, and the design of new clinical facilities.
- Evaluating and taking appropriate action on incidents attributed to equipment malfunction or misuse. The clinical engineering director summarizes the technological significance of each incident and documents the findings of the investigation, subsequently submitting a report to the appropriate hospital authority and, according to the Safe Medical Devices Act of 1990, to the device manufacturer and/or the Food and Drug Administration (FDA).
- Selecting departmental staff and training them to perform their functions in a professional manner.
- Establishing departmental priorities, developing and enforcing departmental policies and procedures, and supervising and directing departmental activities. The clinical engineering director takes an active role in leading the department to achieve its overall technical goals.

Therefore, the core functions of a clinical engineering department can be summarized as follows:

1. Technology management
2. Risk management

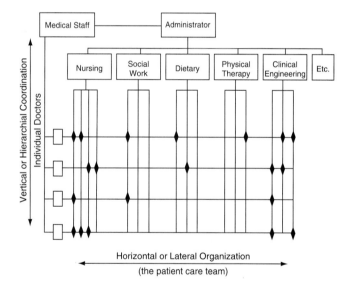

Figure 1-3 Matrix diagram illustrating the bi-directional interdependence and degree of teamwork required to maintain effective interaction between the members of the health care delivery team.

3. Technology assessment
4. Facilities design and project management
5. Quality assurance
6. Training

Professional Status of Clinical Engineering

Upon careful review of our definition of clinical engineering and the responsibilities and functions that clinical engineers assume within the hospital, it is clear that the term *clinical engineer* must be associated with individuals who can provide engineering services, not simply technical services. Clinical engineers, therefore, must be individuals who have a minimum of a four-year bachelor's degree in an engineering discipline. They must be well versed in the design, modification, and testing of medical instrumentation—skills that fall predominantly in the field of engineering practice. Only with an engineering background can clinical engineers assume their proper role working with other health professionals to use available technological resources effectively and to improve health care delivery.

By clearly linking clinical engineering to the engineering profession, a number of important objectives are achieved. First, it enables hospital administrators to identify qualified individuals to serve as clinical engineers within their institutions and to understand better the wide range of functions that clinical engineers can perform, while making it clear that technicians cannot assume this role. Second, from this foundation, it is possible for the profession of clinical engineering to continue to mature. It has been pointed out that professional activities exist if "a cluster of roles in which the incumbents perform certain functions valued in the society in general" can be identified (Parsons, 1954; Courter, 1980; Goodman, 1989). Clearly, this goal has been achieved for the clinical engineer.

One can determine the status of professionalization by noting the occurrence of six crucial events: (1) the first training school; (2) the first university school; (3) the first local professional association; (4) the first national professional association; (5) the first state license law; and (6) the first formal code of ethics (Wilensky, 1964; Goodman, 1989).

Now let us consider the present status of professionalism of clinical engineering. Consider the following: (1) there is continued discussion about the educational needs of existing, as well as beginning, professionals; (2) there is a professional society, the American College of Clinical Engineering, which is effective in establishing the knowledge base on which the profession is to develop; (3) there exists a credentials process that reflects the needs of this new profession; and (4) there is a code of ethics, which the American College of Clinical Engineering has developed for use by all in the profession.

This process toward professionalization will certainly continue in the years ahead as this new professional society continues to seek to define and control certification activities, to define the educational process that is required for these new professionals, and to promote the status of clinical engineering to hospital administrators and society as a whole.

Future of Clinical Engineering

From its early days—when electrical safety testing and basic preventive maintenance were the primary concerns—to the present, the practice of clinical engineering has changed enormously. Yet, it is appropriate to use the cliché, "The more things change, the more they stay the same." Today, hospital-based clinical engineers still have the following as their primary concerns: patient safety and good hospital equipment management. However, these basic concerns are being supplemented by new areas of responsibility, making the clinical engineer not only the chief technology officer but also an integral part of the hospital management team.

In large part, these demands are due to the economic pressures that hospitals face. State-of-the-art, highly complex instruments, such as MRI systems, surgical lasers, and other sophisticated devices, are now used as a matter of course in patient care. Because of the high cost and complexity of such instrumentation, the institution needs to plan carefully—at both a technical and a managerial level—for the assessment, acquisition, and use of this new technology.

With these needs in mind, hospital administrators have begun to turn to their clinical engineering staffs for assistance in operational areas. Clinical engineers now provide assistance in the application and management of many other technologies that support patient care (e.g., computer support, telecommunications, facilities operations, and strategic planning).

Computer Support

The use of personal computers (PCs) has grown enormously in the past decade. PCs are now commonplace in every facet of hospital operations, including data analysis for research, use as a teaching tool, and many administrative tasks. PCs are also increasingly used as integral parts of local area networks (LANs) and hospital information systems.

Because of their technical training and experience with computerized patient record systems and inventory and equipment management programs, many clinical engineers have extended their scope of activities to include personal computer support. In the process, the hospital has accrued several benefits from this involvement of clinical engineering in computer servicing. The first is time: Whenever computers are used in direct clinical applications or in administrative work, downtime is expensive. In-house servicing can provide faster and often more dependable repairs than an outside group can. Second, with in-house service, there is no need to send a computer out for service, thus reducing the possibility that computer equipment will be damaged or lost. Finally, in-house service reduces costs by permitting the hospital to avoid expensive service contracts for computers and peripheral equipment.

With all of these benefits, it might seem that every clinical engineering department should carry out computer servicing. However, the picture is not so simple. At the most basic level, the clinical engineering program must be sure that it has the staff, money, and space to do the job well. To assist in making this determination, several questions should be asked: Will computer repair take too much time away from the department's primary goal of patient care instrumentation? Is there enough money and space to stock needed parts, replacement boards, diagnostic software, and peripheral devices? For those hospitals that do commit the resources needed to support computer repair, clinical engineers have found that their departments can provide these services very efficiently, and they subsequently receive added recognition and visibility within the hospital.

Telecommunications

Another area of increased clinical engineering involvement is hospital-based telecommunications. In the modern health care institution, telecommunications covers many important activities, the most visible of which is telephone service. Broadly speaking, however, telecommunications today includes many other capabilities.

Until the 1970s, telephone systems were basically electromechanical, using switches, relays, and other analog circuitry. During the 1970s, digital equipment began to appear. This development allowed the introduction of innovations such as touch tone dialing, call forwarding, conference calling, improved call transferring, and other advanced services. The breakup of the Bell System also changed the telecommunications field enormously by opening it to competition and diversification of services.

Data transmission capability allows the hospital to send scans and reports to physicians at their offices or at other remote locations. Data, such as patient ECGs, can be transmitted from the hospital to a data analysis system at another location, and the results can be transmitted back. Hospitals are also making increased use of facsimile (fax) transmission. This equipment allows documents, such as patient charts, to be sent via telephone line from a remote location and reconstructed at the receiving site in a matter of minutes.

Modern telecommunications equipment also allows the hospital to conduct educational conferences through microwave links that allow video transmission of a conference taking place at a separate site. Some newer equipment allows pictorial information, such as patient slides, to be digitally transmitted via a phone line and then electronically reassembled to produce a video image.

Clinical engineers can play an important role in helping the hospital administrators to develop plans for a continually evolving telecommunications system. They can provide technical support during the planning stage, can assist in the development of requests for proposals for a new system, and can help to resolve any physical plant issues associated with installing the new system. Thereafter, the clinical engineer can assist in reviewing responses to the hospital's requests for quotations and can lend a hand during installation of the system.

Facilities Operations

Recently, some hospital administrators have begun to tap the high-level technical skills that are available within their clinical engineering departments to assist in other operational areas. One of these is facilities operations, which includes heating, ventilation, and air conditioning (HVAC), electrical supply and distribution (including isolated power), central gas supply and vacuum systems, and other physical plant equipment (Newhouse et al., 1989). This trend has occurred for a number of reasons. First, the modem physical plant contains microprocessor-driven control circuits, sophisticated circuitry, and other high-level technology. In many instances, facilities personnel lack the necessary training to understand the engineering theory underlying these systems. Thus, clinical engineers can effectively work in a consultation role to help correct malfunctions in the hospital's physical plant systems. Another reason is cost; the hospital might be able to avoid expensive service contracts by performing this work in-house.

Clinical engineers can assist in facilities operations in other ways. For example, they can lend assistance when compressed gas or vacuum delivery systems need to be upgraded or replaced. While the work is taking place, the clinical engineer can serve as the administration's technical arm, ensuring that the job is done properly, that it is in compliance with code, and that it is done with minimal disruption.

Strategic Planning

Today's emphasis on health care cost control requires that clinical engineers assist in containing costs associated with the use of modern medical technology. To accomplish this objective, clinical engineers are increasingly becoming involved in strategic planning, technology assessment, and purchase review. During assessment and purchase review, the clinical engineer studies a request to buy a new system or device and ensures that the purchase request includes (1) needed accessories; (2) warranty information; and (3) user and service training. This review process ensures that the device is, in fact, needed (or whether there exists a less costly unit that meets the clinician's requirements) and that it will be properly integrated into the hospital's existing equipment and physical environment.

Clinical engineers can provide valuable assistance during the planning for, and financial analysis of, potential new services. If one considers the steps that are involved in planning for a new patient-care area, many design questions immediately come to mind: What is the best layout for the new area? What equipment will be used there? How many suction, air, oxygen, and electrical outlets will be needed? Is there a need to provide for special facilities, such as those that are required for hemodialysis treatment at the bedside? With their knowledge of instrumentation and user needs, clinical engineers can help select reliable, cost-effective equipment to help ensure that the hospital obtains the best possible plan.

In the future, clinical engineering departments will need to concentrate even more heavily on management issues by emphasizing the goals of increased productivity and reduced costs. By continually expanding their horizons, clinical engineers can be major players in ensuring high-quality patient care at a reasonable cost.

References

AHA. *Hospital Administration Terminology, ed 2,* American Hospital Publishing, Washington, DC, 1986.
Bauld TJ. The Definition of a Clinical Engineer. *J Clin Engin* 16:403-405, 1991.
Bronzino JD. *Technology for Patient Care.* St. Louis, C.V. Mosby, 1977.
Bronzino JD et al. A Regional Model for a Hospital-Based Clinical Engineering Internship Program. *J Clin Engin* 7:34-37, 1979
Bronzino JD. Clinical Engineering Internships: A Regional Hospital-Based Approach. *J Clin Engin* 10:239, 1985.

Bronzino JD. *Biomedical Engineering and Instrumentation: Basic Concepts and Applications.* Boston, PWS Publishing Co, 1986.

Bronzino JD. Biomedical Engineering. In Trigg G (ed): *Encyclopedia of Applied Physics.* New York, VCH Publishers, 1991.

Bronzino JD, Hayes TP. Hospital-Based Clinical Engineering Programs. In *Handbook for Biomedical Engineering.* New York, Academic Press, 1988.

Bronzino JD, Smith V, Wade M. *Medical Technology and Society.* MIT Press, Cambridge, 1990.

Bronzino JD. *Management of Medical Technology: A Primer for Clinical Engineers.* Philadelphia, Butterworth-Heinemann, 1992.

Bronzino JD. Clinical Engineering: Evolution of a Discipline. In *Biomedical Engineering Handbook, ed 1, 2.* Boca Raton, FL, CRC Press, 1995, 2000.

Courter SS. The Professional Development Degree for Biomedical Engineers. *J Clin Engin* 5:299-302, 1980.

Goodman G. The Profession of Clinical Engineering *J Clin Engin* 14:27-37, 1989.

International Certification Commission's (ICC) Definition of a Clinical Engineer, International Certification Commission Fact Sheet, Arlington, VA, ICC, 1991.

Jacobs JE. The Biomedical Engineering Quandary. *IEEE Trans Biomed Engin* 22:1106, 1975.

Newhouse V et al. The Future of Clinical Engineering in the 1990s. *J Clin Engin 1989* 14:417-430, 1989.

Pacela A. *Bioengineering Education Directory,* Brea, CA, Quest Publishing Co, 1990.

Pacela A. Career "Fact Sheets" for Clinical Engineering and Biomedical Technology. *J Clin Engin* 16:407-416, 1991.

Painter FR. *Clinical Engineering and Biomedical Equipment Technology Certification.* World Health Organization Meeting on Manpower Development and Training for Health Care Equipment, Management, Maintenance and Repair. WHO/SHS/HHP/90.4, 1989, pp 130-185.

Parsons T. *Essays in Sociological Theory, rev ed.* Glencoe, IL, Free Press, 1954.

Schaffer MJ, Schaffer MD. The Professionalization of Clinical Engineering. *Biomed Instr Technol* 23:370-374, 1989.

Schaffer MJ, Schaffer MD. What Is a Clinical Engineer? Issues in Definition. *Biomed Instr Technol* 277-282, 1992.

Wald A. Clinical Engineering in Clinical Departments: A Different Point of View. *Biomed Instr Technol* 23:58-63, 1989.

Wilensky HL: The Professionalization of Everyone. *Am J Sociol* 69:137-158, 1964.

2

History of Engineering and Technology in Health Care

Malcolm G. Ridgway
Sr. Vice President, Technology Management and Chief Technology Officer, MasterPlan, Inc.
Chatsworth, CA

George I. Johnston
Dybonics, Inc.
Portland, OR

Joseph P. McClain
Gilbert, AZ

Modern medical care is centered upon the use of a broad range of equipment and medical devices, resulting in an increasing demand for workers with technological expertise. This chapter emphasizes the interdependence of medical device technology and those who manage that technology. Highlighting the significant milestones in medical device development over the past century serves to illustrate the increasing symbiotic relationship between medical device technology and those who must support that technology, the clinical engineers, biomedical equipment technicians, physicists, and equipment maintenance personnel.

The Early 1900s

A minor hospital building boom significantly increased the number of hospitals in the United States. However, the principal means of providing health care was still physicians making house calls and carrying their "little black bags." In general, physicians seemed to be in short supply, but so was the demand for their services. Many of the services that the physician provided also could be obtained from experienced amateurs in the community. The home was typically the site for treatment and recuperation, and relatives and neighbors constituted an able and willing nursing staff. At the time, the average life expectancy was 47 years, and 95% of all births were in the home, although only 15% of the homes had a bathtub. The leading causes of death were pneumonia, influenza, tuberculosis, and dysentery, followed by heart disease and stroke.

The first X-ray machines were introduced following Roentgen's discovery of X rays in 1895. Early radiographs involved large doses of radiation. An image of the human head could require up to 11 minutes of exposure time. The first use of contrast media occurred between 1906 and 1912. Dutch physiologist Willem Einthoven introduced the first ECG recording device, a string galvanometer, in 1901. Because it used a very large magnet, it was large and heavy, weighing about 600 lbs. The first "telecardiogram" was recorded on March 22, 1905, when Einthoven transmitted ECG signals from a healthy volunteer in a hospital to the galvanometer in his laboratory, 1.5 kilometers away by way of pole-mounted telephone cables. The volunteer's elevated R-waves were attributed to his cycling from the laboratory to the hospital for the recording. The latest technology at that time was the radio which, following Marconi's first successful transmissions in 1895, was used primarily for ship-to-ship communications.

Pre-1940

During the period between World Wars I and II (1919–1939), the development of medical instrumentation benefited from electronic amplification and the adaptation of technologies that had been developed for other applications.

The use of X rays for diagnosis continued to progress during this period, as did the use of radium needles for the treatment of tumors. Hospitals where radiation treatments were performed began to employ physicists to assist in the treatment planning and safe use of the radiation sources. The development of instruments and devices for diagnosis or medical research was generally undertaken through the collaboration of physicians and private entrepreneurs.

The use of electronic amplification techniques developed following Lee de Forest's invention of the triode valve in 1906 led to more compact ECG machines. Sanborn and Cambridge Instruments were early manufacturers of these table- or cart-mounted devices. In 1928, Frank Sanborn's company (later acquired by Hewlett-Packard) converted its table model ECG machine into its first portable version, which weighed 50 lbs. and was powered by a six-volt automobile battery. Al Grass contributed to the development of electroencephalography by adapting a low-frequency, high-gain recorder that he had first used to monitor seismic events to charting very small EEG potentials. Arnold Beckman adapted a pH-sensing electrode, which he had developed as a "ripeness indicator" for the citrus industry in southern California, to medical uses in the clinical laboratory. In a slightly different type of collaboration, W.T. Bovie, a physicist employed by the Harvard Cancer Clinic, developed a spark-gap electrosurgical machine that was first used in 1925 by prominent neurosurgeon Harvey Cushing.

Other medical devices commonly found at that time in the hospital included instrument sterilizers and, after 1927, large pressure chamber-type respirators known as "iron lungs" that were used to keep chest-paralyzed polio patients alive. The inventors, Harvard medical researchers Philip Drinker and Louis Shaw, used an iron box and two vacuum cleaners to create the device, which was about the length of a subcompact car. At that time, maintenance of these devices was largely in the hands of the facility's on-site plant engineer and the equipment manufacturer. The United States army had one of the first centrally managed groups of hospitals and was certainly the first on record to take an intelligent interest in maintaining its equipment. On July 11, 1919, the surgeon general of the army sent a letter to the quartermaster general stating:

"It is believed to be better policy to have a central repair establishment in the medical department to which surgical instruments and delicate laboratory equipment can be sent for repairs."

In 1922, the army established a central maintenance facility at its St. Louis Medical Depot, presumably to provide a more proficient capability to repair the more delicate instruments. Later, during the early years of World War II, the army's hospital managers recognized a need for an organized training program, and on January 10, 1943, the surgeon general authorized a three-month biomedical equipment technician training course and requested the adjutant general to publish quotas for a school to be conducted at the St. Louis Medical Depot. This was the beginning of the army's Biomedical Equipment Training Program, the first such program in the United States with the requirement to cover a wide variety of medical equipment.

In 1918, a worldwide flu epidemic killed 20 million people, including 500,000 Americans. The battle against infectious diseases was aided considerably during the latter part of this period by the development in Germany of the sulfa class of drugs, which became the first pharmaceutical tools for use against infectious diseases. Although British bacteriologist Alexander Fleming discovered penicillin in 1928, that drug's potential for use as an antibiotic was not immediately recognized, and the drug was not produced in quantity and made generally available until after World War II. The passage of the Food, Drug and Cosmetic Act in 1938 initiated the requirement that all drugs be approved by the FDA prior to being marketed in the United States.

The 1940s and 1950s

The post-World War II years (1945–1959) can be described best as the "golden age of medical electronics." Innovation was spurred by new technologies that had been developed for military applications and by medical electronics advances that had been pioneered in university-based research-and-development (R&D) labs. The teaching hospital-based model shops, often associated with a medical physics department, collaborated with creative, research-oriented physicians to develop many new aids and began to expand their area of interest beyond mechanical gadgets and toward electronic devices.

In 1947, the Hill-Burton Act ignited another hospital building boom. Over the next 25 years, the federal government subsidized state building programs with more than $3.7 billion. Hospital budgets were also fattened by a growing number of employer-paid group insurance programs. The first group program was provided by the Montgomery Ward retail chain in 1910, and the first Blue Cross plan was established for municipal employees in Dallas in 1929.

Several new, military-developed technologies provided a boost to this new area of medical electronics. Anti-submarine sonar countermeasures gave birth in Japan to A-scan ultrasound, and the development of microwave radar technology accelerated the formation of innovative electronic design teams in the U.S. McGraw-Hill published a series of textbooks called the *Radiation Laboratory* series, in which many clever pulse, waveform, and amplifier circuits were described.

In the larger medical institutions, medical physics departments were staffed by physicists and technicians who assisted with radiotherapy and dispensing isotopes for both therapy and treatment. In these departments, small R&D groups were formed to assist research-oriented physicians with the fabrication of specialized research tools. Initially, these R&D projects were mechanically oriented, with some of these efforts resulting in successful prototypes of devices such as dialysis machines and artificial heart valves. During the post-war years, these groups diversified into electronic gadgeteering, focusing at first on recording bioelectric signals such as nerve- and muscle-action potentials, and then analyzing them using various electronic techniques. Interesting work was published on topics such as vectorcardiography and ballistocardiography. In 1950, at the University of Manitoba in Canada, electrical engineer John Hopps unexpectedly invented the first external cardiac pacemaker while conducting research on hypothermia and experimenting with radio-frequency heating to restore body temperature.

In time, more advanced technology allowed for the development of instrumentation to study intracellular activity. Publication of these designs further stimulated this activity and resulted in a small commercial market. Grundfest was known for his physiological amplifier that was fabricated on a Plexiglas chassis to minimize ground currents and suspended by springs to minimize microphonics. Tony Bak marketed a negative capacitance, unity gain microelectrode amplifier as well as glass microelectrodes (micropipettes). Earl Sandbek began marketing ear oximeters, hand-machined and -wired in his home basement shop. Harry Johnson produced a two-beam oscilloscope using Dumont's pioneer two-gun cathode ray tube, the first such oscilloscope available other than Dumont's until Tektronix introduced its Model 565, which is still a research staple today.

On the diagnostic imaging side, the subspecialty of angiography followed the introduction of the first image intensifier and the cut-film changer around 1955. Radionuclide scanning, the precursor to nuclear medicine, was first performed during the 1950s using special gamma cameras.

The prestigious Institute for Radio Engineers (IRE) which had formed special-interest groups to address such diverse applications as antennas, audio, and microwaves, formed a Professional Group on Medical Electronics and, in collaboration with the Instrument Society of America, initiated a regular series of annual conferences in 1948. The 1958 Conference on the Topic of Computers in Medicine and Biology drew more than 300 attendees to hear more than 70 papers. In 1961, the attendance exceeded 3000, and over 250 papers were presented. At the 1968 conference, 478 abstracts were presented. The multisociety Alliance that organized these meetings also published detailed proceedings that helped to disseminate this information further.

New leadership sprang up in the electrical engineering departments of many academic centers. One particularly noteworthy example was R. Stuart Mackay, a professor of electrical engineering at the University of California at Berkeley who, in 1951, established an R&D laboratory at what is now the University of California San Francisco Medical Center. He was primarily responsible for a prolific series of advances in electron microscopy, medical telemetry, closed-chest ventricular fibrillation, guard-ring tonometry, the application of ultrasound to the study of diver decompression, and early work on color displays for X-ray information.

On the therapeutic side, Dutch physician Willem Kolff introduced the first artificial organ (an artificial kidney) at the end of World War II. Kolff had developed the prototype units in his native land during the time of German occupation, using readily available natural sausage skins as the dialyzing membrane. At this time, the first organ transplants were being performed, and the surgical teams were wrestling with the practical nursing difficulties that resulted from the need to virtually destroy the body's natural immune system, leaving a "defenseless" patient who needed to be kept in near-total aseptic isolation. Interestingly, one of the earliest forms of patient monitoring was the "barrier nursing" systems that were constructed during this period to enable the nursing staff to measure the standard bedside indicators of temperature, pulse rate, and respiration from the nonsterile side of a window in the isolation barrier.

In 1948, Brattain and Bardeen, working under William Shockley at Bell Labs, codiscovered the physical phenomenon that led to the development of the transistor. In 1951, the Joint Commission on Accreditation of Hospitals (JCAH), the predecessor to today's JCAHO, was established. In 1955, Jonas Salk's oral vaccine for poliomyelitis became generally available, as did the birth-control pill. The first patient was treated using a heart-lung machine. Cournand, Richards, and Forssman shared a Nobel Prize in 1956 for their contributions to the advancement of cardiac catheterization. Werner Forssman had passed a urological catheter from a vein in his right arm into his right atrium when he was a surgical intern in 1929 to first demonstrate the feasibility of this technique. In 1956, heart-lung bypass technology received a significant boost with the development of the membrane oxygenator, which allowed extracorporeal circulation of the blood for days, or even weeks, without toxicity or hemolysis. In 1958, physician-engineer Forrest Bird introduced his Universal Medical Respirator, a little, green box known far and wide as "the Bird." It was the world's first highly reliable, low-cost, mass-produced medical respirator. The "Baby Bird" respirator, introduced in 1970, would quickly reduce worldwide mortality for babies with respiratory problems from 70% to 10%.

The 1960s

During the 1960s, the era of the first "medical arms race," hospitals scrambled to be the first to have the most glamorous new equipment and could not get enough of it. The amazing promise of transistorized electronics ignited research into a whole new range of medical devices. Dr. William Chardack, assisted by electrical engineer Wilson Greatbatch, implanted the first internal cardiac pacemaker. The battery life for these early units was only 12–18 months. Other, well publicized "medical marvels" included ingestible endoradiosondes that radioed diagnostic information from the human alimentary tract and fetal heart monitors that promised to reduce or eliminate many of the risks to the unborn child. In 1964, Dr. Bernard Lown introduced the DC defibrillator to replace the less effective AC defibrillator that first had been used by Zoll in 1956.

In 1961, the nation's growing "Medical Technology Imperative" was given a boost by President John F. Kennedy's call for the U.S. to put a man on the moon before the end of the decade. Creation of a federally funded (National Heart Institute) Artificial Heart program in 1964 drew upon the strong technological capabilities of the nation's aerospace contractors. Controversy over the relative promise of artificial organs and transplanted organs continued long after this funding was cut off in the 1980s.

Initiation of the federally funded National Institutes of Health Program of Grants to Universities for Biomedical Engineering Training in 1965 was supplemented by a series of nationwide conferences on a wide variety of biomedical engineering topics such as "multiphasic screening, sponsored by the National Academy of Engineering (NAE). During this period, there was also a considerable amount of activity in many academic centers in areas more related to biophysics and biomathematics. The leaders of this research often had credentials in medicine as well as engineering. According to articles written at the time, there were more than 255 biomedical engineering projects that were supported by research grants from the National Institutes of Health. The Institute of Radio Engineers (IRE) changed the name of the Professional Group on Medical Electronics (PGME) to the Professional Group on Biomedical Engineering (PGBME).

The specialty of clinical engineering arose in the late sixties. It was intended to differentiate the new device safety-related activities from the maintenance and repair activities of in-house biomedical engineering groups.

At this time an amazingly broad range of projects was underway in in-house biomedical engineering R&D laboratories. The contents of a textbook on advances in biomedical engineering, published in 1967, listed such diverse topics as lasers in medicine, dialyzers, hyperbaric oxygen therapy, cardiopulmonary resuscitation, cryogenic surgery, orthopedic prostheses, silicone encapsulation for implants, heart valves, pacemakers, echocardiography, medical telemetry, medical thermography, fiber-optics for endoscopes, fetal heart monitoring, and medical computing. In-house biomedical engineering groups were also beginning to provide limited, on-request, maintenance services for many of the items in the proliferating inventories of medical instruments. They did not provide the elaborate documentation of these services that would be required later. Some in-house groups even provided repair services to equipment manufacturers.

The first multichannel clinical laboratory analyzers provided the opportunity for the hospital to profit from a medical device. Marketing materials provided with the equipment encouraged hospitals to invite the local press to witness and videotape the fascinating technology of bubbles moving through the multiple glass tubes of the new Technicon multichannel auto analyzers. This footage of a medical marvel would then be presented on the local six o'clock news. The financial arrangements were also attractively structured, in a manner similar to a Coke machine lease. The hospital paid for the analyzer on a simple pay-per-use basis and added a fee to guarantee excellent profitability. The more tests the hospital did, the more money the hospital made!

In 1968, the nation's first investor-owned hospital was opened in Nashville by the Hospital Corporation of America (HCA), the company that eventually, under the leadership of Rick Scott, would own more than 800 of the nation's hospitals. Other for-profit hospital companies soon followed HCA's lead. Other events taking place in the background during this period included the following:

- 1964: Dr. Michael DeBakey performed the first coronary artery bypass (CAB) operation
- Mid-to-late 60s: Preliminary work was already being performed on the risk of fibrillation to patients undergoing cardiac catheterization
- 1967: Christian Barnard performed the first human whole-heart transplant in Cape Town, South Africa.
- 1968: Dr. Denton Cooley implanted a short-lived, non-FDA-approved artificial heart.
- 1969: A four-university network of large computers, known as the Advanced Research Projects Area Net (Arpanet) was created by the Department of Defense. This would later expand and develop into the Internet.

The 1970s

Around the beginning of the 1970s, an ingenious and plausible, but only theoretical, electrical safety scare jump-started a new specialty called "clinical engineering." AAMI had already established a certification program for biomedical equipment technicians in 1970 and in response to these widely publicized electrical safety concerns, decided in 1973 to create a similar certification program for this new type of specialist who was being called a "clinical engineer."

Community hospitals boomed during this period. The number of hospitals across the nation peaked during this decade at 5875, about 16% higher than the number of hospitals existing today. There was a concerted effort to establish networks of specially equipped trauma centers in the major metropolitan areas. There also was growing concern about the runaway cost of health care and a variety of new restraining concepts were introduced, including Regional Medical Programs (RMPs), Certificate of Need (CON) requirements, and the increasing use of Health Maintenance Organizations (HMOs).

A series of professional meetings was held to discuss increasing concern about the theoretical possibility that catheterized patients who have external pacing leads or other conductive pathways directly to the inner wall of the heart are, theoretically at least, vulnerable to potentially lethal ventricular fibrillation induced by the very small "leakage" currents that are sometimes present in the various pieces of equipment that often surround them in the intensive care areas of the hospital., Consumer activist Ralph Nader dramatically described this possibility in an article in the 1971 issue of the *Ladies Home Journal*. He quoted several authoritative sources from the various preceding professional meetings, but in the article extended Dr. Carl Walter's original estimate of the number of deaths per year actually occurring from this theoretical scenario from 1200 to 5000.

Largely as a result of this adverse publicity, a proposal to amend the National Electric Code was presented at the Annual Meeting of the National Fire Protection Association (NFPA) in San Francisco in 1972. This proposal would have required that all power provided within a health care facility be provided by way of special, expensive isolated power systems (IPSs). The proposal is voted down in a dramatic showdown on the floor of the meeting after a concerted effort was made to rally opposition to the proposal by a coalition of representatives of the American Hospital Association, the staff of the newly active, not-for-profit Emergency Care Research Institute (ECRI), and a few biomedical engineers from several large teaching hospitals. The intense debate about the real value of isolated power systems would drag on for six or seven more years before it was finally laid to rest by a ruling from the board of directors of the NFPA. In the meantime the originally-thought-to-be-vulnerable catheterized patients were effectively protected by the simple expedient of shielding the exposed ends of the external conductors. A byproduct of this episode, however, was the discovery that the existing quality of the maintenance of the typical hospital's continually expanding inventory of electronic equipment was inadequate. As a result, a new high-intensity focus on equipment maintenance and safety checking was born.

Although the electrical safety scare was quickly shown to be only theoretical, new regulations addressing this potential threat appeared. In 1974, the JCAH issued new standards that prescribed quarterly documented electrical safety tests for all of the facility's patient care equipment. In California, the Department of Health issued new requirements as part of its new Title 22 regulations introducing the soon-to-be-obsolete concept of the "electrically susceptible" patient, and a host of related tests. Several other jurisdictions followed suit. The California regulation and many of the others created at that time persist today and are still enforced in only slightly modified form. This issue created the new specialty of documented electrical safety testing which, in turn, lead to the establishment of several new independent biomedical service organizations.

This same issue did not immediately lead to the widespread creation of more in-house biomedical engineering groups in the smaller hospitals, but it did prompt the W.K. Kellogg Foundation to award a series of seed money grants which over the next five or six years resulted in the formation of about 20 not-for-profit clinical engineering shared service programs. The shared service concept is based on time sharing a critical mass of trained technical staff among a geographical cluster of the smaller community hospitals that otherwise would not be able to afford an onsite technical capability.

The 1970s also saw the beginning of the widespread introduction of the so-called high-tech (i.e., digital, computer-controlled) medical devices. Tomographic scanning had been a part of basic nuclear medicine for some time when the British company EMI introduced the first commercial tomographic scanners for radiography in 1972. Gordon Hounsfield's original computed tomography (CT) scan took hours to acquire a single slice of image data and more than 24 hours to reconstruct these data back into a single image. CT scanners were quickly recognized as an enormous step forward in the area of radiographic imaging and by 1977, when the General Electric Company introduced its line of CT scanners, every hospital wanted one, in spite of the $1 million-plus price tag.

Other examples of new technologies that were being incorporated in a more or less continuous stream to the hospital's continually expanding clinical equipment inventory during this period included anesthesia machines that had been revolutionized by the incorporation of electronic monitors and more reliable vaporizers, more sensitive image intensifiers (developed for use in the jungles of Vietnam) that allowed smaller exposure levels during fluoroscopy examinations, and long-life nuclear batteries for use in implantable cardiac pacemakers.

Other events taking place in the background during this period included the passage in 1973 of the Williams-Steiger Act creating the Occupational Safety and Health Administration (OSHA); the Medical Device Amendments of 1976 to the Food, Drug and Cosmetics Act, which extended the FDA's jurisdiction to medical devices and lead to the formation of the Bureau of Medical Devices; and in 1977 the first *in-vitro* fertilization which resulted in the birth of Louise Brown in 1978. The end of this decade also saw the availability of the first commercial, off-the-shelf personal computers.

The 1980s

During the 1980s, the widespread adoption of high-tech medical devices accelerated and collided with increased pressure to control health care costs. In-house programs encounter the outsourcing challenge. A landslide election victory for the Republicans in 1981 signaled a more determined effort to control runaway health care costs, and the federal government introduced the concept of fixed-price reimbursement for Medicare and Medicaid patients. The capitation arrangement used a complex price list that is related to the diagnostic-related group (DRG) associated with the patient's complaint when admitted to the hospital. In spite of this effort health care cost inflation continued and insurance premiums rose by 90% between 1981 and 1984.

Hospitals that had purchased high-tech devices such as CT scanners began to experience "repair-ticket shock" as they had to replace high-cost parts such as array processors. The manufacturers began to push fixed-price, full-service contracts as a way of "insuring" against these financial risks. They typically portrayed the high cost of these contracts as good value relative to the revenue lost when the device goes "down."

U.S. Counseling Services introduced "maintenance insurance" in 1982 as a lower-cost alternative to manufacturer full service contracts. The cost was typically at least 20% less than a comparable manufacturer contract. Similar offerings by other insurance companies quickly follow. This concept would bring about significant changes in the medical equipment-maintenance aftermarket.

The "time and materials" feature of maintenance insurance, which unlocked the user from using the services of a predetermined service provider, also opened up a significant market opportunity for independent (i.e., non-manufacturer) service. Several companies specializing in maintaining the high-cost, high-tech devices – particularly CT scanners – were formed and quickly became nationwide competitors to the manufacturers.

During this period, more and more hospitals followed a general nationwide business trend and turn to "outsourcing" contractors for many of their nonclinical services. A number of service companies began to offer hospitals resident on-site clinical engineering services on a contract basis, which is perceived by existing in-house clinical engineering groups as a significant threat. Methods of measuring and enhancing productivity subsequently became a major topic of discussion at professional meetings.

On another front, growing concerns about the environment gave rise to ever more zealous regulations concerning the possible overexposure of health care workers to workplace chemicals such as ethylene oxide, formaldehyde, and several other disinfecting agents. In California, the passage of Proposition 65 extended this concern to the possible exposure to these substances of communities adjacent to hospitals, and all hospitals were finally driven to installing emissions-control devices on their exhaust fans. In New Jersey, the discovery of used syringes and other discarded injection paraphernalia on a local beach resulted in the Medical Waste Tracking Act of 1988. Even though the beach debris was eventually attributed to illegal drug users, hospitals remained saddled with elaborate and costly requirements for the packaging, transporting, and offsite disposal of their infectious waste.

Most health care providers experienced financial difficulties throughout this decade. HMOs were the exception and were growing in a spectacular manner. Hospital closures reached an all-time peak in 1988, during which 88 hospitals closed their doors. Medical insurance premiums rose again in the four-year period between 1988 and 1992 by 74%. Nevertheless, the pace of medical technology acquisitions continued. Prototype magnetic resonance imaging (MRI) machines, which had been pioneered by Dr. Raymond Damadian, founder of Fonar, were cleared by the FDA and introduced to the hospital marketplace in 1984. They received the same favorable response that CT scanners had received earlier. Other high-tech devices also continued to be well received. Even though the release of Cyclosporin, the anti-rejection wonder drug, in 1983 put new life into the organ transplant programs, the artificial heart program continued. A new model, the Jarvik 7, had been implanted in 1982 by Dr. William DeVries.

In 1985, Dr. Sam Maslak, working with the Acuson Corporation, developed the 128-channel computed sonography platform which quickly became the "gold standard" for a growing list of ultrasound applications. Later advances during the 1990s dramatically improved the quality and diagnostic value of ultrasound images.

The 1990s

In the early 1990s, the American College of Clinical Engineering (ACCE) was founded [see Chapter 130]. It was then, and remains today, the only professional organization devoted solely to representing the interests of clinical engineers. Its mission statement reads as follows:

1. To *establish* a standard of competence and to promote excellence in Clinical Engineering Practice.
2. To *promote* safe and effective application of Science and Technology to patient care.
3. To *define* the body of knowledge on which the profession is based.
4. To *represent* the professional interests of Clinical Engineers.

During the 1990s, hospitals begin to look at information technology and technology management in general, but continued pressure to reduce labor costs eliminated many of the middle managers who would have been instrumental in making these strategic decisions. Marvelous new, but expensive, prescription drugs began to consume even more of the health care budget. Doing more with less continued to be the dominant theme in hospitals.

In 1993, health care reform became an important political issue but the process stumbled badly and the plan conceived by Hillary Clinton, the wife of then-President William Clinton, never came to fruition. Other "political" factors come into play. There is a widespread public outcry against HMO-related denials of service. Many of the alliances among hospitals began to fall apart because of disappointing cost-saving benefits. Erosion of working conditions in stressed hospitals contributed to shortages of skilled staff.

The assimilation of glamorous new technologies such as PET scanners, SPECT scanners, "Gamma Knives," and PACS networks continued unabated. Few hospitals appeared willing to change their past practices and institute-realistic technology assessments or to create an authoritative "technology manager" or "chief technology officer." The acceptance of some cost-saving technologies, such as endoscopic surgery and telemedicine, reduced the pressure somewhat.

The general trend to outsourcing continued and was reinforced in the clinical engineering area by Columbia's decision to outsource its large corporate in-house clinical engineering program to GE Medical. Some salesperson in one of the national outsourcing companies coined the indefinable term "asset management" and created yet another level of confusion in the marketplace and the clinical engineering community.

After many years of hostility to independent service organizations based on an assertion that only the manufacturer can adequately service high-tech medical devices, several of the manufacturers of medical imaging equipment declared themselves to be in the "multivendor service" business and offered contracts to service other manufacturers' products.

Significant consolidation took place in the independent service marketplace, resulting in fewer choices in some local areas and regions. Consolidation in the medical equipment manufacturing industry also began to accelerate, reducing the hospital's competitive choices and further diminishing the probability that the manufacturers will be willing to sell parts and technical information to their competitors in the after-sale-maintenance marketplace.

Toward the end of the decade, the prescription-drug manufacturers launch an effective campaign of advertising their new "wonder" drugs directly to the consumer. Their success in this area put further pressure on health care costs.

The FDA expressed concern about the quality of refurbished medical devices and in 1997 issued an advanced aotice of proposed rulemaking that reveals some confusion about their understanding of the role of equipment maintainers. It was rumored that this action was precipitated by the equipment manufacturers' desire to further handicap their independent competitors for the after-sale-maintenance business. The FDA stated that it intend to encourage an alternative voluntary registration program that would enable purchasers of used medical devices to clearly differentiate properly restored devices from those in the "spray and pray" category.

The Twenty-First Century

The public's perception of the health care system appears to be at an all-time low. There has been an extended round of whistle-blower suits that have resulted in a number of health care executives being convicted of defrauding the government's Medicare program. The number of Americans who have no medical insurance coverage is at an all-time high in spite of the fact that the nation's expenditures on health care as a percentage of the country's gross domestic product is also at an all-time high. This trend shows absolutely no promise of a change in direction.

The health care industry has recently been rocked by two devastating reports from the Institute of Medicine that are extremely critical of the overall quality of the health care services delivered by the industry, and its failure to date to incorporate modern business methods. Allegations of very high death rates and associated additional costs estimated to be as high as $29 billion per year resulting from avoidable medical errors are very troubling. At the same time, the public is being bombarded with promises of a medical revolution based on almost unbelievable technological breakthroughs such as diagnostic DNA chips, surgical robots, and a revolutionary, new class of clinical-information systems. The FDA has recently approved a completely self-contained artificial heart that will reportedly cost no more than $25,000 when manufactured in quantity and an ingestible capsule that will transmit back pictures of the patient's entire, intact digestive system.

While there still are important contributions to make in the traditional areas of medical physics, biophysics, bioengineering, biomedical engineering, clinical engineering, and maintenance of the medical equipment and devices, there are clearly major opportunities to contribute by creating a new specialty area that might be called "clinical systems engineering."

The experts tell us that a large percentage of medical errors occur because the systems that we have in place to deliver the various segments of health care are too complicated. We need help from professionals who have interdisciplinary skills and some kind of training in systems analysis. We also need technology that will simplify these systems rather than make them more complex. Both of these needs seem to represent a major opportunity for the technological disciplines to continue to make significant contributions.

Reference

Dyro JF. Focus on University Hospital & Health Sciences Center SUNY at Stony Brook Biomedical Engineering Department. *J Clin Eng* 18(2):165-174, 1993.

3

The Health Care Environment

Lee O. Welter
Health Care Consultant
Sacramento, CA

The health care environment can be viewed from multiple perspectives. A historical perspective and philosophical approach improves our sense of intellectual balance and helps us to progress without recreating many mistakes of the past.

A quote from Sir William Osler's *Aequanimitas and Other Addresses* is cited in the preface to Richard Gordon's book, *The Alarming History of Medicine* (Gordon, 1993):

"The philosophies of one age have become the absurdities of the next, and the foolishness of yesterday has become the wisdom of tomorrow, through long ages which were slowly learning what we are hurrying to forget, amid all the changes and chances of twenty-five centuries, the profession has never lacked men who have lived up to the Greek ideals." Despite dramatic changes in the science of medicine, the ethics and art of medicine have been largely sustained.

Ancient and Prehistoric Medicine

Evidence of primitive medical care extends as far back as our knowledge of human existence. Ice-Age cave drawings depicted witch doctors. A body from that era, found frozen in a receding glacier, was carrying pouches of medicinal herbs. Stone Age skulls revealed healed trephine holes (Duin and Sutcliffe, 1992).

King Asoka of Buddhist India, of several centuries BC, established medical schools and hospitals, even a veterinary hospital. Hospitals were first evidenced in ancient Egypt, with this practice later adopted by the Greeks the Romans, whose military hospitals ultimately came to be used for the destitute. A standard of professional ethics that currently applies was developed by Hippocrates, "the father of medicine." In ancient Greece, medical science tended to displace magic and witchcraft. Religion has continued to be a factor in medical practice.

Around 335 AD, the Church and its nursing orders took up this enterprise. Hotel Dieu in Paris, founded in 660 A.D., still survives and reveals architectural attention to vital features such as lighting and ventilation, ease of cleaning, and the placement of patients for ready nursing observation beyond the view of other patients. Early Arabic hospitals provided good nutrition, music, and even storytellers, as well as nurses and physicians. Bladder stones had been surgically treated by the ancient Arabs and Hindus. Due to the lack of adequate anesthesia and perioperative care, these and other surgeries were performed only when urgently needed. Until well into the nineteenth century, such operations were similarly conducted as speedily as possible since only about one third of those patients survived the stress and pain. After general anesthesia came into use, about two-thirds of anesthetized surgery patients survived.

Science Struggles Against Superstition and Tradition

Advocating a scientific basis for medicine, rather than traditional dogma, Roger Bacon, physician and scientist, was not well received by those in power. Among his many ideas was the development of corrective lenses (eyeglasses) to aid persons suffering presbyopia. He envisioned the invention of the powerboat, the flying machine, the mechanical hoist, and the horseless carriage. For his heresy, Bacon was punished by the Dominicans of the Inquisition with imprisonment for nearly the rest of his life. From the mid-fourteenth through the mid-eighteenth centuries, such persecution reigned, finally fading after 1836, when a young woman was drowned for practicing witchcraft. However, the attitudes and underlying human behavior that prompted and condoned these witch hunts have remained part of the human condition.

Social and political issues have remained prominent throughout the ages. Having discovered oxygen, Joseph Priestly had been experimenting with the potential of nitrous oxide gas as an anesthetic approximately a century before it was widely adopted for that purpose. His English neighbors, intolerant of such radical behavior and his unorthodox religious views, torched his home laboratory and forced him to flee in 1791 for the United States, where he became interested in Benjamin Franklin's experiments with electricity and neglected his medical developments.

Resistance to Change Is Human Nature

Resistance to change remains a constant of human nature. This has been particularly dramatic in the sphere of medicine. One striking example was the English discovery in 1601 of citrus fruit (containing vitamin C) to prevent scurvy, which had been killing the majority aboard long sea voyages. Rediscovered in 1747, this practice was not instituted by the British navy until 1795 and by its merchant marine in 1865. To this day, resistance to change is a factor that must be considered in every attempt to progress in practically every human endeavor.

Science and Humanity Gain Ground on Ignorance and Superstition

During the nineteenth century, the world and its medical domain began its accelerating progress. Treatment of the mentally ill, previously by torture to drive out the evil spirits, gradually became more humane. Johann Peter Frank wrote about the concepts of disease prevention and public health welfare near the beginning of the nineteenth century. Around that time, Edward Jenner fostered the innovation of vaccination to prevent smallpox. Xavier Bichat discovered the tissues of which the body is composed, and he recognized the unique characteristics of physiology. Physical diagnosis was advanced by Leopold Auenbrugger and Jean Nicolas Corvisart, and further extended by Rene Theophile Hyacinthe Laennec's development of the stethoscope and its clinical application.

Public Health versus Contagion

Public health regulations were introduced in Great Britain during the mid-nineteenth century. Dr. John Snow served a dramatic and vital public health need during the 1854 "Broad Street Pump Outbreak" cholera epidemic by identifying a particular public well as an infection source. He broke off the well's handle to prevent its use and published *On the Mode of Communication of Cholera* the following year. Around the same time, New York and Massachusetts made similar public-health and urban-planning recommendations. These initiatives considered fundamental physiologic and psychological needs and protection from contagion and accidents. Many of these recommendations had been advocated over many centuries as essentials for a healthful life.

The inflammatory and infectious nature of many diseases began to be recognized. Ignaz Philipp Semmelweis recognized the contagious basis for puerperal fever and advocated that handwashing by clinicians could largely prevent this deadly scourge. Embarrassed by Semmelweis, his contemporaries terminated his employment. Progress was often not cheerfully received.

Medical Science and Technology Progress Despite Opposition

For example, in the last half of that century, even though adequate anesthesia was available for surgery, some surgery continued to be performed without it. When Harvey Cushing advocated the routine use of blood-pressure measurement during anesthesia and surgery, the Harvard Medical School study committee concluded that it was unnecessary. Fortunately, anesthesia practitioners readily adopted Cushing's innovation. Other medical technology, such as the microscope, permitted further research and advances. Louis Pasteur discovered the existence of cells and developed the germ theory of disease. Joseph Lister extended this theory into the practice of antisepsis for surgery, acknowledging Semmelweis's contribution. Johannes Mueller, a professor at Berlin's University, promoted extensive medical research, encouraging many others to pursue such development in physiology, histology, pathology, the extensive application of the microscope, and the invention of the ophthalmoscope. His colleagues and successors advanced the scientific basis of medicine. From there, later devices such as the electron microscope and recording kymograph permitted advances in laboratory and medical science. Ultimately, the disciplines of genetics, immunology, antibiotic and pharmaceutical therapy, endocrinology, neurophysiology, neurology, and psychiatry evolved.

Special surgical tools represent the earliest medical instruments. Their initially primitive nature was matched by the relative crudeness and poor success of early surgery, which had prompted physicians to distance themselves from the practitioners of surgery. This gap between physicians (i.e., internal medicine) and surgeons gradually dissolved as the science and accomplishments of surgery advanced.

By the early nineteenth century, thermometers, anesthesia and blood pressure measurement devices, surgical instruments and furniture, and sterilizers were in use. More complex instruments such as the electrocardiogaph were developed and gradually introduced to clinical practice. The finger-on-the-pulse technique was extended by devices for measuring and for displaying the pulse wave-form. Refined microscopes and further laboratory measurement techniques promoted the development of clinical pathology and laboratory medicine. The power and importance of accurate, reliable medical instruments was emphasized in an 1890 *JAMA* editorial: "successful medical practice (requires) precision in diagnosis," which is dependent upon medical instruments. Instruments that were not easy to use did not become popular.

Hospitals Become Medical Technology Resources

Another powerful innovation evolved as progress in their science led surgeons to be accorded equal status to physicians (internists). Rather than remaining merely a place for sick people waiting to die, the hospital became a focus of medical technology and increasingly specialized practitioners. Joel D. Howell's book, *Technology in the Hospital – Transforming Patient Care in the Early Twentieth Century*, offers the following examples and comments (Howell, 1995). An apparently healthy patient with a broken leg was seen at the Pennsylvania Hospital in 1900, but no X ray was done, and the only laboratory test was a cursory urinalysis. Fifty-one days later, he was "cured" and left the hospital. His entire hospital record was a single page. About 25 years later, a similar patient required a hospital stay of only 21 days, but received an X-ray diagnosis and four urinalyses, and grew a chart eight pages long. This transformation in care is attributed to a greater belief in the scientific basis of medical practice and an attendant faith in objective data. In part, it reflected a greater social acceptance of the hospital's role in medical care.

Wilhelm Roentgen first described X-ray images in 1895, the same year that Frederick W. Taylor first published his writings about scientific management. Hospitals came to be recognized as candidates for the same improved management and information-handling systems as other businesses. Increasing patient loads led physicians to begin expecting payment for their hospital care. Hospitals also devised accounting methods.

Medical Informatics Is Initiated

Advances in hospital organization promoted such advances and triggered new ideas. Boston surgeon, Ernest Amory Codman, Harvey Cushing's medical school colleague, argued for outcomes studies to improve surgical management. Dr. John Shaw Billings, physician director of the Army Surgeon General's Library (later the National Library of Medicine) assisted with the 1880 and 1890 census. He encouraged Herman Hollerith to develop a machine for tabulating population and similar statistics. Invented in 1882, these machines processed the data for 62 million people in the 1890 census. Calculators and then Hollerith card systems were used initially in 1887 for health and vital statistics in Baltimore and New York City; then by the Surgeon General's Office. Hospitals began using this system for accounting and clinical data. Hollerith's Tabulating Machines company, started in 1886, later became International Business Machines (IBM).

Concurrently, the length and complexity of patient charts grew, with increased specialization and departmentalization leading to fragmented, discontinuous records. Subsequently, careful design of specialized forms and charts was undertaken. Other infrastructure improvements improved the availability of medical care: Improved transportation has permitted doctors to see ten times as many patients as they could 80 years ago. The patients could now travel to the doctor (or hospital) rather than requiring the doctor to travel to the patients.

During the 1920s, life insurance companies began using blood pressure, spirometry, urinalysis, to supplement tradition medical assessments of risk. Standards were developed for such measurements and for the differentiation of normal-versus-abnormal values. The importance of precise measurement and documentation became recognized, and the concept of self care was promoted. The environmental and industrial contributions to disease gained recognition.

Medical and Hospital Management Evolve

The Mayo Clinic and Hospitals, in Rochester, Minnesota, which were founded in the late 1880s by Dr. William Worall Mayo and his two sons, became a prototype multispecialty clinic. Its fame is well deserved, with many firsts to its credit, including iodine treatment for goiter, the first blood bank, and the discovery of cortisone. Among other successes, it emphasizes quality of nursing care as well as educational programs for its staff.

The most dramatic changes occurred in surgery, for which hospital developments created an apt environment that internists also adopted. Even though medical care and surgery still could have been provided in patients' homes, hospitals provided good electric lighting and specialized equipment, as well as trained staff. These improvements promoted an increase in the volume of surgery. For example, tonsillectomy could be justified, sometimes "just because the tonsils were there."

Management guru Peter Drucker recognizes that, despite its shortcomings, the evolving hospital organization itself was an innovation that improved medical care (Drucker, 1980). The Great Depression gave management, or business management, a negative connotation, so that Raymond Sloan's study of hospital management led to the discipline called *hospital administration*. Drucker notes that the science of management cuts across every enterprise (Drucker, 1999). He sees favorable evolution in this discipline, as outlined below.

Cost and Quality Issues Arise

During the 1930s, health care accounted for about 1% of the gross domestic product (GDP) in the Unites States; today that figure is about 15%. The hospitals of the 1940s were very labor-intensive, but hospitals now have tremendous investment in expensive high-tech equipment as well. They have become capital-intensive, but specialized personnel are needed to operate and manage these systems. Concentration on special skills is essential for performance improvement, but "only massive increases in hospital productivity can stem the health care cost explosion."

The United States' National Academy of Sciences publication, *Organizational Linkages —The Productivity Paradox*, examines the lack of improved productivity commensurate with the massive expenditures on information technology (IT) in the late twentieth century (Harris, 1994). One of its conclusions is that operations improvement is the most effective IT benefit. The greatest opportunity for this in medicine lies in streamlining the capture of data and in providing timely, relevant information to aid decisions at the point of care.

Medical care's focus upon fighting disease is shifting to the maintenance of physical and mental health. Accompanying this is a shift in emphasis on information technology from the technology to the information. The site of care is increasingly migrating outside of hospital walls.

Historically, our current Information Revolution is the most recent of four. The first was the invention of writing in Mesopotamia, more than 5000 years ago. The second was the invention of the written book in China, around 1300 BC, and in Greece, around 500 BC. The third, between 1450 and 1455, was the adoption of the printing press, engraving, and movable type.

Drucker's observations above are fascinating, but they neglect to mention one important factor. A popular assumption is that improving medical capabilities and quality must lead to increasing costs. Even though medical innovation may account for about a third of the large increase in the United States' medical expenditures, another major factor has not been realistically addressed.

Economic Consequences of Regulation and Good Intentions

During World War II, the U.S. Government imposed wage controls in an attempt to counteract the market forces of high labor demand and low supply. One loophole was that fringe benefits were not limited. Consequently, the practice of employer-provided prepaid medical benefits (with favored tax treatment) became widespread. Patients and doctors were consequently freed from cost constraints in delivering medical care. The inevitable economic consequences were only accelerated in the mid-1960s, when the introduction of Medicare and Medicaid further severed the direct link between consumption and payment.

Economic laws of supply and demand cannot be repealed; rationing is the alternative when price controls prevent supply from increasing in response to demand. This fundamental issue is still not being addressed realistically. In the United States, government and corporate payment intermediaries have been using clumsy forms of rationing since the mid-1980s Regardless of the specific payment mechanisms, improved costs as well as improved quality must both continue to be sought because available resources are finite.

The Need for Modern Management

The lack of cost constraints had other adverse consequences. V. Clayton Sherman, a hospital management consultant, prefaces his book (Sherman, 1999): "in the early 1980s . . . . Health care as an industry was in a time warp, notably behind the advance of the management profession and seemingly oblivious to its direction. . . . we're still catching up to the state of the management art. . . ." Unlike the growth era of the medical device industry from 1965 to 1976, current constraints, beginning in 1983 with limited payments, require new devices to improve costs and quality.

Some such constructive movements have already commenced. Hospitals and other businesses are "unbundling" noncore services. Outsourcing of noncore services is beneficial: The contractor can specialize, develop superior performance, and provide suitable career ladders for its employees. More clinical or core services are also being contracted. These include contracts with physician multi- and single-specialty groups, perfusion services, clinical engineering, and turnkey infusion systems.

Drucker writes, "innovation matters because ours is a knowledge-based society," and health care is heavily dependent upon information management (Drucker, 1999). His example of innovative hospital reengineering is that some hospitals are managing patients as they would emergency patients (who are unable to fill out lengthy forms) – eliminating most of the laborious and expensive admissions process.

Drucker notes that in order to increase knowledge-work productivity, the questions to ask are the following: "What is the task? What do we try to accomplish? Why do it at all?" The answers permit redefining the task and eliminating unnecessary work. Drucker credits the late Karl Bays, who made American Hospital Supply (later acquired by Baxter Labs) the industry leader in the 1970s by listening to customers and determining their needs.

Some institutions have already undertaken such efforts. American hospitals' claim of a "nursing shortage" ignores that over half of the nurses' time is spent with non-nursing tasks, such as paperwork for third-party payors. Some are relegating such tasks to ancillary personnel and are applying appropriate technology.

Medical Informatics Continues to Evolve

Gerhard Venzmer envisioned future extensions of medical informatics based upon his observation that ". . . in the Mayo Clinic . . . case histories of patients are . . . electronically assessed. . . . such a vast amount of data—there are forty thousand different diseases—that no doctor in the world could keep all their symptoms in his head (Venzmer, 1968). The computer, however, stores them all up."

Dr. Morris Collen sketches the recent development of computer-based medical informatics (Collen, 1995). In 1946, the first electronic digital computer was produced at the University of Pennsylvania. Computer biomedical applications began in the late 1950s and have gradually become more pervasive. By the late 1960s, significant progress had been made in the development and availability of computers and of their medical applications. Hospital and office information systems, clinical support systems, automated patient monitoring, and computer-based consulting began.

A 1968 issue of *Bio-Medical Engineering* broached this topic in its *Comment—Can We Avoid Computers?*: "Here almost everything will be new, and only active participation by doctors can ensure that what is devised bears any resemblance to what is really wanted.... Whether we like it or not, doctors are going to be confronted by computers in the very near future. Responsibility for whether this confrontation leads to progress or to frustration for both doctors and patients rests largely with the medical profession."

That challenge still applies. In 1991, the United States' Institute of Medicine published *The Computer-Based Patient Record,* outlining the essential features of an ideal system and encouraging its widespread adoption (Dick, 1997). While this development has tremendous implications for the quality and cost of medical care, the general adoption of such a system is not occurring as quickly as had been hoped.

Specialization and Regulation Bring Progress and Complexity

Progress in the business of medicine and in improving its quality also have increased its complexity, as reflected in a couple of publications. *Medical and Health care Marketplace Guide* (1997-98), names approximately 5000 entities that sell products and services to health care institutions. The FDA has reportedly registered over 16,000 medical device companies. Government regulations may comprise an appreciable portion of development costs for new devices and for delivering health care. *Health care Standards* (2000) lists over 500 entities that produce standards, laws, and regulations for medical systems and care.

The Health Care Environment as an Ecological System

More recently, NASA has catalogued the vital needs of its self-contained space station environment, necessarily incorporating many of the same environmental requirements of health care institutions. Some exceptions are the considerations for microgravity, extreme accelerations and unusual atmospheres, as well as the implications of those factors for certain other activities.

Hospitals must confront many conflicting needs and paradoxes to achieve a productive, safe, and healthful environment. Patients might be incapacitated by the effects of age, disease, treatments, or disorientation in strange surroundings. Those with compromised immune function must be isolated from those who might harbor virulent disease organisms. Mechanical ventilation support of respiration could predispose patients to pneumonia. Surgical and traumatic wounds and skin punctures could permit the growth and entry of bacteria.

Patients are at risk of infectious diseases that once were thought to be totally under control. Antibiotic-resistant organisms and reservoirs with compromised immune function have created hazardous strains of micro-organisms. They can cause diseases such as tuberculosis, meningitis, pneumonia, and other devastating infections that do not respond readily to commonly used medications. Since the 1970s, some hospitals have been transient reservoirs of deadly organisms which are extremely difficult to treat. Commonplace international travel has served to spread some of these diseases worldwide.

Clinical Engineering Emerges from the Tradition of Bioengineering

Numerous successful medical inventions have arisen from the work of individual scientists, many of whom were physicians. Some fascinating accounts include the development of cardiopulmonary bypass by surgeon Dr. John H. Gibbon, Jr. and his wife Mary Hopkinson Gibbon; cardiac catheterization by Dr. Werner Forssmann, further developed by Drs. Dickinson Richards and Andre Cournand; artificial organ development by Dr. Willem Kolff and colleagues; Dr. Paul Zoll's external cardiac pacemaker and its subsequent refinement for transvenous internal pacing. With the advances of science and enabling technologies in the late 20th century, engineers were more frequently a part of these developments.

While such inventions have continued to arise, another series of events influenced the role of engineering in medicine. A somewhat exaggerated fear of hospital electrocution arose during the 1960s, prompting hospitals to welcome knowledgeable engineers and technologists to help resolve this perceived institutional problem. In the process, additional benefits of including such personnel in the hospital management and medical care team came to be recognized. Many pioneer clinical engineers began to write their own job descriptions: This is an ongoing process with many opportunities and challenges to meet.

Effective Medical Systems Planning Requires Profound Knowledge

A sound knowledge of basic science and clinical concerns is essential to evaluating, planning, implementing, and maintaining medical diagnostic and therapeutic technologies effectively and safely. For example, despite thorough warnings, many individuals and institutions have been surprised by the combination of extremely powerful magnetic resonance imaging (MRI) magnets and ferromagnetic objects in close proximity.

Specialized clinical facilities might have particular needs, such as the filtration, temperature and humidity, and high air flow of surgical suites. At times, regulations might be out of synchronization with realistic current needs. Many operating rooms reveal residuals from the time when flammable and explosive anesthetics required special equipment, furnishings, and supplies. The fairly recent adoption of convective patient heaters and other technology helps to protect surgical patients from hypothermia in the cool operating room. At the same room temperature, barrier gowns worn by the surgical team could cause the professionals to be uncomfortably or dangerously warm: A bead of sweat teeming with bacteria could drip into a wound.

Therapeutic oxygen or anesthetic gases could become contaminated or substituted by another gas in error. The stored energy of high-pressure gases could cause barotrauma or could be released explosively to create deadly projectiles. The "fire triangle" of fuel, oxidizer, and ignition source could combine unexpectedly with devastating consequences in health care environments.

The Surgery Suite is a Major Opportunity for Improved Systems

Other facets of the operating room environment need improvement. The typically high noise level of surgical suites is a source of stress and an impediment to clear communication. While lighting is important and generally adequate, further improvement would be welcome. Natural light is not sufficiently used. Most operating-room devices are not designed to fit compactly or to operate well in systems with the commonly found multitude of such equipment.

The disposal of wastes is a pressing environmental concern, especially when medical waste could harbor virulent disease organisms or toxic materials. Vermin and other pathogens can be particularly dangerous within a hospital or clinic.

Safety and Security are Increasingly Vital Issues

Hospital planning and personnel training must consider the special concerns of security and electrical and fire safety. All personnel must be alert to the vulnerability of patients to these hazards. Many such considerations are addressed in *Environmental Health and Safety in Health-Care Facilities* (Bond et al, 1973).

Current controversies include a concern for variations in the incidence of particular treatments or surgeries, and in outcomes. Inefficiencies and errors in medical records and treatment have prompted programs for studying and improving these problems.

Science and Humanity Continue to Battle Superstition and Tradition

John C. Burnham, in *How Superstition Won and Science Lost* (Burnham, 1987), recognizes that ignorance and even well-intentioned social influences can adversely affect the progress of science and medicine. He writes of the ongoing propaganda crusade that began during the nineteenth century: "Like the leaders of superstition in bygone centuries, they (the media) were attempting to assert their authority against the authority . . . of the popularizers of a reductionistic-rationalistic science."

Ignorance, misinformation, and misconception will continue to be integral to the human condition, despite our best efforts and intentions. Often, superstition, ignorance, and intentional misinformation are used for unsavory purposes. The web site "Junkscience.com" and similar publications serve as helpful contemporary resources on this topic. Junkscience.com—"All the junk that's fit to debunk"—debunks bad science that is used by personal injury lawyers, social and environmental activists, power-hungry government regulators, politicians, cutthroat businesses, and overly ambitious scientists.

Ongoing Challenges

Previous beneficial innovations have not been quickly adopted, and those now being promoted are likely to face similar obstacles. Among current challenges are those promoted in 1968 by Venzmer (1968), who concluded, ". . . the most urgent task confronting medical science . . . should be the application of the counterbalance to the increasingly unnatural mode of life to which modern man is condemned in this technological age. . . ." Our lifestyle and personal behavior choices are major determinants of health.

The identification of population disease trends can now be supplemented by increasingly detailed knowledge of individual genetic influences. The tremendous benefits to be gained by aggregating individual health data in data warehouses for analysis must be balanced with the need to protect individual privacy.

Mechanisms must be implemented to better link the demands for medical resources with their supply. Feedback mechanisms regarding consequences (or their likelihood) are needed so that individual can make behavior choices.

Conclusion

Regarding the conflicts among social needs, professional ethics, realistic economics, and political regulation, Gordon (1993) concluded his book fittingly: "The intelligent reader will have grasped that the potential of medicine is infinite, the demands on medicine must be unrestrained, but the resources for medicine are limited. Unless a fearless politician strikes a non-political compromise between all three, the history of medicine, like the history of the world in *1066 and All That*, will come to an abrupt end. The history of medicine will not come to an abrupt end, but, like most of our predecessors, we must continually face the challenge of obstacles to our bettering of the human condition.

References

Bond RG, Michaelson GS, DeRoos RL. *Environmental Health and Safety in Health care Facilities,* New York, Macmillan Publishing, 1973.
Burnham JC. *How Superstition Won and Science Lost—Popularizing Science and Health in the United States,* New Brunswick, NJ, Rutgers University Press, 1987.
Collen MF. *A History of Medical Informatics in the United States—1950 to 1990,* Bethesda, MD, American Medical Informatics Association, 1995.
Dick RS, Steen EB, Detmer DE, (eds) *The Computer-Based Patient Record: An Essential Technology for Health Care, Revised Edition,* Washington, DC, National Academy Press, 1997.
Drucker PF. *Managing in Turbulent Times,* New York, Harper & Row, 1980.
Drucker PF. *Management Challenges for the 21st Century,* New York, Harper Business,1999.
Duin N and Sutcliffe J. *A History of Medicine—From Prehistory to the Year 2020,* New York, Simon & Schuster, 1992.
Gordon R. *The Alarming History of Medicine,* New York, St. Martin's Press, 1993.
Harris DH (ed). *Organizational Linkages: Understanding the Productivity Paradox,* Panel on Organizational Linkages. Washington, DC, National Academy Press, 1994.
Health care Standards, Plymouth Meeting, A, ECRI, 2000.
Howell JD. *Technology in the Hospital—Transforming Patient Care in the Early Twentieth Century,* Baltimore and London, The Johns Hopkins University Press, 1995.
Medical and Health care Marketplace Guide, 1997-98, 13th ed, Dorland's Biomedical—sponsored by Smith Barney Health Care Group.
Sherman C. *Raising Standards in American Health Care—Best People, Best Practices, Best Results V.,* San Francisco, Jossey-Bass Publishers, 1999.
Venzmer G. *Five Thousand Years of Medicine,* translated by Marion Koenig, New York, Taplinger Publishing Company, 1972.

Further Information

Bronzino JD. *The Biomedical Engineering Handbook, 2nd ed.* Boca Raton, FL, CRC Press, 2000.
Caceres CA, Yolken HT, Jones Piehler HR, Schick JW. *Medical Devices: Measurements, Quality Assurance, and Standards, ASTM Special Technical Publication 800,* West Conshohocken, PA, ASTM, 1983.
Fisher JA. *The Plague Makers—How We Are Creating Catastrophic New Epidemics and What We Must Do to Avert Them,* New York, Simon & Schuster, 1994.
Gelijns AC (ed). *Medical Innovation at the Crossroads, Volume III, Technology and Health Care in an Era of Limits,* Washington, DC National Academy Press, 1992.
Kotter JP. *Leading Change,* Boston, Harvard Business School Press, 1996.
Pickstone JV. *Medical Innovations in Historical Perspective,* New York, St. Martin's Press, 1992.
Poynter FNL, Keele KD. *A Short History of Medicine,* London, Mills & Boon, 1961.
Rogers EM. *Diffusion of Innovations, 4th ed.* New York, Free Press, 1995.
Weisse AB. *Medical Odysseys—The Different and Sometimes Unexpected Pathways to Twentieth-Century Medical Discoveries,* New Brunswick, NJ, Rutgers University Press, 1991.

4

Enhancing Patient Safety: The Role of Clinical Engineering

American College of Clinical Engineering
Plymouth Meeting, PA

Opportunities for enhancing patient safety exist within the health care delivery system. Individuals and groups throughout the system are actively pursuing these opportunities. The profession of clinical engineering has a unique role to play in enhancing patient safety, particularly with regard to medical technology as it is applied in our health care delivery system.

The American College of Clinical Engineering was established in 1990 to represent and advance the profession of clinical engineering. ACCE has prepared this white paper to share its vision of the role of clinical engineering in enhancing patient safety. We address it to our colleagues within the health care delivery system.

Our message is that clinical engineers, individually and collectively, will continue to work to improve patient safety. Our message is also that ACCE is taking a leadership role in pursuing opportunities for clinical engineering to contribute even more effectively in the area of patient safety. We seek the cooperation and active support of all our clinical, technical, and administrative colleagues in this pursuit.

Patient Safety

Over the past decade there has been steadily increasing concern regarding patient safety. Results from *the Harvard Medical Practice Study* focused attention on the incidence of adverse events in health care delivery (Brennan et al., 1991; Leape et al., 1991). Subsequent studies provided additional quantitative data (Thomas et al., 2000). In a landmark report, "*To Err Is Human: Building a Safer Health System,*" the Institute of Medicine (2000) estimated that medical errors cause 44,000 to 98,000 deaths annually in U.S. hospitals.

One response to the concern for patient safety has been an effort to improve reporting systems for medical errors–including "near misses" in which errors could have caused death or injury but, fortunately, did not. The objectives are to learn from these errors and to find ways to prevent them from recurring. Such efforts have been hampered by controversies regarding mandatory (as opposed to voluntary) reporting, confidentiality of data, and professional liability.

Another response has been to identify existing knowledge and resources that can be applied to the problem. In some cases, the relevant knowledge and resources lie outside the health care delivery system. For example, past improvements in the safety of commercial aviation may serve as a model for health care delivery. In other cases, however, valuable knowledge and resources may be found within the health care delivery system itself. In this regard, improved cooperation and coordination among health care professionals is considered vital.

Of immediate impact on hospitals and other components of the health care delivery system are new patient safety standards promulgated by the Joint Commission on Accreditation of Healthcare Organizations (2001). The new patient safety standards cut across disciplinary boundaries in an attempt to make safety a fundamental principle of patient care. These standards are linked to long-standing safety-related standards regarding infection control, the environment of care, and other disciplines.

In a follow-up to its earlier report, the Institute of Medicine (2001) has published "Crossing the Quality Chasm: A New Health System for the 21st Century." This report places patient safety within the broader context of quality and calls for a health care delivery system that is safe, effective, patient-centered, timely, efficient, and equitable. Underlying many of the report's specific recommendations is a concern that the health care delivery system is fragmented in ways that produce the opposite of these qualities.

What can be done now to enhance patient safety? First, we can improve our ability to learn from mistakes. Skillful investigation of incidents and responsible sharing of data will help us to identify the root causes of error. Second, we can improve our ability to anticipate mistakes. Cooperation among health care professionals of all disciplines will augment our capacity to probe the weaknesses in the system.

Finally, we can improve the health care delivery system itself. All of us within the health care delivery system can work with other stakeholders to reconfigure the structure of the system, to realign the incentives that guide the system, and to build the resources the system draws on to produce high quality patient care.

Medical Technology and Patient Safety

A defining characteristic of the modern health care delivery system is the ubiquitous use of medical technology. Broadly speaking, medical technology includes not only medical devices, drugs, and biologics, but also the medical and surgical procedures that they enable and the organizational and support systems within which they are used. Diagnosis, monitoring, treatment, and rehabilitation all rely on complex and sophisticated medical technologies.

In today's health care delivery system, the patient is at the center of an intricate network of clinicians, medical devices, and other elements of the system. Every interface between a human being and a machine contains opportunities for error: Information might not be accurately acquired, recorded, or communicated; necessary actions might not be safely and effectively carried out; and adverse events might occur.

The health care delivery system has been criticized for its "culture of blame" in which culpability for failure has been attributed to the human elements of the system: People make errors; therefore people must change their behavior to reduce errors. However, numerous researchers (e.g., Bogner, 1994; Cook, 1998; Perrow, 1999) have found that human errors are more generally associated with latent causes that are hidden within systems and processes. Current thinking places the responsibility for "human error" squarely on the shoulders of latent (i.e., root) causes that can be prevented only by adjustments to systems and processes.

Simplistic models of adverse events involving medical technology have been based on a dichotomy between "device failure" and "user error." However, more sophisticated taxonomies have been developed that recognize numerous sources of error with the potential for complex interactions among them (ECRI, 1991). To emphasize the multifaceted nature of medical technology application some authors employ the term "use error" rather than "user error" (CDRH, 2000; Hyman, 1995). The critical point is that errors associated with the *use* of medical technology should not be automatically ascribed to the *user*. More importantly, efforts to eliminate such errors should not focus on the user in isolation from the system in which he or she works.

In a recent report to Congress (Gardner and Flack, 1999) the U.S. Food and Drug Administration (FDA) stated that under the requirements of the Safe Medical Devices Act medical device manufacturers reported a total of 980 device-related deaths in 1998. In a presentation to the Association for the Advancement of Medical Instrumentation (AAMI, 2000), a representative of the FDA Center for Devices and Radiological Health stated that one-third of the 80,000 incident reports that it receives annually might involve medical equipment "use error." Medical technology is an integral component of the health care delivery system. Efforts to improve patient safety and the quality of health care delivery must take into account the omnipresence of medical technology.

The Clinical Engineering Profession

The American College of Clinical Engineering defines a clinical engineer as "a professional who supports and advances patient care by applying engineering and management skills to health care technology." Clinical engineering became a distinct profession during the 1960s in response to increasing use of medical technology in health care delivery. Since that time, clinical engineering has become a vital component of the health care delivery system, providing leadership in the safe and effective application of medical technology.

Throughout its history, clinical engineering has focused on medical devices as they are used in health care delivery settings: acquiring the appropriate equipment; inspection, maintenance, and repair; regulatory compliance; and related technical issues. Over time, clinical engineering has assumed a leading role in management of medical equipment during its entire life span. As a result, clinical engineers have become deeply involved in quality-improvement and risk-management activities.

Clinical engineers are essential members of multidisciplinary hospital teams investigating incidents in which a medical device might have contributed to injury or death. The clinical engineering perspective can be instrumental in identifying root causes and solutions. An understanding of equipment design principles can produce insights that go beyond the standard behavior of the device in question. An understanding of equipment operation and maintenance can draw attention to likely failure modes and the effect of support systems of device performance. An understanding of systems theory and human-factors engineering can shed light on the interaction between machines and humans.

Clinical engineers have also made contributions to patient safety beyond their own institutions in fields as diverse as anesthesia mishaps (Cooper et al., 1984), radio-frequency interference with medical telemetry (American College of Clinical Engineering, 1998), and remarketing of medical devices (Hatem, 1999). Clinical engineers have advanced the literature of patient safety (Hyman, 1994; Shepherd, 2000) and incident investigation (Bruley, 1994; Dyro, 1998; Shepherd and Brown, 1992). Building on this history, clinical engineers have an important and unique role to play in efforts to improve patient safety.

In the future, individual clinical engineers will continue to act within their own health care facilities to enhance patient safety on a day-to-day basis. Beyond this, the clinical engineering profession will continue to work at all levels within the health care delivery system to improve patient safety and the quality of health care delivery.

The American College of Clinical Engineering

The American College of Clinical Engineering was established in 1990 to represent and advance the profession of clinical engineering, both in the United States and internationally. It has acted both independently and in cooperation with other organizations to achieve its objectives.

For example, ACCE has worked with the U.S. Federal Communications Commission and the American Hospital Association to resolve radio-frequency allocation conflicts that threatened the performance of medical telemetry. Similarly, ACCE has worked with the U.S. Food and Drug Administration and the Association for the Advancement of Medical Instrumentation to establish voluntary standards for the remarketing of medical devices.

Since 1991, ACCE has also conducted numerous Advanced Clinical Engineering Workshops to educate clinical engineers in developing countries on the fundamentals of technology management. In addition, ACCE has recently assumed management of the clinical engineering certification program established in 1977 by the International Certification Commission on Clinical Engineering and Biomedical Technology.

The ACCE Board of Directors has established an ad hoc committee on patient safety. The committee advises the board of opportunities to advance patient safety in health care delivery. The board has identified several strategies that it will adopt as opportunities occur:

- To seek partnerships at national and international levels to pursue common goals for enhancing patient safety
- To provide formal endorsements of new strategies and methodologies designed to enhance patient safety
- To prepare a directory of publications and other resources for reference and application by health care professionals in enhancing patient safety
- To develop educational programs that improve the skills of health care professionals and their ability to participate in efforts to enhance patient safety
- To review the body of knowledge for clinical engineering certification for inclusion of human factors engineering, root-cause analysis, and other fields of study related to patient safety

The first tenet of the ACCE Code of Ethics calls upon clinical engineers "to strive to prevent a person from being placed at risk of injury due to dangerous or defective devices or procedures." The American College of Clinical Engineering pledges itself to that duty in the far-reaching context of patient safety and high-quality health care delivery.

References

Association for the Advancement of Medical Instrumentation (AAMI). Risk-management process growing more important to manufacturers. *AAMI News* 35(5):1-2, 2000.
American College of Clinical Engineering. ACCE partners with telemetry manufacturers: AAMI, ASHE, and the AHA to recommend spectrum allocation to the FCC. *ACCE News* 8(6):11-12, 1998.
Bogner MS (ed). *Human Error in Medicine.* Hillsdale, NJ, Lawrence Erlbaum Associates, 1994.
Brennan TA et al. Incidence of Adverse Events and Negligence in Hospitalized Patients: Results of the Harvard Medical Practice Study I. *N Engl J Med* 324(6):370-376, 1991.
Bruley ME. Accident and Forensic Investigation. In Van Gruting CWD (ed). *Medical Devices: International Perspectives on Health and Safety.* Amsterdam, Elsevier, 1994.
Center for Devices and Radiological Health (CDRH). *Medical Device Use Safety: Incorporating Human Factors Engineering into Risk Management.* U.S. Food and Drug Administration, Washington DC, 2000.
Cook R. *How Complex Systems Fail.* Chicago, Cognitive Technologies Laboratory, University of Chicago, 1998.
Cooper J et al. An Analysis of Major Errors and Equipment Failures in Anesthesia Management: Considerations for Prevention and Detection. *Anesthesiology* 60:34-42, 1984.
Dyro J. Methods for Analyzing Home Care Medical Device Accidents. *J Clin Engin* 23(5):359-368, 1998.
ECRI. *Medical Device Reporting under the Safe Medical Devices Act: A Guide for Health Care Facilities.* Plymouth Meeting, PA, ECRI, 1991.
Gardner S, Flack M. Designing a Medical Device Surveillance Network. Washington, DC; U.S. Food and Drug Administration; 1999.
Hatem MB. From Regulation to Registration: Safety and Performance Needs Drive Industry Consensus on Voluntary Servicing, Remarketing Controls. *Biomed Instr Technol* 33(5):393-398, 1999.
Hyman WA. Errors in the Use of Medical Equipment. In Bogner MS (ed). *Human Error in Medicine.* Hillsdale, NJ, Lawrence Erlbaum Associates, 1994.
Hyman WA. The Issue Is 'Use' Not 'User' Error. *Med Device Diagn Ind* 17(5):58-59, 1995.
Institute of Medicine: *To Err Is Human: Building A Safer Health System.* Washington DC, National Academy Press 2000.
Institute of Medicine. *Crossing the Quality Chasm: A New Health System for the 21st Century.* Washington DC, National Academy Press 2000.
Joint Commission on Accreditation of Health Care Organizations. *Revisions to Joint Commission Standards in Support of Patient Safety and Medical/Health Care Error Reduction.* Chicago, JCAHO, 2001.
Leape LL et al. Incidence of Adverse Events and Negligence in Hospitalized Patients: Results of the Harvard Medical Practice Study II. *N Engl J Med* 324(6):377-384, 1991.
Perrow C. *Normal Accidents: Living with High-Risk Technologies,* Princeton, NJ, Princeton University Press, 1999.
Shepherd M, Brown R. Utilizing a Systems Approach to Categorize Device-Related Failures and Define User and Operator Errors. *Biomed Instr Technol* 26:461-475, 1992.
Shepherd M. Eliminating the Culture of Blame: A New Challenge for Clinical Engineers and BMETs. *Biomed Instr Technol* 34(5):370-374, 2000.
Thomas EJ et al. Incidence and Types of Adverse Events and Negligent Care in Utah and Colorado. *Medical Care* 2000; 38(3):261-271.

For Further Information

For further information regarding the American College of Clinical Engineering and its efforts to enhance patient safety, contact:
American College of Clinical Engineering
5200 Butler Pike
Plymouth Meeting, PA 19462-1298
Phone: 610-825-6067
Web: www.accenet.org

5

A Model Clinical Engineering Department

Caroline A. Campbell
ARAMARK/Clarian Health Partners
Indianapolis, IN

Whether creating a clinical engineering department or managing an existing program, the clinical engineer must continuously consider and plan for the resources that are needed in order to perform the clinical equipment management services. Some of the resources that are fundamental to the clinical engineering department's success are staffing, space, test equipment, tools, communications equipment, training, and a computerized maintenance and management system. Because the ability to obtain some or all of these resources will likely be constrained, the clinical engineering manager must think creatively to fill the needs within those constraints. This chapter provides suggested approaches to these situations.

Staffing

The most critical of all resources is adequate and appropriate staffing. Unfortunately, a shortage of clinical engineers and biomedical equipment technicians has created a significant challenge for the clinical engineering managers. Enrollment in biomedical programs is down, and the sophistication of technology is up. These factors have created a high demand for a resource that is in short supply. To attract employees in this climate, the manager must survey the market to verify that the salaries and benefits being offered are competitive (Campbell, 2000). Several journals publish annual salary-survey data that can be useful in establishing appropriate salary structures for each position (Baker, 2000).

The exercise of staffing a clinical engineering department begins with the development of job descriptions, job titles, pay scales, and certification requirements (Pacela and Brush, 1993; Dyro, 1989a, b). Based on the job description, clinical engineering managers must determine the qualifications for the position. Next, the employer must find qualified applicants. The most common way to find qualified applicants is to use "outside" contacts, such as help-wanted advertising in newspapers, journals, web sites, specialized job-placement agencies, schools, colleges, military resources, regional biomedical societies, and nationwide societies. An "inside" search involves limited internal advertising of the position and the use of personal referrals for candidates. Finally, the clinical engineering manager must screen the applicants. Comparison of candidate qualifications to the position description is the obvious first step in this process, but there are other prescreening techniques such as employment testing. Interviewing is the most common way to screen candidates for job positions. Postinterview screening is a final step to help determine the best job-person match. It includes checking references and validating previous employment.

The mix and quantities of employees must be derived from the clinical engineering department's responsibilities. Typical clinical engineering departments employ clinical engineers (CEs), biomedical equipment technicians (BMETs), and administrative support staff. In departments where both CEs and BMETs are employed, the latter typically perform preventive and corrective maintenance, and the former performs pre-purchase evaluation, management of service contracts, risk management, quality control, education and training, and research and development (Glouhova et al., 2000). Administrative support staff is typically employed to answer telephones, bill for services, perform data entry, and maintain files. If the clinical engineering department has responsibilities related to radiation safety, employment of a medical physicist might be warranted. In general, one full-time equivalent is needed in order to to support 590 clinical devices or $2.5 million worth of equipment. Of course, applicability of these metrics depends on the experience level of the staff and the responsibilities of the department (Glouhova et al., 2000).

Training

Because of the tight labor market, the best job candidates will likely not be the perfect candidates. Therefore, the new employee's orientation and training program have become more important than ever. Additionally, provision of training has become an important staff-retention tool (Dyro, 1989).

Training can be negotiated into new equipment purchases to minimize the associated expense. It also can be obtained at cost from original equipment manufacturers or from independent training centers. Technical training seminars are also often offered in conjunction with professional society meetings. Service manuals must be obtained in order for the training to be beneficial. Teleconferences and Internet-based training provide a cost-effective methodology for training on certain topics.

Space

Space is often a difficult resource to obtain. Successful health care facilities generally allocate the most square footage and the most attractive real estate to patient-care areas. Space may be allocated in one large footprint or scattered throughout the institution. Whatever the case, consciously planning for utilization of the available space will always result in a more efficient environment than will allowing a space plan to develop haphazardly.

Periodic review of the clinical engineering department's short- and long-term objectives is critical to the department's success. All goals need to be reviewed to determine the impact of the environment on the department's ability to fulfill the objectives. Long-term goals might be directly related to space, such as creating the ambience within the department of a world-class service. Other goals can be directly related to provision of a service but can have space implications. For example, a short-term objective of establishing internal support of portable radiology units will require consideration of the width of pathways that will be traveled to get to the point of service. Controls also will be needed in order to minimize unnecessary radiation exposure. These space considerations are integral to fulfilling the objective of establishing internal support of portable radiology units.

Space is not a limitless resource, and hence space planning can include redesign of work processes to better fit the existing space. For example, a shortfall in storage space can be addressed by eliminating hard-copy paper documentation through use of a computerized maintenance-management system (Rice, 1997). A shortage of workspace also might be addressed by pushing certain work out into the patient-care environment. For example, ultrasound scanners frequently can be inspected in patient-care areas after hours. Without this constant balancing of space resources with business practices, the space inevitably will become inadequate and subsequently will impact the efficiency and effectiveness of the department.

If funding permits, an architect should be engaged to help plan and design the clinical engineering space. Usually an architect will be engaged only for a major renovation of space. If an architect is not available, those individuals who are responsible for renovation of space within the institution can provide some guidance for approaching space design. For a successful design, the clinical engineering staff will think about the workflow in terms of its functional needs that are related to space. This will include an exhaustive listing of physical resources, such as gas and water, that are needed for various tasks, as well as quantification of space needed, such as the number of linear feet required for storage of equipment manuals and equipment files. Table 5-1 lists a number of suggested physical resources to be considered in this planning exercise. Using this approach, the space can be designed in a methodical manner to match the needs while fulfilling all applicable codes. For example, if there is only one water source in the department, that is the location where servicing of dialysis machines will take place unless funding is available for additional plumbing. Other equipment requiring a water source, such as humidifiers or lasers, will also be serviced in that location. The workspace surrounding that water source then can be designed around the needs of that same equipment, i.e., parts-storage area and counter space for appropriate test equipment.

Storage space must be designed to accommodate storage of chemicals and test gases, parts, equipment, office supplies, and service documentation. In considering space arrangement, think about what items are always used in one work location, and consider permanent storage of those items in that location. Also consider the items, such as office supplies, that essentially are utilized throughout the workspace. These items might be stored in a central storage depot or interspersed throughout the workspace. Some storage areas may need to be secured. For example, the department's practice may be to restrict access to service documentation. Storage space for staging equipment is necessary. Although not uncommon, the practice of storing pallets of new equipment in a hallway likely violates local fire safety codes. The equipment-staging area ideally will be supplied with ample power and will contain adjustable shelving.

Table 5-1 Infrastructure resource considerations

Equipment work area	Resources
Anesthesia/respiratory therapy	• Phone and data outlet • Ample power outlets • Gas outlets (nitrous oxide, oxygen, air, vacuum) • Large sink and cup sinks • Parts storage
Dialysis	• Phone and data outlet • GFI outlets • Compressed air • Recessed water spigots with raised drains • Parts storage • Sealed-membrane floor with floor drain
Radiology	• Phone and data outlet • 50A, 220V single-phase power • Portable lead shield, lead-lined doors • 10-ft. ceiling • Wide-access door (48 in.) • Incandescent light on dimmer switch • Beam and hoist with 1-ton load capacity • Deep-utility sink • "Room in Use" light • Power for wall-mounted view box
General workbench	• Ample power outlets • Static mat • Electrostatic-free outlet • Durable countertop • Lockable drawers • Manometer mount • Vacuum • Air • Phone and data outlet

Since the core work of the clinical engineering department takes place at the workbench, this area deserves the most investment of attention and resources. The workbench design optimally will incorporate sufficient countertop space, adequate storage space for test equipment and tools, and appropriate resources such as vacuum, grounding mats, and access to sufficient quantities of electrical outlets. To make efficient use of the workbench space, shared resources can be mobilized by mounting on a cart. For example, a mobile solder station can be conveniently brought to the workbench, where a device is already disassembled, instead of bringing printed circuit boards to the stationary solder station.

The needs of a reception area are distinctly different from the equipment service area. The reception area will be used to receive the customers and business partners of the clinical engineering department. Comfortable seating in good repair is needed to accommodate these guests. Because the prestige of the department is partly dependent on the impressions of these individuals, the reception area must be a more traditional office space that is shielded from the noise of drill presses and equipment alarms. The design of the reception area typically supports the administrative work processes of the department and hence typically includes such things as computers, printers, fax machines, copiers, filing cabinets, and desk furniture. Much of the office equipment can be hidden in well-designed closets that are opened as needed for access. In this manner, the clutter of the workspace is minimized while accommodating convenient access.

Critical staffing shortages are facing many clinical engineering departments, and the impact of the work environment on recruitment and retention therefore must be considered. Although a functional space design is necessary for a successful department, creation of the desired ambience must be woven into that functional design. Lighting is a critical factor in creating a desirable ambience. If funding is available, artwork can be very effective in humanizing a technical environment. Mounting of personalized name plates on the workbenches of employees is an example of an inexpensive approach to creating an environment of mutual respect. Personalization of workspace is also a work-satisfaction issue that can be accommodated easily. Provision of small bulletin boards for the posting of family pictures allows for this personalization while avoiding the taping and tacking of things such as pictures and calendars to the walls. No matter how much effort and money is invested in creating the ambience, a dirty environment will always be seen as a dirty environment. Therefore, making arrangements to clean equipment before entering the department and for maintaining order within the department is critical.

Computerized Maintenance and Management System

The foundation of the successful clinical engineering program is the computerized maintenance and management system (CMMS). This database will house the information that the clinical engineering staff will use to make informed decisions and recommendations relating to equipment management. Careful selection of a computerized maintenance and management system will permit access to this information using a variety of queries and the ability to format the information in a useful manner. The CMMS also has evolved into an important process tool for organizing the work of the department.

Although some clinical engineering systems develop their own CMMS, there is a variety of products available on the market. Some of these products are designed specifically for clinical engineering and interface with common test equipment; others are more generic service-industry packages that can be adapted to suit the purposes of clinical engineering. At a minimum, the selected CMMS should permit keeping an equipment inventory with a variety of information including a unique numeric identifier, manufacturer, model number, and serial number, and it should have the capability to generate work orders on a scheduled basis and on demand. Other attractive features include the capability to track a parts inventory and parts utilization, financial tracking capability, and payroll functions. Often these various features are arranged in modules that interact with one another. A high degree of integration between modules is a desirable feature of a CMMS. For example, some systems contain a module for definition of device specific preventive maintenance procedures that can be married to specific devices in the inventory module. When a scheduled preventive maintenance work order for that specific device is generated, the appropriate preventive maintenance procedure appears in the work order module screen.

Access to the CMMS by the clinical engineering staff is desirable wherever work with equipment is performed. Therefore, transportability of the CMMS into the clinical care environment is also a desirable feature. This can be accomplished through connection to the hospital's backbone—wireless communications—or through upload/download of the database (or some portion) onto portable devices.

Test Equipment and Tools

The contents of the test equipment inventory will depend upon the kinds of clinical equipment supported by the clinical engineering department. However, every test equipment inventory will contain certain multipurpose items that will be used to support a variety of clinical equipment. At a minimum, the test equipment inventory will include a digital multimeter, oscilloscope, rheostat, electrical safety analyzer, and patient simulator with ECG and pressure capabilities. Large departments will find it necessary to acquire common test equipment such as multimeters and electrical safety analyzers for most, if not all, BMETs.

Many test devices are specific to clinical devices, e.g., electrosurgical unit tester, defibrillator output tester, and hence a thorough review of the clinical equipment inventory is required in order to determine the appropriate test-equipment inventory. Additionally, as clinical equipment is added to the support responsibilities of the department, a review of the requisite test equipment to fulfill those responsibilities is needed. Table 5-2 lists common test equipment that is needed to support various medical devices.

A set of hand tools also will be necessary to fulfill equipment-support responsibilities. Sharing of tools is not recommended as individuals tend to be particular about the order and condition of the tools that they use. Hence, the completeness and condition of the tool set tends to be a work satisfaction issue. The wise manager will budget annually for replacement of worn and lost tools. Specialized tools that are not frequently needed are more appropriate for sharing.

Communications Equipment

Communications is an integral part of every service industry (Maddock and Hertzler, 1999). The clinical engineering manager will be wise to make a list of the various types of communications that will occur, and then will incorporate appropriate communications equipment to facilitate that exchange. For example, telephone access at the workbench will facilitate conversation with a manufacturer's technical support group with the device under repair within easy reach. A wireless telephone will facilitate this same conversation for devices that cannot be removed from the clinical environment. Communications

Table 5-2 Suggested test equipment

	• *Oscilloscope*
General	• Function generator • Electrical safety analyzers • Multimeters • Capacitance meter • Transistor tester • Pressure gauges • Tachometer • Rheostat • DC power supply • Patient simulator • Frequency counter • Flowmeter • Calibrated weights
Specialty	• Ventilator tester • Respirometer • Gas analyzers • Electrosurgical unit tester • Defibrillator tester • Spectrum analyzer and antennas • Oxygen monitor • Laser sensor and power meter • Audiometric analyzer • X-ray phantoms • Radiation detector • Collimator test tool • Ion chambers • Resolution test tool

devices that will be considered include telephones, fax machines, and computers with Internet access. In this electronic age, the clinical engineering manager also must remember that sharing hard-copy documents continues to play an integral role in communications and therefore will remember to provide bulletin boards and mailboxes.

Communications equipment also permits locating a particular staff member when there is an urgent need requiring that staff member's attention. This can be accomplished utilizing a pager, a two-way radio, or an intercom system. Each of these approaches has limitations including coverage area and contribution of noise (i.e., sound and/or electromagnetic energy) to the clinical environment.

Conclusion

Each clinical engineering department will require some customization based on the services that it will provide and the resources that will be provided to it. However, every clinical engineering manager will need to consider and incorporate staffing, space, test equipment, tools, communications equipment, training, and a computerized maintenance-management system into the department's plan. This chapter has provided a template for considering the resource needs of the clinical engineering department; however, the clinical engineering manager is encouraged to approach each of this with creativity. Other ideas for utilizing resources in the development and implementation of a clinical engineering department often can be obtained from the work of other clinical engineers (Dyro, 1993; Soller, 2000; Hughes, 1995; Gupte, 1994).

References

Baker T. Survey of salaries and responsibilities for hospital biomedical/clinical engineering and technology personnel, *J Clin Eng* 25(4):219-234, 2000.
Campbell S. Attracting and retaining qualified workers in 'today's hot job market, *Biomed Instr Tech* 34 (6), 2000.
Dyro JF. How to recruit and retain staff: Part 1, *Biomed Instr Tech* 23(2):92-96, 1989a.
Dyro JF. Focus on: University Hospital & Health Sciences Center SUNY at Stony Brook Biomedical Engineering Department. *J Clin Eng* 18(2):165-174, 1993.
Dyro JF. How to recruit and retain staff: Part 2, *Biomed Instr Tech* 23(3):230-232, 1989b.
Glouhova M, Kolitsi Z, and Pallikarakis N. International Survey on the Practice of Clinical Engineering: Mission, Structure, Personnel, and Resources. *J Clin Eng* 25(5):269-276, 2000
Gupte PM, Tsunekage T, Ma WP, Adadjo FK. Focus On: Westchester County Medical Center Division of Biomedical Engineering. *J Clin Eng* 19(4):310-323, 1994.
Hughes JD. Focus On: Washington Hospital Center, Biomedical Engineering Department. *J Clin Eng* 20(2):127-134, 1995.
Soller I. Workplace profiles: Biomedical/Clinical Engineering Department at SUNY Downstate Medical Center—University Hospital of Brooklyn. *ACCE News* 10 (6):10-11, 2000.
Maddock K, and Hertzler L., Building a Clinical Engineering Department from the Ground Up, *Biomed Instr and Technol* 33(6), 2000.
Pacela AF. and Brush LC, How to Locate and Hire Clinical/Biomedical Engineers, Supervisors, Managers, and Biomedical Equipment Technicians, *J Clin Eng* 18(2):175-179, 1983.
Rice JD. Using Laptop Computers as a BMET Field Service Tool, *Biomed Instr Technol* 27(6), 1993.

6

Clinical Engineering in an Academic Medical Center

Ira Soller
Director, Scientific and Medical Instrumentation
SUNY Health Science Center at Brooklyn
Brooklyn, NY

The State University of New York, Health Science Center at Brooklyn—widely known as Downstate Medical Center—is located in the heart of urban Brooklyn, New York. Downstate traces its roots back to 1860, when the nation's first hospital-based medical school, Long Island College Hospital (Collegiate Division), was founded in Brooklyn under a New York State-granted charter. For the first time, bedside teaching was included in the curriculum, forever changing the course of medical instruction. The school was renamed the Long Island College of Medicine in 1930, and in 1950 it merged with the State University of New York system and was called the College of Medicine at New York City. In 1985, the institution was renamed the State University of New York Health Science Center at Brooklyn (SUNY HSCB) (Guide to the Archives, 1966).

Today, Downstate is a multifaceted institution providing patient care, education, research, and community services. It serves as a regional resource for Brooklyn, Staten Island, and the surrounding vicinity. It is Brooklyn's premier center for the education of health professionals and one of the leading such centers in the New York metropolitan area.

Downstate is one of New York's leading centers of biomedical research. In 1998, it received worldwide recognition when Dr. Robert Furchgott, distinguished Professor Emeritus, was awarded the Nobel Prize in Medicine for his discovery of the role that nitric oxide plays in the contraction and dilation of blood vessels. Downstate is also well known for the development of magnetic resonance imaging (MRI) by Dr. Raymond Damadian (who produced the first human images in the world using MRI), as well as for the establishment of the nation's first federally funded dialysis program by Dr. Eli Friedman.

Downstate Medical Center consists of the following:

- College of Medicine
- School of Graduate Studies
- College of Nursing
- College of Health Related Professions
- University Hospital of Brooklyn
- Research Foundation

University Hospital of Brooklyn (UHB) is the teaching hospital of Downstate Medical Center. This comprehensive 376-bed facility with several primary-care satellite clinics provides a full range of services from primary and preventive services to tertiary care.

Scientific and Medical Instrumentation Center (SMIC)

History

The Biomedical/Clinical Engineering Department serving Downstate is one of the oldest clinical engineering programs in the nation. SMIC was established in 1963, under the leadership of Dr. Seymour Ben-Zvi (Ben-Zvi and Gottlieb, 1980), several years before Dr. Caceres coined the term "clinical engineering." SMIC's current Director, Ira Soller (Figure 6-1), appointed in 1993, has been on staff since 1977, previously serving as assistant, associate, and acting director.

SMIC has historically supported activities that include the design and development of equipment and devices (usually associated with biomedical engineering research, as well as involvement with medical equipment applied to patients in a clinical setting, a clinical engineering function. Having such versatility, SMIC is capable of providing services in support of all of the goals expressed in Downstate's mission statement, (i.e., patient care, education, research, and community service).

From its inception, SMIC was acknowledged to be one of the leading clinical engineering departments in the country. On April 22, 1970 the W.K. Kellogg Foundation awarded SMIC a $129,216, three-year grant to develop and conduct the nation's first experimental in-house preventive maintenance (PM) program for hospital medical equipment, thus addressing the need to improve equipment maintenance throughout the industry. This PM program proved so successful that it served as a prototype for hospitals throughout the country. Interestingly, this award was made prior to Ralph Nader's article

Figure 6-1 Ira Soller, MSEE, PE, SMIC, Director of Biomedical Engineering.

Figure 6-2 Medical Equipment Management. Associate Directors, Leonard Klebanov and John Czap, are key players in SMIC's efforts to assure that safe, efficacious, medical equipment is acquired, properly tested, and cost-effectively maintained at UHB. Together they have more than 50 years of experience working in the clinical engineering field.

in the *Ladies Home Journal* (March 1971) that raised national consciousness about accidents that could occur in hospitals as a result of poorly designed or faulty medical equipment, and recommended that hospitals hire engineers to eliminate these dangers. This helped to spur the growth of the clinical engineering profession (Curran et al., 1974; Soller, 1999).

Throughout the years to satisfy the needs of the dynamic Downstate environment, SMIC expanded its equipment-support capabilities to provide a diverse range of services to researchers and clinicians. Centralization of hospital equipment management and instrumentation technical services under one department made sense both economically and logistically. It has been demonstrated repeatedly that such services could be provided effectively and in a more timely fashion by in-house staff.

Creative Exposure

Life within a large academic institution, and a teaching hospital in particular, presents many challenges. One must be able to change focus and to prioritize as situations and emergencies dictate. This dynamic environment requires the delegation of tasks and necessitates supervisory clinical engineering staff to have the ability to make independent decisions.

Talented technical personnel are drawn to SMIC because of the Downstate environment, as it offers them exposure to the latest technologies and provides greater opportunity for professional growth than that offered in a nonteaching hospital setting. The depth and breadth of knowledge necessary to meet Downstate's challenges, however, requires SMIC clinical engineers to have expertise in a multitude of disciplines including the basic sciences (e.g., chemistry, physics, and optics), physiology, engineering (e.g., thermodynamics, electronics, and mechanics), and information technology, in addition to having knowledge of regulatory agency requirements, health care codes, and medical equipment standards.

Unique Involvement

Downstate's diverse academic medical center environment has allowed SMIC to have many unique involvements throughout the years. SMIC folklore includes fond tales of Dr. Damadian "scrounging" some parts from SMIC to assist in his development of an MRI machine at Downstate. Over the years, other researchers in addition to Dr. Furchgott have use SMIC services. These include Dr. Henri Begleiter, who studied the genetics of alcoholism, Dr. Mimi Halpern, who studied the sensory system in vertebrates, Dr. Alan Gintzler, who worked on pain perception and addiction (Baron et al., 1983), and Dr. Madu Rao, who developed direct calorimetry in preterm infants (Rao et al., 1995). Likewise, Dr. Randall Barbour requested SMIC's assistance in the development of a near infrared, clinical imaging system that has the potential of becoming as important as MRI (Barbour et al., 2000). As a result of these and other activities, several SMIC staff members have been acknowledged in clinicians' and researchers' technical papers and presentations and have received numerous awards.

SMIC Logo

SMIC's support services to the hospital, academic, and research communities were captured in SMIC's logo, designed by Ms. Tobey Soller, graphic designer, in 1993. The logo consists of a white circle (representing a medical or scientific instrument display screen) containing the green letters SMIC (representing the vitality of an ECG trace), all contained within a blue square (representing Downstate Medical Center and University Hospital of Brooklyn). Taken as a whole, the logo represents a dynamic biomedical/clinical engineering department providing diverse services to the Downstate community.

SMIC Program Services

A summary list of SMIC's biomedical/clinical engineering program services in support of Downstate's mission follows (Wilkow and Soller, 1995). These services are later discussed in detail.

Patient Care

- Acceptance testing (initial checkout, incoming inspection, incoming testing) of new equipment for safety, proper operation, and adherence to manufacturers technical specifications
- Bid evaluation for the purchase of medical equipment and related services
- Centralized patient care instrumentation history files
- Centralized instrumentation technical manuals library
- Clinical engineering consultation
- Clinical engineering participation on Hospital and Health Science Center committees
- Clinical engineering staff and departmental development activities
- Computerized medical equipment inventory
- Coordination of outside medical equipment repair/maintenance services
- Dedicated clinical engineering support to the cardiothoracic open-heart program
- Emergency clinical engineering support, all patient-care areas and operating rooms
- Expert witness testimony
- Hazard and recall alert notification
- Medical equipment asset management
- Medical equipment defect resolution
- Medical equipment evaluation library
- Medical equipment installation supervision and coordination.
- Medical equipment modification
- Medical equipment planning for clinical areas and new programs
- Medical equipment pre-purchase evaluation
- Medical equipment preventive maintenance (PM, scheduled maintenance)
- Medical equipment relocation (as patient areas are refurbished)
- Medical equipment repairs (unscheduled maintenance)
- Medical equipment upgrades
- Medical equipment user in-service education
- No fault found/user error/cannot locate, tracking
- On-call/recall support for critical-care areas and the operating rooms
- Oversight and evaluation of equipment-service contracts
- Patient-related equipment-incident investigation
- Preventive maintenance test procedure generation and update
- Purchase requisition review
- Quality assurance and risk management
- Regulatory agency inspection support (American Association of Blood Banks (AABB), College of American Pathologists (CAP), Department of Health (DOH), Food and Drug Administration (FDA), Joint Commission on Accreditation of Healthcare Organizations (JCAHO)

- Report generation to hospital administration (including clinical capital equipment purchase tracking), the safety committee, and other hospital committees
- Representation of the hospital on the University Healthcare Consortium (UHC) Clinical Engineering Council, and at Greater New York Hospital Association (GNYHA) meetings
- Representation of the hospital in the New York City Metropolitan Area Clinical Engineering Directors Group
- Request-for-quotation (RFQ) generation
- Technology assessment
- Testing of rental, loaner, demonstration, patient-owned, and physician-owned equipment for hospital use
- Vendor and manufacturer interface

Research

- Design, development and construction of specialized instrumentation and devices including electronic, electromechanical and mechanical
- Engineering consultation
- Institutional Review Board (IRB) approval assistance
- Laboratory measurement assistance
- Patent and grant-writing assistance
- Scientific equipment repairs, modification, and upgrade

Educational and Community Involvement

- BMET internship program
- Electrical Engineering Technology Advisory Committee (SUNY Farmingdale)
- Lectures (medical equipment-specific and safety issues)
- Volunteer training

SMIC Support for Patient Care

SMIC's primary focus, to which almost all of its resources are allocated on a day-in day-out basis, is in support of University Hospital of Brooklyn (UHB) and its patient population.

Environment of Patient Care

Providing engineering expertise to ensure that the environment of patient care is safe for both patient and clinician is at the heart of SMIC's clinical engineering program in support of UHB. Medical equipment used on or for patient care comprises a large part of this environment. It must be safe, efficacious (i.e., performing the function for which it was intended), and cost-effective. SMIC's clinical engineers provide technical-support services to ensure this, thus reducing risk to the institution, its patients, and staff. Their diverse body of knowledge and experience enables them to have a unique understanding of the ways that technology and patient care interact within the clinical setting (ECRI, 1989) and is vital when medical equipment is integrated into the environment of patient care. They are able to interface medical systems to the patient and to other medical systems including those that are used for patient data collection (Shaffer, 1997). Their knowledge has proven invaluable to UHB.

Medical Equipment Management

Scope

Medical equipment management (David and Judd, 1993), a cornerstone of SMIC's clinical engineering program, concerns not only equipment maintenance (PM and repair) but also encompasses services that cover the entire life cycle of the patient-care and patient-related medical instrumentation used at UHB. SMIC services start with the planning stages prior to equipment purchase and, as required by the equipment-tracking requirements of the Safe Medical Devices Act, conclude with the formalities of retirement and disposal of older units from service—after their usefulness (i.e., functional or technological) is over (Shepherd, 1996).

SMIC's medical equipment management program is compliant with JCAHO guidelines and regulatory requirements of the Department of Health (DOH), the College of American Pathologists (CAP), and the American Association of Blood Banks (AABB).

The program is involved with equipment acquisition including equipment assessment (pre-purchase evaluation), equipment specification, purchase requisition review, RFQ generation, bid analysis, and vendor selection (See Figure 6-2). It includes installation planning, acceptance testing, defect resolution, user in-service education, selection of service provider, product recall and alerts, incident investigation, and equipment replacement cost analysis and disposal. It strives to save money by reducing equipment downtime and resultant lost patient revenue, and it minimizes the need to pay for equipment rentals. It also allows equipment replacement decisions to be made based not only on equipment capabilities but on past equipment failures and expenses as well. Standardization of equipment and consumables is also encouraged (Cohen, 1994).

Device Responsibility

SMIC is responsible for approximately 10,000 active medical devices. The number changes daily as new equipment is acquired and older equipment is retired from service. The equipment runs the gamut from simple electronic thermometers to sophisticated ultrasound imaging systems, located throughout the institution. To bring order to this vast array of medical equipment and to simplify the gathering and distribution of data relating to these medical devices, SMIC follows the medical device classification nomenclature found in *Health Devices Sourcebook* and *Medical Device Register*. At present, 500 different equipment nomenclatures are used to describe this equipment, the large number suggesting the breadth of knowledge that SMIC staff must have.

SMIC's mandate does not include X-ray or ionizing radiation devices, which are managed by the Radiation Physics Department, nor devices such as stretchers, hospital beds, and wheelchairs, which are maintained by the Facilities Management and Development Department.

Figure 6-3 Administrative Support. SMIC's Administrator, Marcia Wilkow, tracks clinical capital equipment purchases and clinical engineering's purchases. She and SMIC secretary, Emma Johnson, are reviewing purchase requisition information.

Procurement of Medical Devices

UHB acquires medical equipment for a multitude of reasons. These range from the need to replace obsolete equipment to the acquisition of new technology and specialized equipment for new clinical programs, allowing enhanced services to be provided to the community.

SMIC plays a role in the acquisition process by providing input to hospital administration and the Clinical Capital Equipment Acquisitions Committee. Through a purchase requisition review process SMIC ensures that everything required (e.g., peripheral items and supplies) is ordered and that all items are compatible with each other and with existing equipment. This process also ensures that the physical plant is made ready to accept the new equipment, thus expediting equipment installation and use (Soller and Klebanov, 1992). SMIC's Administrator maintains a Clinical Capital Equipment Tracking Spreadsheet Report of all clinical capital equipment purchases, and SMIC works closely with Purchasing and Expenditures Processing to ensure that medical instrumentation is neither ordered unless the purchase requisition has been reviewed nor purchased until the device has successfully passed SMIC acceptance testing. (See Figure 6-3.)

SMIC is also involved in technology management issues relating to proposed new acquisitions. For example, SMIC served on an MRI Acquisitions Committee, visited the vendors manufacturing site, and generated reports for UHB's medical director concerning the acquisition of an Open-MRI device (AAMI, 1997).

New and Renovated Area Planning

SMIC's involvement extends to technical issues relating to medical equipment planning for renovated and new areas within the Hospital. SMIC provides assistance to the clinical staff in determining the location of the medical devices; guidance in connecting new systems to existing systems; and oversight of vendor or manufacturer equipment installation including adherence to contractual requirements (e.g., timeliness of installation). During renovation of existing areas, SMIC assists in safeguarding existing equipment to prevent damage.

Turnkey Installations

SMIC is called upon during request for quote (RFQ) generation, bid evaluation, purchase requisition generation, and acceptance testing for such systems. Although RFQ documents may include equipment and environmental specifications and legal requirements that address issues of noncompliance and penalties, it has been SMIC's experience that large turnkey systems require extensive vendor or clinical engineering interaction and problem solving as installation progresses as documentation for all contingencies is not always possible. Also, several layers of acceptance testing may be required; e.g., bench

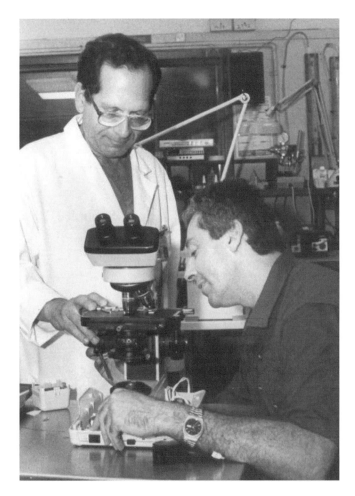

Figure 6-4 Acceptance Testing. Senior biomedical engineers, George Martin and Semyon Ulis, verifying correct operation of a newly purchased microscope to ensure that it is safe for use and that it performs according to manufacturer specifications, and rectifying any defects prior to its being released for clinical use. Payment is not authorized until the unit has successfully passed the acceptance-testing process.

testing of components prior to, as well as after, inclusion in subsystems followed by final global acceptance testing.

Acceptance Testing of Medical Devices

SMIC staff perform acceptance testing, also known as "initial checkout" or "incoming inspection," prior to new devices being placed into service. SMIC also performs an appropriate level of acceptance testing for physician-owned, rental, loaner, donated, or demonstration instrumentation prior to an item being placed into service.

Acceptance testing tends to be more in-depth than PM testing and entails more than just electrical safety testing, e.g., leakage current and grounding resistance. (See Figure 6-4.) This process ensures that all equipment to be used on or for patients functions properly and is safe, both for the clinical user and the patient. Acceptance testing uncovers equipment defects including those not readily apparent, allowing for correction prior to the equipment being placed into service, thus reducing liability to the institution. Sometimes design errors are uncovered as well. Acceptance testing also provides a practical training ground for SMIC's staff, thus allowing people to become familiar with new equipment prior to the occurrence of an emergency clinical situation or patient incident involving this equipment that would require their assistance (Polaniecki, 1989).

Equipment Maintenance and Repair

Proper equipment maintenance and repair, a core element of equipment management, is paramount for an institution to be able to provide high-quality medical care. (See Figure 6-5.) A clinical engineering director must be flexible and fully aware of the skills of the clinical engineering staff and the budgetary constraints of the institution in order to select the proper cost-effective mix of services, the intent always being to provide quality service and to reduce overall costs (Gordon, 1995).

SMIC avails itself of several methods of providing equipment service. Although SMIC provides in-house maintenance and repair for a full range of sophisticated, state-of-the-art medical equipment, which includes anesthesia machines, ventilators, and dialysis units, outside OEM vendor repair and service support is sought out for specialized equipment, as required to supplement in-house capabilities; i.e., equipment that must be con-

Figure 6-5 Preventive Maintenance. Senior biomedical engineer, Lenworth Howe, performing preventive maintenance on a cell saver. Preventive maintenance can be more involved than just performance of a functional and safety test. Problems uncovered must be rectified, and some instrumentation requires that components be replaced due to wear or number of hours of use.

tinuously operational in order to minimize revenue loss to the institution and for which there is little or no back-up equipment. (See Figure 6-6.) SMIC monitors the outside services to ensure work quality, receipt of appropriate documentation for entry into the equipment history files and computerized system, and correct charges (Cohen, 1994).

Providing service to UHB's off-campus facilities (i.e., three clinics, a dialysis center, and a school based clinic) presents challenges that require careful coordination, particularly in transporting equipment and staff between locations, exchanging equipment, and assisting in regulatory inspections.

Emergency Support

SMIC engineers provide on-site troubleshooting during normal working hours to all patient-care units, the operating rooms, and offsite facilities (See Figure 6-7). After hours, on weekends, and on holidays, SMIC provides on-call and recall support to the operating room and critical care areas. SMIC engineers also assist during physical plant emergencies (e.g., water leaks) to safeguard medical equipment.

Cardiothoracic Surgery

One unique SMIC service is the support of UHB's Open Heart Cardiothoracic Surgical Team. A SMIC clinical engineer is dedicated to the team and is present during each case. (See Figure 6-8.) SMIC's engineer assists in equipment setup prior to the operation, equipment troubleshooting during the surgical procedure, patient transport to an appropriate intensive care unit, and bedside physiological monitor connections (Soller, 2000).

Although SMIC clinical engineers assist in the operating room, they do not independently operate the medical equipment or select levels of treatment, e.g., time balloon pump inflation and deflation, prepare medications, heparinize saline solutions, or give fluids to or take fluids from patients by way of, for example, a cell saver or rapid infusion system. Trained clinical specialists, perfusionists, and licensed technicians perform these tasks. However, clinical engineers do provide instrumentation troubleshooting expertise during these procedures.

Risk Management

SMIC's clinical engineering program provides an important oversight function whose purpose is to reduce risk of injury to patients and staff as well as liability to UHB. UHB's Quality Assurance and Risk Management Departments routinely look to SMIC for assistance with regulatory issues, agency surveys (e.g., JCAHO, FDA, DOH, AABB, and CAP), patient- and equipment-incident investigation, and hazard and recall alerts. (See Figures 6-9 and 6-10.) SMIC staff also have been called upon to give testimony as expert witnesses. In addition, user error, equipment abuse, and no-fault-found occurrences are tracked to assist Risk Management in determining whether additional clinical staff in-service training is required.

Figure 6-6 Repair. Senior biomedical engineers, Simon Gerecht and Michael Salts, utilize their extensive electronic troubleshooting skills during repair of an infant warmer to ensure that the thermostat controls are functioning properly.

Clients/Committees/Meetings

SMIC's clinical engineers interface daily with staff from various departments and entities within our health care institution. All of them are considered to be clinical engineering clients and are treated as such. A partial list includes facilities engineering and planning, hospital administration, cardiothoracic surgery, central sterile, expenditures, contracts, risk management, OPD administration, clinics, clinical labs, ambulatory surgery, pharmacy, surgery, anesthesiology, ophthalmology, orthopedics, offsite satellite clinics, clinical areas, nursing units, personnel department, purchasing, and cardiology. Interaction between researchers and educators also occurs.

Such open interaction is essential for a successful clinical engineering program. SMIC maintains that clinical engineering departments must continuously prove themselves, especially as an institution's staff changes. New employees, both administrative and clinical, might be unfamiliar with the benefits of in-house clinical engineering services. Continuous advertising promotion, such as by means of flyers and bulletin boards and attendance at meetings, is required in order to educate them to clinical engineering's vital role within the institution. Report generation serves a similar public relations function.

Committee participation is particularly important. It enables clinical engineering to provide timely feedback to the clinical user on equipment-related issues and questions. This maintains and improves channels of communication between SMIC and its clients. It also allows trends to be spotted such as equipment that is inoperable but not yet reported to clinical engineering.

SMIC staff coordinated the Y2K process at the Hospital and Health Science Center, obviating the need for outside consultants and thus considerably reducing hospital expenditures. SMIC's involvement on various Y2K committees within the Greater NY Hospital Association and SUNY Central in Albany, NY, benefited the institution by raising conscience levels, which resulted in sound cost-effective business decisions for Downstate and University Hospital (Soller, 1999).

SMIC Support for Research

Research support includes consultation assistance in selection of experimental equipment, laboratory setup, measurement technique, grant writing, patent applications, and IRB approval. This support also entails the design, prototype development, and final construction of special-purpose devices that are not readily available commercially or that can be more cost-effectively built in-house.

Involvement in research and educational activities is beneficial. These challenging opportunities help to keep SMIC staff technically competent and involved at the forefront of technology. Such activity also brings prestige to the department, enhancing its professionalism and reputation. To be effective, a team with diverse expertise in electronics, mechanics, and physiology is often required.

SMIC Support for Education and Community Involvement

Education

SMIC supports the Continuing Education Department by coordinating in-service training for the clinical staff on specific medical instrumentation and provides tutorials on safety issues within the clinical environment. SMIC also presented a Post-Baccalaureate Medical Monitoring Course for Downstate's College of Health Related Professions Cardiovascular Perfusionist Program.

SMIC's involvement with SUNY Farmingdale's Electrical Engineering Technology Department led to the establishment of a Biomedical Engineering Technology Internship program. Recently, SMIC was requested by the City University of New York, New York City Technical College to advise in establishing a biomedical engineering technology program.

SMIC has also made clinical engineering presentations to outside institutions; e.g., Brooklyn Museum's *Big Adventure Program* for children; Columbia University's Bioengineering Conference; and Touro College's Levine School of Health Sciences. (See Figure 6-11.)

Community Involvement

Community involvement, provided through on-the-job training programs, has included UHB's volunteer program and the New York Association for New Americans (NYANA) internship program for newly arrived immigrants. When properly supervised, SMIC has found intern and volunteer interaction to be beneficial. For example, qualified volunteers were utilized when the department was faced with the task of multiple device equipment acceptance testing for 300 infusion pumps.

Staff

Overview

SMIC's Director, Ira Soller, MSEE, PE, a recipient of the SUNY Chancellor's Award for Excellence in Professional Service (SUNY Focus, 1996), plays an active role in the clinical engineering community. He presented at AAMI Expositions on clinical engineering legal testimony (Soller, 2001), the impact of health care changes on clinical engineering (Soller, 1996; Soller, 1999), clinical engineering department restructuring (Soller, 1997), and the philosophy of clinical engineering (Soller, 2001). He also contributed to the University Health care Consortium *Compendium of Cost Saving Projects* (UHC, 1995; UHC, 1997) and was an invited panel member at their benchmarking roundtable discussion (Soller, 1995). As a member of the Greater New York Hospital Association (GNYHA) Y2K Workgroup, he was active in defining appropriate policy. He is Coordinator of the

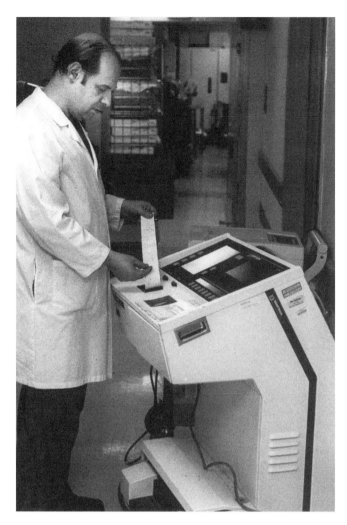

Figure 6-7 Emergency Support. SMIC plays a key role in providing emergency support to troubleshoot and answer questions relating to equipment operation and capability. Associate Director, Leonard Klebanov, is shown testing to ensure that a balloon pump is functioning properly. This device provides support to critically ill patients by reducing their heart's workload until it strengthens. Balloon-inflation and -deflation timing and mode of triggering ensure optimal patient assistance.

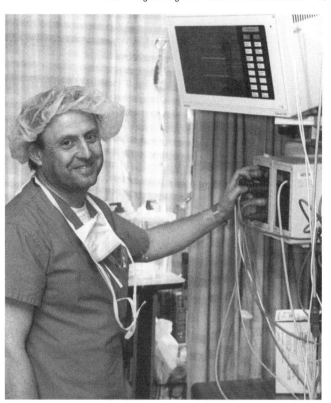

Figure 6-8 Cardiothoracic Surgery Support. SMIC's specially trained engineer is available in the operating room to check the physiological monitoring equipment prior to patient connection as well as to troubleshoot equipment problems should any develop. Biomedical engineer, Boris Shamrakov, is shown checking a patient bedside monitoring setup in the Cardiothoracic Surgical Intensive Care Unit to ensure its functionality and rapid connection upon patient arrival following surgery.

New York City Metropolitan Area Clinical Engineering Directors Group (see Chapter 132) and a representative to AAMI's Biomedical Organizations Committee (Soller, 1999d). Mr. Soller serves on SUNY Farmingdale's Electrical Engineering Technology Advisory Committee and served on its Biomedical Engineering Technology Advisory Committee assisting in establishing an internship program at SMIC. For several years he taught a post-baccalaureate medical monitoring course for Downstate's College of Health Related Profession's Cardiovascular Perfusionist Program and served on its Program and Curriculum Review Board. In 1999, Mr. Soller received an American College of Clinical Engineering Advocacy Award for authoring his article on clinical engineering that was published in the *Wiley Encyclopedia of Electrical and Electronics Engineering* (Soller, 1999). He has been called upon to serve as expert witness of patient incidents and is listed in Strathmore's *Who's Who* (1998-99).

Among SMIC's clerical and administrative support staff of 20 are the following individuals:

John Czap, associate director, has participated in seminars for OSHA and FEMA and has been involved with the New York City Mayor's Office of Emergency Management Special Needs Assessment Program. He is a member of the NYC Emergency Medical Services Council regional faculty. Mr. Czap is SMIC's lead person for patient incident investigations, regulatory agency inspections, and response to environmental emergencies that could affect UHB's medical equipment. His expertise in RF communications is invaluable in matters concerning electromagnetic interference (EMI).

Leonard Klebanov, associate director, is routinely involved in medical instrumentation issues including purchase requisition review for new sophisticated electronic equipment and systems, their acceptance testing, and coordination of installation activities. He oversees the replacement of the hospital-wide physiological monitoring equipment. Mr. Klebanov also supervises SMIC's on-call/recall team, including the support that it provides to cardiothoracic surgery, and serves on the IRB committee. Staff both within and outside SMIC rely on his expertise on technical issues requiring cost-effective solutions.

Administrator Marcia Wilkow adds financial and administrative expertise which is vital to department functioning. Her desk serves as command center, especially during times of crisis. Ms. Wilkow's responsibilities include personnel issues, management of data

Figure 6-9 Incident Investigation. Equipment–related incident investigation is conducted by SMIC whenever a medical device might have caused injury to a patient or clinician. SMIC staff also may anticipate and resolve equipment problems before they result in an incident. Associate Director, John Czap, is shown with an examination lamp bracket that did not meet original equipment-manufacturer (OEM) specifications, which could result in the lamp being easily dislodged and falling. The bracket manufacturer and supplier worked with SMIC to resolve the issue and supplied newly designed brackets.

Figure 6-10 Regulatory Inspection. SMIC staff provides technical assistance during regulatory and accrediting inspections. Here, Lab Supervisor, Sophia Zikherman, discusses an upcoming Department of Health inspection with senior biomedical engineer, Stanley Wei, and biomedical engineers, Yefim Brayman and Michel Chery.

collection and recordkeeping activities relating to SMIC's equipment management program, purchase order tracking, new capital equipment purchase tracking, documentation of SMIC policies and procedures, annual budget preparation, monthly financial analysis, report generation, and interaction with anyone who seeks SMIC services.

Lab supervisors, Luis Cornejo and Sophia Zikherman, contribute important technical expertise. Mr. Cornejo, an expert on sophisticated electromechanical equipment (e.g., ventilators, dialysis equipment, and anesthesia machines), reviews purchase requisitions for such devices and supervises their acceptance testing, PM, and repair. Ms. Zikherman supervises PM and repair services for electronic equipment and serves on SMIC's on-call/recall team.

Other SMIC staff members, with many valuable years of experience, contribute unique talents and expertise to the department. These include senior biomedical engineers Michael Doron, Simon Gerecht, Lenworth Howe, George Martin, Michael Salts, Semyon Ulis, Stanley Wei; biomedical engineers Yefim Brayman, Michel Chery, Ruck Gilles, Boris Shamrokov; and precision machinist Randall Andronica. SMIC secretary Emma Johnson cheerfully keeps everyone on track.

Training

Continuing Education

The environment in which clinical engineering functions changes daily as new technologies such as telemedicine, robotics, and wireless-LAN are introduced into the clinical setting. In this dynamic field, continuing education is the rule. This is as true for SMIC's director (Soller, 2001) as it is for the most junior member of the technical staff. For this reason, SMIC staff are encouraged to participate in continuing education. Training is provided by SMIC (see Fig. 6-12), other departments within UHB, external sources such as independent companies, manufacturers, and vendors, in addition to self-study. When possible, supplemental funding is obtained through the union's continuing-education grants program.

Service Training

As new medical equipment is acquired, employee technical knowledge concerning operation, PM, and servicing must be updated. Whenever possible, service training is included in purchase requisitions andRFQs for new equipment. Service training benefits not only clinical engineering, the equipment maintainer, but also the equipment user each time SMIC is called to assist with equipment-related problems. SMIC staff also attend equipment-operator training provided by vendors for clinical users of medical equipment within the institution.

Safety Training

Employee right-to-know and safety training addressing the hazards encountered in the workplace is also provided. This subject matter includes bloodborne pathogens, hazardous materials, universal precautions (e.g., proper protection when entering patient-care areas, such as gloves and masks), environmental hazards, fire hazards, and a patient's bill of rights.

On-call/Recall Training

Formal on-call/recall training is provided. In addition, regular departmental meetings consider the latest operating-room practices and changes of which on-call staff must be aware to provide emergency services.

Figure 6-11 Educational/Community Support. Senior biomedical engineer, Michael Doron, shown here constructing a test fixture, provided Brooklyn youth with a lecture on blood pressure monitoring as part of the Brooklyn Museum's *Big Adventure Program* for children, thus introducing them to the profession of clinical engineering.

Figure 6-12 In-service Education. In-service education can be provided by manufacturers and vendors, as well as by clinical engineering staff. Here, Lab Supervisor, Luis Cornejo, provides training to other clinical engineers in the use of an automated ventilator tester. Such devices reduce the number of individual test instruments required by integrating several test functions into one device, thus reducing service time leading to more rapid equipment turnaround. Such education also helps to satisfy JCAHO training requirements for clinical engineering staff.

Facilities/Equipment

SMIC maintains electronic and electromechanical laboratories, a machine shop, an administrative area, and storage facilities. Adequate space is allocated for SMIC to perform its myriad functions SMIC and includes limited space to store new equipment, equipment awaiting servicing, equipment awaiting delivery back to the clinical user, and equipment sequestered because of suspected involvement in a patient incident.

SMIC's administrative area acts as the hub of activity. It directs visitors and clients to appropriate personnel, and houses all equipment-management files.

The laboratories for acceptance testing, PM, and repair, contain electrical power, suction, compressed air, secure gas tank storage, sinks, workbenches, test equipment, tools, and storage cabinets.

The machine shop, which is equipped with a drill, lathe, grinder and milling machine, and other useful devices, enables SMIC to perform emergency mechanical repair of damaged medical equipment; customize and modify equipment including crash carts (see Fig. 6-13); mount equipment, e.g., physiological monitors onto anesthesia carts; construct test fixtures; and provide research assistance.

SMIC is the central repository of all regulatory-related medical instrumentation history files, both active and inactive, for equipment that falls within its province. Its library maintains instrumentation technical manuals (e.g., operator, service, and applications) and equipment evaluations. Hence, SMIC is the first place to which one goes for information on patient care or patient-related equipment.

Ethics

SMIC adheres to the ethical code of the ACCE (see Chapter 130). Among other things, this code strives "to prevent a person from being placed at risk of injury due to dangerous or defective devices or procedures" to "work toward the containment of costs by the better utilization of technology"; and to "protect the confidentiality of information from any source." If it does not adhere to confidentiality, a department's credibility is soon lost. Confidentiality relates not only to patient information but also to medical equipment information.

Summary

The challenges of providing a full-service, in-house, clinical engineering program within a dynamic academic medical center and university teaching hospital setting such as that of Downstate Medical Center are formidable. The biomedical and clinical engineering services that are expected far exceed those of a nonacademic community hospital. To meet the challenges, sufficient clinical engineering and administrative staff who possess diverse expertise and flexibility must be available. The challenges are worthwhile and personally rewarding.

SMIC is proud of its past accomplishments, the example it sets, and the input it provides to the clinical engineering community at large. It looks forward to meeting future challenges within the exciting and stimulating environment of SUNY Downstate Medical Center and University Hospital of Brooklyn.

Acknowledgements

I wish to thank all members of SMIC's staff, both past and present, with whom I have worked over the past 24 years. It is through our daily interaction that the concepts presented here have been better defined.

In particular, thanks to: SMIC's associate directors John Czap (for his many years of dedicated service to Downstate) and Leonard Klebanov (for his expert technical consultation); SMIC's administrator Marcia Wilkow (who deserves special praise, often acting as a valued sounding board, and for suggestions to enrich publications such as this one); and SMIC's laboratory supervisors Luis Cornejo and Sophia Zikherman (for keeping SMIC running on a day-to-day basis despite the many challenges faced). By relying on each other over the years of our association, we have managed to push the boundary in the correct direction. I am proud to be part of their team.

Thanks also to: Senior associate administrator William Gerdes, who understands and values SMIC's services; Ernest A. Cuni, Biomedical Communications, SUNY Downstate Medical Center, for the photographs; my daughter Tobey for design of SMIC's Logo; and my wife Helaine for suggestions that improved the readability of this manuscript.

Figure 6-13 Equipment Modification. Lab Supervisor, Luis Cornejo, and biomedical engineer, Ruck Gilles inspects a medication cart that SMIC modified to accept an electrical-outlet strip and retractable reel, oxygen cylinder, defibrillator, and an aspirator so that it can be used as a cardiac-arrest crash cart, which is brought to the patient's bedside for resuscitation during cardiac or respiratory emergencies.

References

AAMI. *Recommended Practices for a Medical Equipment Management Program, Final Draft Standard.* AAMI EQ56-1997. AAMI Arlington, VA, 1997.

Barbour R, Andronica R, Soller I. *Instrumentation and Calibration Protocol for Imaging Dynamic Features in Dense-Scattering Media by Optical Tomography,* Optical Society of America, 6466-6486, 2000.

Baron, Jaffee and Gintzler. Release of Substance P from the Enteric Nervous System: Direct Quantitation and Characterization. *J Pharm Exper Therap* 227(2):1983.

Ben-Zvi S, Gottlieb W. Inspection and Maintenance of Medical Instrumentation. In Feinberg B (ed): *Handbook of Clinical Engineering, vol. 1,* Boca Raton, FL, CRC Press, 1980.

Cohen T. *Computerized Maintenance Management Systems for Clinical Engineering,* Arlington, VA, AAMI, 1994.

Curran WJ, Stanley PE, Phillips DF. *Electrical Safety and Hazards in Hospitals,* New York, MSS Information Corporation, 1974

David Y, Judd TM. *Medical Technology Management,* Washington, SpaceLabs Medical Inc, 1993.

ECRI. Special Report on Technology Management: Preparing Your Hospital for the 1990s. *Health Technol* 3(1):1989.

Gordon G. *Breakthrough Management: A New Model for Hospital Technical Services,* Arlington, VA, AAMI, 1995.

Guide to the Archives, State University of NY Health Science Center at Brooklyn, Medical Research Library of Brooklyn, 1966.

Polaniecki S. *SMIC Guide for the Performance of Initial Checkout,* SUNY Health Science Center at Brooklyn, University Hospital of Brooklyn, August 1989.

Rao M, Koenig E, Song-Li, Klebanov L, Marino L, Glass L, Finberg L. Direct Calorimetry for the Measurement of Heat Release in Preterm Infants: Methods and Applications, *J Perinatology* 15(5):1995.

Shaffer MJ. The Reengineering of Clinical Engineering, *Biomed Instrum Technol* 31(2):177-178, 1997.

Shepherd M. SMDA '90: User Facility Requirements of the Final Medical Device Reporting Regulation, *J Clin Engin* 21(2):1996.

Soller I, Klebanov L. *SMIC Checklist for The Purchase of Scientific and Medical Instrumentation,* SUNY Health Science Center at Brooklyn, University Hospital of Brooklyn, September, 1992.

Soller I, AAMI Conference & Expo. 2001 Preliminary Program, June 2001b, p 9.

Soller I. *Biomedical/Clinical Engineering Department Restructuring, A Practical Management Guide,* AAMI's 32nd Annual Meeting & Exposition Clinical Engineering Best Practices Session, June 7-11 1997. Abstract published in the 32nd Annual Meeting & Exposition Proceedings, June 7-11, 1997.

Soller I. Clinical engineering. In *Wiley Encyclopedia of Electrical and Electronics Engineering, ed 1, vol 2,* New York, John Wiley & Sons, Inc., 1999, pp 451-474.

Soller I. *How to Establish and Maintain a Local Biomedical Organization,* Appendix E Communications, AAMI, 1999.

Soller I. Legal Testimony—The Do's and Don'ts from a Clinical Engineering Perspective, Biomed Instr Technol 35(1):61-62, 2001a.

Soller I. *Medical Equipment Should Only Be Designed By Those Over 50; Bring Back the Canary,* Technical Iconoclast Business & Management Session presentation, AAMI Conference & Expo, 2001.

Soller I. *The Changing Role of Biomedical/Clinical Engineering as a Result of Health Care Reform & Funding Cutbacks,* Management of Clinical Engineering Practice Poster Session, AAMI 31st Annual Meeting & Exposition, June 3, 1996. Abstract published in the 31st Annual Meeting & Exposition Proceedings, June 1-5, 1996, PS-2. p 103.

Soller I. *The Health Care Industry Has Changed—Now What?* Business & Management Poster Presentation, AAMI 34th Annual Meeting & Exposition, 6/7-8/99. Abstract published in the AAMI 99 Session Guide & Abstracts, 1999, p 36.

Soller I. University Health Care Consortium Clinical Engineering Group Round-Table Discussion "Benchmarking pros and cons, and its effect on Clinical Engineering Departments," 1995.

Soller I. Workplace Profiles: Biomedical/Clinical Engineering Department at SUNY Downstate Medical Center—University Hospital of Brooklyn, *ACCE News* 10(6):10-11, 2000.

Soller I. *Y2K Problem Resolution—Medical Equipment.* Presented at Downstate Center–Wide Forum: Will Downstate Be Y2K—Ready? 1999.

Soller I. *Year 2000 Compliance Problem as It Impacts UHB—A Medical Equipment Perspective,* presented at UHB Department Head Meeting. 1998.

Soller I. *Year 2000 Problem—Tools to Assist in Determining Impact on Medical Equipment and Y2K Final Report,* provided to UHB Hospital Administration, 2000.

Strathmore's Who's Who, 1998-99: Small Business Source Book, ed 11, Gale Publishers, 1998, p 1258 (item 15795).

SUNY FOCUS, vol XII, no 18, 1996, p 3.

UHC. *Compendium of Cost Savings and Operations Improvement Projects: Clinical Engineering and Facilities Management,* 1997.

UHC. *Operation Improvement Compendium of Cost Savings Projects: Clinical Engineering,* 1995.

Wilkow, Soller I. *The Scientific and Medical Instrumentation Center,* SUNY Health Science Center at Brooklyn, University Hospital of Brooklyn, 1995.

7 Regional Clinical Engineering Shared Services and Cooperatives

J. Tobey Clark
Director, Technical Services Program
University of Vermont
Burlington, VT

Regional shared service programs, which started in the 1970s, led to safer medical devices and health care technology application and lower support costs for small, community hospitals that did not have the volume to have an in-house department. Many of the clinical engineering programs in developing countries are being set up under the nonprofit, shared services model which has proved to be an effective entry system for clinical engineering.

Although many of the programs evolved over time into for-profit companies or cooperatives in the United States, the general principles of sharing specialized services in a region still are practiced. Programs that continue to offer shared services have been successful because of proactive actions adapted to changes in the health care environment. The Technical Services Program (TSP) described in this chapter is one organization providing shared services as a nonprofit, independent organization. Features of this program are discussed in this chapter.

Introduction

The concept of sharing resources to serve health care in a state or region for the common good of the group fits in with the cooperative efforts that are necessary to provide patient-focused, integrated health care. A shared service is provided through the cooperation of multiple institutions typically serving a certain geographic area. The purpose of shared service organizations is to meet the needs of the regional health care entities in the most cost-effective manner. Shared services allow the economy of scale that individual institutions might not have for low volume, large expense, or highly technical services (particularly regional rural-community hospitals). In the past, many specialty services have been administered in a shared service manner including laundry, blood banking, mobile imaging, and telemedicine technology.

A not-for-profit shared services entity may be part of a hospital cooperative, a hospital association, or university. For-profit shared services also exist. The for-profit kind can be truly a cooperative only when the profits go back to the members. Advantages of the not-for-profit is the enhanced credibility within the hospital community, achieved in part by the involvement of hospital administration via a governing board or advisory group that seeks improved quality and cost-effectiveness—not profits—as the motive.

Historical Perspective

Many factors led to the startup of regional shared service clinical engineering services. They included the electrical safety scare of the late 1960s; Joint Commission on Accreditation of Healthcare Organizations (JCAHO) standards changes; the general lack of expertise of hospitals in the area of medical equipment safety and support; the prohibitive cost of developing full clinical engineering programs in most community hospitals; and the fragmentation of medical equipment support services.

Most importantly, clinical engineering is one area of health care services that are lucky enough to have a major benefactor to fund the start-up of cooperative programs across the United States. The W. K. Kellogg Foundation (Battle Creek, MI) funded regional clinical engineering shared services in the early to mid 1970s (Dodson and Latimer, 1975). The foundation was not new to funding shared hospital services as it had had success in funding management engineering programs in the past. The Foundation funded universities and hospital associations with grant money to start clinical engineering shared services in their state or region. Beginning in 1972, over $3 million in funds were distributed, first to the Northwest Ohio Clinical Engineering Center (Toledo) and later to organizations in Arizona, California, Arizona, Hawaii, Illinois, Maine, Massachusetts, Michigan, New Hampshire, New Jersey, North Carolina, Pennsylvania, Texas, and Vermont, along with the Emergency Care Research Institute (ECRI) in Plymouth Meeting, PA. Some of the early leaders of these programs include Burt Dodson of Carolinas Hospital Engineering Support Services, Malcolm Ridgway of Shared Biomedical Engineering Service, Frank Kane of Medical Instrumentation Systems, and Alex Schwan of Northwest Ohio Clinical Engineering Center. These programs primarily serve small to mid-sized hospitals. (See Figure 7-1.)

Other shared service programs developed from a variety of other funding sources. Initially, most programs were nonprofit tax-exempt corporations. The American Society for Hospital Engineering (ASHE) was the base for the Committee on Shared Clinical Engineering, which fostered networking between groups. Mike Brinkman of Hospital Maintenance Consultants was a primary coordinator who worked with the shared service organizations. Similar shared programs developed in Canada notably the Regional Clinical Engineering Service in Moncton, New Brunswick, which provided services for seven hospitals in the southeastern region of the province.

Today, clinical engineering programs starting in developing countries often model themselves after the shared service model. The Advanced Clinical Engineering Workshops (ACEW), conducted under the sponsorship of the Pan American Health Organization (PAHO) (see Chapter 70) and the World Health Organization (WHO), are presented by faculty from the American College of Clinical Engineering (ACCE) (see Chapter 130). Many of the participants in these programs are part of regional, multi-hospital or university-based shared service organizations in their countries.

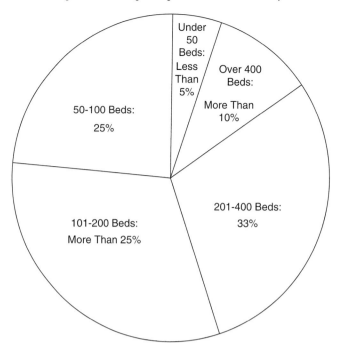

Figure 7-1 Hospitals served by clinical engineering programs as categorized by bed size. From Dodson and Latimer, 1975.

Successful Implementation of Shared Service Clinical Engineering Programs

Although over 20 shared service clinical engineering programs were in existence in the 1970s, few remain, and almost none remains under its original charter. Some were turned into successful private companies. The shared service program organized under the Massachusetts Hospital Association is today Technology in Medicine http://www.techmed.com/. The Hospital Council of Southern California's Shared Biomedical Engineering Service is now MasterPlan, Inc. http://www.masterplan-inc.com/. Other programs ceased operating for a variety of reasons, and a few were bought out by biomedical maintenance companies.

New forms of shared programs operate as for-profit subsidiaries of health systems or cooperatives. Examples would include the following:

- Premier Inc., Clinical Technology Services http://www.premierinc.com, is collectively owned by more than 200 independent not-for-profit hospitals and health systems in the United States and operate under, or are affiliated with, more than 1,800 hospitals and other health care sites.
- Technical Dynamics, Inc. (TDI) http://www.tdibiomed.com provides biomedical and radiology equipment management, maintenance, repair, and clinical engineering support services in Virginia, Delaware, Maryland, and Washington, DC, operating under the auspices of Inova Health Systems.
- TriMedx Health Care Equipment Services http://www.trimedx.com/ is part of Ascension Health, the largest not-for-profit health care system in the United States.

The shared service programs that remained were viable because of several key factors:

- A strong relationship with the state or regional health care organizations. Examples would include state hospital associations, hospital engineering societies, hospitals affiliated with tertiary care centers, and allied health groups;
- Regular communication with customers via meetings, educational events, contributions to nursing and other regional forums, and other interaction. Sponsoring and conducting seminars is one method. E-mail Listservs and Web-based newsletters are good tools to employ;
- Leadership to guide the programs through the hard economic times;
- Proactive program changes to embrace health care technology management—not just maintenance and safety. With the increased safety and reliability of equipment, limited reimbursement for health care services, and less prescriptive regulations, fewer resources could be allocated for maintenance and safety of biomedical devices. Clinical engineering shared service organizations needed to focus on assisting hospitals with the high-expenditure areas of procurement, technology planning, and liability;
- Focusing services on the areas where the dollars were spent in health care technology—radiology, clinical laboratory, and information technologies. A high percentage of the clinical equipment maintenance costs is devoted to expensive, high-tech imaging, laboratory, and clinical computer system);
- Adapting to fewer prescriptive requirements for inspections;
- Moving to fix price service in the age of managed care. Managed care contracts for health care services based on a fixed cost per patient. Many hospitals want the same arrangement for their medical equipment service. Budgeting takes the risk and other unknowns out of financial planning;
- Justifying their existence and resources routinely and when large biomedical maintenance companies proposed to take over clinical engineering services. It was essential for shared services to know their costs on a unit/device basis, to be able to justify services, and to show cost savings for the institutions that they served; and
- The ability to use technology, creativity, and other resources to overcome the logistical problems associated with providing clinical engineering services in a typically rural region. Important elements include the availability of courier services for same-day shipping of parts and equipment, the use of the Internet for communications, interaction with the equipment-management database, education, and customer reporting, as well as specialty staffing such as imaging or dialysis service providers, clinical engineers and account managers located for fast response.

Technical Services Program

The Technical Services Program (TSP) http://its.uvm.edu/ at the University of Vermont is the only shared service clinical engineering program that is still under the same governing body as it was when it was created by the Kellogg grant in 1973 (Clark, 1992). It operates as a nonprofit, independent organization. TSP, which is part of the Instrumentation & Technical Services Department at the University, is a group that is focused on providing community service through clinical engineering services and education—two key elements in the university's mission.

The program serves most hospitals in Vermont, 11 hospitals in northeastern New York, and four in northwestern New Hampshire through contractual arrangements. Many of these sites have resident biomedical equipment technicians who provide direct maintenance, service management, and communicate with users. A staff of three certified clinical engineers, 28 biomedical equipment technicians (BMETs) and additional specialists, managers, and support personnel make up the group. One clinical engineer (CE) supports the teaching hospital, Fletcher Allen Health Care; a second CE consults with community hospitals in New Hampshire and Vermont, and the third covers New York member hospitals. Twenty-two members of the BMET staff are in residence at hospital sites. A fleet of ten vehicles is used to travel to sites in a geographic radius 150 miles to the east, west, and south of the main office in Burlington. A Vermont Hospital Association-sponsored customer board oversees the program and meets quarterly to discuss services, costs, needs, issues, and future planning. This is a key element in the continued development of the program.

TSP has focused on all aspects of health care technology management. A strong effort is made to assist and participate with hospitals in the area of technology planning and acquisition to allow TSP to show its value as professionals assisting hospitals in assessing new technologies, providing guidance on equipment replacement, and in acquiring cost-effective and supportable equipment. In the latter function, TSP can provide significant input into including service training, tools, and documentation with purchase equipment to allow quality support regionally.

Another important element is education and communication with health care staff. Quarterly seminars (Figure 7-3), newsletters, educational participation in regional clinical and facilities engineering society meetings, and, of course, vocational training for university biomedical engineering students in the area of clinical engineering are important parts of continuation of this program (Figure 7-4).

Initiatives of TSP have included:

- Procuring a second grant from the Kellogg Foundation to develop the first IBM-PC-based equipment management system (Hospital Engineering Management System (HEMS));
- Partnership with local industry (Bio-Tek Instruments, now Fluke Biomedical) to develop biomedical test equipment;
- Using the service volume to develop contract-discounting and warranty service arrangements with major medical equipment manufacturers;
- Development of an independent, cost-capped maintenance insurance/group contract arm, Capital Asset Protection Partnership, for reining in service costs for imaging, laboratory, and other high-tech equipment; and
- Participating in a pilot program to replace the FDA's MedWatch reporting system with an Internet-based system.

Although the continued viability of TSP and other shared services depends upon the proactive measures the organizations take to adapt to the changing environment, the regional, nonprofit shared services base of operation will always be an advantageous position for clinical engineering services.

References

Clark JT. Regional Shared Service Clinical Engineering Programs, AAMI Annual Meeting presentation, 1992.
Dodson B, Latimer B. *Strategies for Clinical Engineering through Shared Services.* Battle Creek, MI, W. K. Kellogg Foundation, 1975.

Further Information

Leaders bring expertise to Latin America. *Biomed Instr Technol 36(3):156, 2000.*

Nationwide Clinical Engineering System

Diego Bravar
Research Area of Trieste, CIVAB
ITAL TBS SPA
Trieste, Italy

Teresa dell'Aquila
Research Area of Trieste, CIVAB
ITAL TBS SPA
Trieste, Italy

In a modern hospital, most diagnostic, therapeutic, and rehabilitation activity is based on the extensive use of medical technologies. Between 1983 and 1987, in order to improve the quality of health care services, the Italian National Research Council (CNR) instituted a special project to develop new medical devices and new clinical engineering services. The project had three components, one of which, named ACMAGEST, concerned the assessment of in-house clinical engineering departments (CEDs) by means of the guidelines for their establishment (Donato and Bravar, 1987).

At the end of a widely coordinated activity, that over the years has interested a growing number of operating units (23), the analysis of clinical engineering services development was contradictory. In fact, if from one side, to the existing four services were added six more, from the other the implementation rate of CEDs in Italy was definitely too low to allow Italy to come alongside of the other developed countries. Moreover, when the subproject, ACMAGEST, was completed, CEDs implementation rate suffered another slowdown. Therefore, during the 1990s, because of a drastic financial cut requested by the Italian government, it was impossible to use the internal resources to optimize, the management of biomedical equipment. For this reason, outsourcing the management of the biomedical equipment to private companies became a viable alternative.

In spite of the events of the past, Italy is now slowly on the way to transforming clinical engineering from the sporadic, experimental, and episodic to a reality that will be numerically significant, consistent, and pervasive throughout the national health care scene. Although the number of Italian public hospital beds served by clinical engineering activities has been increasing over the last five years (+80.9%), the number of CE departments within Italian public hospitals has remained low. (Only 20% of total hospital beds are managed by CE.) Today, CE services performed by external private companies are increasing, and the amount of hospital beds served by internal CE departments are about 8%, as compared to 12% served by external private companies.

ITAL TBS activities in the National Health Care Service

ITAL TBS SPA is an independent, third-party service organization established in Italy in the early 1980s after the conclusion of the ACMAGEST subproject. Its main aim is to provide and promote nationwide clinical engineering (CE) services in public and private health care organizations. The company has been ISO 9002- and EN 46002-certified since July 1998.

ITAL TBS began its activities in the Childrens Hospital IRCCS Burlo Garofolo of Trieste in 1992, assigning full-time technical personnel and giving qualified advice for procurement activities. Since then, the company has expanded its CE service throughout the Friuli-Venezia – Giulia Region. Since 1995, its activities have spread nationwide as it supplies CE services to more than 70 hospitals (Fig. 8-1) and manages over 80,000 medical devices.

The laboratories and headquarters of ITAL TBS are located in the Area Science Park of Trieste. The company's Center of Information and Evaluation of Biomedical Equipment (CIVAB) is located here and centers its activity on the collection and distribution of technical and financial information on biomedical technologies. The company has developed and enlarged the range of its offered services. It started with many contracts for the management of biomedical equipment in hospitals and it developed through the impetus of two projects funded by the Italian Ministry of Health. One project, carried out at the CIVAB Laboratories in Trieste, entailed the assessment of the costs of medical device technology. The other, called the SPERIGEST project, was done at the CREAS-IFC Hospital (which specializes in the field of cardiopulmonary care) of the Italian National Research Council (CNR) in Pisa. ITAL TBS always has cooperated in research project shifting the outcome of these in real services (Bravar et al, 1997).

To perform its technical services alone, ITAL TBS employs a staff of about 150 technicians with specific experience in the maintenance of biomedical equipment (Fig. 8-2). The staff is coordinated by 30 engineers, some of whom hold Ph.D.s in clinical engineering. In addition, about 50 technical and graduate personnel are employed in other sectors of the company.

In all hospitals where ITAL TBS operates, technical staff operates on a permanent basis inside the facility to ensure timely intervention and uninterrupted technical support of clinical work, which often is difficult or impossible to achieve with classical maintenance contracts with manufacturers or vendors of the technology. The company offers support services to improve the management of technology, either in terms of the safe operation of biomedical equipment already installed or in relation to aspects connected with the optimization of the maintenance and/or operating costs of these instruments. The range of services offered includes global maintenance of all biomedical equipment installed in

Diego Bravar and Teresa dell'Aquila

Figure 8-1 Geographical distribution of clinical engineering services, located inside hospitals, provided by ITAL TBS.

hospital facilities, assurance of compliance with safety regulations, acceptance testing of new devices, strategic planning with hospital administration for equipment acquisition, preparation of budget specifications for the procurement of biomedical technologies, equipment evaluation and selection, technical evaluation of contract bids, and analysis and monitoring of operating costs.

Preventive Maintenance. All maintenance is generally performed by ITAL TBS staff who are permanently based in the health care facilities involved in the service. Preventive-maintenance servicing is carried out on a regular basis according to the type of appliance and the specific servicing schedules drawn up by ITAL TBS, with extensions where the service and maintenance manuals require. The company technicians carry out planned maintenance according to a detailed schedule established upon acceptance of the equipment. The service is based on operational requirements of health care facilities using the same equipment and on tried and tested schedules.

Unscheduled Repair. Since this service is usually carried out by ITAL TBS technicians who are permanently based in health care facilities, it guarantees prompt interventions and the almost immediate correction of a large percentage of malfunctions, which minimizes machines' down time. Intervention is guaranteed to take place within four hours of the receipt of a service request (unless otherwise specified in the contract). The computerized management of spare parts provides a complete, updated picture of the situation in general and available parts in particular.

Electrical Safety Testing. Electrical safety testing consists of either visual inspection or measurements of biomedical electrical equipment to verify compliance with the requirements of the CEN/CENELEC European standards. The objective is to inspect all electrical equipment found at the client hospital or clinic.

Quality Control. Normal use of biomedical equipment can lead to perceptible deterioration in performance, caused by wear and tear on mechanical and electronic components or, more frequently, by improperly calibrated or aligned equipment components. From the acceptance test onward, it is therefore advisable to conduct a series of tests, based on standardized criteria, designed to ascertain the performance level of a medical device.

Inventory Management. Inventory management establishes the numerical and financial value of the stock of biomedical equipment held by the health care facility. Data are analyzed according to specific statistical processes, such as type of appliances installed, distribution per purchasing year, and obsolescence index.

The Clinical Engineering Consultation and Acquisition Plan. For local health services, the supply of medical equipment that is needed to benefit from the new technologies must be acquired on the basis of a careful analysis of the many factors involved in the use of the technologies themselves. This analysis ensures the correct allocation of resources and avoids the purchasing of unnecessary or even nonproductive equipment. To make cost-effective acquisitions for health care organizations, any new medical technology must be chosen based on the knowledge of its performance, value, and availability. The determining of the correct ratio of equipment stock is carried out by ITAL TBS S.P.A. by means of a specific operational procedure that involves the comparison of total equipment stock with that identified in other developed countries in terms of ratio of this to their respective gross domestic product (GDP). The analysis is also based on the preliminary gathering of specific technical, clinical, and administrative information regarding the requested equipment. In order to ensure the appropriateness and sustainability of the proposed

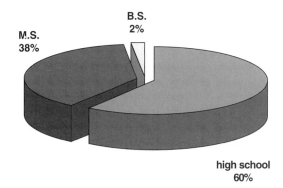

Figure 8-2 Personnel of ITAL TBS.

equipment, the selection of the make and model takes into consideration factors such as operating and running costs, system requirements connected with the installation of equipment, and analysis of alternatives.

Software for the Management of Biomedical Equipment

Since 1988, ITAL TBS S.p.A. has been producing information systems in the field of clinical engineering. Over the last ten years, the company has created a large database, from the activities it has monitored, which includes 80,000 medical devices, over 100,000 maintenance tasks, and about 40,000 electrical safety and quality assurance surveys. Starting with this large and valuable information base and utilizing the CIVAB coding system of medical devices that ITAL TBS S.p.A. developed for the Italian Ministry of Health, the company built an information system integrating technical, economic, and performance data.

ITAL TBS successfully built a complex information system that began operations in late 1998. Since then, it developed an installed base that now serves 30 sites, ranging from complex, multihospital public sites to medium-sized public and private clinics.

The databases are functionally and geographically separated according to the needs and communications costs constraints, while highly integrated by means of distributed and replicated database techniques, to fully achieve the goal of global activity management. The user has the ability to update and access local and global data with the most appropriate user interface ranging from a Windows-based client/server application for data entry to Internet-based data publishing for management at different levels and for customer analytical and synthetic reporting. All available legacy data was taken into account in order to build a system that would be able to supply useful steering and to control pieces of information from the beginning of operation.

The integrated informative system SI^{3}C (Figure 8-3) performs the following functions:

- The recording of each action of corrective maintenance. For each medical device, a computer version of the service manual is available.
- The recording of each action of preventive maintenance. For each medical device, the maintenance protocol is available, with information about the necessary instrumentation and the average time necessary for the action. The system promotes the preparation and performance of a plan of preventive maintenance for a given device.
- The management of the electrical safety tests according to the CE guidelines
- The management of the instrumentation calibration

The Italian Observatory of Medical Equipment and Devices

In order to improve the quality of health care services, the Regional Health Authority of Friuli-Venezia Giulia (a northeastern region of Italy) began an original, three-year program called "Observatory of Medical Devices" (ORT) in 1992. The program's aim was to assess and monitor the existing biomedical equipment in all regional hospitals and to monitor the flow and the expenditure for the most important categories of medical devices (e.g., diagnostic reagents and X-ray films). The ORT program was operated by the CIVAB of ITAL TBS at the Area Science Park of Trieste (Bravar et al., 1995). Concurrently, some Italian regional hHealth authorities (Friuli-Venezia Giulia, Veneto, Emilia-Romagna, Lombardia, Lazio, Sicilia, Trento, and Bolzano) made an agreement with the Area Science Park of Trieste in order to offer advanced services in the field of biomedical technology to each of their public health structures.

In 1997, the Italian Ministry of Health financed a three-year project of the Regional Health Authority of Friuli-Venezia Giulia called "National Observatory of Prices and Technology" (OPT). The aim was to increase the available instruments to all public health structures providing technical and economic information for the more efficient management of medical devices (Bravar et al., 1997). The OPT program, which has been operated by ITAL TBS CIVAB in the framework of an agreement between the Regional Health Authority and the Area Science Park, has developed the following products:

1. Technical bulletins on 36 different categories of biomedical equipment available on the national market (18 issues/ per year)
2. A national database of medical equipment and devices and their suppliers, storing technical and commercial data
3. A national coding system (CIVAB Code) of medical equipment and devices, based on the standard ACMAGEST code
4. A monitoring system of medical equipment and devices purchase prices, based on data collected from a network of 20 Italian regions
5. Remote consultation (via the Internet) of technical information and purchase prices of medical devices in the National Health System

The device and supplies cost-monitoring system has been activated for some of the most important groups of medical devices: 40 different categories of biomedical equipment, representing more than 60% of the market of medical device products in Italy, including diagnostic reagents, X-ray films, hemodialysis filters, implantable pacemakers, angiographic catheters, and knee and hip joint prostheses.

At the end of the activity, the national coding system (CIVAB code) for medical devices and equipment and the 36 different technical bulletins of biomedical equipment, including a data sheet on more than 1500 devices, has been sent out to about 350 Italian Public Administration of National Health Services. The national database of medical equipment and devices and their suppliers (Figure 8-4) consists of more than 26,000 different models of biomedical equipment, 35,000 different models of medical devices, and more than 3,500 addresses of suppliers. The OPT database contains technical and economic data for about 19,000 biomedical devices purchased in 40 Italian Public Hospitals. A password-protected web site, where OPT data and code and analysis of price data are available for all of the Italian Public Hospital Administration. At the end of the project, detailed and up-to-date information about technical and economical parameters of biomedical products is available via a network involving regions and their public health structures.

The Portal for E-Management of Medical Devices in Italy

ITAL TBS, thanks to its wide experiences as a Clinical Engineering Services Company and as manager of the Italian Observatory of Prices and Technologies, is developing a web portal to offer high-quality information about biomedical devices. The model of the "Electronic Office for Medical Devices Management and Procurement" is based on the transfer of the offline core competencies of ITALTBS in an online context: The customer will be able to obtain the technical specifications of equipment, after answering a set of questions about his clinical needs and, more generally, about his activity. In particular, the available services are free-access information, online consulting, and online auction. (See Figure 8-5.)

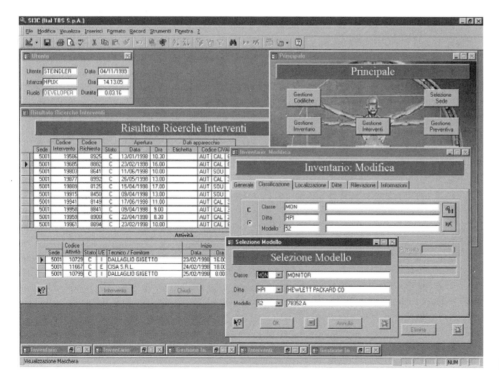

Figure 8-3 SI^{3}C main application user.

Nationwide Clinical Engineering System

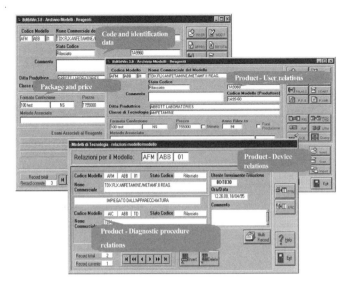

Figure 8-4 The national database of medical equipment and devices.

Available at the first level are the principles of operation for the most important medical equipment, the most interesting news from the health care field, discussion forums on topics of great interest in this field, workshop and congress, and news (product and actions) from manufacturers. To access the second level, the customer must be registered as a client. At this level, he finds detailed technical features and data for each product as well as informative sheets on biomedical equipment. He is guided by a list of questions to the technical specifications of the device needed. The portal shows all of the commercial products and the median price for a particular configuration, and it lists the minimum technical-specifications requirement. On demand, the specifications can be included in a specific document that is necessary for a tender of public administration.

The third level is the organization of inverse auctions, where the technical specifications could serve as a pre-qualification and the bidders could make their economic offers online, lowering the price in order to win the bids. The complete model of the web portal for the "Electronic Office for Medical Devices Management and Procurement" is described in Figure 8-5.

The SPERIGEST Project of the CNR

Since 1995, within the activities of a research project called SPERIGEST of the Italian Ministry of Health and the Italian Authority of the Region Tuscany, the CNR Hospital of Pisa (CREAS-IFC), which is a Hospital Cardiovascular Institution (including the Clinical Physiology Institute of Pisa and the Apuano Pediatric Hospital), implemented the integrated management of its services. The goal of SPERIGEST was to establish the guidelines for the optimized management of a cardiovascular hospital by means of a new model for the management of biomedical equipment and the implementation of integrated software systems for the management of clinical data, signals, and images. Furthermore, in order to optimize the management of biomedical equipment and information systems belonging to the National Health Care Service, a subproject of SPERIGEST, dedicated to clinical engineering, had the goal of ensuring cost-effectiveness and safe and appropriate technology management of all equipment involved in the Hospital's operation. The CREAS-IFC outsourced to ITAL TBS and to the software company INSIEL a subpart of the project including the clinical data, signals, and images; administrative data integration; and the management of biomedical equipment.

The Clinical Engineering subproject of "SPERIGEST." It was a unique opportunity to define the rules within which private companies can promote the diffusion of clinical engineering services at the national level and to redefine the role of CE services in the field of information technology. To ensure cost-effectiveness, safe and appropriate technology management of all the equipment involved in hospital administration, during 1995 the CNR-IFC enlisted ITAL TBS to established CE services. Working out of 8 laboratories, it covers 11 hospitals located in Northern Tuscany for a total amount of nearly 2000 beds and 8000 devices with a capital value of about $60 million.

The ITAL TBS staff of the SPERIGEST project included: 1 biomedical engineer as Project Director, 3 clinical engineers, 16 biomedical technicians, 1 biomedical engineer for economic analysis, 3 computer technicians, and 2 administrative and other biomedical engineers for medical computer science. During this period ,the CNR budget for corrective maintenance (full risk) of biomedical equipment decreased from 7% of the substitution value of installed medical equipment in 1994 to 5% in 1998, mainly because of the clinical engineers intervention and extensive accessibility to integrated information (Fig.8-6). SPERIGEST has been useful in analyzing the set of activities to be carried out by specialists working inside hospitals. Table 8-1 shows the percentage of effort dedicated to the various project tasks.

SPERIGEST has shown how a CE service can dramatically reduce the time for resolving problems with medical equipment, as shown in Figure 8-7.

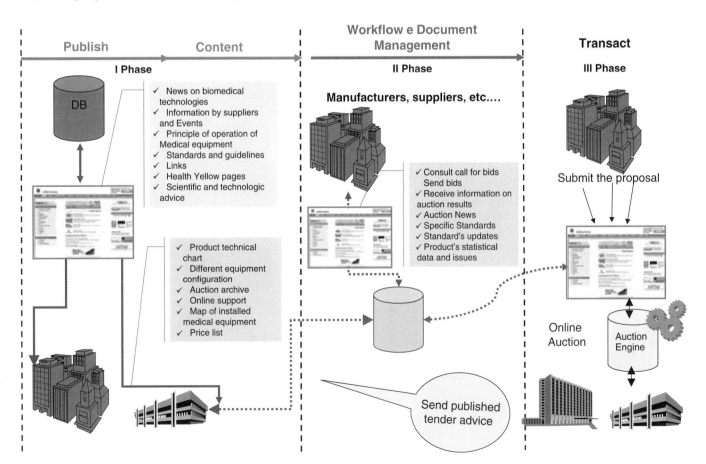

Figure 8-5 Electronic office for medical device management and procurement.

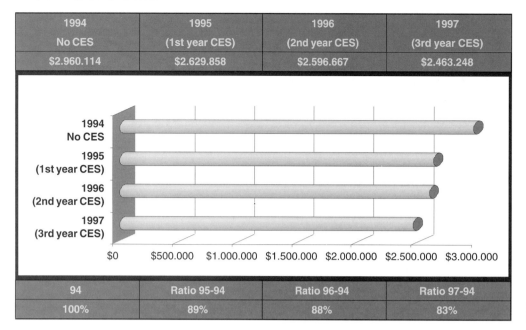

Figure 8-6 Budget for Italian CE service in SPERIGEST.

In conclusion, the outsourcing (total or partial) for clinical engineering services is the better solution for the health care system. The clinical engineering service company represents the better choice in terms of cost-benefit ratio.

"SPERIGEST" software for data, signal and image integration. The principle that has inspired the software-systems structure of SPERIGEST is the maximum integratibility and flexibility with different networks and software. The data flow for the integration is based on 3 archiving levels: (1) administrative, (2) clinical, and (3) multiplicity of local archive named *functional island*. See Figure 8-8.

The first two are central. The functional island is a more defined subsection with its own archive and specific software for diagnostic laboratory as hemodynamic, nuclear medicine, CT scanners, Holter, and digital radiology. The integration of these archives is made by two different methods: middleware and Web. Using standard Internet browsers the web permits the view of clinical and administrative data. The total system connected 7 different sites (both clinic and administrative) managed by 14 nets (Ethernet, Token-Ring, MAN, Serial), 8 routers, and 60 clients.

Inside the SPERIGEST project, ITAL TBS has integrated the software of the cardiology departments of the Hospital (made by CREAS-IFC) with administrative software (made by INSIEL). This integration makes possible the online analysis of health care activities and their costs for each patient. At the end of the project, the SPERIGEST programs are working routinely for hospital admission and discharge, ambulatory reservation, Electronic Medical Record integrated with 6 different diagnostic laboratories and the administrative system, and web tools for hospital management.

Medical Information Technology and Telemedicine Project

ITAL TBS, in the Cardiovascular Hospital of the CNR of Pisa, by using the SPERIGEST project, data, signals, and images have been integrated only for the cardiology departments. Furthermore, ITAL TBS acquired the Austrian company PCS that has already developed and installed, in about 50 Austrian Hospitals, SW (PATIDOK™) for data, signals, and images integration for all hospital departments.

PATIDOK™ is a state-of-the-art hospital information management system that has proven its worth over the past 12 years as highly specialized, user-friendly software that can be configured to satisfy any individual requirement. As a universal program for processing the full range of internal and external workflow processes of health institutions, it embodies intelligent software with impressive gains in application. The clear and simple graphical user surface, the user-specific configuration/limitation of the system, and the tried-and-trusted user guide through the application of the system ensure short training times, a high degree of user acceptance, and the avoidance of application errors.

Many institutions, such as hospitals and clinics, want to organize their internal procedures and their communication with extramural institutions as efficiently as possible. IT systems are meant to organize and accelerate procedures, to rapidly provide needed information, and to collect desired data completely and easily. The application of these systems should be as easy as possible, should guide the user, and should be perfectly adaptable to internal procedures.

PATIDOK™ meets all of these everyday organizational requirements. It is a workflow system that can be individually configured and thus has been able to control workflow difficulties in the health care sector for many years. Starting from the experience of the SPERIGEST project along with PCS's experience in the Austrian market, ITAL TBS aims to develop integrated services of clinical engineering and medical information technology nationwide.

ITAL TBS founded the Second Opinion Italy, a telemedicine company in partnership with Second Opinion BV, a worldwide leader in these services. The service enables the reception of an additional medical opinion provided by specialist physicians in leading American hospitals such as Stanford University and UCSF (University of California San Francisco), just few days from the request.

Table 8-1 Distribution of workload dedicated to various project tasks

Compliance with safety standards	29%
Corrective maintenance	60%
Preventive maintenance	8%
Quality control	2%
Purchase consulting	1%

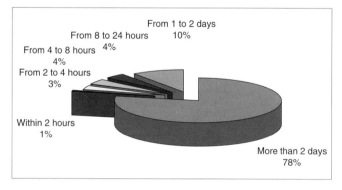

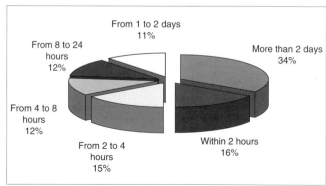

Figure 8-7 Resolution time for manufacturers (top) and for CE services (bottom).

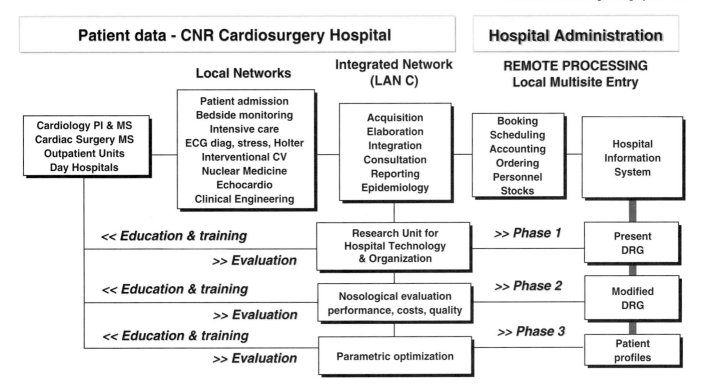

Figure 8-8 SPERIGEST: Clinical–Technical—Administrative integration of hospital resources.

In order to access the service, the user must forward just the medical record to the operational center, located in Milan. The record is digitalized there and forwarded to the specialist. With Second Opinion Italy, ITAL TBS aims to make available a very simple, but innovative, telemedicine service in order to deliver remote care to patients, combining communication technology along with medical expertise.

in this way. The Clinical Engineering Service Company represents the best choice in terms of cost/benefit ratio since the process of integration starts from the biomedical equipment. Then, it is possible to join the know-how and the expertise of clinical engineering with information technology expertise.

Conclusion

The quick evolution of biomedical equipment and information technology together with the raising need of integration of data, signals and images in the hospital produce the need of specialized expertise in computer science and clinical engineering. The health care process is complex. The management tools to make strategic decisions depend on the quality of information. The information for health care service management is generated from two different systems: the administrative system and health care medical record. Information technology should play a large part in the integration of these two systems. The partial outsourcing of the integration process and further operations, including clinical engineering services, is the solution of choice to let health care institutions deal with the more appropriate clinic activities only. A full custom solution can be easily achieved

References

Donato L, Bravar D, "Ruolo e Attività dei Servizi di Ingegneria Clinica," C.N.R. Progetto Finalizzato Tecnologie Biomediche e Sanitarie, Sottoprogetto ACMAGEST., rapporti finali sulle problematiche di acquisizione, manutenzione, e gestione delle tecnologie biomediche, 1987, vol. IV, Roma, pp 1-194.

Bravar D, Benassi P, Ginghiali A, Marcaccioli G, Niccolai M, Rizzo V. "CNR Sperigest Project: Italian Ministry of Health guidelines for developing clinical engineering services." *Proceedings of World Congress on Medical Physics and Biomedical Engineering* 1997, p 1220.

Bravar D, Giribona P, Iacono M, Giuricin C, Carbi N. "A Regional approach for the management of medical technology in Italy: The Oservatory of the Friuli Venezia Giulia Region." *Proceedings of International Conference on Clinical Engineering* 1995, pp 197-210.

Bravar D, Buffolini F, dell'Aquila T, Giribona P, Giuricin C. "A national approach for the management of medical devices in Italy: the Observatory of Prices and Technology." *Proceedings of World Congress on Medical Physics and Biomedical Engineering* 1997, p 1214.

Clinical Engineering and Biomedical Maintenance in the United States Military

Joseph P. McClain
Gilbert, AZ

The biomedical maintenance business area presents significant opportunities to reduce costs through regional cooperation and exploration of alternative maintenance management strategies. Maintenance activities present a substantial cost center for most medical facilities directly through labor and maintenance contract costs and indirectly through equipment acquisition and life-cycle management. Maintenance operations are predominantly facility oriented, with regional or area maintenance activities limited to within service (i.e., Army, Navy, Air Force, Marines, and Coast Guard) lines. Maintenance services that exceed in-house capabilities routinely are outsourced through local contracting.

A regional approach to biomedical maintenance services is consistent with other significant changes in this business area. Maintenance functions are being broadened and deepened into the discipline of clinical engineering. The field of technology assessment and management is being formalized and integrated as a core competency of clinical engineering. Clinical engineering and medical equipment management functions, systems, and processes are assuming renewed significance in the overall management of health care operations as the entire medical logistics field becomes more synergistic and complementary.

Opportunities for Regional Partnerships

Regional opportunities for clinical engineering and biomedical equipment management include the following:

- Shared services and expertise
- A unified approach to technology assessment and equipment life-cycle management
- A unified approach to levels of service
- Shared training;
- Regional contracting for maintenance services;
- Equipment standardization; and
- A unified approach to alternative maintenance strategies, such as risk-based maintenance and *Prime Maintainer* contracts.

These opportunities come with notable challenges. Previous efforts to standardize equipment at the military health services (MHS) level have not been successful in achieving clinical acceptance. Centralized contracting for maintenance services have had uneven success in reducing costs. Differences in service culture and business processes, as well as concerns over the effect of change of jobs and independence, present further challenges that need to be addressed.

TRICARE, the U.S. military's health care system, is patterned after civilian-sector health maintenance organizations, making it in effect the only nationwide managed care system in the U.S. Each TRICARE region or regional business office should establish a subcommittee or process action team of clinical engineering leaders to develop cooperation and business process improvements. Results of these team efforts should be tied directly to capital-equipment acquisition, developed and implemented by the command and local regional contracting. One approach is for teams to collect data on maintenance cost centers, identify high-cost activities, and develop management strategies to achieve the goals of the program. Typical high-cost centers include the following areas:

- X-ray systems
- Laboratory equipment
- Anesthesia equipment
- Patient-monitoring systems
- Fiber optic scopes
- Radiology Digital Imaging Network-Picture Archiving and Communications Systems (DINPACS) or Medical Diagnostic Information System (MDIS)
- Area maintenance support
- Training
- Annual maintenance contracts
- Repair parts

Regional Management Strategies

Regional clinical engineering offers various improvements over the current model of facility-centered maintenance operations. National trends indicate that savings are likely in the following three areas:

1. Consolidation of maintenance services into hub-and-spoke service relationships
2. Regional consolidation of contract maintenance services
3. Alternative approaches to traditional maintenance strategies

Some categories of regional opportunities are emerging as the Department of Defense (DoD) further refines its logistics-support capabilities. For example, Army, Navy, and Air Force activities across the MHS are beginning to share responsibilities for training and developing a tri-service cadre of biomedical equipment technicians. In some cases, Army and Navy activities are using Air Force assessment standards. This cross-pollination is strongly encouraged—it will build a stronger clinical engineering community and will enhance the contributions of that community to the MHS at large.

Regional/National Contracts

Contractor services are an essential adjunct to DoD's in-house maintenance capabilities. Contractors provide more specialized maintenance capability than DoD can affordably train and retain. Contractors allow DoD to acquire and support a broader range of equipment types than otherwise would be possible. Contractors permit increased flexibility in the types and frequencies of maintenance services. Not all contracted services are efficient, however. When separate maintenance contracts are negotiated and administered locally for each DoD treatment facility, service prices might be higher, contractor responsiveness may be reduced, and internal overhead pertaining to award-and-administer contracts is much greater. Individual facilities do not always have the expertise to develop, award, and administer the sophisticated contracts that are needed to obtain today's high-tech biomedical engineering services, but within most DoD TRICARE regions, the expertise does exist.

Maintenance service contracts are developed through statistical averages and expected mean time between service activities for particular equipment types. Regional contracts allow vendors to spread risks across a larger equipment pool and result in more cost-effectiveness than is the case with local contracts. Proven cost savings can range from 10% to 30%, according to prototype programs underway in DoD. Regional maintenance contracts also may allow consolidation of exchange items to provide float while services are performed and permit the consolidation of contractor-owned repair parts and test, measurement, and diagnostic equipment (TMDE) regionally.

Managing Levels of Service

Service response standards can be a significant contributor to maintenance contract costs. The cost of 4-hour service response 24 hours per day, 7 days per week will be significantly more expensive than more liberal standards. A regionally coordinated approach to service levels may reduce costs by providing a more sophisticated approach to the development of Statements of Work and applying a more analytical risk-benefit assessment for equipment to be covered. In other cases, additional risk may be tolerable in an area if equivalent services or equipment are available in the region. Opportunities may exist for shared or back-up support for first response, allowing contract-service-response requirements to be reduced.

Prime Maintainer Contract

Some service organizations are broadening their offerings to provide Prime Maintainer services in which they provide maintenance services that otherwise would be outsourced to multiple equipment maintainers. Under the Prime Maintainer concept, a single service organization provides turnkey support, usually under a negotiated fee schedule, prospective warranty program, or on a capitated basis. Prime Maintainer programs usually allow the in-house clinical engineering organization to enhance its own technical expertise by giving them first-look rights to service equipment or even an opportunity to be reimbursed by the Prime Maintainer for services provided in-house.

For remote sites with low equipment density or with little or no maintenance staff, credit cards, rather than in-house or annual contract support, may be a cost-effective solution. Credit card use must have well-defined controls, including maximum allowable ceilings, prohibitions on certain equipment items, and assurances that they will not be used when other services are available.

The most effective maintenance operation combines the benefits of regional consolidation with the flexibility and responsiveness that is available through other means. It is important to find the right balance among the various options on the maintenance continuum.

Time and Material versus Annual Contracts

As a variation of the Prime Maintainer concept, the in-house management of preventive maintenance and repair services on a time-and-materials (T/M) basis (rather than an annual service contract) may provide cost savings. This approach may be advantageous particularly when there is potential for in-house maintenance support from elsewhere in the region, or when the equipment or service is available.

Risk-Based Preventive Maintenance

Significant savings can be achieved by accepting fewer preventive maintenance or calibration services than are traditionally used. This approach is supported by guidance that indicates health care systems may reduce preventive services by evaluating each generic type of equipment, performing a risk analysis, and matching an appropriate maintenance response. Risk analysis is ongoing and involves continuing adjustment of maintenance frequency, based on gradually accumulating maintenance experience, changing maintenance practices, and technology insertions and improvements. Much of today's equipment is highly reliable and engineered for increased time between preventive services and inspections. In addition, self-diagnostic or remotely diagnosable equipment permits diagnostics or repairs to equipment components without on-site service. Preventive maintenance services–risk adjustment requires more detailed analysis than simple scheduled services. Preventive maintenance is enhanced when equipment density is sufficiently high to spread risk across a larger number of equipment items.

Department of Veterans' Affairs Repair Services

Some DoD facilities (Walter Reed Army Medical Center (WRAMC) in particular) have realized significant savings by using the Department of Veteran's Affairs (DVA) repair facility in Chicago to repair fiber-optic endoscopes. These devices traditionally have been sent to their original equipment manufacturers (OEMs) for repair to ensure quality. Clinicians at WRAMC have been unable to discern a difference between repairs performed by the DVA and the OEM. The DVA cost has been 30% less, with excellent service turnaround times. Regional economics may provide even more advantage for such services. In this example, fiber optic scopes account for a large amount of costs. A regional evaluation of alternative sources, or a regional approach to contracting for such services, can bring these costs down. Another factor that affects costs for scope repair is the cyclic arrival of new medical students or residents in larger facilities. Experience has shown that specific training targeted to these individuals can significantly reduce damage to these instruments, thereby reducing repair costs.

Walter Reed completed a pilot program to evaluate the DVA Durable-Repair Program, using the DVA repair facility in Chicago for the repair of fiber-optic scopes. During fiscal year 1999, documented savings were $124,000, with the average individual repair cost reduced by 35%. Using the DVA improved average turnaround time by 44%.

Equipment Standardization

The standardization of medical equipment may provide significant cost savings throughout the equipment life cycle. A regional approach to equipment acquisition may provide more favorable terms and prices and also may reduce costs for training and maintenance. A prerequisite for effective equipment standardization is a thorough, broadly accepted equipment-requirements determination process, such as the technology assessment and requirements analysis (TARA) process that is used to develop Army requirements for radiology and laboratory equipment. Regions are encouraged to review the current TARA process and to find ways to use or adapt their equipment standardization.

Certain equipment categories lend themselves especially well to standardization because of their high cost, high density, relative criticality, and performance according to pre-existing commercial medical standards or degree of integration. These categories will provide the primary standardization targets for DoD health care regions during the next round of regional operations. Candidates for standardization include the following:

- Equipment approved for use on Air Force medical evacuation aircraft (patient-movement items (PMI))
- Imaging equipment, including digital imaging, ultrasound, and other diagnostic imaging product lines
- Endoscopy equipment and other minimally invasive surgical or diagnostic items
- Physiological monitoring systems, especially integrated patient-monitoring systems; and
- Laboratory and diagnostic equipment, especially HL7-compliant computer integrated systems.

Hub-and-Spoke Services

By forming regional clinical engineering activities, logistics leaders can create hub-and-spoke maintenance organizations that pool technical expertise, TMDE equipment, repair parts inventories, and other assets that otherwise could be distributed unevenly among a region's facilities. The regional, tiered approach to clinical engineering concentrates highly technical and increasingly scarce clinical engineering staffs at the regional level. It permits implementation of the most effective training, cross-utilization, and professional development patterns within and between regions. It will permit the substitution of in-house, hub-owned, clinical engineering staffs instead of expensive, seldom-used services contracts at spoke locations. Finally, it provides the most effective regional visibility of equipment density, utilization, repair record, and modernization needs, all of which are consistent with the shift to technology-assessment roles for DoD's emerging clinical engineering organizations.

More collaboration is possible when DVA facilities are included in regional hub-and-spoke or other collaborative arrangements. DoD's facilities have a much different geographic distribution, with a tendency toward southern states and a concentration of larger medical centers in or near urban centers. DVA facilities, on the other hand, tend to be concentrated in the northeastern United States, and most are not located in urban centers, thus suggesting opportunities for mutual benefit to DVA and DoD. Each organization potentially can provide maintenance coverage for the other in areas where assets are spread thinly.

Technology Assessment and Requirements Analysis

In an environment of constrained resources, it is imperative to apply sound commercial business practices to capital-investment equipment programs (procurement dollars, equipment costing greater than $100,000). The decision makers at the Lead Agent and Military Treatment Facilities (MTF) level must have viable means of acquiring the management information to effectively balance dwindling resources against clinical requirements. The ultimate goal of the TARA program is to establish a standardized methodology for assessing, planning, and pursuing the acquisition of technology within Health Affairs.

The TARA mission is to provide decision makers with the management information that is necessary to making informed decisions on the technology resources that are required to accomplish business-plan missions and to optimize clinical outcomes. This goal is accomplished through an unbiased assessment of radiology and laboratory functional departmental operations. A TARA consists of an assessment of requirements, current equipment, and operations as they relate to the equipment and clinical operations. Clinical issues are reviewed by the service clinical consultant/advisor or representative, and equipment issues are assessed by clinical engineers, medical physicists, and maintenance officers. The results of this assessment are provided only to the requesting facility and their respective lead agent. Trends and command-wide management issues discovered during these assessments will be presented to Health Affairs to assist in policy development and strategic planning. No specific reference will be made to issues at individual facilities except under extenuating circumstances, such as serious safety or risk management issues, or with the permission of the facility.

With Lead Agent visibility by tri-service regions, the return on investment is expected to be substantial. Such a program will assist in standardizing Health Affairs equipment and sustainment regionally, allow for seamless integration of technology, augment the tri-service equipment budget for high-dollar items (i.e., saving money with multi-system buys), and attempt to increase clinical availability and costs.

The Army has identified a cost avoidance of approximately $70 million over the life of the program and expects that this figure will double (at a minimum) by conducting TARAs on a tri-service regional basis. This cost avoidance is recognized by confining high-dollar equipment purchases to equipment that is needed and technologically relevant in military hospitals and clinics.

In addition, by conducting TARAs regionally, other costly issues such as clinical resource sharing can be resolved. For example, consider Army Hospital A and Air Force Hospital AF, which are located within 45 miles of one another. Based on a TARA assessment, Hospital A has three radiologists with a workload for only 2.2, and Hospital AF has two radiologists with a workload for 2.5. Hospital AF sends remaining exams to a civilian group at a cost. The TARA would recommend that one of the Army radiologists share time with the Air Force facility, in return for support in other areas, thus decreasing costs and overhead for both facilities. The Army's TARA team is working with the DVA and plans to incorporate it into regional issues where relevant and cost-effective.

The Transition from Medical Maintenance to Clinical Engineering

The MHS has been in the lead on medical maintenance issues for decades, and continues to set the standard as the profession is transformed from a maintenance orientation to a clinical engineering orientation. As defined by the American College of Clinical Engineering (Bauld, 1991), a clinical engineer is "a professional who supports and advances patient care by applying engineering and management skills to health care technology. Clinical engineers manage personnel, finances, instrumentation, and projects to promote the safe and cost-effective application of technology."

Clinical engineering is a necessity for the complex, regionally based health care systems of the future. The growing complexity of medical technology means that sophisticated equipment, software, and telecommunications systems must be integrated and interoperable. Clinical engineers will participate in the assessment and management of all health care technology throughout its acquisition life cycle. TARA of equipment utilization and needs will ensure deliberate management of clinical technologies and match technology capabilities with health care needs. Clinical engineering operations will be tightly linked with capital and expense equipment acquisition, facility design and management, and maintenance.

As with other logistics disciplines, the emergence of clinical engineering heralds an important change for biomedical equipment maintenance and its place in the MHS.

Increasingly, equipment maintenance is an integral part of a regional, tri-service, managed care strategy to obtain the maximum possible yield from all of a health care system's assets. Clinical engineering is one of many new tools available to logisticians as they seek to define and enhance their role in the MHS and the TRICARE system. The many new directions that are available for clinical engineers fit seamlessly with the other management tools now being used to transform military health care.

References

Bauld TJ. The definition of a clinical engineer. *J Clin Eng* 16:403, 1991.
Thomas Hughes, Concepts in Health Care, Inc., Waltham, MA: 1997.

Functional Economic Analysis for Defense Medical Logistics Standard Support, Office of the Assistant Secretary of Defense (Health Affairs), 1996.

Further Information

http://www.tricare.osd.mil/ The Military Health System Web Site is the Official Web Presence of the Office of the Assistant Secretary of Defense (Health Affairs) and the TRICARE Management Activity. Skyline 5, Suite 810; 5111 Leesburg Pike; Falls Church, VA 22041-3206
Based on earlier work done in the Army's Southeast Regional Medical Command, which has been exploring regional operations for the past two years, the effort will require an investment in hardware, software, telecommunications, personnel, and office space, although marginal costs will decline as the number of sites increases.

10

Careers, Roles, and Responsibilities

Nicholas Cram
Texas A&M University
College Station, TX

The primary driving forces of change in health care are technology, social and political referendum, and market economics (Bronzino, 1992). Changes in available and applied technology in turn affect the responsibilities of persons maintaining and acquiring that technology (Babcock, 1992). In health care, clinical engineers have duties and responsibilities covering a broad range of activities that relate to medical equipment and medical equipment technology (Feinberg, 1986). The clinical engineer is in a position to be the gatekeeper of medical technology. The health care landscape is definitely changing with some areas left barren and some areas of new growth. This environment of change provides several opportunities for career and personal growth in the clinical engineering profession. This chapter will examine traditional, recently acquired, and future duties and responsibilities of clinical engineers.

Traditional duties and responsibilities are related to equipment maintenance. Historically, these duties are a result of the microshock scare of the early 1970s. With large megahospital systems emerging or consolidating in the early 1990s, the need for medical-technology management arose creating several opportunities for clinical engineers. Telemedicine, virtual instrumentation, and the Internet offer still more opportunities for clinical engineers in the near future.

Duties and Responsibilities

Clinical engineers have responsibilities in two broad categories: technical expertise and management. Clinical engineering departments in smaller medical facilities typically do not have the resources to support medical technology assessment programs or functions associated with a chief technology officer. Therefore, clinical engineers in smaller facilities generally have more traditional duties and responsibilities. The author encourages clinical engineers to pursue all opportunities relating to the maintenance, management, and acquisition of medical technology, regardless of the size of the facility. One should not commit to responsibilities that one will not be able to successfully sustain. The traditional duties associated with the hospital clinical engineer involve equipment maintenance, equipment acquisition, consulting for renovation or new construction, and supervision of personnel (see Table 10-1). Equipment maintenance is associated with duties that include unscheduled corrective maintenance of medical equipment and scheduled preventive maintenance (PM) of medical equipment. The documentation of equipment maintenance and risk ranking (i.e., determining the frequency of preventive maintenance) is an extremely important duty relating to these activities.

Appendix 10-A (see Appendices) is a typical position description and list of performance evaluation criteria for a clinical engineer.

Equipment Acquisition

Equipment acquisition involves a planning phase, negotiating phase, and actual receipt of the equipment. Clinical engineers provide valuable technical expertise in the acquisition of medical equipment, which is acknowledged by hospital chief executive officers (CEOs). Medical devices are considered capital budget items i.e., durable items. The purchase of medical equipment should coincide with the mission and vision statements of the organization and should take into consideration current and future needs of the organization.

Equipment Maintenance

Equipment maintenance involves all activities relating to providing an adequate level of service and limiting down time of medical devices in the facility. Traditionally, equipment maintenance is categorized as preventive maintenance and corrective maintenance.

Preventive maintenance (PM) is a scheduled event. PMs are scheduled according to the risk ranking of the particular medical device. Risk ranking is a unitless number derived from a formula that incorporates two factors: (1) relative maintenance frequency and (2) the effect on patient care if the device fails. Another duty associated with the responsibility of equipment maintenance is incoming inspection. All equipment entering a medical facility must be inspected prior to use. The inspection procedure should include performance and safety tests. The College of American Pathologists (CAP), which certifies clinical laboratory operations, has added computer safety testing to its list of device inspection requirements. It could be more cost-effective or a matter of necessity to contract the maintenance of a portion of the medical equipment in the facility's medical equipment inventory. In those instances when outsourcing is required, contract review will be a responsibility of the clinical engineering department. Clinical engineers review service contracts to ensure that PMs are being performed as stated in the contract and that associated corrective repairs are consistent with the work orders sent to the vendors providing service.

Fiscal Accountability

Clinical engineers with managerial responsibilities are accountable for expenditures related to medical equipment (capital equipment) and personnel (operating budget). As health care continues toward consumerism in the United States, fiscal accountability and technology management will become more of a focus for clinical engineers. Fiscal accountability falls into three broad categories: operating budgets; capital budgets; and project budgets. Each of these categories could be a responsibility of the clinical engineer. This will be discussed in detail in Section IV (Management).

Building and Renovation Projects

Clinical engineers have technical expertise and basic engineering skills; therefore, health care facilities rely on clinical engineers during renovation and building projects. It is not uncommon for the clinical engineer to be the only degreed engineer in the health care facility. Clinical engineers assist architects in the design and layout of special systems relating to medical devices during building and renovation projects. The specific section on blueprint drawings concerning nurse-call systems and other electronic or electromechanical devices is referred to as *special systems*. In many instances, clinical engineers may be

Table 10-1 Clinical engineering responsibilities—traditional

Clinical Engineering Responsibilities—Traditional

Equipment Acquisition
Capital Budgets
Strategic Planning (5-Year Plan)
Equipment Maintenance
PMs and Corrective Maintenance
Incoming Equipment Testing
    Computer-Safety Testing
    Other Nonmedical Device Safety Testing
Service Contracts and In-House
Fiscal Accountability
Personnel
Capital Equipment
Operating Budget
Building and Renovation Projects
Writing RFPs
State and Federal Codes
Regulatory Compliance
Risk Management
Safe Medical Devices Act
Joint Commission on Accreditation of Healthcare Organizations
Management—Vision and Growth

Table 10-2 Clinical engineering responsibilities—contemporary

Clinical Engineering Responsibilities—Contemporary

Academic affiliation/teaching
Applications research and design
Consulting
Information Systems Support
In-service training
Technical/clinical investigation—Clinical Trials Support
    IDE compliance
    IRB participation
Technology Management
Technology Assessment

---

required to write a request for proposal (RFP), which outlines specifications for building projects. Appendix 10-B (see Appendices) is a sample RFP for a telemetry renovation.

State and Federal Regulatory Compliance

All state or federal regulations that relate to nurse-call requirements are the responsibility of the clinical engineer. State departments of health have health care facility requirements relating to radiation and laser safety. Clinical engineers are commonly assigned the role of laser-safety officer and training officer for radiation safety. This role involves conducting compliance surveys, distribution of safety-education materials, and ensuring that all required documentation is correct and current. Clinical engineers must have expertise in a broad range of regulatory areas to comply with national standards and codes such as the Safe Medical Devices Act (SMDA) and the National Electric Code (NEC) and regulations promulgated by such federal agencies as the Occupational Safety and Health Administration (OSHA), the National Fire Protection Agency (NFPA), and the Joint Commission on Accreditation of Health care Organizations (JCAHO) as well as local and state standards and codes.

Risk Management

The clinical engineering responsibilities for risk management can be categorized as quality-related or litigation-related. Documentation of all repairs and preventive maintenance as well as monitoring the scheduling and completion of these tasks are important roles in clinical engineering risk management. Clinical engineers are responsible for Food and Drug Administration (FDA) requirements relating to the SMDA. The clinical engineer may act as the chief investigator for medical device incidents that cause death, serious injury, or risk thereof, or he may coordinate investigation activities in accordance with SMDA requirements. Clinical engineers also may serve as expert witnesses in litigation involving medical device incidents.

Department Management—Vision and Growth

At the department level, activities are task-oriented. It is important that clinical engineers, as members of the management team, ensure that the focus of all activities reflect the larger picture of the mission and vision statements of the organization. All activities should be accomplished in an environment that fosters the core values of the system. Technology forecasting and an assessment of current and future needs are integral parts of department management. All successful department managers develop a department mission statement that correlates with the organization's mission and vision statements. An example of proactive technology forecasting and department vision is the role played by clinical engineers who are already anticipating the service opportunities provided by the Internet and telemedicine.

The primary role of the clinical engineer in departmental management is providing documentation for regulatory agencies and supervision of personnel. Although management styles vary according to personality and training, all successful CE management will have the qualities of consistency, respect, and ethics. Management of technical personnel requires an approach that differs from that of management of nontechnical personnel. Young, newly graduated clinical engineers will require closer supervision than will seasoned clinical engineers. In general, technical personnel should be allowed some independence in performing their tasks.

Clinical Engineering Responsibilities—Contemporary

Academic Affiliation/Teaching

Clinical engineering is a hands-on, application-oriented profession. Clinical engineers have a unique opportunity to bring their technical and clinical applications expertise into the classroom in a vibrant and meaningful fashion. The answers to the many challenges that will face future clinical engineers and other health care professionals cannot be found in the textbook. Anecdotal accounts of actual clinical engineering experiences create a visual impression that makes the learning environment interesting for students as well as demonstrating a link between the knowledge of the classroom and the applications in the real world. Although teaching opportunities are somewhat limited, all clinical engineers are encouraged to seek out opportunities of traditional classroom instruction or mentoring a group of clinical engineering students in a structured clinical engineering hospital-internship program.

Applications Research and Design

The patient/device interface is the domain of the clinical engineer. Possibly no other role in clinical engineering can have such a dynamic impact on improving medical devices as that of applications engineering. Clinical engineers understand the clinical environment in which medical devices must function. With more complex devices being introduced into the health care arena, the technology can be overwhelming for physicians, nurses, and clinicians. Clinical engineers understand the medical device customer. Subtle human factors elements, which might be overlooked by other members of the design team, will be obvious to the clinical engineer. Clinical engineers find employment with medical device companies and in applications research for academic institutions.

Consulting

This is a relatively new area of opportunity for clinical engineers. Given the broad range of knowledge and various areas of expertise possessed by clinical engineers, it is not surprising that a niche market has developed for such talents. The range of consulting opportunities spans technology management and maintenance expertise to all of the responsibilities mentioned in this chapter. If a particular health care facility or medical device manufacturing company does not have the resources to accomplish a medical device-related task, whether technical or managerial, clinical engineers can provide a plethora of expertise to any challenge. The federal government also provides opportunities for clinical engineering consulting in its patent office and within the FDA medical device compliance sector.

Information Systems Support

Microprocessor components have become commonplace in electromedical devices. In addition to embedded microprocessor systems, several medical devices have signal-transmission and database interfaces. Several monitoring systems are actually specialized personal computers (PCs). This trend in medical device design provides opportunities for clinical engineers to provide technical expertise of clinical interfaces for information systems (IS).

In-Service Training

As medical devices become more complex, the need for facility user training becomes a necessity. Clinical engineers serve a role as clinical educators for physicians, nurses, and clinicians. Demonstrating the proper use and application of medical devices serves to improve the quality of care received by patients as well as providing more efficient use of medical devices and reducing the clinical staff's fear of technology. This is a role that will continue to expand and to provide future opportunities for clinical engineers.

Technical/Clinical Investigation

Clinical engineers may serve as the coordinator or primary investigator for clinical trials support of medical devices. A clinical trial involves research of leading-edge technology or new applications of proven technology. Clinical engineers will have responsibilities for selecting the team required for the clinical trial and coordinating all of the activities and required FDA documentation. The clinical trials area provides an opportunity for clinical engineers to act as independent consultants or to develop a niche area within their health care facility that could be a source of revenue. The management section of this handbook (Section IV) discusses the categories of revenue-producing and non-revenue-producing departments within a health care facility. The clinical engineering department is traditionally considered in the non-revenue-producing or cost-avoidance category. Any opportunity that allows the clinical engineering department to provide revenue is definitely a value-added service for the department and the facility as a whole.

Two roles associated with clinical trials are investigation drug exemption (IDE) compliance and participation in the institutional review board (IRB). An IDE is given to a medical device for use on a single patient when no alternative devices can provide alternative support. Obviously, these devices would be used in extreme patient-morbidity situations or situations when a proven technology is used in a different application from what the original patent intended. (This does not necessarily imply higher risk for the patient.) In each of the roles mentioned, clinical engineers can provide technical support and expertise that is not available through any other engineering discipline.

Technology Management

Technology management is perhaps the area that will provide the greatest opportunity of employment for clinical engineers. Clinical engineers have both the technical expertise and management tools required for successful management of technology. Technology is the primary driving force for change in health care. Management of technology will continue to be a necessary element of successful health care systems from both a financial and quality-of-care perspective. Technology management in health care involves the audit, planning, and acquisition of medical devices. It is not uncommon for medical device technology, capital, and operational budget items to encompass 35% of a health care system's total budget. Technology assessment has become an important tool for technology management. Technology assessment is a continual process of evaluating the technology within a health care facility with consideration of current and future needs. The technology assessment activity may derive information from primary sources such as manufacturers or in-house evaluations of medical devices being considered for purchase or from secondary sources such as the Internet, consulting firms, FDA sources, or newsletters. Considering the impact of this component, it is easy to recognize why the position of chief technology officer (CTO) has evolved. In some organizations, the CTO will have responsibilities for both medical device technology and information services technology. A more common scenario in larger facilities finds the CTO overseeing medical device technology management and the chief information officer (CIO) overseeing information technology management. The two positions have areas of overlapping responsibilities. Since software programs comprise an innate organ of several medical devices, the CTO and CIO must collaborate if achievement of improved technical quality and fiscal control is to be realized for the health care facility as a whole. A Master of Business Administration (MBA) degree specializing in technology management in addition to a Master of Science degree in biomedical or clinical engineering provides the ultimate resource for a candidate seeking a position as a CTO.

Clinical Engineering Responsibilities—Beyond Contemporary

Clinical engineering has begun—and will continue—to encompass the areas of telemedicine, virtual instrumentation, and web-enabled devices and systems.

Telemedicine

Telemedicine is an electronic-transmission medium that provides the exchange of medical information between at least two persons in geographically separate locations. Essentially, one can consider the telemedicine system as a transmission tube with either special monitoring devices or adapted medical devices connected to each end. Telemedicine activities require knowledge of telecommunications technology, networking technology, and medical device technology. The most intriguing development in telemedicine involves the ability to bring devices into the home for self-monitoring or monitoring by a professional staff. This concept has come to be known as *telehome-health*. With the large numbers of baby boomers approaching retirement age, telehome-health could see a dramatic increase in demand and acceptance as a medical standard. Several issues involving reimbursement need to be resolved before widespread use of telehome-health can be expected. Opportunities abound for clinical engineers in this arena. Several clinical engineers have already made an impact there (see Chapter 101).

Virtual Instrumentation

Virtual instrumentation has evolved in response to a need for more user-friendly and customized medical devices. A virtual instrument is a personal computer application that utilizes existing sensors, transducers or actuators to process the signals and present the information in a format desired by the user. This allows a system to be reconfigured as the demands of the medical equipment environment change while utilizing the same hardware and software (see Chapters 100 and 136).

Web-Enabled Devices and Systems

The Internet will change health care delivery. Exactly what those changes will be are difficult to determine at this time. The Internet has already influenced the manner in which patients—the customers—obtain health care information (see Chapter 73). Web-enabled devices that provide monitoring and signal-transmission capabilities with software interfaces residing on the Internet are a technical reality that may soon become standard practice. This new and modern arena will provide exciting opportunities for clinical engineers. As one might conclude from the sections pertaining to telemedicine, virtual instrumentation, and web-enabled systems, these modalities are intertwined and interdependent. This is the future of clinical engineering. Further insight into the future of clinical engineering can be found in Section XIII of this handbook.

A Day in the Life of a Clinical Engineer (CE)

The Rookie

Jean Smith recently graduated with a degree in biomedical engineering with an emphasis in clinical engineering. Finally, after a month of no response to some 50 resumes, Jean received an offer of employment in the Clinical Engineering Department at City Hospital. Today, Jean starts the orientation process. First, there is the issue of what to wear. (*What do clinical engineers wear, anyhow?*) Jean decides not to overdress and goes business-casual. Jean had completed a three-month, summer clinical engineering internship at the neighborhood hospital clinical engineering department. That summer, she found the passion needed to commit to the profession of clinical engineering. Today would be a different atmosphere. Physicians and nurses might ask questions about a myriad of topics. During Jean's internship, there was always a CE or biomedical equipment technician (BMET) nearby to provide an answer. *"What if I don't know the answers?"* Jean pondered with anxiety. As Jean entered the building, she was met by the director of clinical engineering and escorted to the conference room, where about 30 other new employees were seated. The anxiety was broken temporarily when several three-ring binders were shoved in Jean's general direction. The orientation speakers (there seemed to be a dozen) each introduced themselves and passed out more binders and papers to be signed.

Finally, the general orientation session was over. The director of clinical engineering met Jean at the door of the conference room and exchanged pleasantries, indicating that the entire department was excited about the prospect of adding a clinical engineer to the staff. There was something mentioned about PMs that needed to be finished by the end of the month, but Jean did not remember exactly in what context that related to any assignments that might be given that day. When Jean and the director arrived at the CE department, there seemed to be a flurry of activity. Each BMET and CE that passed Jean's way gave a greeting and a smile.

"I noticed from your resume that you've completed an internship," stated the director.

"Yes," Jean replied, "at the neighborhood hospital in my hometown."

"Excellent," responded the director with a cheery tone.

"I guess we can skip training on the safety, defib, ESU, and NIBP analyzers?" added the director. Jean nodded in agreement but with some hesitation.

Her work area consisted of an electronic workbench inside an 8-foot, square cubicle. Jean was issued a basic electronic tool kit and became familiarized with the oscilloscope attached to a swing arm of the workbench. *"All seems somewhat familiar,"* Jean thought, realizing that the internship training was becoming invaluable. The director was off to a budget meeting and introduced Jean to the clinical engineering supervisor, Jan Jones. Jan had been with the CE department since graduation from engineering school 10 years ago. Jean spent the rest of the day with Jan being introduced to nurse managers and department directors, trying to keep mental notes on the hospital floor plan as well as all of the new names and faces. *"I'll never learn how to get around in this place,"* Jean thought.

That evening, on the drive home, Jean contemplated the first day on the job and realized that, at this point, worrying about not knowing the answers to questions that were never asked by the physicians and nurses was the least of concerns that day. Jean also marveled at the seemingly effortless manner in which Jan navigated the halls and elevators of the hospital and knew everyone's name. She was also impressed with the recognition Jan was given by each member of the hospital staff. *"Maybe I'll get there someday,"* Jean thought. *"Oh well, I survived Day One."*

The Senior Professional

As Jan Jones sat in traffic en route to City Hospital, thoughts of the morning's agenda in the clinical engineering department began to surface. Jan was doing the mental to-do list, which had become a Monday-through-Friday ritual developed over the past 10 years. It seemed to make efficient use of time when the morning rush-hour traffic slowed to a crawl and made the all-too-common situation more tolerable. Two-thirds of the way down the mental to-do list, Jan remembered item 14: *Escort the rookie around the hospital and make him/her feel comfortable on the first day.* Jan's first day at City Hospital 10 years ago seemed a distant memory. Things at City in the CE department were a lot different now. More personnel responsibilities as the supervisor of the Clinical Engineering Department, the staff had doubled in CE, more complex equipment, and a more visible presence among the other hospital co-workers due to the numerous planning meetings involving equipment acquisition, building and renovation projects, and the research partnership with the biomedical engineering department at City University. Traffic was moving again, and Jan's car seemed to find its own way into the now familiar parking stall at City Hospital. *"OK, morning ritual; grapefruit juice (Jan was trying to cut out caffeine), check the work orders, anybody call in sick today?"* The director had already left with the rookie for morning orientation. *"Glad I don't have to go through that process again,"* Jan thought. The rookie would be coming back for a tour of the department and the facility after lunch. Jan decided to work on the capital budget planning and review the month's operational expenses. Before she could start, a BMET had a question about a power supply repair for the monitor in the intensive care unit. Jan asked the obvious, but necessary, question: "Have you checked all the fuses and power from the outlet?" The BMET gave a nod about the fuses, but the outlet had not been checked. Jan was about to return to the budget reports when the process was interrupted by a phone call. It was the nurse manager (NM) for the neonatal intensive care unit (NICU). The NM's voice had an urgency and anxiety that was common to most NM phone calls. "Have the fetal monitors been tested recently?" the NM asked. Jan pulled up some information in the computerized maintenance management system.

"Yes, just last month," Jan replied.

"Well, I've got ECG artifact on three monitors, and these are all critical infants," the NM roared.

"OK, we'll check it out right away," Jan responded calmly. She approached the BMET with the power supply problem. "Can you swing by the NICU and check out the monitors?" Jan asked. "Take some new ECG leads with you. We've got an artifact problem up there." Jan stated matter-of-factly. OK, back to the budget reports. After a few more phone calls Jan was able to make some progress on the presentation for tomorrow's budget meeting. "I've got enough time to grab a bite to eat before the rookie is finished with orientation," Jan thought. Now it was the dilemma of cafeteria food or running across the street to the deli. She decided to take a chance on the cafeteria. *"Hey, not bad,"* Jan thought as the new food-court style cafeteria was surveyed. *"Not like the old days,"* she mused. As she finished lunch and headed back to the clinical engineering department the director of cardioulmonary services rushed by to ask a question. The director quizzed Jan on the status of replacing some of the older ventilators. Jan assured the director that it was on the agenda for the next morning's budget meeting. She made it to the back to the clinical engineering department shortly after the director had arriving with the rookie. "Jean meet Jan, our CE supervisor," the director stated as the traditional handshake process began. "Jan will be your mentor for the first few months until you feel comfortable with your roles and assignments," the director stated. Jan and Jean exchanged some personal background information and then started off on a tour of the hospital. Jan identified each clinical area as they passed them by, introducing Jean to different members of the staff as they became available. Jan noticed the wide-eyed look on Jean's face about a third of the way through the tour. "Don't worry" Jan said, "You'll get used to it. All this will become second nature after your first month." Jean nodded but without any reassurance. Back in the CE department, Jan introduced Jean to fellow CE coworkers. It wasn't exactly a typical day for Jan—a little slower-paced than usual. Tomorrow, Jan could catch up on the dozen e-mails that went unattended due to the rookie's tour. *"Jean will find that niche and become comfortable with the CE environment."* Jan thought. The curiosity and enthusiasm that Jean displayed were key indicators that she had chosen a career field that would fill a personal passion required in health care. *"I'll wait another week before I start assigning Jean any PMs,"* Jan decided. *"No reason to tarnish the academic idealism until the appreciation for our purpose sets in."*

References

Babcock, D., *Managing Engineering and Technology,* Englewood Cliffs, NJ, Prentice Hall, 1992.
Bronzino, Joseph, ed. *Management of Medical Technology: A Primer for Clinical Engineers* Boston, Butterworth-Heinemann, 1992.
Feinberg, Barry N., *Applied Clinical Engineering,* Englewood Cliffs, NJ, Prentice Hall, 1986.

Further Information

Cram, Nicholas, *BMEN 310 Manual: Clinical Engineering.* College Station, TX, Biomedical Engineering Program, Texas A & M University, 2001.
Cram, Nicholas, *BMEN 410 Manual: Clinical Engineering.* College Station, TX, Biomedical Engineering Program, Texas A & M University, 2000.
Ellis, David, *Technology and the Future of Health Care: Preparing for the Next 30 Years,* Chicago, Jossey-Bass Publishers, 2000.
Umiker, William, *Management Skills for the New Health Care supervisor, 3rd ed,* Gaithersburg, MD, Aspen Publishers, 1998.

11

Clinical Engineering at the Bedside

Saul Miodownik
Director, Clinical Engineering
Memorial Sloan-Kettering Cancer Center
New York, NY

For the past several years, the practice of clinical engineering has shifted from the patient's bedside to the office suite and meeting room. Fundamental components of a clinical engineering department's activities should include bedside consultation in the midst of clinical procedures for the purpose of problem solving, interpretation of medical information, and clinical staff education.

The field of clinical engineering has undergone several evolutionary cycles since it emerged as a profession in the 1970s. In many instances, our *raison d'etre* derived from fears of imperceptible, but potentially lethal, 60-Hz leakage currents that might have been taking an untold annual toll in lost patient lives. The reports of dangerous leakage currents in the medical literature were anecdotal in nature and often followed an adverse clinical manifestation that was associated with the use of electrical equipment. A careful engineering investigation of medical equipment and its use in the clinical environment eventually revealed the true nature and scope of this hazard. Although the perceived liability failed to materialize, many clinical engineering jobs and even entire departments owe their existence to this scare. It was inevitable that many clinical engineers were thus directed to the task of tracking medical equipment and their associated leakage currents as their primary function. As the issue of leakage current hazard was gradually put into a more reasoned perspective, clinical engineering departments directed their efforts toward a broader spectrum of medical technology support, management and acquisition. Along with this, systems and policies were established to assure timely inspection intervals, repairs, and compliance with required standards. With the widespread use of computerized systems for medical equipment management, a cornucopia of data became easily available to enable a host of departmental management tools such as performance appraisal, quality assurance (QA), and benchmark comparisons.

Clinical engineering presence in the scope of health care delivery continues to be ubiquitous. Addressing issues of safety, technology acquisition, scheduled and unscheduled maintenance, incoming inspection, regulatory compliance, quality assurance, and facilities management consume larger proportions of available time.

Nevertheless, it seems that the ultimate recipient of medical technology, the patient, has become an increasingly remote component in the delivery of clinical engineering services. Clinical and biomedical engineers have always been in the unique position of acting as a bridge between the medical and technical environments. Theoretically, knowledge in both fields would allow the safer and more efficacious application of a diverse technological armamentarium over a broad spectrum of clinical situations. However, the current trend in clinical engineering appears to be drifting ever farther from its clinical source.

One component of this trend is the "queasy factor." The raw clinical environment is decidedly messy. Many people would choose to avoid the presence of blood, body fluids, secretions, smells and human beings in various states of distress, disassembly, and mortal danger are places. that The nursing and physician communities encounter these as matters of course. Clinical and biomedical engineers and technicians can often selectively avoid the most distressing aspects of the clinical environment.

I am reminded of a poster session I once presented at a meeting of the Association for the Advancement of Medical Instrumentation (AAMI). Meeting attendants would stroll down the poster aisle until a topic of interest caught their eye. My poster was about the development of an extracorporeal gastric-assist device that would transfer a patient's stomach contents into the jejunum, thus bypassing an obstruction or region of paralysis. The visual materials were graphic, showing the gastric and jejunal stomae, the interconnection of the assist device to various tubes entering and leaving the patient's body, and views of stomach contents and bile being transferred by the pumping device. This device was a productive collaboration between the departments of gastroenterology and clinical engineering and required the cooperation of several patients. It directly and immediately improved the patients' quality of life. The clinical engineering department installed these bypass pumps on these patients and consulted with them regarding its care. A few attendees were genuinely interested and stopped to engage in lively discussions. Most, however, accelerated the pace of their stroll when the nature of the poster became apparent. I walked down the poster aisle to see that other topics being presented included such sterile matters as bench testing of mechanical thermostats, various simulators and biomedical test equipment, medical equipment management programs, and a few *in vitro* studies on biocompatible materials. If my poster was the proportional representation of clinical engineering in the clinical environment, then little was being done in this arena.

When clinical engineers actively participate at the bedside, there is the potential for great benefit for the patient, the medical staff, and for the clinical engineering community. The insight, perspective, and problem-solving capabilities that the clinical engineer brings to the scene are often different from those of other members of the medical staff. Once this type of expertise is developed and recognized, it will be called upon with increasing frequency and beneficial results.

Case Studies of CE at the Bedside

Following are several examples of clinical engineering at the bedside.

Case 1: Chin Up! The Lesion Generator Really Does Work

During a computed tomography (CT) guided procedure involving nerve ablation to address chronic facial pain, the clinical engineering department was called to go to interventional radiology to determine why the neural lesion generator was developing insufficient heat. The neurosurgeon demonstrated the temperature rise as measured by the needle probe depositing the energy at the lesion sight. At first, the temperature rose, then reached a plateau. After a few seconds, it dropped. Increasing the output setting of the lesion generator did not improve this situation. Clinical engineers verified the radiofrequency (RF) output of the generator and found it to be nominal. After pondering a while, the clinical engineers entered the CT area and asked to be shown the site of the return electrode. A drape was pulled off of the sedated patient's head where, to the horror of the engineers, an 18-gauge needle with the return wire attached was seen protruding from the patient's chin. There was a significant area of desiccated tissue surrounding this needle. I quickly obtained an electrosurgical return electrode and modified it to be used with lesion generator. The procedure was then able to proceed normally. The neurosurgeon was delighted that the procedure could be completed. In the absence of our intervention, the patient otherwise might have left with even greater pain than before. We explained to the medical staff the nature of electrosurgical return currents and the need to distribute the current density over a wide area. In this instance, the heating at the needle return site dried out the tissue and decreased the overall current produced by the lesion generator and, consequently, the energy available at the desired site for heating. The surgeon was grateful to learn this information (he had been performing this procedure for years with no apparent adverse outcome until now), the patient was able to benefit from the relief of his chronic pain, and the clinical engineering section demonstrated its understanding and quick thinking in remedying the situation.

Case 2: Argon Beam Coagulator Fails under Pressure

Following the introduction of argon beam coagulation (ABC) devices, it was brought to our attention that, under certain conditions, the argon "beam" failed to ignite and to properly deposit the electrosurgical energy at the desired site. The operating room (OR) technical people were constantly swapping out generators in hopes that one unit might perform better than another. All ABC devices performed flawlessly during testing in our clinical engineering laboratory. We went to the OR to observe various ABC surgeries to determine whether a pattern that would help us determine the problem could be identified. It became apparent that problems arose when the ABC handle was being used deep in a body cavity. At first, ignition proceeded normally, but after a few moments the argon beam became intermittent and sometimes ceased completely. Proper operation of the ABC was verified with our electrosurgical unit (ESU) tester, and the argon beam was successfully generated on an open field. Entering the body cavity quickly reproduced the problem. I suggested that perhaps a small amount of back pressure was being developed in the deep cavity that could interfere with directed flow of the ABC. A suction catheter was placed a few centimeters above the ABC handpiece, and its flow was adjusted to approximate that of the argon flow of the ABC unit. That did the trick. Surgery proceeded normally.

Case 3: Clenched Jaw Signals Swamp EEG

During a rather complex neurosurgical procedure to remove an egg-sized tumor from the right side of a patient's brain, a problem was encountered with an electroencephalograph (EEG) data-acquisition system. Following a craniotomy, this device is used to map the functional regions on the surface of the brain to the patient's various motor and sensory areas. In order to do this, the patient must remain conscious. The EEG operator provides a sequence of electrical, optical, acoustic, and mechanical stimuli in order to identify a surgical approach to the tumor that will cause the least amount of damage to the patient's neurological function. In addition to the EEG device, other medical devices were involved, including an ultrasound imaging system, radiofrequency position generator, bipolar coagulator, ultrasonic tissue ablation, stimulators, and irrigators. The EEG operator, who was quite experienced and had done dozens of similar procedures, was unable to record acceptable signals because of an intermittent high-frequency noise on all eight EEG channels. It was initially thought that this noise originated from one of the many other instruments in the room, but this could not be confirmed. Filtering of the signals did not work, as the noise appeared to be in the same frequency domain as the EEG signals. To complicate matters, the patient was able to hear fragments of the conversation between the surgeon and the EEG operator discussing the reason for the delay in the surgery. He readily expressed his anxiety.

At that point, we were called to assist. Every instrument, in turn, had been turned off to determine whether it was the cause of the EEG interference. Indeed, after a few moments of observation, the extraneous "noise" would almost entirely disappear and then, suddenly, completely mask the desired EEG signal.

A closer look at the electrode site was made. The common reference electrode was positioned in the scalp fascia near the craniotomy. The location of the reference electrode is usually not an issue. In this instance, however, the needle electrode appeared to have penetrated some muscle tissue as well. We observed that when the patient was questioned or stimulated, the noise would reappear and sometimes subside as well. We hypothesized that the noise was an electromyographic (EMG) signal caused by tension in the scalp musculature when the patient was aroused or when he felt anxious. A reference electrode site that avoided muscle tissue was carefully chosen. The problem was resolved.

Case 4: Air Today; Gone Tomorrow?

A patient in the intensive care unit (ICU) required the use of a transthoracic pacer. A defibrillator/pacer, normally kept in the unit, was used. Disposable pacing electrodes, which can double as defibrillator paddles, were placed in the classical transthoracic position (over the heart and on the patient's back). However, the pacer would not capture. The pacing current was increased but would still not capture. The defibrillator-pacemaker was changed to a second unit with no better results. Clinical Engineering was called in to assess the "failure" of both pacers as the staff began preparing the patient for the insertion of a temporary transvenous pacing catheter. Our records indicated that both units had been recently inspected and were working properly. A quick test in the ICU confirmed this. I returned to the patient's cubicle and looked around. I noticed that the patient had a chest-drainage tube in place. I spoke to the medical staff who informed me that the patient was being treated for a pneumothorax that had occurred a short while before the pacer was required. The pneumothorax would allow air to enter the chest cavity. If some of this air had been present between the heart and chest wall, where the anterior pacing electrode was placed, it might account for the lack of capture by the transthoracic pacer. Even at high pacing currents, the insulating air prevented enough current from crossing the myocardium and causing capture. The increasing pacing current found a route around the chest wall with very of the current passing through the heart. The electrodes were repositioned to sites that reliably allowed the pacer to capture the patient's heartbeat. This clinical engineering intervention might have averted a catastrophe had the need for defibrillation arisen. The catheter insertion in this patient in the midst of an ongoing episode of bradycardia might have triggered arrhythmias or even ventricular fibrillation. If an attempt had been made to defibrillate the patient, the same air pocket over the heart may also have prevented successful defibrillation.

Case 5: Pump Puts the Squeeze on Plastic Tube, Sending It into V-Tach

The ICU cubicle can become an extremely complex environment in a short period of time. Within 24 hours, patients who start with simple monitoring of ECG, noninvasive blood pressure, and pulse oximetry can require such things as a mechanical ventilator, invasive arterial and venous access, and multiple infusion devices. As the patient's condition deteriorates, additional interventions, therapies, and medications are used.

We were called to the bedside of a very sick patient who apparently had gone into what appeared to be a ventricular tachycardia (V-tach). Two attempts at cardioversion and the administration of potent cardioactive drugs appeared to have been ineffective. To further complicate matters, a 12-lead ECG taken independently of the ECG monitor did not support the observation of V-tach. A view of the central station monitor seemed to confirm the presence of V-tach on the ECG trace. Surprisingly, the invasive pressure and pulse oximeter tracing indicated a relatively normal set of cardiogenic traces, inconsistent with the ECG. Further analysis of the situation was made all the more difficult by the infection-control precautions posted on the door of the patient's cubicle. I looked into the room and was confronted with what appeared to be most of the ICUs portable equipment and about half of its staff. In addition to the normal, full-function physiologic monitoring system and the usual three or four infusion pumps, two extra pumps had been rolled in to administer cardioactive drugs; the defibrillator-pacer was standing by to convert the V-tach rhythm; the 12-lead interpretive ECG machine was in place; and, at the foot of the patient's bed, a portable continuous arteriovenous hemofiltration/dialysis unit (CAVHD) was in use. Except for the defibrillator and the additional IV pumps, the entire assembly had been in use for a number of days with no consequence. The only other piece of information was that the V-tach had begun about 45 minutes earlier following the patient being cleaned and his ECG electrodes changed. The diagnosis of V-tach was identified by the physicians and verified by the arrhythmia computer within the patient monitor.

My line of sight placed the CAVHD alongside the monitor. Without much thought, I alternately scanned back and forth between the various roller pumps on the CAVHD unit and the ECG trace on the monitor. The blood pump was rotating at a rate that seemed to correlate with the ECG trace. Proceeding on a very good hunch based on experience, I asked that the blood pump be stopped. It slowed to a stop after a couple of seconds. When it did, the patient's ECG trace also returned to "normal," to the amazement of the nursing staff and residents and to the chagrin of the attending physician who ordered the cardioversion. What had happened?

There is a published, but little-known, phenomenon that relates to the piezoelectric characteristic of polyvinyl chloride (PVC) plastic commonly used in medical tubing. I had encountered this early in my clinical engineering experience, and a medical journal at a later date published a letter to the editor that I had sent regarding this effect on ECG tracings. PVC tubing will generate a small, but measurable, voltage when it is rapidly squeezed or released. Voltage can propagate along conductive fluid contained within the tubing. The amplitude of the voltage so propagated is dependent on a host of factors (e.g., tubing age, durometer, temperature, composition, and pressure change) The voltage that can arise on the patient surface depends on the tubing penetration site on several factors including the patient, electrode impedance, and intrinsic ECG voltage. Depending on the pinch rate and the other factors mentioned, this piezoelectric effect can mimic or mask a wide range of ECG anomalies, from atrial fibrillation to V-tach. Our patient was somewhat obese and had a fair degree of edema. His intrinsic ECG voltage was < 0.5 mV in lead II. The ECG electrodes were replaced after he was cleaned, with no particular consideration give to skin prep or placement. The entry point of the CAVHD tubing was within 10 cm of the LA ECG electrode and was precisely in line with the RA-LL electrodes for lead II. The roller pump rate for the patient's blood produced one compression and one release of the tubing within the roller guide about every 0.3 seconds. The triangular voltage waveform generated by the pump and seen on lead II of the monitor was in the order of 1.5 mV, virtually masking the patient's own ECG. When skin resistance is high and ECG voltage is low, there is a greater susceptibility to external interference.

When the ECG was taken with the 12-lead automated machine, the electrodes, which were approximately 1-inch square, were placed on the patients limbs significantly removed from the entry point of the CAVHD tubing. In fact, the piezoelectric signal could be regarded as common mode interference considering its relative distance from any but the V-lead chest electrodes. This would allow the high common mode rejection ratio (CMRR) of the ECG amplifier to reject this interference. The large, square area of the electrodes used by the12-lead ECG machine also assured a lower impedance connection to the patient than was possible with the monitoring electrodes. This situation would also tend to reduce the apparent amplitude of the interfering signal. After much judicious skin prep and repositioning of the ECG electrode, we were able to safely monitor the patient.

Conclusions

The health care technology managed by clinical engineering departments should not be viewed merely as entries in a database and subject to an occasional periodic maintenance and repair. Although procedures exist to verify a medical device's proper operation within a set of technical specifications, there is a clinical context to their daily use that must be intimately understood.

In all of the aforementioned cases, the following assumption was made: "Something is wrong with the machine, call Clinical Engineering." However, in each case, the hardware performed flawlessly. A solution to the problem required a comprehensive understanding of the immediate clinical situation, anatomy and physiology, the physics of the instruments in use, and knowledge of fundamental instrumentation characteristics. The practice of clinical engineering at the patient bedside requires the integration of this knowledge, experience, and quick thinking on one's feet.

Structuring these concepts to assure clinical engineering involvement at the bedside is not difficult. At all staffing levels, from the BMET to the section's head it is possible to encourage and participate in this form of activity. Following courses in anatomy and physiology, a basic requirement of a clinical engineer's education and experience should include a familiarity with the kind of activities that occur at the patient's bedside. Ideally, this can be learned or experienced by observation in the intensive care unit, operating room, or emergency room. Following the medical staff on rounds is an effective method of observation. Consistent exposure to the multitude of procedures, terminology, pace of activity, and use of equipment in these areas is a mandatory prerequisite for those who wish to apply their abilities at the bedside. An intimate understanding of principles of the instrumentation used and the clinical situation can often lead to solutions to otherwise intractable problems.

12

The Clinical Engineer as Consultant

J. Sam Miller
Clinical Engineering Consultant
Buffalo, NY

Clinical engineers as consultants play an important role in the health care industry. All practicing clinical engineers serve in some capacity as consultants because they have unique training, knowledge, and expertise that are of value to those in ancillary health care fields. Many of the tasks routinely performed by a clinical engineer in a hospital in support of administration, physicians, nurses, and clinics are actually consulting tasks. This chapter, however, is devoted to consulting in the sense of rendering engineering services to clients for a fee. The topic is presented here with an overview of the present clinical engineering consulting field and answers the dual question of "Who is doing what for whom, and how?" Credentials and qualifications are briefly presented, followed by practical business considerations for both the full-time and part-time independent consultant.

General Overview

Who Are the Consultants?

Clinical engineering consultants are those clinical engineers (CEs) who have found that their level of experience and expertise places them in demand for technical advice in the field of medical equipment. Some have specialty niches, but the majority simply have a broad working knowledge of the use of clinical equipment and management of technology.

There are probably fewer than 100 clinical engineers serving as consultants at the present time, but there is no central list or directory available to confirm that figure. The *Membership Directory* of the Association for the Advancement of Medical Instrumentation (AAMI) lists 5,587 persons as members, with 151 of those listed with the job title of *Consultant*. However, of those, only 20 have a Ph.D., M.S., or certification in clinical engineering. Most of the others listed are consulting as accountants or marketing managers, or with a technical specialty such as sterilization or prostheses. AAMI also offers a *Directory of Medical Technology Consultants* (for a fee to be listed) that currently lists 12 consultants (2002); however, only three of those offer general CE services. The American College of Clinical Engineering (ACCE) membership directory (2003) lists over 150 CEs but does not designate consulting involvement. (See Chapter 130.)

Further insight into who the consultants are can be obtained by searching the web using key words such as *clinical engineering consulting*. There are only a few consultants who currently have web sites to promote their business and to list their qualifications; e.g., http://biomedeng.com and http://www.hrbiomedical.com/page4.htm. This form of marketing consulting services will grow in the future.

How Do They Consult?

The better-known consultants in this field practice on a full-time basis, some as members of consulting firms or groups and some as self-employed independent consultants. There are also well-known consultants who practice on a part-time basis because they are otherwise employed full-time by hospitals, universities, or industry. In this latter case, they usually have an agreement with their employers that allows them to do some consulting on a noninterfering basis. Conflict-of-interest agreements are recommended in these cases to protect both parties; however this usually places limits on the types of consulting projects that can be undertaken.

Who Hires Them to Do What?

There are many different activities in health care that have a need for independent clinical engineering consulting, usually to give technical advice, solve problems, or provide specialty information. The following is a partial list of these activities:

- Assisting the in-house CE whenever expert advice or corroboration is needed
- Assisting hospital administration whenever expert technical advice is needed
- Assisting physicians in new technology or research areas
- Assisting medical equipment entrepreneurs, manufacturers, or vendors with the introduction of new technology (or upgrading old technology)
- Assisting the legal field in product liability/malpractice cases involving medical equipment
- Assisting with the education of medical or engineering professionals
- Assisting hospitals in other countries with the introduction of new programs or technology

Consultants for the first two roles above can be CEs with a particular expertise in a problem area, but they also could be non-CE consultants in a specialty field adjunct to CE, such as emergency power systems, environmental engineering, radiation safety, or telecommunications. Most practicing in-house CEs are naturally reluctant to retain an outside consultant, but if they are able to do so, they find that their own image as an expert is bolstered and that doing so exhibits good management skills.

However, in-house CEs may find that they themselves can serve others as experts in any of the areas described above. Although most in-house CEs do not seek out these consulting opportunities, they are often recruited for such tasks based on their relative experience level and reputation in the community. The legal field often needs to use a CE consultant for technical advice and as an expert witness (see Chapter 13) in malpractice cases involving use or misuse of medical equipment or for product liability cases. CEs who work in local hospitals are commonly used in these instances.

Hospital administrators, particularly those from smaller facilities, can use independent CE advice in reviewing their clinical engineering support services and activities. In some cases, such assistance is requested by the in-house clinical engineering to give credibility to its operations. Consulting for hospitals to advise on programs to ensure compliance with the recommendations of the Joint Commission on Accreditation of Healthcare Organizations (JCAHO) or with state health department requirements. Consultants also assist administrators in the development of technology management programs and policies.

Physicians often need technical assistance with new equipment purchases, grant writing, or research protocols. Human research committees or institutional review boards (IRB) in teaching hospitals and universities can benefit by adding an independent clinical engineer as a consultant to the group. Physicians or other health professionals in state and federal agencies also use CE consultants to review rule making or new medical device applications, or to assist with audits.

Manufacturers of medical equipment—especially smaller manufacturers—often do not have engineers with practical clinical experience on their staff, and they need help in bringing new products to market. Practicing CEs are well-suited to assist with functional engineering design details, with premarket clinical trials, with U.S. Food and Drug Administration regulations (see Chapter 126), and with standards promulgated by organizations such as the American National Standards Institute (ANSI), the National Fire Protection Association (NFPA), and AAMI. (See Chapter 118.) A related consulting area involves technical reviews for venture capital firms of financial proposals from entrepreneurs who need funds to introduce new technology.

To support the field of education, CEs who consult often find themselves invited to participate in the training of students in the health professions, which sometimes leads to part-time faculty positions. Assisting with grant writing and instrumentation used for medical-research projects are other typical university needs that can be filled as a consulting service.

Finally, assisting health care facilities in developing counties with the general area of technology management is an important need at the present time. There are programs funded by the U.S. government that will retain consultants for assistance, but there is also a need for volunteer CE efforts to assist in training programs in these countries. The ACCE routinely conducts Advanced Clinical Engineering Workshops (see Chapter 70) several times a year using a volunteer faculty of CE experts from the consultant ranks.

Qualifications and Credentials

One of the quirks of the consulting field in general is that there are no set standards for being a consultant. CEs can print up business cards and can call themselves consultants. In the field of clinical engineering, however, there are some credentials that help to establish credibility:

- an advanced engineering degree, preferably a Ph.D., from a well-known university
- a Professional Engineering License or certification as a clinical engineer
- membership in professional scientific, technical and engineering societies; e.g., ACCE, AAMI, the Institute of Electrical and Electronics Engineers (IEEE), the American Society of Hospital Engineering (ASHE), the National Fire Protection Association (NFPA), and the American Society of Mechanical Engineering (ASME)
- membership in local professional societies or organizations (e.g., the Chamber of Commerce or health care, engineering, or consulting groups)

There are also certain factors that help to qualify a person as an expert and that are related to years of experience in the field. In a local community, people become known over time and develop a reputation based on the quality of their work. Speaking at local health care society meetings on new technology advances or innovative projects that one has been involved with is an excellent way to enhance one's reputation locally. Referrals from colleagues in related health care fields then often lead to new consulting opportunities.

In the national and international arena, professionals become known as experts over time primarily by their publications and participation in national or international technical meetings. By voluntarily participating in the work and activities of a professional society, a person becomes known by their peers and develops a reputation. In the field of clinical engineering consulting, the most authoritative qualification as an expert is by recognition from peers.

Business Aspects of Independent and Part-Time Consulting

This section considers the factors that are important for a practicing clinical engineer to look into when deciding to offer consulting services on a part-time, or moonlighting, basis. Although most practicing clinical engineers are employed by health care providers, these same considerations will apply to those CEs who work for industry, universities, or the government. Part-time consulting usually begins as follows:

As CEs become established in their work settings, they are soon recognized as technical experts by the physicians, nurses, administrators, and other nonengineering professionals with whom they interact in their day-to-day activities. And they are expected to provide expert advice, as needed, to these coworkers as part of their job responsibilities. However, ethical considerations arise when someone who is not a coworker (e.g., an attending physician or a university researcher) asks the CE for some free advice about some technical problem that they are facing. Most of the time, the CE can provide the needed information in a few minutes of conversation, and nothing much is thought about it from a business-ethics standpoint. But what if the physician, for example, needs to have an interface designed and built so that he can use his PC to view ultrasound images, and he offers to pay the CE for the work?

Another common situation occurs when an in-house CE is contacted by an attorney who works for the law firm used by the CE's employer, and the attorney needs some assistance (for a fee) about a case that does not even involve the CE's employer. If the CE is properly established as a part-time consultant, with their employer's permission, these kinds of events can become win-win-win situations. The CE wins with a bit more income and prestige; the CE's client wins because the client found help without feeling that it was an imposition on the CEs employer; and the CE's employer wins because the employer has become better-known as an organization with recognized experts on its staff and also as a good neighbor to the business community.

How Can Such a Part-Time Service be Arranged?

For the CE, the first consideration should be whether or not the extra work can be done without serious negative impact on the employer, the family, or other personal interests. If the CE is already routinely putting in sixty- to eighty-hour workweeks or attending graduate school in the evenings, then the prospect would be ill-advised. But if the CE is willing to put in the extra time and to give up some holiday and vacation time to work on a project for a client, and if the CE's employer allows some flexibility with work hours, then establishing a part-time consulting service is a definite possibility.

One key to avoiding problems with the employer is to have a written agreement, even as simple as a memo or letter, signed by both the CE and the CE's administrative supervisor, detailing the ground rules for the use of the CE's time for the outside work. The agreement should also detail the CE's use of the employer's resources (e.g., computers, work areas, test equipment, parts, and libraries). In addition, a formal conflict-of-interest statement should be generated by the CE and accepted by the employer, calling out the specific nature of the CE's consulting work and clients. For example, the employer may wish to exclude competitors or providers of goods and services.

Basic Business Elements to Consider

Part-time consulting activities should not impact your normal duties to your employer and should remain as separate as possible. One way to achieve this is with a home office, a place to work independently from the nine-to-five job. But if this option is not practical, then it will be important to keep good written time records of all your consulting activities.

Your initial clients probably will be physicians, lawyers, or manufacturers whom you know from your involvement with them at work. To avoid any conflict of interest, the consulting work should be limited to tasks that are clearly separate from any dealings that your client may have with your employer. It also would be advisable not to accept any consulting work from manufacturers who are competing for business with your employer.

Promoting Your Services

Retaining your first client is usually not an easy task, unless they come to you for help with a specific problem. One recommendation for getting started would be to make a list of potential clients and the tasks that you might perform for them. Then select the one that you would be most comfortable doing work for, and ask that one to have lunch with you to discuss a business idea. Explain that you are now available for consulting work, and let it be known how you could be of assistance. Have an outline of your ideas and an up-to-date copy of your curriculum vitae (CV).

Before you approach the next potential client, do a self-critique of that first meeting to see what you should do next time to improve your approach. It may take only two or three contacts like that to win your first consulting job, and then your efforts should go toward performing that work in an expert manner. Agree to your fee at the same time that the scope of the task is defined, and do 'not underprice your professional fee.

Details to Consider

After one or two successful consulting assignments, you should know whether this part-time, self-employed approach meets your needs and whether you want to continue. If so, here are some ideas to make your efforts more productive:

- Set up a home office and a computer to keep track of contacts, billable and nonbillable time, accounting (including taxes), and to provide Internet access for e-mail and web information searches.
- Have stationery and business cards printed, and consider a simple brochure listing your particular areas of expertise.
- Sharpen your public-speaking skills by way of seminars or participation in the local Toastmasters organization.
- Consider joining local professional or fraternal organizations and attend meetings.
- List yourself as a consultant in any directories that are publicly available. Search the Internet for opportunities to promote your services (e.g., http://www.a2zmoonlighter.com).
- As business increases, ask an accountant for assistance with taxes (you might have to file quarterly estimated-tax payments), and consider errors-and-omissions insurance.

Business Aspects of Independent Full-Time Consulting

This section expands on the part-time business considerations, adding items that are important for starting and operating a full-time clinical engineering consulting service.

Business Basics

Every business should start by developing a written business plan. (See Chapter 140.) This may seem at first to be unnecessary because most engineers think that they know what they want to do and how to get there without needing to write it down as a plan. However, most failed businesses identify the lack of a formal business plan as a major reason for failure. Developing a written business plan forces one to systematically consider details that are otherwise overlooked and provides a practical roadmap to get you where you want to be. A few hours at a public library will provide one with many resources (e.g., Cuppett, 1966) for developing a plan for a relatively small venture such as a consulting service. This plan, if reviewed and updated every 6 months, can become a great working tool for measuring progress and continuously improving quality.

Key Elements of the Plan

- First, clearly state the business philosophy (basic beliefs and important values) and mission (the primary focus of the business).
- Define short- and long-term goals with measurable objectives.
- Develop a strategic plan defining the following elements: present status, marketing plan, operations plan, administrative plan, financial plan, timeline for the above, and tactics.

Development of this strategic plan and tactics should include analysis of strengths, weaknesses, opportunities, threats, and conflicts in a systematic and realistic manner. Marketing should include a list of the needs that your various types of clients have, and which of those needs match your abilities. Operations should detail office space, equipment, and support needs, both currently and for future growth. Financial and administrative details go hand in hand. One needs to determine what type of business one has (e.g., "doing business as" (DBA), incorporated (Inc.), limited liability corporation (LLC), and sole proprietorship); the structure of the business (e.g., officers, owners, survivors); start-up funds and contingency funds, fees, accounting system, tax status and tax filing requirements. Finally, the plan should be projected to cover 5 years. Historically, most new consulting businesses require that length of time to become stable and profitable.

Business Resources

Starting a new business, especially if it is a one-person operation, has the potential to be very rewarding personally and financially. But there are many pitfalls to avoid. There are no bosses, work companions, or secretaries to help keep you going in the right direction, so it is important to develop a support-and-resource network. Many localities have professional consultant membership organizations that meet regularly for the purpose of improving their business skills, sharing problems and leads, networking, and socializing. Although their members generally are from business fields that are not related to clinical engineering, most problems that they face and issues that they address are generic to any consulting field. And the meetings are often a forum for presenting plans or ideas for group critique. This path would be an excellent way to obtain peer review of a new business plan or marketing approach. Check the newspaper listings for calendars of local business meetings to locate such a group, as they generally are not listed in the telephone directory. If there is not a group that is specific to consultants in your locality, there are other business-related, self-help membership groups that operate in a similar manner, such as Toastmasters, Rotary, and the Chamber of Commerce.

Spend some time studying the general field of consulting services. Again, the public library has considerable resources for this kink of information (e.g., Cimasi, 1999; Holtz, 1989). It is also important to maintain professional health care and engineering ties. Organizations such as ASHE, IEEE, and ASME usually have local chapters that meet regularly. Local and/or state health care providers and clinical engineering groups have membership meetings. (See Chapters 131 and 132.) The information and exposure gained by networking at such meetings is invaluable to the new consultant.

Summary

The topic of consulting is of interest to every clinical engineer because it holds the promise of exciting and rewarding work, reasonable compensation, and flexible work hours. The successful consultants, however, are those with many years of experience in a clinical setting, excellent credentials, and, most importantly, a reputation as an expert.

References

Cimasi RJ. *A Guide to Consulting Services for Emerging Health Care Organizations.* New York, John Wiley & Sons, 1999.

Cuppett WT. *Developing Business Plans.* New York, American Institute of Certified Public Accountants, 1966.

Holtz H. *Choosing and Using a Consultant: A Manager's Guide to Consulting Services.* New York, Wiley, 1989.

13

The Clinical Engineer as Investigator and Expert Witness

Jerome T. Anderson
Principal, Biomedical Consulting Services
San Clemente, CA

With the expanding use of technology, the volume of medical devices and capital equipment has greatly increased over the years. This has been accompanied by an increase in injuries and untoward outcomes attributed to medical devices and equipment (see Chapter 84). The trend came to the forefront with Congress's passage of the Safe Medical Devices Act (SMDA) of 1990 (see Chapter 126). Prior to this act, patient injuries were investigated internally for purposes of defending a claim or preparing for potential litigation. While manufacturers were required to investigate injuries reported to them prior to passage of the law, the inclusion of health care facilities in the requirement greatly increased the requirement for investigation of injuries or death thought to be related to the failure or misuse of a medical device.

The call for investigation of medical device-related injuries has grown stronger through the inclusion of the investigative requirements in revisions of the standards of the Joint Commission on Accreditation of Healthcare Organizations and the recent movement in the U.S. to reduce medical errors that result in harm to patients (see Chapter 55).

The clinical engineer is a natural choice to perform investigation of device-related injuries (see Chapter 64). Knowledge of the applications, operation, and potential problems of medical devices is an important factor in the ability to analyze a situation and to reach conclusions about the manner in which a device might (or might not) have contributed to an injury. The clinical engineer is the primary person who can combine an understanding of the device with the clinical situation in which the injury occurred, to provide information on whether the injury was attributable to the device or other factors.

In many cases, there is a propensity to blame the equipment or device for an injury, without understanding the mechanism by which the injury occurred. In 70 to 80% of the cases of reported malfunction of a medical device, the item is found to be operating as designed when tested. The ability of a clinical engineer to unravel the variety of circumstances surrounding a device-related injury is an important contribution to the investigative process, which may also involve risk managers, nursing staff, manufacturers, and others.

In the event that a severe injury or death leads to litigation, the investigation performed by the clinical engineer may become an important part of the litigation process. If this occurs, the investigator may be called upon to function as an expert witness. In this context, the information gained during the investigation, and the conclusions reached, will be subjected to scrutiny by attorneys, opposing experts, and the court system. The clinical engineer will then be exposed to an entirely different system than the objective analysis performed in the engineering arena. The person's experience, qualifications, methodology, results, and conclusions will be examined and questioned. The expert also will be expected to analyze the opinions of the opposing side and to assist in developing positions that will be taken in an attempt to reach a settlement, or, failing that, to win the case at trial.

In order to be successful as an investigator of device-related injuries and to function as an expert witness when called upon to do so, objectivity and neutrality are necessary traits, in addition to the expertise and experience in the field of medical devices. This chapter will assist the reader in understanding what is required to function successfully in this role.

Qualifications to Perform Investigations

The investigation of medial device injuries is a complex task that requires a multidisciplinary approach. The investigator must have knowledge of the theory of operation of the device being investigated, an understanding of the clinical aspects of the event, knowledge of the facilities and systems that may affect the operation of the device, and the ability to work with others to obtain the information necessary to carry out the investigation.

To understand the theory of operation of a device requires knowledge of the physics and engineering principles of the operation of the device. Knowledge of electronics, electricity, pneumatics, hydraulics, optics, digital technology, thermodynamics, human engineering, and other fields may be required in various situations. One must understand how to analyze the normal and abnormal operation of a device in order to determine the ways in which a failure may (or may not) occur and contribute to an injury.

An awareness of the ways in which circumstances outside a device can result in injury is an important part of the investigative process. One common example in modern medical technology is the abnormal operation of a device caused by power-line disturbances. The investigator must continuously consider the role that failure or abnormality of a support system could play in an injury. The crossing of medical gas lines can result in a fatal injury if the wrong gas is delivered undetected through gas-administration equipment. In such a case, failure of the primary device is not the issue; abnormality of the support system is the underlying cause. It often requires a great deal of analysis and attempts at duplicating an event to determine that a support-system problem is the underlying cause of the injury sustained by the patient. Such situations require knowledge of the complex utility systems that are used in health care facilities, not merely the ability to test a piece of equipment on the bench (see Chapter 59).

It is necessary that the investigator understand basic medicine and physiology as well as medical technology. In some cases, the event being investigated has been brought on by the patient's medical condition and has no relation to the equipment and devices in use for clinical care and diagnosis. One example of this situation is the case that the author investigated, in which a patient was originally diagnosed with an electrosurgical burn and an investigation of the equipment used during the surgery was launched. Upon closer examination of the lesion and the circumstances of the patient's care, it was determined that the lesion in question was a pressure injury that had been developing over a period of time and was exacerbated by the extended time of pressure exposure on the operating table. The investigation of the electrosurgical equipment was then abandoned, and the mechanisms by which the patient sustained the pressure injury were pursued. Without knowledge of physiologic and clinical factors, the investigator would not have been able to differentiate between the symptoms and etiology of the lesion, and would have wasted time and money investigating a moot point.

The successful investigator must have the ability to work with clinical staff, both as a member of the investigative team and in the capacity of interviewing staff members involved in the event, in order to gain information about the event. In many cases, the clinical engineer will be looked upon as the leader of an investigation involving an injury that is thought to be related to a medical device. He or she will be asked to assist in making a determination as to whether the device or an error in using the device was the cause of the injury. This challenge may mean performing an independent investigation that is then reported to others, or participating on a team that may include physicians, nurses, risk managers, purchasing staff, administrators, manufacturer representatives, and outside consultants. In many cases, the results of an initial investigation will be used to determine whether or not the incident must be reported under the requirements of the United States Safe Medical Devices Act (see Chapter 126).

In this capacity, it is important that clinical engineers combine their knowledge and findings with those of others on the investigative team. Often there will be cases in which the perspective of someone from another discipline (e.g., nursing) will complement the technical investigator's findings to enable a different or more accurate description of the cause of the event to be developed. It is important that the investigator consider input from members of the team, or that information be obtained from people with expertise in other disciplines prior to drawing final conclusions. Simply saying that the equipment did not fail is not an adequate conclusion to an investigation in which the objective is to determine the underlying reasons for a patient injury.

The Investigational Process

An investigation should be a methodical process that consists of distinct steps. While some investigations will not require or not lend themselves to all of the steps described here, a general framework that will apply to the majority of medical device injury inquiries is presented here. Some of the steps will not be under the control of the investigator, and it is important that the relevant personnel be educated in their roles to ensure that the investigation is timely and thorough.

Gathering of information at the time the event occurs is critical. In most cases, the clinical engineer will not be present at the time of the event. Therefore, clinical staff must be trained beforehand on what to do when a device-related injury occurs. It is important that all settings on the suspect device be recorded, as well as the device identification and patient parameters (where applicable). This should be done as part of the confidential event report (also known by such terms as *incident report* or *variance report*) that is prepared for the facility's risk manager. Settings to be recorded can include such things as output power, rates, alarms that occurred prior to or during the event, the mode in which the device was used (e.g., assist vs. control mode on a ventilator), pressures, and temperatures.

If the device has an alarm or parameter memory, the memory contents should be interrogated prior to sequestration or commencement of any testing. If possible, it is preferable to record, print out, or download memory contents immediately after the event and before the machine is turned off, as some devices use volatile memory that will be cleared of stored information when the power is removed.

The device must be immediately identified so that it can be produced for future testing or other investigation. The manufacturer, model, and serial number should be identified in the event report. Hospital identifiers such as inventory control numbers or department identifications (e.g., ECG #4) also should be included in the identification information. Where appropriate, the most recent maintenance date should be recorded because inspection labels can become illegible or can be removed.

A request to check the device should be made by clinical department staff. The decision as to whether the check should be made by in-house staff or an outside, uninvolved third party should be made by the facility risk manager. It should be clearly stated on a service request form, attached to the device, that it was involved in a patient injury that and is not to be repaired. A brief statement of the suspected malfunction should be included in the inspection request.

Arrangements to sequester devices involved in suspected injuries or deaths should be made in advance of any untoward event. Any supplies that are needed to use the device must also be preserved, such as electrodes, cables, tubing, transducers, and solutions. These frequently provide an insight that cannot be attained by simple performance testing of a device. It is most common for sequestration to be done by the facility's clinical engineering service, but also may be done through risk management or security. In any event, the device must be secured in an area where access is limited to those who have a legitimate reason to inspect or to test the device to protect the legal position of the facility and other interested parties. It is advisable for a facility to have an "evidence locker" where devices can be secured during an investigation. Under no circumstances may a device be released to a manufacturer for evaluation without the considered consent of the facility's risk manger and legal counsel. Release of devices to a manufacturer frequently results in disappearance or repair and return of an item under investigation because of inadequate communication within the manufacturer's organization. Such an event can seriously jeopardize the facility's legal position in the event of litigation. It may be advisable to obtain the services of a disinterested third party to control the device during the investigation, in order to prevent allegations of spoliation of evidence.

The investigator must have the ability to perform information-gathering interviews in a nonjudgmental and nonthreatening manner. It is extremely important that the investigator obtain as much information as possible from those who witnessed the circumstances of the injury. In some cases, these people will be afraid of possible recriminations, simply because they were present at the time the injury occurred. In conducting interviews, the investigator must be viewed as an ally rather than a threat to the staff member. It is also essential that an air of trust be built with the staff member being interviewed. The investigator must have the ability to make the interviewee comfortable with the information gathering process. Conversely, the investigator must be objective–not interjecting personal opinions and not leading the interviewee's responses. An investigation must be handled as a fact-finding mission; all elements of blame must be eliminated if the process is to be successful. It is often best simply to ask the interviewees to describe in their own words what happened, without making any preliminary statements that might color their responses, and then to ask questions based on the event description provided. It will be common to get different descriptions from different staff members. This will depend on their respective function during the event, educational level, experience, and anxiety over the event. The investigator's job is to synthesize all information into a reasonable description of the event so that it can be used in the causation analysis.

Obtaining information on the event is a crucial step in the investigation, but it is often minimized or overlooked. It is not sufficient simply to test a device to determine whether it meets its operational specifications and to report that the problem could not be reproduced. Information on the event will often reveal circumstances that must be considered in order to obtain a meaningful simulation of the event in question. Arrangements should be made to interview all staff members who were present at the time of the event. It is best that the interview of an individual staff member be performed by the investigator with no other people present. The interviewee must be made to feel at ease and assured that the process is not meant to establish fault or to assign blame. Initially, the interviewee should be asked simply to give their account of the event from their perspective. This account then may suggest questions that can be asked in order to elicit additional information. The investigator should have questions in mind to ask to provide information needed in the event that this is not brought out in the initial stage of the interview. The interviewee should be asked which other people were present at the time of the event. It is useful to ensure that such people are included in the interview process if they have not previously been contacted. The staff member should be assured that their information will be kept confidential and used for analysis of the event, not to assign blame. They also should be reminded not to discuss the event or the interview with other parties, in order to preserve confidentiality and to ensure that other parties are not influenced by the interviewee's account and opinions.

Device identification is critical to the investigation. There should be no doubt that the device tested is the actual unit involved in the event under investigation. The device must be photographed on all sides and identification labels (e.g., manufacturer model and serial numbers, owner identification tags, and inspection tags) should be photographed and documented.

The next step in an investigation will be the inspection and testing of a device. It is important that any initial testing be nondestructive. Prior to testing, the unit and all accessories must be visually inspected for damage or contamination that would affect proper operation. The unit and accessories should be photographed, with attention paid to any abnormality found through visual inspection. Equipment covers should be removed where feasible, and the interior inspected for physical damage, contamination, or evidence of abnormal operating conditions (e.g., dirty contacts, overheating, leakage of fluid, or loose connections).

Testing should begin with a functional and safety test of the device to determine whether it meets its operating specifications. The unit should then be tested in the mode in which it was being used at the time of the event, and an attempt should be made to reproduce abnormal conditions to determine the way the item responds. It should be noted whether appropriate alarms occur, the unit switches into a backup mode, or other changes in normal operation of the unit occur. Such testing often will require that staff interviews, event reports, and medical record reviews have been performed prior to testing to determine the circumstances of the event for reproduction during testing.

Equipment and facilities required for inspections can vary greatly. Often the only items needed are those found in a typical, basic clinical engineering laboratory. These will include simulators for ECG, arrhythmia, respiration, blood pressure, temperature, and pressure/vacuum. Basic test equipment for checking the most commonly encountered equipment should include a digital multimeter with interchangeable probe leads, electrosurgical power output meter, digital pressure/vacuum gage, defibrillator output meter, test lung, analytical grade graduated cylinders (10, 100, and 250 ml are useful sizes), stopwatch, digital thermometer with surface and immersible thermocouples, and oxygen monitor. Where more specialized equipment is required, it can be rented from medical test equipment manufacturers, electronic test equipment rental services, or other rental services.

Items used in inspections are the basic tools that one would find in a biomedical service shop. Typical inspection tool sets will include English and metric socket wrench sets; an interchangeable blade screwdriver set with multiple sizes of straight slot, Phillips, and Torx blades. Needle-nose chain nose, pump and slip joint pliers should be included along with a wire cutter/stripper, small hammer, crescent wrenches (a #6 wrench to quite useful), jewelers loupe, hex and spline key wrench sets, jewelers screwdriver set, tape measure, English/metric ruler, hemostat, dental pick, and scalpel handle with disposable #11 blades. A lab coat, safety glasses, supply of exam gloves, and paper towels should be carried for protection from, and cleaning of, potentially contaminated items. Resealable-zipper closure plastic bags in various sizes are useful for packaging and preserving items, and biohazard bags should be available for transporting and storing biohazardous or infectious items.

The investigator also must ensure that test equipment is easily portable as it is often desirable to perform the inspection on-site in conjunction with staff interviews. Custody of items should be retained by the facility and its risk management staff, or the attorneys might not be willing to allow the subject equipment to be removed. It is advisable to have separate photography equipment and tool/equipment cases that can easily be transported to an inspection site as some inspections are visual only and will not require the use of tools or test equipment.

Good-quality photography equipment is necessary for documenting the facility and equipment involved in an investigation. It often will be necessary to have specialized photography capability available, such as macrophotography to obtain the detailed views necessary to properly document an inspection or to photograph small items. One of the biggest problems in properly documenting an inspection is lack of proper photographic equipment. Whether film or digital equipment is used, it must be versatile and be capable of high resolution. Cameras that have interchangeable lenses are recommended so that the best lens for a particular application can be selected. Typical lenses used are a 35- to-70-mm zoom lens and a 60-mm macro lens with ring flash. A wide-angle, 28-mm lens is useful for photographing entire rooms or equipment setups. This latter will also require a suitable flash, as most built-in flashes do not have the power or coverage angle required to obtain good-quality, wide-angle, interior photos. Videotaping equipment is useful for recording a simulation of an event or where a detailed continuous record of an inspection is needed. Depending on what is being done, additional sound and lighting equipment may be necessary to obtain a good-quality videotape.

There will be times when specialized equipment is necessary to perform the appropriate testing of the device. As stated above, it may be necessary to rent equipment from medical or test equipment suppliers and to ensure that the use of the equipment is understood to obtain accurate results. At times, the manufacturer is the only source of specialized jigs or test equipment unique to the device, so they may need to participate in the testing. When this is the case, the testing should be observed by the investigator and an agreed-upon protocol designed to preserve the integrity of the device. The device should remain under the control of the investigator at all times and should be returned to the health care facility when testing is completed. In cases where destructive testing is necessary, all parties involved must agree to it via a signed document with each party's representative assenting to the agreed-upon protocol and the fact that the testing will damage or destroy the item.

Under some circumstances, it will be necessary and desirable to have analytical tests performed by outside laboratories or other experts. In such cases, it is important to ascertain the qualifications of the individuals performing the tests as well as their experience in forensic testing. Examples of such third-party testing would be metallurgical laboratories, analytical chemists, mechanical engineers, and electrical/electronic engineers. When such resources are used, it is important to have a test protocol submitted and agreed upon prior to authorizing such work. Charges of spoliation of evidence can severely weaken a case, and it is important to know the state laws that will be applicable in litigation, in order to ensure that a device is handled in such a way as to avoid these charges. It also must be remembered that the results of outside, third-party testing must be handled properly by the expert, in order to preserve confidentiality and to prevent premature discovery by opposing parties.

Where the inspection and testing process involves removal of the item from the premises where the event occurred, the chain of custody must be preserved. This involves documenting every event in which possession of the item is taken by another party. For example, if the expert delivers the item to an outside testing service, then stores it at the expert's office, and then returns it to the facility that owns it, each step must be recorded. The person releasing the item and the person receiving the item must both sign a document that indicates the description and identification of the item, the date and time it was released into custody of the individual receiving it, and the purpose for the transfer of custody. Such information is necessary for both tracking the item, to ensure it is not lost, and to be able to determine all of the parties who have been involved in handling the item in the event of damage or unauthorized repair.

The final phase of the investigative process is the information-gathering phase. This involves the research necessary to determine whether the event in question has been observed and reported before, or is a unique event. There are a variety of resources available to the investigator for this information. The Manufacturers and Users Device Experience (MAUDE) system, which is the FDA Center for Devices and Radiological Health's medical device reporting database, is an excellent starting point. In addition, access to the Health Devices user reporting system provides a great deal of information on events that do not meet the reporting criteria for health care facilities and manufacturers under the Safe Medical Devices Act. MedLine searches for articles and abstracts relating to problems with the subject device are additional sources of information. If it is not feasible to perform this research in-house, a local database-search firm can provide these services. In using such firms, it is important that all of the search criteria be well defined and that all of the possible search parameters be provided to the service. There have been many incidents of information not being retrieved because the search firm was not given appropriate criteria, thus resulting in a search that failed to reveal information that was immediately available if the proper search criteria had been entered.

In addition to researching external information sources, a basic part of the investigational process is reviewing internal documentation. Such documentation includes the operating policies and procedures of the facility, product literature, manufacturer processing and service instructions, professional processing and sterilization guidelines appropriate to the devices' manufacturer warnings, and the medical records and event reports relating to the incident being investigated. Care must be taken to gather all information and staff opinions before starting to formulate conclusions about the event. It is important that the investigator review all information available before forming opinions. Credibility of the investigator can be severely compromised if new information is revealed that significantly contradicts opinions and conclusions expressed by the investigator.

Reporting

Many investigations are compromised or fail because of inadequate reporting. The device investigation report is described here. The report should be written in a style that can be understood by both nontechnical and nonmedical personnel. The report, whether oral or written, is the best tool by which to educate other parties involved in the incident under investigation. Written reports should include an introduction to the event being investigated, the device description, inspection and functional test results, staff-interview results, observations, conclusions, and recommendations for prevention of future events and injuries. A recommendation as to whether the event should be reported under the requirements of the Safe Medical Devices Act should be made. Under some circumstances, the party requesting the investigation will not desire written reports, and none should be prepared unless specifically requested.

The introduction should summarize the reason for the investigation, a brief description of the device(s) involved, a summary of the event and outcome, the date of the event, the date the item in question was sequestered, and the date of the investigation. The equipment identification section should follow, identifying all devices and supplies by description, manufacturer, model number, serial/lot number, inventory number, equipment-control number, and expiration date as applicable. Where inspection tags are affixed to the device, the date of the most recent inspection should be documented. The device and facility evaluation (as applicable) should include descriptions of the devices and tests performed. Results of visual inspections and performance tests should be reported objectively, with no comments or conclusions made in this section. The observations section will include the investigator's comments on the test results.

The staff-interview section is provided in order to summarize the information obtained by discussing the event with staff members. Ideally, information obtained from staff members who were present at the time of the event should be presented in this section. Where this is not possible, third-party descriptions or accounts of information obtained from those involved in the event should be obtained and reviewed. Ancillary personnel involved in any internal inspections or investigations should also be interviewed in order to obtain any information that they might be able to provide that would allow for a more complete report. It is important to identify the persons interviewed, their positions, and their functions at the time of the event, in order to provide perspective on their information and to permit follow-up if and when needed.

The investigator should summarize findings and relate the event to test results and staff-interview information in the observations section of the report. This section is used to synthesize the information obtained through the inspection, testing, and interview processes. The observations section is also used to comment on test results, clinical conditions related to the event, and the possible mechanism by which the event occurred.

If possible, the conclusions section of the report should state the reason for the event and the cause(s) of the event. The conclusion might be that a particular device failed in a certain manner; that there was an error in setting up or using the devices; that the device design contributed to the problem; or that the event was related to a patient condition and that it was not the result of a device-related problem. At times, it may be necessary to state that there are insufficient data or inconclusive results, and that a final conclusion concerning the reason for the event cannot be reached with the information available.

It is often helpful to provide recommendations for preventing future repetition of the injury or conditions that, under different circumstances, could result in injury to the patient. Whether to include this section depends on the nature of the original request for the investigation. Where an investigation is requested by a facility's risk manager as an internal matter, it is often helpful to provide suggestions for changes in procedure, training of staff, or changes or modifications in devices or equipment. Where the investigation is requested as a result of litigation, such a section generally would not be appropriate, as all that is desired is an objective report of the facts of the situation that occurred and the expert's opinion as to the cause of the event or injury.

When a written report is not created, it is important that the investigator maintain all notes, documents, and photographs in a manner that will enable the presentation of an oral report. An oral report must be based on documented findings. Good notes are essential, in order to prevent misquotation at a later time or having one's information taken out of context in the event that litigation ensues. The investigator will usually be called upon to provide an oral report to the hospital's administration, the risk manager, or defense attorneys representing the hospital in litigation. Under such circumstances, the information obtained during the investigation process must be revealed only to parties who are authorized to communicate with the investigator.

In providing a report, either written or oral, the investigator must consider various issues. A disclaimer should be provided on a written report that protects the confidentiality of the document and guards against unauthorized discovery in jurisdictions where the law protects such reports. The report may have "Confidential" stamped on each page. Another approach is to place the statement "Prepared on Request of Counsel in Anticipation of Litigation" on the title page of the report. Identification of the patient involved must be discussed with the person requesting the investigation, as policies vary widely on this matter. Some facilities request full identification of the patient by name and medical record number, while other facilities want patient identification withheld from the report. Not being aware of the policies of the institutions for which the report is being prepared can be a major pitfall for the unsuspecting reporter.

First and foremost, the investigational report must be objective. There is no place for editorializing or subjective criticism that is not based on established fact in an event analysis. Any criticisms included in a report must be nonjudgmental and written in a constructive manner. No effort to place blame should be made, as to do so could unfairly jeopardize an individual or the institution. Efforts should be made to view the found problems from a systems approach rather than establishing blame. Political considerations also must be eliminated, to the degree possible, from any objective investigational report, as they interfere with performing an objective analysis and forming conclusions that will be meaningful to the institution. For further information on reports, see *Writing and Defending Your Expert Report* (Babitsky and Mangravitti, 2002).

Activities Resulting from the Investigation and Reporting Process

The investigator often will be requested to perform various types of follow-up activities subsequent to generating the investigation report. These may range from simply answering questions to participating in litigation. In the majority of cases, the investigator will not become involved in litigation. (This is a separate issue that will be addressed later in this chapter.)

The investigator may be called upon to make a presentation to an administrative group or medical committee concerning the findings. This usually will involve providing a summary of the event, explaining the theory of how the event happened, and reviewing ways in which a recurrence can be prevented. Frequently there will be pointed questions from the audience. These should be anticipated, and responses should prepared, to the extent possible. The presenter should be aware of who will be at the meeting and their possible issues in order to provide meaningful information and suggestions.

Implementation of recommendations frequently will require the assistance of the investigator and may include training of staff by way of in-service presentations, which may be requested. Recommendations on the modification or development of policies and procedures in conjunction with department management may be an important follow-up action. The investigator may be requested to attend meetings with manufacturer representatives and to participate in manufacturer inspections. In such situations, the role of the investigator should be primarily that of observer, making individual notes and gathering any additional information that may result from the manufacturer's activities.

Interaction with governmental and enforcement agencies is a frequent product of medical device investigations. This may involve compliance with regulations under the Safe Medical Devices Act, if the event occurred in the United States, or similar legislation in other countries. (See Chapters 125 and 126.) Assistance in making reports to governmental agencies may be requested either in the form of helping to fill out documentation or participating in inquiries initiated by a licensing or enforcement body. On rare occasions, an investigator may be requested to participate in formal government activities such as advising on modification or implementation of laws or participating in agency meetings or public hearings.

The Clinical Engineer as Expert Witness

In the event that litigation ensues over a medical device-related injury or death, the clinical engineer may be called upon to act as an expert witness in the matter. In this capacity, the clinical engineer's training and experience will be relied upon to provide analysis and opinions concerning the event being litigated. It will require research to support or refute positions taken by parties to the litigation, working with attorneys preparing the case, and possible testimony at depositions and trial.

The clinical engineer may be designated as an expert to represent an employer or he may be retained as a third-party expert via an insurance carrier or a law firm. While there is an assumption that the clinical engineer will support the position of the retained party, the expert nevertheless must maintain neutrality. An expert witness should function as a disinterested third party whose responsibility is to provide opinions based on the facts of the case and research and testing performed by him. The expert is the only person who is allowed to express opinions in court rather than being limited to simply reporting facts. However, these opinions must be based on theory and information that is accepted practice in the person's field of expertise. Courts rely on experts to educate the court and the jury on the circumstances of a case and to provide opinions on the matter that will be considered in the jury's development of a verdict.

Qualifications to Be an Expert Witness

An expert witness is a person who must demonstrate that he or she is competent to research the circumstances of a case and to reach meaningful conclusions and opinions that the court can rely on. In order to meet this requirement, an expert must pass court tests of competency, which examine the expert's background and credibility. The expert is the person who has special knowledge of an area that is not considered to be in the body of knowledge of the average citizen or that is not discernible by common sense.

Education and special training comprise one of the main criteria used in qualifying prospective experts. The person being considered must have education and training in their field, and it must be significantly greater than that of the common man. In general, the person must have at least an undergraduate degree in, or related to, their field of expertise. In addition, an expert must show evidence of specialized training in the field in which testimony will be offered. This may consist of additional educational credits, attendance at specialized training courses, participation in professional associations related to the field, and certification or licensure in the field. In the case of clinical engineering, this would mean having an undergraduate degree in clinical engineering or related engineering fields with additional training in medical and biological sciences, at a minimum. It also might be possible to qualify with a degree in life sciences or medicine, augmented by training in engineering and technology.

An expert witness must have demonstrated experience that is relevant to the area of proposed testimony. As an example, a person with experience in design of equipment could be expected to testify concerning design defects, but not about the clinical use of a medical device in the operating room, as it is unlikely that a design engineer has spent much time working on surgical cases in an operating room. The amount of experience also must be sufficient to satisfy the court that the person has sufficient knowledge of the field. It is unlikely that a person who has completed their degree, but has no working experience in a field, would be accepted as an expert. Conversely, a person who has only a high-school education but many years of practical experience, such as a machinist or electrician, could qualify.

Credentials are another aspect of the expert's qualifications. A person who is certified as a clinical engineer, a registered nurse, a registered professional engineer in a relevant field, or a person with medical board specialty certification is favored as an expert witness in the field over a person with a similar background who lacks such credentials. Many attorneys require that their experts have a higher degree and/or professional certification of some type to ensure credibility with the court and the jury.

Another basic requirement to function as an expert witness is the ability to communicate. One of the most important functions of an expert witness is to teach the jury so that it can understand the expert's testimony. While clinical engineers are accustomed to working with peers who have knowledge and education in medical and technical fields, as an expert they will be working with people who generally have little or no understanding of engineering or medicine. In addition, it must be considered that the average educational level of most juries is a high-school education. Therefore, it is important that the clinical engineer who wishes to work as an expert witness be able to simplify and to explain medical and engineering concepts to those who have an advanced education but little technical training(e.g., judges and attorneys) and to those whose educational level might not include high-school graduation (i.e., jurors). This must be done as an educational effort at the level appropriate to the audience, and it must not be perceived as technical jargon or as the expert speaking down to the audience. Lack of this ability has been the downfall of many highly educated theoretical experts when juries have been polled after the conclusion of a trial.

The expert also must be able to function psychologically and emotionally in the environment of the legal system as it operates in most jurisdictions. He will be alternately supported and criticized by the various parties to a lawsuit. He must be able to interact objectively with attorneys who are defending their clients' positions by attempting to undermine or disqualify the expert. Many people who have been educated in engineering are not oriented to working in the adversarial atmosphere that characterizes a legal case. During deposition and cross-examination at trial, the experts' background, research, findings, and opinions are challenged, and the expert must be well prepared to defend against these attempts to disqualify them as experts. (See *Cross Examination: The Comprehensive Guide for Experts* (Babitsky and Mangravitti, 2003).) It is important that the prospective expert not take personally any criticism from opposing parties. Conversely, part of the expert's function is to find and exploit the weaknesses of the opposing expert's experience, background, and opinions. Many otherwise well-qualified clinical engineers simply do not have the personality to operate within the emotional atmosphere of litigation.

One of the most important characteristics of a successful expert is neutrality. It is of utmost importance that the expert not be swayed by the emotional issues that arise in medicolegal cases. The expert is probably the only figure involved in the litigation who does not, and should not, have a vested interest in taking sides in the action. The expert must be able to see both sides of a case and to consider them objectively. If the facts dictate a position that is not in favor of the party who has retained the expert, this must be brought to the attention of the attorney with whom the expert is working. One of the services that the expert provides is to be the devil's advocate and to point out weaknesses in the case of the side that has retained him or her. This will assist the attorney in preparing to plug any holes in his arguments and to develop approaches to strengthen the weak points in the case. In some cases, this information may speed settlement of the matter and thus avoid the expense and time associated with a trial that could result in an unfavorable verdict.

Under no circumstances must the expert have any vested interests in the outcome of the case at hand. All offers of contingency fees must be rejected, although compensation is appropriate for the expert's assistance. The expert should have no investment or other monetary interest in any of the parties involved. For example, if the opposing counsel determines that an expert owns stock of a company involved in the case, this fact will be brought out in depositions and court testimony in attempts to disqualify the expert or to undermine the expert's credibility with the jury.

Developing Opinions

The opinions of an investigator or a retained expert are sought either to resolve a problem at the health care-provider level or to assist in litigation. The primary reason for retaining an expert in a legal case is to present opinions to the jury that support the contentions of the side that has retained the expert. In order to express such opinions in court for consideration by the jury, a process must be followed by which valid opinions are developed. If the bases for the opinions expressed do not meet the criteria of the court, the expert's opinions will lose credibility and the expert can be disqualified by the court. Therefore, opinions developed by an expert must be based on provable facts, test results obtained by established methods recognized in the field, and information researched from credible sources.

Among the sources for information upon which to base an opinion are published reports and journal articles in the field. Reports of findings by other experts in the field can be used to back up the opinions developed by the expert, or they can provide reasons to modify them. Such information can be obtained by reviewing technical articles that are pertinent to the area of concern and compiling relevant information. Results of research in a field, government studies, and reports by users, such as the medical device-related reports submitted to the U.S. Food and Drug Administration (FDA), can all provide information to support or refute an expert opinion.

Reviews of the deposition transcripts of other parties involved in the litigation are an important part of developing and supporting expert opinions. It is important to obtain information from witnesses' descriptions of events in order to ensure that the event is fully understood. It is quite possible that one line in the deposition of a seemingly unrelated witness can confirm or disqualify an opinion that an expert has arrived at prior to reviewing case material. It is also important to determine the opinions of opposing experts and their bases for such opinions. While deposition transcript review is often a long and arduous process, it is a valuable source of information for the expert.

Current news reports and magazine articles also can be sources of information that are relevant to the expert's opinions. Reports of successes or problems with a device that is under litigation may act as additional contemporary supporting material for the expert. While it is important to avoid sensationalism in using such reports, they can be valuable supporting material if used properly.

The expert is allowed (and expected) to rely on his or her own experience in formulating opinions concerning a case. Part of the reason why an expert is retained is to provide information based on experience in the field. It is perfectly acceptable to say, "In my twenty years of experience in the field, I have seen this happen often", and then proceed to express opinions on the cause of the event. It is important, however, for the expert not to express opinions in areas where experience is marginal or the expert is not fully competent. A skilled opposing attorney will quickly discern the expert's inexperience and/or weakness in knowledge and will turn it against him. The expert also must guard against being led into an area where he has little or no expertise because he could be asked questions that he could not answer, or he could express opinions that he could not support.

The Litigation Process

In the event you are requested by an employer or retained as a third party to act as an expert witness in litigation, it is important that you understand the general process. This section provides a brief outline of the litigation process as generally followed in the United States and many other countries from the viewpoint of a practicing expert witness.

When an action is filed, both sides begin to prepare their cases. To obtain the information that is necessary to support each side's contentions, the "discovery process" is initiated. This process involves the gathering of information from a variety of sources, including the opposing party. The discovery process may include requesting information and reports from the opposing side, setting up inspections and testing by all parties witnessed, performing searches for information that is relevant to the case, and interviewing all parties involved in the case. There are rules and time limitations concerning discovery that vary among jurisdictions, which the expert should become familiar with in order to ensure that time constraints are observed and that no rules of discovery relating to expert witness activities are violated.

Reports may be prepared at the request of attorneys to assist them in supporting their position in the case. In many cases, the plaintiff's attorneys will request written reports from expert witnesses to use in obtaining information from the defense. In some jurisdictions, experts must prepare reports including supporting facts and opinions that are submitted to the court prior to trial. These reports are then distributed to the litigants for use in trial preparation. In many cases, an expert's reports of investigations will be protected from discovery until the person is designated as an expert for trial. Protection of reports from discovery varies greatly among jurisdictions, and the experts must always assume that their reports will be read by the opposing parties at some point in the litigation process, and that they also could be shown to the jury.

Designation of experts is the formal process by which each side reveals those whom they plan to present as experts at trial. The designation permits the opposing side to review the qualifications and background of the person designated. In many cases, research is done on the expert to determine what cases they have worked on previously as well as their success rate, and to read reviews of their performance. In a process that is related to expert designation, attorneys may work with the experts whom they have retained, to develop a declaration of the expert's opinions concerning the case in order to support efforts to reach a settlement of the case before going to trial.

In most jurisdictions, depositions are taken in which each side is allowed to question the opposing expert(s). This is done under oath, usually in the opposing attorney's office. The proceedings are recorded by a court reporter, and a formal transcript of the deposition is produced and provided to all sides. The expert is required to bring all notes and materials reviewed in forming opinions to the deposition for inspection by the opposing attorney. During the deposition, the expert is questioned on his background, qualifications to provide opinions on the case in question, his opinions, how he has arrived at his opinions, and the facts supporting those opinions. The attorney who retained the expert will accompany him and will have the option of objecting to questions or comments made by the deposing attorney. In some cases, the expert will be instructed not to answer certain questions posed by the deposing attorney. Depositions vary significantly in time, with some as short as 30 minutes, while others have lasted for ten days or more, spread out over several weeks or months. After the deposition is completed, a printed transcript will be sent to the expert by the court reporter for review and correction. At that time, it is important that the expert review the transcript for technical and factual errors, and correct them. It should be noted that if there are major changes made to answers or opinions, this may be used in court by opposing counsel to attempt to embarrass the expert and impugn his integrity. For further information on depositions, see *How to Excel During Depositions: Techniques for Experts That Work* (Babitsky and Mangravitti, 1999).

Many cases (80 to 90%) settle before going to trial. However, the expert witness will testify at trial in those cases that have not been settled. When called to trial, the expert's testimony will revolve around the same items that were covered in the deposition. Sufficient time should be reserved for pretrial preparation to review the notes made in the case and the deposition transcript. It is important that the expert spend time with the attorney retaining him, to discuss what the expert can expect to be confronted with at trial, as well as any new information about the case. At trial, the expert must be careful in understanding the questions asked, before responding. It is not uncommon for opposing counsel to ask a question in several different ways, to elicit different answers and thereby impugn the expert before the jury. Prior to actual questioning, the opposing side might attempt to disqualify the expert at trial by asking a series of questions concerning the expert's background and qualifications.

When testifying in court, the expert should always remember that the jury, and not the questioning attorney, is the audience. Dress and appearance should be conservative and professional. Excessive jewelry should not be worn. It is important for one to look at the jury members and acknowledges their presence. Experts must explain their positions and information in terms that the jurors can understand, as few if any of them will have a technical or medical background. The expert must make an effort to develop a rapport with them so that they will pay attention to the testimony while not violating the court's rules governing interaction with the jury that could result in a mistrial. When testifying, it is important that the expert take time before answering. It is perfectly acceptable in most courts to bring the case file to the witness stand for reference prior to the expert answering a question. This process also gives the expert time to formulate answers and to control the pace of the questioning to a comfortable rate. Normally, an expert will be excused after the testimony is completed. However, it may be necessary for him to be available for recall if instructed by the court or retaining attorney.

When the trial is completed, the retaining attorney generally will advise the expert of the outcome. If the attorney does not communicate within a reasonable time, it is perfectly acceptable for the expert to contact the attorney to learn the verdict. The expert may ask for a critique of his work on the case, as a guideline to areas that can be improved.

The Business of Being an Expert Witness

While many clinical engineers will be involved as investigators, and some may testify as part of their function as employees of health care organizations or manufacturers, there are some adventurous souls who desire to work as third-party experts on a part-time or full-time basis. To do so successfully requires knowledge of the business-management aspects of acting as a third-party expert. (See Chapter 12.)

Establishing a fee structure is an important first step in the process of establishing a practice. Fees are generally charged on an hourly basis for the work performed, and they are dependent on the qualifications of the expert, the general fee structure in the area where the expert is practicing, and rates charged by the competition. There is no specific rule for the hourly rate, which can range from $100 to over $500. Most experts charge a higher rate for trial and deposition testimony, commonly 25-50% more, due to the stress of examination by the opposing side. It is not uncommon to specify a minimum time to be billed for a deposition, such as two hours or half of a workday. If there are other expert witnesses or forensic organizations in the expert's area, they can provide guidelines. Attorneys who use expert witnesses can also provide information on the fees they are accustomed to paying. It is important that a detailed time log be maintained in order to make billing simpler and to justify charges in the event that billings are questioned. It is common practice to charge in quarter-hour increments. Local travel is charged at the hourly rate, with some experts charging for round-trip travel, and others only for one-way. Long-distance travel via air or other carrier is often negotiated with the client. Many experts take work with them, so the travel time is charged as regular work time. Alternatively, work can be done on other cases during travel time to provide income while traveling, as most clients are unwilling to pay an expert to sit on a plane.

It is important that the expert have a signed agreement defining fees and terms of engagement prior to starting work. A contract or letter of agreement should be presented to the client when the client indicates an intent to engage the expert. This document should reference the case assigned to the expert, include terms and conditions, and specify the hourly rates for research and preparation and testimony at depositions and trials. Payment terms must be specified and should require that the expert be notified in advance of any cap or limit on fees, time frame for completion of services, and changes in the nature of services required. There must also be a "stop-work" provision, which specifies that the expert will discontinue work on a case if the client's account falls in arrears. It should be a condition of engagement that all amounts due be paid prior to the expert's appearance at depositions or trial, as a settlement that eliminates the need for further work from the expert may be reached.

When dealing with plaintiffs, it should be stated that payment for the expert's services is not contingent on funding by a third party. This is necessary to protect the expert, as plaintiff's attorneys often work on a contingency basis and the expert could be placed in the position of trying to collect from a third party when working on a plaintiff case. Another item to consider in the agreement is the requirement that all retainers be paid in advance when working on plaintiffs' cases because a plaintiff's attorney could refuse to pay the expert's fee after an unfavorable verdict, claiming that they did not get their contingency, that they were unable to collect the monetary award, or that the client has no money left, due to the cost of pursuing a case that resulted in no monetary judgment being awarded. A significant advance (at least $2,000) retainer should be collected prior to reviewing a plaintiff's case. This acts as a screening device to ensure that the attorney has faith in the case. It also protects the expert from being unable to collect for initial work performed if the case settles or the attorney decides not to pursue it after receiving the initial comments from the expert review of the merits of the case.

Invoices to clients should be done as monthly progress billings unless mutually agreed to by the expert and the client. For short projects such as equipment inspections, or where a lump-sum billing has been agreed to, this may require only one invoice. However, most legal cases will last a year or more, and progress billings will be necessary. It is advisable to use a timekeeping system that can be easily incorporated into the billing process, and there are a number of software programs that prepare billings from input to project time sheets. This will minimize the time and effort involved in billings and will reduce lost time due to forgetting to record time or making transcription errors. The invoice must include the dates of service; a brief summary of the work performed; the hourly rate charged and the total hours billed by each person working on the case; any out-of-pocket expenses incurred on the project; and the total cost. Payment terms should also be stated (e.g., 2%; 10 days; Net 30 days; a service charge of 1.5% monthly to be charged on all past due amounts).

If a client is overdue more than 30 days on payment, an overdue notice should be sent requesting payment within ten working days. In some cases, it becomes necessary to suspend services to a client. One reason for taking this action is due to nonpayment. If there is no response within that time frame, the expert should send a letter warning the client that work on the case will be suspended if payment is not received within 60 days after the bill was rendered. In order to prevent this fate, it is important that the expert track all accounts in order to ensure that they are current.

Another possible reason for suspension of services is that the expert has exceeded the amount of work authorized. The expert must guard against performing more work than an advance retainer will cover in a plaintiff's case, or more than was authorized by an insurance carrier or law firm in a defense case. It may be very difficult to collect after the fact for work done in excess of the amount authorized because of the efforts to control costs of litigation on both sides. The expert cannot afford to think that the client has given him *carte blanche* when signing the retainer agreement.

Business resources will become more important as the income generated grows. The expert should have the support of a bank, a small-business accountant, and an attorney if he or she develops a significant volume of business. To avoid some of the pitfalls of retainer agreements and the possible legal problems associated with a consulting practice, it is advisable to develop an association with an attorney who is familiar with collections and/or contract law. There will be times when a well-worded letter from an attorney will bring about payment that has not been forthcoming. If the business becomes a major source of income, accounting and tax preparation should be done by an accountant who specializes in small business. This can be a valuable resource in preparing statements that may be needed when applying for credit lines or business loans. In addition, business-related tax deductions can be maximized to reduce taxable net income through the services of a knowledgeable accountant. Bank affiliations can be a valuable asset. Establishing credit with a small bank is a valuable asset for an expert who depends on his or her practice for all or a significant portion of annual income. Because the cash flow in this type of business is sporadic, a full-time consultant should have a small-business credit line that can be drawn against when cash flow is poor and that can be repaid when income improves. One major warning is to not use a home mortgage to finance your small business. In the event that the business is unable to generate the funds necessary to make loan payments on time, it is possible that the lending institution would foreclose on the house. Small-business owners have lost their homes because of such circumstances.

Marketing of the expert witness can be done in a variety of ways. One effective method is to purchase listings in the expert directories published by the local bar associations in the area where the expert practices. The expense of such listings can be repaid by obtaining one good case through the listing. The local or regional bar associations can be contacted to determine whether they publish such directories. Professional speaking engagements with local risk management and attorney groups are another excellent way of gaining exposure to potential clients. Such organizations are constantly looking for speakers for their meetings.

Referral services are another aspect of the expert-witness business that must be considered. These services match attorney inquiries with experts in the field, as the case requires. In return, they often require the expert to discount the fee charged to the referral service. They then bill the market rate to the attorney, with the difference being their income. In some cases, the referral service will accept the full rate of the expert and add their percentage to that rate. In the first case, the expert is giving up 25 to 50% of fee income in exchange for the referral. Where the service adds to the expert fee, it creates a danger of reducing the number of cases that will be received if potential clients perceive the total fee as excessive. However, use of referral services can be a good way of building an initial practice. The expert should review the terms and conditions carefully before enlisting in such a service. There is no guarantee that the service will obtain any cases for the expert, and cases referred to the expert will not necessarily be appropriate. Payment of the expert will occur only after the referral service has been paid by the client, so the expert's income is dependent on the referral service's billing and collection practices. Many services also prevent the expert from accepting work directly from a client for a period of several years after a referral. This practice can limit the expert's ability to develop an independent repeat-client base.

When entering into an agreement to become an expert witness, an individual must consider the possible conflicts that may affect his or her effectiveness. Some people are not comfortable working for both plaintiff and defense attorneys and thus limit their practice to serving one or the other category of litigant. This can be used by an opposing attorney to question the expert's neutrality and must also be considered a possible factor in limiting the expert's practice. Prior commitments that will limit the expert's ability to perform work within the schedule set by the attorney or the court must be considered before accepting a case. The ability of an expert to respond to the client's needs will be considered in the decision to retain an individual. In addition, failure to meet the client's schedule can often mean that no further business will come from that client. Inability to respond to litigation issues because of lack of results from the expert places the attorney in a compromised position and could jeopardize the case. Conflicts can arise when a plaintiff asks the expert to participate in a case for a previous client whom the expert has defended on multiple occasions, such as an insurance company or corporate entity with whom the expert has a longstanding relationship. The converse also can occur. It is best to reveal these facts to the inquiring attorney so that they are ware of the potential conflict. As a general rule, the expert should engage the first party who requests his services unless the expert has a history of working for the opposing side on multiple cases. It is important to maintain a balance between defense and plaintiff work so as to avoid a reputation of being too closely allied with either category. If simultaneous inquiries occur on the same case from opposing sides, it is best to work for the side that initiated an inquiry first, unless the expert feels strongly that they wish to work for the opposing side because of previous relationships or the merits of the case.

Whether simply performing an inspection or functioning as a key expert in an extended litigation, the investigator and expert must always be conscious of confidentiality and protection of client information. No investigation or ongoing litigation should be discussed with any person who is not directly involved in working with the expert and representing the interests of the expert's client. Divulging any information about a case to other parties must be strictly avoided. There is no way to know whether the person at the next table in a restaurant, or a new acquaintance at a party, has some relationship to the litigation, or whether they are a relative of a patient whose injury is being investigated. In addition, private investigators have been known to use false credentials to gain credibility with an institution's staff or a person involved in a legal case in order to obtain information for the opposing side. All reports to clients should be marked as confidential, and no information should be revealed to other parties unless the permission of the client is obtained or in compliance with a court order which has been discussed with the attorney for whom the expert is working.

References

Babitsky S, Mangraviti JJ. *Cross-Examination: The Comprehensive Guide for Experts*. Falmouth, MA, SEAK, Inc., 2003
Babitsky S, Mangraviti JJ. *How to Excel During Depositions: Techniques for Experts That Work*. Falmouth, MA, SEAK, Inc., 1999.
Babitsky S, Mangraviti JJ. *Writing and Defending Your Expert Report*. Falmouth, MA, SEAK, Inc., 2002.

Further Information

Babitsky S, Mangraviti JJ, Todd CJ. *The Comprehensive Forensic Services Manual*. Falmouth, MA, SEAK, Inc., 2000.
Malone DM, Zwier PJ. *Expert Rules*. Notre Dame, IN, National Institute for Trial Advocacy, 1999.
Poynter D. *The Expert Witness Handbook*. Santa Barbara, CA, Para Publishing, 1987.
The National Directory of Expert Witness. 2002. Esparto, CA, Claims Providers of America, 2002.

14

Careers in Facilities

Bruce Hyndman
Director of Engineering Services,
Community Hospital of the Monterey Peninsula
Monterey, CA

In the early days of hospital-based clinical engineering programs (usually then called "biomedical engineering departments"), there were usually several levels of positions and most departments were headed by a clinical engineer with a bachelor's or advanced degree or someone with many years of experience. Job descriptions and minimum qualifications for department heads often included a bachelor's degree in biomedical engineering or a preference for a master's degree in the same field because there were more graduate programs in the field than bachelor's level.

At the same time, requirements for chief engineers in hospitals often were based upon experience only, with emphasis on hands-on experience with major building systems.

Today, advertisements for positions in clinical engineering are typically for generalist biomedical engineering technicians (BMETs) or specialists in such areas as radiology- and lab-equipment maintenance and repair. There are few positions available for degreed department heads in the clinical engineering field today. Advertisements for chief engineers, on the other hand, have all but disappeared and have been replaced by position postings for "Facilities Director" or "Facilities Manager," which usually require a minimum of a 'bachelor's degree and often a preference for professional registration such as Professional Engineer (PE) in mechanical or electrical engineering. PE is a licensure granted by state consumer affairs departments. It is earned through experience and testing and is intended to reflect minimum competency to practice as a consultant or designer in the specified field of engineering. It is interesting to note that there is no PE category for biomedical or clinical engineering. This fact has contributed to the desire by some to create a certification process for clinical engineering (see Chapter 133). A certification program for clinical engineers is currently administered through the ACCE Healthcare Technology Foundation (see Chapter 130).

Clinical engineers (CEs) might well consider expanding their horizons to a more diverse application of their engineering and management skills. In fact, there is good reason to conclude that the best thing for hospitals and medical centers to do is to install clinical engineers as facilities directors and plant managers. This career path makes sense for many reasons but may seem controversial or even distasteful to many in the clinical engineering profession. This chapter suggests why this should not be so.

First, most in the clinical engineering field know that many who work in facilities operations lack the sensitivity to, and real understanding of, the clinical environment. This is not to brand all facility managers with the same reputation. However, most clinical engineers will recognize the notion that nurses and physicians often call upon clinical engineers to assist in situations that might be the purview of the facilities department but from whom they have had poor response. An even more frequent current example is the perception of users that information technology (IT) departments really do not meet the service and support expectations of clinical professionals. This is because most of the line staff in a clinical engineering department have more education and training than the typical maintenance or plant engineer and have pursued their career because of an interest in working closely with health care providers in a clinical setting. This puts the BMET or clinical engineer on the clinical side of any activity much more than other maintenance or support departments. In fact, most clinical engineering department staff members crave exposure to the clinical environment and staff on a daily basis, while many facilities staff would just as soon wait until patients and clinical staff are gone before they enter an operating room or other patient-care area. BMET's and clinical engineers sought out their work because they wanted to be in the hospital environment, and their careers are unique to that setting. For the most part, facilities staff might as well be working in an automobile factory, hotel, or high-rise office building. Again, this is not intended as a judgmental statement, but simply a demonstration of the fact that the clinical engineers are uniquely qualified to provide all types of engineering support to the hospital setting.

Second, it is often doctrine to clinical engineers and BMETs that supporting facilities and plant equipment and systems is not nearly as sophisticated and important an activity as supporting medical devices. One still hears the admonition of clinical engineers not to extend department responsibility to include the maintenance of hospital beds because it could demean the reputation of the members of the clinical engineering department. In the last 20 years, the sophistication of plant equipment and systems, and hospital beds for that matter, has come to match that of much of general medical devices. Hospital beds now incorporate electronic scales and microprocessor-based controls for air-pump systems, positioning, and safety features. Operating the air conditioning, energy-management systems, fire alarm systems, automatic doors, isolation-room air supply, and many other systems now requires skills and understanding of the same operating systems, networks, and hardware found in patient monitoring and other medical equipment. The skills that are used to test and maintain complex, sophisticated medical device systems are directly applicable to these other systems now found in the physical plant.

Another reason for this strategy is that many hospitals have incorporated, and more will incorporate, the clinical engineering department under the larger umbrella of a facilities department because the latter will probably have a degreed professional in the directorship of the group. This will eliminate the past argument that clinical engineers should not report to a nondegreed chief engineer. If the facilities department is headed by a clinical engineer, the clinical engineering areas will not receive merely secondary support and interest as part of a larger group, but will receive the primary support and understanding that it deserves.

What, then, of the concerns and misgivings of those clinical engineers who might not see this as an attractive career path choice? The clinical engineer may ask a series of questions as listed below. Answers to these questions are also provided along with their rationales.

Question: *Am I really qualified to manage a facilities engineering department?*
Answer: *Almost certainly.*

- Most clinical engineering degrees have the underpinnings of the same undergraduate basic engineering classes and topics as electrical and mechanical engineering.
- Managing budgets, personnel, and business relationships in the hospital environment is the same for both kinds of departments. In fact, the facilities department budget can be ten times the size of the clinical engineering budget, a fact to be revisited in the advantages of making this career move.
- Although clinical engineers might not know the difference between a 500-ton chiller and a cooling tower, with minimal exposure and some reading clinical engineers will recognize the same principals of physics in the operation of air conditioning that they apply to understanding hypothermia devices and other heat-exchange equipment.
- Clinical engineers are already considered technical experts and will be accepted quickly after demonstrating a vocabulary in the facilities area.

Question: *Will I lose the respect of other professionals by involvement with the maintenance of what may be more mundane equipment and systems?*
Answer: *No.*

- This will require managing your own posture and behavior appropriately. From the standpoint of the chief executive officer (CEO), chief operating officer (COO), and chief financial officer (CFO), the importance of your decisions will be greater because you will affect a much larger part of the operating budget and will contribute more to the planning phases of new facilities. By staying involved at some level with the clinical engineering operation, you will not lose your level of respect.
- Your clinical engineering staff probably will be surprised at your level of understanding of plant equipment and operations.

Question: *Will I enjoy my daily activities and efforts as much as I now do?*
Answer: *Yes and perhaps more so.*

- Contributing to the treatment of patients is an interest of most clinical engineers. The more you can influence the mission of the hospital, the more you can contribute. The attention to one medical system may affect one or a few patients—plant operations and systems affect almost everyone in the hospital simultaneously.
- Confronting contingencies is a commonly quoted activity that makes clinical engineering an exciting endeavor. The number of contingencies in a facilities operation will include all of those that you now encounter, plus many, many more, including power failures, communications failures, and water outages.

Question: *What are the advantages and disadvantages of this career change?*
Answer: *See below.*

Advantages:

- The average salary for a clinical engineering director in 2001 was $60,200, while the average salary for a facilities director in 2001 was $69,700 (Baker, 2002).
- If the clinical engineering operating budget is a portion of an overall facilities budget under your control, your ability to provide adequate funding for the clinical engineering activities is much greater.

Disadvantages:

- The time requirements will be greater. There will be more committee memberships and project management. Clinical engineering activities will be reduced to less than half of your personal effort.
- Mistakes in the management of life safety and other systems can put your hospital on the front page, and you out the door.

It may be that the traditional position of clinical engineer or clinical engineering director will remain in some hospitals and medical centers, but in a recent survey (Cohen, 2002) responding clinical engineers who were employed in hospitals reported their median age to be 45 to 50 years. In the same survey, younger median ages were reported only for those employed in nonmedical industries or for students. Older median ages were reported for CEs in private practice or consulting. If this survey is truly representative, it would appear that there are few clinical engineering positions in hospitals occupied by a new generation.

References

Baker TM. Survey of Salaries and Responsibilities for Hospital Biomedical/Clinical Engineering and Technology Personnel. *J Clin Eng* 27(3):219-236, 2002.

Cohen T. ACCE Body of Knowledge Survey. *J Clin Eng* 27(4):298-299. *Health Facilities Management*, December 2002.

Section II: Worldwide Clinical Engineering Practice

Enrico Nunziata
Health Care Technology Management Consultant
Torino, Italy

Health is an issue of concern for all residents of this planet. No country has the solution for addressing the health care needs of all, since each country, region, and locality has its own particular needs, problems, and resources. Health care technology being applied increasingly in an attempt to address health problems has resulted in wonderful successes and dismal failures. "In 50 years $1 Trillion in aid to poor countries ... failed spectacularly to improve the lot of its intended beneficiaries." (*The Economist* June 26, 1999, p 24).Wealth does not produce immunity to failure, however, as the Institute of Medicine in the United States showed in its report *To Err Is Human*, in which it was revealed that "... at least 44,000 American die each year as a result of medical errors" and that "...the number may be as high as 98,000." Clinical engineers (CEs) bring sound skills and experiences to bear as they strive to more effectively manage the utilization of this technology. Across the globe, the basic principles of clinical engineering can be applied to special situations and particular circumstances.

Clinical engineering is a discipline that is taught, studied and applied throughout the world. National organizations of clinical engineers are found in Mexico, Australia, South Africa, and Italy to name just a few. The American College of Clinical Engineering (ACCE) membership list includes clinical engineers from Albania, Pakistan, Jordan, Poland, India, and many more. These organizations not only are concerned with the local practice of the profession but they also typically take a global view as they realize that information exchange is ultimately beneficial to all. The ACCE, for example, through its International Committee promotes such activities at the Advanced Clinical Engineering Workshop (ACEW), where international faculty present advanced concepts of clinical engineering and health technology management to nations that are intent on improving their health care infrastructure such as China, Russia, Lithuania, Ecuador, Nepal, and Mexico. Information exchange through publications such as *ACCE News* has brought to the world reports of clinical engineering activities in Mozambique, Kyrgyzstan, Mongolia, Namibia, Vanuatu, South Africa, Nepal, and even the United States. This author is editor of the column *CE Around the Globe*, a regular feature of *ACCE News* that describes CE program experience primarily in developing countries.

The International Federation for Medical and Biological Engineering (IFMBE), through its Clinical Engineering Division, has promoted clinical engineering activities among member nations and has sponsored international clinical engineering conferences in France, Italy, and Germany. This Handbook's Editor-in-Chief confided to this author that his participation in these forums and the encouragement, suggestions and advice received from colleagues in attendance from around the world inspired him to assemble papers on all the diverse aspects of clinical engineering under one umbrella text, i.e., this handbook, which would serve as a comprehensive information resource and an educational vehicle for all members of the health technology community and for all aspiring practitioners of clinical engineering.

The Engineering in Medicine and Biology Society of the Institute of Electrical and Electronics Engineers has found a resurgence of interest in clinical engineering and incorporates clinical engineering tracts in its annual international conference. The World Congress on Medical Physics and Biomedical Engineering brings together clinical engineers from around the world. It was on such an occasion in Texas in 1989 that a small group of CEs planted the seed that grew and flowered into the ACCE through hard work and dedication of its members and from the encouragement of the international community. For decades, the IFMBE, the World Health Organization, the Pan American Health Organization, the European Union and nongovernmental organizations such as Project Hope, FINNDA, GTZ, and FAKT have promoted international clinical engineering research and development programs. The contributions have positively affected the development of clinical engineering in many of the countries whose CE practices are presented herein.

This handbook addresses all the issues of which the clinical engineer and, to some extent, all health professionals should be cognizant. By dedicating an entire section on worldwide clinical engineering practice, the editor-in-chief is emphasizing that there is a great deal to be learned from the problems being faced and the solutions being developed everywhere that the clinical engineering profession is practiced.

In this section of the handbook, starting with a World Clinical Engineering Survey (Glouhova and Pallikarakis), the common threads of clinical engineering practice around the world are revealed as well as the sharp differences in such matters as space allocation, staffing, training courses, and medical device inventories. The authors of the following chapters present in-depth perspectives of the origin, development, implementation and expectations of clinical engineering in their respective countries. The countries included are the United Kingdom (Meldrum), Canada (Gentles), Estonia (Aid and Golubjatnikov), Germany (Dammann), Brazil (Brito), Colombia (Gutiérrez), Ecuador (Gomez), Mexico (Velásquez), Paraguay (Galvan), Peru (Vilcahuaman and Brandán), Venezuela (Silva and Lara-Estrella), Japan (Kanai), and Mozambique (Nunziata and Sumalgy), and countries in the Middle East (Al-Fadel). This handbook's international perspective on clinical engineering is not limited to this section, however. The reader also can find examples of clinical engineering practice in other chapters throughout the handbook. Areas of clinical engineering practice, including health care technology management, safety, education and training, medical device manufacture, facilities, standards, and certification, are written by CEs from Australia, Brazil, Cameroon, Canada, Costa Rica, Germany, Greece, Finland, Hungary, India, Japan, South Africa, Sweden, and Switzerland.

15

World Clinical Engineering Survey

Mariana Glouhova
Clinical Engineering Consultant
Sofia, Bulgaria

Nicolas Pallikarakis
Department of Medical Physics
University of Patras
Patras, Greece

Cost-effective management of biomedical technology within the health care systems is vital to the improvement of health care outcomes. Clinical engineering departments (CEDs) have been created in response to this need. The first international survey was conducted in 1988 (Frize, 1990a; Frize, 1990b), to study the degree of involvement of these departments in equipment management. This study, which included the same participants, was followed by a second one three years later that revealed only minor changes in the majority of the factors affecting the development of the clinical engineering field, except for the workload indicators, i.e., the number of devices and equipment value supported by the CEDs (Bronzino, 1992). Over the coming years, their role, organization and responsibilities have increased and have evolved because of the health care reforms occurring in nearly every country. In order to map the current situation in the field–the changes and the trends in the profession–a world clinical engineering survey has been conducted between 1998 and 2000. The results from this study are presented and discussed here.

The Survey

This study was performed by launching a survey, in two successive stages, using two structured questionnaires. The two stages of the survey aimed, respectively, to achieve the following: (1) to map the current situation in the field of clinical engineering in the different countries; and (2) to identify common practices and trends.

The first questionnaire intended to register the structure, personnel, responsibilities, and resources of the departments in different countries. It was distributed to more than 1000 hospitals all over the world. The questionnaire was intentionally short and easy to answer, within 20 minutes. It was authored in English and was translated into French, German, Greek, Hungarian, and Japanese.

The second questionnaire had the objective of investigating trends in current practices and addressed selectively only those departments–gentrified from the first phase–at well established clinical engineering services. It was authored in English. The estimated time required for completing it was about 30 minutes.

Both questionnaire forms were composed with the use of Pinpoint™ software package. Neither one requested confidential information. Prior to mailing, they were peer-reviewed by a small number of experts from the Clinical Engineering Division of the International Federation of Medical and Biological Engineering (IFMBE). The first questionnaire was distributed by both regular mail and e-mail. The second was mainly sent via the Internet, starting six months after the beginning of the survey.

In total, 178 valid responses were received to the first questionnaire (Glouhova et al, 2000). They were grouped in six regions as follows: North America (NA), incorporating the U.S. and Canada; Nordic countries (NC), including Norway, Sweden, Finland, Iceland, and Denmark; Western Europe (WE), consisting of Germany, the Netherlands, and the UK; Southern Europe (SE), involving Italy, Greece, Spain, and Cyprus; Australia (AUS), and Latin America (LA), comprising Argentina, Brazil, Cuba, and Mexico. The samples collected from Eastern European countries, Asia, and Africa were considered inadequate and therefore was excluded from the present analysis. Table 15-1 shows the level of responses for each region. A range of hospital sizes from about 100 to more than 1500 beds was obtained. Most of those surveyed were teaching institutions. The majority of responses came from departments that exist as separate units. The samples from Southern Europe and Latin America include only small (up to 500 beds) and medium sized (5001-1000 beds) hospitals, while the samples from all other regions include large (more than 1000 beds) hospitals as well.

The second questionnaire yielded 82 responses out of 143 sent, which equates to about a 57% response rate. They were grouped according to the first questionnaire scheme in six groups (see Table 15-1). The sample collected from Western and Southern Europe was too small and was excluded from the analysis. Consequently, only the responses from North America, the Nordic countries, Australia, and Latin America were further analyzed. The sample is too small to allow definite conclusions, but the obtained data reveal interesting trends.

Results from the First Stage

Structure

Clinical engineering has a long tradition in North America, the Nordic countries and Australia. Most of the reporting departments had been functioning for more than 30 years. In Western and Southern Europe the majority of departments responding were established in the late 1970s and early 1980s, whereas in Latin America, in the mid-1980s. In the few cases when answers came from hospitals where clinical engineering services function as part of another department, these departments were found to be plant operation, facility management, or engineering services in North America; technical department, medical physics or medical physics, and bioengineering in Australia; clinical physics or technical service in Europe; and engineering or maintenance department in Latin America. Their reporting authorities vary greatly in different countries. Some departments report to the hospital administration, but alternatively may be administrated by the director of engineering, plant or materials management, or the medical director.

Personnel

The educational level of engineers and biomedical equipment technicians (BMETs) in the departments varies considerably. Figure 15-1 represents the skill mix in the departments per region. Table 15-2 shows the variation in the skills per region. It is a general trend in all regions that CEDs consist predominantly of BMETs. However, in some departments in the Nordic countries and Latin America, the number of engineers is sometimes equal to, or higher than, the number of BMETs. CEDs in North America and the Nordic countries are large, with many personnel.

Furthermore, it was found that in Latin America all departments have engineers. In some departments in Latin America and in Europe there are no BMETs at all, while in all other areas there are departments without engineers but not without BMETs. In a number of hospitals, there are also medical physicists within the CEDs. In Europe generally, and especially in the Nordic countries, many departments have engineers holding a Ph.D. In Australia and Latin America, clinical engineers are predominantly at the B.Sc. level, while in North America there is an equal distribution of B.Sc. and M.Sc. graduates.

The educational background of the BMETs was further analyzed and compared to the previous survey in Figures 15-2 and 15-3. A clear trend toward longer educational programs for BMETs is noted. The survey indicates that this is predominantly true for the three- and four-year technical education following high school. This is also seen in Figure 15-4, where the educational backgrounds for all regions are studied. Nevertheless, there are BMETs with B.Sc. degrees and some in the U.S. and in Western Europe even hold M.Sc. degrees.

Figure 15-5 shows the number of technicians per engineer per region. In Australia and in North America the number of technicians per engineer is quite high, i.e., 8 and 7, respectively. This ratio drops to 3 in Western and Southern Europe, while in Latin America data show that on average the number of engineers equals the number of technicians.

Table 15-3 summarizes data on age and experience of engineers and BMETs. Clinical engineers appear to be older and with more years of clinical experience, on the average, as compared to BMETs. The situation is different only in Latin America. Therefore, this table reflects a younger profession of clinical engineering in Southern Europe and Latin America.

Table 15-1 Level of responses per region

Tools/Regions	NA	NC	WE	SE	AUS	LA	Other	Total
Q1 Distributed	495	40	137	101	17	43	167	1000
Q1 Returned	57 (12%)	20 (50%)	36 (26%)	14 (14%)	13 (76%)	16 (37%)	22	178
Q1 Analyzed	57	20	35	14	13	16	0	155
Q2 Distributed	54	20	29	11	13	12	4	143
Q2 Returned	38	9	8	3	10	11	3	82

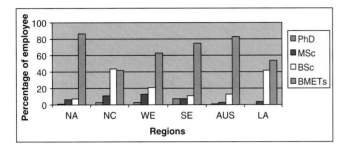

Figure 15-1 Skill mix in the departments per region.

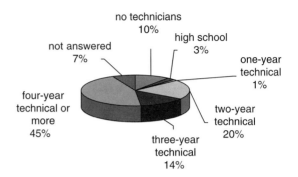

Figure 15-2 Pie chart illustrating the educational background of the BMETs (equivalent regions with the previous survey—NA, NC and WE): 1998-2000 survey.

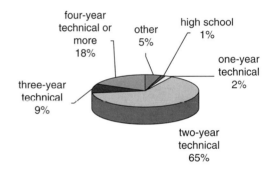

Figure 15-3 Pie chart illustrating the educational background of the BMETs: 1988 survey (Frize, 1990a).

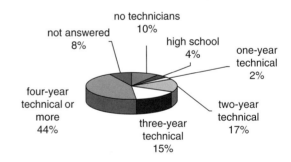

Figure 15-4 Pie chart illustrating the educational background of the BMETs for all regions studied.

Although certification for clinical engineers is required only in Iceland, Norway, the Netherlands, Cuba, and the UK (from October 2000), 22% of the reporting engineers are certified in North America, 23% in Australia, 37% in the Nordic countries, 22% in Latin America, 30% in Western Europe, and 35% in Southern Europe. From the responses, it appears that certification for BMETs is required only in Cuba. Nevertheless, 18% of BMETs in North America, 10% in Australia, 13% in the Nordic countries, 19% in Western Europe, and 29% in Southern Europe are certified and hold licenses.

Training and continuous professional development of clinical engineering personnel appear to be common practice in most regions, with the exception of Latin America. The highest frequency of training courses, attended by CED personnel, is in the Nordic countries, where engineers and BMETs attend a course on an average of every 6 months. In North America and Western Europe, the technical personnel take training once or twice per year. In Southern Europe, the majority take courses once per year, while in Australia they do so once or twice every two years. Few departments in North America and Southern Europe reported no training at all. This is also the situation reported by the majority of the departments in Latin America.

Responsibilities

The number of devices and the equipment value supported per full-time equivalent employee (FTE) are shown in Figures 15-6 and 15-7, respectively. Technical personnel in Australia support the largest number of devices, followed by Western Europe and North America. In the first four regions, FTE support between 550 and 600 devices. In the Nordic countries and North America, however, the highest equipment value supported per FTE is found. The highest numbers of devices supported per hospital bed are observed in North America, the Nordic countries, and Australia (Figure 15-8).

Generally, CEDs in small hospitals support medical devices of lower value and of smaller number. Although there are some exceptions, hospital size has always been considered to be directly correlated with the number of devices and equipment value supported by the CED, which is confirmed with the present survey as well. In the majority medium-sized hospitals in North America and the Nordic countries, as well as in the large hospitals in all regions, CEDs support more than 4000 devices, representing more than $20 million in equipment value. In some large hospitals, CEDs support even more than 10000 devices. More than half of the large hospitals in North America, the Nordic countries, and Western Europe support equipment valued at more than $40 million. Therefore, in terms of absolute number of devices and equipment value supported by the CEDs, a remarkable increase is observed since the previous survey. However, it is important to note that the technology base in the hospitals in many cases includes equipment not supported by the CEDs. Table 15-4 summarizes data on this, for hospitals with a separate department. In some cases, the departments support all equipment in the hospitals, and on average they have taken this responsibility to quite a high level. In Latin America this process is still not well established.

Table 15-2 Variations in the skills within regions (average number and range per department)

	NA		NC		WE		SE		AUS		LA	
	Av.	R	Av.	R	Av.	R	Av.	R	Av.	R	Av.	R
Ph.D.	0,07	0-1	0,55	0-2	0,2	0-2	0,36	0-1	0,15	0-1	0	0
M.Sc.	0,75	0-5	2,2	0-8	0,91	0-7	0,36	0-1	0,46	0-1	0,38	0-4
B.Sc.	0,93	0-9	8,9	0-26	1,54	0-5	0,57	0-2	1,69	0-12	3,63	0-9
BMETs	11,77	1-60	8,4	0-34	4,63	0-27	3,86	0-11	11	1-27	4,69	0-24

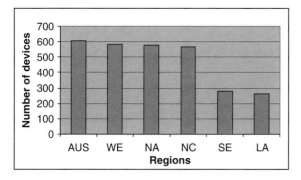

Figure 15-5 Number of technicians per engineer per region.

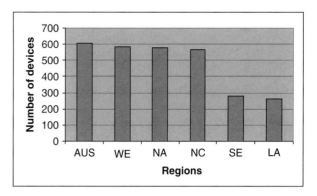

Figure 15-6 Number of devices supported by FTE, on average, per region.

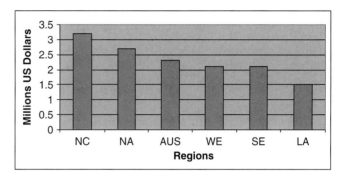

Figure 15-7 Equipment value supported by FTE, on average, per region.

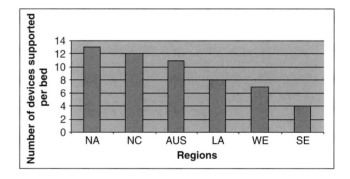

Figure 15-8 Number of devices supported per bed.

Table 15-5 contains the responses regarding the services provided by the departments. Traditional tasks of the departments, such as preventive and corrective maintenance, equipment inventory, acceptance testing, and repurchase consultation (including market survey, development of specifications, evaluation of tenders, and final selection), management of service contracts, and education and training, are performed by a large portion of the departments. However, tasks such as risk management, quality control, and research and development are still not so widely offered. Under "other," departments reported such responsibilities as government activities, project management, administration, consultation, and parts sourcing. An attempt was also made to identify the distribution of tasks between engineers and BMETs. It was found that in the majority of departments, where both engineers and BMETs exist, repurchase consultation (with the four sub stages), management of service contracts, risk management, quality control, education and training, and research and development, are mainly performed by engineers, while preventive and corrective maintenance are predominantly responsibility of BMETs. Equipment inventory and acceptance testing are shared between both professionals. It is important to note however that there is overlapping of roles in performing the tasks to a very high degree.

Resources

Table 15-6 shows the availability of resources other than personnel, which was discussed earlier. The respondents were invited to state the value of their test equipment, spare parts inventory, and budget, as a percentage of the value of biomedical equipment under CED's responsibility. Enormous variations in the reported data are observed. These questions were also not always answered. Therefore, regional analysis and comparisons for these options were not made.

According to the responses, a high percentage of departments in all regions have inadequate staffing levels. Figure 15-9 captures opinions of the respondents on the need for engineers and BMETs. The call for BMETs is quite high in all regions surveyed; moreover, it appears to be higher than the need for engineers. The smallest need of engineers is reported in North America and Western Europe.

Figure 15-10 shows opinions on adequacy of test equipment, spare parts, and space allocation. A level of test equipment, which amounts to at least 1% of the equipment value supported, seems to be adequate for the majority of respondents. It appears, as well, that the same level of at least 1% provides a good basis on which to build spare parts inventory. The value of biomedical equipment under CED management has increased appreciably since the last survey. This percentage reflects as well the significant increase of the value of test equipment and spare parts inventory.

A space allocation of at least 20 square meters per person is considered necessary for departments. The trend from the previous survey of more generous space allocations in the Nordic countries is also seen in the present survey, where 75% of the respondents report having more than 20 square meters per person, followed by 72% in Australia. Departments in Latin America are the least spacious with 94%, with less than 15 square meters per person; the majority are not satisfied with the situation.

Managerial issues

More than 70% of the departments in North America, Australia, and the Nordic countries have implemented quality assurance programs, while the percentages for Western Europe, Latin America and Southern Europe are much smaller, i.e., 43%, 38%, and 29%, respec-

tively. This shows a significant increase for all regions compared with the previous survey. The use of a productivity index as a measure of staff performance was recorded in 56% of respondents from North America, 69% from Australia, 45% from the Nordic countries, 26% from Western Europe and just 14% from Southern Europe. In Latin America, the majority of departments use a productivity index (56%), but few have quality assurance programs. Comparison with the previous survey shows an increase in departments evaluating staff productivity using a productivity index. This has been facilitated by computerized systems for equipment management as more than 90% of the departments in all regions reported their use, with this figure being 69% for Latin America.

Table 15-3 Age and clinical experience of clinical engineers and BMETs

| | Clinical engineers | | | | BMETs | | | |
| | Age | | Clinical experience | | Age | | Clinical experience | |
	Range	Average	Range	Average	Range	Average	Range	Average
Australia	35-55	42	8-30	16	33-46	38	7-15	10
North America	30-62	41	1-38	15	28-52	37	1-23	11
Nordic countries	34-63	43	1-30	14	27-50	39	3-20	12
Western Europe	31-54	41	5-30	14	29-55	37	1-32	10
Southern Europe	29-50	38	4-24	10	25-40	36	3-12	8
Latin America	23-38	31	1-13	7	24-47	32	4-16	8

Table 15-4 CED involvement in the overall management of the equipment in the hospitals

Regions	Equal value supported by CEDs towards equal value in the hospital (percentages)	
	Range	Average
Australia	62-100	92
North America	10-100	83
Nordic countries	6-100	78
Southern Europe	7-100	72
Western Europe	10-100	69
Latin America	7-100	59

Table 15-5 Extent of provision of services by CEDs

Services provided	Departments (percent)
Equipment inventory	95
Preventive maintenance	97
Corrective maintenance	97
Repurchase consultation	92
Participation in the market survey	79
Development of specifications	85
Evaluation of tenders	88
Final selection	86
Acceptance testing	94
Management of service contracts	88
Risk management	66
Quality control	75
Education and training	89
Research and development	48
Other	23

Table 15-6 Availability of resources (range and average)

Resources	Range	Average
Test equipment	0.001–17%	1.8%
Spare parts	0.001–15%	2%
Budget	0.006–20%	4.5%
Space allocation	1.5–77.7 m²/person	20.9 m²/person

Results from the Second Stage

Equipment Inventory

All responding hospitals from North America, the Nordic countries, and Australia use a software package for equipment management, which in the majority of the cases is a commercially available product. This facilitates the support for up-to-date inventory and the availability of complete individual files for each item. The situation is different in Latin America, where a relatively small number of hospitals (45%) use software management packages.

The age of life support and therapeutic devices was indicated to be smaller than the age of all other equipment categories in all regions. Hospitals in Latin America, however, are equipped with older equipment, as compared to the other regions.

Preventive and Corrective Maintenance

The majority of the participants maintain written procedures for safety, performance, and calibration checks, while the percentages referring to the availability of written procedures for corrective maintenance (CM) are much lower. Preventive maintenance (PM) procedures and protocols are set in the majority of the cases for each equipment type and are based on manufacturers' requirements, guidelines and standards of associations and institutes, and individual experience, while CM procedures are mainly based on the manufacturers' recommendations. More than 80% of the participants from North America, the Nordic countries, and Australia state that the test equipment used is calibrated according to standards and that it is checked regularly for proper calibration, while this is the case in only about 55% of the participating hospitals from Latin America. More than 80% of the respondents from all regions report that clinical departments and laboratories notify the CED in case of need for repair, even when the contractor is an external service provider. The majority of participating hospitals have a policy requiring authorization for (estimated) expensive repairs before the repair is made, while the replaced parts are only sometimes kept for additional inspection. In rare cases, PM and CM for clinical laboratory and imaging and radiation therapy equipment are the responsibility of CEDs. By contrast, in most cases, PM and CM of patient diagnostic, life support and therapeutic and other biomedical equipment are performed for more than 75% of the devices by the CEDs, as shown in Figure 15-11 and Figure 15-12.

The Survey revealed that the practice of frequently informing the clinics and laboratories about the results of PM checks and CM costs of their equipment is lacking, while PM and CM records are well documented.

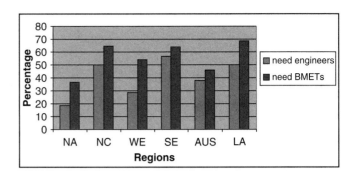

Figure 15-9 Percentage of departments needing additional engineers and BMETs per region.

Acquisition of Biomedical Equipment

More than 65% of hospitals in all regions have written procedures for the acquisition of biomedical equipment, which are based mainly on guidelines and standards of associations and institutes and on individual experience. In the majority of cases, CEDs are always informed when new equipment is to be purchased. They are consulted in almost all purchases of patient diagnostic, life-support and therapeutic and other biomedical equipment, and, to a much lower degree, for the purchases of clinical laboratory and imaging and radiation-therapy equipment. CEDs in the Nordic countries and Australia have the highest level of involvement in the pre purchase consultations, including clinical laboratory and imaging and radiation therapy equipment. CEDs usually participate in the definition of technical specifications, the evaluation of tenders, and the final selection process. In most cases, the maintenance contracts are negotiated with their active involvement.

Acceptance Testing

Acceptance tests (incoming inspections) of the patient-diagnostic, life-support, and therapeutic and other biomedical equipment are mainly overtaken by the CEDs, while for the clinical laboratory and imaging and radiation therapy equipment the tests are conducted in collaboration with the manufacturers. With the exception of Latin America, the great majority of hospitals in all regions have written procedures for acceptance testing of the incoming equipment, and more than 75% of the new equipment is checked before commissioning. Acceptance tests are sometimes performed after major repairs from outside service providers in North America, the Nordic countries, and Australia. Incoming inspections are well documented.

Management of Contracts

Almost all of the responding hospitals in North America, Australia, and Latin America have established procedures for evaluation of the adequacy of biomedical service contracts, while in Latin America about 80% of the respondents have written procedures on how to proceed when a contract is breached.

Risk Management

Risk management is a well-established service within the CEDs' responsibilities in North America, the Nordic countries, and Australia. Almost all hospitals from these regions responded that they have an internal system for reporting all accidents or equipment-related hazards and that they have a person assigned to address the hazard notifications. Therefore, users are immediately informed in case of hazard notification with a device similar to the one they use, and the hazards are well documented.

Education and Training

More than 70% of the hospitals in all regions have a program for training the medical and nursing staff on the proper and safe use of biomedical equipment. In North America, the Nordic countries, and Australia, the majority of the departments report that they can meet less than 50% of user education and training needs. In Latin America, however, in half of the responding hospitals, the CEDs perform more than 75% of the user's education. There is continuing education for the technical staff as well, which is mainly performed on-the-job or in training courses organized by the manufacturers.

The highest frequency of training courses, attended by CED personnel, is in the Nordic countries—every 6 months; in North America—once or twice per year; and in Australia—once or twice every two years. Few departments in North America reported no training at all, while this is also the situation reported by the majority of the departments in Latin America. However, when inquired if this frequency of training courses is satisfactory, 67% from the participants from the Nordic countries answer positively, with the respective numbers for the other regions–63% for North America, 60% for Australia, and 27% for Latin America.

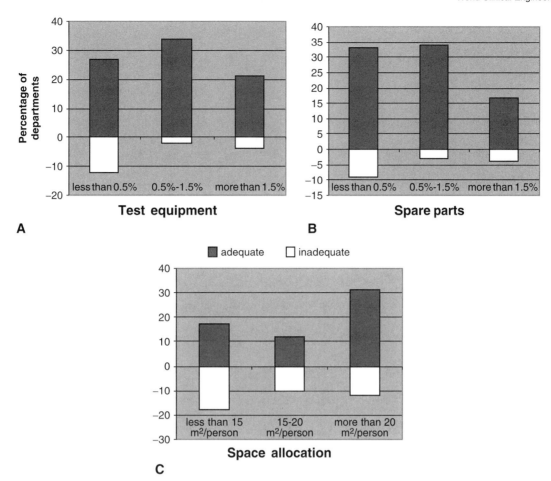

Figure 15-10 Percentage of departments reporting adequacy or inadequacy: (a) test equipment–less than 0.5%, 0.5%-1.5%, more than 1.5%; (b) spare parts–less than 0.5%, 0.5%-1.5%, more than 1.5%; (c) space allocation–less than 15 m²/person, 15–20 m²/person, more than 20 m²/person.

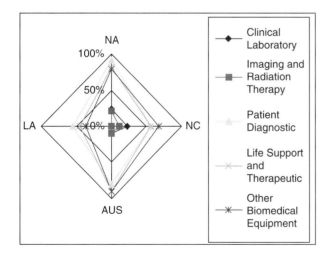

Figure 15-11 PM performed by CEDs to more than 75% of the equipment base.

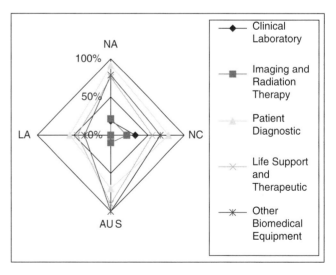

Figure 15-12 CM performed by CEDs to more than 75% of the equipment base.

Research and Development

Research and development (R&D) seems an important activity of the CEDs only in Australia and in the Nordic countries, since 80% and 67%, respectively, of the hospitals from these regions reported that they are involved in such activities. In the Nordic countries, hospitals involved in R&D have more than 5% of their staff assigned to such duties. In Australia, the respective number is about 2%. In some hospitals there is even additional staff hired only for R&D. It is, however, interesting to mention that all hospitals involved in R&D from these regions are teaching institutions.

Quality Assurance

More than 70% of the departments from North America, Australia, and the Nordic countries have implemented quality assurance programs, the majority being in practice for more than 5 years. This shows a significant increase, which comes to justify that the process, which was in its initial phase in 1988, is being more and more implemented and used in practice in 2000.

Finally the survey showed that almost all participating hospitals from North America, the Nordic countries, and Australia have an adequate assess to information related to current codes and standards and have adequate technical libraries. This is not the case, however, for the majority of hospitals in Latin America.

Discussion and Conclusions

The international survey of clinical engineering departments performed during 1998-2000 came in response to the need to update the available information about clinical engineering worldwide and is the largest ever conducted. The study was designed and performed using a two-step approach. Respectively, to the aims and objectives of each step, two specifically designed questionnaires were distributed by regular mail and e-mail. Questionnaires were sent to hospitals from all continents, and responses were received from all continents.

The received responses produced the largest database that is currently available and revealed that clinical engineering as a profession is better established and recognized today than it was a decade ago. More than 80% of the departments feel that they are well accepted in their institutions. The main problems are lack of highly qualified personnel (because of the lack of quality academic programs); limited funding for technical training to maintain staff competencies for all equipment types; and continuous pressure to reduce costs by increasing department efficiency. However, despite the efforts at harmonization, the management of biomedical technology in hospitals all over the world remains nonuniform, with great variations in terms of structure, personnel, responsibilities, and resources.

In the regions with long tradition like North America, the Nordic countries, and Australia, departments are involved in the maintenance of usually more than 75% of the available equipment value in the hospital. There is also a trend, especially in North America and Australia, for one department to cover the management and maintenance of the equipment of several hospitals, striving in this way for cost-justification and efficiency. In Western and Southern Europe, CEDs maintain, on the average, about two-thirds of the equipment value available. The departments in Latin America are responsible to less than 60% of the equipment value available in the institutions.

In Southern Europe, and especially in Latin America, the profession is still relatively young. However, significant investments are in progress in hospitals in Latin America. Data show that as a general trend, the value of equipment in hospitals has increased drastically over the last ten years, reaching levels of over $100,000 per hospital bed in North America and the Nordic countries.

References

Bronzino JD. Management of Medical Technology—A Primer for Clinical Engineers. Butterworth-Heinemann, 1992, p 413.
Frize M. Results of an International Survey of Clinical Engineering Departments – Part 1: Role, Functional Involvement and Recognition. *Med Biol Eng Comput* 28:153-160, 1990.
Glouhova M, Kolitsi Z, and Pallikarakis N. International Survey on the Practice of Clinical Engineering: Mission, Structure, Personnel, and Resources. *J Clin Eng*: 25(5): 205, 2000.

16

Clinical Engineering in the United Kingdom

Stuart J. Meldrum
Director of Medical Physics and Bioengineering
Norfolk and Norwich University Hospital
NHS Trust
Norwich, England

For many years, scientists and engineers in the UK have made major contributions to biomedical engineering. Early emphasis was on research, with more managerial roles emerging only later. Discussion between biomedical and hospital engineers has generally resulted in the most appropriate group in each hospital providing equipment management services. The recent emphasis on risk management and quality of care has been of significant benefit to clinical engineering, with hospitals obliged to establish structures to meet the requirements of clinical governance and controls assurance. A training scheme, introduced in the 1990s, provides theoretical knowledge and practical skills in clinical engineering. Those who successfully complete the training scheme are eligible for professional registration.

Introduction

There has been much discussion about roles and definitions in clinical engineering, but the majority of published work relates to the situation in North America (Goodman, 1989; Bauld, 1991; Shaffer and Shaffer, 1992). In spite of the fact that engineers in the United Kingdom (UK) have a long history of involvement in health care, the term "clinical engineering" has only recently come to be used in the internationally accepted sense of the phrase. The terms "bioengineering," "biomedical engineering," and even "medical engineering" are more commonly used in UK hospitals to describe the range of services considered to be the domain of clinical engineering (Webster and Cook, 1979). In this discussion of clinical engineering in the UK, the term will be used as defined by Schwartz (1984): The application of biomedical engineering knowledge in a health care context.

The late development of the concept of clinical engineering in the UK can be attributed to a number of factors. Initially, engineers entered the field of medicine to work in research and development and only later took on managerial roles. Secondly, biomedical engineering was seen as a scientific, rather than an engineering, discipline, with many of its practitioners working in medical physics departments. For reasons of perceived status, staff preferred to refer to themselves as "medical physicists" rather than "clinical engineers" in spite of the fact that they performed an engineering function. Thirdly, for many years there has been a centralized health care system in the UK that, until recently, showed little awareness of the wider contribution that clinical engineering could make.

Fortunately, with the recent emphasis on the quality of health care, the situation has changed. Clinical engineering is now receiving greater recognition and is seen to be able to contribute more than just equipment maintenance. In recent years, a national training scheme for scientists and engineers in health care has been introduced. Training is provided across a range of subjects from radiation physics to rehabilitation engineering, thus serving to blur the distinction between the roles of medical physicist and clinical engineer.

Early Beginnings

Scientists and engineers in the UK have made a major contribution to the development of biomedical engineering from its earliest days. In 1958, leading scientists and engineers from around the world met in Paris for the first world conference on medical electronics. At that meeting, it was suggested that an international society should be established to serve the needs of the rapidly expanding field of engineering as applied to medicine. The following year, a second international conference was held (again in Paris), at which a decision was taken to form the International Federation for Medical and Biological Engineering (IFMBE). Of the 41 founding members of this organization, three were from the United Kingdom (Mito et al., 1997).

A year later, the world conference was held in London. It served as the catalyst for the formation of a society to serve the needs of biomedical engineering in the UK. Consequently, the Biological Engineering Society (BES) was established in 1960 as a multidisciplinary society, the aim of which was to foster cooperation among engineers, scientists, technologists, and others engaged in the life sciences. Note that the chosen name was "biological engineering" as the society was intended to encompass the application of engineering to biological systems in the widest sense. In subsequent years, many biologists and physiologists played prominent roles in the 'society's affairs. In 1963, the BES was one of the first national societies to affiliate with IFMBE.

From its formation, the BES provided a focus for biomedical engineering activity in the UK. This situation continued until 1995, when the BES merged with the society representing medical physicists to create what is now the Institute of Physics and Engineering in Medicine (IPEM). Thus there is now a single professional body representing the interests of physicists and engineers in medicine, providing an opportunity to integrate the education, training and methods of service delivery.

Emphasis on Research

During its early years, biomedical engineering in the UK was seen essentially as an academic discipline. Many of its contributors were employed directly by clinical departments, with the development of new diagnostic and therapeutic techniques as their primary role. Biomedical engineering departments were established in a few of the major teaching hospitals, and while providing a focus for bioengineering activity, they were often dependent on research funding for their continued existence.

By the 1970s, accounts of departmental research activities began to appear in the literature. Roberts has described how the department at King's College Hospital, London, was established in 1967 (Roberts et al., 1971). Research was carried out in collaboration with vascular surgeons, and prominence was given to the development of novel techniques for the measurement of blood flow. Roberts stresses the importance of carrying out such research in a clinical environment, where there is day to day awareness of clinical needs.

Watson has described the origins and early years of the Department of Medical Electronics at St Bartholomew's Hospital (Watson, 1975). Initial funding was provided by the hospital-research board. Within 10 years, groups were working on problems as diverse as blood flow, incontinence, dialysis, speech aids, and radiotelemetry. Watson was one of the first to acknowledge the importance of providing adequate management of equipment and the bearing that this had on the safe delivery of services.

Considering its population, Scotland has always made a disproportionately large contribution to bioengineering activity. This has been attributed to both the academic traditions of the country and to the close relationship between the universities and the providers of health care. Services in Scotland tend to be based on large, regional departments, a structure that has encouraged the development of bioengineering in each of the major centers of population. By the late 1960s, a wide range of bioengineering activity was taking place (Forwell, 1970).

Equipment Maintenance in Hospitals

As already mentioned, the early emphasis in bioengineering was on research. During the 1960s, however, health care became increasingly dependent on technology. By the early 1970s, hospitals were becoming aware of the lack of appropriate arrangements for managing this technology. Around this time, a group was established to address the issue, with membership drawn principally from regional hospital engineering departments. The outcome was a proposal that a network of properly structured groups should be established across the country, with a remit to supervise the management of all medical equipment (DHSS, 1971). It was envisaged that these groups would be located within hospital works departments and that they would provide maintenance services that were largely independent of the equipment manufacturers.

These proposals—principally that the maintenance of medical equipment was the preserve of the hospital engineer—caused some consternation in bioengineering circles (Anon, 1972a; Anon, 1972b). In response, the medical physicists issued a counterproposal suggesting that clinical equipment was so complex that its maintenance required a scientific, rather than a hospital engineering, background (Perry, 1973). The hospital engineers' proposal was never implemented, but did serve to bring the issue of equipment maintenance to the attention of the wider bioengineering community. Ultimately, services were developed and provided by the most appropriate group in each hospital.

Garrett has described the setting-up, funding, and organization of the highly respected department in Bristol (Garrett, 1984). The initial remit was to identify equipment that could be economically and effectively maintained in-house. It was estimated that this could be done at a cost one-third less than that of the external contractors, who were generally the equipment suppliers. No new funding was provided for the project, as the scheme was expected to create its own funding from cancellation of maintenance contracts. The Bristol development was based on the bioengineering group within the medical physics department as it had the necessary range of expertise.

In addition to providing support for the normal range of electromedical equipment, the Bristol group was also responsible for the high-energy accelerators and cobalt machines used in radiation therapy. The high value of this equipment, and the related savings from transferring to in-house maintenance, played a major part in establishing the viability of the service. Unfortunately, few clinical engineering departments are able to benefit from the inclusion of such high-value equipment in their inventories. Garrett also discussed the difficulties of comparing in-house with external costs but concluded that the simple measure of saving the cost of external contractors' travelling time can be of significant benefit to the hospital. Stamp has described the development of the medical equipment management service in Sheffield (Stamp, 1984). There, the service, which is also based on the expertise of the local medical physics department, caters to a number of adjacent health districts, thus providing greater economy of scale.

Guidance Documents

A centrally managed health care system, such as that in the UK, tends to develop in response to guidance published by the Department of Health. Equipment management services are no exception, and over the years a number of guidance documents have been published on what constitutes good practice in medical device management. The first of these, *Health Equipment Information 98,* was published in 1982 in response to a growing awareness that many accidents in hospitals were caused by poor maintenance and repair (DHSS, 1982). These guidelines provided a comprehensive cradle-to-grave approach to equipment management and were promptly implemented by many hospitals. Guidance was offered under five headings: Procurement, acceptance, maintenance, training, and disposal. Whelpton has described the manner in which these recommendations were implemented across a district-wide service in Nottinghamshire (Whelpton, 1988).

In 1998, the guidance was updated to take account of changes that had occurred in legislation, particularly that relating to CE marking of medical equipment (MDA, 1998). The revised document also acknowledged the increased use of medical devices in patients' homes, and the problems associated with their management and support.

Thus although the components of a clinical engineering service had been in place for a number of years, clinical engineering service has not been referred to as such until recently. There has been relatively little written about the provision of clinical engineering services in UK hospitals, with the work that does exist generally confined to descriptions of how central guidance has been implemented in various centres. While North American authors have frequently written on the economics of health technology management (Furst, 1986; Betts, 1989), this is rarely discussed in papers from the UK. It has been suggested that the anomalies of Health Service accounting are largely to blame, with few clinical engineering departments having sufficient budgetary control to demonstrate the benefits of their services (Charles and Woolley, 1984). In the UK system, large items of equipment are frequently purchased with central funding, with little or no provision to meet revenue costs or end-of-life replacement. Equipment management budgets are often underfunded, and consequently grossly overspent, and equipment runs beyond the end of its economic life.

In the UK today, the clinical engineering function is generally provided by an in-house team. In teaching hospitals and larger, district hospitals, these teams tend to be led by a graduate engineer, while in smaller hospitals the team is lea by a biomedical equipment technician. Private hospitals in the UK tend to be small and purchase clinical engineering services from neighboring state hospitals or from private sector maintenance organizations. While a few independent service organizations do exist, they have made little impact on the overall provision of clinical engineering services.

Clinical Engineering Certification

Medical equipment management services developed in the UK with little acknowledgement of the ways in which clinical engineering had developed in North America (Bronzino, 1995). By the mid-1980s, however, members of the BES had become aware of the clinical engineering certification scheme that was being run in the U.S. and decided to introduce a certification scheme of their own. The UK scheme was loosely based on the one that was operating in North America, but no written examination was involved. Applicants were assessed on the basis of a written account of their work experience and an in-depth interview.

The UK clinical engineering certification scheme was never successful, however, as there was always an underlying uncertainty about its aims and objectives. Unlike the U.S., the UK has no separate professional organization representing rehabilitation engineers. The BES certification scheme, therefore, had to accommodate both groups. Engineers working in rehabilitation claimed that they were the only true "clinical" engineers as they were the only ones playing a direct role in patient care. Consequently, the criteria for certification were modified to include the requirement that applicants spend a certain percentage of their time in direct patient contact. The effect of this was to close off certification to many senior members of the profession, who were exclusively involved in administrative and managerial duties. The certification process attracted too few senior members of the profession to ensure its success and was closed in favor of encouraging biomedical engineers to seek registration as professional engineers.

The Changing Health Care Climate

A new discipline, such as clinical engineering, can develop and grow only in a health care system that values the contribution it makes. State-funded health care has been available in the UK since 1948 and was intended to be comprehensive and free at the point of delivery. Over the intervening years, many organizational changes have been made with the aim of providing a more responsive and cost-effective service (Webster, 2000). The early 1990s saw the most fundamental of these changes with the introduction of the concept of the "internal market." Central to these reforms was the purchaser provider split. Health authorities purchased services on behalf of patients, while hospital and community services were split into about 450 independent self-governing trusts. It was envisaged that, by encouraging hospitals to compete with each other, efficiency of care would be improved, and costs forced down. A more financially aware health care system might be expected to recognize the wider contribution that clinical engineering could make, but this was not the case.

The Quality Agenda

It was only during the late 1990s, with the introduction of a quality agenda, that the health care climate became more conducive to the development of clinical engineering. While formal systems for clinical risk management had been commonplace in some countries for many years, it was only during the 1990s that such arrangements became commonplace in the UK (Walshe, 2001). One factor driving this change was the increased tendency toward litigation and the related increase in cost to health care providers. In the mid-1990s, a national clinical negligence scheme was established to help defray the costs for individual hospitals. The scheme involved the introduction of national standards for risk management, including appropriate arrangements for medical equipment. Hospitals are audited against these standards, and they benefit from a reduced premium if certain standards are met. Thus, there now are clear financial benefits to having access to an effective clinical engineering service.

Clinical Governance and Controls Assurance

Despite these developments, which raised the general awareness of risk management, there was no unifying philosophy upon which to build (Secker-Walker and Donaldson, 2001). This was only introduced in 1998, with the publication of the Government's White Paper on the future of the health service (HMSO, 1997). Trust management boards were required to monitor the quality of the care that they provided, with the same rigor as applied to financial matters. In practice, this requirement was met by introducing the concept of "clinical governance."

Contemporaneously with the introduction of clinical governance came the concept of "controls assurance." While clinical governance was intended to improve the quality of care, controls assurance was intended to improve the environment of care (Figure 16-1). Controls assurance is intended to provide evidence that health care providers are doing their reasonable best to manage their affairs in a way that protects patients from harm. Hospitals are required to audit a range of activities, one of which was the management of their medical equipment. Furthermore, a key component of control assurance is the ability to demonstrate that an appropriate delivery framework exists within the organization. Clinical engineering has an obvious part to play in demonstrating that this is the case.

Controls assurance places new emphasis on involving clinical engineers in the more strategic aspects of equipment management. There is also a recommendation that clinical engineers should be more involved in user training. Until controls assurance was introduced, few clinical engineers had been involved in this aspect of medical device management (Fouladinejad and Roberts, 1998). Involvement in user training is also important in demonstrating that clinical engineering departments are more than just repair shops (Frize, 1994).

National Audit Office Report

The profile of clinical engineering was further raised in 1999 with the publication of a report on the management of medical equipment, prepared by the National Audit Office (NAO) (NAO, 1999). One of the reasons why the government commissioned this report was an awareness of the fact that the medical equipment used in the Health Service represented a considerable national asset. Ensuring good practice in maintaining this asset would save the nation large sums of money. The NAO report discussed the role that clinical engineering could play in coordinating manufacturers' service provision and the benefits that would result from providing at least part of the maintenance program using in-house biomedical engineering staff. The other major area covered in the report was that of device safety and the role of clinical engineers in reporting and investigating device-related incidents. The authors of the report also noted the wide differences in service provision that they had encountered across the country. The report suggested that by benchmarking costs and management practices many financial savings and quality improvements would result.

Training and Registration

Basic Training

In the early 1990s, the British government introduced formal training schemes for a number of scientific disciplines working in health care, including medical physicists and clinical engineers. IPEM was duly charged with developing the scope and content of the program. One of the strengths of the scheme is that it is centrally funded, with trainees employed for the duration of the two-year training period. Training is demand-led, with the number of training places tailored to fit the needs of the various professional groups. Thus, all trainees would expect to find a post somewhere in the country at the end of their training.

All training is carried out in accredited centers that are able to demonstrate that they possess the necessary resources. These are both material resources, such as equipment and books, and human resources, such as staff who have the appropriate seniority and experience. Training centers were initially accredited for a period of three years, but for most centers this has now been extended to five years.

Trainees generally enter the scheme as new graduates in either physics or engineering. During the first two years, emphasis is on the acquisition of basic knowledge and practical skills. Basic knowledge is gained by attending a master's course in an appropriate subject. A list of accredited M.Sc. courses is maintained by IPEM (www.ipem.org.uk). Of the 20 courses currently listed, eight are biased toward biomedical engineering, with the remainder concentrating on medical radiation physics.

Practical skills are gained during hospital placements, when trainees generally opt to study three areas of medical physics or clinical engineering. The range of subjects is shown in Figure 16-2. From the top down, the list starts with subjects that might appeal to a trainee who is interested in medical physics. From the bottom up, the list contains subjects of more interest to the clinical engineer. Presenting the subjects in this way serves to indicate the continuum of subjects in medical physics and clinical engineering, rather than presenting them as separate disciplines. Trainees in clinical engineering will tend to opt for medical electronics and instrumentation, physiological measurement, and possibly information management and technology.

Practical skills are gained during three placements in hospitals, with each subject placement lasting for a period of six months. During this period the trainee is expected to produce a portfolio, describing the work that has been carried out and the manner in which the various competencies have been acquired. In the remaining six months of the two-year training period, trainees study core material such as the structure of the health care system, medical ethics, and relevant legislation. At the end of the two-year training period, trainees undergo an in-depth interview in each of the specialist areas covered. Successful trainees are awarded the Diploma of the Institute of Physics and Engineering in Medicine.

Higher Training

Upon successful completion of training, candidates apply for junior staff posts in hospitals. During basic training, a trainee may choose to study a mix of subjects, including aspects of both radiation physics and clinical engineering. At this point, however, they tend to specialize in one or the other. The first two years in these new posts include further compulsory training, with trainees learning more specialized skills in their chosen field. This is followed by a further two years of supervised responsibility before the trainee can take the examination to become a full-fledged corporate member of IPEM. Upon completion of training, members are expected to follow a balanced program of continuing professional development (CPD). This program requires that a minimum of 50 CPD points are acquired each year, with each point corresponding to one hour of development activity. A record of achievement, which is open to audit, must be maintained.

Engineering Registration

One of the reasons for discontinuing clinical engineering certification in 1998 was that few employers understood its significance, and consequently few posts were advertised with certification as a requirement. When considering the future of certification, IPEM took the view that professional engineering registration was more widely recognized and that it should be the standard to which all engineers aspire.

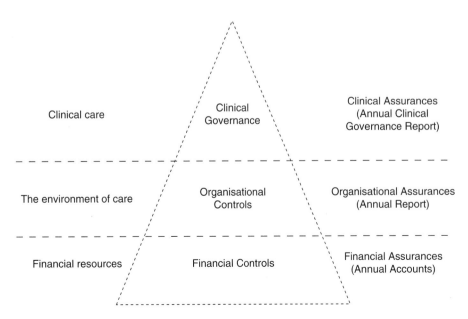

Figure 16-1 Relationship between clinical governance and controls assurance and the various reporting requirements.

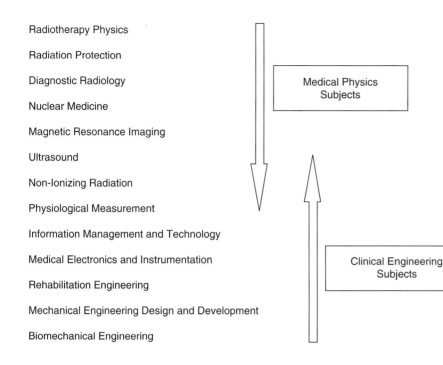

Figure 16-2 Major subject areas in IPEM training scheme for clinical scientists and engineers.

In the UK, the Engineering Council is charged with maintaining the register of professional engineers, who are referred to as Chartered Engineers and use the designatory letters CEng after their names (www.engc.org). The council does not accept direct applications for registration but relies on one of approximately 35 approved engineering institutions to process applications and to uphold its standards. IPEM, as a nominating body, is able to accredit courses and to approve individual development programs and IPEM's corporate-membership interview can also serve as the professional-review interview for registration. Thus clinical engineers who pass through the IPEM training scheme can progress to full professional registration with a minimum of inconvenience. Table 16-1 describes the structured training for clinical engineers leading to professional engineering registration.

State Registration

In 2000, the profession of "clinical scientist," which includes the practice of clinical engineering, was formally recognized with the establishment of a registration board under the Professions Supplementary to Medicine Act (1960). The board, as a statutory body, is the means by which the state ensures that the profession is formally regulated. The board has three main functions: To maintain a register of clinical scientists; to regulate the education and training leading to eligibility for registration; and to cancel registration in cases of infamous conduct. State registration has one aim: The protection of the public. All clinical scientists and clinical engineers working in the UK health service must now be state registered, and loss of registration will mean that the person may no longer work in the public sector. Further details are available from the registration Council (www.cpsm.org.uk).

Conclusions

Since its academic beginnings, clinical engineering has now established its own professional identity. Recent guidance documents from the department of health have stressed the importance of the contribution that clinical engineering can make to health care, particularly in areas of quality assurance and risk management. Recognition of clinical engineering in the basic training scheme, the introduction of state registration for clinical engineers, and a more concise route to professional engineering registration have helped to create a greater sense of identity.

Table 16-1 Structured training for clinical engineers leading to professional engineering registration.

Activity	Duration	Award
Undergraduate course in physics or engineering	3 years	First degree
Basic training in 3 areas of medical physics or clinical engineering	2 years	Higher degree plus IPEM Diploma
Higher degree in medical physics or clinical engineering		State registration
Higher training	2 years plus 2 years responsible experience	Corporate Membership of IPEM Registration as Chartered Engineer

References

Anon. Equipment Maintenance *Biomed Eng* 7:219, 1972
Anon. Equipment Maintenance: Physicists or Engineers? *Biomed Eng* 7:465, 1972.
Bauld TJ. The Definition of a Clinical Engineer. *J Clin Eng* 16:403, 1991.
Betts WF. Using Productivity Measures in Clinical Engineering Departments. *Biomed Instrum Technol.* 23:120, 1989.
Bronzino JD. Clinical Engineering: Evolution of a Discipline. In Bronzino JD (ed). *The Biomedical Engineering Handbook.* Boca Raton, Fl, CRC Press, 2499-2506.
Charles L, Woolley A. The Financial Management of Hospital Equipment. *J Med Eng Technol* 8:254, 1984.
Department of Health and Social Security. Engineering Inter-Board Study Group on EBME Maintenance. *Report EY10.*
Department of Health and Social Security. Management of Equipment. Health Equipment Information No 98.
Forwell GD. Biomedical Engineering in Scotland. *Biomed Eng* 5:434, 1970.
Fouladinejad F, Roberts JR. Analysis of Training Activities of Clinical Engineering Departments in the United Kingdom. *Biomed Instrum Technol* 32:254, 1998.
Frize M. Longitudinal Study of Clinical Engineering Departments in Industrialised Countries 1988 to 1991. *Med Bio Eng Comput* 32:331, 1994.
Furst E. Productivity and Cost-Effectiveness of Clinical Engineering. *J Clin Eng* 11:105, 1986.
Garrett JA. Equipment Management in Practice. *J Med Eng Technol* 8:256, 1984.
Goodman G. The Profession of Clinical Engineering. *J Clin Eng* 14:27, 1989.
HMSO. The New NHS - Modern and Dependable: A National Framework for Assessing Performance. London, HMSO.
MDA. Medical Devices and Equipment Management for Hospitals and Community-Based Organisations. MDA DB 9801. Medical Devices Agency.
Mito K, Richter N, Saito M, et al. Meeting Challenges in Medicine and Health Care Through Biomedical Engineering: A History of the IFMBE. Amsterdam, International Federation for Medical and Biological Engineering.
NAO. The Management of Medical Equipment in NHS Acute Trusts in England. HC 475. National Audit Office, 1999.
Perry BJ. Equipment Maintenance: Hospital Physicists' Association's Views. *J Biomed Eng* 8:395, 1973.
Roberts VC, Sabri S, Cotton LT, et al. Biomedical Engineering at King's College Hospital. *Biomed Eng* 6:199, 1971.
Schwartz MD. The Emerging Field of Clinical Engineering and Its Accomplishments. *IEEE Trans Biomed Eng* 31:743, 1984.
Secker-Walker J, Donaldson L. Clinical Governance: The Context of Risk Management. In Vincent C (ed). *Clinical Risk Management: Enhancing Patient Safety.* London, BMJ Books, 6173, 2001.
Shaffer MJ, Shaffer, MD. What is a Clinical Engineer? Issues in Definition. *Biomed Instrum Technol* 26:277, 1982.
Stamp JM. Multi-District Equipment Management. *J Med Eng Technol* 8:259, 1984
Walshe K. The Development of Clinical Risk Management. In Vincent C (ed). *Clinical Risk Management; Enhancing Patient Safety.* London, BMJ Books. 45-60.
Watson BW. The Departments of Medical Electronics, St. Bartholomew's Hospital—The First Ten Years. *J Biomed Eng*; 10:98, 1975.
Webster C. The History of the National Health Service. In Merry P (ed). *Wellard's NHS Handbook 2000/01* Wadhurst, JMH Publishing, p 1-4.
Webster JG, Cook AM. Clinical Engineering: Principles and Practice. Englewood Cliffs, NJ, Prentice Hall, 1979.
Whelpton D. Equipment Management: The Cinderella of Bio-Engineering. *J Biomed Eng* 10:499, 1988.

Further Information

Institute of Physics and Engineering in Medicine, Fairmount House, 230 Tadcaster Road, York, YO24 1ES (www.ipem.org.ul,).
The Engineering Council, 10 Maltravers Street, London WC2R 3ER (www.engc.org).
Registration Council for Scientists in Health Care, 2 Carlton House Terrace, London SW1Y 5AF (www.cpsm.org.uk).

17

Clinical Engineering in Canada

William M. Gentles
BT Technology Consulting
Toronto, Ontario, Canada

Canada's health care system, its geography, and its political landscape all have had an influence on the way that clinical engineering has evolved as a profession in this country. Like many other countries, Canada has seen dramatic changes in health and health care over the past 50 years. The Canadian health care system is distinguished by the fact that it is funded by tax revenues and that it is free for all residents of Canada. These traits lead to efficiencies, but they also mean that the system is subject to political pressures.

The Canada Health Act (CHA, 2001) is Canada's federal health insurance legislation. The aim of the national health insurance program is to ensure that all residents of Canada have reasonable access to medically necessary insured services, without direct charges. The legislation only permits the establishment of private, for-profit hospitals to provide services that are not defined as medically necessary. The legislation does not cover dental care, and the list of services that are defined as medically necessary is constantly evolving.

In the year 2000, total health expenditures in Canada were approximately $2000 per capita. Health expenditures accounted for 9.3% of Canada's gross domestic product.

Although the Federal government sets standards for health care, the provincial governments have the responsibility to manage and to pay for health care. Tax dollars collected by the federal government are handed over to the provinces and territories according to a complicated formula. In each of the ten provinces and four territories, health care is managed in a slightly different way. In the last decade, most provinces have made major structural changes in their health care systems. The number of hospitals and hospital beds in Canada is in a state of constant flux as provincial and territorial governments attempt various schemes to manage costs, such as regionalization and restructuring. Between 1995 and 2000, approximately 275 acute care hospitals were closed, merged, or converted to other types of care facilities. This consolidation was a natural consequence of the trends to shorter hospital stays, increased day surgery, increased number of minimally invasive procedures, and increased use of home care.

Canada's single-payer heath insurance system results in hospitals that tend to have high occupancy rates, with equipment that is heavily used. Clinical engineering departments have been under pressure to control costs and, in many cases, to downsize as the health care system has been subjected to repeated funding constraints by governments.

Canada is a large country, with a relatively small population. Most of the population is concentrated along the Canada-U.S. border. Health services are spread thinly in the northern regions, and clinical engineering services to support health care in northern regions are often based in larger centers to the south. New technologies, collectively known as "telehealth," are beginning to offer innovative ways to deliver health care services over large distances. Examples are teleradiology and telemonitoring of home-care patients.

Clinical Engineering Departmental Funding Sources and Economic Pressures

For the most part, Canadian clinical engineering departments are funded from hospital global budgets. They must compete with other services within the organization for limited funds. Funding levels are inconsistent, as no national workload measurement system exists for clinical engineering in Canada. One common trend is for departments to grow over time as they prove their value to the organization, and then to get cut back when there is a change in leadership in the department. Although many departments are able to track and to report on the cost of providing services, most are globally funded for their core services and do not charge customers directly for their services.

Many hospitals in Canada have recently begun to reinvest large sums in capital equipment, resulting in an increase in the workload for the clinical engineering service. In some cases, the clinical engineering service has been successful in tying an increase in capital-equipment inventories to an increase in operating funds for the department. At Sunnybrook & Women's College Health Science Centre, for example, the biomedical engineering department receives an increase in its operating budget of 5% of purchase cost of new equipment related to program expansions. Of this amount, approximately 80% is allocated to salaries, and 20% to parts and supplies to support the expansion.

Typical Department Programs and Services

During the development of the Clinical Engineering Standards of Practice for Canada (See Chapter 123), a consensus emerged as to what the basic programs and services that any clinical engineering service in Canada should provide. The following list is based on these Standards of Practice.

Medical Device Technology Management

- Device tracking and inventory
- Acquisition
- Unscheduled maintenance

Scheduled Maintenance Technology Assessment and Planning

- Planning and pre-purchase evaluation
- Assessments of safety, efficacy, and cost-effectiveness
- Long-range device planning

Risk Management

- Reuse policy development
- Managing hazard alerts and recalls from Health Canada, FDA, ECRI, and vendors
- Incident investigation.
- Education of device users
- Modification of devices

Management Models Example of a Nonregionalized Service

Sunnybrook & Women's College Health Sciences Centre (SWCHSC) in Toronto is an amalgamation of three founding hospitals. The founding hospitals were Sunnybrook Health Science Centre, Women's College Hospital, and the Orthopaedic and Arthritic Hospital. These hospitals amalgamated under a government directive in 1998. There was a comprehensive clinical engineering service led by a director, Bill Gentles, in the largest of the three hospitals, Sunnybrook, and a smaller service at Women's College. The Orthopaedic & Arthritic Hospital relied on outside vendors for clinical engineering services (see Figure 17-1).

After the amalgamation, the Department of Biomedical Engineering was one of the first departments to be consolidated. There were no staffing changes—the manager at Women's College simply reported to a new person. The department was given a mandate to consolidate services and to provide service to the Orthopaedic & Arthritic campus. The amalgamated organization has approximately 1200 beds, with trauma, cardiac surgery, and neonatal intensive care generating the majority of the work load for the biomedical engineering department.

The department had developed a growing range of nontraditional services at the Sunnybrook site. These included PC hardware repair, maintenance insurance, a central equipment pool for infusion pumps and other small equipment, and surgical-instrument repairs. These services were funded on a chargeback basis, and other departments were charged by the mechanism of "cost transfers" for services provided.

Total staff in the department is 19 full-time equivalents (FTEs), of whom five are funded by cost-recovery programs, and the remainder by the global budget of SWCHSC. Technologies supported by the department include patient care medical devices, personal computers, lasers, surgical instruments, and some laboratory equipment. Technologies not supported include medical imaging, large lab analyzers, and

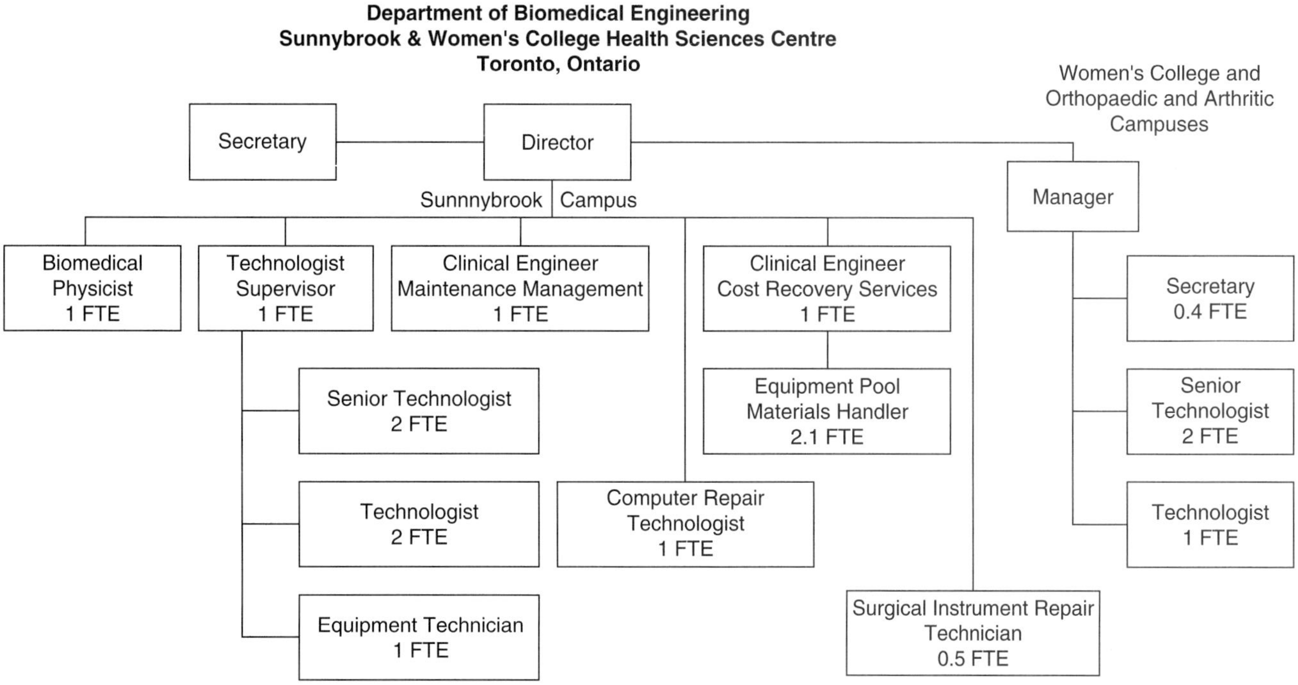

**Department of Biomedical Engineering
Sunnybrook & Women's College Health Sciences Centre
Toronto, Ontario**

```
                          ┌───────────┐
                          │ Director  │
         ┌────────────────┤           ├────────────────────────┐
         │                └─────┬─────┘                        │
    ┌────┴────┐                 │                              │
    │Secretary│         Sunnybrook Campus        Women's College and
    └─────────┘                                  Orthopaedic and Arthritic
                                                      Campuses
                                                   ┌─────────┐
                                                   │ Manager │
                                                   └────┬────┘
  ┌──────────┬──────────┬──────────────┬──────────────┐ │
┌─┴────┐ ┌───┴────┐ ┌───┴────┐ ┌───────┴──────┐       │
│Biomed│ │Technol.│ │Clinical│ │  Clinical    │       │
│Physi.│ │Superv. │ │Engineer│ │  Engineer    │       │
│1 FTE │ │1 FTE   │ │Maint.  │ │Cost Recovery │       │
└──────┘ └───┬────┘ │1 FTE   │ │1 FTE         │       │
             │      └────────┘ └──────┬───────┘       │
             │                        │               │
     ┌───────┴─────┐          ┌───────┴──────┐   ┌────┴────┐
     │Senior Tech. │          │Equipment Pool│   │Secretary│
     │2 FTE        │          │Materials     │   │0.4 FTE  │
     └─────────────┘          │Handler 2.1FTE│   └─────────┘
             │                └──────────────┘   ┌─────────┐
     ┌───────┴─────┐          ┌──────────────┐   │Senior   │
     │Technologist │          │Computer Repair│  │Technol. │
     │2 FTE        │          │Technologist  │   │2 FTE    │
     └─────────────┘          │1 FTE         │   └─────────┘
             │                └──────────────┘   ┌─────────┐
     ┌───────┴─────┐          ┌──────────────┐   │Technol. │
     │Equipment    │          │Surgical Inst.│   │1 FTE    │
     │Technician   │          │Repair Techn. │   └─────────┘
     │1 FTE        │          │0.5 FTE       │
     └─────────────┘          └──────────────┘
```

Figure 17-1 Department of Biomedical Engineering Sunnybrook and Women's College Health Sciences Centre.

renal dialysis. As of March 30, 2001, the director reported to the vice president of clinical support services. This reporting relationship has subsequently changed with the retirement of the director.

Regional Services

As of the year 2000, most provinces and territories (with the exceptions being the province of Ontario and the territory of Nunavut) have reorganized hospitals into regional groups, usually with a single governing board for each region. The clinical engineering services in these regional groups of hospitals tend to thrive as they demonstrate their cost-savings potential.

Prior to this trend in the 1990s, clinical engineering departments were, most often, stand-alone services providing service to only one hospital. Smaller hospitals (i.e., fewer than 200 beds) would typically outsource clinical engineering services to vendors or third-party service providers.

With the trend toward regionalization, clinical engineering departments in larger hospitals found themselves responsible for providing service to a number of smaller institutions within their geographical area. In many cases, they evolved into large, regional groups providing service over broad areas, with funding coming from the regionalized hospital group. In other cases, a large department servicing a single site took on the responsibility for managing groups of technicians in smaller hospitals within the same region, where the technicians were paid by the individual hospitals. In general, the trend to regionalization has led to the growth of in-house services at the expense of for-profit service groups.

The following Organizational Chart (Figure 17-2) shows the staffing levels and management structure for a regional service based in Edmonton Alberta. This service supports six acute-care hospitals. The number of staff in this service totals 63 FTE's, of which about 50 are funded by the region, and the remainder are funded by individual hospitals. Nontraditional services include a central equipment pool, a clinical system IT coordinator, and a maintenance management specialist. Technologies supported by the service include patient care medical devices, lasers, renal dialysis, and some laboratory equipment. Technologies not supported include medical imaging and large lab analyzers. The regional manager reports to the director of equipment management in the capital equipment planning portfolio.

Outsourcing and Third-Party Services

Although there is nothing in the Canada Health Act that prohibits the outsourcing of support services by hospitals, the penetration of for-profit clinical engineering service groups in Canada has been limited. The landscape of third-party service providers is in a state of flux in Canada, as it is in other countries. Mergers and acquisitions are always occurring. The number of independent groups providing for-profit clinical engineering services decreased during the 1990's. GE Medical (Canada) acquired one of the larger independent Canadian groups, formerly known as BESSI, as part of the GE strategy to offer a full range of multivendor services. ServiceMaster Canada acquired a group formerly known as ESNA Biomedical. Consequently, in 2001, there was only one large independent service organization offering clinical engineering services in the country—ServiceMaster Canada. Their penetration of the Canadian market has been limited.

GE Medical has had some success in marketing its multivendor service in Canada. Other large vendors of imaging equipment such as Siemens and Philips do not offer multivendor service in Canada.

Sources of Trained Staff

Managers in clinical engineering commonly come from the clinical engineering program at the University of Toronto Institute of Biomaterials and Biomedical Engineering (http://www.utoronto.ca/IBBME/), which offers a Master of Health Science degree that incorporates three hospital internships. Technologists come from a number of postsecondary programs that offer two- or three-year programs in electronics engineering technology with a biomedical focus: British Columbia Institute of Technology (http://www.bcit.ca), Northern Alberta Institute of Technology (http://www.nait.ab.ca/), Fanshawe College in Ontario (http://www.fanshawec.on.ca/), and College of the North Atlantic (http://www.northatlantic.nf.ca) in Newfoundland.

New Program or Service Initiatives

One growing trend in many departments is to develop enhancements to the service that are funded by charging the end-users directly. These services are run on a business model and compete with outside services that the hospital is already purchasing. If the in-house service is cheaper and of a higher quality, it is usually successful in selling the concept to customers and stakeholders. Examples of such services are computer repairs, medical imaging service, maintenance insurance, and surgical instrument repairs.

Departments are also looking outside their organizations for sources of revenue. In Ontario, the ambulance services are organized across Base Hospitals, and it is common practice for the Biomedical Engineering group in the Base Hospital to have a contract to maintain the defibrillators and other life-support equipment for a fleet of ambulances. Other initiatives include providing contracted services for clinics and labs that are not affiliated with the hospital.

With the growth of complex technologies used in the operating room and intensive care units, there has been a growth in the number of equipment technicians whose function is to operate these complex devices during routine procedures. In many organizations, the clinical engineering service is seen as the logical home for these technicians. In the past, the participation of clinical engineering staff in routine clinical procedures has been avoided by many clinical engineering groups because of staffing limitations. There appears to be an opportunity for growth in this area that many clinical engineering services in Canada have actively pursued.

With the move toward a comprehensive electronic patient record (EPR), clinical engineering services are working more and more closely with information technology (IT) services. There is growing involvement of the IT service in the specification of patient-monitoring equipment that will meet the needs of the EPR. In the case of the Regional Clinical Engineering Service in Edmonton, it can be seen on the organization chart that a distinct position has been created, called Clinical System IT Coordinator. The need for clinical engineering services to have staff trained in computer networking is widely recognized. Positions have been created in many clinical engineering groups to meet this need.

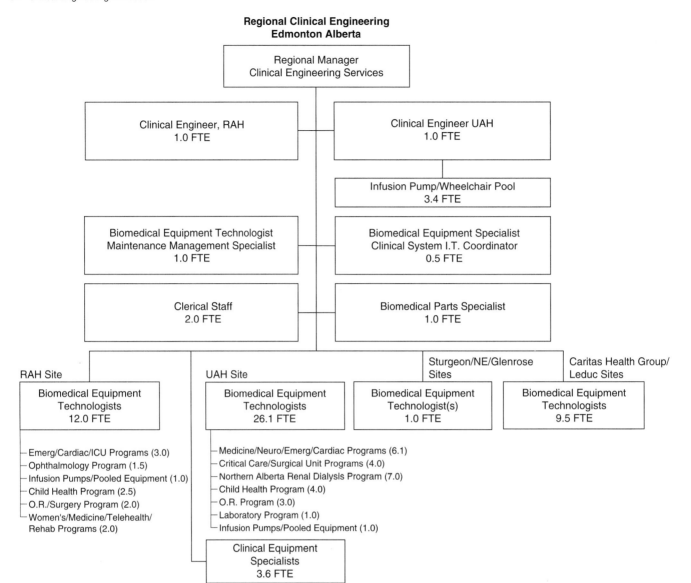

Figure 17-2 Regional clinical engineering: Edmonton, Alberta.

Accreditation and Peer Review

The Canadian Council on Health Services Accreditation (CCHSA) (HSA, 2001) is a nonprofit, nongovernmental organization that is similar to the Joint Commission on Accreditation of Healthcare Organizations (JCAHO) in the United States.

The CCHSA provides health services organizations with the opportunity for voluntary participation in an accreditation program based on national standards. In recent years, these standards have moved away from prescribing what a hospital should look like, to emphasizing patient outcomes. Sections that address equipment management have been weakened, and quality measurement and quality improvement have been given much greater emphasis.

This situation was the stimulus for the development of the Clinical Engineering Standard of Practice for Canada and an associated Peer Review Process, under the sponsorship of the CMBES ((Easty and Gentles, 2003)).

Communication in the Clinical Engineering Community

The members of clinical engineering profession in many of the larger provinces have organized local interest groups in order to facilitate communication. The primary mode of communication is e-mail, but most of these groups also hold regular meetings, either monthly or quarterly. Table 17-1 is a list of the associations and information on how to contact them:

In addition, the Canadian Medical & Biological Engineering Society (CMBES) sponsors a list-server, which can be joined by posting the message "Join CMBES" in the body of an e-mail sent to majordomo@messenger.sbgh.mb.ca. This list-serve is an active and informative communication medium for anyone practicing clinical engineering in Canada. The CMBES URL is www.cmbes.ca. The CMBES holds an annual conference in a different city each year. Information about upcoming conferences can be viewed on the society's web site. Canadian clinical engineers also participate actively in international groups such as the Association for the Advancement of Medical Instrumentation (AAMI), the American College of Clinical Engineering (ACCE), and the International Federation of Medical and Biological Engineering (IFMBE).

Table 17-1 Canadian clinical engineering associations and contact information

Province	Group name	Contact information
British Columbia	Institute of Biomedical Engineering Technologists	www.asttbc.org/ibet
British Columbia	Clinical Engineers Committee - B.C.	Jack.Hauzeneder@ex.thr.bc.ca
Alberta	Alberta Clinical Engineering Society	www.aces.ab.ca
Ontario	Clinical Engineering Society of Ontario	www.ceso.on.ca
Quebec	Association des physiciens et ingénieurs biomédicaux du Québec	www.apibq.org

Summary

The clinical engineering profession in Canada shares many similarities with the profession in other countries. The unique aspects have been determined by the geography, the politics, and the Canadian health care system. The future holds many challenges and much promise as the profession continues to evolve to meet the changing needs of the Canadian health care system.

References

CHA. Overview of the Canada Health Act. Government of Canada. www.hc-sc.gc.ca/medicare/chaover.htm, 2001.
HAS. Canadian Council on Health Services Accreditation. www.cchsa.ca/english/indexeng.html, 2001.

Additional Information

Canadian Medical and Biological Engineering Society, Clinical Engineering Standards of Practice for Canada. www.cmbes.ca, 1998.
Canadian Institute for Health Information, Health Care in Canada. www.cihi.ca, 2001.

18 Clinical Engineering in Estonia

Siim Aid
Department of Medical Devices, State Agency of Medicines
Tartu, Estonia

Ole Golubjatnikov
Estonian American Fund
Syracuse, NY

The goal of this chapter is to demonstrate, using Estonia as an example, some general and country-specific aspects of a medical-technology management challenge, characteristic to a country that is transitioning from an Eastern European (Soviet Bloc) economy to Western technology and the free market. The paper is based on implemented and proposed activities to meet the challenge and provides conclusions and recommendations for transition. It expresses the results of medical equipment surveys and questionnaires in health care facilities of Estonia, practical work in the field, and available statistical data from Estonia and other sources.

Estonia and Its Health care System

Demographics

Geography: Area: 45,000 sq. km. (17,340 sq. miles). Population: Around 1.4 million.
Neighbous: South: Latvia; East: Russia; North: Finland (60 km across Bay of Finland); West: Sweden (120 km.across the Baltic Sea).
Language: Estonian, which belongs to the same language family as Finnish and Hungarian.
History: Together with two other small Baltic states, Estonia has had a rich and turbulent history, mainly because of its location on one of the major contest lines between Western and Eastern Christianity and cultures.

Until the thirteenth century AD, the territory of Estonia was a weak, monolingual union of independent counties with traditional rules and religion based on nature. The country was conquered by northern Crusaders and converted to Catholic Christianity. It has been ruled by noblemen from the descendents of the Crusaders, with additions from other occupants throughout its history (i.e., Denmark, Germany, Sweden, Poland, and Russia). Estonia was under the occupation of the Czar's Russia from the early eighteenth century until 1918. The country gained its freedom during its War of Independence (1918-1920). Seized by the Soviet Union in 1940, occupied by Hitler's Germany during WW II, and reoccupied by the Soviet Union in 1944. It regained its independence in 1991 as a result of the *singing revolution*. Since independence, Estonia has established a Western constitutional democracy. It is a member of NATO and is in the process of becoming a member of the European Union (EU).
Prevailing religion: Lutheran.

Health Care System

Estonia's diverse health care system is rapidly changing because of major reforms. Quite recently, Estonia gained 60 hospitals with 10,000 acute care and long-term beds and a number of outpatient clinics. The goal of the ongoing health care Project 2015 is to convert these facilities to 13 acute-care hospitals with 3300 beds, and the other beds and hospitals are to be converted for nursing care. The goal will be achieved by closing some smaller hospitals and merging a number of hospitals into four to five chain-like facilities consisting of three to eight previously separate health care facilities.

Outpatient facilities (polyclinics) are converted mainly into group practices of family practitioners, with the addition of some specialized care. Private outpatient hospitals and specialist practices are emerging, especially in surgery (e.g., plastic surgery; orthopedic; urological; gynecology; ophthalmology; and ear, nose, and throat (ENT).

Primary care is based on recently the established institution of family practitioners. There are 800 practices each with an average of 1700 patients. Practices are private by statute. Dental care facilities number 460 and consist mainly of private practices.

Health care is mainly financed by social tax (13% of salary), a state budget (e.g., ambulance service), and, increasingly, additional out-of-pocket payments by the general population. Total medical expenditure is estimated at 6 to 7% of the gross domestic product (GDP).

Total yearly medical device purchases by hospitals are estimated at $30 million (including disposables and accessories). Out of this total, the serviceable equipment is valued at approximately $10 million. To this sum, purchases by dental and family practices, emergency service, devices for handicapped persons, over the counter (OTC) directly to the population add approximately $10 to 15 million, mainly for nonserviceable devices. The total installed base of serviceable equipment in all of Estonian hospitals is estimated at $120 to 180 million (based on the purchase price). The annual outside-contract equipment service is estimated at $1.5 million.

Challenges of Health Care Technology Management in Transition

Health care system reform, transition to Western technology, and conversion from Russian language–oriented and centralized Eastern European common market (EEM) to

English language–oriented, free and diverse Western market (WM) cause many challenges. Some of these are listed below:

The Need to Convert Technical and Language Skills of Technical and User Personnel

The EEM had an equipment base where the number of manufacturers was limited to two or three for each kind of device. Therefore, it was easy for a technician to become proficient in servicing this limited and homogeneous device population. The sophistication was modest and lagged behind mainstream WM equipment by 5 to 15 years on the average. Because of this lag, only relatively proven design and performance solutions were implemented and entered the market. User manuals, technical manuals, documentation, and related materials were predominantly in Russian. The training courses were also held in Russian, with the exception of general training, which was provided in national languages. The content of technical documentation was strictly standardized and well covered. Compared to WM devices, however, the documentation frequently was not user-friendly.

Conversion and Decentralization of Decision making

In EEM, devices were distributed, rather than sold, by governmental institutions to healt care facilities. Because of strict standardization, the decision makers had no involvement with the devices. Therefore, they did not care about the content and performance characteristics of acquired devices and accompanying documentation. In the WM, the responsibility for specifying, vendor qualification, procurement, and maintenance rests with the individual health care facility within the legal and regulatory environment of the government.

Four- to Ten-Fold Increase in Device, Hand Tool, and Test-Equipment Prices

In EEM countries, the labor cost was, and still is, many times lower than in WM. In addition, standardization, limited product selection, dictated prices, and retarded technology further reduced the cost of EEM devices. The transition to Western technology and WM caused a four- to ten-fold increase of device prices for health care facilities in the transitioning countries. Hand tools and test devices purchased on the WM present a similar price challenge. Therefore, considering local purchasing power, the relative cost is four- to eight-fold as high as for a buyer in the West.

Steep Decrease in Availability of Specialized Training

Because of the limited size of the local language market, the training for technical staff is not economically feasible for the WM vendors. At the same time, the cost of one day of training abroad is comparable to a hospital technician's monthly salary. Both hospitals and private companies suffer from the lack of trained technical staff because of the high price of courses and training available in English.

The Need to Cope with Conversion from Fixed Prices of the EEM to Floating and/or Negotiated Prices of the WM

In the EEM, the prices were negotiated between the manufacturer and the centralized governmental institution. For WM purchases, all of the Soviet Union acted as a single, centralized group purchasing organization. The hospitals had to accept the result at a fixed price. As a result, hospitals had no experience at negotiating prices. Consequently, during the initial seven to ten years of the transition, facilities typically paid the full list price. Another related problem was the economic loss that stemmed from poorly drafted procurement specifications and the lack of internal standardization

Diverse Installed Equipment Base

Currently, three subgroups of devices can be identified in Estonian hospitals:

1. Old (and outdated) EEM devices – degraded in technology and quality, but well supplied with documentation and well known by technicians who were trained during the Soviet time
2. Equipment installed between 1990 and 1995 through intensive humanitarian aid – mostly used devices of newer technology and quality but poorly supported because of inadequate technical documentation
3. New WM devices purchased after 1995, which now carry most of the workload in hospitals – latest technology, improved quality, and well supported by private companies, poorly supported by in-house technicians who have limited access to technical documentation

Efficacy of Service Resources

The above-described diverse equipment base in the hospitals, combined with limited budgets and lack of trained staff, presents a significant maintenance challenge. There are two ways to reduce the device service costs: Increase the efficacy of service resources or reduce service of the device.

During the first phase of transition to the free market, the majority of hospitals lacked the technology management experience and tended to overutilize the second of these options. This resulted in rapid deterioration of quality and performance of both used and recently purchased equipment.

The Need to Convert the Legal Environment and Market Surveillance Systems Related to Medical Devices

In the European Union, of which Estonia stands as a candidate for membership, the safety-vigilance system relies mainly on customer activities and complaints. (See Chapter 125) It depends on re active, rather than proactive, market control. Expulsion of ineffective or useless devices from the market is expected to be performed mainly by market forces. In order to be effective and to avoid heavy losses of scarce health care resources, such a system calls for technologically competent hospital personnel as customers.

The Need to Have Reliable Device-Related Statistics to Support Legislative and Decision Making Processes

The EEM was a highly standardized and centralized health care system. The local healt care facilities in Estonia had no involvement with the data collection or decision making of the central authorities. Equipment maintenance in hospitals was centrally controlled. Therefore, Estonia's independence brought with it a large collection of legislative, regulatory, and decision making challenges. Data collection, equipment databases, and statistics were necessary to support local hospital decision making, regulatory activities, and legislative action. A number of other, mainly temporary, problems resulting from the rapid transition process exist and need attention.

Conclusions and Recommendations

The majority of the above-described problems will be mitigated or solved, even though sometimes the process is agonizingly slow, particularly under the pressure of a diverse set of market forces. During transition, special attention must be given to the following five factors:

1. Policy, legislation and regulatory action

The interrelation of safety, performance, and availability of devices needs to be addressed during the development of legislative and regulatory systems. Policy and legislative work in Estonia is guided by the European Union requirements. A number of government laws and Social Ministry regulations have been issued. Estonia has made considerable progress during its first five-year Health care Master Plan (i.e., 1995 to 2000). The current healt care reform activities are conducted within the framework of the Health care Project 2015.

2. Efficient management of resources

The transition from centralized to decentralized decision making at the country and health care facility levels represents a major cultural change in Estonia. The four– to ten-fold increase in relative cost of equipment from EEM to WM prices creates a major management and budgetary challenge. Management must provide considerable additional attention to technology management and acquisition practices. Budgetary and manpower planning is necessary to ensure that medical devices are adequately maintained and that they operate safely and effectively.

3. Equipment database and statistics

The transition from EEM to WM represents a revolutionary change in the way equipment is procured. Efficient and effective acquisition of equipment by hospitals requires access to actual (independent) price, safety, and performance information. (See Chapters 33 and 34.) This is especially critical for a small economic area with a unique language. Appropriate resource allocation to medical economics and improvement of data collection methods to guide the transition period are essential. However, some initial data collection will be needed before conclusions and projection of trends can be made.

4. Education and training

Specialized education, training, and retraining of the medical and technical staff in medical device technology management require concentrated efforts and resource allocation by health care, governmental, educational, and international organizations. Isolated private-sector vendors are not well-suited for such education and training. Because of the limited market size, education and training are of questionable profitability for the local provider. The importance of language and terminology issues are frequently underestimated, but they should not be. The Medical Technology Plan of Project 2015 estimates the country's medical physics (MP) and biomedical engineering (BME) staff requirements in the order of 150 to 200 specialists. The plan recommends a ratio of 1:4:10 among physicists, engineers, and technicians. During the Soviet period, there was no formalized education in the MP and BME fields. To address the educational, training, and retraining needs, the Tartu University established in 1996 the Training Center of MP and BME, offering degrees at the master's level. It provides quality control and support to hospitals across the country. The Tallinn Technical University established the Biomedical Engineering Center in 1994, offering degrees at the masters and doctoral levels. High-school education in these fields is in the planning phase.

5. Coordinated and integrated action.

Development of the legislative system, medical devices industry, medical gement in hospitals, and research and development work in medical technology as separate systems is not effective, especially in a small economic area with a unique language. These should develop together and should be kept in mind constantly and equally, to ensure effective allocation of limited financial and human resources.

Summary

Reforms in the health care system of a transitioning country can be successful only as long as it is realized that the reforms, by their very nature, are technology-driven. Health care technology and its associated cost in all countries advances faster than their GDP. To meet their population's increasing expectations, the transitioning countries must conduct major reforms with highly limited resources. Therefore, effective technology management is particularly critical for these countries.

19

Clinical Engineering in Germany

Vera Dammann
Department of Hospital and Medical Engineering,
Environmental, and BioTechnology
University of Applied Sciences Giessen-Friedberg
Giessen, Germany

More than 6000 clinical engineers are working in industry, hospitals, and service companies in Germany today. The legal framework for clinical engineering work in the hospital is the Medical Devices Law of 1994, which is based on the European Medical Devices Directives (MDD,1993; see also Chapter 125) and accompanying directives and regulations concerning ownership, operation, maintenance, and calibration of medical devices. Clinical engineering services in German hospitals are usually carried out by clinical engineers and medical technicians, integrating external services of the manufacturers and service companies. German clinical engineers and biomedical engineering technicians are affiliated in the Fachverband Biomedizinische Technik e.V., fbmt.

Professional Field of Clinical Engineers in Germany

Education of clinical and biomedical engineers started in East Germany in 1953, in West Germany in 1970. Thus, today it is estimated that more than 6000 specialized professionals are working in the field of clinical engineering (Dammann, 1994; Dammann, 2000). The majority found jobs in the medical device industry; and just a minority is working in hospitals. Clinical engineers are supported by skilled craftsmen and biomedical engineering technicians (in Germany, these are simply called "medical technicians"), who attended a special two-year technician college after completion of a three-year apprenticeship in electrical or mechanical technology.

Employers and Jobs of Clinical Engineers

Owing to their interdisciplinary education, clinical engineers are very flexible and thus are able to find adequate jobs. Table 19-1 lists, in descending order of size, the number of CE jobs available.

Survey of Hospitals

Provoked by acute financial problems in the German health insurance system, several laws and directives were revised. The consequences for hospitals are enormous, as the following list shows:

- Budgeting instead of full refinancing of expenses
- Proof of all costs related to special patient treatment procedures and to single departments
- Quality assurance in diagnostics, therapy, and nursing
- strengthening of ambulatory care and reduction of nursing days
- Introduction of a diagnostic-related group (DRG) system and the concept of evidence-based medicine

Many hospitals have closed or reduced their number of beds as a consequence of financial pressures. Today (BFG, 2000), the situation is as follows:

- There are 2200 hospitals, about 10 % of which have more than 500 beds, and among these are 35 university hospitals with up to 2500 beds (e.g., "Charité" in Berlin)
- The average number of beds per hospital is 250
- There are 1400 diagnostical and/or rehabilitation clinics with an average of 135 beds per clinic.

Increasingly more hospitals, including university hospitals, have been converted from public institutions to foundations or private enterprises, with the advantages of better management possiblities, better payment of staff, and, especially, improvement in financing and budgeting. Many hospitals have given up in-house service departments such as catering, laundry, maintenance, clinical chemistry laboratory services, and even clinical engineering services. Instead, they use external service providers.

Legal Requirements in Clinical Engineering

The Law

In the course of harmonization with the EEC Directive on Medical Devices (MDD, 93/42 EEC [MDD, 1993]), the German government signed the German Law on Medical Devices (Medizinproduktegesetz, MPG [MPG, 1994] on August 2, 1994. Like the EEC directive, it is not directed primarily to the owner and operator of a medical device, but to the manufacturer and vendor. It especially defines the prerequisites and procedures for the CE-mark and for a national vigilance system (Pallakarakis, 2003). In addition to the MDD, however, it requires the following at the levels of the manufacturer, importer, and vendor:

- A special person who is responsible for the handling of safety information on medical devices; and
- Special knowledge of the salespersons, advisers, and instructors, who are now called "medical device consultants" (*Medizinprodukteberater*).

The law also authorized the Ministry of Health to publish and enforce special ordinances concerning the following:

- Clinical evaluation of medical devices
- Storage, purchase, sale, information, consulting, and instruction on use and operation, including
  - documentation
  - personal requirements
  - premises
  - hygienics
  - labeling of containers for medical products
  - securing supply
  - handling of medical products that are unsuitable for sale and use
  - regular checks of medical products and documentation of these checks (in the sense of storage checks)
- Education and qualification of medical device consultants
- Instruction of the owner and user of medical devices
- Technical safety checks
- Calibration of measuring devices
- Function checks
- Inventory
- Recordkeeping for dedicated medical devices

All owner-related duties (Böckmann,1994; BGBl, 1998) are typically the responsibility of the in-house clinical engineer.

System of Governmental Authorities

As Germany is a federal republic, the states (*Länder*) are responsible for supervision of safety in health care. The federal Ministry of Health (http://www.bmgesundheit.de) enacts the law and the regulations. Together with the relevant department of the federal Ministry of Health, the state Ministries for Occupation and Social Affairs formed a medical device committee in order to harmonize all procedures. In a particular state, the local trade-supervisory boards (*Amt für Arbeitsschutz und Sicherheitstechnik*), which are the next governmental level below the Ministry of Occupation and Social Affairs, are responsible for the supervision of hospitals and medical device manufacturers and traders. They carry out inspections of documentation and procedures and investigate accidents and incidents. Because maintenance and, especially, safety checks

Table 19-1 Jobs available to clinical engineers by employer and field of specialization

Employer	Field
manufacturer of medical devices	marketing and sales
	support, technical service
	application, training
	product management
	project management
	regulatory affairs
	quality management
	software development
	hardware development
hospital	central clinical engineering
	radiation therapy (as medical physicist)
	hospital engineering (technical director)
	technical support in other clinical department
service company*	maintenance, especially safety checks, repairs, and calibration
engineering office*	hospital and hospital equipment planning
	software development
	assessment of devices (for certification or in case of accident)
	certification ("notified body")
	advisory service (especially quality assurance and regulatory affairs)
	authoring of manuals and tutorials
	training
public administration, ministry	supervision
	regional planning

*Including self-employed in own enterprise.

and calibration of medical devices need adequate knowledge, experience, and equipment, the boards also may check workshop equipment and the qualifications of staff.

The German federal Office of Drugs and Medical Devices (*Bundesinstitut für Arzneimittel und Medizinprodukte* (BfArM) (http://www.bfarm.de) was assigned to medical device matters such as central recording and assessment of observed and reported risks associated with medical devices as well as coordination of measures where necessary. The databases of the governmental vigilance system are maintained at the German Institute for Medical Documentation and Information (DIMDI) (http://www.dimdi.de).

Clinical Engineering Services—Tasks and Structures

Because of legal demands, nearly every hospital employs at least one medical technician. The minimum list of duties includes maintaining the inventory, device records, and schedule of safety checks and calibrations. At the next level, the medical technician or clinical engineer carries out first-line service, manages external orders, controls service contracts, and checks the bills.

A full clinical engineering service covers the entire life cycle of a medical device within a quality assurance system (Böckmann and Frankenberger, 1994; Nippa and Fritz, 1995, 1998, 1999). This entails the following:

- Equipment and investment planning
- Technology assessment and specifications development
- Safety of combination of devices and networks
- Call for tenders
- Ordering
- Contracting with external services
- Installation preparation, including cooperation with architects and hospital engineers
- Acceptance tests
- Installation
- Final check and placing in operation
- Consulting and training of users
- Functional and safety checks
- Calibration
- Preventive maintenance
- Repairs
- Cooperation with purchasing officer for the purchase of consumables
- Logistics and environmental protection concerning the supply, operation, and recycling or disposal of consumables
- Documentation (e.g., inventory and device history) [VDI 2426]
- Analysis of incidents and accidents
- Contact with supervising authorities
- Budgeting
- Quality assurance
- Reselling, disassembly, or disposal of device

The clinical engineer may be involved in clinical trials with new devices, either as the person responsible for technical service or at the application level.

As Germany is a country of rather short distances, external services compete strongly with in-house clinical engineering groups.

In-house Structures

On average, in-house clinical engineering structures have the following:

- Clinical engineer (with a university degree) per 500 beds
- Medical technician or further-qualified craftsman per 100 beds
- Clerk per 1000 beds

Various models exist for integrating the clinical engineering service into a hospital's organizational structure.

Large and Medium-Sized Hospitals

The clinical engineering service may be central or decentralized, with technicians in small, individual workshops in the medical departments; e.g., anesthesia and intensive care unit (ICU). The head of clinical engineering is usually an experienced clinical engineer or physicist. Increasingly more clinical engineering (CE) groups have their own budget, develop a business plan, and make decisions on outsourcing tasks; e.g., regular electrical safety checks, full-service MRI contracts, or single repair orders.

The CE group may be an independent department, with its head reporting directly to the hospital director. If the hospital has a directorium, the head of CE may even be member of the directorium. In other cases the CE group is just a sub-department of the technical services, reporting to the technical director of the hospital.

Along with CE groups with hospital-wide responsibilities, some medical departments employ one or more clinical engineers or medical technicians. For example, many clinical engineers work in radiotherapy departments as experts in medical physics and are responsible for dosimetry, therapy planning, and radiation protection.

Small Hospitals

General hospitals with fewer than 200 beds employ one or two medical technicians. They are responsible especially for documentation according to legal requirements, carrying out regular functional and safety tests, some first-line service, and hiring external services for more complex or extensive work.

Highly specialized hospitals such as cardiac surgery or oncotherapy hospitals with just one medical department usually have a clinical engineering group, offering full CE service, including, in some cases, the operation of complex medical devices.

External and Mixed Services

Clinical engineering services are offered externally by the following models:

- Manufacturers traditionally, offering such services as
  - single-maintenance action
  - part or full maintenance
  - cooperative service with in-house staff
  - consulting and/or full clinical engineering service by a dedicated subcompany
- Small, locally operating service companies that
  - specialize in single tasks such as safety tests or calibration according to medical device regulations
  - act as privatized service centers that are responsible for one large hospital, which may be a shareholder of the service company, or for several hospitals in the region, in some cases selling medical devices and/or comsumables
- Small, supra-regionally operating service companies, that could include
  - distributors who import, sell, and service medical devices
  - consultants in hospital equipment planning, IT technologies, quality assurance, organization engineering, and surveying or certified experts
- Large, nationally or internationally operating service companies that offer
  - consulting services
  - partial or complete clinical engineering service with or without engineers or technicians, permanently delegated to a single hospital (so-called external in-house service).

In contrast to the outsourcing actions of the early 1990s, hospitals now tend toward mixed services, with at least one clinical engineer or technician who keeps the records and has the knowledge of the hospital.

Affiliation of Clinical Engineers and Biomedical Engineering Technicians

In 1984, the *Fachverband Biomedizinische Technik e.V.*, (*FBMT*) (http://www.fbmt.de) was founded by 40 clinical engineers. Today, the FBMT has more than 500 members who cooperate in regional and working groups. The association offers continuous education in seminars and workshops and holds a large congress in Wuerzburg every year. (Proceedings are available in print and on CD-ROM.) The scientific journal *Medizintechnik, mt*, is their forum and is distributed to all members. New initiatives include a European Alliance of Clinical Engineering, together with the national CE associations of the neighbor countries and a certification board for medical technicians.

References

BFG. Statistisches Taschenbuch Gesundheit. Bundesministerium für Gesundheit, Berlin, 2000.
BGBl. Verordnung über das Errichten, Betreiben und Anwenden von Medizinprodukten (Medizinprodukte-Betreiberverordnung-MPBetreibV) BGBl. I vom 29.6.1998, S.1762 ff. 1998.
Böckmann R.-D., Frankenberger H. Durchführungshilfen zum Medizinproduktegesetz. TÜV Verlag, Loose leaf edition, 1994.
Dammann V. Blätter zur Berufskunde Dipl.-Ing. (FH) Biomedizinische Technik. 2-IR 42.. 4th ed. Bertelsmann Verlag, 1994.
Dammann V. Ergebnisse der Berufsfeldrecherche 2. Halbjahr 1998. Medizintechnik 120:25, 2000.
DVMT. Dachverband Medizinische Technik: Qualitätsmanagement der Medizintechnik des Krankenhauses–ein Leitfaden. Stresemannallee 15, D-60596 Frankfurt, DVMT, e-mail: service@dvmt.de, 1999.
MDD. Medical Devices Directive 93/42/EWG / Richtlinie 93/42/EWG des Rates über Medizinprodukte; Amtsblatt der Europäischen Gemeinschaften, 36. Jahrg., L 169/1 bis L169/43, 12. Juli 1993.
MPG. Gesetz über Medizinprodukte (Medizinproduktegesetz) vom 2.8.1994 (BGBl. I S. 1963) Zuletzt Geändert Durch das Erste Gesetz zur Änderung des Medizinproduktegesetzes (1.MPG-ÄndG) vom 6.8.1998 (BGBl. I S. 2005), 1994.
Nippa HJ, Fritz B. Medizintechnischer Service in Krankenhäusern Deutschlands. Medizintechnik 115:184, 1995.
Nippa HJ, Fritz B. Mit Qualitätsmanagement auf dem Wege zur Kostensenkungen und Motivation. Medizintechnik 118 Teil 1: S. 93 ff, Teil 2: S. 133 ff, Teil 3: S. 173 ff, Teil 4: S. 223 ff, 1998.
Nippa HJ, Fritz B. Mit Qualitätsmanagement auf dem Wege zur Kostensenkungen und Motivation. In mt Medizintechnik 119 Teil 5 S. 21 ff, 1999.
VDI. VDI Standard 2426: Kataloge in der Instandhaltung und Bewirtschaftung der Medizintechnik. Blatt 1: Allgemeines, Blatt 2: Blatt 2: Standardisierter Gerätekatalog, Blatt 3: Fehlerkataloge, Blatt 4: Instandhaltungsmaanahmen, Blatt 5: Anwendung von Katalogen in der Instandhaltung und Bewirtschaftung der Medizintechnik

Additional Information

Richtlinie des Rates vom 20 Juni 1990 zur Angleichung der Rechtsvorschriften der Mitgliedstaaten über aktive implantierbare medizinische Geräte; Amtsblatt der Europäischen Gemeinschaften, vom 20.7.90 Nr. L 189 S. 17-36. Council Directive of June 20, 1990 on the approximation of the laws of the Member States relating of active implantable medical devices (90/385/EEC). OJ No. L 189, 20.7.90, 17 ff
Richtlinie des Europäischen Parlaments und des Rates vom 27.10.1998 über In-vitro-Diagnostika. Abl. Nr. L 331 vom 7.12.1998, 1ff

20 Clinical Engineering in Brazil

Lúcio Flávio de Magalhães Brito
Certified Clinical Engineer
Engenharia Clínica Limitada
São Paulo, Brazil

Brazil: The Country

In describing the current practice of clinical engineering in Brazil, it is helpful to have some background information about the country.

Demographics

Brazil is a country with a population of over 169 million inhabitants distributed over an area of over 8.5 million square kilometers. That results in a population density of 19.9 inhabitants/sq. km. In 1980, the infant-mortality rate was 68 per 1000 births. In 1999, this index fell to 34. Adults are living longer. The life expectancy of those born today in Brazil is 68 years. The data of the Brazilian Institute of Geography and Statistics (IBGE) reveals that the population growth is decreasing and that the migration toward the cities is increasing (IBGE, 2001). New economic opportunities are appearing throughout the country, but social inequality continues.

Geography

Brazil has 26 states and one federal district where Brasília, the capital, is located. The map of the country is shown in Figure 20-1.

Hospitals

The number of hospitals in Brazil and its distribution in federate units (FU) is presented in Table 20-1.

Medical Device Market

In December of 1998, the market for medical devices was $13 billion. Of this, $500 million (38.46%) corresponds to diagnostic imaging equipment. The ultrasound market, for instance, began to grow in 1994 and in 1998 the annual market was about $80 to $104 million. The annual growth rate was around 20% and represented 50% of the volume of the medical device market in Latin America.

Medical Device Industry

The Brazilian national industry of medical devices is represented by the Brazilian Association of Industry of Medical Devices and Equipment (ABIMO). ABIMO was created in 1962 by 25 companies out of a total of 40. ABIMO is divided into five groups: Medical hospital equipment; radiology and medical imaging; dentistry; laboratory; and implant and supplies.

In 1996, ABIMO member companies had the following profile:

- 80.6% of the companies were of domestic capital, and 19.4% foreign capital
- 39.7% were small in size, 45.5% were medium, and 14.8% were large

Between 1991 and 1996, participation in ABIMO of small companies decreased from 47.1% to 39.7%, while the participation of large companies increased from 4.4% to 14.8%. The revenue of the market was the following:

- 1994: $1.31 billion
- 1995: $1.44 billion
- 1996: $1.96 billion
- 1997: $2.30 billion

In 1997, 93.9% of sales were domestic, and the rest were exports. In 1998, ABIMO had 230 members.

ABIMO currently has 243 members. They comprise 80% of the national market. Altogether, they manufacture the products to equip about 80% to 85% of a general hospital. Many of the ABIMO members have been in existence for over 100 years.

In the year 2000, the ABIMO stood as follows (Rodriques, 2001):

- The section billed R$ 3,450,000,000 and it exported FOB $149,901,439.
- Composition of the capital of the companies of ABIMO was of 80% national and 20% foreign and mixed.
- 28% of the companies were small; 68.7% were medium; and 3.3% were large.

Figure 20-1 Brazil.

Table 20-1 Brazilian Hospitals

	Quantidade De Municípios		Number Of Hospitals			Population		Market
FU (1)	Cities with hospitals (2)	Total of cities per state (3)	Qtde (4)	% (5)	% Accum distribrib. (6)	Cities with hospitals (7)	Total per state (8)	IPME (9)
SP	419	645	936	12,2	12,2	32.903.041	34.120.886	28,07250
MG	443	853	791	10,3	22,6	14.463.278	16.673.097	10,24670
PR	317	399	674	8,8	31,4	8.561.168	9.003.804	6,19443
BA	262	415	544	7,1	38,5	10.538.569	12.541.745	5,62585
RJ	74	91	510	6,7	45,1	13.104.249	13.406.379	10,87072
GO	181	242	481	6,3	51,4	4.079.789	4.515.868	2,67037
RS	293	467	435	5,7	57,1	8.891.582	9.637.682	7,11841
MA	127	217	385	5,0	62,1	4.299.092	5.222.565	2,09216
PE	158	185	364	4,8	66,9	7.088.179	7.399.131	3,98924
CE	168	184	345	4,5	71,4	6.623.891	6.809.794	3,21637
PA	90	143	272	3,6	74,9	4.815.658	5.510.849	2,38713
PB	115	223	255	3,3	78,2	2.698.279	3.305.616	1,50264
SC	174	293	246	3,2	81,5	4.252.081	4.875.244	3,06380
RN	124	166	205	2,7	84,1	2.346.961	2.558.660	1,26285
MT	91	126	198	2,6	86,7	2.078.779	2.235.832	1,34123
PI	106	221	193	2,5	89,2	2.154.038	2.673.176	1,12000
MS	63	77	143	1,9	91,1	1.830.001	1.927.834	1,25585
RO	40	52	120	1,6	92,7	1.152.202	1.231.007	0,65761
ES	58	77	118	1,5	94,2	2.611.095	2.802.707	1,78319
AM	52	62	105	1,4	95,6	2.269.583	2.389.279	1,22192
AL	43	101	91	1,2	96,8	2.002.662	2.633.339	1,13558
TO	64	139	87	1,1	97,9	828.207	1.048.642	0,37316
SE	30	75	55	0,7	98,6	1.247.582	1.624.175	0,76932
DF	1	1	39	0,5	99,1	1.821.946	1.821.946	1,51172
AC	12	22	25	0,3	99,5	419.617	483.726	0,20957
AP	9	16	21	0,3	99,7	338.153	379.459	0,15205
RR	8	15	20	0,3	100,0	209.236	247.131	0,15561
TOTAL	3.522	5.507	7.658	100,0		143.628.918	157.079.573	100,00

(1) relationship of federative unit
(2) number of cities with hospitals in federative unit
(3) number of cities in federative unit
(4) amount of hospitals in federative unit
(5) percentile distribution of hospitals in the union
(6) accumulated percentile distribution of hospitals in union
(7) total population in cities with a federative unit hospital
(8) total population per federate unit
(9) index of potential market

- The customer base was 48% private sector; 44.3% government; and 7.7% Unified System of Health (SUS).
- 46 (22.5%) of the companies made products certified by the NBR-IEC standards; 40 (19.6%) CE/UC; 33 (16.2%) FDA; 8 (3.0%) UL; 83 (40.7) GMP, and 23 (11.3%) other standards.
- 125 companies had no ISO certification; 71 (34.80%) had ISO 9000 certification; and 8 (3.93%) had ISO 14000 certification.
- ABIMO generated 37,679 direct jobs, of which 21.4% were for professionals with college degrees; 48.2% were for those with high-school degrees; and 30.3% were for those with other qualifications.

Problems

One of the fundamental problems found during the 1980s was the lack of clear policies for planning and administration of medical equipment, especially for the more innovative technologies. Some of the problems that were most difficult to solve concerned the following:

- Ways to manage medical equipment and devices, ranging from simple ones, such as thermometers, to CT scanners
- Ways to present data on safety and efficacy on each new technology
- Ways to obtain premarket evaluation of some technology
- Ways to obtain data on life expectancy of equipment
- The amount of money that a country should invest in health care technology
- The kind of technology (e.g., drugs, procedures, equipment, and administrative technology) and information that a country should invest in
- The costs to operate medical equipment within an entire country
- Replacement costs of each technology

Between 1980 and 1990, Brazilian authorities began to study and to understand this problem of planning and administration of medical equipment. The strong participation in international meetings that considered this theme in a global way was helpful (WHO/PAHO/FDA, 1986). Brazil started to promote the exchange of information, closer communication, and cooperation with other countries (USDC, 1990). Brazilian authorities understood that the lack of specialized manpower was a serious problem and that a corrective action should be taken (Wang and Calil, 1991; MoH, 1992; WHO, 1989). The following recommendations emerged:

– Training, alone, cannot solve the problems related to health care equipment management.
– Clinical engineering manpower development should be a national priority.
– Particular attention should be paid to policy formulation, manpower development physical-infrastructure strengthening, and information support.
– Highly specialized training should be conducted at the global level.
– Training always should be linked to a service workshop in a nearby hospital in order to give trainees practical exposure.
– A wide and serious gap in availability of training materials existed.
– In Brazil, positive conditions exist for developing a program to minimize and to start to control these problems.

From 1991 to 2000, the results of the efforts of the previous decade began to appear. In 1991, six Brazilian engineers were trained at the First Advanced Clinical Engineering Workshop in Washington, D.C. (See Chapter 70.) In the same year, a Health Care Technology College opened in Brazil. It has a 3-year program that trains professionals called "health care technologists," who are able to operate and maintain medical equipment. Since 1994, this college has trained approximately 28 new health care technologists each semester.

In 1992, the Brazilian Health Ministry published a Term of Reference for technicians training in maintenance of medical devices. That document described the fundamental importance of the biomedical equipment technician to the proper maintenance of equipment and proposed a strategy to qualify professionals in the Brazilian health care field. It presented the following scenario:

"The world market for medical equipment was, in 1998, $36.1 billion. Assuming that around 5% to 10% of the value of the equipment needs to be spent on maintenance, one arrives at an amount of approximately $2.7 billion in annual expenditures. As a consequence of the growth of the installed base and the increase of the technological complexity of the equipment, the expenses in maintenance in these countries grew by 50% in the last five years (1988-1992). In Brazil, the installed base of equipment operating in the public health institutions is nearly $6 billion, representing annual expenditure for maintenance on the order of $450 million, in other words, 3.5% of the Budget of the Union for Health in 1991." (DOU, 1991)

In 1993 and 1994, four federal and state universities began a program to train clinical engineers to practice in Brazil. These courses are full-time programs that lasted 12 months. Approximately 160 clinical engineers graduated in two years. During the 1990s, many other initiatives were undertaken, including the following:

- Master's and doctoral programs in biomedical engineering were strengthened and consolidated.
- Some new technical schools started while others consolidated.
- The International Certification Commission certified nine Brazilian clinical engineers, and a Brazilian Board of Examiners for Clinical Engineering Certification was created.
- Many meetings relating to clinical engineering were organized.
- At least four books relating to clinical engineering were published.

Reinforce to the Reorganization of the Unified System of Health (REFORSUS), created at the end of 1996 through a $650 million loan agreement with the Brazilian government, the InterAmerican Bank Development (BID), and the World Bank (BIRD), has been investing in the recovery of Brazil's health care physical infrastructure. REFORSUS invested in the purchase of medical equipment, hospital installations, and movable units; the execution of reform works; the enlargement and conclusion of establishments of health; and in projects for the improvement of the administration of the national health system. In order to operate properly, REFORSUS organized a team of health professionals, including clinical engineers, to evaluate needs and to specify and select medical equipment and hospital facilities in the institutions selected for this project.

The National Agency of Sanitary Surveillance (ANVISA) was created by Law No. 9782 on January 26, 1999. It is a regulatory agency characterized by administrative independence, job stability for its managers, and financial autonomy. In the structure of the Federal Administration, it is linked to Ministry of Health. ANVISA began an intensive program to develop agents specializing in sanitary surveillance to work in the medical device and equipment area. Many of these agents were engineers. Clinical engineers participated as teachers in these programs.

Employment

Brazil has been doing its homework. Consequently, it is better prepared than it was 20 years ago. Today, the job market for Brazilian clinical engineers is broad and diverse.

Hospitals

Clinical engineers who work in hospitals are engineers or health care technologists that have taken specialized courses in clinical engineering. Their duties vary according to the institutions where they work but, in general, include those defined by American College of Clinical Engineering. (See Chapter 130.) They usually report directly to the administration of the hospital, where their responsibility is to provide support decisions that involve medical equipment or hospital facilities. Few clinical engineers have the opportunity to act exclusively with medical equipment.

Industry

Usually, industry offers clinical engineers opportunities to serve in the maintenance area. Some engineers who have managerial abilities also act in administrative areas of these organizations. The development of new equipment sometimes is done by clinical engineers, although it is more common to have this activity performed by biomedical engineers. Besides maintenance, the area of sales is also available for clinical engineers and for health care technologists.

Academia

Another option for clinical engineers, biomedical engineers, and health care technologists is to work in academia. There, among other functions, they support and conduct research leading to new procedures. It is a highly sought-after atmosphere among health care technologists. About 35% of recent graduates, along with 30% of those who already have graduated and are active in the job market, seek postgraduate degrees in biomedical engineering and specialization in clinical engineering.

Consulting

Consulting activities in clinical engineering experienced a considerable increase over the last five years. This fact demonstrates that hospitals, health care industries, teaching institutions, health care companies, and the professionals involved in administration are much more conscious of the inconveniences and waste of resources that will ensue if they make the wrong decisions. Clinical engineers, in this market segment, act as mediators in the decision processes. They serve as objective interfaces, translating the needs of the suppliers for the consumers of these technologies. In short, the clinical engineer offers support for the right decisions and investment of resources.

Companies

Many clinical engineers, observing the need of hospitals, meet those needs (Blumberg, 1987). Some have worked independently as consultants. Others have formed companies. These clinical engineers are currently engaged in consulting, maintenance of biomedical equipment, validation of processes and equipment, and sales. Today, Brazil has only a few professionals in clinical engineering companies. However, because of the specialization and the emergence of this new profession, a number of companies have been established in several Brazilian states. With the increase in competition, the search for excellence and quality will result in patients receiving a higher quality of care.

Sales

Selling medical equipment has become highly specialized, to the point where industries are incorporating many clinical engineers to perform this function. The clinical engineer in sales should understand the technical aspects of the product, equipment operation, physiologic interactions, cost analysis (e.g., depreciation, amortization, and leasing), and clinical applications. Many specialists in clinical engineering are practicing in this area.

Conclusion

The role of clinical engineers in the Brazilian health care community has changed considerably over the last two decades. This profession, which not long ago was unrecognized, is becoming more highly respected and appreciated. Clinical engineers are now in demand. Many hospitals and manufactures have a clinical engineering team that is composed of more than a clinical engineer. The team acts to minimize loses of resources in health care technology management. In this context, clinical engineers play a fundamental part in the team as it decides and recommends particular technology to buy. Today, clinical engineers are working with doctors, nurses, and administrators.

Brazil is changing, and clinical engineers in the country are changing with it. These changes are accelerated by the increased interaction with the world at large. In Brazil, clinical engineers have much work to do.

References

Blumberg DF. Strategic Opportunities for Independent Equipment Maintenance Repair Service in the Hospital Market. *Second Source* 2(3):6, 1987.
DOU. Budget of the Union. Official Diary of the Union, 1991.
IBGE. Brazilian Institute of Geography and Statistics, 2001.
MoH. Training in Clinical Engineering. Ministry of Health. 1993.
MoH. Training of Technicians in the Maintenance of Hospital Equipment. Ministry of Health, 1992.
Rodrigues. Brazilian Association of the Industry of Goods and Medical and Odontology Equipment (ABIMO) Annual Report, 2001.
USDC. Medical Instruments U. S. Industrial Outlook. U.S. Department of Commerce, 1990.
Wang B, Calil SJ. Clinical Engineering in Brazil: Current Status. *J Clin Eng* 16(2):129, 1991.
WHO. Manpower Development for Health Care Technical Service. World Health Organization, 1989.
WHO/PAHO/FDA. Proceedings of the First International Conference of Medical Device Regulatory Authorities (ICMDRA). World Health Organization/Pan American Health Organization/Food and Drug Administration, 1986.

21 Clinical Engineering in Colombia

Jorge Enrique Villamil Gutiérrez
Independent Technology Consultant for Health Projects
Bogota, Colombia

Colombia's law 100/1993 defines all matters related to the Social Security System in Colombia. This law resulted from modifications in existing health services regulations. The National Social Security System was designed to guarantee every citizen's access to the health system. The guiding principles (CoC, 1993) of the system are as follows:

- **Efficiency:** The best social and economic utilization, in an adequate, timely, and competent way, of administrative, technical, and economic resources for the security of the society.
- **Universality:** Guaranteed protection for all people without discrimination.
- **Solidarity:** Mutual help from the strongest to the weakest among people of all ages, economic groups, regions, and communities.
- **Integrity:** Coverage of all contingencies affecting health, economic situations, and life conditions.
- **Unity:** Politics, institutions, regimes, and procedures working together.
- **Participation:** Active participation of the community, which benefits from the system, in organization, control, supervision, and audit of the system.

The system has two major components known as the "contributive regime" and the "subsidiary regime." In the contributive regime, workers with higher incomes are affiliated with an administrative and financial company known as EPS. EPS controls the contributions of workers and has contracts with several IPS centers (institutions that deliver health services, such as hospitals, health centers, and diagnostic centers). These institutions can be either public or private. EPS makes contracts with several IPS centers to guarantee that all affiliated workers and their families will have access to health services. For their contributions, workers and their families receive a basic package known as POS [Mandatory Health Services Plan (Contributive Regime)]. For additional services, the worker must pay an additional fee. A percentage of the contributions of these workers goes to a National Fund known as FOSYGA, the solidarity and guarantee fund. Its resources are used to help to pay for health services for low income workers.

Low income workers belong to the subsidiary regime. Municipalities pay for the affiliation of these workers. Every person is affiliated with an administrative and financial company known as ARS [Administrative Enterprise (Subsidiary Regime)] and is classified in a social and economic table of reference. The citizen receives a card containing personal data and showing the classification. With that classification, they have the right to receive basic health services known as POSS (Mandatory Health Services Plan Subsidiary Regime). The subsidiary regime pays the IPS centers for services provided to these people according to the POSS rate schedule. The citizen pays a percentage of the cost according to his or her classification. The subsidiary regime covers all the costs of the poorest workers.

The National Social Security System provides participants with certain rights, as follows: The right to select the EPS and the IPS centers; the right to receive the health services included in the POS; the right to include the family as beneficiaries of the health services; and the right to select the physician or specialist. Table 21-1 and Figure 21-2 show how many people have Social Security and how the system works (Correa, 2001).

In 1991, Colombia adopted a decentralized government system that gives autonomy to the regions of the country. For every aspect of public administration, the central government gives the political orientation that is mandatory for governors, mayors, and directors of public institutions. But the priorities are set, and the decisions made, by the departmental authorities or the mayors of cities. For example, health and education social services are the responsibility of the mayors. Mayors are the heads of the local health organizations and define, with the councils, the development plan priorities. The mayor selects the managers of the hospitals from a group of three candidates selected by the community, health authorities, and local authorities. The managers serve for at least three years.

To develop and deliver health services efficiently, cost-effectively, and with opportunity, the IPS centers (hospitals) are considered to be companies with autonomy to manage their resources and must compete with other health companies (other hospitals) in their area of influence. To sell their services through contracts with the EPS, the hospitals must fulfill mandatory requirements relating to physical infrastructure, equipment, human resources, administrative organization, controls, audit, and maintenance. These conditions, known as "Essential Requirements," are defined for each health service that the hospital offers. The hospital's manager certifies to the local or regional health authority by means of an official declaration that the services offered comply with the essential requirements and that the hospital is able to operate and provide those services. The National Superintendent of Health and regional and local health authorities assure that the hospital does, in fact, comply with the essential requirements. The mission of the Superintendent and the authorities is to guarantee that the health services provided to their citizens are of excellent quality and are cost-effective and adequate.

Hospitals and Health Centers

Public Sector

In Colombia, health services are provided at three levels: Primary, secondary, and tertiary. The primary level comprises health posts, the health centers (with or without beds), and the local hospitals. Health posts are located in the counties; the health centers could be located in counties, small towns or big cities; and local hospitals are normally located in small towns or, if of a high level, in big cities. Local hospitals normally offer outpatient medical and dental consultation, normal obstetric care, basic surgical procedures, and basic clinical laboratory and X-ray services. Some of them offer basic emergency care. The number of beds varies but average between 10 and 25 per hospital. Health services

Table 21-1 Population with Social Security (Thousand)

Concept	1995	1997	1999	2000
Population Insured	36,558	40,018	41,439	42,299
Contributive Regime	6,708	13,065	13,652	13,454
Subsidiary Regime	4,800	7,026	9,325	9,365
Speical Regime	1,000	1,150	1,250	1,250
Total Insured	12,509	21,242	24,228	24,070
Converted (%)	32.44%	53.38%	58.33%	56.90%

in these establishments are provided by general professionals in medicine, dentistry, infectious diseases, and nursing. At the primary care level, there are 560 local hospitals, 2000 health centers, and 2705 health posts.

Secondary health services are provided by specialists with access to better infrastructure and equipment for diagnoses and treatments. There are 134 regional hospitals located in towns that have more than 15,000 inhabitants, and the number of beds available varies between 30 and 160, depending on the locations and services provided. Some hospitals offer specialized services like intensive care units. Some secondary level hospitals are located in big cities.

Tertiary care is provided in approximately 36 specialized public hospitals located in the main cities of Bogotá, Medellin, Barranquilla, Cali, Bucaramanga, Ibague, Pereira, and Armenia. These hospitals have the most modern infrastructure and equipment needed for treatment and diagnosis, e.g., computerized axial tomography, automated clinical laboratories, nuclear medicine units, intensive care units, cancer treatment units, linear accelerators, and neonatal units. These hospitals have between 250 and 800 beds and have training programs for health sciences professionals.

Social Security Institute

The Social Security Institute (ISS) is a national institution, classified as an "Industrial and Marketing State Company" within the Labor Ministry. Its mission is to provide health, pension, and insurance services to workers. In the health sector, ISS has its own EPS and a network of IPS centers located in several cities. ISS has approximately 40 hospitals and a network of hundreds of small health facilities where affiliated workers and their families receive health services. For many years, ISS was the main social security provider in the country, but recently it has met with strong competition from private EPS and, as a consequence, has lost a significant percentage of higher income workers to EPS. Today, ISS has major problems that can be solved only by the strong intervention of the central government, which must make radical political and administrative decisions. ISS provides health services to a population estimated at between five and six million people and continues to be Colombia's largest health services provider.

Private sector

Law 100/93 brought significant change in the private health sector. Many health enterprises were established and compete with public hospitals to sell services to the different EPS. Table 21-2 shows Colombia's health organization.

Clinical Engineering Facts

The clinical engineer's roles and responsibilities have changed dramatically over the past several decades. In the 1940s and 1950s, hospitals were so elemental that engineers were unnecessary. Problems could be solved by people who had elemental knowledge of electromechanics. Equipment was so simple and of such excellent quality that it could be kept operational for years with only basic maintenance. In the 1960s, electronics arrived, and a fast evolution began. The universities in Colombia had only classical engineering programs such as civil, mechanical, industrial, and electrical. No one studied any subject relating to modern clinical engineering. Health institutions were developed with financing from central government, religious communities, local communities, and philanthropists.

In 1968, under the administration of President Carlos Lleras Restrepo, a national institution was established, called Fondo Nacional Hospitalario (FNH) (FNH, 1991). FNH was

Table 21-2 Colombia Health Organization

Institution	Quantity
Departmental Health Authorities	32
District Health Authorities	4
Local Health Authorities (Municipalities)	1086
Public Health Centers Plus Health Posts*	4705
Public IPS (Local Hospitals I Level)	560
Public IPS (Regional Hospitals II Level)	134
Public IPS (Specialized Hospitals III Level)	36
EPS	27
Private IPS **	2672
Private IPS ***	27328

Gestion Official Report, National Superintendency of Health, year 2000 (Gestion, 2000).
* Includes health centers with beds
** Includes private hospitals and clinics with hospitalization service
*** Includes diagnostic centers, consulting offices, blood banks, and other entities.

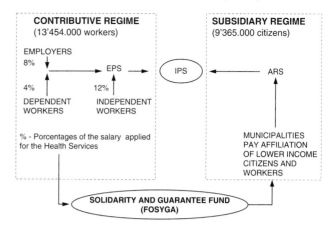

Figure 21-1 Colombian social security system.

charged with the responsibility of improving public hospitals, which were obsolete, inadequately equipped, and generally dysfunctional. This sad state was blamed on unwise investments and lack of proper planning. The contract made between the Health Ministry and the Social Security Colombian Institute (ICSS) establishing FNH defined the mission of managing the investments that were necessary to develop the public hospital network using economic resources from the central government and the ICSS. From 1970 to 1972, FNH began designing a national program targeted at 54 priority hospitals involving every aspect of planning, design, construction, equipment acquisition, and training programs for health professionals, technicians, and engineers. These hospitals, developed with modern ideas of infrastructure development, resulted in better health services for every citizen. To finance the program and to complete the construction and acquisition of equipment, the government increased taxes on beer sales by 8% and authorized international loans with banks and governments. The government also developed a methodology to define investment priorities.

From 1973 to 1992, FNH took out several international loans for new infrastructure development and new equipment acquisition in public and some private hospitals. The Engineering and Maintenance Division of the FNH developed training programs for the hospitals' technicians and engineers. Technicians and engineers from Colombia and throughout Latin America benefited from these programs. In the 1970s and the first half of the 1980s, training programs addressed only traditional engineering, none involving medical equipment. Not until the second half of the 1980s did a few universities and technical institutes begin training programs in electromedicine and bioengineering. During the years of FNH's existence more than 1500 technicians and 200 engineers received technical and administrative training. This training improved their skills and technical knowledge in hospital maintenance. FNH was closed in December 1993, as a consequence of the decentralization process that transferred the activities of the FNH to regional and local authorities. Since January of 1994, all processes relating to construction of new hospitals, equipment acquisition, and maintenance of the infrastructure and equipment have been decentralized and now are the responsibility of regional and local health authorities. In accordance with the priorities that the authorities define, hospital managers must plan, obtain, and expend resources on maintenance, acquisition of equipment, and human resources.

Maintenance Legislation

Law 100/93, which has resulted in better maintenance services in public and private hospitals, states in Article No. 189 the following:

> Public and private hospitals, in which the value of contracts with National or Territorial Entities is more than 30% of their total income, must assign at least 5% of their total annual budget to maintenance of the physical infrastructure and hospital equipment."

As a direct result of this article, every hospital must invest in maintenance every year to protect the physical infrastructure and equipment. Before 1993, investments depended only on the common sense of directors. Now, maintenance is an obligatory activity and is implemented based on strategies chosen by hospital administration.

In 1997, in accordance with Law 100/93, the National Superintendent of Health, as the governing authority of the General Social Security Systems, issued Internal Circular No. 29, which defined the need to develop in every hospital an annual maintenance plan, for which the authorities of each hospital (i.e., manager, chief of maintenance, and fiscal auditor) are responsible (NSH, 1997). The circular states, in part:

> Hospital maintenance is understood to be the technical and administrative activities required to prevent accidents and to repair and maintain the infrastructure and equipment and activities that improve the functionality of equipment (Article 70, Decree 1769 of 1994).

The circular defines the maintenance objectives as follows:

- To guarantee the security of patients and the personnel who administer and use the physical resources of the hospital
- To support the health service in meeting the quality objectives ordered by law
- To ensure availability and efficient functioning of physical resources required to deliver health services and to help reduce the institution's cost of operations

The annual maintenance plan is seen as the management tool that provides guidelines to the maintenance departments in every hospital. The plan must include the objectives; programming activities; and physical, human, technical, and economic resources to meet every hospital's objectives. It must include at least two chapters—one on physical infrastructure and the other on hospital equipment.

Physical Infrastructure

"Physical infrastructure" refers to the buildings; water supply and distribution systems; electrical distribution systems; communications; medical gases; and areas adjacent to the buildings, such as parking lots and grounds.

Hospital Equipment

Hospital equipment includes the following:

- Industrial equipment (e.g., electrical generators; boilers; water pumps; refrigeration and air conditioning systems,; elevators; laundry; and kitchen and other similar equipment
- Biomedical equipment
- Furniture
- Communication and information systems
- Equipment used for processing and duplicating information

"Biomedical equipment" is understood to be every machine or apparatus that uses one or a combination of electrical, electronic, or mechanical means for prevention, diagnosis, treatment, and rehabilitation in health services.

Every year before January 30, hospital managers and maintenance chiefs should send to the regional authorities their annual plan of maintenance, which should specify the priorities of the year, including their costs and action plan. The annual plan is an official document that must be signed by the manager, the maintenance chief, and the fiscal auditor. The authorities responsible for supervision will audit the plan to ensure adequate investment of resources. However, this process requires more control. Some hospitals do not obey the law because they either do not have an annual maintenance plan, do not make the investments detailed in the plan, or do not have adequate technical staff to manage the medical equipment technology in the hospital. The Health Ministry and the National Superintendent of Health must urgently attend to these lapses because the "Essential Requirements" refer directly to maintenance services and define actions that must be taken and information that must be available at any time. According to the essential requirements, every maintenance unit must do the following to comply:

- Have clear rules about the functioning of the service
- Plan the needs of spare parts and consumables
- Have the internal administrative forms to control the maintenance activities
- Plan and develop the annual plan of maintenance
- Have an inventory of spare parts and materials for urgent maintenance
- Plan the daily activities
- Supervise and explain the infection control policies of the institution
- Maintain an updated inventory of spare parts and tools required for service work
- Have a daily control of services performed
- Have maintenance records for each device in the hospital
- Have up-to-date blueprints of every network installed in the hospital
- Keep records of the down time of the equipment
- Make monthly evaluations of the preventive and corrective maintenance done
- Keep records of the response time to requests for maintenance services
- Control and evaluate the maintenance activities made through external contracts

Major Problems of the Maintenance Administration

Although the Essential Requirements define a list of responsibilities and administrative controls that must be available at any time, many hospitals do not comply with them. This reality needs to be addressed and corrected as soon as possible. Some of the most common problems are the following:

- In Colombia, clinical engineering is so new that many health professionals do not know what it is.
- In general, only in the tertiary level hospitals are the technical or maintenance units directed by an engineer or other professional. Only a few secondary level hospitals have engineers, with the majority of personnel having low or no technical profiles.
- The Health Ministry determines the salaries of every job in the national health organization. The maintenance staff salaries are so low, and the responsibility so high, that many engineers or technicians work only as outside consultants on specific projects. In many hospitals, the chief of maintenance does not fulfill the requirements for the job.
- Some managers do not understand the importance of the technical activities and do not pay enough attention to them.
- The technical or maintenance unit, when existing inside the hospital's organization, normally occupies one of the lower positions and, as an indirect consequence, the hospital manager does not rely on his maintenance personnel.
- A large chasm divides law from reality. Although the law adequate prescribes technical work, local or temporal circumstances govern inside the hospital. For example, developing a long-term planning process is not possible because the hospital managers want immediate results. These results must be obtained in the normal, three-year period of the manager's tenure in the hospital. For some managers, what could be done after he leaves the position is of no concern.
- In the public sector, it is almost impossible to efficiently manage the financial resources required for maintenance. Equipment should be repaired as soon as possible, but administrative and financial processes dictate different priorities.
- In many hospitals, the technical unit is considered only for maintenance activities; other important tasks, like acquisition of new technology, are done only by the hospital manager and administrative staff. Sadly, the engineer learns of this only when the new equipment arrives.
- In relation to maintenance human resources, the Essential Requirements refer only to the availability of electric or mechanical engineers in the maintenance unit, but availability does not mean that these professionals are required to be on the hospital staff. By contracting, many hospitals avoid the necessity of having their own maintenance unit and, hence, their own engineers. One commonly finds only carpenters or masons at work. Herein lies an explosive situation. Advanced, complex medical device technology demands the attention of a trained, skilled professional. One cannot create such a person in just a few hours.
- Organization inside hospitals normally considers technical activities to comprise only maintenance activities, and does not consider the other important technology management issues, such as the acquisition processes, the evaluation of the equipment's operational costs, and training programs.
- Several Colombian clinical engineers have recently begun to form a professional association, which could be a first step toward solving the problems mentioned above. The clinical engineering association should approach the Colombian Engineers Society, which is the official consultant to the central government, and through this entity, study, analyze, and propose strategies enabling the government to correctly actualize the legislation covering clinical engineering activities. This new association should interface with International Clinical Engineering Associations to raise the quality of its professionals to international standards. (See Ott and Dyro, 2003; Easty and Gentles, 2003; Hertz, 2003; Pallikarakis, 2003; Goodman, 2003; Nicoud and Kermit, 2003; and Grimes, 2003.)
- The National Superintendent of Health must exert better supervision and control to ensure that hospitals comply with every obligation in the norms relating to clinical engineering.

Maintenance Investments

Since 1994, the central government has made some investments through the Health Ministry to help the regional and local authorities to develop maintenance services. These include training programs and acquisitions of tools and instrumentation to update the technology required, in a first step program, to ensure quality control in maintenance services. A $50 million investment was made tools through an InterAmerican Development Bank (BID) loan. The project was made through a Program known as *Mejoramiento de los Servicios de Salud* (Improvement of Health Services) between 1996–1997.

Maintenance Human Resources

In the early 1990s, several Colombian clinical engineers attended several Advanced Clinical Engineering Workshops organized by the American College of Clinical Engineering with the support, in part, of the Pan American Health Organization and the World Health Organization (Dyro et al., 2003). During 1997 and 1998, through the program mentioned above, approximately 23 engineers who worked in the main and more advanced hospitals in the country had the opportunity to take a clinical engineering course designed and developed by The Andes University—one of the most recognized universities in the country–that enabled them to improve their skills and abilities in performing technical work. Other education institutions have begun training programs in the area. During the 1990's, this resulted in trained technicians who gradually were incorporated into hospitals and began to do more technical work than was done in the 1980's. Some of the universities that now offer training programs in bioengineering, electromedicine, clinical engineering, maintenance, acquisitions, supervision, and control of activities are the following:

It is important to know that the increased number of Colombian technicians and engineers who are working now in Colombian hospitals are products of normal training programs of universities and technical schools. Only recently have these new engineers and technicians had the opportunity to begin work in hospitals.

Physical and Functional Hospital Census

Between 1994 and 1998, the Health Ministry, using resources of BID, and through the Improvement for the Health Services Program, made a Physical and Functional Hospital Census that covered 170 public secondary and tertiary hospitals. The census updated all aspects relating to hospital architecture: Physical infrastructure; engineering systems (including water supply installations, mechanical and special installations, electrical and communication installations); and the medical equipment inventory. For each component, a group of architects and engineers designed the required methodologies to gather the information, and systems engineers developed the required software to process the information collected in the hospitals.

The most important results of the census were the following:

- **Architecture.** Every hospital covered has updated architectural blueprints and a detailed analysis of the architecture style and the ways in which the hospital connects with other health institutions and with the community it serves.
- **Physical Infrastructure.** A detailed evaluation of the physical infrastructure of each building from external areas to the interior spaces. All materials used in doors, windows, floors, and walls were classified and their condition, appearance, and maintenance evaluated. Each service and functional unit was counted and classified producing a unique, national classification with updated blueprints of the distributions of services and functional units inside the hospitals.
- **Engineering Installations.** For each type of installation, the Census provided up-to-date blueprints and detailed information of each system, including mechanical installations, electrical and communication networks, water supply systems, fire protection systems, and sanitary sewer systems.
- An **up-to-date equipment inventory** that covered all medium- or high-technology equipment.

Clinical Engineering in Colombia

Table 21-3 Universities offering training programs on technical subjects

Institute	Area	City
Los Andes	clinical engineering	Bogotá
Francisco José de Caldas (Distrital)	electronic with specialization in electromedicine	Bogotá
Antonio Nariño	biomedical engineering	Bogotá
Manuela Beltran	hospital engineering and biomedical engineering	Bogotá, Bucaramanga
Pascual Bravo	management acquisition processes	Medellin
Escuela Colombiana de Carreras Industriales	electromedicine and biomedical engineering	Bogotá, Cali
Sena (National Apprenticeship Service)	electromedicine, electricity, electronics, machine tools, welding, painting, refrigeration, air conditioning, boilers, out-of-board maintenance motors	Bogotá, Medellin, Manizales, Bucaramanga, Barranquilla, Cali
Sergio Arboleda	hospital architecture and maintenance	Bogotá
El Bosque	electronic with specialization in electromedicine	Bogotá

The information collected by the Census in databases files was delivery to the regional health authorities and to the managers of each hospital. This information enables the hospital staff to improve planning processes and to manage the investments that are required in order to offer health services of the highest quality. This information is also fundamental to maintenance services, and it is the responsibility of the technical team to keep it up to date.

As an example of the information collected, Figure 21-2 shows a national summary of the distribution of equipment of medium- and high-technical complexity by hospital medical and administrative units (Villamil, 1998a). The distribution of medical equipment by functional unit is shown, as well as the convention codes to identify each functional unit. Clinical laboratory and surgical units have the highest concentration of equipment. Such information enables technical departments to define priorities and to plan maintenance actions.

Figure 21-3 contains additional census information. The first graphic contains information about the technological nature of the equipment found; the second, information about the kinds of maintenance services employed; and the third, information on the functionality of the equipment (Villamil, 1998b). It can be seen that electronics is the prevailing technology, followed by electromechanical. This information is important for defining the profiles required by maintenance technicians and engineers working in hospitals. Also apparent is that corrective maintenance (external or in-house) is the most common alternative used in the country, while preventive actions lag far behind. This finding emphasizes the need to develop better preventive maintenance programs to increase the quantity of functioning equipment.

Medical Device and Equipment Technology Legislation

Recently, the Health Ministry signed the Official Resolution No. 00434 (2001/03/27) (MoH, 2001) that defines norms to evaluate and import biomedical technologies and

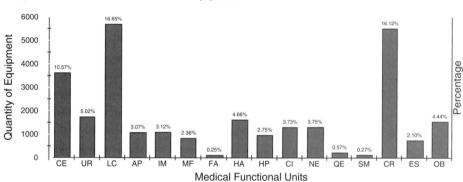

Republic of Colombia — Distribution of medical equipment

REPUBLIC OF COLOMBIA DISTRIBUTION OF EQUIPMENT MEDICAL AREA															
SERVICES	AMBULATORY		DIAGNOSTIC			THERAPEUTIC		HOSPITALIZATION						SURGICAL AND OBSTETRIC		
FUNCTIONAL UNITS	CE	UR	LC	AP	IM	MF	FA	HA	HP	CI	NE	QE	SM	CR	ES	OB
Quantity of Equipment	3592	1705	5659	1042	1062	802	86	1583	934	1268	1273	194	93	5479	713	1508
Percentage	0.1057	0.0502	0.1665	0.031	0.031	0.024	0.003	0.047	0.027	0.037	0.037	0.006	0.003	0.161	0.021	0.044

ADMINISTRATIVE AREA														
SERVICES	HUMAN RESOURCES		ADMINISTRATIVE	FINANCE			GENERAL				COMPLEMENTARY			
FUNCTIONAL UNITS	PE	BS	AU	PR	TP	CC	CO	LA	AG	MA	BI	CP	CA	DO	VM	AO
Quantity of Equipment	76	13	105	15	46	46	667	1245	1888	2581	0	1	9	103	119	84
Percentage	0.2236	0.0382	0.3089	0.044	0.135	0.135	1.962	3.663	5.554	7.593	0	0.003	0.026	0.303	0.35	0.247

Hospitals included: 167 (second and third level).
Total equipments: 33991.
April 1998,

Medical area, convention codes:

CE	Outpatient Consultation	IM	Imaging	HP	Pediatric Hospitalization	SM	Mental Hospitalization
UR	Emergency	MF	Physical Medicine	CI	Intensive care	CR	Surgery
LC	Clinical Laboratory	FA	Pharmacy	NE	Neonatology Hospitalization	ES	Sterilization
AP	Pathological Laboratory	HA	Adult Hospitalization	QE	Burnt Hospitalization	OB	Obstetrics

Administrative area, convention codes:

PE	Personnel	TP	Treasury	AG	General Store	CA	Cafeteria
BS	Social Well-being	CC	Accounting and costs	MA	Maintenance	DO	Classroom
AU	Attention to users	CO	Kitchen	BI	Library	VM	Medical home
PR	Budget	LA	Laundry	CP	Chapel	AO	Auditorium

Figure 21-2 Distribution of medical equipment.

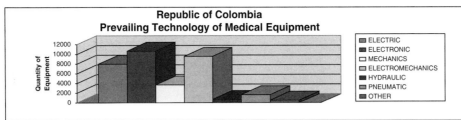

EQUIPMENT TECHNOLOGY	Total Equipment	Percentage
ELECTRIC	7987	22.73
ELECTRONIC	10,666	30.36
MECHANICS	3699	10.53
ELECTROMECHANICS	9577	27.26
HYDRAULIC	697	1.98
PNEUMATIC	1655	4.71
OTHER	523	1.49

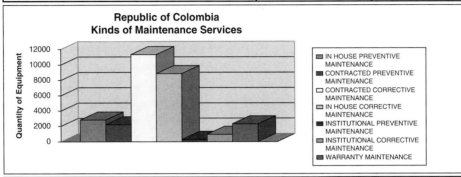

EQUIPMENT MAINTENANCE	Total Equipment	Percentage
IN-HOUSE PREVENTIVE MAINTENANCE	2818	8.02
CONTRACTED PREVENTIVE MAINTENANCE	2199	6.26
CONTRACTED CORRECTIVEMAINTENANCE	11,353	32.31
IN HOUSE CORRECTIVE MAINTENANCE	8872	25.25
INSTITUTIONAL PREVENTIVE MAINTENANCE	300	0.85
INSTITUTIONAL CORRECTIVE MAINTENANC	940	2.68
WARRANTY MAINTENANCE	2297	6.54

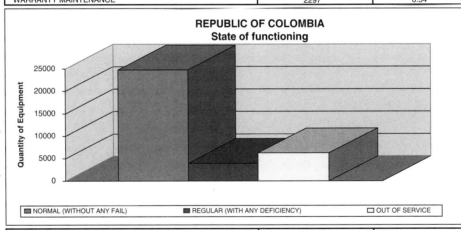

EQUIPMENT STATE OF OPERATION	Total Equipment	Percentage
NORMAL (WITHOUT ANY FAILURE)	24,759	70.46
REGULAR (WITH ANY DEFICIENCY)	3920	11.16
OUT OF SERVICE	6284	17.88

Colombia[10]

Brief Resume of the Country

Official name: Republic of Colombia
National holiday: July 20 (ndependence Day)
Government form: Unitary presidential republic; two legislative chambers
Actual constitution: July 4, 1991
Population: 40,000,000 estimated year 2000

Capital: Santa Fe de Bogotá
Geographic location: From 66° 50' 54" Greenwich West Longitude, to 70° 01' 23 Greenwich West Longitude; from 4° 13' 30" South Latitude to 12° 27' 46" North Latitude
Colombia is situated in the northwestern part of South America and bounded on the north by the Caribbean Sea (1600 km), on the east by Venezuela (2219 km) and Brazil (1645 km), on the south by Peru (1626 km) and Ecuador (586 km) and on the west by Panama (266 km) and the Pacific Ocean (1300 km). Colombia is the only country in South America with coasts on both the Caribbean Sea and the Pacific Ocean.

Total Area: 2,070.408 km
Terrestrial Area: 1,141,748 km
Marine Area: 928,660 km
Political division: 32 departments and a district capital (Santa Fe de Bogotá)
Largest cities:
  Santa Fe de Bogotá 6,000,000
  Cali 2,500,000
  Medellín 2,,500,000
  Barranquilla 2,500,000
  Bucaramanga 2,200,000
Ethnic groups:
  Mestizo: 58% mixed European and Native American ancestry
  White: 20%

Mulatto:	14% mixed white and black ancestry	
Black:	4%	
Mixed black and Native American: 3%		
Native AmericanL 1%		

Urban/rural breakdown
Urban: 73%
Rural: 27%
Official language: Castilian (Spanish);
Other languages: Native American languages (e.g., Arawak, Karib, Chibcha, and Tukano)
Religions: Roman Catholicism (95%); others (5%), including Protestantism and Judaism
Economy: gross domestic product (1998) U.S. $102,896 million

- **Agriculture:** Coffee, sugar cane, bananas, tobacco, cotton, cut flowers, cacao beans, potatoes, cassava, and plantains
- **Mining:** Petroleum, natural gas, coal, nickel, gold, and emeralds
- **Manufacturing:** Food products, textiles, beverages, transportation equipment, and chemical products

- **Employment breakdown:**
  Services 46%
  Agriculture, forestry, and fishing 30%
  Industry 24%
- **Major imports:** Food, machinery and transportation equipment, chemicals, minerals, metals, computers, and electronic and communication equipment'
- **Major exports:** Coffee, agricultural products (especially flowers and bananas), petroleum, coal, gold emeralds, chemicals, and textiles
- **Major trading partners:** United States, Venezuela, Germany, Japan, Netherlands, Brazil, and Peru
- **Currency:** Colombian peso (U.S.$1 = 2350 Colombian pesos -2001)
- **Education:**

Major universities:
National University (Bogotá, Medellin, Manizales, San Andres, Palmira)
University of Antioquia (Medellin);
University of El Valle (Cali)
University of Los Andes (Bogotá); University Javeriana (Bogotá)
University of El Rosario (Bogotá); University of Cartagena (Cartagena)
Eafit (Medellin)

Figure 21-3 Prevailing technologies, kinds of maintenance services, and equipment functioning in Colombia.
Note: The Census detected that there are a total of 10,580 devices that do not receive maintenance services of any kind, and that quantity is 30.11% of the equipment evaluated. For other types of equipment it was not clear if they received any maintenance.

defines the technologies that the authorities must control. The resolution establishes the methodologies and evaluation procedures for technical and economic analysis and methodologies to allocate the technology throughout the country. It also defines the criteria for technology acquisition, importation, and incorporation into the health services, enhancing quality and cost-effectiveness.

The resolution covers biomedical equipment and devices, surgical instrumentation, surgical procedures, drugs, and information systems used in the health services. For the first time, a classification of medical equipment by risk, physical state, and use has been established. The risk classification is based on international risk classifications (high-risk, classes IIB and III; medium-risk, class IIA; and low-risk, class I). The physical state classifies the equipment as like-new (i.e., not used or less than a year from the date of manufacture); old (i.e., less than four years from the date of manufacture); and prototype (i.e., equipment or devices in an experimental stage of development, without authorization to be used to provide health services). Moreover, the resolution defines the process of technological evaluation and finally defines types of controlled technologies, which require a special process to be imported into the country.

References

CoC. Congress of Colombia. Social Security Law No. 100. Chapter I, Article 2. 1993.
Correa CJ. Un Recorrido de Infarto. *El Tiempo* 2:15, 2001.
Dyro JF, Judd TM, Wear JO. Advanced Clinical Engineering Workshops. In Dyro JF (ed). *The Handbook of Clinical Engineering.* Burlington, MA, Elsevier, 2003.
Easty T and Gentles W. Standards of Practice. In Dyro JF (ed). *The Handbook of Clinical Engineering,* Burlington, MA, Elsevier, 2003.
Fondo Nacional Hospitalario. Una Respuesta de Fondo. 23 Años de Gestión Institucional. Fondo Nacional Hospitalario. 1:24, 1991.
National Superintendent of Health, Colombia. *Gestion Official Report, Year 2000:* 7, 2000.
Goodman G. Hospital Facilities Safety Standards. In Dyro JF (ed). *The Handbook of Clinical Engineering,* Burlington, MA, Elsevier, 2003.
Grimes SL. Clinical Engineering Future. In Dyro JF (ed). *The Handbook of Clinical Engineering,* Burlington, MA, Elsevier, 2003.
Hertz E. Recommended Practice for a Medical Equipment Management program. In Dyro JF (ed). *The Handbook of Clinical Engineering,* Burlington, MA, Elsevier, 2003.
MoH. Official Resolution No. 4252. Colombia Ministry of Health. 1997.
MoH. Official Resolution No. 0434. Colombia Ministry of Health. 2001.
Nicoud T, Kermit E. Certification. In Dyro JF (ed). *The Handbook of Clinical Engineering.* Burlington, MA, Elsevier, 2003.
NSH. Circular 29. National Superintendent of Health. 1997.
Ott J, Dyro JF. The American College of Clinical Engineering. In Dyro JF (ed). *The Handbook of Clinical Engineering.* Burlington, MA, Elsevier, 2003.
Pallikarakis N. European Union Directives. In Dyro JF (ed). *The Handbook of Clinical Engineering.* Burlington, MA, Elsevier, 2003.
Santos A, Muñoz A, et al. Casa Editorial El Tiempo. *Colombia Viva.* 2000.
Villamil J. Improvement of the Health Services Program. Official Report of Medium and High Technology Equipment. 30-33, 1998.
Villamil J. Improvement of the Health Services Program. *Official Report of Medium and High Technology Equipment.* 15-25, 1998.

22

Clinical Engineering in Ecuador

Juan Gomez
Director, Hospital Maintenance Division
Ministry of Health
Quito, Ecuador

The Republic of Ecuador, located in the "belt" of Latin America, has an area of 275,830 square kilometers and a population of 12,174,628 (1998 census).

The Ministry of Public Health is the steering agency for health services and manages the largest number of hospital beds in the country, as shown in by level of complexity in Table 22-1 and by number of beds in Table 22-2.

The approximate cost of the equipment for that investment (approximately US$50,000/bed) is US$4.35 billion, while $304.5 million (7%) is required for its annual maintenance. The Ministry of Health's 2001 budget for hospital maintenance is $3,212,622, i.e., one-hundredth the amount needed to keep this technology infrastructure in optimal condition and working efficiently, not including physical infrastructure and facilities.

Background

With the oil boom in the 1970s, the country's physical hospital infrastructure began to be broadly expanded and modernized. In the Ministry of Health, the Department of Hospital Maintenance was created and later transferred to a state agency, the Ecuadorian Institute of Sanitary Works (IEOS), which began providing technical assistance to health care facilities and an ongoing training program for operating and maintenance staff at those facilities nationwide. Technical cooperation agreements were established with PAHO/WHO and friendly governments to train technical maintenance staff.

The 1980s were marked by a national economic crisis that contracted financial resources for maintenance, which led to the rapid deterioration of the health services technology infrastructure. This situation was aggravated by the acquisition of new medical and industrial equipment with technologies that demanded greater financial, human, and administrative resources.

In the 1990s, the country instituted a health sector reform process involving decentralization and modernization of health care facilities. As a result, the IEOS disappeared, and its technical assistance, training programs, and maintenance activities were transferred to the hospitals. Accordingly, in the new millennium, maintenance of the technology infrastructure, particularly infrastructure related to knowledge and the application of clinical engineering, must be developed and adapted to the new role brought on by these changes.

Technology Infrastructure Maintenance Status

In order to establish criteria with respect to the new role that maintenance activities should play in the ministry, it is important to analyze the following related elements:

- Organization
- Medical/industrial equipment
- Human resources
- Budget
- Information

Organization

With the advent of health sector reform, maintenance programs have not been properly defined and identified; therefore, there is no duly structured national maintenance policy. The organizational development of local maintenance systems has been weak; this is further complicated by the presence and activities of unions, as well as low wages, which affect normal operations at the local level. The rise in the market for used and refurbished equipment, as well as greater donations of equipment, is leading to serious operating and maintenance problems, because of the lack of manuals and parts and the absence of regulations and standards controlling the quality and efficiency of these devices. Private-sector involvement in equipment maintenance is inadequate, providing only weak technical service support, due to existing economic constraints and the lack of adequate execution and control mechanisms for such interventions. The lack of coordination among institutions in the health sector often leads to a duplication of equipment in a single city or region. Some of this equipment is quite expensive, resulting in a waste of the limited national funds available for hospital equipment.

Medical/Industrial Equipment

The mix of brands, models, and origins of existing equipment is an important factor in the inefficient delivery of maintenance services. Approximately 50% of the equipment is out of service, boxed up, or operating poorly. Much of the equipment was manufactured 15-25 years ago and therefore has fulfilled its service life. Physical plant and utility systems have an average useful life of 40 years, with a similar percentage of deterioration. The ongoing adoption of the latest technologies in the area of medical devices places greater demands on medical, operating, and maintenance staff and new demands on sites and facilities. It also generates high operating and maintenance costs. Waste and inefficient use of energy resources such as water, electricity, steam, and medical gases leads to high consumption costs and hence, high operating costs. The lack of hospital safety programs directed at all personnel in medical units often causes problems related to the proper use of equipment and facilities, which affects the life of the equipment and the physical integrity of the people who work with it.

Human Resources

There is a shortage of professional and technical maintenance staff. Approximately 76% of staff have received only on-the-job training; 19% of the staff are technical; and 5% professional. The state-of-the-art technology of the new equipment has not been updated because of the limited professional staff, and there is a lack of human resources in these new technical areas, as illustrated in Figure 22-1.

Since the dissolution of IEOS, there has been no ongoing training program geared both towards equipment operators and maintenance staff, and ministry officials very rarely conduct training courses for such personnel. Universities and polytechnic schools are not involved in training these resources. Mid- and lower-level technical personnel have strong commitments to existing unions in hospital units; this often leads to weak collaboration on attempts to establish maintenance programs in the units. The structural adjustment demanded by the reform process will have a considerable impact on existing human resources for maintenance in each hospital.

Budget

Low percentages of the budget are allocated to maintaining each unit and replacing equipment. Cost analyses are not used as a maintenance management tool. The lack of technical and economic inventories of hospital equipment in each unit makes it impossible to determine real budgetary needs for maintenance and equipment replacement. The budget allotments for maintenance and equipment replacement have historically been based on the number of beds, not real needs. Nonreimbursable loans can be used as a source of financing.

Information

In most hospitals, economic and technical equipment inventories are insufficient. The little information available on equipment is not utilized by management or included in health information systems. There is limited access to the scientific and technical information already available or that could be obtained inside and outside the hospital. Computer systems in each hospital are not sophisticated, and the vast majority of hospitals do not have access to current technologies, such as information networks, e-mail, or the Internet.

New Role

In the new framework of health sector reform, and in the process of acquiring and incorporating state-of-the-art technologies into the health services, the role of engineering,

Table 22-1 Hospitals in Ecuador by level of complexity

Hospitals	No. Of Units	%	No. Of Beds	%
Specialty	2	1.6	700	8.0
Provincial	22	17.5	3,422	39.1
Specialized	13	10.3	2,031	54.2
Cantonal	89	70.6	2,593	29.6
Total	126	100.0	8,746	100.0

Table 22-2 Hospitals in Ecuador by number of beds

Hospitals	Number Of Units	%	Number Of Beds	%
a) More than 200 beds	18	14.3	4,882	55.8
b) More than 100 and fewer than 200	12	9.5	1,531	17.5
c) More than 50 and fewer than 100	13	10.3	840	9.6
d) Fewer than 50 beds	83	65.9	1,493	17.1
Total	126	100.0	8,746	100.0

maintenance, technology management, and equipment regulation programs must be redefined by the following means:

- Strengthening the steering capacity of the Ministry of Health in the area of maintenance
- Organizing and providing training courses, workshops, seminars, and internships that make it possible to train technical equipment maintenance and operating staff
- Developing a new concept of maintenance engineering at both the central and local levels

Factors Involved in This Redefinition

The main factors that will affect the proposed redefinition are the following:

- The deterioration in the physical and technology infrastructure
- New health care delivery modalities
- Changes in the epidemiological profile of the population
- Lack of regulatory programs for medical devices
- The cumulative deficit in maintenance services
- Weak selection processes and the acquisition of technology
- Lack of maintenance service options
- Lack of formal and ongoing training for maintenance staff
- No development of new professional programs
- Development of information technology and communications
- Efficiency, safety, and environmental impact of health care facilities

Purpose

Efforts must be directed toward having a physical and technological health services infrastructure that makes a positive contribution to quality assurance; risk management; and health care facilities accreditation.

General Objectives

The general objectives of health care system restructuring is to develop, operate, maintain, and replace the physical and technology infrastructure in health facilities, to guarantee the delivery of services with equity, quality, efficiency, and safety, in order to protect the investment in this infrastructure.

Specific Objectives

Specific objectives are summarized in the following:

- Undertake an inventory of hospital equipment (technical and economic inventory)
- Strengthen engineering, maintenance, technology management, and equipment regulation policies and programs
- Train operating and maintenance personnel in accordance with the technologies incorporated
- Maximize the use of resources allocated to health facility operations and maintenance
- Improve efficiency and safety in operating the equipment, physical plant, and facilities
- Search for alternatives and mechanisms for delivering maintenance services
- Identify project financing mechanisms

Of these objectives, the following warrant special attention:

- Conducting the equipment inventory to provide a baseline for future interventions for repairs and/or scheduled equipment maintenance and replacement
- Intervention in the different stages of technological equipment management: Planning, procurement, management, evaluation, discarding, and replacement
- Regulating medical equipment and devices and their main activities related to: Registry of entry into the national market; after-sale surveillance; control and inspections of manufacturers and importers; safety, efficacy, and operating evaluation; and standardization of regulatory registries
- Establishment of ongoing training programs

Basic Aspects of Maintenance Engineering

In the new role that Ministry of Health maintenance departments should play at both the central and the local levels, the basic activities that must be accorded priority are as follows:

- Maintenance management: Organizing the departments, employing business criteria to help achieve the basic objective of hospital modernization
- Training: Developing ongoing training programs on the use and maintenance of the technology infrastructure
- Safety and biosafety: Establishing, conducting, and monitoring hospital safety programs that provide for the safe, reliable use of equipment and facilities
- Energy efficiency: training all hospital personnel to properly use energy inputs (e.g., electricity, steam, water, and medical gases) in order to save on operating costs
- Environmental sanitation: Establishing control programs for the unit's liquid and solid waste
- Vulnerability and disasters: Organizing and testing emergency evacuation plans and developing studies on the units' vulnerability to disasters

It should be recalled that the maintenance department is part of a larger engineering process whose purpose is to ensure that the hospital has a continuous, safe supply of equipment that runs efficiently and well, at low operating costs, as illustrated in Figure 22-1.

Activities

The activities required to achieve the proposed objectives are organization, training, information, efficiency and safety, research, and mobilization of resources.

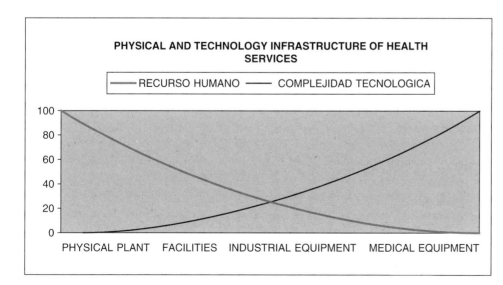

Figure 22-1 Physical and technology infrastructure of health services: human resources and complex medical device technologies.

Organization

- Preparation of national maintenance policies
- Program to regulate medical devices
- Systems for technological management of equipment
- Service maintenance engineering systems
- Programs for preventive maintenance at the local level
- Preparation of codes, standards, and guidelines

Training

- Organization of ongoing training programs
- Internships in technical centers and programs
- Support for graduate students in clinical and hospital engineering
- Technical cooperation with PAHO/WHO and countries in the Americas
- Participation in international congresses and events
- Holding of advanced workshops on clinical engineering

Information

- Technical, economic inventory of medical and industrial equipment
- Organization of management information systems
- Production, compilation, and distribution of technical materials
- Strengthening technical documentation centers
- Links to scientific associations

Efficiency and Safety

- Maintenance management schedule
- Energy resource administration program
- Safety and vulnerability programs in the hospital
- Sanitation program

Summary

Because of new trends toward the modernization of states—particularly in Latin America—that seek to decentralize activities relating to programming, developing, and preserving the physical and technology infrastructure of hospital units in the countries, the individuals who are responsible for preserving such infrastructure, particularly medical equipment, are taking on a new role in their activities.

A technical/economic survey of the current status of hospital equipment, both medical and industrial, must be conducted (equipment inventory) to determine the operating condition of every piece of equipment in each health facility and to schedule the resulting rehabilitation, preventive maintenance, and repair activities; there also must be ongoing training for operators and maintenance staff, using facilities in institutes of higher education and cooperation from international organizations, such as PAHO/WHO, and friendly governments.

National policies on maintaining and resupplying these units must be established, together with programs to regulate equipment and donations, technology management systems for them, and maintenance programs that employ a management perspective, similar to those found in the private sector. The execution of energy-resource administration, hospital safety and vulnerability, and basic sanitation programs plays an important role in this process.

The use of information technology is critical to this management style and is also a source of information for producing, compiling, and distributing technical materials. Research and mobilization of resources must be directed to develop. New maintenance service models, promote operational research, and pursue resources and sources of financing.

When properly managed, all of these elements will make it possible to achieve the general objective of developing, maintaining, and renewing the physical and technology infrastructure of health facilities to ensure equity, quality, efficiency, and safety in service delivery in order to protect the investment in infrastructure.

23

Clinical Engineering in Mexico

Adriana Velásquez
Clinical Engineering Consultant
Mexico City, Mexico

Clinical engineering has developed extensively and rapidly in Mexico over the past two decades. This chapter contains information about the country of Mexico and its health care infrastructure, biomedical engineering education, and clinical engineering activity. The results of a recent survey regarding the practices of hospital based clinical engineering departments are presented. Mexico has a board of examiners for the certification of clinical engineers and a Mexican biomedical engineering society. Finally, the work regarding medical device regulation in Mexico is described, and the challenges that clinical engineers have to face are summarized.

Mexican Health Care System

Mexico is a country of 100 million inhabitants and an area of slightly more than 2 million square kilometers. Unfortunately, almost 40% of the population lives in extreme poverty, although there is an important financial and industrial sector, which supports a middle and upper class. This disparity is reflected in the health statistics (Figure 23-1) that show the causes of mortality to be a mixture of first- and third-world illnesses. Infectious diseases and chronic and degenerative diseases are the major causes of death. Resources need to be allocated to address both ends of the health care spectrum.

Regarding morbidity, the five most important causes are acute respiratory infections, intestinal infections (e.g., diarrhea and amibiasis), diabetes mellitus, high blood pressure, urinary tract diseases, and nutritional deficiencies.

Hospital Infrastructure

Concerning infrastructure (Figure 23-2), Mexico has 21,285 health service units in both the private and public sector. Public sector facilities include 17,496 outpatient care facilities, from which just 171 are tertiary or specialized-care hospitals, and 705 are secondary-care general hospitals. In the private sector, 2841 hospitals have fewer than 40 beds, and 72 hospitals—23 of which are in Mexico City—have more than 50 beds.

The Mexican Social Security Institute (IMSS) provides health services to its 40 million participants. The Social Security and Services Institute for Government Employees (ISSSTE) serves 10 million people. The public sector health state services, Health Ministry, Secretaria de Salud (SSA), provides services to another 40 million. About 10 million people are served by the private sector, such as PEMEX, and state services, such as the army and the navy (Figures 23-3 and 23-4).

Technology Problems in Mexico

Almost 90% of all medical device technology is imported. As a consequence of the lack of national production, some of this imported technology is not as efficient as it would be if it were produced in the native country. This increases the costs of health care delivery.

As it is elsewhere in the world, medical technology (i.e., devices, procedures, and drugs) in Mexico is advancing rapidly and constantly. The absorption of this technology by the health care system places an increased burden upon the users, nurses, physicians, and technicians to receive training and to become specialized.

Unfortunately, financial limitations prevent the purchase of all of the equipment, medical devices, and consumables that are needed for all health care units. However, in many hospitals, equipment is underused or is so antiquated that it no longer can be used safely.

Medical equipment is not equitably distributed across the country. The Y2K program showed that 45 % of all electronic medical equipment is in the capital city, Mexico City. Thus, much of Mexico's population does not have access to high technology services as it does not have the means to travel long distances to the city.

Most hospital administrators or medical directors buy state-of-the-art equipment. However, this equipment is not necessarily the most appropriate to meet the needs of each

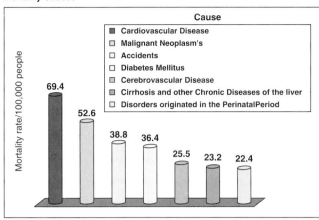

Figure 23-1 Mortality causes in Mexico, first- and third-world-country diseases.

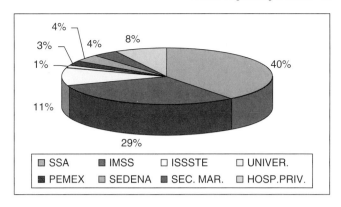

Figure 23-3 Hospital infrastructure in Mexico: Public (SSA 40%; IMSS 29%; ISSSTE 11%; army and navy (SEDENA), SEC MAR 4%, university 1%; and PEMEX 3%) and private 8%.

Figure 23-2 Hospital distribution in three levels of complexity in Mexico.

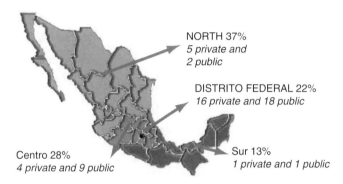

Figure 23-4 Health service units distribution in Mexico and the number of clinical engineering departments in each area.

of the health care units. Usually there are not enough financial resources to purchase maintenance or to acquire equipment consumables. In only a few institutions is the technology evaluated extensively before it is purchased.

Biomedical Engineering Education in Mexico

In 1973, biomedical engineering formally started developing in Mexico with a bachelor's degree program at Universidad Iberoamericana. In 1974, another program started at Universidad Autonoma Metropolitana Unidad Iztapalapa. Today, there are four universities offering bacherlor's degrees and two offering master's and/or doctoral degrees in Biomedical Engineering. One program that started in the 1960s is dedicated to bioelectronics and research. The four universities are in Mexico City, but now there are three electronic engineering bachelor's degree programs, which have specialty courses in medical electronics or medical instrumentation, in Guadalajara, Morelia, Tijuana, and Puebla.

These programs have produced more than 1000 biomedical engineers who are now working in the following areas:

- In 65 public and private sector hospitals, tending to maintenance, acquisition, training, and use of medical equipment.
- At more than 20 companies, as vendors or distributors of medical technology, providing more professional and informed sales and ongoing support
- In research at more than 10 universities, or in the health sector developing new technologies to better serve the sector
- In information technologies in other sectors

In 2002, the universities offering biomedical engineering and clinical engineering were the following:

- Universidad Iberoamericana, founded in 1972
  http://www.uia.mx/ibero/prog/carreras/biomedica/
  Clinical engineering
  Medical electronic instrumentation
  Rehabilitation engineering
  A postgraduate program is planned
- Universidad Autonoma Metropolitana, founded in 1974
  Has a postgraduate program for M.Sc. and PhD.
  http://www.uam.mx/licenciaturas/pdfs/22_5.pdf
  http://www.iztapalapa.uam.mx
  htttp://www.infocbi.uam.mx/planbiom.html
  http://itzamna.uam.mx/posgrado/
  Digital signal processing
  Medical instrumentation
- Universidad Profesional Interdisciplinaria de Biotecnología,
  http://www.upibi.ipn.mx/page40.html, founded in 1987
- Universidad Nacional Autonoma de Mexico, which started its postgraduate program in 2001
  http://inbio.fi-p.unam.mx

Biomedical Engineers in Hospitals

In 1978, the first clinical engineering programs in hospitals started. By 1984, four private and five public hospitals had opened clinical or biomedical engineering departments. Today, there are 65 clinical engineering departments working in secondary and tertiary care hospitals; i.e., general care hospitals and highly specialized care hospitals. These cover barely 10% of the need. Where there are no biomedical or clinical engineers, mechanical or electrical engineers and technicians from plant engineering service the medical equipment. These individuals do not have the proper training to manage these health care technology resources effectively, however.

National inventories of medical equipment technology were taken as part of the country's response to the Y2K problem (see Chapter 105). In the course of the inventory, the following were revealed:

- Medical equipment not working properly because of lack of parts or consumables
- Equipment that had not yet been installed because of lack of coordination among maintenance, purchasing, end user, and vendor
- Equipment that was not needed or used
- Lack of medical equipment management programs, which could result in better use of currently available technology in public and private hospitals

A proposal was made for biomedical engineers at regional levels to help with technology management by performing evaluations and general problem solving. Unfortunately, this has not happened yet at a national level.

The results of a Mexican clinical engineering survey made in 2002 show that the 65 clinical engineering departments in Mexico are located primarily in private hospitals with more than 50 beds and at the Health Ministry (Secretaria de Salud) hospitals.

The education of biomedical or clinical engineering department heads is as follows:

- 91% bachelors in biomedical engineering
- 9% bachelors in other engineering fields such as electronics.

Of those surveyed, 9% have a master's degree and 20% have a management diploma.

Within the hospital organization, biomedical and clinical engineering departments report in the following way (Figure 23-5):

- 15% report to the medical director
- 70% report to the management director or hospital administrator
- 15% report to other departments such as maintenance or research

In several hospitals, the clinical engineer coordinates all technological areas and serves as either technology officer, director, or assistant director in charge of the following:

- Biomedical or clinical engineering departments
- Plant engineering (the machine room and all systems including electrical, hydraulic, and pneumatic)
- Electronics, audiovisual, and communications (e.g., networks, phones, fax, copy machines, and audiovisual equipment)
- Systems engineering (clinical and administrative computers)
- In-house maintenance (painting, furniture, signage)
- Laundry and pharmacy
- Special projects like new construction or technology assessment

Monthly salaries are displayed in Figure 23-6. As is evident, no standard salary for biomedical and clinical engineers exists. The low salaries are inappropriate for jobs that entail high responsibilities. Advocacy for higher salaries is a priority. It is suggested that one avenue for salary increases is to tie increased productivity to increases.

The work done in typical clinical or biomedical engineering departments involves the whole cycle of technology:

- Establishing the need for a new technology
- Planning and research on new products and the status of those already in use
- Pre purchase evaluation
- Participating in the acquisition committee (sometimes merely as observers)
- Transferring devices to the hospital location where they will be used
- Verifying the installation
- Coordinating training for technicians, doctors, or nurses, on different shifts, according to the need
- Calling the vendor for service during the warranty period and performing preventive or corrective maintenance with or without the presence of the vendor
- Reviewing safety issues
- Evaluating the performance, productivity, and quality of service rendered
- Replacing planning for old equipment
- Providing consultation on problems and operation, if involved with technology design, development, production, and sale
- Selecting appropriate technology and vendors

The survey data show that clear distinctions among the work of the CE supervisor, CE, other engineers, and BMETs do not exist (see Table 23-1). In some of the hospitals, the work of the BMET job is done by clinical, electronic, or mechanical engineers or students because there is a lack of BMETs.

A survey was made of hospitals with and without clinical engineering departments (CED). It was found that where there is no CED, there is a considerable amount of equipment that is left without service, unless it is urgent or very important to the institution. In public hospitals, the percentages of medical devices at three levels of complexity were as follows: High complexity, 7%; medium complexity, 36%, and low complexity, 57%.

Medical Device Regulation in Mexico

Mexico has five biomedical engineers at the state level making decisions involving medical equipment: one in the federal unit for regulation of medical equipment; at least 20 at the 10 National Institutes of Health; one making medical equipment purchasing decisions for IMSS (which has 40 million participants); two at the Health Ministry in charge of medical equipment central purchases; and four in the Technology Assessment and Management Unit at the Health Care Ministry advising on the e-Health National Program and the cost-effectiveness of certain appropriate medical equipment for special populations. This unit has programs to promote better use of medical technology and a program to regulate donations.

Two biomedical engineers work in the Mexican regulatory agency, COFEPRIS. This federal commission focuses on sanitation and is responsible for the following:

- Sanitation risks evaluation
- Issuance of regulations, authorizations, registrations, permits, and licenses
- Verification of compliance with regulations
- Law enforcement
- Decentralization and wide coverage
- International affairs
- Management and support systems
- National Public Health Laboratory's Net

Clinical Engineering Certification Program

The Mexican Clinical Engineering Group was established in July 1991. This followed the certification of the first Mexican clinical engineer by the International Certification Commission for Clinical Engineering and Biomedical Engineering Technology (ICC) and the first Advanced Clinical Engineering Workshop (ACEW) in Washington, D.C. (see Chapter 70). The Mexican Clinical Engineering group became the clinical engineering chapter of the Mexican Society of Biomedical Engineering in October 1993.

With the support of the ICC and the U.S. Board of Examiners (see Chapter 133), the first clinical engineering certification exam took place in Mexico. Three more Mexicans became certified, and three students passed the first part of the exam. In July 1994, the ICC approved the Mexican Board of Examiners and accepted the affiliation of the Mexican Certification Commission (MCC). The MCC administers the exam in Spanish.

Clinical engineering certification is not a valid license to practice clinical engineering. According to the laws and rules regarding the education of individuals, Mexico does not recognize it as a legal registration. Of primary importance is that certification demonstrates professional achievement and that it is highly regarded by peers in the profession. The clinical engineering chapter of the Mexican Society of Biomedical Engineering provides information regarding certification. The exam has three parts: A general knowledge exam, a written essay exam, and an interview by three proctors. More information is available at the address below:

SOCIEDAD MEXICANA DE ING. BIOMÉDICA
Atención: Ing. Claudia Cárdenas
Querétaro 210 2° piso, Col Roma. México, 06700 D.F. MEXICO
Tel: (52 5) 574 45 05
Fax: (525 55 574 39 28) de México, D.F.
E-mail: ingclinica@somib.org.mx

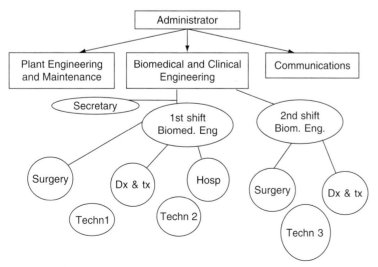

Figure 23-5 Typical organizational structure in Mexican hospitals.

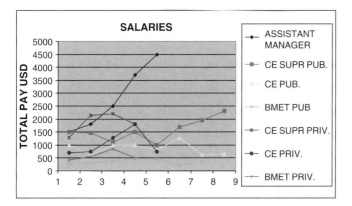

Figure 23-6 Salaries for BMETs, clinical engineers, clinical engineers' supervisors, and biomedical engineers working as assistant managers in private and public Mexican hospitals (July 2002).

Table 23-1 Job responsibilities of clinical engineering supervisors, clinical engineers, BMETs, and others (percentage of respondents indicating responsibilities in listed tasks)

Tasks	Job responsibilities (%)			
	CE Supervisor	Clinical Engineer	BMET	Other
Preventive maintenance	21	33	80	15
Repairs	20	32	40	34
Design & Modifications	33	11	32	
Coordination & Supervision of Outside Services	24	21	10	23
Purchase Evaluation	29	22	27	
User Training	33	18	70	42
Productivity Evaluation	20	28	10	60
Research Protocols	6	2	10	10
Service Administration	35	60		
Outgoing Equipment Evaluation	39	5		
Teaching	10	6		

Mexican Society of Biomedical Engineering

Since 1991 an active group of clinical engineers has been meeting monthly. This group, the Clinical Engineering Club, was later accepted as the Clinical Engineering Chapter of the Mexican Biomedical Engineering Society (SOMIB). SOMIB was formed in 1976 by the first teachers of biomedical engineering. SOMIB now has 450 members, and for 25 years it has presented annual conferences and has published a journal twice per year. Its web site (www.somib.org.mx,ingclinica.somib.org.mx) offers information on events, academic issues, and job advertisements. It has a virtual library, national and international biomedical engineering links, and clinical engineering information. The Biomedical Engineering Society participates in the development of federal regulations and standards.

SOMIB has the following ten chapters:

- Diffusion (e.g., web pages)
- Industry, vendors, and distributors
- Clinical engineering
- Academic
- Instrumentation
- International
- Four provinces

SOMIB is a member of several international organizations, including the International Federation of Medical and Biological Engineering (IFMBE) and the International Union of Physical Sciences and Medicine (IUPSM). It is also a member of the Regional Council of Biomedical Engineering in Latin America (CORAL). SOMIB also has memberships in national federations and is a partner of other societies such as the Mexican Hospital Association (AMH) and the Mexican Federation of Scientific Societies (FESOCIME).

Clinical Engineering Challenges in Mexico

Although Mexico started biomedical engineering programs almost 30 years ago, much remains to be done. Some of the priorities are the following:

- To strive for a more financial and academic recognition for technicians and engineers
- To increase the participation of clinical engineers in social security institutions and in the health secretariats of the states
- To have a larger participation in the regulation of medical devices to ensure safe and effective devices commercialized in Mexico
- To promulgate ideas for the improved use of resources
- To demonstrate the importance of selecting appropriate technologies according to specific needs, the cost-benefit ratio, increase technical quality
- To obtain institutional collaboration for the better use of resources, to increase accessibility, equity, quality and efficiency in the Mexican health services to provide better services to the population

Acknowledgment

The author thanks the following colleagues for their assistance in preparing this chapter Ametia Patiño, Laura López, Sandra Rocha, Beatriz Hernandez, and René Rodríguez.

24

Clinical Engineering in Paraguay

Pedro Galvan
Departamento de Mantenimiento, Instituto de Investigaciones en ciencias de la Salud, Asuncion
Paraguay

Health care technology development plays an essential role today in promoting health and developing health systems and services. Traditionally, in developing countries, nonfunctional equipment and facilities impede the adequate delivery of appropriate health services. Furthermore, a proper equipment management system, which includes health care technology policy, sufficient financial resources, adequate human resources, and a comprehensive maintenance system, is crucial to guaranteeing the development of a sustainable health service.

This chapter provides an overview of the clinical engineering practice in Paraguay and its impact on the care health care technology (HCT) management system for the public health sector. A feasible and sustainable HCT development model is included in the management plan, which is based on a diagnosis of the health care infrastructure that was made few years ago. The HCT management plan focuses on issues such as HCT policy, financial resources, human resource development, and maintenance system to promote a sustainable clinical engineering practice in the country in the framework of an adequate health service delivery policy. Together, a well-structured and organized HCT management with the experiences made in this field can help to establish a comprehensive clinical engineering practice in health care systems.

Introduction

One of the results of globalization and efforts to improve health care systems worldwide has been the recognition that increasingly complex health care technology plays a vitally important role in all health systems.

In the past decade, the majority of countries in the Americas, including Paraguay, have been reforming or considering reform of their health systems and services to promote equity in health and access to health services through improved HCT management (HCTM). Improved HCTM will result in increased efficiency in the allocation, use, and maintenance of resources; improved effectiveness and quality of care; ensured financial sustainability, and encouraging community participation and intersectoral action (PAHO, 1998).

In Latin America and the Caribbean, there are approximately 16,566 hospitals that comprise more than 1,100,000 hospital beds; 44.5% of these hospitals are located in the public sector. Approximately 50% of the medical equipment in public hospitals is out of service or is functioning at a level that is out of compliance with manufacturer safety specifications (PAHO, 1998).

The problem of malfunctioning equipment is complex, however, and it involves issues such as the capital and recurrent cost of equipment; the low level of development of maintenance systems; the lack of standardization; the donation of medical equipment; weak after-sale support service; uncoordinated processes for infrastructure development; a lack of technical capacity; and a shortage of professional and technical staff (75% self taught; 19% technician; and 6% professional) in public sector hospitals.

In Paraguay, the hospital infrastructure lists 1207 hospitals with 6966 hospital beds. Of these hospitals, 68.6% are located in the public sector and comprise 62.2% of the hospitals' beds.

The state of the physical and technological infrastructure in the health sector is similar to that in Latin America. A survey of HCT maintenance practices taken in 11 of the existing 18 regions in the country (Galvan and Isaacs, 1999) showed that 48.9% of existing medical equipment and devices were out of service or malfunctioning, mainly due to a lack of managerial capacity and a shortage of professional and technical maintenance staff.

Finally, as in most developing countries, technology development plays an essential role in promoting health and in developing health systems and services. A well-conceived and well-implemented biomedical or clinical engineering department can make an important contribution to the proper distribution of resources, to the selection of cost-effective technology, to greater efficiency and more effective services, to quality assurance in HCT, and to the facilitation of decision making regarding HCT policy in hospitals and other health care services. This chapter expands the information about the status of clinical engineering in Paraguay and its contribution in improving regulation and the appropriate use and maintenance of HCT.

Diagnosis of the Health Care Infrastructure

As in most developing countries, technological development in the health care system of Paraguay has been based on the transfer of technologies designed for developed countries (PAHO, 1998). In many cases, this technology transfer was incomplete, as it was not adapted to the organizational, economic, social and cultural situation of the country.

In the 1980s and early 1990s, some of the common problems in the health care system were the following:

- Scarcity of many basic technologies
- Excessive and indiscriminate use of expensive HCT
- Lack of policies and standards to regulate the introduction and use of HCT
- Underdevelopment of support technologies
- inequalities in access to available HCT
- Scarce resources
- A shortage of professional and technical staff (i.e., clinical engineers [CEs] and biomedical equipment technicians [BMETs])

Problems such as a lack of HCT policies and standards; uncoordinated donor programs; shortages of, and inadequately qualified, human resources; and suboptimal managerial capacities are consequences of scarce resources, institutional weaknesses, and insufficient capacity to absorb and maintain new technology in the country (Galvan and Isaacs, 1999).

Regarding the field of HCT and physical infrastructures management, it is common to find that equipment and facilities are frequently out of service or malfunctioning (Galvan and Isaacs, 1999) for various reasons, such as the following:

- Lack of infrastructure, equipment, and human resources for maintenance
- Lack of resource planning and management
- Low efficiency and weak after-sale support service
- Low levels of standardization, leading to a high degree of diversity of equipment and physical plant
- Inadequate training of equipment operators and maintenance technicians
- lack of managerial and financial capacity to improve HCT maintenance

Survey reports (Galvan and Isaacs, 1996; Galvan and Isaacs, 1997) showed that only 51.1% of basic biomedical equipment of selected health facilities, 10 of the secondary care hospitals (second referral level), and 48 of primary care hospitals (first referral level), were in proper working condition. This situation is a direct consequence of the absence of a maintenance program and shortage of, as well as inadequate training, of maintenance technicians. For performing maintenance in the selected hospitals, there were available only BMETs (13%) and self-taught technicians (87%) but no clinical or biomedical engineers.

Sustainable Clinical Engineering Practice

A sustainable and strengthened organization and structure of clinical engineering departments in the health care services is of vital importance and implies a challenge for all stakeholders and decision makers (particularly for HCT, as its rate of change has increased considerably faster, in recent years, outpacing the evolution in the management and organization of national health care).

To develop a comprehensive clinical engineering practice in Paraguay, the Biomedical Engineering Department of the "Instituto de Investigaciones en Ciencias de la Salud-UNA" (Health Sciences Research Institute) proposed a feasible and sustainable strategy based on survey results (Galvan and Isaacs, 1996, 1997), with the key focus on HCT policy, financing and human resource development, as well as maintenance system development.

The final outcome will only be as successful and strong as the quality of effort and skill applied to the key issues.

Health Care Technology Policy

In December 1994, the Summit of the Americas reaffirmed the interest of the governments of the region in promoting reforms in their health systems that would guarantee equal access to basic health services. The summit charged Pan American Health Organization (PAHO), the International Development Bank (IDB), and the World Bank with organizing a special meeting on health sector in the Americas (PAHO, 1998).

In this regard, and within the framework of transformation of the health sector, the Ministry of Health of Paraguay has been working since 1994 (Vidovich et al 1998) on the reorganization of health systems, including health technology, to achieve a higher level of equity, quality, efficiency, and universal access to health care.

In the past, decisions on new construction, the purchase or replacement of HCT, and the approval of new applications were made through specialists like physicians and nurses but without participation of clinical or biomedical engineers. Consequently, in most cases, it is common to find that, for complex HCT, what was chosen was what is technically possible instead of what is really necessary and useful for each health service (PAHO, 1998). This unsatisfactory situation contributed to inadequate procurement and operational strategies and impacted adversely on the health care system.

In order to improve the overall situation, in particular HCTM health technology assessment (HTA) issues, the Ministry of Health defined a strategy (Galvan, 2001) wherein a basic health technology policy (HTP) is an integral part of its overall national health policy and development plans (i.e., contingency, short-, mid-, and long-term plans).

One essential part of the strategy is the assumed health technology development framework, as show in Figure 24-1, which includes two important components:

1. Health Technology Assessment (HTA) (PAHO, 1998)
2. Technology Management (TM)

In order to guarantee ownership and sustainability of the HTA policy, the Ministry of Health includes all national stakeholders in the formulation and implementation of the policy. In this regard, it created alliances with the following:

- Health sciences research institutes (basic and applied)
- Scientific organizations
- Multilateral organizations
- International agencies

In addition, in April 1996, a meeting of ministers of health of MERCOSUR (the commercial alliance among Argentina, Brazil, Paraguay, and Uruguay) was held to address quality and health care technology.

Since 1997, MERCOSUR has had technical subgroup 11, which is in charge of promoting HCT and HTA issues, among other things. In this regard, the technical subgroup is working on multilateral cooperation, facilitating cooperation with international agencies and networks (e.g., PAHO/WHO, ISTAHC, INAHTA, ECRI, and CCOHTA), identifying relevant groups and national institutions in HCT management and HTA fields, and emphasizing that the clinical engineer is pivotal in the proper implementation of such issues.

More recently, the Ministry of Health of Paraguay began creating a critical mass of personnel trained in HTA methodology and practice (Galvan, 2001) who have appropriate access to national and international information sources.

To establish an appropriate HCT system and its management, the Ministry of Health created an HCT management department with three specific objectives:

1. strengthening policy and programs in clinical engineering, maintenance, technology management, and regulation
2. optimal use of the resources assigned to the clinical engineering and maintenance programs
3. improvement in quality, efficiency, and safety of the operation of equipment, utilities, and physical plant, adapting to economic realities

The HCT management department has implemented the divisions of planning, procurement, and management (see Figure 24-1) to achieve the specific objectives and to guarantee health services delivery with equity, quality, efficacy, and safety, and to protect the HCT investment.

A short evaluation report (Galvan, 2001) showed that the impact of the implementation of the HCT development plan was focused on acceptable improvement of the health care referral system, giving access to curative services with appropriate technology to deliver a basic package of essential diagnosis, treatment, and rehabilitation, with increased delegation of responsibilities at the district level.

Furthermore, viewed from a practical perspective, this preliminary country analysis showed that, over the implementation period of HCTM, the evidence gathered demonstrates that 85% of the contingency plan of the health technology policy for the period 1999-2003 has been achieved with the general improvement in the state of HCT. In general, improvement of planning system focusing on an appropriate HCT system and its management was reached at the Ministry of Health through the implementation of the HCT development plans.

Financial Resources

According to World Bank reports (World Bank, 1996), it is recommended that up to US$12 per capita is required to comply with current health care demands. The reality of the developing countries demonstrates that such targets are difficult to achieve and that they must make tremendous efforts to comply and meet their health care demands (Heimann et al, 2001).

In the case of Paraguay, the *per capita* spending on health in 1999 was US$112.9, and the public spending was 7.9% of the gross domestic product (GDP) (Vidovich et al, 1998).

The poor financial resources for public health are mirrored in the area of health care technology, with less than 1% of the overall national health budget dedicated to upkeep and maintenance. Despite this alarming situation, the World Bank and the Inter-American Development Bank (IDB) is financing programs with over US$77 million to construct and rehabilitate (organization, HCT, and training of human resources) many health care facilities at primary health level for the next five years.

Consequently, the financial needs far exceed the resources available. In order to improve the available financial resources, a process of decentralization with increased delegation of responsibilities at the district level could prove to be one of the key responses.

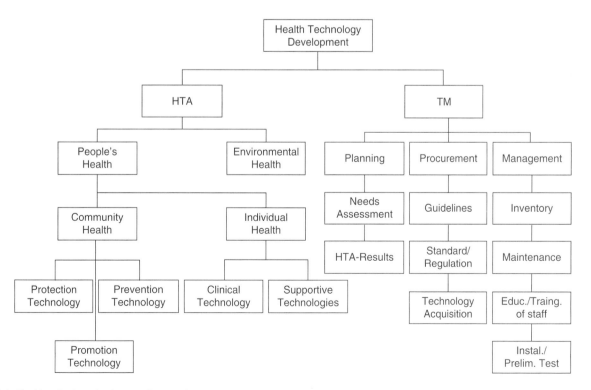

Figure 24-1 Health technology development framework.

Human Resource Development

A sustainable and strengthened HCTM can be achieved only if an appropriate human resource development plan is included in the health technology policy. To establish an appropriate HCT system and its management, human resource availability based on clinical engineers, BMETs, managers, and secretaries as staff members is required.

Such qualified staff should perform activities as listed in Table 24-1 (Bauld, 1987). Task distribution by job classes and the total work effort by all of the staff must be included to account for productivity.

The clinical engineer is the manager of a clinical or biomedical engineering department. The department will be responsible for technology management and at least part of the safety management program. With this variety of responsibilities, several different job roles are required, and each of these may have different educational requirements.

In Paraguay, the main problem with HCTM in the 1980s and 1990s was defined as inadequately qualified human resources and suboptimal managerial capacity. Survey reports (Galvan and Isaacs, 1999) showed that, regarding human resources, the main institutional weaknesses were focused on the following for upkeep and maintenance of HCT:

- Absence of a clinical or biomedical engineer to manage the clinical engineering department
- Shortage of technical staff (2.5% CE, 13% low-qualified BMET, and 84.5% self-taught)
- Inadequate training of maintenance technicians
- Low managerial capacity
- Lack of staff in new technical areas

To improve the HCTM and, in particular, human resource development, a training strategy for BMET and CE was defined. Additionally, another key goal of professional development will be the formation of certification programs for CEs and BMET, with the cooperation and support of the American College of Clinical Engineering (ACCE) (see Chapter 130).

Regarding the training of BMETs, a two-year polyvalent basic training program will be instituted at the National Institute of Health of the Ministry of Health. Such qualified staff should receive further one-year training on medical electronics of biomedical equipment to enable them to fulfill their roles.

The clinical engineering training program includes a five-year polyvalent training at the Polytechnics University of Asunción. Furthermore, a one-year training practicum at hospitals (in clinical engineering departments) or in industry is necessary to enable the clinical engineers to fulfill their roles.

Regular, continuous training courses for staff and clinical equipment users should be implemented, and career opportunities should be explored. Furthermore, specialized courses have been designed and instituted for specific topics, e.g., X ray, CT scanners, lab analyzers, surgery equipment, and ventilators in the annual training program of the Ministry of Health with the cooperation and support of the Senior Experten Service (SES) of Germany.

Moreover, managerial skills are also required at the administrative level (Ministry and Regional Administration) to comply with the needs of a maintenance system and future advances and developments in HCT.

Maintenance System

Maintenance infrastructure should be designed and established at all hospitals and reference centers that have at least 100 beds. Then, in general, at this level there are operating theaters, intensive care units, radiology departments, maternity services, and functional laboratories, all of which need in-house maintenance services.

In Paraguay, the maintenance system situation is similar to that in Latin America. Survey reports (Galvan and Isaacs, 1999) showed that 48.9% of the existing medical equipment and devices of selected hospitals were out of service or malfunctioning, and that one of the main reasons for this situation was lack of infrastructure, tools, equipment, and information systems for maintenance.

To overcome this situation and to improve the HCT maintenance system in the country, the Ministry of Health has designed, and is seeking financial resources to implement, an HCTM Network for national hospitals as well as all regional/district hospitals as reference centers. The HCTM Network comprises a central workshop, located in Asuncion; three regional workshops distributed according the demand of the involved hospitals in main provincial towns; and three mobile workshops to maintain hospitals in rural and remote areas.

According to the assigned capabilities, the HCTM network should cover up to 85% (depending on volume of complex repair) of the daily scope of maintenance activities. Further, the more complex repair and maintenance work (15%) should be performed by private services (i.e., manufacturers and agents). However, to achieve an acceptable level of efficiency in the HCTM system, a good collaboration should exist among the administration, hospital managers, and workshop personnel.

Recommendations

Clinical engineering departments should be installed close to the problem areas in a sustainable and cost-effective manner. This implies that such departments must have competent human resources (CEs and BMETs), adequate infrastructure, sufficient financial resources, and adequate management for the scope of responsibilities. Furthermore, such departments would not be available at rural and remote levels, given the lack of critical mass of HCT. In Paraguay, the majority of rural and remote hospitals have between 5 and 20 beds, with a lack of critical mass of HCT. To compensate for the absence of such departments in rural and remote areas, the use of mobile workshops incorporated in to the clinical engineering maintenance system is recommended.

In order to achieve a well-organized and well-structured clinical engineering department, an HCTM plan developed according the criteria of the Joint Commission for Accreditation of Healthcare Organizations (JCAHO) called "equipment management" (EM) should be established to comply with the HCTM's mission and vision.

Conclusion

Clinical engineering practice in Paraguay does not differ significantly from the practice made in other Latin America countries. Proper HCTM, including performing maintenance, is crucial for adequate health service delivery. But local settings differ from region to region in the same country, and appropriate solutions are not easy to find.

In order to guarantee health service delivery with equity, quality, efficacy, and safety, and to protect the physical plant investment, governments need to improve the availability and management of HCT at hospitals in rural and remote areas for the more deprived population.

Sustainable maintenance systems at an affordable cost contribute substantially toward improving the health system. The establishment of an adequate HCT policy, improvement of financial resources, investment in local human resources, and alliances with relevant groups and national institutions to create common positions to solve problems are of vital importance to guarantee a sustainable system.

World experiences show that in-house maintenance services at the local, regional or national level can save the scarce financial resources if driven by a clinical engineering concept and cost-effective considerations.

Investment in continuing education courses for clinical engineers and BMETs is important but is only beneficial if the other key components like sufficient financial resources, adequate infrastructure, and appropriate supply of spare parts are available.

In conclusion, well-structured and well-organized clinical engineering departments to manage and maintain health care technology are absolutely necessary at health care facilities where a critical mass of HCT and utilities is available.

References

Bauld T. Productivity: Standard Terminology and Definitions. *J Clin Eng* 12(2):139, 1987.
Galvan P, Isaacs J. Evaluación del Estado de los Equipos y Sus Respectivos Programas de Mantenimiento en las 5 Regiones Sanitarias del Proyecto BID "Reforma de la Atención Primaria de Salud." Pan American Health Organization/World Health Organization (PAHO/WHO), Asunción, Paraguay, 1996.
Galvan P, Isaacs J. Evaluación del Estado de los Equipos y sus respectivos Programas de Mantenimiento en las 6 Regiones Sanitarias del Proyecto BIRF "Salud Materna y Desarrollo Integral del Niño/a." Pan American Health Organization/World Health Organization (PAHO/WHO), Asunción, Paraguay, 1997.
Galvan P, Isaacs J. Evaluación del estado de los equipos y sus respectivos programas de mantenimiento en 11 Regiones Sanitarias de Paraguay. Instituto de Investigaciones en Ciencias de la Salud, Asunción, Paraguay, 1999.
Galvan P. Technology Development in Health Systems and Services in Paraguay. Instituto de Investigaciones en Ciencias de la Salud, Asunción, Paraguay, 2001.
Heimann P, Porter D, Schmitt R, et al. Sustainable Health Care Technology Management Systems for the Public Health Sector in Developing Countries. *Medizintechnik* 1:22, 2001.
Pan American Health Organization/World Health Organization (PAHO/WHO). Developing Health Technology Assessment in Latin America and the Caribbean. Pan American Health Organization/World Health Organization (PAHO/WHO), Asunción, Paraguay, 2001.
Vidovich A, Avila R, Echeverría A. Análisis del Sector Salud del Paraguay. Pan American Health Organization/World Health Organization (PAHO/WHO), Asunción, Paraguay, 1998.
World Bank. Health Report. Washington, DC, 1996.

Table 24-1 Task distribution by job classes (Bauld, 1987)

Task	Job			
	BMET	Clinical engineer	Manager	Secretary
In-service training	X	X	X	
Equipment evaluation	X	X	X	
Periodic Inspection	X		X	
Preacquisition planning		X	X	
Repair	X			
Design services	X	X		
Rounds	X	X	X	
Parts ordering	X	X		X
Data analysis	X	X	X	X

25

Clinical Engineering in Peru

Luis Vilcahuaman
Coordinado Bioingenieria, Pontificia Universidad Católica del Perú
Lima, Peru

Javier Brandán
Jefe de la Unidad de Bioingenieria, Pontificia Universidad Católica del Perú
La Victoria-Lima, Peru

The attempts at organizing maintenance service in the health facilities of the Ministry of Health in Peru have been rather varied. Within a very general historical framework, some of the efforts made are enumerated here:

Maintenance Service—1966

The proposal, *Organizing the Maintenance Service of the Hospitals of the Ministry of Public Health and Social Welfare of Peru*, was presented as Item X of the Technical Meeting on Hospitals, held in Lima in December 1966. It called for information on the current state of the maintenance service of Ministry of Health hospitals and health centers; planning of what the service should be in the hospitals, including the aspects of staffing, repair shops, standards, preventive and corrective maintenance, safety, costs, performance, links with other services; and recommendations for the creation of a national Office for Maintenance Supervision.

The presentation to the participants concluded with a proposed Organizational Chart for the Maintenance Service, in which the chief of maintenance reports to hospital management, and a Table of Economic Resources comparing health care facilities in the country.

It is important to recognize the conceptualization of maintenance, as well as its proposed organization, at that time. Also deserving of special attention is the establishment of the category "economic resources for maintenance" as a specific line item for maintenance activities.

However, this proposal was implemented in just one or two hospitals in the country, and only as a result of its author's tenacity. There was nocontinuity in this process.

National Plans for Coordinated Action in Health (PNACS 82/85)

National Plans for Coordinated Action in Health (PNACS 82/85) was implemented in 1981, with financial support from the World Bank, with the purpose of developing the National Health Services System. Here, it is worth mentioning the comprehensive vision of the health services linked with the maintenance services.

The pyramid of care in health services formulated by this plan consisted of four levels parallel with the maintenance service, as follows:

Care in the health services		Maintenance services	
1st Level:	Community Shelter Health post Health center	1st Level:	Maintenance Re-equipping Remodeling Increase
2nd Level:	Local hospital	2nd Level:	Maintenance
3rd Level:	Regional hospital		
4th Level:	Highly specialized Hospital		

This national plan remained only in the study phase, and its maintenance provisions were never implemented.

National System for the Maintenance of Physical Health Services Infrastructure May 1983

This proposal was presented at the first Meeting on Health Services Architecture and Engineering, which was held in Lima from May 2–10, 1983. It began with an overview of the state, as of 1983, of the Office of Maintenance and Reequipping (OMR), organized into the following units: The Medical Equipment Unit, consisting of the following departments: Electromedical, electromechanical, dental, diagnostic imaging, and telecommunications; the Heavy Equipment Unit, consisting of the following departments: Boilers, electrical generators, general services, and sterilization; the physical plant unit, consisting of the departments of civil, sanitary, and electrical engineering; and the Agreements Unit, for the reequipping to be done during that period. It is important to note that at that time, the OMR had staff with different levels of training: 47 professionals (14 with university degrees, 33 with high school diplomas), 82 technicians, 126 auxiliaries, and 50 service people, for a total of 260.

The proposal itself rests on three supports: Research and norms, training, and economic resources for the purpose of conducting a situational diagnosis, post analysis, and operations research on the demand for services, and designing the maintenance system.

As in the preceding case, this proposal did not obtain the necessary support from the decision makers in the Ministry of Health.

Consultancy on a National Maintenance System

The purposes of the Consultancy on a National Maintenance System provided to the Ministry of Health of Peru 15.10.83–15.12.83 were to design the National Maintenance System, to prepare a complete timetable, and to detail the process, as well as the scope, of each activity involved in the development of the Maintenance Plan, based on the established criteria.

This work concluded with the following recommendations:

- That the Pan American Health Organization/World Health Organization (PAHO/WHO) continue its support of the Ministry of Health of Peru to implement a Maintenance System that is more complete and efficient than the existing one.
- That PAHO/WHO participate and actively collaborate in a plan to present the social and economic benefits of programmed maintenance, g.. d to directors of hospitals and offices that support the senior management of the Ministry of Health, indicating indexes and the proportion of capital required by the maintenance.
- Insofar as possible, ensure recognition of the financial burden represented by hospital maintenance for governments such as that of Peru, since the country does not produce hospital equipment.
- That a permanent commission be established to execute the activities according to the approved timetable of the National Maintenance Plan.

In fact, practical action was never taken to establish a maintenance system; thus, this work remained in the consultancy phase.

Situational Diagnosis of Maintenance Services in the Hospitals of Metropolitan Lima, 1984

Situational Diagnosis of Maintenance Services in the Hospitals of Metropolitan Lima, 1984 was conducted by the Office of Physical Infrastructure of the Ministry of Health, under the project known as Development of the Physical Infrastructure of the Ministry of Health (DIF), with joint support from the United Nations Development Program (UNDP) and the Pan American Health Organization (PAHO). This project lists, in the form of technical files, the hospital services, organization of maintenance, ratio of maintenance staff, and area available for maintenance shops. This diagnosis did not lead to concrete results in terms of changes in the country's maintenance system.

National System for the Rehabilitation of Infrastructure and Equipment (June 1986)

The National System for the Rehabilitation of Infrastructure and Equipment (June 1986) was sponsored by the National Service for the Rehabilitation of Infrastructure and

Equipment (SENARINEO) and with the knowledge of the German GTZ Project, described below, which defines the above-mentioned National System with three levels:

1. Central level: Regulates, supervises, provides technical assistance, serves as comptroller, and administers the system at all levels, requiring multidisciplinary, specialized human resources
2. Intermediate level: Responsible for providing services to rehabilitate infrastructure and equipment; it would consist of technical service centers (TSCs) throughout the country, which would have material resources, specialized technical personnel, and rehabilitation workshops
3. Local level: Made up of the maintenance departments of hospitals and health centers, and workers in general

This system was not implemented, despite a ministerial resolution to create 12 technical service centers in the country.

Hospital Maintenance Service Project–Ministry of Health (MdS) and German Technical Cooperation Agency (GTZ), 1982–1994

The Hospital Maintenance Service Project was implemented under a technical cooperation agreement between the Republic of Peru and the Federal Republic of Germany (D.L. No. 21086 of 28 January 1975), based on the previous experience of the German government, in which loans had been made for the construction of new hospitals during the 1960s and 1970s. However, because of ignorance about the concept of maintenance and the activities related to it, the infrastructure, facilities, and equipment deteriorated prematurely; new loans therefore were requested for the construction of new hospitals, or in their absence, for the purchase of new equipment, without considering the necessary expenditures for maintenance of the investments already made.

The first period, 1982–1987, had the following goals:

- Establish a Training Center for Health care Facilities Maintenance Technicians in the city of Chimbote (currently named CENFOTES)
- Establish a regional center in Chimbote to offer maintenance services to hospitals belonging to the former Health Region IV: Ancash-La Libertad (at the present time, the technical service center (TSC), located at the Hospital de La Caleta in Chimbote, is not in operation)
- Promote short training courses in hospital equipment maintenance.

The second period, 1987–1988, was marked by donation from the German government:

- Acquisition of heavy equipment, tools, vehicles, and photocopying machines for four technical service centers (TSCs) in Puno, Cuzco, Huancayo, and Chimbote
- Acquisition of heavy equipment and tools for five regional hospitals: Iquitos, Tarapoto, Trujillo, Chiclayo, and Ayacucho.

The third period, 1989–1993, had the following goals:

- Develop logistic system to supply parts and materials for maintenance
- Reorganize the administration of maintenance
- Establish a technical information system
- Upgrade the education and training of staff involved in the maintenance and operation of equipment and facilities
- Set up a planning preventive maintenance system (PPM)

The project's goals were not fully met, due to a lack commitment of Ministry of Health authorities, especially the programs involved with maintenance at the national level, namely: The Office of Physical Infrastructure (OIF); the Office of Maintenance and Reequipping (OMR); the National Investment Service (SNI); the National Service for the Rehabilitation of Infrastructure and Equipment (SENARINEQ); and the Bureau of Equipment Conservation (DIRECONEQ). All of them today are under the National Maintenance and Equipment Program (PRONAME), which translates into the fact that the project had an official national counterpart for only six months throughout the entire period.

In 1991, as the result of an evaluation sent by the headquarters of the German Technical Cooperation Agency (GTZ), which reported little progress in the programmed activities, the project was deactivated and its remaining economic resources reoriented to providing support for the hospitals of the southern region of the country; e.g., a donation of boilers and the piping system in Tacna and Arequipa.

Hospital Management: The Hospital as a Business

The traditional concept of the hospital, understood as a place where people go to heal, and as a result, where only physicians, nurses, and drugs are found, is used increasingly less, as there is a better understanding that the hospital is a conglomeration of human, material, and technology elements, adequately organized through effective management to provide health services (preventive, curative, and rehabilitative) with maximum efficiency and reliability and optimal profitability.

It is these criteria that govern the concept of the hospital as a business, where management (i.e., the rational use of the available economic, physical, human, and technology resources) occupies a pre-eminent place within the health institution in which it is found. This is why we stated at the beginning of the study that hospitals and, in general, the different types of health facilities, either public or private, will tend to manage their resources on the basis of the variables that influence the production of services—either to boost productivity in the case of public facilities, or to boost profits in the case of the private sector.

The Sphere of Maintenance

As stated above, in this study we introduce the modern concept of the business, where the structure of production costs is increasingly a concern. This is why it is of supreme importance to try to minimize costs in order to ensure greater profits. This criterion is employed to the utmost in the competitiveness business model, where the five major pillars of the business environment are the following:

1. Strategic planning in management
2. Quality systems
3. Monitoring and administration of energy
4. Management of maintenance
5. Staff development

As a result, it is clear from the beginning of the study that the management of maintenance is not an island in the complex organization of the Ministry of Health and the health sector in general. Thus, high levels of interaction are needed in order to achieve the great objective of viewing the hospitals of the Ministry of Health and of the health sector in general as businesses, whose objective of providing health care services should be measured by their quantity and quality.

Definitions of Maintenance

Maintenance is the complex of activities carried out to keep goods or property (e.g., real estate, furniture, equipment, facilities, and tools) in efficient, economic, and safe operating condition. The technical objective of maintenance is to have the goods when and where they are needed. The economic objective is to contribute to the less costly production of goods or services within the business. Hospital maintenance, in turn, has an additional objective, known as a "social objective," which is to prevent death or a worsening of illness when hospital goods are in sound condition, the product of adequate maintenance.

The two types of maintenance defined at the beginning of the study are the following:

1. Preventive maintenance: Actions taken as a result of planning rather than demand
2. Corrective maintenance: Actions taken as a result of demand rather than planning

In principle, predictive maintenance (which is an extension of preventive maintenance) has been discussed, as the need for having diagnostic equipment and adequate computer systems would not correspond to a reality like that of the Ministry of Health. The same holds true for the criterion for rehabilitation or refurbishing, which corresponds to activities outside the maintenance budget (such as investment), but within the sphere of maintenance, which will be described under the heading of "economic resources."

The Management of Hospital Maintenance

By the very nature of its actions, which involve hospital goods and human, economic, physical, and technology resources, the management of hospital maintenance has ceased to be viewed as a complex of technical staff, tools, and parts. It is now seen as a group of activities supported by the four major pillars listed below, requiring a manager who makes rational use of these resources:

1. Administration
2. Economic resources
3. Physical resources
4. Human resources

The Maintenance Organization

With a view to establishing a maintenance organization within a modern concept of management, consideration should be given to the following components:

- Maintenance objectives, which will relate to a reduction in deterioration rates, increases in the availability of equipment, reduction in the high operating costs, and increases in effectiveness and production, which together contribute to an improvement in health service delivery
- Mmaintenance policies, understood as the intention of achieving the objectives in specific periods, known as the short, medium, and long term
- Structuring and execution of the system, defining the functions and location of each system component and prioritizing the criteria for decentralization and autonomy
- Administration, including the criteria for organization, planning, programming, execution, supervision, and control of maintenance activities

In order to achieve the proposed goals of health facility maintenance and to increase production in the services provided, the health facility itself must have adequate infrastructure, facilities, and equipment, as well as repair shops, tools, and parts for maintenance, supported in both cases by sufficient human and economic resources to achieve the optimal levels of care for users of the health services.

Social Security System

While all public hospitals have departments of maintenance and general services, that operate today in Level IV hospitals at the secondary level of management on the Organizational Chart, the first bioengineering unit in Peru was created by general management resolution on May 12, 1995 in the Hospital Guillermo Almenara Irigoyen (Peruvian Social Security Institute, known today as EsSalud). This unit operates in accordance with its manuals on organization and functions and on norms and procedures.

Emphasizing that clinical engineering and technology management represent the administrative structure in the organization of the unit, their human resources are limited in number but have levels of specialization and training that are consonant with the master's level in biomedical engineering. Since its creation, the unit has focused on providing training and assistance in the technology aspect of critical area equipment, as the effort to change a years-old system, with all its inherent external and internal problems, has produced conflict with various interests. Accordingly, our current strategy and policies focus on education at the different levels, with priority given to the goals of clinical benefit to the patients that will result from the fulfillment of institutional programs and policies.

By way of background, during its first year of existence, our unit established internships that were initially intended for residents specializing in anesthesiology, but currently they encompass five specialties. The following year, we were successful in instituting a bioengineering course within the graduate specialty of intensive care medicine at the Universidad Nacional Mayor de San Marcos.

In an effort to consolidate our status as an organization, the Project for Strengthening the Bioengineering Unit was introduced in 1997, supported by the Office of International Cooperation of what was then the Peruvian Social Security Institute (IPSS), and the Representative Office in Peru of the Pan American Health Organization (PAHO/WHO).

That same year, a cooperative agreement was signed between the Pontificia Universidad Católica del Perú and our hospital in the area of bioengineering.

Since the signing of the agreement, the joint studies that have been conducted include the following:

- In the field of electromagnetic interference, a study of the prohibition of cellular telephony in critical areas of public hospitals (1998)
- Energy saving: A study of the distribution and diagnosis of oxygen loss in the Hospital Guillermo Almenara Irigoyen (1999)
- Warning and advice on the Y2K computer problem (PIAS) (1999)

Apart from the agreement, our staff support the research group GIDEM, also connected with the Universidad Católica. The principal tasks of this group of professionals are technology research, development, and innovation, with the fundamental objective of educating human resources in order to achieve a multiplier effect for the benefit of the country. Also noteworthy are the awards won by staff at the national and international levels.

26 Clinical Engineering in Venezuela

Ricardo Silva
The Pennsylvania State University
State College, PA

Luis Lara-Estrella
Simon Bolivar University
Caracas, Venezuela

The Bolivarian Republic of Venezuela is a federal nation organized into 23 states and a district capital. It has an area of 916,446 square kilometers with an estimated population of 24,169,744, 85% of whom live in urban areas. Life expectancy for the period of 1995-2000 was 75.7 for females and 69.9 for males. These and other health care-related data may be found on the Internet (OPS-OMS Venezuela: "Analisis Preliminar de la Situacion de Salud de Venezuela," http://www.ops-oms.org.ve/site/venezuela/ven-sit-salud-nuevo.htm).

According to the Constitution of the Bolivarian Republic of Venezuela (Constitución de la República Bolivariana de Venezuela, Artículo 83), health care is a fundamental social right, and the government has the obligation to guarantee it. In order to do this, there is a National Public Health System, controlled by the Ministry of Heath and Social Development, based on the principles of free service, universality, integrity, equity, social integration, and solidarity (Artículo 84). Finally, the constitution states that financial support for the National Public Health System is a responsibility of the state (Artículo 85), and that everyone has the right to social security as a public, nonlucrative service that warrants health and protection against different contingencies (Artículo 86).

There are 296 public hospitals; 214 of which are integrated into the National Public Health System, and the rest are integrated into several different public organizations. There are 344 private hospitals, of which 29 are nonprofit organizations. By the year 2000, there were 40,675 hospitalization beds integrated into the National Public Health System (17.6 beds per 10,000 inhabitants), with more than 50% of those in the five most developed states (http://www.ops-oms.org.ve/site/venezuela/ven-sit-salud-nuevo.htm).

Origins of Clinical Engineering

The first attempts at organizing clinical engineering-related activities were begun during the 1960s, when the Centro Nacional de Mantenimiento (National Maintenance Center) was created within the Ministry of Health with a grant from WHO-PAHO (VEN 24/U.N.D.P/P.A.H.O.-4862) (Lara-Estrella, 1991). This center was a pioneer in the development of maintenance standards and guidelines for hospital engineering in Latin America. Although it was a good start, the economic welfare of the nation was about to change, and these initiatives where short-lived.

During the 1970s, Venezuela had an impressive economic boom; oil prices were high, and the government had money to spare. The money seemed limitless, and everything could be acquired new. Major investment projects were begun, new hospitals were built, new equipment was bought, everything was overpriced; and there was a lack of government control to avoid excess. At this time it was felt that there was no need for maintenance. Therefore, maintenance organizations were dismantled and clinical engineering was considered unnecessary.

The National Maintenance Center was the basis for what later became known as the Dirección General de Mantenimiento de Infraestructura Física y Equipos (DIFE) (General Direction for Physical Plant and Equipment). DIFE's functions were planning, construction, and maintenance of the physical plant and equipment for all of the institutions within the Ministry of Health. In 1987, in conjunction with DIFE, the Fundación para el Mantenimiento de la Infraestructura Médico-Asistencial para la Salud Pública (FIMA) (Foundation for Maintenance of Medical Infrastructure for Public Health), was created. FIMA was created with the aim of organizing and optimizing the response time for the maintenance of medical infrastructure and equipment. The headquarters were located in Caracas, but FIMA had offices in each of the 23 states. FIMA was a civil nonprofit organization, overseen by the Ministry of Health, with direct income from the central government (Lara-Estrella, 1991).

Both FIMA and DIFE were centralized structures in charge of all of the technological responsibilities for the whole country. The area of competence for the first one was hospitals, while that of the second was small clinics and ambulatory care units. These organizations selected and acquired technology, determined technical specifications, provided maintenance, installed equipment, and supervised service contracts. This proved to be a great failure because acquired equipment did not necessarily respond to a real technological need; response time for repair of damaged and defective equipment was high; and preventive maintenance was almost nonexistent. Moreover, the amount of corruption in both these institutions was high, and customer (hospital clinical personnel) satisfaction was very low.

Redefining Clinical Engineering

Since 1976, clinical engineering has been studied at the Simon Bolivar University (USB) as part of the bioengineering studies program; however, it was not until 1996 that clinical engineering activities were established in a Venezuelan hospital. Clinical engineering is commonly associated with the management of medical equipment, while hospital engineering concerns with the physical plant. In Venezuela, this distinction of activities was impossible because many of the equipment-related problems are the direct result of physical plant problems and vice versa. Therefore, in 1997, Mijares and Lara-Estrella redefined clinical engineering for Venezuela as "the sum of all the engineering and management processes that, as a whole, allows the optimization of the hospital's technological aspects, guaranteeing an overall efficient technological management, with high

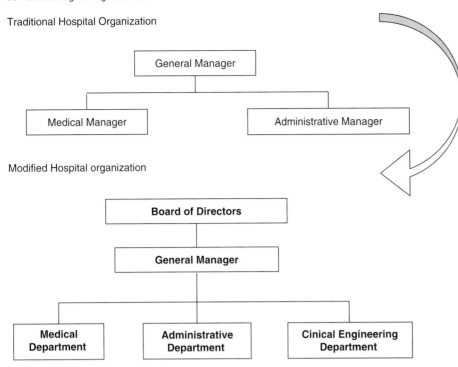

Figure 26-1 Traditional and modified hospital organization models. Under the traditional model, technology management is nonexistent. Consequently, decisions concerning technology are not normally considered within the hospital's management. The new proposed structure includes clinical engineering, in which the technological aspects of health services are considered fundamental elements, due to their important impact on health attention costs.

availability and to the satisfaction of physicians, paramedics, and patients"(Seminario and Lara-Estrella, 1997).

In 1996, the masters program in biomedical engineering was established within Simon Bolivar University and the Unidad de Gestión de Tecnología en Salud (UGTS) (Health Technology Management Group) was created to provide research and funding for clinical engineering-related activities. That same year, a technical assistance agreement was signed between the Simon Bolivar University and the J.M. de Los Ríos Children's Hospital. The objective of the project was to endow the institution with an integral technological management system through the establishment of a clinical engineering department that allowed effective management of all technology available in the institution (Seminario and Lara-Estrella, 1997). The project was begun in March 1996 and lasted ten months. A new organizational structure that considered the technological aspects was proposed. Consequently, the clinical engineering department, structured at a staff level, modified the traditional organization of the hospital in the following way (Figure 26-1).

Successful Implementation of Clinical Engineering

As a consequence of the installation of the clinical engineering department in the Children's Hospital, a generalized increase in productivity could be verified. For example, before the establishment of the department, the hospital generated monthly financial resources of about US$5263; nine months later, this amount rose to US$31,578, essentially due to the fact that most of the hospital's installations and equipment were operative (Seminario and Lara-Estrella, 1997). This project went on for several months, but a change in the political scenario brought the project to an end by September 1997; however, the basis was established and the door was opened for new projects to follow, and the modified hospital organization structure (see Figure 26-1) proved to be the most efficient way of addressing technology management.

Two years later, in 1998, Centro Médico Loira, a private hospital in the city of Caracas, requested the establishment of a clinical engineering department in a way similar to that of the Children's Hospital. There were some major upgrades in this project. A mechanism for the evaluation of progress was established, and the concept of the final user was incorporated into the objective of clinical engineering. Clinical hospitality was incorporated as an essential part of the clinical engineering department (Bylaws of Hospitality Financial and Technology Professionals, http://www.hftp.org/). In our case, clinical hospitality could be defined as the management of infrastructure and accommodation in order to increase customer (patients, medical doctors, and staff) satisfaction, within a clinical/hospital environment. This concept proved to be significant improvement in clinical engineering management, as technical priorities do not necessarily improve customer satisfaction, and a dirty hospital does not provide a favorable impression even if it has the best medical-imaging department. Thus, taking into account the technical aspects, as well as the hospitality issues and splitting the difference, simultaneously improves clinical practice and customer satisfaction.

Centro Médico Loira's project was a success, and its clinical engineering department has been working ever since. The hospital has steadily been growing, customer satisfaction has increased, and the image of the hospital has improved from a second-level hospital to one of the best private hospitals in Caracas. Following Centro Médico Loira's example, some other private hospitals in Venezuela also adapted and incorporated the concept of clinical hospitality into their technological management.

Nationwide Networking

In 1999, a project was begun to establish a clinical engineering department within Hospital Militar, Dr. Carlos Arvelo, the major military hospital. This project incorporated training of a clinical engineering staff to serve as clinical engineers within the other military hospitals and introduced the concept of clinical engineering networking and coordination at a national level.

That same year, the Ministry of Health decided to change their Technology Management Practices and Procedures, and a New Technology Management General Direction (TMGD) was created within the Ministry. FIMA and DIFE were disassembled, and their technological, and some human, resources were incorporated into this new structure. The purpose of this new structure is to become a policy manager and supervisor of Good Clinical Engineering Practice and to transfer operative processes to the regional level. The TMGD will be responsible for establishing and controlling usefulness, accessibility, and quality of new technology introduced into the country, and for supervising regional activities. However, the operation of technology needs assessment, maintenance, and operative procedures will be managed locally by clinical engineering departments created in every hospital. Simon Bolivar's Health Technology Management Group was hired by the Ministry of Health to manage the transition to the new Health Technology Management scheme. Today, personnel are being trained, structures are being created or modified, and policies are being discussed. If everything goes well, by the year 2004, there should be an organized Health Technology Management network within Venezuela's public hospital system. All of the major hospitals should have a Health Technology Management Department, and the central government should have with a structure responsible for coordinating and integrating technological policies and managerial practices. This should provide the country with the tools required to maintain and improve a quality health initiative within the Venezuelan public health system.

References

Lara-Estrella LO. Estudio Critico Sobre la Gestión Tecnológica en el Sector Salud Venezolano. Speech Given at the Universidad Simón Bolívar, 1991.

Seminario RM, Lara-Estrella LO. Establishment of a Clinical Engineering Department in a Venezuelan National Reference Hospital. J Clin Eng 22(4):239, 1997.

27

Clinical Engineering in Japan

Hiroshi Kanai
Institute of Applied Superconductivity,
Tokyo Denki University
Tokyo, Japan

This chapter will address the clinical engineering system and national certification of clinical engineering technicians (CETs) in Japan. The present medical device management situation, the challenges that clinical engineering faces, and the Clinical Engineering Technician laws of Japan are discussed.

Clinical Engineers and Clinical Engineering Technicians in Health Care

Various kinds of sophisticated equipment and apparatus have been applied to modern medicine in many countries. Such sophisticated medical equipment should be managed by specially trained personnel who work alongside physicians and nurses. The need to appropriately apply such medical devices has given rise to the need for new members of the health care team, the clinical engineer (CE) (Bauld, 1991; Bronzino, 1995), the biomedical equipment technician (BMET), and the clinical engineering technologist (CET). CEs are specially educated and trained engineers who manage and otherwise support medical device technology in most hospitals in the Americas and in European countries. (See Chapter 15.) Biomedical equipment technicians (BMETs) and clinical engineering technologists (CETs) work in clinical fields such as open heart surgery, hemodialysis, patient monitoring, and assisted ventilation, and perform maintenance and inspection of a wide range of medical devices. Modern health care, with its increasing use of a wide range of medical devices varying in all levels of complexity, has come to rely upon an efficient and effective technology management system to ensure that this technology is safe and cost-effective. Such a system depends upon the contributions of engineers and technicians working in harmony.

In 1991, the World Congress on Medical Physics and Biomedical Engineering met in Kyoto, Japan, and one of the most important subjects was clinical engineering. The topics featured in the clinical engineering sessions at the Congress were the following:

- Clinical engineering role
- Clinical engineering finance
- Clinical engineering technicians
- Education and career of CEs and CETs
- Standards and regulations
- New technologies and their impact on clinical medicine
- Clinical engineering programs in various countries
- Clinical engineering in developing countries

Ten years have passed since the Congress, but most of the problems discussed there have yet to be solved.

Present State of Clinical Engineering in Japan

Japan has only 130 CEs—30 who manage clinical engineering departments and teach clinical engineering in hospitals and 100 who perform clinical research and development of clinical equipment. Japan has about 16,000 CETs certified by the Japanese government. Most work in hospitals, where 2000 maintain clinical equipment, 12,000 operate clinical equipment, and 500 repair clinical equipment. Others work for medical device manufacturers. The current number of paramedical personnel working in hospitals who operate and maintain medical devices is estimated to be about 15,000.

In Japan, clinical engineers who work in clinical research and academia are considered not to need certification. Moreover, only clinical engineers who manage clinical engineering departments or those who maintain and operate medical equipment, should be certified. However, the current number of CEs who fall into this latter category is only about 50 and, despite the expected increase to 200 in near future, it is felt that this number is too small to justify the establishment of the certification system for CEs. Therefore, instead, there is hope that an international certification system and clinical engineering certification guidelines can be established.

Because well over 10,000 CETs were working in hospitals in Japan, a CET certification system for was established. In Japan, medical doctors and nurses have had to solve equipment-related problems in the hospital, despite having little, if any, relevant education and training to do so. Because they lack the knowledge to operate this life-support equipment safely, this role is performed by the CET. A small number of CETs also perform maintenance and repair of clinical equipment.

Clinical Engineering in Societies and Committees

To improve the clinical engineering system in Japan, clinical engineering committees have discussed what would constitute the most suitable clinical engineering system for the Japanese hospital. Three Japanese scientific organizations are concerned with clinical engineering issues: (1) the Japan Society of Medical Electronics and Biological Engineering (JSMEBE); (2) the Medical Instrument Society of Japan (MISJ); and (3) the Executive Committee for Dialysis Therapy in Japan (ECDTJ). JSMEBE is the affiliated society of the International Federation for Medical and Biological Engineering (IFMBE) and is in involved mainly with basic medical engineering research. MISJ is rather clinically oriented. Each of these three organizations had their own Committee of Clinical Engineering until 1982, when the three combined to form the Joint Committee on Clinical Engineering. They now meet several times a year to discuss the many matters concerning clinical engineering in Japan.

Because their histories and current medical systems are different, each country has its own policy regarding the organizations of paramedical personnel and respective educational systems. In Japan, for example, those who want to work as paramedics in the clinical field must obtain national licenses, which are certified by the Ministry of Health and Welfare. Prior to 1987, the Japanese Government authorized eleven kinds of paramedical occupations: (1) nurse; (2) clinical laboratory technician; (3) clinical radiation technician; (4) physical therapist; (5) occupational therapist; (6) prosthesis technician; (7) dental hygienist; (8) eye therapist; (9) massager; (10) acupuncturist; and (11) bonesetter.

After several studies and meetings, the Joint Committee proposed a structure for the clinical engineering system, which was suitable for Japan. To safely use modern sophisticated equipment, all medical personnel should have a minimum knowledge of electromedical safety and other medical engineering problems. In addition, clinical engineering specialists are needed for this purpose. They are divided into two groups: Clinical engineers and clinical engineering technicians. The CET is equivalent to the BMET in the United States, and the CE is equivalent to the CE in the United States.

Clinical Engineering System Suitable for Japan

In the 1980s, the Joint Committee understood the main purpose of clinical engineering in the United States and in many other countries simply to be keeping medical devices in good condition. They did not see the role of clinical engineering encompassing the operation of medical devices. Therefore, the main responsibilities of clinical engineering in these countries were considered to be preventive maintenance, safety checks, and repair of equipment. The Joint Committee, on the other hand, concluded that clinical engineering in Japan should include all matters concerned with medical engineering equipment; i.e., health care technology management (David and Judd, 1993), from evaluation for purchase of new equipment to the decision to retire old equipment, as well as the operation of some equipment such as hemodialysis machines and heart-lung bypass units. The Joint Committee also considered that clinical engineering should contribute to medical research in hospitals from the engineering point of view.

The Joint Committee proposed a clinical engineering system. CETs must be certified by the Ministry of Health and Welfare. Educational requirements for the CET are at the same level as the clinical laboratory technician and the clinical radiation technologist, but also include a rather basic and broad knowledge of engineering. This engineering knowledge is required because the CET most likely will encounter a wide array of medical device technologies on the job. They also should have enough basic knowledge to understand new clinical equipment as it is developed and introduced into the health care system. If more advanced knowledge is required for the operation of complicated therapeutic equipment, such as a hemodialysis unit, it is recommended that special clinical engineering technicians or those CETs who were specially trained and certified by an appropriate nonprofit society (in addition to the national certificate) should operate the equipment. Clinical engineers, those who supervise all paramedical technicians directly concerned with clinical equipment, should be certified by a nonprofit society.

The Joint Committee of Clinical Engineering had long discussions with the Ministry of Health and Welfare about the certification of clinical engineers and clinical engineering technicians. At the end of 1986, the Ministry decided to improve the situation in clinical engineering in Japan in order to keep sophisticated medical equipment in a safe and reliable condition and to operate them safely. The Ministry discussed with the Joint Committee the clinical engineering system that was suitable for Japan. The Ministry and the Joint Committee agreed that the system, proposed by the Joint Committee, was the most suitable system for Japan. It drafted a bill for the Clinical Engineering Technician law, and the bill was passed by the Japanese congress in June 1987. With the bill's passage, the position of clinical engineering technician became the 12th paramedical occupation authorized by Japanese government. Recently, emergency and ambulatory technicians were nationally authorized, as well.

Clinical Engineering Technician Law

The purpose of the Japanese Clinical Engineering Technician law is the qualification of CETs for the improvement of medical services. The definition of CET is a paramedical person who has a license from the Minister of Health and Welfare and operates, maintains, and inspects clinical life-support and control systems under the direction of medical doctors. A clinical life-support and control system substitutes or assists human life functions of respiration, circulation, and metabolism. A person who wants to work as a CET must pass a national clinical engineering technician examination after graduating from a three-year special college for CETs or an equivalent educational system, and must be certified by the Minister of Health and Welfare.

Education

There are many educational paths that prepare one to be eligible for the national CET examination. The most common is a three-year special college that provides the necessary CET education and training. One alternative is the completion of studies at a one-year special CET college after graduation from a two-year college. Other ways to obtain eligibility exist as well. The law prescribes the CET curriculum of the three-year special college, as shown in Table 27-1.

Because clinical engineering technicians are destined to work on a wide range of clinical equipment problems in the hospital, twice as much time is devoted to basic engineering than to basic medicine and clinical medicine combined. The 33 special CET colleges in Japan graduate about 1500 students each year.

Job Responsibilities

The tasks of the CET are as follows:

- Operation, maintenance, and inspection of clinical life-support and control systems
- Ventilators
- Heart lung bypass units
- Hemodialysis units
- Intensive care units
- Hyperbaric chambers
- Therapeutic equipment (e.g., intra-aortic balloon pumps (IABP) and defibrillators)
- Maintenance and inspection of clinical engineering equipment

National CET Certification Examinations

When the national CET examination program first began, technicians who had been working in Japanese hospitals for at least five years were eligible to sit for the exam. This five-years experience criterion was in effect from 1988 to 1993. During that period, about 13,700 technicians applied for this annual examination, and about 10,000 technicians passed. Since 1994, the national CET certification examination is given each year to applicants who have graduated from the special three-year colleges or from four-year special universities for clinical engineering technicians. Since its inception, the national CET examination process has resulted in the certification of about 17,000 CETs.

Table 27-1 Three-year CET curriculum

Subjects	Time (hours)
Cultural subjects	420
Basic medicine	345
Clinical medicine	240
Basic engineering	765
Clinical engineering	900
Others	330
Total	3000

Distribution of Certified Clinical Engineering Technicians

The distribution of nationally certified CETs is quite different from that of other countries. Of the certified CETs currently employed, 11,000 are working mainly in the operation of hemodialysis units (dialysis techs), 700 in extracorporeal circulation (perfusionists), and 600 for intensive care units (monitoring technicians). Others are working in respiratory therapy (respiratory therapists), and still others in the maintenance of medical devices (biomedical equipment technicians). The Joint Committee recommends that all certified CETs should increase their skills such that they are able to do all of the above jobs. However, it is difficult to change the clinical engineering system in Japanese hospitals.

The most urgent problem facing CETs is how best to work within the organizational structure of the hospital. Special paramedical personnel have long been working in many Japanese hospitals because of the need for those with the engineering knowledge necessary for dealing with modern, sophisticated medical equipment. These paramedical personnel—engineers as well as technicians—are attached to various hospital departments. Most Japanese hospitals do not have special departments for clinical engineering. Only 15% of these paramedical personnel belong to clinical engineering departments. The other clinical engineers and technicians work in other areas of the hospital because doctors find it convenient to have their own clinical engineering technicians in their departments. Most of these technicians have passed the examination and hold a license for working as a CET. Today, throughout Japan centralized clinical engineering services by way of a clinical engineering department are the exception. The rule or common practice on the other hand is the decentralization of CE and CET services in disparate department throughout individual hospitals. With such a decentralized model and, consequently, the lack of a critical mass within a centralized department, health care technology management and other higher-level clinical engineering functions become increasing difficult. With the lack of these services, the hospital, in turn, suffers by inefficient acquisition and use of technology, increased costs, and lower quality of patient care. (See Chapter 30.)

Urgent Problems of Clinical Engineering in Japan

The most urgent problem is the need to revamp the curriculum in the three-year special college for clinical engineering technicians. The Ministry of Health and Welfare has prescribed the clinical engineering technician curriculum in all 33 three-year special colleges and the four-year special universities. It is felt that the time allotted is insufficient to adequately teach the students what they need in order to be an effective CET. Therefore, a new systematic method is needed, to give students a broader and more relevant education in engineering, medicine, and clinical engineering.

An educational system for clinical engineers also should be established. The Joint Committee of Clinical Engineering recommended that the education level for clinical engineers should be at least two years of graduate school in clinical engineering. In Japan, a suitable educational system for clinical engineer does not exist. Only a small number of clinical engineers who have enough ability to manage and to supervise clinical engineering technicians are working in hospitals in clinical engineering. Therefore, in order to increase their numbers, it is urgent that an education system for clinical engineers be established. (See Chapters 66 and 68.)

Japan does not have a certification system for clinical engineers. The Joint Committee of Clinical Engineering is discussing the systems that might be most suitable for the certification of clinical engineers, and it looks to the international community for possible models or affiliation. IFMBE has a Clinical Engineering Division (CED). Over the years, the IFMBE CED has held two meetings per year, in which various problems on clinical engineering and aspects of the international certification system for clinical engineers were discussed. Unfortunately, the expected contribution of CED in promoting the establishment of a world certification system for clinical engineer did not materialize. The American College of Clinical Engineering (ACCE) Clinical Engineering Certification Program (see Chapters 130 and 133) is worth studying for its application to addressing the certification needs of clinical engineers in Japan.

CETs in Japan do not have a society. The author recommends that the Joint Committee assist with the organization of a society for them. Most clinical engineers are members of JSMEBE and contribute their expertise through their work on the JSMEBE clinical engineering and education committees.

Unfortunately, the rapid progress of medicine has brought with it many social problems. These include structural change of disease, increasing numbers of elderly people, economical problems, ethical problems, and safety and reliability of medicine (Bronzino et al, 1990; U.S. Congress, 1987). Clinical engineers have the necessary education, training, and experience to solve these social problems. Japan recognizes the valuable contribution that clinical engineers must make in solving the problems relating to the use of the health care technology of today and tomorrow.

References

Bauld TJ. The Definition of a Clinical Engineer. *J Clin Eng* 16:403, 1991.

Bronzino JD. Clinical Engineering: Evolution of a Discipline. In Bronzino JD (ed). *The Biomedical Engineering Handbook*. Boca Raton, FL, CRC Press, 1995.

Bronzino JD, Smith VH, Wade ML. *Medical Technology and Society: An Interdisciplinary Perspective*. Cambridge, MA, MIT Press, 1990.

David Y, Judd TM. *Medical Technology Management, Biophysical Measurement Series*. Redmond, WA, SpaceLabs Medical, 1993.

U.S. Congress. Assessment, Life-Sustaining Technologies and the Elderly. Office of Technology Assessment, OTA-BA-306. Washington, DC, US Government Printing Office, 1987.

28 Clinical Engineering in Mozambique

Enrico Nunziata
Health Care Technology Management Consulting
Torino, Italy

Momade Sumalgy
Director, Department of Maintenance, Ministry of Health
Mozambique

Clinical engineering in developing countries, particularly in sub-Saharan Africa is different in many aspects from that in developed countries. Often, there is no history of clinical engineering in the health services of developing countries, and external forces (usually bilateral or multilateral donors) often introduce inappropriate maintenance concepts. The major factors affecting the development of clinical engineering in developing countries are the following:

- A lack of a maintenance culture at all levels in the health service
- Extremely scarce resources (financial and material)
- Low productivity and a lack of skilled manpower
- The transfer and use of inappropriate technology

Clinical engineering in Mozambique should cover not only the maintenance of infrastructure and vehicles but also the management of all of its physical assets (also known as health care technology). Physical assets include equipment (medical and general), vehicles, infrastructure, and other installations (e.g., central medical gas supplies). The direct transfer to Mozambique of clinical engineering models developed in western countries and, more recently, in Latin America, is not appropriate, due to technical difficulties and deep cultural differences. This chapter gives a description of a physical-assets management model being developed in Mozambique, along with its strengths and weakness.

Mozambique

Table 28-1 provides some background information on Mozambique. More detailed information is available at the official National Institute of Statistics website: www.ine.gov.mz.

Brief History of the Maintenance System in Mozambique

Prior to independence in 1975, maintenance in the health service had been provided by outside contractors. On the eve of independence, the contractors left, and consequently there was no maintenance being carried out. A centralized national maintenance center was slowly created in the capital city of Maputo, with the support of donors. Staff, from plumbers to X-ray technicians, were introduced and trained to different levels. Maintenance centers were also introduced in the provincial capitals. They were entirely dependent on the National Maintenance Center in Maputo for their supplies—everything from the paper they wrote on to spare parts. The connection between the provincial maintenance centers and the local health service was negligible and sometimes nonexistent.

In 1987, the Government of Mozambique and multilateral donors agreed to implement a Structural Adjustment Plan, and the health service started on a recovery program. At the beginning of the decentralization process in the early 1990s, three specialized maintenance services were opened in the three major hospitals in the country (in the southern, central, and northern regions). More maintenance staff were introduced and trained. Some of the medium-level technicians gained work experience with the help of technical advisors funded by donors. Little formal training was done outside the country, except for occasional short courses provided by medical companies for newly acquired equipment. Centralized planning enabled equipment (i.e., medical and supporting hospital equipment) to be standardized. Consequently, the same types and models of machine were found across Mozambique's health facilities, and, therefore, spare parts planning and acquisition (or donation) of new equipment were simplified.

As the economy continued opening, the private market developed, and some public institutions, like the state-owned electrical and communication companies, were partially privatized. The search for skilled technical manpower drained the already minimal human resources acquired by the health service, and the most capable technicians left. At the end of 1996, the National Maintenance Center in Maputo was stagnant. Few technicians from the center were traveling the country to perform maintenance on medical equipment. Infrastructure maintenance was nonexistent, and most of the government budget for maintenance was absorbed by the repair of vehicles. The National Maintenance Center was totally disconnected from the rest of the health care system. There was no leadership, no control system, no established priorities, and no recorded measurement of the work performed. A new leadership team took over the Directorate of Administration and Management and the Ministry of Health (MOH) as a whole. The maintenance sector underwent a profound change with the support of external technical assistance grants and the willingness to access donors' or soft-loan funds. This change is still ongoing. The reforms were policy document, strategic planning, information system, and Training.

The Policy Document

Around the end of 1996 and the beginning of 1997, the then National Maintenance Center proposed a policy document to reorganize the maintenance systems in the MOH. The document underwent several discussions in the ministry, and after a consensus was reached the MOH Coordination Council approved the new policy, which was signed in July 1997. The policy document was very simple and brief, outlining the major concepts of management and maintenance of physical assets in the health service. It detailed the most important steps that were necessary to create an efficient and sustainable system

The New Organization and the Strategic Plan

Following the adoption of the new policy, a new organizational structure was introduced. Under the new structure, the old National Maintenance Center became the Department of Maintenance, whose responsibilities changed from intervention to regulation and monitoring. In the provinces, the old maintenance centers were supposed to be transformed into management and operational services within the Provincial Directorate of Health. They were to provide professional advice as well as intervening in physical assets management and maintenance. With these changes, the new system was born, but the system was far from operational. There were many constraints, such as a limited skill base and limited material resources available, especially in the provinces. Therefore, a strategic and operational plan, defining the steps to take in the fields of human, material, and financial resources, was prepared in order to complement the policy document. The strategic plan included, among other issues, the following information:

- The typical structure in term of space, human, and material resources of the maintenance unit for each level of the health service pyramid
- An estimation of the financial resources (recurrent costs) needed for maintaining the physical assets, based on a percentage of the estimated replacement cost
- An estimation of the financial resources (capital cost) needed to improve and develop the maintenance sector in terms of infrastructure, material, and human resources

Table 28-1 Basic statistics for Mozambique

Population (1997 Census)	16,099,246
Surface area	799,380 km
Infant mortality rate (IMR)	145.7/1000
Life expectancy at birth (average for men and women)	42.3
Gross domestic product (GDP) per capita (1998)	236.9 USD
Minimum salary (1998)	20.5 USD

The strategic plan was finalized in late 1998, and implementation started during the first quarter of 1999.

The Information System

No system can be managed without information, especially when the system to be managed is not homogeneous and is distributed over a large area, as in Mozambique. Therefore, one priority was the development of an information system (IS) and the regular collection of data. In late 1997, a nationwide information system was introduced, and data collection on assets and maintenance activities started. The IS did not include any preventive maintenance procedures, but the work order form allowed one to differentiate between preventive maintenance and corrective maintenance. The data collected produced a set of databases containing information on equipment around the country. It allowed the Department of Maintenance to produce an annual Retro-Information Bulletin with analysis of the raw data and to produce tentative comparisons among the different maintenance centers around the country. This first IS contained only indicators regarding medical equipment maintenance (e.g., down time and total number of corrective maintenance hours), but no indicators were available to evaluate the system as a whole.

The Implementation Process

In late 1998, one and a half years after the policy became official, the implementation of the policy lagged behind what was expected. It was felt that in the provinces, the directors of health did not know about the policy and its implementation. Probably, one reason for this was that Provincial Directors usually stayed in place for no more than two or three years. Indeed, by the end of 1998, most of the directors who had approved the policy document in 1997 had left for further education or different careers. It became apparent that the administrative system was the key to the successful implementation of the new policy. It was the administrators who, since the early 1980s, really had been managing the provincial health service. The directors—always young doctors who had just finished university—were taking these appointment as a "must do" before proceeding with his or her specialization. They were not interested in management, but only in clinical work. Therefore, the administrative system was not keen on accepting the new modus operandi and was delaying introduction of the changes. Consequently, it was decided that the Department of Maintenance in Maputo needed to monitor and to follow up the policy implementation. This decision was backed up by the Directorate of Administration and Management, who decided to host a National Meeting of the Administration and Management Sub-Sector each year. During these meetings, the development of the administrative and management system (of which the Department of Maintenance is a part) is discussed, and constraints are examined in order to be resolved. These meetings were a great help in boosting the whole policy implementation process.

Monitoring

In order to evaluate the development of a system, it is necessary to measure its changes and to create a monitoring system that is able to detect problems and to allow for adjustments. When the policy document, information system, and strategic plan were prepared, no target, and few indicators, were defined. The only indicators used were those relating to the equipment maintenance and inventory activities. During 1999, the Department of Maintenance performed a study. Its objective was twofold. First, the Department of Maintenance wanted to have a better picture of the cost that would be incurred to expand its services and maintenance activities. Second, it wanted to define a benchmark of indicators that could provide a measurement base for the maintenance sector as a whole. The study, based on known literature, provided a series of possible scenarios for the future of the sector and their related costs. It also defined a benchmark of 10 indicators, taking into account the following factors:

- Productivity (single and multi factors)
- Efficiency
- Administrative management capacity
- Assess life-cycle status
- Qualitative evaluation

It defined a "quality indicator" as a weighted sum of the 10 other benchmark indicators.

Training

The skill level of the technicians was, and still is, low. A system works only if enough of the right human resources are in place, and if their responsibilities and hierarchy are defined and respected. Finally, incentives and stimuli are needed, to retain the personnel. To face these challenges, the Department of Maintenance is preparing a comprehensive proposal for the Ministry of Health that defines the human resource requirements at all levels, from basic biomedical technicians to postgraduate health technology managers. As the salary for technical staff is extremely low, the Department of Maintenance is proposing that the Ministry should create a special career for technical health care personnel, similar to that which exists for nurses. It would allow for a raise in government salaries, due to the special role provided by technical health care personnel in the civil service.

1999–2000: The Preliminary Results

The results of the process started in 1997 with the signature of the policy document, and then with the definition of the strategic plan and its first two years of implementation can be summarized in Figure 28-1 and Tables 28-2, 28-3, and 28-4.

Figure 28-1 provides a graphical representation of the organizational changes to the system as a whole. The vertical structure of the previous system changed into a horizontal structure and allowed integration of the Provincial Maintenance Services into the Directorate of Health.

Table 28-2, provides the Department of Maintenance, in particular, and decision makers in the Ministry of Health, in general, with a graphical representation of the development of the system. To create this table, the Department of Maintenance collected data on the following criteria:

- Policy application and organization
- Human resources
- Infrastructure and material resources
- Information systems

For each criterion, scores were defined for the level of policy implementation. The maximum score for each criterion was 1.00. Looking at Table 28-3 an improvement of almost 100% between 1999 and 2000 can be observed. The number of good, adequate, and bad cells increased from 4, 4, and 12, to 7, 10, and 22, respectively.

Finally, Table 28-4 reproduces the preliminary results of the evaluation of 5 difference maintenance services in Mozambique, based on the benchmark indicators previously described.

The Future

After the first two years of implementation and evaluations (internal and external), the Department of Maintenance initiated a revision of its strategic plan. The process of revision included the following steps:

- Analysis of the 1999–2000 implementation and preparation of the 2001–2204 Strategic Plan
- Revision and improvement of the information system with the inclusion of the benchmark indicators and the inventory and management of the infrastructure
- Revision of the priority areas of the maintenance sector, aligning them with those of the health service as a whole
- Continuous monitoring of the maintenance sector to evaluate its performance and its integration into the health service as a whole

Some of these activities are already under development, and by the second half of 2001 all the revisions should be finalized and the new plan initiated. One major step will be the final separation of the management and the operational responsibilities of the Department of Maintenance.

Conclusion

The change of a system, especially when it has been in place for more than a decade, is a complex, barrier breaking, multidisciplinary effort. It should include a deep analysis of the entire sector of which it is a part, and not only of the one under consideration. Today, the maintenance sector in Mozambique, three and a half years after the introduction of the policy, is still at the beginning of its changing phase. Indeed, even if some major results were achieved, the overall objective of efficient and cost-effective physical asset management is still far away. The difficulties in reaching these objectives are multifarious and different from those encountered when setting up a maintenance service in a hospital in a western environment. In the process, the management core of the Department of Maintenance learned several lessons:

- Culture and social realities are major factors to take into account when planning countrywide physical asset management within the public sector
- Multidisciplinary aspects, such as the administration and logistic sectors, upon which most of the maintenance activities depend, also must be considered
- Donors' and community interests and priorities, which are not necessarily the same as those of the Ministry of Health, must be evaluated and accounted for
- Priorities and target-setting must be addressed as early as possible and must correspond to those of the health service as a whole. It is important to maintain a comprehensive mix of activities and not to skew the interventions in one area to the detriment of another (e.g., the primary health care level versus the tertiary health care level)
- Monitoring and continuous evaluation activities must be in place from the very beginning, so that situations requiring immediate intervention can be identified and acted upon.

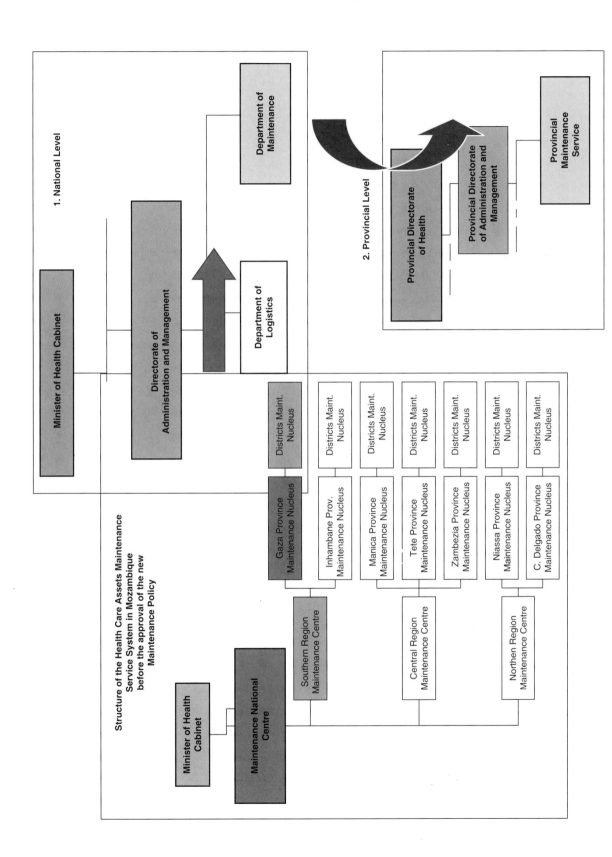

Figure 28-1 The changing from a vertical organizational structure to a horizontal one shows the situation with respect to the inventory process at the end of 1999. The assets inventoried were only medical equipment and hospital supporting equipment such as washing machines or power generators. Only in one province did the inventory include furniture (hospital and general).

Table 28-2 Data of the inventory of medical and general hospital equipment in Mozambique

Maintenance Service	Districts			Health Units[1]			Medical and Hospital Equipment Inventoried			
	No. of Districts ('99)	No. Visited ('99)	% Visited	No. of Health Units ('99)	No. Visited ('99)	% Visited	1993	1998	1999	% Inventoried ('99)[4]
SMH HC-Maputo				1	1	100%	760	0	563	28%
Maputo City	5	5	100%	20	13	65%	227	0	0	65%
SPM Maputo[2]	8	7	88%	16	10	63%	212	438	953	68%
SPM Gaza	12	9	75%	18	13	72%	145	488	488	73%
SPM Inhambane	14	12	86%	56	15	27%	219	439	439	27%
SMH HC-Beira				1	1	100%	351	350	780	92%
SPM Sofala	13	10	77%	65	58	89%	0	1566	1231	89%
SPM Manica	10	10	100%	15	15	100%	108	1790	1769	100%
SPM Tete	13	13	100%	46	43	93%	162	884	1124	94%
SPM Zambézia	17	6	35%	55	12	22%	172	327	327	22%
SMH HC-Nampula				1	1	100%	249			0%
SPM Nampula[3]	21	15	71%	52	28	54%	0	1253	1333	58%
SPM Niassa	16	10	63%	20	14	70%	106	867	867	64%
SPM Cabo Delgado	17	11	65%	23	16	70%	126	811	982	78%
Total	**146**	**108**		**388**	**239**	**62%**	**2.837**	**9.889**	**10.856**	**63%**

(1): **1999 Data. Health Posts are not included.**
(2): Data on the Inventory for 1998 and 1999 of Maputo City and SPM Maputo (Maputo Province) are combined since the work is performed by the personnel available at the Department of Maintenance
(3): The data includes the Central Hospital of Nampula
(4): The amount of existing equipment is estimated based on the average amount of equipment in the Health Units inventoried
HC: Central Hospital
SPM: Provincial Maintenance Service
SMH: Central Hospital Maintenance Service

Table 28-3 Graphical and score summary of the evolution of the maintenance system in Mozambique, 1999-2000

	Scores							
	1999				2000			
Maintenance Service	Policy/ Organization	Human Resources	Infra Structures/ Material Resources	I.S.	Policy/ Organization	Human Resources	Infra Structures/ Material Resources	I.S.
SPM Niassa	0.290*	0.160*	0.250*	0.100*	0.435†	0.160*	0.250*	0.100*
SPM Cabo Delgado	0.580†	0.160*	0.250*	0.000*	0.725‡	0.160*	0.250*	0.500†
SPM Nampula	0.145*	0.334†	0.750§	0.500†	0.580†	0.334†	0.750§	0.600†
SMH HC-Nampula	0.000*	0.334†	0.500†	0.000*	0.500†	0.334†	0.500†	0.000*
SPM Zambéia	0.000*	0.000*	0.000*	0.000*	0.435†	0.160*	0.500†	0.000*
SPM Tete	0.145*	0.160*	0.750§	0.000*	1.000§	0.334†	0.750§	0.600†
SPM Manica	0.435†	0.501‡	0.750§	0.500†	0.870‡	0.501‡	0.750§	0.600†
SPM Sofala	0.580†	0.501‡	0.750§	0.700‡	0.870‡	0.501‡	0.750§	0.700‡
SMH HC-Beira	0.290*	0.334†	0.500†	0.000*	1.000§	0.501‡	0.500†	0.800‡
SPM Inhambane	0.145*	0.160*	0.000*	0.000*	1.000§	0.000*	0.400†	0.600†
SPM Gaza	0.145*	0.160*	0.000*	0.000*	0.870‡	0.334†	0.000*	0.500†
Maputo City	0.145*	0.334†	0.000*	0.000*	0.580†	0.334†	0.000*	0.000*
SPM Maputo	0.000*	0.501‡	0.000*	0.400†	0.580†	0.501‡	0.000*	0.400†

*very poor; †poor; ‡adequate; §, good.

Table 28-4 Benchmark and quality indicators score for 5 maintenance units

NO	Indicators	Scores				
		SMH HC-MAP	SPM MAP	UM MACH	SPM SOF	SPM MAN
1	Activities Performed / Maximum Possible No. of Activities (multifactor productivity)	1.5	1.2	1.4	3.3	2.5
2	Interventions Done / Requested	7.9	1.9	3.9	9.0	11.0
3	Downtime	na	na	na	na	5.0
4	Repeat Repairs	na	na	na	na	5.0
5	Response Time	0.4	0.2	0.2	0.7	0.4
6	Budget's utilisation rate	12.5	3.1	10.0	12.5	12.5
7	Staff Productivity	2.5	0.0	2.5	0.0	5.0
8	Qualitative Score	6.3	5.5	3.5	4.8	7.0
9	Maintenance Cost / Replacement Cost	na	na	na	na	na
10	Activities Performed / Needed	na	na	na	na	na
	GLOBAL QUALITY (Y) SCORE (max = 100)	**31.6**	**12.0**	**21.6**	**30.4**	**48.6**

HC-Maputo: Maputo Central Hospital
SPM Maputo: Maputo Provincial Maintenance Service
UM MACH: Maintenance Service of the General Hospital of Machava (Maputo City)
SPM SOF: Sofala Provincial Maintenance Service
SPM MAN: Manica Provincial Maintenance Service

Further Information

Daft RL, Marcic D. *Understanding Management,* 2nd edition. International Thomson Publishing, 1998.
Newbrander W, Barnum H, Kutzin J. *Hospital Economics and Financing in Developing Countries,* Geneva, World Health Organization, 1992.
Black N. Quality Assurance in Medical Care. *J Public Health Med* 12(2): 97-104, 1990.
Rainer W, Menegazzo E, Wiedmer A. Quality in Management of Biomedical Equipment. *J Clin Eng* 21(2):108-113, 1996.
American Society for Health care Engineering of AHA. Methodologies of Providing High-Quality Customer Services: A Facility Management Approach. Health care Facilities Management Series, Number 055989, October 1999.

Bracale M, Cesarelli M, Rutoli G. Evaluation of Activities in the Health Services. *Med Biol Eng Comput* 25(6):605-612, 1987.
Lodge D. Productivity, Efficiency & Effectiveness in Management of Health Care Technology: An Incentive Pay Proposal. *J Clin Eng* 16(1):29-34, 1991.
Cohen T, Bakuzonis C, Friedman S, et al. Benchmark indicators for medical equipment repair and maintenance. *Biomed Instrum Technol* 29(4):308-321, 1995.
World Bank. Essential vs. Comprehensive Primary Health care, World Development Report, Washington DC, 1993.

Clinical Engineering in the Middle East

Hashem O. Al-Fadel
Managing Director, ISNAD Medical Systems,
Amman, Jordan

With the existence and influx of many types of sophisticated medical equipment, the management of technology is considered one of the important factors in maintaining a high standard of health care (see Chapter 30). Clinical engineering departments monitor, inspect, and ensure that medical equipment is appropriately functioning and properly acquired for its intended purposes. Management of technology is the primary role of any clinical engineering department in a hospital. Such management begins before the time of purchase and lasts through the life of the equipment, and it includes the disposal process.

Maintenance of medical equipment is one of the main functions that attract most of the attention by users and administrators in any hospital (see Chapter 37). This paper reviews and describes the major problems and concerns facing clinical engineering departments in general and in developing countries in particular (Al-Fadel, 1995). In addition, some solutions to existing problems are suggested, and a perspective on the future is discussed.

Problems and Concerns

Some of the common problems or concerns of clinical engineering departments in the Middle East include the following:

- Long lead-times are required, to obtain needed spare parts. Although most departments try to stock some recommended parts, often the parts needed cannot be found in stock. To obtain a particular part, it may take from three months to a year, as sometimes it is not even in the vendor's stock or in the manufacturer's warehouse.
- Sometimes service manuals are not provided or are not even adequate, thus resulting in longer times to repair.
- Adequate support from some vendors, or even from manufacturers, is often not provided. For example, a vendor might not be prompt in solving the problem. Some local vendors might not even employ qualified service personnel to solve the problems.
- Lack of full awareness of some clinical staff about the roles of clinical engineering departments could result in lack of proper communication between users and the clinical engineering staff.
- Some types of equipment might have intermittent problems that could be caused by age, design problems, or the environment. This problem may occur in any health care institution and is addressed by technical departments throughout the world.
- Difficulty in finding well-qualified technicians or engineers often occurs.
- Difficulty in maintaining some equipment by in-house personnel because of the lack of standardization often occurs. Having many one-of-kind devices, each of which requires that the staff obtain special training for servicing, places an onerous burden on the training resources that are available to the department.
- Sometimes there is a lack of competent in-house training programs offered for users and clinical staff.
- Not enough trained local staff are necessarily available to meet hospital needs.
- The process of training staff is long.

Discussion

These problems are common not only in the Middle East, but in hospitals worldwide—even in many hospitals in the industrialized countries (see Chapter 15). Cooperation among hospitals, and proper selection of equipment from cooperative vendors and manufacturers are crucial for long-term equipment support. Other factors, such increasing clinical engineering awareness by clinical engineering departments presenting seminars to the clinical staff, training programs for local clinical engineering graduates, and emphasis on standardization in the purchasing process, will contribute positively toward reducing these problems. These efforts are made in few clinical departments but are not consistent from one hospital to another.

In Saudi Arabia in 1990, the Clinical Engineering Club was founded at the King Faisal Specialist Hospital and Research Center in Riyadh. The club has started the process of cooperation among hospitals in terms of technical consultations, seminar presentations, and expertise exchange. This process can strengthen the cooperation among the technical departments to a certain extent and can result in mutual benefit for all hospitals. Similarly, the Egyptian Clinical Engineering Society, when it was active in late 1989, has contributed some assistance to this field. Exchange of information at the international level benefits all parties. In 1991, the King Faisal Specialist Hospital hosted the week-long International Conference on Health Technology Management, which featured the leading clinical engineers and technology managers from around the world. Several clinical engineers from The Kingdom of Saudi Arabia were among two dozen participants at the 1st Advanced Clinical Engineering Workshop, held in Washington, DC, in 1991 (see Chapter 70). Other examples of international cooperation and information exchange can be found in non-governmental programs to improve health systems. For example, Project HOPE provided financial and technical assistance in Egypt between 1979 and 1986 (Weed and Gellert, 1995).

The problem of acquiring spare parts in a prompt manner needs a great deal of attention and research by all departments in order to find the optimum solutions. This may include setting guidelines on agreements of sharing services, experience, and, possibly, spare parts among hospitals. Another possibility is for hospitals to have comprehensive service agreements with experienced major original equipment manufacturers (OEMs) for optimum service. These subjects could be open for discussion toward any suggested practical solution.

One cannot deny the fact that local vendors will not be able to stock spare parts for equipment other than that covered by service contracts. In the meantime, the suggested or recommended spare parts to be purchased by customers are often exaggerated. Accordingly, neither vendors nor customers can maintain sufficient stocks of spare parts. This situation can be nearly corrected by rationalizing the lists of recommended spare parts and by offering them at lower profit margins.

Future Outlook

Clinical engineering departments should continue striving to keep up with the rapid future technology demands and to increase the awareness of hospital administrations to support their continuous staff development programs. More meticulous planning for stocking a reasonable number of spare parts, especially for vital equipment, in order to minimize equipment down time, should be implemented. Careful selection of equipment for purchase with an emphasis on after sale service is a necessity, to avoid withdrawal or lack of service support. Relationships with clinical engineering departments in neighboring local and regional hospitals should be strengthened by way of activities such as clubs, seminars, and workshops, to promote cooperation in technical support.

Conclusion

Managing health care technology has become an important aspect of the operation and maintenance of hospitals. Local experience by clinical engineering departments in hospitals indicates that problems encountered do not differ widely from those experienced by similar departments worldwide. Maintaining stocks of spare parts, having the correct service documents, staff development programs, and recruiting qualified technicians are the main concerns of the clinical engineering departments in hospitals. Finally, the high-technology equipment always should be maintained by major OEMs or sole agents, to minimize down time and to increase availability and productivity of technology.

Mutual cooperation among regional hospitals is one way to overcome most of the major problems encountered. Hospital administration support is needed more than ever, to aid the management of health care technology. The successful transfer of high technology demands recognition and support for CE training and practice from health care policy makers and administrators (Weed, 1989). As the cost of this technology continues to increase and the technology's complexity grows, the need for effective management will grow commensurately. To avoid being overwhelmed by the advances of technology, a systematic and innovative approach is needed.

References

Al-Fadel HO. Clinical Engineering Issues in Developing Countries. In Bronzino J (ed). *The Biomedical Engineering Handbook*. Boca Raton, FL, CRC Press, 1995.

Weed HR, Gellert GA. Advancing Biomedical Engineering in Developing Nations: Project HOPE and the Potential Impact on Nongovernmental Organizations. *J Clin Eng* 20(5):394, 1995.

Weed HR. Clinical Engineering and Biomedical Equipment Technician Education in the Developing World. 11th Annual IEEE/EMBS International Conference, Seattle, WA, 1989.

Section III

Health Technology Management

Thomas M. Judd
Director, Quality Assessment, Improvement and Reporting,
Kaiser Permanente Georgia Region
Atlanta, GA

Health care systems everywhere face the *STEEP* test of being Safe, Timely, Effective, Efficient, Equitable, and Patient-centered. Meeting these current and perceived future challenges to improve health care content and delivery are often associated with using increasingly sophisticated technologies for diagnosis and treatment. Health care technology management activities offer a range of solutions to address these requirements and to improve quality while reducing cost. These activities include, but are not limited to, setting technology priorities based on the disease burden of the population served, allocating resources (physical, financial, and human) matched to these health needs, and managing these resources through their full cycle of potential use. This section will review the full spectrum of those activities.

The 2001 U.S. Institute of Medicine (IOM) Report regarding 21st century health care in the United States suggests *highest quality care* would be achieved when:

- All preventive, acute, and chronic care services are delivered accurately and correctly.
- All indicated services are delivered at the right time.
- Services not helpful to the patient or reasonably cost-effective are avoided.
- Safety hazards and errors that harm patients and employees are avoided.
- The patient's unique needs and preferences are respected.

The definition of *health technology* used by the World Health Organization (WHO) includes drugs; devices; medical and surgical procedures; the knowledge associated with these in the prevention, diagnosis, and treatment of disease as well as in rehabilitation; and the organizational and supportive systems within which care is provided. This section of the Handbook does not include focus on management of drugs, technology users, or evidence-based clinical procedures. However, it does include management of other health technology *assets,* such as medical devices and supplies; physical infrastructure, such as health buildings and associated services and utilities; and logistics support and information systems. These latter items are important technology components in ensuring quality and cost-effective outcomes for patients.

Health technology management (HTM) is defined as a systematic process in which qualified health care professionals, typically clinical engineers (with their unique ability to visualize a wide range of systems issues and to determine important linkages and solutions), in partnership with other health care leaders, plan for and manage health technology assets to achieve the highest quality care at the best cost. HTM activities in classical terms begin with strategic planning as well as technology assessment and facilities planning, proceed with technology procurement, and conclude with service or maintenance management.

Nearly 20 esteemed colleagues have made major contributions to this section, describing the impact of a wide range of HTM activities, as conducted in both developed and developing countries. They and others have contributed to a collection of case studies that WHO analyzed in 1998,[1] demonstrating the quantitative evidence of HTM impacts on the provision of appropriate *STEEP* and *highest quality care*, which are noted below.

Medical Devices	Cost (or Time) Savings
● Maintenance	20-30%
● Reduced investment through planning	10-20%
● Reduced development time for acquisition specifications	(2-4 weeks)
● Appropriate technology introduction	10-90%
● User training, reducing maintenance	10%

Health Facilities

● Reduced investment through planning	10-20%
● Reduced time for planning	30%
● Utilization	
● Ambulatory care	20%
● Inpatient care	20%
● Diagnoses and treatment	50%

Health Delivery System Processes

● Chronic disease treatment	
● Appropriate use of technology	20%
● Supplies and logistics process redesign	20%

Because HTM activities typically consume 20% or more of all health care capital and operational investments, using the strategies outlined in this section is a must for any health care delivery system in the 21st century.

Reference

WHO's Informal Consultation on Physical Infrastructure, Technology and Sustainable Health Systems for Developing Countries, Geneva, Switzerland, December 1998.

30

Introduction to Medical Technology Management Practices

Yadin David
Director, Center for TeleHealth and Biomedical Engineering Department, Texas Children's Hospital
Houston, TX

Thomas M. Judd
Director, Quality Assessment, Improvement and Reporting, Kaiser Permanente Georgia Region, Atlanta, GA

Raymond P. Zambuto
CEO Technology in Medicine, Inc.
Holliston, MA

The quest of every society is to continuously improve the quality of its members' lives, through promotion of health, prevention of disease, and access to an efficient health care delivery system. Many different methods and strategies for pursuing efficient delivery systems have been tried, and others will be experimented with in the future, but it is evident that we have not yet found the optimal approach. Health care ranges from the fight against diseases to the maintenance of physical and mental functioning, and its delivery largely depends upon technology, especially medical technology. Therefore, medical technology management is one of the most important segments of the health care system, and it is the segment that carries the best potential for clinical engineers (CEs) to demonstrate their unique expertise and leadership excellence.

Medical technology contributes to the advancement of health care in many ways. It contributes to screening of abnormalities and their risks. It contributes to the diagnosis of clinical signs that identify the nature or the cause or the extent of the pathology. It contributes to treatment in the restoration, improvement, and replacement of bodily function as well as preventing further deterioration or pain sensation. It contributes to rehabilitation by restoring, replacing, improving, or maintaining physical or mental function impairment. Technology is expected to reduce the risk of a disease, shorten illness duration, improve the quality and accuracy of care, increase access to care, and replace or limit the decay of a person's functions so and return that person to a state of quality life. In addition, technology is expected to contain cost, to enhance healthy behavior, and to reduce intervention risks. In summary, acquisition of medical technology is accomplished primarily for the following five reasons:

1. To improve diagnostic, therapeutic, or rehabilitation efficiency.
2. To increase the health system's cost-effectiveness or reimbursement.
3. To reduce risk exposure and eliminate errors.
4. To attract high-quality professionals.
5. To expand the service area or to better serve the beneficiary base.

Health care delivery systems around the world are going through major transformations. While knowledge is continuously being created and disseminated at an accelerating rate, the allocation of resources for implementation of preferred solutions is lagging behind, creating a gap that could overwhelm the system if left unchecked. This chapter addresses technology management practices that close this gap by achieving an efficient and effective methodology for the assessment and deployment of medical technology.

Technologies in general, and medical technology in particular play a significant role in the health care transformation. To ensure that technology is safe and effective, there is a need to understand adequately the potential of technology and the importance of its associated management methodology and tools. Without such management methodology and tools, technology function and patient outcomes will be impaired. Forward-looking managers recognize that properly constructed medical technology management methodologies and tools provide objectives and guideline protocols for efficient practice and decision making processes in the following stages in the technology life cycle:

- Strategic technology planning
- Technology assessment
- Technology acquisition and implementation
- Technology risk management and quality improvement
- Technology utilization and servicing
- Technology value or cost/benefit ratio analysis

The management of assessment and deployment of safe and effective medical technology lags behind both the knowledge and practice patterns of management in general. In the highly complex environment of the health care delivery system, the challenge to invest in management methods and practices has diminished such that the consequences of medical technology decisions are inadequately factored into the larger strategy. While this varies from one patient population to another, and from one hospital type to another, these management tools, where they are used, have a direct impact on patient care outcomes, hospital operations, and financial efficiency. Only by applying these tools and methodologies can the system optimize the development of medical technology and the facilities that house it.

There are three types of managers: Those who make things happen, those who watch things happen, and those who wonder what happened. This chapter describes the managerial tools that can facilitate the transformation of a "watcher" into a "maker".

Strategic Medical Technology Planning

The health care delivery system is going through a transition that is driven by four major forces: Budget, structure, technology, and social expectations. The impact of any one or combination of these forces may change from time to time, as does their relative significance, creating a result that is the subject of public debate. It is clear, however, that health care is being subjected to mounting pressure by the needs to (1) identify its goals; (2) select and define priorities; (3) allocate resources more effectively; and (4) achieve system-wide integration.

The health care delivery system presents a complex environment wherein policies, facilities, technologies, drugs, information, and a full range of human interventions interact. It is in this clinical environment that patients in various conditions, skilled staff, contract labor, and a wide variety of technologies converge. The dynamics of this swirling environment, as they relate to medical technology management, include leadership, resources, competencies, risk exposure, regulations, rate of change, and the ability to demonstrate impact on outcomes.

Care providers are faced with the ubiquitous presence of medical technology at the vortex of changing provider and patient roles compounded by system accessibility and integration challenges. Society demands, in addition to user competency, improved quality of care, reduction in error rates, and containment of expenditures. Without a systematic approach, this scenario often leaves hospitals without a clear direction for meeting these expectations. Short-term cost pressures can drive hospital decisions that conflict with the other factors.

One apparent solution that would bring a sense of order and reason to this volatile environment is to seek ways for hospitals to more effectively manage their available technology resources and to do more with less available capital by only selecting "appropriate" technologies that have longer and more reliable life cycles. Proven technologies that fit well into their budgets and operations can be supported and relied upon to provide safe and effective care. Health care delivery organizations have begun to combine strategic technology planning with other technology management activities in programs that effectively integrate decisions about adoption of newer technologies with the hospital's existing technology base—a process that has resulted in better care outcomes at higher efficiencies. Well-integrated medical technology programs will steer hospitals through these transition times by improving performance, eliminating preventable errors, and reducing operational costs.

The Scope of Technology to Be Managed

Technology, as defined by David and Judd (1993), means merely the use of "tools," that is, the involvement of any agent that assists in the performance of a task. In this context, the technology that has been developed for, and deployed in, the health care delivery system ranges from the "smart" facilities within which care is being provided to the products that are used in and around the provision of health care services. Technology "tools" have been introduced at an increasing rate during the past 100 years and include the use of techniques, instruments, materials, systems, facilities, and information. Of all the factors and resources that will shape the future of the health of humankind, the one that most often stretches the imagination is medical technology. However, medical technology is often blamed for contributing to the escalation of health care costs without receiving recognition for improving access to the system and the quality and efficiency of the system.

The past decade has shown a trend toward increased legislation in support of more regulations in health care. These and other pressures will require technology managers to be familiar with the regulations and to be able to manage a program that demonstrates compliance with these requirements throughout the life cycle of the technology. If you subscribe to the saying, "You cannot manage what you do not measure, and you cannot measure what you do not define," then the need for the development of a systematic and comprehensive planning process for technology adoption is obvious. In terms of defining the scope of technology to be managed, the health care organization must develop a rationale for adoption. Without this most basic tool, the process becomes increasingly randomized over time until no consistent system of management can survive. One example of a ranking of rationale for technology adoption is the following list:

Clinical Necessity

- Contribute to meeting/exceeding standard of care
- Positively impact care quality or level
- Impact life quality
- Improve intervention's accuracy, specificity, reliability, and/or safety
- Reduce disease longevity/length of stay

Operational Support

- More effective care/protocol/decision making
- Impact operational efficiency and effectiveness
- Impact development or current service offering
- Impact liability exposure, contribute to reduction in errors
- Increase compliance with regulations
- Reduce dependence on user skill level
- Impact supporting departments
- Increase utilization rate and reduce maintenance load

Market Preference

- Impact access to care
- Increase customers' convenience and/or satisfaction
- Impact organization or service image
- Improve return on investment (ROI) or revenue stream
- Lower the cost of adoption and ownership
- Impact market share

Strategic Planning Process

The strategic planning process is the road map for the introduction and development of technology and services, and their related policies into the core business of the hospital to maximize the value outputs of the program. The outputs of this process are measured as changes in cost, quality, performance efficiency, or quality of life. The road map is an important guideline because it identifies a common vision for timely response to fundamental needs. The following key components must be present in the plan to ensure the optimal allocation of funds needed:

- Regional planning, coordination, and technology assessment
- Strategic technology planning and priority setting
- Budget development and approval processes
- Technology management and service planning
- Technology acquisition
- Technology audit and risk management

A technology strategic plan is derived from, and supports, well defined clinical objectives. The ability to contribute to this process and the development of these components and their interaction with budgeting processes require a unique set of skills and technical management expertise that is consistent with the characteristics of a mature clinical engineering professional. This expertise facilitates the integration of clinical objectives with management and technical threads that permeate the organization. This aspect of the planning process must include the following elements:

- Creation of a plan to support the facility's vision and communicate its process to staff
- Periodic review of the alignment between the vision and strategy
- Identification of areas/topics where changes are needed
- Determination of priorities and creation of a plan to meet the objectives
- Inclusion in the plan of the details of specific expectations from information technology, medical technology, and building spaces—transforming experts' knowledge into service strategy
- Delineation of clinical goals for road map planning, interaction with operations and capital budgeting processes, acquisition and deployment timing, equipment asset management, and monitoring and evaluation

In order for the planning process to maximize the value it adds, it must include standard elements of analysis and must be somewhat predictive in several areas where trends may change over the course of implementation of the plan. The planning process must include the following elements:

- Assess changing clinical goals. The clinical goals are updated annually. For a given year, key hospital participants, through the strategic planning process, determine the clinical services that the hospital should be offering in its referral area. These must be projected with accuracy at the outset.
- Take into account health care trends, demographic and market-share data, and space and facilities plans.
- Analyze the facility's strengths and weaknesses, goals and objectives, and opportunities and threats.
- Conduct an audit of the existing technology base, including its condition, life expectancy, and utilization rate.
- Audit and project the costs of health care providers using the existing technology, considering turnover of personnel as well as technology.
- Integrate assessment and prioritization of new and emerging technologies.
- Ensure strong compliance with and support of anticipated technological and utilization standards.
- Review technological trends and their operations impact.

If all of these areas are considered, the outcome of this process will be the following:

- A coherent plan that supports the objectives outlined in the organization's vision for the coming year
- A predictable level of technology that is capable of meeting requirements for a standard level of operation in the referral area
- Offerings of better and more efficacious and consistent health care services
- Effective use of limited resources and provision for growth of the organization's intellectual property
- A strategic technology plan that helps technology managers to match available technical abilities (both existing and new) with clinical requirements and financial capability
- A definition for the level of service expected
- Priorities in budgeting for technological adoption and acquisitions

To accomplish this goal, clinical engineers and technology managers must understand why their institutions' values and mission are as they are; must pursue knowledge and collect information that supports their institutions' strategic plans; and must be able to translate their operations according to the strategic planning process utilizing the often limited resources allocated to them. Although a technology manager might not be assigned to develop an institution's overall strategic plan, he or she must understand and be ready to offer logical and informative input to the hospital management. The clinical engineer will be prepared to provide this input in the following ways:

- By committing to a professional involvement with, and understanding of, all the hospital services
- By understanding technology assessment methodology and equipment life cycle functions
- By determining the ways in which the hospital's technological deployment is best evaluated
- By articulating justifications and provisions for adoption of new technologies or enhancement of existing ones
- By assisting in providing a review of emerging technological innovations and in determining the impact that they can have on the hospital. (A good rapport with the research and development industry facilitates this.)
- By visiting the sites of technology development—research or manufacturing—as well as the exhibit areas at major scientific and medical meetings, because tomorrow's clinical devices are in the research laboratories today
- By being familiar with the institution and its equipment users' ability to assimilate new technology

The past decade has seen a trend toward increased customer expectations, legislation, and regulation in health care. These developments and financial pressures require that additional or replacement medical technology be well anticipated and justified. Proper planning will provide the rationale for sound technology adoption. Today's marketplace demands cost effectiveness, competitiveness, and flexibility from every hospital if it is to survive and grow. Such demands require that the effective clinical engineer be able to articulate the differences among factors such as clinical necessity, code compliance, management support, market preference, and arbitrary decision.

Technology Assessment

As medical technology continues to evolve, so does its impact on patient outcomes, hospital operations, and financial resources. The ability to manage this continual evolution and its subsequent implications has become a major challenge in all health care organizations. To be successful, it must be an integral part of hospital operations that address the needs of the patient, and it must smoothly mesh people and technology. The manager who commands knowledge about the organization's culture, the equipment users' needs, the existing environment within which equipment will be applied, equipment engineering, and emerging technological capabilities will be successful at implementing and managing technological changes.

In the technology assessment phase, the clinical engineering professional needs to wear two hats in order to lead the team and to contribute to the decision making process. The team should incorporate representatives of equipment users, equipment maintainers, physicians, purchasing or reimbursement managers, administration, and other members from the institution, as applicable.

Technology Audit

With a coherent clinical strategic plan in place, the hospital can conduct a credible audit. Each major clinical service or product line must be analyzed to determine how well the existing technology base supports it and supports the conditions of that technology. A Medical Technology Advisory Committee (MTAC), consisting of hospital management, physicians from major specialties, nurses, program managers, and clinical engineers should be appointed to conduct this analysis. The key steps that should be taken during the audit are as follows:

1. Develop a hospital-wide complete inventory (i.e., quantity and quality of equipment included), and compare the existing technology base against known and evolving standard-of-care information, patient-outcome data, and known equipment problems
2. Collect and review information on technology utilization and assess appropriate Use, opportunities for improvement, and reduction of risk level
3. Review technology users' (physicians, nurses, technologists, and support staff) educational needs as they relate to the application and servicing of medical equipment
4. Determine appropriate credentialing of users for competence in the application of new technologies, assess needs, determine whether requirements are being met, and assess risks involved (credentialing committees will be the primary group to match clinician skills with evolving clinical treatment procedures or protocols)
5. Keep current with published clinical protocols and practice guidelines using available health care standards directories
6. Utilize clinical outcomes data for quality assurance and risk management program feedback

The audit will allow the gathering of information about the existing technology base and will enhance the capability of the MTAC to assess the need for new and emerging technologies as well as the impact of these technologies on their major clinical services. In this assessment, the following issues should be considered:

- Needs
- Value of the technology
- Technical validity
- Ability to assimilate the technology
- Ability to integrate with existing technological platforms
- Medical staff satisfaction
- Impact on staffing and delivery of care
- Impact on facilities
- Impact on standards of care and quality
- Economic considerations (e.g., reimbursement, life cycle costs)

The committee will then set priorities for equipment replacement and implementation of new and emerging technologies, which, over a period of several years, will guide the acquisitions that provide the desired service developments or enhancements. Priorities will be set based on need, risk, cost (acquisition, operational, and maintenance), utilization, and fit with the clinical strategic plan.

Budget Strategies

All of the information collected above will bear on the developing of budget strategies. Strategic technology planning requires a 3- to 5-year long-range capital spending plan. The MTAC, as appropriate, will provide key information regarding capital budget requests and make recommendations to the capital budget committee each year. There is a three-fold purpose for the capital budgeting process:

1. To develop procedures to solicit and review technology requests
2. To coordinate capital expenditures with available resources
3. To determine optimal financing methods for acquisition

The MTAC should review the final capital budget listing in order to recommend when the items should be purchased during the next year and, if possible, to determine if there should be centralized, coordinated acquisition processes planned for similar items from different departments.

Long-term capital equipment budgets are derived from the analysis of replacement life cycles, organization financial conditions, annual operations support costs (including service, upgrades, and repairs) and true needs justification coupled with a 3-year budget cycle. Each item of equipment listed on the budget is highlighted as either a replacement or a new requirement for an existing or new program. The replacement life cycle is modified from standard tables by factors such as average duty cycles and utilization and escalating repair and service history. Economic justifications for clinical services revolve around a "make or buy" decision—whether the service should be performed by the clinical services in-house or should just be purchased from the commercial market. The needs justification usually centers on the capabilities of the clinical staff.

Prerequisites for Medical Technology Assessment

Medical technology has a major strategic factor in positioning the hospital and its perception in the competitive environment of health care providers. Numerous dazzling new biomedical devices and systems are continuously being introduced. They are being introduced at a time when the pressure on hospitals to contain expenditures is mounting. Therefore, forecasting the deployment of medical technology and the capacity to continuously evaluate its impact on the hospital require that the hospital be willing to make the commitment and to provide the support such a program. An in-house "champion" is needed in order to provide the leadership that continuously and objectively plans. This figure might use additional in-house or independent expertise as needed. To focus the function of this program in large, academically affiliated, and government hospitals, the position of a chief technology officer (CTO) is becoming justifiable. While executives have traditionally relied on members of their staffs to produce objective analyses of the hospital's technological needs, they nevertheless are too often subjected to the biases of various interest groups, including marketing and vendor appeals. More than one executive has made a purchasing decision for biomedical technology only to discover later that some needed or expected features were not included with the installation or that those features were not yet approved for delivery. These features have come to be known as "futureware" or "vaporware". Or, alternatively, it may be discovered that the installation has not been adequately planned, ending therefore as a disturbing, unscheduled, expensive and long undertaking.

Most hospitals that will be providers of quality care will be conducting technology assessment activities in order to be able to project needs for new assets and to efficiently manage existing assets within the limits of the available resources. In order to be effective, an interdisciplinary approach and a cooperative attitude are required because the task is complex. The ability to integrate information from disciplines such as clinical, technical, financial, administrative, and facilities in a timely and objective manner is critical to the success of the assessment.

Medical technology includes medical and surgical procedures, drugs, equipment and facilities, and the organizational and supportive systems within which care is provided. This definition focuses on equipment, systems, facilities, and procedures (but not drugs). There are considered to be two tiers of investigation in medical technology assessment, given that it is the evaluation of the effectiveness of equipment, systems, and procedures in treating or preventing disease or injury:

1. Primary: Clinical safety and effectiveness in terms of physical indicators of patient care outcome
2. Secondary: Synthesizing the results of clinical impact to project financial outcome and reimbursement decisions for payers.

This chapter also emphasizes medical equipment management as an essential element of medical technology management, including the notion of the skills to forecast medical equipment changes and the impact of those changes on the hospital market position. While most consideration is usually given to capital asset management (see Chapter 35) when it comes to medical equipment, one should not exclude the accessories, supplies, and disposables from the medical equipment management program. Another often-overlooked factor in medical equipment management is the impact of the maturity of the technology on education and training as well as on servicing. Equipment that is highly innovative, in development or in clinical trials, will have a far different learning curve for users as well as maintainers than equipment based on more mature technologies.

Technology Assessment Program

Increasingly more hospitals are faced with capital or equipment requests that are much larger than the capital budget. The most difficult decision, then, is the one that matches clinical needs with financial capability. In that process, the following questions are often asked: How can a hospital avoid costly technology mistakes? How can a hospital wisely target capital dollars for technology? How can a hospital avoid medical staff conflicts as they relate to technology? How can a hospital control equipment-related risks? How can a hospital maximize the useful life of the equipment or systems while minimizing the cost of ownership?

As mentioned earlier, technology assessment is a function of technology planning that begins with the assessment of the hospital's existing technology base. Technology assessment is, rather than an equipment comparison, a major, new function of a clinical engineering department. It is important that clinical engineers be well prepared for the challenge. They must have a full understanding of the missions of their particular hospitals, a familiarity with the health care delivery system, and the cooperation of the hospital administration and the medical staff. To maximize their effectiveness, clinical engineers need access to database services and libraries; the ability to visit scientific and clinical exhibits; the capability to establish an industrial network; and a relationship with peers throughout the country.

The need for clinical engineering involvement in such a program is evident when one considers the problems typically encountered:

- Recently purchased equipment, or its functions, is underused
- Users experience problems with equipment.
- Maintenance costs are excessive.
- The facility is unable to comply with standards or guidelines (e.g., JCAHO requirements) for equipment management.
- A high percent of equipment awaits repair
- Training is inefficient because of a shortage of allied health professionals

A deeper look at these symptoms using a proper technology assessment analysis likely would reveal the following:

- The lack of a central clearinghouse to collect, index, and monitor all technology-related information for future planning purposes
- The absence of procedures for identifying emerging technologies for potential acquisition
- The lack of a systematic plan for conducting technology assessment, and thus an inability to maximize the benefits from deployment of available technology
- The inability to benefit from the organization's own previous experience with a particular type of technology
- The random replacement of medical technologies, rather than a systematic plan based on a set of well-developed criteria
- The failure to integrate technology acquisition into the strategic and capital planning of the hospital

The following scenario suggests one way to address these problems and symptoms.

To address these issues, efforts to develop a technology assessment plan are initiated with the following objectives:

1. To accumulate information on medical equipment
2. To facilitate systematic planning
3. To create an administrative structure supporting the assessment program and its methodology
4. To monitor the replacement of outdated technology
5. To improve the capital budget process by focusing on long-term needs relative to the acquisition of medical equipment

This program, and specifically the collection of up-to-date, pertinent information, requires the expenditures of certain resources and active participation in a network of colleagues who practice in this field. Membership in organizations and societies that provide such information should be considered, as should subscriptions to computerized databases and printed sources.

The Director of Clinical Engineering (DCE) chairs the MTAC while another CE from the same department serves as the committee's designated technical coordinator for a specific task force. Once the committee accepts a request from an individual user, it identifies other users who might have an interest in that equipment or system, and it authorizes the technical coordinator to assemble a task force consisting of users who the committee has identified. This task force then serves as an ad hoc committee that is responsible for the establishment of performance criteria that will be used during the assessment of the equipment described on a Request for Review (RR) form. During any specific period, there might be multiple task forces, each focusing on a specific equipment protocol.

The task force coordinator cooperates with the material management department in conducting a market survey, in obtaining the specified equipment for evaluation purposes, and in scheduling vendor-provided in-service training. The scheduling of the in-service training for the users can be highly frustrating at times, as the shortage of allied health professionals reduces availability for training while increasing the need for training due to higher staff turnover rate. It is highly recommended, therefore, that this activity be well coordinated with the users' group training coordinator.

After establishment of a task force, the committee's technical coordinator analyzes the evaluation objectives and devises appropriate technical tests, in accordance with recommendations from the taskforce. Only equipment that has successfully passed technical tests will proceed to a clinical trial. During the clinical trials, the clinical coordinator collects and then reports to the task force the summary of experiences gained. The technical coordinator then combines the results from the technical tests and the clinical trials into a summary report and prepares the task force's recommendations for MTAC approval. In these roles, the CE serves as the technical coordinator and as the clinical coordinator bridging the gap between the clinical and the technical needs of the hospital.

The technology assessment process, begins with a department or individual filling out two forms: (1) an RR form (Figure 30-1), and (2) a Capital Asset Request (CAR) form (Figure 30-2). These forms are submitted to the hospital's product standards committee, which determines whether an assessment process is to be initiated, and the priority for its completion. It also determines whether a previously established standard for this equipment already exists.

In the RR form, the originator delineates the rationale for acquiring the medical device. For example, the way the item will improve patient care; generate cost savings, support the quality of service; and provide ease of use, as well as who the primary user will be. In the CAR form, the originator describes the item, estimates its cost, and offers some justification for its purchase. The CAR is then routed to the capital budget office for review. During this process, the optimal financing method for acquisition is determined. If funding is secured, the CAR is routed to the materials management department where, together with the RR, it will be processed.

The rationale for having the RR accompany the CAR is to ensure that pricing information is included as part of the assessment process. The CAR is the device by which the purchasing department sends product requests for bid. Any request for review that is received without a CAR, or any CAR involving medical equipment that is received without a request for review is returned to the originator without action. Both forms are then sent to the clinical engineering department, where a full-time employee designated as a coordinator reviews and prioritizes various requests for the committee to review.

Both forms must be sent to the MTAC if the item requested is not currently used by the hospital or if it does not conform to previously adopted hospital standards. The committee has the authority to recommend either acceptance or rejection of any request, based on a consensus of its members. If the request is approved by the MTAC, then the requested technology or equipment will be evaluated using technical and performance standards. Upon completion of the review, a recommendation is returned to the hospital's product standards committee, which reviews the results of the technology assessment, determines whether the particular product is suitable as a hospital standard, and decides whether it should be purchased. If approved, the request to purchase will be reviewed by the capital budget committee (CBC) to determine whether the required expenditure fits within the available financial resources of the institution, and whether or when it might be feasible to make the purchase. To ensure coordination of the technology assessment program, the chairman of the MTAC also serves as a permanent member of the hospital's CBC. Accordingly, there is a planned integration between technology assessment and budget decisions.

As a footnote to this example, it is important that those involved in the process understand fully the way that standards are developed, the way they are used and modified, and, most significantly, the effect of these activities on the entire spectrum of health-related matters. Some standards address, for example, protection of the power distribution system in the health care facility; protection of individuals from radiation sources, such as lasers and X-rays; and protection of the environment from hazardous substances (see Chapter 117). The practicing professional should fully appreciate the intent of standards in general and should participate in their development and use.

Technology Assessment and Clinical Engineering

Clinical engineering departments are at the threshold of a revolution toward the comprehensive management of all health care technology. Increasing pressures for greater attention to the quality, fiscal containment, and risk mitigation and error reduction should be matched with skillful and competent management focusing on the characteristics of health care technology. A well-organized program will have a significant impact on the hospital's bottom line, which is a highly desirable outcome in today's financial climate. Hospitals and vendors that operate with organized asset management programs are already benefiting from the involvement of clinical engineering professionals. The role of clinical engineers is threaded throughout the program as it relates to medical equipment and systems. Clinical engineers contribute to, and participate in, every phase of the equipment life cycle, from the capital budget planning, the equipment evaluation, and the performance validation, to the acceptance testing, user training, inventory control, repair and maintenance services, and incident investigation. Their involvement improves the planning for the new (and the management of the existing) equipment inventory, thus impacting integration, quality, finance, and risk.

Device Evaluation

One of the best methods of ensuring that the contribution a technology makes is valuable to the hospital is to analyze carefully each medical device in preparation for its assimilation into the hospital operations. This process of equipment evaluation provides information that can be used to screen unacceptable performance, by either the vendor or the equipment, before it becomes a problem for the hospital (see Chapter 33).

The evaluation process consists of technical, clinical, financial, and operational aspects. These aspects were evaluated earlier, as described in the MTAC function; however, the emphasis here is on the clinical engineer's responsibility. It is assumed that in order to fulfill these duties, the CE is familiar with the emerging and evolving technologies and can translate the clinical needs of the users into an effective and comprehensive bid specification document. The document should be clear, facilitating a competitive bidding environment and comparison of vendors and their wares. This document sets the whole equipment evaluation and selection into motion. Validation criteria for key elements, such as system configuration, extent of facility preparation and operation disturbance, performance requirements, users and maintainers training, warranty, documentation, delivery schedule, and implementation plan, should be spelled out. Cost of service support and price for future upgrades need to be locked.

After the vendor has responded to the informal request or the request for proposal (RFP) information, the clinical engineering department will be responsible for evaluating the technical responses, while the materials management department evaluates the financial responses.

In translating clinical needs into a specification list, key features or "must have" attributes of the desired device are identified. In practice, clinical engineering and materials management develop a "must have" list and an "extras" list. The "extras" list contains features that could tip the decision in favor of one vendor, all other factors being equal. These specification lists are sent to the vendor and are effective in a self elimination process that results in a time savings for the hospital.

Once the "must have" attributes have been satisfied, the remaining "candidate" devices are evaluated technically and the "extras" are considered. This is accomplished by assigning a weighing factor, e.g., 0–5, to denote the relative importance of each of the desired attributes. The relative ability of each device to meet the defined requirements is then rated. Consider the following examples of attributes:

- Accuracy and repeatability
- Ease of use
- Reliability
- Expected user's skill level
- Serviceability and warranty
- Performance
- Compatibility and interchangeability
- Ability to be upgraded
- Safety
- Cost

Each of these attributes is important, but some are more important than others. In assigning the weighing factors, the CE must take into account the relative importance of each of these attributes. He or she should create a bidding environment that will enable a direct comparison of vendors. Therefore, the RFP should provide details of delivery training and installation, a detailed description of the "must haves" and the "extras," and the cost of service and upgrades, as well as identifying recourse for vendor deficiencies.

The performance of the acceptance testing accomplishes the following:

- Verifies by incoming inspection that each medical device received is capable of performing its designed function
- Obtains baseline measures that can be used later to resolve specified problems
- Assures compliance with the equipment management program, the relevant factors of which include:
  - Verification that the chosen vendor has delivered a complete system with all of the accessories and other needed supplies
  - Documentation of full compliance with terms that were prescribed in the Conditions of Sale and agreed upon when the bid was awarded
  - Initiation of an asset control record by the clinical engineering department

This last item is the point where the equipment enters into the equipment maintenance program, the warranty period is initiated (if applicable), and testing criteria are documented.

The review of each of the vendors' responses, the performance of comparative tests and value analysis, and the performance of acceptance testing are the steps that will reduce procurement costs and problems. They will prevent problems such as dissatisfaction, cost overrun, postimplementation surprises, unplanned service costs, slow resolution or delayed response, prolonged startup, performance gaps, overcharges, and unauthorized promises.

Request For Review by the
MEDICAL TECHNOLOGY ADVISORY COMMITTEE

(Complete all pertinent information)

☐ New product ☐ New equipment ☐ Replacement item ☐ Single user ☐ Other
Submitted by _____ To _____ Date _____
Your position (title) _____ Department name & number _____
Brief description of the item (manufacturer & catalog no.) _____
Item used for _____
Item unit cost _____ Anticipated annual usage _____
Current item being replaced (if applicable) _____
What is the annual utilization of the item being used? _____
Unit cost of the item currently being used (if not stock) _____
Manufacturer & catalog no. _____
Please give your assessment of the proposed item over the current item:
    ☐ Better patient Care   Explain: _____
    ☐ Cost savings   Explain: _____
    ☐ Better quality   Explain: _____
    ☐ Easier to use   Explain: _____
    ☐ Other   Explain: _____
Who will be the main user of this item? _____
What other facilities are using this item? (if any) _____
List current user of this item (if any) _____
Any other pertinent information (attach items/literature)

Authorized signature _____ Date _____

FOR MEDICAL TECHNOLOGY ADVISORY COMMITTEE USE ONLY

Request received by _____ Date _____
Does this item comply with standards regulations? (if applicable) _____
Additional action taken: _____
Presented to Product Standards Committee on: _____
Was not presented to the Product Standards Committee due to: _____
Action taken: _____

From: Chairman of the Product Standards Committee   To: _____

Please accept our appreciation for your recommendation. The item you suggested was:

☐ Considered and approved for evaluation   ☐ Considered and not approved
☐ Referred for further evaluation   ☐ Pending consideration due to lack of information
☐ Other: _____

Figure 30-1 Request for Review (RR) form.

Existing inventory utilization should be monitored periodically. The utilization level can be measured and compared with the budgeted level. The utilization rate of existing inventory is a good indicator in justifying additional capital requests.

Risk Reduction

Significant progress in controlling risk has been achieved with the early implementation of an equipment management program. With the development of the dynamic equipment risk factors and associated failure analysis techniques, proactive techniques to contain risk can now be implemented (see Chapter 56). These techniques should be used for assessing new equipment as well as for the management of existing inventory. Error avoidance and lessons learned from analysis of near-miss events are useful tools for further reduction of potential risk (see Chapter 55).

An organization may have a variety of objectives, such as profit, growth, and the performance of a public service. However, a fundamental management commitment to minimizing the adverse effect of accidental loss to an organization is the founding principle of a risk management program. Risk management is the process of making and carrying out decisions that will minimize adverse incidents. Such a program requires development of criteria, identification of problems, and action to reduce those problems.

The medical technology management program participates in the organization effort early on and throughout the equipment life cycle by assessing equipment performance. Impact of risk and quality are monitored prior to purchase decision; during installation, maintenance and repair; and as indicators for disposition or replacement. Faulty design,

CAPITAL ASSET REQUEST

This form is required for all fixed asset purchases with a cost of $500 or more.
Please type and complete all requested information.

I. ORIGINATOR

Date   Department   Cost Center   Requested By   Extension

(Includes model #, mfg., name, accessories, color, size, style, etc.; attach any brochures, pamphlets, spec. sheets, etc.)

Item description

Estimated unit cost   Quantity   Total cost   Date required
$ $

Is item in the approved capital budget?   Yes   No   If yes, what is the budget number?

Suggested Vendor(s)

Are building modifications needed?   If yes, have you contacted facilities management?
Is installation required?
Will there be a maintenance contract?   If yes, have you contacted Biomed?
Inhouse
Vendor   Estimated annual cost?
Comments

Signature of department manager   Date

II. Purchasing

Quotes obtained from:
Vendor #1 _____   Terms _____   F.O.B. _____
Recommended _____   Contact _____   Delivery _____
                                         Phone _____   Total price $ ____
C.O.S. Status _____

Quotes obtained from:
Vendor #2 _____   Terms _____   F.O.B. _____
Recommended _____   Contact _____   Delivery _____
                                         Phone _____   Total price $ ____
C.O.S. Status _____

Quotes obtained from:
Vendor #3 _____   Terms _____   F.O.B. _____
Recommended _____   Contact _____   Delivery _____
                                         Phone _____   Total price $ ____
C.O.S. Status _____

Signature of department manager   Date

III. General Instructions

Routing	Initial	Date Rec'd	Initial	Date Fwd
1. Facilities(furniture only)				
2. HIS(computer only)				
3. Biomed(all items except furniture)				
4. Purchasing				
5. Originating department				
6. Accounting				

Step 1: Department completes Section 1
Step 2: Furniture only forward to Facilities
Step 3: Computers only forward to HIS
Step 4: All items excluding furniture forward to Biomed
Step 5: Forward to purchasing
Step 6: Purchasing complete Section 2
Step 7: Return to department
Step 8: Forward requisition to Accounting
Step 9: Secure proper approvals
Step 10: Forward requisition to Purchasing

Figure 30-2 Capital Asset Request (CAR) form.

poor manufacturing, lack of compatibility with existing technology, and mismatch with users' skills or needs can be corrected during the equipment selection and incoming inspection. On the other hand, incorrect operating procedures, the lack of a (or an inadequate) maintenance program, or faulty repair work can be corrected by failure analysis and corrective action based on information collection and an evaluation system that has been described in the JCAHO Plant, Technology, and Safety Management publication. The collection of equipment failure analysis information over several years indicates that equipment risk factor is dynamic. The dynamic equipment risk factor is a modification of a static factor that is assigned to a medical device when it first enters the equipment management program. This static factor is being modified, continuously, over its life cycle by risk factors that derive from information collected about the equipment performance experiences.

Periodically, a summary report of significant equipment-related performance is prepared. The report's data comprise elements that show

1. The ratio of completed to scheduled inspections
2. The number and percentage of devices that fail to pass the prescribed inspection
3. The number and percentage of devices for which a user's complaint was registered, even if no problem was found
4. The number and percentage of devices that show physical damage
5. Devices that were involved in an unusual event, i.e., an accident. Each element is counted as an event, and thresholds above which unsafe conditions may exist can be determined.

Each individual device has its own history and thus its own risk level. The failure elements report allows the structured progression from considering an isolated device performance to clustering equipment users' behavior and their interaction with the devices. In essence, this change is a translation of an equipment repair service into a technology management function that aids the hospital in selecting better equipment, establishing more effective users training that is proportional to measured risk, scheduling maintenance more efficiently, and prioritizing capital replacement. The trending of this information over time will guide the annual review of the effectiveness of the clinical engineering program.

The program should be complemented with a professional communication between the clinical engineer and the various manufacturers. It will result in the availability of better and safer products, less complex operation and maintenance instructions, more effective in-service training, and rapid resolution when action is needed.

A well-managed equipment program provides a systematic approach to controlling technology-related risks in all of its phases, from the needs analysis to equipment disposition. Equipment-related data elements provide qualitative criteria for evaluating equipment and users performance in relation to equipment use. Through the development of a failure analysis program, continuous improvement in equipment performance and simultaneous reduction of risk potential in the clinical environment is achievable.

Technical Asset Management

An accountable, systematic approach will assure that cost-effective, efficacious, safe, and appropriate equipment is available to meet the demands of quality patient care. Such an approach requires that resources committed to the acquisition and the management of medical equipment will be monitored. It is assumed that the financial group manages cost accounting. The medical equipment management program's purpose is to ensure that a process is dedicated to the management of technology.

The MTAC provides a comprehensive and integrated approach to the analysis, implementation, and management of new or additional medical technology. It will turn a fragmented and unpredictable decision making process into a new technique that is well conceived and that supports the hospital's mission. Bold action is required in order to achieve this, including gathering knowledge regarding the trends in medical technology; the development of decision criteria and analytical techniques; the interaction with budget strategies and financial alternatives; the implementation of the capital assets management program; the determination of facility design and long-range services impacts; and the coordination of technology and assets information into hospital operations. This technique will fill many gaps in database reports that are critical to effective operation and hospital performance. Once it becomes integrated, this process may impact a broad range of parameters, including monthly profit and loss, employee productivity, cost accounting, departmental utilization, the effect of physicians' practice patterns on resources, use of hospital resources in relation to patient outcome, analysis of charges, comparative data from other hospitals, profitability forecasts, and procedures pricing.

Asset Management

The attributes of ideal asset management are demonstrated through the continuous availability of robust and reliable equipment and systems at the lowest possible life cycle cost, whenever and wherever needed (see Chapter 35). Asset management attributes are outlined below.

1. Acquisition and equipment life cycle
   (a) Involvement in the process of determining the need for equipment (both short- and long-term needs)
   (b) Preparation of bid specifications and supporting negotiation
   (c) Careful and detailed pre-purchase evaluation and selection
   (d) Development and performance of acceptance testing
   (e) Technical support over the equipment's life cycle
   (f) Recommendations for, and assistance in, its disposition by replacement, refurbishment, upgrade, or declared obsolescence

2. Technical support
   (a) Establishment of complete equipment inventory with control records, files containing operating and service manuals, and testing and quality-assurance indicators
   (b) Incoming equipment acceptance testing and application of a control-number tag
   (c) Hazard and recall notification and incidents handling system
   (d) Periodic, as well as preventive, maintenance of all equipment, performed by either hospital personnel or outside vendors
   (e) Equipment repair, including management and integration of service vendor activities
   (f) Day-to-day assistance to equipment users promoting improvement in clinical use of equipment (e.g., periodic 'equipment rounds' in the diagnostic-imaging department)

3. Information and Training
   (a) Dissemination of user's of manuals and other labels;
   (b) Processing and tracking hazard and recall data
   (c) Initial and ongoing training of all clinical personnel in the safe and effective use of patient care equipment on at least an annual basis
   (d) Investigation of incidents, prompt reporting as appropriate, equipment-related incidents, hazards, and problems. Methods to avoid learned errors should be discussed during staff training

4. Monitoring and evaluation
   (a) Development of, implementation of, and participation in quality assurance and risk management activities
   (b) Periodic assessment of the equipment management program's effectiveness with the combination of objective and subjective data
   (c) Ensuring of effective communication and feedback between relevant personnel in the hospital, (e.g., clinical staff, purchasing, clinical engineering, hospital administration, and equipment vendors). It is important to focus all service-related communication between hospital departments and vendors in the clinical engineering department.

5. Documentation of program activities described above to meet regulatory, accreditation, and problem solving requirements and to minimize liability

A clinical engineering program, through outstanding performance in equipment management, will win the support of the hospital and will be asked to be involved in the full range of technology management activities, including:

- An equipment control program that encompasses routine performance testing, inspection, periodic and preventive maintenance, on-demand repair services, incidents investigation, and actions on recalls and hazards
- Multidisciplinary involvement in equipment acquisition and replacement decisions; development of new services; and planning of new construction and major renovations, including intensive participation by clinical engineering, materials management, and finance departments
- Training programs for all users of patient care equipment
- Quality improvement (QI), as it relates to technology use
- Technology-related risk management

Clinical Engineering Needs

Because medical assets, the technology, the information, and their interaction with users are mission critical, professional management review is expected to guide this process. Clinical engineering professionals have the skills and the competency to provide this service. However, an effective program requires that administrative and clinical personnel have a clear vision of the program deliverables and return on investment. The deliverables must be well documented and periodically reported, highlighting changes in medical-assets characteristics and performance, clinical engineering personnel development and turnover rate, risk mitigation results, cost containment achieved, customer satisfaction, and participation in scientific publications.

To deliver all of these items, a clinical engineering program requires strong and capable leadership, commitment of budget for personnel, test equipment, and appropriate space. Clinical engineering leadership must be able to identify the needs for a quality program and to determine the impact if the expected level of fiscal support is not obtained. To accomplish this, workload and budget allocations per unit of service must be developed and established for the organization and for clinical engineering functions. Because medical assets management consists of a variety of tasks, individual impact and alternatives should be studied and presented to management.

A successful clinical engineering program is largely dependent on adequate budgetary support for training, administrative overhead, subscription to technical services, and access to supplies. Strong relationships with peers in other organizations, including professional societies and vendors, should be encouraged, and information technologies such as computer hardware and software programs are necessary.

Within the organization there should be a demonstration of strong support for the clinical engineering program through clear and immediate communications and involvement of members of the program in space and equipment planning, purchasing decisions, and service contract review. Organizations that have adopted this approach have harvested the benefits of planting and nourishing the seeds of optimal medical technology management.

Reference

David Y, Judd TM. Medical Technology Management. Biophysical Measurement Series, Redmond, WA, SpaceLabs Medical, 1993.

Further Information

Andrade JD. *Medical and Biological Engineering in the Future of Health Care.* Salt Lake City, University of Utah Press, 1994.

Bronzino JD. *Management of Medical Technology: A Primer for Clinical Engineers.* Boston, Butterworth-Heineman, 1992.

Bronzino JD, Smith VH, Wade ML. *Medical Technology and Society: An Interdisciplinary Perspective.* Cambridge, MA, MIT Press, 1991.

Reisner SJ. *Medicine and the Reign of Technology.* New York, Cambridge University Press, 1978.

31

Good Management Practice for Medical Equipment

Michael Cheng
Health Technology Management
Ottawa, Ontario, Canada

Joseph F. Dyro
President, Biomedical Resource Group
Setauket, NY

Establishing essential elements for good medical device technology management practices (Cheng, 1996) is desirable and is in keeping with the development of concepts of best practices or good practices standards and recommendations. In the government-regulated fields, Good Manufacturing Practice (GMP) has long been established as a quality assurance requirement for drugs, and later it was extended to medical devices regulated by the United States Food and Drug Administration (see Chapter 126). More recently, quality systems have become mandatory regulatory requirements for manufacturing medical devices in an increasing number of countries worldwide, following the recommendations by the Global Harmonization Task Force (see Chapter 124). Quality assessment and performance improvement regulations have recently been promulgated by the United States, requiring hospitals to develop and maintain a quality assessment and performance improvement (QAPI) program.

In North America and many other parts of the world, the quality assurance of professional practices is delegated to the professionals themselves (whereas it is delegated normally to their associations). Standards of practice for clinical engineers exist in Canada (see Chapter 123). Similarly, there are good distribution practice, good laboratory practice, good clinical practice, and many other "good practices." Clinical engineers (CEs) and biomedical equipment technicians (BMETs) are professional managers of medical devices in health care institutions. Basic guides to Good Medical Device Technology Management Practices (GMDTM) or Good Management Practices (GMtP) are clearly needed.

The need for GMtP is even more urgent in developing countries for the following reasons: In developing countries, medical equipment maintenance remains an unresolved problem. Over fifteen years ago, the World Health Organization reported that 25%-50% of all equipment in hospital inventories is out of order; the situation has not improved significantly since then. In fact, the situation is complicated by the rapid global diffusion of medical devices. A GMtP could serve as a guide for CEs worldwide. Medical device maintenance is considered one component of several essential elements that constitute medical technology management (see Chapter 30). Placing maintenance within the framework of a total medical technology management system will facilitate the development and sustenance of maintenance programs. Recently, recommended practice for a medical equipment management program was developed in the United States under the aegis of the Association for the Advancement of Medical Instrumentation (AAMI) and was accepted as an American National Standards Institute (ANSI) (see Chapter 122). This was a step in the right direction, and it is a document upon which one could build a more comprehensive approach to medical technology management.

This chapter presents a proposal for a GMtP that incorporates all of the essential elements of technology management (see Chapter 30) and discusses the practical applications of such a good practices document.

Proposed Essential Elements for Management

Starting with the essential element of medical device maintenance, one can readily grasp the fact that many interrelated factors at all of the stages of the equipment life span, if left unmanaged, greatly complicate the maintenance problem. However, it is clear that if maintenance capabilities are considered during the initial stage of planning and acquisition, maintenance problems can be minimized. For a comprehensive and more effective way to manage medical equipment, one can learn from GMP by using the system approach to consider the management of the different aspects within the life cycle of a medical device. The typical life cycle of a medical device has the stages shown in Figure 31-1.

Information gathering for the planning phase normally starts during the last three phases, but it can occur at any phase of the life cycle.

Planning

Maintenance is just one element to be managed. Proper management of each of the other elements has an impact on the maintenance function, but it is especially important in the planning phase. For example, in the planning phase, one can specify the following conditions that should be met in order to aid the decision-making process:

1. Demonstrated needs and benefits
2. Available qualified users
3. Approved and reassured source of recurrent operating budget
4. Confirmed maintenance services and support
5. Adequate environment support
6. Regulatory compliance

If all of these conditions are met before the acquisition of the device, problems (including maintenance) that can occur later will be reduced. If some conditions are not met, potential problems can be anticipated rather than coming as surprises that cause wastes of precious resources as well as obvious frustration for the maintainer. The previous six conditions are simple and should be applied to any routine acquisition of a medical device. A policy on medical device acquisition that includes meeting these conditions as prerequisites to acquisition will reduce medical device problems later in the life cycle of the device. For example, a consideration of the skill level that is required for operating a device will ensure that only the appropriate technology is acquired. Similarly, assessing the financial resources of the acquiring enterprise will determine whether the total cost of operation of a device can be maintained. Items such as cost of disposables, training costs, and maintenance costs can be identified, and appropriate financial planning can ensue. Figure 31-2 shows the "Acquisition Iceberg" with most of its life cycle costs hidden from view. Planning for a large amount of medical equipment or the adoption of

Stages in the Life Cycle of a Medical Device

- Planning
- Acquisition
- Delivery and Incoming Inspection
- Inventory and Documentation
- Installation, Commissioning, and Acceptance
- Training of Users and Operators
- Monitoring of Use and Performance
- Maintenance
- Replacement or Disposal

Figure 31-1 Essential elements for life-cycle management of medical devices.

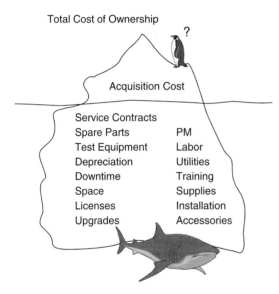

Figure 31-2 Medical device acquisition iceberg.

new technology should also include a broad range of enriched considerations, as indicated in Table 31-1. The significance of managing the other phases of the life cycle is briefly introduced in the following sections, with respect to minimizing the maintenance-related problems.

Acquisition

The acquisition phase can be subdivided into evaluation (see Chapter 33) and procurement (see Chapter 34). Evaluation of devices includes such factors as safety, performance, and ease of maintenance. Such items as the need to standardize on models or manufacturers of equipment should be considered. Furthermore, in the procurement process, conditions can be included in the purchase order to specify that the supplier must supply operating and service manuals, operation and service training, and essential spare parts. Other special requirements also can be specified here, such as withholding payment if specified conditions are not met.

Delivery and Incoming Inspection

Incoming equipment should be carefully checked for possible shipment damages; compliance with specifications in the purchase order; and delivery of accessories, spare parts, and operating and services manuals (see Chapters 6, 95, and 122).

Inventory and Documentation

Medical device inventory and documentation systems can provide information to support different aspects of medical equipment management (see Chapters 35 and 36). One important aspect is consideration for standardization. Inventory entries should include accessories, spare parts, and operating and service manuals. It is advisable to make copies of the manuals for distribution to the users, while the original copies of the manuals should be stored at the technical document library for safekeeping.

Installation, Commissioning, and Acceptance

Installation and commissioning can be carried out by in-house technical staff if they are familiar with a given item of equipment. If installation and commissioning by the suppliers are needed, in-house technical staff should monitor the process so that any technical matters can be noted and recorded on the equipment service history. The occasion also provides an excellent opportunity for in-house technical staff to gain familiarity with the new item.

Training of Users and Operators

Training of users and operators is important in ensuring the safety and effectiveness of medical devices. Operator error is a leading cause of device malfunction. In developing countries, expatriate physicians frequently change equipment, and retraining before the use of complex devices should be mandatory. Incorrect use of devices also will greatly increase maintenance problems.

Monitoring of Use and Performance

One common mistake is to believe that the warranty period is covered by the supplier, so no in-house technical attention is necessary. In-house technical staff should become the link between user and supplier and should observe any supplier's technical services. Such warranty services should be recorded in the equipment service history. This also will provide a learning opportunity for the in-house technical personnel. Vigilance and the regulatory requirements of postmarket surveillance must be implemented.

Maintenance

Proper maintenance of medical equipment is essential in order to obtain sustained benefits and to preserve capital investment (see Chapter 38). Medical equipment must be maintained in working order and must be calibrated periodically for effectiveness and accuracy (see Chapter 37).

Replacement or Disposal

Given that the majority of medical equipment in developing countries is old and that spare parts are often in short supply, some old units can be dismantled to provide spare parts for similar units. This process also will provide an opportunity for cultivating technical innovation using local resources. Disposal of equipment must follow safety procedures (see Chapter 61) in order to protect people and the environment. The equipment inventory must be updated to reflect the fact that devices have been retired.

The preceding nine management functions are proposed as essential elements for good medical device technology management practices. From the simplest one-technician workshop to an elaborate national network of workshops and from the poorest country to the richest, GMtP requires the appropriate management of each phase during the life span of medical equipment.

Enrichments of the Essential Elements

The management of each phase can be enhanced when needs arise and when the resources are available. For example, the basic maintenance service is curative repair. Preventive maintenance can be added whenever time and resources become available. Further enhancement would add risk management and quality improvement. Planning, in addition to the six basic items listed previously, would include health technology assessment, national health policy, and socioeconomic considerations (see Chapter 32). Information technology can enhance one or all of the phases. Table 31-1 offers a few suggestions, but ultimately this table should include all topics associated with the management of medical devices.

Table 31-1 Enriched elements in life cycle management

Essential element	Basic application	Enriched application
Planning	Basic planning questions before acquisition	Enriched tools
	Demonstrated needs and benefits	Health technology assessment
	Available qualified users	National health policy
	Approved and reassured source of recurrent operating budget	Health reform policy
	Confirmed maintenance services	Socioeconomic elements
	Adequate environmental support	Political-cultural elements
	Regulatory compliance	Alternative health care
		Other considerations
Acquisition		
Delivery, incoming inspection		
Inventory and documentation	Paper or cards	Computers
Installation, commissioning, and acceptance	Risk management, quality improvement, and information technology	
Training users and operators		
Monitoring use and performance		
Maintenance	Corrective, preventive	
Replacement or disposal	Personal safety and environmental procedures	

Applications

Table 31-1 provides a framework that integrates all topics relating to medical equipment management. "High profile" topics (e.g., regulations, health technology assessment, acquisition strategy, risk management) are linked to the shop level operational elements. This holistic framework provides a "big picture" for everyone, and it can serve as a simple reminder to senior management that planning is an integral part of GMtP. In the training of new technicians, this framework is also very useful for seeing the basic elements in relation to advanced topics.

Good management practice using the life cycle approach could offer an effective management system in developing countries, to meet the challenge of the rapid global diffusion of medical technologies. It also should improve the recognition by decision makers of the tasks and skills needed by medical equipment professionals to manage medical equipment properly. The life cycle management concept was reported in a health care policy mission to Mongolia (Cheng, 1996). In the Republic of Vanuatu, national policy and guidelines for the management of medical equipment based on the life cycle management concepts were established in 2001 by the country's Ministry of Health. (Cheng 1999; MoH, 2000). Again, based on the life cycle management framework, Lao (2003) has recently established a medical equipment management policy and has presented this national policy at an international forum (World Bank, 2003).

Some medical equipment professionals in industrialized countries may claim that managing these simple essential elements is already their daily routine and there might not be a need to state the obvious. While this may be true, it further emphasizes the importance of documenting what basic good practice is. Furthermore, many CEs and BMETs have expressed their lack of involvement in the planning and acquisition aspects of medical equipment. This good management practice guide should help to legitimize their role and to gain support from senior management to include CEs and BMETs in the activities that are important to good management of medical devices.

One suggested reason for the unsatisfactory development of medical equipment management capacity in developing countries was the lack of an international standard to guide this development (World Bank, 2003). The GMtP framework can be developed into an international standard, with detailed application tools for international applications. This standard also can serve as a guideline for monitoring, evaluation, or audits in a way that is similar to the use of other quality standards. Weakness in certain areas of service can be systematically identified and strengthened in this manner.

Developing countries have expressed a need to know "how to fish" for themselves in managing medical technologies (World Bank, 2003). For many developing countries, the life cycle management framework described in this chapter and the practical maintenance strategy (see Chapter 38) can be used together systematically to manage medical equipment. No significant external resources are needed for the application of this methodology. They can serve as tools to "fish for oneself."

References

Cheng M. Ulaanbaatar, Mongolia Mission Report. RS/96/0416. Western Pacific Regional Office, World Health Organization. Geneva, Switzerland, 1996.

Cheng M. Vanuatu Mission Report. MR/1999/0446, Western Pacific Regional Office, World Health Organization. Geneva, Switzerland, 1999.

Medicare and Medicaid Programs; Hospital Conditions of Participation: Quality Assessment and Performance Improvement; 42 CFR 482; Federal Register 68(16):3435-3455.

Lao PDR. Medical Equipment Management Policy. Ministry of Public Health, 2003.

Republic of Vanuatu Ministry of Health. National Policy and Guideline for the Management of Medical Equipment. Ministry of Health, 2000.

World Bank. International Forum for Promoting Safe and Affordable Medical Technologies in Developing Countries. World Bank, May 2003.

32

Health Care Strategic Planning Utilizing Technology Assessment

Nicholas Cram
Texas A&M University
College Station, TX

The effective and efficient management of medical technology has become a matter of financial survival for health care facilities. Coordinating purchases of medical equipment and medical equipment service contracts requires both technical and managerial expertise. Regardless of the size of the health care facility, the process of successfully maintaining and acquiring medical equipment remains the same. Only the economies of scale differ with regard to size. Medical technology assessment provides a management tool linked to the specific needs of a specific health care facility, as well as the mission and vision statements of that health care facility. This linking process of assessment and acquisition of medical technology is vital to long-range strategic planning. In such planning, existing and projected additions of technology within the health care system, must be evaluated to determine necessity, efficiency, safety, and cost versus benefit. Existing and newly acquired medical technology must also be evaluated with respect to the current and future needs of the health care organization. Developing a technology evaluation system that incorporates technology forecasting and off-the-shelf technology with specific needs and future goals is the most efficient way to provide a decision making template for strategic planning. Medical technology purchase decisions are often made with respect to a single department or a single event, without a systems approach or multidepartment involvement. A well-defined technology assessment program can incorporate provisions for medical device system interfacing and future growth, which is usually not a priority with the single-department approach to medical technology acquisition.

The ability to interface disparate devices with a central monitoring area and/or database is an ever-growing requirement among health care systems. Both in-house interfaces and telemedicine applications need to be considered, even if they are not currently required. Coordination of purchases through a technology development plan based on technology assessment will prevent disappointments and costly retrofittings.

Standardization and economies of scale can be achieved with a system-wide approach to technology acquisition and can provide a better bottom line for the organization. The development of medical device vendor standards must include clinical and technical criteria. A lower cost product that is of poor quality and not accepted by the clinical staff could potentially jeopardize patient care. The selection of a standard product, therefore, must involve a combination of quality, effectiveness, and cost factors, all of which can be achieved with a medical technology assessment program.

The Medical Technology Assessment Process

The most robust and comprehensive method of technology assessment is known as randomized controlled clinical trials (RCTs). Although this method provides extremely useful information, the process is complicated, expensive, and time consuming. Clinical engineers (CEs) and chief technology officers (CTOs) need to have access to a continuous flow of information concerning the technology baseline and future needs of their specific organizations. Medical technology assessment (MTA) is a continuous process of evaluating the medical equipment in use, planning for future technology needs, and acquiring medical equipment. The most important and essential step in beginning a medical technology assessment program is the existence of an accurate equipment inventory. The most efficient review of equipment is provided with a computerized maintenance management system (CMMS) (see Chapter 36). A physical inventory should be taken annually and should be verified against the inventory reflected in the CMMS database.

The technology baseline is accomplished using a medical equipment audit, which consists of both an inventory and a utilization level measurement of all medical devices that the facility is currently using. An inventory listing alone is not useful as an audit tool to establish the technology baseline. If the facility has 100 devices listed on the inventory, and 50 are never used due to either maintenance problems or user preference, the actual technology baseline should reflect this fact. As an example, during a preventive maintenance check of infusion pumps, the CE finds several of the items stuffed in the corner of the soiled hold room. Upon investigation, the CE learns that the nursing staff has not used the pumps in six months because they did not understand the programming features of the pumps. The solution is to determine whether operator training in-service will solve the problem or whether the technology is so complicated that it is not user friendly. In either case, the actual utilization of the equipment must be reflected in the technology baseline, which is accomplished with the technology audit.

Planning for the future technology needs of the organization is based on patient surveys, mission and vision statements, and demographic information of the medical population served. An integrated, interdepartmental committee with representation from clinical engineering, nursing, medical staff, purchasing, and administration provides the best results for defining and implementing the future technology needs of the organization. As with all programs, support of the medical technology assessment process must come from upper administration in order to receive the resources needed for it to be successful. Because of the importance of technology management in relation to competitive positioning, and because of the effect of technology costs on the bottom line, many health care organizations have created the position of Chief Technology Officer (CTO). In organizations with a CTO, medical technology assessment is a primary focus of one individual. In larger organizations, the chief cechnology officee (CTO) is typically the technology committee chair.

Acquisition of medical technology (i.e., medical devices) will be determined by the technology baseline of the facility and the needs of the organization, which are determined by the technology planning committee. Medical technology assessment (MTA) of new equipment can be accomplished through in-house resources of the clinical engineering department and clinical staff or through secondary sources. Certain devices, such as those used in imaging, are not appropriate candidates for in-house evaluations due to size and portability restrictions. Site visits and secondary sources provide the best means of MTA for devices in this category.

Technical and clinical in-house evaluations are limited in scope but can provide valuable hands-on assessment of specific devices, that cannot be achieved through any other methods. The in-house assessment process can be accomplished using a weighted matrix or ranking system for decision making. A weighted matrix is a comparison of a vendor's ability to meet certain preselected criteria. The weighting occurs due to valuing the criteria at different levels ("points"), depending on their priority. A weighted matrix is valuable in selecting the best equipment to purchase for the specific area of priority. This process is similar to that provided by *Consumer Reports* magazine in comparing various manufacturers' products. An example of a weighted matrix is illustrated in Table 32-1, in which a point system (0-5) is used to evaluate the various features of four telemetry systems considered for purchase. Table 32-2 shows an example of a report to a technology committee, based upon the results of the clinical evaluation shown in Table 32-1, in which advantages and disadvantages of the four systems under consideration are summarized.

A ranking system helps to focus on a particular area of technology. For instance, if a certain amount of capital budget money is set aside for technology acquisition in the upcoming year, a ranking system can set priorities for how much and where that money is spent. Secondary sources of information include published data, peer group shared information, group purchasing organization (GPO) flyers, manufacturer inquiries, and subscriptions from private and governmental organizations. A full discussion of MTA is presented elsewhere (David and Judd, 1993) and in other chapters of this handbook (see Chapters 30 and 33). But a brief overview of the in-house evaluation process and secondary MTA resources will be provided to allow continuity of the topic.

In-House Clinical and Technical Evaluations

An in-house clinical evaluation of new medical equipment should never be confused with a clinical trial. A clinical trial involves medical devices, that have not been cleared for commercialization by the Food and Drug Administration (FDA) under paragraph 510(k) of the Medical Devices Amendments of 1976 (see Chapter 126). These devices may be part of a study to obtain FDA premarket approval (PMA). In-house clinical evaluations, on the other hand, are performed with medical devices already in production by medical device manufacturers. The in-house clinical evaluation process uses nursing and/or clinical staff members, for evaluation of the new technology over several days as a part of their normal clinical practice or nursing protocol. As mentioned earlier, criteria are selected for the weighted matrix, and the clinical staff will evaluate each vendor against the others, based on the criteria. Clinical criteria should relate to clinical issues and user-friendliness of the equipment. Examples of clinical criteria include monitor visibility, range and clarity of signal, ability to program the device, and human-factors elements. Human-factors elements are related to features on the medical device that are related to user-friendliness. For example, if a bedside monitor requires three steps in order to print an ECG strip, is the three step process intuitive? Does it require that a blue button be pressed, and then a red button? Is the system Windows®-based? Is there a point-and-click process similar to those of personal computer (PC) processes? How quickly did the nursing staff learn how to use this new medical device after the initial presentation? The clinical evaluation is designed for selection of the medical device that best suits the clinical (i.e., patient care related) needs of the clinical staff, regardless of the price. Therefore, relative vendor pricing never should be disclosed to the clinical staff until after the evaluation process is completed.

The in-house technical evaluation should involve maintenance and interface related criteria such as mean time between failures, nonproprietary features of hardware and software, cost of parts, cost of the system, cost of ancillary parts (e.g., cables, pads, electrodes), installation requirements, ability to be upgraded, scalability, and the cost and availability of technical and clinical training. It should be noted that the criteria for in-house technical and clinical evaluations also could serve as selection criteria for information received from secondary sources of MTA.

Table 32-1 Prepurchase evaluation – Clinical Engineering department Technology Assessment: Clinical and technical evaluation of telemetry systems

Clinical Evaluation
Telemetry System–Budget Item #7

Vendor	Company 1	Company 2	Company 3	Company 4
CRITERIA				
Water-resistant	4.1	3.5	3.6	4.0
Training	4.3	3.3	4.2	3.9
Ease of use	4.2	3.5	4.4	4.5
Monitor visibility	4.2	3.2	4.0	4.6
Range & clarity of signal	4.2	3.7	4.0	4.4
Upgrade	4.0	3.4	4.5	3.0
ST segment analysis	4.1	3.6	4.1	0.0
72-hour trending	4.0	3.6	4.0	4.0
TOTAL POINTS	33.1	27.8	32.8	28.4

Criteria scored 0-5 points, with 5 as best score. Company scores reflect the average score given

Technical Evaluation

Vendor	Company 1	Company 2	Company 3	Company 4
CRITERIA				
Water-resistant	5.0	3.5	3.6	3.5
Training	5.0			3.0
Nonproprietary leads & cables	5.0	1.5	3.0	4.0
Nonproprietary boards & parts	5.0	1.0	3.0	4.0
PC-based	5.0	1.0	5.0	4.0
Upgrade	5.0	3.0	5.0	0.0
Installation costs	5.0	2.0	2.04.0	
Reputation of vendor	3.0	9.0	3.0	7.0
TOTAL POINTS	38	26	27.6	29.5

Criteria scored 0-5, except "reputation of vendor" 0-10, with 5 and 10 as best scores respectively

Planning Strategies

The format for MTA presented in this chapter is an expanded version of the approach of ECRI, a long-time advocate of medical technology management and strategic technology planning (ECRI, 1989; ECRI, 1992). All planning strategies must have the support of upper management in order to be successful. This support must include budgetary resources required to carry out the plan. As in the case of MTA, the medical technology strategic plan is long-range (i.e., 5 or more years) and is continually evolving. The plan should be updated at least annually.

Two key elements of the technology strategic plan are the technology audit and the community needs assessment. The technology audit provides a technology baseline for a specific facility. In essence, this is the technology snapshot of where a facility is today. The community needs assessment provides essential information regarding potential opportunities for expanded health care services among a specific patient population. The community needs assessment provides a *technology vision* of where a

Table 32-2 Technology assessment: Final report to technology committee

Company 1 ($198,000)	Company 2 ($217,000)
Advantages:	Advantages:
1. Price	1. Price includes two portable monitors for transport
2. Best score for both clinical and technical evaluations	2. Not user-friendly
3. Committed to research and development	Disadvantages:
4. Technical and clinical training provided	1. No ST segment analysis
5. Full disclosure system	2. Upgrades depend on agreements with another company
Disadvantages:	3. No technical or clinical training
1. Lesser-known company; no track record	4. Not an OEM; reseller of device
2. Not an OEM; reseller of device	

Company 3 ($225,000)	Company 4 ($349,000)
Advantages:	Advantages:
1. Excellent vendor reputation	1. Best score on clinical evaluations
2. System operates in UHF range for improved signal quality	2. User-friendly
Disadvantages:	Disadvantages:
1. Not user-friendly	1. Price
2. Not PC-based	2. No clinical or technical training provided
3. Worst overall score on clinical evaluations	3. Unknown vendor reputation

health care facility should position itself. Combined with demographic information, the community needs assessment and the technology audit are powerful tools for strategic technology planning.

The sole purpose of the medical technology strategic plan is to support the clinical strategy of the health care facility. The medical technology strategic plan therefore must have balanced input from both clinical and technology sectors of the health care facility. Valuable first-hand information can be obtained from the technology audit. Additional clinical input can be obtained through a committee setting or face-to-face interviews with key decision makers in your facility.

The medical technology strategic plan is a marriage of convenience. The clinical goals and clinical strategy are compared to the baseline technology provided by the technology audit, and the basic question is posed: How do we get there from here? It may require the purchase of new medical devices, or there may be a need to support a completely new technology. An awareness of emerging technology should always be incorporated into the medical technology strategic plan. There will be an increasing trend in health care technology for software-driven applications, including remote diagnostic procedures for medical devices, robotics, and electronic records automation. Although the exact impact of biomedical research might not be predictable, clearly noninvasive and minimally invasive procedures are on the rise. Changes in your health care facility's infrastructure must be considered today in order to avoid costly renovations or missed opportunities in the future. Since all planning involves futuristic targets, there will always be a risk of unknown events or circumstances, which have an impact on the planning process.

Planning must be a fluid process that is flexible enough to allow for change. Risks can be minimized by thorough research, and specific contingency plans should be provided for suspect critical elements of any strategic plan. Attending professional conferences and subscribing to professional journals will provide adequate information on emerging technology and health care trends. The Internet is a rich source of emerging technology information. All facility strategic plans should support and complement the mission and vision statements of the organization. Figure 32-1 provides an illustration of a flow chart for a technology strategic plan.

Utilization of Automated Systems

The flow of information for any clinical of engineering process can be overwhelming. Coordinating, reviewing, and implementing of process changes are best accomplished using automated systems. Manual methods of reviewing and sorting mountains of information are inefficient and not practical. Most computerized maintenance management systems (CMMS) provide user-defined fields. The CMMS provides a database of all medical equipment in the system. CMMS reporting and trending capabilities provide complete maintenance histories of all medical devices in the system. Cost benefit and equipment life cycle calculations can be derived from user-defined fields in the CMMS. A user-defined field allows the system user to create analysis reports, which automatically update the status of priority areas. The priority areas are flagged in the CMMS according to indicators identified by the user. An indicator is defined as a quality characteristic of a process that can be measured and contributes positively or negatively to an outcome. Indicators are best identified when a process is detailed in a flow chart. An example of a process flow chart with identified indicators is illustrated in Figure 32-2.

Defining Terms

Chief technology officer (CTO): An organizational position that is responsible for the management, application, and strategic planning of all technology within the organization.

Community-needs assessment: A survey or questionnaire that provides essential information regarding potential opportunities for expanded health care services among your specific patient population. The community-needs assessment provides a technology vision of where your health care facility should position itself.

Indicator: A quality characteristic of a process that can be measured and contributes positively or negatively to an outcome.

In-house clinical Evaluations: An assessment process used in medical technology assessment that uses facility patient care staff for hands-on analysis of medical equipment in a pre-purchase scenario.

In-house technical evaluations: An assessment process used in medical technology assessment that uses facility technical (Clinical Engineering) staff for hands-on analysis of medical equipment in a pre-purchase scenario.

Medical technology assessment (MTA): A management tool linked to the specific needs of a specific health care facility, as well as the mission and vision statements of that health care facility. A continual process of assessment and acquisition of medical technology is vital to long-range planning. Existing and projected additions of technology within the health care system must be evaluated to determine necessity, efficiency, safety, and cost/benefit. Existing and newly acquired medical technology also must be evaluated with respect to the current and future needs of the health care organization.

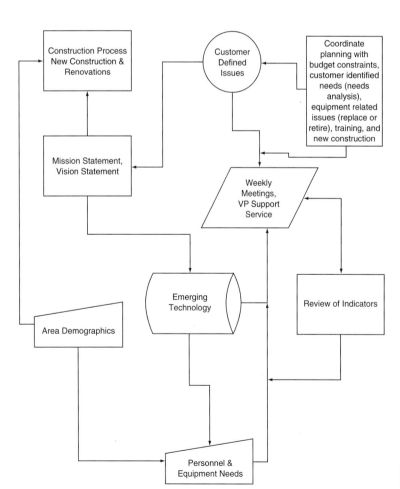

Figure 32-1 Clinical engineering and telecommunications strategic planning process.

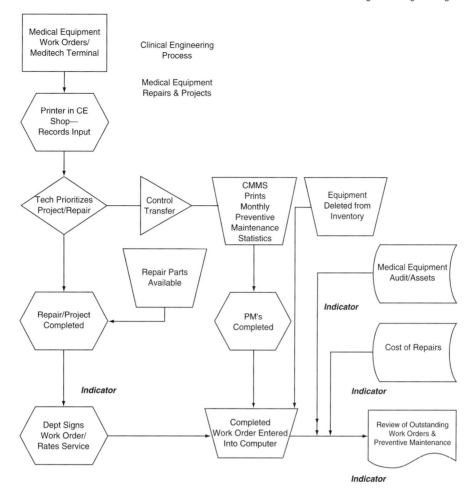

Figure 32-2 Flow chart of a clinical engineering process, medical equipment repairs and projects, showing four indicators: Repair/project completed, medical equipment audit/assets, cost of repairs, and review of outstanding work orders and percentage of completed preventive maintenance procedures.

References

Bronzino D. *Management of Medical Technology: A Primer for Clinical Engineers*. Boston, Butterworth/Heinemann, 1992.

Carey RG, Lloyd RC. *Measuring Quality Improvement in Health care: A Guide to Statistical Process Control Applications*. New York, Quality Resources, 1995.

Cram, N. Using Medical Technology Assessment as a Tool for Strategic Planning. *J Clin Eng* 24(2): 113-123, 1999.

David Y, Judd T. *Medical Technology Management*. Redmond, WA, SpaceLabs Medical, Inc, 1993.

ECRI. Technology Management: Preparing Your Hospital for the 1990s. *Health Technol* 3(1):1-43, 1989.

ECRI. Beginning Strategic Technology Planning; Capital, Competition, and Constraints: Managing Health care Technology in the 1990s. Plymouth Meeting, PA, ECRI, 1992.

33

Technology Evaluation

Gary H. Harding
Greener Pastures, Inc.
Durango, CO

Alice L. Epstein
CNA HealthPro
Durango, CO

Participation by clinical engineers in the evaluation of technology can directly contribute to and affect the quality of patient care and patient outcomes and financial situation of the facility. The extent of training required to effectively participate in the process and the workloads of various department required to contribute will need to be analyzed. The impact of new or reengineered technologies varies from the time the technology is introduced through its obsolescence. Technology evaluation is most often associated with complex devices, but clinical engineers should be aware that what appear to be simple, noncomplex devices should be subject to the same technology evaluation processes; otherwise, there can be dramatic and negative impacts. Technology evaluation is not merely the review of devices or supplies, but the review of those devices or supplies for particular clinical need(s). If the technology selected does not meet the clinical needs of the patient or cannot be successfully used by those who are responsible for direct patient care, then needed or optimum procedures could be impossible to provide. If the technology overlaps the utility of existing equipment or is rapidly made obsolete by other technology expected to be introduced onto the commercial market, there can be a substantial negative financial impact. If user and service support are not considered, frequent training and retraining in correct use of the technology and mitigation of accidents, as well as constant down time for repair or maintenance, can result.

Clinical engineering's contribution to careful and complete evaluation of technology can ensure that the introduction of new technology meets the medical goals of the organization in the present and, to the extent possible through forecasting, in the future. Examining ways in which technology would integrate system-wide into the organization, or what alternatives there are or might be in the near future, can help to control costs, thereby improving the organization's financial situation. Determining whether and how technology can be rapidly assimilated by clinical staff and whether service expectations are within reason and acceptability can reduce training requirements, accidents, and down time.

This chapter will discuss many of the critical aspects of technology evaluation. Evaluation of technology is a team effort. Members of the clinical staff, administrative management, facilities management, and many others should (and do) participate in this process. As with all team efforts, there will have to be concessions because of such things as cost and user limitations. While there are many acceptable and necessary reasons to make concessions, clinical engineers must be careful to identify and to evaluate requested concessions that do the following:

- Impact patient and/or staff safety
- Are not truly required, as the limitation is perceived and not real
- Have potentially severe utility or financial implications
- Fail to meet laws, regulations, or acceptable practice

Evaluating technology can be a challenging and rewarding experience for the clinical engineer—one in which others can learn from the engineer, and the engineer can learn from others.

Strategic Technology Planning

Where does (or should) an organization start in the technology evaluation process? Because of the importance of technology to the overall patient care and financial goals of the organization and the risks of haphazard acquisition, the first step in technology evaluation should be development and implementation of a strategic technology planning program (or process, depending on the size and resources of the organization) (Cram 2003). The main purpose of strategic technology planning is to:

Gain a complete understanding of the existing technology-related equipment and services within the organization and the expected progression of and changes in services that can be forecast with an intent to provide a base on which to consider and time acquisition of future technology.

The clinical engineer has the primary responsibility for identifying existing technology within the organization. This duty often takes the form of the clinical engineering inventory, although it is not uncommon to supplement this with inventories from specialty departments such as radiology and clinical laboratory, and from capital equipment financial bases, e.g., accounting. For the purposes of strategic technology planning, a complete tabulation of current technology should be developed and should include not only the name and manufacturer, but also the age and expected useful lifetime. Other information (e.g., location in the facility and progressive annual costs to use and maintain) can be useful as well, although more detailed information can be acquired later in the process as specific technologies are considered. The importance of this tabulation is primarily to assist in determining the status of current technology within the organization that may or may not be in actual clinical use, vendors that have supplied technology, and technology approaching obsolescence and need for replacement.

An inventory and overview of technology-related clinical services delivered by the organization is typically the responsibility of administrative management. It might include departmental administrators and/or clinicians. It is important that an overview of current technology-related services be compiled so that (1) future requests for new services can be compared against services already provided; and (2) future requests for new technology acquisition can be examined for ability to meet current needs and to expand utilization in areas other than that of the requester. It would be unwise to purchase a new surgical laser for one department when there was an existing, acceptable surgical laser in use by another department that was underutilized, for example. It would be important to know that a request for new service by one medical specialty is a service already provided by a different medical specialty, or that a technology requested for one procedure could be used in many more. Technology relating to imaging illustrates this point. Efforts by the Agency for Health care Research and Quality (AHRQ) in 2001 included evaluating Positron Emission Tomography not only for its application in diagnosis of breast cancer, but for myocardial viability, soft tissue sarcoma, thyroid cancer, and Alzheimer's disease. Once the existing technology and services are compiled, requests for replacement or acquisition of new technology can be pursued.

In order to ensure that the strategic technology planning process is supported, it is important to seek participation by affected parties. It is not essential that all parties be represented in, for example, a strategic technology planning and acquisition committee, but some mechanism(s) to ensure that the voices of affected parties are heard and that rationales for decisions are distributed to these parties must be provided. There will be difficult, and often political, decisions about replacement, supplementation, and introduction of technology that be necessary to make. In today's economy, financial resources are limited and must be distributed carefully. Unrealistic requests may be made; powerful participants may not understand the relative importance of their need or desire versus the relative importance of another participant's need outside their expertise. One value of the strategic technology planning process is that participants can be advised of all of the needs and the rationales for decisions. Experience shows that these highly trained and motivated participants typically endorse decisions or offer constructive alternatives when provided with valid information. Organizations might want to consider utilizing independent consultants for initial process development and periodic participation in lieu of developing their own structure for strategic technology planning.

Technology and Alternatives

Once the strategic technology planning program has been instituted and identification of technologies for consideration for acquisition occurs, the process of technology evaluation can proceed. The first step in the process is to identify and assess the requested technology and alternative technologies already in use within the organization or available commercially that could meet the identified clinical need. This process will formalize the following:

1. Identification and detailing of the clinical procedure(s) for which the requester intends use of the technology
2. Collection of information about the specific technology being requested
3. Identification of other clinical procedures for which the technology could be used
4. Collection of information about alternative technologies that could be used for the same clinical procedures
5. Comparison of the requested and alternative technologies to the existing technologies within the organization
6. Determination of the risks and hazards associated with the use of the requested and alternative technologies
7. Examination of the clinical efficacy of the requested and alternative technologies
8. Performance of a conceptual needs analysis based on all of this information

Clinical Procedures

It is typical for there to be an official or unofficial clinical sponsor or advocate for acquisition of the technology. The sponsor should provide a narrative and supporting literature (e.g., authoritative published reviews of the procedures and use of the technology) for the technology being requested. The narrative should provide the sponsor's understanding of the clinical procedures for which the technology may be used now and in the future; their analysis of the clinical literature that clearly supports the use of the technology in these clinical procedures; the technology they are currently using or could use to perform these clinical procedures; and their reasons why the specific technology should be acquired. Complete copies of authoritative published reviews are desirable. A review is an article that examines experience derived from many studies; these reviews often can save those who are responsible for making decisions from having to seek out, select, obtain, and read volumes of separate articles—all of which may have individual limitations in their performance and application. Reviews often provide detailed bibliographies that provide additional references in one place and also provide the reviewer with an idea of how well established and analyzed the technology is. For example, if no reviews exist, or only three articles exist in the world, or all articles are foreign-based, the reviewer could gather important information on which additional questions to ask.

The sponsor also might be aware of, or might seek out, information about other clinical procedures for which current technology, requested technology, and alternative technology may be used. Resources that are readily available for the sponsor in this endeavor include peer-reviewed literature, the medical director and medical staff, and the manufacturer(s) of the technologies.

Comparing Requested and Alternative Technologies

The clinical engineer is well prepared to direct the effort to collect information about the requested technology and alternative technologies. A literature search should be performed emphasizing the specific technologies and include related medical literature, as well as engineering, financial, and governmental databases (see Chapter 86). Resources that identify manufacturers of the technologies (e.g., group purchasing lists of acceptable equipment, Medical Device Register (MDR, 2003), the Health Devices Sourcebook (ECRI, 2003), and MD Buyline (MD Buyline, 2003)), should be consulted. Requests for information should be made to each manufacturer of the requested and alternative technologies, emphasizing that the manufacturer should provide all information, e.g., brochures, user manuals, sales and marketing documents, support documents, and published articles that the manufacturer believes the organization should consult before making a decision about the technology acquisition. Organizations could specifically ask each manufacturer to provide a comparison of their product to that of its competitors. Such a comparison should be readily available or easily compiled, as it is important for sales and marketing professionals to differentiate their product from those of competitors. It is convenient at this time to request data that the manufacturer either submitted to the FDA (see Chapter 126) or identified to the FDA as available to support the manufacturers' contention that the technology is safe and effective for its intended purposes, as well as a compilation of the clinical procedures that the manufacturer identifies as an intended purpose of the device. In new or developing technologies, it is worthwhile to ask specifically which clinical procedures normally encountered within the medical specialties for which the technology is used are not currently within the intended purposes of the technology.

Comparing Owned vs. Requested or Alternative Technologies

Once information about possible clinical procedures and technologies has been received, and before any other extensive investment of time or money, the information should be reviewed and compared against comparable information on technology that is already owned. All affected parties should be consulted, and their viewpoints should be considered. While much of the discussion might have occurred as part of the strategic technology planning process, this effort provides an opportunity for parties to become aware of the latest information about requested and alternative technologies and how they could impact on, or be utilized by, all parties. Identification of potentially unacceptable financial impact or equipment compatibility issues could arise. Participants from other medical specialties can identify an opportunity and interest in supporting and using the technology if, for example, necessary options for their specialty are purchased; this can be particularly important to the original sponsor if the sponsor's utilization projections are marginal. A rational decision to forgo acquisition of the requested or alternative technology, to use or upgrade existing technology, or to proceed with the process as existent or modified can be made.

Risks, Hazards, and Clinical Efficacy

It is important to remember a basic premise in health care: First, do no harm. In order to meet this basic premise, potential risk must be identified. While adverse clinical events occur, and some might be inevitable given the complex nature of medicine, unnecessary complications and accidents should be avoided when possible. It is imperative that risks be minimized and eliminated where possible through the proactive identification of risks and hazards to identify whether and how they can be mitigated (see Chapter 56).

Process

Most engineers are familiar with standard design engineering practices, such as Failure Modes and Effects Analysis (FMEA) (Stamatis, 1995). One can consider technology risk and hazard analysis to be analogous to a system-wide process applied to the technology rather than the device (Stalhandske et al., 2003). The points are the same: Eliminate unnecessary failure modes; guard against those that cannot be eliminated for some reason; and warn about those that cannot be eliminated and where guarding alone is insufficient to mitigate the risk or hazard. Elimination of risk or hazard could mean redesigning a feature of the product while in device design. In technology evaluation, elimination could mean discontinuing consideration of a product from a specific manufacturer. Similarly, while in device design, a fail-safe limit switch may be provided to guard against the potential for limb entrapment; in technology evaluation, however, the guard might be accepting products only from vendors who have a commercially available system that has been in place in other facilities for some time, without resultant adverse occurrences. Furthermore, while a caution label could serve as the warning in device design, alerting medical staff to limitations in use of each of the technologies under consideration could be an analogous technology evaluation warning. Hence, it is important that a systematic, unbiased process of risk and hazard identification and mitigation be used for technology evaluation. While the process can be identified by many names and the features vary somewhat from process to process, the goals of the risk and hazard analysis process should be the same.

Financial Risk

Risk, for the purposes of technology evaluation, means not only the risk of physical injury or facility damage typically recognized, by clinical engineers, but also financial risks that the technology may pose as well, e.g., underutilization, unacceptable dedication of financial resources, loss of clinical specialists, increase in liability premiums or potential losses, or negative publicity. While addressing certain risks would be beyond their ability or control, clinical engineers can ensure that a participant in the process who can address them is responsible to identify and review risks and their extent. While any aspect of technology evaluation can be a "make or break" item, more often decisions are based on a comparison of many aspects, with financial risk being significant.

Resources

Unfortunately, the governmental agency that we often think of as being responsible for ensuring the safety and efficacy of medical products, the United States Food and Drug Administration (FDA) Center for Devices and Radiological Health (CDRH), has a history of poor performance in this regard, as demonstrated by the number of patient and staff injuries caused by medical devices. While physicians and other medical specialists often believe that if a product has gone through the FDA process it is safe and effective, the reality is that for many device types, and in many cases, the FDA process does not determine safety and efficacy. Due to human and financial resource limitations primarily imposed by Congress, the FDA often relies on the manufacturer of the product to determine safety and efficacy. In the 510(k) premarket clearance process, the FDA often does not even require the manufacturer to provide the data on which the manufacturer made its determination that the product was safe and effective. Hence, potential purchasers of technology cannot, and should not, rely on a manufacturer or a technology having received FDA clearance to market the technology as any indication that the product is safe and effective. Greater assurance is provided if that product has gone through the premarket approval (PMA) process, usually reserved for Class III, complicated, or other technologies identified in the past as particularly problematic.

The FDA process can be abused by manufacturers, distributors, and others as a means, for example, to avoid cost, dissemination of information, or appearance of acceptance of responsibility to determine safety and efficacy. For example, a large manufacturer who held and still holds major market share in injection and blood-collection needle products did not perform unbiased, statistically significant clinical studies prior to, or for years after, placing its safety products on the market to determine that these safety products actually reduced the occurrence of accidental needle sticks. Instead, the manufacturer made it a de facto responsibility of each organization to determine on its own whether the safety products were "safe and effective."[1] CDRH cleared such products from this and other manufacturers again and again, for years, without review of safety data or even requesting that the data be provided to them.

While organizations should request this information from the manufacturer and should examine FDA and/or CDRH databases,[2] a healthy degree of skepticism in findings, or the lack thereof, could be warranted. If the manufacturer did not submit clinical data to the CDRH and does not provide this data in response to the request by the organization, it is difficult to offer a valid reason to continue to consider that device. It is reasonable to expect that the manufacturer is most knowledgeable (i.e., expert) about the product and its safety and efficacy. If a manufacturer cannot, does not, or will not provide substantiating information that is convincing, it is reasonable to exclude that manufacturer's technology. If a manufacturer transfers the responsibility for determining safety and efficacy to the organization, it is reasonable to exclude that manufacturer's technology from

---

[1]This manufacturer introduced a safety-needle product in 1988. However, it was not until Younger et al. (1992), the CDC (1996), and Mendelsohn (1998) performed studies that clinical safety and efficacy of safety devices were evaluated.

[2]The CRDH website has databases relating to registration of the manufacturer, listing of devices, 510(k)s, PMAs, and occurrence reports, which can be accessed at http://www.fda.gov/cdrh/databases.html.

further consideration. Document the situation so it can be conveyed to the clinical sponsor in an unbiased, straightforward manner.

There are other governmental and private agencies that can offer information on technology evaluation. For example, the Agency for Health care Research and Quality (AHRQ) performs technology assessments on behalf of the Center for Medicare & Medicaid Services (CMS) for the purpose of CMS determining whether Medicare/ Medicaid will provide financial reimbursement for application of the technology. The AHRQ has performed a number of technology assessments since 1990.[3] AHRQ states that it (or its evidence-based practice centers) uses state-of-the-art methodologies for assessing the clinical utility of medical interventions. Reportedly, their assessments are based on a systematic review of the literature, along with appropriate qualitative and quantitative methods of synthesizing data from multiple studies. While useful information can be gained from studies performed by, or on behalf of, AHRQ and CMS, it is important to differentiate the interests and motivation of the organization from the interests and motivation of a governmental agency impacting on government spending. If third-party or Medicare reimbursement is an important financial aspect of the technology acquisition, and the technology is not covered, this would be important information to acquire, document, and distribute.

The International Network of Agencies for Health Technology Assessment (INAHTA) currently includes 39 member agencies from 20 countries around the world. It offers a database of technology assessments, a technology assessment checklist, and links to member technology assessment sites.[4] A number of private World Health Organization (WHO) collaborating agencies (e.g., ECRI) are included. If a technology under consideration is new or "cutting edge" technology in the United States, experience gained by use in other countries can be obtained by accessing INAHTA information.

Private agencies, such as ECRI, can offer technology assessment information of varying detail (Keller, 2003). While this information is typically fee-based (e.g., cost per report and cost per annual subscription), it can be a wise investment if the resource provides specific, detailed information about the requested or alternative technologies under consideration that is correct and unbiased. In some instances, fees cover telephone consultation with agency experts to whom specific questions can be addressed. On-site and comprehensive consulting services also can be useful, especially where organizational expertise does not exist or where organizational internal dynamics challenge the ability for the system to operate properly otherwise.

Local, regional, national, and international organizations of clinical engineers can provide a mechanism for obtaining advice from peers who may have encountered the same technology evaluation issue. (See Chapters 130, 131, and 132).

Conceptual Needs Analysis

The purpose of the conceptual needs analysis is to capture and synthesize clinical procedure and technology information that is pertinent to decisions making. That is, it is not typically reasonable to provide participants with a package of original articles, manufacturers' brochures, or technology assessments. A small, experienced taskforce should preview the information and present only the information that is necessary to do the following:

1. Educate the participants
2. Inform participants why preliminary procedures, technologies, or even specific vendors have been excluded from further consideration
3. Briefly identify aspects that are similar to all technologies under consideration
4. Identify in greater detail specific aspects that are dissimilar and that are important to decision making
5. Identify known risks and hazards and methods to successfully mitigate them[5]
6. Detail the clinical efficacy of the technologies

Some organizations may direct this task force to make specific recommendations based on their analysis of these considerations.

The preliminary analysis should be provided to all participants in anticipation of their individual or collective (i.e., committee) review. Ample time for participants to examine and consider the information must be provided. Participants may pose new questions that will have to be explored. Collective review allows an opportunity for discussion and the back and forth consideration of important items, be they financial, technical, or administrative. The desired outcome of the conceptual needs analysis is to determine whether existing, requested, or alternative technologies should receive further consideration. If the outcome is not to proceed, it is important to document the reasons for discontinuing pursuit; this may be especially important when the issue was related to replacing existing equipment that is nearing obsolescence. Documenting the decision can simplify future consideration if and when it arises; this is not to say future consideration will be avoided or result in the same outcome, since reasons for discontinuing pursuit may have been overcome at a later date. It can help the organization's litigation position if there are valid reasons for discontinuing pursuit and if an incident occurs with the current equipment.

Several other important decisions can be made at this juncture:

- Which top three to five vendors should be considered further?
- Should the technology receive further evaluation of only the preliminary clinical procedures that were identified, or should a broader or narrower application be pursued?
- Are there other issues (e.g., potential physical location, staffing, volume of disposables, and medical professional certification) that must be considered, and how critical are they to ultimate acquisition and safe introduction of the technology?

[3]A list of, and access to, these reports can be found at http://www.ahcpr.gov/clinic/techix.htm

[4]Go to http://www.inahta.org/

[5]It is important to recognize that mitigation may not mean elimination of the risk, but, for example, implementation of risk, control actions or purchasing insurance to cover the risk.

- How well does the technology fit into the current user environment? (Can the technology be expected to fit easily, safely, and reliably into current experience and practices of clinical users, service support professionals, and the patient population without undue resistance or extraordinary requirements for training?)

Assuming that a decision is made to pursue the technology further, the next step involves determining desired specifications and starting the bid process. As the bid process is an important and complex step, it is discussed in this handbook separately in the next chapter (Chapter 34). This technology evaluation model assumes that desired specifications have been identified and that a reasonable bid process has been achieved. The remaining discussion assumes that the bid process was performed and that it resulted in a further reduction of vendors and identification of a more detailed compilation of device features and options, service support offerings, acquisition alternatives, and training requirements.

Testing Laboratory and Engineering Evaluation

The most desirable next step is to have the remaining vendor products brought into the facility for further evaluation. If possible, the technology or a representative sample thereof (e.g., a single patient monitor) should be delivered by each vendor to the facility. In some cases, where the equipment is extremely large or costly, or requires special utilities or facilities, this may not be possible. In such cases, the testing laboratory and engineering evaluation should be performed either at the manufacturer's location or in a representative installation at a time when the technology is not expected to be in use. It is important that those who are responsible for the laboratory and engineering evaluation receive adequate training from the manufacturer prior to the actual evaluation. It is also essential—especially when the testing of equipment is performed in a representative installation—to advise the vendor of tests that are expected to be performed prior to their performance, in order to allow the vendor to identify tests that could be potentially destructive to the equipment or harmful to staff.

Laboratory and engineering evaluation prior to introduction of the technology serves several purposes. Of primary importance is the keeping of potentially unsafe equipment out of clinical areas. It is important to examine the quality, features, performance, possible failures, and service requirements of the equipment as well.

Incoming Inspection

Technology entering a facility on a trial or evaluation basis should be subjected to full incoming inspection testing (see Chapters 6 and 122). Because the clinical engineering components of the evaluation are significant, the tests performed during incoming inspection of a trial/evaluation device should be even more comprehensive. Clinical engineers are expected to gain information that will allow decision makers to differentiate and decide among competing products. While incoming inspection of a purchased product can focus in part on verifying that operation specifications are met, incoming inspection of a trial/evaluation product could justifiably focus on determining the full range of operation and the comparability of operation from one trial/evaluation product to another.

Clinical engineers have a multitude of incoming inspection checklists available,[6] including the manufacturers' recommendations on how to examine electrical and mechanical safety of the products. Most address classic aspects of electrical and mechanical safety through inspection and testing. All accessories required for proper operation should be present, as should the operators' manuals and technical service manuals and schematics. Some manufacturers might attempt to limit this information, but the information is important to understanding exactly how the product works and the limitations that are inherent. User/operations and service manuals, like the product and options, can be returned to the manufacturer if their product is not selected, thereby avoiding cost to the manufacturer or release of information that the manufacturer considers proprietary and sensitive.

Proper operation of the equipment as specified in the performance specifications in the manufacturer's literature should be confirmed. However, if possible, clinical engineering should develop and perform tests that are specifically intended to differentiate competing products in important areas. Tests that differentiate equipment but that are not clinically significant or not significant in any other way (e.g., safety-wise or financially) should be avoided.

The clinical engineer should note whether the product has been evaluated in any manner by nationally recognized testing entities; e.g., Underwriters Laboratories (UL), Canadian Standards Association (CSA), or City of Los Angeles. The clinical engineer should acquire and peruse all applicable standards that address the technology under evaluation (see Chapters 117, 118, and 125). Further, it should be noted whether the product does or does not comply with governmental labeling and performance requirements; e.g., use on physician prescription only and OSHA-compliant for reducing accidental needle sticks.

Gaining the Experience of Others

Clinical engineers should require the manufacturer to provide references, including customers who currently utilize the product. They should contact these references directly, to discuss the reference's experience with the product and the vendor. Clinical engineers should contact their counterpart(s) within the reference organization(s) to seek their input on their experiences, whenever possible. If available to them, clinical engineers should contact others not included in the manufacturer's reference list who have experience with the product. Such lists might be available through group purchasing organizations, professional societies, and shared service organizations. Clinical engineers should be willing to contact representatives from other organizations who have noted use of the product in, for example, the medical literature. When gaining the experience of others, however, it is

[6]One example is the ECRI Inspection and Preventive Maintenance Manual.

important to remember to compare, to the extent possible, "oranges to oranges." Experience with a product that might be years older than the product under evaluation can result in misinformation if the clinical engineer is not careful. Similarly, comparing a vendor's historical performance to its current performance might not be appropriate, in either a positive or a negative way. The clinical engineer should examine the information and its limitations, with an emphasis on determining exactly what can be said and supported with regard to positive and negative aspects of the experience of others.

Technology and the Facility

Will the facility be a good fit for the technology? The clinical engineer has the user's manual, the service manual, the schematics, and, most importantly, the product. Most clinical engineers have a working knowledge of the physical facilities in which the product can be used and of the utilities necessary to support the technology. Certainly there might be other engineering professionals who have greater expertise and responsibility for the facilities and utilities. However, laboratory and engineering evaluation of the products is a perfect opportunity for clinical engineers to examine and discuss, with their internal counterparts who are responsible for the facility, what the perceived needs are for the product. For example, there might be special power or shielding requirements. There might be special water, gas, or other requirements. Systems (e.g., fire suppression systems, to protect the physical product) might be required. The device might be too large to fit into the facility, or it might be too heavy for the support structure. The facilities engineer should be invited to participate in the laboratory and engineering evaluation. More than one strong alliance between clinical engineers and facilities engineers has been created by clinical engineers recognizing the facilities engineer's expertise and encouraging participation.

Maintenance and Service Requirements

The clinical engineer can examine and perform manufacturer-identified maintenance and service procedures to determine human and financial resources that are required. The ease of disassembly and the need for special tools can be examined. The clinical engineer can evaluate whether special training is required and whether the manufacturer can and will provide such training. But there are other issues that might be more difficult to evaluate. For example, how reliable is the product? What are the expected maintenance and service requirements? While the results of manufacturer testing and experience in the field can be requested, the actual applicability of laboratory-simulated testing of a component might have no relationship to the reliability of the system in actual clinical use. The clinical engineer can request that the manufacturer provide a complete list of spare parts recommended by the manufacturer. The actual service experience of organizations that use the same technology should be examined as well.

It is important to remember that some issues relating to maintenance and service can be transferred to others (e.g., the manufacturer or a shared service) by contract (see Chapters 7, 8, and 39). The bid could include, for example, provisions for the manufacturer to guarantee replacement parts availability, defined service response time and up time, the provision of "loaner" equipment, and training.

Laboratory Evaluation by Clinicians

Clinical evaluation of the product is an important part of the technology evaluation process. But clinicians should first evaluate the product in the testing laboratory before allowing it to be placed into the clinical environment. While there is an obvious safety aspect to laboratory evaluation, laboratory (i.e., simulated use) evaluation of the technology by clinicians offers an opportunity for clinicians to be trained in the use of the product, to be advised of limitations or differences noted by clinical engineering review, and to systematically examine the operation and human factors aspects of the product without the interference of a patient's clinical need. Human factors with the interference of clinical need of a patient will be examined during the clinical evaluation. If simulated use identifies a shortcoming before use on a patient, the clinical engineer should not allow a product to progress to the clinical evaluation phase, where an accident might ensue (e.g., from a defeated alarm that staff thought was active).

Laboratory evaluation of the product by a clinician offers an opportunity for the collaboration and development of a relationship between clinical staff and clinical engineering staff. It can offer an opportunity to identify special training needs and requisite warnings that clinical staff who are responsible for the clinical evaluation should receive prior to initiation of the process. Current clinical practice on the existing product in use might not be recommended, or could even be hazardous, if used with the technology under evaluation. Staff need to know, prior to use of the product, that a change in practice not only can make the product more attractive to them, but also that it can enhance patient and/or staff safety. Including clinicians in the testing laboratory evaluation of the product can facilitate the inclusion of the clinical engineer in the clinical evaluation of the product. Such inclusion can have a number of benefits, including verification of laboratory findings, identification of new issues, and strengthening of the engineer/clinician relationship.

The results of the laboratory evaluation should be compiled, and a recommendation provided regarding whether and how to proceed further. The results might eliminate a product from further consideration. The recommendation might identify previously unidentified needs or costs. It should provide details of training and warnings that clinical staff who are responsible for clinical evaluation should receive prior to actual clinical use of the product. It might also identify additional questions to be answered prior to proceeding further.

Clinical Evaluation

Assuming that the product has not been eliminated by previous processes, it is important to assess the product in the area of expected use by representative expected users. Just as certain device design and facility limitations could have precluded the introduction of a product into the facility for laboratory testing, so too, for various reasons, examining the product within the organization on a trial/evaluation basis might be impossible. In such instances, clinical evaluation could be limited to simulated clinical use, observation of clinical use by others, or use on, for example, volunteer, if performed within the confines of laws and regulations. In some cases, there might be special requirements for Institutional Review Board (IRB) approval of a protocol in order for the technology to be used on a patient or volunteer.

Clinical evaluation has many of the same features (e.g., confirming operations) as the laboratory and engineering evaluation of the product. Clinical evaluation extends laboratory and engineering evaluation by interjecting the clinical environment, including the patient, the user, and the facility. Clinical evaluation includes examining the performance, ease of use, human factors (see Chapter 83), and safety aspects of the product, but also should emphasize examining the differences between or among competing products. Clinical engineering can participate in the clinical evaluation process as an observer; although it is not paramount that the observer be a clinical engineer, it is paramount that an impartial observer participate. Often, a user does not realize exactly how they are using, or misusing, a product at the actual time in question. The user's attention might be diverted to the patient, to another patient, to another product, or to another activity that is occurring but that does not include them. They might misinterpret user instructions and might believe they are using the product in accordance with manufacturer instructions. In fact, the device might contribute to such "user errors," and this contribution is important to identify and note. They might misinterpret device presentations (e.g., visual warning or alarms), or correctly interpret them, but not be able to act accordingly without resetting the product.

An impartial, trained observer can watch how the products are being used and misused. Difficulties in use, product-related efforts that are particularly time consuming, and frequent need for replacement of accessories or options can be noted. Depending upon the specific issue and any safety ramifications, the observer can advise the clinical user of the situation or can withhold this information until it can be analyzed and evaluated. An argument can be made that the clinical engineer is well positioned to be an observer as he or she already will have received training on the device and will have operated it in the course of the laboratory and engineering evaluation. A strong argument can be made for the observer to be a member of the clinical staff that ultimately will make use of the product, because that professional eventually will need to be trained anyway and also has direct, hands-on, clinical experience. Regardless of who is chosen to observe, it is important to recognize that constant observation is neither recommended nor possible. Clinical use during the evaluation should be as natural as possible and should not compromise patient or staff safety. Observers who have tried to maintain constant vigilance for hours on end have been found to miss many important events.

Features of a clinical evaluation should encompass objective and subjective observations. Performance results (e.g., correct capture and identification of an arrhythmia that has occurred) can prove out objectively. Set up might be judged difficult by one evaluator, but easy by another. Such subjective assessments deserve further analysis if possible. It is important that the objective and subjective assessments of clinical staff be captured during the clinical evaluation. Objective assessments should be compared against, for example, actual performance specifications provided by the manufacturer and results of laboratory and engineering tests. Inconsistencies should be examined. Subjective assessments that might impact positively or negatively on final selection should be examined carefully to determine whether any objective tests can be performed to prove out or supplement the subjective results. One benefit of collecting subjective assessments is that staff misunderstanding can be addressed at that time and need not wait until the product has been purchased and placed into actual use.

The results of clinical evaluation should be compiled, and a recommendation provided, regarding whether and how to proceed. The results might eliminate a product from further consideration. They might identify two or more products that clinical staff would find acceptable for acquisition. They might identify training or staff issues or costs previously not identified. They might identify issues that remain to be clarified and, for example, require reexamination in the laboratory.

Vendor Evaluation

Certain aspects of vendor performance are part of other evaluations (e.g., parts availability in testing laboratory and engineering evaluation and past experience with a vendor's product by clinicians in the clinical evaluation). But there are additional aspects to consider and to include in the complete process of technology evaluation. For example, how long has the vendor been in business, and how stable is it? Does the vendor have the manufacturing capacity and expertise to provide the product, training, and support? Does the vendor have an acceptable reputation within the industry, and has it behaved responsibly?

Final Evaluation and Selection

The process of final evaluation and selection should include all affected parties. A small group of experienced evaluators can cull through all of the information and results on the various evaluations in order to identify specific information that is important to the final decision, but inclusion of all parties and their interests in the final determination is desirable. The small group might seek additional information from others, (e.g., manufacturers), who will clarify remaining outstanding issues to the extent possible. They might seek to determine whether vendors are willing to provide additional financial or other incentives, or to provide additional training that testing has determined would be necessary for staff based on the results of the laboratory and engineering evaluation or the clinical evaluation. Where more than one product has successfully passed all of the internal evaluations with no clear preferences identified, vendors should be advised that the organization is seeking ways in which to differentiate one opportunity from another, to facilitate a final decision. Vendors can offer concessions at this point. In the absence of concessions, the organization simply will make the best possible business decision.

Once all of the results and supplemental information have been compiled, the small group can develop a cost evaluation of the alternatives. The cost evaluation should capture a myriad of factors, including, but not limited to, the initial cost of the product purchase, the cost of accessories and options, freight cost, and installation cost (see Chapter 30). The cost of such things as spare parts and consumables should be identified. Costs related to staffing and training should be provided. Costs for operating expenses, such as electric or water cooling, should be provided if they are significant. If a life cycle cost analysis can be developed and depended upon for the purposes of comparison of one product to another, it should be provided as well. Armed with the results of the various evaluations, a final decision can be made. The decision and the rationale for the decision should be provided to all affected parties.

Conclusion

Technology evaluation is an important part of heath care operations. Throughout the process, it requires the cooperation of many professionals, including the clinical engineer. It offers opportunities for the development of alliances and sharing of information for the benefit of patients and staff alike. If performed properly, it can provide a means for rational decision making that is acceptable to all stakeholders. It can help make the best use of human and financial resources in an industry where both are limited.

References

Centers for Disease Control and Prevention. Evaluation of Safety Devices for Preventing Percutaneous Injuries Among Health Care Workers During Phlebotomy Procedures. *Morb Mortal Wkly Rep* 46(2):17, 1996.
ECRI. *The Source Book.* Plymouth Meeting, PA, ECRI, 2003. Also MD Buyline. www.MDBuyline.com, 2003.
MDR. *The Medical Device Register.* Los Angeles, Canon Communications, 2003. www.cancom.com
Mendelsohn MH. Efficacy of a "Safety" Winged Steel Needle in Preventing Percutaneous Injuries (PIs) in Health care Workers. Presented at the Eighth Annual Meeting of the Society of Health care Epidemiology of America, 1998.
Stalhandske E, DeRosier J, Patail B, Gosbee J. How to Make the Most of Failure Mode and Effect Analysis. *Biomed Instrum Technol* 37(2):96, 2003.
Stamatis DH. Failure Mode and Effect Analysis. Milwaukee, WI, ASQC Quality Press, 1995.
Younger B, Hunt EA, Robinson C, et al. Impact of a Shielded Safety Syringe on Needlestick Injuries Among Health Care Workers. *Infect Control Hosp Epidemiol* 13(6):349, 1992.

Further Information

The FDA CRDH website has databases relating to such details as registration of the manufacturer, listing of devices, 510(k)s, PMAs, and adverse occurrence reports. Accessible at http://www.fda.gov/cdrh/databases.html.
Agency for Health care Research and Quality (AHRQ) reports can be found at http://www.ahcpr.gov/clinic/techix.htm.
International Network of Agencies for Health Technology Assessment (INAHTA), http://www.inahta.org/.

34

Technology Procurement

Gary H. Harding
Greener Pastures, Inc.
Durango, CO

Alice L. Epstein
CNA HealthPro
Durango, CO

Technology procurement is an important element of the overall hospital technology management program (see Chapter 30). Procuring technology is a complex, detail focused process, although some members of the health care team and vendors believe that it is merely the act of writing the check. Unless careful attention to procedure and detail is maintained throughout the procurement process, significant and even catastrophic events may occur. For example, costly unbudgeted and unexpected "required options" might need to be acquired after equipment has been ordered, or unexpected installation changes could be required. In a worst case scenario, the technology might not be usable by the facility at all. It might be severely underutilized, thus having a dramatic financial impact, or use could result in injury to patient or staff.

It is imperative that an organized framework be developed to ensure that the actual procurement of technology results in the organization's receiving the proper technology under the terms it believes appropriate. A flexible approach will be important to process the variety of technologies to be acquired. However, it is important to note the use of some basic, standardized tools; e.g., a request for quotation, which can be modified when need indicates, can result in better performance of the procurement process, thereby saving time and money. Procurement comprises the three following essential processes:

- Organizing the presentation of initial important requirement, once a decision has been made to consider a technology for acquisition
- Determining final specifications, options, terms, and conditions under which technology will be acquired
- Ensuring that technology and services expected are provided and/or insured before payment

The clinical engineer (CE) can utilize many of the skills learned in didactic training and through experience to assist the organization in methodically and safely procuring technology. It is imperative that clinical engineers remember at all times that they are team members, and not solely responsible for either the technology procurement itself or the process. Technology procurement and the overall hospital technology management program require cooperation among and participation of professionals from a variety of disciplines. The clinical engineer can be a strong team member who can help to ensure that the actual clinical needs of the user are identified, addressed, and, to the extent possible, met. The clinical engineer can serve as the communication bridge among professionals of various disciplines, both inside and outside the organization. Important attributes that contribute to the effectiveness of the procurement process include the following:

- Attention to detail and process
- Quick study of technical and health care issues, both clinical and administrative
- Ability to create and manage relationships, inter- and intradepartmentally as well as externally
- Willingness to ask questions and request clarifications
- Excellent communication skills, both written and oral
- Ability to compile and manage information with an emphasis on determining comparability.

Breakdowns in communication (e.g., between clinicians and vendor personnel) or lapses in process (e.g., specification of a term without a written commitment or agreement by the vendor in its response) can create significant and even insurmountable problems later in the technology acquisition process. The clinical engineer can facilitate procurement and the process by listening, interpreting, evaluating, and teaching. By asking the right questions, communicating the defined need, and monitoring whether expectations can be or are being met, the clinical engineer can contribute to the quality and scope of patient care as well as to the financial well-being of the organization.

The remaining portions of this chapter discuss several areas of specific importance to technology procurement and provide additional resources for the clinical engineer to access.

Technology Evaluation Process

Procurement follows evaluation. However, benefits accrue from the overlap of the two processes. The purpose of the technology evaluation process is to determine the value and need for existing, new, and alternative technologies that the organization must consider in order to provide clinical care to the patient population it serves. (see Chapters 32 and 33). As the technology evaluation team and those responsible for technology management assessment proceed through the process, the need to integrate information and actions with the technology procurement team becomes apparent.

Request for Information

The acquisition of initial information about requested or alternative technology precedes procurement. While the technology evaluation team could seek this information independently of the technology procurement team, there are several advantages to the procurement team's collaboration in such steps as the development of the initial request for information (RFI). The procurement and evaluation teams can begin the cooperative process of understanding the scope of the technology to be considered, the way the evaluation is to proceed, and the terms and conditions under which the technology will make its way into use. Specific advantages of close collaboration for evaluation and procurement include the following:

- Identification of basic specifications, options, terms, and conditions to be discussed in the early stages of the technology evaluation process
- The opportunity to advise vendors early in the process and in a consistent manner as to what the "ground rules" will be
- Vendors who are experienced in responding to standardized organizational requests and needs and who will appreciate the opportunity to respond in a thorough and cost-effective manner
- Reduction of duplication of efforts (tasks generally will be performed only once during the process)
- Errors and omissions that might arise from lack of (or miscommunication over) differences between the technology evaluation process and the technology procurement process can be minimized, controlled, and addressed expediently

The initial RFI is not intended to present all questions or even to acquire all important information. It is intended to seek adequate information from potential vendors of requested and alternative technologies, and to allow the technology evaluation team to compare information and prioritize some initial decisions about the technology(Ies), clinical procedures, and vendors that should be considered further. A specific RFI can be prepared in cooperation with the technology evaluation team and distributed to potential vendors. The two teams should work together to detail the current specific clinical need identified within the organization and should request related information from potential vendors. Elements of the RFI should include the following:

- A comparison, by each vendor, of the vendor's and the competition's products, including past, current, and alternative products that are intended to (or otherwise might) meet the clinical need
- A discussion of other clinical specialties that might make use of the products
- A discussion and description, with references, of the clinical utility of the products, including all of the clinical procedures within the intended use of the product, those that are not currently included within the intended use but may be included at a later date, and those that are not included
- The current status of device listing with the United States Food and Drug Administration (FDA) and the manner in which the vendor received clearance to market the product
- The data that the vendor relied upon in its medical vendor submission to the FDA to assert that the product was safe and effective
- A description of the actual equipment, options, and accessories that will be necessary for the organization to acquire in order to perform the clinical procedures identified
- A description of the special considerations (e.g., for installation, utilities, service, or training) that the organization should consider during the technology evaluation process
- All user, service, and installation documentation (e.g., manuals)

Many other questions can be asked or requests for additional information made; the extent of these questions and requests will vary based on the complexity of the technology, the characteristics and timing structure of the technology evaluation process, and the status of the technology (e.g., experimental vs. widely accepted).

Information that is important to the procurement process might already have been gathered by the technology procurement team. Relevant information should be requested from this team as well. It is often worthwhile to consider proactively the need for information from all teams so that the process for acquiring information can be standardized.

Technical Specifications and Other Requirements

The technology procurement team should be identifying questions to be asked regarding specifications, as well as evaluating responses to the questions asked. Incomplete, unacceptably limited, or incomparable specifications often can be identified on initial inspection. Mistakes that might be made by comparing "apples to oranges" can be reduced. Important comparable differences can be identified and can result in either discontinuation of consideration of that vendor or a request for further clarification. Initial efforts can assist in the development of a detailed Request for Quotation (RFQ) and can identify specific attributes that deserve closer inspection during laboratory and clinical evaluations.

Standardization

There is intrinsic value in standardizing technology to the extent practicable. Vendors, models, characteristics of similar (but not the same) equipment types, and service responsibilities can be standardized. The standardization can be within the products and services offered by a specific vendor, or include characteristics (e.g., communication buss) and services (e.g., guaranteed availability of loaner units across different vendors, service providers, and clinical areas of use). Members of the technology procurement team can do the following:

- Identify in which places and in what ways standardization is currently in place
- Evaluate whether existing or new standardization can apply to the technologies under consideration
- Identify the specific standardization of vendors, features, or services that should be addressed
- Compare offerings to what is available or currently in place within or throughout the organization

Standardization of specific elements should be placed in perspective with respect to other issues for the specific technology and decision under consideration; for example, the opportunity to rate the value of standardization against clinical utility.

Long-Term Relationship

The value of a long-term relationship with a vendor can be important to some decisions. If the vendor and the organization have experienced the procurement process together over prior acquisitions, and with great success, there is good reason to believe that additional successful mutual processes are achievable by the organization and the vendor. However, it is also important to be able to place a balanced perspective on long-term relationships with respect to all other aspects of the evaluation. For example, while an organization may have a successful long-term relationship with a vendor of patient monitoring equipment, other factors, such as the need to work with a different division, or the offering of first-generation equipment in another device type, could call into question the value of the long-term relationship for the specific technology under consideration. The clinical engineer can help the organization to assess the value of long-term relationships for the specific technology under consideration.

Service and Training

Service capabilities of the vendor and the clinical engineering department must be evaluated. The vendor's present and past ability to provide service and to train the hospital's clinical engineering department in servicing the equipment are important procurement considerations. Training needs of the user may encompass educational services as well as clinical user training. Aspects of safety training, such as those required with laser equipment, should also be considered.

The vendor's response to the RFI can provide a basic understanding of the service and training needs from the perspective of the vendor. The vendor is not the end user; the vendor might understate those needs or might not understand the expected environment of use within the organization. The vendor might describe the needs and its ability to provide support as it would like things to be, or as it perceives things to be, rather than as things actually are, from the user's perspective. CEs who are experienced in procurement can examine information submitted from all parties and can compare it to their experience with these and other vendors. An "apples to apples" comparison is more likely to result, and differences in service and training can be kept in the proper perspective among vendors and with other aspects of the evaluation.

Subjective Bias

Subjective bias is decision making or evaluation based on personal, poorly measurable, and unverifiable data or feelings that are improperly weighted against objective, unbiased data. Subjective bias can cloud the issues, impede the ability of a participant or group to make valid decisions, and result in costly mistakes. Participants in the technology evaluation and procurement processes might or might not recognize subjective bias on their part or on the part of other participants. As the relative power of the participant rises, subjective bias on their part and its potential negative effect on the process also rise. In an extreme case example, subjective bias on the part of the technology's hospital advocate who is already determined to select one vendor's product over all others can make it almost impossible to perform a valid assessment to acquire the best technology for the organization. If the technology from the "preferred because of an invalid bias" vendor happens to be the best technology for the organization, the outcome might be fruitful; if not, the organization risks acquiring the inappropriate technology, or the organization could lose the services of a talented clinician if decisions go against the clinician's bias.

The CE can help to identify and eliminate subjective bias simply by asking questions and requesting references in support of a position. An inability or unwillingness to answer questions or to provide references could be the result of subjective bias, although there might be other reasons as well, e.g., lack of competency in a new technology. If resources within the organization or the vendor are unable to answer questions or to provide references, the CE can seek an answer elsewhere, e.g., in the literature, from noted authorities, or from experienced users. Remembering that the technology evaluation and procurement processes are cooperative efforts, the CE can point out where there is bias or subjectivity and the importance of objectivity in decision making whenever possible. In some cases, it can be beneficial to utilize an unbiased, independent, expert consultant to facilitate consideration.

Procurement Financial Alternatives

It is often highly desirable to consider the financial impact and financial alternatives early in the technology evaluation and procurement processes. One reason to do so is to find alternatives for consideration of equipment that the organization simply cannot afford. Another reason is to avoid eliminating equipment that the organization could not afford under lump sum cash outlays, but that it might be able to afford under long-term, periodic payment structures that might be available.

There are three typical financial methods for acquiring the use of technology: Cash purchase, rental, or lease-purchase. Cash purchase typically requires the large, up-front expenditure of capital with periodic and ongoing maintenance, as well as service and upgrade costs. A portion of the value of the product can be written off each year until the product is fully depreciated. The purchaser typically owns the product as soon as it is accepted.

Rental usually requires a lesser initial capital expenditure, if any, and the organization does not own the product. There is no depreciation of the product because it is not owned, but there are periodic (e.g., annual or monthly) rental payments. The lease-purchase alternative is somewhat of a cross between purchase and rental; the terms of the lease and transfer of ownership at some point during or at the completion of the lease are negotiable and usually vary dramatically.

Reasons for choosing one alternative over another include cash flow considerations, source of funding, attractiveness of offers, expected product life cycle, expected frequency of upgrades required, and effects of each alternative on the overall financial position of the organization. One newer alternative, which is useful for especially costly technology (e.g., a proton beam therapy system), is development of a joint venture among, for example, a health care provider, a technical vendor, and a commercial partner. Up-front investments and responsibilities for various aspects of product use, upgrade, and support are split among participants, with the financial risk spread among them.

While scrutiny of the financial alternatives often rests with financial specialists within the organization, the CE can contribute to the validity of the data upon which these professionals will rely; for example, examining and evaluating the projected costs for maintenance, upgrades, and service within each alternative. That is, if the projected cost data and expected life expectancy of the product are incorrect, the financial assessment might be performed correctly, but the conclusion reached will be incorrect.

All of the above may be considered in the initial stages of technology evaluation through a conceptual needs analysis. The CE responsible for procurement can contribute to the decision of whether to move forward and with which vendors and technologies to do so.

Each recipient of the RFI should be advised of all decisions affecting that vendor. If a vendor has been eliminated from further consideration, the rationale for doing so can be conveyed to its representatives so that there is closure; the vendor can improve its sales efforts in the future if it has the ability and desire to do so. The vendors eliminated from the process might request the opportunity to resubmit or to supplement the submission of information. Permitting a vendor to resubmit should be a joint decision made by all team members.

Request for Quotation

Request for quotation (RFQ) is also known as request for bid, request for proposal, and request for tender. Assuming that the technology evaluation processes have been performed and that a decision to move forward has been made, a much more complete and detailed effort is necessary in order to identify specific equipment, services, and support that can be provided by each qualifying vendor. Extraneous applications and options can be eliminated from further consideration. Each vendor who is still under consideration can be provided with a much more complete idea of the nature and scope of the clinical need and technology interest. The RFQ is the organization's document; it is an opportunity for the organization to ask for whatever it wants. CEs who have been involved in prior procurements are well qualified to lead the effort to convey the interests and expectations of the organization to the vendor because they were involved in the initial RFI and some parts of the decision making process. As it is likely that much of the information eventually will be restated in the form of an acquisition contract, it is expedient for there to be consistency in the RFI, RFQ, and purchase contract processes. It should be understood that the RFQ, while an important step in the overall process, is not a legal transaction. It is an evolving process; i.e., it is subject to change as, for example, new information or opportunities are identified. Information that is gleaned from other subsequent tasks (e.g., laboratory and clinical evaluation) and discussions might require clarifications or changes in the RFQ.

Standard Template

It is unwise to attempt to reinvent the wheel every time a technology is evaluated. While the RFQ document should be flexible to evaluate varying technology, starting with a standard template and making necessary modifications to fit the specific technology have distinct advantages. There is a greater likelihood that the request will be complete and not miss a crucial item and that it will be compiled more rapidly and efficiently. Vendors can become experienced in what is expected of them and can "standardize" their responses if applicable, thereby reducing their costs. Those who are responsible for evaluating vendor responses can become experienced in responses and their implications. Often, a standardized comparison tool (e.g., a tabular program) can be more easily utilized if the RFQ is based on a standardized format. The process can expediently drive the comparison of "apples to apples." Often, the simplest way to begin developing a standardized request is to review and modify samples of RFQs that the organization has utilized successfully in the past.

One purpose of the RFQ is to convey the organization's interests and expectations to the vendor. In order to do so, the communication must be clear and understandable. The structure of the RFQ might include sections devoted to specific topics; e.g., introduction, equipment specifications, terms and conditions, installation, service, and training. Similarly, questions and requirements should be clear, concise, and easy for the vendor to understand.

There is a variety of templates available for consideration (ECRI, 1976; Mandell, 1986; ASHE, 1991), but most include some or all of the following elements:

- A narrative description of the organization's interest in the technology(ies), the contacts within the organization, and the period in which responses, presentations and procurement is expected to take place. It is not necessary to fully detail, for example, the clinical significance or efficacy of the device, as the vendor theoretically is expert in these areas or would not be providing the technology.
- Technical specifications that are particularly important to the intended clinical use of the technology and to the specific installation in the physical facility. Often, the RFQ provides the vendor with sufficient information on the basis of which to ask for and to gain more detail (from, for example, the clinical sponsor and the facilities engineer within the organization).
- Options that are mandatory and those that would or should be considered. The RFQ is an opportunity for the organization to ask for and to receive sufficient information on the basis of which to acquire and to use technology without the immediate need for additional options or costs subsequent to this acquisition, as well as to plan for clinical service changes and technology upgrades in the future.
- Consumables that must or should be acquired in order to use the technology. Cost is not the only issue related to consumables; availability also might be important if consumables are subject to back order, recall, or replacement.
- Documentation that will be provided. Typically, user and service manuals are mandatory. Descriptions of diagnostic software and specialty tools (e.g., test boards) are helpful.
- Installation requirements, including space, power, cooling, and weight. Typically there are two approaches to examining installation requirements. The organization can detail what it expects will be necessary, and it can require the vendor to advise the organization if this is correct and complete, or it can require the vendor to advise it as to what is necessary for a complete and correct installation. In either event, it is reasonable to require the vendor to perform a site analysis as part of the RFQ process so that unexpected and unnecessary problems (e.g., the floor's inability to carry the weight) do not occur at the time of actual installation.
- Costs over the expected lifetime of the technology (life cycle costs) for the various acquisition options (i.e., cash purchase, rent, lease, or joint venture) that are available through the vendor or that have been utilized by others acquiring the technology. Vendors are often aware of methods and sources of financing, as well as reimbursement;
- Training courses and materials that are available from the vendor, their agents, or others that are mandatory or desirable for clinical staff and service/repair professionals. Disclosure of cost, if any, should be required. In some cases (e.g., lasers), special physician training might be strongly recommended. In most cases, training of clinical staff may need to be performed in multiple locations throughout the 24-hour period, or the vendor will need to train the trainers. The need for service training should not be overlooked, even if vendor or third-party service is expected. At least one professional within the organization should have a working knowledge of the technology. Turnover in clinical and service staff occurs, and some mechanism to address training in the event of turnover is desirable.
- Identitites of ten most similar clients, including contact names and information. It should be clear to the vendor that the organization intends to contact these clients to discuss their experience with the technology and the vendor, including use, service, and costs.
- Service options and resources available. The RFQ is an opportunity for the organization to learn everything that it can about what to expect in the way of service after procurement and installation are complete. Important service topics that can be examined include:
  - Service contracts and cost, as well as reduction in cost if first screening is performed by the organization
  - Guaranteed up time
  - Employer or independent dealer service
  - Location of service personnel
  - Number of personnel at the service location trained on the specific equipment, and location of backup service and contact options (e.g., hotline)
  - Availability and attributes of guaranteed response time
  - Parts availability, location, shipping schedule, and price list
  - Labor rate(s), standard hours, and travel time
  - Frequency and content of preventive maintenance provided
  - Availability of loaner equipment
  - Indemnification and liability insurance coverage
- Terms and conditions under which the technology will be acquired. This is the opportunity to do the following:
  - Require the vendor to install and acceptance test the technology
  - Identify how and when the vendor will receive payment
  - Specify penalties if the technology does not perform reliably or is not installed according to schedule
  - Detail how future versions of product hardware and software will be made available, performed, and charged
  - Require evidence of compliance with local, state, and federal codes and regulations

If the organization has unreasonable or unrealistic expectations or asks for information that is impossible to obtain, responses may be minimal without any need for them to be so. Hence, it is important for the organization to think about the aspects of the technology and vendor support or other offerings that are important to and desirable for the specific procurement. The organization might even direct (in the RFQ) vendors to contact the organization to discuss any aspects of the RFQ that the vendors find confusing or objectionable. Although obtaining the best technology at the best price is always foremost on the minds of the organization, the organization and the vendor must work together effectively throughout the lifetime of the technology. If the RFP and subsequent interactions are fair, a solid working relationship can result.

Aspects of the RFQ and the successful vendor responses eventually will be reduced to a procurement contract. If the RFQ or responses are incomplete, the procurement contract might be expected to be incomplete. If, for example, a vendor does not respond and agree to provide service manuals, this issue probably will arise later, perhaps unexpectedly.

Vendor Presentation

Some organizations find that a formal presentation by the vendor is beneficial. Vendor presentations are opportunities for learning, both for the professionals within the organization and for the vendor. Clinical professionals can learn new information about the attributes of the technology and gain an understanding of the extent of costs that will be assumed and the support that will be required. Vendors can learn more about the relative importance that the organization places on cost versus performance, or on some other attribute of the procurement.

In some instances, it might be possible to eliminate a vendor based on preliminary analysis or on the vendor's response and presentation. to do so it is clearly indicated and supported objectively. It would be unwise to eliminate a vendor where a more complete analysis of the vendor's response or subsequent questioning might result in that vendor as the best, not the worst, alternative. Rebidding should be discouraged as it costs both the organization and the vendor time and money.

Response Analysis

Once the vendors have responded to the RFQ, the responses must be compiled and compared. If the RFQ is standardized, the responses are complete, and a standardized analysis tool can be utilized, then the process may be fairly straightforward. Many organizations find a tabular format easy to use. One major responsibility of those who are responsible for the analysis is to be sure to compare "apples to apples." If a like comparison is not possible, contact with the vendors should be made in order to acquire information that is more comparable.

The analysis should focus on information that is important to comparing procurement options. While it is important, for example, to ensure that all of the vendors' technology meets the basic required technical specifications, once this determination has been made, the information can be eliminated from the comparison table, and just a note need be included that all vendors meet the specifications. It is important to identify differences among vendor responses (e.g., in price); however, it is just as important to remember that a vendor's response could be subject to change if it becomes a potential reason for eliminating that vendor and they are advised of that potential. Providing vendors with the opportunity to provide the rationale for questionable items may identify a miscommunication and subsequent clarification.

Comparison and tabular presentation of differences in responses with regard to price, payments, and terms are typical. However, it is also important to capture crucial information such as omissions and errors in the responses, a vendor's special attributes, and special technology attributes. Omissions and errors should be discussed with the vendor, to determine whether additional information can be acquired from the vendor. Special attributes of the vendor often relate to the vendor's history in the industry and their reputation with customers. For example, if the vendor has a history of unsatisfied customers, and the dissatisfaction is based on objective information, this is important information to include in the analysis. Still, it is important to remember that some causes of dissatisfaction (e.g., service response time) can be addressed, to some extent, by terms and conditions within the purchase contract. Special attributes of the technology often relate to such issues as a small installed base, a brand new product line, or an older system, compatibility, or special design.

Ultimately, the analysis may include a formal tabular comparison, additional objective findings, subjective findings, and points raised in the analysis to develop recommendations for proceeding. The results of the analysis can range from discontinuation of the procurement and technology evaluation process, to elimination of some of the responding vendors, to consideration of all responding vendors in subsequent steps. Many organizations find it helpful to stress within the analysis that the results, findings, terms, and conditions are subject to change prior to procurement, either because of changes in need or as the result of negotiations.

Laboratory and Clinical Evaluation

Once the analysis is completed, decisions related to laboratory and clinical evaluation can be finalized. Which vendors to consider, what technology and options to request for evaluation, and the timetable to be utilized to perform the evaluations will have to be decided (see Chapter 33). It is advisable to keep the CE who is responsible for procurement in the information loop during the evaluation period. Decisions made and information learned may directly affect the procurement process subsequent to the evaluation. Once these important evaluation aspects of the process are completed, tasks that are related to the procurement process can be expected to accelerate.

Revision of Technical Specifications and Other Requirements

Revision of specifications is often necessary from what is learned during the evaluation process. Revision may include identification of the preferred vendor(s), specific options, accessories, features, training, or service. It is important that procurement professionals work closely with the evaluation team to identify any desirable changes or additions that have become apparent during the evaluation process. It is common for questions to remain and to require addressing by the vendor(s).

Unless there is overwhelming reason to do so, it is not typically wise to reduce the field of vendors for final procurement to one vendor. Doing so can minimize the organization's ability to negotiate such items as cost and terms. In addition, it is reasonable and ethical to explain to each vendor the identified differences and shortcomings of the response and equipment, along with an offer to each vendor to explain any rationale for the difference or the ability to compete. Questions about cost, performance, and options are ethically appropriate for the vendor and organization alike.

It is also important to capture and communicate to team members and decision makers requirements for installation, training, and service of the vendors who remain in competition. In some instances, a recompilation and re-analysis of information might be necessary. Occasionally, a rebid of the entire procurement may be necessary. It is important for procurement professionals to analyze the information received from the initial RFQ process, the information received from the results of the evaluation processes, and any clarifying information received by the vendors to determine whether sufficient information exists on which to finalize the award. If additional information is needed, or if insufficient information exists, the procurement professional should take the time to rectify the situation or to ensure that the proper professionals within the organization are aware of the issues.

Assuming that adequate information on which to base the award has been collected, and that the original RFQ was complete, the process of revision can be expedited. If the revision document is complete, the process of moving to award with a procurement contract can be expedited.

Selection and Payment

RFQs, evaluations, and revisions have been completed. Thus, the procurement team moves to "award" and the job of the CE in the procurement process is done, right? Wrong! It is important to ensure that the technology, options, services, and other "promised items" are delivered. The first step in assuring that promised items are delivered is to provide an award document that clearly spells out what these promised items are.

Contract

A procurement or purchase or lease contract, by whatever name, is desirable. The language of the document should be tailored to the specific aspects of the technology and should have undergone prior review by legal counsel. If standardized, transferring information into the award procurement contract from the preceding procurement documents can be expedited.

Standard features of a procurement contract include a detailed description of the technology, options, accessories, documentation, and software to be provided by the vendor. The contract usually also includes the specific conditions of sale, such as the following:

- Installation requirements, responsibilities, and timetable
- Acceptance testing requirements, responsibility, and timetable
- Warranty details, spare parts availability, and service requirements
- Clinical, service, and other training to be provided, assignment of cost for same, and schedule for doing so

Of course, the contract will incur costs. Often, it is wise to identify not only the costs for which the organization will take responsibility, but also the costs for which the vendor or others involved in the procurement will take responsibility. Doing so can avoid subsequent questions of payment responsibility and the unexpected need to assign more dollars than expected. The contract also typically identifies payment schedules (e.g., 10% withheld until final acceptance testing complete), ability to assign the contract, grounds for and methods of cancellation, and price protection.

The CE who is responsible for the procurement can protect the organization by taking the time to identify all of the technology features, options, accessories, services, and other acquisition issues that are important to the organization, and for ensuring that those items are contained within the procurement contract. Sufficient detail as to each, in order to identify what is required, to ensure that each is delivered, and to specify penalties can minimize misunderstandings and future problems.

Sometimes overlooked in the process is the need to advise vendors that are not receiving the award that an award has been made. The vendors typically have devoted time and resources to their response, and each vendor deserves to be advised. The organization should document the reasons for the selection. Depending on the situation, there may be merit in advising the vendor(s) of the reasons. Vendors might be able to improve their marketing and presentation response in the future and might decide to improve their technology and offerings. In some cases, the organization might subsequently need to consider acquisition of technology from a vendor that it did not select for a particular procurement.

Incoming Inspection and Acceptance

Once the award has been made, the process of monitoring and verifying the installation and delivery of the technology will begin. While this is an important aspect of the procurement process, it is discussed in detail elsewhere in this handbook (see Chapter 6 and 122). In short, acceptance testing, monitoring installation, and verifying delivery of specified items ensures the following:

- The technology performs its intended function
- Initial (baseline) performance characteristics are measured and recorded
- Incoming inspection and acceptance testing is performed and met
- The remaining terms and conditions of sale are met

Payment

Assuming that payment conditions have been detailed in the procurement contract, the CE need advise the appropriate personnel in the organization only when milestones are achieved or in the event that problems have arisen. In the event that a problem arises, the CE who is responsible for procurement may assume primary responsibility for determining the way that problems are to be resolved, and he may negotiate costs based on the effect of the problem. While CEs may not have the final authority on these aspects of the process, they can and should contribute significantly, identifying the ramifications of problems and the adequacy of proposed resolutions.

Monitoring

The procurement process is not complete upon final payment. Often, monitoring of medical staff satisfaction, equipment performance (e.g., uptime), provision of subsequent services (e.g., preventive maintenance and quality control), and calculation of life cycle costs must be performed. Along with addressing direct promises and needs, such monitoring can contribute to the organization's knowledge base and expertise for future acquisitions. In some instances, it can help to contribute to meeting accreditation; (e.g., JCAHO) and to satisfying regulatory requirements (e.g., CLIA).

Conclusion

The clinical engineer has a solid background on which to assume primary responsibility for the procurement process. It is important that the clinical engineer recognize the clinical and financial experts within the organization as critical participants in the process. Careful attention to detail throughout the process with the intent to capture important information and to advise the organization's and vendor's professionals of this information helps to improve the process. Developing a request for proposal and following through to the procurement contract, payment, installation, and use of the technology is truly a lesson in communication. Expediting communication through the assurance that adequate information is passed along to all parties and that responsibilities are clear can make the process a win-win one for all participants.

References

ASHE. Appendix B–Bid Specification. ASHE/AHA, Chicago, IL, 1991.
ECRI. Proposal Rating System: Health Devices. Plymouth Meeting, PA, 1976.
Mandell SF. The request for proposal (RFP). *J Med Syst* 10(1):31-39, 1986.

35

Equipment Control and Asset Management

Matthew F. Baretich
President, Baretich Engineering, Inc.
Fort Collins, CO

The number of critical medical devices used in a hospital may range from something on the order of 1000 devices for smaller community hospitals to well over 10,000 devices for large, academic medical centers. In this context, "critical" refers to a medical device that requires individual tracking and management to insure safe, effective, and economical use.

The importance of this task calls for a carefully designed equipment control and asset-management program. In practice, the magnitude of the task effectively requires the use of a computerized maintenance management system (CMMS). A well-crafted, CMMS-based equipment control and asset management program (see Chapter 36) represents the fundamental information resource that supports many of the medical technology management activities described in this handbook.

This chapter describes basic principles that can be used to determine the medical devices that should be included in an equipment control and asset management program. Inappropriate exclusion of a device can compromise its safety and effectiveness; inappropriate inclusion of a device is a waste of maintenance resources. Thus, well-designed inclusion criteria are necessary for optimization of safety, effectiveness, and economy.

Acknowledgment

Research for this article was supported by Hospital Shared Services of Denver, Colorado.

Background

The Joint Commission on Accreditation of Health Care Organizations (JCAHO) requires hospitals to "establish and maintain an equipment management program to promote the safe and effective use of equipment" (JCAHO 2002). Efficient and economical compliance with this standard is a challenge facing most hospitals in the United States.

A fundamental aspect of medical equipment management is a program of scheduled maintenance. As reflected in JCAHO Standard EC.1.6, the essential components of a scheduled maintenance program are the following:

- *Inventory.* Definition of medical devices to be included in the program
- *Procedures.* Definition of maintenance activities to be performed on devices in the program inventory
- *Scheduling.* Definition of when maintenance activities are to be performed on devices in the program inventory
- *Monitoring.* Continuous measurement and periodic review of program performance

Prior to 1989, JCAHO standards required hospitals to include all medical devices in the equipment management program inventory. Since that time, JCAHO standards have allowed hospitals the flexibility to exclude certain devices so that maintenance resources can be focused where most needed. Current JCAHO standards allow hospitals to establish "risk criteria for identifying, evaluating, and taking inventory of equipment to be included in the management program." These criteria must address "equipment function (diagnosis, care, treatment, and monitoring), physical risks associated with use, and equipment incident history."

Thus, the concept of a "risk-based inventory" was established. In pursuit of operational cost-effectiveness and compliance with JCAHO standards, most United States hospitals have implemented some form of risk-based inventory system.

Literature Review

Since 1989, there have been numerous publications in the clinical engineering literature proposing risk-based inventory systems for equipment management programs.

Of particular influence have been recommendations published by recognized professional organizations, e.g., ECRI, the Association for the Advancement of Medical Instrumentation (AAMI), and the American Society of Hospital Engineering (ASHE); highly regarded journals (e.g., *Journal of Clinical Engineering* and *Biomedical Instrumentation & Technology*); and JCAHO itself.

In the literature and in practice, two broad approaches to designing a risk-based inventory for medical equipment are followed. The first approach is to identify "critical" or "high-risk" medical devices and to use this information as the primary factor in deciding which medical devices to include in the equipment management program. An important example of this approach is the *Health Devices Inspection and Preventive Maintenance System* that is published by ECRI (1995). This system divides medical devices into "high-risk," "medium-risk," and "low-risk" categories on the basis of "risks of injury caused by device failure or user error." ECRI suggests that devices in the first two categories are typically included in an equipment management program. However, ECRI also emphasizes that "although risk level is an important indicator, it alone does not define a device's support requirements"

Another example of this approach is the medical device classification system, which is run by the United States Food and Drug Administration (FDA) (www.fda.gov/cdrh/prodcode.html). (See Chapter 126.) The primary purpose of this classification system is to establish the level of regulatory control required for the marketing of medical devices. Class I devices require the least regulatory control, while Class II and Class III devices require increasing levels of regulation. The FDA notes that "the risk the device poses to the patient and/or the user is a major factor in the class it is assigned."

The FDA does not represent its device classification system as being applicable to designing a risk-based inventory system for medical equipment maintenance. However, some publications do refer to the FDA classifications in this regard. A prominent example is the consensus standard, Recommended Practice for a Medical Equipment Management Program (American National Standard ANSI/AAMI EQ56:1999), which states that "most health care organizations will want to include devices falling into Class II or Class III within their equipment inventory list" (AAMI 1999). (See Chapter 122.)

The second basic approach is to use a numerical algorithm to determine the medical devices that should be included in an equipment management program (Fennigkoh and Smith, 1989).

The Fennigkoh and Smith algorithm scores medical devices on three factors:

1. Function (2-10)
2. Risk (1-5)
3. Required maintenance (1-5)

The sum of these scores yields an "equipment management" (EM) number. Devices for which EM ≥ 12 are included in the equipment management program. The Fennigkoh and Smith algorithm and its many derivatives have been incorporated into commercial CMMS software, implemented as online calculators, (e.g., www.currentpath.com) and applied by numerous health care organizations. American Society for Health care Engineering (ASHE 1996) developed one of the most widely distributed derivatives, which includes additional factors and a different method of calculation. Wang and Levenson (2000) provided a detailed review of algorithm-based systems. These authors also recommended a modification of the Fennigkoh and Smith algorithm to incorporate a "mission criticality" score reflecting the importance of a medical device to the mission of a health care organization.

Although the Fennigkoh and Smith-based algorithms were valuable responses to the 1989 changes in JCAHO standards, there now is reason to question their conceptual underpinnings. For example, although these algorithms generally include a "maintenance requirement" factor, it is possible for a medical device with recognized maintenance requirements to be excluded from the equipment management program if it has low enough "scores" on other factors in the algorithm (a consequence of the additive nature of these algorithms). Also, some Fennigkoh and Smith derivatives ASHE (1996) use the result of a single algorithm to determine both inclusion in the program and frequency of maintenance, two fundamentally distinct concepts that should be based on different considerations.

Ridgway has proposed a different conceptual approach (2001) in which medical devices to be included in a "monitored maintenance program" are those that are "critical devices in the sense that they have a significant potential to cause injury if they do not function properly" and that are "maintenance sensitive in the sense that they have a significant potential to function improperly if they are not provided with an adequate level of PM." Noncritical devices are excluded, as are devices for which there is no evidence of benefit from "planned maintenance."

Principles

Health care organizations vary widely in their implementations of JCAHO-compliant equipment management programs. They have adopted a variety of approaches to designing risk-based inventory systems, developing maintenance procedures and schedules, and measuring performance. A significant variation in terminology exists with respect to scheduled maintenance programs. Recognizing the wide range of practice in the profession, this chapter offers general principles that health care organizations can use as guidelines in developing their equipment control and asset management programs. These guidelines are based on best practices described in the literature and observed in practice. They should be adapted to the policies, procedures, and terminology of each unique health care organization.

What are the objectives of a scheduled maintenance program?

Using ECRI (1996) terminology, the following are the key objectives of a scheduled maintenance program for medical equipment:

- To reduce the risk of injury or adverse impact on patient care
- To decrease equipment life cycle costs
- To comply with codes, standards, and regulations

What Categories of Procedures Are Part of Scheduled Maintenance?

Adapting ECRI's (1996) and Ridgway's (2001) terminology, scheduled maintenance activities may be categorized as follows:

- Procedures to reduce the likelihood of premature failure (true preventive maintenance)
- Procedures to determine the performance of the device relative to its specifications
- Procedures to determine the safety of the device relative to applicable standards

What devices should be included in a scheduled maintenance program?

A medical device should be included in a scheduled maintenance program if any of the program objectives can be achieved by any of the maintenance activities described above. Here are some examples of this principle:

- If a device can fail in a manner that may cause patient injury, and if the likelihood of that failure can be reduced by one or more categories of scheduled maintenance activities, the device should be included in the scheduled maintenance program.
- If the life cycle cost of a device can be reduced by one or more categories of scheduled maintenance activities, the device should be included in the scheduled maintenance program.
- If applicable standards require a device to receive periodic inspection (falling into one or more categories of scheduled maintenance activities), it should be included in the scheduled maintenance program.

What devices present a risk of injury or adverse impact on patient care?

Two widely recognized resources regarding the "risk level" of medical devices are ECRI (1995) and FDA (www.fda.gov/cdrh/prodcode.html). These resources provide an initial assessment of "risk level" that can be adjusted on the basis of experience within a particular health care organization.

What devices are "maintenance sensitive"?

Following the definition in Ridgway (2001), a device is "maintenance sensitive" if it can be expected to function improperly in the absence of appropriate scheduled maintenance. Ridgway provides conceptual guidelines for this concept. Device-specific information can be gathered from manufacturers' recommendations; guidelines published by recognized organizations (e.g., ECRI, AAMI, and ASHE); experience within a particular health care organization; and professional knowledge regarding medical device design and operation.

Summary

These broad principles may be used to determine the medical devices that should be included in an equipment control and asset management program. After this determination has been made, these devices can be entered into a computerized maintenance management system. The CMMS can be used as a basis for planning and documenting scheduled and unscheduled maintenance activities. The CMMS also can be used for performance monitoring of the equipment control and asset management program. The result is a management tool for achieving safety, effectiveness, and economy in equipment maintenance.

References

AAMI. Recommended Practice for a Medical Equipment Management Program. American National Standard ANSI/AAMI EQ56, 1999.

ASHE. *Maintenance Management for Medical Equipment*. Chicago, American Society for Health Care Engineering, 1996.

Collins JT, Dysko J. Risk Assessment in a Medical Equipment Management Program. *ASHE Health Care Facilities Management Series* 055369, 2001.

ECRI. Health Devices Inspection and Preventive Maintenance System. Plymouth Meeting, PA, ECRI, 1995.

Fennigkoh L, Smith B. Clinical Equipment Management. *JCAHO Plant, Technology & Safety Management Series* (2)5-14, 1989.

Fennigkoh L, Lagerman B. Medical Equipment Management. *JCAHO Environment of Care Series* (1):47-54, 1997.

Gullikson ML, David Y, Blair CA. The Role of Quantifiable Risk Factors in a Medical Technology Management Program. *JCAHO Environment of Care Series* (3):11-26, 1996.

JCAHO. Medical Equipment Maintenance: Medical Equipment Management Plan and Inventory Selection. *JCAHO Environment of Care News* 3(1):2-3, 2000.

JCAHO. Comprehensive Accreditation Manual for Hospitals. Oakbrook Terrace, IL, Joint Commission on Accreditation of Health Care Organizations, 2002.

Ridgway M. Classifying Medical Devices According to Their Maintenance Sensitivity: A Practical, Risk-Based Approach to PM Program Management. *Biomed Instrum Technol* 35(3):167-176, 2001.

Stiefel RH. Medical Equipment Management Manual. Arlington, VA, Association for the Advancement of Medical Instrumentation, 2001.

Wang B, Levenson A. Equipment Inclusion Criteria: A New Interpretation of JCAHO's Medical Equipment Management Standard. *J Clin Eng* 25(1):26-35, 2000.

36
Computerized Maintenance Management Systems

Ted Cohen
Manager, Clinical Engineering Department, Sacramento Medical Center
University of California,
Sacramento, CA

Nicholas Cram
Texas A & M University
College Station, TX

How do clinical engineers (CEs) ensure that the medical systems installed in their customers' health care organizations are functioning optimally? What do they do when those medical systems malfunction? What is the quickest route to restoring performance of the medical system to the customer's satisfaction, and at what cost? How do they help to determine when to replace a major system? How do they interact with the health care providers and all other customers in a modern, complex, health care organization such as accrediting organizations, the FDA, device manufacturers, risk managers, and attorneys? Most use some type of computerized record keeping system to help them to manage their business.

Computerized maintenance management systems (CMMSs) have evolved into a useful tool for providing technology support. Whether supporting a three-technician shop or an international service organization. almost all medical equipment support organizations are using some type of CMMS in their operations. A CMMS can be classified broadly as internally developed (typically using commercial off-the-shelf (COTS) personal computer hardware and database software); commercial CMMS applications; or the newest approach—application service providers (essentially a web-based software rental service). This chapter reports on the current status of computerized maintenance management systems (CMMS) in use in hospitals and health care systems today and some of the future technologies that the future leaning providers of medical technology support services are starting to use.

Computerized maintenance management systems can provide the technology management staff with a wealth of information to help manage many technology support-related functions. Examples include the following:

1. Quantitative equipment reliability assessments can be made, based on failure rate, down time, and repair and maintenance costs. These assessments can be used to determine equipment that should be replaced and that should assist in the subsequent vendor selection for the new product being purchased.
2. User/operator training needs can be identified based on trends in use error problems (e.g., problem not repeatable, incorrect settings, liquid spills, and physical damage).
3. Scheduled maintenance can be prioritized based on the risk to the patient of an equipment failure and the maintenance needs of the device. CMMSs can be used to balance and manage this often large workload better.
4. Scheduled maintenance program effectiveness can be measured by the rate of problems identified (yield), parts replaced, equipment not found/not available as compared to the total number of inspections performed by risk priority, and again, used to manage this often very large workload.
5. Workorder systems can be used to prioritize repair requests and better manage down time of critical systems.

This chapter addresses the following areas: Medical equipment management fundamentals, the CMMS core, CMMS modules, data integrity, reports, utilities, network and multiuser issues, and new technologies.

Medical Equipment Management Fundamentals

Technology plays a major role in the delivery of today's health care. Keeping the medical devices that help to provide that care in good working order is the role of equipment service management. In 1993, U.S. hospitals spent over $3 billion ECRI, 1993) to service their medical equipment. Appropriate equipment service, whether performed by hospital personnel, equipment manufacturers, or contracted third-party organizations, is essential to modern, high-quality, cost-effective health care.

Modern health care organizations must carefully manage all equipment service independent of the service provider, who may be an employee, a contractor, a manufacturer representative or an employee of another organization affiliated with the health care institution, e.g., group purchasing organization (GPO). Comprehensive equipment service management includes all medical technology and all service providers. In many cases there will be multiple service providers that may include multiple in-house groups (e.g., clinical engineering, radiological engineering, multiple manufacturers, and multiple independent contractors). In other cases, there may be one in-house or contracted organization coordinating all of the service regardless of who is the actual service provider for a specific device or repair.

In addition to the actual equipment maintenance and repair, the technology management team should participate in all phases of a device's lifespan, including capital budget planning, equipment needs analysis, specification and request for proposal/bid authoring, vendor selection, service provider evaluation and monitoring, service contract evaluation and monitoring, installation planning, installation, acceptance, user education, product recalls and alerts, incident and accident investigation, replacement analysis, de-installation, and salvage. (See Chapter 30.) Overall program cost management is another key function.

Regardless of the service provider model, service is coordinated by a service manager who may be on-site in larger organizations or who may be shared among several sites for smaller hospitals. The service manager could be an employee of the health care organization, a contracted employee, or an employee of a vendor. There are many different cost models for the payment of medical device service, and it is the responsibility of the service manager to coordinate such service in an efficacious and cost-effective manner within the defined description of the manager's job as directed by senior leadership.

Equipment Management Implementation

The first, and one of the most critical, steps in implementing an equipment management system (computerized or noncomputerized) is to complete an accurate inventory of all equipment that will be under the equipment management program, including devices that will be serviced by other organizations but whose service needs to be tracked. Each device that needs to be tracked must have an equipment control number assigned to it and labeled on the device. In some hospitals, the hospital asset number or property number already assigned to the device may be used as an equipment control number. If the asset tracking system is incomplete or if the format of the asset number is not compatible with the equipment management system, then an independent equipment control numbering system might need to be developed. Without an effective inventory system, it is impossible to track maintenance and repairs, alerts and recalls, and most of the other equipment-management functions accurately.

The inventory must be kept accurate and must be frequently updated as new equipment is added to the hospital's inventory and old equipment is removed. This task involves policies and procedures that include the equipment management function as a critical part of the hospital's new equipment-receiving function and equipment-removal and -salvage operations.

Important parameters to be tracked in association with each device in the inventory are the model, serial number, warranty expiration date, risk of the device, type of device, ownership information, maintenance scheduling information, and purchase information. A sample list of data fields and their definitions is included Cohen (1994).

JCAHO

The Joint Commission on Accreditation of Healthcare Organizations (JCAHO) (JCAHO, 2000) has considerable impact on the equipment management program. The following is a discussion of some of the equipment management-related JCAHO requirements.

Standard EC.1.6: The hospital plans for managing medical equipment.

The intent statement for EC.1.6 starts as follows: "The hospital identifies how it will establish and maintain an equipment management program to promote the safe and effective use of equipment"

The equipment management program is documented via the equipment management plan. This plan includes processes for selecting and acquiring equipment and for establishing criteria for identifying, evaluating, and taking inventory of equipment to be included in the management program before the equipment is put into clinical use.

A hospital has the option of including all patient care equipment in its program or developing written-assessment criteria and applying those criteria on a device-type by device-type basis to determine whether a device is to be included in the equipment-management program. All patient care equipment must be assessed, independent of acquisition source, including rentals, leases, donations, and physician-owned equipment. The written assessment criteria should include the following:

- Equipment function (e.g., Does the device provide life support or a resuscitation function?)
- Physical and clinical risks to the patient: What is the likelihood that the device will fail in such a way as to cause serious physical injury? How is the device used, and how is data from the device used? What would happen if the device displayed or reported clinical data erroneously?
- Maintenance requirements: Does the device require periodic parts replacement, lubrication, recalibration, or other routine maintenance in order to perform properly and safely? What maintenance is performed by the user, and how frequently is it performed? What additional maintenance must be performed by the service provider?
- Incident history: Does the device have a history of incident reports or other serious reported problems?

Many different schemes have been developed to determine the risk associated with a given device or type of device. The simplest provide a three-level categorization (high-risk, moderate-risk, and low-risk) for each device included in the equipment management program. Some systems also include the ability to define and identify critical systems. Moreover, low-risk devices with no maintenance requirements might not need to be included in the plan as they are dependent on local codes, standards, and policies.

In addition to the JCAHO, various other state, local, building code-related authorities (e.g., National Fire Protection Association [NFPA]), and professional organizations (e.g., College of American Pathologists [CAP] and the American Blood Bank Association [ABBA]), have accreditation requirements and licensing laws, regulations, and standards that vary by locale and type of institution.

Data Collection

The primary incoming data to the equipment management program are repair information, scheduled maintenance workorders, new equipment receipts, product recalls and alerts, and incidents and other hazard information. This large amount of information needs to be collected in a user-friendly yet controlled and organized manner. Nonautomated means of collecting these data have been by way of paper forms for work orders (both scheduled and unscheduled), telephone requests, and incoming equipment documents (internal documents or vendor packing slips).

Equipment data management schemes typically are based on each device's unique equipment control number as assigned by the health care institution or its contracted service-management organization. A file is established for each device and activity repair and maintenance activities logged in that file. This file should include all maintenance and repair activities performed on the device, regardless of the service provider.

With nonautomated equipment management systems, it is difficult to implement statistical analysis of the data including cost center accounting information, device-type problem analysis, and reporting and other statistical analysis for trending and reporting problems. When the inventory includes large numbers of devices and therefore large numbers of scheduled and unscheduled workorders, statistical analysis by noncomputerized methods become overly cumbersome. Many large hospitals have more than 10,000 devices and complete more than 20,000 work orders in one year. Statistical analysis of these data must be performed in order to reveal useful trends of problems categorized by department, by equipment model and by type of equipment. These trends then become information that can be used to identify equipment requiring replacement, user-training requirements by department, and user-training requirements by type of device or model. Obviously, management of this large amount of data is one of the reasons that computers are used for modern equipment management programs.

Clinical engineers and hospital administrators are recognizing the cost and technical advantages of centralization in equipment management and maintenance services, as can be seen by the increasing responsibilities that many clinical engineering departments and contractors are receiving. These include increased maintenance and repair responsibilities for clinical equipment in radiology and the clinical laboratory and for nonclinical equipment involving information technology, telecommunications, and material-moving robots. In addition, as large hospitals purchase or jointly manage neighboring smaller hospitals, clinics, free-standing surgery centers, and other healthcare-related facilities, the repair and maintenance of medical equipment in those facilities is frequently coming under the responsibility of the parent organization's clinical engineering department. As growth in responsibility, technical complexity, and volume continues, it becomes clear that computerized databases are required in order to manage the large amount of data and to convert those data into useful management information.

The Computerized Maintenance Management System Core

The core of an equipment management system comprises equipment inventory, repair and maintenance history, and work order control. The equipment inventory is an automated file of all of the equipment that has been included in the CMMS. The repair and maintenance history is a record of each repair and maintenance event, independent of who initiated the event and who provided the service. Work order control is used to dispatch and to prioritize requested work, to schedule periodic inspections and preventive maintenance (PM), and to track the status of pending scheduled and unscheduled workorders.

Equipment Inventory

In the typical CMMS, when new equipment is received, a biomedical equipment technician (BMET) ensures that the order is complete; inspects and tests the device in accordance with the service manual that is provided as part of the order; and, based on the type of device, the organization's inclusion criteria, and the policies of the clinical engineering organization, determines whether the device needs to be included in the equipment management program. If it does, the BMET then enters the new item onto the database (or completes a form so that a data entry clerk can enter it) as well as completing an incoming inspection work order.

Device descriptions and other fields should be made as consistent as possible by using the ECRI device nomenclature (ECRI, 1994) and relational database techniques where, for example, each unique model entry has one entry that is referenced by all equipment entries of that model. Each model and device type also can include various defaulted fields (e.g., scheduled maintenance information such as inspection frequency and maintenance procedure reference). A similar construct may be used for the owning department (cost center) with defaults for minimum scheduled maintenance frequency and location. This building block approach, with references to standard tables (e.g., departments and type descriptions) and built-in default values, allows equipment records to be built quickly with maximum data integrity and flexibility.

Cohen (1994) lists fields, and their definitions, that typically are stored as part of the equipment inventory.

For those new devices whose model and/or type description are not yet in the CMMS, the equipment type record must be built first, and then the model record, before the new equipment record can be generated.

The Integrated History Record

The second part of the core of the CMMS is the integrated history record. Whereas equipment records are fairly standardized across most CMMSs, the maintenance and repair records are not standardized, nor is there a consensus on the data that are necessary to collect. There is a large variation among CMMS regarding the particular data that are collected, how they are collected, and how they are used.

The integrated history record concept provides a service provider with independent date-and-time-tagged repair and maintenance history associated with each service event. The fields for the integrated history record are listed in Cohen (1994).

The typical history record contains the following information:

- Original problem or request: Text of original problem request
- Work order type: Category of work (e.g., scheduled maintenance, repair, incoming inspection, project, and recall/alert)
- Open date and time: Origination date and time
- One or more tasks where each contains the following information:
- Start and end dates, and times for each task
- Status of the work order at the end of the task (e.g., complete or awaiting parts) and referred to a vendor
- Service provider, technician, engineer identification: Who performed the task (e.g., vendor, CE, or BMET)
- Labor hours for the task, including travel and overtime
- Parts and materials cost
- Reference to a parts purchase order, vendor repair order, or stock parts sales order
- Down time: Especially important for high-cost medical equipment (e.g., imaging equipment) where down time significantly adversely affects patient throughput and thus revenue

Associated with each task is a list of zero or more specific actions taken as part of the task. There is no consensus in the clinical engineering community as to whether these additional important data should be encoded or free text. Certain special fields that cause other actions to occur must be encoded (e.g., codes that indicate the status of the work order and codes that are used to update the scheduled due dates). These include electrical safety pass, inspection complete, and PM complete. Other fields can be either free text or encoded. Free text generally provides more comprehensible information; encoded data are easier to analyze statistically.

Work Order Subsystem

The work order subsystem of the typical CMMS consists of the following modules: An unscheduled (requested) work order manager and technician dispatcher and an inspection and PM scheduler. Some systems also include an inspection preventive maintenance procedure library.

The unscheduled work order manager documents incoming requests for repair services and keeps track of the work order until completion. Typical tracked information includes request for name and phone number, equipment identification, equipment problem and/or service requested, equipment location, type of work order (i.e., repair, new inspection, or product recall/alert), and priority of the work order (i.e., how soon the customer needs the work completed). See Cohen (1994) for a list of fields and their definitions.

The maintenance scheduler is used to initiate and to manage scheduled inspections, scheduled parts replacement, and scheduled preventive maintenance work orders.

Scheduled inspections are defined as periodic inspections that verify that the product meets manufacturer's specifications by using externally calibrated sources and measuring devices. Generally, scheduled inspections will be completed on site and will not require removing equipment covers. Periodic parts replacements are replacements of specific, predetermined parts at predetermined intervals (e.g., replace nickel-cadmium defibrillator batteries every two years).

PM are actions that are necessary or desirable in order to extend the operational intervals between failure, to extend the life of the equipment, or to detect and correct problems that are not apparent to the user. PM could include scheduled parts replacement and inspection activities, but PM is generally more invasive, typically requiring the equipment to be brought to a shop location and requiring access to the internal portions of the equipment. For example, lubricating the moving parts of an electric bed would be preventive maintenance.

In the typical CMMS, each device has associated with it scheduled inspection data that include the following for each type of scheduled inspection: Provider responsibility (i.e., the service provider that is responsible for completing the inspection and PM), an interval (e.g., 12 months), fixed or floating flag, a synchronization date for fixed scheduling, the most recent completion date, and the next due date.

The next due date is calculated based on either a fixed or a floating basis. A fixed schedule results in the work being scheduled at the same time each year, based on the interval, and a synchronization date, regardless of when the last inspection was completed. A floating schedule results in a new due date determined by the inspection interval and the date the last inspection was completed. Upon request or at fixed times (e.g., weekly or monthly), the inspection scheduler initiates and prints, based on the parameters listed above, a summary of detailed work orders that are the scheduled work for the following period.

The inspection and preventive maintenance procedure library consists of a set of equipment type and model-specific procedures itemizing the tasks and parts that are required in order to complete periodic inspections, preventive maintenance, and periodic parts replacement. Some organizations use references to service manuals. Others develop typed procedures using word processors and reference those, and still others use sophisticated CMMS-based procedures that use database fields for common tasks and build the procedures as sets of tasks. There is no consensus on the data format for procedures other than those procedures that are model-specific and required for most high-risk devices (e.g., ventilators and complex systems, such as imaging systems).

It is required that technicians performing scheduled maintenance tasks consistently and appropriately test the device being inspected. Most CMMSs allow scheduled maintenance work orders to be generated as individual work order forms with the itemized procedure on the printed work order, or as a list of work orders with a reference to a printed procedure. Many commercial systems provide both options and leave to clinical engineering management the decision to print procedures for each work order.

Parts and Service Provider Management

One of the underlying philosophies of a quality CMMS is that all services be tracked and costs recorded regardless of the service provider. This sections describes various ways to track and manage repair parts and vendor services.

Parts One of the ways to categorize repair parts is by the way that they are obtained. Repair parts typicallyare obtained from one of the following types of sources: Stock parts, parts purchased directly from a vendor and shipped to the hospital for specific use on a specific repair, or parts supplied by the vendor as part of vendor service that includes installation labor.

Stock parts are stored locally for future use. These parts may be stored in a stock room, at a technician's workbench, or in other locations. All clinical engineering departments maintain some stock of electronic components, batteries, wire, nuts, bolts, and other common hardware. Many departments also stock circuit boards and other expensive medical device-specific parts.

Determining the parts to stock or to order on an as-needed (i.e., just-in-time) basis is based on how critical the device is and the ability of the clinical engineering department to obtain specific parts from the source quickly. As it is impossible to predetermine all failures, and as it is cost prohibitive to stock all parts for all devices, major high-cost, low failure-rate parts would not be stocked.

Some clinical engineering departments have purchasing authority and issue their own purchase orders. Purchasing authority, combined with a direct receiving function in clinical engineering and the availability of Federal Express, UPS, or other overnight carriers, frequently allows the ordering and receipt of parts to occur within 24 hours, thus minimizing the need to stock high-cost, low failure-rate parts. However, some clinical engineering departments are still burdened by institutional purchasing protocols that require the purchasing department to preapprove and/or to place parts orders, thereby adding significant delays to the ordering and receipt of parts. Computerized parts management systems are nonstandardized and typically differ a great deal from one department to another and from one institution to another.

CMMS-based parts management software usually requires that all parts entered have a previously issued and unique clinical engineering part number. Alternatively, some systems do not require a unique clinical engineering-generated part number and instead use the manufacturer's part number as the index to the parts management system. Other fields that typically are collected include a part description, price(s), manufacturer, manufacturer part number, vendor, and vendor part number.

Most hospital purchasing and accounting systems are much more complex and require more complex systems to manage parts. For example, some clinical engineering departments stock large numbers of parts and/or high-value parts and require that pricing includes a handling markup to pay for parts stockroom overhead. Other departments may require pricing calculations based on a rolling average price paid when several parts (bought at different times and at different prices) are resold to the clinical engineering customer as part of a repair. Other institutions that rely on purchasing parts quickly and on an as-needed basis may require a reference to a unique purchase order number or purchase order tag number for each purchase.

Service provider management No clinical engineering department can provide 100% of the equipment service that is required, 100% of the time, on 100% of the equipment inventory. In order to control the cost and quality of vendor medical equipment repair and maintenance services, it is appropriate for clinical engineering departments to control and to coordinate vendor equipment services.

Vendor services can consist of nonbillable warranty work, service performed on a fee-for-service basis, service performed under a prepaid service contract, billable services performed under a prepaid service contract but outside the prepaid terms and conditions of the contract, and other billable and nonbillable services (e.g., product recalls and installation work). All costs (e.g., prepaid, billable, freight, tax, parts, and labor) should be tracked.

All vendor service work needs to be coordinated and tracked in a similar manner to the data collected for in-house work. Typically, vendors complete service reports when they have completed a task or when they leave the site even if the task has not been completed. These service reports are among the key documents used for data collection. Service reports should be required by and provided to clinical engineering for all vendor work, both billable and nonbillable.

In addition, billable services typically require a purchase order prior to the vendor providing the service. Billable services also will eventually yield an invoice itemizing pricing. Although estimated pricing for vendor services can be used to track costs, it is much better business practice for clinical engineering departments to receive an invoice, match the invoice with the purchase order and service report, review the charges to make sure that the cost is appropriate for the work that was completed, and then complete the documentation by entering appropriate technical and cost data into the CMMS.

The CMMS requirements become more complex if maintenance insurance, partnership contracts, and other risk-sharing, cost-sharing schemes (e.g., parts caps) are implemented. These typically require keeping track of various terms and conditions that trigger additional costs or rebates, and sometime these triggers accumulate across a large number of devices over several years.

Data Accuracy and Integrity

New equipment entries and equipment service history entries comprise the majority of data collected in a CMMS. Service history data are typically collected by biomedical equipment technicians and then either directly keyboarded into the CMMS or a paper form filled out and data entry clerk keyboards the data into the CMMS. Equipment data can be entered by clerks or technicians from purchase orders or new equipment forms filled out by technicians for later entry by clerks. Service reports from other service providers are entered into the CMMS by technicians or clerks, or they are scanned.

Data entry requirements are intended to optimize data accuracy while minimizing data entry times. Data entry times can be minimized by using defaults and not requiring any redundant entries or keystrokes. Data accuracy and data integrity can be optimized by making the CMMS easy to use correctly, enforcing data integrity rules when appropriate, and establishing operational policies and practices that require every employee to enter accurate and complete data.

Data integrity issues can be further subdivided as follows (Barta, 2001):

- Data integrity enforced by the CMMS
- Data integrity enhanced (but not enforced) by the CMMS
- Data integrity based on department-wide, standardized definitions and operating procedures

Enforced data integrity can be used where flexibility in practice is not required and where an absolute relationship always occurs. This occurs most often in customized CMMSs, where the business rules can be built into the system and the flexibility of being able to match many different institutions' rules are not required. One example of enforced data integrity would be that every parts purchase order must have a valid purchase order number, a valid department, and a valid vendor associated with the order. In every case where there is an absolute requirement for a valid reference field, it is appropriate to enforce data integrity.

Some fields require an entry but do not necessarily reference another table or file. Examples of these required fields are a model and manufacturer for each equipment control number, and start and end dates for each work order. Other types of absolute data enforcement include the correct type of data (e.g., numeric, string, or date), range checking (e.g., valid date checking), and inter-field relationship checking, sometimes called "sanity checks." For example, a work order end date cannot fall before the work order start date.

When data integrity cannot be absolutely enforced by the CMMS because of exceptions or situations that require more flexibility than absolute data integrity allows, it is appropriate to use the same data integrity concepts but with defaults, warnings, or workarounds that allow the user to bypass the data integrity system legally. For example, when a new model is added, the CMMS may display a list of similar models and ask the data entry person to select one of the displayed models as the correct model nomenclature or indicate that the new model entered is indeed a valid new entry. This helps decrease the number of duplicate models inadvertently entered because of subtle differences in the model spelling and the punctuation used; however, this does not preclude the entry of new model names.

Every department that operates a CMMS should have a department-wide set of standardized definitions and operating procedures describing system use, data formats, data measurements, required fields and management's clarification of the ways they want certain tasks documented parts. For example, are transit time and wait time documented as part of a repair work order? This set of standard operating procedures establishes a consistency that cannot be programmed into a CMMS (even a custom CMMS) and makes the CMMS a much more valid and useful management tool.

Reports

The reporting capabilities of a CMMS are the heart and soul of a vendor's product from a management perspective. The ability of the CMMS to produce relevant, informative, concise reports transforms the software into a management tool. When justifying the purchase of a CMMS, a demonstration of reporting capabilities to senior administration is one key to whether or not the deal is consummated (with considerations for price). All successful technology management processes are dependent on the ability to gather data, to aggregate the data to convert it to useful management information, and then to produce reliable decisions based on the information. Reporting is the fundamental communication and decision support tool for the CMMS.

The ability to make the CMMS work toward your specific needs centers on the reporting capabilities of the system. The reporting access structure should be easy to use and intuitive yet offer sufficient reporting power to allow any data field to be aggregated and reported in any manner that the end user feels is appropriate. Typically, a tiered approach is used with canned reports for commonly used reporting requirements (e.g., monthly reports to customers), ad hoc reports for selection of various fields commonly reported (e.g., equipment history reports with date ranges), and custom reports for more complex reporting requirements (e.g., cost benchmarking) that may require some programming skill to generate.

Point-and-click structures should be the reporting baseline with common graphics or symbols that are easily recognizable and understood regardless of the computer literacy of the staff. Microsoft Windows®-based structures are one way to provide familiarity and user friendliness. More advanced reporting tools also may use a structured query language (e.g., SQL) to extract data from the database and additional (often third-party) reporting tools (e.g., Crystal Reports®) to format and write the report. The CMMS must have a reliable query process that allows an interface with the main database for all required reports. Reports should have the capability not only to print out on to paper but also to view on screen and to publish them as computer-transmittable files (e.g., pdf files) or web pages (e.g., html files).

All CMMS products will have generic reports. A review of the vendor's generic reports may reveal a need for reports that are specific to your health care facility. Specific reports can be generated with user-defined fields or through custom reports written by the vendor, a third party, internally by information technology (IT) or clinical engineering department staff depending on resources available, and the report writing tools that are available in the CMMS. The authors recommend that an agreement with the original vendor be included in the contract price for all initially required custom reports. Third-party software vendors charge as much as $50,000 for each customized report. Pop-up displays that are triggered by indictor limits provide a dashboard for decision support and management control. (See Chapter 100.)

The CMMS must be capable of satisfying regulatory reporting requirements, particularly those of the Joint Commission for the Accreditation of Healthcare Organizations (JCAHO), the Food and Drug Administration (FDA) (see Chapter 126), and state and local health agencies. The basic reporting structure for clinical engineering should include an inventory list for all medical equipment, a work order generation and completion process, risk-ranking of medical devices, preventive maintenance scheduling, personnel time allocation, and resource allocation (both physical and personnel). Table 36-1 lists many commonly used reports produced by a CMMS.

Utilities

A utility is a computer program that is not contained in the main operating system (OS) or the core CMMS and that provides an essential function or feature. In most cases, as customers demand similar utilities, the programs are incorporated into the OS and/or CMMS, but during selection of the CMMS for your facility it is wise to have a checklist of required features.

Features that are required but not contained in the OS or core CMMS should be available from the vendor as utilities. Examples of utilities are database validation, database diagnostics, security features, global data changes (e.g., manufacturer A merges with manufacturer B), user-defined fields, initial installation, setup, database conversion from a previous vintage or foreign vendor's CMMS, and configuration programs.

The system administrator has access to the utilities programs and will become familiar with their functions. General CMMS users most likely will not be aware of, or have access to, utilities programs. As the needs of the facility change, available utilities programs allow the CMMS to become flexible and scalable. The addition of a utilities program allows a system to grow as needs change, without the expense of replacing the entire system. This is an important purchase consideration that should be explored thoroughly.

Network and Multiuser Issues

All but the smallest of clinical engineering departments will require more than one user with simultaneous access to the CMMS. The connectivity and throughput capabilities of the CMMS will be determined by the existing network, or by the funding that is available to purchase and install the recommended network infrastructure. Prospective CMMS vendors should provide network bandwidth requirements for their systems. The vendor should be asked for complete technical specifications, and these specifications should be reviewed with the IT department.

The optimal CMMS vendor selection process should involve a committee that includes a staff member from the IT department. Begin with a list of multiuser needs and expectations. Is there a need for telecommuter system access? Does the enterprise consist of a single facility or multiple local or remote facilities with varying telecommunications capabilities? Who are its customers? Who are the system users? How many workstations will be required? After answering these questions, compare those needs to the system capabilities and the institution's current network infrastructure. Can the desired system be achieved within the existing infrastructure? If not, what will it cost to provide a network that is capable of meeting the needs? Will funding to upgrade the network be available?

Assuming that the funding and/or the network is suitable for the needs of the organization, review the available networking technology. Are there areas where a wireless interface is more cost-effective, or will the entire network be hardwired? Will the system be run on an intranet corporate enterprise network, a web-based network, a local area network (LAN), or a combination of these technologies?

Most CMMSs will use a client/server, local area network (LAN) structure for multiuser access. The client/server system consists of a main server (or multiple servers if the CMMS is operating across multiple facilities) and several workstations that typically are interconnected via a cabled network. The typical workstation will be the personal computers (PCs) in your work areas and will be used to run the CMMS client and other desktop-based applications (e.g., word processor, spreadsheet, and calendar). The server is a more robust computer with the power and memory to host and drive the entire system. All CMMS data will be stored on the server and will be sent to the client workstations on an as-needed basis. The server may be located in the clinical engineering or information technology departments. It is managed by the system administrator.

For those mobile users who cannot often access a LAN-connected workstation, most CMMS vendors offer a laptop, standalone version that can be periodically, intermittently connected to and resynchronized with the server.

Data Telecommunications Technology Primer

Most hospitals will have existing network infrastructure (e.g., cabling, routers, switches, and hubs). If applications are required for remote workstation interfaces (i.e., those outside the LAN framework), an understanding of the available Local Exchange Carrier (LEC) telecommunications infrastructure is essential. LEC is a local telephone company or competing company that can provide the desired connectivity.

For connectivity over longer distances than the internal wired LAN/WAN, there are a variety of telecommunication options including microwave, optical, and various wired technologies. Microwave connections can be considered only for short (line-of-sight) distances of less than three miles. Rain, snow, fog, birds, and bats can cause network interruptions of microwave transmissions. Infrared transmission, known as Freespace Infrared Local Area Network (FIRLAN), also can be used for remote connection to the LAN. FIRLAN's advantages and disadvantages are similar to those of microwave, and installation costs can run above $20,000. These types of remote interconnections to the LAN are known as point-to-point connections. Figure 36-1 illustrates a generic point-to-point telecommunications system.

Table 36-1 Common CMMS reporting capabilities

Complete equipment inventory listing	Productivity report by technician by date	List of vendor sources for repair parts by inventory number
Print scheduled PMs	Incomplete work orders by assigned technician, department, and date	List of equipment manufacturers by equipment type and model
List of open, suspended, closed work orders by department, technician, and date	Estimated PM hours by facility by date	Graph bar chart corrective repair costs vs. PM costs by date
Total work orders by facility by date	Life-cycle costing by inventory number by date	Graph bar chart corrective repair costs vs. PM costs by date by inventory number
Total work orders completed by technician by date	FTE estimate based on estimated PM hours by facility by date	Personnel listing technician wage/salary, start date, certifications, OEM schools attended
PM completion percent by date	FTE estimate based on current corrective repairs by facility by date	Personnel listing vacation and sick hours available
PM completion percent by cost center by date	Total dollar amount for equipment in inventory	Risk management SMDA reportable events by department by date
Total parts cost by inventory number by date	List of equipment by equipment type, category	Risk management SMDA reportable events by date
Total corrective hours by inventory number by date	Weekly productivity by technician by date	Risk management SMDA reportable events by inventory number by date
Incomplete PMs by technician by date	Graph line chart of total inventory by date (from date to date)	Risk management SMDA reportable events by equipment category
Incomplete PMs total number by date	Graph line chart of total inventory asset dollar value by date (from date to date)	Graph pie chart of resources by PM hours and corrective hours by facility by date
Performance indicator reports (e.g., down time, response time)	Ratio of repair and maintenance cost to acquisition cost by type of equipment	Purchase order by vendor by date

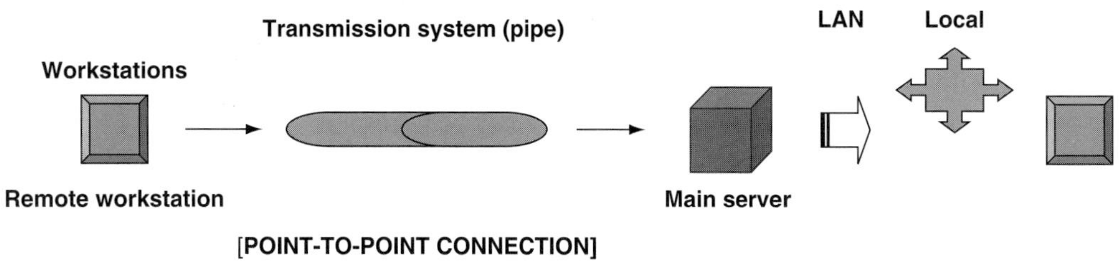

Figure 36-1 Point-to-point remote communications.

The rate of uninterrupted flow of information is known as "throughput." Throughput is directly related to available bandwidth and router capacity. Table 36-2 illustrates the various LEC services and their bandwidth limitations.

The primary network issues involving the installation of a CMMS will involve the local area network (LAN) and access to a server. A network consists of nodes (i.e., workstations), hubs, media access unit (MAU), bridges, routers, and gateways that are linked together for the purpose of transmitting data from one location to another. LANs provide a robust, flexible, and scalable means to interconnect all users of the CMMS. Although LAN standards are beyond the scope of this chapter, further in-depth information concerning LAN specifications can be found in the Institute of Electrical and Electronics Engineering (IEEE) 802 family of standards.

LANs are constructed in specific layouts or connection pathways known as topologies. The major LAN topologies are mesh, star, ring, and bus. Figure 36-2 illustrates each of these topologies, which provides a visual concept of the way that various networks are linked together.

The bus and ring topologies are the most common LANs. Star and mesh topologies are more expensive because of the increased number of links in the network. Bus and ring topologies are well-suited for CMMS applications. Bus topology with standard Ethernet (10Mb/sec) or high-speed Ethernet (100Mb/sec) are the most commonly used LAN infrastructure for CMMS applications.

The principle of ring topology involves a "token," which contains requested information and is passed around the ring until it is accepted by the originator of the request (i.e., the workstation). Ring topology ensures uniform distribution of requested information and guards against bandwidth saturation. Both shielded and unshielded twisted pairs can be used at transmission rates up to 16 Mbps, which is somewhat higher than standard Ethernet bandwidth. Bus topology has an advantage of simplicity for maintenance, but access depends on the position of the workstation on the bus.

Each LAN will have one access method and one or more specified physical media and topology. The most typical LAN uses bus topology and Ethernet. Ethernet uses carrier sense multiple access with collision detection (CSMA/CD) as the access method. Ethernet refers to the physical link and data link protocol for LAN interconnection. The physical connection may be a twisted pair or a coaxial cable. The maximum transmission rate of an Ethernet link is 14,800–64 byte packets per second (PPS) or more commonly 10 Mbps. Ethernet connections which use twisted pair phone wires will have an RJ 45 connection at the terminal end and are known as "10base-T." High-speed Ethernet, which has a transmission speed of 100 Mbps, is known as 100base-T. If the facility has sufficient network infrastructure to meet the bandwidth requirements of the CMMS, then the clinical engineering department should use that infrastructure. Table 36-3 provides a comparison of physical media and LAN topology characteristics.

In some instances, it may be cost effective for the facility to purchase one CMMS, which can be utilized by multiple service departments such as IT, physical plant, and clinical engineering. Difficulties arise in this scenario because of multiple departmental regulatory reporting requirements and variations in project priorities. If a single server is used, the drive should be segmented by department in order to allow easy access to information pertaining to a specific department. Without segmentation, all data are lumped into one large pool. The queuing factors must sort through several lists before the specific required information is retrieved. Although it is more expensive, a dedicated server for each department eliminates several software conflicts; is simpler to install, customize, upgrade, and maintain; and reduces problems with throughput.

Additional challenges of a multiuser network involve access levels and access security. The system administrator is required to have access to all features and functions and to manage security and access for the other staff. The manager and supervisors, administrative assistant/clerical/dispatch, and technician staff all may have access to some common features and some others that are dependent on their access group. If a nurse manager is concerned about a shortage of vital signs monitors, a read-only access level provides information on the repair status for a specific piece of equipment. Allowing clinical staff read-only access to CMMS inventory and work order status screens is an efficient communications and management tool. End user access to enter new repair work order requests also can be useful.

The system administrator will have the highest access level that is capable of performing all CMMS functions. A supervisor access level just below the system administrator should be available for managers. This level of access provides reporting capabilities and sensitive personnel information concerning salary, date of hire, home telephone number, and other personal information.

General access should be available to all technicians and engineers in the CE department. Personnel outside the department should be given read-only access. Personnel without a knowledge of the way a department functions or the way the CMMS operates might unknowingly modify or delete important data fields or files and corrupt the database. Senior administration might wish to have access to some of the reporting capabilities of the system. Again, a read-only access level is the prudent approach for system entry.

Although most CMMS systems will not contain patient information, system security must be included (see Chapter 104). As mentioned earlier, there could be sensitive personnel information in the database, and users who are unfamiliar with the system might induce errors that would require reentering information. General security methods using password protection are sufficient safeguards to prevent unauthorized entry into the CMMS. Firewall protection must be considered if Internet access is linked to the system or if the CMMS interfaces with another network.

New Technologies

Innovations and new applications of existing telecommunications and computing technology are emerging at a rapid pace. To state that new technology is being implemented is relevant only when describing the comparison to a facility's existing technology and, then, only in a specific and short time period. The Internet and Computer Telephony Integration (CTI) are perpetual, cutting-edge concepts that continually infuse new technology into the mainstream of networking.

When electronics nanotechnology is added to these rapid transformations, the world of networking takes on new meaning almost daily.

The most dramatic impact in telecommunications has come from, and will continue to come from, deregulation ushered in by the Telecommunications Act of 1996 and the resulting impetus for development of additional infrastructure. Satellite communication, fiber optics, ISDN, DSL, and a host of other, faster broadband communication technologies are bringing more and more data to homes and businesses at a faster and faster rate. Web-based programs offer flexibility and accessibility unparalleled by any existing hard-wired LAN system. Virtual Private Networks (VPNs) offer the opportunity to have LAN types of applications on a worldwide scale. The availability of wireless network links (e.g., 802.11 Ethernet and Bluetooth) undoubtedly will be included in the networks of the future. Remote and distributed computing technologies will overcome problems with scalability, performance, and cost, the bane of hard-wired networks.

Technologies that are currently available but not widely utilized in CMMSs include personal digital assistants (PDAs), CMMS-like software embedded in test and analysis equipment, bar-coding interfaces, remote diagnostics, service manuals and operator manuals residing on a server or web site, e-mail interfaces, video, technical and clinical in-services, and web-based CMMS programs.

Several CMMS vendors have interfaced PDAs and similar handheld or wearable workstations to their applications. These new portable devices can provide additional features with their built-in and add-on CMMS and other common computer applications (e.g., word processing, calendar, and spreadsheet). The CMMS applications for the PDAs and similar devices range from the full CMMS inventory and maintenance data to selected records determined by the CE staff using the PDA. Touch screen technologies, speech recognition, and voice-in and -out add to the user-interface features. Synchronization is

Table 36-2 Comparison of communication network technology utilized for point-to-point connections

Service	Carrier/medium	Bandwidth	Data Integrity	Application(s)
PSTN (standard phone cable network)	4-wire dedicated	56Kbps	Moderate to very good	Remote PC/workstation to LAN PC
Full T-1 24 digital channels	T-1	64 Kbps/channel	Excellent	Supports video capabilities. Only fractional T-1 required for most CMMS applications
Switched 56	Fractional T-1	56 Kbps	Moderate to very good	Remote PC/workstation to LAN
BRI	ISDN (2 B + D)	128 Kbps	Excellent	Must have an ISDN interface for legacy equipment
Frame Relay	ISDN 2-wire	64-128 Kbps	Excellent	Also supports wireless applications
Broadband	Optical fiber & ATM	>1.5 Mbps	Excellent	Supports video applications. Not required for general CMMS applications

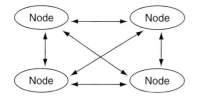
Interlinking system of the mesh topology

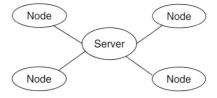
Interlinking system of the star topology

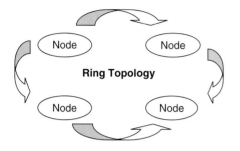

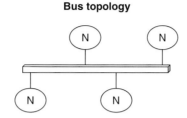

Figure 36-2 Mesh, star, ring, and bus topology.

by intermittent, ad hoc connection to the CMMS LAN, but wireless network capability (e.g., 802.11 wireless Ethernet) for these portable CMMS devices is on the horizon. Challenges for these devices continue to be the following: Data management (e.g., sending extracts of the CMMS database or the entire database to the PDA and the concomitant memory size issues); nonstandard display size and resolution; touch screen keyboard; nonstandard keyboards and keyboard size; cost versus durability (e.g., the ability to withstand a drop to the floor); and battery life, which is typically shorter than an eight-hour work day.

Several CMMS vendors have interfaced PM software into test and analysis equipment for paperless maintenance inspections, work orders, and electrical safety testing. Historically, these systems have focused on electrical safety testing and storing electrical safety leakage? current values, which has limited, if any, legal or practical application. Automated integration of test equipment with a CMMS application may have some application where there are large numbers of devices that can be automatically tested and certain limited, critical data values stored.

Bar coding has been used by CMMSs for many years. Bar code systems and readers have improved to the point at which they are useful for equipment and parts inventory control and may have some utility in work order management.

Various paging systems have been interfaced to CMMSs to allow automatic paging based on user entry of work orders and directed paging based on staff availability schedules and other data. As cellular phones, pagers, PDAs and other devices become integrated, there is an opportunity to develop a "technician communication tool" that combines several of the features discussed here as individual products. To be most useful, such a product would need to have seamless wireless network access to the full CMMS (e.g., IEEE 802.11 wireless Ethernet).

Another new technology is the application service provider (ASP). ASPs are software applications running remotely at a vendor's site that are accessed by the end-user in a fashion that is similar to the way applications are accessed by a LAN. A CMMS ASP would look little different from a LAN CMMS, with a web browser as the graphical user interface (GUI) to the CMMS. The vendor of the ASP takes care of all access and security issues as well as backups, upgrades, and regular data maintenance. Fee structures are under development and include fees based on the amount of data stored, the number of users, bandwidth requirements, and other parameters. Security enhancements for these shared remote networks are under development.

Conclusions

Information and communication technologies are rapidly changing. The CMMS technology market is relatively small and somewhat mature. Accordingly, changes occur much more slowly. CMMSs most likely will not drive the wireless communication infrastructure development in health care institutions, but once the wireless infrastructure is in place, clinical engineering departments and CMMS vendors will be able to take advantage of them and offer much more robust portable applications. Until then, new developments will be incremental improvements in software as well as and new decision support tools to help convert the large amount of data collected to useful information for managing technology.

Defining Terms

Asynchronous Transfer Mode: A high-speed network that provides optimal multimedia transmission at prohibitive costs

Bridge: An electronic device that interconnects two disparate LANs

Bus Network: A network topology where all workstations are attached to the same main physical medium (wire or cable)

Coaxial cable: A shielded cable for linking nodes within the network

Database: An accumulation of information stored in an orderly protocol, which can be accessed through a query application, updated, and viewed

DSL: Digital Subscriber Line.

Ethernet: The physical link and data link protocol for LAN interconnection. One of the oldest and most cost-effective network links

Graphical User Interface (GUI): A graphical tool that provides the user with point-and-click, intuitive characters, or icons to control the actions of the computer

Hub: A network node, which serves as a multiport repeater to other links in the network

Hypertext Transfer Protocol (HTTP): A computer standard, which allows computers operating in different environments to transfer data

ISDN: Integrated Services Digital Network

Intranet: A private network utilizing Internet-based technologies within an organization or enterprise

Local Area Network (LAN): A group of computers sharing a common database and/or server for a common purpose, forming nodes connected by a physical medium

Local Exchange Carrier (LEC): The local telephone companies or competing companies that can provide a user with the desired connectivity. In telephony jargon, they are divided into "incumbent LECs" (ILECS–the companies that have a majority of the local business) and "competitive LECs" (CLECs–the companies that compete for more of the ILECs business)

Network: A system of nodes (i.e., workstations, hubs, routers, bridges, and gateways) that are linked together for the purpose of transmitting information (e.g., voice, data, and video) from one location to another

Server: A more robust computer that has the horsepower to drive the entire system

Throughput: The rate of uninterrupted flow of information

Topology: The construction or physical geometric pattern of specific layouts (connection pathways) for a LAN

User-defined field: An open field that allows the user of a CMMS program to determine the characteristics of the queuing requirements and display outputs

Table 36-3 Physical media vs LAN topology comparison

Topology	Bus, star, or ring	Bus or ring	Bus or ring	Bus, star, or ring
Physical medium	Twisted pair	Baseband coaxial cable	Baseband coaxial cable	Fiber
Maximum nodes	About 250	About 1000	3,000-4,000	3,000–4,000
Available channels	Single channel	Single channel	Multichannel	Multichannel
Maximum transmission rate	100 Mbps	100 Mbps	400 Mbps	Several Gbps
Disadvantages	Lesser bandwidth, susceptible to noise	Susceptible to noise	Expensive, requires headend set up	Expensive, specialized training
Advantages	Low cost	Ease of installation	Transmission of voice, data, and video	Transmission of voice, data, and video

Utility: A small computer program, not contained in the operating system (OS), that provides an essential function or feature

References

Barta, RA., A Computerized Maintenance Management System's Requirements for Standard Operating Procedures, Biomedical Instrumentation and Technology. 35, 2001.

Cohen T, et al. *Computerized Maintenance Management Systems for Clinical Engineering,* Association for the Advancement of Medical Instrumentation. Arlington, Virginia, 1994.

ECRIHealth. Technology Management. ECRI, Plymouth Meeting, PA, 1993.

JCAHO. Comprehensive Accreditation Manual for Hospitals. Joint Commission on Accreditation of Health care Organizations. Chicago, 2000.

Further Information

Conrad JW, ed. *Handbook of Communications Systems Management,* New York, CRC Press, Auerback Publications, 1998.

Cram N. Computerized Maintenance Management Systems: A Review of Available Products. *J Clin Eng* 23:369-179, 1998.

Held G. *Handbook of Communications Systems.* New York, John Wiley & Sons, 1999.

Held G. *Ethernet Networks.* ed 2. New York, John Wiley & Sons, 1996.

Held G. *Practical Data Communications.* New York, John Wiley & Sons, 1995.

Sloan JP, ed. *Handbook of Local Area Networks.* Boston, Auerbach, RIA Group, 1997.

Stiefel R, ed. Medical Equipment Management Manual: How to Be in Complete and Continual Compliance with JCAHO Standards. AAMI, 2001.

37

Maintenance and Repair of Medical Devices

James McCauley
Director, Biomedical Engineering,
Central Coast Hospital
Gosford, New South Wales, Australia

This chapter looks at ways in which maintenance and repair activity may be managed. No hard and fast rules are set; rather, a more general approach is adopted. The reason for this is to encourage those who are responsible for managing or providing maintenance and repair services to think in a wider context so that they may have confidence to develop and explore different strategies or ideas. Furthermore, this chapter makes no attempt to be either specific or comprehensive nor to define any existing requirements, for to do so would ensure obsolescence in a rapidly changing world.

Maintenance and repair activity is required in order to ensure that devices are kept functioning within the limits imposed by the test criteria and to return devices to the required level of functioning after breakage or other failure. Additionally, safety and performance testing is required in order to identify unsafe or incorrectly performing medical devices that could pose a risk to either patients or staff. Such testing is usually performed at acceptance and then on a routine basis. Together, safety and performance testing and maintenance and repair form the basis for any medical device management program. (See Chapter 122.) However, the two are often seen as quite separate activities, as safety and performance testing is often mandated (typically every 12 months), while maintenance and repair activity is usually performed in reaction to device breakage or failure. In practice, maintenance and repair are generally managed differently from the way safety and performance testing are managed. Although maintenance and repair are viewed as basic "bread and butter" activities, they are in fact more like "chalk and cheese." They are opposite sides of the same coin and almost the antithesis of each other. "Maintain" comes from the Latin "to hold in one's hand," meaning to protect and look after. "Repair," on the other hand, means to return things to the way they were, to make better, or to fix. The primary goal of maintenance activity is to reduce or if possible to eliminate, the need for repairs.

Unfortunately, it will never be possible to reduce repair activity to zero, despite effective maintenance programs that eliminate foreseeable problems and that keep devices functioning within acceptable safety and performance criteria. Random component failure will never be entirely eradicated, and the practical reality ensures that devices will be broken in use, despite staff training or attitude. On the other hand, it is possible to reduce maintenance activity to zero, in which case all available resources are allocated to repairs (and to required routine safety and performance testing). In fact, once resources are limited, many organizations tend to slide toward a low-maintenance and high-repair outcome. Once repairs reach a certain level, it could be difficult to continue any maintenance plan or balance repairs with routine safety and performance testing. If it were not for mandatory or professional obligations, some organizations could end up doing little safety and performance testing and end up operating on "run 'til failure" mode. While such a strategy is clinically unacceptable, it may be the most efficient in terms of equipment management cost and adequate resource allocation, for medical device management is often difficult to achieve in the health care environment. Only when clinical service delivery or statutory obligations cannot be maintained—possibly leading to income reduction or increased exposure to liability—does funding for medical device management take priority over clinical needs.

Medical device management is funded by many methods. These include historical funding, funding based on the number and type of devices, funding based on the number and type of beds and services in a health care organization, or a combination of these and other factors such as various "insurance" schemes. (See Chapter 39.) Regardless of whether an organization performs maintenance and repair "in-house" or contracts some or all of this service out by whatever means, it is essential to understand the way in which funding is obtained and allocated. (See Chapter 48.) This perspective helps adequate resources to be sourced, distributed, and retained to meet the ongoing changes in medical device management requirements.

Framework Guiding Maintenance and Repair

Regulation

Before maintenance and repair work is carried out, it is important to establish the regulatory framework under which such work is to be undertaken, to ensure compliance with local requirements. (See Chapter 124.) Requirements vary according to location, the type of equipment being managed, the nature of the health care organization's operation, and possibly among different manufacturers. In some countries there is little regulation, and work might be carried out by any capable person, regardless of qualifications, experience, or training. To some extent this situation exists almost everywhere, as some organizations employ (or contract out to) repair personnel who have had limited training, opting instead to provide skilled supervision so that people who have minimal qualifications can do the job. (See Chapter 38.) Some countries regulate the registration of technical personnel, including supervising personnel, and might further regulate repair and maintenance to the extent that these activities can be audited to ensure that quality control is maintained. (See Chapters 19 and 27.) Auditing can extend beyond the country of manufacture such that all repairs undertaken by the manufacturer or supplier are open to being audited, regardless of where they are performed (particularly if there is a chance that the device could re-enter the country of manufacture), with penalties if quality standards are breached. (See Chapter 126.) In such instances, manufacturers will seldom take responsibility for maintenance and repair performed by unknown personnel. (See Chapter 13.)

In addition, there are various standards (see Chapter 117) that govern medical device management, and most countries adopt at least one such standard, either as part of the overall quality process or more rigorously as a legislative requirement. Typically, such standards define the nature and frequency of safety and performance testing but do not define maintenance requirements other than in the most general terms. These standards seldom if ever define repair quality issues.

The personnel who are responsible for medical device management programs should be well informed as to the regulatory requirements in their area. Further, they should know whether there are requirements covering the qualifications of technical personnel and whether standards exist to cover maintenance and repair procedures. If all requirements covering device management can be met, then the recipient health care organization should meet reasonable public expectation for quality health care delivery, given local circumstances, and should have its capability assured.

Risk Management

The reality of many medical device management programs is that they are relatively under-resourced, particularly with the adoption of more, and more complex, technology. This presents a difficult situation. If scarce resources are allocated to yearly safety and performance testing, perhaps for mandatory or accreditation reasons, then the repair backlog may increase to a level where clinical service delivery becomes affected. Also, if the repair backlog increases, then maintenance activity will reduce, thus compounding the problem.

In order to operate with limited resources, various techniques have been adopted to justify reduced safety and performance testing frequencies and to define maintenance requirements more precisely (Fennigkoh and Smith, 1989; Fennigkoh and Lagerman, 1997; Gullikson, David and Blair, 1997). These techniques have included classifying the maintenance needs of items or protocols for justifying the extension of safety and performance testing intervals. These techniques rely either on some measure of device "criticality" risk analysis or on sound statistical reliability data. Usually, insufficient data are available for valid statistical analysis, and these techniques then must be based on risk-management techniques. Risk management typically involves developing a "matrix" with the probability of failure along one axis and the consequence of failure (from minor to major) along another. Devices can be placed in the matrix according to where they sit on each axis and ranked, allowing resources to be targeted at the highest risk items (i.e., those where both the probability and consequence of failure are highest). The rationale behind any technique adopted should always be documented. (See also Chapter 56.)

Maintenance Strategy

Regardless of the technique adopted to balance available resources against "risk," fundamental maintenance and calibration strategies will help to improve device management efficiency over time. Further innovative strategies should always be sought out and considered, particularly if they bring increases in efficiency, as any increase in efficiency can reduce the overall risk.

One fundamental step is to move toward "standardization" of the devices being managed and to reduce the different types of devices found across a health care organization. This may be done without compromising service delivery, while still allowing a degree of clinical choice. If implemented carefully, it will benefit an organization because there will be less variation in maintenance and repair requirements, thus allowing limited resources to be more effectively applied. Staff training (both technical and clinical) can become more focused. Such a strategy seldom needs justification. For the maintenance and repair service provider or medical device manager to have the most valuable insight into this process, they need clinical knowledge as well as technical familiarity with the devices being used.

Pre-purchase estimation of projected maintenance costs needs to be considered, particularly as these costs often can exceed the capital cost of the equipment over its lifetime. (See Chapter 30.) Although often overlooked by clinical staff and management, it should be an essential consideration in any procurement process. If the maintenance requirements can be paralleled with existing equipment (e.g., by using the same type of batteries or other components), then such factors should be investigated because it might be possible to reduce overall long-term labor and parts costs.

"Ownership" of the equipment is another issue to be considered. Is equipment "pooled," or does it belong to a particular ward or other location? Often there is a mix of ownership across a health care organization. Whatever ownership model is employed, it will have implications when it comes to locating equipment and to the funding of maintenance and repair. No single strategy has emerged as the best. Ownership is often defined as part of the health care organization's broader business process. This reality and staff sentiment often make it hard to change ownership patterns. With vigilance, areas where the current ownership practices are impeding efficient device management can be identified, and different practices possibly can be adopted. Any maintenance and repair service provider should be prepared to adapt to existing conditions.

Other strategies to consider include balancing the level of support against available technical capability, both internal and external. In some countries, health care organizations find it almost impossible to obtain any technical support for their medical devices, while in other countries the capability and availability of such services are high. Scrutiny of all technical resources available to a given locality is essential in order to ensure that the highest quality and most cost-effective service are provided to the health care organization.

There always will be new and different strategies to consider. Some of these will be brought about by changes to technology and improvements in device management techniques. Others will be brought about by changes in financial or business requirements. No single strategy should be considered sacrosanct, and new strategies always should be considered or sought out.

Linking Maintenance Requirements to Test Schedules

One major problem in managing medical equipment is difficulty in obtaining the equipment when needed. Often equipment is in use, or a particular device might be impossible to locate. Different organizations have adopted various strategies to help ensure that all equipment is tested when required, with differing degrees of success. The best strategies often depend largely on local knowledge and contacts. Nevertheless, obtaining access to the equipment can be a time-consuming exercise and is often a major and unaccounted for drain on resources.

The simplest and possibly least developed option is to link maintenance activity to safety and performance testing. Safety and performance testing of all devices is usually required at least every 12 months. More recently, this period has been extended for some devices based on reliability and failure analysis. Certain maintenance procedures for particular devices are required more frequently than every 12 months and some might be required less frequently. For certain types of equipment, maintenance is based on occasions of usage or on hours in service.

With thought, most maintenance can be included in the routine testing schedule regardless of the way the test schedule is managed. It is not whether, but how, maintenance is included that leads to increases in efficiency without increasing risk. Battery replacement is a good example to illustrate this point. Typically, devices with batteries require some form of battery capacity check for battery deterioration. One way to perform such a check as part of routine testing is to ensure that the devices are first fully charged and then left functioning for a set period. If during this time the battery voltage drops below a certain level, usually indicated by the low-battery alarm, the battery is deemed to have insufficient capacity and must be replaced. This is a time consuming process for testing personnel, who must fully charge the battery, discharge it (perhaps over several hours), and then recharge the battery before returning the device to service. One alternative strategy would be to replace batteries beyond a certain age. Assuming that good quality batteries are available, which has become harder to guarantee, then instead of testing the capacity of each battery, a policy of changing batteries that are over 18 months old could be adopted. In this case, given yearly testing, no battery in service would be over 30 months old. While there are arguments for and against both strategies based on risk, cost, and efficiency, this example serves to illustrate that there are different ways to incorporate maintenance into the routine testing schedule. Different strategies could be developed for virtually all maintenance activities. The astute maintenance and repair service provider or device manger should consider adopting different maintenance strategies if this leads to increased efficiency, provided that risk does not increase unacceptably, as any increase in efficiency can be used to improve overall device management.

Maintenance Planning

Forward planning of maintenance procedures is important. This process requires detailed knowledge of maintenance requirements and the resources that are required in order to perform maintenance. The resources required include labor, parts, materials, and tool costs. If maintenance is contracted, then all of these costs are often rolled into one by the contractor, although with some contracts parts costs might be kept separate.

Not all computer databases used for medical device management allow for the forward planning of maintenance requirements. Industrial maintenance databases, possibly designed for the production industry, are often good at forward maintenance planning and estimating forward resource requirements, but they tend to lack the required functionality of dedicated medical device management databases. The database system in use will determine how sophisticated forward maintenance planning can become. (See Chapter 36). Nevertheless, regardless of the database or methodology used, even gross attempts at forward planning of maintenance requirements can benefit both the provider of the service and the recipient health care organization. These benefits include better budget forecasting, the opportunity to identify items that might no longer be cost-effective to maintain, and the potential to rationalize maintenance expenditure across an organization. To establish maintenance requirements, the current condition and maintenance history must be determined for the devices to be managed. If no reliable maintenance history is available, then inspection and condition monitoring, or testing, can help to establish the status of a device. In order to obtain an overall assessment of maintenance requirements in an organization, a "walk through" review of all areas with medical devices is beneficial. It helps to establish the conditions under which devices are being used or stored. This is important because many devices have particular usage or storage requirements, such as needing to be left on charge or in an upright position; if these requirements are not being observed (for whatever reason), then this will have implications for maintenance and repair management. Areas where limited "user maintenance" of devices is possible should be encouraged or instigated if confidence in the capability of users can be assured. This will free up resources, give a certain degree of ownership and responsibility to users, and reduce the need for repairs.

Maintenance planning should be realistic and achievable. Planning initially should focus on major generic maintenance requirements, such as battery replacement, and also on the most critical items, such as ventilators. By tackling the major maintenance requirements first, any increases in efficiency will result in improved availability of resources, allowing smaller (and perhaps more difficult) issues to be tackled. It also will have the added intangible benefit of increasing confidence. With time and confidence, maintenance and repair providers or managers will move from prescriptive and reactive strategies to more creative and proactive management techniques. In doing so, capacity and capability will increase, and the ability to adapt to changing conditions will improve.

Repair Strategy

Fault Diagnosis

Accurate fault diagnosis is critical and may be simple or extremely complicated. (See Chapter 59.) Damage to a device or lack of functionality may be apparent. However, the root cause of many faults is not readily apparent, leaving repair personnel reliant on their skills and experience to accurately identify the cause. (See Chapter 55.) As the actual fault might have been caused by disconnected factors, it is important to keep a broad view when trying to identify causes. For example, a blown fuse could be the result of an internal component failure but also could be the result of a transient spike on the power supply, beyond control. Furthermore, with experience, the cause of certain faults becomes more apparent, or certain strategies become more effective. The old adage "If no fault is found in the power supply, then check the power supply" sums up the value of experience when trying to diagnose faults correctly. Should the cause of a fault not be readily apparent, then a strategy needs to be followed. The purpose of such a strategy is to identify accurately the cause of the fault in the least possible time so that the appropriate repair can be made.

If possible, information from a user about the fault should be obtained. This information may be communicated directly by the user or it may be written on some sort of fault tag. One should carefully evaluate the information provided by the user because it may be invaluable in accurately identifying the cause of the fault, or it could be confusing and incorrect and could lead repair staff in the wrong direction.

One should try to replicate the fault if it is safe to do so. The fault might be simple to replicate (e.g., "the device won't turn on") or more complicated (e.g., "the external pacer injector does not appear to be in synch with this patient's ECG"). Complicated faults might be difficult to identify, as the fault may only present under certain clinical conditions or with particular types of patients. Nevertheless, if the fault is linked to a clinical procedure (for example, "this thermometer reads high"), then one should be wary, because the device actually might be functioning correctly and clinical staff are relying on preconceived perceptions of the patient's condition rather than the device. A real fault must never be discounted.

Should any type of fault be unable to be replicated, then one may assume that user error is to blame. Such faults may have their origin either in inadequate knowledge of the operation of the device or accidental user error. To jump to either conclusion might be premature, and many service personnel will have experienced the embarrassment of having blamed the user only to later identify a genuine fault with the device. If user error is to blame, then it is important to try to prevent this from recurring. Often a strategic label may be the best way to eliminate the problem, particularly if certain functions of a device are ambiguous or not easy to identify. (See Chapter 83.) Furthermore, simple modification could prevent the problem from recurring. If user error is to blame, then issues surrounding staff training need to be considered.

Most modern microprocessor-controlled equipment may contain service menus, error logs, or diagnostic modes of operation. These may be accessed either directly through the device or by some link to a diagnostic program contained in a remote computer. Repair staff should be familiar with, and should have access to, all possible methods of fault identification. Often the only way to identify intermittent faults is by these means, particularly if there is a device history that might contain an error log.

There might be more than one fault. Perhaps there was fluid ingress or further mechanical damage? Maybe one fault will not become apparent until another has been fixed. The permutations and combinations are limitless, but with experience and by relying on objective analysis, most faults eventually can be found.

Fault Rectification: Repair

The level to which any repair is undertaken is a combination of technical skill, parts availability, clinical need, and cost. Repairs should be performed only when personnel have sufficient training, technical information, tools, test equipment, and parts. Should any of these factors be lacking, then the repair should be performed elsewhere or delayed until these factors become available. It is important to establish boundaries as to what type of repairs will be attempted. Should an overly ambitious repair be attempted, then time and spare parts could be wasted if the repair is unsuccessful, or the safety of the device could be compromised. With modern, multilayer circuit boards and programmable surface-mount components, often the only option is to swap the board and to have the faulty one repaired by the original equipment manufacturer.

Technical skill is a combination of training and experience. All training will improve technical skill, and both general and specialized training should be made available. Experience comes with time, but the experience of others may be used. Establishing good relationships with suppliers and other repair organizations can be invaluable. Attempts at either formal or informal "networking" will allow exchange of ideas, experience, and information. Staff members should be encouraged to "network" because it will increase their knowledge and improve their skill.

Obtaining the required technical information can be difficult. Should diagrams, schematics and component information be unavailable, it may be impossible to effect a repair, regardless of technical skill. Nevertheless, all devices will have comprehensive service information somewhere. Unfortunately, this is not always readily available; and even if service information is required to be supplied as part of the procurement process, it might lack sufficient detail to perform a repair in the future. Some manufacturers provide technical information on CD or make it available over the Internet. These sources may be used to good effect.

Having the correct, high-quality tools and the proper test equipment is imperative. While a poor workman may blame his tools, a good workman will not use poor-quality tools. Simple things like good quality screwdrivers prevent damage to screw heads, thus allowing the device to be maintained and repaired many times without causing worry about extracting damaged screws in the future. Electrical static protection is also important because static can either weaken or damage electronic components, thus resulting in unforeseen or premature device failure.

Parts availability and storage is a complicated issue, for consideration must be given to supply time, quality, and stock levels. Repair time is minimized by keeping common parts on hand. However, this may lead to devices of considerable capital value sitting idle, with some parts becoming obsolete in time. Whether to use parts supplied by the original equipment manufacturer or by "second sources" always presents a dilemma. So does the level to which parts are held as stock, or obtained as required, particularly as some parts may be difficult to obtain rapidly or may deteriorate with time. Parts salvaged from devices destined for disposal, or reconditioned components, could be used for repairs. Such a strategy could improve availability of parts, but quality and long-term reliability could be compromised. A parts strategy that minimizes repair turnaround time but that also is financially efficient and safe must be developed.

Testing

After any maintenance procedure or repair, testing is required in order to ensure that the device is safe and performing up to specifications. Whether a full safety and performance test should be run is a matter for professional judgement, dependent on the level of repair. Furthermore, consideration should be given to "soak" testing repaired devices by leaving them functioning for a considerable time, particularly if the original fault was intermittent.

Conclusion

This chapter reviews some of the more general aspects of maintenance and repair management. This will lead those who are responsible for maintenance and repair of medical devices to look to different or innovative strategies or options, rather than being introspective, when trying to address the considerable and crucial demands of managing medical devices.

References

Fennigkoh L, Lagerman B. Medical Equipment Management. *JCAHO Environment of Care Series* (1):47-54, 1997.

Fennigkoh L, Smith B. Clinical Equipment Management. *JCAHO Plant, Technology & Safety Management Series* (2)5-14, 1989.

Gullikson ML, David Y, Blair CA. The Role of Quantifiable Risk Factors in a Medical Technology Management Program. *JCAHO Environment of Care Series* (3):11-26, 1996.

38
A Strategy to Maintain Essential Medical Equipment in Developing Countries

Michael Cheng
Health Technology Management
Ottawa, Ontario, Canada

Basic medical equipment is now widely used in the district health facilities of developing countries. This large volume of essential medical equipment is supporting primary health care to the general population. District facilities are numerous and widespread in most developing countries. Maintaining this large volume of basic, essential medical equipment presents a special challenge.

Status of Maintenance in Developing Countries

Large cities in most developing countries now have maintenance workshops, often built on the grounds of major hospitals. Many of these workshops can repair complex medical equipment and offer adequate services to the major hospitals with which they are affiliated. They are fully occupied, however, in coping with the ever increasing numbers and complexity of medical equipment in major hospitals; therefore, their services cannot be easily extended to district health facilities. The growth of maintenance services simply is not keeping pace with the rate of deployment of medical equipment in these countries.

The ministries (or departments) of health in developing countries, often lacking in equipment maintenance expertise, may believe or expect that the few established workshops have the capacity to resolve the equipment maintenance problems of the nation as a whole. In contrast, the technical staff in those workshops, traditionally confined to lower-ranking positions, rarely have the opportunity to develop a national perspective. Their needs and objectives are usually to further their training in order to advance their high-technology skills; these goals are often justified, considering the complex equipment that they service. These conflicting expectations do not lead to an effective solution for the national needs in medical equipment maintenance. What, then, is an appropriate strategy to tackle this problem?

Priority needs to be given to large-scale training of "basic" technicians and the establishment of smaller workshops in the districts to maintain essential medical equipment. This training task will take less time and cost to accomplish as compared with training technicians to maintain complex items or equipment, and it has a number of additional advantages described below. These issues are analyzed with the aid of an empirical model.

An Empirical Model

A country's medical equipment inventory can be represented by the pyramid shown in Figure 38-1 (Cheng, 1993).

In Figure 38-1, height represents, in approximate order, equipment complexity, while width represents equipment quantity. The pyramid shape indicates that simple equipment greatly outnumbers complex equipment. For example, there are obviously more weights, scales, stethoscopes, and sterilizers than ultrasounds, lasers, and CT scanners. As the complexity of the devices increases, the numbers available decrease. The bottom half of the pyramid comprising devices such as sterilizers, microscopes, water baths, scales, and stethoscopes can be maintained by technicians with a minimal amount of training and cost.

The maintenance of the wide range of equipment identified in Figure 38-1 requires a correspondingly wide range of technician skill levels, and the cost or time required to train a technician increases dramatically with the level of skills required. This situation is illustrated by the curve (oc) in Figure 38-2.

In Figure 38-2, the inventory of equipment is divided into complexity categories A and B; ob represents the cost or time to train a basic technician to maintain the simple category B; ba represents the cost or time required to train a mid-level technician to maintain category A. This graphical comparison suggests that at a much smaller cost, or in a shorter period of time (ob compared with ba), technicians can be trained to maintain a larger quantity of basic essential medical equipment (B compared with A). The actual relationship between complexity and cost or time (curve oc in Figure 38-2) will not be exactly as illustrated, but in general, the idea of a rapid increase of cost or time with complexity is valid. If special attention is given to the maintenance of essential medical equipment in support of primary health care facilities, the population at large can benefit comparatively quickly in many developing countries.

Development and Implementation

Given the current situation in many developing countries, the pyramid model suggests an appropriate strategy to attack the problem of medical equipment maintenance. This strategy calls for a priority in the training of technicians to maintain the relatively simple but plentiful essential medical equipment commonly found in district health facilities. This strategy requires less time, costs less, and delivers benefits to a larger population.

Here lies a happy coincidence in which the easier way of doing things is also more economical and beneficial. The following further advantages can be derived from this strategy:

- Mid-level technical expertise local trainers in existing workshops can be utilized.
- Because of less stringent prerequisites for selection, a larger number of candidates can be recruited for training, enabling a relatively rapid multiplication of technical human resources.
- Because of the large amount of similar equipment in use, it may be possible to carry out apprentice training on the job so that the trainees can provide actual services to health facilities.
- This strategy suits well the market economy that is now found in practically all countries around the world. Complex repairs can be done more cost-effectively by company enterprises. Companies are obliged to have trained technicians to support the equipment they supply. In-house and company services can be used to best advantage.

On the other hand, if the alternative of advanced training for staff of existing workshops is pursued without a larger backup workforce, it is necessary to select candidates who already have mid-level skills and to send them to service schools for training. Because there are usually few such technicians in developing countries, existing services would be interrupted. In some cases, suitable technicians would go from one overseas training program to another, with little time to deliver actual services in the home country. Moreover, upon completion of a training program, some trainees would choose to work for another organization, or to emigrate to a country where prospects of a more lucrative career are better. The loss of highly trained technicians could be devastating to a country that is striving to build up a maintenance service while coping with a shortage of technical personnel.

By initially concentrating on training a large number of technicians to maintain basic, essential medical equipment, a pool of technically skilled people will be made available for services and for higher training. This can provide a more stable base to develop maintenance services further.

No doubt, successful development of medical equipment maintenance services depends on critical factors such as political will, local cultural attitudes (Taylor et al, 1994), and financial support. In several countries that the author has visited, all necessary supports were available, but the development programs were still getting nowhere because actions were not effective in solving the problems. Both the national executing office and the international aid agency were dissatisfied and were searching for a better approach. It was from these countries that the strategy elaborated in this chapter was developed.

The situation in these countries can be likened to that of a ship. Given a captain and crew (i.e., political will and operating support) and fuel (i.e., financial support), the ship can move along. If the captain has no clear direction to steer the ship, however, the ship could be moving around in circles in the middle of an ocean. The strategy presented in this chapter is like giving the captain a direction to steer the ship toward a productive destination. When one loses the way, a compass can provide guidance. When a particular issue is important to a country, a strategy can help to coordinate all national efforts toward a national goal.

The simplicity of the pyramid model can help nontechnical decision makers to grasp better the technicalities of equipment maintenance. Such a strategy would call for clinical

133

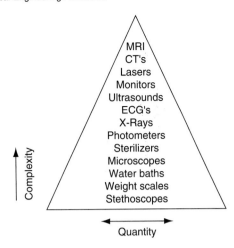

Figure 38-1 Diagrammatic inventory of medical equipment.

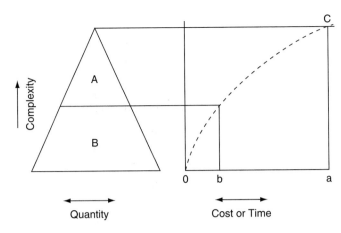

Figure 38-2 Relationship between complexity of equipment and cost and duration of training technicians.

engineers in developing countries to acquire management and training skills in order to lead or manage national medical equipment maintenance programs.

Conclusion

Giving a priority to maintaining essential medical equipment to support primary health care facilities has the advantage of meeting the urgent, current needs in many countries. It can decrease the required time to bring benefit to the broader population. Furthermore, this approach can utilize skilled personnel as local trainers in existing workshops to multiply the workforce efficiently. Adequate technical human resources and a network of basic workshops across a country can provide a foundation to create a culture and an infrastructure to facilitate higher development in medical equipment maintenance, thus helping to overcome some chronic obstacles in countries that lack a strong industrial base.

This also can be used as a general strategy to promote cost-effective maintenance for medical equipment in the market economy where private companies would do most specialized repairs while the in-house staff would confront problems on the front line.

References

Cheng M. A Priority in Maintaining Medical Equipment in Developing Countries, Proc. of the Joint Conference of COMP and CMBES, Ottawa, pp 314-315, 1993.

Cheng M. Medical Device Regulation and Policy Development, International Forum for the Promotion of Safe and Affordable Medical Technologies in Developing Countries. The World Bank, May 19-20, 2003.

Taylor K, Frize M, Iverson N, Paponnet-Cantat C. The need for the integration of clinical engineering and sociological perspectives in the management of medical equipment within developing countries; J Clin Eng 19(4):291-296, 1994.

39

Outsourcing Clinical Engineering Service

Peter Smithson
QA & Organizational Development Manager, Principal Clinical Scientist, Medical Equipment Organization,
Bristol, England, UK

David Dickey
President & CEO,
Medical Technology Management, Inc.
Clarkston, MI

Clinical equipment ownership results in a financial liability that extends far beyond the cost of equipment purchasing, installation, testing, and user training. Yet when properly defined and aggressively managed, a consolidated medical equipment service management program can create substantial cost savings for any health care organization, especially given the continued increase in new and replacement high-technology equipment purchases. Considering the fact that many capital acquisitions are of assets that have a useful life of seven or more years (especially when considering the asset's longevity through planned upgrades), it is not uncommon to expect an annual ownership (service) expenditure of 5% to 12% of the equipment's original purchase price, depending upon the mix of equipment modality. Choices as to the best way to provide for a consolidated medical equipment maintenance management program can typically be categorized into two types of program models: (1) in-sourced (in-house) and (2) out-sourced.

Outsourcing

What is out-sourcing? Essentially, it is the transfer of any defined business operation or responsibility to another organization. In terms of clinical engineering, it could involve the transfer of responsibility for selected portions of equipment service, procurement, or program management to an external business entity or organization, all for one agreed upon or "not to exceed" price. It could involve transfer of the financial risk, with or without service staffing, as well. Typically the company providing the outsourced service makes all service decisions and assumes responsibility for all outcomes (i.e., good and bad).

Two Service Models

One key differentiating factor between the two program models of outsource and in-house relates to identification of the financial risk. As is seen in Table 39-1, all programs can be divided into two main categories, depending on who is taking on the program over-budget risk.

It is important to realize, however, that most in-house (in-sourced) clinical engineering programs typically continue to utilize the services of external vendors, as it would be unusual for any single program to have sufficient internal resources to be able to perform 100% of required services. Given the high number of inventoried devices typically contained within a clinical equipment management program, it would not be unusual for the in-house staff to service only 70%–80% of the equipment base, relying on external vendors, with their specialized expertise, to service the remaining items. Therefore an in-house (in-sourced) program still contains some elements of an outsourced program. For purposes of this discussion, an outsourced program is defined as one that is fully provided by non-hospital staff and whose management (the entire service function) is the responsibility of an external provider. Under a fully outsourced program, all staff are on the out-sourced vendor's payroll, and the vendor is fully responsible for all aspects of program management, performance, and financial outcomes. The hospital pays one price to one vendor, who then assumes all responsibility for paying for all subcontractors, staff salaries, parts, supplies, and overhead.

While there are no requirements or regulations stipulating who can (or must) provide repair services on most medical equipment, certain countries and/or individual states may have guidelines relating to service, testing, or calibration of specific device types such as high-energy radiation treatment or mammography systems. In the United Kingdom (UK), the Medical Devices Agency (MDA) has issued guidelines to hospitals and community-based health facilities that state that any servicing organization must maintain a properly trained staff, the correct manuals and tools, access to spare parts approved by the manufacturer, and adequate quality control procedures (MDA, 1988; MDA, 2000). These could apply to in-house as well as outsourced program providers.

Why Outsource?

The reasons why many organizations consider use of an out-sourced program typically include the following:

- Desire to reduce the number of employees who are directly on the hospital payroll. Many organizations have management goals to reduce the total number of employees (full-time equivalents [FTE]) to meet required staffing benchmark levels based on total number of employees per occupied bed.
- Cost savings. Many outsourced programs are sold and justified on the total program cost, which is shown to be less than the actual or estimated cost of the organization's current in-house programs. This is especially true when the majority of the equipment base, such as for a radiology service program, comes from one vendor or manufacturer who has access to parts and labor at their internal cost.
- Access to resources that are not readily available to the organization (e.g., trained staff and parts). Many outsourced programs are provided by equipment manufacturers who have direct access to parts, supplies, and software products directly or indirectly related to the equipment base of the program being outsourced. In addition, most outsourced programs have developed custom policies, procedures, and software systems that can be rapidly installed at a low cost.
- Short-term solution to a problematic in-house program. Many in-house programs are short staffed or have employees with limited expertise and service capabilities. The outsourced vendor may have underutilized staffing resources in the local geographical area who can be assigned full- or part-time to assist in implementing or providing routine support to the organization.
- Reduced internal overhead related to invoice processing. Given the high cost of processing individual purchase orders for parts, supplies, and individual vendor repair services, an outsourced program can have an indirect impact on the hospital's cost if the internal overhead need to process these invoices and related paperwork can be eliminated or reduced.

OEM Full-Service Contracts

One form of outsourcing involves contracting with numerous vendors for defined capped cost services on individual equipment. Contract costs from the OEM have come down over the years and can range from 2% to 12% of the equipment acquisition cost, depending on equipment modality. Service staff provided are typically well trained, and they have access to all of the tools, software diagnostics, and spare parts designed by the OEM for supporting the equipment.

During the sales process, the equipment vendor may offer discounts to its multiyear agreement and may hope that the one-time offer of the discount will persuade the purchaser to sign up. Many multiyear agreements do not have an out clause or have a high cancellation penalty should the purchaser elect to cancel early. Signing a multiyear service agreement with no "out clause" might not be advisable, as it severely limits future cost savings opportunities, especially if the equipment failure rate is determined to be lower than expected. However, if you bought an item that has a high failure rate, locking in a reduced-rate service agreement would have been advisable. Doing an adequate amount of homework on expected failure rates prior to signing the purchase agreement is obviously advisable.

Given the fact that most of today's medical equipment is software driven, OEM vendors have the ability to bundle new software upgrades into their contract pricing, thus making the software upgrades appear to be "no charge." However, not all equipment will need routine software upgrades, and the astute purchaser should consider negotiating a "not to exceed" price for future upgrades, independent of any full-service contract options. While there is nothing inherently wrong with the full-service contract approach, the main concern is that if you sign a multiyear, full-service agreement, you exclude yourself from all options and future cost saving opportunities should the equipment have a low failure rate or should you wish to use an alternate service vendor or to convert to an in-house program.

Table 39-1 Two models for hospital medical technology service

Hospital transfers risk	*Hospital carries risk*
(outsourced program)	(in-sourced program)
OEM full-service contracts*	OEM T/M**
ISO/OEM asset management***	ISO T/M
Maintenance insurance	In-house self-insured

* OEM—Original Equipment Manufacturer; **T/M—Time and Materials; *** ISO—Independent Service Organization

OEM Asset Management/Multivendor Service Programs

It is difficult to identify exactly when the first OEM advertised multivendor program was marketed, but it seems to have coincided with the onset of "asset management" programs. While most advertisements tried to market these programs as something new, the concept of having a dedicated or shared technical staff who are trained to service a multitude of products has been one of the key features of hospital-based clinical/biomedical engineering programs going back as far as the early 1970s.

What is new for the OEMs is the offering of their ability to have their employees service equipment not manufactured by them, hence "multivendor services." The interesting twist here is the fact that for years, many OEMs had continued to claim that only they could service their equipment, as is was so complex, requiring parts that only they could supply. Then, almost overnight, they suddenly had the expertise to be able to service anyone's equipment, regardless of manufacturer or model.

The reality is that the majority of equipment problems do not always require specialized training and can be resolved by a generalist (biomedical or radiology service) engineer who has basic knowledge and experience on modality-based products. When specialized expertise is needed, it is not uncommon for the multivendor service provider to call in their competitor for assistance, paying for the work under a negotiated time and material (T/M) basis.

Total program cost for a consolidated multivendor asset (service) management program typically will be less than the cost of individual service contracts. Customer satisfaction for these types of programs varies by facility, with most problems relating to irregular staffing, increased equipment down time, and lack of value-added services (to support basic biomedical equipment management functions not specifically outlined in the contract agreement). One primary concern relates to the concern over conflict of interest, based on the questioning as to how the on-site multivendor equipment service staff could possibly work as hard on making their competitors equipment work as well as their own?

ISO Maintenance Service Programs

The Independent Service Organization (ISO) has been in the multivendor service business since the early 1960s. It offers a cost-effective alternative to the OEM or total in-house service program. As outsourced maintenance service providers, ISOs have been providing asset management-type services well before the entry of OEMs into this market, and they typically function at a cost that is less than that charged by the OEM provider. One key feature of an ISO program is that it typically does not manufacturer or sell any vendor's equipment, thereby leading to the potential claim that they can truly be unbiased on the issue of equipment-replacement selection. Many of the ISO programs are geographically based, with ability to offer cost-effective services only to a client population within their geographical proximity.

As with an in-house program, the ISO is generally able to service and support a wide range of equipment, but it may lack the specialized training that is necessary to perform extensive troubleshooting and maintenance of complex systems such as newer computerized tomography (CT) and magnetic resonance imaging (MRI) scanners. ISO programs generally call in the OEM vendor for assistance, either on T/M service or under a limited contract to support such systems when needed. A well-trained ISO program should be able to provide first call maintenance services on 60%–75% of the equipment base, with the balance serviced primarily by the OEM.

Over the years, the number of qualified ISO programs has diminished, with many having been bought up by the large OEMs as a means to expedite their entry into the multivendor service market. The main concern over the use of ISO problems is business longevity.

Maintenance Insurance

The concept of maintenance insurance works on the premise that, by bundling together the service contract budgets of multiple items, one can lower the overall program cost by averaging the T/M service cost needs of all items (Tran, 1994). Using actuarial service cost data on multiple types of equipment, the maintenance insurance program establishes an estimate of T/M costs for each item, including preventive maintenance (PM) costs, then adds in an overhead and profit margin. Under the ideal program, the equipment user then continues to call in their previously used OEM for service when needed, then either pays the bill itself and waits for insurance reimbursement or submits the bills directly to the insurance carrier for payment.

Over the years, variations on the insurance program concepts have been implemented, such as:

- Inclusion of a rebate or shared savings component to reward the hospital in working with the insurer to minimize the annual service costs
- Reimbursement to the hospital when allowing their in-house staff to perform a portion of the maintenance (at in-house labor costs of $35–$70 per hour instead of paying the OEM rates of $120–$230 per hour)
- Provision of a first-dollar versus stop-loss limit-based program
- Provision of on-site clerical and/or technical staff to assist with management of the program, and its paperwork
- An arrangement whereby the insurer pays the service vendor directly
- Multiyear agreements whereby the annual program premium cost increase is capped
- Ability to add and delete individual items via a pro-ration schedule
- Telephone support assistance in locating alternate parts and labor sources

Some programs have restrictions and other program requirements, however, that could limit the maintenance insurance program applicability to many health care organizations, such as:

- The mandate that all equipment items are included in the program rather than allowing the hospital to select the items that are included for coverage
- Selective coverage, or the situation where the insurance carrier drops coverage for a specific item should its maintenance costs greatly exceed the itemized premium
- Timing restrictions, whereby field service reports and invoices must be submitted within a predefined number of days after the work event, or else they will be excluded from coverage
- The requirement that a work event that is predicted to result in a cost that exceeds a predefined limit (e.g., $5,000 or $10,000) must be called in to the insurance program for authorization prior to having the work initiated
- The requirement that the insurance program can mandate the source of parts and labor for certain equipment items. Repairs performed by non-authorized sources (usually at a higher cost) may not be covered by the program, even if the equipment owner has determined that their source (in many cases, the OEM) is better able to perform the repair.
- Scheduled preventive-maintenance procedures that take longer than what the insurance carrier has estimated may be excluded from coverage
- If the equipment is scheduled for preventive maintenance, or if a noncritical repair is needed but the equipment is unavailable, the service staff may have to wait for the equipment to become available, which might be billable time from the service vendor, but might not be covered by the insurance program
- Certain glassware components such as X-ray tubes, image intensifiers, gamma camera crystals, ultrasound transducers, and CRT displays might not be covered if they need to be replaced because of poor image quality (rather than catastrophic failure), as the failure might be categorized as a planned-obsolescence replacement rather than a covered maintenance event.
- Because of cash flow limitations, the insurance carrier might elect not to pay the maintenance vendors within previously stated time frames, thus resulting in the service vendor putting the hospital on "credit hold," meaning that no additional maintenance services will be provided until all owed payments are made.
- Many marketed maintenance insurance programs are not endorsed by regulated insurance carriers. If not held to the same standards as the insurance industry, they could cancel their contract (or drop selected equipment coverage) with the hospital at any time should their profitability targets not be met. Should they go out of business or have cash shortfalls, the hospital could become liable for having the overdue invoices paid, even though they have already paid the maintenance insurance vendor.

Maintenance insurance, when properly selected and used, can be a useful tool for a growing in-house service program, especially if it results in the taking over of service and support of items previously under warranty or OEM full-service agreement where no maintenance cost data were kept or monitored. This transfers the risk of going over budget while the hospital collects maintenance cost data for a year or two, thus allowing the hospital to be in a better position to self-insure their budget. Ideally, maintenance insurance should not be used for a period greater than one year, or two at a maximum. If annual maintenance costs exceed the program premium, the insurance carrier could raise their rates in order to recover their losses.

Larger hospitals (i.e., those with 300 beds or more) may be positioned to self-insure their maintenance budgets, assuming that they have the infrastructure to manage the paperwork and vendor-services aspects of the program. Cost savings by going to a self-insured program typically cost less than the cost of the maintenance insurance program. However, if a maintenance insurance program is desired, be sure that it is underwritten by a regulated insurance carrier that has a high rating. Programs that the provider "self-insures" should be considered with caution! In the UK, official guidelines advise National Health Service users not to use insurance-based schemes, as their experience has shown that the promised savings are not as high as predicted.

Summary

When selecting an outsourced service provider, it is important that all program components be identified in writing and made part of the contractual agreement. Issues related to on-site, dedicated staffing, parts sources, scheduled preventive maintenance, backup support, external OEM vendor assistance, data ownership, reporting, equipment relocations, upgrades, software, test equipment, user training, committee participation, and total budgetary responsibilities also must be clearly defined.

The cost effectiveness and quality impacts of out-sourced clinical engineering programs versus in-house programs have been the subject of much debate, with many organizations having tried both models. One given fact is that a service program that is outsourced to a for-profit company has two masters to serve: The hospital client, and the business entity shareholders. One master wants the lowest cost and the other wants maximum profit margin. The hospital mandates quality and the out-sourcer is obligated to deliver it. The debate over cost-effectiveness and long-term impact on equipment quality continues to make the choice over outsourcing versus in-sourcing a case-by-case determination.

Even with a fully outsourced program, the hospital still holds the ultimate responsibility for the impact of its equipment on patient care outcomes. The outsourcing of the management of the service program does not mean that the hospital will not be held harmless for patient injuries because of malfunctioning equipment. An outsourced program still needs to be audited, monitored, and held to the same standards to which an in-house program would be held.

References

MDA. Medical Device and Equipment Management for Hospital and Community-Based Organizations, Medical Devices Agency Device Bulletin 9801, January 1988.
MDA. Medical Devices and Equipment Management: Repair and Maintenance Provision, MDA DB2000 (02) Medical Devices Agency, June 2000.
Tran TG. Use of Maintenance Insurance to Minimize Costs. *J Clin Eng* 19(2):143-147, 1994.

Further Information

The Management of Medical Equipment in NHS Acute Trusts in England, report by the comptroller and auditor general for the NHS executive. National audit office H. C. 475 session 1988–99, The Stationery Office, London, June 10, 1999.
Insurance in the NHS: Employers/Public Liability and Miscellaneous Risk Pooling, Health Service Circular HSC 1999:021, Department of Health, http://www.open.gov.uk/doh/coinh.htm, February 3, 1999.
Medical Devices Management, NHS Executive Controls Assurance Standard, Rev. 01, November 1999.
Bendor-Samuel P. Turning Lead into Gold. The Demystification of Outsourcing, http://www.turning-lead-into-gold.com.
http://www.outsourcing-journal.com/.
http://www.outsourcing-exchange-center.com/.
An Insider's Look at Increasing Role of Service Outsourcing in the Health Care Market D. F. Blumberg Associates, Inc. http://www.dfba.com/hc_service.htm, 2003.
http://www.echohealth.org.uk.

40

New Strategic Directions in Acquiring and Outsourcing High-Tech Services by Hospitals and Implications for Clinical Engineering Organizations and ISOs

Donald F. Blumberg
President, D.F. Blumberg & Associates, Inc.
Fort Washington, PA

The growing costs involved in the support of medical electronics and diagnostic imaging technology and other high-technology equipment and systems found in hospitals and health care facilities today require an objective and professional analysis and evaluation in order to understand fully the underlying causes of these rising costs and to provide a logically defined set of options and alternatives for the industry. Hospitals and health care facilities use, and therefore must service and support, a wide variety of high-tech equipment on their premises, including, but not limited to, the following:

- Computers
- Office automation technology
- Telecommunications, including voice and data networks
- Medical diagnostics and instrumentation
- Other medical instrumentation and technology
- Building and infrastructure control systems

In addition, the very nature of the hospital technology base is changing as hospitals acquire outlying professional offices and health care facilities, moving to a geographically dispersed wide area network configuration. (See Figure 40-1.) The introduction of network distributed diagnostics systems, (e.g., picture archiving and communication systems (PACS) is changing service requirements. In essence, full service in hospitals now requires a broad base of technical capabilities, central control and coordination of calls and logistics, and increasing network service and support. Simple service of diagnostics imaging and medical electronics in one building or small campus will no longer suffice.

In broad terms, the general options that are available to hospitals and health care facilities, the end users of medical electronics and other high-technology, for service and support, are as follows:

- To maintain all equipment through an internally organized and operated biomedical engineering and management information system (MIS) support organization
- To outsource all or a portion of these maintenance, repair, and technical support services to the original equipment manufacturer (OEM) or independent third-party maintenance (TPM) organizations
- To expand services offered to other hospitals; in essence, to move in to the provision of third-party maintenance services in an attempt to increase economies of scale

To the extent that all or a portion of maintenance, repair, and related support services is outsourced, the following three options are available for outsourcing:

- The OEMs that are beginning to offer service on an array of different products and technology on a multivendor service basis
- Independent service organizations (ISOs) providing third-party maintenance services for one or several OEMs' products
- A full turnkey or facility manager or site manager to manage and/or service all or a portion of the equipment at the hospital/health care site.

Until recently it has been extremely difficult for the hospital and health care users, the OEMs, and the ISOs to fully understand and make use of these options and alternatives. Recent extensive benchmarking studies by the author and D.F. Blumberg & Associates (DFBA) of the service of both general high-technology equipment used in the hospitals and specific medical electronics technology, as well as research into the size and dimensions of these market opportunities, has enabled an assessment and evaluation of these issues to be made. This information is presented below.

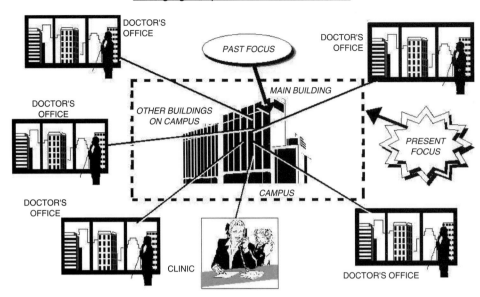

Figure 40-1 Emerging hospital service environment.

Total North American Hospital / Healthcare High-Tech service Market
- BY EQUIPMENT / SERVICE CATEGORY -
($ IN BILLIONS)

EQUIPMENT / SERVICE CATEGORY	1999	2000	2001	2002	2003	2004	CAGR
DATA / PROCESSING & OFFICE AUTOMATION EQUIPMENT	4.73	5.30	5.95	6.68	7.49	8.37	12.1%
DATACOM NETWORKS & PERIPHERALS	5.13	5.90	6.77	7.75	8.86	10.10	14.5%
TELECOM / WIRELESS EQUIPMENT	3.32	3.60	3.91	4.25	4.59	4.96	8.4%
MULTI USE NETWORKS AND HOSPITAL AUTOMATION SYSTEMS	10.67	12.70	15.05	17.75	20.86	24.51	18.1%
SUBTOTAL: INFO TECH	23.85	27.50	31.67	36.44	41.80	47.94	15.0%
BUILDING AUTOMATION EQUIPMENT	0.30	0.33	0.36	0.40	0.44	0.48	9.9%
MEDICAL ELECTRONICS	13.89	15.78	17.90	20.36	23.21	25.99	13.3%
SUBTOTAL: MEDICAL & BUILDING EQUIPMENT	14.19	16.11	18.26	20.76	23.65	26.47	13.3%
SOFTWARE SUPPORT	5.90	7.08	8.52	10.32	12.65	15.17	20.8%
FACILITIES MANAGEMENT	0.59	0.62	0.65	0.68	0.72	0.75	5.0%
DESIGN & ENGINEERING	0.68	0.78	0.89	1.03	1.18	1.34	14.5%
LOGISTICS SUPPORT	0.19	0.22	0.25	0.29	0.34	0.40	16.0%
SUBTOTAL: VALUE-ADDED SERVICES	7.36	8.70	10.32	12.32	14.89	17.67	19.1%
TOTAL SERVICE MARKET	45.40	52.31	60.26	69.52	80.34	92.08	15.2%

Figure 40-2 Total North American hospital/health care high-tech service market by equipment/service category (in $ billions).

Market Size and Benchmark Evaluation

The general market for service and support for high-tech equipment in hospitals in North America is quite large (exceeding $10 billion in 1996) and is growing, as shown in Figure 40-2. This evaluation, based on an extensive survey of over 500 small, medium, and large general and specialized hospitals showed that the largest single specific area of expenditures in hospitals was for the service and support of medical diagnostics and imaging (electronics) equipment.

In order to clearly understand the underlying economic dynamics of high-technology service and support in hospitals, it is of interest to examine key results of a recent study carried out by DFBA, utilizing a variety of sources, to examine existing cost elements and components of the typical 300–400-bed general hospital located in major metropolitan areas in the United States. The largest single functional component of cost is direct labor and benefits. Other major cost items include direct overhead administration and infrastructure costs; outside material, equipment, and supply purchases; and all other costs. If trends in these costs over the 1992-1994 timeframe are examined, as shown in Figure 40-3, the direct labor costs can be seen to continue to rise along with the incremental costs of maintenance and supply. It is important in this analysis to note that the significant portion of the total costs of high-technology maintenance and support is buried in direct labor, overhead, and other related costs of the hospital. If the internal and external costs of high-tech maintenance and support along with the building infrastructure, environmental technology, and support were to be computed, the costs of this *function* would represent the second largest cost component in hospital operations.

KEY TRENDS IN TOTAL HOSPITAL COSTS FOR TYPICAL 300–400 BED HOSPITAL

KEY FACTORS	YEAR			% CHANGE
	1992	1993	1994	
DIRECT LABOR AND BENEFITS	27.5	29.0	30.9	12.4
DIRECT OVERHEAD, ADMIN., & INFRASTRUCTURE COSTS	2.9	3.0	3.0	3.4
OUTSIDE MATERIAL, EQUIPMENT, & SUPPLY PURCHASES	3.1	3.3	3.4	9.7
TOTAL MAINTENANCE & SUPPORT COST*	17.1	18.3	20.0	18.1
INCREMENTAL MAINTENANCE AND SUPPLY COSTS**	6.3	7.0	8.0	27.0
ALL OTHER COSTS***	3.3	2.9	2.7	-18.2
TOTAL COSTS OF OPERATION	43.1	45.2	48.0	11.4
OPERATING PROFIT	1.2	1.2	2.6	116.7
PROFIT AS A % OF TOTAL REVENUE	2.7%	2.5%	3.5%	29.6

*   ALL MAINTENANCE AND SUPPORT COSTS FOR HIGH TECHNOLOGY, BUILDING AND ENVIRONMENTAL INFRASTRUCTURE, ETC. (INCLUDES DIRECT LABOR & OUTSIDE PURCHASES)
** ALL MAINTENANCE AND SUPPORT COSTS, LESS INTERNAL DIRECT LABOR & BENEFIT COSTS & OTHER PURCHASES
*** DEPRECIATION, G&A, ETC.
SOURCE: ESTIMATED BY DFBA, BASED ON AHA, OTHER STUDIES AND REPORTS, AND INDEPENDENT DFBA STUDIES AND SERVICES OF HOSPITALS

Figure 40-3   Key trends in total hospital costs for typical 300–400-bed hospital.

This analysis also indicates that the typical medium to large size hospital's operating profit, though small, has been increasing, primarily because of a significant decrease in other support costs and a significant slowing in the rate of growth associated with direct labor and benefits. An analysis of the total costs of high-technology and building infrastructure maintenance and support as compared to the operating profit provides a clear indication of the potential impact of getting increased economies of scale through outsourcing, downsizing, or growing the service business base. In essence, a 10% savings in high-technology and building maintenance and support costs could lead to a doubling of operating profit. This represents a strong incentive for hospital executives, managers, and administrators to look for all paths to improving the efficiency and cost effectiveness of high-technology, infrastructure, support costs and services.

In order to assess and evaluate the subcontract buy, sell, or outsource paradigm for medical electronics and other high-technology service from the hospital (administrator user), OEM, and ISO perspective, it is necessary to accurately benchmark and evaluate the cost, profitability, performance, and prices of these three major classes of service organizations in the industry. Such benchmarking studies to develop industry averages and company-specific information have recently been done. Probably one of the most important OEM companies, from the perspective of medical electronics and diagnostic imaging service is GE Medical, which is a major player in the medical electronics industry and one that serves as a standard with respect to diagnostic imaging and related medical electronics technology service base. GE Medical has now moved into health technology multivendor equipment service (MVES) and support.

As indicated in Figure 40-4, key benchmark parameters have been estimated for General Electric Medical, the medical technology OEM industry as an average, and typical ISO/TPM organizations for revenue per field engineer, profitability, call response, and repair completion time parameters. Where comparable, performance information with respect to internal biomedical engineering organizations on the average, and for the largest and most efficient internal biomedical engineering organizations, is provided.

Based upon this benchmark data, further analysis and evaluation were made to examine the true cost of support on a per unit basis and levels of performance in order to gain an understanding of both the economics and service-performance capabilities of these different options for a number of specific product examples. The data suggest that while the

BENCHMARK COMPARISONS OF ALTERNATIVE MEDICAL SERVICES ORGANIZATION

	TYPE OF SERVICE ORGANIZATION				
	MOST PROFITABLE (GE MEDICAL)	TYPICAL OEM	TYPICAL TPM	TYPICAL INTERNAL BIOMEDICAL SVCS.	MOST EFFICIENT BIOMEDICAL SVCS.
REVENUE/BUDGET PER FIELD ENGINEER ($THOUSANDS)	385	301	265	136	101
PROFIT PER FIELD ENGINEER AS A % OF REVENUE	42%	31%	16%	NA	NA
MEAN TIME TO RESPOND (HOURS)	1.5	2.0	1.4	.9	.4
TYPICAL TOTAL MEAN TIME TO REPAIR (HOURS)	2.3	2.3	2.5	2.2	2.1
AVERAGE CALLS PER DAY	2.8	1.9	2.2	2.8	4.2

SOURCE: DFBA BENCHMARKING ANALYSIS BASED ON 1993-1994 DATA

Figure 40-4   Benchmark comparisons of alternative medical services organizations.

typical internal biomedical organization is obviously more responsive because it is based on-site, it also can be somewhat more costly than the typical ISO/TPM organization unless it is efficient. The in-house biomedical engineering department can be inefficient because of low density, lack of an effective management system, and lack of strong incentives to improve productivity. It is not at all clear that the more responsive on-site capabilities of the typical biomedical department offset its higher costs. On a cost-response basis the well run ISO/TPM can offer a better solution, from the viewpoint of the user, than can an inefficient in-house service organization.

An in-depth review for CT scanners and most other products, for example, shows that density affects productivity and cost; i.e., the denser the equipment base served, the more efficient the service. Thus in general, the larger service organization supporting a large base will be more efficient and cost-effective as density increases. In examining outsourcing options, the management's desire for high target margins also affects the equation. Typically, ISO/TPMs, for example, are more cost-effective than the OEM organizations for end users, primarily because of the high profit margins charged by the OEMs to offset low product margins and profitability. GE Medical for example shows profitability estimated in the range of 35% to 42% pretax, whereas the ISOs are accepting a lower level of profitability and therefore, offer a lower price option.

The primary reason for the typical small, in-house biomedical engineering organization being somewhat more expensive and less efficient is that its utilization rates are much lower because of lower density. In essence, both the OEMs and ISOs, by servicing a broad array of customers, can keep their service engineers busier and at a higher level of training and skills than can the internal organization, which spreads its service force across only its own internal installed base. Clearly, one way that hospitals could improve this utilization rate would be to either expand into TPM service for other hospitals or to establish an ISO, which could be utilized by a number of hospitals in one chain or by multiple hospitals in a given geographic area as their central source of service. This has been done in the case of Kaiser Permanente Health Services, for example, where one internal biomedical engineering department serves several hospitals. In essence, under this concept an ISO/TPM built from an internal biomedical engineering department may become the accredited service organization for one or more hospital chains or hospital groups on a negotiated price basis. With the high density of the served base, productivity and cost-effectiveness increase.

Change in Competitive Factors Affecting Efficient Use of Service in the Health Care Market

In the past, OEMs have attempted to counter this trend by creating barriers to service competition through the following tactics:

- Withholding required parts
- Withholding or denying access to required diagnostics
- Withholding or denying training or documentation to others who desire to service the equipment
- Withholding or denying software support or service
- Other actions to deny access to needed information, materials, and software or to tie one service to another, thus making it difficult if not impossible for competitive service providers to operate effectively

In the medical field, OEMs engaged in the above practices and had informal understandings that one OEM would not service the equipment of a competing OEM.

These anticompetitive practices recently came under attack. Through antitrust and other legal actions, the right of OEMs to raise such anticompetitive barriers was challenged. The typical general defense by OEMs engaged in these practices is that they are forced to do so in order to maintain product quality. However, in a case involving Etek (a small TPM) against Picker (a large diagnostics imaging OEM), the court found differently. Typical defenses by OEMs in antitrust actions,(e.g., that they could not be monopolistic in the service market when they were not monopolistic in the product market and that the customer's perfect knowledge of the market would predict price escalation) also have been weakened. A decision by the U.S. Supreme Court in the case of ITS versus Eastman Kodak (to remand the case for review, and providing a new framework for examining these issues) has begun to erode this defense (Blumberg, 1992; Blumberg and Quinn, 1992). However, the OEM anticompetitive (and potentially monopolistic) practices did lead to the creation of real inefficiencies in the market, particularly in selected segments such as diagnostic imaging and instrumentation.

The author has determined that the effect of these monopolistic or anticompetitive type practices is to generate a surcharge or profit increase of 20%, and in some cases 40%, over a true competitive price. In the area of medical diagnostics and imaging service and support, this amounts to a surcharge to the hospital and health care industry of over $500 million annually. Thus, increased enforcement of existing antitrust rules and regulations requiring OEMs to be more open and supportive of all service providers could yield savings of over $1 billion annually to the health care industry. This strong market incentive is beginning to create demand for new MVES offerings to health care.

Emergence of New Competitive Integrated Service Organizations in the Health Care Market

In general, competition in the health care technology service market is somewhat fragmented and narrowly focused (see Figure 40-5). As noted above, traditionally, almost all medical technology OEMs operated under an informal, unstated "gentlemen's agreement" in which they would not service a competitor's equipment and would not sell parts to or support ISOs. These policies, which were not consistent with the general direction of OEMs in the data processing and office automation industries, tended to erect real or imagined barriers to the entry of integrated and TPM service providers and significantly reduced the service options available to health care organizations.

However, the following two factors recently have caused a major reversal of this situation:

1. A decline in the demand for the purchase of new capital equipment for medical technology, resulting in a significant decrease in revenues and profits associated with product sales by OEMs and great pressure to reduce the service organization size in light of the reduction in the demand for service of the OEMs products. This, in turn, has led several medical technology OEMs—e.g., GE, Picker, Phillips, and Siemens—to cast off the gentlemen's agreement and to move toward providing multivendor service and support of medical electronics technology in general.
2. As indicated above, certain service-based antitrust cases (including the Eastman Kodak case, which was remanded by the U.S. Supreme Court for district-level review, with guidance) have begun to focus on the monopolistic practices of withholding parts, diagnosis, and documentation support, with the allegation that this creates a monopoly. Both the Supreme Court remand and recent Federal Circuit decisions seem to suggest that the manufacturer's position of withholding parts sales from independent service organizations was at increasing risk. In fact, a recent request by Kodak to restrain action requiring Kodak to sell parts to independent service organizations has been overturned. In essence, with the movement of major medical technology OEMs into multivendor service and support, and an increase in the risk of OEMs withholding parts sales from independent service organizations because of antitrust legal action, the industry providing integrated service and support to the health care market has begun to open up.

As indicated in Figure 40-6, several new or existing service organizations have begun a dialogue with hospital administrators to discuss the prospect of full or partial outsourcing of high-technology service and support on a multivendor basis. Existing medical OEMs, such as GE Medical, Picker, Siemens Medical; independent service organizations who have focused on the medical technology field, such as Innoserve Technologies, COHR, and AMSCO; and relatively new service providers from other market segments, such as Olivetti and Marriott, are now beginning to create the infrastructure to support outsourcing by hospital operators.

Key Trends and Factors Affecting Health Care Technology Service and Support Outsourcing

Managed care is the most critical factor affecting health care technology service and support today and is changing virtually every aspect of health care delivery. Under the traditional fee-for-service contract, health care practitioners were prepared to do whatever was necessary to care for and to cure ailments in their patient population. The role of the health care practitioner was to relieve pain and to focus on caring for the needs and requirements of individual patients. Managed care, with its glass ceiling on fees and expenses, has forced the health care community to manage the outcome of the patient population rather than to attend to the needs of the individual patient.

Because of this paradigm shift, health care community is becoming increasingly more cost-conscious and business-minded. The health care community must constantly weigh and balance the alternatives of providing the highest quality of care against the cost of potential outcomes this care provides. While this has a negative effect on suppliers of health care products and services, some vendors have found an increase in sales because they have demonstrated that their products and services improve the bottom line. Other vendors who have not been able to adjust effectively to the new realities of the health care industry have experienced a decrease in sales.

The new reality of health care is that the industry is becoming increasingly more market driven. Health care organizations (HCO) and payers (i.e., insurers) are thinking and acting more and more like major corporations. Their primary focus is on increasing revenue while controlling costs. However, capitated payment plans of insurers severely affect the level of revenue a health care establishment can obtain. As a result, a hospital or a physician's practice can no longer purchase new products and technology and then place the costs on the patient population by raising fees. Today, new equipment can be purchased only if it can help to win new health maintenance organization (HMO) contracts or to improve the productivity and efficiency of health care delivery. As a result, the life of existing equipment is extended, while increasing the utilization of the equipment in order to recover the full cost of ownership. In order to achieve this end, the hospital itself is functioning more like a factory, attempting to increase the number of patients who go through the system, while shortening the length of time the individual patient is in the system. In essence, hospital administrators are concentrating on developing and implementing practices that decrease the cycle time and increase the throughput of the patient population in order to improve efficiency, productivity, and ultimately, the profitability of the HCO.

This market-driven corporate focus has led to a number of new and emerging trends, which have a significant impact on the effects of hospital operations and in turn impact requirements for technology service and support. These trends include the following:

- A movement toward greater consolidation, which results in the creation of economies of scale and reduces the overall size of the purchase opportunity for new equipment, but increases the overall opportunity for services while decreasing the number of potential buyers of service.
- Hospitals acquiring offices of general practitioners and specialists to increase control of the geographic market. As a result, the remote offices and clinics must be integrated with the central hospital campus and combined with a greater emphasis on increasing utilization of the existing facilities and technologies. The technology service organization is no longer responsible for merely equipment within one building or campus, but for that of the entire hospital network. Furthermore, increased utilization of equipment

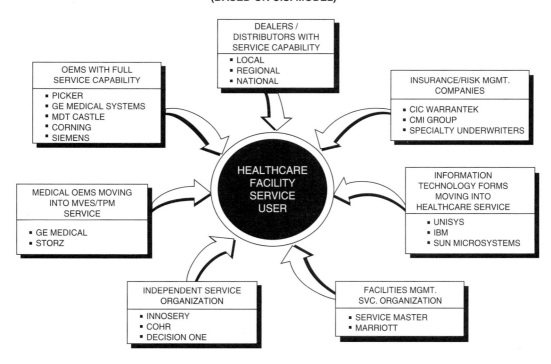

Figure 40-5 Classes of competition in the health care service market based on the United States model.

results in more frequent equipment failure. As a result, service response and repair become highly critical.
- Deployment of advanced networking technology, such as PACS technology, integrates clinical functions such as diagnostic imaging and laboratory analysis into a distributed network that is shared by all facilities and locations that are part of the health care delivery system. (See Chapter 102.) The increased deployment of the network technology results in the requirement of sophisticated network service and support to be available on a distributed basis.
- Increased cost considerations result in the use of a cost control committee and the placement of more purchasing responsibility on the materials managers, thus, increasing the emphasis on equipment and facilities planning (see Chapters 30 and 32) and enhancement of efficient asset management and control (see Chapter 35). As a result, equipment and service purchases are scrutinized for cost efficiency and justification. In addition, equipment and service are purchased on a shared basis for more efficient utilization of capital and operating costs, thus increasing utilization rates, needs for service coverage, and higher equipment up time requirements. (See Chapter 47.)
- Collaboration is more commonplace and results in an increased need for integration and shared reward programs, which leads to a decrease in equipment service purchases and consolidation of service contracts.

In essence, key paradigm shifts in the health care industry infrastructure have had a large impact on the purchase of high-tech equipment and supporting services. On one hand, there is an increased demand for highly reliable equipment and related support services creating a major opportunity to those vendors who can meet the needs and requirements of hospital purchasers. On the other hand, the downward pressure to control costs and to improve operating efficiency has resulted in hospital decision makers placing greater scrutiny on the value and benefits that the acquisition of new equipment and services have on the health care delivery system. The ability of vendors and suppliers to meet these conflicting requirements is a key factor in determining future success in the health care market.

Implications to the Health Care Technology Service Market

The new paradigm shift in health care has resulted in hospitals becoming part of a large corporate organization. The primary motives of the health care organization are no longer medically driven but financially driven, as hospital decision makers behave more and more like corporate executives. Decisions for equipment service purchases are made centrally, at an administrative or executive level. Furthermore, the decision maker focuses on the requirements of the entire organization rather than those of one department or floor within the hospital. Like most corporations, the chief executive officer of the hospital delegates decision making responsibility through the organization and relies on recommendations from subordinate managers regarding strategic decisions with respect to health care technology service and support. Accordingly, in-house clinical engineers are in a key position to impact and to influence the decisions regarding questions of whether to outsource service or to expand it internally.

In order to make informed decisions on the issue of outsourcing or expanding service internally, clinical engineers now must look at how they are going to meet the service needs and requirements of the entire hospital network. (See Figure 40-1.) The total installed base, which must be managed, includes not only medical devices inside the hospital facility itself but also the devices within doctors' offices and clinics, as well as technology that is shared or integrated between facilities. In essence, the focus of the clinical engineer is no longer on supporting a medical device on a floor of the hospital but on all the facilities that constitute the hospital infrastructure. The clinical engineer is responsible for delivering support on a densely populated, somewhat dispersed, installed base of medical technology. This installed base geography can be as large as a major metropolitan city and thus it can have the characteristics resembling those of a branch office of an OEM or ISO.

In-house clinical engineers must manage this total installed base of technology in the most efficient and effective way possible. (See Chapter 30.) The key issues revolve around providing the necessary support to extend the technology life cycle and providing the necessary service to maintain high end user requirements for service response and repair times. In order to make the optimum decision, the in-house clinical engineer should analyze all components of the service delivery value chain (See Figure 40-7.) This chain can be defined as all activities, tasks, and procedures that are required in order to respond to and to resolve requests for service. The value chain analysis is a new approach, which many clinical engineers are using to manage optimally the entire service operation. The components of the value chain include the following:

- On going and continuous evaluation of service strategy and tactics
- Implementation of this strategy vis-à-vis current service operations, which include:
  ○ receipt of service request
  ○ isolation of problem
  ○ determination of probable causes and symptoms accepting ownership for problem resolution
  ○ determining corrective action and identifying and utilizing service resources (i.e., parts, skills, and documentation)
  ○ implementing corrective action and problem resolution
  ○ updating experience and fine-tuning service strategy tactics

The clinical engineering manager or administrator can improve service efficiencies and costs by placing emphasis on managing one or all of the individual components of the

NEW OR EMERGING COMPETITIVE INTEGRATED SERVICE ORGANIZATIONS IN THE HEALTHCARE MARKET

FIRM	FOCUS	GENERAL STRATEGY
GE MEDICAL	• SERVICE OF DIAGNOSTIC IMAGING TECHNOLOGY • NOW MOVING INTO MVES	NOW EXPANDING INTO FULL SITE SERVICES TO OFFSET PRODUCT SALES DECLINING
INNOSERV TECHNOLOGIES, INC. (R-SQUARED SCAN SYSTEMS AND MEDIQ)	• SERVICE OF DIAGNOSTIC IMAGING • SERVICE OF INSTRUMENTATION • FULL SERVICE OF ALL HIGH TECHNOLOGY IN HOSPITALS	• CONTINUING EXISTING STRATEGY OF TECHNOLOGY/MODULAR SERVICE • ATTEMPT ACQUISITION AND EXPAND
PHILLIPS MEDICAL	• MULTIVENDOR SUPPORT SERVICE	• EXPANDING INTO MULTIVENDOR SUPPORT THROUGH PARTNERING WITH IBM, ETC.
OLIVETTI	• FULL SITE SERVICE • MULTIVENDOR SERVICE	TOTAL SERVICE COMMITMENT BASED ON SUCCESSFUL BANKING INDUSTRY MARKET APPROACH
COHR	• TOTAL BIOMEDICAL SUPPORT	ATTEMPTING TO EXPAND SERVICE FOCUS
MARRIOTT	• FOOD SERVICE • CLEANING SERVICE • BIOMEDICAL SUPPORT	EXPANDING SERVICE OFFERING
SERVICE MASTER	• CLEANING SERVICE • MAINTENANCE SERVICES	HAS ATTEMPTED EXPANSION IN MARKET
AMSCO	• SURGICAL TECHNOLOGY • STERILIZERS • DIAGNOSTIC IMAGING	• HAS PARTNERED WITH OTHER VENDORS TO OFFER BROADER SERVICES • USED ACQUISITION OF TRW MEDICAL TO EXPAND INTO MVES • DIRECTIONS NOT CLEAR DUE TO MAJOR MANAGEMENT CHANGES & SALE TO NEW COMPANY
PICKER	• MULTIVENDOR SUPPORT SERVICE	NOW EXPANDING TO FULL TPM SERVICES

Figure 40-6 New or emerging competitive integrated service organizations in the health care market.

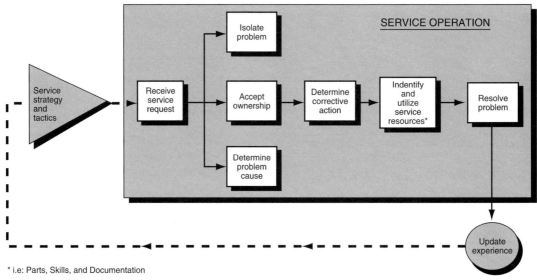

* i.e: Parts, Skills, and Documentation

Figure 40-7 Service delivery value chain.

service delivery value chain. In essence, these components can be performed internally or they can be outsourced to an external vendor in order to realize gains in service productivity, quality, and efficiency. By focusing on the value chain, it becomes increasingly clear that there is a high demand for ancillary support services such as remote diagnostics, help desk, and parts management. These ancillary support services enable the manager and administrator to manage optimally the value chain. The focus is no longer on determining how many people will be required in order to support the technology but on determining the functions and processes that will be required in order to maintain medical devices effectively. From the supplier's perspective, the service delivery value chain creates new opportunities for value-added ancillary support services, which have a high value in use, thus creating new service offering and the potential for unbundling service contracts and offering some services at a premium. The acceptance of the value chain management concept ultimately will result in more opportunities for health care technology service and support.

Current and Emerging Service Requirements of Hospitals

The author has conducted extensive market research to validate and quantify the basic assumptions about health care technology service and support as well as measure the impact of key trends on managing health care technology service and support requirements. Conclusions are based on market research designed to quantify attitudes and perceptions of hospitals, clinical engineers, biomedical equipment technicians (BMETs), and administrators. The research, conducted over a 12-month period (1996-1997), utilized six focus groups comprised of clinical engineers, biomedical equipment technicians (BMETs) and administrators and entailed telephone surveys of 350 small, medium, and large hospitals.

The focus group results confirmed that hospitals are continually exploring the possibility of outsourcing services. Focus group participants believed that managed care had caused, and that it would cause, many hospitals to eliminate their in-house service staffs. In addition, with more hospitals merging, they believed that it made sense to streamline service by either consolidating in-house clinical engineering departments or utilizing an outside service provider. Focus group participants suggested that the culture of a hospital determined whether the hospital would outsource the clinical engineering department or allow that department to grow. A hospital would be less likely to outsource the department if the administration has a high level of confidence in the department manager and has made a commitment to keep the department in-house. However, often because of a lack of knowledge at these administrative levels, the biomedical department can have a substantial influence upon decisions.

Many of the focus group participants indicated that the level of service that they were getting from outside providers did not meet expectations, particularly in the areas of telephone technical assistance and parts availability. To that end, focus group participants indicated that they desired better service from outside providers. Many participants confirmed that their hospitals had a high value in use for specific ancillary support services. For example, focus group participants expressed a willingness to pay a premium if the level of support that they received would help to lower turnaround time, lower costs, and improve quality—particularly those technologies and modalities where in-house support capabilities were limited to nonexistent. This comment fully validated the position that CEs and BMETs are now utilizing ancillary support services to manage effectively the service delivery value chain. Most importantly, these focus group findings confirmed that CEs and BMETs were managing service and support strategically, utilizing outside vendors, and emerging technology to manage the service value chain.

These focus group findings were further substantiated by the results of a second large-scale survey of over 350 decision-makers and hospital organizations. The survey results indicated that the decision making had become more centralized and that hospitals had tight requirements for service quality. (See Figure 40-8.) The most important factors when evaluating service performance appear to be technical knowledge, ability to resolve problems over the telephone, and first time fix right. The cost of maintenance services is less important than service quality-related issues. Survey research has shown that a high percentage of hospitals have an interest in an integrated service contract from a single-source provider. (See Figure 40-9.) In addition, hospitals desire a broad array of ancillary support services (see Figure 40-10) and are willing to pay a premium over and above the basic service contract rate to receive these value-added ancillary support services, which are vital to managing the service delivery value chain.

Strategic Evaluation

In summary, the hospital health care technology service market is large and growing. The American market for health care technology service was $20 billion in 1996 and is expected to grow to $41 billion by the year 2001. The hospital market offers some unusual features, which can become the basis for an extremely profitable growth business and service. Much of the growth will come from new ancillary support services, which are generally in demand for health care technology, including network service and support, technology assessments, and telephone technical support hotlines. Furthermore, service requirements are changing dramatically because of a number of key trends of an economic, managerial, and technical nature, including increasing cost-containment pressures, hospitals' desire for control of the total market by acquiring outpatient facilities and doctor's practices, new network technology, and increasing criticality of services.

The health care technology service market can be segmented into two types of organizations: Hospitals without internal biomedical service organization that require basic and value-added services and hospitals with internal service or service management (i.e., total asset management providers). (See Figure 40-11.) These types of organizations still require added-value ancillary support services in order to manage the service delivery value chain, although they do not require basic services. Competition, which is still fragmented because of hospital OEM strategy, is now changing. Of the major classes of competition, the OEMs and ISOs will have the most significant impact because of their ability to offer a fully integrated service portfolio. While there are a number of OEMs with nationwide service organizations, it is clear that the strategic direction of the organization will support two extremely different patterns:

- OEMs expanding services in a strategic line of business and/or expanding their service portfolio
- OEMs moving out of service

The traditional ISOs will consolidate and continue to become more aggressive. Other ISOs in the information technology service market will penetrate the hospital technology service market. The ISO market will continue to grow and expand into more areas of equipment service, and manufacturers will be less of a force in the service of their own equipment.

An emerging class of competition for both the OEMs and ISOs are hospitals forming their own for-profit service companies or shared service organizations. A number of hospital clinical engineering departments are pursuing this strategy in an attempt to expand the service capabilities while combating the threat of outsourcing.

Not only are the market and competition changing rather rapidly in health care, but also new service management systems and technology (Blumberg, 1996) to improve the efficiency, productivity, cost effectiveness and delivery of service are becoming available. These new technologies focus mostly on the optimization of service resource allocation. The new technology is taking a number of forms, including improved service planning and forecasting models, new technology for call analysis, help desk support, technical assistance and call avoidance, new software for optimized scheduling and dispatch, wireless communications, and mechanisms to improve interaction between headquarters call management activity and the service engineering using a laptop or personal digital assistant (PDA). The new technology is enabling service providers to manage the value chain and to generate service quality improvement more efficiently. The technology is becoming a basic requirement for competing in the health care technology service market. However, investment in this technology is significant, making the decision to enter or to expand into service a strategic related question.

Strategic Steps in Determining Optimum Action by Hospitals

It is evident that hospitals and OEMs will continue to consider seriously three possible alternatives for managing the service operation:

- Significantly grow or expand
- Outsource subcontractor or downsize
- Joint venture or divest

This decision is clearly not a simple one. It requires a strategic assessment and evaluation where a number of considerations must be made, including, evaluating the following factors:

- The importance of service to the service organizations' customers and users
- The observed perception of service quality and responsiveness on the part of the market or user community to individual company considering outsourcing
- The current level of service efficiency and productivity as compared to other equivalent service organizations in the market

These key issues will determine whether a service organization will be best advised to make any appropriate strategic changes in order to significantly reduce its costs and to improve flexibility, efficiency, and performance. The basic decision making process is similar to the typical make-or-buy decision in manufacturing. It involves a matrix (Figure 40-12) relating to the critical importance of service and the perceived service image of the service organization versus the current level of service efficiency and productivity, benchmarking against industry standards and patterns. The key decision and recommendation will vary as a function of the interaction of these key parameters.

This decision making process has proven to assist health care service providers to arrive at the optimum decision and solution with respect to cost reduction and efficiency and profitability improvement that may result in a decision to grow the service organization and improve significant economies of scale; outsourcing or subcontracting of key service functions and processes to reduce cost; forming a joint venture partnership or relationship with another service organization to gain both economies and scale; and the ability to sell excess or utilized time for an increase in the direct business base and service portfolio. An in-house clinical engineering department utilizing the same methodology could arrive at one to four possible conclusions:

- Maintain and service all equipment using a facilities-based clinical engineering and MIS service department
- Use a combination of facilities-based service and subcontract service
- Outsource a significant portion of all service requirements to an independent service organization or multivendor equipment supplier, where such organizations are reliable to improve service cost-of benefits
- Consider creating an independent service organization that would be utilized by one or more hospital chains, by either organizing a new independent biomedical engineering service organization or joint venture merger with one or more OEMs

In summary, hospitals and health care delivery organizations should closely examine their present and emerging service needs, requirements, and costs in order to determine the best way to provide service and support over the next decade. These decisions ultimately will move the health care organization into one of two directions: (1) growth and expansion of their internal service organization or (2) outsourcing and downsizing. In order to meet the emerging needs and requirements of health care organizations, OEMs and ISOs need to assess and to analyze the market needs and requirements, taking into account the full range of technology and the need for quality service levels in light of current perceptions of health care providers.

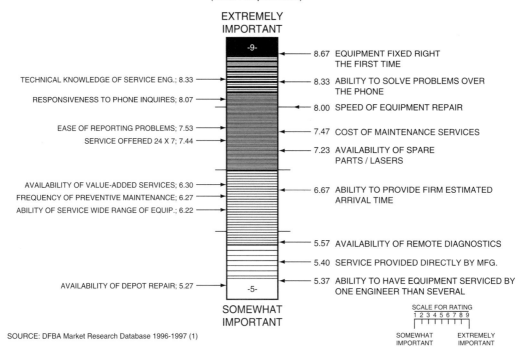

Figure 40-8 Overall rating of importance of service vendor selection criteria.

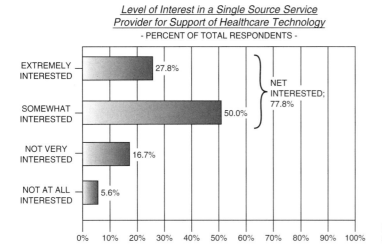

Figure 40-9 Level of interest in a single-source service provider for support of health care technology.

Final Analysis and Evaluation

In summary, as indicated above, hospitals and health care organizations and their clinical engineering departments should seriously consider the following two possibilities:

- Outsource all or a significant portion of their service to ISO/TPMs where such organizations are reliable and can provide cost-effective service. This is particularly valid for small and medium-sized health care organizations where the density is low or highly skilled internal personnel cannot be maintained and supported efficiently.
- Consider creating an ISO that is utilized by one or more hospital chains by either organizing a new independent biomedical engineering service organization created by outsourcing existing biomedical engineering groups into a new, self-supporting service unit or entering into a contract with a national ISO that is willing to offer the required level of service for the broad array of the installed base at a minimum profit.

Under any circumstances, hospitals and health care organizations should recognize the potential interest in these new options to reduce significantly their health care costs to the ultimate user, the patient. Hospitals and health care facilities that utilize efficient multi-vendor equipment, TPMs, or broad-based clinical engineering service organizations could cut annual costs by 45% or more. Based on a general rule of thumb, the cost of contracted-out maintenance and support of all high-tech medical, information technology, building, and environmental equipment in a typical 300–400-bed hospital is about $5.5 million. Thus, outsourcing to an efficient ISO/TPM or an efficiently run coordinated biomedical engineering services organization or OEM, providing a fully integrated service portfolio, could produce a net annual savings of $500,000.

INTEREST/WILLINGNESS TO PAY A PREMIUM FOR VALUE-ADDED SERVICES AS A PERCENTAGE OF EQUIPMENT PURCHASE/LEASE
- PER YEAR -

VALUE-ADDED SERVICE (AS PART OF EQUIPMENT PURCHASE/LEASE)	WANT IN SERVICE AGREEMENT	RATING OF IMPORTANCE	WILLINGNESS TO PAY PREMIUM	% PREMIUM WTP
REPLACEMENT PARTS	84%	8.8	56%	11%
LABOR	21%	7.9	50%	20%
PREVENTIVE MAINTENANCE PROGRAM AVAILABLE	30%	6.8	48%	3%
OFF-PEAK HOUR SERVICE	56%	7.2	44%	6%
7-DAY, 24-HOUR SERVICE	68%	7.7	47%	5%
GUARANTEED SAME DAY ON-SITE RESPONSE TIME	32%	8.1	50%	55
GUARANTEED 2-HOUR ON-SITE RESPONSE TIME	25%	7.2	60%	8%
UNLIMITED FIELD SERVICE CALLS	42%	6.4	12%	10%
LOGISTICS/INVENTORY MANAGEMENT	79%	6.8	33%	6%
SOFTWARE OR APPLICATIONS SUPPORT	94%	6.5	50%	3%
TELEPHONE TECHNICAL SUPPORT	80%	8.4	51%	8%
SELF-MAINTENANCE TRAINING AND DOCUMENTATION	90%	8.2	67%	15%
QUICK-FIX SPARE-PARTS KIT AVAILABLE	70%	7.5	78%	2%
REMOTE DIAGNOSTICS	42%	6.7	25%	3%
3-5 DAY DEPOT REPAIR TURNAROUND	63%	6.5	33%	4%
DROPPED UNIT COVERAGE (DAMAGE INSURANCE)	22%	6.4	75%	8%
NETWORK INTEGRATION	81%	6.8	62%	8%
NETWORK SUPPORT	62%	6.9	56%	4%
ON-LINE DOCUMENTATION/INTERNET SUPPORT	79%	7.3	54%	5%
PARTS CONSIGNMENT	65%	6.4	57%	9%

○ INDICATES FIRST HIGHEST WTP
□ INDICATES SECOND HIGHEST WTP
△ INDICATES THIRD HIGHEST WTP

SOURCE: DFBA MARKET RESEARCH DATABASE 1996-1997

Figure 40-10 Interest/willingness to pay a premium for value-added services as a percentage of equipment purchase/lease per year.

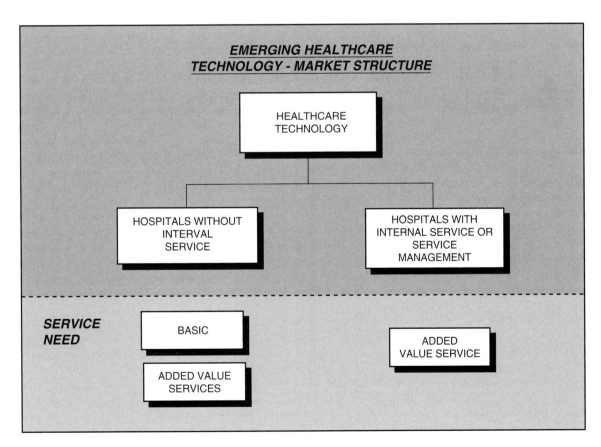

Figure 40-11 Emerging health care technology–market structure.

Key Decision Model for use in Outsourcing & Reducing Service Costs

		RESULTS OF EXTERNAL MARKET SURVEY			
		SERVICE CRITICAL		SERVICE NOT CRITICAL	
		STRONG POSITIVE PERCEPTION	WEAK NEGATIVE PERCEPTION	STRONG POSITIVE PERCEPTION	WEAK NEGATIVE PERCEPTION
RESULTS OF INTERNAL PRODUCTIVITY AND BENCHMARK SURVEY	MORE PRODUCTIVE & EFFICIENT THAN INDUSTRY	EXPAND SERVICE AGGRESSIVELY TO CUSTOMER BASE	PARTNERING WITH MORE EFFICIENT VENDOR	TO BE DETERMINED BASED ON RESULTS	OUTSOURCE SERVICE THROUGH JOINT VENTURE OR PARTNERSHIP
	SAME AS INDUSTRY STANDARDS	EXPAND SERVICE	TO BE DETERMINED BASED ON RESULTS	TO BE DETERMINED BASED ON PARTS	OUTSOURCE
	BELOW INDUSTRY STANDARDS	EXPAND SERVICE, BUT SUBCONTRACT CERTAIN FUNCTIONS	OUTSOURCE TO EFFICIENT SERVICE VENDOR	SUBCONTRACT CERTAIN SERVICE FUNCTIONS	OUTSOURCE AGGRESSIVELY

Figure 40-12 Key decision model for use in outsourcing and reducing service costs.

For OEMs and ISOs, the best option would be to develop strategies for significantly increasing the density of the base served. Given the significant technological similarity among medical electronics, computers, office automation, telecommunications, data networks, and building automation (all using microprocessor circuit boards and LAN-type communications), a significant payoff could be realized in terms of reduced costs, improved efficiency, and improved response time by expanding the portfolio of equipment served. Experience also shows that a significant improvement in call-handling efficiency and effectiveness using advanced call diagnostics and call avoidance tools can reduce the number of on-site calls significantly. Improved logistics control to reduce certain calls (e.g., when something is broken because of lack of parts) also will significantly improve productivity.

Summary

Based on extensive benchmarking comparison and market estimates, the issues of make, buy, or outsourcing are neither trivial nor easy to resolve in the health care field. However, based on the different perspectives involved, the following general guidance should apply:

- Hospitals and hospital-based clinical engineering service organizations: Small to medium hospitals should consider outsourcing of service to MVES TPMs and ISOs or OEMs who are willing to service the broader array of technology installed in order to achieve economies of scale, lower total price, and reduced response time (by positioning one or more technicians on site). Large hospitals and hospital chains should consider either outsourcing to a large multi service TPM or an integrated major OEM offering a broad service portfolio or facilities management capability. Larger hospitals and chains should also consider expanding their existing clinical engineering support groups into full TPM/MVES, service to increase their served base density.
- OEMs: OEMs should consider expanding their portfolio of served and supported technology to include a broader range of equipment as found in the new metropolitan hospital complex (Figure 40-1) and increasing use of computerized call management and call avoidance and integrating logistics control to eliminate up to 35%–40% of on-site service calls.
- ISO/TPMs: Independent TPMs should explore partnering, joint venturing, mergers and acquisitions to increase economies of scale and served base density. A critical opportunity for ISOs would be to move into fourth party depot repair to rehabilitate and refurbish older but still useful medical diagnostics and imaging equipment, selling at a significant discount from OEM *new* products, and generating profits from service and support. ISOs should also expand their product support portfolio and increase use of infrastructure and systems for call diagnostics and avoidance and logistics control.

Finally, the government itself could help significantly in reducing costs associated with high-tech service by more aggressively enforcing antitrust laws, particularly with respect to medical diagnostics, imaging, and instrumentation technology service. This alone could reduce health care costs by $1 billion or more annually.

References

Blumberg DF. The Strategic Implications for Field Service Industry of Eastman Kodak. *The Computer Lawyer,* 1992.

Blumberg D. Strategic and Tactical Optimization. The New Paradigm of the Service Executive. *AFSMI Journal,* September 1996.

Blumberg D, Quinn T. Leveling the Playing Field in Medical Technology Service and Support/ *AFSMI Journal,* December 1996.

41
Vendor and Service Management

Joseph F. Dyro
President, Biomedical Resource Group
Setauket, NY

This chapter describes ways in which the clinical engineer can foster excellent relationships with the medical device vendor and can effectively manage the vendor as service is provided for that device. A good working relationship with manufacturers is necessary to ensure proper after-sales technical support by way of such items as documentation, service, parts supply, upgrades, and recalls. On the other hand, it is wise not to be limited to only one or a very few suppliers nor to become too reliant on the manufacturers. The proper relationship starts at the equipment planning stage and not when service problems first appear. Ways in which the clinical engineer can relate to the vendor (i.e., the original equipment manufacturer [OEM]) at the planning stage will be described. Relationships with OEMs throughout the life of the equipment will be optimal if terms and conditions are clearly delineated at the procurement stage. The way to write an effective purchase agreement containing all the necessary conditions will be presented. Throughout the life cycle of the equipment, the responsibilities of the clinical engineer, such as performing thorough incoming inspection, and the responsibilities of the OEM, such as providing clear and comprehensive service manuals, will be described. If the OEM is to play a major role in equipment service, agreements for this should be arrived at during the procurement stage as part of maintenance planning and implementation. Examples of the OEM's expectations of hospitals and health authorities will be given.

After device acquisition and expiration of the warranty, the hospital must arrange for ongoing service for that device. This chapter will describe ways in which the clinical engineer can manage that service provider, whether that be the original vendor or a third-party service organization. Emphasis will be placed on managing outside service rather than in-house service. A concise review and critique of the different methods available to provide maintenance services to equipment purchased will be presented. The advantages and disadvantages of independent (nonmanufacturer) service organizations (ISOs) will be described. Possible ways to structure a maintenance program will be explored, with consideration given to the role played by the in-house maintenance department. The concept of types and levels of required service will be explored. Maintenance management techniques such as auditing existing agreements, identifying equipment serviced, identifying costs, and identifying services provided will be provided (ASHE, 1988). The way to select the most efficient and cost-effective overall management approach will be described. Suggestions will be made as to the way to communicate with and manage service vendors in order to obtain the best possible service at the lowest possible cost. Evaluation of program options should include, for example, identifying all possible sources of maintenance and evaluating advantages and disadvantages. Various types of service contracts, such as pay-as-you-go arrangements, shared risk and managed maintenance plans, and single contractor for all equipment will be explored. Effective cost control techniques will be presented.

Purchase Agreement Requirements

When procuring medical device technology, the time during which the buyer has the maximum leverage to negotiate the most satisfactory agreement is before the purchase. One's bargaining position is severely weakened after the acquisition. Therefore it behooves the clinical engineer to know what the vendor can supply. This would minimize the overall long-term cost of ownership of the acquired technology. The clinical engineer, as a participant in the technology acquisition process, must ensure that purchasing agreements contain, in writing, the following items, at a minimum:

Documentation: Two copies, including such articles as operation manuals, video tapes, service manuals (with theory of operation, detailed electrical and mechanical schematics, parts lists, and troubleshooting and preventive maintenance guidelines and procedures.

Software updates for the life of the equipment: Updates inform the buyer of the availability of upgrades at a minimum. Negotiations before purchase should strive to obtain these updates free of charge or at a reduced charge. Updates should be accompanied by all pertinent documentation.

Parts available for the life of the equipment: Inevitably, even the most reliable devices will fail at some point, for a variety of reasons including user error, physical abuse, adverse operating conditions, inherent design flaws, and random component failures. The useful life of any technology is extended when repairs can be made with replacement parts. Some of these parts may be proprietary and unavailable from other companies (i.e., "second source" companies that manufacture and sell parts, usually at substantially lower prices than the OEMs'). As a rule, second sources always should be considered for replacement parts in order to reduce the life cycle cost of the equipment. Not only should the parts be available from the OEM but they also should be readily available (i.e., able to be shipped overnight by any of a number of common carriers). Terms for maximum allowable shipping delays should be established as part of the purchasing agreement. Depending upon the criticality of the equipment, the time for shipment will vary from one day to one week.

Parts on consignment: The OEM should supply critical parts so that repairs can be made without delay, with the hospital paying for these parts when used. Working with the OEM can establish what these critical parts are and the quantities recommended. The inventory of consignment parts will depend upon the number of devices in the system and their critical nature.

Spares: For large systems such as physiological monitoring systems, the buyer should insist on acquiring spares. For example, if fifty bedside monitors are purchased, then several spare monitors should be acquired, either through purchase or through a loan agreement. Having a spare monitor to substitute for a failed monitor will ensure same-day remedy and will reduce down time significantly. This lack of critical devices could mean that a bed or an operating room or examination room remains unavailable, resulting in lost revenue for the hospital.

Exchange policy: Inquiries should be made concerning the OEM's policy on exchange of a device that has failed. As discussed above, in the case of large systems with multiple identical devices, spare devices will obviate the need to ship a replacement device. For one-of-a-kind devices of a critical nature, exchanging the item for another in the event of a failure will ensure maximum up time and will reduce the pressure to troubleshoot and to repair a failed device on an emergency basis.

Service reports provided by manufacturer's service representative: The purchasing agreement should stipulate that all interventions by the OEM's service representative be documented and that these reports be given to the hospital, to be kept with the device's service records. Legible service reports should detail the identification of the device serviced (e.g., model, serial number, and equipment control number), the full name of the service personnel, the time necessary for the intervention, the problem addressed, the solution implemented, the parts or subassemblies replaced or repaired, and the reasons for failure. In this way a complete history can be developed and can assist in troubleshooting future failures, or it can serve to document inherent design flaws that cause premature failures.

Warranty: An explicit statement of warranty is necessary. Negotiating with the vendor can result in extended warranties that will reduce the life cycle cost of operation of the technology. The systems, devices, subassemblies, and parts that are covered by any warranty should be explicitly stated, as should the conditions of the warranty. For example, in the event of obvious physical abuse of the equipment, the OEM might be reluctant to exchange or repair a failed device. Determining the contribution of physical abuse to the failure is affected by the conditions under which the device is expected to operate, however. When it is foreseeable that a device would be subjected to shock and vibration (e.g., a pole-mounted infusion pump colliding with a door jam) in the course of a typical hurried emergency application, the OEM should be responsible for the repair or replacement. It is the OEM's responsibility to anticipate extremes of operating conditions in the hospital environment when designing and manufacturing medical devices. Adequate environmental testing in the research and development stages of device manufacture should reflect the conditions that will obtain when the device is actually used.

The warranty period must begin after final acceptance of the device or system by the clinical engineering department. This is particularly important in the case of large installations in which implementation may extend over several months. Not until all systems are in place and are functioning according to previously agreed specifications should the warranty period for the whole system commence. Extending the warranty period for as long as possible is an effective way of reducing total life cycle equipment costs.

Environment: The OEM should supply documentation detailing the environmental requirements for the device, such as operating and storage ranges of temperature and humidity, electrostatic discharge withstood, electromagnetic interference susceptibility, and electromagnetic interference capability.

Facilities: Facilities requirement also should be stated, such as required electrical supply (i.e., voltage, frequency, and power consumption); water or gas pressures and flows; and air quality.

Service training: The buyer should require that hospital-based service personnel be trained in the troubleshooting and repair of the device or system. For large systems such as physiological monitoring systems, it is likely that clinical engineers (CEs) and most biomedical engineering technicians (BMETs) will be involved in service and therefore they all should be trained. On-site training is typically the most cost-effective for the OEM and for the hospital. For one of a kind device, such as an ultrasound imaging unit, it would be more advantageous for the hospital to send at least one (preferably two) CEs or BMETs to the manufacturing facility for service training. All training materials should be given to the trainees so that they, in turn, can serve as resource people for the further training of BMETs within the CE department. Redundancy in service capability is highly desirable.

Negotiations should strive to obtain such training as part of the overall purchase. At this time, commitments for ongoing training should be obtained from the OEM in the case of staff turnover or major upgrades to the original equipment purchased. The level of training should be equivalent to that which the vendor provides to its field service engineers and technicians and other representatives. The hospital should reserve the right to determine the individual who receives training and when such training should occur. The OEM should provide a schedule of available training courses and site locations. Obtaining this training as part of the purchase package will avoid training costs at a later date and will further reduce the life cycle cost of the device.

Specialized test equipment: The identification, cost, availability, and training requirements for specialized test equipment must be stated. The price of acquisition of this equipment at the time of the purchase agreement should be negotiated.

Diagnostic software: The identification, cost, availability, training requirements, and licensing agreements of diagnostic software must be stated. Negotiation is advisable because the cost of diagnostic software can be substantial.

Technical support line: A technical support line must be stipulated, indicating hours of availability and minimum response time to inquiries. Methods for communicating technical support should be ascertained, such as phone, facsimile device, e-mail, on-site presence.

Operator training: Provisions for training the operator of the device must be established in the purchasing agreement. The training schedule must accommodate the work schedules of the operators and must be done as soon as the device or system has been inspected and is readied for deployment or installation. It is preferable to begin training during the course of installation for those large systems that have a lengthy installation period. For such systems, a trainer might need to be present for several weeks during the time of system implementation and available for refresher training at least throughout the warranty period.

Vendor service considerations: The various options for vendor service must be understood. Any offer of service contract should detail the response time guarantees and whether response is on site or whether devices must be returned to the vendor. Costs should be established for service contracts typically to begin after the warranty period is over. If a service contract is the desired route to follow, establishing a price and locking in that price at this time would guarantee that a price quotation after the warranty period does not escalate. The service contract should state at a minimum the number and frequency of PM procedures, up time guarantees, and response time. The CE department might wish to take over all service after the warranty period. Time and materials rates (e.g., standard, overtime, and travel) should be established.

Service technician qualifications: Guaranteed presence of OEM technical support is particularly desirable in the case of a large-scale system. It is not unreasonable to require that a manufacturer's service representative be on site for a year if the system is particularly extensive, if the hospital has adequate physical space to accommodate the representative, and if the OEM and hospital agree that the representative can use the hospital presence as a base of operations for services other systems in the immediate locality. This is particularly helpful for the hospital and the manufacturer, especially in a large metropolitan area with many installed systems in area hospitals.

Incoming Inspections

Final payment must be withheld until all conditions are satisfied. The following is a checklist of items to consider at incoming inspection.

- Check for all items listed on the purchase agreement.
- Check for manuals (service and operators) and other documentation as described above.
- Check for accessories such as cables, probes, and remote controls.
- Check for safety of the device or devices and for safe installation (e.g., adequacy of mounting brackets for overhead lights or wall-mounted monitors). Do not limit incoming inspection to an electrical safety check on the laboratory bench. All features of a device's operation should be verified. Appropriate device operation should be confirmed in the environment in which the device is intended to function. Adverse device device interaction from, for example, electromagnetic interference or adverse environmental interaction such as excessive ambient light or noise that could deleteriously affect alarm visualization or auditory perception can be detected at this time.
- Check for proper operation by following the manufacturer's test procedures at a minimum. These may be adapted and incorporated into the CE department's procedures for periodic testing after installation. Many clinical engineering departments have procedures for incoming inspections that are more rigorous than those of some manufacturers. Special attention should be paid to the incoming inspection of technology that is just beginning its life cycle and that is being installed for the first time.
- Ensure that all users are trained by the manufacturer before first use on patients. In some cases, it is sufficient for training to be done by in-house personnel, such as clinical engineering or nursing education.

Only after the above have been done should payment be authorized.

Service Agreements

Provided that they are properly negotiated and carefully monitored, the use of a service agreement (i.e., a service contract) can be an effective and appropriate tool in the ongoing support of medical devices. It is to the vendor's advantage to enter into a service-contract agreement with the hospital. The vendor then would be in a position to have increased hospital exposure and contact with hospital personnel, thus increasing the chance of selling more equipment, upgrades of existing equipment, and recommending replacement of a competitor's equipment that also might be in use. With a regular presence at the hospital, the vendor can enhance customer relations and ensure a future revenue stream.

Service Contract Competition

Medical equipment maintenance is a major industry. In the United States, over $2 billion is spent on clinical equipment maintenance each year. Manufacturers look to their service divisions for making corporate profits. Many hospitals feel that equipment service is already too expensive, and they look for ways to reduce the cost of this service. As a consequence, support services are under increased scrutiny. All service providers are under increased pressure to deliver services at the best possible price for the hospital. Clinical engineering departments feel this economic pressure and must compete with original equipment manufacturers (OEMs) and independent service organizations (ISOs) for the business of medical equipment service. Equipment is in general becoming more reliable and less likely to fail, and it requires less preventive maintenance. Regulatory bodies, quasi-regulatory organizations, and standards organizations have recognized this increased reliability and have tended to relax requirements and recommendations for maintenance.

Determining the Required Service

The hospital must determine the service or maintenance that is needed on a case-by-case basis. Service comprises the following actions:

- *Corrective maintenance,* usually called repair, returns the device to its original function.
- *Preventive maintenance* is used to prevent failures.
- *Upgrades* typically advance the function of the device.

Maintenance comprises overhauls and refurbishing, corrective maintenance or repair, preventive maintenance, calibration, performance testing, quality assurance, safety testing, visual inspections, and user maintenance.

When assessing service requirements for a particular medical device, the following criteria must be considered:

• Maintenance requirements	• Manufacturer recommendations	• Regulations
• Device history	• Utilization	• Environment
• Device function	• Device risk	• Resources

Rationale for Purchasing a Service Contract

The hospital may be inclined to enter into a service contract under the following conditions:

- The hospital lacks the expertise to provide in-house service.
- The hospital lacks sufficient in-house resources to provide support.
- Users prefer the OEM (common with one-of-a-kind, high-tech devices such as clinical laboratory analyzers, catheterization laboratory units, and ultrasound diagnostic units).
- Support is included or may be required with purchase or lease of the equipment, especially clinical laboratory equipment.
- Product would be more expensive to support with in-house resources after considering the cost of training, travel, and special test equipment.
- The in-house service capability cannot guarantee the equivalent service response, particularly when only one CE or BMET is trained and available for service. Dependency on one person is less desirable than dependency on an entire company.

Types of Providers

Many options are available to the hospital for what provider can supply the above service including the OEM, distributor, third-party, or independent service organization (ISO) and in-house department. The following is a summary of the various organizations that can provide service, along with their respective advantages and disadvantages:

Manufacturer (OEM) Service

Table 41-1 summarizes the main advantages and disadvantages of service contracts offered by the original equipment manufacturer.

Independent Service Organization

There are many independent service organizations, varying sizes, locally, regionally and nationally. The advantages and disadvantages of the ISO are listed in Table 41-2.

Hospital-Based Service Organizations

Service organizations that are hospital-based or in-house are typically the clinical engineering departments. These organizations may be known within the hospital as "biomedical engineering," "medical instrumentation," "medical equipment service," or simply "biomed." Advantages and disadvantages of this service support option are shown in Table 41-3.

Payment/Coverage Options

Service providers (e.g., ISOs or OEMs) offer a wide range of options for service of medical devices. These include full-service agreement (FSA) contracts; limited hours (e.g., 8-5, Monday through Friday); limited service calls; preventive maintenance only; labor or parts only; depot; anticipatory (hospital clinical engineering department screening); and risk sharing.

Full Service Contracts

FSA contracts account for 5%–20% of the equipment-acquisition cost. FSAs may be purchased up front (i.e., capitalized). Long-term contracts are offered. They can be included in a lease or in a disposable (reagent) agreement. Table 41-4 lists the advantages and the disadvantages of a full-service contract.

Hospital Time and Materials

A hospital may chose to "go bare"; that is, not to have any service contracts. When service is needed, the hospital pays for the time and materials that are consumed in effecting a repair. Such services can be obtained from a wide variety of sources including OEM and ISOs. The advantages and disadvantages of time-and-materials coverage are shown in Table 41-5.

Table 41-1 Advantages and disadvantages of the manufacturer service contract

Manufacturer Service Contracts	
Advantages	Disadvantages
Resources (service tools, documentation, parts)	Cost
Arrangements can be built into the purchase	Agreements
Once on-site fast service	Coverage limited to manufacturer's products
	This is changing
	Service lock-ins due to technical design or company policy

Table 41-2 Advantages and disadvantages of independent service organizations, service contracts

Independent Service Organizations Service Contracts	
Advantages	Disadvantages
Lower cost	Quality variable
May be local	Stability
May be able to cover multiple brands or device types	Reliability
May be specialists in a niche area	Training and parts resources may be most limited

Table 41-3 Advantages and disadvantages of hospital-based service organizations

Hospital-Based Service Organizations	
Advantages	Disadvantages
Should be lowest cost	Staff usually are generalists
Fast response; no travel time or cost	Limited technical support
Mission is the same as the customer's	Access to documentation and parts is more limited
Might manage all institutional technology	Might not have skills to "fix" the customer

Table 41-4 Advantages and disadvantages of a full-service contract

Full-Service Contracts	
Advantages	Disadvantages
Fixed cost for good budget control	Usually the most expensive approach; up-front payment
Minimal paperwork; limited management effort after negotiation	Additional hidden costs
Perception of preferred customer status	Locked in/little flexibility/cannot second source
On occasion, service personnel are the best trained and most skilled	Difficult to quantify service received; little or poor documentation
Obviates need for possibly more expensive in-house service support	No incentive for users to control repair calls
May include free upgrades	Exclusions: OT, certain parts, limited calls, terms...
Complex contract may be difficult to negotiate	

Single Service, All Inclusive Contractor

Service agreements made with a single contractor offer the advantages and disadvantages listed in Table 41-6.

Traditional Maintenance Insurance

Maintenance insurance is a scheme for providing service for all hospital equipment at a fixed cost. While it has several advantages, it has disadvantages as well (see Table 41-7).

Managed Risk-Shared Savings

Shared savings programs are structured to differ from maintenance insurance primarily by their offering an incentive to the hospital to employ good maintenance practices. As in all of the other schemes described, this too has both its advantages and disadvantages as shown in Table 41-8.

Service Contract Terms and Conditions

The following is a checklist of terms and conditions to look for in the service contract:

- Cancellation clause
- Payment
- Access to equipment
- Exclusions
- Parts kits/stock
- Guaranteed prorated rebate
- Renewal
- Contract length
- Indemnification
- PM specification

Table 41-5 Advantages and disadvantages of time and materials coverage

Time and Materials Coverage	
Advantages	Disadvantages
Best documentation	Difficult to predict cost–pay as you go
Maximum flexibility	High ongoing management–considerable paperwork
Information allows management to understand service patterns and to take action; incentives	Vulnerable to vendor practices and the "Big Hit"
Self-insurance (bank account) helps	

Table 41-6 Advantages and disadvantages of single-contractor service agreements

Single-Contractor Service Agreements	
Advantages	Disadvantages
One-stop shopping	Limited flexibility
Good budget control	Many contract exclusions
Minimal paperwork and ongoing management	Difficult to address quality issues that occur because of lack of specialization and generalist approach, i.e., "Jack of all trades"
	Obtaining parts in a timely and cost-effective manner

Table 41-7 Advantages and disadvantages of maintenance insurance

Maintenance Insurance	
Advantages	Disadvantages
Good budget control	Reimbursement cash-flow delays
Good documentation	Rejected claims
Single contract can cover all clinical (and non-clinical) equipment	Administrative effort is significant
	No incentive to control cost

Table 41-8 Managed risk–shared savings programs advantages and disadvantages

Managed Risk–Shared Savings Programs	
Advantages	Disadvantages
Good budget control	Hospital must buy into program
Incentives for good maintenance practices	Shared savings returns
Single contract	Operations can be challenging
Documentation good	
Flexible	

Negotiating

Everything is negotiable. The following is a summary of the types of items that one should consider as being open to negotiation when entering an equipment purchase or a service contract agreement. One might find it helpful to create one's own contract. A contract addendum can be attached to the equipment purchase contract or the vendor's service contract that all must sign.

- Try to negotiate future service contract pricing at equipment purchase. (This allows for future budgeting, should you choose to sign a service contract after the warranty.)
- Choose payment terms that meet your needs, such as paying up front or in installments.
- Be sure to evaluate different service contract options (e.g., parts only, labor only, or preventive).
- Maintenance, normal business hours, full service.
- Where possible, consider several service contractors (not just the manufacturer).
- Prepare for vendors threatening that you will "have to wait longer for service if you do not have a contract."
- Beware that vendors might try to go directly to the device users in order to try to pressure you to sign a service contract.
- Consider the use of maintenance insurance.
- Examine closely what is included in the contract (for example, parts, hours, consumables, and up time).
- Negotiate your terms and conditions:
  - Vendor service representatives must sign in with clinical engineering upon arrival at the hospital.
  - Service representatives must leave copies of all service reports, including labor hours, travel hours, and parts used.
  - Establish and enforce vendor response times (by telephone and on-site).
- Cancellation clause
- Avoid automatic renewal
- Carefully consider committing to multiyear service agreements

Support Resources to Be Negotiated for In-House Service

The following are support resources that an in-house service organization requires in order to properly provide service to medical devices. During the procurement process, negotiations should result in these items being supplied along with the equipment purchased.

- Theory of operation/schematics
- Parts lists
- Troubleshooting guide
- PM procedures
- Diagnostic software
- Training: live, computer-based, video, or teleconference; on-site or factory
- Training (should be the same as vendor service representatives receive and should be available for the individual specified by the hospital)
- Service tools
- Spare parts
- Technical support assistance

Negotiate to reduce cost of service contracts by utilizing first-response screening and troubleshooting by in-house clinical engineering or other in-house personnel and by entering into a shared risk contract.

Manufacturer Service Information Requirements

When negotiating a service contract, the vendor must produce the following information:

- Call back, on-site response, and repair time
- Labor rate
- Standard, overtime, and travel rates
- Service staff
- Qualifications, location, number
- Contract
- Coverage attributes (e.g., number of PMs, uptime guarantees, cancellation, exceptions)
- Shared risk, insurance, biomedical screening and other creative arrangements

Partnerships with Vendors

A good relationship with the vendor will enhance the cost-effective utilization of the technology acquired. Understanding the company's dynamics and knowing its products and personnel yields more successful assessment activities and improves the overall acquisition process. Good relationships make it easier to resolve safety, installation, reliability, and maintenance problems. After the useful life of the product is over, a maintained good relationship invariably will improve the equipment replacement planning process. For much of the life of the product, the vendor is the primary support resource. Establishing a good relationship is the precursor to establishing a partnership with the vendor, which can benefit the hospital by increasing vendor support during the life of the equipment.

How to Establish Relationships

The clinical engineer begins to build a relationship with the vendor at the beginning of the acquisition process. To do this, the CE must be an active participant in the technology management process; i.e., the CE must be involved at the first stage, which is technology assessment (David and Judd, 1993; Bronzino, 1992). At that stage, the CE should come to know all the vendors of a particular technology and should begin to establish a good rapport with those representatives assigned to the hospital's account.

The representatives will come to regard the CE as a principal player in the acquisition process and will be more responsive to requests for information and could positively influence the course of future negotiations. Be cordial to the manufacturer's representative. Communicate regularly and expect the same behavior in return. Ask the right questions, express your needs, and look for the manufacturer to do the same. Establishing a good relationship will benefit both parties.

Actions to Avoid

The vendor must not be alienated. If possible, one should not limit oneself to a single supplier for all needs in a modality. A single supplier equates to no competition, which in turn means higher prices. When acquiring a large hospital-wide system such as a physiological monitor system, it is best to cultivate relationships with three of the prime contenders. While standardization of devices has many advantages, it carries with it the disadvantage of having only one vendor for the device type. Standardization within units or areas of the hospital (rather than hospital-wide standardization) will permit the acquisition of devices from different manufacturers and will maintain the hospital's competitive edge. Do not reveal all information to the vendor, and (especially) do not indicate that decisions have been made to purchase a vendor's equipment before a formal agreement has been reached, because this will adversely affect the negotiation process.

Manufacturer Relationship

Assessment Stage

During the technology assessment phase, the clinical engineer should learn the companies' strengths and weaknesses and should obtain valid assessment data of the companies' technology.

If the hospital has already decided at this stage to purchase from a single vendor (i.e., sole source) it is particularly important for the CE to know and trust the vendor. In order to interact well with the vendor, it is necessary to communicate with a good scientific background on the technology, participate in clinical trials, and thoroughly analyze the vendor's manufacturing capabilities and field service support.

Acquisition Stage

During the acquisition stage, quotations are submitted in response to a request for proposals (RFP). The manufacturer's response should be based on well-defined needs in the specifications established by the hospital and should not be based on extraneous features (sometimes referred to as "bells and whistles.") The manufacturer should reply to the RFP with accurate and complete responses. If the vendor perceives the possibility of a long-term financial relationship with the hospital, negotiations will be easier and could yield such items as software and hardware upgrades and additional units to retain as replacements in the event of device failure.

A good relationship with the vendor will enable a clearer view of attributes such as manufacturer support resources. The clinical engineer must know the equipment support requirements and the mean time between failures (MTBF). Good relationships promote compliance with requests for service documentation, service tools, diagnostic software, parts, technical support, and direct access to key service personnel and managers.

Service support planning begins prior to purchase. Building a relationship with the vendor early in the acquisition process improves the chances of favorable negotiations.

Vendor Characteristics

The CE and vendor should be comfortable in a give-and-take relationship. Vendors who have the hospital's interest and success in mind are preferable for long-term relationships. Vendors come in all sizes and shapes. Some are good and some are bad. One should cooperate with and help to build up the good ones that exhibit the following characteristics:

Low down time	Good technical skills	Good communication
Respect for the customer	Good reputation	Dependability
Information provided when requested	Few callbacks and problems	Quick repairs

Bad vendors have the following characteristics and should be avoided:

No (or inadequate) service documentation	Remote support only
Manufacturer parts only	Will not work with maintenance insurance, third parties, or in-house support
Noncancelable service contracts	Upgrade requirements for support
Sole vendor for department	Do not return telephone calls

Techniques for Establishing Vendor Partnerships

The first step in establishing a vendor partnership is to agree that a partnership is in the best interests of the vendor and the hospital where both can achieve financial advantages. A good partnership begins with both parties gaining respect for each other. Good communication skills will foster each understanding the other's needs and the identification of what is to be accomplished through the partnership. When committing to shared success, it is advisable to make outcomes measurable, to establish problem-resolution methods, and to meet commitments.

One should understand that the manufacturer typically needs to earn the profit margin and to obtain the hospital's business. While obtaining the list price is the goal, the better, more flexible, and creative vendors will be willing to negotiate. Be aware that although service and sales may be separate lines, vendors will be willing to derive less immediate profit from sales in the face of greater profits through a service contract. Some vendors are providing multivendor service, so their interest in service goes beyond the equipment under consideration and includes other hospital equipment that might require a service contract that the vendor can provide.

After the equipment purchase, good manufacturer-CE department communication is needed for planning, coordination, and implementation of the installation, training, incoming inspection and acceptance testing, and ongoing support planning for the operational system.

Several examples of partnerships are illustrated in the following:

Case Study Number One

A rental equipment company partnered with a hospital in the following way: The company established a presence at the hospital. The company owns equipment. It tracks, processes, and charges on the hospital's computer. The company follows the procedures established by the hospital CE department and performs all inspections, preventive maintenance, and repairs; documents work performed; and reports results to the CE director. This scheme benefits both hospital and company. While they both share the risk, they both benefit from administrative and technical service cost savings. Their on-site presence further benefits the patient as it fosters appropriate utilization of the technology.

Case Study Number Two

A major manufacturer of physiological monitoring equipment had an installed base valued at $2 million in 30 hospitals that are under service contracts to a large, regional clinical engineering shared service organization in the northeastern United States. The equipment had been purchased by the various hospitals over the previous three years. The manufacturer had adopted a new philosophy of outsourcing service in order to reduce its service staff and costly support resources. The manufacturer and the ISO struck an agreement whereby the manufacturer would provide on-site service training for all CE staff and would place parts kits on consignment at the ISO site. The ISO, for its part, charged the manufacturer a reduced rate.

Case Study Number Three

A leading California-based service organization entered into a long-term partnership with a large Vermont-based service organization. The Vermont organization required cost-effective answers to maintenance problems on large, sophisticated, complex imaging and laboratory systems. The California group had the expertise and infrastructure to help them accomplish this and desired an expansion to the eastern United States as well as increased income. The Vermont group was able to continue to provide comprehensive management, sales support, and reporting for the California group.

Managing Service Contract Vendors

Managing service contract vendors and managing one's own in-house clinical engineering department share many similarities. Both entail negotiation, knowledge of services to be provided, monitoring and measuring those services, and analysis and reporting of the results in order to obtain the most cost-effective service.

The goal of any service organization that manages an equipment maintenance program is to provide equally good or better service at a lower cost than that previously or currently provided. The service organization, whether in-house or ISO, must support the hospital's mission; equipment must be available for patient care; service must be cost effective; customers such as physicians, nurses, and administrators must be satisfied; and safety concerns and regulatory requirements must be met. It is typical for one hospital in developing a system of service support to mix a variety of service options such as combining an in-house program with several OEM service agreements.

Centralized Clinical Equipment Maintenance Management

A centralized clinical equipment maintenance management system is recommended for several reasons. Such centralization achieves compliance with standards of the United State Joint Commission on the Accreditation of Health care Organizations and the regulations under the U.S. Food and Drug Administration Safe Medical Devices Act. Typically, over 30% savings over decentralized CE and service contracts are realized. Centralization is a concept that is well understood by bottom-line-orientated administrators who have the perception that equipment service is a commodity.

Over the last decade the trend in hospitals toward outsourcing the service of all clinical and nonclinical technology while maintaining management control gave CE departments (which were in the best position to manage technology) an excellent opportunity to assume this responsibility while achieving significant cost avoidance. CE departments, which already were in synchrony with the hospital's mission, thus added imaging, clinical lab, and other areas to their scope of responsibilities, thus demonstrating that the in-house organization is in the best position for cost avoidance. The utilization of a computerized database system for overall management as well as other administrative and technical tools has enabled on-site service organizations to provide cradle-to-grave management of technology. The success of in-house programs of this sort owes a great deal to the clinical engineers trained as problem solvers and equipped with the skills and knowledge to provide total technology management and service support.

Implementation Basics

The clinical engineer must make key decision makers in the institution aware of the opportunity for cost reduction. To begin, the clinical engineer must perform an audit to determine inventory, expenditures, and coverage. After analyzing the data, the plan should show the changes in operations that will yield improvement in clinical equipment support management. The plan must clearly delineate the mechanisms of monitoring and reporting, the value obtained, the improvements over the current process, the appropriate

service levels for each device in the inventory, the hours of service, the type of services provided (e.g., PM, full-service, or parts only), the response time and completion time, and compliance with regulations and liability issues.

Management Data

The basic rule of management is that one cannot manage what one cannot measure. Therefore, accurate records of such things as time worked, travel time, parts used, and parts costs are essential. An accurate inventory is always the starting point. The maintenance source then must be matched with each item in the inventory. The cost, coverage, exclusions, term, and uptime provisions of all service contracts must be determined. Accounts payable data must be obtained in order to capture expenses for parts, external labor, and other details. All other arrangements, such as leases, must be considered. The result of this analysis is the quantification of in-house costs.

Breakdown of Medical Equipment Maintenance Expenses

A typical management report would include such a graphic as shown in Figure 41-1, the expenditures of the principal users of medical device technology by percent of the total costs.

Several guidelines that will help the clinical engineer to know whether costs are in line with national averages have been developed. For example, the following are estimates of the maintenance costs ($/bed) for hospitals of various sizes:

0–75 beds = $1200/bed
76–200 beds = $1500/bed
201–400 beds = $2250/bed
>400 beds = $2800/bed

Cohen (2002) maintains that the best measure of service cost is a percentage of the total technology expenditure acquisition cost. This averages to about 5% for large, U.S. hospitals with CE departments.

Additional factors must be taken into account when managing equipment service, such as customer satisfaction and the quality of service. Such information can be obtained by means of customer surveys. Current service arrangements that have flexibility are targets for additional cost reductions.

Inventory Audit

An inventory audit must be comprehensive and must include all technologies. Data must be obtained for any item that must be managed. Data must be presented in a standardized nomenclature and must include, at a minimum, such items as manufacturer, model, serial number, acquisition cost, location, status and condition of the technology, maintenance provider, type of maintenance (from full-service to electrical safety testing), and a program identification number.

Service Contract Analysis

Service contracts may be obtained for analysis from various areas, including facilities, purchasing, clinical departments, and accounts payable. The type of data should include service provider, term of contract, purchase order, annual cost, type of coverage, special notes, and exclusions and inclusions. Cruz et al. (2002) have developed methodologies for assessing the quality of service of ISOs and for reducing the costs of these services.

Time and Materials Analysis

Accounts payable is typically the source of data about time and materials expenditures. Electronic databases and filing cabinets are the usual repositories of such reports. All items that are not on contract require knowledge of the maintenance account code or vendor code. This process can be time-consuming because one must look at each record to confirm the data. One will find many non-maintenance items such as training, accessories, and rentals. Some expenditures will be missed.

Other In-House Costs

An accurate assessment of the CE department's costs would include cost of personnel, benefits, equipment, operations, space, utilities, advertising, and training. Services should be

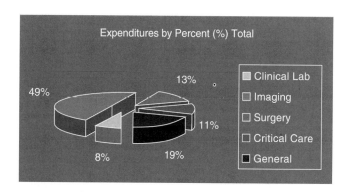

Figure 41-1 Expenditures of various clinical departments as a percentage of total hospital cost.

matched to specific devices. Other costs would include such items as maintenance insurance, administrative costs (e.g., the cost to generate a purchase order), and management costs.

Analysis and Proposal

The following are many of the options that are available to the clinical engineer who manages the hospital's medical technology service:

- Review all data and determine the best mix of service providers.
- Eliminate service contracts where possible.
- Change service providers.
- Research ISOs in your locality or region.
- Move service contracts to maintenance insurance.
- Keep items under service contracts or add.
- Critical equipment, service resources not available, proprietary, long-term, cost
- Add and delete items to be maintained.

What is Best Method of Providing Medical device technology Service?

In general, the best option is a mix of in-house service, vendor time and materials, service contracts, and managed maintenance insurance. Annual records of the relative costs of each of the service options selected should be kept. Such data enable analysis in order to determine the optimum mix of service providers.

Monitoring Vendor Performance

Managing service-contract vendors requires the monitoring of vendor performance by such vehicles as auditing of service reports and sign-in logs. Ongoing communication with service representatives is mandatory, especially when devices fail repeatedly. Monitor device-user and service-user satisfaction through surveys and communication. Best vendor service management is obtained when clinical engineering has the authority to sign vendor payments. Contract expiration should be anticipated, as it will provide an opportunity to review the past service in preparation for negotiating a new contract or canceling the old one in favor of another vendor or using the current vendor under a different contract agreement, such as time and materials (cost per hour, plus parts cost). All of this requires keeping accurate records of time worked, travel time, parts used, parts costs, and periodic summaries.

Effective Cost Control Techniques

To effectively control the cost of services provided by any organization, whether it is the OEM or the in-house clinical engineering department, the guidelines listed below should be followed.

Determine the Role of the In-House CE Department

First the CE department must establish that it is the logical choice for managing medical device service contracts and managing the vendors from whom the hospital has purchased medical devices. This starts with an audit of the current activities and resources in CE. Determine the CE department's strengths, weaknesses, obstacles, and threats (SWOT analysis) See Chapter 138 for a discussion of SWOT analysis.

Technical, management, and information systems resources are required along with the ability to form partnerships with vendors. A proposal should be in the form of a business plan (see Chapter 140). The plan should include the following elements:

- Current activities
- Proposed changes
- Investments and return
- Risk-shared contracts/guaranteed costs
- Implementation plan
- Why the CE department is the best choice to manage
- Potential problems with outside vendors
- Ongoing monitoring
- Quality assurance

To gain the hospital's support in obtaining resources, the CE must have a record of quality service, must demonstrate good relationships with clinical and administrative staff and manufacturers, must show value in the acquisition process by defining requirements in the RFP, and must provide valuable decision making information. The CE must define the quality and cost advantages of building CE department support into the purchase.

The CE department should consider an internal contract which defines the requirements for CE and clinical departments, objectives, measurement tools, methods of communication and problem resolution, and reporting costs to user departments. Such a plan will reduce the tendency to point the finger or assign blame for problems and put more attention on patient care.

Cost Control by Vendor Management

The following are various processes and tools for controlling cost by vendor management:

- Service requests
- Who will call vendor? (Department or CE?)
- Logging call: Get all information, open a work order
- Flag high-cost service events and look for options
- Alternate parts, the device worth fixing
- Effect of service on clinical practice
- Reviewing service reports and invoices
- Do not pay for training the vendor's staff
- Parts changes: Only pay for the one that fixes the problem
- Was the problem fixed the first time? Take special note of recurring problem reports or service requests because they indicate that inadequate or incomplete service was performed.
- Timely and accurate submission of insurance claims
- Researching and keeping current on alternative service sources
- Communication with peers to develop new strategies
- Group contracts
- Managing in-house technical staff in their role
- Procedures

Monitoring and Reporting

Service management requires regular monitoring and reporting. The communication process is an effective way to demonstrate the value of clinical engineering and to demonstrate managerial and technical skills. Such indicators as those listed below are typically included in the monitoring and reporting process.

- Cost-based reports
- Budgeted expenses, actual expenses, and variance
- Phase-in of changes
- Additional opportunities
- Quality reports
- Annual or semi-annual customer survey
- Down time or the time taken to return to service

The Maintenance Consultation

In an attempt to establish the value of vendor services, one must be capable of asking good technical questions. Establishment of a friendly relationship with the vendor enhances the flow of information. Questions such as "will the vendor install a hospital-furnished part?" and "is equipment worth fixing?" should be among those posed in the course of attempting to control the cost of vendor service. Vendors might need to be reminded about replacing swapped boards. They also might need to be asked to provide an exchange price for part(s).

Reviewing Service Reports

On occasion, a vendor's service report is poorly written, illegible, inaccurate, missing information, and unsigned by the recipient of the service. Reviews of service reports have revealed duplicate billing; invoices that do not match service reports; inconsistent travel charges; billing for parts not replaced; billing for unnecessary reports; overcharging for parts; tax added on labor; and billing for time on site, inconsistent with the standard time actions performed required. Effective cost control requires that the vendor comply with some basic requirements when completing a service report such as legibility and accuracy.

Reviewing Maintenance Histories (Trends)

The clinical engineer will find ways of reducing costs by reviewing maintenance histories. For example, the need for user training might be indicated when the same problem recurs with atypical frequency. Should problems recur and continue to go unsolved, the clinical engineer should suspect an inherent design fault that even the factory-trained technician is unable to detect. This is cause for requesting warranty credits or a reduction in service charge or replacement under extended warranty of the failing device.

Other Cost-Containment Strategies

Several other cost-containment strategies are listed below:

- Avoid "lock-in" contracts (i.e., those that cannot be canceled without a substantial penalty).
- Perform an end-of-year trend analysis.
- Utilize the proper skill level for the problem.
- Encourage and help the equipment users to help themselves by basic troubleshooting and an understanding of the theory of operation of a device.
- Plan for maintenance at the time of purchase.

Conclusion

This chapter has detailed the essential elements and requirements of a purchase agreement. Incoming inspections have been shown to be a necessary step to ensure that all equipment meets with specified requirements and that authorization can be given to pay the vendor for the delivered product. The various types of service agreements and equipment service providers have been described. The advantages of partnerships with vendors have been presented, along with techniques for fostering those partnerships. The role of the clinical engineer as the appropriate individual to manage service contracts and vendors has been emphasized. The essential elements of an equipment-control program have been reviewed. Finally, effective cost-control techniques have been described.

References

ASHE. Maintenance Management for Medical Equipment, American Society for Hospital Engineers, Chicago, IL. 1988.

Bronzino J. *Management of Medical Technology: A Primer for Clinical Engineers.* Butterworth-Heinemann, Burlington, MA. 1992.

Cohen T, Cruz AM, Denis ER, Puñales EP. Management of Service Contracts Using an Independent Service Provider (ISP) as Support Technology. *J Clin Eng* 27(3):202-209, 2002.

David Y, Judd T. *Medical Technology Management.* Spacelabs, Redmond, WA. 1993.

42

Health Care Technology Replacement Planning

J. Tobey Clark
Director, Technical Services Program
University of Vermont
Burlington, VT

The health care technology planning process includes assessment of new technologies, acquisition protocols to purchase and introduce effective devices and systems to the hospital, and methodology to objectively and proactively replace equipment. All equipment reaches the point in its life where the cost-benefit ratio goes to the negative because of decreased reliability, increased downtime, safety issues, compromised care, increased operating costs, changing regulations, or simply obsolescence. At that point, replacement must be considered.

Many hospitals respond to health care technology replacement requests with "knee-jerk" reactions, thus causing immediate, unplanned, and unbudgeted replacement of expensive technology. Examples of reactive replacement requests include the following:

- When the device fails at a critical time, even though it might have been reliable in the past
- When the physician returns from a conference and states, "The current technology is obsolete and must be replaced."
- When the department manager complains, "My equipment never works right" at the capital budget meeting and then "politics" for new equipment
- When undergoing repair, it is found that parts and support are no longer available

Replacement under these conditions can be avoided through a systematic plan to replace equipment in a prioritized fashion (Clark and Forsell, 1990). Health care organizations have limited funding for capital purchases, with many under strict budgeting guidelines. Replacement based on anecdotal and subjective statements and politics creates havoc related to finances, morale, and operations. Clinical engineering departments are impacted if they do not have a replacement plan when major repairs occur in older equipment.

The ideal health care technology replacement planning system would be facility-wide, covering all clinical equipment; would utilize accurate, objective data for analysis; would be flexible enough to incorporate nonequipment factors; and would be futuristic by including strategic planning relating to clinical and marketplace trends and hospital strategic initiatives relating to technology. The plan should encompass many factors relating to cost-benefit analysis, safety, support, standardization, and clinical benefit.

Methodology

Data

One key to developing a robust equipment replacement plan is accurate data for analysis. The equipment inventory is the starting point. A correct listing of the medical device inventory in a hospital is essential. Work orders performed on equipment should reference the unique identifying number in order to tract and report on medical device history.

External factors also should be available to help the equipment planning process. Table 42-1 is a comprehensive list of factors to evaluate in determining equipment replacement.

Analysis

An analysis program can be used to interrogate the equipment management database to list a "first pass" listing of items for replacement and scoring. The analysis should cover all clinical equipment on the health care systems inventory. Factors, sample scoring, and weights are shown in Table 42-2. Some are absolute factors calling for urgent replacement, such as serious safety issues or unavailability of parts. Other factors might have formulas associated with the data, which are used to determine whether a threshold has been exceeded. Examples would include the following:

- Maintenance costs over the past three years exceeding 25% of the purchase cost of the device
- Age >125% of AHA equipment life (Arges, 1998) or U.S. Army (U.S. Army, 1992)
- Failure rate outside of one standard deviation for like equipment

Once the first pass list is generated, qualitative factors shown in Column 3 of Table 42-1 would be added to the analysis. Factors including regulatory prohibition or significant cost benefits might call for immediate action. Scoring or adding in these factors requires professional clinical engineering staff making assessments and adjusting the scoring based on their knowledge and communication with administration, clinicians, BMETs, and others. Reduced costs or improved revenues, strong user preferences, and clear clinical advantages are some factors that would put a device or system into a high-priority category for replacement. Various algorithms have been developed for this analysis, some using a simple scoring system and formulas (Fennigkoh, 1992), and others using more complex systems that employ predictive cost, reliability, and parts-availability data and/or analysis (Ridgway, 2002). For the most expensive items (>$250,000), such as imaging systems, laboratory equipment, and radiation therapy, a life cycle cost (LCC) analysis should be performed to assist with the replacement decision. (See Chapter 30.)

Replacement of an item with a score above the threshold might not be the only option. For some items, an upgrade might solve safety or performance issues, which would allow the hospital to keep the equipment in service. A second option would be the transfer of the device to a less critical application.

The equipment item removed from the department can be disposed, sold to a used-equipment dealer, donated for humanitarian purposes, used in a research lab, or "cannibalized" for parts. Also of note is that not all items removed from service must be replaced. If utilization or need is low, the item might not be replaced, thus saving the hospital the cost of device testing, maintenance, space/utilities utilization, and training.

To be effective, the equipment replacement planning recommendation should prioritize items for replacement into the following categories:

- Urgent—immediate
- Priority 1—next fiscal year
- Priority 2—two fiscal years from current
- Priority 3—three fiscal years from current
- Advisory—product-line changes, published end of support, regulatory (e.g., medical telemetry frequency change)

Communicating Results

The data are best presented in a report showing a facility-wide replacement plan to administration and a separate report used to show a "hit list" for individual departments. The reports should be easy to read and should show inventory item, justification for replacement, and replacement cost. The department report that lists recommended replacement should be presented a month ahead of budget submissions by department heads, to allow review and discussion. It should be clear that the equipment replacement report is a recommendation that provides guidelines and generates discussion and that never should be considered a mandate by clinical engineering.

Once the replacement plan has been finalized, the agreed-upon replacement date should be loaded into the equipment management database to allow proactive actions, should maintenance be required, or should safety issues occur. For example, if an item were listed as being replaced the following fiscal year, a "red flag" would alert BMET staff when a work order was opened that major repairs should not be undertaken.

Conclusion

Equipment replacement planning is an important part of the technology planning process. Immediate benefits include dramatically reducing emergency purchases of replacement equipment and improving the safety and effectiveness of clinical technology. Administration (including finance) and clinicians have a better appreciation of the value of clinical engineering through the objective and rational recommendations made in the equipment replacement report.

Table 42-1 Information requirements for equipment replacement planning

Equipment inventory data	Work order data	External information
• A unique identifying number • Standardized device type, manufacturer, model, and serial number • Device risk and function • Safety features • Purchase date/age, P.O. number and cost • Upgrade status (e.g., software and hardware) • Service provider • Service/support documentation, tool and software availability • Equipment replacement schedule based on national data (e.g., AHA or Armed Forces) or depreciation • Replacement cost • Manufacturer end-of-support date • Condition assessment by technical staff	• Incoming inspection parameters including date accepted, safety and performance values, and pass/fail results • Failure rate and trends • Failure type • Unresolved failures • Down time • Costs for parts, labor, contracts, and other work on the equipment • User problems such as user errors, reports where no problem was found, and abuse • Recalls, alerts, incidents, and other safety-related information	• Standards of care for techniques where the technology is utilized • Technological status: emerging, current, mature, or obsolete • Utilization data for the equipment • Standardization in the facility by equipment type • Reference data and studies documenting cost advantages of replacement of existing technology • Marketing, competition, and strategic planning information • Regulatory changes • Alternative service sources and capabilities • Vendor data on upgradability and production • Backup equipment availability

Table 42-2 Analysis of replacement factors

Factor	Scoring	Weight
Age	>125% AHA life = 1 Depreciated = 1	Low
Risk/function	Life support/critical/income producing Non-critical	Medium
Support status	Ended/will end = 1 (unless reliable alternative exists or spare parts)	Absolute (date of end of support = year for replacement)
Condition assessment by BMET staff	Fair = .5 Poor = 1	Medium
Safety issues	Unresolved safety issues = 1	Absolute
Failures	# Events > mean plus standard deviation = 1 Unresolved failures = 1	Medium High
Down time	>10% = 1	Medium
Maintenance costs	>25% of purchase cost over past 3 years = 1	Medium
User problems	# Events > mean plus standard deviation = 1	Medium
Recalls and alerts	# Recalls/alerts > mean plus standard deviation = .5 Unresolved safety problems = 1	High
Standard of care	Does not meet = 1	High
Technological status	Obsolete = 1 Mature = .5	Medium
Utilization	Little or low utilization = −1	Medium
Standardization	Does not conform to standard/cost savings available = 1	Medium
Cost savings/revenue increase	Both = 2 One factor = 1	High
Regulatory	Does not meet regulations = 1	Absolute (date of regulation = year for replacement)
Product status	No longer made/no upgrades available = 1	Low
Backup equipment	Sufficient backup equipment available = −1	Low

References

Arges GS. *Estimated Useful Lives of Depreciable Hospital Assets.* American Hospital Association. American Hospital Publishing, Chicago. 1998.

Clark JT and Forsell R. Clinical Equipment Replacement Planning. Biomedical Instrumentation and Technology, Hanley & Belfus, July-August 1990, pp 271-276.

Fennigkoh L. A Medical Equipment Replacement Model. *J Clin Eng* 17(1): 43-47, January-February 1992.

Ridgway M. *Preliminary Equipment Replacement Planning (PERT) Report.* Presentation at the Advanced Clinical Engineering Workshop. San Jose, Costa Rica, February 28, 2002 U.S. Army. Maintenance Expenditures Limits for Medical Material—TB MED (http://www.armymedicine.army.mil/usamma/maintenance/tech-man.html), June 1992.

43

Donation of Medical Device Technologies

Joseph F. Dyro
President, Biomedical Resource Group
Setauket, NY

Most charitable organizations donate medical equipment that is unsuitable for its intended use (ECRI, 1992). Most recipients receive donated medical equipment that is unsuitable for the institution. Donors fail to ensure that donated equipment is functional, meets standards of safety and performance, has adequate spare parts and accessories available, and is the type of technology that the recipient wants, needs, and is able to operate and support. Most recipients do not adequately describe their medical requirements, available financial support, facilities preparation, or available utilities. By following checklists that accompany guidelines for donation and receipt of medical equipment, substantial waste and inefficiencies can be eliminated. Guidelines for the donation of medical equipment are described in this chapter. The American College of Clinical Engineering (ACCE), the World Health Organization (WHO), and several other health care organizations have been leaders in the development of donation guidelines (ACCE, 1995; WHO, 1997; FAKT, 1994).

The Current State of Medical Equipment Donation

Rationale for Donations

For decades, developed countries have donated used medical equipment to less-developed countries. This recycling of goods accomplishes several important objectives. In developed countries, it helps to keep hospital corridors clear of obsolete equipment and reduces the amount of waste that ordinarily would go to landfills. Frequent equipment turnover also increases the rate of introduction of new technologies. Not insignificantly, donations give a benefactor a good feeling and a belief that that needy people in some faraway country are being helped. Financial advantage through tax deductions also contributes to a donor's sense of well-being. In developing countries, physicians gain quicker access to sophisticated technologies and, most importantly, less-privileged patients gain wider access to better health care (Muschlitz, 1996). In some countries, nearly 80% of health care equipment is funded by international donors or foreign governments (WHO, 1987).

Rationale for Guidelines: Frequently Encountered Problems

Unfortunately, not all donations achieve their goals. Much donated equipment is not or cannot be used by the recipients. It is estimated that as much as onethird of all donations do not achieve their eventual goals, wasting precious time and resources. In Sub-Saharan Africa, up to 70% of equipment lies idle due to mismanagement of the technology acquisition process, lack of user training, and lack of effective technical support (Bloom and Temple-Bird, 1990). Many factors contribute to this reality. Some donors are so anxious to get rid of their unwanted hardware that they pay little attention to the equipment's condition, availability of parts, documentation, supplies, and operator training in the recipient country. In a typical hospital's obsolete equipment storage area, only one in ten devices might be functional. Perhaps more ominously, in some cases, donated equipment has been recalled in the donor country by the medical device regulatory body of that country. Some international relief organizations are forced to concentrate on volume rather than on the quality of donated goods in order to gain publicity and to please corporate donors.

On the other side, many recipients do not carefully screen what they ask for nor invest enough time and resources to plan and support what they get. Sometimes they are spoiled by the notion that they always can ask for another one and can discard what they do not want or failed to maintain. Finally, the lack of communication between the donor and recipients before the shipment of goods is probably the single most important reason why many donations do not work out well.

In most cases, donations circumvent the selection and procurement systems of the recipient country, where such systems exist. As a result, little consideration is taken of actual local requirements, the number of user staff and their capability, and the level of technical expertise of available maintenance personnel. Further problems relating to equipment calibration and operation, purchase of consumables, and availability of spare parts can transform the donated equipment into a liability rather than an asset to the recipient institution.

Some used equipment is repaired or refurbished before it is donated. Care must be taken, however, and it must be realized that beauty often is only skin-deep (Wang, 1996). Figure 43-1 shows what is called the "Spray and Pray" refurbishing method of painting a device to make it look new without doing adequate maintenance or repair of the item.

Guideline Development

First Efforts

Over the last decade, many organizations in industrialized and developing countries developed guidelines on equipment donations. Some of these are listed below:

- Christian Medical Association of India (CMAI): *Study Report on Medical Equipment and Supplies and the Role of CMAI* (1989).
- International Medical Device Group (IMDG): *Policy Position on Donating and Selling Medical Equipment* (1992).
- Association for Appropriate Technology (FAKT), Germany, and Churches' Action for Health (CMC) of the World Council of Churches: *Guidelines on Medical Equipment Donations* (1994).
- Ministry of Health of Tanzania: *Guidelines on Donations of Drugs and Medical Equipment to the Health Sector* (1995).

The above guidelines and other similar efforts, although addressing the same problem, treated it from different angles, going into varying degree of detail, focusing on different levels (i.e., national or international), and targeting different audiences (i.e., donors or recipients, governments or non-governmental organizations (NGOs). Hence there was a common concern expressed on numerous occasions over the fragmentation of efforts, lack of uniformity and comprehensiveness in presenting the issue, and unmet need of developing countries to have a managerial tool to properly address donations getting the maximum benefit from them (PAHO, 1997).

In 1995, the American College of Clinical Engineering (ACCE) published the first comprehensive managerial tool to aid in the donation process: *Guidelines for Medical Equipment Donation* (ACCE, 1995). Drawing largely upon the work of ACCE, the World Health Organization (WHO), in collaboration with several other organizations listed below, drafted guidelines: *Guidelines for Health Care Equipment Donations, 4th Draft* (WHO, 1997), which include additional management tools.

American College of Clinical Engineering

In 1993, the American College of Clinical Engineering formed a committee to discuss and to find means to improve the effectiveness of donations of medical equipment. The committee members, all of whom had significant international experience, were: Alfred G. Jakniunas, Chairman; Joseph F. Dyro, Ph.D., CCE; David Harrington, CCE; Thomas Judd, ME, CCE; Denver Lodge, ME, CBET, CCE; Robert Morris, CCE; and Binseng Wang, Sc.D., CCE.

By adopting these guidelines, donor and recipient organizations can expect to realize the following benefits:

Figure 43-1 Spray and Pray refurbishing method.

- Better matching of need and availability
- Improved pre-donation planning and preparation
- Assurance of completeness and quality of donated goods
- High likelihood that received equipment will be installed and used
- Assurance of maintainability of donated equipment
- Continuous quality improvement through follow-up evaluations

These guidelines, described below in detail, are divided into five sections. The first section, "Suitability for Donation," helps the potential donor to screen out equipment that should not be made available for donation. The second section, "Evaluation of Potential Recipients," is designed to help the donor to find the right recipient through a careful evaluation of the recipient's clinical need and its resources for operating and maintaining the donated equipment. The third section, "Pre-Donation Planning," helps the donor and the recipient to plan and prepare for the donation. The fourth section, "Donation Implementation," provides reminders for ensuring a successful transfer of goods, including assembly, packaging, shipping, documentation, customs clearance, unpacking, and installation. The fifth and final section, "Follow-Up Evaluation," is devoted to an evaluation of the each donation so that both parties can learn how to avoid past mistakes and how to improve future transactions.

These guidelines are focused only on medical equipment and do not address other medical devices such as disposables, *in-vitro* diagnostic kits, or reusable accessories. As special precautions are needed in handling those devices, potential donors and recipients should seek expert advice from clinical staff or manufacturers.

World Health Organization

The World Health Organization, under the leadership of Dr. Andrei Issakov, reviewed its field experience, existing guidelines, and related documents to develop a globally acceptable document that is flexible and adaptable to a variety of country situations and conditions (Issakov, 1994). The document is a result and culmination of joint efforts of several international and national organizations and individuals who are concerned with problems of health care equipment management in general and health care equipment donations in particular. Collaborating with WHO were the following organizations:

- African Federation for Technology in Health Care (AFTH)
- American College of Clinical Engineering (ACCE), USA
- Association for Appropriate Technology (FAKT), Germany
- Centre for Health Technology, Cameroon
- Churches' Action for Health (CMC), World Council of Churches
- International Medical Device Group (IMDG)
- Medical Research Council (MRC), South Africa
- Technical Cooperation Agency (GTZ), Germany

WHO propounds the following core principles for equipment donations:

- Health care equipment donations should benefit the recipient to the maximum extent possible. This implies that all donations should be based on an expressed and validated need and that unsolicited donations are to be discouraged.
- A donation should be given with full respect for the wishes of the recipient and their authority within the health system and should be supportive of existing health policies and administrative arrangements.
- There should be no double standards in quality: If the quality of an item is unacceptable in the donor country, it is also unacceptable as a donation.
- There should be effective communication between the donor, the recipient authority, and whenever possible the end user before, during, and after the donation.

Suitability for Donation

This section on suitability for donation and the following sections comprise a description of the donation process, taken directly from the ACCE guidelines.

Prior to making equipment available for donation it is crucial that the potential donor make a critical evaluation of it. It is not only a waste of precious resources to move useless and unsafe equipment from one place to another, it also undermines the good will and trust that everyone is trying to build.

General Quality

The donor should ensure that donated medical equipment is fully operational at the system and subsystem levels. All essential accessories and supplies should be available. The donor should follow a checklist to ensure that all subsystems, components, accessories, and supplies (for initial operations) are included, and they should supply the recipient with such a checklist. (Checklists are often found in manufacturers operating manuals and can be prepared by the former operators.) Documentation, especially operating and service manuals with part lists, is critical to the eventual usability of the donated equipment.

Safety, Specifications, and Standards

All medical equipment should meet or exceed existing safety and performance specifications provided by the manufacturer. In addition, they should meet standards promulgated by the American National Standards Institute (ANSI), Association for the Advancement of Medical Instrumentation (AAMI), National Fire Protection Association (NFPA), Underwriters Laboratories (UL), Canadian Standards Association (CSA), or international bodies such as International Standards Organization (ISO) and International Electrotechnical Commission (IEC). Equipment that has not been approved by the United States Food and Drug Administration (FDA) Center for Devices and Radiological Health requires special export licenses. Equipment that is subject to FDA or manufacturer recalls or hazard alerts should not be donated. Equipment that has nonfunctional subsystems can be donated, provided that those subsystems are clearly identified and labeled.

Obsolescence

A minimum of two years of manufacturer's sales and technical support should be available. This support should include repair parts; accessories (either reusable or disposable); and troubleshooting, repair, and maintenance assistance. Obsolete equipment or equipment for which replacement parts are unavailable should be shipped only if they are designated for parts only.

Appropriate Technology

In considering the provision of medical equipment to developing countries, potential donors should favor the following desirable characteristics in such equipment:

- Simplicity of operation
- Minimal number of accessories required
- Availability of necessary operating supplies (particularly disposable) in the recipient country
- Standardization with other equipment in the locale
- Ease of maintenance
- Tolerance to hostile environment

Evaluation of Potential Recipients

The most important prerequisites for a successful donation are that the potential recipient truly needs the equipment being requested and that it has the means to operate and provide maintenance for it. Although equipment supplied may be completely functional, it will be ineffective if it is not appropriate for the services provided at the recipient site and if it cannot be financially supported through its remaining life cycle. Previous recipient experience with donations is a plus but not essential for the success of the operation. Figure 43-2 is a form and Figure 43-3 is a checklist to assist potential donors and recipients in addressing the issues discussed below.

Clinical Need

To properly justify the need for the goods being requested, the donor should demand from the potential recipient(s) the following information: Which, and how many procedures, will be performed using the requested equipment; why the resources currently available (attach a list with descriptions) are not satisfactory; and an analysis on how the requested equipment will help meet the expected clinical demand.

Readiness to Absorb the Technology

The recipient should be required to provide information showing that it is ready to use and maintain the equipment being requested. Such readiness includes trained operators, appropriate environment, ancillary equipment, maintenance capability, and financial viability. It is acceptable that the recipient demonstrates that it has plans and means to address any shortcomings.

Human Resources

The recipient must employ properly trained physicians, nurses, and/or technicians who will be operating the requested equipment. If none is available currently, an explanation of the way the training is going to be achieved is required.

Environment

The recipient should describe available facilities such as physical space, electrical and pneumatic power, heating, ventilation, and air conditioning (HVAC) to install and operate the requested equipment. Particular care should be devoted to the availability of a stable electrical supply, air conditioning, and humidity control, which are vital to the performance of sophisticated medical equipment.

Material Resources

If the donated equipment is not accompanied by required ancillary equipment, the availability or means to acquire the latter should be described. The availability of supplies that are needed to operate the equipment locally must be researched and confirmed.

Maintenance Resources

The recipient should describe human, material, and financial resources that are available to service and to maintain the requested equipment within its institution. Information about service available from local manufacturer representatives and independent service organizations also should be included.

Financial Feasibility

The recipient must demonstrate that it has the financial ability to install, operate, and maintain the requested equipment.

MEDICAL EQUIPMENT DONATION REQUEST FORM

Note: Fill a form for each TYPE of equipment requested even when several units are needed. Attach sheets with more additional information if there is not enough space on this form.

1. **REQUESTER IDENTIFICATION**
   - NAME OF THE INSTITUTION:
   - NAME OF THE DEPARTMENT:
   - STREET ADDRESS:
   - CITY, STATE:
   - COUNTRY, POSTAL CODE:
   - PHONE & FAX NUMBERS:
   - CONTACT PERSON & TITLE:
   - DATE & SIGNATURE:

2. **EQUIPMENT IDENTIFICATION**
   - EQUIPMENT NAME:
   - IMDC CODE (ECRI):
   - CLINICAL APPLICATION (S):
   - QUANTITY REQUESTED:
   - SAMPLE BRANDS AND MODELS:
   - ACCESSORIES NEEDED:

3. **REQUEST JUSTIFICATION**

   3.1 Procedure(s) that will be performed using the requested equipment, with estimated number per month:

   3.2 Explain why the resources (equipment, methods, procedures, etc.) presently available are not satisfactory:

   3.3 When equipment is being requested to complement or replace existing equipment or services, please describe the resources presently available:

   3.4 Compare the expected demand and the production capacity of the equipment being requested:

4. **READINESS TO ABSORB THE TECHNOLOGY**

   4.1 Human resources available (indicate additional training if necessary)

   4.2 Material resources (additional equipment and devices needed)

   4.3 Space and special installation available or planned

   4.4 Maintenance requirements (in-house service, external contracts, etc.)

   4.5 Financial considerations (for installation, operation and maintenance)

Figure 43-2 Medical equipment donation request form.

MEDICAL EQUIPMENT DONATION ACTION CHECKLIST

ACTION	DONOR RESPONSIBILITY	RECIPIENT RESPONSIBILITY
Suitability for Donation • General Quality • Safety, Specifications and Standards • Obsolescence • Appropriate Technology	Evaluate unnecessary equipment prior to offering it for donation	

Evaluation of Potential Recipients
• Clinical Need
• Readiness to Absorb the Technology
    Human Resources
    Environment
    Material Resources
    Maintenance Resources
    Financial Feasibility
	Request information from potential recipient and evaluate it to determine likelihood of success for donation	Submit information to potential donor using request form

Pre-Donation Planning
 	
• Installation, Operation, and Maintenance Requirements
    Installation Requirements
    Operation Requirement
    Maintenance Requirement
    Special requirements
    Pre-Donation Recipient Preparations
	Provide data to recipient	Use donor's data to prepare personnel and infrastructure

Donation Implementation • Assembly, Packaging, and Shipment • Customs Clearance, Unpacking, Installation, and Maintenance	Donor's responsibility	Recipient's responsibility
Follow-Up Evaluation	Analyze process and improve procedure	Provide feedback to donor

Figure 43-3 Medical equipment donation action checklist.

Pre-Donation Planning

Assuming that the donor is reasonably convinced that the potential recipient really needs the equipment and can support it for the remainder of its useful life, the two parties should start planning for the donation. First, the donor must provide the recipient with detailed information regarding the installation, operation, and maintenance of the equipment. This information will enable the recipient to begin pre-installation tasks, including the training of personnel for operation and maintenance. In a few extreme cases, the recipient may cancel the donation after realizing that it cannot support the equipment. After all requirements have been satisfied, the recipient will notify the donor to assemble and package the equipment for shipping.

Installation, Operation, and Maintenance Requirements

Installation Requirements

The donor should specify the installation location, accessibility, floor-loading capacity, space and power requirements (i.e., voltage, frequency, phase, and power consumption) and environmental conditions. Care should be taken to identify any unusual extremes of temperature, humidity, dust, and power fluctuations that could adversely affect the equipment's operation. If available, the donor should ensure that detailed installation instructions are provided. Most if not all of this information is available in the equipment's operating or service manual.

Operation Requirement

The donor should inform the recipient of all the necessary subsystems such as cables, reagents, filters, electrodes, and recording paper that will be required in order to operate the equipment to be donated. Test equipment and calibration standards often are required to ensure performance and accuracy of the equipment. Availability of these items throughout the remaining useful life of the equipment should be ascertained. Again, any required operator training should be stated clearly.

Maintenance Requirement

The donor should seek guidance from its own service personnel so it can provide detailed maintenance requirements such as technician training, special tools, preventive maintenance frequency and materials, and test and calibration equipment needed.

Special Requirements

Any special requirements should be identified and communicated to the recipient. These include air or water cooling; electrical power; water quality; mechanical, layout or radiation or acoustic shielding requirements. Specialized software may be required to install, operate, or maintain equipment.

Pre-Donation Recipient Preparations

Assuming that the donor is reasonably convinced that the potential recipient really needs the equipment and can support it for the remainder of its useful life, the two parties should start planning for the donation. First, the donor must provide to the recipient detailed information regarding the installation, operation, and maintenance of the equipment. This information will enable the recipient to begin pre-installation tasks, including the training of personnel for operation and maintenance. In a few extreme cases, the recipient may cancel the donation after realizing that it cannot support the equipment. After all requirements have been satisfied, the recipient will notify the donor to assemble and package the equipment for shipping.

Implementation and Evaluation

Assembly, Packaging, and Shipment

Prior to packaging the equipment to be donated, the donor should ensure that it is safe and performing within manufacturer's specifications. This can be accomplished by performing an operational verification procedure that is found in most operating manuals. In addition, all accessories and supplies should also be checked. All software that is necessary for equipment operation should be included. Training aids such as slides, books, and videotapes should be supplied if available. The checklist (Figure 43-3) should be used to verify that all subsystems, components, and accessories and supplies are included. This checklist also will be helpful in the preparation of the shipping documents.

Equipment that may contain patient material should be properly decontaminated prior to packaging and shipment. Radioactive sources should be removed and properly packaged in special shipment containers (with radioactive marking on the outside). Fluids should be drained and fragile parts packaged with great care. International shipments are often handled roughly by people who lack the proper training and equipment, and therefore they are subject to a high probability of damage.

All equipment should be shipped with operation and service manuals. Software version numbers and significant hardware updates, if applicable, should be noted. The operation manual should contain detailed operating instructions and should list all necessary subsystems, accessories, user-replaceable parts, reagents, and other supplies such as chart paper, gases, coolants, and chemicals. The service manual should contain specifications, schematics, operating instructions, troubleshooting, repair and maintenance procedures, cleaning and/or sterilization recommendations, and a replacement parts list. The supplier should provide, if available, procedures or recommendations for periodic inspection, maintenance, and calibration to ensure that the equipment is maintained in a safe and effective operating condition. If the documentation is not available, the donor should consider purchasing it to ensure the eventual usability of the donated equipment.

Donated equipment should be packaged in accordance with the method of shipment to minimize damage in transit. For surface shipment, waterproof wrapping and wooden crates are necessary. Air shipment requires less sturdy packaging, but limitations in size and weight are more severe.

Shipping documents should list everything inside the respective packages and should clearly indicate that the shipment is a donation. Sample shipper's export declaration forms and instructions can be obtained from the Department of Commerce through the Bureau of Census, Washington, DC 20233.

Donors who are not familiar with packaging, shipping, and documentation might consider seeking the assistance of freight forwarders and companies that specialize in assisting exporters in transferring goods to other countries.

Customs Clearance, Unpacking, and Installation

Customs clearance is the sole responsibility of the recipient. If special documentation is needed, the recipient should request it prior to the shipment.

The recipient should inspect all containers and contents for damage and should verify that the contents are intact and that nothing is missing. Any irregularities should be reported immediately to the donor and to the shipping company for insurance claims. The manuals received with the equipment should be distributed to the appropriate personnel: Operating manual to the operator and service manual to the clinical engineer or biomedical technician. If a centralized library exists, the manuals should be forwarded to that location.

Installation should be performed according to the instructions received from the donor. After the installation, verification of proper and safe operation must be performed prior to clinical use.

After verifying that the equipment received is working properly, the recipient should implement a program of periodic inspection, maintenance, and calibration to ensure that the equipment is maintained in a safe and effective operating condition for its remaining useful life. If an in-house maintenance department does not exist, the recipient should recommend such a department to the institution's administration.

Evaluation

After installation and operation, the donor and the recipient should assess the level of operational success or failure of the medical equipment donated. This assessment fosters communication between donor and recipient, encourages the continued support of the donor, and allows both parties to learn to improve from previous experience.

The recipient should not hesitate to identify mistakes made by the donor. Of equal importance, the donor should demand honest and timely response from the recipient. The success of future donations will be enhanced as a result of such assessments.

Forms and Checklists

The Medical Equipment Donation Request Form and the Medical Equipment Donation Action Checklist are shown in Figure 43-2 and Figure 43-3, respectively.

References

ACCE. American College of Clinical Engineering. *Guidelines for Medical Equipment Donation*, Plymouth Meeting, PA, 14, 1995.

Bloom G, Temple-Bird C. Medical Equipment in Sub-Saharan Africa: A Framework for Policy Formation. IDS Research Report No. 19, WHO/SHS/NHP/90.6, WHO, Geneva, 1990.

ECRI. IMDG, Donating and Selling Used Medical Equipment. *Health Devices* 21(9):295-297, 1992.

FAKT. Association for Appropriate Technology (FAKT), Germany, and Churches' Action for Health (CMC) of the World Council of Churches. Guidelines on Medical Equipment Donations. Technical Library, FAKT. In CMC Publication. *CONTACT*. World Council of Churches, Geneva, October, 1994.

Issakov A. Service and Maintenance in Developing Countries. In Van Gruting CWD (ed). *Medical Devices: International Perspectives on Health and Safety*. Amsterdam, Elsevier, 1994.

Muschlitz L. Donated Equipment: From Trash to Treasure. *Health care Technology Management*, 36-38, April 1996.

PAHO. Pan American Health Organization. *Creating Non-Governmental, Government and Private Sector Networks for Effective Coordination and Delivery of Donated Medical Supplies and Equipment*. 1997.

Wang B. The Facts on Used Equipment Export. *Biomedical Technology Management*. 64, 1996.

WHO. World Health Organization. *Guidelines for Health Care Equipment Donation, 4th Draft*, WHO/ARA/97.3. 1997.

WHO. World Health Organization. Report of the Interregional Meeting on Maintenance and Repair of Health Care Equipment. Nicosia, Cyprus, November 1986. WHO/SHS/NHP/87.5, WHO, Geneva, 1987.

Further Information

Christian Medical Association of India (CMAI). *Study Report on Medical Equipment and Supplies and the Role of CMAI*. 1989.

International Medical Device Group (IMDG). *Policy Position on Donating and Selling Medical Equipment*. 1992.

Ministry of Health of Tanzania. *Guidelines on Donations of Drugs and Medical Equipment to the Health Secto*. 1995.

44

National Health Technology Policy

Thomas M. Judd
Director, Quality Assessment, Improvement and Reporting,
Kaiser Permanente Georgia Region
Atlanta, GA

Many countries do not have the health technology (HT) resources that they need in order to improve their populations' health. As defined by the World Health Organization (WHO), HT resources include human resources, pharmaceuticals, equipment and supplies, and facilities. A rational process for identifying, acquiring, and managing needed resources in care delivery requires the development of an explicit policy. The policy should address HT use in all levels of a national health care system. Clinical engineers (CE) around the world have contributed to the development of health technology policies (HTP), particularly as it relates to the medical equipment component. These and other, wider experiences will be discussed in this chapter.

Rationale for a National Health Technology Policy

All countries grapple with limited resources for health. The central issue concerns the best way to use available resources or to plan optimally for future needs. HTPs seek to have the following affects:

- Maximize limited financial investment in health.
- Minimize quality waste.
- Maximize loans and donations.
- Ensure rational use of health resources.

Recommended HTP Elements

A literature search was conducted to identify several models for HTP in developing countries for which the global clinical engineering community has played a significant role. As a result of this survey, the following ten recommended HTP elements emerged as most important:

Political

- HTP is linked to national health policy and health-reform initiatives through the Ministry of Health.
- HTP has a Ministry of Health-led communications plan, internally for the health system and externally for technical aid supporters and donors.

Planning

- HTP is focused on national health priorities, based on their population's epidemiological data.
- There is a rational HTP methodology for planning and management, with multidisciplinary expert panel consensus.
- HTP integrates planning and management for all HT resources.

Assess, Acquire, Manage

- The HTP establishes and supports a technology assessment program, including ongoing input of evidence-based medicine.
- The HTP ensures that the country conducts a health technology management (HTM) process, from planning to acquisition through the full life cycle of resource use.
- The HTP ensures development of HTM capacity in the country.
- The HTP ensures that ongoing activities for quality improvement, risk reduction, and patient safety for HT are conducted in the health delivery system.
- The HTP ensures that a regulatory and legal framework for HT resource research and development (RandD), appropriate business ethics, and resource quality assurance is developed.

HTP Models and Key Features

- Brazil 1989: Binseng Wang advocates a comprehensive medical equipment management approach, integrating RandD and regulation through the equipment life cycle.
- WHO Geneva 1999: Authored by Andrei Issakov and others, this model focuses on equipment, facilities, and related support systems. It provides an extensive "Country Situation Analysis" as well as an HTP framework, guidelines, process flows and checklists that address all elements of HTP.
- WHO Africa Region 1999: Authored by Yunkap Kwankam and others, this model notes that HTP should address health technology assessment (HTA) and HTM, human resource (HR) development, research and communication, and access to information.
- South Africa 2000: Authored by Peter Heimann, Mladen Poluta, and others, this HTP model focuses on equipment, facilities, related support systems, and key clinical services. It advocates a country-wide HT audit, use of Essential Health Technology Package (EHTP) HT resource planning and management methodology, development of HT regulatory framework, HTM capacity building, and recognition that "highly specialized systems" such as transplants require unique management. (See Chapter 45.)
- PAHO 2000: This model by Antonio Hernandez highlights HTP focuses on improving facility and equipment infrastructure in Latin America through intercountry and interinstitutional coordination, and the development of HR and HTM capacity to improve equity, quality, efficacy, and safety.
- England 2000: Authored by Caroline Temple-Bird, this model recommends HTP elements for developing countries that focus on improving facility and equipment infrastructure; addresses HTA, acquisition, life cycle management, and developing HR and HTM capacity, research and local production.
- World Bank 2001: Authored under Yolanda Taylor's oversight, this model of HTP development links all health technology resources, emphasizes life cycle cost analysis and management including HTA, linkage with macro policy and regulatory mechanisms, acquisition and capacity building.
- Mexico 2001: Authored by Adriana Velazquez, this model of HTP development links all health technology resources, provides a focus on telemedicine for the e-health sector, use of HTM in public institutions to assist HTA, acquisition, and overall resource management.
- WHO-EHTP 2002: Authored by Heimann, Issakov, Kwankam, Thomas Judd, and others, this model of HTP development is based on integrated health resource planning and management and on meeting health priorities through resource optimization based on evidence-based clinical practice guidelines.

Case Study 1: HTP Model, South Africa 1999

The following statistics were taken from the remarks made by Nonkonzo Molai at the Advanced Health Technology Workshop held in Cape Town, South Africa, November 1999. (See Chapter 71.)

Health Problems

- Population of 40 million with following 1998 health data/1000 (WHO Region average):
- Infant mortality – 59 (57)
- Maternal mortality – 230 (430)
- Malnutrition under age 5 – 23%
- Health expenditures – 7.9% of GDP (5.2%)
- Leading causes of mortality (Africa) – 8.2% acute lower-respiratory infection (LRI); 19% HIV/AIDS; 7.6% diarrhea conditions; 5.5% perinatal complications; epidemics of HIV/AIDS, malaria, and tuberculosis in region.

Country's Needs, Priorities, and Resources

The level of health service delivery has not improved much in spite of considerable investment in importing new technologies and operational expense of supporting these technologies. For example, the lack of equipment maintenance and repair is a major stumbling block in the effective provision of care and is related to poor planning and acquisition and procurement.

The most important intervention is ensuring a stable supply of competent health technology managers. The Workshop is intended to assist decision-makers and planners in addressing complex issues of planning, life cycle, transfer, management, and utilization of health technology so that plans for education, training, and placement of HTM professionals can be developed.

The Workshop also is intended to impart skills to the HTM community to take actions to ensure and enhance the safety of health technologies and facilities.

HTM is one of the major cost drivers in our health care system within the area of resource mobilization. The broad framework for comprehensive HT policies has been created. Our top priority is the creation of an HT system composed of acquisition, management, planning, and HTA subsystems. The major short-term strategic action is to conduct a comprehensive national HT audit. A second action is to field-test the Essential Health Technology Package (EHTP). Findings from these two activities will be considered in order to determine the level of intervention from government that will be required.

Framework for HTP

Macro Policies

- HT Acquisition Subsystem: This area focuses on reducing the unnecessarily high HT cost that results from the unsystematic and unplanned acquisition and inappropriate use.
- HT Management Subsystem: This is aimed at ensuring safety, efficiency, and effectiveness of HT, which in turn will enhance the delivery of health care services. HTM systems audits will determine needs and capacity for HTM maintenance, HT HR training and development strategies, and further development of HTM systems and structures.
- HT Planning Subsystem: This area focuses on ensuring rational and optimal distribution of HT, at improving equity in access through appropriate use of the EHTP methodology for planning and management of primary, secondary, and tertiary care.
- HTA Subsystem: This is aimed at ensuring appropriate introduction, adoption, and continued use of appropriate HT in the country.

Operational Policies

- Review existing and prioritize policies that need development based on concrete examples; e.g., budget figures and waste incurred in HT due to wrong practices that require prompt attention.
- Develop HT-related policies for HR, intersectoral collaboration, and HT information systems.

Mission and Desired Outcome

The mission of health care planners and policymakers is to ensure that health technology is harnessed to its fullest extent as one of the tools to improve the delivery of health services and to devise a strategy that facilitates appropriate utilization of health technology for the South African health system. The desired outcome is a unified and harmonious HT system that ensures optimal distribution of the limited HT resources and facilitates equity in access, with the ultimate aim of improving the quality of health services.

Case Study 2: HTP Communication Plan – Example of WHO African Region, 1999

WHO convened Region member states and established a HT Task Force to provide assistance to member states. It produced two important documents (Guidelines for Donation of Health Care Equipment and Guide for the Formulation of National Health care Equipment Policy) as key communication tools. The policy tool is summarized below:

Guiding Principles for Health care Policy Development

- To give HT policy high priority as an essential component for the comprehensive development of the health system and for ensuring the improved equitable access of the population to affordable and sustainable quality care.
- To plan the introduction of HT and to manage it properly, taking into account the needs and aspirations of the population, the environment and its trends, and available resources.
- To systematically give preference to the technological options that, for the same life cycle cost, have proved their effectiveness in the region and in other countries in similar contexts.

Executive Summary of Report and Communication Plan to Region Member States

- Health for all through primary health care, proclaimed in Alma Ata in 1978, will remain a major goal in the coming years and century.
- To achieve this goal, the majority of countries in the region are reforming their health systems and services in order to ensure equity in health and access to health care.
- Technology development has played, and continues to play, an essential role in promoting health care delivery.
- The scope of health technology is very wide. This document focuses on medical equipment, one of the major concerns of the region.
- The policy statement and strategy for its implementation proposed in this document aim to help member states to formulate HTP and plans that support the achievement of national health care policy.
- Priority interventions for HT should address: (1) technology management; (2) human resources development; (3) research and communication; and (4) access to information.
- The Regional Committee is invited to examine the proposed document, make suggestions for its improvement, and adopt it in support of the health-for-all policy in the region.

Financial, Medical, Public Health Benefits

Technology can play a key role in achieving the objectives of health sector reform, particularly better access to health care by the population, greater equity, improved quality of care, and increased cost-effectiveness of health services. A rational health technology policy for each country can solve many common problems. Purchase price of equipment is usually only a small fraction of life cycle costs. In all procurements, provision should be made for coverage of recurrent costs that reach the annual rate from 3%-8% of the purchase price.

Target Groups and Communication Strategies

The ministries of health (MoHs) in the member states should be the main targets and conduits for information on health technology policy. Partnerships should be encouraged among the respective MoHs and other government bodies, nongovernmental organizations (NGOs), consumer groups, the private sector, and industry.

WHO should focus its activity at the country level by embarking upon advocacy vis-à-vis policymakers and health officials to create the political will to include HT policy in the list of national policy priorities and to identify and mobilize institutional partners so as to benefit from their support. This political will should result in the establishment of appropriate mechanisms and tools for the implementation, monitoring, and evaluation of the HT policy. At the regional level, measures will focus on endorsement by member states, leadership to support of HT policy in national health development priorities, and implementation of relevant national plans. A recommended HT-awareness campaign and communication plan for South Africa should include the following actions:

- Debunking myths (e.g., that HT is synonymous with expensive HT such as MRIs and CT scanners
- Showing relevance and appropriateness
- Tailoring policy to a country's objectives and strategies
- Ensuring that HT is seen as a tool and a resource
- Not losing sight of the end objective, which is to improve patient care

Case Study 3: The Experience of Mozambique

Enrico Nunziata, clinical engineering consultant to the Ministry of Health of Mozambique from 1997 to 1999, reported the following (Nunziata, 2003): After 10-15 years of technical assistance, the maintenance system was stagnant after externally funded projects ended. In 1997, an HT policy was developed and approved within the MoH "Modernization Program" for the Administrative and Management Sector of the health care system. In 1998, the new administrative structure was introduced at all levels. A new medical equipment management system (MEMS) was developed and introduced, and a sub-Sector General Action Plan (GAP) was developed, approved, and presented to participating donors. In 1999, the GAP became known and accepted and became a tool for the local HT manager and a reference point for donors. Data collection for MEMs became more regular, and an economic study was performed to evaluate real maintenance cost to correct the GAP for introducing future options. By 1999, a new breed of engineers was attracted to enter the system. HT policy was a first step, but providing appropriate HT was still dependent on foreign aid and technical assistance. The development of HT maintenance capacity remained strongly dependent on other MoH sectors.

Case Study 4: The Experience of Brazil

Context and Structure

The 1990s were a time of national health system reform in Brazil with decentralization and unification of public and private health services. A new HTP was needed for Sao Paulo State's 560 hospitals (approximately 85,000 beds), 2100 clinics, and 10 research and manufacturing facilities providing care for 33 million people. Toward this end, the Office of Equipment Advisor to the Secretary of Health was established. Binseng Wang, advisor from 1987–1990, implemented a structure that addressed all aspects of the equipment life cycle listed below:

- Planning
- Procurement
- Acceptance
- Commissioning
- Maintenance
- Refurbishment
- Decommissioning

Regulation and research and development were incorporated in the plan in an effort to control the safe and effective use of medical device technology and to encourage the growth of a native medical device industry.

Opportunities and Critical HTP Success Factors

Opportunities

- Insufficient financial resources required careful planning with input from all stakeholders.
- Inadequate infrastructure required use of creativity to use efficiently what is available.

- Lack of trained human resources required an investment in education and training.
- Difficulty in acquiring HT equipment spare or repair parts through alternative sources.
- Missing equipment documentation led to inclusion in acquisition agreements and assistance from global colleagues.

Critical HTP Success Factors

- Keep decision-makers well informed.
- Create a strong link between health policy and HTP.
- Secure wide stakeholder support.
- Use transparent and rational methodologies for planning, management, and support of technology.
- Develop a strong evaluation system that includes feedback *from* equipment users, equipment service staff, and patients and *to* policy makers and stakeholders.
- Emphasize education and training to build HTM capacity for the long term.
- Plan for equipment replacement and continuous equipment service support.

Case Study 5: The Experience of Kenya – Late 1990s, John Paton, GTZ/MoH

- MoH of Kenya initially wanted standardization of equipment for management and spare-parts reasons.
- MoH adopted international standards for accurate and detailed equipment specifications.
- MoH modeled development of HT Policy on existing drug-policy document approved by Parliament and in wide use in Kenya.
- A draft policy was developed by an MoH working team. All relevant stakeholders in the health delivery system were asked for comments. A workshop was convened to develop a consensus-based final document.
- The HTP policy was seen as a first step. It was kept broad and is subject to ongoing review and update regarding goals, requirements, and responsibilities.

Case Study 6: EHTP–A Rational Planning and Management Framework for Kyrgyzstan, 2000–2004

Overview

The Essential Health Care Technology Package (EHTP) is a new methodology and set of tools that was specifically developed to strengthen health resource planning and technology management. These resources, termed "health technologies," are, primarily pharmaceuticals, human resources, medical devices, and facilities. Figure 44-1 shows the main elements in the EHTP. (See Chapter 45 for a detailed description of the EHTP.)

The link between resources and the delivery of quality care is achieved by identifying all health technologies that are needed to provide well-delineated promotive, preventive, and curative health interventions. Health interventions in turn are linked to the WHO International Classification of Diseases (ICD9, ICD10) and other disease classifications (e.g., ICD9-CM). This mapping between HT and health care interventions is based on internationally accepted and recognized clinical practices. It is expressed in the form of a "matrix," or generic template, whose elements include economic, technical, and other information typically associated with acquisition of HT and with their utilization in providing the interventions. This EHTP matrix is the basis of the EHTP methodology and tools, which are implemented in a suite of software packages. Each intervention is decomposed into progressively smaller components, all the way down to procedures and associated techniques. The decomposition uses the Current Procedure Terminology (CPT) of the American Medical Association. Over 350,000 links thus have been created between international classification of diseases and the CPT.

One unique and powerful feature of EHTP is the integration of epidemiological profiles and recognized clinical practices (such as clinical practice guidelines [CPGs]) with the EHTP matrix. This allows advanced modeling and simulation for identifying economic and other resource requirements given varying scenarios of disease incidence and prevalence and CPGs. Conversely, simulation can indicate the epidemiological conditions that can be covered with available resources by using specific CPGs.

Implementation in Kyrgyzstan

EHTP was developed over several years and provided by the WHO Department of Health Service Provision and the Medical Research Council, South Africa, to WHO member states beginning in 2000. In 2001, the Kyrgyz Republic (Kyrgyzstan) Ministry of Health (MoH) requested implementation. Goals and objectives include:

- Applications in the broad context of health systems development and reform and thus aligned with and possibly able to accelerate present health reform efforts.
- Methodology and computer-based tool expected to measurably improve effectiveness, efficiency, performance, and quality of health delivery.
- Ability to overcome past fragmented and often untimely efforts to plan for and manage health resources through WHO-designated essential lists.
- Simplified coordination of resources by central, district and local health leaders and integrated management of these resources to allow more efficient distribution.
- Focusing on the country's disease burden, as defined by in the Kyrgyz Republic by MoH and the National Drug Council, who direct the national CPG development process.
- Use of best medical evidence to direct care at primary, secondary and tertiary levels by selecting interventions or CPGs best fit to available financial resources and existing constraints.
- Methodology and tool may be dynamically updated with advances in medical technology.

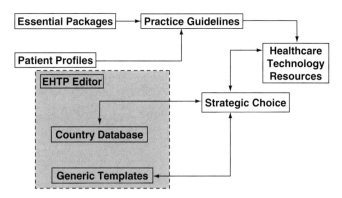

Figure 44-1 Essential Health Technology Package simulation.

- Piloted at three sites for primary and secondary level care in 2003, with nation wide implementation in 2004.
- Pilot sites initially used EHTP to better manage five target conditions selected by MoH and pilot leaders: Hypertension, acute respiratory illnesses in children, iron deficiency anemia in pregnancy, brucellosis and tuberculosis (TB).
- Quality, cost and accountability indicators were developed and measured.
- The implementation process is overseen by a MoH national commission that includes a national coordinator, expert resource working group leaders, coordinating office team, and WHO technical advisors.

Initial Findings at Pilot Sites

- Three secondary level hospitals and their primary level clinics have significant CPG noncompliance:
- Pharmaceuticals: Problems featured overuse of recommended drugs, use of obsolete drugs, or drugs not priced affordably in the pilot area.
- Equipment: Recommended diagnostic lab and imaging tests were not completed because of lack of appropriate equipment or lack of supplies to operate equipment (e.g., lab reagents, chest films for X-ray, or appropriate X-ray and ultrasound equipment).
- Human Resources: Overuse of physicians occurs, notably at the secondary level; nonphysician staff competencies are not defined, resulting in a wide variation of staff utilized at both levels.
- Facilities: Because pilot sites chosen at locations whose facilities already had been modified to meet new health reform standards based on "case rates" instead of population size, facilities were adequate to meet requirements of CPGs.
- Resource running costs were measured for all resources (e.g., initial and ongoing training of staff; maintenance and operational costs for equipment and facilities; and variable costs for drugs in a given area).

Key Definitions and Applications

EHTP "Simulation Tool" software: allows for the integration of linked interventions to CPGs, health and patient profiles and health packages. Definitions and explanation of terms as they apply to EHTP are presented below.

Scenarios

- Scenarios are graphical interface and/or representation of CPGs.
- MoH-approved CPGs are used to direct care providing best clinical outcomes and optimal use of limited resources for clinical conditions that are most prevalent in country.
- The pilot had 30 primary and secondary CPGs programmed in EHTP Computer Tool.
- The scenarios can be used to map published CPG and clinical/nonclinical pathways based on HT-linked interventions (procedural steps, involving a specific clinical technique, each with known health resources required, called the "technology lists" [described below]).
- The graphical interface allows easy interpretation of CPGs and allows easy modification and editing of CPGs. Scenarios are saved in unique files, thus allowing easy sharing and Internet distribution. CPGs can be based on existing evidence-based CPGs from other countries, thus reducing development time.

Technology Lists

- Centrally validated previously for each of four resources needed for each procedural step of CPGs; or each box/health intervention in CPG flowchart mapped into EHTP Computer Tool. In pilot, Technology Lists were validated at primary and secondary levels.

Patient Profiles

- Patient profiles are separate modules that can be used to graphically display patient utilization of scenarios.
- Patient profile data can be based on epidemiological data.
- Historical data can be recorded. Statistical mapping of future data based on epidemiological data can be used to plan for future interventions.
- Pilot locations mapped profiles for each of 30 conditions for 2002 to use as a baseline.

- When combined with the Scenarios and validated Technology Lists for each CPG, will allow local health leaders to calculate resource requirements to treat that condition throughout a given year.
- Profiles are a unique method of quantifying magnitude to CPGs. Based on timing of scenarios and interventions, resource scheduling is an integral part of patient profiling.

Scheduling

- Intervention scheduling forms the basis for the calculations of resources within scenarios.
- Resources scheduling may be based on actual intervention's time periods or resource utilization within the scenarios, taking into account known resource constraints. (For example, consumables, surgical instruments and linen are calculated on procedure length.)

Resource Constraints

- Scheduling of resources can be calculated within predefined resource constraints.
- Constraints that can be predefined are resources availability (e.g., human resource and competency availability, medical device and infrastructure availability, and the availability of pharmaceuticals).
- Time constraints for human resource availability, holidays, and working hours can be set and included in the model. Other constraints, such as effectiveness of certain support services, can be set to be included in the calculation of the scheduling process.

Health Packages

- Health packages are a graphical interface used to link scenarios and patient profiles.
- Multiple linked scenarios and health profiles represent a set of services delivered at a point of care. For example, caesarean section, antenatal care, and pre-eclampsia would be part of a mother-and-child health package for a specific level of health care intervention, or all of the services delivered at a specific pilot site could be a health package.
- Health care packages can be saved in an EHTP-specific file format, thus allowing easy sharing and distribution on the Internet. Any number (system dependent) of scenarios and patient profiles can be associated to define the health package.
- Resource requirements derived from the EHTP health packages are calculated based on the actual time schedules, patient profiles and loads, and their applicable CPGs.
- Health packages were developed for each pilot site; when combined with the Scenarios with validated Technology Lists for several CPGs, allows local health leaders to calculate aggregated resource requirements to treat the many conditions throughout the year and to develop resource-sharing strategies as appropriate.

Country Database

- Country-specific databases are derived from the Generic Template provided by WHO to the national coordinating office team, under the oversight of the EHTP national coordinator and commission. The EHTP Generic Template provides a single reference database with relatively complete information on all known health resources and health interventions in the world. This comprehensive database is vital in ensuring a manageable updating and synchronization process for country-specific information.
- The country databases are a subset of the generic database and usually reflect the country-specific needs and practice, as does the Kyrgyzstan country-specific database.
- The subset of the country databases are then modified (over time) to reflect the specific country conditions.
- This process includes the modification of the macro properties (i.e., procedure timing and actual health care technologies that are needed to perform the intervention).
- Micro properties are also changed. These include details such as quantities; technical details and specifications; costs and medical device maintenance; and management details.
- The ability to change details at both the micro and macro level of care, scenarios, patient profiles, and health packages provide complete flexibility and allows users to adapt their planning scenarios to virtually any conditions applicable in their target setting.

Costing of Interventions

- Costing of clinical and nonclinical interventions within the health sector is complex. The unique association and interaction of health care technologies required to provide services is not well defined and changes from country to country.
- The EHTP methodology of unbundling all services and interventions into their respective health care technologies provides a unique opportunity to attach costs, scheduling, and quantities to all elements.

EHTP Costing Capabilities

- Unit costs for all HT, including recurrent, installation and training costs.
- Intervention costs are determined by compounding the costs attributed to all HT for a single intervention.
- Stringing health care interventions into CPGs allows the costing of interventions based on practice and takes into consideration medical ethics and quality considerations.
- The combination of CPGs further refines the costing module, as resource-sharing based on clinical practice can be determined.
- The costing module can be used to determine unit and marginal costs for any area of health service delivery and planning. New and additional costs can be determined for services.
- The costs of certain HT within a bundle of services can be determined. Cost factors associated with increasing the level of demand, quality, and service provision can be modeled.
- Increased costs associated with the introduction of new techniques, medical devices, or drugs (and any combination thereof) can be quantified and evaluated.
- Cost evaluations of providing a new service within an established service portfolio also can be appraised and quantified in terms of its base resources.

Other Applications

- EHTP provides a unique platform of mapping the current services' portfolio complete with actual resource requirements and costing data.
- EHTP thus provides a mechanism to set the virtual target and to compare the effort and cost required, thus improving the current health care system to a planned and envisaged system.
- The costs and additional effort required can be determined and can form a significant evidence base for decision-makers in their approach of improving health.
- EHTP thus provides an appropriate tool to establish both a current and projected list of services that can be used by donor and technical assistance agencies, nongovernmental organizations, and consultants, to plan their support.
- This planning process can be used primarily to determine the direct needs, such as medical devices and pharmaceutical and human resource needs, but it also quantifies the extra support that is required, to cover the indirect requirements, such as maintenance and other recurrent costs.
- EHTP also can be used to determine the donor agencies that can support a given area within the complex set of requirements. EHTP provides a useful tool for donor agencies to make more effective interventions and also provides a guide to recipient countries to gauge and prioritize their requirements, as well as to base new requests on evidence.

Conclusions

There are several models to assist developing countries for creating effective health technology policies. Some of the case studies highlight medical equipment management (HTM) best practices and others demonstrate a broader view of HTP and its implications. Clinical engineers continue to contribute significantly to these efforts. The reader is encouraged to contact those individuals noted in the reference section to discuss ways in which this work can assist the reader's efforts in this regard.

Reference

Nunziata, E. EHTP in Mozambique, *ACCE News* 2003; 13(3):7-8.

Further Information

Central Asia & Africa, WHO EHTP Materials, Peter Heimann, Andrei Issakov, Tom Judd, 2001–2002, pheimann@mrc.ac.za.
Mexico, A. Velazquez, 2001–02, avelazquez@ssa.gob.mx.
PAHO, A. Hernandez, 2000–02, 1hernana@paho.org.
Kwank A. HTP in the African Region. WHO, 1999.
Molai N. HTP Development/Implementation. South Africa, 2000–2002.
Temple-Bird C. Practical Steps for Developing HTP (in Developing Countries). UK, 2000.
Sakova I. HTP Development Materials, Country Situation Analysis. WHO, Geneva, 1999.
Wang B. Brazil HTP and World Bank. 1989 and 2001.

45

The Essential Health Care Technology Package

Peter Heimann
Medical Research Council
Cape Town, South Africa

Andrei Issakov
Program Manager, WHO, Health System Development,
Geneva, Switzerland

Yunkap Kwankam
Health Systems Development, WHO
Geneva, Switzerland

Global economic, political, and social changes, as well as an increased and changing disease burden, have generated widespread efforts to reform and adapt health care systems to improve their efficiency, equity, and effectiveness. Health authorities worldwide are increasingly concerned with defining policies and strategies to contain growing costs of care within the context of serious economic constraints while trying to preserve the health system's social imperatives of equity and solidarity. At the same time, health care systems in all countries, rich and poor, become increasingly more technology dependent, and as health care technology continues to evolve, so does its impact on patient outcomes, hospital operations, financial efficiency, and overall performance of health services and systems (Heimann and Issakov, 2000).

Technology has a paradoxical role in health care systems. It is frequently cited as the most significant contributor to the unacceptable cost increases while at the same time it equips health care providers with indispensable tools to perform their functions more effectively and efficiently. The promise of new technologies is becoming increasingly familiar to populations. Rising public expectations for better health services combined with growing market and professional pressures for introducing the latest technological advances into practice have commonly led to the perception that the quality of care is directly linked to the sophistication of technologies used in providing that care. In many instances, however, the high expenditure related to the introduction of costly and complex new technologies is not justified by the overall benefit achieved in terms of access, quality, and health outcomes (Issakov, 1994a). Moreover, expansion of health care infrastructure and proliferation of technologies have far outpaced the capacity of many countries to effectively absorb, adequately support, and fully utilize these technologies, thus creating serious imbalances in health services provision.

Within this paradigm, health care technology (and in particular, related management strategies), have repeatedly come under the spotlight as health care providers seek to identify cost-effective methods for providing equitable and quality health care (World Bank, 1993). It is commonly accepted that only rationally planned health care technology interventions based on a sound needs assessment and ensured life cycle support provision can maximize their potential benefits, contain costs, and improve health outcomes rather than become a long-lasting burden on scarce health sector resources. However, the ability to realize this is a major challenge faced by health care decision makers and other health professionals in many countries. It is compounded by the lack of appropriate management tools for optimizing the deployment of health care technology and facilities that house it and those for matching the country need and capacity with available technology options.

Responding to this urgent need of the World Health Organization (WHO) member states for a comprehensive tool to assist in better planning and management of health care technology interventions, WHO and its Collaborating Centre for Essential Health Technologies at the South African Medical Research Council (MRC) in Cape Town embarked in 1996 on a major research and development initiative. This work has resulted in developing the concept, methodology, and software-based tool called Essential Health Care Technology Package. EHTP ensures that health care technology interventions are properly assessed and successfully implemented and helps to make informed decisions on acquisition, deployment, and utilization of health care technologies, thus contributing toward cost-effective, quality and equitable health services delivery. It also helps in defining the essential services that are affordable and sustainable by identifying and quantifying all kinds of essential resources needed to perform required procedures for the majority of common and prevalent diseases and health-related conditions. Broadly, EHTP is instrumental in translating strategic plans into operational realities.

The Essential Health Care Technology Package

The EHTP, therefore, is both a methodology and a tool aimed at strengthening and optimizing health care technology planning and management. The methodology is based on the hypothesis that effective health care delivery can be realized only if all health care technologies (e.g., medical devices, drugs, human resources, and physical infrastructure) are available and correctly managed to deliver that care.

The Context

There have been significant changes in the scope and complexity of health services provision in the last few decades, and the recent introduction of "high-tech" and other sophisticated health care technologies has necessitated the establishment and strengthening of complete support systems and services. Unfortunately, the rate of health care technology deployment has far outpaced the rate of implementation of these support services, which often has resulted in health care technology management inefficiencies, as clearly demonstrated by the high percentage of inoperable and unused medical devices.

These constraints have forced health care providers and decision-makers to refocus their attention on optimizing existing health care systems in particular, at improving health care technology management, and on reassessing the important role that health care technology plays in the provision of quality health care. The identification of appropriate health care technology interventions and the optimization of the delivery process within the framework of financial constraints and changing technology environments have become paramount in any health care and delivery planning process. Globalization and the need to be competitive have placed a further demand on health care providers to ensure that the real needs of the people are addressed and optimized.

Identifying and assessing the real needs of any health care system and ensuring that the health care technology base in a country adequately supports the delivery of health care have become aspects of any health care planning process. The need to base decisions on sound evidence has also gained significant momentum and has become an integral part of the health care planning process. Linked to this process is the development of tools that are flexible and responsive to changes in burden of disease and changes of complexity and availability of health care technologies.

The transfer of health care technologies further adds to the complexity of the health care planning process and often results in significant technology transfer mismatches, particularly in the process of technology acquisition through donor aid (World Bank, 1994). One consequence of this mismatch is the inability of less-developed countries to absorb and fully utilize health care technologies effectively, especially new and sophisticated technologies. This problem is compounded by a lack of adequate technical support structures and skilled users and by increasing financial constraints. While the rate at which technologies are being developed and introduced into the regional markets is increasing, the ability of health care systems in these countries to utilize these technologies fully is likely to decrease. For these reasons, it has become widely accepted that what is needed is improved management and planning of health care technology throughout its life cycle.

Health Care Technology Management

The goal of health care technology management (HTM) is to optimize the acquisition and utilization of technology to achieve maximum beneficial impact on health outcomes at the national level. At the organizational level, HTM is an accountable, systematic approach to ensuring that cost-effective, efficacious, safe, and appropriate equipment is available to meet the demands of quality patient care. HTM is a complex multidisciplinary process that can have substantial positive impact on health care outcomes provided that proper attention is given to political, organizational, economic, cultural, ethical, and legal aspects of technology interventions and their management. As part of this process, extra care must be exercised to ensure that those technologies that are acquired and deployed have maximum positive impact on health outcomes and that they can be utilized effectively and supported to ensure sustainability of the interventions.

To ensure effective HTM, therefore, a framework within which these aspects can be incorporated must be identified and quantified. While aspects of HTM have been considered in detail, these contributions often apply to a specific context and timeframe and are presented with varying degrees of observer bias (Medical Alley, 1995; Webster, 1979; Free, 1992; Barnes, 1990; Banta and Andreasen, 1990; Drummond, 1994; Scenario Commission, 1987; Jennett, 1988; McKie, 1990; Goodman, 1992). One further shortcoming of current approaches is that most if not all have been formulated in the setting of an industrialized country. These approaches–explicitly and implicitly–make assumptions relating to many components of HTM, from skills of individuals and resources available to the technical and organizational infrastructures. Moreover, most of the current approaches focus on technical issues without sufficient consideration of the complex and dynamic human factors in the management process and the differing HTM environments in less-developed countries (Issakov, 1994b). This is being realized increasingly in many areas of health care technology management.

HTM research in many developing countries is limited, and even experience and knowledge that have been gained have not been widely disseminated. In particular, the link between the research community and decision makers is tenuous, and in many instances it does not exist at all. This is not unique to the degree of development, but the implications of this shortcoming are perhaps more serious than in the industrialized countries. Moreover, the positive impact that continued research could have in the decision making process is not fully realized because many countries in the region continue to rely on intermittent, ad hoc information sources for decision making.

It has been shown that better HTM is urgently required as an integrated component of the health care delivery process to ensure both sustainability of the health care system and an improvement in the quality of care (Commonwealth Regional Secretariat, 1995). It will be especially important to be able to predict, measure, and assess the effectiveness and impact of health care technology interventions and thereby to optimize their implementation and utilization. In order to achieve this, the elements of the management process–and their interaction–must be identified, and this can only be done through effective and appropriate research.

Justification/Rationale

In the process of health sector reform, special care is being taken to ensure that the implementation of reforms does not result in unwanted or unnecessary consequences. Changes in the health care sector generally are preceded by a comprehensive analysis of the prevailing situation, and medical audits and needs assessments have been conducted to identify real and perceived needs and to ensure that the health sector reforms are effective (WHO, 1995a; WHO, 1995b; WHO, 1996).

Due to the seriousness of the situation in many countries and the shortage of time available in which to achieve the desired changes, much of the data that have been used to justify the reform process have been extrapolated from few studies. Similarly, information on health care technology assessment and management practices is limited, and thus reforms in these fields are also based on few studies (ISTAHC, 1994). However, important studies covering many WHO member states have indicated the existence of a management crisis in most phases of the technology life cycle.

Historically, the methods of technology procurement and the organizational structures for this process have resulted in a general lack of accountability for the capital investments made. Few facilities have an effective inventory system, particularly for technology items, and therefore health care technology-related information is not available at the district, regional, or national levels. Stock taking and asset management are not performed (indeed, they are not possible) at any level of the health care system in many countries. Assessment of health care technology interventions is therefore not possible, and this situation severely hampers effective overall management.

A significant investment must be made in order to correct the problem. Unfortunately, there is no "quick fix" solution, and the problems currently experienced are merely symptoms of complex and interrelated shortcomings of the HTM process within the constraints imposed by the health system and general environments.

Health care technology is not necessarily linked to health outcomes, as is clearly demonstrated by the lack of appropriate health care technology policy and planning in many countries. Health care technology is often purchased on an ad hoc basis and generally is linked to the prevailing availability of funds rather than to a planned acquisition strategy. One typical consequence of this approach is the large proportion of medical equipment that is unused and/or nonfunctional; as much as 70% of the equipment does not function in certain countries.

These basic issues and their relationship to the efficiency of technology interventions are thus key elements in the understanding of the HTM process. Research is needed to link these to the entire management cycle and thereby to create a tool for effective decision making (Attinger and Panerai, 1988).

Research from the health systems paradigm also must be incorporated into an integrated assessment of the health care technology intervention process. This implies that health care technology must be linked to health outcomes and that the HTM process must be defined and quantified to ensure that decision makers are able to predict and manage health care technology interventions as an integrated element of health care provision.

Although health care technologies are often blamed for the high rise in costs associated with the provision of health care, very little evidence-based research has been offered to support this assertion. In fact, very little research has been done to establish the impact of health care technology on health outcomes. This is evident from the difficulties that researchers and planners have in linking technology with health care outcomes and in measuring quality-of-care indicators such as QALYs (quality adjusted life years) or disease burden indicators such as DALYs (disability adjusted life years) (Murray and Lopez, 1994; Mooney and Creese, 1995; Prost and Jancloes, 1995).

What is needed therefore is an initiative to address these issues in a global manner. It is proposed that the primary step was to develop an EHTP that would provide a planning tool for health care planners in terms of the management of health care technology, contribute to greater efficiency of health care technology resources, and facilitate improved management and support of the technology interventions (Hanmer et al. 1994; WHO, 1994; AFTH/MRC, 1995).

The EHTP, through its formulation and implementation, will help to ensure that technology interventions in the health sector reforms process can be properly assessed and successfully implemented. The package will provide governments, donors, policy formulators, and decision-makers with much needed tools to facilitate improved management of health care technology and thereby to ensure that cost-effective and equitable interventions can be provided at all levels of health care delivery. The EHTP will be equally useful for developing and developed countries alike.

Package Development

Methodology

The EHTP is a methodology and a tool aimed at strengthening and optimizing health care technology planning and management. The methodology is based on the hypothesis that effective health care delivery is possible only if all health care technologies (e.g., medical devices, drugs, human resources, and physical infrastructure) that are needed to support the health care delivery process are available and correctly managed. This implies that all health care interventions (which can either be promotive, preventive, or rehabilitative) that comprise the health care delivery process must be supported by the correct ratio and nature of medical devices, drugs, human resources, and physical infrastructure. For example, an appendectomy must be supported by the correct combination of medical equipment, surgical instruments and support equipment (medical devices), physician and nursing staff (human resources and knowledge), medical gases and anesthetic agents (drugs), and an appropriate surgical setting (physical infrastructure) in order for it to be successfully implemented and executed.

The ratio and magnitude of the combination of health care technologies will depend on the skills and clinical practice that are applicable for each technology used in each stage. The lack of one or more of these health care technologies clearly may have an impact on the effectiveness and/or on the ability to implement the desired intervention. In many cases the omission of critical and essential health care technology elements would impede the viability of the specific intervention. However, there are many interventions (in particular, promotive health care interventions) that require only one or two health care technology elements in order to be implemented effectively.

The objective of the EHTP therefore is to identify and quantify the above-mentioned nature and ratio of health care technologies for all medically recognized health care interventions. Figure 45-1 represents the relationship between health care interventions and their corresponding technologies. Health care interventions (i.e., promotive, preventive, or rehabilitative) that are represented on the x-axis can be linked to a corresponding medical device, human resource, and/or physical infrastructure, thus forming the EHTP matrix. The EHTP matrix is developed for all levels of health care delivery, thus reflecting the different nature of services and clinical practice, technological complexity, and sophistication of health care technologies that will be applicable to and realistic for the different levels of health care.

A three-dimensional matrix is therefore developed reflecting the inter relationship among health care technologies, health interventions (x and y-axes as in Figure 45-1) and levels of health care delivery (z-axis). This relationship provides the basis for a powerful management and planning tool since it enables health planners and decision makers to quantify and qualify similarities and differences in health care interventions (with respect to health care technologies), thus providing an unique platform to integrate services into more efficient and effective health care packages.

Linking the health care technologies to their corresponding interventions is based on current clinical practice and therefore is time and knowledge dependent, so the EHTP matrix is not universally applicable but must be adapted to become country-specific. Alternatively, it does not prescribe clinical practice but merely provides a basis for currently available methodology.

Logical Framework

The EHTP logical framework is based on linking existing and proprietary information sources such as the International Classification of Diseases (ICD) database and the Current Procedure Terminology (CPT) database with medical equipment, human resources, facility databases and drugs databases (Figure 45-2). The changes in extent and rate at which the contents of database change necessitates that updated proprietary databases (e.g., ICD and CPT) are incorporated into the EHTP package to ensure that up-to-date information is always available.

As shown in Figure 45-3, the ICD and CPT databases are cross-linked. In addition, three versions of the International Classification of Diseases database (ICD 10, ICD 9 and ICD CM) have been cross-referenced to expand the user base of the EHTP package.

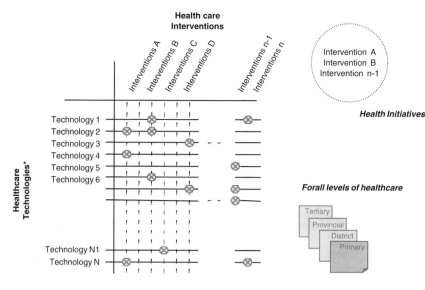

Figure 45-1 Two of the three dimensions of the EHTP matrix are shown above. The matrix represents the relationship of health care interventions and health care technologies. Health care technologies include medical equipment, human resources, drugs, and facilities and represent the mix of these health care technologies that are needed to perform a specific intervention. The interventions may be preventive, promotive, and rehabilitative health interventions. The EHTP matrix is developed for all levels of health care. A particular health initiative requires specific interventions, as indicated in the circle.

Currently there are more than 225,000 established links between the ICD and CPT databases.

The health care technology databases (i.e., medical equipment, drugs, human resources, and facilities) are then linked to the procedures database. This process is guided by the expertise of medical societies, experts, and individuals using their respective fields of knowledge. The process described above is used in the development of the EHTP template database, which will be described in detail below.

Figure 45-3 represents the logical framework of the linked database of the EHTP package. Each rectangle represents a proprietary database. The International Classification of Diseases Databases, ICD 9 and 10 (shown in orange) are the work product of the WHO and have been cross-referenced as described above. The link to ICD 9CM has been provided in order to ensure that countries using the Clinical Modification (CM) of the ICD 9 also will be able to use the EHTP software package.

The cross-referenced ICD databases have been linked to clinical procedures via a template that contains more than 225,000 links to the CPT procedure database, which is the copyright of the American Medical Association. This link of the procedure databases to proprietary health care technology databases (UMDNS, HR, DRUGS, and FACILITY) forms the EHTP template.

One standard feature of the EHTP database is the essential equipment (EEL), human resources (EHRL), drugs (EDL), and facility (EFL) lists, as shown in Figure 45-3. Limited querying and analysis tools are incorporated into the EHTP template software to provide limited functionality and analysis capabilities.

The EHTP package contains two main components. The first is the generic template (Figure 45-4) that has been described above and is used to develop the EHTP generic template database. In addition, the generic template forms the basis, which is modified by countries to reflect their specific conditions and realities. The second component of the EHTP package is the development of simulation tools, as shown in Figure 45-4. This package incorporates the generic template database and allows access to the generic template. The simulation tool is the actual package that provides the front end to health care planners to simulate and plan health care technology interventions as part of health care delivery.

The simulation tool contains four main components. The first incorporates the economic analysis of health care technology. Basic economic information, such as medical equipment fixed and recurrent costs, human resource costs, drug costs, and facility costs are stored in the generic EHTP template database and are accessed by the Simulation Package. Here the stored costs are applied to the simulation tools, and costing analysis becomes available.

The other elements of the simulation tools are structured in the same way. They include analysis tools for clinical, epidemiological, and related medical equipment information such as maintenance requirements. Accordingly, health care planners can simulate various possible health care delivery scenarios by varying cost, human resource and medical equipment availability without changing the EHTP template database and without much effort.

The template database information can be updated whenever new versions of the propriety database become available. This incorporation of the template database thus provides an up-to-date database for planners and also provides a conduit for them to sample both the latest developments of medical equipment and classification of diseases and preventive, promotive, and curative procedures.

Figure 45-4 further shows that the entire EHTP package is made up of two separate components that share a common core or template database. As will be discussed below, the separation of the two systems assists in the implementation of the EHTP as the generic template must be modified to produce the country-specific core database before the implementation of the simulation tools.

Additional modifications have been undertaken on the generic EHTP template database. In Figure 45-5, the incorporation of the EHTP classification database for the Integrated Management of Child Illness (IMCI), Integrated Management of Pregnancy and Childbirth (IMP AC) and Adult Lung Health Initiative (ALHI) is shown. These classifications of interventions, as defined in these special health initiatives, are directly linked to the CPT database. The WHO identified this requirement because many health interventions are performed without identifying the underlying disease, and this applies in particular at lower levels of health care delivery (e.g., clinic and health center).

The incorporation of this modification of the EHTP methodology does not alter the development of the EHTP generic template database. Substantially more effort will be required in the development of the EHTP simulation tools. The link between the diagnostic codes (the ICD databases) and the CPT procedures database needs further elaboration, however. In Figure 45-6, a breakdown of the complete linking process is given. The ICD classification has been divided into its root classifications, which are represented by Category, Block, Sub-block and Intervention. This representation of the disease classification codes follows the WHO coding structure. Therefore the category is represented by an alphanumeric code (e.g., 000-099 for pregnancy, childbirth, and puerperium). The Block is a subset of the category classification and is represented by the subset of the coding (e.g., 080-084 for delivery). The Sub-block and Intervention are represented by 080 for single delivery by forceps and vacuum extractor and "080.1" for low forceps delivery.

The next element is an artificial step called "Step," which was included to differentiate between diagnostic and therapeutic "Interventions." Many interventions performed at

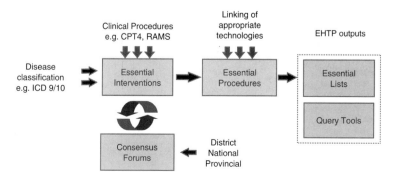

Figure 45-2 The point of origin for the EHTP is disease classification. The disease classifications are linked to clinical procedures which in turn are linked to appropriate technologies, which include medical devices, human resources, drugs, and facilities. The essential interventions are selected by modifying the EHTP templates through a process of country-specific consultations and consensuses. EHTP output includes Essential Lists for Human Resources, Drugs, Physical Infrastructure and Medical Equipment. In addition, powerful query and simulation capabilities ensure maximum benefit for health care planners.

166 Clinical Engineering Handbook

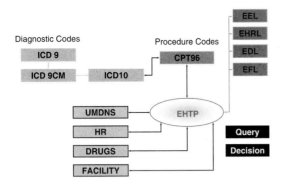

Figure 45-3 The EHTP logical framework. The disease classifications databases (ICD 9, 10 and 9CM) are linked to the clinical-procedures database (CPT), which in turn is linked to its appropriate technologies through linking medical devices (UMDNS), human resources (HR), drugs (DRUGS), and facilities (FACILITIES) databases. This link provides the basis for the EHTP templates. The EHTP output includes Essential Lists for Human Resources, Drugs, Physical Infrastructure, and Medical Equipment. In addition, limited query and simulation capabilities ensure maximum benefit for health care planners.

lower levels of health care delivery address only the diagnostic component of a disease classification, but due to practical and other reasons involve a requirement to refer the patients to higher levels of care to address diagnostic and/or therapeutic components of clinical practice. In order to capture this difference of therapeutic and diagnostic elements for health care delivery, this artificial step has been included. No special codes are attached to this step.

Next is the CPT and/or ICPC coding. As mentioned above, a template of codes linking CTP to ICD is provided in the EHTP template database for guidance. The CPT codes and nomenclature (e.g., "62278, vaginal delivery only") are given as shown in Figure 45-6. Multiple CPT procedures are again possible for a single Step and ICD Intervention. Again, linked to each procedure is another artificial element of the tree called the Techniques. This step is incorporated to ensure that all possible techniques may be captured to perform a single Procedure. Linked to the Techniques are the health care technologies (e.g., medical devices, drugs, human resources, and health facility), as described previously.

For each of the health care technology propriety databases, collateral information is collected. For example, for medical equipment, data on life expectancy, user training, complexity, criticality for the procedure, and maintenance requirements are collected. For the purposes of this chapter, not all details of this information will be discussed.

Implementation

Scope

The nature of implementation is slightly different for the different levels of health care delivery.

At the national level, the EHTP mainly will be used as a planning tool aimed at identifying the relationship between health care technologies and health care interventions. At lower levels of health care, the EHTP will focus on identifying resources with particular emphasis on simulation and comparing different health care delivery scenarios. Economic evaluation and cost analysis can identify hidden and recurrent costs, which typically include medical equipment costs such as installation, maintenance, user training, commissioning and decommissioning, consumables and spare parts, and purchase costs.

Adaptability of EHTP

The EHTP was developed specifically with the intention of being applied and implemented globally. The development of the EHTP template covering the entire spectrum of the ICD 10 codes contributes toward the country-specific implementation. Language adaptability has been provided, and most of the proprietary databases are available in various languages. For example, the UMDNS is available in English, Spanish, French, German, and Russian, while the ICD 10 is available in English, French, and Spanish, and the CPT in English and Spanish. (A French version of the CPT will be available within two years.) The EHTP database management system has been fully developed and tracks changes to the EHTP generic template ensuring proper management of the data-capturing process.

Implementation Process

Key activities for the EHTP implementation and information dissemination are given below and may be divided into three phases.

Phase 1

The first phase includes the training of trainers for EHTP implementation. Experts in the field of health care technology will be identified and will be invited to attend two workshops. These experts, once trained, will become the core group of trainers who will be responsible to support and guide the implementation, also monitoring and evaluation of the EHTP implementation both nationally and regionally. The curriculum for the training workshops includes hands-on training of the EHTP software, its methodology, and implementation strategies required to ensure successful implementation at the national level. Part of the first phase will be the development of training materials, guidelines, and facilitator notes. This documentation also will act as the reference material for the facilitators and trainers at the national level.

Phase 2

The second phase includes the identification of national counterparts from ministries of health or national institutions who ultimately will manage and facilitate the implementation of the EHTP at the national level. These individuals may be selected from within national ministries of health and finance and/or appropriate national institutions and will be responsible for all local management associated with the EHTP implementation. These national counterparts will attend two training workshops to familiarize themselves with and strengthen their understanding of the EHTP methodology and use of the EHTP software. In each participating country, infrastructure will be provided in order to support and manage the EHTP implementation. It will include the EHTP software, computer system and peripherals, modem, and documentation.

Phase 3

The third phase includes modification of the EHTP generic template database.

The EHTP software is provided to each participating country with a generic EHTP template containing data on such details as medical device costs, human resources, and drug requirements. These templates must be modified to reflect the requirements and

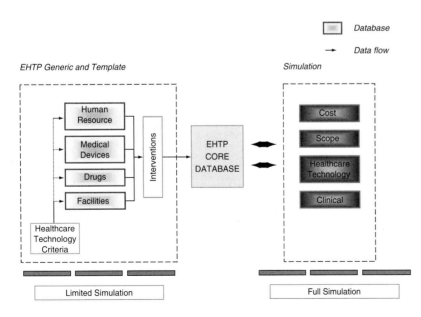

Figure 45-4 The EHTP system overview. The EHTP generic template database (shown on the left) is used to modify, edit, and append changes to the EHTP providing the basis for country-specific databases. These changes, which may include country-specific modifications, are stored in the EHTP core database (shown at the center). This core database is accessed by the simulation tools (shown on the right), which embrace all of the simulation and querying functionality of the EHTP package.

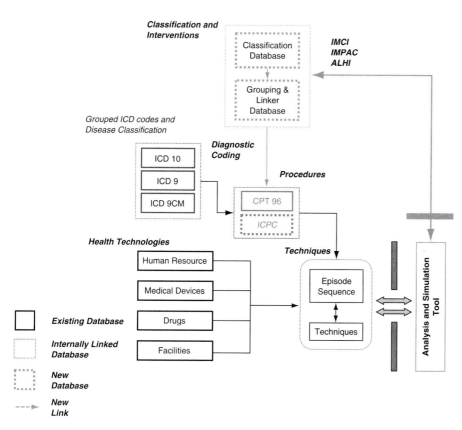

Figure 45-5 The EHTP template system overview. The EHTP health care technology databases (classification of diseases (ICD) and interventions of the special health initiatives) are linked to the CPT and International Classification of Primary Care (ICPC) procedures databases. A procedure sequence capability has been added to ensure logical flow of procedures. The limited simulation and tools component is shown. An additional link between the simulation tools and the special health initiative database has been implemented to ensure that grouped interventions, as defined in the special health initiatives packages by the WHO, can be simulated.

realities of the specific country of implementation. This process of modifying the EHTP templates will be managed by the national counterparts, as described in Phase 2, above.

The national counterparts will organize numerous workshops during the first year of implementation. Each workshop will address a specific clinical specialty (for example, anesthesiology, pediatrics, or mother-and-child health care), and local medical specialists and experts will be able to modify the EHTP generic template to reflect local conditions. The modified EHTP template then will be incorporated into the country-specific EHTP package at the national level and will form the basis of the health care technology management planning tool for decision-makers.

Information Promotion and Dissemination

This phase will be an ongoing activity for each participating partner country and its collaborating partners. Results and other outputs of the EHTP implementation will be disseminated at scientific meetings and meetings of regional stakeholders and will be provided to regional governments for their consideration.

In collaboration with the WHO, a number of countries have been selected for EHTP implementation. These include Namibia, Mozambique, South Africa, Estonia, Russia, Nepal, Sri Lanka, and Cuba. Numerous presentations have been made to the appropriate authorities in the selected countries. The benefits and implications of the EHTP have been presented and highlighted. The identification and selection of the Trainers of Trainers has been completed, and the first workshop was presented in Geneva and South Africa in 2000. The implementation schedule for the selected countries has been finalized, and resource material (e.g., facilitators' manuals, implementation notes, and guidelines) have been developed.

EHTP Justification

The burden of disease has become an integral component in the setting of health priorities for many countries. The concept of disease burden is well established and has been documented extensively. Many organizations have assisted governments in identifying their burden of disease, and the concept is now well entrenched and linked to economic, quality, and disability aspects qualifying health care delivery.

The processes of priority setting, health policy formulation, and the development and implementation of health initiatives that are specifically tailored to address disease burden are vast (in their own right) and therefore are not addressed in this chapter. The formulation of many health initiatives aimed at addressing specific diseases and illness (i.e., IMCI, IMP, and AC) and promoting well-being (i.e., Essential Programme on Immunization [EPI]) have been well documented and supported internationally. Guidelines and significant support infrastructure have been developed at the country level with the support of the United Nations agencies and governmental and nongovernmental organizations. Globally, the majority of governments have adopted and incorporated these health initiatives into their own action plans to address the prevailing national disease burden.

It is the realization and implementation of the above-mentioned health initiatives into existing governmental structures that has proven to be most difficult. The vertical nature of these health initiatives and the rigid hierarchy of government services indeed prevent effective integration and highlight the precarious situation that health care planners and decision-makers face.

The difficulty of implementing these health initiatives is partly explained by the nature and application of these health initiatives. Figure 45-7 shows a representation of the relationship between the health systems' requirements for various levels of health care delivery. At the managerial level (health systems-orientated), substantial managerial and planning component are required in order to achieve the objectives. However, as we descend through the layers of health care delivery, the technical (case management or patient-orientated) component becomes more important and the managerial aspects are reduced.

At the regional level, patient-orientated aspects of health delivery are often addressed exclusively by clinical experts while planners and decision-makers in dedicated but separate institutions address managerial and planning issues. This aspect of separating the patient-orientated and health systems-orientated planning and service provision elements is generally observed up to the district level. For health care delivery levels below the district level, both health systems and patient-orientated functions are usually allocated to one individual (typically a nurse or an appropriate health care professional with a clinical background), often resulting in the omission of management and planning components at levels below the district level. This fact significantly impacts on the effectiveness of service delivery and on the overall implementation of the above-mentioned health initiatives as management information and planning detail at the technical level (case management) is not incorporated into systems planning. The practical consequences of this are often reflected by the absence of sufficient drugs, medical devices, and human resources at clinics, health care centers, and, to a lesser degree, at district-level facilities.

This failure at higher levels of service delivery to provide direct support and guidance with regard to the management component concerning the allocation of resources is one of the most important shortcomings of the implementation strategy for most health packages and other initiatives.

Improvement of resource allocation cannot be addressed effectively if, in addition to the above-mentioned shortcomings, no mechanism exists to ensure that the planning process at the managerial level knows the requirements at lower levels of health care delivery. This feedback loop in Figure 45-2 is essential for the planning process.

The EHTP provides a unique method by overcoming this shortcoming concerning the management and technically components of health care delivery. The EHTP matrix provides a unique mechanism enabling the management level to identify resource requirements at the technical level (case management), particularly as it provides a platform to simulate patient-load and disease-profile scenarios.

One further shortcoming is the assumption that resources are allocated with specific reference to the health initiatives (e.g., the allocation of resources for health care technologies, drugs, infrastructure, and staff for special health initiatives like the IMCI

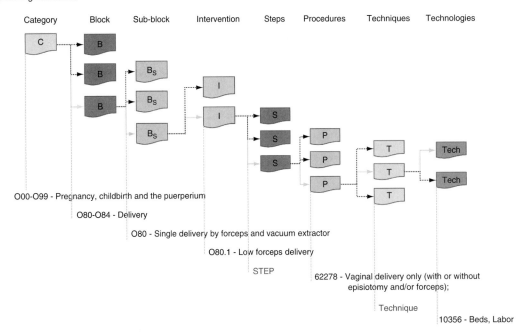

Figure 45-6 Linking of the International Classification of Diseases (blocks C, B, Bs, and I) to procedure (P), and technologies (Tech). An example from pregnancy and childbirth is given below the tree structure. C represents the disease classification; B represents the disease block; Bs represents the block subset; I represents the intervention; S represents the step; P represents the procedure; T represents the alternative techniques per procedure, and Tech represents the technology (e.g., medical devices, drugs, human resources, and health facility).

program). In reality, however, government ministries are funded at the functional and departmental levels, and specific and dedicated health initiative resource allocation is thus very difficult and often virtually impossible.

This process of priority setting, health policy formulation and implementation, and the development and specification of the Essential Health Packages often receives attention at the highest national level. Often more resources and political support ensuring some form of success support this process. The implementation of the Essential Health Packages, however, does not always receive the same importance from both the political and financial level, and therefore the implementation often fails or is implemented suboptimally.

Not only are political and financial resources required for the successful implementation of the Essential Health Packages, but also accurate and specific information is needed. The inability of lower levels of health care delivery (e.g., the district level) to provide meaningful information feedback to central and national ministries also contributes toward this suboptimal implementation of the Essential Health Package. Often the management and service provision details, which are applicable at the district level, are not known at the central level, which might lead toward the inappropriate design and implementation of the Essential Health Packages. Other reasons for the suboptimal implementation include:

- Lack of the implementation funds
- Shortage of skilled personnel
- No monitoring and evaluation of implementation process
- Omission of guidelines
- Inability to detail the realities of lower health care delivery levels

It is specifically this barrier that the EHTP aims to overcome by providing detailed information to decision-makers, thereby ensuring that the Essential Health Package is translated into a realistic package, permitting practical application.

The operational focus of the EHTP is shown in Figure 45-8. The EHTP specifically focuses on overcoming the barrier, as described above, and ensuring that a benefit is derived both at the district level and at the central level via the feedback loop, as indicated.

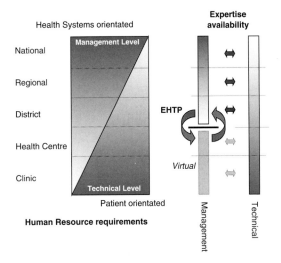

Figure 45-7 The EHTP matrix provides a "virtual" management environment by providing detailed information to health managers at a technical level. The EHTP also ensures that management information from a higher level of health care provision is reflected within this virtual management environment and thus ensures that the information exchange between the technical and management levels is maintained. (See horizontal arrows.) The figure on the left indicates the management/technical requirements associated with each level, of health care. Below the district level, however, the management capacity of health care providers in many countries is significantly reduced due to the exclusively technical nature of staff. The black line represents this barrier whereby the circular arrows represent a bridging of the EHTP gap.

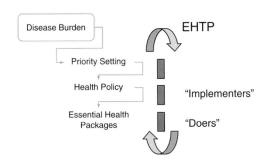

Figure 45-8 In the delivery of health care, the development of Essential Health Packages through prioritizing interventions to address disease burden is well known. However, the integration of these "vertical" Essential Health Packages into existing health care programs is extremely difficult due to the rigid resource-allocation mechanism. Therefore a barrier exists between the planning and the implementation of these Essential Health Packages. The EHTP provides a mechanism to bridge this barrier and ensures that information is both provided to and received from the implementation process.

EHTP Matrix in the Support of Health Care Technology Management and Planning

The development of the EHTP matrix provides health care planners and decision makers with a unique tool supporting the process of resource allocation. In addressing the burden of disease and the subsequent development of Essential Health Packages (through priority setting), the issue of allocating resources is paramount. Although resource allocation may be categorized and even associated at a macro level to disease burden, the allocation of actual resources within executing agencies such as ministries of health are exclusively linked to human resources, capital, and recurrent expenditure. The decision making process therefore becomes confronted with a complex problem of allocating resources to existing and known health care technologies (e.g., medical devices, human resources, drugs, and infrastructure) and trusting that these health care technologies when bundled into action and together with health care programs will address and alleviate the prevalent disease burden. The absence of specific information of the effect that these health care technologies have on the management of disease burden and how this comes about might be the weakest link in the current health care technology planning process. The EHTP, through application of its matrix, therefore provides a fundamental component in the planning process and uniquely addresses the shortcomings mentioned above. In addition, the EHTP also provides a qualitative method through which resource allocation (specifically health care technologies) can be evaluated and assessed.

The identification and application of this matrix therefore supports decision makers in prioritizing which health care technologies (medical devices, human resources, drugs and facilities) must be supported to guarantee the effective implementation of health care interventions to address identified diseases. In particular, the ratio of recurrent cost versus capital investment within any given health care system can be determined within the context of addressing the identified diseases.

Link to Other Health Initiatives

One of the criteria in the formulation of the EHTP concept was that the point of origin should be able to tie into existing health planning and -management tools and indicators. This requirement was seen as fundamentally important in ensuring continuity between macro and micro aspects of health care delivery.

Disease burden has been recognized as a fundamental and important indicator in any health planning process and has found universal application. The inherent macro nature of the disease burden indicators clearly makes this indicator unsuitable for the process of linking health care technologies to known and applied health service indicators. However, disease burden indicators have been linked to disease classifications and diagnostic codes, thus making the disease classification and codification an ideal point of entry for the EHTP.

The ICD, in their own right, have found universal application, and many nations have standardized on one of the versions available. The latest version, ICD 10, is a revision of the widely distributed ICD 9 codes, which classified the majority of curative diseases. These codes, which have been developed under the auspices of the WHO, now also include specialty areas such as neurology, mental health, dentistry, and anesthesiology. International codes on primary health care (ICPC) have been developed as well; however, they have not found the same acceptance as the ICD 9 and 10. This has resulted in many nations developing their own primary health care classifications and codes, preventing accepted classification. This issue makes the link between the EHTP matrix and the primary health care classification far more difficult.

EHTP Advantages

EHTP Application

The EHTP finds its principle application in the planning and management of health care technologies. The management of medical devices, human resources, drugs, and facilities is multidisciplinary and requires attention at various levels of the health care system. The scope and complexity of the planning and management process of health care technologies are different for the various levels of health care. For example, the range and even the level of complexity of medical devices and skills required at the district level is vastly different from that required at the regional or central level. The degree of effort to manage these technologies therefore also differs from level to level and requires a unique approach for each level of health care delivery.

The management of these health care technologies can best be described by its life cycle. The life cycle of health care technologies can be categorized into three distinct phases: Provision, acquisition, and utilization. The EHTP supports all three phases by providing both quantitative and qualitative information, but also it provides a unique opportunity to simulate various conditions and effects for different health care scenarios. Within this context, the EHTP has found a unique application for which the main areas of use are summarized below.

The EHTP:

- Provides a map and link of all international disease classifications to promotive, preventive, and rehabilitative procedures
- Provides the essential health care technology elements (e.g., medical devices, human resources, drugs, and facilities) required to reduce the burden of disease
- Provides a list of all health interventions and typical time requirements per intervention for each disease classification
- Provides dynamic Essential Lists for specific levels of health care delivery and/or disease classification. These include medical equipment lists, human resources lists, essential drug lists, and essential facility lists
- Provides a qualitative method of determining capital and recurrent costs of medical devices per disease classification
- Provides management data on medical devices, including technical classifications, maintenance, and utilization information
- Provides economic data on human resources, facilities, and drugs
- Simulates the costs of medical devices, human resources, and drugs for all disease classifications, interventions, and clinical procedures for various levels of health care
- Simulates the interdependencies of medical devices, human resources, drugs, and facilities for various health interventions and scenarios
- Provides a reference to health care technology audits, existing standard equipment lists, and staffing requirements

Essential Equipment Lists

The static nature of the standard equipment lists makes them less attractive in the planning and allocation of resources. Many standard equipment lists have been developed before and have been modified repeatedly, but the shortcomings of these lists are not really addressed. The development of the EHTP provides a number of advantages (given below) as compared to the standard equipment lists.

EHTP provides a dynamic Essential Equipment List. The link between use of medical devices and health interventions is uniquely identified and is therefore modifiable. Changes to either the health intervention (which includes changes to clinical practice) or health technologies (which includes medical devices, human resource skills, or techniques) can be easily incorporated into the Essential Equipment Lists, which are based on the EHTP methodology.

The fact that all health interventions have been mapped to their respective health care technologies also provides an additional benefit. For example, the addition of new health care technologies to the Essential Equipment Lists can be evaluated by assessing their utilization rate within a package of health care interventions and determining the other health interventions that could benefit from the introduction of this new health care technology. Similarly, shared human resources and skills can be identified and easily evaluated.

Changes in disease patterns and epidemiological differences within a regional or country can be incorporated into specific and focused equipment lists that are derived from the EHTP. In addition, the development process and effort associated with formulating standard equipment lists is greatly reduced, thus allowing valuable resources time to be applied to other activities.

A fundamental shift in the development of the Essential Equipment Lists has also been achieved through using the EHTP methodology. The static nature of the standard equipment list determines that a limited number of experts normally formulate these lists. In comparison, the Essential Equipment Lists derived through the EHTP methodology have a much larger expertise base because each disease specification and its resulting health intervention is addressed by an expert or panel of experts in that field. The advantage of this process is that each medical specialty has input in the development of an appropriate health care technology list without prescribing or influencing the lists formulated by other medical specialties.

Requirements at any given health facility as a function of human resources, medical devices, and drugs also support improved decision making. This issue in particular is important because facilities in the past have been treated independently and as being quite distant from health interventions and the prevailing disease burden. The development and implementation of facilities has a significant budgetary implication for any health care system. Independent and dissociated physical-infrastructure planning clearly has a negative impact on the effectiveness of health care delivery. In particular, recurrent and long-term liabilities of the physical infrastructure planning process are ignored in the formulation of standards and norms associated with facility planning. The availability of the Essential Facility Lists based on the inter relationship among facilities, medical devices, drugs, and human resources therefore provides a significant improvement in the health care technology planning process and assists decision-makers in gauging and defining an appropriate level and scope of physical infrastructure.

Scientific Basis of EHTP

One major concern in the development and implementation of the Essential Health Package was that much of the planning required quantitative data. Rapid changes in health care technology trends, clinical practice, disease burden and profiles, and economic realities have made health care technology planning more challenging than ever before.

Evidence has shown that planning and decision making based on past trends seldom correlates accurately with future situations for systems as complex as those in the health sector. Planning for the future therefore requires significant emphasis to be placed on providing solutions that are based on evidence and that are scientifically defendable. The EHTP methodology has incorporated this trend by providing scientifically sound methods of relating health care technologies to health interventions without losing sight of the overall objective.

The need to determine the impact of health care technologies on health care delivery also forces us to consider a scientific basis that can quantify this impact. During the last decade, many donor agencies have been confronted with the fundamental question of what the impact is of the health care technologies that they have provided on the recipients' health care system. The lack of any proper scientific basis to determine this impact forced these agencies to use empirical evaluations and assessments to determine the effect within a focused and time-limited survey.

Promotive, Preventive, and Curative Health Care

The EHTP methodology allows the linking of any health care intervention to its relevant technologies. Currently the curative health care component has been incorporated into the EHTP, based on the immediate availability of the ICD. However, promotive and preventive health interventions form an important component of the overall health delivery package of any nation, and provision therefore has been made to incorporate these into the EHTP.

Dynamic System

Changes in disease profiles at all levels of health care suggest that the Essential Health Package needs similar changes to address such changing disease profiles. The static nature of Essential Health Package seldom is in the position to address any rapid changes–especially those in disease profiles.

Due to the static nature of the specialized health initiatives as part of the Essential Health Packages and the enormous effort that is required in order to implement these packages, even minor changes to the Package require enormous effort to ensure that these changes are applied throughout the health care system. These changes and the untested and unevaluated impact and long-term effects have made the implementation of the Essential Health Packages such a difficult task.

The EHTP addresses this static nature of the Essential Health Packages, by providing the platform of relationships between service interventions and their required health care technologies. Therefore an addition or omission of any health intervention from the Essential Health Package can be evaluated and assessed rapidly without having to re-evaluate the entire EHTP package.

Simulation of Interventions

The EHTP also provides a useful platform to simulate the cause and effect of health care technologies concerning health care interventions and disease burden. The nature of the EHTP matrix determines that all health technologies are linked to the associated health interventions, providing a useful tool to determine the effect of health technologies upon each other and upon health care interventions and the burden of disease. For example, the omission of a medical device will reveal the health technologies (e.g., other medical devices, drugs, human resources and facilities) that will be affected by this omission. In addition, all medical and nonmedical procedures and interventions that will be affected will then become known.

Clinical Procedures and Clinical Practice

The rapid development of new medical devices, knowledge, and clinical practice has resulted in significant changes in the delivery of health care. The enormous scope of new medical devices reaching the markets each year, the impact that they have on clinical practice, and the burden that they have created for health care planners and decision-makers alike have necessitated a reinvestigation of current planning procedures and policies. It is apparent that the magnitude and complexity of new technologies that have been introduced into the health sector no longer can be managed by traditional, existing managerial capabilities.

Coupled with the increasing rate of technological advances, the increased application to specialized health and medical areas and the increased need for support technologies all have increased the need for a new health care technology management paradigm. Existing health care technology management tools, such as the static equipment lists, no longer are able to incorporate these changes fast enough to provide a useful management tool.

In addition, the multidisciplinary nature of the technical advances, which span the spectrum of curative, preventive, and promotive health care, requires the input of a team of experts in the planning process, to ensure that all areas of health care provision are covered. Reality dictates that a continuous input from experts in the planning process cannot be maintained. The EHTP partially addresses this problem through its multidisciplinary approach. The EHTP matrix also allows simultaneous and independent modification and alterations without affecting the availability of the management tool.

National Implementation

In addition, national implementation has the advantage that local experts can oversee and perform modifications of EHTP template, thus ensuring that local conditions that reflect economic realities, clinical practice, skills base, and social and political considerations can be incorporated into the EHTP. Basic and essential services can be determined within the constraints of the budget and may be included as part of the EHTP implementation.

Definition and Delineation of Terms

The definition of "health technology," used by WHO, includes devices, drugs, medical procedures, and surgical procedures, and the knowledge associated with these, used in the prevention, diagnosis, and treatment of disease, as well as in rehabilitation, and the organizational and supportive systems within which care is provided (Broomberg and Shisana, 1995).

Health care technology therefore includes all medical equipment, devices, and consumables, as well as the organizational, support, and information systems used at all levels of health care delivery. Medical and surgical procedures, especially those that make substantial use of health care technology as defined, are included by implication.

Health care technology is therefore an umbrella term for a family of systems, techniques, and physical items used in health care delivery.

This definition also incorporates:

Organizational and physical infrastructure, including health facilities and buildings; their installations and plant; energy sources and water and gas supplies; and supportive and logistical systems, whose components are supply systems, information systems, communication and transport systems, and waste disposal systems.

The health care technology life cycle is considered to consist of three phases:. Provision, acquisition, and utilization, each of which can be subdivided as follows: Provision includes assessment of needs, research and development, testing, manufacture, marketing, technology transfer, and distribution. Acquisition includes technology assessment and evaluation, planning, procurement, installation, and commissioning. Utilization includes operation, user training and support, maintenance and repair, risk management, assets management, and decommissioning. The first phase falls within the domain of industry, while the last two are generally the responsibility of national health systems and health care providers.

References

AFTH/MRC. Proceedings of the AFTH/MRC Regional Workshop on Health Care Technical Services in the Sub-Saharan Region. In Poluta M (ed). Tygerberg, South African Medical Research Council, Midrand, South Africa, April 1995.

Attinger EO, Panerai RB. Transferability of Health Technology Assessments with Particular Emphasis on Developing Countries. *Int J Technol Assess Health Care* 4(4):335-335, 1988.

Banta HD, Andreasen PB. The Political Dimension in Health Care Technology Assessment Programs. *Int J Technol Assess Health Care* 6(1):115-123, 1990.

Barnes DE. Value of Technology Assessment in Developing Countries. *Int J Technol Assess Health Care* 6(3):359-362, 1990.

Broomberg J, Shisana O. *Restructuring the National Health System for Universal Primary Health Care: Executive Summary.* Pretoria, South African Department of Health, June 1995.

Commonwealth Regional Secretariat. Health Reform Initiatives in East, Central and Southern Africa. Proceedings of the Directors Joint Consultative Committee Meeting. Nairobi, Kenya, August 1995.

Drummond M. Evaluation of Health Technology: Economic Issues for Health Policy and Policy Issues for Economic Appraisal. *Soc Sci Med* 3(12):1593-1600, 1994.

Free MJ. Health Technologies for the Developing World: Addressing the Unmet Needs. *Int J Technol Assess in Health Care* 8(4):623-634, 1992.

Goodman C. It's Time to Rethink Health Care Technology Assessment. *Int J Technol Assess Health Care* 8(2):335-358, 1992.

Hanmer L, Heimann P, Issakov I, et al (eds). Executive Report on the Regional Workshop on Health Care Technology in the Sub-Saharan Region. Somerset West, South African Medical Research Council, Tygerberg, April 1994.

Heimann P, Issakov A. *Integration of IMCI, IMPAC and ALHI and the EHTP*. Proposal to WHO, February 2000.

Issakov A. Health care equipment: A WHO Perspective. In van Gruting CWD (ed). *Medical Devices: International Perspectives on Health and Safety.* Elsevier Science BV, Burlington, MA, 3–5, 1994.

Issakov A. Service and Maintenance in Developing Countries. In van Gruting CWD (ed). *Medical Devices: International Perspectives on Health and Safety.* Elsevier Science BV, 21–38, 1994.

ISTAHC. Technology Assessment in Health Care for Developing Countries. *Int J Technol Assess Health Care (Special edition reprint)*. Cambridge University Press, 1994.

Jennett B. *High Technology Medicine–Benefits and Burdens.* Oxford University Press, 1988.

McKie J. Management of Medical Technology in Developing Countries. *J Biomed Eng* 12(5):259-61, 1990.

Medical Alley. *Measuring Cost Effectiveness: A Road map to Health Care Value. Document Two: A Technical Guide.* Minneapolis, 1995.

Mooney G, Creese A. *Priority Setting for Health Service Efficiency: The Role of Measurement of Burden of Illness.* Geneva, WHO, 1995.

Murray CJL, and Lopez AD. *Global Comparative Assessments in the Health Sector: Disease Burden, Expenditures and Intervention Packages.* Geneva, WHO, 1994.

Prost A, Jancloes M. *Rationales for Choice in Public Health: The Role of Epidemiology.* Geneva, WHO, 1995.

Scenario Commission on Future Health Care Technology. Anticipating and Assessing Health Care Technology, Vol 1: General Considerations and Policy Conclusions. Martinus Nijhoff Publishers, 1987.

Webster JG, Cook AM (eds). *Clinical Engineering–Principles and Practices.* Prentice-Hall, 1979.

WHO Health Systems Workshop. Arusha, Tanzania, November 1995.

WHO Health Care Technology Experts' Working Group Meeting. Brazzaville, Congo, April 1996.

WHO Intercountry Meeting on Achieving Evidence-Based Health Sector Reforms in Sub-Saharan Africa. Arusha, Tanzania, November 1995.

WHO. Selection and Development of Health Technologies at District Level. Resolution AFR/RC44/R15 Adopted at the 44th Session of the WHO Regional Committee for Africa. Brazzaville, September 1994.

World Bank. *World Development Report: Investing in Health.* Oxford University Press, 4, 1993.

World Bank. *Better Health in Africa: Experiences and Lessons Learned.* World Bank, Washington, DC, September 1994.

46

Impact Analysis

Thomas M. Judd
Director, Quality Assessment, Improvement and Reporting,
Kaiser Permanente Georgia Region, Atlanta, GA

There are many issues contributing to health in developing countries. Availability of appropriate health technology (HT) resources is one principal factor. The World Health Organization (WHO) defines HT resources as including human resources, pharmaceuticals, equipment, and facilities. Proper management of HT can assist in meeting health needs by addressing supportability and sustainability of HT resources. Integrating HT resource planning and management as they support best clinical practices remains a challenge. This has been demonstrated through several relevant case studies in which clinical engineers have directed or assisted HT management (HTM) activities providing cost benefits, and improved quality of care and patient outcomes.

Effects of Health Care Technology Management

Economics and Health

Developing countries typically spend about 7% of their Gross Domestic Product Purchasing Power Parities (GDP PPP) on health. For example, U.S.$2 billion out of U.S.$28 was spent in recent years in a Latin American country, or $240 annually per person with per capita annual income of about U.S.$2,000 (Judd, 1997). Ideally, the investment in health would address the major health needs, would assist in improving health status, and would improve life expectancy (LE). Major needs and disease prevalence for the example include cardiovascular disease and cancer for adults, and intestinal, nutritional, and respiratory conditions for infants and children under five years of age. Interventions typically require the use of various health facilities and medical devices. Economic growth also affects LE. This growth is influenced strongly by availability of physical resources, knowledge resources (physical, technology, and human), and economic liberty (sociopolitical systems).

Impact of HT on Health

Proper use of HT resources can reduce risk of disease, shorten duration of disease, improve access to care, improve patient functional status, contain health care costs, and lower intervention risk. Technical innovation in health care is a relatively new phenomenon, providing a wide range of positive treatments in the prevention and cure of illness.

Health organizations as "socio-technical systems" are highly dependent on the human factors that bridge the people and other HT resources used in treatment. The discipline of clinical engineering (CE) has emerged in the last quarter of the 20th century, to assist health professionals in various health settings with the effective use and management of a vast array of medical devices, processes, and health facilities that are currently available. CE is a subspecialty of biomedical engineering that is devoted to the application of engineering methods and technologies in the delivery of health care. HTM addresses challenges in HT planning, procurement, and management, as shown in Table 46-1.

Health System Structure Issues

Health care typically is delivered through both the public and the private health sectors. Local, provincial, and national health authorities interact in any country. Empowering these authorities should be one objective of health care reform. Typically, the central government's role in improving health is largely in standards-setting and fiscal control through, for example, a ministry of health (MoH). As a new health care model is formulated in a country that implements reforms entailing increased investment in health technology, development of increased HTM capability is required. HTM can assist in overcoming waste and the "iceberg" syndrome, as shown in Figure 46-1. This effort requires the MoH's support in recognizing and implement HTM to support best clinical practices.

A health care system that falls victim to the iceberg syndrome, largely as a consequence of inadequate HTM, typically experiences problems and corresponding wastages, as shown in Table 46-2.

Table 46-1 Elements of health technology management

Health Technology Management of Medical Equipment		
Planning	*Procurement*	*Management*
Needs analysis	Specifications	Receiving
Strategic technology planning	Life-cycle cost analysis	Staging
		Architectural, engineering, and construction coordination
Technology assessment	Terms and conditions	
Facility evaluation	Tendering	Installation quality assurance precommissioning inspection
Financial evaluation	Bid analysis	
Equipment planning	Selection	Management systems and software
Architectural	Purchasing	Warranty monitoring
Coordination	Financing	Service contract monitoring
Engineering support		Training management
		Risk management and quality assurance
		CE department procedures
		ISO 9000 quality documentation

Assessment of Impact

In order to improve performance of health systems, health indicators are needed in order to measure the effect of HTM practices. Such effects of HTM practices include those listed below:

- DALYs (disability adjusted life years, global burden of disease) assess patterns of illness and injury with units that take into account premature death and impact of disabilities. Better HTM improves DALYs.
- Better HTM improves HT resource planning for unmet regional health and resulting HT resource needs.
- Patterns of health expenditures on inappropriate, wasteful, or potentially harmful services, reducing overuse.
- Cost-effective care providers and practices (e.g., evidence-based medicine (EBM)), and increasing underused practices.
- Interventions that improve safety and quality-of-care processes and outcomes in all health care settings.

Quantitative Evidence of HTM Impacts

The following quantitative evidence of the impact of HTM has been observed with medical devices, health facilities, and health delivery system processes (Judd, 1998).

HTM Case Studies

USA—Medical Equipment Management

HTMs for equipment resources were conducted in seven geographic areas with 21 hospitals, whose average size was of 225 beds (David and Judd, 1993). The seven HTM teams of 10 technicians each provided equipment management services to a wide range of medical equipment including general biomedical, laboratory, and radiology equipment for each three of these hospitals. Three years after program start-up, there was a 30% cost savings for internal equipment management versus external manufacturer support, primarily due to a 50% reduction in radiology maintenance costs. For example, there was $540,000 in support costs at baseline of which $340,000 was for radiology equipment; after three years, there was a $170,000 reduction after all costs were taken into account.

Success for internal equipment maintenance led to involvement in other HTM activities such as the following, accomplished by the staff:

- Moved and reinstalled X-ray rooms.
- Upgraded CT scanner systems.
- Were key decision-makers in new acquisition, and assisted in improving equipment-replacement decisions.
- Led a process that resulted in standardization of medical imaging equipment.
- Improved quality and cost-effectiveness of high-cost radiology equipment consumable items (e.g., glassware).
- Developed processes to identify risk-related trends in equipment use and maintenance.
- Became more widely involved in related facilities planning issues.
- Improved technology planning based on equipment utilization and uptime indicators.

HTM Workshops

The American College of Clinical Engineering (ACCE), World Health Organization (WHO), and Pan American Health Organization (PAHO) have conducted Advanced Health Technology Management Workshops (Judd et al., 1991). The effects on medical equipment HTM demonstrated in the United States above have been partially extended to national health care technical services (HCTS) in developing countries. The process of transferring these kinds of results has been led by WHO Department of Health Services Provision (WHO/OSD), which has worked from the mid-1980s to the present with the global clinical engineering community through a series of international workshops. (See Chapter 70.) ACCE began international HTM workshops in partnership with WHO and PAHO in 1991, which, to date, have resulted in training key equipment HTM leaders on five continents and in 75 countries. Principal global areas that had been served prior to that time include Latin America (1991, 1994, 1997, and 1998), Eastern Europe, and the former Soviet Union (1993), the People's Republic of China (1995), the Middle East (1996), Russia (1999, 2000), and Sub-Saharan Africa (1999). The ACCE website http://www.accenet.org/InternationalCommittee/index.html describes past workshops and locations dating back to 1999.

ACCE developed a syllabus for these workshops, under contract for WHO (Dyro et al., 2003). The following outline demonstrates HTM training issues for equipment and facilities in developing countries.

Section 1 Introduction
1.1 Advanced Clinical Engineering Workshops
1.2 Health Care Technical Services
1.3 Guide for the Formulation of a National Health care Equipment Policy
1.4 WHO Essential Health care Technology Package (EHTP)
1.5 Maintenance in Remote Areas and Mobile Maintenance Units
1.6 Clinical Engineering Department Management
1.7 Organization of Equipment Maintenance Program or Department
Section 2 Maintenance And Service Management
2.1 Elements of an Equipment Control Program
2.2 Determining and Organizing the Technical Workload
2.3 Unscheduled Service
2.4 Improving Service Quality
2.5 Managing a Maintenance Department
2.6 Equipment Replacement Planning
2.7 Maintenance Procedures
2.8 Preventive Maintenance Management
Section 3 National Health Care Equipment Planning
3.1 National Health Care Equipment Policy and Communication Plan
3.2 Needs Assessment and Planning
Section 4 Impact Analysis
4.1 Technology Management in Preventive, Primary, and Tertiary Care
4.2 Strategic Planning
4.3 Asset Management
4.4 Technology Management
Section 5 Selection and Procurement Strategies
5.1 Planning
5.2 Selection and Procurement
5.3 Donated Medical Equipment
5.4 Managing Installation and Service Agreements
Section 6 Information Systems
6.1 Principles of Database Management
6.2 Implementing Data Collection and Record Keeping Technology
6.3 The Equipment Control Master File
6.4 Documenting Service Work Performed
6.5 Parts Management
6.6 Contract Management
6.7 Information Systems
6.8 Computerized Inventory and Records Management Systems
Section 7 Technology Assessment
7.1 ABCs of Technology Assessment
7.2 Technology Assessment in the Local Health care Facility
Section 8 Resource Allocation
8.1 Setting Up a New Program
8.2 Developing a Budget

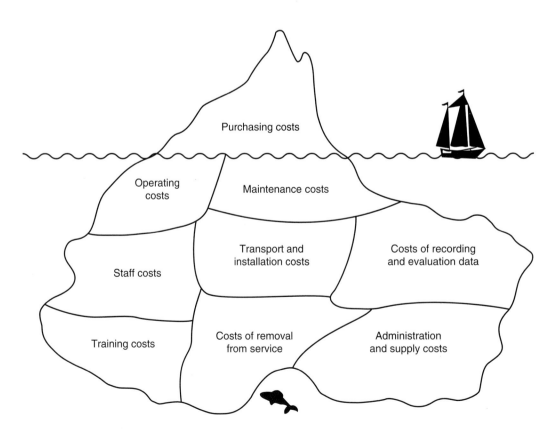

Figure 46-1 The iceberg syndrome in life-cycle costs of health care equipment.

Table 46-2 Problems and corresponding wastages as found in many developing countries

Problem	Waste
Inability to correctly specify and foresee total needs when tendering and procuring equipment	10–30% extra cost
Purchase of sophisticated equipment that remains unused due to lack of operating skill and technical staff	20–40% of equipment
Extra modifications or additions to equipment and/or buildings unforeseen at the initial tender stage due to lack of staff expertise	10–30% of equipment
Maltreatment by operating and maintenance staff	30–80% of lifetime
Lack of standardization	30–50% extra spare parts cost
Down time due to inability to use or repair, or due to a lack of spare parts or accessories.	25–35% of equipment

Table 46-3 Quantitative evidence of HTM impacts (Judd, 1998)

System	Cost (or time) savings
Medical devices	
Maintenance	20–30%
Reduced investment through planning	10–20%
Reduced development time for acquisition specifications	(2–4 weeks)
Appropriate HT introduction	10–90%
User training reducing maintenance	10%
Health facilities	
Reduced investment through planning	10–20%
Reduced time for planning	30%
Utilization	
• Ambulatory care	20%
• Inpatient care	20%
• Diagnoses and treatment	50%
Health delivery system processes	
Chronic disease treatment	
• Appropriate use of technology	20%
Supplies- and logistics-management process redesign	20%

8.3 Developing a Business Plan
8.4 Centralized versus Charge-Back Budgeting Strategies
8.5 Marketing and Selling Your Services
8.6 Ongoing Relationships with Manufacturers
8.7 Financial Management for an ISO
Section 9 Human Resources Development
9.1 Defining Job Roles and Responsibilities
9.2 Staffing Levels
9.3 Roles of Supervisors and Managers
9.4 Setting Standards and Measuring Candidates
9.5 Defining and Measuring Productivity, Efficiency and Effectiveness
9.6 Training
9.7 User Error—Discovery and Response
9.8 Stress Management
9.9 Professional Development
Section 10 Safety, Risk Management and Quality Improvement
10.1 Safety
10.2 An Integrated Safety Program
10.3 Safety Training Programs
10.4 Risk Management
10.5 Total Quality Management Principles
10.6 Quality Improvement
Section 11 Utility Systems
11.1 Electrical Distribution Systems
11.1 Heating, Ventilation and Air Conditioning
11.3 Medical Gas and Vacuum Systems
11.4 Water Systems
11.5 Fire, Smoke and Heat Detection Systems
11.6 Sterilization
11.7 Waste Management
11.8 Utility Problems in Rural Areas
Section 12 Technical Lectures on Medical Devices
12.1 Troubleshooting of Medical Equipment: Concepts and Applications
12.2 Clinical Laboratory Technology
12.3 Radiology Equipment Technology
12.4 Physiological Monitoring Equipment
12.5 High-Tech Equipment

Section 13 Summary of Workshop Goals and Objectives
13.1 Summary of Workshop Goals and Objectives

Africa–Medical Equipment Management

The following case studies were presented at the ACCE HTM Workshop for Sub-Saharan African countries in Cape Town, South Africa, in November 1999.

Ghana–Juergen Clauss, FAKT, 1999

- Forty of 124 hospitals are assisted by Ministry of Health (MoH) HCTSs, and another 15 hospitals are assisted by two nongovernmental organizations' HCTSs and three private companies.
- Health sector reform is underway and includes contracting of health services to public and private providers for a variety of health resources, including equipment and facilities.
- Equipment management and technical training courses are provided by US NGO International Aid.

Tanzania–Juergen Clauss, FAKT, 1999

- Seventy-five of 195 hospitals are assisted by the MoH HCTSs, and another 60 hospitals are assisted by NGO HCTSs. There are several MoH HCTSs in different regions, as well as three private companies and an equipment HTM training school.
- Health sector reform is underway, however there is no HT policy process at this time. There are various donor investment programs at work.

Senegal–Roger Schmitt, GTZ, 1999

- There are six HCTSs in six cities, with a focus on regional and national hospitals (between 150 and 400 beds in size).
- Conclusions of several studies in the 1990s: HCTSs with up to two technicians and two craftsmen are economical. However, equipment preventive maintenance is irregular; there are equipment procurement problems; health-reform decentralization complicates improvements in HTM programs; and often donor activities are not helpful to the development of HCTS.

Kenya–John Paton, GTZ, 1999

- There was a study of medical equipment spare parts that focused on specifications, procurement, storage, distribution, and finance issues. The study initially identified 10–20 more urgently needed parts that were purchased with EU funding and distributed to HCTSs, where they were quickly used. Effectiveness of repairs was to be studied next.
- A comprehensive list of needed parts is being provided as a result of establishing equipment standards, including parts distribution and storage; establishing a procurement system; and ensuring long-term financing to promote sustainability.

HTM in Preventive Care

Proper preventive screening with appropriate HT resources leads to better patient outcomes (Judd and Painter, 2000). Equipment management should be focused in the areas that are known to support the best clinical practices, as shown in Table 46-4. Resources and their impacts are cited in evidence-based prevention guidelines, as shown below:

Breast Cancer Screening

Impacts

- Breast cancer is the most common cancer and major cause of cancer death in women aged 40–69; the American Cancer Society estimates over 180,000 new invasive cases annually and over 40,000 deaths.
- Appropriate mammography can reduce mortality by as much as 30%.
- Approximately 75% of women with breast cancer have no identifiable risk factors.

Key Aspects of Evidence-Based Practice Guidelines

- Ages 40–50: Offer yearly, with discussion of potential risks and benefits
- There is a higher false-positive rate (lower positive predictive value) for screening mammography for women under age 50. This results in calling patients back to repeat films; additional studies, including biopsies; and anxiety for the patient.
- Ages 40–65: Conduct mammograms yearly; little risk from accumulated radiation exposure due to calibrated low doses from equipment from ACR accreditation.

Table 46-4 Clinical practices and primary technology for screening for breast and cervical cancer

Clinical impact	Primary technology involved (devices, drugs, processes)
Breast cancer screening	Radiology Mammography
Cervical cancer screening	Laboratory cytology

Cervical Cancer Screening

Impacts

- Cervical cancer affects approximately 13,000 women in the United States annually, with approximately 4,600 deaths annually, and is the leading cause of cancer death among women aged 35–54.
- Between 1955 and 1992, the number of deaths from cervical cancer declined by 74%. The main reason for this change has been the use of the Pap test to detect cervical cancer early.
- The five-year survival rate for early invasive cervical cancer is 91%. The overall (for all stages combined) five-year survival rate is 69%.

Key Aspects of Evidence-Based Practice Guidelines

- Ages 21–69: Repeat Pap smears every year until patient has three negative smears; subsequently, all women with an intact uterus should be screened every three years.
- For high-risk patients, or after total abdominal hysterectomy, screen annually.

HTM in Primary Care

Proper diagnosis and treatment with appropriate HT resources leads to better patient outcomes (Judd and Painter, 2000). Equipment management should be focused in the areas that are known to support the best clinical practices (see Table 46-5). Resources and their impacts are cited in evidence-based clinical practice guidelines as follows:

CAD

Impacts

- CAD is the leading cause of death and leading cause of premature, permanent disability in the labor force in the United States due to acute myocardial infarction (AMI).
- Use of revascularization procedures for treating CAD is increasing rapidly, such as PTCA and CABG.

Key Aspects of Evidence-Based Practice Guidelines

- Patient education focuses on cholesterol control, diet, exercise, and smoking cessation.
- Risk factor management focuses on smoking cessation, diabetes control, appropriate diet and exercise, treatment of depression as appropriate, and receiving appropriate immunizations, such as influenza and pneumococcal shots during the appropriate season of the year.
- Medications include antiplatelet therapy (aspirin), lipid-lowering agents, beta-blockers after AMI, possibly ACE Inhibitors for CHF symptoms, and antihypertensives for blood-pressure control.
- Goals: For patients with known CAD, have LDL cholesterol <130 mg/dl; ensure patients on beta-blockers immediately after AMI; and reduce variation in use of cardiac procedures; that is, reduce rates of repeat revascularization following angioplasty.

CHF

Impacts

CHF is the most frequent cause of hospitalization for people aged 65 and older, and it is rare among persons aged 45 years and younger. About half of patients diagnosed with CHF die within five years.

Key Aspects of Evidence-Based Practice Guidelines

Patient education focuses on managing symptoms and weight.

- Outpatient care management focuses on weight control, ensuring timely follow-up care depending on severity, ongoing laboratory screening, encouragement regarding smoking cessation, and appropriate immunizations.
- Medications include ACE Inhibitors or other vasodilator therapy.
- Goals include ensuring use of ACE Inhibitor therapy and ensuring outpatient follow-up after hospitalization to minimize re-admissions.

Asthma

Impacts

In 1998, approximately 83,000 Kaiser-Permanente (KP) members in the United States (5–44 years of age) had mild, moderate, or severe persistent asthma; prevalence was 2.8% among KP's pediatric population (5–17 years), U.S.-wide; 6.34% for ages 0–17; and 1.5% among KP's adult members (ages 18–44); U.S.-wide 4.49% for ages 18–44.

- National guidelines call for use of inhaled anti-inflammatory medications (preventers) in a ratio of at least 1.0 to use of inhaled beta-agonists (relievers) for optimal therapy to prevent asthma symptoms and flare-ups.
- Asthma-related hospital admissions or observation stays are generally considered to be preventable through effective outpatient management; well-informed providers and

Table 46-5 Clinical practices and primary technology for screening for coronary artery disease, congestive heart failure, asthma, and diabetes

Diagnosis and treatment	Primary technology involved
Coronary artery disease (CAD) secondary prevention	Laboratory -lipid profile; radiology–angiography Pharmaceuticals -beta blocker (use post-heart attack) (AMI)
Congestive heart failure (CHF)	Process: Multidisciplinary team approach with patient self-management tools; general devices–ECG; Laboratory–various; radiology–chest X-rays; Pharmaceuticals ACE Inhibitor use
Asthma	Devices: Peak flowmeters and spirometry; patient self management education and treatment plan; Pharmaceuticals: relievers/preventers
Diabetes	Laboratory: blood glucose, lipids, renal; patient self-management (glucometer) and education; Pharmaceuticals–various; outpatient management

patients are crucial for proper management; primary care providers need better education and support to ensure that patients have the knowledge and confidence to manage their conditions.
- Despite the fact that exposure to smoke is known to increase symptoms and to precipitate attacks, 13% of patients who responded to a national KP survey smoked.

Key Aspects of Evidence-Based Practice Guidelines

- Treatment schedules are stratified to severity of disease, from Level 1, intermittent (lung function greater than 80% of predicted), to Level 4, severe persistent (lung function less than 60% of predicted).
- Routine visits to primary care range from level 1, annually, to level 4, three per year.
- Routine screenings should include: Flu vaccine annually; office review of peak flows each visit; review of environmental triggers each visit; and update treatment plan and home management skills each visit.
- Diagnostic tests should include: Laboratory tests generally not helpful; radiology: chest X-ray at initial evaluation (pediatrics possible sinus X-ray); peak flow, spirometry and/or pulse oximetry tests.
- Treatment plan should include: Provider establish and agree on treatment goals with patient; ensure patient education; ensure patient self-management plan; review ongoing drug therapy with step downs if possible; patient conduct home peak flow monitoring three times daily, depending on severity, during flare-ups will allow patient to determine adjust drug use; care manager for sicker patients for closer monitoring and control.

Diabetes

Impacts

In the United States, 15.7 million people (5.9% of the population) have diabetes. An estimated 10.3 million have been diagnosed, but 5.4 million are not aware that they have the disease. Each day, approximately 2,200 Americans are diagnosed with diabetes. About 800,000 Americans will be diagnosed annually.

- Diabetes is the seventh leading cause of death in the United States, contributing to 198,140 deaths in 1996. This chronic disease has no cure.
- By afflicting 3%–5% of KP members, this condition typically has several comorbidities and is responsible for increased rates of AMI, stroke, kidney disease, and limb amputations, among other serious ailments.
- Health care costs are typically three times other than for nondiabetics. Optimal management of diabetes requires a system to identify patients, to deliver needed services, and to track consequences of care.
- An ongoing set of outcomes' measures can clarify which systems of care have the greatest benefits and which are the most cost-effective.

Key Aspects of Evidence-Based Practice Guidelines

- Treatment Team equals patient and the primary-care health care team (HCT). The primary-care HCT should address diabetes at every visit for any clinical problem. Diabetes is typically diagnosed by glucometer for two fasting plasma glucose tests or one random plasma glucose plus classic symptoms, and should have ongoing glycemic control laboratory testing. Care is stratified into three levels: Level 1: Glycemic control (HbA1c<7.1) and complications controlled; Level 2: HbA1c (7.1 – 8.1); Level 3 (>8.1).
- Initial-care activities and ongoing routine care by level with primary-care practitioner and other diabetes practitioners for ongoing glycemic control; as well as lipid management through ongoing laboratory tests; retinal screening; practitioner interactions for blood pressure control, foot exams, initial/ongoing education; renal screening through laboratory tests; behavioral health screening (25% have depression that affects adherence to guidelines); medication strategies; and outpatient-management plans.

HTM for Integrated HT Resources

The Essential Health Care Technology Package (EHTP) is a new methodology and set of tools that has been specifically developed to strengthen resource planning and technology management in health care (Heimann et al., 2003). It integrates management of all HT resources, including human resources, pharmaceuticals, equipment, and facilities. (See Chapter 45.)

The link between resources and the delivery of quality care is achieved by identifying all HTs that are needed to provide well-delineated promotive, preventive, and curative health interventions. Health interventions in turn are linked to the WHO International Classification of Diseases (ICD9, ICD10) and to other disease classifications (e.g., ICD9-CM). This mapping between HT resources and health care interventions is based on internationally accepted and recognized clinical practices. It is expressed in the form of a "matrix" or generic template whose elements include economic, technical, and other information, typically associated with acquisition of the HT resources and with their utilization in providing the interventions. This EHTP matrix is the basis of the EHTP methodology and tools, which are implemented in a suite of software packages. Each intervention is decomposed into progressively smaller components, down to procedures and their associated techniques. The decomposition uses the Current Procedure Terminology (CPT) of the American Medical Association. Over 350,000 links have been created between international classification of diseases and the CPT.

A unique and powerful feature of EHTP is the integration of epidemiological profiles and recognized clinical practices (such as clinical practice guidelines-CPGs) with the EHTP matrix. This allows advanced modeling and simulation for identifying economic and other resource requirements given varying scenarios of disease incidence and prevalence, and CPGs. Conversely, simulation can indicate epidemiological conditions that could be covered with available resources by using specific CPGs.

EHTP was developed over several years and was provided by the World Health Organization (WHO), Department of Health Service Provision (OSD), and Medical Research Council (MRC), South Africa, to WHO member states beginning in 2001, with the Kyrgyz Republic (Kyrgyzstan) Ministry of Health (MoH) requesting implementation beginning in 2002. Strengthening resource planning through effective management of health care technologies, EHTP achieved the following goals and objectives in Kyrgyzstan:

- Applied in the broad context of health systems development and reform, and thus aligned with (and possibly able to accelerate) present Health Reform efforts.
- Methodology and computer-based tool expected to measurably improve effectiveness, efficiency, performance, and quality of health delivery.
- Simplifies coordination of resources by central, district, and local health leaders and integrates management of these resources to allow more efficient distribution.
- Focuses on the country's disease burden, as defined by in the Kyrgyz Republic by MoH and the National Drug Council, who direct the national CPG development process.
- Uses best medical evidence to direct care at primary, secondary, and tertiary levels by selecting interventions or CPGs that are best fit to available financial resources and existing constraints.
- Methodology and tools may be dynamically updated with advances in medical technology.
- It has been piloted at three sites for primary- and secondary-level care in 2003 and is scheduled to be implemented nationwide in 2004. The pilot sites initially use EHTP to better manage five target conditions selected by MoH and pilot leaders: Hypertension (HTN), acute respiratory illnesses in children (ARI), iron deficiency anemia in pregnancy (IDA), brucellosis (BRU), and tuberculosis.
- Quality and cost indicators were developed; baseline results were measured in 2002, and follow-up results in 2003. Accountability measures to ensure that pilot understanding of key tool features will be conducted in 2003.
- Preliminary quality outcomes measured include those shown in Table 46-6.

EHTP Impacts

- Quality indicators show improvement across several key measures, both clinical process and clinical outcome related.
- Cost data are not comparable, as there has been no unified method for calculating costs within MoH hospitals and clinics. An MoH work group is addressing this issue.
- Use of EHTP as a resource planning through effective management of health care technologies has potential for significant improvement in quality in the Kyrgyz Republic and in many other developing countries.
- There are several other significant quality indicators that EHTP can address that have been identified in WHO literature; a sample of these indicators is shown in Table 46-7, with recommendations of ways in which the MoH in Kyrgyzstan could use them. Other indicators are noted at http://www.who.int/health-services-delivery.

Conclusions

Several models to assist developed and developing countries in creating effective health technology resource management show demonstrable impact on improved costs and quality. Some of the case studies highlight medical equipment management (HTM) best practices, and others demonstrate a broader view of HTM and its implications. Clinical engineers continue to contribute significantly to these efforts.

Table 46-6 Preliminary quality outcomes from EHTP implementation in kyrgyzstan

No.	Indicator	Baseline 2002				Re-measure 2003				Comments
1.	Clinical Practice Guidelines (CPGs) approved by MoH & programmed in EHTP tool (have MH-validated lists of technologies used)	• 32 Primary level CPGs • 17 Secondary CPGs • 3 Tertiary CPGs				• 13 more Primary CPGs • 13 more Secondary • 3 more Tertiary				These 45 conditions & 81 CPG (at 3 levels) account for approx. 80-90% of all care delivered and resources used at pilot sites
2.	% of patients treated in compliance with CPGs at pilot sites (n=10 treatment per CPG per site; observed by MoH CPG expert for 5 target conditions)		Primary	Secondary			Primary	Secondary		Baseline higher CPG non-compliance at secondary level is caused by later EHTP-facilitated MoH approval of these CPGs. Data show care improvement in pilot sites in 2003
		Issyk-Ata	70%	10%		I-A	68%	91%		
		Ak-Tala	58%	27%		A-T	95%	93%		
		At-Bashy	50%	39%		A-B	95%	58%		
3.	Primary level (treated for 5 key conditions)		I-A	A-T	A-B		I-A	A-T	A-B	
	• Number of diagnostic tests done:	ARI	-	-	-	ARI	-	-	-	
	-ARI - None required	HTN	90%	100%	40%	HTN	78%	90%	0%	
	-HTN - % of patients referred ECG	IDA	100%	100%	100%	IDA	83%	97%	100%	
	-IDA - % pts referred to CBC/hemo		8%	5%	6%		9%	7%	8%	
	• % referred for hospitalization									
	• Treatment outcomes:									
	-Full recovery	ARI Full recovery				ARI Full recovery				
	-Clinical improvement of condition	HTN Clin. improvement				HTN Clin. improvement				
	-Clinical worsening of condition	IDA Clin. Improvement				IDA Clin. improvement				
4.	Secondary level									
	• Number of diagnostic tests done:		I-A	A-T	A-B		I-A	A-T	A-B	There has been some improvement in ARI testing and in HTN testing;
	-ARI - % of patients referred to CXR	ARI	0%	0%	0%	ARI	10%	50%	0%	
	-HTN - % of patients referred ECG	HTN	100%	70%	80%	HTN	95%	83%	100%	
	-IDA - % pts referred to CBC/hemo	IDA	100%	100%	100%	IDA	100%	100%	100%	
	• Re-hospitalizations in 1 month		0%	0%	0%		0%	0%	0%	
	• Treatment outcomes:									
	-Full recovery	ARI Full recovery				ARI Full recovery				
	-Clinical improvement of condition	HTN Clin. improvement.				HTN Clin. improvement				
	-Clinical worsening of condition	IDA Clin. improvement.				IDA Clin. improvement				

Table 46-7 Quality indicators with potential for improvement through EHTP in Kyrgyzstan

No.	Indicator	How EHTP Can Help	Baseline Measurement Recommended	Re-measurement Period Recommended
1.	**Population health (gains)** • Desired process outcomes • Desired care outcomes • Overall level raised • Distributed appropriately • Health system-related outcomes' accountability, e.g., timing of care, sustainability of delivery system performance • Provide performance and incentives re outcomes (see 16 below) • Health expectancies & gaps	• Quality can be improved either by better use of existing resources or through contribution of added resources. EHTP models an integrated solution here based on best science that can be used in light of existing constraints. • EHTP can assist in measuring outcomes, competence, appropriateness, safety, timeliness, preventive care, access, availability, and continuity of care.	Selected process & care outcomes are currently being measured. Others could be chosen based on other key epidemiological conditions, with baseline in 2003.	2003 and beyond
2.	**Responsiveness of the health system** to legitimate expectations of the population (excludes health improving dimensions noted in 10.) • Increase average level • Reduce inequalities in responsiveness • Respect for the person: dignity, autonomy, need for confidentiality and information; patient's demand & rights • Respect for customer (client) orientation: prompt attention, provision of basic amenities, access to social support networks, and choice	• Satisfaction surveys could be added to measurements by the EHTP central Coordinating Office Team at pilot sites and other sites as country-wide implementation proceeds to measure and begin to improve existing responsiveness. This information could be brought back to monthly EHTP National Commission meetings for discussion and referral to appropriate branches of MoH for follow-up action. • Measures could include patient choice, patient experience, acceptability, respect & caring, availability of information, and timeliness.	2003	2004
3.	**Fairness in financing** (financial risk protection for households – should not impoverish patients; poorer patients should pay lower proportion of disposable income) Note evidence that prepaid financing that are publicly financed are more successful in containing costs & making health gains. Balance risk by capitation with providers and fair co-payment by patients. Stimulate competition between providers by giving feedback on provider performance to patients; must be in context of strong government regulation & publicly financed system.	• Extensive MoH efforts ongoing regarding health finance reform to support fairness. EHTP cost indicators could continue to inform this process at the local, district and central levels. • Measures could include affordability, access, availability, and continuity. • Could also include Provider Performance, see 16 below.	2002	2003 and beyond
4.	**Health system inputs** (amount and quality of resources, typically health expenditures per capita) • Recurrent costs of service provision • Physical availability of inputs • Skill mix of health care personnel • Utilization of medical equipment and facilities	• These are currently standard EHTP measures that could be used in a variety of ways by MoH.	2002	2003 and beyond
5.	**Health system organizational structure and processes** • Level and type of autonomy and integration • Incentive structures	• Understanding how health resources are individually managed and then integrating these activities through EHTP is a key way it assists the MoH. • Using EHTP to develop data to assist in establishing provider and local, district and central health system incentives is a logical next step.	2003	2004
6.	**Coverage with Effective Interventions** (the probability of receiving a necessary health intervention conditional on the presence of a certain health problem or need) • Access (availability, accessibility, affordability, acceptability) • Utilization (access, personal health behavior) • Effectiveness (efficacy, inputs, QA mechanisms, patient compliance, & external factors) • At individual level, also measure inequality of effective coverage	Criteria for how interventions are identified for measurement of effective coverage: • Able to produce health gain in relatively short time period • Size of health problem at global and country level • Evidence on effectiveness of interventions, and inherent credibility • Aligns with national health priorities, needs, policies • Balance between different modalities of health care • Cost-benefit ratio of obtaining information at country level • Able to link global processes with country priorities	2003 These criteria are nearly exactly matched with EHTP's existing capabilities. WHO's global emphasis regarding effective coverage is currently on communicable diseases such as HIV/AIDS, TB, and chronic diseases like ARI, asthma, hypertension, depression, angina, etc.	2004
7.	**Provider Performance** • Individual provider quality performance & competency • Individual provider safety performance • Provider motivation & incentives based on performance	• EHTP provides the tools to assess and improve these measures, in partnership with MoH.	2003	2004
8.	**Improving Health System Efficiency** Varies according to the choices a country makes about the mix of interventions purchased with available health expenditures, includes both technical and allocative efficiency.	• WHO recently completed this measure this for all of its 191 member countries. • WHO studies indicated that more outcome goals can be met without always increasing health expenditures; EHTP provides the means for this assessment.	2003	2004
9.	**National Quality Assurance (QA) / Quality Improvement (QI) Programs through Accreditation** EHTP processes for improving health system resource planning and management provide the framework to assist in QA/QI accreditation activities.	• EHTP's national implementation is a natural link for establishing accreditation programs for inpatient and ambulatory facilities. The United States models of JCAHO accreditation for inpatient facilities and NCQA for ambulatory facilities links well with EHTP's methodology and tools.	2003	2004
10.	**Patient Safety** Health care interventions are intended to benefit patients but they also cause harm. The complex combination of processes, technologies & human interactions in a health delivery system can bring benefits but also cause adverse events.	• EHTP directly addresses many of these issues, could assist national efforts to emphasize patient safety as a concern in health system performance and quality management, and help establish the evidence base of adverse events.		

Table 46-7 Quality indicators with potential for improvement through EHTP in Kyrgyzstan—cont'd

No.	Indicator	How EHTP Can Help	Baseline Measurement Recommended	Re-measurement Period Recommended
	Developing countries often need particular attention for adverse events due to poor state of infrastructure and equipment, unreliable supply and quality of drugs, shortcomings in waste management and infection control, poor performance of personnel because of low motivation or insufficient technical skills, and severe under-funding of essential operating costs.			
11.	**Cost –Effectiveness Analysis CEA** CEA of a wide range of interventions can be undertaken to inform a specific decision-maker. This person faces a known budget, and a series of other constraints. A decision-maker may be able to allocate an entire budget or only a budget increase; he may be a donor, a MoH, a district medical officer, or a hospital director. Choices available are typically limited by some combination of health resources or the current mix of interventions.	• In partnership with a health economist who understands all of the constraints involved, the information generated from and able to be modeled by the EHTP tool can potentially be a powerful tool in various forms of CEA.	2003	2004

References

David Y, Judd TM. *Medical Technology Management.* SpaceLabs (Medical, Inc. Biophysical Measurement Series). Redmond, WA, 1993.

Dyro JF, Wear JO, Grimes JL, et al (eds). *Advanced Clinical Engineering Workshop Syllabus.* Plymouth Meeting, PA, ACCE, p 232, 2003.

Heimann P, Issakov I, Kwankam Y, et al. *Quality and Management Applications of EHTP. Kyrgyzstan District Health Leaders.* December 2002, May 2003.

Judd TM et al. *Empirical and Anecdotal Evidence from WHO's Informal Consultation on Physical Infrastructure, Technology and Sustainable Health Systems for Developing Countries.* Geneva, Switzerland, 1998.

Judd TM, David Y, Wang B, et al. *Advanced Clinical Engineering Workshop (ACEW) for Latin America–Executive Summary.* 1991.

Judd TM. HTM Example. Health in the Americas (Developed from this PAHO document). *Health Care Technology: Current Status and Trends Presentation at Ministry of Health Seminar.* Cartagena, Colombia, South America, 1997.

Judd TM, Painter FR. *The Linkage of Quality Management and Technology Management in Health care.* HealthTech Annual Conference, 2000.

Section IV Management

Alfred M. Dolan
Samuel Lunenfeld Associate Professor in Clinical
Engineering, Associate Director, Institute of Biomaterials and
Biomedical Engineering, University of Toronto
Toronto, Ontario, Canada

The field that is now referred to as "clinical engineering" needs to be considered in the context of the development of biomedical engineering and in turn, engineering.

A number of years ago the following definition of engineering served to illustrate that the profession has long recognized its responsibility in applying the application of scientific techniques, theories, and technology for the solution of societal needs.

"In the past engineering was defined as the facility to direct the sources of power in nature for man's use and convenience (Dolan et al, 1973)."

That definition presents issues which all engineers, and clinical engineering professionals in particular must address. Those issues are of course the need to maintain an appropriate level of scientific knowledge and technological skills and a commitment to define and meet the real societal needs in health care. These issues are addressed as the different aspects of clinical engineering practice are discussed in the chapters of this book.

Biomedical engineering origins can be traced to ancient times through the work of such people as Alcmaeon, Plato, and Galen who investigated, observed, and systematized the world they studied around them including the human body. Galen's work on hemodynamics persisted for more than 1200 years to the time of Maimonides. Leonardo da Vinci, who was arguably the greatest engineer in history, also applied physical principles and experimental analysis to the study of physiology and medicine. Helmholtz built on his own early interests in physics and mathematics to undertake a career in medicine in 1838. He made prodigious contributions to the understanding of physiology and invented the ophthalmoscope he needed to investigate the physiology and psychology of hearing,

Investigators, scientists, and engineers like Helmholtz understood that the same laws apply to studies of physics, engineering, or biology—the unity of the sciences and the essence of biomedical engineering. That understanding represents the great challenge and potential for the applied side of biomedical engineering, which has come to be called clinical engineering, where all aspects of science, engineering and technology come to play directly in the health care field.

Clinical engineering has developed in health care facilities around the world over the last four decades of the 20th century. There was broadly held recognition of the technological explosion that had affected society in general and health care in particular, and a series of workshops held in 1972 provided a forum for the discussion of the need for an engineering approach to effect some control on this technology (Hopps, 1972; Craig, 1972), That and subsequent work serves to illustrate that there was recognition of the problems that existed in health care in that time period but further that there was recognition of the potential for biomedical engineering to solve them. It was, however, not only engineers who realized how broad that problem solving role could become. That broader understanding came from visionaries like Robert Rushmer, a cardiovascular research scientist, and Cesar Caceres, a clinical cardiologist, whose recognition of what the range of engineering involvement in health care could be defines, in essence, the role of clinical engineering.

Cesar Caceres offered perhaps the most insightful description of clinical engineering: "An engineer who is trained for and works in a health service facility where he or she is responsible for the direct and immediate application of engineering expertise to health and medical care (Caceres, 1980)."

While this definition unnecessarily constrains the practice of clinical engineering to that being within the precincts of the hospital, it still accurately outlines the role of clinical engineering and assumes both appropriate engineering expertise, and direct benefit to patient care or health care. Some three decades later the American College of Clinical Engineering emphasizes both patient care and management by defining a clinical engineer as "a professional who supports and advances patient care by applying engineering and management skills to health care technology (ACCE, 2001)."

Health care and the science and technology supporting health care will continue to develop. This accelerating trend in scientific developments in health care will in turn, drive the development in clinical engineering. In keeping with our understanding of clinical engineering as supporting and advancing patient care through application of engineering management and technology, it is important to recognise that the field of clinical engineering will need to continue to develop as health care develops. This will include developments both in sophistication, such as physiological functional imaging or highly integrated information systems, and in scope, such as in wellness care or in distributed clinical care. Active development in clinical engineering will be required if the field is to continue to thrive and it is toward that end that this Handbook endeavours to encourage the profession.

This section addresses management issues from the perspective of engineers who have applied management principles in their professional careers in the health care environment. Seaman (Industrial/Management Engineering in Health care) with case studies makes a convincing case for utilizing the special skills and insight of clinical engineers to improve the utilization of and maximize the investment in health care technology. Seaman, working with the Editor-in-Chief of this Handbook in a tertiary-care hospital, demonstrated that sound industrial/management techniques applied in the clinical setting yielded enhanced medical device utilization. Wang (Financial Management of Clinical Engineering Services) provides comprehensive description of the financial concepts the clinical engineering must embrace to manage effectively with available resources. He gives concrete examples of best financial management principles and offers templates for putting the principles into practice. Fennigkoh (Cost-effectiveness and Productivity), in providing the tools with which the clinical engineer can measure the output of the clinical engineering department, shows the CE how to maximize on available human resources to yield optimal medical device technology management. In order to improve on a department's performance, one must understand its objectives, appreciate its resources, and measure its accomplishments. Autio (Clinical Engineering Program Indicators) recommends ways by which the clinical engineer can use indicators of performance to enhance overall departmental operations.

The human aspect of the clinical engineer's role in management is emphasized in Personnel Management (Wear), Skills Identification (Cram), and Management Styles and Human Resource Development (Epstein and Harding). The precious resource of the individual must be appreciated, nurtured, developed, guided and encouraged to function at the highest levels. Wear gives an overview of the responsibilities of the clinical engineering director in recruiting and retaining personnel. Cram affirms that identifying the skills of each employee is at the foundation of effective work performance. That there is not one simple formula for managing and for developing human resources is emphasize by Epstein and Harding as they trace the development of management techniques over time and space. Judd completes the picture with an insightful chapter on quality. He draws upon his broad and extensive experience in putting into practice methods for achieving the best results. He shows how to assess present levels of performance and recommends techniques and tools to achieve higher levels of quality.

References

ACCE. *Enhancing Patient Safety: The Role of Clinical Engineering*, American College of Clinical Engineering, White Paper, 2001.
Caceres CA (ed). *Management and Clinical Engineering*. Artech House Books, 1980.
Craig JL (ed). *First International Biomedical Engineering Workshop Series. I. Biomedical Equipment Maintenance service programs*. The American Institute of Biological Sciences, Bio Instrumentation Advisory Council, 1972.
Dolan AM, Wolf HK, Rautaharju PM: Medical, Engineering—Past Accomplishments and Future Challenges, *Journal APENS* 1973.
Hopps JA, (ed). *First International Biomedical Engineering Workshop Series. I: Biomedical Equipment Maintenance service programs*. The American Institute of Biological Sciences, Bio Instrumentation Advisory Council, 1972.

47

Industrial/Management Engineering in Health Care

George Seaman
President, Seaman Associates
Northport, NY

If it moves, it's Mechanical;
If it doesn't, it's Civil;
If it smells, it's Chemical;
If it is invisible, it's Electrical;
And if you can't even imagine it, it's INDUSTRIAL!

The IE page at the University of Nebraska.
Attributed to Professor Schmuel S. Oren, UC -Berkeley.

Industrial/management engineering is probably the broadest of all the engineering disciplines. Industrial/Management engineers (I/MEs) integrate human, information, material, monetary, and technological resources to produce goods and services optimally. Major areas of specialization applied in the health care sector include:

- Engineering economics and decision analysis
- Human factors engineering
- Process/service delivery systems
- Operations research/resource optimization/decision science
- Logistics, information, and materials management
- Statistical analysis/quality management
- Stochastic systems/simulation/modeling

Typically I/MEs are people-oriented problem solvers whose assignments address management decision making. They tend to be diverse, often requiring interaction with health care administrators and specialists such as those who have particular expertise in biomedical technologies with the underlying guideline of providing optimal care to patients.

According to the ACCE definition (Bauld, 1991), a clinical engineer (CE) is a professional who supports and advances patient care by applying engineering and management skills to health care technology. As health care has become increasingly dependent on more sophisticated technology and associated complex biomedical equipment, the clinical engineer's role has risen to the level of the technology officer in supporting health care organizations with technology planning, implementation, and ensuring a safe and efficient utilization during the life span of the technology.

Both skill sets complement each other in the areas of general engineering knowledge, and both require a unique set of management skills and knowledge specific to their areas of specialization. Aligned toward a specific goal, a synergistic model can emerge that can result in a rationally planned and well-executed implementation of technological change. This chapter facilitates the understanding of what the I/ME does, by demonstrating problem solving strategies and the way I/MEs can work effectively with the CE in working with the new and old frontiers of technology in health care organizations.

Definition of an Industrial/Management Engineer

"Efficiency is doing things right; effectiveness is doing the right things."–Peter F. Drucker

I/MEs design the optimal combination of human and natural resources to make systems perform at their best. The integration of people, materials, capital, equipment, and energy into production systems is the I/ME's main concern. An I/ME may be involved in scheduling staff and systems resources, designing health care delivery systems, or building information systems to support organizational development and decision making.

The U.S. Bureau of Labor Statistics has described a typical I/ME's function as follows: Industrial engineers determine the most effective ways for an organization to use the basic factors of production-people, machines, materials, information, and energy to make or process a product. They are the bridge between management and operations. They are more concerned with increasing productivity through the management of people, methods of business organization, and technology than are engineers in other specialties, who generally work more with products or processes.

The I/ME's Role in Decision Making

"The purpose of a model is not to fit the data but to sharpen the questions"–Samuel Karlin, 11th R A Fisher Memorial Lecture, Royal Society 20, April 1983

Decision science is an essential part of an I/ME's training in the development of modeling skills. A model is an abstraction of a real-world process, such as transport-delivery system, customer service, or behavior of health care markets. Sound analysis of a model's output can help to improve an organization, service, or program's performance. In order for such models to be beneficial, their results must be clearly communicated to the company. In order to identify the problem and to utilize the information that the models generate, the ability to ask the right questions and good acumen in listening and communication skills are necessary.

The Bureau of Labor Statistics has also noted that to solve organizational, production, and related problems most efficiently, industrial engineers carefully study the product (or process) and its requirements, design logistics and information systems, and use mathematical analysis methods (such as operations research) to meet those requirements. They develop management control systems to aid in financial planning and cost analysis, design production (or process) planning and control systems to coordinate activities and to control product (or service) quality, and design or improve systems for the physical distribution of materials and services. Industrial engineers conduct surveys to find facility locations with the best combination of resources, transportation, accessibility, and costs. They also develop wage-and-salary administration systems and job-evaluation programs. Many industrial engineers move into management positions because the work is closely related to the art of management science.

Decision-Making in Health Care

"Success is relative: It's what we can make of the mess we have made of things."
– T. S. Eliot

The health care industry is the largest single industry in the United States, comprising over 700,000 physicians and 5000 hospitals. Total expenditures exceed $1.3 trillion dollars, and according to some experts, the figure is expected to double by the year 2007 (HCFA, 1997). This increase in costs in delivery of health care services has caused the federal government, employers, and consumer groups to pressure the health care industry into seeking more cost-efficient methods for servicing a growing number of health care consumers, e.g., Baby Boomers. Concomitant with rising costs, health care organizations are under pressure to respond to increased regulations, e.g., the Health Insurance Portability and Accountability Act (HIPAA), (see Chapter 104), increasingly complex and cumbersome managed-care constructs, e.g., precertifications, bundling of services in response to optimize HMO reimbursement structuring, and to contend with aging technology and infrastructure.

On the horizon is the new world of technology that is growing at a geometric rate and promising fantastic medical strides in genetics and molecular biology, pharmacology, medicine and surgery, diagnostic tools, physiological monitoring, and robotics. These and the informatics that cross-section all of these technologies promise momentous advantage and value. These new developments often put the chief executive officer (CEO) in an investment quandary between managing costs and buying something that will improve quality and reduce (or contain) future costs (improving value). If the investment is large, as almost all health care investments are, there may be federal, state, and local regulatory impediments that may take so long as for the investment to lose its utility as a "competitive" instrument or, worse, render the technology nearly obsolete. This is particularly critical to a smaller hospital that might not have the funding capability of a large medical center or network that can absorb the risk; however, it must invest to survive.

There are occasions when the CEO must make an investment to attract or retain a prominent physician, respond to physician or consumer-community pressure, or, more imperatively, respond to an urgent or emergent situation to avoid a disastrous outcome. Regardless of the investment quagmire, hospitals must invest in technology to thrive and to continue doing the "right thing." But in doing so, how does the CEO know in the decision to invest is the right way? Finance can work out the break-even, the physician or nurse can tell him or her how great it performed at XYZ hospital or conference, or a board member can provide a nudge. But the CEO is accountable and, therefore, must rely on other factors that effect his or her decision and the future of the organization.

Planning

"Whatever failures I have known, whatever errors I have committed, whatever follies I have witnessed in private and public life have been the consequence of action without thought."–Bernard M. Baruch

Technology Acquisition

"When the only tool you have is a hammer, then every problem begins to look like a nail."–Abraham Maslow

How does the acquisition of technology start? It should start at the strategic planning level. (See Chapters 30 and 32.) Because technology is extremely dynamic, planning at the micro level should be near the endpoint, and not the beginning! Often, the capital budgeting process serves as a surrogate for the strategic planning process. It requires uninformed staff members to provide a guesstimate of their needs and desires, usually in short order, and often without a knowledge cross-connect to other services. This is disastrous in the sense that using a capital budget as the strategic plan will result in suboptimal utilization of technology, such as creating an implementation-connectivity nightmare; exorbitant service contract and inventory costs; or buying soon-to-be-obsolete technology and leaving the disconnected I/ME and CE to contend with a plethora of patchwork and other unnecessary operational implementation and management challenges that otherwise could have been avoided.

Engineering Economics and Decision Analysis

A Tale of Two CT's A major driving force behind many management decisions and fundamental to Business 101 is turning a profit. In non-profit health care facilities it is a contribution to margin. The I/ME often plays a major role in providing the analytic work to support decision making through a variety of techniques. For example, consider a radiology service that has purchased a new CT scanner in addition to their current machine to meet increasing demand. Each unit takes a different length of time to scan the patient. Because CT_A is new, it averages 16 minutes per scan, while the used scanner, CT_B, averages 24 minutes per scan. The total time to process a patient averages 12 minutes per patient, net of either scanner's processing time. Previous cost analysis determined that 24 patients per day had to be processed for the service to break even.

To determine a solution, we must look at the processing rates. The rate for CT_A is 2.1 patients per hour (60 min/hr/ 16+12 minutes per patient) and 1.7 patients/hr for CT_B. If the facility operates for seven hours, a total of 26.6 patients/day can be processed, assuming a steady state operation, so the service easily breaks even. However, delays such as patients arriving late, staff breaks, absenteeism, lost paperwork, and machine delays all contribute toward undermining that ideal state. If five minutes per patient added for delay variances or the unit incurred X minutes of down time were factored in the process daily volume would be reduced to 23 patients per day, thus sustaining a perpetual loss. The reality is that there are stochastic processes at work that cause delays that are variable in duration and not evenly distributed on a per patient basis. This can result in as much as a 20% reduction in utilization, or 18–20 patients per day in this case.

The I/ME in applying a particular approach can conduct a time study analysis and construct a flow process chart to identify ways to reduce the patient processing time by eliminating or combining steps, or removing certain steps in the "critical path" of the process to reduce the direct processing time. Other tools might include a simulation of the process where a dynamic flow of sequenced of events emulating the work process can be constructed, and specific events or subprocesses varied to determine service-time sensitivity while incorporating those stochastic processes such as patient arrivals and resource down time.

If performed within a rational planning process, this analysis could be performed prior to purchasing equipment with predetermined time standards, logistics, and monitoring in place. A note of interest: The hospital decided to utilize all inclusive service contracts that totaled $92,000 per year. If the hospital had utilized the in-house clinical engineering department, not only would they have saved money, but there would have been added value in utilizing during the CE's non-CT dedicated time as well.

Project Management

"Everything is okay in the end. If it's not okay, then it's not the end."–Unidentified

Managing Large, Complex Projects

Gantt Chart

A Gantt chart is a horizontal bar chart developed as a production-control tool in 1917 by Henry L. Gantt, who was an American engineer and social scientist. Frequently used in project management, the Gantt chart provides a graphical illustration of a schedule that helps to plan, coordinate, and track specific tasks in a project (Figure 47-1). Gantt charts can be simple versions created on graph paper, or more complex, automated versions created using project management applications such as Microsoft Project or Excel. (See, e.g., http://goanna.cs.rmit.edu.au/~geoff/280/cgi/sample/timing/section1.htm)

PERT Chart

The limitation of a Gantt chart is that it does not show interdependency. A PERT chart, ("Program Evaluation Review Technique") was developed by the U.S. Navy in the 1950s to manage the Polaris submarine missile program to overcome that limitation. These are often combined with critical path method (CPM), which was developed in the private sector around the same time. CPM identifies the path sequence that is most critical to completing the project.

A PERT chart presents a graphic illustration of a project as a network diagram consisting of numbered nodes (either circles or rectangles) that represent events or milestones in the project, linked by labeled vectors that represent tasks in the project. The direction of the arrows on the lines indicates the sequence of tasks. In the PERT chart shown in Figure 47-2, for example, the tasks among nodes 1, 2, 4, 8, and 10 must be completed in sequence. These are called dependent or serial tasks. The tasks between nodes 1 and 2, and between nodes 1 and 3 are not dependent on the completion of one to start the other, and they can be undertaken simultaneously. These tasks are called "parallel" or "concurrent" tasks. Tasks that must be completed in sequence, but that do not require resources or completion time, are considered to have event dependency. Represented by dotted lines with arrows, these are called "dummy activities." For example, the dashed arrow linking nodes 6 and 9 indicates that the system files must be converted before the user test can take place, but that the resources and time required to prepare for the user test (e.g., user training and writing the user manual) are on another path. Numbers on the opposite sides of the vectors indicate the time allotted for the task

These techniques are useful in organizing, managing, assigning responsibility, and updating status on large, complex projects. Typically, the I/ME will create a list of tasks working with specialists such as architects, other engineers, and programmers and will develop and organize a more detailed list of activities, sequence the necessary events, and determine the interdependencies between activities and events. This particular specialty of I/ME falls under project management, which is an entire field in itself. There are certified project managers (PMs), special tools used for developing schedules, and international societies devoted to this important field. (For more information on project management, see http://www.allpm.com/static.html.)

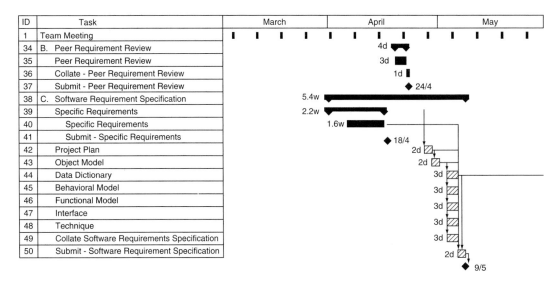

Figure 47-1 Example of a Gantt chart.

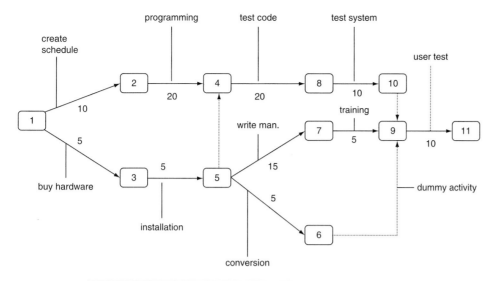

Figure 47-2 Example of a PERT chart.

Reusable vs. Disposable Blood Pressure Transducers: A Case Study

"An approximate answer to the right question is worth a good deal more than the exact answer to an approximate problem."–John Tukey

The following is one example of how the areas of specialization of an I/ME and CE can be combined in analyzing cost/benefit use of alternative medical devices in a hospital: The cost of reusable transducers had been on the increase over the previous year. At the behest of a sales representative, a hospital administrator wanted to evaluate the use of disposable transducers. The I/ME reviewed the problem with the CE and began to structure an approach. Looking at the numbers in a "make/buy" analysis, the reusable transducer cost $275 while a disposable one cost $2.38. Total inventory was 149 transducers, with 12 in storage. Random sampling indicated that, on average, 82% of the transducers were in use on a daily basis. The average repair rate was 26 per month at a cost of $63 in parts, labor (CE cost based on time-study analysis), and overhead, and a replacement rate of 37 per month.

On the surface, this is a simple problem. Extending the replacement cost on an annual basis comes out to $141,756/year: (37 replacements/mo. × $275/txd 12/mo/yr) + (26 repairs/mo × $63/Repair × 12 mo.). Using disposable units would cost $114,686/yr: (161 transducers × 82% utilization × $2.38/transducer). Using disposable units would yield a savings of $27,000/year. Clearly, the choice would be to use disposables.

The clinical engineering director, however, had the intuition and foresight to have previously enacted development of a comprehensive medical equipment management system (MEMS) (Dyro, 1984, 1993). Given the wealth of data available through the system, the I/ME decided to look more closely at Mean Time Between Failure (MTBF) rates by user group.

The results were startling. The I/ME identified two units that had the shortest MTBF and consequently the highest replacement and repair rate. The culprits were an intensive care unit (ICU) and the operating room (OR). Upon interviewing the staff in these locations, the I/ME determined that they often mishandled these transducers–sometimes just tossing them on a table, sometimes dropping them on the floor. The answer became crystal clear: Training, sensitizing, storage accessibility, and monitoring of staff who handle the transducers. By reducing the incidence of breakdown by 20%, the cost was shifted in favor of reusable transducers. Of course, this course of action was transmitted to all areas, thus resulting in a sharp decrease in repairs and significantly increasing the MTBF and financial savings.

Resource Optimization/Operations Research

"Mathematicians do not study objects, but relations between objects. Thus, they are free to replace some objects by others so long as the relations remain unchanged. Content to them is irrelevant: They are interested in form only."–Jules Henri Poincaré (1854–1912)

The field of operations research (OR) took root in the years just prior to World War II. In 1938, experiments began to explore ways in which information provided by radar could be used to direct aircraft. A multidisciplinary team of scientists working on this radar/fighter plane project studied the actual operating conditions of these new devices and designed experiments in the field of operations, and the new term "operations research" was born. The team's goal was to derive an understanding of the operations of the complete system of equipment, people, and environmental conditions (e.g., weather and time of day), and then to improve upon it. Their approach was later paralleled in the U.S. with the first team working on antisubmarine tactics. The U.S. group developed a series of mathematical models, "entitled search theory," that were used to develop optimal patterns of air search. Today, each branch of the U.S. military has its own operations-research group that includes both military and civilian personnel. OR moved into the industrial domain in the early 1950s and paralleled the growth of computers as a business planning and management tool. As the field evolved, the core moved away from interdisciplinary teams to a focus on the development of mathematical models that can be used to model, improve, and even optimize real-world systems. They include deterministic models such as mathematical programming, routing or network flows, as well as probabilistic models such as queuing, simulation and decision trees. The following are simple examples in linear programming:

Production Planning

"How can it be that mathematics, being after all a product of human thought independent of experience, is so admirably adapted to the objects of reality?"– Albert Einstein (1879–1955)

Example 1

A gynecologic service performs two diagnostic tests (X and Y) using two machines (A and B). Each patient who has an X diagnostic test procedure requires 50 minutes processing time on machine A and 30 minutes processing time on machine B. Each patient who requires a Y diagnostic test procedure requires 24 minutes processing time on machine A and 33 minutes processing time on machine B.

At the start of the current week, there are 30 patients for of X and 90 patients for Y on the schedule. Available processing time on machine A is forecast to be 40 hours and on machine B it is forecast to be 35 hours (an additional 1hr/day for setup and QC).

The average demand for X in the current week is forecast to be 75 patients, and for Y it is forecast to be 95 patients. Service policy is to maximize the combined sum of the patients of X and of Y scheduled during the week. The following approaches can be taken:

- Formulate as a linear program the problem of determining how many of each group of patients to process in the current week.
- Solve this linear program graphically.

Let

- x be the number of diagnostic test procedures of X patients in the current week
- y be the number of diagnostic test procedures of Y patients in the current week

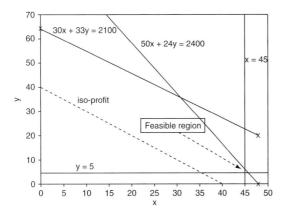

Figure 47-3 Graphical solution to determine the maximum the number of patients processed each week.

Table 47-1 Times for diagnosis, repair, PM, and profit (i.e., marginal cost contribution [MCC]) for each of four variants

	Diagnosis (min)	Repair (min)	PM (min)	MCC ($)
Variant 1	20	30	20	1.50
Variant 2	40	20	30	2.50
Variant 3	30	30	20	3.00
Variant 4	70	40	50	4.50

Then the constraints are:

- $50x + 24y \le 40(60)$ machine A time
- $30x + 33y \le 35(60)$ machine B time
- $x \ge 75 - 30$
- i.e., $x \ge 45$ so production of X >= demand (75)–initial schedule (30), which ensures that we meet demand
- $y \ge 95 - 90$
- i.e., $y \ge 5$ so production of Y >= demand (95)–initial schedule (90), which ensures we meet demand

The objective: Maximize $(x+30-75) + (y+90-95) = (x+y-50)$, i.e., to maximize the number of patients processed by the end of the week, Figure 47-3 shows that the maximum occurs at the intersection of $x = 45$ and $50x + 24y = 2400$.

Solving simultaneously, rather than by reading values off the graph, we have that $x = 45$ and $y = 6.25$ with the value of the objective function being 1.25.

Example 2

Obviously, in the real world there are many problems that exceed two-dimensional states. Variables with greater than 10 variants are not unusual. Consider the following example:

A CE department provides four variants of a service, i.e., in-house, contract-out, a hybrid of the two, OEM service. In the final part of the process, there are diagnostic, repair, and preventive maintenance. For each variant, the time required for these operations is shown in Table 47-1 (in minutes), as is the marginal cost contribution (MCC) per unit repaired (i.e., profit).

Given the current state of the professional workforce, the department estimates that each year they have 100,000 minutes of diagnostic time, 50,000 minutes of repair time, and 60,000 minutes of checkout time available. What apportionment of each variant should the department service per year, and what would be the associated MCC?

Suppose now that the department is free to decide how much time to devote to each of the three operations (diagnosis, repair, and PM) within the total allowable time of 210,000 (= 100,000 + 50,000 + 60,000) minutes. How many of each variant should the department make per year, and what would be the associated profit?

Production-planning solution:
Variables

Let: x_i be the number of units of variant i (i=1,2,3,4) made per year
T_{diag} be the number of minutes used in diagnosis per year
T_{rep} be the number of minutes used in repair per year
T_{chk} be the number of minutes used in PM per year

where $x_i \ge 0$ i=1,2,3,4 and $T_{diag}, T_{rep}, T_{pm} \ge 0$

Constraints

(a) operation time definition

$T_{diag} = 20x_1 + 40x_2 + 30x_3 + 70x_4$ (diagnosis)
$T_{rep} = 30x_1 + 20x_2 + 30x_3 + 40x_4$ (repair)
$T_{pm} = 20x_1 + 30x_2 + 20x_3 + 50x_4$ (pm)

(b) operation time limits

The operation time limits depend upon the situation being considered. In the first situation, where the maximum time that can be spent on each operation is specified, we simply have:

$T_{diag} \le 100,000$ (diagnosis)
$T_{rep} \le 50,000$ (repair)
$T_{pm} \le 60,000$ (pm)

In the second situation, where the only limitation is on the total time spent on all operations, we simply have:

$T_{diag} + T_{rep} + T_{pm} \le 210000$ (total time)
Objective: Maximize MCC–thus, the objective function is:

maximize $1.50x_1 + 2.50x_2 + 3.00x_3 + 4.50x_4$

There are a variety of software packages and mathematical modeling techniques that can be used to solve these and other problems. This is but the surface of the types of problem-solving techniques used by I/MEs and OR scientists. One Windows®-based version of these packages is called WinQSB®, and there are several other packages that employ spreadsheet methodologies for solving complex OR problems.

Decision Trees

"The roads we take are more important than the goals we announce. Decisions determine destiny."– Frederick Speakman

In many problems, chance (or probability) plays an important role. "Decision analysis" is the general name that is given to techniques for analyzing problems that contain risk/uncertainty/probabilities. Decision trees are one specific decision-analysis technique.

Example 1:

A company faces a decision with respect to a medical instrument, Model M997, developed by one of its research laboratories. It must decide whether to test market M997 or to drop it completely. The company estimates that test marketing will cost $100K. Past experience indicates that only 30% of products are successful in test markets. If M997 is successful at the test-market stage, then the company will face a further decision relating to the size of plant to set up to produce M997. A small plant will cost $150K to build and to produce 2000 units a year, while a large plant will cost $250K to build and to produce 4000 units a year. The marketing department has estimated that there is a 40% chance that a competitor will respond with a similar product and that the price per unit sold will be as follows (assuming that all production is sold):

	Large plant	Small plant
If competition respond	$20	$35
If competition do not respond	$50	$65

Assuming that the life of the market for M997 is estimated to be seven years and that the yearly plant-running costs are $50,000 (for both sizes of plant), should the company proceed with test marketing M997?

To view this problem, consider the decision tree shown in Figure 47-4.

- Decision nodes represent points at which the company must make a choice of one alternative from a number of possible alternatives (e.g., at the first decision node, the company must choose one of the two alternatives: "drop M997" or "test market M997.")
- Chance nodes represent points at which chance, or probability, plays a dominant role and that reflect alternatives over which the company has (effectively) no control.
- Terminal nodes represent the ends of paths from left to right, through the decision tree.

Solution–Step 1 For each path through the decision tree, from the initial node to a terminal node of a branch, one works out the profit (in dollars) involved in that path. Essentially, in this step, one works from the left-hand side of the diagram to the right-hand side.

- Path to terminal node 2–we drop M997.
Total revenue = 0
Total cost = 0
Total profit = 0

Money that already has been spent on developing M997 is a "sunk cost" (i.e., a cost that cannot be altered, no matter what our future decisions are) and has no part to play in deciding future decisions.

- Path to terminal node 4–We test market M997 (cost $100,000) but then find that it is not successful, so we drop it.

Total revenue = 0
Total cost = 100
Total profit = –100 (all figures in $K)

- Path to terminal node 7–We test market M997 (cost $100K), find it is successful, build a small plant (cost $150K), and find that we are without competition (revenue for seven years at 2000 units per year at $65 per unit = $910K).
- This continues for each path to terminal node.

Solution–Step 2 Consider chance node 6 with branches to terminal nodes 7 and 8 emanating from it. The branch to terminal node 7 occurs with probability 0.6 and total profit

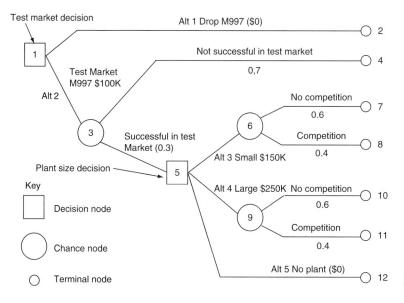

Figure 47-4 Decision tree to determine whether a company should test market a medical device.

of 310K, while the branch to terminal node 8 occurs with probability 0.4 and total profit of −110K. Hence, the "expected monetary value" (EMV) of this chance node is given by:

0.6 × (310) + 0.4 × (−110) = 142 ($K)

Essentially this figure represents the expected (or average) profit from this chance node (60% of the time we get $310K and 40% of the time we arrive at −$110K so on average we get (0.6 × (310) + 0.4 × (−110)) = 142 ($K)).

The EMV for any chance node is defined by "sum over all branches, the probability of the branch multiplied by the monetary ($) value of the branch." Hence, the EMV for chance node 9 with branches to terminal nodes 10 and 11 emanating from it is given by:

0.6 × (700) + 0.4 × (−140) = 364 ($K)
node 10   node 11

A decision node relating to the size of plant to build now can be developed as in Figure 47-5, where the chance nodes have been replaced by their corresponding EMVs. The plant-decision node now presents three alternatives:

Alternative 3: Build small plant EMV = 142K
Alternative 4: Build large plant EMV = 364K
Alternative 5: Build no plant EMV = −100K

It is clear that, in dollar terms, alternative number 4 is the most attractive alternative, so we can discard the other two, giving the revised decision tree shown in Figure 47-6.

Repeating the process carried out above:

The EMV for chance node 3 representing whether or not M997 is successful in test market is given by

0.3 × (364) + 0.7 × (−100) = 39.2 ($K) plant decision node node 4

Hence, at the decision node representing whether or not to test market M997, we have the two alternatives:

Alternative 1: Drop M997 EMV = 0
Alternative 2: Test market M997 EMV = 39.2K

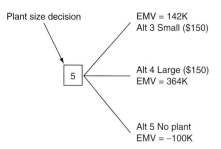

Figure 47-5 Decision tree to determine the size of a manufacturing plant to build.

It is clear that, in dollar terms, alternative number 2 is preferable, so we should test market M997 given that this decision has an expected monetary value (EMV) of $39.2K. If M997 is successful in test market then we anticipate, at this stage, building a large plant (reference the alternative chosen at the decision node relating to the size of plant to build). However, in real-life situations another review should take place after test marketing has been completed.

Example 2:

A company is trying to decide whether or not to bid for a certain contract. It estimates that merely preparing the bid would cost $10,000. If the company were to bid, then it is estimated that there is a 50% chance that the bid would be put on the "short list" of competitors, but otherwise their bid will be rejected.

Once "short listed," the company will have to supply further detailed information (entailing costs estimated at $5,000). After this stage, this bid will be accepted or rejected.

The company estimates that the labor and material costs associated with the contract are $127,000. It is considering three possible bid prices: $155,000, $170,000, and $190,000. It estimates that the probabilities of these bids being accepted (once they have been short listed) are 0.90, 0.75, and 0.35, respectively.

What should the company do, and what is the expected monetary value of your suggested course of action?

Solution Refer to the decision tree for the problem shown in Figure 47-7.

Below, step 1 of the decision tree solution procedure is carried out, which (for this example) involves calculating the total profit for each of the paths from the initial node to the terminal node (all figures in $).

Step 1

path to terminal node 7–the company would do nothing Total profit = 0

- path to terminal node 8–the company would prepare the bid but would fail to make the short list

Total cost = 10 Total profit = −10

- path to terminal node 9–the company would prepare the bid and would make the short-list, and their bid of $155K would be accepted

Total cost = 10 + 5 + 127 Total revenue = 155 Total profit = 13

- path to terminal node 10–the company would prepare the bid and would make the short list, but their bid of $155K would be unsuccessful

Total cost = 10 + 5 Total profit = −15

- path to terminal node 11–the would company prepare the bid and would make the short list, and their bid of $170K would be accepted

Total cost = 10 + 5 + 127 Total revenue = 170 Total profit = 28

- path to terminal node 12–the company would prepare the bid and would make the short-list, but their bid of $170K would be unsuccessful

Total cost = 10 + 5 Total profit = −15

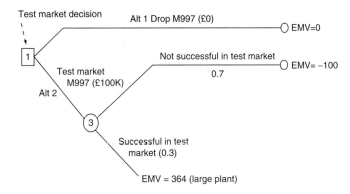

Figure 47-6 Revised decision tree to determine plant size.

- path to terminal node 13–the company would prepare the bid and would make the short-list, and their bid of $190K would be accepted

Total cost = 10 + 5 + 127 Total revenue = 190 Total profit = 48

- path to terminal node 14–the company would prepare the bid and would make the short-list, but their bid of $190K would be unsuccessful

Total cost = 10 + 5 Total profit = –15

- path to terminal node 15–the company would prepare the bid and would make the short-list and then would decide to abandon bidding (an implicit option available to the company)

Total cost = 10 + 5 Total profit = –15

Hence one arrives at Table 47-2 indicating for each branch the total profit involved in that branch from the initial node to the terminal node.

One now can carry out the second step of the decision tree solution procedure whereby one works from the right-hand side of the diagram back to the left-hand side.

Step 2

Consider chance node 4, with branches to terminal nodes 9 and 10 emanating from it. The expected monetary value for this chance node is given by 0.90(13) + 0.10(–15) = 10.2
Similarly, the EMV for chance node 5 is given by 0.75(28) + 0.25(–15) = 17.25
The EMV for chance node 6 is given by 0.35(48) + 0.65(–15) = 7.05
Hence, at the bid-price-decision node, we have four alternatives:

1. Bid $155K EMV = 10.2
2. Bid $170K EMV = 17.25
3. Bid $190K EMV = 7.05
4. Abandon the bidding EMV = –15

Hence, the best alternative is to bid $170K, leading to an EMV of 17.25.
Hence, at chance node 2 the EMV is given by 0.50(17.25) + 0.50(–10) = 3.625.
Hence, at the initial decision node. we have the two alternatives:

1. Prepare bid EMV = 3.625
2. Do nothing EMV = 0

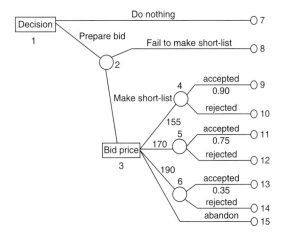

Figure 47-7 Decision tree to determine whether or not a company should bid for a certain contract.

Hence, the best alternative is to prepare the bid leading to an EMV of $3625. In the event that the company is short-listed, then (as discussed above) it should bid $170,000.

Queuing Theory and Simulation

"Mathematicians do not study objects, but relations between objects. Thus, they are free to replace some objects by others so long as the relations remain unchanged. Content to them is irrelevant: They are interested in form only." Jules Henri Poincaré (1854–1912)

The Hotline Problem

The use of queuing theory is used in a variety of ways to solve waiting-time problems. Typically there is the multiline/multiservice model that is similar to several lines of cars lining up at a gas station or information going to a printer queue. Other types, such as single-line/multiservice or multiline/sequential multiservice, can be described as well. Other factors, such as fixed or variable service times or random or fixed arrival times, enter into these problems. There are a number of basic mathematical algorithms that can be used to determine the probability distribution of waiting time (e.g., 5% wait >20 min., 10% wait <2 min.) The current method used in solving these and other problems utilizes software not unlike simulating electronic circuit configurations or flight simulators. Following is an example of setting up a telephone hotline center for a health care institution. Other applications will be intuitively obvious.

A hospital is planning to set up a cancer hotline center and needs to know the required staffing level to process calls. The I/ME conducts a series of trial calls to determine the expected mean time (and standard deviation) that it would take to process each call (service time). Research in gathering data at other similar centers reveal that, on average, 67% of the calls come in the a.m. and 23% in the p.m. and that calls arrive randomly.

In simulating this problem, the operation was simulated varying the daily volume and the number of operators. The service time was 3.8 minutes and considered exponential in that some calls may be extremely long. The results of the simulation are presented in the following table:

Table 47-3 represents relationships among daily call volume, the number of nurses answering the phone, the number of rings before the call is answered, and the productivity. Within each cell, under "Pickup," is the percentage of the time that someone will pick up the call, at a 95% confidence level, within the corresponding number of calls. For example, at the 80-calls-per-day level, one could be 95% sure that, with two nurses, a call would be picked up within 5 rings 87% of the time, or within 7 rings 91% of the time. The productivity level with two nurses would be 36%. Now, the manager must make a decision based on service level and productivity. The manager could decide to forego productivity for better service, or vice versa. These simulation techniques can be applied to queuing models and are valuable in determining quality-service levels.

Table 47-2 Total profit for each branch from the initial node to the terminal node

Terminal node	Total profit $
7	0
8	–10
9	13
10	–15
11	28
11	–15
13	48
14	–15
15	–15

Table 47-3 Relationships among daily call volume, the number of nurses answering the phone, the number of rings before the call is answered, and the productivity

		95% Conf. pick-up within				
Calls/day	Nurses	3	5	7	<=10	Productivity
60	1	88	93	96	>98	54%
60	2	94	97	99	>99	27%
60	3	97	98	99	>99	18%
80	1	82	86	88	>92	72%
80	2	84	**87**	**91**	>95	36%
80	3	89	90	93	>99	24%
100	1	75	78	81	>83	90%
100	2	77	79	84	>87	45%
100	3	80	84	89	>94	30%

Cost-Benefit Analysis

"Informed decision making comes from a long tradition of guessing and then blaming others for inadequate results"–Scott Adams

As budgetary issues grow in complexity and pressures increase for medical organizations to remain fiscally sound in the face of increasing regulation, cost-benefit analysis offers decision-makers a potentially potent source of counsel. This decision support tool provides a format for enumerating the range of benefits and costs surrounding a decision, aggregating the affects over time using an approach called "discounting," and arriving at a dollar-denominated "present value" that in concept is comparable with other uses for scarce financial resources.

In order to reach a conclusion as to the desirability of a project, all aspects of the project–positive and negative–must be expressed in terms of a common unit, that is, there must be a "bottom line." The most convenient common unit is money, which means that all benefits and costs of a project should be measured in terms of their equivalent monetary value. A project might provide benefits that are not directly expressed in terms of dollars, but there might be some amount of money that the recipients of the benefits would consider to be just as good as the project's benefits. For example, a project might result in fewer job injuries or in better patient access. In the grossest sense, the value of that benefit to a worker who risks injury is the minimum amount of money that that recipient would take instead of the medical care or that the organization would lose in a negligence lawsuit. This value could be either more or less than the market value of the medical care provided. It is assumed that more esoteric benefits, such as preserving aesthetic space or historic sites, have a finite equivalent monetary value to the organization.

A simple, systematic, and structured approach to apply in supporting business decisions should include the following considerations:

- Which vendor's outsourcing proposal is the best choice?
- Are we better off building or buying the application?
- If we can't pay for all the projects, which ones should we select?
- Are we better off implementing the only option or doing nothing?

Example

A materials management director is deciding whether to implement a new computer-based contact-management and purchasing-processing system. The department has only a few computers and his staff are not computer literate. He is aware that computerized work forces are able to process more requests and to give a higher quality of reliability and service to "customers." They also will be more responsive to their commitments and regulatory requirements and can work more efficiently with inventory control, warehousing, and logistics. The financial cost/benefit analysis is shown below:

Costs:

- New computer equipment:
- 10 network-ready PCs with supporting software @ $1225 each
- 1 fileserver @ $1750
- 3 printers @ $600 each
- Cabling and installation @ $2300
- Support software @ $7500
- Training costs:
- Computer introduction–8 people @ $ 200 each
- Keyboard skills–8 people @ $ 200 each
- Support System–12 people @ $350 each
- Other costs:
- Lost time: 40 man days @ $ 100/day
- Lost requisitions through disruption: Estimate: $10,000
- Lost requisitions through inefficiency during first months: Estimate: $10,000

Total cost: $55,800

Benefits:

- Tripling of mail capacity: Estimate: $20,000/year
- Ability to sustain telepurchase campaigns: Estimate: $10,000/year
- Improved efficiency and reliability of follow-up: Estimate: $25,000/year
- Improved customer service and retention: Estimate: $15,000/year
- Improved accuracy of customer information: Estimate: $5,000/year
- Increased ability to manage purchasing effort: $15,000/year

Total Benefit: $90,000/year

Payback time: $55,800/$90,000 = 0.62 of a year = approx. 7 months

Inevitably, the estimates of the benefit given by the new system are quite subjective. Despite this, the materials management director is likely to introduce it, given the short payback time.

Cost/benefit analysis is a powerful, widely used, and relatively easy tool for deciding whether to make a change. To use the tool, first work out how much the change will cost to make. Then calculate the benefit you will derive from it. Where costs or benefits are paid or received over time, calculate the time that it will take for the benefits to repay the costs. Cost/benefit analysis can be carried out using only financial costs and financial benefits. However, one might decide to include intangible items within the analysis. As one must estimate a value for these items, this process inevitably brings an element of subjectivity into the process.

The I/ME Education

"Students today depend too much upon ink. They don't know how to use a pen knife to sharpen a pencil. Pen and ink will never replace the pencil."–National Association of Teachers, 1907

The industrial engineering curriculum prepares engineers to design, improve, install, and operate the integrated systems of people, materials, and facilities needed by industry, commerce, and society. Industrial engineers solve problems that arise in the management of systems, applying the principles of engineering science, product/service and process design, work analysis, human factors principles, and operations research.

The typical industrial engineering curriculum combines four professional areas of practice: Product and production process design, work analysis, decision sciences, and engineering-management science. Students also are offered exposure to the more specialized areas of automated manufacturing systems, information systems, quality management, operations research, and safety engineering. In the freshman and sophomore years, a two-year foundation in basic sciences, engineering science, mathematics, the humanities, and the social sciences is taught to provide an adequate background for the courses presented in the following years. The last two years focus on developing expertise in statistics, operations research, information systems, systems analysis, organizational management, manufacturing, and industrial engineering methods. Courses stress fundamental principles and concepts that culminate in a system design dealing with real engineering and management situations in an industrial, commercial, or public-service enterprise. Students take required and elective courses that are relevant to their chosen major. Some of these courses are listed below:

- Introduction to Telecommunication Service and Systems
- Computer Simulation
- Decision Analysis
- Engineering Aestho-physiology
- Engineering Economy and Cost Analysis
- Engineering Statistics I
- Engineering System Design
- Facilities Planning and Design
- Management Organization Theory
- Operations Research I
- Operations Research Software Studies
- Product Development and Product Information Management
- Production Management I
- Quality Control
- Regression Analysis and Experimental Designs
- Statistical Data Analysis
- Work-Cell Programming for Factory Automation
- Human-Computer Interaction Design

Professional Organizations

A number of professional organizations help I/MEs to maintain their professional skills and to keep up with the latest problem-solving techniques that their peers have learned and applied in a particular specialty. One major web-access page is:

TechExpo–Directory of Hi-Tech Organizations for the Engineering and Medical/Life Sciences: http://www.techexpo.com/tech_soc.html

Other societies that are specific to the area of industrial/management engineering include the following:

- Society for Engineering & Management Systems
- Society for Health Systems
- Society for Work Science (Ergonomics & Work Measurement)
- Aerospace & Defense Division
- Energy, Environment & Plant Engineering Division
- Engineering Economy Division
- Facilities Planning & Design Division
- Financial Services Division
- Logistics Transportation & Distribution Division
- Manufacturing Division
- Operations Research Division
- Quality Control & Reliability Engineering Division
- Utilities Division

- Computer & Information Systems Interest Group
- Consultants Interest Group
- Electronics Industry Interest Group
- Engineering Design Interest Group
- Government Interest Group
- Process Industries Interest Group
- Production & Inventory Control Interest Group
- Retail Interest Group

Conclusion

"Coming together is a beginning, staying together is progress, and working together is success."–Henry Ford

This chapter has acquainted the clinical engineering professional with the field of industrial/management engineering. Collaborating with a management engineering professional yields the benefits of developing a comprehensive approach to planning, problem-solving, and decision making. Many professionals fear loss of ownership for a particular assignment and are apprehensive about multidisciplinary partnerships that can lead to conflicts in leadership, approach strategy, and solution determination. However, these partnerships are necessary to effectively meet a progressively complex, technology-driven, austerity budgetary environments. Further, it challenges the engineering professional to approach these issues as part of a knowledge-expert team.

Having graduated from college and having been hired as a clinical engineer at a hospital, one day the CE is seated next to the chief operating officer (COO) at a business lunch. The CE knows that the COO started as a transporter 30 years ago and that he rose through the ranks of the organization to his current position. The COO believes that a person only needs (innate) ability and experience in order to succeed in business. What arguments would the CE use to convince the COO that the decision making techniques performed by a CE and an I/ME are valuable?

The points to consider should include:

- Engineering analyses are obviously valuable in tactical situations where data are well-defined.
- The advantage of explicit decision making is that it allows assumptions to be examined explicitly.
- That an "analytical" approach might be better (on average) than a person.
- CE and I/ME techniques combine the ability and experience of many people.
- Sensitivity analysis can be performed in a systematic fashion.
- CE and I/ME techniques enable problems that are too large for a person to tackle effectively to be dealt with.
- Constructing an engineered model structures thought about what is and is not important in a problem.
- Training in engineering teaches a person to approach problems in a logical fashion.
- Using standard CE and I/ME techniques prevents the need to "reinvent the wheel" each time a person meets an appropriate problem.
- Modeling techniques enable computers to be used with (usually) standard packages and, consequently, all benefits of computerized analysis (e.g., speed, rapid [elapsed] solution time, and graphical output).
- CE and I/ME techniques complement ability and experience, not substitute for them.
- Other companies use engineering techniques: Do we want to be left behind?
- Ability and experience are vital, but CEs and I/MEs to use these effectively in tackling large problems.
- CE and I/ME techniques free executive time that could be spent on more creative tasks.

References

Bauld TJ. The Definition of a Clinical Engineer. *J Clin Eng* 16:403-404, 1991.
Dyro JF. Focus On: University Hospital & Health Sciences Center SUNY at Stony Brook Biomedical Engineering Department. *J Clin Eng* 18(2):165-174, 1993.
Dyro JF. Management Applications Utilizing Downloaded Mainframe Data. In Semmlow JL and Welkowitz W (eds). *Frontiers of Engineering and Computing in Health Care*. New York, IEEE, 141-145, 1984.
HCFA. The Health Care Financing Administration (CMS). National Health Expenditures (1997).

Acknowledgment

Examples courtesy of JE Beasley, who earned his undergraduate degree in mathematics at Emmanuel College, Cambridge (1971–1974) and M.Sc. (1974–1975) and Ph.D. (1975–1978) in management science at Imperial College, London, where he is currently a member of the faculty. Dr. Beasley is also a senior lecturer in operations research in the management school.

48

Financial Management of Clinical Engineering Services

Binseng Wang
Senior Director, CE & QA, MEDIQ/PRN Life Support Services, Inc.
Pennsauken, NJ

One of the intrinsic and fundamental components of the successful management of clinical engineering services is financial management. Unfortunately, many managers have overlooked the importance of financial management. For the last decade, American health care organizations have been under increasing pressure to contain and reduce costs in all areas, including the support services for medical equipment. This pressure has resulted in the contraction or even the demise of many in-house clinical engineering departments that were not able to prove their cost-effectiveness or competitiveness against independent service organizations or multivendor service divisions of major manufacturers. In reality, the financial management skills required are fairly elementary and well within the grasp of persons with strong technical background and mathematical proficiency. To illustrate the way an in-house clinical engineering department would manage its finances, a hypothetical example for a small community hospital is provided here. This example starts from an annual budget that includes all of the typical expenses and overhead and ends with an analysis of the principal components of the cost of service. Through this analysis, the manager will be able to determine the best opportunities for improving the department's cost-effectiveness and competitiveness. On the other hand, if the clinical engineering services are part of a for-profit enterprise, the financial management skills required are substantially more involved but still within reach of those who are willing to invest some time to learn. A second example is provided, to demonstrate the way a financial plan can be drafted for the first five years of of a start-up service company. This example shows the main planning tools (i.e., projected income, cash flow, and balance sheets) as well as an analysis of the viability and performance of the enterprise. Using these tools, the company's management can plan and monitor progress and can seek funding from venture capital firms and lending organizations to transform the business from a "garage operation" into a *Fortune* 500 company (assuming a little bit of luck!).

Introduction

"Just about everything that we do involves allocation of resources."
–Marvin Minsky, professor and director, Artificial Intelligence Lab, Massachusetts Institute of Technology, Cambridge, MA (Minsky, 2001).

Although the statement above was made in the context of research on the human thinking process, it holds true for most human endeavors and certainly for clinical engineering (CE) activities. Among the various resources that must be allocated and therefore properly managed, financial resources are among the most prominent. Unfortunately, financial management is often poorly understood and frequently ignored by CE practitioners, especially in the United States. Multivendor service companies successfully exploited this weakness in the last decade, which led to the contraction and demise of many in-house CE departments. In developing countries, the need to justify their existence in financial terms is generally better understood by the clinical engineers, although only a few of them have published their achievements (see, e.g., Wang & Bellentani 1986).

During its infancy in the U.S. (i.e., the 1960s and 1970s), CE growth was stimulated primarily by a fear of liability and a desire to improve health care quality. Money was of little concern. The quality of equipment manufactured was often poor, so hospitals were willing to invest in CE to ensure quality and to reduce risks. The CE departments were managed by persons who had strong technical backgrounds but little or no business management training or experience. Some of the clinical engineers learned how to manage the finances of their department or company, but most had been reluctant to meet this challenge head-on. This scenario changed significantly in the later part of the 1980s and profoundly in the 1990s because of the nationwide cost-containment and reduction efforts. Except for a few prestigious hospitals, most health care organizations were under enormous pressure to reduce their costs while improving their quality. CE departments had to prove their cost-effectiveness in order to survive in this new environment. Initially, benchmarking was used as a way to compare efficiency among hospitals and against some outsourcing companies. One of the common financial benchmarks was the cost of the CE department as a percentage of the total cost of the equipment. As discussed by Cohen et al. (1995), this parameter depends very much on the types of equipment covered. For example, the percentage cost is much higher for the CE departments that do not cover imaging equipment than it is for those that provide maintenance to big-ticket items such as X-ray machines and CT scanners. For this benchmark to be meaningful, it is mandatory to compare the total equipment-support cost with the total equipment-inventory cost (Cohen et al., 1995). Unfortunately, because many CE departments do not cover imaging equipment, there is no way for them to use this benchmark to gauge their efficiency.

Over the last decade, multivendor service companies have begun to emphasize hourly rates and total maintenance costs when they have pitched their services to health care organizations. Even though these parameters are also imperfect benchmarks, they are easier for the hospital administrators to understand because of their MBA training. Therefore it is essential for the CE managers to know how to calculate their hourly rates or total department costs in order to remain competitive (or to keep their jobs!). One of the benefits of learning how to calculate these values is in understanding what the most important factors of a department's costs are and therefore how they can be reduced or eliminated.

This chapter addresses the principles and practices of financial management by analyzing two fictitious, but realistic, examples. The first case is that of a small, in-house CE department that is responsible for the management of all of the medical equipment—including imaging and clinical laboratory—but not the facilities. This in-house team performs repairs and preventive maintenance on equipment for which it has technical competence, and supervises service contracts for equipment on which it lacks expertise. By analyzing its annual budget and ways in which the institution can fund this department, insight into ways to manage its finances is gained. Furthermore, whether this department has any chance of fighting off an assault from outsourcing companies is discussed.

The second case study analyzes the finances of a small CE service company. This company is a for-profit business that was started to in order provide maintenance and consulting services to the hospital mentioned above, and that subsequently grows to provide the same services to other health care organizations in its vicinity. The goal here is to see how much profit this company must generate in order to survive its first five-year period.

By comparing these two case studies, we are able to understand the challenges and opportunities that each organizational structure has. Furthermore, it is hoped that this comparison will help those who want to progress from an in-house team into a shared-services organization, or even into a for-profit independent service organization (ISO).

Although the values used in these case studies are realistic, many details were ignored, and some simplifications were made in order to reduce the complexity of the calculations. All of the illustrations were originally calculated using a computerized spreadsheet. Readers are encouraged to construct their own spreadsheet models and to change the values to resemble their own local realities. This helps to improve understanding of the relative importance of each variable and, therefore, helps in finding opportunities to improve management of finances.

Financial Management for CE Departments

"If you're not in the hunt on cost, you're not going to have the opportunity to demonstrate quality."–Harley Warren, Equipment Asset Manager at Baptist Memorial Hospital, Pensacola, FL (Forrest, 1997).

One of the basic duties of any department manager is to produce an annual budget and then try to control the expenditures within that budget. In this case study, we start by examining the budget of a small CE department composed of the following full-time staff:

- One clinical engineer who spends a portion (60%) of his/her time managing the department while still performing hands-on maintenance services
- Three biomedical technicians (BMETs) whose primary responsibility is to support technology by such means as performings inspections, repairs, and preventive maintenance (PM) and by providing user training
- One administrative assistant who provides all administrative support

This team is responsible for all technical support in the hospital (called "General Hospital"), servicing whatever it can, and managing what is serviced by others (e.g., OEMs and ISOs). The CE department also supervises the service contracts that have been made for some equipment (e.g., imaging and clinical lab). Although it is not really necessary, it is assumed that their employer is a nonprofit, community hospital with approximately $10 million worth (in terms of replacement cost, not original acquisition cost) of medical equipment (including imaging, clinical laboratory, and biomedical), with a total unit count of 2,500 medical devices. As these values are totally hypothetical, no attempt is made to use them for comparison with actual hospitals.

It is further assumed that this CE department has been in existence for some time, so it is not necessary to purchase all test and measurement equipment. However, it will be necessary to acquire some items, either to replace those that are worn out or to gain additional technical capabilities.

Budgeting

With the assumptions stated above, the budget for the CE department is discussed. A typical expense budget is divided into two parts: One for capital investment, and another for operating expenses.

To assist future analysis and discussion, the concepts of "direct costs" and "overhead" are introduced. Direct costs are those that can be traced to a particular product or service, such as repair parts used for a specific medical device, a maintenance contract for a particular system, hours of time that a BMET spent on performing certain PM functions, and repair fees paid to an ISO for a certain device. On the other hand, overhead includes money spent on tools and supplies that are used for multiple medical devices, training courses for the BMETs, calibration of test equipment, and the time that the clinical engineer and BMETs spent in CE department meetings. Overhead is often confused with fixed costs because the latter are included in the former, but overhead also includes variable costs, such as those for lubricants and medical gases used in equipment service, that increase with the amount of equipment serviced. In other words, overhead is the sum of fixed costs and the portion of variable costs that cannot be easily (or is not worthwhile to be) attributed to individual medical devices. This distinction is shown to be important later in this chapter, when funding alternatives are discussed.

Table 48-1 shows the annual budget for the CE department for the fiscal year of 2001. The values in this table are explained below in detail. It suffices to point out that the total budget is $454,530, of which $285,300 corresponds to direct costs, and the remaining $169,230 to overhead.

Capital Investment

As stated before, it is assuming that this CE department has been in existence for some time, so it is not necessary to purchase a complete set of new test and measurement

Table 48-1 Sample annual budget for a clinical engineering department

General Hospital Clinical Engineering Department Annual Budget For Fiscal Year 2001

	Total (us$)	Direct	Overhead
1. **Capital Investment**	$8,480	$0	$8,480
1.1 Equipment	$1,200		$1,200
1.2 Information Systems	$2,150		$2,150
1.3 Tools & Instruments	$330		$330
1.4 Furniture	$600		$600
1.5 Books & Publications	$3,400		$3,400
1.6 Other Capital Investments	$800		$800
2. **Operating Expenses**	$446,050	$285,300	$160,750
2.1 Personnel	$169,300	$120,300	$49,000
2.2 Benefits	$71,100	$0	$71,100
2.3 Supplies	$24,500	$20,000	$4,500
2.3.1 Replacement Parts	$20,000	$20,000	
2.3.2 Maintenance Supplies	$4,000		$4,000
2.3.3 Office Supplies	$500		$500
2.4 Third Party Services	$152,700	$145,000	$7,700
2.4.1 Maintenance Contracts	$120,000	$120,000	
2.4.2 Time & Materials (OEMs & ISOs)	$25,000	$25,000	
2.4.3 Training	$1,500		$1,500
2.4.4 Consulting	$3,000		$3,000
2.4.5 Test Equipment Calibration	$2,000		$2,000
2.4.6 Form Printing	$1,200		$1,200
2.5 Freight & Travel	$3,200	$0	$3,200
2.5.1 Freight	$700		$700
2.5.2 Travel	$0		$0
2.5.2.1 Food & Lodging	$1,500		$1,500
2.5.2.2 Tickets	$1,000		$1,000
2.6 Other Operating Expenses	$25,250	$0	$25,250
2.6.1 Communication	$1,200		$1,200
2.6.2 Energy	$450		$450
2.6.3 Medical Gases	$250		$250
2.6.4 Mail	$200		$200
2.6.5 Professional Associations	$600		$600
2.6.6 Physical Space (rent)	$7,500		$7,500
2.6.7 Miscellaneous	$15,050		$15,050
Total	**$454,530**	**$285,300**	**$169,230**

equipment. On the other hand, one assumes that it will be necessary to acquire some items, either to replace those that failed or to gain additional technical capabilities.

Table 48-2 shows the capital-investment budget. Note that the list includes many medical devices that already are in the inventory (i.e., those with quantity equal to zero). These existing items and their respective values will be used later (in section 3.1.3) for depreciation calculations. Most of the items are self-explanatory and low-cost. One exception is the allowance for magazines and periodicals. This is because some subscriptions are rather expensive (e.g., ECRI's *Health Devices* costs about $2000 per year).

Operating Expenses

Like other organizations within the service industry, CE departments typically have operating expenses that are much higher than the capital investment. There are many ways to classify operating expenses, each with its own advantages and disadvantages. For simplicity and didactic reasons, as few classifications as possible are used–only enough to visualize the most important components and to avoid missing important expenditures.

Table 48-3 shows an example of an operating budget. Here, again, not only the values are provided but also the breakdown between direct expenses and overhead. Each of the main items is explained below.

Personnel and Benefits As explained above, the CE department is composed of the following full-time staff:

- One clinical engineer
- Three biomedical technicians (BMETs)
- One administrative assistant

For simplicity, it is assumed that the three BMETs earn the same annual salary, such as $31,000. An allowance is made for 10% overtime for the hourly-paid employees; i.e., all except the clinical engineer. In addition to their salaries, the costs of the benefits mandated by law (e.g., state unemployment contribution by employers) and voluntarily provided by the employer (e.g., partial coverage of health, dental, and life insurance premiums) must be considered. Furthermore, it is important to make allowance for overtime and, often, on-call charges. For simplicity, these two charges are bundled into a single value. Table 48-4 shows an example of the personnel-expense budget. The values on this table were arbitrarily chosen and are probably low for major metropolitan areas, but reasonable for small cities. Readers are welcome to recalculate the values using numbers that are more realistic for their individual cases.

Supplies Included in this budget category are replacement parts, maintenance supplies, and office supplies. The distinction between replacement parts and maintenance supplies is somewhat arbitrary but necessary to distinguish between parts e.g., printed circuit boards (PCBs) and pneumatic subassemblies, used for specific medical devices and those that can support multiple medical devices (e.g., a tube of lubricant or a package of nuts and bolts). Often, many items that could be traced to individual devices are simply accounted for in bulk as maintenance supplies, simply because their low costs do not justify the labor associated with detailed accounting. Included in the office-supplies category are such items as paper, notebooks, pens, and clips. Again, the values chosen in Table 48-3 are probably on the low side, but they suffice for this case study.

Third-Party Services Included within this category are all the expenses that are paid to an entities outside of the hospital for delivering such items as maintenance contracts, maintenance services, training of BMETs, consultant services, calibration of test equipment, and printing of service-report forms used to record services performed. Naturally, there could be more or fewer expense items in individual cases. Again, the values in Table 48-3 were chosen arbitrarily but they reflect the reality of a community hospital.

Because the service contracts and time and materials expense can be traced to individual medical devices, they have been classified as direct expenses. The rest should be spread out evenly among all of the equipment covered by the CE department and, therefore, it is considered overhead.

Freight & Travel This category includes allowances for the incoming shipment of maintenance materials (i.e., parts and supplies) and outgoing shipment of equipment for service elsewhere. Also included in this category are the travel expenses associated with training. In this case, two trips are being planned. Note that the training cost (tuition) was already included in the previous category. The values chosen for the items of this category are shown on Table 48-3.

Other Operating Expenses Included in this category are all the other expenses that were not included in the previous categories and for which specific categories were not desired. Table 48-3 shows some common examples. As most of them are self-explanatory, no attempt is made to describe each one. However, facilities (e.g., physical space, energy, and gases), provided by the health care organization are included. It is important to

Table 48-2 Sample annual capital-investment budget for a clinical engineering department.

General Hospital Clinical Engineering Department Capital Investment Details

	Quantity	Cost (us$)	Extension (us$)
1. Capital Investment			
1.1 Equipment			$1,200
oscilloscope	0	$2,000	$0
digital multimeter	0	$150	$0
electrical safety analyzer	0	$350	$0
patient simulator	1	$1,200	$1,200
pneumatic calibration system	0	$5,000	$0
telephone system	0	$300	$0
fax equipment	0	$500	$0
intercom system	0	$600	$0
1.2 Information Systems			$2,150
computers	0	$2,500	$0
peripherals	1	$400	$400
local network	1	$500	$500
software	5	$250	$1,250
1.3 Tools & Instruments			$330
stopwatch	0	$25	$0
thermometer	0	$35	$0
mechanical tools	3	$50	$150
electronic tools	3	$60	$180
1.4 Furniture			$600
desks & chairs	0	$450	$0
workbench & stools	0	$900	$0
antistatic mats & straps	0	$120	$0
shelves & cabinets	2	$300	$600
1.5 Books & Publications			$3,400
1.5.1 books & manuals	1	$400	$400
1.5.2 magazines & periodicals	1	$3,000	$3,000
1.6 Other Capital Investments			$800
1.6.1 Miscellaneous	1	$800	$800
Total			**$8,480**

Table 48-3 Sample annual operating budget for a clinical engineering department.

Clinical Engineering Department Operating Expenses Details

	Total (us$)	Direct	Overhead
2. Operating Expenses			
2.1 Personnel	$169,300	$120,300	$49,000
2.2 Benefits	$71,100	$0	$71,100
2.3 Supplies	$24,500		
2.3.1 Replacement Parts	$20,000	$20,000	
2.3.2 Maintenance Supplies	$4,000		$4,000
2.3.3 Office Supplies	$500		$500
2.4 Third-party Services	$152,700		
2.4.1 Maintenance Contracts	$120,000	$120,000	
2.4.2 Time & Materials (OEMs & ISOs)	$25,000	$25,000	
2.4.3 Training	$1,500		$1,500
2.4.4 Consulting	$3,000		$3,000
2.4.5 Test Equipment Calibration	$2,000		$2,000
2.4.6 Form Printing	$1,200		$1,200
2.5 Freight & Travel	$3,200		
2.5.1 Freight	$700		$700
2.5.2 Travel			
2.5.2.1 Food & Lodging	$1,500		$1,500
2.5.2.2 Tickets	$1,000		$1,000
2.6 Other Operating Expenses	$25,250		
2.6.1 Communication	$1,200		$1,200
2.6.2 Energy	$450		$450
2.6.3 Medical Gases	$250		$250
2.6.4 Mail	$200		$200
2.6.5 Professional Associations	$600		$600
2.6.6 Physical Space (rent)	$7,500		$7,500
2.6.7 Miscellaneous 5%	$15,050		$15,050
Total	**$446,050**	**$285,300**	**$160,750**
Percentages	**100%**	**64%**	**36%**

Table 48-4 Sample of annual personnel-expense budget for a clinical engineering department.

General Hospital Clinical Engineering Department Personnel Expenses Details

	Qty	Salary	Extension	Direct	Overhead
2.1 Personnel					
Clinical Engineer	1	$45,000	$45,000	$18,000	$27,000
Biomedical Technicians	3	$31,000	$93,000	$93,000	
Administrative Assistant	1	$20,000	$20,000		$20,000
Overtime 10%			$11,300	$9,300	$2,000
TOTAL			$169,300	$120,300	$49,000
2.2 Benefits					
Mandatory 30%		$47,400	$47,400		$47,400
Voluntary 15%		$23,700	$23,700		$23,700
TOTAL			$71,100		$71,100
TOTALS			$240,400	$120,300	$120,100

account for these "invisible" costs, to make any calculation and comparison realistic. An allowance for small variances and oversights is included as "miscellaneous," which is assumed to be 5% of the total operating cost.

Financing Alternatives

There are several ways to cover the total annual budget of $454,530 for the CE department. Each of these methods is briefly described and discussed below.

Cost Center

One of the most common and traditional methods is to consider the department a cost center within the organization's administrative budget. In other words, the CE department is part of the overhead that all the revenue-generating departments must support. Examples of other administrative departments are finance, information technology, and human resources. The head of CE department—in this case the clinical engineer—negotiates the department's annual budget with his/her supervisor and then tries to "live" within this constraint. The budget is composed of capital investment and operating expenses. (See the top section of Table 48-5.) Section 2.3.1 discusses the way the clinical engineer should analyze the budget in order to become more cost-effective and to increase the value of the department to the employer.

Cost-Sharing

Another method is to make each user department (e.g., ICU, surgical center, and cardiology,) pay a certain percentage of the CE department's budget. The percentages contributed by different departments are pre-established and typically are commensurate with the amount of services that they receive from the CE department.

The user departments (e.g., psychiatry) that have fewer medical devices prefer this method because they pay less than they would in the cost center method, where all revenue-generating departments contribute equally, regardless of the usage. The departments that have more equipment (e.g., imaging, cardiology, and surgery) will have to contribute more because of their frequent use of the services provided by the CE department.

While this method may be "fairer" than the previous one, the head of the CE department must be careful not to bias his attention toward the heavier users ("bigger customers"), ignoring the less frequent users. To do so could create a politically dangerous situation.

Reimbursement for Service

Yet another financing method is reimbursement (or "chargeback") for service performed. The CE department "charges" each user department monthly (or quarterly) for the material and labor provided to that department. Typically, the CE department computes the hourly labor rate it that will need in order to cover all of its direct costs and overhead, and it then multiplies the number of hours provided to each department by this labor rate. To this rate, the CE department adds the "pass-through" costs, i.e., those costs that are spent for specific medical devices (or department).

The middle section of Table 48-5 shows the reimbursement values for our case study. In this case, the "pass-through" costs are the maintenance contracts, time and materials, third-party services, and replacement parts. The difference between the annual budget and these "pass-through" costs are called "non-pass-through expenses" that must be recovered through reimbursement. To these expenses must be added the depreciation of capital investment (assumed at the rate of 10% per year), to build reserve for replacing test and measurement equipment in the future. Therefore, the total to be reimbursed is $286,482. In order to derive the hourly labor rate, the total number of productive hours that the CE department is capable of generating must be determined first.

Table 48-6 shows the way the total number of productive hours is calculated. First, the holidays, vacations, and sick days and other paid absences (e.g., jury duty, bereavement leave, and training) must be subtracted from the total number of paid working hours. The number thus calculated is called "hours available for work." Each worker must take some breaks for refreshments, toilet use, walking to user departments, or helping a user or colleague, so not every available hour can be used to service equipment. The productivity for BMETs is typically in the range of 70%–85% (Furst, 1997). In this case study, a productivity of 75% is assumed. Furthermore, it is assumed that the clinical engineer can contribute only 40% of his or her total time to hands-on work; the rest is spent on meetings and administrative duties. Therefore the total number of productive hours per year for the entire CE department is 4671.6 hours.

Returning to the middle section of Table 48-5, with this value, the hourly labor rate to charge for reimbursements is $61.

This financing method is more laborious than those previously presented, but it has the advantage of providing user departments with an accurate accounting of the services provided. This helps to reduce the number of unnecessary service calls, but the labor involved in the accounting ultimately may reduce the productivity of the CE department.

It is possible to carry the reimbursement model further and to compute the actual cost associated with each activity. This type of accounting is known as "activity-based costing" (ABC) (Marchese, 2001). For example, in our case study, one could separate the PM activities from the repairs. If the three BMETs have different levels of expertise and are paid at different rates, the CE department manager might assign the PM duties to the least experienced person and might let the more experienced BMETs perform the more challenging repairs. In this case, the cost of PM activities could be billed at a lower rate than repairs. This accounting method is valuable when there is a large enough difference in staff pay to justify the accounting time invested.

Profit Center

An astute CE department director can actually use the reimbursement model to his advantage and to become profitable. By calculating precisely the hourly labor rate and carefully controlling the expenses, it is possible to generate some budgetary surplus, which then

Table 48-5 Examples of three financing methods for a clinical engineering department: (1) as a cost center; (2) through reimbursement per service; and (3) as a profit center

General Hospital Clinical Engineering Department Financing Methods	
	Value
1. Cost Center	
Capital Investment	$8,480
Operating Expenses	$446,050
TOTAL	**$454,530**
2. Reimbursement Per Service	
Pass-Through Costs	
Maintenance Contracts	$120,000
Per-Call Maintenance Services	$25,000
Replacement Parts	$20,000
Operating Expenses	
Non-Pass-Through Expenses	$281,050
Depreciation of Capital Investments (10%/year)	$5,432
TOTAL	$286,482
Total Number of Productive Hours (from table 6)	4671.6 hours
Hourly Labor Rate for Reimbursement	**$61**
3. Profit Center	
Pass-Through Costs	
Maintenance Contracts	$120,000
Per-Call Maintenance Services	$25,000
Replacement Parts	$20,000
Operating Expenses	
Non-Pass-Through Expenses	$281,050
Depreciation of Capital Investments (10%/year)	$5,432
TOTAL	$286,482
Profit	**$71,621**
Desired profit margin (as a % of cost)	25%
TOTAL	$358,103
Total Number of Productive Hours	4671.6 hours
Hourly Labor Rate to be Charged to Customers	**$77**
Management Fees over Pass-Through Costs	**$24,750**
Management margin	15%
Total Profit (before taxes)	**$96,371**

can be used to buy more tests tools or training courses or to improve the work environment. However, this approach must be used with great care, preferably with explicit approval of the organization. Furthermore, the CE department must be attentive in not overemphasizing its profit because this could prevent it from being competitive when it is compared to the outside service providers, especially the ISOs. Moreover, other departments could become jealous and create political troubles.

Alternatively, a real for-profit service operation can be set up (even under the umbrella of a nonprofit organization). The bottom section of Table 48-5 shows the way this enterprise would have to calculate its hourly labor rate. Assuming, arbitrarily, that the desired

Table 48-6 Sample calculation of the total number of productive hours per year for a clinical engineering department

General Hospital Clinical Engineering Department Productivity Analysis		
	Values	Hours
Hours Paid per Employee		
Hours/week	40	
Weeks/year	52	
Total		2080
Paid Holidays		
Holidays/year	12	
Holiday hours/year		96
Paid Vacation		
Vacation days/year	15	
Vacation hours/year		120
Paid Absence/Leave/Sick Days		
Absent days/year	4	
Absent hours/year		32
Hours Available for Work		
= paid hours − holiday hours − vacation hours − absent hours		1832
Productivity		
Typical productivity of biomedical technicians	75%	
Range	70-85%	
Productive Hours per Year		
For each biomedical technician		1374
As a percentage of paid hours	66%	
Total Number of Productive Hours for the Clinical Eng. Dept.		
Number of biomedical technicians	3	
Percentage of hands-on time for the clinical engineer	40%	
Total number of productive hours/year		4671.6

profit margin is 25% (over the cost), the hourly rate for the same CE department would rise from $61 to $77. In addition, it is likely that it will also add a management fee to the pass-through costs (after all, someone has to spend time to supervise what the third parties are doing). For our case study, we will assume that 15% will be added. Therefore the total profit that this service center will generate per year would be $96,333. The next section examines whether the assumptions of a 25% profit margin and 15% cost markup are enough to keep this business model afloat.

Naturally, a combination of the methods mentioned above also can be used in a single organization. It also should be stressed that the same financing method does not necessarily work best for all organizations of similar size and complexity. Each one might have a different financing method that is preferred within that institution, and the CE department is required to follow that model.

Discussion

Now that budget creation has been reviewed, the most important factors, and the ones that we can manipulate to improve the CE department's competitiveness, are to be identified and analyzed.

Budget Analysis

First, the overall budget (Table 48-1) is analyzed in two ways: By comparing capital investment with operating costs, and by comparing direct costs and overhead. As stated above, in being a service organization, the CE department spends most of its budget on operating expenses rather than on capital investment. Table 48-7 shows that, in our case study, the operating expenses form 98% of the budget, while capital investment accounts for only 2%.

More interesting is to look at the relative proportions of the direct costs versus overhead. In the budget shown in Table 48-1, direct costs are 63% of the total budget and overhead is 37% (see item 2a of Table 48-7). However, if the user departments were to pay directly for the service contracts and third-party services (which happens in many hospitals in the United States), the overhead would jump to 55% (see item 2b of Table 48-7). If the user departments also were to pay for the replacement parts, the overhead would become almost 50% larger than the direct costs (item 2c of Table 48-7), even though nothing has changed in reality. This explains why the CE departments are often perceived as being inefficient; they spend most of their budgets on nonproductive work. The CE department director must be able to explain that this number manipulation hides the fact that the CE department spends time and energy supervising service contracts and third-party services and replacing failed parts with new ones. In other words, the CE department is adding value to those services and replacement parts and, therefore, should be entitled to include these expenses when computing the overhead.

Another way to visualize the relative importance of different components of the budget is to put them into a pie chart. Figure 48-1 shows that the main components of the budget are labor (26%), overhead (37%), and maintenance contracts (26%). The rest, third-party services (6%), parts (4%), and supplies (1%), add up to a total of only 11%. Included in the overhead are capital investment, freight, travel, and other operating expenses, in addition to the indirect personnel expenses and benefits.

Cost Analysis

The hourly labor rates that the CE department charges the user departments are now addressed. From Table 48-4 (personnel budget) we can see that the total direct cost of labor is $120,300. If this value were divided by the total number of productive hours, the hourly rate for productive labor would be obtained. Table 48-8 shows that the result is $26 per hour. Subtracting this value from the reimbursement hourly rate of $61, the hourly rate for overhead is $36 (rounded up from $35). In other words, only 42% of the hourly rate of $61 is actually used to pay for work that is directly related to maintenance, while the remaining 58% is for overhead. Again, this is a somewhat distorted way of viewing the facts, as the amount of $120,300 cannot buy all of those productive hours. For example, without the benefits, one simply cannot get the employees to work. It is also illegal not to pay mandatory benefits.

Nonetheless, it is clear from this simple analysis that one must carefully control the overhead. Any activity that is not directly related to the production of services, i.e., overhead, will increase the hourly labor rate that the CE department needs to charge to the user departments without bringing tangible benefits to the latter. A closer look at Table 48-4 will show that the situation should improve significantly if the ratio of hands-on personnel versus administrative personnel increases. This fact explains why larger CE departments, OEMs, and ISOs have advantage over smaller CE teams. This is the reality of economies of scale.

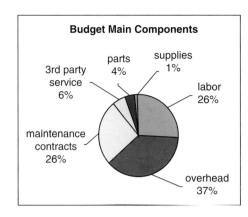

Figure 48-1 Main components of the annual budget for a clinical engineering department

Competitiveness Improvement

One natural way to improve competitiveness would be to improve the quality of services by, for example, reducing the amount of "rework" (e.g., repeated repairs). In spite of its obvious importance, this does little to decrease the hourly labor rate. There are at least three possible ways to decrease this rate. As shown in Table 48-9, the first possibility is to increase the productivity to 80%, which would increase the amount of productive hours and would reduce the labor rate to $57 per hour. The second option would be to decrease the overhead by 10% by for example reducing the amount spent on external consulting and travel and by negotiating the training costs into the initial equipment-acquisition price. This would reduce the labor rate to $58 per hour. If one could simultaneously increase productivity and reduce overhead, it would be possible to reduce the labor rate to $54 per hour. The eternal challenge for all CE department managers is in decreasing costs while improving the quality of services.

Financial Management for CE Service Providers

"Know yourself and your enemy, and you will win one hundred percent of your battles." –Sun-Tze, ancient Chinese war strategist

As it is often said, business is war. In order to survive and thrive in any business, one must know how to control its finances, especially its costs. In other words, know your costs and those of your competitors, and you will win and prosper.

Unfortunately, it is not enough to know how to prepare a budget and to operate within its limits. Quite a few additional financial concepts are needed. Readers might want to consult some textbooks on business accounting (e.g., Siegel et al, 1987; Anthony, 1997; Fridson, 1996; and Brigham and Houston, 1998) in order to appreciate fully the case study presented below. There are also a number of business-accounting software programs, such as Peachtree™ Complete Accounting and QuickBooks,™ by Quicken.™ These programs allow a small business to manage its finances without a full-time accountant, and they have some examples and tutorials that are worth examining.

Financial Planning

One of the fundamental differences between managing a cost center versus a business enterprise is the need to look at the relations between income and expenses for a certain period of time (typically five years). This long-range planning is necessary in order to visualize the ultimate self-sustainability of the enterprise as well as the potential return on the investment of capital and time. If the long-term prospect is not good, or if the return on investment (ROI) is not promising, there is no reason to waste time and money. It would be better to put the money in a bank and to enjoy life on the beach.

The following case study concerns a service company called "Fix-All Biomed Company," which is composed of the people who were in the CE department described in the previous case study:

- One clinical engineer, who spends a portion (60%) of his/her time managing the department while still performing hands-on maintenance services
- Three BMETs, whose primary responsibility is to support technology (i.e., to perform inspections, repairs, PM, and user training)
- One administrative assistant, who provides all administrative support

Table 48-7 Analysis of the overall budget by comparing capital investment and operating costs, and by comparing direct costs and overhead in various scenarios

Budget Analysis		Capital	Operating
1. Comparison between capital investment and operating costs		2%	98%
2. Comparison between direct and overhead	**DIRECT**	**OVERHEAD**	
2a. With everything	$454,380	$283,800	$170,580
percentage	100%	63%	37%
2b. Without service contracts & 3rd-party services	$309,530	$140,300	$169,230
percentage	100%	45%	55%
2c. Without replacement parts	$289,530	$120,300	$169,230
percentage	100%	42%	58%

Table 48-8 Analysis of the labor cost, comparing the direct cost and overhead components of hourly rate

Cost Analysis		
Total cost of productive labor (direct costs in figure 4)	$120,300	
Total Number of Productive Hours	4671.6	
Hourly labor cost for productive labor	$26	42%
Hourly labor cost for overhead	$36	58%
Total cost of overhead within operating expenses	$160,750	

Table 48-9 Alternatives for improving the hourly labor rate for a clinical engineering department

Alternatives For Improving Competitiveness		
1) Improve quality	The costs will not change, but less rework may save money.	
2) Decrease costs		
a) Increasing productivity to		80%
Total number of productive hours increase		4983
Hourly labor rate for reimbursement decrease		**$57**
b) Reduce the overhead by 10%		10%
Total cost of overhead decreases to		$144,675
Hourly labor rate for reimbursement decrease		**$58**
c) Doing both		
Total number of productive hours increases		4983
Total cost of overhead decreases to		$144,675
Hourly labor rate for reimbursement decrease		**$54**

Table 48-10 Sample five-year financial plan for a clinical engineering service company

Fix-All Biomed Company Clinical Engineering Department Five-Year Financial Plan

	Year I	Year II	Year III	Year IV	Year V
1. Sales/revenue	$420,280	$494,521	$577,889	$671,661	$791,779
1.1 Service [time & material]	$297,200	$339,117	$386,982	$441,642	$504,067
1.2 Contracts	$120,000	$151,200	$185,220	$222,264	$277,135
1.3 Consulting	$3,080	$4,204	$5,688	$7,755	$10,576
2. Cost of Sales	$264,300	$301,729	$315,878	$398,699	$519,579
2.1 Pass-through Costs	$144,000	$177,820	$188,252	$267,244	$384,181
2.2 Own Expenses	$120,300	$123,909	$127,626	$131,455	$135,399
3. Gross Income/profit	$155,980	$192,792	$262,011	$272,962	$272,200
4. Administrative Expenses	$186,348	$195,731	$205,714	$216,367	$228,490
4.1 Operating Expenses	$160,750	$165,573	$170,540	$175,656	$180,926
4.2 Capital Depreciation	$4,584	$5,432	$6,280	$7,128	$7,976
5. Financial Expenses	$6,250	$6,644	$5,775	$4,218	$3,365
5.1 Interest Received	$250	$356	$225	$782	$635
5.2 Interest Paid	$6,500	$7,000	$6,000	$5,000	$4,000
6. Net Income/profit (loss)	($36,618)	($9,582)	$50,522	$52,377	$40,344
7. Taxes 35%	$0	$0	$17,683	$18,332	$14,120
8. After-tax Net Income (loss)	($36,618)	($9,582)	$32,839	$34,045	$26,223

This company has a comprehensive service agreement with the General Hospital described above, with the obligation of servicing whatever it can, and of managing the rest that is serviced by third parties, i.e., OEMs and ISOs. The company also supervises service contracts that it makes for the hospital to maintain certain medical devices (e.g., imaging and clinical lab). Although it is not really necessary, it is assumed that the General Hospital is a community hospital with approximately $10 million worth (in terms of replacement cost, not original acquisition cost) of medical equipment (including imaging, clinical lab, and biomedical), with a total unit count of 2,500 medical devices. After the first year the company will try to obtain business from other health care organizations within the same geographical area, thereby expanding its revenue base in order to be less dependent on a single customer.

Table 48-10 shows a typical five-year financial plan for the Fix-All Biomed Company. This spreadsheet is organized quite differently from the annual budget shown on Table 48-1. First, it includes income, known as "revenue," as well as expenditures. The expenses are divided into three categories: Cost of sales, administrative expenses, and financial expenses. The difference between revenue and cost of sales is called "gross income." From the gross income, one must subtract the administrative expenses and financial expenses to arrive at the net income. The net income could be negative (loss) or positive (profit). If the net income is positive, the company must pay income tax. After the deduction of income tax (assumed to be 35% of the net income), one can see what the company has earned, known as "after-tax net income." This calculation is projected (or estimated) for a period of five years. The format that is used to present each year's information is known as an "income statement in accounting." The other classical accounting reports are the "balance sheet" and "cash flow statement," which are discussed below.

Each line item on Table 48-10 is explained in detail below, including the way these values change from year to year. As in the case of the CE department budgeting, the values are calculated using many assumptions. Because projections must be made for five years, some of the assumptions could become invalid and would need to be revised fairly soon, or even could need to be totally replaced. Good financial planning depends not only on experience and foresight but also on luck.

Revenue

The revenue shown on Table 48-10 is detailed in Table 48-11. First let us examine the revenue that the Fix-All Biomed Company expects to generate in the first year from its repair services (billed in the form of time spent and material used). In the first year, it expects to generate 3600 billable hours at the rate of $77 per hour, which is the value previously calculated in Table 48-5 (for a for-profit CE department). The total billable service revenue is then $297,200. Customers will be billed $20,000 in replacement parts that can be traced to individual medical devices serviced.

It is assumed that Fix-All will have 10 contracts with individual values of $12,000 each, totaling $120,000 of contract revenue. Finally, it is estimated that 40 hours of consulting work (e.g., advising hospitals about the brands and models of equipment that are most appropriate for their next purchase) will be provided. At the same rate of $77 per hour, this consulting work will generate $3,080 in the first year.

For the subsequent years it is assumed that the number of billable technical hours will increase at the rate of 10% per year and that the cost of replacement parts will increase proportionally to the number of billable technical hours (i.e., more parts replaced due to more work performed). It is also assumed that the hourly rate will increase at the rate of 4% (i.e., 1% above the rate of inflation, which is assumed to be 3% per year). However, the value of service contracts is projected to increase at the rate of 5% because more equipment will be added to each contract. The number of service contracts is assumed to grow at the rate of 20% per year (by contracting with other health care organizations in the same geographical area). Furthermore, Fix-All expects to be able to increase the hourly rate of its consulting services at 5% per year because it believes that its reputation will improve with time; i.e., its customers increasingly appreciate its advice. The number of consulting hours will grow at the rate of 30% per year.

Cost of Sales

Table 48-12 shows the cost of sales that Fix-All expects to have in the first five-year period. These costs are called "cost of sales" because they are a consequence of sales. In other words, if no sales were made, these costs would not exist. They are often known as "variable costs" or "increment costs." The costs that exist regardless of sales are called "administrative costs" or "fixed costs" and are discussed below. The cost of sales is divided into two main categories: (1) pass-through costs and (2) internal expenses.

The first-year pass-through costs are examined. The first item (2.1.1) is what Fix-All must pay others to perform the work that its own people cannot perform because of the shortage of manpower. From Table 48-6, it can be seen that the maximum number of productive hours is 4672. These hours can be used for repair services, fulfilling contracts, and perform consulting. Any time beyond 4672 hours must be covered by external help (or overtime from internal staff). Table 48-12 shows that the total number of

Table 48-11 Sample five-year sales/revenue projection for a clinical engineering service company

Fix-All Biomed Company Clinical Engineering Department Five-Year Revenue Projection

	Year I	Year II	Year III	Year IV	Year V
1. Sales (revenue)					
1.1 Service [time & material]	$297,200	$339,117	$386,982	$441,642	$504,067
# billable hours	3600	3960	4356	4792	5271
hourly rate	$77	$80	$83	$87	$90
billable service [= hours × rate]	$277,200	$317,117	$362,782	$415,022	$474,785
replacement parts	$20,000	$22,000	$24,200	$26,620	$29,282
1.2 Contracts	$120,000	$151,200	$185,220	$222,264	$277,135
# of contracts	10	12	14	16	19
value/contract	$12,000	$12,600	$13,230	$13,892	$14,586
1.3 Consulting	$3,080	$4,204	$5,688	$7,755	$10,576
# billable hours	40	52	67	87	113
hourly rate	$77	$81	$85	$89	$94

Table 48-12 Sample five-year cost of sales projection for a clinical engineering service company

Fix-All Biomed Company Clinical Engineering Department Five-Year Cost Of Sales Projection

	Year I	Year II	Year III	Year IV	Year V
2. Cost Of Sales					
2.1 Pass-through Costs	$144,000	$177,820	$188,252	$267,244	$384,181
2.1.1 Third-party maintenance service	$0	$0	$0	$40,123	$114,593
#available hours for repairs	4672	4664	4461	4445	4328
# payable hours	0	0	0	347	943
hourly rate	$100	$105	$110	$116	$122
2.1.2 Replacement Parts	$20,000	$22,000	$24,200	$26,620	$29,282
2.1.3 Third-party maintenance contracts	$120,000	$151,200	$158,760	$194,481	$233,377
# of contracts	10	12	12	14	16
value/contract	$12,000	$12,600	$13,230	$13,892	$14,586
2.1.4 Third-party consulting service	$4,000	$4,620	$5,292	$6,020	$6,928
#billable hours	40	44	48	52	57
hourly rate	$100	$105	$110	$116	$122
2.2 Internal Expenses	$120,300	$123,909	$127,626	$131,455	$135,399
2.2.1 Personnel	$120,300	$123,909	$127,626	$131,455	$135,399
2.2.2 Benefits *[see Admin. Expenses]*	$0	$0	$0	$0	$0

hours available for repair services drops progressively with time as Fix-All absorbs more and more third-party contracts and consulting services. The hourly rate that Fix-All must pay for temporary help is assumed to be $100 per hour, and it increases at the rate of 5% per year. This rate is higher than the rate Fix-All charges its customers because of the temporary nature and the fact that the company is not paying benefits to the subcontractors.

The second item is simply the cost of replacement parts for which Fix-All expects to be reimbursed by its customers. The third item is composed of maintenance contracts that Fix-All must maintain with OEMs and ISOs because of the lack of expertise or specialized equipment. It is assumed that initially all contracts are simply passed through to third parties (i.e., with no administrative charges from Fix-All). After the first two years, Fix-All's staff learns how to perform some of the work itself and gradually takes over some of the contracts. Each contract is assumed to take 96 hours of labor per year (equivalent to one day per month).

Similarly, the consulting services provided by Fix-All are initially delivered by outsiders. Gradually, Fix-All's staff takes over and keeps the dependence on external experts to a minimum. This keeps the growth of the number of hours to be paid to outsiders at 10% per year.

The second part of the cost of sales is the direct personnel costs, shown in Table 48-4. All overhead is included in the administrative expenses described below. Like the budget for the CE department, it is assumed that the payroll will increase with inflation at the rate of 3% per year.

Looking back at Table 48-10, one can see that the difference between revenues and costs of sales, called "gross income" (or "profit" if it is positive, or "loss" when it is negative), increases steadily with time, giving the impression that the company is growing well.

Administrative Expenses

Table 48-13 shows the administrative expenses anticipated by Fix-All for its first five years. Like the budget for the CE department, these expenses are divided into three main categories:

- Operating costs
- Bad debt
- Capital depreciation (note that this is not capital investment)

Comparing Tables 48-12 and Table 48-3, one can see that all of the operating costs of Table 48-3 that were not included in the costs of sales are included in Table 48-13. For example, the overhead for personnel and all of the benefits from Table 48-3 are transcribed into Table 48-13, as are the costs of supplies, third-party services, travel and shipping, and other operating costs.

Never seen before, however, are the bad debts. These values correspond to the amount of money that Fix-All believes that it might not be able to collect from its clients. While a CE department never has to worry about the hospital not paying for its services (i.e., salaries and benefits to the employees), a private company must make allowance for the possibility that some customers might dispute some of its bills or not pay for its services for whatever reason. In this case, bad debts equal to 5% of account receivables are assumed.

Also included in Table 48-13 are capital depreciation costs that were not considered in the analysis of the CE department. From Table 48-2, the initial investment that was necessary to buy all tools and test and measurement equipment needed to equip the Fix-All Biomed Company (or the CE department) can be computed. This total is $45,840. Assuming that the company continues to add $8,480 worth of test and measurement equipment every year, line 4.3.3 shows the growth of the cumulative capital investment over the five-year period. For simplicity's sake, linear depreciation in 10 years is assumed. Therefore, each year, the depreciation of capital investment is 10% of the cumulative investment shown in line 4.3.3 of Table 48-13.

Cash-Flow Projection

Before examining the financial expenses summarized in financial planning (Table 48-10), cash flow projection is explained here. Financial expenses are composed primarily of interest that will be owed on money that is borrowed in order to keep a company solvent. Through cash-flow analysis, one is able to determine how much money the Fix-All

Table 48-13 Sample five-year administrative expenses projection for a clinical engineering company

Fix-All Biomed Company Clinical Engineering Department Five-Year Administrative Expenses Projection

	Year I	Year II	Year III	Year IV	Year V
4. Administrative Expenses	$186,348	$195,731	$205,714	$216,367	$228,490
4.1 Operating Costs	$160,750	$165,573	$170,540	$175,656	$180,926
4.1.1 Personnel *[Adm. Personnel Only]*	$49,000	$50,470	$51,984	$53,544	$55,150
4.1.2 Benefits *[All employees]*	$71,100	$73,233	$75,430	$77,693	$80,024
4.1.3 Supplies	$4,500	$4,635	$4,774	$4,917	$5,065
4.1.4 Third-party services	$7,700	$7,931	$8,169	$8,414	$8,666
4.1.5 Travel & Shipping	$3,200	$3,296	$3,395	$3,497	$3,602
4.1.6 Other Operating Costs	$25,250	$26,008	$26,788	$27,591	$28,419
4.2 Bad Debt	($21,014)	($24,726)	($28,894)	($33,583)	($39,589)
4.2.1 Noncollectable receivables	($21,014)	($24,726)	($28,894)	($33,583)	($39,589)
4.3 Capital Depreciation	$4,584	$5,432	$6,280	$7,128	$7,976
4.3.1 Initial Investment $45,840					
4.3.2 Annual Capital Expense $8,480					
4.3.3 Cumulative Investment	$45,840	$54,320	$62,800	$71,280	$79,760
4.3.4 Annual Depreciation [@10%/year]	$4,584	$5,432	$6,280	$7,128	$7,976

Table 48-14 Sample five-year cash flow projection for a clinical engineering company

Fix-All Biomed Company Clinical Engineering Department Five-Year Cash Flow Projection

	Year I	Year II	Year III	Year IV	Year V
2. Inflows	$414,516	$475,151	$549,220	$638,860	$752,825
2.1 Accounts Receivable	$420,280	$494,521	$577,889	$671,661	$791,779
2.2 Bad Debt [assumed at 5%]	($21,014)	($24,726)	($28,894)	($33,583)	($39,589)
2.2 Interest Received	$250	$356	$225	$782	$635
2.3 Additional Investment (stocks issued)	$15,000	$5,000	$0	$0	$0
3. Outflows	$477,390	$482,782	$523,581	$616,167	$742,105
3.1 Capital Expenses	$45,840	$8,480	$8,480	$8,480	$8,480
3.2 Operating Expenses	$425,050	$467,302	$486,418	$574,355	$700,505
3.3 Interest Paid	$6,500	$7,000	$6,000	$5,000	$4,000
3.4 Taxes Paid	$0	$0	$17,683	$18,332	$14,120
3.5 Dividends Paid	$0	$0	$5,000	$10,000	$15,000
4. Net Increase (decrease)	($62,874)	($7,630)	$25,639	$22,693	$10,719
5. Loans & Payments	$65,000	$5,000	($10,000)	($10,000)	($10,000)
5.1 Loans	$65,000	$5,000	$0	$0	$0
5.2 Payments	$0	$0	$10,000	$10,000	$10,000
6. End-of-year Balance	$7,126	$4,496	$15,639	$12,693	$719

Biomed Company will have to borrow each year and then to calculate the interest on the borrowed money.

Table 48-14 shows the five-year projection for cash flow. At the beginning of the first year it is assumed that the owners of the company have only $5,000 left after making the needed capital investment to buy the test and measurement equipment and an initial stock of supplies and other consumables. The cash inflows comprise projected revenue, known as "accounts receivable," for repair services, contracts and consulting work (from Table 48-11), interest received (discussed below), and additional investments that the owners (or other investors) are willing to make each year. Deducted from the inflows are the bad debts discussed above. The cash outflows consist of capital investments, operating expenses, interest, taxes, and dividends (i.e., money paid to the stockholders).

At the end of the first year, Fix-All must raise $15,000 from its owners (or other investors, collectively called "stockholders") and must borrow $65,000 from lenders (e.g., banks) to cover its shortfall of $62,724, to end up with a positive year-end balance of $7,276. In the second year, the situation is still difficult, because the company still must obtain $10,000 (i.e., $5,000 from owners and $5,000 from lenders) to close the year with a positive balance of $4,808. The financial outlook for Fix-All improves in the third year, when it is able to pay $5,000 in dividends to its owners and $10,000 in principal to its lenders. The balance at the end of this year is projected at $15,753. In the fourth year, the situation continues to improve. Fix-All is not only able to pay $10,000 in dividends but it also pays back $10,000 of principal to its lenders. However, in the fifth year, the situation degenerates. Fix-All is able to close the year with a positive balance of only $833 after paying $15,000 in dividends and $10,000 on loans.

A closer look shows that in the fifth year, the cash outflows are almost as high as the inflows. Several reasons contributed to this bad situation. First, the steady growth of the business created a labor shortage (see Table 48-12), which had to be covered at high cost by hiring temporary workers at $122 per hour. Second, the growth of the hourly rate charged by Fix-All (4% per year) was simply not fast enough to keep up with the cost of outside services and service contracts. Finally, by adopting the in-house initial rate of $77 per hour, Fix-All started in a barely profitable condition and just could not earn enough profit to cover the financial costs and dividends. In other words, $77 per hour would have been fine for an in-house CE department but it was unrealistic for a for-profit enterprise. Had a sixth year been projected, Fix-All would have ended up in deep trouble (i.e, with a negative cash balance).

Table 48-14 clearly shows that the financial planning based on projected income statements (Table 48-10) alone is insufficient if not misleading. In the first two years, if the company were not able to land additional investments from its owners (or new investors) and to borrow money from lenders, it probably would fail. More significantly, even when things seemed to improve in the third and fourth years, Fix-All still would get itself into cash flow trouble starting in the fifth year. This discrepancy is probably easier to visualize by comparing Figures 48-2 and 3. Figure 48-2 shows the after-tax net income from Table 48-10 in the first five years, while Figure 48-3 shows the cash flow in the same period of time. In spite of the ascending trend of the income, the dip in cash flow in the fifth year is a troubling sign. In fact, had the projected numbers for the sixth year been computed, negative end-of-the-year balance would have appeared, showing that the company might become insolvent at that time. This example illustrates well the accounting axiom, "Cash is king."

It should be stressed that, for simplicity's sake, projecting the cash flow monthly and including the delay between invoicing and receipt of cash was not attempted. Although most invoices are issued with 30-day allowances, it is not unusual to receive the money as many as 60 or 90 (sometimes even 120) days later in the current U.S. health care market. This practice makes cash-flow management even more challenging and forces many companies to delay their payment to their vendors, which could retaliate by increasing the price or even refusing to extend credit.

Financial Expenses

With the loans calculated in Table 48-14, the financial expenses projected for the first five years can be computed now. It is worthwhile to call attention to the fact that financial expenses were totally absent from the in-house CE department budget, even though the hospital's finance department is required to consider these costs.

Table 48-15 shows the progression of financial expenses over the first five-year period. The interest that the company receives is simply the interest to be earned on its bank accounts (e.g., a money-market checking account), assuming the rate of 5%. Instead of attempting to calculate exactly the interest to be earned on a daily or monthly basis, it is assumed that the average bank balance is the same throughout the year (obviously an oversimplification, but not material in this case study because of the low value). The values shown in line 5.1.1 are actually those derived earlier in the cash-flow projection. The interest to be paid depends on the net amount of the debt (line 5.2.4) and the interest rate (assumed to be 10%).

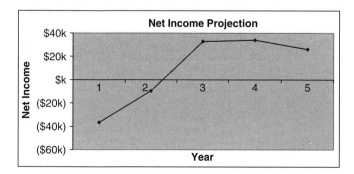

Figure 48-2 Projected after-tax net income for a clinical engineering service company in its first five years.

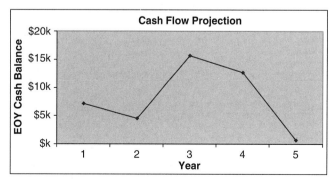

Figure 48-3 Projected cash flow for a clinical engineering service company in its first five years.

Table 48-15 Sample five-year financial expenses projection for a clinical engineering service company

Fix-All Biomed Company Clinical Engineering Department Five-Year Financial Expenses Projection

	Year I	Year II	Year III	Year IV	Year V
5. Financial Expenses	$6,250	$6,644	$5,775	$4,218	$3,365
5.1 Interest Received	$250	$356	$225	$782	$635
5.1.1 Average Bank Balance	$5,000	$7,126	$4,496	$15,639	$12,693
5.1.2 Interest Rate	5%	5%	5%	5%	5%
5.2 Interest Paid	$6,500	$7,000	$6,000	$5,000	$4,000
5.2.1 Prior Debt	$0	$65,000	$70,000	$60,000	$50,000
5.2.2 New Debt	$65,000	$5,000	$0	$0	$0
5.2.3 Debt Payment	$0	$0	$10,000	$10,000	$10,000
5.2.4 Net Debt	$65,000	$70,000	$60,000	$50,000	$40,000
5.2.4 Interest Rate	10%	10%	10%	10%	10%

Balance-Sheet Projection

Now that the two primary financial reports (income statement in Table 48-10 and cash flow in Table 48-14) have been discussed, attention is turned to the third and last primary financial report, the balance sheet. Table 48-16 shows the projected five-year balance sheet for the Fix-All Biomed Company. First, notice that we have a column prior to the first year called "initial." This column records all of the investments made prior to the beginning of operations. All of the labor that went into the organization of the company and miscellaneous expenses such as attorney's and accountant's fees have been lumped together and are called "organization costs." The purpose of this column is to enable the calculation of the return on the money invested in the company.

The balance sheet follows the classical division of assets, liabilities, and equity. In the assets category are included the typical line items of cash, accounts receivable, inventory, property and equipment (also known as "fixed assets"), and organization costs. In the liability category are accounts payable, short-term debts, taxes payable, long-term liabilities, and dividends paid (i.e., to the owners). By definition, equity is the difference between assets and liabilities and therefore it can be positive or negative. Equity is composed of total investment or paid capital (i.e., capital stock at the beginning of the year plus additional stocks issued that year) and retained earnings, which represents money that the company has earned but has not distributed to the owners or shareholders. Because total investment is always positive, retained earnings can be positive or negative, depending on how well the company performs that year.

Prior to the start of operations of Fix-All, the owners invested $60,840. In the first year, an additional investment of $15,000 was made, elevating the total invested to $75,840. Because the liabilities are now higher than the assets, the company ended the year with negative retained earnings of $126,718. Therefore the value of the company (equity) is negative $50,878 (= $126,718 -$75,840) at the end of the first year (in spite of the fact that a total of $75,840 was invested).

In the second, third, and fourth years, the situation improved, but the liabilities were still higher than the assets. The retained earnings are negative on all five years, but the equity turns positive in the third year and grows steadily afterwards (Figure. 48-4). However, at the end of the fifth year, the equity is still lower than the total investments made by the owners/stock holders. In other words, if Fix-All were to shutdown at the end of the fifth year, its owners would end up losing some of their money after paying all the debt.

At first glance it would seem that the investors would have been better off putting their money into a bank account and enjoying their lives on the beach! This is not necessarily true, however. First, the business has not failed and the owners may still be able to sell the company with some small profit (i.e., for a price that is higher than their accumulated investment). The selling price of a company is not solely related to its current revenue, but a reflection of the future cash earning power. In other words, the trend is more important than the current financial status.

Financial Performance Analysis

Many indices and ratios have been developed by analysts to evaluate the financial performance of companies. Each index (or "ratio") provides some idea about the company's financial status, but there is no universally agreed upon set of indices or ratios that will precisely predict the success of any enterprise. Each industry sector typically has a set of benchmarks that are used by analysts to judge the performance of individual companies.

Some of the indices commonly used for profitability analyses are shown in Table 48-17. The first one, gross income margin, is the ratio between gross income and revenue (both available from Table 48-10, Financial Plan). The next two indices, net income margin and after-tax net income margin, show the net income and after-tax net income as percentages of the revenue, respectively. These three indices show how profitable the company is when compared to others in the same industry. For example, supermarkets typically have low margins (< 5%) but have high turnover of inventory. Stores that sell major home appliances, on the other hand, command much higher margins but have lower sales volumes. The service industry, including CE service companies, often have intermediate margins (5%–10%) and moderate volumes when compared to other industries.

Also shown in Table 48-17 are four indices used to measure how successful or efficient the company is in generating return from the investments made in the form of money (investment), capital (equity), material goods (assets), and total capital employed (assets without considering current accounts receivable and payable). They are, respectively,

- Return on investment (ROI)
- Return on equity (ROE)

Table 48-16 Sample five-year balance sheet projection for a clinical engineering service company

Fix-All Biomed Company Clinical Engineering Department Five-Year Balance Sheet Projection

	Initial	Year I	Year II	Year III	Year IV	Year V
1. Assets	$60,840	$445,522	$521,225	$599,995	$701,573	$813,243
1.1 Cash on hand	$5,000	$5,000	$7,126	$4,496	$15,639	$12,693
1.2 Accounts receivable -Bad debt	$0	$399,266	$469,795	$548,995	$638,078	$752,190
1.3 Inventory	$0	$0	$0	$0	$0	$0
1.4 Property & equipment	$45,840	$41,256	$44,304	$46,504	$47,856	$48,360
1.4.1 Buildings & land	$0	$0	$0	$0	$0	$0
1.4.2 Equipment & furniture+fixtures	$45,840	$45,840	$54,320	$62,800	$71,280	$79,760
1.4.3 Accumulated depreciation	$0	$4,584	$10,016	$16,296	$23,424	$31,400
1.5 Organization costs	$10,000					
2. Liabilities	$0	$496,550	$544,302	$570,101	$647,687	$758,625
2.1 Accounts payable	$0	$425,050	$467,302	$486,418	$574,355	$700,505
2.2 Short-term debts	$0	$6,500	$7,000	$6,000	$5,000	$4,000
2.3 Taxes payable	$0	$0	$0	$17,683	$18,332	$14,120
2.4 Long-term liabilities	$0	$65,000	$70,000	$60,000	$50,000	$40,000
3. EQUITY	$60,840	($51,028)	($23,077)	$29,894	$53,886	$54,618
3.1 Initial capital stock	$60,840	$60,840	$75,840	$80,840	$80,840	$80,840
3.2 Additional equity [new stocks issued]		$15,000	$5,000	$0	$0	$0
3.3 Total investment		$75,840	$80,840	$80,840	$80,840	$80,840
3.4 Retained earnings		($126,868)	($103,917)	($50,946)	($26,954)	($26,222)

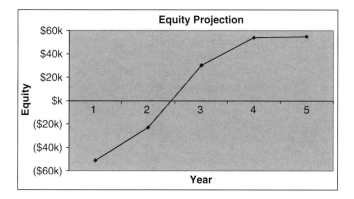

Figure 48-4 Projected equity for a clinical engineering service company in its first five years.

- Return on assets (ROA)
- Return on total capital employed (RTCE)

The data needed to calculate these indices are available from Table 48-16, Balance Sheet. As the investments and returns are made and realized at different times, the net present value (NPV) of all investments and returns has been computed so that they can be properly compared at a single time instant (i.e., at the end of each year). A detailed explanation of NPV can be found in most accounting or finance textbooks, and the calculation can be made by using formulas built into spreadsheets or financial calculators. For this example, a general business-loan interest rate (including the inflation rate) of 12% is assumed. Because in the first three years, the accumulated after-tax net incomes are negative, all four indices are not meaningful. However, they become positive in the fourth and fifth years, and both ROI and ROE are above expectations when compared to other businesses. This is not surprising, as service companies typically require few assets and investments in order to start and operate. The expertise and talent of their owners and employees are their most important asset, but they are not quantified until the company is sold (i.e., they are included in the good will). The reason that ROA seems to be below expectations is that it includes a large amount of accounts receivable (see Table 48-16). To eliminate the influence of accounts receivable, a more appropriate parameter, the total capital employed (TCE), has been created. TCE is defined as the difference between assets and current liabilities, which is shown calculated in Table 48-18 by subtracting the long-term liabilities from all liabilities. In other words, TCE is essentially all of the assets without accounts receivable and accounts payable. Again, TCE for Fix-All is above expectations because this is a service company.

Table 48-18 shows two indices, current and quick ratios, which are designed to measure the liquidity of the enterprise (how capable is the company to meet its short-term financial obligations). The current ratio is calculated by dividing the values of all current assets (i.e., excluding fixed assets) by all current liabilities (i.e., excluding long-term debts), whereas the quick ratio also excludes the inventory from the assets (because it is not certain that the existing inventory can be sold at the value estimated by the company). The quick ratio is also known as the "acid test" because it measures the company's capability to meet its immediate obligations, whereas the current ratio provides an estimate of this capability for the near term (i.e., 3–6 months).

Break-Even Analysis

Although not always required by lenders and venture capital companies, the break-even analysis is a useful tool for the company managers and owners. This analysis is used to determine the level of sales (i.e., break-even sales [BES]) that must be attained in order for the company to meet cash obligations, such as operating expenses and debt payments. Table 48-19 shows how this analysis is made.

BES is defined by the following equation:

break-even sales (BES) = fixed costs/(1-variable costs/revenue)

To determine BES it is necessary to segregate all costs into fixed and variable categories. In the fixed-costs category are the administrative expenses and the financial expenses. In the variable costs are the cost of sales. All of these values can be obtained from Table 48-10 (Financial Plan). Comparing the BES thus calculated with the projected revenue (also from Table 48-10), BES is higher than the projected revenue in the first two years but then becomes lower, starting in the third year. This calculation confirms the fact that the net income in the first two years is negative.

Besides this obvious conclusion, the BES equation helps us to visualize two statements made previously:

Table 48-17 Sample performance (margins and returns) analysis for a clinical engineering service company

Fix-All Biomed Company Clinical Engineering Department Viability/performance Analysis

Source		Initial	Year I	Year II	Year III	Year IV	Year V	Expected
FinPlan	3. GROSS INCOME		$155,980	$192,792	$262,011	$272,962	$272,200	
FinPlan	1. REVENUES		$420,280	$494,521	$577,889	$671,661	$791,779	
Gross Income Margin			37%	39%	45%	41%	34%	
FinPlan	6. NET INCOME		($36,618)	($9,582)	$50,522	$52,377	$40,344	
FinPlan	1. REVENUES		$420,280	$494,521	$577,889	$671,661	$791,779	
Net Income Margin			NA	NA	9%	8%	5%	
FinPlan	8. AFTER-TAX NET INCOME		($36,618)	($9,582)	$32,839	$34,045	$26,223	
FinPlan	1. REVENUES		$420,280	$494,521	$577,889	$671,661	$791,779	
After-tax Net Income Margin			NA	NA	6%	5%	3%	>5%
FinPlan	8. AFTER-TAX NET INCOME		($36,618)	($9,582)	$32,839	$34,045	$26,223	
	NPV AFTER-TAX NET INCOME		($32,695)	($7,639)	$23,374	$21,636	$14,880	
	Accum NPV AFTER-TAX NET INCOME		($32,695)	($40,334)	($16,959)	$4,677	$19,557	
Balance	Investments	$60,840	$15,000	$5,000	$0	$0	$0	
	NPV Investments	$60,840	$13,636	$3,986	$0	$0	$0	
	TOTAL NPV Investments		$74,476	$78,462	$78,462	$78,462	$78,462	
Return on Investment -ROI			NA	NA	NA	6%	25%	>10%
FinPlan	8. AFTER-TAX NET INCOME		($36,618)	($9,582)	$32,839	$34,045	$26,223	
	NPV AFTER-TAX NET INCOME		($32,695)	($7,639)	$23,374	$21,636	$14,880	
	Accum NPV AFTER-TAX NET INCOME		($32,695)	($40,334)	($16,959)	$4,677	$19,557	
Balance	3. EQUITY		($51,028)	($23,077)	$29,894	$53,886	$54,618	
	NPV EQUITY		($45,561)	($18,396)	$21,278	$34,246	$30,992	
Return on Equity -ROE			NA	NA	NA	14%	63%	>15%
FinPlan	8. AFTER-TAX NET INCOME		($36,618)	($9,582)	$32,839	$34,045	$26,223	
	NPV AFTER-TAX NET INCOME		($32,695)	($7,639)	$23,374	$21,636	$14,880	
	Accum NPV AFTER-TAX NET INCOME		($32,695)	($40,334)	($16,959)	$4,677	$19,557	
Balance	1. ASSETS		$445,522	$521,225	$599,995	$701,573	$813,243	
	NPV ASSETS		$397,788	$415,517	$427,064	$445,862	$461,456	
Return on Assets -ROA			NA	NA	NA	1%	4%	>9%
FinPlan	8. AFTER-TAX NET INCOME		($36,618)	($9,582)	$32,839	$34,045	$26,223	
	NPV AFTER-TAX NET INCOME		($32,695)	($7,639)	$23,374	$21,636	$14,880	
	Accum NPV AFTER-TAX NET INCOME		($32,695)	($40,334)	($16,959)	$4,677	$19,557	
Balance	1. ASSETS		$445,522	$521,225	$599,995	$701,573	$813,243	
	Current Liabilities (see below)		$431,550	$474,302	$510,101	$597,687	$718,625	
	Total Capital Employed (TCE)		$13,972	$46,923	$89,894	$103,886	$94,618	
	NPV Total Capital Employed		$12,475	$37,407	$63,985	$66,022	$53,689	
Return on TCE -RTCE			NA	NA	NA	7%	36%	>10%

Table 48-18 Sample liquidity (ratios) analysis for a clinical engineering service company

Fix-All Biomed Company Clinical Engineering Department Liquidity Analysis

Source		Year I	Year II	Year III	Year IV	Year V	Expected
Balance	2. Liabilities	$496,550	$544,302	$570,101	$647,687	$758,625	
Balance	2.4 Long-term liabilities	$65,000	$70,000	$60,000	$50,000	$40,000	
	Current Liabilities	$431,550	$474,302	$510,101	$597,687	$718,625	
Balance	1. ASSETS	$445,522	$521,225	$599,995	$701,573	$813,243	
Balance	1.4 Property & equipment	$41,256	$44,304	$46,504	$47,856	$48,360	
	Current Assets	$404,266	$476,921	$553,491	$653,717	$764,883	
Current Ratio		**94%**	**101%**	**109%**	**109%**	**106%**	**>40%**
	Current Liabilities	$431,550	$474,302	$510,101	$597,687	$718,625	
	Current Assets	$404,266	$476,921	$553,491	$653,717	$764,883	
Balance	1.3 Inventory	$0	$0	$0	$0	$0	
	Current Assets -Inventory	$404,266	$476,921	$553,491	$653,717	$764,883	
Quick Ratio/"Acid Test"		**94%**	**101%**	**109%**	**109%**	**106%**	**>30%**

- To increase profit (or value, in the case of an in-house CE department), one should try to reduce the fixed costs (or overhead) as much as possible, as it reduces the amount of sales needed to break even.
- It is acceptable to increase variable costs, as long as they help to increase revenue by at least the same percentage. Better yet would be to reduce variable costs as a percentage of the revenue.

Discussion

From the foregoing analysis it is clear that the transition from an in-house CE department to a for-profit company is quite challenging in financial terms. Although initially the profit margin seems to be adequate, the cash-flow projection shows that the Fix-All Biomed Company could falter because of cash-flow troubles. One of the problems detected in the fifth year is the lack of internal labor to cover all of the service and contract needs. One possible solution that was not explored here would be the hiring of an additional biomedical technician. Because an employee's rate is less costly than that of temporary help (or an overtime rate), this option could improve the bottom line. If, however, a more global solution were desired, the possibility of raising the hourly rate from $77 to, for example, $85 should be considered. With a spreadsheet one can see that this small (10.4%) increase will improve the whole financial picture significantly. Obviously there are other possible solutions. The purpose of financial planning is exactly to allow the managers to visualize potential problems and to simulate different scenarios and alternatives before making a decision.

As stated before, numerous assumptions and forecast have to be made in the projection of income, cash flow, and balance. Consequently, the financial plan can be only as good as the assumptions and forecasts made. As it is impossible to foresee all of the possibilities and challenges, management will need to make adjustments and corrections as the company evolves. This does not mean that the initial plan is useless or that it would not be worthwhile to go through the trouble of performing financial planning. The value of financial planning is actually not in the plan achieved but in the planning process itself (thus the saying "Having a plan is nothing, planning is everything"). This is because during the planning process one is forced to think carefully about every step of the business process before taking it. This exercise helps managers to identify the critical steps of the process and the assumptions that might have to be revised when the circumstances change.

Although the financial-performance analysis presented above seems to provide a simple and powerful set of tools to analyze and compare companies. These tools (ratios and indices) should be used with great care. In reality, these tools produce better questions than answers, i.e., they will suggest when and where to look deeper for signs of trouble or poor performance. If properly used, financial-performance analysis can provide information and insights for at least three critical tasks. First, it will assist in the allocation of capital in order to maximize return and to minimize risk. Second, it will identify a company's financial and operating strengths and weaknesses (two important components of a financial strengths, weaknesses, opportunities, and threats [SWOT] analysis). Finally, it will help to evaluate managerial performance in order to determine appropriate compensation and promotions. Typically, this kind of analysis is performed by comparing companies within the same industry (cross-sectional analysis). Occasionally, the analysis also can be made across various industries to determine the overall performance of each industrial segment. Sometimes, analysis can be executed to identify trends and exceptions over time (time-series analysis) and to compare financial position and results from period to period.

Conclusions

The two case studies presented above illustrate ways in which CE managers can manage the finances of their departments or companies, especially ways to control and reduce the costs without affecting the quality and thus to remain competitive. Certainly there are many omissions and inaccuracies in these case studies, but they provide enough details to demonstrate the basic principles of financial management for an in-house CE department and for an independent service company. Starting from these case studies, readers should be able to build more complete and accurate financial models of their own operations in order to assist them in successfully growing their department or enterprise.

Although the in-house CE department seems to be more cost-effective than the for-profit company, one should not conclude that this is always the case. The reality may be quite different. Often, for-profit companies are less generous than health care organizations in terms of compensation and benefits. The former are often subject to contracts negotiated by labor unions. Service companies also tend to have a much higher ratio of BMETs to clinical engineers than do hospitals, especially when they sustain multiple health care organizations, effectively reducing the overhead costs (an advantage provided by economies of scale). Finally, multivendor service companies are typically less involved in activities that are not strictly maintenance and repair (e.g., user training, incident investigation, and pre-acquisition assessment), thereby improving their productivity when compared to in-house CE departments. After all, for-profit companies would not remain in business if they could not be profitable.

A cost benchmark that has been used and preferred by some CE managers is the total cost of services as a percentage of the equipment inventory (Cohen et al., 1995; Cohen, 1998). In the case studies presented here, the hourly rate was used as the main cost benchmark. It is indisputable that the hourly rate depends significantly on the sophistication of the equipment covered. (Typically, the imaging technicians command higher pay than those who service biomed equipment.) However, a smart CE manager utilizes a mixture of internal capabilities and outside service contracts to satisfy the need of his employer, no matter whether it is a health care organization or a service company. The goal of every CE manager should be to maximize the service impact while reducing costs rather than building a CE empire. Therefore if the CE department or company were properly managed, the hourly rate benchmark would be more accurate and easier to compute than the percentage-of-inventory benchmark, as the latter requires constant revision of current equipment replacement costs. The original acquisition cost should not be used because it does not consider inflation and accumulated interest, or possible deflation of equipment prices.

Another benchmark often used by CE managers is cost avoidance or money saved (Cohen et al, 1995; Gordon, 1995). This benchmark is calculated by comparing the costs of internal labor versus outside services and/or the reduction of breakdowns (down time) because of preventive maintenance. Unfortunately, the money saved is not tangible (it is known as "soft money") and thus it has little credibility with administrators and financial analysts.

Irrespective of the benchmark(s) used, it is clear that CE managers must know how to manage their finances at least as well as–and probably better than–they know how to service the equipment. The case studies shown here demonstrate that the mathematics needed is not that advanced and the calculations can be accomplished with a stan-

Table 48-19 Sample break-even analysis for a clinical engineering service company

Fix-All Biomed Company Clinical Engineering Department Break-even Analysis

Source		Year I	Year II	Year III	Year IV	Year V
Break-Even Sales (BES) = Fixed Costs/(1-Variable Costs/Revenue)						
	Fixed Costs	$192,598	$202,374	$211,489	$220,585	$231,856
FinPlan	4. ADMINISTRATIVE EXPENSES	$186,348	$195,731	$205,714	$216,367	$228,490
FinPlan	5. FINANCIAL EXPENSES	$6,250	$6,644	$5,775	$4,218	$3,365
	Variable Costs	$264,300	$301,729	$315,878	$398,699	$519,579
FinPlan	2. COST OF SALES	$264,300	$301,729	$315,878	$398,699	$519,579
	BES	$518,945	$519,100	$466,459	$542,780	$674,426
FinPlan	1. REVENUES	$420,280	$494,521	$577,889	$671,661	$791,779
Excess (Shortfall)		**($98,665)**	**($24,579)**	**$111.430**	**$128.881**	**$117.353**

dard spreadsheet program or basic business-accounting software. Accordingly, one should start immediately to collect the numbers and to enter them into the computer to calculate current hourly labor cost unless, of course, one plans to retire in the next six months or intends to visit the unemployment office to fraternize with the in-house CE managers who lost their jobs because they were unable to justify their costs. Worse yet would be if one stood in the bankruptcy court queue with the CEOs of independent or shared-services companies who "lost their shirts" through improper management of cash flow.

References

Anthony RN. Essentials of Accounting, 6th ed. Reading, MA, Addison-Wesley, 1997.
Brigham EF, Houston JF. Fundamentals of Financial Management, 8th ed. Orlando, FL, Dryden/Harcourt Brace, 1998.
Cohen T, Bakuzonis C, Friedman SB, et al. Benchmark Indicators for Medical Equipment Repair and Maintenance. *Biomed Instr Tech* 29:308, 1995.
Cohen T. Validating Medical Equipment Repair and Maintenance Metrics, Part II: Results of the 1997 Survey. *Biomed Instr Tech* 32:136, 1998.
ECRI. Device and Dollars, Health Devices–Special Issue. Plymouth Meeting, PA, ECRI, 1988.
ECRI. Technology Management: Preparing Your Hospital for the 1990s. *Health Tech* 3:1, 1989.
Forrest W. Meet the New Technology Manager. *Medical Imaging* 12(11):108, 1997.
Fridson MS. Financial Statement Analysis—A Practitioner's Guide. 2nd ed. New York, John Wiley & Sons, 1996.
Furst E. Budget; Basic and Productivity and Other Performance Measures; and Advanced Productivity and Additional Performance Measures. *Clinical Engineering Improvement Tools.* Chicago, IL, American Society of Hospital Engineers and IMPTECH Improvement Technologies, 1997.
Gordon G. *Breakthrough Management–A New Model for Hospital Technical Services.* Arlington, VA, Association for the Advancement of Medical Instrumentation, 1995.
Minsky M. Moody Computers–An Interview with Tom Steinert-Threlked. *Interactive Week*, 8(8): 47, 2001.
Marchese MS. The ABCs of Costing. *24x 7* 6(2):32, 2001.
Siegel ES, Schultz LA, Ford BR, et al. *The Arthur Young Business Plan Guide,* New York, NY, John Wiley & Sons, 1987.
Wang B, Bellentani IF. Maintenance of Medical Equipment: Experience from a Developing Country, Proc. 9th. International Congress of Hospital Engineering, Barcelona, Spain, 1986.

Cost-Effectiveness and Productivity

Larry Fennigkoh
Associate Professor, Electrical Engineering and Computer Science Department, Milwaukee School of Engineering, Milwaukee, WI

Throughout most of clinical engineering, the maintenance, support, and management of health care technology occupy most of the profession's time, talent, and thought. Increasingly, and somewhat ironically, these functions also are receiving considerable financial scrutiny by hospital administrators and quality-improvement consultants. The irony is that some of the very services that exist in order to save money often are being eliminated, supposedly to save money. Calls for increased productivity, accountability and staff and budget reductions abound. Clinical engineering, like many other health care service functions, continues to struggle with the challenge of answering such calls. Despite the efforts and early pioneering work of many of the profession's founding fathers (Shaffer, 1974; Ridgeway, 1980; Johnston, 1983; Furst,1986; Bauld, 1987), a consensus as to the best way to characterize and monitor clinical engineering's cost-effectiveness and productivity has yet to evolve. As a result, in part, many hospital-based clinical engineering departments have been caught somewhat off-guard when the challenges to justify their existence are made.

While this chapter does not attempt to summarize the work of the past 30 years, it does offer (1) a brief review of the underlying difficulties associated with the development and use of performance measures (acquiring an appreciation and awareness of these difficulties is crucial before implementing any form of productivity-measurement program), (2) a summary of some reasonably well-accepted definitions and performance metrics, and (3) some suggestions for improving clinical engineering cost-effectiveness and productivity through a more proactive approach to department management.

Why Measure Cost-Effectiveness and Productivity?

The implicit benefits of being able to measure cost-effectiveness and productivity comes from an underlying–and perhaps somewhat flawed–assumption that service activities in general, and clinical engineering functions in particular, may be managed like closed-loop feedback-control systems. Here, and as illustrated in Figure 49-1, this assumption maintains that if and only if outputs from these functions can be appropriately measured, then outputs can be controlled and improved. While this notion is particularly true in the manufacturing sector and works extremely well at optimizing and maintaining product quality, it has not transferred well to clinical engineering, where the outputs tend to be services, not manufactured goods. Admittedly, this author was also an early and strong proponent of this classical method (Fennigkoh, 1986, 1987).

Specifically, the fundamental differences between services and products make use of this model in a service environment so difficult. As characterized by Dunn (1985) and Fennigkoh (1987), service functions differ from manufactured products in the following, unique ways:

- **Intangibility:** "Services are intangible and can seldom be tried out, inspected, or tested in advance." This vagueness also increases as the services become more sophisticated and technical. Here the buyer's perception of what is being "bought" could be entirely different from what is actually being "sold".
- **Perishability:** "Services are also perishable, meaning that they cannot be inventoried for later sale. An unsold service is lost forever." In essence, time is the product. (Williams, 1986).
- **Nonstandardization:** While service-delivery mechanisms can be standardized, e.g., the manner in which orders are taken, processed, the manner in which the services are provided depends largely on the highly variable and complex behavior of the human delivering the service.
- **Inseparability and buyer involvement:** "Services are generally inseparable from the source that provides them or from the buyer who benefits from them." This inseparability places a considerable time constraint on the classical feedback-control model. The service already has been delivered before any feedback is obtained (i.e., the service function is effectively operating in an open-loop, rather than closed-loop, mode).

These unique differences between services and products become even more profound and problematic when combined with the simplified closed-loop model of clinical engineering service delivery (Figure 49-1). While the feedback model appropriately acknowledges the need for management to monitor CE department function and performance and to use this information to further improve such performance, it also presumes that the service outputs are being measured appropriately, consistently, and accurately. Errors or flaws in obtaining these measures result in management responding to, and attempting to control, the wrong things. Skepticism, disagreement, and mistrust result.

Problems with the Measurement of Service Productivity

The following concepts embody the fundamental difficulties associated with the measurement of service cost-effectiveness and productivity:

1. Any closed-loop service model is primarily reactive in nature. That is, only after the service has been rendered and its performance measures presumably collected and analyzed, can the service be improved. In many institutions, this may be months after the service has been delivered. Granted, the availability of meaningful measures of service output can be extremely useful in monitoring trends and in developing

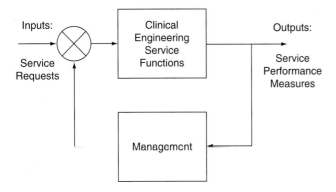

Figure 49-1 Classic closed-loop model of clinical engineering service delivery.

projections. However, the inherently reactive nature of this model places an often unintended emphasis on counting (e.g., productivity ratios, hours of service, number of repairs, and PMs) rather than on serving. This distinction is precisely what Peter Drucker (1993) noted in an article in the *Wall Street Journal* in which he encouraged the service industry to develop measures that "give us effective business control" rather than those that are means for simply counting things.

2. The "good" or "bad" judgments that tend to accompany such counting-based programs often only encourage the service staff to make the numbers "look better" rather than to deliver better services.

3. Perhaps the most insidious of the productivity-measurement problems is related to the reliability and validity of such measures. These terms have very specific and crucial meanings within the statistical and scientific community. Weaknesses in either of these characteristics weaken the use and value of the measure. Most of the problems experienced with the development and use of clinical engineering performance measures can be traced to weaknesses in their reliability and validity. Specifically, the validity of any given measure is defined as "the extent to which an instrument measures what it is intended to measure" (Portney, 2000). Validity also implies that a measurement is relatively free from error; i.e., that it is also reliable. In this context, reliability is the extent to which a measurement is consistent and free from error. The differences and relationship between these two concepts are illustrated Figure 49-2. Here, the target offers a useful analogy of what clinical engineering should be "shooting for" in its continuing development of appropriate service measures.

Many of the performance measures discussed below suffer from weaknesses in both their validity and reliability. Consider, for example, the classic measure of productivity, which often is expressed as a percentage of the ratio of recorded labor hours to paid hours. As such, it is simply a measure of time that was recorded, not necessarily how effectively the service staff did their jobs. In this regard, a productivity ratio would not be a valid indicator of service effectiveness. Similarly, if the service staff were not consistent and honest in their recording of labor hours, such a measure also would not be reliable. Consequently, managers who now attempt to "improve" service performance based on this single, unreliable, and invalid measure of "productivity" often create more problems (and less productivity).

The development of valid performance measures, particularly for service-based functions, is especially problematic because service quality is largely subjective in nature. It is, quite literally, in the perceptual eyes of the customer. Consider the many factors that influence a customer's level of satisfaction with any service function (Figure 49-3). Now try to determine a way to measure these factors. The difficulty in doing so (and, often, the inability to do so) also represents issues of measurement validity. Any measure that does not truly measure what it is intended to measure is not a valid measure.

In the typical health care environment, clinical engineering service quality and hence its perceived level of cost-effectiveness are consistently evaluated by the composite mix of all of these factors. The intrinsic difficulty in trying measure these service attributes is precisely why the proactive approach to clinical engineering productivity is offered later in this chapter.

Even if the clinical engineering profession were to derive some valid measures of cost-effectiveness, the second crucial criterion for this measure is its reliability. Again, reliability refers to the relative freedom from error and the consistency of the measure. Because virtually all of the published metrics rely on the accurate and consistent reporting of service labor hours, quantities, or costs, all of these are subject to considerable variability in their capture and recording (i.e., they can lack reliability). Reliable time recording is especially problematic if it is perceived that these records will be used in subsequent employee performance reviews and productivity reports. Here, the tendency is for the service staff simply to record more time than they actually have expended, to make

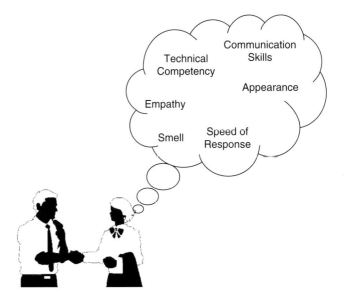

Figure 49-3 Perceptual factors influencing customer satisfaction with clinical engineering service delivery.

their numbers "look good". As a graph of any inverse function, e.g., $y=1/x$, would reveal, any performance measures involving labor hours in the denominator of a ratio (e.g., service costs/hour, repairs/hour, and PM inspections/hour) are profoundly sensitive to variations or error in these labor-hour measures. Here, relatively small inconsistencies in reported time can have profound effects on the derived measure. Unless management has some assurance that the reported labor hours are reasonably accurate, these labor-based hours should be used judiciously.

Despite these problems of measurement validity and reliability, a variety of derived measures and techniques have been reported. While care needs to be exercised in the interpretation and use of these measures, they may offer clinical engineering managers some insight as to the way their departments are functioning. In this regard, they may be particularly useful in revealing performance trends—e.g., growth, stability, or decline—and thus their use is would be encouraged. All such measures, however, remain reactive, or retrospective, in that they can be calculated and evaluated only after clinical engineering services have been rendered.

Before reviewing such measures, some basic definitions upon which these measures are based are reviewed.

Definitions

Despite some institutional variations in productivity nomenclature and definitions, a few reasonably standardized terms and concepts should be established before any measures are developed and implemented.

Effectiveness: "The capability of producing the desired results, i.e., doing the right things at the right times" (Bauld, 1987). Within the service industry in general and health care in particular, effectiveness is a subjective assessment made by the customer, often based on perceived quality and speed of delivery.

Efficiency or Productivity: "The ratio of output per unit of input or the ratio of production to the capital and resources invested" (Bauld, 1987). More specifically, productivity has been defined as:

Productivity (%) = [(Chargeable hours)/(Total worked hours)]100

Lodge (1991) makes a further distinction between productivity and efficiency and with measures of the latter for any given task being expressed as below:

Efficiency (%) = [(actual job hours)/(industry standard hours)]100

Other derived measures that have been particularly useful for trending service functions on a monthly basis are summarized in Table 49-1 (adapted from Fennigkoh, 1986).

One of the most informative and intuitively meaningful measures, however, remains the effective hourly labor rate or service cost per hour. Simply, how much does it actually cost the institution to support a clinical engineering service function? As originally reported

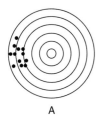
A

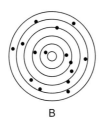
B

C

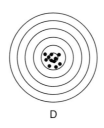
D

Figure 49-2 Representative targets illustrating the differences between reliability and validity.
(A) Measures are highly reliable, but not valid.
(B) Measures are neither valid nor reliable.
(C) Reliability has improved but is still low.
(D) Measures are both valid and reliable. (Adapted from Portney and Watkins, 2000).

Table 49-1 Derived measures for trending service functions

Performance metric	Clinical engineering service functions		
	Unscheduled service requests	Scheduled maintenance inspections	Failure-mode assessments
Work volume-based	Volume/month Volume/device class Volume/technician Volume/FTE % completed (1 day)	% completed PMs/month	Volume or % of: No problem Operator error
Labor (hours)–based	Hours/request Hours/device Hours/technician % productive	Hours/PM % PM labor	Labor hours or % due to: No problem Operator error
Cost-based	Costs/request Cost/device % acquisition cost Average labor costs	Costs/PM Costs/PM/device	

by Johnston (1983), measures of hourly labor costs can be determined with the following expression:

Cost per hour = (total department budget)/(total productive hours)

In order for such a measure to be valid, department costs should reflect all costs that the department incurs for the institution. Conventionally these are referred to as "variable" and "fixed" costs. These would include, at a minimum, salaries, benefits, travel, parts and service supplies (even estimates of parts-inventory carrying costs should be included here), telephone, depreciation, and fixed-cost estimates associated with the space that the department occupies. In essence, what would be the department's total costs if it were forced to become totally self-sufficient.

In the denominator of the cost-per-hour equation should be a measure of the total productive, or billable, hours that are available within the department. This includes adjustments for vacation, sick time, training days, holidays, and some reasonable estimate of productivity, e.g., 70%–75%. Effectively, this is a measure of time during which the service staff is actually doing something that it legitimately could charge a paying customer. As such, and as mentioned earlier, this measure of labor cost per hour is extremely sensitive to variations (or errors) within the denominator of this equation. Given that,

Total productive hours = (total paid hours − vacation productivity)

the complete equation becomes:

$$\text{Costs per hour} = \frac{[(\text{variable costs} - \text{parts costs}) + (\text{fixed costs}) + (\text{administrative costs})]}{[(\text{No. of FTE's}) \times (\text{total productive hours})]}$$

Again, with the productivity factor in the denominator of this equation, it should be apparent why productivity has such an impact on the effective costs to the organization.

A Proactive Approach to Clinical Engineering Productivity

Rather than view clinical engineering functions as a set of traditional, closed-loop systems and processes, a somewhat open-loop–and, admittedly, somewhat abstract–view of the service function is offered. Such a view, however, is from the perspective of the customer and not that of clinical engineering department management. As illustrated in Figure 49-4, this view is of what the customers, the users of health care technology, consciously or unconsciously expect from providers of clinical engineering support. These expectations include the following:

1. Speed or a reasonably quick response to their service requests
2. Quality work, which, in the eye of the beholder, includes a pleasant attitude and appearance, and empathy, as well as work that correctly solves the original problem(s)
3. Reasonably low cost or transparent cost

As shown in Figure 49-4, customers are the most satisfied when the clinical engineering department consistently operates in the zone of high-quality, quick response time, and low cost. In doing so, the department tends to become increasingly more invisible to the end-customer. In many ways, such a level of invisibility becomes the epitome of customer service and cost-effectiveness. Departments that achieve such levels of invisibility can

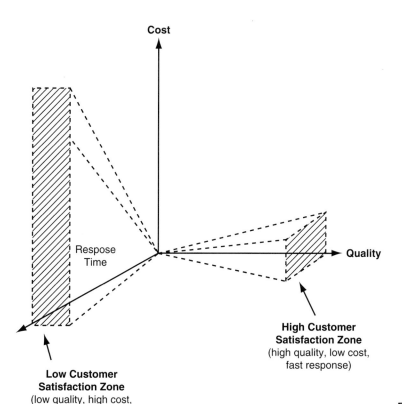

Figure 49-4 Factors influencing customer satisfaction with clinical engineering services.

References

Bauld TJ. Productivity: Standard Terminology and Definitions. *J Clin Eng* 12(2), 139-145, 1987.
Drucker P. We Need to Measure, Not Count. *The Wall Street Journal*, April 13, 1993.
Dunn DT et al. Marketing High Tech Services: Target Your Sales. *Business* 35:3-11, 1985.
Fennigkoh L. Medical Equipment Maintenance Performance Measures. *ASHE Technical Document Series*, September 1986.
Fennigkoh L. *Management of the Clinical Engineering Department: Converting a Cost Center into a Profit Center*. Brea, CA, Quest Publishing, 1987.
Furst E. Productivity and Cost-Effectiveness of Clinical Engineering. *J Clin Eng* 11(2):105-113, 1986.
Johnston GI. Analysis of In-House Costs. *IEEE Frontiers of Engineering and Computing in Health Care* 473-476, 1983.
Lodge DA. Productivity, Efficiency, & Effectiveness in the Management of Health care Technology: An Incentive Pay Proposal. *J Clinl Eng* 16(1):29-34, 1991.
Portney LG, Watkins, MP. *Foundations of Clinical Research–Applications to Practice*, 2nd ed. Prentice-Hall, Upper Saddle River, NJ 2000.
Ridgeway M. Part II: Measuring the Performance of the Hospital's clinical or Biomedical Engineering Program. *J Clin Eng* 1980, Oct-Dec; 5(4): 287-298.
Shaffer MI et al. *A System Analysis Approach for Estimating the Costs of a Clinical Engineering Service*. Proceedings of the 27th ACEMB, 1974.
Williams JF. Making Clinical Engineering a Business–and Improving Communications with Administration. *Biomedical Technology Today* 1(4):132-136, 1986.

50

Clinical Engineering Program Indicators

Dennis Autio
Sr. Clinical Engineer, Department of Veterans Affairs
Portland VA Medical Center
Portland, OR

The role of the clinical engineering department evolved from a task-oriented repair shop to a complex business incorporating a full range of clinical engineering services (Bronzino, 1992). This change has been rapid over the past 15 years, often motivated by economic factors affecting health care delivery. To continue to exist during these changing times, clinical engineering departments had to demonstrate performance (Keil, 2000) and value-added services to their parent organization if they were to be successful in competing for limited financial resources. Difficult decisions were made. Some programs were eliminated, and the previous services were contracted out to other organizations. Other programs were significantly reduced in attempts to reduce costs by reducing services. Successful clinical engineering programs embraced this change, often struggling initially to develop tools and practices to help them to create a new, business-focused, bottom-line-oriented support organization.

As new business practices were integrated into existing clinical engineering programs, it became important to develop objective, reliable indicators to document performance and to measure improvements in the services provided (Fennigkoh, 1986; Hertz, 1990; JCAHO, 1990; Audio and Morris, 1995; Keil, 1998a Audio and Morris, 2000;). This became a powerful tool when departments could demonstrate high-quality service at a competitive cost (Stiefel, 1991). Clinical engineering programs now could use this tool to strategically plan for the future using high-performance work teams that used continuous quality improvement methodologies (JCAHO, 1994). This empowering process took time to understand and to identify the ways in which it could be implemented in the workplace. With training and time, the clinical engineering staff was able to implement unique solutions for its facilities. Some of these solutions could be integrated into programs at other facilities. Throughout this process, one common factor was the use of indicators to monitor program performance. When coupled with threshold measurements and trend/pattern-analysis techniques, clinical engineering departments could better compete with the marketplace to provide timely and cost-effective proactive services.

Developing a clinical engineering program that uses indicators to manage performance requires several important planning steps. The services provided by the program must be defined. The process to document these services and to acquire data must be standardized into a usable database. The department's philosophy must be understood by all staff members and aligned with those of the organization. Staff must understand the process to develop program indictors with appropriate thresholds and how use objective data to measure the department performance. These indicators also can be used to help assess problems and to identify opportunities for quality improvement (Sherwood, 1991; AAMI, 1993a; Keil, 1998b; Al-Fadel and Crumley, 2000). When compared with similar indicators used at other facilities, they can be used to identify best practices for integration into your program.

Clinical Engineering Program Services

Clinical engineering programs have evolved from equipment maintenance (e.g., repairs, periodic inspections, and modifications), to include equipment management (e.g., installation, incoming inspections and acceptance testing, hazard notification, user-error identification, incident investigation, contract management, training, and database management) and technology management (e.g., assessment of new technology, pre-purchase equipment evaluations, and specification development). (See Chapter 30.) Not all clinical engineering departments provide the same type or level of services. Therefore it is important to define the clinical engineering services that are specific to a particular department. A broad range of clinical engineering services is listed below:

- Corrective Maintenance: Any services that involve medical equipment repair would be included in this category. Specific services include repairs performed by in-house personnel or vendors, repairs completed during the warranty period, repairs completed as a result of a hazard notification, repairs resulting from user error, and repairs performed under a service contract (see Chapter 37).
- Inspections and Periodic Maintenance (IPM): This includes various types of services, including electrical safety inspections, operations checks, scheduled IPMs performed by staff or vendors, and unscheduled IPMs performed by staff or vendors. Some of the IPM services may pass inspection, some may fail the inspection, and some inspections may be canceled.
- Education: Various educational services are provided by clinical engineering departments including training users on procedures and proper equipment use. In addition, clinical engineering staff training must be provided and documented (see Chapter 51).
- Equipment Management: Various services are specific with managing the equipment supported including developing and managing the database, pre-purchase equipment evaluations, specification development, incoming inspections, acceptance testing, and removing equipment from service (AAMI, 1993b).
- Contract Management: Contracts with various vendors that are managed by your clinical engineering department may be in effect. Services here include developing the specific contract specifications, coordinating services with the vendor, reviewing vendor services provided, and documenting services into your equipment management program (Hyman and Cram, 1999).
- Risk Management: This category includes assessing equipment for inclusion into the equipment management program, identification of user errors, hazard-notification review, incident investigation, Safe Medical Device Act-Medical Device Tracking, and Safe Medical Device Act-Device Incident Reporting (David and Judd, 1993; Wang and Levenson, 2000).
- Technology Assessment: Depending on the focus of the department, different clinical engineering services can be provided including problem investigation, equipment modification, instrumentation design, planning for implementation of new technology, and planning for the replacement of existing technology (see Chapter 32).

Clinical Engineering Program Database

Once one is able to define the clinical engineering services provided by one's department, the next step is to identify the way to capture the data associated with providing these services. Various equipment management programs that provide this service are available commercially (AAMI, 1994; Selsky et al., 1991). Data entry and data retrieval are important issues to assess with any of these programs because one will want to be able to compile the data quickly and present them in a manner that allows for quick analysis. (See Chapter 36.)

Specific equipment data are needed to uniquely define each equipment item included in the equipment management program. General equipment information includes manufacturer, model number, and serial number. Equipment management information includes facilities asset management tracking number, department using the equipment, equipment category type, and risk assessment. Additional helpful equipment information includes acquisition date, acquisition value, warranty period, and warranty expiration date. One should capture this basic information for all equipment that is supported in this program.

There is a core set of data that are helpful in managing the performance of a program. In addition to providing an equipment history for review, these data allow the categorization of the services provided, documentation of the amount of time spent performing these services, identification of the timeliness of service, and definition of the cost for this service. At a minimum, these data sets should enable the following data to be captured:

- Work Categories: Definition of the specific work tasks that categorize the services provided by the department is necessary. These tasks may include repair, IPM, incoming inspection, acceptance testing, equipment design/modification, hazard notification, user error, training, evaluation, and specification development. It may be important to further define some of these services into additional categories. For example, one may want to break down IPMs into "IPM-Complete," "IPM-Uncompleted," or "IPM-Failed." This allows easy monitoring of the IPM services provided.
- Equipment Identification: Identify the specific equipment in which this service was provided. Some services will not be associated with an equipment item, such as equipment evaluations, specification development, or review of hazard notifications.
- In-house Support: This includes both the labor cost and the number of hours spent performing the service. Labor cost should reflect salary, benefits, space, utilities, productivity, and administrative overhead. Hours should include travel time, research time, coordination time, and the actual time required to provide the service.
- Vendor Support: This information is included on the service report. It includes the hours spent providing the service, including travel hours and repair hours. The costs may include a travel rate, repair rate, and possible *per diem*.
- Contract Vendor Support: The actual hours and cost incurred when equipment is supported by a vendor under a service contract must be identified. While important to include in the equipment history, this information often is not captured. In addition to the total cost of the contract, one should capture the individual vendor's hours spent supporting the equipment and the appropriate labor cost.
- Parts: Identify the description of the parts used and the cost of each part. The actual cost of a part is not simply the purchase price but may include the costs of acquiring and storing it. The descriptions and costs for the parts provided by the vendor and the staff should be included in the equipment history.
- User Identification: Identify the person who is bringing the problem to your attention, including their name, phone number, department, and location.
- Date/Time Indicators: Identify the date and time when the problem was first identified, when it was responded to, and when it was resolved.
- Problem Description: Define the nature of the problem encountered.
- Solution Description: Describe the solution identified for solving the problem.

In addition to capturing the above data, this program also should allow you to schedule and document IPMs and user training. Merely defining the services provided and the database is not enough. The clinical engineering staff must be trained on the department's philosophy, policies, and procedures. This ongoing process must be constantly reviewed and updated as a program evolves. Each staff member must be able to define the specific clinical engineering services provided and the way that this information is captured in the database. This staff understanding of practice and procedures will provide standardized data for future analysis.

Clinical Engineering Program Management

After determining the services provided and defining the database, the management practices must be assessed. Successful clinical engineering departments define their mission, vision, values, and goals, making sure that they are aligned with those of the parent organization. A mission statement describes the services that the department provides to the organization. The vision statement describes the direction in which the department is going. A value statement and definition of excellence help to define what is important to the department. Goals then can be developed, integrating the mission, vision, and values.

Strategic planning is an important process for any successful clinical engineering department. It allows assessing the current position, identifying immediate problems to be addressed, and defining future directions. This is often not an easy process. It requires balancing available resources with the needs and priorities of the organization. Effective planning requires access to data that describe the department's performance and the way its resources are being utilized. Indicators are measurement tools that are useful in quantifying these services. They can be used to identify opportunities for quality improvement and to compare services with those that are provided by other organizations to identify opportunities for improvement. Each of these techniques allows the development of strategic plans, including department goals to improve performance.

Although a department may define generic goals, it is important to define specific goals to focus their attention and efforts. These goals will change periodically, often being driven by other issues that affect the organization. It is important that staff members understand the specifics of how these goals were developed and how they can meet them. In order to determine whether progress is being made to meeting goals, an indicator and measurement tool to monitor performance is required. Effective indicators can be used to monitor department performance and quality improvement opportunities and to allow comparison with other organizations.

Monitoring Department Performance: Indicators can be used to monitor the services provided by staff members, teams, and the department. Indicators should be accurate and objective. They can range from simple tallies (e.g., the number of jobs completed), totals (e.g., the number of hours spent doing specific tasks), or percentages (e.g., the percentage of available time spent performing specific tasks). There is a tight linkage among individual, team, and department performance. As indicator data are made available, they can become a tool to prioritize and schedule pending activities, keeping everyone focused on what is important.

Quality Improvement Opportunities: While managing the department, problems will be defined. These often are identified when program-indicator thresholds are not met or when trends and patterns are identified. At this time, the department can implement a quality improvement process to define the problem, assess the options, develop a solution, and implement an action plan to correct the problem. Program indicators are then used to monitor the effectiveness of the solution.

External Comparison: On occasion, the opportunity may arise to compare program indicators with those that are used in other facilities. These benchmarking opportunities allow the incorporation of best practices of other organizations. Care must be taken in carefully defining the program indicators used for this comparison. If one is able to define comparable program indicators among different departments, the resulting gap analysis (i.e., differences in program indicator values obtained among facilities) can identify opportunities for improvement.

Clinical Engineering Program Indicators

Once services have been defined, the database developed, and the management practices put into effect, developing a process to review and assess program performance should begin. It is important to recognize the importance of valid data. By identifying a minimum data set required to define the services provided, one can reduce the variability of data acquired by different staff members. After appropriate training, every staff member should be able to effectively identify the clinical engineering services provided and to document it properly. Without complete data the analysis becomes problematic. Over time, the information in the database becomes important to help assess department and staff performance and in relation to the services provided. This information increases in importance as problems are resolved. In order to use these data effectively, a program that utilizes indicators to monitor clinical engineering services performance must be designed and implemented.

An indicator is a reliable, valid, quantitative process or outcome measurement that relates to performance quality. This objective measurement is reliable if different people can derive the identical measurement from the same data set. It is valid if the information obtained from analysis presents an opportunity for quality improvement. Keep in mind that an indicator can be positive (e.g., the number of IPMs completed in a month) or negative (e.g., the number of IPMs not completed in a month). In either case, it provides information for further analysis. An indicator can assess different aspects of performance, including timeliness, efficiency, productivity, efficacy, safety, or customer satisfaction. With time, the reliability and validity of program indicators should evolve to the highest level possible.

An indicator can be used to measure and monitor different steps of a process. It can be used to assess the outcome of a specific process, such as the number of anesthesia-machine IPMs completed in a year. It also can be used to assess specific steps or results obtained during a process. One example would be the number of infusion pump repairs in a year that took over two weeks to complete because parts were not available.

Indicators can be fall into one of two categories. A "sentinel event" indicator documents an undesirable event, often relating to safety issues, and does not occur often. One example could be an equipment failure that results in patient injury. An "aggregate data" indicator documents performance based on many events. One type is a "continuous variable" indicator, where the value is measured over a continuous scale. One example of this type of program indicator would be IPMs scheduled per year. The other type is a "rate-based variable" indicator that is represented as a proportion or ratio. As a proportion, the numerator is a subset of the denominator, such as the number of IPMs completed per total number scheduled. As a ratio, the numerator and denominator measure different parameters, such as the number of repairs per 100 beds.

There are many program indicators that can be used to monitor clinical engineering performance. It is easy to fall into the trap of trying to measure everything. Identifying appropriate, selective indicators requires careful planning and assessment. The indicator is only one step in the performance management process. The indicator should evolve with time as an understanding of the processes is gained.

Managing Clinical Engineering Program Performance Using Indicators

The design of a program that uses program indicators to manage clinical engineering performance requires careful attention to detail and must be integrated with the overall program goals. The flowchart of this process is shown in Figure 50-1 and is explained below.

Identify Indicator and the Threshold: A specific part of the program to monitor performance should be selected carefully. This selection process may result from a professional assessment of the program, user-satisfaction surveys (Keil, 1999), choices made staff members, suggestions from administration, and identification of a quality improvement opportunity or benchmarking with another organization (Coopers and Lybrand, 1994). One should identify the specific part of the program being monitored, describe

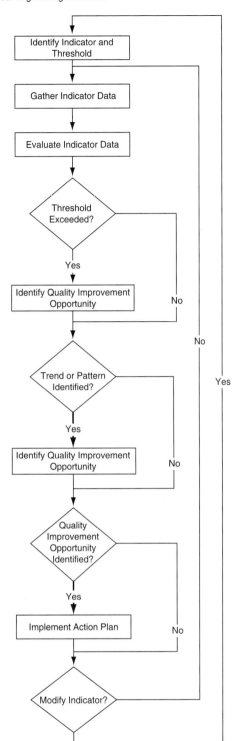

Figure 50-1 Indicator management process.

Gather Indicator Data: Once an indicator is defined, the various data sources and data elements must be defined. A standardized database assists greatly in this process because the staff continues to use the same process as before to collect data. Timeliness of data collection is important. Staff should be encouraged to document these activities in a timely manner to allow for rapid data collection.

Evaluate Indicator Data: As data are gathered it becomes important to be able to quickly compile the relevant data and present them in a manner that is easy to interpret. Using thresholds often allows the use of a single number to define performance. This can be easily reported in tabular or graphical format.

Determine Threshold: One first should determine whether the threshold of the indicator data was met. If so, then one should proceed to the next step of looking for trends and patterns. If the threshold was not met, then a quality improvement opportunity possibly exists and must be explored. This requires a more detailed analysis of the indicator data. Sometimes a department decides to change the threshold measurement based on historical data and a desire to better document the performance provided by a department.

Identify Trends or Patterns: A trend is the general direction that an indicator takes over a period of time. It can be positive, negative, or neutral. A pattern is a distribution of indicator measurements. A pattern analysis often takes place after a threshold is crossed or a trend is identified. Additional information is often required in order to answer questions that may arise, to identify trends or patterns. Again, staff proactive participation is important during this process.

Identify Quality Improvement Opportunity: During this time, one should carefully reviewed the data with the staff to determine reasons why the threshold was not met or to identify trends or patterns. This is an important process that must be approached in a proactive manner. One must carefully look at the process to identify problems. Difficult questions often must be asked. Were enough resources allocated to perform this service within the timeframe required? Was appropriate priority given to provide this service within the timeframe identified? When obstacles are identified, options for removing them must be considered. Careful review with staff will identify opportunities that can improve the current process. These improvements must be identified, and then an action plan must be developed in order to implement them. An action plan identifies the problem, the data that were assessed, the analysis of the data, and recommended actions to be taken. This documentation is an important part of this process. It starts to tell the story of what the department is doing, what was found, and the way to plan to improve it. One important step in this process is to review this action plan to document whether the solution was successful in improving performance. This can be accomplished by using the same indicators to monitor performance after the action plan was implemented. This is an iterative process, constantly undergoing improvement.

Modify Indicator: When a quality improvement opportunity is identified, the indicator often will evolve. This is the time to redefine the indicator and the threshold, and then begin the process again. At some point one may decide that there is no longer the need to manage this indictor as intensively as in the past. At this point the indicator may be retired or it may be used to report performance. The only time when one would take any further action is if a threshold is crossed.

Case Study: A Project to Enhance IPM Services

In the following example, a case study reviews the way some of these concepts can be integrated into a complex process to monitor and improve scheduled IPM services.

Problem: Current monthly IPMs completion rate is 52%. Some IPMs took several months to complete. Some IPMs are not completed. IPMs included a variety of tasks, including electrical safety inspections, operation checks, and periodic maintenance procedures. Procedures are not well documented.

Goals: Identify timetable to complete monthly IPM services. Identify critical equipment to be included in this program. Deploy resources proactively to prioritize and complete IPMs in a timely manner.

Project team: All staff members participate in this project.

Initial Assessment: Reviewed the scheduled IPM for the past two years and identified several trends and patterns. All IPM services are treated the same–there is no prioritizing of services based on equipment criticality. There was no proactive process to distribute IPM workload. The initial indicator identifies only the number of IPM services completed by the end of the month. There is no process to help the staff to assign, prioritize, and close out scheduled IPMs.

Phase 1 Action Plan

1. Continue with the current process with the same indicator. It is noted that no threshold was determined. Additional information and time are needed in order to define a proper solution.
2. Review equipment management program to standardize equipment categories and assign a risk assessment based on equipment function, physical risk, and maintenance requirements. Define the population of equipment with high- and medium-risk, based on standardized risk assessment criteria. This was the first step to standardizing the program.
3. Verify that appropriate IPMs are scheduled, define the specific procedures to be used, and identify estimated time to complete. This would provide an opportunity to project specific resource requirements for each month.
4. Provide staff training on the new risk assessment process and how IPMs are processed.

Second Assessment: Once the first phase was completed, the projected IPM schedule for the year was reviewed. With a standardized database, monthly workload projections could be determined based on risk category. IPM schedules will need to be modified to distribute workload better (Downs and McKinney, 1991). The indicators need to be revised and appropriate thresholds developed based on risk assessment. Proactive IPM assignment to staff and monitoring of IPM status is required.

any specific terminology used, and define the specific indicator to be used to monitor performance. Once the indicator is identified, an associated threshold must be defined. A threshold is a specific data point that, once reached, identifies the need for further review to analyze the data and to assess why the threshold was reached or exceeded. A threshold should be considered a goal for the department, which can change with time and understanding or the processes. If a threshold is not met, then a possible quality-improvement opportunity exists to improve performance. One example of a threshold could be 100% of the scheduled IPMs for equipment with risk score greater than 12 is completed within four weeks. The threshold (100%) is set for the indicator (completed IPM's) monitoring a population of equipment (with risk score greater than 12) for a period of time (four weeks).

Phase 2 Action Plan

1. After assessing the existing IPM schedules from an equipment category perspective, these inspections are distributed equally throughout the year. Peak workloads that would require significant resources to be allocated for only IPMs, such as anesthesia machines, were identified throughout the year. In this case, these IPMs were scheduled quarterly–all of them on the same month–and required a significant amount of staff time to complete. By doing a third of these each month, the monthly workload and allocation of resources could be better managed while still providing quarterly inspections.
2. Program indicators with thresholds were determined for high- and medium-risk equipment categories. All of the high-risk equipment and 80% of the medium-risk equipment were to be completed within four weeks. Indicator data are posted at the beginning of each week, identifying the number of IPMs completed during the last week (per risk category). Weekly completion goals were established at 25% of the scheduled high-risk equipment and 20% of the scheduled medium-risk equipment.
3. IPMs are no longer assigned by management but are selected proactively by the staff at the beginning of the month. Staff develop a proactive plan to schedule this work to be completed weekly.
4. Provided opportunities for the staff to meet and review this process, make observations, and recommend future actions.

Third Assessment: After the second phase, the staff felt that they had a better understanding of what was required, and they were provided with tools to monitor this task. Additional procedures were needed to define ways to close scheduled IPMs if they were not completed during the four-week interval. Monthly review meetings provided opportunities to assess the services provided and to identify opportunities to improve the services.

Phase 3 Action Plan

1. Define procedure to close out an IPM at the end of the month. High-risk equipment had to be completed. Medium-risk equipment could be canceled if there were no equipment repairs within the past three months and the last scheduled IPM was completed or if the equipment could not be located.
2. Continue to use the program indicators and thresholds to monitor performance on a weekly basis.
3. Quick monthly meetings were held to review the IPM indicators for the past month, identify any trends or patterns, and review the IPM scheduled for the next month.

Process Review: This process took over a year to implement. When it was completed, there was a defined process to manage IPMs within the department, and all staff were trained as to its use. Workload was better distributed throughout the year. Standardized procedures were identified defining what was to be done and how long it took to perform these tasks. Workload data were identified proactively and distributed by staff for completion. Many opportunities for quality improvement were identified based on defining the services provided and reviewing the data collected, thus resulting in improved quality, productivity, and timeliness of service.

The Future

Successful clinical engineering departments have adopted many business practices to identify high-quality and cost-effective performance (Furst, 1986; David and Rohe, 1986; Bauld, 1987; Betts, 1989; Lodge, 1991; Mahachek, 1987; Mahachek, 1989). Effective utilization of program indicators with appropriate analysis of thresholds, trends, and patterns is an important tool in this process. When thresholds are exceeded or trends and patterns identified, a potential quality improvement opportunity exists. Comparing comparable indicators with other programs will provide opportunities to benchmark performance and to identify best practices for implementation. This process will continue to evolve, especially as tools become available to provide complex analysis of the data gathered in your database. One should not be surprised if in the future, software applications will assist in the identification of equipment for replacement, based on different indicators assessing multiple data points captured from information included in the equipment-management program.

Acknowledgement

The author had the opportunity to work for seven years with Robert L. Morris, PE, CCE, at the Department of Clinical Engineering at Oregon Health Sciences University. During that time they collaborated to develop many clinical engineering management practices, some of which are described in this chapter. Although Bob is no longer with us, his spirit lives on in the knowledge and enthusiasm for clinical engineering that he passed on to others.

References

AAMI. *Management Information Report MIR 1: Design of Clinical Engineering Quality Assurance and Risk Management Programs.* Arlington, VA, Association for the Advancement of Medical Instrumentation, 1993.

AAMI. *Management Information Report MIR 2: Guideline for Establishing and Administering Medical Instrumentation Maintenance Programs.* Arlington, VA, Association for the Advancement of Medical Instrumentation, 1993.

AAMI. *Management Information Report MIR 3: Computerized Maintenance Management Systems for Clinical Engineering.* Arlington, VA, Association for the Advancement of Medical Instrumentation, 1994.

Al-Fadel H, Crumley R. Auditing the Performance of a Clinical Engineering Department for Quality Improvement. *J Clin Eng* 25(1):50, 2000.

Autio DD, Morris RL. Clinical Engineering Program Indicators. In Bronzino JD (ed). *The Biomedical Engineering Handbook.* Boca Raton, FL, CRC Press LLC, 1995.

Autio DD, Morris RL. Clinical Engineering Program Indicators. In Bronzino JD (ed). *The Biomedical Engineering Handbook, Second Edition.* Boca Raton, FL, CRC Press LLC, 2000.

Bauld TJ. Productivity: Standard Terminology and Definitions. *J Clin Eng* 12(2):139, 1987.

Betts WF. Using Productivity Measures in Clinical Engineering Departments. *Biomed Instrum Technol* 23(2):120, 1989.

Bronzino JD. *Management of Medical Technology: A Primer for Clinical Engineers.* Stoneham, MA, Butterworth-Heinemann, 1992.

Coopers and Lybrand International, AFSM. *Benchmarking Impacting the Bottom Line.* Fort Myers, FL, Association for Services Management International, 1994.

David Y, Rohe D. Clinical Engineering Program Productivity and Measurement. *J Clin Eng* 11(6):435, 1986.

Downs KJ, McKinney WD. Clinical Engineering Workload Analysis: A Proposal for Standardization. *Biomed Instrum Technol* 25(2):101, 1991.

Fennigkoh L. ASHE Technical Document #055880: *Medical Equipment Maintenance Performance Measures.* Chicago, American Society for Hospital Engineers, 1986.

Furst E. Productivity and Cost-Effectiveness of Clinical Engineering. *J Clin Eng* 11(2):105, 1986.

Hertz E. Developing Quality Indicators for a Clinical Engineering Department. *Plant, Technology & Safety Management Series: Measuring Quality in PTSM.* Chicago, Joint Commission on Accreditation of Healthcare Organizations, 1990.

JCAHO. *Primer on Indicator Development and Application, Measuring Quality in Health Care.* Oakbrook, IL, Joint Commission on Accreditation of Healthcare Organizations, 1990.

JCAHO. Framework for Improving Performance. Oakbrook, IL, Joint Commission on Accreditation of Healthcare Organizations, 1994.

Hyman WA, Cram N. In-Source, Out-Source, On-Site, Off-Site: A Checklist of Clinical Engineering Services. *J Clin Eng* 24(3): 172, 1999.

Keil OR. Performance Measurement. *J Clin Eng* 23(4): 236, 1998.

Keil OR.. The Challenge of Building Quality into Clinical Engineering Programs. *Biomed Instrum Technol* 23(5): 354, 1998.

Keil OR. Data Driven Survey Process. *J Clin Eng* 24(6): 339, 1999.

Keil OR. Telling Your Performance Management Story. *J Clin Eng* 25(1):6, 2000.

Lodge DA. Productivity, Efficiency, & Effectiveness in the Management of Health care Technology: An Incentive Pay Proposal. *J Clin Eng* 16(1):29, 1991.

Mahachek AR. Management and Control of Clinical Engineering Productivity: A Case Study. *J Clin Eng* 12(2):127, 1987.

Mahachek AR. Productivity Measurement: Taking the First Steps. *Biomed Instrum Technol* 23:16, 1989.

Selsky DB, Bell DS, Benson D, et al. Biomedical Equipment Information Management for the Next Generation. *Biomed Instrum Techno* 25(1):24, 1991.

Sherwood MK. Quality Assurance in Biomedical or Clinical Engineering. *J Clin Eng* 16(6):479, 1991.

Stiefel RH.. Creating a Quality Measurement System for Clinical Engineering. *Biomed Instrum Technol* 25(1):17, 1991.

Wang B, Levenson A. Equipment Inclusion Criteria: A New Interpretation of JCAHO's Medical Equipment Management Standard. *J Clin Eng* 25(1):26, 2000.

51
Personnel Management

James O. Wear
Veterans Administration
North Little Rock, AR

Personnel management is an important role of a clinical engineering director. Although a clinical engineering department might have a maximum of 10–15 employees, the personnel management can comprise a significant portion of the director's job.

The director needs to establish an appropriate staffing level and to monitor this staffing level. Appropriate job descriptions must be written and modified as the technology of the hospital equipment that is being maintained changes. The recruitment of staff is a function of the director, as is selection from the pool of recruits. Having a good retention program so that there are no vacancies to fill over a period of time can reduce this part of the job.

Supervision of the staff is an important function to ensure that the engineers and technicians are motivated to provide good customer service and maximum productivity. A fair evaluation of the individuals, including their customer service and productivity, is an important aspect of the retention. At the same time, there must be appropriate compensation.

One important part of any clinical engineering program is that there is some pathway for career development and a career ladder that employees can climb. Without a career ladder, it is more difficult to retain employees because they will seek promotions and more challenging job opportunities elsewhere. Where technicians and engineers may belong to a union, the director must maintain appropriate management-labor relations.

Staffing Levels

Most hospitals are understaffed in their clinical engineering departments. This means that the work of the department director as a manager is important and that scheduling and establishing priorities is critical. In the clinical engineering department there will be clinical engineers, biomedical engineering technicians (BMETs) and clerical support. BMETs may be called "biomedical equipment technicians" or "biomedical electronics technicians." A department may not have individuals in all of these positions but all of these functions must be carried out. Instead of a clinical engineer there may be a senior-level technician as director of the department and instead of clerical staff, the BMETs may be doing their own clerical work. The roles and responsibilities of each of these will be discussed below.

When reviewing a staffing situation for a clinical engineering department, there are at least four situations. The first one may be staffing a brand new hospital, which has just been built and equipped. This will require developing a staffing program over a period of time because the staff requirements will not be as great initially (with new equipment) as it will be as equipment ages.

The second situation is for a hospital that has no in-house program. This is similar to a new hospital except there already is equipment in place that needs to be maintained, and clinical engineering needs to develop inputs on new purchases. This requires staffing a full department as quickly as possible but also has the opportunities to show real savings as the equipment is being maintained with service contracts or on-call service.

The third situation is for an existing program that is a poor clinical engineering program. This is the worst case that a manager can face from a staffing standpoint. Generally it is not possible to go in and eliminate employees and then replace them. This situation requires analyzing the staffing needs and matching that with the existing staff and determining how the gaps can be filled with training or the recruitment of staff as that opportunity becomes available.

The fourth situation is staffing in an existing program that is a good program. This program probably already will be well staffed, and therefore the situation becomes one of maintaining the staffing level and increasing it as needed.

The hospital size is a factor in the staffing required but probably more important are the amount of equipment and the sophistication of the equipment. Some of this can relate to the type of hospital. Most hospitals are general medical and surgical hospitals. They range from very small to large, tertiary-care units, but they primarily provide acute care. There are also long-term care facilities, which are primarily psychiatric hospitals, and which may have a large number of patient beds and not a large amount of instrumentation. Most large, general medical and surgical hospitals also have a number of psychiatric beds. If a hospital is a teaching hospital, then it will have more equipment and more maintenance needs because a lot of students and interns are using the equipment. This is where different people and more users of the equipment increase the wear and tear, and no user is actually responsible for keeping the equipment in good condition. A teaching hospital will also have a significant amount of research, which will involve some unique equipment as well as some needs for modification of equipment. A hospital with significant research activity has some definite clinical engineering needs above what a general hospital will have, in or outside of the maintenance area.

One of the first steps in ideally developing the staffing levels, as well as numbers of staff, is the use of a skills assessment. There are six steps in developing the skills assessment: (1) to identify the competency required; (2) to build profiles of staff; (3) to assess the existing employees; (4) to identify gaps between the profiles and the employees; (5) to identify development options; and (6) to follow through to meet the competency needs.

The first step is to decide on the equipment that will be maintained and the other functions (e.g., equipment evaluation, service training, and contract management) will be performed. Next, one breaks these functions down into the competencies required. If individuals are going to maintain radiology equipment, they need to have the following competencies: 1) a knowledge of X-ray theory; 2) a knowledge of the operation of X-ray equipment; 3) good digital electronics; 4) computer operating systems; 5) networking; 6) general mechanical skills; and 7) radiology quality control. Similar competencies will be required for any category of equipment that is to be maintained.

Some of the same competencies are required for contract management, even though the people are not actually performing maintenance on the equipment. In addition, they must have good oral and written communication skills. They must have basic contract knowledge, including knowledge of how to write the appropriate specifications into a contract, and a sense of what to do whenever the contractor does not meet the terms of the contract.

To provide support to the clinical staff in evaluating equipment requires a general overall knowledge and an ability to read and interpret specifications on equipment. It requires good communication skills for dealing with the clinical staff and gaining their confidence.

In-service training is an important function of the clinical engineering program and requires special communication skills and knowledge of educational techniques. It requires learning how to use the appropriate visual aids and being able to evaluate whether or not people have learned the subject. It requires an ability to gain the confidence of the clinical staff or any other staff members that are being trained. This requires a broad knowledge of equipment if staff are being trained on the use of equipment, as well as an ability to read manuals and to develop meaningful training.

After the competencies are developed, one must develop profiles of staff that would be desirable to have those competencies. For instance, if someone is going to maintain intensive-care equipment and will be responsible for training the staff in the use of the equipment and electrical safety, then the individual would need the appropriate competencies for maintaining the equipment and for training. It should not be assumed that if someone is good at maintaining equipment, he is a good trainer or has good communication skills for working with clinical staff.

The next step is to assess the current employees to see whether any of them meets the profiles that have been developed. After looking at the competencies of employees, one might adjust the areas that they work in so that they come closer to the profiles needed. This could be done with a matrix of competencies needed with the competencies of the existing staff. One detail that should be considered here is whether personnel have competencies that are not currently being utilized either from previous work or previous training. It could well be that existing staff is being underutilized with the skills that they have.

After assessing the existing employees, one will find gaps in what exists and what is needed among the staff. For instance, there may be someone who has all of the technical skills for maintaining equipment in the clinical lab but does not have the skills to train people in that area. A person may be very good at monitoring a contract but might not have the skills for developing the contract. In a maintenance area, someone might be capable of maintaining analog X-ray equipment but might not have the required skills in digital electronics to maintain the newer digital equipment.

After looking at the gap analysis, one must identify what options are available for developing the competencies. If there are vacancies on the staff, and thus the opportunity to hire new people, then one can recruit to fill these gaps. The recruitment must be combined with profiles so that individuals are hired for the profile desired and not to fill gaps that have no relationship to each other. With an existing staff, one must consider ways in which personnel can be trained to fill the gaps. For instance, with a person who is good at maintaining clinical lab equipment but has no experience in teaching, one must assess whether they have the appropriate personality for teaching. (If so, they can be trained in teaching techniques and evaluation.) The individual who is good at monitoring a contract can be trained to write statements of work and can work with the contracting office to develop a meaningful contract. The X-ray technician who has good analog experience can be trained in digital electronics and then can trained to maintain newer X-ray equipment.

After each of the options has been looked at and developed, the critical factor is in following through with filling these gaps from recruitment or training existing staff. Priorities will have to be established because everything cannot be done at once. Establishing priorities will depend on the equipment that must be maintained, including the sophistication and amount of the equipment. The sophistication of the equipment will have more of a bearing on the skill level of the staff required than the number of staff. In fact, some of the more sophisticated equipment might have fewer maintenance problems due to high reliability and built-in diagnostics. However, it will require a more skilled technician to maintain it.

The amount of equipment will have the biggest bearing on the staff required. This can be considered in terms of the total dollar value of the equipment as well as in terms of the number of equipment items. The best measure is probably the number of items of equipment; the more pieces of equipment that must be maintained, the more technician time will be needed. This will require more time for inventory–finding the equipment as well as training users.

The external resources that are available are another factor that must be considered. At times it is more cost-effective to contract for maintenance of equipment.

There are some rules of thumb for the numbers of staff members required. They can be useful when there is no other information available. One BMET for every 100–150 beds was a good ratio in the early 1990s. But as more and more outpatient services are performed, this number is less reliable. Keeping one BMET for every $1,000,000 in equipment value is another rule of thumb. Of course, that depends on what kind of equipment a hospital has. Furthermore, as hospital equipment has become more expensive, that number is probably higher today. Another ratio that has been used is one BMET for every special service area, such as intensive care, clinical lab, or radiology. Again, this depends on the amount of the equipment or the size of the services. Another figure that has been used is one BMET for every 400 items of equipment. This works out reasonably well in a large facility that has several thousand medical devices.

In looking at staffing from this standpoint, one supervisor for every 8–10 BMETs and one clerical person for every 8–10 BMETs would be appropriate.

It is tempting to say that any size hospital needs to have one technician to maintain equipment or someone who oversees contracts. However, if a hospital has one technician, it must have some method for backup because that technician will have vacation, sick, and training days. Therefore a facility must have two technicians or an arrangement to have outside coverage by service from vendors or possibly a sharing agreement with another hospital.

The best tools for determining staffing needs are workload data and historical maintenance information. Historical work order information provides types of equipment that need to be repaired as well as estimates of the amount of time required for different repairs. One can consider preventive maintenance, or so-called scheduled work, to determine the number of staff hours that are required. However, the preventive maintenance must be viewed realistically because no hospital can accomplish all preventive maintenance that manufacturers recommend. Appropriate preventive maintenance program must be developed to avoid too little or too much expenditure of human resources. The hospital's vacation policy must be considered with projections for sick leave and training, to account for the amount of hours that are available for technicians.

The Department of Veterans Affairs in the U.S. has 172 hospitals with about 50,000 beds, about 100 clinical engineers, about 1000 BMETs, and about 100 clerical staff. They have approximately $3 billion in medical equipment. Most of these facilities are teaching hospitals. Some are relatively small–fewer than 100 beds–but most are in the 250–400-bed range. Looking at these data, there is one engineer for about 10 BMETs; there is one engineer for 500 beds, and one engineer for $30 million worth of equipment. At the same time, there is one BMET for 50 beds, and one BMET for $3 million worth of equipment. The VA maintains over 90% of this equipment in-house at an annual cost of about 4% of the acquisition cost of the equipment.

In a clinical engineering program there are several levels of positions. Table 51-1 lists and describes these different positions, based upon annual salary surveys published in the *Journal of Clinical Engineering*. The biomedical engineering technician starts as a BMET I and can progress up the career ladder to BMET II and BMET III, and to BMET specialist, which involves working in a particular area of medical equipment, and to BMET supervisor.

An engineer may start as a clinical engineer (junior level) and, with experience, may become a senior clinical engineer. A clinical engineer also may be a supervisor. In the top management position will be a clinical engineering director who could be a clinical engineer or, especially in smaller hospitals, a senior biomedical engineering technician.

Table 51-2 is the summary of the work done by BMETs and clinical engineers in hospitals, based on a recent *Journal of Clinical Engineering* survey. It lists the percentage of time that someone in each of these categories devotes to particular job functions.

Figure 51-1 is a representation of a career ladder for a clinical engineering department. It shows the way a BMET can progress up the levels of BMETs or can, with additional education, become a clinical engineer.

Education

A BMET might have military experience. The United States has the largest training program for these technicians. They might come from a one-year certificate program either electronics or biomedical equipment technology. Increasingly, they have a two-year associate's degree either in electronics or in biomedical engineering technology. Many of those who have military experience have achieved the associate's degree through affiliations that the military has with colleges, or through the Community College of the Air Force. A few BMETs will have bachelor's degrees, but typically those will be in nonrelated fields.

Recruitment

Two of the best areas for recruitment of biomedical engineering technicians have been the U.S. Army and Air Force. They had separate training programs, but a few years ago they were combined into one training program in Wichita Falls, Texas. Their technicians receive extensive electronics training and extensive training on medical equipment, including X-ray.

Another source has been community colleges that offer an associate's degree with specialization in biomedical equipment or biomedical electronics technology. These typically have been small programs with 10–20 graduates each year. There is considerable variation in the quality of the programs because there is no accreditation, as such, for a biomedical equipment or electronics program. Some of the programs are accredited as electronics programs. The primary emphasis of these programs has been on electronics, and the depth of the biomedical equipment depends upon the availability of equipment for the students to work on. The better programs have internships (of a semester or more) with hospitals so that the students obtain real-life, hands-on experience. At any given time, there are 30–40 of these programs in the U.S.

The recruitment of clinical engineers is more difficult because most programs in the U.S. that train biomedical engineers train primarily in the areas of research rather than in hospital clinical engineering. Some programs that have had hospital-based internships have conferred master's degrees in clinical engineering. Many clinical engineers are at the bachelor's-degree level and have been recruited from electronic engineering programs. One of the best sources of recruitment has been to find a junior engineer or a second engineer in a hospital. They will have gained the clinical experience from work and will have learned to apply their engineering skills in the clinical setting.

One detail that makes the recruitment of technicians and engineers difficult for hospitals is compensation. In general, these individuals can obtain higher-paying jobs outside of the hospital, with the exception of some specializations such as imaging technicians. Medical equipment companies generally will pay more than hospitals, but candidates frequently are not interested in jobs with the former because of the travel that is required. In fact, sometimes good sources of recruitment are companies where a technician has decided that he no longer likes to spend the time traveling yet enjoys working with the medical equipment.

One of the problems of recruiting good technicians and engineers is the hospital working environment. Frequently, individuals cannot stand the smells, sights, and sounds of the hospital setting. The sight of blood, and the fact that patients die, can be unsettling to some, as can the psychological stress of working on a medical device and realizing that someone's life may depend on how well that device is repaired. Some people, interested in helping others, gravitate toward the medical field but are not interested in being a nurse or a physician. Instead, they chose the psychologically rewarding fields of biomedical engineering technology and engineering.

Retention

Retention is important in any field, but particular emphasis must be placed on retention when compensation is not as high as in other places and when there is a significant investment of on-the-job training. This is true with the clinical engineering staff. One of the best areas for retention is the continuous training of technicians and engineers to help them meet the challenges of changing technology. Being exposed to, and working with, some of the newest technology in the medical field helps with retention. Technicians especially like the freedom they are given in a well-managed program to work as part of the medical-care team with given departments.

Motivation

Motivation should be one of the criteria in hiring clinical engineering staff. A good program depends upon individuals who are self-starters and well-motivated to help the clinical staff in the delivery of patient care. Motivation is enhanced by developing a relationship where the clinical engineering staff is part of the health care delivery team and where caregivers can depend upon them to do whatever it takes to keep the equipment functioning. One of the motivations for BMET and clinical engineers is that they perform most of their work in the clinical setting, rather than in a shop and therefore that they are able to see the results of their work and its impact on patient care.

Labor and Management Relationships

Most hospitals are not unionized, but where they are, supervisors and management must pay more attention to issues in which the union might become involved. Where there is a union, there will be a union contract, and all levels of supervision must be knowledgeable about the contract. There should be no difference in the management of a program where the staff belongs to a union and a clinical engineering department where the staff does not belong to a union.

Table 51-1 Generalized Job Descriptions

Position	General	Direction received	Education level	Certification	Average years of experience	Supervisory responsibilities
BMET I: Biomedical equipment technician I (Jr. BMET)	An entry level or Jr. BMET. Has no or minimal related or equivalent experience. Performs skilled work of routine difficulty under close supervision. Fills out report forms. Primarily performs preventive maintenance, repairs, and safety testing.	Works under close supervision. Receives specific and detailed instructions. Work is checked during progress and is reviewed for accuracy upon completion.	Associate of science degree (58%) No degree (31%) Bachelor's degree (11%)	Not certified	2.2 years	Not a supervisor
BMET II: Biomedical equipment technician II (BMET)	Under general supervision, performs skilled work of average difficulty. Has at least several years of related or equivalent experience. Has good knowledge of schematics. Maintains records and makes reports. Primarily does preventive maintenance, repairs and safety testing. Can give staff training on equipment safety and proper operation. Selects and requisitions needed parts; coordinates repairs with outside firms.	Works independently in repair and preventive maintenance programs. Receives general supervision.	Associate of science degree (68%) No degree (24%) Bachelor's degree (8%)	About 19% are certified	6.5 years	Might be assisted by BMET I or other technicians, but is not a supervisor.
BMET III: Biomedical equipment technician III (Sr. BMET)	Works in repair, preventive maintenance and incoming inspection. Has a significant amount of experience or training. Performs highly skilled work of considerable difficulty. May specialize in certain types of equipment. Skilled with test equipment, schematics, service manuals. Can give staff training on equipment principles of operation and safety. May participate in selection of new clinical equipment.	Works independently under minimal supervision.	Associate of science degree (56%) No degree (30%) Bachelor's degree (14%)	About 41% are certified	11.6 years.	Works independently and is not a supervisor. Might be assisted by other technicians.
Equipment specialist: RES or LES	Works in repair, preventive maintenance, and incoming inspections. Has specialized training and experience on Radiology (RES) or Laboratory (LES) equipment. Performs highly skilled work of considerable difficulty. Has comprehensive knowledge of practices, procedures and specialized equipment. Highly skilled with specialized test equipment, schematics, service manuals. Can perform staff in-service training on principles of operation and safe use of specialized medical equipment. Might participate in selection and management of new clinical equipment or systems.	Works independently under minimal supervision.	Generally an associate of science degree or higher	CRES and CLES certifications available from ICC		Works independently under minimal supervision. Might be assisted by technicians or other specialists, but is not a supervisor.
BMET supervisor (A BMET with supervisory responsibilities)	Supervises peer or subordinate BMET workers. Has a significant amount of education or training or equivalent experience. Performs highly skilled work of considerable difficulty. Coordinates, schedules and assigns work for others. Might specialize in certain types of equipment. Has comprehensive knowledge of practices, procedures, and types of equipment. Skilled with test equipment, schematics, and service manuals. Continues to perform highly skilled repairs but also has a wide variety of professional level responsibilities.	Usually reports to a department head.	Associate of science degree (62%) No degree (21%) Bachelor's degree (17%)	About 46% are certified	13.5 years	Supervises other BMETs and support personnel. In some smaller hospitals or departments, might be the department head.

Role	Description	Supervision	Education	Certification	Experience	Supervisory responsibilities
Clinical engineer (CE)	Works in design, modification, analysis, equipment selection, evaluation, planning, performance testing, or R&D. Performs engineering-level work of considerable difficulty. Can manage the process of high-tech medical equipment selection and acquisition. Has the capability to design and modify medical devices and to perform analysis of devices and systems. Manages medical equipment within the patient-care setting. Performs a broad range of professional level tasks. Can provide extensive training and in-service education.	Works independently under minimal supervision.	Is usually a degreed engineer. Bachelor's (61%) Master's (35%) Ph.D. (2%) Note: a few non-degreed or associate degree individuals are pioneers in the field or have been certified by the ICC as clinical engineers.	About 21% are certified.	8.5 years	Works independently and is not a supervisor.
Clinical engineer supervisor (A CE with supervisory responsibilities)	A Clinical Engineer that supervises BMETs or peer and subordinate CEs. Is usually a degreed Engineer. Has the capability to design and modify medical devices and to do analysis of devices and systems. Performs a broad range of professional level tasks including pre-purchase planning and evaluations, in-service programs, coordination of services, and a significant amount of supervision.	Reports to a department head	Bachelor's degree (52%) Master's degree (42%) Ph.D. (6%)	About 16% are certified	13.9 years	Supervises BMETs, CEs and support personnel, but is not the overall department manager
Manager or director of clinical biomedical engineering department	The department head of a clinical or biomedical engineering department is usually designated a "manager" or "director." (Most such departments report directly to hospital administration.) Most of these department heads are educated and experienced as clinical engineers or BMETs. Others are educated in business administration or hospital administration or may have extensive health care supervisory experience gained in other hospital management positions. The department head must have a significant amount of technical or management education or equivalent experience. They have the skills necessary to manage the process of high-tech equipment selection and acquisition; manage the maintenance and repair of all hospital high-tech equipment; participate in or direct an equipment-related incident and risk-management program; monitor and control department cost-effectiveness and productivity; conduct planning and budgeting; coordinate with other departments; plan and direct department development; and supervise others.	Usually receives direction directly from hospital administration, but in some cases reports to the head of a larger department.				Supervises BMETs, CEs, and support personnel. Participates in hiring process, reviews and disciplinary actions. Is the overall department or group head.

Table 51-2 BMET and Clinical Job Responsibilities by Percentage of Time

Position	BMET I	BMET II	BMET III	BMET specialist	BMET supervisor	Clinical engineer	Clinical supervisor	Clinical engineering director
Repairs/PM/Safety testing	73%	69%	66%	68%	29%	24%	13%	12%
Clinical support	3%	5%	6%	6%	6%	14%	9%	7%
Incoming testing	5%	5%	5%	3%				
Receive training	4%	3%	3%					
Provide training			4%	4%	4%			
Coordinate outside services				4%	5%		7%	7%
Purchasing	4%	4%				5%	7%	
Pre-purchase evaluation						10%		6%
Supervision of others					25%	6%	25%	27%
Department development					9%	9%	14%	18%
Other	11%	14%	16%	18%	22%	28%	25%	23%

Among issues that supervisors must be cautious about in a union shop are work assignment and ensuring that the workload is equitable and within everyone's job description. The fairness issue extends to training and promotional opportunities. As management is open to a grievance under a union contract, all of these areas must be well-documented. When disciplinary action is taken against an employee, he will have the right to union representation, which must be honored appropriately.

Another major area detail about a union contract, over which the shop supervision has little control, is the periodic negotiation of the contract. It occurs with the hospital management; major issues include compensation and benefits. In most cases, this is a contract for all clinical employees or, at least, those who perform engineering functions. In most cases, as long as supervision and management treat all employees equally and fairly, there should be no issues with regard to labor and management relationships.

Management

Supervisors and managers must be able to measure what happens in their organization. If a function cannot be measured, then it cannot be managed. Managers are responsible for planning the workflow of an organization, organizing ways in which the work can be accomplished, and determining the staff needed. However, their main function is team-building in order to achieve the needed results.

Managers cannot do all of the work, so they must work through people and with people, so they must develop the people who work for them. They must be able to realize the human potential of each employee they have and to recognize the desires of those employees. By developing personnel for better work and new tasks, they will shape the organization to accomplish what is needed.

As managers cannot do all of the work, one of their more important functions lies in delegating. Delegation is the way that one works through people. At the same time, this does not mean that they delegate something and forget about it. Delegation is an art that a manager must learn. When an assignment is delegated, it must be explained thoroughly, and a manager must ensure that the employee understands the explanation. If there are any potential problems with the assignment, they must be identified at the time of the assignment or as soon as they are recognized. Clearly defined checkpoints should be agreed to by the manager and the employee; next, the assignment should be summarized so that both the manager and the employee will understand it. The manager must not forget the checkpoints for the assignment because these are opportunities to measure the work that is accomplished and to encourage and evaluate the employee.

One of a manager's important roles is to build a team and then to coach and maintain that team. The team must learn to be dependent on each other. Team members must be aware of the fact that they are a team and of ways in which they can work together. This awareness will give them the power to work and to achieve the best results. Team members might need to be selected for their willingness to work as a team. This is where the manager must have the insight into the personalities and desires of the employees.

The manager must establish a working environment for the team. In other words, one of management's roles is to remove obstacles that would prevent the team from functioning. A team must know and understand the environment. They must trust the manager, and both the team and the manager must be committed to teamwork. All of the team members and management must cooperate for the team to develop and to produce maximum results.

A manager should expect high productivity from a team. The team should be expected to be more creative as it works together. The team should be able to be better at problem solving as members build on each other's strengths. As a team accomplishes work that members probably could not have accomplished individually (or at least not in as good or as timely a manner), the members should gain personal satisfaction. By working on the team and building on each other's strengths, each member will have increased personal growth and will have become a stronger member of the team.

The manager becomes the coach of the team, like the coach of an athletic team. He gives the team and individual members credit. The manager will provide constructive criticism, and at the same time they must be able to take criticism from the team. The manager must have good listening skills so that he can listen to the team (especially to the individual members) and can understand their personal needs. The manager must set a good example and must develop open communication with the team so that he listens and discusses the work with them. He must care for each one of the team members and must be able to mediate disputes between the members of a given team or between the members of different teams.

The bottom line comes down to the manager taking the blame if the team does not accomplish the work, in the same manner that a coach takes the blame for an athletic team's loss. At the same time, if the work is accomplished well, the team receives the credit.

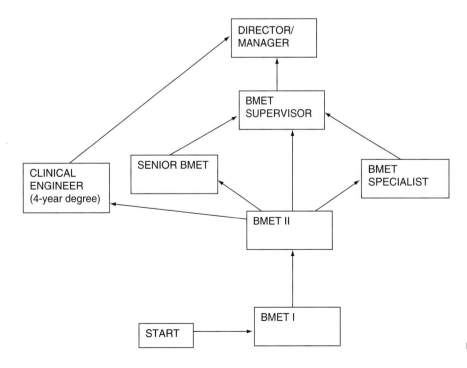

Figure 51-1 BMET Career Paths.

Performance Evaluation

In order to evaluate an employee's performance, there must be some understanding and agreement with the employee about what is expected from him. Evaluations can be performed in several ways. Peer evaluation is one way that is especially useful in the case of teams; team members can evaluate each other. A hospital can use customer evaluation. This is a critical part of the overall evaluation of the clinical engineering department as well as of the individual employees. A customer evaluation can be conducted by a periodic survey of the departments.

Productivity

The manager must be knowledgeable about productivity and might even need to provide management with reports on productivity. However, productivity is defined in different ways and is difficult to measure accurately. One can measure changes in productivity once it is defined and how it is measured.

One of the necessary details is the amount of time that is available for the technicians. Table 51-3 shows a breakdown of an employee's time, based on five eight-hour days per week, for 52 paid weeks. Accounting for holidays, vacation, average sick leave, and two 15-minute breaks per day, one has 204 days available, or 78.5%.

This example is for a senior technician working for the Department of Veterans Affairs, where the employee has 26 paid days of vacation. A more typical example might be 10 days of vacation, which would change the calculation to 220 days available, or 84.6%. Productivity, therefore, is based on time paid. The absolute maximum that an employer could realize from an individual would be this percentage.

Table 51-4 shows data on nonproductive or lost time for the average industrial worker and for maintenance workers in the Department of Veterans Affairs' Medical Centers (VAMC). This is not just for biomedical engineering technicians but for maintenance workers in general. Data were taken from surveys of supervisors who provided estimates. The VAMC supervisors did not report any idle time. Lost time, or nonproductive time, is any time during which the employee is not making repairs, consulting with staff, or in maintenance-related training. This is an area where a manager can work at reducing some of this lost time, and at increasing productivity.

Productivity relates to efficiency. The more efficient a program is, the more productive it will be. Output divided by the input is a simple definition of productivity. Of course, the final results depend on the ways that output and input are defined.

One of the details that must be considered in discussing productivity is what one considers to be a productive task. The following are some definitions of tasks:

- *Productive direct* are items like repair, installation, inspection, and in-service training.
- *Productive nondocumented* are the same items as productive direct, but there is no record of them. These might be small repairs and informal discussions.
- *Productive indirect* are items that are not directly related to repair. These include attending meetings, reading journals, and technical training.
- *Nonproductive* are vacation time, sick time, and breaks.

Going a level deeper into definition, *repair time* is the time required to diagnose a problem, to get parts, to make the repair, and to document the repair. *Travel time*, which is one area where productivity can be increased by reducing travel time, is included in this.

Table 51-3 Available time

Paid days 52 × 5	260
Holidays (U.S.)	10
Vacation	26
Sick leave	6
Breaks (1/2 hour per day)	14
204 days available or 78.5%	

Table 51-4 Maintenance worker lost time

	Industry	VAMC
Company practices	16.1%	17.4%
Idle	9.2%	—
Traveling	16.0%	6.0%
Personal	7.3%	3.8%
Early quits	4.4%	3.2%
Waiting	4.6%	4.4%
Instructions	4.4%	4.2%
Clean up	5.2%	5.6%
LOST TIME	67.2%	44.4%

Periodic maintenance time or *preventive maintenance time* is that time that is required to check the medical device for any defects and to perform any tests that are required, including electrical-safety tests. *Travel time* is the time required to go to and from a repair site. Because many devices in a hospital setting are repaired in the place where they are used, travel time can be a significant factor. *User down time* is the time when equipment could be used to perform work in the clinical setting but it is not available because it needs repairs. This does not include the time that equipment might be down at an off-hour when preventive maintenance might be being done. *User up time* is the time when the user is working and the equipment is available to be used. This is an important time to write into service contracts.

One method for presenting maintenance productivity is by showing that the maintenance percentage cost is equal to the cost of maintenance, divided by the inventory value.

Conclusion

This chapter indicates the importance of personnel management to a successful clinical engineering program. Various topics discussed emphasize areas in which the director of the department must develop skills. Communication, coaching, and team-building skills are the most important skills the director must have. Through personnel management, the director can create a working environment for a successful and highly productive department.

52

Skills Identification

Nicholas Cram
Texas A & M University
College Station, TX

Academic Knowledge

Certain academic institutions have earned banners of excellence based on years of developing a turnkey product for the clinical engineering or health care marketplace (Baker, 2002). Employers are more familiar with the scholastic product of these institutions and rely on the premise that certain basic concepts will be part of the engineer's or technician's basic skills. It is not always the case, but there is a comfort level for the employer. An employer generally waits six months before making a basic skills assessment of a recent graduate, whether that graduate is from a two-year associate's program in applied science or a four-year bachelor of science program in engineering.

The management and maintenance of biomedical equipment is a dynamic and expanding career field, but it is a relatively small arena compared to almost any other profession. In this small field, the grapevine and the Internet rapidly convey information regarding academic institutions. If entrance standards and course requirements become diluted, an academic institution's long-held reputation can tarnish quickly.

Military-trained biomedical equipment technicians (BMETs) sometimes find themselves at a disadvantage if the employer has ties to other civilian academic institutions. Care must be taken, as well, in evaluating the military-trained BMET's skills because of the many variables involved in military BMET training (Wear, 2003). For example, the military emphasizes radiology training as its ultimate goal. If the BMET received that training but is assigned to a field-post aid station, BMET skills in radiology repair and maintenance will not have an opportunity to blossom. On the other hand, if a military BMET is assigned to a major medical center, that BMET will receive excellent experience and possibly additional training. Therefore, a wide variation in skill level is quite possible to encounter in military-trained BMETs. (See Chapter 9.)

Hands-on Skills

Generally, BMETs will receive more practical, hands-on training than will engineers, and they will continue to excel in bench work particularly if they receive training by way of original equipment manufacturer (OEM) or skills seminars. Clinical engineers (CEs) typically are more inclined to take advantage of opportunities to develop their public-speaking, communications, and management skills. They may also volunteer for OEM training and may tend to learn more rapidly than BMETs. Simply put, CEs are oriented toward problem-solving, while BMETs are task-oriented. Both skills are useful, and their value to the employer depends on the employer's needs, which ultimately are to address the needs of the clinical engineering department's customers. Skills requirements vary from facility to facility and from project to project within the same facility. Different skill sets are required, depending upon the mission of the health care facility. A clinical engineering department in an academically affiliated facility may support research, whereas a department in a specialty hospital may support devices used in cardiac procedures and organ transplants. The demands of the customer change the skills that are required in order to meet that customer's expectations. An experienced clinical engineering manager will seek opportunities to provide service to customers who are not currently satisfied. This could entail the identification and re-engineering of an inefficient process.

As new technology is introduced, BMETs and CEs must adapt accordingly by improving their skills and understanding of the technology. The CE manager is constantly engaged in the skills-identification process, striving to make the department more efficient. The core skills taught in academic institutions are basic building blocks upon which additional skills are placed. A manager who is a mentor is invaluable. The experience obtained from a good mentor can increase the opportunity for professional advancement and standing within a department.

As the health care facility expands its services to the community, the CE department must be versatile and must maintain pace. It must increase its visibility to clinical and administrative personnel. Developing communications skills will help with bridging the gap between clinical and administrative worlds and will facilitate the accent of the career ladder.

The manager who keeps the work environment challenging finds that personnel enjoy the challenge and stimulation and that they derive job satisfaction as a consequence. Technical personnel, while performing enjoyable, challenging tasks, develop new skills and a sense of accomplishment and competency. Babcock (1991) states, "The professional's view of what is challenging must be reconciled with the needs of the organization, and the challenge to the supervisor is not just making wise decisions, but structuring them as much as possible to provide the desired challenge and then persuading the individual of their importance."

Accepting Risks and New Challenges–Developing Your Skills

Clinical engineering duties and responsibilities are generalized in the job description. Successful clinical engineering departments commit to tasks that other technical departments may consider an overcommitment of resources or beyond the scope of their technical expertise. Telemedicine (see Chapter 101), picture archiving and communications systems (PACS) (see Chapter 102), and computerized maintenance management systems (CMMS) (see Chapter 36) are just a few examples of the neohealth care technology without specific departmental boundaries. If the challenge of these new technologies is accepted, one should make a full commitment to ensure success. If one is concerned that the department will fail to demonstrate competency, one should not undertake the project. However, if the manager never ventures beyond the norm, the department and associated skill levels will reflect that lack of commitment.

Contacts and Networks

BMETs and CEs should be involved in their respective professional organizations and should attain certifications offered within the profession. (See Chapter 133.) A certification or registered license conveys a base of knowledge to fellow associates in the same profession. To those who engage in the debate over whether certification or licensure makes one a better technician or engineer, it must be stated that improving professional skills is not the goal of certification or licensure. It is, rather, a professional recognition that the certified or licensed individual has a proven core knowledge required of the profession. Numerous salary surveys have affirmed that those CEs and BMETs who are certified receive higher compensation (Baker, 2002).

Attending professional meetings and becoming involved in professional societies provides an opportunity to exchange professional and managerial information, discuss work environments, and explore standards of living. If an inspection by the Joint Commission on Accreditation of Healthcare Organizations (JCAHO) is impending, discussions with fellow BMETs and CEs who recently have endured the process will be of value. Insight gained from these discussions can be shared with the hospital's team leaders.

Networking provides a link to best practices and benchmarking without having to reinvent the wheel. In short, it is an efficient and reliable management pipeline that often leads to lifelong professional associations. It can serve as a career-development tool and can provide insight for professional, management, and personal decisions. When utilizing a professional network, adhere to your professional ethics and avoid gossip channels. Always be considerate of the time constraints of those contacted, and strive for a *quid pro quo* relationship.

Resumes

Perusing a resume in connection with the hiring process provides valuable information about an individual's recorded achievements. It provides little information, however, about the magnitude of those achievements and how well any needed skills would augment your department's performance. A personal interview and a resume provide a manager with about 50% of the information that is needed to make a wise hiring decision. There are several intangibles that determine the long-range success in the hiring process that cannot be analyzed with a resume and interviews. For example: Will there be a common chemistry that is required for all interactive technical partnerships to be successful? Will your service or product be enhanced? The best that a manager can hope for is that the skills have been identified for a probability of success in the hiring process. The other significant aspect of whether an employee will enhance the CE department's technical expertise and customer satisfaction is the work environment itself. This depends largely on the kind of a department that the manager has created or inherited, the manager's ability to motivate, and the proficiency with which the manager can bring out the best talents of every employee.

References

Babcock DL. *Managing Engineering and Technology.* Englewood Cliffs, NJ, Prentice-Hall, 1991.
Baker T. Survey of Salaries & Responsibilities for Hospital Biomedical/Clinical Engineering & Technology Personnel. *J Clin Eng* 27(3):210-236, 2002.
Cram N. *Clinical Engineering Manual.* Biomedical Engineering Department, Texas A&M University, 2001.

53

Management Styles and Human Resource Development

Alice L. Epstein
CNA HealthPro
Durango, CO

Gary H. Harding
Greener Pastures, Inc.
Durango, CO

Management of engineers and technological staff presents distinct challenges. Staff are highly trained, focused, and intense professionals with high degrees of self-motivation and direction. Managing the egos, vast skills, and individuality of such staff requires skilled, trained management who understand the way to mold staff into a cohesive team. The primary tasks of a technical manager will be to provide support and to facilitate, rather than to maintain a close hold on direction.

Technical managers need not be the individuals with the greatest intellect or ability to perform technical functions. While the manager must have a firm grasp on the concepts and science, it is more important for the manager to have the necessary skills to ensure that tasks are assigned appropriately and completed. It is more important that the manager facilitates support, such as ensuring that the group and individuals within it receive the training and support that are needed to progress and improve. Staff often can develop pockets of expertise that would remain isolated unless a technical manager were to facilitate dissemination of the expertise to others who might make use of it in their tasks.

The following sections provide specific information on subjects that are important to technical management and human resources.

The Evolution of Management Theory

The early twentieth century brought to the attention of industry the study of time and motion as a method to improve work-practice efficiency. Frederick W. Taylor introduced the concept of scientific engineering, in which he recognized that each task could be examined scientifically and rationally optimized to improve productivity. Taylor is a controversial figure in management history because although his innovations in industrial engineering paid off with dramatic improvements in productivity, he has been credited with destroying the "soul" of work and with dehumanizing factories. Taylor recognized the importance of training an employee to the highest level of his capabilities and the responsibility of management to prepare systems to enable the worker to work better and more quickly than he otherwise could. Taylor believed that workers should be taught by and that they should receive the "most friendly help" from their managers instead of being driven or coerced by their bosses or left to their own unaided devices.

The Ford Motor Company introduced production lines and injected monetary incentives (piece work and higher wages). Production improved, although it became clear that an excessive pace wore out workers in a short number of years.

Henri Fayol and Luther Gulick introduced variations of scientific management in the 1920s. They argued that the use of principles such as specialization of labor and hierarchical leadership structures would result in optimal organizational performance. Fayol's components included planning, organizing, commanding, coordinating, and controlling. Gulick's components were similar.

In the 1920s, Elton Mayo explored human relations by studying changes in productivity at the Hawthorne Works of the Western Electric Company. He noted that changes made to lighting, processes, and work breaks (up or down) resulted in an increase in productivity. He concluded that involving people in ideas and changes (even irrational or noneconomic ones) improves morale and productivity. Conclusions showed that:

- Complaints are not always about what they seem to be, but may be the result of some personal disturbance.
- All actions in an organization are given a social meaning. Perceived and intended meaning may be very different.
- The behavior of an individual in a job is the result of more than instructions or job descriptions. It is affected by personality, values, and attitudes expressed by co-workers.
- Worker status is very important and is determined by the way the worker perceives his working environment.
- There is an informal organization within any formal one. This is made up of relationships, communications patterns, and perceptions.

The scientific management perspective was attacked by Herbert Simon (1946), who argued that the principles of scientific management were vague and contradictory. Widely accepted, scholars turned toward the importance of specialization, chain of command, unity of command (one boss), span of control (optimum subordinates), and minimum authority levels.

Motivation was introduced as a management concept by A. H. Maslow in the 1950s. Maslow determined that all individuals have a hierarchy of needs:

1. Biological (the most basic): Hunger, warmth, rest
2. Safety: Protection from danger
3. Socialization: Love, affection, affiliation
4. Self-esteem: Autonomy, dignity, respect
5. Self-actualization: Realization of one's potential through competence, creativity, and achievement

E.H. Schein introduced the theory of the complex man in the 1960s. Schein believed that the study of management theory was too simplistic and that no single management style can succeed in improving the performance of all workers. The motives of an individual can be extremely complex and liable to change over time. He believed that a high level of satisfaction does not necessarily lead to increased productivity.

In the 1960s and 1970s, Peter Drucker introduced "management by objectives" to include fundamentals of strategic management by setting objectives for staff and assessing their achievement. He believed that decentralization was imperative, in order to confront sudden changes and to take advantage of new situations.

Peter Senge, in the early 1990s, purported that those organizations that are capable of decentralizing leadership to enhance the capacity of all people to work productively toward common goals results in learning organizations being capable of rapid change. Only organizations that are flexible, adaptive, and productive excel. All people have the capacity to learn, but the structures in which they function are often not conducive to learning because they lack the tools and guidance for employees to make sense of situations they face.

Empowerment has always been perceived to be highly motivating and extremely efficient, but it is little used because leaders often desire and control power and efficiency. Throughout history, empowerment leadership was implemented when "getting the job done" took priority over control. Empowerment as a management style authorizes employees to perform their work without the need to seek approval from supervisors. However, it also holds employees accountable for their actions and the outcomes of projects they assume. Traditional management does not embrace empowerment because managers are reluctant to relinquish control. Effective empowerment programs require restructuring and relearning of responsibilities by managers and employees.

West Meets East

Management style, like society, generally varies from U.S. and European organizations to Far Eastern organizations. If we simplify styles, Western management style may be described as open, direct, and confrontational, stressing short-term profitability. Eastern management style stresses seniority, relationships, and family ties, stressing long-term profitability and employment. As globalization progresses and financial challenges arise, differences may shrink—perhaps taking the best of both styles and merging them into one, but more likely taking the features of both styles that improve the opportunity for economic survival.

Arthur Andersen and the Batey Research and Information Centre performed a survey in 1997 that examined the differences in management styles from the perspective of senior business executives (405) themselves.[1] While the results support the position that management styles continue to differ, the results also point to styles converging. For example, "respect by business community" and "commanding a good salary" were ranked first and second as indicators of a good manager by the eastern and western executives. While western managers place a great deal of emphasis on initiative and often make decisions unilaterally as "the boss," eastern managers often prefer personal meetings and development of consensus. Western managers tend to stay within their bounds and focus on results rather than the way a task is completed, while eastern managers typically provide detailed goals, instructions, and schedules and look past their group assignment to the way their work might interact with other groups.

Most would agree that much of the globalization of the market is resulting from the expansion of western organizations into previously untapped or limited opportunities (markets and production capability alike). Eastern executives, while they may not embrace western methods, have studied these methods meticulously, and, as noted above, some have embraced particular features of them. However, western executives rarely receive training in eastern management styles, although there are exceptions. The challenges for eastern and western managers are to learn and to continue to be open to those practices in each of their respective regions that not only work within their primary domain but also in domains in which they may expand or otherwise contact. By coming to understand the history, rationale, and specific features of each style, engineering managers can adopt practices that do not conflict with accepted business practices to an extent that will result in project failure, loss of strategic alliances, or loss of technological and human resources. A simple difference in, for example, a western manager directly confronting an eastern participant in public about a disagreement (versus taking the time to discuss the issue on a personal level in private) can lead to irreparable damage to a relationship or alliance. Sometimes even the fact that damage has been done is not recognized by the manager until the long-term effects (e.g., dissolution of the alliance, drop in production, or loss of employees) become evident. Similar negative results could be expected if an eastern manager holds a tight rein on a western engineer rather than "letting him do his job." Inclusion of women in the work environment remains highly divergent among some cultures. As the western management cultures have come to embrace and promote (by law) equal employment between the sexes, sociological differences in other regions of the world have a far different perspective on women in business and in religious activities in terms of what equality might mean. If you are a woman, these differences most certainly can affect your opportunities to work or to manage. From a western perspective, there currently seems to be no good answer to what we perceive to be a problem and an unacceptable inequality. Suffice it to say that new opportunities in the east (and elsewhere) must be carefully examined with respect to this issue, and they must be discussed among male and female executives to construct the organization's approach to maintaining harmony within the organization.

The use of experienced mentors, external consultants who are familiar with the area and styles and training and education on these subjects can help to make the experience successful and enlightening.

Cultural Diversity

Engineering in medicine in the United States is a multicultural professional discipline. Trained engineering professionals and engineering professionals-in-training bring with them diversity in virtually every aspect of personal (e.g., language, religion, and home life) and professional (e.g., education and exposure to technology level) experience. Add to this the globalization of industry and technology-related humanitarian efforts that are being made around the world; it becomes obvious how important it is for engineering managers to develop an understanding of the cultural differences and strategies for not only overcoming impediments but also for taking maximum advantage of these differences to promote project success. Major impediments can include:

- Poor quality of communication and understanding, leading to errors in technical and/or project development
- Inability to match personal sociological understanding and experience with that of the community, either when a foreign engineer enters the U.S. community or when a U.S. engineer enters a foreign community
- Application of incorrect risk/cost/benefit analysis to that community
- Inability to provide adequate support services (e.g., cultural food stuffs, religious services, and time within a day for purposes outside the basic U.S. business practice) to meet the needs of the whole professional for the cultural situation in which he has been placed

Typically, the engineering curricula have not set aside a substantial amount of time to address these issues. Similarly, not much time has been set aside within industry to prepare managers or working professionals for the sometimes difficult and sensitive tasks of addressing cultural differences. Even so, it has been our experience that the mutual intellectual basis of engineering, regardless of cultural diversity, can provide a solid background on which to seek and understand the nature of our colleagues, their needs, and the needs of their communities.

The first step in addressing cultural diversity is to promote the interest among engineering professionals and students in investigating and experiencing cultural diversity firsthand; it might come in the form of personnel exchanges, required learning, or side-by-side work experience. While the nature of our cultural differences can change from relationship to relationship, the willingness to seek out information on our own initiative and to ask questions when a matter may be unclear, and remaining nonjudgmental and receptive, can go a long way toward promoting understanding. Assigning mentors who already have experienced and successfully addressed what might be expected by the less experienced engineer is also an excellent method for addressing issues of cultural diversity. Such mentors not only can tutor the culturally diverse but also the less-experienced colleague who is experiencing cultural diversity for the first time.

The value of making use of cultural diversity also should also not be minimized. Professionals who are involved in a culturally diverse project should seek out experienced colleagues who are likely to be able to contribute to the specific issues that could face management and the working professional. While a language barrier is an easily identified issue where the value of collegial intervention is understood, similar review of project goals and conceptual design can avert misunderstanding, errors, or failure of the project and personal or professional discontent.

The Creative Process

Many organizations make their daily profit on "meat and potatoes," doing exactly what they have done best and routinely. There is nothing wrong with this as long as a predisposition to routine does not (1) impact on the ability of the organization to do things better, cheaper, or safer; (2) stifle innovation that could help to ensure future success or; (3) create an employment environment that stifles the creative nature of its employees so that an important part of their basic personality needs is not met. If the organization does not accept the challenge to do things better, cheaper, or safer, a competitor might arise and take that meat and potatoes off the table altogether. If the pipeline for new products, new projects, or new clients shuts down, the organization could slowly fade away or become a takeover target. If creative employees are not provided with an avenue for release, employees might leave or become ineffective when the organization needs them most.

The creative process is all about team building, teamwork, problem-solving, and productive creative thinking. If a group jumps into problem-solving or creative thinking without having built the team or having cultivated teamwork, many participants might feel stifled or might prefer not to express themselves freely if there would be a resulting negative impact on the creative process. Others might not know how to participate. Engineers need to have a concept and perception of team, and a recognition of their place within it, in order to feel routinely comfortable sharing ideas that have not yet been thought through fully. They also need to know that the process itself is not flawed; for example, that the process will catch an idea that seems to have potential but that is ultimately flawed before someone thrusts it into use without proper consideration. Conversely, they need to know that those ideas that have potential will not be ridiculed, shut down, or pushed aside because of one person's viewpoint or shortcomings.

Some creative-process aficionados suggest that brainstorming should be a scheduled event, with limitations on group size and time allowances. While there is no reason not to hold such meetings, creative thinking actually occurs all throughout the day, the month, and the year. Cultivating an environment where brainstorming has a release that is supported at all times–even when the employee is at home or on vacation–may be a more effective way to capture those ideas that occur to special individuals throughout their employment. This might be in the form of cultivating an environment in which every employee is encouraged to write down their ideas when they arise and to provide various methods by which they can convey and discuss their ideas with others. Some creative thinkers do not create well on demand, but they can let their creative thinking flow when the environment is right for them.

Much has been said about the value of meetings, and certainly the increasing dependence on, and comfort with, e-mail and other communication venues impacts the need for face-to-face meetings. However, the arguments against face-to-face meetings must be weighed against the intrinsic value of face-to-face confrontation, reading a person's body language for signals, and immediate and unguarded response. Many argue that face to face meetings are too costly. For example, we saw one argument citing that an employee

1 Asia 21, 16-19, September 1997.

paid at $40,000 per annum cost $300–$400 per hour in a meeting. Suffice it to say that the arguer was not the CFO, nor very good at math, in any event. Similarly, we have found that engineering professionals typically will go out of their way (e.g., attend a nighttime dinner meeting not on company time, although the company might purchase dinner) to participate in the creative process if the environment for the process is appropriate, and if viable creative outcomes are introduced.

Facilitators have developed some basic structures for brainstorming, problem solving, and creative thinking. For example, some programs have a first section where all ideas are taken, without any negative comment allowed. A second section allows for a period of private consideration by smaller groups. The third section brings all of the groups back together to examine their findings and to promote further discussion. It should be noted that not all groups or individuals will respond well (i.e., creatively) to a specific structure. The groups and individuals should help to determine the structure, with no "canned program" arbitrarily accepted without this input. It is like saying, "Hey, let's all think creatively, but within this box." We have seen some recommendations that managers should maintain a confidential file on each and every employee. The purpose of doing so purportedly is for the manager to monitor productivity for time-management purposes. While such activities might work well for operations that just "turn the crank," they work poorly for the creative process. Usually it is an accounting/time-management professional who is interested in eking out greatest the profitability from existing operations; unfortunately, there often is no place within their models for creativity, although their salaries and those of all within the company might ultimately depend upon that creativity and the basic operations that might have resulted from past creativity.

In brief, the creative process should be cultivated in every organization. While more attention to it might be provided in certain groups (e.g., research and development), it is important to recognize that creativity is part of human nature. Some more, some less, some in one manner, some in another, some that may be useful to the organization in its mission, and some that may not be useful to the organization, other than in cultivating a good work environment for all.

Communication Skills

As we recognize the importance of communication in bridging cultural-diversity gaps, communication affects all aspects of a business. Communications skills are essential to leadership. The ability of a manager to communicate and to relate organizational goals to the individuals to whom leadership is afforded is directly tied to the success of that manager. The manager of a technical staff must ensure that communications flow in both directions for success to be attained.

It is important to realize that communication can be oral, written, or interpreted by action – action of the company (e.g., lack of funding of what otherwise has been interpreted and projected to others as an important program or sector) or action by the individual (e.g., body language). If you provide a message that has one meaning but convey a different message by one of the other methods of communication, do not be surprised if staff become confused, outspoken in support or lack of support thereof, or frozen in place.

The discipline of risk management almost always regards written messages and documentation as more acceptable and indicative of desired practice than spoken messages; i.e., unless written messages and documentation are repeatedly not followed by actual practice over time—an unfortunate occurrence that repeats itself much too often. The actual message conveyed to different staff members might be mixed and very different as a result of the total communication package that one manager presents to staff, versus what that staff member might then impart to others, or versus what another manager presents. One simple example might be a senior management executive conveying to his research-and-development director the need to complete a project within budgetary and time constraints; the director might convey the message, but mixed, by averting his eyes from his engineering staff potentially interpreted by staff that one or both goals is unrealistic or that the result of an inability to meet the goals could result in unemployment.

Communication can be impaired by many different types of barriers. Physical barriers to good listening are often environmental; too hot, too cold, or distracting. Mental barriers exist in the form of stress, prejudgment of the material or speaker, or rambling by the speaker. A manager should realize the existence of these barriers and should cultivate an environment in which such barriers are not likely to exist.

There are a number of counterproductive communication errors to avoid, such as:

- Failure to greet another
- Too little or too intense eye contact
- Making a person wait
- Accepting phone calls or working on other projects during a meeting
- Interrupting
- Using meaningless or confusing words or expressions
- Obfuscating or embellishing message
- Poor listening
- Constructing a physical or psychological environment that is not conducive to promoting honest discussion

These errors apply not only to discussions and communications within an organization, but also to communications (e.g., customer-service contacts) that the organization might have with others outside the organization. Some of these communication errors can be addressed within the organization if managers receive adequate training, understand ways to mitigate the errors, and receive constructive criticism based on actual practice. Confrontational-based criticism (during, for example, a salary review) without prior provision of adequate training and experience should be avoided; such activities indicate more of a need for the manager to be retrained in communication. A manager who is practicing the art of effective communication will never think or assume that staff members have all of the information that is essential to their satisfactory performance of job duties.

Mentors can make an important contribution to improvement of communication. If the mentor can be seen by the mentee as interested first in the professional development of the mentee, and perhaps only second in the success of the organization in the particular matter, the message of criticism may be communicated more effectively because the mentee might be better prepared to listen. Where it is clear by review or occurrence that communication systems have problems, outside consultants can be helpful.

It is important to understand and to support that listening is the responsibility of all parties–including very senior management, who might resist criticism of their abilities or systems that they have developed, or new hires, who might have had limited experience in listening. Listening requires concentration. Poor communication often results in suboptimal performance.

The Written Word

Much of the writing that an engineer prepares may well be related to technical issues; for example, conceptual-design review, project-resource allocation, meeting details, service recommendations, or user instructions. Unlike personal, written communications that we may make to friends and family, business documents can be viewed as permanent records long after the projects for which they were written are completed, or long after the individuals who have composed or read them have departed from the organization. Similarly, unlike verbal discussions we may have with co-workers, clients or others, where a dialogue including an exchange of views is provided, written documents do not provide for an immediate exchange of information or concerns or for an opportunity to utilize the other methods of communication (e.g., body language or facial expressions) to confirm understanding or confusion. As a result of these and other factors, it is important that engineers be skilled in the art of business writing. Every engineering curriculum should teach technical writing, should address all manner of typical business documents, and should identify potential errors and methods for avoiding them. In addition, the employer should provide focused training on methods for and policies and procedures concerning business writing; mentoring to provide ongoing, unbiased criticism of actual examples of business writing can be very helpful.

One of the first questions to ask is "What type of document is it, or should it be?" For example, after receiving a user report from the field, regarding their perceived identification of a limitation in safe use of a product, it is important to consider which document(s) will best respond to address the issue(s) raised. A customer-service letter to the reporting client might be warranted—perhaps a memorandum to regulatory affairs, maybe a new user instruction, emphasizing the proper methods for the use of the product and the potential result(s) of misuse.

A second question to ask is "Who is the audience?" Perhaps the writing will be of a sensitive nature to in-house legal counsel, or to the user who is not an engineer or who is not a college or high-school graduate, or to a technical professional who must make a change in material or physical feature.

A third question to ask is "What is the goal of the written document?" It might be to help the user, who has seventh-grade reading skills, to understand how to safely utilize a product. It might be to place an attorney on notice that serious repercussions might be about to occur, or to convince senior management to fund a project, or to convey technical information or direction to a colleague or staff.

Once these three initial questions have been asked and answered, the tone and content of the writing can become much clearer. The writing can be direct, courteous, and sincere, can provide the appropriate emphasis, and can be written at a level that is appropriate for the audience. Emphasis can be in the form of word treatment (i.e., bolding, underlining, capitalization, and italicization) or word-smithing. It is important to avoid discriminatory language, obfuscation or embellishment of the issue, and overemphasis (e.g., yelling—CAPS in e-mails) when not indicated or necessary. It is important to identify the issue, to detail what is expected, and to stress the benefits to the reader. If you are confident in your viewpoint, it is helpful to stress that confidence.

Never forget that the document might appear where you least expect it. It might be read and interpreted by someone to whom you did not direct it. If, on second reading (a habit that all should embrace before sending a document), specific information included in the writing does not contribute to the goal of the document, it should be considered for deletion or revision until all that is left does contribute to the goal. If it is recognized that a document is particularly significant or sensitive, a "second opinion" might be worthwhile; sometimes this second opinion might even come from the primary audience to whom the document is addressed.

Remember that written documents might not replace the need for interpersonal discussions, meetings, or additional presentation vehicles (e.g., slides, videos, and training sessions). It is also important to be fully familiar with your organization's position on company documents. Documents prepared for business purposes might be defined as business products, wholly owned by the business, and you might not be permitted to use or maintain these documents, even within your office. Many organizations have document-retention policies, and it is important to be familiar with them.

Mentoring

Mentoring is the provision of guidance from an experienced party to one who might be less experienced generally or who might be less experienced in specific areas only. The purposes of mentoring include:

- Orientation and policy and procedure introduction
- Technical training and development
- Professional development and diversification
- Error/issue management
- Unbiased, nonconfrontational and routine critique

Mentors might need to set aside management responsibilities; that is, if the business situation is such that the boss must also be the mentor, the professional mentor must be able to set aside many of the responsibilities of a boss in order to provide an opportunity for the mentoring process to work. While the process of mentoring is intended to assist the mentee, one oft overlooked aspect of the process is that mentors must be trained and must understand how to be an effective mentor. Not all experienced or exceptional engineers can be, without training, even an adequate mentor. Analogies include:

- While a person may be a great individual of high standards and knowledge, they might be a terrible parent.
- While a student might have expert potential, a poor teacher can obstruct the student's realization of their potential, perhaps for life.

Mentoring can be either informal or formal, and typically some degree of mentoring (which may be as simple as showing mentees that this is the way we do things) is better than none. Mentoring need not be provided by engineers solely within the mentee's group; mentoring by professionals within, for example, a local professional society, a separate division, or a separate group is often more acceptable because the relationship is outside the political constraints of employment. Most experts agree that compatibility and respect are the foundation of a mentoring relationship.

Providing attention to development of the mentor program and supporting all participants in the program can provide important advantages in job satisfaction, job performance throughout the management hierarchy, and employee retention.

Motivation

Managers might believe that employees only work for money or tangible rewards. But when asked which rewards make a difference, employees rarely list money at the top, instead identifying recognition, involvement in decision making, or interpersonal rewards. People perform better when their efforts are rewarded, although rewards need not always be financial. The manner in which you treat your employees can be an effective reward. Nonfinancial rewards include public recognition and praise, providing the opportunity to expand job responsibility, kind words in a performance review, and notification to upper management of an accomplishment.

One significant form of motivation is feedback. Feedback, positive or negative, should always be constructive. Praise is a nonexpense method to increase morale, motivation, and productivity. Most people respond to praise by working harder. Those who feel unappreciated are likely to cut back their efforts, figuring that management does not care.

The manager should ensure that criticism can be clearly interpreted as criticism of behavior or performance and not of the person. Managers must take full responsibility for communicating to the employees their perception of the work. Meaningful feedback is immediate. One of the first things that managers should ask themselves before criticizing is determining whether the employee should be able to perform the requisite task with the current skills. If the answer is no, it is the responsibility of the manager and the organization to see that appropriate training is provided to that employee.

Educational opportunities and training can be effective motivators. They demonstrate that the manager recognizes the employee's potential, that they are interested in the employee's growth, and that they expect to expand the employee's participation within the organization. It is imperative that the value of the opportunity be recognized by the manager and understood by the employee. Time might need to be allotted in schedules and project assignments in order to get the most out of the education without the employee regretting being away from assigned work.

Motivated employees are those who are comfortable with their assignments and who believe that their managers value them. It is essential that performance standards be clearly specified in advance. If the manager and employees are not in agreement about what is expected, confusion and frustration are likely to result. The manager must clearly communicate individual and organizational goals. When assigning projects, it is important to spell out assignments clearly. Employees often get incomplete information or conflicting messages about what they are expected to accomplish.

When a work assignment is imparted, create an atmosphere that fosters employee questions. Employees might be embarrassed to ask questions because they might fear they would appear incompetent. Provide the option to ask further questions in various venues, and of various participants. Let employees know that you value them and that you are providing them with an assignment because you truly believe that they will be able to accomplish the task with excellence. Take a risk and give the assigned person full responsibility for the successful execution of the assignment. An effective organization is built of employees who want to do their assignments and take on more responsibility.

Some organizations have discovered ways to demonstrate their appreciation and concern for employees, such as management staff spending several days per year working on the "line" or "floor." Few things motivate and impress employees more than seeing their manager pitching in when the pressure is on. Extra days or afternoons off, which are not counted against their paid leave, have been successful as well. Check with a human resources specialist before awarding any noncompensatory/benefit-type rewards. Run out of new ideas? Ask your employees what would be meaningful to them, and ask them annually. No one knows better than the employee what will incentivise, reward, or motivate them.

Employees work harder and more effectively in a humanized environment that treats employees as individuals working in a team environment. Career ladders are important to motivate employees toward self-improvement. They can be created to offer technical professionals the opportunity to transition from functional specialists to managers. In engineering, classical career ladders (e.g., new titles, staff-management responsibilities) can be difficult to construct. Selection for assignment by project complexity and importance might well replace the classical ladder. It is recommended that opportunities for advancement be made whenever possible. It is difficult for employees to be satisfied working in cubicles when upper management are not, especially if the differences are excessive. While most employees understand the financial implications of workspace and the need for managers to have space allowing confidentiality, it is important to provide quiet, private workspaces for cubicle-bound employees.

Financial rewards definitely motivate employees. Most staff are familiar with what their peers are being paid and with comparable benefit packages elsewhere. It is absurd to think that an employee who can, for example, calculate the increase in the height of the Empire State Building that occurs when the moon is directly overhead, will not know how his colleague is compensated in a competitor organization. Managers need to be aware of salary and benefit-package surveys, and they need to be prepared to impart the information to employees and to senior management.

Organizations employing technical professionals face a common dilemma: How to establish reward systems and work environments that are both stimulating to the professional and productive for the organization. Standard management strategies for motivating, rewarding, and leading nontechnical employees frequently fail to work for those who are involved in research, development, and engineering, thus resulting in performances that are suboptimal for the individual and the organization. Yet if financial compensation is not available (for good reason), staff might be receptive to creative motivational alternatives.

Performance Management

Performance management requires the participation of both parties to the process; the manager/supervisor and the employee. It can be defined as the constant communication between the parties of status of work, needed revisions or changes, and reflection on performance to improve future efforts. Performance management is not solely the annual review. The effectiveness of the performance-management process suffers from long periods where there is no communication. The best processes provide ongoing, almost immediate review and feedback, timely corrections, flexibility, and training focusing on addressing recognized weaknesses. Immediate, ongoing performance management can take the "financial" aspect out of the performance-management process except for, perhaps, the scheduled annual and biannual review. Unfortunately, it is recognized that managers and employees alike can be uncomfortable and even defensive at annual reviews. Eliminating the financial aspect can allow both the manager and the employee to focus on the actual events that are occurring and on determination of the support that the employee may need in order to learn, improve, and progress.

The basis of the performance-management process includes clear definition of the job, expectations, ways in which performance will be measured, and the process that will be used. Managers and employees alike can become truly frustrated by constant changes in the process, obfuscation of expectations, inclusion of unclear or poorly defined expectations, or inability to measure performance. While some flexibility in the process is necessary, organizations that have haphazard, poorly coordinated performance-management processes or that lack senior support for same usually contribute to problems that develop between managers and employees. Organizations that do not clearly define the manager/employee relationship and that do not promote immediate, ongoing feedback can find themselves with ineffective managers and stagnant or lost employees.

The Performance Review

If the performance-management process is immediate and ongoing, the performance review need not dwell on specific events that occur around the date of the review. Rather, it can focus on broader issues such as salary, promotion, training, discipline, and termination. While many engineering managers would like to avoid discussion of discipline and termination, sometimes the financial aspects of the organization or a specific project require a reduction in human resources that may well include good, productive engineers. The performance review, if performed correctly and consistently, can provide one measurement of value and contribution to the success of the organization. Unfortunately, if the process is flawed (e.g., if expectations are inflated beyond reason, or if measurability is impossible), performed inconsistently, or lacking in objectivity, the measurement is useless. There is no wonder that employees become defensive, elect to leave employment there, or are so disheartened that it affects productivity to the extent that no one throughout the entire organization meets, for example, a performance expectation.

Performance Review Criteria

Criteria should be based on the actual job—the duties and responsibilities of the position. Nothing can be seen as more ridiculous to an employee than a performance evaluation based on criteria that are outside their control. Inclusion of such criteria outside the control of the professional often causes employees to wonder whether senior management or their direct supervisor actually knows (or cares to know) what the contribution of the particular professional actually is.

One method of classifying criteria is measurable and behavioral. Measurable criteria can be evaluated in terms of quantitative and qualitative terms. For example, all projects were completed on time and within budget; 20 client visits were performed out of 20 expected; profitability rose 5%, reaching the set goal; or projects were completed in a manner that is acceptable to the organization. Behavioral criteria are actions that can be observed or described (e.g., an employee fails to get along with others at times, always contributes to creative processes, or consistently conveys the organization's message).

One simple way to improve the effectiveness and utility of performance reviews is to train supervisors and employees in the process. Training in giving and receiving constructive criticism can improve not only the process, but also attitude. Training managers in ways to observe and interpret behavior or common factors that distort reviews can improve the value of the process. Unfortunately, organizations that place little emphasis on the performance-management process can be expected to place even less emphasis on training because the process itself is so confused.

It is important that the review include open dialogue; a monologue by the supervisor could be expected to ensure defensiveness and to promote misunderstanding. Documentation of performance and behavior should be a mainstay of the process. All too often the manager throws together uncharacteristic examples of both, at the last minute, in order to meet a perceived need for a performance review to occur, without implementing a process that includes immediate and ongoing performance management and documentation of same throughout the preceding year.

Discussion of the results, not necessarily ending in agreement or consensus, should occur. Objective review of these discussions and remaining differences can often result in identification of invalid criteria, trends in employee dissatisfaction or misunderstanding, or managers requiring assistance.

Development of an action plan to move forward for the next period should not only be developed but agreed upon. If a feature of an action plan cannot be agreed upon, it is best to address that detail immediately. This may require arbitration, mediation, or discussion with others. It is rare, in our experience, for employees to take a stand against a particular aspect of a performance-review action plan. When they do, often there is either something truly wrong with the feature, or the discussion has led them to misunderstand the nature of the plan. While performance review often has a negative connotation and promotes negative images to the employee, it can become an opportunity to promote team building, to enhance the opportunity to coach or mentor an employee, to motivate or to encourage an employee to improve their position, or to emphasize an employee's security and value to the organization.

Stress Management

Relax. Take a rest. Go fly a kite, or just calmly sit and watch the world go by. Easy to say, but often difficult to do—maybe even stress-producing. Stress exists in everyone's life, but not all stress is necessarily bad. What stresses one individual might not stress another. Managers might have their own stress, plus inherited stress of others. How stress manifests itself, how we react; and how we manage it can be different for each of us. For example, the opportunity to present the outcome and success of a team project to upper management might be an agonizing prospect for one employee, but an honor, a challenge, and great fun for another.

Our response to stress can take different forms:

- Physiological: Increased heart rate, muscle tension
- Behavioral: Talking faster, social isolation, decreased or increased productivity
- Cognitive: Negative thinking, unwanted recurrent thoughts, new ideas
- Emotional: Excitement, anxiety, depression

Stress can undermine our effectiveness. Agitation and short fuses discourage employees or colleagues to approach each other in an open, nondefensive manner. If you are stressed, others know it, and the effect cascades. Calm, controlled leadership can inspire calm, controlled employees. Ability to identify stressors, their effects, and methods to harness the energy that stress produces can make managers more effective. It would be ideal if we could eliminate all stressors, but that would be unrealistic in the complex world where we live and work. You can remove some stressors from your life—a process that, in and of itself, can prove stressful. Some people actually take comfort in knowing that they are stressed–they are in demand, or they consider the causes of their stress to be important. Any method that you find effective in approaching stress is acceptable as long as the stress does not negatively impact your life or those around you.

People are different; they want or need different things, they behave differently (even under the same circumstances) and may believe differently. Some people like to be around other people, while others prefer to be by themselves—going to a party is actually stressful for some. Some make decisions quickly, while others prefer to keep their options open, perhaps collecting more information before making a decision. Some enjoy working from checklists, laying things out in an orderly fashion and systematically accomplishing items one at a time. Others find their creativity stymied if they are forced to work under such requirements, and they find it humorous to infuriating that more time might be spent documenting the performance of activities than in actually performing them. Some are highly rational, logical, and orderly, while others are reflective and intuitive. Some go "by the book," doing things consistently and predictably every time. Others seek creativity and innovation; they thrive on chaos and are bored by routine.

Experience shows that some personal styles are generally better suited to technical endeavors, scientific activities, and engineering. Certainly not all engineers are alike, but there are many similarities in personality traits that lead to similarities in managing stress. There are many personality-testing instruments available today. None comes close to measuring all the nuances of human personality, but they can help us to obtain a glimpse of who we are. The Meyers-Briggs Temperament Indicator is one of the best known. The full test consists of several hundred questions, although a shorter adaptation was developed by David Kiersey and Marilyn Bates. The tests examine the way people perceive the world and the way they determine what to do. Human reactions and perceptions have been analyzed, and the tests are designed to identify an individual's traits; i.e., general personality styles. People often straddle more than one category, and categories can overlap. The tests' designers believe that it is important to pay attention to those personality elements (i.e., psychological preferences and needs) that have the most significance, while acknowledging the overlaps.

Both tests score along the same four scales below and are available on the Internet or through management consultants. Remember that the tests are not infallible but are merely indicators of personality and temperament. Test instruments can be fooled if not answered directly and truthfully. In addition, temperament on the day of the test can moderately affect results. The results can identify strengths and weaknesses in leadership as well as stressors that need to be mastered in order to be more productive and effective.

1. E/I = Extrovert/Introvert: This scale measures the ways in which use of perception and judgment is affected by yourself and those around you. The person who prefers to be with other people probably tests more E than I. The person who prefers his own inner world of ideas and reflection probably tests more I than E. Approximately 75% of the population tests with an E preference, and 25% with an I.
2. N/S = Intuition/Sensing: This scale measures the ways in which one gathers data for decision making. The person who believes that he is practical and fact-oriented probably tests S. The person who relies on unconscious perceptions, "gut" reactions or "hunches" probably tests N. Experts say that differences on this scale are the greatest source of miscommunication and misunderstanding of the four scales. The S person perceives through facts and believes that experience is the best teacher. The N person is concerned about what could be. He can be bothered by reality, and he prefers to speculate about possibilities. Engineers and scientists most often test out as N's. Approximately 75% of the population tests as S while 25% tests as N.
3. T/F = Thinking/Feeling: This scale measures the ways in which one chooses or decides how to make decisions. T type people are more comfortable with impersonal, objective judgments. F-type people are more comfortable with personal, value-driven judgments and often are put off by rules. Formal schooling tends to develop our T side, and engineers and scientists tend to score as T's. This is the only scale that is influenced by gender: 60% of women test as F, while 60% of men test as T.
4. J/P = Judging/Perceiving: This scale measures the ways that we respond to what is happening. Persons who test J generally prefer a great deal of closure with a significant need for things to be settled, including deadlines. P-type people prefer to keep their options open, often wishing to delay their judgments, including deadlines. The populations tests approximately 50% J and 50% P.

Stress-Management Techniques for Introverts: Find comfort in your inner peace. Recognize your strengths and knowledge. Draw on your strengths to increase communication and interaction with peers. Listen quietly to others' input. Take action when you are comfortable with the action. Remember that more high-technology professionals are introverts than extroverts.

Stress-Management Techniques for Extroverts: Use your inner focus to assess relevance of ideas before sharing them with others. Use that inner focus to assess appropriateness before speaking, and reflect before acting.

Stress-Management Techniques for Sensors: Focus on the bigger picture to prioritize the details, and proceed accordingly. Consider new solutions that would be appropriate to the problem at hand.

Stress-Management Techniques for Intuitives: Focus on relevant details, concrete realities, and the particular problem at hand. Consider existing solutions.

Stress-Management Techniques for Feelers: Recognize that some conflict is inevitable and that you are not going to be satisfied with the outcome of all situations. When comfortable, assert your own interests. Try to understand that you will never fully understand others and their situations.

Stress-Management Techniques for Perceivers: When you are comfortable and ready, make a decision, even if all information that you believe is important is not available. Build in a structure for obtaining the information that you need.

Stress-Management Techniques for Judgers: Consider significant information that is relevant to a decision. Recognize that organization is only a means to an end–the method for arriving at the decision. Evaluate, for relevance and completeness of work, your decisions that were made without all relevant information.

The goal is not to eliminate all stress. If you know your stressors, you can tackle them in a proactive manner and can create a more balanced, flexible approach to future stressors. Stress-management models can be helpful. Other helpful methods include relaxation training, cognitive restructuring, assertiveness training, vacations, and exercise. If you are experiencing a significantly increased level of stress that is resulting in a significantly decreased level of function, you might want to consult with a professional.

Interviewing

Match the applicant to the job, situation, and environment! Assuming that the organization is participatory and team-oriented, the purpose of an interview is to find the best person to supplement or replace a team member within a team that already has goals. Maintaining an emphasis on the goals of the team and ways in which an applicant might contribute or (fail to contribute) to attaining those goals can keep the interview process in the correct perspective. Interviewers essential try to gain sufficient information with which to forecast whether and how an applicant will work within the team. Forecasting is fraught with problems, and forecasting based on incorrect or incomplete information is even more problematic. Keeping the above in mind, some of the objectives of the interview process include:

- Fleshing out the details of data on the application
- Obtaining additional data that is not on the submission
- Probing the applicant to determine how he might respond in situations that are likely to occur within the team and its projects
- Monitoring and assessing the applicants' behavior (e.g., motivation, appearance, character, or forthrightness)

Common failures within the interview process include:

- The interviewer being unfamiliar with the actual needs of the specific job or team
- The interviewer being unprepared (e.g., untrained in the interview process; failing to have a structure for the interview; or failing to know and focus on the goals of the interview
- Bias (e.g., conscious or unconscious application of interviewer biases or preferences toward the interviewee or the position)

- Unreasonable reliance on first impression or intuition
- Creating an interview environment that is not conducive to the task (e.g., noise, constant interruptions, or insufficient time allotted)
- Failure to listen (e.g., propensity to talk, leaving little time for the applicant to convey data).

The interviewer(s) can take some initial steps to properly prepare for interviews and to avoid, to the extent practicable, these problems. Interviewers should receive training in ways to conduct and analyze interviews and should remember that there are laws (e.g., Title VII of the Civil Rights Act) that prohibit discrimination. Interviewers should carefully review submissions prior to the interview as well as aspects of the specific job and environment into which the successful applicant may be inserted. It can help to develop, in advance, questions that can elicit new information about gaps or ambiguity in the submission; how the applicant has responded to challenges in the past; how they might approach a hypothetical problem-solving situation that could arise within the environment in which they will need to work, and work and personal work habits that might meld well or poorly within the current team.

Gathering of the interviewers, post-interview, to review, compare, and discuss information gathered is usually a good idea; new information and impressions can be shared among all interviewers; misinformation can be identified (both given by the interviewee and collected by the interviewer); and bias can be identified and resolved. It is important that the purpose of an interview (i.e., find the best hire to supplement or replace a team member within a team that already has goals) should be re-emphasized at this time.

Problem Employees

One of the most time-consuming and difficult situations that a manager can face is in having a problem employee. A problem employee can be a new hire who brings along "baggage" that slips through the interview process and probation period. A long-term employee can become a problem employee due to personal or other factors.

First, a manager should try to determine the extent and cause of the problem (e.g., tardiness, inability to meet deadlines, or undermining the authority and effectiveness of management). Second, a manager can examine whether the organization contributes in any way to the problem. Third, the manager should evaluate whether or not the situation can be remedied realistically and the resources that would be necessary to do so.

Replacing employees can be costly, especially when considerable resources have been expended in recruiting, time, and training. If the employee is receptive to discussing and addressing the problem, the situation can improve. If the employee is not receptive to discussing and addressing the problem, the problem could spread and undermine the morale and productivity of the entire staff.

When a manager suspects or is advised of a problem, the approach should be direct, discussed with all parties in confidence, and begin the documentation process. Always get all sides of the story, from the involved employee(s), from bystanders, or from others outside the organization. Be sensitive to the fact that all of these parties eventually might need to interact together again, and promote mutual respect where possible. Once the facts are known, confront the problem employee if it is appropriate. Never criticize an employee in front of other employees. Do not cause someone to "lose face." Take the sting out of criticism by focusing the discussion on the task, and not on the person.

Some managers find it helpful to allow the problem employee to develop a remedial program that can be followed, to address the problem and to set milestones to meet to review progress. Unfortunately, it must be accepted that not all problems can be resolved, and that termination of problem employees might be best for all parties. The manager must be aware of, and must follow, the organization's human resource policies relating to steps to follow under these circumstances. Most termination policies do not begin at the point of termination. Rather, they begin weeks or months earlier, to ensure that adequate attempts are made to rehabilitate the problem employee and that the organization's liability is minimized. Hence, managers should seek counsel with, and the guidance of, human resources professionals as soon as they suspect that a problem exists.

Employment Laws and Regulations

All managers should have a basic understanding of the important elements of the employment-related laws and regulations. Human resource departments and attorneys specializing in employment law are your best resources when an action or issue arises. Equal-employment laws are enforced by a variety of federal agencies, including the Office of Civil Rights and the U.S. Department of Labor. The U.S. Department of Labor administers and enforces more than 180 federal laws. These mandates and regulations cover many workplace activities for about 10 million employers and 125 million employees. Following is a brief description of many of the applicable statutes, but it should not be considered complete or extensively detailed. It is intended to identify the major labor laws, not to offer a detailed explanation of them. Consult the statutes and regulations for details.

Wages & Hours: The Fair Labor Standards Act (FLSA) delineates standards for wages and overtime pay. The act is administered by the Wage and Hour Division of the Employment Standards Administration (ESA). It requires employers to pay covered employees who are not otherwise exempt at least the federal minimum wage, and overtime pay of one-and-one-half-times the regular rate of pay. The Wage and Hour Division also enforces the labor standards provisions of the Immigration and Nationality Act that apply to aliens who are authorized to work in the U.S. under certain non-immigrant visa programs.

Workplace Safety & Health: The Occupational Safety and Health Act is administered by the Occupational Safety and Health Administration (OSHA). Safety and health conditions in most private industries are regulated by OSHA or OSHA-approved state systems. Employers have a general duty under the OSH Act to provide work and a workplace that is free from recognized, serious hazards. OSHA enforces this statute through workplace inspections and investigations.

Employee Protection: Most labor and public-safety laws, and many environmental laws, mandate whistleblower protections for employees who complain about violations of the law by their employers. Remedies can include job reinstatement and payment of back wages. OSHA enforces the whistleblower protections in most laws.

Uniformed Services Employment and Reemployment Rights Act: Certain persons who serve in the armed forces have a right to re-employment with the employer they were with when they entered service. This includes those called up from the Reserves or National Guard. These rights are administered by the Veterans' Employment and Training Service (VETS).

Employee Polygraph Protection Act: This law bars most employers from using lie detectors on employees, except under limited circumstances. It is administered by the Wage and Hour Division.

Garnishment of Wages: Garnishment of employee wages by employers is regulated under the Consumer Credit Protection Act, which is administered by the Wage and Hour Division.

The Family and Medical Leave Act: Administered by the Wage and Hour Division, the law requires employers of 50 or more employees to give up to 12 weeks of unpaid, job-protected leave to eligible employees for the birth or adoption of a child or for the serious illness of the employee or a spouse, child, or parent.

A number of federal laws and regulations prohibit discrimination on the basis of race, color, religion, sex, or national origin, both by the federal government itself and by organizations receiving various types of federal financial assistance or government contracts. These include:

- Title VI and Title VII of the Civil Rights Act of 1964, as amended in 1972 and 1991
- The Equal Pay Act of 1963
- The Age Discrimination in Employment Act of 1967 (Public Law (PL) 90-202) (applies only to people at least 40 years of age; for participants in programs or activities receiving federal financial assistance, the relevant law is the Age Discrimination Act (ADA) of 1975, which applies to people of all ages)
- The Americans with Disabilities Act of 1990 (P.L. 101-336)
- Section 503 and 504 of the Rehabilitation Act of 1973, as amended

The EEOC's regulations are published annually in Title 29 of the Code of Federal Regulations (CFR). The CFR is available online through the U.S. Government Printing Office. (http://www.gpoaccess.gov/cfr/index.html)

Resources

Management Theory

McNamara C. Various Styles of Management. http://www.mapnp.org/library/mng_thry/styles.htm

Croft C. The Evolution of Management. University of Western Australia, Department of Electrical and Electronic Engineering. http://www.ee.uwa.edu.au/~ccroft/em333/leca02.html

The Creative Process

MindTool.com. Brainstorming: Generating radical new ideas. http://www.mindtools.com/brainstm.html

Communication Skills

Hirsch H. Essential Communication Strategies: For Scientists, Engineers, and Technology Professionals. Wiley-IEEE Press. Indianapolis, IN. 2nd Ed., 2002.

Building Customer Loyalty. Engineers International. http://www.engineers-international.com/customer_loyalty.html

The Written Word. Online Writing Lab, Purdue University, http://owl.english.purdue.edu/handouts/pw/

Vilmi R. Writing Help. http://www.ruthvilmi.net/hut/LangHelp/Writing/

Mentoring

Yao JTP. Guidelines on Career Paths and Mentoring. ASCE Zone III Younger Members Conference. Department of Civil Engineering, Texas A&M, Houston, TX, February 6, 1999. http://lohman.tamu.edu

Guidelines for Mentoring Relationships. The Institution of Electrical Engineers. http://www.iee.org/EduCareers/Mento/guidelines.cfm

Motivation

Heathfield SM. http://humanresources.about.com/cs/moralemotivation/

HR Web Guide: Employee Motivation. The University of Vermont. http://www.bsad.uvm.edu/hrm/Motivation/motivationhome.htm

Business Bureau, UK. http://www.businessbureau-uk.co.uk/growing_business/employee_relations/motivation.htm

Topics in Motivating People, BusinessTown.com. http://www.businesstown.com/people/motivations.asp

Croft C. Motivating Job Performance. University of Western Australia, Department of Electrical and Electronic Engineering. http://www.ee.uwa.edu.au/~ccroft/em333/lectures/Lec%20L/Notes.htm

Performance Management

McNamara C. Basics of Conducting Performance Appraisals. Management Assistance Program for Nonprofits. http://www.mapnp.org/library/emp_perf/perf_rvw/basics.htm

HR WebGuide. The University of Vermont.

Croft C. Performance Appraisals. http://www.bsad.uvm.edu/hrm/PerfAppraisals/perfhome.htm
Human Resource Management. University of Western Australia. Department of Electrical and Electronic Engineering. http://www.ee.uwa.edu.au/~ccroft/em333/lectures/Lec%20J/Notes.htm

Stress Management

Sime WS. Stress Management: A Review of Principles. University of Nebraska Lincoln. http://www.unl.edu/stress/mgmt/
Kiersey DM. Temperament and Character. http://keirsey.com/
Stress Management Techniques. MindTool.com. http://www.mindtools.com/pages/article/newTCS_00.htm

Interviewing

Hiring & Firing. BusinessTown.com. http://www.businesstown.com/hiring/index.asp

EEO Guidelines for Interviewing Applicants. Wake Forest University. http://www.wfu.edu/hr/forms/eeo-guidelines.pdf
Recruiting & Hiring. CareerLab. http://www.careerlab.com/hiring.htm

Problem Employees

DelPo A, Guerin L. Dealing with Problem Employees: A Legal Guide., Nolo Press. Berkley, CA 2001.
Managing People: Problem Employees, BusinessTown.com. http://www.businesstown.com/people/employees.asp
Dealing with Problem Employees. HRTools.com. http://www.hrtools.com/HREssentials/P05_7320.asp

Employment Laws and Regulations

U.S. Department of Labor, Office of the Secretary. http;//www.dol.gov
U.S. Equal Opportunity Employment Commission. http://www.eeoc.gov

54

Quality

Thomas M. Judd
Director, Quality Assessment, Improvement and Reporting,
Kaiser Permanente Georgia Region
Atlanta, GA

This chapter on quality introduces the reader to the main concepts of clinical quality improvement (QI) and ways in which they can be applied to clinical engineering activities. Clinical engineers (CEs) need to understand and to apply basic quality principles in their work, such as having the right people doing the right thing at the right time (Deming,1986), and they need to learn how their health technology planning and management work can best contribute to improved quality of care.

Techniques are presented for improving the quality of health care delivery and technology management practice, in particular. Quality can be measured and improved through:

- Appropriate quality culture and QI infrastructure
- Prioritized QI initiatives based on impact (disease burden), improvability (quantified gap between current and evidence-based best practice), and inclusiveness (broad relevance and reach)
- Appropriate QI indicators and tools
- Performance feedback and other methods
- Using best science for care (evidence-based medicine) and CE practices

Some applications of these techniques include medical equipment quality-assurance and improvement programs, regulatory-compliance programs, customer-satisfaction surveys, patient safety, and improving cost-effective use of technology resources. These applications will be demonstrated through several case studies.

Quality Culture and Infrastructure

Business Principles

- Leadership implies establishing unity of purpose, direction, and the internal environment of an organization, creating the environment in which people can become fully involved in achieving organizational objectives.
- Factual approach to decision making: Effective decisions are based on the logical or intuitive analysis of data and information. "You can't manage what you can't measure."
- Involvement of people: People at all levels are the essence of an organization, and their full involvement enables their abilities to be used for the organization's benefit.
- Customer focus means knowing whom your organization serves, understanding their current and future needs and requirements, striving to exceed customer expectations, and acknowledging other stakeholders.
- Process approach: Desired results are achieved more efficiently when related resources and activities are managed as a process. Process-performance examples include accuracy, timeliness, dependability, cycle time/throughput, staff effectiveness and efficiency, utilization of technologies, and cost reduction.
- "Systems approach to management" implies identifying, understanding, and managing a system of interrelated processes for a given objective that contributes to the effectiveness and efficiency of an organization.
- Continuous improvement is a permanent objective of the organization.

- Mutually beneficial supplier relationships enhance the ability of the organization and its suppliers to create value.

Health Care Delivery Principles

- Preventive activities are stressed. Prevention or early identification of harmful events extends resource life.
- Reduce inappropriate variation in diagnoses and treatment. Find the best science and implement it; avoid overuse and underuse. Activities should have compelling evidence, also known as "doing the right thing."
- Deliver care in a patient-centered manner. Provide all the information they need in order to participate in decisions.
- Those who know how to deliver care (or service) must lead the improvement process.
- QI is most efficiently done through multidisciplinary teams.
- Measure and give feedback, and intervene to improve. Reduce waste in all forms.
- Good medicine is good business, characterized by physicians ("the right people") making health care decisions and "doing the right thing" for patients. Leadership in their marketplace for "quality of care, service, and affordability" is a core value for health care organizations and increasingly is a key business strategy and goal.
- Quality of care is a measure or indicator of the degree to which health care is expected to increase the likelihood of desired health outcomes and to which it is consistent with the standards of health care (United States National Institutes of Health).
- Quality Indicators for an integrated health care delivery system:

• Population health	• Community benefit	• Satisfaction	• Research
• Episode characteristics	• Financial performance	• Efficiency	• Capacity
• Episode prevention	• Quality of care	• Education	

- *Health technologies (HT) impact on quality of care*: The concept is based on the premise that effective delivery of quality health care services depends on the availability and proper management of appropriate resources. These resources are primarily pharmaceuticals, human resources, medical devices, and facilities.
- Essential Health Technology Package (EHTP): The link between HT resources and the delivery of quality care is achieved by identifying all HTs needed to provide well-delineated promotive, preventive, and curative health interventions (Heimann et al., 2003).

Quality Improvement Priorities and Infrastructure Model

The strategic goals in 2002 in the health care organization that this author represents (Kaiser Permanente [KP]) were to have profitable growth, to hold or increase market-leading quality of care and service performance, and to improve cost position with competitors (Judd et al., 2003). KP pursued these goals through its physicians and employees; improving the work environment is key to success and is another organizational goal. Care and service (care experience) quality priorities in 2002 included focused objectives in disease-state management, patient safety, continuity and coordination of care, and national report-card-quality measures; and maintaining high national-accreditation status for the services that KP delivers.

The health care organization annually sets quality priorities and evaluates the performance of the QI program to assist in meeting strategic targets, to ensure compliance with all appropriate regulatory and accreditation requirements, and to meet business imperatives (e.g., risk, contractual, and financial). Goals, initiatives, structure, and/or responsibilities are changed or aligned to ensure an effective, ongoing program. Responsible managers and physicians evaluate performance against targets, identify opportunities to improve (and constraints against that improvement), and submit evaluations and proposed work plans to their sponsoring operations leaders. Various committees and work teams (the quality infrastructure) evaluate their efforts in their areas of expertise and submit evaluations and proposed work plans to the organization's key quality committee.

The KP's key quality committee oversees performance against stated priority goals and quality themes throughout the year through regular reports and updates, and it reviews progress toward goals annually. There are three levels of review, with increasing intensity—quality monitoring, QI projects (QIPs), and QI activities (QIAs). Quality monitoring is the systematic measurement of key processes and outcomes used to ensure consistent performance and to detect opportunities. QIPs use a template to describe activities relating to the systematic improvement of a measured process or outcome and may be clinical- or service-oriented. QIAs are long standing QIPs that will meet the rigor of scrutiny for accreditation. A quality theme is an area of broad organizational interest that cuts across many structural areas. It represents an issue that should be a consideration of some, if not all, QIPs. In 2002 our quality themes were patient safety, continuity and coordination of care, member care experience, elder care, health-services research, and issues of clear community interest (e.g., minority and women's health). Quality themes and QIAs are our clearest organizational quality priorities and are displayed in the Quality Projects Matrix, a tool to assist the key quality committee in its QI program oversight.

Throughout the year, new action plans are concurrently incorporated into established work plans in response to identified issues and recommendations received from internal and external sources. Committees in our quality structure document these new plans in committee minutes. At the end of each year, action plans are evaluated, and actions that are not completed are rolled forward, as appropriate, to the following year.

Quality Improvement in Health Technology Management

Historically in clinical engineering, this has meant conducting those QI activities that keep desirable outcomes with use of medical equipment occurring and preventing the occurrence of undesirable outcomes. A QI process was developed by the U.S. hospital-accrediting body, the Joint Commission on Accreditation of Healthcare Organizations (JCAHO), in order to provide a patient care environment that is continually safer and that provides higher-quality patient outcomes. A typical array of QI indicators that may be used in HTM are in Table 54-1. This is typically an ongoing, hospital-wide process that is the responsibility of the entire staff at all levels. It empowers employees to make changes that improve care and service outcomes. It is a 10-step method that allows assessment and improvement of performance, through structure, process analysis, and outcome measurement. Its application is noted in the first case study below.

Case Studies

A Clinical Engineering Project to Enhance IPM Services, Dennis Autio, OSHU. (See Chapter 50.)

The following example reviews ways in which some of these principles can be integrated into a complex process in CE to monitor and improve scheduled inspections and preventive maintenance (IPM) services.

Problem

The current monthly IPMs completion rate is 52%. Some IPMs take several months to complete. Some IPMs are not completed. IPMs included a variety of tasks, including electrical-safety inspections, operation checks, and periodic maintenance procedures. Procedures are not well-documented.

Goals

Identify timetable to complete monthly IPM services. Identify critical equipment to be included in this program. Deploy resources proactively to prioritize and complete IPMs in a timely manner.

Project team

All staff members participate in this project.

Initial Assessment

The project team reviewed the scheduled IPMs for the past two years and identified several trends and patterns:

- All IPM services are treated the same; there is no prioritizing of services based on equipment criticality.
- There is no proactive process to distribute IPM workload.
- The initial indicator identifies only the number of IPM services completed by the end of the month.
- There is no process to help the staff to assign, prioritize, and close out scheduled IPMs.

Table 54-1 Clinical engineering quality indicators.

Inspection and preventive maintenance	Repair
Type/number of devices scheduled for service	Down time (up time)
Type/number devices inspected	Specific equipment failure
Type/number of devices that failed an inspection	Number of repairs
Type/number: on-demand service	Average time per repair
Type/number found with physical damage	Down time due to repairs
Type/number of no problem found	Repair turnaround time
Type/number serviced more than once in 7 days	Response time for repairs
Type/number involved in incident	Repeat repairs
Type/number requiring abnormal labor or parts	Repairs delayed due to parts orders
Inspections failed	Down time associated with parts orders
No inspection-equipment not located or in use	Mean time to repair
Users	Miscellaneous complaints
User-related problems	Incident investigations
Percentage of user errors associated with high-risk devices	Equipment recalls
Number of user errors	
Number of repairs caused by user misuse or abuse	
Frequency of repairs by user errors	
Frequency of user errors on same shift or same unit	

Phase 1 Action Plan

- Identify the current process with the same indicator. It is noted that no threshold was determined. Additional information and time is needed in order to define a proper solution.
- Review equipment management program to standardize equipment categories and assign a risk assessment based on equipment function, physical risk, and maintenance requirements. Rate equipment as high or medium risk, based on standardized risk-assessment criteria, the first step toward standardizing the program.
- Verify that appropriate IPMs are scheduled, define the specific procedures to be used, identify estimated time to complete, and provide an opportunity to project specific resource requirements for each month.
- Provide staff training on the new risk-assessment process and ways in which IPMs are processed.

Second Assessment

Once the first phase was completed, the projected IPM schedule for the year was reviewed. With a standardized database, monthly workload projections could be determined based on risk category. IPM schedules will need to be modified to better distribute workload. The indicators need to be revised, and appropriate thresholds need to be developed based on risk assessment. Proactive assignment to staff and monitoring of IPM status are required.

Phase 2 Action Plan

- After assessing the existing IPM schedules from an equipment category perspective, these inspections are distributed equally throughout the year. Peak workloads that would require significant resources to be allocated for only IPMs, such as anesthesia machines, were identified throughout the year. In this case, these IPMs were scheduled quarterly, all of them on the same month, thus requiring a significant amount of staff time to complete. By doing one third of these each month, the monthly workload and allocation of resources could be better managed while still providing quarterly inspections.
- Program indicators with thresholds were determined for high- and medium-risk equipment categories. All of high-risk equipment and 80% of medium-risk equipment were to be completed within four weeks. Indicator data are posted at the beginning of each week, identifying the number of IPMs completed during the last week (per risk category). Weekly completion goals were established at 25% of the scheduled high-risk equipment and 20% of the scheduled medium-risk equipment.
- IPMs are no longer assigned by management but selected proactively by the staff at the beginning of the month. Staff develops a proactive plan to schedule this work to be completed weekly.
- This provided opportunities for the staff to meet and review this process, make observations, and recommend future actions.

Third Assessment

After the second phase, the staff felt that they had a better understanding of what was required, and they were provided with tools to monitor this task. Additional procedures were needed to define the way to close scheduled IPMs if they were not completed during the four-week interval. Monthly review meetings provided times to assess the services provided and to identify opportunities for improvement.

Phase 3 Action Plan

- Define procedure to close out an IPM at the end of the month. High-risk equipment had to be completed. Medium-risk equipment could be canceled if there were no equipment repairs within the past three months and the last scheduled IPM was completed, or if the equipment could not be located.
- Continue to use the program indicators and thresholds monitor performance on a weekly basis.
- Quick monthly meetings were held to review the IPM indicators for the past month, identify any trends or patterns, and review the IPM scheduled for the next month.

Process Review

- This process took over a year to implement. When it was completed, there was a defined process to manage IPMs within the department, and all staff were trained as to its use.
- Workload was better distributed throughout the year.
- Standardized procedures were identified, defining what was to be done and how long it would take to perform.
- Workload data were identified proactively and distributed by staff for completion.
- Many opportunities for QI were identified based on defining the services provided and by reviewing the data collected, resulting in improved quality, productivity, and timeliness of service.

Novamed's QI-Focused Approach (Painter, 1999)

- Keep it simple, using only two or three indicators.
- Differentiate between quality and performance measurements, such as speed of resolution of problem versus number of inspections performed. Be consistent in measurement.
- Communicate the process to everyone, including the administrator, the appropriate clinical staff, and your own staff.
- Publish results in the work area, and review them in meetings.
- Put a committee of senior CE department staff together to review data and report results. Use your internal committee to evaluate data frequently enough to make a difference and to make your follow-up steps actionable (e.g., review monthly, report every six months).
- Expect improvement. Change your practices to improve. Look for steady and incremental changes, not major changes. Change the system, and not the people, to accomplish improvement. The people will follow.
- If indicator is not valid (e.g., not as good a measure of quality as you would like), change indicators.
- Never stop trying to improve. Keep changing the system to get better results.
- Make this process easy, and it will work. Make it complicated, and it will fail.

Mediq PRN's QA-Focused Approach (Wang, 1999)

This program is designed to provide safe, effective rental medical equipment to customers. All equipment complies with or exceeds JCAHO, United States Food and Drug Administration (FDA), and most original equipment manufacturer (OEM) recommendations. There are five program components, as follows:

Equipment Acquisition

- All equipment is approved for safety and efficacy through FDA through clinical studies.
- It is virtually all tested and/or listed by Underwriters Laboratories (UL), ECRI, and Canadian Standards Association (CSA). (See Chapter 118.)
- Equipment acquisition is based mostly on customer demand and return-on-investment considerations.
- Pre-purchase testing and evaluation is performed on more critical equipment, and there are factory visits for equipment purchased in large quantities.

Inspection and Maintenance

- Categories include: Safety and performance inspection (at delivery or equipment transfer); preventive maintenance (according to established schedules); and reconditioning (extensive replacement of critical parts after thousands of hours of use).
- The inspection and maintenance schedule is automated based on OEM recommendations, risk analysis, and experience.
- Inspection, maintenance, and repair procedures are modeled after ECRI's IPM system using OEM's recommended checks (ECRI, 1979).
- The service-records archiving system accepts daily service reports and has data stored on write once read many (WORM) optical disks for legal purposes.
- Test equipment calibration is conducted at least annually.

Quality Monitoring

- For service center (i.e., central location) inspections, separate staff inspects equipment on scheduled and random bases.
- At branch (i.e., remote) locations, central staff inspects twice annually with reports to senior management.
- Unsatisfactory equipment transfer reports allow feedback between branches, and provides with central management oversight.
- Overdue inspection reports are provided monthly.
- Customer complaints are aggregated centrally, with branch follow-up, and finally are closed centrally.
- Each technician's service reports are reviewed for accuracy and completeness through service-data audits.
- A sample of each technician's service report data is compared with the computer system to ensure accuracy and completeness through service data entry audits.

Regulatory Compliance

- Medical device recall information is collected and analyzed, and appropriate follow-up is conducted.
- Medical device tracking reports acquisition of tracked devices to OEM.
- Device-incident reporting includes investigation of every equipment-related death or serious injury.
- Upcoming regulations are continually watched.
- OSHA regulations compliance is ensured for blood-borne pathogens and hazardous chemicals.

Staff Management and Quality Improvement

- Staff qualifications and training includes clear requirements for background, initial, and ongoing training.
- Customer surveys are conducted for external and internal customers.
- The QA program is being revised constantly as new science of HTM emerges.

Service-Quality Improvement

Overview

Health care organizations should recognize that many sources of customer-related information exist and should establish processes to gather, analyze, utilize, and deploy this information. The organization should identify sources of customer and end-user information in written and oral forms from internal and external sources. Customer information might include feedback on all aspects of product, customer requirements and contract information, market needs, and information related to competition. Sources of information on customer satisfaction might include customer complaints, direct communication with customers, questionnaires and surveys, focus groups, reports from consumer organizations, and reports in various media.

The clinical engineer strives to deliver quality service. In order to know and to improve the level of the service provided, and to understand the expectations and the perceptions of the customer of that service, the CE must understand and use various sources of information on service quality or customer satisfaction. Survey instruments are an effective way of gaining this information. By studying other surveys, the CE can create a customized survey tool to measure customer satisfaction for a particular organization.

Measurement of service quality typically involves the following considerations:

- Asking customers for feedback on their care experience
- Focusing on a few satisfaction measures, and understanding key drivers of service quality
- Developing a survey based on customer needs and changing the surveys when the customer's needs change
- Understanding that how the survey is delivered matters
- Repeating surveys to make the results actionable; trending results; and intervening to improve
- Inviting comments with specifics related to individual customer experiences

Survey Examples

Several forms of satisfaction surveys are shown for health care and HTM customer satisfaction with key features as noted in Table 54-2.

Table 54-2 Health care and health care technology management customer-satisfaction survey elements

Common features	CAHPS* Managed Care Member Survey (NCQA, 2003)	CE Internal Customer Satisfaction (John Dempsey Hospital, 1999)	Hospital External Service Provider Survey (Novamed, 1999)	OEM Customer Satisfaction (GE Medical Systems; Varian Oncology Systems)
Timeliness to start	Getting care quickly	Response time to arrive, prioritize services	Response time to arrive	Response time to arrive, prioritize services
Timeliness to finish	Getting needed care	Response time to complete	Response time to complete	Response time to complete
Communicate results	How well doctors communicate	Service provider communicated results		Service-provider communicated results
How service provided	Customer service	Service provider demeanor		Service-provider demeanor
Staff assistance	Courteous and helpful office staff	Service demeanor by support staff		Support staff ease of contact, style, parts availability
Business aspects	Claims processing		Equipment particulars	Sales support: style, responsiveness
Provider competence	Personal care provider rating	Technical competence	Technical competence	Technical competence
Results	Health care rating	Overall satisfaction	Overall satisfaction	Overall satisfaction
Management rating	Health plan rating	Management responsiveness and competence		Management responsiveness and competence
Other	Comments	Comments; Evaluator	Comments; Evaluator	Comments; Evaluator

*Consumer Assessment of Health Plan Satisfaction (CAHPS) used by the National Committee for Quality Assurance (NCQA).

Overview

Rosow and Grimes (2003) state in their *Nursing Administration Quarterly* article on "Technology's Implications for Health care Quality: A Clinical Engineering Perspective" the following views:

As hospitals are motivated by the need to reduce medical and medication errors, improve patient safety, create effective processes and cut operating costs, they look to and draw from quality management processes used by manufacturers and other industries. Four of the most popular processes used by health care institutions are the following:

1. ISO 9000 series quality management systems standards, first adopted in 1987 by the International Organization for Standardization (ISO). Originally intended as a quality systems tool for industry and manufacturers, the standards have been used by health care institutions to assess and improve their systems and processes.
2. Six Sigma, which is a rigorous, data-driven, decision making process. It utilizes a systematic five-phase, problem-solving process called DMAIC: Define, Measure, Analyze, Improve, and Control. DMAIC helps ensure that teams stay on track by establishing specific deliverables for each phase. Six Sigma is designed to eliminate deviations in operations. A company that adheres to Six Sigma standards will have 3.4 defects per 1 million opportunities, by using statistical models to measure performance in various categories. The key is to eliminate variation and give customers consistent, reliable service.
3. The Malcolm Baldrige National Quality Award employs seven categories that include leadership, strategic planning, customer and market focus, information and analysis, human resource focus, process management and business results. Managed by the National Institute of Standards and Technology (NIST), in cooperation with the private sector, the Baldrige award is given annually to businesses in the manufacturing, service, and small business sectors as well as to education and health care organizations. Education and health care categories were added in 1999. In 2002, there were 17 applicants in the health care category, the highest total so far.
4. Failure Mode Effect Analysis (FMEA), which incorporates a prospective assessment that identifies and improves steps in a process thereby reasonably ensuring a safe and clinically desirable outcome. In effect, FMEA is a systematic approach to identify and prevent product and process problems before they occur (Stamatis, 1995).

Our ability to achieve the level of quality necessary for the health care delivery system of the future requires the adoption and integration of technology into the health care process.

Recommended Clinical Engineering Involvement

Over the past decade, the medical device and information technology industries have made significant progress in designing and creating "smart" products and systems.

Figure 54-1 illustrates various levels of technology adoption and integration through the health care delivery process. Only by advancing more of the health care processes to Level 3 can we achieve the degree of quality that the system needs. Through adoption, standardization, and integration of technology, the quality, safety, availability, and cost of health care will improve.

In this era of rapid change and diffusion of medical and information technology, clinical engineers are essential members of multidisciplinary hospital teams. The clinical engineering perspective can be instrumental in identifying root causes and solutions with regard to patient safety and process improvement. An understanding of equipment and system design principles can produce insights that go beyond the standard behavior of the device or system in question. The ability to utilize best practices and quality-management systems allows clinical engineers to contribute to the improvement of management processes, efficiencies, and patient safety measures.

Linking Health Quality and Health Technology

Overview

Essential Health care Technology Package (EHTP) (see Chapter 45) is a new methodology and set of tools specifically developed to strengthen resource planning and technology management in health care (Heimann et al., 2002, 2003). As noted before, these resources predominantly include pharmaceuticals, human resources, medical devices, and facilities. The link between resources and the delivery of quality care is achieved by identifying all health technology (HT) resources that are needed in order to provide well-delineated health interventions. Health interventions are linked to the World Health Organization (WHO) International Classification of Diseases. This mapping between HTs and health interventions is based on internationally accepted and recognized clinical practices.

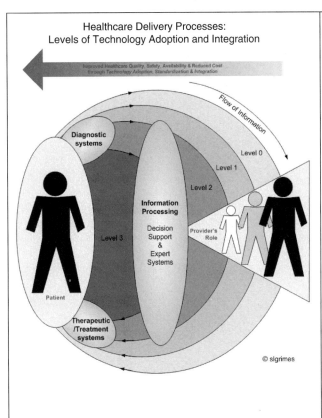

Figure 54-1 Recommended clinical engineering involvement in improvement of management processes, efficiencies, and patient safety measures.

Table 54-3 Key health conditions, clinical practice guidelines, and targeted medical device technologies linked

Key national health conditions	Primary level: Critical step(s) in CPGs for screening—treatment and medical device technology used	Secondary level: Critical step(s) in CPGs for screening—treatment and medical device technology used
Communicable diseases		Complications of condition diagnosed at primary care level
1. HIV/AIDS	Screening: Rapid HIV-1 tests via blood with lab equipment	Chest X-ray to rule out TB/pneumonia: Lab tests for TB and pneumonia if abnormal
2. Tuberculosis	Screening: Skin test, then sputum-slides-microscope if abnormal	Chest X-ray, lab tests to assess sputum and to grow TB cultures
3. Malaria	Screening: Rapid diagnostic tests, microscope-slides-blood	Lab tests to assess complications with liver and kidney
Key chronic diseases		
4. Hypertension	Screening: Assess BP with manometer-stethoscope; lab tests for cholesterol and kidney function	Cardiac status–ECG and chest X-ray; lab tests for cholesterol and kidney function (urine and creatinine)
5. Diabetes	Screening and tracking: Glucometer	Eyes–ophthalmoscope, kidneys–lab tests (urine and creatinine)
6. Asthma	Screening and tracking: Spirometer	Treat with aerosol chamber
Mother and infant health		
7. Prenatal care	Screening: Lab urine tests (or dipstick) to determine whether any infections; HIV-1 test	
8. Peri-partum care	Fetalscope	Fetalscope, forceps, oxygen delivery to mother
9. Cholera and diarrhea child health	Screening: Cholera packs; capability for IV resuscitation if diarrhea severe	IV therapy
10. Acute respiratory infections	Ears: Otoscope	Chest X-ray to rule out tuberculosis, asthma, malaria
Trauma		
11. Various	Suture materials	Orthopedic treatment: radiographic room (long bone X-rays)

From Heinemann et al. 2003.

Table 54-4 Assessment of optimal health care technology management for targeted health technologies

QI indicators by HTM category	QI indicators by title for targeted HTs	For 5 key conditions at each site, assess questions about HT "critical" items noted on Table 54-3
Planning (1-5)	1. Percentage selected	Are noted HT typically/sometimes/seldom selected through a formal assessment process? Why or why not?
	2. Percentage with appropriate facility plan	Are the noted HT typically provided with the appropriate facility planning when selected? Why or why not?
	3. Percentage with appropriate human resource (HR) Plan	Are the noted HT typically provided with the appropriate HR planning when selected? Why or why not?
	4. Clinical process and outcomes measures for key conditions	Are typical clinical process and outcome data collected for key conditions noted? Why or why not?
	5. Percentage timely access to care goals met	Are the noted HT typically available for appropriate follow-up care for these conditions? Why or why not?
Acquisition (6-9)	6. Percentage procurement meeting appropriate criteria	Are the noted HT typically acquired with appropriate criteria noted and met? Why or why not?
	7. Percentage needs met appropriately by donors	What percentage of noted HT needs are met by donors?
	8. Percentage meeting appropriate regulatory standards	Appropriate regulatory standards typically met?
	9. Percentage acquired with supplies/consumables plan	Are noted HT typically acquired with appropriate ongoing supplies and consumables plan? Why or why not?
Training and Support (10-13)	10. Percentage with appropriate User Training	What percentage of noted HT users receive user training?
	11. Percentage targeted HT with timely availability of supplies/consumables	What percentage of noted HT typically have timely availability of supplies and consumables? Why or why not?
	12. Percentage with timely information support during operation or use	What percentage of noted HT typically have timely information support during operation or use? Why or why not?
	13. Percentage with appropriate reuse of disposables	What percentage of noted HT typically have appropriate reuse of disposables? Why or why not?
Asset Management (14-15)	14. Percentage on appropriate HT inventory	What percentage of noted HT typically is on an appropriate HT inventory? Why or why not?
	15. Percentage with appropriate	What percentage of noted HT typically have appropriate HT inventory data analysis? Why or why not?
Maintenance (16-21)	16. Percentage on appropriate PM program	What percentage of noted HT typically is on appropriate PM program?
	17. Percentage on appropriate repair program	What percentage of noted HT typically on appropriate repair program?
	18. Percentage receiving timely repair	What percentage of noted HT typically is receiving timely repair?
	19. Percentage with appropriate communications between Users and HCTS	What percentage of noted HT typically have appropriate communications between users and HCTS?
	20. Percentage with timely availability of spare/repair parts	What percentage of noted HT typically have timely availability of spare/repair parts? Why or why not?
	21. Percentage with appropriate performance review and decommissioning process	What percentage of noted HT typically has an appropriate performance review and decommissioning process? Why or why not?
Risk Management (RM) and Quality Improvement (QI) (22-25)	22. Percentage receiving timely RM review and follow-up (FU) actions	What percentage of noted HT typically receives timely RM review and follow-up (FU) actions? Why or why not?
	23. Percentage receiving timely QI review and FU actions	What percentage of noted HT typically receives timely QI review and follow-up (FU) actions? Why or why not?
	24. Percentage receiving timely regulatory recall and product feedback communications	What percentage of noted HT typically receives timely regulatory recall/product feedback communications? Why or why not?
	25. Percentage utilized in context of a quality-accreditation program	What percentage of noted HT typically is utilized in context of a quality-accreditation program? Why or why not?
Other (26-28)	26. Sustainability of HTM Services	Do you expect HTM Services to be stronger in 5 years or absent?
	27. Customer expectations met	For noted HT use, are client expectations typically met?
	28. Service impact of equipment malfunction	Can clinical services typically be delivered when noted HT malfunctions?

Key Health Conditions, Clinical Practice Guidelines, and Targeted Medical Device Technologies Linked

One unique and powerful feature of EHTP is the integration of epidemiological profiles and recognized clinical practices such as evidence-based clinical practice guidelines (CPGs) or clinical pathways. If best clinical practice is known, then all HT resources to optimally deliver that intervention may be specified. This allows CE practitioners and health planners to potentially conduct optimal HT planning and management by meeting CPG requirements. Priority CPGs can be selected based on addressing the most significant disease burden for the population served and can be implemented in a manner that has been demonstrated to result in the highest quality of patient outcomes. (See Table 54-3).

Assessment of Optimal Health Care Technology Management for Targeted Health Technologies

As a result, a new technique has been developed to link medical equipment HTM techniques with their use in CPGs The following case study illustrates this linkage. The technique is based on identifying targeted medical devices or HTs that are critical to successful implementation of the CPGs and their related interventions (see Table 54-4).

Conclusions

Case studies have demonstrated ways in which quality can be measured and improved through:

- Appropriate quality culture and QI infrastructure
- Prioritized QI initiatives based on impact (disease burden), improvability (quantified gap between current and evidence-based best practice), and inclusiveness (broad relevance and reach)
- Appropriate QI indicators and tools
- Performance feedback and other methods
- Using best science for care (evidence-based medicine) and CE practices

Applications of these techniques have shown models of medical equipment quality-assurance and -improvement programs, regulatory-compliance programs, customer-satisfaction surveys, patient safety, and linkage of quality and technology for improving cost-effective use of technology resources.

References

Deming WE. *Out of the Crisis*. Cambridge, MA, MIT University Press, 1986.
ECRI. A New Concept in Inspection and Preventive Maintenance. *Health Devices* 8(4).71. 1979.
GE Medical Systems. *Customer Evaluation*. October 1992.
GE Medical Systems. *Varian Oncology Systems Customer Satisfaction Survey*. 1999.
Heimann P, Issakov I, Kwankam Y, et al. Quality and Management Applications of EHTP. Kyrgyzstan District Health Leaders, December 2002; May 2003.
John Dempsey Hospital. *Service Vendor Performance Evaluation*. 1999.
Judd TM et al. 2002 QI Program Evaluation. Kaiser Permanente. Georgia, May 2003.
Judd TM. Linking Health Quality and Technology Management. Health Technology Scoping Study. Republic of South Africa Department of Health, August 2003.
NCQA. CAHPS National Member Satisfaction Survey. http://www.ncqa.org re
NovaMed. Biomedical Engineering Department Customer Satisfaction Survey. 1999.
Painter F. NovaMed QI Program. Sub-Saharan Africa Advanced Clinical Engineering Workshop, 1999.
Rosow E, Grimes SL. Technology's Implications for Health Care Quality: A Clinical Engineering Perspective. *Nursing Administration Quarterly*, 2003.
Stamatis DH. Failure Mode and Effect Analysis. Milwaukee, WI, American Society for Quality, p 494, 1995.
Wang, B. Mediq QA Program, Sub-Saharan Africa Advanced Clinical Engineering Workshop, 1999.

Further Information

Berwick D et al. *Curing Health Care*. Jossey-Bass, 1990.
CMMS. Hospital Conditions of Participation: Quality Assessment and Performance Improvement. 42 CFR Part 482, Medicare and Medicaid Programs. Department of Health and Human Services, Centers for Medicare and Medicaid Services (CMMS). *Federal Register* 68(16):3435-3455, 2003.
Donabedian A. The Definition of Quality and Approaches to Its Assessment. Ann Arbor, MI, University of Michigan Administration Press, 1980.
Institute of Medicine. *Crossing the Quality Chasm*. Washington, DC, National Academy Press, 2001.
Kohn LT, Corrigan J, Donaldson MS (eds). *To Err Is Human: Building a Safer Health care System*. Washington, DC, National Academy Press, 1999.
Millenson ML. Demanding Medical Excellence. University of Chicago, 1997.
Simmons D. Health Care Quality and ISO 90001:2000. In Dyro JF (ed). *The Handbook of Clinical Engineering*. Chapter 119. Burlington, MA, Elsevier, 2004.

Section V Safety

Marvin Shepherd
DEVTEQ Consulting
Walnut Creek, CA

A medical device is just one component of a minisystem that delivers a clinical benefit to the patient. As device-related minisystems increase in number, the hazards associated with their use become more varied, and the clinical environment, more complex. Devices can become nonfunctional because of electromagnetic interference (EMI), or they can become fire-ignition sources for patients who are undergoing treatment in oxygen-enriched environments. Such medical device-related events require a clinical engineering investigation of the event and recommendations to prevent similar, future events. Because corrective recommendations frequently involve hospital professional staff and processes, they must be integrated into the total hospital safety program. The authors of the chapters in this section on safety address many of these issues and make recommendations for assuring a safe clinical environment.

Methodologies for making the clinical environment safer are described in the first four chapters. Patail (Patient Safety and the Clinical Engineer) gives the perspective of an experienced clinical engineer working within the National Center for Patient Safety (NCPS), showing that the clinical engineer is ideally suited a leadership role in promoting patient safety. Such systematic techniques as root cause analysis and failure mode and effects analysis, and tools such as process-flow diagrams, hazard-scoring matrices, and decision trees have enabled the NCPS to make measurable positive strides in a short time. Epstein and Harding (Risk Management), with their extensive experience in advising health care organizations on risk-management issues, present a comprehensive overview of this subject and present guidelines for adoption of effective techniques and programs. Vegoda and Abramson (Patient Safety Best Practices Model) bring their formidable expertise in information technology (IT) to bear on the patient safety issue as they outline a model system for patient safety best practices. Baretich (Hospital Safety Programs) provides an expanded view of hospital safety going beyond safety as applied only to medical devices. He describes the safety structure and requirements of a complete hospital safety program as required by the Joint Commission on Accreditation of Healthcare Organizations (JCAHO). The program encompasses the safety of staff, patients, and visitors from the various hazards in a health care environment. He emphasizes that, in order to be most effective, the safety-related aspects of clinical engineering practice must be integrated into this hospital-wide safety program.

Shepherd (Systems Approach to Medical Device Safety) identifies the five fundamental components of a medical device-related minisystem, a system that delivers at least one clinical benefit. In addition, he discusses the way these components can fail in such a manner as to prevent the clinical benefit from being delivered and may result, instead, in an injury or death. By means of this generic model, one can understand ways in which a patient might experience a particular hazard as well, and one can employ methodology to trace the fundamental causes of an injury back to the latent causes that were present in the minisystem. As the number and complexity of medical devices have increased, so have reports of interactions between various minisystems. Miodownik (Interactions between Medical Devices) explores some of the interactions among device-related minisystems when they are connected and operating simultaneously on or around a patient. Through case studies, he shows that the patient-selection criteria might not always identify those within a population who might be harmed by a diagnostic or therapeutic intervention. Device–device and device–patient interactions might directly result in an injury or a malperforming minisystem. His engineering analysis gives warning that clinical engineers must remain vigilant in order to detect unexpected minisystems interactions.

Cheng (Single-Use Injection Devices) details the safety aspects of needles and syringes, with an emphasis on their use and safety in developing countries. Recognizing that reuse of single-use injection devices is a leading cause of infection, Cheng recommends alternatives to this practice, such as auto-disable syringes, safety boxes, and the disposal of used sharps. He applies life cycle management concepts to injection devices to ensure safety in health-program planning and delivery.

Tan and Hinberg (Electromagnetic Interference with Medical Devices) provide a review of international EMI standards and an overview of EMI issues, including the effects of wireless telecommunication, wireless LAN, metal detectors, and article-surveillance equipment on medical devices is also included. Their chapter, developed from both the regulatory and the practical viewpoint, includes suggestions on managing the risks of EMI. Health care facilities are experiencing an increasingly hostile EMI environment. To ensure that a clinical environment is safe from EMI-provoked disturbances, clinical engineers must proactively manage the environment through detection, correction, and prevention of EMI. Paperman, David, and Hibbetts (Electromagnetic Interference in the Hospital) describe components of such a management program and present case studies from their own experiences to illustrate its value.

The role of the clinical engineer as medical device-safety officer or as independent forensic engineer requires specific skills in accident investigations. Dyro (Accident Investigation) provides a comprehensive overview of device-related accident investigations and describes the knowledge, skills, and investigative techniques that are necessary to a competent investigator. He emphasizes that the intended result of any investigation is to identify the latent (root) causes of an event and the minisystem modifications that are necessary to prevent similar events in the future. Accident investigators estimate that as many as 70% of all device-related, adverse events have some contribution from the human operator, often from the limitations of human abilities but frequently from inadequate human-factors-design considerations of the device manufacturer. Dyro discusses the fundamentals of human error and human factors design and their contributions to accidents. Clinical engineers, who should direct accident and incident investigations, must become familiar with the fundamentals of human error and human factors designs as they affect minisystem processes.

Finally, Ridgway (Electrical Safety in Perspective) describes the history of "microshock," a device-related hazard that appears to have been more imagined than real but nevertheless resulted in many unnecessary and expensive corrective actions in hospitals. In the early 1960s, small electrical currents that caused microshock were alleged to have killed thousands of hospital patients annually. Over the following twenty years, the concern for microshock emerged through both misinformation and a lack of information. Ridgway's insightful retrospective on microshock and the ensuing preoccupation with electrical safety describes the effort and resources that were expended and largely wasted in addressing this relatively minor hazard.

55

Patient Safety and the Clinical Engineer

Bryanne Patail
Biomedical Engineer, Department of Veteran's Affairs,
National Center for Patient Safety
Ann Arbor, MI

The clinical engineer (CE) is ideally suited for a leadership role in promoting patient safety. The safety of patients has gained widespread attention with reports of unusually high incidents of errors occurring in the administration of therapeutic and diagnostic interventions. Some of the reasons for the apparent compromising of patient safety explored in this chapter include financial constraints, fear of legal liability, and reluctance to admit to commission of errors. The Veterans Administration National Center for Patient Safety is in the vanguard of the patient safety movement as it attempts to change the culture of blame in its system of hospitals to the culture of patient safety. With proven systematic techniques such as the root-cause analysis and failure mode and effects analysis and the expertise of the CE, the NCPS has made measurable positive strides in a short time. Some of the tools, such as process-flow diagrams, hazard scoring matrices, and decision trees, used by the Center are described in this chapter.

The Clinical Engineer: Well-Suited for Patient Safety

The concept of patient safety is not new to CEs. In the past three decades since one of the first in-house clinical engineering departments was conceived and developed, safety has been the mission of such a department. A mission statement of a typical clinical engineering department is: "To assure a cost-effective, safe, quality environment for patients, employees, volunteers, visitors, and medical staff of the hospital, relative to patient care, diagnostic, therapeutic, and life-support medical devices, instruments, and systems." The American College of Clinical Engineering (ACCE) defines a *clinical engineer* as: "a professional who supports and advances patient care by applying engineering and managerial skills to heath care technology." One of the missions of ACCE is "to promote safe and effective application of science and technology in patient care."

Clinical engineers are uniquely positioned to address patient safety issues for the following reasons:

By Law

The minute one claims to be an engineer, one is bound by fiduciary responsibility for the safety of the public. Webster's Dictionary defines *fiduciary* as "a person who stands in a special relation of trust, confidence, or responsibility in his/her obligation for public safety."

By Education

In any bona fide engineering school, core courses such as statics, dynamics, thermodynamics, mechanics of materials, physics and mathematics (calculus) prepare one to better understand the cause and effect relations of most problems and incidents. Engineers are known for their problem solving abilities. Most CEs hold undergraduate degrees in the calculus-based engineering discipline such as electrical engineering, mechanical engineering, or computer engineering, and a graduate degree in biomedical/clinical engineering. Some also hold a doctoral degrees. In addition, many CEs are certified as certified clinical engineers (CCE).

By Experience

Over the past 30 years, CEs have participated in prospective approaches to ensure patient safety vis-à-vis technology assessment (T/A) projects. They have developed and published consensus standards (as members of AAMI) for almost all critical medical devices and systems, and many are now ANSI standards. Some CEs work as company consultants or collaborate with companies to conduct alpha or beta site testing of medical devices. Some spend enough on devices from one manufacturer to cultivate the influence to provide feedback and/or participate in the design of fault-tolerant medical devices and systems.

These activities are possible only if the CEs are also involved in retrospective assessments to ensure patient safety vis-à-vis incident investigations and since 1991 because of the Safe Medical Devices Act of 1990, CEs have been reporting all medical device vulnerabilities to the manufacturer and/or a central database (FDA MAUDE). The knowledge gained from reading investigations of incidents that occurred elsewhere allows CEs to plug back into the prospective risk-assessment process.

By Reputation

In addition to being credentialed with degrees and certification, CEs are well respected by their peers, appreciated by their co-workers, and trusted by their employers and customers. They usually have excellent track records in participating, educating, and consulting activities to improve patient safety. A high percentage of CEs participate in professional activities.

By Position in the Institution

Staff vs. Line?

Most staff positions protect the institution from exposure/liability; CEs work closely with the Legal Affairs department and lawyers' fiduciary responsibility is to protect their client the institution, which is sometimes in conflict with a CE's fiduciary responsibility to protect the public. Therein lies the problem. Solutions to this problem are presented later in this chapter.

On the other hand, aline position, or in some cases called operations, is responsible to produce the actual product (crank out the widgets). Someone in this position is expected to produce tangible, cost-saving, production support for the caring of the patients. Clinical engineering departments in most hospitals are relegated to repairing medical devices (mostly electronic devices). Therein lies another problem.

Leadership

Most reputable CEs also hold leadership positions in their hospitals. A CD is usually a department head, chairperson of a committee or committees, process owner, educator, or consultant.

Despite the unique position the CE might hold in a hospital or a heath care organization to ensure safety, it has been estimated that close to 98,000 deaths occur annually in U.S. hospitals due to medical errors (Kohn et al., 2000). One is compelled to ask:

- What happened?
- Why did it happen?
- What are the root causes?
- What can we do to prevent it from happening again?

Factors Contributing to Medical Errors

There are many dynamics that played a part to reach to this point, and several are listed here.

Heath Care Reimbursement

What Happened?

The changing landscape of heath care reimbursement process in the U.S. played an important role in reaching the current state of affairs. Over three decades reimbursement went from a cost, plus a certain margin contractual fee-for-service arrangement, to a Diagnostic Related Groups (DRGs) capitated payment system, to HMOs and PPOs. The whole issue of treating heath care as a business is the culture change that had devastating long-term impacts on patient safety. Strictly bottom-line orientation is the modus operandi of most hospital administrators (MBAs). Some nonprofit heath care facilities are spinning off profit making arms and the COO of one of these service companies, LLC, told me "Patient safety initiatives are diametrically opposed to the objectives of this company." Quality of care and safety of patients took the back seat. The theme of hospitals these days is to do more with less. Is this the best way management can save money?

Why Did it Happen?

Many hospitals are struggling to stay afloat. When one is fighting to survive, one forgets the real mission of the hospital. Core values take the back seat and/or are missing.

What Are the Root Causes?

Our business schools must stop teaching MBA students the notion that "all is fair in love, war, and business." Business leaders and CEOs must understand that they have an obligation to the community that provides them with the infrastructure, services, and other resources to conduct their businesses.

A CE, who is usually the department head, is bogged down with the daily tasks of managing the department, working with budgets, operations reports, meetings, and employees' needs and demands. He or she does not have the time and resources to attend to the RCAs and technology/systems-related prospective and retrospective assessments. Therefore, in order to survive, many CEs manage to carve out their scope of responsibility to only repairable medical devices (mostly electronic devices) because the technicians in the department came from an electronics technician educational program or the ranks of electronic technicians.

What Can One Do About It?

Identify a need to reevaluate the core curriculum of MBAs. Teach them that the community provides a certain infrastructure, services, and other resources to allow them to conduct business in the community, and they have certain obligations to the community and the patients they serve. Core curricula of BMETs and CEs need to be reevaluated also. Teach and share with them the different kinds of failures/incidents/injuries that can occur in hospitals. Include not only the repairable medical devices arena, but also problems with the application of these devices, processes that have failed with nonrepairable devices, and the systems failures that have occurred so far. Teach them how to conduct root cause analysis and some of the successful forcing functions and other robust, fault-tolerant solutions that have been implemented to solve these problems. If the expectation is to have these people get involved in improving patient safety, consider some of the unsolved problems and how to conduct HFMEA™ to look at solutions and use new technologies and new processes to solve these problems. The CE's job functions must meet the mission statement of his or her department.

Administrators Need to See Tangible Results

What Happened?

Myopic: Most administrators see tangible results of their investment in a few full time equivalents (FTEs) to repair medical devices as opposed to contracting out with the manufacturer for a maintenance program to save money. Did the administrators lose sight of the real mission of hospitals? CFOs literally control the direction of the business enterprise. The business people have done an excellent job to change the culture of the hospitals to think in terms of budgets and bottom lines. We are in the business culture. Although we are in the patient care business, patients are not number one–*business* is number one. However, the quality and safety people have not yet done an adequate job. If all the constituents in the hospital are properly represented, and their voices are heard equally, then there should not be a skewing of emphasis toward business.

Why Did it Happen?

Businesses complain about the spiraling cost of heath care. The automotive industry estimated that heath care cost accounts for $300 of each car's cost. So what? The American public demands the latest and best technology for its care. It is more glorious to have an MRI than an X-ray.

The ratio of Biomedical Equipment Technicians (BMETs) to CEs is 100 to 1. If you mention Biomed to a heath care worker, the first thing that pops into his or her head is the repairman. Why? Because they see and meet the BMETs more often than they see and meet the CEs. Since the CE department is relegated to repairs, the CEs are also looked upon as repairmen.

Litigious Society

What happened?

Medical malpractice suits are on the rise. The U.S. public, in general, has become enamored with lawsuits and expects to become rich from the inconveniences and/or injuries sustained as a result of medical care. They perceive the physician, the hospital, and the medical device manufacturers as people with deep pockets and unlimited funds to pay for damages. The cover story in the June 9, 2003, issue of *Time* delineated how the soaring cost of malpractice insurance is driving some physicians out of certain regions, out of certain high-risk specialties, or out of medicine completely. And so the pattern continues.

Why did it happen?

Personal injury lawyers may be encouraged by the recent malpractice award of $140 million in New York, chronicled in an article in *The New York Times* titled "New York Hospitals Fearing Malpractice Crisis" by Richard Perez-Pena.

What happened?

CONFIDENTIAL: CEs are not privy to all the patient safety incidents in a facility. Everything is confidential. The pattern repeats itself. In order to avoid litigations, physicians, nurses, health care workers, and hospitals treat all medical adverse events as confidential and do not discuss them with the front-line care providers. Many facilities take meticulous steps to avoid discovery of such events.

Why Did It Happen?

A punitive environment is a factor as well. Many surveys have shown that a high percentage of heath care workers chose to work in heath care to make a difference in curing the sick, healing the injured, and saving lives. They are dedicated, compassionate, intelligent, and motivated individuals. However, when a mistake is made and/or a patient is injured, the first question that most bosses ask is "Who did it?" Disciplinary action is often the first recourse. Mistakes or are not tolerated. Root causes are not identified, and eliminating an individual from a position seems to be the norm to solve problems.

What Are the Root Causes?

- Pharmacies conduct their own investigations on their own employees.
- Labs conduct their own investigations on their own employees.
- Radiology labs investigate their own employees.
- Operating rooms investigate their own employees.
- Nurses investigate their own employees.
- Cardiologists investigate their own employees.

What Can One Do About It?

Create a non-punitive environment, form multidisciplinary teams, invite subject-matter experts and process owners to the table, and discuss all patient safety issues. Make the discussion a learning experience, with the focus on preventing adverse events in your facility.

Solutions

Creating a Culture of Safety

Only by viewing the health care continuum as a system can truly meaningful improvements be made. A systems approach that emphasizes prevention, not punishment, can create patient safety success stories. Other high-risk businesses such as airlines and nuclear power plants have used this approach to accomplish safety goals. To make the prevention effort effective, we use methods of gathering and analyzing data from the field that allow the formation of the most accurate picture possible. People on the front line are usually in the best position to identify issues and solutions, so both root cause analysis teams and heath care failure modes and effects analysis teams formulate solutions, test, and implement strategies, and measure outcomes in order to improve patient safety. Findings from the teams are shared with other facilities in the system. This is really at the core of what we mean "by building a culture of safety." It is portrayed as the engine that propels the system toward the goal of maximum safety. This kind of cultural change does not happen overnight. It can only happen as a result of effort on everyone's part to take a different approach to the way we look at things. We must constantly ask whether we can do things in a better, more efficient, and safer manner. We must never let "good enough" be good enough. We must be relentless in our pursuit of finding ways to improve our safety systems. We do not believe that people come to work to do a bad job or to make an error, but given the right set of circumstances any of us can make a mistake. We must force ourselves to look past the easy answer–that it was someone's fault – to answer the tougher question of why the error occurred. There is seldom a single reason. Through understanding the real underlying causes of errors, we can better position ourselves to prevent future occurrences. Although the saying goes, "Experience is the best teacher," it is one of the most expensive teachers as well. One of the best ways to reduce the expense is to take advantage of lessons present in close calls, where things almost go awry, but no harm is done. Establishing a culture of safety where people are able to report both adverse events and close calls without fear of punishment is the key to creating patient safety.

Challenges

Historically, accident prevention has not been a primary focus of medicine. Hospital systems were not engineered to be prevent or absorb errors; they changed reactively without being proactive. The hospital system had misguided reliance on faultless performance by heath care professionals. Medical culture rewards perfection and punishes errors in a complex system not engineered for risk.

Unrealistic expectations discourage openness and honesty, and preclude learning from close calls. Mistakes may be made by capable, conscientious, and compassionate individuals trying to do the right things.

Tradition and values are also important; they were important to me. (that is the way I used to do it ... now you can do the same)

Human performance is fallible.

Create a Culture of Safety the VA Way

The following is a list of the steps that the Veterans Administration took to create a culture of safety:

- Assured a nonpunitive environment, except for intentional unsafe acts.
- Received a public commitment from heath care leadership tied to the performance appraisal process of every health care facility director.

- Dedicated resources: 200 Patient Safety Managers (PSMs) and 22 Network Patient Safety Officers (NPSOs)
- Established Special Patient Safety Centers of Inquiry (SPSCIs)
- Provided Incentives for the VHA workforce to promote safety
- Directed safety efforts
- Created the National Center for Patient Safety
- Trained 1000 patient safety managers, patient safety officers, and others from 163 VA facilities and several private sector health care facilities over 3 years at a 3-day course on Patient Safety 101 and about 200 safety professionals at a 3-day course on Patient Safety 202.
- Developed and published the *VA Patient Safety Handbook*
- Created the Patient Safety Website at www.patientsafety.gov
- Developed tools such as RCA Software (SPOT) and the RCA database, HFMEA,™ and distributed them to every hospital CEO in the United States through the American Hospital Association
- Developed and distributed of cognitive aids
- Disseminated patient safety alerts and advisories
- Participated as consultants in the procurement and standardization of medical devices, products, and services
- Established a patient safety work group in the information systems (Office of Information) support team.

Since taking these actions, close-call reports in the VA have gone up by 900%, an indication that the blameless reporting system is working well.

Vehicles to Change the Culture

Root Cause Analysis (RCA)

Root cause analysis RCA is a process or technique used to identify the most fundamental reason or contributing causal factor as to why a problem occurred. It is a retrospective assessment, focused on finding vulnerabilities in the system and developing countermeasures. The process of RCA addresses four basic questions. 1. What happened? 2. Why did it happen? 3. What are the contributing causal factors? 4. What can we do to prevent it from happening again? The emphasis must be on developing effective countermeasures. The RCA team must be interdisciplinary in nature, involving experts from the front line who are closest to the safety process and who have the best ideas to solve the problem. RCA is a process that continually digs deeper by asking, "Why, why, why?" at each level of an event. It is a process that identifies changes that need to be made to systems, is as impartial as possible, and moves beyond blame. The RCA process must consider human factors and other factors, related processes, and systems, and analyze the underlying cause-and-effect relationships. Relevant literature must be reviewed during the process and internal consistency must be achieved. The RCA must identify risks and their potential contributions to safety errors, and must determine potential improvements in processes or systems. To be credible, an RCA must include the participation and support of the leadership of the organization and those most closely involved in the safety process and systems. Figure 55-1 is a flow diagram of an RCA team process.

The following are five rules of causation adapted for patient safety from a publication by David Marx on the NCPS website, www.patientsafety.gov:

Rule 1–Causal statements must clearly show the cause and effect relationship.

This is simplest of the rules. When describing why an event occurred, show the link between the root cause and the bad outcome, and each link should be clear to the RCA team and others. Focus on showing the link from the root cause to the undesirable patient outcome under investigation. Even a statement such as "resident was fatigued" is deficient without a description of how and why this led to a close call or mistake. The bottom line is that the reader needs to understand the logic in linking the cause to the effect.

Rule 2–Negative adjectives such as poorly or inadequate are not used in causal statements.

As humans, we try to make each job we have as easy as possible; unfortunately, this human tendency works its way into the heath care documentation process. We may shorten our findings by saying, "Maintenance manual was poorly written" when we really have a much more detailed explanation in mind. To force clear cause and effect descriptions and avoid inflammatory statements, do not use negative descriptors that are merely placeholders for more accurate, clear descriptions. Even words such as "carelessness" and "complacency" are bad choices, because they are broad, negative, judgments that do little to describe the actual conditions or behaviors that led to the mishap.

Rule 3–Each human error must have a preceding cause.

Most of our mishaps involve at least one human error. Unfortunately, the discovery that a human erred does little to aid the prevention process. Investigate to determine WHY the human error occurred. It can be a system-induced error, such as a step not included in the medical procedure or an at-risk behavior, such as doing task by memory, instead of with a checklist. For every human error in your causal chain, there must be a corresponding cause. The cause of the error, not the error itself, leads to productive prevention strategies.

Rule 4–Each procedural deviation must have a preceding cause.

Procedural violations are like errors, in that they are not directly manageable. Instead, we can manage the cause of the procedural violation. If a clinician is violating a procedure because it is a local norm, address the incentives that created the norm. If a technician is missing steps in a procedure because he is not aware of the formal checklist, improve education.

Rule 5–Failure to act is only causal when there was a preexisting duty to act.

We can all find scenarios in which our investigated mishap would not have occurred – but this is not the purpose of causal investigation. Instead, we need to find out why this mishap occurred in our system as it is designed. A doctor's failure to prescribe a medication can only be causal if he is required to prescribe the medication initially. The duty to perform may arise from standards and guidelines for practice or from other duties involving patient care.

Health Care Failure Mode and Effect Analysis (HFMEA)™

To optimally meet the needs of a prospective risk assessment of heath care processes, the National Center for Patient Safety (NCPS), with assistance from Tenet Health Systems of Dallas, Texas, developed a hybrid method, the Healthcare Failure Mode and Effects Analysis (HFMEA)™, that combines concepts of FMEA from industry with the Hazard Analysis and Critical Control Point (HACCP) from food safety, as well as tools and concepts that are integral to the VA's RCA process (Stalhandske et al., 2000).

The HFMEA™ is a systematic approach to identify and prevent problems with products and processes before they occur. It is a prospective assessment that identifies and improves steps in a process, reasonably ensuring a safe and clinically-desirable outcome. HFMEA(streamlines the hazard analysis steps found in the traditional FMEA process by combining the detectability and criticality steps of the traditional FMEA into an algorithm presented as a decision tree. It also replaces calculation of the risk priority number (RPN) with a hazard score that is read directly from a Hazard Matrix Table developed by NCPS specifically for this purpose.

Prologue

During a midnight shift on an intensive care unit, a patient with an infectious disease was monitored by a physiological monitor while on a ventilator in an isolation room with an anteroom. While the patient was asleep, the assigned caregiver decided to help transfer her other patient, who was recovering well, back to a regular floor. While the caregiver was out of the isolation room, the first patient managed to extubate himself. No one heard the ventilator alarm or the physiological monitor alarm. When the caregiver came back to the room she immediately called a code and then attempted to resuscitate the patient, but the patient could not be revived.

What is the relevance of this story to proactive risk assessment and HFMEA™? A team examining this high-risk situation might have identified a number of vulnerabilities that could have been mitigated without the harm and tragedy that occurred. The following example depicts how this outcome may have been avoided by using the HFMEA™ proactive risk-assessment model.

Basics of HFMEA™

HFMEA™ is a five-step process that uses a multidisciplinary team to proactively evaluate a health care process. The team uses process-flow diagramming, a Hazard Scoring Matrix™ (Table 55-1), accompanying Severity Rating System (Table 55-2) and Probability Rating System (Table 55-3), and the HFMEA™ Decision Tree (Figure 55-2) to identify and assess potential vulnerabilities. The HFMEA™ Worksheet is used to record the team's assessment, proposed actions, and outcome measures. HFMEA™ includes testing to ensure that the system functions effectively and new vulnerabilities have not been introduced elsewhere in the system.

STEP 1 DEFINE the HFMEA™ TOPIC

- Define the topic of the HFMEA along with a clear definition of the process to be studied. Think about narrowing the scope so that the review is manageable and the actions operationally sound (see Figure 55-4).

STEP 2 ASSEMBLE the TEAM

- The team should be multidisciplinary including subject matter expert(s) and an advisor (see Figure 55-4).

STEP 3 GRAPHICALLY DESCRIBE the PROCESS

- Develop and verify the flow diagram (this is a process vs. chronological diagram).
- Consecutively number each process step identified in the process flow diagram (see Figure 55-5).
- If the process is complex, identify the area of the process to focus on (i.e., take manageable bites).
- Identify all subprocesses under each block of this flow diagram. Letter these subprocesses consecutively under each block (i.e., Under block 1 as A, B, ..., D, under block 2 as A, B, ... E) (see Figure 55-5).
- Create a flow diagram composed of the subprocesses (see Figure 55-6).
- Transfer these to the HFMEA™ Worksheet, Line 1 (see Figure 55-7).

A Helpful Hint: It is important that all process and subprocess steps be identified before proceeding.

STEP 4 CONDUCT a HAZARD ANALYSIS

- List all possible/potential failure modes under the subprocesses identified in Step 3. Transfer the failure modes to the HFMEA™ Worksheet. (Hint: Failure modes include anything that could go wrong that would prevent the subprocess step from being carried out; they describe what could go wrong. For example: If logging onto a laptop computer is the process step, two possible failure modes are (1) not being able to log in and (2) delayed login. Use various methods including the NCPS triage/triggering questions, literature reviews, and brainstorming to identify potential failure modes.)

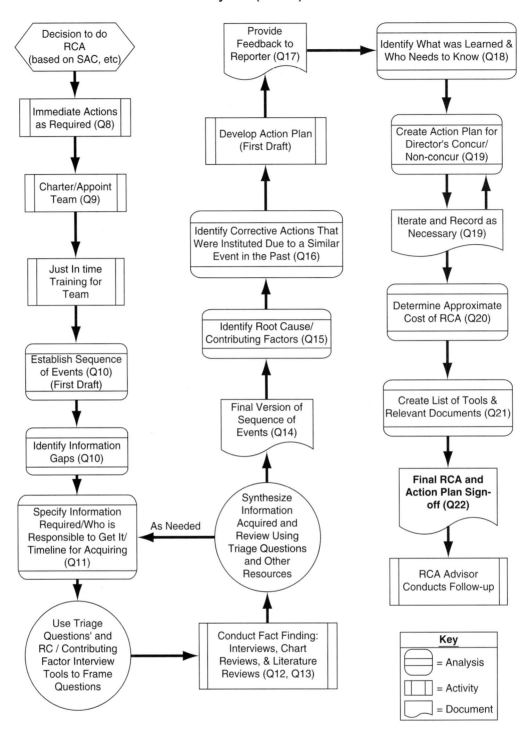

Figure 55-1 RCA team process.

- Determine the severity (Table 55-2) and probability (Table 55-3) of the potential failure mode and look up the hazard score on the Hazard Score Matrix™ (Table 55-1). Go to the HFMEA™ Decision tree (Figure 55-2). Use the decision tree to determine whether the failure mode warrants further action. Record the action to "proceed" or to "stop" on the HFMEA™ Worksheet (Figure 55-3).
- List all of the failure-mode causes for each failure mode where the decision is to "Proceed" and record them on the HFMEA™ Worksheet, Line 3. *(Hint: Remember that failure-mode causes are the reasons why something could go wrong. Each failure mode may have multiple failure-mode causes. For example, possible failure-mode causes for not being able to log in and delayed login with a computer would include the computer not being available, no power, and no log-in ID for the operator.)*

STEP 5 ACTIONS and OUTCOME MEASURES

- Identify a Description of Action for each failure mode that will be eliminated or controlled.
- *(Hint: Place the control measure in the process at earliest feasible point. Multiple control measures can be placed in the process to control a single hazard. A control measure can be used more than once in the process. Solicit input from the process owners if they are not represented on the team. Try to simulate any recommended process change to test them before facility-wide implementation.)*
- Identify outcome measures that will be used to analyze, and test the redesigned process.

Table 55-1 Hazard scoring matrix

Probability	Severity of effect			
	Catastrophic	Major	Moderate	Minor
Frequent	16	12	8	4
Occasional	12	9	6	3
Uncommon	8	6	4	2
Remote	4	3	2	1

To use this matrix: (1) Determine the severity and probability of the hazard based on the definitions included with this matrix. (NOTE: These definitions are the same as those used in the RCA safety assessment code.) (2) Look up the hazard score on the matrix.

- Identify a single, responsible individual by title to complete the recommended action.
- Indicate whether top management has concurred with the recommended action.
- Test to ensure that the system functions effectively and that new vulnerabilities have not been introduced elsewhere in the system.

How to Develop Reasonable and Concrete Failure Modes

There are several techniques besides brainstorming that should be used to develop reasonable and concrete failure modes once process diagrams are complete and the focus areas are chosen. Reviewing databases, such as the United States Food and Drug Administration (FDA) Manufacturer and User Device Experience (MAUDE), could provide malfunctions and user-interface design issues such as inadvertent shutdown of external pacemaker machines. Usability tests that are done by the HFMEA™ team or found in a literature search can be useful (Welch, 1998). Usability testing is a human-factors-engineering technique that can be done on devices, work areas, or larger processes (Gosbee et al., 2001).

Another technique is patient safety rounds. For example, upon sitting down with ICU nurses and residents, the CE may find that the nurses are worried about missing alarms several times a week because of distractions or noise levels and that they have nearly missed important alarms because of distraction or noise level. Another approach is to use the findings from routine safety-assessment tool "audits." For example, the CE may find that many of the hospital's compressed air wall outlets have green Christmas tree light adaptors attached; the VA's approach to this issue has been to send out an alert that all facilities should switch to a clear adaptor. This thereby avoids the potentially hazard and vulnerability caused by matching the wrong color adaptor and thereby providing the wrong color cue.

In a complex and difficult HFMEA,™ video documentation can be done on high-hazard areas. The video analysis provides data on close calls or adverse events, and is one of the most concrete development tools to list failure modes. For example, the University of Maryland has used this for research into safety issues during respiratory arrest resuscitation (Xiao and Moss, 2001).

Those that have FMEA experience and skills may find themselves in demand to respond to the new JCAHO standards and need for proactive risk assessment within heath care.

JCAHO Standards

The Joint Commission on Accreditation of Healthcare Organizations (JCAHO) drives much of the activity within heath care through its standards that must be met to gain accreditation. While voluntary, almost all hospitals choose to apply for JCAHO accreditation. The new LD 5.2 JCAHO patient safety standard reads as follows: "Leaders ensure that an ongoing, proactive program for identifying risks to patient safety and reducing medical/health care errors is defined and implemented." The intent section clarifies that annually at least one high-risk process be examined. For this process it is required to identify failure modes, and for each failure mode identify the possible effects. For the most critical effects, conduct analysis of the systems issues that allow this to occur, and mitigate the serious issues.

Table 55-3 Probability rating

Frequent–Likely to occur immediately or within a short period (could happen several times in one year)
Occasional–Probably will occur (could happen several times in 1 to 2 years)
Uncommon–Possible to occur (could happen sometime in 2 to 5 years)
Remote–Unlikely to occur (could happen sometime in 5 to 30 years)

Procurement Using Proactive Risk Assessment Model

CEs are called upon to fix, repair, maintain, and update equipment throughout the hospital and clinic environment. Skills gained in engineering schools and practical knowledge developed through work experience are applied in these traditional activities. However, CEs can broaden their impact by influencing procurement activities, thereby increasing the visibility of the CE department, improving the safety of the procurements, and improving the bottom line.

Engineers are taught to think in terms of systematic, logical, objective-based, well-supported conclusions. Consider researching the ways procurement decisions are reached and selling the capital acquisition committee on the added benefit of applying CE skills to the procurement process.

- Provide a proactive approach to procurement; think about what might go wrong with the different models under consideration and provide a modified failure mode effect analysis to compare different models
- Encourage end-user input into the procurement decision process
- Include consideration of past experience (good or bad) with the particular manufacturer
- Use existing data from databases of fellow CEs, ECRI, and MAUDE to support the safest purchase
- Develop explicit criteria to enhance an objective evaluation of the equipment

The CE department will be involved downstream in maintaining and fixing equipment brought into the facility. Why not become involved initially to make the most prudent purchase considering full life cycle costs, safety, and usability in your decision process?

Definitions

Effective Control Measure: A barrier that eliminates or substantially reduces the likelihood of a hazardous event occurring.

Health Care Failure Mode and Effect Analysis (HFMEA™): (1) A prospective assessment that identifies and improves steps in a process, thereby reasonably ensuring a safe and clinically desirable outcome; (2) a systematic approach to identify and prevent product and process problems before they occur.

Hazard Analysis: The process of collecting and evaluating information on hazards associated with the selected process. The purpose of the hazard analysis is to develop a list of hazards that are of such significance that they are reasonably likely to cause injury or illness if not effectively controlled.

Failure Mode: Different ways that a process or subprocess can fail to provide the anticipated result.

Probability: See the Probability Rating System (Table 55-3).

Severity: See the Severity Rating System (Table 55-2).

Table 55-2 Severity rating

Catastrophic event (Traditional FMEA rating of 10-Failure could cause death or injury)	*Major event (Traditional FMEA rating of 7-Failure causes a high degree of customer dissatisfaction.)*
Patient outcome: Death or major permanent loss of function (sensory, motor, physiologic, or intellectual), suicide, rape, hemolytic transfusion reaction, surgery/procedure on the wrong patient or wrong body part, infant abduction, or infant discharge to the wrong family **Visitor outcome:** Death; or hospitalization of 3 or more. **Staff outcome:*** A death or hospitalization of 3 or more staff **Equipment or facility:** **Damage equal to or more than $250,000 **Fire:** Any fire that grows larger than an incipient	**Patient outcome:** Permanent lessening of bodily functioning (sensory, motor, physiologic, or intellectual), disfigurement, surgical intervention required, increased length of stay for 3 or more patients, increased level of care for 3 or more patients **Visitor outcome:** Hospitalization of 1 or 2 visitors **Staff outcome:** Hospitalization of 1 or 2 staff or 3 or more staff experiencing lost time or restricted duty injuries or illnesses **Equipment or facility:** **Damage equal to or more than $100,000 **Fire:** Not applicable – See Moderate and Catastrophic
Moderate event (Traditional FMEA rating of "4" – Failure can be overcome with modifications to the process or product, but there is minor performance loss.)	*Minor event (Traditional FMEA rating of "1" – Failure would not be noticeable to the customer and would not affect delivery of the service or product.)*
Patient outcome: Increased length of stay or increased level of care for 1 or 2 patients **Visitor outcome:** Evaluation and treatment for 1 or 2 visitors (less than hospitalization) **Staff outcome:** Medical expenses, lost time or restricted duty injuries or illness for 1 or 2 staff **Equipment or facility:** **Damage of more than $10,000 but less than $100,000 **Fire:** Incipient stage‡ or smaller	**Patient outcome:** No injury, nor increased length of stay nor increased level of care **Visitor outcome:** Evaluated and no treatment required or refused treatment **Staff outcome:** First aid treatment only with no lost time, nor restricted duty injuries nor illnesses **Equipment or facility:** **Damage of less than $10,000 or loss of any utility? without adverse patient outcome (e.g. power, natural gas, electricity, water, communications, transport, heat/air conditioning) **Fire:** Not applicable – See Moderate and Catastrophic

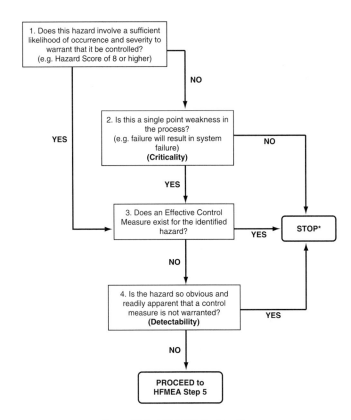

Figure 55-2 HFMEA™ Decision Tree.

Failure Mode: First Evaluate failure mode before determining potential causes	Potential Causes	Scoring			Decision Tree Analysis				Action Type (Control, Accept, Eliminate)	Actions or Rationale for Stopping	Outcome Measure	Person Responsible	Management Concurrence
		Severity	Probability	Haz Score	Single Point Weakness?	Existing Control Measure?	Detectability	Proceed?					

HFMEA™ Step 4 Hazard Analysis | HFMEA™ Step 5 - Identify Actions and Outcomes

Figure 55-3 HFMEA™ Worksheet.

Patient Safety and the Clinical Engineer 233

Step 1. Select the process you want to examine. Define the scope (Be specific and include a clear definition of the process or product to be studied).

This HFMEA ™is focused on _____

Step 2. Assemble the Team
HFMEA™ Number _____
Date Started _____   Date Completed _____
Team Members 1. _____   4. _____
2. _____   5. _____
3. _____   6. _____
Team Leader _____
Are all affected areas represented?   YES   NO
Are different levels and types of knowledge represented on the team?   YES   NO
Who will take minutes and maintain records? _____

Figure 55-4   HFMEA™ Process Steps 1 and 2.

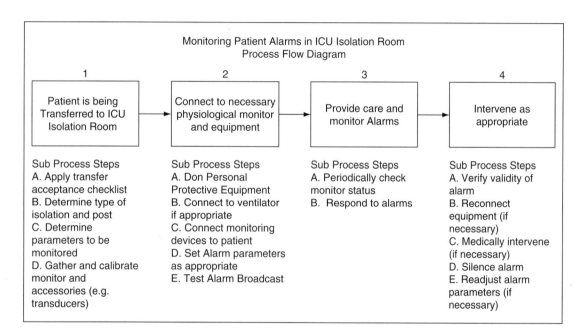

Figure 55-5   Monitoring patient alarms in ICU isolation room: Process flow diagram.

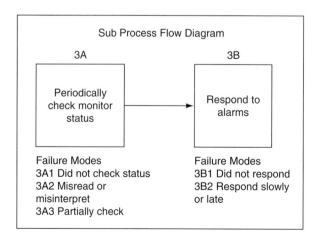

Figure 55-6   Sub Process flow diagram.

Solutions

Staff vs. Line? Be neither. If one is serious about patient safety and is truly passionate about it, the ideal chain of command is an autonomous arm reporting directly to the board. The National Center for Patient Safety serves as a good model. The IOM report also recommends forming a National Center for private-sector hospitals using NCPS as a model. This will overcome two problems, delineated earlier in the document.

Conclusion

CEs are uniquely positioned to address patient safety issues. They have the proper education, the proper mindset, and solid experience, entrusted by the public and trusted by their peers. They are performing prospective and retrospective risk assessment on certain groups of medical devices. They need to expand their horizons and get involved with the multidisciplinary teams to conduct these risk assessment in all heath care technologies and processes. Close to 90% of the hospitals in the United States are looking at the Practitioner Order Entry Systems (POE) and the Bar Code Medication Administration Systems (BCMA) to mitigate well-documented Adverse Drug Events (ADEs). These are truly complex, high-cost technologies with complicated process issues to which CEs should apply their expertise. CEs must play the role of Stewarts to

Failure Mode: First Evaluate failure mode before determining potential causes	Potential Causes		HFMEA™ Step 4 Hazard Analysis							HFMEA™ Step 5 - Identify Actions and Outcomes			
			Scoring			Decision Tree Analysis			Action Type (Control, Accept, Eliminate)	Actions or Rationale for Stopping	Outcome Measure	Person Responsible	Management Concurrence
			Severity	Probability	Haz Score	Single Point Weakness?	Existing Control Measure?	Detectability	Proceed?				
Don't respond to alarm			Catastrophic	Frequent	16	→	N	N	Y				
	Ignored alarm (desensitized)	3B1a	Catastrophic	Frequent	16	→	N	N	Y	C	Reduce unwanted alarms by: changing alarm parameter to fit patient physiological condition and replace electrodes with better quality that do not become detached.	Unwanted alarms on floor are reduced by 75% within 30 days of implementation.	Nurse Manager	Yes
	Didn't hear; caregiver left immediate area	3B1b	Catastrophic	Occasional	12	→	N	N	Y	C	Alarms will be broadcast to Central Station with retransmission to pagers provided to care staff.	Alarms will be broadcast to the central station within 4 months; complete by mm/dd/yyyy	Biomedical Engineer	Yes
	Didn't hear; alarm volume too low	3B1c	Catastrophic	Occasional	12	→	N	N	Y	C	Set alarm volume on isolation room equipment such that the lowest volume threshold that can be adjusted by staff is always audible outside the room.	Immediate: within 2 working days; complete by mm/dd/yyyy	Biomedical Engineer	Yes
	Didn't hear alarm; remote locatin (doors closed to isolation room)	3B1d	Catastrophic	Frequent	16	→	N	N	Y	C	See 3B1b	See 3B1b		
	Caregiver busy; alarm does not broadcast to backup	3B1e	Catastrophic	Occasional	12	→	N	N	Y	C	Enable equipment feature that will alarm in adjacent room(s) to notify caregiver or partner(s).	Immediate: within 2 working days; complete by mm/dd/yyyy	Biomedical Engineer	Yes

Figure 55-7 HFMEA™ sub process step: 3B1—Respond to alarms.

change the culture of safety in health care facilities. Make patient safety everybody's business and create that "engine that continues to propel the system toward the goal of maximum safety."

References

Gosbee JW, Arnecke B, Klancher J, et al. The Role of Usability Testing in Heath Care Organizations. In *Proceedings of the Human Factors Society 40th Annual Meeting*. Santa Monica, CA, Human Factors Society, pp 1308-1311, 2001.

Kohn LT, Corrigan JM, Donaldson MS (eds). To Err Is Human: Building a Safer Health System. Institute of Medicine. National Academy Press. March 2000.

Stalhandske E, DeRosier J, Patail B, et al. How to Make the Most of Failure Mode and Effects Analysis. *Biomed Instrum Technol* pp 96-102, March/April 2000.

Welch DL. Human Factors Usability Test and Evaluation. *Biomed Instrum Technol* pp 183-187, March/April 1998.

Xiao Y, Moss J. Practices of High Reliability Teams: Observations in Trauma Resuscitation. Proceedings of Human Factors and Ergonomics, 44th Annual Meeting. Santa Monica, C, HFES, p 395, 2001.

56

Risk Management

Alice L. Epstein
CNA HealthPro
Durango, CO

Gary H. Harding
Greener Pastures, Inc.
Durango, CO

Risk management, the organized effort to identify, assess, and reduce physical and financial risk to patients, staff, and business, has deep roots both inside and outside of health care. Sweeping new initiatives by the United States and international regulatory agencies to inject risk management techniques into the development and use of medical devices have surfaced and are receiving increased attention. However, few new, innovative changes to the risk management process have occurred over the last 35 years, although some would argue otherwise (ASHRM, 1999). The authors concur with this statement on the basis of more than 20 years experience in health care risk management (Epstein and Harding, 2003). Surely, the names of processes, inventors, and regulations have changed with the introduction and remodeling of old concepts into newly titled ones, but the actual need to identify, assess, and manage risk overshadows the sparkle of the new name one applies to the process or the honor bestowed upon its inventor.

The following applications each include aspects of risk management:

- Enterprise risk
- Total quality management (TQM)
- Performance improvement (PI)
- Quality inspection system technique (QIST)
- Failure mode and effect analysis (FMEA)
- Corrective and preventive action (CAPA)
- Risk analysis–ISO/IEC 14971 Clause 7
- Root cause analysis (RCA)
- Six Sigma
- Title 21, Chapter 1, Part 820 (Quality System Regulation)

Risk management is not about complying with the latest concept or approach, i.e., complying precisely with the letter of the risk management technique in vogue. Rather, it is about taking the steps necessary to effectively identify, assess, and address risk.

Risk management is not about placing blame. Multiple studies have found that accidents and incidents do not decrease in an environment of blame. Rather, accidents and incidents most often occur because of system failure. The adverse occurrence is rarely caused by the actions of one individual. Usually there is a series of actions and resultant interactions that allow the event to occur. Most failures have resulted from the lack of attention to one or more aspects of the basic risk management techniques that have been available for over 35 years. Blind adherence to a process in vogue may be the greatest impediment to successful risk management. A process may be accepted by one group, but it may conflict with the processes and viewpoints of another group and stifle communication.

Risk management is all about communication–communication of information among health care professionals and sometimes with others, for example, the patient, the alternate caregiver, and the insurer. It is all about ensuring that the proper risk management techniques are in place and that they are in use consistently. While there are differences in actual operational application of risk management techniques, the basic tenets of risk management are the same for health care clinical processes, health care, clinical engineering device management, and medical device manufacturers. This chapter discusses risk-management basics, pertinent regulatory authorities, and specific risk management issues from the perspective of the hospital clinical engineer. Resources are provided not only to address specific clinical engineering issues but also to provide the experiences of industries that have successfully applied the basic tenets of risk management from which others can and should learn.

Definitions of Risk Management

Kavaler and Spiegel define risk management as "...an organized effort to identify, assess, and reduce where appropriate, risks to patients, visitors, staff, and organizational assets" (Kavaler and Spiegel, 2003). The American Society of Health care Risk Management (ASHRM) defines risk management as "The process of making and carrying out decisions that will assist in prevention of adverse consequences and minimize the adverse effects of accidental losses upon an organization. Making these decisions requires the five steps in the decision process.... A systematic and scientific approach in the empirical order to identify, evaluate, reduce, or eliminate the possibility of an unfavorable deviation from expectation and, thus, to prevent the loss of financial assets resulting from injury to patients, visitors, employees, independent medical staff, or from damage, theft, or loss of property belonging to the health care entity or persons mentioned. The definition includes transfer of liability and insurance financing relative to the inability to reduce or eliminate intolerable deviations." (ASHRM, 2003), in a contract for the Canadian government, defined risk as follows, "For the most part, risks are perceived as any thing or event that could stand in the way of the organization achieving its objectives" (Treasury Board of Canada Secretariat, 1999).

Hence, for these organizations, risk management is not about being "risk averse." Risk management is not aimed at avoiding risks. It has as its focus the identification, evaluation, control, and mastering of risks. Risk management includes taking advantage of opportunities and taking risks based on an informed decision and analysis of the possible outcomes. Snow related that risk management in medical device design and production is composed of five core activities – planning, risk analysis, risk evaluation, risk control, and postproduction control (Snow, 2001).

History of Risk Management in Health Care and Related Industries

The concept of risk management developed from within the insurance industry. Primarily related to the theories of legal liability (e.g., malpractice, product liability, errors and omissions, and completed operations); the insurance industry had and still maintains a direct interest in reducing the occurrence and impact of adverse events. Simply put, if the insurer received a premium to provide insurance for accidental loss, and actual losses were below the value of the premium attained, a profit might be made by the insurance provider. If a situation developed wherein the losses due to adverse events (e.g., an insured ship sank) exceeded premium attained, not only was the insured in danger of going out of business, but so was the insurer. For self-preservation, insurers developed and applied the concept of risk management many years ago.

While Kavaler points out that the principles of risk management in medicine date back to the Code of Hammurabi, under which a surgeon might be bound to lose a hand if a patient lost an eye as a result of medical treatment, the emphasis in this review is more recent. With respect to the practice of medicine, the introduction and application of risk management in health care was designed as a response to the medical malpractice crisis in the 1970s, due in large part to significant jury awards and a decrease in investment returns for large insurance carriers. Decreased investment returns led to increased medical malpractice insurance premiums, and caused some malpractice insurance carriers to cease selling insurance policies to physicians and health care organizations.

For many years, the practices related to risk management for medical device manufacturing developed along with other manufacturing sectors, for example, heavy equipment and automotive, without special emphasis on health care. In the past few decades (late 1990s and 2000s), a heightened recognition of and emphasis on the application of risk management to medical devices outside of the insurance industry has occurred with the introduction of the United States Medical Device Amendments of 1976 and the continuing development of the Food and Drug Administration Center for Devices and Radiological Health (formerly known as the Bureau of Medical Devices and the Bureau of Radiological Health). One driving force was the Institute of Medicine (IOM) report on medical errors, with an emphasis on taking the steps necessary to reduce the occurrence of these errors (Kohn et al, 2000). Another driving force could be the implementation of *caps*, that is, limits of liability, by state governments for hospitals and health care providers, which limit their financial exposure to, for example, $250,000. Such caps make any other party to a lawsuit, e.g., surgeons and medical device manufacturers, the only remaining "deep pockets," i.e., those parties with substantial financial resources, from which the injured party can extract the potential multimillion dollar awards for costs of medical care, pain and suffering, and the loss of consortium.

Health Care Operational Risk Management Overview

Health care risk managers have various training and backgrounds, including health care administration, nursing, insurance, risk management, law, quality improvement, medical records, facilities management, and even (although a rare occurrence) clinical engineering. While there are universities and colleges offering bachelor's and master's degrees in risk management, none specialize in health care risk management at present. The majority of risk managers in the health care field have been trained in a discipline within health care, either clinical or administrative, from which they grew into the roles of risk managers. Certification in health care risk management through testing and experience is available from ASHRM, an individual membership group of the American Hospital Association. A broader, insurance-industry certification is available from the Insurance Institute of America, which offers the designation of Associate in Risk Management (ARM) after successfully completing three courses and accompanying examinations.

Risk Management Process

To apply risk management to the department of clinical engineering, one must understand the basic components of the risk management process. The process consists of five steps:

1. Identify and analyze exposures.
2. Consider alternative risk treatments techniques.
3. Select the best technique to manage and treat the risk.
4. Implement the selected technique.
5. Monitor and improve the risk management program risk management program.

Risk Management Program Structure

Risk management structures and operations may differ from health care organization to organization, even within the same community, city, state or region. While there are certain elements common to most risk management programs, specific functions vary according to the complexity of the organizations and interests of the governing body and chief administrative officer. As the responsibilities defined under various risk management program risk management programs differ, so will the role of the clinical engineer in risk management. All risk management programs are based on patient safety and the preservation of assets, and have risk identification and loss prevention as their fundamental components. ASHRM has suggested that there are six functional areas within a comprehensive risk management program:

- Risk identification and risk treatment through loss prevention and reduction
- Claims management
- Risk financing
- Regulatory and accreditation compliance
- Risk management departmental operations
- Bioethics

Risk Identification

Risk identification is the process through which the engineer becomes aware of the potential for a loss exposure. An exposure can include patient injury, loss of an asset, plaintiff verdicts and malpractice payments, loss of reputation, or governmental fines. Methods often implemented to identify potential and actual risk include incident reporting, generic occurrence screening, patient complaints, patient satisfaction surveys, accreditation survey reports, insurance carrier risk assessment reports, and state licensure surveys. Clinical engineering problem reports, unexpected or unacceptable equipment downtime, and MDR reports could also be used to identify potential or actual adverse events. All too often, risk managers are unaware of these engineering databases and do not integrate them into an organization's loss prevention reports.

Risk Treatment

Risk treatment refers to the options and choices available to handle a specific risk. Risk can be controlled internally through risk avoidance/prevention or risk reduction/minimization. Risk can be controlled financially through risk acceptance/retention or risk transfer. Risk avoidance is used when the risk is considered significant enough to avoid the risk by avoiding the action that would create exposure to it. For example, an organization in a rural setting may not be able to hire adequate staff for a neonatal intensive care unit. Therefore, the organization would only perform well-baby deliveries and transfer all high-risk deliveries.

Risk reduction is the most common risk management treatment. For example, an organization without a neonatal intensive care unit may have emergency equipment such as a radiant warmer, an on-call system for staff with basic neonatal resuscitation training, and a transfer agreement with an NICU in case of high-risk deliveries that an organization can not foresee or avoid.

Risk acceptance involves assuming the potential losses associated with a given risk, such as precipitous high-risk deliveries, and making plans to cover the potential financial costs of the risk. In order to cover the cost of a high-risk delivery with complications, an organization may decide to pay for insurance with low deductibles (typically a high-cost option as the insurance company essentially pays almost the first dollar on every loss) or assume high deductibles within the insurance program and pay for many of them out-of-pocket.

Risk transfer means recognizing that medical care can not be provided without assuming some risk. Therefore, and organization may insure for possible misadventures, which transfers the risk to the insurer, or assign responsibility to another service provider (e.g., independent clinic or surgicenter), which transfers the risk to the service provider or its insurer.

Evaluation and monitoring of the implemented risk-treatment technique selected is essential to determine the effectiveness of the choice. Evaluation of the treatment should be multidisciplinary whenever possible, as other disciplines bring varying insights into the success or failure associated with the chosen techniques. Benchmarks or thresholds should be established prior to the implementation of a technique so that measurement can be made against defined parameters.

Claims Management

Apart from the actual physical injury that an adverse event can cause to patients or staff, the risk to the organization may be a lawsuit. Claims management includes activities associated with managing potential and actual claims or lawsuits. Serious adverse events are often treated as potential litigation, and as such, information must be gathered and protected. Sometimes the protection of information may seem contrary to the clinical engineer's professional tendency to share information in order to optimize an investigation and minimize the likelihood of another similar injury. There is a balance that must be struck. Appropriate professionals both inside and outside the organization should discuss the issue of protection of information versus the common good. Protection may include sequestering medical devices thought to be involved in a serious incident, deciding whether a third-party independent investigation is necessary, interviewing involved staff members, and locating user manuals, policies, and procedures.

Risk Financing

Risk financing consists of activities to finance losses. This strategy includes maintaining and coordinating exposure data and values, for example, capital inventory, disposable values, replacement costs, and utility interruption costs. The most common method of financing the cost of risk is through the purchase of insurance. Organizations must make decisions regarding what to insure, the justification for the insurance, coverage limits, deductibles, and attachment points to ensure that all exposures are adequately covered. The legal liability of clinical engineers has been discussed in the chapter by Richards and Walter. Although insurance does not protect an engineer from being found guilty of faulty practice or negligence, it can provide financial assistance in the form of defense costs and payment of jury awards if the engineer is found guilty. The insurance company can eliminate the tribulations associated with a court case if the decision is made to settle a case with payment to the injured party before the case goes to court. Insurance polices rarely, if ever, will cover the cost of fines imposed by the court. The health care organization can purchase insurance for the professional actions of the engineer, product liability exposures, completed operations, errors, and omissions, and employee on-the-job injuries.

Contract Review

Contract review is integral in evaluating the adequacy of insurance, since contracts can contain both harmless and indemnifying language. In addition, an organization must ensure that outside contractors, such as installers, are adequately insured for work-related injures and the liability of their errors and omissions. While the risk manager may be aware of the need to finance these risks, the clinical engineer can participate in the technical oversight in order to be sure that outside contractors actually perform up to the expected standards.

Data Collection, Analysis, and Reporting

Risk managers must develop reports to demonstrate trends and indicators for the governing body and administration relative to adverse event reporting, loss prevention activities, and investigations. The risk management department may be responsible for developing policies and procedures related to patient safety activities such as informed consent, adverse event reporting and disclosure, product recall, and confidentiality. The department may develop and provide educational programs for these topics. In some hospitals, risk managers find themselves responsible for developing a program for management of adverse event exposures resulting from contracts, affiliation agreements, leases, purchase agreements, and warranties.

Governmental and Professional Organizations' Contributions to the Importance of Health Care Risk Management

The American College of Surgeons (ACS) was one of the first medical professional organizations to recognize the importance of a systematic approach for improving patient safety. In 1979, the ACS published the first edition of the patient safety manual as a response to the malpractice insurance crisis of the mid-1970s. The ACS created the patient safety system as a program that coordinated related functions of quality assurance, risk management, medical staff credentialing, and peer review. The goals of the patient safety system were to improve the quality of patient care, reduce preventable patient risks, and manage losses due to professional liability. The system's creators envisioned collaboration between hospital governance, management, and the medical staff. Using quality of care as a guiding ethic, they found four common interests involving patient safety (American College of Surgeons, 1985).

1. Minimize risks of patient injury.
2. Minimize financial losses due to malpractice awards and settlements.
3. Use hospital resources appropriately and cost-effectively.

4. Meet JCAH/JCAHO requirements. In 1985, what is now known as the Joint Commission on Accreditation of Healthcare Organizations (JCAHO) was the Joint Commission on the Accreditation of Hospitals (JCAH). See Chapter 121 for a description of the JCAHO.

Safety engineering was considered an important element in the risk management function of the patient safety system. The ACS believed there should be an organized safety engineering or clinical engineering program and a systems approach to the safe use of medical equipment that included the development of a relationship with quality assurance and risk management professionals.

The ACS recognized that although hospitals rely heavily on modern medical technology, the human factor was the key to patient safety. "The patient's safety depended not on a piece of equipment, but rather, on the health professional responsible for purchasing equipment, checking it prior to use, operating it properly, assuring correct settings, reporting problems to the engineer, and documenting preventive maintenance" (American College of Surgeons, 1985). The risk management aspects of a safety engineering department included incident reporting and follow-up for equipment-related occurrences, prepurchase review, preoperation testing, staff training, preventive maintenance, and documentation. In addition, the ACS recognized the need for an in-house medical device recall system to ensure that staff members were aware of confirmed equipment hazards related to equipment in use or considered for purchase. The system required monitoring of product recalls and service notices from manufacturers, government agencies, and independent reviewers, as well as reporting hazardous equipment to the Food and Drug Administration (FDA) and manufacturers. See Chapter 126 for more information on this subject. The inclusion of these elements brought clinical engineering into the mainstream of risk management. was The ACS's concern that a malfunctioning device could jeopardize the safety of future patients was a driving force for including product information in risk management plans.

Integral risk management elements of the clinical engineering program included the procedures for investigation following an equipment-related incident, the need for formal preventive maintenance procedures, and clinical staff training in the safe and effective use of medically-related equipment. The ACS recommended that the program elements be audited every 3-5 years to identify the strengths and weaknesses of the existing programs and to target priority areas needing improvement.

Another aspect of the patient safety system was the development of a response plan for equipment-related incidents. The ACS stated that equipment involved in an adverse event should not be returned to service without careful inspection and repair if warranted. They noted that the reason for the device failure may be an important defense in a malpractice claim, especially if the hospital seeks to transfer some or all of the liability to the equipment manufacturer or service vendor. The response to the incident required completion of the incident report form and prompt submission to the risk manager, tagging of the suspect device, and removal of the device from service. In coordination with the risk manager, the device should tested by the hospital's engineering department or a qualified outside party. The ACS emphasized that the device should not be tested by the manufacturer. Only after testing should the hospital decide whether the device should be repaired and returned to service, held for evidentiary purposes, or returned to the manufacturer. The defect found in the device should be documented and any similar devices in the facility should be examined. Hospitals should report any incidents to the FDA national Problem Reporting Program (termed the PRP and one predecessor to today's MedWatch program). Only then should the manufacturer be notified that the device was potentially involved in a patient injury and that the injury may result in a claim.

The National Patient Safety Foundation (NPSF), was founded in 1996 by the American Medical Association (AMA) as a collaborative initiative involving all members of the health care community. It was developed to stimulate leadership, foster awareness, and enhance the knowledge, dissemination, and implementation of patient safety protocols. The NPSF attracted attention in 1999 when the November release of the patient safety report by the Institute of Medicine (IOM) of the National Academy of Sciences put patient safety in the national spotlight. The IOM report defines safety as freedom from accidental injury. Distinguishing between medical accidents and errors was a significant step in framing accident investigations. The word *error* has a judgmental quality, while *accident* recognizes the complexity of the event. The study reported that preventable adverse events are a leading cause of death in the United States. The results of two studies imply that at least 44,000 and perhaps as many as 98,000 Americans die in hospitals each year as a result of medical errors when extrapolated to the approximately 33.6 million admissions to U.S. hospitals in 1997. Although the majority of these accidents are related to medication errors, one study cited in the IOM report found that 44% of preventable adverse events have been attributed to technical errors, which include equipment and systems failures (Kohn et al, 2000). The CDRH estimates that approximately one-third of the 80,000 device reports they receive each year may involve medical equipment "use error" (Campbell, 2001).

Following the dissemination of these reports, several agencies began studying the best methods to improve patient safety. In 2001, the Agency for Healthcare Research and Quality (AHRQ) released an Evidence Report/Technology Assessment, entitled *Making Health Care Safer; A Critical Analysis of Patient Safety Practices*, which reviewed evidence from 79 patient safety practices (AHRQ, 2001). The report listed 73 practices that could improve patient safety and described 11 practices that the researchers considered workable but are not performed routinely in the nation's hospitals and nursing homes. The following patient safety practices are relevant to clinical engineers:

- Preanesthesia checklists to improve patient safety
- Recognizing the impact of intraoperative monitoring on patient safety
- Using human factors to reducing device-related medical errors
- Refining the performance of medical device alarms (e.g., balancing sensitivity and specificity of alarms and ergonomic design)

In 2003, the National Quality Forum (NQF) released a consensus report entitled *Safe Practices for Better Health Care* (NQF, 2003). The report details 30 health care practices that the NQF considers appropriate in applicable clinical care settings to reduce the risk of harm to patients. The report focuses on practices that meet the following criteria:

- They have strong evidence that they are effective in reducing the likelihood of harming a patient.
- They are generalizable (i.e., they may be applied in multiple clinical care settings and/or multiple types of patients).
- They are likely to have a significant benefit to patient safety if fully implemented.
- They are useful for consumers, purchasers, health care providers, and researchers.

Practices that may be of importance to clinical engineers who work with electronic data and information management include:

1. Creating a health care culture of safety.
2. Preparing patient care summaries or other similar records from documented information, not from memory.
3. Ensuring that health care information, especially changes in orders and new diagnostic information, is transmitted in a timely and clearly understandable form to all of the patient's current health care providers.
4. Implementing a computerized prescriber order entry system.
5. Implementing a standardized protocol to prevent the mislabeling of radiographs.

The American College of Clinical Engineering (ACCE) emphasized the role of the clinical engineer in patient safety in its White Paper, *Enhancing Patient Safety; The Role of Clinical Engineering* (ACCE, 2001).

Regulatory and Accreditation Compliance as Risk Management Tools

Risk managers are often tasked with overseeing regulatory and accreditation compliance, since compliance is a fundamental risk management tool. If the organization meets regulations and standards, they are better able to control risk and demonstrate to a jury that they have met the standard of care. Clinical engineers may be responsible for compliance with the following:

- State and federally required reporting of certain types of adverse events and incidents
- Life Safety Codes
- Occupational Safety and Health Administration Act
- Needlestick Safety Prevention Act
- Patient Self-determination Act
- Safe Medical Devices Act
- Health Care Quality Improvement Act
- Emergency Medical Treatment and Active Labor Act
- Health Insurance Portability and Accountability Act

The Clinical Engineer's Role in Risk Management

Risk management is an art, not a science. While there have been attempts to measure risk and the effectiveness of risk treatments, such measurements are based on experiences, not hard and fast scientific equations. Management of equipment-related patient injury risk is a combination of equipment assessment, appropriate training, proper clinical technique, preventive maintenance, and failure analysis (Maley and Epstein, 1993). The selection of the safest equipment may be ineffective in controlling risk if preventive maintenance, software upgrades, and overhaul requirements do not take place. "Gradual deterioration without maintenance may bring the safety level below an acceptable level of manageable risk" (Gullikson, 2000). A number of clinical engineers (Gullikson, 2000; Ridgway, 2001; Dolan, 1999) have attempted to address proactive risk management through the identification of preventive maintenance risk factors as occurrences, such as the return of equipment for service within nine days. However, there is no consensus or hard, well-supported data upon which to assess and accept such factors. While some engineers may identify failure to pass a preventive maintenance procedure as an occurrence, there is no well-supported information to indicate that such a device is any greater risk to patients than a device that passed the preventive maintenance procedure. The use of algorithms to identify critical devices or risk factors for medical devices that constitute actual risk of injury may be the clinical engineering profession's best attempt to proactively address risk, but clinical engineers should recognize the limitations of the practice and not rely solely on such algorithms.

Clinical engineers can contribute to the risk management function, both within their respective departments and in the larger scope of facility-wide risk management. The primary methods all hospitals use to facilitate risk identification are incident (also known as adverse event) reporting, medical record review, and patient complaints. It is easy to visualize the role clinical engineering should play in the reporting and investigation of adverse events, since patient equipment is in use at the time of most adverse events.

Risk managers are often responsible for developing and maintaining collaborative relationships with other hospital departments related to the identification and treatment of risk. If the risk manager is unaware of the capabilities of the clinical engineering department with respect to problem identification and investigation (a limitation all too common in the authors' experience), the risk manager may not involve the clinical engineer in the process until he or she understands how a device can play a role in an adverse event. For example, a device may seem to operate properly in performance testing after an adverse event, but radiofrequency interference may have caused erratic operation of the device in the actual environment of use.

Other chapters in this handbook emphasize the importance of clinical engineering in the acquisition of new technologies (See Chapters 30-34). Due diligence, as a risk-management tool, should be completed for all new acquisitions of facilities and equipment. Many health care organizations are merging or purchasing complete facilities or existing services. Preacquisition risk analysis can help administrators make a more educated decision on whether or not to proceed with the acquisition. Such analysis can also help administrators determine terms that should be included to control risks and steps that may need to be taken after the acquisition to address risks. An interdisciplinary team including the clinical engineer can review the clinical, facility, and equipment issues on a preacquisition basis to offer information on ways to control risk that may contribute to a better acquisition opportunity from a financially or other perspective.

Proper risk management dictates basic requirements for all hospital departments. Policy and procedure manuals must reflect the actual operations of the department. Whether the policy and procedure manual is borrowed from another facility, purchased from a commercial vendor, or provided by the manufacturer in-house professionals must review the information to ensure that each policy or procedure accurately reflects the functions performed and is appropriate for the department and the facility. For example, does the document represent the standard of care as practiced at the facility? Can the staff meet the requirements of the policy or procedure?

Each policy should be identified with a policy number and the date of inception. As policies and procedures are removed from service or revised, a copy of the retired policy should be archived for a minimum of seven years. Each policy and procedure manual should be reviewed at least biannually and as needed. Policies and procedures can be updated and revised at any time. However, as old policies and procedures are retired or new policies and procedures are introduced, all manuals should be updated, and all affected staff should be apprised of the change. If there is a policy that affects more than one department, e.g., nursery equipment preventive maintenance schedules, all related departments should incorporate and understand the identical policy.

Although seemingly obvious, all policies and procedures should reflect the name of the health care organization using the document. If borrowed documents are used, the name of the originating facility should be removed. Obtain permission in writing from the originating facility to utilize a policy within another organization. If there are blank spaces included in a commercial policy or procedure, the blanks should be completed with the information required.

Policies and procedures should reflect the national state of practice. If there is an injury and a related policy and procedure does not reflect the current accepted standard of practice, a jury would be hard-pressed to understand why the health care organization could not or would not meet that standard.

User manuals should be present at the location in which equipment is used and clinical staff should be made aware of the availability and location of the manuals. The manuals should not be locked away for safekeeping.

Risk management dictates that staff be adequately trained in their job functions and understand their individual contributions to risk management and patient safety. In-service education, continuing education, and manufacturers training should be well-documented. Documentation should include names of attendees, course content, program sponsor, evidence of attendance, and date of attendance.

Medical device training is important for clinicians as well as for the clinical engineer. Clinicians' user errors and improper use of medical devices have been linked to multiple patient injuries. The FDA MedWatch and Manufacturer and User Device Experience (MAUDE) data indicate that user error and improper use are the two most often-cited reasons for patient injury. Since most of the entries in the FDA MedWatch and MAUDE are manufacturer-generated, it is not surprising if their reports may appear to shift responsibility and risk to the clinician and health care organization. It is important to note that plaintiff's legal counsel may use these reports to assert responsibility by the clinician and health care organization. Therefore, clinicians must have adequate training specific to the medical devices being used. Clinical competency checks are imperative, especially for high-tech equipment or where experience shows a need for special training. Refresher and remedial clinical training may be indicated.

Clinical engineers should be aware that manufacturers of medical devices are under increasing regulatory pressure to measure and address risk associated with their devices. While a discussion of medical device manufacturer risk management is beyond the scope of this chapter, clinical engineers should know that information about the risk character of a medical device may be available from the FDA or the manufacturer. Alerting the risk manager to the potential availability of such information may assist the risk manager in properly transferring risk and loss to a medical device manufacturer when appropriate.

Clinical engineers must be cognizant at all times of the potential discoverability of information by litigants and others. Routine documentation, memoranda, laboratory test notebooks, and investigation data may all be discoverable by litigation opponents. It is important to learn appropriate methods of writing and maintaining documents. While document control is not intended to alter the truth, it is intended, to the extent possible, to document only factual information and minimize unnecessary conjecture or opinion.

Statutory Risk Management Requirements

The Emergency Medical Treatment and Active Labor Act (42 USC 1395dd; 42 CFR 489, et al.) requires that anyone who presents at a hospital for what he or she believes is an emergency condition is entitled to a screening assessment and any needed stabilizing treatment. For the clinical engineer, this indicates the need to ensure that the equipment necessary to stabilize patients in an emergency is readily and always available and operational.

The Patient Self Determination Act of 1990 (42 USC 1395cc) requires that health care organizations inform patients about their rights regarding decisions about the course of their treatment, even if they are incapable of expressing those decisions at a later date. For the clinical engineer, this infers that inpatient health care organizations must be capable of preserving, terminating or removing artificial life support. Ventilators, feeding pumps, and other life-support equipment must be available and in working condition.

The Safe Medical Device Act of 1990 (21 USC 360; 21 CFR 803) requires that the FDA gather and health care organizations report information regarding the safety of medical devices. See Chapter 126. Sometimes, the clinical engineer is the primary operational department, reporting directly to the FDA. However, the risk management department may assume this role. Regardless of the reporting department, the information and responsibilities must be shared between the clinical engineer and the risk manager.

The Mammography Quality Standards Act of 1992 (42 USC 263; 21 CFR parts 16, 900, 1308, 1312, 42 CFR part 498) requires annual facility compliance inspections by an FDA-approved accrediting body, adherence to specified equipment testing and maintenance protocols, and implementation of quality control programs.

The Protection of Human Subjects Act (45 CFR part 46, 21 CFR parts 50, 56, 312, 812) addresses issues related to clinical trials of drugs and devices. Clinical engineers may become involved in the approval and clinical trial process any time a clinical trial of a device is being considered at the healthcare facility (See Chapter 87). Participation on the Institutional Review Board (IRB) on an ad hoc basis is recommended whenever a device is involved.

The Needlestick Safety Prevention Act (P.L. 106-4) requires that safety devices be used whenever possible. Clinical staff may resist changing their practices, even though the new device or technique is safer and does not affect clinical efficacy. Clinical engineers can identify clinicians in other facilities who are successfully implementing safety products and ensure that device trials are properly structured to examine clinical utility and safety without bias. See Chapter 61.

OSHA has many provisions that affect clinical engineering:

- Control of exposure energy to hazardous energy standard, commonly referred to as the Lockout/Tagout Standard, applies to any situation in which an unexpected energization of equipment or the release of stored energy could occur and injure an employee. The standard requires that equipment be locked out, i.e., rendered unusable unless a lock is removed during repair work. If a piece of equipment cannot be locked out, it must be tagged to warn other employees not to turn it on. OSHA requires a detailed educational program and plan to comply with this regulation.
- Regulation of mercury exposure may involve clinical engineers in evaluating the effectiveness of mercury-free devices such as sphygmomanometers, thermometers, and gauges.
- Regulation of exposure to methyl methacrylate, used in orthopedic surgery, should involve clinical engineers in the selection and maintenance of mixing hoods and laser exhaust and smoke evacuator systems (See Chapter 108).
- Additional chemicals with exposures governed by OSHA o include acetone, alcohol, benzene, toluene, and xylene.
- OSHA also governs exposure to airborne contaminants and tuberculosis. Clinical engineers may be involved in the design of special rooms and the selection of respiratory-protection equipment (See Chapter 114).
- OSHA does not directly regulate exposure to antineoplastic and chemotherapy drugs; however, it issued guidelines on these subjects in its technical manual in 1995. If these drugs require reconstitution before use, they should be mixed in a Class II Biological Safety Cabinet, preferably one that is vented to the outside of the building. HEPA filtration cabinetsalso may be used, but they are less efficient.
- Although OSHA does not regulate exposure to laser and electrocautery plumes, these plumes have been shown to carry intact, viral DNA, bacteria, chemical vapors and gases, mutagenic materials, carcinogens, cyanides, formaldehyde, and smoke particles (West, 2001).

Accreditation as a Risk Management Technique

Clinical engineers are usually familiar with the accreditation requirements that are specific to their department; they should also be familiar with organization-wide standards that may affect their operations. For example, JCAHO utility-management requirements may not be the responsibility of clinical engineers, yet the requirements may affect proper operation of equipment.

The JCAHO has included patient safety and clinical engineering-related standards since its first published standards in 1953. Acute care hospitals are not the only organizations accredited by the JCAHO. Accreditation is also available through the JCAHO for the following enterprises:

- Psychiatric, children's, critical access, and rehabilitation hospitals
- Home care organizations, including those that provide home infusion and durable medical equipment services
- Nursing homes and other long term care facilities
- Behavioral health care organizations
- Ambulatory care providers, outpatient surgery facilities, rehabilitation centers, infusion centers, group practices as well as office-based surgery
- Clinical laboratories, including blood-transfusion and donor centers.

The JCAHO is supported by the American Hospital Association, American Medical Association, American College of Physicians, American Society of Internal Medicine, American College of Surgeons, and American Dental Association. Nurses are have an at-large representative.

The JCAHO has integrated patient safety and elements of risk management into its performance improvement standards. The 2004 JCAHO risk management-related standards include the requirement for health care organizations to manage safety by identifying risks, and planning and implementing processes to minimize the likelihood of those risks causing incidents. The standards further require that the organization conduct proactive risk assessments that evaluate the potential adverse affects of equipment on the safety and health of patients. In addition, organizations must define and implement an ongoing, proactive program for identifying and reducing unanticipated adverse events and safety risks to patients. The commission believes that "such initiatives have the obvious advantage of preventing adverse events rather than simply reacting when they occur. This approach also avoids the barriers to understanding created by hindsight bias and the fear of disclosure, embarrassment, blame, and punishment that can happen after an event."

The performance improvement process is composed of the following activities:

- Collection of data to monitor performance
- Systematic aggregation and analysis of data, with analysis of undesirable patterns or trends in performance
- Implementation of processes for identifying and managing sentinel events
- Use of information from data analysis to make changes that improve performance and patient safety and reduce the risk of sentinel events

Organizational performance improvement requirements for the collection of patient safety data include gathering staff perceptions of risks to patients and suggestions for improving patient safety, as well as staff willingness to report unanticipated adverse events.

JCAHO outlines the following proactive performance improvement activities to reduce risks to patients:

1. Select a high-risk process (i.e., a process that if not planned and/or implemented correctly, has a significant potential for impacting the safety of the patient, to be analyzed).
2. Describe the chosen process (e.g., through the use of a flowchart).
3. Identify the ways in which the process could break down (i.e., the failure modes or fail to perform its desired function).
4. Identify the possible effects that a breakdown or failure of the process could have on patients and their potential severity.
5. Prioritize the potential process breakdowns or failures.
6. Determine why the prioritized breakdowns or failures could occur, this may include performing a hypothetical root cause analysis.
7. Redesign a risky process and/or underlying systems to minimize the risk patients.
8. Test and implement the redesigned process.
9. Monitor the effectiveness of the redesigned process (JCAHO, 2004).

The 2004 risk management standards also include the requirements that all equipment is maintained appropriately and that the organization ensures responses to product safety recalls. The two standards written specifically for the clinical engineering department are short, concise, and to the point: "The organization manages medical equipment risks," and "Medical equipment is maintained, tested, and inspected" (JCAHO, 2004). There are, however, many requirements to meet these two standards.

Management of equipment risks requires the following:

1. Development of a written management plan describing the processes to manage the effective, safe, and reliable operation of medical equipment.
2. Identification and implementation of a process(es) for selecting and acquiring medical equipment.
3. Establishment and use of risk criteria for identifying, evaluating, and creating an inventory of equipment for inclusion in the medical management plan before the equipment is used. Criteria to address include equipment function (diagnosis, care, treatment, and monitoring), physical risks associated with use, and equipment incident history.
4. Identification of appropriate strategies to achieve effective, safe, and reliable operation for all equipment on the inventory.
5. Defined intervals for inspecting, testing, and maintaining appropriate equipment on the inventory (i.e., those medical devices on the inventory benefiting from scheduled activities to minimize the clinical and physical risks) that are based upon criteria such as manufacturers' recommendations, risk levels, and current organizational experience.
6. Identification and implementation of processes for monitoring and acting on equipment hazard notices and recalls.
7. Identification and implementation of processes for monitoring and reporting incidents in which a medical device is suspected in or contributed to the death, serious injury, or serious illness of any individual, as required by the Safe Medical Devices Act of 1990.
8. Identification and implementation of processes for emergency procedures that address:
   (a) What to do in the event of equipment disruption or failure
   (b) When and how to perform emergency clinical interventions when medical equipment fails
   (c) Availability of backup equipment
   (d) How to obtain repair services

Maintenance, testing, and inspection of medical equipment requires documentation of the following:

1. Current, accurate, and separate inventory of all equipment identified in the medical equipment management plan, regardless of ownership.
2. Performance and safety testing of all equipment identified in the medical management program before initial use.
3. Maintenance of equipment used for life support that is consistent with maintenance strategies to minimize clinical and physical risks identified in the equipment management plan.
4. Maintenance of non–life-support equipment on the inventory that is consistent with maintenance strategies to minimize clinical and physical risks identified in the equipment management plan.
5. Performance testing of all sterilizers used.
6. Chemical and biological testing of water used in renal dialysis and other applicable tests based upon regulations, manufacturers' recommendations, and organization experience.

Review of these standards demonstrates the importance of clinical engineering processes in improving patient safety and reducing the likelihood of patient injury resulting from a medical device.

JCAHO introduced the concept of sentinel event reporting in 1998. The sentinel event requirements of the JCAHO are found in the performance improvement standards. Accredited organizations are required to investigate sentinel events, which are to be analyzed from a systems perspective by the health care organization. "A sentinel event is an unexpected occurrence involving death or serious physical or psychological injury, or the risk thereof. Serious injury specifically includes loss of limb or function. The phrase, 'or the risk thereof' includes any process variation for which a recurrence would carry a significant chance of a serious adverse outcome." The sentinel event is then subject to a root cause analysis.

"Root cause analysis is a process for identifying the basic or causal factors that underlie variation in performance, including the occurrence or possible occurrence of a sentinel event. A root cause analysis focuses primarily on systems and processes, not individual performance. It progresses from special causes in clinical processes to common causes in organizational processes, and identifies potential improvements in processes or systems that would tend to decrease the likelihood of such events in the future, or determines, after analysis that no such improvement opportunities exist." JCAHO requires an action plan that identifies the strategies the organization intends to implement to reduce the risk of similar events occurring in the future. (See http://www.jcaho.org/accredited+organizations/hospitals/sentinel+events/se_pp.htm)

In addition to the self-investigation requirement, the JCAHO has a sentinel event advisory group that reviews data and decides whether to issue an alert and develop a related patient safety goal. If an alert or patient safety goal is developed and published, all accredited facilities are expected to review the information and take the necessary action to improve patient safety.

Of the 25 JCAHO safety reports available, seven have a direct impact on the activities of clinical engineers:

1. *Preventing Surgical Fires*, June 24, 2003
2. *Bedrail-related Entrapment Deaths*, September 6, 2002
3. *Preventing Ventilator-related Deaths and Injuries*, February 26, 2002
4. *Preventing Needlestick and Sharps Injuries*, August 2001
5. *Medical Gas Mix-ups*, July 2001
6. *Fires in the Home Care Setting*, March 20, 2001
7. *Infusion Pumps: Preventing Future Adverse Events*, November 30, 2000

In addition to the accreditation requirements, the JCAHO has begun setting national patient safety goals. Two of the seven 2004 JCAHO National Patient Safety Goals are relevant to clinical engineering:

1. "Improve the safety of using infusion pumps by ensuring free-flow protection on all general use and patient controlled analgesia intravenous infusion pumps."
2. "Improve the effectiveness of clinical alarm systems by implementing regular preventive maintenance and testing of alarm systems. Assurance that alarms are activated with appropriate settings and are sufficiently audible with respect to distances and competing noise within the unit."

Although JCAHO is the primary accrediting body for health care organizations, there are additional accrediting bodies that address patient safety and medical device safety. Other significant accrediting organizations include:

- The Accreditation Association of Ambulatory Health Care accredits all types of ambulatory care providers, including office based surgical practices, urgent care centers, and lithotripsy centers.
- The Commission on Accreditation of Rehabilitation Facilities accredits rehabilitation facilities, adult day care services, assisted living facilities, behavioral health organizations, community service organizations, and employment training and service organizations.
- The Community Health Accreditation Program-accredits community nursing centers, community rehabilitation centers, home care services, home infusion therapy programs, and home medical equipment providers.

Summary

Risk managers are, by nature, a skeptical group, assuming the worst. That is, if something can go wrong from a design, systems, staffing, training, or misuse perspective, it will. There are many activities in which clinical engineers participate that can help the risk manager sleep at night, if these activities are managed and communicated to the risk manager consistently. Clinical engineering risk issues that have come to the forefront of health care risk managers in recent years include the following:

- Device recalls
- Devicetracking requirements
- Device-incident investigations using root cause analysis
- Device procurement using Failure Modes and Effects Analysis (FMEA)
- Clinical department technology risks, such as:

   1. Critical alarms
   2. Neonatology
   3. Lasers
   4. Hemodialysis
   5. Anesthesia

- Reuse of devices
- Alternate parts and supplies
- EMI interference
- Disposal of used equipment
- In-house repair and device modifications

- Off-label, off-shelf, and unapproved uses of devices
- Patient- or physician-owned equipment
- Borrowing and lending equipment

Other areas where clinical engineers can contribute to risk management include lasers and anesthesia and other gases. The clinical engineer can make significant contributions to the selection, safety procedures in use, and continuing safe operation of surgical and ophthalmologic lasers. While some lasers may be handled by laser technicians or the manufacturer, it is important for a trained professional within the organization to have a close working knowledge of the devices in case an adverse event may be related to them.

In addition, problems in administrating the proper gas and in eliminating known waste gases have resulted in significant risk and losses to health care organizations. In some organizations, the responsibility for such issues rests with facilities engineering. In other organizations, partial or complete responsibility for medical gases and their waste elimination rests with the clinical engineer. Initial and routine testing of medical gases and the effectiveness of their elimination methods are recommended. Testing gas elimination methods following any work performed on the systems can help control risk.

If clinical engineers acknowledge these significant risks and communicate with the risk manager to identify the policies and procedures for minimizing risk in these areas, the risk manager will sleep better at night and patients will be safer.

Risk management is every employee's responsibility. Each employee performs risk management tasks every day, although they may not often be aware of this important function, nor may they given credit for their contributions to patient safety. From a clinical engineering perspective, identification of potential or actual problems with equipment, adherence to regulations, and reporting and investigation of incidents are all risk management functions.

References

Agency for Health Care Research and Quality. *Making Health Care Safer; A Critical Analysis of Patient Safety Practices.* Evidence Report/Technology Assessment, No. 43. July 2001. http://www.ahrq.gov/clinic/ptsafety/

American College of Clinical Engineering. *Enhancing Patient Safety; The Role of Clinical Engineering.* White Paper. Plymouth Meeting, PA, American College of Clinical Engineering, 2001.

American College of Surgeons. *Patient Safety Manual.* ed 2, Bader & Associates, Inc. 1985.

American Society of Health Care Risk Management. *Barton Certificate in Health Care Risk management Program.* American Society for Health Care Risk Management of the American Hospital Association., 2003.

Campbell S. Exploring Ways to Reduce Errors. *Biomed Instrum Technol* 243-248, July/August 2001.

Carroll R (ed). *Risk Management Handbook for Health Care Organizations.* ed 4. Jossey Bass, 2003.

Dolan A. Risk Management and Medical Devices. *Biomed Instrum Technol* 33:331-333, 1999.

Epstein AL, Harding GH. Risk Management in Selected High-Risk Hospital Departments. In Kavaler F, Spiegel A (eds). *Risk Management in Health Care Institutions.* Boston, Jones Bartlett Publishers, p 325-363, 2003.

Gullikson ML. Risk Factors, Safety, and Management of Medical Equipment. In Bronzino JD (ed). *The Biomedical Engineering Handbook.* ed 2, Vol 2, Sec 17, Clinical Engineering, Subsec 169. CRC Press, LLC, 169-1–15, 2000.

Joint Commission on the Accreditation of Health Care Organizations. *Hospital Environment of Care Standards.* 2004. http://www.jcaho.org/accredited+organizations/hospitals/standards/new+standards/ec_hap.pdf

Kavaler F and Spiegel AD. *Risk Management in Health Care Institutions: A Strategic Approach.* Ed 2., Jones and Barlett Publishers, 2003.

Kohn LT, Corrigan JM, Donaldson MS (eds). *To Err is Human: Building a Safer Health System.* Institute of Medicine, National Academy Press, 2000.

Maley RA, Epstein AL. *High Technology in Health Care: Risk Management Perspectives.* Chicago, American Hospital Publishing, Inc., 1993.

Ridgway M. Classifying Medical Devices According to their Maintenance Sensitivity: A Practical, Risk-Based Approach to PM Program Management. *Biomedical Instrum Technol,* p 167-176, May/June 2001.

Snow A. Integrating Risk Management into the Design and Development Process. *MD & DL,* p 99-113, March 2001.

NQF. The National Quality Forum. *Consensus Report: Safe Practices for Better Health Care.* 2003.

Treasury Board of Canada Secretariat. Best Practices in Risk Management: Private and Public Sectors Internationally. Part 2. April 27, 1999.

West JC. Occupational and Environmental Exposures for Health Care Facilities. In Carroll R (ed). *Risk Management Handbook for Health Care Facilities.* ed 3. American Society for Health Care Risk Management. Jossey-Bass, 2001.

Further Information

The following is a list of government and professional organizations that include patient safety issues of interest to clinical engineers. This list is not all-inclusive nor does inclusion in this list connote endorsement by the authors of this chapter.

Agency for Health Care Research & Quality (AHRQ). www.ahrq.gov/qual/errorsix.htm
This site shows AHRQ's involvement with patient safety through links to press releases, documents, speeches, statements, and hearings.

Morbidity and Mortality Rounds on the Web: This web-based patient safety resource and journal features monthly expert analysis of five cases of medical errors in the areas of medicine, surgery/anesthesia, obstetrics-gynecology, pediatrics and "other." The site also links to interactive learning modules, forums for online discussion, and resources links.

American Academy of Orthopaedic Surgeons (AAOS). http://orthoinfo.aaos.org/
Association site contains information on orthopaedic conditions and treatments, injury prevention, and wellness and exercise, along with fact sheets on patient safety surgery issues.

American Association of Nurse Anesthetists (AANA). www.anesthesiapatientsafety.com
Patient safety site focuses on general nurse-anesthetist information and includes articles relating to patient communication, conscious sedation, and standards for anesthesia safety.

American Hospital Association (AHA). http://www.aha.org/patientsafety/safe_home.asp
Includes information on AHA's initiatives, and focuses on medication safety and safety culture. Resources include links to successful practices, reporting information, the Strategies for Leadership series in PDF format, and links to major reports.

American Medical Association (AMA). NPSF's Web site (www.npsf.org)

Includes information on patient safety and AMA guidelines.

American Nurses Association (ANA). Occupational Safety: www.ana.org/dlwh/osh
Focuses specifically on occupational health, including articles on needles sticks and links to OSHA and similar organizations.

Anesthesia Patient Safety Foundation (APSF). www.apsf.org
Devoted to patient safety, this site provides a video library, newsletters, research grants, links to other web sites, questions and answers, and a protocol for adverse events.

Association for the Advancement of Medical Instrumentation. www.aami.org
Dedicated to increasing the understanding, safety, and efficacy of medical instrumentation, this site includes industry standards, forums, and newsletters.

Association of Health care Risk Management (ASHRM). www.ashrm.org

Devoted to health care risk management, links to related sites, forums, newsletter
Association of Perioperative Registered Nurses (AORN). www.patientsafetyfirst.org
Information for nurses; patient section of this site provides materials on safety before, during, and after surgery.

Association of Professionals in Infection Control and Epidemiology (APIC). www.apic.org/safety This site includes information on a patient safety training course titled Patient Safety: Tools for Implementing an Effective Program, an overview resource entitled the *Patient Safety Backgrounder*, and a tool entitled *Effective Strategies for Improving Patient Safety*.

American Society of Anesthesiologists (ASA). www.asahq.org
This site links to several potential resources in anesthesia safety (including videos).
Department of Defense (DOD). www.afip.org/PSC (Patient Safety Workgroup)
This site contains information on the center for patient safety in the DOD, which is geared toward identifying systemic errors in the military and includes patient safety newsletters..

Emergency Care Research Institute (ECRI). www.mdsr.ecri.org (Medical Device Safety Reports) ECRI provides a range of tools and resources related to patient safety and health care quality, particularly focusing on devices and technology.

Food and Drug Administration (FDA). www.fda.gov/cder/drug/MedErrors/default.htm (Center for Drug Evaluation & Research) Associated with medical device errors, directions for reporting an error (www.fda.gov/medwatch/), and federal guidelines.

Institute of Medicine (IOM). http://www.iom.edu/IOM/IOMHome.nsf/Pages/Quality+ Initiative Includes current and past reports and studies related to quality of care and patient safety.

Joint Commission on Accreditation of Healthcare Organizations (JCAHO). www.jcaho.org
Information regarding standards, standards development, sentinel events, and legislative issues if available on this site.

Leapfrog Group. www.leapfroggroup.org Includes information on the Leapfrog Group's approach and information and fact sheets on three initiatives, including computer physician order entry, evidence-based hospital referral, and ICU physician staffing.

National Quality Forum (NQF). www.qualityforum.org Includes information regarding NQF's on-going projects and reports in patient safety, as well as links to completed reports.

Patient Safety Institute (PSI). www.ptsafety.org Includes information on this organization, which strives to work with hospitals, providers and patients to reduce medical error through the use of technology and the building of relationships.

Premier, Inc. http://www.premierinc.com/safety/ Includes resources for patients and health care professionals; Safety Share electronic newsletter is archived on website and an online store offers many free safety resources.

Risk Management Foundation (RMF). www.rmf.harvard.edu./patientsafety (Center for Patient Safety) Includes information on risk management, patient safety, and other related resources.

The Institute of Electrical and Electronics Engineers. www.ieeeusa.org Includes standards and provides a forum for medical device engineers.

Veteran's Administration (VA). www.patientsafety.gov (National Center for Patient Safety) Includes resources in patient safety, information on culture of safety, root cause analysis, a glossary of terms, a discussion room, and a library.

57

Patient Safety Best Practices Model

Paul Vegoda
Manhasset, NY

Carl Abramson
Malvern Group
Malvern, PA

Patient safety has become the watchword of patient care in the first years of the 21st century. This is the result of the often-quoted report of the Institute of Medicine (IOM) on patient injuries (Kohn et al., 2000). According to another report (AHRQ, 2001), 700,000 patients are victims of medical errors every year.

Patient safety does not just happen. It takes careful analysis, planning, and effort. (See Chapter 55.) Improvement requires will because durable improvement is not an accident; it takes effort (Berwick, 2003).

Many health care organizations depend on analyzing reported incidents to determine what caused the incident and how to prevent that type of incident from reoccurring. (See Chapter 64.) Correcting reported problems, which have occurred at the patient's expense, is considered an acceptable approach to decreasing incidents. This retrospective approach presumes that a strong patient safety foundation is in place. It also assumes that corrective steps for making the equivalent of last minute adjustments to a close-to-perfect system results in an environment with acceptable risks.

A better approach however is to complete a proactive review of potential problem processes and then implement redesigned processes to prevent reportable incidents from occurring (Dolan, 2003).

Silver Bullet

The primary emphasis in health care has been to apply technology to the problem of patient safety. The authors believe that technologies such as bar coding, computerized physician order entry (CPOE), and clinical data repositories are excellent tools to support patient safety initiatives and also improve the quality of patient care, but they are not the silver bullets that will eliminate the medication errors that lead to adverse drug events (ADEs).

The causes of medication errors can be attributed to the following (Ten Key Elements of Medication Safety, http://www.ismp.org):

- Critical patient information missing
- Critical drug information missing
- Miscommunication of drug order
- Drug name, label, or, packaging problem
- Drug storage or delivery problem
- Drug delivery device problem
- Environmental, staffing, or workflow problem
- Lack of staff education
- Lack of patient education
- Lack of quality control or independent check systems

Although miscommunication of a drug order can be virtually eliminated by CPOE, robust clinical information systems must be tightly integrated with the CPOE use to reduce the occurrence of missing patient and drug information that is critical in determining an appropriate therapy (Bates and Gawande, 2003).

The first three causes of medication errors listed previously can be reduced by the use of CPOE and associated clinical information.

Automating order entry provides only a partial solution to reducing medication errors. Automation of a poorly designed process compounds the problem.

Process Analysis

The premise that technology alone is sufficient to eliminate medication errors must be replaced by the plans to conduct an analysis of existing processes and redefinition of workflow prior to the implementation of new technologies.

The analysis of existing processes will reveal the extent to which technology can support the reduction of medication errors and expose the need for critical remediation in areas not impacted by technology.

Modeling both the existing processes and published standards for best practice to determine potential high-risk processes (existing processes that can cause harm to a patient), allows the development of new processes to replace the high-risk processes. At this point, the new processes can take advantage of the technology tools available to support them.

As is evident from the Institute for Safe Medication Practices (ISMP) error categorization list previously described, errors can include problems in practice, products, procedures, and systems. A process is a combination of practice, products, procedures, and systems that completes a purpose (e.g., delivers a medication dose).

Patient safety is impacted by each of these hospital domains:

- Inpatient
- ICU
- ED
- OR
- Outpatient
- Ancillary services
- Administration

In each of these domains, individuals provide information, make decisions, and take actions to carry out medication-administration processes. These individuals include the following:

- Patients (provide history)
- Attending physicians (determine therapies)
- House staff (monitor patient progress)
- Nurses (administer medication doses)
- Pharmacists (check medication order for appropriate medication)
- Ancillary technicians (record test results)

Given the interaction of many elements, the processes that ultimately affecting the medication of a patient must be modeled and analyzed systematically for potential risk. Special attention must be given to variations in processes throughout different domains.

The methodology described by the authors is based on process-modeling techniques that were developed for Health Insurance Portability and Accountability Act (HIPAA) assessments, which provided significant findings about process problems far beyond the scope of HIPAA requirements. (See Chapter 104.) Those findings are discussed below.

Methodology

The methodology developed is based on three fundamental premises:

- An organization must know how its processes work..
- Variability in processes must be reduced..
- Repeatable processes must be implemented.

While these premises may seem fundamental, they are not necessarily practiced by many organizations. The methodology creates a baseline model of an organization's existing processes, which can be measured against the Patient Safety Best-Practices Model developed by the authors.

The Patient Safety Best-Practices Model was developed based on Joint Commission on Accreditation of Healthcare Organizations (JCAHO) patient safety goals and accreditation requirements, National Coordinating Council for Medication Error (NCCMERP) and American Society of Health System Pharmacists (ASHP) guidelines, ISMP, and others. These guidelines and requirements were combined to provide a best practices model that will provide the basis for the assessment of current practices.

The Patient Safety Best-Practices Model is an evolving model, which is updated as new standards, and requirements are published. The model represents the best thinking of the analysts, physicians, nurses, and pharmacists who are working with the authors.

The initial step in the methodology is data collection. Data collection is a multistep process that involves two distinct sub-processes; direct interviews of nurses, pharmacists, and physicians and an anonymous staff survey on patient safety. The use of an anonymous survey is critical to the success of the patient safety data gathering, since some punitive health care cultures tend to discourage the reporting of medication errors. While these unreported errors might not necessarily lead to an ADE in any particular instance, there could be a fatality if the processes that lead to the particular practice are left unchanged. In the experience of the authors, some health care organizations report extremely low medication error rates have a low rate of internal reporting due to a culture that punishes honesty and discourages the reporting of errors to management.

Concurrent with the data collected by the interviews, models of the processes described by the interviewees are built using ProcessMapper©, a proprietary model-building software system. Figures 57-1 and 57-2 show examples of ProcessMapper© output in both text format (Figure 57-1) and graphical representation of the process (Figure 57-2).

Concurrent with the development of the organization's current process model is a review of current policies, procedures, and forms. This review is necessary to determine whether current policies meet best-practice and accreditation requirements.

If policies meet requirements, a review of how the processes are implemented is critical since implementation of processes may differ from policy requirements.

Patient safety committee minutes and incident-reporting logs are studied to determine the types of incidents that were reported in the past and the processes that were developed to prevent reoccurrences of the incident. The anonymous survey will fill in the gaps in the patient safety committee minutes and the incident-reporting logs.

After the data are gathered and the current process models are built, an analysis of the gaps between the Patient Safety Best-Practices model and the current process model is completed. This analysis determines potential risks from the continuation of current practice and provides a series of recommendations, which will bring the current processes into compliance with the recommended best practices.

A report is prepared which lists the recommended best practices, the current processes in place, the risk, if any, associated with current processes, and recommendations for compliance.

Patient Safety Best-Practices Model

4.	Improve Medication-Administration Process
4.1	Establish policy for identifying and managing sentinel events
4.2	Establish policy for internal reporting
4.3	Establish policy for medication-administration process
4.3.1	Establish policy for enforcement of medication-administration standards
4.3.2	Establish policy for staff education about medication administration
4.3.3	Establish policy for labeling medications dispensed to patients
4.3.3.1	Ensure that medication-labeling policy promotes safe use of medication
4.3.4	Establish policies for preparation of drugs

Figure 57-1 Sample of patient safety best-practices model text.

The methodology can be summarized as follows:

- Build a model of current medication-administration processes.
- Review any process changes implemented due to past events.
- Compare the model of current processes to the Patient Safety Best-Practices model.
- Assess the gaps between current processes and the recommended standards.
- Develop a plan to close gaps.

Results of Process Modeling

The process-modeling methodology described above was originally developed to provide a cost-effective approach to assess the impact of the HIPAA regulations on the operations of health care providers (Vegoda and Abramson, 2003). Process models of the HIPAA Privacy

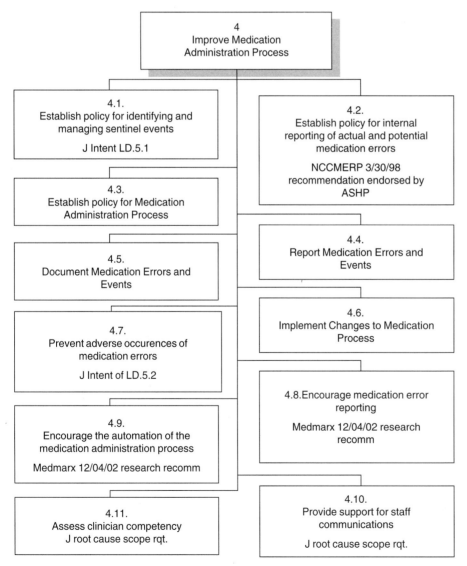

Figure 57-2 Graphic representation of patient safety best practices model.

and Security regulations were built. (See Chapter 104.) The use of the ProcessMapper© system facilitated the update of the regulations models as the original requirements changed.

Staff members were interviewed so that a model of the processes that were currently in place could be built. After the staff members who had been interviewed confirmed the correctness of the model, it was assessed against the privacy and security models and the gaps relative to HIPAA compliance were determined and documented. Compliance recommendations were developed based on the gaps. These recommendations suggested changes to current processes or the implementation of new processes required to comply with the regulations.

Policy models were built based on HIPAA requirements since HIPAA compliance requires that every process be documented with policies and procedures. The current policies were compared to the models to determine policy gaps. Modifications to existing policies and implementation of new policies were recommended as the result of the comparison.

Concurrent with the development of the process models, the ProcessMapper© provides for the documentation of observations and recommendations.

Numerous non-HIPAA related observations and recommendations resulted from the process modeling methodology. In a number of instances, management had no idea of the processes that were being followed by the staff until they reviewed the process models.

Time Bombs

Some examples of problems that were uncovered during proactive process data collection for process-model building are listed here.

A pharmacist prepared the morning delivery medication carts in the evening before leaving the hospital. The next morning, the pharmacist added any orders to the cart that came in during the night. The pharmacist delivered the carts to the nursing units by the required time of 8:00 a.m. As part of the model building interview, the pharmacist was asked if the orders that came in during the night were checked for drug–drug interactions or drug allergies. The reply was that there was no time to do the checks if the carts were to arrive on time. The checks could be completed post-administration.

A review of the hospital's policies showed that there was a requirement for the pharmacy to complete drug–drug interaction and drug allergy checks prior to dispensing any medication. Although the policy requirement was in place, the process being performed by the pharmacist was not the process that management believed was in use. The use of proactive process modeling unveiled this very high-risk process.

A second example of how process modeling can uncover potential high-risk processes involved the modeling of nursing practice. Many hospitals have implemented 12-hour nursing shifts for 3 days. During the modeling process, it was determined that the hospital being assessed had its nurses work three 12-hour shifts while the nursing supervisors worked five 8-hour shifts so that there was continuity of supervision over the work week.

As a result of the difference of shift hours, two different supervisors supervised a nurse during the nursing shift. Building the process model led to the discovery that the different shift supervisors required different processes for a given practice. This caused confusion among the nursing staff and a huge variability of practice. One supervisor followed approved hospital policies and procedures, while the other established processes that were not known to the hospital clinical management.

While either of the processes modeled might accomplish the required outcome, the risk lay in the variability demanded by the nursing supervisors.

These examples demonstrate that proactive analysis of processes can reveal hidden features of the medication administration process that may harm patients. Also, a proactive analysis of the medication administration process comparing actual processes to accreditation requirements and best-practice recommendations can go a long way toward identifying the process gaps that can cause harm.

References

Agency for Health Care Research and Quality (AHRQ). Reducing and Preventing Adverse Drug Events to Decrease Hospital Costs. AHRQ, March 2001.
Bates MD, Gawande MD. Improving Safety with Information Technology. *NEJM* 348:2526-2534, 2003.
Berwick MD. Errors Today and Errors Tomorrow. *NEJM* 348:2570-2572, 2003.
ISMP Classification System. *Ten Key Elements of Medication Safety*. http://www.ismp.org.
Kohn LT, Corrigan JM, Donaldson MS. *Err is Human: Building a Safer Health System*. National Academy Press, Washington, DC, 2000.
Vegoda P, Abramson C. An Integrated Business Approach to Process Improvement and HIPAA Compliance. *JHIM* 17:59-62, 2003.

58 Hospital Safety Programs

Matthew F. Baretich
President, Baretich Engineering, Inc.
Fort Collins, Co

Safety and effectiveness have long been watchwords for clinical engineers. Medical devices should do what the health care practitioner wants them to do (effectiveness) and not do what the practitioner does not want them to do (safety). These are the two sides of the coin of clinical engineering. This chapter looks at the safety side of the coin. However, it does not focus directly on medical device safety—those topics are thoroughly addressed in other articles in this handbook. Instead, it describes the broader context of safety programs throughout the hospital. To be most effective, the safety-related aspects of clinical engineering practice must be integrated into these hospital-wide safety programs.

JCAHO Environment of Care Standards

The *Comprehensive Accreditation Manual for Hospitals* (CAMH) published by the Joint Commission on Accreditation of Healthcare Organizations (JCAHO) includes a chapter entitled, "Management of the Environment of Care." The standards contained in this chapter require hospitals to develop management plans in seven areas:

- Safety management (Standard EC.1.1)
- Security management (Standard EC.1.2)
- Hazardous materials and waste management (Standard EC.1.3)
- Emergency management (Standard EC.1.4)
- Life safety management (Standard EC.1.5)
- Medical equipment management (Standard EC.1.6)
- Utility systems management (Standard EC.1.7)

Each of the standards cited has associated standards that provide details regarding implementation of the management plans. For example, Standard EC.1.6 requires hospitals to develop a medical equipment management plan. Standard EC.2.6 requires hospitals to implement the medical equipment management plan. Standard EC.2.10.3 requires medical equipment used in hospitals to be maintained, tested, and inspected. Table 1 illustrates the relationships among the JCAHO Environment of Care standards.

Also illustrated in Table 58-1 are three other components of the JCAHO Environment of Care standards:

- Orientation and education (Standard EC.2.8)
- Other environmental considerations (Standards EC.3.1 through EC.3.4)
- Measuring outcomes of implementation (Standards EC.4.1 through EC.4.3)

The Safety Committee

Each JCAHO Environment of Care standard includes a set of intent statements that describe critical aspects of the standard. For example, Standard EC.1.1 requires hospitals to establish a safety-management Plan that includes specific components listed in the intent statements. One of the intent statements in Standard EC.1.1 requires a hospital's safety management plan to identify processes for examining safety issues by appropriate hospital representatives. Details regarding the role of this group, which has traditionally been referred to as the safety committee but is now often called the environment of care committee, are included in Standard EC.4.2.

Table 58-1 JCAHO Environment of Care Standards

	Introduction (not scored)	Safety	Security	Hazardous and waste materials	Emergency management	Fire prevention	Medical equipment	Utility systems	Orientation and education
Planning	EC.1	EC.1.1. EC.1.1.1 EC.1.1.2	EC.1.2	EC.1.3	EC.1.4	EC.1.5 EC.1.5.1	EC.1.6	EC.1.7 EC.1.7.1	
Implementation – general	EC.2	EC.2.1	EC.2.2	EC.2.3	EC.2.4	EC.2.5	EC.2.6	EC.2.7	EC.2.8
Implementation – drills	EC.2.9				EC.2.9.1	EC.2.9.2			
Implementation – maintenance	EC.2.10	EC.2.10.1				EC.2.10.2	EC.2.10.3	EC.2.10.4 EC.2.10.4.1	

	Introduction (not scored)	Application
Other environmental considerations	EC.3	EC.3.1 EC.3.2 EC.3.2.1 EC.3.3 EC.3.4
Measuring outcomes of implementation	EC.4	EC.4.1 EC.4.2 EC.4.3

The JCAHO refers to the safety committee as a multidisciplinary improvement team. The activities of the safety committee, as described in Standard EC.4.2, include the following:

- Analyze of safety issues in a timely manner
- Develop and approve of recommendations
- Communicate of safety issues to the leaders of the hospital, to individuals responsible for performance-improvement activities, and, when appropriate, to relevant components of the hospital-wide patient safety program
- Communicate of recommendations for one or more performance improvement activity to the hospital's leaders at least annually

Typically, the clinical engineering department of the hospital is represented on the safety committee. It is appropriate, but less common, for the clinical engineering representative to serve as chairperson of the committee.

The safety committee serves as a forum for reports from the seven environment of care areas. For example, the safety committee will receive reports as specified in the hospital's Medical Equipment Management Plan, including performance monitoring data and incident investigation reports. Serving as a reporting forum is an essential role of the safety committee. However, the great value of the safety committee is found in its ability to subject these reports to multidisciplinary analysis. Careful analysis from a range of perspectives will produce communications, recommendations, and performance-improvement activities that enhance safety throughout the hospital.

The Safety Officer

The JCAHO Standard EC.1.1 requires a hospital's safety management plan to designate "a qualified individual(s) to oversee development, implementation, and monitoring of safety management." To carry out these responsibilities, most hospitals formally designate an individual to serve as safety officer. Standard EC.1.1 also mandates an "individual(s) to intervene whenever conditions pose an immediate threat to life or health or threaten damage to equipment or buildings." This authority is typically granted at least one member of the hospital staff, including the safety officer.

In many hospitals, the safety officer serves as chairperson of the safety committee. However, it is also common to separate these functions, so the safety officer serves as a member of the safety committee but not as its chairperson. In smaller organizations, the safety officer may be given responsibility for not only the safety management plan but also other environment management plans such as those for life safety management, management of hazardous materials, and waste management. In a relatively small number of hospitals, clinical engineers serve as safety officers. This represents an appropriate opportunity for career development for hospital-based clinical engineers.

A critical aspect of the safety officer's role is the collection and reporting of information regarding the hospital's environment of care. Details regarding these activities are included in Standard EC.4.1. It is important to note that JCAHO expects the safety officer to review information from all environment of care areas. For example, the standard states that the safety officer "reviews summaries of deficiencies, problems, failures, and user errors related to . . . managing medical equipment." Thus, the safety officer is in a position to coordinate information across the entire range of JCAHO environment of care issues.

Environment of Care Management Plans

Management plans serve as executive summaries of how a hospital meets the intent statements for each environment of are standard. Each management plan should include a statement of objectives. As a hospital defines the objectives for a particular management plan, it is important to be consistent with the objectives that JCAHO incorporates into its environment of {See hyphenation query} care standards. For example, Standard EC.1.6 states that the medical equipment management plan should "promote the safe and effective use of equipment." Thus, the medical equipment management plan is *not* a description of how the clinical engineering department functions; it is a description of how the hospital meets the intent statements under Standard EC.1.6.

To continue this example, the standard requires the management plan to establish emergency procedures regarding "when and how to perform emergency clinical interventions when medical equipment fails." This is generally a responsibility assigned to patient care components of the organization, such as nursing services, rather than to clinical engineering. However, this responsibility should be addressed in the medical equipment management plan and it should be considered as a component of the hospital's environment of care program.

Each management plan also should include a statement of the scope of the plan. In general, the scope of a hospital's environment of care program should cover all facilities in which the hospital's patients receive care, and all facilities in which the hospital's employees work. The scope statements for particular management plans may build on this fundamental definition. For example, the scope statement for a hospital's medical equipment management plan might specify the places in which medical equipment is used (e.g., main hospital, offsite clinics, home care) or categories of medical equipment that the hospital manages in different ways (e.g., imaging equipment maintenance managed by radiology, general medical equipment managed by clinical engineering). However, the scope statement is written for a particular management plans, it must cover the full range of responsibilities for the hospital and not only those for an individual department.

Following the statements of objectives and scope, each environment of care management plan should address all intent statements in the relevant standards. It is useful to include the actual intent statement and to follow it with a brief description of ways in which hospital meets the intent. Describe who does what with regard to the intent statement in a few sentences. This description should be followed by references to detailed policies, procedures, and documentation that the reader may pursue for further information.

As a hospital develops its environment of care management plans, it should carefully examine each activity for completeness. First, does the activity fully meet the intent of the standard? Second, does the activity contribute to achieving the stated objectives for the management plan? Third, does the activity address the entire scope of the management plan? The management plan is complete only when all of these questions can be answered affirmatively.

The Performance Improvement Cycle

Performance improvement is a fundamental concept underlying all JCAHO standards. This is a continuous process of monitoring and evaluating performance to identify and realize opportunities for improvement (JCAHO, 2001). Figure 58-1 illustrates the performance improvement cycle for the environment of care standards.

Each environment of care management plan is required to establish a process for ongoing monitoring of performance. Typically, each management plan will specify a small number of "performance monitors" that measure critical aspects of performance under the plan. For example, the medical equipment management plan may specify periodic monitoring of the percentage of scheduled maintenance procedures that are completed on schedule (See chapter 50). When specifying a performance monitor, it is important to define how it is calculated (e.g., mathematical definitions of the numerator and denominator), how often it is calculated (e.g., monthly), how it is reported (e.g., quarterly to the safety committee), and whether a target value has been determined (e.g., 95% based on JCAHO scoring standards for Standard EC.2.10.3).

Each environment of care management plan is also required to describe "how an annual evaluation of the . . . program's objectives, scope, performance, and effectiveness will occur." When describing the annual evaluation process, it is important to define who conducts the evaluation (e.g., a task force of the safety committee), what process is followed (e.g., review of performance monitoring data for the preceding 12 months), and how the findings are reported (e.g., through the safety committee to administration). Reporting requirements are addressed in Standard EC.4.2.

The results of the annual evaluation may be used to modify the environment of care management plans. For example, if a particular activity does not adequately address the

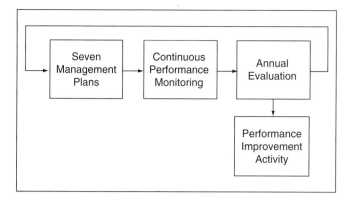

Figure 58-1 Performance improvement cycle for the environment of care standards.

entire range for which it is applicable, the activity may be expanded to cover the full scope as defined for that management plan. The results also may be used to modify the performance monitors that are specified in the management plans. For example, review of medical equipment management data may suggest the need to monitor the failure rate of a particular type of medical device.

Components of Hospital Safety Programs

The performance-improvement cycle links the safety committee, the safety officer, and the managers of the seven environment of care management plans into a framework for coordinating the many separate activities that constitute a hospital-wide safety program. Key components of the safety program are described as follows.

Safety Management

One of the most widely-recognized activities under the safety-management program is the performance of hazard-surveillance rounds on a regular basis (Standard EC.2.10.1). Although these rounds may be conducted by one or two people (e.g., the safety officer and the manager of the unit under surveillance), it is often more effective to broaden the objectives of the process and include other personnel in a small hazard-surveillance team. Additional participants may be drawn from infection control, environmental services (housekeeping), facilities engineering, and clinical engineering. Although these additional participants may focus primarily on their own areas of responsibility, they also can provide complementary perspectives for other members of the team.

Hospitals are also required by JCAHO to establish worker-safety programs (Standard EC.1.1.1). Typically, such programs are based on Occupational Safety and Health Administration (OSHA) regulations and other related government regulations. Protection from hazardous materials is a prominent issue for all health care workers (Standard EC.1.3). Protection from blood borne pathogens (e.g., hepatitis B virus) and airborne pathogens (e.g., tuberculosis bacteria) is of particular concern to clinicians. Other requirements, such as lockout-tagout programs, are applicable to workers in clinical engineering and facilities engineering.

Emergency Management

JCAHO recently expanded the scope of its emergency management standards to include four phases of activity: Mitigation, preparedness, response, and recovery (Standard EC.1.4). The involvement of clinical engineers is valuable in all phases, particularly with regard to the role of medical and other technologies in managing emergency situations.

Life Safety Management

JCAHO and most fire-safety jurisdictions require hospitals to comply with the life safety code NFPA 101 of the National Fire Protection Association (NFPA), published in 1997. This standard provides detailed specifications for design, construction, and operation of buildings to protect occupants from fire. JCAHO has developed a Statement of Conditions (SOC) document that hospitals must use to verify compliance with the life safety code (Standard EC.1.5.1). When deficiencies are identified, hospitals must document each deficiency and specify a plan of corrective action in Part 4 of the SOC referred to as a plan for improvement, and, as appropriate, implement interim life safety measures (Standard EC.1.5) to mitigate the deficiencies.

Hospitals are also required to conduct and evaluate fire drills at regular intervals (Standard EC.2.9.2) and to prohibit smoking in hospital buildings except under strictly-controlled circumstances (Standard EC.1.1.2).

Medical Equipment and Utility Systems Management

In the JCAHO standards, many parallels exist between the medical equipment management program (Standard EC.1.6) and the utility systems management program (Standard EC.1.7). In both cases, critical devices and systems are identified for inclusion in a program of scheduled inspection, testing, and maintenance. A well-designed program contributes to both the safety and effectiveness of patient care.

JCAHO has continued to expand the scope of utility systems management to include infection control (Standard EC.1.6), particularly for waterborne pathogens (e.g., legionella bacteria) and airborne pathogens (e.g., tuberculosis bacteria). Infection control has also become a prominent issue in facility design and construction (Standard EC.3.2.1) with the publication of the 2001 edition of *Guidelines for Design and Construction of Hospital and Health Care Facilities* by the American Institute of Architects (AIA). These infection control–related requirements have promoted greater cooperation between infection control professionals on one hand, and clinical and facilities engineers on the other.

Patient Safety

In 2000, the Institute of Medicine published *To Err is Human: Building a Safer Health System* (Kohn, 2000). This landmark report estimated that medical errors cause 44,000 to 98,000 deaths annually in hospitals in the United States. Although there is some controversy regarding the actual number of deaths caused by medical errors, there is a strong consensus that patient safety can and should be improved. There is also general agreement that improvements in patient safety will *not* come from blaming health care providers and exhorting them to do better. Instead, hospitals need to improve the systems in which patient care is provided.

At the heart of a systems approach to patient safety is coordination of all systems components—the patient, patient care providers, support personnel, medical technologies (e.g., devices, drugs, and medical and surgical procedures), and the environment in which patient care is provided (See Chapter 59). The systems approach includes proactive analysis to anticipate ways in which a system can fail. It also includes root cause analysis of actual failures. The objective of both activities is to improve the systems in order to make future failure less likely.

As of July 2001, JCAHO extensively revised its hospital accreditation standards to directly address patient safety. The fundamental revisions were made in the leadership chapter, which now mandates that "leaders ensure implementation of an integrated patient safety program throughout the organization" (Standard LD.5). Building on earlier requirements to conduct retroactive root cause analyses of sentinel events (Standard PI.4.3), JCAHO now requires hospitals to conduct proactive analyses to identify potential risks to patient safety (Standard LD.5.2). This approach to risk assessment is explicitly based on the principles of systems engineering.

Additional revisions to JCAHO standards in January 2002, included integration of Environment of Care activities into the hospital-wide patient safety program. Standard EC.4.1 assigns the responsibility for integration to the safety officer. Standards EC.4.2 and EC.4.3 address the flow of information from the environment of care program to the patient safety program. This is a clear indication of the vital role that environment of care activities have played in advancing patient safety.

Thus, improving patient safety has become a dominant theme for JCAHO and for the entire health care delivery system. The area of patient safety offers with vast potential for clinical engineers to use their skills in medical and nonmedical technology management, their close association with clinical and nonclinical professionals throughout the hospital, and their solid grounding in general and specialized principles of engineering science. (See Chapter 55.)

As described in *Enhancing Patient Safety: The Role of Clinical Engineering,* a White Paper published by the American College of Clinical Engineering, clinical engineers are in a unique position to make valuable contributions to patient safety (ACCE, 2001). Through the paper, ACCE committed itself to promoting the principles of systems engineering as part of the body of knowledge for clinical engineering practice. Development of such skills will augment the ability of clinical engineers to adopt prominent roles in this new era of patient safety.

Summary

Hospital safety programs—designed to protect patients, visitors, and staff—are practical applications of the medical dictum "First, do no harm." They are fundamental in creating an environment in which effective patient care can occur. Since its earliest days, the profession of clinical engineering has addressed safety in the clinical environment. Working in cooperation with professional colleagues throughout the health care delivery system, clinical engineering can extend their admirable record of contributions to hospital safety.

References

American College of Clinical Engineering. *Enhancing Patient Safety: The Role of Clinical Engineering.* Plymouth Meeting, MA, American College of Clinical Engineering, 2001.

American Institute of Architects. *Guidelines for Design and Construction of Hospital and Health Care Facilities.* Washington, DC, American Institute of Architects, 2001. Washington, DC.

Kohn LT, Corrigan JM, Donaldson M S (eds). *To Err is Human: Building a Safer Health System. Institute of Medicine.* National Academy Press, March 2000.

Joint Commission on Accreditation of Healthcare Organizations. *Comprehensive Accreditation Manual for Hospitals.* Oakbrook Terrace, IL, Joint Commission on Accreditation of Healthcare Organizations, 2001.

Joint Commission on Accreditation of Healthcare Organizations. *Performance Improvement in the Environment of Care.* Oakbrook Terrace, Illinois, Joint Commission on Accreditation of Healthcare Organizations, 2001.

National Fire Protection Association. *NFPA 101 Life Safety Code.* Quincy, MA, National Fire Protection Association, 1997.

59

Systems Approach to Medical Device Safety

Marvin Shepherd
DEVTEQ Consulting
Walnut Creek, CA

Health care is delivered through an assemblage of systems and minisystems. If a device-related minisystem of an assemblage fails to deliver its clinical benefit or delivers a physical insult to a patient, an incident investigation will begin. Information will be gathered, devices will be tested, and personnel will be interviewed. Some conclusions will be reached, and corrective actions will be taken. However, when the investigative process is discussed with in-house investigators, the process is frequently found to be incomplete, flawed, or biased. These deficiencies appear to be related to the infrequency with which most in-hospital investigators perform investigations, a lack of education and training in investigative techniques, and a lack of understanding of analytical methods for determining the cause(s) of an event.

A generic system's risk model has been developed for analyzing the performance of a minisystem. It provides the investigator with a mental model of the interacting components of the mini-system and provides a logical pathway toward the root causes of an adverse event. In addition, at the time of a minisystem failure, the model can be used as a checklist to assure that information about each component has been gathered and analyzed for its contribution to the event. Although errors still will be made in causal determinations, they will not be due to a forgotten or neglected component or condition.

Background

The medical community has only recognized that appropriately designed health care delivery systems, subsystems and processes are the best approach to reducing medical errors (Moray, 1994; Reason, 1990; Senders and Moray, 1991; Norman and Draper, 1986). In 1976, a system safety model was developed that could be applied to the failure of medical devices from the viewpoint of equipment maintenance (Shepherd et al, 1976; Shepherd and Shaw, 1976). Only later was the model expanded to identify factors that prevented a clinical benefit from being delivered to a patient or that delivered a physical insult (Shepherd, 1983; Shepherd and Brown, 1992). In 1989, a device-education model for nurses, based on this model, was presented at an FDA conference (Abbey and Shepherd, 1989). As risk and safety are opposite sides of the same coin, and as the major players in health care were moving away from safety concepts and toward risk concepts, the name of the model was changed from "systems safety model" to "systems risk model" (SRM) in 1997 (Shepherd, 2000). With recent advances in human error theory, systems analysis, and increased concerns for patient safety, the model was modified again in 1999 to better address the search for the root causes of adverse events and to suggest more effective methods to prevent recurrences of such events (Shepherd, 2000; Shepherd, 1999).

The Systems Risk Model (SRM)

The systems risk model is illustrated in Figure 59-1 and is presented from the viewpoint of the health care provider.

Description of the SRM

The large envelope of Figure 59-1 represents the health care provider, i.e., hospital, ambulatory care center, nursing home, or other patient care location. In the center of the large envelope is a gray, shaded area containing the terms "operator," "facility," "device," "environment," and "patient." This represents a generic, device-related minisystem that delivers a clinical benefit. Each of the five components leads to smaller envelopes. These smaller envelopes contain the subcomponents of the five major components. There are a total of 14 subcomponents in the SRM. The five major components are at the end of the delivery system where patients receive care. The subcomponents are part of the end of the system that supports the minisystems at the patient care end (Cook and Woods, 1994). The flaws found within the subcomponents could lead to a component failure, that is, lead the minisystem outside its expected performance boundaries, or they could cause a physical injury to the patient. More complex SRMs can be constructed by adding more devices or more operators.

Flaws located within the subcomponents frequently can be traced to their origins, before they entered the clinical environment. For instance, a flaw in a device component may originate with its design and construction; a flaw in a facility component may originate during the planning and preconstruction phases; and a flaw in the operator component may originate during basic education and training. These factors are shown in Figure 59-1 as being outside the large envelope. The health care provider has an opportunity to minimize these flaws through careful examination of a component before it enters the care delivery environment. Once the component becomes active in the clinical environment, the provider has lost this opportunity and must manage the risks associated with any flaws that passed through the initial scrutiny.

Health care providers organize their facilities to deliver care to a specific patient population; heart transplants generally are not performed at substance abuse centers. The facility component is designed, the device component is selected, and the operator component is trained to accommodate the selected population. Patients who fall outside this population are placed at a different risk level, which must be considered when analyzing minisystem failures.

Definitions for the Subcomponents

The definitions for the subcomponents of the systems risk model are provided in Table 59-1, columns 1 and 2. Column 4 contains some general questions to pose during an investigation that may lead an investigator to flaws of the system, often referred to as root causes. Column 3 contains the person or group administratively responsible for that subcomponent at the time of an adverse event. The designation of a responsible group is useful to the health care provider. Such a group can assist in determining the root causes and in deciding where to establish defense barriers, i.e., the special shielding or relocation of EMI-sensitive equipment away from community transmitters, the use of focused prepurchase device evaluations to reduce unpredictable device failures, and the use of patient data entry forms that screen for unusual patient characteristics.

The Direct and Root Cause(s) of Failure

Determining the direct cause of an adverse event provides a doorway to the root cause.

Direct Cause

The direct cause of failure as it applies to a device-related, minisystem is a deviation in expected the performance of a component of the SRM that results in the minisystem performance falling outside of specified boundaries or that delivers a physical insult to the patient. If the operator component fails, it is not only considered a direct cause, but also an active cause. Identification of the operator component as the direct cause is not intended to suggest blame, but is simply a modeling mechanism to establish a doorway to the subcomponents and to root cause(s). The failure of a component is just the beginning of an investigation.

Root Cause

Flaws in a minisystem are the root causes of adverse events. Flaws may be obvious, but managed or ignored, or they may go unrecognized until an adverse event occurs. Prior to an adverse event, the identification of these flaws is part of a quality improvement program. Following an event, identification of flaws is the goal of the accident investigator.

Finding the Cause(s) of Failure

The performance boundaries of a minisystem are established by many factors, including the design of the device, clinical trials, professional standards, operator expertise, organi-

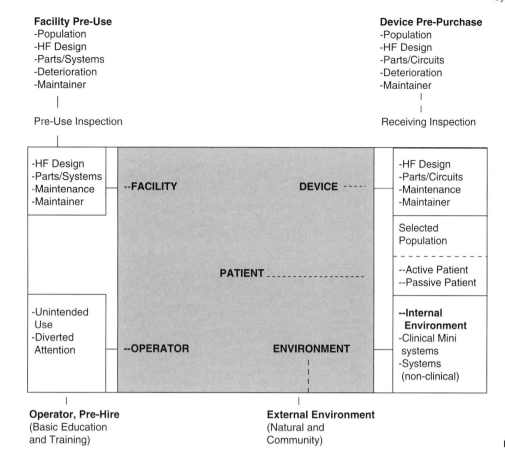

Figure 59-1 A systems risk model (SRM).

zational goals, policies and procedures, laws, and others. Finding the causes of failure requires the investigator to identify the component or components that exceeded the boundary limits. The search must begin at the site of injury, death or device malfunction, traced back to the SRM component(s), on to the subcomponent(s), and finally to the root cause(s). The fundamental mechanisms of injury and device malfunctions must be well understood by the investigator to begin the process properly.

Each component is examined to determine whether it deviated from expectations or can be excluded based on the conditions surrounding the event. It is typical to begin with an examination of the device to determine whether it meets the manufacturer's performance and safety specifications. The same is true of the facility component. Both have objective specifications that can be readily measured. Any interactions between systems and minisystems must be assessed, and any characteristic of the patient that may have contributed to the event needs to be identified. The operator component is evaluated last and except for education and training, is it more difficult to assess.

Case Study of a Monitor That Failed to Alarm

A female adult patient with end-stage cancer was admitted to an emergency room (ER). She was subsequently placed in a private room, where vital signs were taken and a cardiac monitor was connected. The monitor was about 10 years old, and the heart rate alarms on this monitor had to be activated by pushing a button. The registry nurse did not activate the rate alarm because the nurse believed that default rate alarms activated automatically when the monitor power was turned on. The patient was visually monitored every 10-15 minutes. Approximately 90 minutes after the patient arrived and 10 minutes after being last examined, the patient was found dead. No heart rate alarms had alerted the nurse.

Investigating the Event

The following information was acquired during the investigation from numerous sources, including patient records, interviews, device examination and testing, and manuals.

- Device Component. The heart monitor itself and the 4-bed system were examined and found to function to the specification of the manufacturer. In this system, no default alarms occurred automatically when the power was on. Alarms would only function with a single, push-button activation.
- Facility Component. Although a noisy power line was considered as a possible source of noise that might turn off the alarms, it was rejected without any testing of the line.
- Patient Component. The patient was in the end stages of cancer and barely alert. Neither active nor passive characteristics were identified as contributing to the event. A "no code" had not been placed in the patient's record.
- Environment Component. Other systems operating at the time, such as a radio system that was intermittently active and located within 6 feet of the foot of the bed, were considered and rejected as the source of the alarm failure.
- Operator Component. The operator of the monitor was a registered nurse (RN). Her education, training, and credentials were excellent for working in an emergency room. However, she was trained to use newer monitors that automatically set default limits when turned on. She had never worked with older monitors, such as the one in this case, that required a push-button to set the alarm.

Determining Direct Cause

In this instance, the hazard was the failure of the alarm to function in a timely manner. Although the patient would have died anyway, since a "no code" had not been entered into the patient's record, a Code Blue team would respond if the alarm sounded. In evaluating each component of the system, all functioned normally except for the operator component. The operator simply failed to press the default alarm button. Using the SRM analysis, the operator was considered to be the direct cause as well as the administrative cause of the adverse event.

Determining Root Cause

The root causes of this event are rather interesting. One cause relates to the use of a mixture of monitoring systems of different designs. In some systems, default alarms are set automatically. In others, they are set with a push of a button. Neither the registry nurse nor the ER knew of these conditions.. If either had known, the adverse event might not have occurred. If the ER had known the nature of the alarm and had failed to train all nurses to use it, the ER would have been negligent.

During investigational interviews, we encountered one regular ER nurse who said she knew that the monitor alarms were set automatically. She attempted to prove this by applying ECG electrodes to her body and inserting the ECG cable into a monitor. When her ECG was visible on the monitor, she lifted one of the electrodes from her body and an alarm sounded, but it was a leads-off alarm. The nurse did not know the difference between a leads-off alarm and a heart rate alarm.

TABLE 59-1 Definitions of Subcomponents and Administrative Responsibility

Component/Subcomponent	Subcomponent definition	Administrative Responsibility at time of event	Generic questions related to root causes
Device			
Human factors design	Design features defined by accepted human factors (HF), checklist guidelines or standards, or by the manufacturer.	Operator	Was the HF feature unusual? A new feature on a new device? A common feature on other devices? Different from devices with similar functions? Was it mentioned in the education and training curriculum? Mentioned in operator's manual? Was knowledge gained by the provider prior to hire? Was the human factors feature related to a device modification?
Parts/circuit design	Sudden and unpredictable failure of equipment, parts, or circuits	Manufacturer	Was unpredictability known prior to purchase? After purchase? Can one buy more reliable devices? Did the purchasing process examine other databases, such as FDA or ECRI, for similar problems? Does the maintenance database contain similar events with similar devices? Was there any indication that the unpredictable failure might be related to maintenance actions?
Deterioration	Slow and predictable changes in equipment accuracy, precision, and reliability	Device maintainer	Was device listed in inventory for PM? Was it scheduled for PM? Was PM performed? Did maintainer forget PM detail? Not enough maintainers to do PM? Budget cut prevented PM?
Maintainer	Action or inaction of maintainer resulting in equipment failure	Device maintainer	Could not repair device? Repaired same device multiple times? Maintainer never trained or educated to repair/PM this device? Received wrong information for repair? Not allowed enough time for repair? Distracted from repair? Under-trained in equipment repair (i.e., soldering or troubleshooting)?
Facility			
Human-factors design	Design features defined by human-factors (HF) guidelines or standards, or by the architects and engineers	Operator	Was HF feature unusual? New feature on new utility? Common feature on another utility? Different from utilities with similar functions? Was it mentioned in the ed/train curriculum? Mentioned in P&P orientation? Knowledge gained by provider prior to hire?
Parts/systems design	Sudden and unpredictable failure of facility parts or systems	Architect & engineer	Was unpredictability known before or after construction? Can one buy more reliable parts or systems? Did design process examine any databases for similar problems? Own? FDA? Colleagues? Does maintenance database contain similar events on similar devices? Any indication that the unpredictable failure might be related to maintenance actions?
Deterioration	Slow and predictable changes in the facility systems	Facility maintainer	Was utility/equipment listed in inventory for PM? Was it scheduled for PM? Was PM performed? Did maintainer forget PM detail? Not enough maintainers to do PM? Budget cut prevented PM?
Maintainer	Action or inaction of maintainer resulting in facility parts or systems failure	Facility maintainer	Could not repair system or equipment? Repaired same system or equipment multiple times? Maintainer was never trained or educated to repair/PM this system or equipment? Received wrong information for repair? Not allowed enough time for repair? Distracted from repair? Under-trained in basic repairs (e.g., soldering or troubleshooting)?
Patient			
Active			
educated	Patient who is knowledgeable of a hazard performs a harmful action	Patient	Is patient normal for this treatment area? Was patient knowledgeable of hazard? Was provider? Could device have been placed beyond patient's reach?
uneducated	Patient who is ignorant of a hazard or uncontrollably performs a harmful action	Operator	Should patient have been restrained? Sedated? Placed where they could be watched more carefully?
Passive	Patient condition or sensitivities	Operator	Was patient screened for sensitivities? Was sensitivity unusual? Should provider have known? Should this patient be treated at this provider's facility?
Operator			
Unintended use	Action or inaction due to lack of education or training that provides unintended results	Operator	Knowledgeable of this minisystem? Under trained/educated for this minisystem? Received orientation? Mentored? In-serviced? Operator manual missing?
Diverted attention	Action or inaction of educated/trained operator under certain circumstances that provides unintended results	Health care provider	Long hours? To many patients? Inadequate staff? Personal problems? Budget cuts? Excessive meetings? Did operator exceed the physical limits of an accessory or part?
Environment			
External	Unsafe or inefficacious device due to external systems outside the control of the health care provider	Nature or the community	Was the failure due to a natural disaster (i.e., earthquake, hurricane, or flood)? Normal environment of heat, sun, wind, and rain. Local disasters, (i.e., plane crash or utility failure)?
Internal	Unsafe or inefficacious device due to interaction with other systems or minisystems controlled by the health care provider	Operator	Caused by other clinical minisystems? Caused by utility systems? Telecommunications? Internal disaster, (i.e., a power failure that affects multiple patients)?

In this hospital, intensive care unit (ICU) nurses would occasionally work in the ER. We went to the ICU, which had only the new style of heart monitors, and interviewed four ICU nurses. Two of those nurses said that the default heart rate alarms on ER monitors were automatically set when the monitor power was turned on. This suggests that the nurse in question was simply the operator of a system that failed but the mistake could have been made by an ICU nurse as well. A literature search for similar events yielded at least two publications that warned of the possibility of errors when two systems that had with different methods for setting alarms were mixed. However, these warnings never filtered down to the registry nurse or the ER staff.

To prevent a future event, the ER added a label under each of the alarm buttons that directed nurses to push the button to activate the alarm. The hospital also updated nurse orientation training material and to the operating manual for the monitor itself with a note explaining the need to push a button to activate alarms. The hospital also developed a long-term plan to replace the older monitors.

One interesting aspect of this case is that if a "no code" had been placed on the patient's chart, no Code Blue response would have been necessary. The death of the patient would have been a non-event, and the failure of the nurse to set the alarms might have gone unnoticed. In addition, when the nurse checked the monitoring system to determine why the alarm had failed, she noted that the monitor clock was an hour behind real time. She said she believed that if the clock was wrong, the system must not have been functioning properly.

Summary

The SRM has been used for over 20 years to analyze device-related adverse events. It ensures that surrounding factors and pertinent conditions have been considered for their roles in the adverse event. It provides a logical pathway for pursuing patient injuries or system malperformance from an event to its root causes. The SRM identifies process locations where the health care provider can evaluate components to identify obvious flaws and prevent them from entering the clinical environment. Following an investigation, the SRM can also be used as a checklist to suggest changes in the minisystem that might prevent a recurrence of the event.

References

Abbey J, Shepherd M. The Abbey-Shepherd Device Education Model. Nursing and Technology: Moving into the 21st Century. Annapolis, MD, HHS/FDA, 4189-4231, 1989.

Cook R, Woods D. Operating at the Sharp End. The Complexity of Human Error. In Bogner MS (ed). *Human Error in Medicine*. Hillsdale, NJ, Lawrence Erlbaum Associates, pp 255-310, 1994.

Moray NP. Error Reduction as a Systems Problem. In Bogner MS (ed). *Human Error in Medicine*. Hillsdale, NJ, Lawrence Erlbaum Associates, pp 67-91. 1994.

Norman D, Draper S. *User-Centered System Design*. Hillsdale, NJ, Lawrence Erlbaum Associates,1986.
Reason J. *Human Error*. Cambridge, UK, Cambridge University Press, 1990.
Senders JW, Moray NP. *Human Error: Cause, Prediction and Reduction*. Hillsdale, NJ, Lawrence Erlbaum Associates, 1991.
Shepherd M. *A Systems Approach to Hospital Medical Device Safety*. (Monograph) Arlington, VA, AAMI, 1983.
Shepherd M. Managing the Risks of Device-Related Adverse Events; Identifying Direct & Root Cause(s). Durham, NC, Duke University Medical College, 1999.
Shepherd M. *Medical Device Incident Investigation & Reporting*. Walnut Creek, CA, Devteq Publishing, 1.1-1.39. 2000

Shepherd M, Brown R. Utilizing a Systems Approach to Categorize Device-Related Failures and Define User and Operator Errors. *Biomed Instrum Technol* 26:461-751, 1992.
Shepherd M, Collard J, Collins M. *Some Preliminary Results From a Shared Engineering Database*. Boston, MA, 10th Annual AAMI Conference & Expo, l976.
Shepherd M, Shaw D. *Managing Patient Care Equipment: Analyzing the Maintenance Database*. Plant, Technology & Safety Management Series. Chicago, IL, Joint Commission on the Accreditation of Hospitals, 1976.

Interactions Between Medical Devices

Saul Miodownik
Director, Clinical Engineering, Memorial Sloan-Kettering Cancer Center
New York, NY

The increasing complexity of medical instruments eventually causes conflicts between devices that are normally used on the same patient. Fundamental understanding of the technology and its relationship to specific aspects of the patient's anatomy and physiology is required to address potential interference with other devices.

As medical devices are deployed into various settings, including health care, their initial control and operation are often the responsibility of technical experts who are familiar with their nuances and quirks. By the time any technology is mature enough to be run effectively by appliance operators, that detailed knowledge is no longer needed and is often lost. Indeed, this is the normal evolution of most technology in society. Everyone uses televisions, DVD players, computers, and cellular phones with impunity, but few people are even vaguely conversant with the underlying technology that drives them.

Medical technology is no different in this regard. However fascinating the scientific and engineering underpinnings of a medical device may be, the end-user has no need to be aware of them. In isolation, this may be totally acceptable and, perhaps, preferable. Unfortunately, there are many instances when medical devices conflict with each other and can interfere with the correct diagnosis of a patient's condition or the delivery of the desired therapy. The responsibility increasingly falls on those who manage and purvey technological matters to analyze, anticipate, and ultimately prevent these interactions from negatively affecting the patient. In most settings, this vigilance and required knowledge become the responsibility of clinical and biomedical engineers.

Unanticipated medical device interactions have occurred at various levels for more than a century. Since the early days of the X-ray image, it has been known that some materials (often metallic) would affect the proper exposure of a diagnostic X-ray film. This was the direct consequence of ionizing radiation. Electrosurgical devices were in widespread use for decades before routine electrocardiogram (ECG) monitoring became a common anesthetic practice. The interaction between these necessary but conflicting technologies (much noise and, sometimes, temporary loss of the ECG signal during an X-ray) was immediately understood and accepted. The1960s and 1970s raised a great deal of conjecture about, and fear of, lethal microshock as an undesirable consequence of the use of medical equipment and electrical appliances at the patient bedside. Although this danger never materialized, it did demonstrate that technology could interact in unexpected and potentially dangerous ways. (See Chapter 65.) The deployment of increasingly esoteric instrumentation can cause interactions between totally unrelated systems and can yield unpredictable consequences.

Perspectives

Very early in my experience as a clinical engineer, I was called to one of our intensive care units to check a cardiac monitor that was presenting an incorrect heart rate. From the central station, I observed the bed in question and what appeared to be a fairly reasonable low-voltage ECG. I confirmed the patient's heart rate reading of 160 BPM. The nursing staff told me that this could not be possible. After launching into an elaborate discussion of double-triggering and elevated P-waves mimicking R-waves, one of the intensive care unit (ICU) nurses said, "Excuse me, but he's been quite dead for over an hour!" I went into the patient's cubicle and looked around. The monitoring electrodes were still connected to the patient, as were the intravenous (IV) lines. I noticed that one of the IV lines ran through an infusion controller that was still turned on and delivering fluid to the patient. When the IV controller was turned off, the patient's ECG signal disappeared. At first I thought a radiated pulse from the controller might be disrupting the ECG leads. Although I had no good explanation for this phenomenon at that time, the nurses thanked me for fixing the problem as I left. It was an episode that I never forgot (Miodownik, 2003).

EXAMPLE 1: The Van de Graff Generator Treadmill

In one case, the clinical engineering department was asked to configure a treadmill-based cardiac stress test system. There was an available automated treadmill that stepped through a preprogrammed elevation and speed sequence. We provided a spare ECG monitor that had a digital storage screen, heart rate meter, alarm limits, and strip chart recorder. The system was quickly assembled, configured, tested, and delivered to the ECG department in the autumn to everyone's satisfaction. Early in January of the very next year, we began receiving intermittent complaints of huge bursts of noise that distorted the ECG signal and ruined the test. The noise seemed to be related to the higher speeds of the treadmill, as it never occurred early in the stress test protocol. We investigated and discovered that the noise was the result of missing or incorrect pieces in the ECG A/D converter system. A direct recording from the analog ECG signal revealed no appreciable noise. We could find no defect in the monitor, or anything obviously wrong with the treadmill or its controller. Internal leads to the treadmill motors were fitted with the appropriate RF filter, as was the ECG cable. When the system was operated with a simulator, no noise was heard at any treadmill speed or elevation. Changing the monitor and the treadmill with different units that worked fine elsewhere still resulted in noise. The problem persisted, and led to the repetition of one or two stress tests per month.

By March of the same year, the complaints had subsided and finally disappeared. Throughout the summer, no further complaints came from the ECG Department. However, by winter they returned in force. This time, an able runner in our section volunteered to run on the treadmill for a stress test under our close scrutiny. Electrodes were placed, and the stress test protocol was conducted as for any patient. At a speed of about six miles per hour, dropped bits were in evidence on the screen. By the time the speed reached eight miles per hour, the ECG was unreadable, and the test had to be terminated. We repeated the problematic segments of the test but still could offer no explanation. We returned to the drawing board and pulled out the schematics of all of the affected instruments, and looked at them for clues. None appeared. We also wondered why the problem appeared to be seasonal.

Our hospital HVAC is particularly well-known for providing desert-like heat in the winter and a soggier form of cooling in the summer. At other times of the year, ambient climatic conditions usually prevail. We therefore wondered whether the static electrical charges that have a tendency to build up under very dry conditions might be related to this problem. In the final analysis, sudden discharge of static electricity was the proximal cause of the monitor problem. The etiology, however logical, was not obvious.

Patients who are scheduled for treadmill stress tests are asked to report for their test in comfortable running clothes and to wear sneakers or running shoes. After being prepped and outfitted with electrodes, they mount the treadmill and grab onto a handrail in front of them. The handrail has an insulating grip. This position was then maintained throughout the test. The interface between the treadmill running belt and the patient's sneaker sole behaved in the same manner as a Van de Graff generator. As the treadmill speed increased, electrons were stripped off the belt at an increasing rate. Since the patient was essentially electrically isolated from grounding (isolated ECG amplifier, insulated sneakers, and insulated handrail on the treadmill), this system would charge the patient to ever increasing voltages. This continued unless the charge could leak away or until a sudden discharge of the accumulated electrons occurred. On more humid days, the charge-leakage rate through the air was sufficient to limit this voltage to low levels. On very dry days in the winter, the path of discharge was either through dielectric breakdown in the treadmill or through exceeding the isolation limits of the front-end ECG amplifier. Close inspection of the ECG amplifier schematic revealed a spark-gap discharge voltage limiter across the isolation barrier with a breakdown voltage of about 4000 volts. It was hypothesized that the sudden discharge of the spark gap sent enough voltage into the digital storage section of the monitor to cause the dropped bits. We revisited the treadmill with our runner and opened the ECG monitor to observe the activity when it was in use. Sure enough, at sufficient treadmill speed the spark gap was arcing and dropped bits obliterated the ECG trace. The solution to the problem was as simple as the etiology of the problem was complex. A 75-megohm resistor was placed across the spark gap to slowly drain off charge buildup while maintaining sufficiently high impedance for safety.

In this instance, a number of advanced design concepts failed because of the device interaction and potentially put the patient at risk. The isolation system was designed to ensure patient electrical safety; the digital storage and display of the ECG waveform was designed to provide easier viewing of the signal for immediate and accurate determination of ECG changes related to exercise; the spark gap was to protect the isolation system from potentially damaging voltages. It was precisely at the higher levels of exercise, when the most accurate reading of the ECG was needed, that the stress test system failed. However, this was no real failure; it was an interaction between medical devices (Elliott and Gianetti, 1995).

EXAMPLE 2: Blood Pressure High on Bubbles

Invasive pressure monitoring became common at the patient's bedside in the early 1970s. Pressure transducers required special domes that allowed coupling to the patient's vasculature via fluid-filled tubing. The nursing staff often complained that systolic pressure readings seemed high. They questioned these hypertensive pressure measurements and checked them against their pressure cuffs and stethoscopes. We would check the calibration of the pressure transducer and amplifier and pronounce them accurate. We would explain how the pressure module's digital display was surely more accurate than the nurse's meager attempt at blood pressure measurement by auscultation. However, their complaints persisted, and eventually were proven legitimate.

Invasive pressure measurement has now been in common use in critical care units and operating rooms for many years. With the advent of reliable, inexpensive, disposable pressure transducers and tubing sets, questions of accuracy, linearity, and repeatability have largely disappeared. However, the issue of fluid-filled tubing interacting with the pressure transducer and the monitor remains a thorny problem.

Most invasive pressure monitoring instruments extract diastolic, systolic, and mean pressure information by observing the waveform's minimum and maximum values and integrating the area under the waveform curve to extract its mean value. Transmitting a pressure waveform from the blood vessel, from which it is derived, to the transducer appears trivial. One simply connects a fluid-filled, non-compliant tube from the vessel to the transducer. However, when this is done, a new system is inadvertently created with a life of its own. The arrangement of fluid-filled tubing and associated stopcocks creates an ideal environment for trapping gas bubbles, whose quantity and volume vary unpredictably. This causes a second-order under-damped resonant system that is easily excited by the patient's arterial pressure waveform. This is analogous to an under-damped mass, spring, and resistance system. The mass is represented by the fluid mass in the tubing, the spring is represented by the compliance of the tubing and any trapped bubbles, and the resistance is represented by resistance to the movement of the fluid. The combination of these elements exhibits a characteristic tuning frequency, or resonance, determined by the element values. On a theoretical basis, when excited by a step function, amplification of up to 100% is possible. When this happens in the clinical arena, systolic pressures are amplified to varying degrees as high as 75-100 mm Hg. Diastolic values are usually not affected, since they represent a slower part of the pressure waveform. At high heart rates, under some conditions, falsely lower diastolic pressures may also appear. Mean pressure, however, does not change and is independent of this resonance. Although this phenomenon, called "resonant overshoot," has been reported in the literature for more than 70 years, solutions that consistently work are as yet unavailable. Such techniques as filtering and gas debubbling do not properly address this issue. It is not unusual to review invasive pressure waveforms in an intensive care unit and find that more than 50% of the systolic readings are suspect. The problem is twofold. First, the left ventricular rate of pressure increase (dp/dt) is crucial to the way the second-order tubing system is excited. Second, the resonant frequency of the tubing system changes as bubbles are added and removed and as temperature variations cause the bubbles to shift in volume. The resonant frequency cannot be sufficiently determined to apply an inverse filter to correct the resonance of the tubing system. Although the pressure waveform may contain obvious markers of overshoot (e.g., additional dicrotic notches), this is not always the case. Identification of a suspect waveform containing systolic overshoot remains difficult.

This is a direct example of the interaction of medical devices (the transducer and its tubing set) that is often ignored to the possible detriment of the patient. No contraindications are mentioned in information insert for the tubing set that warn against relying on the systolic pressure measurement to diagnose or treat the patient. If the systolic reading, in the presence of serious resonant overshoot, is used to titrate a vasoactive drug that is controlling the patient's blood pressure, changes in the tubing environment could cause the clinician to make incorrect adjustments that could result in lower or higher patient blood pressure than desired.

MRI Interactions

Magnetic resonance imaging (MRI) systems pose a host of potential problems when used with medical devices. The powerful static and dynamic magnetic fields required to image the patient, high power radio frequencies, and the sensitivity of the MRI detection system to radio frequencies from other devices, provide ample opportunities for troublesome interaction. In anticipation of this possibility, most MRI procedures have been contraindicated for patients with in-dwelling pacemakers due to their metallic wires.

Fundamentally, the MRI magnetics can physically displace metallic objects that are near, implanted, or attached to the patient. There have been reports of surrounding object becoming projectiles, with the displacement of aneurysm clips leading to the deaths of patients undergoing an MRI. MRIs also cause wires in devices in, on, or near the patient to overheat. Sufficient energy can be coupled with the MRI high power radio frequency generator and create currents in nearby wires. The quantity of that energy depends on a number of factors, including the wire length and its proximity to the patient. In any event, the patient or someone else must remove the additional thermal burden. This has led to patient burns in a number of instances. In addition, wires moving through the static magnetic field as the patient is moved into the magnet will also generate voltages that can interfere with the proper operation of medical devices. Special considerations must be given to wires that may enter the MRI in the process of monitoring or providing life support to the patient (e.g., ECG lead wires, pulse oximeters, or ventilators).

The increasing use of automated, implantable cardiac defibrillators (AICDs) and pacemakers in the population presents a host of potential interaction issues. Although patients with these implants were initially excluded from MRI procedures, newer devices and lower magnetic field strength MRIs permit this usage albeit with caution.

Conclusions

The explosive growth of wireless devices, such as cellular phones and wireless access points increases the potential for interference with medical devices. The literature contains ample evidence for and warnings of these effects. However, comprehensive solutions remain wanting. In addition, the proliferation of automated external defibrillators (AEDs) in the nonhospital setting raises concerns. Will an AED recognize a fibrillating heart in the presence of an implanted pacemaker? Will the discharge of an external defibrillator have an adverse effect on an implanted device and its lead wire system?

The previously mentioned examples indicate how certain types of medical devices can interact in unexpected though not unexplainable ways. The combination of instruments or accessories was not unusual or contraindicated by the manufacturers involved. The technologies that conflict are often assembled as a natural extension of improved diagnostic and therapeutic capability, but interactions are difficult to assess and problems are difficult to predict prior to their actual occurrence. As medical instruments and the environments in which they are used continue to upgrade with new technologies, it will be necessary to analyze and to understand their fundamental aspects in order to anticipate conflicts.

References

Achenbach S, Moshage W, Diem B, et al. Effects of Magnetic Resonance Imaging on Cardiac Pacemakers and Electrodes. *Am Heart J* 134:467-73, 1997.

Agarwal SK. Infusion Pump Artifacts: The Potential Danger of a Spurious Dysrhythmia. *Heart Lung* 9:1063-1065, 1980.

Campbell WB. EKG of the Month: Left Atrial Enlargement, Intraventricular Conduction Delay, Myocardial Injury, and Electromagnetic Artifact due to an Intravenous Infusion Pump. *J Tenn Med Assoc* 76:586-587, 1983.

Chen WH, Lau CP, Leung SK, et al. Interference of Cellular Phones with Implanted Permanent Pacemakers. *Clin Cardiol* 19:881-886, 1996.

Elliott WR, Gianetti G. Electrostatic Discharge Interference in the Clinical Environment. Brief Cold Snaps or Humidification Disruptions can Cause ESD Problems. *Biomed Instrum Technol* 29:495-499, 1995.

Groeger JS, Miodownik S, Howland WS. ECG Infusion Artifact. *Chest* 85:143, 1984.

Keeler EK, Casey FX, Engels H, Lauder, et al. Accessory Equipment Considerations with Respect to MRI Compatibility (Review). *J Magn Reson Imaging* 8:12-18, 1998.

Lauck G, von Smekal A, Wolke S, et al. Effects of Nuclear Magnetic Resonance Imaging on Cardiac Pacemakers. *Pacing Clin Electrophysiol* 18:1549-1555, 1995.

Miodownik S. Clinical Engineering at the Bedside. In Dyro JF (ed). *The Handbook of Clinical Engineering*. Burlington, MA, Elsevier, 2004.

Nakamura T, Fukuda K, Hayakawa K, et al. Mechanism of Burn Injury During Magnetic Resonance Imaging (MRI)–Simple Loops Can Induce Heat Injury. *Front Med Biol Eng* 11:117-129, 2001.

Niehaus M, Tebbenjohanns J. Electromagnetic Interference in Patients with Implanted Pacemakers or Cardioverter-Defibrillators. *Heart* 86:246-248 2001.

Price RR. The AAPM/RSNA Physics Tutorial for Residents. MR Imaging Safety Considerations (Review). *Radiographics* 19:1641-1651, 1999.

Pride GL Jr, Kowal J, Mendelsohn DB, et al. Safety of MR Scanning in Patients with Nonferromagnetic Aneurysm Clips. *J Magn Reson Imaging* 12:198-200, 2000.

Pruefer D, Kalden P, Schreiber W, et al. In vitro Investigation of Prosthetic Heart Valves in Magnetic Resonance Imaging: Evaluation of Potential Hazards. *J Heart Valve Dis* 10:410-414, 2001.

Ridgway M. The Great Debate on Electrical Safety—In Retrospect. In Dyro JF (ed). *The Handbook of Clinical Engineering*. Elsevier, Burlington, MA, 2004.

Rothe CF, Kim KC. Measuring Systolic Arterial Blood Pressure. Possible Errors from Extension Tubes or Disposable Transducer Domes. *Crit Care Med* 8:683-689, 1980.

Sakiewicz PG, Wright E, Robinson O. Abnormal Electrical Stimulus of an Intra-Aortic Balloon Pump with Concurrent Support with continuous Veno-Venous Hemodialysis. *ASAIO J* 46:142-145 2000.

Schenck JF. Safety of Strong, Static Magnetic Fields (Review). *J Magn Reson Imaging* 12:2-19, 2000.

Shan PM, Ellenbogen KA. Life After Pacemaker Implantation: Management of Common Problems and Environmental Interactions. *Cardiol Rev* 9:193-201, 2001.

Todorovic M, Jensen EW, Thogersen C. Evaluation of Dynamic Performance in Liquid-Filled Catheter Systems for Measuring Invasive Blood Pressure. *Int J Clin Monit Comput* 13:173-178, 1996.

Zegzula HD, Lee WP. Infusion Port Dislodgement of Bilateral Breast Tissue Expanders After MRI. *Ann Plast Surg* 46(1):46-48, 2001.

61

Single-Use Injection Devices

Michael Cheng
Health Technology Management
Ottawa, Ontario, Canada

This chapter details the safety aspects of needles and syringes–medical devices used for injecting medications into the human body and withdrawing fluid from the body. Although the chapter focuses on developing countries, many aspects also apply to industrialized countries.

The majority of injection devices are now produced for single use. Safety problems associated with the use of needles and syringes include infections caused by the reuse of contaminated devices and accidental pricks by needles (needle sticks). Pricks by contaminated needles also lead to infections. The serious impact of injection safety in global health has led to an active group called the Safe Injection Global Network (SIGN, www.injectionsafety.org) and a SIGN Internet forum (e-mail:sign@uq.net.au).

This chapter describes technical reasons not to reuse devices, auto-disable syringes to prevent reuse, safety boxes to contain sharps immediately after use, and the ultimate disposal of used sharps. The life cycle management concept proposed for major medical equipment is applied to injection devices to ensure safety in health program planning and delivery. (See Chapter 31.)

Injections and Safety

Injections are one of the most common health care procedures. Between 12 billion and 16 billion injections are administered worldwide each year. Most injections (90% to 95%) are given for therapeutic purposes and are often unnecessary. About 5% of injections are given for immunizations.

In developing countries, often because of insufficient number of injection equipment or inaccessibility to more (see injection equipment security details following) up to two-thirds of the injections in some countries are reported to be unsafe. Although immunization injections are thought to be safer than therapeutic injections, nearly 30% of immunization injections are still considered to be unsafe. Unsafe injections are responsible for millions of cases of Hepatitis B and C and an estimated 250,000 cases of HIV annually (WHO, 2000).

An injection is considered safe when it does no harm to the recipient, does not expose the health care worker to any risk of infection, and does not result in waste that is dangerous to the community or harmful to the environment.

In many developing countries and some sectors of developed countries, the reuse of injection equipment without effective sterilization may cause cross-infection. Used and often contaminated sharps (i.e., needles or syringes) are often handled in a way that exposes health care workers to needle-stick injuries. Unsafe management and improper disposal of sharps waste can also cause contamination. Sharps waste is often carelessly thrown away where waste pickers and other people can be pricked and infected.

Preventing the Reuse of Single-Use Injection Equipment

The reuse of single-use medical devices is a worldwide problem. In industrialized countries, the reprocessing and reuse of medical devices for a range of applications has raised serious concerns, and regulatory agencies are establishing measures to control syringe and needle reuse. In the United States, the Food and Drug Administration (FDA) now requires the reprocessing of a single-use device to follow the same regulatory requirements as those followed by the original manufacturer of the device. In developing countries, international health agencies find the reuse of injection equipment to be a major contributor in the spread of spread of hepatitis, HIV, and other infections. In some regions, up to 75% of single-use syringes have been reused.

Manufacturers design devices that are labelled "single-use" with the intention that they will not be reused. The technical reasons against reuse include the following:

- Devices may not be taken apart for proper cleaning.
- Single-use devices may not be properly resterilized.
- The mechanical integrity and/or functionality of some single-use devices may not stand up to rigorous reprocessing.
- The cleaning chemicals or sterilizing agents could affect the reprocessed devices, the medication, or the patient.

Sterility, functionality, and possible hazards caused by chemical reactions after reprocessing are important issues.

Most disposable injection syringes are made of plastics that can not withstand high-temperature sterilization. With ethylene oxide gas sterilization, the bio-burden after syringe use varies a great deal from syringe to syringe, which renders reliable resterilization impractical. Other points of concern include the invasive needle that critically contacts blood, the piston rubber seal, and the lubricant, each of which could harbor pathogens. In addition, the deformity of the needle after use is likely to degrade its functionality. The risks of infection associated with the reuse of single-use injection equipment are extremely high. The seriousness of the spread of diseases caused by the reuse of single use injection equipment led to a WHO-UNICEF-UNFPA joint statement on the use of auto-disable (AD) syringes in immunization services (http://www.injectionsafety.org/).

The AD syringe (PATH, 2000) has a built-in mechanism that is designed to give a single dose of vaccine, after which the syringe is permanently locked or disabled. Such a mechanism prevents the reuse of contaminated syringes and needles, and eliminates their unauthorized packaging and resale. Currently, many types of AD syringes are commercially available. Increasingly, AD syringes are used in therapeutic and other injections. In many models, it is essential to completely eject the medication in the syringe in order to render the syringe nonreusable.

Preventing Sharps Injuries

To avoid needle-stick injury after an injection, the used needle should be immediately deposited into a safety box or container to prevent direct access, not recapped. Safety boxes must be made of puncture-proof material. While single-use syringes and needles should never be reused, it is possible to consider the recycling of safety boxes. However, special boxes and decontamination procedure must be strictly followed. (See http://www.noharm.org/.)

The use of safety boxes should not be regarded as the end of the safe disposal of used needles or syringes. Final appropriate and safe disposal requires systematic considerations.

Ultimate Safe and Appropriate Disposal of Sharps Waste

The ultimate disposal of injection equipment can be part of the institution-wide health-waste disposal system. Safety boxes containing used sharps must be disposed in a manner that is consistent with protecting the environment.

The Appendix on p.253 provides a comparison of various methods for the disposal of sharps waste. It should be noted that each method has its strengths and weaknesses, and no single waste disposal solution is best. Health care facility administrators must identify the most appropriate disposal technologies depending on the local regulations, the size of health care facilities, the location, (e.g., urban versus rural) and the availability of construction material for building incinerators or pits or to encapsulate waste. Geographical factors are also important and include the ease of access to final disposal sites and the nature of the soil where the health facility is located (e.g., sandy, muddy, flood-prone, or rocky ground.)

Some of the current options, such as incineration, are considered unacceptable by a number of environmentalist groups. Health waste disposals are important issues and

Table 61-1 Life Cycle Management Applied to Injection Equipment

Essential element for major equipment (life cycle management)	Injection equipment application	SIGN Toolbox references* (8 May 2003)
Planning	Situational and needs assessment	1.1, 1.2, 1.3, 1.4, 1.5, 1.6, 1.7, 1.8, 2.5, 2.6, 3.1
	Qualified health care workers?	
	Real need for medication?	
	Oral alternative?	
	Behavioral change?	
	Injection equipment security	
	Adequate storage?	
	Disposal:	
	Safety boxes	
	Disposal facilities	
	Communication strategies	
Acquisition	Procurement steps, price negotiation	1.6
	Quality assurance criteria	
Delivery	Shipment damage?	1.6
Incoming inspection	Specified goods?	
Inventory and documentation	Master inventory	1.6
	Distribution records	
Commissioning	Proper storage and safeguard	
	Good distribution practices to health facilities	
Training of users	Appropriate use	1.2, 1.4, 1.5, 1.8, 2.1, 2.2, 2.3, 2.4, 2.6, 2.7
	Safe use	
Monitoring of use and performance	Monitor risks, such as:	
	Breakage	
	Needle-stick	
	Incomplete ejection to disable syringe	
	Splash of blood or medication	
	Reuse	
	Safety boxes for used sharps	
Maintenance	Adequate resupply	1.1, 1.6
	Injection equipment security	
	Maintain number matched to medications	
Replacement disposal	Appropriate disposal	2.5, 2.7
	Safe disposal	

*The SIGN Toolbox references are available on the web site: www.who.int/injection_safety/toolbox

alternative technologies are being researched and developed. (See http://www.noharm.org/.)

Application of Life Cycle Management

In health projects such as an immunization program or any therapeutic program involving injections, ensuring product quality of vaccines and injection equipment is not enough to ensure safety. All aspects of immunization, including product quality, storage, handling, administration of injection, and disposal must be considered. In fact, the life cycle management concept developed for major medical equipment can be useful here. This concept calls for the inclusion of other essential elements for good management. (See Chapter 31.) For example, introducing an injection safety component at the planning phase of health projects involving injections is a proactive way to ensure safety. The different essential elements in the life cycle management concepts, from planning to disposal, can apply.

In Table 61-1, major management items are cross-referenced to various publications available in the Toolbox of the Safe Injection Global Network (SIGN). These publications are neither standard nor exhaustive, but they contain detailed descriptions of the management items to illustrate the applicability of the concepts.

Planning for safe disposal in various countries includes assessment of the disposal situation in each country. Once the supply of injection equipment has been determined, injection equipment security, safety boxes, and disposal facilities are considered as follows:

- Injection equipment security. Matching the quantity of medications or vaccines with the supply of injection equipment to ensure enough devices for all medications or vaccines without the need to reuse equipment. The lack of access to safe injection equipment frequently leads to reuse.
- Safety boxes. Sufficient number of safety boxes must be supplied to contain the sharps immediately after use. The use of the safety box is not the end of the disposal responsibility but a part of the safe use phase.
- Final disposal. The means and facilities for the final disposal of the safety boxes containing used sharps must be determined at the planning stage. The disposal procedures must ensure personal safety and protect the environment.

Discussion

The life cycle management table provides a simple, one-page checklist for key items of the whole management system. (See Table 61-1.) Details of key items that are specific to a certain health program or project can be added for customized consultation. This one-page picture also allows any health worker to see the relevance of his or her role in the overall system, an important first step toward increasing the system efficiency and effectiveness.

The life cycle management approach, originally developed for major medical equipment, also applies to nonmajor, but essential, medical devices and may be extended to additional devices.

References

PATH. Safe Injection Manual. PATH Technologies for Immunization, http://www.path.org/technos/ht_safe_injection_manual.htm, PATH, 2000.

WHO. Injection Safety. World Health Organization Secretariat Executive Board 107th session, EB107/23, 2000.

WHO/UNICEF/UNFP. An Online Joint Statement on the Use of Auto-Disable Syringes in Immunization Services. http://www.injectionsafety.org/, 2000.

Appendix Comparison of Methods for the Disposal of Sharps Waste

Waste burial pit or encapsulation	Simple	Potential of being unburied
	Inexpensive	No volume reduction
	Low-tech	No disinfection of wastes
	Prevents sharps-related infections or injuries to waste handlers or scavengers	Pit may fill quickly
		Not adapted for nonsharp infectious wastes
		Presents a danger to community if not properly buried
		Inappropriate in areas of heavy rain or if water table is near the surface
Burning (< 400°C degree), including:	Brick-oven burners	Incomplete combustion
	Drum burners	May not completely sterilize
	Pit burning	Results in heavy smoke
	Relatively inexpensive	May require fuel or dry waste to start burning
	Minimum training required	Potential for toxic emissions (i.e., dioxins or furans) if waste stream is not properly managed
	Reduction in waste volume	
	Reduction in infectious material	
Incineration (> 800 degrees Celsius)	Almost complete combustion and sterilization of used injection equipment	Relatively expensive to build, operate, and maintain
	Reduces risk of toxic emission	Requires trained personnel to operate
	Greatly reduces volume of sharps waste	May require fuel or dry waste to ignite
	Compliant with local environmental laws	Potential for toxic emissions (i.e., dioxins or furans) if waste stream is not properly managed.
Needle removal/Needle destruction	Reduces occupational risks to waste handlers and scavengers	Potential needle-stick injuries during needle removal
	Plastic and steel may be safely recycled for other uses after treatment (e.g., for buckles and coat hangers)	Fluid splash back and needle manipulation ma create opportunities of blood borne pathogen transmission
	Manual technologies available	Used needles/syringes need further treatment for disposal
		Safety profile is not established
Melting in industrial ovens	Greatly reduces volume of sharps waste	Expensive
		Requires electricity
		High capital cost
Autoclave steam sterilization followed by shredding	Sterilizes used injection equipment	Requires electricity
	May reduce waste volume	High operational costs
		High-maintenance

From the SIGN toolbox.

62

Electromagnetic Interference with Medical Devices: *In Vitro* Laboratory Studies and Electromagnetic Compatibility Standards

Kok-Swang Tan
Medical Devices Burea,
Therapeutic Products Directorate, Health Canada
Ottawa, Ontario, Canada

Irwin Hinberg
Medical Devices Bureau
Therapeutic Products Directorate, Health Canada
Ottawa, Ontario, Canada

Electromagnetic interference (EMI) from radiofrequency sources may cause a variety of medical devices to malfunction and compromise patient safety. An overview of the following EMI issues is presented: Interference with medical devices, including implantable cardiac devices, from wireless telecommunication systems, wireless local area network systems, electronic article surveillance systems, and metal detectors and assessment of the susceptibility of medical devices to EMI from these sources. This chapter includes recommendations on solutions to minimize risk and a review of international standards on electromagnetic compatibility (EMC) requirements.

Electromagnetic Interference

Electromagnetic interference (EMI) is a phenomenon that may occur when an electronic device is exposed to an electromagnetic (EM) field. Any device that has electronic circuitry can be susceptible to EMI. With the ever-increasing use of the electromagnetic spectrum and the more complex and sophisticated electronic devices, issues of EMI are attracting attention. When addressing EMI issues, consider a source, a path, and a receptor. The electromagnetic energy from the source propagates through the path and interferes with the operation of the receptor. All three must exist to have an EMI problem. The path can be conducted, radiated, inductive, or coupled with a capacitor or with electrostatic discharges, or a combination of any of the above. Therefore, to understand the effects of EMI, consider two factors: Emissions and immunity (also known as susceptibility). Emissions are a measure of electromagnetic energy from a radiofrequency source. Immunity concerns the degree of interference from an external electromagnetic energy source on the operation of the electronic device. The device will be immune below a certain level of EMI and become susceptible above that level. The three most common EMI problems are radio frequency interference, electrostatic discharge, and power disturbances. This chapter will focus on radiated interference from various radiofrequency sources.

Interference with Medical Devices

The operation of components of modern medical electronic devices such as sensitive analog amplifiers and sophisticated microprocessors can be disrupted by EMI when electromagnetic field strengths in the surrounding environment are greater than the immunity of the medical device. Electromagnetic interference has been responsible for many life-support and critical care medical device malfunctions, which raises concerns about the safety of patients who depend on these devices. Numerous incidents involving EMI with medical electrical equipment have been reported to regulatory agencies in Canada, the United States, the United Kingdom, and Australia, and published in scientific journals (Silverberg, 1993; Hayes et al., 1997; Carillo et al,, 1996). Sources of interference include wireless telecommunication systems, electronic security systems, electrosurgical devices, electrostatic discharge, and electrical disturbances from main power supply systems. However, the incidence of unreported EMI malfunctions is unknown. Between 1984 and 2000, Health Canada's Medical Devices Bureau (MDB) received 36 reports of medical device malfunction attributed to EMI (MDB, 2000; Tan and Segal, 1995). These included four reports of medical device malfunctions caused by wireless cellular phones, two cases of EMI interference from electronic article surveillance (EAS) systems on implantable cardiac pacemakers, and possibly one case of decreasing battery voltage of a pacemaker and activating the premature failure error message. The MDB also investigated reports of interference from other radiofrequency sources. These included interference of an electrosurgical device with the electrocardiogram signals displayed on the monitor of an automated external defibrillator; complete inhibition of the pacing signal of a pacemaker by a pulsating magnetic field from a video display terminal; failure of the R-wave detection circuitry of a cardiac defibrillator in the presence of a simulated muscle artifact signal from an electrocardiogram simulator; and interference of the line-isolation system in an intensive care unit with the performance of a defibrillator. These reports highlight the need for guidelines on the management of EMI within hospitals, especially in critical care areas.

The United States Food and Drug Administration Center for Devices and Radiological Health (FDA, CDRH) has also evaluated the potential EMI risks to medical devices based on the CDRH MedWatch incident report system. Silberberg published a report that cites 101 incidents the FDA received from 1979 to 1993 (Silberberg, 1992; Silberberg, 1993). Witters et al. reported in 2001 that 576 out of 150,000 incident reports received between 1984 and 1995 were found likely to be related to EMI. About 79%, or 456, of these reports involved implantable cardiac pacemakers and defibrillators, and indicated that intervention was needed to protect patients. They also reported that, in the last 10 years, the FDA has received more than 80 EMI-related reports related to security systems, such as EAS systems and metal detectors.

Hospital staff, including biomedical and clinical engineers, clinicians, nurses, and administrators, as well as the general public, have expressed concerns about EMI on incident reports and other publications. Frequently asked questions include: What is the safe distance between a cell phone and a medical device or a pacemaker? Should hospital ban the use of cell phones? Will the cell phone affect cardiac pacemakers? Will the EAS system affect the pacemaker? Can hospital use the use new wireless LAN system? Will any wireless communication systems or security systems cause inappropriate inhibition or triggering of a pacemaker pacing output, or activate asynchronous pacing on a cardiac pacemaker? Will such systems reprogram the pacemaker to a backup mode, permanently damage the pacemaker, or inappropriately trigger an implantable cardioverter defibrillator (ICD)?

Electronic circuits of cardiac pacemakers resist EMI only up to a low level of radiation. At higher levels, pacemakers react by switching to a fixed-rate mode. Strong, amplitude-modulated, low- or high-frequency EM fields can cause asynchronous pacing as well as an inhibition of the pacemaker's stimulation function. Since the introduction of wireless hand-held cell phones, the user carries not only a receiver, but also a transmitter close to the body. The transmitter emits relatively strong EM fields into the body. The emitted EM waves, with their analog, digital, or both kinds of modulation, penetrate the body, where they can influence existing electronic implants, such as pacemakers. Several research groups have investigated the interference on implantable cardiac pacemakers in the proximity of cell phones (Tan and Hinberg, 1998; Hayes and Wang, 1997; Carillo et al., 1996; Barbaro et al., 1995; Eicher et al., 1994; Imich et al., 1996; Schlegal et al., 1995; Hayes et al., 1996). In addition, researchers have reported effects of other commercial radiofrequency instruments, such as security systems, on implantable medical devices (Tan and Hinberg, 1998; Tan and Hinberg, 1999; Dodinot et al., 1993; Lucas et al., 1994; McIvor, 1995; Mathew et al., 1997; Copperman et al., 1988).

To identify emerging EMI issues, concerns, and risk management strategies, the authors investigated the susceptibility of critical care medical devices to various radiofrequency sources and assessed the risks. The sources were the following: Wireless telecommunication devices, such as analog and digital cellular phones, two-way radios, personal communication service (PCS) systems, companion technology systems, family radio services (FRS) systems wireless local area network (LAN) systems, medical telemetry systems, electronic security systems, such as EAS systems, walk-through metal detectors (WTMDs) and hand-held metal detectors (HHMDs). Medical devices used in these studies included 29 therapeutic devices, 13 monitoring devices, 10 diagnostic medical devices, 22 implantable cardiac pacemakers, and 2 implantable defibrillators. The degree to which medical electrical devices are affected by radiofrequency sources depends on the device's susceptibility to the EMI, the power and the operating frequency of the radiated sources, and the distance of the medical electrical equipment from the radiated sources.

Electric and Magnetic Field Strengths from Various Radiofrequency Sources

Before any EMI testing was done on medical devices, the researchers measured electric and magnetic field strength from various radiofrequency radiated sources as a function of distances or operating frequencies. An adjustable dipole antenna, EMCO model 3121C and an Amplifier Research log periodic antenna model AT1000, were used to measure the electric field intensity radiating from the source. The output of the antenna was connected to a spectrum analyzer (Hewlett Packard models E8541 and model E4407B). EMCO passive loop antennas, models 7604 and 6509, were used to measure magnetic field intensity generated from the source. In either case, electric and magnetic field strengths were calculated from the values obtained by the spectrum analyzer, to which both the frequency-dependent antenna factor and cable loss were added.

Wireless Telecommunication Devices

Figure 62-1 shows the electric field strengths as measured from four two-way radios and one cell phone as a function of distances. Four models of cell phones were tested with similar results. At a distance of 1 meter (m), the electric field strength measured from two-way radios and cell phones ranged from 7–15 volts/meter (V/m) and 3–5 V/m respectively, depending on their maximum output powers and operating frequencies. The two-way FRS systems produced less electric field strength than cell phones. At a distance of 1 m, electric field strengths of 2–2.5 V/m were measured. A wireless LAN system generated an electric-field strength of about 0.1 V/m at a distance of 1 m from the antenna on the system (Lucas et al., 1994).

Security Systems

Figure 62-2 shows the magnetic field strengths measured from three EAS systems, six WTMDs and 13 HHMDs. Two of the three EAS systems generated fields stronger than any metal detectors, approximately 1–5 Gauss. Walk-through metal detectors generated stronger magnetic fields than hand-held metal detectors. When compared to the Canadian Safety Code No. 18 for metal detectors (HWC, 1977), all metal detectors tested complied with the existing safety code.

Medical Devices Testing

Methodology for the Susceptibility of Medical Devices to EMI from Wireless Telecommunication Devices

Portable medical devices were tested in a vacant hospital room, while permanently mounted devices were tested on-site. Normal operating conditions for each device applied. Four models of analog cell phones with maximum power of 0.6 watts (W) and four models of two-way radios varying from 2-5 W were tested. The susceptibility of the medical device to EMI from wireless telecommunications devices was tested at a distance of 3 m. If EMI was not observed, testing was repeated at shorter distances until an effect was observed or until the telecommunication device was within 2 cm of the medical device. If EMI was observed, the distance was moved back until no effect occurred. At each distance, tests were done, first with the antenna in a vertical position and at different heights ranging from ground level to the height of the medical device. All tests were repeated with the antenna in a horizontal position and at the back, top, and both sides of the medical device. All distances were measured parallel to the ground, from the base of the antenna on the telecommunication device to the face of the medical device. After each test, the medical device was checked to confirm that it had not been permanently damaged or reprogrammed by EMI. Any observation was repeated at least three times to rule out coincidence. If it appeared that the effect changed over time (e.g., fluctuating values), the transmitting time was extended to determine whether the effect would increase, disappear, or remain within certain boundaries. A similar test procedure was incorporated into an American National Standards Institute (ANSI) recommended practice guideline (1997).

Results of EMI Testing

Wireless Telecommunication System

The results of EMI testing are shown in Table 62-1. The most obvious result is that more than 60% of the medical devices tested were affected by the presence of a wireless telecommunication device. Of these, 64% (32 of 50) were affected by two-way radios and 26% (13 of 50) were affected by analog cell phones. The two-way radios caused the greatest interference at the largest distances. The cell phones had a much smaller effect, which often occurred only at very close distances. Two-way radios caused some devices to malfunction at distances of 3 meters, while analog cellular phones did not cause malfunction at distances greater than 1 meter.

To qualitatively and quantitatively characterize the EMI-induced medical device malfunctions, each effect was assigned one of the following fault codes:
A. Visible and/or audible alarm with device stoppage
B. Visible and/or audible alarm with no device stoppage
C. Readout errors with no change in operation
D. Measured readout variation causing change in device operation

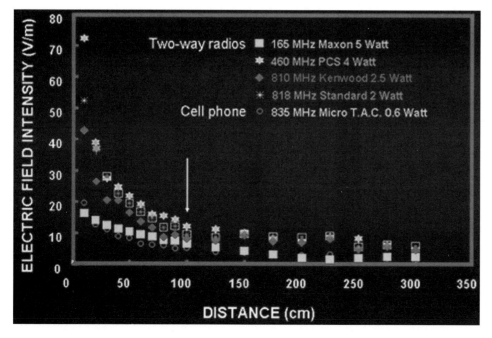

Figure 62-1 Electric field intensity measurement of wireless telecommunication devices.

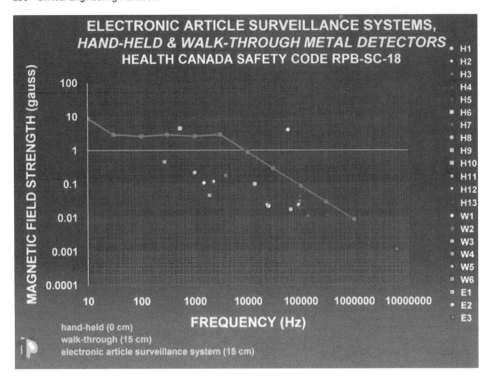

Figure 62-2 Magnetic field strength vs. frequency of electronic article surveillance systems and hand-held and walk-through metal detectors.

E. Audio indicator distortion
F. Device reboots or powers down by itself
G. Loss of input being measured
H. Distortion present in device measurement display (waveforms)
I. Device malfunction, no alarms
J. Display malfunction
K. Device needed to be manually reset to continue proper operation
L. Device changes operating mode
M. Recorder malfunction
N. Alarm malfunction
O. No observed effect
P. Device could not be approached from that direction

Figure 62-3 summarizes the percentage distribution of each fault code. The most common faults were readout errors with no change in operation (C), measured readout variation causing change in device operation (D), and device rebooting or powering-down by itself (F).

Figure 62-4 shows the results obtained at different distances. At any given distance, two-way radios caused more malfunctions and more severe malfunctions than analog cell phones. At distances less than 30 cm, two-way radios operating at 810 MHz caused more malfunctions than those operating at 165 MHz, 460 MHz and 818 MHz. Two-way radios operating at 165 MHz and 460 MHz caused malfunctions at distances greater than 3 m while the other radios had no effect. Analog cell phones did not cause malfunctions at distances greater than 1 m.

Table 62-1 shows that the wireless telecommunication devices interfered with the operation of the majority of medical devices. However, some of the devices showed an effect only at very close distances and specific frequencies.

The severity of EMI effects depended on the following factors:

- Distance from the wireless telecommunication device to the medical device
- Distance from the wireless telecommunication device to the transmission tower base station that determines the power output of the wireless telecommunication device
- Frequency of the wireless telecommunication device;
- Transmitting time, in some cases
- Shielding of the medical device and cables

These test results implied that cell phones caused a larger effect when initially transmitting a call. Once the connection was established, the severity of the effect often diminished slightly. The reason could be that cell phones adjust their power outputs depending on the distance from the operating site to the base station.

To analyze the EMI risk, all observed malfunctions were categorized into five classes (lowest risk to highest risk):

I   Change in output (e.g., distortion of waveform) but operation of device and safety of patient is unaffected
II Transient error message display, no reset of device required
III False output readings, which may affect treatment of patient
IV Error message display, manual reset of device required
V   Operation of device is affected, but there is no visual indication.

Figure 62-5 shows the distribution of malfunctions by risk category obtained with the five wireless telecommunication devices. The analog cell phones, which had lower power outputs than the two-way radios, produced fewer device malfunctions than the radios. The 165 MHz and 818 MHz two-way radios generated fewer device malfunctions than the 460 MHz and 810 MHz radios. Of all the malfunctions, 29% were Class IV and V (the two higher risk categories). Of these, 28% were caused by two-way radios, while 1% was caused by the analog cell phones.

Table 62-1 Effects of Wireless Telecommunication Devices on Medical Devices

	Two-way radios				Cell phones
Frequency (MHz)	165	460	810	818	828-846
Power (W)	5	4	2.5	2	0.6
Electric field strength measured at 1 m from the dipole antenna	7	15	11	10	5
# of Medical Devices Tested			50		
# of Medical Devices Malfunctioned			32		18

Distances (m)	Number of Malfunctions					Total
< 0.3	26	32	39	34	19	150
0.3-1.0	7	17	10	13	4	51
1.0-3.0	10	9	5	3	0	27
3.0-5.0	2	4	0	0	0	6
> 5.0	1	1	0	0	0	2
Total	46	63	54	50	23	236

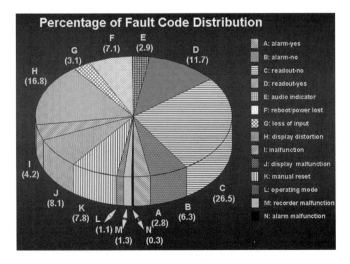

Figure 62-3 Percentage fault code distribution.

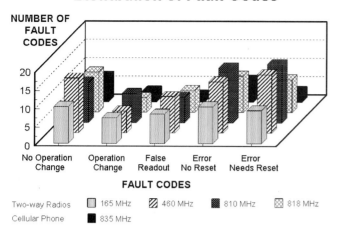

Figure 62-4 Distribution of fault codes.

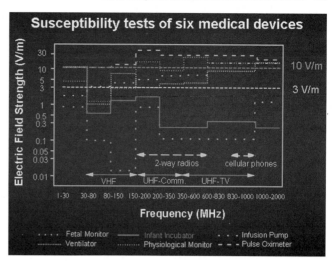

Figure 62-6 Susceptibility tests of six medical devices.

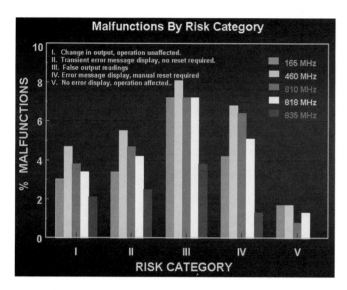

Figure 62-5 Malfunctions by risk category.

Since some medical devices showed an effect only at short distances from the telecommunication devices and only at specific frequencies, the researchers conducted susceptibility tests for a selected number of medical devices. Figure 62-6 shows EMI susceptibility curves for the six medical devices that were the most susceptible to EMI from wireless telecommunication devices. Four of the devices failed to meet the current International Electrotechnical Commission (IEC) electromagnetic compatibility (EMC) requirements of 3 V/m (IEC, 1993).

The results indicate that cell phones create a greater EMI effect when establishing a connection. Once the connection has been established, the severity of the effect often diminishes. The reason could be that cell phones adjust their power outputs depending on the distance from the operating site to the base station.

These results are generally consistent with the ad hoc testing results reported by other investigators (Segal et al., 1996; Tan and Hinberg, 1995a; Tan and Hinberg, 1995b; Rice et al., 1993; Malach, 1993; Lange, 1993; Bostrom, 1991; Tan and Hinberg, 1995c; Baba et al., 1998; Hietanen et al., 2000). Segal et al. (1996) found that two-ways radios caused malfunctions in 50-70% of medical devices tested. One device malfunctioned when the two-way radio was 3 m away. Baba et al. (1998) reported that 53% of medical devices (57 of 108) were affected by cell phones with maximum power of 0.6 W-0.8 W and concluded that these effects were low-risk. A low-power personal hand-held phone with a maximum output power of 0.08 W interfered with only 4% of the devices. Hietanen et al. (2000) studied the effects of cell phones on 23 medical devices in one hospital and concluded that GSM 1800 MHz phones were safe for use in hospitals, while GSM 900 and NMT900 phones should be restricted to areas with no sensitive medical devices.

Wireless LAN and Medical Telemetry Systems

A wireless LAN system operating at 2.42 GHz generated an electrical-field strength of 0.1 V/m at 1 m from the antenna of the system. The background electrical field strengths at each test site were below 0.1 V/m, except near the corridor of the elevator outside the operating rooms, where the electrical field strength varied from 0.15 V/m-0.25 V/m.

The authors tested 106 medical devices located throughout one 300-bed and two 800- to 900-bed hospitals in Canada to assess their susceptibility to EMI from two wireless LAN systems and one medical telemetry system (Tan and Hinberg, 2000a). When the wireless LAN system was in standby mode and within 10 cm of three models of hand-held ultrasound Doppler heartbeat detectors, the Doppler units produced periodic high-pitched beating sounds, which could be misinterpreted as normal heartbeat sounds from a patient. The periodic sounds changed to random static noise when the wireless LAN system was transmitting data. In addition, the quality of the data transmission from the wireless LAN system and its base station deteriorated in the colonoscopy room, possibly because the room was lead-shielded. A hard-wired connection from the wireless LAN would prevent deterioration of transmission quality. None of the devices was affected by a medical telemetry system operating at 466 MHz. This study indicated that a wireless LAN system, employing very low field intensities, could be used safely in close proximity to medical devices.

FRS Two-Way Radio Systems

Two models of FRS systems, operated at 461.8 MHz and 467.8 MHz, could cause malfunctions on four out of seven models of infusion pumps from four manufacturers. All four pumps either prompted an occluded message, that would require the operator to push the flow button to continue, or a failure message, that would require the operator to reset al. l settings. The interference occurred when the FRS two-way radio was about 10 cm closer to the pump. FRS systems also caused one model of infant incubator to increase its temperature display readings from nearly 50 cm away. The heating circuitry was affected and would not turn on the heater to maintain the incubator's preset temperature. In addition, FRS systems affected the trace display of two models of electrocardiograph monitors. These observed malfunctions were similar to the regular two-way radios or cell phones, both of which had higher power than FRS systems. FRS systems did not interfere with the two models of external defibrillators or two models of pulse oximeters tested.

Implantable Cardiac Device Testing

Methodology for *In Vitro* Testing

The setup for testing the susceptibility of implantable cardiac pacemakers to EMI included a human torso simulator (Tan and Hinberg, 1998). The simulator consisted of a plastic box filled with 0.18% saline solution (0.03 M). The plastic box was placed horizontally on top of a table to test wireless telecommunication devices and vertically on a table with moving wheels to test electronic security systems. The pacemaker and its leads were mounted on a grate and totally submerged in saline. The leads formed a semi-circle with a diameter of about 20 cm in the shape of a brass clef, and were fastened onto a grate. Figure 62-7 shows the schematic diagram of the experimental setup. Two stainless steel plates were used to monitor the electrical activity of a pacemaker. Signals obtained from these plates were amplified with a differential amplifier and the signal from the amplifier was displayed on a storage oscilloscope. Simulated electrocardiography (ECG) signals from a patient simulator were displayed on an ECG monitor. The amplified ECG signals from the monitor were applied to two other stainless steel plates that were mounted perpendicular to the first two and used to apply simulated ECG signals that would inhibit the pacemaker. Implantable defibrillators were tested with the same setup.

Inhibition

Any changes in the pacing signals, such as the amplitude and the pulse duration of the pacing signals, the pacing intervals between the atrial and ventricular pulses, and the pacing frequency, were recorded. This was done to determine whether the signal from the radiofrequency sources could cause the pacing outputs to induce a total, partial, or prolonged inhibition.

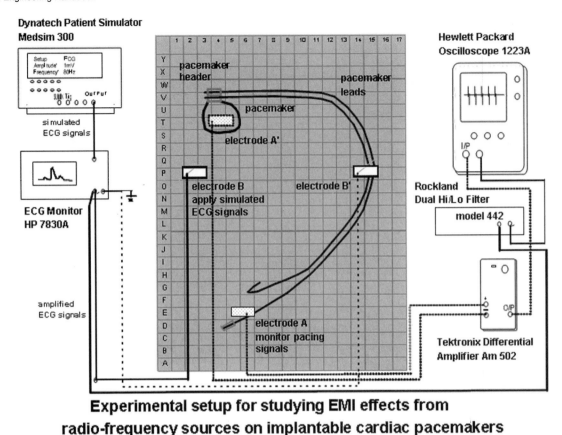

Figure 62-7 Experimental setup for studying EMI effects from radiofrequency sources on implantable cardiac pacemakers.

Reactivation

After the threshold amplitude of the amplified ECG signal was established, the output amplitude of the amplified ECG signal was set to twice its threshold amplitude. The effect could be to change the pacemaker's operating mode to an asynchronous mode with a rate different from the simulated ECG signals, or to a rate synchronous with the pacing frequency of the simulated ECG signal. When the reactivation took place, the pattern of pacing, either regular or irregular, was recorded to determine whether EMI could reactivate the pacing signal of the pacemaker after it had been inhibited by the simulated ECG signal.

Wireless Telecommunication Devices

Each wireless telecommunication device was tested at different points on the grate with the antenna of the wireless device oriented either parallel or perpendicular to the long axis of the simulator. In the case of the analog and digital cellular phones, and FRS systems, actual communications were carried out either by keying out or calling in to the wireless device. Only keyed-out communication was programmed for the PCS system.

Walk-Through Security Systems: Static and Dynamic Tests

The simulator was placed on a turntable mounted on a wooden table with rigid rubber wheels. The header of the pacemaker was 122 cm above the ground. Static tests simulated a person standing with his shoulder against the transmission panel of a security system (perpendicular orientation), or with his back against the panel (parallel orientation) by turning the turntable 90° (see Figures 62-8 and 62-9). Tests for inhibition and reactivation were conducted at various locations along the transmission panel and various turntable angles. Dynamic tests were conducted by slowly rolling the wooden table at a speed of 20 cm/s to simulate a person slowly walking through the security system. Tests were repeated for the parallel orientation with the front surface of the simulator placed parallel to the transmission panel of the security system.

Linear Positioning

For the parallel orientation, the surface of the simulator was kept parallel to the surface of the transmission panel of the security system. (See Figure 62-8.) The distance from the surface of the security system transmission panel to the header of the pacemaker was set between 8.5 and 9.4 cm to keep the distance from the mid-point of the panel to the header of the pacemaker at about 11 cm, since the thickness of the panel of the security systems varied. Figure 62-6 shows the orientations and locations relative to the EAS panel.

For the perpendicular orientation, the turntable holding the simulator was rotated 90 degrees perpendicular to the transmission panel. (See Figure 62-9.) The distance from the surface of the transmission panel to the header of the pacemaker was 12 cm. Both inhibition and reactivation tests were conducted at 5 cm intervals along the transmission panel, with the last interval varied because the width of each security system changed. Tests were repeated by rotating the turntable 90 degrees in the other direction perpendicular to the transmission panel. The distance from the surface of the transmission panel to the header of the pacemaker was changed to 23 cm. This was done to simulate a person with an implantable pacemaker at 12 cm and 23 cm from the transmission panel of the security system when the person walked through the system in opposite directions and touched the transmission panel while walking. The locus for the header of the pacemaker formed a parabolic curve with various distances from the transmission panel of the security system.

Rotation

Additional tests were carried out with the header of the pacemaker 122 cm from the ground, by rotating the torso simulator at 15-degree intervals until the simulator was perpendicular to the transmission panel. (Figure 62-9.) The simulator was kept at each position for 10 seconds. Inhibition and reactivation tests were performed for each position at three locations: at the middle and at both ends of the panel.

Hand-Held Metal Detectors

The HHMD was slowly swept up and down five times in front of the simulator to simulate a pacemaker patient being searched by security personnel. Three models of EAS systems from two manufacturers were tested. Six WTMDs from three manufacturers and 13 HHMDs from nine manufacturers were also tested.

Results of *In Vitro* Testing

Wireless Telecommunication Systems

Nine of the 20 pacemakers tested were susceptible to EMI from nearby digital cellular phones (Barbaro, 1995). Interference effects were generally not observed when the cell phone was more than 15 cm from the pacemaker. (See Figure 62-7.) Dual-chamber (DC) pacemakers were more susceptible to interference than single-chamber (SC) models. All four models of digital cell phones used in the tests decreased the output pulse rate of two SC pacemakers and two DC pacemakers. All four phones also induced rhythmic pacing in three SC pacemakers and four DC pacemakers when the output pacing of the pacemaker was inhibited by an external ECG signal. There was a ten-fold increase in the monitoring peak amplitudes of the pacing output from one SC pacemaker and one DC

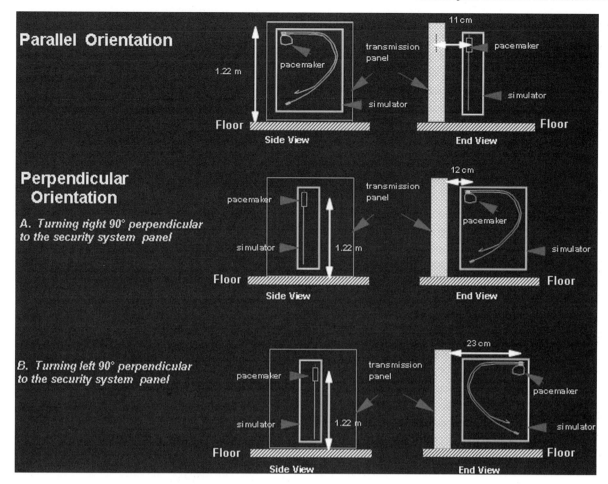

Figure 62-8 Orientations for the testing of security systems on cardiac pacemakers.

pacemaker from the same manufacturer. The 3.4% rate of interference is consistent with the results reported in other *in vitro* and *in vivo* studies. (See Table 62-2.) The EMI generated by the wireless devices did not reprogram the pacemakers. The interference effects ceased when the phones were turned off, and the pacemakers returned to their normal operations. Analog cell phones, digital PCS systems, and FRS systems did not cause any interference with pacemakers. These findings suggest that pacemaker-dependent patients should use analog or PCS phone systems. This is consistent with the findings by Hayes et al. (1997) that analog cellular phones are safe for patients who have pacemakers.
The study showed that some pacemaker designs appeared to be immune to EMI. Two models of DC pacemaker, with and without filters, were tested to investigate the effectiveness of filtering circuitry. No EMI effects were observed in the model with filters, and transient inhibition was observed in the model without filters. A clinical study by Carillo et al. (1996) confirmed that there were no interference effects on pacemakers with filters. These findings suggest that improvements in filtering technology to reduce the susceptibility of pacemakers to radiofrequency EMI should be encouraged.

To determine the vertical distances needed to induce EMI effects between the wireless system and a pacemaker, additional testing was done on the pacemakers found to be susceptible to EMI. A digital cellular phone was placed on wooden blocks above the grate of the simulator and the distance was measured when an inhibition or a reactivation was observed. Figure 62-10 shows the average 3-D EMI effects on nine pacemakers. The average maximum vertical distance at which interference was observed was 3.4 cm above the pacemaker. However, in one case interference was observed at a distance of 40 cm from the pacemaker. In all cases, EMI occurred only when the phone was in the transmission mode.

In the test of FRS two-way radio systems, 21 out of 22 models of cardiac pacemakers and two models of implantable defibrillators were not affected. FRS systems caused one model of dual-chamber cardiac pacemakers to increase the pacing amplitude by 1.5-fold and keep the same pacing frequency. When the FRS transmission stopped, the interference diminished.

Two wireless LAN systems were tested in close proximity of 22 models of cardiac pacemakers and two models of implantable defibrillators; none of the pacemakers or defibrillators was affected.

Electronic Article Surveillance Systems

The authors also studied the susceptibility of 22 pacemakers and two implantable defibrillators to EMI from three types of EAS systems: Pulsed magnetic field (UM), continuous magnetic field (SM), and sweep-frequency magnetic field (CS) (Tan and Hinberg, 1998 and 1999). During static tests, UM systems decreased the electrical output pulses of eight of the pacemaker models, while SM systems affected five pacemaker models, and CS systems did not affect any pacemakers (Table 62-3). The decrease in the pacing frequencies of some pacemakers may be of concern to patients who are fully dependent on their pacemakers. Pacemaker-dependent patients may experience dizziness or collapse if pacing ceases for more than 3 seconds. When the output of the pacemaker was inhibited by an external ECG signal, UM systems induced rhythmic pacing on 15 pacemakers at distances up to 33 cm from the transmission panel, while SM systems affected 12 pacemakers, at distance up 18 cm. This asynchronous pacing could compete with the normal sinus rhythm to induce arrhythmia. Dodinot et al. (1993) and Lucas et al. (1994) reported

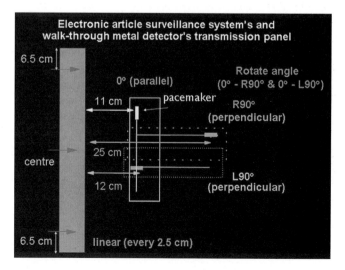

Figure 62-9 Linear and rotation modes for the testing of security systems on cardiac pacemakers.

Table 62-2 Effects of Various Wireless Telecommunication Devices on Pacemakers: *In Vitro* and *In Vivo* Studies

	Power (W)	Frequency (MHz)	in vitro studies			in vivo studies
			Health Canada	Univ. of Oklahoma	US FDA	
# of Pacemakers			20	29	30	975 patients
Incidence of interference			3.4 %	4.7 %	-	20 %
Analog cellular phone	0.6	828	0 %	0 %	0	0.5 %
TDMA-50	0.6	835	3.4 %	4.7 %	10 %	4.2 %
TDMA-11	0.6		-	-	36.7 %	10.5 %
CDMA	0.6		-	2.8 %	-	3.1 %
PCS	0.6	1810	0 %	0.6 %	-	0.2 %
GSM	0.6		-	-	0	-
FRS	0.1	468	0 %	-	-	-

Zero indicates no interference.

interference effects from other EAS systems. In addition, the bipolar lead configuration appeared to be less susceptible to interference from SM systems than the unipolar lead configuration. This finding is consistent with those of Dodinot et al. (1993) and Lucas et al. (1994). Furthermore, the same EMI effects occurred inside and outside the transmission panels of the EAS gate. There was no interference when the simulator was rolled through any EAS system during the dynamic tests. The two defibrillators were also unaffected by any EAS systems. However, McIvor (1995) and Mathew et al. (1997) have reported that patients with implantable defibrillators experienced inappropriate firing while in close proximity to an EAS system. None of the EAS systems caused any permanent damage or reprogramming to the pacemakers and defibrillators. These findings suggest that pacemaker patients should not stop within 33 cm of either side of the transmission panel of an EAS.

Walk-Through and Hand-Held Metal Detectors

None of the 13 HHMDs interfered with the 22 pacemakers and the two implantable defibrillators tested (Hayes et al., 1996). One WTMD decreased the pacing frequencies of five SC and three DC pacemakers during the inhibition tests, and 10 SC and four DC pacemakers induced reactivation. The other five WTMDs caused interference with only two SC pacemakers during inhibition and reactivation tests. No interference was observed when the simulator was rolled through the WTMDs during the dynamic tests. These findings indicate that WTMDs may affect some pacemaker models if the patient remains near the detector for longer than 2 seconds. However, these findings are not consistent with the report by Copperman et al. (1988) that WTMDs have no effect on implanted pacemakers. None of the WTMDs interfered with the implantable defibrillators except for one model of WTMD, which stopped defibrillation on one model of defibrillator when a ventricular fibrillation was simulated. In no case did the WTMDs or HHMDs cause reprogramming or permanent damage to any pacemakers or defibrillators. Asch et al. (1997]) reported that airport security could affect cardiac pacemakers and other implanted metallic devices. These studies suggest that pacemaker and defibrillator patients should not stop within 33 cm of either side of the transmission panel of the WTMDs.

Discussion

Although many device manufacturers have addressed EMC, there still appears to be insufficient awareness of EMI problems. The need to educate users, manufacturers, and regulators about EMI and to develop EMC standards persists. Preventing all EMI malfunctions is difficult given the challenge of making medical devices immune to all sources of EMI. The exposure to an EM environment, and the frequency, location, orientation, and design of a device can influence whether and how a device will be affected by EMI. In practice, it is impossible to completely stop EM energy at its source. Modern society has become too dependent on the convenience of instant communication. Some medical devices themselves often emit EM energy, and simultaneous use of several medical devices in a hospital room can also cause EMI problems.

All ad hoc studies of EMI susceptibility of medical devices have consistently reported that wireless telecommunication devices can cause a high percentage of medical devices to malfunction. In contrast, a relatively small number of EMI incidents have been reported to regulatory agencies in Canada and the United States. Under-reporting may be due to a lack of awareness of EMI as the cause of device malfunction. Based on the studies mentioned above, researchers (Tan et al., 1995; Siegel et al.,1996) have suggested that hospitals should develop policies for handling the use of any wireless telecommunication devices. Participants in the Round Table on EMC convened by Health Canada in September 1994 unanimously agreed that cell phones should not be banned in hospitals, but that policies should be established to govern their use (Tan et al., 1995).

The results presented here confirm that the very low field intensities generated by wireless LAN systems do not interfere with medical devices in the hospital. These findings sug-

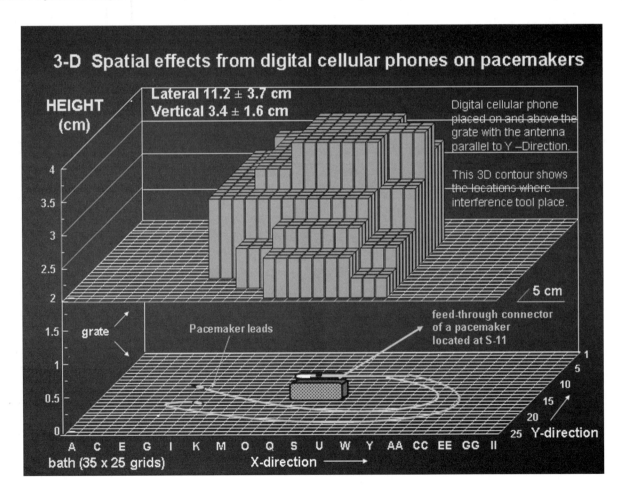

Figure 62-10 3-D spatial effects from digital cellular phone on pacemakers.

Table 62-3 Effects of security systems on pacemakers

Type	Mode	Carrier frequency	Magnetic field strength (ΦT)*	Effects on pacemakers	
				Inhibition	Reactivation
EAS	Continuous	535 Hz	450	23 %	55 %
	Modulated pulse	Carrier: 58.4 KHz modulation: 60 Hz	400	36 %	68 %
	Sweep	7.4-9.1 MHz	0.1	0	0
WTMD	Repetitive pulse	250-500 Hz	4.5 -10	5 %	9 %
	Repetitive pulse	89 Hz	45	36 %	64 %
	Modulated Pulse	250 -909 Hz	18 -22	5 %	9 %
	Modulated pulse	210 Hz	12	9 %	14 %
HHMD	Continuous	14 kHz -1.8 MHz	0.2 -10	0	0

*Measured at 15 cm from the transmission panel of EAS and WTMD systems, and 2.5 cm from HHMDs; zero indicates no interference effects.

gest that these wireless systems are acceptable for use in hospitals in view of the benefits of obtaining real-time access to patient medical information. Nevertheless, hospitals should test each new wireless communication system on potentially susceptible devices before putting them into general use. One of the conclusions of the Canadian Task Force on EMC in Health Care, established in 1994 after the Round Table Discussion on EMC in Health Care (Tan et al., 1995), was that "the potential for EMI is minimized by using RF sources having the lowest possible transmission power" (Siegel et al., 1996). The task force recommended replacing high-power sources with low-power ones. The findings described in this chapter support the use of low-power wireless communication devices in hospitals.

Furthermore, these studies confirm that EMI from nearby digital cellular phones may affect the operation of some implanted cardiac pacemakers under worst-case conditions. The greatest risk of interference occurs when the antenna of the phone is in close proximity to the pacemaker. The resulting decrease in the frequencies of the output pulses or transient responses of some pacemakers may be of concern to patients who are fully dependent on their pacemakers. The unwanted rhythmic pacing induced in some pacemakers during reactivation is a concern, since it could compete with the normal sinus rhythm of the heart and induce arrhythmia. In all cases that produced EMI effects, if the phone is in stand-by mode and is placed in close proximity to the pacemaker, an incoming call will switch the phone to its transmission mode and may interference with the pacemaker. When the phone is turned off, the interference stops, and the pacemaker resumes its normal operation.

EMI did not occur with the analog cellular phone or the PCS system, probably because of the low power output of these communication devices. These findings suggest that pacemaker-dependent patients should use an analog cellular phone or a PCS system.

Strategies to Reduce EMI Risk

Managing EM Fields Within and Outside Hospitals

Hospital administrators and biomedical engineers or clinical engineers should understand radiofrequency and its sources. Hospital administrators should consider developing hospital policies for controlled use of radiofrequency sources and the frequency spectrum. Administrators should also consider integrating radiofrequency sources and systems with their existing communication systems to eliminate any uncontrolled communication sources in hospitals.

Managing Medical Device Immunity

Hospital administrators and clinical and technical staff should be educated about EMI-related risk. It is important for biomedical engineers, clinicians, and nurses to be aware of EMI-related incidents and report those incidents to the correct authority. If any staff member finds any sensitive medical devices susceptible to EMI, they should relocate them or separate them. Biomedical or clinical engineers should determine the immunities of their life-support and critical care medical devices. Due to financial constraints, many hospitals continue to use older medical devices that were not designed or tested for electromagnetic compatibility. Biomedical and clinical engineers should advise hospital administration that all medical devices considered for purchase must at least comply with existing EMC standards.

Hospital Policies

ECRI recommends against the use of mobile telephones or hand-held radio transmitters (walkie-talkies) in health care institutions because of the potential for their interference with the safe operation of life-support or critical monitoring equipment (ECRI, 1993).

The Regina Health Board has prohibited the use of cellular phones and other transmitting devices in certain areas of hospitals (Regina Leader-Post, 1993). The Royal Alexandra Hospital in Edmonton has a policy that only emergency service personnel who require constant communication during emergency situations, may use their cellular phones and portable two-way radios in patient rooms and critical patient care areas. These telecommunication devices must be used as far away from any nearby medical devices as possible (e.g., 3– 4 m). If possible, users of portable two-way radios should leave the patient area to make the transmission. The policy also requires signs at all public entrances to the hospital to warn visitors to turn off cellular phones or two-way radios, and signs at all entrances to critical care areas to restrict the use of cellular phones and two-way radios. Individuals who need to use cellular phones and two-way radios are advised to use them in places where medical equipment is not used, such as lounges, waiting rooms, offices, and cafeterias.

Many hospitals in Scandinavian countries have prohibited the use of mobile telephones (and radio transmitters) within the entire hospital area (Bostrom, 1991). Some hospitals prohibit the use of mobile telephones within certain distances (e.g., 8 m) from a medical device.

Section 5.3.6 of the Canadian Standards Association (CSA) publication CAN/CSA Z32.2-M89, "Electrical Safety in Patient Care Areas" (CSA, 1989), states that "Patient-owned radio transmitters of any kind, (e.g. citizen band radios, walkie-talkies, portable or cellular telephones, amateur radios, etc) should not be permitted in or near intermediate and critical care areas."

Recommendations for Implantable Cardiac Devices

The chances that EMI from digital cellular phones would produce life-threatening situations are low. However, data presented here support the recommendations in the Health Protection Branch's Medical Device Alert No. 108 (MDA, 1995) that cautions patients with pacemakers as follows:

1. The use of a digital cellular phone very close to the pacemaker may cause the pacemaker to malfunction.
2. It is advisable to avoid carrying the cellular phone in a breast pocket directly over the pacemaker because an incoming call will switch the phone to its transmission mode and may cause interference.
3. In using a cellular phone, patients should hold it to the ear farthest from the pacemaker.

Wireless Technology Research (WTR) and the Health Industry Manufacturers Association (HIMA) Pacemaker Interest Group recommend 15 cm as an adequate separation distance between a pacemaker and a wireless phone (WTR, 1996). This distance appears to be valid for minimum risk with most pacemaker designs.

Interference with pacemakers by security systems such as EAS systems and WTMDs may pose a potential risk, depending on the model of pacemaker, the programmed mode, and the type of the security system. The greatest risk of interference occurs when the pacemaker is close to the security system and the individual is stationary. This condition is unlikely to be encountered during normal exposure to an EAS system or walk-through metal detector, since pacemaker patients do not remain stationary within these systems.

Interference with pacemakers occurs on both sides of the transmission panel of an EAS system or a walk-through metal detector and at distances as far as 33 cm for UM, 18 cm for SM, and 33 cm from WTMDs. Patients may incorrectly assume that EMI effects occur only when they are inside the gate. Data described in this chapter suggest that pacemaker patients should not stop and linger within 33 cm of either side of the transmission panel as measured from the patient's shoulder to the transmission panel of the security system. Instead, they should walk through the system at a normal pace without stopping. Manufacturers should consider placing a sign on the transmission panel to identify the presence of a security system and advise pacemaker and defibrillator patients to walk through the system at a normal pace. In addition, EAS systems should not be installed in the vicinity of cash registers in such a way that pacemaker-dependent patients must stand within 33 cm of the EAS system while completing their transactions.

EMI Awareness

Many EMI malfunctions could be prevented through education. Health Canada places a high priority on providing information to the public. *Medical Device Alert* and *It's Your Health* bulletins are sent to health care professionals to warn them of problems and provide recommendations for management of the associated risks (Tan and Hinberg, 2000b). Health Canada scientists have also published scientific papers on EMI and have given many lectures on the topic at scientific conferences and workshops (Tan et al., 1995; Tan and Hinberg, 1999; Tan and Hinberg, 1995a; Tan and Hinberg, 1995b; Tan and Hinberg, 1995c; Tan and Hinberg, 2000b).

Electromagnetic Compatibility (EMC) Standards

The U.S. FDA issued a voluntary standard MDS-201-0004 in 1979 specifying that radiated immunity of medical electrical equipment should be 7 V/m from 450 MHz to 1000 MHz. The international collateral standard IEC 60601-1-2 on electromagnetic compatibility for medical electrical equipment states that an immunity of 3 V/m shall apply for the frequency range from 26 MHz to 1000 MHz. However, a higher level of immunity was proposed in recognition of risks to life-support equipment. The draft second edition of the IEC 60601-1-2 proposes 10 V/m for life-support equipment and 3 V/m for other medical electrical equipment for the frequency range from 80 MHz to 2.4 GHz. However, immunity of 10 V/m does not guarantee that EMI effects will not occur. In addition, IEC 60601-1-2 (1st edition) requires testing with 1 kHz amplitude modulation at a single modulation frequency. The draft 2nd edition proposes testing at 80% amplitude modulation at 2 Hz if the physiological simulation frequency and operating frequency of the medical device under test is less than 1 Hz or greater than 3 Hz, and it is no longer required to test at 1 kHz.

The need for establishing specific EMC standards for medical electrical equipment and systems is well recognized. Designing medical devices to be immune to electromagnetic fields up to 10 V/m for life-support medical electrical equipment and 3 V/m for non-life-support medical electrical equipment, as proposed in international standards, would greatly reduce the hazard from EMI. The ISO TC150SC6 Working Groups 1 and 2 are drafting specific EMC standards for implantable cardiac pacemakers and defibrillators.

Conclusion

In conclusion, management of the use of radiofrequency sources and attention to the electromagnetic compatibility of medical devices in health care facilities will provide some assurance as to the safe use of medical devices and may prevent EMI problems.

References

Asch M, Liu D, Mawdsley G. Detection of Implanted Metallic Devices by Airport Security. *J Vasc Interv Radiol* 8:1011-1014, 1997.

Baba I, Furuhata H, Kano T, et al. Experimental Study of Electromagnetic Interference From Cellular Phones with Electronic Medical Equipment. *J Clin Eng* 23:122-134, 1998.

Barbaro V, Bartolini P, Donato A, et al. Do European GSM Mobile Cellular Phones Pose a Potential Risk to Pacemaker Patients? *Pacing Clin Electrophysiol* 18:18-24, 1995.

Bostrom U. Interference from Mobile Telephones: Challenge for Clinical Engineers." *Newsletter of the International Federation for Medical & Biomedical Engineering*, November 10, 1991.

Canadian Standards Association. Electrical Safety in Patent Care Areas." CAN/CSA-Z32.2-M39, Toronto, Canadian Standards Association, 1989.

Carillo RG, Williams DB, Traad EA, et al. Electromagnetic Filters Impede Adverse Interference of Pacemakers by Digital Cellular Phones. *J Am Coll Cardiol* Abstract 901-922, 15a., 1996.

"Cell Phone Warning. Regina Leader-Post, February 16, 1993.

Copperman Y, Zarfati D, Laniado S. The Effect of Metal Detector Gates on Implanted Permanent Pacemakers. *Pacing Clin Electrophysiol* 11:1386-1387, 1988.

Dodinot B, Godenir JP, Costa AB. Electronic Article Surveillance: A Possible Danger for Pacemaker Patients. *Pacing Clin Electrophysiol* 16:46-53, 1993.

ECRI. Guidance Article: Cellular Telephones and Radio Transmitters: Interference With Clinical Equipment. *Health Devices* 22, 1993.

Eicher B, Ryser H, Knafl U, et al. Effects of TDMA-Modulated Hand-Held Telephones on Pacemakers (Abstract I-1-10). *Proceedings of the Bioelectromagnetics Society Conference*, Stockholm, Sweden, June 1994.

Hayes DL, Wang PJ, Raynolds DW, et al. Interference With Cardiac Pacemakers by Cellular Phones. *N Engl J Med* 336:1473-1479, 1997.

Hayes DL, Carillo RG, Findlay GK, et al. State of the Science: Pacemaker and Defibrillator Interference from Wireless Communication Devices. *Pacing Clin Electrophysiol* 19:1419-1430, 1996.

Hietanen M, Sibakov V, Hallfars S, et al. Safe Use of Mobile Phones in Hospitals. Helsinki, Finnish Institute of Occupational Health, Technical Research Centre of Finaland, 2000.

IEEE. *American National Standard Recommended Practice for an On-Site, Ad Hoc Test Method for Estimating Radiated Electromagnetic Immunity of Medical Devices to Specific Radio-Frequency Transmitters ANSI C63.18-1997*. Institute of Electrical and Electronic Engineering, New York, 1997.

Irnich W, Batz L, Muller R, et al. Interference of pacemakers by mobile phones. *Proceedings of the Bioelectromagnetics Society Conference*, pp 121-122, 1996.

Lange S. Cellular Phones and Transmitting Devices in Critical Care Areas. Regina General Hospital, Regina, Saskatchewan, Private Communication, 1993.

Lucas EH, Johnson D, McElroy BP. The Effects of Electronic Article Surveillance Systems on Permanent Cardiac Pacemakers." *Pacing Clin Electrophysiol* 17:2021-2026, 1994.

Malach T. "Preliminary Testing of the EMI susceptibility of medical equipment," University of Alberta Hospital, Edmonton, Alberta, Private Communication, 1993.

Mathew P, Lewis C, Neglia J, et al. Interaction Between Electronic Article Surveillance Systems and Implantable Defibrillators: Insights from a Fourth-Generation ICD. *Pacing Clin Electrophysiol* 20:2857-2859, 1997.

McIvor ME. Environmental Electromagnetic Interference from Electronic Article Surveillance Devices: Interactions with an ICD. *Pacing Clin Electrophysiol* 18:2229-2230, 1995.

Health Canada. Medical Device Alert No.108: Digital Cellular Phone Interference with Cardiac Pacemakers. Health Canada, Ottawa, Canada, 1995.

Health Canada. *Medical Devices Incident Report Database*, Medical Devices System, Medical Devices Bureau, Therapeutic Products Programme, Health Canada.

IEC. Medical Electrical Equipment Part 1: General requirements for safety. 2. Collateral standard: Electromagnetic compatibility -Requirements and tests; IEC 60601-1-2. Geneva, International Electrotechnical Commission, 1993.

Rice ML, Smith JM. Study of Electromagnetic Interference Between Portable Cellular Phones and Medical Equipment. *Proceedings of the Canadian Medical Biology Engineering Society Conference* 17, 1993.

Health Canada. Safety Code Recommends Safety Procedures for the Selection, Installation and Use of Active Metal Detectors RPB-SC-18, Health and Welfare Canada, January 1977.

Schlegal RE, Raman S, Grant FH, et al. In-Vitro Study of the Interaction of Cellular Phones with Cardiac Pacemakers. Proceedings of the Workshop on Electromagnetics, Health Care and Health, Montreal, Canada, pp 33-36, 1995.

Segal B, Retfalvi S, Pavlasek T. Sources & Victims: The Potential Magnitude of the Electromagnetic Interference Problem. In Sykes S (ed). *Electromagnetic Compatibility for Medical Devices: Issues and Solutions*. Anaheim, CA, Food & Drug Administration and Association for the Advancement of Medical Instrumentation, 1996.

Segal B, Retfalvi S, Townsend D, et al. Recommendations for Electromagnetic Compatibility in Health Care." *Proceedings of the Canadian Medical Biology Engineering Conference* 22:22-23, 1996.

Silberberg, JL. Medical Device Performance Degradation Due to Electromagnetic Interference: Reported Problems. *Proceedings of the EMI Control in Medical Electronics*, Interference Control Technologies, June 15-17, 1992.

Silberberg JL. Performance Degradation of Electronic Medical Devices Due to Electromagnetic Interference. *Compliance Engineering* 10:25-39, 1993.

Tan KS, Hinberg I. Investigation of Electromagnetic Interference with Medical Devices in Canadian Hospitals. *Proceedings of the 17th IEEE EMBS Annual Conference and 21st CMBEC* 17:105, 1995.

Tan KS, Hinberg I. Malfunction in Medical Devices Due to EMI from Wireless Telecommunication Devices. *Proceedings of the 30th AAMI* 30:96, 1995.

Tan KS, Hinberg I. Can Wireless Communication Systems Affect Implantable Cardiac Pacemakers? An In Vitro Laboratory Study. *Biomed Instrum Technol* 32:18-24, 1998.

Tan KS, Segal B, Townsend D, et al. *Proceedings of the Round Table Discussion on Electromagnetic Compatibility in Health Care*. Ottawa, Canada, 1995.

Tan KS, Hinberg I. Can Electronic Article Surveillance Systems Affect Implantable Cardiac Pacemakers and Defibrillators? *Pacing Clin Electrophysiol* 21:960, 1998.

Tan KS, Hinberg I. A Laboratory Study of Electromagnetic Interference Effects from Security Systems on Implantable Cardiac Pacemakers. *Proceedings of the XXVI General Assembly, International Union of Radio Science on Electromagnetic Interference with Medical Devices* 868, 1999.

Tan KS, Hinberg I. Effects of a Wireless LAN System, a Telemetry System and Electrosurgical Devices on Medical Devices in a Hospital Environment. *Biomed Instrum Technol* 34:115-118, 2000.

Tan KS, Hinberg I. Susceptibility of Medical Devices to Radiofrequency Electromagnetic Fields. *Proceedings of the Advanced Clinical Engineering Workshop*, Santo Domingo, 2000.

United States Congress: Wireless Technology Research, Final Report: Evaluation of Interference Between Hand-Held Wireless Phones and Implanted Cardiac Pacemakers. Washington, DC, 1996.

Witters D, Portnoy S, Casamento J, et al. Medical Device EMI: Analysis of Incident Reports: Recent Concerns for Security Systems and Wireless Medical Telemetry. *Proceedings of the IEEE EMC Symposium*, 2001.

63

Electromagnetic Interference in the Hospital

W. David Paperman
Clinical Engineering Consultant
Cut 'N Shoot, TX

Yadin David
Director, Center for TeleHealth and Biomedical Engineering Department, Texas Children's Hospital
Houston, TX

James Hibbetts
Texas Children's Hospital
Houston, TX

The density of occupancy of the electromagnetic spectrum is increasing because of the expansion of wireless services (the culprits). Despite the efforts of manufacturers to harden clinical devices to the effects of electromagnetic interference (EMI), reports of new incidents of interference to previously unaffected medical devices (the victims) appear in medical literature and anecdotally (CBS, 1994; Paperman et al, 1994). The role and the knowledge base of the clinical engineer must expand to understand and manage these ever-increasing challenges.

Electromagnetic Radiation

Electromagnetic radiation occurs when an alternating current is generated. An electromagnetic field is created in the vicinity of the source. The range, or distance of the radiated electromagnetic field can be increased when coupled to a conductor. The magnetic field is further radiated by the flow of the current along this conductor. Even at a reduced amplitude, a corresponding current will be generated when another conductor is subjected to the field of the radiating conductor. The radiated field will have many characteristics. The most important of those characteristics are amplitude, periodicity, and waveform.

The amplitude of the impressed alternating current defines the energy of a radiated field. This field is generally expressed in volts per meter (V/M). Amplitude at the source point (transmitting conductor) will define, in conjunction with frequency and distance, the amplitude of the electromagnetic field induced in the secondary (receiving) conductor subject to the Inverse Square Law.

The periodicity or frequency of alternation is expressed in Hertz per second, which allows the calculation of the wavelength of the radiated electromagnetic field. Knowing the wavelength versus the physical length of both the radiator and the secondary conductor into which the radiated energy is induced provides an estimate of the potential amplitude developed by or within a device at risk. Waveforms other than linear (sinusoidal) can produce multiple and variable frequencies. The foremost example of a nonlinear waveform is the square wave. The square wave produces many multiples, or harmonics, which have a range many times that of the fundamental frequency. The majority of digital devices generate square waves.

Depending on the amplitude at the point of generation, the efficiency of the auxiliary radiator (antenna), the generated frequency, and the waveform, the potential for interference between digital devices and clinical devices can exist for many kilometers from the source of electromagnetic radiation.

EMI threats come from a multitude of sources. These are broadly divided into two categories: Devices that emit intentional electromagnetic radiation for communications and control, and devices that, as a byproduct of their operation, emit unintentional (incidental) radiation (Figure 63-1). Some of the major and more common sources of EMI encountered in the clinical environment follow.

Intentional Radiators:

- Television broadcast stations: Analog and digital
- Commercial radio stations: Analog and digital
- Land mobile radio: Fixed base (FB), mobile, and portable two-way radio sources (walkie talkies)*
- Paging: One-way and two-way wireless messaging service transmitters
- Cellular telephones and sites*
- Wireless personal digital assistants (PDAs)
- Wireless networking devices (an increasing threat in the 2.4 GHz industrial, scientific, and medical ISM band)
- Unlicensed and unauthorized users of two-way radio communications equipment (Pirates)*

Unintentional Radiators:

- Lighting systems (especially florescent), including energy saving electronic ballasts*
- High energy control systems (HVAC controllers), especially variable-speed controllers*
- Malfunctioning electrical services*
- Universal-type electric motors*
- Pulse oximeters*
- Displays (CRT and plasma), computer, television, and instrumentation
- Wired computing networks
- "Smart" fire detection and alarm devices*
- Electrosurgical units (ESU)*
- Defibrillators

These sources of electromagnetic radiation, as seen from the previous lists, can be intentional or unintentional. The challenges and the institutional responses imposed by the presence of these EMI sources can be divided, for all intents and purposes, into four parts: Mitigation, detection, correction, and prevention. Although the mitigation and prevention phases appear to be similar, they differ in their approaches and applications.

Controlling the Effect of EMI

Mitigation

In the mitigation phase, the principal concern is reducing risk to, or victimization of, clinical devices by devices emitting unintentional or intentional electromagnetic radiation. The sources of the devices may be internal or external to the institution. The devices may be intentional or unintentional radiators of electromagnetic fields, and the electromagnetic energy may be radiated or conducted. Figure 63-1 illustrates the electromagnetic interaction among intensional and unintentional radiators and intentional and unintentional targets.

Mitigation is accomplished in part by the careful analysis of a device, including the accessories to be used with it, and the electromagnetic environment in which it will be used. A successful approach has been implemented at Texas Children's Hospital. This approach is accomplished in two phases: Thorough analysis of the environment in which the device is to be placed, called "footprinting," and an equally thorough analysis of the clinical device itself, called "fingerprinting." The procedures of quantifying potential EMI and mitigating the effects through environment analysis are described in greater detail later in this chapter.

---

*Sources (the culprits) of incidents attributed to EMI encountered at Texas Children's Hospital

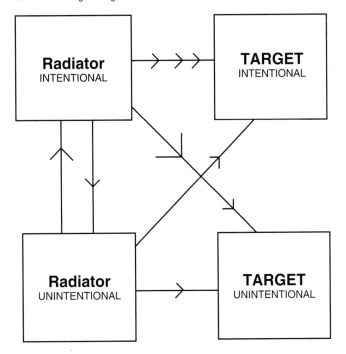

Figure 63-1 The complexities of EMI-radiators and targets: Culprits and victims.

Another vital part of the mitigation process occurs during facility planning. Whether planning a new facility, an expansion of an existing facility, or remodeling of a facility, clear and concise communications between affected departments, architects, contractors, and the clinical engineering department are vital at all stages of the project. For example, one institution installed expensive, screened rooms at great cost. When the clinical engineer, acting on complaints of erratic EMG operation, determined that the room played a role in the culprit/victim relationship, the rooms were disassembled, also at great cost. In this case, the relocation of the affected department was not an option. The clinical engineering department had not been made aware of these rooms, or of their intended application that involved the use of known culprit devices.

Examples of proactive mitigation include consulting services provided by the Biomedical Engineering Department and the Television Services Group during the latest expansion of Texas Children's Hospital. These services, which involved the expertise of all groups within the biomedical engineering department, included not only space design, but the design, implementation, and expansion of wireless paging and communications systems. Through the use of a distributed energy antenna systems, (i.e., leaky coax-radiax) radio frequency power levels were maintained at levels deemed safe for clinical devices while providing the required coverage area for a nurse call–specific paging system. Safe wireless telephonic communications were implemented using low-power microcellular systems. Infant abduction systems were reviewed for specifications including emission characteristics, and a system was selected based in part on its low electromagnetic radiation levels.

The use of portable radio communication devices in a large campus environment is a basic requirement for security, engineering, and guest services. As part of the mitigation process, instruction in the safe use of two-way radio equipment in the clinical environment is mandatory for all personnel who use this equipment (ECRI, 1993).

Detection

Detection is implemented upon the report of a device malfunction. A preliminary analysis of the incident may indicate that the cause of the malfunction, intermittent or permanent, may have been electromagnetic interference. In the case of continuing or continuous device malfunction not otherwise attributable to defects within the device itself (e.g., no problem found [NPF] service reports), a careful investigation begins, using various test equipment, some of which may require specialized construction. This section presents a more detailed description of the basic equipment necessary to locate and identify sources of EMI and some of the basic techniques, referred to as "ghost hunting." Once the type and source of the interference is detected and analyzed, the next step is to reduce or eliminate the interference to the victim device. However, the victim device is removed in some cases, because of an intensely hostile electromagnetic environment. Replacement of the victim device with an equivalent device that may have other, less sensitive responses is an option.

The detection phase can eliminate the possibility of electromagnetic interference as the culprit and cause of improper operation of a specific device. Indeed, subsequent investigation can reveal technical problems within a device that were previously attributed to electromagnetic interference.

Correction

Correction of victim/culprit relationship(s) can take many forms, some practical, some impractical. Correction may be part of a process to ensure that the victim device meets all specifications that can affect its susceptibility to electromagnetic radiation. For example, hospital staff should ensure that proper case-to-case contact is made in coated conductive coatings and that the coatings have not been worn or abraded. In some instances, the victim device can and should be removed from the environment where it is at risk. If the culprit is local, removal may be one practical solution to the problem. Instances of electromagnetic radiation emanating from abandoned wiring (passive reradiation) have occurred that dictated removal of the wiring. Interference to clinical devices originating from active (nonfiberoptic) wiring, such as networking trunk conduits, mandates rerouting of the wiring. The same can apply to modern in-plant telephone systems, usually digital in nature. Experiences at Texas Children's Hospital support the work of researchers (ECRI, 1988) in finding that shielded rooms rarely correct problems when the source of electromagnetic radiation is contained within the local environment. In extreme cases, the existing shielded room must be disassembled or the victim equipment (and department), moved to other quarters within the institution.

Prevention

Sustainable EMI prevention must include institution-wide compliance with guidelines and with those policies created within an institution to limit possible sources, EMI as well as the selection and deployment of medical devices that offer more effective immunity from EMI. Overall, observance of this two-pronged policy will have the effect of reducing the risk of electromagnetic interference to the proper operation of clinical devices and therefore reducing the risk to the institution. At Texas Children's Hospital, EMI prevention takes several forms (David, 1993). A proactive policies and procedures manual that, guided by the biomedical engineering department, defines allowable sources of radiation within the institution and mandates training procedures for employees of the institution that are required by job necessity to carry sources of electromagnetic radiation, i.e., intentional radiators. As an example, signage mandated by the policies and procedures manual directs that all cellular telephones be turned off. Their use within the institutional campus is not allowed. The policies and procedures manual further mandates that all employees using radio equipment, especially handheld devices, be trained in their safe use within the institution. Additionally, a stipulation that clinical devices must be EMI compatible is incorporated into the Condition of Sale, issued and agreed to by to vendors, to define the technical conditions that must be met under the purchase contract.

Case Histories

Role Reversal

An unusual example occurred when the victim (i.e., a pager), normally considered a culprit (i.e., a physiological monitor), and the culprit, normally a victim, exchanged roles. Pagers had ceased receiving calls in a cardiovascular intensive care unit. Field intensity measurements, interrogation of employees, and other investigative procedures conducted by the clinical engineer yielded a scenario in which some pagers were not able to receive calls when near a well-known intensive care monitoring system.

Standards and practices relating to on-site testing for the presence of electromagnetic radiation were reviewed. A modified Open Antenna Test Site (OATS) procedure (ECRI, 1992; ECRI, 1988), was used as the measurement guideline (Bennett, 1993). As for equipment, an Empire NF-105 field intensity metering system (loaned by a staff member previously engaged in field site radio frequency [RF] measurements), was used as the measurement device. An area characterization (i.e., *footprint*) in the area of interest was taken as well as a characterization of a representative culprit device (i.e., *fingerprint*). The field intensity measurements of a representative unit of the monitor system showed the existence of an unintentional radiofrequency close to the operating frequency of the paging system. Once the source of the EMI was identified, the results were presented to the manufacturer. Initial responses from the manufacturer's media representatives were not encouraging. During one conversation, the manufacturer's electromagnetic interference expert said that the solution was to increase the radiated power of the paging transmitters. This solution was immediately dismissed from consideration, because increasing the level of intentionally radiated electromagnetic energy to compensate for the effect of existing unintentional electromagnetic radiation would increase the potential for risk to other clinical devices. Only when the biomedical engineering department was ready to persuade the client department to cancel the purchase order for the remaining devices did the manufacturer discover a modification to reduce the severity of the interference. This modification, when installed, did not remove all of the unintentional radiation but did reduce the level of the interfering electromagnetic radiation sufficiently to relieve interference with pagers. The manufacturer essentially redistributed and dispersed the radiated energy to other points in the spectrum, which reduced the amplitude at the specific pager operating frequency.

This incident was an interesting introduction to the practical effects of device compatibility (or lack of) and electromagnetic interference in the clinical environment. After a technician from the Texas Children's Hospital Biomedical Engineering Department installed the modification provided by the manufacturer, a fingerprint of the modified device was taken again.

This incident was concluded so successfully that two mutually beneficial results were obtained: The problems of EMI in the clinical environment were clearly and graphically demonstrated within the institution, and the hospital obtained funding for the purchase of modern signal analysis equipment.

An Unusual Source

Another "ghost hunt" occurred at one of the biomedical engineering departments' client hospitals. The wireless medical telemetry there began intermittently displaying error

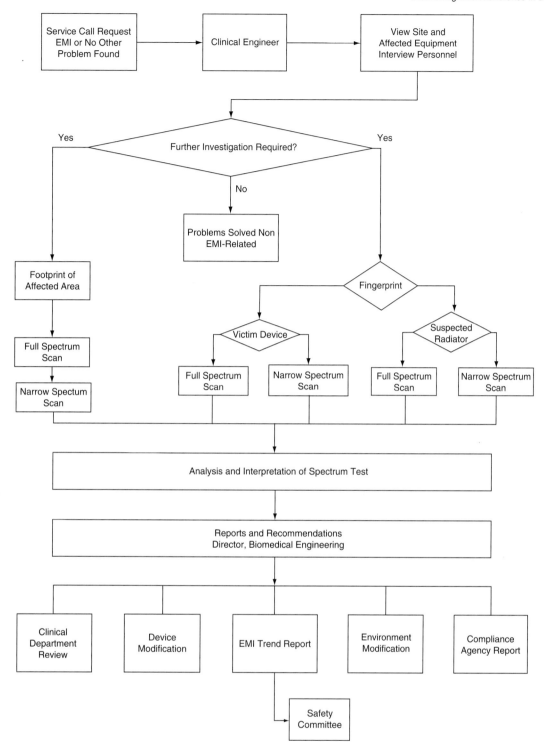

Figure 63-2 EMI problem resolution flowchart.

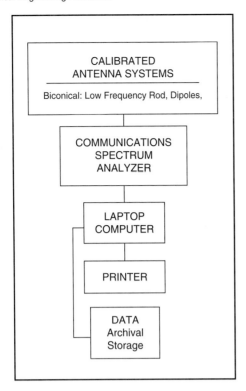

Figure 63-3 Typical equipment configuration for fingerprinting and footprinting.

codes (i.e., loss of data). Staff members were carefully questioned. The times when the interference occurred were established. Further probing revealed that a recently upgraded fire alarm system was undergoing acceptance tests during the periods of interference. As this would be an unusual source of interference, cooperation was sought from plant engineering personnel and the representatives of the fire detection company. Questions posed to the fire detection engineers yielded no prior indications of interference. A spectrum analyzer and broadband antenna system were set up at the site. The spectrum analyzer showed the presence of a recurrent but not time repetitive pulse, indicating that the interference source was radiating a quasi-random pulse.

Working with the fire alarm and plant engineering personnel, circuits controlling the various alarm devices were isolated and shut down individually. When no results were obtained on the affected floor, the same shut-down procedure was implemented on the floor below. When the enunciators—the audio-visual (A/V) units that proved audible and visual warnings of a fire—were shut down one floor below the affected area, the pulse and the interference effects disappeared. The floor below the affected area was a mechanical plant floor, and so the fire alarm company had installed high-powered A/V units. These A/V units used strobe lights. On discharge, these devices emitted a fast-rising, short-duration pulse. The telemetry antennas were receiving these pulses. However, the preamplifiers in the antennas and in the receivers themselves were not saturating or being desensitized by these pulses because the pulse was of such short duration. The pulses passed right through the RF portions of the telemetry system and corrupted data bytes. Open junction boxes with excess wiring hanging in circular loops contributed to the radiation of the pulses. Properly closing the junction boxes and replacing the strobe lights with lower intensity devices within fire code guidelines resolved the situation.

Risk Prevention

The following is an example of how the application of footprinting and fingerprinting may have prevented an EMI incident. The diagnostic imaging department bought new telemetry equipment. Footprinting revealed that the level of incidental emissions in that department was approximately sixty-seven microvolts. Fingerprinting a representative telemetry transmitter showed that the transmitted energy level at the standard one-meter distance was 12 microvolts above the background level.

The defined risk was that there was not sufficient signal-to-noise ratio to preclude intolerably long periods of loss of useable signal throughput. Testing the telemetry transmitter (fingerprinting) is done under controlled conditions at a fixed distance that can not be maintained under real world conditions. A patient wearing a telemetry transmitter can not be expected to maintain a 1-meter distance from the receiving antennas. As the distance between a patient and the receiving antennas increases, the received signal degrades because of the various topographical conditions and the inverse square law. As the signal degrades, the level of acceptable data degrades accordingly. If the degradation is significant, there is a chance that the telemetry system will not be able to recognize the emergency if a patient is in cardiac distress.

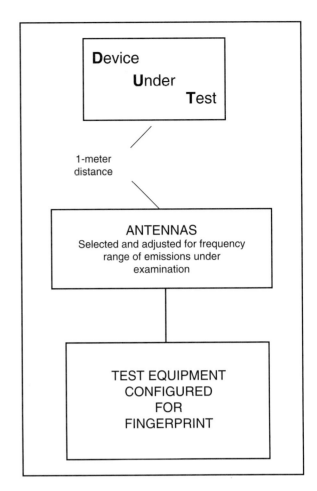

Figure 63-4 Fingerprinting.

Interference of Another Type

But not all ghosts are due to radiated or conducted electromagnetic fields. For example, a report was received of intermittent interference to an EMG device in the physical therapy department of the hospital. The department felt that the culprit was the MRI system located directly above the area containing the victim device. During footprinting, measurements entailing more than the normal broad spectrum procedures were indicated. In this application, the fingerprinting equipment was set up to measuring any radiated electromagnetic fields at the resonant frequency of the MRI that might leak from the shielded room environment that contains the MRI system. During a period of several hours, no leakage was detected. A broader spectrum scan showed that this department was in a remarkably quiet location with respect to radio frequency energy.

Further interviews with the doctors and staff led the clinical engineer to perform a somewhat unorthodox series of tests. The filtration on the EMG was broadened and the leads laid out, unterminated, on the couch on which patients were placed. During this procedure the clinical engineer observed that, when pressure was applied to the couch cushion, the baseline of the EMG machine would vary synchronously. Several additional adjustments to the sensitivity (gain) of the device and the time base were made. It then appeared that any motion in the immediate vicinity of the device would cause this baseline shift. Based on the results of these tests and of an investigation that determined the material composition of the environment (vinyl cushions on the couch and highly waxed vinyl floor), the engineer concluded that the cause of the interference with the device was electrostatic, not electromagnetic. Furthermore, the intermittent appearance of the problem was attributed to the fact that hydrotherapeutic baths were located two doors away from the EMG room. Their intermittent operation would raise the humidity sufficiently to reduce the potential for the generation of intense electrostatic charges. This was a decidedly different ghost hunt.

Interference Not Caused by EMI

One of the recent trends related to electromagnetic interference is that of a manufacturer's technical problems being attributed to EMI. Although EMI may play a significant role in the performance degradation in certain clinical devices, it is not always the cause. One example involved radiological equipment. X-ray films from one of the radiology labs displayed an artifact described by the radiologists as "chicken scratches." This artifact was present primarily during a Temporo-Mandibular Joint (TMJ) procedure. This artifact was initially attributed to either conducted or radiated electromagnetic interference, specifically conducted EMI from the power lines. An outside consultant measured radiation from the power lines, but there was insufficient indication that the artifact was due to the power system. After several months, the biomedical engineering department was called in for consultation.

Initial investigation, including timing the artifacts as they appeared on the film, indicated some degree of synchronicity (timing repeatability). This implied that the interference was time-locked (synchronized) within the radiology system. Due to the characteristics of the interference, such as the timing (frequency) of the interference in relationship to the scan rate, a probe was designed, constructed, and tuned to resonate within the frequency range of the suspected interfering signal. A thorough investigation of the areas surrounding the radiology suite included probing of the three-phase electrical service panels that provided the distribution of power to the suite. No high levels of radiated electromagnetic energy were encountered within the frequency of interest. The investigation was then conducted within the radiology suite. Despite intensive probing of all devices within the suite, including video monitors, computers, and control systems, and examination of the system involved in the TMJ procedure during which these artifacts most often appeared, the source could not be attributed to electromagnetic interference.

A detailed report was filed with the radiology department. The manufacturer's representative was called in. Based on the inconclusive findings, a greater in-depth analysis of the problem was conducted by the representative, the clinical engineer, and technicians from the biomedical engineering department radiology group. During the analysis, the investigators discovered a disconnected bonding strap within the camera head. Reconnecting this strap improved the performance of the system and reduced the artifact.

This incident further illustrates the benefits of cooperation to resolve interference issues, irrespective of the sources of interference. It also demonstrates the effectiveness of a proactive, competent, and supported electromagnetic interference program in the clinical environment. Placing blame rarely mitigates risks.

Variability: A Demonstration of the Problem

Variability in the repeat cause and effect (culprit/victim) device relationship poses a problem when instituting proactive programs designed to reduce the risks attributable to EMI with the support of management.

In 1996, at the request of NHK (Japan Educational Television), the television services group of the biomedical engineering department at the Texas Children's Hospital in Houston was asked contribute a segment to an educational program about electromagnetic interference issues in the hospital environment. The hospital administrators decided that a demonstration of EMI using actual medical equipment would meet the program's objectives of demonstrating the variability of effects of a common source radiator on identical clinical devices.

Two of the same model of hemodialysis machines from the same manufacturer were used in the demonstration. Under the supervision of the clinical engineer, the machines were prepared in such a way that accessories, calibrations, wiring positions within the devices, and location within the demonstration area were as identical as possible.

An intentional radiating source was placed in transmit mode at a distance of 1 meter from the clinical devices. The source was a walkie-talkie commonly encountered in the environment (151.625 MHz frequency band, measured power output of 4 watts at the transmitter, antenna efficiency of approximately 40%, frequency modulation at 5-KHz deviation). The results illustrated of the issue of variable susceptibility. One device failed repeatedly in a noncatastrophic condition, while the other device, located nearby, was unaffected.

Demonstrations such as this, performed on demand, tend to support the many anecdotal EMI-related reports that our department receives daily from other institutions. Regrettably, many EMI incidents go unreported due to a lack of specific programs to address them.

The varied nature of EMI-related equipment malfunctions and the associated risks mandate a proactive program of EMI identification and methodology for risk reduction. An effective program relies on the cooperation of all parties potentially affected by EMI; medical staff, plant engineering, information services, biomedical engineering, and device manufacturers. Due to the highly diversified knowledge and experience required to coordinate detection and mitigation of EMI, the clinical engineering department and its personnel experienced in RF must take responsibility.

An operational protocol must be developed to address EMI issues. (See Figure 63-5.) This provides a defined structure to process requests for EMI investigations and a structure for processing and reporting the results of the investigations.

Programs and Procedures

Testing for electromagnetic compatibility (EMC) in the clinical environment introduces a host of complex conditions not normally encountered in laboratory situations. In the clinical environment, various RF sources of EMI may be present anywhere. Isolating and analyzing the impact from the sources of interference involves a multidisciplined approach based on training in and knowledge of the following:

- Operation of medical devices and their susceptibility to EMI
- RF propagation modalities and interaction theory
- Spectrum-analysis systems and technique (preferably with signature analysis capabilities) and calibrated antennas
- Established methodology of investigating suspected EMI problems, which includes testing protocols and standards

Both standard test procedures adapted for the clinical environment and personnel trained in RF behavior increase the odds of proactively controlling EMI in the clinical environment, thus providing a safer and more effective patient care environment. The methods employed in the following procedures are variations of the OATS (Open Antenna Test Site) technique (ECRI, 1992; ECRI, 1988), a standard for open site testing (Southwick, 1992; Bennett, 1993; ANSI, 1991) and ANSI C63.4-1991 (ANSI, 1991) and ANSI C63.18-1997.

The selection of the spectrum analyzer and the options installed in it were influenced by several factors. A spectrum analyzer of the communications system test type was deemed desirable because there is no better way to characterize devices that emit RF—intentional or incidental—and the environment in which those devices operate. Broadband devices indicate relative RF activity but do not indicate the operating frequencies or modulation types. Both characteristics, independently and together, affect the susceptibility of clinical devices. Table 63.1 shows the test equipment used for EMI testing by Texas Children's Hospital's Biomedical Engineering Department.

A digitally based communications analyzer was needed to archive the results of the EMI tests; both fingerprints and footprints. The flexibility of performance requirements of the device was important, and as a cost-saving benefit, Texas Children's Hospital also uses it to maintain of the hospital radio communications systems.

Since no two modified OATS environments are identical, no two results obtained under the same testing parameters will be identical. Many factors affect the detailed test results, including complex absorption and reflection variables that are totally site-dependant.

Footprinting

At Texas Children's Hospital, the EMI testing program has evolved from years of experience and analysis. This program is not static. Sources of EMI and susceptibility characteristics of devices are in constant change. As new threats arise, the plan is periodically reviewed and modified to contend with them. Further modifications to the plan and procedures are based on a continuous review of wireless industry trends. They represent a proactive response to perceived future threats. At present the program consists of the procedural plan outlined above and three series of tests: (1) area characterization, footprinting (Figure 63-3), (2) device characterization, fingerprinting (Figure (63-4), and *ad hoc* susceptibility testing according to IEEE guidelines (Knudson and Bulkeley, 1994).

Footprinting an area means performing a series of spectrum/amplitude scans for electromagnetic radiation in a defined or designated area. Footprinting is an ongoing technique that defines electromagnetic radiation in a clinical facility or facilities in a multibuilding campus. Footprinting is also done upon request from a department experiencing performance degradation of clinical, diagnostic, or therapeutic devices when EMI is the suspect. The procedure involves a series of 20-MHz-wide spectrum sweeps, beginning at 2 MHz and ending at 1 GHz with antenna(s) in the horizontal plane. This procedure is repeated with the antenna(s) vertically polarized. Tunable standard antennas are adjusted for the correct resonant length for the center of each 20-MHz window.

The footprinting procedures yield two results: An overview of the radiated electromagnetic fields present at a specified location in the environment and the amplitudes and types of emission of those fields. Several incident investigations have been successful using this information. In cases where EMI has been site-originated, the source has been removed and the problem has been corrected. In some cases, the affected device required maintenance to correct the problem. An added value of footprinting is that the data obtained during the process meets the basic requirements for a site search similar to the OATS procedure for fingerprinting individual devices. The results of the footprinting scans are transferred to a storage medium and filed. They form a comparison database that is used to evaluate new devices before introducing them into a specified area.

Fingerprinting

The process of fingerprinting a device has several steps. Again, the spectrum analyzer and calibrated antenna system are the primary tools used to analyze the device under test (DUT) (see Figure 63-4). To minimize the loss of information from the DUT due to masking by other sources of electromagnetic radiation, areas within the clinical physical plant should be tested using the footprinting procedure until a relatively quiet area is found.

For the fingerprinting procedure, as in the footprinting procedures, the standards antenna should be located as far as possible from any conductive material. The standards antenna should be located at a height equal to the center of the DUT. Due to constantly changing EM fields and frequencies, they execute the footprinting procedure in the selected area immediately prior to fingerprinting the DUT. Once the data from the latest footprint has been stored, place the DUT on a nonmetallic stand located as far as possible from any conductive material. The standard antenna(s) is then mounted on a tripod and, in a horizontally polarized mode, placed 1 meter from the front of the DUT. As in the footprint procedure, a series of 20-MHz-wide spectrum sweeps are performed and the results are recorded. This procedure is repeated with the antenna(s) vertically polarized. When using tunable standard antennas, adjust them to the correct resonant length for the center of each 20-MHz window. Upon review of the data collected, any emissions attributable to the DUT should be rescanned. To increase detail in the frequency range in which the emissions were observed, the spectrum analyzer window is narrowed to a 200-KHz/division or smaller, (e.g., 5 KHz/division) sweep width.

Fingerprinting tests are conducted for two reasons. First, compliance with hospital policy on acceptance testing requires that representative samples of all devices be tested prior to entering the clinical environment for the first time. Second, both theory and experience have demonstrated the need for device testing. A device that radiates unintentionally can victimize other devices. It may also be susceptible to victimization by ingress at their egress frequencies and modulation parameters. Those devices already in the environment are fingerprinted if there is reason to believe that they have the potential to be an EMI victim or a culprit.

The procedure used to initiate and track an EMI investigation is shown in Figure 63-2. Members of the clinical staff are instructed to call the biomedical engineering department and television services group in any case of a device malfunction not attributable to a routine failure. In cases of a new device entering the clinical environment for the first time or a new application for an existing device, a request to test the device for compatibility is generated. This call is referred to the clinical engineer responsible for EMI investigations. When an incident might be attributable to EMI, the engineer visits the site as a part

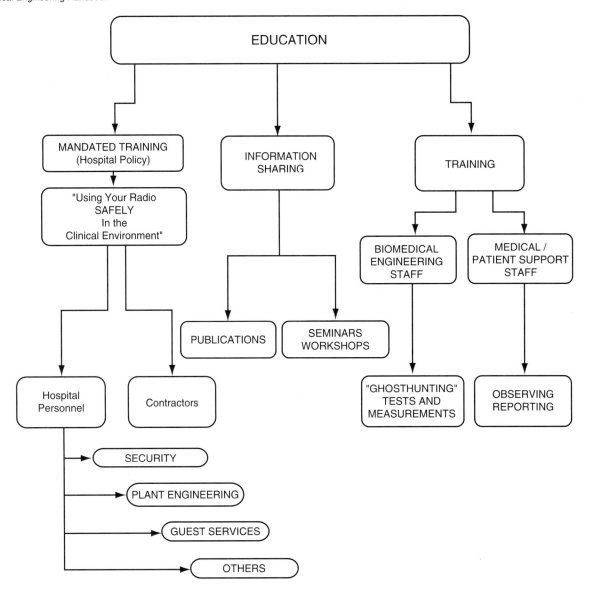

Figure 63-5 Flow diagram of hospital educational program in electromagnetic compatibility and interference relating to medical device technologies.

of the initial investigation. The possible victim device is viewed and tested to determine whether it might have been, or is being, affected by EMI. Interviews with the personnel responsible for the area and the operation of the device(s) must be conducted. Based on the results of this preliminary investigation, a decision is reached as to the desirability or feasibility of further investigation.

The decision to continue the investigation is based on several factors: Is there a high probability of operator error? Is this a very rare occurrence? Is this either a very old device that might be reaching the end of its reliable life cycle or a new device experiencing infant mortality. As part of the preliminary investigation process, maintenance histories of the device are reviewed. Another component involves the review of equipment added to the environment that might have increased the overall RF hostility in the area enough to cause interference. Footprinting records of the area are reviewed as are fingerprints of the victim devices(s).

The mode of device failure (table 63-2) is an important part of the evaluation. Is the victim device alarming? Is it operating erratically? Is it changing its operational parameters either temporarily or permanently? Is it shutting down? Is there a latching change that created the alarm? The answers to these questions can all point to the criminal device or devices. If the failure mode can be duplicated and if the failure appears to have been caused by an intentional or incidental radiator, the culprit device can usually be identified. If it is within the clinical environment, it is silenced or removed. Many times, no further investigation is required. An incident report is generated and filed. A copy is provided to the department initiating the service request. A copy, if deemed necessary, is given to the appropriate reporting agency, such as the Food and Drug Administration Center for Devices and Radiological Health (FDA, CDRH).

If a complete testing procedure is required and a recent footprint of the area exists, a new set is acquired and compared to the older set. Changes in the area environment are noted and analyzed. This is also compared with any existing fingerprints of the victim device. If no fingerprints exist for a representative victim device, or when a device shows signs that its ability to resist ingress may be compromised, it will be fingerprinted. After a cursory footprint of the quiet area, to ensure that no significant changes have occurred in that environment, the victim device, if practical (size and weight can affect the test location), can be moved to this area for fingerprinting.

Generally, the fingerprinting procedure depends on the operating characteristics of the device. For example, intentional radiators are tested for emissions at their operating frequencies, modulations, and at second and third harmonic frequencies. Unintentional radiators (i.e., microprocessor controlled devices) are tested from 1/4 clock frequency to 300 MHz. There are exceptions to this guideline, such as when relatively strong emissions continue to the 300 MHz point. In these cases, readings are continued to 1 GHz. The new digital cellular telephones (GSM) are now in service. Third Generation (3G) wireless communications devices should be in service in the near future. The Biomedical Engineering Department has already received reports of interference to clinical devices by digital telephones.

As both the fingerprint and footprint tests proceed, the records presented by the spectrum analyzer are reviewed. Any RF emissions exceeding a predetermined value are noted, especially frequencies not easily correlated to known sources of intentional radiation. The window of observation is then narrowed to obtain a greater resolution, or magnification, of the emission of interest. This typically identifies the type of modulation of the intentional radiator. After the series of footprinting and fingerprinting tests have been completed, both the clinical engineer and the technician performing the tests analyze the results and generate reports and recommendations. These are discussed with the Director of Biomedical Engineering and possible resolutions of the problem are analyzed and reviewed.

There is not always an ideal solution to an interference problem. Many factors may contribute to device victimization, some of which are not easily or practically resolved. For example, if a recently added high-power paging transmitter were installed on an adjacent building, the additional energy radiated could affect devices that exhibited no previous EMI-related reactions. Historically, if the licensed transmitter is installed on a building over which the institution has no control, it will be difficult to remove. Such transmitters are licensed by the Federal Communications Commission and the medical device has no statutory protection; a one-way street (see Chapter 103).

Modifications to the environment are recommended if they are practical. These might include screening an area or relocating a device within an area. If modification of a medical device appears to be the only solution to the problem, then the manufacturer of the device is advised of the problem. The biomedical engineering department can and does advise the manufacturer. Ad hoc modifications of medical devices to mitigate susceptibility or egress are improper and should not be attempted. Performing such modifications would, in most cases, be a violation of FDA regulations and would increase the institution's risk for liability.

Summary

Despite gains in spectrum protection for some categories of patient monitoring devices through the development of standards, the overall electromagnetic environment is becoming increasingly hostile to the safe operation of clinical devices (see Chapter 103). The establishment and maintenance of a safe environment for the operation of clinical devices requires a multidisciplined approach. There must be a program of education involving technical, clinical, maintenance, and management personnel and staff. A clinical engineer experienced in RF and EMI, as well as in appropriate test and measurement equipment, must be available

A proactive testing and evaluation program that includes ongoing measurements of plant and incoming devices must be established.. Maintained devices—especially those that have demonstrated a potential for electromagnetic wave be spot-tested periodically, using the fingerprint method.

This chapter has presented the need for a proactive EMI management program designed to limit the destructive effects of EMI on clinical devices. Some of the previously mentioned causes of EMI and methods used to test both the environment and the devices within it are based on experiences at Texas Children's Hospital.

References

ANSI. American National Standard for Methods of Measurement of Radio-Noise Emissions from Low-Voltage Electrical and Electronic Equipment in the Range of 9 KHz to 40 GHz. C63.4. New York, American National Standards Institute, 1991.
Bennett WS. Making OATS Measurements Reproducible from Site to Site. *EMC Test & Design* 4:34, 1993.
CBS. Haywire. *Eye-to-Eye with Connie Chung*. CBS, December 1, 1994..
US Government. CFR 47; Part 15.103 (c,e). Code of Federal Regulations of Telecommunications. Code of Federal Regulations, 1991.
David Y. Safety and Risk Control Issues: Biomedical Systems. In Dorf RC (ed). *The Electrical Engineering Handbook*. Boca Raton, FL, CRC Press, 1993.
ECRI. Patient-Owned Equipment. *Health Devices* 17:98, 1988.
ECRI. Ventilators, High Frequency. *Health Device Alert* p. 2, December 18, 1992.
ECRI. Guidance Article: Cellular Telephones and Radio Transmitters-Interface with Clinical Equipment. *Health Devices* 22(8,9):416, 1993.
Knudson T, Bulkeley WM. Stray Signal, Clutter on Airwaves Can Block Workings of Medical Electronics. *The Wall Street Journal*, p. 1, June 15, 1994.
Paperman WD, David Y, McKee KA. Electromagnetic Compatibility: Causes and Concerns in the Hospital Environment. *ASHE Health care Facilities Management Series*. Chicago, IL, ASHE, 1994
Southwick R. EMI Signal Measurements at Open Antenna Test Sites. *EMC Test & Design* 3:44, 1992.

64

Accident Investigation

Joseph F. Dyro
President, Biomedical Resource Group
Setauket, NY

The increased dependence of caregivers on medical devices, the increased complexity of medical device design and capabilities, the increased pressure on medical staff to operate efficiently and cost-effectively, the decrease in training opportunities, and human error contribute substantially to medical device–related accidents in the health care setting.

Accidents involving medical devices have occurred and will continue to occur for many reasons, including device design, human error, and lack of training. Accident investigation addresses several concerns: improved patient safety, elimination and reduction of risk, prevention of recurrence, compliance with regulatory imperatives, and discovery of facts that support the litigation that often follows a serious injury.

An accident response system with policies and procedures, resources to implement them, and administrative support should be in place in all institutions so that staff members and departments are aware of their respective responsibilities if an accident occurs. The clinical engineer (CE) must take a leadership role in any accident response system. This chapter introduces the clinical engineer to accident investigation and recommends that the CE develop the skills necessary to perform investigations either as an independent consultant or as an employee of the hospital or independent service provider. This chapter describes the methodology and tools that enable the CE to put accident investigation into practice.

Forensic clinical engineering examination and analysis, either as medical device safety officer or as independent forensic engineer, requires skills in accident investigation. Accident investigation yields best results when the hospital administrators and staff cooperate with a sound understanding of policies and procedures pertaining to accidents. Techniques are described for addressing the issues of evidence, documentation, reconstruction, investigation tools, interviews, root cause analysis, reports, and recommendations.

This chapter also provides methods for the CE who investigates accidents involving the use of medical devices. A systems analysis technique guides the investigator through five factors that often contribute to accidents: The facility, the operator, the patient, the device, and the environment. Several case studies are presented in which the methods described enabled the identification of failure mechanisms contributing to injury and death. The methods are applicable to a wide range of medical devices used in the hospital and in the home setting. The chapter also offers an accident investigation checklist to aid the clinical engineer in acquiring and documenting relevant details important in forensic examination. Education in accident investigation and available resources are also presented.

While this chapter focuses on accidents, it is equally applicable to the critical incident technique. The investigative techniques described apply to the identification of preventable incidents that could have led, or did lead, to an undesirable outcome. The occurrence rate of preventable incidents may be reduced through the investigation of "near misses" and "no-harm events" as well as adverse events (Cooper, 1978).

Rationale for Medical Device Accident Investigation

Accident investigation is done for many reasons, including public pressure to reduce medical errors and improve the quality of care, regulatory imperatives, and legal and ethical considerations. A principle objective of accident investigation is to discover the cause of the accident (Shepherd, 2000). Once identified, the cause can be addressed, eliminated, or controlled, with a reduced likelihood of recurrence. Improvement in quality of care follows a comprehensive accident investigation that reveals the root-cause and yields recommendations for improvements in the health enterprise, personnel, maintainers, and medical device manufacturers and distributors (Driscoll, 2003; Patail and Bruley, 2002; Berry and Krizek, 2000; ECRI, 2001; JCAHO, 2000).

Patient Safety Movement

The public demands an appropriate response from health care providers in the wake of reports of deaths and serious injuries that occur in hospitals (Kohn et al., 2000). Organizations of health care professionals are unified in their support of patient safety initiatives (ACCE, 2001; JCAHO, 2001; Shojania et al., 2001; Shepherd, 1999; Dyro, 2000). The Institute of Medicine (IOM, 2001) in its report, "Crossing the Quality Chasm: A New Health System for the 21st Century," calls for a health care delivery system that is safe, effective, patient-centered, timely, efficient and equitable. Patient safety can be enhanced by improving the ability to learn from mistakes. Skillful investigation of incidents and responsible sharing of data will help in the identification of root causes of error (ACCE, 2001). Accident investigation alone will not address the issue of risk due to medical technology. Investigation will, however, provide valuable insight into the root causes of a specific injury or a class of injuries that, if addressed, will minimize the odds of recurrence.

Effective risk management (see Chapter 56), a vigorous patient safety program (see Chapter 58), and a strong clinical engineering presence are essential in enhancing patient safety (Dyro, 1988). Attention to accident investigation will heighten the hospital staff awareness of the existence of an active patient safety and risk reduction program. Shifting the emphasis from blaming the individual to seeking the root cause will encourage disclosure of accidents that otherwise might have gone unreported for fear of blame, retribution, and punishment (Shepherd, 2000a).

Investigation techniques that use the preferred systems approach (Shepherd, 1993) will increase the hospital employee's awareness of the elements of the "system" and of element interactions, such as device-device interference, that lead to inappropriate patient care. Hospital employees will recognize that accident investigation by the clinical engineer is only one of the many facets of technology management and that they can be active participants in the safety team by being alert to device malfunctions, such as unusual odors, unexpected or erratic functioning, or electrical sparks.

Regulatory Imperative

The United States Food and Drug Administration (FDA) is charged under the 1976 amendments to the Food, Drug, and Cosmetics Act to regulate medical devices for the public good. Since then, many initiatives that place responsibility for post-market surveillance on manufacturers, distributors, and health care providers (hospitals and individuals) have been launched. The Safe Medical Devices Act (SMDA) requires that a hospital report any incident in which the manufacturer's medical device may have contributed to the injury to the manufacturer within 10 business days. If the manufacturer is unknown, the occurrence must be reported to the FDA. The FDA follows post-market surveillance with policies and procedures to gather information to fulfill its mandate to protect the public. (For more on the FDA, see Chapter 126.) The Joint Commission on Accreditation of Healthcare Organizations (JCAHO) requires that hospitals have a mechanism to analyze and reduce through corrective action sentinel events, i.e., an unexpected occurrence involving death or serious physical or psychological injury, or the risk thereof (JCAHO, 2001). OSHA must be alerted when an employee's health has been compromised by an injury in the workplace. The state health department, and local, regional, and national authorities have reporting requirements for accidents that result in injuries. In the event of any workplace injury, it is in the hospital's best interest to launch its own investigation quickly.

Litigation

A death or serious injury that occurs unexpectedly in a hospital may lead to a lawsuit. A medical device-related death or injury may result in the injured party, the plaintiff, suing several entities and individuals including the hospital, doctors, nurses, technicians, support staff, the device maintainer, the distributor and the manufacturer of the device. The hospital that immediately conducts a thorough investigation can better assemble facts and evidence that could shift liability from the hospital and its personnel to another party, such as the manufacturer of a defective device.

Ethics

The American College of Clinical Engineering states that the first tenet of the ACCE Code of Ethics calls upon clinical engineers "to strive to prevent a person from being placed at risk of injury due to dangerous or defective devices or procedures (ACCE, 2001)."

Role of the Clinical Engineer as Investigator

The CE is the logical choice as the investigator of medical device-related accidents (see Chapter 55, Patient Safety and the Clinical Engineer, and Chapter 13, The Clinical Engineer as Investigator and Expert Witness). By education, training, and experience, the CE is the one hospital employee who has the greatest knowledge of the principles of operation of medical devices. The American College of Clinical Engineering Health Technology Foundation (AHTF), acknowledging that clinical engineers are engaged in accident investigation, has drafted guidelines for incident investigations (AHTF, 2003).

The clinical engineer is a logical choice to serve as the medical device safety officer in a hospital. The CE is skilled in safety awareness education and training, accident investigation and analysis of the causes of accidents, and documenting and reporting findings. The CE can serve as the coordinator of investigations. Hospital CEs should recuse themselves from the role of primary investigator when the device involved in the accident is one serviced, maintained, and repaired by the clinical engineering department. In such cases, the services of an independent CE should be retained. An independent clinical engineer may work in another hospital, as an independent consultant (see Chapter 13), or working in an organization that specializes in medical device testing and evaluation.

Response to Accident

An accident should be responded to immediately. Resources must be available on a 24-hour, 7-days-per-week basis. Policies and procedures must be in place to guide the accident response and the subsequent investigation. Adequate training must prepare staff members to react appropriately in an emergency situation. An accident in one department ultimately involves many parts of the hospital, such as administration, risk management, legal affairs, clinical engineers, staff development, safety officer, medical staff, and nursing department. All must know the appropriate documentation required by hospital policy and by local, state and federal regulatory bodies. It may be necessary to notify outside authorities and organizations such as the local fire department, environmental protection agency, OSHA, the FDA, the device manufacturer or distributor. The hospital must be prepared to notify and interact with these entities.

General Guidelines

The first response to an accident is to protect the injured party or parties from further harm, e.g., removing a patient away from a fire. The injury should be assessed and treated by the medical team immediately. Attention should then be directed to controlling the source of the danger, for example, extinguishing the fire. Clinical engineering should come to the scene. It may be necessary to call for assistance or to activate alarms at this point, such as calls to maintenance if the problem involves the facility's utilities or physical plant, calls to the fire department in case of fire, calls to the infection control officer if the accident involved the release of an infectious agent. Once the source of danger is contained, the facility should be inspected for changes that could adversely affect other patients, such as loss of power or smoke in the air handling system. The environment should be checked so as to ensure that other devices were not adversely affected by the accident. For example, if a power surge caused a device to injure a patient, that surge may have adversely affected the operating characteristics of devices powered by the same circuit.

Next, an accurate description of the occurrence should be entered into the medical record and an incident report should be completed. Notify other hospital personnel—risk management, public affairs, and chief operating officer—of the hospital policy and procedures.

Policies and Procedures

Policies and procedures must be in place to guide the accident response and subsequent investigation.

Administration, legal affairs, risk management policies and procedures should be crafted for the following:

- Policy: Incident Reports
- Procedure: Preparation and Processing of Incident Reports
- Policy: Safe Medical Device Act Reporting
- Procedure: Preparation and Processing of Incident Reports Affected by Safe Medical Devices Act

The clinical engineering department should have the following:

- Policy: Incident Investigations
- Standard Operating Procedure: Incident Investigation

This procedure should incorporate the details of incident investigation described in this chapter, e.g., preservation of evidence, data collection, photography, interviewing witnesses, and report preparations.

A Sample Clinical Engineering Policy Regarding Incident Investigations is shown below:

Clinical Engineering Policy Regarding Incident Investigations

Clinical engineering has the responsibility to investigate and report on all incidents at John Endall Hospital that involve medical devices. This responsibility is detailed in Management Manual Policy 008. The purpose of these investigations is to discover the facts surrounding an incident, analyze them, and propose recommendations that will lessen the possibility of recurrence. Additionally, Clinical Engineering must report these incidents to the Hospital's Equipment Quality Performance Committee, to device manufacturers, and the FDA according to the regulations imposed by the Safe Medical Device Act of 1990.

Individuals contacted by clinical engineering in the course of an incident investigation should be as candid as possible to facilitate fact-finding. The purpose of these investigations is to improve the quality of patient care, not to fix blame. Except as necessary for reporting, information contributed will be kept confidential.

Clinical engineering strives to maintain objectivity in all its incident investigations. The actions of clinical engineering are subject to comment and recommendation, as are those of other departments and medical staff. In cases where maintaining objectivity is difficult, an outside party may conduct an investigation after legal affairs is consulted.

Clinical engineering's years of experience in investigating variances at John Endall Hospital have repeatedly shown individuals to be helpful and cooperative in supplying information and in working together to find solutions to problems. We look forward to continuing to serve you and our patients in this capacity.

John B. Booté
Director, Clinical Engineering

Team Approach: Medical Staff, Nursing, and Risk Management

Adequate training ensures that staff members are prepared to react appropriately in an emergency. An accident in one department ultimately involves many parts of the hospital organization such as administration, risk management, legal affairs, clinical engineers, staff development, safety officer, medical staff, and nursing department.

Clinical Engineering

Clinical engineers should respond immediately to the scene of an accident. Steps to be taken include the following:

- Preventing or minimizing further injury or damage to patients and staff
- Establishing a perimeter and cordon off with to minimize loss or alteration of evidence
- Checking for damage to ancillary devices and facilities
- Photographing, sketching, recording, and otherwise documenting the scene
- Obtaining first-hand witness accounts
- Beginning to establish a spatial frame of reference and a time line
- Preserving and sequester evidence
- Launching investigation

Documentation and Reporting

It may be necessary to notify outside authorities and organizations, such as the local fire department, Environmental Protection Agency (EPA), OSHA, FDA, the device manufacturer or distributor about an accident (FDA, 1996; Kessler, 1993). The clinical engineer must be prepared to not only notify but also interact with these entities. The hospital must be familiar with the appropriate documentation required by their own policy and by local, state and federal regulatory bodies.

Skills Requirements

Effective accident response and investigation requires that the clinical engineer possess technical expertise, knowledge of human factors (Guyton, 2002), and good communications skills. Expertise, knowledge, and communication skills are typically gained through formal and informal training and education, and from practical experience. ACCE in its clinical engineering certification process stresses that knowledge is required in human factors engineering, root cause analysis, and other fields of study related to patient safety (ACCE, 2001). ACCE Teleconference Series has featured conferences on these subjects (Shepherd, 1996; Dyro, 2000a; Patail and Bruley, 2002). A survey of clinical engineers established that accident investigation falls within the "body of knowledge" expected of a clinical engineer (Cohen, 2001).

Technical

A sound knowledge of clinical engineering in addition to good analytical ability and problem solving techniques is required for accident investigation. The knowledge must include the principles of operation of medical devices, the use of those devices to treat and diagnose patients, the fundamentals of physiology, and an understanding of the interaction of physical forces with the human body. For example, the effective accident investigator will understand the operation of a hypothermia machine and thermoregulation of the body. The CE will understand such phenomena as skin lesions and their causes, physiological response to and injury mechanisms of electricity, and mechanical forces and their effects upon the body.

Human Factors

The clinical engineer will find knowledge of human factors useful whether the CE is working in a hospital (see Chapter 83), as an independent forensic clinical engineer (Bogner, 1994), or as a design engineer (Gosbee, 1997; Gosbee et al., 2001). Cooper (1978; 1984) has shown that human factors analysis of adverse events and equipment design leads to a safer clinical environment.

A device involved in an accident could be found to be performing within its design specifications. However, the specifications may not have adequately taken into account the good human factors design and substituted a poor design feature, for example, a switch or a knob that could inadvertently be activated by a person brushing by causing a change in diagnostic measurement sensitivity or therapeutic energy output.

Communications

A medical device is one element of a system that includes the physical environment and its influences, plus the human elements—the patient and the device operator. Other hospital personnel or visitors may witness an accident, and the information they contribute can assist in determining the cause. Personal fact-finding interviews require good communication skills including socio-political awareness, clear speaking, and active listening skills. Documentation of the investigation and its findings, conclusions, and recommendations requires good writing skills.

Systems Approach

The systems approach to medical device accident and incident investigation (Shepherd, 1983) describes a system composed of five elements: Device, operator, facility, environment, and patient (see Chapter 59). For example, Figure 64-1 shows a diagram of a neonatal intensive care unit (NICU) containing the five elements. A neonate (the patient) is in an infant radiant warmer (the device) that is electrically powered from the wall receptacle (the facility). The operator (the nurse) prepares to connect the skin temperature sensor to the warmer's control panel. A nearby examination lamp and external influences of a bright sun and electromagnetic waves emitted by a local station (the environment) are shown. All five elements of the system can interact and cause an accident resulting in patient injury. The brief summary of the systems approach to accident investigation that follow stresses the importance of using the technique

Medical Device

A medical device can contribute to patient injury as a result of defects and deficiencies that can be introduced into the device at a number of stages of its life cycle, including design, manufacture, transportation, inspection, maintenance, repair, post-market surveillance, and obsolescence.

Design

Device performance, accuracy, reproducibility, and safety require sound design and function of appropriate parts and components, the correct circuit diagrams and mechanical drawings, the optimal routing of wires, and adequate connectors. The theory of operation of the electrical and mechanical aspects must be based on sound principles, and the design must also consider mechanical shock and vibration, fluid ingress, electromagnetic interference, and electrostatic discharge (ECRI, 1974).

Sudden or unpredictable failure of a part or circuit is classified under this system subcomponent. Device design should incorporate failure mode and effect analysis (FMEA) (Stamatis, 1995; Stalhandske et al., 2000; Willis, 1992). The effect of a sudden component failure on the operation of the device can be predicted. If the results endanger the patient, fail-safe design must be incorporated. Bruley (1994) described an incident in which a patient was crushed by a descending motorized radiotherapy gantry that did not stop in response to its normal operating control, its emergency stop, or its automatic limit switch. All three controls operated through the same electrical relay, which failed. FMEA would have predicted this failure mode and would have resulted in an alternate design in which relay failure would stop gantry movement.

Human factors design is that aspect of device design that operators, maintainers, and others use when working with the device (Christoffersen and Woods, 1999). A device may work as designed and may be safe and effective when used properly, but its design may confuse the operator or the maintainer such that it is utilized incorrectly or repaired improperly (Welch, 1998). Controls and indicators can be difficult to see. Examples include numbers on a control knob or display that are too small to be seen clearly or, if illuminated, are too dim in high ambient light conditions to be visible. Parts should not be able to fit into a device in ways other than those that are intended for proper device operation. For example, check valves that can be reversed in a breathing circuit have cut off air supply to patients.. Attention to human factors design should be directed to evaluation and incoming inspection of medical devices (Hyman and Cram, 2002).

Connectors are device components with a long history of deficiency in human factors design. Proper markings, use of mechanical interlocks, keying, proper placement of connectors on the medical device, and color-coding are some of the means to minimize human error due to deficient human factors design. Figure 64-3 shows the use of a temperature sensor from another manufacturer's radiant warmer inserted into the connector of a warmer used to treat an infant. The radiant warmer overheated, causing brain damage to the infant.

The device-patient connection is often a weak link in the system. These connections involve such devices as tracheal tubes, needles for intravenous or intra-arterial lines, ECG electrodes, and temperature sensors. In a case study described below, a partially dislodged tracheal tube resting against the patient's neck, deprived air from entering the patient's lungs while preventing the low airway pressure alarm from activating because of the obstruction of the breathing circuit.

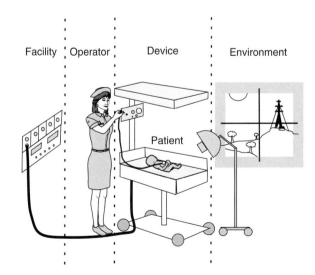

Figure 64-1 Systems approach to medical device accident investigation: The neonatal intensive care unit. Adapted from *A Systems Approach to Hospital Medical Device Safety* (Shepherd, 1983).

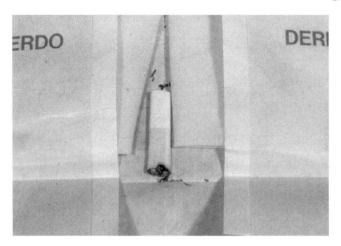

Figure 64-2 Cigarette butt packed with surgical gloves

Manufacture

The medical device can be rendered defective during the manufacturing process. For example, a package of surgical gloves could be contaminated if a cigarette butt discarded by an assembly line worker is packed in with them. (see Figure 64-2). In another example, a misassembled PEEP valve resulted in the blocking of the exhalation port resulting in barotrauma and death of a ventilator-dependent patient. The packaging error in the former example was relatively easy to detect and resulted in no harm; the misassemble PEEP valve in the latter example was not checked, and led to grave consequences. Both examples underscore the importance of inspection prior to use of any medical device. Such inspection can often be as simple as visualization.

Transportation

The transport of a device from one location to the next, (e.g., from manufacturing plant to hospital loading dock or from an equipment storage room to a patient care area,) may subject the device to shock, vibration, temperature, and humidity extremes that the device was not designed to tolerate. One in three monitors of 120 purchased by one hospital failed incoming inspection in the author's experience because in transport a particular circuit board dislodged from its connector. The mechanical design was inadequate; it did not contemplate the typical stresses placed upon the monitor during shipping and, hence, did not provide for adequate retention mechanisms. Devices particularly susceptible to damage in transportation include surgical lasers, heart-lung machines, dialysis units, and incubators.

Inspection

Defects may be introduced during manufacture, assembly, or transportation. However, any device that arrives at a hospital must be inspected before it is used for patient care. Failure to do so permits the use of defective devices to the detriment of the patient.

Maintenance

Circuits and parts can fail as a result of use, aging, and the environment. Maintainers of medical devices can make mistakes that lead to adverse events and injury to patients (Shepherd, 1998). For example, the maintainer can introduce a device defect by replacing a part with an incorrect part or by incorrectly installing a modification or upgrade. The author investigated an adverse occurrence in which the input and output lines on an anesthesia machine manifold were reversed by a serviceman, resulting in an anesthetic overdose and brain damage to a 13-month-old infant. Further investigation showed that the designs of the input and output lines were faulty from a human factors perspective—they were easily reversible and did not have adequate markings to indicate the proper assembly. The author procured the modification kit from the manufacturer and with little difficulty was able to cross-connect the input and output lines to the manifold. The instructions for the modification were poorly written and poorly illustrated, adding to the confusion of the serviceman, who failed to check the results of his work with an anesthetic gas analyzer.

Recall and Modification Notices

Failure to address a recall or modification notice allows a defective device to continue in use at the patient's peril. On average, only 4% of devices subject to recalls are ever located and returned. Undoubtedly, this is due in large part to ineffective or non-existent recall and hazard alert systems in healthcare organizations.

User

The user is under increased pressure to perform work in as little time as possible to maximize profits for the enterprise. The burden of complicated technology is heavy and misuse or abuse is likely to occur (ECRI, 1993). Errors in the use of a medical device can stem from misuse, abuse, and inattention (Shepherd and Brown, 1992).

Misuse

Two examples of misuse are described as follows. Figure 64-3 shows the control unit of a radiant warmer with an incorrect temperature sensor attached. The wrong sensor was used and a nurse who noticed that the connector was loose in the jack secured the sensor to the control unit with white tape. Lethal hyperthermia resulted from this misuse. This error was in part attributable to the defective human factors design of the temperature sensor connector that enabled it to be plugged into the wrong jack. The neonatal intensive care unit (NICU) used warmers made by several manufacturers. The operator was unaware of the difference between the available connectors, and temperature sensor interchange was likely to occur. The nurse erred in attempting to remedy a situation by making a technical adjustment with the use of white tape (Kermit, 2000; Shepherd, 2000b; Shepherd and Dyro, 1982) rather than calling for clinical engineering service.

An infant suffered lesions to the groin area while cared for in an intensive care unit of a major children's hospital. The infant warmer's high temperature alarm activated frequently. When the alarm sounded, the heating stopped, and the reset button needed to be pressed to resume heating. The simple solution for the nurse was to tape over the reset button, which meant that the warmer operated in continuous heat mode (see Figure 64-4). Betadine that had pooled in the groin area, combined with the irradiance from the warmer, resulted in skin lesions. Investigation of the incident soon after its occurrence found that

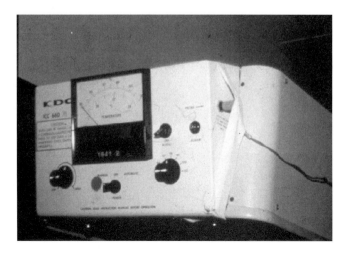

Figure 64-3 A skin temperature sensor is taped to the control unit. The wrong sensor was used, and the sensor was secured to the control unit with white tape.

Figure 64-4 The high temperature alarm and heater output reset button shows the residue of white adhesive tape remaining on the surface of the control panel around the reset button.

the radiant warmer was operating within its design specifications. Visual examination of the control panel revealed that the reset button had been taped down, as indicated by the residue of white tape left on the panel around the button.

Abuse

Abuse is typically unintentional, but may be intentional. For example, a nurse, hurriedly moving an incubator from an equipment storage room to the NICU, crashes the incubator into a wall or doorframe damaging controls or internal components. Figure 64-11 below describes this kind of device abuse.

Inattention

Long hours without sleep have adverse affects on the level of alertness of nurses and doctors. Sleep deprivation, ignorance, and distraction are factors in inattention to both the patient and the device. Studies have shown that accidents are most likely to occur on the late night to early morning, a.k.a. the graveyard shift (Abramson et al., 1980). Undoubtedly all these factors combined to result in the self-defibrillation incident that occurred in one hospital's emergency department (Iverson and Barsan, 1979).

Facility

The same factors that influence the performance of a device influence the adequacy of the facility to provide the necessary services, e.g., utilities.

Human Factors Design

The location of a wall receptacle in a location that results in the line cord running along the floor may result in the nurse tripping or the disconnecting the cord from the receptacle, or the cord being damaged by a wheel running over it (see Figure 64-1).

Part/Systems Design

The design of the facility includes such factors as the location of a window (see Figure 64-1) or the location of an electrical receptacle. See Chapter 95 for a case study in which a patient received a near-fatal electrical current because a line cord was routed in a way in which it could be damaged.

Deterioration

Facilities must be maintained and will deteriorate if they are not. The wall unit shown in Figure 64-1 contains medical gas outlets and suction in addition to electrical outlets. Allowing these utilities to deteriorate could adversely affect patient care. For example, the continual insertion and withdrawal of plugs into and from receptacles will eventually degrade the receptacle, resulting in poor electrical contact. Gas and suction outlets can leak over time, affecting pressures and flows. A leaking oxygen outlet is particularly dangerous as it could result in an oxygen-enriched atmosphere that, with fuel and an ignition source, could result in an intense fire.

Maintainer

The maintainer could cause a hazardous situation by, for example, replacing a defective receptacle with the wrong receptacle, e.g., replacing a regular power receptacle with a red one. If power subsequently fails, a critical life-support device such as a ventilator may fail to operate if it was connected to what a nurse thought was an emergency power outlet, but was instead a regular outlet.

Environment

The environment can be separated into two categories: The internal environment and the external environment.

External

In Figure 64-1, the examination lamp illuminates the internal environment. The lamp's output, while illuminating the neonate with visible light, also heats with infrared radiation. This internal environmental factor is an example of a device-device interaction that can contribute to such an occurrence as neonatal hyperthermia or burn.

Internal

The sun illuminates the external environment, and its radiance in the room may influence thermal balance. In addition, the television transmission tower located on a nearby hill can cause electromagnetic interference with the devices in the NICU (see Figure 64-1).

Patient

The patient can contribute to an accident in several ways (Kenney, 1983). An active patient can adversely affect the functioning of a device or interfere with a device's operation. A passive patient may be more or less susceptible to an injurious element of a device's function simply by virtue of age, weight, height, or physical condition.

Active (Educated; Uneducated)

An active patient may, for example, attempt to get out of bed even though bed rails are up. Sliding out of bed and entrapment of the head and neck can occur. The size, weight, age, and level of awareness of the patient are all factors that can influence the odds of entrapment. The active, uneducated patient may, for example, change the controls of a ventilator on which he or she depends.

In one example, a side panel of an infant radiant warmer was left in the down position as shown in Figure 64-1, either from failure of the nurse to reposition it or because of failure of latches to secure the panel. This action contributed to an active neonate falling off the mattress surface and onto the floor.

Passive

A passive patient, simply by virtue of the patient's physical condition, height, weight, age, and mobility, may contribute to an accident. For example, a passive critically ill, low-weight, premature neonate whose hand is exposed to an elevated heat source is less likely to tolerate the thermal load because of impairment of peripheral circulation that can disperse the thermal energy absorbed by the hand.

Investigation Procedures

While no two accidents are exactly alike, investigation procedures include several well-defined steps, described as follows. Variations in each step may be necessary depending on the particular circumstances (see Shepherd, 1993; Bruley, 1994).

First Response

Clinical engineers should respond immediately to the scene of the accident, and should take the following steps:

- Prevent or minimize further injury or damage to patients and staff.
- Establish a perimeter.
- Check for damage to ancillary devices and facilities.
- Photograph, sketch, record, and otherwise document the scene.
- Obtain first-hand witness accounts.
- Begin to establish a spatial frame of reference and a time line.
- Preserve and sequester evidence.
- Launch investigation.

Examining the Scene

Arrive at the scene as soon as possible. Time is of the essence as environment changes. There is often a haste to clean up, bring in replacement equipment, and dispose of packaging material that could establish the identity of a disposable such as an electrosurgical pencil or a grounding pad. Search the trash for discarded wrappers and package inserts.

Establish initial conditions and record these by means of photographs, drawings, and notes. Photographs of the injury immediately after occurrence and subsequent photographs over time can be helpful in determining the nature of the injury especially in the case of skin lesions of unknown origin. A laboratory notebook or equivalent means of recording data should be used. Notes should be taken in ink rather than pencil. A medical device incident investigation form serves well as a checklist for obtaining information. The investigator is encouraged to develop a form, based on the following list of items that would typically be acquired in the course of a site inspection:

Incident Investigation Checklist

- Device information (manufacturer, brand name, model/catalog number, serial number, lot number, hospital control number, and labels)
- Maintainer information (service history)
- Event information (date, injury, date and time of awareness of event, date of reports to manufacturer/FDA, location, use as intended by manufacturer, and description of the event)
- Patient information: Name, address, phone, classification (inpatient, etc.), patient ID number, room number, age, weight, sex, physicians who treated them, and their medical condition before and after the event
- Injury information: Description, time of injury discovery, time of device application, , locations of device and patient, treatment
- Incident investigation procedure: Initial conditions, e.g., control settings, environmental conditions, device history, relative position of device, patient, disposables, personnel, furniture (sketches and photographs); witnesses; documents (e.g., labels, service records, manuals and photographs of scene, devices, patient injury; tests performed; and findings).
- Investigation conclusions: How the device contributed to the event, and to what extent

Notes and sketches made at this time will aid in later reconstruction efforts. Space may be made available on the accident investigation form, or the notes and other documents may be attached to the form.

Note locations of all items in the vicinity, e.g., instruments, medical devices, disposables, carts, beds, furniture, personnel, visitors, documentation, manuals, labels, and package inserts. If items were moved after the occurrence, ascertain their location prior to moving. Failure of a device to operate in a safe mode may have been caused by its proximity to another device (see Chapter 60 for interactions between medical devices) that may have emitted electromagnetic interference. (see Chapters 62 and 63). Observe and note all instrument settings, control knobs, meter readings, and connections. Settings should not be changed unless necessary in order to prevent further injury to the patient. For example, the author investigated one accident in which several of a newborn's fingers were burned and eventually lost. The first response of the nurse and doctor was to remove the source of heat (excessively high temperature humidified air delivered by a hose next to the fingers) from the fingers and to lower the temperature setting on the heated humidifier. These actions, recorded in the medical record and memorialized in witness statements, assisted the author in determining the cause of the injury.

Determine the identities of all witnesses, including doctors, nurses, technicians, physical plant maintenance personnel, housekeeping staff, visitors, respiratory therapists, and students. These individuals may have altered the position of devices, damaged line cords or connectors, or altered settings. See Chapter 95 for a discussion of a line cord entrapment caused by the hospital housekeeper who moved a motorized parallel bar in order to clean the floor and entrapped the cord in the process. Later activation of the bars resulted in a near-lethal electric shock. Also contact personnel who might not have witnessed the accident, but who might have been in attendance at some time or who might have other information to contribute (e.g., a person who performs the same kind of work in other areas of the hospital).

Note the environmental conditions such as time, temperature, and relative humidity. Note the conditions of the facilities upon which the device operation depends, such as line voltage, medical gas pressures and flows, water pressure and flows, and suction. Certain environmental conditions could adversely affect the operation of devices.

During this initial response, one should begin to establish a spatial-temporal relationship of significant events involved in the accident. (See Table 64-1.) Police reports, medical records, witness statements, depositions, and interviews may provide data for the time line. Typically, different observers have varying perceptions and recollections of time and space. These perceptions tend to be less clear with the passage of time. Obtaining evidence and facts as early as possible will minimize the effect of fading memories.

Evidence

The investigator must identify, document, examine, and preserve all evidence at the scene. A thorough search must be conducted, including a search of the trash for items that may have been discarded. The evidence should be labeled, photographed, placed in a suitable container, and impounded. Access to the evidence should be controlled and all parties who subsequently examine the evidence should be recorded to establish a chain of custody. Testing of all evidence may not be possible at the time of the initial response to the event. Proper identification and preservation will permit examination at a later date by the investigator or other parties.

In cases of serious injuries or deaths, examination of the device in question should be performed in the presence of all interested parties, e.g., the hospital representative, device manufacturer, and independent investigator.

Preservation

The investigator must take care not to do destructive testing or to damage the evidence in any way. The outcome of legal proceedings that may occur subsequent to the accident can hinge on the issue of spoliation, i.e., alteration of the evidence. Impounded evidence, such as a medical device, should not be placed back into operation until all parties have had an opportunity to examine it. The personnel who are responsible for routine maintenance and repair should not inspect the device. They may attempt to conceal errors committed previously during servicing visits.

Once notified, a manufacturer will typically ask that the device be returned for examination or that a manufacture's represented be permitted to examine it in the hospital. The manufacturer may repair or replace a defective component, and issue a report that the device was found to have no defects. Manufacturers have occasionally claimed that the returned device never reached its destination. One should not send a suspect device to the manufacturer without a written agreement as to the actions to be taken during a device inspection. Such an agreement should address the following points at a minimum:

- Description of product (e.g., serial number, hospital equipment control number, lot number)
- Description of malfunction
- Request to examine for defects
- Letter to notify sender upon receipt of device
- Request for report within 30 days of a finding, with conclusions and recommendations
- Description of test methods
- No destructive testing without prior permission
- Prompt return of device upon completion of testing or sooner if necessary
- Chain of custody preservation such as transportation records

Should destructive testing be necessary by the manufacturer, it should be done in the attended by a representative of the hospital and the injured party.

Requests from the FDA to test the device should not be granted unless approved by the hospital's legal counsel. The ability to have the device tested by an independent party may be adversely affected and evidence could be lost.

Documents

All pertinent standards, manuals, incident report, medical records, notes, maintenance records, test results, and labels should be obtained. Investigators should also obtain instructions, technical communications, and warnings, an analyze any warnings Product liability lawsuits often allege failure to warn and provide appropriate safety information (Peters and Peters, 1999).

Documents upon which the investigator may rely upon include standards and guidelines related to performing investigations (ASTM, 1995; ASTM, 1996; ASTM, 1997 and 1997a; ASTM, 1998: IEEE, 1985) and other standards such as the Standard for Health Care Facilities, NFPA 99 (NFPA, 2002), and National Electrical Code. NFPA 70 (NFPA, 2002a), and the National Electrical Safety Code (IEEE, 1997). See Section XI of this Handbook for more information on standards and regulations. Other documents include hazard alerts, recall letters, manufacturer specifications, design and testing documents, and post-market surveillance databases, e.g., the FDA Manufacturer and User Facility Device Experience Database (MAUDE).

The reliance upon standards becomes especially important should an investigator's findings be challenged in a court of law. Recent court decisions beginning with *Daubert v. Merrill Dow Pharmaceuticals* have placed increased responsibility upon judges to scrutinize the bases for the opinions of experts (Babitsky et al., 2000). Whether or not standards exist for an investigative technique's operation is one criterion applied. The peer review to which a method or technique is subjected during the standards development process and the existence of the standards themselves can be the key to admissibility of evidence (Lentini, 2001).

Examination of Evidence

Measurement techniques, inspection and measurement tools are described as follows.

Measurement Techniques

Use the human senses: Sight, hearing, smell, taste, and touch to examine evidence. If one possesses a sixth sense, use that as well.

Testing of devices should not be done in isolation on a laboratory bench, but ideally at the location where the accident occurred, or in the context of a reasonable reconstruction of the scene. A device might perform satisfactorily on the test bench in the laboratory but not when tested in the environment where the injury occurred. Factors such as electromagnetic interference from another device or from an outside source, a power line irregularity in the patient's room, or a gas source providing inadequate pressures and flows could contribute to a device's inadequate performance.

All available devices involved in the accident should be inspected, along with all disposables, accessories, packaging, and peripheral devices. The facility should be inspected as well, and the output of electrical receptacles and medical gas outlets should be measured if these were used. Weather conditions should be noted—a lightning strike may have caused a momentary interruption in power or a power line transient that adversely affected devices. If weather details are unknown, the United States Meteorological Survey can provide the weather conditions from the time in question.

During a site inspection, take photographs of everything examined. Photographs of the scene should be taken from all directions. The front, back, sides, top, and bottom of the device or devices should be photographed. Take close-ups of relevant components such as controls, meters, and connectors. Dimensions of the object photographed can be seen if a ruler is placed in the field of view. Record the subject matter in each photograph in a laboratory notebook. Record a sketch of the scene in the notebook and note the photographic frame, using an arrow to show the camera angle. Pay special attention to damage, which could be a sign of physical abuse or of unusual connections, improper settings, or incompatible accessories.

Physical measurements include temperature, relative humidity, time, electrical parameters, dimensions, and weight. Note a device's construction materials. Use a magnifier—at times, the smallest bit of evidence, such as a tiny metallic whisker that causes a relay to fail, provides the greatest clue. Take macro photographs to capture such details.

While it is beyond the scope of this discussion to present all testing procedures, functional verification of device performance includes the following:

- Therapeutic devices should be operated to determine and measure their output, e.g., energy, waveform, frequency, irradiance, direction, flow, pressure, and volume.
- Diagnostic devices can be tested for performance by using appropriate simulators for such physiological parameters as blood pressure, temperature, and QRS waveform.

Testing a device is best done in an environment that replicates the actual environment at the time of the accident as closely as possible. Failure may have occurred due to an interaction between the device and another device in the environment or to some abnormality in the facility such as line voltage variation or electromagnetic interference. One must always be on the alert for the possibility of an intermittent failure that occurs only under particular circumstances.

Inspection and Measurement Tools

An accident investigation kit should be available at an instant's notice. Figure 64-5 shows the "yellow safety box" used by the author in his investigations (Dyro, 1995a). Figures 64-6 to 64-9 show the items stored in the box. Other items may be added depending upon the nature of the event to be investigated.

Reconstruction

When possible, attempt to reconstruct the accident. Experiments can be performed to simulate the event if the appropriate exemplars can be obtained and if enough is known about the spatial and temporal relationships of the event. Use simulators if, for example, a physiological signal such as ECG or temperature is required. Photographs and video recording of the simulation will help interpret the results.

As an example, the steps to take in accident reconstruction for electrical injury cases is as follows (Fish and Geddes, 2003):

- Determine current flow
- Determine expected medical effects of the electrical current
- Determine the nature of the injury actually sustained
- Compare the expected effects of current flow on the actual injury
- Consider that the person's condition is due to disease or other processes unrelated to the electrical injury
- Cite references from the scientific literature that support and illustrate the consultant's opinions

Figure 64-5 The author with his yellow safety box entering a client's home prepared to perform an accident investigation (Dyro, 1998).

Figure 64-6 The yellow safety box, shown open with its two trays removed to display their contents.

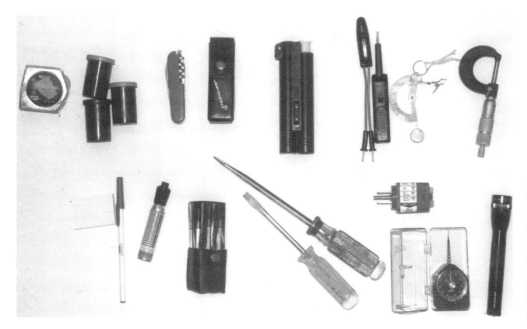

Figure 64-7 The contents of the upper tray. Upper row from left to right: tape measure, photographic film, Swiss army knife, Leatherman® tool, 30X illuminated magnifier, electrical circuit voltage tester and live circuit checker, letter scale (0-100 gm), micrometer. Lower row from left to right: Post-It notes, ball point pen, marking pen, jeweler's screwdrivers, screwdrivers (large and small flathead), receptacle tester, force gauge, flashlight.

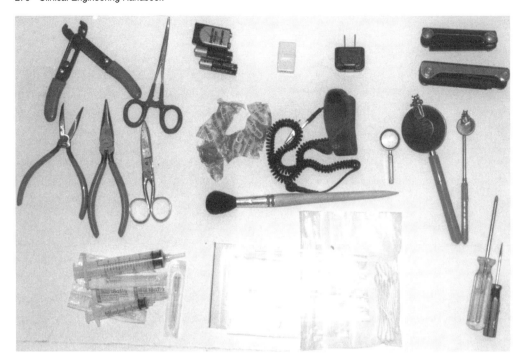

Figure 64-8 Contents of the lower tray. Upper row from left to right: wire strippers, forceps, batteries, adapter, electrical flasher, and Allen wrenches (metric and English). Middle row (l to r): curved tip pliers, needle nose pliers, scissors, Atomic Fireballs (candy to distract curious children during home investigations), camel hair brush, electrical grounding strap, 10X magnifier, inspection mirrors (large and small). Lower row from left to right: assorted syringes, plastic collection bags, cotton swabs, and screwdrivers (Philips head and small flathead).

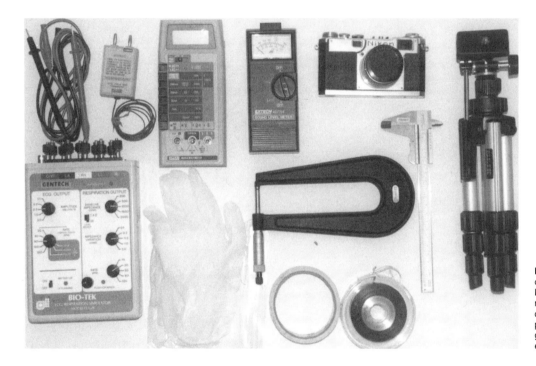

Figure 64-9 Contents of the bottom of the box. Upper row from left to right: leads, thermocouples and adapter, multimeter, sound level meter, and camera. Lower row from left to right): physiological simulator, examination gloves, micrometer, masking tape, electrician's tape, calipers, tripod.

The established spatial-temporal relationships aid in the reconstruction. These relationships are first documented in the initial response to the accident, sometimes in a spreadsheet format. An example of how such a document can be formatted appears in Table 64-1.

Interaction with and Interviews of Others

Interview Techniques

The CE should also be aware that during a time of an accident witnesses are typically under increased stress and may not have accurate recollections of events, times, and other personnel who may have been present. Some personnel may also attempt to conceal information that could implicate themselves or others in contributing to the injury. Ask open-ended questions, allowing the person interviewed to answer in his or her own words. Details such as movement and location of personnel, and observations such as smells, sounds, and sights should be ascertained and explored. The clinical engineer should adhere as closely as possible to the following guidelines for conducting a fact-finding interview:

- Question one person at a time, to avoid uncomfortable interpersonal relationships (peers, subordinate/employer).
- If several people are involved in an equipment setup, interviewing them together may be helpful.
- Understand how the position of the person interviewed may influence answers to questions.
- Be prepared with technical details and a list of questions.
- Remember to determine who, what, where, why, and how.
- Take notes while maintaining eye contact, refrain from audio or video recording.
- Ask questions to put the interviewee at ease.
- Keep an open mind.
- Do not jump to conclusions: Note discrepancies, and defer judgments until all interviews have been done.
- Do not interrupt an answer to ask for more details.
- Ask open-ended questions so that the person tells a story; even seemingly irrelevant details may prove to be important after all the data have been collected.
- Ask questions to elicit observations rather than judgments.
- Rephrase response to ensure that the answer is understood.

Table 64-1 Accident reconstruction spread sheet. The format shown in this spreadsheet may be changed depending on the nature of the event under investigation

Person	Source	Facts and statements	Date & Time	Significance
Buckley	NE Weather Science	Relative humidity (RH)=90%; Temperature (T)=98°	1/16/2 11-12 PM	Perspiration of foot in shoe, sweat on hands
Cragin	Deposition page 10, line 5	"The floor is wet all the time."		Increased conductivity
Allen	Affidavit 5/12/2	"The lights flickered."	1/16/2 11:05 PM	Power line disturbance
Svenson	Service Report #8098	Short circuit repaired	1/17/2	Faulty device

- End each interview by emphasizing the role of improving quality through root cause analysis.
- Summarize notes and sign and date the document upon conclusion of interviews.

Witnesses and Other Hospital Personnel

All eyewitnesses should be interviewed, in addition to any personnel that may have relevant information, such as technicians and any service personnel responsible for cleaning, sterilizing, inspecting, and maintaining the items used at the time of the injury.

Manufacturers

The device manufacturer must be notified in accordance with FDA medical device reporting (MDR) regulations (Guyton, 2002; FDA, 1996).

FDA

Discussions with the FDA should be coordinated with the hospital administration's FDA contact person.

Attorneys

All discussions between hospital staff and attorneys must take place with the advice and consent of the hospital legal affairs office.

Media

The hospital clinical engineer should not speak to the media, e.g., radio, television, and newspapers. Any contact with the media should be limited to the official hospital spokesperson. The clinical engineer should be aware that information transmitted to the media may not be disseminated in a way that reflects all the information deemed importance or significant (Dyro, 1998a).

Reports

Upon completion of an investigation, a written report is prepared. Reports should be well written, concise, complete, and accurate. The report should include of the following parts: Abstract, introduction, methodology, results, discussion, conclusion, opinions, root cause analysis, and recommendations. Use advanced techniques for report writing to make your report come to life. For example format it well, and use photographs, sketches, figures, and tables to help tell the story (Dyro, 2003).

Case Studies

Studies of several cases which the author has investigated follow. These illustrate one or more of the five elements in the system and the investigation techniques described above.

Infant Incubator

The author investigated a death at a large university teaching hospital. An infant died of hyperthermia. The temperature of the infant incubator (Figure 64-10) was 108° F when checked by a nurse at 2:00 a.m..
Investigation revealed that the operator, environment, and the device itself (both from a manufacturer and maintainer perspective), contributed to the death.

Operator

User abuse (abuse) is illustrated in the four views of the incubator shown in Figure 64-11. The figure shows obvious signs of rough treatment at several points of the incubator. Bent control knobs, a cracked meter face, and a bent carrying handle are clearly shown. The nurses were both inattentive and uneducated. They failed to observe these signs of mechanical shock and failed to recognize the possible consequences of such physical damage.

Environment

The environment in which this event occurred was a nursery and the event occurred during the night shift about 3:00 a.m. At this time, the nurse was inattentive to her duties and failed to check on the neonate.

Device

The maintenance of the incubator was deficient because the hospital had no clinical engineering department, and the incubator was not under any kind of a service agreement with an outside party.

No incoming inspection or periodic inspection had been performed that may have detected the design inadequacy and certainly would have detected the signs of user abuse.

The device design was defective. The forces to which the control unit and incubator are typically subjected were not considered, and the thermostat system was inadequately equipped to handle these forces. The backup, high temperature, mercury-in-glass thermostat failed because the design did not consider the typical rough handling of the control unit. Mechanical shock and vibration separated the mercury column. Because the column was broken, it did not form a conductive pathway between the two circumferential electrodes on the exterior of the glass thermostat capacitively coupled to the column (Narco, 1982; Harding, 1982; Haffner, 1982; Dyro, 1977) (see Figure 64-12).

As a result of recommendations made by the author, the hospital instituted a clinical engineering department.

Ventilator and Apnea Monitor

A young child died at home in bed while on a ventilator and while monitored by a heart rate and respiration monitor (Dyro, 1998). The breathing circuit became dislodged from the tracheal tube. The ventilator low airway pressure alarm failed to activate because the patient connector of the breathing circuit was occluded by the child's neck. The ECG/respiration rate monitor failed to alarm because it detected artifact ECG and respiration signals generated by the mechanical force of the ventilator tubing against the child and the bed (Sahakian et al., 1986). With each ventilation, pressure generated in the patient circuit was conducted to the bedside monitor arrangement, causing movement of the transthoracic electrodes. This motion prevented the monitor from alarming. The monitor and ventilator were plugged into an old power strip that the parents had found in their garage. The power strip was faulty, and its neon light "on" indicator circuitry produced electromagnetic interference that resulted in artifact signals detected by the monitor. Reconstruction of the event was possible because all of the components of the system or appropriate exemplars were available.

All five elements of the system described above were involved in this event: The ECG/respiration rate monitor (device), ventilator (device), patient circuit (device), child (patient), facility (power strip), environment (bed), and the parent who assembled the equipment (operator).

Skin Lesions

The investigator will on occasion be asked to investigate an event that resulted in a skin lesion. Often the investigator will be told that the patient received a burn. Care must be taken in the interpretation and investigation of injuries that are described as burns. It is natural to think of a burn as an injury to the skin caused by excessive heat. However, the investigator must consider all the mechanisms that can produce a skin injury (Gendron, 1988). Table 64-2 provides a list of several of several devices that led to skin lesions as a result of several different injury mechanisms. Device, injury, and mechanisms are shown in the table.

Serious burns have occurred in at least two incidents that the author has investigated in which neonates in incubators crawled from the mattress surface to hot air vents (Figure 65-13). Missing barriers and excessive temperatures emanating from air vents

Figure 64-10 The infant incubator under test. A monitor with air temperature sensors measures the thermal characteristics of the incubator.

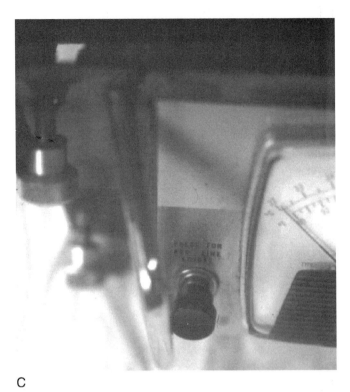

Figure 64-11 A–D Four views of control panel of incubator showed many signs of rough handling such as bent knobs (A, B) and handle (C) and cracked meter face (d).

presented a hazard. The active infants were able to move over the vents. Inattentive nurses failed to observe the movement of the infant to the areas of the incubator at dangerously high temperatures. In one instance, the manufacture issued an advisory letter indicating that a modification kit was available to isolate the infant from the source of heat, but the hospital failed to receive the letter, and the modification was not made.

Accident Investigation Education

Learning about accident investigation techniques can occur as part of college courses, or during workshops, seminars, or teleconferences, and in the workplace.

Formal Courses

Formal courses in the clinical engineering curriculum should be developed to include accident investigation education. This chapter can be used as an outline around which to

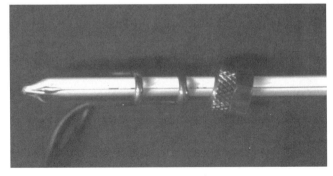

Figure 64-12 The backup thermostat is shown with the broken mercury column.

Table 64-2 Several examples of incidents involving skin lesions investigated by the author are listed in terms of devices and types and mechanisms of injury.

Medical device	Injury	Mechanism
Heating pad	Infant whole body burns	Thermal (conduction)
Infrared lamp	Adult skin burns	Thermal (infrared Radiation and conduction)
Operating room table mattress	Whole body lesions	Chemical (ethylene oxide sterilant) burns
Heated humidifier	Infant face burns	Thermal (conduction and convection)
Electrosurgical unit in oxygen enriched atmosphere	Face and upper body burns	Oxygen enriched OR fire thermal (fire)
Infant radiant warmer	Lower body burns (white tape defeated alarm)	Thermal (infrared radiation)
Electrosurgical unit	Burns to chest; Burns around mouth; Internal burns	Thermal (electrosurgical current)
Portable oxygen unit	Burns to legs	Thermal (fire)
Pulse oximeter skin electrode	Skin lesions	Thermal (conduction)
Surgical retractor	Skin and muscle damage	Pressure necrosis

develop a course on accident investigation for the clinical engineer. Examples follow for classroom techniques, case studies, demonstrations, examinations and reports that should be included in a formal college course. The course length and depth can be adapted to the educational requirements of the class. Undergraduate and graduate courses in clinical engineering should include at least six hours of class lecture and six hours of preparation.

Classroom Techniques

The instructor should develop systems approach skills by creating a hypothetical situation involving a specific medical device. The students should be directed to describe the system with a drawing. The student should strive to incorporate all five elements (device, operator, facility, environment, and patient) into the drawing and analysis. Each element should contribute in some way to the cause of the accident. The class should be given a list of devices from which to choose, such as infant incubator, ventilator, anesthesia machine, EKG machine, electrosurgical unit, hyperthermia unit, infusion pump, defibrillator, and physiological monitor. After completion of the exercise, the student should present the hypothetical case study to the class, using the graphic he or she created.

Case Studies and Demonstrations

Lecture presentations of actual case studies are more relevant and hold student interest better than hypothetical cases. Classroom and laboratory demonstrations with actual medical devices reinforce understanding of device interactions with other elements in the system.

Examinations

The examination for the certification of clinical engineers administered by the American College of Clinical Engineering includes questions on accident investigation. The candidate for certification is expected to answer multiple choice, essay, and oral questions during the examination process.

Reports

Good communication skills include writing skills. Students should learn how to write a technical report that is accurate, concise, and complete and includes the basic elements: abstract, introduction, methods and materials, results, discussion, and conclusions. Reports should contain references to supporting documents, such as hospital standards and operator and service manuals. Techniques to make a report more effective, such as good formatting and the use of photographs, diagrams, figures and tables should be emphasized (Dyro, 2003).

Informal Courses and Independent Study

Everyday occurrences in the home provide the opportunity to practice accident investigation. If a home appliance, e.g., radio, television, toaster or dishwasher, operates in an unexpected fashion, the clinical engineer should use this event as an opportunity to practice and refine investigation techniques. In addition to technical and scientific texts on subjects related to accident investigation. He or she can obtain additional insight from television and film in which plots involve forensic examination techniques. The CE will find, however, that all the clues necessary to establish the cause of the adverse event may not be present in reality as they always seem to be in novels, movies, and TV.

Resources to assist in accident investigation education and training are available from many sources, including the following:

- Standards

American Society for Testing and Materials
Association for the Advancement of Medical Instrumentation
National Fire Protection Agency
Institute of Electrical and Electronics Engineers
Chapters 117 and 120 of this handbook

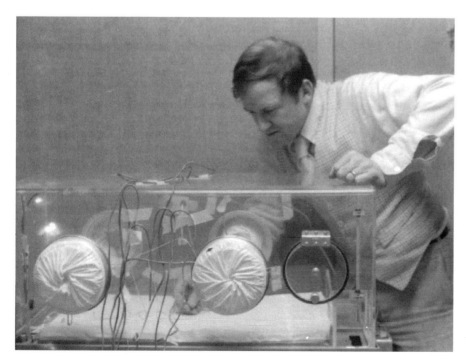

Figure 64-13 The author positions thermistors to measure the temperatures of surfaces and air within an infant incubator.

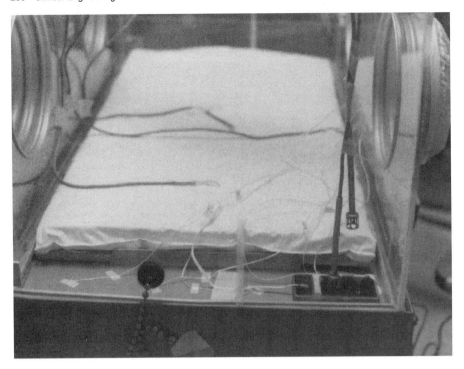

Figure 64-14 Surface temperature thermistors and an air temperature thermistor are shown affixed to surfaces and suspended above the air inlet vent of infant incubator in which infant suffered third-degree facial burns after contact with heated, humidified air at excessive temperatures. The hospital had not installed the plastic shield that the manufacturer made available as a retrofit.

- Regulatory Agencies

Food and Drug Administration (FDA, 2002); (Dyro, 2003a); (See Chapter 126)
Joint Commission on Accreditation of Healthcare Organizations (See Chapter 121)

- Books

Medical Device Incident Investigation & Reporting (Shepherd, 2000c)
Unexplained Patient Burns (Gendron, 1988)
Medical Device Accidents and Illustrative Cases (Geddes, 2002)
Human Error in Medicine (Bogner, 1994)
The Portable Poe (Poe, 1973)
Accident and Forensic Investigation (Bruley, 1994), chapter in Medical Devices: International *Perspectives on Health and Safety* (Van Gruting, 1994)
Writing and Defending Your Expert Report (Babitsky and Mangraviti, 2002)
Handbook of Electrical Hazards and Accidents (Geddes, 1995)
Electrical Injuries: Engineering, Medical, and Legal Aspects (Nabours et al., 1999)
Warnings, Instructions, and Technical Communications (Peters and Peters, 1999)

- Courses at local, regional, and national conferences

Annual National Expert Witness and Litigation Seminar (Dyro, 2003b)
Annual Meeting of the Association for the Advancement of Medical Instrumentation (Dyro, 1995)

- Organizations

American College of Clinical Engineering (ACCE) Teleconference Series (Dyro, 2000a; Shepherd, 1996; Patail and Bruley, 2002).
Human Factors and Ergonomics Society
National Forensic Center
National Center for Patient Safety (http://www.va.gov/ncps/)

- Manufacturers
- Television and Cinema

References

American College of Clinical Engineering. *Enhancing Patient Safety: The Role of Clinical Engineering.* Plymouth Meeting, PA, American College of Clinical Engineering, 2001.
American College of Clinical Engineering Healthcare Technology Foundation. *Guideline for Incident Investigations.* Plymouth Meeting, PA, American College of Clinical Engineering Healthcare Technology Foundation, 2003.
American Society for Testing and Materials. *Standard Practice for Collection and Preservation of Information and Physical Items by a Technical Investigator.* ASTM E1188-95. West Conshohocken, PA, American Society for Testing and Materials, 1995.
American Society for Testing and Materials. *Standard Practice for Reporting Incidents.* ASTM E1020-96. West Conshohocken, PA, American Society for Testing and Materials, 1996.
American Society for Testing and Materials. *Standard Practice for Examining and Testing Items That Are or May Become Involved in Products Liability Litigation.* ASTM E860-97. West Conshohocken, PA, American Society for Testing and Materials, 1997.
American Society for Testing and Materials. *Standard Practice for Reporting Opinions of Technical Experts.* ASTM E620-97. West Conshohocken, PA, American Society for Testing and Materials, 1997.
American Society for Testing and Materials. *Standard Practice for Evaluation of Technical Data.* ASTM E678-98. West Conshohocken, PA, American Society for Testing and Materials, 1998.
Abramson N, Wald KS, Grenvik ANA, et al. Adverse Occurrences in Intensive Care Units. *JAMA* 244:1582-1584, 1980.
Babitsky S, Mangraviti JJ. *Writing and Defending Your Expert Report.* Falmouth, MA, SEAK, Inc., 2002.
Babitsky S, Mangraviti JJ, Todd CJ. *The Comprehensive Forensic Services Manual.* Falmouth, MA, SEAK, Inc., 2000.
Berry K, Krizek B. Root Cause Analysis in Response to a "Near Miss." *J Healthc Qual* 22:16-18, 2000.
Bogner MS (ed). *Human Error in Medicine.* Bethesda, MD, United States Food and Drug Administration, 1994.
Bruley ME. Accident and Forensic Investigation. In Van Gruting, CWD (ed). *Medical Devices: International Perspectives on Health and Safety.* Amsterdam, Elsevier Science, 1994.
Christoffersen K, Woods DD. How Complex Human-Machine Systems Fail: Putting "Human Error" in Context. In Karwoski W and Marras W (eds). *Handbook of Occupational Ergonomics.* Boca Raton, FL, CRC Press, 1999.
Cohen T. ACCE Body of Knowledge Survey. *ACCE News* 11:13-14, 2001.
Cooper JB, Newbower RS, Long CD, McPeek B. Preventable Anesthesia Mishaps: A Study of Human Factors. *Anesthesiology* 49:399-406, 1978.
Cooper J, et al. An Analysis of Major Errors and Equipment Failures in Anesthesia Management: Considerations for Prevention and Detection. *Anesthesiology* 60:34-42, 1984.
Driscoll C. Conducting the Right Investigation. *Biomed Instrum Technol* 37:65-66, 2003.
Dyro JF. The Development of a Standard for Infant Warmers and Incubators. FDA/BMDDP-77/43, PB-263 250/3WV. Springfield, VA, National Technical Information Service, 1977.
Dyro JF. Hospital Safety Program. In JG Webster (ed). *Encyclopedia of Medical Devices and Instrumentation.* New York, Wiley, 1988.
Dyro JF. Building Awareness: Educating Health Care Providers and the Public. In Sykes S (ed). *Electromagnetic Compatibility for Medical Devices: Issues and Solutions.* Arlington, VA, Association for the Advancement of Medical Instrumentation, 1995.
Dyro JF. The Safety Box. *Proceedings of the AAMI 30th Annual Meeting, Anaheim, CA.* Arlington, VA, Association for the Advancement of Medical Instrumentation, 1995.
Dyro JF. So You Want to Be in Pictures? Seventh Annual National Expert Witness and Litigation Seminar, Hyannis, MA, June 19, 1998.
Dyro JF. Methods for Analyzing Home Care Medical Device Accidents. *J Clin Eng* 23:359, 1998.
Dyro JF. Get on Board the Safety Train. *ACCE News* 10:3, 2000.
Dyro JF. Investigating an Equipment Incident. ACCE Teleconference. Plymouth Meeting, PA, American College of Clinical Engineering, October 19, 2000.
Dyro JF. Advanced Report Writing Techniques: Bringing Your Report to Life. 12th Annual National Expert Witness and Litigation Seminar, Hyannis, MA, June 26, 2003.
Dyro JF. JCAHO Patient Safety Goal #6: Improving the Effectiveness of Clinical Alarm Systems. US Food and Drug Administration, Medical Product Surveillance Network (MedSun) Teleconference, April 10, 2003.
Dyro JF. Accident/Incident Investigation: Advanced Techniques, 12th Annual National Expert Witness and Litigation Seminar, Hyannis, MA, June 26, 2003.
ECRI. Hazard: Monitors and Static Electricity. *Health Devices* 3:275, 1974.
ECRI. Investigating Device-Related Incidents. *Medical Device Reporting: A Guide for Health care Facilities.* Plymouth Meeting, PA, ECRI, 1991.
ECRI. Case History: The Hazards of New Technologies. *Accident Investigator* 2:1-2, 1993.
ECRI. Root Cause Analysis. *Health care Hazard Control: Safety and Behavior* 2.2.1:1-8, 2001.
Food and Drug Administration. Medical Device User Facility and Manufacturer Reporting, Certification and Registration. 21 CFR Part 803. July 31, 1996.

Food and Drug Administration. Integrating Human Factors Engineering into Medical Device Design and Development: An FDA Q&A. *J Clin Eng* 27:123-127, 2002.
Fish RM, Geddes L. *Medical and Bioengineering Aspects of Electrical Injuries.* Tucson, AZ, Lawyers & Judges Publishing Company, 2003.
Geddes LA. *Handbook of Electrical Hazards and Accidents.* Boca Raton, FL, CRC Press, 1995.
Geddes LA. *Medical Device Accidents and Illustrative Cases, 2nd Edition.* Tucson, AZ, Lawyers & Judges Publishing Company, 2002.
Gendron FG. *Unexplained Patient Burns: Investigating Iatrogenic Injuries.* Brea, CA, Quest Publishing, 1988.
Gosbee J. The Discovery Phase of Medical Device Design: A Blend of Intuition, Creativity, and Science. *Medical Devices & Diagnostic Industry* 79-82, November 1997.
Gosbee JW, Arnecke B, Klancher J, et al. The Role of Usability Testing in Healthcare Organizations. *Proceedings of the Human Factors Society 40th Annual Meeting.* Santa Monica, CA, Human Factors Society, 2001.
Guyton B. Human Factors and Medical Devices: A Clinical Engineering Perspective. *J Clin Eng* 27:116-122, 2002.
Haffner ME. Malfunction of a Neonatal Incubator (Letters). *JAMA* 247:2372, 1982.
Harding GH. Malfunctioning Neonatal Incubators (Letters). *JAMA* 248:2835, 1982.
Hyman WA, Cram N. A Human Factors Checklist for Equipment Evaluation and Use. *J Clin Eng* 27:131-3, 2002.
Institute of Electrical and Electronics Engineers. IEEE Recommended Practice for an Electromagnetic Site Survey (10 kHz to 10 GHz), 473-1985. New York, Institute of Electrical and Electronics Engineers, 1985.
Institute of Electrical and Electronics Engineers. National Electrical Safety Code. C2-1997. New York, Institute of Electrical and Electronics Engineers, 1997.
Institute of Medicine. *Crossing the Quality Chasm: A New Health System for the 21st Century.* Washington, DC, National Academy Press, 2001.
Iverson K, Barsan W. Accidental Cranial Defibrillation. *JACEP* 8:24-25, 1979.
Joint Commission on Accreditation of Healthcare Organizations. *Root Cause Analysis in Health Care: Tools and Techniques.* Oakbrook Terrace, IL, Joint Commission on Accreditation of Healthcare Organizations, 2000.
Joint Commission on Accreditation of Healthcare Organizations. Revisions to Joint Commission Standards in Support of Patient Safety and Medical/Health Care Error Reduction. Oakbrook Terrace, IL, Joint Commission on Accreditation of Healthcare Organizations, 2001.
Kenney RJ. Comparative Negligence: The Patient's Duty to Use Care. *Forum* 4:15-16, 1983.
Kermit E. Medical Adhesive Tape Should Be a Controlled Substance! *ACCE News* 10:15, 2000.
Kessler DA. Introducing MEDWatch: A New Approach to Reporting Medication and Device Adverse Effects and Product Problems. *JAMA* 269:2765-2768, 1993.
Kohn, LT, Corrigan JM, Donaldson MS. *To Err Is Human: Building a Safer Health System.* Washington, DC, National Academy Press, 2000.
Lentini JJ. Standards Impact the Forensic Sciences. *ASTM Standardization News* 16-19, February 2001.
MDR Report. FDA DEN M54439. January 24, 1992.

Narco Scientific, 1982. Warning Letter. July 1982.
National Fire Protection Association. Standards for Health Care Facilities. NFPA 99. Quincy, MA, National Fire Protection Association, 2002.
NFPA. National Electrical Code. NFPA 70. National Fire Protection Association. Quincy, MA, 2002.
Nabours RE, Fish RM, Hill PF. Electrical Injuries: Engineering, Medical, and Legal Aspects. Tucson, AZ, Lawyers & Judges Publishing Company, 1999.
Patail B, Bruley ME. Investigation with Root Cause Analysis: ECRI and VA Approaches. ACCE Teleconference. Plymouth Meeting, PA, American College of Clinical Engineering, November 21, 2002.
Peters GA, Peters BJ. *Warnings, Instructions, and Technical Communications.* Tucson, AZ, Lawyers & Judges Publishing Company, 1999.
Poe EA. The Murders in the Rue Morgue, *Graham's Magazine*, April, 1841. In Stern PVD (ed). *The Portable Poe: Selected Works of Edgar Allan Poe.* 1973.
Sahakian AV, Tompkins WJ, Webster JG. Electrode Motion Artifacts in Impedance Pneumography. *IEEE Trans Biomed Eng* 32:448, 1986.
Shepherd M, Dyro JF. The Whimsical Use of White Tape. Slide/Audio Program. University of California, San Francisco, 1982.
Shepherd M. *A Systems Approach to Hospital Medical Device Safety.* Arlington, VA, Association for the Advancement of Medical Instrumentation, 1983.
Shepherd M, Brown R. Utilizing a Systems Approach to Categorize Device-Related Failures and Define User and Operator Errors. *Biomed Instrum Technol* 26:461-475, 1992.
Shepherd M. The Systems Technique for Analyzing Device-Related Failures. In Shepherd M (ed). *Systems for Medical Device Incident Investigation and Reporting.* New York, Raven Press, 1993.
Shepherd M. Incident Investigation. ACCE Teleconference. American College of Clinical Engineering. Plymouth Meeting, PA, March 21, 1996.
Shepherd M. Device Servicer Error: An Under-Reported Hazard? *J Clin Eng* 23(3):215-222, 1998.
Shepherd M. National Patient Safety Movement. *ACCE News* 9:9, 1999.
Shepherd M. Mending the Way of Our Errors. *ACCE News* 10:3-4, 2000.
Shepherd M. Eliminating the Culture of Blame: A New Challenge for Clinical Engineers and BMETs. *Biomed Instrum Technol* 34:370-374, 2000.
Shepherd M. *Medical Device Incident Investigation & Reporting.* Walnut Creek, CA, Devteq Publishing, 2000.
Shojania KG, Duncan BW, McDonald KM, et al. Making Health Care Safer: A Critical Analysis of Patient Safety Practice. *Evidence Report/Technology Assessment* 43, 2001. Rockville, MD, Agency for Healthcare Research and Quality.
Stalhandske E, DeRosier J, Patail B, Gosbee J. How to Make the Most of Failure Mode and Effects Analysis. *Biomed Instrum Technol* 2:96-102, 2000.
Stamatis DH. Failure Mode and Effect Analysis. American Society for Quality Press, 1995.
Van Gruting CWD (ed). *Medical Devices: International Perspectives on Health and Safety.* Amsterdam, Elsevier Science, 1994.
Welch DL. Human Factors Usability Test and Evaluation. *Biomed Instrum Technol* 2:183-187, 1998.
Willis G. Failure Modes and Effects Analysis in Clinical Engineering. *J Clin Eng* 17:59-63, 1992.

65

The Great Debate on Electrical Safety—In Retrospect

Malcolm G. Ridgway
Sr. Vice President, Technology Management and Chief Technology Officer, MasterPlan Inc.
Chatsworth, CA

The 1960s saw both the advent of open-heart surgery and the increasing use of cardiac catheterization procedures. More and more patients with externalized transarterial catheters, usually enclosing leads that could be quickly connected to an external cardiac pacemaker, were appearing in the new cardiac care or special care units of hospitals. Once physicians realized that this highly conductive pathway not only bypassed the usually protective layers of relatively resistive body tissues, but that it also directed current to the most electrically sensitive areas of the inner walls of the heart, concerns arose about the possibility that these patients could be electrocuted by currents much smaller than those that would affect the exterior of the body.

Patients with externally accessible conductive pathways leading directly to the heart came to be known as electrically susceptible (ES) or electrically sensitive patients (ESPs). The theoretical phenomenon, in which an ES patient might be induced into fatal ventricular fibrillation by the passage of a small level of current through the transarterial catheter, became known as "silent electrocution" or "microshock."

Concerns about this potential scenario were published as early as August 1961 in an editorial in the journal *Circulation* titled, "*Hidden Hazards of Cardiac Pacemakers*." Laboratory experiments indicated that the levels of current that could trigger potentially fatal ventricular fibrillation were indeed significantly lower than the levels associated with conventional electrocution—on the order of tens of microvolts. Tests also established that the now-familiar phenomenon of current "leakage," as well as the process of relatively large currents passing through low resistance grounding conductors, could easily send high magnitude currents into the exposed conductive pathways.

The concept of equipotential grounding, in which substantial (green) grounding conductors are used to connect all exposed conductive surfaces to a central grounding point in a star configuration, was developed as a prime defensive measure against this new hazard. Others championed the use of isolation transformers as the best way to reduce leakage current in the ground circuits of the hospital's electrical distribution system. Several regulatory and standards-setting organizations began taking notice.

In April 1968, the division of medical sciences of the National Research Council (NRC) held a two-day workshop on "Electrical Hazards in Hospitals" that was attended by more than 100 people. The proceedings of this workshop were edited by Dr. Carl Walters and later published by the influential National Academy of Sciences (Walter, 1970). Carl Walter was a renowned surgeon at the Peter Bent Brigham Hospital in Boston, a member of the faculty at Harvard Medical School, and chairman of the committee on hospitals of the National Fire Protection Association (NFPA). Dr. Walter has been credited with establishing one of the world's first blood banks in a basement room at Harvard in 1934, and later (1949) with the invention of the blood bag, which ended the cumbersome and dangerous procedure of pumping blood directly from donor to patient via paraffin coated glass tubes. In addition, his insight and pioneering work with the Castle company led to the introduction of high pressure steam sterilizers (sometimes called autoclaves) for reprocessing surgical instruments. Before the autoclave, surgical instruments were simply "sterilized" in boiling water.

At the 1968 NRC-sponsored workshop, Dr. Walter first speculated on the probable incidence of death by "microshock" in US hospitals. During a discussion of national statistics on electrocution that were available at the time, he claimed that an insurance actuary, whose statistical hobby is electric shock and electrocautery injuries, had assured him that there were 1200 misdiagnosed electrocutions annually in hospitals during 1964 and 1965. That would have amounted to one "misadventure" annually in every seven hospitals in this country. Walter said these misadventures were classified as cardiac arrest, but the deaths occurred during resuscitation efforts unrelated to the patient's primary disease or during application of electric appliances. He also said the statistics were culled to demonstrate the prevalence of the problem and to show why the medical profession has not recognized the problem (i.e., during a lifesaving venture, those involved may not perceive what is going on). Walter claimed to have encountered three such instances in a recovery room himself and further said that when the situation was recreated, it was readily apparent exactly which device caused the trouble. Finally, Walter went on to assert there were at least a dozen analyses in existing literature of the patient electrocution risk and claimed that is why so many doctors were becoming interested in electric shock.

On January 27, 1969, a report titled, "Accidental Electrocutions Claim 1200 Patients a Year" was published in *Electronic News* (Electronic News, 1969). The report quoted microshock statistics obtained from Dr. Walter during a telephone interview. These same statistics were repeated again during presentations made by Dr. Walter and others at the 71st Annual Meeting of the American Hospital Association in Chicago in August 1969. After the proceedings of the NRC workshop were published in 1970, these statistics were repeated at a press conference and widely reported throughout the national press.

In June 1970, a report was distributed by the UPI wire service that Ralph Nader, an attorney and consumer activist, had alleged in a speech that 5000 deaths attributable to microshock occurred each year in the nation's hospitals. To this day Mr. Nader has not provided any independent substantiation for his figure.

In March 1971 the *Ladies Home Journal* ran an article quoting Ralph Nader titled, "Ralph Nader's Most Shocking Expose," which stated that, *"at the very least, 1200 Americans are electrocuted annually during routine diagnostic and therapeutic procedures,"* and that *"medical engineers such as Professor Hans von der Mosel, co-chairman of the Subcommittee on Electrical Safety of the Association for the Advancement of Medical Instrumentation and safety consultant to New York City's Health Services Administration, believe that the number might be ten times as high as the conservative estimate of 1200"* (Nader, 1971). This is the source of the sometimes quoted "estimate" of 12,000 deaths per year. Interestingly, this same article states that, *"Only three hospitals in the country have biomedical engineers on their staffs to supervise the operation and maintenance of complex machines: Downstate Medical Center in New York City; Sinai Hospital in Baltimore; and Charles S. Wilson Hospital in Johnson City, N.Y."*

By the mid-1970s, the NFPA's Committee on Hospitals had developed and distributed for public comment some proposed amendments to Article 517 of the 1971 edition of the National Electric Code (NEC) that would require all hospitals to have isolation transformer-based "Safe Patient Power Centers" in all special care areas of the nation's approximately 6000 hospitals. The potential financial impact of this proposal shocked the health care community. The technical inadequacy of the proposed solution also shocked the embryonic clinical engineering community.

In the spring of 1971, shortly before the Annual NFPA Meeting in San Francisco at which the Committee on Hospitals' proposed amendment would be voted on, the Hill-Burton Program Committee convened a private meeting in Rockville, Maryland, at which ten experts in "electronics in hospitals" were invited to debate with Dr. Walter and his technical advisors about the merits of the proposed new requirements. In a follow-up report, one of the ten experts concluded Dr. Walter arrived at his "estimate" of 1200 deaths per year due to microshock by noting one patient death in his hospital he suspected as having been the result of microshock and then extrapolating that to 1200 microshock electrocutions per year on the basis of his hospital caring for about one in 1200 of all US patients yearly.

Participants at the meeting indicated that the proposed solution was technically inadequate because the isolation monitor, that the National Electric Code required to be used with an isolation transformer, injects far more current into the circuit than the "safe" level of 15 microamps. Dr. Walter's team was unable to rebut the criticism. The report goes on to state that, *"The Hill-Burton Program Committee's findings were never publicized, but the committee did inform the NFPA that if isolated power in all special care areas was required by the NFPA in its forthcoming standards, the committee would terminate its long-standing requirement that hospitals receiving its funds comply with NFPA's standards."* This was a substantial threat—at that time virtually all new hospital construction and renovations were subsidized with federal funds from the Hill-Burton Program.

When the proposed amendments to the NEC were presented at the NFPA Annual Meeting in San Francisco in May 1971, they prompted a very lively floor debate, after which adoption was deferred and they were returned to committee by a 106 to 38 vote of the membership of the electrical section.

In spite of this period of spirited discussion about the reality or non-existence of this new, perhaps life-threatening hazard, and the uncertainty about whether or not the various proposed countermeasures and elaborate safety tests could eliminate or reduce the threat, a battery of new electrical safety requirements appeared. Many of these requirements persist today in only slightly modified form as part of various regulations. The Joint Commission on Accreditation of Hospitals issued new standards that prescribed quarterly documented electrical safety testing for all of a facility's patient care equipment. In California, the State Department of Health issued stringent electrical safety measures as part of its new requirements for general acute care hospitals. Title 22 of the State Administrative Code introduced the soon-to-be-obsolete concept of the "electrically sensitive patient" and a host of related tests. It was a time of absurdities, such as festoons of green grounding wires connecting every piece of exposed metal surface within the vicinity of any special care bed (even in the adjacent bathrooms) to substantial central grounding posts. However, diligent, dedicated investigations over the next several years for possible occurrences of microshock failed to turn up any credible evidence that this ingeniously conceived but still theoretical hazard was claiming any lives.

In September 1973, Dr. Joel Nobel, Director of the Emergency Care Research Institute, made a statement during hearings before a Senate subcommittee on the proposed Medical Device Amendments of 1973:

"The issue of microshock electrocution, its real versus claimed incidence, its widespread publicity, the enactment of codes and laws to combat it, and the economic fortunes of the electrical transformer industry, are inextricably intertwined. Phony statistics have been used to promote the sales of safety equipment and manipulate the National Electric Code to require the use of specific products. We are not suggesting that the microshock electrocution issue was fabricated by the industrial and code making camps and consumer advocates. Each, however, capitalizing on the issue, has distorted both the technical problems and the priorities rather badly. The result is that many millions of dollars have been diverted from more critical areas of health care. This electrical safety issue has, however, performed a useful catalytic function in drawing attention to other problems associated with the use of technology for health care. It has helped hospitals to understand the broader needs for engineering support of patient care, including the judicious purchase, inspection, and preventive maintenance of medical equipment.

In later testimony, he added:

"Our information and priorities are sometimes distorted by special interest groups, however, and this is acceptable. By way of example, consider how much attention has been devoted to the problem of electrical safety in hospitals during the last 5 years, especially by the engineering community and the manufacturers of safety devices and equipment. Speculation is often translated into reality, or at least belief, by the very fact of statement or publication. Bogus statistics on electrocution in hospitals have been proclaimed and republished without end or confirmation, for 5 years. Many millions of words have been written about microshock and many millions of dollars spent to avoid it. It is obvious, however, that we still know nothing of its real incidence. Is it a widespread problem or a phantom ? We are not suggesting that the electrical safety problem is nonexistent. Our data show that it does exist, and it is significant; but its characteristics and magnitude are rather different than is generally believed. Our biggest problem is not electrocution by microshock but, instead, inadequate or unreliable power. Not too much electricity but too little.

And in August 1975, a report appeared in the journal *The Medical Staff* under the heading, "The Myth of Iatrogenic Electrocution: Its Effect on Hospital Costs." That report discussed Walter's widely publicized claim that there were 1200 electrocutions a year in US hospitals and how that charge was reported regularly thereafter in the lay press. The report went on to point out that evidence supporting Walter's claim had never been produced and that John Bruner, MD, assistant professor of anesthesia, Harvard Medical School, flatly stated to an AMA annual scientific assembly in 1975 that there had been no documented death due to electricity in a US hospital in more than a decade. While Bruner did admit he was among those first concerned with iatrogenic (physician-caused) electrocution, he reported having subsequently developed little evidence to support concerns that it might be a common occurrence. On the contrary, Bruner said that unwarranted concerns about microshock likely lead to the scrapping of useful equipment in favor of high priced "safety-featured" gadgetry. While acknowledging the potential for injury wherever electricity is used and where haste, stress, and moisture and other environmental factors combine to increase that risk, Bruner nonetheless saw that much of the effort and cost incurred in addressing microshock had little real benefit since the magnitude of the hazard was largely imaginary in the first place.

In the meantime, someone had—with a stroke of genius—realized that this entire threat could be completely eliminated by simply protecting the exposed conductive ends of the patient's catheter. Proper terminations for transarterial catheters providing low impedance pathways to the heart and great vessels became the order of the day, and electrically sensitive patient's need for special environmental consideration disappeared almost overnight. Articles on electrical isolation of the patient appeared. See, for example, Guidelines for Clinical Engineering Programs; Part I: "Guidelines for Electrical Isolation" (Ridgway, 1980).

A byproduct of this extended episode, however, was the discovery that the existing quality of maintenance of the typical hospital's ever-expanding inventory of electronic equipment was inadequate. A new high intensity focus on equipment maintenance and safety was born.

Another interesting sidebar is the parallel, then subsequent, debate about the rationale for perpetuating the isolated power requirement in operating rooms where the use of flammable agents had been prohibited. The original requirement for isolated power had been introduced into the NFPA standards governing anesthetizing locations in 1941, along with other antistatic measures intended to reduce the number of accidents due to the ignition of flammable agents such as cyclopropane.

In 1970, the standard addressing anesthetizing locations (NFPA 56–Code for the Use of Flammable Anesthetics) had been renumbered as NFPA 56A and given the title "Standard for the Use of Inhalation Anesthetics (Flammable and Nonflammable)." According to this new document, anesthetizing locations where the use of flammable agents was prohibited did not have to install or use any previously required antistatic safeguards, except the isolated power system. In retrospect this might appear strange, until one considers the other issues faced by the NFPA's Committee on Hospitals at that time. The committee was advocating the use of isolated power systems (IPS) in other special care areas of the hospital as a safeguard against microshock. The, often acrimonious debate continued throughout the 1970s and well into the next decade, documented in *Guidelines for Clinical Engineering Programs; Part IV: Isolated Power in Anesthetizing Locations? History of An Appeal*, (Ridgway, 1981). The IPS advocates finally settled for permitting isolated power in anesthetizing locations, but not requiring it. The debate was particularly interesting because advocates of the less stringent approach required considerably more professional courage and belief in their analyses than those advocating the "safer," more extravagant solution. One approach that proved useful in bringing some uncertain observers around to the more radical position was the use of a probabilistic illustration to semi-quantify the level of risk, documented in *Guidelines for Clinical Engineering Programs; Part III: The Risk of Electric Shock In Hospitals* (Ridgway, 1981).

There have been no significant adverse trends in electrical accidents in operating rooms over the past 20 years. The predominant categories of equipment-related misadventures in the operating room continue to be patients accidentally burned by poorly implemented electrosurgical procedures, and patients injured by pressure sores resulting from extended contact with the unyielding surface of the surgical table. Both of these problems are often misdiagnosed as accidental burns.

References

Accidental Electrocutions Claim 1200 Patients a Year. *Electronic News* January 27, 1969.
National Fire Protection Association. Standard for the Use of Inhalation Anesthetics (Flammable and Nonflammable), NFPA 56: Code for the Use of Flammable Anesthetics. Quincy, MA, National Fire Protection Association, 1970.
Nader R. Ralph Nader's Most Shocking Exposé. *Ladies' Home Journal* 3:98-179, 1971.
Nobel J. *Testimony before a Senate sub-committee on the proposed Medical Device Amendments of 1973*. Washington, DC, September, 1973.
Ridgway M. Guidelines for Clinical Engineering Programs. *J Clin Eng* 5:287-298, 1980.
Ridgway M. Guidelines for Clinical Engineering Programs. *J Clin Eng* 6:287-298, 1981.
Walter CW. *Electrical Hazards in Hospitals.* National Academy of Sciences Workshop Proceedings. Washington, DC, 1970.

Section VI: Education and Training

James O. Wear
Veterans Administration
North Little Rock, AR

The terms *education and training* are frequently used synonymously, but they have two very different meanings. Both require teaching and result in learning—but one is very broad and general, while the other is very specific.

Education develops a general competence in a field by fostering expansion of knowledge and wisdom. The best example of education is formal schooling—that is, a course or series of courses of formal study or instruction. Education consists of learning theories, philosophies, and how to obtain and use information. Synonyms for the word "educate" include develop, enlighten, illumine, enlarge the mind, fill with new ideas, and edify.

Training, on the other hand, is very specific. It exercises someone in a profession or directs them in obtaining a skill. Training involves giving instructions and drilling individuals in habits of thought or action. Training is used to shape or develop character of people by discipline or by precept. Synonyms for the word "train" include coach, drill, exercise, practice, and familiarize (e.g., we train soldiers, athletes, and animals).

A third word that is sometimes used synonymously with educate or train is teach. In teaching, one is providing lessons, lecturing, interpreting information, indoctrinating individuals, or influencing their behavior. In the process of teaching, one is attempting to cause a change in behavior; if the individual learns, then a change in behavior will have been accomplished. Teaching actually involves both education and training. It could involve only one of the two, such as a course called *Introduction to Clinical Engineering* or a course on *How To Use a PDA*. *Introduction to Clinical Engineering* would be education since it probably would not contain any specific skills; *How To Use a PDA* would probably be entirely training. In most cases, however, teaching does involve both education and training.

A college education is general education in a pure sense because most of what is learned, as far as specific information, is obsolete at the time it is learned or shortly after one finishes college. The result of a college education on a long-term basis includes only two things: An individual learns 1) how to think and 2) where to find information. With this educational background, one can then be trained to do the specific skills and jobs that must be accomplished.

The college education of an engineer contains more training than that of someone graduating from a College of Arts and Sciences. An engineer goes through many courses in which he learns specific skills. He learns how to make specific calculations so that when he graduates he is immediately able to perform some specific tasks. However, in our rapidly changing technology, even this ability is lost within approximately 3 years as he becomes obsolete without further education and training.

A clinical engineer may have additional training and education through an internship, either as part of his curriculum or after graduation. An internship is probably more training than education, but it does contain some of both. Once on the job, the clinical engineer must, as must any other engineer, go through lifelong continuing education or, one should say, continuing education and continuing training.

With rapidly changing technology, much that clinical engineers learn becomes obsolete in short periods of time and as new technologies come along. They may also find themselves going in new career directions. Clinical engineers attend professional meetings, short courses, and university courses, which may lead to advanced degrees. Training on specific equipment will be required, as well as learning new procedures and techniques. Continuing education activities may be formal or very informal, but both are necessary for a clinical engineer to continue to be well educated and well trained.

Academic programs in clinical engineering are at different levels. The Associate's Degree programs that train Biomedical Engineering Technicians vary greatly in program content. Some are electronic technician programs with a course or two in biomedical equipment added, while others are more intense programs with internships. US undergraduate and graduate programs are mostly biomedical engineering programs rather than clinical engineering programs. There are very few clinical engineering programs in the U.S. Biomedical Engineering programs are more theoretical and equipment-design oriented whereas clinical engineering is an application hospital-based type of training activity. The key to any clinical engineering training is an internship. A clinical engineering department is typically hospital-based, like the one at the University of Connecticut, which is a general clinical engineering program with a one-year internship. Several programs, which are more commonly called rehab engineering, could be considered a subspecialty of clinical engineering. They have internships in rehab programs and within hospital settings.

Clinical engineering programs are more highly developed at the international level, such as in Germany, Ireland, and Australia. The American College of Clinical Engineering, in conjunction with the World Health Organization and international clinical engineers, has developed programs in advanced clinical engineering workshops around the world primarily for developing and transitional countries, for engineers and administrators to learn about health care technical management. Dyro, Judd, and Wear summarize the programs that have been presented. Judd, Dyro, and Wear present an overview of an advanced clinical engineering workshop presented in Cape Town, South Africa and specifically tailored to encompass the area of health technology management. These workshops, in addition to providing educational opportunities for engineering and health care administrative personnel in target countries, have provided cross-fertilization in the international engineering field with involvement of faculty from different countries.

Continuing education of clinical engineers is important since they must keep current with engineering developments that impact equipment and other health care technology. They must also keep up with the changes in the health care delivery system. Some of the clinical engineers have become so involved in continuing education that they have actually changed their career and have become physicians. Gilchriest describes an academic program on the retraining of clinical engineering personnel in some of the developments in biomedical engineering.

Distance education has become the way many clinical engineering staff receive continuing education. This has partially been brought about due to the cost of travel and the changes in technology. Of course, distance education is not new. Correspondence courses have been a form of distance education that has been around for many, many years. For a few years, audio teleconferencing has been used by the American College of Clinical Engineering to provide quality continuing education programs for engineers, and has been well received by both presenters and participants. The presenters do not have to travel and can make a presentation from the office. At the same time, the participants can receive the training during their lunch hour or at some other time at their local medical center.

Another method of continuing education utilizing technology has been video conferencing, which initially only involved "talking heads," but has become a more real educational experience where skills and techniques can be demonstrated. Video conferencing takes two modes. One is two-way video conferencing, either over high-speed phone lines or through a satellite broadcasting system where people make a presentation and participants can ask questions. A more common method is a satellite broadcast, which is a one-way communication, but participants can ask questions through phone lines, fax systems, or e-mail. Video conferencing is considerably more expensive than audio conferencing, but if something needs to be shown, this is a methodology that can be utilized very efficiently. Audio teleconferencing and video teleconferencing allow for presentations to be made anywhere in the world and allow clinical engineers access to presentations by international experts.

The Internet with the World Wide Web has also become a technology that can be used for college education or continuing education. This technology can be used on a very low level, where material is put on a site and downloaded similarly to a handout in a classroom. It can be made interactive with an instant session through chat rooms or various messenger services. The Internet provides a method that instructors can interact with and answer questions from students. The Internet and the World Wide Web have provided an opportunity for clinical engineering personnel anywhere in the world to receive instruction from international experts.

In addition to clinical engineers participating in various continuing education activities, they also are the source of continuing education with in-service training for the clinical engineering staff and health care delivery staff such as nurses and physicians. Clinical engineers will provide health care providers with information on the latest technology with regard to medical instrumentation. They may also provide the education on codes and standards that relate to the equipment at a facility. Clinical engineers are heavily involved in education and training both for their own professional development and for the professional development of the health care providers in the clinical setting.

66

Academic Programs in North America

Tim Baker
The Journal of Clinical Engineering
University Heights, OH

The Journal of Clinical Engineering surveyed 106 universities, colleges, and technical schools that offer biomedical instrumentation technology training. Fifty-five of these institutions offer associate's degrees and/or certificates in biomedical instrumentation. The remaining 51 institutions offer undergraduate, master's, medical, and/or doctoral degrees in clinical engineering or with a focus on bioinstrumentation. This article describes the trends reported by the staff at these programs regarding curriculum, staff size, and student numbers and quality.

Survey of Academic Institutions in the United States

The Journal of Clinical Engineering surveyed 106 universities, colleges, and technical schools that offer biomedical instrumentation technology training. See Appendix A for the list of questions asked. Appendices B and C provide complete lists of institutions that received surveys. There was a 15 percent response rate to the survey, which was mailed and/or e-mailed in July 2001. A complete list of institutions that received surveys can be found at the end of this article. Of the 51 graduate and undergraduate programs, 32 offer an undergraduate degree, 43 offer a master's degree (one offers an MD/MA), and 39 offer doctoral programs, including seven that offer MD/PhD degrees.

Bachelor's degree programs are available in 20 states, including Arizona, Massachusetts, New York, Pennsylvania, Ohio, North Carolina, Maryland, Wisconsin, Louisiana, Georgia, Illinois, Texas, Connecticut, Alabama, California, Iowa, Michigan, Virginia, Tennessee, and Missouri, as well as the District of Columbia. (See Figure 66-1.)

Master's programs are available in 22 states, including Arizona, Massachusetts, California, Ohio, Texas, South Carolina, New York, North Carolina, Georgia, Maryland, Wisconsin, Illinois, Indiana, Louisiana, Alabama, Connecticut, Iowa, Tennessee, Washington, Utah, Virginia, Missouri, and the District of Columbia. (See Figure 66-2.)

PhD degrees are offered by schools in 20 states, including Arizona, Ohio, Massachusetts, South Carolina, New York, Pennsylvania, North Carolina, Georgia, Illinois, Wisconsin, Maryland, Indiana, Texas, Louisiana, Alabama, California, Connecticut, Washington, Tennessee, Missouri, and the District of Columbia. (See Figure 66-3.)

Associate's degree programs and certificate programs in biomedical engineering technology (BMET) are offered by schools in slightly more than half the states, including Alabama, Arkansas, California, Connecticut, Delaware, Florida, Georgia, Illinois, Indiana, Massachusetts, Virginia, Washington, Wisconsin, Tennessee, Texas, South Carolina, South Dakota, Pennsylvania, Ohio, Oklahoma, Missouri, New Jersey, New York, North Carolina, Maryland, Michigan, and Minnesota—a total of only 28 states. (See Figure 66-4.) The Canadian provinces of Alberta, British Columbia, and Ontario boast institutions that offer AA/certificate BMET programs; however, they were not included in this survey.

Texas and Pennsylvania are the states with the most options for students looking to earn an associate's degree or certificate in biomedical equipment testing, maintenance, and repair; they offered four programs each at the time of the survey. However, Penn State, which offered clinical engineering at both its Wilkes-Barre and New Kensington campuses, discontinued the Wilkes-Barre program at the end of 2001 due to a lack of student interest.

Other AA/certificate programs recently discontinued include an associate's degree in biomedical engineering offered by Baker College in Flint, Michigan and an associate's degree program in biomedical equipment technology from the Northwest Technical College in Detroit Lakes, Minnesota.

States with three clinical engineering programs include Illinois and Washington. States with two programs include Alabama, California, Florida, Georgia, Indiana, Massachusetts, South Carolina, Ohio, Wisconsin, Tennessee, and North Carolina. States with one program include Arkansas, Connecticut, Delaware, Virginia, South Dakota, North Carolina, Oklahoma, New Jersey, Missouri, New York, Maryland, and Minnesota.

States that do not offer BMET AA/certificate schooling include Oregon, Nevada, Idaho, Montana, Utah, Arizona, New Mexico, Colorado, Wyoming, North Dakota, Nebraska, Kansas, Louisiana, Iowa, Louisiana, Mississippi, West Virginia, Maine, Vermont, New Hampshire, and Rhode Island.

Canadian schools that offer BMET AA/certificate programs include the Northern Alberta Institute of Technology in Edmonton Alberta, the British Columbia Institute of Technology in Burnaby, and Fanshawe College in London, Ontario.

Associate's Degree/Certificate Program Trends

Results from the survey reveal that information technology and management are not high priorities in the curriculum of most trade schools and associate's degree programs. This finding was particularly interesting given the growing importance of the microchip and networking in the day-to-day use of many devices used at the patient interface within hospitals and clinics. Many biomedical equipment technicians (BMETs) and clinical engineers (CEs) have expressed concern that their departments will eventually become divisions of hospital information technology (IT) departments unless BMETs and clinical engineers master these internet-age skills. However, US schools that train BMETs do not seem to be addressing the issue.

As seen in Figure 66-5, about half of the training for BMETs focuses on equipment repair and preventive maintenance. Thirty percent of the typical curriculum focuses on electrical engineering and electrical safety. Only 15 percent of the course of study focuses on computer science, including database management, LAN devices, and networking. About 10 percent is dedicated to human physiology and medical terminology.

One disturbing result of the survey is that the quality of incoming BMET students has declined. (See Figure 66-6.) About 70 percent of those who responded said that student quality had declined in the ten years from 1991 to 2001.

While this finding contains a degree of subjectivity, student enrollment does not. Half of the BMET training programs that responded said student enrollment declined from 1991 to 2001. Half reported that it has remained steady, or that it has slightly declined. Two-thirds of these respondents reported that they are worried about the future viability of their programs.

When asked if these educators are worried about the future of the industry, the answers were almost always indicative of the local environment, not national trends. Where there is strong hospital support and outreach to local BMET students, educators reported that they were not worried about the future employment prospects of their graduates. Educators in smaller towns or rural regions, where one might expect job opportunities to be scarce, seemed to be the most confident. Many of these programs reported having strong ties to local hospitals and good reputations for producing high-quality BMET graduates.

Most BMET training programs have three or fewer staff members. About 15 percent have six or more. (See Figure 66-7.)

The BMET student population reported by survey respondents ranged from two to 110. The average was 34 students, the mean was 56 students, and the median was 18 students. Seventy percent of the respondents reported having 20 students or fewer enrolled as of 2001. Twenty percent reported having more than 100 students. Thirty percent had fewer than 10 students.

Undergraduate and Graduate School Trends

Universities that offer undergraduate and graduate coursework in biomedical engineering reported that the interest in their programs is on the rise. Most reported a 20 to 50 percent increase in the student population from 1991 to 2001. Nearly all of these programs have staff sizes of fewer than 10 people. Only a few reported having a staff of more than 25 members. (See Figure 66-9.) Two-thirds of these universities reported that the quality of entering students is better today than it was in 1991. (See Figure 66-10.)

The focus of study at US undergraduate and graduate clinical engineering programs is equally split between electrical engineering and computer science (see Figure 66-8); the latter group includes computerized medical instrumentation, telemetry, web-enabled devices, database management, LAN devices, and general information technology. The University of North Carolina at Chapel Hill devotes nearly 70 percent of its curriculum to computer science.

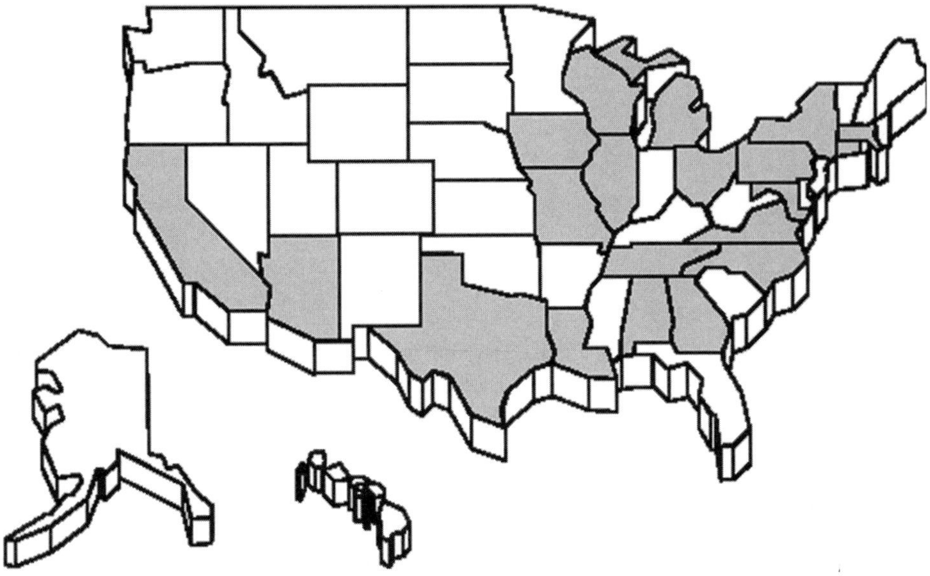

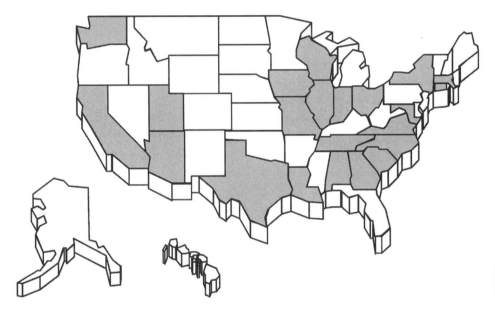

Figure 66-1 Twenty states plus the District of Columbia offer bachelor's degree programs in Clinical Engineering.

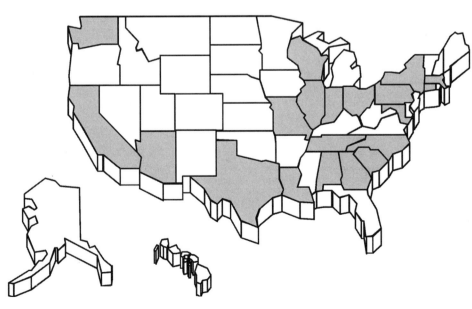

Figure 66-2 Twenty-two states plus the District of Columbia offer Clinical Engineering master's degree programs.

Figure 66-3 Twenty states plus the District of Columbia offer Clinical Engineering PhD programs.

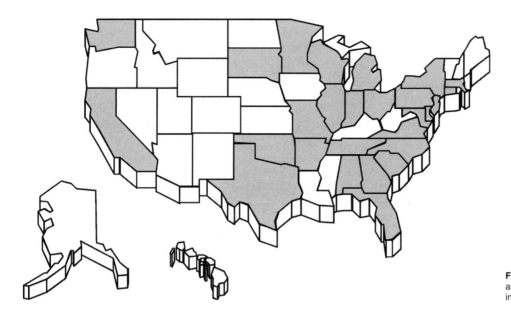

Figure 66-4 Twenty-eight states offer associate degree and/or certificate programs in Biomedical Equipment Technology.

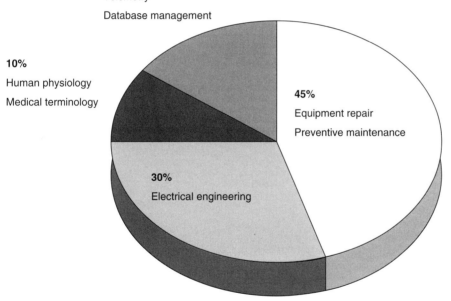

Figure 66-5 Typical curriculum at associate's degree schools that responded to the survey.

As expected at this level of study, less than five percent of total instruction time is spent on electrical safety testing. None of the respondents from undergraduate and post-graduate schools reported teaching infection control.

The survey questionnaire failed to ask for student enrollment broken down by undergraduate and graduate student populations, so it is impossible to provide this information. Overall, graduate and undergraduate clinical engineering student populations ranged from 10 to 320. The average was 147 students, the mean ws 165 students, and the median was 110 students. Sixty percent of the respondents reported having 100 or more students. Only 15 percent reported having 10 or fewer students as of 2001.

Why Programs Fail

In the last 18 months, several BMET associate's degree or certificate programs have been discontinued, primarily because of a lack of students. The deans and program directors at several schools expressed genuine concern about the future viability of their BMET training programs.

An interview with Albert Lozano, an assistant professor of engineering at the School of Engineering Technology and Commonwealth Engineering, Pennsylvania State University at Wilkes-Barre, provided some insight into why established BMET training programs with good reputations are discontinued. The Penn State Wilkes-Barre BMET program was discontinued in 2001 due to a sharp decline in student interest.

"Enrollment in the 2-year program continued to decrease over the last several years–two to three years–before we ended the associate's degree in Biomedical Equipment Technology program," Lozano said. "In fact, there were only two students in the last class. The interest on the part of the biomed kids just dropped."

Fortunately for the students of northeastern Pennsylvania, Penn State continues to offer associate's degrees in electrical engineering technology and telecommunications engineering technology as well as a baccalaureate degree in electrical engineering technology at Wilkes-Barre. "The interest in these has declined as well, but not to the extent of the biomedical associate degree," Lozano said.

The biomedical equipment technology program existed at Wilkes-Barre for more than 15 years and had been accredited by the Technology Accreditation Commission (TAC) of the Accreditation Board for Engineering and Technology (ABET) until it was discontinued. Penn State still offers the program at the New Kensington Campus, near Pittsburgh.

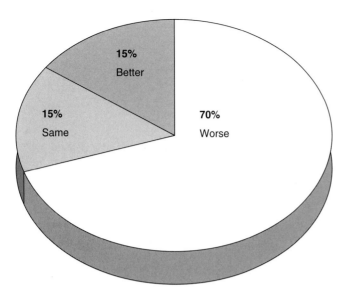

Figure 66-6 Response by associate's degree/certificate programs to the following question: "How would you describe the quality of incoming students compared with students ten years ago?"

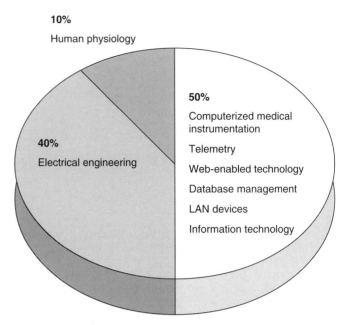

Figure 66-8 Typical curriculum at US undergraduate and graduate clinical engineering programs.

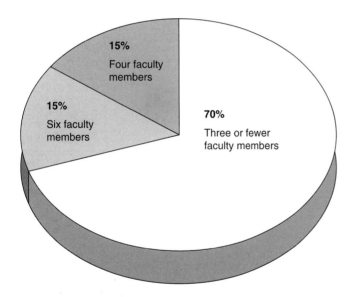

Figure 66-7 Typical staff size at responding associate's degree/certificate programs.

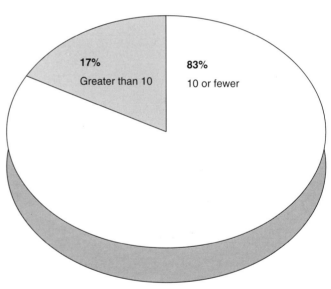

Figure 66-9 Typical staff size at U.S. undergraduate and graduate clinical engineering programs.

Lozano said he believes that several factors contributed to the demise of the Wilkes-Barre program. First, he said he thinks there is a stigma attached to all two-year degree programs that discourages students from pursuing them and parents from supporting them. Second, at the time the program ended, the economy was booming. "Students were finding it easy to make money without too much education," he said. Many students who were technically gifted found work in information technology. This leads to Lozano's third theorized reason: The lure of IT and computer engineering. That work is far more enticing to high-school graduates who are interested in technical careers, Lozano believes. Lastly, Lozano suspects that the curriculum at Penn State, where associate's degree graduates must take physics, extensive mathematics, English, and other humanities, hurt recruitment. "Students realized they could earn a degree in biomedical maintenance from a technical school without having all of this extra coursework and could still find good jobs. Of course, we feel that we produce a different, more global and competent kind of graduate, but students may feel they did not need the humanities and other courses."

Lozano also said that the low profile of the biomedical equipment profession also made recruiting young students difficult. Many potential students and their parents simply do not understand what a BMET does, which adds an extra hurdle to the recruitment process. "Society doesn't have a good understanding of what a biomedical professional does. It's an unknown profession," Lozano said.

Lozano also believes that there is an overall decrease in interest among US high-school graduates in technical subject matter, which hurts all areas of science. Programs such as the Wilkes-Barre BMET associate's degree, which were not wildly popular to begin with, are particularly impacted by this trend.

Based on his recruitment experience within high schools, Lozano has learned that many parents discourage their children from working toward a two-year degree, which is a problem with most associate's degree programs no matter what the topic. "Parents often ask if their child with a two-year degree will get a good job with a salary that is comparable to that of a graduate with a four-year electrical engineering degree." Parents

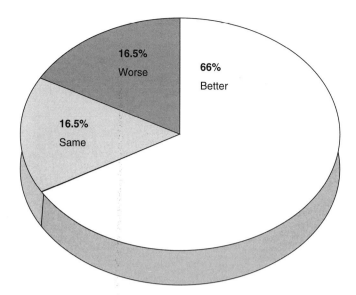

Figure 66-10 Ranking of student quality at US undergraduate and graduate Clinical Engineering programs compared to students ten years ago.

generally want their children to have four-year degrees, even though this is not necessary to thoroughly train qualified BMETs.

The termination of the Wilkes-Barre program has not gone unnoticed by local hospitals, Lozano said. There are approximately 15 hospitals in the Wilkes-Barre region and in nearby Binghamton, New York that relied on the program for new BMETs. Lozano said that 80 percent of his associate's degree graduates were placed at local hospitals. Support from the hospitals was strong. Many offered internships and donated equipment for training purposes. "Hospitals still often call saying 'we need more graduates'; we just have to say 'sorry, the program has ended'."

Lozano said student quality was not an issue, just student interest. In fact, the employers praised the quality of the graduates. Those who were eager to enter the biomedical field and work in hospitals were high-caliber students. "There were just too few of them," he said.

Lozano feels that the Wilkes-Barre program offered a curriculum that trained students for the realities of the modern BMET profession. The training focused less on component-level troubleshooting in favor of networking, risk management, safety, and technology integration. The program relied on biomedical professionals from local hospitals who worked on the program's curriculum committee. Based on the recommendations of these professionals, the program continued to revise the curriculum to fulfill the current demands of the profession.

Lozano is still in contact with many of his previous students, most of whom are happy in their chosen profession and are doing well. One student who graduated in the late 1990s began working at a New Jersey hospital at a salary of $35,000 plus benefits including an automobile. There are not many entry-level positions that offer such generous compensation, he added. Within about two and a half years, that student was making $50,000 at the same hospital.

Specific Comments

"In the last five years, the school has made an effort to improve the program and work with area hospitals to provide up-to-date learning experiences," said one BMET program director. "The program has also worked to provide some educational materials on-line for students who are located throughout the country. Student interest is increasing as the school has put more effort into recruiting and publicity. Student quality is improving as admission requirements to the university have tightened over the last five years. Our program continues to thrive given the high demand for graduates and the on-line environment attracting a wider geographic population. Two years is only enough time (to train BMETs) to provide the basics. However, we find that employers would prefer to customize additional education opportunities as they see fit."

One professor at a BMET associate's degree program is not worried about the future of the profession. "We have steady enrollment, and there are always jobs to be found," he said. "There are not many other programs in the area, and we have good hospital support." Two years of training is enough for BMETs, especially "with comprehensive internships and continuing education from manufacturers. Our grads know how to keep learning."

On the same topic, a department chair of a BMET and radiology-technician program said, "For a general BMET program, two years (training) is more than enough. It's the quality and efficiency of the training program that counts, not so much the length."

One BMET instructor said that he is worried about the future of his program in particular, and the profession in general. "Starting salaries and working conditions are not attractive to today's students."

One director of an associate's degree program described the curriculum and goals of the program as follows: "The Biomedical Equipment Technology curriculum prepares individuals to install, operate, troubleshoot, and repair sophisticated devices and instrumentation used in the health care delivery system. Emphasis is placed on preventive and safety inspections to ensure that biomedical equipment meets local and national safety standards. Course work provides a strong foundation in mathematics, physics, electronics, anatomy and physiology, and troubleshooting techniques. People skills are very important, as well as the ability to communicate both in written and oral form. A biomedical equipment technician is a problem solver. Graduates should qualify for employment opportunities in hospitals, clinics, clinical laboratories, shared-service organizations and manufacturers' field service. With an AAS degree and two years experience, individuals should be able to become a certified biomedical equipment technician."

One staff member at an associate's degree program said, "I am very concerned about the program. One program in (our state) has already folded because of a lack of interest. Student quality is much worse than ten years ago. Students can't do algebra and can't write on a college level."

"Our biomedical electronics major teaches students how to repair, calibrate, and evaluate the performance of electro-mechanical equipment used in the medical fields." said one professor at an associate's degree program. "Students receive a strong background in electronics and the sciences, as well as training in various areas of biomedical equipment."

This individual is worried about the future. "Due to the reforms in health care, there is a negative perception to working in the health care field, and with the new areas of computer information systems and networking, there is difficulty in recruiting students into biomedical engineering technology." "With all of the changes in health care technology, two years is still sufficient time for the fundamental training that a BMET requires; then it becomes on-the-job training and specialized training after employment."

Appendix A: Survey Questions

1. Does your school offer a degree/certification?

   - Certificate
   - Biomedical equipment technician
   - Radiology equipment specialist
   - Clinical laboratory equipment specialists
   - AA/AS degree
   - Bachelor's of science (or arts) degree
   - Master's degree
   - PhD
   - ScD

2. Please give the formal name of your program(s) and degree(s).
3. If yours is a certificate program, how many quarters of study are required?
4. How long has your program(s) existed?
5. Describe your curriculum.
6. How much of the curriculum focuses on (provide %):

   - Hands-on maintenance and PM? (%)
   - Medical terminology? (%)
   - Management? (%)
   - Infection control? (%)
   - Electrical safety testing? (%)
   - Electrical engineering? (%)
   - Human physiology? (%)
   - Computerized medical instrumentation? (%)
   - Web-enabled medical devices? (%)
   - Telemedicine? (%)
   - Telemetry? (%)
   - Database management? (%)
   - Information technology? (%)
   - Microsoft NT training? (%)
   - Unix training? (%)
   - Computer OS certification preparation? (%)
   - LAN devices? (%)
   - Computer languages, such as C++? (%) (If so, please list languages.)

7. State your name and title.

8. How many people are on your staff, and what is their training?
9. What is the student population today?
10. What was the student population ten years ago?
11. Is student interest declining?
12. Is student quality

- the same as ten years ago?
- better than ten years ago?
- worse than ten years ago?

13. Are you concerned about the future viability of your program? (If so, please elaborate.)
14. With all of the changes in health care technology, are two years sufficient training for a BMET? (Please elaborate.)
15. Are you considering offering more specialty training? (Please elaborate.)
16. Do you require internships?
    ☐ Yes
    ☐ No

Appendix B: BMET Certificate and Associate's Degree Programs by State

State	Institution	Degree
Alabama	Jefferson State Community College	College associate's degree
	Northwest-Shoals Community College	Certificate in biomedical equipment technology
Arkansas	University of Arkansas for Medical Sciences	Associate's degree
California	Los Angeles Valley College	Two-year electronics technology certificate
	Santa Barbara City College	Associate's degree/certificate
	Napa Valley College	Associate's degree
Connecticut	Gateway Community College	Associate's degree
Delaware	Delaware Technical and Community College	Associate's degree
Florida	Broward Community College	Associate's degree
	Santa Fe Community College	Associate's degree
	Florida Community College at Jacksonville	Associate's degree
	Keiser College of Technology-Fort Lauderdale	Associate's degree
Georgia	Central Georgia Technical College	Certificate in biomedical electronics
	Chattahoochee Technical College	Associate's degree
Illinois	Richland Community College	Associate's degree in biomedical electronics
	South Suburban College	Associate's degree/certificate in biomedical electronics technology
Indiana	Indiana University–Purdue University at Indianapolis	Associate's degree
Kentucky	Madisonville Community College	Associate's degree
Maine	Southern Maine Technical College	Associate's degree
Massachusetts	Franklin Institute of Boston	Medical electronics
	Quinsigamond Community College	Engineering program
		Associate's degree
	Bunker Hill Community College	Associate's degree
	Springfield Technical Community College	Associate's degree
Maryland	Howard Community College	Associate's degree
Michigan	Baker College	Associate's degree (discontinued)
	Schoolcraft College	Associate's degree
Minnesota	Northwest Technical College	Associate's degree (discontinued)
Missouri	St. Louis Community College-Florissant Valley	Associate's degree
New Jersey	Thomas Edison State College	Associate's degree
New York	Erie Community College	Associate's degree
North Carolina	Stanley Community College	Associate's degree
	Caldwell Community College and Technical Institute	Associate's degree
Ohio	Cincinnati State Technical and Community College	Associate's degree
	Owens Community College	Associate's degree
Oklahoma	Tulsa Community College	Associate's degree in (electronics, with biomedical equipment technology option)
Pennsylvania	Penn State, Wilkes-Barre	Associate's degree (discontinued)
	Penn State, New Kensington	Associate's degree
	Community College of Philadelphia	Associate's degree
	Johnson Technical Institute	Associate's degree
South Carolina	Greenville Technical College	Certificate in biomedical equipment Technology (after completion of an AS in electronics)
	York Technical College	Certificate program
South Dakota	Southeast Technical Institute	Certificate program
Tennessee	Southwest Tennessee Community College	Associate's degree
Texas	Grayson County College	Certificate program
	Texas State Technical College-Harlingen	Associate's degree
	St. Philip's College	Associate's degree
	College-Waco	Associate's degree
Virginia	ECPI College of Technology	Associate's degree
Washington	North Seattle Community College	Associate's degree
	Spokane Community College	Associate's degree
Wisconsin	Western Wisconsin Technical College	Associate's degree
	Milwaukee Area Technical College	Associate's degree
Alberta, Canada	Northern Alberta Institute of Technology	Two-year diploma
British Columbia	British Columbia Institute of Technology	Two-year diploma
Ontario	Fanshawe College	One-year post-graduate diploma

Appendix C: Undergraduate and Graduate Programs by State

State	Institution	Degree
Alabama	University of Alabama	MS, PhD
Arizona	Arizona State University	BS, MS, and PhD
	University of Arizona	MS and PhD
California	California State University-Sacramento	MS
	University of California at Berkeley	PhD
	University of California at San Diego	BS, BA, MS, PhD, MD/MS, and MD/PhD
	University of Southern California	BS, MS, and PhD
Connecticut	Trinity College	BS
	University of Connecticut	MS and PhD
	University of Hartford	BS
Georgia	Georgia Tech	MS and PhD
	Mercer University	BS and MS
Illinois	Northwestern University	BS, MS, and PhD
	University of Illinois-Chicago	BS, MS, and PhD
	University of Illinois-Urbana/Champaign	BS, MS, PhD
	Southern Illinois University-Carbondale	BS in electronic management
Indiana	Purdue University	MS and PhD
	Indiana University–Purdue University at Indianapolis	BS
	Indiana State University	BS
Iowa	University of Iowa	BS, MS, and PhD
Louisiana	Tulane University	BS, MS, and PhD
Maryland	Johns Hopkins University	BS, MS, and PhD
Massachusetts	Boston University	BS, MS, PhD, and PhD/MD
	Worcester Polytechnic Institute	BS, MS, and PhD
Michigan	University of Michigan	BS and MS
Missouri	Washington University	BS, MS, and PhD
New Jersey	Thomas Edison State College	BS
New York	Columbia University	BS, MS, and PhD
	Cornell University	BS, MS, and PhD
	Rensselaer Polytechnic Institute	BS, MS, and PhD
North Carolina	Duke University	BS, MS, and PhD
	University of North Carolina at Chapel Hill	BS, MS, PhD, and PhD/MD
Ohio	Case Western Reserve University	BS, MS, PhD, and PhD/MD
	Ohio State University	MS, PhD, and MD/PhD
	University of Akron	BS, MS, and PhD
	Wright State University	BS and MS
Pennsylvania	Carnegie Mellon University	BS, MS, and PhD
	Drexel University	MS and PhD
	University of Pennsylvania	BS, MS, and PhD
	University of Pittsburgh	BS, MS, PhD, and MD/PhD
South Carolina	Clemson University	MS and PhD
Tennessee	Vanderbilt University	BS, MS, and PhD
	East Tennessee State University	BS
Texas	Rice University	MS and PhD
	University of Texas	MS and PhD
Utah	University of Utah	MS and PhD
Virginia	University of Virginia	BS/ME, MS, and PhD
	Virginia Polytechnic Institute	BS
Washington	Walla Walla College	BS
	University of Washington	MS, PhD, and MD/PhD
Wisconsin	Marquette University	BS, MS, and PhD

# 67

Clinical Engineering Education in Germany

Vera Dammann
Department of Hospital and Medical Engineering,
Environmental, and Biotechnology
University of Applied Sciences Giessen-Friedberg
Giessen, Germany

A member of the technical staff in a clinical engineering department in Germany has a formal technical education as a skilled craftsman, with or without specialization or additional higher qualification as "Meister," as a technician, or as an engineer with a university degree. All formal education is controlled by the government with regard to level of quality, duration, and final certificate. Education, entry requirements, and final examination with certificate at every level are regulated by law. Traditionally, universities issue the diploma in engineering after four or five years of study. There is no advanced degree except the doctorate. Since the late 1990s, there has been a trend at universities towards the bachelor/master system.

Most clinical engineers have a university diploma, usually from a university of applied sciences.

German System of Technical Education

Qualified Craftsman Level

Entrance qualification for the three- or three-and-half-year vocational training is a qualified school-leaving certificate of 10 years. The law states that young persons up to 18 years of age may be trained only in a recognized training occupation. Attendance at vocational schools is compulsory (up to 12 hours per week). The setup of the vocational schools is largely standardized. The on-the-job training takes place in a hospital or related enterprise and is characterized by its systematic and casuistic nature, as defined by training regulations (the federal syllabus for in-plant training) and the presence of the trainer and Meister[1] specialized in training, to guide the apprentice. Public vocational schools do not charge study fees; the apprentice gets a small salary from the enterprise.

There is no formal vocational training for a medical technology craftsman. Craftsmen working in the clinical engineering field go through an apprenticeship in electronics or mechanics and obtain experience through several years of on-the-job training in a maintenance workshop and/or by attending special seminars offered by institutions of continuing education or by the manufacturers of medical devices. These seminars qualify the apprentice for single, usually rather simple, tasks; e.g., maintenance of a specific type of infusion pump.

The head of an electronics or mechanics workshop in a clinical engineering department is usually an experienced master craftsman, i.e., "Meister."[1] The Meister certificate is also necessary to run a private workshop offering maintenance service in Germany. This regulation is effective in all commercial fields.

Technician Level

An experienced biomedical engineering technician (BMET) in Germany is called "medical technician" and is allowed to carry out all maintenance, instruction, and documentation tasks in clinical engineering. The medical technician is aware of the legal requirements concerning his or her work, and at any time must be able to demonstrate the way in which necessary special knowledge for dedicated maintenance tasks was gained.

Ten technical colleges/technician schools offer medical technician courses for a fee, and the entrance requirement is relevant vocational training. A course takes 2400 to 2900 contact hours (i.e., two years full-time or four years part-time) and ends with an examination acknowledged by the state.

In contrast to the designation "engineer", (Ingenieur), the general term "technician" is not protected by law in Germany. The title "certified technician" (staatlich geprüfter Techniker), however, is reserved for technicians who passed the exam at a certified technical school.

University Level

Clinical engineers hold a university degree, the diploma in engineering. There are two types of universities offering courses in (bio)medical engineering/clinical engineering: the traditional-style university, "Universität," and the university of applied sciences (UAS), "Fachhochschule (FH)." The diplomas of the two university types are equivalent under the law.

Clinical engineering is considered an applied science, and is taught in an interdisciplinary program at UASs based on physics, electronics, or precision engineering. All curricula comprise mathematics, physics, chemistry, data processing, electrical engineering, electronics, material sciences, mechanical engineering, human biology (e.g., physiology), physiological measurement, imaging, quality assurance, and safety. Most of the subjects are obligatory. Subjects are taught for up to half a semester. The amount of classical engineering sciences depends on the basic specialization (see Tables 67-2 and 67-3) of the biomedical engineering (BME) department. For example, curricula based on mechanics could include computer-aided design. All new revisions of the curricula show an increase in so-called nontechnical subjects like economics, quality assurance, management, law, and foreign languages.

Traditional universities offer biomedical engineering courses for course specialization within electrical or mechanical engineering or physics in the third or fourth year of study.

All graduates can work as clinical engineers directly after graduation, without any additional certificate. Since every university and every course of study must be accredited by the government (ministries of science and education), the quality is defined and the level of examination is high. It is up to the employer to decide whether to hire a "greenhorn" or an experienced engineer.

Traditional System

As all the German institutions are public and lie within the formal and financial responsibility of the respective state governments, the framework of all studies is similar to the following:

- Admission requirements: At least 12 years of elementary and secondary school (comparable to A-level) plus practical workshop experience of two to six months
- Six semesters (at universities of applied sciences) or eight semesters (at traditional universities) of full studies (25 to 30 lesson hours per week, including exercises and lab work)
- One or two half-years of internship in industry or hospital with supervised project work at universities of applied sciences
- An individual final examination project (three to six months in a university laboratory, in research and development, or at a hospital under the supervision of a professional) with a thesis at the end
- A degree in engineering at graduation: "Dipl.-Ing." The UAS adds the abbreviation *FH*: "Dipl.-Ing. (FH)." Law protects the designation "Ingenieur".
- The diploma of a UAS is at a level between bachelor's and master's degrees, while the diploma of a traditional university is master's level.
- Only the traditional universities can offer a doctorate, but graduates of both traditional institutions or UASs may enter a doctorate program. Graduates of a UAS, however, may be required to pass an additional year of advanced theoretical studies.
- Students are not charged a study fee. This is a form of scholarship that the German government grants to every enrolled student, including students from foreign countries.

Development of New Structures

Since the late 1990s, most German universities have redesigned existing courses of study to create modularized courses, leading consecutively to the bachelor's, the master's, and the doctoral degree. In addition to the approval by the respective ministry, such new courses should be accredited by an independent agency. The modules in the new courses

---

[1] The qualification "Meister" requires a successfully accomplished vocational training plus some years of experience on the job and the attendance of a "Meister course" of about one year. The Meister course and final exam cost about 20,000 Deutschmarks (DM). The certificate is issued by the respective chamber or guild.

are credited by the European Credit Transfer System (ECTS). The number of credit points for a module depends on the amount of work the student invests. A full year of studies equals 60 ECTS points. The ECTS[2] was created in order to acknowledge exams passed at another university within the European Union.

Harmonization of university courses and degrees in Europe is now carried out according to the "declaration of Bologna 1999."[3]

Promoted by a conference of all university presidents, financially supported by the federal ministry of education and science, and assisted by the German Academic Exchange Service (DAAD),[4] several dozen international courses of study at the bachelor's and master's level have been created. These new courses offer first-year lessons in a foreign language (mainly English) and integrate a compulsory student exchange with an institution abroad. The "International Studies in Clinical Engineering" program at the UAS in Giessen[5] is part of this promotional program. It leads to the bachelor's degree after three years and the bachelor's degree with honors after four years.

Continuous Education in Clinical Engineering

In addition to the degree courses at universities and product-related courses offered by manufacturers, many universities and private institutions offer seminars and workshops in the following areas:

- Legal aspects of medical devices for manufacturers, distributors, owners, or maintainers
- Maintenance, safety tests, or calibration of special kinds of medical devices
- Health care technology management
- Quality assurance, quality management, and auditing

Regional groups of clinical engineers organize meetings with experts for special discussions and for the presentation of new technology. Clinical engineers attend medical congresses, including the annual congress of clinical engineers held by the Fachverband Biomedizinische Technik e.V. in Wuerzburg.[6] The MEDICA,[7] the world's greatest exhibition of medical devices, is another forum for seminars and workshops that takes place each November in Duesseldorf.

Lists of Technician Schools and Universities Offering Courses in Clinical Engineering

Table 67-1 lists certified technician schools. Table lists 67-2 UASs. Table 67-3 lists traditional universities.

Table 67-1 Certified Technician Schools

City	Address
Ansbach	Maschinenbauschule Ansbach Eyber Str. 73 D-91522 Ansbach
Dortmund	Robert-Bosch-Berufskolleg Brügmannstr. 29 D-44135 Dortmund
Esslingen	Medizinische Technische Akademie Esslingen Kesselwasen 17 D-73728 Esslingen
Hannover	Schulen Dr. W. Blindow Baumstr. 20 D-30171 Hannover
Heidelberg	Carl-Bosch-Schule Mannheimer Str. 23 D-69115 Heidelberg
Kiel	Techniker-Fachschule Kiel e.V. Schleusenstr. 1 D-24106 Kiel
Köln	Rheinische Akademie e.V. Vogelsangstr. 295 D-50825 Köln
Neustadt	Berufsbildende Schulen Neustadt Bunsenstr. 6 D-31535 Neustadt
Regenstauf	Private Technische Lehranstalt Eckert Bayernstr. 20 D-93128 Regenstauf
Stadthagen	Schulen Dr. W. Blindow Hüttenstr. 15 D-31655 Stadthagen

Table 67-2 Universities of Applied Sciences (for actual URLs of the institutions please refer to www.fbmt.de → Information → Studium)

City	Address	Course, Degree	Basing on
Berlin	Technische Fachhochschule Berlin Luxemburger Str. 10 D-13353 Berlin	"Medizinisch-Physikalische Technik" Diploma	physics
Bremerhaven	Hochschule Bremerhaven An der Karstadt 8 D-2568 Bremerhaven	"Medizintechnik" Diploma	physics
Gelsenkirchen	Fachhochschule Gelsenkirchen, Neideburger Str. 43, D-45897 Gelsenkirchen	Mikrotechnik und Medizintechnik" specialising in "Gesundheitstechnik" Diploma	physics
Giessen-Friedberg	Fachhochschule Gießen-Friedberg, Bereich Gießen Wiesenstr. 14, D-35390 Gießen	"Medizintechnik" Diploma (planned: bachelor's and master's) or International Studies Clinical Engineering" Bachelor or "Computer Science in Medicine" Diploma (planned: bachelor's and master's)	physics
Hamburg	Fachhochschule Hamburg Lohbrügger Kirchstr. 65, D-21033 Hamburg 80	"Medizintechnik" Diploma (planned: bachelor's and master's)	physics
Jena	Fachhochschule Jena, Tatzendpromenade 1b, D-07745 Jena	"Biomedizinische Technik" Diploma	physics
Jülich	Fachhochschule Aachen Abteilung Jülich Ginsterweg 1 D-52428 Jülich	"Physikalische Technik" specialising in "Biomedizinische Technik" Diploma	physics
Köthen	Fachhochschule Anhalt in Köthen Bernburger Str. 52-57 D-06366 Köthen	"Elektrotechnik" specialising in "Biomedizinische Technik" Diploma	electrical engineering
Lübeck	Fachhochschule Lübeck Stephensonstr. 3 D-23562 Lübeck	"Medizintechnik" Diploma	physics
Mittweida	Fachhochschule Mittweida Technikumsplatz 17 D-09648 Mittweida	"Physikalische Technik" specialising in "Medizinische Technik" Diploma	physics
München	Fachhochschule München Lothstr. 34 D-80335 München	"Feinwerk-und Mikrotechnik" specialising in "Medizintechnik" Diploma	precision mechanics
Münster	Fachhochschule Münster, Abteilung Steinfurt Stegerwaldstr. 39 D-48565 Steinfurt	"Physikalische Technik"specialising in "Medizintechnik" Diploma	physics
Remagen	Fachhochschule Koblenz,RheinAhrCampus in Remagen Südallee 2 D-53424 Remagen	"Medizintechnik und sportmedizinische Technik" Diploma	
Stralsund	Fachhochschule Stralsund Zur Schwedenschanze 15 D-18435 Stralsund	"Medizininformatik und Biomedizinische Technik" Diploma	electrical engineering and computer science
Villingen-Schwenningen	Fachhochschule Furtwangen in Villingen-Schwenningen, Jakob-Kienzle-Str. 17 D-78054 Villingen-Schwenningen	"Medical Engineering" Diploma	mechanical engineering
Ulm	Fachhochschule Ulm, Albert-Einstein-Allee 55 D-89081 Ulm	"Medizintechnik" Diploma	precision mechanics
Wilhelmshaven	Fachhochschule Wilhelmshaven, Friedrich-Paffrath-Str. 101 D-26389 Wilhelmshaven	"Medizintechnik" Diploma or Biomedical Engineering Diploma	precision mechanics
Zwickau	Fachhochschule Zwickau Dr.-Friedrichsring 2A D-08056 Zwickau	"Physikalische Technik" specialising in "Biomedizinische Technik" Diploma	physics

* "Specialising" means full-time advanced studies of 2 to 4 semesters.

---

[2] see http://europa.eu.int/comm/education/socrates/ects.html
[3] see http://www.unige.ch/cre/activities/Bologna%20Forum/Bologne1999/bologna%20declaration.htm
[4] see www.daad.de. DAAD publishes all relevant information about the German university system, including courses, prerequisites, and applications.
[5] see kmubserv.tg.fh-giessen.de/pm/CE
[6] see www.fbmt.de
[7] see www.medica.de

Table 67-3 Traditional Universities (for actual URLs of the institutions please refer to www.dgbmt.de Studienmöglichkeiten BMT)

City	Address	Course	Amount of Specialization
Aachen	RWTH Aachen Helmholtz-Institut für Biomedizinische Technik Pauwelsstr. 20 D-52074 Aachen	Elektrotechnik" or "Physik" or Informatik" or "Maschinenbau"	single facultative subjects
Berlin	TU Berlin Institut für Feinwerktechnik und Biomedizinische Technik Dovestr. 6 D-10587 Berlin	"Biomedizinische Technik"	specialization after basic studies in mechanical engineering or electrical engineering
		"Maschinenbau" or "Elektrotechnik"	single facultative subjects
	TU Berlin Institut für Elektronik und Lichttechnik Einsteinufer 17 D-10587 Berlin	"Elektrotechnik" or "Technische Informatik" or "Physikalische Ingenieurwissenschaften"	single facultative subject: medical electronics
Bochum	Institut für Hochfrequenztechnik Ruhr-Universität / IC 6 D-44780 Bochum	"Elektrotechnik"	specialization in last year
		"Maschinenbau" or "Physik" or "Medizin"	single facultative subjects
Braunschweig	TU Braunschweig Insitut für Elektrische Meatechnik und Grundlagen der Elektrotechnik Hans-Sommer-Str. 66 D-38023 Braunschweig	"Elektrotechnik"	specialization
Dresden	TU Dresden Institut für Biomedizinische Technik Mommsenstr. 13 D-01062 Dresden	"Elektrotechnik"	specialization
Darmstadt	TU Darmstadt Institut für Automatisierungstechnik Landgraf-Georg-Str. 4 D-64283 Darmstadt	"Elektrotechnik"	single facultative subjects
Erlangen	Universität Erlangen-Nürnberg Zentralinstitut für Biomedizinische Technik Turnstr. 5 D-91054 Erlangen	"Physik" or "Werkstoffwissenschaften" or "Elektrotechnik" or "Fertigungstechnik" or Chemie und Ingenieurwesen"	specialization in last year
Hannover	Medizinische Hochschule Hannover Institut für Biomedizinische Technik und Krankenhaustechnik Postfach 610180 D-30625 Hannover	"Biomedizintechnik" (just certificate)	postgraduate course in BME
Heidelberg / Heilbronn	Universität Heidelberg / Fachhochschule Heilbronn Fachbereich Medizinische Informatik Max-Planck-Str. 39 D-74081 Heilbronn	"Medizininformatik"	full undergraduate course in computer science in medicine
Ilmenau	TU Ilmenau Institut für Biomedizinische Technik und Informatik Gustav-Kirchhoff-Str. 2 D-98684 Ilmenau	"Elektrotechnik"	specialization after basic studies
Kaiserslautern	Universität Kaiserslautern Fachbereich Elektrotechnik und Fachbereich Physik Erwin-Schrödinger-Str. (Geb.58) D-67663 Kaiserslautern	"Medizinische Physik und Technik" Certificate	postgraduate distance learning course of studies
Karlsruhe	Universität Karlsruhe Institut für Biomedizinische Technik Kaiserstr. 12 D-76128 Karlsruhe	"Elektrotechnik"	specialization
Lübeck	Medizinische Universität Lübeck Ratzeburger Allee 160 D-23538 Lübeck	"Informatik"	single facultative subjects
München	TU München Lehrstuhl für Technische Elektronik Arcisstr. 21 D-80333 München	"Elektrotechnik"	specialization in last year
		"Maschinenwesen" or "Physik"	single facultative subjects
Rostock	Universität Rostock Medizinische Fakultät Institut für Biomedizinische Technik Ernst-Heydemann-Str. 6 D-18055 Rostock	"Maschinenbau"	specialization in biomechanics / BME
Saarbrücken and Homburg/Saar	Fraunhofer-Institut für Biomedizinische Technik Ensheimer Str. 48 D-66386 St. Ingbert	"Informatik"	minor subject
Stuttgart	Universität Stuttgart Institut für Biomedizinische Technik Seidenstr. 36 D-70174 Stuttgart	"Maschinenbau" or "Verfahrenstechnik" or "Technische Kybernetik"	specialization in last year (major or minor)
Ulm	Universität Ulm Zentralinstitut für Biomedizinische Technik Albert-Einstein-Allee 47 D-89069 Ulm	planned	planned

68

Clinical Engineering Internship

Izabella A. Gieras
Clinical Engineer, Clinical Engineering and Technology Management Department,
Beaumont Services Company, LLC
Royal Oak, MI

Frank R. Painter
Biomedical Engineering Graduate Program, University of Connecticut
Trumbull, CT

Education is a fountain of knowledge. It is through this knowledge that one gains experience and the indispensable preparation to go forth in life. People choose to pursue educational disciplines including arts, law, medicine, engineering, and business. However, some brave souls choose to pursue the magical and mysterious world of biomedical engineering. Biomedical engineering concentrates on the use of engineering principles and practices to search for new knowledge of life processes. The field includes the search for applications to cure and control diseases and ways to provide life support to humans and animals (Bronzino, 1995).

Clinical engineering is the newest and most diverse subspecialty of biomedical engineering. It combines the application of engineering and managerial skills with health care technology for ultimate support and advancement in patient care. Clinical engineering encompasses such areas as biochemical engineering, bioinstrumentation, biomaterials, biomechanics, biomedical imaging, ergonomics, and neurobiology (Bauld, 1991).

The University of Connecticut (UCONN) at Storrs, Connecticut is one of several academic institutions that have instituted programs in clinical engineering. UCONN offers an excellent graduate degree in biomedical engineering with a clinical engineering internship. Students complete their master's degree while they work in a hospital environment acquiring skills and an understanding of clinical engineering in action. This chapter describes the origins, present curriculum, hospital affiliations, and typical activities of the UCONN clinical engineering internship program.

Internship History

The clinical engineering internship was established in the greater Hartford area in 1974 and moved its academic affiliation to the University of Connecticut in 1997. Students who pursue this program earn a Plan A Master of Science (MS) in biomedical engineering that requires a minimum of 24 credit hours of course work and a thesis. A Bachelor of Science (BS) degree in either engineering, physical sciences, or mathematics is a prerequisite for admission to the biomedical engineering program at UCONN. (See www.bme.uconn.edu).

University of Connecticut MS Curriculum

The MS in biomedical engineering at UCONN consists of a total of eight graduate courses (3 credits each), the thesis project (9 credits), and intern experience (6 credits). There are three courses that form the foundations for the degree and for the specialized clinical engineering internship program associated with it. The following courses are required for all clinical engineering interns:

- Physiological Systems I
- Clinical Engineering Fundamentals
- Clinical Instrumentation Systems

The rest of the curriculum consists of biomedical engineering graduate courses, which should be from engineering disciplines related to the intern's background, interests and future career plans. The typical student's workload consists of two classes per semester. Students spend their summer vacation researching master's thesis projects. During the two academic years of this program, students work at affiliated hospitals while taking courses and working on their theses.

Internship Affiliations

The UCONN clinical engineering internship program relies upon local hospitals with which it has formal affiliations. Those in Connecticut include Hartford Hospital, the University of Connecticut Medical Center, Yale/New Haven Hospital, West Haven VA Medical Center, and St. Francis Hospital. The program is also associated with Bay State Medical Center in Springfield, Massachusetts. Students are required to spend approximately 20 hours per week working in the hospital. Each hospital is unique in terms of the research and projects available to the interns and the ways in which clinical engineering is practiced. For example, the Hartford Hospital internship primarily provides technical support and consultation to all engineering and maintenance departments throughout the hospital (Rosow, 2002). Students have the opportunity to get involved in repair and maintenance of clinical and medical equipment while being exposed to numerous projects in research and development applications using programming software such as LabVIEW and BioBench (see Chapter 100). The biomedical engineering department at Hartford Hospital provides an enriching experience in terms of its versatile engineering and clinical projects, such as the machine vision project (see Chapter 92).

The University of Connecticut Medical Center's program supports all medical equipment, including clinical laboratories and radiology. The clinical engineering department provides students with skills in electronics, reading and comprehension of service and maintenance manuals, repairs, inspections, and special departmental projects. Due to its affiliation with the University of Connecticut Medical School, the hospital provides a fertile environment for student research. Thesis projects often stem from such research opportunities.

The Yale/New Haven Hospital offers exposure to a wide range of hospital specialties, including clinical laboratories, diagnostic imaging, and hemodialysis. Interns work on repairs, inspections, and clinical education program development.

The West Haven Veterans Administration (VA) Hospital is closely affiliated with the Newington, CT VA Hospital and is part of the Veterans Service Integrated Network 1 (VISN 1), VA New England Health care System. VISN1 is perhaps the largest health care network in New England. The VA New England Health care System is comprised of eight medical centers and 37 community-based outpatient clinics throughout the six New England states. Students are exposed to a variety of technology projects related to the requirements of the VA patient population. Students can become involved with the medical research activity that the Yale School of Medicine conducts at the West Haven VA.

St. Francis Hospital, in downtown Hartford, provides the student with exposure to a well-organized clinical engineering department with a focus on the practical aspects of technology management, a wide variety of current medical equipment, and a progressive clinical environment.

The Bay State Health System possesses modern test equipment and computer facilities and supports all hospital medical equipment. Interns are exposed to report preparations, oral presentations, asset management, database management, and medical device inspections.

Undoubtedly, this clinical engineering internship program offers indispensable, in-depth clinical and engineering opportunities through exposure to the hospital environment and the different academic courses offered. The two years that students spend in their respective hospitals provide them with excellent opportunities to pursue their personal interests while learning from numerous projects offered by each hospital.

First Year of the Internship

The clinical engineering internship program is structured in such a way that in the first year, clinical engineering interns dedicate their time to rigorous clinical rotations throughout the hospital and participate in various projects undertaken at their respective hospitals. These rotations may include areas such as:

- Clinical engineering
- Respiratory therapy
- Operating room
- Physical therapy
- Anesthesiology
- Lab medicine/pathology
- Emergency room
- Diagnostic imaging
- ICU, adult and neonatal
- Nuclear medicine
- OB/GYN and labor & delivery
- Radiation oncology

Depending on individual interests, these rotations can vary from two weeks to one month. Interns are encouraged to interact with all clinical and medical personnel, and to learn not only the function of the equipment used in each area, but also the overall operation of each department through which they rotate.

In the first year, interns search for their potential master's thesis projects and subsequently start the necessary background research. Some responsibilities of the intern during the first year may include the following:

- Assisting in the equipment management program by performing corrective maintenance and preventive maintenance on specific medical equipment
- Learning and becoming proficient in the use of the equipment management software that is used in running the department
- Performing incoming inspections on new equipment
- Performing electrical safety testing and medical gas outlet testing
- Providing in-service training for medical equipment users
- Preparing technology assessment studies and product comparisons
- Providing administrative support to the director of clinical engineering with budgeting and Joint Commission on Accreditation of Healthcare Organizations (JCAHO) recommendations.

Second Year of the Internship

The second year of the internship program is dedicated to completing a clinical engineering–oriented thesis project. Depending on the hospital, the intern may continue to have various departmental responsibilities, using the remaining internship time on thesis project research. Students working on complex engineering and clinical projects gain valuable knowledge to enhance their clinical and engineering expertise. The overall success of this program is based on the following criteria (Gieras, 2000):

- Application of engineering techniques to patient care and hospital-based research
- Completion of diverse clinical rotations in different departments of the hospital
- Interaction with hospital personnel, including administrators, nurses, technicians, and medical staff

Professional Organization Involvement

During the clinical engineering internship, students become involved in a variety of societies, such as the American College of Clinical Engineering (ACCE) (see Chapter 130), the Association for the Advancement of Medical Instrumentation (AAMI), the American Society for Health care Engineering (ASHE), the New England Society of Clinical Engineering (NESCE) (see Chapter 131), and the Institute of Electrical and Electronics Engineering/Engineering in Medicine and Biology Society (IEEE/EMBS). Students are also encouraged to attend conferences, symposia, seminars, and workshops. The annual conference and exhibition of AAMI provides multidisciplinary programs that enrich the attendees' knowledge and understanding of the development, management, and use of medical instrumentation and related medical technologies. (See www.aami.org).

ACCE (see www.accenet.org) sponsors a wide range of educational programs, including a teleconference series (see Chapter 72), an annual symposium (Cohen, 2002), and Advanced Clinical Engineering Workshops (ACEW) in the United States (Dyro, 1999) and worldwide (see Chapters 70 and 71). ACCE members constitute the majority of speakers on clinical engineering at annual and regional conferences held by ASHE, AAMI, and IEEE. The workshops (ACEWs) provide an opportunity to learn from other clinical engineers about different aspects of clinical engineering, such as evaluation and acquisition of medical devices and systems, technology management, medical equipment inspections, medical device research activities, and medical device incident investigations.

NESCE cosponsors the annual Northeastern Biomedical Engineering Symposium, which provides a forum for interaction with other clinical engineering professionals and an opportunity to enhance technical skills and knowledge through numerous lectures, seminars, and courses (Francoeur, 2002). A large exhibit area offers an opportunity to inspect the latest medical device technologies and clinical engineering support tools.

Clinical engineering interns can also attend the annual IEEE EMBS Northeast Bioengineering conference, held at various universities throughout New England. This provides an excellent opportunity for all second-year interns to present papers about their thesis projects and to learn about research undertaken by their peers.

Students are also encouraged to become involved in local societies, such as the UCONN IEEE BMES/EMBS Society, where some hold positions as the society's president, vice-president, treasurer, and secretary. Students organize guest speakers, field trips, and social events in order to sustain the biomedical engineering tradition at UCONN. Participation in local societies is one way to maintain contact with peers and achieve professional advancement (see Chapter 132).

Life After Internship

The clinical engineering internship program endows students with excellent skills and knowledge. This leads to further academic studies, research activities, or professional work. Graduates from this program have been successful in taking the following paths:

- Clinical engineering departments of hospitals
- Clinical engineering departments of independent service organizations
- Medical device companies
- Medical schools
- PhD Programs
- Research
- Other academic disciplines related to biomedical engineering

Many graduates accept positions as clinical engineers or directors of clinical engineering departments. Some choose paths that take them into industry, working for companies like Phillips Medical, Zoll, and Baxter. Other students, driven by a passion for research and the search for further knowledge, choose academia. In summary, the UCONN clinical engineering internship program prepares students for career challenges in life.

References

Bauld TJ. The Definition of a Clinical Engineer. *J Clin Eng* 16: 403–405, 1991.
Bronzino JD. *The Biomedical Engineering Handbook*. Boca Raton, FL, CRC Press, 1995.
Cohen T. The ACCE Symposium: Perspectives for Successful Leadership in Clinical and Information Technology Services. *ACCE News* 12:5, 2002.
Dyro JF. Advanced Clinical Engineering Workshop, Hartford, Connecticut. *ACCE News* 9:13, 1999.
Francoeur D. 20th Northeastern Biomedical Symposium. *ACCE News* 12:11-12, 2002.
Gieras I. Clinical Engineering Internship. *ACCE News* 10:12-13, 2000.
Rosow E. Workplace Profiles: Hartford Hospital, Hartford, Connecticut. *ACCE News* 12:15-17, 2002.

Biomedical Engineering Technology Program

Bruce J. Morgan
Clinical Engineering Consultant
Riverside, CA

Developing a biomedical engineering technology curriculum is fraught with problems and compromises. This chapter examines the successful program at the State University of New York College of Technology at Farmingdale to see how one college dealt with the problems. The program was designed to be a two-year terminal program. That is, the majority of the graduates are expected to go directly to work in the health care field as biomedical equipment technicians (BMETs), although a small percentage continue with full-time education. As such, the program emphasized the specific skills the graduates would need to work in hospitals.

Overview

The electrical engineering technology courses concentrate on basic electrical theory and its relation to analog and digital systems. The lecture courses are analytical and problem-oriented to foster logical thinking and an understanding of the relations between various topics.

The electrical technology laboratory courses reemphasize the aspects learned in basic theory classes and show the theoretical principles applied to actual circuits. The laboratories are equipped with modern test equipment. All the EET courses have associated laboratory experiments, on which students work individually or in pairs. The biomedical technology laboratories are an integral part of the lecture courses and stress the development of analytical, mechanical, and inspection skills, which are essential in the industrial setting. The laboratories also provide an application of the material learned in lecture.

The biomedical engineering technology courses concentrate on the fundamentals of biomedical equipment technology, emphasizing safety concepts, operation of the health care system, and the application of technology to clinical problems. In order to keep the student current in the years following graduation, the program stresses the development of the individual's initiative and ability to learn independently.

There is a four-semester limit for an associate's degree, including several college requirements, such as two semesters of English and two social sciences courses. Also, it is the college's policy to have all degree courses accredited by the appropriate agency. The curriculum is based on the electrical engineering technology (EET) program due to the heavy emphasis on electronics in medical equipment.

Curriculum

Please note that one semester hour of credit consists of one hour of lecture or two to four hours of lab study per week for a fifteen-week semester.

Course Descriptions

BIO 165–Basics of Human Function

This course surveys the major physiological and morphological relationships of the human body's organ systems. The course emphasizes the integrative pathways and regulatory processes that reflect holistic concepts.

BME 202–Medical Instrumentation I

An introduction to the principles of operation and the clinical applications of medical instrumentation. Emphasis is placed on the electronic principles used in health devices, including amplifiers, signal conditioning, and power supplies. The course examines the integration of physics, chemistry, and electronics in medical devices, as well as hospital organization and the legal aspects of medical devices. Laboratory work includes experiments in electronics, medical equipment operation, preventive maintenance, and repair.

	Lecture Hours	Lab Hours	Credit Hours
First Semester			
EET 101 Electric Circuits I	4	3	5
EET 102 Basic Instrumentation	1	3	2
EET 108 Computer Applications I	1	2	2
EGL 101 Composition: Rhetoric	3	0	3
MTH 129 Technical Math A	4	0	4
	13	8	16
Second Semester			
EET 103 Electric Circuits II	4	3	5
EET 107 Basic Electronics	4	3	5
EET 109 Computer Applications II	1	2	2
EGL 102 Composition: Literature	3	0	3
MTH 130 Technical Calculus A	4	0	4
	16	8	19
Third Semester			
BIO 165 Basics of Human Function	3	0	3
BME 202 Medical Instrumentation	3	3	4
CHM 140 Introduction to General, Organic And Biochemistry	3	2	4
EET 233 Digital Electronics	4	3	5
Social Science Elective	3	0	3
	16	8	19
Fourth Semester			
BME 203 Medical Instrumentation II	3	4	5
BME 254 Electronic Health Care Systems (BMET Clinical Internship)	1	2	2
EET 251 Microprocessors	2	3	3
PHY 131 College Physics I	3	2	4
Social Science Elective	3	0	3
	12	11	17

BME 203–Medical Instrumentation II

A continuation of Medical Instrumentation II. Emphasis is placed on the operation of more sophisticated medical devices and systems, particularly those that are computer- or microprocessor-based. Laboratory work includes inspection and repair procedures for various types of equipment similar to that commonly found in operating rooms, intensive care units, clinical laboratories, and various clinical diagnostic units.

BME 254–Electronic Health Care Systems (BMET Clinical Internship)

Application of the principles learned in BME 202 and BME 203. Students work in a local hospital under the direct supervision of an experienced BMET or clinical engineer (Hines and Dyro, 1990). Emphasis is on the application of principles learned in lecture and laboratory to actual clinical situations.

CHM 140–Introduction to General Chemistry, Organic Chemistry, and Biochemistry

A one-semester course with laboratory work designed primarily for dental hygiene students and biomedical engineering students. Basic principles of general chemistry, organic chemistry, and biochemistry are presented with emphasis on their applications to health science. Topics include measurement, states of matter, bonding theory, solutions, acids, buffers, and pH, and the structure and function of carbohydrates, lipids, sterols, amino acids, and proteins, as well as a molecular approach to enzymatic action, digestion, metabolism, and nutrition.

EET 101–Electric Circuits I

A basic course in direct current circuit theory, including Ohm's law, Kirchoff's law, analysis of series, parallel and combination circuits, mesh analysis, nodal analysis, superposition, Thevenin's, Norton's, and maximum power transfer theorems. Course work also includes study of electric fields and capacitance, magnetic fields and inductance, and analysis of R-C and R-L networks. The laboratory work is coordinated to support the theory course.

EET 102–Basic Instrumentation

The objective of this course is to familiarize the student with commonly adopted techniques and procedures of electronic instrumentation in an industrial setting. The curriculum includes measurement of current, voltage, power, frequency, and time using electronic instruments. The course also covers principles of transducers to extend instrument ranges and conversion of physical quantities to electrical signals.

EET 103–Electric Circuits II

This is the second of a two-course sequence to provide the background needed to analyze electric networks. Topics covered in this course include sinusoidal waveforms and nonsinusoidal waveforms. Also discussed are Fourier series and harmonics, the phasor representation of sinusoidal signals, steady-state conditions, series and parallel resonance, average power calculation, simple passive filters, frequency response (dB magnitude and phase) and its relations to the step response of simple RC, RL, and RLC networks, transformer principles, and types of transformers.

EET 107–Basic Electronics

The fundamentals of semiconductor diodes, bipolar junction transistors, and field effect transistors are discussed. The course analyzes basic diode, circuits, rectifiers, RC filters, zener diodes, and elementary zener regulated power supply, and describes three terminal IC regulators. Small signal single-stage bipolar and field-effect transistor amplifiers are analyzed in terms of voltage, current and power gains, and input and output impedance at midband frequencies. In addition, ideal operational amplifiers are discussed and analyzed in noninverting, summing, and difference amplifier configurations at midband. The laboratory portion of the course requires abridged and formal laboratory reports.

EET 108–Computer Applications I

An introduction to the application of computers in electrical technology. Topics covered are C programming as applied to electric network analysis and program control of electronic instrumentation using the standard IEEE-488 bus. Controlled instruments include switchers, power supplies, and multimeters.

EET 109–Computer Applications II

A continuation of EET 108, Computer Applications I. Topics covered include advanced programming in C, further use of the IEEE-488 bus, programmable instrumentation to perform automated testing of both electronic devices and networks, and the use of software for the analysis of electric networks.

EET 233–Digital Electronics

Analysis and design of combinational and sequential logic circuits. The curriculum addresses number systems, coding, Boolean algebra, Karnaugh map minimization technique, cod converters, SSI and MSI circuits, flip-flops, counters and shift registers, integrated circuit families, multiplexers, semiconductor memories, and D/A and A/D converters.

EET 251–Microprocessors

This course covers fundamental microprocessor concepts including theory of operation, circuitry, programming, signals, timings, and I/O interfacing, with laboratory work on microprocessor trainers. The student is required to interface input and output devices to the microprocessor and to quantify the associated hardware/software trade-offs.

EGL 101–Composition: Rhetoric

This expository writing course emphasizes the use of acceptable patterns of English and the application of rhetorical principles and research. Students will gain experience in the writing process, including revision. A research paper is required, with assignments in library research, note taking, outlining, and incorporation of sources into a final draft.

EGL 102–Composition: Literature

An introduction to plays, poetry, short stories, novels, and essays. Papers are written on forms, techniques, and themes of literature.

MTH 129–Technical Math A

This is a pre-calculus course with applications from various disciplines including technology, science, and business. Topics include families of functions, mechanics of functions, exponential and logarithmic functions, trigonometric functions, and complex numbers. The emphasis is on application and problem solving, and students must have a graphing calculator.

MTH 130–Technical Calculus A

This is a calculus course for those not majoring in mathematics, engineering science, or computer science. Topics include the derivative and differentiation of algebraic, trigonometric, exponential, and logarithmic functions, and applications of the derivative and the definite integral. Applications are taken from technology, science, and business. Problem-solving is stressed, and a graphing calculator is required.

PHY 131–College Physics I

This course integrates the theory and laboratory work of a general college physics course without calculus. Topics include fundamental concepts of units, vectors, equilibrium, velocity, and acceleration in linear and rotational motion, as well as force, energy, momentum, fluids at rest and in motion, and oscillatory motion. Students complete laboratory problems, experiments, and reports associated with the topics studied.

Biomedical Engineering Courses

All of the clinical engineering content is contained in the three biomedical engineering courses, BME202, BME203, and BME254. The courses are designed to give the students all the skills necessary to function in a hospital setting with a minimum of supervision. The curricula allow students to use available resources to fill in gaps in their knowledge.

The lecture courses, BME202 and BME203, given in the third and fourth semesters, respectively, contain some electrical engineering technology topics that are not covered in other courses. These topics include subjects from large signal amplifiers, integrated circuits, and communications. Medical terminology is also included. Students use a self-instruction text (Dennerll, 1998) and take a weekly quiz given to ensure their satisfactory progress.

The first course, BME202, starts with an overview of hospital organization and BMET job descriptions that stimulates the students' interest. Students learn how to solder and desolder in the lab, and spend several hours learning electrical safety principles. In addition, the students learn to read large schematics for a single-channel ECG machine and are introduced to the twelve lead ECG configurations. Laboratory work consists of performing electrical safety tests on the classroom and an ECG machine. Next, the students perform a performance test on the ECG machine, and they learn how to find components on both the schematic and the equipment. This is followed by troubleshooting the equipment in a laboratory setting. A malfunction is placed in the machine and the entire class is taken step-by-step through the process of finding and correcting the problem. The students practice finding various problems for the remainder of their first semester's laboratory time. The remaining lectures cover subjects from transducers and electrodes, differential amplifiers, and Wilson networks to power supplies. By the end of the first semester, the student is thoroughly familiar with patient monitors and techniques for inspection and troubleshooting. Towards the end of the semester the students learn universal precautions to prepare them for their clinical internships.

BME203, given in the following semester, addresses typical biomedical equipment other than monitors, including defibrillators, electrosurgical instruments, infusion pumps, chart recorders, telemetry, central station monitors, and incubators. In the lab, each student is assigned a medical device on a round-robin basis and must test and repair it as necessary. Typical troubles are put in each device to give the student real experience. Lectures cover all the equipment used in the lab; however, the students must often work on a device before it has been discussed in a lecture. The students must then find the information they require from the available documentation and other resources, such as *Health Devices* (see www.ecri.org). The course also includes a few hours of lecture on the care and maintenance of batteries, including sealed lead acid, nickel cadmium, nickel metal hydride, and lithium ion.

Textbooks are a problem, since no single book addresses the majority of the subject matter covered in the two biomedical engineering lecture courses. Two textbooks have been tried (Aston, 1990; Carr and Brown, 1998) and each is adequate, although neither one is complete.

The clinical internship occurs during the last semester of the program. This is the capstone course and was described in detail in an article in the *Journal of Clinical Engineering* (Hines and Dyro, 1990). In this course, the students apply what they have learned during the previous three semesters in an actual clinical setting. The internship consists of one-and-a-half to two hours of lecture and eight hours of work per week. Topics covered include computerized medical equipment service and inventory systems, enteral feeding devices, infusion devices, syringe pumps, ECG monitors and machines, temperature measurement, invasive blood-pressure monitors, physiological monitoring equipment, hypo-hyperthermia devices, blood warmers, electrosurgical machines, surgical lasers, infant warmers, incubators, transport systems, pulse oximetry, oxygen monitors, and noninvasive blood-pressure monitoring. Lectures also address patient data management systems, electron microscopes, suction and vacuum devices, magnetic resonance imaging, radiology equipment, diagnostic ultrasound, hemodynamic monitoring, intra-aortic balloon pumps, defibrillators, nuclear medicine, radiation safety, pacemakers, service careers, and the moral and ethical aspects of technology in medicine.

Facilities

The program has a dedicated classroom/laboratory of approximately 1000 square feet. Equipment includes typical electronic instrumentation such as oscilloscopes, multimeters, frequency counters, function generators, and power supplies. The facility also features specialized biomedical instrumentation such as patient simulators, safety testers, defibrillator testers, electrosurgical testers, and blood pressure calibrators, as well as typical clinical instrumentation such as ECG machines, patient monitors, defibrillators, infusion pumps, recorders, and telemetry units. The total value of the equipment is approximately $200,000, the bulk of which was purchased with three government grants. Also, the facility obtained significant donations of slightly dated equipment from local hospitals that upgraded, which is still useful for teaching.

Faculty

College faculty members teach all courses. Senior faculty members have a master's or doctoral degree and substantial teaching experience. Newer faculty members typically have doctoral degrees. Clinical engineers with at least fifteen years of industrial experience teach the biomedical engineering technology courses. One technician is assigned primarily to the biomedical engineering technology program to maintain the equipment.

Advisory Committee

Input is solicited from the advisory committee in order to ensure that the academic program produces graduates with the knowledge and skills necessary to work successfully in a clinical environment. This committee consists of representatives from hospitals, shared service companies, and medical device manufacturers. Committee members are kept abreast of any proposed changes in the curriculum, and meetings are held at least annually to discuss these changes and to obtain input on any matters that the members consider important.

Accreditation

The Technology Accreditation Commission Accreditation Board accredits the program for Engineering and Technology (TAC of ABET). This organization publishes Criteria for Accrediting Programs in Engineering Technology annually. This publication contains general requirements for all engineering technology courses, as well as specific requirements for particular specialties. The curriculum previously described meets both the general and the specific criteria. The general criteria address such topics as adequacy of college resources and requirements for technical courses. The specific criteria cover requirements applicable only to the biomedical engineering technology curriculum. These criteria can be obtained from the ABET at the address, below:

Accreditation Board for Engineering and Technology, Inc.
111 Market Place, Suite 1050
Baltimore, MD 21202
Telephone: 410-347-7700
Fax: 410-625-2238
E-mail: accreditation@abet.org
Or, the information is available on their website at http://www.abet.org.

References

Aston R. *Principles of Biomedical Instrumentation and Measurement.* Merrill, 1990.
Carr JJ, Brown JM. *Introduction to Biomedical Equipment Technology,* 3rd ed. Upper Saddle River, NJ, Prentice Hall, 1998.
Dennerll JT. *Medical Terminology Made Easy,* 2nd ed. Albany, NY, Delmar Publishers, 1998.
Hines EW, Dyro JF. The BMET Internship Program at University Hospital, Stony Brook, New York. *J Clin Eng* 15:309, 1990.

70

Advanced Clinical Engineering Workshops

Joseph F. Dyro
President, Biomedical Resource Group
Setauket, NY

Thomas M. Judd
Director, Quality Assessment, Improvement and Reporting,
Kaiser Permanente Georgia Region, Atlanta, GA

James O. Wear
Veterans Administration
North Little Rock, AR

Clinical engineering education programs are underdeveloped or nonexistent in most developing countries and even in some industrialized countries. The World Health Organization (WHO), the Pan American Health Organization (PAHO), and other international bodies concerned with health care infrastructure development have collaborated with the American College of Clinical Engineering (ACCE) in the development and implementation of an Advanced Clinical Engineering Workshop (ACEW) program to address educational needs. This chapter presents a ten-year retrospective of the ACEW. Approximately 1000 participants, including faculty, represented over 50 countries in planning the program, and this chapter also includes the mechanics of the ACEW development, implementation, and review. Participants at the 18 ACEWs thus far include health care decision makers, hospital administrators, health ministry officials, nurses, allied health professionals, physicians, and clinical engineers. Results have been positive. Programs in clinical engineering education have arisen, participants have advanced in positions of leadership, governmental health technology management policies have been developed, and clinical engineers have successfully completed requirements for clinical engineering certification.

ACEW Rationale and Origins

Recognition of Need

In May 1987, the 49th World Health Assembly adopted a resolution on economic support for national "health for all" strategies, in which it called upon member states to establish programs for better management and maintenance of equipment through appropriate procedures and training of personnel, and called upon WHO to provide the technical support they required to establish such programs, (WHO, 1987). This call for action reflected the growing concerns of the international community about the worldwide unsatisfactory situation regarding the management, maintenance, and use of health care equipment. Additional concerns included the deteriorating quality of health care delivery and the resulting waste of national and international resources.

The need for a workshop to train and update the leading clinical engineers from Latin America was perceived after discussions with representatives of the WHO division of strengthening of health services and the PAHO.

American College of Clinical Engineering

The American College of Clinical Engineering accepted the challenge of organizing the Latin American workshop and sought the help of the International Federation for Medical and Biological Engineering (IFMBE) and PAHO. Many members of ACCE travel extensively on volunteer assignments to provide clinical engineering support to developing countries. Recognizing the need to improve the level of medical device technology management and clinical engineering in the Americas and throughout industrialized and developing countries, the ACCE agreed to spearhead the development of advanced clinical engineering workshops in collaboration with other motivated organizations.

First ACEW

An organizing committee included Dr. Gloria Coe and Mr. Angel Viladegut (PAHO), Dr. Robert Nerem (IFMBE), and Mr. Frank Painter, Dr. Yadin David, and Dr. Binseng Wang (ACCE). The first ACEW was held in Washington, DC, in June 1991, at the Pan American Health Organization building (Advanced Clinical Engineering Workshop, 1991). A faculty of 19 taught 22 participants, most of whom hailed from 14 Latin American countries. Two weeks of didactic lectures were followed by two weeks of practical, hands-on technical management training in selected technologies at various institutions throughout the United States. Seven participants received funding from the PAHO Central American Project (Guatemala) while others used personal funds and institutional support. The balance of the revenue was derived from nine companies: Hewlett-Packard, SpaceLabs, General Electric, ReMedPar, Physio-Control, Ohmeda, J&J/Critikon, Siemens, and DIAL. Faculty and participants stayed at Georgetown University student dormitories.

Design, Mechanics, and Implementation

Host Needs

ACEWs are tailored to the needs of the host country or region and to the needs of the participants.

Planning

Budget

A budget is prepared that includes faculty and participant transportation, lodging, and food, as well as the cost of facilities, handout materials, translators, and advertising. Any registration fees, plus any grants and contributions from workshop sponsors, are added to the revenue column.

Agreements

Formal written agreements ensure that the workshop is financed. Advanced funding for faculty travel and *per diem*s are recommended, to reduce the odds of delayed reimbursement.

Sponsors

The WHO and PAHO have provided the bulk of the financial support over the course of the ACEW program. In addition, medical device companies, publishers such as Aspen Press and the Journal of Clinical Engineering, and research organizations such as ECRI, have contributed in cash or in kind. Medical device manufacturers have underwritten the costs of lunches, dinners, and receptions and have provided staff specialists to lecture at ACEWs. Nongovernmental organizations and private foundations have given financial support to workshops.

Local Support

Local support includes translation, audio-visual, transportation, social events, photography, public relations, media relations, advertising, registration, photocopying, and meeting faculty at the airport and transporting them to workshop venues and lodging. In addition, some local organizers have arranged for a professional photographer to take pictures during the workshop that could be purchased for a nominal cost.

Design

The needs and desires of the host country or region are considered as the workshop coordinator and faculty assemble the ACEW program. The draft program is sent to the hosts, who typically will make minor adjustments.

Length of Workshop

A typical day begins at 8:30 a.m. and concludes at 5:30 p.m. It is not unusual for sessions to go beyond 6:00 p.m. and to continue into the night on topics of special interest. Certification exams occur in the evenings, which entails extra work for the faculty as they distribute, proctor, and grade the examinations. Breaks and lunch are scheduled. The time period for the didactic lecture portion of ACEWs has ranged from three days (Brazil, 2001) to two weeks (Boston, 1993). Practicum periods have ranged from one to two weeks. Faculty members report a high level of interaction and commitment to learning at all ACEWs, and that participants pay attention to and respect the faculty.

Participants

In some instances, participants are selected by a formal process, but with mixed success. Participants add much to the success of ACEWs. Although participants may be reluctant to speak up, ask questions, and interact with the faculty early in the workshop, this reticence often vanishes after the first day's welcome reception, which facilitates social interaction and intimacy. All participants enjoy a lively interaction by the end of the ACEW. This interaction creates a strong learning environment, increases the level of receptivity of the participants, and allows the faculty to learn a great deal about the problems faced by participants in their day-to-day work. Faculty members often leave these discussions enlightened by the unique and creative solutions to problems in the face of limited resources.

Faculty

Composition

Core faculty are drawn from the ranks of ACCE membership. Adjunct faculty includes selected experts on clinical engineering and medical device technology management, as well as other select fields, such as human resource development, performance improvement, motivation and morale when dealing with change, and measuring and improving client satisfaction. Faculty members from local host countries and countries in the immediate region contribute substantially to the workshop, as these individuals are usually native speakers and relate well to the participants. ACCE faculty members agree that there is much to learn from the experiences of these local faculty members. ACCE encourages all of its members to participate in the ACEW program, and it maintains a list of interested speakers. A typical composition of an ACEW faculty is three to four experienced members and one or two first-time faculty members.

Compensation and Finances

Faculty members are compensated for travel expenses, food, and lodging, but do not receive a salary, honorarium, or other compensation for their time and effort. Expenses incurred during the preparation of materials may be reimbursed, but faculty members usually rely upon the good will of their companies or businesses to defray the expense of paper, typing, reproduction, and phone and fax services. The ACCE ACEW pen is a much appreciated expense item. All workshop participants receive a pen embossed with the place and the date of the ACEW. In addition, faculty members often distribute items to the participants, including pens, small tools, books, catalogs, pocket protectors, hats, and t-shirts.

Workshop Coordinator

The ACEW coordinator maintains communication with local hosts, sponsors, and faculty. The coordinator assigns lecture responsibilities and ensures that faculty members complete their lectures and handout materials in a timely fashion. The coordinator assembles all lectures into one document, "ACEW Proceedings," for distribution to all participants. The coordinator keeps the workshop on schedule and accommodates last-minute adjustments. The coordinator gives the welcoming remarks and introduces fellow faculty members.

Faculty

Faculty members are responsible for preparing their lectures, slides, overhead projections, videos, handouts, outlines, and other workshop materials. While they are responsible for their own lectures, faculty members also draw upon material presented at previous ACEWs that is outlined in the ACEW Syllabus, described later in the chapter. Dress is business-style on the first day and casual later in the week. Jackets are quickly shed in the tropical climes, whereas ample clothing is donned in the cold, damp September in Moscow.

Venues

Hotels, hospitals, banks, universities, convention centers, retreat houses, libraries, and government buildings have all served as venues for ACEWs, as have lecture halls, classrooms, laboratories, and conference rooms.

Facilities

Workshop facilities should include the following: convenient access to lunch and break areas, adequate lighting, photocopy machines, communications (e.g., e-mail, fax, and phone), adequate seating, sound-level control, and public address system with lavaliere and lectern microphones. A table should be placed in or near the classroom for last-minute handouts or papers too large to include in the ACEW Proceedings.

Presentation Format

Format of presentations varies considerably and often reflects an individual faculty member's preferred teaching style. Presentations may be individual lectures with question-and-answer periods, panel discussions, lectures with guided group discussion, lectures with exercises, break-out groups for problem solving, or role-playing, where participants play the roles of purchasing committee members, such as a hospital director, clinical engineer, nurse, physician, or finance director., Group discussions where faculty members are not present have proven to be effective for critical review of workshops. Such reviews have been a part of each ACEW and contribute substantially to quality improvement.

The lecturer may stand in front of participants at a lectern, or use a lavaliere microphone and move about the room, with participants assembled about tables in a horseshoe arrangement so they can see each other's faces (Dyro, 2000).

Materials

Necessary materials for lectures might include a blackboard, whiteboard, flip chart, overhead projector, 35mm slide projector, videotape player, video projector, or laptop computer. Other helpful materials include paper and overhead transparencies, colored marking pens for overhead transparencies, a laser pointer, and a watch or timer.

Practicum

The days or weeks immediately following a one- or two-week classroom presentation are spent working and learning at a clinical engineering facility, hospital, research lab, or manufacturing site.

Site Visits

Participants and faculty visit local hospitals, manufacturing plants, and academic centers during the course of a one- or two-week ACEW. Such visits are preferably arranged and scheduled before the ACEW, but they can be arranged during the ACEW as need and interest dictate.

Evaluation

Evaluation forms are completed on the last day of the ACEW, and open discussion elicits comments and criticisms from the participants.

Archiving

ACEW activities are routinely documented through published reports. Handouts and lecture materials are kept by faculty members for use in future ACEWs. Materials are often in the form of technical and scientific article reprints, photographs, CDs, slides, tapes, and overhead projections, but PowerPoint® presentations have gained favor with faculty, and they are encouraged to use the program, since electronic storage of lecture presentations facilitates dissemination of ACEW content.

Typical ACEW

Planning Sessions

Either the WHO or the PAHO identifies the need for an ACEW. Individual countries often request ACEWs when they recognize the impact they have on the health care systems of neighboring countries. Once budgetary constraints are analyzed and feasibility is determined, WHO or PAHO contacts the ACCE workshop coordinator for scheduling. The coordinator, host country, and sponsoring agency determine a date, and planning begins. The coordinator identifies the faculty and begins to plan the ACEW.

Lecture Preparations

Faculty members complete their lectures and handout materials before the ACEW, and they are encouraged to present them using PowerPoint®.

Arrival and Organizational Meeting

Upon arrival at the ACEW site, the faculty members assemble to discuss the upcoming workshop. Particular concerns include last-minute changes in the program; requests for additional modules; requests for local participation; site visits; and schedules of dignitaries, health officials, and the press. At this time, last-minute adjustments are made to the orientation and emphasis of the lectures. Special requests are considered and incorporated if possible. No ACEW schedule has been followed to the letter, because changes always arise and demand flexibility on the part of the faculty.

First-Day Ceremonies

A typical first day consists of the following: The workshop coordinator, representatives of host organizations, local sponsors, dignitaries, politicians, and health officials give introductions. Faculty introduce themselves with a brief biographical sketch and a description of the subjects to be covered. The coordinator gives a presentation of the workshop objectives and describes the succeeding days' schedules in detail. The schedule for the week is distributed, along with the compendium of lectures and their supporting documentation. Other handouts are also distributed at this time. After the end of formal lectures on the first day, faculty and participants enjoy a welcoming reception. Establishing rapport with the participants at the reception promotes increased information exchange throughout the remainder of the workshop. Such social functions result in long-term associations, to the benefit of faculty and participants.

Typical Workshop Day

Workshops begin approximately at 8:00 a.m. and end at 5:00 p.m., with morning and afternoon breaks and lunch breaks taken during the day. Lectures should not last beyond 1½ hours without giving participants an opportunity to stretch. Lectures can include any of a number of topics contained in the ACEW Syllabus (Dyro and Wear, 2001), and detailed descriptions of the contents of past ACEWs have been published (Advanced Clinical Engineering Workshop, 1991; Wear, 1995; Dyro, 1996; Dyro, 1999a; Dyro, 1999b; Judd and Dyro, 1999; Dyro, 2000c; Dyro, 2000a; Dyro, 2001; Clark, 2002).

Social Programs

Social programs form an integral part of most ACEWs. These might include a simple reception with coffee, tea, and snacks, or a formal banquet. Local hosts often arrange tours to scientific, medical, historic, cultural, architectural, and environmental sites. While most faculty members are present during each other's presentations, arrangements can be made for an individual faculty member to spend some free time visiting the local attractions (see Figure 70-1).

Closing Ceremonies and Evaluation

Certificates of participation are awarded to participants, and faculty may also receive them. It is not atypical for participants to recognize the contributions of the faculty with presentations of small gifts such as handicraft items or books describing the city or country. Closing ceremonies also include group photographs.

Misadventures

Visas

Moscow's Sheremetyevo Airport has two terminals: Terminal One and Terminal Two. A faculty member, Bob Morris, flew from China to Moscow, where he was to meet fellow faculty member Al Jakniunas, who had Bob's visa. Al waited at Terminal 2 and Bob arrived at Terminal 1. With no visa, Bob went directly to a detention cell and was booked on the next flight back to China (Figure 70-2). Hours later, after scouring the airport, the ACEW search party spotted Bob peeping through the window of his cell. Visa in hand, the workshop went on.

Medical Emergencies

Faculty member Al Jakniunas became sick with an intestinal problem en route to Moscow (Figure 70-3). The chief surgeon at the top military hospital in Moscow, who was a workshop participant, ordered him to the hospital for emergency surgery. After four hours under the knife, Al's intestines were repaired, and he recovered completely during his two-week recuperation in the VIP suite (Dyro, 1999b).

Gastrointestinal afflictions from inadequately cleaned food (e.g., oranges washed in the fetid tributaries of the Nile) are not uncommon during travels to foreign lands.

A pre-occurring emergency, such as when illness struck one faculty member days before the Mexico ACEW, or during an ACEW, means that healthy faculty pick up the slack and make the presentations in place of the afflicted. The author had a close call, but, fortunately, the local host facilitated the services of a skilled endodontist to treat an abscessed tooth.

Participants are not immune to emergencies—an asthma attack found a Panama ACEW participant in the hospital.

Health insurance is highly recommended. In fact, WHO contracts include insurance coverage for all faculty.

Air and Ground Travel

During the practicum phase of the first ACEW, 12 participants packed into Frank Painter's station wagon for a site visit to New York City. On the way from Trumbull, CT to their destination, the overloaded station wagon's suspension system collapsed.

Tom Bauld relates his account as a first-time faculty member at the Panama ACEW (Bauld, 2001). Airplane maintenance delays meant a missed flight and an overnight at the airport hotel. A gang that was headquartered in a hotel not far from the ACEW venue, was arrested and charged with plotting to assassinate Fidel Castro, who was attending a Latin American summit. While walking through the national park under a giant palm, a branch fell, narrowly missing us.

Lost or delayed suitcases are not uncommon. Bob Morris was clothed entirely in apparel borrowed from four others on his team.

Watch those potholes and burros and racecar drivers! Local roads, traffic conditions, and drivers may differ in quality from those to which faculty are accustomed.

Unexpected and Memorable Events

Terrorists bombed apartment complexes in Moscow during the week of the ACEW. The bombing near the American Embassy in Lima kept the faculty in its hotel, thus severely curtailing sight-seeing (Painter, 2002).

A flash fire during the ACEW in Quayaquil, Ecuador tested the participants' accident/incident investigation and root cause analysis skills (Dyro, 2001). During a presentation on the systems approach to accident/incident investigation, a fire broke out in a food pantry immediately outside the classroom. Acrid black smoke filled the area, sending gasping faculty and participants in search of fresh air. After the fire was extinguished and the smoke cleared, the presenter assigned teams of participants to put into practice what they had just learned. The ensuing investigations yielded results that were then presented to the workshop class. Faculty served to critique the presentations and to guide the class in contributing their insights into the analysis.

The Beijing ACEW faculty nearly disappeared on the Great Wall of China. Fortunately, one member of the faculty, Dr. Binseng Wang, was somewhat familiar with the native language and obtained the necessary directions back to the workshop.

The timing of breaks during the Nepal ACEW was often determined by power outages, which prevented projection equipment from working (Wear, 2001).

Everyone would do well to use the philosophy practiced by veteran world traveler and educator Bob Morris. During his illustrious career, he often found himself in difficult situations about which he could do nothing. He maintained serenity and tranquility, for to do otherwise (e.g., to become angry or depressed) would not change the situation.

ACEW Retrospective

A total of 19 ACEWs have been presented since the first ACEW in Washington, DC in 1991. Following is a list of those ACEWs.

List of ACEWs and Published Workshop Reports

1. Washington, DC—June 1991 (Advanced Clinical Engineering Workshop, 1991)
2. Boston–May–June 1993 (Second International Advanced Clinical Engineering Workshop Executive Summary, 1993; Wear, 1995)
3. Beijing–November 1995 (Dyro, 1996)
4. Washington, DC–June 1997 (Advanced Workshop, 1997)
5. Mexico City–November 1998 (Dyro, 1999c)
6. Hartford, Connecticut–June 1999 (Dyro, 1999a)
7. Moscow–September 1999 (Dyro, 1999b)
8. Cape Town, South Africa–November 1999 (Judd and Dyro, 1999)
9. Santo Domingo, Dominican Republic–March 2000 (Dyro, 2000c)
10. Chicago, July 22-23, 2000 (Dyro, 2000b)
11. Vilnius, Lithuania–September 18-22, 2000 (Dyro, 2000a)
12. Panama City, Panama–November 13–17, 2000 (Hernández, 2001; Bauld, 2001)
13. Guayaquil, Ecuador–March 26-30, 2001 (Dyro, 2001)
14. Nepal–April 9-13, 2001 (Wear, 2001)
15. Havana, Cuba–May 2001 (David, 2002)
16. Brazil–June 4-6, 2001 (Wear, 2001)
17. Costa Rica–February 25-March 1, 2002 (Clark, 2002)
18. Lima, Peru–March 18-22, 2002 (Painter, 2002)
19. Guayaquil, Ecuador–September 9–13, 2002 (Gentles, 2002)

ACEW Syllabus

Having realized the value of ACEWs, WHO supported the development of the ACEW Syllabus (Dyro et al, 2003). No two ACEWs are exactly alike, because educational needs and health care development efforts vary from one country to the next. Advances in technology and practices continue to change with time. The Syllabus includes a detailed description of the many models that constitute an ACEW.

WHO Specifications

In order to best match the needs of the host country or region to the educational content of the ACEW, WHO determined that the Syllabus should contain a shopping list of various topics that address all aspects of clinical engineering and health care technology management. Topics include those suitable for the educational needs of all levels of participants in the health care infrastructure; i.e., health ministry officials, hospital administrators, directors of clinical engineering, clinical engineers, educators, and allied health professionals. The Syllabus then can be referenced during the course of ACEW planning between host region or country and ACEW faculty.

Contributors

Under WHO direction and specification, the Syllabus contains a representative outline of a two-week ACEW. The following faculty members have served on ACEWs and have provided the bulk of the material for the Syllabus: Tobey Clark, Yadin David, Joseph Dyro, Jonathan Gaev, David Harrington, Alfred Jakniunas, George Johnston, Thomas Judd, Robert Morris, Frank Painter, Mladen Poluta, Henry Stankiewicz, Binseng Wang, and James Wear.

Contents

The 232-page Syllabus comprises some 65 modules, all of which have been presented at least once at one of the 19 ACEWs presented to date. Each individual module comprises the following: learning objectives, summary of the lecture, lecture outline, and references. The modules are grouped under the following areas: introduction, maintenance and service management, national health care equipment management policy, health facility needs assessment and planning, impact analysis, selection and procurement strategies, information systems, technology assessment, resource allocation, human resources development, safety, risk management and quality improvement, utility systems, troubleshooting, technical lectures on medical devices, and summary of workshop goals and objectives. The Syllabus addresses virtually all aspects of clinical engineering and health care technology management, and therefore represents a valuable guide for any educational program in these fields.

Value Analysis

As a direct result of the ACEWs presented to date, programs in clinical engineering education have arisen; participants have advanced in positions of leadership; governmental health technology management policies have been developed; and clinical engineers have successfully completed requirements for clinical engineering certification.

Indicators

Number of ACEWs 19

Number of Participants: Over 1000

CCE Exams: Over 20

Career-Ladder Ascent

Several participants have risen in their institutions' hierarchies or have forged initiative in the health care infrastructure, such as new medical device management companies. One participant is now special advisor to the president of Mexico; another has founded a successful shared services program in Brazil.

Academic Programs

Intent to establish academic programs in clinical engineering has intensified, and discussions between ACEW faculty and academic institutions in host countries for program development are underway.

Evaluations

Participant evaluations of the ACEW have been consistently excellent.

Clinical Engineering Advocacy

While difficult to measure, promotion of the profession of clinical engineering occurs. Name recognition is increasing. Governmental bodies are incorporating recommendations made during the course of ACEWs into laws and regulations.

Advantages and Disadvantages

Faculty

While traveling, the faculty can visit with relatives in foreign lands, gain appreciation for local customs, cuisine, and culture, improve dancing and language skills, foster cordial relations with professionals around the globe, and learn from participants and faculty from host and other countries. Although the work is hard, and the travel long, and although time is lost when one is away from busy projects at home, the overall consensus of the faculty is that participation in ACEWs is a broadening experience, a wonderful change of pace, and an opportunity to visit other parts of the world. The visits are distinctly different in that the hosts provide personal insight rarely obtained when one travels as a mere tourist.

Participants receive world-class clinical engineering education in their own countries. This minimizes their expenses. Often, the costs of air travel, accommodations, and living outside of their countries are prohibitive. Education now becomes affordable.

Infrastructure in Developing Countries

Most developing countries have rudimentary, at best, or nonexistent, at worst, health care technology management policies.

Networking

Participants are encouraged to maintain contact with faculty. Internet access enhances networking. Faculty have hosted participants who visit the United States subsequent to ACEW attendance.

Resource Expenditure

ACEWs are a cost-effective way of providing world-class education to those who are most likely to benefit from it.

Conclusion

ACEWs have demonstrated a positive influence on the development of clinical engineering education and practice around the globe. ACEWs have heightened the awareness that health care technology management is a fundamental building block in a nation's health care system. The ACEW has shown itself to be an effective means to transfer information in a relatively short period of time at a modest expenditure of financial resources. The ACEW will continue to be utilized well into the future. Among those countries in which the need and desire for ACEW workshops has been identified are the following: Jamaica, Mexico, Venezuela, Kazakhstan, Kyrgyzstan, Mongolia, Tajikistan, Turkmenistan, Uzbekistan, Moldova, India, Ukraine, and Armenia.

References

Advanced Clinical Engineering Workshop. *ACCE News* 2(1):8, 1991.
Advanced Workshop. *ACCE News* 7(4):1, 8, 1997.
American College of Clinical Engineering. *Guidelines for Medical Equipment Donation.* ACCE, Plymouth Meeting, PA, 1995.
Bauld T. Bauld Reflects on Panama ACEW: A First-Time Faculty Member Opens His Diary to the Reader. *ACCE News* 11(1):7-8, 2001.
Clark T. Costa Rica Welcomes ACCE Team at Workshop. *ACCE News* 12(3):13-14, 2002.
David Y. Advanced Clinical Engineering Workshop: Havana, Cuba. *ACCE News* 12(1):15, 2002.
Dyro JF, Wear JO, Grimes SL, Baretich MF. *Advanced Clinical Engineering Workshop Syllabus*, Plymouth Meeting, PA, ACCE, 2003.
Dyro JF. ACEW Baltics 2000. *ACCE News* 10(6):6-7, 2000.
Dyro JF. ACEW Chicago. *ACCE News* 10(5):6, 2000.
Dyro JF. Advanced Clinical Engineering Workshop–Hartford, Connecticut. *ACCE News* 9(4):13, 1999.
Dyro JF. Advanced Clinical Engineering Workshop in Ecuador. *ACCE News* 11 (3,4,5):4-5, 2001.
Dyro JF. Advanced Clinical Engineering Workshop in Russia. *ACCE News* 9(5-6):12-17, 1999.
Dyro JF. Dominican Republic Advanced Clinical Engineering Workshop. *ACCE News* 10(3):9-10, 2000.
Dyro JF. Mexico ACEW. *ACCE News* 9(1):4-5, 1999.
Dyro JF. Report on Advanced Clinical Engineering Workshop, Beijing, People's Republic of China. *ACCE News* 10–11, Spring 1996.
Gentles W. Advanced Clinical Engineering Workshop, Guayaquil, Ecuador. *ACCE News* 12(6):9, 2002.
Hernández A. ACEW Panama 2000. *ACCE News* 11(1):6-7, 2001.
Judd T, Dyro JF. Advanced Health Technology Management Workshop, Cape Town, South Africa. *ACCE News* 9(5-6):22-25, 1999.
Painter F. ACEW in Peru. *ACCE News* 12(3):14, 2002.
Second International Advanced Clinical Engineering Workshop Executive Summary. *ACCE News*, August 1993; 3(4):7-9.
Wear J. Advanced Clinical Engineering Workshop, Health Care Technology Management, Kathmandu, Nepal. *ACCE News* 11(6):14-15, 2001.
Wear J. Advanced Clinical Engineering Workshop, Saó Paulo, Brazil. *ACCE News* 11(6):15, 2001.
Wear JO. Second International Advanced Clinical Engineering Workshop, May/June 1993. *J Clin Eng* 20(2):92-96, 1995.
World Health Organization. Report of the Interregional Meeting on Maintenance and Repair of Health Care Equipment, Nicosia, Cyprus, November 1986. WHO/SHS/NHP/87.5. Geneva, WHO, 1987.

71

Advanced Health Technology Management Workshop

Thomas M. Judd
Director, Quality Assessment, Improvement and Reporting,
Kaiser Permanente Georgia Region, Atlanta, GA

Joseph F. Dyro
President, Biomedical Resource Group
Setauket, NY

James O. Wear
Veterans Administration
North Little Rock, AR

Health technology management leaders from twenty sub-Saharan African countries participated in an advanced health technology management workshop in Cape Town, South Africa, from November 8–12, 1999 (Judd and Dyro, 1999). Eight American College of Clinical Engineering (ACCE) members were among the faculty presenting to approximately 65 attendees. This workshop was one in a series of Advanced Clinical Engineering Workshops (ACEWs) organized by the American College of Clinical Engineering. Thomas Judd (ACEW coordinator), James Wear, Frank Painter, Bob Morris, Binseng Wang, and Joseph Dyro (United States) were joined by fellow ACCE members Andrei Issakov (Geneva, Switzerland) and Enrico Nunziata (Mozambique).

On the last day of the workshop, faculty, participants, and sponsors critically reviewed the five-day event. The workshop focused on advanced health care technology management: planning, funding and management of health care technology and the required infrastructure to achieve optimal outcomes. The evaluation reviewed workshop goals and objectives, collected an accurate list of participants and faculty with their contact information, and summarized key messages from each workshop presentation. The evaluation meeting also solicited recommendations from participants for improving future workshops as well as recommended next steps for follow-up by the participants, their home countries, and the World Health Organization (WHO), and gave participants an opportunity to offer feedback on workshop messages and logistics. This chapter presents both a detailed description of the Cape Town workshop itself and the methodology for the critical review of all ACEWs and similar workshops.

Organization and Participants

The workshop was organized by the American College of Clinical Engineering (ACCE), World Health Organization (WHO), African Federation for Technology in Health care (AFTH), Department of Health of South Africa, and South African Medical Research Council (SAMRC) in association with the International Federation for Medical and Biological Engineering (IFMBE) and the International Federation of Hospital Engineering (IFHE). The event was cosponsored by the Association for Appropriate Technologies (FAKT), European Union (EU), Finnish International Development Agency (FINNIDA), German Technical Cooperation Agency (GTZ), and Oregon Health Sciences University (OHSU).

Welcoming and opening remarks were made by the following dignitaries: Peter Heimann (SAMRC), director of the WHO Collaborating Centre for Essential Technologies in Health and secretary-general of AFTH, as well as Andrei Issakov (WHO), Robert Morris, past president, ACCE, Mladen Poluta, director of the health care technology management program at the University of Cape Town and chairman, IFMBE working group on developing countries, Nonkonzo Molai, director of health technology policy, on behalf of A. Ntsaluba, director-general of the department of health of the Republic of South Africa; Nico Walters (MRC) on behalf of M.W. Makgoba, MRC president, and Bernard Shapiro, IFHE secretary-general.

Participants included 12 instructors, eight of whom were ACCE members, and 65 attendees who were health technology management leaders from 20 sub-Saharan African countries.

Goals and Objectives

The purpose of the critical review is to enable the participants to understand clearly workshop goals, objectives, and key messages. In this case, the workshop goal was to understand how to partner with other participants in the sub-Saharan Africa region and other supporting organizations to implement the key health care technology management (HTM) principles, practices, and messages learned during the workshop.

The regional WHO director determined that health technology policy in Africa should address technology management, human resources development, research and communication, and access to information. The overall goal of the workshop was to build strength and capacity for health technology management (HTM) in Africa. The prime objectives were to raise awareness of HTM for health decision makers at policy level in the region, to provide the latest information and to expand awareness of fundamental and advanced concepts and methods of HTM, and to share positive HTM experiences with decision-makers from other countries in the region. The workshop addressed planning, funding, and management of health care technology and the infrastructure required to achieve optimal outcomes.

In February 1999, the WHO regional director stated that priority interventions for developing health technology in Africa should address the following issues:

1. Technology management
2. Human resources development
3. Research and communication
4. Access to information

The opening address by the department of health at the Cape Town ACEW in South Africa on November 8, 1999 highlighted the following:

The level of health service delivery has not notably improved in spite of considerable investment in importing new technologies and the operational expense of supporting these technologies. For example, the lack of equipment maintenance and repair is a major barrier to providing effective care, and stems from poor planning and acquisition and procurement. The most important intervention is ensuring a stable supply of competent technology managers. It is the desire of the department of health that the workshop teach decision makers and planners to deal with complex issues of planning, life cycle, transfer, management, and utilization of health technology, so that organized plans for education, training, and placement of HTM professionals can be developed. This workshop is intended to impart skills to the HTM community to ensure and enhance the safety of health technologies and facilities. HTM is one of the major expenses in our health care system within the area of resource mobilization. The broad framework for comprehensive health care technology (HT) policies has been created. An HT system composed of acquisition, management, planning, and assessment as subsystems is top priority. The major short-term strategic action is a comprehensive national HT audit. Another action is field testing of the Essential Health Technology Package (EHTP) software. The results of these two actions will determine the level of government intervention required.

Major Topics Presented

Technology Trends and Impacts

The link between health outcomes (life expectancy) and a country's economic system is clear. To understand and improve health system performance, health data are needed for analysis. Disability adjusted life years (DALYs) are an effective measure of disease burden (Murray and Lopez, 1996). There is quantitative evidence of the positive impact of HTM on health system performance. Various technology trends can impact future delivery systems for devices and supplies, logistics, medical procedures and techniques, and sites of appropriate care.

National Health Technology Management Policy and Communication Plan

Major changes in the sophistication and complexity of medical technology are expected in the next decade. Management capacity will determine a country's ability to absorb technology. Health care technology (HT) policy is indispensable and will provide the framework for health technology management (HTM); it drives the entire system and its ongoing improvement.

WHO Guidelines for Formulation of National Health Care Equipment Policy

Generic guidelines and implementation documents covering both process and content have been drafted to assist countries in developing and monitoring national HT policies (WHO, 1999). The HTM process is complex; its policy must address all issues of the HT life cycle.

Managerial Competency Requirements

The workshop participants discussed managerial competency required for the effective use of technology and health care planning for district health management in Africa. Active learning is the best method for training technology managers, with incremental changes expected over time. Leadership needs an effective district-level information system to manage by fact. The book entitled *Yenza* (Human, 1998) served as a blueprint for management transformation.

Macro Technology Assessment and Strategic Planning

Macro technology assessment includes evaluation of safety, efficacy, and cost-effectiveness of new procedures and equipment. It is appropriate in developing countries, and needs to be done to ensure "evidence-based" decision making. It creates a bridge between science and decision making and is the only real contribution towards sustainability. Questions to consider are when new technologies should be adopted and older ones replaced, which clinical services should be offered to address the needs of the patient population, and what changes should be introduced to the existing clinical services. Good sources of information on technology assessment include ECRI (www.ecri.org) and the International Society of Technology Assessment in Health Care (ISTAHC).

Micro Technology Assessment

Micro technology assessment includes the evaluation of equipment to be purchased for a single hospital. For example, one hospital developed a multidisciplinary committee for capital equipment assessment, including the evaluation of information available from clinical engineers. The content of technology audits was described in *Medical Technology Management* (David and Judd, 1993). Examples of techniques for conducting the audits came from Senegal, South Africa, and Namibia.

Planning, Selection, and Procurement Strategies

Other key issues discussed at the workshop included the importance of planning, life cycle cost or cost of ownership, pros and cons of the purchase process, evaluation of the need, impact, costs, and benefits of new technology, evaluation of available products on the market, and acquisition of the best product to meet the facility's needs. Alternatives such as lease, rental, and donation were also discussed.

WHO Essential Health Care Technology Package (EHTP)

The EHTP represents a new and unique decision making tool for effective HTM. The EHTP links human resources, medications, equipment, and facilities for diagnostic and therapeutic procedures. It is adaptable and appropriate for developing countries and is available to them at no cost from WHO. The package focuses on interventions and addresses all the necessary requirements; it allows for various scenarios and can be used to determine the criticality of equipment in the provision of care (see Chapter 45).

Health Care Equipment Donations

Donation guidelines proposed by FAKT (1994), ACCE (1995), and others have been combined into draft WHO guidelines (1997), which can be adapted to each country's (or organization's) circumstances. The WHO guidelines formalized the communication link between donors and recipients. Forms requiring essential information are provided. Donations play an important part in providing needed services; up to 80% of the equipment in some countries arrives through donations. Donations must be managed for appropriateness and effectiveness.

Provider–Vendor Relationships

Health care organizations should seek out manufacturers and vendors who have their interests in mind. Everything is negotiable; a good relationship benefits both parties. Health care organizations can improve the quality of the service received from the suppliers by communicating clearly, documenting expectations, and supporting good vendors.

Maintenance and Service Management

A comprehensive inventory is the foundation of a good management program. Gathering financial data along with service activity helps hospitals evaluate actual service costs. Data and the ability to act in response to data are needed for effective service management. Departments that demonstrate cost savings can create resources in an environment of fixed resources. Examples of resource creation and service management were provided through two case studies. Technology management is a developing process; one can start small and then grow.

Budgeting and Financial Reporting

Key issues presented were budgeting for an in-house clinical engineering department, cost recover methods, and financial planning for a for-profit service company (see Chapter 48).

Human Resources Development

Planning and Management

Key issues discussed included the roles and responsibilities of biomedical equipment technicians and clinical engineers, ways managers can improve the clinical engineering program, ways to improve the productivity of maintenance personnel, and ways to determine the personnel requirements of a health care technical service (HCTS).

Training, Continued Education, and Use Training

Professional development of clinical engineering personnel through education and professional organizations was discussed. Appropriate user training can enhance health outcomes.

Health Care Technical Services (HCTS)

HCTS provide a structure for a decentralized, countrywide provision of HTM services. Representatives from Kenya, Malawi, Cameroon, Senegal, Nepal, Mozambique, and Tanzania presented case studies of various models. Key contributors to development of these models in the region include German Technical Cooperation (Deutsche Gesellschaft für Technische Zusammenarbeit (GTZ)), FAKT GmbH, European Union (EU), and FINNADA. Decentralization has necessitated a rethinking of older models and has allowed greater opportunities for public and private mix of ways to provide services.

Safety, Risk Management, and Infection Control

Additional issues presented included infection control, along with maintenance of medical equipment, protection of the staff, the identification of risks and hazards by incident review, risk reduction, and sharing. The use of a systems approach and personal vigilance for device problems, frequent assignment of technology managers as radiation safety officers, an increase in training related to user errors, a review of international and US safety standards and regulations, and a discussion of liability insurance were included.

Quality Assurance and Improvement

Issues of medical equipment quality assurance and improvement concepts were presented, including user and service training and updating and regulatory compliance (e.g., recalls, tracking, reporting, and patient incident investigations). Quality of care improves through development of appropriate infrastructure, use of evidence-based medicine to drive care, provision of data feedback to practitioners on the effectiveness of their care, and creative use of the EHTP.

Computerized Management Systems

A computerized equipment management system is essential to handle a large inventory. The workshop speakers described the design principles and philosophy and the basic elements of a countrywide system for developing countries. Service costs, productivity, and quality-improvement initiatives require a computerized data-analysis system, and the purchase price of hardware and software is only part of the commitment for use of this system.

Utility Systems

HTM procedures should include utility systems requirements. Clinical engineering should be involved in the design of both facilities themselves and their utility systems to increase awareness of utility challenges and possible solutions. Infection control is important with regard to all utilities, particularly HVAC and water.

- Electrical distribution
- Heating, ventilation, and air conditioning systems
- Medical gas and vacuum systems
- Water systems
- Sterilization
- Maintenance procedures

Professional Registration and Certification

Certification is the only professional standard available to evaluate practicing clinical engineers and biomedical equipment technicians. Certification is a voluntary, peer-reviewed process, not a mandatory countrywide registration. Developing countries have used certification to increase recognition of clinical engineers and to measurably raise the level of health care provided.

Outsourcing of Service Management

The workshop speakers also discussed the extent of outsourcing, including full, partial, and management services only. Elements of successful outsourcing include thorough knowledge and information, good initial data, and continuous monitoring of the relationship.

Medical Equipment Troubleshooting

A centralized approach with readily understood, practical techniques was presented, and participants received a guide to troubleshooting medical equipment.

Recommendations

Points Workshop Organizers Should Consider

- Participants' stated workshop objectives:

  1. Training in the basic principles of health technology management used in the United States
  2. Exchanging information and experiences with colleagues from other countries
  3. Promoting the growth of the profession

- Content issue: Provide an introduction on preparing action plans (2- to -3year short-range plans), operating plans (6 months to 1 year), detailed activity plans to monitor work, and reference terms for studying the sector.
- Faculty meetings: Conduct these during the workshop to ensure that presentations are on track.
- Notebooks: Use tabs so that presentation materials are clear.
- Notebooks: Ensure that presentation materials are spell-checked.
- Notebooks: Whenever possible, provide presentation materials prior to the presentation.
- Faculty preparation: Recognize where different presentations will cover the same materials to discern the advantages and disadvantages of overlap.
- Faculty preparation: Consider ways to have faculty better prepared to understand the regional situation prior to the workshop.
- Presentation style: Consider using case studies and group exercises to reinforce the presentation materials.
- Presentation style: Demonstrations are extremely helpful.
- Participant assessment: Set aside specific times for participant feedback to assess how well participants are learning and how to improve; conduct evaluation at the conclusion of the workshop.
- Participant follow-up: Ensure that contact information is provided for all participants.
- Participant follow-up: Ensure that photographs taken at the workshop are provided for all participants.
- Faculty follow-up: Prepare a handbook of standard workshop presentation materials for participants and others to use.
- Clarify order of presentation of the workshop earlier for faculty convenience.
- Recognize how learning can diminish with long workshop days.
- Ensure that participants have time to network with the speakers and each other.
- Ensure that presentation scheduling leaves time for participants to ask questions and share their experiences.
- ACCE and WHO should consider development of a communications/publications/document retrieval service for African health technology managers of clinical engineering with information from 1980 to the present.
- Recognize that participants appreciated and benefited from the presentations, and from the wealth of experience of the faculty. "We have been given a vision of how we can find a way to our solutions," one participant said.
- Include more discussions of participants' experiences, not only faculty experiences, to add balance.
- Participating countries' Ministries of Health (MOH) should receive objectives prior to the workshop to aid in the selection of the most appropriate participants.
- The WHO Essential Health Technology Package (EHTP) review should be expanded.
- Consider using a professional facilitator in workshop to widen participant discussion; continue to bring in an outside management expert/faculty member.
- Provide a participant list with key job responsibilities and local objectives related to the workshop.

Participants and Countries in Africa

Participants in African workshops should consider the following:

- Join the African Federation for Technology in Health care (AFTH)
- Short-term goals:
  - Formulation of a national policy through a task force
  - Evaluation of the WHO EHTP
  - Organization of local workshops (AFTH/MRC/UCT)
- Development of a plan for collaboration with the WHO through the AFTH.
- Sign up for INFRATECH@LISTSERV.PAHO.ORG, worldwide e-mail distribution.
- Distribution of the ACCE News, the newsletter of the ACCE, by e-mail and otherwise.
- Distribute comments, feedback, and follow-up to participants, via Infratech e-mail, regular mail, and AFTH website.
- Share workshop information within participating countries and with those who were unable to attend.
- Note that participants in the African workshop expressed concerns including: "Will countries take action on what we learned and recommendations at this workshop?" "What outside pressures will there be on monitoring and enforcement from the WHO or from Africa?"
- Develop strategies for the training of health technology managers at the "grass roots" level; build upon concepts discussed in *Yenza* (Human, 1998).
  - Actual requirements
  - How to form partnerships, as the Pretoria, South Africa, Technicon program has done
- Establish Senegal regional training sites and a web site, as well as a center to collaborate with the WHO.

World Health Organization (WHO)

After the Africa workshop, the WHO considered the following:

- EHTP: Encourage audits and integrate audit technology
- Consider strategies for how countries are monitored against resolutions:
  - By individuals
  - By WHO representatives at the WHO regional planning meetings
- Facilitate distribution of workshop participant feedback and country situation analysis 6 to 9 months after the workshop

Evaluation

Survey Questions

A survey was distributed at the end of the fourth day of the Africa workshop and collected at the end of the workshop the next day, with the assumption that all participants knew enough English to understand and respond to the questions.

The following eleven questions were asked:

1. What were the best three parts of the workshop program?
2. What three parts of the workshop program were the least useful to you?
3. Rate the presentations in general by checking one of the following: 1 = poor to 5 = excellent
4. Check how useful the workshop presentations will be to your job: not useful/useful/very useful
5. How will you use this information when you return to your job?
6. Rate the usefulness of the handout materials: poor/fair/good/very good/excellent

7. How would you improve the handouts?
8. List some topics that you would have liked to discuss in the workshop.
9. Check the categories of people that you would recommend to attend this workshop.
10. Would you recommend this workshop to others?
11. What is your role in the health care delivery system in your country?

Survey Results

The overall results of the evaluation are very good. There is no doubt that the participants enjoyed the workshop, which is the criterion for a level-one evaluation. The participants were not evaluated on knowledge gained, but the survey suggests that they gained useful information. Thirty-one of the participants returned the surveys, and the following summary includes percentage responses where possible.

1. What were the three best parts of the workshop program?

The following is a summary of the topics in the order presented. Any topics not listed did not receive any rating. If a topic is not rated, it does not mean that it was not good, but only that no one rated it as the best.

10%	Technology Trends and Impacts
23%	National Health Technology Management Policy and Communication Plan
6%	WHO Guidelines for Formulation of National Health Care Equipment Policy
23%	Managerial Competencies Required for the Effective Use of Technology
6%	Y2K Issues Panel
29%	Macro Technology Assessment and Strategic Planning
3%	Micro Technology Assessment
23%	Planning, Selection, and Procurement Strategies
13%	WHO Essential Health Care Technology Package
16%	Health care Equipment Donations
10%	Manufacturer–Vendor Relationships
35%	Maintenance and Service Management
13%	Budgeting and Financial Reporting
16%	Human Resources Development (Planning and Management)
6%	Safety, Risk Management, and Infection Control
6%	Quality Assurance and Improvement
3%	Utility Systems Panel

The topic of greatest interest was the maintenance and service management. Bob Morris was the only faculty member mentioned for his presentation, which related to this topic. The policy and planning topics were rated the next most important to the participants.

2. What three parts of the workshop program will be the least useful to you?

The following is a summary of the least useful topics in the order presented. Any topics not listed did not receive any comments. The fact that a topic is listed does not mean that it was a bad topic or poorly presented, but that it was the least useful to one or more of the participants. In some cases, the same topics appear here that were listed above as best parts of the workshop.

3%	National Health Technology Management Policy and Communication
6%	Managerial Competencies Required for the Effective Use of Technology
6%	Health Care Planning for District Health Management in Africa
10%	Y2K Issues Panel
3%	WHO Essential Health Care Technology Package
6%	Manufacturer–Vendor Relationships
26%	Budgeting and Finance Reporting
3%	Human Resources Development (Planning and Management)
6%	Safety, Risk Management, and Infection Control
6%	Quality Assurance and Improvement
10%	Computerized Management Systems
32%	Utility Systems Panel
13%	Professional Registration and Certification

The utility systems panel was, by far, the least useful, and it probably was the weakest topic in the program. Budgeting and financing was the next least useful topic mentioned. Because most of the participants were government representatives, they had difficulty relating to some to the concepts.

3. Rate the presentations in general by checking one of the following:

Poor 0%   Fair 0%   Good 42%   Very Good 48%   Excellent 10%

All participants thought that the presentations were good to excellent, with 58% rating the presentations either very good or excellent. While the survey did not distinguish among the presenters, no presentation received a rating of "not good."

4. Check how useful the workshop presentations will be to your job.

Not Useful 3%   Useful 55%   Very Useful 42%

The overwhelming majority of the participants (97%) felt that the workshop presentations would be useful in their jobs.

5. How will you use this information when you return to your job?

35%	Increase decision makers' awareness of the importance of HCTS.
19%	Formulate and implement an HCTS program policy.
16%	Share the information with counterparts to see whether it applies in my country.
10%	Lecture students and conduct my own workshop.
6%	Apply the guidelines in planning, procurement, outsourcing, and equipment policy.
6%	Share information with medical equipment and health care technologists.
6%	Improve my weak areas in the field.
3%	Integrate all the district maintenance units in a coherent system.
3%	Set up a central unit for procurement strategies.
3%	Link up with the Health Infrastructure Division on ways in which it fits into the National Health Policy and Plan under SWAP.
3%	Develop a workshop of training and equipment activities.
3%	Advocate for training and continued education in HCT.
3%	Encourage purchase of better equipment management in my area.
3%	Change some work procedures to improve work safety.

The percentages add up to more than 100% because some participants indicated more than one planned use of the workshop information on returning to their jobs. Similar responses were combined to assess the overall impact of the workshop. More than one-third of the participants planned to use the information to encourage decision makers to implement a Health Care Technology Policy.

6. Rate the usefulness of the handout materials.

Poor 0%   Fair 13%   Good 45%   Very Good 42%   Excellent 0%

None of the participants rated the usefulness of the handout materials as poor or excellent, and only 13% rated them as fair. Most of the fair ratings also commented on the need for better organization and more timely availability. Since 87% rated the handout materials as good or very good, the participants should have been able to use these materials to accomplish the items they listed in question 5 when they returned to their jobs.

7. How would you improve the handouts?

35%	Make handouts available at beginning of workshop
19%	Organize handouts with dividers
19%	Make handouts more condensed, and include references
3%	Edit the overheads into pictures and captions
3%	Make handouts available electronically
3%	Add more detailed material
3%	Provide more tools
3%	Provide a list of information services and web sites
3%	Print double-sided to reduce volume of paper

The participants obviously wanted the handout materials at the beginning of the workshop, and organized so they could follow with the lectures. Nineteen percent wanted fewer materials, but more references so they could research additional information on their own. Some participants noted specifically that material that was not covered in the lectures should not be in the handouts. These percentages do not add up to 100% since some participants made more than one comment and some made no comments.

8. List some topics that you would have liked to discuss in the workshop.

19%	Situation analysis of the health technology management in each country
6%	Action plans and workshop resolutions
6%	Strategies to convince the government that we are a big part of health care in South Africa
3%	WHO's health technology policy in the African region
3%	Human resources development for specific countries
3%	More practical issues on topics like quality assurance or risk management
3%	More of the policy and planning role, less on managing and implementing
3%	Generic specifications to assist in the procurement of medical equipment
3%	Practical setting-up of maintenance/engineering workshops
3%	In-house maintenance vs. outsourcing
3%	Case studies on productivity and effectiveness of health care provinces
3%	Southern Africa Development Community training using pooled resources
3%	Training of technicians and users
3%	How to interface with decentralized and privatized health facilities
3%	Medical technology management as part of the whole health-district system
3%	Presentations by health equipment manufacturers
3%	Information on certification of clinical engineers and BMETs in the United States
3%	WHO's Essential Health Care Package

Most of the additional topics that participants wanted focused on individual countries' experiences. This would have allowed for more sharing and might have given them better ideas for specific solutions to their local problems. The percentages do not add up to 100% since some did not list any topics and some listed more than one.

9. Check the types of individuals to whom you would recommend this workshop.

Manager 71%   Engineer 77%   Planner 58%   Health care Administrator 58%
Technician 35%
 Others–Politician, Advisor, Nurse, Decision Maker, Policymaker, Manufacturer

These recommendations probably reflect the jobs of the participants. Some thought that the attendees should be in pairs of a technical person and an administrative person. Others said that they would have liked to have the workshop agenda beforehand, to help in the selection of the attendees.

10. Would you recommend this workshop to other people?

Yes 97% No 0% Maybe 3%

This is probably the most significant result of the evaluation. If 97% of the participants would recommend the workshop to others, they must have found the information useful.

11. What is your role in the health care delivery in your country?

The jobs of the participants were grouped below into "technical or engineering" and "administrative or planning" roles. However, this grouping may not be accurate, since engineers may function in both administrative and planning positions. Conversely, participants who are not engineers might also have HTM roles.

Technical or Engineering

Clinical engineering coordinator (private hospital)
Electronic engineer in medical equipment maintenance
Engineer, provincial maintenance section
Head of biomedical engineering services in the ministry of health
Health technology management
Health care technology manager for secondary hospitals
Hospital engineer
Hospital medical equipment management and repair
HTM advisor
Manage equipment in my province
Medical support manager (provincial general hospital)
Specialist in disability and assistive services
Technical advisor at ministry level

Administrative or Planning

Advisor in health care delivery programs
Advisor to hospital management
Advisor to provincial directorate of health
Consultant
Decision maker
Draft health-related policies and plans
Head of planning division
Health economist planning health services
Health system consultant
Manage health laboratory services at the ministry of health
Manage health care technology for country
Management of health technology in ministry of health
Manager in radiography
Planner
Provincial health care manager
Provincial manager of public health

Conclusions

Faculty, participants, and organizers considered the Advanced Health Technology Management Workshop to be a success. Workshop evaluations indicated overall satisfaction with both the content and the method of delivery. Areas of improvement were identified that will be incorporated in future Advanced Clinical Engineering Workshops. Impact evaluations with the participants at some point after the workshop are useful to determine whether the information from this workshop made a difference in health technology management in the participants' respective countries. The Cape Town AHTM Workshop serves as a model for future health technology management workshops presented for developing nations.

References

American College of Clinical Engineering. *Guidelines for Medical Equipment Donation*. Plymouth Meeting, PA, American College of Clinical Engineering, 1995.
David Y, Judd TM. *Medical Technology Management*. Redmond, WA, SpaceLabs, 1993.
Dyro JF, Wear JO. *ACEW Syllabus*. Plymouth Meeting, PA, American College of Clinical Engineering, 2001.
FAKT (Association for Appropriate Technologies). *Guidelines: Medical Equipment Donations*. Geneva, CMC, 1994.
Human P. *Yenza*. Oxford, 1998.
Judd T, Dyro JF. Advanced Health Technology Management Workshop, Cape Town, South Africa. *ACCE News* 9:22-25, 1999.
Murray JL, Lopez AD. *WHO: The Global Burden of Disease*. Cambridge, MA, Harvard University Press, 1996.
World Health Organization. *WHO Guidelines for Formulation of National Health care Equipment Policy*. Geneva, World Health Organization, 1999.
World Health Organization. *Guidelines for Health Care Equipment Donation, 4th Draft*. Geneva, World Health Organization, 1997.

72

Distance Education

James O. Wear
Veterans Administration
North Little Rock, AR

Alan Levenson
MediqPRN
Pennsauken, NJ

Distance education is not new. It has been around for decades, but recently it has received more publicity with the advent of the Internet. The first distance-education courses were correspondence courses, back before the turn of the 20th century.

With the reduction in hospital budgets, continuing education is becoming more difficult for clinical engineers (CEs) and biomedical engineering technicians (BMETs) to obtain. Education costs for all staff are being reduced because they are not easily charged back to patient care. At the same time, CEs and BMETs are required to have continuing-education credits (CECs) to maintain their certifications or licenses as professional engineers. Distance education provides an economical way to obtain this continuing education.

Many electronic technicians were trained with extensive correspondence courses in general electronics or radio and television repair. When introduced 40 years ago, audio teleconferencing became a useful adjunct to distance education. In audio teleconferencing, lectures are presented over a telephone, and students are able to ask questions. A dialogue between instructor and student can occur. Another form of distance education has been the use of videotapes and audiotapes, where lectures have been put on videotape or audiotape. With the addition of written material and an exam, while not interactive, this distance education program is an enhancement over textbook correspondence courses. The introduction of video teleconferencing was yet another improvement in distance education; lectures could be seen, not just heard. Video teleconferencing is an enhancement over a videotape in that the presentation is live and, in some cases, two-way video technology can render the teleconference interactive. Correspondence courses changed with the advent of computer technology. These computer-based courses are simply correspondence courses utilizing computers. While actually similar to some of the correspondence program instruction courses, computer-based courses are more interactive.

The Internet and courses on the World Wide Web can be interactive. Frequently, some of these courses that have been computer-based or videotaped are now placed on the web so that they are available at any time. Interactivity can be added to these courses. With the advent of all of the new technology and the interest in distance education, degrees can be offered either partially or entirely by distance education.

Correspondence Courses

Correspondence courses have been popular for generations. The enrollee can study, do exercises, and take tests at a remote site of convenience. Educational material, typically textbooks or lecture notes, is sent to the student by way of the US Postal Service or common carrier. Upon their completion, assignments and exercises are sent in for grading. Generally, one examination is given at the end of the course. In some programs, the examination is monitored, or taken at a school, library, or other remote location. If the program is considered a self-study program, it might have a self-administered examination, which is then returned to the program for grading.

College credit is given for correspondence courses offered by many colleges and universities. One can complete general education requirements for an associate's or baccalaureate degree by this method. The completion rate for these programs is 40%–50%.

Some schools provide only correspondence-course programs. These schools typically bestow certificates of successful completion of a course and, in some cases, offer courses for which the student can receive college credit. Over the years, various electronics programs have been instituted in this way. These have extensive correspondence courses, including physical experiments. These programs have been accredited, and program expenses can be covered by the GI Bill. Many programs were established after World War II and trained people for radio and television repair. Some of these courses include the actual construction of a radio or a TV, or, more recently, a computer. Most of these courses originated after World War II as a result of the GI Bill passed by Congress and signed by President Franklin D. Roosevelt to provide benefits for soldiers returning from conflict. The United States GI Bill Act of June 22, 1944 (NWCTB-11-LAWS-PI159E6-PL78 (346)) put higher education within the reach of millions of veterans of World War II and later military conflicts. Veterans' benefits could be used for these courses if they met certain standards. As technology changes, correspondence schools change as well. They now feature such subjects as computer repair and communications networks.

The International Correspondence School has been in business for many years and has offered a variety of subjects by correspondence. It is now known as Thomson Education Direct. (See www.educationdirect.com.) TPC Training Systems (www.tpctraining.com) has a variety of courses that can be used by an institution as correspondence courses, with TPC grading the exams. Scientific Enterprises, Inc. (www.scientificenterprises.com) has a large number of self-study courses in biomedical engineering, hospital engineering, and hospital safety.

In addition to companies, several government agencies have developed their own correspondence programs. The Department of Veterans Affairs has had an extensive correspondence program for over 30 years for hospital engineering and biomedical engineering. The military as well has had extensive correspondence programs over the years.

Some correspondence programs have gone through transitions from strictly textual material to self-paced material, which was partially interactive and allowed people to go through a study activity at their own pace. Computer-based instruction is now available in which the text material and video clips are put on CD-ROMs. Exams can be given at particular places throughout the course. By these means, the material is interactive. If a question is missed, the program can refer the student back to the section for restudy. Prior to the CD-ROM, some of these programs were available on floppy disks, but they did not have the capability or as much of the interactivity of video clips because of the medium's insufficient storage capacity.

Audio Teleconferences

Audio teleconferencing, which has been in existence for decades, remains a viable teaching modality. The two main programs in clinical engineering, which have been in existence for several years, are sponsored by the Department of Veterans Affairs (DVA) and the American College of Clinical Engineering (ACCE). (See Chapter 130.)

An audio teleconference is a relatively inexpensive form of distance education that can provide many participants with the same information. It does not require any sophisticated equipment except the bridge that connects the phone lines. In the United States, commercial sites can furnish the bridges and phone lines for less than $25 per hour per site. All that is required by the speaker and the participants is a speaker phone with a mute button. The audiovisual support can be provided either by handout materials or by slides, transparencies, computer files such as PowerPoint® presentations, or even videos that can be shown at the local site.

A facilitator is desirable at each site with half a dozen or more participants, in order to encourage maximum interaction. This allows for local discussion and the possibility of workshop exercises. It is possible to go offline in the course of a presentation to conduct a workshop or discussions, or to watch a video, and then to go back online for discussions with the speaker.

In order to achieve maximum interaction, having no more than 50 sites online at a time is advisable. The number of participants at a site can be as few as one or as many as 20, with the most desirable number being between four and ten. Enough participants must be at a site in order to have some discussion, but too many participants make hearing or participation in the local discussion difficult, if not impossible.

The prime advantage of an audio teleconference is that the speaker can be at any site where he has access to a speaker phone; likewise, the participants can be anywhere as long as they have a speaker phone with a mute button. The mute button is critical to preventing noise and feedback over the lines. Another advantage of the audio teleconference is that it can be a series of conferences, rather than just one. Because two hours is about the maximum time that one can maintain the interest of a group, an extensive topic such as codes and standards is best presented with sessions extending over several days.

Department of Veterans Affairs Experience

The Department of Veterans Affairs (DVA) has been using teleconferencing for each distance education since 1974. Because the DVA has 172 hospitals, it typically does each audio teleconference four times, one for each of the four time zones of the contiguous 48 states. With this modality, the DVA can train 1000 or more biomedical engineering and safety staff on updates to codes and standards.

A committee of three clinical engineers develops the training material and serves as faculty. A pre-test, handouts, case studies, evaluation, and post-test are developed. A site facilitator is selected for each site. This facilitator must have some content knowledge and is usually the clinical engineer or a senior BMET. Before the teleconference training, there might be some benefit in a conference call with all of the facilitators, to discuss the procedures.

Case studies are developed from questions that are asked of the DVA clinical engineers in DVA headquarters. The participants at a site discuss these case studies, and they come to a consensus solution. The case studies are then discussed online with the faculty. At the end of the teleconference, each participant receives a copy of the correct solutions and the rationale.

The DVA has used this modality to teach people how to teach. Students watch videotapes and then come together to discuss what they have seen. They also make presentations between audio teleconferences and discuss their experiences, both successes and failures.

The DVA conducted a series of audio teleconferences to prepare people to take the clinical engineering certification exam (see Chapter 133). First, it surveyed the DVA clinical engineers to determine the topics that they needed to review. They arrived at the following topics: clinical laboratory, respiratory therapy, radiology, physical plant, and utilities. Then, the DVA education division located speakers for these topics and organized one or more audio teleconferences on each topic. Clinical staff and a clinical engineer presented each session. Most of the clinical staff speakers were not associated with the DVA but had experience teaching by audio teleconference. The faculty prepared handout materials, including audiovisuals, and provided appropriate reference articles. In the final session, the speaker was the chairman of the clinical engineering certification board, who reviewed the examination application form with the participants.

The DVA conducts a six-month follow-up evaluation of the effect of the audio teleconferences. These generally receive excellent ratings and demonstrate cost savings. The cost savings might be in the form of reduced maintenance time, better equipment evaluation, or canceled or reduced service contracts. Depending on the program, improved clinical customer response also might be an outcome. Of course, one outcome is the improved understanding of codes and standards as well as the survey process of the Joint Commission on Accreditation of Healthcare Organizations (JCAHO).

American College of Clinical Engineering Experience

In an effort to provide education opportunities for its members at a reasonable cost, the American College of Clinical Engineering initiated an annual series of audio teleconferences on clinical engineering topics in 1995. The format for these teleconferences is a 45-minute presentation followed by a 15-minute question-and-answer period. At least one week prior to the teleconference, the teleconference coordinator distributes handout material, including copies of computer-graphic presentations, at the participant sites. Participants complete evaluation forms on each session, and the University of Arkansas for Medical Sciences gives continuing education units to the participants.

The number of sites participating has varied from 10 to 20, with one to 10 people at each site. There have been 50 to 100 participants on each of the programs, and evaluations have been very good. The speakers receive modest honoraria. The sites have been charged $125 per teleconference for up to four people, with an additional charge of $10 per person. The teleconference series has been profitable for the American College of Clinical Engineering, providing funds that are used for further educational activities and ACCE member benefits.

Table 72-1 shows a list of some of the ACCE Audio Teleconference topics.

When the program first began, handout materials were mailed, but now these materials are typically sent to the participating sites electronically. At some sites, the PowerPoint® presentations are shown as the speaker talks. In any case, each participant receives a copy. The sessions are also recorded and are available for sale with a copy of the handouts. Copyright releases are obtained from the presenters for the material to be made available in this format.

Audio teleconferencing also has been used to teach courses for college credit in biomedical instrumentation technology. A program at the University of Arkansas for Medical Sciences has taught an introductory biomedical instrumentation course by audio teleconference. The lectures were presented for two hours each week. A student visited particular departments in hospitals to see and talk to the operators about equipment. In the following week, students could ask questions and discuss what they had seen. Codes-and-standards courses, including NFPA 99 and 101 and the JCAHO Environment of Care Standards, were also taught by audio teleconference.

Video Teleconferencing

The next highest level of distance education is videoconferencing or satellite broadcasts. Government agencies and commercial enterprises have used it often. Among the several technologies for audio and video communication that can be utilized are point-to-point high-speed phone lines or bridges to connect several points. This method is interactive and allows the instructor to see the participants and allows the participants to see each

Table 72-1 Topics of ACCE Audio Teleconferences

Benchmarking
Building Teamwork Between CE Staff & Maintenance Staff
Business Plans
Can Clinical Engineering Departments REALLY Do Anything about Human Error?
Contract Management
Equipment Management Inclusion Criteria: An Improved Method of Including Equipment into the Program and Determining Inspection Frequency
FDA Issues
Financial Analysis of Operations
HIPAA's Impact on Clinical Engineering
How to Justify (to Your Administration) Additional Manpower, Increased Salaries, and Attracting and Retaining Good Help
Incident Investigation
JCAHO's Changing Perspective and Its Impact on Clinical Engineering
Managing Electromagnetic Compatibility in the Hospital
New Opportunities for Clinical Engineers
Non-Profit to Profit
Remote Diagnostics–Where Are We Today?
Technology Assessment
Tools for Technology Managers: Strategic Technology Planning
Understanding the Health care Marketplace
What Does It Take to Perform In-House Radiology Maintenance?
What Is the Difference between CEs and BMET Managers?

---

other. Among the many systems available are V-Tel™ and PicTel.™ Again, the bridge time can be rented from a third party (see Chapter 73).

Satellite presentations can be made through an uplink and then received at sites with a downlink, which is not too expensive. However, this does not allow any interaction except by a telephone call. True two-way interaction requires a full uplink at both ends of a program, which is very expensive. With the advent of digital systems, a good-quality satellite broadcast can be presented and picked up at the downlink sites. This gives a much higher quality presentation than has been possible in the past, but it still is only interactive through telephone calls.

The DVA has been using satellite broadcasts for their biomedical engineering and hospital safety programs. They have four studios around the U.S. with broadcast capability, and all of the 172 medical centers have downlinks for the digital signal. At the same time, the DVA has four full-time channels on a satellite. Broadcast programs are typically one to two hours and are presentations by individuals or panels. A toll-free number enables participants to call in with questions. However, as compared to an audio teleconference where people simply push a mute button for an open line, just the small effort in making a telephone call decreases interactivity. A satellite broadcast is a good way to present information, especially when participants need to visualize material rather than merely hear a lecture.

The next highest level of technology involves bringing the satellite broadcast down to the desktop, where it can be received through the network on personal computers. The DVA is currently installing this technology.

The Internet

Another area of distance education is the use of Internet and web sites. Universities are using this method to present material to students on campus, but it can be used to present material to anywhere that people have access to the necessary technology. (See Chapter 73.)

The Internet, with e-mail, can be utilized for teaching courses that might have been taught strictly by correspondence. This allows a student access to the instructor in a much shorter time frame than that which elapses in correspondence by letter. One additional advantage is that it is much less expensive than a telephone call. Exams can even be presented by e-mail. Web-based courses can actually be interactive and can include video clips. These can be useful, but they can be expensive to develop appropriately. A student must have a computer with a high-speed modem or some other rapid access to the Internet, to make efficient use of their time with a web-based course. Again, some of the web-based courses are used at universities or at hospitals on their local area networks. These networks typically have significantly increased transmission speeds. Because they can be taken at almost any time, web-based technology is particularly useful for short courses (e.g., code compliance) required for employees.

A web site's availability 24 hours per day, 7 days per week, is a significant advantage, enabling participants to work on courses at any time. The courses can be taken on demand, on the student's time, and not on the instructor's time. Because students can be from any country in the world, language barriers can increase the challenge for the instructor.

The one disadvantage of web-based courses is that they generally will not meet US Occupational Safety and Health Administration (OSHA) requirements as stand-alone programs. OSHA requires that training be localized, which means that one must have local names and phone numbers in the program. OSHA also requires that someone be present during the course at all times to answer questions. A question cannot be asked by way of e-mail, to be answered later. This means that if someone is taking a web-based course at 2:00 a.m., someone else must be reachable by telephone to answer a question. OSHA requires a great deal of hands-on training. Therefore, web-based courses are similar to those that utilize videotapes and that can be used in support of training, but they generally cannot be used exclusively. While the presence of someone who can answer questions can be arranged, the need for hands-on training remains a limitation of web-based courses.

Netmeeting, a specialized system that can be used over a network, allows all the participants to see the instructor's screen on their computers. This is useful for teaching the use of computer programs.

It is possible to obtain a college degree from accredited universities with a combination of residency and distance education. In some programs, a degree can be obtained without having to meet residency requirements. These programs can be composed entirely of distance education, including some online courses, experience credit, and company training. One must approach this area with caution, however, because some schools are accredited by regional accreditation agencies, and some are not.

Distance education has become a way of life for many professionals as technology, medical equipment, and the means of delivery of health care change. In order to keep abreast of key developments, CEs and BMETs must continue their educations. The only economical way to accomplish this is by distance education, which allows individuals to learn on their own time, at their own pace, at an affordable cost. Fortunately, the change in technology has also provided newer modalities with which distance education can be provided. Many find that, as adult learners, they can learn as much or more through distance education as they can in a traditional classroom on the college campus.

73
Emerging Technologies: Internet and Interactive Video Conferencing

Albert Lozano-Nieto
Assistant Professor of Engineering, The Pennsylvania State University, Wilkes-Barre Campus
Wilkes-Barre, PA

The main goal of clinical engineering academic programs is to train future professionals in the maintenance, repair, and overall management of clinical and medical instrumentation. The previous chapters in this section have discussed several academic approaches to clinical engineering education, either at a degree level or for continuing education.

The clinical engineering profession of today is undergoing major changes. Clinical engineers' main role is shifting from the traditional approach of performing preventive maintenance and repairs on equipment, toward management of technology in the whole health care system (Zambuto, 1997). The clinical engineers of today play a multifaceted role in the clinical environment as the result of their continuous and growing involvement with the management of clinical equipment. They regularly interact with the clinical staff, hospital administrators, vendors and manufacturers, clinical engineers at other facilities, and regulatory agencies.

It is in this constantly changing scenario that one must evaluate and modify approaches to the education and training of these future professionals. Educators must incorporate new and more efficient education techniques to keep students in consonance with the changes occurring in the industry. In the classrooms, they must anticipate the future needs of industry while maintaining the high levels of quality that traditionally have characterized clinical engineering education. Finally, educators must remember that employers demand not only a high level of technical skill from prospective employees, but also some experience in the clinical setting and teamwork and, especially, the outstanding communication skills that are so critical in the work environment (Xu et al, 1997).

The traditional components in clinical engineering education have consisted of lectures delivered by instructors and laboratory work focusing on troubleshooting and repairing diverse medical devices at the component and subsystem levels, as well as performing preventive maintenance tasks on the equipment. Lectures in the classroom address the basic principles of electricity and electronics, physiological transducers, the origin of biopotentials and biological signals, and the interaction between living tissues and electricity. The experimental portion in the laboratory normally is used to visualize the concepts previously developed in the classroom as well as stimulate the student's intellectual curiosity in areas that are not covered in lectures. Some academic programs also require students to successfully complete an internship in a health care institution, thus giving their graduates the opportunity to acquire experience in the workplace before graduating, which constitutes a definite asset to their future employers (Lozano, 1998). This approach is perfectly valid today inasmuch as students still need to know the basic working principles of medical equipment and need to be able to identify possible causes for its failure or malfunction, as the majority of in-house service calls are caused by operator error or erroneous equipment configuration (Westover, 1998). However, this approach is not enough by itself. Given the size and infrastructure of current academic programs in clinical engineering, the instructors alone cannot provide the breadth of perspective and in-depth coverage for all the issues that affect clinical engineering, from both a technical and a professional point of view. Furthermore, it is also widely recognized that educators will be able to increase their critical-thinking abilities and consequently become more effective once in the workplace by exposing future professionals to a variety of different opinions and points of view (Newport, 1997).

Clinical Engineering Education through Internet Resources

It is irrefutable that, in the last few years, different Internet services have played a major role in changing the ways in which one accesses, transmits, and shares information. These new resources can, and should, be used to improve the way one approaches the education of future clinical engineers, as the Internet and its related services constitute dynamic media with strong possibilities of continuous adaptation to the necessities of the user. Furthermore, the same platform supports a broad range of informational resources that was previously disseminated among a large number of locations and media. The Internet also makes available to the students information that traditionally has been difficult for them to obtain, such as the technical data on medical devices. The growing number of commercial electronic medical equipment manufacturers that distribute their information through the Internet allows students to access quickly all of this information from the classroom. The Internet must be viewed as it was conceived: as a powerful and valuable tool to exchange ideas and information among individuals located at geographically distant points.

Developing Basic Skills Using Electronic Mail

Although virtually all of the students who are enrolled in academic programs in today's colleges and universities have computer accounts with Internet access, not all students will use them on their own initiative, nor will they explore the different and more complex possibilities that these resources offer. In the author's particular experience, e-mail was heavily used by the instructor to send and to review assignments and homework, as well as for other communication purposes. This exercise was based on a gradual approach of increasingly using more complex functions such as composing and editing, replying to the sender and to all other recipients, forwarding messages, sending attachments, and working with different fonts and layouts. The goal of this e-mail usage was to enable students to use e-mail, with all of its possibilities, on a regular basis.

Participating in Clinical Engineering Listservs

After the students become familiar with the use of e-mail, they are required to subscribe to a clinical engineering distribution list (or listserv). The primary goals of these distribution lists are to discuss and to share information among professionals working in the different areas relating to clinical engineering. Although the students do not feel comfortable enough to participate actively in the discussions, they gain important experience by reading the messages interchanged in the listserv. They feel as though they are a part of this professional group, learning not only about technical details of equipment, but also about their goals as professionals and the current problems and concerns that clinical engineers face today. This feeling of belonging to a distinct professional group is unique and cannot be achieved in the classroom, making this an invaluable experience. In addition, when the instructors are also monitoring the listserv, they can bring some of their current topics of discussion into the classroom. The author's personal experience shows that some of the most interesting discussions during the lectures were brought up by the students based on topics that were being discussed in the listserv. Table 73-1 lists some of the listservs carrying discussions that are interesting to the clinical engineering community. Because of the ephemeral life of some of these distribution lists, and their possible changes of electronic address, they should be considered only as examples, and readers are advised to search for an active list that meets their interests.

The World Wide Web as a Tool for Clinical Engineering Education

The World Wide Web (WWW) has become an indispensable resource for clinical engineering students to have closer contact with industry as well as a tool to search for employment opportunities. Almost all manufacturers, vendors, and distributors of medical equipment currently use the WWW as a vehicle to disseminate information about their products. Instructors in clinical engineering programs can use it by assigning work to the students in which they need to access and process this information, for example, asking students to conduct research on a particular medical device or to analyze the market characteristics and trends of specific equipment. The use of the WWW has tremendous advantages over the traditional media of distributing technical information, such as technical literature and brochures from manufacturers. These are costly items to produce and distribute, and they must be updated frequently as new models are introduced in the market. Manufacturers might then limit the distribution of these brochures to potential buyers, while the academic world might be overlooked. Even in those cases of good cooperation with academic programs, the delays between the requests for technical literature and delivery to the students could be too long to be incorporated effectively into the academic program. By posting all of this information on the WWW, manufacturers can update their catalog of products quickly and at a minimal cost, thus making it accessible to a wider audience. Students can access all of the information (from different manufacturers) that they need for their projects

TABLE 73-1 Clinical Engineering distribution lists

List Name	Address	Notes
BIOMEDTALK	listserv@listserv.aol.com	General discussions, mostly by US professionals. Average of about 20 messages per day.
BmList	lisproc@nor.com.au	International: Australian clinical engineering.
Cmbes	majordomo@clineng.sbr.umanitoba.edu	International: Canadian professionals.
Electromedicina	electromedicina@valme.sas.junta-andalucia.es	International: For Spanish and Latin American clinical engineers. In Spanish.
SDbiomeds	SDbiomeds@yahoogroups.com	Regional: San Diego area.
biomedchat_Missouri	biomedchat_missouri@yahoogroups.com	Regional: Missouri area.
BMETs-Houston	BMETs-Houston@yahoogroups.com	Regional: Houston area.

from the classrooms, which allows clinical engineering instructors to incorporate these examples of industry uses into their academic curricula.

As the WWW technology evolves, fewer usage problems can be expected. Even though an Internet address (URL) is not permanent, the widespread use of custom domain names has solved the problem of having to remember complex URL addresses. Furthermore, these custom domain names offer an added warranty that the user need not know the address of the host where the information is stored, but only the URL for a particular manufacturer or distributor of equipment. The use of higher-bandwidth connections, even from homes, allows the possibility of incorporating stronger multimedia content without excessively increasing the amount of time needed to download a web page. This allows students to view, for example, realistic views of medical equipment, and to watch them in working action, and thus to increase the ways in which they assimilate this information. However, some of the old problems in using the WWW persist. Manufacturers of medical equipment do not acknowledge the existence of similar products from their competitors, leaving students without a complete and realistic understanding of the real market share of a specific manufacturer or the extent to which the claims of uniqueness of a particular technology made by manufacturers are shared by products from their competitors. One also must take into account that the vastness of the WWW (one of its great assets) can work against its best pedagogic use. Most of the students, especially in their first academic years, are used to working with only a limited number of resources and may feel overwhelmed by all of the possibilities of information offered on the WWW. Students also may experience problems finding and filtering information that is of interest to their projects if they are not completely familiar with the different search engines and searching strategies and the nonlinear structure of the WWW.

In addition to the technical information provided by the manufacturers and distributors of medical equipment on the WWW, the web pages from the clinical engineering departments of different hospitals offer the students a realistic view of the profession. Students and instructors can discuss the roles and mission of these departments, their structure, and the policies in use at specific facilities, and can take virtual tours of these departments to learn about the equipment that they service and the clinical staff with whom they must interact. Because the WWW is (mostly) not constricted by geographical boundaries, students also can compare ways in which different countries approach clinical engineering, and thus can get a broad vision of clinical engineering beyond what the educator might discuss in the classroom.

The WWW is also useful for introducing students to different professional organizations relating to clinical engineering. Clinical engineering instructors can use these organizations, whose URL addresses are shown in Table 73-2, as another tool to help students become closer to the reality of the profession and to understand the current issues and concerns in the field. For students who are about to graduate, the WWW has become an extremely important resource for seeking employment. Students can find job openings in different clinical engineering areas using the web pages from professional organizations, manufacturers, and independent service organizations as well as clinical engineering and biomedical engineering technology professional societies. Students can easily select the type of employment that they believe will best suit their personal preferences, while targeting specific geographic areas to which they would prefer to relocate. Faculty also can play a major role at this stage of the students' learning experience by discussing with them strategies for resume writing, interviewing, and negotiating compensation packages. In addition to these web pages, the previously discussed e-mail distribution lists also contain announcements of job openings at different levels, which contributes to an immersed learning experience.

Bridging the Gap Between Industry and Academia: Interactive Videoconferencing

Interactive videoconferencing is a useful and cost-effective way to bring students closer to clinical engineering professionals and industry. At the author's institution, interactive videoconferencing is used in the Guest Lecture program, a series of presentations given to the students by professionals from the clinical engineering field for two hours per week during one semester. This series of lectures gives the students the opportunity to learn about the different kind of industries and employment opportunities that they might consider, as well as the opportunity to learn about the current important topics in the field. These lectures, which are summarized in Table 73-3, are balanced between those with highly technical content, which gives the student an in-depth view of a specific piece of medical instrumentation, and those aimed at providing them with a breadth of perspective on the profession. The technical lectures describe in great detail the technical and functional aspects of different medical devices, basic working principles of which have already been covered in the classroom. Their purpose is to analyze the equipment in detail, taking into account its functional aspects and different operational modes. The breadth-of-perspective lectures focus on different aspects of clinical engineering that traditionally have been omitted from the formal curricular plans in academia but that will have a strong impact on their careers (Elder and Corrin, 1996).

Because the two campuses at Penn State that offer this degree are more than 300 miles apart, the guest speakers come to one of the campuses (normally the one closest to them), and their lectures are transmitted to the other campus by interactive videoconferencing. This technology allows for access to a higher number of students, while reducing the traveling costs and the inconveniences associated with traveling. The classrooms at both ends, where the lectures are originated or sent, are equipped with the necessary hardware to send and receive interactive videoconferences using a PicTel™ system. At the reception end, one TV screen shows the guest lecture being transmitted, while a second screen shows the image of the classroom that is sent to the remote end. This allows the interaction between the near-end students with the speaker. The speaker has a view of the local students, the remote-end students on the TV monitor, and a view of the signal sent to the remote end.

The assessment of this Guest Lecture program indicated that the majority of the students were satisfied with the program and that they welcomed the opportunity to interact with industry professionals. This analysis also showed that most of its perceived drawbacks were caused by the technology used. PicTel™ uses compressed video to limit the amount of information transmitted through the phone lines. The compression algorithm makes it so that, for example, fast movements in the near end appear noncontinuous at the far end, and introduces a delay between both ends. This relatively small delay (around two seconds) can appear significantly long to the students at the remote end. For example, when one of these students asked a question of the guest lecturers, he had to wait two seconds until the lecturer acknowledged this question. Because the students are used to interacting with live instructors, this delay is erroneously perceived as the guest lecturer not paying attention to the students at the remote end. Once the students at the remote end become used to the specifics of this technology, they become more comfortable and interact at a higher level with the speakers after only a few lectures.

Figure 73-1 summarizes the major benefits, as selected by the students at the remote end, resulting from incorporating the Guest Lecture series into the clinical engineering program. The students positively valued the possibility of acquiring more in-depth technical information on specific medical devices, as well as the interaction with professionals who are currently working in the field. Although all the students indicated that the program was well balanced between technical lectures and more general lectures, almost all of them showed a preference for the technical lectures, as compared to the breadth-of-perspective lectures. This last result is not surprising, as these students have chosen to

TABLE 73-2 Clinical Engineering Organizations on the WWW

Organization	URL Address	Comments
AAMI	http://www.aami.org	Association for the Advancement of Medical Instrumentation
ACCE	http://accenet.org	American College of Clinical Engineering
ASHE	http://www.ashe.org/	American Society for Health Care Engineering
ECRI	http://www.ecri.org/	Health care technology assessment
JCAHO	http://www.jcaho.org/	Joint Commission on Accreditation of Healthcare Organizations

TABLE 73-3 In-depth and breadth-of-perspective guest lectures delivered by interactive videoconferencing

Guest Lecture Program Contents	
In-Depth Technical Lectures	Breadth-Of-Perspective Lectures
Career opportunities in clinical engineering	Consequences of ineffective technical writing
Defibrillation technology	Preparing effective resumes
Clinical practice of electrosurgical unit	Interviewing and job prospects
Vital signs monitoring	Independent service organizations (ISOs) opportunities
Y2K issues in health care	Writing business letters
Blood borne pathogens – universal precautions	Developing customer relations skills
Patient lead wires	OSHA regulation: lockout/tagout procedures
Introduction to radiology	BMET and CE relationships with other health care professionals
Introduction to anesthesia gas machines	Medical device investigation

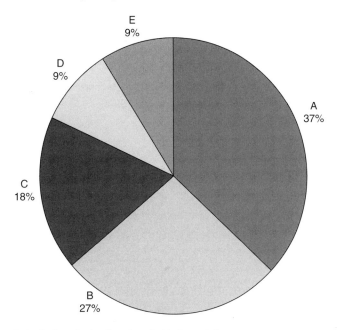

A. In-depth understanding of medical instrumentation
B. Interaction with industry professionals
C. Introduction of new equipment to students
D. Dialog with clinical engineering professionals
E. Introduction of new topics

Figure 73-1 Summary of benefits from the Guest Lecture program as perceived by students at the remote end.

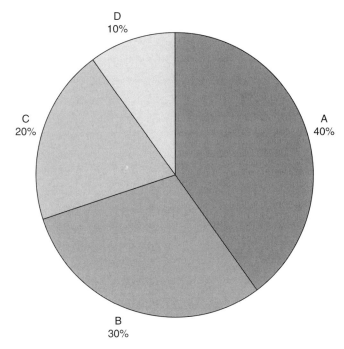

A. Lecture was balanced between information and technical content
B. Speaker had a good sense of humor
C. Speaker had good communication skills and used good audiovisual aids
D. The equipment under discussion was very interesting

Figure 73-2 Summary of good characteristics of interactive videoconference lecturers as perceived by students at the remote end.

pursue careers in a specialized and highly technical field. The major problem identified by the remote-end students was the lack of human interaction between the guest speaker and themselves. Human contact and personal interaction still is perceived as the single most important aspect of education. In this context, the speaker's personality, the interaction with both groups of students, the careful use of audiovisual tools, and especially the realization that there are two groups of students with different needs become critical to ensuring that the audience at the remote end is reached successfully.

Figure 73-2 summarizes the characteristics that made the most interesting lectures, as perceived by the students at the remote site. Figure 73-2 shows that the speaker's personality (in particular, their enthusiasm and interest in involving their audience in the topic being discussed and creating a relaxing atmosphere) is valued as one of the most critical parameters for a successful lecture. The students also positively valued loud and clear speaking, as well as the effective use of audiovisual aids. These characteristics are not new, as they are known by those who routinely speak before audiences. However, they become critical during videoconferencing because all of the other communication channels that one has with a near-site audience are absent, and they become the only human contact between the speaker and the remote audience.

Conclusions

The new goals and approaches in clinical engineering education, which are needed in order to keep up with the ever-changing professional scenario, demand the use of new technological tools that, combined with the traditional tools and approaches, can bring a new dimension to clinical engineering education, creating a learner-centered approach. This learner-centered approach shifts the traditional paradigm of instructor-to-student information delivery to one of making the students the center of their educational experience. Internet tools, as well as videoconferencing, are two of the different techniques that can be used in conjunction with more traditional approaches, to increase learning experiences and to better educate future clinical engineers.

The different Internet services constitute an exciting and promising aid in a clinical engineering educational program to enhance the students' learning experience. These tools provide a vast array of resources to complement what the students have learned during the traditional lectures and laboratory experiences. Professional distribution lists also play a unique role in enhancing their education. These lists open a window to the professional field of clinical engineering while the students are still in school, giving them the opportunity to have a sense of belonging to a distinct professional group while learning about their problems, rewards, and other issues that clinical engineers face daily. The wise incorporation of some Internet services in a clinical engineering curriculum will enhance their learning experience and will prepare them to become better professionals.

Another emerging technology with a bright future for clinical engineering education is the use of interactive videoconferencing. This approach brings industry into the classroom, allowing students to become closer to the different facets of clinical engineering. The students who have participated in the Guest Lecture program at Penn State University have recognized this fact as the most important outcome of this program. Students also have recognized that despite their preference for a certain type of lecture, there is a need to be well-versed in all of the areas that may pertain to clinical engineering. Clinical engineering educators can help students to understand this need by stressing from the beginning of the program the need for future professionals to acquire a breadth of exposure to all of these areas, which will play major roles in their future career development and possibilities for advancement as clinical engineers (Kearney, 1996).

The use of these technological tools is not exempt from potential risks. The dynamic media and the immense capacity for information storage and retrieval that drives the Internet resources can become their major drawback if faculty do not carefully design its use as an educational tool. The benefits of the use of videoconferencing systems can be reduced by the absence of human contact between the speaker and the students, as the technology does not allow all of the communication modes of face-to-face lectures. Given these limitations in communication modes, the personality and skills of the speaker to reach the remote audience become critical to ensuring that students at the remote end obtain the most benefits from the videoconference. Prospective distance speakers need to evaluate their communication techniques and to realize that the students at the remote end will see them only through a television screen. These speakers must pay constant attention to the remote-end students who are trying to engage them in interactive discussions during the entire session, in order to have a productive atmosphere.

The use of Internet resources and videoconferencing with industry professionals can become excellent tools for clinical engineering educators, as they bring students closer to the reality of the profession. Like the introduction of any other tools in the classroom, they require an adaptation period in which instructors and students need to work together and explore the ways in which these tools can be used in order to obtain the maximum benefits (Lozano, 1998b). Undoubtedly, all of these new approaches impose additional burdens on the faculty, as they must create new ways of instruction, delivery, and experimentation with these tools. However, the positive effects that these methods will have on the professional performance of graduates clearly counterbalance any inconveniences. With the aid of these technologies, clinical engineering faculty will be able to develop newer academic approaches, moving toward a learner-centered paradigm. This will allow students to be immersed in the reality of the profession at an earlier stage and, consequently, to become more interested and involved in the clinical engineering academic program, thus working synergistically to produce better professionals.

Summary

New technologies, particularly the use of Internet resources and interactive videoconferencing, can become extremely important educational tools for academic programs in clinical engineering. The roles and responsibilities of clinical engineers are undergoing major changes that must be reflected in the way that future professionals are being trained. The Internet, with its vast array of resources, can be used effectively by faculty in these programs to bring industry closer to the students. Students can use Internet resources to

References

Elder S, Corrin N. Biomedical Engineering's Role in Hospital Health Technology Assessment. *Proceedings of the 17th IEEE Eng in Med & Biol Conf*, paper 477, 1995.

Lozano-Nieto A. Internship Experiences in Biomedical Engineering Technology: An Overview of Students and Prospective Employers. *Proceedings of the 1998 ASEE Conference.* Session #1148, 1998.

Lozano-Nieto A. How Technology Teaches the Way We Teach: Benefits and Risks. *Proceedings of the IEEE International Professional Communication Conference* 75-83, 1998b.

Newport C, Elms D. Effective Engineers. *International Journal of Engineering Education* 13:325-332, 1997.

Kearny BJ. Developing High-Quality Biomedical Equipment Technicians: A Tech Prep Baccalaureate Degree. *J Clin Eng* 21(5): 402-406, 1996.

Westover AR, Moog T, Hyman WA. Human Factors Analysis of an ICU. *J Clin Eng* 23:110-116, 1998.

Xu Y, Wald A, Cappiello JC. Effective Communication and Supervision in the Biomedical Engineering Department. *J Clin Eng* 22(5): 328-334, 1977.

Zambuto RP. Health Care Trends and Clinical Engineering. *Biomedical Instrumentation and Technology* 31(3): 228-236, 1997.

74

In-Service Education

James O. Wear
Veterans Administration
North Little Rock, AR

One important role in a clinical engineering department in a hospital is to provide in-service education for the clinical staff (Dyro, 1988). The Joint Commission on Accreditation of Healthcare Organizations (JCAHO) requires continuing education for all of the hospital staff, especially in the clinical areas (see Chapter 58). This continuing education ensures that the staff maintains competency, and it helps to reduce errors and to improve patient safety.

In-Service Topics

The clinical engineering staff is in the best position to provide in-service education on topics that relate to medical equipment and to certain areas of safety. In addition to maintaining medical equipment, the clinical engineering staff should be aware of the way the equipment operates; therefore, the staff is frequently in the best position to teach people the proper use of equipment as well as any required operator maintenance. All employees must be particularly aware of the JCAHO Environment of Care standards and the medical equipment management plan, in particular. The clinical engineering department should be responsible for providing in-service training on the implementation of the medical equipment management plan to the clinical staff. Despite the fact that the electrical safety concern–one of the cornerstones upon which clinical engineering was built–has been exaggerated and is but a small part of the clinical engineer's concern (see Chapter 65), it is nevertheless advisable that the clinical staff receive electrical safety in-service training. Electrical safety remains a part of the patient safety program and remains a concern of individual employees.

Below is an outline for a typical electrical safety program for nursing personnel. Its elements should be in any course outline. The objectives of any training activity must be well defined and articulated, as they serve to keep the presentation focused. The outline, which includes the objectives, provides a direction for the presentation and lets the participants know what will be covered.

When to Conduct In-Service Training

One could say that any time is the right time to conduct in-service training, but there are times when it definitely must be provided. When new equipment arrives, there must be in-service training for the operators. If the vendor is providing this training, then the clinical engineering staff also should receive this training because they need to know how to operate the equipment that they are going to maintain. Clinical engineering staff will provide future in-service training on the equipment. The initial in-service training is critical to ensuring that the clinical staff understand how to properly use the equipment; this understanding can reduce, if not eliminate, operator error, a major factor in medical device related incidents and accidents (see Chapter 55). Thus, in-service is a fundamental component of a hospital safety program (see Chapter 58).

In-service training must be conducted for new users who are unfamiliar with the operating characteristics of a device. If the turnover rate of the clinical staff using the equipment is high, the in-service training burden increases, especially in a teaching institution where students and interns who might be using the equipment come and go with regularity. Frequently, the service in which the device is used will have knowledgeable users who are capable of providing in-service training. These services might depend on the clinical engineering staff to provide some parts of the training, however. Clinical engineering staff should welcome the opportunity to be part of the in-service training team, because knowledgeable users lead to increased patient safety and a low level of maintenance associated with user error.

When a pattern of user errors is identified, in-service training can be one way to reduce the incident of errors. The clinical engineering staff, with the aid of such tools as a computerized maintenance management system (see Chapter 36) is often in the best position to identify patterns of user error. One indicator of the need for training is a pattern of frequent calls for medical device maintenance in which no problem is found. This typically means that the users do not know how to use the equipment properly. Another symptom of user error is the recurrence of a particular maintenance problem for which there does not appear to be any explanation. Obvious equipment abuse or misuse indicates the need for in-service training. The general maintenance record on a medical device can suggest user error. User error should be considered whenever a device has a higher than expected frequency of service.

In-service training where user error is involved should be conducted in a sensitive manner so as not to polarize nursing and clinical engineering (Dyro, 1983). The user generally does not want to admit that he does not know how to use a medical device properly. This in-service training could be no more than just reinforcing for the doctor, nurse, or technician ways to use the equipment, and what might have caused an equipment failure. If conducted properly, training is accomplished without the operator losing face. In a more serious situation, such as accident investigation (see Chapter 64), the clinical engineering department might need to discuss with the supervisor of the using department the need for in-service training. In any case, the training must be done in a positive manner and must not place blame on the user for improperly utilizing the equipment.

Table 74-2 lists the considerations that the clinical engineering staff should use in establishing an in-service training program. These can be used to develop a single presentation, a series of presentations, or a complete training curriculum.

Table 74-1 In-service outline for electrical safety for nursing

Objectives:
1. Nursing staff will understand basic electrical concepts.
2. Nursing staff will be able to distinguish between microshock and macroshock.
3. Nursing staff will be able to discuss leakage current.

I. Basic electricity
   a. DC electricity
   b. AC electricity
   c. Ohm's law
II. Electricity in the hospital
III. The patient as an electrical conductor.
IV. Macroshock vs. microshock
V. Leakage current
VI. Electrical safety policy
VII. Electrical safety testing

Table 74-2 Considerations for a training program

I. Why Have a Training Program?
   A. Employees are unable to do something.
   B. Employees are unable to do something well enough.
   C. Employees are doing something wrong.
II. Prerequisites For a Training Program
   A. Determine knowledge requirements.
   B. Determine skills requirements.
   C. Appraise individual capabilities.
   D. Decide on training methods to be used.
   E. Investigate availability of appropriate training materials.
   F. Establish proper environmental conditions for training.
III. Objectives of a Training Program
   A. Provide the knowledge and skills needed.
   B. Provide knowledge that is relevant to work experience.
   C. Provide continuing education opportunities.
   D. Encourage individual effort.
   E. Remain flexible to suit varying work situations in your hospital.
IV. Questions to Be Answered Prior to Start of Training
   A. What will be the size of the training group?
   B. How frequently will sessions be held?
   C. What should be the length of each training session?
   D. Where will the training sessions be located?
   E. What time of day are study sessions to be scheduled?
   F. What are the objectives of the program?
V. Establishment of Administrative Guidelines
   A. Method of selecting trainees
   B. Specific schedule for training
   C. Scheduling work experience
   D. Reviewing or testing of the trainees' progress
   E. Certificates of completion, or formal recognition at the completion of training
VI. Review Points for Evaluation of the Training Program
   A. Make the evaluation an integral part of the training program.
   B. Let trainers and trainees know the evaluation methods and standards.
   C. Were objectives met?
   C. Was the training valid?
   D. Were the mode of the instruction and the administration of the program satisfactory?
   E. Determine trainee behavioral change–job satisfaction.
   F. Base evaluation on demonstrated performance.

How to Train

Training can be accomplished in many different ways, and the modality that is used will depend upon the trainer as well as upon the participants. The method used will also depend on whether someone is trying to teach cognitive material or skills. In Table 74-3 are some important things that a trainer should consider, to create a good learning environment for trainees. These are some of the basics of adult learning:

One method of training is by discussion–an informal method that can be a one-on-one discussion or with a small group. This method allows for the greatest amount of interaction between the instructor and the learners. In many cases, this is the least threatening training tool and is useful to train someone when training is needed. For example, when a clinical engineer (CE) or biomedical equipment technician (BMET) sees a user improperly using a medical device, proper operation can be discussed, and demonstrations made. The user is in a highly receptive mode at this point, and the chance of learning and retention is enhanced as a consequence. One could call this method "just-in-time learning."

The classical method of training is the lecture. It is particularly good for teaching groups of people cognitive material. It is a more formal method of teaching and requires time to prepare handouts, audiovisuals, and other support material. Typically, two or three hours outside of the lecture are required to prepare a one-hour lecture. For a lecture to be appropriately presented, the instructor must find ways to hold the attention of the learners. If possible, the instructor must get the learners involved with some interaction. This involvement becomes more difficult with a larger class.

Table 74-3 Some important things that a trainer needs to do to help trainees learn

1. Learners control learning; instructors control teaching.
2. Learning might not result from teaching.
3. Motivating learners is the first task of instructors (trainers).
4. Information to be learned must be organized logically.
5. Teaching must provide for learner participation response and experience.
6. Organize detail into structured patterns to increase memory retention.
7. The way a learner feels influences his or her ability to learn.
8. Learning is a perceived change in behavior.
9. Behaving and learning are products of perceiving.
10. Practice, alone, is not adequate for learning to occur.
11. Retention increases when learners can self-correct their work.
12. Varying the teaching methods improves learning.
13. Information to be learned must be presented to the learner at his or her level of sophistication.
14. Learning is increased when new information is presented in a fashion that closely simulates the real world.

One important form of in-service education is the demonstration. Because much of the in-service education provided by the clinical engineering department is the proper use of medical equipment, this is best taught by demonstrating usage. This training is actually performed where the equipment is used and may be done for one or several learners. This technique is utilized when someone is found to be misusing a medical device. In order for the instructor to demonstrate the use of the equipment, he needs to know how to use it. The CE and the BMET need to know how to use the equipment they manage. It is also a good idea, when doing the demonstrations, to have a systematic job aid to leave with the learner.

Videotapes are frequently used as part of a training program. They never should be used as the only part of a training program. The keys to any training activity are the opportunity for the learner to ask questions, and for the instructor to be sure that the concepts being taught are learned. The videotape might do an excellent job of the presentation, but it does not allow for any interaction or feedback. It can be shown and then discussed with the participants. One good technique is to use a series of short questions that cover the highlights of the videotape as a self-graded quiz. Then, the answers to the questions can be discussed. This actually provides a significant amount of reinforcement of the concepts that are being taught.

A slide and audio tape or a computerized slide presentation using such software as PowerPoint® can be used as part of a lecture presentation. Students can view the presentation on their own. However, as in the caution given concerning learning by watching videotapes, the presentation is not sufficient on its own. A well designed slide presentation can be used effectively as a support for training. Participants can go through it, take a quiz, and then discuss the material. It is possible that the discussion arising from either a computerized slide show or a videotape could occur later.

Modalities such as CD-ROMs or web-based training can be utilized, but these are typically a mix of PowerPoint® slides or video clips (see Chapter 73). If well done, they can provide some participant interaction. However, for the most effective training, the learner must have the opportunity to ask questions and to discuss the material with the instructor. With some training, such as that required by OSHA, there is a requirement that there be the opportunity at the time of the training for the participant to ask questions. This can be satisfied by having someone be available by telephone.

An examination is an important training tool. Unfortunately, most people look at an exam as a threatening situation, especially adults who have been out of school for many years. However, the exam can be used in a nonthreatening manner and does provide assurance that someone has gained some knowledge or has learned a skill. An examination can entail writing answers in response to specific questions, demonstrating the proper use of a medical device, or both. These methods provide reinforcement to the learner, who must recall the information that has been presented. Discussing the exam and the correct answers thus becomes another method of reinforcement. In general, the exam need not be graded. However, it is important for the instructor to have some feedback, in order to know whether the learners are able to recall the material presented.

In an exam in which an individual must demonstrate how to use the equipment, only a passing or failing grade is possible. It is unacceptable and unsafe for medical personnel to know how to use the equipment only 70% of the time, or to know how to use only 70% of the features. They either must know how to use the equipment or not use the equipment. This would be like having a pilot of an aircraft know how to take off, but not how to land.

All of these training techniques are in the instructor's toolbox, and many times they are used in combination with each other. The instructor must determine the best training modality for the material presented and for the type of learners being taught.

Lecture Improvement

For in-service education and training, the lecture is probably the modality that is used most of the time. Videotapes and computers might support the lecture. A discussion is really a form of lecture. Even though it is informal, it should have some of the elements of a lecture. The lecture is prepared with an outline like the example in Table 74-1 and with the considerations listed in Table 74-3. Table 74-4 lists ways to present effective lectures. One must be flexible with the presentation and must use feedback. Students become bored quickly if they already understand the material being presented. Watching the students provides feedback. Bored expressions, wandering eyes, fidgeting, doodling, and yawning are all signs that one might be able to move on to new material that would bet-

Table 74-4 Guidelines for an effective lecture

1. Do not attempt to teach too much material.
2. Focus on student performance instead of on your own.
3. Tell students your objective.
4. Tell students why the objective is important to them.
5. Make your talk personal.
6. Begin with simple, basic, or familiar tasks and move to complex, sophisticated, or unfamiliar information.
7. Break the material into short sections:
    -Ask questions.
    -Ask for judgments.
    -Let them know whether they were successful.
8. Change the pace–stop often, get feedback, and give feedback.
9. Use visual images and concrete aids whenever you can.
10. Find out how effective you were at the end, with a short post-test.
11. Ask your students to evaluate your lecture anonymously by answering interview-type questions.
12. Analyze the data you receive from the post-test and interview questions.

ter hold their interest. This will allow the instructor to meet the objectives of learning new skills and concepts. One should not be surprised if a one-hour lecture is completed in 30 minutes. On the other hand, one may find that the lecture material was above students' knowledge and skill base. In this case, additional remedial material must be covered. Extending an in-service lecture is not advisable, however, because the participants must return to their jobs. Should remedial material be added, it is preferable to cut some of the prepared material, thus ensuring timely completion and maximum comprehension.

Take time to make good visuals, such as slides, overheads, or computer graphics. A good presentation can be wasted with poor visuals. The visuals should enhance the presentation, not diminish it. The following are guidelines for creating effective slides:

1. Use horizontal format.
2. Keep text and graphics within a safe central area.
3. Use big, bold lettering, 24-point or larger.
4. Use only six lines per slide.
5. Use only six words per line.
6. Check for readability. (If displayed on a 17-inch monitor, can it be read at 17 feet?)
7. Do not use vertical lettering.
8. Present one main idea per slide.
9. Keep the focus on key points.
10. Keep the text simple.
11. Illustrate difficult concepts.
12. Use visuals to enhance text.
13. Do not use fancy backgrounds.
14. Use contrasting colors.

References

Dyro JF. Educating Equipment Users: A Responsibility of Biomedical Engineering. *Plant, Technology & Safety Management Series.* Chicago, Joint Commission on Accreditation of Health Care Organizations, 1988.

Dyro JF. Depolarizing Nursing and Biomedical Engineering. *Device Techniques* 4(1):9-10, 1983.

Ridgway M. The Great Electrical Safety Debate In Retrospect. In Dyro JF (ed). *The Handbook of Clinical Engineering.* Burlington, MA, Elsevier, 2004.

Wear JO, Levenson A. Distance Learning. In Dyro JF (ed). *The Handbook of Clinical Engineering.* Burlington, MA, Elsevier, 2004.

75

Technical Service Schools

Manny Roman
President, DITEC, Inc.
Solon, OH

Knowledge Transfer

We need to transfer information and knowledge from subject-matter experts to others, to continuously maintain and expand the body of available knowledge. In the traditional, formal education system, the "taking of students from the known to the unknown" has been performed by the "talking head" method. Students sit in a classroom and listen intently while the teacher makes the presentation. Sometimes training aids are used, but the information transfer is accomplished via the talking head at the front of the room. This method has many limitations, such as the expertise of the educator, presentation skills, limited audience, time constraints, and varying student learning abilities.

Early "distance education" methods continued to use the talking head in videotape or slides. Often, the educator was not visible, and the student saw the training aids most of the time. With the widespread availability of computers, the Internet, and the World Wide Web, present-day distance education has proliferated in all aspects of education and training.

We should point out that there is a difference between education and training. Education is broad in scope, general in nature, and intended to provide a basis for future application of the knowledge. As an example, think of the study of the production of X-rays in machines without application. Training is focused on a particular topic and is intended to provide the student with specific information regarding a task; for example, the way that a particular machine generates X-rays, or the requirements for keeping that machine in proper working order.

Distance-education techniques, specialized education software, and advanced delivery techniques have allowed educators to reach many more individuals. Training and education packages cover a myriad of topics to varying degrees of complexity, functionality, cost, and usefulness. The more familiar the student is with the subject matter, the better the chance that the package will be useful. One unfortunate effect of the ease of development and distribution of distance education is that some packages are full of "fluff," without much "meat." Some packages are nothing more than a book on a CD, without any real curriculum.

Another concern with "training packages" is that the information is presented in only one way, and no matter how many times you view it, it still says the same thing, the same way. Still another concern is that you cannot ask the package questions for clarification or understanding. Some packages will provide for some type of feedback from the student and even a forum for communication with other students and the subject-matter expert. More sophisticated, and expensive, packages contain checkpoints, quizzes, and examinations to ascertain the level of knowledge transfer.

So, is the talking head *dead?* Many feel that most education issues, and even some training issues, can be adequately addressed using distance education and training packages. Some examples are the way X-rays are generated within the X-ray tube, the way the Doppler effect is used in ultrasound, the purpose of a flood in nuclear medicine, and the way a gradient affects the MRI magnetic field. These types of knowledge do not really require the presence of the subject-matter expert or even a machine. With the proper design and delivery, a training package could present the requisite information for understanding of the concept involved. This type of knowledge is relatively easy to transfer with a package because it involves a type of memorization. The actual installation, calibration, and optimization of an X-ray tube are different matters, however.

It is appropriate here to speak of the ways in which people learn. Some people are visual learners–they learn more easily when they can see the subject matter. Others are aural learners–they learn better by hearing. The third group is kinesthetic learners–the doers, the hands-on learners. In the attempt to take people from the known to the unknown, the curriculum developer/presenter must take into consideration the particular learning characteristics of the target audience.

Our dilemma as trainers is to choose the appropriate curriculum and delivery method for the subject matter and the learner. In diagnostic imaging service training, the student is more likely to be a hands-on type of learner. Their choice of profession, and our extensive experience in training them, clearly punctuates this. Add to this the inescapable and inherent danger in servicing high-voltage, radiation-producing machines, and we come to the conclusion that a student should have live access to the machine.

The access must comprise controlled and supervised laboratory exercises and troubleshooting lessons as well as checkpoints, quizzes, exams, and the timely critique of individual progress. The student must develop a strong respect for, and the appropriate fear of, the dangers presented by these types of systems. These components of the training process cannot be provided by a distance-training package that is developed and delivered by present technology. The subject-matter expert must be available to ensure maximum transfer of knowledge and to observe the learning process. Because learning is "a relatively permanent change in behavior," the student must be monitored for the requisite behavioral pattern. The proper and efficient maintenance of these complex and dangerous machines requires a behavioral process that is best provided and supervised by the expert "talking head."

A quality curriculum design and presentation expert will provide the student with the greatest opportunity to acquire the requisite behavioral skills to fully service complex and dangerous equipment. One often-made mistake is that an organization will ask the local "service expert" to provide on-the-job training to coworkers in an attempt to save training dollars. Knowing a system does not equate to having the capabilities to teach the system properly. Good talking heads don't grow on trees.

Quality education and training of complex and dangerous equipment such as diagnostic imaging systems requires the proper mix of expertise, knowledge, curriculum design, delivery method, checkpoints, quizzes, exam, hands-on laboratory and troubleshooting exercises, instructor and student critiques and follow-ups, and, yes, a "talking head."

Training Sessions

DITEC is one of several organizations that provide training and other educational opportunities such as conferences, seminars, and publication primarily in the field of high-tech medical technologies. DITEC presents training sessions throughout the year on diagnostic imaging service and management training. Figure 75-1 is a typical schedule of programs. Modules are presented from two to seven times annually. Participants travel to the DITEC training facilities in Solon, Ohio. DITEC also takes its modules to groups throughout the United States. For example, in 2003, a program on diagnostic radiology systems was presented for the members of the New England Society of Clinical Engineering. (See Chapter 131.)

Diagnostic Imaging Continuing Education Conference

For a decade, DITEC has sponsored an annual Diagnostic Imaging Continuing Education Conference. Conference sessions typically include the following areas:

- Management
- Multivendor service
- New technology
- Regulatory and performance
- Professional development

The following is a list of the courses presented at one of these conferences.

Financial Management for Clinical Engineering

In today's environment, an understanding of financial reports, accounting systems, and reimbursements for services is important to all managers. Clinical engineering managers must have a clear understanding of the ways in which these systems work, to be able to make informed decisions on capital and operating budget development and analysis, equipment selection, variance analysis, future and present values of money, cost analysis, and other functions.

The True Cost of Lowest Price: Thinking Beyond the Current Budget

Managers often are faced with such rigid budget constraints that they are unable to think past their current budget. The focus should be on the price/cost relationship, and understanding that a lower price may mean much higher cost over the asset's lifetime, and significantly higher over the institution's life. Dynamic asset management must take many factors into account in order to provide an institution with the best solution.

Equipment Planning: Sanity vs. the Lost Opportunity

Come and see how you can develop and execute a comprehensive Equipment Planning Strategy addressing these critical components: Health care technology, capital budget, strategies procurement/installation coordination, equipment application vs. technology advancement, and building interdepartmental relations. Understand the approach for the simple process of planning current and future capital equipment expenditures.

Image Systems—Upgrades Without Fear

This session will cover interfacing and servicing the ideal CCD Camera and Image Processor combination, the technical aspects of the host equipment requirements, image intensifiers, optics, and pulsed fluoroscopy. The applications covered are for R/F, DSA, and Cine. With flat panels, image intensifiers, optics, CCD Cameras, PC-based image processors, how do I make sure my investment will not be obsolete in a few years?

Increase Your Worth Through Effective Reporting

This presentation is designed to show how a department manager needs to keep senior management aware of the effectiveness of the department through various reports. Learn some of the "dos and don'ts" of reporting that will make your report show your best side and help gain the respect of your senior management and meet the requirements of the JCAHO.

Getting the Most Life from Your CT Tubes

Hospitals are dramatically cutting their overall operating costs of CT glassware by using the patented TLS Oil Process to extend tube life. Find out how this process works and why it *now* makes financial sense to put your expensive glassware on a regular PM schedule!

Marketing Yourself in the New Century

This seminar will discuss elements to consider when seeking employment or promotions in health care management, engineering, or a technical field. Emphasis will be on the professional growth value in a health care career path. Participants will learn marketing venues within the Internet and other resources, powerful resume formats that are best for today's fast-paced environment, proven successful interviewing techniques, and acceptable attire.

Who Moved My Cheese?

Change can be a blessing or a curse, depending on your perspective. Our lives and belief systems are mostly built around our "cheese"—our jobs, our career paths, the industries we work in. We have to be alert to changes in the cheese, and be prepared to go running off in search of new sources of cheese when the cheese we have runs out. This session will discuss some of the ways we can improve our "cheese."

Defining Your Performance Indicators

This session will discuss how members of the service profession can measure their service programs with tangible measuring tools. It will help establish guidelines for customer satisfaction up to the executive level. If your organization already has indicators, this session will help you improve them.

Spiral/Helical CT Scanners

The state of the art in CT scanning is called spiral/helical scanning. The session will cover this expanding and changing modality. The emphasis will be on the clinical and technical advantages of spiral/helical CT imaging and how it enhances 3D images.

Ten Common RF Studies

Good communications with the radiologist and the technologist requires a strong understanding of the RF studies routinely performed. This session will cover the 10 most common RF procedures, with an emphasis on the technologist's viewpoint. Positioning and technique selection required for good image quality will also be discussed.

FDA Regulations on ISO and In-House Servicers

This session will discuss the new responsibilities and accountabilities of ISO and in-house servicers of medical equipment under FDA guidance and enforcement. Practical methods of using the Code of Federal Regulations to improve safety, reduce costs, and eliminate restrictions on servicing medical devices will be thoroughly discussed. An exchange of questions and comments is encouraged.

In-House Technology Management

Health care managers face a wide variety of issues while trying to operate in an increasingly cost-containment–focused environment. In-house clinical/biomedical managers face even greater challenges with all the options and threats presented by others. This session will cover the options and resources a technology manager needs to be effective in today's health care market.

Flat Panel 2001—The Litmus Test

This session will provide an overview of the flat-panel imaging devices on the market and how they are holding up under the scrutiny of the critics. It will include the various application reviews and updates on future engineering projects and a prospectus of the new designs and the latest product developments by the industry leaders. Come prepared, and bring your best imaging questions.

Networking Today's Computers

This session begins with basic networks, their topology, and their cabling. This leads into a discussion of Ethernet and token ring technology, the ISO-OSI Model, TCP/IP, DICOM 3.0,

Category	Sub	Course	Training Type
X-RAY LEVEL SERIES	DITEC'S LEVEL SERIES	FUNDAMENTALS OF SERVICING DIAGNOSTIC IMAGING SYSTEMS - LEVEL I	Service Level Training
		ADVANCED CONCEPTS OF RADIOGRAPHIC IMAGING MAINTENANCE - LEVEL II	
		ADVANCED CONCEPTS OF FLUOROSCOPIC IMAGING MAINTENANCE - LEVEL III	
		ADVANCED CONCEPTS OF DIGITAL IMAGING MAINTENANCE - LEVEL IV	
	PACS	ADVANCED CONCEPTS OF PACS, ICOM, AND TELERADIOLOGY SYSTEM MAINTENANCE - LEVEL V	
X-RAY IMAGING PRODUCTS	GENERAL ELECTRIC	ADVANTX LFX (12 PULSE) X-RAY CONTROLS MAINTENANCE	Manufacturer Specific Training
		ADVANTX MPPU (MID FREQUENCY) X-RAY CONTROLS MAINTENANCE	
		ADVANTX SCPU (HIGH FREQUENCY) X-RAY CONTROLS MAINTENANCE	
		ADVANTX R&F IMAGING, SPOTFILMER, TV CAM, LI., ABC, & TABLE MAINTENANCE	
		MVP 60/80/100 X-RAY CONTROLS MAINTENANCE	
		R&F IMAGING - (L-500) 8835 SPOTFILMER, MS89 TC CAMERA, & TABLE MAINTENANCE	
		MPX/SPX X-RAY CONTROLS MAINTENANCE	
		AMX PORTABLES (110,2,3,4) MAINTENANCE	
	SIEMENS	POLUDOROS 50/80SX X-RAY CONTROLS MAINTENANCE	
		POLYDOROS 50/80S/100S X-RAY CONTROLS MAINTENANCE	
		R&F IMAGING - EXPLORATOR ML/MR FILMER, VIDEOMED N/H CAMERS, SIRESKOP 5 TABLE	
	PHIL	SUPER 50/80/100 CP X-RAY CONTROLS MAINTENANCE	
		R&F IMAGING - SCOPO 76 FILMER, XTV11 DIAGNOST 76/66/56 TABLE MAINTENANCE	
	PICK	MTX 340/360/380/3100 X-RAY CONTROLS MAINTENANCE	
		R&F IMAGING - 1720/1721 SPOTFILMER, BETA TV CAMERA, VECTOR/ELITE 9000/4500 TABLE	
	TOSH	TOSHIBA - DC & KX0 30/50 HIGH FREQUENCY X-RAY CONTROLS MAINTENANCE	
	OEC	OEC 9000 SERIES (9000/9400/9600) C-ARM SYSTEM MAINTENANCE	
CRES	CERT	ICC - CERTIFICATION FOR RADIOLOGICAL EQUIPMENT SPECIALIST (CRES) PRETESTING	
NETWORKING	MGR	NETWORKS, DICOM, AND PACS FOR SERVICE MANAGERS AND ADMINISTRATORS	
		IMPLEMENTING AND MANAGING IN-HOUSE RADIOLOGY	
MANAGERS		PRINCIPLES OF DIAGNOSTIC IMAGING FOR MANAGERS AND SALES PROFESSIONALS	
OTHER MODALITIES	US	FUNDAMENTALS OF DIAGNOSTIC ULTRASOUND ACUSON 128XP / ATL ULTRAMARK / HP SONOS	Modality Specific Training
	NUC	FUNDAMENTALS OF NUCLEAR MEDICINE MAINTENANCE	
	LAB	FUNDAMENTALS OF SERVICING LABORATORY EQUIPMENT	
	MRI	PRINCIPLES OF SERVICING MAGNETIC RESONANCE IMAGING SYSTEMS	
	CT	PRINCIPLES OF SERVICING CT SYSTEMS	
	MAMMO	GE DMR/700/800T MAMMOGRAPHY SYSTEM MAINTENANCE	

Figure 75-1 Typical courses presented at the DITEC diagnostic imaging service and management conference.

and network software. The key issue of DICOM Conformance is highlighted. Fundamentals of network planning, implementation, and troubleshooting finish the session.

Mammography Quality Standards Act (MQSA)

The Mammography Quality Standards Act of 1992 requires that all facilities producing, processing, or interpreting mammograms have been certified by the Secretary of the Department of Health and Human Services by October 1, 1994. Failure to comply could result in civil penalties of $10,000 per day. This seminar discusses the requirements for certification.

Image Systems Evaluation (Hands-On Demo)

The fluoro-imaging chain has many variables and components that must complement each other to ensure an acceptable image. This session will present and demonstrate the techniques used by imaging experts to evaluate and optimize fluoro images. Subjects will include image optimization, resolution, image specifications, test equipment use, and common imaging problems. It will be held at the DITEC Training Center.

Strategic Planning in the New Century

The need for strategic planning in today's business environment cannot be overemphasized. As business systems become bigger and more complex, proper planning is essential for organizational survival. This seminar will cover the fundamentals of planning, as well as advanced, new techniques that can be utilized to ensure success and growth in future endeavors.

Asset Management–Data Tracking Reporting

A very important function of asset management is the data tracking and reporting piece. This session will discuss JCAHO requirements, what hospitals must do and how they must do it, as well as provide a live demonstration of an actual database. Emphasis will be on the information reports generated for hospital management.

Principles of Diagnostic Ultrasound

Due to its relatively compact design, advanced digital circuitry, and ease of maintenance, ultrasound is proving to be another source of revenue for ISOs and a cost-savings factor for in-house programs. This session is designed to introduce the skills necessary to service ultrasound with an emphasis on image analysis and operation.

Picture Archiving and Communications Systems (PACS)— Hands-On Demonstration

Advancements in electronics, communications, and computer hardware and software, as well as the high cost of film, have caused hospitals to consider filmless radiography. Picture archiving and communication systems are now available from several manufacturers. This session will introduce the fundamentals of PACS and will include a hands-on demonstration of the concepts using a PACS workstation simulation.

Mammography Physicist Testing (Hands-On Demonstration)

The current trend of regulation and accreditation in mammography requires all those who are associated with this modality to upgrade their skills and knowledge. This session will cover the physicist tests mandated by the Mammography Quality Standards Act (MQSA) to ensure the lowest dose to the patient as well as the highest-quality image. It will be conducted at the DITEC Training Center.

Industry Future—Panel Discussion

Today's environment of "health care reform," managed care, consolidation, and multi-vendor service presents challenges and opportunities never before encountered in our profession. Guest panelists from organizations that have taken on the challenges of leading the industry will guide the discussion as we head into the new century.

Pitfalls of Purchasing Refurbished Equipment

Refurbished equipment provides for considerable savings over new equipment in many cases. Some organizations claim to produce "refurbished" equipment when, in reality, the equipment is only repainted. This session will discuss how to avoid trouble when purchasing refurbished equipment by getting it in writing, getting references, visiting facilities, and determining whether it "feels right."

Principles of MRI

Magnetic resonance imaging has been used in a clinical setting for years. This technology is another opportunity for in-house groups and ISOs to expand into servicing. This session will cover the MRI principles of resonance, gradients, RF, image analysis, and algorithms, as well as an overview of the hardware needed to produce an image.

Imaging Glassware—Generation to Display

This comprehensive session will describe the fluoroscopic X-ray system from the inception of the X-ray beam to the television monitor output. This detailed description includes the X-ray tube, both glass and metal technology, image intensifier/amplifier, camera, tube or solid state, and the monitor. Monitor discussions will include everything from normal 525-line systems to 5-megapixel PACS units.

Principles of Diagnostic X-Ray Imaging

This session is intended for managers, sales professionals, and service professionals who need an introduction or refresher course in the technical aspects of radiology and fluoroscopy. Topics will include diagnostic imaging system components and factors that affect image quality.

Pulsed Fluoroscopy

The pulsed fluoroscopy technique offers advantages over conventional continuous fluoroscopy and is becoming a popular option. This session will cover the advantages of pulsed fluoro as well as the equipment changes needed to control the exposure in X-ray systems.

Digital Diagnostic Imaging

Advancements in electronics as well as their reduced price have permitted an increase in the use of computers to perform digital imaging. This session will cover the most widely used methods of producing digital images, including mask subtraction, road mapping, temporal averaging, ejection fraction, wall motion, and many other imaging algorithms.

Imaging Technology for the New Century

Medical imaging has come a long way since Roentgen's first X-ray in 1895. Armed with such techniques as MRI, CT, and ultrasound, physicians now peer inside the human body in ways that were previously done only through surgery. But even that technology is changing. DT-MRI, OCT, VCSLs, T-Scan, and a host of other techniques loom on the horizon. This seminar brings you to the leading edge of these exciting technologies.

Adding Style to Your Reports

The Microsoft Office Suite provides a lot of capability to produce good-looking and effective reports. Armed with some basic tips and tricks, you can create easy-to-read custom reports for the equipment you service.

ISO 9001:2000 VS ISO 9001:1994

ISO 9000 is sweeping the world, rapidly becoming the most important quality standard. Thousands of companies in over 100 countries have adopted it, with many more in the process of doing so. Why? Because it controls quality, it saves money, customers expect it, and competitors use it. If you are currently ISO9002/3:1994 certified, you will now need to become ISO9001:2000 certified. This seminar explains how to make the transition.

Principles of Nuclear Medicine

This session will provide introductory knowledge in the principles of servicing nuclear medicine systems. Topics will include nuclear physics, gamma detectors, pulse-height analyzers, x–y–z coordinates, flood correction, and image formation, as well as an overview of the hardware needed to produce an image.

High-Frequency Principles & Troubleshooting

During the 1980s, high-frequency KV circuits became technically feasible as the switching times of high-power SCRs got faster. The concepts of high-frequency KV and mA control have become the state of the art in today's X-ray systems. This session will cover the concepts and troubleshooting techniques needed to maintain these systems.

Non-Invasive Measurements: Past, Present, & Future

This session will discuss some of the advances in non-invasive X-ray measuring equipment. Included in this explanation are techniques associated with standard as well as specialized measurements (e.g., cine, non-invasive mA, fluoro, portables, etc.) In addition, a new method and device for focal-spot measurements will be presented.

Resource Services Program

DITEC offers the DITEC Resource Services Program (DRSP), which is aimed at servicers of diagnostic imaging technologies. Below is a listing of services provided through that program.

- Telephonic support assistance
  - Management
  - Technical
- Management information reports of DRSP usage
- Specially discounted on-site technical assistance
  - Provided by DITEC or other appropriately trained individuals
- Specially discounted management assistance
  - Provided through our network of management consultants
- Special 10% tuition discount on "closed door" on-site training by DITEC
  - For organizational and corporate members only
- 50% discounted attendance at yearly DITEC Conference
  - Technical and support services exhibits from national suppliers
  - Over 60 special seminars presented by industry leaders
  - Special panel discussions on the industry, in-house, and independent service
- One Call Parts Sourcing Assistance saves money, time, and aggravation:
  - Help to troubleshoot and identify the fault

- Help with the part identification
- Have an inventory of surplus systems and parts
- Research suppliers of the best price and quality
- Get the item for you or send you directly to the supplier
• DITEC Parts Sourcing Directory on the World Wide Web—a listing of suppliers and contacts for:
  - Equipment
  - Parts
  - Glassware
  - Other services

• DITEC INK Quarterly Newsletter
  - DITEC news
  - Industry news
  - Featured organization—featured articles—technical tips
  - FDA compliance updates
  - Special test equipment loans

Further information

www.DITECnet.com

76 Clinical Engineering and Nursing

Thomas J. Bauld, III
Biomedical Engineering Manager, Riverside Health System,
ARAMARK/Clinical Technology Services
Newport News, VA

Joseph F. Dyro
President, Biomedical Resource Group
Setauket, NY

Stephen L. Grimes
Senior Consultant and Analyst, GENTECH
Saratoga Springs, NY

Impact of Technology on Clinical Engineering and Nursing

Advanced technology has been a mixed blessing. Its positive impact on health care has been the prolongation of life and the improvement of the quality of life. It also has created immense problems, particularly by influencing illness patterns to shift from the acute to the chronic, engendering a complex medical instrumentation environment where stress, bewilderment, intimidation, and fear can abound. The expansion of technology has increased the complexity of nursing practice and has raised ethical and legal questions never before confronted (Stephens, 1992).

The resolution of most of these issues requires an interdisciplinary approach; many of these issues are of direct concern to clinical engineering and nursing, and they face the rest of the health care team as well.

Some of the consequences of advancing technology include the following:

- The failure to utilize a technological advance can increase liability exposure.
- An assessment must be performed to determine when it is appropriate to use an innovative technology.
- Technology's advances will enable more home care of the chronically ill, raising the issues of ultimate responsibility for the safety of home-use devices such as ventilators, infusion pumps, dialysis machines, and apnea monitors.
- Advancing technology may increase or decrease the incidence of medical malpractice litigation. Technology that requires excessive nursing attention fosters an impersonal climate because less time is available to spend interacting with the patient. Technology that is less nurse-dependent promotes a more personal experience, and a patient's favorable impression of his hospital stay often mitigates the legal consequences of complications.

Changing legal attitudes in response to technological advances have resulted in additional responsibilities for nurses. Recognition of the day-to-day condition of devices, the need for in-service training, and signs of incipient device failures are essential components of a sound medical device safety program required by legal and regulatory authorities. Diligence in reporting device-related incidents and accidents ultimately results in the reduction of liability exposure. Full cooperation is recommended in incident investigations. The systems approach to incident investigation identifies the contribution to the incident of the device manufacturer, the maintainer, the facility environment, the patient, the family members, or the clinical users.

Today, microprocessor-based monitors provide more information and improved accuracy. Unless this information is assimilated and appropriate action taken, nurses could be liable for rejecting it. Advancing technology places significantly more responsibility upon the nurse. Efforts to reduce liability exposure must begin with education. Education can help to minimize the likelihood of improper use of state-of-the-art medical devices (Dyro, 1983a).

Nursing Education in Technology

The lack of medical instrumentation courses in nursing curricula is often cited as an explanation for difficulties that a nurse might experience in the use of a medical device (Harton, 1982; Schultz, 1980; Shaffer, 1983; Laing, 1982). However, some educators question whether heavy emphasis on the latest technological advances is appropriate in the nurse curricula. These educators call for a softening of the technical–mechanistic approach to care, particularly in the critical care setting (Zemaites, 1982).

Nursing education in this country began with the Nightingale schools, which stressed a highly disciplined, apprenticeship approach to the education of the "trained nurse" (Notter and Spalding, 1976). Early in this century, it was recognized that the hospital-based diploma program's educational function was compromised by the hospital's manpower needs. As nurses sought autonomy, the locus of educational preparation shifted from the hospital-controlled-diploma nursing program to the collegiate-degree-granting institutions. Three educational routes exist, i.e., associate's degree, diploma, and baccalaureate in nursing, for entering nursing practice and for attaining eligibility status for registered nurse licensure examination.

In 1965, the American Nurses' Association (ANA) declared that education for those who work in nursing should take place in institutions of higher education. Professional nursing-practice education should consist of a minimum of a baccalaureate degree in nursing, and technical nursing practice education should consist of a minimum of an associate's degree in nursing (ANA, 1965). The ANA contends that diploma and associate's degree nurses should be recognized as "technical nurses" and baccalaureate degree nurses should be recognized as "professional nurses."

Technical nursing practice is the direct nursing care in areas of physical comfort and safety of patients who have health problems. The technical nurse performs nursing functions with patients who are under the supervision of a physician and/or professional nurse and assists in planning the day-to-day care of a patient and supervising other workers in the technical aspects of care. Technical nursing practice involves coordination of functions with other health services and the provision of quality nursing care under the leadership of the professional nurse.

In contrast, the professional nurse is prepared to meet the health care needs of individuals, families, and groups in any setting. The focus for health care involves emphasizing the promotion of health and prevention of illness, as well as caring for clients who are already manifesting health problems. Despite the ANA position, graduates of baccalaureate and associate's degree and diploma programs are still licensed, hired, and expected to function as professional nurses. Graduates of the three different programs must be clearly identified because their education has prepared them for different roles and levels of responsibility (Lunn, 1982).

Integral to the development of clinical proficiency in critical care nursing is the nurse's acceptance and use of technology. However, Harrington (1983) polled 19 medical, nursing, and medical technologist schools and found none that offered courses in medical device technology. A recent study of the way operating room nurses learned to use electrosurgical units showed that most received the required knowledge via instruction from a staff member (94.3%) followed by on-the-job training (85.2%) and orientation (56.6%). On the other hand, only 15.7% received instruction in nursing school (McConnell and Hilbig, 2000).

There is always intense competition for new curricula items, and the clinical topics typically outrank the medical device ones. Because few, if any, basic nursing education programs include either theory or clinical practice related to medical devices, particularly those used in critical care nursing, the burden of preparing clinicians rests with the hospital in-service training department (McConnell and Hilbig, 2000). This is accomplished through specific education strategies that include the clinical preceptor, methods for expanding the knowledge base, self-evaluation, computer-based training, and resource personnel.

The clinical preceptor is selected from a corps of seasoned clinicians who are clinically competent, caring, willing, and able to share their expertise with an orientee. Success depends on the clear identification of preceptorial functions, the support from nursing administration and middle management, the assumption by the preceptor of the teacher/counselor role, and a system to reward and recognize the preceptors for their willingness to share.

In expanding the knowledge base, it is necessary to teach the basics, such as the jargon of bioinstrumentation. Orientees require the basics, while seasoned clinicians, because they are problem-centered, are interested in troubleshooting techniques. Multimedia self-learning modules are a useful adjunct to didactic lectures. Such modules must contain objectives, a pretest, a slide, tape, or filmstrip, and a posttest. Ongoing self-evaluation relies on the identification of essential competencies. Essential members of the resource base include the clinical engineer (CE) and the instrumentation technician, who are members of the critical care team.

The clinical engineer can play an important part in nurse education in medical instrumentation. In the presentation of orientation lectures on medical device safety matters, the clinical engineer can stress the need for the new employee's participation in patient-equipment management and, at the same time, can enhance the identity of the clinical engineering profession (Shaffer, 1983).

Developing Understanding Between Clinical Engineering and Nursing

In the discussion of roles of the CE and nurse, the wide variation in position descriptions and job classifications must be recognized. In both disciplines, a wide spectrum exists. For many, the term "clinical engineer" or "biomedical engineer" means the physical plant mechanic or carpenter who carries a leakage current meter and a roll of inspection tags. Others view the clinical engineer as the designer of artificial hearts. To a large extent, the responsibility and position within the hospital hierarchy depends on the engineer's educational preparation, motivation, and interpersonal skills.

Erroneous Perceptions

The nurse and the CE usually have a clear and accurate perception of themselves (Dyro, 1983b). They have a good sense of themselves, their work, and their relationship to other health care professionals. However, others usually do not perceive these professionals as they view themselves.

The characters illustrated in these two figures are composite sketches of a nurse and a clinical engineer. The caricatures depicted are not those of any particular individuals these authors have encountered. However, they do portray one character's mind's eye view of the other when communication between the two professionals is lacking and when perceptions become distorted. A composite sketch is a drawing based upon the observations of a number of individuals who have a certain perception of a person. The artist gleans certain details from people who have encountered the subject being portrayed—details such as a sloping forehead or bushy eyebrows. There are several limiting factors in the generation of a composite sketch that prevent it from being a perfect representation of reality. Some observers have photographic memories, but many do not. Ability to recall varies from one individual to the next. The length of time over which the observer has encountered the subject and the quality of the encounter affect the resultant perception of the subject. Observers vary in their abilities to listen and to understand what they are told. If one never has an occasion to see or to talk to the subject, one may form images based upon ways that person has altered the environment. For example, one knows that there is a Sasquatch, Yeti, or Abominable Snowman because one can see the big footprints in the snow.

The two figures illustrate perceptions of individuals when an observer has little or no contact time with the person being described. The illustrator interviewed several CEs and biomedical engineering technicians (BMETs) whose exposures to nurses were brief and who based their descriptions on the results of their perceived nursing interactions with medical devices. The BMET, who had been called upon to repair a monitor that had been smashed to pieces, imagined that the nurse had great strength and possessed a stout frame and muscular arms and legs. The BMET imagined that the nurse transported physiological monitors by dragging them across the floor by the line cord because several monitors under repair had damaged strain reliefs and broken insulation at the plug-to-cord junction. A CE felt that a nurse was always equipped with a gasket scraper for fine-calibration adjustments and a generous supply of white tape (which is used to fix a vast array of medical devices) because the CE frequently observed white tape liberally applied to loose connectors and broken parts. Other imagined features included a decorative necklace of paperclips, which were used to hold things together, and hair containing an ample supply of bobby pins, which had been found jammed into switches holding down alarm-reset buttons so that nuisance alarms would not sound. Improper settings on devices led the CE to believe that the nurse had extremely poor eyesight and that she must wear very thick glasses. The CEs and BMETs interviewed assumed that nurses have great difficulty with their vision because on a number of occasions they responded to calls from nurses who reported not being able to find an ECG cable, only to find that the cable lay on the patient's bedside stand. On the basis of these observations from the CEs and the BMETs, the illustrator composed the Gorilla Nurse.

Several nurses were asked to describe the clinical engineer. They responded with a description of a troglodyte or a caveman-like person who lives somewhere deep in the bowels of the building and belongs to the biomed tribe, which includes carpenters, biomedical engineers, biomedical engineering technicians, biomedical equipment technicians, repair men, maintenance, and clinical engineers. Accordingly, the artist rendered the sketch of the Caveman Clinical Engineer. He gave the Neanderthal CE a sloping forehead. The CE's brain case is of diminished proportions because in his capacity he would not possess the higher-level intelligence required to understand the principles of hemodynamic monitoring or electrocardiograph measurement. The CE has with him several essential items, such as inspection stickers. These are all prenumbered and preapproved. The nurses observed these tags being hurriedly placed on equipment just prior to Joint Commission on Accreditation of Health care Organizations (JCAHO) inspections. The CE carries an electrical leakage current tester because, as far as the nurse is concerned, the only inspection that the CE performs is a check of the stray electrons lurking behind control knobs, ready to spring upon the unwary. He has an ample supply of tools for fixing the laundry chute, the car, the hi-fi, and the radio.

The sketches depict the nurse and CE as they are sometimes seen by each other. Effective communication between members of the two professions is a good first step toward correcting some misconceptions.

Accurate Perceptions

In 1982, the Association for the Advancement of Medical Instrumentation (AAMI) 16th Annual Meeting featured a technical session on a systems approach to medical device safety, i.e., the identification of the five elements affecting medical device malfunction: the user (often the nurse), the device (and the maintainer, often the clinical engineer), the facility, the environment, and the patient. During that session, a clinical nurse specialist, in addition to showing ways that nursing influenced patient safety, stressed the need for close cooperation between the two disciplines of nursing and biomedical engineering (AAMI, 1982). Her colorful images of the nurse "air-head" and the clinical engineer "Maytag repairman" were attempts to illustrate that members of both disciplines are often stereotyped by each other and by those outside of their respective fields (Dyro, 1983c). She showed that the images were not accurate, that the critical care nurse is technologically sophisticated, that the nurse often contributes to the advancement of medical device technology, that clinical engineers complement nurses on the critical care team, and that benefits could be derived from interaction between the two disciplines. One of the first clinical engineers the world knew was the nurse Florence Nightingale, whose establishment of hospitals to treat the sick and wounded during the Crimean War demonstrated that engineering the environment to improve sanitation and infection control reduced morbidity and mortality.

Clinical Engineer as Educator

There are few endeavors where clinical engineers can have a greater beneficial effect on health care than in their role as educators. Knowledgeable individuals are naturally limited by time and location in what they can accomplish with their knowledge. Teachers dramatically extend their influence by sharing their knowledge with others. With their understanding of technology and its application in the clinical environment, clinical engineers are in a unique position to judge the technology-related information to share, and with whom to share it, in order to make the most profound impact on the quality and efficacy of health care.

A clinical engineer's most important teaching role is in the instruction of clinicians on the proper use of medical devices. While some staff might have extensive training in the operation of specific medical devices and systems (e.g., medical imaging technicians, laboratory technicians, and respiratory therapists), most physicians and nurses operate medical devices incidentally to their primary roles. As a consequence, their expertise regarding the effective use of medical devices is often uneven. Coordinating their efforts with the organization's educational team, CEs should assess the need for staff training and should focus on the development of programs that will be most effective in meeting the organization's needs. Service histories can be an effective tool for identifying areas in which training would be beneficial. A significant number of service reports on a particular model or type of device where there was "no problem found" or "use related" problems a strong indication of the need for staff training. Other means of identifying areas where medical device training is needed include staff surveys, incident reports, and anecdotal reports.

Generally, staff training on medical devices by CEs should include proper set-up procedures, setting alarm parameters, indications and contraindications for use of the device, problems to expect during routine use, and basic troubleshooting techniques. A review of clinical application procedures by the nursing educators should cover step-by-step use of the device's various operating modes. There should be a general discussion of the indications and contraindications for use of a device and the device's effect when used in these circumstances. Finally, staff training should address relevant troubleshooting techniques that will help to ensure that the patient benefits from the best application of the device while reducing the likelihood of unnecessary clinical down time and service calls.

Clinical Engineer in Nursing Support

Clinical engineering typically supports nursing in a number of ways, several of which are listed below:

- Introduction to the use of new systems, and refresher training on existing systems
  - Indications and contraindications
  - Proper use
  - Basic troubleshooting
- Provide "help desk" services; i.e., answer technical questions regarding operation of medical devices, including use and basic troubleshooting
- Education based on analysis of "user-related" problems
- Consultation on "technology-related" issues

Customer Satisfaction

In assessing nurse satisfaction with CE support, the CE should ask the nurse the following questions:

- Do you feel that you have been adequately trained in the appropriate use of the medical technology you employ?
- Do you know whom to call with biomedical device repairs or technical issues?
- Are you generally satisfied with the quality of technical advice you receive?
- Are you confident that the medical devices you use are adequately maintained and operating properly?
- Do you know the steps to take in case an incident involving a medical device occurs?

Taking annual user-satisfaction surveys, using the same questions from year to year to track improvements, is recommended. Questions can be changed if there are new areas to explore or if prior results were ambiguous. The most recent version of the survey used at the Riverside Health System in Virginia is shown below.

On a 1 to 5 scale, please rate the performance of the Biomedical Engineering department

Scale:
5 = Very Satisfied   4 = Satisfied   3 = Fairly Satisfied   2 = Poor   1 = Very Poor   N/A

Questions:
1. Ease of contacting biomedical engineering
2. Prompt response to service requests
3. Timely completion of equipment maintenance
4. Effective communication on the status of equipment requiring service
5. Technical competence of biomedical engineering personnel
6. In-service training provided on equipment user maintenance requirements
7. Support received from your technology manager, Tom Bauld
8. Assistance with the selection of new or replacement equipment items
9. Professionalism of the staff
10. The overall effectiveness of the biomedical engineering program in meeting your needs

Following the survey process, the results are tabulated. Any questions with a score of 3 or less should be addressed with an action plan, and pertinent users contacted for follow-up.

Teamwork for Risk Reduction

Together, the CE and the nurse form a powerful team that is able to identify and eliminate potential medical device problems that could lead to patient injury. The CE is reasonably certain that when a medical device is evaluated, purchased, inspected, and introduced into use with adequate education of the user, the medical device will perform safely and adequately for its intended purpose. Nevertheless, random component failure may occur, or a weakness in design and manufacture might have gone undetected in the course of incoming acceptance testing. In addition, the clinical environment can be a rather hostile one for a device, and such an environment can be deleterious to proper operation. Device–device interaction, mechanical shock and vibration, fluid incursion, inadvertent and excessive strain, and intentional or unintentional improper use can occur. The nurse is on the frontline of device use. With appropriate knowledge, training, and experience, the nurse can detect adverse device operation before patients are injured. The nurse as problem-spotter is crucial in identifying incipient device failure and bringing this to the attention of the medical device expert, the clinical engineer. Nurses typically are well trained, observant, communicative, persistent, and determined when patient care and safety are at issue.

The Role of Nursing and CE in Technology Management

The CE's role as technology manager depends upon close cooperation of all members of the health care team, not the least of whom is the nurse. Pre-purchase device evaluation, service, utilization, incident and accident investigation, training, and eventual obsolescence are points in the life cycle of a medical device where the nursing input to clinical engineering is needed. For example, utilization of a device might reveal a human factors design flaw that might have escaped detection during incoming inspection because not all clinical conditions or device–device interactions can be predicted and simulated during pre-purchase evaluation and acceptance testing. The nurse is ideally positioned to observe such flaws that could lead to what is often described as user error. User error often results from improper ergonomic device design or lack or inadequacy of warnings and instructions. Alerting the clinical engineer to such observations can lead to equipment being designed or modified to minimize the potential for user error.

Training and education is one of the many hidden elements of the life cycle cost of medical technology (Sweeney, 1992). This cost must be factored into any new technology acquisition decision.

Incident Reporting and Investigations

Should an incident occur that can or did adversely affect a patient or anyone else, such as a nurse, doctor, or patient visitor, the prompt reporting of this incident to risk management and clinical engineering can minimize the risk of further device problems and can aid in the investigation of problems that did occur. The CE must train the nurse to follow the correct procedures when an accident occurs and a medical device is involved in the diagnosis or treatment of the patient. For example, the nurse should observe device settings; placement of the device with respect to the patient and other devices; and unusual sounds, smells, or sights, and should list the accessories used; environmental conditions (e.g., temperature, humidity, and electrical power fluctuations); other personnel in attendance at the time of the incident; and the condition of the device immediately prior to the event. The nurse must be trained to preserve evidence (e.g., disposables, leads, connectors, packaging, and package inserts). Complete and accurate documentation of the facts surrounding the incident, recorded as close as reasonably possible to the time of occurrence (attention to the injured party may delay such immediate documentation) will be more accurate than a recollection of events made a day, a week, or a month from the time of occurrence.

Collaboration on JCAHO Activities

The changing requirements for JCAHO accreditation are among the immutable constants in health care. The Environment of Care Standards have patient and staff safety as their foundation. Collaboration among many disciplines is required to meet the standards, with a matrix-management approach becoming more important. Regardless of other disciplines, the two that interact most in ensuring patient safety are nursing and clinical engineering. Working on JCAHO committees, doing environmental and safety rounds, and working together on specific equipment management tasks (such as the scheduled inspection process) all benefit from the close working relationship of staff from each department.

JCAHO Six Patient Safety Goals for 2003

To improve practice in several main areas, the JCAHO has required each institution to develop responses to initially six, and now seven, National Patient Safety Goals. The goals were prompted by Sentinel Events and the realization that the circumstances leading to the Sentinel Events probably existed widely throughout the health care system. Implementation of plans to address the majority of the goals are the responsibility of the clinical staff, but two of them require input and activity by clinical engineering. Unveiled in July 2002, hospitals had until July 1, 2003 to complete the implantation of their plans.

Goal # 1 Improve the Accuracy of Patient Identification
JCAHO Recommendations
 A Use at least two patient identifiers whenever taking blood samples or administering medications or blood products.
 B Prior to the start of any surgical or invasive procedure, conduct a final verification process, such as a "time-out," to confirm the correct patient, procedure, and site, using active, not passive, communication techniques.

Goal # 2 Improve the Effectiveness of Communication among Caregivers
JCAHO Recommendations
 A Implement a process for taking oral or telephone orders that require a verification. "Read-back" of the complete order by the person receiving the order.
 B Standardize the abbreviations, acronyms, and symbols used throughout the organization, including a list of abbreviations, acronyms, and symbols *not to use*.

Goal # 3 Improve the Safety of using High-Alert Medications
JCAHO Recommendations
 A Remove concentrated electrolytes (including, but not limited to, potassium chloride, potassium phosphate, and sodium chloride >0.9%) from patient care units.
 B Standardize and limit the number of drug concentrations available in the organization.

Goal # 4 Eliminate Wrong-Site, Wrong-Patient, and Wrong-Procedure Surgery
JCAHO Recommendations
 A Create and use a preoperative verification process, such as a checklist, to confirm that appropriate documents (e.g., medical records and imaging studies) are available.
 B Implement a process to mark the surgical site and to involve the patient in the marking process.

Goal # 5 Improve the Safety of using Infusion Pumps
JCAHO Recommendation
 A Ensure free-flow protection on all general-use and patient-controlled analgesia (PCA) intravenous infusion pumps used in the organization.

Goal # 6 Improve the Effectiveness of Clinical Alarm Systems
JCAHO Recommendations
 A Implement regular preventive maintenance and testing of alarm systems.
 B Assure that alarms are activated with appropriate settings and are sufficiently audible with respect to distances and competing noise within the unit.

Goal # 7 Reduce the Risk of Health care–Acquired Infections
JCAHO Recommendations
 A Comply with current CDC hand-hygiene guidelines.
 B Manage as Sentinel Events all identified cases of unanticipated death or major permanent loss of function associated with a health care–acquired infection.

For clinical engineers, most attention has been directed to National Patient Safety Goals #5 and #6 because they work with instrumentation that is routinely part of the institution's equipment management plan and clearly are within the nursing/clinical engineering collaboration arena. The issue of effectiveness of clinical alarms is an area that has needed attention for a long time, and one sees a wide variety of approaches being developed nationwide that will lead to improvements in patient care because of the collaboration of nurses and clinical engineering professionals.

Clinical Alarm Audits: A Research Opportunity

One way to evaluate the effectiveness of clinical alarms is to perform an audit of what occurs on a nursing unit when device alarms are activated. At the Riverside Regional Medical Center, the alarm audit has evolved into a long-term, quality improvement research project involving the School of Professional Nursing and the Clinical Engineering department. A multidisciplinary team initially proposed an audit process and developed an audit tool, which were piloted with nursing managers and staff-development educators. After consideration of the results, a lack of consistency, difficulty in tabulating data, and difficulty assigning staff resources to the task were found. Discussions with the director of the School of Professional Nursing revealed that one of the senior courses had a requirement for the students to perform an audit. The instructors were enthusiastic, as were the nurse managers. The audit tool was refined to provide data in a more measurable format so that trends of the audit outcomes could be developed. As an added benefit, the opportunity for the Clinical Engineering department to collaborate with senior nursing students provided an excellent environment in which to develop working relationships and understanding of each discipline's abilities.

Education Opportunities and Techniques

Rationale for Educational Programs in Medical Device Technologies

Experts in equipment-related injuries believe that the vast majority of incidents are due to user error, not equipment malfunction. As the complexity and sophistication of equipment grow, regular training of equipment users becomes ever more important as an element of technology management and an integral part of technology-related quality assurance and risk management. The scope of equipment training should include not only hospital personnel but also physicians and any others who regularly use equipment, e.g., residents, medical students, and allied health students.

In particular, the nurse, as operator of many sophisticated and complex medical devices, needs in-service education and training to keep pace with the rapid advances in health care technology. A strong education program is an indispensable means of maximizing the use of medical device technology within a hospital (Dyro, 1988).

What the Nurse Must Know

CEs have a unique perspective not only on the benefit of technology, but also on its risks. The use of technology in the health care environment where there is a constant staff turnover and procession of patients, many made more vulnerable because of their situation, introduces significant risks. There are risks in the ways in which each new medical device interacts with other devices and systems, with a complex environment containing a diverse range of electric, electromagnetic, radiological, thermal, chemical gas/fluid, biological, auditory, and material factors, and with people.

CEs can assist in the developing educational programs that acquaint nurses with relevant risks and can provide them with the knowledge and tools that they need in order to manage and minimize those risks. Such training should emphasize the importance of preempting problems by reporting anything out of the ordinary with a medical device (e.g., signs of physical damage, inadequate cleaning, missing components, loose controls, or unusual odors or sounds), a description of situations in which use of a device should be terminated (e.g., when alternate devices are available, when the potential risk of continuing use outweighs the benefit of continuing), and the procedure for reporting (e.g., to report service needs to clinical engineering). Training also should include appropriate precautionary steps when using a device. This typically includes a brief pre-operational physical inspection and operational check, thus ensuring a safe environment (e.g., safe electrical and gas supply systems and operating away from combustibles or devices that could interfere with proper operation).

Nurses must be trained as to what to do when a device failure or incident occurs. This should include noting circumstances of use, device settings, other devices, and accessories used. Their training should include the procedure for reporting this information, as it is critical in determining remedial actions that are necessary to prevent any reoccurrences.

Equipment users must be trained before new equipment is introduced into patient care, and they also should receive annual training in equipment use. However, the annual in-service training should *not* simply cover fundamentals of equipment use, because most of those receiving the refresher training use the equipment regularly and can be presumed to know the basics. Rather, the annual refresher should concentrate on life-support or problem-prone types of equipment, particularly problems that actually have occurred over the past year, ways in which they can be prevented, and ways in which the users can troubleshoot and resolve them.

Education material for training of nurses in medical instrumentation is available from a number of sources (Webster, 1998; Carr and Brown, 1998; Aston, 1990). The nursing literature contains some articles on medical instrumentation (Miller and Zbilut, 1983). Medical device manufacturers are also abundant sources of technical information, especially as it relates to the products that they manufacture (Sweeney, 1992).

Educational Venues

The clinical engineer is involved in nurse education in the following situations:

- New employee orientation
- Formal in-service continuing education programs
- Equipment fairs
- Informal, as-needed education during medical device problem reports and requests for service
- Training programs presented when new equipment is introduced
- Participation of clinical engineering in patient care areas, such as by clinical instrumentation specialists

Orientation

An orientation lecture by clinical engineering on medical device safety acquaints the new employee with the essentials of medical device safety and the role and function of the clinical engineering department, and ways to request service assistance. At orientation, the informal educational role of CEs is emphasized.

Formal In-Service Programs

CEs should make a multicomponent educational effort to train hospital staff in medical device operation and safety. The staff development department, together with the clinical engineering department, should develop orientation programs that consist of lectures presented by nursing staff educators and self-study modules on subjects such as ECG, cardiac-arrest response, IV therapy, and respiratory care. Training in equipment use is usually best conducted by someone from the same discipline as those being trained, such as a staff development nurse for nursing personnel or a technologist supervisor for radiology or respiratory therapy. However, CEs should participate in the development of the content, especially when problems of equipment use are to be discussed. There is an increasing emphasis on computer-based self-training, especially for a set of mandatory topics done on an annual basis to ensure competency. Clinical engineering input into these computer-based courses is beneficial.

Situational Education and Training

Most training and education is on an individual basis and occurs in answer to a nurse's request for service, a time at which the nurse has the highest level of attention and interest in the problem occurring with the medical device. Analysis of computerized medical equipment management system failure categories such as operator error and equipment abuse can influence the course of educational efforts. Focused training to particular groups or on particular device types can be given if failure analysis indicates that it is warranted.

New Equipment Introduction

During the acquisition and capital review process, the CE must ensure that all educational training tools are acquired along with newly acquired medical devices. A new device must not be placed into use until the users are adequately trained. Working with the nursing staff development and the manufacturer, the CE can aid in ensuring that an adequate education and training program that will enable all users to be properly trained is put into place.

Clinical Instrumentation Specialists

Clinical instrumentation specialists (CISs) not only deliver medical device service at the point of care, with minimal response time, but also foster improved communication between care-giver and clinical engineering support provider. A CIS group within a clinical engineering department initially formed to provide technical assistance to the open-heart surgical suite and other operating rooms expanded its activities into other critical care areas such as the recovery room, intensive care units, and the labor and delivery suites (Lauria et al., 1986). Expansion of the CIS activities improved the clinical staff's perception of the CE department and provided growth and flexibility for CE personnel involved with this program.

Many hospitals, especially those that regularly perform open-heart and neurological surgery, have a small group of technicians who are dedicated to providing technical assistance and service to the operating room (OR). This type of group typically works directly for anesthesiology or surgery and is responsible for the calibration, repair, and maintenance of OR medical instrumentation. Clinical instrumentation specialists work effectively in areas such as the recovery room, intensive care units, and the labor and delivery suites. Their typical day-to-day activities are the following:

- Providing assistance and guidance to users of medical instrumentation
- Interfacing with staff and supervisory personnel in clinical areas to discuss needs for increased user in-servicing, expediency of CE services, quality of CE services, equipment allocation problems, and overall reliability of the instrumentation
- Providing feedback to CE, staff development, and other support groups within the hospital
- Screening out unnecessary requests for equipment repair service so that BMETs are not needlessly overloaded and so that equipment is not taken out of service without good reasons
- Modifying or developing new test procedures for use by BMETs and clinical staff
- Performing pre-purchase evaluations and assisting in clinical studies involving medical instrumentation
- Assisting clinical and CE staff in performing in-house research and development

Expansion of the CIS program resulted in a more positive perception of the role of CE by the clinical staff. Users reported that equipment problems were solved more quickly and that their understanding of the equipment improved. The program effectively remedied instrumentation problems before the clinical staff lost trust in either the equipment or the CE staff.

Teaching Methods

The training must utilize formal and informal methods, taking into account principles of adult learning and instructional design. All available tools, such as videos, audiotapes, slides, books, manuals, posters, and simulators should be obtained. Pre- and post-tests are vital to the assessment of the real effectiveness of the training.

The following is a set of questions that will assist the clinical engineer and nurse staff development team in determining the extent and nature of training:

- What are the training requirements for this equipment?
- Can the suggested training program be customized?
- Who will deliver the training?
- How many users will need to be trained?
- Will a train-the-trainer method be used?
- Can clinicians be relieved from their duties during training?
- Can cost-effective personnel arrangements be made for off-shift training?
- Is there a financial commitment from management for an effective training program?
- Is there a hospital or nursing policy that mandates training for all new staff and new equipment?
- Do staff changes, technology changes, regulatory requirements, and other factors make an ongoing original equipment manufacturer (OEM) training agreement a financially sound investment?

Information about proper equipment use must be explained in a manner that those receiving the training can understand in order to be effective. Discussions between the clinical engineering representative and the staff development nurse or other experienced trainer, in advance of the training session, can be helpful in identifying the best way to approach the material to be covered. Adult learners absorb the material in a variety of ways, but the most effective method has the learners demonstrating the required tasks under the observation of the instructors. This has significant implications for the quantity of equipment and the number of instructors needed for in-services.

Information must be transmitted not only accurately but also in a way that captures the interest of the audience. The educator must seek out ways in which the presentations captivate, rather than overload, the student (Berens, 1988). Recent developments in simulation technology offer the potential to enhance training to a new level that involves not only operational features, but also assessment and prediction of the patient's clinical changes resulting from changes in equipment settings.

Audits of many clinical engineering departments have shown that the approach taken by clinical engineering personnel is crucial to the effectiveness of their training (ECRI, 1989). Those who conduct the training must be sensitive to the fact that most equipment users have little or no knowledge of electronics or engineering.

It is vital that technical staff involved in educating clinicians avoid the use of jargon and acronyms. While these forms of communication can facilitate conversations between peers, they will impede understanding when used in an interdisciplinary setting.

Indicators of Education Efficacy

Equipment-related training should be monitored and assessed regularly for its effectiveness, either by the clinical engineering department or by the individual departments conducting it. Training provided by manufacturers should be included in such reviews. Camplin (1988) recommended the following ways in which effectiveness of education can be measured: incident or accident reports, feedback from equipment tests and inspections, findings from hazard surveillance, critiques of drills and exercises, surveys, unit rounds, and inspections. Risk management, quality assurance, and safety management are all means of obtaining and processing data. Other measures include pre- and post-test and follow-up post-tests several months after the training. One objective assessment that can be monitored and trended over time as a performance indicator is the number of use-related work orders for the specific device. During CE rounds, observation of nursing practice in utilizing medical devices can yield evidence of device misapplication, misuse, or abuse.

Evaluation of Effectiveness of In-Service Training

Table 76-1 shows a training impact assessment form that is used to gauge the effectiveness of medical device training. In this form, corrective work requests are tabulated for six months. The period needs to be at least 6 and, in some cases, 12 months following the training.

Observation During CE Rounds

Some examples of improper device use that a clinical engineer might observe during rounds, indicating that nurse training, including the following, is required:

- Physical damage, missing screws, missing components, missing accessories
- Battery-operated device not plugged into AC outlet between uses
- Devices with "broken" notes taped
- Devices reported as broken when subsequent testing reveals "no problem found"

Examples of what CEs have observed in the course of rounds are shown in Figures 76-1 through 76-5. On occasion, the CE might witness improper use of medical devices.

Organizational Considerations

Administration

Those who are responsible for the administration and management of hospital affairs must ensure that full support is given to the CE department's role in education and training of clinical personnel with regard to medical device operation, utilization, and safety. Policies and procedures should describe this educational component of the CE department. The relationships with nursing staff development should be described.

Policies and Procedures

Policies and procedures that pertain to the role of clinical engineering in nurse education should include the following elements:

- Staff members must be trained prior to their initial use of medical equipment, and periodically as necessary thereafter. Training should include the following:
  - Indications and contraindications
  - Proper operation
  - Basic troubleshooting
  - Appropriate "backup" procedures when the device fails
- Procedures to follow when incidents and accidents occur involving medical devices

Clinical Engineering Department Roles and Responsibilities

The following are typical roles and responsibilities of the clinical engineering department with regard to nursing and nursing education and training in particular:

- Equipment life cycle cost: nurse (user) training is a component of cost
- Incident investigations
- Accidents: evidence preservation and sequestering techniques

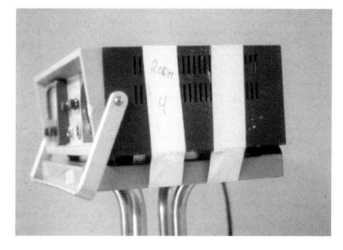

Figure 76-1 White tape, a.k.a. "nurse tape," used to fasten monitor to stand.

Table 76-1 Training impact assessment form

Device Type	Pre-training G			Post-training G			Diff		
	# WRs	Repair hours	Materials cost	# WRs	Repair hours	Materials cost	WRs	Repair hours	Materials cost
PCA pump									
Intensive care									
General care									
Labor & delivery									
TOTAL									

326 Clinical Engineering Handbook

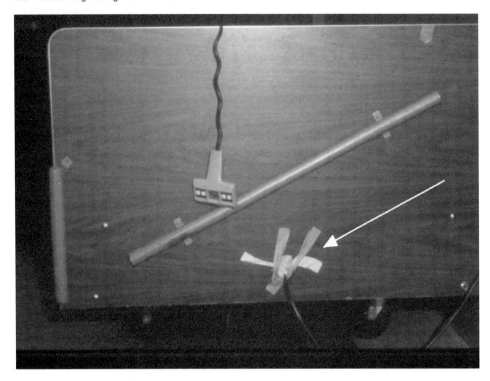

Figure 76-2 White tape used to secure ill-fitting electrical power cord to bed receptacle.

Figure 76-3 Improperly stacked monitor.

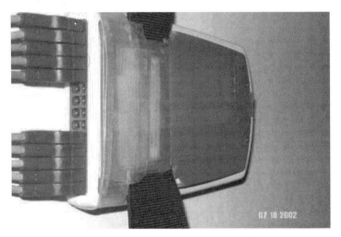

Figure 76-4 A 12-lead EKG interface secured with multiple layers of clear tape.

- Recall and hazard-report management
- Equipment selection and clinical trials
- New-staff orientation to electrical safety issues and department operations
- User training based on tracking instances of user error by device type and department
- Development of competencies for equipment use
- Risk management
- Obtaining replacement user manuals

Accountability and Appreciation

Equipment control records should include detailed records of the equipment types and the departments where training was provided, as well as documentation of persons who have

Figure 76-5 Multiple cracks observed in the plug of a nebulizer's power cord.

provided the training. In a similar fashion, nursing and other departments are responsible for maintaining records of persons who were trained on particular equipment. Computerized tracking of employee attendance and program evaluation facilitates JCAHO documentation and reporting requirements (Franklin and Landureth, 1988). Nursing policy indicates that there is a formal process to address training, and documentation demonstrates that the training has been done. However, JCAHO surveyors go beyond merely verifying that personnel files document training. They assess whether or not the clinical staff can demonstrate and use the knowledge acquired.

While the health care environment is serious and professional, the ability to celebrate milestones and recognize co-workers in their profession's special week or month is a golden opportunity to show appreciation and to build relationships. Many clinical engineering departments host an annual food event for colleagues in nursing, radiology, and the laboratory. At Riverside Health System, the CE department held an ice cream social during Laboratory Week and provides multiple flavors and all of the toppings for personnel from all three shifts at all three of our hospitals. For Radiology Week, the department provided pizza lunches. During Virginia's Biomedical Equipment Technician Week, the department hosted open houses with snacks, popcorn, slushies, and a random drawing for a dinner certificate. Each of these events was fun and enjoyable for all involved. Some of the benefits include an opportunity to have the service staff interact with the customers in a nonservice environment, a chance to show appreciation and recognition for the work of the clinical staff, and a way to have departmental or institution-wide announcements that build recognition of clinical engineering as part of the health care team.

Summary

The advancement of medical device technologies has resulted in greater numbers of complicated devices. Nurses, clinical technicians, and physicians have more effective tools for caring for the patient, while at the same they bear an increased burden of responsibility to use devices properly.

The partnership role of clinical engineers is nowhere more evident and needed than with the largest group of health care workers, the nurses. CEs and nurses are beginning to perceive each other's profession with more clarity and appreciation. The nurse perceives the clinical engineer as a skilled professional who has expertise in the design, utilization, and management of medical devices, and looks to the CE as an ally in ensuring that the proper devices are introduced into the clinical arena and that they are maintained in a safe and reliable condition throughout their life cycles. Nurses seek out and welcome the educational support provided by CEs. CEs, for their part, understand the particularly demanding work that nursing entails, and they strive to support that aspect of the work that entails the utilization of technology in support of patient care. Customer satisfaction surveys are effective in increasing the perception of nursing needs and CE responsiveness to those needs.

By close CE–nurse collaboration in many aspects of the equipment management life cycle, such as pre-purchase evaluation, postmarket surveillance, and training, a significant improvement in medical device utilization and patient safety can be achieved. Capital-equipment acquisition committees, accident/incident investigation teams, and educational programs are among those elements in the technology management process where such close teamwork is particularly beneficial.

The use of performance indicators such as repeat problems, no problem found, and user error can assist the CE in assessing the effectiveness of medical device training. Walking around and observing ways in which technology is utilized yields further insight into equipment use and enables refinement of educational programs and risk-reduction efforts.

Administrators must understand that an investment of clinical engineering time in training users of medical device technologies is a wise use of resources, and they must promote such efforts through policies and procedures and adequate financial and managerial support. By training the users to be competent in the effective use of those technologies and to report any perceived risks, incidents, or close calls associated with the utilization of equipment, clinical engineers can greatly extend their ability to improve the quality and efficacy of health care delivery.

References

American Nurses Association. *A Position Paper: Educational Preparation for Nurse Practitioners and Assistants to Nurses*. Washington, DC, ANA, 1965.

Aston R. *Principles of Biomedical Instrumentation and Measurement*, Merrill, 1990.

Bauld TJ. *Planning, Execution and Evaluation of In-Service Training Programs*. Arlington, VA, Association for the Advancement of Medical Instrumentation, 1983.

Berens W. How to Inspire Students Suffering from Information Overload. *IIE Transactions* 20(1):119–120, 1988.

Camplin CR. Education: Measuring Effectiveness. *Plant, Technology & Safety Management Series*. Chicago, Joint Commission on Accreditation of Health care Organizations, 1988.

Carr JJ, Brown JM. *Introduction to Biomedical Equipment Technology, 3rd Edition*. Upper Saddle River, NJ, Prentice Hall, 1998.

Dyro JF. The Changing Legal Implications of Advancing Technology. *Proceedings of the AAMI 18th Annual Meeting*. May 22–25, 1983.

Dyro JF. Depolarizing Nursing and Biomedical Engineering. *Device Techniques* 4(1):9–10, 1983.

Dyro JF. *Impact of Technology on Biomedical Engineering and Nursing*. Arlington, VA, Association for the Advancement of Medical Instrumentation, 1983.

Dyro JF. Educating Equipment Users: A Responsibility of Biomedical Engineering. *Plant, Technology & Safety Management Series*. Chicago, Joint Commission on Accreditation of Healthcare Organizations, 1988.

ECRI. The Transition From Equipment Control to Today's Clinical Engineering Department. *Health Technology* 3(1):10–19, 1989.

Franklin RK, Landureth L. Tracking Employee Attendance at Safety Programs: A Computerized Approach. *Plant, Technology & Safety Management Series*. Chicago, Joint Commission on Accreditation of Health care Organizations, 1988.

Harrington D. Technology for hospitals. *Med Electronics*, 6:60–83, 1983.

Harton HR. Iatrogenic Complications: The Nurse Device User. *Proceedings of the AAMI 17th Annual Meeting*. Arlington, VA, Association for the Advancement of Medical Instrumentation, 1982.

Laing G. The Impact of Technology on Nursing. *Med Instrum* 16(5):241–242, 1982.

Lauria MJ, Kintas S, Dyro JF. A Bridge Between BME and the Clinician–The CIS Program. Proceedings of the 39th Annual Conference of Engineering in Medicine and Biology, 28:294.

Lunn RL. Impact of Educational Preparation on Entry into Nursing Practice. *Med Instrum* 16(5):234–236, 1982.

McConnell EA, Hilbig J. A National Study of Perioperative Nurse Education in Two Technologies. *AORN J*, 72(2):254–264, 2000.

Notter L, Spalding EK. Professional Nursing: Foundations, Perspectives and Relationships (9th ed). Philadelphia, LB Lippincott, 1976.

Schultz JK. Nursing and Technology. *Med Instrum* 14(3):211–214, 1980.

Shaffer MJ. Applicability of Personnel Management Concepts to Clinical Engineering Practice. *Med Instrum* 17(1):34–37, 1983.

Sweeney L. Controlling Technology Costs: How Two Hospitals Save Money by Looking at the Whole Picture. *Advances*. Palo Alto, CA, Hewlett-Packard. 1992.

Webster JG. *Medical Instrumentation: Application and Design*, 3rd ed. New York, Wiley, 1998.

Zemaites M. The Development of Critical Care Nursing. *Proceedings of the AAMI 17th Annual Meeting*. Arlington, VA, Association for the Advancement of Medical Instrumentation, 1982.

Retraining Programs

James Gilchriest
Touro College School of Health Sciences
Bay Shore, NY

Program Description

Program Summary

The Institute for Biomedical Engineering and Rehabilitation Services (IBMERS) of Touro College, Bay Shore, New York, offers a Master of Arts (MA) in clinical engineering. In particular, the program attempts to create clinical engineering research as a research-oriented service provider discipline on the graduate level. The program capitalizes on manpower by building onto the Institute's presently offered graduate-level certification program in Clinical Engineering and Biomedical Technology. This program is presently retraining individuals released from the aerospace and defense industries.

IBMERS is uniquely positioned to develop clinical engineering as a discipline, inasmuch as it is a technology-driven, highly productive, biomedical engineering "think tank" with a critical mass of 34 interdisciplinary scientists and engineers who have worked together since 1986. The collective experience includes *Fortune* 500 CEOs, physician-engineers, behavioral scientist-engineers, rehabilitation specialists, computer scientists, and physicians. These individuals represent IBMERS' departments of biomedical engineering, neurocybernetics, neuroscience, behavioral sciences, instructional technology and applied bioscience, and medical humanities. All of the faculty work collectively beyond disciplinary boundaries.

The application areas for employment of graduates include hospital engineering functions such as quality assurance, equipment safety standards, and management of the equipment maintenance function, as well as intra-operative equipment monitoring and the support of highly sophisticated functions such as positron emission tomography (PET); industrial environments including equipment design, manufacturing processes in pharmaceutical and equipment manufacturing environments, FDA regulatory affairs, clinical trials, and beta-site testing; government applications including policymaking for technology, and product evaluation; and rehabilitation environments, including the design of orthotics/prosthetics, wheelchairs, and assistive aids for individual patients. Here, individual patients would consult with the clinical engineer to produce for the individual patient a less restrictive environment. The pool of students included a significant number of displaced defense workers who had more than ten years of experience and many graduate degrees among them. There are in excess of 20,000 of these individuals in the New York City area, thus allowing the opportunity for highly selective admissions standards.

The Faculty

The program faculty members are interdisciplinary active researchers and provide a broad knowledge base for the students. They have experience in the following biomedical engineering specialties: Medical imaging, brain blood flow, Alzheimer's disease, genetic engineering, nuclear medicine, neurophysiology, human information processing, neurocomputing, electrophysiology, systems theory, neuropsychology, clinical engineering, communications systems, magnetic and solid-state physics, electrodiagnosis, acoustics, neurocybernetics, neural networks, automata, medical physics, and continuum neurophysics.

Neuropsychology and Sensory Communication

Curriculum and Course Outlines

Students in the program are trained to gain as complete an understanding of engineering applied to biomedicine and health service delivery as possible. They are prepared in anatomy and physiology as applied to systems sciences and engineering; instrumentation design and equipment safety; sensors and transducers; signal processing; materials sciences; regulatory affairs; rehabilitation engineering and human factors psychology applied to biomedicine; imaging sciences; health information management systems; clinical laboratories; biomedical ethics; statistics; and research design.

The program gives students significant exposure to patent law, forensic examination, manufacturing, hospital inventory control, and FDA regulations. Students hone patient-management skills by research and service in human clinical laboratories in the institute's clinical facilities as well as in the clinical laboratories of affiliated field sites. Furthermore, students demonstrate the ability to work independently in their chosen field of research and index scientific literature. They make an original contribution of significance to the advancement of knowledge in their chosen field of application in the form of a dissertation.

During the first year of graduate study, all students enroll in a graduate statistical and research design sequence as well as in a physiology sequence. There is also a first-year research requirement that starts the student on an active program of research. The student also must enroll in a sequence of courses designed to give exposure to various topics and methods. In addition, the student must demonstrate competence in a major field of study within the institute. Before admission to candidacy for the master's degree at the end of the first year of study, the student must pass a comprehensive examination in the field of concentration and related fields. This examination tests the student's mastery of a broad field of knowledge, not merely the formal coursework.

A variety of advanced research seminars are taught on a regular basis. Upon completing the comprehensive examination, students engage in dissertation research that culminates in an oral defense.

Two opportunities are provided to pass the qualifying examinations at the end of the second and fourth semesters. Those who fail this examination, or for some other reason do not meet the assessment criteria for entrance into dissertation advisement, are not permitted to proceed further. Students may be awarded the advanced certificate in clinical engineering if they have completed the requirements for that program. The comprehensive examination is held upon completion of the first 33 credits.

Two opportunities are provided to pass both the first qualifying examination at the end of the second trimester and the comprehensive examination at the end of the final trimester. Those who fail the first examination are not permitted to proceed further.

Program Schedule

The program can be completed on a part-time or full-time basis. Full-time matriculated students require a sequence of a minimum of 9 credits per semester to maintain their full-time status for each trimester. Full-time students enrolled in an approved internship require a commitment of at least 20 hours per week for two semesters of 15 weeks each. Part-time students may complete their internship experience by completing the same number of hours as those students with full-time status (i.e., a minimum of 600 hours).

Program scheduling is in a year-round, two-year, three-semester format. Classes typically meet once per week for the required amount of time, depending on the class. Most courses are scheduled at night (some on weekends) to accommodate the working professionals in the program. The following is a sample course schedule for the Master of Arts (M.A.) in clinical engineering:

First Trimester:	Hours
IBMB 200: Introduction to Research Methods and Methodology	3
IBMN 500: Human Physiology for Engineers	4
IBME 705: Research Advisement	1
IBME 500: Biomedical Instrumentation	1
IBME 510: Microcomputer-Based Biomedical Instrumentation	1
IBMI 500: Business/Management Skills in Biomedicine I	2
	14 credits

Second Trimester:	Hours
IBMB 210: Multivariate Analysis	3
IBMN 510: Physiology for Engineers II	4
IBME 706: Research Advisement	1
IBME 540: Medical Device Safety and Design	3
IBMI 510: Business/Management Skills in Biomedicine II	2
	13 credits

First qualifying examination

Third Trimester:	Hours
IBME 550: Biomedical Engineering Laboratory	1
IBMH 500: Biomedical Ethics	3
IBME 560: Electrophysiology and Signal Analysis	3
IBME 710: Dissertation Advisement	2
IBME 750: Internship in Clinical Engineering/Technology	0
	9 credits

Fourth Trimester:

IBME 750: Independent Study	1

One of the Following Three Courses:	Hours
IBME 520: Medical Imaging Systems, or	3
IBME 530: Principles in In Vivo NMR Imaging/Spectroscopy, or	3
IBMC 540: Digital Image Processing	3

One of the Following Courses:	Hours
IBME 610: Speech Synthesis Recognition and Acoustics, or	3
IMBC 500: Computer Models of Nervous System Function, or	3
IBMC 510: Analysis of Computation/Systems in Biomedicine, or	3
a specialty course in the student's area of research interest, and	3
IBME 751: Internship in Clinical Engineering/Technology	0
IBME 710: Dissertation Advisement	2
	9 credits

Program and Course Requirements

Requirements for Admission

The following are the criteria for admission to the program:

- A baccalaureate degree in any engineering discipline from an ABET-accredited program or
- A baccalaureate degree in any applied science discipline from an accredited college or university or
- A degree from a foreign university deemed to be the equivalent of any engineering discipline as evaluated by a college-approved credentials service, such as the International Education Research Foundation, Inc.
- Submission of transcripts of undergraduate and all prior graduate work.
- Submission of three letters of recommendation, one of which must be from an individual who will address the matter of the applicant's research potential.
- A clearly written statement of professional goals and how one plans to achieve them.
- By appropriate record of achievement as well as one or more interviews, applicants must satisfy the admission committee that their preparation indicates a high potential for successful advanced study and the requisite skills necessary to work intensively with patients in great need.

Objectives

The program prepares broadly trained clinical engineering professionals who will be competent in a variety of settings and applications including hospitals, industrial design, rehabilitation, government, and the pharmaceutical industry.

Program Requirements

Courses are divided into five areas of specialization for a total minimum number of 45 program credits; these areas are progressively developed over the five levels of the program. The following is a breakdown, by area of specialization, of the required courses in the program.

1. Research Methods and Methodology
2. Systems Physiology for Engineers
3. Clinical and Biomedical Engineering
4. Medical Humanities
5. Professionalism

Research Methods and Methodology

The offerings are designed to teach skills that are useful in understanding, critiquing, designing, implementing, and applying research to help solve and understand clinical problems. (Minimum 12 credits.)

IBMB 200: Introduction to Statistics and Research Design (3 credits)
IBMB 210: Inferential and Multivariate Analyses (3 credits)
IBME 705–706: Research Advisement (1 credit each)
IBME 710–715: Dissertation Advisement (2 credit each)

Systems Physiology for Engineers

The offerings are geared toward providing a solid intellectual grounding in fundamental and applied human physiology from a systems and designers' viewpoint. Courses offered are from the Department of Neuroscience and currently are part of the IBMERS course offerings. (Minimum 8 credits)

IBMN 500: Human Physiology for Engineers I (4 credits)
IBMN 510: Human Physiology for Engineers II (4 credits)

Clinical & Biomedical Engineering (Minimum 24 credits)

The following courses are required (18 credits)
IBME 500: Biomedical Instrumentation (3 credits)
IBME 510: Microcomputer-Based Biomedical Instrumentation (1 credits)
IBME 540: Medical Device Safety and Design (3 credits)
IBME 550: Biomedical Engineering Laboratory (1 credits)
IBME 510: Microcomputer-Based Biomedical Instrumentation (1 credits)
IBMB 200: Introduction to Statistics and Research Design (3 credits)
IBME 750–752: Internship in Clinical Engineering/Technology (0 credits)
IBME 560: Electrophysiology and Signal Analysis (3 credits)
IBME 580: Mechanics of Materials in Biomedicine (3 credits)

The student must take a minimum of 3 credits from any of the following courses:

IBME 520: Medical Imaging Systems (3 credits)
IBME 530: Principles of *In Vivo* NMR Imaging/Spectroscopy
IBMC 540: Digital Image Processing (3 credits)

Students also must take a minimum of 3 credits from any one of the following courses:

IBME 600: Biomechanics of Movement and Gait (3 credits)
IBME 610: Speech Synthesis Recognition and Acoustics (3 credits)
IMBC 500: Computer Models of Nervous System Function (3 credits)
IBMC 510: Analysis of Computation/Systems in Biomedicine (3credits)
Any specialty in the student's area of research interest (3 credits.)

Medical Humanities (Minimum 3 credits)

Students are required to take the following course:
IBMH 500: Biomedical Ethics (3 credits)

Professionalism (Minimum 4 credits)

Students are required to take the following courses:

IBMI 500: Business/Management Skills in Biomedicine I (2 credits)
IBMI 510: Business/Management Skills in Biomedicine II (2 credits)

Students may elect to take the following courses:

IBMI 520: Introduction to Forensic Examination (3 credits)

Comprehensive Written Examination and Oral Presentation of Thesis

A comprehensive examination is given on some of the coursework taken during this program. The examination is required for successful completion of the program. The comprehensive examination includes anatomy and physiology, electric and electronic fundamentals, safety in health care facilities, and medical equipment problem-solving. The examination is scheduled at the end of the student's fourth trimester in the program. Upon successful completion of the coursework, the comprehensive examination, and the successful completion of the thesis, the student is awarded the degree of Master of Arts in Clinical Engineering and Biomedical Technology.

A variety of advanced research seminars are taught on a regular basis. Upon completing the comprehensive examination, students engage in dissertation research that culminates in an oral defense. The defense is on work that the student has published, either alone or with a faculty mentor in the peer-reviewed indexed scientific literature.

Course Descriptions

Department of Biomedical Engineering

IBME 500: Biomedical Instrumentation

Origins and characteristics of bioelectric signals, recording electrodes, amplifiers, chemical pressure and flow transducers, noninvasion monitoring techniques, and telemetry are covered.

(3 credits)

IBME 510: Microcomputer-Based Biomedical Instrumentation Laboratory

Provides hands-on laboratory experience with common biomedical transducers and instrumentation used in physiological and clinical evaluation. Laboratory experiments cover electronic circuit design and construction, analog/digital signal acquisition and processing, principles of hardware and software designs for interfacing biomedical sensors to general-purpose PCs. Co-requisite: IBME 500: Biomedical Instrumentation (or equivalent).

(1 credit)

IBME 520: Medical Imaging Systems

Provides an overview of the physics of medical imaging analysis. Topics covered include X-ray tubes, fluoroscopic screens, and image intensifiers; nuclear medicine; ultrasound; computer tomography; and nuclear magnetic resonance imaging. The imaging quality of each modality is described mathematically, using linear systems theory, Fourier transforms, and convolutions.

(3 credits)

IBME 530: Principles of In Vivo Nuclear Magnetic Resonance Imaging and Spectroscopy

Emphasizes the applications of Fourier-transform nuclear magnetic resonance (FTNMR) imaging and spectroscopy in medicine and biology. Course topics include review of the basic physical concepts of NMR (including Block equations); theoretical and experimental aspects of FTNMR; Theory of relaxation and relaxation mechanisms in FTNMR instrumentation for FTNMR imaging techniques (point, line, plane, and volume methods); and *in vivo* NMR spectroscopy (including volume-localization techniques). Prerequisites: Differential and integral calculus; ordinary differential equations.

(3 credits)

IBME 540: Medical Devices, Safety, and Design

Provides an introduction to biomedical engineering in the hospital. The basic theory of operations of a wide range of medical devices (from heating pads to intensive care monitors and ultrasonic diagnostic units), equipment hazards, and evaluation are stressed.
Co-requisite: IBME 550: Biomedical Engineering Laboratory; IBNM 510: Physiology for Engineers.

(3 credits)

IBME 550: Biomedical Engineering Laboratory

Covers biomedical measurements: Instrumentation analysis and calibration, frequency response of biomedical instrument sources, and reduction of noise in biopotential recordings. Laboratory work includes noise reduction in biopotential amplifiers, circuits, and damping techniques in biomedical recorders. Co-requisite: IBME 540: Mechanical Devices, Safety, and Design; IBNM 510: Physiology for Engineers I.

(1 credit)

IBME 560: Electrophysiology and Signal Analysis

The general theory of decomposition, the Fourier transform theorems, examples of singularity functions, line spectra, sampling, line systems, filters, and spectral analysis are covered. The relationship between Laplace transforms and Fourier transforms, Hilbert transforms, and analytical properties of realizable systems are also covered. Applications to EEG, evoked potentials, event-related potentials, electrocardiography, and electromyography are stressed. Prerequisite: IBME 500: Biomedical Instrumentation.

(3 credits)

IBME 580: Mechanics of Materials in Biomedicine

Materials used for implantable devices are studied. Selected topics include sheer center, asymmetrical bending, curving beams, Castigliano's Theorem, the theory of strengthening mechanisms with emphasis on dislocation theory, and physical ceramics.

(3 credits)

IBME 600: Biomechanics of Movement and Gait

Covers isokinetics, gait simulation, CAD/CAM, force velocity relationships, and biomechanical analysis techniques. Computerized cinematography, electrogoniometry, electromyography, and McDonnell Douglas B 200 systems are used.

(3 credits)

IBME 610: Speech Synthesis Recognition and Acoustics

Hardware and software aspects of speech synthesis and recognition are handled. Speech synthesis, recognition, system design, linear predictive coding, synthesis and recognition applications, acoustics, and noise reduction are covered.

(3 credits)

IBME 700: Independent Study in Biomedical Engineering

The topic is agreed upon between the engineering program student and an IBMERS faculty member. Approval of the director of IBMERS is required.

(1 credit)

IBME 750–752: Internship in Clinical Engineering

A supervised field experience for the student is provided in either a hospital engineering department, a hospital-based clinical diagnostic or treatment facility, a biomedical industry design or field-engineering application; or a private, technology-intensive, clinical group practice.

(0 credits)

In addition to taking courses in the Biomedical Engineering department, students take courses in the following departments:

- Department of Neurocybernetics
- Department of Neuroscience
- Department of Humanities in Medicine
- Department of Applied Biosciences
- Department of Biobehavioral Science
- Department of Rehabilitation Science

Department of Neurocybernetics

IBMC 500: Computer Models of Nervous System Function

Covers continuum neurophysics, artificial networks, linear and nonlinear models, bifurcation, and chaos theory as vehicles for simulating nervous system function and dysfunction.

(3 credits)

IBMC 510: Analysis of Computations and Systems in Biomedicine

Probability, combinatorics, recurrence relations, and the establishment of asymptotic bounds are studied in relation to biomedical systems.

(3 credits)

IBMC 540: Digital Image Processing

Reviews fundamental concepts of digital-image processing, analysis, and understanding, including modeling, algorithms, hardware components, and system architecture. Theoretical topics include image acquisition (sampling) and mathematical characterization, linear operations on images, image compression, enhancement and restoration, image matching and tracking, and feature extraction. Applied topics: CCD imaging and camera calibration.

(3 credits)

Department of Neuroscience

IBMN 500: Physiology for Engineers I

Defines the functions required of multicellular organisms in order to sustain life in terms of physical, chemical, and system principles. Topics covered prepare the student for more advanced work on a particular organ system. Included are diffusion; osmotic pressure; membrane potential; cellular transport; body fluid compartments; circulation; and the heart as a pump, respiration, blood-gas transport, urine formation, and acid–base balance.

(4 credits)

IBMN 510: Physiology for Engineers I

Defines the functions required of multicellular organisms in order to sustain life in terms of physical, chemical, and system principles. Topics covered prepare the student for more advanced work on a particular organ system. Detailed systems analysis is performed on the nervous system, its organization, self-organizing properties, feed-forward and feed-back systems, relax systems, disordered states, communication system principles, and tonic–sensory interactions. Prerequisite: IBMN 510: Physiology for Engineers I.

(4 credits)

Department of Humanities in Medicine

IBMH 500: Biomedical Ethics

Explores ethical issues generated by the application of scientific and technological advances to the preservation, destruction, and programming of human life. Topics covered include ethics in medical research; abortion; euthanasia; behavior control; allocation of medical resources and ethics of the patient physician interaction; and the ethics of technology application.

(3 credits)

Department of Applied Biosciences

IBMI 500: Business/Management Skills in Biomedicine I

Consists of lectures and discussions of topics essential to biomedical technology business development and interaction of practitioners with the biomedical industry. Topics include Concept identification and validation, business aspects of biomedical technology, finance, marketing, product development, good manufacturing practices, regulatory affairs, and distribution networks.

(2 credits)

IBMI 510: Business/Management an Skills in Biomedicine II

Instruction and student participation in interactive setting. Practical application and development in oral and written communication skills required in business and management in the biomedical/hospital industry. Topics include interpersonal communication, sales, formal and informal presentations of technology, interdepartmental relations, public relations, advertising, correspondence, and business and budget planning.

(2 credits)

IBMI 520: Introduction to Forensic Examination

Survey of the applications of clinical engineering, biobehavioral science, neurosciences, and biomechanics to forensic examination. Topics include the nature and analysis of slip-and-fall injuries, pain management, product design and liability, disability estimation, analysis of residual work function after injury, accident reconstruction, and the process of giving court testimony.

(3 credits)

Department of Biobehavioral Science

IBMB 200: Introduction to Research Methods and Methodology

Three hours of lectures and one two-hour lab per week. Introduces descriptive and inferential statistics and their roles in clinical and technical practice and evaluation. Topics include correlation, regression, t-test, Chi square, ANOVA, nonparametric methods, sample-size estimation, and probability. Also included are topics in experimental design and correlation analysis, including multiple correlation and multiple regression, curvilinear relationships, missing data, interactions, ANOVA and its generalization, logistical regression, selected complex factorial designs, multiple comparisons, and meta-analyses; causal models, construction estimation, and testing of causal models for correlational data. Particular attention is paid to models with unobserved variables.

(3 credits)

IBMB 210: Multivariate Analysis

Scientific concepts, matrix theory, and computer techniques of multivariate analyses for clinical research and evaluation. Topics include cluster and factor analyses, multiple regression, and discriminant functions. Emphasizes research and clinical technology rather than mathematical theory. Prerequisite: IBMB 200.

(3 credits)

Coursework

Program Literature

The majority of literature in the clinical engineering master's degree programs comes in the form of handouts. The handouts are from a variety of sources that include textbooks, scientific journals, product literature, government publications, industry magazines, the Internet, society meetings, and newsletters. While most courses include handouts, the classes that utilize the most handout material include IBME 510: Microcomputer Based Biomedical Instrumentation; IBMH 500: Biomedical Ethics; IBMI 500 and 510: Business/Management Skills in Biomedicine I and II; IBME 540: Medical Devices, Safety and Design; IBMB 200: Introduction to Research Methods and Methodology; IBME 500: Biomedical Instrumentation; IBME 550: Biomedical Engineering Laboratory; IBME 580: Mechanics of Materials in Biomedicine; IBMB 210: Multivariate Analysis; and IBMI 520: Introduction to Forensic Examination.

The following is a sample of some of the textbooks that are required for the clinical engineering master's degree program:

Bronzino JD, editor. *The Biomedical Engineering Handbook*. The *Handbook* is utilized throughout the program as a general reference text or to investigate a specific area of biomedical specialization. Classroom lessons, projects, and homework from the majority of courses listed in this program rely heavily on the *Biomedical Engineering Handbook* for source and reference material.

Curry TS, Dowdey JE, Murry RC. *Christen's Physics of Diagnostic Radiology, 4th Edition*. This textbook is used for a number of courses but mainly for IBME 520: Medical Imaging Systems.

Webster JG, editor. *Medical Instrumentation Application and Design, 2nd Edition*. This textbook is used primarily for IBME 500: Biomedical Instrumentation; IBME 540: Medical Devices, Safety and Design; and IBME 560: Electrophysiological Signal Analysis.

Guyton CA and Hall JE. *Textbook of Medical Physiology, 9th edition*. This textbook is used mainly for IBMN 500: Human Physiology for Engineers I and IBMN 510: Human Physiology for Engineers II.

Tompkins WJ, editor. *Biomedical Digital Signal Processing*. This textbook is used for the course IBMC 540: Digital Image Processing.

Nordin M and Frankel VH. *Basic Biomechanics of the Musculoskeletal System, 2nd edition*. The textbook is used in the course IBME 600: Biomechanics of Movement and Gait.

Computer Hardware and Software

Touro College provides a computer laboratory to all students. Computer software applications are used for the following courses.

IBME 510: Microcomputer Based Biomedical Instrumentation uses the Dataq data-acquisition hardware and software; IBMC 540: Digital Image Processing uses software that comes with the class textbook; IBMB 200: Introduction to Research Methods and Methodology and IBMB 210: Multivariate Analysis both use Systat Statistical Software, and C++ and Visual Basic software are reviewed. All courses require work to be performed with a computer using word-processing programs, spreadsheet programs, and presentation programs, and using the Internet for literature and subject searches.

Class Lessons and Projects

In general, most lecture courses come in the format of lectures, textbook and handout work, class discussions, assigned homework, and class projects. The laboratory courses are hands-on. For the courses IBME 500: Biomedical Instrumentation; IBMI 520: Introduction to Forensic Examination; and IBME 550: Biomedical Engineering Laboratory, the instructor brought in many medical devices for examination and evaluation. Medical devices are operated on simulators. Defibrillators, electrosurgical units, incubators, wheelchairs, pulse oximeters, heart rate and respiration monitors, endoscopes, and orthopedic implants are among those devices utilized in the laboratory setting.

Class projects take the form of hands-on diagnostics and operation of medical devices, oral presentations, and written projects. All projects and written reports are geared toward "real-world" applications, requiring scientific and industrial literature searching, acquiring product literature, and sometimes product samples. In addition, multiple hospital-site visits, society meetings, and guest speakers help to introduce students to the "real world" outside the classroom.

Present Program Status

The following dissertations were written by students who have been awarded the MA in Clinical Engineering:

Robert C. Counts Sr., "A comparative study of an annual quality control report as it relates to computer aided tomography vs. digital fluoroscopy," November 2000.

James A. Gilchriest, "Orthokinetic sensor cuff for the precision measurement of hand flexion; proof of concept study," June 2001.

Masha Lorenz, "The effect of the accessibility of recent data through the Internet on the time an article is accepted for publication in scientific journals," July 2001.

William McConlogue, "Proof of theory, new methods for high speed sorting of solid oral dosage forms by weight," December 2002.

The following programs are currently offered at Touro College:

Clinical Engineering and Biomedical Technology at Touro College School of Health Sciences
  Certificate programs
    Biomedical equipment technician (undergraduate)
    Clinical engineering (postgraduate)
  Degree programs
    Master of Science in clinical engineering and biomedical technology
      Track 1: Clinical Engineering—Degree awarded: Master of Science in clinical engineering and biomedical technology
      Track 2: Biotechnology and Bioprocess Engineering—Degree awarded: Master of Science in clinical engineering and biomedical technology (bioprocess engineering track)

78
Techno-Bio-Psycho-Socio-Medical Approach to Health Care

T. G. Krishnamurthy
Chairman, Paramedical Discipline Board, CEP-AICTE,
UVCE Campus
Bangalore, India

Clinical Engineering Practices in Developing Nations

It is a universally known fact that two-thirds of the world's population lives in developing and underdeveloped nations. The affluence level in these nations is around 15 percent. About 80% live in semi-urban and rural areas. Socioeconomic, traditional, and cultural parameters effect the preventive, diagnostic, and curative aspects of medical/rehabilitation (medi/rehab) care to a large extent. Developing strategies in a realistic, feasible, need-based, and time-bound plan is necessary. Optimization of the large manpower pool (training and untrained), reasonable material resources, and very limited funds appear inevitable and essential.

The challenge is an integrated interdisciplinary center involving a techno-bio-psycho-socio-medical teamwork approach (Taylor, 1994). Urban, semi-urban, and rural needs differ widely with regard to preventive health aspects. One gets a glimpse of the magnitude of the task only when one actually visits the rural and semi-urban areas apart from the slums and dwellings in the urban areas. The education and training of various levels of medical, paramedical, and nonmedical personnel involving medi/rehab care programs is a formidable task. Inspection, approval, and accreditation of various levels of centers are intricate tasks. The authorities concerned must have a clear concept of the basic infrastructural requirements for establishing these centers.

It is not difficult to regulate, standardize, and accredit at higher levels, namely, doctoral, master's, graduate, and diploma levels. However, in these underdeveloped and developing nations, the human resources potential of the less affluent, who form 80% of the population, poses the challenge to those who provide leadership in integrated interdisciplinary areas like clinical engineering, rehabilitation engineering, health-rehabilitation care (HRC) technology, and paramedical education. Many intelligent teenage boys and girls living in the rural and semi-urban areas after their 10 or 10+2 courses cannot afford migration to cities and towns to study in hostels and continue education. Many of them have aptitude and are intelligent. Various levels and types of necessary paramedical personnel have been identified. Some of these can be located in the semi-urban and rural areas to the benefit of humanity and society. In the last decade, attempts have been made to regulate the education-training aspects of this cadre in a planned way by regulating such items as the curricula, course duration, and fee structure.

The Need for Health Sciences Technology Education and Training

- Human resources potential must be developed based on the needs in urban, semi-urban, and rural areas.
- Socioeconomically backward children who are talented, after their 10 or 10+2 courses, must get an opportunity to nurture and harness their potential based on their aptitude.
- It becomes mandatory in all democratic nations to provide education and training at reasonable costs.
- It is most desirable to locate these centers in the semi-urban and rural areas as far as is feasible.
- Considering the affluence of 10–15%, medical engineering education is beyond the reach of the majority of talented youth. Even merit seats are not accessible in view of the escalating fees.
- Capitation/management seats in engineering, medicine, nursing, pharmacy, and dental are almost exclusively meant for the very highly affluent.
- Considering the sad and pathetic situation after more than 50 years of Indian independence, strategies must be evolved to improve the knowledge base and placement potential of those who are less fortunate.
- Health sciences technology education and training at the graduate level (6 semesters + 2 of internship) as initiated by the Sri Ramachandra Medical College and Research Institute (SRMC & RI) at Chennai could serve as a model for similar programs at selected medical institutions all over the nation. Such a course is within the reach of some in the middle-income group.

Role of Clinical Engineering Centers

Clinical engineering centers throughout the nation share the goals listed below:

- To train various levels of personnel to operate, maintain, and repair all types of hospital equipment
- To encourage innovation, initiate design, and fabricate simple-use diagnostic, therapeutic, and bio-analytical devices and rehabilitation devices and appliances
- To enlighten and educate concerned medical, paramedical, and nonmedical personnel about various systems of medicine and techniques
- To initiate planned scientific steps to optimize manpower material utilization utilizing the latest management techniques
- To promote closer interdisciplinary interaction among clinical, biomedical, paramedical, and nonmedical personnel by periodic seminars and short-term courses in various clinical areas
- To maintain close relations and strong interaction with the Clinical Engineering Division of the International Federation for Medical and Biological Engineering (CED IFMBE)
- To initiate and introduce concepts of health care and rehabilitation care technology planning, coordination, management, and administration, and to attempt their application at all levels of hospitals and medical institutions under governmental and voluntary agencies
- To set up the National Clinical Engineering Certification Program in association with the All India Council for Technical Education–Continuing Education Programme (AICTE-CEP) in Bangalore
- To plan and organize awareness programs at various centers all over the country with the cooperation of local medical, paramedical, and nonmedical personnel under the guidance of biomedical specialists sponsored jointly by AICTE-CEP and the Clinical Engineering Centre (CEC) in Bangalore
- To promote joint projects with other global nationals in the fields of clinical and rehabilitation engineering, and sponsor younger colleagues to update their knowledge
- To promote wider utilization of the wide range of useful therapeutic techniques whose clinical efficacy is proven
- To help set up feasible and economically viable integrated therapy centers in the rural and semi-urban areas under governmental and voluntary agencies
- To help improve the quality of care at all levels using the latest interdisciplinary advancements, in cooperation with state Medicare authorities

Technology Management in Health Care Delivery Systems

A dynamic, innovative, interdisciplinary, integrated-planning, coordination, management, and administration approach is essential, in order to manage the health care delivery systems in a cost-effective manner in the urban, semi-urban, and rural areas. Improving the quality of medical rehabilitation care of 1020 million people with 100 million handicapped and 50 million elderly along need-based, time-sensitive, realistic, feasible lines poses a formidable techno-bio-psycho-socio-medical challenge to all who are genuinely committed (see Table 78-1).

Obviously, in a unique, multilingual, multireligious nation like India, it involves very close interaction among specialists of various disciplines in the fields of medicine, engineering, sciences, and technology. The actual needs in the urban, semi-urban, and rural areas must be evaluated depending on the local situation. A wide range of parameters must be considered in order to optimize the utilization of the large manpower pool, reasonable material resources, and limited funds. Human and medical ethical aspects must be considered while adapting automation and semi-automation techniques. Humanism and optimism are to be integrated with action. There is an urgent need to evaluate in a planned scientific way the clinical diagnostic and therapeutic potential of various systems of medicine and the wide range of techniques. As part of continuing medical

Table 78-1 Health rehabilitation care data for India

Population	1020 million
Handicapped	100 million
Elderly	50 million
Medical colleges (attached hospitals with 500-1000 beds)	175
Major hospitals (500 and above, up to 2000 beds)	1500
Referral centers of clinical excellence (100–500 beds)	200
Specialist medical institutions	100
District level hospitals (150–300 beds)	550
Rural hospitals/community health centers (30–100 beds)	2,500
Primary health centers (5–10 beds)	30,000
Community health sub-centers	150,000
Bed-to-population ratio	1:1300
Total number of hospital beds	800,000

education (CME) programs, awareness workshops, short- and long-term training and education, refresher courses must be planned and organized in medical colleges, hospitals (district level upward), and in centers of clinical excellence. Medical, paramedical, and nonmedical personnel at various levels must be exposed to these programs in order to update their knowledge base and to improve their placement opportunities. Funds alone can never improve the quality of health care delivery systems. Dedication and commitment, coupled with optimism and humanism, are the basic fundamental prerequisites to improve the quality of care at various levels. Some of the topic areas are identified in Table 78-2. These topics deserve consideration for integration into awareness/training programs.

The devices and systems listed in Table 78-3 integrate low-cost, medium-cost, high-cost, and low-tech, medium-tech, and high-tech equipment and systems used in day-to-day work in all levels of hospitals (i.e., from primary health centers to centers of clinical excellence). The above exclude high-cost, high-tech systems used in medical research and development.

Optimizing Equipment Utilization

Optimizing the utilization of medical devices requires an understanding of the principles of operation and the underlying medical condition being diagnosed or treated. Table 78-4 lists the subjects covered in the course *Optimizing Equipment Utilization in Medical Institutions*.

Paramedical Education and Training

Overview of Centers of Certificate Programs

1. The prerequisite for one-year certificate programs is Pass in 10 or 10 + 2 who have aptitude.
2. The majority cannot afford medical, engineering, dental, nursing, pharmacy, and physiotherapy courses.
3. The development of human resource potential is a challenge to all who are involved in this significant task.
4. Inspection, approval, and accreditation aspects have been regulated and standardized for higher levels, namely doctoral, master's, degree, and diploma.
5. The cadre of clinical assistants/health care technicians form the second line to assist those such as trained nurses, physiotherapists, ECG and X-ray technicians.
6. This cadre is meant for those students who are intelligent and show aptitude, but who are not affluent enough to enroll in higher-level courses.
7. Development of this cadre is bound to improve the quality of medical/rehabilitation care all over, especially in the semi-urban and rural areas, making it cost- effective and accessible to the nonaffluent millions.
8. Certain minimal infrastructure in the form of classrooms libraries, laboratories, and technical staff (both full-time and part-time) are needed. These centers can be graded depending on such factors as location and training facilities.

Table 78-2 Health care technology management topics

Health care delivery systems development	Health care delivery systems management
Predictive, preventive, curative maintenance aspects	Safety, standardization, calibration
Hospital communication systems	Critical care areas management
Air- and water-quality monitoring	Blood bank and clinical laboratory services
Imaging-techniques systems	Trauma-center management
Organ-transplant techniques management	Computers in medicine
Impact of internet and multimedia	Yoga, naturopathy, meditation
Integrated-therapy approach	Dental-care systems
Energy management and audit	Hospital waste management and disposal
Essential hospital engineering services	Human resources management
Basic hygiene and cleanliness	Prevention of infection
Rehab management of elderly and handicapped	Mobility, transport, accessibility aspects
Human engineering management	Intra-ocular lens implantation
Holistic systems management	

Table 78-3 Equipment and systems routinely used in hospitals for medical diagnosis, analysis, and therapy

Diagnosis	Bio-analysis	Therapy	General purpose
Electrocardiographs	Microscopes	Pacemakers	Cold chain
Electromyographs	Centrifuges	Defibrillators	Refrigerators
Electroencephalographs	Autoclaves	Ventilators	Sterilizers
Audiometers	Incubators	Heart-lung machines	Heaters
X-ray units	P.E. Calorimeters	Resuscitators	A.C. Units
Sonographs	Flame photometers	Stimulators	Hospital communication systems
CAT Scan	Spectrophotometer	X-ray units	UPS systems
MRI Scan	pH Meter	Cobalt therapy unit	RP systems
Blood pressure apparatus	Electrophoresis	Linear accelerator	Emergency
Stethoscope	Chromatographs	Physiotherapy	Generators
ICU Setup	Gas analyzers	Dialysis units	General and OT
CCU Setup	Ovens	Hearing aids	Lighting
Polygraphs	Water quality	Speech correction	Fans
Transducers	Analyzers	Devices	Computers
Recorders	Auto analyzers	Ortho-Pros implants	
CCTv	Tissue processors	IO Lenses	
Catheterization lab	Radiation safety	Anesthesia apparatus	
Bio-telemetry	Setup	Radio isotopes	
Endoscopes	Glassware	Behavior therapy	
Computerized TO	Reagents	Electronarcosis	
Diagnostic systems	Blood banks system	Traction units	
		Rehab. aids and appliances	

9. About 60 of the 125 centers are located in Kerala, where missionaries pioneered paramedical education in Asia. Some of them have been functioning for the last two to three decades. Most of those trained are now employed all over the world.
10. Any new center needs a gestation period of at least 3–5 years to overcome the initial teething difficulties, bottlenecks, handicaps, barriers, and hurdles for its survival and eventual development.
11. There is an urgent need to regulate and standardize the duration, curricula, fee structure, staff pattern, and laboratory facilities. Thus, under AICTE-CEP, Ministry of Human Resource Development, the Paramedical Council of India should be instituted along the lines of the Nursing Council, Pharmacy Council, Dental Council, Rehab Council, and Medical Council for ensuring quality of education and training.
12. The twin purpose of improving the knowledge base and enhancing placement potential can be achieved for all of those younger colleagues who are not able to afford higher levels of education.

Interdisciplinary Approach

Interdisciplinary areas involve a close integration of a wide range of diverse disciplines. One outstanding example is the field of health rehabilitation care technology, which encompasses biomedical engineering, clinical engineering, rehab engineering, and complimentary medical systems and techniques. Training of competent faculty required for the certificate, diploma, degree, post-graduate, and doctoral levels is a formidable challenge. Various branches in the fields of medicine, engineering, science, technology, sociology, and psychology are integrated to meet the preventive, diagnostic, therapeutic, and research aspects that pertain to the well-being of the individual. Safety, standardization, calibration, operation, and maintenance are parameters that must receive appropriate attention. Yoga, naturopathy, and meditation play a vital role in augmenting the efficiency of various systems of medicine and a wide range of techniques utilized to relieve pain and suffering. Treating elderly and handicapped patients involves mobility, transport, and accessibility aspects. Dynamic leadership is essential to conceiving and evolving realistic and feasible programs to suit the needs of the medical fraternity at various levels. One can visualize the teamwork that is required in order to achieve time-sensitive targets. Approval and accreditation of these need an unbiased approach to ensure quality care at all levels. Courses must be developed in accordance with the needs of their participants.

Table 78-4 Optimizing Equipment Utilization in Medical Institutions course content

Basics of medicine–clinical anatomy	Maintenance aspects
Basics of medicine–clinical physiology	Safety aspects
Basics of medicine–clinical biochemistry	Standardization/calibration aspects
Medical diagnostics	Anesthetic techniques
Bioanalytical techniques (e.g., blood bank)	Imaging techniques
Medical therapeutics	Health rehabilitation care technology
Cardiological techniques	Clinical engineering overview
Neurological techniques	Institutional wastes disposal
Nephrological techniques	Institutional energy management
Oncological techniques	Refrigeration and air conditioning
Otolaryngological techniques	Institutional communication systems management
Prosthetics–orthotics	Operation theater planning
Biomaterials and biotransducers	

Table 78-5 Certificate-level technician courses at the All India Council for Technical Education–Continuing Education Programme in Bangalore (AICTE-CEP)

Medical laboratory technology	Clinical engineering
Physiotherapy	Rehabilitation engineering
Integrated therapy	Anesthesia
Blood bank	Hospital waste management
Imaging technology	Energy management
Dialysis	Communication systems
Operating theater	Computer operation
Electrocardiography	Biomaterials
Electroencephalography	Sonography
Ophthalmology	Medical transcription
	Audiometry

In the final lap of any interdisciplinary program, due weight must be given to the aptitude of the trainee, namely whether the trainee opts for (1) research; (2) development; (3) operation and maintenance; or (4) safety standardization and calibration. It is obvious that all of these involve integration of CME, continuing education programs (CEP), and quality-improvement program (QIP) aspects. It will be impossible for training institutions to equip with a very wide range of diagnostic and therapeutic systems. They will need to find professional support from medical institutions and hospitals to expose trainees to the day-to-day use of equipment and systems. The faculty must be unique, service-minded, humanistic, and optimistic in their approach. In addition to regular programs, various levels of workshops, seminars, and symposia must be organized in medical institutions and hospitals, to update knowledge. Improving quality of care involves such topics as waste-management, safe blood supply, energy audit, and accident emergencies. This note highlights salient features involved in developing curricula for courses of various levels.

Health Scenario

India is the largest democratic nation in the world. Its population exceeds one billion and includes 100 million handicapped and 50 million over age 62. India is also unique because it integrates a multilingual, multi-religious population. Affluence is around 10–15%, and nearly 80% live in the rural areas and in slums and dwellings located in the urban and semi-urban areas. One can visualize the major challenge facing all of those genuinely involved interdisciplinary intellectuals who obviously must provide the much-needed innovative initiative and dynamic leadership under such circumstances. It is bound to be a feasible, realistic, need-based, and time-sensitive approach involving optimization of the large pool of trained and untrained manpower. Reasonable material resources, and very limited monetary resources, are available for health rehabilitation care programs. A techno-bio-psycho-socio-medical teamwork approach appears mandatory if the quality of health rehabilitation care is to be improved at all levels, especially in rural areas, where the majority of the socioeconomically deprived persons live.

Health Care Technique

Preventive, diagnostic, analytical, and curative aspects of medi/rehab care involve a wide range of techniques that, in turn, involve various levels of instrumentation. The recurring and nonrecurring costs involved depend on the level and quantity of the high-tech, medium-tech, and low-tech instrumentation required. A wide range of parameters like location, manpower, material facilities availability, disease pattern, and required funds play decisive parts. Health/medicare delivery is the responsibility of concerned state governments, ably augmented by voluntary organizations. Broadly, a three-tiered approach starting from the grassroots level, i.e., primary health center through district-level hospitals to major hospitals or specialist referral centers. Complementary systems of medicine, namely Ayurveda, Unani, homeopathy, Siddha, and a wide range of techniques under energy medicine, including yoga, meditation, and naturopathy, are slowly but surely finding wider acceptance, on both national and global levels. Safety, standardization, quality, and reliability are to be evaluated and assessed periodically in a planned scientific way, to ensure optimized medi/rehab care at low to moderate costs to the nonaffluent millions.

Human Resources Development–Education and Training

Human resources development is the basic and fundamental objective of all CEP and QIP programs initiated by the AICTE, International Society for Technology in Education (ISTE). Agencies like University Grants Commission (UGC), Council of Scientific and Industrial Research) (CSIR), Indian Council Of Medical Research (ICMR), Department of Science & Technology (DST), Medical Council of India (MCI), New Anglican Missionary Society (NAMS), World Health Organization (WHO), United Nations Education, Scientific and Cultural Organization (UNESCO), United Nations Children's Fund (UNICEF), UNDP, World Bank, Ford Foundation, Voluntary Health Association of India (VHAI), Christian Medical Association of India (CMAI), and The Institution of Electronics and Telecommunication Engineers (IETE) augment the efforts of AICTE-CEP and ISTE. Medical diagnosis and therapy is mainly based on qualitative and quantitative analytical laboratory clinical examination results. Paramedical or surgical intervention, followed by medical or surgical intervention, followed by a medical or surgical procedure, followed by post-medical/surgical corrective procedure, involves utilization and application of paramedical techniques provided by trained paramedical specialists and technicians. All of these depend on instrumentation that is low-tech, medium-tech, and high-tech, of low-cost, medium-cost, or high-cost, depending upon the clinical parameters to be analyzed. Paramedical technicians form the fundamental base of any medical or surgical diagnosis therapy. Education and training of this cadre is essential. Autopsy of 50 years of human resource center development (HRCD) programs, and biopsy of the present will be the pointer for overcoming past barriers, bottlenecks, and hazards in planning for the future. Technology needs at various levels are to be assessed and developed. Certain mandatory measures might be inevitable, to improve the quality of medi/rehab care on a cost-effective basis. Accountability is the basic and fundamental necessity. Implementation of recommendations at various levels is most urgently needed and must be a top priority.

As in the other areas, trained-manpower needs require a realistic, need-based, feasible, time-senstitive, and cost-effective approach. Integrated interdisciplinary education forms the basis of any paramedical education or training program. The mode of delivery is by way of organizing awareness, seminars, symposia, and workshops to be followed by short-term and long-term courses, six-month or one-year certification courses on selected and relevant topical areas. The number needed depends on the type of center, namely primary (PHC), secondary (district level) or tertiary (major and referral centers). Referring to Table 78-7, at the PHC, one can utilize 1–4, at the secondary center, 1–7 and 12–17, and at major hospitals and referral centers, 1–18. Nursing assistants and pharmacy assistants are needed in larger numbers to assist trained nurses and pharmacists who complete courses (degree or diploma). It is estimated that every trained nurse or pharmacist needs three or four assistants to assist them in their work. Therefore, placement-oriented, knowledge-based courses, which form a major part of AICTE-CEP objectives, will be fulfilled. It is also part of the strategy for the development of human resources, especially in semi-urban and rural areas. It also helps to narrow the widening urban and rural knowledge and placement gap. Governmental and voluntary organizations can make use of such technicians. The AICTE-CEP is approving paramedical courses after due inspection of such centers that are offering these courses. To date, nearly 100 centers have been approved. It is most heartening that a majority of them are located in semi-urban and rural areas to benefit the socio-economically needy, who are educated up to 10th standard or 10 + 2 level and could not continue because they are unable to pay the capitation or donation demanded by nursing schools and colleges and colleges of pharmacy. Physiotherapists also are needed in large numbers. A broad-based, open-minded, integrated, interdisciplinary teamwork approach is needed at all levels. The course-fee structure must be flexible to allow the training institutions located in rural areas to charge less for the needy and female trainees. Lack of funds should never block talented rural students from rising up and helping to close the urban–rural technology gap. Paramedical courses play a vital role in improving the quality of life of the nonaffluent millions in rural areas, by promoting human resource development (HRD) and utilizing CEP of AICTE. A review of the past 50 years of education and training in this field is a worthwhile exercise, followed by a study of the present, to apply corrective measures. One can get a meaningful glimpse of the reality of the situation only when one visits rural areas. Such visits leave one feeling sad but with hopes for a better decade by educating and training rural youth.

Outstanding medical specialists, along with equally reputable health rehabilitation care (HRC) technologists, are part of the Board of Examiners, which is headed by an internationally renowned humanist neurosurgeon and assisted by a dynamic secretary, who plans and conducts theoretical and practical examinations at centers approved by AICTE-CEP. Currently, courses on medical lab technology, nursing assistants, physiotherapy, X-ray, and ECG have been approved depending on the institutions' requests, after due inspection of their facilities. More courses listed might become acceptable as the national and global placement potential improves. It is most heartening that quite a number of those trained have found, and are finding, gainful openings all over the nation, and overseas. It augurs well for both the trainees and approved institutions, and they can be optimistic.

Holistic Approach to Health Rehabilitation Care–Challenges in the Millennium

The ever-escalating modern Medicare costs make it mandatory and imperative to explore alternate approaches to evolve strategies to develop an integrated interdisciplinary holis-

Table 78-7 The Need for Paramedical Education and Training)

1. India's population exceeds one billion, with nearly 10% handicapped and 3% elderly. Affluence is around 10–15%, and nearly 80% live in rural areas.
2. In the above context, one must conceive and evolve strategies for education, training, and development of an enormous human resource potential, to improve the quality of life, especially in semi-urban and rural areas.
3. Unfortunately, prevalent socioeconomical conditions force a large number of intelligent students to seek paramedical education and training after their 10 or 10+2 courses. Paramedical courses began almost 30 years ago in India, especially in Kerala.
4. About 3 years ago, AICTE-CEP initiated steps to constitute a board of examiners in this area, with a view to standardizing a syllabus, duration, and basic infrastructure facilities. The normal procedure is to inspect and approve those who meet the basic conditions.
5. Broadly, about 20 types of technicians have been identified, considering present and futuristic placement potential.
6. Three levels of technicians are needed, starting from the primary, grassroots level and moving to the secondary district level and final tertiary level of city hospitals. Twin targets of improved knowledge base and better placement can be achieved.
7. Approval, assessment, accreditation, and QIP, have received attention at the post-doctoral, doctoral, master's, graduate, and diploma levels.
8. It is only apt, as part of HRD, CEP-AICTE develops human resource potential in semi- urban, rural areas by educating and training youth. It must be a missionary effort.
9. The 26 nodal CEP Centers can initiate steps to plan, coordinate, promote, and propagate paramedical education throughout the nation.
10. By these measures, the poor can have an opportunity to harness their potential, based on aptitude. A gestational period of three years is inevitable before one can assess, evaluate, standardize, and accredit these need-based, cost-effective, paramedical training centers.

Table 78-7 Paramedical personnel clinical assistants required (ranked according to demand)

1	Nursing
2	Medical laboratory
3	Physiotherapy
4	Electricians
5	ECG
6	X-ray
7	Blood bank
8	Imaging
9	Dialysis
10	Cardiac care
11	Neurological care
12	Ophthalmic care
13	Dental care
14	Anesthesia/operating room
15	Pharmacy
16	Dietary
17	Audiometer
18	Medical transcription

NOTE: 1 to 4 PHC Level (grass roots); 1 to 7 & 12 to 17 (district level); 1 to 18 major hospitals & referral centers.

tic approach. This must be cost-effective, need-based, realistic, feasible, time-sensitive, and target-oriented. Optimizing utilization of the large manpower pool, limited material resources, and very limited monetary resources is the most urgent need. The health rehab care of nearly one billion people, with 10% handicapped and 3% elderly, is a formidable techno-bio-psycho-socio-medical challenge.

Integrated therapy centers, using an open-minded, broad-based, holistic approach, are the only hope for improving the quality of medi/rehab care of the nonaffluent millions in India. Urban, semi-urban, and rural health rehabilitation care requires a planned, scientific approach. Fortunately, a wide range of noninvasive techniques and alternate systems of medical care, namely Ayurveda, homeopathy, Unani, and Siddha, are available for wider utilization. Yoga, naturopathy, and meditation care augment any system of medicine, thus reducing the medical care costs. A holistic medical care approach with integrated therapy appears to be the only hope for the nonaffluent millions in third-world nations, where two-thirds of the world's population lives.

One sincerely hopes that the coming decade will see the emergence of the holistic care to benefit humanity. Karmayogis who are optimists with humanism as the base are the answers.

Conclusion

Attempts must be made to modify and to adapt the techniques of yoga, naturopathy, and meditation into everyone's day-to-day life whether in the urban, semi-urban, or rural areas. Selected yogasanas, pranayama, and meditation for about 30 minutes daily, coupled with modified naturopathy practices to suit the practical daily life, can go a long way in enabling wider utilization by all sections of the people in society. There is a most urgent need to evaluate the percentage of people who practice them every day. Agencies like National Institute of Naturopathy (NIN), Central Council for Research in Yoga & Naturopathy (CCRYN), and Indian Systems of Medicine & Homeopathy (ISMH) can initiate steps to get a glimpse of the present status, so that they can plan for the future. Table 78-8 in the Appendix lists leading yoga, meditation, and naturopathy research/training/education institutions in India. One must evolve strategies to develop a feasible, holistic approach. Outstanding yogacharyas, naturopaths, and meditators have undergone modern medical/surgical procedures when cardiac, neurological, and orthopedic intervention becomes inevitable to provide relief from pain and suffering. Birth and death are accepted realistically.

In addition to those institutions listed in Table 78-8, scientists in the National Institute for Mentally Handicapped, Secunderabad (NIMHANS), Defence Institute of Physiology Allied Sciences (DIPAS), Institute of Nuclear Medicine & Allied Sciences (INMAS), All India Institute of Medical Science (AIIMS), MAMC, Jawaharlal Institute for PG Medical Education & Research (JIPMER), Post Graduate Institute for Medical Education & Research (PGIMER), India Institute of Technology (IITS), India Institute of Science (IISc), and physiology departments of some of the medical colleges are actively involved in planned scientific evaluation of the technique. Some university yoga departments are conducting research. UGC plans to help universities to set up yoga centers. Agencies like the Department of Science and Technology of India (DST), Indian Council of Medical Research (ICMR), CCRYN, Defence Research & Development Organisation (DRDO), and UGC are supporting these projects.

"Think globally, act locally, coupled with simple living and high thinking" should be our motto.

Reference

Taylor K, Frize M, Iverson N, et al. The Need for the Integration of Clinical Engineering & Sociological Perspectives in the Management of Medical Equipment Within Developing Countries. *J Clin Eng* 19(4):291-296, 1994.

Appendix

Table 78-8 Leading Yoga, meditation, naturopathy research/training/education institutions in India

Bihar School of Yoga, Munger, Bihar
Kailvalya Dharma, Lonavala, Maharashtra
Divine Life Society, Rishikesh, Uttar Pradesh
V.K.Yoga Research Foundation, Bangalore
Central Research Institute of Yoga, New Delhi
International Center for Yoga Education and Research, Kottakuppam
National Council for Education, Research and Training, New Delhi
Verman Yoga Research Institute, Secunderabad
Himalayan International Institute, Dehradun
S.V. Institute of Yoga and Allied Sciences, Bangalore
Yoga Institute, Santacruz, Mumbai
RMM Institute of Yoga, Pune
Aurobindo Centre, Pondicherry
Brahmakumaris International Centre, Mount Abu
AU Institute of Yoga and Consciousness, Vishakapatnam
CCRYN, New Delhi (Supporting Body)
Indian Council for Research in Yoga, Bhopal
Haryana Institute of Alternate Medicine, Panchkula
BM College of Naturopathy and Yogic Sciences, Ujire
R.K. Institute of Moral and Spiritual Education, Mysore
Adhyatma Sadhana Kendra, New Delhi
National Institute of Naturopathy, Pune
Bapu Nature Care Hospital, Delhi
Jindal College of Naturopathy and Yogic Sciences, Bangalore
JSS College of Naturopathy and Yogic Sciences, Bangalore
Vyakthi Vikas Kendra, Bangalore
Prakruthi Jeevan Kendra, Bangalore
Sadharana Sangam, Bangalore
Institute of Naturopathy and Yogic Sciences, Bangalore

Table 78-8 Some leading design, research, and training centers in India

Advanced Training Institute for Electronics and Process Instrumentation, Hyderabad
All India Council for Technical Education–Continuing Education Programme, Bangalore
All India Institute of Medical Science, New Delhi
All India Institute of Physical Medicine and Rehab, Mumbai
All India Institute of Speech and Hearing Handicapped, Mysore
Artificial Limbs Manufacturing Co. of India, Kanpur
BMS College of Engineering, Bangalore
Bhaba Atomic Research Centre, Mumbai
Blind Men Association, NREI, Ahmedabad
Central Scientific Instruments Organisation, Chandigarh
Centre for Cellular & Molecular Biology, Hyderabad
Centre for Electronic Design and Technology, Mohali
Christian Medical College Hospital, Vellore
Clinical Engineering Centre, Bangalore
Coimbatore Institure of Technology, Coimbatore
Defence Electromedical and Bio-Engineering Laboratory, Bangalore
Defence Institute of Physiology Allied Sciences, New Delhi
Electronics Regional Testing Laboratory, Trivandrum
Indian Institute of Science, Bangalore
Indian Institute of Technology, Bombay, Delhi, Chennai, Kanpur Kharagpur and Gauhati
Institute of Aviation Medicine, Bangalore
Institute of Nuclear Medicine & Allied Sciences, New Delhi
Jawaharlal Institute for PG Medical Education & Research, Pondicherry
Madras Institute of Magnetobiology, Chennai
Manipal Institute of Technology, Manipal
National Institute of Hearing Handicapped, Mumbai
National Institute for Mentally handicapped, Secunderabad
National Institute for Ortho Handicapped, Calcutta
National Institute for Rehab Training & Research, Otapur
National Institute of Visually Handicapped, Dehradun
National Institute of Nutrition, Hyderabad
National Institute of Occupational Health, Ahmedabad
National Institute of Mental Health & Neurosciences, Bangalore
Nizam's Institute of Medical Sciences, Hyderabad
OU College of Engineering, Hyderabad
Post Graduate Institute for Medical Education & Research, Chandigarh
SCTI Medical Sciences & Technology, Trivandrum
Shri Bhagubhai Mafatlal Polytechnic, Mumbai
Sri Jayachamarajendra College of Engineering, Mysore
University of Roorkee, Roorkee
Vikalanga Kendra, Allahabad

Section VII

Medical Devices: Design, Manufacturing, Evaluation, and Control

Gary H. Harding
Greener Pastures
Durango, CO

Alice L. Epstein
CNA HealthPro
Durango, CO

The authors of the next chapters provide historical and contemporary insights into many aspects of the history, design, manufacturing, costs, and control of medical devices from a clinical engineering perspective. These chapters begin with the evolution and growth of medical device technology, from the early healer's use of sharpened stones to endoscopy, and through the intricate financial aspects of technology acquisition, software development, and the excitement of research and design of new, cutting-edge technology. These chapters examine issues that face medical device manufacturers, as well as hospital-based engineers, innovators, designers, testers, and evaluators. They share interests, concerns, and issues and address them with an interest in providing insight into the information that will prove valuable to a clinical engineer, regardless of the base from which the design, manufacturing, or control is performed.

Richter provides a thoughtful review of the evolution of medical device technology, noting the contributions of Hippocrates, Santorio, Galileo, and others, with specific emphasis on the importance and interface of temperature measurement, bioelectric signals, blood pressure, pain relief, imaging, and patient monitoring, Richter provides support for the concept that man-made technical innovation influences human biological evolution and that the physician-engineer relationship will continue to strengthen and expand.

Gaev notes that with more than 5000 different types of medical devices on the market, it becomes almost impossible to have medical intervention without the application of a device. He provides a basic overview of the leading clinical maladies that affect the world population and the expenditures that are made worldwide to address these maladies. Gaev poses the question of allocation of health care spending and the relative success in treating illnesses and injuries. His examination of modalities to treat same clinical morbidities includes an example of treating diabetic patients to help understand the various ways in which diagnostic and therapeutic devices come into play as a result of the clinical course of a disease.

Canlas et al. provide a glimpse into the actual clinical engineering practices at William Beaumont Hospital (Michigan) with respect to technology management within the hospital setting. The authors bring to the forefront the importance of a systems approach to technology management, including ensuring medical device safety through participation in equipment selection and evaluations. They review the need for continued vigilance in the evaluation of technology after product selection, including cost-effectiveness review, anticipation of future technology needs, evaluating designs for efficacy and reliability, and occasional prototype design and redesign.

Oberg discusses the nuances of device innovation, beginning with research and design, providing insight into invention, patents, product life cycle, publishing, and actual device development. Oberg believes new product ideas should be put through a rigorous assessment process when evaluating effective product development. Market analysis, technology assessment and economic assessment should preface concept testing and prototype development. Oberg examines the way that commercial exploitation through licensing and starting new companies for commercialization of a product typically progresses.

Hyman and Wangler examine the ways in which human factors considerations affect safe and effective use of medical devices in the actual health care delivery setting. They discuss human factors in the context of all devices in use, physical environment of use, and the personnel using the device. The authors acknowledge the importance of the differences in people (e.g., vision, hearing, and height.) device interactions, the multiple patient environments relating to the physical environments, and the incalculably important issue of personnel. They note that all human factor-related interventions might not be successful unless there are adequate management and resources, and they believe that this should include a nonpunitive system of adverse and near-miss event reporting with a focus on investigations to provide safeguards within the health care delivery system.

Geddes sets a paradigm of failure modes by types of medical devices; diagnostic, therapeutic, and assistive/rehabilitative. A brief explanation of product liability that is specific to product defect, misuse, negligence, and design defect is presented. Accidents relating to the three types of medical devices are explored through specific examples.

Fries and Bloesch address the burgeoning issue of software-intense medical devices. They provide lessons and an organized approach to specifying software requirements, creating software design through implementation. Elements of software development include identification and review of existing software standards, evaluation of design alternatives, and tradeoffs to determine the software architectural model. The authors note the importance of selecting effective design methodology and programming language from a realistic framework. Fries and Bloesch explore the relationship of software risk analysis and metrics based on the types of hazards identified, speculation as to the way the equipment can fail, and what failure is tolerable. Design review, performance predictability, design simulation, verification, and validation all lay the groundwork for implementation.

Keller details the ECRI (Plymouth Meeting, PA) method of comparative evaluations of medical devices. He notes the importance of identifying equipment needs, technology alternatives, and any other specialties that might benefit from the proposed equipment prior to the device evaluation. ECRI's model uses a rating system of acceptability, based on essential features. The model includes a detailed comparative evaluation that is based on a pre-established set of comprehensive criteria. Keller lists examples of critical criteria, including performance and safety, human factors, construction quality, vendor service, and cost. He provides a discussion of ways to address findings, noting the importance of the evaluator's consideration of the relative importance of various aspects (e.g., performance, safety, and ease of use) of the evaluation.

Balar examines the process of evaluating investigational devices in the clinical setting by providing insight into the ways in which clinical engineers at William Beaumont Hospital (Michigan) evaluate such devices for their Institutional Review Board (IRB). The author explains the role of the clinical engineer's in evaluating the safety and efficacy of investigational devices used in research studies. There are discussions of the evaluation process, from initiation through research, risk analysis, and reporting. Tips for overcoming obstacles in this arena are presented.

After reading this section, clinical engineers will gain some understanding of specific issues that must be considered in medical device design, manufacturing, and control as it relates to clinical engineering. Useful personal insights by professionals who are the clinical engineer offer the benefit of experience and the lessons learned through applied research. The complexity of medical device design, manufacturing, and control becomes apparent through the eyes of these experts, all of whom understand that the future might be even more complex. Stressing the importance of knowledge, systems, and clinical engineering participation and support, medical device design, manufacturing, and control can be performed safely and effectively to the benefit of patients, medical professionals, and others.

79

Evolution of Medical Device Technology

Nandor Richter
ORKI
Budapest, Hungary

The Origins

In the early phase of civilization, instruments were used in medical treatment. Healers used sharpened obsidian stones for skull trepanation, and metal knives later became common tools. Thousands of years ago doctors sought to cast a glance into the human body. They tried to observe the inner structure and the functioning of the living organ through the mouth, the rectum, and the vagina. Observation through these parts caused minimum functional disorder, and minimal tools were needed. Around 400 BC, Hippocrates was the first to mention the endoscopy when he described a rectum examination. At that time, a light source was not yet available. Around AD 1000, the well-known Arab doctor Abu al-Quasim used a glass mirror for vaginoscopy. Because of inaccurate notions and the lack of measuring instruments, observations could lead to only rough conclusions.

The Enlightenment brought with it a significant impetus as the dogmatic, retarding forces of the Middle Ages influenced scientific activities less and less. During the seventeenth century, the exact sciences developed vigorously. The curious human examined and understood nature better and more objectively. Progress in healing required the increasing application of tools and equipment. For example, doctors realized rather early the importance of body temperature and its fluctuation. Again, it was Hippocrates who considered body temperature as the most important sign in the case of acute illnesses. However, it was not until around 1612 that the Italian doctor Santorio made the first clinical thermometer. He is also credited with the introduction of the weight balance and hygrometer. With these devices, the desire of Galen (131–201 AD), that not only the type of illness be identified but also the "quantity" of the illness should be measured, could now be realized. Galileo stated "measure the measurable and make measurable what was not measurable till now." His words are equally important and basic even today. One can apply this to all diagnostics and to most medical equipment.

Temperature Measurement

Thermometers have been enormously helpful to physicians. Time series of temperatures were made possible, providing information on the progress of an illness and the trend of the patient's condition. One good indicator of the importance of the thermometer is that over one billion temperature measurements are carried out yearly in the hospitals in the United States.

Physicians looked for symptoms that could increase their knowledge about the condition of the patient, the nature of illness, and its seriousness. Besides temperature, the color of the face, the color and odor of urine, and the rhythm of the heart were all peculiar and perceptible to the patient. It was desirable to measure these parameters quantitatively and not simply to observe the color of the patient's face. Quantifying color, heart rhythm, and chemical composition of urine had to wait for further developments in the chemical and technical sciences.

Bioelectrical Signals

The discovery of bioelectrical signals was of paramount importance in that it led to diagnostic and therapeutic applications. Today, we know that bioelectrical phenomena are characteristic of all living organisms. Long before bioelectricity was well understood, much was already known about the electrical activity of nerves and muscles. Platon (427–347 BC) mentioned electric rays (torpedinidae) found in the Mediterranean Sea. Aristotle (384–322 BC) also wrote about these marine creatures. He mentioned that touching these rays could cause deafness. In the eighteenth century, Aloisius Luigi Galvani (1737–1798), a professor at the University of Bologna, performed experiments on frog nerve-muscle preparations. He stimulated with electrical charges the nerve that led to the frog's femur and found that during stimulation the femur muscle contracted. In one of his experiments, he demonstrated the cell membrane potential, the polarizing potential of cells in rest. For his achievements, Galvani was named the father of electrophysiology. From that time onward, an increasing number of scientists started to study the action potentials generated in living organisms.

These electrical signals are primarily the cell membrane potentials, action potentials of heart and muscles, and the action potentials of the brain. Studying action potentials requires sensitive equipment and thorough knowledge in measuring technique. Depending on the source of their origin, the amplitude and frequency domains of these signals are significantly different. From the point of view of measurement technique, the heart action potentials, the ECG waves, can be considered as the easiest to measure. Voltage potentials of these signals are in the millivolt (mV) range, their useful frequency domain is 0.1—100 Hz. Brain action potentials are on the order of microvolt (µV) and the frequency domain is 0.1–10,000 Hz. Evoked potentials are even lower with a frequency range of approximately 10–10,000 Hz.

Around 1856, electrocardiographic signals of a frog's heart were measured for the first time. In 1903, Willem Einthoven (1860-1927) introduced his string-galvanometer and recorded ECG signals. Hans Berger started to study the electrical signals of the brain by electroencephalograph measurements in 1924. However, his first good-quality electroencephalogram was not ready until 1929. In 1943, Weddel succeeded in registering muscle action potentials with an electron beam oscilloscope. But the first commercial electromyograph did not enter the market until around 1960.

Blood Pressure

Aware of the importance of blood circulation, studying blood pressure was a natural requirement. In one of the first experiments, a tube was inserted into the neck artery of a horse, and the pressure variation generated by the heart was measured. For human application, the bloodless measurement of blood pressure was a necessity. However, one had to give a measurable, practical definition of the blood pressure. There was no simple, bloodless way of measuring the instantaneous value of blood pressure until the introduction of the systolic-diastolic measurement. Still accepted in daily routine, this detection technique is based on the Korotkoff sounds. Digital technology made small blood pressure meters possible with a concomitant surge in their use in the home as well as the hospital.

Anesthesia and the Relief of Pain

Horace Wells, an American dentist, performed the first painless dental operation in 1844. He used nitrous oxide, known also as "laughing gas." Therefore, he is considered the inventor of narcosis. Anesthesia, the painless operation on patients, was an immense development in surgery. It gave enormous push to the development of surgical instruments and apparatuses used in operating rooms and beyond.

It is also interesting to draw attention to a different medical advance, acupuncture anesthesia. Western medicine is learning more and more about this practice of traditional Chinese medicine. While its mechanism is still not completely known, it does prove useful in certain areas of anesthesia.

X-ray and Nuclear Medicine

The desire of physicians for centuries to be able to look into the human body was finally realized in 1895 with Conrad Roentgen's discovery of X-rays. In the following century, the application of X-rays to diagnosis and therapy gave strong momentum to the advance of medicine. In hospitals and clinics, the medical applications of X-rays became a separate professional field. With the development of the engineering sciences and, later, informatics and computer techniques, application of the images generated by X-ray increased. Of course, this went along with the construction of more sophisticated equipment. Therefore, the demand increased for experts who are familiar with the operation and maintenance of these devices.

Imaging and Image Processing

X-ray images, with their increasingly fine details, provided increasingly more information. Image intensifiers and video monitors rendered possible the manifold applicability of X-ray equipment. High-power and fine-resolution X-ray tubes expanded further the application possibilities. X-ray equipment, such as the tomograph, angiograph, angio-cardiograph, and the urograph, was developed for special applications. Imaging and image-processing methods provided the information-enhancing possibility for physicians.

CT and MRI

In the field of imaging, the computer tomograph (conceived by William Oldendorf and developed by Godfrey Hounsfield and Allen Cormack) brought revolutionary change. The first successful clinical experiments occurred in 1972. The Nobel Prize was bestowed upon the two inventors in 1979, and Godfrey Hounsfield was knighted in 1981. The basic idea of computer tomography went beyond X-ray imaging. The principle is also becoming important in fields where the source is non-ionizing radiation.

The basic phenomenon of magnetic resonance imaging was first observed in 1946. Certain atomic particles in strong magnetic fields absorb very high-frequency electromagnetic waves selectively. Particularly those nuclei in which at least one proton or one neutron is unpaired show this phenomenon. This absorption can be measured and evaluated. In the human body, water and lipid (fat) molecules, which contain hydrogen, show the effect in measurable magnitude. This noninvasive, relatively hazard-free diagnostic method is based on this selective absorption phenomenon. The first MRI equipment was put on the market during the early 1980s. Since then, its application has become broad including, among others, the study of the brain, breast, heart, kidneys, liver, pancreas, and spleen.

Nuclear Medicine

The use of radioactive isotopes as tracers by George C. de Hevesy in 1912 was a great epoch-making discovery for which Hevesy was recognized with the Nobel Prize in 1943. Wide-ranging application of radioactive tracers became possible after World War II. Radioactive isotopes came to the market after the 1950s. Around this time, development of the equipment necessary for measurements was completed as well. Importance of manmade scintillation crystals was extraordinary. These crystals make possible the detection, counting, and measurement of the radiation emitted from disintegrating nuclei. The development of synthetic crystals, the electron multiplier, and spectrum analyzers altogether resulted in the general application of nuclear-measuring technique in health care. In the following decades, the evolution of imaging and image-processing technology further broadened the field of nuclear measuring techniques. The introduction of the gamma camera made possible the imaging of larger parts of the body.

Ultrasound

In 1880, Madame Maria Sklodowska Curie and her husband Pierre discovered radium and subsequently received the Nobel Prize. Pierre's brother Jacques discovered the piezoelectric effect. When high-frequency electric fields are imposed on a piezoelectric material, mechanical vibration results. Because the frequencies of these vibrations are above the audible frequency domain, the vibrations produced are termed "ultrasound." Mulwert and Voss reported the first ultrasound therapy intervention in 1928 when they tried to cure certain deafness with ultrasound irradiation. Pohlmann, in 1939, made the first ultrasound therapy equipment for treating humans. In 1942, K. Tr. Dussik was the first to attempt to apply ultrasound for diagnostics. After World War II, ultrasound was used in different fields of medicine. The one-dimensional method (A-scan) was followed by the two-dimensional method (B-scan). Essentially, these are imaging methods with which place and form of tissues and organs can be examined.

By the application of the Doppler effect, ultrasound was expanded to dynamic measurement areas. Reflection of ultrasound from certain moving parts of the body makes the measurement of radial velocity of the reflecting surface possible. This phenomenon is applied, for instance, in measurement of the blood flow velocity and in the examination of fetal heart movements.

Considerably higher power must be applied in the case of therapy. Accordingly, the problem of regulating and making accurate measurements of ultrasound dose and energy had to be solved. Today, medical application of ultrasound is expanding rapidly, and equipment with new features appears on the market continually. As a result, quantity and diversity of knowledge required for the application of ultrasound is growing.

Microscope and Endoscope

Most likely, Zacharis Jansen, a Dutch optician, discovered the compound microscope in 1590. Its general use gave enormous impetus to the development of medical sciences. With its help, otherwise invisible elements of the body could be studied. Since this discovery, many varieties of microscopes have been developed. At the far end of the spectrum are sophisticated devices, such as the electron microscope and the atomic-force microscope.

For centuries, physicians wanted to directly visualize the inner parts of the functioning body. Various endoscopes were developed to assist in achieving this goal. The major improvement in endoscopes started with the invention of Bozzini, a physician in Frankfurt. Bozzini succeeded in introducing a beam of light into a hollow organ and directing the reflected light to the eye of the observer. The technician, Nitze, constructed the first truly practical endoscope. These instruments were used for rectoscopy and cystoscopy. From that time onward, the endoscopes have been continuously improved. Nowadays, endoscopes and laparoscopes furnished with flexible fiber optics or with cylindrically shaped lens systems (the Hopkins-optic) are in use. With the help of these modern devices, minimally invasive diagnostics and surgery progressed rapidly. The diagnosis and therapy of the digestive system, rectum, bladder, respiratory tract, joints, and abdominal cavity can be carried out with less trauma than with the conventional methods, which typically required large incisions. Introduction of catheters, biopsy forceps, electrocoagulation instruments, and lasers can be done with devices evolved from endoscope.

Laboratory Devices

Together with the development of biology, medical science, and measurement techniques in chemistry and physics, the variety of laboratory devices has increased enormously. With the help of the modern laboratory automatons, the number of different measurements done today is on the order of hundreds of millions per day. Of course, this practice also raises the number of the companies that supply the reagents and disposables.

Surgical Instruments

These instruments are usually not of concern to clinical engineers. However, for the sake of completeness, they should be mentioned. The number of surgical instruments ranges in the thousands. Many of them are engineering marvels, the products of cutting-edge fabrication technologies and advances in metallurgy and materials science.

Patient Monitoring Devices for Intensive Therapy

Modern technology and engineering opened the door for the application of individual equipment in systems. Two good examples are the complex array of equipment in the operating theater or the installations in the intensive care units. Patient monitoring systems, respirators, defibrillators and others constitute an integrated system. The measurements of one device might affect the operating parameters of the other. The harmonized functioning of these devices has required the attentive presence and intervention of the physician or intensive therapy nurse.

One ongoing development is the automation of processes and the involvement of information technology. The patient will become a part of a feedback loop, in which the system automatically sets and maintains the optimum parameters. Such parameters might include pulse, respiration, ECG, blood pressure, and blood sugar level. This, to some extent, is the vision of the future; but considering the pace of development in the last few decades, one may safely say that this is not the illusion of the far future. The science of robotics is enabling machines to replace some of the operations of the surgeon. Will the human ever be completely replaced by machines?

Understanding the Physical World

In the previous paragraphs the development of medical device technology was introduced in a nutshell, through examples. These brief episodes demonstrate well that, in the course of development, combination and interaction of different scientific disciplines have played an important role. Formerly, usually one person facilitated this interaction. In the time of Hippocrates, for example, science-minded people had wide interest and philosophical inclination. Because of the lesser quantity of the available scientific knowledge at that time, surveying it was easier for one person. This helped to develop those statements and discoveries that demanded multidisciplinary knowledge from the inventor. Such was the situation up until the last few centuries.

Taking a great leap in historical terms to the Enlightenment, the progress in understanding the law's of nature was rapid and substantial. The seventeenth century is considered to have been the century of geniuses, and the eighteenth century is considered to have been the period of brightness and the mind. The Bernoulli brothers (Jacob and Johann), Herman von Helmholtz, Gottfried Wilhelm Leibnitz, and Isaac Newton towered above the geniuses. Newton, in his *Principia*, who also built upon the observations of Galileo, Copernicus, and their contemporaries, gave the best description of mechanics of their time. With the infinitesimal calculus, and with Newton's mechanics as a basis, Leonhard Euler, a Swiss mathematician, described the movement of solid bodies and developed the basic equations of hydrodynamics. Galvani and Volta laid the foundation for knowledge of contemporary electricity.

During the eighteenth century and beyond, the bulk of work went into observation and understanding natural phenomena, the qualitative recognition of the laws of nature, and their formulation in statements and equations. The growing amount of knowledge and the immersion into the details necessarily steered scientists toward specialization. The epoch of polymaths began to fade away. Due to the prompting effect of the industrial revolution, demand on practical application of sciences strengthened. A technical revolution followed the scientific revolution. Application of discoveries accelerated. The number of inventions grew. The steam engine and the principle of the dynamo, followed by the general use of electricity, had great effects on almost every aspect of life. Medical science provided its share, too. Physicians discovered the potential of electrical phenomena. For instance, they introduced electrical shock therapy for a variety of purposes. The rapid development of basic sciences, physics, chemistry, and biology, and the engineering sciences, together with the demands emerging from the medical side, gave impetus to the developers of medical devices. The pace of development during the nineteenth century became significant. Increasingly, humans began to understand the phenomena of their surroundings, the laws of nature, and their effects on the individual as well as on society. Discoveries and inventions followed each other.

During the first part of the twentieth century, significant milestones included the discovery of X-ray, the introduction of ECG, and the development of anesthetic and respiratory equipment. Then came World War II and the harnessing and release of atomic energy, with one of its practical applications being nuclear medicine. Discovery of special semiconductors and transistors in the 1950s also gave a large thrust to the development of electromedical equipment.

During the first half of the last century, pharmaceuticals dominated medical advances. But since the 1960s, developments in medical engineering have been unprecedented. In the second half of the last century, a great variety and number of medical devices entered hospitals and consulting rooms. Today, the number of different medical devices is around 10,000. If one considers the variations of the equipment manufactured by different companies for the same application, then this number increases to tens of thousands.

It is no exaggeration to state that health care and health care organizations are among the most complex and complicated structures of our society. Physicians must select from an extensive arsenal of methods, materials, instruments, and from combinations of them, and they must do so with the intention of providing the best care for the patient. Evidently, in such a complex system, various specialists must work together. In the course of time, pharmacists appeared in the hospital, joining the physician and the nurse. They helped with the proper handling and preparation of the vast number of pharmaceuticals available. As with pharmaceuticals, the introduction of a multitude of increasingly complicated medical devices urgently emphasized the importance of the support of specialists in physics and engineering. By the 1930s, medical physicists began to appear in hospitals. Their tasks were to handle and control X-ray equipment, to perform dose measurements, and, later, to work with other ionizing radiation sources. Dose planning and control and quality assurance also became their tasks. Nowadays, they are often found working with non-ionizing radiation, such as ultrasound and lasers, as well.

Since the 1970s, complicated equipment such as patient monitoring systems and heart-lung machines have made indispensable the presence in hospitals of those specialists called clinical engineers. Soon it became evident that besides requiring technical knowledge, these clinical engineers needed experience in other fields. Apart from becoming knowledgeable about the human body, these engineers needed to know how to communicate with the medical staff and to have experience in hospital management and administration. Step by step, the concept of clinical engineer was developed. In our time, clinical engineering has become an accepted profession in many countries.

The number, variety, and complexity of medical devices, combined with the growth of informatics, made it important and possible to think and work with systems. Health informatics gradually gained ground all throughout health care. The great quantity of patient' data can be organized and managed only with the help of informatics. Informatics finds particular importance and applicability in acquisition, processing, storage, archiving, retrieving, transmission, and display of diagnostic images.

The Future

The current trend will continue. The fields of medicine, engineering, and science will develop independently but will help each other, a phenomenon known in electronics as "bootstrapping." Progress in medical science makes new demands upon engineering, which in turn are reflected in the development of new methods and equipment. New engineering technologies provide fresh possibilities for physicians. Cellular and tissue engineering come closer to the routine application of their results. The introduction of nanotechnology will soon yield practical applications. Robotics will open up new opportunities in surgery. The combination of these technologies will result in spectacular achievements. As an example, new techniques in heart surgery combine robotics and endoscopy, thus enabling an operation without the need to fully open the thorax, without the need to use a heart-lung machine, and without having to lower the patient's temperature. The field of gene technology shows promise of progress and advancement.

Biomedical engineering was literally created and developed during the twentieth century. Considering the accelerated pace of development of the present, it is impossible to predict developments that will occur over the next few decades. However, it is possible to affirm with confidence that the physician-engineer relationship and the importance of clinical engineers in health care will remain and grow in prominence. Besides engineers who are directly involved in hospitals, clinical engineers will also play an important role in research and development. Equally important is the role of the engineers who are employed in medical equipment factories. In summary, relatively little is known about the forthcoming development in this century; however, even this modest foresight hints at enormous technical progress.

Social Effects

What are the impacts on the society of the rapid technical progress taken place this century? Colossal resources of energy are at the disposal of mankind. The bulk of physical work has been taken over by machines. Comfort of life has increased. Initially, humans saw only the pleasant side of this progress, that living conditions improved. Only in the last few decades has mankind started to feel and get to know in more detail the harmful effects of the environmental hazards that accompanied the technical revolution and which affect the entire biosphere.

In addition to environmental damages, the technical revolution has affected the practice of medicine, which now must face new maladies. As illnesses shift in their nature and significance, adaptation of health care services requires continuous change and development.

The extended life span of humans brings into focus health problems of the aged. Attending to the health problems of the aged will impact medical device development in two ways: (1) requirements for innovative medical devices for home use and (2) increased number and variety of devices for hospital use. Those in the aged population are not necessarily ill, in the general sense, but they require special support structures and medical devices which may significantly differ from those found in a general hospital.

In the long run, man-made technical means influence the trend of human biological evolution. The so called "homo technomanipulatus" is coming into being. Abilities, once so important, like physical fitness, the subtle function of the sensory organs–the anatomy that was so well matched to extensive movement are all gradually diminishing. In place of these, new abilities required for intellectual work are augmenting. In the longer-term, negative effects of these will be significant in health care. For instance, the incidence of the diseases of the locomotor, cardiovascular, psychic, and neural systems will increase. The results of a recent research project demonstrated the likelihood of manipulation of the human genome. As a consequence, deliberate or accidental modification of the human genes will be possible. The "homo genomanipulatus" may come into being. Impact of this on humans is unimaginable today. The consequences of gene manipulation for our descendants' are difficult to foresee. We do not know the future of the biosphere. However, genetic manipulation of humans will soon become reality. But we do not know to what extent the new features created in this way will be advantageous or rather disadvantageous in the biosphere that is also changing beyond our control and in an unknown way. Because of the different and long time constants of the processes, one cannot survey the whole system (man + biosphere).

It is clear that health care must be vigilant and ready to respond to new challenges. While some of the impact of advances in technology may be deleterious, it is some comfort to know that predicting and reacting positively to this impact is within mankind's reach. It is imperative that those engaged in health related technical activities must continuously progress by advancing to appropriate knowledge levels and developing technological means so as to enable the generation of solutions to complex and rapidly mutating immerging problems. Taking all of this into consideration, the knowledge and efforts of clinical engineers will be appreciated in the future, as they are in the present.

Further Information

Borst C. Operating on a Beating Heart. *Scientific American* 283(4):46, 2000.
Csaba G. Quo Vadis Homine? *Természet Világa* (World of Nature) 125(1):12, 1994.
Csaba G. Homo Biomanipulatus. *Természet Világa* (World of Nature) 131(4):167, 2000.
Csaba G. Homo Technomanipulatus. *Természet Világa* (World of Nature) 131(8):357, 2000.
Csaba G. Homo Genomanipulatus. *Természet Világa* (World of Nature) 132(1):2, 2001.
Encyclopedia Britannica 1997
Katona Z. Brief History of the Detection of Bioelectric Phenomena. *Kórház és Orvostechnika* (Hospital and Medical Engineering) 26(3):70, 1988.
Katona Z. History of Medical Technique. *Kórház és Orvostechnika* (Hospital and Medical Engineering) 26(4):108, 1988.
Katona Z. Thermometer. *Kórház és Orvostechnika* (Hospital and Medical Engineering) 26(6):161, 1988.
Katona Z. History of Instruments for Temperature Measurement. *Kórház és Orvostechnika* (Hospital and Medical Engineering) 27(1):19, 1989.
Katona Z. Short History of the Application of Ultrasounds in Medicine. *Kórház és Orvostechnika* (Hospital and Medical Engineering) 27(4):97, 1989.
Katona Z. Early History of Surgical Anesthesia. *Kórház és Orvostechnika* (Hospital and Medical Engineering) 29(4):97, 1991.
Katona Z. Short History of Endoscopy. *Kórház és Orvostechnika* (Hospital and Medical Engineering) 29(5):141, 1991.
Olshansky SJ, Carnes BA, Butler RN. If Humans Were Built to Last. *Scientific American* 284(3)42, 2001.
Sorid D, Moore SK. The Virtual Surgeon. *IEEE Spectrum* 37(7):26, 2000.

80 Technology in Health Care

Jonathan A. Gaev
International Programs, ECRI
Plymouth Meeting, PA

Medical technology encompasses a wide range of medical devices. Practicing clinical engineers often focus on a few areas of medical device technology, such as defibrillators, electrosurgical units, and physiological monitors. This chapter addresses the entire spectrum of medical devices with which the clinical engineer should be concerned, the medical devices that are required for almost all diagnostic and therapeutic medical interventions. These devices have different lifetimes because they may be disposable, reusable, or implantable. They are made of a range of materials, such as plastics, ceramics, metals, wood, and biologic products, and they rely upon all types of physical principles for their functioning (e.g., electronic, hydraulic, mechanical, chemical, optical, and radiation). People use them to improve patient health. Medical devices, as distinguished from drugs, achieve their action without directly entering metabolic pathways.

Like the systems of the body, devices are specialized to perform specific tasks. Over 5000 different types of devices are on the market today, and the clinical engineer must be comfortable working with this great variety of devices. The main features to consider when working with a medical device in a health care setting are the clinical condition to be addressed, the user of the device, and the requirements for use, including power, training, storage, maintenance, and cost. Reusable devices, disposable devices, accessories, and consumables must be considered in the effective management of medical device technology. Devices are usually part of a chain of equipment. For example, an X-ray unit might require film and film processors in order to produce an image for a radiologist to read.

It is almost impossible to find a medical intervention that does not involve medical technology. Simple devices, such as a scale, stethoscope, thermometer, latex gloves, and sphygmomanometers, are part of almost all medical examinations. Complex devices include imaging equipment, clinical laboratory equipment, and implants. Some devices are used just once, and some are used for 20 years or more. The tremendous range of application, sophistication, cost, life span, and functionality of medical devices and their intimate relationship with people set them apart from many other technologies. Training, maintenance, selection, and use must be considered to ensure that the device will do its job of helping people to stay well. This chapter discusses the role of medical device technology in health care.

The Health Care Market

Epidemiology

The goal of the health care provider is to solve clinical problems (not to buy, use, or maintain medical devices). One must understand the relationship between the clinical problem and the devices used. For example, coagulation of blood can be achieved by using a bandage costing ten cents, an electrosurgical unit costing $3000, or a laser costing $120,000.

The causes of morbidity and mortality will vary. In wealthier populations and countries, people tend to live longer and tend to have more chronic diseases, including coronary artery disease, cancer, stroke, and chronic obstructive pulmonary disease. Poorer populations suffer more from infectious diseases, including diarrhea and malaria (Coe and Banta, 1992). The world's 10 leading killer diseases in 1996 (MHMG, 1998a) are listed below:

Coronary heart disease: 7.2 million
Cancer (all sites): 6.3 million
Cerebrovascular disease: 4.6 million
Acute lower respiratory infection: 3.9 million
Tuberculosis: 3.0 million
Chronic obstructive pulmonary disease: 2.9 million
Diarrhea (including dysentery): 2.5 million
Malaria: 2.1 million
HIV/AIDS: 1.5 million
Hepatitis B: 1.2 million

Health Care Spending and Health

Most of the world delivers the majority of health services through the public sector. The United States is an exception, delivering almost all services through the private sector. In 1998, in the developed world, over $2,000 was spent per person each year, representing on average, 8% of a nation's GDP.

For comparison, Table 80-1 shows a sample of countries with the highest per capita expenditures in their respective regions of the world. Although the United States spends more on health care than any other country in the world, its population does not enjoy the longest lifespan. Therefore, one must ask, "What does money buy in terms of health?"

In wealthier countries, demand for drugs and devices is strong. Long-term care for the aged represents a large and growing expenditure. In less-wealthy countries, spending is focused on drugs, which generally are quite cost-effective to treat diseases. Less funding is available for medical devices and for device maintenance (Table 80-2). In many developing countries, the majority of medical devices do no function, because of lack of maintenance, training, disposables, power, or other requirements.

Devices represent only 4.4% of the total spent on health care. Because of the size of the market, this money is spent differently in each part of the world. In the larger markets of the developed world, purchasers have direct relationships with device manufacturers. In the smaller markets of the developing world, manufacturers work through local distributors. Information such as hazard warnings and recalls can be lost because of these intermediaries, and spare parts and repairs tend to take much longer when working through distributors.

Devices are a relatively small part of worldwide health expenditures (Table 80-3). Much more is spent on drugs than is spent on medical devices. About one-half of the drug expenditures are for treatment of cardiovascular, respiratory, and central nervous system problems and infectious diseases.

Medical Device Sales

In the United States, imaging equipment and other "big ticket" items represent a smaller portion of the medical device market than do disposable devices (Table 80-4). In 1996, of the medical devices sold, X-ray equipment comprised 7%; general electromedical equipment, 14%; for surgical appliances and supplies, 30%; surgical and medical instruments,

Table 80-1 Health Care Expenditures and Life Expectancy

Area	Health care spending % GDP $USA	Health care spending Per Capita $USA	Life expectancy male Born 1999 Years	Life expectancy female Born 1999 Years
United States	14	$4,270	73.8	79.7
Switzerland	10.1	$3,564	75.6	83
Israel	8.2	$1,385	76.2	79.9
Argentina	8.2	$676	70.6	77.8
South Africa	7.1	$268	47.3	49.7

From WHO Report 1997 estimates, Table 8 pp 192–195; Table 2 pp 156–163.

Table 80-2 Estimated health care expenditures in 1996

	Developed world	Developing world
Total ($ billion)	$2,374	$552
Hospital care	37%	42%
Physician services	19%	22%
Drugs and medical Nondurables	12%	19%
Long-term care	6%	3%
Other	26%	14%

From Medical Health care and Marketplace Guide (MHMG, 1998b)

Table 80-3 Medical Device Expenditures 1996

Worldwide ($ billions)	$129
United States	42%
European Union	27%
Japan	15%
Rest of the world	16%

From MHMG, 1998c.

Table 80-4 Medical Device Sales 1999

Product	1999 Revenue $million
Incontinence supplies	$2010
Home blood glucose monitoring products	$1710
Wound-closure products	$1500
Implantable defibrillators	$1334
Soft contact lenses	$1159
X-ray equipment	$1140
Orthopedic fixation devices	$1122
Pacemakers	$1112
Examination gloves	$1088
Coronary stents	$1086
Ultrasound equipment	$1070
Arthroscopic accessory instruments	$919
Magnetic resonance	$890
Computed tomography	$790

From MDDI, 2000; MHMG, 2000.

30%; diagnostic products (mostly in the clinical laboratory),15%; and dental equipment and supplies, 4%.

Clearly, most of the money spent on medical devices does not go toward "big ticket" items. In hospitals, labor is the largest expense, averaging 53.8% in 1998 in the United States (Health Forum, 2000). Capital medical equipment is a small part of the budget of a typical hospital. Disposables are also a significant expense for most hospitals and may exceed the cost of capital medical equipment (Table 80-4).

One sees a similar pattern when analyzing individual procedures. For example, in one institution's costs for a thoracotomy for lung cancer, salaries accounted for 54%, supplies and medication 27%, and capital equipment only about 7% (Marrin et al., 1997).

Definitions

Although medical devices do not represent society's largest health care expense, they remain crucial to the delivery of health care. In all societies, government often is involved in use and sale. The United States Food and Drug Association defines a medical device as: "an instrument, apparatus, implement, machine, contrivance, implant in vitro reagent, or similar or related article, including any component part, or accessory, which is:

1. Recognized in the official National Formulary, or in the United States Pharmacopoeia, or any supplement to them;
2. Intended for use in the diagnosis of disease or other conditions, or the cure, mitigation, treatment, or prevention of disease, in man or other animals; or
3. Intended to affect the structure or any function of man or other animals, and which does not achieve its primary intended purpose through chemical action within or on the body of man or other animals and which is not dependent upon being metabolized for the achievement of any of its intended principle purposes." (MDA, 1976)

Devices that meet this definition include disposables (some are reused for a single patient, and some are thrown away after a single use), reusables, and implants. Medical devices are part of the elements of health care that include medical procedures, surgical procedures, and drugs.

The medical device technology area of medical information–based devices, including information systems, diagnostic tools, picture archiving and retrieval systems (PACS), and electronic patient records is new. The trend certainly is to include more information-processing capability in all medical devices (see Chapters 97, 98, 100, and 102).

Attitudes

The intimate relationships among the technology, the practitioner, and the patient are critical distinctions in the medical field. Devices are used to affect the health of the patient. The practitioner feels responsible for achieving the best possible result, so he must feel comfortable using the devices. Practitioners need to have experience using the devices and must believe that the devices work properly. Familiarity with devices and the results that can be obtained with them are important to the health care professional. Clinical effectiveness is hard to prove, and practitioners can be skeptical about acceptable results being obtainable via a new technology. This need for reliability and consistency and the need for training in the use of technology lead many medical professionals to resist changes in technology unless a clear benefit can be demonstrated in clinical practice. If the current device is adequate for a practitioner's art, the practitioner will resist acceptance of new technology, especially if it will require more training, or if it will change the way the clinician works.

For the clinical engineer to perform effectively, the attitudes and the environment of the user must be understanding. Health care professionals are trained in the use of devices, but not necessarily in the general principles of engineering, mathematics, or physics relating to the devices. Many health care professionals see devices as tools; they prefer not to spend too much time on the tool, so that they can focus on the clinical needs of their patients. Time is usually at a premium in health care settings, so the health care professional does not always have the necessary amount of time to learn how to use a new device. Clinicians might be unaware of alternatives to the devices they are using. Clinical engineers can bridge the gap between technical knowledge and clinical needs and thus can help the clinician to deliver the best possible patient care.

Nomenclature and Codification

Nomenclature systems are essential to managing the great variety of medical devices. A unique reference for each device enables one to organize information related to that device, such as work orders, service history, service contract information, hazards, recalls, in-service training, and other device-related expenditures. The problem is similar to that facing a library. Having millions of books is a useless resource if they cannot be located. Just as a library uses a classification system to organize its books, the clinical engineer needs to have a nomenclature system to uniquely identify medical devices. Maintaining the nomenclature system is quite time-consuming, so most institutions use a standard system, such as the Universal Medical Device Nomenclature System™ (UMDNS™), which is available without charge from ECRI (www.ecri.org).

It is no surprise that there are standards, rules, and regulations for the use of technologies that impact health. In ECRI's directory of national and international standards, there are over 43,000 medical equipment standards, prepared by over 1400 agencies throughout the world. Clinical societies also publish guidelines for their members regarding medical technology. The National Guidelines Clearinghouse (http://www.guideline.gov) is a good source. Some hospitals require the practitioner to complete a training program before using certain technologies, such as surgical lasers or endoscopes.

Safety

Safety is also related to standards, guidelines, and, mostly, common sense. Maintenance and device selection are important, but the majority of the problems come from the way the equipment is used and the relationships between the device and other systems (see Chapter 60). Energy-producing devices, such as lasers and radiological equipment, have special engineering- safety requirements.

Device classification systems depend on perspective. If one were classifying devices for safety, one might use high risk for life support devices; medium risk for devices whose failure would affect patients, but not cause serious harm; and low risk for devices whose failure would not be likely to harm patients.

The Environment of a Device

Devices do not help people all by themselves. They help only when they are part of a medical intervention. An intervention requires a patient, a practitioner, and it might require a drug, device, and medical facility. One must understand this context to ensure that devices contribute appropriately to clinical outcomes. More specifically, one must take a systems approach (see Chapter 59) and must consider the four following interfaces (Figure 80-1) (see Bruley, 1994):

- Device–user
- Device–patient
- Device–environment/facility (hospital, home, or other)
- Device–accessories/disposables/consumables

Device–User

Although one thinks of the doctor as the user of the device, most of the time a nurse or technician is the actual user. One must be aware of the differences in perspective, training, and education between the person who specifies the device and the person who uses it. Users must be trained to operate the device and to interpret the results. Most of the medical device problems in the hospital are related to user error. Frequently, the clinical engineering department can help to reduce errors by training users (see Chapter 74).

Device–Patient

Almost all devices touch patients or draw samples from the patients' bodies. The wide variation in the patient population must be understood for an adequate understanding of ways in which the device will accommodate that variation. It is crucial to understand exactly how a device interfaces with a patient and how it can accommodate a range of patients. Certain groups of patients, such as newborns, children, very large or small adults, and elderly patients, might have special needs. Some devices require special accessories so that they can accommodate groups of patients.

Device–Facility

All devices have conditions and requirements for their storage and use, including temperature, humidity, electrical power, water, pH, shielding from electromagnetic fields, and connection to specialized gases. Be especially careful when devices are designed to be used in one environment, such as a major hospital, but are used in another environment, such as a home (Dyro, 1998). Similarly, caution must be taken when devices designed for adults are used with children. Even simple devices, such as sharps containers, have contextual implications. In a children's hospital, they must be out of the reach of the child because the child might consider the device to be a toy and thus might want to put his hand inside to play with it.

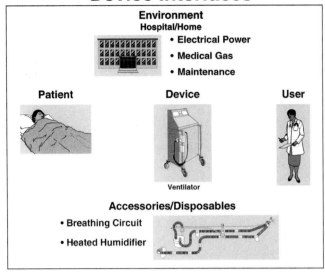

Figure 80-1 Device interfaces.

Accessories, Disposables, and Drugs

Device-Accessories/Disposables/Consumables

Most devices have reusable and disposable components. An infusion pump has a disposable infusion set; a defibrillator may have disposable electrodes; and an X-ray unit typically has a film cartridge. The entire group has to function correctly. The manufacturer of the device might not produce or sell the disposables or consumables. Conversely, the manufacturer of the disposable or consumable might not sell or produce the medical device. Sometimes, only one manufacturer makes the disposable or consumable. The relationship between the disposables and the consumables affects the use and cost of the device. The profit from consumables is so high for some clinical lab equipment that the manufacturer will offer the device for free if the hospital commits to purchasing the consumables from the manufacturer.

Disposable Devices

Disposable devices are designed to be discarded after one use, or after multiple uses on the same patient. Some of these devices are quite expensive (catheters used in catheterizations can cost $400 or more), so there are powerful financial incentives to reuse devices, even though the devices were designed for a single use. Some institutions reuse single- use devices. There is a great controversy about the risks to the patient when single-use devices are reused. In many cases, there are alternatives to a reusable and disposable device.

Some associate the following characteristics with reusable devices: Strength, difficulty to clean, and low cost (i.e., if labor costs to reprocess the device are low). Disposable devices are thought to be cleaner, easier to keep sterile, less robust, and, ultimately, more expensive. When making a choice, it is important that one be certain as to the reason for deciding on a reusable or disposable device, and to review assumptions as time goes on. The cost of reprocessing can change as labor costs change and as new technologies to help in reprocessing and sterilization become available. Both types of devices, reusable and disposable, still need to be inspected before use.

Differences Between Devices and Drugs

Drugs and devices sometimes come together directly, as in an intravenous (IV) pump, and indirectly, as in the case of clinical lab analyzers, often determine what drugs are prescribed by the clinician. Devices such as stents are now impregnated with drugs during manufacture. The line between drug and device is changing, so it is important to appreciate the differences between them.

Devices and drugs are used to help patients, require a trained person. Drugs, however, do not require periodic maintenance, do not break, and do not require as many people to keeping them working. Drugs do not have a "work history," nor do they require service contracts and the same type of record keeping. Teaching a doctor how to prescribe a drug is quite different from teaching a doctor, nurse, or technician how to use a medical device.

Example: Treating Patients with Diabetes

The following discussion details the requirements for treating diabetes, one of the world's most common diseases. Diabetes is expected to affect over 200 million people worldwide by 2005. In the United States, about 10 million people have been diagnosed, and another seven million have the disease and do not know it. Annually, over 600,000 new cases are diagnosed in the United States, where the disease afflicts men and women in equal proportion. Diabetes is the seventh leading cause of death in the United States (NDIC, 2001).

Diabetes and its associated complications are among the most prevalent, costly diseases in the world. The worldwide market for glucose-monitoring products exceeds $4.7 billion. In the United States, direct costs of diabetes care are estimated at about $50 billion, almost 6% of the total personal health care expenditures, while the complications of uncontrolled diabetes result in nearly $100 billion in annual medical costs.

Diabetes is a chronic disorder of blood glucose regulation that commonly results in the development of cardiovascular, ophthalmic, neuropathic, and nephropathic complications. It is classified by two major types: Type 1 (lack of insulin production and release) and type 2 (resistance to insulin's actions). In type 1diabetes, islet cells have been destroyed by an autoimmune response, and insulin production is reduced to insufficient levels for maintenance of blood glucose regulation; thus, insulin deficiency is the main cause. In type 2 diabetes, the body cannot properly respond to the insulin produced by the pancreas. Glucose remains in the blood instead of supplying the body with the fuel it needs.

In the Unites States, about 500,000 people have type 1 diabetes, which typically begins in childhood. The rest have type 2 diabetes, which usually develops after age 40. Type 2 diabetes is often found in elderly people and is also associated with obesity. Several recent studies have shown that keeping glucose levels close to normal was associated with a major reduction in the secondary long-term complications of diabetes. For a patient with type 1 diabetes, this management requires blood glucose testing several times per day, three to four daily insulin shots, and lifestyle changes. Tight control is recommended as an important way to delay the onset and to dramatically slow the progression of long-term complications from diabetes, such as retinopathy, neuropathy, nephropathy, and cardiovascular disease. About 40% of the patients who are diagnosed with type 2 diabetes will require insulin therapy and will benefit from tight control of their blood glucose levels. The disease must be managed during the life of the patient because there is no cure.

The diagnosis and treatment of an adult patient with type 2 diabetes is discussed below. The equipment used, the location of the equipment used, and the practitioners involved in treating the patient are described. The equipment ranges from simple to complex devices and includes disposables, reusables, and home care equipment (i.e., a wide range of technology involved in treating a patient). The relevant equipment, assuming that the disease progresses significantly, is described. Fortunately, not all patients will suffer all symptoms of the disease.

Diagnosis

During a routine physical examination, the patient might report symptoms such as excessive eating, thirst, frequent urination, or weight loss. Blood tests confirm the diagnosis. Because many patients have no symptoms, screening is performed by measuring the glucose level as part of general blood tests performed on patients over age 45. Urine tests for glucose are no longer used to diagnose diabetes in the United States.

Blood is drawn in a physician's office and is contained in either a sealed test tube that has a preservative to inhibit glycolysis, or in a special tube with a barrier to keep the serum/plasma separate from the red blood cells. To prepare the sample so that it can be read on the clinical chemistry analyzer, it must be spun on a tabletop centrifuge (acquisition cost: $2000) at 3200 to 3,500 rpm to separate the serum/plasma from the rest of the sample. (Most clinical laboratory equipment does not accept whole blood—only serum/plasma). The sample should be stored cold and should be sent to the clinical laboratory within a few hours of being drawn.

Once received by the clinical laboratory, the blood sample is tested on a clinical chemistry analyzer (acquisition cost: $100,000) by a laboratory technician. Glucose levels outside of predetermined values indicate that the patient has diabetes. The tests should be repeated on another day, to confirm the diagnosis.

Diet and Exercise

Many patients with type 2 diabetes can be treated with diet and exercise. Counseling is important. Ideally, the patient meets with a dietician or a nurse educator to obtain advice regarding changes in diet and activity. The patient's weight is monitored at home and during office visits, using a patient scale.

If diet and exercise do not control the levels of blood glucose, the patient will be prescribed oral medications, or, in some cases, insulin. He or she will need to monitor blood glucose levels at home, using a blood glucose monitor.

Blood Glucose Monitors

Blood glucose monitors are portable, battery-powered devices (ECRI, 2000a). The patient places a drop of blood from the finger onto a paper test strip, which is impregnated with a glucose-specific enzyme that reacts with the glucose in the blood. The strip is inserted into the blood glucose meter and is read using either reflectance photometry or electrochemical technology to determine the glucose level in the blood. These monitors are also used to monitor blood glucose levels in clinical settings.

The average cost for home blood glucose monitors cost is $55. They last about three years. (Hospital units, which have additional capabilities to store and transfer information, cost about $,000) The strips cost about $0.35 apiece, for home use ($0.70 for hospital use). Assuming two readings per day, the test strips cost about $260 per year. The worldwide market for glucose meters and strips is about $3 billion dollars per year and is expected to double by 2008.

Persons with diabetes need to have their blood glucose levels measured by a clinical laboratory two to four times per year, depending on the severity of their disease. A hemoglobin A1c test is performed in the clinical laboratory using manual or automated equipment at least twice per year. Other tests, including kidney-function and urine-microalbumin tests, can be performed as needed.

Patients can be prescribed medications to help the body to use the insulin that it produces more efficiently. Most of these patients monitor their blood sugar at home by using a blood glucose monitor.

Insulin Injections

Some type 2 diabetes patients will require insulin injections in order to maintain appropriate blood glucose levels and thus to decrease their risk of complications, which can include blindness, kidney damage, nerve damage, and circulatory problems. Home monitoring using a blood glucose monitor is a critical part of the treatment.

These patients must inject themselves two or more times per day with insulin, using an insulin syringe. Insulin must be stored in a refrigerator and must be protected at all times from temperatures greater than 86° F (30° C) or less than 36° F (2° C). Injections are given at room temperature to facilitate absorption, so the patient may keep a personal supply at room temperature. Insulin is stable for one month at room temperature. The patient always needs to have an extra vial of insulin on hand in case of emergency.

The syringes cost about $28 per box of 100. They are sold as single-use devices, although some patients choose to reuse them. The annual cost for two syringes per day is about $200 per year.

Blood Testing

Patients with Type 2 diabetes normally will have a blood glucose test performed every three months, and a hemoglobin A1c test performed every six months. Glucose monitoring represents 19% of all *in vitro* diagnostic tests performed annually in the United States.

Because the patient uses a blood glucose monitor at home to check their glucose levels and to adjust their insulin dosage, it is important to compare the results using home machines to those obtained by the clinical chemistry analyzer. Agreement should be within 15%.

Because diabetics are at greater risk of heart disease, cholesterol testing (or, more precisely, testing for lipids and triglycerides) is performed one or more times per year. Some diabetics will develop kidney disease, which may progress to the point where they require dialysis.

Dialysis

Dialysis is the removal of toxins from blood by a machine (ECRI, 2000b). Some patients can treat themselves at home by flushing their abdominal cavity (peritoneal dialysis—typically performed daily). More often, hemodialysis is required. It takes place during three sessions of five hours, using a dialysis machine in a hospital, nursing home, or other facility. The equipment required includes a water-purification system, dialysis machine (acquisition cost: $25,000), dialyzer, and disposables. The patient is connected to the machine by a dialysis technician. The machine filters the patient's blood, toxins are removed, and the blood is returned to the patient. A medical specialist supervises the operation of the facility.

Eye Examination

To prevent blindness, an ophthalmologist performs detailed eye exams several times per year. Retinal surgery is often required and is performed using an ophthalmic laser (ECRI, 2000c) (acquisition cost: About $55,000).

Heart

Because diabetes can lead to coronary artery disease, the cardiovascular condition of the patients is carefully monitored using EKG and stress tests. Prevention includes control of weight, exercise level, and diet. If coronary artery disease develops, bypass surgery, stents, angioplasty, and laser treatments may be performed.

Circulatory complications can lead to the need for amputation of the feet and legs. More than half of the amputations that take place in the Unites States are performed on diabetics. Wounds do not heal well for diabetics, and wound-healing on extremities is difficult.

Summary of Devices used in Diabetes

Diabetes is one of the most costly diseases in the world. The direct cost of diabetes in the U.S. represents about 6% of the total personal health care expenditures. Treating the complications of this terrible disease costs nearly $100 billion each year.

The devices that are used to manage diabetes are found in home, doctor's offices, hospitals, and specialized centers. Patients, caregivers, doctors, nurses, laboratory technicians, and medical specialists use the equipment. The equipment chain includes syringes ($0.28) scales, sphygmomanometers, stethoscopes, fundoscopes, blood glucose monitors, refrigerators, and clinical chemistry analyzers ($100,000). The annual cost for syringes and test strips for home glucose measurements is $460. More sophisticated equipment is required when complications occur. All of the equipment in this chain is required in order for the patient to receive proper treatment.

References

Briggs J. Diagnostics Industry Overview. *Clinical Laboratory Products Magazine*. http://www.clpmag.com, 2001.

Bruley M. Accident and Forensic Investigation. In Van Gruting CWD (ed). *Medical Devices, International Perspectives on Health and Safety*, Amsterdam, Elsevier Science, 1994.

Coe GA, Banta D. Health Care Technology Transfer in Latin America and the Caribbean. *International Journal of Technology Assessment in Health Care* 8:2 255–267, 1992.

Dyro JF. Methods for Analyzing Home Care Medical Device Accidents. *J Clin Eng* 23(5):359-368, 1998.

ECRI. Portable Blood Glucose Monitors (Update Evaluation). *Health Devices*. 29(6):200–237, 2000a.

ECRI. Hemodialysis Units. Health care Product Comparison System, 2000b.

ECRI. Ophthalmic Lasers. Health care Product Comparison System. 2000c.

MDA, 1976. (21 USC 321h.) Medical Device Amendments of 1976.

Health Forum, LLC. Table 8, 1988 Utilization, Personnel and Finances. *Hospital Statistics 2000*, 164-165.

Marrin CAS, Johnson LC, Beggs VL, Batelden PB. Clinical Process Cost Analysis. *Ann Thorac Surg* 64:690-94, 1997.

Medical Device and Diagnostic Industry (MDDI). *Industry Snapshot* 12:47–56, 2000.

Medical Health Care and Marketplace Guide (MHMG), 16th Ed., Vol. 1 2000–2001. October 2000, Dorland Health care Information, Philadelphia, PA USA.p I-1010

Medical Health care and Marketplace Guide, 14th Ed., 1998, Dorland's Biomedical, Philadelphia, PA USA; 1998a, p I-21.

Medical Health care and Marketplace Guide, 1998, Dorland's Biomedical, Philadelphia, PA, 1998b, p I-28.

Medical Health care and Marketplace Guide, 14th Ed. Ed., 'Philadelphia, Dorland's Biomedical, 1998c, p I-495.

National Diabetes Information Clearinghouse (NDIC). Diabetes Statistics. http://www.niddk.ni h.gov/health/diabetes/pubs/dmstats/dmstats.com, February 2, 2001.

World Health Organization. *The World Health Report 2000 Health Systems: Improving Performance*. Geneva, World Health Organization, 2000.

World Health Organization. *The World Health Report 1997 Health Systems: Improving Performance*. Geneva, World Health Organization, 1997.

Further Information

Association for the Advancement of Medical Instrumentation
1110 North Glebe Road, Suite 220
Arlington VA 22201-4795
Tel: 703 525 4890
Fax: 703 276 0793
http://www.aami.org

American Hospital Association
One North Franklin
Chicago Illinios 60606-3421
Tel: 312 422 0300
http://www.aha.org

ECRI
5200 Butler Pike
Plymouth Meeting, PA 19462
Tel: +1 610 826 6000
Fax: +1 610 834 1275
http://www.ecri.org

Journal of Clinical Engineering
Aspen Publishers, Inc.
7201 McKinney Circle
Frederick, MD 21704
Tel: 800 638 8437

Canon Communications LLC
11444 W. Olympic Blvd.
Los Angeles, CA 90064
Tel: 310 445 4200
Fax: 310 445 4299
http://www.devicelink.com/mddi

National Guidelines Clearinghouse (http://www.guideline.gov).

Serpa-Flórez F. Technology Transfer to Developing Countries: Lessons from Colombia. *International Journal of Technology Assessment in Health Care* 9:233–237, 1993.

81

Medical Device Design and Control in the Hospital

Joel R. Canlas
Clinical Engineering and Technology Management
Department, Beaumont Services Company, LLC
Royal Oak, MI

Jay W. Hall
Manager, Clinical Engineering, St. John Health
Detroit, MI

Pam Shuck-Holmes
Product Development Engineer
Detroit, MI

Clinical engineers in the hospital setting use modern technical, scientific, engineering, and management training. Specialization, knowledge, and experience of clinical engineers serve humankind by contributing significantly to the provision of safe and effective medical technology. This view is in harmony with the American College of Clinical Engineering (ACCE) definition of a clinical engineer as "a professional who supports and advances patient care by applying engineering and managerial skills to health care technology" (Bauld, 1991). In this chapter, the authors, reflecting on over 40 years of collective experience as clinical engineers in a hospital setting, describe their role in designing, manufacturing, evaluating, and controlling medical devices to ensure their safe and effective application in health care.

The extent to which a clinical engineer (CE) is able to perform a responsible function in the hospital setting depends to a large degree on the acceptance and understanding of their roles and capabilities by the hospital administration and staff. Without the full support of the hospital, from the lowest to the highest levels, clinical engineers cannot function to the best of their abilities. Thus, when empowered, clinical engineers not only follow a course of action specified in good clinical engineering practice guidelines but also go beyond the call of duty. The CE's clients (i.e., doctors, nurses, respiratory therapists, imaging specialists, other hospital staff, patients and their families, health maintenance organizations [HMOs], and preferred provider organizations [PPOs]), are usually well informed about medical technologies, which is a result in part, of the Internet and the accessibility of up-to-date information that once lay only in the books and journals stored in the deep recesses of libraries.

As clinical engineers, the authors have striven, consciously and subconsciously, to change the staff's perception of clinical engineers as simply repairmen armed with a cell phone and three pagers dangling from their belts. At William Beaumont Hospital, CEs have focused on a systems approach to medical technology management. For example, hospital beds are not treated as simply a support surface for patients. The CE considers a hospital bed from the perspective of mechanical design, safety features, operating mechanisms, interfaces with the patient, intelligent nurse-call and alarm systems, and electrical power requirements. The CE analyzes the risks that the bed or its accessories pose to a patient, a nurse, or other equipment. Similarly, a robotic system includes the control device, all of the system components, the maintainers, the users, the storage environment, the physical-facility requirements, the specialized test equipment, and the training requirements. If all devices and systems were so perfect as to be predictable with no possibility of causing injury to the patient and user, then the need for clinical engineers would diminish. As this is not the case, the clinical engineer must identify and utilize available resources, examples of which include the purchasing department, which can provide financial analyses and assess the manufacturers' financial health, test equipment, education and training in the scientific and engineering method, the manufacturers' engineers and technical staff, the Internet, and device-user experiences.

As in the health care industry at large, in the hospital there are reactive and proactive elements in managing medical technology. The proactive, or preventive, component in medicine staves off the possibilities that can cause disease in a patient; e.g., preventing obesity can prevent the various diseases known to be associated with it such as diabetes (see Chapter 80). The reactive, or treatment, component addresses the patient who is already afflicted with a disease and the means for making them well again. This similar approach works well with the hospital's overall goal of providing world-class customer service.

By describing clinical engineering experiences in this chapter, the authors provide practicing and prospective clinical engineers with a window into the clinical environment, showing ways in which CEs can help the hospital and patients and can advance health care. Engineers are never satisfied with the status quo. They constantly explore various solutions to a problem, seeking the optimal outcome.

The following anecdote captures the sense of the engineer's fervent desire to seek engineering solutions to a problem. A long time ago, three folks were about to lose their heads on the guillotine. The first to go, a peasant, was asked by the executioner whether he wanted to have his head cut off while facing the sky, or downwards. He chose to look downwards. The blade was released but stopped just short of his head. Everyone took it as a miracle, and the peasant, to his extreme joy, was pardoned of his crime and was set free. The same thing happened to the second person, a lawyer, who chose to face the sky. The last person, an engineer, also choosing to face the sky, saw a knot in the rope and, just before the guillotine was released, cried out, "Hold everything—I see the problem!"

Design and Modification of Medical Devices in the Hospital

Clinical engineers thrive on change. They must, because medical technology is constantly evolving and changing. What seems to work well today could be superseded by something better tomorrow. Keeping up with the technology, and managing current and future needs are two services that clinical engineers can provide adequately. This goes well with the hospital's goals of customer satisfaction by providing a holistic approach to medical care in a timely and effective manner. By helping to ensure an environment of safe and effective medical devices through good technology management, CEs enable clinicians to have the armamentarium that is needed to treat patients effectively.

Managing medical technology requires that both the medical device and the human factor component be understood. Ensuring that a medical device is safe and effective requires consideration of a host of factors, such as device design, construction materials, performance history in the clinical setting, the time in the market, FDA allowances and approvals, the manufacturer's reputation, parts availability, and the vendor's responsiveness to its clients' needs. The human factor component of technology management looks at the human interface, which includes the patient, the clinician, the clinical/biomedical engineer, the service provider(s), and the manufacturer's representatives. The hospital relies on the clinical engineer to facilitate communication among medical, nonmedical, and technical people. In addition to a working knowledge of the engineering process and specifications, the CE must be able to translate this information to such diverse professionals as doctors, nurses, financial managers, administrators, and technicians. At the same time, the CE needs to be able to transmit the clients' feedback on a medical product to the manufacturer in a constructive fashion in order to resolve product problems in a timely fashion. Such feedback often results in product improvements as well.

Ensuring Medical Device Safety by Clinical Engineering

Various tools are available to the CE in the hospital setting to ensure the safety of medical devices. These tools, tried and tested by the authors, are listed below.

Equipment Selection and Evaluations

Clinical engineers conduct short equipment evaluations in-house to provide an educated estimation of the product's usefulness, safety, and effectiveness (see Chapter 33). The focus at this stage is on the product's suitability as a hospital standard. Such factors that must be considered include the manufacturer's financial health, which could determine whether the company will even exist in 10 years to support its products, and the product's record of performance.

The manufacturer's financial health can be assessed by analyzing information gleaned from annual reports, the Internet, and information services such as ECRI and the MDE Group. Purchasing and finance departments, by the nature of work that they do, are excellent resources. A product or a technology's track record, i.e., performance history, can be assessed through various sources, like interviewing user hospitals (vendors should be able to provide you with a list of hospitals and specific departments using their products), researching websites such as the FDA's Manufacturers And User Device Experience (MAUDE) database at www.fda.gov/, and ECRI at www.ecri.org.

Hospital technology management requires the monitoring of the life cycles of medical devices and systems (see Chapter 82). Systems comprise not merely the device, but also the support group of equipment and services that ensure its proper functioning and the competency of the user group. As a device approaches the end of its rated life expectancy, or if it has required inordinately high maintenance, this information must be conveyed to the user group. However, as experience shows, through constant communication, the clients (i.e., the hands-on staff members (users)) inform the CE of changes in a device such as deterioration, inaccurate measurement, or intermittent operation. When equipment has been identified for obsolescence in at least two years, work begins on selecting replacements.

At times, if the technology has not changed significantly, and the hospital is satisfied with the equipment, only a financial analysis of the pricing and service quotations from several vendors and manufacturers is required. Rapid changes in medical technology, however, dictate that even the most mature, most often used, and most accepted technologies, despite small generational improvements, should be compared with newer competing products and technologies through evaluation and, where necessary, clinical trials. Purchases that exceed a certain monetary value require a formal bid process entailing a request for proposal (RFP). The RFP is addressed later in this chapter.

A medical device becomes a hospital standard after the process of engineering evaluation, clinical trials, financial analysis (based in large part on the responses to the RFP), and selection as the most cost-effective product.

Clinical Trials

Participation in clinical trials is often an enjoyable, rewarding, and challenging aspect of the CE's work. The rapid growth in the biomedical engineering and its application to medical device technology market results in new and varied devices coming to market (see Chapter 80). This, in turn, presents the CE with more devices to evaluate, and more clinical trials to manage. Clinical trials find CEs becoming increasingly involved with the clinical and medical staff. The following is a list of many of the steps taken by CEs in managing and running a clinical trial:

- Obtain user department cooperation. (This tends not to be a problem because the department usually has requested the trial.)
- Arrange with the vendors and manufacturers for use of their products on a clinical trial of a duration of a few days to several weeks. Vendors are generally quite helpful, even supplying necessary, and often costly, disposables. Occasionally, the hospital supplies the disposables, especially if these are stocked items.
- Arrange with the vendors and manufacturers for in-service training of the staff involved with the trial. Obtain user and service manuals and make them readily available to staff. Schedule the trial so that most staff members have an opportunity to use and evaluate the product.
- Prepare a survey instrument that lists the desired specifications and evaluation criteria (e.g., ease of use). It is recommended that the survey instrument be on only one page, that the scoring be succinctly explained, that all terms used be understandable by the evaluators, that it states where to return completed forms, and that it contain a space for the name and mailing address of the evaluator.
- Clearly explain and document test methodology to ensure scientific validity of the study.
- Facilitate the monitoring of the devices for proper operation. This monitoring is usually the responsibility of the vendor or manufacturer. Because the vendor cannot be at the hospital at all times, a person in the department who is properly in-serviced should be designated as the resource person.
- Prepare a schedule lasting over a reasonable time that covers every possible step needed for the clinical trial. This is essential to ensure timely completion of the trial.

Negotiating for Safety Features

Sometimes certain features that help to improve the safety of the device are incorporated or are even sold as options. One should ascertain that the products provided for the trial have all the desired specifications and that the competing brands have the same features.

Note the willingness of the manufacturer to provide training of in-house maintainers to the same level as that of the manufacturer's field-service technicians. While in most cases, the manufacturer is quite willing to provide this level of training, the manufacturer occasionally claims that the proprietary nature of its technology militates against this. In these cases, one should relinquish servicing responsibility to the manufacturer or should negotiate a plan for utilizing in-house maintainers. The reasons for this are threefold: (1) immediate response to any equipment problems; (2) control over the safety, turnaround times, and costs; and (3) an immediate familiarity with the equipment and user group.

Education

In-Servicing of Clinical Users

The hospital should demand in-servicing for all new devices, including new standard devices that are upgrades of the old standard being replaced. Every department has an assigned education specialist, who will need to be trained well so that that specialist can train other staff members, i.e., a train-the-trainer program. (Preferably, every user should obtain hands-on training initially.) That department should provide periodic in-servicing to users, either on demand or on a scheduled basis. Training should entail the theory of operation. A knowledgeable user is a prerequisite for efficient and safe use of medical devices.

The operation of complex devices is not easily mastered by typical users. It is, therefore, imperative that staff members continuously receive in-servicing that includes theory of operation, device features, and available resources with the goal of achieving staff familiarity with, and respect for, devices.

Training the Maintainers

Training and in-servicing should be provided to the servicing department. In particular, the in-house department should designate an education specialist, who will be the department's resource when questions arise.

Surveillance

Monitoring Medical Device Reports for Indicators

Researching the products' safety record can be done from several sources, including the hospital's medical equipment database; other hospitals; professional colleagues at local, regional or national conferences; the FDA; and ECRI. Maintaining a healthy communication line with every departmental client in-house helps tremendously because bits and pieces of information gleaned from users can prove invaluable in fine-tuning requirements to help keep the device in constant and safe operational condition.

Variance Reporting Process

Once the device has been purchased, has passed all safety inspections, and has been deployed in the hospital, surveillance of device performance and safety must occur. Any variance in device operation must be documented by means of a variance, incident, or adverse-occurrence report. (The title of such a report varies from institution to institution.) Any equipment that malfunctions or fails while in use on a patient must be investigated, and the variance, findings, and resolution must be recorded. The hospital has legal responsibilities under the Safe Medical Devices Act of 1990 (Alder, 1993; JCAHO, 1993; 21 CFR, 1995).

Cradle-to-Grave Management

This is a circular process of ensuring the continuity of making available at all times the necessary patient care equipment. Key to this process is the collaborative interaction among the client clinical staff, equipment maintainers, clinical engineers, and purchasing and financial analysis departments. Having responsibility for certain steps in this process and interacting with members of the project team, the clinical engineer stands to achieve a relatively safe, reliable, and cost-effective medical equipment inventory. The various components in this process include the following:

- Identifying the equipment to be purchased based on (1)
  clinical need (newer technologies have a way of rendering some older technologies obsolete);
- Planned obsolescence (the device is nearing the end of its rated useful life);
- Maintenance cost (the cost of maintaining a device exceeds depreciated value).
- Conducting pre-purchase evaluation of equipment. This process consists of two components: (1) performing engineering evaluations of competing products, and (2) conducting clinical trials, after narrowing the field down to a few candidates, to rank and select the best fit. Where the cost of purchase exceeds a certain amount, a more formalized pre-purchase evaluation is conducted, the RFP that involves the competing vendors and manufacturers to submit competitive bids that include the desired hospital specifications.
- Justification of the selected product and vendor and manufacturer is based on results of the pre-purchase evaluation or the RFP, cost-benefit analysis, life cycle costing, and the anticipated cost of ownership that includes, among others, expected cost of maintenance, depreciation, and obsolescence.

Clinical engineers, bridging the communication gap between clinicians and design engineers, can positively influence product design. Frequently, the CE explains to the manufacturer those things that can improve a product's efficiency and safety. Consider an example that the authors encountered in which foot supports used on an operating room (OR) bed repeatedly loosened and fell, but only on one side. Clinical engineering analysis revealed that the clamps' unidirectional thread enabled them to tighten down on the bed rail when a load (the foot) was placed on it but to loosen on the other side because the load applied a loosening force. Working with the manufacturer produced a clamp that works on either the right or the left side of an OR bed, without loosening.

At some time in the life of a medical device, it is likely that failure will occur. If the clinical engineering department provides in-house service, the CEs and BMETs must be trained at the same level as that of the manufacturer's service providers. Based on the manufacturer's recommended preventive maintenance schedules, the equipment is included in the scheduled maintenance program.

Maintaining an up-to-date database is a prerequisite to being able to predict early on a device's maintenance schedule. Also, including the projected useful life span of every device in this database enables the clinical engineer to begin the replacement process before frequent breakdowns occur, which would force the hospital to take costly measures like renting similar equipment to make up for the shortage.

Ensuring Efficacy and Effectiveness

Specifying Medical Devices for Cost-Effectiveness

Cost-effectiveness means that the product meets all of the desired specifications at a price that is competitive with, or even lower than, that of other vendors. Clinical engineers must work closely with the client to prioritize specifications, listing them as (1) those that are necessary; (2) those that are desired but that the organization can live without; and (3) those that are not needed, based on current and anticipated use. It is also best to work closely with the manufacturers, who can recommend the best devices to suit the hospital's needs without including unwanted or nonessential features that would drive up the price.

Anticipating Future Needs During Selection

Health care is a constantly evolving process, and, at times, it is even revolutionary. The instrumentation that is required to deliver current state-of-the-art technologies could become obsolete in a short time. For example, robotics is making great strides in the hospital setting. William Beaumont Hospital's medical laboratories are major users of robotic systems to perform various tests efficiently and cost-effectively, utilizing small specimens. It allowed the hospital laboratory to expand its capacity and capabilities without increasing skilled and educated labor. Robotic surgical systems have also revolutionized the way surgeries are performed in house. A robotic arm system was quickly superseded by a more sophisticated one that is a precursor of one enjoying more widespread use in surgical telemetry.

The clinical engineer, with various information sources available, is becoming an unofficial bio-informatics officer. With our participation in the institutional review board (IRB) process in-house; the technology acquisition process; hands-on familiarity with the standard equipment used in-house; communication on almost a daily basis with our clients regarding equipment performance; keeping up with medical research and current trends; consultations with clients, manufacturers, and their representatives, the in-house clinical engineer can provide technological guidance to the future.

An engineering professor's favorite question to his students is "When is software considered obsolete?" The answer is "As soon as it is marketed." Clinical engineers need to be alert to the possibility that the product that is recommended to the hospital today will be obsolete in a short time. Clinical engineers do not need to have extrasensory perception to be able to look into the future. Being well-read and scientifically curious enables CEs to unearth information that, by itself or in various combinations with other technologies, can help to paint a picture of what is to come.

Evaluating Designs for Efficiency and Reliability

When assessing efficiency and reliability, the CE should have the actual device, its user, service manuals, and an in-service from the manufacturer's representative. A block of time should be set aside to operate the device, while becoming acquainted with its various features and capabilities. This opportunity to learn new technology is quite often stimulating and enjoyable for the CE. Assessing a product's efficiency and reliability, long before it is used, need not be a guessing game. If the product's design and technology are relatively new and untested in the market, the CE looks at the manufacturer's reputation and record of accomplishment for similar products. The authors prefer devices that have been on the market for at least one year. Problems occurring in this one-year introductory period would appear in manufacturers' recalls, warning letters, ECRI Hazard Alerts, and the FDA MAUDE database.

The manufacturer's representatives usually can provide the CE with a list of user hospitals in the region or in the entire country. Checking with these users usually provides one a good insight into what to expect from the product, especially if one communicates with all levels of users and maintainers in the hospital. This is where membership at the local biomedical engineering society or any similar professional groups can be of great help.

Minimizing User Errors through Ergonomics in Device Design

Elegant design often relies upon the KISS ("Keep It Simple, Sweetheart!") principle. The design should make the equipment intuitive, simple, and easy to use. The fewer controls for staff to manipulate, especially during procedures, the better. All critical controls should be on the front panel and not susceptible to fluid spills, which is a constant risk in the hospital setting. Operating the equipment should eliminate, as much as possible, steps that are unnecessary and consolidating those that are similar and that may require a modified sequence of operation. Going beyond the equipment itself, make sure that the supporting infrastructure is available in the areas in which it will be used. A simple oversight can lead to terrible delays, not to mention costs and frayed nerves, as well as having to install or customize the facility to accommodate the equipment's needs.

Criteria for Involvement

Clinical engineering services are provided in response to the following conditions:

- Requests by human investigation committee (HIC) or institutional review board (IRB)
- Requests by medical staff
- Requests to meet a "critical" patient need
- Humanitarian use devices (humanitarian device exemptions)

Requests by HIC or IRB Investigational devices are evaluated for safety and efficacy prior to use in the hospital as part of a study, usually as part of the FDA premarket approval (PMA) process. Less frequently, the CE department assesses devices that designed and constructed by in-house staff.

Requests by Medical Staff Often, doctors return from conferences brimming with ideas and requests to introduce new products into the hospital. Where the products show promise of being more cost-effective without compromising on safety and efficacy, they become the subject of extensive CE evaluation and clinical trials. Through this mechanism, beneficial, innovative technology can be made available to the patient population.

Meet a "Critical" Patient Need At times, the caregiver or the clinical engineer conceives of ways to customize or to modify a device to expand its capabilities, thus providing more benefit to the patient and possibly to the user. The clinical engineer possesses the education, intellect, skill, and training to perform such customization or re-engineering of a device. Such modification typically requires that the CE expend time in research. The following example illustrates the way the CE does engineering. For certain procedures, OR patients must be positioned on their side, with their 'arms on the side in parallel fashion. Standard arm rests that clamp onto the OR table do not allow this positioning. Alerted to this clinical need, the CE working with the OR staff conceived of a double-decker design. Some research revealed that the local armrest manufacturer sold modular components. A double-decker armrest was created from modular components. In-house manufacture of a specialty device such as this would not have been cost-effective and would have required extensive time to design and prepare mechanical drawings. Even with the aid of computer-aided design (CAD) software, preparing the necessary drawings for an instrument maker to follow would have been prohibitively expensive. In this case, the use of standard modular hardware in a novel configuration accomplished the objective at a minimum expense.

Humanitarian Use Devices (Humanitarian Device Exemptions) Humanitarian use devices are exempt from normal FDA 510(k) and PMA processes. The FDA allows the use of such equipment on a limited number of patients where the product might be their best hope (hence the term "humanitarian device exemptions"). The hospital usually treats devices of this sort as investigational and processes the introduction of this technology by way of the IRB. Under the HDE, such devices need not go through the rigorous evaluation that the clinical engineer, as a technical consultant to the IRB, must perform. However, an assessment of its safety and efficacy must be weighed against the possible benefits that the product can bring to the patient. In most cases, the risk of patient injury is far outweighed by the alternatives; e.g., permanent disability or death of the patient if he is left untreated.

Redesigning or Customizing Medical Devices

Improving Ergonomics

Not all products come customized to a user's personal specifications. By helping to improve the ergonomics of equipment, the CE can help the clinician to perform a better job more safely. Because medical devices are regulated by the FDA, care must be exercised to ensure that specifications are not altered in such a way as to conflict with these regulations. Modifications are typically limited to helping improve a product's ergonomics and do not necessarily change its specifications or performance. Examples of such modifications include the following:

- Mounting an electrosurgical unit (ESU) on a wheeled cart at waist-level height, so that a 5′2″ doctor to operate comfortably
- Providing a padded armrest and a comfortable chair for a surgeon who is reconstructing a patient's middle ear
- Selection of foot-pedal controls to prevent accidentally stepping on two pedals at once
- Placing foam pads to protect a surgery patient from pressure necrosis and decubitus ulcer formation

Some in-house designs have been patented, but for the most part, the CE makes these adjustments based on need, without regard to commercialization (see Chapter 82).

Prototyping Designs to Fill Specific and Specialized Needs

After producing mechanical drawings with CAD, the next step is to render the design into a working prototype. Many times, theory does not readily match reality. Engineers should be prepared to encounter the familiar *Murphy's Law*: Anything that can go wrong will go wrong. However, with a bit of persistence, a final design is rendered, and the finished product is executed in collaboration with the machine shop or instrumentation shop. After the device is deployed, care must be taken to monitor closely the product's safety and effectiveness. Close consulting with the clients greatly helps to perfect the finished product.

Typically, the cost of such device-design activity is built into the general operating fund for the hospital. Where the project involves a research or an activity with funding other than by the hospital, the CE department bills the grant or source of funding through the client department. This funding mechanism will differ among institutions.

Should a device design progress to the stage of patenting, the department could realize some financial advantage by selling the rights to manufacture the product. The transition from device design, to patent, to manufacturing, to marketing and sales is long and costly, requiring substantial financial resources and device development infrastructure. Nevertheless, patenting can be an effective way to preserve the rights of the inventor to some remuneration from the invention.

Design/Ergonomics

Increased attention is being paid to ergonomics and human factors in medical device design (see Chapter 83). Increased efficiency and reduction in human error are achievable using an ergonomic approach in equipment design. For example, minimally invasive surgical procedures made possible by ergonomically designed endoscopic instrumentation has reduced (or, in many cases, eliminated) the need for an open incision, thus reducing surgical time and expense and improving patient outcome (e.g., less anesthesia and reduced chance of infection). Robotic arms now extend the surgeon's precision beyond unaided physical capabilities. Lengthy surgeries lead to physicians' tired muscles, but when translating motions into action, robot helpers can eliminate the tremors of a tired surgeon's hands. Years can be added to a surgeon's career by enhancing his senses. The patient population ultimately benefits from the prolonging of a skilled surgeon's longevity.

Documenting

A well-written report can be effective in impressing upon the hospital and clients the clinical engineer's thoroughness, knowledge of the subject, and grasp of the issues involved. Writing reports helps the CE to understand more thoroughly what was accomplished in the evaluation, clinical trial, or device design. Putting thoughts into words forces one to consolidate thoughts and to communicate in a logical, understandable fashion for the target audience. On occasion, in the process of writing, various issues, options, and alternate solutions emerge that were not considered during the progress of the work. The CE should never underestimate the power of the written word. A clinical engineer's responsibility in this endeavor is to be careful and deliberate in arriving at conclusions and recommendations and to be astute enough to modify them as new information becomes available. Preparing a well-written report can induce a customer-satisfaction response that exceeds financial reward. A satisfied customer will regard the engineer highly and will seek help in the future. Furthermore, the customer is likely to recommend the engineer to colleagues.

A clinical engineer's good work is only helpful when it is communicated properly to the correct audience, who can act on the findings and recommendations. Good communication skills require that the author know the intended audience and phrase words accordingly. In the hospital, the CE's reports are usually directed to the administrators, managers, the nursing staff, and the medical staff. These different audiences require slightly different approaches that match their expectations, training, education, and experience. A layperson's approach works well with nonmedical persons. The scientific approach appeals to the clinical professional. The nursing staff seems to respond well to reports that address protocol or standard procedures for its areas of responsibilities. Doctors, by virtue of their extensive scientific education and training, respond well to a reporting style used in medical textbooks and journals. With a mixed audience, if recommendations particularly affect one group, the reporting style should be slanted in that direction. Another option is simply to write two or three different reports to send to the different groups.

Product Improvements

Clinical engineers are uniquely positioned and qualified to recommend to manufacturers device modifications that help to improve the safety and effectiveness of their products. Such modifications could be simply adding warning labels or changing existing warnings to make them more effective. Many product improvement ideas stem from work performed during engineering investigation of medical device variance reports, engineering evaluations, and clinical trials.

Clinical engineering in the hospital should not be limited to recommending and implementing changes in medical devices and systems. The clinical engineer has much to offer in helping clinical staff to modify protocols and other standard procedures to improve safety, efficiency, and the effectiveness of the equipment used. For example, recommending the use of split return pads when using ESUs and advocating for smoke evacuators in the OR where ESU is used, are two measures that enhance patient and staff safety.

Conclusion

The hospital-based clinical engineer is in a unique position to advance patient care through engineering skills. Engineering is the process of synthesis and design, the creation of something that was not there before—something that fills a need. The need for good clinical engineering abounds in today's health care organizations. The clinical engineer should not be content with the role so often cast by well-intentioned, but inadequately informed, fellow health care professionals. Although some engineers think of device modification and design as merely offshoots of the job of maintaining hospital medical equipment (a view more in keeping with the education), experience and skills of an engineer is that analysis and synthesis is the main stem. Caceres (1980) would be delighted to know that clinical engineering has emerged from the shadows of the distant repair shop to the bedside where engineering skills can be most effectively applied.

Acknowledgment

As staff clinical engineers assigned to different hospital department clients, the authors did a lot of collaborative work, which helped to forge a warm friendship. We acknowledge each other's support and encouragement during this professional affiliation. We have supported each other through many rough bumps and have celebrated each other's career successes. As with most everything else, all good things must eventually end. Although two of the group have left for positions in other institutions, we thought it fitting to assemble in this chapter some collective thoughts on the role that clinical engineers can play in the hospital, with an emphasis on the design and redesign of medical devices.

References

21 CFR Parts 803 and 807; Medical Devices; Medical Device User Facility and Manufacturing Reporting, Certification and Registration. *Federal Register* 60(237):63578-63607, 1995.

Alder H. *Safe Medical Devices Act of 1990: Current Hospital Requirements and Recommended Actions; Health care Facilities Management Series.* Chicago, American Society for Hospital Engineering of the American Hospital Association, 1993.

American College of Clinical Engineering. *Enhancing Patient Safety: The Role of Clinical Engineering.* Plymouth Meeting, PA, American College of Clinical Engineering, 2001.

Bauld TJ. The Definition of a Clinical Engineer. *J Clin Eng* 16(5):403-405, 1991.

Caceres CA. *Management and Clinical Engineering,* Norwood, MA, Artech House Books, 1980.

JCAHO. Medical Equipment Safety: Meeting the Requirements of the Safe Medical Devices Act. *Plant, Technology & Safety Management Series* 2. Chicago, Joint Commission on Accreditation of Healthcare Organizations, 1993.

82

Medical Device Research and Design

Åke Öberg
Linköping University
Linköping, Sweden

Engineers who are active in the health care sector have great opportunities to contribute to long-term quality development by developing new techniques or improving existing ones. By working daily with commercially available medical devices and observing ways in which products are used in practice, the clinical engineer (CE) gains valuable knowledge that, when coupled with keen insight and creativity, can lead to ways to improve existing techniques or to solve long-standing problems. The CE has a perspective on developmental requirements, which is an extremely valuable attribute in order to function as an innovator. Furthermore, the CE's proximity to the point of health care delivery provides ample opportunities for the testing and trial of new products in the end-user environment. In the pursuit of improvements in health care, many with expertise in the medical sciences (e.g., doctors, nurses, and therapists) collaborate well with engineers, for they understand the engineer's gifted ability to analyze problems and synthesize solutions. The beneficial result of such analysis and synthesis, the warp and weft of the engineer's mantel, is invention, the creation of something useful; e.g., an object, a machine, or a technique, that did not exist before.

Clinical engineering was born from the concept that engineering attributes (i.e., the analysis and synthesis) are needed to improve health care (Caceres, 1980). Over the last two decades, however, the pendulum has swung and paused over the repair-shop and financial spreadsheets. Mercifully, it is swinging back to its original position, the bedside, the point of care delivery (see Chapters 11 and 138), with no small impetus imparted by recent revelations of inadequacies in the so-called health care delivery system (Kohn et al., 2000). Regulator demands and financial pressures have absorbed and redirected time and talent, thus denying clinical engineering departments the resources for time-intensive product-development activities. Available personnel must resolve the immediate daily tasks such as keeping the hallway clear of equipment, fixing broken infusion pumps, and entering no-problem-found codes in the computer. Inventions and technical development work fell to the lowest rung on the departmental priority ladder. Sadly, hospital administrators underutilized or spurned the talents of even highly competent, well-educated, and skilled clinical engineers, often out of ignorance of their enormous potential, but also in part because of the inability of CEs to articulate their value and to advocate their profession.

The need for improved or new products for health care is the major driving force for innovations and industrial production of medical devices. New medical products are born in the light of new clinical requirements and new technical possibilities. To be able to identify the new needs and possibilities requires a high degree of competence in both the engineering and medical fields.

Developing new products takes time. A 5- to 10-year period from inception to commercialization is not unrealistic. The interpretation of marketplace trends is important in successful product development. Two main questions arise: Where will health care be 10 years from now, and which new techniques will emerge in the meantime? The needs that one sees today will not necessarily be the same 10 years from now (see Chapter 142). To correctly assess future trends and needs is of the utmost importance in the development of new, commercially successful products. The world market for medical and health care products is worth approximately $150 billion, of which pharmaceuticals represent $100 billion; the remaining $50 billion represents instrumentation and medical devices (OECD, 1992).

Today health care and the biomedical industry use highly sophisticated technology in their products. It is reasonable to assume that technological progress will find early and advanced applications in the health care field. Therefore, it is of interest to discuss personal qualities and knowledge that are prerequisites for the CE who wants to engage in, and to be dedicated to, innovation activities and product development.

From Inventor to Innovator

An innovation is defined as an invention that has been successfully marketed and has reached a stage of commercial success. Usually, a distinction is made between an inventor and an innovator. The role of an innovator is much more demanding than that of an inventor. An inventor is strongly focused on the technology and the technical development of the invention; the product. The inventor seldom finds invention commercialization exciting and avoids creating the contacts necessary for further business development of the product.

An inventor must learn to become a successful innovator and entrepreneur. Knowledge in the following fields is particularly important to gain:

- **Marketing**, to perform market assessments and to understand the essentials of marketing new products
- **Economy**, to do project budget plans
- **Law**, to the protection that a patent gives

The Innovator as a Person

Concepts like innovations, innovators, inventors, product ideas, and creativity are commonly used when the importance of product development is discussed. Some of these concepts are considered to be synonymous, but sometimes one needs a short and practically oriented discussion to define these concepts.

"Creativity," for example, has hundreds of definitions in the literature. In most cases, they can be summarized as "the ability to generate something new and/or useful." For others, the word "useful" can be exchanged for "interesting," "attracting" or even "selling." In the case of technical innovations, the "degree of utility" is defined as something that can be sold on the market or can be exploited financially. Within the Organization for Economic Cooperation and Development (OECD), the concept of innovation has been taken to mean a technical idea that has been realized in the form of a product, which has been successful on the market. Creativity is the willingness and ability to produce something new that is closely related to curiousness, an inherent human attribute. In addition, the inclination and the necessity to invent and to explore are deeply satisfying and rewarding activities for the inventor. Also satisfying are creating and developing new ideas and seeing solutions to problems. One reward is to see the ideas realized, a satisfaction that is similar to that of an artist over a finished painting or sculpture. Creativity is also a necessity, even a compulsion. The inventor is often obsessed by thoughts surrounding a problem. It can seem impossible not to come back to the problem repeatedly.

Creative persons might have many unique qualities, which are easily identified such as the following:

- Flexibility
- Sensitivity to problems
- Originality
- Motivation to create
- Endurance
- Concentration on the task

A visionary mind is crucial, e.g., a talent to foresee scenarios in the future. But a visionary mind is not enough. The abilities to get things done to control project development are important. The innovator must constantly draw knowledge from the wells of marketing, economics, visualization, and patent law, all of which are important in the process of developing an idea into a marketable product.

The Life Cycle of a Product

Every product has a life cycle. It is born, matures, reaches a maximal sales figure, and finally disappears from the market. The life cycle curve of a product is a well-known concept within marketing, product development, and market research (Figure 82-1). It shows the sold quantity of a product as a function of time from its introduction on the market to the time when the product no longer is marketed, and it is usually divided into phases (Ohlsson, 1992):

- **Introduction**–when the product is first introduced on a market and the first sales occur.
- **Sales growth**–when the product has been on the market for some time and the awareness of its existence spreads and increasing numbers of the product are sold.

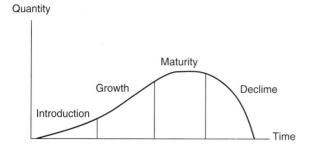

Figure 82-1 Product life cycle curve from introduction to obsolescence.

- **Maturity**–where the consumers interested in the product have already bought it and sales no longer increase.
- **Decline**–where the demand for the product decreases.

The duration of the life cycle can vary from a few months to many decades. The life cycle curve is a useful instrument by which the future market potential for a product can be analyzed. If the sales of a particular product are studied, one usually can tell which of the phases the demand will be in, at a given moment. Usually, the introduction and development phases are more interesting than the saturation and decline phases. The future total consumption can be calculated also by means of the life cycle curve. Using today's data, the future market potential can be forecast.

Patenting and Publishing

Good patent protection is essential for anyone who wishes to exploit an invention industrially, to invest in it, and to manufacture it. The patent is a guarantee that no one else, without risking some form of legal action, can exploit the original idea for commercial profit. In order to secure a patent for an invention, it must be "new". If one places the idea before the public by means of a publication or a lecture, then patenting possibilities are voided, thus creating a high hurdle to jump in the path to product innovation.

At university hospitals and other research-oriented hospitals, there are often visible differences in outlook between the academic researchers who will publish, or give lectures on, their new results, and the innovators who will patent new ideas as a base for industrial development. Often, both groups can be satisfied by patenting first, and then publishing somewhat later. By this procedure, the publishing is marginally delayed while at the same time the great commercial value such a patent can be utilized.

Strategic Assessments for Effective Product Development

A preliminary assessment of a new product idea must always include the following considerations:

- Market analysis
- Technology assessment
- Analyses

The results from these preliminary assessments are important cornerstones in the management of product development projects leading to a commercial success. Figure 82-2 is a flow diagram showing the steps in product development, starting with assessment and moving toward patenting and marketing and production.

Market Analysis

Early market assessments will answer two important questions (Committee of Science and Technology, 1980):

1. Is there a need (i.e., a "market") for the product that is under study?
2. Is the market large enough that an investment in a development project would be profitable?

Whether developing simple or complicated technology, one should aim to gather a good overview of the market. An early and preliminary view of a potential market helps the innovator to plan for the long-range development of the invention.

Technology Assessment

Technology assessment includes analysis of problems relating to product technology as well as production technology. Product technology assessment is particularly important for medical products in the clinical environment. Of course, safety considerations are greatly important. The technical design of a medical device must comply with existing national and international standards and regulations. Accepted design principles must be used.

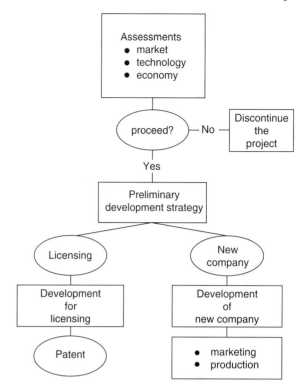

Figure 82-2 Product development flow diagram.

The involvement of external persons who are not familiar with the project is an important part of technology assessment. Such individuals are valuable for conducting tests and giving user feedback. Such an involvement often results in major improvements in the design and functionality of a device. Technology assessment of a small number of devices (e.g., 5 to 10) in "the real world" of hospitals and clinics often will affect the final design of the product in a decisive way.

Involvement of "external" persons in product assessment must include some type of protection in terms of a secrecy declaration or contract (nondisclosure or confidentiality agreement) in which the involved person is prohibited from revealing or utilizing the knowledge obtained during their evaluation work.

The assessment of production technology problems is important before mass production starts. The choice of proper production technologies can strongly affect the market price of a product. If new production equipment must be set up, then large investments might be necessary, and they would affect the economy of the whole project. If a long production run of many devices is contemplated, a production technology-oriented design will result in large savings, particularly for an inventor in a university or hospital environment, who usually does not have extensive experience in industrial production techniques. The initiation of close and early collaboration between the inventor and the final (industrial) producer of the device is strongly recommended. Such collaboration often shortens the route to the market considerably and can reduce production costs.

Economic Assessment

According to its definition, an innovation must involve financial success on the market. Thus, it is essential that the project costs are monitored regularly and that, as early as possible, an assessment be made of the feasibility of the product being a technical and a commercial success. At an early stage, only rough calculations can be made. They should be based on the cost of product sales and an estimate of the size of the product's market share that is necessary to the product being a financial success.

Concept Testing

Before starting to exploit an idea technically and commercially, an investigation should be launched to ascertain whether it really is new. A patent search often reveals that the same, or a similar, idea has been invented already. Concept testing is important for answering the following questions:

- Is the idea new?
- Does the invention infringe on another patent upon commercial exploitation?
- Can the idea be patented?
- What has been done in this area lately?

Concept testing can be carried out partly by screening patents in areas in the proximity of the current invention. A good conception of the "state-of-the-art" also can be gained by searching databases of different types. Libraries have large databases where correctly formulated questions in the form of search profiles can give adequate guidance on innovations in the current area of work. Published scientific literature can provide interesting information on the news value of an invention. The literature can suggest the diagnostic value that is inherent in the utilization of the invention. This, in turn, is an indicator of market size.

Prototype Development

Prototype design is an important step of the innovation process. Prototype development enables one to test the strength of an idea from a technical and an economic point of view, to evaluate the response of the market, and to calculate the production cost. It is usually a long way from the prototype stage to mass production, and often it is practical to proceed in several steps:

- Sketches: Show the idea from a functional and technical point of view in terms of drawings or technical diagrams.
- Models: A three-dimensional presentation of the product idea in wood, metal, or plaster, without any functional demands. This form is especially good for marketing purposes.
- Functional model: A three-dimensional model, as the one mentioned above, but one that also can demonstrate how the idea will work.
- Prototype: The prototype is an exact model of the product as it will function and appear in mass production.
- O-series: A smaller series of the product, often used for demonstration to prospective customers, or for evaluation by experts
- Mass-produced products: Products from the production line

Forms of Exploitation

Commercial exploitation of an invention can be accomplished in many different ways. The two most usual are licensing and the starting of new companies. Licensing entails transferring the rights of the invention to an entrepreneur, who is responsible for continued development and commercial introduction. Generally, when choosing the form of licensing, the inventor has strong patent protection as a basis for negotiations with the entrepreneur. Without strong patent protection, the licensing negotiations are difficult, as it generally is the patent that, at this time, is the only meaningful feature in the negotiation. In some cases, it can be advantageous to start a new company that conducts the development work and is responsible for market launching. If the invention to be exploited concerns completely new areas where few or no established companies are active, then a start-up company for commercialization can be a quicker route to the market. After successful product development and market launching, the company can be sold. Upon starting a company, the patent issue is generally less important, while matters concerning marketing and manufacturing are more important. After the invention has been assessed from market, technical, and economic perspectives, it is time to decide on the form of full-scale exploitation.

Table 82-1 Comparison of activities on exploiting an invention

Activity	Demand	
	New company	Licensing
Patenting	less	more
Prototype design	same	same
Marketing	more	less
Manufacturing	more	less
Financing	same	same
Organization	more	less
Time span	better	worse
Risks	same	same

Often, it is the inventor's own disposition that determines the form of exploitation chosen. If the inventor is more interested in the continuous development of a new technique, then licensing (which amounts to transferring the marketing and financial problems to someone else) is the choice. If, however, the inventor finds it stimulating to work with the entire production process from technical development work through to sales, then starting a company is the choice. A clinical engineer who is employed by a hospital will find that the hospital's policy also influences the choice of form of exploitation. Table 82-1 describes differences between the two forms of exploitation:

References

Caceres CA. *Management and Clinical Engineering.* Artech House Books, 1980.
OECD: *Innovation Policy-Trends and Perspectives.* Paris, OECD, 1992.
Kohn LT, Corrigan JM, Donaldson MS. *To Err is Human: Building a Safer Health System*, National Academy Press, 2000.
Ohlsson L. *R&D for Swedish Industrial Renewal* (DS 1992:109). Stockholm, Utbildningsdepartementet, 1992.
Committee of Science and Technology. *Small, High Technology Firms and Innovation.* Washington, D.C., U.S. Government Printing Office, 1980.
Fölster S. *The Art of Encouraging Invention.* Stockholm, IUI, 1991.

Further information

Brown KA. *Inventors at Work: Interviews with 16 Notable American Inventors*, TEMPUS Books, 1988.
This book gives interesting perspectives on the invention process.
In addition, the serials *Journal of Medical and Biological Engineering and Computing, Physiological Measurements, Medical Engineering,* and *Medical Physics* publish papers on new instrument ideas. National authorities give advice and recommend ways to start new companies.

83
Human Factors: Environment

William A. Hyman
Biomedical Engineering Program
Texas A&M University
College Station, TX

Valory Wangler
Biomedical Engineering Program
Texas A&M University
College Station, TX

Human factors considerations in clinical engineering address the ability of personnel to effectively use and service medical devices in a safe and consistent manner under the actual conditions found in the medical setting. The application of human factors begins with the design of individual medical devices. This area of human factors has received considerable attention, including for specific areas of the hospital, e.g., anesthesia and intensive care (Cooper, 1978; Westover et al. 1998), by individual manufacturers, and from the U.S. Food and Drug Administration (FDA) and professional organizations (ANSI/AAMI, 2001). The attention of manufacturers and the FDA is primarily on the individual device because that is what is being designed and regulated. Extending beyond the individual unit, human factors considerations also must be applied to the assemblage of equipment in the clinical setting. Thus, the first component of the environment to be considered is that of multiple medical devices being placed near each other and simultaneously used. The next level is the multiple patient environment, where the use of several different sets of devices on patients in close proximity to each other adds to the overall equipment complexity. Further extending the perspective, the physical environment also has human factors issues, as does the personnel environment. Ultimately, the management system and associated policies and procedures also have human factors components. Thus, human factors must be applied at several levels of design and management in order to achieve the ultimate goal of medical devices being consistently used in a safe and effective manner.

The System Problem

The need for the effective application of human factors is consistent with the Institute of Medicine report's (Kohn et al. 2000) emphasis on medical errors being an important issue in the delivery of quality of health care. This report, and both earlier (Shepherd and Brown, 1992; Bogner, 1994; Leape, 1994) and subsequent (Becher and Chasin, 2001) work, emphasizes that errors are usually a result of a combination of factors, often involving multiple instances of less-than-adequate human or system performance in a complex world, rather than a single error by a single person. Such errors have been called "organizational accidents" (Reason, 1997) to distinguish them from truly individually induced accidents. While the majority of such attention has been focused on medication errors, surgical errors and medical device use have also received attention. Thus, it is not just one device that requires attention to human factors but also that device in the context of all of the devices in use, in the actual physical environment of use, as used by a combination of personnel, most of whom also have multiple other duties.

Human factors Issues for Individual Devices

The underlying principles of the application of human factors to the design of individual medical devices are that humans have a variety of physical, psychological, intellectual, and motivational attributes and that devices must be designed so that challenges to the capacity these attributes is minimized or eliminated. Thus, vision, perception, hearing, reach, strength, fatigue, common expectations, and other realistic performance capabilities must be considered during design. These must be considered for normal operating conditions and a variety of abnormal conditions and emergencies. They also must be considered for all of the users including installers, clinical and centralized set-up and maintenance, repair, and direct patient care. People make mistakes, and many of these mistakes are predictable based on general and documented experience. Similarly, there is a common group of contributing causes to such mistakes. Based on this knowledge, it is inappropriate, and potentially dangerous, to design equipment assuming perfect user performance at all times. Herein lies the basic contribution of effective human factors design—it is the bridge between a device that is technically capable of achieving the desired outcome, and the actual and consistent achievement of that outcome under the actual conditions of use, by the real users.

Device Interactions

Medical devices have the potential for adverse technical interaction (Miodownik, 1993). Well-known examples of this are the early vulnerability of monitoring equipment to defibrillator discharge, laser-induced tracheal tube fires (de Richemond and Bruley, 1993), and the incompatibility of oxygen and aluminum regulator valves (NFPA, 1999). In addition to technical interaction, there is a human factors interaction. A single problematic device can adversely affect the use of other devices because of the problematic device distracting or occupying the user because of the excess attention it requires. In addition, frustration and related adverse responses can generally degrade overall user performance. The reverse situation also occurs in which a "user-friendly" medical device allows the user to accomplish quickly, efficiently, and accurately tasks associated with its use, increasing time available for other duties, including the use of the other equipment. Well-designed equipment also positively affects the user's overall experience and sense of being in control.

Another aspect of device interaction is general clutter presenting the opportunity for multiple displays, tubes, and cables to be confused with, or masked by, each other. The general principal that displays should be logical and user-friendly in order to facilitate the finding of needed information is well established (Wickens et al. 1998). Some actually take pride in the complexity of the intensive medical environment, as opposed to recognizing that a collection of various devices from various venders that are physically placed with, in some cases, limited thoughtfulness would not be acceptable in many other high-tech industries. Even the use of a single vendor for multiple devices does not ensure a reasonable final assemblage of devices. The ability to stack monitoring devices, while useful, does not automatically provide reasonable system integration. Those individuals should ask themselves whether they would want to fly on a commercial airliner if they saw that the cockpit was a conglomeration of individually acquired and placed instruments with cables strewn between them.

The lack of standardized color-coding (or any color-coding at all) of the function and connection points of various tubes and cables has been surprisingly resistant to improvement. Even the physical placement of multiple devices is a challenge at the clinical level with respect to location, access, and view. Also problematic when multiple devices are used together is the increasing number of displays and user-adjustable parameters, each coupled with a variety of display formats, e.g., analog, digital, and strip chart, and a variety of user interfaces, e.g., rotating knobs, sliders, and up/down buttons. Problems with multiple alarms have also been identified, including too many identical or similar sounds and inconsistent hierarchies of the criticality of sounds, combined with the challenge of locating some kinds of sounds in space, or otherwise identifying which parameter of which device (let alone what condition of the patient) is causing the alarm. Multiple simultaneous alarms compound the challenge. The psychophysiology of sound detection in terms of volume, frequency, and other parameters has been studied (Wickens et al. 1998; Malkin, 1991) but not adequately applied to the medical environment.

Look-alike devices that perform different functions, or that are not as interchangeable as they appear to be, provide another form of potentially dangerous device interaction. Various patient cables that have the same (or not quite the same, but physically interchangeable) connectors at the device end, but that use the conductors in different ways, and endoscopic dissection balloons that look essentially identical outside the body but that provide a different inflation field are two examples of such look-alike devices. Programmable devices also look identical but can perform different functions, depending on the way they are set-up. Control of these potential problems includes rigorous identification of such devices, avoidance of their presence in the hospital through purchasing decisions, post-purchase labeling to add further device identification, and focused training of personnel who are likely to encounter both devices. Training is generally considered the least desirable way to address a hazard that requires control.

The Multiple Patient Environment

Human factors challenges with respect to the use of multiple devices on a single patient are compounded when several patients are receiving care in close proximity to each other. Here, the physical and visual complexity is compounded, and the setting can be so crowded that the devices are physical barriers to care. In addition, as the number of devices used increases, the numbers of parameter settings, displays, and alarms increase proportionally. However, the number of personnel does not generally increase in direct proportion to the number of patients, so that the ability of a nurse or other direct caregivers to perform optimally with respect to each individual patient can be compromised, especially during one or more emergency conditions.

The Physical Environment

Physical environment issues include those associated with the placement of one or more medical devices, and their various patient and other interconnections as discussed above. In addition, lighting, sight lines, noise, temperature, and other environmental parameters can aid or inhibit the safe and effective use of medical equipment and, therefore, the quality of care. Lighting can be too bright or too dim (Wickens et al. 1998; Bridger, 1995). Bright lights can interfere with the ability to read displays effectively, and the homemade glare-reducing shade is still to be found. Continuous bright lighting has an adverse effect on patients. Dim lighting obviously can interfere with the ability to observe unlit or inadequately lit displays or to detect other adverse medical device or patient situations. The color spectrum of patient lighting can affect visual detection and interpretation of the patient's appearance. The general aspects of lighting requirements and effects can be found in a variety of resources (e.g., Bridger, 1995).

Sight lines are important with respect to the ability to read some kinds of displays, especially in the case of screens that have a narrow useful-viewing angle or that are partially obscured by antiglare shades. Displays at substantially different distances from the viewer can further challenge those using different glasses for different tasks and those whose focal distance no longer adjusts as rapidly as it once did. Even if technically viewable, wide or inconvenient placement of multiple displays requires a scan strategy that is rarely, if ever, taught or discussed. The arrangement of multiple displays with respect to relative importance and need to access should be practiced as well (Wickens et al. 1998). The default condition is that displays will not be routinely scrutinized if they are not easily viewable. Another sight-line issue is the ability to see a patient, or multiple patients, from one or more workstations. Even specific periodic visual observation of a patient cannot ensure that serious adverse events will not occur. While one-on-one, continuous, direct observation of all patients is not feasible, or even indicated, there must be a deliberate system in place to make specific observation-requirements decisions for each patient. Moreover, violations of such guidelines because of inadequate staffing are not acceptable. The use of monitoring is not a panacea for the problem of patients who are out of sight. To be effective, the mode of monitoring must be specifically relevant to the medical condition of the patient and must cover all likely or anticipated adverse events. The provision of less-than-adequate monitoring has the psychological risk of leading the caregiver to believe that the monitoring will bring their attention to a wider, and generally unspecified, range of adverse conditions. Therefore, vigilance can be degraded, rather than enhanced, by monitoring. The question of when, if ever, monitoring can be a substitute for direct attention is an open one, although manufacturers tend to advise that their monitoring equipment not be used in this fashion. Thus, risk management attempts by the manufacturer can shift to the hospital the responsibility for monitoring related adverse events.

Noise pollution has two components. One is the direct effect of environmental noise on the ability to hear or distinguish important sounds emanating from medical devices, or even from patients. As discussed above, distinguishing significant device sounds, such as alarms, can be difficult even under ideal conditions and is surely degraded by excessive noise. The second component of noise pollution is its generally disturbing nature, at least to some people, including both staff and patients. Annoyed people are not likely to perform at their best, and therefore a particular real environment might not produce the same level of performance as other actual or simulated environments.

The effect of temperature and other environmental annoyances and discomforts is hard to document, but under routine circumstances, people are believed to work best when their comfort level is enhanced, especially over long durations, as opposed to their ability to respond to crises, which might be unaffected by adverse external influences. Environmental air and chemical hazards must be addressed, of course, for both direct safety and human performance.

Personnel

Ultimately, it is the staff in the hospital who determines the quality of care, including the safe and effective use of medical devices. Similarly, medical error is essentially a human problem and therefore is within the domain of human factors. Any hospital's staff reflects a range of education, training, experience, motivation, distraction, and personal stress. Inadequately trained personnel cannot be expected to function consistently at a high level, especially when confronted with unusual or stressful situations. Of particular relevance to medical devices is the distinction between knowing generically how to use a type of medical device (e.g., an infusion pump) and knowing how to use a particular manufacturer's brand and model of that type of device (e.g., brand X, model Y infusion pump). Even devices that were originally identical can perform differently if there are user- or technician-selectable options. This issue can be difficult to address when temporary nurses are employed, either from outside the hospital or from a different department. Here, the lack of standardization of important features between devices from different manufacturers, which might be reflected even within a single hospital, makes such distinctions particularly important. Motivation is also a variable factor. At the worst extreme is deliberate adverse behavior. Presumably, this is not a common event. The broad middle ground can include a range of personal and institutional disgruntlement that results in degraded performance, as compared to an individual's underlying ability. This is perhaps a perfect example of the human factors issue. The individual is generally capable of achieving the desired outcome, but, for one or more reasons, does not.

Stress, whether personal or work-related, is another adverse factor that can interfere with one's ability to achieve what one otherwise would be capable of achieving. Ideally, people would be stress-free, or at least they would be able to isolate stress from their on-the-job performance. However, the world is not ideal, and any tacit or explicit assumptions that one makes provides an opportunity for adverse consequences. In hands-on health care, both the demands of the job and the potential for disaster are high, in contrast to many office jobs, for example, where an individual under self-identified stress can chose to do relatively little on any particular day, and any work that is attempted has little potential for personal injury. A systems approach first must design the work to minimize stress-related failures (O'Neil, 1998), must have a means to consider whether staff members are not capable of fully performing their duties, and must have a nonpunitive means to encourage such self-identification, at least on an occasional basis.

The adequacy of hospital staffing (i.e., the number of people and their skill levels) has been a particular issue during the managed care cost squeeze. Many hospitals have downsized their staffs and/or have replaced highly skilled positions with less-skilled, less-costly, personnel. Having more patients to care for, more equipment to manage, and more paperwork to do has a limit before performance is degraded. In most organizations, the staff rises to the challenge of doing more with less, thus appearing to reward the decision makers. In addition, a period of downsizing is not an opportune time to complain about any potential, but not realized, negative impacts. Furthermore, systematic studies of the impact of such decisions are not routine. However, selected studies have revealed the obvious. Patients do better when the staff treating them have the highest skill and experience level and are not so overworked that they cannot apply these skills in a conscientious and timely manner.

Clinical engineering departments have faced these same pressures with the additional threat, and reality, of the entire department being outsourced to reduce cost. Such cost reduction implies either that the old organization had bloated salaries or that the new organization is more efficient than the old one because of higher technician productivity and less waste, or that the new organization will provide fewer services than the old one did. The latter explanation often can be attributed to the fact the total services provided by an effective clinical engineering department cannot be explicitly found on a list of functions and duties, yet provide real, value-added service (Hyman and Cram, 1999). When only the enumerated duties are outsourced, those that were not enumerated are likely to disappear.

The Management System

It is sometimes possible for good people to overcome inadequate management and inadequate resources. However, it is best if the management system is a good one that has clear knowledge of the challenges and that assures resources and controls that address those challenges. While it also might be possible for good management to overcome a weak staff, good management combined with good personnel is the key to success. This includes an understanding of human factors issues across the spectrum outlined above. It also requires strong support for the need to include human factors considerations in all aspects of the selection, use, and maintenance of medical equipment. Finally, adverse events with use-related errors must be considered from the systems perspective, rather than from a blame-the-error maker perspective (see Chapter 55), and there must be procedures in place that require an adequate investigation of the underlying causes of device-related and other problems (see Chapter 64). At the FDA, and elsewhere, this perspective has lead to the term "use error" over "user error" because the latter implies a conclusion as to what caused the event, while the former more factually describes the event.

Another important risk management requirement in this regard is an open, encouraged, and nonpunitive system of adverse, near miss, untoward, and uncomfortable event reporting. It is always disturbing to discover that one or more people were aware of device or system malperformance before it was involved in a patient (or other) injury. Reasons for not reporting might be related to fear of being labeled a complainer, or unsympathetic responses from support personal when a "silly" question or complaint occurs. The former can be addressed by anonymous reporting mechanisms, or by providing rewards for valuable comments. Inappropriate responses to reports are a human factors issue in themselves and can be addressed by having buy-in through training as to the importance of open and uninhibited communication.

The FDA is also encouraging voluntary reports under its "Use Difficulty Feedback" program (FDA, 2003a), and reporting of certain types of medical device events to the FDA is required under the MDR regulations (FDA, 2003b). External reporting of adverse experience is also encouraged by various nongovernmental organizations such as the Joint Commission on Accreditation of Healthcare Organizations (JCAHO) (JCAHO, 2003). Whether voluntary or mandatory, external reporting has important risk management implications with respect to the possible use of such information in adversarial legal actions associated with allegations of negligent patient injury. Therefore, it is appropriate for such disclosures to be made under suitable guidance and control.

The public, or membership-based, accessibility of the disclosures of others provides a valuable resource for identifying problems before they occur at a particular institution. Therefore, there should be a system to actively monitor such reports, to determine their local relevance, and to take steps to implement preventive actions. Merely bringing such reports to the attention of relevant staff members is generally not effective and will serve only to demonstrate that the problem was known before it manifested itself. There must always be an assignment of specific responsibility to assess and to resolve the issue, and a systematic review of that assessment and resolution.

Summary

Human factors cover a broad range of issues, from the design of individual medical devices to the design of management systems. It addresses the fact the people are less-than-ideal performers at their best, and variable performers over time. Consequently, people make mistakes or otherwise perform on occasion inadequately. Fortunately, most lapses are benign, but on occasion they can lead to patient, user, or bystander injury and/or damage to valuable equipment. It has always been tempting in these situations simply to blame the individual, and to admonish them to focus and do better, perhaps to send them to a retraining session, to terminate them if the situation was grievous, and then to consider the event resolved. The more modern systems-oriented view requires, at a minimum, an investigation of why the individuals were in a situation of not performing to their capacity, and, in particular, what it was about the situation, the equipment, the environment, and the system that might have contributed to degraded performance, and what steps can be taken to correct the causative chain or to provide safeguards against the causative chain reaching the adverse endpoint. Of course, it is also necessary to implement these steps.

References

ANSI/AAMI. *Human Factors Design Process for Medical Devices.* ANSI/AAMI HE74:2001. Arlington, VA, Association for the Advancement of Instrumentation, 2001.

Becher EC, Chasin M. Improving Quality, Minimizing Error: Making it Happen. *Health Affairs* 20(3), 2001.

Bogner MS. *Human Error in Medicine.* Hillsdale, NJ, Lawrence Erlbaum Associates, 1994.

Bridger RS. *Introduction to Ergonomics.* New York, McGraw-Hill, 1995.

Cooper J. Preventable Anesthesia Mishaps: A Study of Human Factors. *Anesthesiology* 49:399-406, 1978.

De Richmonde AL, Bruley ME. Head and Neck Surgical Fires. In Eisele DW (ed). *Complications in Head and Neck Surgery.* St. Louis, MO, Mosby, 1993.

FDA. Reducing User Error. U.S. Food and Drug Administration, http://www.fda.gov/cdrh/useerror/New.html, 2003a.

FDA. Reporting Problems with Medical Devices. U.S. Food and Drug Administration, http://www.fda.gov/cdrh/mdr.html, 2003b.

Hyman WA, Cram N. In-Source, Out-Source, On-Site, Off-Site: A Checklist of Clinical Engineering Services. *J Clin Eng* 24(3):172-174, 1999.

JCAHO. *Joint Commission on Accreditation of Health care Organizations Sentinel Event Reporting*, http://www.jcaho.org/index.html, June 12, 2003.

Kohn LT, Corrigan JM, Donaldson MS. *To Err is Human: Building a Safer Health System*, National Academy Press, 2000.

Leape L. Error in Medicine. *JAMA* 23:1851-1857, 1994.

Malkin J. *Hospital Interior Architecture: Creating a Healing Environment for Special Populations.* New York, Van Nostrand Reinhold, 1991.

Miodownik S. Interaction Between Medical Devices; Medical Equipment Safety: Meeting the Requirements of the Safe Medical Devices Act. *Plant, Technology & Safety Management Series* 2. Chicago, Joint Commission on the Accreditation of Health care Organizations, 1993.

O'Neill MJ. *Ergonomic Design for Organizational Effectiveness.* Boca Raton, FL, Lewis Publishers, 1998.

NFPA. *Recommended Practice on Materials, Equipment, and Systems Used in Oxygen-Enriched Atmospheres.* Quincy, MA, National Fire Protection Association, 1999.

Reason J. *Managing the Risks of Organizational Accidents.* Aldershot, UK, Ashgate Publishing, 1997.

Shepherd M, Brown R. Utilizing A Systems Approach to Categorize Device-Related Failures and Define User and Operator Errors. *Biomedical Instrumentation and Technology* 26(6):461-475, 1992.

Westover AW, Moog T, Hyman WA. Human Factors Analysis of an ICU. *J Clin Eng* 23(1):110-116, 1998.

Wickens CD, Gordon SE, Liu Y. *Introduction to Human Factors.* New York, Longman, 1998.

84

Medical Devices: Failure Modes, Accidents, and Liability

Leslie A. Geddes
Showalter Distinguished Professor Emeritus of Bioengineering, Purdue University
West Lafayette, IN

This chapter addresses the three types of medical devices and the types of malfunction that can occur. It gives a short explanation of product liability. Several examples of medical device accidents are presented. It is the objective of the chapter to assist the clinical engineer in identifying foreseeable accidents and to provide guidance on obtaining medical device standards and historical information on medical device accidents. In one way or another, clinical engineers work with three types of medical devices: Diagnostic, therapeutic, and assistive or rehabilitative.

The diagnostic device acquires information for presentation to the human senses. In essence, the diagnostic or data-acquiring device is an extender of the human senses. It is in this area that transducers and display devices play key roles. The therapeutic device is used by the physician, or the physician's delegated representative, to arrest or control a physiological process that has gone awry because of disease, trauma, or some other agent. Although chemical agents (drugs) are most prominent in the therapeutic area, physical devices, such as the high-voltage X-ray (to kill cancer cells), the pacemaker (to initiate rhythmic heartbeats), or the defibrillator (to arrest the lethal cardiac arrhythmia, ventricular fibrillation), are but a few prominent examples of therapeutic devices. Assistive or rehabilitative devices are used to make up for diminished function or to provide for a lost function, often when a disease process has been arrested but a deficit remains. Some of these devices might be life-supporting or life-sustaining. The objective of many rehabilitative devices is to make the subject capable of being self supporting, thereby allowing him or her to do things without reliance on the help of others.

Failure Modes

Although well designed, well made, and closely regulated, the three types of medical devices can fail and cause injury, property damage, or death to a patient. Examples and the legal aspects (product liability) will be discussed below; but first it is useful to identify at least some of the failure modes and consequences that can accompany the use of these three types of medical devices.

Diagnostic Devices

- Provide inaccurate information leading to inappropriate medical decision and action
- Malfunction or give erroneous information due to environmental effects
- Are sometimes used for other than the intended use
- Are hazardous or injurious to the patient due to component failure or due to excessive leakage current because the device is not grounded properly
- Are hazardous due to multiple devices connected to the patient (device interaction)
- Fail when alarm is turned off
- Can cause an allergic response when in contact with the body

Therapeutic Devices

- Failure to deliver adequate or delivery of excessive therapeutic agent
- Lack of indication that therapeutic agent is being delivered
- Delivery of inappropriate therapy
- Inadequate training or experience of the operator
- Failure to heed warnings in instruction manual
- Failure to read instruction manual or package insert
- Allergic response to therapeutic agent or delivery system
- Interaction between multiple therapeutic agents or multiple devices connected to same patient
- Susceptible to malfunction due to environmental agents (e.g., EMI, noise, ultraviolet rays, heat, or humidity)

- Drug synergism or antagonism
- Adverse interaction between two or more therapeutic devices connected to a patient

Assistive or Rehabilitative Devices

- Patient inadequately informed or trained
- Failure to report for checkup
- Device inadequately protected from environmental agents (e.g., EMI, noise, vibration, heat, moisture, and ultraviolet rays)

Product Liability

The term "product liability" describes the litigation in which a complaining party (the plaintiff) alleges that a product caused personal injury, death, property damage, and compensation is sought for the loss from one or more parties (the defendants), who usually include the manufacturer, contractor, assembler, distributor, vendor, and/or practitioner (and often assistants) who used the product. Product liability falls within the purview of civil law, and to prevail requires that proof be established by a "preponderance of the evidence." The dictionary definition of "preponderant" is "superior in weight, number, or power." One loose definition of adequate proof, under this standard, would be "more likely than not."

Product liability extends into three areas: Manufacturing, design, and misrepresentation. To prevail, a plaintiff must prove that the defect existed when the product left control of the manufacturer or vendor; the provider knew (or should have known) of the existence of the defect; there were feasible alternatives that would have eliminated the risk; and the product caused the injuries or damages complained of. Misrepresentation involves making unmet claims about a product. See Chapter 127 for a detailed discussion of product liability.

Product Defect

Legally, a product is defective when it fails to meet the provider's own specifications and applicable industrial and/or governmental standards. However, meeting the requirements of such standards is not enough; it must be shown that the product is safe for any reasonable and foreseeable use. For a product to be found to be defective, the plaintiff must prove that the defect was present when the product left the manufacturer's premises. Because a product might pass through one or more vendors, proof of when the defect was induced can be difficult. It is for this reason that the vendor(s) and the manufacturer are often named as codefendants in a product liability lawsuit.

Inspection and maintenance records are very useful in showing that a product was or was not defective. Manufacturer's instruction manuals often specify the type and frequency of these activities. However, if the maintenance is performed by unauthorized personnel, the manufacturer's warranty becomes void, and the manufacturer cannot be held responsible for any harm done by a defective product.

An improvement (however useful) in a product by the purchaser renders the product not what the manufacturer intended the product to be. This constitutes tampering, and the manufacturer cannot be held responsible if an accident occurs.

The instruction manual identifies the features, operation, and proper use of a product. It often identifies risks and hazards and cites warnings associated with use of the product. This information is often valuable in explaining the cause of an accident.

Some medical devices are leased, and it is the responsibility of the lessor to provide a properly functioning and safe product. Therefore, the lessor's inspection, testing, and maintenance records and proof of adherence to the manufacturer's specifications are important. Such information can be useful in explaining the cause of an accident.

All products have a finite life, and despite proper inspection, testing, and maintenance, a product can malfunction because of a component failure. In such a case, it can be difficult to establish liability. However, maintenance records can show whether a particular component has failed frequently. In many cases, reports of medical device related accidents or incidents are maintained by the FDA in a Manufacturer and User Device Experience (MAUDE) database. MAUDE should be consulted to investigate whether the same event has been reported by other users of the same product. If it has been, and if a competitor's product does not have the same defect, then the issue of a design defect arises.

Although accelerated life testing is possible with some products, with others it is not. The best information on expected product life comes from actuarial data, i.e., the past history of a product.

One important aspect of a product defect is in identifying the consequence(s) of the defect. In other words, it is essential to show that the presence of the defect caused or contributed to the harm, injury, or death. It should be pointed out that, in some cases, harm was done that was unrelated to a product defect. One should be alert to the practice of blaming a device when improper use or negligence may have been the cause.

Misuse

When harm results from the misuse of a product, it is usually assumed that the product was without a manufacturing or design defect. Therefore, the issue becomes one of foreseeable user behavior. The U.S. Supreme Court has held that foreseeability is significant in product liability cases when the product is what it is intended and known to be, but injury was suffered because the product was misused. A product is not misused merely because the manufacturer intended that it be used in a different manner; the manufacturer must show that the use that caused the injury was not reasonably foreseeable. Therefore, the manufacturer is responsible for designing a product with reasonable, foreseeable misuse in mind. It is in this area that the instruction manual can play an important role by identifying the best procedures, intended use, and potential for injury in the event of misuse.

There are two types of misuse: Contributory negligence and assumption of risk. In the former, the test of reasonableness applies. In other words, did the user act as a reasonable person in the same or similar circumstances? In the latter, the assumption of risk is evaluated on the basis of what the user knew, just before the incident. In assessing the assumption of risk, there are three elements that the plaintiff must show, on the part of the user: That the user knew of the risk before the incident; acted voluntarily; and acted unreasonably.

One should always bear in mind that tampering or sabotage could have occurred, and thus might have been the cause(s) an accident. "Tampering" is defined as meddling for such purposes as altering, damaging, or misusing something. "Sabotage" defined as is malicious injury to work, tools, machinery, or other items in order to cause malfunction.

Negligence

Negligence is the failure to conform one's conduct to that of a reasonably prudent person under similar circumstances. If such conduct failure results in harm to a person (or property), a lawsuit can be filed. The purpose of negligence law is protection from unreasonable risks or harm, which are foreseeable and therefore preventable. To prevail in a negligence lawsuit, five elements must be proven: The existence of a duty of care; breach of that duty; proximate cause (or "legal cause"); actual cause (or "cause-in-fact"); and actual harm or injury (i.e., "damages").

"Duty of care" refers to behavior as a reasonable and prudent person; breach of that duty involves showing that something has happened and that the defendant acted unreasonably under the particular circumstances. "Proximate cause" refers to identification of the reason for the accident. "Damages" refers to details like the value of personal injury, property damage, emotional stress, or lost wages that resulted from the negligent behavior.

Design Defect

A product does not have a design defect when it is safe for any reasonably foreseeable use. Such a safe product also has met all of the applicable functional specifications. Before a design defect can be alleged, it is necessary to establish that the product has met all applicable manufacturing, industrial, and governmental requirements, and that despite the manufacturer having met these requirements, the harmful incident occurred.

To establish that a design defect exists involves the following elements:

1. Identification of the design defect
2. Establishing a link between the design defect and the harm done
3. Identification of alternate designs that would have prevented the harmful incident
4. Comparing the product performance with like products offered by other manufacturers

One good example of a design defect is a product that functions normally in one typical environment but fails in another. In such a case, the injured party can raise the issue of a design defect. For example, if an apnea monitor operates properly in one environment (e.g., in a hospital ward), but fails to alarm cessation of breathing in another environment (e.g., the home), the issue of design defect can be raised. If the reason for failure were electromagnetic interference (EMI), it would be fruitful to inquire whether similar apnea monitors provided by competitors also have failed in the same environment or a comparable environment. If they have not failed in both environments, it shows that technology that can eliminate this hazard is available. If the manufacturer of the monitor that failed was unaware of, or chose to ignore, the available technology, a case for a design defect can be made.

Medical Device Accidents

Accidents can be associated with the use of diagnostic, therapeutic, and rehabilitative devices, and the clinical engineer might be called upon to investigate. The investigative procedure is embodied in the saying "When you hear hoof beats, think of horses, not zebras." In other words, look for the simple and obvious things first. A few examples of device malfunction are presented below. Additional information on accident-investigative techniques for the clinical engineer can be found in Bruley (1994), Dyro (1998, 2003), and Geddes (2002).

Diagnostic Device Accidents

As stated earlier, the diagnostic device presents information to the human senses. One frequently used diagnostic device is the pulse oximeter, which measures the redness of the blood. The reading that it provides is termed "oxygen saturation" (S_aO_2); however when this quantity is measured with the pulse oximeter, the notation used is S_pO_2. Typically, the finger, toe, or ear is the site for measurement.

Murphy (1990) describes an unusual accident that occurred with a pulse oximeter: "A 4.4-kg, full-term infant was born by vaginal delivery with the aid of vacuum extraction. A subcapsular occipital hematoma developed, and the patient was transferred to the newborn nursery. All the nursery's pulse oximeter probes (manufacturer A) were in use, and a probe (manufacturer B) was borrowed from the anesthesia department. The nursery personnel connected the probe on the infant to the A machine. Initially, a wrap-around finger probe was applied to the right ring finger, but no saturation (S_pO_2) reading was obtained. The problem was assumed to be a faulty probe. This probe was removed after approximately 1 minute and noted to be 'a little warm.' A second type of probe (clip-on type) was then applied to the superior aspect of the right ear pinna. After approximately 3 minutes without obtaining any S_pO_2 or pulse rate readings, the tip of the ear was noted to be acquiring a red color, the probe was removed, and further attempts to monitor S_pO_2 were abandoned. Over the next several hours, it became apparent that the infant had been burned in two locations, suffering from a second-degree burn to the right ring (and adjacent middle) finger and a third-degree burn to the tip of the right ear. The infant subsequently recovered uneventfully from the occipital hematoma. The burns were initially treated with topical therapy, and the patient was seen in consultation by a surgeon prior to discharge. Reconstructive surgery will likely be required at sometime in the future for repair of the right ear."

Subsequently, the authors measured the temperature at the tip of the clip-on probe while connected to the machine used in this incident. Within two minutes, the temperature at the tip of the probe was 200° F. This information was reported to the manufacturer and to the Food and Drug Administration. Was this a product failure?

The surprising thing about the foregoing incident is that the connector of one manufacturer's probe could engage the receptacle on the other manufacturer's instrument. No mention of any warnings that this could happen appeared in the manuals of manufacturers A or B. The foregoing incident raises several questions: Were the nurses adequately trained? Was there a design defect? Had the manufacturer of the probe made warning labels that indicated that the pulse oximeter probes should only be used with the instrument?

Clearly, if the nurses had had adequate training, they would not have thought that oximeter probes were interchangeable among different manufacturers' instruments. Regarding a design defect, this is unlikely. However, it would be prudent for a designer to use a connector that is different from that used by a competitor. In retrospect, warning labels should have been placed on the pulse oximeter probes and machines, and in the instruction manuals.

Therapeutic Devices

Therapeutic devices take many forms and are used by physicians or their delegated representatives. As stated earlier, chemical agents feature prominently in this category, and obvious errors can occur. However anesthesia machines can be considered therapeutic devices because they allow surgeons to perform their tasks effectively. Two examples of malfunction of therapeutic devices (infusion pump and anesthesia machine) follow.

Infusion Pump Mishaps

Drugs are often injected into the vascular system by way of a programmable infusion pump. Typically, a sterile syringe containing the drug is inserted into the pump, and the delivery program is entered by a keyboard. Such devices are typically microprocessor-controlled.

Silberberg (1993) cited the following cases of infusion pump malfunction: (1) An infusion pump changed rate when a cellular phone was placed on the instrument stand; and (2) an infusion device on an isolated circuit (high source impedance) behaved erratically because of conducted EMI at 33 MHz. Obviously, the drug infusion pump was inadequately protected from environmental electromagnetic interference (EMI).

Anesthesia Machine Mishaps

An anesthesia machine delivers a gaseous anesthetic and oxygen to the patient. The exhaled carbon dioxide (CO_2) is removed by a CO_2 absorber (soda lime) in the breathing circuit. The CO_2 absorber is in an optically transparent cylinder and contains an indicator (ethyl violet), which turns purple when the absorber is exhausted. Carbon dioxide is a potent respiratory stimulant that causes the breathing rate and depth to increase (hyperventilation). An increase in the concentration of CO_2 in the blood is known as "hypercarbia." If hyperventilation occurs when a spontaneously breathing patient is connected to an anesthesia machine, several undesirable consequences can occur, not the least of which is the inhalation of more and more anesthetic, thereby deepening the level of anesthesia, which can lead to respiratory arrest and death.

Andrews et al. (1990) reported a case of a patient hyperventilating spontaneously during anesthesia; yet the CO_2 absorber did not turn blue indicating a functioning CO_2 absorber. Fortunately, an accident was prevented because blood pCO_2 was measured routinely. It was decided to change the CO_2 absorber. Later, it was found that the CO_2 absorber failed because of deactivation of the color indicator by radiation from operating-room fluorescent lights (which emit a small amount of ultraviolet radiation), which can bleach the indicator.

Andrews et al (1990) carried out a series of carefully controlled experiments and stated that after 24 hours of fluorescent-light exposure with a received flux density of 46 mwatts/cm^2 at 254 nm, the concentration of functional ethyl violet remaining in pulverized Sodasorb was 16% of the baseline value. Furthermore, using multiple light sources of various intensities, the greater the intensity of light, the more rapid the rate of decline of the ethyl violet concentration. Finally, the authors pointed out that photodeactivation is more likely to occur in operating rooms (ORs) that are brightly lit or in those that utilize ultraviolet germicidal bulbs. Intensely lit ORs are the trend, and many existing ORs contain germicidal UV bulbs. This stated that end-tidal CO_2 monitoring might be more beneficial than color change of absorbent to diagnose rebreathing in those operating rooms.

It is obvious that the CO_2 absorber should have been contained in a chamber that filtered out ultraviolet radiation and allowed visible light to pass, to identify the color change when the CO_2 absorber is exhausted. Clearly, this is a design defect because technology was available to prevent the accident.

Rehabilitative Device Accidents

Rehabilitative devices are designed to make a patient more self-supporting. Some devices are life-sustaining and must not fail. One common type of rehabilitative device is the electrically powered wheelchair.

Powered Wheelchair Mishaps

Motor-driven wheelchairs and scooters used by patients contain electronic controls that have malfunctioned. Silberberg, of the FDA Center for Devices and Radiological Health (CDRH) (1993) reported that a quality assurance manager, who had previously worked for a large wheelchair manufacturer, inquired in a letter dated June 26, 1992 about the status of medical device electromagnetic compatibility (EMC) standards. As a side issue, he stated that in his previous position he had received reports of powered wheelchairs driving off curbs and piers unintentionally when a police or fire vehicle, harbor patrol boat, or CB or amateur radio transmitter was operating in the vicinity. Silberberg had reproduced the phenomenon under controlled conditions and had observed powered wheelchairs "go by themselves" within 15 or 20 of a police or fire radio. The CDRH databases were checked for reports of electromagnetic interference problems with powered wheelchairs. While several reports of unintended motion were found (several involving serious injury), none in the report had been attributed to electromagnetic interference.

CDRH engineers investigated the EMI susceptibility of the motion controllers of several makes of powered wheelchairs and scooters. The powered wheelchairs tested were found to be susceptible to field strengths in the range of 5 to 15 V/m. At the lower end of the susceptibility range, the electric brakes would release, which could result in a hazard of rolling if the chair were stopped on an incline. As the field strength at a susceptible frequency was increased, the wheels would begin turning, with the speed of rotation being a function of the field strength.

CDRH engineers also measured the emissions from police and fire radios. A 100-watt state police radio transmitting at 39 MHz was found to have a field strength of 41 V/m at a distance of 0.9 m. CDRH also learned that a small number of manufacturers of wheelchair controllers supply much of the powered-wheelchair and scooter industry. It is obvious that the controllers used in these wheelchairs were inadequately shielded to exclude such electromagnetic interference (EMI).

Information Sources

FDA MAUDE Database

In 1990, the U.S. Congress passed the Safe Medical Devices Act (SMDA), which went into effect on May 28, 1992 and was amended with the passage of the Medical Device Improvements Act in 1992. This act, as amended, amendments mandated the reporting of all incidents associated with the use of medical devices that result in serious injury or death. Such incidents must be reported to the FDA under Medical Device Reporting regulation (MDR) Part 803 of Title 21, Code of Federal Regulations (CFR). The FDA maintains a database of problem reports called "Manufacturer and User Device Experience" (MAUDE). Such reports are available from the FDA and can be useful in establishing facts in a lawsuit. Recent FDA regulations relating to 21 CFR parts 803 (i.e., MDR) and 807 (i.e., pre-market approval) clarify and expand requirements for medical device manufacturers and health care facilities (60 FR 63597, Dec. 11, 1995, as amended at 62 FR 13306, Mar. 20, 1997; 65 FR 4118, Jan. 26, 2000).

The reporting requirements mandated by the Safe Medical Devices Act are highlighted in the following:

Health care facilities must submit a report whenever they receive, or otherwise become aware of, information, from any source, that reasonably suggests that a device has, or may have, caused or contributed to a death or serious injury.

The term "serious injury" is defined as an injury or illness that:

1. Is life threatening; or
2. Results in permanent impairment of a body function or permanent damage to a body structure; or
3. Necessitates medical or surgical intervention to preclude permanent impairment of a body function or permanent damage to a body structure

The following are required to report:

1. Hospitals, ambulatory surgical facilities, nursing homes, home-health agencies, ambulance providers, rescue groups, skilled nursing facilities, rehabilitation facilities, hospices, psychiatric facilities, and all other outpatient treatment and diagnostic facilities that are not a physician's office
2. Manufacturers of medical devices
3. Distributors of medical devices (at present, covered by a different regulation)

It is always useful to know whether the accident under investigation is common or rare, or whether there are precedents that could affect a judgment. Such background information is not always easy to locate. However, there are many sources of information on medical device accidents. MAUDE reports, institutional medical records, the incident report and depositions obtained from those present at the time of the accident contain valuable information.

Another useful source of FDA information is the 510(k) application for permission to sell a device that is substantially the same as one on that is the market. The IDE (Investigational Device Exemption) and the PMA (pre-market approval) are FDA documents that ensure that safety and efficacy considerations were met so that limited clinical trials of new medical devices are allowed. See Chapter 126 for a more detailed discussion of the FDA Center for Devices and Radiological Health.

Textbooks, Publications, and Organizations

The Handbook of Electrical Hazards and Accidents (Geddes, 1995) contains much information on the response of the body to the passage of low-and high-frequency current. A book by Pearce (1985) contains many examples of electrosurgical accidents. A chapter in a book by Fish (2000) describes many electrical accidents and includes many medical device accidents.

Several journals describe medical accidents; among these are *Medical Malpractice, Verdicts, Settlements and Experts,* and *The Forensic Examiner.* Reports of accidents also appear in *Health Devices* and *Health Devices Alerts* (see ECRI). *Health Devices Alerts* provides abstracts of selected technology assessments and contains a summary of reported medical device problems, hazards, recalls, and updates. The selections derive from extensive review of the medical, legal, and technical literature; reporting networks; and government sources worldwide. Most abstracts are taken from reports received within one month of the publication date of the issue.

SEAK, Inc. hosts many expert witness conferences that address medical malpractice and accidents. (See, e.g., Dyro, 2003b.)

Information on anesthesia-related accidents is available from the American Society of Anesthesiologists, which maintains a closed-claims database. A wide variety of accidents that occurred during anesthesia are described in it. In addition, the Wood Library–Museum of Anesthesiology also maintains a database of anesthesia accidents.

Published papers on accidents can be accessed through a library search of key words. The National Library of Medicine and the Web of Science are good sources.

Medical Device Standards

Before the FDA was granted authority to regulate medical devices in 1976, various professional societies had made recommendations for minimum performance of devices. In parallel with these activities, several standards-promulgating groups, notably the National Fire Protection Association (NFPA) and Underwriters Laboratories (UL), concerned themselves with the safety of medical devices. Now, there are standards for efficacy and safety for medical devices that have resulted from the activities of the FDA device classification panels. The American National Standards Institute (ANSI) serves as the national coordinating body for all standards-setting organizations. Organizations that publish medical device standards include the Association for the Advancement of Medical Instrumentation (AAMI), the American Society for Testing Materials (ASTM), and the Electronic Industries Association (EIA). The Canadian Standards Association (CSA) and the International Electrotechnical Commission (IEC) each promulgate medical device standards. The IEC has a wider range of performance standards than is required in the U.S., and when a U.S. standard is absent, the IEC standard is often cited. A wide variety of standards documents can be obtained from Global Engineering Documents and the Guide to Biomedical Standards.

Legal Documents

Two enormous databases cover the legal aspects of medical mishaps, as well as a wide variety of other subjects, including patents and patent infringement cases. The Lexis-Nexis database serves 60 countries. It has 7300 databases and adds 9.5 million documents each week. Both legal and nonlegal information is available. The West Publishing Company operates the Westlaw database and contains federal and state case law entries. Attorney editorial services can be provided, and a nominal charge is levied for each case, there being no subscription fees.

References

Andrews JJ, Johnston RV, Bee DE, et al. Photodeactivation of Ethyl Violet: A Potential Hazard of Sodasorb. *Anesthesiology* 72:59-64, 1990.
Bruley ME. Accident and Forensic Investigation. In van Gruting CW (ed). *Medical Devices: International Perspectives on Health and Safety.* Amsterdam, Elsevier, 1994.
Dyro JF. Methods for Analyzing Home Care Medical Device Accidents. *J Clin Eng* 23(5):359, 1998.
Dyro JF. Accident/Incident Investigation: Advanced Techniques. 12th Annual National Expert Witness and Litigation Seminar, Hyannis, MA. SEAK, Inc. 2003.
Nabours RF. *Electrical Injuries.* Tucson, AZ, Lawyers & Judges Publishing, 2000.
Geddes LA. *Handbook of Electrical Hazards and Accidents.* Boca Raton, FL, CRC Press, 1995.
Geddes LA. *Medical Device Accidents,* 2nd ed. Tucson, AZ, Lawyers and Judges Publishing, 2002.
Murphy KG, Secunda JA, and Rockoff MA. Severe Burns from a Pulse Oximeter. *Anesthesiology* 73(2):350-352, 1990.
Pearce JA. *Electrosurgery.* London, Chapman Hall, 1985.
Silberberg JL. Performance Degradation of Electronic Medical Devices Due to Electromagnetic Interference. *Compliance Eng* 3:1-8, 1993.

Further Information

AAMI, Association for the Advancement of Medical Instrumentation. 3330 Washington Blvd. Arlington, VA 22201-4598.
American National Standard Institute. 1430 Broadway, New York, NY.
American Society of Anesthesiologists. 520 N. Northwest Highway, Park Ridge, IL 60068-2573. http://www.asahq.org/.
Canadian Standards Association. 235 Montreal Rd., Ottawa, Ontario, Canada K1A 0L2
ECRI, 5200 Butler Pike, Plymouth Meeting, PA 19462.
EIA, Electronic Industries Association Health Care Electronics Section. 2001, Eye Street, N.W., Washington, D.C. 20006.
FDA, Food and Drug Administration. 5600 Fisher Lane, Rockville, MD 20857.
Food and Drug Administration. Center for Devices and Radiological Health, 12720 Twinbrook Parkway, Rockville, MD 20857.
Forensic Examiner. American College of Forensic Examiners, 2750 E. Sunshine St., Springfield, MO 65804.
Global Engineering Documents. Clayton, MO
International Electrotechnical Commission (IEC). Centre du Service Clientele (CSC), Commission Electrotechnique Internationale, 3 rue de Varembe, Case postale 131, CH1211-Geneve 20 (Geneva) Suisse (Switzerland).
Lexis-Nexis. 9443 Springboro Pike, P.O. Box 933, Dayton, OH 45401.
Library Searches: Web of Science. http://www.webofscience.com (fee required).
Medical Malpractice, Verdicts, Settlements and Experts. 901 Church St., Nashville, TN, 37203-3411.
National Electrical Code: National Fire Protection Association (NFPA)< P.O. Box 9145, Quincy, MA. 02269-9959
National Library of Medicine. http://www.nlm.nih.gov/.
SEAK, Expert Witness Conferences. SEAK Inc. Falmouth, MA. www.seak.com.
The Guide to Biomedical Standards. Gaithersburg, MD, Aspen Publishers.
Underwriters' Laboratories Inc., 207 East Ohio Street, Chicago, IL 60611.
West Publishing Company. 620 Opperman Drive, P.O. Box 64779, St. Paul, MN 55164-0779. Webmaster.westdoc@westpub.com
Wood Library. Museum of Anesthesiology, 520 Northwest Highway, Park Ridge, 60068-2573. ASA Web page: http://www.asahq.org and e-mail to WLM@asahq.org.geddes.pap. techexp.5/27.

85
Medical Device Software Development

Richard C. Fries
Manager, Support Engineering, Datex-Ohmeda, Inc.
Madison, WI

Andre E. Bloesch
Datex-Ohmeda, Inc.
Madison, WI

Many kinds of medical devices are rapidly becoming software-intensive. Software controls their operation, collects and analyzes information to help make treatment decisions, and provides a way for users to interface with the medical device. In these devices, the software transforms a general purpose computer into a special-purpose medical device component. As in hardware design, specifying the software requirements, creating a sound software design and correctly implementing it are difficult intellectual challenges. Good software development is based on a combination of creativity and discipline. Creativity provides resolution to new technical hurdles and the challenges of new market and user needs. Discipline provides quality and reliability to the final product.

Software design and implementation is a multi-staged process in which system and software requirements are translated into a functional program that addresses each requirement. Software design begins with the work products of the Software Requirements Specification. The design itself is the system architecture, which addresses each of the requirements of the specification and any appropriate software standards or regulations. The design begins with the analysis of software-design alternatives and their trade-offs. The overall software architecture is then established, along with the design methodology to be used and the programming language to be implemented. A risk analysis is performed and then refined to ensure that malfunction of any software component will not cause harm to the patient, the user, or the system. Metrics are established to check for program effectiveness and reliability. The Requirements Traceability Matrix is reviewed to ensure that all requirements have been addressed. Peers The review the software design for completeness.

The design continues with modularizing the software architecture, assigning specific functionality to each component and ensuring that internal and external interfaces are well defined. Coding style and techniques are chosen based on their proven value and the intended function and environment of the system. Peer reviews ensure the completeness and effectiveness of the design. The detailed design also establishes the basis for subsequent verification and validation activity. The use of automated tools throughout the development program is an effective method for streamlining the design and development process and assists in developing the necessary documentation.

The key to success in verification and validation is planning. Verification and validation planning encompasses the entire development life cycle, from requirements generation to product release. The initial planning of verification and validation is documented in a Software Verification and Validation Plan (SVVP). The SVVP describes the verification and validation life cycle, gives an overview of verification and validation, describes the verification and validation life cycle activities, defines the verification and validation documentation, and discusses the verification and validation administrative procedures. An excellent guideline for verification and validation planning can be found in IEEE Std 1012, with an explication of Std 1012 found in IEEE Std 1059.

Software Standards and Regulations

There are a myriad of software standards to assist the developer in designing and documenting a software program. IEEE standards cover documentation through all phases of design. Military standards describe the way that software is to be designed and developed for military use. There also are standards on software quality and reliability to assist developers in preparing a quality program. The international community has produced standards that primarily address software safety. In each case, the standard is a voluntary document that has been developed to provide guidelines for designing, developing, testing, and documenting a software program.

In the United States, the FDA is responsible for ensuring that the device utilizing the software, or the software as a device, is safe and effective for its intended use. The FDA has produced several drafts of reviewer guidelines, auditor guidelines, software policy, and good manufacturing practices (GMP) regulations addressing device and process software. In addition, guidelines for FDA reviewers have been prepared, as have training programs for inspectors and reviewers. The new version of the GMP regulation addresses software as part of the design phase.

The United States is ahead of other countries in establishing guidelines for medical software development. However, there is movement within several international organizations to develop regulations and guidelines for software and software-controlled devices. For example, ISO 9000-3 specifically addresses software development in addition to what is contained in ISO 9001. CSA addresses software issues in four standards covering new and previously developed software in critical and noncritical applications. IEC has a software-development document currently in development. They also have a risk analysis document (IEC 601-1-4).

Design Alternatives and Tradeoffs

The determination of the design and the allocation of requirements is a highly iterative process. Alternative designs are postulated that could, or are candidates to, satisfy the requirements. The determination of these designs is a fundamentally creative activity, a "cut and try" determination of what might work. The specific techniques used are numerous and call upon a broad range of skills. They include control theory, optimization, consideration of man-machine interface, use of modern control test equipment, queuing theory, communication and computer engineering, statistics, and other disciplines. These techniques are applied to factors such as performance, reliability, schedule, cost, maintainability, power consumption, weight, and life expectancy.

Some of the alternative designs will be quickly discarded, while others will require more careful analysis. The capabilities and quality of each design alternative is assessed using a set of design factors that are specific to each application and to the methods of representing the system design.

Certain design alternatives will be superior in some respects, while others will be superior in different respects. These alternatives are "traded off," one against the other, in terms of the factors that are important for the system being designed. The design ensues from a series of technology decisions, which are documented with architecture diagrams that combine aspects of data and control flow. As an iterative component of making technology decisions, the functionality expressed by the data flow and control flow diagrams from system requirements analysis is allocated to the various components of the system. Although the methods for selection of specific technology components are not a part of the methodology, the consequences of the decisions are documented in internal performance requirements and timing diagrams.

Finally, all factors are taken into account, including customer desires and political issues to establish the complete system design. The product of the system design is called an "architecture model." The architecture includes the components of the system, allocation of requirements, and topics such as maintenance, reliability, redundancy, and self-test.

Software Architecture

Software architecture is the high-level part of software design, the frame that holds the more detailed parts of the design. Typically, the architecture is described in a single document called the "architecture specification." The architecture must be a prerequisite to the detailed design, because the quality of the architecture determines the conceptual integrity of the system. This, in turn, determines the ultimate quality of the system.

A system architecture first needs an overview that describes the system in broad terms. It also should contain evidence that alternatives to the final organization have been considered and the reasons why the organization used was chosen over the alternatives. The architecture should also contain:

- Definition of the major modules in a program. What each module does should be well defined, as well as the interface of each module
- Description of the major files, tables, and data structures to be used. It should describe alternatives that were considered and should justify the choices that were made
- Description of specific algorithms or reference to them
- Description of alternative algorithms that were considered and indicate the reasons why certain algorithms were chosen
- In an object-oriented system, specification of the major objects to be implemented. It should identify the responsibilities of each major object and ways in which the object interacts with other objects. It should include descriptions of the class hierarchies, of state transitions, and of object persistence. It also should describe other objects that were considered and should give reasons for preferring the chosen organization.
- Description of a strategy for handling changes clearly. It should show that possible enhancements have been considered and that the enhancements most likely are also the easiest to implement
- Estimation of the amount of memory used for nominal and extreme cases

Software architecture contemplates two important characteristics of a computer program: The hierarchical structure of procedural components (modules); and the structure of data. Software architecture is derived through a partitioning process that relates elements of a software solution to parts of a real-world problem implicitly defined during requirements analysis. The evolution of a software structure begins with a problem definition. The solution occurs when each part of the problem is solved by one or more software elements.

Choosing a Methodology

It seems there are about as many design methodologies as there are engineers to implement them. Typically, the methodology selection entails a prescription for the requirements analysis and design processes. Of the many popular methods, each has its own merit, based on the application to which the methods are applied. The tool set and methodology selection should run hand-in-hand. Tools should be procured to support established or tentative design methodology and implementation plans. In some cases, tools are purchased to support a methodology already in place. In other cases, an available tool set dictates the methodology. Ideally, the two are selected at the same time, following a thorough evaluation of need.

Selecting the right tool set and design methodology should not be based on a flashy advertisement or a suggestion from an authoritative methodology guru. It is important to understand the environment in which it will be employed and the product to which it will be applied. Among other criteria, the decision should be based on the size of the project (i.e., the number of requirements), the type of requirements (i.e., hard or soft real-time), the complexity of the end-product, the number of engineers, the experience and skill level of the engineers, the project schedules, the project budget, the reliability requirements, and the future enhancements to the product (i.e., maintenance concerns). Weight factors should be applied to the evaluation criteria. One way or another, whether the evaluation is conducted in a formal or informal way, involving one person or more, it should be done to ensure a proper fit for the organization and product.

Regardless of the approach used, the most important factor to be considered for the successful implementation of a design methodology is software-development-team buy-in. The software development team must possess the confidence that the approach is appropriate for the application and must be willing and "excited" to tackle the project. The implementation of a design methodology takes relentless discipline. Many projects have been unsuccessful as a result of lack of commitment and faith.

The two most popular formal approaches applied to the design of medical products are the Object Oriented Analysis/Design and, the more traditional, (top-down) Structured Analysis/Design. There are advantages and disadvantages to each. Either approach, if done in a disciplined and systematic manner along with the electrical system design, can provide for a safe and effective product.

Choosing a Language

Programming languages are the notational mechanisms that are used to implement software products. Features available in the implementation language exert a strong influence on the architectural structure and algorithmic details of the software. Choice of language also has been shown to have an influence on programmer productivity. Industry data have shown that programmers are more productive using a familiar language than an unfamiliar one. Programmers who work with high-level languages achieve better productivity than do those who work with lower-level languages. Developers who work with interpreted languages tend to be more productive than those who work with compiled languages. In languages that are available in both interpreted and compiled forms, programs can be productively developed in the interpreted form, and then released in the better-performing, compiled form.

Modern programming languages provide a variety of features to support development and maintenance of software products. These features include:

- Strong type checking
- Separate compilation
- Used-defined data types
- Data encapsulation
- Data abstraction

The major issue in type checking is flexibility versus security. Strongly typed languages provide maximum security, while automatic type coercion provides maximum flexibility. The modern trend is to augment strong type checking with features that increase flexibility while maintaining the security of strong type checking.

Separate compilation allows retention of program modules in a library. The modules are linked into the software system, as appropriate, by the linking loader. The distinction between independent compilation and separate compilation is that type checking across compilation-unit interfaces is performed by a separate compilation facility, but not by an independent compilation facility. User-defined data types, in conjunction with strong type checking, allow the programmer to model and to segregate entities from the problem domain using a different data type for each type of problem entity. Data encapsulation defines composite data objects in terms of the operations that can be performed on them, and the details of data representation and data manipulation are suppressed by the mechanisms. Data encapsulation differs from abstract data types in that encapsulation provides only one instance of an entity.

Data abstraction provides a powerful mechanism for writing well-structured, easily modified programs. The internal details of data representation and data manipulation can be changed at will and, provided that the interfaces of the manipulation procedures remain the same, other components of the program will be unaffected by the change, except perhaps for changes in performance characteristics and capacity limits. Using a data-abstraction facility, data entities can be defined in terms of predefined types, used-defined types, and other data abstractions, thus permitting systematic development of hierarchical abstractions.

Software Risk Analysis

Software risk analysis techniques identify software hazards and safety-critical single- and multiple-failure sequences; determine software safety requirements; including timing requirements; and analyze and measure software for safety. While functional requirements often focus on what the system shall do, risk requirements must also include what the system shall not do, including means of eliminating and controlling system hazards and of limiting damage in case of a mishap. An important part of the risk requirements is the specification of the ways in which the software and the system can fail safely and extent the extent to which failure is tolerable.

Several techniques have been proposed and used in for conducting risk analysis, including:

- Software hazard analysis
- Software fault tree analysis
- Real time logic

Software hazard analysis, like hardware hazard analysis, is the process whereby hazards are identified and categorized with respect to criticality and probability. Potential hazards that must be considered include normal operating modes, maintenance modes, system failure or unusual incidents in the environment, and errors in human performance. Once hazards are identified, they are assigned a severity and probability. Severity involves a qualitative measure of the worst credible mishap that could result from the hazard. Probability refers to the frequency with which the hazard occurs. Once the probability and severity are determined, a control mode (that is, a means of reducing the probability and/or severity of the associated potential hazard) is established. Finally, a control method or methods will be selected, to achieve the associated control mode.

Real-time logic is a process whereby the system designer first specifies a model of the system in terms of events and actions. The event-action model describes the data-dependency and temporal ordering of the computational actions that must be taken in response to events in a real-time application. The model can be translated into Real Time Logic formulas. The formulas are transformed into predicates of Presburger arithmetic with uninterpreted integer functions. Decision procedures are then used to determine whether a given risk assertion is a theorem that is derivable from the system specification. If so, the system is safe with respect to the timing behavior denoted by that assertion, as long as the implementation satisfies the requirements specification. If the risk assertion is unsatisfiable with respect to the specification, then the system is inherently unsafe because successful implementation of the requirements will cause the risk assertion to be violated. Finally, if the negation of the risk assertion is satisfiable under certain conditions, then additional constraints must be imposed on the system to ensure its safety.

Software Metrics

Software must be subjected to measurement in order to achieve a true indication of quality and reliability. Quality attributes must be related to specific product requirements and must be quantifiable. These aims are accomplished through the use of metrics. Software-quality metrics are defined as quantitative measures of an attribute that describes the quality of a software product or process. Using metrics for improving software quality, performance, and productivity begins with a documented software development process that will be improved incrementally. Goals are established with respect to the desired extent of quality and productivity improvements over a specified time period. These goals are derived from, and are consistent with, the strategic goals for the business enterprise.

Metrics that are useful to the specific objectives of the program, that have been derived from the program requirements, and that support the evaluation of the software consistent with the specified requirements must be selected. To develop accurate estimates, a historical baseline must be established, consisting of data collected from previous software projects. The data collected should be reasonably accurate, collected from as many projects as possible, consistent, and representative of applications that are similar to work that is to be estimated. Once the data have been collected, metric computation is possible.

Metrics that can be used to measure periodic progress in achieving the improvement goals are then defined. The metric data collected can be used as indicators of development process problem areas, and improvement actions identified. These actions can be compared and analyzed with respect to the best return on the business's investment. The measurement data provide information for investing wisely in tools for quality and productivity improvement.

A feedback mechanism must be implemented so that the metrics data can provide guidance for identifying actions to improve the software development process. Continuous improvements to the software development process result in higher-quality products and increased development team productivity. The process-improvement actions must be managed and controlled so as to achieve dynamic process improvement over time.

Requirements Traceability

It is becoming increasingly apparent how important thorough requirements traceability is, during the design and development stages of a software product, especially in large projects with requirements that number in the thousands or tens of thousands. Regardless of the design and implementation methodology it is important to ensure that the design is meeting its requirements during all phases of design. To ensure that the product is designed and developed in accordance with its requirements throughout the development cycle, individual requirements should be assigned to design components. Each software requirement, as might appear in a software-requirements specification, for example, should be uniquely identifiable. Requirements that result from design decisions (i.e., implementation requirements) should be uniquely identified and tracked along with product functional requirements. This process not only ensures that all functional and safety features are built into the product as specified, but also drastically reduces the possibility of requirements "slipping through the cracks". Overlooked features can be much more expensive when they become design modifications at the tail end of development.

The Requirements Traceability Matrix (RTM) is generally a tabular format with requirement identifiers as rows, and design entities as column headings. Individual matrix cells are marked with file names or design-model identifiers to denote that a requirement is satisfied within a design entity. A Requirements Traceability Matrix ensures completeness and consistency with the software specification, which can be accomplished by forming a table that lists the requirements from the specification versus the way each is met in each phase of the software-development process.

Software Reviews

Timely and well-defined reviews are integral parts of all design processes. Each level of design should produce design-review deliverables. Software project development plans should include a list of the design phases, the expected deliverables for each phase, and a sound definition of the deliverables to be audited at each review.

Reviews of all design material have several benefits. First and foremost, authors are more compelled to elevate the quality of their work when they know that their work is being reviewed. Second, reviews often uncover design blind spots and alternative design approaches. Finally, the documentation generated by the reviews is used to acquire agency approvals for process and product.

Software reviews can take several different forms:

- Inspections, design, and code
- Code walk-throughs
- Code reading
- Dog-and-pony shows

An inspection is a specific kind of review that has been shown to be extremely effective in detecting defects, and to be relatively economical as compared to testing. Inspections differ from the usual reviews in several ways:

- Checklists focus the reviewer's attention on areas that have been problems in the past
- The emphasis is on defect detection, not correction
- Reviewers prepare for the inspection meeting beforehand and arrive with a list of the problems they have discovered
- Data are collected at each inspection and is fed into future inspections to improve them

The general experience with inspections has been that the combination of design and code inspections usually removes 60%–90% of the defects in a product. Inspections identify error-prone routines early, and reports indicate that they result in 30% fewer defects per 1000 lines of code than walkthroughs do. The inspection process is systematic because of its standard checklists and standard roles. It is also self-optimizing because it uses a formal feedback loop to improve the checklists and to monitor preparation and inspection rates.

A walkthrough usually involves two or more people discussing a design or code. It might be as informal as an impromptu bull session around a whiteboard; it might be as formal as a scheduled meeting with overhead transparencies and a formal report sent to management.

Following are some of the characteristics of a walkthrough:

- The walkthrough is usually hosted and moderated by the author of the design or code under review.
- The purpose of the walkthrough is to improve the technical quality of a program, rather that to assess it.
- All participants prepare for the walkthrough by reading design or code documents and by looking for areas of concern.
- The emphasis is on error detection, not correction.
- The walkthrough concept is flexible and can be adapted to the specific needs of the organization using it.

When used intelligently, a walkthrough can produce results that are similar to those of an inspection; i.e., it typically can find between 30% and 70% of the errors in a program. Walkthroughs have been shown to be marginally less effective than inspections, but under some circumstances they can be preferable.

Code reading is an alternative to inspections and walkthroughs. In code reading, source code is read for errors. Readers also comment on qualitative aspects of the code, such as its design, style, readability, maintainability, and efficiency. A code reading usually involves two or more people reading code independently and then meeting with the author of the code to discuss it. To prepare for a meeting, the author hands out source listings to the code readers. Two or more people read the code independently. When the reviewers have finished reading the code, the code-reading meeting is hosted by the author of the code and focuses on problems that the reviewers have discovered. Finally, the author of the code fixes the problems that the reviewers have identified.

"Dog-and-pony shows" are reviews in which a software product is demonstrated to a customer. The purpose of these reviews is to demonstrate to the customer that the project is viable, so they are management reviews rather than technical reviews. They should not be relied upon to improve the technical quality of a program. Technical improvement comes from inspections, walkthroughs, and code reading.

Performance Predictability and Design Simulation

The effort to predict the real-time performance of a system is a key activity of design that some software developers often overlook. During the integration phase, software designers often spend countless hours trying to fine-tune a system that has bottlenecks "designed in". Execution estimates for the system interfaces, response times for external devices, algorithm execution times, operating system context-switch time, and I/O device access times in the forefront of the design process provide essential input into software design specifications.

For single processor designs, mathematical modeling techniques such as Rate Monotonic Analysis (RMA) should be applied, to ensure that all required operations of that processing unit can be performed within the expected timeframes. System designers often fall into the trap of selecting processors before the software design has been considered, only to experience major disappointment and "finger pointing" when the product is released. It is imperative to a successful project that the processor selection comes after a processor-loading study is complete.

In a multiprocessor application, up-front system performance analysis is equally important. System anomalies can be very difficult to diagnose and resolve in multiprocessor systems with heavy interprocessor communications and functional expectations. Performance shortcomings that appear to be the fault of one processor are often the result of a landslide of smaller inadequacies from one or more of the other processors or subsystems. Person-years of integration phase defect resolution can be eliminated by front-end system design analysis and/or design simulation. Commercial tools are readily available to help perform network and multiprocessor communications analysis and execution simulation. Considering the pyramid of effort that is needed in software design, defect correction in the forefront of design yields enormous cost savings and increased reliability in the end.

Coding

For many years, the term "software development" was synonymous with coding. Today, for many software-development groups, coding is now one of the shortest phases of software development. In fact, in some cases, although very rare in world of real-time embedded software development, coding is actually done automatically from higher-level design (mspecs) documentation by automated tools called "code generators".

With or without automatic code generators, the effectiveness of the coding stage is dependent on the quality and completeness of the design documentation generated in the immediately preceding software development phase. The coding process should be a simple transition from the module specifications, and, in particular, the pseudocode. Complete mspecs and properly developed pseudocode leave little to interpretation for the coding phase, thus reducing the chance of error.

The importance of coding style (how it looks) is not as great as the rules that facilitate comprehension of the logical flow (how it relates). In the same light, in-line code documentation (comments) should most often address "why" rather than "how" functionality is implemented. These two focuses help the code reader to understand the context in which a given segment of code is used. With precious few exceptions (e.g., high-performance-device drivers) quality source code should be recognized by its readability, and not by its raw size (i.e., the number of lines) or its ability to take advantage of processor features.

Design Support Tools

Software development is very labor-intensive and, is therefore, prone to human error. In recent years, commercial software-development support packages have become increasingly more powerful, less expensive, and more available to reduce the time spent doing things that computers do. Although selection of the right tools can mean up-front dedication of some of the most talented resources in a development team, it also can bring about significant long-term increase in group productivity.

Good software-development houses have taken advantage of CASE tools that reduce the time spent generating clear and thorough design documentation. The advantages of automated software-design packages are many. Formal documentation can be used as proof of product development procedure conformance for agency approvals. Clear and up-to-date design documents facilitate improved communications between engineers, thus lending to more effective and reliable designs. Standard documentation formats reduce learning curves that are associated with unique design depictions among software designers, thus leading to better and more timely design formulation. Total software life cycle costs are reduced (especially during maintenance) due to reduced ramp-up time and more efficient and reliable modifications. Finally, electronic forms of documentation can be easily backed up and stored off-site, thus eliminating a crisis in the event of an environmental disaster. In summary, the adaptation of CASE tools has an associated up-front cost that is recovered by significant improvements in software quality and development-time predictability.

Design as the Basis for Verification and Validation

Verification is the process of ensuring that all products of a given development phase satisfy given expectations. Prior to proceeding to the next-lower level or phase of design, the product (or outputs) of the current phase should be verified against the inputs of the previous stage. A design process cannot be a "good" process without the verification process ingrained—they naturally go hand-in-hand.

Software project management plans (or software quality assurance plans) should specify all design reviews. Each level of design will generate documentation to be reviewed or deliverables to verify against the demands of the previous stage. For each type of review, the software-management plans should describe the purpose, materials required, scheduling rules, scope of review, attendance expectations, review responsibilities, what the minutes should look like, follow-up activities, and any other requirements that relate to company expectations.

At the code level, code reviews should ensure that the functionality implemented within a routine satisfies all expectations documented in the "mspecs." Code also should be inspected, to satisfy all coding rules.

The output of good software designs also includes implementation requirements. At a minimum, implementation requirements include the rules and expectations placed on the designers to ensure design uniformity as well as constraints, controls, and expectations placed on designs to ensure that upper-level requirements are met. General examples of implementation requirements might include rules for accessing I/O ports, timing requirements for memory accesses, semaphore arbitration, inter-task communication schemes, memory-addressing assignments, and sensor- or device-control rules. The software-verification and -validation process must address implementation requirements as well as the upper-level software requirements to ensure that the product works as it was designed to.

Verification and Validation Life Cycle

An SVVP lays out the framework from which the verification and validation activities proceed. IEEE 1012-1986 is very detailed about development life cycle and the verification and validation products that are generated at each phase. Specifically, the standard calls out the following phases:

- Concept phase
- Requirements phase
- Design phase
- Implementation phase
- Test phase
- Installation phase
- Operation and maintenance phase

In industry, it is more typical for software verification and validation to follow a simpler model, as shown in the Figure 85-1.

This model is simpler in that it focuses verification and validation activities on the software that is generated, rather than on output of every phase of development. Another document, called the Software Quality Assurance Plan (SQAP), is used to cover other verification and validation aspects of product development, such as software requirements reviews, hazard and risk analysis, and design reviews.

Using the model as shown in Figure 85-1, software validation testing covers validation of the software product by testing against the requirements generated in the requirements-generation phase. Software validation is analogous to black box testing. Integration testing covers verification of features defined from the requirements-generation phase, the concept/high-level design phase, and the feature-interface design phase. Integration testing tests the collection of software modules, where the software modules are joined together to provide a product feature or higher level of functionality. Finally, unit testing covers verification at the lowest level, which includes the feature interface design phase and the low-level design phase. Unit testing typically covers the verification of smaller collections of software modules.

Verification and Validation Overview

The verification and validation overview describes many of the project management details that must be addressed in order to perform the verification and validation functions. They include the verification and validation organization; a schedule; resources; responsibilities; and any special tools, techniques, and methodologies.

The organization describes who is responsible for carrying out the various verification and validation efforts. In practice, this section is used to describe who will be responsible for maintaining the laboratory, who will manage the validation testing, who will manage verification testing, and who will be responsible for anomaly resolution.

Verification and validation scheduling is very important for planning activities in order to support the overall product schedule. Often, a higher-level schedule is entered into the software verification and validation plan, and a detailed schedule is maintained apart from the software verification and validation plan. This is generally due to different software applications for generating a schedule and for generating documentation.

The resources section describes all of the equipment, software, and hardware that are required, to perform verification and validation activities. A subsection also should be included on the validation of using any software tools, where the software tool is used as part of the verification and validation process. Per the FDA's regulations (FDA, Quality

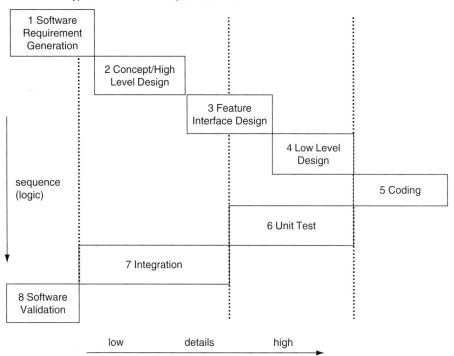

Figure 85-1 Typical software development model.

System Regulation, 21 CFR Part 820, June 1, 1997), any device that is used in a company's quality system must be validated. A test fixture or a customized software application clearly falls under this definition. Thus, it is necessary to validate the software tool.

Verification and Validation Life Cycle Activities

In the SVVP, this section typically includes the criteria, inputs/outputs, reviews, testing approach, and training for each verification and validation phase. The criteria describes the goal that each phase is defined to achieve. The inputs/outputs phase defines what things are needed as inputs into the phase and what the product of each phase are.

It is also helpful to add a subsection on the general functionality that will be tested. The IEEE standard does not call this out. However, adding this subsection is helpful to reviewers of the software verification and validation plan. It provides information on the functionality that is scheduled for testing.

Verification and Validation Documentation

Each verification and validation phase has its activities and documentation that is associated with it. For unit, integration, and software validation test phases, typically there are test plans, test procedures, and test results and reports.

For software validation, there usually are some additional documents. The Software validation group typically has a requirements-to-test cross-reference. The cross reference ensures that all the requirements defined in the SRS are covered in some test. It is also helpful in the cross-reference document to include a test-to-requirement cross-reference. This backward reference helps to ensure that a requirement has adequate coverage. The software validation group also generates a procedure-response document, which serves as a means to close out any issues or comments that occurred on during an official validation pass. Finally, the software validation group has been responsible for writing a final report called the software verification and validation report (SVVR), which summarizes the activities from all of the previous verification and validation phases.

Verification and Validation Administrative Procedures

The verification and validation administrative procedures section provides additional guidance on ways that the verification and validation will be conducted. IEEE Std 1012-1986 defines this section as including anomaly reporting/resolution, task iteration policy, deviation policy, control procedures, and standards/practices/conventions.

It is also useful to include section metrics. IEEE Std 730.1-1989, IEEE Standard for Software Quality Assurance Plans, includes metrics as part of a SQAP. The following subsections describe in more detail anomaly reporting, metrics collection, and reporting, and they provide some additional topics that are important to verification and validation.

Anomaly Reporting and Resolution

One result of verification and validation is the generation of anomalies found during testing. A process for reporting anomalies must be in place to record anomalies. Some institutions have used a lab notebook to document anomalies. More frequently, a PC-based database program is used to log and to store anomalies electronically.

In addition, a process must be in place for resolution of anomalies found against the product software. It can be a committee of people who are responsible for the product or simply the software test coordinator who is working with the primary software developer. The responsible parties determine the risk of each anomaly.

Validation Metrics

There are several aspects of verification and validation in which it is helpful to generate metrics. The first and most obvious one is anomaly metrics. An example of anomaly metric tracking is shown in Figure 85-2. This chart shows the tracking of open severitized anomalies as a function of time.

The second aspect of test metrics is those involving test development. These metrics are used to monitor the test development process and to monitor overall test parameters for all projects. Monitoring the test development process ensures that the test development is proceeding according to schedule. An example is shown in Figure 85-3.

Configuration Management

Just as it is important to maintain a configuration for product development code, it is equally important to practice configuration management on test protocols. Configuration of test materials includes:

- Script libraries
- Test scripts
- Manual protocol
- Test results

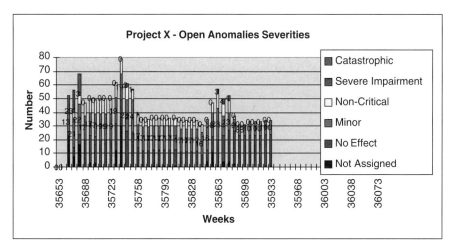

Figure 85-2 Metric chart showing open anomalies sorted by severity

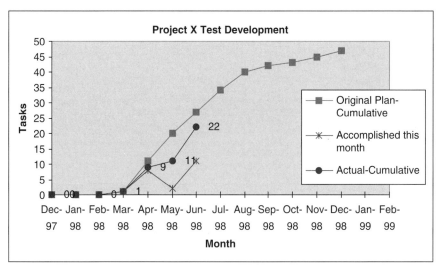

Figure 85-3 Monitoring the test-development process.

- Test plans and procedures
- Test system design

Protocol Templates

To make the development of tests more uniform and efficient, it is advisable to have templates for the test protocol. If test scripts are written, then a set of development guidelines is suggested.

Test Execution

This section discusses approaches that are used when executing manual protocol and automated scripts.

Process Improvement

No test process is perfect. There are always ways to improve the test-development process to do things more efficiently. Thus, some time should be set aside for review of the current processes that are in place. If the test team is large enough, a process team can be created to address processes and process changes in the test group. Discussions should occur among test team members to determine those processes that worked well and those that did not.

Test Development

Requirements Analysis and Allocating Requirements

Today, it is assumed that a software requirements specification (SRS) is written. The SRS describes the behavior and functionality of the software in the medical device. It should obtain its inputs from the following sources:

- Derivation of the product specification
- Control procedures from the risk and hazard analyses
- Regulatory requirements, such as safety-relevant computing

A large part of the verification and validation effort will concentrate on proving that these software requirements are fulfilled. Therefore, the requirements in the software requirements specification must be enumerated to permit proper allocation to the different test phases and the specific tests.

The initial analysis of the software requirements specification involves determining where to test a particular requirement. Where a requirement is tested is often based on the capabilities of the test fixture used in validation. Usually, most requirements can be tested at the validation level. The balance of the requirements is then tested during integration or unit testing. In general, a requirement that has external stimulus (to the central processing unit (CPU)) and the ability to monitor externally are done at the validation level. Exceptions to this are conditions that might require very specific timing, and that timing is calculated internally to the CPU.

Requirements that are tested during integration or unit testing are typically those that cannot be tested sufficiently during validation testing. Testing a "software watchdog error" requirement is good example of a requirement tested during integration or unit testing. In this case, the software is monitoring itself, and there is no external means to cause the fault.

The second step of the requirements analysis phase is to determine the best way to group the requirements into tests. One relatively straightforward approach to doing this is to create tests based on sections in the software requirements specification. Typically, a section in an SRS will describe a feature almost fully. By following the software requirements specification format, the tests then are developed by feature, which is a logical approach. Of course, there will be requirements in the software requirements specification where this approach will not apply. In such cases, further analysis must be done to assess whether additional tests should be created, or whether one of these "special" requirements can be allocated into those tests that follow the software requirements specification.

As mentioned at the beginning of this section, control procedures from the risk and hazard analysis should be written into the software requirements specification. If this is done, then those risk/hazard requirements simply will be analyzed and allocated to tests just like all other requirements.

Storing the allocation matrix is a final aspect of requirements allocation to test. . It is simple enough to keep the matrix in a spreadsheet or an electronic document. However, it can also be stored in a database. The advantage of storing it in a database is that a user can track test development and also can generate numerous reports.

Testing Phases and Approaches

Unit Testing

Unit testing is performed on a smallest amount of code. What comprises a "unit" is often the subject of lengthy discussion and debate. No matter how a unit is defined, the intent of unit testing is to ensure that the lowest-level software modules are tested.

There are several approaches to unit testing, such as branch path analysis or module interface testing. In branch-path analysis, the developer uses a development tool to step through each path within a unit. In module-interface testing, the developer tests the unit from a "black box" perspective. Thus, the unit is only tested by varying the inputs into the software unit.

The type of unit testing that will be required will depend on many factors. In medical devices, the detailed branch path analysis testing is usually done against software units that are considered to be critical to the safety of the patient or the user. Other, lesser critical software units can be justifiably tested using the less thorough module interface testing.

Integration Testing

Integration testing has several definitions. The definition covers the integration of two or more software units together. A broader definition covers the integration of physical subsystems (each with its own embedded software) to ensure that they work together.

Either type of integration testing brings separately developed entities together to ensure that they work together. The type of testing that is usually performed at these levels involves writing protocols to exercise the interface.

Validation Testing

Validation testing is the process of proving that the product meets the product specification. It also can mean going the extra step and trying to ensure that the product does not do what it is not supposed to do. Thus, several approaches to testing are employed to exercise the software product. These approaches are discussed below.

Requirements-Based Testing

Requirements-based testing is the primary approach that is used to validate software. Essentially, the requirements from the software requirements specification are analyzed and allocated to specific tests. Different approaches to testing, such as threshold testing or boundary testing, are used in developing these tests. The test steps are the sequential actions that must be taken to prove that the requirement has been met.

Threshold Testing

Threshold testing is the process of proving that an event will occur when a specified parameter exceeds a certain value. An alarm, such as a low-battery alarm, is a good example of this on a medical device. If the device is running on a battery, and the voltage level in the battery falls below a threshold level, then an alarm is enunciated. In these types of tests, a tolerance band is placed around the threshold level. The tolerance is determined by the requirements in the software requirements specification and also can be influenced by the accuracy of the test system. The test is then designed to vary the battery voltage to cross through the threshold and to see that the alarm is tripped within the tolerance band. Once the alarm has been tripped, the next part of the threshold test is to vary the parameter (in this case, the battery voltage) in the reverse direction to ensure that the alarm no longer enunciates.

Boundary Testing

Boundary testing is the exercise of testing a parameter at its limits, and of trying to exceed those limits. A measured value of O_2 that is defined as not to exceed 110% would be a good example of this. (An O_2 reading can exceed 100% if the O_2 sensor is out of calibration or calibrated incorrectly.) The test then would be designed to prove that 110% O_2 could be displayed. Additionally, the test then would see how the device would react if the measured O_2 value were to exceed 110%.

Stress Testing

Stress testing is the process of subjecting the unit to a bombardment of random inputs—often keypad presses or knob turns—to try to cause a software failure. This can be done manually or with an automated test system.

Manual stress testing and automated stress testing each have their own advantages. The advantage of manual stress testing is that the tester can test a certain aspect of the device can and observe anomalous behavior in other areas of the device. The advantage of automated stress testing is that the test system can provide inputs into the device much more quickly than a human tester can.

Volume Testing

Volume testing is the process of exercising the unit for an extended period of time. In our medical applications, we have found that running a volume test for a 72-hour period can uncover problems not found elsewhere.

Volume testing is almost always done with an automated test system. Volume testing does not necessarily use random fast, slam-bang key presses like stress testing does. Volume testing uses a logical approach to testing all of the paths in a software program. Once the device is in a certain state, probabilistic algorithms like Probability Density Function (PDF) are used in determining the next state.

Scenario Testing

Scenario testing is the act of writing a test that emulates the actual use of the software from the user's perspective. For example, in anesthesia machines, a test could be written to cover a clinical situation. Scenario testing is helpful because it can uncover problems in the system, or it can detect flaws in the overall use of the product.

System Testing

System testing occurs to validate the entire medical device, not just the software that is embedded in the device. Depending on the way a test station is instrumented for software testing, there can be significant overlap between software and system testing. For example, software embedded in a ventilator usually requires other mechanical-electrical subsystems in order to function. It is difficult to write sophisticated simulations to trick the ventilator software into thinking that all of the other subsystems of the ventilator are present. Therefore, a software test system might resemble a complete-system. It is up to the responsible parties to ensure that the degree of overlap is made known. More importantly, it is vital that everything is tested at some point in the product development.

Test Execution and Reporting

Test Plan

For software validation, a test plan is written prior to executing a test pass. The format follows that defined in ANSI/IEEE Std 1012. The key aspects of the test plan are the description of the software changes, and the tests that will be run to prove that those software changes are implemented correctly. The test plan is written whenever a full validation or a regression test is run.

Test-Configuration Form

Because it is very important to be able to repeat a test, it is necessary to document the configuration of the equipment that is used to run a test against. This documentation is written on a test configuration form. The contents of these forms can vary. However, the primary fields are date, item, and item serial number, and there is a place for the test engineer to sign his name. The test configuration form becomes part of the test-documentation suite.

Executing Manual Protocol

Executing manual protocol for a formal test involves printing out the test procedures and going through the steps in the procedures. Because these printouts are a part of the official documentation, the test engineer must sign and date the documents. If the tester finds problems during the test, he will document the issue on the test procedures.

It is important that the handwritten notes on the paper procedures be addressed. The notes on the procedures could be problems found or procedure deviations. All of these markings must have closure. This closure is formally documented in the procedures response document. The procedures response document is covered later in this chapter.

Executing Automated Protocol

Typically, executing automated protocol amounts to setting up a batch job to submit a series of tests to be run. A test engineer fills out a test configuration form to have a piece of paper to start off the test. The batch job is then run overnight, and the results are analyzed the following day.

Experience has shown that a test in the batch occasionally can fail. This failure usually has been attributable either to timing issues or to the target-device feature having changed but the test not having changed. When this happens, it is usually a practical matter simply to correct the test script and to rerun it. However, in instances where correcting the test script would consume a large amount of time, it is sometimes more efficient to run that test manually to complete the test pass.

Test Results

Once test results have been generated, they must be analyzed. The results from the automated tests are reviewed by running searches for keywords that indicate whether the test had problems. Additional analysis is performed by reviewing samples of the automated test results. This ensures that the tests executed as expected. Other analysis might need to occur, such as when a large amount of data are generated during the test, after which data reduction and analysis must be performed to assess correctness.

Test results must be managed along with the other test documentation. The results from the manual test procedures must be included with the other paper documentation and stored in the formalized location. The electronic results files also must be put under configuration control. This is done with a configuration-management tool or a process of backing up the files to controlled media.

Test Reports

Software Validation Test Report

The key items in this report are a summary of what was tested, how it was tested, any problems that were encountered, and a recommendation for release.

Requirements to Test Cross Reference

This document covers the ways in which the requirements were allocated to tests.

Procedures Response Document

As mentioned earlier, all handwritten marks on the manual test procedures must be reviewed and provided a closure. The results of this closure are documented in the procedures response document. Typically, each mark-up in a test procedure is a software problem, a procedure error, a procedure deviation, or a comment.

Further Information

Bass L, Clements P, Kazman R. *Software Architecture in Practice*. Reading, MA, Addison Wesley Longman, 1998.
Bloesch A. Overview of Verification and Validation for Embedded Software in Medical Devices. *Handbook of Medical Device Design*. New York, Marcel Dekker, 2001.
Boehm BW. A Spiral Model of Software Development and Enhancement. *Computer* 5, 1988.
Boehm BW. *Software Engineering Economics*. Englewood Cliffs, NJ, Prentice Hall, 1981.
Booch G. *Object-Oriented Analysis and Design with Applications, 2nd Edition*. Redwood City, CA, Benjamin Cummings, 1994.
Booch G, Jacobson I, Rumbaugh J. *The Unified Modeling Language User Guide*. Reading, MA, Addison Wesley Longman, 1998.
Deutsch MS, Willis RR. *Software Quality Engineering—A Total Technical and Management Approach*. Englewood Cliffs, NJ, Prentice Hall, 1988.
Dyer M. *The Cleanroom Approach to Quality Software Development*. New York, Wiley, 1992.
Fairley RE. *Software Engineering Concepts*. New York, McGraw-Hill, 1985.
Food and Drug Administration, *21 CFR Part 820*. Washington, DC, U.S. Government Printing Office, 1997.
Fries RC. *Reliable Design of Medical Devices*. New York, Marcel Dekker, 1997.
Fries RC, Pienkowski P, Jorgens J. Safe, Effective, and Reliable Software Design and Development for Medical Devices. *Med Instrum* 30(2), 1996.
Hatley DJ, Pirbhai IA. *Strategies for Real-Time System Specification*. New York, Dorset House, 1987.
Humphrey WS. *Managing the Software Process*. Reading, MA, Addison-Wesley, 1989.
Institute of Electrical and Electronics Engineers. *IEEE Standard 730–IEEE Standard for Quality Assurance Plans*. New York, Institute of Electrical and Electronics Engineers, 1989.
Institute of Electrical and Electronics Engineers. *IEEE Standard 829–IEEE Recommended Practice for Software Requirements Specifications*. New York, Institute of Electrical and Electronics Engineers, 1998.
Institute of Electrical and Electronics Engineers. *IEEE Standard 830–IEEE Standard for Quality Assurance Plans*. New York, Institute of Electrical and Electronics Engineers, 1989.
Institute of Electrical and Electronics Engineers. *IEEE Standard 1012–IEEE Standard for Software Verification and Validation*. New York, Institute of Electrical and Electronics Engineers, 1986.
Institute of Electrical and Electronics Engineers. *IEEE Standard 1016–IEEE Recommended Practice for Software Design Descriptions*. New York, Institute of Electrical and Electronics Engineers, 1986.
Institute of Electrical and Electronics Engineers. *IEEE Standard 1059–IEEE Guide for Software Verification and Validation Plans*. New York, Institute of Electrical and Electronics Engineers, 1993.
Kan SH. *Metrics and Models in Software Quality Engineering*. Reading, MA, Addison–Wesley, 1995.
Leveson NG. *Safeware*. Reading, MA, Addison-Wesley, 1995.
McConnell S. *Code Complete*. Redmond, WA, Microsoft Press, 1993.
McConnell S. *Rapid Development*. Redmond, WA, Microsoft Press, 1996.
Page-Jones M. *The Practical Guide to Structured Systems Design—2nd Edition*. Englewood Cliffs, NJ, Prentice Hall, 1988.
Pressman R. *Software Engineering*. New York, McGraw-Hill, 1987.
Putnam LH, Myers W. *Measures for Excellence*. Englewood Cliffs, NJ, Prentice Hall, 1992.
Rakos JJ. *Software Project Management for Small to Medium Sized Projects*. Englewood Cliffs, NJ, Prentice Hall, 1990.
Rumbaugh J et al. *Object-Oriented Modeling and Design*. Englewood Cliffs, NJ, Prentice Hall, 1991.
Rumbaugh J, Jacobson I, Booch G. *The Unified Modeling Language Reference Manual*. Reading, MA, Addison-Wesley, 1998.
Sommerville I, Sawyer P. *Requirements Engineering*. Chichester, England, Wiley, 1997.
Storey N. *Safety-Critical Computer Systems*. Harlow, England, Addison-Wesley, 1996.
Thayer RH, Dorfman M. *Software Requirements Engineering—2nd Edition*. Los Alamitos, California, IEEE Computer Society Press, 1997.
Yourdon E. *Modern Structured Analysis*. Englewood Cliffs, New Jersey, Yourden Press, 1989.

Comparative Evaluations of Medical Devices

James P. Keller
Director, Health Devices Group, ECRI
Plymouth Meeting, PA

Evaluating medical device technology is a complex process. Many factors contribute to its complexity, ranging from the wide variety of devices purchased and used in hospitals, to the various settings and applications where devices are used, to the many different types of users of the equipment. Many hospitals do not have a good process in place to address these complexities effectively or to evaluate their choices for medical technology. They subsequently make inappropriate, and sometimes unsafe, equipment procurement choices. The purpose of this chapter is to present a process that health care organizations can use to effectively evaluate medical technology. Many of the steps in this process are modeled after a 30-year-old, comparative brand-name evaluation program and processes used by ECRI, a research institute in Plymouth Meeting, Pennsylvania, to advise hospitals on medical device selection. Examples from ECRI's evaluation program will be used throughout this chapter.

Complexities of Hospital Purchase of Medical Equipment

Hospitals buy a large number of medical devices. $155.5 billion was spent on medical devices worldwide, and $66.5 billion was spent in the United States in 1999 (MHMG, 2000). The types of devices include capital equipment, computer-based systems, equipment integrated with computer-based systems, drug/device combinations, disposable/reusable products, and supplies. Most applications of the devices are for use within the hospital. However, many hospitals also acquire equipment and supplies for distribution by a durable medical equipment (DME) provider; for placement directly in a home setting by a hospital (e.g., without using the DME as a "middle man"); for research or a combination of research and clinical use; and for use in clinics, doctors' offices, nursing homes, patient-transport vehicles, same-day surgery facilities, and other settings. Depending on the setting, the hospital will have one user to a wide range of users for any given type of equipment. In most cases, the users of the equipment are health care professionals with some level of degree- or certificate-based training. However, many devices are used by untrained individuals and even by patients.

There are often several different types of devices that can be used for the same, or some of the same, types of applications. Hospitals typically have many vendors to choose from when selecting a class of medical equipment to purchase. Often, one manufacturer will offer several choices of equipment within the same device class, or several different configurations of the same device. Many hospitals rely on vendor information and specifications to make their purchasing decisions. However, vendor specifications are not presented in a standardized format from vendor to vendor. Specifications, product descriptions, and general product information (e.g., summaries of clinical studies) are also presented in the most positive light for each manufacturer. Clinical literature about the utility of products also might be biased (e.g., funded by manufacturers of the products being evaluated), incomplete, or of poor quality, and typically it is not comparative. This all results in a confusing, and often misleading, equipment purchasing data set for the hospital.

Medical equipment purchasing is rarely centrally controlled within an institution. Purchasing typically takes place independently in many departments throughout the same institution. In many cases, the different departments are purchasing equipment for the same or similar purposes. This results in a wide range of criteria and purchasing practices within the same institution, often with significantly different standards of care. When one hospital is part of a larger health care system, the number of variables in system-wide purchasing can become even more overwhelming.

Hospitals that participate in, or that are members of, group purchasing organizations (GPO) theoretically avoid some of the complexities described above. GPOs typically take advantage of the economies of scale provided by their membership, to negotiate substantial group-rate-type discounts for a large list of devices. GPO member hospitals that buy devices from the GPO lists will receive the negotiated pricing. The GPOs will list one-to-several choices for members within each device category on the list. However, GPO member hospitals usually have not had a direct say in selection of the GPO-selected products. These products have not been selected with the unique needs of each GPO member in mind. In fact, they might not have been selected for their clinical utility, effectiveness, safety, or cost-effectiveness. The GPO's selection criteria sometimes can be solely based on the lowest negotiated price. Although the GPO member is not typically locked into only buying the products on the GPO contract list, it is in its best interest to strongly consider the discounted prices that are available. In fact, some GPO contracts are based on the GPO member committing to purchasing a certain number of devices from the GPO list in order to obtain the negotiated prices. However, similar or better bargains might be available elsewhere, and in order to be thorough, hospitals need to investigate whether the GPO-listed devices actually meet their specific needs.

The regulatory status of devices can impact their purchase. Most devices used for clinical applications are clearly labeled as medical devices according to FDA definitions. However, some devices, such as clinical information systems, do not have a formal regulatory status or others are in the process of becoming approved or cleared for sale within certain jurisdictions. Hospitals that are interested in purchasing this type of equipment have an additional set of factors to consider before purchasing, relating to liability, insurance coverage, government reimbursement, competence of users, patient consent, and other issues.

The complexities discussed above are but a few of those that hospitals are confronted with when purchasing their wide range of medical devices. The best way to address these complexities is for hospitals to have a process in place to objectively evaluate its equipment needs and the true value of, and need for, the equipment that the institution is considering. This process will be discussed below and will be modeled, in large part, after a 30-year-old comparative brand-name evaluation program and processes used by ECRI to advise hospitals on medical device selection.

Identifying Equipment Needs

Medical devices are, in theory, purchased to meet one or more specific needs for the health care facility. Before any type of device evaluation takes place, however, those needs must be clearly defined. The need could be as simple as a measuring a patient's blood pressure. On the surface, the solution to the problem is to buy enough sphygmomanometers and stethoscopes to meet the needs of the clinicians who will be taking the measurements. However, "taking blood pressure measurements" is a general criterion. The hospital's true need, at least in some clinical departments, might be to record blood pressure readings automatically; to display the data in a way that can warn clinicians when blood pressure reaches dangerous levels; to store the readings in some form of database for trending of blood pressure fluctuations; and to integrate the blood pressure data with other clinical parameters to manage patient care more effectively. Handheld sphygmomanometers and stethoscopes clearly cannot meet these needs.

As with the blood pressure measurement example, most equipment purchase decisions are not as simple as the number and brand of device to buy. Each equipment purchase consideration must answer a series of questions before the purchasing process gets to the point of considering the number and brand to purchase. The first question to ask is whether you need equipment for an application. Sometimes, a better choice for an application might be drug therapy, rather than a device solution. If a device solution is appropriate, the next question concerns the type of equipment needed. Equipment evaluators should keep in mind that there might be several types of equipment choices, such as lasers, electrosurgical units, and harmonic scalpels for surgical dissection and coagulation. Each alternative must be considered with the specific needs of the institution in mind.

Equipment evaluators also should determine the clinical specialties that would need the equipment or that have similar needs. For example, orthopedic surgery might request a certain type of laser for dissecting tissue during joint surgery. However, the laser might not be used often enough merely by orthopedic surgeons, to justify such a costly acquisition; lasers for orthopedic surgery typically cost at least $100,000. However, if it is determined that other surgical specialties will benefit from use of the laser as well, the acquisition could be justified.

Determining true clinical need is a critical step in the evaluation process. Unfortunately, many hospitals do not carefully consider this issue and have paid the high price of buying medical devices that are not really needed; the devices become expensive dust collectors. In some cases, the equipment is utilized, but not often enough for staff to become proficient with its use. Each time the device is used, the clinical staff must relearn how to use it. When this happens, the cost of poor purchase decisions also becomes a safety concern. Patients and staff are at risk of being injured by medical devices that operators do not know how to use proficiently. ECRI has operated a medical device problem-reporting and investigation program for over 30 years. It has investigated many cases of serious patient or staff injury because use of medical devices that clinical staff did not fully understand how to use.

Careful research is required, to determine the true clinical need for medical devices. It should include interviews with staff who will be using or affected by the equipment; review of relevant clinical and technical literature on the benefits and drawbacks of the technology; review of the hospital's past history with similar equipment; surveys of other organizations' experience with the equipment being considered; discussions with manufacturers; review of manufacturer literature; and consultation with outside experts. In some cases, there will be overwhelming evidence that a certain type of technology is a standard of care and totally appropriate for the application(s) being considered. However, with new and emerging technologies, the information gathered must be examined critically for bias and quality of data. ECRI has observed many examples of new and emerging technologies being prematurely implemented because of over-reliance on poor-quality clinical data or overly vigorous promotion of the technology by proponents who will receive financial gains from its success.

Establishing Basic Selection Criteria

Once it has been determined that a certain class of equipment is needed by the hospital, the actual device-evaluation process can begin. The first step is to determine the features needed from the device. As mentioned above, there are often a wide variety of devices offered within one device class. Many of the features that are available for the different models are absolutely necessary for the devices to serve their main purpose. However, as with a car, many devices have features that might offer little more than cosmetic benefits that come at a high cost, the often-termed "bells and whistles." Other features might be clinically useful, but they are not needed for the specific application being considered. Conversely, some devices lack a key feature in order to work as designed. One of the first steps of a device evaluation is to identify the "must have" features for the devices being considered for purchase. Once these features have been identified, the universe of devices available for purchase can be narrowed down to a manageable number.

ECRI's brand-name comparative evaluation program uses a rating system that is based on an essential-features analysis. The rating system includes the rating categories of "preferred," "acceptable," "not recommended," and "unacceptable." Devices rated "unacceptable" have one or more features that prevent them from performing their intended purposes; create an unacceptable risk; or are prohibitively expensive. For example, infusion pumps that do not have automated protection from free-flow delivery of medications and other intravenous fluids can deliver a fatal overdose. ECRI rates infusion pumps that lack this feature as "unacceptable" and recommends that hospitals remove these devices from their purchase-consideration list. In fact, ECRI recommends that these devices be removed from use and replaced with the safer, free-flow protected devices. Other device features that have resulted in "unacceptable" ratings by ECRI have included needle-stick-prevention devices that lacked automated needle-stick protection (ECRI, 1999) and thoracic aspirators that lacked sufficient specified volumetric airflow to overcome large air leaks in the thoracic cavity and thus provide poor suction (ECRI, 1998a). In each example, the devices were not able to perform their intended purposes, at least according to ECRI's expectations, and they were unsafe.

The essential-features analysis from the ECRI evaluation examples cover generic problems or characteristics that would apply to any institution's situation. However, some issues will come up in a hospital's device selection that will be unique to its circumstances or at least a small sampling of similar hospitals. Part of the essential-features analysis is to identify the features that the devices under consideration must have, or should not have, in order to meet the hospital's unique needs. For example, if the hospital is operating under a specific software platform, with which a medical device based computer system is not compatible, the medical device should be ruled out of the hospital's purchase consideration unless the hospital is in the process of also upgrading the software platform to a compatible system. In another example, many hospitals are installing picture archiving and communications systems (PACS), systems used for archiving and viewing radiographic images. In order for these systems to work, they must be compatible with the radiographic imaging devices in the hospital (i.e., the PACS should be able to receive and read data from the imaging devices). If a significant number of devices in the institution are not compatible with the PACS, then the PACS will not be appropriate for consideration. In the PACS example, an extensive analysis, typically based on each system's level of compliance with the American College of Radiology (ACR) and the National Electrical Manufacturers Association (NEMA) Digital Imaging and Communications in Medicine (DICOM) standard, might be required in order to determine compatibility.

Detailed Comparative Evaluation Criteria

After narrowing down the device-selection options using an essential features analysis, the real comparative nature of the evaluation process can begin. This process involves comparing the devices or systems being considered in a variety of general categories. Within each category, specific criteria must be established in order to effectively compare the systems to one another. The general categories include overall performance and safety, ease of use and other human factors issues, quality of construction, service and support, and cost. Depending on the type of device or system being evaluated, one category might need to receive more attention than others in the evaluation process. Another factor that determines the category that receives a higher level of attention will be the setting for the device under consideration. Devices intended for home use, especially those used and operated by patients, will have significant ease-of-use concerns. Devices intended for transport use, especially in helicopters, should receive a greater emphasis on quality-of-construction factors, such as the ability to withstand shock and vibration.

In order to conduct a thorough evaluation, a comprehensive set of criteria must be established for each device being evaluated in each of the general categories mentioned above. This can be a daunting task for any hospital or other health care institution, especially considering the large number and variety of medical devices purchased every year. However, despite the confusing set of device-selection data with which hospitals are confronted, resources do exist to make this process much more manageable. One of the most well-known and respected resources is the Health Devices Evaluation program operated by ECRI, a nonprofit corporation that engages in the independent evaluation of technology. For over 30 years, ECRI has been conducting evaluations comparative to those of a similar institution, Consumers Union, publisher of *Consumer Reports*. ECRI has established rigorous conflict-of-interest rules and evaluation methods that support its unequaled reputation as a totally objective and comprehensive medical technology information source. Hospitals routinely refer to ECRI's many comparative device evaluations for guidance on establishing their own evaluation criteria, or the ECRI criteria are fully adopted into the hospital's process.

Criteria for ECRI's comparative evaluations are developed through rigorous review of related clinical and technical literature, standards, and previous ECRI evaluations, and from ECRI's experience in working with and investigating problems involving the evaluated technology. ECRI technical staff members also spend many hours interviewing experts on, and users of, the technology and observing the technology in clinical use. Manufacturers of the evaluated products are invited to ECRI's laboratories to demonstrate their products and for questioning by ECRI technical staff. Through this process, important performance characteristics are identified, user problems are noted, and other evaluation-related factors are established, from which the details of the criteria are developed. Examples of device areas where evaluation criteria are established include allowable turning rotation on control knobs, alarm volume thresholds on anesthesia machines, flow control accuracy on critical care ventilators, radiation exposure dose limits on CT scanners, puncture resistance for surgical gloves, and level of acceptable display-screen clutter on physiologic monitors.

The ECRI criteria are packaged in a well-organized document and critically reviewed by ECRI technical staff, outside experts, users, and manufacturers whose products are being evaluated. This results in a comprehensive, but practical, set of expectations for the evaluated device or system. Hospitals must examine each of ECRI's criteria to verify that they apply to their unique situation. In most cases, the criteria will be completely applicable. Where a criterion does not apply or has not been established by ECRI, evaluations of similar technology by ECRI can be used to develop new, or to revise existing, criteria. The hospital conducting the evaluation can also use ECRI's process of reviewing relevant clinical and technical literature and standards, and consultations with experts (including ECRI staff) and users to develop the criteria.

Critical Evaluation Criteria

Because of the wide range of devices used in hospitals and other health care settings, it is impossible in the context of this chapter to describe critical-evaluation criteria for all of these devices. However, a number of examples based on recent ECRI evaluations and the experience of the author are presented below. Examples are provided for each of the general categories mentioned above.

Performance and Safety

Arthroscopic irrigation and distension systems are designed to distend a joint cavity with fluid to provide an orthopedic surgeon with a workable operating space within the joint during arthroscopic surgery. High pressures from the arthroscopic irrigation and distension systems are known to cause extravasation of the distension fluid into the tissue surrounding the joint cavity. At a minimum, extravasation can cause irritation and delayed healing in and around the surgically treated joint. More serious complications include nerve palsy, arterial compression, and even cardiac arrest. Clinical studies have shown that joint pressures of as low as 180 mm Hg can cause extravasation. To minimize the risk of extravasation, arthroscopic irrigation and distension, systems should have a limit so that they cannot deliver fluid in the joint at pressures higher than 180 mm Hg and joint pressure should be monitored by the arthroscopic irrigation and distension systems so that users can be alerted, should pressure levels reach the 180 mm Hg limit (ECRI, 1998b).

Ease of Use and Other Human Factors Issues

Blood glucose monitors are used by diabetic patients and clinicians to monitor and to help control a patient's blood glucose level. Most of these devices are used by patients outside of the hospital setting. Many of these patients are elderly, with limited manual dexterity and impaired vision. Blood glucose monitors have a reading port into which patients place a small calibrated test strip, onto which they have applied blood, for analysis. In order to accommodate users with limited manual dexterity, the test strips cannot be so small as to make the strip difficult for the user to handle. The display image must be large enough to present the glucose measurement in an easily legible format, clearly viewable under a variety of lighting conditions (ECRI, 1997).

Quality of Construction

A relatively new type of electrosurgical return electrode has become available. It uses capacitive coupling to return electrosurgical energy used for cutting and coagulating of tissue to an electrosurgical unit from the surgical site on a patient. The return electrode is designed to disperse the electrosurgical current over its large surface area and to eliminate high-current densities in the electrosurgical current return path, which can cause a patient burn. This electrode is unique because of its large size (almost 2 feet by 3 feet) and is designed to be reused many times. The electrode consists of a large sheet of conductive fabric that is covered by a urethane insulating material. Because this device is reusable, it likely will be exposed many times to sharp objects and other abuse in the operating room over its expected useful life. If penetrated by sharp objects or otherwise damaged, the conductive fabric could be directly exposed to the patient's skin, at which point electrosurgical current could concentrate and cause serious burns. This product, therefore must be resistant to tearing and other damage that might occur during normal use in the operating room (ECRI, 2000a).

Service and Support

PACS are complex, software-intensive systems that interface with many devices and systems throughout the hospital (see Chapter 102). These systems have many users, also throughout the hospital and sometimes from remote locations. They cannot be installed, operated, or maintained without significant support from the manufacturer. Many extensive software upgrades to these systems are expected throughout their useful lives. One critical criterion for an evaluation of these systems is for the manufacturer to have the ability to provide comprehensive, responsive, and reliable support from installation, day-to-day operations, routine maintenance and repairs, system upgrades, and user training (ECRI, 2000b).

Cost

Rarely is the cost of a medical device limited to the purchase price of the device itself. The additional costs could include accessories, service and support, installation fees, additional staff needed to run the equipment, reagents, annual software license fees, and financing fees. In order to do a true comparative analysis of the costs for several choices of medical devices or systems, all of the applicable costs must be considered over the expected life of the equipment, typically for at least five or seven years. When reagents or accessories are involved, evaluators must estimate the annual use of these elements in the equation and must come up with a projected annual cost for their use. The general cost criterion for any device evaluation is that a device should have a favorable life cycle cost as compared to other evaluated devices. The life cycle cost analysis must include all expected costs for the equipment, taking into consideration expected annual inflation rates and a discount factor to account for the changing value of money over the time of the analysis.

Testing for Compliance with Evaluation Criteria

Verifying compliance with essential criteria is relatively easy. Evaluators simply need to check with manufacturer specifications and to determine whether the devices being considered are specified to meet the essential criteria. If they do not, they can be immediately ruled out of the evaluation process. However, verifying the more specific performance and safety, ease of use, or other criteria is not that easy. In order to get it right and to avoid inappropriate or more seriously ineffective, overly expensive, or unsafe choices, a comprehensive effort is required. In some cases, physical testing is required. In other cases, qualitative assessments and comparative analyses must be used. With life cycle analysis, all costs must be identified, and projected volumes must be carefully estimated. The data then must be entered into analytical tools such as life cycle cost calculators to obtain results for comparison.

The person who does all of the research, testing, and analysis depends on the technology being evaluated and the hospital's previous history with the technology being considered. Older, more established, or simple technology might not require substantial hands-on analysis by the hospital. Consultation with an outside organization like ECRI for its experience with, and purchase advice for, the technology might be sufficient. Newer or more sophisticated (especially highly customized) technology like PACS or hospital information systems likely will need a dedicated team, including outside experts and nonbiased analyses, to determine how effectively the systems comply with the criteria, especially those that address the evaluating hospital's unique circumstances or customization needs.

It is important to include potential users on any assessment that occurs at the hospital. Users can provide valuable insights on any assessment of performance, device usability, or other characteristics and will be much more willing to embrace new technology that is eventually acquired if they believe that their opinion was sought and headed in the selection process. On many occasions, ECRI has observed technology being thrust upon clinical staff without its input. These situations create ill will between equipment selectors and users and often result in poor utilization of the new technology.

When ECRI conducts a comparative evaluation of a medical device, it uses detailed test methods and presents these and its findings in its monthly publication, *Health Devices*. Hospital equipment evaluators should look to this publication for guidance on ways to perform their analyses. In many cases, ECRI will have tested and presented results on the specific products being considered by the hospital. In this case, all that the hospital must do is to verify that ECRI's findings apply to the hospitals' unique circumstances. ECRI technical staff also can be used by the hospital evaluators for advice on ways to perform analysis and testing, which testing is most important to consider, and whether testing by an independent body is warranted for a particular device evaluation.

Dealing with the Findings

No technology is perfect. Some devices or systems might work well but are somewhat difficult to use. Another might work well yet might have a safety problem that is correctable with routine user intervention. Another might be the best performer, might be the easiest to use, and might possess the most desirable feature, yet it might be significantly more costly than the alternatives. These are the types of findings that equipment evaluators are likely to be confronted with at the end of the evaluation process. The evaluator's role is to assess the importance of each finding, then apply a weight or some other type of judgment scale to the finding, and tally the results to come to a final conclusion. The challenge for the evaluators is to determine the appropriate weight for each finding. And, unfortunately, one weighting system cannot be used for every class of device. For example, image quality is a critical factor when assessing digital radiography system. However, it is important, but not critical when assessing a physiologic monitoring system.

Equipment evaluators must look at the individual features of each evaluated technology and determine the features that are most important. Findings relating to the more important features can be assigned greater weight in the analysis. Evaluators then need to look at individual results in every category to determine whether any finding could affect the safety or overall effectiveness of the equipment significantly. Depending on the nature of the problem, a finding relating to an infrequently used feature could end up receiving a high weight in the device's final analysis. As an example, consider an infusion device evaluation where the device was found to work well, except for over a range of infrequently used flow settings. The accuracy at those flow settings was such that patients could receive a dangerous overdose of medication. Despite doing well on all of the frequently used and important settings, the system should be given an overall low rating for the evaluation. How low the rating is set would depend on how likely the infrequently used flow settings actually would be used at the hospital. This also could depend on the type of instructions that would have to be provided to the clinical staff on the problem, how likely the staff would follow the instruction over time, and whether there are other acceptable alternatives to the device with the accuracy problem.

Once the analysis is complete, evaluators should review their findings and conclusions carefully. The review should conclude that the hospital would end up acquiring a safe, effective, easy-to-use, clinically needed, and cost-effective device or system. If not, the appropriateness of the acquisition decision should be seriously considered.

Summary

Evaluating medical technology is a complex process. Many factors make the process so complex, ranging from the wide variety of devices purchased and used in hospitals, to the various settings and applications where devices are used, to the many different types of users of the equipment. Many hospitals do not have good processes in place to effectively evaluate their choices for medical technology, and subsequently they make inappropriate, and sometimes unsafe, equipment procurement choices.

In order to have an effective evaluation program, a process must be put in place to identify the true need for medical equipment. Once a true need for equipment has been identified, the features required by the hospital for the equipment must be identified. Only equipment with the required features should be considered for further evaluation. Detailed criteria then must be established for the remaining equipment under consideration. Resources like ECRI's Health Devices evaluations are valuable tools to use in establishing the evaluation criteria. ECRI's evaluation criteria often can be wholly incorporated into the hospital's evaluation process.

Once the evaluation criteria have been established, a careful process of assessing the equipment for compliance with the criteria can begin. ECRI and other outside resources can be utilized for this type of information. Hospitals likely will need to do some analysis on their own or with the assistance of outside experts, especially for highly customized equipment, such as clinical information systems. Once results have been obtained, a weighting system must be established to help the evaluating committee come to a final conclusion. Factors such as safety, overall performance, ease of use, likelihood of the equipment being used, clinical utility of the equipment, and cost should be important considerations in establishing the weighting system.

References

ECRI. Needlestick Prevention Devices (Update Evaluation). *Health Devices* 28(10):395, 1999.
ECRI. Thoracic Aspirators (Evaluation). *Health Devices* 27(12):434, 1998.
ECRI. Arthroscopic Irrigation/Distension Systems (Evaluation). *Health Devices* 28(7):247-248, 1998.
ECRI. Portable Blood Glucose Monitors (Evaluation). *Health Devices* 26(9-10):347, 1997.
ECRI. Megadyne Mega 2000 Return Electrode (Evaluation). *Health Devices* 29(12):451-453, 2000.
ECRI. Picture Archiving and Communication Systems (Evaluation). *Health Devices* 29(11):406, 2000.
MHMG. *Medical and Health Care Marketplace Guide 1999-2000*, p. I-653. Philadelphia, Dorland's Biomedical, 2000.

# 87

Evaluating Investigational Devices for Institutional Review Boards

Salil D. Balar
Beaumont Services Company, LLC
Clinical Engineering and Technology Management Department
Royal Oak, MI

Most U.S. hospitals have Institutional Review Boards (IRBs). The United States Food and Drug Administration (FDA) defines an IRB as "an appropriately constituted group that has been formally designated to review and monitor biomedical research involving human subjects." The FDA grants IRBs the authority to approve, require modifications in, and disapprove all research activities covered by FDA regulations. The department of Clinical Engineering and Technology Management (CE & TM) at William Beaumont Hospital (WBH) evaluates the safety and efficacy of investigational devices used in research studies before the devices enter into the stream of commerce. The role of the CE department in evaluating these devices varies from hospital to hospital. Too often, the IRB does not take advantage of the CE department's expertise, thus posing a safety risk to the patients and medical device users involved in the studies. This chapter describes how CE & TM initiated, developed, and sustains a productive relationship with the Beaumont Hospital's IRB, known at the hospital as the Human Investigation Committee (HIC). The CE & TM outlines the scope of the clinical engineer's responsibilities, and describes a uniform procedure for evaluating and approving investigational medical devices used in IRB applications. It also suggests ways in which clinical engineers can adapt these concepts to their own situations.

A Cross-Functional Team

William Beaumont Hospital has been in compliance with IRB guidelines since the FDA began publishing them in 1981. However, it was not until 1988 that CE & TM began working with the hospital's IRB. The CE & TM began service to a single research department regulated by the IRB and copied the IRB chairperson on reports to that department. At every opportunity, the CE & TM communicated the results of its evaluations for both patient and medical device user safety.

Gradually, the IRB began to recognize the value of the CE & TM to researchers, based on the following attributes:

- Extensive knowledge of the medical device industry's problems and trends
- The ability to obtain objective information from the FDA and other regulatory resources
- High engineering skills and standards in device examinations and data analysis
- Expertise in envisioning the clinical use of medical devices in the context of clinical protocols

Over time, the IRB suggested that all researchers consult with it about the safety and efficacy of devices used in their studies. Recently, the William Beaumont Hospital IRB made the decision to require CE & TM approval of investigational devices before a trial begins, and it enforces compliance to CE & TM recommendations throughout the trial.

The CE & TM works with many of the hospital's research departments, including cardiology, radiation oncology, radiology, vascular surgery, orthopedics, pediatrics, urology, and the eye institute. During the course of an evaluation, the CE & TM interacts with the hospital's IRB chairperson and staff members. Information is exchanged with principal investigators and co-investigators, physicians, physicists, clinical research coordinators, and administrative managers in various research departments. CE & TM also communicates with manufacturers' representatives, including their regulatory affairs experts and product-development engineers. Periodically, the director of CE & TM has an in-depth meeting with the IRB chair and vice-chair. At these conferences, the director requests feedback on the services provided and suggests new ways to expand the working relationship for mutual benefit.

The Scope of Evaluation Activities

It is within the scope of CE & TM responsibilities to evaluate investigational medical devices used in IRB projects. However, CE & TM does not assist the IRB with drug-related studies. The following is a summary of the five most common circumstances under which CE & TM evaluates medical devices:

1. The IRB requests evaluation of a device for the first time.

CE & TM typically prepares a full evaluation report, described below,

- If the device needing evaluation is investigational (subject to IDE)
- If the device is being used for the first time at WBH in an investigational study for an IRB project

2. The IRB requests an evaluation update when there is a minor modification or upgrade to a previously evaluated device.

CE & TM evaluates the device and completes a short update report:

- If the device is investigational
- If the device has already been evaluated once
- If the device has a minor modification or upgrade

3. The IRB requests an evaluation of either an FDA 510(k) cleared device or a humanitarian use device (HUD), a device allowed by the FDA to be used under the Humanitarian Device Exemption (HDE).

CE & TM provides a memo in a format similar to an update report:

- If the device has 510(k) clearance
- If the device is being used in an investigational study for the first time at WBH
- If the chairperson of the IRB needs a memo addressing the risk of using the device

4. The IRB requests an evaluation of an FDA-approved device to be used for a different clinical application.

CE & TM provides a memo in a format similar to an update report:

- If a device is FDA-approved
- If a device has been used for one kind of clinical application in an investigational study at WBH, but will now be used for an entirely different clinical application
- If the chairperson of the IRB needs a memo addressing the risk of using the device.

5. The IRB requests investigation of an adverse event involving a device used during an investigation study.

CE & TM provides an adverse-event investigation report summarizing its findings and follow-up recommendations regarding the device.

The Evaluation Process

CE & TM's evaluations focus on the safety and efficacy of the devices. It has found that the evaluation process can be divided into four steps: Initiation, information gathering, risk analysis, and preparation of a written report.

Initiation

When a research department needs a device evaluation, it provides CE & TM with the following items:

- A copy of the research protocol and a copy of the researcher's application to the IRB
- A sample of the device

- The trial's anticipated start-up date
- Device packaging inserts
- Device specifications with operator and service manuals, if available
- Literature references
- A project number (to account for our time)

Research

The CE & TM gathers additional information from interviews, written records, and hands-on examination of the device. After reviewing the initial materials provided by the research department, CE & TM generates a list of questions for the investigator and manufacturer. For example:

- What testing is performed on the product?
- What material comes in direct contact with blood and what biocompatibility standards are followed for compliance?
- Have researchers experienced problems with the device to be used in the trial, or with similar products?

Next, CE & TM conducts its own information search. First, it reviews the Food and Drug Administration's Manufacturer and User Device Experience (MAUDE) database. The database contains reports of adverse events involving medical devices. The data consist of all voluntary reports since June 1993, user facility reports since 1991, distributor reports since 1993, and manufacturer reports since August 1996. The ECRI Health Devices database is searched next. ECRI provides information about product evaluations, alerts about technology-related hazards, and advice about technology acquisitions, staffing, and management (see Chapter 86). CE & TM then searches all other sources for any published product evaluations and adverse-event-investigation reports.

CE & TM clinical engineers closely examine the medical device under investigation. Step-by-step, they attempt to anticipate the modes of failure for each feature, using such techniques as Failure Mode and Effect Analysis (FMEA) (Stalhandske et al., 2003). Observations are compared to those described in published product evaluations and adverse-event-investigation reports.

Risk Analysis

Evaluating the product against established quality criteria and gathering relevant information is essential for an insightful analysis and a correct approval decision. In CE & TM's analysis, the quality of the manufacturer's instructions is assessed. Information about previous generations of the device is compared with what is claimed or known about the new version of the product. The manufacturer's compliance with common standards, such as those for sterilization and biocompatibility, are determined. Assessment also includes the risk associated with the circumstances or the conditions in which the device is made. The regulatory status of the device is confirmed. The suitability of the device for the particular proposed clinical application is verified. Only after the above tasks are completed does CE & TM formulate conclusions about the product's safety.

Report

In a written report, CE & TM does the following:

- Summarizes the facts gathered from interviews with researchers and manufacturers and from reviews of published information
- Describes the examination of the product
- Describes the other factors considered to reach conclusions
- Indicates approval or disapproval, and whether the approval is "full" or "conditional"
- Concludes with recommendations and/or stipulations

Recommendations for approved devices usually focus on user cautions. For example, it might be explained that the user should monitor temperature and pressure, check the device for damage and integrity of sterilization before use, or be aware of electrical, chemical, and fire hazards.

Recommendations for disapproved devices usually focus on a contingency plan. For example, the principal investigator might be requested to write a "failsafe" mode, to write patient instructions if the device fails during use, or to obtain more information from the manufacturer. If CE & TM cannot approve a device, the department works with the investigator and manufacturer to have the device or protocol modified as needed, so it can change the disapproval to an approval if possible.

The report is addressed to the IRB chairperson, and copies are made for the principal investigator, the director of the department that is responsible for the project, and other key personnel.

Challenges

The clinical engineer might encounter obstacles while developing a relationship with the hospital's IRB and an evaluation process that works for all involved. The following suggestions may mitigate these obstacles.

- Anticipate that the hospital's IRB may be uninformed or misinformed about the clinical engineering profession.
- Take the initiative to meet IRB members and researchers.
- Explain the benefits that clinical engineering would provide to the IRB and to research departments.
- Be prepared to suggest a procedure for working together but be flexible enough to incorporate the IRB's input.
- Expect the number of requests for your involvement to be limited at first.
- Anticipate that the investigators will sometimes have unrealistic expectations of the speed with which the evaluation and approval process can be performed.
- Encourage investigators to send all relevant information early to facilitate a timely completion of the evaluation. Anticipate that some manufacturers will be reluctant to provide a sample device. Explaining the need for the device and the method of evaluation usually will be sufficient to overcome the resistance.

Occasionally, a manufacturer insists on signing a confidentiality agreement before CE & TM can see a new device. Obtain the advice of your hospital's legal department in responding to such requests.

Summary

Clinical engineers are particularly suited for working with a hospital's IRB. Suggestions incorporated in this chapter provide ideas for initiating, developing, and sustaining a productive relationship with the IRB. Guidelines are presented to assist clinical engineering departments in determining the scope of their involvement and developing an evaluation procedure that will work for their particular hospitals. The clinical engineer who follows the recommendations described in this chapter will contribute to minimizing risks to patients and medical device users involved in research.

Defining Terms

Humanitarian Use Device (HUD): A device that is intended to benefit patients by treating or diagnosing a disease or condition that affects fewer than 4,000 individuals in the United States per year.

Humanitarian Device Exemption (HDE): An application similar in both form and content to a Premarket Approval (PMA) application, but is exempt from the effectiveness requirements of a PMA.

Investigational Device Exemption (IDE): An exception from the 510(k) clearance process under special conditions, such as when the device has a well-defined protocol or consent form, or can only be used for research purposes.

Premarket Submissions: For a medical device to be marketed in the United States, it must have clearance from the FDA. There are two types of FDA premarket submissions: A premarket notification 510(k) and a premarket approval application (PMA). Unless the FDA classification regulation specifically exempts a device, a premarket notification 510(k) submission is required for Class I and Class II devices. Depending on its type, a Class III device may require either a premarket notification 510(k) submission or a premarket approval (PMA) application.

Acknowledgements

The author thanks Paula Stachnik, medical writer, for her editorial support. He also thanks Michael Tanner, director of clinical engineering and technology management, for his insightful direction and guidance regarding regulatory-affairs compliance issues.

References

Stalhandske E, DeRosier J, Patail B, et al. How to Make the Most of Failure Mode and Effect Analysis. *Biomedical Instrumentation & Technology* 37:96-102, 2003.

Further Information

FDA website: www.FDA.gov
ECRI website: www.ECRI.org
Standard Operating Procedure for IRB Evaluations, developed by the Clinical Engineering and Technology Management Department of Beaumont Services Company, for use in William Beaumont Hospital, Royal Oak, Michigan.

Section VIII

Medical Devices: Utilization and Service

Joseph F. Dyro
President, Biomedical Resource Group
Setauket, NY

Technology management entails medical device assessment, evaluation, selection, acquisition, inspection, service, education and replacement. The scope of medical device technology is broad, encompassing devices ranging from relatively simple, low-cost, disposable ECG electrodes to highly complex and costly picture archiving and communications systems.

To effectively manage this wide array of health care technology, clinical engineers must understand the rationale for medical device utilization, the constraints placed upon the technology by the facility, the patient, the operator and the environment, and the effect that those constraints have on device performance and safety. They must know what, why, how, where, and when devices are used in order to manage effectively. What was only a short time ago a biomedical engineering specialty concerned only with electronic monitors and defibrillators has returned to the original concept espoused by those early pioneers who saw clinical engineering as a profession that applied engineering skills to analyze and solve problems in the clinical environment, whether that entails purchasing the most cost-effective examination gloves or designing a new ambulatory care pavilion. Whether complex or simple, all medical devices fall within the purview of the clinical engineer, and they must be managed to ensure that the technology is performing satisfactorily at all times and without compromising patient or personnel safety.

This section places emphasis on medical devices that are used throughout various departments of primary- to tertiary-care hospitals, and some mention is made of technologies that might find use in the ambulatory, long-term-care, and home setting. While all medical specialties could not be covered with the limits of this handbook, this section contains a representative sampling of typical medical and surgical services. Rather than address in detail the design, theory of operation, specifications, and performance of a specific medical device, the authors give an overview of the types of devices that are commonly encountered in their clinical engineering practice in all areas of the hospital. Throughout this section, the reader's attention is directed to those particular aspects of a device's performance and safety that the author deems most significant in daily practice.

Miodownik (Intensive Care) describes the historical development of the intensive care unit (ICU) from facilities that treated casualties of war to those today that treat all life-threatening illnesses and injuries. Described as the CEs most challenging environment, the ICU accommodates patients often on the brink of survival intimately linked to a concentration of complex, expensive diagnostic and therapeutic devices utilized by dedicated, hard-working and occasionally overworked, overstressed, and inadequately trained personnel. Financial pressures and proliferation of medical technologies militate against utilizing the available technology in a thoughtful, deliberate, unhurried manner. Misuse, misapplication, and abuse are sometimes the consequences of overworked, overstressed personnel. Miodownik describes the significance of those devices that are typically encountered in the ICU, including monitors of physiological parameters such as temperature, blood pressure, heart rate, respiration, end-tidal carbon dioxide, arterial blood oxygen saturation; intraaortic balloon pumps; left ventricular assist devices; infusion devices; ventilators, and dialysis units. Clinical engineers are looked to as repositories of technical knowledge that is beyond that of the user nurse and physician. They must maintain a substantial level of expertise in the physics and physiology of measurements and therapies, knowledge of computer operating systems, networks and communications protocols, and the peculiarities of many different instruments, as well as an appreciation of way to keep all of these systems in operation.

Smith, Rane, and Melendez (Operating Room) describe the medical technology, surgical specialties, personnel, and the physical setting of a conventional operating room. The complex and dynamic interaction of these facets of the operating room is thoroughly illustrated, thus enabling the clinical engineer support this environment to more effectively. They begin with by describing the role of the operating room, typical operating room floor plan, surgical suite layout, and operating room infrastructure. They then list common and specialized device technologies such as electrosurgical units, anesthesia machines, surgical instruments, defibrillators, monitors, thermal support devices (hyperthermia machines), headlamps, tourniquets, lasers, imaging technologies, minimally invasive technologies (e.g., endoscopes and endoscopic instruments), and heart-lung machines. Lastly, they describe the CE's role in ensuring safety in the OR.

Melendez and Rane (Anesthesiology) expand upon the role of anesthesia in the OR as they give a thorough accounting of the role of the clinical engineer in supporting the delivery of that service by ensuring safe and efficacious medical devices. Starting with an overview of anesthesia, amnesia, and pain control, they move to a detailed description of typical technologies, such as the anesthesia machine and ventilators with associated humidifiers, vaporizers, patient breathing circuits (carbon dioxide absorbers, scavengers, gas supplies, airway pressure, temperature and oxygen-monitoring systems). Airway-management tools, capnography, agent analysis, and principles of assisted ventilation are discussed.

The discussion of imaging technologies in the OR (Smith, Rane, and Melendez) is enlarged by Harrington (Imaging Devices), who describes the basics of radiation physics and the theory of operation of the X-ray machine (tube and housing, collimators, filters, and grids, film and film cassettes, and power supply). An overview of fluoroscopy and the video chain is given. Harrington then describes a typical table, overhead tube, control console, processors, multiloaders, and dry imaging. The film processor is explained, as are dark room films and daylight films. Specialty units are described, including mammography, tomography, bone-density analyzers, interventional radiology (special procedure rooms), and the cytoscopy room. He concludes with the portable C-arm and a general discussion of the techniques for effective management of imaging devices.

Machine vision can be defined as the "acquisition and processing of images to identify or measure the characteristics of objects." Rosow and Burns (Machine Vision) provide useful information on the three steps required for successful machine-vision applications: Conditioning, acquisition, and analysis. They illustrate this with two case studies: (1) the EndoTester—a virtual instrument-based quality control and technology assessment system for surgical video systems, and (2) a VabVIEW-based wound measurement system. This chapter underscores the application of clinical engineering in the research and design of diagnostic technologies.

The specialties of perinatology (DeFrancesco) (i.e., the care of the newborn before, during, and after birth), has made tremendous strides over the last few decades, owing largely to the rapid and simultaneous and synergistic advances in medical devices, pharmaceuticals, and clinical procedures. The flow of patients through a typical facility is described from prelabor, labor and delivery, and postpartum support areas. A typical floor plan for a labor and delivery area and individual room layouts are described. Changes in the patient environment (i.e., the building infrastructure) that have resulted in a safer and more efficient environment are noted. The tertiary-care facility often incorporates neonatal intensive care units and pediatric intensive care units to provide support for premature infants and critically ill infants in outlying hospitals without adequate support facilities. The special needs of security, data communications, climate control, medical gases, and vacuum-power distribution to support perinatal care is described. Explanations are given of most of the typical technologies such as maternal/fetal monitors, neonatal monitors, incubators, radiant warmers, apnea monitors, phototherapy units, ventilators, ultrasonic scanners, and the electronic medical record. The new therapeutic system of Inhaled Nitric Oxide (INOvent) delivery and monitoring is described.

Goodman (Cardiovascular Techniques and Technology) begins with some anatomy and discussion of cardiovascular disease. He then explains diagnostic technologies for cardiovascular disease, including noninvasive static and stress tests in diagnostic offices (e.g., electrocardiograms, phonocardiograms, and Holter monitors) and invasive diagnostic tests in the cardiac catheterization laboratory (e.g., cardiac output monitors and cardiac catheters). Hospital-based surgical interventions for cardiovascular disease in the operating room and cardiac catheterization laboratory, such as angioplasty, are explained. Cardiovascular care medical devices, including intra-aortic balloon pumps, cell savers, pacemakers, and defibrillators, are addressed.

The next chapter, Beds, Wheelchairs, and Other General Hospital Devices, explains that all medical devices must perform satisfactorily and must not pose a threat to patient or operator safety. Clinical engineering departments have shied away from such low-tech medical devices as beds, wheelchairs, and stretchers, perhaps because these items are not sophisticated enough and because the skill level of the department's staff members exceeds that required to service these items. Case studies demonstrate that the failure or misuse of the simplest general hospital medical devices can cause, and has caused, death and serious injury. Dyro shows that by matching skills and tasks, a clinical engineering department can cost-effectively manage these oft-forgotten devices to the benefit and safety of the patient population.

In the last chapter, a methodology is presented for determining the cause of a medical device malfunction. The generalized and systematic techniques recommended are effective in troubleshooting all medical devices, whether large or small, costly or inexpensive, mechanical or electrical, complicated or simple. The methodology applies to all causes of apparent medical device failure, including random component failure, device-device interaction, human factors design errors, operator misuse, and abuse.

88

Intensive Care

Saul Miodownik
Director, Clinical Engineering,
Memorial Sloan-Kettering Cancer Center
New York, NY

There has always been the perception that those patients closer to nursing stations fare somewhat better than their counterparts who are further away. The concept of the "intensive care unit" evolved from a series of medical needs as varied as the treatment of large numbers of shock casualties during World War II, nursing shortages in the late 1940s, and the polio epidemics of 1947–1948. The care and monitoring of such patients was more efficient, and survival improved when patients were grouped in a single location. By 1958, approximately 25 percent of community hospitals with more than 300 beds reported having an intensive care unit (ICU). By the late 1960s, most United States hospitals had at least one ICU. In 2002, there were approximately 6000 intensive care units in the U.S., treating 55,000 patients daily, with an annual budget of approximately $180 billion. Several types of intensive care units exist, including surgical ICU, medical ICU, neonatal ICU, pediatric ICU, and Nero (burn) ICU. The training of the staff accommodates the specific needs of the each population and the specialized equipment used. Many of the techniques and interventions are common to all types of ICUs (Marino, 1991).

A patient's illness or trauma can rapidly become a life-threatening situation that demands immediate and continuous attention. ICUs are repositories of technology and expertise where patients can receive the treatment required for survival. In this arena, clinical engineers (CEs) are presented with their most complex and challenging environment. It is useful to categorize these technologies and equipment in the manner shown in Figure 88-1.

All four elements are required, to manage effectively any patient's care. However, a patient who has been admitted to an ICU presents with a set of abnormal physiological conditions of varying severity that can suddenly change and become life threatening. Consequently, physiological parameters such as blood pressure, heart rate, blood oxygen saturation, and respiration are continuously monitored; laboratory results (e.g., blood chemistry, gases) and other information are frequently, though intermittently, obtained, integrated, and interpreted; and a therapeutic intervention is put in place, adjusted, or removed. Although this "closed loop" process is by no means automatic or unique to the intensive care setting, it represents an algorithmic approach to patient management that requires tight control and rapid response to emergent conditions. Consequently, the technologies that are required to implement these functions are often specialized for the ICU setting.

Monitoring and Diagnostics

Physiological monitoring in the ICU setting closely follows the patient's cardiac, hemodynamic, and respiratory statuses. While early physiological monitors could view one lead of the ECG and, perhaps, an invasive blood pressure, today's systems integrate a vast array of parameters into a relatively small monitoring device. A typical, full-featured, physiological monitor might contain a set of built-in monitored parameters or might have the ability to add plug-in modules to configure the unit as the clinical situation demands. Monitors located at the bedside are typically connected to monitors at a central station where one may observe the physiological status of many patients at once (see Figure 88-2). ICU monitors commonly include the ability to display the waveforms, values, and trends for the following parameters:

- 12-lead electrocardiogram (ECG)
- Arrhythmia monitoring and interpretation
- ST segment analysis
- Impedance based respiration
- One to three invasive blood pressures
- Noninvasive blood pressure
- Pulse oximetry (Sao_2)
- Two temperatures
- Thermal dilution cardiac output

The following additional monitoring functions are also available as separate units or modules that can be relevant to the management of ICU patient:

- End tidal CO_2
- Continuous cardiac output (CCO)
- Impedance based cardiac output
- Metabolic monitoring
- Real-time blood gas monitoring
- Mass spectrometry

Although provided as part of the therapeutic aspect of the intensive care unit, the life-support ventilator provides a full range of pulmonary monitoring parameters as well.

The admission of a patient to an ICU is often (but not always) prompted by compromises or threats to cardiac and/or respiratory function. Insufficiencies in these areas must be addressed rapidly. The delivery of oxygenated blood to body tissues is therefore the first concern of the ICU staff. ECG, pulse oximetry, and noninvasive blood pressure (NIBP) are physiological measurements that can be obtained immediately and noninvasively upon connection of the patient to the monitoring equipment.

Monitoring and therapy at this point become progressively more invasive with the addition of vascular access to measure internal blood pressures on both the arterial and venous side of the circulatory system. Catheters can be inserted into the pulmonary artery to obtain thermodilution cardiac output information (Trautman, 1988) as well as measurement of patient fluid load. Vascular access also provides a convenient site for obtaining blood samples for a variety of laboratory tests. Although attempted for many years, a practical and economical indwelling blood gas analysis catheter remains to be developed.

Additional diagnostic tools are often available in the ICU as well. It is common to find ultrasonic scanning and imaging systems as part of the unit's armamentarium. Although 12-lead ECG acquisition is often part of the bedside monitor, a separate ECG machine is also available. Depending on the ICU size and census, the ICU also can own items such as dedicated, portable X-ray machines and blood gas analyzers. Because of the volatility of the patient's medical condition, and the need to respond quickly to sudden changes, point-of-care blood chemistry machines, which are small enough to be positioned at a patient's bedside, have been developed.

Bedside physiological data is often routed to a central station for display, printing, and alarm monitoring. In recent years, the importance of the central station as a location to be carefully watched for adverse patient events has diminished. Staffing in the ICU environment usually involves a high nurse-to-patient ratio. Most alarm events are recognized at the bedside, with the central station serving more as workstation for archival retrieval of alarm history and analysis of trends. The availability of inexpensive computer hardware has made possible the implementation of full disclosure systems that permit the archiving of all waveforms for a period of up to several days for all of the beds in the ICU. These are often packaged as software within the central station monitor. A consequence of the capabilities has been a reduction in the need and use of strip-chart recorders at the bedside or central station. There is little need to sift through yards of paper strips to find the event of interest because the full disclosure or alarm history record can be printed at will. The laser printer has become the *de facto* hardcopy device of choice.

Information Collection and Clinical Information Systems

Most, if not all, health care facilities in the 21st century have installed an information systems (IS) infrastructure that permits the dissemination of all manner of information to the hospital staff (see Chapter 97). Computer workstations are inexpensive and widely available, and they tap into the institution's high-speed networks. Thus, ICU staff accessibility of such data as patient medical records and test results is assured.

As the patient is admitted to the ICU, information from his medical history and ongoing monitoring are collected and correlated. In today's modern facility, a vast array of information is available to assess the patient's condition. To this end (in addition or as part of the bedside monitoring), high-speed data networks route a large volume of information to the bedside for analysis. Past ECGs are retrievable from archived records and can be compared with the latest 12-lead ECG available at the bedside. Blood chemistry values are obtained, and relevant radiological and ultrasound images are retrieved electronically by way of a picture archiving and communications system (PACS) that is available for viewing near the patient's bedside (see Chapter 102).

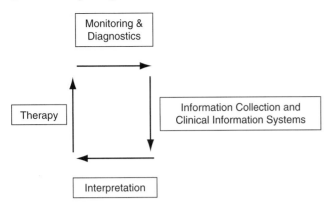

Figure 88-1 Four elements required for effective management of patient care.

The actual display devices for physiological monitoring and patient lab data are often identical. Increasingly, there has been a movement away from proprietary CRT monitors or thin screen displays in physiological monitoring. Many manufacturers have begun to use components and interfaces that are commonly and inexpensively available in the IS industry. Additionally, digital information is transacted using standard Ethernet protocols. This allows a certain degree of compatibility between physiologic monitoring systems and the patient's medical records. In many instances, parametric information from physiologic monitoring system passes to clinical information networks and workstations. To varying degrees, laboratory data are viewable on the physiologic monitors (see Chapter 98).

Interpretation

To address conditions that are life threatening and that require immediate intervention, several algorithms have been developed that, to a certain degree, address a given acute clinical situation. These algorithms can be taught to newer ICU staff and compensate to some extent for the experience that otherwise would be required. This distills the array of information into a simplified set of required parameters to treat a specific situation. A decision tree is followed with feed back information that is derived from the monitoring and diagnostic equipment.

Not all ICUs are managed the same way. There are two prevailing schools of thought regarding this issue. In some hospitals, when a patient is admitted to the ICU, most, if not all, of the patient's subsequent medical management is in the hands of a dedicated staff of intensive care specialists (IC specialists) whose responsibilities lie primarily in the ICU environment. This team of specialists includes physicians with specialties in such areas as pulmonary, cardio-vascular, infectious diseases, and internal medicine. Additional medical specialties are consulted as needed. Control of the patient's care is usually relinquished by the referring physician or surgeon. In other institutions, the referring physician might continue to monitor and adjust the therapy to the patient with the assistance of specialized nursing staff and technicians. The prevailing literature indicates that there may be benefits in terms of improved ICU survival rates, reduced length of stay, and a decrease in operating costs in an IC specialist-controlled ICU model.

The availability of this vast array of data and diagnostic tools does not necessarily translate into better outcomes for ICU patients. The complexity of a particular patient's condition makes the interpretation of the data a less-than-automatic or intuitive process. Additionally, clinical studies that challenge certain prevailing ICU practices are constantly emerging in the literature. The relatively wide latitude of patient response to, and tolerance of, interventional procedures further clouds the issue. It cannot be stated with any clear degree of certainty that the introduction of advanced monitoring, therapies, and data management technology has resulted in a definitive decline in patient mortality. What advanced information technology has provided, however, is a streamlining of the information assembly process. This simplifies and, to some extent, reduces, the time required to assess and begin treating the patient.

Therapy

Patients who have been admitted to the ICU present with a host of problems that might or might not be life threatening. Severe dehydration, chest pain, and shortness of breath are common symptoms that dictate ICU entry. Upon admission, monitoring of basic vital signs is begun immediately and includes the measurement of electrocardiogram (ECG) (Plonsey, 1988), noninvasive blood pressure (NIBP) (King, 1988), and arterial blood-oxygen saturation (S_aO_2) (Welch, 1990). The resulting parametric information points to the first therapeutic interventions. Disturbances in the heart rhythm and rate are immediately visualized, blood pressure in the abnormal range identified, and low levels of S_aO_2 addressed. A peripheral intravenous (IV) line is started for the administration of medications, fluid support, and withdrawal of blood samples for diagnostic evaluation. Oxygen is applied by facemask to raise saturation levels. These are usual starting points, and many patients will require not much more therapy than this throughout their ICU stay. The patients are monitored, and the levels of therapeutic support are correspondingly adjusted, until the underlying disease process or condition is resolved. This type of patient represents a sizable percentage of the short-stay ICU population that is primarily being observed and requires minimal or moderate support during their stay. The more severely ill patients will undergo more radical and invasive therapies. Following are major categories of ICU therapeutic intervention.

Respiratory Care: Intubation, Mechanical Ventilation

Respiratory failure can be described as the inadequacy of the patient's intrinsic respiratory efforts to produce normal blood oxygenation and carbon dioxide clearance, as verified by arterial blood gas measurement. It can be caused by a host of factors, including infection, trauma, and paralysis. The primary consideration of mechanical ventilation has been to restore the patient's blood gas value to nominally normal levels. In spontaneous ventilation, the lungs inflate when negative pressure is created in the thoracic cavity by the downward displacement of the diaphragm. Until the mid 1950s, this was done with devices such as the iron lung, a container that surrounded the patient with his head protruding by way of an airtight collar (Mörch, 1985). This subjected the patient's body to a negative pressure, relative to the outside air, allowing the patient to inspire ambient air through his mouth as the iron lung cycled. This form of ventilation placed little stress on the patient's pulmonary system. He could eat, drink, and talk because his airways were otherwise unencumbered.

After the polio epidemics of the 1950s, positive-pressure mechanical ventilation became more widespread. In this mode, the lungs inflate when positive pressure is generated by the ventilator, forcing the lungs open and causing downward displacement of the diaphragm. Most often, this is done by delivering a volume of humidified, blended air and oxygen mixture by way of an endotracheal tube. Modern mechanical ventilators (see Figure 88-3) have a wide range of settings over a range of pressure limits such as O_2 percentage, tidal volumes respiratory rates, and positive-end expiratory pressure (PEEP) (Behbehani, 1995). This type of technology has been in place for more than 35 years, with various improvements in machine size, intelligence, display, modes of ventilation, interfacing, and data collection. However, there is a danger, long recognized, of ventilator-induced lung injury (VILI) or ventilator-associated pneumonia (VAP) with the indiscriminate use of positive-pressure mechanical ventilators. Overdistension of the alveoli, either by excessive pressures or volumes, can cause dangerous lung damage especially in the presence of high-oxygen concentration. This is often evidenced in the longer-term patient requiring higher levels of respiratory support. In addition, injury to the lung can produce a cytokine cascade effect that can cause damage to organs throughout the rest of the body. One approach to alleviate this problem is the use of permissive hypercapnia. Smaller tidal volumes are used to prevent lung distension and high airway pressure. This produces less CO_2 clearance, and the blood pH will be lower (more acidic), but this method can protect the lungs from permanent injury.

As an alternative, other ventilator technologies are available. High-frequency jet ventilation (HFJV), high-frequency positive-pressure ventilation (HFPPV), and high-frequency oscillation (HFO) all limit the volume and, hence, the pressure provided to the patient on each breath by operating at supranormal respiratory rates (Hamilton, 1988). HFJV and HFPPV operate in the realm of 100–200 breaths per minute, and HFO in the domain of 10–30 Hz. This is far in excess of the 10–30 breaths per minute encountered in spontaneous or mechanical ventilation. These high-frequency breaths produce lower peak airway pressures, possibly reducing lung injury.

Cardiac Care

Among some of the more frequently utilized devices in cardiac care are the defibrillator (see Figure 88-4), pacemaker, intra-aortic balloon pump (IABP) (Jaron and Moore, 1988), left ventricular assist device (LVAD) (Rosenberg, 1995), and extracorporeal membrane oxygenator (ECMO). For decades, life-threatening cardiac rhythm disturbances (e.g., ventricular fibrillation, ventricular tachycardia, and atrial fibrillation) have been successfully addressed utilizing defibrillators (Tacker, 1988). These devices have been standard in the ICU since their inception decades ago. Little has changed in their basic operation. They all deliver energy over a wide range of output energy and have had the ability to be synchronized (i.e., to perform ECG-synchronized cardioversion). Recently, however, the use of biphasic waveforms has been revisited. The assumption (not yet accepted in wide-ranging studies) is that more successful cardioversion can be achieved at lower energies, thus

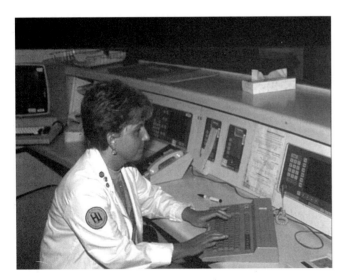

Figure 88-2 Physiological monitoring at the central station.

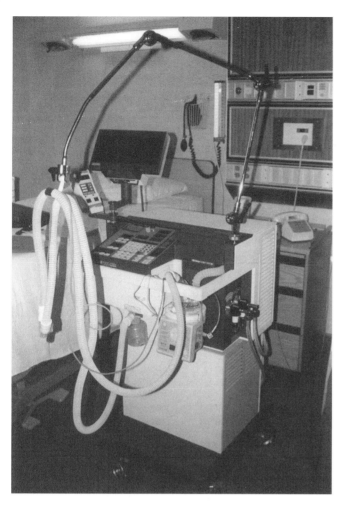

Figure 88-3 Puritan-Bennett model 7200 ventilator.

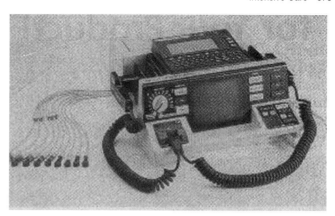

Figure 88-4 Defibrillator.

avoiding potential damage to the myocardium. Additionally, the refinement of arrhythmia detection has allowed these devices to be automated, detecting dangerous cardiac events and firing the defibrillator without the need for operator intervention. Additionally, many of these units can be configured to perform transthoracic pacing.

Because of myocardial infarction, or for other reasons, patients sometimes develop temporary or permanent heart block, preventing the heart's conduction system from beating at an appropriately consistent rate. A pacemaker is typically utilized to correct this condition. Depending on the acuity of a patient's condition, the first-order intervention probably would entail the use of a transthoracic pacemaker. Pacing also can be done by inserting a pacing catheter in the right heart, which is then connected to a small, external pacemaker.

The aforementioned interventions effectively treat conduction or electrical defects that can arise spontaneously or in response to infection or myocardial infarction. The heart muscle is, for the most part, largely intact. When the heart muscle fails, additional measures must be taken, to sustain the patient's life. At the ICU bedside, this is accomplished by the use of an intra-aortic balloon pump (IABP) (Jaron and Moore, 1988). A special catheter is inserted into the aortic arch containing a balloon that can be rapidly inflated and deflated by a low-viscosity gas, usually helium. The timing of the balloon inflation is synchronized to either the patient's ECG or blood pressure waveform. Its primary purpose is to improve perfusion of the heart muscle by inflating the balloon as the aortic valve opens, providing additional backpressure to the carotid arteries and rapidly deflating it so that the rest of the body can be perfused. This procedure sustains the patient as the myocardium can recover sufficiently to pump on its own. In the event that a replacement heart is required, the procedure will extend a patient's life until a donor heart is available.

Alternatively, there are a number of LVADs that enhance the pumping action of the left ventricle until a donor organ is found (Rosenberg, 1995). They are surgically implanted in the patient's chest or abdomen and are managed by the ICU staff. Less widely used, though available also, is the extracorporeal membrane oxygenator (ECMO), or heart lung machine (Dorson and Loria, 1988). This device, used during coronary-artery bypass graft, can provide both respiratory and circulatory function, allowing the heart and lungs to be mechanically inactive for a time. Efforts such as this are extreme and require significant technical support during use.

Infusion Devices

Most ICU patients arrive with some form of vascular access and usually receive additional IV lines during their stay. Much of the therapeutic intervention and monitoring occurs via vascular access. Fluid management and drug delivery can be accomplished using gravity as the source of positive pressure to push liquids into the patient. Infusion rates are monitored visually, and adjustments can be made to achieve the required goals. This approach is not viable when six to eight infusions are running simultaneously, with rates varying from 0.1 cc/hour, for certain pain medication, to hundreds of cc/hour, for rapid fluid replacement. This is handled now with a variety of programmable infusion pumps (Figure 88-5) and controllers (Voss and Butterfield, 1995).

Infusion pumps and controllers employ a variety of electromechanical mechanisms to regulate fluid flow to the patient. These include peristaltic or diaphragm pumps, syringe devices, and variable pinch clamp mechanisms that control gravity-driven flow. Usually, the tubing sets used by these devices are dedicated and not interchangeable. Infusion pumps with a wide range of rates (1–999 cc/hour) have been available for decades. Modern infusion devices have a wide range of programmable infusion rates and extensive alarm capabilities. Features include proper set placement detection, pressure limits, proximal and distal occlusion alarms, end of infusion alarm, and programmable keep-vein-open (KVO) rates. Additional safety features to prevent unrestricted free flow of IV fluid to the patient are incorporated as well. In order to avoid medication errors, systems that will associate pharmacy requisitions, the infusion device, and patient identification are being developed. Additional intelligence in such systems will identify patient sensitivities, incompatibilities between infused medications, and violation of nominal drug concentrations and infusion rates.

Dialysis: Kidney and Organ Support

The sickest ICU patient will begin to manifest failure in one or more organs. Acute renal failure (ARF) is not uncommon in the ICU setting. ARF occurs in approximately 5% of all hospitalized patients and, historically, has been associated with a high risk of mortality. Following the introduction of dialysis therapy more than 50 years ago, patient mortality dropped from 90% to 50%. However, mortality in the intervening years has not improved from the 50% level. In ICU patients presenting with other co-morbidities, the mortally rate

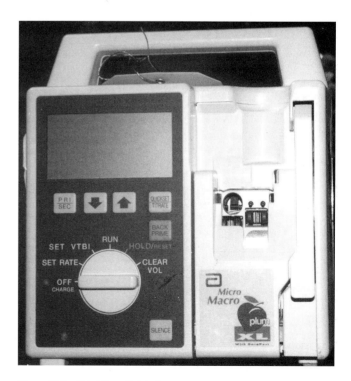

Figure 88-5 Abbot Plum XL infusion pump.

is between 50% and 80%. Although the etiology of ARF varies, it is often associated with a reduction in urine output and a drop in systemic blood pressure. Waste products accumulate in the blood, electrolyte levels become abnormal, pH usually drops, and edema sets in.

Initially, ARF was treated in the same manner as chronic renal failure, with intermittent hemodialysis. This approach quickly removes the metabolite, electrolyte, and fluid imbalance. However, these levels vary widely because the body continuously generates their production between dialysis sessions. Consequently, there has been a move toward continuous renal replacement therapy (CRRT) (Galletti et al., 1995). In this mode, renal function is supplied continuously while the patient recovers. There are several different CRRT modalities. Historically, arterial circulation was used to provide the force to move blood across a dialyzer cartridge. Because of the need for arterial access and inherent complications and the technical improvement in venovenous therapies, these methods have been largely abandoned. The venovenous modalities differ primarily in the methods of clearance. Slow continuous ultrafiltration (SCUF) is a method employed to remove volume by taking off fluid filtered by a dialysis membrane by convective force. Continuous venovenous hemodialysis (CVVHD) provides clearance using diffusive clearance by running dialysate across the membrane. Continuous venovenous hemofiltration (CVVHF) removes fluid by convection (as in SCUF) and then provides replacement fluid back to the patient. Finally, continuous venovenous hemodiafiltration combines the clearance properties of CVVHD and CVVHF. These approaches are provided by bedside machines incorporating the required pumps, control instrumentation, filters, and fluids.

Clinical Engineering and the ICU

Perhaps the most complex and technically challenging hospital environment is the ICU. Traditionally, CEs have been involved in medical device evaluation, inspection, maintenance, layout, design, and integration of the various instrumentation systems. As the devices' and systems' complexity has increased over the years, clinical engineers have been looked to as repositories of technical knowledge that is beyond that of the users (i.e., the nurses and physicians). This requires maintaining a substantial level of expertise in the physics and physiology of measurements and therapies; knowledge of computer operating systems; networks and communications protocols; and the peculiarities of many different instruments as well as an appreciation of ways to keep all of these systems in operation. The many interactions and safety issues of the various monitoring and therapeutic instruments must be understood and translated so that the CE can provide answers to questions posed by the clinical staff, whose responsibility is the care of the patient, not the technology.

References

Behbehani K. Mechanical Ventilation. In Bronzino JD (ed). *The Biomedical Engineering Handbook*, Boca Raton, FL, CRC Press, 1995.

Dorson WJ, Loria JB. Heart-Lung Machine. In Webster JG (ed). *Encyclopedia of Medical Devices and Instrumentation*, New York, Wiley, 1988.

Galletti PM, Colton CK, Lysaght MJ. Artificial Kidney. In Bronzino JD (ed). *The Biomedical Engineering Handbook*, Boca Raton, FL, CRC Press, 1995.

Hamilton LH. High-Frequency Ventilators. In Webster JG (ed). *Encyclopedia of Medical Devices and Instrumentation*. New York, Wiley, 1988.

Jaron D, Moore TW. Intraaortic Balloon Pump. In Webster JG (ed). *Encyclopedia of Medical Devices and Instrumentation*. New York, Wiley, 1988.

King GE. Blood Pressure Measurement. In Webster JG (ed). *Encyclopedia of Medical Devices and Instrumentation*, New York, Wiley, 1988.

Marino L. *The ICU Book*. Philadelphia, Lea and Febiger, 1991.

Mörch ET. History of Mechanical Ventilation. In Kirby RR, Smith RA, Desautels DA (eds). *Mechanical Ventilation*. New York, Churchill-Livingstone, 1985.

Plonsey R. Electrocardiography. In Webster JG (ed). *Encyclopedia of Medical Devices and Instrumentation*. New York, Wiley, 1988.

Roa RL. Clinical Laboratory: Separation and Spectral Methods. In Bronzino JD (ed). *The Biomedical Engineering Handbook*. Boca Raton, FL, CRC Press, 1995.

Roa RL. Clinical Laboratory: Nonspectral Methods and Automation. In Bronzino JD (ed). *The Biomedical Engineering Handbook*. Boca Raton, FL, CRC Press, 1995.

Rosenberg G. Artificial Heart and Circulatory Assist Devices. In Bronzino JD (ed). *The Biomedical Engineering Handbook*. Boca Raton, FL, CRC Press, 1995.

Tacker WA. Electrical Defibrillators. In Webster JG (ed). *Encyclopedia of Medical Devices and Instrumentation*. New York, Wiley, 1988.

Trautman ED. Thermodilution Measurement of Cardiac Output. In Webster JG (ed). *Encyclopedia of Medical Devices and Instrumentation*. New York, Wiley, 1988.

Voss GI, Butterfield RD. Parenteral Infusion Devices. In Bronzino JD (ed). *The Biomedical Engineering Handbook*. Boca Raton, FL, CRC Press, 1995.

Welch JP, DeCesare R, Hess D. Pulse Oximetry: Instrumentation and Clinical Applications. *Respir Care* 35(6):584–601, 1990.

Further Information

Kirby RR, Banner MJ, Downs JB. *Clinical Applications of Ventilatory Support*. New York, Churchill Livingstone, 1990.

Webster JG. *Medical Instrumentation: Application and Design*, 2nd ed. Boston, Houghton Mifflin, 1992.

89

Operating Room

Chad J. Smith
Product Development Engineer, Kensey Nash Corporation
Exton, PA

Raj Rane
Massachusetts General Hospital
Boston, MA

Luis Melendez
Massachusetts General Hospital
Boston, MA

The operating room is an exciting location in which to apply the principles of engineering. The surgical environment encompasses an extensive arrangement of medical technologies and processes. The clinical engineer must apply the principles of several engineering disciplines to support an efficient, productive, and safe operating room.

This chapter describes the medical technology, surgical specialties, personnel, and physical setting of a conventional operating room. The complex and dynamic interaction of these facets of the operating room is thoroughly illustrated, thus enabling the clinical engineer to support this environment more effectively.

The Role of the Operating Room

Surgery is the diagnosis and treatment of medical injuries, diseases, and deformities by manual or operative means. It is a fundamental service that the health care facility provides to a community. A vast assortment of surgical procedures is performed to treat the medical conditions within the scope of health care.

A surgical procedure is a delicate and complicated undertaking. It must be performed in a controlled environment with numerous technical resources and a staff of medical professionals and support personnel. Depending on the patient and the nature of the surgical procedure, surgical cases are characterized in one of three ways: Inpatient, outpatient, and same-day. Inpatient procedures are performed for patients who have been admitted to the hospital. Outpatient procedures are generally for those undergoing minor surgical procedures. In those cases, the patient typically will undergo a local anesthetic and will be admitted and released in the same day. Same-day surgery consists of more extensive cases than outpatient procedures and might include general anesthesia. However, the patient is also discharged in the same day. The operating room is the location in a health care facility that is equipped for the performance of surgery. Patients who require invasive treatment are transported to this department to undergo induction of anesthesia, the surgical procedure, and resuscitation. It is a technologically advanced area filled with complex equipment, highly trained professionals, and stringent environmental controls. For the scope of this chapter, the term "surgical suite" will be used to refer to the individual room where a surgical procedure is performed. The "operating room" will signify the entire department, within the health care facility, in which the operating suites and supporting areas are located.

A typical hospital comprise numerous departments that provide a wide range of services. The operating room (OR) plays an important role in a hospital and in the overall process of providing patient care. The following lists outline the typical steps for a patient who is undergoing a scheduled surgical procedure, as well as an unscheduled procedure. It is meant to demonstrate methods in which a surgical procedure is arranged, and the relationships among hospital departments.

Scheduled Surgery

- The patient consulted a primary care physician.
- The primary care physician made a diagnosis and directed the patient to consult a surgical specialist.
- A surgical specialist diagnosed the condition and confirmed the need for surgical treatment. Images from the radiology department and test results from the medical lab aided in the diagnosis.
- The surgery date was scheduled, and OR time was booked.
- The patient was admitted to a pre-admitting test center to review the medical history and to undergo various tests.
- An anesthesiologist studied the patient's medical history and planned the method of anesthesia.
- The patient was sent to the pre-operating area (induction room), where she was prepared for surgery, and anesthesia was initiated.
- Surgery was performed in a surgical suite.
- Following surgery, the patient was delivered to the post-anesthesia care unit (PACU) to recover from the effects of anesthesia.
- The patient was delivered to a care unit for complete recovery.
- The patient was discharged from the hospital.
- The patient made periodic visits to the hospital for physical rehabilitation.

Unscheduled (Emergency) Surgery

- The patient placed an emergency call, and an ambulance was dispatched.
- The patient was delivered to the hospital's emergency department.
- The ER physician ordered the patient be transported to the radiology department for imaging and sent samples to the medical lab for analysis.
- A surgical specialist was consulted to aid in the diagnosis.
- The patient was transported to the pre-operating area (induction room), where she was prepared for surgery, and anesthesia was initiated.
- Surgery was performed in a surgical suite reserved for emergency cases.
- Following surgery, the patient was delivered to the PACU to recover from the effects of anesthesia.
- The patient was delivered to the intensive care unit for complete recovery.
- The patient was discharged from the hospital.
- The patient made periodic visits to the hospital for physical rehabilitation.

A wide range of surgical procedures can be performed in an operating room. The surgical specialties of a particular hospital are based on the resources and medical scope of that health care facility. The following is a list of some surgical specialties that are commonly found in a hospital's operating room. A brief description of each specialty is included.

General Surgery

This form of surgery includes a broad spectrum of surgical care that involves largely the surgical management of diseases of the bowel, gallbladder, stomach, and other digestive organs.

Thoracic and Cardiovascular Surgery

Thoracic and cardiovascular surgery concerns the diagnosis and treatment of disorders of the thorax. The thorax is the upper part of the trunk between the neck and the abdomen, containing the chief organs of the circulatory and respiratory systems, such as the lungs and heart. Two major types of thoracic surgery are classified as pulmonary (i.e., pertaining to the lungs) and cardiovascular. Cardiovascular surgery treats diseases and conditions of the heart and the blood vessels of the entire body. Common cardiovascular procedures include coronary bypass surgery, aortic or mitral valve replacement or repair, and aneurysm repair.

Neurosurgery

Neurosurgery is the specialty of surgery that addresses the diseases and disorders of the nervous system. Within the realm of neurosurgery are the brain, spinal cord, and associated vascular supply. Common procedures include cervical fusion. Disorders commonly treated by neurosurgeons include intracranial tumors, vascular malformations, carpal tunnel syndrome, spinal cord injury, and stroke.

Orthopedic Surgery

Orthopedic surgery treats and corrects deformities, diseases, and injuries to the skeletal system, its articulations, and associated structures. Some examples of orthopedic surgery include hip and knee replacement, cartilage repair, and fracture repair.

Plastic Surgery

Plastic surgery is concerned with the repair, restoration, or improvement of lost, injured, defective, or misshapen parts of the body, caused by congenital defects, developmental abnormalities, trauma, infection, tumors, or disease. It is generally performed to improve functions, but it also is done to approximate a normal appearance.

Gynecology

Gynecology involves the physiology and disorders primarily of the female genital tract, as well as female endocrinology and reproductive physiology.

Urology

Urology is the study, diagnosis, and treatment of diseases of the urinary tract in both sexes, and the genital tract in the male.

Burn Treatment

Burn treatment is surgery performed to treat and manage burn injuries by reconstructing damaged tissues and improving functionality of the damaged area. A burn team might include plastic and general surgeons.

Pediatric Surgery

Pediatric surgery is concerned with the health of infants, children, and adolescents.

For the purpose of this chapter, the setting for any of the following procedural areas can be characterized as a surgical suite. In many cases, the equipment and facility requirements that are present in these areas are similar to those found in a conventional operating room environment.

- Labor and delivery
- Endoscopy
- Ophthalmology and otolaryngology procedure rooms
- Dental surgery
- Radiology suites, including interventional radiology and radiation oncology
- Interventional electrophysiology
- Treatment rooms in which minor procedures are performed

Operating rooms, large and small, rely heavily on medical personnel and support staff to carry out a wide variety of tasks. The following is a list of some of the employees who might be found working in, or providing a service to, the operating room.

- Physicians and anesthesiologists
- Registered nurses, circulating nurses, scrub nurses, nursing assistants, and surgical technicians
- Clinical engineers, biomedical engineering technicians, and anesthesia technicians
- Schedulers, record keepers, patient transporters, turnover team, purchasers, and administrators

Operating Room Floor Plan

The floor plan of a well-organized and safe OR involves a great deal of research and planning. It is a complex environment with numerous design criteria. The layout of the facility directly effects user productivity and satisfaction. The OR must be arranged in a manner that is conducive to the flow of patients, staff, and equipment. It also must include space for functional support areas that are common to most ORs, such as instrument processing, technical support, and equipment storage areas. It is necessary for clinical engineers to have knowledge of the OR layout and design issues. Clinical engineers, as technical experts of medical technology and the environment in which it is used, also may serve as a valuable resource in the design of an OR.

The arrangement of surgical suites often reflects a hospital's available space and surgical requirements. Several OR floor plans have been developed to utilize available space and maximize productivity most ideally. The L- or T-shaped single-corridor layout is common in smaller ORs. Larger facilities typically utilize the multiple-corridor layout, with the clustering of surgical suites by surgical specialty. (Bronzino, 1992). Grouping surgical specialties together aids in the efficiency of sharing common resources.

The location of functional support areas that serve the OR is another important consideration in a floor plan. The efficiency of these services is enhanced when the distances that must be traveled are minimized. The following is a list of some functional support areas that are commonly found in, or adjacent to, the OR.

Induction and Pre-Operative Holding Areas

Induction and pre-operative holding areas are utilized to prepare the patient for surgery, and the administration of anesthesia. These services are located either immediately adjacent to the operating room or within the department. The procedures and services performed in these areas will vary from hospital to hospital. The holding areas at some facilities are used simply to prepare the patient for delivery to the surgical suite, and other hospitals might utilize this area to fully anesthetize the patient.

Scrub Stations

Scrub stations used for hand washing contain sinks and cleaning supplies and are common to all ORs. Two scrub positions must be located adjacent to the entrance to each surgical suite.

Control Desk

The control desk is the communication center of the operating room. Arrangement of the case schedule is a primary function of the control desk. This is vital for the efficiency and productivity of the OR. Several things must be considered when planning the schedule. The scheduler must coordinate the urgency of surgical cases, surgical location, time considerations, patients, surgeons, anesthesiologists, support staff, equipment and supply availability, and post-anesthesia care. The control desk also might be involved in patient billing and recordkeeping.

Technical Support Services

Technical support services are vital to the productivity of the OR. Support for operating room technology must be immediately available to ensure the efficiency and safety of the surgical environment. Support groups serving the OR include anesthesia technical support and clinical engineering (see Figure 89-1). Anesthesia technical support is located within, or in the immediate vicinity of, the OR. The anesthesia workroom is primarily used for cleaning, testing, and storing anesthesia equipment. In larger facilities, clinical engineering support workrooms also might be situated within the OR. Later in this chapter, the responsibilities of the clinical engineering department will be discussed in detail.

Patient Support Service Areas

Patient support service areas include personnel, equipment, and supplies used to provide assistance to a variety of operating room functions. Space must be incorporated into the operating room floor plan to accommodate services such as equipment or patient transporters, orderlies, and nurse's aides.

Housekeeping Areas

Housekeeping areas are the workspaces for housekeeping and turnover team personnel, equipment, and supplies. Turnover team members are responsible for cleaning and setting-up surgical suites between cases.

Pharmacy

The pharmacy is included in the overall design of a health care facility to provide medications for patient care. Hospitals of considerable size can include a pharmacy within the operating room.

Instrument Processing

Instrument processing can be managed in a variety of ways, depending on the resources and preferences of a particular institution. Regardless of the sterilization method, a process must be present for delivering contaminated instruments to the processing area and redistributing the appropriate instruments to the surgical location. Sterilization methods can be divided into two categories: Centralized and decentralized.

A centralized processing department is designed to serve a large volume and can be located within the operating room. Smaller hospitals sometimes utilize a centralized processing department that serves the entire facility. Centralized instrument processing can employ an assortment of sterilization processes. The two most common methods include steam sterilization and ethylene oxide (EtO) sterilization. Autoclaves are devices that sterilize instruments using steam, with temperature, exposure time, and pressure as variables. Variables in EtO sterilization are EtO gas concentration, moisture, temperature, and time. Dry-heat sterilization and chemical sterilization are other methods of instrument processing that can be utilized in health care facilities.

Decentralized instrument processing is the practice of sterilizing instruments inside, or within the immediate proximity of, a surgical suite. Perhaps the most common method involves the use of autoclave technology. An autoclave can be designated for each operating suite or shared among a cluster of suites. Table-top sterilizers are used for flash sterilization located in common areas in the operating room. They are autoclavable at around 270° F for 3 to 5 minutes

Materials Management

A hospital also must allocate space for the delivery, storage, and distribution of supplies to the operating room. The process of waste disposal also must be accommodated. Trash and biological waste must be properly stored, handled, and eliminated. In addition, the operating room must establish a manner for processing surgical drapes, scrubs, and other linens used in the surgical environment.

Equipment Storage Areas

A common dilemma is the lack of sufficient storage space. The operating room layout must include areas for the storage of medical devices. Large devices, such as stretchers, spare anesthesia machines, and X-ray equipment, must be stored in accessible locations where they will not clutter or obstruct hallways.

Other areas

Functional areas that also can be located in or near the operating room include dark rooms, conference areas, and locker rooms.

Operating Room Facility Infrastructure

The operating room is a complex environment that requires precise climate control and numerous utility services. The following is a list of utilities services that are present in an operating room. A description of each utility is included.

Climate Control

The purpose of a heating, ventilating, and air-conditioning (HVAC) system is to provide and maintain environmental conditions, including proper airflow, heating, and cooling within a certain area or the entire hospital (Hyndman, 2003). Types of HVAC systems include variable air volume, multizone systems, displacement ventilation, and water-loop heat pumps. An HVAC system mainly consists of an air/water supply system, filters, heating and cooling coils, compressors, fans, motors, exhaust and evacuation systems, air ducts, vents, and control mechanisms. It can provide heat and ventilation to the entire hospital with provisions for controls for individual zones (e.g., the OR), or each zone can have their own HVAC system and associated control. The control mechanism is an integral part of a highly efficient HVAC system. The controls monitor, display, and allow the user to regulate various parameters, including temperature, pressure, and humidity. Precise control over temperature and humidity is important in the operating room. For example, a clinician must be able to adjust the temperature of the operating suite for surgical specialties such as pediatrics and orthopedics, where higher room temperatures are desired in order to help maintain patient temperature. Some cases, such as burn treatments, require a higher humidity.

The OR requires specialized criteria for the ventilation system as well. The surgical area must be replenished with fresh air at regular time intervals time to maintain a safe and sterile environment. A minimum of three air changes per hour of outdoor air, and 15 total air changes per hour, is required (AIA, 1996). A positive pressure must exist between the operating room and adjacent areas, to prevent the inflow of contaminated air. The ventilation system also should be designed so that there is no recirculation of air from room to room.

Adequate filtering is another important criterion of the ventilation system. A hospital-grade high efficiency particulate air (HEPA) filter is installed at the location where the air is drawn in, to filter debris and to control airborne infection. Hospital-grade HEPA filtration is constructed with an air-intake grill to catch larger debris such as leaves, seeds, and insects. Next is a layer composed of fibers to remove smaller objects like dirt and dust. Finally, a corrugated fine-mesh removes particles such as pollen, dust, and viruses.

Electrical Power Distribution

The prevalence of electrical devices for the diagnosis and treatment of patients has made electrical power distribution one of the more important systems in modern medical facilities (see Chapter 109). The nature of the operating room environment demands a particularly elaborate set of requirements (NEC, 2003). Electrically powered devices surround the patient and perform a wide range of vital functions. Therefore, proper design and maintenance of the electrical power system are essential, to ensure the safety of both the patient and the clinician. As the need for a continuous power is essential throughout a hospital, electrical systems must comprise at least two sources of power. The typical primary power sources are public utility lines. To provide for the continuous supply of power to the critical areas of the hospital, an alternate, or emergency, source of power is necessary. In the event of an interruption, hospitals are required to have an in-house alternate source of power, such as a generator.

The operating room is densely populated with electrically powered devices. Individual surgical suites must be supplied with enough electrical wall outlets to accommodate the equipment, and these outlets must be in locations that are conveniently accessible to the clinician. Wall outlets in a health care facility deliver power to equipment using a three-wire configuration of hot, neutral, and ground wires. The ground wire protects patients and staff from electrical hazards by providing a low resistance pathway to channel fault or leakage currents away from an electrically powered device to ground. Some surgical techniques and instrumentation bypass the patient's body resistance, thus increasing the

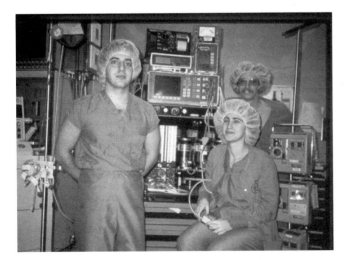

Figure 89-1 Clinical engineering department engineers and technicians support the OR.

patient's susceptibility to electrical energy from external sources. Minimizing the electrical energy to which a susceptible patient is exposed is prudent. For additional information on electrical safety, see Ridgway (Chapter 65) and Baretich (Chapter 109).

Although most equipment is designed to operate on the same power requirements, the facility also must be prepared to accommodate devices that require varying power consumption. While most wall receptacles provide 115 volts, some electrical devices, such as lasers and X-ray equipment, require 230 volts to operate (see Chapter 91).

In the past, the use of combustible anesthetic gases (e.g., cyclopropane and ether) created a volatile environment that demanded additional safety measures. It was essential to reduce the likelihood of a static discharge (spark) that could lead to an explosion in the operating suite. Although the risk of explosion is not typically an issue today, it has played an important role in the design of electrical power systems in many existing operating rooms. For example, power outlets have typically been installed approximately four feet above floor level because volatile explosive agents tend to settle in the air. Special plugs, termed "explosion-proof," were also commonly installed on power cords that, when plugged in, created an airtight seal. Conductive flooring is also commonly found in operating rooms. This technology was installed to eliminate the electrical isolation of equipment and personnel in an operating suite. Electrical isolation between objects is conducive to the build-up and potential discharge of static electrical charge. Isolated ungrounded power systems, line isolation monitors (LIM), and ground-fault circuit interrupters (GFCI) are further examples of technologies employed in operating rooms to reduce the risk of explosion in the presence of combustible anesthetic agents. These technologies also serve to reduce the risk of explosion from electrical shocks resulting from stray (leakage) current.

An isolated ungrounded power system is an electrical power distribution system in which all of the current-carrying conductors are isolated from ground (and earth) by a high impedance (Feinberg, 1980). The most common and economical method of isolation is to use an isolating transformer. In a properly installed system, no hazardous current will flow from either conductor to ground, but the two conductors will function as though they were connected directly to a ground. This is especially useful in a "wet environment" such as the operating room, where liquid spills and standing water increase the risk of electrical hazards.

An LIM displays the degree of electrical isolation in an isolated ungrounded power system. It provides an early warning system of possible leakage or fault currents from either of the current-carrying conductors to ground. A GFCI also serves to protect patients and staff members. This technology monitors the ground fault current and interrupts the power to the electrical receptacle when the current exceeds a preset limit. Although LIM technology is more expensive, one advantage over the GFCI is that the LIM will alarm when a fault is detected, but will not interrupt electrical service.

Gas Pipelines

Pressurized gases must be distributed to locations throughout the health care facility (see Chapter 110). Medical gases such as oxygen, nitrous oxide, and medical air are particularly important for the administration of anesthesia in the operating room. Medical gases are stored either in metal cylinders or in the reservoirs of bulk gas storage and central supply systems. The central supply system is the source of medical gases that are distributed via the facility pipeline system. The number and location of wall outlets in the operating room varies, depending on facility design and surgical specialty. Figure 90-2 shows a wall outlet for nitrogen commonly used to power pneumatic tools such as saws and drills.

The method of storage for the central supply of oxygen depends on the demands of the facility. Smaller facilities tend to store oxygen in a series of cylinders connected by a manifold or high-pressure header system (Ehrenwerth and Eisenkraft, 1993). Larger facilities store their bulk oxygen in pressurized liquid form, which enables the hospital to store more oxygen in a smaller space. The central supply of medical air is most commonly generated by air compressors, stored in a reservoir, and delivered to the piping system. Other sources include pressurized cylinders or a proportioning system that mixes the appropriate amount of oxygen and nitrogen from central sources. The nitrous oxide supply systems include the cylinder manifold system or a bulk liquid storage system similar to the one used for oxygen system (Ehrenwerth and Eisenkraft, 1993). All central supply systems for medical gases must be designed with a separate reserve system. Details concerning the use of medical gases in anesthesiology are discussed elsewhere in this handbook (see Chapter 90).

Nitrogen and carbon dioxide can be piped into the operating room, as well. Nitrogen, primarily used for gas-powered equipment, is stored as a series of pressurized tanks connected by a manifold. Carbon dioxide is also typically stored as a cylinder manifold system.

Vacuum

The central vacuum system is another essential utility that must be piped into the operating room. The system is vital to both surgery and anesthesia. It provides pressure for drainage, aspiration, suction during surgical procedures, and the scavenging of waste gases from the anesthesia machine breathing circuit. A health care facility must have two suction pumps, each individually capable of handling the overall demand.

Lighting

Lighting is an essential utility in a health care facility. The operating room requires special attention to lighting because of the small-scale delicate tasks that must be performed. Although the choice of a surgical light is subjective, several characteristics must be addressed. The criteria include light intensity, light color, focusing capability and range, degree of shadow production, heat production and dissipation, choice of mounting, lamp head maneuverability, and ease of cleaning (Bronzino, 1992).

Communication

The speed and reliability of communication in the operating room is vitally important. Communication systems are becoming increasingly complex with the continual advances in networking and wireless technology. Fundamental modes of communication such as telephones, paging systems, and intercoms are common in operating rooms.

Advances in computer technology and network systems have had a profound effect on the communication systems within the operating room environment. E-mail, Internet access, online scheduling, and patient databases have vastly improved the quantity and quality of exchanged information. See the section on information, in this handbook.

Fire Protection

All health care facilities must be equipped with a fire protection system. The fundamental elements of such a system include smoke detectors, alarms, a sprinkler system, and automatically closing doors to contain the fire. The hospital staff must have an established evacuation plan for patients and staff. Several operating room fires occur each year. The combination of an oxygen-enriched atmosphere, fuel source (e.g., drapes, endotracheal tubes, and the patient) and ignition source(e.g., electrosurgical units and lasers) have produced catastrophic fires (de Richmond and Bruley, 1993; NFPA, 1994).

Surgical Suite Layout

A well-designed surgical suite provides ample space for staff and equipment. It encourages a smooth flow of people and materials within the room. While the layout of an individual surgical suite varies by surgical service, all suites share some characteristics, as described in below.

The size of the operating suite should reflect the space required by the patient, staff, and equipment to carry out the surgical procedure. The standard surgical suite is at least 400 square feet. Surgical suites that require specialized equipment or additional personnel can be significantly larger. It is important to provide space so that sensitive medical equipment is not positioned in high-traffic areas. The shape of an operating suite can take on a variety of forms, (e.g., square, rectangular, round, or oval), depending on design preferences and overall operating room layout.

The surgical table is typically oriented in the center of the room, with the surgeons positioned along the side of the patient. The scrub nurse is positioned in an area where both the instrument table and the surgeon's outstretched hand are within reach. The anesthesiologist and anesthesia equipment are generally located at the head end of the table. This allows the anesthesiologist to have access to the patient's airway. The anesthesia equipment also must be positioned near medical gas, vacuum, and power outlets that can be located on a wall or column. Other devices, such as defibrillators, electrosurgical units, video towers, and bypass machines, are located further outside the surgical field. Figure 89-3 shows a typical layout of an operating room.

The surgical suite also must provide accommodations for storage. Commonly, cabinets, shelving, and carts will line the walls of a surgical suite. In addition, surgical suites are usually equipped with communication equipment. Computer, intercom, and telephone access are standard in modern surgical suite layouts.

Common Operating Room Technologies

The following is a list of typical medical technologies commonly found in the OR. A brief description of each technology is included.

Furniture

Furniture is utilized extensively in the operating room and serves a wide range of functions. Patient beds are employed for transporting patients and as an adaptable platform for performing surgery. Transport beds, or stretchers, are designed with side rails and large casters to safely transport patients within the hospital (see Chapter 95).

Figure 89-2 OR wall outlet for nitrogen gas.

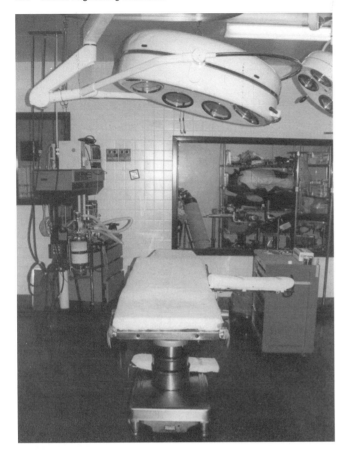

Figure 89-3 Typical OR layout showing OR table, overhead lights, anesthesia machine and monitors, medication and supplies cart, storage room, and wall outlets for gases.

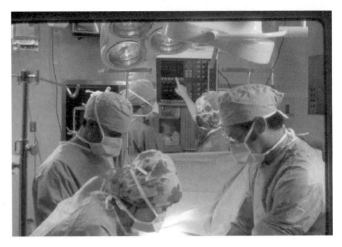

Figure 89-4 The surgical team at work on a draped patient.

Surgical beds are valuable tools for patient positioning during surgery. In order to accommodate various procedures, surgical beds are more substantial and complex than transport beds. Modern beds are equipped with mechanical, electronic, and hydraulic systems that allow clinicians to position a patient in numerous orientations, such as raising or lowering the height, tilting to the left or right, Trendelenburg (head down, legs up), and reverse-Trendelenburg. Surgical bed accessories include arm support boards, leg supports, foot extensions, restraints, and padding. They also can be outfitted with a wide variety of specialized surgical positioning devices, such as neurosurgical headrests, radiolucent tops, and the Andrews Frame used for spinal surgery.

Other standard items of furniture in the OR include instrument tables, intravenous (IV) poles, waste storage bins, stools, and chairs. Racks, carts, and shelving, used to store medical supplies, surgical supplies, and instruments, are also common in a surgical suite.

Surgical Drapes

Surgical drapes are used in the OR to protect the patient, clinicians, and equipment. Drapes can be made of cloth or paper, and reusable or disposable. Figure 89-4 shows a surgical procedure underway with the patient draped. Important characteristics include barrier protection effectiveness, resistance to ignition, and durability. Surgical drapes are employed to provide a physical barrier that protects the surgical field from contamination. An "ether screen" is the wall of drapes set up in order to provide a barrier between the anesthesia work area, at the head of the patient, and the surgical field. Drapes are also placed in the surgical field around the incision site to cover the patient and to collect fluids. They also can be used to wrap sterile surgical instruments and to cover equipment in the surgical suite.

Anesthesia Equipment

The delivery of anesthesia requires a gas delivery system, as well as continuous and detailed monitoring of patient physiology. The anesthesia machine is used to deliver a known mixture of gases to the patient. Three main sections of the anesthesia machine are the gas supply and delivery system, the vaporizer, and the patient breathing circuit (see Chapter 90; Calkins, 1988; Paulsen, 1995; Dorsch and Dorsch, 1984).

A variety of technologies are employed by the anesthesiologist to monitor the physiology of the anesthetized patient. Physiological monitoring equipment includes electrocardiograph monitors, pulse oximetry, invasive blood pressure monitors, noninvasive blood pressure monitors, temperature monitors, respiratory gas monitors, and electroencephalograph monitors (see Chapter 98). Other devices utilized by the anesthesiology department include infusion devices and fluid warmers.

Surgical Instruments

Surgical instruments are hand-held tools or implements used by clinicians for the performance of surgical tasks. A vast assortment of instruments can be found in an operating suite. Scalpels, forceps, scissors, retractors, and clamps are used extensively. The nature of certain surgical procedures requires a more specialized set of instruments. For example, bone saws, files, drills (Figure 89-5), and mallets are commonly utilized in orthopedic surgery. Surgical instruments are generally made of carbon steel, stainless steel, aluminum, or titanium, and are available in a range of sizes.

Electrosurgical Units

The electrosurgical unit, or Bovie, is a surgical device used to incise tissue, destroy tissue through desiccation, and to control bleeding (hemostasis) by causing the coagulation of blood. This is accomplished with a high-powered and high-frequency generator that produces a radiofrequency (RF) spark between a probe and the surgical site that causes localized heating and damage to the tissue (Gerhard GC, 1988). An electrosurgical generator (Figure 89-6) operates in two modes. In the monopolar mode, an active electrode concentrates the current to the surgical site and a dispersive (return) electrode channels the current away from the patient. In the bipolar mode, both the active and return electrodes are located at the surgical site.

Defibrillators

A defibrillator is a medical device that is used to deliver an electrical shock to the heart (Tacker, 1988). The shock is intended to correct irregular electrical activity of the heart and to establish an organized rhythm. A shock of adequate power and duration will cause the cells of the heart to simultaneously repolarize and allow a normal rhythm to return. The defibrillator uses a capacitor to store the required energy, measured in joules (i.e., watts per second), to deliver the shock. A DC power supply charges the capacitor to the selected energy level.

Electrodes are used to deliver the electrical shock to the patient. Electrode types include reusable paddles and adhesive electrodes. External defibrillation is applied to the chest of the patient with external electrodes or paddles. An internal paddle set is used when defibrillation is delivered directly to the heart. Factors governing the set-up and performance of a defibrillator include patient impedance, energy waveform shape, and electrode type and placement.

Most defibrillators are designed with technology to monitor the patient's ECG signal and allow for synchronized cardioversion. Synchronized cardioversion is the delivery of energy to the heart during ventricular depolarization, or upon the detection of the QRS

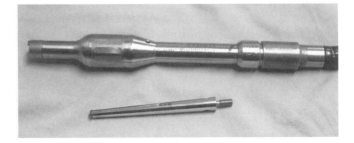

Figure 89-5 Surgical drill.

Figure 89-6 Aspen Labs MF 380 electrosurgical unit.

complex. This feature serves to protect the patient by preventing the inadvertent delivery of energy during the ventricular refractory period.

Temperature Regulation Devices

Temperature monitoring and regulation are crucial to ensure the safety of a surgical patient (Vaughan, 1988). Although heating and air conditioning controls in the operating room help to maintain a safe surgical environment, a patient is at risk of suffering from the effects of hyperthermia, and more commonly, hypothermia. Heat loss can be attributed to contact with conductive surfaces, exposed body cavities, cold irrigation solutions, and convective heat loss due to the considerable flow of air in operating suites. In addition, the effects of anesthesia can impair the body's natural mechanisms for maintaining proper temperature.

Several methods are employed to equalize a patient's body temperature. Blankets, sheets, and clothing are common methods of preventing heat loss. A blanket warmer is also utilized to prevent hypothermia by delivering warmed air to an inflatable blanket surrounding the patient. A hyper-hypothermia unit circulates heated or cooled water through a blanket, to raise or lower a patient's temperature. A fluid warmer is a device used for warming intravenous or irrigation fluids prior to contact with the patient.

Headlamps

Headlamps are used to supplement the facility lights in the surgical field. The headlamp apparatus, worn by the clinician, is connected to a light source by a fiber-optic cable. This versatile source of light is particularly useful when overhead lights are insufficient or obstructed.

Tourniquets

A tourniquet is a surgical device that is used primarily to temporarily occlude blood flow to a part of the body and to obtain a nearly bloodless operative field. A pneumatic tourniquet uses pressurized air to restrict blood flow and comprise an inflatable cuff, connective tubing, pressure source, pressure regulator, and a pressure display. Tourniquets are commonly used in amputations and various other orthopedic surgical procedures (see Figure 89-7).

Specialized Operating Room Technologies

Lasers

A laser is a device that directs an intense beam of radiation to the surgical site to cut, coagulate, or vaporize tissue (Powers, 1988; Judy, 1995). The major components of a laser are the lasing material, mirrors, a cooling system, an optical or electrical pump source, and a

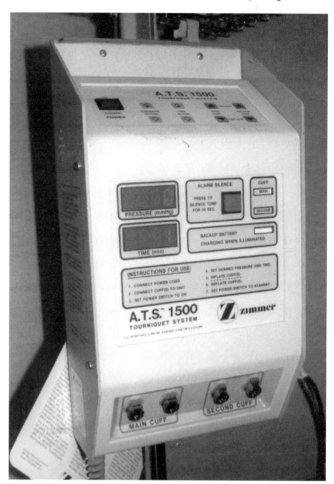

Figure 89-7 Zimmer ATS automated tourniquet system.

delivery system. The lasing material fixes the output wavelength of the laser and primarily comprises a gas-filled tubular cavity (gas) or a solid-state medium. Lasers with differing lasing material, wavelengths, beam shapes, and guidance methods, are used in the operating room for various physiological effects. Four laser types that are typically used in the operating room to coagulate, ablate, or remove soft tissue are carbon dioxide (CO_2), argon ion (Ar-ion), neodymium-yttrium-aluminum-garnet (Nd:YAG), and gallium-aluminum arsenide (GaAlAs).

Robotics

The use of robotics has provided extensive benefits to many industries, and the medical field is no exception. Specifically, surgery is a field wherein the application of robotics can lead to advantages in providing health care. The inexhaustibility, repeatability, and precision of robotics are favorable in surgery.

Imaging Technologies

Body organs, structures, and tissues are studied using imaging techniques (Siedband, 1992). Several technologies are available to the clinician. Hard-tissue imaging is done using X-rays. Soft-tissue contrast and functional information cannot be sought using X-rays. There are many techniques, including computed tomography and magnetic resonance imaging, that are effective in fairly accurate determination of damaged cells or soft tissue.

X-rays

Radiography utilized electromagnetic waves to produce two-dimensional images of anatomy, captured on a photographic film. The intensity of the image on the film is determined by the intensity of the rays emerging from the anatomy. Tissues with different densities will exhibit varying degrees of X-ray absorption. This is the most common form of imaging technique used in clinical practice.

Angiography

Angiography is a radiographic technique where a radio-opaque (i.e., visible on an X-ray) contrast material is injected into a blood vessel for the purpose of identifying its anatomy on X-ray. This technique is used to image arteries in the brain, heart, kidneys, gastrointestinal tract, aorta, neck (carotids), chest, limb, and pulmonary circuit.

Fluoroscopy

Fluoroscopy is a form of diagnostic radiology that enables the radiologist, with the aid of a contrast agent, to visualize an organ or the area of concern in motion, via X-ray. This contrast agent allows the image to be viewed clearly on a television monitor or screen. Contrast agents, also known as "contrast media," can be introduced into the body by injection, swallowing, or enema.

Computed Tomography

Computed tomography is a diagnostic technique that uses X-rays to acquire detailed information of soft tissue structures, muscles, bones, and organs. It uses an X-ray source that revolves around the object to be imaged. The detector captures the rays (raw data) that penetrate through the organs. The raw information is processed and reconstructed using a computer algorithm to form images. The images are in the form of cross-sectional slices. This technology is particularly useful for producing images of the brain.

MRI

Magnetic Resonance Imaging (MRI) is a noninvasive method of imaging structures and soft tissues inside the body. MRI imaging is very detailed and provides a high degree of diagnostic accuracy, as compared to other modes of imaging. It is based on the principle that hydrogen nuclei in a strong magnetic field absorb pulses of radiofrequency energy and emit them as radio waves, which can be reconstructed into computerized images. The images produced are of a high-quality and give a good indication of the properties of internal body parts. Magnetic resonance imaging is widely utilized by many surgical specialties. It is particularly valuable for imaging soft tissue, the brain and spinal cord, joints, and the abdomen.

MRA

Magnetic Resonance Angiography (MRA) is a noninvasive method of vascular imaging and determination of internal anatomy without injection of contrast media or radiation exposure. The technique is used especially in cerebral angiography, the radiography of the vascular system of the brain, as well as for studies of other vascular structures.

Ultrasound

Ultrasound imaging is technique in which high-frequency sound waves are reflected from internal organs, and the echo pattern is converted into a two-dimensional picture of the structures beneath the transducer.

Minimally Invasive Surgical Devices

For many surgical procedures, the method of choice has shifted from traditional open surgery to the use of less-invasive means. Minimally invasive surgery is performed with the aid of a viewing scope and specially designed surgical instruments (Garrett HMS, 1994). The scope allows the surgeon to perform major surgery through several tiny openings without making a large incision. These minimally invasive alternatives usually result in decreased pain, scarring, and recovery time for the patient, as well as reduced health care costs.

Endoscopy is an examination of the interior organs and body cavities, through a natural body opening or a surgical incision, using a light and a rigid or flexible viewing instrument called an endoscope. The viewing component of an endoscope is made up of hundreds of light-transmitting glass fibers bundled tightly together.

Laparoscopy has become a common surgical technique in the operating room (Brooks, 1994). This procedure is the examination of the interior of the abdomen with a slender endoscope, called a laparoscope. The laparoscope is inserted through an incision in the abdominal wall in order to perform surgery. Laparoscopic techniques are performed to remove gall bladders, to perform antireflux operations on the esophagus, and to remove organs such as the adrenal glands. Although minimally invasive surgery is beneficial for the patient, it is technologically more demanding than traditional surgery. Special training is required for clinicians, and the associated medical technology is more advanced. A laparoscopic procedure is usually performed with a mobile cart that is outfitted with equipment for visualization, instruments for exposure and manipulation, and equipment and instruments for cutting and coagulation.

The equipment for viewing internal tissues and organs include a light source with fiberoptics to transmit the light, a high-resolution camera that can withstand sterilization, a video processor, a high-resolution monitor(s), a video recorder, and a printer. Several devices are employed for exposure and manipulation in laparoscopic procedures. Trocars are sharp-pointed surgical instruments, used with a cannula to puncture a body cavity and to provide intra-abdominal access. A high-flow carbon dioxide (CO_2) insufflator is used to expand the abdominal cavity, to make internal organs more accessible. Graspers are devices, generally with two movable and serrated jaws, that are used to grasp and retract organs (Figure 89-8). Figure 89-9 shows an endoscopic instrument for applying clips. Irrigators (or aspirators) are also commonly used. Instruments used for cutting and coagulation in laparoscopic procedures include microscissors, electrocoagulating dissectors and graspers, heater probes, and lasers.

Heart-Lung Machines (Bypass)

A heart-lung machine (Figure 89-10) is an apparatus that does the work both of the heart (i.e., pumps blood) and the lungs (i.e., oxygenates the blood) during, for example, open-heart surgery (Galletti and Colton, 1995). The basic function of the machine is to oxygenate the body's venous supply of blood and then to pump it back into the arterial system. Blood returning to the heart is diverted through the machine before returning it to the arterial circulation. Some of the more important components of these machines

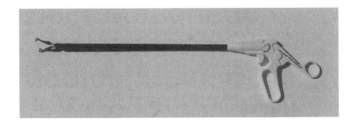

Figure 89-8 Endoscopic instrument: Autosuture Endo Babcock graspers.

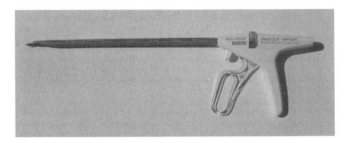

Figure 89-9 Endoscopic instrument: Autosuture Endoclip applier.

include pumps, oxygenators, temperature regulators, and filters. The heart-lung machine also provides intracardiac suction, filtration, and temperature control.

Transesophageal Echocardiogram

The transesophageal echocardiogram is a devices that utilizes ultrasound technology for imaging cardiovascular anatomy and physiology (Schluter and Hanrath, 1984). This device uses an ultrasound probe that is inserted in the esophagus and stomach to image the heart and its associated blood vessels. The internal perspective avoids the interference from the anatomy of the chest encountered with external echocardiograms. Because the esophagus is located just behind the heart this specialized view is clear. The transesophageal echocardiogram uses the same sound-wave technology as a regular echocardiogram. The ultrasound probe emits sound waves of a certain frequency upon the object to be imaged, and the return wave frequency is detected by a transducer. The characteristics of the structure are determined by analyzing the return frequency, and the signals are converted into computerized images.

Transesophageal echocardiograms are frequently used to monitor the heart during surgery. Other common uses include searching for an abnormality in the heart or major blood

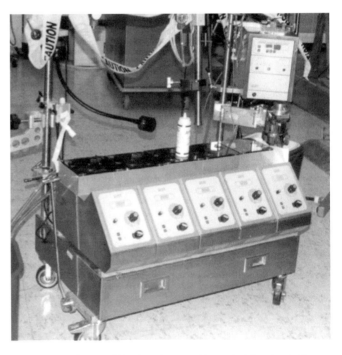

Figure 89-10 Stockert-Shiley heart-lung machine under laboratory testing.

vessels that could be responsible for causing a stroke; looking for infections on the heart valves; and evaluating the aorta for a tear in its wall.

Cell Savers

A cell saver is a device that collects and returns a patient's blood that otherwise would be lost during a surgical procedure. This process is referred to as "auto transfusion." The cell saver receives salvaged blood that has been suctioned from the surgical site. The blood passes through filters to remove surgical and cellular debris. Centrifuge technology is employed to separate the heavier oxygen-carrying red blood cells from the other components of the blood. The red blood cells are processed and stored for auto transfusion.

Pacers and Pacemakers

A pacemaker is an electrical pulse generator designed to support the electrical activity of the heart (Greatbatch and Seligman, 1988). A pacemaker can be implanted within a patient or used externally. External temporary pacemakers are commonly used in cardiovascular procedures to treat heart conditions such as bradycardia, atrial and/or ventricular arrhythmias, and cardiac arrest.

The basic pacemaker comprises a pulse generator, a programmer, and a lead. The pulse generator consists of the power source and circuitry, which senses the heart's electrical activity and generates the output. The programmer allows the clinician to adjust pacing variables such as pulse rate, amplitude, duration, and the sensitivity of pulse detection. The lead is an insulated wire that carries the stimulus from the pacemaker to the heart and delivers an ECG signal back to the pacemaker.

Ventricular Assist Devices

The primary aim of a ventricular assist device (VAD) is circulation support. When the myocardium is damaged, the heart is unable to maintain the required cardiac output and blood pressure to maintain blood flow. A VAD is implemented to relieve the workload of the myocardium. Another use of the VAD is ventricular assistance. Patients with heart failure, which can be reversed if the heart is given sufficient time to recover, are candidates for a VAD. Using a VAD is also common after patients undergo cardiopulmonary bypass or any other traumatic heart surgery. A VAD can assist in the recovery of the right or left ventricle (RVAD or LVAD) or both ventricles (BIVAD). An LVAD diverts blood from either the left atrium or left ventricle, sends the blood through a pump, and then returns the blood to the aorta. The RVAD operates in the same manner, with the blood diverted from the right atrium and returned into the pulmonary artery.

Microscopes

Due to the spread of microsurgical procedures through the various surgical disciplines, the surgical microscope has become a valuable tool in the operating room. Microscopes are used to provide magnification and illumination in order to view objects during minute and intricate surgeries. These devices are generally employed when performing procedures, such as neurosurgery, where structures are often very small and can not be viewed clearly with the naked eye. Surgical microscopes generally contain binocular lenses, light sources, optic fiber cables, and focusing mechanisms. Surgical microscopes must be equipped with a mounting or support system in order to be utilized in the surgical field. A variety of support systems are available, including mobile floor stands and ceiling mounts.

Safety in the Operating Room

The hazards associated with medical devices, clinical processes, and human error are of primary concern to the clinical engineer (see Chapter 55). A prevalence of medical equipment and staff in the OR, along with the vulnerability of the surgical patient, creates an environment that requires additional precautions. Potential hazards associated with operating room devices and technologies are vast and can be encountered in many forms. Medical equipment, for example, can pose electrical and/or mechanical hazards to patients and staff. A patient whose natural resistance to current flow has been compromised (perhaps due to an invasive connection to a medical device) is particularly vulnerable to electrical hazards. Devices that are mechanical in nature, such as transport beds and surgical tables, also pose a risk to patients and staff if they are not used and maintained properly. It is a responsibility of the clinical engineering department to implement a program to ensure medical device safety and to manage risk. The program must include the preventive maintenance and inspection of medical devices and systems. It also must conform to codes standards set forth by regulatory agencies such as NFPA, Underwriters Laboratories (UL), Occupation Safety and Health Act (OSHA), and the Joint Commission on Accreditation of Healthcare Organizations (JCAHO).

Other hazards associated with technology in the operating room include environmental hazards, such as the spread of infection due to poor filtration of operating room air; biological hazards, resulting from poor sterilization practices; and radiation hazards from diagnostic machines and therapeutic devices that release radiation (Bronzino, 1992).

A lack of knowledge and communication within the clinical staff significantly increases the likelihood of injuries in the operating room. Familiarity with medical devices, technical processes, and awareness of emergency procedures are key factors in avoiding injury. To meet these requirements, a clinical engineer must provide, or participate in, the in-servicing of new equipment, continual user training, and incident reporting and evaluation.

Adequate preparation requires that the medical staff be properly trained to respond to the following emergencies (see Chapter 116):

- External (community) disaster
- Fire
- Facility-system failure (e.g., power, medical gases, communication systems, elevators, sewer, steam supply, water loss or leak, or HVAC)
- Hazardous-material contamination (e.g., chemical, biohazard, or radiation)
- Infection control hazard

Clinical Engineering Roles

Clinical engineering is the application of technology to improve the quality of health care in health care facilities. It is the responsibility of a clinical engineer to apply engineering principles to understand, develop, control, and maintain medical technologies, systems, and processes. Ensuring the safety of patients and staff is also at the core of a clinical engineer's responsibilities. The role of clinical engineering departments can differ among institutions of varying resources and services. Some clinical engineering departments have an active role in the application of technology for patient care, while others are mainly a "fix-it" shop. Smaller institutions tend to have a centralized clinical engineering department that services the entire facility. On the other hand, larger institutions might allocate groups to specific departments within the hospital.

The operating room is particularly challenging to a clinical engineer. It is an area of abundant and complex technologies. In order to provide proper support to the operating room, the engineer must understand surgical processes and terminology, in addition to providing technological expertise. Ethics and professionalism are also vital characteristics in this field. Members of a clinical engineering department must be conscious of their impact on the patient care process and overall safety in the operating room.

The evaluation and introduction of new technology is a major function of a clinical engineer. The process of medical equipment procurement includes identification of equipment needs, selecting the equipment specifications and vendor, demonstrating the equipment to the users, purchasing the equipment, preparing it for use, staff training, and installation. Medical devices and technologies are provided to promote quality patient care, yet ensure cost effectiveness. Hospitals generally operate with limited funds with a growing demand for medical services. Therefore, it is vital for clinical engineers to plan investments in medical technology carefully and to maintain the equipment following procurement. The clinical engineer also must be mindful of the codes and standards that regulatory agencies have applied to health care technology.

The operating room's dependency on technology for patient care demands that medical equipment be reliable and available. The clinical engineer is responsible for ensuring that equipment is properly maintained. Preventative maintenance and scheduled inspections are necessary to minimize malfunctions, to verify functionality, and to prevent device-related injuries. Spare devices and replacement parts should be maintained for all vital equipment.

Inventory records must be kept in order, to manage the assets of the operating room. A computer database is an essential tool for managing data such as equipment inventories, parts inventories, and inspection schedules. A device's individual performance and repair histories also should be stored for future reference. Other responsibilities of a clinical engineer can include device design, project management, budgeting, staff training, incident investigation, and facility design.

Conclusion

The operating room is a complex department that serves an essential role within a health care facility. This environment comprises a wide range of clinical professionals, medical technologies, and complex systems. The clinical engineer, as a technical expert, is essential to the efficiency and safety of the health care facility. To support the OR effectively, the clinical engineer must maintain a thorough knowledge of medical devices, clinical practices and procedures, facility systems, and safety guidelines.

References

AIA. *Guidelines for Design and Construction of Hospital and Health Care Facilities*. Washington, DC, The American Institute of Architects Press, 1996.

BMET Certification Course. Mercer Island, WA, Morse Medical, Inc.

Bronzino JD. *Management of Medical Technology: A Primer for Clinical Engineers*. Boston, MA, Butterworth-Heinemann, 1992.

Bronzino JD. *The Biomedical Engineering Handbook*. Boca Raton, FL, CRC Press, 1995.

Brooks DC. *Current Techniques in Laparoscopy*. Philadelphia, PA, Current Medicine, 1994.

Calkins JM. Anesthesia Machines. In Bronzino JD (ed). *The Biomedical Engineering Handbook*. Boca Raton, FL, CRC Press, 1988.

De Richmond AL, Bruley ME. Head and Neck Surgical Fires. In Eisele DW (ed). *Complications in Head and Neck Surgery*. St. Louis, MO, Mosby, 1993.

Dorsch JA, Dorsch SE. *Understanding Anesthesia Equipment*. Baltimore, Williams & Wilkins, 1984.

Eichorn JH, Ehrenwerth J. Medical Gases: Storage and Supply. In Ehrenwerth J, Eisenkraft J (eds). *Anesthesia Equipment: Principle and Applications*. St. Louis, MO, Mosby, 1993.

Feinberg BN. *Handbook of Clinical Engineering Volume 1*. Boca Raton, FL, CRC Press, 1980.

Galletti PM, Colton CK. Artificial Lungs and Blood-Gas Exchange Devices. In Bronzino JD (ed). *The Biomedical Engineering Handbook*. Boca Raton, FL, CRC Press, 1995.

Gardner TW. *Health Care Facilities Handbook, 6th Ed*. Quincy, MA, National Fire Protection Association, 1999.

Garrett HMS. *Surgeon's Reference for Minimally Invasive Surgery Products*. Montvale, NJ, Medical Economics Data Production Co.

Gerhard GC. Electrosurgical Unit. In Webster JG (ed). *Encyclopedia of Medical Devices and Instrumentation*. New York, Wiley, 1988.

Greatbatch W, Seligman LJ. Pacemakers. In Webster JG (ed). *Encyclopedia of Medical Devices and Instrumentation*. New York, Wiley, 1988.

Judy MM. Biomedical Lasers. In Bronzino JD (ed). *The Biomedical Engineering Handbook*. Boca Raton, Florida, CRC Press, 1995.

Laufman H. Surgical Management: Developments in Operating Room Design and Instrumentation. In Ray CD (ed). *Medical Engineering Year Book*, Year Book Medical Publishers, 1974.

NEC. *National Electrical Code.* Quincy, MA, National Fire Protection Association, 2003.

NFPA. *Manual on Fire Hazards in Oxygen-Enriched Atmospheres.* NFPA 53. Quincy, MA, National Fire Protection Association,

Powers SK. Laser Surgery. In Webster JG (ed). *Encyclopedia of Medical Devices and Instrumentation.* New York, Wiley, 1988.

Schatt S. *SAMS Understanding Series, 3rd ed.* Prentice Hall Computer Publishing, 1992.

Schluster M, Hanrath P. The Clinical Application of Transesophageal Echocardiography. *Echocardiography* 1:427, 1984.

Siedband MP. In Webster JG (ed). Medical Instrumentation: Application and Design, 2nd Ed. Boston, Houghton Mifflin Company, 1992.

Square D. *Hospital Isolated Power Systems.* Product Data Bulletin No. 4800PD9701, June 1997.

Tacker WA. Electrical Defibrillators. In Webster JG (ed): *Encyclopedia of Medical Devices and Instrumentation.* New York, Wiley, 1988.

Vaughan MS. Temperature Measurement in the Clinical Setting. In Webster JG (ed). *Encyclopedia of Medical Devices and Instrumentation.* New York, Wiley, 1988.

90

Anesthesiology

Luis Melendez
Massachusetts General Hospital
Boston, MA

Raj Rane
Massachusetts General Hospital
Boston, MA

To excel as a clinical engineer or biomedical engineering technician supporting, the use of equipment in anesthesia requires equally strong understanding of both the clinical and technological components of the job. This is an important distinction that anyone must face when involved with anesthesia. Physiology and technology meet—the two cannot be separated. If you know the technology well but have less than able comprehension of medical terminology, human anatomy, and physiology, you will have difficulty making efficient decisions to correct a problem in the middle of surgery or to choose the best technology to be used for the next 15 years. This chapter is primarily dedicated to existing and present technology. Future machines are not discussed in detail, as it would be purely speculation as to what might work. Most discussions are limited to principles and not specific detail.

This chapter is best introduced and summarized with a case report. Events similar to the one described below have happened, and will happen, at most any institution. Suppose that in the middle of a total hip replacement in an operating room (OR) where ultraviolet (UV) lights and laminar flow ventilation systems are used to keep infection rates to a minimum, the clinical engineer (CE) is asked to diagnose a large leak that has developed in the breathing circuit during the case. The resident and attending physicians are concerned about the course of events, and the CE can feel the tension upon entering the room. Alarms are sounding, and it appears that the doctors are having a difficult time ventilating the patient. The physiological monitor is indicating poor oxygen saturation; it tells the CE that the capnogram shows a poor waveform. The ventilator is flashing a number of alarms. The attending physician wants a new anesthesia machine, now. The CE is told that the patient is paralyzed and that they will use an Ambu bag (i.e., a manually operated resuscitator) while the CE gets that broken machine out of the room.

The CE wants to perform a couple of tests on the machine but is concerned that asking for the time to do it will add more tension. Wanting to solve the problem as fast as possible, the CE does as told. Quickly, the CE is out of the OR and off to find another machine. Meanwhile, they disconnect the patient from the machine, start manual ventilation, and rearranging things in the operating room so that the CE can wheel the large device out and roll in another in its place. The CE returns and leaves the new machine just outside of the room, helps the physicians move all items and drugs from the existing machine, disconnects patient monitoring, gas-supply hoses, and electrical connections. The CE wheels the machine out of the room (careful not to touch sterile tables and drapes, or else there would be even more upset people), and moves the new machine in. With great care, things are reconnected, and every detail is put back in its place. The new machine and then the ventilator are turned on. The very same problem is still there. The bellows do not fill properly, even with a fresh gas-flow rate of 10 lpm. Tension is building. The CE has done everything asked of him, and there is nothing else that the he can think of. The physicians are growing even more upset. Not wanting to be in the way, the CE decides that it is time to leave.

Back in the clinical engineering department, the CE finds a co-worker and discusses the recent occurrence. The co-worker convinces the CE that they both should go to the OR. The first machine is still in the hallway. They plug it in and turn on the oxygen cylinder. A quick breathing circuit leak test and a functional ventilator test indicate that nothing is wrong with that machine. In the room, everyone is near an uproar because they cannot ventilate adequately. The co-worker asks the physicians to turn off the ventilator and to manually ventilate the patient. They must use the oxygen-flush key just to keep enough volume in the breathing circuit. The CE notices that the breathing circuit bag is actually collapsing between breaths. There is something pulling vacuum in the breathing circuit. The CE knows that there is now a different scavenger connected, so that cannot be the problem. Remembering the nasogastric suction catheter (used to empty the esophagus and stomach contents), the co-worker asks for it to be pulled out a little. The physician pulls back on the tube, and everything comes back to normal. The CE is amazed that a doctor missed the cause of such a leak.

Humans make mistakes. Safety is the responsibility of everyone involved. There must be redundancies to help prevent a number of events that would result in an adverse outcome. Multiple factors played into this case. Tunnel vision, a lack of comprehension in the application of technology, simply following orders without thinking or communicating, striving to be the one to solve a problem, and not seeking help from others all contributed to the confusion of both clinical and technical staff while putting the patient at greater risk and increasing the possibility for serious infection.

What Is Anesthesia?

Pre-Operative Assessment and Plan

Anesthesia is more than the manual labor and technology involved in relieving patients of pain that they would otherwise endure during surgery. To determine whether a patient is ready to undergo anesthesia, anesthesiologists must assess a patient's current state with a physical examination, evaluate lab results and other relevant tests (e.g., EKG and renal and pulmonary function) and medical history while taking into consideration personal preferences, and even religious beliefs, on occasion. A perioperative management plan must be developed. The physician and patient will discuss options and risks of local, regional, and general anesthesia as well as backup plans in case unexpected events occur. In doing so, they must take into account any risks to which the patient will be subject. A neonate with poor lung function, a healthy adult, and a hemodynamically unstable elderly patient all pose different challenges. In anticipating what could happen during the surgical procedure and what the level of difficulty any one individual patient may be, the anesthesiologist must decide among many things, such as level of monitoring, vascular access (replenish fluids and sample blood gases) and airway management.

Amnesia and Analgesia

Anesthesia is an induced, controlled state combining amnesia and analgesia during surgery, obstetric, therapeutic, and diagnostic procedures. Overall, the safest approach is to leave the patient as self-sufficient as possible. Under proper management, a patient who is spontaneously ventilating (breathing unassisted) is inherently at less risk than a patient requiring mechanical, or assisted, ventilation. For minor procedures, local anesthesia, as with the use of peripheral nerve blocks, is performed by an anesthesiologist and monitored by less-trained personnel during the procedure. This is known as monitored anesthesia care (MAC) and often involves nurses without extensive anesthesia training. For more invasive cases, spinals and epidurals (regional blocks), combined with sedation, can be used to perform many surgical procedures. There are significant risks involved. Regional anesthesia frequently requires the immediate availability of appropriately trained personnel and general anesthesia equipment as backup. Epidurals can be used for surgical needs and postoperative pain management.

Conscious sedation is often used in intensive care units. It can be used as a means to restrict patient movement and to aid relief in highly stressful times. It is also used for imaging child patients or irritated individuals. There is a hazy, but fine, line (often misunderstood) between where conscious sedation ends and general anesthesia begins. Anesthesiologists tend to have a clear understanding of the boundaries but others who are not as experienced in the practice might not.

General anesthesia (GA) affects the entire body, rather than any one specific area. GA shuts down the body's reactions to noxious stimuli most often also involving paralysis. The minimum alveolar concentration (also referred to as MAC) is the minimum anesthetic agent to prevent 50% of the population from moving because of noxious stimuli (i.e., incision). By definition, one MAC is not enough to treat patients, as half would react to an incision. A value of 1.3 MAC will prevent most any patient from moving. One MAC varies by intravenous drug and inhalation agent. It is a measuring tool to standardize comparisons. For example, one MAC is accomplished at 1.68% of the volatile liquid anesthetic ethrane and at 1.15% of the anesthetic forane. Because GA affects the entire body, there is total loss of consciousness and the ability to communicate, thus requiring physiological monitoring to best determine the patient's condition. During GA, the patient is often paralyzed to the point where autonomic control of the diaphragm is shut down, thus requiring the use of positive-pressure (mechanical) ventilation and related monitoring.

Critical Care

Many anesthesia departments are deeply rooted and involved in the day-to-day operations of intensive care units (ICUs). Training in anesthesia is a solid foundation for management of critically ill patients (Stoeling and Miller, 1994). Patient care during surgery is, in many ways, similar to that of other critical care areas, but changes happen in terms of minutes or hours, versus days or weeks. Because of their specialty and experience, anesthesiologists must understand intravenous administration of rapidly acting drugs, fluids, and volume control. Critical care calls for in-depth knowledge of mechanical ventilation, airway management, and cardiovascular monitoring, including invasive pressure lines of hemodynamically unstable patients. Anesthesiologists are also responsible for cardiac and pulmonary resuscitation (code response) in the ORs and ICUs.

Pain Management

Anesthesia departments are also responsible for pain management during diagnostic, therapeutic, and obstetric procedures. Pain units that are staffed by anesthesiologists diagnose and treat painful syndromes when patients suffer from chronic issues. Obstetric floors often have anesthesiologists present, to help reduce the pain involved with labor and delivery.

Considering all of this, do not overestimate anesthesiologists. They are not superhuman. No individual will master all of these subspecialties. As a clinical engineer or biomedical equipment technician (BMET) supporting the needs of the department and its staff, it helps to understand the breadth of responsibilities and expectations of the physicians as a whole.

Where Is Anesthesia Performed?

Operating rooms, treatment rooms, and intervention radiology suites are beginning to look more similar than ever. Historical functions of the different areas and physician specialties are blending, making formerly clear lines quite hazy. Patient flow through the hospital's multifaceted care units must be carefully planned. Preparing and recovering a surgical day-care patient (rather than one from general surgery) will put significantly different demands on support systems and staffing. The hospital cannot afford bottlenecks that result with unnecessary delays in surgical, imaging, or other expensive care areas.

For various reasons, doctors and patients alike express interest in providing anesthesia at remote locations (out of the operating room) of the hospital. This poses significant risks and unique challenges in planning, supporting, and actually meeting the demand. These remote sites most often were not designed with anesthetizing-location-facility requirements in mind. Details of facility criteria are discussed elsewhere in this handbook (see Smith et al., Chapter 89). Reliable supplies of oxygen and suction, along with adequate electrical power and lighting, are key components. The lack of any of these can negate further discussion. If facility requirements are met, but physical space is an issue, the anesthesiologist, machine, and supplies are often crowded into an uncomfortable corner. From a user's perspective, they are away from the space to which they have grown accustomed. People must be extra vigilant when they are out of their normal element, as supplies and other things are not where they are expected to be. More importantly, support personnel are not immediately available to help in emergencies, as in they are in the operating room. One example is the need of imaging pediatric patients. Small children tend not to stay still enough for adequate results, and therefore need sedation. Unless properly planned, pediatric supplies are not commonly found throughout the hospital. Delays in procuring critical care items could lead to disastrous results.

The special needs of magnetic resonance imaging (MRI) and radiation oncology (e.g., proton beam accelerators) present unique challenges to all involved in their use. The most significant are magnets and invisible forces. Both technologies involve strong forces that cannot be seen, and catastrophic events can result if mistakes occur. Specialized equipment is needed, to detect the energy field generated by both types of devices.

MRI requires highly specialized equipment: Nonferrous magnetic materials can be used, CRTs are a problem, radiofrequency (RF) sensitive environment, physiological monitors are less than optimum, the same level of monitoring is simply not available, application of MRI is expanding and it is being used more frequently during surgery. Radiation oncology requires leaving anesthetized patients by themselves, which is not something that people take lightly. Anesthesiologists still need to be able to monitor the patient and machine while they are out of the room

Safety Concerns

The single most significant difference in supporting anesthesia technology, as distinguished from any other clinical engineering function, is that there is a susceptible patient connected to life-support equipment, and they are given medications that bring them relatively close to death. Anesthesiologists are keenly aware of this. As a whole, stress levels of people in operating rooms are exacerbated because the patient has undergone two forms of injury when on an OR table. They have suffered the initial illness or trauma that has brought them to the OR, and they must undergo the trauma of the surgical procedure itself. It can be difficult to believe that the patient often must experience additional trauma in order to get better.

Patient safety is a team effort. No single department bears the responsibility for overall patient safety. Everyone must evaluate their individual responsibilities and must look for ways to prevent and improve their environment, to prevent mishaps from occurring. One example of a relatively simple sounding task, but often complicated by various factors for the OR team, is patient positioning. A number of injuries have occurred because of positioning. Different departments must take into account their own prospective needs. Surgeons must have access to certain areas for sterile prep and the surgical site. Nurses must make sure that the items for which they are responsible are accessible and do not cause potential pressure points, while anesthesiologists are looking out for potential nerve injuries. Because the patient is unconscious, he is unable to tell them that the position is uncomfortable, and staying that way for six hours can hurt.

Anesthesia safety is frequently compared to flying an airplane. On airplanes, most of the problems (and, therefore, most hazards) are associated with take-off and landing. In anesthesia, this is equivalent to induction and emergence. Both have multiple systems working in unison to maintain function. On a commercial plane, fuel is delivered precisely to engines that are attached via structural members to a fuselage, with a crew using radiofrequency communication and a global positioning system to move the plane from one location to another. During anesthesia, oxygen is mixed carefully with an inhaled anesthetic agent and delivered through a breathing circuit driven by a ventilator, controlled by a physician who watches a display to monitor electrical and physical changes in a patient who cannot communicate. As long as things are going well, everything appears to be relatively simple.

Unfortunately, things do not always go as planned. A number of factors can contribute to undesired events during anesthesia, including noise, fatigue, and boredom (Weinger and Englund, 1990). Studies have shown that anesthesiologists can be idle 40%–70% of the time during a surgical procedure (Drui et al., 1973; Boquet et al., 1980), which could further affect their vigilance at their primary responsibility of monitoring the patient, procedure, and equipment. When something out of the ordinary occurs and takes anesthesiologists off their planned course, it can lead to moments of high activity, where many things must be accomplished in little time, potentially in a state of concern. These periods are often associated with high task density, where people average less time on individual tasks in contrast with the time spent during less busy periods (Herndon et al., 1991). In both aviation and anesthesia, there are lengthy, intensive training periods and highly educated and skilled personnel. Each field relies on a person who has a widely varying workload, to maintain order. It does not take much time for one or two additional missed warning signs to result with significant demands on the individual at the controls. For a pilot or an anesthesiologist, an uneventful day is a good day. No one likes unpleasant surprises on an airplane or in anesthesia.

Redundant systems help to prevent surprises. Studies in the aviation industry show that adverse outcomes frequently happen when a number of undetected smaller events occur, involving different factors (e.g., human error, equipment failure, and supply mishaps) cumulating in an undesired result (Billings and Reynard, 1984). There always should be additional resources for people to turn to for help. This can be difficult for some, because they may feel that asking for help shows a weakness and that they should be able to figure it out on their own, lest they be seen as less than fully capable. Patient care environments must try to foster a setting in which looking for help when an individual is not fully comfortable with a condition is a perfectly respectable option and not cause for punitive action. Solutions can be relatively simple but not visible when an individual is fixated on something else. Another viewpoint might be all that is needed. Physicians frequently have enough to worry about and might not always see the solution; this state of affairs can be more prevalent at teaching institutions. Clinical engineers can back physicians up by understanding the demands placed on them and on the clinical environment.

To minimize the possibility of equipment failures during use, the aviation industry implemented a preflight inspection. Similarly, those who are concerned with anesthesia safety developed the United States Food and Drug Administration (FDA) Anesthesia Apparatus Check-out Recommendations (FDA, 1993). A team of people of varied backgrounds and interests created the FDA procedure. When followed correctly, this comprehensive procedure can identify most any problem. Although it is a simple procedure, it does require the user to know proper technique. It can be simple to make a mistake (e.g., negative pressure leak test with the machine turned on) that results in false positives or negatives. Most anyone with a reasonable understanding of an anesthesia machine can perform the checkout. A technician or engineer with more in-depth knowledge of the components that are actually being tested can use it as a useful troubleshooting tool to easily identify system failures within the machine.

The aviation industry has used simulators for years. Their use is expanding medicine to put physicians in stressful situations without putting patients at risk. People need to be taught crisis management. One needs to conceptualize and understand how to work in a team when encountering stressful and challenging situations. The human mind does not function rationally when in a panic.

Safety at off-site locations is worth noting. A number of undesired events have prompted review of office-based anesthesia. The associated risk versus cost is an example of the many challenges faced in health care, particularly at large teaching hospitals. In an effort to stay price-competitive with the entire medical industry, procedures are being performed out of the operating room and in doctors' offices. No one should ignore the fact that infrequent events with potentially catastrophic outcomes do happen. Key components can be as simple as proper and adequate supplies, additional personnel, and available telephones. Other components depend on the procedure and level of anesthesia performed. Even though some procedures appear to be relatively simple, the requirements for anesthesia can call for a machine to be present at all times. The emergent need for mechanical ventilation and use of volatile agents is quite possible. These machines might be underutilized, but they are subject to greater wear because they are moved more frequently than machines based in the operating room.

There are supply concerns, as well. Additional stock is not as readily available when users are not in the main supply area, which normally is the operating room. If items stored differently, individuals are out of their normal surrounding, and this contributes to disorientation and difficulty locating the items. Delays in supplies can have catastrophic results, as well. If the correct endotracheal tube is not readily available, minutes might as well be hours when there is a problem with an airway.

Infrastructure

Smith et al. described facilities' infrastructural needs (see Chapter 89). Much of an OR's evolution has involved the application of anesthesia and some of its historical limitations. Conductive flooring, other electrostatic discharge protection, and explosion-proof electrical connections were directly and indirectly employed because of former anesthetic technology. Currently, there is ongoing debate about the continued use of the isolated power supply (IPS) in new construction. The use of ground fault interrupts (GFI) meets code requirements for wet locations at a lower cost than isolated power. Some facilities elect to continue with the more expensive option because it does not shut off power to an outlet at the first fault. If their application is poorly understood, GFIs could be hazardous when life-support equipment such as anesthesia or cardiopulmonary bypass machines is used. Shutting of life support equipment when one plugs a faulty device into an outlet on the same circuit is not a wise alternative.

Following are a number of other systems that are needed, to support specific needs for anesthesia:

- Immediate access to the hospital's pharmacy and a means to control narcotics with potentially high throughput for busy centers
- A well-organized means to purchase, receive, and stock single-use supplies
- Facilities and associated documentation to reprocess, decontaminate, and sterilize multiple-use devices
- Centralized storage of equipment that is not used every day but must be available
- An area for preadmissions testing
- Space for an interview and simple testing

Patients and physicians need to interact preoperatively to assess and plan the anesthesia. This may be done in general-care areas if the patient has been admitted in the hospital.

Induction (or holding) rooms offer a number advantages and disadvantages. They require more patient movement and additional equipment but can reduce turnover time and can increase throughput. Primarily, they provide a location to help set up lines and epidurals without occupying time in the operating room itself. They are no longer true induction rooms, as anesthesia is not induced in them, but they are helpful prep areas.

In the ORs, anesthesia machines must connect to existing infrastructure for gas supplies. It is useful to have standardized connections so that machines can be used in any location with adequate infrastructure. Individual surgical anesthetizing teams might have varying needs and preferences for machine and monitoring layout, depending on the surgical service that they support. There also are specialized sites like radiation oncology and MRI that will require the machine to be tailored to meet their requirements.

Post-anesthesia care units (PACUs) have established clinical personnel requirements for emerging patients (ASA, 1994). Patients can be unstable; so immediate critical care, airway management, and possibly anesthesia equipment must be available. Their workload varies. At one moment, the space could be used primarily for pre-op holding; at other times, things can be relatively calm. Then, suddenly, three or four cases finish at once. The PACU is an ICU and must have similar monitoring requirements with centralized alarms. It must accommodate varying demands for stretchers used in patient transport.

Airway Management Tools

One of the most critical tasks in caring for the ventilated patient is in securing an airway. If there is not a reliable means of moving gases in and out of the lungs, cardiac resuscitation follows respiratory arrest. Among the many tools available to anesthesiologists, none is more important for airway management than suction. Reliable and adequate sources of oxygen and suction are vital. Anesthesia cannot be performed safely without both. Yankaur catheters are the most typical device used to aspirate airways. Nasal-gastro tubes are used to empty stomach contents that otherwise could interfere with the airway.

Probably the device with which most people are familiar is a facemask. Masks are relatively simple devices that come in numerous sizes, both disposable and reusable. They can leak if not fitted correctly, and they require a hand or strap to hold them in place. Their greatest drawback is that they do not prevent possible aspiration of stomach contents. They are not the best option for longer cases, and excessive pressure can cause physical injury to the patient.

To use most any other airway management tool requires direct visualization with a laryngoscope. There are standard airway classifications, depending on patient anatomy. Laryngoscopes are available in various configurations to best meet the needs of different airway anatomy. Common blades that are used to obtain direct visualization of the vocal cords are straight, straight with curved tip, or curved (Jackson-Wisconsin, Miller, and MacIntosh, respectively). When patient anatomy or trauma is such that use of a laryngoscope is difficult or impossible, a fiber optic scope is used to help intubate the patient. The two services that most often require these tools are thoracic and plastic surgery. Thoracic teams use them to visualize airways more easily, evaluate tube placement, and aspirate secretions. Reconstructive plastic surgery requires the use of fiber optic scopes, as significant percentage of these patients have disfigurement or trauma that has altered normal anatomy. Video equipment can help teach their proper use by enabling two people to visualize the same image at once.

Endotracheal (ET) tubes are the most common item used to maintain an airway. They are available in numerous sizes and in cuffed (Figure 90-1) and uncuffed configurations, although cuffed tubes are more common. They have a balloon-like outer section at the distal tip that inflates to seal with the inner walls of the trachea to prevent leaks and inhalation of gastric contents or other secretions. They are nearly always used on adults, and they pose other potential problems if the patient is intubated for periods over 48 hours. Uncuffed tubes do not put pressure on the inside of the trachea that can be more problematic with pediatric patients but can contribute to airway leaks. The most frequent problem associated with intubation is a sore throat from the pressure exerted on the inner tracheal mucosa. There are specialized ET tubes with two lumens used most frequently during lung surgery, enabling ventilation of one lung or both. Because they are in the immediate surgical vicinity, these tubes are subject to greater external forces and, therefore, are often reinforced. Another option is the laryngeal mask airway (LMA). Because of its seal design, its use is limited to ventilation pressures of about 20 cm H_2O, and it does not prevent aspiration of gastric contents. The LMA is most efficient in environments where surgical procedures are generally short, and it is a helpful tool for emergent needs.

Services that pose unique challenges are pediatrics and oral surgery. Children are smaller, potentially making tasks more challenging. In oral surgery, scavenging can be challenging because surgery takes place in the immediate area where gases are flowing.

Anesthesia Machines

Anesthesia machines (see Figure 90-2) are constructed of a number of systems assembled as one device. There are standards developed by the American Society for Testing and Materials (ASTM) (ASTM, 1989) for many of the subassemblies used on or with the machines. Its major systems can be broken down to gas delivery (frequently referred to as the "machine," itself), vaporizer(s), breathing circuit, ventilator (including related monitoring), physiological and CO_2 and agent monitors. One standard does not cover all aspects of the machine; for example, there are standards for machine, ventilator, oxygen monitor, and breathing circuit. Unfortunately, they can be vague and interpreted in different ways, making them somewhat difficult to read and understand.

Gas Supplies

The machine's primary function is to reduce supply-line pressures, mix a number of gases (most typically oxygen, nitrous oxide, and air), and deliver a controlled output to the breathing circuit. Primary gas supplies feed the machine 50 psi. A pressure-relief valve opens above 75 psi in case of infrastructure system failure. Technicians and engineers need to be familiar with a number of pressure-measurement units. The most common are pounds per square inch, millimeter of mercury, and centimeters of water (psi, mmHg, and cmH_2O, respectively). A rough equivalent is that one psi is about 50 mmHg and about 70 cmH_2O. In the same units of measure, an anesthesia machine needs to safely reduce and control gases fed at 3500 cmH_2O and to supply them to the breathing circuit normally operating at about 35 cmH_2O. In other words, the supply pressures are 100 times that of the breathing circuit. Machines are constructed with a high-pressure side and a low-pressure side. The high-pressure side is primarily the supply, and the low-pressure side is any part operating near breathing-circuit pressure.

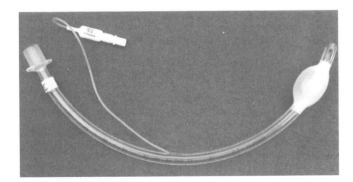

Figure 90-1 Cuffed endotracheal tube.

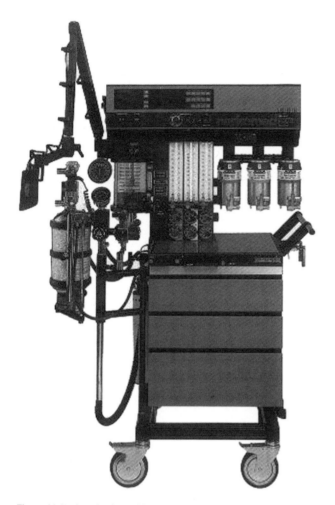

Figure 90-2 Anesthesia machine.

There are a number of items with designed incompatibility in anesthesia machines. Pipeline gas-supply connections are available in a few different configurations. The diameter indexing safety system (DISS) is a common configuration. The thread and inner diameters vary to prevent accidental connections to a wrong supply. Virtually all machines use E cylinders as backup supplies. Backup oxygen is vital for patient safety and must always be available. Cylinders use the pin index safety system (PISS) to prevent wrong connections. Holes in cylinders work in conjunction with pins in the yokes. For anesthesia-supply gases (oxygen, nitrous oxide, and air), one pin location is fixed, and the other varies depending on gas type. One inherent weakness in this system is that if the wrong pin pulls out or breaks, it could lead to a misconnection.

Oxygen Pressure Detecting System and Distribution

To detect the presence of oxygen pressure, there is a pressure-controlled, normally off, valve called the "fail safe," which is intended to protect the patient from a hypoxic mixture. Its function is to shut off secondary and tertiary gases in case of oxygen loss. The name is a misnomer because the device does fail. It is ineffective when other gases are delivered in the oxygen pipeline or when flow-control valves or hypoxic mixture interlock systems are out of adjustment. Machines also have a pressure-sensing alarm that indicates when supply pressure has fallen below a threshold to inform the user to take evasive action (e.g., to turn on cylinder supply).

Oxygen-pipeline supply is plumbed to at least five locations at full pressure within the machine. The oxygen-flush valve is always active and capable of delivering 50–65 lpm. It incorporates a safety-pressure relief at about 120 mmHg for catastrophic failures that far exceed normal breathing-circuit pressures. The flush valve is dangerous to the patient in the hands of an uninformed user. Activating the flush for one second will increase the tidal volume delivered by one liter, which is easily enough to cause barotrauma to a susceptible patient. The fail-safe and pressure-sensing alarms, as described above, require line pressure to operate correctly. Flowmeters combined with the flow-control valves feed the oxygen supply to the patient via vaporizer(s) and breathing circuit. They are the interfaces between the high- and low-pressure sides of the machine. Ventilators need a working gas. The United States has standardized on oxygen, but other parts of the world could use air. Auxiliary flow meter and other power outlets are fed from accessory connections.

The anesthesia machine has two limits for the minimum oxygen-flow rate. One is fixed and sets at the absolute minimum deliverable, while the other varies in proportion to the nitrous oxide flow-rate setting. Most machines are designed for a minimum of 200 ml, but some go as low as 50 ml. This is intended to reduce the possibility of a hypoxic mixture because there is always the minimum patient-uptake requirement supplied. An average adult will metabolize about 2-300 ml/min of oxygen.

Nitrous oxide flow control is linked directly or indirectly to the flow of oxygen, also to prevent delivery of a hypoxic mixture. This introduces significant associated construction costs and complexity. As technology and practice evolve, people are questioning the long-term use of nitrous oxide and wonder whether it still should be made available in all machines. To compensate for potential supply pressure changes that could alter flow settings from one machine to the next connected to a common supply line, some manufactures use two-stage pressure reduction to minimize the effects.

Gas Mixing

Traditionally constructed (mechanical) machines use a needle valve combined with a rotameter for each supply gas to control flow. The rotameters all feed a common output, thereby regulating gas composition fed to the vaporizers. After mixing with the inhalation agent delivered by the vaporizer, the combined mixture is called the "fresh gas flow" (FGF), delivered at the common gas outlet. Some electronic machines have settings for total gas flow and oxygen percentage that control gas-flow rates independently.

There is a check valve built into the common gas outlet on Datex-Ohmeda machines to prevent breathing-circuit pressures from back-pressurizing the vaporizers. This valve requires a negative pressure, to open and correctly perform a leak test on the low-pressure system. If a positive-pressure test is performed, it will not open this valve, and the machine will not be tested properly. Drager machines do not have this check valve.

Machine and Space Layout

Much of the machine consists simply of shelf space, essentially a convenient location to store monitoring equipment and supplies needed when performing anesthesia. A working surface and tabletop space is an important location to place medications and other items needed within hands' reach of the physician. There is also a chest of drawers to store supplies and other items needed for use with the machine.

People adjust the layout of most anything they use regularly. They become accustomed to what they have, and grow to like where things are positioned. If the things with which one interfaces change radically, it affects one's comfort level. One is more likely to make mistakes by turning the wrong knob, leading into additional confusion because of rituals and habits of looking for things in one location. Consistency, particularly at large institutions, has its advantages and disadvantages.

The physical space layout and the way the machine is positioned in the room can be significant. If the user is forced, due to an awkward machine orientation, to use their right hand for holding the mask, and their left for manual ventilation, it is very difficult to adapt. To help understand this, one need only try brushing one's teeth with the nondominant hand. It can be done, but would one want a person's life in one's hands when doing it the first few times? The patient's orientation affects access to the endotracheal tube and ease of inspecting lines, leads, and electrodes. The patient could be prepped and draped so that it is very difficult to reach most anything.

Vaporizers

Vaporizers (Figure 90-3) add and control the concentration of volatile inhalation anesthetic agent in the gas delivered to the machine's common gas outlet. In order to understand the way a vaporizer functions, the reader first must be familiar with a few terms and principles of fluid dynamics. A fluid is anything (liquid, gas, or both) that takes the shape of its container. Materials have three phases: Gas, liquid, and solid. Vapor is the gaseous phase of a liquid at room temperature and atmospheric pressure. Pressure, volume, and temperature are related. A gas is fully saturated when it contains the maximum amount of vapor possible without precipitating out to a liquid. If two or more gases are mixed in one container, the total pressure exerted is made up of the sum of partial pressures created by each individual gas.

Partial Pressure

An understanding of partial pressure is required, to support anesthesia technology effectively. Partial pressures are an absolute measurement defining the total number of molecules where percentages are relative to total gas mixture. Partial pressure is used and applied in a number of technologies within anesthesia (e.g., vaporization, ventilation, and respiratory gas monitoring). One example to help clarify the term is dry (versus moist) air. Air is made of 21% oxygen. The remainder is mostly nitrogen and trace gases (negligible partial pressure for this example). Atmospheric pressure is approximately 760 mmHg. On a dry, warm day (98°F or 37°C) of 0% relative humidity, oxygen partial pressure is 21% of 760 (0.21 × 760), or 159.6 mmHg. The remaining pressure must be from nitrogen (760-159.6), or 600.4 mmHg. Then, on the following day, the temperature and atmospheric pressure are the same as the day before, but the relative humidity is 100% (the air is fully saturated with water vapor and cannot contain any more without rain-out). At 37°C, fully saturated water-vapor pressure is 47 mmHg. On the second day, atmospheric pressure contains three gases (nitrogen, oxygen, and water vapor). Oxygen concentration remains constant at 21%, but its partial pressure is a function of the three gases [0.21 × (760-47)] or 149.7 mmHg. The partial pressure of nitrogen is 0.79 × (760-47) mmHg, or 563.3 mmHg.

Because volatile inhalation agents have varying partial pressures that affect concentration output, modern vaporizers are made agent-specific. The most common agents currently in use are forane, ultane and suprane (manufacturer trade names are isoflurane, sevoflurane and desflurane, respectively). Vaporizers are available as either funnel- or key-filled. Funnel-filled vaporizers can be more convenient and reduce the possibility of

Figure 90-3 Vaporizer.

environment in which is it stored when in use. Mainstream vaporizers control output by injecting measured amounts of agent directly into the main stream of carrier gas.

Future machines might incorporate new designs that are more closely related to the mainstream vaporizers for other agents. Vaporizers are also used on heart-lung machines to control the anesthetic agent delivered through the oxygenator during cardiopulmonary bypass surgery.

Breathing Circuits

Breathing circuits serve as the interface between patient and machine. Because people breathe in volumetric flow and pressure cycles and anesthesia machines deliver a unidirectional stream of gas (at a specific oxygen concentration and controlled anesthetic agent), an interface between a person and the anesthesia machine is required. Breathing circuits convert the machine's steady gas output to a flow and pressure cycle that is consonant with the human breathing cycle.

In keeping with the Ideal Gas Law, $PV = nRT$, to maintain a constant baseline pressure, one must also maintain a constant baseline volume. If there is excessive (or a growing) volume in a closed system, pressure will pressure. Respiratory baseline pressure is measured at end-tidal expiration when there is no longer significant flow. This measurement is known as "positive end-tidal expiratory pressure" (PEEP). A second ventilation indicator is "peak inspiratory pressure" (PIP), the maximum pressure attained during the inspiratory cycle.

When a patient is paralyzed to the point where they no longer spontaneously ventilate, positive-pressure ventilation is used. This involves delivering pressurized tidal volumes of gas to the patient's lungs for oxygen uptake supporting their required metabolism. Under normal circumstances (no leaks), an anesthesia machine's gas output has two places to go: Patient uptake or its scavenging system.

A primary component of breathing circuits is the tubing and its configuration to the patient. Some breathing circuits rely on high fresh gas flow rates to prevent rebreathing of respiratory gases while others involve absorbent to neutralize expired carbon dioxide. A thorough discussion of all the breathing circuit (open, semi-open, semi-closed, and closed) options is too great for this text. The two most common circuits used in the Unites States, the Bain Circuit and Circle System, are described below.

Bain Circuit

A Bain Circuit is a semi-open breathing circuit that does not recirculate respiratory gases and relies on high fresh gas flow rates to prevent rebreathing. It is most often used for neonatal and thoracic applications. Its greatest advantages are that it creates little dead space, it is lightweight, and gases can be scavenged easily. The most significant drawbacks to this circuit are that it requires high fresh gas flow (delivering cold dry gases to the patient) and can be cost-prohibitive with costly volatile agents.

Circle System

By far, the most common is the circle system, where respiratory gases move through a housing incorporating unidirectional valves to recirculate gases, minimizing waste and maximizing warmth and humidity in the circuit. The system's greatest drawback is its level of complexity. However, nearly all anesthesia machines are designed to operate with a circle system.

Carbon Dioxide Absorbers

Recirculating breathing circuit gases require the patient's expired carbon dioxide to be removed, to prevent hypercarbia. This is accomplished by means of a CO_2 absorbent contained in a housing, called an "absorber." The breathing circuit absorber must allow the use of both automated (machine ventilator) and manual (breathing circuit bag) ventilation. Most present-day absorbers are stand-alone subassemblies that tie into related functions of the machine. They are modular and can be exchanged easily if needed. They are the union between patient and machine and are where respiratory gases mix with fresh gas flow from the machine. Delivered oxygen concentration is measured in the absorber's inspiratory limb that is most proximal to the patient and in a relatively safe location where it is unlikely to be damaged or disconnected. One advantage of modern absorbers is that they measure PIP after the inspiratory check valve. Older style absorbers require the use of a second gauge when a PEEP valve is used for clinical reasons (discussed below). Gages on older designs only display interior absorber pressure. When the inspiratory check valve closes, they do not display true airway pressure and do not reflect PEEP.

In an effort to address some of the limitations of the relatively simplistic time-cycled, volume-controlled ventilators that are typically found in anesthesia, manufacturers are introducing a new generation of absorbers that are more fully integrated into the machine. This design approach has advantages and disadvantages. New designs can be sterilized. While not required in the United States, parts of the European Community require that breathing circuit components be sterilized between patients.

One significant point to consider is that absorbers require daily service and therefore have greater potential for problems such as leaks. Although they are designed with this in mind, personnel who are not technically trained might partially disassemble the absorber. The system is subject to repeated changes of disposable breathing circuit tubing, spills, daily cleaning, and absorbent changes, and is otherwise exposed to a demanding environment.

Absorbent

Various commercially available absorbents are used to absorb carbon dioxide. For practical purposes, the CE and BMET must be aware and concerned with the absorbent's systematic application and its use. By-products of the chemical reactions between the

vapor lock when filling, but they are susceptible to being filled with the wrong agent. Key-filled vaporizers are more cumbersome and require use of a filler, but they virtually eliminate the possibility of cross-contamination because the agent is not poured directly into the vaporizer as in the case of funnel-filled units.

Operating Principles

Side-Stream

All of the mixed gas flows through the inside of a side-stream vaporizer. The main stream of this gas (when flowing through a vaporizer it is also known as the "carrier gas") has a fraction diverted to the wick/sump assembly. The amount of flow of this side-stream diverted to the wick assembly is controlled by the setting on the vaporizer output dial. The higher the output setting, the greater the amount diverted. This side-stream flow becomes fully saturated with anesthetic vapor in the wick assembly and then returns to the main-stream controlling agent concentration output in the fresh gas flow.

Two physical principles cool this style of vaporizer, requiring temperature compensation. Forced convection (flow of gas) and latent heat of vaporization (energy required to vaporize a liquid) cool the vaporizer while in use. Cooling an anesthetic agent reduces its partial pressure, which in turn lowers the vaporizer's output. To compensate for this cooling, vaporizers incorporate a bimetallic temperature-sensitive diverter that adjusts the amount of flow fed to the side-stream from the main carrier gas. As temperature drops, the side-stream of gas increases to maintain a constant output over varying temperature during normal operation. Side-stream vaporizers, which are relatively simple in design, are the most commonly used and are available for all of the currently used agents, with the exception of Suprane.

Mainstream Injection

Suprane is a volatile agent with high partial pressure that makes it difficult to deliver with a traditional side-stream vaporizer. It can be delivered only using a mainstream injection vaporizer. Suprane vaporizers are heated and pressurized to control more precisely the absorbent and CO_2 are beneficial to the patient and to the proper function of the absorber. However, one reaction might prove to be a hindrance in excess. Removal of carbon dioxide is clearly beneficial and a primary function of the absorbent. In the process, a chemical reaction produces a color change in the absorbent, which acts as a visual indicator of the absorbent's activity. The color of absorbent changes from off-white to violet as its ability to absorb carbon dioxide diminishes. Under certain, but rare, conditions, the color change reverts to the original. A more common problem is that users overlook, or forget to notice, the color change. In addition, optical properties of the housing can change, thus making the color change difficult to see. Channeling occurs when respiratory gases take the path of least resistance and channel through a relatively narrow cross-section. This can result in unwanted inspired carbon dioxide.

Heat and water are by-products of CO_2 absorption. The heat generated is not excessive and is beneficial to the patient because it helps to warm cool supply gases, which could irritate the lung lining. Water produced is also beneficial because supply gases are dry, and added moisture reduces lung irritation. If water production is neglected, it will build to the point where it could contribute to problems in the breathing circuit. The most frequently noted problems are sticking expiratory check valves from surface tension between valve disc and housing, and water accumulation in hoses and tubing to the ventilator, affecting proper function by inhibiting proper pressure sensing and gas flow.

Gas Scavengers

When the machine is set for an FGF rate of anything higher than the patient's uptake (normally 200–300 ml of oxygen per minute for most adults), any excess must be removed to avoid a build-up of volume as discussed above. Although PEEP is used for certain clinical indications, its uncontrolled application is disastrous for the patient. The scavenging system plays a vital role in conjunction with the breathing circuit to prevent this from happening. During automated ventilation, the bellows pop-off valve opens at the end of tidal expiration to divert excess volume to the scavenging system. During manual ventilation, the user must open and close (as necessary) an adjustable pressure-limiting (APL) valve to set the upper limit for breathing-circuit pressures. If the APL valve is open, or any pressure occurs in excess of its setting, gas flows to the scavenger. The physical construction of scavengers used on the machines varies and can require user interaction. The various designs offer their own risks and benefits; some put greater demands on the hospital infrastructure (vacuum pumps), while others require the user to make adjustments according to FGF rate.

PEEP Valves

During spontaneous ventilation, the body maintains a residual volume in its lungs to prevent them from collapse. During positive-pressure ventilation, the ventilator maintains a slight PEEP (2–3 cm H_2O) to accomplish the same. Clinical indications (e.g., adult respiratory distress syndrome [ARDS]), PEEP valves are used to increase the resistance that a patient encounters during exhalation, increasing the residual volume in their lungs. Different models of PEEP valves can be permanently installed either on the machine/absorber or temporarily on the expiratory limb. For the latter, the design is a variation of a check valve, making it dangerous if installed backwards or in the inspiratory limb, as it would result in little gas flow and inadequate oxygenation (Figure 90-4).

Humidification

As discussed previously, supply gases are dry and can irritate the lung's lining, particularly with susceptible patients who are on the machine for extended periods. Although the use of active humidifiers has declined over the years, they may still be used for certain patients. Two techniques are used to maintain moisture in the gas that the patient breathes: The heated active humidifier and the heat moisture exchanger (HME).

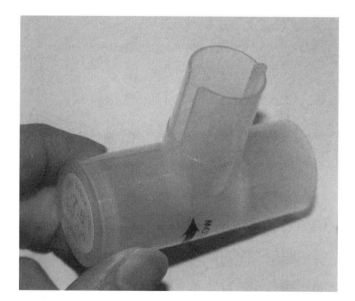

Figure 90-4 Example of a PEEP valve.

Active heated units are generally more effective in adding both humidity and heat to the breathing circuit. The operator must connect the unit to an electrical power source, must monitor its temperature, and must clean and fill them. Many units incorporate check valves to prevent inadvertent delivery of fluid to the patient if they are installed incorrectly. Backward installation can result in significant flow restriction and inadequate oxygenation. These units are installed in the inspiratory limb and have a predetermined "in" and "out."

HMEs are single-use, disposable units. They are connected to the distal end of the breathing-circuit tubing that is proximal to the endotracheal tube. Their primary function is to maintain as much heat and humidity as possible in the patient while preventing heat and humidity from entering the absorber. HMEs are occasionally known as "artificial noses."

Ventilators

Ventilators (Figure 90-5) free the users' hands so they are able to perform other tasks. It is a way to apply cyclical pressure to the equivalent of the breathing circuit bag at a controlled rate and frequency. An expiratory valve inside the ventilator closes during the inspiratory cycle to direct drive gas into the bellows housing, creating a positive pressure forcing breathing circuit gases to flow. At end inspiratory cycle, the expiratory valve opens, releasing drive gases from the bellows housing, and returns the patient to atmospheric pressure. The working gas that is used to drive the bellows varies by country, as discussed previously. Comparable components between manual and automatic ventilation are the bag and bellows and APL and bellows pop-off valve. The bag and bellows offer a means to buffer a volume as gases move in the breathing circuit. The APL and bellows pop-off valves control removal of excess gas from the breathing circuit to the scavenging system, as discussed previously. Both valves require a small amount of PEEP for preferential flow into the bellows (rather than the scavenger) but the bellows pop-off is set a fraction higher, at about 2–3 cm H_2O.

Pressure versus Volume-Controlled

Historically, anesthesia ventilators have been relatively simple, time-cycled, and volume-controlled which might or might not meet the changing health care environment. Patient care can be classified as acute or noncritical. Acute care is a growing population at larger teaching hospitals, as sicker patients are moved out of smaller community hospitals by modern health management organizations. This patient population is challenging to manage and requires equipment that is more sophisticated. Noncritical care machines should be simple to use, with relatively little user-interface complexity.

One example of the limitations of traditionally simplistic anesthesia ventilators is in the function of the bellows pop-off valve. It closes completely during inspiration so that FGF mixes directly with the set tidal volume delivered. Changes in FGF will affect actual volume delivered as a function of inspiratory time; the greater the flow, the larger the tidal volume without making any changes to settings. In response to the growing market of acute care ventilated patients, manufactures have begun to incorporate more complex pressure-support/controlled ventilators into their anesthesia machines.

Airway Pressure Monitoring

Most volume-controlled ventilators are pressure-limited and will not deliver settings if pressure limit is triggered. Some have an adjustable pressure limit, and others are preset. The transducer, which senses patient breathing-circuit pressures, is physically located in the ventilator fed from a tube taped into the breathing circuit. As mentioned above, this should be in the inspiratory limb after the check valve, but its location can vary depending on machine vintage (Figure 90-5).

Oxygen Concentration

The American Society for Testing and Materials (ASTM) standards require an oxygen concentration measured in the inspiratory limb that turns on with the machine and has a battery backup. Most utilize galvanic fuel-cell technology, which involves a chemical reaction between an electrolyte and two poles (anode and cathode), similar to those of a battery. The reaction produces an electrical potential relative to the oxygen concentration that is measured and displayed. It has a relatively long time constant and cannot provide breath-to-breath oxygen concentration.

Paramagnetic technology is incorporated in some CO_2 and agent monitors. It has a fast enough response to overcome the limitation of galvanic fuel cells but is more expensive. Using a switched magnetic field, it exploits oxygen's natural magnetic properties to create a pressure differential between gas streams (sample and reference), which is detected and displayed as a concentration (Ehrenwerth and Eisenkraft, 1993).

Volume Measurements

Given the limitations of relatively simplistic volume-controlled ventilators, volume measurement is the most accurate device currently in use to provide users with feedback of what tidal and minute volume is actually delivered to the patient. Because of compliance of gas, breathing circuit tubing, and additional dead space in the circuit, the measurement is close, but not exact. A vane anemometer measures a volume of gas by means of monitoring the mechanical rotation of a vane or blade. These devices can be either analog or digital. Technology in newer machines includes ultrasound using Doppler acoustical properties between two points to measure gas flow. Hot wire systems measure energy requirements and changes of a heated wire within a housing as gas volume cools the surface. Pitot tubes in opposing directions working in conjunction with differential pressure transducers are also used to measure flow.

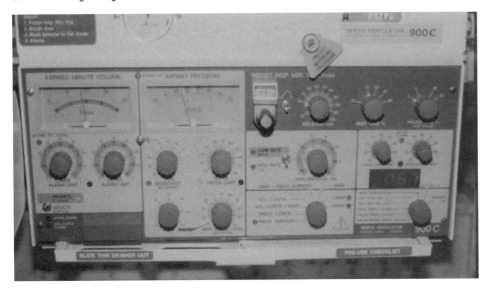

Figure 90-5 Siemens-Elema 900C ventilator.

Associated Alarms

Many of these measurements are, primarily, a means to inform the user of the possibility of a problem, such as disconnect or leak. Oxygen monitors identify potential hypoxic mixtures and catastrophic events and cannot be set below 18% without alarming. Pressure alarms can be used to indicate significant changes in the patient's condition or restriction in the airway. Subatmospheric alarms indicate the possibility of an active vacuum in the airway, or whether the patient is drawing a negative pressure by fighting the ventilator.

Monitoring

The American Society of Anesthesiologists (ASA) (ASA, 1998) has set minimum requirements for monitoring during anesthesia. Depending on the patient's acuity and other factors, such as the surgical procedure and anesthetic plan, the level of monitoring chosen by the anesthesiologist can exceed minimum requirements to include invasive pressures (arterial, pulmonary artery, and central venous), neuromuscular blockade, and consciousness.

The first requirement of the ASA monitoring standard is "qualified anesthesia personnel..." The care provider is ultimately the most important monitor and must be able to concentrate on the patient without considerable distraction. Secondarily, but no less significantly, the standards specify monitoring for oxygenation, ventilation, circulation, and temperature. Continuous physiological monitoring is used during anesthesia, utilizing plethysmographs, electrocardiograms, and oscillometric devices. Figure 90-6 shows a pulse oximeter used for monitoring the patient arterial blood-oxygen saturation, Sa_aO_2.

Studies have shown that although a large percentage of mishaps during anesthesia were due to human error, a significant number were related to equipment failures related to leaks, misconnects, gas-flow-control errors, and circuit disconnects (Cooper et al., 1984). Two of the ventilation monitors (oxygen and volume) were discussed previously. The ASA standard requires that there be a device to detect a breathing-circuit disconnect and that it must have an audible alarm. Along with the use of pulse oximetry, the ASA specifies that there be adequate lighting and exposure to assess patient color and proper oxygenation. This serves as a reminder that one cannot rely solely on monitors. Taking care of a patient is ultimately the responsibility of the provider, not of the monitor.

Capnometry and Agent Analysis

ASA standards originally stated that capnometers would be used to verify the presence of carbon dioxide to ensure tracheal, versus esophageal, intubation. The latest revision specifies continuous CO_2 measurement. There are clinical indications that can be detected using capnometers. Apnea caused by disconnects, ventilator failure, and complete obstruction in the breathing circuit or the scavenging system can be detected, possibly early enough before arterial blood-oxygen desaturation occurs. CO_2 waveforms during controlled ventilation display characteristics that help the physician to identify potential patient-management issues. An abnormally slow rise can indicate a restricted airway or kinked endotracheal tube, while a baseline drift can indicate a faulty expiratory limb valve, consumed CO_2 absorbent, or channeling. BMETs and CEs should learn typical and atypical capnograms and their causes. This knowledge assists in field troubleshooting and in communication, especially when a physician states that the CO_2 waveform looks peculiar.

Monitors can be divided into two categories: Side stream or mainstream measurement devices. Side stream monitors remove a small sample from the breathing-circuit gases and delivers it to a measurement chamber. The effluent can be either scavenged or returned to the expiratory limb to be recirculated. Mainstream monitors measure the patient's expired carbon dioxide concentration through an optical window in a tube connected to the breathing circuit, most often at the end of the endotracheal tube.

There are a number of different technologies employed in side-stream monitors, including mass spectrometry, infrared (IR) analyzers (single and dual beam), and Raman spectroscopy. The advantages of a side stream monitor are the simplicity of items connecting to the breathing circuit and the ability to read multiple gases. The ancillary items do not need to be reprocessed, are disposable (thus minimizing risk for cross-contamination), do not need optical properties or cleanliness, and therefore are easier to use on a daily basis. The components added to the breathing circuit are physically smaller—an added advantage when working with limited access and sterile drapes. Their largest drawback is in working with expired water. The sampling tube can condense water vapor, and if the monitor is unable to clear the droplets on its own, it requires user interaction. Many monitors incorporate a water trap that must be emptied on occasion, and hydrophobic filters that can occlude.

Molecular asymmetry is required for IR light absorption, resulting in vibration or rotation of dipole bonds. Nonpolar molecules, such as oxygen and nitrogen, do not absorb IR energy and cannot be measured with infrared spectroscopy. Absorption correlates with the number of molecules, so it is an absolute, and not a relative, measurement. Carbon dioxide absorbs IR light at about 4.3 μm, nitrous oxide (N_2O) about 4.5 μm, and volatile anesthetic agents range from 9–12 μm. Because CO_2 and N_2O absorb IR closely to one another, there is signal bleeding between the two. As a result, some CO_2 monitors need N_2O compensation when calibrated. In the monitor, the sampled patient gas runs through water-permeable tubes to dry the moist sample prior to delivery to the measurement bench. Water vapor absorbs IR at the same spectrum and therefore is a contaminant for CO_2 measurements. To measure different gases, dual-beam infrared spectrometers use a spinning wheel with band pass filters to tune its spectrum for the particular gas of interest for maximum signal and a reference. The reference is a zero or minimal response point to which to compare the measurement. Absorption is detected and displayed as a waveform and number.

Volatile anesthetic agents can be measured using infrared spectroscopy and have individual signatures at different wavelengths. Some monitors require simple software updates to read newer agents, while others require more extensive replacement of hardware. Unfortunately, there also are models that are cost-prohibitive to update.

End-user calibration of side-stream monitors involves a zero and span setting. Newer-style monitors have automatic zeroing where room air sample is taken internally. Older monitors require the person completing the calibration to remove the sampling line from the breathing circuit, and then to complete a sequence of actions (i.e., turn knobs or activate keys). Span requires the person calibrating to use a known calibration gas that can be specific to the model of monitor. The calibration gas is sprayed into the sampling port, and the upper measurement range is established either by software or physically setting a potentiometer.

Existing mainstream monitors are more common for use outside of anesthesia due to technology constraints as they are limited to measuring only one gas. They are frequently found as options on physiological monitors.

Figure 90-6 Ohmeda Biox model 3740 pulse oximeter.

Temperature

Temperature is the last physiological parameter mentioned in the ASA monitoring standard. Patients undergoing anesthesia frequently experience hypothermia caused by the mechanics of anesthesia and skin exposure to the cold environment of an operating room. Induction of anesthesia suppresses the body's ability to regulate core temperature at the most fundamental state. An individual who is awake will make behavioral changes (e.g., in terms of dress or shivering) when sensing a change in environmental temperature. The hypothalamus responds to temperature changes and induces vasodilation or restriction as necessary to control and redistribute blood volume regulating core and peripheral body temperatures. Most volatile agents impair vasoconstriction, potentially altering blood-volume distribution. Greater blood volume in extremities increases heat loss, thus lowering core body temperature. Muscle relaxants and anesthesia otherwise clearly inhibit shivering and reduce heat production by resting muscles (Morley-Forster, 1986).

Accidental hyperthermia during anesthesia is not common in the United States, as most all operating rooms are air-conditioned. Hyperthermia is more common in tropical climates, in operating rooms without air conditioning. Malignant hyperthermia can be detected by monitoring the patient's temperature, but an increase in expired carbon dioxide is an earlier indicator.

Thermistors are most commonly used to measure the patient's temperature. Their use poses little risk to the patient; they are reliable and relatively simple to use, and they provide an accurate measurement when used correctly. The most frequent measurement location is nasopharyngeal. The most common problems are cooling from respiratory gases from leaking and misplacement. These devices generally consist of probes that connect to a temperature-sensing device, either a stand-alone box or integrated into a cardiovascular monitor. Thermistors are also used in pulmonary artery catheters, which are used for invasive pressure monitoring and provide a core-temperature measurement.

Infrared tympanic membrane measurement can provide an accurate core temperature but is susceptible to user error. An infrared scanning device with disposable protective cap is inserted in the ear canal, where the energy radiated from the tympanic membrane is measured, converted, and displayed as a temperature.

Liquid crystal thermometers are available to measure skin temperature. However, because skin temperature does not correlate well with core temperature, their use is not practical in anesthesia.

Peripheral Nerve Stimulators

Peripheral nerve stimulators are used to provide an indicator of neuromuscular blockade or otherwise general paralysis. During anesthesia, muscle relaxants are most frequently used to relieve natural muscle tension to facilitate intubation and surgery. In an intensive care unit, muscle relaxants can be used to aid the use of mechanical ventilation. The drugs used are characterized as depolarizing or nondepolarizing. They work by blocking neuromuscular transmission, thereby reducing the muscle's ability to contract, and paralyzing the patient. Paralysis can be achieved also at high doses of volatile anesthetic agents. One important distinction is that a patient can be paralyzed but may not necessarily be anesthetized. Failure to render the patient unconscious results in a most undesirable situation in which the patient can experience excruciating pain of surgery yet, because of paralysis, is unable to cry out or otherwise signal to the surgeon.

Peripheral nerve stimulators (also known as "twitch monitors") are relatively simple pulse generators. Most stimulators are battery-powered devices that connect to the patient by means of a cable and electrodes. Most often, the electrodes are similar to those used for ECG monitoring but can be specific for the application. Most monitors provide information to the physician by pure observation. The user must feel and observe the patient's reaction to the electrical stimuli. There are units available that can monitor the patient's reaction electrically and/or mechanically and can display patient characteristics (electronic or paper chart recorders). These units are used less regularly, often for teaching purposes.

Most frequently, the electrodes are placed over the ulnar nerve near the wrist. The location preference is primarily related to muscle innervation. The ulnar nerve solely innervates the adductor pollicis muscle. Stimulating the ulnar nerve minimizes the possibility of cross talk with other muscles, allowing the user to monitor for a response only on the patient's thumb. Less frequently, the electrodes can be placed on a facial nerve if access to extremities is difficult.

Twitch monitors generate a single-phase DC pulse, normally of a fixed duration and adjustable current. Preprogrammed units are available that deliver pulses in sets to help the physician to identify the state of neuromuscular blockade. The most common pulse sets are:

- Single twitch—upon demand, every ten seconds or one per second
- Train of four—a set of four pulses at half-second intervals or can be set to repeat every twelve seconds
- Double burst—a set of three pulses, 20 milliseconds apart, followed by another similar burst 750 milliseconds later
- Tetanus—Repetitive, single pulses at 50 Hz or greater

References

ASA. Standards for Basic Anesthetic Monitoring. Park Ridge, IL, American Society of Anesthesiologists, 1998.

ASA. Standards for Postanesthesia Care. Park Ridge, IL, American Society of Anesthesiologists, 1994.

ASTM. Minimum Performance and Safety Requirements for Components and Systems of Anesthesia Gas Machines, F1161-88. Philadelphia, PA, American Society for Testing and Materials, 1989.

Billings C, Reynard W. Human Factors in Aircraft Incidents: Results of a Seven-Year Study, *Aviat Space Environ Med* 55:960-965, 1984.

Boquet G, Bushman JA, Davenport HT. The Anesthesia Machine: A Study of Function and Design, *Br J Anaesth* 52:61-67, 1980.

Cooper JB, Newbower RS, Kitz RJ. An Analysis of Major Errors and Equipment Failures in Anesthesia Management: Considerations for Prevention and Detection. *Anesthesiology* 60:34-42, 1984.

Drui AB, Behm RJ, Martin WE. Predesign Investigation of the Anesthesia Operational Environment. *Anesth Analg* 52:584-591, 1973.

Ehrenwerth J, Eisenkraft J, eds. *Anesthesia Equipment Principles and Applications*. St. Louis, MO, Mosby, 1993.

FDA. Anesthesia Apparatus Check-Out Recommendations. Rockville, MD, United States Food and Drug Administration, 1993.

Herndon O, Weinger M, Paulus M, et al. Analysis of the Task of Administering Anesthesia: Additional Objective Measures. *Anesthesiology* 75:A487, 1991.

Morley-Forster PK. Unintentional Hypothermia in the Operating Room, *Can Anaesth Soc J* 33:515-527, 1986.

91

Imaging Devices

David Harrington
President, SBT Technology, Inc.
Medway, MA

Darkroom

Medical imaging predates the recording of an electrocardiogram (ECG) by about eight years. X-rays were discovered on November 8, 1895, by Wilheim Roentgen. Seven weeks later, Roentgen conducted additional research and published his findings. He did not apply for patent protection on his discovery. He was awarded the Nobel Prize for Physics in 1901. Many of the top scientists of the era became involved with X-rays, including Maria Skladowska, Pierre Curie, and Thomas Alva Edison. Clarence Daley, who worked with Edison, provided the next important milestone in the development of imaging when he died in 1904 of radiation poisoning. Progress was slow for many years as other applications of X-rays were developed, such as treating acne and tonsillitis and viewing feet in shoes at the local department store to ensure a good fit. After World War II, mobile vans were equipped with X-ray units for screening the population for tuberculosis.

In the late 1940s, sonar, which used to detect underwater objects, was adapted in Japan for use on patients to view internal organs and structures. This technology arrived in the United States during the early 1950s, mostly as a research method, and did not enjoy wide clinical use until the late 1960s. The 1980s brought color to ultrasonic imaging, and in the 1990s three-dimensional images were developed.

In the 1970s, computerized tomography (CT) was introduced, but its wide spread use was hindered by high costs and low reliability. As electronics advanced, so did CT. It is now rare that a hospital does not have at least one CT scanner. In addition, size has decreased to the point where portable CTs are now manufactured. Now, the scanner can be brought to the patient, instead of bringing the patient to the scanner.

The 1980s brought the introduction of a technology known as "nuclear magnetic resonance (NMR)." For obvious marketing reasons, the name quickly changed to magnetic resonance imaging (MRI). The nuclear reactor accidents at Three Mile Island and Chernobyl made anything containing the word "nuclear" a concern to the general population. With patient-movement problems and long scan times in the magnet field reduced or eliminated, MRI is now one of the primary diagnostic tools in health care.

The 1990s brought positron-emission tomography (PET) and single-proton emission computed tomography (SPECT) imaging. These are just starting to become available in more than a few hospitals. As with CT and MRI before that, costs, reliability, and reimbursements have limited the widespread use of PET and SPECT technology.

Now, several imaging technologies are combining to diagnose and treat patients.

X-Ray Generation

X-rays are electromagnetic energy at short wavelengths, generally 0.1 to 1 angstrom. (An angstrom is 10^{-10} meter.) Because the wavelengths are so short, they behave like particles rather than waves, so the names "photon" or "quantum" are used to indicate small quantities of energy. As the frequency of the wave doubles, so does the energy. This means that for the same voltage and current settings, by increasing the frequency (reducing the wavelength) the energy (radiation) doubles.

Photon (quantum) energy is measured in electron volts (eV). An electron volt is the amount of energy that an electron gains when it is accelerated over a voltage difference of 1 volt. Clinically useful X-rays are in the range of 45,000 eV. As electrons from a source (filament) are accelerated and strike the target (anode), heat is generated, and two types of X-rays are produced: General and characteristic radiation.

General radiation, also known as "breaking radiation," or *"bremsstrahlung,"* is the X-ray that is generated when the accelerated electron passes near the nucleus of an atom in the target where it is deflected and decelerated. When this occurs, a small amount of the total energy (1%) is emitted as X-ray, and 99% is in the form of heat.

When the accelerated electron, passing near the nucleus of an atom in the target, collides with and ejects an electron from one of the inner rings of the atom, characteristic radiation is generated. An electron from one of the outer rings will move to replace the ejected electron, and as it moves, it emits an X-ray.

The "intensity" of the X-ray beam is the number of photons in the beam, multiplied by the energy of each photon (both general and characteristic), measured in units of electron volts or Roentgens per minute (R/min).

The X-Ray Machine

X-Ray Tube and Housing

CEs and BMETs working in the field should never replace an X-ray tube, as this requires considerable skill and specialized equipment. What CEs and BMETs do replace in the field is the tube housing, which contains a new tube that has been aligned in an oil-filled housing. The first step in understanding X-ray technology is to know what comprises a tube and what some of the problems associated with the tube are.

All tubes contain three basic parts:

1. An enclosure, generally made from borosilicate glass, but some high-power tubes may have metallic or ceramic enclosures.
2. A filament made from a tungsten alloy.
3. An anode, or "target," that is made with tungsten or other metals and backed with copper.

Most tubes will have several filaments, one for each focal track on the anode; a "focusing cup," or cathode, to better direct the electrons moving from the filament to the target; and a target, the anode, that has one or more focal tracks machined in and that rotates at up to 20,000 rpm. The various parts are assembled in the enclosure, which is sealed under vacuum. The material that is used to seal the enclosure must have the same expansion characteristics as the enclosure so that the vacuum is not broken when the tube heats up. If the enclosure is not adequately evacuated, internal arcing can occur resulting in inconsistent output (gassy tube). Sometimes this can be corrected by a process called "seasoning," which is performed when the tube is installed or reactivated after not being used for a long time.

Filament voltages range from 2.5 to 15 volts, and each filament in the tube could have a different voltage. The filament current range is generally between 3 and 6 amperes. Tube current is measured in milliamperes (mA). When the unit is energized and in a keep-warm state, the filaments are constantly on. Because the filaments are kept warm, they release some electrons that are kept confined by the focus cup (cathode). The focus cup also aims the released electrons at the target when the system is generating X-rays.

The anode (target) of a tube is a disc about 10 centimeters (cm) in diameter and 2 cm (or less) in thickness. The disc surface is a tungsten alloy over a copper base. Copper is used for its heat conduction properties. Electrons, accelerated from the filament, hit one or more of the angled surfaces on the disc, called "focal surfaces," or "tracks," to generate X-rays. In all but the simplest tubes, mostly those in dental X-ray units, the anode rotates. The rotating anode allows the heat (recall that only 1% of the energy is in X-ray and 99% is heat), to be distributed over the entire target. This also keeps the focal surfaces from being distorted from the heat, which keeps the focal spot consistent and increases the life of the tube. The bearings on the anode are sealed and self-lubricating and cannot be serviced in the field.

The tube is placed into a tube housing, where connections are made between the filaments and anode and the exterior connection on the housing. In most cases, the housing is then filled with oil, which acts as both an electrical isolator and a thermal conductor to conduct the heat away from the tube. The tube housing is cooled by way of conduction in low-use application, by convection with a fan in higher-use settings, and with cooling systems in the highest-use systems. Tubes are rated in heat units (HU) and will shut down if the rating is exceeded. Tube housings with fans must be cleaned on a regular basis, to ensure good cooling. Fluid levels in the tube housings requiring cooling systems also should be checked on a regular basis. On one side of the tube housing will be a "window" where the X-ray beam exits the tube housing. This can be plastic or glass and might have tape or some protective material over it for shipping. It must be removed before mounting the collimator to the housing. Several recent studies have indicated that the oil loses its ability to conduct heat after prolonged use and may contribute to shortened life of the tube. An indication of the problem is in tube "sputtering" (non-linear output or radiation), below the average HU units at which sputtering occurred in the past. HU problems are rare with X-ray film studies but are common with fluoroscopic studies. One is advised to document carefully the procedures that were being performed when the tube started to "sputter."

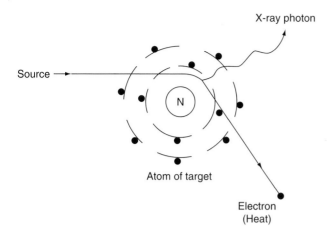

Figure 91-1 Production of general radiation.

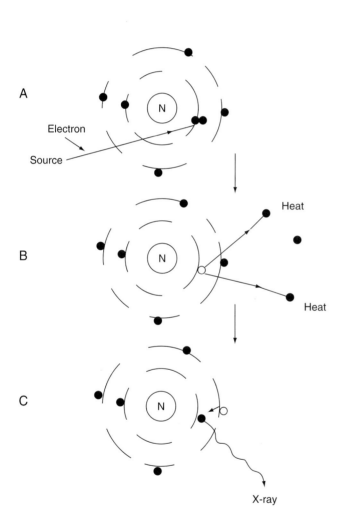

Figure 91-2 Production of characteristic radiation.

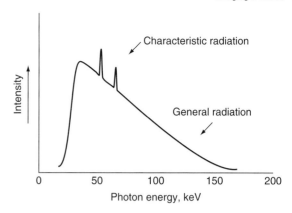

Figure 91-3 X-Ray output.

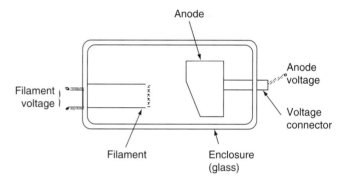

Figure 91-4 Simple tube with stationary anode, low duty cycle.

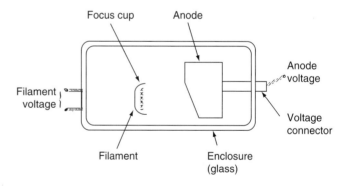

Figure 91-5 Same tube with focus cup (cathode).

The focal spot of a tube is a function of the geometry of the target, target material, texture of the target material, size of the filament and the focus cup (cathode), plus any wobble that might occur in the anode as it rotates. Figure 91-7 illustrates how these factors typically interact. Focal spots are stated in mm, typically 0.3, 0.5, 1.0 and 1.5 mm. Measurement of focal spots is made using a pinhole camera or a star pattern. With a pinhole camera, the image is measured and divided by the amplification factor. The star pattern requires calculations that are more mathematical and it is sometimes difficult to establish where the line blur occurs. These measurements are made at the factory and by a physicist during validation testing. The focal spot size, as built, is listed on the tube housing and serves as a reference point during future validations.

As the tube ages, the focal spot will increase in size. Some of the growth in size can be compensated for by changing "techniques," i.e., voltage, current, and time settings, by the technologists. The smaller the focal spot, the better the resolution of the image. While that is the "official" line of thinking, the consistency of the X-ray generated also affects the resolution. This is evident when edges of objects are not sharp but blended. The use of filters and grids can increase sharpness, also called "clear image" or "central area," and can decrease the blurred/blended area, also called "edge gradient" or "penumbra." The testing of a new tube at installation should include documenting not only the size and shape of the focal spot, but also any tilt, blurred edges, or uneven radiation. A hard copy of this information should be kept on file as long as the tube is in service.

The focal spot geometry produces an uneven beam of X-ray called the "heel effect." This means that there is more energy at one end of the field than at the other. Ideally, aligning the tube so that the radiation caused by the heel effect is directed toward the thickest part of the object being studied helps to maintain the density (i.e., how dark the film is) and contrast (i.e., darkness relative to surroundings) of the film. If complaints about density and contrast arise suddenly, one should check to see that the tube has not been rotated.

Most of the preceding material in this chapter has concerned areas into which BMETs cannot, or should not, venture, or it was simply background information. The remaining

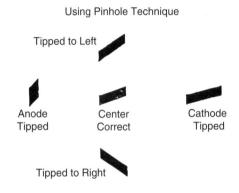

Figure 91-6 Tube with rotating anode.

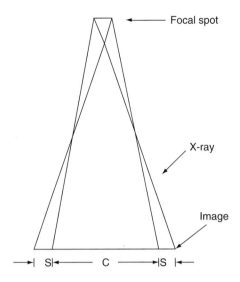

Figure 91-9 Effect of focal spot size: c = clear image; s = blurred area or penumbra.

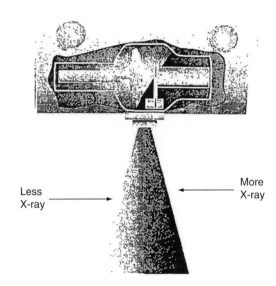

Figure 91-10 Heel effect.

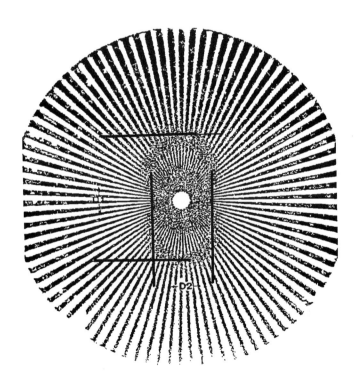

Figure 91-7 Focal spot, using a pin hole camera, when alignment is 15° off in any direction.

Figure 91-8 Star pattern test, look for missing lines.

material addresses items on which BMETs or clinical engineers actually work. Because of this, an attempt is made to use clearer language.

Collimation of the X-ray beam starts with the tube design, specifically the focal spot track, which focuses the majority of the X-rays generated in one direction. The tube housing further reduces stray radiation and continues the collimation process. The beam then exits the tube housing through the "top hat," a lead cone with a rectangular hole in it that sets the maximum film size that can be exposed at a given distance, also called "source-image distance" or "focal-film distance." This also shapes the beam and helps to reduce stray radiation. The collimator mounting block (which is made from aluminum) and the rotation ring secure the collimator to the tube housing. These items should be checked regularly to ensure that the hardware is secure. One old trick is to use a dab of nail polish on the screw head to housing surface. Cracked nail polish will indicate that the hardware is starting to loosen.

On some collimators, there is a slot in the housing, near the mounting block, that allows for the placement of filters in the X-ray beam. These are thin sheets of copper or aluminum (but other material could be used) that are inserted to remove spikes in the radiation intensity that are generated by characteristic radiation, and to smooth the remaining radiation so that it is consistent. This is also referred to as "hardening the beam" or removing soft radiation. These filters should be added or removed only as directed by the physicist doing the radiation certification on the unit. The certification might need to be redone after major repairs to the generator, but otherwise it is done as required by local and federal regulations. The maximum length of certification is three years. Commonly, technologists readjust their techniques after filtration is added or removed. In addition, it is common to experience an increase in "retake" films immediately after changes are made in the system, based on the requirements of the physicist. CEs and BMETs need to work closely with the department supervisors and technologists to ensure that the retake rate returns to, or decreases below, the previous level.

The upper-level shutters are below the filters. These are flat, lead alloy strips of metal that have beveled mating edges to totally block X-ray passage when closed. The upper-level shutters are mechanically linked with the lower-level shutters, so they work together. On some units, there is a feature called "automatic collimation," where the unit senses the size of the film cassette installed and automatically adjusts the shutters to that size. When a problem occurs with the automatic collimation, it is generally traceable to the sensors in the Bucky tray (see below). With manual collimation, problems generally are based in the adjustment knobs or the linkage between the knob and the shutters.

A lamp and a radiolucent mirror are in the space between the upper and lower shutters. This is called the "aiming light," "centering light," "collimation light," or "field light." The light beam is focused through the protective lens on the bottom of the collimator, which has cross hairs to align properly the tube with the area of the patient to be X-rayed. The alignment of the centering light to the film cassette must be checked during each preventive maintenance (PM) inspection to ensure proper alignment. The lamp is replaced from the side of the collimator, never by removing the protective lens with the cross hairs and reaching through the shutters. The light should only remain on for less then a minute.

As part of a PM, the shutters should be fully closed, and a film should be exposed. There should be no exposure on the film if the collimator is working properly. If the shutters do not fully close, the exposure will indicate the shutter that is not closing or that is damaged. A damaged or malfunctioning shutter can adversely affect the quality of patient films and can result in retakes. Retakes are costly and expose the patient to additional ionizing radiation.

At the bottom of the collimator on some units are channel slides that are used to mount cones, cylinders, or other shaped metal units to focus the beam further to an even smaller spot on the patient. These are mechanical devices, and must to be checked only when purchased. The security of their mounting does require checking at PMs.

As the X-ray beam passes through a body striking hard or soft tissue, some of the beams can be deflected, and some will trigger the release of "characteristic" radiation, all of which can cause the lack of definition of the object being studied. By placing a grid between the patient and the film, deflected and Compton scatter radiation is reduced, thus giving a sharper image on the film. Compton scatter is caused by X-ray beams reacting with body parts and creating secondary X-rays, similar to chromatic radiation. Two basic grids are used: A parallel grid, and a focused grid. Focused grids are limited in that they are designed for a single source image distance (SID) or focal film distance (FFD).

Grids come in various ratios of the height of the strip to the space between strips. One common grid is 8:1. where the lead strip is 8 times as high as the space between it and the next strip. The common ratios in use are 5:1 for low-voltage work, 8:1 for mid-range voltages, 12:1 general purpose (the most common), and 16:1 for high-voltage work. The next item to consider is the spacing of the strips. The common choices are 30 strips per centimeter for general use, 45 strips for skull studies, and 60 strips for vascular studies. In most cases, the technologists can change grids as needed, without clinical engineering involvement. When the technologist has a problem with a grid, it generally involves poor films. Two common problems are the cases where the grid is not installed flat or is installed upside down, especially with a focused grid. Occasionally, they will be off-center, thus resulting in film edges that are not clear, and contrast that is off on one or more edge.

If the films come out with noticeable grid lines, there is a problem with the "Bucky." Dr. Hollis Potter and Dr. Gustav Bucky, working independently, developed this system in the 1920s. The grid is moved 2–3 cm in one or more directions so that the grid lines do not appear. The movement of the grid is the "clunk" that you hear during an X-ray exposure. The drive mechanism to move the grids can be springs, a motor, solenoids, or some other simple drive. The film cassette is placed in the Bucky tray, which sets the collimator to that particular film size. The grid is mounted over the tray and moves during the exposure. It generally takes more time just to get to the grid to fix the Bucky mechanism than it does to fix it.

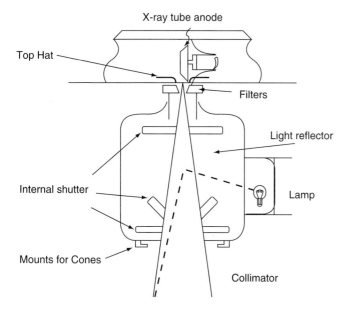

Figure 91-11 Collimator.

The OFD (Object-Film-Distance) is the amplification factor of the imaging system. The greater the distance, the larger the object appears on the film. This is rarely used, but sometimes it is useful in locating breaks in small bones.

Film and Film Cassettes

For most X-ray examinations, dual-emulsion films are used, which means that both sides of the film take the image. These films are more durable then the single-emulsion films that are used for mammography studies (see below). Double-emulsion films tend to be brighter, but they give up some detail.

In the film cassette, two intensifying screens emit light when exposed to the X-ray. This light, along with the X-rays, exposes the film, creating an image. This allows for the patient to absorb a lower dosage, but there is a small loss in clarity of the image on the film. Sometimes the screens become damaged, and the resulting images are not of a high-quality. Sometimes problems with the film-cassette screens are reported as problems with the X-ray unit. One indication of a screen problem is that only one of a group of films has a problem. Finding the bad cassette can be time-consuming. Users also should monitor cassettes for light leaks. This is usually evident by the clear edges of the developed film. When troubleshooting X-ray problems, one always should inspect the last films taken before doing anything with the equipment. The films will provide an unbiased report of what, if anything is wrong with the generator.

Power Supply

Because the maximum voltage of the electrical power entering a hospital from the power grid is 13,800 volts (well below the voltage that is needed to create an X-ray) transformers are used to obtain the needed higher voltages. The highest voltage on hospital floors is 480 volts in a "Y" configuration of four wires.

Simple X-ray units, such as a dental unit, can work on single-phase line voltage, 117 VAC, with a step-up transformer, and with the tube acting as a rectifier. This is a single-pulse system and, as such, is inefficient; and the quality of the X-ray is not great. This is a half-wave rectifier system where the X-ray is produced only for the short time when the half-wave reaches the critical voltage, (45,000 volts). These are reasonably small and inexpensive units, with a fixed anode, and they rarely require service. One thing to remember is that these units are not well collimated, and stray radiation is common. Patient and user exposure levels are higher than in most radiographic rooms.

In some very old units, the power supply is a single-phase line with full-wave rectification. The rectifiers are in the same "tank" as the transformer. Either vacuum tubes or solid-state components and capacitors smooth the ripple of the full-wave rectification. Most of these units have been replaced in modern health care facilities, but they still can be found in use in a physician's office. Repair parts are hard to find, and some tanks might contain polychlorinated biphenol (PCB) based oil, even though all PCBs were supposed to be replaced in the early 1980s. These were known as "two pulse" wave systems and were found mostly in dental offices.

The majority of medium and high power units installed up to the early 1990s used 480-volt, three-phase power supplies. These were also called "12-pulse systems." When the three legs of the output side of the transformer are combined into DC, the ripple is so small that 95% of the voltage will produce the X-ray, as compared to less than 50% in a single-pulse system.

For voltages of 150,000, as used in angiography suites, step-up transformers are needed. Because of line fluctuation and inefficiencies in transformer designs, a turns ratio of about 1:350 would have to be used. The size and weight of these transformers might require special floor braces and additional room. There could be several in the same area if more than one tube is used. As these units can become hot, especially if constant use is required, heat generation is a consideration. The transformers require regular inspection to check the oil level. Darkened oil in the transformer tank is an indicator of arcing or excess heat. The units also must be examined for leaks, dents, or bulges in the sidewall. All can indicate impending failure of the system.

Most modern X-ray systems use what are called "high-frequency" power supplies. The incoming frequency is amplified to approximately 100 kHz, which allows smaller transformers to obtain the required voltage gain. At these high frequencies, there is no need to rectify the waveform. The X-ray intensity generated is quite constant, which allows the patient to be exposed to lower doses. Most high-frequency power supplies are smaller, lighter, and less prone to failures and can be put in the console, thus freeing up floor space.

One should have service manuals before starting any repairs on the high-voltage systems. "Lockout–tagout" procedures must be used at the location for safety of the person and staff. More than one incident has occurred in which a unit was not properly locked out and tagged out, only to be inadvertently turned on. Not only is there an electrical-shock hazard, but serious damage can be done to the device. It is generally a good idea to have a second person in the area when working on high-voltage systems, for the protection of all.

It is also important that one communicate what one is doing, when one expects to be done, and whether the physicist must be called to validate the system after the repair is completed. This physicist should be notified when any changes are made in the high-voltage power supplies or timing circuits, and when a tube, image intensifier, or video chain (see below) is modified.

Fluoroscope

The original screen fluoroscope was invented in 1895, by Thomas Edison, who promptly patented it. The basic design remained constant for close to 60 years before being replaced by video systems. Fluoroscopic (fluoro) exams view dynamic events in the body,

396 Clinical Engineering Handbook

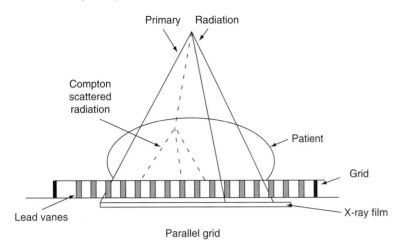

Figure 91-12 Parallel grid.

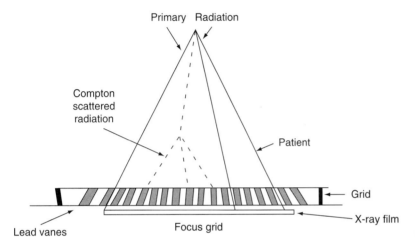

Figure 91-13 Focused grid.

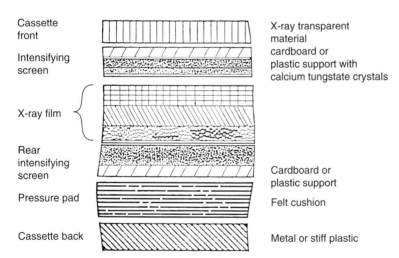

Figure 91-14 Cassette/film cross-sectional view.

such as movement in the intestinal track, or blood flow. It is real-time imaging. If a record is needed, a "spot film" is taken. Some units can be equipped with a rapid film changer, also called a "puck." This unit allows for the quick sequencing of film during a procedure. Some units can have a "multiformat" camera attached to them, in which case up to 32 images can be placed on one sheet of film. The same information that would be recorded by the puck on many sheets of film is done electronically from the fluoroscopic image received by the video system and is transmitted to the multiformat camera. Because the image is captured by a video camera, it can be digitized and manipulated. This is sometimes called "digital subtraction" or "digital subtraction imaging." The fluoroscope works at a lower power level than that required for the typical X-ray film. Therefore, if a film is required, the system goes into "boost" mode to expose the film, and then automatically drops back to its previous exposure level. Generally, patients receive a higher dose of radiation during a fluoroscopic study than they would if only films were taken. Although the power is lower, the exposure time is longer, thus giving the higher dosages of radiation. Hanging from the sides of the I/I are strips of lead shielding, which must be inspected on a regular basis for secure mounting and adequate coverage.

With the exception of C-arm units (see below), the X-ray tube is located under the table on which the patient is positioned. The tube is in the same type of housing as previously described. However, it does not rotate. The collimator is simpler as it has no targeting light or cross hairs. After the X-ray beam passes through the patient, it is detected by the image intensifier (I/I). At the back end of the I/I is a video camera that amplifies the light generated by the X-rays striking the I/I and displays the light patterns on a CRT. That is the simple explanation of a fluoroscope.

The I/I comes in various sizes, from 3 to 15 inches in diameter; 12- and 15-inch units are not common in newer units as the radiation dose for these is high, and modern techniques allow the radiologist to better target the areas being examined. Some units are

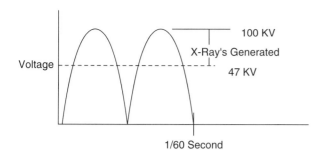

Figure 91-15 Single-pulse wave.

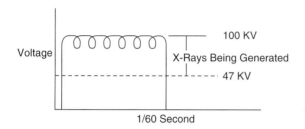

Figure 91-16 12-pulse waveform.

multisized in that either electronically or mechanically they can be set for 6- or 9-inch diameter receiving areas. The I/I is a vacuum chamber, usually made of glass, with a phosphorous coating on the inside of the large end. As X-rays excite the phosphors to emit light, that light is focused at the small end of the tube. The light beam can be linked directly to a camera, split with mirrors to two cameras, or split between a camera and direct viewing ports, which is not common in modern designs. The most common configuration is the direct link to a camera, followed by a beam splitter that links the video camera and a cinematography (*cine*) camera to the light beam coming from the I/I, as in a catheterization installation.

The camera can be a Vidicon, Plumbicon, CCD, or other construction, and is matched to the rest of the video chain. Cameras come in a wide variety of sizes and mounting styles, so one must be careful to specify the correct camera when selecting replacements. On some systems, there may be a photo detector between the I/I and camera that is used to provide the feedback for automatic brightness controls, sometimes called "automatic gain control." In other systems, the feedback is generated after the camera.

Monitors come in a many sizes, formats, and resolutions. On older units, 450–700 lines are used, with the most common being the 525-line monitor. Some newer units use 1024-line monitors. Many units use 1024 × 768 interlaced or 1280 × 1024 or 1600 × 1280 non-interlaced monitors. The system must be matched from camera to monitor for best results. One should not spend the extra money for a high-resolution monitor if the camera is not capable of the higher resolution. When replacing a monitor, one also must consider its weight if it is "hung" on a floating arm. One might have to rebalance the arm to ensure that it remains stationary when repositioned and that does not float upward or downward.

As previously mentioned, spot film systems, rapid film changers and multi-format cameras are often used for "still" recordings of body structures. For motion studies, videotape and cine film have been replaced with dynamic memory—a computer with a large memory that can document several studies before they must be transferred to a CD, DVD, or optical disc. It is, essentially, a picture archiving and communications system (PACS) unit. Videotape is inexpensive and requires no film processing, but it does not have the capability of high-resolution of small objects that film does. Videotapes are often used as backups for cine and are becoming more widely accepted by the medical profession. These tape systems are not off the shelf, from the local video-supply store, but are professional grade units with "frame grabber," super slow motion, and other features. Cine requires a special developing unit and an editing console to view the films. These are high failure and high maintenance devices that require good PM procedures to reduce costs and down time.

Most fluoro systems have automatic gain or brightness controls. These allow for the automatic adjustment of voltage (KV) and current (mA) during the study so that the contrast and density of the image remain constant. This system can be a source of problems that are often more user-centered than technology-centered. Hard failures are uncommon with these circuits, but often they need adjustments as the tube ages. When a tube is replaced, it is a good practice to reset the gain control to the new baseline of tube output. On some units, this is done automatically.

The Table

The simplest X-ray table is a horizontal surface with limited motion. The film cassette tray—the Bucky—is stationary, as is the overhead tube, and the patient is positioned using the "floating top." The head-foot movement is generally two feet or less, with side-to-side motion of about half the head-foot motion. Once the table and patient have been properly aligned, the top is locked in place using electrically powered brakes. The tabletop is made of a composite material that allows X-rays to pass through it with little or no loss, and no additional scatter. Many of these tops have weight limits of 175–200 kilograms. Damaged tops should be replaced when found, as they could compromise patient safety by contributing to lock failures, poor film quality, infection, and patient injury.

Powered tables have the head/foot and side/side motions along with multicassette trays, plus the ability to move the patient from a prone (i.e., horizontal) position to perpendicular positions. The I/I moves with the angular position so that it is always aligned with the tube under the table. Because there are motor drives that are used to move the table and patient to selected angles, it is common for linens to catch in the drive mechanisms. These should always be covered. When they are not covered, they can present a major potential for a patient injury. In addition, it can put an unreasonably heavy demand on the motors and can lead to failures. The locking systems in tilting tables are critical and should be a major point on any PM program. As with the simple table, the tabletop is subject to damage from heavy patients and should be inspected closely after a heavy patient has been on the table. When performing annual PM inspections on the table, the gearing should be checked for missing teeth and lint accumulation. These two hazards can contribute to premature failures of the drive motors. When working on tables that are used for reflux or voiding studies, one should follow universal precautions procedures for personal protection.

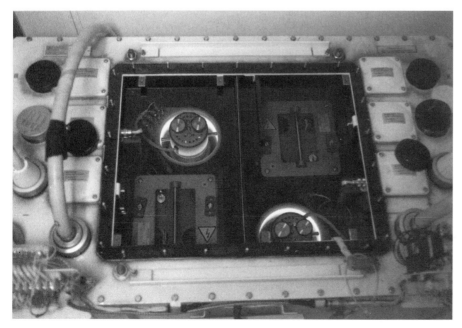

Figure 91-17 High voltage tank with cover removed.

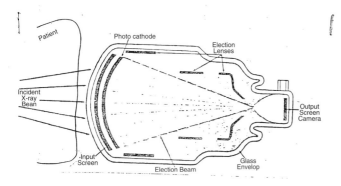

Figure 91-18 Image intensifier.

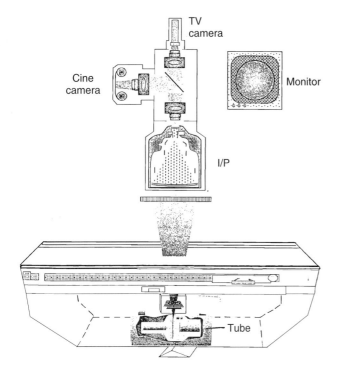

Figure 91-19 Video chain.

The Overhead Tube

When an overhead tube is present, it is mounted on a track system, similar to an operating room (OR) light, except that the amount of movement is much greater. The overhead tube assembly will move in four horizontal directions. Detents (locking positions) and stops on the tracks are required, to ensure safety and proper alignments. The security of the detents and the stops must be verified on a regular basis. Keeping other devices from infringing upon the area of movement of the overhead tube is difficult. Many devices have been damaged by being hit during the movement of the tube. The end stops on the tracks need to be inspected on a regular basis, to ensure that the tube assembly cannot fall off the tracks. In many installations, a counter-balance system using cables and weights is used to allow easy and smooth movement of the tube. These cables fray over years of use and occasionally need to be replaced, especially with systems that brake or lock by clamping on the cable. Frayed cables on the up-and-down movement of the tube over the patient can be particularly dangerous.

With any locking system, locks will stick open, making it impossible to lock the tube in position. Troubleshooting lock problems is a two-person function. While one person observes the locks, often while standing on a ladder, the other energizes the lock from the tube head handle. Broken wires in the locking system are common. A 10-foot length of wire with alligator clips on each end can be used to locate a broken wire. One end is clipped to the switch, and the other to the lock. If the lock energizes when the switch is activated broken wire is indicated. Replacing lock wires is not easy, as it can be difficult to route the wire properly. Temporary repairs to get through the case or, at worst case, the day should not be considered permanent repairs. Rubber-jacketed wire *should* not be used in any area where radiation exposure is expected. Radiation drives the rubber molecules into the metal, creating high resistances over years of exposure. Use only plastic or Teflon-insulated wires.

The Console

The control unit, or console, is located in a room adjacent to the main room. A leaded glass window protects the technologist from scattered radiation and affords the technologist a full view of the patient (Strzelczyk, 2004). On the console, the technologist selects the technique; i.e., the KV, mA, and time for the exposure. Depending upon the age of the unit, selections are made via rotary switches, push-button switches, paddle switches, or up-and-down rocker switches. On some very old units, the exposure time is set with a mechanical timer (such as an egg timer). While older units might have only an analog meter displaying KV and mA, modern digital units have displays of KV, mA, time location buttons (for skull, thorax, abdomen, pelvis, or extremities) that automatically set the KV, mA, and time for the technologist. Override controls enable the technologists use other settings as needed.

Most problems associated with the console are the same as with any control surface on any device and include loose knobs, stuck switches, and cuts in the membranes that protect panel switches.

The panels may display elapsed-time and dose information when in the fluoro mode, heat units, and other information. Many of the newer units have an alarm-memory system that logs faults in the system. This always should be viewed when any problem is reported. Generally, alarm memories are available only when a special sequence is entered, when the unit is in the service mode. "Technique" errors indicate that the technologist is overriding the automatically selected settings and that he might not be getting proper exposures. This situation could be difficult to handle because, if the technologist is challenged, he could become more critical of the equipment.

The junction box where the wiring from the console connects to the rest of the room can be a problem point. Coffee spills, paper clips, and heavy dust accumulation can short connections out.

When radiologists are performing a fluoroscopic study, they are in the room with the patient and will use either a hand or foot control to control the unit, make spot films, and perform other adjustments. Both are failure points and need constant work. Foot controls are subject to spills, kicks, and general abuse, while hand controls are often dropped, thus causing severe mechanical shock from sudden stop or rapid deceleration. One solution is to put a rubber protector on the hand control, as one would on a pulse oximeter.

On some systems, there are up to five instrument cabinets, two high voltage tanks, and various other components, all of which are interconnected. It is important to inspect all exposed wiring (at least annually) and to keep the cabinets well ventilated and dust free. The manuals should be kept in the area and secured. Some organizations keep service records in a binder in the area, along with any field notices or updates. In radiology service, information is power, and one must protect that source of one's power. It is uncommon for X-ray systems to be connected to the hospital's emergency power system. Most systems can be manually connected to emergency power for short periods if there is enough generator capacity. Commonly, parts of the system are connected to power conditioners or even an uninterruptible power supply (UPS). It is a good idea to list of how each device in the room is powered, as some fluoroscopic video monitors could be on the same power system as the table or control panel and will only work if those systems are energized.

People have gone the extra mile to make X-ray systems more complicated and expensive than they have with any other medical device. One has two to three names for the same functions; one has interconnections for devices two feet apart that will travel 20 feet into the floor or ceiling to a junction cabinet and then back. In addition, costs are high, and alternate vendors for many items are not available. In imaging service, it is almost better to be a good manager of technology than to be a good repairer of technology.

Processors, Multiloaders, and Dry Imaging

In radiographic systems, film is used to document images, not PACS or Dymanic memories. The most common films are blue or green based. Films can have varying speeds, (i.e., varying light needed to expose them), emulsions, (molecules that accept light), base material, single- or dual-sided emulsions, and sizes. There is daylight film, for which no darkroom is required, and normal film, for which a darkroom is required. Combined with daylight or darkroom cassettes that have intensifying screens in them that match the film being used, the chance of mistakes and mismatches is quite high. The film type (blue or green), the speed, the intensifying screens in the cassettes, and the chemicals in the processor all must be compatible, in order to achieve high-quality results. Combining these factors with technique variations (KV, mA and time settings), it is easy to understand why the waste-film bins are always full.

Darkroom Films

Working in a darkroom under a "safe light," the technologist hand loads sheets of film into cassettes by opening the back of the cassette. The cassette is either hand-carried out of the darkroom or placed into a "pass box" where another technologist picks up the cassette and performs an exam by placing the cassette into a Bucky tray and exposing the film with a patient between the source (X-ray tube) and the film. The technologist brings the cassette back to the darkroom or pass box, where the cassette is opened and the film is manually fed into the film processor.

Some of the problems commonly encountered include:

- Light leaks around the edges of the film, indicating that the cassette did not close properly, or that it is defective (look for indications that the cassette has been dropped and damaged)
- Marks on the film caused by dust or dirt on the intensifying screens
- Cover latches not closing properly

Daylight Films

Under normal lighting conditions, packs of films in various sizes are placed into a multiloader by a technologist, usually 100 sheets at a time. The technologist pushes the cassette into the "loading slot," where mechanical fingers open the end of the cassette while at the same time determining its size. The machine selects the correct size of film and uses push rollers to force the film into the cassette. The technologist then uses the cassette for a procedure. After the procedure, the technologist places the cassette into the "unloader," which is attached to the film processor. Mechanical fingers open the cassette and eject the film onto a roller system that transports the film into the processor.

Common problems include the following:

- Light leaks, generally only of one edge of the film, which could be caused by a bad cassette or light exposure to the film as it was loaded into the multiloader.
- Marks on the film caused by the loading or unloading rollers.
- More than one film loaded into the cassette. This is a common problem if the humidity in the multiloader is allowed to get too high.

Film Processor

Auto film processors have changed little since their introduction in the 1960s. The exposed film moves downward into the developer solution, and then back up to the crossover rack that sends the film downward into the fixer bath, and then back up to another crosser rack and down through the rinse bath, and then into the drying rack before being dropped into the developed-film bin on the front of the processor. With each film that passes through the processor, additional chemicals are added to keep the system in balance. If the processor sits for long periods without films being processed, the chemical balance can be affected. Most processors have a "stand-by" cycle, during which developer and fixer are added to the processor every hour, replacing the evaporated chemicals. This can cause problems with the quality control on films, as the concentrations can get too high and thus cause very dark films. It is wise to discourage placing film processors in low-use areas, as they require more care then those in high-use areas.

The solutions used must be compatible with the film emulsions and must be monitored for quality by performing "densitometry" tests at least ever other day (or, better yet, every day). A film is exposed with a sensiotometer and run through the processor. On the edge of the film is a 21-step exposure that ranges in density from 0.05–3.05 in steps of 0.15. Using a densitometer, three points on the exposed film are measured.

The first point is the film outside of where the gray scale appears from the sensiotometer. This is the base, or fog reading, that should remain constant. Next, two steps on the gray scale are measured, such as 9 and 13; this can vary from hospital to hospital and even within a single hospital. The selected steps are part of the quality assurance (QA) program for the radiology department. The measurements from these points are plotted, and any variations longer than two days generally will require either the cleaning of the processor or adjusting the chemical mix or temperature. If all 21 steps are measured and plotted, one would have a characteristic curve of density and exposure. This is also call and "H & D curve" (developed by Hunter and Driffield). The QA chart for each processor shows the four key points, the base fog of the film, the readings at each selected step, and the temperature of the developer.

The crossover rollers, or rack, should be cleaned each morning to remove any build up of dried chemicals. Rinsing under running warm water for a minute or less is sufficient to remove any accumulation. If not cleaned, the dried chemicals can lead to "pick off" on the film and could indicate the presence of an anatomical change in the patient that is not there. This can be a serious problem, especially in mammography (mammo) studies.

Most hospitals will have a silver recovery system connected to the processors to capture the silver that washes the films as they are developed. These can cause problems if not changed on a regular basis, and they can impede the flow of water from the processor to the drain. Clogged drains, leaking tanks, temperature changes, and stuck pumps are common problems with processors. If the processor rollers and racks are cleaned on a regular basis, and the rollers are replaced as needed, they are not major problem areas. However, they become problem areas if they are not cleaned. A good QA program must be followed, to minimize problems with processors and to ensure consistent results.

In most hospitals, there is a dedicated film processor for mammo films. These films are single-emulsion films and require temperatures that are different from those of regular radiographic study films.

When digital systems are in place for either fluoroscopic or radiographic studies, the signals are transmitted to a device called a "laser camera" or "laser imager." In this device, the images are scanned onto film and transported into the film processor. These systems are prone to failures and require regular servicing, as with any digital system. One must be aware of light leaks between the laser camera and the processor because they can affect the quality of the film.

A verification of this is called the "dry imaging system" or "dry laser camera." This device takes the digital images and "prints" them onto the acetate-base "film." This is not a true film; it is clear acetate that contains printed images. There is no film processor, no cassettes, no film and no chemicals to mix. It is less expensive to operate, and environmentally friendly, but capable of being used only with digital imaging systems. Some radiologists are not comfortable with it. These units are found mostly in the CT and MRI areas. These systems will become more common as PACS are installed in hospitals. PAC systems are expensive and require major changes in department policies and procedures, but pricing is coming down. As more digital systems are installed in procedure rooms, the better the return will be on investments in PAC systems (Cohen, 2003).

Specialty Units

Specialty units might be physically located not within the main radiology department, but in other areas of the hospital. Mammography and bone-density services are often located in the women's health clinics or in outpatient clinic areas. Tomography units could be located in the orthopedics clinics, while the interventional radiology areas could be close to the operating rooms. The remote locations of these services can strain one's servicing capability. For example, when devices to be serviced are not in the main radiographic area, there are no convenient alternate methods of conducting the specialized examinations. These examinations are generally expensive procedures, and the hospital does not want to forgo reimbursement for them. Rescheduling patients can be a problem, so good PM programs are needed, to keep the "up time" high on these devices and systems and to prevent negative affects on the hospital's financial picture.

Mammography

A mammogram is part of regular testing for women over age 40. This is a soft-tissue radiograph, requiring specialized X-ray equipment and accessories. Originally, when a mammo was done, the collimator was removed from the tube, and a cone was mounted in its place. The cone was lowered to the body surface, and an X-ray was taken. The voltage used was between 40–50 kV, with a long exposure period. It was an uncomfortable procedure, and the dose could be as high as 10 rads. Now, specially designed mammography units, while still uncomfortable for the patient, deliver much lower doses. The use of high-frequency generators has further reduced exposures and has increased contrasts on the films because the X-ray beam is more consistent.

In a mammo procedure, the breast is compressed between adjustable "paddles" that move tissue to a more consistent thickness, and the film is taken. This is a single-emulsion film designed for low-contrast images. The film is developed in a dedicated processor for that type of film. Digital imaging is slowly becoming more common, but its progress has been slowed by reimbursement problems, as many insurance carriers have not approved the use of digital systems in mammography studies. The FDA was also slow in issuing its approval of the technology.

Common problems in mammography units include compression paddles that crack or slip, and mechanical positioning. Replacing the tube is common after two to four years of use and can be expensive. The physicist should closely monitor the focal spot size on the tube to ensure that it does not exceed its specified size range. The growth of the focal spot will affect the quality of the exams. One should obtain from state authorities a copy of the regulations for mammography services and should ensure that the installations meet these requirements.

Some radiologist will also have a dedicated viewing panel for the mammo films. The lights might be brighter and might have more output in the blue spectrum of the light. If this is the case, it is a good idea to replace all of the light tubes at the same time when performing PM.

Tomography

This procedure is known by many names (e.g., "tomo," "tomography,"" body-section radiography," "laminography," and "planegraphy.") The basic principle is to blur unwanted images while maintaining the contrast and density of the desired object. The main use of this technique is for examinations of the spine. The operating principle is simple in that the patient remains motionless while the tube and film move in opposite directions. Mechanical problems are the most common, and spare parts are often difficult to obtain.

Another version uses the long cassette and film of up to 30 inches in length. The film and patient remain stationary while the tube moves. Other than the movement of the tube and film, everything else on this unit is standard, including the development of the film.

Bone Density

The loss of minerals in bones had been difficult to diagnose for many years. While tomo exams were used to confirm degenerating vertebrae, they were not practical for mass screening. In the late 1980s, new devices emerged that were specifically designed to measure the density of bones. These system are simple and generally reliable.

Under the table is an X-ray tube with dual energy ranges (70 and 140 KV are typical). By using a high-frequency power supply, the quality and quantity of the X-rays generated is consistent. Instead of a collimator, which limits the output to a square or rectangular shape, a long and narrow "fan beam" is created. These are rotating anode tubes, and some require additional cooling to maintain the ideal operating temperatures of the tube. The beam passes through the patient and is detected by a series of sensors mounted over the patient. The sensors are connected to a computer, which reconstructs the data into a picture. In most cases, the image is printed out on a standard computer printer (laser or inkjet), with computed densities printed next to the image. If a dry imager is available, the image can be printed on that as well.

Scan times can range from a few seconds, to image a small area, to up to 20 minutes, for a full body scan. The dose levels to the patient are about equal to a chest X-ray, even with the long scan time. Because the beam is so confined, the walls do not require lead lining, and the operator needs no shielding.

On most systems, the patient is placed on the table and positioned at the point where the scan is to start. The tabletop moves the patient past the X-ray tube and detectors until the scan is complete. Because the total movement of the table is about 30 inches, the ends of the table should be kept clear of other devices that could obstruct table movement.

Another version of bone densitometry is the heel ultrasonic unit. This is a small unit that is good for mass screenings but not for definitive diagnosis of patients. The patient places a foot onto the positioning platform of the device, and the heel bone is scanned. If the reflected power is below a certain number, the patient is advised to get a scan using the X-ray system.

Interventional Radiology or Special Procedures Rooms

Interventional radiology (or special procedures) rooms encompass a wide range of rooms, including cardiac catheterization laboratories, angiography suites, electrophysiology suites, and neurology suites. All special procedures rooms are equipped with fluoroscopic capabilities but with better video resolutions, smaller focal spots, higher power, and a multitude of support equipment. Many of these units are almost like a separate practice within the hospital. As the equipment is used for twelve or more hours every day, incorporating it into a PM program is difficult. One must be flexible by servicing these areas at times when, for example, staff are having patient conferences or are on breaks.

The support equipment in these areas includes such devices as defibrillators, multichannel monitors, lasers, pacer programmers, electrosurgical devices, and radioactive sources used in angioplasty. One might have problems in maintaining a good inventory of equipment in this area as some of the devices may be on loan or consignment, or simply "borrowed" from other parts of the hospital. Care should be taken to follow universal precautions when working in these areas.

Cystoscopy Room

The cystoscopy (cysto) room is generally located in the OR suite and is specially designed to perform studies or procedures relating to the kidney, urethra, bladder, and prostate. It is a fluoro room with a special table. Films are not taken often, as the surgeon uses the fluoro imaging to guide instruments to remove or crush stones. In other procedures, dye is infused into the kidney, and the movement down the urinary track is observed. Another common procedure in the cysto room is the transurethral resection (TUR) surgery of the prostate.

This room is a wet environment, and close inspection of grounding and measurement of electrical leakage currents is essential for the protection of the patient and staff. Generally, this room is available for inspections and PMs in the mid to late afternoons on most days of the week. One should follow universal precautions when working in this area. Probably more than in any other X-ray installation, the lubrication of mechanical systems must be done on a regular basis. In addition, one should check for rusting on the machines.

Portable C-Arm

Portable C-arms are similar to portable X-ray units except that they are fluoro systems. They use a high-frequency power supply and have an image intensifier and view chain to display images. Most have a storage scope where the first image of that patient is kept to serve as a reference and guide to the impending procedure. The systems come in two basic sizes. The larger size units are used in operating rooms, often on orthopedic cases (such as hip replacements) and provide the surgeon with an accurate indication on joint is aligned and seating. They are also used in other cases in the operating room for placement of drains, catheters, and other devices into the patient. Generally, C-arms have two components: The "C," which contains the tube; and an I/I attached by way of cables to the chassis, which contains the generator and the video-storage electronics.

The smaller units, often called "mini C-arms" are commonly used to assist physicians in resetting broken bones in hands, wrists, and feet. Because the mini C-arm is in one chassis, it is easy to move and takes up less space than the larger units do.

Common problems associated with C-arms include the following:

- Interconnection cables are easily damaged, as are the connectors.
- Positioning locks not holding during procedures allow the head to move.
- Casters and caster locks.
- Mechanical damage from running into walls and other objects.

Although technical discussion of CT, MRI, nuclear medicine imaging, PET, SPET, and ultrasound is not included in this chapter, the technical management problems are similar to those associated with the devices described above. Technical management is discussed in the following section.

Management of Imaging Devices

Several hard and fast rules apply to managing imaging devices. These rules do not vary between hospitals and cannot be appealed to the courts.

No matter how capable one is, in the hospital one will not be able to repair every item, every time, in a timely fashion. One needs access to outside experts, whether they are others in one's service company, an independent service organization (ISO), or the manufacturer of the device. Even the field service organizations of a supplier often have to call the factory for guidance on problems. No one knows everything.

One will be overcharged for repairs parts, time, or travel, on a regular basis. Therefore, one should review all service reports and invoices.

One will not be able to locate a second source of all repair parts, especially tubes on newer units.

There often will be a discrepancy between the real problem and what the physicist reports. Performing good PMs that find and correct minor problems before they become major problems brings the greatest financial savings.

It is important that the chief technologist be kept informed of all repair and PM work done in the department. While one need not to give detailed reports, it is a good idea to meet on a regular basis to review problems that have occurred. Some problems could be caused technologists using poor techniques or attempting tests utilizing a unit that is not suited for those tests.

It is also a good idea to work with the department on any quality issues that effect patient care or costs, especially retakes. When a study or exam must be repeated, extra costs are incurred. More importantly, repeated exams expose the patient to more radiation and, if the test is invasive, to an increased chance of infection. Retake rates can indicate equipment and personnel problems. While these problems are sometimes difficult to separate and to correct, their solution brings with it decreased costs and increased patient safety.

One must be aware of any changes in film, processor, or contrast-media supplies as they may affect the quality of examinations by requiring changes in equipment settings.

One must keep good records of what is done. This is not easy to do, as some of the work will be done "off hours" and the service reports might not be sent to the clinical engineering department. All service reports should be reviewed for completeness and should include, at a minimum, a description of the work performed and the time expended. Reports should be checked also for the "boomerang effect," which occurs in all service organizations. It is not uncommon to have a service call on a device the day after a PM was done, nor is it uncommon to have the same failure within days of a so-called repair.

One should take care in record keeping, which can save the hospital major expenses, particularly on repairs or PMs that were not done or that were done improperly, for which the hospital was billed.

Lastly, when working on imaging equipment, one should "think simple" and should look for the oblivious. As with other medical device technologies, adhering to this philosophy usually results in problem solution. One should ask questions and should take the time required to understand the problem fully, before starting any repair.

References

Strzelczyk J. Radiation Safety. In Dyro JF (ed). *The Clinical Engineering Handbook.* Burlington, MA, Elsevier, 2004.

92

Machine Vision

Eric Rosow
Director of Biomedical Engineering, Hartford Hospital
Hartford, CT

Melissa Burns
Holyoke, MA

The human observer relies on a wide range of cues, drawing from color, perspective, shadings, and a vast library of particular individual experiences. Visual perception relies on a uniquely human capacity to make judgments. However, a machine-vision system does not have an experience base from which to make comparative decisions. Everything must be specifically defined. Simple problems, like locating an object, differentiating top from bottom, and distinguishing a light artifact from a part of the interesting object, become complicated tasks in machine vision. That is why, in order to reduce the number of variables, the vision system must be provided (from the beginning) with the "best image" possible, meaning the image that makes it easiest for the machine vision computer system to do its task.

"Machine vision" can be defined as "the acquisition and processing of images to identify or measure the characteristics of objects." The objective of this chapter is to provide useful information on the three steps that are required for successful machine-vision applications: Conditioning, acquisition, and analysis. In addition, several real-world examples of machine vision and computerized motion control are presented, to illustrate the practical applications of these technologies within the biomedical field.

Steps to Success Using Machine Vision

This section introduces three simple steps (Figure 92-1) that help one to achieve success with machine vision. The three steps to machine-vision success are as follows:

1. **Conditioning:** The process of preparing the imaging environment. One must consider factors such as lighting and motion, as well as the specific feature(s) of the objects being observed.
2. **Acquisition:** The process of selecting image-acquisition hardware, a camera, and a lens. One also uses software to capture and display the image.
3. **Analysis:** Includes all of the image-processing steps one uses to answer questions about the image. Questions such as "How big is this object?" and "Is this object present in the image?" can be answered.

Other practical examples of machine vision and virtual instrumentation are discussed in the following three examples. The first example shows how a computer can read a seven-segment display and can display these "analog images" in the form of a digital indicator, as illustrated in Figure 92-2.

This next example (see Figure 92-3) shows how virtual instrumentation and machine vision can be used to read the position of a needle. These "virtual instruments" (VIs) are useful in many applications, such as the calibration of speedometers and gauges in the automotive and medical industries.

The last example (see Figure 92-4) illustrates how machine-vision applications can read standard barcodes. In this example, the user performs a simple calibration procedure by selecting a "region of interest" (ROI) that crosses the barcode area, and the VI then will automatically interpret standard barcodes into their numeric equivalents.

In the following, several real-world examples of biomedical applications that use machine-vision and/or motion-control technologies are presented.

Case Study #1: The EndoTester™—A Virtual Instrument-Based Quality Control and Technology Assessment System for Surgical Video Systems

The use of endoscopic surgery is growing, largely because it is generally safer and less expensive than conventional surgery, and patients tend to require less time in a hospital after endoscopic surgery. Industry experts conservatively estimate that about 4 million minimally invasive procedures were performed in 1996. As endoscopic surgery becomes more common, there is an increasing need to evaluate accurately the performance characteristics of endoscopes and their peripheral components.

Introduction

The assessment of the optical performance of laparoscopes and video systems is often difficult in the clinical setting. The surgeon depends on a high-quality image to perform minimally invasive surgery, yet assurance of proper function of the equipment by clinical engineering staff is not always straightforward. Many variables in the patient and equipment may result in a poor image. Equipment variables, which may degrade image quality, include problems with the rigid endoscope, either with optics or light transmission. The light cable is another source of uncertainty because of optical loss from damaged fibers. Malfunctions of the charge-coupled device (CCD) video camera comprise yet another source of poor image quality. Cleanliness of the equipment, especially lens surfaces on the endoscope (both proximal and distal ends), is a particularly common problem. Patient factors make the objective assessment of image quality more difficult. Large operative fields and bleeding at the operative site are just two examples of patient factors that can affect image quality.

The evaluation of new video endoscopic equipment is also difficult because of the lack of objective standards for performance. Purchasers of equipment are forced to make an essentially subjective decision about image quality. The authors have developed an instrument, the EndoTester,™ with integrated software to quantify the optical properties of fiberoptic endoscopes. Figure 92-5 illustrates the main menu of this application.

Materials and Methods

The EndoTester™ is a specialized optical bench used for the quantitative testing of the fiber-optic path and the lens system in rigid and flexible endoscopes. In addition to the specialized test station, the EndoTester™ requires the following:

- A high-intensity variable light source
- A flexible fiber-optic cable
- A CCD video camera processor
- An optional video monitor (recommended)
- A standard PC and a video-capture board
- Laser Printer (600 dpi or higher recommended)

A standard PC containing either an 8-bit grayscale or a 24-bit color video digitizing board (MRT Videoport, MRT Micro, Inc, Boca Raton, FL or IMAQ-PCI-1408, National Instruments, Austin, TX) and custom software allows for easy acquisition and analysis of the endoscope's optical properties. Collectively, the optical test bench and specialized software allow the user to perform a series of tests on the endoscope and its peripheral devices. Specifically, these tests include the following:

- Relative light loss
- Reflective symmetry
- Percent of lit (i.e., good) fibers
- Geometric distortion
- Modulation transfer function (MTF)

Each series of tests is associated with a specific endoscope to allow for trending and easy comparison of successive measurements. Specific information about each endoscope (e.g., manufacturer, diameter, length, tip angle, department/unit, control number, and operator), the reason for the test (e.g., quality control and pre/post repair) and any problems associated with the scope are also documented through the electronic record. In addition, all quantitative measurements from each test are automatically appended to

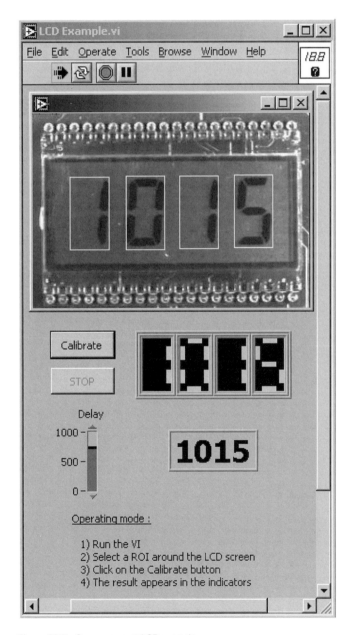

Figure 92-1 Three steps required for successful machine-vision applications. (Courtesy of National Instruments).

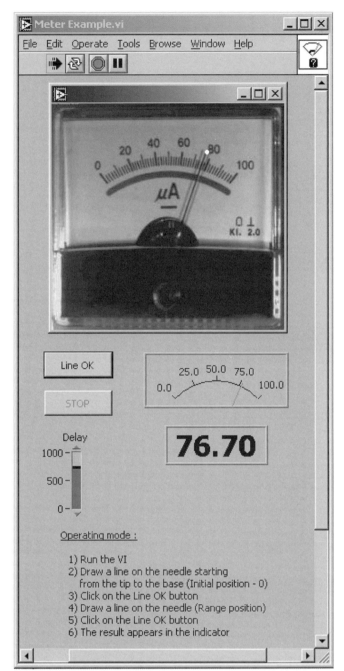

Figure 92-3 Meter example.

Figure 92-2 Seven-segment LCD example.

the electronic record. Figures 92-7 and 8 illustrate the information profile and problem-entry screens of the EndoTester.™ Figure 92-9 illustrates the test sequencer, in which the various tests can be selected either manually or automatically.

Endoscope Tests

Relative Light Loss in Optic Fibers

The relative optical light loss measurements quantify the degree of light loss from the light source to the distal tip of the endoscope. The relative light loss will increase with fiber-optic damage. Changes in the light-source intensity or the condition of the fibers in the fiberoptic cable are normalized out of the relative loss calculation because the relative light loss is determined directly from the endoscope's light output and light input.

Having a simple means to quantitatively measure performance variation with respect to time is desirable for all endoscopes but appears to be particularly valuable for the evaluation of disposable endoscopes. These units can be projected to have a rated life of 30 uses, but measurements of performance change under actual operating, and sterilizing conditions at a given institution support periodic quantitative measurements.

The relative light loss for the optic fibers of the endoscope under test is calculated by the following equation:

$$\text{Relative Light Loss} = 10 \log \left[\frac{\text{Light Out}}{\text{Light In}} \right]$$

The "light in" value is measured by configuring the output-connection end of the fiberoptic cable to a test fixture that holds the cable a fixed 2-inch distance from a photometer (Edmund Scientific, Model INS-DX100). The "light out" value is measured by the photometer illuminated by the endoscope under test.

Reflective Symmetry

"Reflective symmetry" is a measure of light amplitude in the endoscope's field of view. This value is important in that it quantifies the effective distribution of light.

By employing five magnitude comparators, it is possible to transform the continuous illumination pattern into five annular rings of decreasing grayscales. The pattern of gray rings produced by this test should be nearly centered on the image. The pattern is circular for zero-degree endoscope tip angles and sometimes is elliptical for angled endoscope tips. A histogram graphs the number of pixels from each comparator output (light intensity). For each "ring" that is displayed in the filtered image, the user can see graphically how many pixels exist for each band of intensity. In order to pass this test, the "percent bright area" is required to be greater than or equal to 50% of the maximum brightness. The following equation defines how percent bright area is calculated. "Gray 1" and "Gray 2" refer to the two brightest (i.e., innermost) rings.

$$\text{Percent Bright Area} = \left[\frac{(\text{Gray 1} + \text{Gray 2})}{\Sigma \text{ pixels}} \right] \times 100$$

In addition to calculating the percent bright area of the field of view, the "boresight error" can be measured. The boresight error is defined as the difference between the center of the field of view and the center of illumination provided by the endoscope. A grid of 0.2-inch × 0.2-inch squares, whose center square is solid black, is positioned in the center of the field of view. Combining this grid with the ring pattern shows the geometrical distance from the center of illumination to the center of the field of view. By counting the 0.2-inch square grid, a quantitative measure of the boresight error is determined for the endoscope at a 2-inch tip distance.

Lighted Fibers

A close-in reflection (less than 0.25 inch separation) of the distal end of the endoscope from a polished mirror or lucite surface is captured by the CCD camera. This provides a record of the pattern of lighted optical fibers for the endoscope under test. The number of lighted pixels will depend on the endoscope's dimensions, the distal end geometry, and the number of failed optical fibers. New fiber damage to an endoscope will be apparent by comparison of the lighted fiber pictures (and histogram profiles) from successive tests. Statistical data are also available to calculate the percentage of working fibers in a given endoscope.

In addition to the two-dimensional profile of lighted fibers, this pattern (and all other image patterns) can be displayed in the form of a three-dimensional contour plot, as shown in Figure 92-12. This interactive graph can be viewed from a variety of viewpoints in that the user can vary the elevation, rotation, size, and perspective controls.

Geometric Distortion

The geometric distortion test is used to quantify the optical distortion of the rod–lens system. The EndoTester's video frame grabber captures the image of a square grid pattern. "Distortion" is defined as the change of magnification at points around the field of

Figure 92-4 Barcode example.

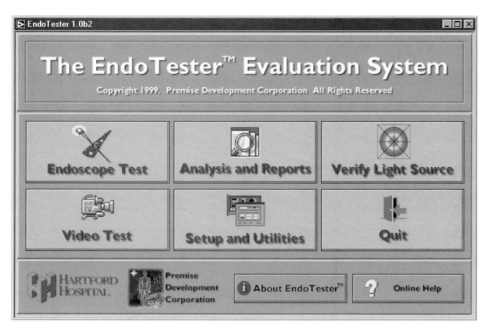

Figure 92-5 The EndoTester™ main menu.

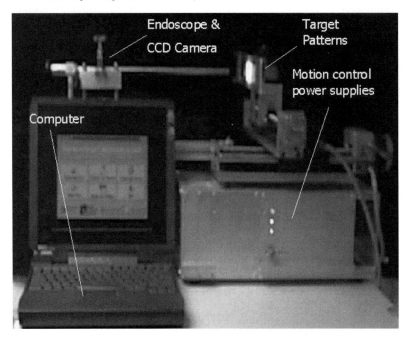

Figure 92-6 The EndoTester™—Basic Test Fixture.

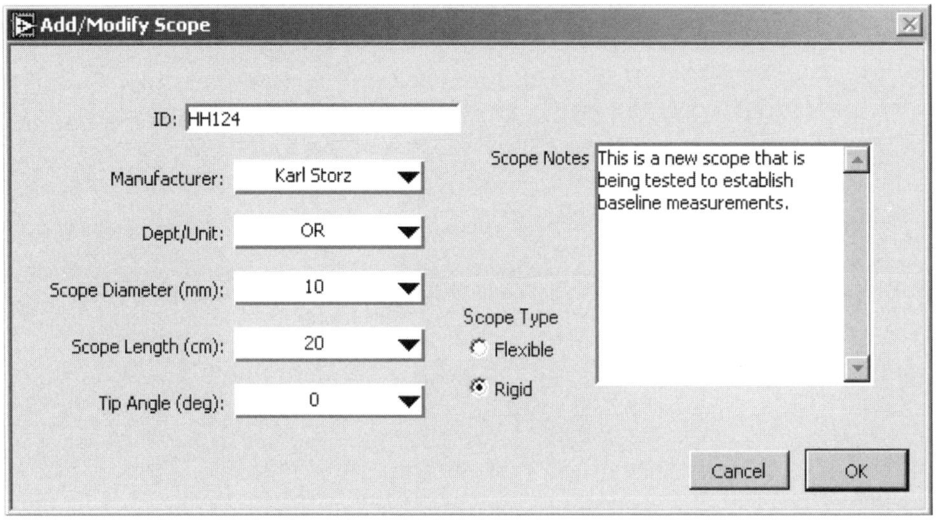

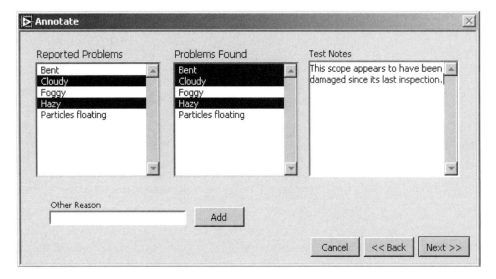

Figure 92-7 Endoscope information profile.

Figure 92-8 Problem-entry screen.

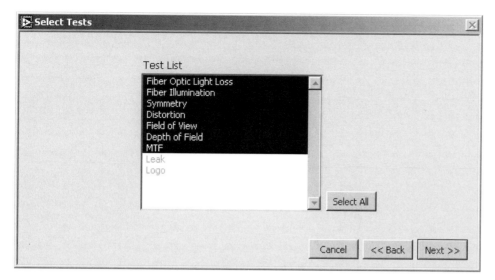

Figure 92-9 EndoTester™ test selection.

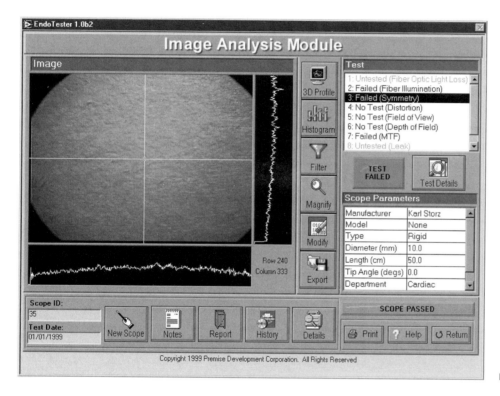

Figure 92-10 Reflective symmetry test.

view with respect to the maximum magnification occurring at the center of the field of view. By measuring the diagonal length of the central square of the pattern, with respect to the diagonal length of any other square, the geometric distortion can be determined. The percentage of distortion is calculated by the following equation:

$$\text{Percent Distortion} = \left[\left[\frac{\text{Other Square Diagonal Length}}{\text{Central Square Diagonal Length}}\right] - 1\right] \times 100$$

Figure 92-13 illustrates the screen that is used to measure the geometric distortion of the endoscope under test.

Modulation Transfer Function

Aperture response is a universal criterion for specifying picture definition and other aspects of imaging system performance. It can be used for film images, camera lenses, television camera imagers, receiver picture tubes, and the human eye. The aperture response is measured as a contrast ratio by a square-wave pattern (contrast transfer function [CTF]) or a sine-wave pattern (modulation transfer function [MTF]).

The square-wave response data can be converted to equivalent sine-wave data by mathematical manipulation; however, the sine-wave method is the most direct approach. Variable density film targets are now available as sine transmission targets; hence, a direct measurement of MTF can be performed. In fact, even a single-space frequency measurement can provide a good index of performance for an optical lens system such as an endoscope. This is the approach used by the EndoTester.™ The endoscope is tested with a one-cycle-per-millimeter sinusoidal transmission target (Sine Patterns, Penfield, NY).

The MTF of the lens system is measured at a spatial frequency of six cycles per degree of apparent field of view. Measurements at this frequency are considered to be accurate indications of good optical instrument performance, when the MTF is high. Thus, a single spatial frequency on the test target provides a good quantitative index of the local spot performance of the endoscope's lens system. Typically, the center region of a lens has the best optical performance, and the edges of the lens have less visual sharpness. Therefore, the MTF test measures performance at the left and right edges, the top and bottom edges, and the center of the lens. Studies have shown that a user's perception of image quality can be correlated to high values of modulation transfer function.

The MTF tests use a film target whose transmittance of white light varies sinusoidally across the X-axis and is constant with respect to the Y-axis. The fiber-optic cable is removed from the endoscope under test and is used to illuminate the sine target. The endoscope views the transmitted sinusoidal light pattern. The frame grabber captures the sinusoidal light variation along the horizontal sweep axis of the television image of the target.

"Modulation" is defined as:

$$\text{Modulation} = \frac{(T_{\max} - T_{\min})}{(T_{\max} - T_{\min})}$$ Where T is transmitted light intensity.

When this measurement is made over a range of spatial frequencies, a plot of the curve of the modulation transfer ratio as a function of spatial frequency is called the "modulation transfer function" (MTF). The MTF of a system is equal to the product of the MTFs

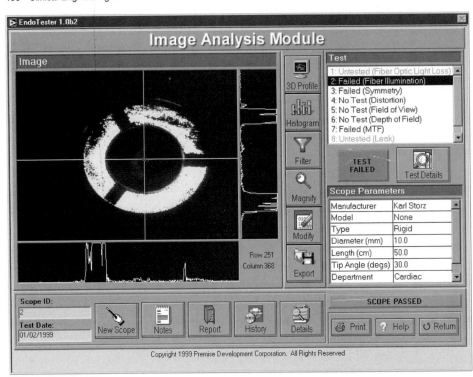

Figure 92-11 Endoscope Fiber Illumination Test.

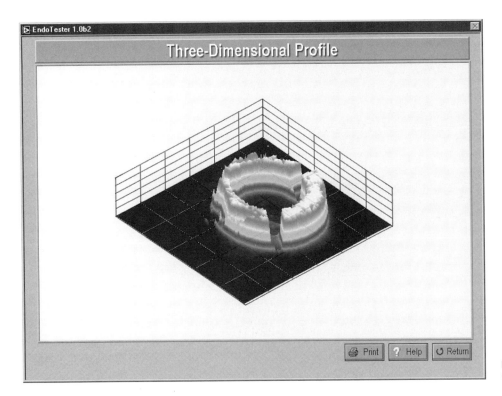

Figure 92-12 Endocope fiber Illumination (3-D contour profile).

of each component of the system. Thus, the MTF of the signal captured by the frame grabber is the product of the endoscope's MTF and the TV system's MTF. Figure 92-14 illustrates the screen that is used to measure the modulation transfer function of the endoscope under test.

Discussion

An easy-to-use optical evaluation system for endoscopes has been developed. This system allows objective measurement of endoscopic performance prior to equipment purchase and in routine clinical use as part of a program of prospective maintenance. Measuring parameters of scope performance can facilitate equipment purchase. Vendor claims of instrument capabilities can be validated as a part of the negotiation process.

The adoption of disposable endoscopes raises another potential use for the EndoTester.™ Disposable scopes are estimated to have a life of 20–30 procedures. However, there is no easy way to determine exactly when a scope should be "thrown away." The EndoTester™ could be used to define this end-point.

The greatest potential for this system is as part of a program of preventive maintenance. Currently, in most operating rooms, endoscopes are removed from service and sent for repair when they fail in clinical use. This practice causes operative delay with attendant risk to the patient and an increase in cost to the institution. The problem is difficult because an endoscope might be adequate in one procedure but might fail in the next, which might more exacting because clinical variables such as large patient size or bleeding. Objective assessment of endoscope function with the EndoTester™ could eliminate some of these problems.

Equally as important, an endoscope evaluation system also will allow institutions to ensure value from providers of repair services. The need for repair can be better defined,

Machine Vision 407

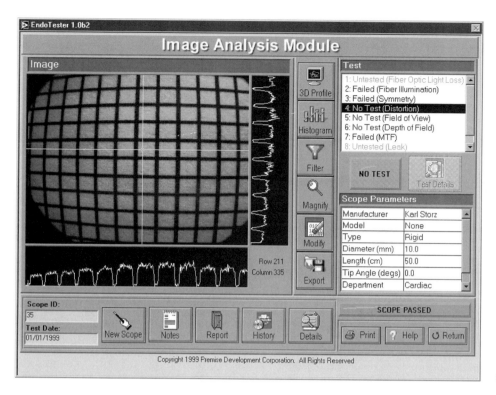

Figure 92-13 Distortion-analysis panel.

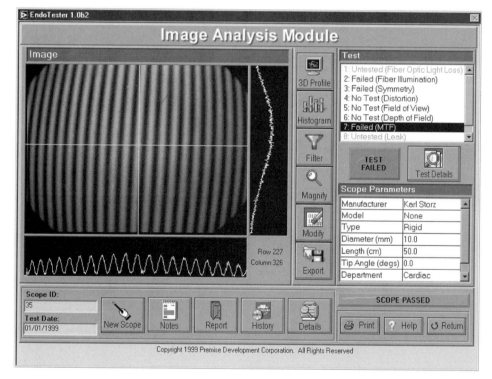

Figure 92-14 Modulation transfer function.

and the adequacy of the repair verified, when service is completed. This ability becomes especially important as the explosive growth of minimally invasive surgery has resulted in the creation of a significant market for endoscope repairs and service. Endoscope repair costs vary widely throughout the industry, with costs ranging from $500 to $1500 or more per repair. Inappropriate or incomplete repairs can result in extending surgical time by requiring the surgeon to "switch scopes" (several times, in some cases) during a surgical procedure.

Given these applications, the authors believe that the Endo Tester™ can play an important role in reducing unnecessary costs, while at the same time improving the quality of the endoscopic equipment and the outcome of its utilization.

It is the sincere hope of the authors that this technology will help to provide accurate, affordable, and easy-to-acquire data on endoscope performance characteristics, which clearly are to the benefit of the health care provider, the ethical service providers, manufacturers of quality products, the payers, and of course, the patient.

Case Study #2: A LabVIEW-Based Wound Measurement System

Description of System

The wound management system is a simple, portable, computer-based approach to wound assessment that implements both digital and infrared imaging technology. The digital image is analyzed to determine wound area and color properties, while the infrared image will be analyzed to determine blood perfusion properties of the wound.

The purpose of the wound management system is to provide a simple and cost-effective method by which health care personnel can quantify the healing rate of a wound, assess the effectiveness of wound care treatment techniques, and document wound healing progress. The computer-based system improves the repeatability and effectiveness of both wound area measurement and wound coloration analysis as compared to current practices.

Current Wound Management Practices

Wound measurements currently involve invasive and inconsistent measurement techniques that utilize manual measurement tools or plastic wound overlays and require estimation of wound size using geometric simplifications (e.g., considering only the length and width of a wound and using a square area approximation of wound size). Current practices also involve qualitative, subjective clinician assessment of wound colorization and inconsistent use of photographs to document patient progress.

Wound care currently encompasses a wide variety of patient types, from acute, simple lacerations seen in the emergency room; to devastating wounds resulting from trauma and surgical interventions; to chronic, lower extremity ulcers and pressure sores. There is fragmentation of wound care personnel and facilities within the medical community. Currently, new treatment modalities are appearing at a rapid rate. It is often beyond the ability of a single individual to comprehend and incorporate the new information into improved and cost-effective techniques for patient care.

Chronic wounds represent a major issue in health care today, particularly among the elderly. Nonhealing wounds can persist for years, causing pain to patients and placing them at risk for secondary infection or loss of limb. The financial burden of the care of these patients is phenomenal: Institutional care of nonhealing wounds is estimated to cost approximately $1000 per day. A recent study estimates that approximately eight million people in the United States currently suffer from nonhealing wounds, resulting in over $10 billion in additional annual health care costs.

As a result, wound care has become a specialty in itself. Over the last 15 years, an enormous number of wound care products and services have emerged, claiming to empower clinicians with the ability to manage the healing process effectively. In addition, several "wound measurement" products have been introduced to the market. These products range from simple rulers and pliable transparent sheets (with "bull's eye" target patterns) that cost less than $20, to high tech, computer-based measurement systems that cost over $80,000.

Several factors recently led to the emergence of wound care as a distinct medical discipline. Dedicated wound care programs are increasingly caring for Americans who suffer chronic wounds, and they offer the potential to produce enhanced outcomes at reduced costs for payers and consumers of health care services. Evidence of this trend can be found in a variety of recent reports and studies. A report by Frost and Sullivan indicates that the United States wound-management products market is now $1.74 billion and that it should grow to $2.57 billion by 2002. The cost of treating the chronic wound is estimated at $5 billion to $7 billion per fiscal year, and these wounds are increasing at a rate of 10%per year, according to the report. (For more information, see http://www.pslgroup.com/dg/daf6.htm.)

The development of the American Academy of Wound Management (AAWM), a national, nonprofit certifying board is another indication that wound care is coming into its own as an industry. Board certification is now available for physicians, nurses, therapists, researchers, and other health care professionals involved in wound care. Over 1000 wound care professionals have requested applications for board certification through AAWM.

Advantages of System

In most hospital settings today, the daily clinical routine for wound care involves little or no quantitative documentation. Measurements are performed generally with the help of a ruler or a disposable transparency onto which the wound is drawn. In some cases, clinicians will periodically take pictures with a digital camera, but no formal protocol has been established, no lighting considerations are taken into account, and the purpose is more educational and qualitative than quantitative. Figure 92-15 illustrates examples of current tools and techniques that are used to measure wounds and healing rates.

The LabVIEW-based Wound Management System eliminates subjectivity and approximations by using computer algorithms to: Calculate the wound area based on the true, irregular wound outline; determine the percentage of wound area that is red, yellow, and black (the colors most indicative of wound healing progress); and evaluate or present information on wound vascularization (via infrared imaging). In addition, because the system is based on obtaining successive images of the wound as treatment progresses, clinicians have access to an image archive of patient progress. Clinicians also can add personal notes to the file for a patient to create an inclusive database of images, analysis, and clinical progress. The system has been designed to be portable and easily used by a typical clinician. The system will be self-contained and will utilize a fixture to address concerns such as lighting, distance from camera to patient, and photographic angle.

Indicated Population

The applicable patient population includes sufferers of acute wounds (e.g., lacerations, trauma, and surgical wounds) and chronic wounds (e.g., arterial, venous, and diabetic ulcers, and pressure sores) Application of the wound healing system will focus on the chronic wound population, which accounts for the most extensive costs to health care and would result in the most patient benefit. Initial proof-of-concept testing occurred at Hartford Hospital's foot clinic, which focuses on diabetic foot ulcers.

Figures 92-16, 92-17, and 92-18 illustrate some of the image acquisition, analysis, and data management modules of the wound measurement system.

Benefits of Wound Management Technology

The benefits of wound management technology can be divided into three primary categories: Benefits to the clinician; benefits to the patient; and benefits to the health care organization.

Benefits to the Clinician

- Consistent and repeatable method for measuring wounds
- Objective and quantitative data for assessing wound care approach
- Access to historical data on wound healing rates based on treatment type and patient-specific variables
- Easily accessible image archive of patient progress

Benefits to the Patient

- Noninvasive wound measurement and assessment
- Reduced likelihood of nosocomial infection
- Enhanced client understanding of and participation in wound management (e.g., measured improvement and visual presentation of wound progress motivates continuation of proactive patient practices)

Benefits to Health Care Organization

- More rapid assessment of wound care effectiveness, thereby decreasing treatment time and related costs.
- Decreased time required by the clinician to document wound condition.
- Remote sharing of information between sites for teaching and/or collaborating.

Discussion

Wound care continues to present financial and human burdens to the health care community. Clinicians recognize the usefulness of wound images and tools for tracking wound progress, but they lack tools that provide valuable information while limiting impact on patient flow and clinical efficiency. Without quantitative methods to assess individual patients and available treatment methods, it is difficult to optimize treatment protocols and to minimize patient suffering. User-interface development tools provide user-friendly and highly configurable means of interfacing with vision-capture devices and turning the resulting images into clinically relevant quantitative data. Algorithms for evaluating wound size and appearance eliminate the qualitative nature of human perception and replace perception with numeric and graphical data that can be quickly assessed by the busy clinician and applied to the treatment of suffering patients. Eventually, an integrated wound-measurement tool can become part of a successful and financially competitive wound care program. These tools will allow doctors, physical therapists, and nurses to become more proactive in their approaches and to limit reactive measures and the associated recovery time with the resulting benefits being shared by the patients, clinicians, and the health care delivery system.

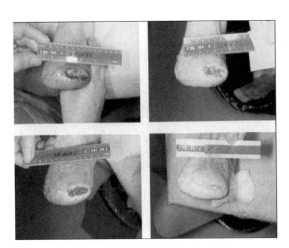

Figure 92-15 Wound documentation as it is ideally performed today. Source: Hartford Hospital, Hartford, CT.

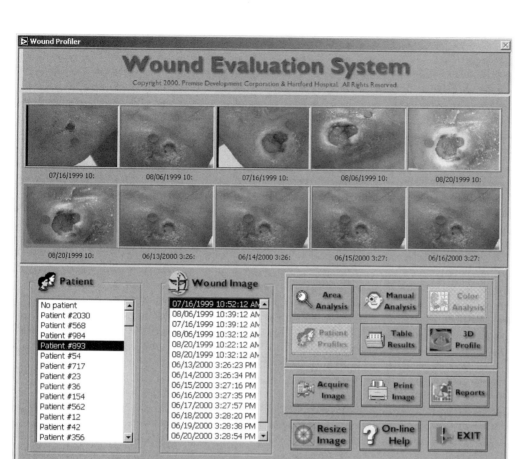

Figure 92-16 Patient and wound selection panel illustrates how any patient (and his or her respective wound images) can be selected for historical and visual comparison. In addition, a particular image can be selected for advanced analyses, as shown in Figure 92-17.

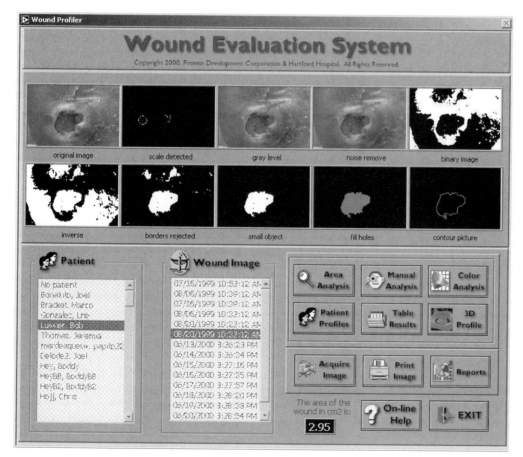

Figure 92-17 Automated wound analysis illustrates how a selected wound can be automatically analyzed in terms of its area and colorization. Computer algorithms are employed to identify the region of interest and to perform specific analyses including wound contour, area, and coloration. Analyses that are being developed include infrared imaging and volumetric analysis.

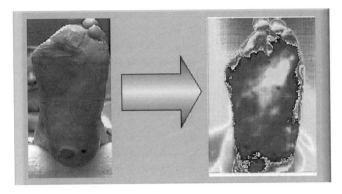

Figure 92-18 Infrared/thermal imaging. Example of thermal imaging on a diabetic foot wound. This technique can be an effective method to monitor wound healing and revascularization, as well as predict future wounds and ulcers.

Conclusions

Virtual instruments and executive dashboards allow organizations to harness effectively the power of the PC to access, analyze, and share information throughout the enterprise.

The case studies discussed in this chapter illustrate ways in which various institutions have conceived and developed "user-defined" solutions to meet specific requirements within the health care and insurance industries. These dashboards support general operations, help hospitals to manage fluctuating patient census and bed availability, and empower clinicians and researchers with tools to acquire, analyze, and display clinical information from disparate sources. Decision makers can easily move from big-picture analyses to transaction-level details while at the same time safely sharing this information throughout the enterprise to derive knowledge and make timely, data-driven decisions. Collectively, these integrated applications directly benefit health care providers, payers, and, most importantly, patients.

References

Campbell J. *Snapshot for LabVIEW® for Windows*. Rochester, NY, Viewpoint Software Solutions, 1994.
Inglis A. *Video Engineering*. New York, McGraw-Hill, 1993.
Kutzner J, Hightower L, Pruitt C. Measurement and Testing of CCD Sensors and Cameras. *SMPTE Journal* 325–327, 1992.
IEEE. Measurement of Resolution of Camera Systems, *IEEE Standard 208*, 1995.
Rosow E, Olansen J. *Virtual Bioinstrumentation: Biomedical, Clinical, and Health care Applications in LabVIEW.*, Prentice Hall, 2001.
Surgical video systems, *Health Devices* 24(11):428–457, 1995.
Walker B. Optical Engineering Fundamentals. New York, McGraw-Hill, 1995.

93

Perinatology

Vinnie DeFrancesco
Clinical Engineer, Biomedical Engineering Department
Hartford Hospital
Hartford, CT

Perinatology is a specialty that involves the medical care and treatment of an infant and mother immediately prior to, during, and following childbirth. This time in most people's lives is the happiest time they will ever spend in a hospital. This chapter will focus on the environments and technologies experienced by mother and child before, during, and after birth.

The hospital labor and delivery (L&D) environment has changed considerably over recent years because of technological advances; however, most of the environmental changes in this area have been aesthetic. The aesthetic changes are the ones considered most comforting for the patients. The environmental changes associated with the neonatal intensive care unit (NICU) and pediatric intensive care unit (PICU) tend to be more general. However, given the focus on patient care, efficiency, and safety, these changes and redesigns are quite effective. In this case, the patient can be a child ranging from a premature infant to a young adolescent.

Gaining an understanding of the area and its function is much like understanding the schematics of a medical system or device. The various areas have defined floor plans. Thoroughly understanding associated patient flow is crucial in the planning, design, and implementation stages of an L&D unit. Once an understanding of the floor plan is gained, it is important to consider the required support areas. Just as important as the floor plan for the functionality of the area, the support areas of the given location are critical to provide effective and timely care to the patient. The individual room layout and configuration (even within the same department, such as L&D) can vary considerably. Furthermore, in addition to differences among individual patient rooms, the various clinical and support areas can vary as well. Within the various room layouts, the equipment orientation and storage is discussed.

The facility should be designed to support the varying services and equipment needs of each clinical area. The important and highly visible technology of infant security systems, their implementation, and integration into the existing facility infrastructure are discussed in this chapter. The complex interrelationship among data communications, clinical facility infrastructure, and medical device technologies is discussed as well. Medical gases and vacuum systems vary in the clinical areas, depending on the procedures performed in the rooms. Other utilities concerns are climate control and electrical power distribution.

The various medical device technologies in perinatology areas range widely in complexity and criticality. Their use depends on the needs of the patient and the clinical services provided (Schreiner, 1981). The medical technologies described in this chapter (e.g., monitors, radiant warmers, and ventilators) require technical support information for their effective utilization. Some emerging technologies discussed include the electronic medical record and advanced apnea-monitoring devices.

The Changing Environment of Childbirth

Until the early 20th century, childbirth was the domain of women. Pregnant women gave birth at home, generally with other women attending the birth. Anesthetics and pain medications were not used. By the 1940s and 1950s, women flocked to hospitals to give birth, under the care of male physicians. In the hospital, the common practice was to anesthetize women during labor to minimize or eliminate any pain during childbirth. The expectant father was relegated to the waiting room to shield him from the "gruesome reality" of childbirth. During that time in France, Dr. Ferdinand Lamaze developed "childbirth without pain" (now known as the Lamaze method) based on his observations of women in the Soviet Union giving birth without anesthesia. The Lamaze method gained popularity in the United States after Marjorie Karmel wrote of her childbirth experience using Lamaze in the 1957 book, *Thank You, Dr. Lamaze*. By the 1970s, "natural childbirth" was in vogue. Although most women continued to deliver in hospital settings, couples were drawn to the Lamaze method. Women began to demand alternatives to the sterile hospital environment with its cold, hard delivery tables and stirrups. Eventually, many women and natural childbirth advocates regarded birth as a normal process that should occur in a warm setting with the family present. The consumer demand for change grew and resulted in changes in hospital practices. As natural childbirth was gaining recognition and popularity, physicians were making advances in pain management during labor. The development of the epidural anesthetic dramatically altered the birth experience for women. Suddenly, a woman could obtain pain control with drugs without losing alertness or awareness of the birth and without delivering a sluggish, "drugged" baby. Even women who required cesarean sections no longer had to have general anesthesia that deprived them of the awareness of the arrival of the baby. Many still regard natural childbirth as the ideal method, and the vast majority of women in the U.S. deliver in hospital settings. In addition, most men remain at their partner's side during delivery.

What has changed most recently in the hospital setting is the L&D environment itself. Hospitals have made great efforts to make the birthing experience occur in a warm setting with the family present. Today, the final stages of labor and the delivery occur in one room. In some cases, that room also serves as a recovery area. The L&D rooms have all of the necessary equipment for the birthing process while maintaining a pleasing appearance. Everything from the color scheme to the furniture creates a warm and peaceful setting, however, there is still the need to adhere to hospital standards. In the home, one does not have medical equipment (usually), medical gas outlets, or a bed with stirrups. Therefore, the challenge was to have these necessary resources available without significantly altering the environment. Interior designers and architects devised ways to hide the medical devices, as well as a method of retrieving them quickly when needed. Most of the equipment can be placed in cabinets located at the bed-head wall. Stirrups can be hidden under the patient's bed. The adverse effects of these aesthetic changes can cause the medical equipment to exceed operational levels, however. For example, a fetal monitor in a cabinet might overheat from lack of ventilation. The specifications on medical devices such as maximum operational temperature and heat output should be regarded well by architects who design the cabinetry.

The environmental changes for the NICU or the PICU have addressed the needs of their patients' normal environment. The general environments for children's hospitals have changed to be more suited for children. Designers and architects go to great lengths to look through a child's eyes, from a child's height. Their efforts focus on keeping children occupied to that they do not view a hospital as a bad or painful place. Great emphasis is placed on creating a quiet and peaceful environment with such devices as acoustical ceiling tiles and incandescent indirect lighting.

Patient Flow and Floor Plan

The patient's experience at the time of labor can be extremely hectic, even though the expectant mother and family have probably been preparing for this moment for the previous eight to ten months. An efficient and effective floor plan can be beneficial for the expectant mother, her family, and the hospital in general. The hospital has specific procedures that it must follow to provide the required support to deliver the newborn and ensure the well-being of the mother.

Prelabor, Labor, Delivery, and Postpartum

When the expectant mother first arrives at the hospital, she most likely will be sent to the admitting office, or to the prelabor area of L&D. The prelabor area generally comprises two separate sections: The admitting area and the patient room. The admitting area has a control desk were the clinical staff can monitor the expectant mother in her room while gathering patient information from the father necessary to complete the admissions process. In the prelabor patient rooms, the clinicians assess the expectant mother's true condition. The patient might experience contractions of the uterus that are actually "false labor" because they do not open the cervix as true labor does. At this point, the clinician can make the decision to admit the patient or to send her home. The physical location of the prelabor area should be relatively close to the emergency department/admitting office, or at least within a short walking distance. In most hospitals, the L&D rooms are adjacent to the prelabor area, because in most cases it would be the location where the expecting mother will give birth. The patient can bypass the prelabor area when the signs of labor are clearly present and can go directly to the L&D rooms, where she will go through labor and will deliver the baby. This area also may have a control desk where a clinical person will be present to perform the same duties as are performed in the prelabor area. The L&D area is usually in the center of all of the maternity areas, and the center of the L&D area is the clinical control area. This area is where most of the clinical staff will record and document the patient's progress through the labor process. The clinical staff also will assess and decide how the labor process is progressing and whether any intervention or pain medications should be administered.

Ideally, a practitioner decides for a cesarean birth if he or she ascertains that a surgical birth would be safer for the mother and the baby. A Cesarean birth is one in which the baby is removed through a surgical incision in the mother's abdomen rather than emerging vaginally. The cesarean birth is a surgical process that typically will occur in an operating room. Most hospitals have dedicated operating rooms for cesarean procedures or other operations that may arise during labor. As with any operation, the patient will be carefully monitored, and the surgical staff will be prepared for any emergency. The operating rooms are usually located adjacent to the L&D rooms for quick transport in case of emergencies. In a non-emergency situation, the preoperative procedures will be initiated calmly.

In an emergency, however, the preoperative period can be quite rushed and can appear chaotic as hospital personnel hurry to complete all of the necessary tasks rapidly. Commonly, some preoperative procedures are initiated in the mother's room before transfer to the surgical suite. Immediately after surgery, a woman typically will be wheeled to the recovery room, which usually is located in, or adjacent, to the operating rooms. Usually, a woman spends about one hour in recovery. Then, she is transferred to a room in the postpartum unit. With changes in insurance reimbursement policies, most women spend only three days in the hospital after a cesarean section, and about two days after a vaginal birth.

Support Areas

Several crucial support areas should be considered when evaluating the patient flow and floor plan of an L&D unit. Theses include pharmacy, anesthesia workroom, clinical engineering, and equipment storage. During labor, contractions can cause much discomfort, which will cause patients to request pain medications. The doctor will select the most appropriate medicines, taking into account the medical condition of the mother and her baby. In the hospital, these medications are controlled and administered by a pharmacist and anesthesiologist. The anesthesia workroom is usually located in the same area as the operating rooms to facilitate anesthetics administration if necessary. A pharmacist will distribute the medications that the doctor chooses; these medications are usually analgesics. They are sometimes administered by injection to help relieve the pain associated with contractions. Typically, a pharmacy on the maternity unit consists of a medication station. Such a station can be a self-contained medicine-dispensing unit, which usually requires a key or code to access. These devices also will record and document the clinician who administers the medication, the time of administration, and the amount administered. In large maternity units, there may be more than one medication station. Clinical engineering's main responsibility is to create and implement an equipment management program to support the operation of the medical devices in the hospital, and it may have a satellite workroom on the maternity floor.

Although most of the equipment is located in the patient rooms, specialized and less frequently used equipment may be needed at times. An area should be allocated for storage of these infrequently used devices. The storage locations should be spread throughout the maternity area so that the clinical staff can gain quick access when they need to.

Individual Room Layouts

Prelabor and Postpartum

The patient room general requirements for the prelabor and postpartum room requirements are quite similar to each other. The requirements are the same for medical and surgical nursing units. The maximum room capacity is only two patients, with a minimum of 120 square feet of clear floor area for a double-occupancy room and 100 square feet for a single-occupancy room. For cesarean delivery suites, the minimum clear floor area is 360 square feet, and 300 feet for delivery rooms. The required clear floor area does not include toilet rooms, closets, lockers, alcoves, or furniture. There must be a minimum of three feet between the sides and foot of the bed, and a four-foot clearance is required in the case of multiple bedrooms. Each patient room that is intended for 24-hour occupancy is also required to have a window or vents that can be opened from the inside. Each room must be configured such that the patient can access a toilet room without entering a general corridor. One toilet room cannot serve more than four beds, and in new construction a hand-washing facility is required if a toilet room serves more than two beds. Hand-washing facilities for patient rooms most likely are located in patient rooms. The specification on medical gases and climate control are discussed below.

Labor and Delivery

In the equipment orientation for a typical L&D room shown in Figure 93-1, a fetal monitor is built into the cabinetry of the head wall unit, which is at the head of the bed. The suction and oxygen regulators are also located at the head of the patient's bed. In new L&D rooms, electronic charting systems are commonly found. With the same cabinetry at the head wall unit, a computer displays the electronic chart of the patient in the room, as well as that of other patients on the unit. Having these medical devices built into the cabinetry gives the room environment a warmer, more home-like look. However, one should be aware of ventilation and monitors' power requirements. Given that the room looks and operates quite differently when the equipment is in use, as opposed to when it is not in use, the flow of clinicians and patients must be evaluated in both scenarios. Moving in a clockwise direction, starting at the bed (Figure 93-1), one will see an infant warmer, in which the newborn is placed. The newborn also may require medical gases and suction. These utilities are built into the wall and typically are hidden by a painting when not in use. An entrance to a toilet room and hand-washing facility is located at the opposite end of the patient bed, as are a TV and VCR, which are built into cabinets. Lastly, a reclining chair that can extended to form a bed is also provided for the patient's husband or any other person accompanying the mother at this time.

Nursery

The postpartum area of a maternity unit has the basic requirements as the Pre-labor and L&D areas. A hospital that has 25 or more postpartum beds is required to have a separate nursery to care continually for infants who require close observation. The minimum requirement for floor area is 50 square feet per bassinet or infant incubator. Most hospitals have replaced traditional nurseries with those in postpartum and labor-delivery-recovery-postpartum (LDRP) areas. These areas have the same minimum floor area, ventilation, electrical power, and medical vacuum and gases requirements as a full-term nursery does.

NICU and PICU

The NICU and PICU care for critically ill neonates and pediatric patients (Klaus and Fanaroff, 1979; Korones, 1981). These specialty intensive care units must provide for

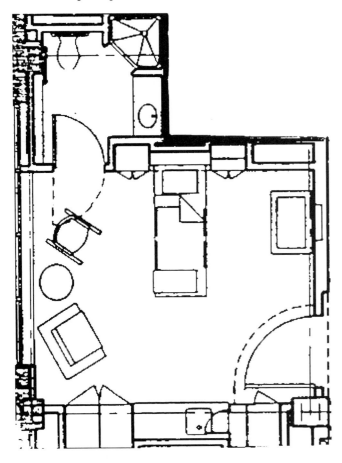

Figure 93-1 A typical floor plan for a labor and delivery unit.

appropriate staffing, isolation, and safe transportation of the patients, along with life support from supporting hospital areas. A minimum of one airborne-infection control room with the ability for nursing observation is required. The number of infection control rooms is dependent on an infection control risk assessment. These rooms must be equipped with only one patient bed and its own toilet and hand-washing facilities. The rooms' walls, ceilings, and floors should be sealed tightly so that air cannot penetrate into the room from the outside. One of the major differences between a standard critical care unit and the PICU is that a sleeping space for parents must be provided in the PICU. In addition, a consultation/demonstration room in or next to the PICU must be available for private discussions between clinical staff and family. An examination and treatment room can be located centrally to serve the PICU or an adjacent area such as the NICU. The floor area must be a minimum of 120 square feet and must contain hand-washing fixtures, storage facilities, and a countertop or desk for writing. Some of the equipment in these treatment rooms might include patient monitors, incubators, and ventilators, all of which will be discussed later in the chapter.

In all these areas, a charting/recording and dictation room for the clinical staff must be allocated. The equipment to support these functions is becoming more sophisticated and integrated with the medical devices. For example, in many institutions, the traditional whiteboard used for charting the patent's progress has evolved to become an electronic whiteboard. These types of devices require additional infrastructure such as a network that communicates with the patient monitors in each of the patient rooms.

Facility Infrastructure

Security

In addition to the security measures that a hospital implements to ensure the safety of patients and visitors, most have installed added security measures in the maternity area. These security systems are specifically keyed to operate with the parents and the newborns, to ensure that the newborn leaves specific areas of the unit with only the parents or appropriate clinical personnel. Most systems operate using an electronic ID bracelet, which transmits a signal from the newborn to receivers located at entrance and exit points. Securing it to the infant or child's limb with the alarm band activates the transmitter. A display panel showing the floor plan of the unit designating the different locations or zones will visually and audibly inform the staff when an alarm has occurred. In addition to a display panel, a graphical display (usually with a print function) collects all transmitter data and displays the location and transmitter number of the infant or child involved. These control devices are placed at the control desk or centrally located room; they also are integrated into a main security panel at the hospital's security office. Some more sophisticated systems also can be connected to the elevators, causing them to bypass the floor, thus preventing egress by completely isolating the floor from the rest of the hospital.

Nurse-call systems are required for the maternity unit and the NICU and PICU. These systems ensure communication between individual patients in their rooms and a centralized location like the control desk. They also are used to communicate to staff from the patient rooms or the entire unit. For emergencies, pull-string–activated call boxes are located in the patient's toilet rooms for activating the nurse-call system at the control desk. For additional security, various areas of L&D install and integrate into the nurse-call system a video camera located at the entrance to the unit, which allows surveillance of those requesting admittance.

These types of systems require little preventive maintenance (PM). To ensure the operation of such security systems, a transmitter tester is required, to verify the operation of any transmitter with the sensors throughout the unit. This type of test can be scheduled annually. It is important to have available space, replacement circuit boards, control panels, and nurse-call boxes for immediate replacement in the event of a malfunction. The nurse-call system should be always active to ensure patient safety.

Data Communications

Data communications can be complex and difficult to implement, depending on the existing infrastructure of the facility. Although it is not a new issue in a hospital, it can be an infrastructure concern with the medical devices in the clinical setting. Most areas in a hospital use networks to integrate hospital/department applications for data collection and access to patient information. Manufacturers of medical equipment and, especially, patient monitors have created a data communication network to display patient data. The communication protocol between the devices has been the priority for the manufacture software application and is transmitted via a serial cable and RS-232 connector. Networks were created to collect real-time patient data from multiple deceives and to display them in a centralized location. These data networks are installed in a hospital infrastructure specifically for these medical devices and exclusive to the rest of the hospital's data networks. Having these exclusive networks resulted in a secure and reliable network for real-time critical patient data. Because these devices are on an exclusive network and usually are proprietary, they cannot communicate with other applications and data unless these are provided by the same manufacture using the same protocols. These systems are limited to a specific area, usually one critical care area with limited distance between the farthest two devices or nodes on the network. Most of the medical equipment manufactures have adopted standard network protocols established by IEEE 802 standards (see Chapters 97, 98, and 102). This allows more integration with standard network hardware and other hospital systems and applications.

The protocol by which the medical devices communicate determines the infrastructure required. More and more, hospitals will need to grapple with the question of how medical devices communicate their data throughout the hospital reliably, rapidly, accurately, and securely, and how that information is to be integrated into the patient's electronic record, which will include such items as patient history, vital signs, and hospital and doctors' bills.

Climate Control

Relative humidity and ideal temperature requirements for most of the hospital are similar. The ventilation and filtration requirements vary throughout specific areas of the hospital. Table 93-1 shows the requirements for the clinical areas discussed in this chapter. For more detailed specifications and information on other clinical areas, please refer to

Table 93-1 Climate and control requirements for various clinical areas within a hospital

Location	Air movement relationship to adjacent area	Minimum outdoor air changes per hour	Minimum total air changes per hour	Recirculated by means of room units	Relative humidity (%)	Design temperature (degrees F/C)
Labor/delivery/recovery		2	2			70–75 (21–24)
Labor/delivery/recovery/postpartum		2	2			70–75 (21–24)
Caesarean/delivery	Out	3	15	No	30–60	68–73 (20–23)
Recovery room		2	6	No	30–60	70 (21)
Newborn nursery		2	6	No	30–60	75 (24)
PICU		2	6	No	30–60	70–75 (21–24)
NICU		2	6	No	30–60	70–75 (21–24)
Infectious isolation Room (airborne)	In	2	12	No		75 (24)

the American Institute of Architects' *Guidelines for Design and Construction of Hospital and Health Care Facilities* or *Health Care Facilities Handbook* by The National Fire Protection Association (NFPA 99, 90A) guidelines.

Medical Gases and Vacuum

For all of the areas discussed in this chapter, such as L&D, postpartum, NICU, and PICU, the requirements for oxygen, medical air, and vacuum are the same. A hospital's piped gas and vacuum systems are categorized as a level-one system (by NFPA 99) because patients served by this system will be at the greatest risk should the system fail. The percentage of oxygen at the outlet should be 100% (+/- 2%). The static pressure (i.e., pressure at the outlet without any flow) for oxygen, air, and nitrous oxide (N_2O) should range from 40–60 pounds per square inch gauge (psig). Gas-flow pressure is drawn from the outlet at an increasing rate until flow reaches the test value of 100 liters per minute (lpm) or the pressure drops to within +/- 5 psig of the range noted earlier. The static pressure for vacuum (suction) ranges from 12 (minimum) to 31 (maximum) inches of mercury (in Hg). The PM for medical gas and vacuum systems consists of annual testing of pressures and flow rates and of vacuum (Frank, 2003). The typical repairs for theses systems involve the outlets for the various medical gasses and vacuum or the device attached to the outlet, such as a suction regulator. Within each area, the required number of outlets for medical gases and vacuum can vary greatly. The required number for outlets is shown in Table 93-2.

Power Distribution

All electric material and equipment, including controls and signaling devices, should comply with the electrical standards specified by National Fire Protection Association (NFPA). (see Chapter 109). Electrical standards should comply with applicable section of NFPA 70 and NFPA 99 (NFPA, 2000). The major concern for electrical standards is to prevent electrical hazards from failure within electrical medical devices or facility's electrical distribution systems.

One of the requirements for electric power distribution in L&D, NICUs, and PICUs is the provision of emergency power supplied by hospital-based generators. Specially marked (by, for example, a red outlet faceplate) emergency power outlets should be provided for life-support medical devices that must remain electrically powered in the event of a power failure. Medical devices integrated with a hospital-wide information system and data processing equipment should be on emergency power. Typically, emergency power circuits are energized within 10 seconds of loss of normal power. Most data processing systems, such as servers, will restart within that time. Systems that cannot tolerate power interruption for as long as 10 seconds must be energized by an uninterruptible power source (UPS).

Technologies

This section describes some of the electronic and electromechanical medical devices that a clinical engineer is likely to encounter in the service of perinatology. Only brief descriptions of the technologies are provided, and the reader is encouraged to delve further by consulting the references cited.

Neonatal Monitors

NICU and PICU monitors have many features in common with those used for adults in the intensive care units (Neuman, 1988a; Brans, 1995). Patient monitors have developed into a computer platform requiring sophisticated microprocessors and software (see Chapter 98). Some important specifications of which to be aware are performance specifications, network capabilities, interface capabilities, operating conditions, and support issues. Evaluating the processing and software capabilities of a monitor can be similar to evaluating those of a computer. One should evaluate the main processor speed and type of chip (e.g., Motorola® MPC860P (PowerPC), 66MHz). Also the graphical capabilities as determined by the graphics processor (e.g., 8-, 16-, or 32-bit). At a minimum, the monitor should have local area network (LAN) communication capability to communicate to a central monitoring device. A more sophisticated monitor will have a LAN Communication Processor (32-bit) that uses normal IEEE 802 standards. Some of the manufacturers are even using standard operating system software, which will make it easier to create comprehensive interfaces from device to device and from device to hospital information system. The typical power requirements are 110±20 VAC, 50/60 Hz, single phase and may average 100 W (max) power consumption. When installing monitoring systems in a new or existing area, one should ensure that the heating, ventilation, and air-conditioning (HVAC) specifications for the room (e.g., ambient temperature, relative humidity, and air exchanges) can support the operation requirements of the monitor(s) (Hynman, 2003). Heat dissipation is a critical monitoring parameter (measured in BTU/hr or watts) to consider when the room draws call for the equipment to be located in a confined space like a cabinet.

Support issues for patient monitors can become much more complex if the monitors are integrated through a network. A patient monitoring network can become another system that requires support for the hardware and software components that make up the network. If the manufacture uses proprietary communication devices and software, it will be more difficult to support the equipment without support from the manufacture.

A PM program for the monitors should follow the manufacture's suggestions. Specialized test equipment may be required, to perform the recommended tests of proper operation of the monitors. Monitors may have special connectors or cables that are proprietary to the manufacturer. The sole-source nature of these accessories usually renders them more expensive than those that can be obtained from second sources. This sort of information should be considered at the time of equipment evaluation and purchase as it can have a sizeable effect on the cost of operation (see Chapter 33 and Figure 93-2, showing bedside monitors on a radiant warmer on which a neonate is lying).

Maternal/Fetal Monitors

Maternal and fetal monitors (Neuman, 1988b) provide information on fetal heart rate, uterine activity, maternal blood pressure, maternal arterial blood oxygen saturation ($MSao_2$), and maternal ECG (MECG). Early in labor or in prenatal examination, indirect measurement of the fetal heart is obtained utilizing ultrasound transducer technology. A beam of ultrasound from a transducer placed on the maternal abdomen is directed to the fetal heart. Motion of the heart causes a Doppler shift of the frequency of the impinging beam as it is reflected back to the transducer. The transducer receives the reflected signal and electronic circuitry detects any differences between the transmitted and received ultrasound frequencies. Differences in frequencies indicate detected heart motion. Periodicity of the motion is displayed as heart rate. This rate is printed on a strip-chart recorder. At the same time, uterine activity is typically detected, also indirectly, by a transducer (i.e., tocodynamometer), which is applied to the maternal abdomen and measures the relative strength of uterine contraction. Uterine activity is also recorded on the

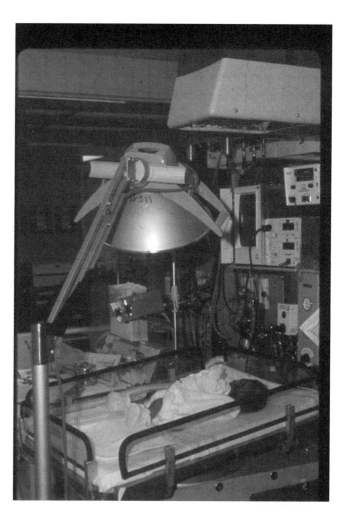

Figure 93-2 A typical array of neonatal monitors and associated cables and leads mounted on a shelf on an overhead radiant heating lamp. Also shown are examination lamp, ventilator tubing, and air-oxygen flow control.

Table 93-2 Medical gas outlet requirements for various clinical areas within a hospital

Location	Oxygen	Medical air	Vacuum
Labor room	1	1	1
Recovery room	1		3
Labor/delivery/recovery	2		2
Labor/delivery/recovery/postpartum	2		2
Caesarean/delivery	2		3
Postpartum bedroom	1		1
Newborn nursery	1	1	1
PICU	2	1	3
NICU	3	3	3
Infectious isolation room (airborne)	1		1

same strip chart. Comparing fetal heart rate variations with uterine activity gives the clinician information that helps him to determine the condition of the fetus (Spencer, 1991).

Ultrasound transducers are sensitive and easily damaged by shock and vibration, and they are not repairable. When projecting costs, the replacement cost or transducers must be taken into account. Many fetal monitors have a telemetry feature, which allows patients in labor to ambulate while the fetal heart rate and their uterine contractions are continuously monitored. This added feature requires installing a telemetry antenna field covering the L&D area. The telemetry transmitters and receivers will require additional PM and repair because they are often damaged by being dropped.

Direct sensing of heart rate and uterine pressures is done immediately before delivery, when direct access to the fetal scalp is achieved. Fetal-scalp electrodes sense the electrical activity of the heart, and intrauterine pressure transducers measure uterine pressure. To aid in the determination of the physiological status of the fetus and the mother, monitoring of the fetal and maternal arterial blood oxygen saturation is performed (see Figure 93-3).

Fetal monitoring systems can include the feature of electronically recording and archiving data. Manufactures have expanded these types of applications to include an electronic whiteboard, where information pertaining to the patient and fetus can be displayed and documented during labor. At the L&D control room, a central station display may be present for staff to survey the maternal waveforms of all patients on the unit. Implementation and support of these types of systems requires knowledge of networks and servers.

Incubators

Infant incubators provide thermal support for the neonate (Perlstein and Atherton, 1988). Most incubators also incorporate means for controlling oxygen levels and relative humidity of the air the infant breathes. Microprocessors incorporated in most modern incubators assist in the accurate control of temperature, humidity, and oxygen levels while enabling such features as graphical data trending of the critical parameters controlled by the incubator. An incubator PM program should take into account the manufacturer's recommendations and should include measurement of sound levels, operating temperatures of humidifiers, and oxygen sensors. Servo-controlled oxygen and humidity delivery systems typically require unique calibrations to be preformed during PM. In addition to calibration, humidifiers require periodic replacement of the air-intake filter.

Inadequate maintenance can result in incubators that are hazardous to the infant. Incubators are used for many years, during which time they are subjected to appreciable shock and vibration because most are mounted on casters and are moved about for cleaning and storage. In the past, mechanical stress has damaged temperature control mechanisms, which, in turn, have overheated infants, causing brain damage or death. Old incubators that have been relegated to storage can be placed back into service during a high census period (Figure 93-4). These semiretired incubators might lack necessary safety features and manufacturer-recommended upgrades or modifications and can pose serious risks to neonates. For example, on some older models, a missing heat shield can permit an infant to crawl over a hot air vent and thus suffer severe burns. Lack of maintenance results in the use of incubators with high ambient-noise levels originating from defective or misaligned air-circulating fans, which can cause hearing loss. Defective door latches can enable an infant to crawl out of an incubator and fall to the floor (Dyro, 1977). Older incubators used mercury-based temperature sensors, which often broke, thus exposing the infant to hazardous mercury vapors (Dyro, 1981). It was not uncommon to observe a pool of mercury from broken thermometers and mercury switches on the floor of the heating compartment directly beneath the infant mattress.

Phototherapy Units

Phototherapy units treat hyperbilirubinemia by irradiating the baby with light in the blue region of the spectrum from 420–500 nm (Neuman, 1988a). This light oxidizes the bilirubin in the blood, thus producing compounds that can be eliminated from the body. The unit consists of a bank of 20-watt fluorescent lamps positioned 30–40 cm above the infant. Of concern is the lamp lifetime (typically averaging 1000 hours). Some phototherapy units incorporate an hour-meter to record bulb life and infant-treatment time. In order to maintain irradiance levels (measured in w/cm^2/nm) at therapeutic levels, periodic measurement with a photometer of irradiance at the infant surface is required. Figure 93-5 shows a stand-alone, moveable phototherapy unit. A bank of fluorescent lamps is housed in the overhead compartment. In use, the unit is wheeled to the neonate's bedside (either bassinet or incubator) and illuminated with the output of the lamps. Many radiant warmers have integral or accessory phototherapy units.

Ventilators

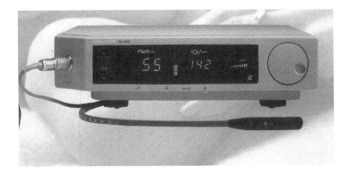

Figure 93-3 Fetal and maternal arterial blood oxygen saturation monitor.

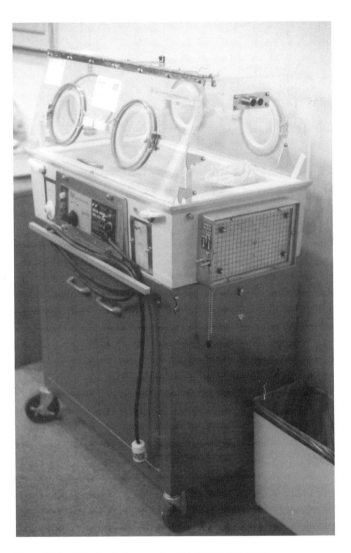

Figure 93-4 Older incubators may be placed back into service.

The ventilator's main purpose is to assist the patient in breathing (Behbehani, 1995; Kirby et al., 1990). When used with neonates and pediatric patients, the ventilator must have sufficient flow range and sensitivity. The flow range is an important consideration for purchasing ventilators for use in an NICU and PICU. Typical inspiratory and expiratory tidal volume can range from 10–399 ml ± 0.5ml for pediatrics patients, and 2.0–39ml ± 0.5ml for neonates. The flow-rate can range can be as sensitive as 0.00–3 l/s ± 1%.

Ventilators are designed to provide the proper pressure flow range (0.1 ml/s – 3 l/s) for all hospital patients. The power supply requirements are standard, as for most medical equipment, although general battery backup devices are essential and should provide battery power for approximately 30 minutes. Additional features to consider with regard to power are hospital-grade twist-lock power plugs to prevent disconnection if the ventilator is accidentally moved. Ventilators require a medical gas supply that is free from water and oil, to ensure the proper operation. The amount of water in the air supply should be less than 5 g/m^3, and the oil content should be less than 0.5mg/m^3. The amount of water in the oxygen supply should be less than 20 mg/m^3. Air and O$_2$ pressures should be in the range of 200–650kPa. The medical gas supply for the hospital should be tested prior to ventilator use. One should evaluate the hospital's standard on medical gas fittings so that the proper connectors for the ventilators can be purchased.

The proper support of these devices can be quite complicated and can require services from various hospital departments. Ventilators require proper cleaning according to the manufacture's specifications between patients for infection control and proper operation. Most ventilators also require some type of calibration or operations check between patient uses. Supporting ventilators can be quite complex and time-consuming. A preventive maintenance program for ventilators should follow the manufacture's suggestions. Test equipment is required for PMs and to ensure the proper operation of the ventilators. Ventilators require PMs to be performed on a time-in-operation basis (typically measured in hours; e.g., every 1000 hours of use). The PM requirements for the various hour intervals can vary depending on the manufacture. In many cases, PMs will require replacement parts. Parts that are required for the PMs can be purchased as a kit, and the price can vary depending on manufacture and model. The most typical parts can range from filters and O-rings to oxygen sensors. These requirements must be evaluated before purchase of ventilators, to properly plan for both infrastructure and operational cost in supporting ventilators.

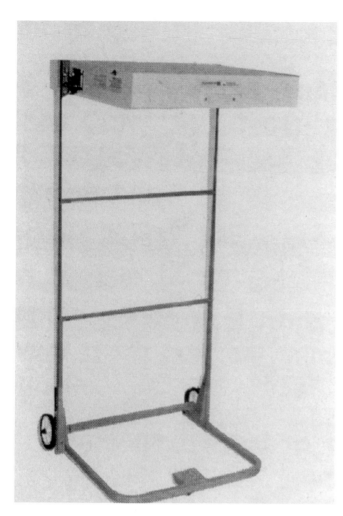

Figure 93-5 Moveable phototherapy unit.

High-Frequency Oscillating Ventilators

The high-frequency oscillating ventilator (HFOV) is one of several high-frequency ventilators shown to be effective in the management of neonates in respiratory failure, especially those with severe barotrauma (Null et al., 1990). Other high-frequency ventilators include high-frequency positive-pressure ventilators, high-frequency jet ventilators, and high-frequency flow interruptors. HFOVs are capable of supporting the ventilatory and oxygenation requirements of neonates, full-term infants, and children who might be failing despite conventional ventilation. This type of ventilator uses a diaphragmatically sealed, electromagnetically driven piston, which is the main mechanical difference between the jet ventilator and conventional ventilators. For this type of ventilator, it is important to follow the manufacture's suggestions on a PM program. Again, like the traditional ventilators, test equipment is required for PMs and calibration, to ensure the proper operation of the ventilators. The drive piston requires changing after it has performed a certain amount of cycles, as well as calibration every six months. The piston requires centering, which requires applying electrical counterforces to the piston coil, which maintains the piston centering. This amount of pressure can range from ± 130 cmH$_2$0, with an accuracy of ± 2% of reading or ± 2 cmH$_2$0. The same specifications for medical gases, emergency power, and backup battery of the conventional ventilator apply to the high-frequency ventilator.

Inhaled Nitric Oxide Delivery Systems

The Inhaled Nitric Oxide (INOvent) delivery system is a delivery and monitoring system for inhaled nitric oxide therapy (Walding, 2003). The INOvent delivery system is not a stand-alone medical unit. It works in combination with a mechanical ventilator (or anesthesia machine) to deliver a constant concentration of nitric oxide (NO). The INOvent has its own comprehensive alarms to ensure safe and reliable operation with continuous monitoring of NO, NO$_2$, and O$_2$. Nitric oxide is supplied to the system by way of a gas cylinder. Both the gas and the system are highly specialized, heavily regulated, and available only through the manufacturer.

The INOvent is a highly specialized medical device. The manufacture recommends that all repairs and PMs be performed by a specified, direct supplier company. The system does require a performance-verification procedure before using the unit on a patient. This performance verification includes bleeding and purging the head unit and the NO tanks prior to operation or testing. Using a pressure gauge and a test circuit, one can perform various calibration tests to ensure proper operation.

Radiant Warmers

Most radiant warmers are integrated for L&D or LDRP areas of the hospital. Some new models of radiant warmers have changed with the environment of most LDRP rooms by incorporating such features as detachable wooden carts, which can blend into the aesthetics of the room. Some radiant warmers have pneumatic modules for resuscitation. Radiant warmers can be equipped with many options for measuring the newborn's temperature. The main function of the device is to providing radiant heat by way of infrared heating elements, which typically are mounted above the infant mattress. The output of these elements is regulated by a control mechanism that responds to infant skin-temperature information obtained from temperature sensors attached to the infant.

Figure 93-6 shows a radiant warmer termed "neonatal care station" by the manufacturer. Radiant warmer technology has advanced over the last two decades, in part through the development of performance and safety standards (Dyro, 1977). Early versions of the radiant warmer had few of the safety features now incorporated into modern radiant warmers. While new, more effective, and safer technology exists, one must remain vigilant to control older technology that might be brought out of storage to address a high-census situation, as discussed above.

Care should be taken to ensure that temperature-sensing mechanisms are calibrated and that compatible temperature sensors are used. Incompatible sensors cause some radiant warmers to emit maximum heat output. Such uncontrolled maximum heat has burned infants (Dyro, 1977). The heater output should have a sensitive incremental adjustment, typically 10% of output. Inspections should include a check of infant containment mechanisms such as side panels. Defectively designed or worn latches have given way, thus enabling infants to fall off the mattress and onto the floor.

The pneumatic mechanisms give clinicians the capability to ventilate the neonate. They provide gas flow for a neonate and can operate with a gas blender to provide specific levels of oxygenation. Source gas pressures and flows affect performance and must be inspected periodically.

Apnea Monitors

Apnea monitors are part of the patient monitoring systems in the PICU and NICU (Neuman, 1988a). Some stand-alone apnea-monitoring systems are frequently used in the home for infants who have had apneic episodes in the hospital or who are at risk of apnea. (See Figure 93-7.) These incorporate heart rate monitoring as well as respiratory monitoring. Monitoring of breaths is typically done by transthoracic impedance monitoring. The same electrodes that are used to detect the ECG for heart rate monitoring are used to conduct a low-level, high-frequency (20-100 kHz) current through the infant. Breathing motions change the impedance, thus resulting in a voltage change detected by the monitor. When the monitor senses that breathing motions have ceased, it alarms after a delay of typically 20 seconds. The alarm alerts the care-giver to intervene. In many

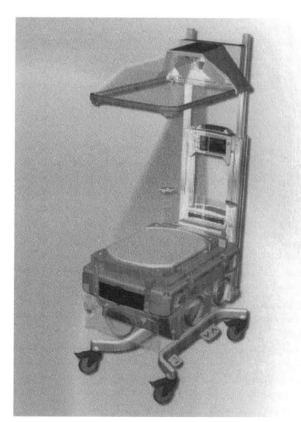

Figure 93-6 Infant radiant warmer.

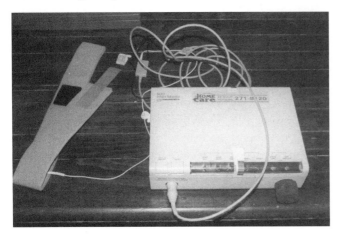

Figure 93-7 Transthoracic impedance apnea and heart-rate monitor.

cases, an apneic infant will resume breathing in response to slight tactile stimulation. In some cases, aggressive cardiopulmonary resuscitation is required.

Apnea monitoring is an imperfect technology in that all techniques that are used to detect breathing (e.g., transthoracic impedance, pressure sensors, airway temperature monitoring, pneumotachography, carbon dioxide sensors, and sound detectors) are susceptible to artifact, which can defeat the alarm system of the monitor. Dyro (1976, 1978) has described the hazards and risks of a wide range of apnea monitoring technologies. Environmental factors such as external vibrations from other devices and electromagnetic interference, and internal factors such as cardiogenic artifacts can cause the monitor to detect breaths when the infant is not breathing (Dyro, 1998).

Ultrasonic Scanners

Ultrasonic scanners use measure fetal age, gestational age, and fetal weight to obtain multiple ratio calculations, fetal biophysical profile, and twin gestational measurements and to generate reports. These are standard obstetric parameters that are usually incorporated into general ultrasound-imaging units (Goldberg et al., 1995). Computerized ultrasound units have specific programs and built-in ultrasound probes. Most machines allow for all fetal-growth data and ratio data to be stored on one 3.5-inch diskette for convenient data storage and retrieval and report printing.

A PM program for ultrasonic scanners should follow the manufacture's suggestions. Required testing phantoms are required, to perform the recommended tests to ensure the proper operation of the ultrasonic scanner and probes. Replacement probes need not be repaired or purchased from the manufacture of the ultrasonic scanner unless they are specialized.

Future Technologies

The L&D, NICU, and PICU areas recognize the need to utilize an electronic medical record (EMR). The patient flow that was highlighted in the beginning of this chapter could be improved if all one had to do at the admitting office were to provide a card containing pertinent patient information. As patients progressed through labor, all vital signs, administered medications, and maternal parameter would be captured. For example, a maternity unit could request the acquisition of fetal monitoring data. If alarm conditions occurred, or if some type of intervention were necessary, (e.g., administration of medication), the information typically would be printed from the fetal monitor and retained. The fetal-monitoring data (e.g., heart rate variations and uterine contractions) also could be captured electronically and displayed on monitors, usually in the control room area. If a documentable event occurred, the clinician could make some type of indication on the fetal strip printout. Clinicians want the event to correlate to the correct time and fetal information on the fetal strip. Even though the fetal strip can be captured electronically, the information that the clinician dictates on the fetal strip cannot be. The department therefore keeps the printed fetal strips and electronic data documentation in the patient's medical record. A playback feature for the fetal monitoring recorded strips is needed, to document accurately the exact time an intervention was performed. The clinician then can play back the fetal strip to the point where a notation was required, without stopping the collection of new fetal-monitoring information.

Clinical engineering research and development has developed closed-loop monitoring systems. For example, traditional apnea monitors, after sensing apnea or low or high heart rate, trigger an alarm, which alerts the nurse to intervene. The nurse's first response would be tactile stimulation by hand contact, which often results in resumption of breathing. Providing a closed-loop monitoring method, a computer system can record EKG, heart rate, and pulse rate from a pulse oximeter to sense apnea. When apnea is detected, a signal is sent to a vibrotactile transducer, which can be placed on the neonate's thorax to stimulate breathing (Pichardo, 2003). The computer also can record the EKG, heart rate, and pulse rate of the neonate and correlated the data in the same time frame the stimulation was administered.

Conclusion

A wide range of varied and complex medical device technologies are utilized in perinatology. Opportunities exist for the clinical engineer to ensure that appropriate technologies are specified and acquired and, after purchase, to ensure that they are maintained in proper operating order. Research and development by clinical engineers can lead to improvements in existing technology.

References

The American Institute of Architects. *Guidelines for Design and Construction of Hospital and Health Care Facilities*, 2nd Impression., The Institute, 1998.
Behbehani K. Mechanical Ventilation. In Bronzino JD (ed). *The Biomedical Engineering Handbook*. Boca Raton, FL, CRC Press, 1995.
Brans YW, Hay WW. *Physiological Monitoring and Instrument Diagnosis in Perinatal and Neonatal Medicine*. Cambridge, Cambridge University Press, 1995.
Dyro JF. Apnea Monitors. In Moore TD (ed). *Iatrogenic Problems in Neonatal Intensive Care*. Columbus, OH, Ross Laboratories, 1976.
Dyro JF. *The Development of a Standard for Infant Warmers and Incubators*. FDA/BMDDP -77/43, PB-263 250/3WV. Springfield, VA, National Technical Information Service, 1977.
Dyro JF. *An Investigation to Determine the Risks and Hazards Associated with Apnea Monitors*. FDA/BMDDP-78/142, PB-284 858/8WV. Springfield, VA, National Technical Information Service, 1978.
Dyro JF. Mercury Hazards Should Be Removed from Incubators. *Device Techniques* 2(4):37, 1981.
Dyro JF. Methods for Analyzing Home Care Medical Device Accidents. *J Clin Eng* 23(5):359-368, 1998.
Frank B. Gases. In Dyro JF (ed). *The Handbook of Clinical Engineering*. Burlington, MA, Elsevier, 2004.
Goldberg RL, Smith SW, Mottley JG, et al. Ultrasound. In Bronzino JD (ed). *The Biomedical Engineering Handbook*. Boca Raton, FL, CRC Press, 1995.
Kirby RR, Banner MJ, Downs JB. *Clinical Applications of Ventilatory Support*. New York, Churchill Livingstone, 1990.
Klaus MH, Fanaroff AA. *Care of the High-Risk Neonate, 2nd Edition*. Philadelphia, WB Saunders Co, 1979.
Korones SB. *High-Risk Newborn Infants, 3rd Edition*. St. Louis, Mosby, 1981.
Miodownik S. Intensive Care. In Dyro JF (ed). *The Handbook of Clinical Engineering*. Burlington, MA, Elsevier, 2004.
Gardner TW. *Health Care Facilities Handbook, 6th Edition*. New York, National Fire Protection Association, 2000.
Neuman MR. Fetal Monitoring. In Webster JG (ed). *Encyclopedia of Medical Devices and Instrumentation*. New York, Wiley, 1988a.
Neuman MR. Neonatal Monitoring. In Webster JG (ed). *Encyclopedia of Medical Devices and Instrumentation*. New York, Wiley, 1988b.
Null D, Berman LS, Clark R. Neonatal and Pediatric Ventilatory Support. In Kirby RR, Banner MJ, Downs JB (eds). *Clinical Applications of Ventilatory Support*. New York, Churchill Livingstone, 1990.
Perlstein P, Atherton H. Infant Incubators. In Webster JG (ed). *Encyclopedia of Medical Devices and Instrumentation*. New York, Wiley, 1988.
Pichardo R, Adam JS, Rosow E, et al. Vibrotactile Stimulation System to Treat Apnea of Prematurity. *Biomed Instrum Technol* 37(1):34–40, 2003.
Schreiner RL. *Care of the Newborn*. New York, Raven Press, 1981.
Spencer JAD. *Fetal Monitoring: Physiology and Techniques of Antenatal and Intrapartum Assessment*. New York, Oxford Medical Publications, 1991.
Walding DL, David YB, Garcia X, et al. Design of a Virtual Instrument to Correlate Tagged Exhaled Nitric Oxide Breaths and Pulmonary Mechanics Measurements for Ventilated Pediatric Patients. *J Clin Eng* 28(2):123-125, 2003.

Further Information

Fletcher MA, MacDonald MG, Avery GB. *Atlas of Procedures in Neonatology*. Philadelphia, JB Lippincott, 1983.

Monitors

Corometrics Fetal 129 Series Monitor Operations and Service Manual, 1998.
HP. *Merlin Service Manual.*, Hewlett-Packard, 1995.

Incubators:

Hill-Rom. *Hill-Rom Air-Shields Isolette C2000 Operations Manual*. www.hill-rom.com.

Phototherapy Units

http://www.medela.com.

Ventilators

Siemens. *Servo 3000 Service Manual.*, Siemens, 1999.
The Sensormetics Oscillating Ventilator Service Manual, 1998.
INOvent. *INOvent Delivery System: Datex-Ohmeda Operations Manual*. 1999. www.ohmeda.com.

Radiant Warmer

www.hill-rom.com.

Apnea Monitors

Corometrics Medical Systems. *Corometrics Service Manual*, 1988.

Ultrasonic scanners:

wwwsiemensmedical.com.

94
Cardiovascular Techniques and Technology

Gerald Goodman
Assistant Professor, Health Care Administration,
Texas Women's University,
Houston, TX

There is perhaps no discipline of medicine that mimics electrical engineering as much as cardiology. From the study of electrical conduction for heart activity, to cardiovascular dynamics, the basic $E = IR$ equation applies. If one substitutes electron flow for blood flow, substitutes electrical resistance for vascular resistance, and views the heart as a power source, then one has the following: Cardiac output = (blood flow) × (vascular resistance). In addition, cardiology may employ more electronic diagnostic technologies than any other medical discipline.

This chapter defines the medical specialty areas grouped under cardiovascular diseases, discusses the history of cardiovascular and cardiac technology as applicable to clinical engineering, and identifies future trends in this technology.

Definitions

In order to define and clarify this diverse field, one must first define the field anatomically. One can then define the medical specialties that typically practice on these organs and systems.

Anatomical Definitions

Thorax typically refers to the chest and lungs. Thoracic surgery can be any procedure in the chest cavity. However, it is typically associated with surgery on the heart or major arteries and veins.

Cardiac refers to the heart itself. Reference may be to heart muscle, heart valves, or other cardiac structures. The term *coronary* is tightly associated with cardiac disease (coronary artery disease).

Coronary is defined as "encircling in the manner of a crown"." It applies to vessels, nerves, ligaments, and other areas. For our discussion, "coronary" refers to that part of the coronary vascular system (coronary arteries) that serves the heart muscle.

Vascular refers to veins or arteries, or more generally, any specialized structure for conducting fluid. Within the terminology are several subcategories:

- **Cardiovascular**—relating to the heart and blood vessels (arteries, veins, and capillaries); the circulatory system.
- **Cerebrovascular**—relating to or involving the blood vessels in the cerebrum.
- **Peripheral vascular**—relating to the large blood vessels of the arms, legs, and feet.
- **Microvascular**—relating to the small blood vessels which lead to organs and nerves.

The adjective "cardiovascular" is typically used all-inclusively to describe diseases and treatments related to the heart and its associated circulatory system.

Cardiovascular Disease Medical Discipline Definitions

From the anatomical definitions come several major medical subspecialties for treating diseases of the cardiovascular system. The two major disciplines are cardiology and cardiovascular, or cardiothoracic, surgery. The distinction between the two is found in the term "surgery." Cardiologists typically provide medical interventions, while cardiovascular surgeons provide surgical interventions.

Not all vascular diseases directly involve the heart. While the cardiologist, by definition, treats diseases of the heart and the cardiovascular surgeon performs surgery on the heart, each specialty must address diseases that can involve any part of the circulatory system. A cardiologist may diagnose and treat vascular disease as well as heart disease. Vascular surgeons may operate on the blood vessels only, while cardiovascular surgeons may operate on both the heart and associated vascular system. Typically, the two surgical specialty names are used interchangeably.

Cardiology has evolved into both a diagnostic and an interventional medical practice. Procedures that typically have been performed in the operating room by a cardiovascular surgeon, such as cardiac pacemaker insertions, may now be performed by the cardiologist in the cardiac catheterization laboratory ("cath lab") (Lin, 1988). The technologies that are available to treat blockages in the coronary arteries, such as angioplasty (DeSciascio and Cowley, 1988), can also be performed in the cath lab. Cardiologists who practice in this area are referred to as "interventional cardiologists."

Cardiology has a subspecialty in the diagnosis and treatment of cardiac arrhythmias, which are variations in the electrical rhythm of the heart. An electrophysiologist specializes in the diagnosis and treatment of electrical conduction problems that cause arrhythmias.

Diagnostic Technologies for Cardiovascular Disease

The technologies used in the diagnosis and treatment of cardiovascular diseases are quite diverse. They include surgical interventions that physically repair diseased components of the heart (e.g., heart-valve replacement surgery), various surgical and nonsurgical techniques to clear artery blockage, and treatment with drug regimens to control such parameters as blood pressure and cholesterol level.

The discussion of technology is organized first by type of intervention, and then by location of use of the technology. Types of interventions are noninvasive and invasive diagnostic and surgical. Locations of use are office-based, cardiac cath lab-based, and operating room-based technologies.

Diagnostic Office–Based Noninvasive Technologies (Static Tests)

Diagnostic technologies for cardiovascular disease have evolved rapidly over the past ten years. The technology has focused on noninvasive diagnostic methods that reduce patient cost and patient risk. The primary classifications of diagnostic testing are static tests, stress tests, and ultrasound-based tests. All are office-based and typically are performed on an outpatient basis.

Static testing for cardiovascular disease refers to those tests that do not require the patient to exercise or otherwise physically alter their cardiac performance either during or immediately before a diagnostic test. Static tests include electrocardiograph and heart sound auscultation.

Electrocardiography

The electrocardiogram (EKG or ECG) is the basic assessment tool of cardiology (Clements and Bruijn, 1990). The technology uses surface electrodes placed on the chest of the patient (Neuman, 1998). Electrical activity of the heart—the electrocardiogram—is then reproduced on an oscilloscope-type display, and in a paper-chart recording.

While the physiology of the ECG is beyond the scope of this article, the essential facts are that the ECG represents the electromotive force generated by heart muscle (Berbari, 1995). These forces are represented as force vectors in the same manner as any force vector in engineering. The standard ECG technology uses multiple leads to record the vectors. Typically, a 12-lead view of heart electrical activity is reproduced simultaneously in a test that lasts less than one minute. Common technical problems in using and ECK machine include poor electrode adhesion, lead wire fractures, and background electrical noise (60-cycle electrical noise).

Phonocardiography

The phonocardiogram (PCG) detects and records heart sounds, the sounds made by the various cardiac structures pulsing and moving blood. The sound is caused by the acceleration and deceleration of blood and turbulence developed during rapid blood flow. The essential technology for listening ("auscultation") to heart sounds is still the basic stethoscope. Electronic stethoscopes are available, but they have had little impact. Heart-sound auscultation is both a science and an art. The science of the sounds that relate to a specific heart problem has been well established. The ability to discern the sounds or to recognize the sounds that are being heard is perhaps the art, although time-frequency techniques have been employed to analyze the phonocardiogram (Bulgrin et al., 1993).

Nuclear Source Testing

There are a number of cardiovascular static tests available that employ a radioactive source that is injected into the patient. These demonstrate damages and evaluate the function of the heart. As the injected radionuclide circulates to the heart, the presence of radioactivity in the ventricles of the heart or the myocardium is measured. Both planar and three-dimensional nuclear medicine are used in these studies. Single-photon emission computed tomography (SPECT) is a nuclear medicine technique that yields a three-dimensional reconstruction of the distribution of the radiopharmaceutical (Fahey et al., 1988).

Technetium-99m pyrophosphate scintigraphy is used to diagnose acute myocardial infarction at least 18 hours post-infarction. Areas of damaged myocardium contain calcium, which complexes with the Technetium-99m. Therefore, infracted areas appear as hot spots.

Persantine thallium can be used for patients who cannot exercise. Persantine™ is the trademark for the drug dipyridamole, which is a coronary vasodilator. Persantine™ dilates normal coronary vessels more than the occluded vessels. Blood would preferentially flow to the normal vessels in a sort of "steal" phenomenon. A thallium scan can then reveal the ischemic areas by demonstrating reduced perfusion.

Diagnostic Office–Based Noninvasive Technologies (Stress Tests)

The diagnosis of cardiovascular disease may require the patient to physically alter their heart activity. The heart under stress from such activities as walking or running can give a clearer diagnostic picture than can a heart at rest. Although low, there are risks to stressing through exercise the heart of someone with suspected heart disease. Mortality was 0.5 per 10,000 in the United States and 0.2 per 10,000 in Europe in the late 1970s. Morbidity, primarily from ventricular tachyarrhythmias, is equally infrequent (Mukharji, 1988). Stress testing is performed under tightly controlled conditions, with continuous cardiac function monitoring, and with full cardiac resuscitation capability available. Conventional and nuclear stress testing are available.

Conventional Stress Testing

Stress testing can be accomplished with the patient walking or running on a standard exercise treadmill, with continuous electrocardiograph and blood pressure monitoring (Mukharji, 1988). Figure 94-1 shows a typical stress test system. Arrhythmias may appear during the exercise period or the resting period immediately after exercise.

Nuclear Stress Testing

Stress testing can also be accomplished in conjunction with the injection of a radionuclide. Thallium stress testing evaluates heart function by the observation of thallium distribution in the myocardium during exercise. Thallium is taken up by normal cells, in which case the scan shows an evenly distributed pattern of radioactivity. Areas of reduced blood flow (ischemic areas) induced by stress will appear as cold spots. In addition to the conventional equipment required for a stress test, the thallium stress test requires the use of a nuclear camera to record the thallium distribution. Because a nuclear source is being used, oversight by a radiation safety office that is certified for nuclear radiography is required (see Chapter 113).

Long-Term Trend Monitoring

The monitoring of cardiac activity during activities of daily living is a technique for capturing subtle arrhythmias (Tompkins, 1988). While not specifically a stress test, long-term monitoring does look at cardiac function under varying conditions of physical activity.

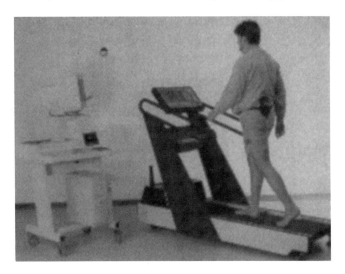

Figure 94-1 Stress-test system.

The patient wears the Holter monitor for a continuous 24- to 48-hour period (Holter, 1961). The monitor and technique are named after the late American biophysicist Norman Jefferis Holter. The monitor records a continuous electrocardiogram over this period. Monitors that can record blood pressure as well as electrocardiograph are also available. Older technology uses analog tapes for the recording. Newer monitors employ digitizing technology. In addition to the monitor, a scanner is necessary to decompress and analyze the recording. The output of the scanner typically includes a summary of events by arrhythmia type, and an actual electrocardiographic recording.

An advance to the continuous recording Holter monitor is an event monitor carried by the patient, and used only when the patient senses a cardiac event. The monitor records a short-duration electrocardiogram, and can store up to six such events. The patient can send the recording to a receiving center via a telephone modem (Thakor et al., 1984). The receiving station records the arrhythmia and forwards a report to the attending cardiologist. The patient might carry the monitor for a month or more.

Diagnostic Radiography

Conventional radiography has been employed in the diagnosis of cardiovascular disease in the form of the basic chest X-ray. Of interest to the cardiologist would be enlarged heart size, fluid between the membrane lining the heart and the heart itself, and fluid in the lungs.

CT scanning was not previously available to evaluate heart disease because the organ of interest—the heart—was in constant motion (Cunningham and Judy, 1995). Fast CT scanning has overcome this problem by employing basic CT scanning technology and enhanced computer technology that can reconstruct a beating heart (Boyd and Lipton, 1983; Ritman, 1990).

While not a radiography technique, positron emission tomography (PET) scanning is also employed to evaluate heart disease. PET involves the tracking of a nuclear source injected into the patient (Budinger and VanBrocklin, 1995). ET scanning is not a stress test like that used in nuclear medicine. Limitations in PET technology are the expense of the hardware, the necessary radiation protection for the facility, and the availability of nuclear pharmacology.

Diagnostic Ultrasound

Also known as cardiac ultrasound, diagnostic ultrasound is used to diagnose, in real time, the movement of valves and the walls of the heart during a cardiac cycle. Valvular stenosis and regurgitation, septal defects between the chambers, ventricular dyskinesia, and growth of myxoma (a type of cancer) can be diagnosed (Yoganathan et al., 1995; Ferrara, 1995). An echocardiogram is a method of studying the heart's ventricular size, function, wall thickness, motion, and valves by using Doppler ultrasound.

Ultrasound is becoming the standard assessment tool for cardiovascular disease. It is completely noninvasive, requires no exercise risk, uses no contrast media, and uses no ionizing radiation. The technology has evolved from simple M-Mode (motion mode) ultrasound, to 2-dimensional ultrasound and color ultrasound (Ferrara, 1995).

As with other diagnostic techniques, the ultrasound is both science and art. The science of what ultrasound can show has been well established. The art is in the ability of the cardiologist to interpret what he sees.

Diagnostic Cardiac Catheterization Laboratory–Based Technologies (Invasive)

The cardiac catheterization laboratory has evolved into a highly sophisticated diagnostic laboratory. Typical technology employed includes an imaging system, physiological monitoring, and radiographic recording technologies for both standard radiographic images and cinematic film, and an injector for radiopaque dye. By its nature, all procedures performed in the cardiac cath lab are invasive.

Coronary Angiogram

The standard procedure performed in the cardiac cath lab is coronary angiography (Van Lysel, 1995). This procedure involves the introduction of a catheter into the right or the left side of the heart to study the pressures in the central vein, across the valves of the arteries and the chambers of the heart. The volumes in the cardiac chamber during the cardiac cycle and the patency of the coronary artery are also measured by observing the flow pattern of radiographic dye injected through the catheter.

Right-heart catheterization measures the right atrial, right ventricular, pulmonary artery, and pulmonary capillary wedge pressures, oxygen saturation, and cardiac output. The catheter is inserted through the femoral vein or subclavian vein. Left-heart catheterization measures the aortic and mitral valve stenosis and regurgitations, global and regional left-ventricular functions, and coronary arteriography. The catheter is inserted through the femoral artery or the brachial artery.

The radiographic system typically employed in the cath lab is a c-arm single plane or bi-plane fluoroscopy system. The c-arm arrangement allows for easy viewing of right-heart or left-heart features. Bi-plane imagery also facilitates the examination of left-heart or right-heart features without the need to reposition the patient. The X-ray table is typically a cradle arrangement that allows the patient to be moved during fluoroscopy. The dye injector can be synchronized with the radiographic unit in order to capture the dye-flow pattern. During the procedure, both conventional radiographic film and cine film of the dye distribution are made.

Physiological monitoring includes surface electrocardiograph recordings as well as intra-cardiac electrocardiograph recording. Because of the very low electrical signals being recorded, electrical background noise must be suppressed. The electrical grounding system in the cath lab is a critical installation issue. The physiological monitor will also typically record multiple intra-cardiac invasive blood pressures.

A variation of the cardiac cath lab is the vascular lab. There, the intent is to examine the vessels other than those associated directly with the heart. The vascular lab would employ similar radiographic equipment in order to evaluate the vessels serving, for example, the brain.

Electrophysiology Study

A separate area of diagnosis performed in the cardiac cath lab is an electrophysiology study. There, the cardiologist seeks to identify aberrant electrical pathways that can cause cardiac arrhythmias. The technology employed includes the standard cath lab radiographic and physiological monitoring equipment. In addition, equipment specifically designed to study or "map" cardiac electrical impulses is employed.

The cardiologists will electrically stimulate the heart and will record the electrical conduction timing from the origination node in the heart, the sinoatrial node, through the electrical pathways in the heart, the HIS Bundle, and finally through the distribution system to the heart muscle itself. These latter conduction structures are the Purkinje fibers. A three-dimensional map of cardiac electrical activity can be developed. Once the pathways have been identified, a radio frequency pulse is delivered to cut the aberrant pathway. The pulse is very similar to that developed in electrosurgery devices. The procedure is called a "radiofrequency ablation." The procedure is both diagnostic and therapeutic in that a cure to the arrhythmia results from the ablation.

As with any cardiac electrical measurements performed in the cath lab, electrical background noise can disrupt the procedure. Electrical grounding, to include isolation transformers on critical diagnostic equipment, is required.

Intra-Cardiac Catheters

The catheters employed in cardiac diagnosis are a technology all their own. There, multi-lumen catheters are employed to measure pressures, inject dye or medications, and measure electrical activity. Measuring electrical activity requires that the catheters incorporate multiple poles spaced along the distal tip of the catheter. The catheter must be large enough to accommodate the lumen and electrodes, yet small enough and flexible enough to be inserted in the heart via a femoral or jugular vein.

A different type of catheter is used for the electrophysiology study. This catheter will incorporate multiple electrodes to measure the intracardiac electrocardiogram, as well as a pole for the delivery of the radiofrequency pulse to ablate the electrical pathway.

Surgical Interventions for Cardiovascular Disease

Operating Room–Based Surgical Interventions

Surgical interventions for cardiovascular disease include, at one extreme, cardiac transplantation, and at the other, replacement of diseased valves and other heart structures. Because the heart is essentially a pump, surgical interventions have been compared to "plumbing repairs". For some congenital heart defects, the surgeon does, in fact, reroute the great vessels serving the heart. Patches can be placed on or in the heart to correct holes or areas where heart muscle has died. Figure 94-2 shows a typical radiology department special procedures room where interventional cardiovascular procedures are undertaken.

The intervention can be limited to the chest cavity or can be an open-heart procedure. For the latter, the technology employed includes advanced physiological monitoring, anesthesia support equipment, and the heart-bypass (extracorporeal) pump (Dorson and Loria, 1988). Because the "pump" supports both circulation and respiration, the blood circulating through the pump may be warmed or cooled. In addition, in order to limit blood loss, a blood scavenger system, or "cell saver," (Haemonetics Ltd.) may be employed. Such a device separates and washes red cells from blood aspirated at the operative site. The washed red cells then may be resuspended in plasma protein solution and autotransfused.

Perhaps the most common cardiac surgical procedure is the coronary artery bypass graft (CABG or "cabbage"). There, the coronary arteries serving the heart are bypassed with a vessel taken from the patient. One or more coronary arteries may be bypassed in this manner. This is one of the most common surgical procedures performed in the United States.

The cardiovascular surgeon also performs vascular procedures. One common procedure is a carotid endarterectomy. In this procedure, the surgeon physically removes plaque from the inside lining of the carotid artery, restoring normal blood flow to the brain.

A number of technologies are used in support of cardiac surgery. Physiological monitoring after cardiovascular surgery may include cardiac output monitoring. Several techniques have been employed (Geddes, 1995). The current state of the art in cardiac output measurement is "thermal dilution." The "gold standard" of any cardiac output testing is the Fick Method. In conjunction with cardiac output monitoring, pulmonary arterial pressure (Gardner, 1988) can be continuously measured using a Swan-Ganz catheter (Swan et al., 1970).

Heart muscle that has been damaged as a result of blocked coronary arteries might need supplemental pumping support postoperatively. Intra-aortic balloon counterpulsation (balloon pumping) is typically used. Intra-aortic balloon pumps (Jaron and Moore, 1988) utilize a balloon placed in the descending aorta. Pulsating counter to the pumping of the heart, the pump reduces the workload on the heart by reducing the peripheral vascular pressure. The pump then provides support pumping to maintain adequate blood pressure for circulation (see Figure 94-3).

Another form of counterpulsation is enhanced external counterpulsation. This noninvasive treatment is designed to lessen the work for the heart, usually for patients with angina (Lawson et al., 2000). Blood pressure cuffs are placed on the legs, and in a coordinated way some cuffs are inflated with air while others are deflated. It helps the blood circulation in the heart. Treatments are observed in a doctor's office, lasting one to two hours each, and then are repeated over time. Figure 94-4 shows an ambulatory care patient receiving enhanced extracorporeal counterpulsation treatment.

Cardiac Catheterization Laboratory–Based Surgical Procedures

It is probably incorrect to classify the following procedures as surgical procedures. However, they are discussed under this heading because they often replace procedures performed in the operating room.

The interventional cardiologist has had a significant impact in moving cardiovascular surgical procedures out of the operating room and into the outpatient setting of the cath lab. The majority of the interventions have been directed at clearing blockages in the coronary arteries, providing a less costly, safer alternative to CABG.

The techniques and technologies used for clearing the coronary arteries include coronary angioplasty (also known as "percutaneous transluminal coronary angioplasty," or PTCA), stenting, and lasers. The technology and techniques have changed over the past 10 years and continue to be an intense area of clinical research. The basic technology employed in the cath lab is the same as that used for a coronary angiogram, but the change has been in the type of catheter used and in the skills of the interventional cardiologists.

In coronary, or balloon, angioplasty, a catheter is threaded into a blocked coronary artery (Walton, 2002). A balloon on the tip of the catheter is then inflated. The inflation of the balloon pushes the plaque on the inside of the artery wall back, reopening the artery. Stenting is a modification of this procedure. There, once the plaque is pushed back, a stent or rigid tube is positioned inside the artery to keep plaque from reclosing the artery.

In both procedures, reocclusion of the artery can occur. Repeat procedures can be performed, although there is a risk of rupture of the coronary artery, or releasing plaque into the circulatory system. Various technologies are under investigation to work in conjunction with balloon angioplasty and stenting to reduce reocclusion.

Other Technologies in Cardiovascular Disease

Three other technologies are used for the treatment of cardiovascular disease: Pacemakers, defibrillators, and drugs.

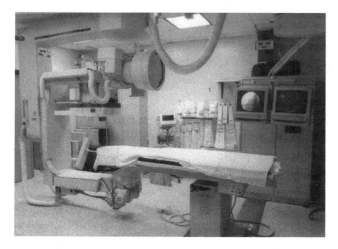

Figure 94-2 Special procedures room.

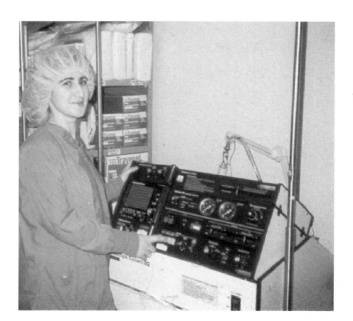

Figure 94-3 Intra-aortic balloon pump being prepared by a clinical engineer.

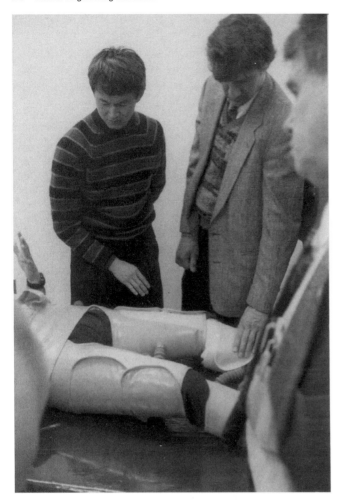

Figure 94-4 Vasogenics enhanced extracorporeal counter-pulsation system treating an ambulatory patient who is being treated on an out-patient basis. Those observing the procedure are clinical engineering interns.

Pacemakers

Pacemaker technology has evolved in both sophistication and reduction in physical size (Geddes, 1995). In addition, improvements in battery technology, such as the lithium battery, now provide pacemakers with multiyear life spans. Battery life has been improved because of new battery technology and because of better pacemaker lead wires and electronics with lower power consumption. Pacemaker technology, itself, has evolved to include pacemakers where a single unit can stimulate either the atrium or the ventricle.

The traditional pacemaker application is to stimulate the heart when the patient's own rate slows below a preset limit, or to detect certain arrhythmias and to override the arrhythmia. The newest technology (as of the time of writing) in pacemakers is a pacemaker intended to be used for patients in congestive heart failure. There, the pacemaker attempts to stimulate the atrium and ventricle in a pattern that is similar to the heart's natural pattern, a bimodal pacemaker.

The pacemaker is typically implanted during a cath lab procedure (Forde and Ridgely, 1995). Previously, the procedure was performed in the operating room. The pacemaker lead wires are installed in much the same fashion as a catheter insertion for a coronary angiogram. The pacemaker "generator" is installed in a pocket under the skin. The pocket is created using a local anesthetic.

Defibrillators

Defibrillator technology has changed dramatically since 1995 (Tacker, 1995). Implantable defibrillators are available, using the same technology and implantation techniques used for pacemakers (Duffin, 1995). They are highly effective and reliable.

External defibrillators are undergoing a metamorphosis of sorts. What has been the tradition in defibrillators (a single, pulse-modified Lown waveform) is being replaced with a biphasic, modified Lown waveform. As a result, the standards for defibrillation are changing, with lower maximum output required to achieve defibrillation. There remains controversy over the effectiveness of the biphasic waveform, yet many equipment manufacturers have changed their product lines to feature only biphasic units.

Drugs

Technology in health care comprises three components: Medical devices, techniques, and pharmaceuticals. In medicine, drugs are effective technologies for the treatment of cardiovascular disease. Drugs that are used to reduce blood pressure, increase blood pressure, increase cardiac muscle contractility, and reduce cholesterol levels are valuable adjuncts to the medical devices and techniques discussed above. While a description of drugs and their applications is beyond the scope of this chapter, the reader should be aware that pharmaceutical interventions in cardiovascular disease are an essential part of the total technology picture.

References

Berbari EJ. Principles of Electrocardiography. In Bronzino JD (ed). *The Biomedical Engineering Handbook*. Boca Raton, FL, CRC Press, 1995.
Boyd DP, Lipton MJ. Cardiac Computed Tomography. *Proceedings of the IEEE* 198, 1983.
Budinger TF, VanBrocklin HF. Positron-Emission Tomography (PET). In Bronzino JD (ed). *The Biomedical Engineering Handbook*. Boca Raton, FL, CRC Press, 1995.
Bulgrin JR, Rubal BJ, Thompson CR, et al. Comparison of Short-Time Fourier, Wavelet and Time-Domain Analyses of Intracardiac Sounds. *Biomed Sci Instrum* 29:465-472, 1993.
Clements FM, de Bruijn NP. Electrocardiography: Monitoring for Ischemia. In Lake CL (ed). *Clinical Monitoring*. Philadelphia, WB Saunders, 1990.
DiSciascio G, Cowley MJ. Coronary Angioplasty. In Webster JG (ed). *Encyclopedia of Medical Devices and Instrumentation*. New York, Wiley, 1988.
Dorson WJ, Loria JB. Heart-Lung Machines. In Webster JG (ed). *Encyclopedia of Medical Devices and Instrumentation*. New York, Wiley, 1988.
Duffin EG. Implantable Defibrillators. In Bronzino JD (ed). *The Biomedical Engineering Handbook*. Boca Raton, FL, CRC Press, 1995.
Ferrara KW. Blood Flow Measurement Using Ultrasound. In Bronzino JD (ed). *The Biomedical Engineering Handbook*. Boca Raton, FL, CRC Press, 1995.
Forde M, Ridgely P. Implantable Cardiac Pacemakers. In Bronzino JD (ed). *The Biomedical Engineering Handbook*. Boca Raton, FL, CRC Press, 1995.
Gardner RM. Hemodynamic Monitoring. In Webster JG (ed). *Encyclopedia of Medical Devices and Instrumentation*. New York, Wiley, 1988.
Geddes LA. Cardiac Output Measurement. In Bronzino JD (ed). *The Biomedical Engineering Handbook*. Boca Raton, FL, CRC Press, 1995.
Geddes LA. Cardiac Pacing: Historical Highlights. In Bronzino JD (ed). *The Biomedical Engineering Handbook*. Boca Raton, FL, CRC Press, 1995.
Holter NJ. New Method for Heart Studies: Continuous Electrocardiography of Active Subjects Over Long Period is Now Practical. *Science* 134:1214, 1961.
Jaron D, Moore TW. Intra-Aortic Balloon Pump. In Webster JG (ed). *Encyclopedia of Medical Devices and Instrumentation*. New York, Wiley, 1988.
Lawson WE, Hui JC, Cohn PF. Long-Term Prognosis of Patients with Angina Treated with Enhanced External Counterpulsation: Five-Year Follow-Up Study. *Clin Cardiol* 23(4):254–258, 2000.
Lin JP. Cine and Photospot Cameras. In Webster JG (ed). *Encyclopedia of Medical Devices and Instrumentation*. New York, Wiley, 1988.
Mukharji J. Exercise Stress Testing. In Webster JG (ed). *Encyclopedia of Medical Devices and Instrumentation*. New York, Wiley, 1988.
Neuman MR. Biopotential Amplifiers. In Webster JG (ed). *Medical Instrumentation: Application and Design, 3rd Edition*. New York, Wiley, 1988.
Ritman EL. Fast Computed Tomography for Quantitative Cardiac Analysis: State of the Art and Future Perspectives. *Mayo Clin Proc* 65:1336, 1990.
Swan HJC, Ganz W, Forrester J, et al. Catheterization of the Heart in Man with the Use of a Flow Directed Balloon-Tipped Catheter. *N Engl J Med* 283:447–451, 1970.
Tacker WA. External Defibrillators. In Bronzino JD (ed). *The Biomedical Engineering Handbook*. Boca Raton, FL, CRC Press, 1995.
Thakor NV, Webster JG, Tompkins WJ. Design, Implementation, and Evaluation of a Microcomputer-Based Portable Arrhythmia Monitor. *Med Biol Eng Comput* 22:151–159, 1984.
Tompkins WJ. Ambulatory Monitoring. In Webster JG (ed). *Encyclopedia of Medical Devices and Instrumentation*. New York, Wiley, 1988.
Van Lysel MS. X-ray Projection Angiography. In Bronzino JD (ed). *The Biomedical Engineering Handbook*. Boca Raton, FL, CRC Press, 1995.
Walton A. Balloon Angioplasty. Medtech News: Technology & Innovation. MedTech1.com, December 2002.
Yoganathan AP, Hopmeyer J, Heinrich RS. Mechanics of Heart Valves. In Bronzino JD (ed). *The Biomedical Engineering Handbook*. Boca Raton, FL, CRC Press, 1995.

95
General Hospital Devices: Beds, Stretchers, and Wheelchairs

Joseph F. Dyro
President, Biomedical Resource Group
Setauket, NY

Wheelchairs, beds, and stretchers are found in abundance in hospitals, short- and long-term care facilities, and the home. These devices are not high technology devices in general and are usually classified as low-risk devices. As a consequence, these devices often received much less attention than high-tech and high-risk devices such as defibrillators, electrosurgical units, anesthesia machines, and dialysis units. Nevertheless, they present a finite risk to patients and must not be ignored. Beds, stretchers, and wheelchairs have caused serious injury and death. This chapter addresses these devices and others usually referred to as "general hospital devices" and "rehabilitation devices," such as patient lifts, parallel bars, and chairs. Recommendations are made in the areas of evaluation, purchasing, training, service, modification, and obsolescence. Guidelines are presented for investigation of incidents and accidents involving these devices. Several case studies are presented to illustrate the types of risks involved and steps to take to minimize or eliminate these risks to patients and staff.

Beds

The bed is often forgotten because clinical engineering departments focus attention on more complex devices such as defibrillators and patient monitors (Poulsen, 2001). Unfortunately, such inattention can result in poorly maintained and hazardous beds. Bed, stretcher, and wheelchair management has been successfully incorporated into a clinical engineering department (Dyro, 1993). Clinical engineers have participated in the evaluation and selection of appropriate technologies. Appropriately trained general mechanics have proven themselves able to work effectively under the supervision of clinical engineers to inspect and maintain these devices.

Patients in the hospital or home care spend most of their time in bed. As technology has advanced, the first beds found in some of the first hospitals (Figure 95-1) have gotten more complex and multifunctional. The bed is a medical device and is found in many configurations; e.g., electric beds and non-electric (i.e., mechanical) beds. The hospital bed today may be hydraulically or electrically powered, enabling the user to position the bed at various levels above the floor, and various parts of the mattress surface to adjust the position of the head, knee, and foot positions. Accessories for beds include head and foot boards, bumpers, mattresses, mattress covers, occupancy alarms, bed-wetting alarms, pads, railing pads, bedrails, trapezes, and scales. Side rails come in a range of styles including removable side rails, adjustable side rails, and half-side rails.

Special beds include flotation-therapy beds to reduce the risk of decubitus ulcer formation, barobeds to support extremely obese patients (Figure 95-2), and circle beds to position and stretch patients. Advanced life-saving and life-extending interventions find more patients staying in bed longer, thus increasing the likelihood of the development of skin ulcers from prolonged contact of the patient with the bed under pressure. Obesity is ever increasing in our society. As more extremely heavy patients are cared for or are treated for their obesity with, for example, bowel resection, medical devices are feeling the strain (BS&S, 2002). Circle electric beds and other devices (e.g., traction tables) that relieve the stress imposed on the skeletal system by gravity are seeing increased use as back problems afflict an increasingly sedentary society. A wide variety of beds accommodate the many special needs of the broad spectrum of patients. Beds include the following types: Air fluidized beds, birthing beds, circle electric beds, cribs, electric beds, obese electric beds, floatation-therapy beds, fluoroscopic beds, hydraulic beds, infant incubators, infant radiant warmers, rocking cradles, rocking beds, bassinets, pediatric beds, labor beds, low-air-loss beds, mechanical beds, metabolic beds, orthopedic beds, and tilt beds.

The numerous considerations taken into account when specifying a bed include patient age, size, weight, and condition. Criteria also include the intended use and the environment in which it will be used. Beds must accommodate a wide range of physical characteristics of the patient, from a small, pediatric patient to an obese adult. Included in this chapter are the bed features of devices including infant incubators, radiant warmers, bassinets, and cribs. These devices are included because in many hospitals they are treated to the same lack of attention given to a hospital bed. In several of the incidents that the author has investigated, lack of medical device management was the root cause of patient injury and death (see Chapter 64).

Figure 95-3 shows a newborn infant (this author's daughter) resting comfortably on a mattress in an infant incubator. At the other end of the life cycle curve, a centenarian, (the newborn's great aunt) rests comfortably in an electric bed in a home care situation (Figure 95-4). Note that many design characteristics of an adult bed are found even in an incubator. For example, the infant is resting on a mattress; the mattress can be inclined; the mattress is appropriately sized so as not to permit gaps and spaces that could entrap limbs; and the infant is restrained from leaving the bed surface and falling to the floor by the incubator enclosure that consists of rigid walls and movable access panels (i.e., side panels). The concerns to ensure patient safety, whether for the infant or the adult, revolve around issues such as patient entrapment, patient containment, patient comfort and support, patient positioning.

Design Features

Figure 95-5 is an illustration of a typical hospital bed, showing several design features. The following are the principal design features of beds:

Frame	Casters and wheels	Brakes
Side rails and latches	Foot and head boards	Electronic controls
Mattresses	Padding and bumpers	Motors
Gears	Positioning mechanisms*	IV pole connectors

*Trendelenburg, reverse-Trendelenburg, Fowler, foot, head, and knee elevation

Bed Management

Clinical engineers should actively participate in the assessment, evaluation, purchasing, maintenance, repair, modification, and obsolescence of beds and bed accessories. Neglecting these medical devices increases the risk of patient injury from a variety of mechanisms, several of which are described below.

Injuries

Several examples follow, illustrating that death or serious injury can occur from the use of adult beds and infant incubators.

Entrapment

Bedrails intended to keep a patient in bed and to prevent the patient from falling out of bed have entrapped patients, causing death and serious injury. This happens when a disoriented patient attempts to leave the bed and becomes entangled in the bedrail or between the bedrail and the bed or does not attempt to leave the bed but simply slips between the mattress and the bedrail (Miles, 2002). About 2.5 million hospital and nursing-home beds are in use in the United States. Between 1985 and 1999, 371 incidents of patients caught, trapped, entangled, or strangled in beds with rails were reported to the United States Food and Drug Administration. Most patients were frail, elderly, or confused (FDA, 2000). Parker and Miles (1997) described 74 entrapment deaths that occurred between 1993 and 1996 and described the mechanisms of these deaths. Miles and Parker (1998) show pictures describing two cases of fatal bedrail entrapment. Figure 95-6 shows several points at which a patient can be entrapped in a bed (FDA, 1995). The incidence of fatal bedrail entrapments have lead some to question their use or to conclude that bedrails should not be used in ambulatory patients (Huffman, 1998; Hanger et al., 1999).

Figure 95-7 is a re-enactment of the scenario described at hazard point 3 in Figure 95-6—a strangulation of a patient who attempted to get out of bed only to become hung up on the side rail. (The volunteer for the re-enactment suffered no injuries.)

The lifetimes of the various components of a bed differ. For beds, it is 25 years; for rails, 10–15 years, and for mattresses, about 5 years (Miles and Levenson, 2002). In this mix-and-match situation, the rail-to-bedframe spacings can vary over the life of a bed. Furthermore, adjustments to the bed are typically made by housekeeping, janitors, or maintenance workers—personnel who have little familiarity with medical devices and the serious consequences of improper bed configuration.

After many years of silence and neglect, the health care community is taking some action to address the problem (Parker, 1997). In 1999, the Hospital Bed Safety Working

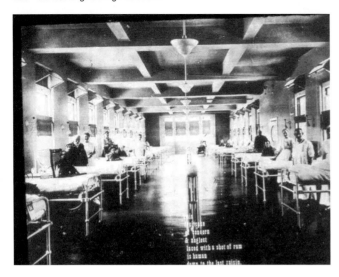

Figure 95-1 Hospital ward in the early 1900s.

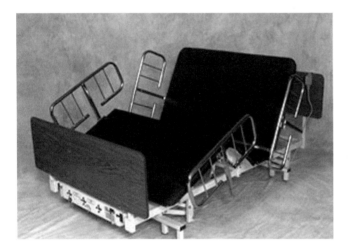

Figure 95-2 Bed for the obese—Stenbarr SB Bari 600 electric bed. 48″ wide, 600-pound-capacity bed frame.

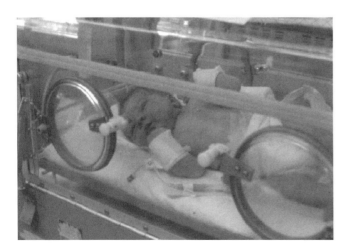

Figure 95-3 Infant incubator bed for a neonate.

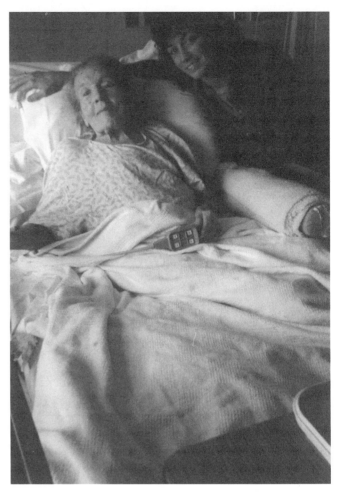

Figure 95-4 Centenarian resting comfortably in a hospital bed in the home care setting.

Group (HBSW) was established at the Department of Health and Human Services in Washington, DC under the leadership of the United States Food and Drug Administration (FDA). Joining the FDA in this effort are representatives from the medical bed industry, national health care organizations, patient advocacy groups, and other federal agencies (Centers for Medicare & Medicaid Services; the Consumer Product Safety Commission, and the Department of Veterans Affairs). The Group's goal is to reduce the risk of entrapment and injuries relating to hospital beds and bedrails through standardization of definitions; standardization of the evaluation of beds, mattresses and side rails; and outreach to providers and patients. Assessment kits supplied by the HBSW enable hospitals to assess the risk levels presented by their beds (BS&S, 2002a). For further information on the activities of the group, see http://www.fda.gov/cdrh/beds/index.html. Other organizations such as the Joint Commission on Accreditation of Healthcare Organizations (JCAHO)(JCAHO, 2002, 2003) have issued alerts to the health care community regarding the hazard of bed rail entrapment. Recalls have been issued because of possible patient entrapment (BS&S, 2003) and latch failures (BS&S, 2003a, 2003b).

The Health Care Financing Administration (HCFA) and the Food and Drug Administration (FDA) differ in their interpretation of restraints especially as it relates to side rail use. The HCFA's requirements pertaining to the use of restraints in nursing homes are at 42 CFR Section 483.13 (a). HCFA defines "physical restraints" under "Interpretive Guidance" in the *State Operations Manual* as: "any manual method or physical or mechanical device, material, or equipment attached or adjacent to the individual's body that the individual cannot remove easily which restricts freedom of movement or normal access to one's body."

The FDA defines "protective restraint" at CFR Section 880.6760 as:

"a device, including, but not limited to, a wristlet, anklet, vest, mitt, straightjacket, body/limb holder, or other type of strap that is intended for medical purposes and that limits the patient's movements to the extent necessary for treatment, examination, or protection of the patient or others."

The HCFA and FDA recently clarified the distinctions that each agency makes with regard to the definition of restraints (HCFA/CDRH, 2000).

Another form of entrapment occurs when a bed is lowered onto patients or visitors who are under the bed. Adults and children who have crawled under beds have been crushed when beds have been lowered, most frequently when beds have incorporated a "walk-away down" feature that enables the bed to be lowered with the activation of a single control switch.

General Hospital Devices: Beds, Stretchers, and Wheelchairs   423

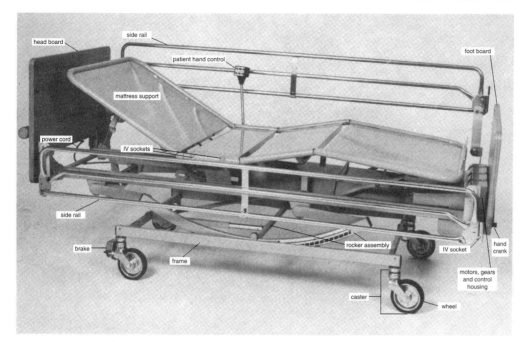

Figure 95-5 InterRoyal Fred Bed showing typical design features.

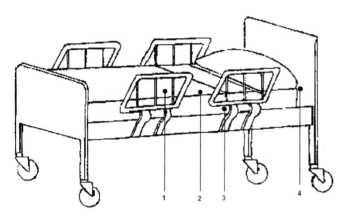

Figure 95-6 The four locations on a bed with bedrails, where patients have most frequently become entangled: (1) head inserted between widely spaced vertical bars in the bedrails; (2) patient slid either on their abdomen or on their back though the slot between head and foot rails; (3) patient slid legs first through the slot until the anterior neck jammed against the rail; and (4) head buried in the triangular space created by the right angle of the bedrail and the headboard with the mattress-corner curves (FDA, 1995). Parker (1997) describes death from asphyxiation when a patient's entire body becomes entrapped in the space between the side of the mattress and the bedrails and when the head of the patient in bed passes over the top of the bedrail and the throat rests on the rail.

Falls

Falling is a common cause of morbidity and the leading cause of nonfatal injuries and trauma-related hospitalizations in the United States (Baker et al., 1992). Falls while getting into and out of beds are too common. Ironically, risk of falling may be increased by the presence of bedrails because some people try to climb over them and fall in the process (Parker, 1997). Improper bed and bed rail height are factors contributing to falls associated with beds (Agostini et al., 2001). Inadequately latched bed rails or deteriorated latches that allow side rails to collapse with no or some assistance from the occupant, or rails that are ineffective in restraining patients may contribute to patient falls (Miles and Levenson, 2002). The button type of latch is particularly subject to deterioration. A spring-loaded button is constrained to a hole in a side rail vertical tube member. To raise or lower the side rail, the button must be pushed in to allow free movement of the tube member. Over time, the button slides up and down, and the tubes that are sliding over the button wear away the button, rendering the button-latch ineffective (Health Canada, 1988).

The author has investigated incidents in which infants have fallen out of incubators by crawling out of open or inadequately closed or locked hand ports, and when they have fallen out of radiant warmer bassinets when side walls have collapsed because of inadequate and defective latches. (Figure 95-8.) The striker plate shown in Figure 95-8 was manufactured from a soft metal that, over time, wore away as shown, rendering the latch ineffective for securing the door.

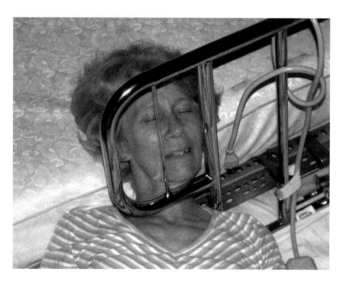

Figure 95-7 Reconstruction of a bedrail entrapment event.

Impalement

While relatively rare, cases of impalement on bed parts have occurred (Parker and Miles, 1997). The author investigated an impalement event in which a nursing home resident had fallen from her bed (Figure 95-9), impaling her throat on the vertical element of the half-bed side rail lying on the floor as shown in Figure 95-10.

Burns

Serious burns have occurred when neonates in incubators crawled from the mattress surface to hot air vents. Lack of maintenance and inspection allowed an infant incubator, missing a barrier separating the infant mattress from the hot air vent, to be used (Figure 95-11). A serious burn resulted when the infant crawled over the hot air vent.

Other Hazards

Electric shock (BS&S, 2003c), fire (BS&S, 2003d), contamination with infectious agents, instabilities (BS&S, 2003e, 2003f), and tipover (MDR, 1988; BS&S, 2003g) are several other hazards that are present in some beds. Power cords of electric beds often have been damaged by being pinched between the bed and the floor, wall, or room furniture. An electrical shock hazard can result from a damaged line cord with missing insulation that is in contact with the bed. Damaged power cords from other medical devices and household appliances such as radios and televisions that are in contact with

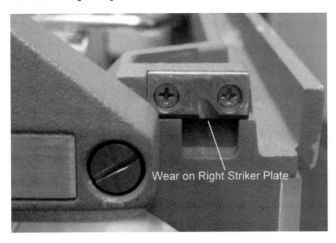

Figure 95-8 Incubator door latch striker plate showing wear to soft metal of plate. Such wear enabled an infant to open the door merely by the leaning on the door. Reproduced with permission from Shepherd (2003).

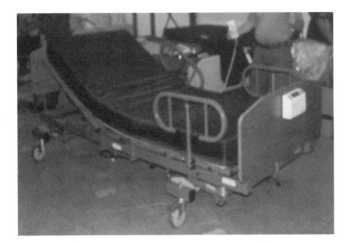

Figure 95-9 Burke Tri-Flex "A" bed.

Figure 95-10 Side rail.

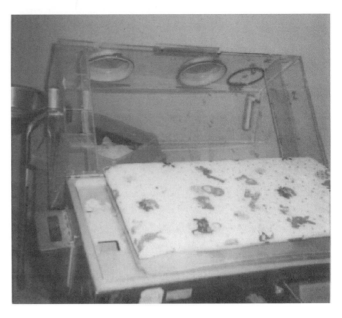

Figure 95-11 Infant burned foot by exposure to heated humidified air entering incubator infant compartment through vent.

a bedrail also pose a hazard (Figure 95-12). Beds can harbor infectious agents and must be kept clean.

Operating Room Tables

Among the wide variety of tables utilized in health care are the following types: Anesthetist's, autopsy, bedside, blood donor, dissecting, examination/treatment, instrument, nuclear medicine, obstetrical, operating, orthopedic, overbed, physical therapy, proctologic, radiographic, radiotherapy, special procedures, tilt, traction, and urological. They range considerably in complexity and cost. Among the more complex of these are operating room (OR) tables, which are used to support a patient while a surgical procedure is performed (Figure 95-13). OR tables are also referred to as "surgical beds" or "operating room beds."

Design Features

The medical device that is utilized to support a patient undergoing a surgical procedure is known by many names (e.g., surgical table, operating room table, OR table, and surgical bed). Holding steady at the desired height and angles of the patient platform is the most important evaluation criterion of an operating table. As with all medical devices, clinical engineering can contribute substantially in the assessment, evaluation, purchasing, maintenance, repair, modification, and eventual disposition of these devices. OR tables are becoming increasingly sophisticated and complex with microprocessor circuitry and voice-activated table-position control. It is not for this reason alone, however, that CE departments should include service of these devices in their programs. Even the least complicated table, if nonfunctional, severely affects the utilization of an operating room. Cancellation or rescheduling of a surgical procedure equates to a loss of revenue, inconvenience to staff, and a danger to the patient in need of surgery. While infrequent, reports of OR table failure indicate that they do occur, such as the recent report of the top of an OR table becoming detached in the middle of a surgical procedure (BBC News, February 27, 2003).

OR tables are used for a wide variety of surgical procedures, including the following: General, ambulatory, neurological, gynecological, urological, ophthalmological, orthopedic (articulated posturing of patient for reconstructive and reparative orthopedic procedures), minimally invasive (such as endoluminal stent-grafting, abdominal aortic aneurysms (AAA), intraoperative angioplasty, cardio/endo vascular, coronary angiography, mid CAB, mini CABG).

The following is a listing of the four main design features of an OR table and evaluation criteria:

- Table top
- Support base
- Controls
- Power supply

All four design features must result in a stable patient platform that is ergonomically suitable for the operating room staff and that can be attained conveniently, quickly, reproducibly, and reliably. Radiographic imaging during surgery places special requirements on such characteristics of the table top as ease of positioning an imaging devices such as a C-arm and radiolucency, scattering and absorption of X-ray radiation.

Tabletop design criteria include surface area, overall dimensions, articulations (head, back, legs), standard Trendelenburg, reverse Trendelenburg, and various tilt orientations.

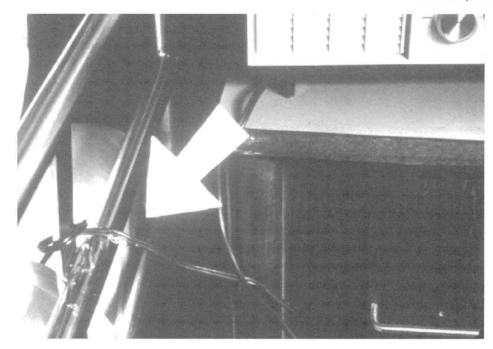

Figure 95-12 The line cord of a radio with worn insulation lying against bedrail.

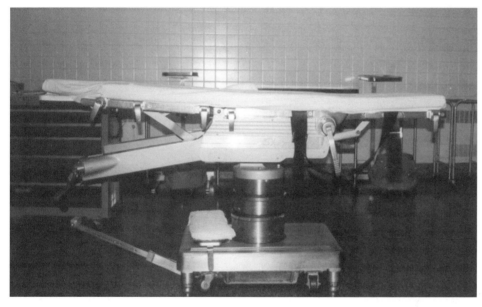

Figure 95-13 Shampaine operating room table.

The versatility of an OR table depends upon the availability of accessories or extensions to accommodate positioning requirements and upon such features as modular design that enables easy reconfiguration of the table for different procedures.

Control of the articulations, orientation, and height of the patient surface should be convenient, quiet, and reliable. Manual positioning, powered positioning regulated by pendant controls that can be accessed at various positions around the table, wall-mounted controls, and voice-command controls are considerations. Power-assisted patient-positioning systems should operate smoothly and quietly throughout the full range of height and tilt positions of a table's patient surface. OR tables can be positioned by manual, hydraulic, electric, or electro-hydraulic power sources. A manual backup system must be incorporated for times of electrical failure.

The structural characteristics will dictate the weight-bearing capacity, stability, and maneuverability of the table. Movable tables must be stable on level and irregular surfaces and typically have wheels and casters that provide for good maneuverability and stability. Some tables consist of permanently mounted OR table supports, such as pedestals, which accept modular table surfaces.

OR Table Accessories

A wide variety of accessories are utilized in conjunction with OR tables, including such devices as pads, clamps, tabletop extensions, head rest, stirrups, Clark crutch sockets, arm boards, arm and hand tables, restraints, retractors, leg and calf support, abductor bars, sacral rest, and perineal posts.

Hazards

The most prevalent complication associated with OR tables is that of skin lesions and underlying tissue trauma associated with ischemia induced by prolonged excessive pressure to parts of the circulatory system of the patient.

Pressure Necrosis

Constant pressure applied to parts of the body affects the underlying blood circulation. If perfusion of a volume of tissue is compromised for a long enough period, ischemia results and can lead to necrosis of the affected tissue. Such accessories as retractors, mattresses, railings, and leg supports, if inappropriately designed, manufactured and utilized, can be that source of excessive pressure that can be injurious to the patient during long operations (Gendron, 1988). For example, the author investigated an incident in which a patient had suffered neurological damage as a result of prolonged, high-pressure contact of a limb with the vertical elements of the Pittman retractors, as shown in Figure 95-14. Surgical staff must be appropriately trained to avoid such situations and to use adequate padding to distribute the force of a hard object over a large area in order to minimize the pressure upon the patient's vasculature.

Chemical Burns

Mattresses have been the cause of serious skin damage to patients undergoing long procedures. Skin lesions can be caused by such mechanisms as excessive temperature,

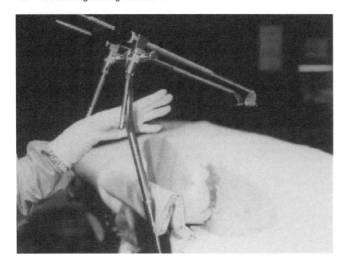

Figure 95-14 Pittman retractors.

pressure necrosis, chemical attack, electrical current, and laceration. The author unraveled a mysterious, full-backside skin lesion that a patient suffered while undergoing a six-hour cardiovascular procedure. Typical sources of insult such as hyperthermia, mattresses, and electrosurgical generators were first considered. It was only when the staff during interview admitted that the OR table mattress had recently been sterilized by ethylene oxide (EtO) and that, because of the OR scheduling demands, was returned to service immediately upon completion of the sterilization, that the author realized that the lesion was caused by EtO that had remained in the mattress after sterilization. The central sterile supply and OR staff had not taken adequate precautions to ensure that the mattress had been sufficiently aerated after EtO sterilization to allow the elution of the caustic sterilant.

Instability

Unstable supports for the limbs have resulted in product recalls (BS&S, 2003h) and interruption and delay in surgical procedures (BBC, 2003).

Stretchers

A stretcher, also referred to as a "gurney" or a "patient trolley," is a medical device upon which a patient lies for transportation. A wide variety of stretchers are available and their design depends upon the use to which the stretcher is put. A stretcher may be as simple as those utilized in military conflicts or emergency rescue—essentially, a piece of canvas affixed to two sturdy poles on which the injured person is carried by rescuers. Stretchers for use in the field can be considerably more complicated, with more features such as the ambulance stretcher shown in Figure 95-15, designed so that its undercar-

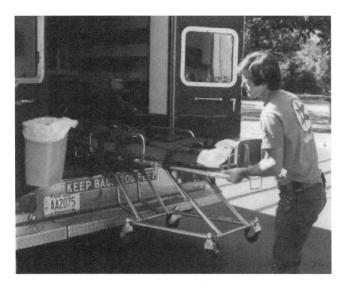

Figure 95-15 Ferno-Washington collapsible stretcher being taken from an ambulance.

riage will fold, thus allowing the stretcher to be slid into the ambulance compartment. This stretcher also has wheels to facilitate transport over smooth terrain. Hospital-based stretchers are used to transport patients between patient rooms, and from a room to diagnostic and therapeutic departments such as X-ray and the operating room. Figures 95-17 and 95-18 show two devices designed for transporting patients to and from the operating room and post-anesthesia recovery area. These hospital-based stretchers range from the simple to the complex, incorporating such features as shelves for physiological monitors and hydraulically powered mechanisms for adjustment of stretcher height and mattress position.

Design Features

In selecting a stretcher, one should evaluate it for the presence of pinch points, entrapment zones, ease of maneuverability, adequacy of locks and brakes, ease of operation of bed-positioning controls, clarity of labeling and indicators (e.g., for height adjustment), weight capacity, surface area, ease of cleaning, and sturdiness of construction. Stretchers are subject to considerable shock and vibration, as they are intended to move about and often inadvertently collide with walls, door jambs, other stretchers, beds, and furniture.

Listed below are a number of design features that should be included in any pre purchase evaluation of stretchers. These features, similarly, should be carefully examined in the course of routine maintenance.

Mattress	Frame	Side rails
Head and foot boards	Wheels and casters	Brakes
Steering guides	Hand holds	Padding
Bumpers	Shelves and storage areas	IV poles
Oxygen cylinder storage	Positioning systems	Labels and indicators

Stretcher Incidents

Defective design, manufacture (e.g., defective welds (BS&S, 2003i)), maintenance, repair, and improper use of stretchers can result in patient injury.

Figure 95-15 shows a stretcher whose undercarriage can fold up, thus permitting positioning in an emergency vehicle. Defects in latching mechanisms and improper use can result in the unexpected collapse of the undercarriage while the patient is being transported outside of the ambulance with legs in the fully extended position.

Figure 95-16 shows a Stryker PACE stretcher involved in an incident in which a nurse allegedly pinched her finger between the corner of the stretcher and the wall. Extensive testing by the author confirmed that the stretcher had excellent maneuverability and that all wheels and brakes were in conformance with specifications.

Figure 95-17 shows a Stryker Model 935 SurgiCare clean-corridor system stretcher. A patient allegedly injured fingers while attempting to sit on the stretcher. The author found that inadequate maintenance and failure to properly install the manufacturer's product modification allowed a pinch point to exist.

Wheelchairs

Wheelchairs have aided the movement of people for centuries. The ancient wheelchair depicted in Figure 95-18 stands in stark contrast to wheelchairs incorporating the most recent technological developments, such as the one shown in Figure 95-19. Patients whose ambulation is impaired are frequently transported about the hospital in wheelchairs. Wheelchairs are found increasingly in nearly every setting as architectural barriers to access are removed in response to federal legislation ensuring equal access to the handicapped. Patients, their relatives, and their caregivers have ample opportunity to see and feel them. Often, the first medical device to which a patient is exposed is a wheelchair. A patient seated in a wheelchair that is dirty, in disrepair, or malfunctioning forms a poor impression of the hospital, which is responsible for the wheelchair's maintenance. That same patient is exposed to increased risk of injury from a poorly maintained wheelchair (Young, 1985). If a simple wheelchair cannot be adequately maintained, what does that say about the safety and performance of more sophisticated devices to which a patient might be exposed? Improper use and poor maintenance and repair contribute to wheelchair hazards (HPCS, 1994). According to one study (Calder and Kirby, 1990), fatalities occur in approximately 0.2% of serious wheelchair-related accidents, or 7.6 fatal accidents per 100,000 users per year. Most of these fatalities were associated with falls and tipping. Active research programs (Kirby et al., 1989a, b), advances in wheelchair design, and adoption of international safety and performance standards will mitigate some of these problems (Brubaker, 1988), but the need for vigilance, proper utilization, and adequate maintenance will continue to exist. For a detailed discussion of wheeled mobility, wheelchairs, and personal transportation, the reader should consult the work of Cooper (1995).

Wheelchair Design Features

Wheelchairs, both motorized and nonmotorized, share many design features. The significant design features of a wheelchair are listed below:

Frame	Wheels and tires	Casters
Brakes	Handles	Arm, leg, and foot rests
Trays and tables	Storage compartments	Seats
Padding	Anti-tip devices	Motors and gears
Drive mechanisms	Batteries and chargers	Electronic controls

Figure 95-20 shows a schematic representative of a typical nonmotorized wheelchair and terminology used to describe salient features of the wheelchair.

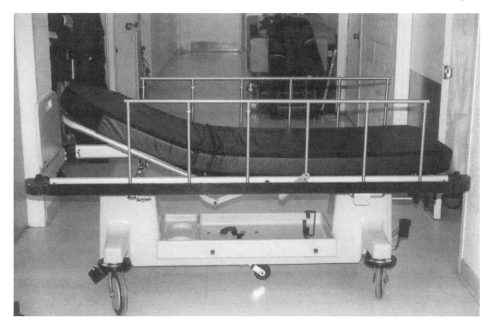

Figure 95-16 Stryker PACE bed.

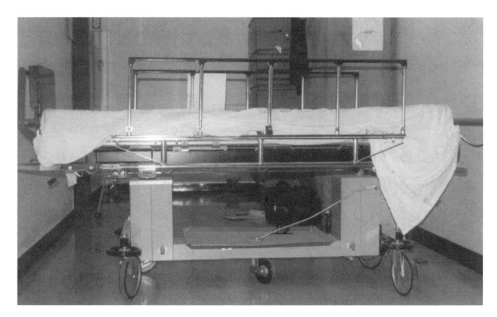

Figure 95-17 Stryker SurgiCare model 935.

Wheelchair Management

Management of wheelchairs entails the elements of assessment, evaluation, purchasing, maintenance, repair, modification. and obsolescence. The numerous considerations taken into account when specifying a wheelchair include patient age, size, weight, and condition. Criteria also include the intended use and the environment in which it will be used (Brubaker, 1986). Wheelchairs must accommodate a wide range of physical characteristics of the patient from a small pediatric patient to an obese adult. Wheelchairs are used for transfer from a patient bed to an examination room, from the discharge desk to the awaiting car, van, or ambulette, from the emergency room to the treatment area and as an alternative to lying in bed. Wheelchairs are used increasingly by athletes in track and field, basketball, and other sports (Brubaker, 1988). Patient conditions dictate the amount of assistance needed in transferring a patient to and from a wheelchair and for movement.

A wide variety of accessories are available for wheelchairs, including trays, seats, and rests for arms and feet. Motorized and nonmotorized wheelchairs (Figures 95-21 and 95-22) are manufactured by scores of companies and used by millions of people throughout the world.

Safety Considerations

Understanding the hazards associated with wheelchairs is beneficial for the manufacturer who must design these devices to eliminate such hazards. A retrospective based upon the literature, the author's accident investigations, and hazard-reporting systems has revealed a number of injuries resulting from wheelchair use. The following describes and lists the most prevalent dangers.

Tipping

Tipping to the sides or forward or backward can cause a wheelchair to tip over. The occupant of such a wheelchair at the time of tipover can be seriously injured as he strikes the floor. Irregular surfaces, rough terrain, inclines, gaps on the surface, or a combination of the former can cause a wheelchair to tilt far enough to tip over. A wheel falling off would also cause the wheelchair to tip over.

Collapse

If the wheelchair frame loses its structural integrity through a broken frame (e.g., defective welds or metal fatigue) or loose or missing fasteners (e.g., screws, nuts, bolts), the wheelchair may no longer be able to support the weight of, and forces exerted by, the patient. The author has investigated incidents in which wheelchairs have collapsed from inadequate and improper repair and maintenance.

Brake Failure

Brakes to prevent the wheels from turning might not adequately engage, thus permitting the wheelchair to move as the patient is transferring. Falls can result. Wheelchairs on inclines may run away, gain momentum, and forcibly eject the patient when the wheelchair comes to a sudden stop. Brake mechanisms, such as the one shown in Figure 95-23, require inspection and maintenance to ensure that fasteners are secure and properly adjusted. Testing the adequacy of breaking force should be done after servicing.

428 Clinical Engineering Handbook

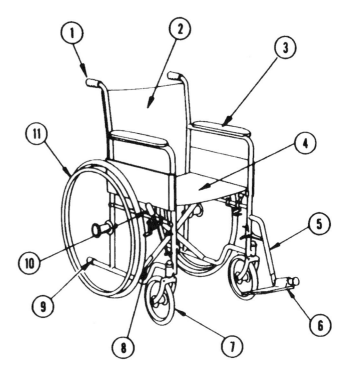

1. Handgrips/Push Handles
2. Back Upholstery
3. Armrests
4. Seat Upholstery
5. Front Rigging
6. Footplate
7. Casters
8. Crossbraces (Serial No.)
9. Tipping Lever
10. Wheel Locks
11. Wheel and Handrim

Figure 95-18 Ancient wheelchair.

Figure 95-20 Wheelchair terminology.

Figure 95-19 Advanced-technology wheelchair: INDEPENDENCE ™ iBOT ™ 3000 Mobility System. This wheelchair is capable of going up and down stairs, over curbs, and rising up on its two rear wheels to put its occupant at eye level with a standing adult.

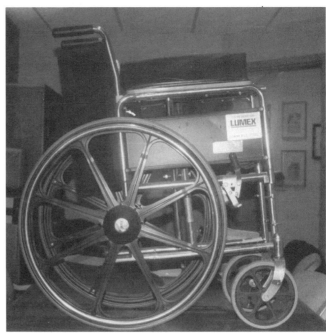

Figure 95-21 Nonmotorized wheelchair.

General Hospital Devices: Beds, Stretchers, and Wheelchairs 429

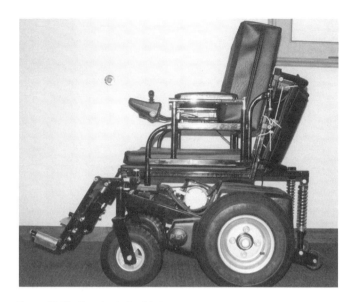

Figure 95-22 Motorized wheelchair.

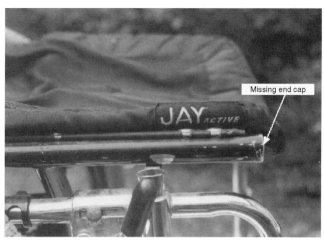

Figure 95-24 Missing a protective cap, the sharp edge of a wheelchair frame lacerated the skin of patient during transfer.

Figure 95-23 Brake mechanism of wheelchair.

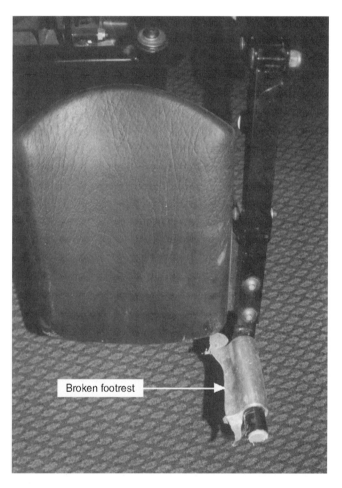

Figure 95-25 A broken footrest left a sharp edge and a cutting hazard.

Pressure Points

Irregular, hard surfaces can result in high-pressure points or areas on the patient's body (e.g., thighs, buttocks, and back). Decubitus ulcer and pressure necrosis can result. Research has resulted in seat-cushion designs that minimize these dangers (Ferguson-Pell et al., 1986).

Sharp Edges and Pinch Points

Patients may have reduced sensitivity to tactile stimulation. Rough surfaces and sharp and jagged edges can cause lacerations of the skin (Figures 95-24 and 96-25).

This is especially possible during transfer and may affect many areas of the body such as the hands, fingers, wrist, arm, and legs. Fingers can be pinched if caught between hard, metallic surfaces (e.g., between frame and foot and leg rest; in the levers that actuate the brake; and between wheel spokes and frame).

Susceptibility to Electromagnetic Interference (EMI)

Numerous cases have been reported of motorized wheelchairs and scooters moving in an uncontrolled fashion when subjected to electromagnetic radiation (Witters and Ruggera, 1994). Serious EMI reactions by these devices over a wide range of radio frequencies (covering many private and commercial transmissions) were observed in CDRH laboratory investigations and testing (Witters, 1995). EMI can occur in the presence of a wide spectrum of electromagnetic energy frequencies and intensities. Sources of EMI include pagers, cell phones, aircraft beacons, radio and television transmissions, and common household appliances and tools such as microwaves, electric light dimmers, electric drills, and saws (Dyro, 1995). While the FDA has done extensive testing of the susceptibility of wheelchairs to EMI and has advised the manufacturers of its findings, clinical engineers should remain vigilant to spontaneous activation of powered wheelchairs, especially those that were introduced on the market prior to FDA advisories in 1994 (Albert, 1994; Morrison, 1994). (See Chapters 62, 63, and 103 for more on EMI.) The FDA requires that a warning sticker be placed on each powered wheelchair or scooter, indicating the risk due to EMI (Cooper, 1995).

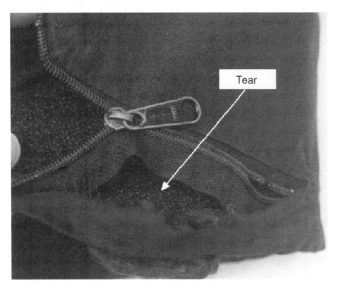

Figure 95-26 Torn seat cushion allows ingress of filth and vermin.

Fires and Explosions

Gases released by batteries have ignited, resulting in explosion and fire. Wheelchairs recently have been recalled by the FDA because motors could catch on fire (BS&S, 2003j). A short circuit with the charger harness can cause heat damage to the units with potential for fire, according to the FDA.

Lack of Cleanliness

Incontinent patients confined for extended periods of time can soil a wheelchair. The wheelchair that is not properly cleaned at regular intervals becomes a safe harbor for dust, dirt, filth, infectious agents, and vermin. Rust and corrosion can degrade the structural integrity of the wheelchair. Figure 95-26 shows a tear in a wheelchair seat cushion, a point of entry for filth and vermin.

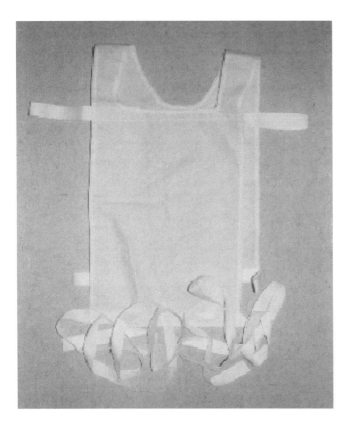

Figure 95-27 Patient vest restraint.

Entrapment

Wheelchairs, stretchers, and beds may be used in conjunction with patient restraints. Restraints can cause death by asphyxiation (Miles and Irvine, 1992; Rubin et al., 1993). The author investigated an incident in which a pediatric patient who had been restrained to a wheelchair with a vests restraint (such as that shown in Figure 95-27) was injured when he attempted to get out of the wheelchair. The wheelchair tipped, and the patient hit his head on the floor.

Other Hazards

Recent recalls of wheelchairs and mobility aids have been made for various reasons, including electrical-grounding defects (BS&S, 2003k), drive wheels coming off axles (BS&S, 2003l), wheels shattering (BS&S, 2003m), and back-support collapse (BS&S, 2003n).

Standards

Voluntary standards for wheelchairs exist (ANSI, 1990a, b, c, and d). Such standards, when adhered to by the manufacturer, assure a minimum level of performance and safety of wheelchairs. While a new wheelchair may have been designed to rigorous standards and may be without defect at the time of incoming inspection, it soon is placed in an often uncontrolled environment, used by uninformed patients and staff, and maintained and repaired by unqualified people. Safety and performance characteristics become degraded, and patient injury results.

Lifts, Parallel Bars, and Chairs

Patient Lifts

Patient lifts are utilized in transferring patients (e.g., from bed to chair). They are relatively simple devices, but they vary significantly in several aspects. For example, heavy-duty lifts accommodate the special needs of extremely obese patients. Figure 95-28 shows such a patient lift. The design features of a patient lift are shown below. When evaluating and purchasing a lift, the clinical engineer should be aware of the following issues: Stability, propensity to tip, ease of movement, swivel bar rotation, sling hook-up strength, load-bearing capacity, strength of welds, caster movement and locking, pinch points, and height of lift range. Accessories such as scales should be examined as well, because they could collapse because of defective design and manufacture (BS&S, 2003o).

Casters	Latches	Brakes
Handles	Swivel bar	Base legs
Sling	Sling hook-up	Hydraulic system
Manual crank	Battery and charger	Low battery alarm

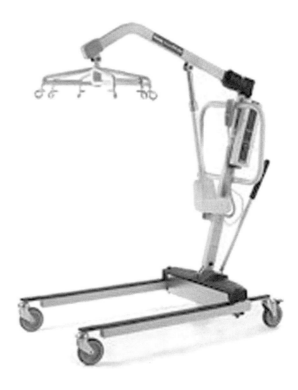

Figure 95-28 Invacare Reliant 600 Heavy Duty Power Lifter. This particular lift incorporates a battery-powered, heavy-duty hydraulic lift to meet the special needs of patients who weigh up to 600 lbs.

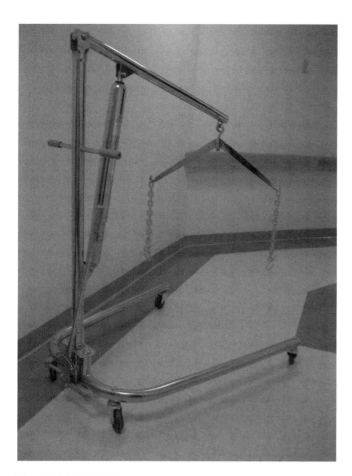

Figure 95-29 Patient lift.

The following clinical engineering study of patient lifts illustrates that this relatively simple mechanical medical device can fail and seriously injure the patient who is being lifted, and can cause injury to the operator (BS&S, 2003p). A clinical engineering analysis of a hospital incident involving a patient fall from a lift found that lack of or improper maintenance resulted in a failure of a locking mechanism (Grimes, 2003). When using the lift, the patient fell back onto the bed, but the two nurses making the transfer pulled their backs and could not return to work until they recovered.

The lift involved in the incident and shown in Figure 95-29 is designed with legs that can be spread (for increased stability) when lifting patients. Spreading of the lift's legs is accomplished by means of a vertically mounted spreader arm that forces the legs apart through mechanical linkage. A spring and bushing are normally mounted on the arm's pivot shaft. They are designed to force the arm into one of several detents. The lift was found to be missing several components (bolt, steel washers, spacer/spring, spacer/bushing) and that rendered the locking mechanism ineffective (Figure 95-30). A survey of patient lifts on the nursing floors of the hospital where the incident occurred resulting in the discovery of a second unit with missing hardware.

Clinical engineering made the following recommendations for remediation:

1. Repair or replace defective patient lifts.
2. Educate staff on the proper use of patient lifts (i.e., how to stabilize the units when using on a patient). An anecdotal report by a patient-lift user suggests that they might not be properly stabilizing the lift before use; i.e., operators might not be spreading lift legs.
3. Teach staff how to identify and report defective devices.
4. Those responsible for maintenance should schedule visual and mechanical inspections as a part of rounds

Parallel Bars

The following case study illustrates that even what appears to be a rather simple device can prove to be lethal to the patient. A patient suffered a near-fatal electric shock while exercising on motorized parallel bars (Figure 95-31). A patient undergoing therapy in the physical therapy department took his position on parallel bars for stepping exercises. The physical therapist (PT) flicked the toggle switch on the control panel to activate the motor that adjusted the height of the bar. Within seconds, the physical therapist heard a groan and observed the patient grimacing and clutching the stainless steel bar tightly. Noise and sparks were noted coming from the control box. The patient was an unwitting part in an electrical circuit through which flowed enough electrical energy (i.e., a level above the let-go current (Dalziel and Lee, 1968)) to cause tonic spasms of his musculature. Fast thinking and response by the PT in pushing the patient away from the bar broke the circuit and avoided a death from respiratory collapse.

Investigation by the author revealed that the line cord leading to the motor control box was caught under the base of the parallel bars. As the parallel bar raised, so did the box and the line cord, but because the cord was caught, it could not travel with the box, and a strain was placed upon it at the point of its emergence from the control box. The inadequate strain relief gave way, and the sharp circular cutout in the chassis cut through the insulation of the line cord, enabling the hot conductor to contact the box that was at the same potential as the stainless steel parallel bars. Current flowed through the patient.

Incoming inspection, had it been performed, would have revealed the inadequate strain relief. Clinical engineering inspection of the installation would have revealed the inadequate access to the electrical supply (facility) and the resultant inappropriate device modification for rerouting the AC power cord. Periodic preventive maintenance and inspection would have revealed strain-relief failure from the obvious signs of damaged strain relief, damage to the line cord, and inadequate routing of the line cord as a result of the device modification.

The evening before the incident occurred, a member of the housekeeping department moved the parallel bars during routine cleaning, inadvertently entrapping the line cord (Figure 95-32) and preventing its movement as the bars were raised. The motor control unit rose with the bars, placing strain on the line cord and resulting in the chassis cut-out in the control unit cutting through the line-cord insulation, thus shorting the 120-volt line to the chassis and to the parallel bars.

Chairs

Patient rooms typically are furnished with chairs for the comfort of visitors. Such a simple item as a chair has been shown to have caused adverse reactions from patients and visitors. The author experienced this as he visited a patient in a local hospital. The room was small and cramped, with barely any room to move about. When doctors, nurses, therapists, housekeeping, and other visitors entered the room, one had to move the chair to make space for those entering and moving about the room. This movement of personnel and chairs was ongoing and was not limited to one room. All the rooms to the sides, above, and below this room were similarly appointed. The movement of the chair was a relatively minor inconvenience. The result of the movement was the problem. The feet of the chair scraped against the floor, causing a terrible din that could be heard several rooms away. The patient, seeking rest and recuperation, was not afforded the quiet needed because a horrible screeching of chairs on the floor filled the air all through the day and night.

The installation of an inexpensive, low-friction pad or similar device to the feet of the chairs would have significantly reduced or eliminated the raucous noise. This anecdote emphasizes that one need not expend large amounts of resources to improve the quality of care. Engineering solutions to large problems can be simple and inexpensive.

Management of General Hospital Devices

Management of any health care device technology entails a cradle-to-grave responsibility. Starting with needs assessment, it continues through evaluation, purchase, service, and retirement. The discussion below pertains especially to the clinical engineering management of general hospital devices used in the hospital. Beds, stretchers, and wheelchairs, as well as many other general hospital devices, are used in extended care facilities, nursing homes, and in home care settings and should be managed as well as those found in the hospital (Dyro et al., 1996; Dyro, 1998).

Evaluation and Purchase

Clinical engineering participation in the evaluation of devices prior to their purchase is prudent. To promote the proper selection of a particular wheelchair, for example, the manufacture may provide detailed selection criteria (Everest & Jennings, 1989a). These criteria, as well as published evaluations, should be consulted when making a device acquisition (HPCS, 1994).

Clinical engineering should provide assistance to the purchasing department to ensure that bid specifications are correct and that necessary terms and documentation requests are incorporated in the bid document. In preparing bid specifications, the clinical engineer should consult applicable standards. For example, numerous standards exist for wheelchairs (ANSI, 1990 a, b, c, and d).

Service

In the case of hospital-based devices, responsibility for obtaining service rests on the shoulders of the facility whether the device is owned by the facility, donated, rented, or on loan. The facility may elect to employ an in-house unit, an independent service organization, the manufacturer, or a rental agency to provide this service. The level of service should be the same, regardless of the choice of support organization. In practice, however, a wide variation in the level of service exists among these organizations.

When general hospital devices are serviced in-house, this service is typically the responsibility of the hospital facilities and maintenance departments, the clinical engineering or biomedical engineering department, or a unit devoted exclusively to these devices only. The author, who has successfully integrated service of these devices into a full-service clinical engineering department (Dyro, 1993), advocates that this responsibility be held by clinical engineering. Where the responsibility falls to facilities or some other hospital unit, it is important to maintain the management principles of inventory control, preventive maintenance, and repair. General hospital devices should not be allowed to fall between the cracks. When these devices are in the home, the home health care agency, rental company, or authorized repair facility is responsible for providing

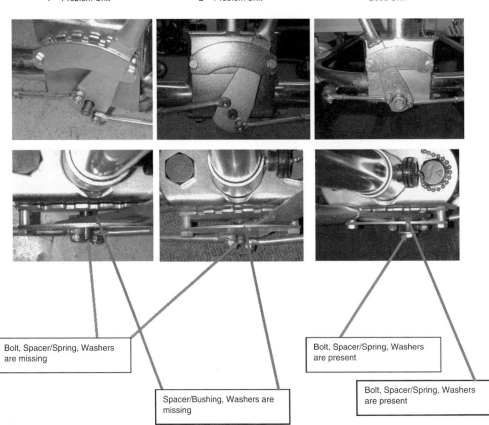

Figure 95-30 Patient-lift locking mechanisms. Mechanisms of three lifts of the same design are shown: 1st problem unit; 2nd problem unit; and good unit.

Figure 95-31 The author in the physical therapy department testing motorized parallel bars.

service. Those responsible for the care of an incapacitated patient are responsible for the coordination of service.

Service of general hospital devices, whether used in the hospital or not, entails the following:

- Incoming inspection
- Preventive maintenance
- Cleaning
- Repair
- Modification

Incoming Inspection

Incoming inspection of all medical devices upon receipt, and prior to initial use, should be performed to ensure that devices are safe and effective for their intended use. The manufacturer (Everest & Jennings, 1989c), professional organizations, the literature, colleagues, and other hospitals can be sources of recommended procedures for this inspection. Table 95-1 is a list of generalized medical device evaluation factors that can serve as a checklist for evaluating medical devices. Knowledge of the evaluation factors will ensure that the evaluation considers all factors affecting safety, performance, and cost. The list is intended to include all facets that are pertinent to evaluations of a broad spectrum of device types. One should include those factors that pertain to the device being evaluated. For example, if the device is a nonelectric wheelchair, then Section 2.4.2 (electrical safety) can be omitted.

The devices should be entered into an inventory management system along with pertinent information such as manufacturer, vendor, date of purchase, purchase price, model number, serial number, and preventive maintenance frequency. Operators and service manuals should be obtained and cataloged at this time. Service staff should receive training in the appropriate service procedures. This may entail attending a manufacturer's training program, an in-house training session, or a self-taught program utilizing the manufacturer's training aids such as books, manuals, and videotapes.

Preventive Maintenance

Preventive maintenance should be performed, and its frequency and intensity should based upon some objective criteria (Stiefel, 2002). A program based upon the level of risk ascribed to a medical device, taking into account the level of risk associated with the device and the class of patient for which the device is used, is one such example (Gullikson, 2000; Ridgway 2001). Inspection and maintenance procedures for most medical devices are available in the literature and from manufacturers (e.g., Dyro, 1978). (See

General Hospital Devices: Beds, Stretchers, and Wheelchairs   433

Figure 95-32 Device modification at installation leads to potentially fatal electric shock. Simulated line cord is shown in white, leading from the motor-control box down to a channel cut into the wooden base of the parallel bars. The line cord passed under the base and led to a wall-mounted electrical receptacle on the far right.

Figure 95-33 White tape fails to secure side panel.

Chapters 37 and 56 for discussions of maintenance programs and risk management, respectively.)

Cleaning

Cleaning occurs during preventive maintenance and during unscheduled repair service. The patient care staff in the case of hospital-based devices are responsible for daily checks for cleanliness of all medical devices and should either perform simple cleaning or delegate this task to the unit that is responsible for this function (e.g., housekeeping). In the case of devices located in the home, it is the responsibility of the patient or the caregiver to maintain the cleanliness of a device.

Repair

Repair of the device whether in the hospital or the home is performed by qualified service personnel. During preventive maintenance, necessary repairs are made.

Required repairs should not be delayed until the designated time for periodic preventive maintenance but, rather, should be performed as the need arises. To this end, if a device is hospital-based, the hospital staff (including doctors, nurses, technicians and housekeeping–those individuals who have frequent day-to-day contact with these devices) should be aware of the need for repairs and should request that repairs be done. Hospital staff should sequester devices in need of repair when defects render them inoperable or hazardous to patients. Patient care staff should not repair devices (e.g., nurses should resist the temptation to do a quick fix by application of white tape [Shepherd and Dyro, 1992]). Figure 95-33 shows an attempt to repair a defective latch on the side panel of a radiant warmer. The white adhesive tape, also known as "nurse tape," lost its adhesive properties when heated by the overhead radiant warmer. The side panel subsequently collapsed, and the infant rolled over the side and fell to the floor. Fortunately, the infant was uninjured by the fall (see Chapter 96 on troubleshooting).

If a device is used in the home, the patient or caregiver is responsible for recognizing the need for repair and for notifying the responsible party (e.g., a durable medical device rental company or manufacturer). The need for repair can be recognized when, for example, the head end of a bed cannot be elevated, the bedside rail binds when raised or lowered, a wheelchair brake does not adequately lock a wheel in place, or screws, nuts or fasteners loosen and cause mechanical instability in the device. As in the case of the hospital-based device above, quick-fix repairs an inadvisable as they often promote a false sense of security. For example, if tape were used to secure an arm pad to a wheelchair frame rather than the proper mechanical fasteners, the arm pad could loosen when subjected to the weight of a patient during a bed-to-wheelchair transfer, and such instability could result in a fall (Figure 95-34).

Modification of a device is deemed appropriate at times when a patient's needs can best be met by altering the design of the device, especially when alternative device designs are not commercially available. For example, the author has authorized the modification of a hospital bed, thus enabling it to support the weight of an extremely obese patient. Such modification should be undertaken with great care and with the advice and consultation of clinical engineering. The performance and safety of the device must not be degraded as a result of modifications.

Facilities for Service

A well-equipped work space is required for the servicing of general hospital equipment such as beds, stretchers, and wheelchairs. The author incorporated a general hospital-equipment service function into a full-service clinical engineering department (Dyro, 1993). The general mechanics workshop has a two-ton hoist, air evacuation system, welding and refrigeration equipment, standard power hand tools, tool kits, and a complete stock of spare parts, mechanical connectors, and fasteners. The workshop has adequate bench space for three mechanics and a spacious work area. Storage is provided for maintenance and service manuals, which most major manufacturers make available.

Post-Market Surveillance

The clinical engineering department must be part of the hospital's hazard-and-recall system. Some estimate that the percentage of recalled devices is as low as 4% of those in the field.

Table 95-1 Generalized Medical Device Evaluation Factors

1. Clinical significance (literature search and historical information)
   1.1 Physiologic rationale for device
   1.2 Statistical incidence and criticality of disease amenable to diagnosis or treatment by the device
   1.3 Relationship of engineering design criteria to clinical need and physiologic rationale
   1.4 Clinical efficacy of the device
   1.5 Safety and reliability of patient-device and patient-operator interface
2. Laboratory and engineering evaluation
   2.1 Description
   2.1.1 Configuration
   2.1.2 Principles of operation
   2.1.3 Special features
   2.2 Physical characteristics
   2.2.1 Dimensions
   2.2.2 Weight
   2.2.3 Materials
   2.3 Operational performance
   2.3.1 Measured engineering performance
   2.3.2 Comparison with manufacturer's specifications, other published standards, and physiologic requirements
   2.4 Operator and patient safety
   2.4.1 Mechanical
   2.4.2 Electrical safety
   2.4.2.1 Inspection
   2.4.2.1.1 Quality of line cord, plug cap, strain relief
   2.4.2.1.2 Accessibility and labeling of fuses or circuit breakers
   2.4.2.1.3 Suitability of protective devices and ratings
   2.4.2.1.4 Fault indicators or alarms
   2.4.2.1.5 Quality of switches, controls, connectors, and cables
   2.4.2.1.6 Power switch(es) in hot or neutral side
   2.4.2.1.7 Appropriate choice compatibility and protective incompatibility of connectors for patient cables and transducers and for connecting cables or line cords
   2.4.2.1.8 Visibility and suitability of warning legends
   2.4.2.1.9 Presence of balanced filter in primary of input power circuit
   2.4.2.1.10 Voltage/current ratings of power supply capacitors, fuses, transformers, rectifiers, and other components
   2.4.2.1.11 Quality of internal components
   2.4.2.1.12 Construction practices
   2.4.2.1.12.1 Mechanical integrity
   2.4.2.1.12.2 Insulation to chassis where indicated
   2.4.2.1.12.3 Identification of trimpots, relays, components, and other items
   2.4.2.1.12.4 Minimization of wire harness used for routing
   2.4.2.1.13 Presence of electrostatic shielding of power transformer
   2.4.2.1.14 Resistance between power supply sources and patient interface
   2.4.2.1.15 Design considerations that inhibit catastrophic failure
   2.4.2.1.16 Fire- and explosion-proofing
   2.4.2.2 Tests
   2.4.2.2.1 High-voltage test-input plug to chassis
   2.4.2.2.2 With combinations of input power polarity, grounding and machine operation, transient and steady-state leakage current (AC and DC) as measured:
   - Between patient leads
   - Between patient interface and ground
   - Between patient interface and case
   2.4.2.2.3 Ground-circuit resistance
   2.4.2.2.4 Static resistance between hot and neutral input power lines and chassis/building ground device
   2.4.2.2.5 Susceptibility to electromagnetic interference (EMI)
   2.4.2.2.6 Radiated and/or conducted electromagnetic radiation
   2.4.2.2.7 Injection currents from fault. Warning system (if applicable)
   2.4.3 Thermal
   2.4.4 Chemical
   2.4.5 Hazard warning and fail-safe features
   2.5 Human factors
   2.5.1 Device-operator interface
   2.5.1.1 Ease or complexity of application and use
   2.5.1.2 Reliability
   2.5.1.3 Operating instructions
   2.5.2 Device-patient interface
   2.5.2.1 Ease of application
   2.5.2.2 Complexity
   2.5.2.3 Reliability
   2.5.2.4 Tissue compatibility
   2.6 Reliability
   2.6.1 Failure
   2.6.1.1 Frequency–MTBF
   2.6.1.2 Failure modes
   2.6.1.2.1 Failure mode analysis
   2.6.1.2.2 Primary and secondary hazards
   2.6.1.3 Self-test and checking
   2.6.1.4 Failure-warning features
   2.6.2 Repeatability
   2.6.2.1 Dependency on energy source
   2.6.2.2 Operator proficiency
   2.6.2.3 Effect of varying ambient conditions
   2.6.2.4 Dependence on patient cooperation
   2.6.2.5 Calibration requirements
   2.6.2.6 Self-calibrating features
   2.6.3 Durability
   2.6.3.1 Quality of components
   2.6.3.2 Mechanical integrity
   2.6.3.3 Manufacturing practices and workmanship
   2.6.3.4 Shock resistance
   2.6.3.5 Impact resistance
   2.6.3.6 Vibration resistance
   2.6.3.7 Abrasion resistance
   2.6.3.8 Flexure and strain resistance
   2.6.3.9 Life testing
   2.6.4 Compatibility of materials and mechanisms with purpose and environment
   2.6.4.1 Pressure or altitude
   2.6.4.2 Temperature
   2.6.4.3 Humidity
   2.6.4.4 Chemical compatibility
   2.6.4.5 Aging and wearing characteristics
   2.6.4.6 Vibration
   2.6.4.7 Sound
   2.6.4.8 Electromagnetic compatibility
   2.6.4.9 Toxicity or toxic by-products
   2.7 Maintenance
   2.7.1 Availability and thoroughness of maintenance manuals
   2.7.2 Scheduled maintenance cycle
   2.7.2.1 Frequency
   2.7.2.2 Complexity
   2.7.2.3 Ease of cleaning
   2.7.2.4 Ease of disassembly and re-assembly
   2.7.2.5 Method of sterilization
   2.7.2.6 Downtime
   2.7.3 Unscheduled maintenance and repair
   2.7.3.1 Ease of disassembly and reassembly
   2.7.3.2 Accessibility to components and mechanical ease of repair
   2.7.3.2.1 Test points
   2.7.3.2.2 Ease to locate components from schematics and illustrations
   2.7.3.3 Specialized parts or tools requirements for servicing
   2.7.3.3.1 Tools
   2.7.3.3.2 Parts
   2.7.3.4 Parts availability
   2.7.3.5 Specialized service-training requirements
   2.7.3.6 Mean time to repair
   2.7.3.6.1 Availability of manufacturer's data
   2.7.3.6.2 User data
   2.7.4 Manufacturer's service capability
   2.7.4.1 Mode of operation
   2.7.4.1.1 Central organization
   2.7.4.1.2 Local representatives
   2.7.4.1.3 Training of service personnel
   2.7.4.1.4 Loaner and exchange policy
   2.7.4.1.5 Spare parts stocking policy
   2.7.4.1.6 Central backup and responsiveness to local service personnel
   2.7.4.2 Responsiveness to customer
   2.7.4.2.1 Timeliness of response
   2.7.4.2.2 Mean time to repair
3. Clinical Testing
   3.1 Operational performance in relationship to clinical and physiological requirements
   3.2 Performance judged subjectively by clinical personnel
   3.3 Safety
   3.3.1 Factors not apparent from laboratory testing
   3.4 Reliability
   3.4.1 Accuracy
   3.4.2 Repeatability
   3.4.3 Durability
   3.4.4 Compatibility of materials and mechanisms with purpose and environment
   3.5 Human factors
   3.5.1 Operator-device interface
   3.5.1.1 Ease of setup and use
   3.5.1.2 Ease of disassembly
   3.5.1.3 Ease of re-assembly
   3.5.1.4 Interpretability and readability
   3.5.2 Patient-device interface
   3.5.2.1 Ease of application
   3.5.2.2 Complexity
   3.5.2.3 Reliability
   3.5.2.4 Tissue compatibility
   3.5.2.5 Physical comfort
   3.5.2.6 Psychological effects
   3.5.3 Training requirements to maintain acceptable performance levels
   3.5.3.1 Initial training-time investment
   3.5.3.2 Effectiveness of training materials
   3.5.3.3 Proficiency retention

Table 95-1 Generalized Medical Device Evaluation Factors—cont'd

3.5.3.4	Retraining cycle and requirements		5.1.4	Installation
4.	Evaluation of manufacturer or supplier		5.1.4.1	Planning
4.1	Corporate history		5.1.4.2	Special construction or power supply
4.1.1	Age		5.1.4.3	Rigging and moving
4.1.2	Economic stability		5.1.5	Training
4.1.3	Size and characteristics of staff		5.1.5.1	Materials and subsystems
4.2	Manufacturing capabilities		5.1.5.2	Staff time
4.2.1	Plant facilities		5.2	Operating Expenses
4.2.2	Quality control procedures		5.2.1	Manning
4.2.3	Dependency on outside vendors		5.2.2	Energy source
4.3	Service Capability (See Section 2.7.4)		5.2.3	Operational upkeep and preventive maintenance
4.4	Distribution		5.2.4	Consumable supplies and parts
4.4.1	Mode of local representation		5.2.5	Unscheduled maintenance and repairs
4.4.2	Continuity of representation		5.2.6	Repair parts
4.4.3	Mode of shipping		5.2.7	Retraining
5.	Cost-Benefit Analysis		5.2.8	Warranty conditions
5.1	Initial Costs		5.3	Value Analysis
5.1.1	Administrative and professional time for selection, specification, and purchasing		5.3.1	Initial and operating costs
5.1.2	Acquisition cost		5.3.2	Lifespan and amortization
5.1.2.1	Basic unit		5.3.3	Cost of money
5.1.2.2	Accessories		5.3.4	Utilization
5.1.2.3	Spare parts and special tools		5.3.5	Labor-saving factor
5.1.2.4	Freight and delivery		5.3.6	Value of increased or new capability
5.1.3	Acquisition		5.	Correlation of data and ranking of attributes in relationship to competitive equipment

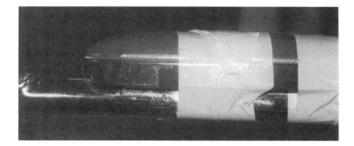

Figure 95-34 Wheelchair armrests taped to frame.

Inattention to recalls and ignorance of recall and hazard notices may result in defective devices remaining in use in the hospital (see Chapter 139). The clinical engineer has a duty to report to the manufacturer and to the FDA medical devices deemed to be hazardous.

User Education

The manufacture is responsible for providing adequate instructional material on the proper use of the medical device. Most major manufacturers of beds, stretchers and wheelchairs publish booklets aimed at user, caregiver, and maintainer education (Everest & Jennings, 1989b). The clinical engineering department should maintain a library and should have audio-visual equipment to present educational material (e.g., slides and videotapes) provided by the manufacturer and should play an active role in a hospital's in-service education program (Dyro, 1988).

Obsolescence

Clinical engineering should ensure that a plan is in place to retire general medical devices when their useful lives are over. Well-maintained devices that are retired in favor of newer technology may be considered for donation (see Chapter 43).

References

ANSI. *Wheelchairs: Determination of Overall Dimensions, Mass, and Turning Space*. RESNA Standard. ANSI/RESNA WC/05. New York, American National Standards Institute, 1990a.
ANSI. *Wheelchairs: Determination of Static Stability*. RESNA Standard. ANSI/RESNA WC/01. New York, American National Standards Institute, 1990b.
ANSI. *Wheelchairs: Nomenclature, Terms, and Definitions*. RESNA Standard. ANSI/RESNA WC/00. New York, American National Standards Institute, 1990c.
ANSI. *Wheelchairs: Type Classification Based on Appearance Characteristics*. RESNA Standard. ANSI/RESNA WC/7930. New York, American National Standards Institute, 1990d.
Agostini JV, Baker DI, Bogardus ST. Prevention of Falls in Hospitalized and Institutionalized Older People. In *Making Health Care Safer: A Critical Analysis of Patient Safety Practices*. Evidence Report/Technology Assessment No. 43, Publication 01-E058. Rockville, MD, Agency for Health Care Research and Quality, 2001.
Albert S. *Dear Powered Wheelchair/Scooter or Accessory/Component Manufacturer Letter*. Rockville, MD, United States Food and Drug Administration Center for Devices and Radiological Health, 1994.
BBC. Woman Dies After Operation Room Problem. *BBC News*, February 27, 2003.
Baker SP, O'Neill B, Ginsburg MJ, et al. *The Injury Fact Book, 2nd Edition*. New York, Oxford University Press, 1992.
BS&S. Obese Patients Strain Equipment as Geriatric Surgery Grows in Popularity. *Biomedical Safety & Standards* 32(21):171–173, 2002.
BS&S. Few Hospital Bed Injuries Result from Entrapment. *Biomedical Safety & Standards* 32(3):17–18, 2002.
BS&S. Possible Patient Entrapment Cited in Bed Recall. *Biomedical Safety & Standards* 33(15):117, 2003.
BS&S. Fluidized Therapy Units Have Flaw. *Biomedical Safety & Standards* 33(15):120, 2003.
BS&S. Side Rails Faulted for Acute Care Bed Recall. *Biomedical Safety & Standards* 33(1):3, 2003b.
BS&S. Beds Pose Electrical Shock Hazard. *Biomedical Safety & Standards* 33(14):107, 2003c.
BS&S. Bed Recall. *Biomedical Safety & Standards* 33(11):87-88, 2003d.
BS&S. Beds May Be Dangerous. *Biomedical Safety & Standards* 33(5):39, 2003e.
BS&S. Electrical Defects Lead to Recall of Neonatal Incubators. *Biomedical Safety & Standards* 33(3):23, 2003f.
BS&S. Cots Could Tip. *Biomedical Safety & Standards* 33(4):29, 2003g.
BS&S. Birth Bed May Be Unstable. *Biomedical Safety & Standards* 33(7):54, 2003h.
BS&S. Weld Failure Leads to Recall of Turning Frame Stretcher. *Biomedical Safety & Standards* 33(3):20, 2003i.
BS&S. Wheelchair Motor Fire Hazard. *Biomedical Safety & Standards* 33(9):68, 2003j.
BS&S. Wheelchairs under recall. *Biomedical Safety & Standards* 2003k; 33(12):94.
BS&S. Firm Recalls Powered Wheelchairs. *Biomedical Safety & Standards* 33(9):68, 2003l.
BS&S. Wheels on Scooters Could Shatter. *Biomedical Safety & Standards* 33(14):108, 2003m.
BS&S. Wheelchair Component Recalled. *Biomedical Safety & Standards* 33(12):93, 2003n.
BS&S. Scales Recalled; Bolts May Break, Causing Injury. *Biomedical Safety & Standards* 32(19):158, 2003o.
BS&S. Patient Lifts Subject of Recall. *Biomedical Safety & Standards* 33(2):10, 2003p.
Brubaker CE. Advances in Wheelchair Technology. *IEEE Eng Med Biol Mag* 7(3):21-24, 1988.
Brubaker CE. Wheelchair Prescription: An Analysis of Factors That Affect Mobility and Performance. *J Rehabil Res Dev* 23(4):19–26, 1986.
Calder CJ, Kirby RL. Fatal Wheelchair-Related Accidents in the United States. *Am J Phys Med Rehabil* 69(4):184–190, 1990.
Cooper RA. Wheeled Mobility: Wheelchairs and Personal Transportation. In Bronzino JD (ed). *The Biomedical Engineering Handbook*. Boca Raton, FL, CRC Press, 1995.
Dalziel CF, Lee WR. Reevaluation of Lethal Electric Currents. *IEEE Trans Ind Appl* 4(5):467–476, 1968.
Dyro JF, Sarkany G, Beltrani F, et al. *Handbook for Home Health Care*. Proceedings of the AAMI 31st Annual Meeting. Arlington, VA, Association for the Advancement of Medical Instrumentation, 1996.
Dyro JF. Inspection and Preventive Maintenance of Infant and Transport Incubators. AAMI 13th Annual Meeting. Washington, DC, Association for the Advancement of Medical Instrumentation, 1978.
Dyro JF. Educating Equipment Users: A Responsibility of Biomedical Engineering. *Plant, Technology & Safety Management Series* 1. Chicago: Joint Commission on Accreditation of Healthcare Organizations, 1988.
Dyro JF. Focus On: University Hospital & Health Sciences Center, SUNY at Stony Brook Biomedical Engineering Department. *J Clin Eng* 18(2):165–174, 1993.
Dyro JF. Building Awareness: Educating Health Care Providers and the Public. In *Electromagnetic Compatibility for Medical Devices*. FDA/AAMI Conference Report. Arlington, VA, Association for the Advancement of Medical Instrumentation, 1995.
Dyro JF. Methods for Analyzing Home Care Medical Device Accidents. *J Clin Eng* 23(5):359–368, 1998.
Everest & Jennings, Inc. *Wheelchair Selection*. Camarillo, CA, Everest & Jennings, 1989a.
Everest & Jennings, Inc. *Safety and Handling*. Camarillo, CA, Everest & Jennings, 1989b.
Everest & Jennings, Inc. *Care and Service*. Camarillo, CA, Everest & Jennings, 1989c.
FDA. *FDA Safety Alert: Entrapment Hazards with Hospital Bed Side Rails*. Rockville, MD, United States Food and Drug Administration Center for Devices and Radiological Health, 1995.
FDA. *A Guide to Bed Safety*. http://www.fda.gov/cdrh/beds/, 2000.

Ferguson-Pell M, Cochran GVB, Palmieri VR, Brunski JB. Development of a Modular Wheelchair Cushion for Spinal Cord Injured Persons. *Journal of Rehabilitation Research and Development* 23(3):63-76, 1986.
Gendron FG. *Unexplained Patient Burns*. Brea, CA, Quest Publishing, 1988.
Grimes SL. 2003. (Personal Communication)
Gullikson ML. Risk Factors, Safety, and Management of Medical Equipment. In Bronzino JD (ed). *The Biomedical Engineering Handbook, 2nd Edition*. Boca Raton, FL, CRC Press, 2000.
HCFA/CDRH. HCFA/CDRH Letter Regarding Physical Restraint Definition. August 1, 2000.
HPCS. Nonpowered Wheelchairs. Health care Product Comparison System. Plymouth Meeting, PA, ECRI, 1994.
Hanger HC, Ball MC, Wood LA. An Analysis of Falls in the Hospital: Can We Do Without Bedrails? *J Am Geriatr Soc* 47:529–31, 1999.
Health Canada. Possible Collapse of Metalcraft Hospital Bed Side Rails, Model 1092. 89, 1988.
Huffman GB. Rates of Deaths Associated with Use of Bedrails. *Am Fam Physician*. January 15, 1998.
JCAHO. *Focus on Five: Disarming the Bed Rail Trap*. Chicago, Joint Commission on Accreditation of Healthcare Organizations, 2003.
JCAHO. *Sentinel Event Alert*. Chicago, Joint Commission on Accreditation of Healthcare Organizations, 2002.
Kirby RL, Kumbhare DA, MacLeod DA. "Bedside" Test of Static Rear Stability of Occupied Wheelchairs. *Arch Phys Med Rehabil* 70:241–266, 1989b.
Kirby RL, Atkinson SM, MacKay EA. Static and Dynamic Forward Stability of Occupied Wheelchairs: Influence of Elevated Footrests and Forward Stabilizers. *Arch Phys Med Rehabil* 70:681-686, 1989a.
MDR. Incubator Tipover. *MDR Report* 8809: , 1988.
Miles S, Parker K. Letters to the Editor: Pictures of Fatal Bedrail Entrapment. *Am Fam Physician*, 1998.
Miles S, Levenson S. Bed Siderails in Long-Term Care. *Nursing Home Medicine: The Annals of Long-Term Care Online*, http://www.mmhc.com/nhm/articles/NHM9806_b/Miles.html, 2002.
Miles SH, Irvine P. Deaths Caused by Physical Restraints. *Gerontologist* 32:762—766, 1992.
Miles SH. Deaths Between Bedrails and Air Pressure Mattresses. *J Am Geriatr Soc* 50:1124–5, 2002.
Morrison J. *Radio Waves May Interfere with Control of Powered Wheelchairs and Motorized Scooters*. (Sent with Dear Powered Wheelchair/Scooter or Accessory/Component Manufacturer Letter and Dear Colleague Letter.) Rockville, MD, United States Food and Drug Administration Center for Devices and Radiological Health, 1994.
Parker K, Miles SH. Deaths Caused by Bedrails. *J Am Geriatr Soc* 45:797-802, 1997.
Ridgway M. Classifying Medical Devices According to Their Maintenance Sensitivity: A Practical, Risk-Based Approach to PM Program Management. *Biomed Instrum Technol* 35(3):167–176, 2001.
Rubin BS, Dube AH, Mitchell EK. Asphyxial Deaths Due to Physical Restraint. *Arch Fam Med* 2:405–408, 1993.
Shepherd M, Dyro JF. *The Whimsical Use of White Tape* (educational slide presentation). University of California, San Francisco, 1992.
Shepherd M. Managing the Risks of No Problem Found (NPF) and Near Miss (NM) Incidents. In *Medical Device Incident Investigation and Reporting Manual*. Walnut Creek, California, Devteq, 2003.
Stiefel RH. Developing an Effective Inspection and Preventive Maintenance Program. *Biomed Instrum Technol* 36(6):405–408, 2002.
Witters D. Medical Device EMI: The CDRH Perspective. In *Electromagnetic Compatibility for Medical Devices*, FDA/AAMI Conference Report. Arlington, VA, Association for the Advancement of Medical Instrumentation, 1995.
Witters DM, Ruggera PS. Electromagnetic Compatibility (EMC) of Powered Wheelchairs and Scooters. In *Proceedings of RESNA 17th Annual Conference* 359–360, 1994.
Young JB, Belfield PW, Mascie-Taylor BH, et al. The Neglected Hospital Wheelchair. *Br Med J* 291:1388–1389, 1985.

96

Medical Device Troubleshooting

Joseph F. Dyro
President, Biomedical Resource Group
Setauket, NY

Robert L. Morris
Oregon Health & Science University
Portland, OR

The goal of troubleshooting is to repair or correct a fault in an instrument system. This chapter provides a sound introduction to logical troubleshooting of medical devices, medical technology systems, and scientific and analytical instruments. The material presented in the chapter is based upon a general theory of troubleshooting, with stress on general techniques and strategies.

General Approach to Troubleshooting

This section provides a general overview of the process of logical troubleshooting. Other sections delve into the subject in more detail.

Systems Approach

An instrument system consists of three major components: The operator, the environment, and the instrument. A failure in, or poor performance of, an instrument system can be the result of difficulties in any one of the system's three major components.

The operator: The end-user typically makes the initial decision that a failure or malfunction has occurred problems can be due to such factors as incorrect or improper operation, or incorrectly set controls.

The environment: The total environment surrounding the instrument and operator problems can be related to environmental and other factors such as temperature, drafts, dirt, vibration, incorrect electrical supply, electrical or chemical interference, input exceeding the dynamic range, bad reagents, or bad electrodes.

The instrument: The device that performs a task such as measurement or control. Failures of the device fall into two general categories: (1) non-electric (e.g., loose or broken connections, dirt, corrosion, or mechanical wear. These are the most probable sources of instrument failures); and (2) electronic (a component or circuit failure). In general., the electronic portion of an instrument is the most reliable part.)

The first problem of troubleshooting is to determine whether the fault is in the operation, environment, or instrument component of the instrument system. The purpose of the methodology of troubleshooting is to gather information about the poor performance or failure of an instrument system in a logical and systematic manner. Questions to ask include the following:

- How does the system normally behave?
- What are the conditions of failure?
- What are the symptoms caused by the failure?
- What is the cause of the failure?

Steps in Troubleshooting

- Determine the symptoms and analyze them.
  - Listen, think, look, smell, and operate.
  - Use all available sources of information such as manuals, maintenance records, other people, and the agent or manufacturer.

○ State the symptoms as clearly and precisely as possible. Stating that a device does not work, while perhaps true, is not a clear, informative statement of symptoms.
- Localize the problem to a functional module
- Isolate the problem to a circuit
- Locate the specific component or problem.
- Determine the cause of the failure.
- Replace or correct the defective component or problem and correct the causes of the failure.
- Check for correct operation and calibration.
- Complete the record keeping.

Proceed by asking questions and finding answers to them. State questions as simply as possible. Select questions that can be answered easily and that will provide the maximum information to help find the fault. Answer the questions by thinking, observing, and testing. Refer to experts, manufacturers, or agents. Look in manuals, text books, catalogs, or other technical data sources.

Make observations and perform tests, proceeding from the simple to the complex. Think about failure, the degree of failure, the causes of failure, the time course of the failure, and the combinations of failure. In other words, given the information at hand:

- What is most likely to have failed?
- What was the most probable cause of the failure?
- How can the fault and the cause be corrected?

Keep records. Good records immediately help to keep you proceeding in a logical sequence and to keep track of what has been done. Records help in the future by providing a reference in case of additional failures in the same or a similar instrument. They can be used to provide feedback to manufacturers and their agents and to provide management information to aid in making repair or replacement decisions. Records can be used to detect a problem in a particular instrument that might require modification to correct.

Logical Steps in Troubleshooting

The troubleshooting process is the effort to find the discrete element or elements that have failed, starting at the system level and working down to the discrete element. A step-by-step procedure is listed below, adapted from Aston (1990). In troubleshooting, one would follow the steps indicated below until the problem is identified, and then would skip to step 25.

1. Assess the equipment environment with the human senses (i.e., eyes, ears, nose, tongue, hands).
2. Consider operator error.
3. Assess at the system level.
4. Check connections.
5. Isolate the problem device.
6. Verify device operation.
7. Check patient cables, connections, and accessories.
8. Utilize equipment troubleshooting devices.
9. Check diagnostic error messages when available.
10. Execute equipment self-testing procedures.
11. Open case and view interior.
12. Make a module-level assessment.
13. Check all connections.
14. Analyze symptoms for board-level problem isolation.
15. Use specialized diagnostic aids, maintenance manuals, and diagnostic software.
16. Analyze equipment history (e.g., age, preventive maintenance compliance, previous repairs).
17. Analyze data for problem isolation.
18. Consult with manufacturer's technical representative.
19. Make test-point measurements.
20. Analyze data for board-level problem isolation.
21. Take measurements using the device's block diagram.
22. Analyze data for board-level and ancillary-component isolation.
23. Make measurements using module schematic diagrams.
24. Analyze data for component-level problem isolation.
25. When the problem is found and corrected, take action to minimize recurrence (e.g., in-service education, device replacement, or increased preventive maintenance).

Troubleshooter Characteristics and Skills

Characteristics

A good troubleshooter has the following characteristics:

- Curiosity about why things are the way they are, and about how things work, both separately and together
- A thorough grounding in fundamentals, not only electrical, but also mechanical, optical, and chemical
- A good understanding of the general principles, functions, and characteristics of instruments, instrument systems, and computers
- Knowledge of transducers and their characteristics
- Knowledge of the basic components from which instruments are built, such as:

Resistors, capacitors, transformers, inductors, diodes, transistors and other semiconductors, analog and digital integrated circuits, motors, switches, wire, cable, connectors, circuit cards, clutches, and fasteners.

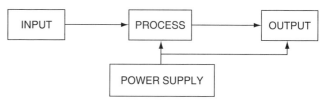

Figure 96-1 General device functional block diagram.

- Knowledge of the failure modes of instruments and components
- Knowledge of their own capabilities (e.g., when to stop and when to ask for help)
- Good manual dexterity
- A systematic, logical approach to problems and problem-solving
- The ability to work well with people

Skills

Troubleshooting requires a broad knowledge base including electronics, physics, chemistry, optics, mechanics, measurement theory, and equipment operation. Troubleshooting requires the following practical skills:

- Careful observation–knowing what to look for, when to look, and where to look
- Effective utilization of instrument service manuals, and other sources of technical data
- Proper use of hand tools and small machinery–using the proper tool or machine for the task at hand, and using them safely and correctly
- Good soldering techniques–using a good, clean soldering iron and following proper practice to prevent damage to delicate components, circuit cards, and insulation
- Knowledge of device assembly, disassembly, and component removal methods–proceeding cautiously and carefully noting the order of things to prevent additional difficulties and to allow proper re-assembly
- Good safety awareness and practices–not only for the trouble shooter but also to note and change those things that could be a hazard to the operator or user of the device

General Description of Instruments

This section presents a general method of describing instruments through the use of functional block diagrams and emphasizing their value in troubleshooting and understanding instruments.

A functional block diagram as a model for any instrument is shown in Figure 96-1.

The functional block diagram can be made more detailed and device-specific, as shown in Figure 96-2.

Further expansion of the functional block diagram is shown in Figure 96-3. This process of expanding a functional block diagram can be repeated to obtain increasing detail and aids in the understanding of the device's operation and possible problem areas. Figure 96-4 shows a device-specific diagram with more detail.

A single block in a functional block diagram of an instrument (e.g., the power supply (see Figure 96-3)) can be broken into its own functional block diagram. The power supply for example would be depicted as in Figure 96-5.

Complete schematic diagrams do not exist for many instruments. There are only functional block diagrams, interconnecting diagrams, and partial schematics. This is true of nearly any instrument that contains integrated circuits. With digital instruments, a logic diagram is furnished. A logic diagram is merely a detailed functional block diagram.

If you understand one instrument, you know something about every instrument. If you understand one type of instrument, such as a spectrophotometer, you know functionally about every instrument that uses or measures light. If you understand functional modules, such as power supplies, amplifiers, you know something about every instrument that includes them. If you understand components, resistors, capacitors, semiconductors, you know something about every instrument that contains those components.

In a functional block diagram of troubleshooting, information is the input; knowledge and understanding is the process; the solution is the output; and the brain is the power supply, as shown in Figure 96-6. As is true with instruments, when the power supply is not working, nothing works well. Instruments and components have more similarities than differences. Knowledge gained by studying one instrument or component can be applied to all instruments and devices.

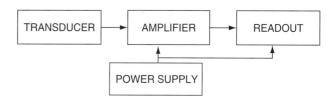

Figure 96-2 Functional block diagram expanded.

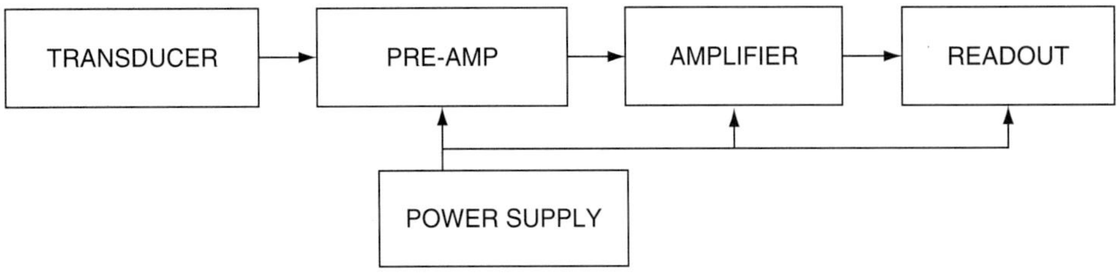

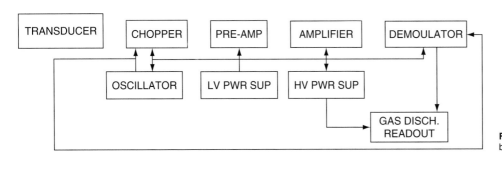

Figure 96-3 Functional block diagram further expansion.

Figure 96-4 A device-specific functional block diagram.

Figure 96-5 Functional block diagram of a power supply.

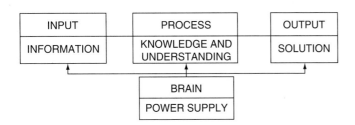

Figure 96-6 Functional block diagram of a troubleshooting system.

Symptoms and Analysis

This section provides detailed information on the first and most important part of the process of troubleshooting: Determining symptoms and their analysis. The determination and analysis of symptoms are repeated continuously throughout the process of troubleshooting. At each step, localization to a functional module, isolation to a circuit, and locating the defective component, the redetermination and analysis of symptoms through the process of listening, looking, smelling, operating and testing is repeated over and over. Thinking and asking questions to guide the thinking process are essential.

Thought Process

Use the following questions help to guide the thought process as the troubleshooter seeks to identify and learn more about the symptoms of the problem.

- What are the known symptoms?
- Can the symptoms be verified?
- Did the symptoms appear gradually or suddenly?
- Are the symptoms always present?
- Are the symptoms only present under some circumstances (e.g., after warm-up, with certain control settings)
- Are the symptoms only present intermittently and in an unpredictable manner? This is the most difficult case and may require the careful application of thermal., voltage, vibration, or other stress to determine the conditions of failure and to reproduce the failure. Record-keeping of the failure conditions is very important. Interference by another device or system often shows up as an intermittent problem, which is present only when an interference-generating system is operating.
- Are there any other symptoms?
- What is the normal or expected behavior of the instrument?
- How does the actual behavior differ from normal?
- Can the symptoms be restated in a different way that will be more exact and can provide more information?
- What are the easiest checks or tests to perform to verify the symptoms?

When inquiring as to the nature of the problem, formulate and pose easy-to-answer questions that require only yes-or-no answers. The questions should be designed so that an answer can eliminate major portions of the system from further consideration as a potential source of the problem. Examples of these types of questions include the following:

- Did the device ever operate correctly under the conditions to which the failure was attributed?
- Is the device plugged in?
- Is it turned on?
- Is the fuse good?
- Is there an output? Is it normal?
- Is there an input?
- Do the symptoms indicate a possible fault in the power supply or other functional module?

Be flexible with questions. If it is not possible or difficult to answer a question, rephrase it or go on to a different question. Use all available sources of information when analyzing symptoms, and when asking and answering questions. Questions and their answers may be temporary or tentative and subject to change. They are only aids to acquire information so a problem may be solved. They are not the real problem. The direction in troubleshooting is always from the general to the specific, from system to instrument to function to component. The direction in testing is from the simple to the complex.

Looking for Clues

Operate the Controls

- Remember to record the original settings of the controls. The problem could be caused by a control in the incorrect position for proper operation.
- Do all indicator lights (e.g., warning alarms) work properly?
- Do the controls behave normally? Observe device performance and all read-outs and indicators.
- Do the controls look, sound, and feel appropriate?
- Do the controls have an appropriate effect on the operation of the instrument?

Listen, Look, Smell, and Feel

Your senses and brain make up the best and most flexible test and analysis system in the world. Listen to the operator or user, and to the device for normal and abnormal sounds such as:

- Information as to the conditions and symptoms of failure from the operator.
- Rattles, squeaks, buzzes. These can indicate wear or lack of lubrication or motion.
- Sizzling or humming. Either can indicate a short circuit, arcing, high-voltage corona, and excessive loading of transformers.
- Excessive loudness or silence. Each can indicate a broken or defective part such as a cooling fan.

Look for the following:

- Physical damage and wear. (Figure 96-7 shows a bent knob on the front panel of an infant incubator.)
- Loose, inadequate, incorrect, or incompatible connections or connectors. Figure 96-8 shows two connectors on a pressure transducer cable, both of which have bent pins. Investigation revealed that the operator was attempting to connect the cable to the incorrect receptacle on the blood pressure monitor.
- Dirt, dust, lint, corrosion, leaks, or vermin. (Figure 96-9 shows lint accumulation about temperature-control elements in an infant incubator.) Signs of fluid staining also can be seen. Figure 96-10 shows a printed circuit board infested with red ants. The ants apparently were attracted by the tempting odors emanating from the board at the site of power transistor mountings.
- Burned, charred, or discolored insulation, components, circuit cards, or chassis. Figure 96-11 shows a hole and scorch marks produced by electricity from a power cord passing through the chassis of height-adjusting control unit of parallel bars. The cord was inadequately protected with grommet and strain relief.
- Physical relationships–use these to build a mental image of the functional relationships
- Proper mechanical operation–no binding or jerking
- Ways to disassemble the device or module

Smell for the following:

- Overheated transformers. This can indicate a defect in the transformer, an external short, excessive load, or a cooling failure.
- Burned resistors or resistors where you cannot read the color code. These can be caused by reduced value due to age effects, failure of another component or function, or cooling failure.
- Burned or charred insulation. This can be caused by a short circuit, excessive current flow, or an overheated nearby component.

Figure 96-7 Front panel of an infant incubator, showing a bent knob.

Tues, Sept 9, 2003

The 3 cables picture here were picked up by clinical engineering staff from Neurology today. Staff should be reminded that if a connector doesn't seem to fit properly, it should not be forced ... and when one cable doesn't fit, forcing a second or third cable is not a proper response (It also can delay and possibly prevent access to diagnostic data on the patient). Please seek help from other knowledgeable staff or contact clinical engineering for assistance.

The replacement cost of these cables is $ ____ and this kind of damage cannot be reliably repaired.

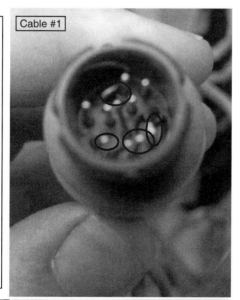

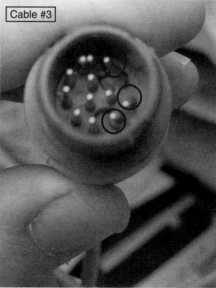

Figure 96-8 Bent pins on a connector, resulting from user error.

Figure 96-9 Lint and fluid accumulation on the temperature-control unit of an infant incubator.

Figure 96-10 Ant infestation of printed circuit board.

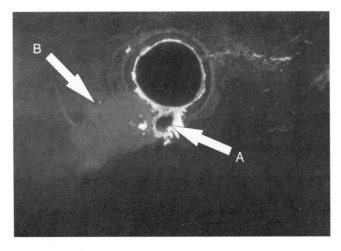

Figure 96-11 Burns on a chassis caused by a short in the power cord.

Feel for the following:

- Hot spots. Be careful not to get burned or shocked.
- Rough edges or worn spots. These can indicate such things as mechanical wear, incorrect alignment or assembly, or broken parts.
- Smooth operation. Few devices are designed to function in a jerky or irregular manner.

Do not look at schematics or begin taking measurements until you have exhausted the preceding process. Most (nearly 80%) instrument system problems can be solved by using your general and specific understanding of instruments and systems and the knowledge to be gained through a careful utilization of your senses of sight, hearing, smell, and touch.

More Clues–White Tape

White tape–medical adhesive tape–has been properly used for many years to hold catheters in position or to fasten bandages. On occasion, white tape is used inappropriately to make a temporary repair on a medical device. The observation of white tape, or simply the residue of tape previously applied, is a clue in troubleshooting as it reveals that location and the nature of the problem.

Shepherd and Dyro (1982) investigated the uses of white tape. In one instance, white tape covered the plug of a thermistor used on a radiant warmer. Discussions with the nurse determined that, without the tape, the plug had been loose and that the heating of the neonate had been intermittent. The nurse had accomplished a temporary repair that had allowed the device to continue functioning. In another instance, white tape covered a bed-control plug where it entered a bed. The purpose of the tape was to keep the plug from falling out of its mating connector. On yet another occasion, several layers of white tape were noted as being wrapped around an ECG transmitter. Discussion with a nurse determined that, in the middle of the night, the transmitter had been dropped and the case had become cracked. The nurse had wrapped white tape around the case, and the transmitter had begun working again and had been returned to service. In another instance, white tape was noted at the rear of a device holding a fuse into its fuse post. The fuse cover had been lost or broken, but the tape allowed continued functioning of the device.

Troubleshooting should not be restricted to the laboratory bench and examination of only those devices reported to have failed and brought to the clinical engineering department for repair. The clinical engineer and biomedical engineering technician must be ever vigilant throughout the hospital. The presence of a cracked connector on an ECG cable, a broken meter face on an electrosurgical unit, or white tape on a radiant warmer bassinet latch are signs that underlying problems exist with the devices that may eventually emerge to the detriment of the patient even though the device may appear to be operating satisfactorily.

Shepherd and Dyro (1982) also found white tape used to reduce loud and irritating alarms. In one instance, tape had been placed over the audible output portion of a neonatal warmer to reduce the sound level. In a second case, tape had been placed over a reset button that was normally used to temporarily silence a radiant warmer alarm. The tape held the button "in" so that no alarm could occur. The alarm circuitry was, in effect, permanently reset. A neonate was severely burned by this warmer while the tape was in use. These uses and others have prompted Kermit (1993) to urge tight restrictions on white tape.

Localization, Isolation, and Location

This section continues the process of troubleshooting to localize, isolate, and locate a fault. The general techniques in the above section on symptoms and analysis will be followed. The general sequence to be followed is localization to a function module, isolation to a circuit, and locating the defective component.

- A "function module" is a part of an instrument that performs some major function. Usually, it would correspond to a block in the functional block diagram of an instrument. Examples include power supplies, amplifiers, oscillators, and pre-amplifiers.
- A "circuit" is a part of an instrument that performs an electronic subfunction. It is a part of a function module. Examples include the input stage of an amplifier, the filter of a power supply, the regulator of a power supply, and the output drivers of an amplifier.
- A "component" controls a circuit, current, or voltage. A group of components connected together form a circuit. Examples of components include switches, resistors, capacitors, transformers, transistors, integrated circuits, and lamps.
- Components are used to make circuits. Circuits make up a function module. Function modules make up a device or instrument. Devices can be connected together to make up a system.

Always do the easiest things first. Consider what is most likely to be the problem, and check it first. If the problem is not what you thought it was, stop and think before proceeding. Always be systematic, and record your results so you will know what has been checked and what is known to function correctly.

Localization to a Function

Before a fault can be localized to a function, it is necessary to know what the functions are. One must be able to draw a block diagram. If little is known about the device, the block diagram will be quite simple. As more is learned about the device, a more complete and detailed block diagram can be created. If the trouble is not found after the above symptoms and analysis steps are followed, then do the following:

- Make a decision as to the most probable cause, based on the symptoms.
- Determine what and how to test or measure to verify your decision.
- Proceed in a logical systematic manner so that you will know what you have done and what portions of the device have been found to be functioning correctly.
- Repeat the above process until the trouble has been located, and then correct it.

General Strategies of Testing and Measuring

All other things being equal., it is usually a good idea to check the power supply voltages first.

Substitution: Substitution is often the fastest means of finding a faulty function module, but for it to be effective, several factors must be taken into consideration:

- Think before substituting. The actual fault may be such that it will cause a failure in the substitute. For example, if the 5-volt regulator had failed and the 5-volt supply was actually 8.5 volts, substituting TTL logic cards or integrated circuits would only ruin the substitutes rather than isolate the fault. It is a good idea to check the power supplies before substituting.
- It is generally not a good idea to substitute high-power, current, and/or voltage functional units with high power, as that could damage the power supply and perhaps the substitute.
- The substitute must be known to be working. It is not helpful to replace one broken unit with another.
- When substituting circuit cards, there may be jumpers, switches, or controls on the card that must be set correctly before the substitute will work properly. One example is a memory board in a computer that often has switches or jumpers that must be set to give the correct address to the CPU.

- Substituting a working component for one that has failed can simply create more failed components if the reason for the original failure is not considered and corrected.

Input-to-Output Method: This method involves measuring or testing the input to a functional module. If the measurement is satisfactory, then look at the output of the module. If the output is satisfactory, then the module under test must be good. Move on to the next module.

Output-to-Input Method: This method is the same as the approach above, except that the output is tested first. The choice between the output-to-input method or the input-to-output method is a matter of convenience. Use whichever is easier and quicker.

Half-Split Method: This is a good method to use if there are a large number of identical functional blocks in series or a number of identical circuit cards in an instrument. Test or measure the output from the middle block or circuit card. If the results are correct, then the first half of the blocks are OK, and the problem must lie in the second half. Split the second half of the blocks in half, and measure the signal. If the signal is incorrect, the problem lies between the half and 3/4 block. Further measurements will isolate the problem to a particular block or circuit card. The half-split method is very efficient in searching for a problem in a large number of identical functions that are connected in series. However, several assumptions are made:

- All circuit blocks are equally reliable.
- Only one fault exists.
- All measurements are similar and take the same amount of time.

Most systems do not consist of only series-connected blocks but possess parallel branches and feedback loops. The connections that complicate fault location are:

- *Divergence:* An output from one block feeds two or more other blocks. Check each output in turn, then continue to search for a faulty block in the area that is common to the incorrect inputs.
- *Convergence:* Two or more input lines feeding a single functional block. First, check the inputs at the point of convergence. If all are correct, the fault is beyond the convergence point. If any input is incorrect, the fault must lie in that input circuit.
- *Feedback:* The output of a block that is connected to the input of the same block or to the input of an earlier block.

Feedback systems are more complicated. There are two types of feedback.

Modifying Feedback: Modifying feedback occurs when the feedback is used to modify the characteristics of a system. One example of modifying feedback is the automatic gain control used in radio receivers and video amplifiers. It is often possible to break the feedback loop and to check each circuit inside the loop separately, without a fault signal being fed around the loop. It is usually best to disconnect the feedback loop at the input block. Care must be taken that bias conditions are not upset. In the case of an AC feedback signal, it can be shorted to ground with a suitable capacitor.

Sustaining Feedback: Sustaining feedback occurs when the feedback is essential before any output can exist. Sustaining feedback is used in many position-control systems where a feedback signal, proportional to the position of some output device, is used to cancel the effect of an input signal. Most servo-type chart recorders fit this description. When the feedback is disconnected from the input, it might be possible to inject a suitable signal in place of the feedback and then to check the circuit blocks within the loop.

Isolation, Location, and Cause Determination

Isolation to a circuit within a function is a continuation of the symptoms-and -analysis process, described above. The only additional factor that comes into use is your detailed knowledge of circuits. When isolating to a circuit, it is often useful to look at the schematics, if they are available. Continue to think, look, listen, smell, and feel, and take measurements.

Locating the Defective Component

- Everything that has been said above applies.
- When looking for a faulty component, keep in mind the components that are least reliable and therefore most likely to fail.
- Check the component that is most likely to have failed first.
- Remember to use a systematic approach and to do the easy things first.

When the Fault is Found

Determine the cause of the fault. If the problem is a failed component, determine the cause of failure before replacing the failed component. Replacing a burned-out resistor that was damaged by a shorted transistor will only consume more resistors until the shorted transistor is replaced. Correct the problem, and test the device for proper operation. Review the entire troubleshooting and repair process. Look back at ways in which the problems were found and corrected.

- Were there symptoms that you ignored but that actually would have helped you to find the problem sooner?
- Given similar symptoms, could you find the problem sooner next time?

With the device operating correctly, measure voltages and signals at critical points in the instrument, and record the readings in the manual or on the schematic for future reference. Review the entire process of troubleshooting and repair as a learning experience to find out the things you did correctly and those things that you should do differently. Complete the necessary paperwork to maintain accurate and complete records.

Test Equipment and Tools

This section reviews basic considerations for measurements and the characteristics of common test equipment. The troubleshooter must be thoroughly familiar with the basic principles of measurements and with the test equipment to be used, before the maximum information can be obtained from any measurement performed. One must know what, where, when, how, and why to measure, and must be able to interpret the results of a measurement. Troubleshooting can require a wide range of tools from the basic human senses (eyes ears, nose, mouth, and hands) to specialized equipment such as ventilator analyzers. Examples of tools that are helpful in troubleshooting are described below.

Measurement

No measurement should be made without knowing, before the measurement, what to expect as a result of the measurement. If you do not know what voltage or signal you should have at a particular point in a circuit, how can you tell whether the results of the measurement are correct or incorrect? The purpose of using test equipment and making measurements while troubleshooting is to gain information about a failure or fault, or about the cause of a failure or fault. The following are definitions of some common terms:

Accuracy: The closeness of a measurement to the "true" or "correct" value. Accuracy is often expressed as a percentage of reading or a percentage of full scale. Digital measuring instruments always have an additional limit on their accuracy of plus or minus one in the least significant digit, due to the nature of digital counting circuits. A known accuracy is the result of calibration.

Precision: The number of significant digits available to describe a measurement. It is usually expressed as a number of digits. For example, the precision of a measurement or instrument can be stated as 5 digits or 3 ½ digits. Precision does not necessarily imply anything about accuracy. It is possible to build or to buy a digital multimeter that has 4 digits of display but only an 8-bit D/A converter. If that were the case, the precision of such an instrument would be 4 significant digits, but the accuracy could not exceed 0.5%.

Resolution: How finely a measurement can be read or a value resolved. A statement of resolution is usually of the form (referring to the specifications of a DMM) "A resolution of 100 microvolts on the 1-volt range." This statement of resolution implies 4 ½ or 5 digits of precision. It says nothing about accuracy. Sometimes (particularly when making differential measurements) it will be more important to have high resolution than an accuracy sufficient to match the resolution.

Reproducibility: If repeated measurements of the same signal are taken, how well do the results of successive measurements agree with each other? Reproducibility is affected by the stability of the measuring instrument and the signal source.

Stability: If a stable, unchanging signal is being measured continuously, how stable or unchanging is the indication on the measuring instrument?

Whenever a measurement is made, the use of a measuring device affects whatever is being measured. Most of the effect is due to a change in circuit parameters brought about by the attachment of the probes of a test instrument to a point in the circuit. All electronic or electrical measuring instruments have finite input impedance. Effects might or might not be significant. The effect might be to change the measured quantity so that the result is not useful. Attaching a test instrument to a circuit to make a measurement can change the way a circuit behaves. One example would be trying to measure or to look at the signal at a sensitive point in an oscillator. The process of trying to make a measurement could shift the frequency or could stop the oscillator from oscillating. The consequences of attaching a test lead to a point in a circuit must be considered and understood before the results of the measurement can be interpreted. The effect of attaching a test lead to a point in a circuit is usually called "loading." Loading does not affect the precision or resolution of a measurement, only the accuracy.

Sometimes a measurement is made and it simply is not appropriate. One example would be looking with an oscilloscope or voltmeter directly at the base of a common emitter AC transistor amplifier. The measurement will tell you something about the bias conditions that are present, but very little about the signal. Because a transistor is controlled by current into the base instead of voltage, one probably will not be able to see any signal. If one does see a signal, it will be much smaller than expected.

Yellow Safety Box

The yellow safety box (Dyro, 1995) contains most of the devices and measurement tools that are required in inspecting medical devices (Figure 95-12). Its chief advantage is its portability, which enables the troubleshooting at remote locations. The yellow safety box is particularly applicable to investigating accidents in which people have been injured by medical devices.

A partial listing of the items in the box and their applications follows: Hand tools (pliers, screw drivers, wrenches, hemostats); photometer; digital voltmeter (DVM), sound-level meter; sample collection bags (swabs, plastic bags, brush); measurement devices (letter scale, force gauge, tape measure, electronic tape measure, ruler, micrometer, calipers, protractor; camera (still, moving); tape (white, black, masking); personal protection (gloves, respirator, dusk mask, safety glasses, ear plugs); batteries (AA, AAA, D, C); camera; visual-examination aids (flashlight, headlamp, loop, magnifying glass 5X-30X); markers (pen, pencil); paper (investigation form, graph, ruled, notes, Post-It notes); film (400 ASA 12-, 24-, 36-exposure; video film); candy; and physiologic simulator. (See Chapter 64 for a detailed description of the yellow safety box.)

General Purpose Test Equipment and Tools

Several companies specialize in the manufacture and sale of general purpose test equipment that performs physical measurements such as weight, dimension, pressure and flow, electrical parameters (voltage, current, resistance, frequency, waveform), temperature, humidity, irradiance, relative humidity, particle count, gas concentration, sound level, time, and force. These companies include Extech, Fluke, Ohaus, Tektronix, Chatillon, and Hewlett-Packard.

A troubleshooting and repair facility should be equipped with a full range of basic hand tools such as screwdrivers, wrenches, pliers, cutters, soldering irons, magnifiers, examination mirrors, drills, hammers, and saws. The list of possible tools is extensive.

Specialized Test Equipment

Many companies (e.g., Metron and DNI Nevada, Inc.) manufacture and sell specialized test instruments designed specifically to test one particular type of medical device or system; e.g., electrosurgical unit analyzers, ventilator testers, defibrillator analyzers, infusion pump analyzers, S_pO_2 analyzers, ECG analyzers, non-invasive blood pressure monitors, fetal monitors, gas analyzers (oxygen, carbon dioxide, nitrous oxide, halogen, ethylene oxide), and electrical safety analyzers.

Several companies manufacture and sell test devices that are specific to their products; e.g., Nellcor (pulse oximeters), Ohmeda (anesthesia machines and infant incubators), and Medtronic (pacemakers and implantable programmable infusion pumps).

Simulators are available to aid in the testing of medical devices. They generally take the place of the patient on whom the medical device is used. These simulators include finger phantoms, test fingers, pulse oximetry, heart rate, blood pressure, ECG, arrhythmia, fetal simulator (ultrasound, maternal ECG, tocodynamometer, beat-to-beat variability), temperature, respiration (impedance), breast phantoms, and lungs.

Many companies are involved in the design, manufacture, sale, and distribution of the above classes of test equipment for use by clinical engineers in the hospital and in the home health care setting. Table 96-1 is a partial listing of some of these companies not mentioned above. Table 96-2 lists most of the test equipment and tools utilized in a full-service clinical engineering department.

Components

All engineers and technicians, whether they are involved in design, troubleshooting, or repair, must know the properties and characteristics of the components used in the electrical and electronic devices with which they are working. In the case of maintenance and repair, a thorough knowledge of components, their characteristics, and their failure modes are important to help in locating and repairing faults also when exact replacement parts are not available and a substitute must be used in its place. Some of the more important characteristics of key components that are relevant to troubleshooting are described in this section.

Resistors

General Characteristics

Power Rating: Common resistors are available in power ratings ranging from 1/8 watt to 10 watts.

- The power rating should never be exceeded.
- The power rating decreases with a rise in ambient temperature. The published power rating usually only applies with an ambient temperature of 20° C and decreases to zero at some temperature, depending on the materials used to fabricate the resistor.
- The power rating of a resistor used in a circuit should be at least twice that of any required power dissipation. This usually will provide a sufficient safety margin to prevent premature failure.

Tolerance: Resistors come in standard tolerances of 3%, 5%, 10%, and 20% (seldom used. anymore). A resistor will only remain within its accuracy tolerance if it has not been placed under stress through exceeding its power, temperature, or voltage limits.

Temperature Coefficient and Temperature Range: The value of a resistor will change with temperature. The amount of change is dependent upon the materials from which the resistor was fabricated. Some resistors will return to (or near) their original value when the temperature is reduced, and some will exhibit a hysteresis effect.

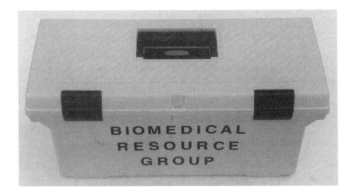

Figure 96-12 Yellow safety box (Dyro, 1995).

Table 96-1 Test equipment manufacturers (partial listing)

Test Equipment Manufacturers		
Alnor	Bio-Tek Instruments	Manley Lung Simulator	Riken
Bacharach	Clinical Dynamics	Mini-OX	TIF
BAPCO	ComfortCheck	Netech Corporation	Timeter
BC Biomedical	Dale	Nuclear Associates	TSI
BC Group International., Inc	Mallinckrodt	Ohmic	UMA

Table 96-2 Clinical engineering department test equipment and tools

Anesthesia test equipment	Anesthetic gas analyzers	Cable and LAN testers
Cases, carts, and dollies	Conductivity meters	Defibrillator analyzers
Dewpoint meters	Diathermy analyzers	Electrical safety analyzers
ESU analyzers	Endoscopy test equipment	Environmental test equipment
ESD/static straps, and mats	Fetal monitor simulator	Flowmeters
Halogen leak detectors	Humidity meters	Infusion pump testers
Laser power meters	Leads and probes	Light meters
LIM/GFI testers	MAS meters	Medical gas analyzers
Miscellaneous testers	Miscellaneous hand tools	Multimeters
Nitrous oxide monitors	NIBP simulators	Non-magnetic tools
Oscilloscopes	Oxygen analyzers	Pacemaker analyzers
Patient simulators	Phototherapy radiometers	Power line and EMF monitors
Power supplies	Pressure meters and gauges	Pulse oximetry simulators
Radiation dosimeters	Sound level meters	Tachometers
Temperature meters	Ultrasound wattmeters	Ventilator testers
Weights and balances	Workbenches and furniture	Imaging phantoms

Voltage Rating: Resistors have a voltage rating which must not be exceeded or permanent damage to the resistor will result. The voltage rating is a maximum even if the power rating of the resistor is not exceeded.

Linearity or Taper (Variable Resistors): How the resistance varies with wiper position is an important parameter. Most potentiometers used in audio circuits as level or volume controls have a log taper because the ear responds logarithmically to sound levels. Power ratings apply when the power is dissipated throughout the entire resistive element. Variable resistors connected as rheostats are easy to burn out if the rated power is dissipated in a small portion of the element.

Fixed Resistors

Carbon Composition: Composition resistors are made of carbon particles in a binder material that is pressed into shape. The most common failure modes are the following:

- High in value. Caused by migration of carbon or binder under heat, voltage, or moisture. Absorption of moisture causes swelling, thus forcing the carbon particles to separate.
- Open Circuit. Caused primarily by excessive temperature. Carbon composition resistors decrease in value with temperature. Some do not return to their original value when cooled and their values slowly decrease with age until they short and then burn open.

Carbon Film Resistors: Carbon film resistors are made by depositing a carbon film on a ceramic tube. The film is usually in the form of a spiral on the outside of the ceramic base. It is covered with a paint to seal it from moisture and the outside world. The most common failure modes are:

- Open Circuit. Caused by disintegration of the film due to high temperatures. Film also can be scratched or chipped.
- High Noise. Caused by bad contact of the end connectors. It is usually the result of mechanical stress caused by poor assembly in a circuit.

Metal Film Resistors: Manufactured the same way as carbon film but using a metal film as the resistive element. Metal film resistors are more temperature-stable and age more slowly than carbon composition or carbon film resistors. The failure modes for metal film resistors are the same as for carbon film.

Wire-Wound Resistors: Wire wound resistors are made by winding a wire on a ceramic tube and covering it with a ceramic paint. Wire-wound resistors are used where high power dissipation is required. Their principle failure mode is open circuit caused by a fracture of the wire, especially when fine wire is used, and by progressive crystallization of the wire due to impurities. The failure also can be caused by corrosion due to absorbed moisture. Wire-wound resistors also open due to a failure in the welded end connection.

Variable Resistors

Partial failures such as a rise in wiper contact resistance causing noise are common. This is usually caused by dust, grease, or other material trapped on the element. Excessive noise also can be caused by wear of the resistive element. Open-circuit failures are due to corrosion, moisture, high temperatures, or wear from the wiper sliding on the resistive element.

Capacitors

Capacitors come in many shapes and sizes. They are classified into two major categories: Polar and nonpolar.

Polar Capacitors

Polar capacitors are sensitive to the polarity of DC voltages applied to them. They are further classified by the metal used for the anode. Polar or electrolytic capacitors have a much higher capacitance-to-volume ratio than do nonpolar capacitors. Polar capacitors have wide tolerances. Their capacitance varies dramatically with temperature, voltage, current, and frequency. Most electrolytics become inductive at frequencies above a few hundred kilohertz.

Aluminum is the most common material used to make polar or electrolytic capacitors, which are relatively inexpensive and are used mostly as power supply filters. Reversing the voltage across an electrolytic capacitor will usually cause a short circuit and will destroy the capacitor, due to overheating the electrolyte. The capacitor can explode when subjected to reverse currents or to AC voltages.

It is a characteristic of aluminum electrolytic capacitors that the DC voltage rating will deteriorate with time. In a circuit, the voltage rating will gradually fall to the value required by the circuit. If a large voltage spike occurs on an old aluminum electrolytic capacitor, it can cause a short and failure of the capacitor even though the amplitude of the spike is less than the voltage rating of the capacitor.

When stored on a shelf, the voltage rating of an aluminum electrolytic capacitor will gradually decrease towards zero volts. The voltage rating can be restored through a process called "forming." A voltage is applied to the capacitor through a resistor (to limit the current) until the current drops to a reasonable value. The capacitor then will have a voltage rating equal to the applied voltage.

Electrolytic capacitors have a liquid electrolyte that forms the cathode or negative side of the capacitor. This electrolyte tends to dry out with age, ruining the capacitor. Overheating the capacitor can cause the electrolyte to leak out through the pressure seal and can cause the electrolyte to dry out.

The best way to check an electrolytic capacitor is by substitution. If the capacitor has failed open circuit or has decreased in capacitance, it can be checked very quickly by putting another capacitor in parallel. Many so-called capacitance checkers cannot effectively check electrolytic capacitors. Electrolytic capacitors often fail or require replacement because of an increase in their equivalent series resistance. Most capacitor checkers do not measure the equivalent series resistance of the capacitors they check..

Tantalum electrolytic capacitors use tantalum as the anode material. Tantalum capacitors have a higher capacitance-to-volume ratio, lower leakage, and a higher reliability than aluminum electrolytics. Tantalum electrolytics are also more expensive than aluminum. They are not made with voltage ratings above approximately 50 volts.

Nonpolar Capacitors

Nonpolar capacitors are not sensitive to the polarity of applied voltage. They are classified by the material used as a dielectric.

- Paper foil capacitors are very common in older instruments but are seldom used anymore. They have been replaced by plastic film capacitors. Paper capacitors tend to fail short circuit or intermittent open circuit.
- Ceramic capacitors have fairly good high-frequency characteristics. They do not have good capacitance stability and are usually used in circuits where the actual value of capacitance is not critical, such as for bypass. Ceramic capacitors can fail open circuit, short circuit, or intermittent open circuit.
- Plastic film capacitors are made using many different kinds of plastic film. Mylar and polystyrene are the most common materials used. Plastic capacitors are the most common type used in building electronic circuits. When they fail, it is almost always as an open circuit. This makes it easy to check by merely tacking another capacitor in parallel to see whether the problem has been corrected.
- Mica capacitors are very stable in value and have excellent high-frequency characteristics. They are not made with capacitance values above 0.01 microfarads. They tend to fail by short circuit or intermittent open circuit.
- Oil-filled capacitors are used primarily for motor start or run, energy storage, and power-factor correction and phase shifting of AC voltages. Oil-filled capacitors are available with capacitance values from 0.001 to 50 microfarads. They tend to be physically large, and they usually fail through leakage of the oil dielectric. The oil used, particularly in units manufactured prior to 1973, might contain polychlorinated biphenyl (PCB), which can cause cancer following prolonged exposure.

Semiconductors

Diodes

Diodes are semiconductor devices that have only one junction. Current will flow in only one direction. Diodes can fail open or shorted. There are many types:

- **Power diodes or rectifiers**: These diodes are characterized by their high current and voltage ratings. They are most commonly used as rectifiers in power supplies or other applications that require relatively high currents.
- **Signal diodes**: Signal diodes are physically smaller than rectifier diodes and have lower voltage and/or current ratings. Signal diodes have a much lower junction capacitance, allowing them to be used for higher frequencies and in switching applications.
- There are many other kinds of diodes, such as light-emitting diodes (LED) and Zener diodes.

Transistors

Transistors are three-terminal devices that contain two or more junctions. They are available in many different packages, with different voltage, current, and power ratings, and with different polarities. They typically fail at the junctions. The junction can become either open circuit or short circuit, although short circuit failures are more common in high-power transistors. Failures can be caused for the following reasons:

- The manufacturing process. Most of these defects will be picked up during the various quality control checks and tests performed during manufacture. Some units will possess defects that will cause premature failure in equipment.
- Most failures are caused by misuse; bad handling practices; excessive power, voltage, or current ratings; or overheating. Many transistors, and most integrated circuits, will be damaged if they are removed from, or inserted into, a circuit with power on.
- Static electricity is an important source of failures in all types of semiconductors, but especially in MOSFET devices. Proper precautions must be followed during shipping, storage, handling, insertion, removal, and soldering.

Integrated Circuits

In general, all of the comments for transistors apply to integrated circuits (ICs). Because of the number of pins, care must be exercised in removing an integrated circuit, to prevent damage to the integrated circuit, the remainder of the circuit, or the printed circuit card. Particular caution is necessary when the IC is not in a socket and must be unsoldered. The IC can be easily damaged by excess heat from a soldering iron while soldering it to, or removing it from, a circuit card. In general, integrated circuits are more susceptible to damage from static electricity than are transistors. MOS integrated circuits are especially vulnerable to damage.

Fuses

Fuses are two terminal devices that are present in nearly every instrument. The purpose of fuses is to protect against fires and/or consequential damage in an instrument due to a circuit or component failure within the instrument. Fuses are unique among electronic components because a fuse can be said to have performed correctly when it fails, blows, or open-circuits. When a fuse has performed its function, it must be replaced.

A number of parameters characterize fuses:

- *Current rating:* The value of current through the fuse, which, if exceeded, will cause the fuse element to melt or to soften, and the fuse to open-circuit.
- *Speed:* How long after the current rating of the fuse is equaled or exceeded will it open-circuit? Fuses are classed into two major speed categories: "fast-blow" and "slow-blow." There are many variations, ranging from "ultra-fast-blow" to "long-delay."
- *Voltage Rating:* Fuses are voltage-rated by the amount of voltage that can be reliably interrupted by the fuse when it opens. When a fuse opens, the voltage that can appear across the fuse can cause an arc to form, which can maintain electrical continuity. The arc must not be maintained because the circuit can open and prevent further current flow.

Circuit breakers are the functional combination of a fuse and a switch. They have the advantage over a fuse of being able to be reset. When the current rating of a circuit breaker is exceeded, a switch is automatically opened which breaks the circuit. The switch can be manually actuated, and the circuit breaker reset to again perform its function. It does need not be replaced after functioning as a fuse does. Circuit breakers are designed with one of two actuating mechanisms; thermal and magnetic.

- *Thermal:* In a thermal circuit breaker, the current through the breaker passes through a heater, which activates the switch portion if a certain temperature is exceeded.
- *Magnetic:* In a magnetic circuit breaker, the magnetic field caused by the current through the breaker will activate the switch part of the breaker if it exceeds a preset value.
- Magnetic circuit breakers operate much more quickly than do thermal circuit breakers. A thermal breaker may require 0.2 to 0.5 seconds to open, while a magnetic circuit breaker will open in 0.05 seconds under the same conditions.

Fuses and circuit breakers will change their ratings and characteristics, depending upon ambient operating temperature, vibration, age, and other environmental factors. Circuit breakers–especially thermal breakers–will change their ratings after repeated actuations. It is important to replace a fuse or breaker with one of the same characteristics. It is particularly bad to replace a fast-blow or regular fuse with a slow-blow type. Serious circuit damage can result because of the longer actuating time.

Microprocessors

Many modern instruments include a microprocessor and associated circuits. It is important that the troubleshooter have some knowledge of how to locate and correct faults in these instruments. It is also important that spare specialized circuits, such as ROMs, be purchased at the same time as the original instrument because such parts might not be available later, or might be manufactured to different specifications. There is an increased emphasis on data communications systems and instrumentation, including short- and long-distance systems. There will be significant improvements in maintainability and built-in diagnostic programs. These will aid the operator and the troubleshooter. Built-in diagnostics are only useful if and when the microprocessor itself is functioning. The following faults can prevent diagnostics from being helpful: No power, or a faulty power supply; faulty clock; faulty ROM; or faulty peripheral integrated circuit or function.

Troubleshooting and Repair Strategies

The optimal strategy for troubleshooting and repairing microprocessor-based instruments depends on several factors, such as:

- How much is known about the instrument, its microprocessor, and associated software
- Available manuals or documentation
- Available test equipment
- Available spare parts
- Available information from the local agent or manufacturer

There are certain rules or methods of troubleshooting and that are always correct:

- Check and correct any problems that may exist in the power supplies.
- Diagnostics that are built into the equipment should be exercised to see whether any problems can be detected.
- Substitution of components or sub assemblies is a good, rapid technique, but it requires spare parts that are known to be good.

If nothing works, check the following:

- Power supplies
- Clock system
- Program memory (ROM)
- Address bits (for toggling)

Signals always should be measured at the IC pins, not on the circuit card or connectors, when verifying that an IC works or when checking power supplies. By following the simple guidelines outlined above, most microprocessor-based equipment problems can be solved. Those that cannot be solved through the above process usually require extensive testing and analysis.

Interference and Noise

This section concerns the problems of interference and noise. No other area of troubleshooting causes so much difficulty for the average troubleshooter or is as difficult to find and correct. One reason for the difficulty is that the source of the noise or interference might be external to the device under test and might not be present in the test and repair area. The source may also be intermittent. Noise and interference problems can be classed as internal or external and are treated in different ways.

Internal

Internal interference and noise occur when the source of the interference or noise is within the instrument. It can be caused by component failure and mechanical factors such as loose connections and improper routing of wires. Many types of component failure can inject large amounts of noise into a circuit:

- Film resistors
- Paper, ceramic, or mica capacitors
- Dust or other foreign material under the wiper of a variable resistor
- Semiconductors (In some instruments, a semiconductor junction is deliberately used as a source of noise)
- Linear integrated circuits sometimes produce what is called "shot noise" or "popcorn noise."
- Connectors age, become corroded, collect dust and dirt, and lose contact pressure.
- Switches suffer from the same problems as connectors.

Mechanical sources of interference and noise include the following:

- Broken or loose wires or connections.
- Loose bolts or screws that hold the instrument together (particularly if there are small differences in potential between the two parts of the instrument or if the two parts rub together)
- Static electricity generated by loose connections and the rubbing contact between two dissimilar materials
- Misrouted signal or power leads. In general, signal leads should be kept separate from power leads, and low-level signal leads should be kept separate from high-level signal leads or sources.
- Dust or sharp points in a high-voltage circuit causing corona discharge
- Missing panels, shields, and grounding connections

Finding the internal source of the interference and noise and correcting the problem is much easier said than done. The following steps should be taken:

- If the source within the instrument is unknown, the best approach is first to go through the instrument very carefully, tightening all loose screws and connections.
- Wiggle cables, connections, and circuit cards and rap on panels while observing the occurrence of the noise. Be careful not to wiggle things too much or to tap them too hard. You could cause more problems than you solve.
- Clean all dirt and dust from the instrument. Pay particular attention to those areas where high voltage is present.
- Carefully inspect, clean, and observe the operation of any switches or relay contacts that are switching high power or voltage.
- Carefully observe the noise or interference with an oscilloscope. Try to define the characteristics of the unwanted signal.
    - Is it related in frequency, repetition rate, period, or in some other way, to a legitimate signal generated within the instrument?
    - Does the instrument have a switching power supply or regulator? Is the noise or interference related to the switching frequency?
    - Is the noise or interference related to any signal with which you are familiar, such as the mains frequency, twice the mains frequency, or the vertical or horizontal rate of a raster scan CRT?
- Try to isolate the source to a function, section, or circuit.

- If the noise is present on the power distribution lines, carefully check the capacitors in the power supply. There should be a ceramic or other capacitor with good high-frequency characteristics, in parallel with the main filter electrolytics because aluminum electrolytic capacitors become inductors at frequencies above a few hundred kilohertz. The impedance rises with further increase in frequency.
- Carefully check the bypass capacitors in any circuit that generates high-frequency or high-amplitude signals.

External

External interference or noise occurs when the source of the interference or noise is external to the instrument under test. The external source could be any of the following:

- Interference or noise conducted through the mains
- Interference or noise radiated through the air
- Interference or noise due to varying electric or magnetic fields in the vicinity of the instrument
- Interference or noise generated by other instruments in the vicinity

Elimination of problems due to external interference can be accomplished using three general techniques:

- Modifying the sensitive devices to make it less sensitive to the interference. This can be done through shielding, filtering, or grounding.
- Modifying the source of the interference to reduce the amount of interference generated or transmitted. This is done through appropriate shielding, filtering, and grounding.
- Moving either the source of the interference or the sensitive device to a location where the problem no longer occurs.

Interference or noise also can be caused by connecting two or more instruments. The most common reason is current flow in a ground loop. Ground loops occur when two or more devices are connected together in such a way that there is more than one connection to ground. The multiple connections form a loop through which current can flow if there is a potential difference between points in the loop. Because the resistance around a ground loop is very small, only small potential differences are needed, to cause large currents to flow. These currents generate magnetic fields that create the interference. Many times, all ground loops cannot be eliminated in actual practice. Actually, the problem is not that a ground loop exists–problems only occur when a current flows in the loop. If the current is reduced in a ground loop, the interference is also reduced. Often it is possible to insert a small-value resistor (15–50 Ohms) in the loop. This usually will reduce the current so that there will be no harmful interference.

Avoiding More Trouble

This section provides information to prevent any additions to the already difficult job of troubleshooting. The areas of concern are primarily mechanical, electrical, and measurement.

Mechanical

The Right Tool for the Job

During disassembly and assembly of an instrument, use the proper tool for the job. Use of an incorrect tool can cause damage that will prevent the removal of a screw or component, or it can cause unsightly damage such as burrs or scratches. For example, there are at least three types of cross-point screws.

- *Phillips:* It is identifiable because the screw head indentation has a flat bottom, and the driver also has a flat tip. This is the most common type of crosspoint screw.
- *Pearson:* It looks just like the Phillips, except that the screw head indentation has a pointed bottom, and the driver has a pointed tip.
- *Posi-Drive:* The tip of the screwdriver has a convex curvature to match the screw head. The screw head is identifiable by the extra lines between the indentations. This type of screw is used on all Hewlett-Packard and Tektronix equipment.

Screwdrivers and screws come in many sizes. Use the proper size driver for the screw. Screws and nuts are often very tight and will not come loose easily. Several techniques can be used to loosen the screw. Clean the head of the screw. Place a proper size screwdriver firmly in the screw head and give a quick snap of the wrist. This often will loosen an otherwise impossible screw. If the screw still will not loosen, tap the end of the screwdriver handle with a hammer, and then try again with a snap of the wrist to loosen the screw. Apply torque to the screwdriver while tapping it with the hammer. Apply a penetrating oil or solvent and allow it to soak, and then try again to loosen the screw. If the screw has been set with a varnish, shellac, or other substance to prevent loosening, heat it with a soldering iron, and then try to loosen it while it is still warm.

Needle-nose pliers are not made for use as a wrench or spanner to tighten or loosen bolts and nuts. In general, pliers are not manufactured to use as spanners. They mark up the nuts and bolts so that a proper spanner will no longer fit.

Wire-cutting diagonal pliers should not be used to cut steel wire or small screws and bolts. The jaws are not made to cut materials harder than copper alloys. Special cutting pliers are made to cut steel.

There are many different type of hammers. The reason is that each type is designed to be used for a particular purpose. Certain hammers are designed to be used to strike steel, while others are made to pound nails. Pounding steel with a hammer that is designed for pounding nails is very dangerous, as the head could shatter or chip, tossing bits of the steel hammer head about. Workers have been blinded by use of the wrong hammer.

Physical Relationships and Alignments

Use caution when taking something apart, especially if you are not familiar with the device. Clear your work bench and surroundings of all non-essential material, tools, and equipment. Observe carefully, and think before you begin disassembly. Determine how it comes apart, and the tools that are needed. If there are springs or other small parts that could be easily lost or misplaced, they should be placed in a small container. Scratch or scribe a mark across the crack between parts before taking them apart. This will allow reassembly keeping the original alignment. It is an especially valuable technique to use when taking apart motors or other devices with cylindrical symmetry. When disassembling a device, it is helpful to lay the parts out on a clear area of your work bench in the same order in which they are removed. Most things will go back together in the reverse order of disassembly and this approach will allow you to remember the order.

Electrical

There are three major causes of electrical problems that can add to your difficulties: Static, transients, and short circuits.

Static

Static electricity causes far more damage than most people realize. The average person has a capacitance of between 250 and 500 picofarads. It is easy to collect sufficient static charge to generate voltages from 5–20 kilovolts equivalent to a stored energy ranging from 3–100 millijoules. That much energy dissipated in a short-duration discharge is sufficient to damage most low- or medium-power semiconductor junctions. Most engineers are aware that static electricity can damage MOS and CMOS devices, but it is important to realize that bipolar devices such as TTL integrated circuits, operational amplifiers, and bipolar transistors also can be damaged.

While static discharge will usually destroy MOS devices, the damage done to bipolar devices is generally not so obvious. The discharge usually results in a weakened junction, which will fail later under normal voltage or thermal stresses. Transistors often exhibit low, unstable, or very non-linear betas after being subjected to static discharge. Static electricity is especially bothersome when the relative humidity is below 40%.

The best way to prevent static damage is to be aware and to prevent build-up of voltage differences due to static electricity. The following guidelines should be followed to minimize the risk of static electricity:

- Store and transport static-sensitive devices in conductive containers. Do not store integrated circuits and transistors in plastic drawers, or stick them into white styrofoam plastic.
- If possible, have an area of your workspace set up with conductive bench surfaces and floormats.
- Provide a wrist strap or other means to refer yourself to a ground potential.
- The conductive bench surface and floormat and the wrist strap should not be connected directly to a ground point. Connection should be made through a 100-kilo-ohm resistor to ground. This will allow enough to bleed off static charge, but high enough to prevent a shock hazard if you come in contact with high voltages in the device under test.
- When handling dual-in-line integrated circuits, grasp them with your fingers on the ends of the package. Do not touch the pins unless it is necessary to do so.
- Use a grounded-tip soldering iron.
- While a device that is mounted on a circuit card is generally less sensitive to static electricity, it is not always so, and semiconductors can be damaged even though mounted in a circuit card.

Transients

Spikes, glitches, or other voltage and current transients can cause component failure. They usually occur as the result of switching or sudden changes in voltages and currents.

- Do not plug or unplug components with power applied to the circuit.
- Do not remove or insert printed circuit cards with power turned on.
- Beware of inductive transients. They can be caused by such things as motors, relays, and transformers. that are either not provided a means to dissipate magnetic-field energy when turned on or off, or the protective devices provided have failed.

Short Circuits

Inadvertent short circuits can be caused by the following:

- Bending together or shorting two points while removing panels, gaining access to a circuit, or replacing a component, or while making measurements
- Using an incorrect tool or using a tool inappropriately
- Slipping with a probe while making a measurement or while attaching the probe
- Spilling liquid on or into an instrument or device

Accidental short circuits can be prevented by observing the following rules:

- Be careful.
- Turn the power off while attaching test probes.
- Use the proper probe tip. Do not attempt to measure at the pins of an integrated circuit with a blunt probe. It is too easy to slip.
- Do not have coffee, tea, coke, or other unnecessary liquids nearby when working on electrical or electronic equipment.

Measurements

Problems caused by measurements fall into three categories: Improper tools, improper measurements, and improper measurement technique.

Improper Tools

When taking measurements, it is important to use the proper probe or probe tip. For example, without the proper probe tip, it is very difficult to take measurements from the back of a wire-wrapped circuit card. On many circuit cards, spacing is very close, and it is almost impossible to prevent an inappropriate test probe from contacting more than one point at a time.

- Use the proper probe.
- Use aids such as a clip to go on dual-in-line integrated circuits to make the pins more accessible.
- In some cases, it may be necessary to solder a lead to a point in the circuit to gain access for a measurement, particularly if you do not have an extender card to use.
- Extender cards and cables can be very helpful. Remember that extender cards usually do not have keys to ensure that the component side of a circuit card is facing the correct direction.

Improper Measurement Technique

In the case of an improper measurement technique, the idea of the measurement is correct, but the measurement is made in such a way that the fault is not detected or apparent. For example, if an electronic regulated power supply output is measured under no-load conditions, the output can appear correct even if there is a base-emitter short in the series-pass transistor. To detect the problem, the power supply must be loaded. Extender cards will change the electrical characteristics of a high-speed, digital, low-level analog, or any other system or circuit where lead length is important. Under these conditions, extender cards might not be very useful.

Safety

This section is about safety: Safety for the troubleshooter, safety for the equipment operator, and safety for the patient in the case of medical equipment. There are many ways in which one could be injured while working on or around instruments and equipment. The sources of hazards can be classified as electrical, mechanical, radiation, burns, or biological.

Electrical Hazards

Physiological Effects

Electrical current does the damage. Currents less than 500 microamperes produce no sensation or feeling. Sensory nerves are stimulated at current levels of 500 microamperes to 2 milliamperes. Above 2 milliamperes, the motor nerves are stimulated, and muscles contract. Above 1 ampere, burns are likely as a result of the power losses and heat generation in tissue.

Electrical sensitivity is a function of frequency. The frequency to which humans and other biological systems are most sensitive lies in the range of 50 to 60 Hertz. At frequencies above and below this range, more current is required before any sensation is felt, and more current is required to stop the heart from beating.

Most injuries from electric shock are trauma injuries or burns. The trauma results from the "startle reaction" or sudden muscle contractions. For an electrical shock to occur, three conditions must be met: (1) there must be a difference of potential; (2) there must be a path for current flow; and (3) sufficient current must flow.

Safety Rules to Prevent Electrical Shock

- If possible, turn off all power while working inside an instrument, when removing covers and panels, when attaching a probe to a high-voltage test point, when working in tight quarters, and whenever you leave a device with exposed high voltage.
- If power must be on, use only one hand. Keep the other hand in your pocket, where it cannot contact another part of the system to provide a path for current through you. There must be a path for current and a voltage difference before an electrical shock can occur.
- Interlocks are designed for a purpose. Carefully consider the safety consequences before defeating interlocks.
- Use insulated tools whenever and wherever possible.

The purpose of the third or "ground" wire in an instrument's power cord is to protect the user or handler of the equipment from electric shock. Using ungrounded devices increases the risk of electric shock. It is important to realize that the equipment will operate perfectly well without the ground wire intact. The only way to tell whether a device is grounded is to measure continuity from the ground pin of the power plug to the chassis.

Mechanical Hazards

Power Tools

Electric power tools such as drills, grinders, and saws can be dangerous. Clamp your work when using a drill press. Protect your eyes from flying metal chips and glass. Beware of sharp edges that can cut you. Do not defeat the guards and shields that are provided for

protection. Use protective clothing, face shields, or other protective equipment. Most importantly, use tools and equipment correctly.

Compressed Gas Cylinders

Compressed gas cylinders are potentially very dangerous devices. When filled at high pressure, a relatively light blow can turn the cylinder into a rocket with enough power to go through a concrete wall. Gas cylinders should always be secured so that they cannot fall over. They also should be protected from sharp blows.

Lifting

Lift carefully and properly. Back injuries can be permanent. Do not attempt to lift too much. Find help. Be especially careful when lifting awkward or odd-shaped items or when lifting any object over your head.

Radiation Hazards

Non-Ionizing Radiation

Radio frequencies, microwaves, and visible and ultraviolet light are forms of non-ionizing radiation. The principle effect is a burn of some type. Radio frequencies and microwaves can cause serious burns that show up as a small hole in the skin but cause a larger damaged volume beneath the skin. Such burns heal very slowly and have a high probability of becoming infected.

Damage to the eyes often shows up as cataracts, which can cause blindness. Cataracts might not form until years after exposure. Many scientific instruments use or contain a light source. The source might be infrared, visible, or ultraviolet; gas discharge, arc, incandescent, or laser. Many of these light sources are very intense or bright and can cause serious and permanent eye damage. Most people are aware of the dangers of ultraviolet light, but the infrared wave lengths are equally dangerous to the eyes. Always protect your and your fellow workers' eyes when working on instruments that include lasers or other light sources.

Ionizing Radiation

Ionizing radiation cannot be seen, heard, smelled, or felt. One must depend upon instruments to detect its presence. Always wear film badges or other means of detection and dose measurement while working around X-ray machines. Do not go into a room while an X-ray is operating without suitable protective clothing. When working with radioactivity, beware of contamination or spills.

Some instruments use potentially dangerous amounts of radioactivity as a part of them. Examples include a cobalt therapy unit and an ionization detector, used in many gas chromatographic instruments. Some high-current, high-voltage electronic switches contain a radioactive source to stabilize their turn-on characteristics. If these switches are broken, hazardous amounts of radioactivity can be released.

Sound Radiation

High-intensity sound can cause permanent hearing loss and tinnitus (ringing in the ears). Many ultrasonic devices (particularly large, ultrasonic cleaners and ultrasonic disintegrators used to disrupt cells in biological research) have sufficient power output to cause headaches or other problems. Use ear plugs. Putting cotton in your ears is better than nothing.

Other Hazards

Burn Hazards

These are mostly self-explanatory. Hot soldering irons, vacuum tubes, heaters, lamps, motors, power resistors, and heat sinks all can cause painful burns. The rule is to think and beware of all things hot.

Biological Hazards

These hazards apply only for some medical equipment, such as dialysis apparatus, surgical devices, and patient monitors, which have been used on patients who have infectious diseases. If you have a medical device that has blood on it, assume that it is potentially infectious. Clean it before doing any troubleshooting or repair. Use gloves and other protective clothing. One common instrument, which often contains a biological hazard, is a centrifuge. The inside of the bowl is often coated with material from centrifuge tubes that have broken while spinning.

It is the responsibility of the repair person to ensure that a device that is repaired and returned for use is safe to operate. This means that all interlocks are operations, that all protective shields or guards are in place and functioning, and that there are no exposed electrical wires to shock the operator.

Patients are a special class of equipment user. They often are not aware of hazards and cannot move themselves. Patients do not know that they are not supposed to be shocked by a medical device. They often are much more susceptible to injury than well persons are. Patients with heart disease are much more likely to suffer damage from an electric shock than are well persons.

Conclusion

There are no deep, dark secrets to effective troubleshooting. Troubleshooting is the process of applying what you already know, in a logical, systematic manner, to acquire the necessary additional information to find and correct faults in scientific instruments. The most important troubleshooting device is the troubleshooter's brain and senses of sight, hearing, smell, and touch. The most important thing to remember in troubleshooting is to be logical and systematic. Stop, think, and plan your approach. The most common cause of electronic failure is heat and high temperatures. The most common cause of overheating is dirt, dust, or other material covering components, circuits, or instruments affecting heat transfer and airflow characteristics.

Keeping instruments clean is the best way to reduce the frequency of failure. Many faults are non-electronic. The failures might be due to mechanical failure; broken wires, connections, or parts; loose or missing screws, bolts, connectors, or fasteners; or worn or misaligned parts. Remember to always look and think and do the easy things first.

Work from the outside of the instrument, in. Learn all you can from the outside. A careful visual inspection, while keeping the description of the fault in mind, is a useful guide. Look, think, and operate. Look for obvious, simple, potential causes of the problem first. Keep the functional block diagram in mind to help guide you in your search for the problem. Your senses and brain are the best troubleshooting combination.

Do not refer to schematics or take measurements until you have used your senses as far as possible. When measurements or other tests are required, do the simplest or fastest ones first. Do not take any measurements if you do not know what the result of the measurement should be. When looking for faults or a faulty component, look for those that are most likely. Keep in mind the failure rates and failure modes of the components, modules, and instruments.

Be sure that all connections are clean and making good electrical contact. In instruments that have external electrodes or transducers, the most likely failures and the parts to check first (after careful visual inspection) are the electrodes, transducers, and cables.

Do not forget to complete the paper work. It is important. After the repair is complete, review the entire troubleshooting and repair process, going over each step that was taken. Ask yourself whether the process you took was the most efficient or the best under the circumstances. Would you attack the problem the same way if you had another instrument with similar symptoms? What did you learn from the entire process?

References

Aston R. *Principles of Biomedical Instrumentation & Measurement.* Merrill, PA, Macmillan Publishing, 1990.

Dyro JF. Methods for Analyzing Home Care Medical Device Accidents. *J Clin Eng* 23(5):359–368, 1998.

Dyro JF. *The Safety Box.* Proceedings of the AAMI 30th Annual Meeting. Arlington, VA, Association for the Advancement of Medical Instrumentation, 1995.

Kermit E. *Medical Adhesive Tape Should Be a Controlled Substance.* Boston, 28th Annual AAMI Meeting, 1993.

Shepherd M, Dyro JF. *The Whimsical Use of White Tape* (educational slide presentation), University of California, San Francisco, 1982.

Section IX   Information

Elliot B. Sloane
Assistant Professor, Department of Decision and Information Technology, Villanova University
Villanova, PA

This section is devoted to the profound and important transformational role of computers and information systems (IS) in health care in general, and on the clinical engineering profession in particular. Clinical engineers, who apply technology to improve patient care, need to redefine their relationship to computers and IS. Prior conceptions of IS as business and scientific technologies must yield to a new appreciation for the unique and valuable contributions that they continue to make to patient care. Clinical engineers who do not understand (or who avoid) this conclusion may find their circles of influence and activities rapidly diminishing. The information in this section is intended to help avoid that fate by briefly reviewing the past and then illustrating why topics like Y2K and the Health Insurance Portability and Accountability Act (HIPAA) are important in the successful evolution of computers from technical, administrative, and financial tools to invaluable clinical technologies.

One of the earliest and most comprehensive reviews of IS was coauthored in 1971 by a clinical engineering pioneer, Dr. Cesar Caceres. In a paper in the *Annual Review of Information Science and Technology* (ARIST), Dr. Caceres identified many of the emerging roles that early IS played, including clinical information management, medical diagnostic support, and general administrative and financial services. Although he enthusiastically predicted a valuable role for IS in health care, his mainframe-based viewpoint could scarcely have anticipated the rate of change that the modern microprocessor would facilitate. Despite the compelling importance of IS in medicine, that ARIST review has not been updated in more than three decades.

Dr. Caceres pointed the way towards the future role of IS in health care, to "implement things useful to the physician in practice and spend medical information dollars in the implementation of service systems for those practicing physicians currently taking care of the increasing patient load." His concern was that the then 6% burden of health care on the GDP (now 14%) increased demands on physician diagnoses. He foresaw IS as providing an important solution to the problem. Though great strides have been taken in this direction, much remains to be done, since few of the available IS tools are easily integrated and many physicians still rely on paper records and personal diagnostic skills. In fact, most of the IS progress has occurred in three areas: Billing and administration, medical imaging, and clinical laboratory work. To date, however, even these applications have suffered from proprietary designs that limit electronic information sharing and integration.

This section examines the key technological elements needed to understand how IS produces dramatic diagnostic and therapeutic advances. The first chapter, Information Systems Management, is a tutorial on information systems. Information systems (IS) in health care have become a standard and fundamental infrastructure, as much as reliable electricity, clean water, and adequate heating, ventilation, and air conditioning. The IS role will continue to expand, spanning virtually all business and clinical care processes. The chapter is written for the clinical engineering manager, who must provide technology leadership in the health care institutions. The topics covered focus on crucial management issues rather than deeply technical topics in order to create a framework for the clinical engineer (CE) and biomedical engineering technician (BMET) to contribute to this growing field.

The next chapter, Physiological Monitoring and Clinical Information Systems (CIS), addresses both the technologies of bedside monitoring and CIS, and their interdependencies. The state of technology in both areas, reasons for the integration of the two, and the groundwork necessary to manage them are discussed. A brief overview of an electronic patient record (EPR) and the impact of physiological monitoring and clinical information systems have on EPR are presented.

The following chapter, Advanced Diagnostics and Artificial Intelligence, shows that the application of advanced diagnostics and artificial intelligence technology in the field service industry has yielded significant improvements, due to intelligent dispatch and the assignment of both parts and service engineers. The author examines both the current state of technology and experiences in the application of problem diagnostics and resolution technology in the health equipment field service industry and explores the successes and failures of artificial intelligence and the advanced remote diagnostics and decision-support methodology.

The next chapter, Real-Time Executive Dashboards and Virtual Instrumentation, shows how multidimensional executive dashboards can empower health care organizations to make better and faster data-driven decisions by leveraging the power of virtual instrumentation and open architecture standards. Through case studies of bed management/census control, operational management, and data mining and business intelligence applications, the author shows how user-defined dashboards can connect to hospital information systems (e.g., ADT systems and patient monitoring networks) and use Statistical Process Control (SPC) to visualize information.

The chapter on Telemedicine: Clinical and Operational Issues reviews the evolution of this technology and the major issues to consider before initiating a telemedicine program. The authors draw upon their experiences at a leading medical center. Telemedicine is a relatively new modality in the health care delivery system, and many providers are still debating the appropriate entry course and the following regulatory and sustainability issues. This chapter provides guidance for those who are contemplating implementation of this technology.

The next chapter gives a detailed description of Picture Archiving and Communication Systems (PACS), systems that electronically process, store, distribute, and retrieve digital medical images in a portion of, or throughout, a health care enterprise. PACS epitomizes the integration of medical, computer, and communication technologies with its requirements for large amounts of storage, sophisticated databases, large-bandwidth, high-speed networks, computational power at the desktop, ease of use, for a variety of people, and specialized medical application software. The author notes how PACS can improve the imaging department workflow compared to film-based technologies and emphasizes that the practical implementation of PACS requires a high degree of integration with the Radiology Information System (RIS) and/or the Hospital Information System (HIS) for scheduling, workload management, and radiology reports. As PACS reduces film use, film processing, and film storage costs, as well as the number of lost films and retakes, it continues to develop, facilitated by the growth of high-bandwidth intranets.

Another chapter, The Changing Face of Wireless Medical Telemetry, provides a brief summary of electromagnetic interference (EMI) problems for the traditional wireless medical telemetry widely used in hospitals and provides information that clinical engineers can use to reduce the risks of EMI. The chapter describes the worrisome interaction between EMI and medical devices and what has been done to address these concerns. It highlights the creation of the new wireless medical telemetry standard (WMTS) and relates the need to make immediate assessments of medical telemetry equipment and plans to address the coming changes to the airways where wireless medical telemetry has been used.

The author of the HIPAA chapter penetrates the labyrinth of legislation and associated rules to show that clinical engineers must implement solutions to problems wrought by broad changes in health care technology. HIPAA's security rule was conceived to protect the integrity, availability, and confidentiality of health information or data associated with a patient. This includes the obvious; i.e., data in medical and billing records found in provider information systems, but it also includes diagnostic and therapeutic data found in biomedical devices and clinical information systems, a fact with broad implications for clinical engineering. Biomedical devices and systems used by health care providers represent a substantial and growing area of risk with respect to HIPAA security rules. The total quantity of biomedical devices and systems in a typical hospital may run 300% to 400% more than the number of IS-related devices and systems in that same hospital. The author reveals two major trends contributing to the significance of the biomedical technology-related risk. First, biomedical devices and systems are being designed and operated as special purpose computers with more automated features resulting in increasing amounts of health data being collected, analyzed, and stored in these devices. Second, there has been a rapidly growing integration and interconnection of disparate biomedical and information technology devices and systems where health information is exchanged.

The chapter on "Information Management Year 2000" (Y2K) describes the genesis, recognition, and remediation of the problem of computers using only two digits to designate a particular year, (e.g., using "17" to mean "1917)". Memory was precious in the early days of computing. During the 1980s and 1990s, medical devices incorporated computers and associated software. During the mid-1990s reports surfaced of an inherent and possibly fatal flaw in many of these devices. Computer experts predicted that many systems would produce unpredictable results and some might catastrophically fail as the calendar passed from the year 1999 to the year 2000. Some experts postulated that these failures would occur because systems were incapable of recognizing that the year 2000 followed 1999. The author describes the benefits of the remediation efforts, completed at enormous expense. The author also provides a review of our reliance on technology and the vulnerability that represents, the recognition that the vulnerability of reliance on technology required contingency planning, the realization that the replacement of obsolete or unsupportable technologies was long overdue, and that recognition that more effective risk assessment and management methods needed to be developed.

The section concludes with the chapter The Integration and Convergence of Medical and Information Technologies, which describes the convergence of clinical and information technologies in the face of two emerging trends the widespread use of commercial off-the-shelf (COTS) hardware and software, and communication technologies that interconnect the office, the enterprise, the community, and the world. COTS technology significantly reduces manufacturers' costs and improves manufacturers' time-to-market for new products. These technologies allow many of the major medical systems that are sold today to operate as computer systems as well as medical devices. The modern hospital can interconnect these medical devices using standard data ports in patient care areas and standard wiring, hubs, switches, and routers in data closets. These systems integrate information and clinical technology, and support of these systems requires an integrated approach by those who are trained and familiar with both clinical and computer technology.

97

Information Systems Management

Elliot B. Sloane
Assistant Professor, Department of Decision and Information Technology, Villanova University
Villanova, PA

In many ways, information systems (IS) in health care have become a standard and fundamental infrastructure essential, just like reliable electricity, clean water, and adequate HVAC. It is likely that the IS role will continue to expand, spanning virtually all business and clinical care processes. This chapter provides a framework for the clinical engineer (CE) and biomedical engineering technician (BMET) to use in managing their careers in the years ahead. Some CEs and BMETs attempted to distance themselves from the personal and business computer systems of the 1990s, citing their irrelevance to patient care. Fortunately, that is no longer a useful debate; most business and clinical systems now share most of the same common hardware and software components, regardless of the application. Today's IS situation has been shaped by three key factors: (1) the rapid price reductions for IS equipment and rapid increases in computing power; (2) hardware and software standardization and reliability; and (3) the stabilization of data communication standards. Today, many complex systems are developed using combinations or variants of commercial off-the-shelf (COTS) components, thus making it far easier to apply common system design, implementation, and support skills throughout the health care system.

The reader who seeks technical training will find many appropriate sources on the Internet and at bookstores, libraries, and community and engineering colleges. Therefore, this chapter is written instead for the clinical engineering manager, who must provide technology leadership in the health care institutions. The topics that follow focus on crucial management issues rather than deeply technical topics.

Convergence of Health Care's Business IS and Clinical Technologies

Because the cost of standardized microprocessors, sensors, memory, and wired and wireless data communication devices has dropped to just a few dollars per component, these elements are being used ubiquitously and pervasively throughout institutions. The time might not be far off when virtually every electrical and electromechanical device has the ability to capture, store, analyze, and communicate data to help enhance its performance and reliability and to allow it to contribute to overall system performance and reliability. The inherent value of this technology is its ability to configure and monitor the performance and safety of the device that it controls and the ability to communicate the device performance, problems, and relevant patient data throughout the institution's IS "nervous system." This ultimately can allow reliability by communicating problems earlier and more comprehensively and can afford more safety and redundancy by sharing computing resources when problems occur.

Operationally, health care institutions are finding that increasingly more departments, whether administrative, operational, or clinical, are able to interconnect or directly share full- or part-time IS resources like computers, memory/storage, communication pathways, and input/output devices. This has brought IS and biomedical technology support teams closer technically, but it also has sometimes created turf conflicts. Successful management of these converging and pervasive technologies requires that all parties collaborate to ensure that the clinical users and patients receive seamless, reliable support in the most efficient manner possible. To do so requires a long-term commitment to learn, master, and, ultimately, manage these technologies professionally. This chapter is intended to help make that possible by enabling clinical engineers and biomedical technicians to understand the IS field better.

The basic building blocks of IS infrastructure today might be captured in the areas of IS design and management methodologies, computing hardware and software, and data storage and communication. The following information about these areas will help clinical engineers and technicians to participate more effectively.

System Analysis, Design, and Management

The first topic that must be discussed is IS analysis, design, and management because the intended purpose and environment of use must, by necessity, define the hardware and software that will be used. Although most IS texts delve deeply into systems life cycle methodologies for the new engineer or business student, this should not be a new area for health care technologists. Health technology assessment and life cycle cost-analysis techniques now include the use of flexible, computerized decision-support-system tools. The reader must assume that virtually all IS projects require several careful analysis phases to understand the feasibility and consequences of an IS decision. It is likely that the health care system will to need to live with each installed IS for many years to come, and the overall systemic expenses, reliability, and efficiency of the institution is likely to be significantly impacted by each IS as well. The key areas of concern for health care IS system designs are not dissimilar to those for any medical technology. The selected system (1) must be evaluated using an accurate set of specifications that describe the necessary features and their relative priority; (2) must be properly configured to integrate efficiently and reliably with current and future IS systems in the institution; (3) must be installed and tested for proper function according to a written specification, prior to acceptance; and (4) must be included in the institution's ongoing technology management program to ensure reliable testing, maintenance, and repairs.

Needs Analysis

Competent IS specification and evaluation is by no means a simple task today, nor is it likely to become easy in the near future. In the absence of explicit FDA standards for medical devices, medical software, or human factors, most vendors have pursued their own, often unique, designs. In every health care market niche, therefore, the available IS choices vary widely in features, total cost of ownership, and overall compatibility. *Caveat emptor* ("Let the buyer beware"), therefore, must be the institution's policy. This must start with a careful analysis of the actual need to be fulfilled by the system and, importantly, the other systems with which it should successfully interface. Even if an institution or user believes that interfacing is not essential, one should be wary of giving up this option. Too often, the need to expand the system and/or to integrate it into a larger mission will emerge, and the cost for custom interfacing at that point can be enormous. One must be certain that the system can be efficiently and effectively learned and used by the intended staff, as well. Despite many decades of research in the human factors area, too many IS developers fail to take the time to understand the people, skills, and environment in health care. Poorly designed IS designs can cause major and unsolvable training problems that lead to unsustainable error rates and/or abandonment because using them "feels worse" than doing without.

Total Cost of Ownership

Careful system evaluation must take total cost of ownership (TCO) into account (see Chapter 30). Many other industries have come to respect the hidden costs as much as, or more than, the obvious purchase costs. In doing a TCO analysis for an IS project, one would be well advised to consider the consequences of successful adoption and growth on the need for additional hardware, software, and support in the three- to five-year period ahead. It is interesting to see how many IS analyses fail to plan for success, even though that is the hoped-for result of the acquisition. In many cases, success will lead to more users, greater dependence on the system, longer hours of use, ever-increasing storage and retrieval demands, greater printout support, remote access and security, and, ultimately, integration with other IS resources in the hospital. Additional hardware and software capacity can be assessed and controlled more easily if the IS uses a significant amount of COTS hardware and software. The larger the non-COTS portion, the more uncertain the costs and risks will be. One must not exclude the inherent application software from this assessment, either; if it is not based on industry standard operating systems, languages, and data management software packages, the sole-source nature of the system can become an increasingly expensive and inconvenient limitation. If the happy outcome of the IS selection and implementation leads to successful use and adoption, one should plan carefully to ensure that the institution's budget will enable the IS to meet the dependability demands of the use. In health care, many successful systems rapidly become "business critical systems" on which many people and health care delivery processes are dependent. This usually requires improved electrical supply, temperature and humidity control, capacity, stability, backup and recovery resources, and increased data communication capability. A TCO analysis that does not address these issues will grossly underestimate the budget required to support the IS project. There is rarely a "perfect" IS product to select; each introduces different, and sometimes complex, trade-offs. Again, *Caveat emptor* should be firmly in mind, as the vendor will rarely bring these issues to light in his own analysis. If the IS project is expensive and likely to become so successful that it will become part of the institution's infrastructure for quite a while, it may pay to use a decision support system like the Analytic Hierarchy Process (Schmoldt et al., 2001) to help document and weigh the pros and cons.

System Installation and Acceptance

System installation and acceptance testing is becoming more and more complex in today's hospital, especially because many IS products must share electrical and data communication infrastructure; in fact, some systems share much more, including terminals, printers, and data storage. Purchasing documents from the institution should specify the vendor's provision of specific acceptance testing methods, ensuring that required interfaces with other existing or planned IS resources are fully detailed. If these details are omitted, finger-pointing is inevitable, and the ultimate satisfaction of the user and the successful on-budget installation can be seriously jeopardized. In this rapidly evolving field, few standards are fully complete, and even fewer vendors are totally compliant with all aspects of the standards that do exist. The best of the IS standards in health care often have gaps, if for no other reason than the field of medicine and the business of health care are rapidly changing. Most standards take years to be written, and by the time they are published and adopted, new major changes in the environment, such as outpatient care, robotic surgery, telemedicine, or hybrid drug/device combinations, have been adopted. If the new IS cannot adapt to the existing or planned system, or if changes to the existing or planned systems are required, then the TCO analysis might need to project significant extra costs to show that impact accurately.

Maintenance

The ongoing management of the new IS project is likely to require a combination of user, in-house engineers and technicians, and outsourced maintenance and repair support. Some responsibilities, such as changing paper, printer toner and ink cartridges; recalibrating printers, charging batteries; and maintaining backups may fall to the users or their department. Even if they do, the costs of training and retraining staff and the necessary supplies belong in an accurate TCO analysis. Some institutions create or subcontract for "help desk" (see Chapter 99) services to assist the users; if so, the pro-rata costs for the new IS project can be allocated. Although complex preventive maintenance is rarely needed for IS technologies, it is wise to allocate some resources for regular vacuuming of dust, cleaning of filters, replacing crucial batteries, and managing and optimizing disk-storage space as required by the system. Another often overlooked, but crucial, maintenance task is the creation of backup disks or tapes and routine periodic testing to ensure that those backups are actually usable. Ensuring usability can be much more complex and time consuming than initially expected because proper testing may require access to a complete, identical system! Otherwise, the testing may be of little use if the original system has a catastrophic failure. The backup copies can be useless if they cannot be read by a different system or if there is no way to locate or to assemble an acceptable replacement system rapidly.

Maintenance and repair have long been in the domain of the CE and BMET. The passage of the Health Insurance Portability and Accountability Act (HIPAA) (see Chapter 104) has lead to increased CE and BMET responsibility because many devices that they maintain hold or communicate patient-specific medical information as a consequence of the pervasive deployment of microprocessor technologies to such devices. In brief, practices and policies need to be created to ensure that patient-specific data are erased before any device is released for outside service and that memory is electrically erased or physically destroyed before it is disposed of or sent out for repair. For further guidance on this topic, contact the American College of Clinical Engineering (ACCE) at its web site, ACCEnet.org; The ACCE HIPAA Taskforce is currently developing a device-by-device HIPAA risk profile to make it easier to identify affected devices. (See Chapter 130.) If in doubt, however, it is probably wiser to err on the side of caution and to take extra precautions to erase patient data whenever possible.

A further maintenance and repair concern is that all devices, software, and communication systems might not be precisely compatible even when industry standards allegedly have been met. This can be a serious problem with software-configurable devices and systems because a small defect, incompatibility, inadvertent misconfiguration, or other inconsistency could be extremely difficult to detect. For example, a display monitor that looks identical to others but is defective or set for lower resolution or contrast could be difficult to detect, but it could cause disastrous diagnostic or therapeutic mistakes by eliminating clinically significant details. Further, a device could cause other systemic failures throughout the IS when it is connected through the network by tying. For example, a device that causes extensive resending or reformatting of digitized data throughout the IS could "tie up" limited bandwidth, memory, and/or computing resources, causing other parts of the system to slow down or malfunction. The only way to prevent such problems is to develop, document, and enforce careful device and system validation at the time of maintenance or repairs. If the hospital's overall IS is a complex mixture of technologies from diverse manufacturers spanning many generations of technology, such validation could be onerous and expensive, thus affecting the TCO.

Privacy and Security

Two additional critical health care IS management topics that need consideration are privacy and security as legally mandated by HIPAA regulations. Virtually every health care institution is legally required to ensure that only individuals who have a genuine need or right to know a patient's medical information can have access to it. Because much of the information is now in electronic form, software and hardware must be used to reliably prevent unauthorized access. Not only are passwords needed, but mandatory password update cycles and documented training to prevent sharing passwords or private information are also needed. To enforce compliance and to rigidly restrict access to confidential information or sensitive systems such as drug prescriptions, institutions also may selectively implement biometric screening (e.g., iris, fingerprint, voice, or facial scanning) and/or personal physical access control devices (e.g., smart cards, magnetic cards, IR tags, or digitally encoded buttons). However, these security measures are only effective if access to each terminal, printer, communication port, and removable disk drive is physically restricted to authorized personnel and for the intended duration of their use.

The aspect of privacy that relates to restricting IS access moves the discussion to one important facet of security ensuring that unauthorized individuals cannot access the IS. Not only is physical protection needed for IS resources within the institution, but it also is needed to prevent potential access by outsiders. The most serious outside risks are via the communications channels to the IS, including unauthorized tapping into wired or wireless networks and Internet and telephone/modem connections. The last two generally can be addressed by a combination of robust hardware and software firewalls, but addressing the network issues can be difficult indeed. At issue are the trade-offs between convenient and inexpensive access and reconfiguration of wired and wireless network stations and the need for security. The penalties of violating the HIPAA law might leave little alternative but to embrace cost and complexity of formal network restrictions, but it probably will take time and audits, to ensure that the necessary steps are not quickly abandoned or forgotten. "Locking down" the network most likely will require investment in encryption technologies and putting network access and control authority in a few key support staff's hands, which could affect the TCO analysis. Ensuring that these individuals are available whenever required, and providing adequate software and hardware for them to use could become a far more visible cost to health care institutions. Two remaining aspects of security bear further mention. They both relate to ensuring the integrity of the patient's medical data, because the data could impact diagnosis and/or treatment directly. Essentially, the burden is on the institution to ensure that every patient's medical data is protected from inadvertent or unseen damage, change, or partial or total loss. The means for doing this are not simple, requiring a judicious mix of hardware, software, and procedures for compliance. Physical system protection, data backups, and disaster-recovery practices, as discussed above, are parts of the solution. In addition, special hardware and software reliability enhancements for the system, as described in the system and communication sections below, could be needed.

IS System Components

The key components of an IS were classically defined by Von Neumann as input/output, memory, and a logic/computational capability. When these three elements are combined, they perform computing tasks by following carefully orchestrated sequences of instructions that represented in programs provided by human developers. Practically speaking, input/output can be broadened to include physical devices (e.g., keyboards, printers, and displays), biologic sensors (e.g., ECG, blood pressure, and pulse oximetry), of all sorts, and communication technologies that move data from place to place. The memory elements can be static (programmable or erasable memory PROM or EPROM), dynamic random access memory (RAM) that may be in the form of separate chips or imbedded within a microprocessor's own packaging, or disk drives of many forms, including CD/DVD optical storage and magnetic floppy or rigid disk drives. The logical/computational capability can be in the form of general purpose microprocessors of widely varying cost and speed or specialized, low-cost, dedicated logic devices with limited roles.

Before discussing these elements in further depth, it is critical to realize that the low cost and miniaturization of electronic components has resulted in many, if not most, products having built-in input/output, memory, and logic/computational that allows it to perform its task more reliably and flexibly. It is unwise, therefore, to oversimplify the complexity of current and future computing technologies because, in fact, they are made up of multiple interconnected computing systems. Successful design, implementation, and repair of such systems require understanding the role of each component as well as its function within the larger system.

As more and more health care technology shifts to multiple interconnected digital platforms (digital convergence), the design, installation, and support challenges expand. Not only must each device perform as designed, but the entire systemic performance must be validated every time a change occurs. Unfortunately, such changes might be invisible, in the form of program changes (e.g., patches and updates) that might not even be documented or detectable. Emerging standards in the health care field may in the management of these problems, but the multivendor, multigenerational interconnection of technologies required in most institutions will keep a significant portion of the burden on its own staff, not on the manufacturers or government agencies.

Input/Output Technologies

The input/output technologies in health care applications play an important and often overlooked role, as they may inadvertently damage the human information that they are used to represent. This problem is a byproduct of the necessary conversion of analog signals that characterize life and the discrete, or digital, nature of the IS components. The human analog signals must be converted from analog to digital (A/D) before they can be managed by the IS, and then must often be converted to another analog representation such as a display monitor or printout, which requires some form of digital-to-analog (D/A) conversion. Although the frequency and rate of change of most human waveforms is quite slow, relative to many other areas of technology (e.g., radio and radar), the fidelity of A/D and D/A processes is a function of providing enough speed, resolution, and faithful storage and computing to ensure that critically important diagnostic or therapeutic details are not lost or distorted. The ECG signal provides a simple example: If A/D conversion is performed either too slowly or with too little resolution, the digital data obtained and used for diagnostic or therapeutic interventions can be forever damaged by losing fine details, right from the outset. Further, even if the A/D process is correct, if any part of the storage, communication, or computing systems that handle the digital data fail to maintain or interpret every bit in the digital representation accurately, permanent loss of information and/or inaccurate analysis and intervention may occur. Finally, even if the digital representation is accurate within the system, inadequate D/A conversion into a graphical image on a display or printer can prevent a clinician from seeing critically important details.

Problems from this later D/A stage can be very difficult to avoid, as they often also depend on the human observer's optical, auditory, or tactile interpretation of the analog representation. The cognitive abilities and limitations of the human-computer output interface are still being studied, but it is clear that there can be wide performance variations of human and technology that greatly affect the result. Using the ECG as an example, a statistically significant number of clinicians might not perceive certain artifacts on a display, due to their inability to accommodate the contrast or intensity limitations–or settings–of that device. A simple systemic failure like this can be difficult to isolate and repair today, as many clinicians might use different displays, both portable and fixed, and both local and remote, during the course of a day.

The above examples are cautions that the clinical engineer and biomedical electronics technician should bear in mind whenever they work with I/O devices. Even though most of these devices might be very simple and reliable by themselves, when integrated into a health care system the overall safety, reliability, and performance of the system might be severely but unobtrusively limited by the weakest link's failings. Another new risk is that many medical devices could become input devices to other processes that were never considered in their initial designs. The data port from an infusion pump could become the data source for another manufacturer's drug interaction warning system or patient record charting system. Newly discovered drugs, novel dose dilutions, or unexpected use of stand-alone pumps without data ports could defeat the expected safety, clinical, or administrative role of the system.

Aside from the systemic issues mentioned above, certain aspects of digital I/O bear cautious consideration as well. We know from classical clinical, electrical, and mechanical engineering experience that analog waveforms often must be amplified and filtered carefully to ensure reliable fidelity. Many of those techniques now have been subsumed by digital signal processors (DSPs), which can provide inexpensive and flexible waveform manipulation using computation rather than capacitors or inductors. The DSP allows complex filter designs that would be difficult, expensive, and perhaps even impossible, with physical components. These products have allowed signal retrieval and enhancement in situations never before possible. At the same time, there is no simple way to anticipate or control the interactions that can occur when multiple devices are interconnected in a system. The ideal outcome is to achieve a cumulative advantage for the system, but that should not be taken for granted. An unexpected induced artifact (e.g., a newly enhanced signal that is larger than expected, or the removal of part of a signal by a filter that was inadvertently turned on) could cause disruptions or serious adverse consequences. The DSP is likely to be more widely deployed for many purposes and should not be ignored in design, installation, or maintenance.

Communications

Telecommunications, digital communications, and networking might be considered a separate field in many texts, but they are included here to suggest to the reader that they could, and should, be viewed as a continuum, extending the distance and/or flexibility of I/O among patients, doctors, devices, and IS components.

Telecommunications

Telemedicine and telehealth are discussed elsewhere in this handbook (see Chapter 101), but their growing deployment offers further support to this chapter's approach. When needed, physicians can remotely download ECGs, reprogram pacemakers, order prescriptions, and share diagnostic X-rays, and it seems that more and more medical care can be, and will be, delivered this way. A decade ago, the clinical engineer might have comfortably ignored telecommunications as "merely" supporting discussions or business faxes that would be like sticking one's head in the sand today. Traditional telecommunications systems used relays and digital switches to create wire-to-wire connections for analog signals (i.e., sounds). Modern telecommunication systems, however, often convert sound to digital data, using D/A and A/D devices like those mentioned above. Even if a standard telephone handset is used for a person-to-person conversation, there is little or no way to know whether digitization is occurring at one or more points along the route. Many medical devices can use telephone connections to share data, and wired or digital telecommunication technologies each can introduce unique distortions that can be difficult to detect, diagnose, or fix. If one is to be responsible for ensuring reliable medical communication via phone lines, one must understand the telecommunications technologies that are available. With that knowledge, there is little to separate the training and ability that are needed to support regular business and personal phone calls.

Facsimile transmissions (faxes) use a combination of digital and analog techniques to transfer images from one point to another, typically via the telecommunication system, although the Internet is sometimes used instead. With the emergence of HIPAA security and privacy laws, the simple FAX machine is being called into service–perhaps as an interim fix–to send patient information to other doctors or institutions without using formal digital pathways like e-mail or web servers. Technically, this might be seen as a step backward, but it sidesteps some of the more rigid rules, and risks, in the HIPAA law. Specifically, HIPAA's regulations excuse discussions or documents being sent by telephone, as they are not stored, nor intentionally transmitted, in digitized electronic form. One presumption of a telephone or FAX communication is that it is a point-to-point connection of wires, in which the recipient's phone number ends at a known location like a doctor's office or a patient's home. Electronic transmission by e-mail or digital document on the Internet is not exempted from HIPAA regulations, however, as those documents must go through many public and private computers en route to their destination. Unless the documents are encrypted in a secure fashion so that only the intended recipient may view them, they may be read by one or more unauthorized parties, violating the patient's privacy. The result is that many physicians have been advised not to communicate by e-mail or web sites, leaving the FAX as one of the few convenient alternatives. Unfortunately, FAX transmissions suffer from numerous limitations, including crude scanning and printing resolution for most inexpensive products, and unpredictable handling of colors. As described in the above ECG example, this may cause unanticipated artifacts and distortions in the received record, and the problem will be exacerbated by repeated faxing from one physician to the next. This issue is only now emerging, and it might take some time before these problems are well documented and understood. Suffice it to say that this is another example of a technology that could affect the quality of medical care, putting it well within the purview of most clinical engineers and biomedical technicians. As an aside, the HIPAA exemptions to using FAX may be short-lived, because today there is not really any way to prevent the FAX from being routed through a digital or Internet network anyway. Many available telephone systems use digital routes and techniques to save money and time. Unless the user can ensure that both the sender and receiver are truly connected to analog phone wiring, HIPAA constraints may continue to reign.

Digital Communications

We now can shift to a somewhat more technical discussion of general digital communications, networks, and the Internet, to provide a working vocabulary and understanding of the basic principles that will be helpful in the field. There have been many communication and network technologies over the past couple of decades but increasingly more companies are using a limited number of industry-proven standards, which makes the discussion a bit easier. We will begin with wired communication technologies and will close with wireless ones.

Digital communication generally requires reliably sending and receiving binary (1/0, or on/off) signals, which may be represented by a pre-agreed combination of (low) voltage, ground, or a (low) negative voltage. Any two devices could be designed to share digital data as long as they shared a pre-agreed protocol, such as "on" or "1" being 5 volts and "off" or "0" being -5 volts. The other element needed for reliable communication is a timing protocol to ensure that the recipient is measuring the signal at the right time. To accomplish this, either synchronous clocking signals are sent in parallel, or asynchronous triggers (timing signals) are embedded within the data stream. With synchronous communication, an extra wire or pair of wires sends a clock signal. Because the clock and data signals would encounter similar distortion along the route, a clean clock signal might serve as a validation that the wiring path is adequate. Each time a clock pulse is detected, the voltage received can be measured, determining the on or off state of that bit. Asynchronous communication depends on both the sending and receiving device having an extremely reliable and precise crystal timing clock working at formally agreed rates. The receiving device waits for an initial signal to tell it that the message is starting, and then uses its internal clock to pace its measurement rate. Depending on the agreed-upon protocol, the message will usually last a fixed number of time intervals, or the message will begin with a numeric value telling the message length before the data itself is sent. Extremely large amounts of data may benefit from sending multiple signals on separate wires simultaneously, which is called "parallel communication." A common multiple is 8 bits, or signal wires. When configured with an electrical ground wire for each clock wire and a clock and ground wire, each such cable will require at least 18 wires plus spares and any external electrical shielding. Because the cost and size of parallel cables is significant, they are used only infrequently, such as for connecting some printers and for high-speed computer connections (e.g., SCSI). Most current communication systems use an asynchronous serial-information format instead of clocked parallel signals, sending one signal after the next using careful timing to separate them. RS232/422, Ethernet, USB, USB-2, and Firewire formats are based on asynchronous serial communication.

Even if all voltage and timing agreements are in place, proper electrical and mechanical designs are needed to ensure that the voltage and signal shape are not degraded (i.e., diminished, delayed, or otherwise harmed) along the way. Even the best designed communication system designs, it is difficult to guarantee that some external electrical, mechanical, or other problem has not adversely influenced the communication. Therefore, most designs include an active exchange of confirmations between the sending and receiving devices to ensure that the intended message got through accurately. Simple communication software within the devices handles this automatically.

The software generally requires that the sending device describe the length of the information being sent as well as one or more quality control parameters. For example, the sending message might inform the recipient that it should receive 256 signals and that, when added up, there should be an odd or even quantity of 1s in the message, called "even or odd parity." This simple format will serve well if a single bit is lost, but it is not adequate for critical data because interference that causes multiples of an even number of 1s to be misinterpreted as zeros still would have the correct parity. For most high-speed communication, therefore, more sophisticated coding systems are used. These systems work by performing more computations on the data being sent, and then sending data about characteristics of the message that should be received at the end. If the recipient's device performs the same computations on the received data and finds that it does not match those sent with the message, it knows that a problem exists. In some cases, the receiving device is given enough information to repair simple mistakes, but a simpler handshaking communication protocol is often used instead.

If the receiving device is satisfied with the quality and accuracy of the message, it can send its own brief confirmation message, "ACK," to the sending device; if the message was damaged, it can send a negative message instead, "NAK." The handshake protocol then may provide for a sequence of remedial steps to resend and recheck the message, including, if necessary, automatically choosing a slower transmission rate or even a different communication channel. In early RS232 signal designs, separate wires were reserved for sending these handshake signals back and forth ("hardware handshake"), but modern designs provide bidirectional data and signal pairs and use coded messages for the same purpose ("software handshake").

The software handshake protocol allows additional message and codes to flow between sender and receiver like a "busy" signal when a printer's buffer is full, or a "toner empty" message when appropriate; these signals may be different for different systems, thus causing additional incompatibilities. For example, older Hewlett-Packard (HP) printers did not use the same codes as other printers did, which created printout problems. This is part of the reason why unique "drivers" or software programs are needed with different printers, scanners, and terminals so that the system can interpret the codes that it receives and take proper action. Today, inexpensive communication chips (UARTs and USARTS) handle much of the timing and handshaking automatically, leaving the higher-level decisions–like setting off an alarm when a color-ink cartridge runs dry–to device software to notify the operator to correct the problem.

Networks

Networks and the Internet generally rely on the asynchronous, software handshake protocols described above but add a few more formalities to the process. Ethernet, for example, has become a ubiquitous asynchronous network protocol widely used throughout hospitals. The nature of a network, however, requires that many devices are able to share the digital signal capacity of a single set of wires that run throughout a building. This minimizes the cost of sending the signals throughout the building but requires that the devices have a way to discern the messages that belong to it, and those it should ignore. To accomplish this, a combination of hardware and software is required.

A great deal of specific detailed networking information is available in textbooks and online sources like wwwcisco.com, so only a brief overview is provided here. At the software level, one fundamental network design concept is that data are sent in finite length packets of bits with a digital address for the recipient at the beginning of the packet. Regardless of the network details, one basic networking rule is that all other devices should ignore messages that are not addressed to them. However, this does not mean that other devices cannot detect and read the messages, which creates significant security and privacy challenges in the health care setting. Every computer on the network has an internal network adapter card (Network Interface Card (NIC)) that handles the electrical design and communication protocols (Network Access Method (NAM)) for the network it connects to. That card also has a unique ID number, known as the "MAC address" (network Method Access Control).

Topologies

Three general cabling layouts, known as "topologies," can be used: The star, token ring, and bus. Each represents a tradeoff of wiring cost and communication speed and reliability. The star configuration uses a separate wire to each device, which offers high reliability and large signal width, but also has a high cost because wires must be run to every device in a building. A bus network can use one wire running all through a building, and each device can connect to the wire as it passes nearby. If the bus wire is broken, however, no messages can pass the break. Furthermore, if all of the devices sharing the bus are trying to send a lot of critical data, there can be many "collisions," in which packets interfere with each other. Although the bus topology can be quite inexpensive, it can be readily overloaded. Without special software and hardware design and components, a simple bus network could prove inadequate for critical applications.

The third method, token ring, uses a continuous wire that runs all through the building and returns to the starting point. Messages spaces, called "tokens," are time-allocated on the wire, and any device can put information into an unused token space with the MAC address of the recipient. As the tokens go past each device connected to the ring, the device reads the contents and makes a decision: If the token is addressed to it, the device reads the contents and either stores its own message to send or clears the token for other devices to use. A token that goes all the way around the ring without being read is erased because the recipient device is assumed to be offline, and the sending device is free to take other corrective action as needed. The token ring has another advantage in that if the ring is broken at any point, the signals still can propagate to all devices.

In actuality, no single topology is universally or exclusively preferred. In most institutions, combinations of all three are common, balancing costs, performance, and reliability as appropriate. Each subnetwork is connected to nearby subnetworks with a gateway that transforms the data signals and protocols as needed. The gateways work with a Network System Management system (e.g., Novell Netware™) to keep track of the location of each device by storing its MAC address and the subnetwork where it is connected. Messages are routed through the appropriate gateways until the desired subnetwork is accessed. When the collection of subnetworks is located in a single building or area of a building, it is usually called a "Local Area Network" (LAN). Larger networks that span multiple buildings on a manufacturing, research, or manufacturing campus are referred to as "Wide Area Networks" (WANs), and city-wide networks are referred to as "Metropolitan Area Networks" (MANs). The longer the distance, and the higher the necessary total bandwidth, the more likely it will be that network wires will be replaced or supplemented with more expensive microwave or optical fiber links. Specialized NICs and gateways are needed to accommodate these enhancements, but the general architecture described above applies, regardless of the medium or the size of the network.

Internet

The Internet is a specialized network that was designed by the Department of Defense's Defense Advanced Research Projects Agency (DARPA) as a reliable global network that can withstand major physical damage without stopping messages from being received. The Internet functions in a similar way to the general networks described above, with two major exceptions. First, each device on the Internet has an Internet Protocol (IP) address that is quite different from LAN/WAN/MAN MAC addresses.

An IP address is a four-part number, know as a "dotted quad" number. A dotted quad number comprises four separate numbers that are divided by a decimal point (e.g., 174.201.33.101). Each unique number functions like a MAC address or a phone number; the dotted quad IP address represents the address for that particular device or user. Some numbers are permanently assigned to a device, such as a company's main web server. Other numbers are dynamic IP addresses, assigned to individual PCs or devices as soon as they connect to the Internet. The IP addresses are maintained in an international collection of Domain Name Server (DNS) computers connected to the Internet for the purpose of keeping several pieces of Internet addressing information in global sync: Each valid IP address, the textual name assigned to that address (such as www.att.com), and the physical location of the device with respect to its nearest access point to the Internet network.

The global matrix of web sites connected by the Internet is known as the "World Wide Web" (hence, the "www" prefix on most web addresses). Information that is sent via the Internet is broken into packets, each of which has an IP address for the sending and receiving device, sequencing information for each packet, and the time it was sent. Each packet is sent to the nearest Internet access point and it is then read by the nearest Internet packet-forwarding station. These stations are run by military, government, academic, and commercial institutions, based on international contracts and agreements.

The packet forwarding stations get DNS information for the recipient and make a decision to send the packet to another nearby forwarding station. If the forwarding stations or communication lines in the general direction in which the message must go are busy, the packet will be sent in another direction. Each forwarding station moves the packet in the best path it can calculate, with the ultimate goal being the delivery of the packet even if it has to go backward via a long journey around the world to its destination. The receiving device must wait until all of the packets expected in the message reach it, and then they are placed in proper sequence. If the message is satisfactory, it can send a message back to the sender confirming the receipt. If there is a flaw, or if one or more packets are missing, after a predetermined wait the receiver will send a request to the sender to resend the necessary information. If that request fails, the receiving device must presume that the sender is no longer working and notifies the recipient that the communication link has failed. In addition, if a packet-forwarding station finds that a packet has been moving around for a long time without making its way to the recipient, it can kill the packet so that further Internet resources are not wasted. This approach has proven very robust and reliable, which is one reason why the Internet has been so successful.

The successful system for forwarding stations used by the Internet is also its main Achilles heel for privacy and security, however. Every one of the forwarding stations can read, store, and/or possibly change the packets that it forwards. Unless each packet is encrypted in a way that prevents all but the sender and receiver from reading or changing it, the information is essentially open to the public. In addition, many government agencies around the world are believed to be scanning the packets for terrorist or other illegal content. The U.S. government, in particular, has required that the keys to any commercial encryption package be given to it, so even encrypted information could be visible to our government. This means that unencrypted, patient-specific health information, including files, e-mail, pictures, videos, and lab results cannot pass the HIPAA privacy and security rules. Furthermore, it would seem that sending FAX documents through the Internet would not be acceptable either, because they are not encrypted and can be printed easily with FAX-viewing software. These HIPAA conflicts are the reason why many doctors are limiting their patient-specific information to regular FAX, telephone, and/or regular mail. There may be ways around this eventually, including encrypted e-mail or Virtual Private Networks (VPNs), which use encryption and password controls to control the access to web sites and the privacy of information being sent through the Internet. While VPNs are not necessarily complex, they add new expenses because such services are run by private companies that charge for this capability.

Another emerging use of IP addressing is the so-called Voice over Internet Protocol (VoIP) telephony. This is an interesting technology because it uses the free and reliable worldwide bandwidth of the Internet for voice communication. This is essentially one specialized type of digitized voice communication that certain companies like Cisco Systems are championing. Within an institution, a VoIP telephone network may offer flexible installation and movement of telephones and may allow microphone-equipped PCs to serve as telephones as well. Investment in a VoIP system in a hospital may also reduce some wiring costs and complexities by eliminating some or all telephone wiring. After all, if digital wiring is going to be installed everywhere, why not consider using it for the telephone needs, too? Such installations still may need separate wiring for LAN applications, but the choice of VoIP wiring may facilitate deployment of wireless products as well.

VoIP allows routing long distance calls through VoIP services, thus reducing long distance calls to the service charges paid to local phone companies and eliminating commercial long distance costs. VoIP has limitations for long distance calls, however. When used within a facility, the VoIP signals may rarely overtax the bandwidth available. Once the digitized voice packets are sent through the Internet, however, uncontrollable delays may occur for each packet. The result might be choppy sound quality, which might not be an acceptable trade-off. If both parties on both ends of the communications have VoIP systems, however, special software and hardware can minimize these problems. Therefore, VoIP might be better suited for regional or national health systems, which can use long distance savings to help offset the cost of installing compatible VoIP systems.

Many network-based computers are also used to access the Internet. This access is facilitated by a special type of network gateway product that has the ability to maintain lists of the active MAC and IP addresses in use. All inbound or outbound IP addresses are routed to or from the correct MAC address by the gateway and the Network Management System (NMS). In most cases, each PC is assigned a dynamic IP address when it first connects to the Internet, conserving the available Internet connections for active users. By contrast, web servers that need to be constantly online would be assigned a static IP address so that the DNS system can properly route inbound requests for service.

Wireless Communication

Wireless communication (see Chapter 103) is becoming more and more standardized, and it often simply becomes an extension of the wired network. The most common protocols for achieving this are the IEEE 802.11 series. The 802.11a protocol is the oldest and slowest; 802.11b allows more bandwidth, and 802.11g is the newest, fastest, and most generally flexibly compatible of the protocols. Depending on the sophistication of the 802.11 components, partial or complete compatibility may be possible between each of these series. From the computer and network's point of view, each computer on a wireless link is addressed just as if it had wires running to it.

This 802.11 standard is offering new connectivity opportunities never before possible. For example, thousands of Starbucks Coffee shops are set up with 802.11 networks. Customers who pay for access can go online through the Internet whenever they are within radio range of a Starbucks, anywhere in the country. Special security challenges exist, however, because suitably equipped receivers can copy all of this wireless data. If adequate encryption software is not used during all confidential communications, anyone can see it. This presents a serious exposure for web sites and e-mail users, as userids (user identifications), passwords, and private information all could be exposed. Further, if the userids and passwords allow access to internal institutional e-mail, databases, and files, the public disclosure can open the institution to intrusion, hacking attacks, and other serious ongoing security breaches. Unfortunately, although most laptop, desktop, and tablet PCs have adequate encryption capabilities, many PDAs (e.g., simple Palm Pilots™ and HandSpring™ units) do not, thus making enforcement and control very difficult, if not impossible. The convenient, portable, reliable, and low-cost advantages of wireless 802.11 technologies for health care cannot be ignored, but each use may require careful planning and encryption controls to ensure HIPAA compliance.

Other wireless technologies are in use, and each has advantages and disadvantages. One, infrared (IR) communication, often used with PDAs and some notebooks, has the advantage of low cost and a limited amount of line-of-sight security control. Unfortunately, encryption is rarely used, limiting IR applications in health care to non-confidential data. Another technology of note is known as "Bluetooth," which is a low-power, limited-proximity radio link that should also be low-cost. Few Bluetooth applications are available as yet, but eventually the technology might allow physical docking areas where doctors, nurses, or family might access select network or Internet resources without creating radio interference for other nearby devices. Again, encryption of such messages might be limited or nonexistent, so HIPAA-regulated information might not be suitable for Bluetooth applications.

A third digital, in-house, wireless technique that is worth mentioning is the Internet's IP addressing scheme. As long as a radio frequency that complies with hospital frequency allocations is selected, some devices may use a wireless form of the IP address instead of a LAN. The fundamental difference is that the devices do not use a LAN MAC address but the four part "dotted quad" address mentioned earlier. The advantage of this method is that the devices can use standard web-enabled tools such as browsers or file transfer programs. In addition, installation of wireless IP capability can facilitate VoIP applications. This may ultimately allow safe wireless intra- and extra-facility telephone calls throughout the hospital, as long as the VoIP limitations mentioned earlier are not a major hindrance.

Privacy and Security Considerations

No discussion of communications in the health care world is complete without a few words about securing the communications to prevent unauthorized access or tampering. The rapid development of wired and wireless communication standards has made it far easier to train staff; acquire hardware and software; and install, test, operate, and test digital-based systems. This has not come without a price, however, in that the very same standardized components and software are available to anyone who wishes to tap into the system. This chapter cannot fully address this deeply technical and rapidly advancing field, but certain points should be made. First, the access to and from the Internet exposes the whole system to hacker attacks of many sorts. Viruses, for example, are readily attached to e-mails, files, and applications such as Instant Messenger (IM). An automatically enforced antivirus installation and updating methodology may be unavoidable. Unfortunately, the available programs are not free, and each one comes with trade-offs, like stealing significant computer speed from each user and causing occasional program and hardware incompatibilities that can be severely disruptive. However, operating without such software cannot be easily rationalized.

Firewalls

Each Internet user is exposed to attacks through programs that can damage files, modify data, or steal confidential information because during their Internet session their IP address is visible to all sites that forward their packets. Malicious programs can use that IP address to send hidden instructions to the user's PC. For most facilities, therefore, a special suite of hardware and software firewalls must be installed to manage the traffic into and out of the facility. Firewalls are not autonomous products, however. They need to have updates systematically applied, and someone must evaluate the IP traffic to detect and counter a large number of potential threats. Although the threats are not solely external, those that are external require careful scrutiny to detect patterns of hackers trying to infiltrate or interfere with the firewall or other resources. The last thing a facility needs is for a hacker to steal disk, computing, or communication resources for some playful or malicious purpose. Some infiltrating programs have built-in time delays that allow them to hide silently once they have been installed.

Multiple isolated, brief intrusions can allow the institution's system to be attacked from within when staffing is low, thus creating significant internal or external damage. One external use is to create denial of service (DoS) or spam attacks on other outside computers. In a DoS attack, a large number of computers can be set up to simultaneously send bulky messages to, or request time and resource intensive services from, a targeted commercial or military site. Similarly, an e-mail spam attack may use one or more institution's email systems to send out thousands of pornographic, political, or advertising e-mails using its own lists and/or hospital e-mail lists. For this reason, some firewall managers often restrict the number of e-mail copies that any internal user may send and may set thresholds on the type, frequency, and quantity of Internet messages that any one user can send. Those limits can vary by day and time.

Further management of IP traffic is possible, including tracking or limiting e-mail use, scanning e-mail content for privacy and harassment violations, connections to child-pornography sites, and other violations of written institutional policies. Similar firewalls and strategies may need to be deployed within the LAN, WAN, or MAN systems to prevent or detect similar internal privacy or security violations.

Physical Access Controls

Physical access controls must be considered as well. Although the IP and MAC addressing schemes often can flexibly allow any user to plug any computer into any available wired outlet or wireless access point, that may not be supportable in most institutions. The alternatives include rigid password policies and enforcement and limited permission for key maintenance and technical support personnel to implement new or changed connections to the hospital's networks. The trade-offs are not simple. Allowing the networks to allow anyone with an Ethernet or 802.11 wireless adapter to access communication or computing resources might not be acceptable from a security and privacy perspective, but it does make it much easier for clerks, nurses, technicians, or even patients to install and move approved devices around the hospital. At the other extreme, restricting the control of such moves to a single person might not work either, as an emergency relocation of a patient and all of the associated network-based equipment to the ICU cannot be delayed because the person cannot be reached. Having available multiple persons on each shift could solve the problem, but clear documentation and communication will be needed in order to avoid problems.

Memory/Storage Technologies

Memory or, more generally, storage technologies present unique challenges and opportunities in health care. The good news is that storage technologies are becoming so inexpensive and reliable that most problems can be solved reasonably.

Reference

Schmoldt D, Kangas J, Mendoza GA, et al. *The Analytic Hierarchy Process in Natural Resource and Environmental Decision Making*. Dordrecht, Germany, Kluwer Academic Publishers, 2001.

98
Physiologic Monitoring and Clinical Information Systems

Sunder Subramanian
Director, Clinical Information Systems, Information Technology Division, University Hospital and Medical Center, SUNY at Stony Brook
East Setauket, NY

The future direction of the clinical engineering profession will be tightly coupled with Information Technology (IT) more than ever before in the history of the profession. Convergence of the two professions is driven by factors such as advances in the state of the art in device technology, the need for integration of bedside data with clinical documentation, existing and emerging standards in the area of data communication, demand for accurate and reliable clinical indicator data by regulatory agencies, caregiver and customer knowledge and understanding of emerging clinical technologies, and consolidation of device manufacturers and clinical information systems vendors in the marketplace. This evolution is compelling the clinical engineering profession and hospital clinical engineering departments to expand their scope, knowledge, and expertise in the area of clinical information systems (CIS) and to become active participants in the selection, implementation, and support of bedside and clinical information systems.

This chapter addresses the technologies of bedside monitoring and CIS and their interdependencies. The state of the art in technology in both areas, reasons for the integration of the two technologies, and the necessary groundwork for managing these technologies are discussed. A brief overview of an electronic patient record (EPR) and the impact that physiological monitoring and clinical information systems have on EPR are presented.

Clinical Information Systems

One definition of a Clinical Information System (CIS) would be of a system that should interoperate with bedside monitors, bedside peripheral devices, and hospital information systems (HIS) in providing an integrated clinical information charting and data management support system in a local area network (LAN) configuration (Dyro and Poppers, 1988a,b). The capabilities of the clinical information system should be available to the user population distributed on this local area network in a multitiered health care enterprise that is part of an integrated health care delivery system. The CIS should be able to support user needs that are expected to be diverse and dynamic over the service life of the system with resultant growth and change in functional capabilities, data processing, and information-transfer volume (Gage et al., 1986).

The CIS should support generation of, and access to, clinical/patient information on a real-time and archival basis, with response times designed to eliminate manual charting burdens on the clinical staff. The architecture should provide a user-customizable, fault-tolerant redundant configuration for efficient and secure acquisition and entry, query processing, sharing and exchange, storage, and retrieval of clinical data. At the minimum level of functionality, the system should automate patient/clinical data acquisition, processing, storage, and retrieval and should provide capabilities analogous to existing capabilities with the elimination of the need for paper. The CIS should support real-time interfaces to bedside devices such as physiologic monitors and ventilators and to the HIS and ancillary systems such as laboratory, radiology, and pharmacy. This will reduce manual data entry and paper documentation by automating the flow of demographics and vital signs data, lab results, and medication orders into customized charts and forms. Clinical information presented this way is legible, recoverable, reproducible, and accessible throughout the enterprise. Providing integrated clinical information will assist caregivers in making informed clinical decisions about the care for their patients. An integrated query processor and a report writer are necessary to support datamining applications. These tools will greatly benefit quality assurance, regulatory compliance, financial, and research applications.

At the system level, the three critical requirements would be reliability, performance, and size. The best designed systems offer reliability through redundant fault-tolerant architectures where every component of the system is duplicated, thus eliminating or greatly decreasing the possibility of system failures that could result in loss of system availability and patient/clinical data. Response time, a criterion for measuring the performance of the system, is a function of the hardware and communication networks. High-speed processors, with large online storage, that are connected to clinical workstations using a fiber-optic network or high-bandwidth coaxial connections can sustain high data throughput and can provide fast response times. The size of the system, defined in terms of available system memory (RAM and hard drive), not only dictates the storage capacity but also affects the retrieval speeds of stored data. A typical clinical information system that is necessary to store text, numerical, and graphical information will have large and high-speed storage subsystems.

CIS Architecture

Traditional systems such as radiology and pharmacy were best-of-breed application systems. These systems, due to their focus on the application, performed very poorly in the areas of data communication and integration. As a result, they became islands of clinical information, disconnected from the rest of the enterprise. In today's environment, it is important for a clinical information system to have the ability to communicate and seamlessly integrate with the rest of the enterprise. One such example of a CIS is the Centricity product from GE Medical Systems Information Technology (Centricity, 2002). Shown in Figure 98-1 is the architecture used by the Centricity Enterprise Clinical Information System product from GE Medical Systems.

Clinical information systems must incorporate industry standards on all aspects of their architecture, and this should apply across the board, from the user interface to operating systems and to clinical workstations. Using PCs as clinical workstations can be economical and user-friendly. A graphical user interface (GUI), supporting the use of a mouse or a track ball for most common transactions, would be necessary for the success of these systems. Using industry standard databases will enable clinical information systems to readily share data with other systems and to become a part of the enterprise's clinical information management solution. Industry standard communication protocols will enhance accessibility, security, and portability of clinical data over a distributed health care delivery system. A web-enabled CIS can provide caregivers access to clinical data over the world wide web using popular desktop browsers. In a distributed health care delivery system, with patients and caregivers separated by large distances in some cases, having e-mail messaging capabilities can be a useful clinical tool.

CIS Components

Components of a clinical information system include a powerful central computer system running the operating system and the application software with online and archival storage subsystems, communication/network, real time, and batch interfaces clinical workstations and bedside devices.

System Hardware

The Computer System

Clinical information systems typically use the client-server architecture with redundant processors and storage subsystems. In addition, the servers also have archival subsystems and communication links to other systems for sharing and exchanging information. System architecture will be optimized for central and distributed processing resources by partitioning processing capabilities to provide high-speed response times, while utilizing redundant, central data processing and datastorage architectures coupled with high-performance, intelligent clinical workstations. The architecture will also provide coordinated central and distributed processing functionality in data acquisition, processing, and retrieval, with response times optimized for transaction and information retrieval requests. In the event of a system failure, the clinical workstations will reconnect to the remaining functional system, and complete CIS capability will be available to the user with continuous access to all patient data. Most CIS configurations will support simultaneous running of training and production systems.

The state of the art in the server technology is to implement the application using the CITRIX server architecture (Citrix Systems). Hospital Information Systems departments, experiencing a decrease in their resources, face the daunting task of managing ever-increasing responsibilities. Heterogeneous computing environments further complicate the situation. The CITRIX approach enables the information systems departments to meet this challenge with streamlined installation and configuration, centralized administration, and security for

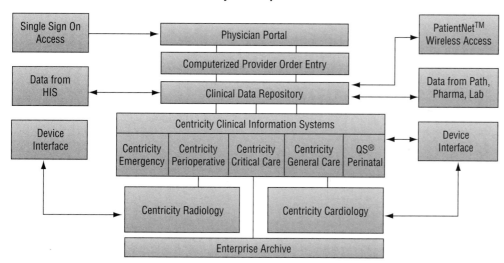

Figure 98-1 Centricity Enterprise Clinical Information System architecture from GE Medical Systems Information Technologies.

information and applications. Fast deployment capabilities and ease of management helps information systems departments to achieve rapid turn-out on upgrades and maintenance. In today's market, some application vendors certify their application to run on dedicated CITRIX servers, while a few others support sharing of servers with other applications.

The servers and clients run well-known operating systems such as Windows NT, Windows 2000, UNIX, and Open VMS. Due to the large installed base for these operating systems, obtaining vendor technical support and internal support is easy. Maintenance, management, and implementation of these systems are relatively trouble free as well.

System security will allow for multiple levels of access that is typically site configurable. At a minimum, a user will need an identifier and a unique password to gain access to the system. In addition, the capability to assign privileges to care givers using a common group profile based on a clinical function or a job title will be supported to enhance the security. Behind-the-scenes capabilities (such as audit trail) to track transactions and administrative tools to restrict read write privileges to the item level will be integral to the application.

Primary On-line Storage

Clinical information system specification for primary storage will include data storage capacity and redundancy, data security and reliability, and access time/response time. The storage capacity will be designed to support response time requirements with the capability of incremental expansion to meet changing clinical and operational needs. Real-time memory (RAM) and hard disk upgrades at the servers and clinical workstations will be supported by the architecture. In today's technology, large-sized disks (gigabyte capacity or larger) in the Redundant Array of Independent (or Inexpensive) Disks (RAID) configuration offer response time, capacity, and redundancy options that easily meet the demands of the clinical environment. RAID5 is one of the common configurations used in application servers that use at least three disk arrays designed for fault tolerance and performance. RAID5 provides data-striping (i.e., spreading out blocks of each file across multiple disks) with higher error detection and correction capabilities. The RAID configuration also allows hot swapping of failed drives, thus significantly reducing the system's down time.

Archive Storage

Clinical information system specification for archival data storage will include media technology, storage capacity and redundancy, data reliability, archival life, archival security, administrative control, and online availability/response time. The archival process, which will be transparent to the user, will be capable of archiving complete patient records without degrading system performance. The archive subsystem will be able to support data life over a number of years and data-availability guarantees, irrespective of changing media technologies and storage software. Access-security functionality will be consistent with the needs of the clinical environment and will meet the stipulation of the regulatory agencies. Write Once and Read Many Times (WORM) technology, once the most commonly used media in CIS configurations, are being replaced by Compact Disc Read Only media (CD-Rs). WORM media offer large-sized storage capacity (gigabytes) and require dedicated drives for reading them. CD-Rs offer portability and use of standard CD ROM drives at the expense of offering smaller storage capacity.

Integration

The communication network subsystem will support CIS requirements for connectivity, response time and data throughput, network maintenance, diagnostics, reconfiguration, reliability and continuous availability. The network will have a bandwidth capable of sustaining throughputs of 100 megabits/second or more, using connectivity protocols that are based on the open systems interconnection (OSI) model. TCP/IP (Kabachinski, 2003) and Ethernet protocols are commonly utilized in CIS communication network architectures. Network topology design and configuration will complement system guarantees in meeting response time criteria. In the area of connectivity, a single point of failure will not bring the system down; alternate routing schemes through redundancy will be part of the architecture. Network availability should be continuous and consistent, with reconfiguration capability in case of a single link failure preventing propagation of network faults. Off-the-shelf components, such as hubs, switches, and routers, will form subsystem building blocks allowing for easy swap out of failed components.

Database

The database is a key component of the CIS architecture that dictates the level of flexibility offered by the system. Ideally, the Database Management System (DBMS) should be based on industry standard models, such as relational, object, or network, that are capable of communicating with other databases and statistical software packages in the health care enterprise. Most commercial CIS products typically use proprietary database architectures that will not allow users to import data items from external databases. Instead, commercial CIS products configure their databases with templates and data items that are user/site-customizable with the help of a configuration tool. The clinical-configuration tool will provide the user/site the ability to create new database items and clinical flow sheets that are closely tailored to their practice. In addition, CIS architectures will support real-time or batch/scheduled outbound interfaces for exporting clinical data to other foreign systems such as a data warehouse or to populate a foreign application such as an electronic patient record (EPR).

A natural query language type interface module will be an integral part of the DBMS and will provide the user with the ability to mine for clinical data for teaching, research, regulatory, and quality assurance objectives. An integrated online audit function that maintains a record of all user transactions (such as additions, deletions, or corrections made to the patient record) is required for risk management, litigation evidence, regulatory compliance, patient confidentiality, and privacy reasons. Most previous generation CIS database architectures did not support robust report writers; instead fulfilled both the database and reporting needs by exporting clinical data in ASCII (Kabachinski J, 2003) format to desktop software packages such as Microsoft Access. In today's market, CIS architectures have to offer an integrated report writer as part of the application. If not, at the very least, these systems should have the ability to interface to industry standard report writers. The state of the art in databases is the use of fourth-generation packages, such as Oracle, that are capable of sharing data with the environment more effectively and easily.

Application Software

The application software, residing on top of the operating system, is run from the server. This software is responsible for managing all clinical transactions, such as data acquisition, display and storage, print, and clinical configuration. The clinical configuration tool and the interface module are part of the application software. The complexity and flexibility of the application software is a function of the clinical data being managed, and also the extent of the clinical charting being done.

Clinical information is a mix of different data types. A patient medical record can consist of discrete laboratory results; continuous analog EKG waveforms; physician and nursing notes and summaries; vital signs trend graphs; radiological images and reports; and patient demographics. A robust CIS should be able to receive and process all of the different data types mentioned above and to integrate them to create a longitudinal electronic patient record. In reality, most systems provide connectivity to other systems that store specialized data. For example, image and waveform data are typically stored in the parent systems (such as radiology and cardiology), and the CIS will provide a pointer to these foreign databases. The pointers enable the clinician to pull up a snapshot of the image or the waveform for the patient alongside other clinical data. Although integration of all clinical data under a single product or single vendor solution is ideal, the fact that a CIS can interface to foreign systems for access to data not present in its database is a useful feature and it can provide a level of integration that is practical, without putting too much burden on the enterprise infrastructure.

Clinical charting is the documentation of clinical events that occurred during the course of a patient's care. One of the objectives of the CIS is to facilitate electronic documentation of these events (Vegoda and Dyro, 1986). CIS databases usually come pre-loaded with most commonly used data items and document templates. Clinical documentation templates are time-based charts and visit-based forms with clinical data and are used by the clinical practice for patient care. These templates are readily usable or site-customizable with the integrated configuration tool. The configuration tool will enable the clinician to create new, site-specific data items and document templates, and also to modify existing templates in the database. A customization tool that is user-friendly and -intuitive will assist in the automation of current charting practices and, at the same time, will enable the user to take full advantage of the power of the computer system. For example, charting on a data item such as infant birth weight is essential for the mother-baby practice. Once the infant weight has been charted, conversion algorithms can be created to run in the background to present infant birth weight in different units of measure.

Clinical information systems should be able to add value to the clinical practice and to the health care enterprise. Reduction in documentation time as a result of automation can enable the clinicians to devote more time for patient care at the bedside (Subramanian et al., 1987). This could contribute to the improvement in patient satisfaction. Electronic charting can improve legibility and recoverability of the data, thus, greatly benefiting information management and reporting for compliance and regulatory needs. Most importantly, CIS can contribute to improvements in patient care, simply because of the availability of complete patient clinical information at all times. For CIS implementations to be successful, they also should be complemented by process redesign. The medication administration record (MAR) and physician order entry (POE) transactions, when computerized, are examples of process redesign that are necessary in order for an EPR implementation to be successful in the health care enterprise.

Medication Administration Record

The MAR is one of the most important tools used by clinicians in the health care setting. It is an abbreviated story of the patient's progress, struggles, and current condition that looks like a table with 31 columns, one for each day of the month and rows that are marked when medication is given (Institute for Safe Medication Practices, 2000). This flow gives all of the important information to safely give medicine. The MAR includes the patient name, name and type of medicine, dosage to be given, time given, name of the ordering physician, and the name of the person or persons dispensing the medicine.

Most vendors offering a clinical information system have an MAR integrated into their product. The very nature of automation enables the clinician to be able to update, review, and modify entries on the MAR from multiple locations and, in doing so, is not hampered by illegible handwriting or nonstandard abbreviations for medications. With a POE module in place, all of the transactions associated with medication orders can be tightly coupled, thus resulting in the reduction of medication errors. Integrating the pharmacy information system into this flow will complete the chain in the medication order-processing transaction.

Computerized Physician Order Entry System

The POE module can be an integral part of the HIS or a separate application that interfaces with the HIS for admission, discharge and Transfer (ADT) and billing information. The product that is most sought after today by health care enterprises in the United States is the Computerized Physician Order Entry (CPOE) for inpatients. This is due to the increase in medication errors reported by health care providers and also due to regulations that are calling for a reduction in medication errors (Kohn et al., 2000). To date, few hospitals have a fully implemented and functional CPOE in place.

CPOE is clinical software designed primarily to automate the physician-ordering process wherein the physician will now create patient orders electronically and no longer use paper (Leap Frog Group, 2003). The main focus in the industry today is implementing an electronic order entry system for inpatients. These systems allow the creation and maintenance of pre-configured physician order sets that are designed to speed up the ordering process. There are 8 to 10 vendors in today's market who offer CPOE solutions. It is important for a health care enterprise to choose the appropriate vendor solution because CPOE is at the beginning of the EPR implementation process. It is critical to ensure that the CPOE can be integrated with other EPR modules, such as clinical documentation, pharmacy, and laboratory. CPOE solutions today offer real-time decision support tools such as appropriateness of the order, ordering process, drug interaction and contra-indications prompts, advisory messages on limits, patient clinical and demographic information, dose calculators, industry standard drug reference database, and the ability to perform edits in real time. The decision making tool is a rules-based engine that can be activated, built on, and modified by the health care enterprise. In operation, the CPOE module is linked with other applications and databases that provide the necessary patient information.

Interfaces

Clinical information systems can no longer survive as dedicated departmental systems with their databases isolated from the rest of the health care enterprise. It has become increasingly necessary for these systems to be able to interface to ancillaries such as the laboratory and pharmacy systems and to the hospital information system. If the enterprise is planning on implementing a clinical data repository (CDR) and an electronic patient record (EPR), it is imperative that the CIS be able to share its data with the CDR and the EPR.

Clinical information systems should support industry standard interface architectures for information interchange with bedside monitoring systems, the hospital information system, and ancillary information systems. A patient medical record is a repository of data originating from disparate systems, and the main objective of an interface is to facilitate the sharing and exchange of data electronically between these disparate systems. For example, patient census (registration) from the HIS, lab results from the laboratory information system (LIS), medication orders from the pharmacy system, and vital signs data from the beside systems, such as the physiologic monitors, ventilators, and infusion pumps, are components of the EPR and are acquired by separate systems. Bedside system interfaces are different from the interfaces to hospital systems and, as such, merit a discussion of their own that is included under the physiologic monitoring system section of this chapter.

Information exchange with HIS and other ancillaries is normally achieved using the industry standard Health Level 7 (HL7) protocol (Health Level Inc., 1998). The HL7 communication protocol should be supported by the CIS interface architecture for information interchange with other systems in the health care enterprise. The HL7 interface uses layer 7 of the OSI (Kabachinski, 2003) model for information interchange, and an interface engine typically will serve as the translator for messages flowing between systems. The interface engine can reside on a separate server using third-party vendor software or can be an integral part of the HIS tightly coupled with other application modules. The CIS interface architecture should support, at a minimum, unidirectional ADT interface and bidirectional clinical orders and results (CORE) interfaces.

During registration, patient demographical and financial information, such as name, medical record number, encounter number for the visit, address, blood type, allergies, insurance eligibility, and referring physician are captured by the HIS. This HIS information is communicated electronically in real-time or batch mode to recipient systems such as the CIS by the ADT interface. The interface eliminates the need for this data to be re-entered as the patient moves across the enterprise. Physician orders for laboratory tests and medications are communicated to ancillary systems using an HL7 orders interface, which enables the clinicians to manage all patient orders electronically. Order information, which is also available to nursing, pharmacy, laboratory and other ancillary services, can populate other clinical tools such as MAR. Studies (e.g., Bates et al., 1999) have shown that there are significant reductions in medication order transcription errors as a result of using an electronic orders interface. A bidirectional-results interface based on the HL7 protocol will enable recipient systems like the CIS to receive results information, such as lab values, from the ancillary systems electronically. Finally, the CIS can send patient charges to the financial system using a billing interface to the HIS. All of the information received from the interfaces become part of the patient's medical record and can be shared with the EPR and the CDR.

Clinical information systems can reside anywhere on the capability curve depending on the functionality offered by the system. Although, applications such as pharmacy and laboratory have been implemented historically on dedicated systems, most vendors are moving away from that model. Vendors are beginning to offer a suite of applications, including laboratory and pharmacy, as part of their enterprise solution for an EPR.

Clinical Workstations

Clinical workstations are the interfaces that allow the users to interact with the system. They must be simple and user-friendly in order be accepted and valued by the clinical staff. Using a personal computer (PC) for clinical transactions can be economical and sensible. The health care enterprise can manage the purchase, upgrade, and maintenance of the hardware for a much lower cost, as compared to deploying dedicated clinical workstations. At the user level, there is a definite improvement in acceptance, simply because of the popularity of PCs and the familiarity of the interface. Vendor applications support most popular PCs that are available on the market today.

Bedside Devices

Bedside devices include physiologic monitors, ventilators, infusion pumps, and others that acquire, display, and manage patient vital signs data in real time. All clinical information systems should support interfaces that are capable of acquiring real-time patient data from bedside interfaces because patient vital signs are an integral part of the patient's medical record (Subramanian et al., 1988a; Subramanian et al., 1988b, c). A more in-depth discussion of their capabilities and architecture is presented in the following physiologic monitoring systems section.

Physiologic Monitoring Systems

The process of monitoring a patient at the bedside and/or a central location can be defined broadly as the method of tracking continuously, in real time, the vital signs of a patient. In today's clinical environment, this is accomplished using a variety of bedside technologies, the most common and sophisticated among them being the physiologic monitor. The term "physiologic monitor," having been through many adaptations reflective of the device capability and functionality, is referred to as the "physiologic monitoring system" in today's world. As will be seen later in this chapter, these devices have grown so much over the years in sophistication, functionality, and intelligence that they have come to be known as "systems," as synonymous with "information systems." This is further proof that there is a convergence of what used to be traditionally a clinical engineering sphere of expertise, and what has come to be called "clinical information systems." The real-time aspect of a monitoring system is the only thing that sets it apart from an information system.

Some of the commonly monitored parameters include electrocardiogram (EKG); arterial and venous blood pressures (invasive and noninvasive); oxygen saturation using pulse oximetry; respiration; temperature; cardiac output; and, on occasion, parameters such as intracranial pressure. The age of the monitored patient population can range from a newborn to an adult, and the acuity of the patient population can vary from critically ill to ambulatory. Clinical settings can include tertiary care hospitals (Poppers and Dyro, 1985), ambulatory care clinics, community health facilities, nursing homes, physicians' offices, and ambulances.

The physiologic monitoring system (PMS) technology has successfully integrated a plethora of device functions into a single powerful bedside clinical tool. The obvious and measurable benefits are in real estate, single point of critical patient information, integrated alarm management, and unified user interface. Due to the single point of connection, patient cable management is simpler. Although there is the possibility of a single point of failure, resulting in the loss of all functionality, there are definite advantages in areas of maintenance and parts and standardization. As there is a single point for data acquisition, communicating this data to other systems in the enterprise is less complicated.

PMS Architecture

PMS architecture can be classified in many ways. A monitor can be networked (wireless or coax cabled) or stand-alone, hardwired or telemetry, portable or fixed bedside, configured or user-configurable. One example of the state of the art in bedside monitoring technology is the Solar 8000M (Figure 98-2) product from GE Medical Systems.

Physiologic monitoring systems today are highly software-driven and, as a result, have become scalable to the type of care, type of patient, and financial constraints of the health care enterprise. At the same time, the hardware technology is moving rapidly toward the PC model with built-in connectivity features aimed at providing ease of integration within the health care enterprise. As health care institutions move toward a CDR and an EPR, bedside systems will be required to coexist and communicate with other information systems in the environment because clinical data acquired at the bedside will become part of the CDR and the EPR.

The monitoring systems should support a scalable architecture to match patient acuity across the continuum of care from the emergency department through the operating room and the intensive care unit to step down units (Poppers et al., 1986). These systems should be configurable to monitor neonatal, pediatric, and adult patient population. At the same time, they should be capable of meeting the needs of different patient care services such as cardiology, medicine, maternal child, mental health, and surgery, as well as dynamic units such as the emergency department, step down telemetry units, and specialty care services such as endoscopy and neurology.

PMS Components

The PMS comprises of the front-end data acquisition and processing subsystems, the user interface with the display and the output devices, and the back-end data and network interface subsystems. Under certain configuration, other components such as slave displays, tickers, and alarm processing units can be part of the PMS.

Vendors make every effort to include standards in their system architecture. Inclusion of standards at all levels of product design directly impacts scalability and post-implementation support, as well as the ability of the monitors to interface to a CIS. Modular design concepts are used to isolate functional blocks wherever possible. This can improve system up times significantly, as it will be easier and quicker to identify and replace failed modules in the field. Another important strategy is the use of off-the-shelf components and modules in the overall architecture. This reduces cost and vendor dependencies for parts and specialized peripherals. As a result of the adaptation of current standards and practices from the computer industry, bedside monitors have become intelligent data-processing and communicating systems—a far cry from the dedicated medical devices of just a few years ago.

Data Acquisition and Processing

The front-end data-acquisition and data-processing subsystems acquire and process the signals received from the transducers attached to the patient. The processed information, consisting of real-time analog traces and associated numerical values of the monitored parameters, are communicated to the display subsystem. Signal- and power-isolation functions (also part of the front-end subsystem) provide a high degree of isolation between the patient and power sources. The state of the art in the data-acquisition and processing subsystems (Figure 98-3) is the TRAM multi-parameter module from GE Medical Systems.

The front-end data acquisition and processing subsystems are a combination of analog and digital electronics and firmware written in machine and high-level languages. Significant benefits have been realized in real estate and power utilized by the hardware due to advances in semiconductor technology and digital signal processing. Very-large-

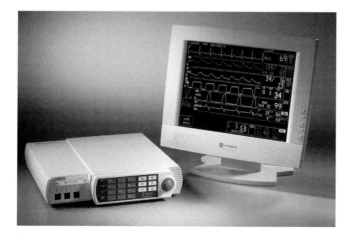

Figure 98-2 Solar 8000M Modular Patient Monitor from GE Medical Systems Information Technology.

Figure 98-3 TRAM multiparameter module from GE Medical Systems Information Technologies.

scale integration has brought about a reduction in the number of discrete functional components and in the energy used by them for their operation. In today's PMS architecture, data acquisition and processing modules (largely hardware-driven in the past) are a combination of digital signal processing components and high-level software. As a result, systems are customizable to match the changing needs of the environment without compromising the functionality. Functional consolidation and technology integration, in addition to scalability, have also made the monitors smaller and modular. It is now much easier to install smaller monitors in tight and cramped spaces in hospital-patient rooms and nurse stations.

User Interfaces

Physiological Monitoring Systems support a variety of user interfaces that enable the clinicians to perform interactive tasks such as managing (e.g., admitting, transferring, and discharging) patients, changing alarm limits, configuring printers, storing and retrieving parameter trends, and monitoring remote patients. The interface should be intuitive and easy to use and should be consistent across all monitoring products used in the environment. The user interface, hardwired and wireless, should support the input of alphanumeric data and should serve as a navigation tool for selecting on-screen functions.

The acceptance and effectiveness of a specific interface technology depends on user education and preference. Most vendors offer a variety of user interfaces, such as standard keyboards, abbreviated keypads and trim knobs, front-panel buttons for the most commonly used functions, and wireless keyboards and keypads, and they try to maintain the same interface across all of the different models of their product line. The standard keyboard is most convenient for performing patient management functions because this process involves typing lengthy fields such as the patient's name and the medicalrecord number. Obviously, keyboards take up the most space. Most care units use a combination of a standard keyboard and the trim knob, touch screen, or keypad for other routine transactions. Some vendors also offer wireless interfaces that are not hardwired to the monitor.

Display

The PMS architecture should be flexible when it comes to the display. The user should be able to choose the size (i.e., small, medium, or large), the technology (i.e., flat-panel or cathode ray tube), and the configuration (i.e., integrated or discrete) of the display that best meets the need of the environment. Trace resolution and trace separation are functions of the quality and size of the display, respectively. Larger monitors, which are capable of displaying color traces and bigger character fonts, are easy to read and more visible from a distance. Flat-panel displays are space savers and would be ideal for small and cramped locations. Finally, the architecture should support the use of off-the-shelf displays.

Output/Print

The PMS should support dedicated digital writers and laser printers. The user should have the flexibility to print alarm events, trend information, and on-demand traces on both printers. Most clinical printing will be done using the dedicated writers that are calibrated and scaled to provide true recordings of all the clinical traces. A laser printer, while providing a report-quality print out of clinical events, can also be shared by other applications when installed on the network.

Integration

A PMS vendor that does not offer network connectivity as a product feature is unheard of in the industry today. Distributed health care delivery systems demand the economies, functionality, flexibility, and interoperability that are achievable using standards for improving and maintaining the quality and competitiveness of their clinical practice and also for meeting their business objectives. A robust communication network is essential and critical for achieving these objectives. The PMS and other information systems in the enterprise are components on this network and, consequently, are required to adhere to industry standards and practices in implementing their architecture. Because physiologic monitors manage critical clinical data, loss of data due to collisions is unacceptable. A typical monitoring system network runs at 10 megabits/second with the network utilization ranging from 25%–30% of the total bandwidth. It is also not uncommon for the monitoring system to reside on a private network, totally isolated from the rest of the enterprise. The network security in this case is virtually unbreakable because the addresses are private and not accessible from outside the network. At the monitor level, all of the configuration functions are layered and are accessible only with a password. All of the functions required for routine management of the patient are at the user level and can be accessed by all of the clinical staff.

Interface

Bedside medical devices connect to the CIS using any one of three distinct configurations. The capability of the bedside system is the determining factor when it comes to selecting the configuration. One way is for these devices to connect directly to the CIS network. High-end devices like physiologic monitors can do this using their built in network adapter and a unique network address. Low-end devices, such as infusion pumps, do not have this capability and require a host device to facilitate the network connection. Typically, physiologic monitors reside on a private network because of their up time and bandwidth requirements, while CIS servers and workstations are normally found on the enterprise network. As a result, although the monitors are network-ready for this configuration, additional dedicated hardware would be required to interface the monitor data to the CIS servers. This might not be desirable and most likely would not be cost-effective. In addition, transferring data across two separate networks poses real challenges when it comes to management.

A second model uses an intelligent multiport device to connect bedside devices to the CIS network. The multiport device is a node on the CIS network with a unique address. This commonly used configuration provides, at every bedside, the ability to connect multiple devices to the CIS network. This architecture also supports the integration of devices at all levels of sophistication.

A third way is for bedside devices, without network-interface adapters, to use a physiologic monitor to connect to the CIS network. The physiologic monitor, in this case, serves as a node on the CIS network using the built-in network adapter and the unique network address. This configuration requires the host device to handle network traffic while continuing to provide real-time patient information at the bedside. As a result, the primary function of the host device can be compromised, and consequently this configuration is less frequently used in connecting to the CIS network.

PMS Configurations

Hardwire

Most patient care units have bedside and central station monitoring capability. In the hardwire configuration, every bedside has a monitor that could be in one of the following modes or a combination of modes: Networked or stand-alone, configured or user-configurable, and portable or station-specific.

All critical care units are typically equipped with monitors at every bedside and at the central nurse station. The bedside monitors and the nurse station's monitor(s) are on a network and enable the clinical staff to manage patient care from both locations. The bedside monitor consists of a processing unit that is a node on the local area network; a color display that is capable of displaying six to eight real-time traces; a data acquisition and processing module with memory that is portable and patient-centric; a user interface such as a trim knob, keyboard or a touch pad; and a dedicated digital writer. The bedside monitors and the central stations form the local area network. These monitors typically are high-end devices that are capable of monitoring all of the vital signs simultaneously in real time with the ability to add or drop parameters to meet the changing acuity of the monitored patient population.

The data acquisition and processing module provides the link to the patient via patient cables and typically plugs into a base unit that has AC power and a link to the main processing unit. The beauty of the acquisition module is its ability to move with the patient while continuing to acquire, process, and display real-time data. This ensures continuous monitoring of patient vital signs during transport, without the clinician having to go through the cumbersome and time-consuming task of disconnecting and reconnecting patient cables for the transport. The acquisition module also saves the zero and alarm settings for parameters such as blood pressure from the bedside and eliminates the need for recalibration during transport and at the new location. The module continues to save patient alarm history during transport and at the new location. Thus, transporting patients between care units that are similarly equipped is quick, easy, and clinically beneficial. All that it takes at the destination location is for the clinician to unplug the module from the transport unit and to plug it into the base unit at the bedside. The storage capacity of the acquisition module is sufficiently large and can store a significant number of events and data trends for prospective review and print.

The enterprise can standardize all of the critical care patient rooms with the high-end monitors. As previously mentioned, in that case, moving patients between these units will be seamless. Their network capability will allow the clinician to view patient alarms and vital signs remotely and, when necessary, print trends and alarm histories. Alarm management in terms of setting limits and acknowledging alarms can be accomplished only locally at the bedside, although temporary alarm silencing can be accomplished from the remote location. The software allows the care unit and the clinical staff to configure the alarm capabilities of the monitor in one of two ways. All of the beds could be configured to the same default alarm strategy (e.g., parameter limits, acuity, audible and visual), based on the historical patient population admitted to the care unit. In this model, alarm settings are changed only for the specific patient, and the monitor will return to default settings after the patient is discharged. For the clinical engineering department, it is quite simple to document one set of default settings for the care unit and to use these settings when replacing a broken monitor. For the nurse in the unit, it is easy to customize the specific bedside alarm settings on demand and not have to worry about keeping track of all of the original settings. Institutions that use this model typically store a copy of the care unit's default alarm settings at the nurse station and also in the clinical engineering department. One alternate method is to tailor each bedside to meet the acuity of the patient. This option is meaningful and beneficial for a floor with multiple care units. Obviously, it is tedious and time consuming and can contribute to potential clinical issues due to improper alarm settings. This model works well in small care units that have designated beds.

Configurability also plays a major role in designing the visual area of the monitor. This is the ability to decide the number of traces displayed by the monitor in real time, the order in which they are displayed, and their colors. This functionality, as well as the alarm configuration, is typically buried below the user level in what is termed the "service mode." The service mode is password-protected and can be accessed only by approved staff. Color designation to specific parameters can be standardized across the enterprise that can greatly improve both the visual comfort of the users and also the clinical separation of the different traces. The color option becomes clinically significant when monitoring parameters such as arterial and venous blood pressures.

Trace size, speed, color, and location on the display are customizable. The user sees, in real time, six to seven seconds of the real-time trace moving from right to left, complemented by numerical values and alarm limits for the monitored parameters. Alarm messages are annunciated audibly and visibly at the patient's bedside and also at the nurse station monitor. The alarm event also triggers the printing of 15 to 20 seconds of the trace, including six seconds of pre- and post-event data. One other feature that is useful clinically is the "view remote unit/bed" function. Because these monitors are on a network, they can be programmed to display traces from a remote bed in addition to the local bed. The local monitor display is on a split-screen mode, providing three to four seconds of real-time trace for the two beds. The remote view capability also can be extended to include the enterprise, allowing a physician to provide consultations for patients without having to go to the bedside.

The bedside monitors can also be configured as a cluster consisting of beds with patients with similar acuity or patients managed by a single nurse. The monitors in the cluster can be configured to share alarm messages with each other. This feature provides the clinician with the flexibility to make decisions with regard to responding to the alarms and can be clinically significant in areas where patients are spread out and also in areas where the patient-to-nurse ratio is high.

Ethernet and TCP/IP protocol are used for messaging by the local monitoring system network. The monitoring system network, running at a 10 megabits/second data transfer rate, uses CAT5 (Kabachinski, 2003) cabling and multiport hubs and switches for establishing connectivity between the bedside and the central station. For performance and response time optimization purposes, the local network architecture is designed to keep traffic within the subnet. Fiber-optic links provide high-speed (100 megabits/second) connectivity between the local network and other care units in the enterprise that are also on the private monitoring system network. Because the monitors reside on a private network, traffic generated by the monitoring system has no impact on the performance of the enterprise backbone. The high-end monitors, when used in a stand-alone configuration, will have no network connectivity to other bedsides or the nurse stations.

Hardwire monitors that are preconfigured with factory-set parameters are well suited for low-acuity care units. Preconfigured monitors are typically single, integrated units consisting of the processor, display, and patient and network interfaces. They are less expensive, as compared to the configurable monitors, and they can be deployed in stand-alone and network modes. With no central station monitoring capability, the stand-alone mode is suitable for care units that have a high nurse-to-patient ratio. In the networked mode, these monitors find wider use where the central-station monitors are capable of displaying vital signs for all of the patients in the unit.

Preconfigured and configurable monitors can reside on the same local monitoring network. Unlike the high-end monitors, the preconfigured monitors do not have the detachable acquisition module that can go with the patient during transport. As a result, during transport, the clinician will have to take the preconfigured monitor from the bedside or use a separate transport monitor for monitoring patient vital signs. Using the preconfigured monitor for transport will necessitate bringing in a new monitor to the bedside. On the other hand, using a transport monitor will require the clinician to swap patient cables and zero transducers before the transport and after the patient is returned to the bedside. There will be a break in patient monitoring during transport using either monitoring modes. In this case, if ever there were a need for retrospective review of the data, the history of events that occurred during transport will not be a part of the continuous record.

Another major advancement in monitoring technology has been the integration of 12-lead ECG measurement function. With the latest physiologic monitors, the clinician can initiate and complete a 12-lead ECG measurement at the bedside, like all other vital signs for the patient. In the past, this function was managed by the EKG department in cardiology and required dedicated EKG technicians and specialized equipment such as the EKG carts. Because EKG carts used a unique set of patient cables and leads, it was necessary to change patient cables and leads whenever a 12-lead ECG was performed on a patient at the bedside. The physiologic monitors, with 12-lead ECG capability, have enabled health care institutions to decentralize the EKG departments and to reduce their staff because nurses and residents are now able to perform this task at the bedside. By standardizing cable, leads, and bedside monitor software version, hospitals can ensure 12-lead ECG capability across the enterprise.

Telemetry

As the phrase suggests, in the telemetry monitoring mode, there are no monitors at the bedside, and patients are managed from the nurse's station using central monitors. Telemetry monitoring is commonly used to manage less-critical patients who are able to (and sometimes required to) ambulate. Current telemetry systems are capable of monitoring in real time, 5-lead ECG, noninvasive blood pressure and oxygen saturation. One example of the state of the art in telemetry monitoring (Figure 98-4) is the Apex Pro Telemetry System product from GE Medical Systems.

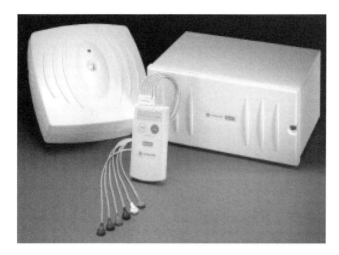

Figure 98-4 Apex Pro Telemetry System from GE Medical Systems Information Technologies.

A telemetry monitoring system consists of transmitters, antenna subsystems, receiver cabinets and central station monitors. The transmitter, attached to the patient, sends real-time digitized vital signs data over a radio-frequency (RF) carrier signal. A network of antenna subsystems, installed at strategic locations in the care unit for maximum coverage, communicates the digitized data to a set of receivers for processing. The central monitor located in the nurse station displays and trends the real time information. All patient transactions, such as admitting, transferring, discharging, alarm configuration, and printing are performed and managed at the central station. The patient also can initiate a print transaction from the transmitter.

The central monitors store alarm events and parameter trends. They can be complemented by dynamic ticker displays and slave monitors placed strategically around the unit. These complementary displays can provide up-to-date patient status information to the clinicians when they are away from the nurse station. The slave monitors display the same information as the central station.

The telemetry configuration requires dedicated staff at the nurse station to monitor the patients and to alert the clinicians when there is a change in patient status. Clinical staff in the telemetry unit will use the hospital paging system to alert the clinicians. Step-down care units typically have a high patient-to-nurse ratio and also a high patient turnover every day. The acuity of the patients in these units, in many cases, calls only for telemetry monitoring. It is also likely that patients from multiple-care units located on different floors are monitored centrally from one nurse station. The total number of patients managed at these nurse stations can be upwards of 40. The biggest challenge in this environment of high patient turnover and inadequate staffing is to keep the clinicians informed of the changing status of their patients.

Fast and reliable methods to communicate alerts to the caregivers are available in the market today that communicate alarms from a monitoring station to the alphanumeric pagers carried by clinicians. These systems reside on the same network as the telemetry receivers and the central monitors, receiving patient alarm information in real time, just like the central monitors. They are programmable to send alarm messages to one or more pagers, depending on the severity of the alarm and the responsibilities of the clinician. The alarm messages include the patient's name and room, as well as the date, time, and type of the alarm. The system can be programmed to send alarm messages from selected patients to a designated pager and a backup pager. Management reports generated by these systems can provide data on frequency and response times for different alarm categories that can be utilized by the institution for implementing changes in clinical practice and adjusting staffing levels in the unit. Note that these alarm communication systems were designed and sold only as adjuncts to the traditional monitoring systems and that they cannot replace the role of dedicated and knowledgeable clinical staff.

The telemetry receiver cabinets and central monitors can reside on the same network with hardwire monitors. Because bedsides can be configured for both telemetry and hardwire monitoring, the clinician can tailor monitoring system capability to the acuity of the patient. Patients on telemetry and hardwire monitoring systems can be viewed on the same central station monitor, and managed from one nurse station. This flexibility significantly reduces the need for moving the patient within the unit and greatly benefits bed-management operation. Because both the telemetry and hardwire systems use the same patient interfaces (e.g., cables and leads), patient discomfort is also greatly reduced when switching between the two systems. Intermediate care units, with varying patient acuities, are usually equipped with hardwire monitors and an antenna network for the telemetry system for this very reason.

Wireless

A third mode that uses wireless communication technology finds use in certain areas of health care enterprise. The wireless configuration lends itself to non-station specific locations such as hallways, bathrooms, waiting rooms, and other overflow areas that otherwise would not be considered monitored locations. A highly fluctuating patient census in a hospital may require that monitoring capabilities be available at these locations. Furthermore, staffing and environmental constraints may require these locations to be monitored centrally as well. The hardwire and telemetry monitoring modes require the installation of permanent connections or an antenna network at these locations, a significant overhead cost for the enterprise because there is no certainty of ever using these locations for patient monitoring. In addition, there is no real way to estimate what would be a sufficient number and location for overflow beds and, as such, in the case of a disaster, there still could be a shortage of monitored locations. The wireless architecture makes the monitoring devices non-station-specific and truly enables the clinicians to bring the monitor to the patient.

Spread spectrum and cell phone technologies are used in implementing the wireless architecture. Every patient monitor and central station is equipped with the transmitter/receiver pair that is capable of sending and receiving information from all devices that are part of the cluster. This configuration does not require connection locations, and, as such, it can be used to monitor patients anywhere in the unit, including the hallways, waiting rooms, and even bathrooms. In the past, the wireless data channel was able to monitor only a limited number of parameters, due to bandwidth limitations. That is no longer a limitation, as these monitors are capable of handling the full range of patient acuity and can be deployed in most care units in the health care enterprise.

Most hospitals use a combination of monitoring modes to accommodate the clinical, as well as financial, needs of their practice. It is imperative that the different monitoring modes and technologies coexist in the environment and also communicate seamlessly with each other.

Other Monitors

Transport Monitoring Transport monitors are compact, portable monitors with limited capabilities, as compared to bedside monitors. The state of the art in this area is small, lightweight displays that are capable of accepting the data acquisition modules from the bedside monitors and displaying up to four real-time traces. Patient transfer is seamless because there is no need to recalibrate transducers and no need to disconnect and reconnect patient cables to get the patients ready for transport. One example of the state of the art in transport monitoring (Figure 98-5) is the DASH 4000 product from GE Medical Systems.

The transport monitors run on batteries and manage alarms in the same way as the bedside monitors. In effect, the transport monitors can serve as bedside monitors when there is a need, and they provide clinically acceptable levels of monitoring and care. Typical battery life is an hour, and under certain configurations the battery life can be increased with the addition of a second battery. All care units that use the transport monitors should have a charger base(s) to plug into, to charge the batteries.

Central Station Monitors Central station monitors are capable of displaying data from multiple patients at a central location such as a nurse station. For this to happen, the bedside monitors and the central station monitors should be on a network. Each central monitor can display two-real time traces per patient for up to eight patients continuously. Patient transaction, such as admitting, discharging, and alarm acknowledgement can be performed at the central station. In fact, in a telemetry unit, all patient transactions are done at the central station because there is no user interface at the bedside. The central station can display all patient alarms and can save and print user-selected patient events. An example of the state of the art in central station monitoring is the CIC Pro Clinical Information Center product from GE Medical Systems.

Windows NT-based workstations that are capable of running multiple applications are the state of the art in central station technology. These stations are capable of launching applications that run in the background without compromising their primary function as real-time patient monitors. Consolidating applications will result in the reduction of hardware clutter in clinical areas and, at the same time, translate to cost savings for the hospital in replacement hardware and technical support. With the help of a couple of key strokes, users will be able to navigate between the real-time patient vital signs information, a 24-hour, full-disclosure software and a HIS terminal emulation session to review patient census and laboratory results. These Windows NT stations support industry-standard laser-jet printers that can be shared between applications.

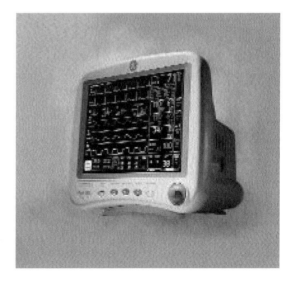

Figure 98-5 DASH 4000 Transport Monitor from GE Medical Systems Information Technologies.

PMS Enterprise Integration

Integration of bedside monitoring systems with the rest of the health care enterprise is rapidly gaining in importance. Patient demographics (registration) such as name, age, and sex are being brought to the bedside monitors from the HIS using industry standard HL7 interfaces. Similarly, 12-lead EKGs performed at the bedside monitor are being transmitted electronically to the cardiology information management system. Once the EKG is read and confirmed by a physician, a charge for the service can be sent to the hospital financial system to generate a patient bill. In both cases above, a working interface will be required that is available in today's market.

There is also the convergence of functionality between traditional noninvasive cardiology functions, such as Holter recording and the full-disclosure feature of the monitoring system. In the full-disclosure mode, a dedicated workstation on the monitoring system network acquires, stores, displays, and prints beat-by-beat vital signs data. These data can be retrieved within the 24-hour window for retrospective analysis of events. Holter recording, a modality where patient EKG is acquired and stored over a period of time for retrospective analysis and reporting, is accomplished using the 24-hour full-disclosure monitoring modality of the physiologic monitoring system.

The architecture of the monitoring system lends itself to better management of patient parameter alarms. Alarm messages are more commonly managed from a central location and, in some cases, such as the intensive care units, are managed at the bedside. In addition, care units that use telemetry to monitor their patients use other remote tools, such as pagers, tickers, and slave displays. Hospitals have also started using PCs for alarm management. A PC, running dedicated software, can be a node on the monitoring system network communicating critical patient alarms to alphanumeric pagers carried by clinicians. The message sent to the pager contains information on patient identity and location and also on the acuity of the alarm. As a result, clinicians can make decisions on responding to the alarm without having to rush to the patient's bedside. This tool is useful and beneficial to care units that are short-staffed and also to units that use telemetry for patient monitoring. The alarm histories and response times tracked by the computer can be used by management for quality assurance and process improvement objectives.

Physiologic Monitors and Clinical Information Systems

Previous generation monitors used proprietary connectivity architecture and as such, extracting data from them by external systems was difficult (Subramanian et al., 1988a). The design focused mainly on pushing data from the bedside for viewing at the central station, and no standards were in place, either for the network architecture or the data-communication interfaces. Over the years, the function of a monitor has changed from a display unit to that of an integrated information management tool. Vendors have been enhancing the capabilities of the monitors to meet the demands of clinical environment to the point where they have become powerful bedside computers ready to communicate with other computers in the enterprise. To some extent, this evolution has also been influenced by the demand for bedside data from regulatory and benchmarking agencies. Physiologic monitoring system vendors have risen to meet this challenge by integrating information technology standards in their architecture. In the health care environment today, data acquired at the bedside, such as the 12-lead ECG, are also impacting the financial status of the health care institution. It is all the more important for the bedside monitors to be able to integrate and interface with other systems in the enterprise for sharing and exchanging of clinical information.

Most clinical information systems are able to interface to bedside monitors to receive vital signs data in real time, and also to track clinical trends. They also interface to the HIS to create a patient medical record with important identifiers, such as name, medical-record number, address, and age, and the vital signs data acquired from the bedside are linked to this patient record. In this model, bedside devices are identified by their physical location, such as room number, and the data association with the medical record is achieved when the patient is admitted to that location. This is functional and workable only if the bedside devices are on the same network as the CIS. In the case of physiologic monitors residing on private networks due to bandwidth requirement and up-time guarantees, HIS patient-identifier data can be routed to the monitors using dedicated gateways. The monitors themselves should have the software capability to receive and populate these data fields in order for this to happen. The state of the art in bedside monitor software is still evolving in the area of field edits for patient demographic information and in interfacing to information systems for demographics data, order processing, results reporting, and billing. The need for this enhanced capability at the bedside is being driven by clinical and financial pressures, and vendors are beginning to address it by incorporating features in the monitors to meet specific applications.

One such application is the acquisition of 12-lead EKG using the bedside monitor. The EKG carts receive the physician's order and patient-demographics information from the HIS, and charges are dropped on the HIS financial systems after EKGs are read and confirmed. As discussed earlier in this chapter, bedside monitors are also capable of transmitting 12-lead EKGs to a cardiovascular information system (CVIS) for interpretation and confirmation. Most bedside monitors today allow users to enter demographic information (e.g., patient name, age, medical record number, care-unit name, and bed location) using the "admit" function. Current software versions for the monitors do not qualify this data as mandatory or required fields by providing screen edits, thus permitting the clinical staff to skip fields or even to enter incorrect information. As a result, EKGs are transmitted from the monitor to the CVIS with missing or incorrect patient identifiers. Patient EKGs without proper identifiers are not billable and result in loss of revenue for the health care institution. There is also the serious matter of noncompliance because patient EKGs without identifiers are not usable for diagnosis and care. A second problem is the inability of the bedside monitor to store real-time 12 EKG traces. Monitors display real-time parameter waveforms continuously, including six to eight seconds of EKGs, and only store alarm events and parameter trends for retrospective review and print. Because 12-lead EKG waveforms are not stored by the bedside monitor, the decision to transmit them to the CVIS must be made by the clinician in real time. If subsequent EKG traces are cleaner and clinically significant, the only option for the clinician is to transmit these traces also to the CVIS. Because every 12-lead EKG requires an order from the physician, this results in multiple EKG traces for one order. Given the volume of 12-lead EKGs done on any given day by a large institution, it is nearly impossible to manually match up the appropriate EKG trace with the order. Unresolved EKGs are not billed because they are not read and confirmed, and they are also not available for clinical use. The last problem is the inability of the interface gateways to queue EKGs when there is a loss in connectivity between the monitoring system network and the CVIS. EKGs are communicated to the CVIS from the bedside monitors by dedicated workstation(s) acting as gateway(s). In most health care enterprises, the CVIS resides on the hospital network, and the gateways provide the connectivity between the private monitoring network and the hospital network. When the gateways go down or lose connectivity to the CVIS, 12-lead EKGs from the bedside are not transmitted to the CVIS. These EKGs are not recoverable even after the link is re-established between the CVIS and the gateways. Gateway software with buffering capability complemented by a redundant system can eliminate this problem.

Vendors have been slow to come up with a solution to these issues. Gateways, with upgraded software, are now capable of storing EKGs over a period of days when there is a loss in connectivity with the CVIS. A typical 12-lead EKG file size is 5 kilobytes, and the gateways are able to store a significant number of EKG waveforms while waiting for connectivity to be re-established with the CVIS. Bedside monitors can now acquire patient demographics information electronically from the CVIS using the "admit" function. Because the CVIS already acquires the ADT information from the HIS for order processing and results reporting, the bedside monitors can be populated with the same information when a clinician requests that information using the monitor's "admit" function. Errors due to manual data entry are no longer a concern, and the likelihood of the CVIS receiving EKGs with proper patient identifiers is high.

Bedside monitors' capability has not been upgraded to store 12-lead EKG traces. This feature will provide the clinician with the ability to select from the stored EKG traces. After reviewing the stored traces, the clinician can select the appropriate trace for transmission to the CVIS. At the CVIS end, as there is only one ECG trace per physician order, reconciliation and billing from that point forward will be easy.

The orders and results interfaces between the HIS and the CVIS complete the circle as far as processing 12-lead EKG transactions are concerned. EKG orders entered on the HIS are electronically sent to the CVIS using the HL7 protocol and are electronically matched by the CVIS with the EKGs sent from the bedside monitors and the carts. After the EKGs are interpreted and confirmed, a message is sent back to the HIS, acknowledging completion of the transaction and authorizing dropping of the charges. While real-time EKG traces are stored in the CVIS, results that consist of measured and calculated parameters and interpretation text can be sent to the HIS, an EPR, and a data repository. Installing the CVIS client application at workstations located throughout the enterprise will enable clinical staff to view and print confirmed 12-lead EKGs. Because all aspects of the 12-lead EKG process are captured electronically and stored in one system or another, medical records departments are no longer be required to store paper traces or to microfilm them.

Clinical Engineering and Clinical Information Systems

Bedside devices today are data and information management systems, and not just bedside monitoring tools. To a large extent, this came about as a direct result of advances in technology. Health care institutions have come to recognize the role of bedside devices as important sources of data that are being sought after by regulatory, provider, and benchmarking agencies. As a result, clinical engineers, traditionally and historically responsible for the management of these devices, have been forced to modify their role in the enterprise. Their focus and skill sets must be retooled to meet this changing environment.

The traditional inspection, preventive maintenance and repair (IPMR) model is not necessarily the only support model, nor is it practical due to the changing technologies and changing dynamics of the health care landscape. Electrical safety is still an important consideration in the design and ongoing maintenance of these devices, although digital technology advances in hardware design have significantly reduced the risk to the patient in this area. Implementation of functions in firmware and software coupled with large-scale integration in semiconductor technology has led to reduction in space and power consumption. Medical devices have benefited from the transfer of this technology, as is evidenced by the shrinking footprint and increasing functionality of the devices. Repair and maintenance, for the most part for many of the devices, should be a matter of identifying the failed module and replacing it. Most medical devices also store error codes in real time to assist in the quick and easy troubleshooting of a failed module or device.

Medical devices have become data and information management systems that are connected to each other (and, potentially, to foreign systems). On many occasions, it will be necessary for clinical engineering departments to look beyond the immediate device to identify, isolate, and resolve problems. This effort calls for additional and, in some cases, different skill sets. It will be necessary for the clinical engineering profession to expand its role and also to hone its skill sets to meet these new challenges offered by the health care enterprise.

One easy way to meet the changing needs is to integrate the clinical engineering department with the information technology (IT) division. In this model, the traditional role of the clinical engineering staff will be preserved because other IT staff will be responsible for resolving all issues behind the device. This model severely restricts the opportunity for CE departments from foraying into cutting-edge technologies (such as network infrastructure) that could provide new career opportunities for CE professionals.

Proactive CE leadership can convince hospital administration of their readiness and willingness to meet these new challenges.

The distinction between a computer system and a medical device can become insignificant in the near future, and the possibility of the two technologies merging is high. One can envision a single workstation at the patient bedside that can pull up, on demand, patient clinical events such as alarm history and trends and that can perform clinical-information transactions such as lab orders and results, vital signs flow sheets, and others. With a single sign-on and context-switching capability, the workstation can be a powerful clinical tool. In addition, this approach greatly reduces the clutter around the bedside by freeing up space.

A great need exists for improving the skill level and overall education of the technical staff within the CE departments to be able to meet these emerging challenges. In addition to the knowledge base, CE departments will be required to find ways to participate actively in the selection, implementation, and management of these systems. Communicating and interacting with other professionals will become increasingly important. Technology will require clinical engineers to work closely with information technology professionals. Communicating effectively with the IT staff will be important.

First, CE professionals must learn about networking and computer systems. Networking fundamentals, such as operation of hubs, switches, and routers, will be useful tools in understanding how information is exchanged on the network. Basic education in network topologies and OSI standard on network layers is a requirement. After hard-device failures, the second most common problem with bedside monitors today is configuration. Frequent operational issues, such as inability to print, missing beds at the central station, inability of the bed to communicate with other beds in the unit and with the central station, and an inability to see remote beds are network- and configuration-related. Today's monitors are highly configurable to the needs of the specific patient and offer flexibility in the area of sharing resources such as printers. Having a firm understanding of software fundamentals will be useful in troubleshooting many of these problems.

The concept of IP addressing is commonly used in the monitoring network communication architecture, and it is imperative that the CE technologist understands how to use the IP function to resolve basic connectivity issues. The central stations are computers that run operating systems such as Windows NT and that are capable of running other applications in the background. The CE technician should have a working knowledge of the operating system environment in order to be able invoke certain transactions at the command level, monitor error logs, configure printers, and reallocate resources. The management of the monitoring system network, for example, will be a shared responsibility between the CE and IT departments. In most cases, the CE staff will get the first call when there is any monitoring system problem. The CE staff should be able to isolate the source of the problem within the scope of their responsibility before triaging it to the IT staff. Lastly, understanding the technology, the concepts, and the jargons will assist in troubleshooting a problem while following instructions from the vendor's technical support over the phone.

In teaching and research institutions, individual researchers will demand real-time data from monitors. CE staff should have a good understanding of the communication protocols and their configurations. Most monitors will support a serial interface that can be used to connect a computer for data acquisition. In some cases, the computer can be assigned a fixed IP address to become a node on the monitoring network, and then can be programmed to listen to communications from a specific device or devices on the network. This can be a problem if the computer uses the same IP address as another valid and live device on the network, such as a monitor or a central station. Because the monitoring network uses dedicated IP addresses, the possibility of duplicate IP address is not unlikely and staff must guarded against it. The CE who has the appropriate education and experience can support the research application without compromising the real-time patient care function of the monitor.

Changing Role of Clinical Engineering Departments

Clinical Engineers and their hospital departments have a new and important role to play when it comes to automation in the clinical environment (Bronzino, 1992). The bedside monitoring systems and other medical devices (such as infusion pumps) are important sources of patient and clinical data that populate the patient's chart, and clinical engineers are the experts within the enterprise on the management of these devices. Integration of the medical device network into the enterprise backbone for data interchange is critical, and in order to be successful there must to be a seamless and cooperative working relationship between the CE and IT departments. Here is a tremendous opportunity for CE staff to educate the IT staff with their expertise in clinical device technologies. To be accepted in the new role, CE departments must enhance their knowledge base and expertise in enterprise-technology solutions, such as system integration, communication protocols and standards, hardware, and software architectures. The trends in the medical device industry and marketplace are on a convergence path with IT, and it is in the best interest of the clinical engineering departments to get onboard at the front end of this change must prepare their staff to recognize and embrace the challenge. A proactive CE leadership will inform and educate its staff about the changing role, will provide the staff with technology based continuing education, and will offer its services in enterprise-wide projects.

Health Insurance Portability and Accountability Act

The federal government's mandate under the Health Insurance Portability and Accountability Act (HIPAA) for compliance in areas of privacy and patient confidentiality when it comes to access to, and communication of, identifiable patient information has a major impact on CE departments. In fact, CE departments will be required to support the compliance initiatives of the health care enterprise because of the medical devices that are managed by them (see Chapter 104). Similar to the Y2K situation, the number of medical devices (stand-alone and networked) vastly outnumbers the computer systems that exist in the health care environment. A number of types of these medical devices display, save, and transfer patient-identifiable data that are covered under the HIPAA requirements. CE departments will have to be proactive in classifying the medical devices by the type of data they handle and assigning levels of risk (Marchese, 2002). Obviously, the level of risk will determine the necessary corrective action for compliance, and clinical engineers will play an important role in this area.

References

Bates DW, Teich JM, Lee J, et al. The Impact of Computerized Physician Order Entry on Medication Error Prevention. *JAMA* 6:313–21, 1999.

Bronzino JD. *Management of Medical Technology: A Primer for Clinical Engineers*. Butterworth-Heinemann, 1992.

Dyro JF, Poppers PJ. Hospital Information System for Intensive Care. *Int J Clin Monit Comput* 3(1):37–38, 1986a.

Dyro JF, Poppers PJ. Improving Patient Care Through Integrated Computing. *Int J Clin Monit Comput* 3(3):217–218, 1986b.

Gage J, Poppers PJ, Dyro JF. A Hospital Information System for Anesthesia Intensive Care. *Proceedings of the AAMI 21st Annual Meeting*. Arlington, VA, Association for the Advancement of Medical Instrumentation, 1986.

Health Level 7, Inc. *Health Level 7 Implementation Support Guide for HL7 Standard version 2.3*. HL7, 1998.

Kabachinski J. Acronyms: The Essence of Modern IT Life. *Biomed Instrum Technol* 37(2):125–127, 2003.

Kohn LT, Corrigan JM, Donaldson MS. *To Err Is Human: Building a Safer Health System*. National Academy Press, 2000.

The Leap Frog Group. *Computer Physician Order Entry Fact Sheet*. The Leap Frog Group, March 2003.

Marchese, MS. Securing Patient Data in Preparation for HIPAA. *24x7*. May 2002:30–35.

Poppers PJ, Dyro JF. Physiological Monitoring in a Tertiary Care Center. In Ruegheimer E and Pasch T (eds). *Notwendiges und Nuetzliches Messen in Anaesthesie und Intensivmedizin*. Berlin, Springer-Verlag, 1985.

Poppers PJ, Dyro JF, Gage JS. Physiological Data Acquisition and Management as an Advancement of Intensive Care. In Bergmann H, Kramar H, and Steinbereithner K (eds). *Beitrage zur Anaesthesiologie und Intensivmedizin, Vol. 16*. Vienna, Verlag Wilhelm Maudrich, 1986.

Subramanian S, Davis GJ, Villez PA, et al. A Medical Device Data Acquisition System Using a Microcomputer-Based Software Interface. *Proc Tenth Ann Conf IEEE Eng Med Bio Soc*. Piscataway, NJ, IEEE, 1988.

Subramanian S, Dyro JF, Poppers PJ. Toward a Totally Integrated Hospital Information System: Infusion Pump Link to Patient Data Management System. *Proc Ninth Ann Conf IEEE Eng Med Bio Soc*. Piscataway, NJ, IEEE, 1987.

Subramanian S, Dyro JF, Poppers PJ. A Microcomputer-Based Software Interface for Automatic Acquisition of Fetal Monitoring Data. *Int J Clin Monit Comput* 5(4):247–250, 1988.

Subramanian S, Dyro JF, Poppers PJ. A PC-based Fetal Monitor Data Acquisition System. *Proc Eighth Ann Conf Comput Crit Care*. Hershey, PA, 1988c.

Vegoda PJ, Dyro JF. Implementation of an Advanced Clinical and Administrative Hospital Information System. *Int J Clin Monit Comput* 3(4):259–268, 1986.

Further Information

Abrami PF, Johnson JE (eds). *Bringing Computers to the Hospital Bedside: An Emerging Technology*. Springer Publishing Company, 1990.

Citrix Systems. Citrix Nfuse Elite Access Portal Server: The Simple, Powerful Access Portal Server That Provides Secure Aggregation of Applications and Information. Citrix Systems, 2002.

Dick RS, Steen EB. The Computer-Based Patient Record: An Essential Technology for Health Care. National Academy Press, 1991.

GE Medical Systems Information Technologies. *Apex Pro Telemetry System Specifications*. General Electric, 2000.

GE Medical Systems Information Technologies. *Centricity Enterprise Clinical Information System Specifications*. General Electric, 2002.

GE Medical Systems Information Technologies. *CIC Pro Clinical Information Center Specifications*. General Electric, 2002.

GE Medical Systems Information Technologies. *DASH 4000 Pro Patient Monitor Specifications*. General Electric, 2002.

GE Medical Systems Information Technologies. *Solar 8000M Modular Patient Monitor Specification*. General Electric, 2000.

GE Medical Systems Information Technologies. *Tram Multiparameter Modules Specifications*. General Electric, 2001.

Institute for Safe Medical Practices. MAR reference and definitions.

Advanced Diagnostics and Artificial Intelligence

Donald F. Blumberg
President, D. F. Blumberg & Associates, Inc.
Fort Washington, PA

Over the last decade, there has been a significant increase in the development, application, and use of advanced diagnostics and artificial intelligence technology in field service. Research[*] carried out in the early 1980s, in fact, showed that service problem diagnostics could result in avoidance of between 30% and 35% of all on-site field service calls, and that in-depth diagnostic evaluation could significantly reduce the number of "broken" field service calls through more intelligent dispatch and assignment of both parts and service engineers (Blumberg, 1984). Since that discovery, there has been a great deal of work carried out attempting to both develop and apply advanced diagnostics technology in the field service industry. It is, therefore, of value to examine both the *current* state-of-the-art and the experience in the application and use of problem diagnostics and resolution technology in the health equipment field service industry, as well as to pragmatically explore both the successes and failures of artificial intelligence and advanced remote diagnostics and decision support methodology in service. This analysis is summarized in this report.

An Overview of the General State-of-the-Art in Remote Diagnostics

The role of artificial intelligence (AI) and remote diagnostics in the service environment is usually part of the overall call-managing process and is based on the strategic use of information and data acquisition methods to identify, isolate, analyze and, ultimately, diagnose and evaluate a fault within a unit of equipment or system. The goal is to increase the efficient allocation and timeliness of service-oriented resources, and to raise the productivity of the service force and the up time of equipment through efficient use and deployment of service personnel and parts to support the up time objective. Improving performance in the service function relies on having the most efficient technology to find, identify, isolate, predict, and repair a fault or potential fault. This requires a service organization of such sophistication as to fully exploit the available and potential benefits of the diagnostic maintenance practice in technological depth.

In general, service productivity can be improved in four ways:

- The anticipation of potential future failures through predictive and preventive maintenance to initiate field work that avoids emergencies and unanticipated failures
- The reduction or elimination of actual, on-site, service calls through more efficient problem and fault diagnosis in response to an emergency call requests and prior to emergency call dispatch.
- The use of improved fault diagnosis to optimize the site-dispatched calls by identifying the required craft and skill levels and parts
- The reduction of on-site repair time through more rapid isolation of the actual problem

The underlying issue is the acquisition and use of information to diagnose and refine the service-call decision as part of the call-handling process, prior to actually initiating the on-site service visit. The data acquisition methods used, and means by which that information is manipulated and presented, represent the technological focus of artificial intelligence and remote diagnostics today in service. Within this context, the maintainability of the database is the most critical issue in terms of utilizing diagnostic knowledge bases; as knowledge bases become increasingly complex, due to the inclusion of ongoing subsequent data, the maintainability of these systems becomes a key factor. Knowledge bases must be constructed to ensure reliable diagnostic information within the limits of a maintenance system.

As such, the more intense and detailed a knowledge base can be transformed, proportional to the cost that will be incurred, to ensure its maintainability. This can be applied in terms of breakdown maintenance/emergency repair, preventive maintenance, and predictive maintenance.

In the traditional practice of breakdown maintenance and emergency repair (Figure 99-1), the information that is used to diagnose the equipment or system fault is usually obtained post facto. Costs associated with breakdown and emergency-repair maintenance practice include unanticipated production losses and related costs of idle production inventory, wasted and wrongfully allocated craft and manpower, and down time and damage of associated machinery resulting from the initial failure.

Preventive maintenance, in which maintenance is scheduled at regular time intervals, provides significant levels of protection from catastrophic equipment failure (Figure 99-1), however, it may be creating unreasonable costs due to assigning parts replacement or overhauls, when the equipment may be operating well within desired operating parameters. Manpower costs can be concurrently overextended by the preventive maintenance approach. Information used in the design of a preventive maintenance schedule is typically taken from the manufacturer's original design and maintenance specifications data. Optimally, the preventive maintenance schedules will be tempered by the personal experience of the responsible maintenance engineer.

Predictive maintenance is condition-restricted. Service provided when the equipment's operating condition is shown to be deteriorating toward catastrophic failure. These condition-monitoring methods are dependent upon the criticality of the equipment and the equipment function, the ability to apply sensors to monitor conditions and to communicate that data, and the volume of information that is needed to be interpreted to make a competent service decision and predictions.

The primary tool for acquiring accurate information in predictive maintenance is the use of condition-monitoring techniques to assess the operating "health" of a particular piece of equipment, or equipment system. Condition-monitoring typically utilizes measures such as:

- Vibration analysis (fast Fourier spectrum analysis)
- Temperature
- Fluid flow/volume
- Pressure
- Electric current fluctuation

Preventive and predictive maintenance can provide service when it is not required, increasing the costs of equipment service. The condition-monitoring approach, while requiring a cost to implement the monitoring apparatus, could optimize the usage of service, providing maintenance and repair only when critically necessary to maintain the equipment's up time integrity. In general, preventive and predictive maintenance works best for mechanical and electromechanical equipment that fails gradually. Demand repair works best for equipment that is either working or not working (i.e., electronic technology).

The actual use of remote diagnostic tools and practices in service has occurred consistently in several places within the full service call process architecture.

- The most common diagnostics mechanism is the technical assistance center (TAC). The TAC typically will have a staff of experts complemented with technical documentation and possibly with access to diagnostic-based systems. The TAC will provide end-users with initial call handling and will implement any call-avoidance routing (e.g., talking the end-user through the solution, sending a user-replaceable part or, when required, beginning the service call escalation procedures).

The overall call management system with remote diagnostic typically contains:

- One or more help desks or call management centers
- Technical assistance centers (TACs) linked to a help desk
- Local area networks (LANs) connecting to individual workstations in both the help desk/call center and TAC

In developing an integrated approach to call management and TAC systems, five general functional modules are necessary, including:

1. The *configuration management* module that is required for overall network control as it provides the ability to identify online and locate hardware/software components
2. The *troubleshooting* module that provides the ability to examine information from the LANs, PC controllers, and other devices through the use of integrated diagnostics
3. The *problem tracking and trend analysis module* that provides the ability to track all supporting activities and responses by configuration as well as identifying long-term trends in problems or service calls

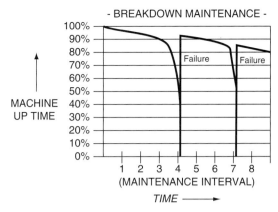
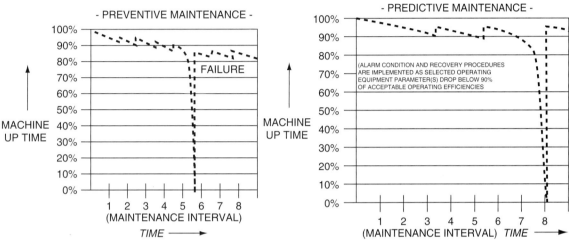

Figure 99-1 Illustrative example of maintenance and repair practices on equipment up time.

4. The *software control module* that provides the ability to support the process for loading new programs, controlling system enhancements, and enforcing software-version control
5. The *shared and coordinated peripheral resources module* maintains complete tracking of the status of reporting and printing in the system

In examining the typical customer configuration and need for service and the business environment in which service operates, one will observe a variety of calls arriving as a function of the stage and the life cycle of support of the system; the extensiveness and sophistication of the system; the size of the deployment; and the education, training, and sophistication of the user group.
The general classes of customer service calls (Figure 99-2) fall into the following categories:

- Type 0—Systems Designs, Specification, and Integration Requests
  - These calls tend to be related to new or emerging requirements and are often random in nature, and are typical of the types of calls that would occur during the very early stages of a system development project.
- Type 1—Installation
  - The actual start of system specific and general customer requests. These calls will occur at the point of installation after system approval, design, and integration. These calls often relate to scheduling of the implementation, installation, initial training, and concerns over workability.
- Type 2—Maintenance and Repair Service Requests
  - The typical customer call in this category involves requirements for maintenance and repair services due to hardware failure, software failure, or a combination of hardware and software failure. These calls are not only very critical but also tend to be those that are perceived by most service organizations as the primary type of service call request. However, the pattern of call arrivals for Type 2 calls is quite different from the call patterns for other types of calls and the mechanisms for response are also different.
- Type 3—Instructional and "How to" Customer Calls
  - A new class of customer requirements and needs, emerging as a result of the increasing sophistication of the system, the fact that increasingly, systems are being used in "real time" by non-technical personnel, and that initial instruction at the point of installation might be either not as comprehensive, or as in-depth as required, relates to user questions concerning the application and use of the equipment, or user inexperience, or lack of user understanding of a specific or general application.

  - These instructional and "how to" calls tend to be critically important in terms of the overall service delivery process to the user, but it is equally important to avoid confusing Type 3 calls with Type 2 calls since the mechanisms, delivery approaches, and optimum-handling mechanisms are quite different. In addition, many service organizations tend to see Type 3 calls, as "non-critical" or, even worse, as "not our responsibility." However, from the user perspective, these calls are just as important and critical as Type 2 calls.

- Type 4—Network Calls
  - A new class of customer calls now occurring, as a result of the increasing deployment of sophisticated local and wide-area networks, are those relating to failure of the network as a whole or elements of the network.
  - Because the general user, particularly in the distributed mode, is unable to determine that the cause of his problem could be a network-type failure, rather than a failure or problem in his own unit, these calls are also often perceived as Type 2 or Type 3 by the customer/user. In addition, the service organization can easily misinterpret these as "Type 2" calls unless some particular technical screening and network-oriented evaluation process is put into place. In essence, the failure of either a local or wide-area network can generate a very large number of Type 2 calls by individual users. An appropriate problem-resolution and self-organizing data system can discern relatively easily whether this rapid series of arriving calls is, in fact, network-related and can "flag" such calls, in order to avoid further work to try to fix the individual box that is failing to operate, due to a network failure, rather than to a box failure itself. The service organization system must be able to recognize this problem as network-related and must direct its focus on the network "fix" if it is to avoid being overwhelmed by inaccurate Type 2 or Type 3 calls from individual users.

- Type 5—Moves, Adds, Changes (MAC) and Field Improvements
  - These calls are usually generated as a result of the change in systems' environments and/or user needs. In that sense, they tend to be very similar to Type 0 calls.
  - Alternatively, they could be self directed as a result of a pattern or statistical analysis and evaluation of the demand associated with Type 2, 3, or 4 calls that tend to dictate the need for a field engineering change or modification in hardware, software, or network, structure and topology in the field, in order to reduce or eliminate certain types of problems in the future

In general, the character of the total demand pattern of these calls is well defined. The total demand pattern for all calls per day is quite predictable as shown in Figure 99-3.

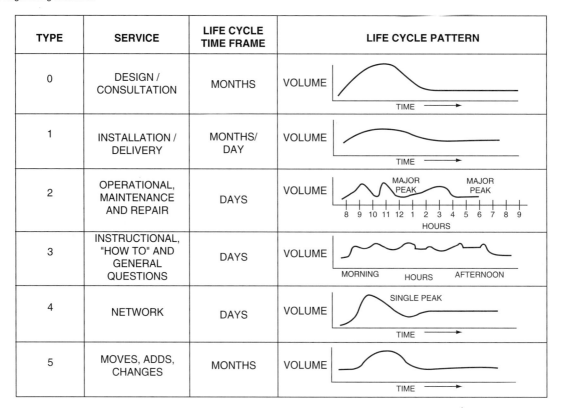

Figure 99-2 Call patterns by type.

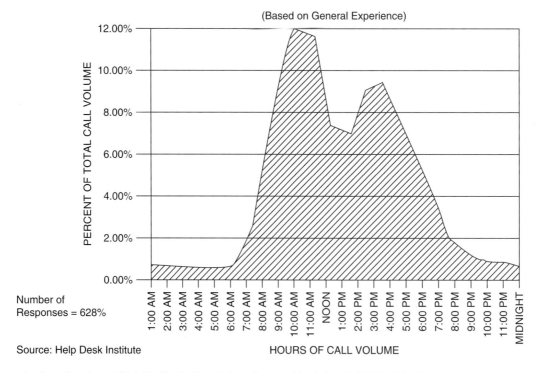

Figure 99-3 General pattern of service and "Help Desk" call volume by hour (in percent) in single rgion Help Desk Institute.

Remote Diagnostics State of the Art

Work in the field of advanced remote diagnostics and artificial intelligence in the field service industry initially tended to focus on the use of "expert" oriented decision techniques based upon either rule or "heuristic" expert models and processes. Within the last few years, there has been a further extension of this array of technology to include self-organizing systems and processes, as well as the application of new intelligent data storage and retrieval approaches. The general process now can be described in the model shown in Figure 99-4.

Within the explicit active remote diagnostic advisory systems are more specific mechanisms, as outlined in Figure 99-4, including: General fault models, troubleshooting models, and deep casual models. The differences between these approaches are outlined below:

- *General fault models* show all the ways that a device can fail and the casual linking among the failures. Information about applicable tests and repairs is attached to each fault situation. This approach has been shown to be not very useful for new products and technologies repair diagnostics.
- *Troubleshooting models* are usually augmented fault models that follow the pattern of the general fault model above but also describe how the fault diagnostics strategy should be modified, under certain conditions. In general, troubleshooting models add "if-then" logic to the standard fault model to support repair analysis.
- *Deep casual models* actually describe the way the unit under repair is structured and assembled. The knowledge base describes the way the device is put together, the way it works, what the relationships are, how likely each piece is to fail, and generates a troubleshooting strategy, based on these relationships. These models are most often used where the equipment being diagnosed is very new or highly complex or where little or no effective troubleshooting strategy currently exists.

In addition to model-based expert systems are rule-based expert systems that involve either flow charts or procedural rules that can be used to diagnose a repair situation, or the use of decision tree technology. The decision tree concept, outlined in Figure 99-5, is based upon the use of a tree-like structure that can relate machines or products to complaints to symptoms to causes and, ultimately, to corrective action recommendations. This type of analysis can be carried out before any type of repair action is completed, (i.e., *a priori*) by reliability and maintainability engineers, to construct the "tree" and "branch" relationships without actual experience. Alternatively, a call handling, dispatch, and call closeout process could be established, making use of the same decision tree structure on a *post priori*, or after-the-fact, basis. This is easily done by simply linking information on the arriving calls for service in terms of identified product or machine related to complaints and symptoms, and then, ultimately, tying this information to a specific closeout call relating the cause to the symptom and the ultimate corrective action that resulted in a fix or repair.

In substance, decision trees could be built before repair actions through reliability analysis or through reverse-engineering analysis. In this case, they are being used in the context of a rule-based expert system. They also could be built-in to the call handling process to construct actual decision trees based upon completed and closed out call data, in the context of a self-organizing approach. Recent developments in the design and application of self-organizing systems also have focused on parallel and sequential processing, utilizing a combination of logic, hypertext, and parallel processing to achieve the same solution as utilized in the *post priori* decision tree-approach described above.

A third major area of technology applied in diagnostics is to make use of newly developed intelligent (data-) retrieval systems. These approaches attempt to use high-speed computer technology to search existing files for case-based situations that are similar in nature to the diagnostics being executed or to make use of neural networks or use very high-speed hypertext search.

All three approaches attempt to make use of historically collected data to present the correct and appropriate information to the repair provider, either centrally or in the field. In summary, the current state of the art involves two general types of structure (i.e., retrieval and diagnostic advisory systems), and two types of explicit diagnostic systems (i.e., expert systems and self-organizing systems).

Expert systems are the other category of advisory systems. Expert systems provide an architecture for recording and using the experience of "expert" human service engineers:

- Rules-based expert systems are produced by recording and linking the heuristic decisions that are made by the human mind when problem-solving or troubleshooting a unit or system's fault, based on "a priori" or before-the-failure analysis. Literally, this approach documents the minute steps that a service technician must go through to remedy a fault. Typically these systems are structured to work on a forward-chaining or a backward-chaining logic depending upon the specific problem for which the system was designed to solve
- Model-based expert systems are also developed on an *"a priori"* basis from a database of the system's or unit's calculated or anticipated failure structure and behavior. This format makes the model-based approach better suited for the fault analysis of new equipment that service personnel might not be familiar with; methods to add historical information as it is developed are desirable in a model-based expert system.

Within this diagnostic state of the art, the basic process model for call resolution and fault diagnostics and isolation, as shown in Figures 99-6 and 99-7, follow these procedural steps:

- Gather and collect information to determine the symptoms and situation in which the problem appears
- Form a hypothesis to select a fault that is most likely to have caused the observed symptoms
- Select a goal or set of goals to resolve a particular call derived from the hypothesis that would have been previously developed this step usually takes the form of

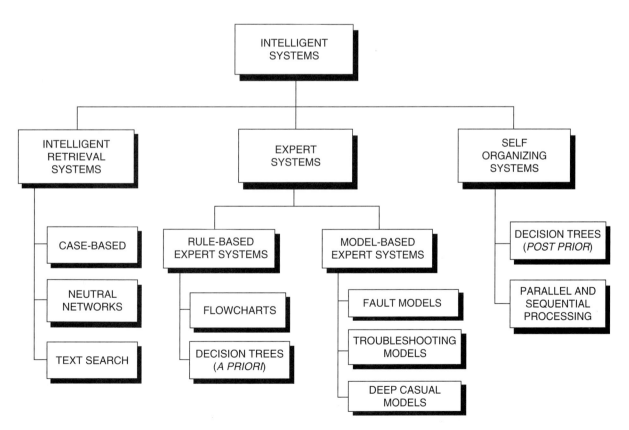

Figure 99-4 Overview of types of data and intelligent diagnostic systems used in help desk/TAC.

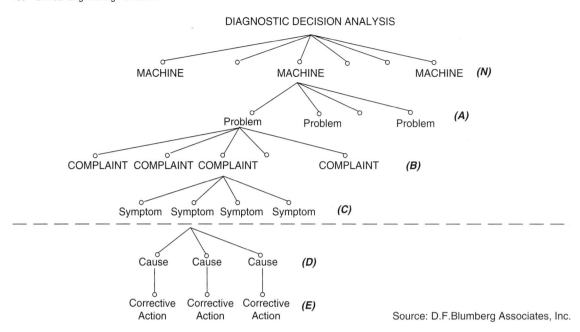

Figure 99-5 Basic diagnostic decision analysis tree structure.

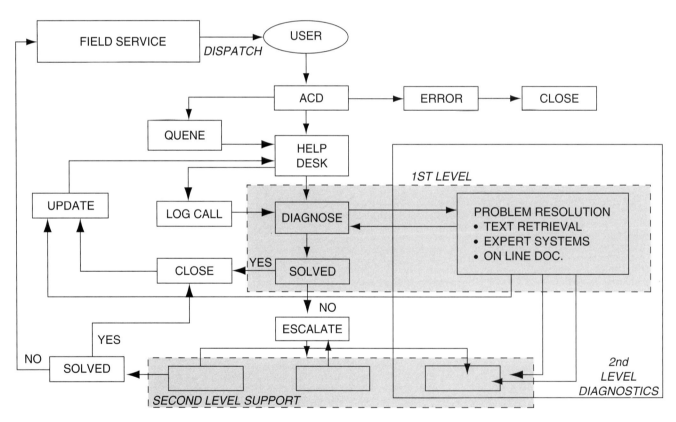

Figure 99-6 Diagnostics process flow model.

"determine whether element x is faulty"," by selecting specific parts out of all the parts to blame, making the process much more manageable and leading to a faster resolution
- Determine which action would most efficiently guide towards achieving the selected goals. The action might be diagnostic, by observing certain values or indicators, or performing a test, or it might be repair-oriented, such as calibration, cleaning, or replacing a part

- Performing the action recommended, and observe its results. If, for example, the action is repair-oriented, observation usually will require repair verification, or retesting functions of the equipment that failed to see whether the repair action caused these functions to work
- Analyze whether the actions' results strengthen or weaken the current hypothesis. If the hypothesis is strengthened and is specific enough, then the problem can be identified and resolved. Otherwise, the hypothesis must be refined, and the process must be repeated

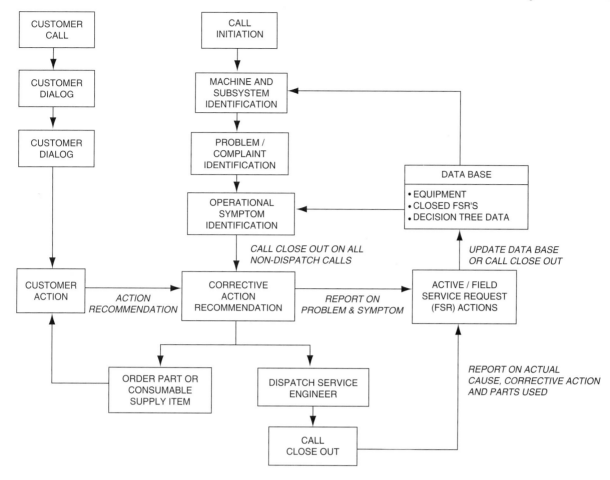

Figure 99-7 Remote diagnostic (decision support) process flow in help desk/TAC application.

The above fault isolation process requires extensive knowledge of the serviced equipment; service engineers obtain such knowledge from formal training, on-the-job training, accumulated experience, and equipment manuals. Several trends, however, make it difficult for service engineers to obtain and to preserve all of the knowledge required to perform their jobs most efficiently. One such trend is the rising complexity of modern equipment. Another is the number of different types of equipment that each engineer must service: Different manufacturers, different models from the same manufacturer, different versions of the same model, each one requiring a large amount of knowledge to be instantly available.

The results of these trends are increasing service costs and decreasing service quality. Because service costs and perceived service quality have immediate and important effects on sales and profits, many companies have decided to search for service support tools.

The application potential will vary as a function of type of diagnostic approach and by the product technology employed, as well as by the application. For example, mechanical and electromechanical equipment tends to fail gradually, and thus expert systems can have value in identifying requirements for predicting the need for advance maintenance. This preventive and predictive maintenance approach has proven to be of some value, particularly for equipment such as heating, ventilation, and air conditioning (HVAC), medical technology, and building automation. Diagnostics and troubleshooting for electronics products tend to be on more of a "go-no-go basis" and therefore self-organizing systems tend to have greater value in these applications. The overall assessment will, obviously, change over time with new breakthroughs in the diagnostics and artificial intelligence state of the art, and as a function of commitment to apply the technology for various types of products and systems. From practical industry experience, it appears to be correct to say that retrieval systems and the self-organizing systems have a role in service and that the applicability of the technology will vary as a function of type of product (e.g., electronic, electromechanical, or mechanical), and the stage of product use (e.g., new product rollout versus mature product versus product being phased back). Specific examples of technology application by company are shown in Figure 99-10.

Current Overview of Diagnostics Technology in Field Service

In field service, a variety of new products to support diagnostics and repair decision making have arisen as a result of extensive developments in the area of remote diagnostics technology and artificial intelligence. These products seek to improve service-based troubleshooting and repair capability for call and help desk-based diagnostics and call avoidance, TAC support, and support in the field. These are shown in Figure 99-8. Fully integrated help desk and technical assistance centers (TAC) systems are also available separately, on a stand-alone basis or as part of an integrated field service management system program.

Direct Experience in the Impact of Diagnostic Technology on Health Technology Service Productivity and Efficiency

As indicated above, there have been significant increases in the commitment of funds by vendors and field service organizations to the development, application, use, and roll-out of advanced diagnostic technology in field service. While the overall demonstration of significant, quantitatively measurable, impact is.

Since 1998 or so, considerable work has been done on developing customized remote diagnostics technology for use within the medical technology field. The key leaders in this area include General Electric Medical, Picker International, Elscint, and Medic Engineering and Maintenance Services (now part of Innoserv). Most use the remote diagnostics in service as a competitive differentiator.

General Electric

General Electric (GE) Medical offers remote diagnostics on CT scanners and MRI through its in-site service program. In-site is targeted to high-end, top-of-the-line equipment including GE9800 CT Scanners, Signa Advantage MRI, Starr Nuclear Imaging, and Advantx Digital X-Ray.

The technology includes a microprocessor, bidirectional modem, and software attached to equipment at the customer facility. The customer telephones the GE customer service center located in Milwaukee, Wisconsin, available through the computer-assisted, repair and engineering services (GE CARES) program, to report a service malfunction. GE CARES alerts and dispatches an in-site technician that patches into the customer's equipment via modem. The technician uses in-site analysis software and database (expert systems) to diagnose the problem. In many instances, a solution can be provided over the telephone.

Service calls that require on-site repair are transferred back to the GE CARES dispatch center by the in-site technician, who schedules a service call and assigns necessary parts. A limitation on in-site is that it is more conducive to identifying software problems than hardware, because GE's equipment is not designed to allow in-site to inspect hardware circuitry. Furthermore, in-site is essentially a reactive diagnostic tool that is primarily geared for operational problem-resolution such as shutdowns created by a power surge.

VENDOR/PRODUCT	PARADIGM	GREATEST STRENGTHS	LIMITATIONS
LOGICA (TEST BENCH™) FIVE PPG PLACE PITTSBURGH, PA 15222 (412) 642-6900 www.logica.com	• FAULT HIERARCHY REASONING • DECISION/HIERARCHY • CASE BASE REASONING • RULE OBJECTS	• MATURE PRODUCT • CAN MEET 95% OF REQUIREMENTS OF INTERNET, EXPERT SYSTEM, AND REMOTE DIAGNOSTIC REQUIREMENT AS PER MODEL • HAS EXPERIENCE IN DEVELOPING HIGHLY COMPLEX DIAGNOSTICS • USES "DETERMINISTIC", LOGIC TO PROVIDE HIGH TEST RELIABILITY FOR COMPLEX FAULTS OR PROBLEMS	• PRODUCT SALES LIMITED
E-GAIN (E-GAIN KNOWLEDGE AGENT™) 455 W MAUDE AVENUE SUNNYVALE, CA 94086 (408) 212-3400 www.egain.com	• CASE BASE REASONING • CAN BUILD IN DECISION TREE LOGIC IF NEEDED, BUT NOT PRIMARY FOCUS	• STRONG HELP DESK CAPABILITY • TOOLS ALLOWS FOR EASY DOCUMENTATION DEVELOPMENT	• FRONT END SPARSE, NEEDS CUSTOMIZATION TO USER REQUIREMENT • WOULD NEED SYSTEM INTEGRATOR TO PROVIDE FOR FULL DIAGNOSTICS • CUSTOMIZATION IS REQUIRED FOR FULL DIAGNOSIS • PRODUCT TYPICALLY GEARED TO ONLY TAC OPERATIONS
ENHANSIV (ECRM™) 7505 IRVINE DRIVE IRVINE, CA 92618 (949) 784-5000 www.enhansiv.com	• FORWARDED RULES BASED MULTI-LANGUAGE MODEL	• STRONG HELP DESK • STATE OF THE ART DIAGNOSTIC PHONE CALL AVOIDANCE SYSTEM	• WOULD NEED SYSTEM INTEGRATION • NOT TYPICALLY INVOLVED IN REMOTE DIAGNOSTIC HOOK-UP WITH INSTALLED BASE
EMERALD INTELLIGENCE INC. (PROCEDURAL™) 405 ALEXANDRIA BLVD., STE 100 OVIEDO, FL 32765 www.emeraldi.com	• DECISION TREE AUGMENTED WITH SESSION FILES	• CAN MEET 90% OF DIAGNOSTIC REQUIREMENTS • HAS EXPERIENCE WITH LARGE COMPLEX DIAGNOSTICS PROBLEM • CAN INTEGRATE WITH VARIOUS EMBEDDED TECHNOLOGIES	• SMALL COMPANY • MAY NOT BE ABLE TO PROVIDE FULL ONGOING PROJECT MANAGEMENT OR CONSULTING
ARMAS SOLUTIONS (ADVANTAGE™) 17950 PRESTON DR. STE 900 DALLAS, TX 75252 (972) 381-2880 www.armassolutions.com	• AUGMENTED FAULT MODEL (AN EXPERT SYSTEM BASED ON MODELS AND RULES)	• STRUCTURE OF KNOWLEDGE DEVELOPMENT (AUTHORING TOOLS) • CAN MEET THE DIAGNOSTIC CAPABILITY • ABILITY TO INTEGRATE INTO WEB	• LIMITED LEVEL OF INTEGRATION NEEDED
CLICKSOFTWARE TECHNOLOGIES, LTD. (CLICKFIX™) 3425 S. BASCOM AVE., STE 230 CAMPBELL, CA 95008 (408) 377-6088 www.clicksoftware.com	• MODEL BASED • CBR	• PROVIDES EASE OF USE THROUGH THE FAULT ISOLATION PROCESS • FAULT DIAGNOSTIC PROCESS COMBINES MODEL BASE LOGIC WITH CBR TO PROVIDE BETTER RELIABILITY THEN JUST CBR	• DOES NOT PROVIDE FOR BILLING, LOGISTICS, OR SERVICE CONTRACT CONSIDERATIONS • NOT A STAND ALONE PRODUCT: DOES NOT INTEGRATE WITH CALL MANAGEMENT SYSTEM
SERVICEWARE (E-SERVICE SUITE™) 333 ALLEGHENY AVENUE OAKMONT, PA 15139 (412) 826-1158 www.serviceware.com	• KNOWLEDGE BASED	• THE INPUT OF CALL LOGS, ON-LINE DOCUMENTATION, PC DATA AND RELATED FORMAT PROVIDE FOR A FULL STAND ALONE KNOWLEDGE SYSTEM • EASY TO DEPLOY • OPEN ARCHITECTURE • INTERFACE WITH THE INTERNET	• SELF ACTIVATION DIAGNOSTIC CAPABILITIES WOULD REQUIRE SYSTEM INTEGRATOR FOR ON-BOARD DIAGNOSTICS

Figure 99-8 Comparative analysis of vendors of diagnostic products vendor capabilities.

In response to market demand for proactive remote diagnostic tools, GE developed quantitative system analysis (QSA) for Signa Advantage MRI customers with access to in-site. QSA essentially monitors equipment, usually after hours, to detect system problems by performing regularly scheduled computer-to-computer checkups, including an automated analysis of each subsystem of the imaging equipment.

QSA enables the operation of two proactive diagnostic tests:

1. A "top-level test" is a weekly evaluation of the entire system. It includes signal-to-noise ratio and other parameters. Results are benchmarked against standards and are compared against other GE Signa MRI units. Data are reported back to original the customer.

2. Small sample test analyzes subsystem performance with the goals of faster calibration and allowing the service professional to troubleshoot on-site.

In-site is available to GE service technicians and service contract customers free of charge. GE maintains "proprietary" rights to diagnostic equipment/system technology attached to the customer's equipment, thereby limiting access to noncontract customers and unauthorized personnel.

The benefit to customers in utilizing in-site is the lowering of downtime costs by lowering repair time from 1.5 hours to 15 minutes. However, the ultimate benefit is realized by GE because in-site lowers costs of labor (dispatch of technician, travel time) and capital (spare parts carrying costs). Note that in-site does not eliminate the role of the field technician. It merely provides a "first line of defense" by performing an initial analytic/diagnostic check.

Picker

Picker's remote diagnostic tool for CT scanners is an expert-based system named Expert. Expert is a portable computer system that field representatives can use to modem into computers at the Picker service center. The interactive link enables step-by-step diagnostic routines. Expert is primarily used by Picker field engineers and support staff. The system is licensed for a fee or sold under exclusive contract to end-users. Picker will train hospital in-house service staff in the use of Expert.

The company resists use of Expert by third-party maintainers (TPMs) because TPM overheads are generally lower. Use of Expert would enable TPMs to gain a significant competitive advantage, which would not be in the best interest of Picker management

Mediq

Mediq, an Arlington, Texas-based independent medical imaging service provider (now part of Innoserv) uses its own proprietary remote diagnostics called Memserv. Memserv (whose name means Mems' evaluation, repair, and verification) utilizes a PC at point-of-service to download the service software to the CT scanner. It collects hard data by interfacing with the hardware and analyzes data to identify problems or help in calibration.

This technology replaces GE advanced diagnostic service software that is not available to independent service organizations (ISOs). It was developed out of necessity and from a strategic decision and investment to be state-of-the-art and on the leading edge by developing its own software.

Elscint

Elscint has applied the Rosh (ServiceSoft) developed expert system shell called CAIS, for computer aided intelligent service, applied to diagnostics of Elscint nuclear imaging technology. CAIS is basically a database that relies on an ongoing feedback loop of knowledge, acquisition, storage, and delivery. This system, marketed by Elscint under the name Mastermind, has been applied to Elscint's Apex nuclear medicine system.

Mastermind requires a service expert to enter data related to operation and service of particular medical equipment. It is user-friendly, and no programming experience is required.

Mastermind, based on Advisor, a diagnostics expert system, identifies problems and recommends appropriate solutions through a PC. The system logs data regarding the service call to the point-of-service allowing CAIS customers the ability to download this experience data into original database.

Benefits that accrue to Elscint from the Mastermind system include enhanced up time, faster service calls, faster PM, expediting entry-level technician training, and reducing loss of expertise from turnover of senior technicians.

Cost Benefit and Impact of Remote Diagnostics

Full quantifiable measurement of remote diagnostics cost benefits has yet to be pragmatically demonstrated in a broad sense. However, it is, clear that diagnostics technology does have some measurable impact. A comprehensive study of the primary benefits and impacts of service diagnostics in equipment service showed an array of primary and direct benefits and impacts, as indicated in Figure 99-9. These benefits can be measured in terms of improvements in service delivery as viewed by the service provider, the field service organization, and the customer.

A survey of well over 100 service providers indicated that service diagnostics can reduce the training time and can reduce or eliminate the effort required to collect and analyze repair and maintainability data. Diagnostics also can help the service organization directly in improving the allocation of craft personnel and parts resources, to reduce costs. Diagnostics technology also has been found to provide a way to deliver better service and product differentiation to the customer from the perspective of the service providers identified above and customers (based upon a survey of over 250 customers). Additional direct benefits include reduction in the overall mean-time-to-repair, including a reduction in the number of "broken" calls due to a lack of availability of parts or skills on-site. Finally, the service providers and the customers observe that service diagnostics could be used to improve the ability to deliver service in remote or inaccessible areas. Quantification of these benefits was more difficult, however.

The primary direct benefit of diagnostics, as shown in Figure 99-10, was the creation of a competitive advantage, reduction in mean-time-to-repair, and reduction of the number of "broken" calls. This was most directly measured in terms of the reduction in the total elapsed time for repair in complex calls, which is typically about 20% of the total number of calls carried out by service organizations in the computer, building and plant automation, and medical electronics service markets. Where measured, diagnostics also tended to increase the up time and the mean time between hard failures. Finally, the survey of service organizations suggested that the best delivery mechanisms were via laptop or portable devices through deployment directly to field service engineers.

In summary, a final area of the D.F. Blumberg & Associates' survey analysis related to the current use and future plans for development of diagnostics in the field service industry and the observable cost savings or benefits used to justify diagnostics products in field service and to evaluate return-on-investment (Figure 99-11). In this respect, the Blumberg studies compare favorably to similar results found in the Serviceware study (1991) of remote diagnostics. Generally, both studies suggest that the use of diagnostics technology varies dramatically by industry with the highest level of implementation being in the industrial control, building- and plant automation markets, telecommunications, and computer industry. There is also a high level of investment occurring across the board to examine the applicability of diagnostics technology for future use.

Actual pragmatic experience in cost savings or benefits justification utilized in diagnostics products tends to vary as a function of the technology serviced and supported as well as the basis of focus (Figure 99-12). In the past, general justification was based upon the concept of gaining a competitive edge over other service organizations, capturing and retaining expertise and knowledge, and improving telephone troubleshooting as part of a TAC operation. These more qualitative justifications tended to be utilized in more cases than direct cost or time savings with respect to reduction in training time, mean time to diagnose and repair, and in the number of calls (e.g., broken due to lack of parts or technical skills). It is quite clear that additional experience will be required before direct quantitative measures of cost benefits can be utilized in justification with a high degree of accuracy and confidence.

In essence, as shown above, the overall justification and impact of diagnostics technology in field service, as of this time, is still based mostly on "soft" issues although there are clearly some specific situations and instances where the savings have been substantial.

COST JUSTIFICATION BASIS	TECHNOLOGY			
	COMPUTERS & OFFICE AUTOMATION	TELECOMMUNICATIONS	MEDICAL ELECTRONICS	BUILDING & PLANT AUTOMATION
GAIN GENERAL COMPETITIVE EDGE	55%	35%	50%	64%
CAPTURE & RETAIN SOME EXPERTISE/KNOWLEDGE	41%	45%	48%	55%
IMPROVED TELEPHONE TROUBLE SHOOTING	35%	42%	28%	31%
IMPROVED CALL AVOIDANCE	31%	43%	21%	34%
REDUCED TRAINING TIME & COSTS	21%	18%	14%	21%
REDUCED MEAN TIME TO DIAGNOSE AND REPAIR IN FIELD	37%	28%	19%	34%
REDUCED BROKEN CALLS DUE TO PARTS/TECHNICAL SKILLS	31%	21%	22%	20%
OTHER (INCREASED REVENUE, JUSTIFICATION OF ADDED PRODUCT VALUE/PRICE)	11%	10%	12%	10%

Figure 99-9 Observed specific cost savings and/or benefits in cost justification based on actual diagnostics projects implemented.

BENEFITS	EFFECTS
IMPROVED CRAFT AND RESOURCE	REDUCED PERSONNEL AND PARTS INVENTORY COSTS
REDUCE THE TRAINING TIME AND SKILL LEVELS OF MAINTENANCE	REDUCED LABOR COSTS PER MAINTENANCE PERSON
REDUCE OR ELIMINATE EFFORT REQUIRED TO ANALYZE DATA	REDUCED PERSONNEL REQUIREMENTS
REDUCE OR ELIMINATE EFFORT REQUIRED TO COLLECT DATA	REDUCED PERSONNEL REQUIREMENTS
REDUCE ELAPSED TIME REQUIRED TO ASSESS PROBLEM; REDUCE MTTR FOR BOTH DEPOT AND FIELD REPAIR	REDUCED PERSONNEL REQUIREMENTS IMPROVE UPTIME
IMPROVE PREDICTABILITY OF FAILURE IN ORDER TO INITIATE PREVENTIVE MAINTENANCE PRIOR TO EMERGENCY FAILURE; INCREASE MTBF	IMPROVE UPTIME INCREASE PERSONNEL REQUIREMENTS
IMPROVE RESPONSE TIME OF SERVICE CALL AND/OR CALL AVOIDANCE ROUTINES	GREATER CUSTOMER SATISFACTION CUSTOMER RETENTION
IMPROVED CRAFT AND RESOURCE ALLOCATION	REDUCED PERSONNEL AND PARTS INVENTORY COSTS
GREATER SERVICE AND PRODUCE DIFFERENTIATION	POTENTIAL TO GAIN NEW MARKETS AND IMPROVE POSITION IN CURRENTLY SERVED MARKETS

Figure 99-10 Benefits of use of advanced diagnostics and AI technology for service operations.

In summary, remote diagnostics technology has been shown to have a direct cost-benefit impact on the service and support organization, and the service user. The impact is, by far, the greatest on the service and support organization.

The primary observed impacts are on the following:

- Creation of a competitive advantage and reduction of direct competitive threat by ISOs
- Reduction in mean-time-to-repair
- Reduction in field diagnostics and troubleshooting time
- Improvement in dispatch assignment and allocation
- Reduction in number of broken calls due to lack of parts
- Improvement in ability to detect the need for field changes and/or field modifications,
- Better control and allocation of high value, critical parts inventory

Remote diagnostics technology is usually implemented via either internal development (as in the case of DEC, Honeywell, Texas Instruments, Johnson Controls, and GE.) or external purchase of software on either an off-the-shelf or customized basis (as in the case of Elscint, Varian, U.S. West, Bailey, and Siemens.).

The effect on the service organization varies primarily as a function of the implementation approach taken. If the technology is based on internal development, R&D is usually maintained and carried out by the corporate R&D organization with liaison support into the service organization. If the technology is purchased, the development and application is usually controlled and coordinated by MIS and/or the service organization. In either case, maintenance of the database and diagnostics algorithms must be maintained by the service organization, typically by the technical and logistics support department.

Summary

In summary, the Blumberg in-depth survey of the state of the art and the results of application, implementation, and rollout of problem resolution diagnostics in control help desks and field service clearly shows the promise and potential of on-line remote diagnostics technology (Figure 99-13) in improving the utilization of service force productivity and efficiency, and making more effective use of service resources (people and parts).

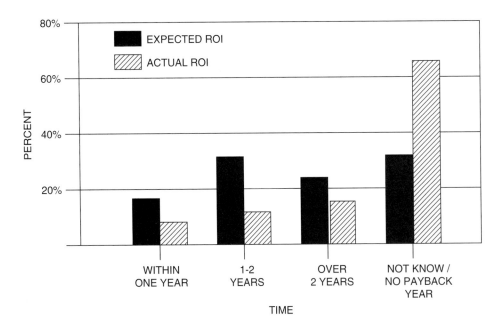

Figure 99-11 Expected and actual return on investment from remote diagnostics.

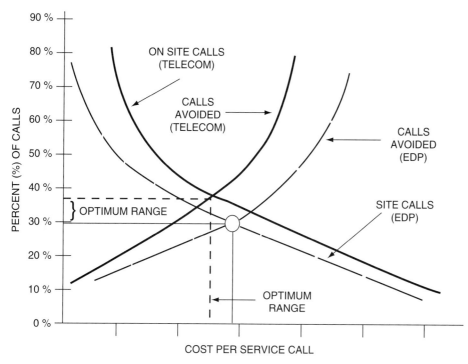

Figure 99-12 Optimum use of TAC and call avoidance.

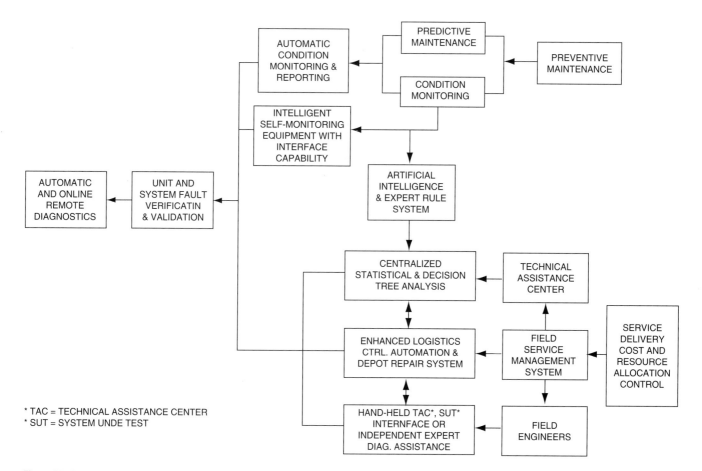

* TAC = TECHNICAL ASSISTANCE CENTER
* SUT = SYSTEM UNDE TEST

Figure 99-13 Illustrative schematic of automatic online remote diagnostics.

The survey of the current and planned future expenditures for diagnostics used in field service clearly indicates that the overall service diagnostics market is currently sizable, in the range of $2 billion as of 1997. This current expenditure and growth primarily will occur in the electronics arena, but there is still substantial development and application investments occurring in electromechanical and mechanical areas. The pace of investment will fall off as the service industry shifts from development and experimental research to the application and rollout of standard off-the-shelf technology. Thus, the overall pattern of the diagnostics market indicates a continuing increase in expenditures by the field-service industry for electronics-oriented service diagnostics and some increase with respect to the application and use of this technology in electromechanical and mechanical application areas.

The application and use of diagnostics technology will also be impacted by the ability of the field service management community to recognize the need to search for an optimum (as opposed to a feasible) solution in dispatch, assignment, call screening, call avoidance, and logistics deployment. The further rollout of affordable wireless communications technology such as Ardis and Ram mobile and the new cellular based CPDP technology, coupled with the increased availability of inexpensive wireless based laptop, portables, and CD ROM-based terminals, will increase the ability to provide an array of powerful tools for diagnostics and repair directly to the service engineer in the field. This ability to augment the diagnostics capability of the service engineer in the field, coupled with the centralized application of artificial intelligence-based remote diagnostics for helpdesk/TAC support that build upon, and make use of, an increasing array of data collected within the integrated field service management system, could result in a significant improvement in the positive impact of problem resolution and diagnostics technology in field service (Figure 99-14). The new, integrated field service management systems provide the basis for automatically collecting problem and symptom data and relating these to the cause and corrective action information. As the availability of accurate and reliable maintainability and repair data increases to meet the data requirements of the diagnostics technology, the scene will be set for further significant improvements in the application and use of diagnostic technology in the field service industry in the 21st century (Figure 99-15). As shown in the above analysis, cost justification of advanced problem resolution and diagnostics technology clearly shows the value of this technology in improving service productivity, efficiency, and profitability.

As the purchase and use of off-the-shelf software and predeveloped knowledge bases prevails over expensive academically and research and development oriented diagnostics applications, this technology will be broadly applied.

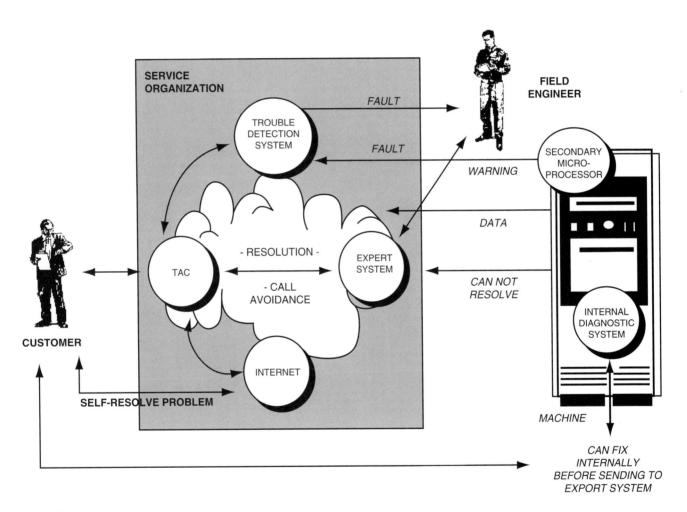

Figure 99-14 Emerging state-of-the-art diagnostic overview.

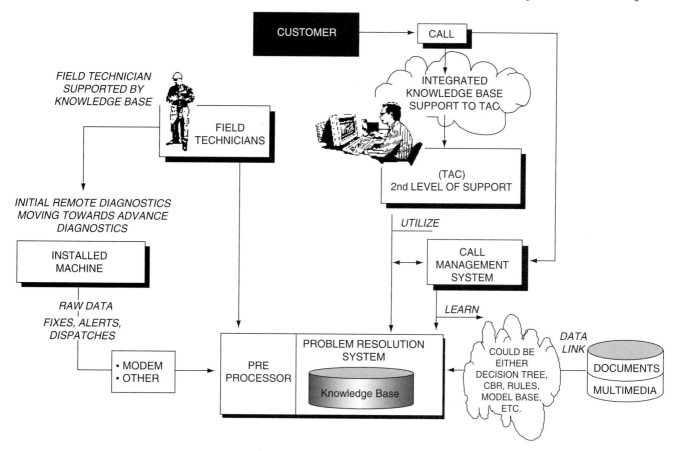

Figure 99-15 Remote diagnostics and knowledge-based support and system architecture.

References

Blumberg DF. Strategies for the Use of Diagnostics Technology in Improving Field Service Productivity. *FSMS Journal*, 1984.
Serviceware. Study of AI and diagnostics technology, 1991.

Further Information

Blumberg DF. Service Problem Diagnostics and Decision Support Technology for Improving Health Technology Service Efficiency and Productivity. *Field Service Management Journal* 1998.
Elity Systems. Expanding the Power of Data. , Elity Systems, 2001.
Blumberg DF. *Managing High Tech Services Using a CRM Strategy.* , St. Lucie Press, 2002.
Ben-Bassat M, Beniaminy I, Joseph D. Can Model-Based and Case-Based Expert Systems Operate Together?. *IET Intelligent Electronics*.
Bearse TM. Deriving a Diagnostic Inference Model from a Test Strategy. *Proceedings of the 1987 IEEE International Workshop on System Test and Diagnosis*.
Ben-Bassat M, Beniaminy I, Joseph D. Different Approaches to Fault Isolation Support Software. *Proceedings of the 1998 IEEE International workshop on System Test and Diagnosis*.
Ben-Bassat M, Ben-Arie D, Beniaminy I, et al. AITEST–A Real Life Expert System for Electronic Troubleshooting (A Description and a Case Study). *Proceedings of the IEEE Conference on AI Applications*. San Diego, IEEE, 1988.
Ben-Bassat M. Expert systems for clinical diagnosis. In Grupta MM, Kandel A, Bandler W, Kiszka YB (eds). *Approximate Reasoning in Expert Systems*. North Holland, 1985.
Panzani MJ, Brunk C. Finding Accurate Frontiers: A Knowledge-Intensive Approach to Relational Learning. The National Conference on Artificial Intelligence. Washington, DC, AAAI Press,
Quinlan J. Learning Logical Definitions from Relations. *Machine Learning* 5 (3), 239–266, 1990.
Simpson WR, Sheppard JW. *System Test and Diagnosis*. Norwell, MA, Kluwer Academic Publishers,
Sheppard JW. Inducing Information Flow Models from Case Data. *Proceedings of the 1998 IEEE International Workshop on System Test and Diagnosis*., 1998.
PCAI Editors. New Technology in a Large Customer Call Center. *Personal Computer Artificial Intelligence*, 6:24–47, 1997.
NEC Research Institute. *Learning Search Engine Specific Query Transformations for Question Answering*. , NEC, May 2001.
Mitsubishi Electric Research Laboratories (MERL). *Learning Hierarchical Task Models by Demonstration*. , MERL, 2001.
MIT Media Laboratory. Building HAL: *Computers that Sense, Recognize, and Respond to Human Emotion*. , MIT Media Laboratory, 2001.
MIT Media Laboratory. *A Family of Algorithms for Approximate Bayesian Inference*., MIT Media Laboratory, 2001.
European Commission. Context Modeling and Transformation for Semantic Interoperability. , European Commission, March 2000.

100

Real-Time Executive Dashboards and Virtual Instrumentation: Solutions for Health Care Systems

Eric Rosow
Director of Biomedical Engineering, Hartford Hospital
Hartford, CT

Joseph Adam
President, Premise Development Corporation
Hartford, CT

Successful organizations have the ability to measure and act on key indicators and events in real time. By leveraging the power of virtual instrumentation and open architecture standards, multidimensional executive dashboards can empower health care organizations to make better and faster data-driven decisions. This paper will highlight ways in which user-defined dashboards can connect to hospital information systems (e.g., ADT systems and patient monitoring networks) and utilize Statistical Process Control (SPC) to "visualize" information. The case studies described will illustrate enterprise-wide solutions for:

- Bed management/census control
- Operational management
- Data mining and business intelligence applications

Background

Virtual Instrumentation Revolution

In the last decade, the virtual instrumentation (VI) revolution has empowered engineers to develop customized systems and solutions, in the same way the spreadsheet has empowered business managers to analyze financial data.

Virtual instrumentation can be defined as:

"A layer of software and/or hardware added to a general purpose computer in such a fashion that users can interact with the computer as though it were their own custom-designed traditional electronic instrument."

Benefits of User-Defined Virtual Instruments

The benefits of virtual instrumentation are increased performance and reduced costs. Because the user controls the technology through software, the flexibility of virtual instrumentation is unmatched by traditional instrumentation. The modular, hierarchical programming environment of virtual instrumentation is inherently reusable and reconfigurable.

Virtual instrumentation applications have encompassed nearly every industry including the telecommunications, automotive, semiconductor, and biomedical industries. In the fields of health care informatics and biomedical engineering, virtual instrumentation has empowered developers and end-users to conceive, develop, and implement a wide variety of research-based biomedical applications and executive information tools.

Case Study #1: A Real-Time Bed Management and Census Control Dashboard

The Challenge

To provide a hospital or an integrated health care delivery network with enabling technology to maximize bed resources, manage varying census levels, reduce emergency department wait times, and avoid patient diversions.

The Solution

The Bed Management/Census Control Dashboard (BMD) is an integrated, easy-to-use application that enables users to access, analyze, and display historical and current patient and bed information easily. This information is continuously updated from the hospital's Admission/Discharge/Transfer (ADT) system. The BMD utilizes real-time monitoring, automation, and communication to achieve a combination of Process Improvement and Decision Support advantages.

- Process Improvement
  The BMD enables more efficient patient placement by reducing and eliminating phone calls and paper processes, and automatically matching patient requirements to available resources. Specifically, the system assists with the clinical and business decision processes that occur when a patient needs to be assigned to a specific bed location. Via continual, "real-time" updates across the hospital's network, BMD users are assured that they have accurate data available to them when and where they need it.
  Current ADT systems are admissions and billing systems geared toward financials. The BMD complements these systems with clinical requirements and was designed to improve operational efficiencies on a unit and/or hospital-wide. Intuitive modules are used to display and provide alerts on quantitative results in real-time, providing an easy way for personnel to know such details as which beds are available and ready for immediate use, which beds have been requested by other units, or which beds have patients with special requirements.
- Decision Support
  The BMD is an easy-to-use Business Intelligence application. It has been designed to allow administrators, clinicians, and managers to easily access, analyze, and display real-time patient and bed availability information from ancillary information systems, databases, and spreadsheets. The dashboard provides on-demand historical, real-time, and predictive reports, alerts and recommendations. Decision-makers can easily move from big-picture analyses to transaction-level details while safely sharing this information throughout the enterprise to derive knowledge and to make timely, data-driven decisions.

An Air Traffic Control Tower for Beds

In many ways, the BMD is similar to an air traffic control tower. Like a real air traffic control tower, this application is real-time and mission-critical. It must handle scheduled and emergency events. The system assists with the clinical and business decision processes that occur when a patient needs to be assigned to a specific bed location. Collectively, this system provides organizations with an array of enabling technologies to:

- Schedule/reserve/request patient bed assignments
- Assign and transfer patients from the emergency department and/or other clinical areas such as intensive care units, medical/surgical units, operating rooms, or post-anesthesia care units.
- Reduce and eliminate dependency on phone calls to communicate patient and bed requirements
- Reduce and eliminate paper processes to manage varying census levels
- Apply Statistical Process Control (SPC) and "Six Sigma" methodologies to manage occupancy and patient diversion
- Provide administrators, managers, and caregivers with accurate and on-demand reports and automatic alerts via pagers, e-mail, phone, and intelligent software agents.

PROBLEM	SOLUTION
ADT systems are not designed to provide sufficient clinical information for appropriate patient placement.	BMD integrates all required information in real-time (e.g.., monitor required or negative pressure room).
Lack of accurate bed availability information can result in lost admissions and excessive wait times.	BMD provides the information that is necessary to significantly improve the efficiency of patient placement and discharge.
Inefficient communication while "searching" for the appropriate bed for a patient.	BMD automates the notification process via dashboard, pager, or e-mail.
23+ hour "observation" outpatients that occupy inpatient beds without payer authorization.	BMD has built-in 24 × 7 automated alerts via intelligent agents and pagers.
Difficulty accessing meaningful historical, current, and predictive data.	BMD provides data mining and decision support tools to create useful information from data.

How It Works

The Bed Management Dashboard is accessible via a web browser or via a client application. The supporting architecture of the BMD system is a standard N-tier server-based system, which, depending on the end-users' needs, consists of one or more of the following: A web server, an application server, a database server, and an ADT interface server.

Key information from the hospital's ADT system is automatically fed to the BMD via a Health Level Seven (HL7) data stream. The system also has been designed to accept inputs from patient monitoring and nurse call networks. Figure 100-1 illustrates the flow of information from the ADT system to the user's desktop. The system typically receives up to 12,000 transaction messages per day that are parsed into appropriate data elements by a HL7 parser and are stored on a database server.

The log-in authentication module accesses a hospital's central user-login repository to validate the user's login information. This module enables single-user sign-on (SSO) capability by providing a user with the same username and password used by the other applications that authenticate to the central user log-in repository.

The integrated data mining and report module consists of a number of standard reports that take advantage of the data-warehouse nature of the BMD. These reports include, but are not limited to, historical census report, real-time census report, bed manager report, physician discharge report, and discharge compliance report. In addition, access to the data by industry standard report writers, such as Crystal Reports and Access give technical users the ability to create complicated or special reports as desired. For example, the BMD offers extensive *ad hoc* reporting capabilities that range from length of stay (LOS) and "Care Day" metrics, to asset management analyses (on beds and patient monitor utilization), to biosurveillance reports that have been requested by state and federal organizations such as the Department of Public Health and the Centers for Disease Control and Prevention..

The utility and configuration module gives system operators the ability to administer BMD users (e.g., add new users and modify new or existing user-security settings); administer service, unit, rooms and beds (e.g., add or modify clinical services, units, rooms, and beds and their inter relationships); define automated alert thresholds; and configure unit floorplan diagrams.

The embedded backup utility module consists of a local version of the bed management database that is constantly updated. Access to this database gives users a self-contained and mobile version of the system that can be used in the event of catastrophic failure of the system hardware or network hardware, or in the event of a crisis that removes the users from direct access to the hospital network.

One of the most important features of the BMD is that it reformats information from the ADT system and presents it to the clinical user in a more "user-friendly" and process-oriented manner. Dynamic and interactive graphical presentations of data are used extensively. Figure 100-2 illustrates how all patients from a given admitting source, such as the Emergency Department, can be displayed and selected from a dynamically sortable "smart table."

Hospital beds are classified as having predefined "attributes," such as being "monitored" or being assigned to the "surgery" service. The needs of patients are similarly described with attributes, such as "monitor required" or "scheduled for surgery." As illustrated in Figure 100-3, the BMD helps to find those available beds in the hospital that meet the specific needs of a patient by guiding the clinical staff through a set of process screens that perform the match.

Decisions for patient placement can be centralized or decentralized. The dashboard allows proper communication between the appropriate parties. Status of decisions is automatically tracked, and a monitoring process can detect and notify key stakeholders of any process delays. Admitting or emergency departments can be notified of automatically decisions, if appropriate. Reporting of information is provided by on-line screen views of data tailored to the needs of a particular class of system user. Unit personnel can view detailed information as well as summary roll-ups of their patients. Figure 100-4 illustrates how patient information can be viewed in a dynamic and interactive floorplan mode.

Administrators and program directors can view data over a wider scope that encompasses multiple units, services, or physicians. An example of a summary report is shown in Figure 100-5.

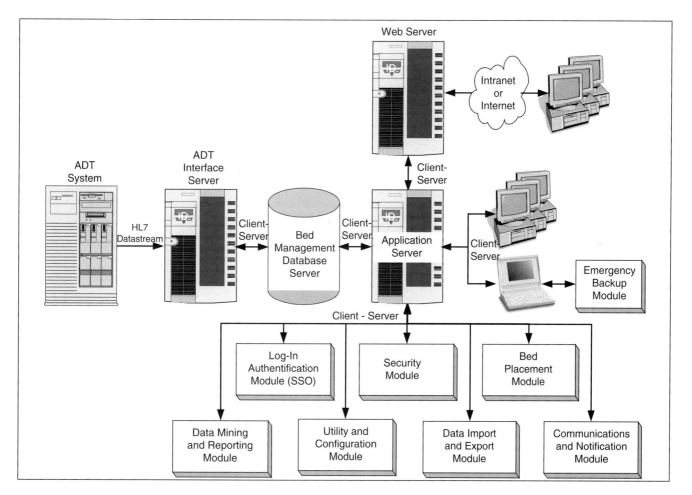

Figure 100-1 System diagram. The BMD consists of four major components: the ADT interface, the database server, an application server, and a web server. The system has been designed to interface with any ADT system that provides an HL7 interface and utilizes TCP/IP as a communication protocol.

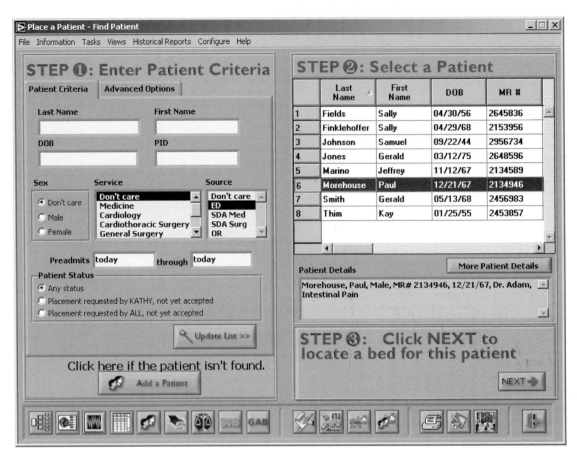

Figure 100-2 Find Patient. This screen is primarily used to request a bed for a patient. The user first enters various criteria to identify the patient. The system then displays all of the patients who meet the specified criteria. Finally, the user selects the patient in question and presses "NEXT" to move to another screen, where an available bed is located and requested.

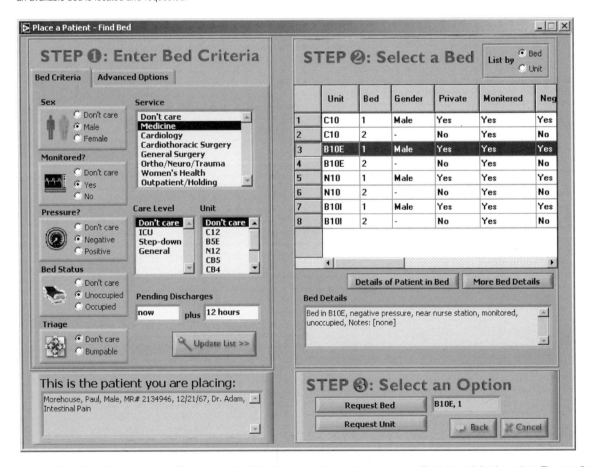

Figure 100-3 Find Bed. Once a patient is selected, this "Find Bed" screen is used to locate a specific bed or unit for the patient. The user first enters various criteria about the type of bed that is needed (e.g., patient gender, monitor required, or negative pressure room required.). The system then displays all of the available beds that meet the specified criteria. Finally, the user selects a particular bed or unit for the previously specified patient.

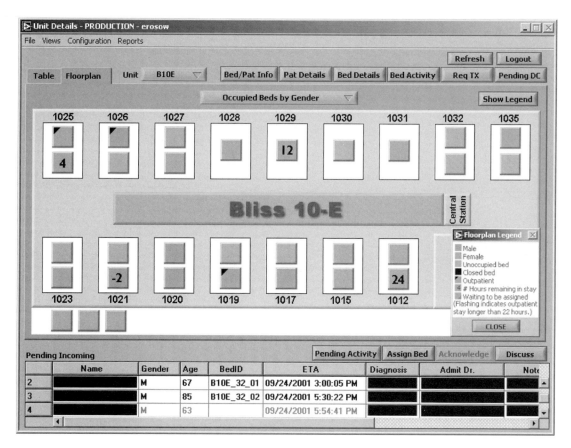

Figure 100-4 Unit Details (Floorplan view). This screen provides a graphical view of an intensive care unit. Each bed is presented as a square using a simplified floorplan view. Colors are used to indicate the selected attribute of the patient or bed. For example, the above screen indicates available beds in green, and occupied beds in red. The gray beds (flashing) are those with pending discharges, and the black bed is closed/inactive. Many other color-coded options are available via the pull-down selector. These include patient and bed attributes such as gender, monitored bed, negative pressure room, and service (e.g., cardiology, surgery, and orthopedics).

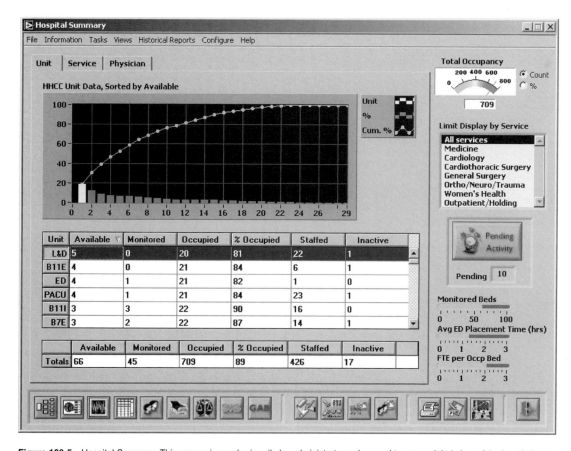

Figure 100-5 Hospital Summary. This screen is used primarily by administrators who need to see a global view of the hospital status. Table and Pareto charts profile the various units and the current status summaries of each unit. These patients also can be rolled up into services, or grouped by physician. In addition, patients can be aggregated in many other ways such as time of admission, length of stay, and admitting diagnosis.

One key feature of this system is its use of "Intelligent Agents." These online agents, as shown in Figure 100-6, are constantly monitoring and analyzing patient and census information, and they have the ability to detect key system situations, such as high census in a unit (i.e., no available beds), excessive ED placement time for a particular patient, or delays in responses to placement requests.

Patient Confidentiality

A full security system is embedded within the dashboard to audit user access and to assign users to definable system roles. These roles are restricted to specified processes and are prohibited from viewing or changing certain data. The system is designed to be fully compliant with the evolving Health Insurance Portability and Accountability Act (HIPAA) regulations.

Case Study #2: PIVIT – Performance Indicator Virtual Instrument Toolkit

The Challenge

Managing multiple data sets and performance indicators can be tedious and time-consuming. This results in a compromised ability to assess effectiveness, detect problems, and capitalize on opportunities for improvement.

The Solution

An easy-to-use dashboard (PIVIT) was developed to integrate disparate data sets and performance indicators to monitor operational performance and to identify areas for performance improvement.

System Overview

PIVIT is an acronym for "Performance Indicator Virtual Instrument Toolkit" and is an easy-to-use data acquisition, analysis, and presentation application. PIVIT applies virtual instrument technology to access, analyze, and forecast clinical, operational, and financial performance indicators. Examples include applications that profile institutional indicators (e.g., patient days, discharges, percent occupancy, ALOS, revenues, and expenses) and departmental indicators (e.g., salary, non-salary, total expenses, expense per equivalent discharge, and DRGs). Other applications of PIVIT include 360-degree peer review, customer satisfaction profiling, and medical equipment asset management.

PIVIT can access data from multiple data sources. Virtually any parameter can be easily accessed and displayed from standard spreadsheet and database applications (e.g., Microsoft Access, Excel, Sybase, and Oracle) using Microsoft's Open Database Connectivity (ODBC) technology. Furthermore, multiple parameters can be profiled and compared in real-time with any other parameter via interactive polar plots and three-dimensional displays. In addition to real-time profiling, other analyses such as statistical process control (SPC) can be employed to view large data sets in a graphical format. Statistical process control has been applied successfully for decades to help companies reduce variability in manufacturing processes. These SPC tools range from Pareto graphs to Run and Control charts, to process capability distributions.

Statistical process control has enormous applications throughout health care. For example, Figure 100-7 is an example of how Pareto analysis has been applied to a sample trauma database of over 12,000 records. The Pareto chart can display frequency or percentage, depending on front-panel selection, and the user can select from a variety of different parameters by clicking on the "pull-down" menu. This menu can be configured to automatically display each database field directly from the database. In this example, multiple fields (e.g., e-code, DRG, principal diagnosis, town, zip code, and payer) can be selected for Pareto analysis. Figure 100-8 further illustrates some of these custom virtual instruments that apply statistical process control to a variety of data sets.

Figure 100-9 shows how equipment selection filters allow the user to restrict data analysis to desired equipment and equipment parameters (e.g., device category, manufacturer, status date, locations, or models). Figure 100-10 illustrates the equipment statistics module that allows users to graphically profile CMMS data in the form of a histogram or a Pareto chart. In this example, the devices are sorted and plotted in a Pareto chart. By clicking on any of the column headings or selecting a different option from the pull-down menu, the graph will dynamically update. Detailed data are rolled up to summary levels and can be drilled down to fine detail, all with a few clicks of the mouse. Users do not need to know how to write queries or program code. The equipment history profile in Figure 100-11 allows users to visualize relationships of work order events for a specific device, as they occur in time. Each vertical line or cursor represents a specific activity (e.g., incoming inspection, preventive maintenance, unscheduled repair, or user error).

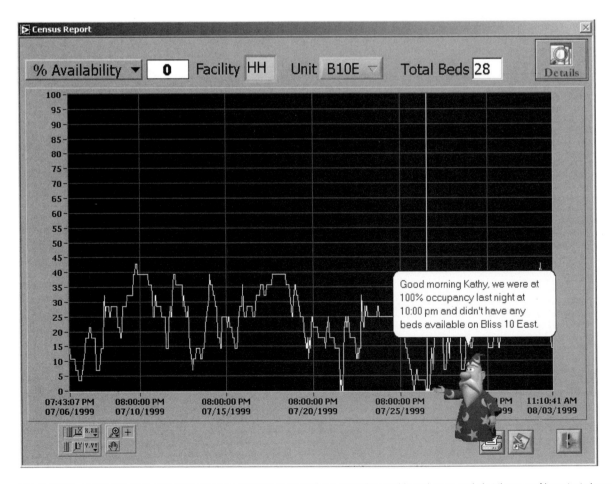

Figure 100-6 Intelligent agents. The BMD can also employ on-line "intelligent agents" to provide assistance and alert the user of important alarm conditions that otherwise might have gone unnoticed. Messages can be in the form of an onscreen agent (such as Merlin the Wizard) or also via e-mail, pager, fax, and/or telephone.

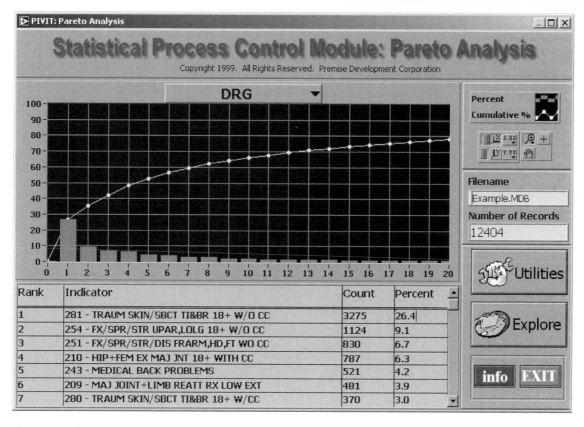

Figure 100-7 Statistical process control–Pareto analysis of a sample trauma registry.

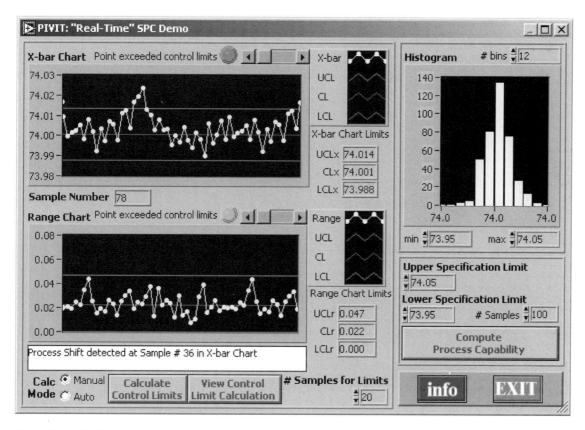

Figure 100-8 "Real time" control chart, range chart, and histogram.

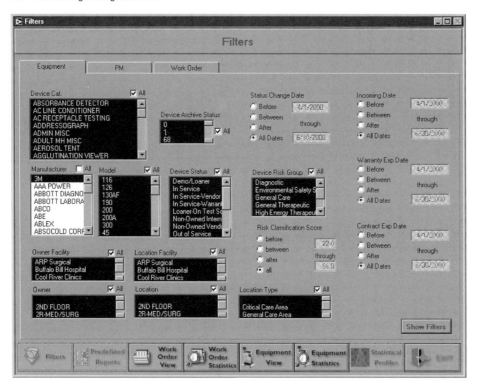

Figure 100-9 Equipment selection filters.

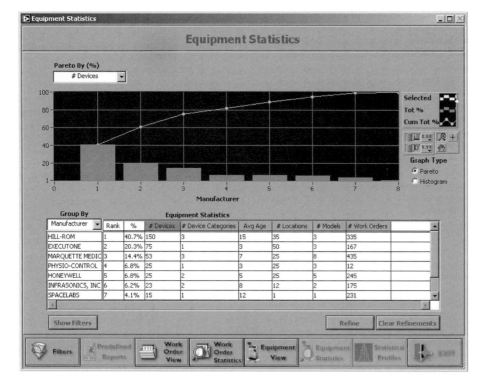

Figure 100-10 Equipment statistics module.

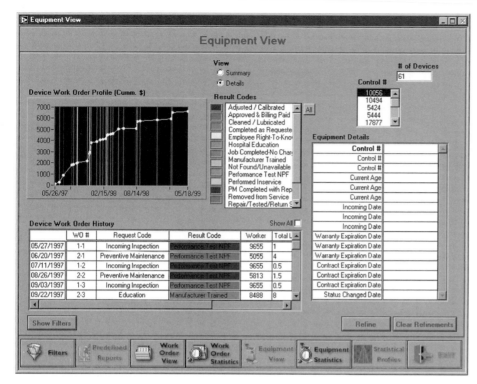

Figure 100-11 Equipment history profile.

Conclusion

Executive dashboards and virtual instrumentation allow organizations to effectively harness the power of the PC to access, analyze, and share information throughout the enterprise. The case studies discussed here illustrate ways in which various institutions have conceived and developed "user-defined" solutions to meet specific requirements within the health care and insurance industries. These dashboards support general operations, help hospitals to manage fluctuating patient census and bed availability, and enable insurance companies to identify best practice patterns, accurately price products, and increase market share. Decision-makers can move easily from big-picture analyses to transaction-level details while safely sharing this information throughout the enterprise to derive knowledge and to make timely, data-driven decisions. Collectively, these integrated applications directly benefit health care providers, payers, and, most importantly, patients.

101

Telemedicine: Clinical and Operational Issues

Yadin David
Director, Center for TeleHealth and Biomedical Engineering Department, Texas Children's Hospital
Houston, TX

In a remote village in Central America, an otherwise unmanaged child's infection could have taken a turn for worse, had it not been for the long distance diagnosis capacity brought about by the recently installed telemedicine system among Zacapa, Guatemala, and Houston, Texas, where subspecialists in the field of Pediatric Dermatology viewed the wound and prescribed the critically needed treatment. Recent developments in the telecommunications and the information technology fields hold the promise of improved access to, and the better utilization of, health care-related resources. In addition to these developments, the deployment of interactive distant training programs offers an opportunity to decrease the knowledge gap between the leading academic medical centers, where new medical knowledge is continuously being discovered, and the remote health care practitioners who find themselves pressed to deliver quality care that meets the needs of their communities in a competitive environment of limited resources. Telemedicine is creating new opportunities for imagining the possibility of a more efficient and accessible health care delivery system. Telemedicine is merely the tool (not the end) by which to negotiate and overcome barriers on the road to the delivery of quality services that the present system cannot deliver as effectively. The developments of modern telecommunications and information technology tools present a radical opportunity to change the health care delivery infrastructure from the ground up. The implementation of telehealth programs in such an environment is supported by enthusiastic approaches, but not with enough planning tools. There is a need to understand how to best adopt the advantages offered through intelligent communication that limitlessly extends the boundary of our senses, records, activities, and outreach. At its basic application, telemedicine can take the distance out of caring. Clinical needs, including the provision for innovative quality continuing education, is one of the programs' prime focuses. However, once moved beyond the clinic, the financial, legal, and engineering implications of telemedicine challenge the *status quo*. If telemedicine is going to become a staple of medical practice on our planet, a massive re-engineering of the national and international telecommunications infrastructure, reimbursement legislation, and clinical practices will be required. There are over 150 telemedicine programs in operation today. It is estimated that 60% of them are less than six years old (Peredni, 2000), representing a relatively new modality in the health care delivery system, where the vast majority of the providers are still debating the appropriate entry course and the following regulatory and sustainability issues. This chapter reviews the evolution and the major issues to consider before embarking on a telemedicine program.

Same Problems: Different Solution

The Center for TeleHealth (CTH), one of the first telemedicine programs to focus solely on the extension of pediatrics subspecialty services, is the result of a collaborative program between the Texas Children's Hospital and the Baylor College of Medicine. While the Center was created for the specific goal of linking the institutions' expertise to remote communities in Texas as well as globally, it would not have enjoyed the executive and medical faculty's support if it had not been for the success demonstrated during its initial applications, the result of using video engineering concepts to solving intrahospital problems. It required the installation of a video network for the transmission of real-time information transmission of echocardiography to the locations where the experts were available (i.e., the cardiology reading area). Soon, other problems were solved with a similar measure of success. The subspecialty coverage of a remote clinical area, the newborn nursery, successfully tested the provision of remote video monitoring of babies by neonatalogists (Adcock et al, 1999). While new knowledge is being discovered and acquired at an accelerated pace, specifically in the field of diagnostic and therapeutic medicine, within the major cluster of the academic medical centers the care provider (who practices in the remote and, at times, rural area, or in the overburdened urban clinics) has yet to realize how best to absorb and deploy this new information. Telehealth is providing a unique opportunity to restructure the dissemination of knowledge methodology used in today's health care delivery system in a way that will distribute specific competencies to the most needed environment—the best practice at the point of care. This focus will allow our society to move from a reactive mode of delivering care toward a life-long health management where individual accountability, preventive medicine, and customized wellness programs are integral components of a new health management methodology.

The Telemedicine Evolution

Telemedicine is not a new entity. It has enjoyed over seventy years of historical evolution. The April 1924 issue of the *Radio News Magazine* showed a drawing of a physician viewing a patient over short-wave radio set that included a television-like display. More realistically, the early programs in the 1950s and 1960s were highlighted with some success by the Nebraska Psychiatric In-state and the Massachusetts General Hospital and Boston's Logan International Airport link (Dakins, 1997). However, these programs lacked the ability to deliver sustained satisfaction to the health care providers or to the patients involved. Following the invention of the color television and launching of satellites in 1965, Dr. Michael DeBakey, the cardiovascular surgeon at the Baylor College of Medicine, started to incorporate video tools into the medical training program. That year, he successfully broadcast the first live cardiac surgery, from the operating rooms at the Methodist Hospital at the Texas Medical Center to Europe, utilizing satellite transmission. Colleagues in Amsterdam viewed and heard Dr. DeBakey mentoring the procedure at the time a new technique for teaching cardiac surgical intervention. The investment by the United States government in the late 1980s and early 1990s in the integration of better telemedicine tools placed a focus on the rural communities and the need to extend care to the internationally deployed military forces. These efforts provided the initial impetus for today's telemedicine programs. The new modality, the Internet, provided another example of ways in which the barrier of distance can be overcome by having access to tools that enjoy a much more ubiquitous communications platform to transmit information between any two distant locations at anytime. In 1983, the first use of the Internet to disseminate medical information between participants took place at the Texas Medical Center and demonstrated the potential for scientific interactions that had been untried before then. Along with the use of satellites and telephone lines to deliver medical information, the Internet now provides communication between two single points, as well as the ability to share information among multiple sites simultaneously, at an affordable cost. The early applications of video engineering were, essentially, an attempt to overcome intrahospital information-flow "bottlenecks." The connection with remote rural communities was not attempted until sufficient experience had accumulated with intra-hospital transmission of video, with the available telecommunications services including the "last mile," the point between the shared infrastructure and user connectivity. Thus, the Texas Children's Hospital/Baylor College of Medicine Medical Center program began in 1993 to take advantage of these opportunities to reach out with its educational and clinical programs beyond the confines of the Center, reaching out 400 miles southwest of Houston to McAllen, Texas, near the Mexican border. The program then, as today, focused on the extension of pediatric subspecialty expertise to communities in need. The growing acceptance of telemedical-based services is changing the way health care is delivered and provides opportunities that had not been available before. In a recent survey of telemedicine service providers conducted by *Telemedicine Journal*, almost 70% of the responders cited the need to increase access to specialty care as the major motivation for initiating remote interactive clinical service while 33% of the responders cited cost saving as the primary methodology for measuring the success of such a service (Field J, 1996). While the service can be administered in a variety of platforms, the needs to accommodate clinical requirements and to appreciate technological limitations are universal. However, these requirements and limitations change from one program to another and from one location to another. Successful development and implementation of telemedicine service is dependent upon the ability to focus on and address specific clinical needs, the integration of infrastructure with existing platforms, and the availability of a reliable and sufficient end-to-end broadband telecommunications service. In addition, sensitivity to cultural preferences and compliance with legal issues has an impact on performance as well.

The Telemedicine System

Interactive computers and advanced telecommunications technologies are transforming every aspect of our life. Medicine is no exception. From virtual classrooms to simulation research, and from pharmaceutical dispensers to artificial organ replacements, advances are generating breakthrough applications that improve the quality of life for an ever-increasing number of people. Health care providers and educators as constituents of this transformation era enjoy new opportunities created for their active participation in this transformation. Reviewing the positive impact that the integration of ostensibly independent patient care services has on the efficient management of the total quality of care, of education, and of collaborative research, it is not surprising that telehealth deployment is on the rise. The forces that drive this phenomenon include the following:

- The need to manage the entire disease encounter
- The desire to apply standard quality of care over a wide geographically distributed community
- The escalation of customer expectations
- Globalization of health care and its support services
- An increase in patient and provider convenience
- User acceptance of the current technological competency

Telehealth can be envisioned as the practice of health care delivery through the use of advanced communication technologies, computers, video instrumentation, and medical devices to exchange information and to deliver services that overcome the barriers of distance, time, and socio-cultural differences. Specific applications include the following:

- Delivering medical services
- Consultation or validation
- Prevention
- Diagnosis,
- Treatment
- Transferring medical data
- Education
- Collaborative research

Current applications are classified as follows:

1. Initial screening/evaluation of patients
2. Triage decisions and pretransfer screenings
3. Medical and surgical follow-up and medication review
4. Consultation for primary care encounters
5. Real-time subspecialty care consultation and planning
6. Management of chronic diseases and conditions
7. Extended diagnostic work-ups
8. Review of diagnostic images
9. Preventive medicine and patient education

In addition, the Medical Center has successfully conducted real-time interactive clinical engineering seminars between Houston, Texas and Riyadh, Saudi Arabia. Many other examples, such the Visions of Health and Peace project, through the use of telemedical tools, demonstrated the promise of global biomedical partnership between Israel and the United States. This is a critical opportunity for engineers to participate in supporting telemedicine programs by demonstrating skill sets that can assert and establish the quantifiable level of service that ensures an appropriate match between the fidelity of acquiring and delivering sound and video levels and the clinical requirements of the specific encounter and that can service and support the operations of this complex system.

The Telemedicine Practice

Telemedicine is the use of electronic information and communication technologies to provide and support health care services when distance and time separate the participants (Field MJ, 1996). In its simplest form, a telemedicine program consists of a site where a consulting physician is located (the "hub") and a remote location (the "spoke") where the referring physician and the patient are present. The consultation between the two peers can take the form of a diagnosis session or a second opinion session. The central hub can serve one or many spokes located at remote sites. The peer-to-peer review can be practiced in two different modalities—in real-time or in delayed time known as the "store-and-forward" mode. The difference between the modalities is the time span associated with the consultation protocol and the requirements of bandwidth.

Telemedicine practices can be categorized into variety of modalities, for example, according to type of services provided. Telemedical encounters can focus on clinical, educational or administrative programs. A program that provides clinical encounters must comply with a variety of regulatory guidelines and must adhere to policies and procedures that are quite different from those required for the provision of educational services. Another categorization can be along the types or the size of network deployed, such as a point-to-point or a hub-and-multiple-spokes system. However, the most popular categorization is done along the span of time within which the telemedical encounter begins, is reviewed, and completed.

Most of the telemedicine programs operate in a synchronous mode, or in what is known as an interactive real-time full-motion encounter. The other modality, store-and-forward, is an asynchronous mode whereby the referring physician and the patient do not have to be present at the time of the consultation together with the specialist. Rather, the referring physician gathers the patient information over a period of time utilizing a battery of diagnostic instruments, ranging from electronic stethoscope and X-ray images to pathological slides or cine fluoroscopy as shown in Figure 101-1. Other patient information, such as echocardiograms, can be incorporated, as can dictation of patient symptoms, family history, vital signs and demographics. This information is then packaged and moved as a completed file, electronic portfolio of the patient, to the consulting physician address at the hub. The transmission is encrypted to protect privacy and to provide confidentiality of information. The consulting physician, after being alerted to the arrival of new electronic file, will open the portfolio, review it, and then render a diagnosis and an opinion as to the management of this condition. The consultant will send it back to the referring physician for action. There are several advantages to the store-and-forward modality. The patient need not be constrained by the schedules of the referring and the consulting physicians; the cost of telecommunications is minimal. On the other hand, the advantage of the real-time interactive mode is the high presentation quality that can be deployed, and the ability to have multiple subspecialists participating simultaneously in management of the condition permits earlier and quicker decision making processes. Accompanying the increase in clinical applications, the appreciation for the minimum features level and fidelity provided by the various types of equipment and communication systems must become clearer in order to efficiently match capability with service needs.

There is a critical opportunity for engineers and clinicians to participate in this process by defining the appropriate fidelity level that will best accommodate the clinical requirement. There still are critical financial, legal, and technical barriers to overcome. These include the following:

- Reimbursement
- Professional component
- Administrative component
- Technical component
- Liability
- Licensure
- Scheduling issues
- Equality/universality of care
- Patient/MD relationship
- Appropriateness of application
- Patient resistance
- Abuse/over-utilization
- Confidentiality/privacy

Transmission Test Conditions

The evaluation of room-based telemedicine systems requires the commitment of many resources. There is a continual need to evaluate new, evolving systems and technologies for an optimal match between patient and physician needs and technological capability in order to reach display-fidelity optimization and to refine and standardize efforts. The Medical Center test was subjective in that its panel of experts was scoring competitors based on their impression and perception of "quality." Seven years ago, it conducted an evaluation test to determine the magnitude of variation in video processing between the systems. The testing protocol provided for a single video source to simultaneously broadcast its signal into two video conferencing systems. The videoconferencing systems were designed to communicate with the video source at different bandwidths. The communication with video source was in real time over a protocol H.320 full T1 carrier service (1.544 Mbps) down to a 1/4 T1 (386Kbps). The video source was changed from an ultrasound examination of a patient cardiac condition to a simulated patient with advanced Parkinson's disease. Other rapid and sudden movements against a large, white background (such as a downhill skiing race) were used as well. A panel of senior medical subspecialists observed the presentations and related their ability to render a medical opinion based on the quality of the displayed video and the fidelity of the audio (Figure 101-2).

For a slow-moving event, such as an examination of skin conditions, a fraction of a T1 line was acceptable for rendering an opinion. However, for ultrasound examination and downhill skiing, the panel opinion was that only presentation-quality 1/2 T1 or better was sufficient to determined minute changes in the conditions presented. Proper lighting and the positioning of the camera during patient examination were also noted as critical for successful/assessment of clinical conditions. Finally, the resolution and the overall size of the video screen were determined to have impact on the clinical acceptance of the telemedical encounter as well. The closer the display area was to a real-life size, the more the panel felt it was acceptable.

Consequently, the Center conducted telemedical encounters where content included significant motion through only 1/2 T1 or better and employed the diagonal measure of 31-inch television sets (Figure 101-3). All video sources were calibrated as to level and were connected via a 1X8-video switcher (selector). This enabled the selection of the various sources utilized in the evaluation at the request of the panel members. The output of the selector was passed to a distribution amplifier (D/A) to compensate for losses resulting from the division of the video signals into two paths. The video signals were then passed to the two transmissions CODECs, (Compressor/DECompressor) "A" and "B." Equal lengths of transmission lines connected the CODECs at the "near end" used for analog-to-digital conversion and subsequent compression and transmission of the signals to the CODECs at the "far end" used for reception, decompression, and digital-to-analog conversion. The output analog information from the two "far end" CODECs were connected to individual multisync monitors ("A" and "B"). The monitors, which were both of the same manufacturer and model, had been calibrated prior to the evaluation, using an NTSC laboratory-grade color-bar generator and photometric color tem-

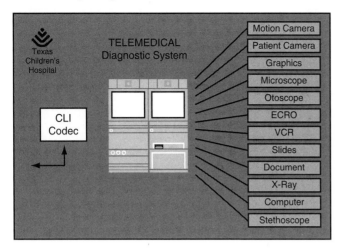

Figure 101-1 Patient information gathered from a battery of diagnostic instruments.

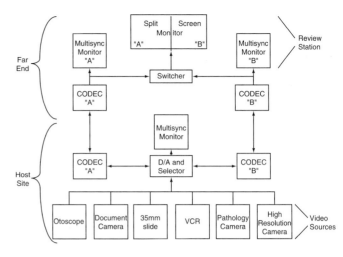

Figure 101-3 Telemedicine evaluation test.

Figure 101-2 Panel of senior medical subspecialists observing and evaluating telemetry presentations.

perature and luminance meters. This calibration ensured that both monitors were objectively as similar as possible in terms of picture quality. In addition, to provide a better side-by-side comparison of the two systems under evaluation, a split-screen monitor was employed. Video information from both CODEC sources was split and sent through a switcher, enabling the side-by-side viewing of video information on a single monitor. Immediately prior to the tests and evaluation, the clinical engineer in charge ensured the integrity of the "double blind" nature of the evaluation by making the final connections from the outputs of CODECs "A" and "B" to the monitor chain. This information was kept confidential until the evaluations were completed.

Resolution of these issues will extend the ability to exchange information and to increase interoperability with legacy systems. This will extend the capacity to engage in this evolution and to determine when it presents sufficient fidelity to deploy this service in the clinical setting. Finally, the legal and regulatory aspects, in addition to the domestic and global aspects, for the test described above must be reviewed.

The Telecommunications Component

Telecommunications, one of the most quickly changing fields, is enabling existing and new platforms to exchange information and to support interaction through unique applications including the medical peripherals. Thus, the management of a telemedicine program must incorporate the understanding of telecommunications technologies and the optimization of this component. The convergence of the digital and the analog transmission methods and the advantages of the various networking technologies require an ability to evaluate and match the network needs with the capabilities of the available infrastructure. Clinical requirements must be matched with technical capacity and supported by financial sustainability to influence the design and the implementation decisions when selecting the platform to be employed. The many modes of communications support speed and bandwidth that ranges from the narrowband used for the traditional telephone system to the wideband application supported by an integrated services digital network and satellite communications. Various selectable bandwidths and availability and level of service should be considered. In selecting the telecommunications technology, one must clearly know how the network will be designed, what protocols are expected to be used, how much and how often data will be transmitted, and, finally, what the budget will allow for this program.

Video compression, a coding technique that is used to reduce the bandwidth required for the transmission of video images, is a developing science. Recent developments are awaiting validation of its clinical acceptance and are still largely unsubstantiated by a conclusive body of research. Opinions vary widely on the topic, but few hard data exist. The science of visual perception is unique in that it is a subjective function of the brain as well as the eye. No two individuals perceive a particular image in exactly the same way. The threshold of persistence of vision and the measurable speed at which a motion-picture or video screen ceases to flicker and begins to "move" may vary significantly among different observers under different lighting conditions. The eye and brain are thought to retain a visual impression for approximately 1/30th of a second. When viewed as a continuum, this collective retention is "seen" by the viewer as uninterrupted motion. It is important, therefore, that the effect of the compression technique chosen for a telemedicine system will not further modify the assessment of the medical conditions being viewed.

Summary

Telemedicine is an emerging industry with the potential to revolutionize the delivery of care to the benefit of health care systems, providers, and patients. The application of electronically connected parties in the provision of medical care and training has already begun to show some very exciting possibilities. The expansion of telehealth systems will deliver services to the benefit of providers' networks, the economy, and patients. Recent developments and improvements in standards in the telecommunications and information technology fields hold the promise of improved access to, and better utilization of, health care-related resources. Telemedical services can be deployed in a variety of platforms. Those platforms keep improving the capturing, transmission, and presentation of medical information. The deployment of any one platform, or a combination of several platforms, is dependent on such conditions as clinical needs, level of staffing training, legal and cultural constraints, limited telecommunications infrastructure, regional economics, and the stage of technological development. Therefore, some applications are better suited than others to support the needs of distant health care providers. Telemedicine systems must be tailored to the needs of the participating health care system. Usually, a tiered approach consisting of a mixture of real-time and store-and-forward technologies, wide bandwidth, and low frame rate is the optimal solution. With proper planning and the anticipated advances in computing power, direct medical digital imaging and telecommunications will continue to increase the cost-effectiveness and will expand the scope of telemedicine services. Two modalities of practice are emerging: The real-time interactive mode and the store-and-forward mode. The combination of these modalities with the capabilities to exchange medical information as supported by the new telecommunications infrastructures places telemedicine at a central position for improving access to quality health care by all. Financial, legislative, and regulatory penalties (Center for Telemedicine Law, 2004) must continue to be removed in order to permit everyone in need to reach their goal by finding the ramp and to get onto the telemedicine highway. The contribution of this transformation to the betterment of health for all must be identified and quantified. This challenge is the next and biggest obstacle for the field of telehealth. Dr. Michael DeBakey put it best: "These technologies will allow us to improve the standards of health care around the world ... providing an important opportunity for distant learning and for the dissemination of medical knowledge and support of this goal."

Sources

1. Adcock L, David Y, et al. *Telemedicine Examination of Newborn Infants: How Accurate Is the Technology?* San Francisco, CA, Society for Pediatric Research, 1999.
2. Dakins DR. Telemedicine in Review, TeleHealth: Steps to Successful Implementation. *U.S. Proceedings of the TeleHealth 1997 Conference.* Dallas, Texas, HIMSS, 1997.
3. Field MJ. *A Guide to Assessing Telecommunications for Health Care.* National Academy of Medicine, 1996.
4. Perednia D. Summary of the 1999 ATSP Report of Telemedicine. *U.S. Proceedings of the TeleHealth 2000 Conference.* Los Angeles, HIMSS, 2000.
5. Reid J. *A Telemedicine Primer: Understanding the Issues.* Iowa, Innovative Medical Communications, 1996.
6. The Center for Telemedicine Law. Washington, DC, www.ctl.org, 2004.

102 Picture Archiving and Communication Systems (PACS)

Ted Cohen
Manager, Clinical Engineering Department,
Sacramento Medical Center, University of California,
Sacramento, CA

A picture archiving and communication systems (PACS) is a system that electronically processes, stores, distributes, and retrieves digital medical images in a portion of, or throughout, a health care enterprise. PACS is most commonly used in radiology departments to process and route radiographs and other diagnostic imaging modalities including MRIs, CTs, nuclear medicine scans, and ultrasound images. PACS epitomizes the integration of medical, computer, and communication technologies with its requirements for large amounts of storage, sophisticated databases, large bandwidth, high-speed networks, computational power at the desktop, ease of use requirements for a variety of end-users, and specialized medical application software. PACS can also be used to store image information from cardiac catheterization procedures, endoscopy, dermatology, radiation oncology, echocardiology, and pathology.

One goal of PACS is to improve the imaging department workflow as compared to film-based technologies. PACS automates and, to varying degrees, improves the workflow among imaging modalities, image acquisition, image processing, image distribution and storage, and the associated personnel within the health care institution. Because 70% of the typical radiology department's workload is plain film images, as radiology departments convert to digital modalities, PACS implementation on a larger scale becomes more practical. In addition, the practical implementation of PACS requires high integration with the Radiology Information System (RIS) and/or the Hospital Information System (HIS) for scheduling, workload management, and radiology reports. PACS also may help to alleviate film, film processing, and film storage costs and to reduce the number of retakes and lost films. The growth of high-bandwidth intranets will also facilitate the growth of PACS.

PACS Technology

PACS are computer network-based image storage, processing, distribution, and retrieval systems that can be used to capture, store, recall, display, manipulate, and print medical images and associated data. Digitally stored images can be accessed at any time from any PACS-enabled workstation, either within the radiology network or remotely (e.g., off-site). Ideally, PACS would facilitate an X-ray filmless environment, and most institutions using PACS see a considerable decrease in film usage (filmless, not film-free). Most PACS are interfaced to hospital information systems or radiology information systems and, to a lesser extent, clinical information systems. However, the ideal, paper-free, fully electronic patient record with demographics, images, clinical data, and other medical records fully integrated is very rare.

PACS are typically made up of the following components: Modality interfaces, a high-bandwidth network, two or more tiers of image-review workstations, two or more tiers of image archives, specialized servers, and interfaces to other information systems and peripherals (e.g., film printers and paper printers). These components are composed of commercial off-the-shelf (COTS) hardware, COTS operating system software and specialized, developer/vendor-specific PACS application software. Each of these PACS components is discussed in detail below.

Modality Interfaces

All radiological imaging devices can be interfaced to a PACS. Digital modalities (e.g., CT, MRI) are usually interfaced directly with DICOM (see DICOM sidebar below) capabilities built in to the modality, as are newer digital modalities (e.g., digital radiography (DR)). Special add-on interface devices can be added to older models of imaging equipment to provide DICOM compatibility.

A traditional X-ray makes up 60% to 80% of all diagnostic radiography examinations. X-ray machines can be interfaced to PACS by using computed radiography (CR). With CR, conventional X-ray acquisition can be performed using a phosphor screen with energy storage capability as the X-ray image receptor. The phosphor screen is contained in a standard-sized radiographic cassette and can be used in existing radiographic tables and wall stands. After exposure, the cassettes are transferred to a reader system. There, the image plate is scanned with a laser device that stimulates luminescence proportionally to the local X-ray exposure. The luminescence signal is converted to an electrical signal and is digitized. The raw digital data representing the image are further processed to optimize the diagnostic content of the visual data. The processed image then can be stored, transmitted to a PACS, and/or printed via laser-printed film.

X-ray rooms with integrated CR devices or solid-state, flat-panel detectors are also available. For small volumes, X-ray film can be individually scanned with a film digitizer to convert them to digital format.

Networks

A PACS is typically configured as its own local area network (LAN) or as a dedicated sub-net in a wide-area-network (WAN) environment. Modalities (or modality interfaces), servers, workstations, and peripherals are interconnected with high-speed computer networks via coaxial, twisted pair, and/or fiber optic cables. Individual networks differ in design, but all have to take into account the very large amount of image data and the speed requirements of PACS. A variety of methods can be used within PACS networks, depending on the institution's legacy network and the geography of the institution. PACS networks have successfully used Ethernet (including 100 megabit and gigabit Ethernet), ISDN, T1 and T3 connections for remote sites, ATM, frame relay, and other high-speed data circuit routing and switching. It is important that the data circuit bandwidth be sized sufficiently large for the amount and speed requirements of the data. Internally, within the PACS, faster data throughput can be obtained using fiber-channel and other new storage and networking technologies (e.g., optical SCSI Storage Area Networks (SAN)).

A centralized PACS has a central archive around which all modalities and workstations are attached. This architecture allows equal access to the entire PACS. However, on a very large system, access can be slowed by one or more slow components (e.g., deep archive search). Although this architecture may include pre-fetching from the deep archive to the short-term archive, it does not require that the pre-fetch go to a specific workstation or distributed server.

Distributed PACS architectures have certain modalities connected to specifically identify distributed servers and/or archives. Attempts are made to optimize the server, workstations, and/or archive used for a particular modality or location. This can result in faster response times when the identified workstation is used, but it reduces flexibility and/or significantly reduces response time when a study must be read from a workstation cluster that is not the primary identified workstation for that modality or location. Both centralized and distributed topologies have been successfully implemented for PACS. However, to overcome some of the limitations of the centralized and distributed topologies, hybrid, "on-demand" systems have been developed. Figure 102-1 shows a sample topology for a large generic hybrid PACS. Images, upon completion of a study, automatically move from the modality to the near-term archive and the web-server. Demographic and image storage parameters, but not the images themselves, are stored on the database server. Any image can be viewed from any workstation. Images also can be automatically or manually routed to specific primary diagnostic workstations based on workflow parameters stored in the database server and/or individual requests (e.g., requests to push an image to a specific location or requests to pull an image from a specific location).

The server database acts as a traffic cop on the on-demand system. The database structure includes information about a particular image (e.g., date and time of acquisition, patient demographics, and modality), but not the image itself. When users select an image based upon a certain criteria, the database locates the image that meets the criteria, identifies and notifies the archive (short or long-term), and requests the archive to pull the image and route it to the workstation. Typically, an additional server is used to connect the database and archives to the web-based workstations. In addition to requiring more sophisticated server databases, the on-demand systems work best when there is a high degree of interoperability between the RIS (Radiology Information System) and the PACS.

Workstations

Primary diagnostic workstations are used by the radiologist to view images for patient diagnosis. These workstations require the highest resolution and speed. Workflow must also be optimized as the radiologist might be reading a large number of studies in one day. Comparisons with prior studies might be required. These diagnostic workstations replace the motorized view boxes that are typically used in a film environment. Primary diagnostic workstations are required to display an image in sufficient resolution, comparable to film for film-based modalities, for the radiologist to make an accurate diagnosis. Images are selected based on a wordlist that comes from the modality or the RIS. Workstation users can review and manipulate (e.g., magnify, rotate, zoom on a region of interest) the images and add, edit, and delete annotations for later reference by themselves or other clinicians. High-resolution CRTs of up to 2048×2560 (2 K $\times$ 2.5 K) pixels are required. Multiple displays (2 or 4) are often used to view multiple images and/or do serial comparisons against prior studies or multimodality comparisons of the same current region of interest. Except for some specialized applications (e.g., nuclear medicine or ultrasound), displays are monochrome, gray-scale monitors.

Images also must be sequenced and displayed in a consistent manner, based on the needs and desires of the radiologist, modality, and type of study. Configuration sets are built so that the appropriate formats for display and sequencing are available based on the specific radiologist user's log-in, the modality, and the type of study (e.g., anatomy). These configuration sets are sometimes called "hanging protocols" based on their historical comparison to hanging multiple films on motorized X-ray view boxes in a certain order and place.

All primary diagnostic workstations offer some image manipulation tools including zoom, pan, magnify, invert, rotate, and measure (e.g., calipers). These workstations also allow image sequencing (sometimes called "stack mode") to dynamically view scan slices. In addition, specialized workstations are available that offer further manipulation

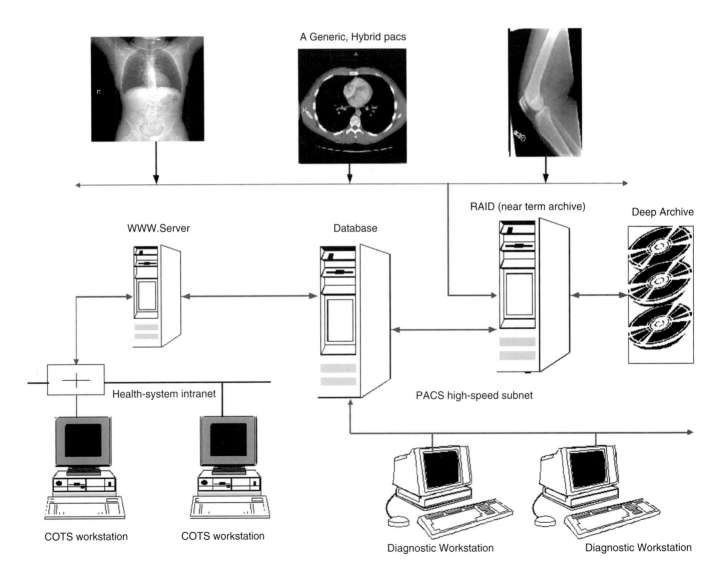

Figure 102-1 Sample topology for a large generic hybrid PACS.

similar to the workstations available with the modalities. Examples include CT-based virtual colonoscopy and 3-D rendering for surgery planning using CT or MRI.

These high-resolution monitors are expensive and require periodic calibration in order that each has consistent brightness and contrast so that a given study will look the same, regardless of which specific monitor is used to display the exam. These high-resolution monitors also decay in brightness over time, and various manual and automated calibration systems are in use to keep these monitors calibrated and to determine when their CRTs need to be replaced.

The primary task of the radiologist is to view the images and to report significant findings to the referring physician. In order to improve the radiologist's efficiency, various schemes for integrating report generation (e.g., speech recognition and typed reports) into the workstations have been developed.

Secondary review workstations, sometimes called "clinical review" workstations, are used by medical personnel (e.g., referring physicians) outside of the radiology department to review images in areas where image information is important and timely, such as intensive care units (ICUs), emergency rooms (ER), and orthopedics. Referring physicians might need to view an image and sometimes might show the image to the patient. This can be accomplished with a less sophisticated, lower-resolution display system. Specialized secondary review workstations are vendor-specific and expensive.

Tertiary-review workstations have even lower resolution and computer requirements. Most tertiary review workstations are web-enabled, commercial-off-the-shelf systems (COTS) with special PACS software and perhaps a special monochrome monitor and video interface/driver. The trend in secondary and tertiary review workstations is to use COTS and standard WWW networks in order to avoid the high cost of PACS workstations. (See cost section below). As the tertiary, COTS, and web-based workstations improve, custom secondary workstations will become obsolete because of their high cost.

Archives

Large radiology departments can generate more than 70 gigabytes of data per day. The ability to collect, store, retrieve, and distribute that large amount of data is one of PACS's primary challenges. Quick retrieval is always an important specification for the radiologist for initial diagnosis. Numerous studies have shown that images are looked at an average of x times within the first week, but only y times with the next z years (x>y>z). Therefore, a tiered approach has been used for image archive; near-term and long-term archives.

All images that are recent and/or otherwise likely to be retrieved in the near-future are typically stored on near- (or short-) term archive, usually using redundant array of independent disks (RAID) technology. RAID can store for a few weeks up to a year or two worth of images recently taken or images queued (i.e., pulled from the long-term archive) for likely retrieval, based on an upcoming appointment or other scheduled event. Unread studies are always stored in the near-term archive.

Long-term archival library systems are automated, robotic devices that index and load various types of media (e.g., tape or optical disks) into readers for retrieval of old studies. These devices can accommodate several different types of media and are expandable to hundreds of terabytes (one terabyte = 1000 gigabytes). As media design changes, and the cost per terabyte goes down, new jukeboxes and other long-term archival systems will be developed. Table 102-1 shows a comparison of long-term archive media. One concern in this rapidly changing technology is that the ability to read current media in several years may be dependent on keeping obsolete technology operational. An analogy would be trying to read eight-inch floppy disks today.

ASP Model

One of the newer solutions offered for the purpose of archiving is off-site storage via an application service provider (ASP). This remote archiving allows a hospital to outsource the hardware, software, and support for its image archive to a third-party. Images are retrieved via a web-browser application, and the ASP provider takes care of all the updates, uptime requirements, backup, and disaster recovery. Speed of retrieval is dependent on the same parameters as an in-house archive in addition to the speed of the wide-area-network connection to the ASP. Some reports have shown speed improvements using an ASP because the ASP was able to use larger jukeboxes and, therefore, to keep all image data online compared to operator intervention for archive magneto optical disk (MOD) loading. A typical cost model for ASP archiving is to charge a fee per image and a fee per retrieval. Definitive, long-term cost studies have not yet been done—only projections to determine the point at which it becomes beneficial to use an ASP rather than in-house storage model, and vice versa, have been made

Another similar idea to the ASP archive is to locate an institution-wide archive in the information technology (IT) department and to archive all information systems including PACS and laboratory information system (LIS), in a centralized location.

PACS Performance

The archive section above discussed the large amount of data that can potentially flow through a PACS system in a period of time. Two of the challenges when handling this large amount of data are retrieval speed and fault tolerance (redundancy). Both add significantly to the cost of any robust PACS.

Speed

Radiologists demand fast retrieval time. They are used in viewing film that is pre-hung on motorized X-ray view boxes and as long as the appropriate prior files were hung, they can view serial comparisons and other relevant past films quickly. PACS technologies that determine retrieval and display speed are the speed of the workstation processor(s), with gigabytes of memory and special fast video interface cards, and the speed of the network (pipes) and associated routers and hubs (100 megabyte Ethernet minimum; gigabyte Ethernet or fiber to the workstation optimum). Pre-fetching optimization (make sure that the retrieval records are available in the near-term archive) and speed of the short-term and long-term archives also impact overall system performance.

New technologies are being developed that address both the speed and cost issues with PACS. One of these technologies is called Dynamic Transfer Syntax (DTS). DTS allows sending image information in a format where the transmitted resolution starts low and increases as the requirements for more resolution are demanded. For example, on a tertiary workstation, an image would only be sent up to the maximum resolution of the workstation, instead of clogging the network and slowing the workstation with a full resolution image. (See DTS sidebar below.)

Redundancy

Modern high-availability, high-cost computer systems can be designed to reach 99.999% up time. If the resultant downtime occurred all at once, it would amount to approximately five minutes per year. Such systems use redundant networks, redundant file servers, hot-swappable disk drives (drives that can be changed with the power on), RAID, hot swappable power supplies, multiple processor computers, and other high-reliability components. As PACS becomes more highly utilized in filmless environments, it becomes a critical system that must be operational at virtually all times. Several of these redundancy options also can increase speed (e.g., storage area networks (SAN) with optical fiber channel connections) but they also can significantly increase costs. Systems need to be designed to meet the up time requirements of the institution while at the same time meeting cost requirements. Five minutes of down time per year is excellent, and one hour of down time (99.99% up time) might be acceptable, but 99.9% up time (~ 8 hours of down time) might not be acceptable, particularly if it occurs all at once. Standard COTS computers running Microsoft NT 4.0 are designed for 99.5% up time (i.e., two days of down time per year).

One promising new computer technology combines redundancy requirements with load balancing to offer speed and advanced fail-safe features. Instead of installing redundant servers that are in a standby (e.g., mirror) mode, these systems use software to manage clusters of servers that look like one host to the end-user but balance the workload between computers. This improves performance while all servers are working properly; senses problems; and then allows continued operation, albeit at lower speed, if one of the servers fails.

Table 102-1 Archive Media Comparison (Anon (b), 2001)

Archive media	Media cost/megabyte	Speed	Advantages	Disadvantages
RAID	High	Fastest	Used for all near-term archives.	High price per gigabyte. Not removable media
DLT and dual spool tape	Low to moderate	Slow to moderate for dual spool	Lowest cost media. Highest capacity media. Newer dual spool systems are improving access speed. Relatively high transfer rate.	Slow seek times
MOD	High	Moderate	Used by most PACS vendors today. Sturdy form factor for long-term, off-line storage. Seek times are good.	Transfer rates slow
DVD	Moderate	Moderate	Commercial/consumer use may drive prices considerably lower.	Commercial/industrial DVD standards are not mature.
On-site data center	varies, but potentially lower	varies, slower than local storage	Consolidates archive.	Needs high-speed network. Institution priorities may not be PACS priorities
ASP	varies, but potentially low cost when all costs are figured in	varies, slower than local storage	Institution does not need to worry about variable costs, system upgrades, backup, disaster recovery, staff support.	Loss of control. Potential speed issues.

Data Compression

With the large storage requirements of PACS and large bandwidth requirements for its networks, the PACS would seem to be a good candidate for compression. There are two primary forms of compression: Lossless and lossy. Lossless compression results in no image degradation and typically can average only a 3:1 compression ratio. Lossy compression results in some image degradation but can reach compression ratios of as high as 200:1.

There are two popular image compression methodologies: Joint Photographic Experts Group (JPEG) and wavelet. JPEG is most common and is included in the DICOM standard. Wavelet allows higher compression ratios and is being investigated by the DICOM committee for standardization and inclusion in a future update of the DICOM standard. ECRI (Anon, 2001) recommends that, for now, only lossless compression be used on any image that will be read by a radiologist, at least until the legal issues surrounding the use of lossy compression are resolved. In addition, ECRI recommends the use of JPEG compression until wavelet compression standards are approved.

Cost Considerations

PACS technology is expensive (see Table 102-2). Four-headed diagnostic workstations can cost upwards of $150,000 per workstation. Infrastructure improvements are frequently required and can be expensive, depending on the size, age, and type of building construction and various regulatory issues that could get in the way of the improvements (e.g., asbestos abatement).

Savings from implementing PACS have been reported to range from $4 per imaging study (savings in film and associated film processing and supply costs) to $12 per study (e.g., film savings plus productivity improvements, file room staff reductions, and film storage space savings,). At 250,000 studies (films) per year, savings would be between one million and three million dollars. A PACS implementation for 250,000 studies per year with 20 diagnostic workstations will cost approximately $5 million dollars in the year 2000. With annual support (software and hardware) estimated at 10% per year, annual upgrade expenses estimated at 10% per year, and a system lifetime of seven years for depreciation purposes, payback is between three years and never, based on the $4- or $12 per-study savings, respectively.

Improvements in technology and hospital infrastructure present opportunities to improve the cost-effectiveness of a PACS purchase and implementation. These include the following:

- Project timing (i.e., coordinating PACS implementation with the installation of high-speed network infrastructure in new buildings and during major remodels).
- Increasing and optimizing the use of (existing) web-based workstations as much as possible
- Using an institution-wide archive
- Reducing the transition time of the simultaneous use of both PACS and film-based technologies for the same modality
- Radiology department productivity improvements from workflow optimization (requires tight integration of PACS and RIS)

Ongoing Support

A PACS is dynamic, and ongoing technical support is extremely important. After implementation, a support team is required to keep the PACS operating effectively. The size of the system and the amount of in-house (vs. contracted) support will determine the size of staff required, and its required skill level. Of course, in the PACS cost analysis, expenses for support staff (contracted or in-house) need to be included. While PACS will allow a reduction in file room support staff, it will cause an increase in IT support staff. Although far less IT staff are required than file-room staff, IT staff have much higher salaries.

Table 102-2 Sample Hardware and Software Initial and Support Costs

Item	Hardware and software cost	Application
PACS software, MOD long-term archive, RAID short term, web-server	$2,000,000 (rough estimate) depending on redundancy, size of RAID, size and media of long-term archive	Main system
(5) 4-headed high-resolution Diagnostic Workstations	150,000 each = $750,000	Radiologists = primary read workstations
(15) 2-headed high-resolution diagnostic workstations	$75,000 each = $1,125,000	Radiologists = primary read workstations
Secondary access (use web-based COTS) with improved video	200 each × $5000 for PC, video and software license $1,000,000	Secondary read
Capital equipment and software subtotal	$5,000,000	
Annual software and hardware support @ 10%	$500,000 per year	
Annual upgrade expenses	$500,000 per year	

PACSs are continually changing, with the typical vendor offering one or two major upgrades per year, and many more minor upgrades for bug fixes. Installing and implementing upgrades probably causes more down time than hardware failure or any other scheduled or unscheduled system work. Validation of upgrades prior to installation is one support concern, particularly in a multi vendor PACS environment (e.g., making sure that new PACS archive version a.0 works with new CR software b.2.).

The following is a short description of two typical support staff jobs:

PACS Director/manager: This person's job is to direct the overall PACS effort; to act as a liaison among the PACS vendor, PACS users, and hospital administration,; supervise the data administrator; and coordinate all support (i.e., in-house and/or vendor purchased support).

Data administrator/system administrator: Any PACS will have routine data support requirements for building data tables, correcting data problems, performing routine backups, and training end-users. For most large systems, provisions must be made to provide some support 24 hours per day, seven days per week.

PACS Installation Planning

Prior to purchasing a system, it is critical to understand how a system will be used. Will the majority of the radiologists be performing primary reads in a modality-specific reading room, a central PACS reading room, or from their office? Will the PACS hardware (e.g., servers, backup jukebox,) be located in the radiology department, the IT department or somewhere else? How should the radiology department workflow change to optimize productivity with a PACS? Where will the CR readers be located? Considering that 70% of a typical radiology department's work is plain film and that the expense of DR is high, in order to digitize this plain film work, CR most likely will be utilized as an integral component of the digital radiology department. Where will the PACS administrator's office be located? Ideally, it should be near the PACS hardware. How and where will referring physicians review images (e.g., PACS secondary workstations, any web-based PC)? What are the PC requirements for PACS web-based image viewing?

Reading Room

PACS installations will have one or more reading areas set up for reading images, either by modality (e.g., MRI reading area, CT reading area, or CR reading area) or by anatomy (e.g., body reading area, neuro reading area, bone reading area). System requirements will determine networking cabling parameters (e.g., fiber or category-5 twisted-pair cable). It is good planning to include additional cable drops whenever cabling is first installed or upgraded. Most systems are flexible enough to allow reading from any primary diagnostic workstation, although some modalities might require specialized workstations. For example workstations for ultrasound and nuclear medicine studies utilize color displays, and some MRI and CT studies require 3-D and other more sophisticated tools.

Lighting is one of the most critical factors that must to be addressed in PACS reading room design. A PACS reading room should have limited extraneous ambient light coming from behind the workstation. Background lighting and glare issues can have a significant effect on the readability of the images on the workstations. Indirect lighting is critical to the room design, including desk lamps for reading paper reports. These simple design criteria become very complex, if not impossible, when mixing film-based reading systems (e.g., motorized X-ray view boxes with their banks of high-intensity bulbs) with PACS workstations in the same room. Obviously, ample electrical outlets and HVAC are also required.

Challenges

All vendors of DICOM-compliant devices will provide a DICOM conformance statement for each device. DICOM conformance statement review continues to be a critical, albeit complex, task that must be performed for each pair of PACS components and modalities that must communicate with each other. This comparison must be performed before initial PACS implementation for existing and new equipment. DICOM compliance review also must take place prior to any upgrades or additions to the PACS or the modalities. Where in-house staff are not capable of performing DICOM conformance statement reviews, institutions should consider hiring outside experts to ensure that the compliance is complete and that the devices will function as planned. This is particularly critical when the pair of devices being investigated is provided by two different manufacturers. One common cause of problems in multivendor PACS is when incompatibilities occur because one of the systems is using private data that the other system does not understand.

Improved workflow can have a significant productivity impact on the radiology department. One technology issue that would significantly improve workflow is improved integration between RIS and PACS. Currently, unless the RIS and PACS are bought from the same vendor (and sometimes not even that helps), integration is minimal. Examples of improvements needed are reducing the need for redundant data entry (e.g., demographics, time of exam start, and exam complete), and automatic sending of messages from the modality to the PACS and/or RIS open exam completion.

In the operating room (OR), surgeons frequently need to refer to X-rays during surgery, and ORs have a high amount of background light. Many ORs are crowded with video and other equipment. Therefore, one PACS challenge is to develop easy-to-use systems (hardware and software) with high-luminescence flat panels or other viewing technologies that can coexist in the high ambient light environment.

Another challenge is the integration of the radiology PACS and its secondary workstations with some of the specialty areas and their need for specialized mini-PACS. These specialized systems are typically purchased separately, and portions of them are redundant in function to the radiology PACS and its archive. Examples of this include modalities where color is essential (e.g., color Doppler studies in ultrasound or nuclear medicine), motion studies (e.g., cine loops for cardiac catheterization labs or echocardiography), physiological data integration with images for cardiac catheterization labs, and treatment planning in radiation oncology.

Summary

PACS technology will continue to be driven by improvements in computer power and lowering of prices for COTS. The specialized applications for servers and archives will use the newest and more expensive computer technology, while the bulk of the workstations will use COTS. The further implementation of higher-speed intranets will allow the secondary viewing of images from any workstation. New standards (e.g., DICOM-2000/JPEG-2000) will improve the speed and interoperability of systems, while new digital modalities (e.g., digital mammography) will also increase the scope of PACS. Such systems will allow the radiologist to perform a primary read from any geographic location, thereby increasing the use of, and lowering the cost of, teleradiology.

Dynamic Transfer Syntax

A University of Pittsburgh team has developed dynamic transfer syntax (DTS), a new protocol for lower-cost and enterprise-wide image distribution. The DTS technique could reduce the need for expensive high-bandwidth networks and high resolution, high-powered primary workstations. DTS was developed to meet the demands of referring physicians who require resolution and image-manipulation features that are similar to those that radiologists have in their own clinics, offices, or operating rooms. Research at the University of Pittsburgh showed that in 60% of the cases, a timely, written radiologist's report (with no images) was sufficient for the referring physician. In another 20% of the cases, a lower resolution set of images was also required. In the final 20% of the cases, clinicians wanted to review at least some of the images at the same resolution, and with similar image-manipulation tools as the radiologist had. Furthermore, the referring physicians wanted access to any and all archived exams for the patient so they could determine which exams are relevant for additional review. Without these capabilities, the referring physicians would still require film and would obviate a major reason for using PACS (i.e., reduction or elimination of the use of X-ray film).

Typical diagnostic-quality PACS workstations can cost upwards of $150,000 each. To solve the above problems without investing hundreds of thousands of dollars, DTS was developed. DTS uses wavelet-based compression to send the requesting workstation data that fits the requirements of the user, the network, and the workstation. The wavelet data scheme allows additional image detail to be sent only when the user needs it. When an image request is made, DTS determines the number of images and the resolution at which to transmit, after assessing the resources available at the workstation. This real-time image delivery scheme enables the use of less powerful networks, such as standard Ethernet (10 megabits per second or 100 megabits per second). In theory, DTS therefore allows full functionality on COTS workstations with standard network infrastructure. DTS is in its infancy and has not been widely implemented yet. It remains to be seen whether it will be as accepted in actual practice as it has been at its developers' sites and whether it can scale up to the demands of the large institutions with hundreds of referring physicians and therefore hundreds of workstations and hundreds of thousands of images.
DICOM Sidebar

DICOM was developed as a digital-image transmission standard in order to enable users to retrieve medical images and associated information from digital imaging equipment in a standard format that would be the same across multiple manufacturers' systems. DICOM was designed to be very flexible. This explanation is based on DICOM version 3.

Standards-based electronic communication is commonly modeled as a set of layers, with each layer performing a defined set of functions such as the Open Systems Interconnection (OSI) 7-layer model promulgated by the International Standards Organization (ISO). This model (see Table 102-3) looks like a stack with the first (i.e., lowest) layer being the physical media (e.g., wire or fiber-optic cable) over which the information will be sent or received, and the top (i.e., seventh) layer being the application software. The middle layers concern character sets that are used to represent information, the rules for making connections over the physical layer, packet breakup and reassembly, communication between layers, and error handling. Each layer in the stack performs a well-defined set of functions and accepts input and provides output to adjacent layers only. Established rules allow communication between different devices at their corresponding layer.

One advantage of layered systems is that a layer can be replaced by a newer layer without affecting other layers. For example, if a new physical medium is 100 times faster than an older medium (e.g., fiber cable vs. standard Ethernet), only the physical layer must be replaced in order to take advantage of this additional speed.

Table 102-3 ISO 7 Layer Model

7	Application layer	Gives user applications access to network services
6	Presentation layer	Encodes and converts user information to binary data
5	Session layer	Opens, manages, and closes communication between computers
4	Transport layer	Sequences data packets and requests retransmission of missing packets
3	Network layer	Routes data packets across network segments
2	Data link (MAC) layer	Transmits frames of data from computer to computer
1	Physical	Defines cabling and connections

DICOM uses services to request communication with another device. The network protocol will indicate that the network is either available or busy. If available, DICOM initiates a series of actions that request an association with the other device. DICOM refers to the devices as "application entities" (AEs) because it is the application-layer software (i.e., the top layer in the stack) that initiates the communication process. The association establishment negotiates the capability of the application entity with which it is attempting to communicate by requesting a list of tasks that it needs to complete. The receiving-application entity then replies with the items on that tasks list that it can do. Subsequent exchanges are based on the capabilities that the two application entities have in common. For example, if the two application entities need to exchange numeric information that is more than one byte long, then they first need to determine how their numeric data is byte-encoded. Is the most significant byte stored and sent first (i.e., "little endian"), or vice versa (i.e., "big endian")? If two devices use the opposite representations to exchange numeric data, such an exchange will be erroneous unless the devices know that the representation is incorrect and they can facilitate conversion. DICOM allows for both methods of representation and resolves the one that is to be used at the association phase of the communication.

DICOM breaks the data that it needs to send, into a series of elements. Applications might have different ways of representing the values contained in any given data element. DICOM has very specific definitions of the different data types allowed, called "value representations" (VR). Examples include text strings of different maximum lengths, binary numeric data, and dates and times. DICOM historically defines the VR in a data dictionary. In DICOM 3 and higher VRs are explicitly imbedded in each data element. DICOM allows both methods, and this is one of the issues negotiated between application entities at association time. The VR and the byte order are part of a set of information required for a successful interchange. This set is called the "transfer syntax." As noted elsewhere in this chapter, some of the newer technologies (and proposed DICOM standards) support *dynamic* transfer syntax, which allows the defined association to change parameters "on-the-fly," dependent on the circumstances of the data requested (e.g., requestor needs low resolution to look at thumbnails, but higher resolution after a particular image is selected for full-screen viewing).

DICOM has also adapted some conventions from other standards, such as the patient name format specified by the HL-7 standard (Health Level 7). DICOM 3.0 also defines the relationship between entities; for example, the way a patient identifier (e.g., name, medical record number) associates itself with a study; the way a study associates itself with a series of images; and the way a series of images is associated with a radiologist's report. This object-oriented model is called an "entity-relationship model." Each of the entities in an entity-relationship model has certain attributes. For example, the patient entity has the attributes of details such as last name, first name, medical record number, height, weight, and date of birth. The DICOM standard defines several information objects (e.g., patient, study and modality) and their mandatory, optional, and conditional attributes. Conditional attributes are required (mandatory) only under certain circumstances. Whenever DICOM needs to uniquely identify an object (e.g., patient), it uses an international standard technique for generating a unique identifier, called a "UID."

DICOM provides standardized services that are used with the information objects. DICOM builds its more common services out of a set of service elements called "DICOM message service elements" (DIMSEs). There are 11 DIMSEs that fall into the categories of operations (e.g., store or query-retrieve) or notifications (e.g., error report or event report). Some DICOM services require more than one DIMSE. For example, query-retrieve is made up of the "find," "get," and "move" DIMSEs. These simple DIMSEs are used to build most of the services in a PACS. Because of the object-oriented nature of PACS, services are referred to as "service classes." It is also important to know whether a device provides a service, or uses a service, or both. As part of the negotiation process discussed earlier, it is communicated whether a device is a service-class provider (SCP), service-class user (SCU), or both. The service classes and information objects are combined to form the functional units of DICOM. This combination is called a "service-object pair (SOP) class."

A DICOM conformance statement follows a standard structure, making comparison between two manufacturers' DICOM conformance documents possible. The contents of the conformance statement include the implementation model of the application, the presentation contexts to be used, the manner in which associations are to be handled, the SOP classes to be supported, the communication profiles to be used, and any extension, privatization, or specialization.

The implementation model of the application is a diagram that shows how the application is associated with local (within the medical device) and remote (across the DICOM interface) activities. Presentation contexts consist of the SOP class(es) and the transfer syntaxes used for that SOP class. The conformance statement must describe the way an activity handles associations and whether or not it will handle multiple associations, and, if so, how many.

The list of SOP classes supported is one of the key portions of the conformance statement. This list describes the service classes and information objects that will be offered and accepted. The communication profile indicates whether any or all of the available communication profiles are used. The available communication standards are point-to-point, OSI, and TCP-IP.

The final section of the conformance statement lists privatization, specializations, or extensions to the standard classes. Extensions are supersets of standard classes and should not cause major compatibility problems. Neither specialized nor private SOP classes use DICOM-defined UIDs and, therefore, they are prone to causing manufacturer-to-manufacturer incompatibilities. This is a common cause of problems in multivendor PACS where unknown incompatibilities are caused by the use of private data.

Consider the following example: Suppose that a technologist is sending ("pushing") an MRI study from an MRI scanner to a workstation using DICOM so that the radiologist can perform a preliminary read prior to the patient leaving the scanner area. The operator selects the specific study to send to a specific workstation. The communication protocol used (typically TCP-IP) will carry the study from the scanner to the workstation. The application software at the MRI scanner console begins assembling a series of SOP instances (i.e., MRI storage instances). The DICOM process begins the communication process by having the MRI scanner application request an association with the workstation. The communication network has previously handled the steps of setting up a communication channel between the MRI scanner and the workstation. During the association, the MRI scanner will provide its presentation context, telling the workstation that it supports the MRI image storage SOP class as a service class user. The MRI also declares that its transfer syntax is implicit VR "little endian." The workstation replies that it supports the MRI SOP class as a service class provider and user. The workstation replies that it can use VR "little endian" transfer syntax. Notification of acceptance of the association is sent to the respective devices.

The MRI sends the request for storage service, along with the images to the software that assembles the DICOM message by putting together the necessary command and data elements. The DIMSEs needed for the storage-service class and the data set are handled by the communication stack, where the message may be broken up into smaller packets. The message is reassembled into the DICOM message in the communications stack (ISO middle layers) of the workstation. The reassembled DICOM message, including image data, is received by the application in the workstation that stores the MRI images in a local cache for display and, if necessary, manipulation.

The speed of this process is dependent on the amount of data, the speeds of the storage media, the workstations, and the network infrastructure. Typically, once the association is established, it can be reused for multiple SOP instances. Associations must be closed when they are no longer needed or when computer resources are being wasted and slowing the system down.

Acknowledgment

Thanks to J. Anthony (Tony) Seibert, Ph.D., medical physicist at the University of California Davis Medical Center, for his review of this chapter.

References

American College of Radiology (ACR) and National Electrical Manufacturers Association (NEMA). Digital Imaging and Communication in Medicine (DICOM), Parts 1 through 15. http://medical.nema.org/dicom.html.
Anon. Picture Archiving and Communication Systems. *Health Devices* 29 (11):3,2000.
Anon (a). DICOM Reference Guide. *Health Devices* 30 (1-2):3, 2001.
Anon (b). Inside PACS: Costs and Benefits of an ASP E- Archive Model. *Medical Imaging* 4:, 2001.

PACS Vendor References

AGFA: http://www.agfa.com/healthcare
General Electric: http://www.gemedicalsystems.com
Kodak: http://www.kodak.com/US/en/health/productsByType/pacs/pacs_Product.jhtml
Stentor: http://www.stentor.com/

103

Wireless Medical Telemetry: Addressing the Interference Issue and the New Wireless Medical Telemetry Service (WMTS)

Donald Witters
Physicist, FDA, Center for Devices and Radiological Health
Rockville, MD

Caroline A. Campbell
ARAMARK/Clarian Health Partners
Indianapolis, IN

The use of wireless medical telemetry in hospitals is expanding as one way to help meet the ever-rising cost of health care. As users of medical devices know from experience, the radio signals transmitted from the patient to the monitoring station are vulnerable to electromagnetic interference (EMI), which can present a real risk to patients being monitored. If these signals are interfered with or lost while the patient is suffering a significant adverse health event (cardiac arrhythmia, for example), the medical response could be delayed, and serious patient consequences are likely to result. Because of the importance of these signals and the likelihood of EMI with these vital transmissions, the Food and Drug Administration, the Federal Communications Commission (FCC), medical device manufacturers, and the health care community banded together to examine the EMI issue with wireless medical telemetry and developed solutions to minimize the risks to patients posed by EMI. These efforts culminated in the creation of the new Wireless Medical Telemetry Service (WMTS), with its separate frequency spectrum and coordination, which is designed to reduce the risk of EMI to the vital patient telemetry signals from other radiofrequency transmitters operating in the same frequency bands.

The purpose of this chapter is to provide a brief summary of the EMI problems for the traditional wireless medical telemetry widely used in hospitals and clinics and provide information that clinical engineers can use to assess their situation and implement solutions to reduce the risks of EMI. This chapter will briefly touch on the concerns about EMI with medical devices and what has been done to address these concerns. It will highlight the creation of the new WMTS and will relate the need to make immediate assessments of medical telemetry equipment and plans to address the coming changes to the airways where wireless medical telemetry has been used. The last section in this chapter speaks to ways of assessing the EMI risks. Each hospital and situation will likely be different, requiring familiarity with the local situation (e.g., numbers of telemetry channels being used or planned) and the risks and other factors that commonly arise with major technology changes.

The rapid changes to the radio spectrum where medical telemetry has traditionally operated make it imperative that hospitals and clinics change their approach for medical telemetry systems in order to minimize vulnerabilities and risks of EMI. Further, the issue of EMI with wireless telemetry must be looked upon as a part of the whole picture of EMI with medical devices. Simply changing the wireless telemetry will not eliminate the threat posed by EMI to other sensitive medical devices. At a time when medical and communications technology is rapidly changing, the clinical user is faced with choices about the deployment of wireless technology and other radio transmitters in their facility that must be balanced with considerations for the numerous other medical devices in the facility that are susceptible to EMI. Only through a basic understanding of the complexities of medical device EMI and the drive toward electromagnetic compatibility (EMC) can the clinical engineer begin to appreciate the risks involved and ways to undertake procedures

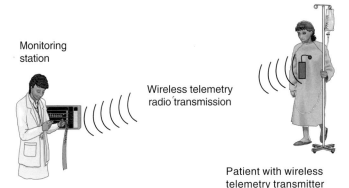

Figure 103-1 Typical wireless telemetry components.

that will minimize these risks (Witters, 1997; Witters and Silberberg, 1997). In the end, all health care facilities must come to grips with the promise of wireless technology and the increasing risk of EMI on patient safety.

Wireless medical telemetry as it is widely used today is a relatively simple way to maintain physiologic monitoring of patients while allowing them to move freely without wires connecting the patient to a stationary monitor, helping to speed their recovery and thus lower costs. Figure 103-1 illustrates a basic configuration for wireless telemetry, consisting of a small, patient attached, limited range radio transmitter sending physiological data in real time to a central monitoring station. Each of the components (patient attached physiological sensor and radio transmitter, radio signal, and central monitoring equipment) is susceptible to EMI. However, because of the changes to the radio airways and allocations used for radio transmission, and the greater use of "wireless" radio transmission links in and around the hospital, the present installed base of wireless telemetry is a greater risk by EMI than ever.

Concerns for EMI and the Drive toward EMC

To understand the drive toward electromagnetic compatibility (EMC), which is essentially the opposite of EMI, the reader must have a basic understanding of these terms and the ways in which they relate to medical devices. EMI is the disruption of the medical device function by electromagnetic energy (EM), which can take the form of any of the following:

- Radio frequency (RF) radiation (e.g., transmissions from radio, TV, radar, and cellular telephones)
- Conducted electrical energy (e.g., electrical surges, sags, and harmonics, and higher-frequency signals that can be induced onto leads and conducted into the device)
- Electrostatic discharge (e.g., lightning and the discharge of static electricity that humans can carry on their bodies to a lower-voltage potential on a medical device)

Each form of EM energy can disrupt medical devices on its own or in combination with one or more of the other forms. In the case of RF radiation, the threat from multiple transmitters can lead to even more complex situations that might be additive or subtractive. In addition to the intentional transmissions from various transmitters, electrical currents flowing through a conductor induce electric and magnetic fields. Thus, all electrically powered medical devices, as well all electrically powered products and systems, create emissions that add to the congestion of EM energy in the environment.

The major two areas of concern for medical device EMC are: (1) device emissions and the effects of these on other devices and (2) the immunity from (or, conversely, the susceptibility to) the EM energy that is expected in the environment where medical devices are used. The key to EMC is to minimize the emissions and to maximize the immunity. In most cases, these two areas interplay. By minimizing the emissions with proper design and shielding, the susceptibility is also changed, and immunity to EMI is improved. The key point to remember is that both emissions and immunity are important parts of the drive toward medical device EMC. The AAMI TIR 18-1997 "Guidance on Electromagnetic compatibility of Medical Devices for Clinical/Biomedical Engineers-Part 1: Radiated Radio Frequency electromagnetic Energy" contains an excellent overview of medical device EMI concerns and ways to address these concerns for the biomedical/clinical engineer (AAMI, 1997). The JCAHO Environment of Care Series: EC/PTSM Series No. 2 contains several articles in this area. Also, see Geddes (1998) for examples of medical device EMI and Kimmel and Gerke (1995) for a good overview of medical device EMI/EMC and how to design, build, and test equipment for immunity to most common EM energy.

Traditional Secondary Status of Wireless Medical Telemetry

Wireless medical telemetry has worked fairly well for many years as the secondary users of the RF spectrum between the high power channels for the Private Land Mobile Radio Service (PLMRS) and on vacant local television channels. According to the FCC rules, as the secondary users of the spectrum, medical telemetry users must accept interference but not cause EMI to the licensed primary spectrum users (e.g., the TV broadcasters). In general, the low radio output power of medical telemetry poses little threat to the high-power broadcasters, and for many years there were adequate numbers of low-power PLMRS channels and unused TV channels (because the older analog TV broadcasts required vacant spectrum between channels). Thus, in a real sense the TV broadcasters of soap operas about hospitals had more rights to use the airways than the hospital transmitting vital signals from actual patients.

However, with the coming of the digital age in the early 1980s, and the ever-changing and increasing needs for radio spectrum, the available channels and frequencies where medical telemetry is able to function have dramatically decreased. In the same space of time, medical technology has increasingly utilized communications improvements to speed the delivery of medicine. Thus, the two worlds of medicine and communications were placed on a collision course, with medical telemetry restricted by with its secondary status for the use of the radio airways and the communications industry rushing to meet ever-increasing demand.

Problems for EMI with wireless medical telemetry and concerns about the changing radio spectrum have long been known (Pettijohn and Larsen, 2001; FCC, 1999). In 1994, a group of wireless medical telemetry device manufacturers petitioned the FCC to allow medical telemetry to use more frequencies. With FDA support, the FCC granted medical telemetry the ability for secondary use of vacant TV channels 14 to 69 (470 MHz- 806 MHz) in 1997. This expanded the frequencies for use by medical telemetry that had been limited to vacant TV channels 7 to 13 (174 MHz to 216 MHz). However, it was not until the initial test broadcasts of the new digital television (DTV) in early 1998 that the potential magnitude of the EMI problem with wireless medical telemetry came to be fully recognized.

A survey done by the American Society of Healthcare Engineering (McClain, 2002) of several hundred hospitals showed that nearly 40% of the wireless medical telemetry used in the U.S. utilized vacant TV channels to carry the radio signals from patient to monitor. Approximately 60% were found to use the PLMRS in the 450 MHz–470 MHz frequency range. A few hospitals were using equipment operating in other frequencies, such as the Industrial, Scientific, Medical (ISM) bands centered around 915 MHz–2450 MHz. Table 103-1 summarizes the various frequencies where wireless medical telemetry can operate.

Table 103-1 Wireless Medical Telemetry Operating Frequencies

Frequency band	Use status	Power output	Limitations	Recent changes
TV channels: 174–216 MHz (Ch. 7–13)	Secondary	1500 μV/m @ 3m transmissions confined to a 200 KHz band	Vacant channels	No new FCC approvals after October 2002
Low Power Radio Service (LPRS): 216–217 MHz	Secondary	100 mW	Nearby transmitters must coordinate with Aricebo Observatory in Puerto Rico	
TV Channels: 174–216 MHz, 470–669 MHz (Ch. 14–46)	Secondary	200 mV/m @ 3m,	Used on health care premises, vacant channels, coordination required for channel 37	No new FCC approvals after October 2002
PLMRS: 450–470 MHz,	Secondary	20 mW without separate license, low power use <2W may need license	Limited use of offset channels, license required for high-power use	No new FCC medical telemetry approvals after Oct. 2002. Freeze on high-power use and channel width changes in 450–460 MHz lifted as of Jan. 2001; freeze on 460–470 MHz lifted October 2003
WMTS: 608–614 MHz	Primary	200 mV/m @ 3m, operation within <1.5 MHz bandwidth	Register with frequency coordinator required, coordination required near radio astronomy	
WMTS: 1395–1400 MHz, 1429–1432 MHz	Primary	740 mV/m @ 3m, operation within <1.5 MHz bandwidth	WMTS is primary on part of this band in certain locations, Itron (utlities) is 150 μ 1/m at the loction in primary on part of band in the WMTS registered health care other location facilities	Government operations cease Jan. 2009 for 1395–1400 MHz; Jan. 2004 for 1429-1432 MHz
WMTS: 1427-1432 MHz 902–928 MHz, 2400–2500 MHz, 5725–5850 MHz	Primary/ Secondary			

Most of the radio spectrum bands used for wireless medical telemetry have restrictions and limitations on these transmissions, and some are being changed in ways that will dramatically increase the vulnerability of present device systems to EMI from other in-band transmitters.

DTV Incidents

The collision between the licensed broadcasters and medical telemetry became clear in March 1998, when the wireless medical telemetry in two Dallas hospitals was disrupted by the initial DTV test broadcasts from a local TV station (FDA, 1998). Quick work by the clinical engineering and nursing staff at these hospitals avoided the situation of unmonitored patients. Fortunately, no patients suffered cardiac arrest or other serious condition while their wireless telemetry was interfered with. The clinical engineers at one institution, Baylor Medical Center, went further and were able to identify the source of the problem as a TV signal covering nearly the entire channel where their telemetry was transmitting. They tracked the TV signal and contacted the local broadcaster to ask that the signal be stopped (FDA, 1998). It seems clear from Figure 103-2, which depicts a typical TV channel over its normal 6-MHz channel width and an idealized DTV signal, that the DTV signals take the entire channel spectrum, leaving no space for low-power radio signals in these frequencies. The result is that the wireless medical telemetry is overwhelmed by the DTV signal on the same channel. As the medical telemetry has no protections from the "in-band" transmissions of the DTV, it simply will cease to get the patient information through to the monitoring equipment and nurses.

However, these incidents brought the situation into focus for the FCC and TV broadcasters, who clearly understood the needs of real patients. With the U.S. acceptance of DTV, the FCC mandated that all TV broadcasters would be allocated an additional TV channel for DTV, with the anticipation of phasing out the older analog technology by the year 2009. This meant that medical telemetry users were left with far fewer vacant TV channels as they had had previously, and the potential for EMI from the DTV signals. Information on the allocations for DTV can be found on the FCC website: www.fcc.gov/healthnet/dtv. The FDA issued a Safety Alert aimed at the hospital and clinical users of medical telemetry, advising them to coordinate with local TV broadcasters (FDA, 1998). It also initiated a meeting with all of these parties to develop ways that would address the immediate issues with wireless medical telemetry, with a goal to develop long-lasting solutions including protected, separate frequencies and coordination. Because of the concern for medical telemetry use in hospitals, the American Hospital Association (AHA) took the lead in organizing a Wireless Telemetry Task Group with the goal of developing specific recommendations for solutions.

Concerns with PLMRS Telemetry

From information in the 1998 ASHE survey of wireless medical telemetry nearly two-thirds of the wireless medical telemetry operates in the frequency range between 450–470 MHz as secondary, low-powered transmitters (see Figure 103-3). Most users are actually operating in this band between 460 and 470 MHz. In the late 1980s, the FCC determined that this band was crowded and should be "refarmed" or changed to allow for increased numbers of channels and users. The plan, adopted in 1995, changed this band from its present 25 KHz bandwidth channels to 12.5 KHz channels with the eventual goal of 6.25 KHz channels. The PLMRS refarming plan also eliminated the low-powered interchannel frequencies that most wireless telemetry used, instead making all users compete for the higher power 6 KHz channels. The narrower channels, combined with newer digital transmissions, will allow several more users access to the PLMRS band. However, it was clear that under this plan all medical telemetry operating in the PLMRS band were immediately vulnerable to EMI from any number of mobile radio transmitters that would use the new allocations. Figure 103-3 and 103-4 illustrate shows the cascade of refarming the PLMRS band from the 25 kHz channels, through the 12.5 KHz channels, into the 6.25 KHz channels. Because of the potential conflicts, a group of the medical device manufacturers petitioned FCC to delay the refarm plan on this band. Fortunately, the FCC recognized the possible conflicts for medical telemetry and placed a freeze on implementation of the refarming. Because of the success in creating the WMTS (see below), the FCC has (announced January 29, 2001) lifted the freeze on high-power applications for the new 12.5 KHz channels in the frequency range 450–460 MHz (FCC, 2001).

Origins of the Wireless Medical Telemetry Service (WTMS)

The AHA Task Group looked at the issues for wireless medical telemetry and developed recommendations to address major areas of concern, including:

- A definition of wireless medical telemetry
- The medical needs for wireless telemetry
- The engineering and RF needs for telemetry
- Ways to raise awareness and educate users

The results of their work developed a clear definition of what wireless medical telemetry is and ways in which it is has been used. This definition was used by FCC in creating the WMTS.

"Wireless medical telemetry is therefore defined as the measurement and recording of physiological parameters and other patient-related information via radiated bi or unidirectional electromagnetic signals. This technology may be contained within a health care facility or extend beyond to other buildings or locations." (FCC, 2000)

The Task Group surveyed several hospital users and found that by using the present technology (transmitting at 0.8 bits per second per Hertz) up to 500 concurrently operated telemetry transmitters within a geographical area would require a minimum of just over 6 MHz of spectrum frequencies. The survey results indicated that the greatest uses for wireless monitoring are for adult electrocardiogram, 12-lead parametric monitoring, pulse oximetry, invasive blood pressures, and respiration (AHA, 1999 and FCC, 2000). Future needs were estimated to make use of at least double this number of concurrent transmitters within the same area and consequently require at least 12 MHz of spectrum.

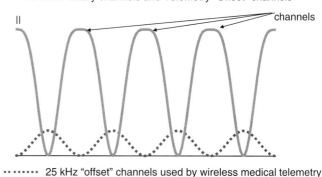

Figure 103-3 Illustration of PLMRS 450–470 kHz radio signals with wireless medical telemetry in offset channels.

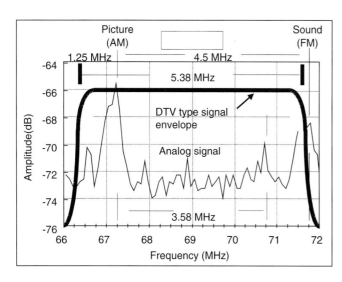

Figure 103-2 Typical 6 MHz TV broadcast channel, with DTV signal overlaid.

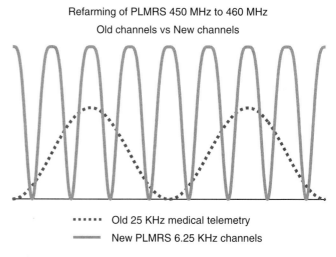

Figure 103-4 New, refarmed 6.25 MHz PLMRS signals with old 25 KHz wireless medical telemetry signals.

The Task Group also identified spectrum candidates, with input by the FCC members, and rated the candidates for several factors, including: Needs for power consumption and transmitter power, vulnerability to EMI, multipath fading, effects of frequencies on humans, bandwidth available, costs for equipment and maintenance, and radiation efficiency. The frequencies chosen as the best candidates totaled 14 MHz and were located in 3 separate frequency bands: 608 MHz to 614 MHz (TV channel 37), 1385 MHz to 1390 MHz, and 1432 MHz to 1435 MHz. This forward-looking assessment allowed for bi-directional communications on separate frequencies, thus potentially opening the future for monitoring patients and treating them via patient-attached medical devices. In addition, the Task Group recognized the continuing need to coordinate use of the frequencies to avoid conflict within and without a hospital. In some urban areas around the U.S., there are several health care facilities using medical telemetry that are located within a short distance of each other. This is especially problematic in the northeastern U.S., where more than 1000 patients may be monitored within a mile of each other. The FCC has commonly used a Frequency Coordinator to oversee the use of various radio-transmission-frequency spectra. Thus, the recommendations included a frequency coordinator for the wireless medical telemetry.

The Wireless Medical Telemetry Service (WMTS)

On June 8, 2000 the FCC adopted the Report and Order ET docket 99-255 creating the WMTS for wireless medical telemetry (FCC, 2000). Under the WMTS, for the first time, medical telemetry has primary status and all of the regulatory protections that go with this status. This means that other transmitters that interfere with WTMS equipment will be in violation of the FCC rules and could face regulatory penalties. In addition, the WMTS will be overseen by the WTMS frequency coordinator, who is responsible for maintaining a database of the use of these frequencies and for facilitating sharing of these frequencies among hospital users and the remaining government operations at protected government sites. Users of WMTS are licensed by rule so that individual licenses are not required; instead, the WMTS users are expected to share the frequencies and to utilize the coordination within their locales. Authorized users of WMTS include authorized health care personnel, health care facilities, and trained technicians under the control of the health care facility. The FCC has chosen to define health care provider facilities as essentially 24-hour-type care facilities, with specific exclusions for mobile or home use.

Unlike the traditional wireless telemetry use of channelized transmission, the AHA Task Group recommended and FCC agreed that there be limited technical specifications and no channelization for the WMTS. This opens the WMTS for many different types of modulation schemes to carry important patient data and allows for the use of bi-directional transmissions, (i.e., from the patient to the monitoring equipment and back to the patient attached device). The minimal technical restrictions enable growth of the medical telemetry technology by allowing, for instance, the advanced communications technologies used in cellular communications to be used for WMTS. By looking toward the future challenges, the lack of technical specifications opens the way toward increased utilization while challenging the designers and users of these systems to maintain system integrity.

Subsequent to the creation of the WMTS, there were FCC actions that affected the 1427–1432 MHz band spectrum where part of the WMTS is located (FCC, 2002). Over the years, public utility companies have been using a radio transmitter system by Itron in this band (as secondary frequency users) to remotely read utility meters. Following negotiations among AHA, Itron, and the FCC, it was decided that the WMTS and Itron and other non-medical telemetry could share the frequencies in the band. Under the final plan WMTS is primary in part of the band and secondary in the remainder of the band, with the primary allocation determined by geographic location.

"Generally, WMTS operations are accorded primary status over non-medical telemetry operations in the 1427–1429.5 MHz band, but are treated as secondary to non-medical telemetry operations in the 1429.5–1432 MHz band. However, there are seven geographic areas in which WMTS and non-medical telemetry operations have 'flipped' the bands in which each enjoys primary status. These seven areas, termed the 'carve-out' areas, are (1) Pittsburgh, PA; (2) the Washington, DC metropolitan area; (3) Richmond/Norfolk, VA; (4) Austin/Georgetown, TX; (5) Battle Creek, MI; (6) Detroit, MI; and (7) Spokane, WA. In these seven areas, in contrast to the rest of the country, WMTS has primary status in the 1429–1431.5 MHz band, but is secondary to non-medical telemetry operations in the 1427–1429 MHz band (FCC WMTS web site, 2003)."

The key features of the WTMS are listed below.

Use of WMTS:
By health care facilities and professionals, licensed by FCC rule

Frequencies bands and technical specifications:
608–614 MHz, 200 mV/m maximum field strength at 3 meters
1395–1400 MHz, 740 mV/m maximum field strength at 3 meters
1429–1432 MHz, 740 mV/m maximum field strength at 3 meters (for important limitations on the use of this band, see Limitations of use below).

- Uni- and bi-directional transmission allowed
- Voice and video not allowed (physiological waveforms are not considered video)
- Limitations on out-of-band emissions
  - 1395–1400 MHz and 1429-1432 MHz bands, no channelization specified
  - 608–614 MHz (TV channel 37) band has limitations of spread spectrum technologies and limitations on assignments of channels so that no facility can have exclusive frequency use. Operations in this TV channel band are not protected from adjacent band interference from channels above and below this TV channel.

Limitations of Use:
- Use of the 1395–1400 MHz and 1429–1432 band is restricted in some geographical locations because of continued government use and a band plan arrangement with other users of the 1427–1432 MHz band. The restrictions related to government will phase out in 2009 and 2004, depending upon the band.
- Use of the 608–614 MHz band is restricted within 80 Km of several sensitive radio astronomy observatories noted in the FCC report and order (FCC, 2000). Written concurrence and frequency coordination near these observatories is required.
- In the 1427–1432 MHz band, WTMS is generally primary at 1427–1429.5 MHz and secondary from 1429.5–1432 MHz, except in cerain geographic locations where the primary and secondary frequencies are reversed. See the FCC web site http://wireless.fcc.gov/services/personal/medtelemetry/

Frequency Coordinator:
- Specific frequencies or range used
- Information required by coordinator
- The American Hospital Association, American Society for Healthcare Engineering Serves as the Frequency Coordinator
- Modulation scheme used and bandwidth
- Effective radiated power
- Number of transmitters in facility and manufacturer and model numbers
- Name of health care provider
- Transmitter location (coordinates, street address)
- Point of contact for authorized health care provider

Other Wireless Medical Telemetry Frequencies

Wireless medical telemetry is allowed to operate over other frequencies, including the ISM bands (915 MHz, 2450 MHz). The ISM bands were intended for the use of industrial, medical, or scientific products and devices that use RF energy for their function. Examples of these products include microwave ovens, RF-wielding equipment, some radars, and medical diathermy. Because few technical restrictions for equipment exist in these bands, and the cost of components for transmissions in these bands have decreased, an increasing amount of communications equipment has migrated to these frequencies. However, the ISM bands, especially the higher bands mentioned above, are rapidly being filled with large numbers of users such as portable phones and wireless LANs. The equipment operating in these bands must be robust and able to withstand significant EMI from the other in-band transmitters. Microwave ovens, for example, are well known to modulate widely over the entire ISM band where they operate and have been seen to interfere (Buffler and Risman, 2000). However, if the equipment is robust enough, it might be able to perform adequately in most environments for use with medical telemetry. Knowing tthere are risks of EMI with wireless medical telemetry, users must perform their own risk assessment and plan their strategy to design and configure their wireless medical telemetry to minimize the risks of EMI. The FDA recognized the need to address these risks and issued a guidance document to the medical device industry with recommendations to assess and appropriately address these risks including helping the device user community (FDA, 2000).

EMI Risk Assessment for Health care Facilities

With the creation of the WMTS it is possible to significantly reduce the risks associated with wireless patient monitoring. As recommended by the FDA, medical device manufacturers and the user community (primarily hospitals) need to assess their telemetry systems for risks of electromagnetic interference and, if warranted, take appropriate action to reduce that risk. The recommendations to users of these telemetry systems focus on review at the institutional level. The institution must balance the factors of operating, such as the cost of altering or replacing the telemetry system, against the risks of harm for the patients.

One way to frame the decision about altering or replacing medical telemetry is based on a risk model from the American National Standard ANSI/AAMI/ISO 14971:2000 Medical Devices—Application of Risk Management to Medical Devices (ANSI, 2000). Under this model, a stepwise progression is taken to plan, assess, and control the risks for the use of medical devices. This approach can be summarized by the following steps:

- Plan the risk management process
- Perform a risk analysis
- Evaluate the risks, and ways to address these risks
- Take action to mitigate and control the risk
- Put into place a program to monitor EMI incidents and proactively reassess telemetry for changes in transmission or device use

Under this model, the need to address risk is based on the seriousness of the hazard, the likelihood that an adverse event will occur, and the impact on the patient if the adverse event actually occurs. The risks for wireless medical telemetry must be continually reassessed as new hazards may arise or the device may be used in new ways that impact the acceptability of known risks. In the end, each facility must decide how it will address the issue, both now and in the future. Figure 103-5 demonstrates a risk assessment and -management process developed specifically for wireless medical telemetry.

Risk Management Planning

At the beginning stage, it is important to establish the responsibilities and scope for the risk management process. Involvement of the appropriate parties in the process is critical to its success. Ideally, the institution's safety committee should have oversight of the risk assessment process and responsibility for implementing the resulting recommendations. In addition, knowledge of radio and communications technology is requisite for the successful risk management process related to wireless medical telemetry. Thus, this expertise must be brought into the process from within the facility (e.g., clinical engineering, telecommunications, information management) or contracted with appropriate EMC experts. The safety committee should designate the point of authority for implementing the spectrum-management program. Such person must have a good working knowledge

Develop Plan to Assess Risk to Wireless Telemetry
1. Assign responsibilities and scope of risk management process, including identification of a point of authority for implementation of spectrum management program.
2. Determine types and numbers of telemetry devices, patient parameters being monitored, device locations, and transmission frequencies used.
3. Identify safety hazards (EMI and patient impact).
4. Define risk acceptability criteria.

Risk Analysis and Evaluation
1. Identify EMI risks.
2. Inventory all radio transmitters (e.g. walkie talkies and wireless communications devices, such as wireless LANs).
3. Determine acceptability of electromagnetic interference risk.

Mitigation and Ongoing Monitoring
1. Analyze options for mitigating risk.
2. Take action to mitigate risk and synchronize and coordinate those actions with spectrum changes timeline.
3. Establish a proactive spectrum management program to manage, monitor, report, and mitigate EMI risks and issues.
4. Establish an education program for caregivers about the potential for EMI, how to report incidents of EMI, and how to react to them.

Figure 103-5 Risk management process for telemetry.

of wireless medical telemetry concerns. Because there is a continual reassessment involved in the process (due to the ongoing changes in medical devices and communications technology) it is highly recommended that the risk assessment and management process include all of the people involved in the purchase, use, and maintenance of telemetry equipment.

Early in the risk assessment planning stage, it is useful to determine the types and numbers of wireless medical telemetry devices, the patient parameters being monitored, device locations, transmission frequencies and number of channels used. The information obtained about the wireless medical telemetry use will be helpful during the risk analysis phase. As described previously in this chapter, wireless medical telemetry can operate in different parts of the spectrum. For this reason, changes in spectrum utilization can pose risks to wireless medical telemetry products operating in the same frequency spectrum but likely will have little or no impact on medical telemetry operating in another part of the spectrum. For example, telemetry operating on channel 37 (608–614 MHz) likely will not be impacted by changes in the ISM band.

The last step in the planning phase is to determine the acceptability of the risk for each health care facility's situation. For example, a tertiary-care facility with a critical cardiac population might have a higher level of concern about interference to the wireless medical telemetry than a community-based hospital. The risk acceptability criteria are a determination of how much interference or loss of wireless patient information is tolerable to the health care facility. These criteria can be categorized to determine the qualitative levels of risk (e.g. occasional, probable, frequent, and certain). Determining the potential safety hazard or impact on patient safety in the event of interference is a prerequisite to determining the risk acceptability criteria. During this step, a decision about what is acceptable to the institution will be used to drive decisions in the risk analysis and evaluation phase.

Risk Analysis and Evaluation

The first step in the risk analysis and evaluation phase is to identify the EMI hazards to wireless telemetry signals. An AHA Task Force (ACCE News, 1998) identified the most immediate potential hazards as changes to the PLMRS, emerging DTV, and another TV broadcast called "low-power television" (LPTV). The LPTV service was established in 1982 by the FCC with the intention of providing opportunities for locally oriented television service on a secondary-frequency user basis. Because of power restrictions on the transmission signals, LPTV is targeted to small localized areas but still may threaten medical telemetry if that telemetry operates on the same frequencies as the LPTV station.

There are other risk factors that can affect all wireless communication systems, including telemetry systems, regardless of the transmission frequency utilized. These risk factors include EMI with the device connected to the patient or the remote monitoring device. The EMI can be a result of other wireless communications transmissions or other electromagnetic disturbances (e.g., AC power problems or electrostatic discharge). All of these sources of electromagnetic disturbances contribute to the potential for medical device EMI, either with the transmission of patient physiological signals or the devices that receive or transmit these signals. Therefore, the risk analysis and evaluation phase should include an inventory of all wireless communications devices used in the health care facility, including transmitters on the roof. (See the AAMI TIR 18-1997 for a thorough explanation of medical device EMI from radio transmitters and how to address this issue. (AAMI, 1997))

The radio transmission frequencies utilized by the wireless telemetry system will determine the applicability of these various hazards for the users. For example, a telemetry system operating in the TV channels in the VHF frequencies typically will not be impacted by the refarming of PLMRS but are likely to be affected by the DTV or LPTV. Analysis of the risks posed by PLMRS changes, DTV, and LPTV involves predicting the probability of interference from these sources and the potential impact of that interference on patient safety. Applying the ANSI medical-device risk management model to this analysis will help determine the acceptability of the risks (ANSI, 2000).

For wireless telemetry systems that currently are operating in the 450–460 MHz frequency band, the probability of interference increased dramatically in January 2001, when the FCC lifted its freeze on PLMRS refarming for these frequencies. The probability of interference to systems operating in the 460–470 MHz frequencies will likewise increase dramatically in July 2003, when the PLMRS is refarmed for these frequencies as well. Although it is difficult to quantify the growth rate of PLMRS use following the lifting of these freezes, there likely will be up to four times more users of the new narrow channels. Because PLMRS users are typically mobile sources, interference with the wireless medical telemetry likely will be intermittent as well as difficult to track and identify. The likelihood of experiencing interference from PLMRS in part depends on how many of these transmitters are used in the vicinity of the medical telemetry system. There likely will be more PLMRS transmitters in use in densely populated areas. Therefore, the risk of interference for wireless medical telemetry currently transmitting on the PLMRS frequencies is more likely in urban areas that have a concentration of commercial PLMRS transmitters. Given these various considerations, the probability of continued or intermittent interference for medical telemetry in the PLMRS frequencies is at least occasional, but more likely is probable or frequent. Furthermore, the resultant risk of injury to the patient due to a disruption in the monitoring is likely at a critically unacceptable level. Thus, actions must be taken to mitigate the interference risk for most of these systems operating at 450–470 MHz frequencies.

The probability for interference to medical telemetry from DTV in different locales can be predicted through review of DTV information posted on the FCC web site (http://www.fcc.gov/) and following FCC announcements and petitions by local TV broadcasters. In general, the local broadcasters are supposed to notify any affected hospitals in advance of beginning DTV transmissions. If a DTV begins station broadcast on a TV channel utilizing the same frequencies as a local telemetry system, then the telemetry system will be rendered inoperable over the entire 6 MHz of the DTV channel (Figure 103-2 shows that the DTV signal takes up virtually all of the particular TV channel frequency band). For this situation, the risk of interference with the wireless medical telemetry signal seems more certain, and the resultant risk of injury to a patient is at a critically unacceptable level.

LPTV broadcasters are also affected by DTV, and must migrate to different vacant TV channels as the DTV broadcasts begin. Thus, the changes brought about by DTV will cascade down to LPTV and possibly affect medical telemetry. When the LPTV changes frequencies, it may produce interference to the local wireless telemetry, even though the clinical telemetry user has made efforts to coordinate with the DTV changes. However, by comparing a list of DTV stations in the area with a list of LPTV stations in the user's local area (found through the FCC website, http://www.fcc.gov/mmb/vsd/lp/lp.html), the reader can determine which LPTV stations will have to change because of the DTV broadcasts. Medical telemetry users should then contact and coordinate with the LPTV broadcaster to plan for an orderly migration to new frequencies that minimize the risks to patients. If there is a direct conflict among the new LPTV frequencies and the frequencies used by the existing medical telemetry, the risk analysis and evaluation for the telemetry is very similar to that for DTV. Because additional LPTV frequency designations are anticipated in the future, continued vigilance in this area is necessary.

Wireless communications transmitting on one of the ISM bands have a recognized potential for interference and thus use technologies to protect themselves from interference. Currently, there is great debate within the clinical engineering community about how successful these protection mechanisms are in ensuring the reliability of ISM medical telemetry systems and their ability to co-exist with other ISM frequency users. At best, the probability for interference in the ISM bands will remain a concern as new ISM technologies emerge and present interoperability issues. When utilizing an ISM band telemetry system, the user must continually manage all of the communications technologies that exist within the health care facility to decrease the potential for interference.

Under the ANSI risk management model, it is necessary to estimate how often the interference to the ISM wireless medical telemetry might occur. As the congestion of unlicensed secondary users in the ISM band increases, the risk of interference grows. For the ISM band telemetry system, the probability of interference could be remote, occasional, probable, or frequent; but as congestion in these bands increase, so too will the probability of interference. The risk of injury to the patient still can be considered as critical, based on the use of telemetry, and hence this risk will need to be mitigated. As the probability of interference increases, the risk likely will become unacceptable.

The potential for interference to wireless medical telemetry from other wireless communications systems used within the facility will be dependent on the relative transmitter output power levels and frequency, as well as the separation of the transmitter from the medical device. An obvious conflict to wireless telemetry systems operating in the PLMRS is posed by use of two-way radios that also operate in this band. Similarly, wireless LANs operating in the ISM bands may pose a risk to wireless telemetry systems that transmit in the same ISM bands. As part of the risk analysis and evaluation, each type of wireless communication device in the facilities' inventory must be analyzed and evaluated. A thorough inventory of radio transmitters must also include both fixed, (e.g., TV) and mobile (e.g., police fire, and commercial) radio transmitters that operate in the general geographic location. As mentioned above, the changes to the PLMRS, DTV, and LPTV are outside the control of the health care facility but must be taken into account when performing the risk assessment and evaluation. Equipment that works today may pose high risks tomorrow as the radio spectrum continues to change.

Mitigation and Ongoing Monitoring

Once the level of risk to the telemetry system is identified, appropriate action must be taken to mitigate or reduce that risk. Mitigation may involve simply retuning medical telemetry transmitters to other frequencies. For example, if a small installation of VHF TV channel frequency medical telemetry is at risk of interference from a DTV station, there may be sufficient spectrum available at a different, vacant, TV channel frequency to retune the telemetry transmitters and thus mitigate the risk. Such an approach can be adequate for a time, but eventually the use of the radio airways will be such that the facility will be left with little alternative but to migrate to a spectrum offering more protection and lower risks. For larger installations in a similar situation, there might be insufficient vacant frequencies to allow continued use of these TV channel frequencies, and thus migration to another spectrum (e.g., WMTS) will be necessary. The actions taken to mitigate the risks for interference will need to be synchronized and coordinated with the spectrum changes that drove the EMI risk decision. For example, telemetry systems operating in the 460–470 MHz frequency band will need to migrate prior to July 2003, and telemetry systems operating in a band to be used by a DTV station will need to migrate prior to the DTV test broadcasts.

A program for ongoing monitoring and mitigation of EMI issues will become more important as hospitals, and the society at large, embrace wireless technology. Such a program must facilitate and continuously coordinate for the reliable functioning of all medical equipment including telemetry. A proactive and continuous spectrum management program will include the following:

- A point of authority for maintaining reliable wireless communications,
- A policy for reporting and evaluating incidents of interference
- An inventory of all existing transmitters used on the premises, including those transmitters that are allowed to reside on the hospital's roof (or anywhere close to the health care facility) for a fee
- An assessment of interoperability of medical device (including wireless telemetry) prior to acquisition of additional wireless technology
- Facilitating the integration of wireless technology through proactive management of all radio transmitters in or around the health care facility
- Methods for minimizing interference
- Reporting interference incidents and issues to the point of authority for such matters
- Reporting patient injury or death related to device malfunction as required under the Medical Device Reporting (MDR) regulation (Part 803 of Title 21, Code of Federal Regulations (CFR)) pursuant to the Safe Medical Devices Act of 1990.

Based on an increased understanding of EMI and its impact on medical equipment functioning, hospital personnel should pay more attention to interference risks than they have in the past. Clinical engineers and hospital personnel in general need to begin proactively addressing incidents of interference. This will involve education of the caregivers so that they will report all medical equipment problems to the clinical engineering department for thorough investigation. Ideally, all patient caregivers will be familiar with the policy for reporting and evaluating incidents of interference. A continuing education effort will likely be needed, to make staff members aware of the changes and to keep them apprised of developments in this rapidly evolving area. Clinical engineering personnel must use the reported information to assess the potential impact of RF interference in reported situations based on the inventory of transmitters used in the hospital. Where interference is identified, action must be taken to prevent recurrence.

When considering acquisition of new wireless technology, the ability of that technology to operate with the existing base of wireless technology should be assessed. This ability can be determined based on a review of the complete inventory of wireless equipment with the incumbent vendor and requiring the vendor to qualify that their new equipment will work with the present systems and in the intended use environment.

Establishment of a point authority for implementing the wireless communications and telemetry management program is highly recommended because there is an appetite in the medical community for wireless technology, which is expected to grow rapidly. Recommended responsibilities for this entity include maintenance of the wireless equipment inventory, rigorous evaluation of reports of potential interference, tracking and trending incidents of interference, evaluation of the interoperability of new equipment, and recommendations for mitigating interference risks.

A survey performed by the AHA Wireless Medical Telemetry Task Force revealed that a wide variety of parameters would be monitored via telemetry if the technology to do so were available. In recognition of these needs, the definition and service rules adopted in the creation of WMTS allows for innovative use. Therefore, the ongoing monitoring phase of the risk management process will include a re-evaluation of the risk based on the changes in the intended use of telemetry. For example, if telemetry were used to monitor fetal heart rate on an ambulating expectant mother, and that monitoring indicated fetal-heart decelerations, the level of risk to the patient could be higher than for a typical ambulatory cardiac patient.

Conclusions

The use of wireless medical telemetry is expected to grow in the coming years as the population ages and the wireless technology becomes more common and less expensive. At the same time, the risks of EMI with these vital signals are increasing for users of the older, secondary frequencies as these frequencies are being changed by the FCC to accommodate more commercial use. Indeed, the changes have begun with the DTV real-locations and the announcement of the changes to the PLMRS–both frequency bands that have been widely used for medical telemetry. Hospital users of wireless telemetry must immediately begin to assess their devices for the risks of EMI and must make changes to address the increased risks for EMI. Recommendations to reduce the risks of EMI with wireless medical telemetry include switching to the new WMTS frequencies where the medical signals are afforded protections and the oversight of a frequency coordinator. With appropriate assessment and proper planning, the EMI risks to wireless medical telemetry should be dramatically reduced, both now and into the future.

References

AAMI. *Guidance on Electromagnetic compatibility of Medical Devices for Clinical/Biomedical Engineers–Part 1: Radiated Radio-Frequency electromagnetic Energy, Technical Information Report AAMI TIR 18-1997.* Arlington, VA, Association for the Advancement of Medical Instrumentation, 1997.

ACCE News. ACCE Partners with Telemetry Manufacturers; AAMI, ASHE, and the AHA to Recommend Spectrum Allocation to the FCC. *ACCE News* 8(6):11–12, 1998.

AHA. American Hospital Association (AHA) Medical Telemetry Task Force Letter to the Chairman of the Federal Communications Commission. , AHA, 1999.

ANSI. ANSI C63.18-1997: Recommended Practice for On-Site Ad Hoc Test Method for Estimating Radiated Electromagnetic Immunity of Medical Devices to Specific Radio-Frequency Transmitters., American National Standards Institute, 1997.

ANSI. ANSI/AAMI/ISO 14971:2000, Medical device – Applications of risk management to medical devices. , American National Standards Institute, 2000.

ASHE. EMC: How to Manage the Challenge. *Health care Facilities Management Series*, Chicago, American Society for Healthc are Engineering, 1997.

Baker SD. Larsen M, Wiley JS. Medical Telemetry Frequency Bands: How Should Hospitals Approach Their Decision on Which Band to Use? *J Clin Eng* 26(2):136–143, 2001.

Buffler CR, Risman PO. Compatibility Issues Between Bluetooth and High Power Systems in the ISM Bands. *Microwave J* 7:126–134, 2000.

FCC. Federal Communications Commission (FCC) Report and Order ET Docket 99-255, PR docket 92-235, Amendment of Parts 2 and 95 of the Commission's Rules to Create a Wireless Medical Telemetry Service, FCC 00-211. FCC, 2000.

FCC. FCC Public Notice, Freeze on the Filing of High Power Applications for 12.5 KHz Offset Channels in the 450–460 MHz Band to be Lifted January 29, 2001, DA 00-1360., FCC, 2001.

FCC. Federal Communications Commission (FCC) Erraum, This Erratum corrects errors in the Report and Order in WT Docket No. 02-8, FCC 02-152, released May 24, 2002. Information on the WMTS is located on the FCC web site at http://wireless.fcc.gov/servces/personal/medtelemetry/

FDA. Public Health Advisory: Interference Between Digital TV Transmissions and Medical Telemetry Systems. , FDA, 1998.

FDA. *Emergency Service Radios and Mobile Data Terminals: Compatibility Problems with Medical Devices.* Medical Devices Bulletin MDA DBI999(02). , FDA, 1999.

FDA. *FDA Center for Devices and Radiological Health (CDRH) Guidance for Industry: Wireless Medical Telemetry Risks and Recommendations.*, FDA, 2000.

Geddes LA. *Medical Devices Accidents with Illustrative Cases.* Boca Raton, FL, CRC Press, 1998.

Hatem MB. Managing the Airwaves: New FCC Rules for Wireless Medical Telemetry. *Biomed Instrum Technol* 34(3):177, 2000.

JCAHO. Managing Electromagnetic Interference. *The Environment of Care Series* 2. Chicago, Joint Commission on Accreditation of Health care Organizations, 1997.

Kimmel WD, Gerke DD. *Electromagnetic Compatibility in Medical Equipment: A Guide for Designers and Installers.*, IEEE Press and Interpharm Press, 1995.

McClain JP. ASHE/AHA Partnership Makes Medical History. *Inside ASHE* 5(8), 1998.

McClain JP. 1998. (Personal Communication.)

Pettijohn D, Dempsey MK, Larsen M. A Review of the Technologies Associated with Indoor Wireless Patient Monitoring. *J Clin Eng* 26(2):155–165, 2001.

Sherman P, Campbell C. Assessing Existing Telemetry Systems for Risk of Electromagnetic Interference. *J Clin Eng* 26(2):144–154, 2001.

Witters D. Medical Devices and EMI: The FDA Perspective. *ITEM Update* 22–32, 1995.

Witters DM, Silberberg JL. Electromagnetic Interference with Medical Devices: FDA Concerns and approaches toward solutions, Managing Electromagnetic Interference. *The Environment of Care Series* 2. Chicago, Joint Commission on Accreditation of Health care Organizations, 1997.

104

Health Insurance Portability and Accountability Act (HIPAA) and Its Implications for Clinical Engineering

Stephen L. Grimes
Senior Consultant and Analyst, GENTECH
Saratoga Springs, NY

What is a chapter on the Health Insurance Portability and Accountability Act (HIPAA) doing in a handbook on clinical engineering? This is an understandable question! Judging solely by the name, most readers would understandably expect to find a discussion on health insurance and finance. But scratch the surface and you will actually find the most significant piece of health care legislation since Medicare was signed into law in 1965. Look further into the labyrinth of legislation and associated rules, and you will find that HIPAA introduces standards that will require broad changes in our health care operations as well in as the way we adopt and implement technology.

The Origins of HIPAA

HIPAA (also known as Public Law 104-191 or the Kassebaum-Kennedy bill) was signed into law by President Clinton in 1996 (for more about Public Law 104-191 see http://www.cms.hhs.gov/hipaa/hipaa2/general/background/pl104191.asp). The original law was conceived during the first Bush Administration and later was included in the Clinton Administration's Health Care Reform Initiative. As the name implies, HIPAA was originally intended to provide a means for consumers to maintain their health insurance coverage regardless of their employment situation. As happens with most legislation between conception and passage, peripheral provisions were added to the bill. Arguably, the most significant of these was called the "administrative simplification provision" (Figure 104-1).

Administrative simplification was added to HIPAA to address a problem brought to Congress's attention in the early 1990s. In 1991, the *New England Journal of Medicine* published an article reporting that 19 to 24 cents of every dollar spent in the U.S. on health care went toward "administrative" costs (Woolhandler and Himmelstein, 1991), as compared with 11% in Canada and 7% in most of Europe (Figure 104-2a). These administrative costs were due in large part to industry inefficiencies: 70% of data being manually keyed into industry computers were output from other computers; at the same time, the industry was bogged down in manual filing, photocopying, faxing, telephoning, and mailing (Moynihan, 2003). The problem would worsen. In 1988, the U.S. spent $558 billion on health care, or 11% of its gross domestic product (GDP) (Heffler et al., 2003). By 2001, those expenditures had risen to $1,425 billion, or 14.1% of the GDP. By 2012, they are estimated to rise further to $3.1 trillion, or 17.7% of the GDP (Figure 104-2b). In 1994, Congress was made aware of at least one possible solution to the problem of escalating health care costs. That year, a report prepared by a respected industry group, the Workgroup for Electronic Data Interchange (WEDI), claimed that the health care industry could save $73 billion annually if it adopted electronic data interchange (EDI) for most administrative and financial transactions, as many other U.S. industries had already done (Duncan et al, 2001). Congress recognized that the nature of electronic transactions made industry vulnerable in ways unlike their paper counterparts and added requirements for privacy and security to the administrative simplification provision of HIPAA.

As the law was written, the Department of Health and Human Services (HHS) was to be responsible for formulating the actual Transaction and Code Sets (TCS) rule and the security rule. The law directed HHS to avail itself of the expertise of various industry organizations, including the National Committee on Vital and Health Statistics (NCVHS) when developing these rules. Within HIPAA, Congress gave itself 36 months to prepare the privacy rule, with that responsibility falling back to HHS only if Congress were to fail to complete the task in the allotted time frame. It did fail to do so and HHS ultimately took on the task of formulating the privacy rule, too.

According to law, HHS was to prepare a draft for each rule and publish it as a notice of proposed rule making (NPRM) for comment in the *Federal Register*. The public then would have 60 days in which to comment, after which HHS would prepare the final rule, taking those comments into consideration. Once the final rule was published in the *Federal Register*, there would be a 60-day review, followed by a 24-month grace period before the rule would take effect (except for small health plans, which were given an additional 12 months to comply). Congress's plan had been to have HIPAA's TCS, privacy, and security provisions in place by the year 2000. However, Congress provided little funding for HHS to begin the process, and the Y2K issue was quickly consuming industry resources (see Chapter 105). Barring other significant delays, the last of the major HIPAA rules will be in effect by April of 2005.

Overview of HIPAA's Administrative Simplification Provisions: TCS, Privacy, and Security Rules

The three major rules associated with administrative simplification are transaction and code sets, privacy, and security. HHS finalized the rules over nearly three years, beginning with TCS in October 2000, and ending with security in April 2003. A brief overview of each of the rules follows:

Transaction and Code Sets (TCS) Rule

The compliance date for the Transaction and Code Sets (TCS) Rule was October 16, 2003. A one-year extension applies to small (i.e., >$5M annual revenue) health plans and to those that have applied for a one-year extension (under ASCA). Covered entities (other than small health plans) that did not apply for one-year extension before October 15, 2002 needed to be compliant by October 16, 2002. The Enforcing Authority is the HHS Centers for Medicare and Medicaid Services (CMS).

The TCS rule is the heart of administrative simplification and the source of any projected cost savings. The rule calls for the adoption of standardized transactions and code sets by health plans, health care clearinghouses, and health care providers who transmit health information in electronic form in connection with any of the transactions defined below.

The standardized transactions are defined by American National Standards Institute (ANSI) X12N specifications and address the following:

- Premium payment /remittance
- Health care claim/encounter
- Plan enrollment /De-enrollment
- Eligibility for health plan
- Coordination of benefits
- Claim payment/remittance advice
- Health care claim status
- Referral Certification and Authorization
- Health care claim attachment (under development~HL7 records)
- First report of injury

The standard code sets are taken from versions defined by the following:

- International Classification of Diseases (ICD)
- AMA's Current Procedural Terminology (CPT)
- HHS's National Drug Codes (NDC)
- Health Care Finance Administration's Common Procedure Coding System (HCPCS)
- ADA's Codes on Dental Procedures and Nomenclature
- National Council for Prescription Drug Programs (NCPDP)

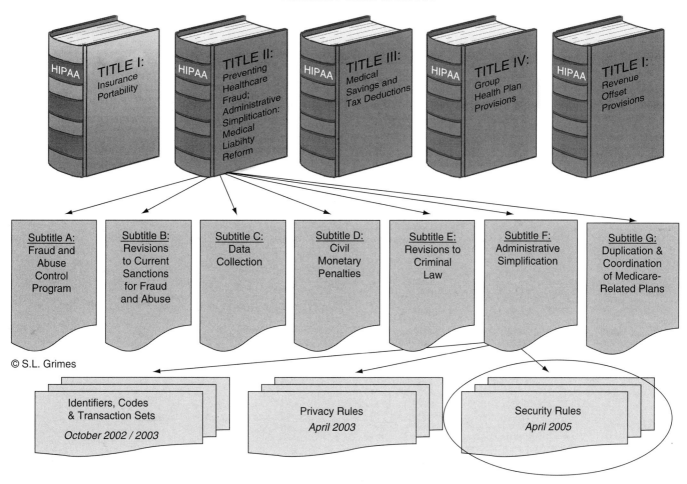

Figure 104-1 Administrative Simplification (Subtitle F) represents a small part of the HIPAA legislation but is responsible for a large part of HIPAA's impact on the health industry.

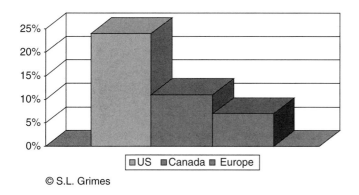

Figure 104-2a Percent of total healthcare expenditure required to cover administrative costs.

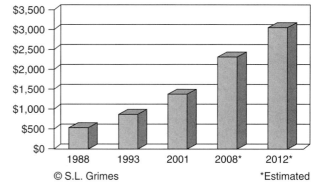

Figure 104-2b Trend in United States healthcare expenditures.

These standardized codes address the following:

- Diagnoses
- Treatments
- Procedures
- Tests
- Services
- Drugs
- Supplies/equipment

Privacy Rule

The compliance date for this rule was April 14, 2003, except for small health plans, which have an additional 12-month grace period. The enforcing authority is the HHS Office for Civil Rights (OCR).

The privacy rule covers health plans, health care clearinghouses, and health care providers who transmit any health information electronically. The rule requires these covered entities to develop and implement policies and procedures as follows:

- Limit access to a patient's individually identifiable health information (IIHI) in any form (i.e., electronic, written, oral) to only that information required by those who have a need based on their involvement in treatment, payment, or health care operations (TPO) unless the patient has explicitly authorized broader access.
- Give patients the right to access to their medical records, to request corrections to those records, and the right to know who has accessed those records.
- Compel business associates (who may, as a consequence of their work, access IIHI) to adopt privacy policies and procedures consistent with the requirements of this rule.
- Provide appropriate sanctions for any person or business associate who inappropriately obtains or uses a patient's IIHI.

Security Rule

The compliance date for the security rule is April 21, 2005, except for small health plans, which have an additional 12-month grace period. The enforcing authority is the HHS Centers for Medicare and Medicaid Services (CMS).

The security rule also covers health plans, health care clearinghouses, and health care providers who transmit health information to any of the covered transactions electronically. The rule applies to electronic protected health information (ePHI) that is transmitted or maintained in electronic media, although HHS warns "that standards for the security of all health information or protected health information in non-electronic form may be proposed at a later date."

The rule requires covered entities to ensure the confidentiality, integrity, and availability of their ePHI by establishing the following:

- Administrative Safeguards
  - Security management processes that include an effective risk analysis and risk management program
  - Workforce security (e.g., authorization, clearance, and sanctions)
  - Security incident (e.g., response and reporting)
  - Education and training of workforce on security issues
  - Contingency plans (e.g., criticality analysis, back-up, and disaster recovery)
  - Evaluation of security program effectiveness
  - Business-associate contracts
- Physical Safeguards
  - Facility access (e.g., contingency, security, access control, and maintenance)
  - Workstation use and security
  - Device and media controls (e.g., disposal, backup, and accountability)
- Technical Safeguards
  - Access controls (e.g., user identification, encryption, and emergency access)
  - Audit controls (e.g., tracking access activity)
  - Integrity (e.g., authentication of ePHI)
  - Transmission security (integrity controls and encryption)
- Organizational Requirements
  - Business associate contracts
- Policies, Procedures and Documentation Requirements
  - Policies and procedures necessary to comply with standards
  - Documentation (e.g., maintain records of compliance actions, activities, or assessments)

Failure to comply with the administrative simplification provisions of HIPAA carries the risk of substantial civil and criminal penalties. There are civil fines of up to $100 per violation and up to a maximum of $25,000 per year possible for identical violations. Because any failure to comply could involve violations of multiple standards, the actual total penalties per incident could far exceed the individual maximums. Criminal penalties range from $50,000 and one year in jail for anyone who obtains protected health information, to $50,000 and 5 years in jail for anyone who obtains protected health information under false pretenses, to $250,000 and 10 years in jail for anyone who obtains protected health information with the intent to use it for commercial advantage, personal gain, or malicious harm.

HIPAA's Effect on Clinical Engineering

Over the next several years, HIPAA compliance efforts will be a high-priority item on the agendas of virtually all of the 1.2 million providers of health care in the US. These providers include hospitals, laboratories, imaging and surgery centers, clinics, pharmacies, physician practices, and other clinical practices.

HIPAA will consume a significant portion of the hospital's resources. It is estimated that HIPAA will cost hospitals over $43 billion to implement (Lageman and Melick, 2000), compared with the $8.5 billion that hospitals spent addressing the Y2K problem (Marietti, 1999).

The strategic interests of provider organizations are closely aligned with HIPAA compliance. Each operational unit within the provider organization (including clinical engineering) is obliged to use any resources that are appropriate and available to assist their organization in achieving compliance. While clinical engineering shares some of the overall HIPAA compliance burden with other operational units, it will have the greatest impact in compliance with the security rule.

HIPAA's security rule was conceived to protect the integrity, availability and confidentiality of health information or data associated with a patient. This includes data in medical and billing records found in provider information systems. It also includes diagnostic and therapeutic data found in biomedical devices and clinical information systems. In fact, biomedical devices and systems used by health care providers represent a substantial and growing area of risk with respect to HIPAA security issues. The total quantity of biomedical devices and systems in a typical hospital may be 300 to 400% more than the number of IT devices and systems in that same hospital. HIPAA-affected systems in both the IT and biomedical inventories can represent a substantial portion of that hospital's total systems. There are two major trends contributing to the significance of the biomedical technology-related risk:

1. Biomedical devices and systems are being designed and operated as special purpose computers with more automated features. This results in increasing amounts of health data being collected, analyzed, and stored in these devices.
2. There has been a rapidly growing integration and interconnection of disparate biomedical and information technology devices and systems where health information is exchanged.

While biomedical technology represents a substantial repository of health information, these devices and systems are often overlooked by what is frequently an IT-centric approach to HIPAA security.

Security is an issue for biomedical technology because any compromise in ePHI integrity or availability can lead to the improper diagnosis or therapy for a patient. Delayed or inappropriate treatment can result in harm, or even death. Any compromise in confidentiality can compromise patient privacy and may result in financial loss to the patient and/or the provider organization.

To successfully address HIPAA security, clinical engineers must be prepared to adopt a different mindset. Clinical engineers must approach the management of technology from a security standpoint, recognizing that ensuring the integrity and availability (as well as the confidentiality) of health information is the heart of an effective technology management program.

A Closer Look at the Security Rule

HIPAA's security rule was designed to ensure the integrity, availability, and confidentiality of all electronic protected health information (ePHI) the provider creates, receives, maintains, or transmits. For purposes of the security rule, integrity, availability, and confidentiality are defined as follows:

- Integrity: Data have not been altered or destroyed in an unauthorized manner
- Availability: Data are accessible and useable upon demand by an authorized user
- Confidentiality: Data are not made available nor disclosed to an unauthorized person

A few other key definitions associated with the Security Rule include the following:

- Electronic Protected Health Information (ePHI) is Individually Identifiable Health Information (IIHI) that is transmitted or maintained in electronic media.
- Individually Identifiable Health Information (IIHI) is information that is a subset of health information, including demographic information collected from an individual, and
  1. is created or received by the healthcare provider and
  2. relates to the past, present or future health or condition of an individual; the provision of care to an individual and
     a. identifies the individual
     b. with respect to which there is a reasonable basis to believe the information could be used to identify the individual
- Electronic media (Figure 104-3) is
  - Storage media including
    - Memory devices
    - Removable/transportable digital memory medium (e.g., magnetic tape or disk, optical disk, digital memory card)
  - Transmission media used to exchange information already in electronic storage media including
    - Internet, Extranet
    - Leased and dial-up lines

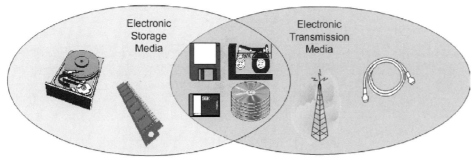

Figure 104-3 Electronic media as defined by the HIPAA security rule.

- Private networks
- Physical movement of removable/transportable electronic storage media

The security rule was written to be flexible. Providers are permitted to use any security measures that enable them to reasonably and appropriately implement the requirements. In selecting the security measures to use, the providers must take into account the following:

- Size, complexity, and capabilities of the organization
- Technical infrastructure, hardware, and software security capabilities of the organization
- Costs of security measures
- Probability and criticality of potential risks to the confidentiality, integrity, and availability of ePHI

The security rule presents its requirements as a series of standards. Standards are required policies, practices, services, or systems that the provider must implement in order to be in compliance with the security rule. The security rule also presents a series of implementation specifications associated with most standards. Implementation specifications are guidelines as to ways to meet a particular standard. The security rule classifies the provided implementation specifications as either "required" or "addressable." A provider must use "required" implementation specifications to meet their respective standards. A provider should use "addressable" implementation specifications to meet their respective standards unless the provider can demonstrate that it meets the standard with a reasonable and appropriate alternative. Factors justifying the use of alternate schemes to meet standards include the results of risk analyses, risk mitigation strategies, the existence of any security measures already in place, and the cost of implementation. When alternate schemes are taken, the decision process and the chosen implementation methods must be well documented.

The security rules standards and implementation specifications are divided into the following categories

1. Administrative safeguards, which include:
   Documented, formal measures (including policies and procedures) to manage the selection, development, and execution of security measures to protect electronic data and the conduct of the workforce in relation to the protection of that electronic data
2. Physical safeguards, which include:
   Measures to physically protect systems storing or transmitting health data and related buildings and equipment from:
   - Natural and environmental hazards (e.g., fire and flood)
   - Intrusion (i.e., use of locks, keys, and administrative measures to control access)
3. Technical safeguards, which include:
   Technical measures to protect systems storing or transmitting health data in the following ways:
   - Protecting information access
   - Controlling and monitoring information access and system activity
   - Verifying data integrity
   - Ensuring availability
   - Ensuring transmission security
4. Organizational requirements, including:
   Use of business associate agreements to compel agents, subcontractors, vendors, consultants, and others who may have access to electronic protected health information to employ appropriate security measures to safeguard ePHI
5. Policies/procedures and documentation requirements, including:
   Measures to ensure that policies and procedures comply with the standards and implementation specifications of the security rule and that all actions, activities, and assessments conducted with respect to these requirements are documented.

A detailed matrix of security rule standards and implementation specifications is provided in Table 104-1.

The Compliance Process

Security Management

HIPAA's security rule requires providers to formulate a security management program and integrate risk analysis and management elements into that process (see Figure 104-4).

To establish an effective security management program, each organization must first assign a security official who will have the overall responsibility and authority to develop, implement, and manage the organization's security measures. After assigning a security official, the provider must progress through the five levels (Figure 104-5) of a security plan in order achieve the goal of a truly effective security program. These levels include the following:

Level 1 Develop necessary policies per security rule's standards.
Each provider organization must establish a set of policies that are designed to meet the specific requirements of the security rule but are also designed to fit the character (i.e.,

Table 104-1 HIPAA Final Security Rule—Standards and Implementation Specifications Matrix

		Administrative Safeguards	
Sections	Standards		Implementation Specifications
164.308(a)(1)	(i) Security management process. Implement policies and procedures to prevent, detect, contain, and correct security violations		(A) Risk analysis (REQUIRED). Conduct an accurate and thorough assessment of the potential risks and vulnerabilities to the confidentiality, integrity, and availability of electronic protected health information held by the covered entity. (B) Risk management (REQUIRED). Implement security measures sufficient to reduce risks and vulnerabilities to a reasonable and appropriate level to comply with §164.306(a). (C) Sanction policy (REQUIRED). Apply appropriate sanctions against workforce members who fail to comply with the security policies and procedures of the covered entity. (D) Information system activity review (REQUIRED). Implement procedures to regularly review records of information system activity, such as audit logs, access reports, and security incident tracking reports.
164.308(a)(2)	(i) Assigned security responsibility. Identify the security official who is responsible for the development and implementation of the policies and procedures required by this subpart for the entity		(REQUIRED)
164.308(a)(3)	(i) Workforce security. Implement policies and procedures to ensure that all members of its workforce have appropriate access to electronic protected health information, as provided under paragraph (a)(4) of this section, and to prevent those workforce members who do not have access under paragraph (a)(4) of this section from obtaining access to electronic protected health information . . .		(A) Authorization and/or supervision (ADDRESSABLE). Implement procedures for the authorization and/or supervision of workforce members who work with electronic protected health information or in locations where it might be accessed. (B) Workforce clearance procedure (ADDRESSABLE). Implement procedures to determine that the access of a workforce member to electronic protected health information is appropriate. (C) Termination procedures (ADDRESSABLE). Implement procedures for terminating access to electronic protected health information when the employment of a workforce member ends or as required by determinations made as specified in paragraph (a)(3)(ii)(B) of this section.
164.308(a)(4)	(i) Information access management. Implement policies and procedures for authorizing access to electronic protected health information that are consistent with the applicable requirements of subpart E of this part. . . .		(A) Isolating health care clearinghouse functions (REQUIRED). If a health care clearinghouse is part of a larger organization, the clearinghouse must implement policies and procedures that protect the electronic protected health information of the clearinghouse from unauthorized access by the larger organization. (B) Access authorization (ADDRESSABLE). Implement policies and procedures for granting access to electronic protected health information, for example, through access to a workstation, transaction, program, process, or other mechanism. (C) Access establishment and modification (ADDRESSABLE). Implement policies and procedures that, based upon the entity's access authorization policies, establish, document, review, and modify a user's right of access to a workstation, transaction, program, or process.
164.308(a)(5)	(i) Security awareness and training. Implement a security awareness and training program for all members of its workforce (including management) . . .		(A) Security reminders (ADDRESSABLE). Periodic security updates. (B) Protection from malicious software (ADDRESSABLE). Procedures for guarding against, detecting, and reporting malicious software. (C) Log-in monitoring (ADDRESSABLE). Procedures for monitoring log-in attempts and reporting discrepancies. (D) Password management (ADDRESSABLE). Procedures for creating, changing, and safeguarding passwords.
164.308(a)(6)	(i) Security incident procedures. Implement policies and procedures to address security incidents . . .		Response and Reporting (REQUIRED). Identify and respond to suspected or known security incidents; mitigate, to the extent practicable, harmful effects of security incidents that are known to the covered entity; and document security incidents and their outcomes.

Continued

Table 104-1 HIPAA Final Security Rule—Standards and Implementation Specifications Matrix—Cont'd

	Administrative Safeguards	
Sections	Standards	Implementation Specifications
164.308(a)(7)	(i) Contingency plan. Establish (and implement as needed) policies and procedures for responding to an emergency or other occurrence (for example, fire, vandalism, system failure, and natural disaster) that damages systems that contain electronic protected health information....	(A) Data backup plan (REQUIRED). Establish and implement procedures to create and maintain retrievable exact copies of electronic protected health information. (B) Disaster recovery plan (REQUIRED). Establish (and implement as needed) procedures to restore any loss of data. (C) Emergency mode operation plan (REQUIRED). Establish (and implement as needed) procedures to enable continuation of critical business processes for protection of the security of electronic protected health information while operating in emergency mode. (D) Testing and revision procedures (ADDRESSABLE). Implement procedures for periodic testing and revision of contingency plans. (E) Applications and data criticality analysis (ADDRESSABLE). Assess the relative criticality of specific applications and data in support of other contingency plan components.
164.308(a)(8)	(i) Evaluation. Perform a periodic technical and non-technical evaluation, based initially upon the standards implemented under this rule and subsequently, in response to environmental or operational changes affecting the security of electronic protected health information, that establishes the extent to which an entity's security policies and procedures meet the requirements of this subpart.	(REQUIRED)
164.308(b)	(1) Business associate contracts and other arrangements. A covered entity, in accordance with §164.306, may permit a business associate to create, receive, maintain, or transmit electronic protected health information on the covered entity's behalf only if the covered entity obtains satisfactory assurances, in accordance with §164.314(a) that the business associate will appropriately safeguard the information. (2) This standard does not apply with respect to (i) The transmission by a covered entity of electronic protected health information to a health care provider concerning the treatment of an individual. (ii) The transmission of electronic protected health information by a group health plan or an HMO or health insurance issuer on behalf of a group health plan to a plan sponsor, to the extent that the requirements of §164.314(b) and §164.504(f) apply and are met; or (iii) The transmission of electronic protected health information from or to other agencies providing the services at §164.502(e)(1)(ii)(C), when the covered entity is a health plan that is a government program providing public benefits, if the requirements of §164.502(e)(1)(ii)(C) are met. (3) A covered entity that violates the satisfactory assurances it provided as a business associate of another covered entity will be in noncompliance with the standards, implementation specifications, and requirements of this paragraph and §164.314(a)...	Written contract or other arrangement (REQUIRED). Document the satisfactory assurances required by paragraph (b)(1) of this section through a written contract or other arrangement with the business associate that meets the applicable requirements of §164.314(a).

	Physical Safeguards	
Sections	Standards	Implementation Specifications
164.310(a)	(1) Facility access controls. Implement policies and procedures to limit physical access to its electronic information systems and the facility or facilities in which they are housed, while ensuring that properly authorized access is allowed....	(i) Contingency operations (ADDRESSABLE). Establish (and implement as needed) procedures that allow facility access in support of restoration of lost data under the disaster recovery plan and emergency mode operations plan in the event of an emergency. (ii) Facility security plan (ADDRESSABLE). Implement policies and procedures to safeguard the facility and the equipment therein from unauthorized physical access, tampering, and theft. (iii) Access control and validation procedures (ADDRESSABLE). Implement procedures to control and validate a person's access to facilities based on their role or function, including visitor control, and control of access to software programs for testing and revision. (iv) Maintenance records (ADDRESSABLE). Implement policies and procedures to document repairs and modifications to the physical components of a facility which are related to security (for example, hardware, walls, doors, and locks).
164.310(b)	Workstation use. Implement policies and procedures that specify the proper functions to be performed, the manner in which those functions are to be performed, and the physical attributes of the surroundings of a specific workstation or class of workstation that can access electronic protected health information...	(REQUIRED)
164.310(c)	Workstation security. Implement physical safeguards for all workstations that access electronic protected health information, to restrict access to authorized users....	(REQUIRED)
164.310(d)	(1) Device and media controls. Implement policies and procedures that govern the receipt and removal of hardware and electronic media that contain electronic protected health information into and out of a facility, and the movement of these items within the facility...	(i) Disposal (REQUIRED). Implement policies and procedures to address the final disposition of electronic protected health information, and/or the hardware or electronic media on which it is stored. (ii) Media re-use (REQUIRED). Implement procedures for removal of electronic protected health information from electronic media before the media are made available for re-use. (iii) Accountability (ADDRESSABLE). Maintain a record of the movements of hardware and electronic media and any person responsible therefore. (iv) Data backup and storage (ADDRESSABLE). Create a retrievable, exact copy of electronic protected health information, when needed, before movement of equipment.

Table 104-1 HIPAA Final Security Rule—Standards and Implementation Specifications Matrix—Cont'd

Technical Safeguards

Sections	Standards	Implementation Specifications
164.312(a)	(1) Access control. Implement technical policies and procedures for electronic information systems that maintain electronic protected health information to allow access only to those persons or software programs that have been granted access rights as specified in §164.308(a)(4)....	(i) Unique user identification (REQUIRED). Assign a unique name and/or number for identifying and tracking user identity. (ii) Emergency access procedure (REQUIRED). Establish (and implement as needed) procedures for obtaining necessary electronic protected health information during an emergency. (iii) Automatic logoff (ADDRESSABLE). Implement electronic procedures that terminate an electronic session after a predetermined time of inactivity. (iv) Encryption and decryption (ADDRESSABLE). Implement a mechanism to encrypt and decrypt electronic protected health information.
164.312(b)	Audit controls. Implement hardware, software, and/or procedural mechanisms that record and examine activity in information systems that contain or use electronic protected health information...	(REQUIRED)
164.312(c)	(1) Integrity. Implement policies and procedures to protect electronic protected health information from improper alteration or destruction	(2) Mechanism to authenticate electronic protected health information (ADDRESSABLE). Implement electronic mechanisms to corroborate that electronic protected health information has not been altered or destroyed in an unauthorized manner.
164.312(d)	Person or entity authentication. Implement procedures to verify that a person or entity seeking access to electronic protected health information is the one claimed...	(REQUIRED)
164.312(e)	(1) Transmission security. Implement technical security measures to guard against unauthorized access to electronic protected health information that is being transmitted over an electronic communications network.	(i) Integrity controls (ADDRESSABLE). Implement security measures to ensure that electronically transmitted electronic protected health information is not improperly modified without detection until disposed of. (ii) Encryption (ADDRESSABLE). Implement a mechanism to encrypt electronic protected health information whenever deemed appropriate.

Organizational Requirements

Sections	Standards	Implementation Specifications
164.314(a)	(1) Business associate contracts or other arrangements ... (i) The contract or other arrangement between the covered entity and its business associate required by §164.308(b) must meet the requirements of paragraph (a)(2)(i) or (a)(2)(ii) of this section, as applicable. (ii) A covered entity is not in compliance with the standards in §164.502(e) and paragraph (a) of this section if the covered entity knew of a pattern of activity or practice of the business associate that constituted a material breach or violation of the business associate's obligation under the contract or other arrangement, unless the covered entity took reasonable steps to cure the breach or end the violation, as applicable, and, if such steps were unsuccessful: (A) Terminated the contract or arrangement, if feasible; or (B) If termination is not feasible, reported the problem to the Secretary	(i) Business Associate Contracts (REQUIRED). The contract between a covered entity and a business associate must provide that the business associate will – (A) Implement administrative, physical, and technical safeguards that reasonably and appropriately protect the integrity, and availability of the electronic protected health information that it creates, receives, maintains, or transmits on behalf of the covered entity as required by this subpart.; (B) Ensure that any agent, including a subcontractor, to whom it provides such information agrees to implement reasonable and appropriate safeguards to protect it; (C) Report to the covered entity any security incident of which it becomes aware (D) Authorize termination of the contract by the covered entity, if the covered entity determines that the business associate has violated a material term of the contract. (ii) Other arrangements. (REQUIRED) (A) When a covered entity and its business associate are both governmental entities, the covered entity is in compliance with paragraph (a)(1) of this section, if – (1) It enters into a memorandum of understanding with the business associate that contains terms that accomplish the objectives of paragraph (a)(2)(i) of this section; or (2) Other law (including regulations adopted by the covered entity or its business associate) contains requirements applicable to the business associate that accomplish the objectives of paragraph (a)(2)(i) of this section. (B) If a business associate is required by law to perform a function or activity on behalf of a covered entity or to provide a service described in the definition of business associate as specified in §160.103 of this subchapter to a covered entity, the covered entity may permit the business associate to create, receive, maintain, or transmit electronic protected health information on its behalf to the extent necessary to comply with the legal mandate without meeting the requirements of paragraph (a)(2)(i) of this section, provided that the covered entity attempts in good faith to obtain satisfactory assurances as required by (a)(2)(ii)(A) of this section, and documents the attempt and the reasons that these assurances cannot be obtained. (C) The covered entity may omit from its other arrangements authorization of the termination of the contract by the covered entity, as required by paragraph (a)(2)(i)(D) of this section if such authorization is inconsistent with the statutory obligations of the covered entity or its business associate.
164.314(b)	(1) Requirements for group health plans. Except when the only electronic protected health information disclosed to a plan sponsor is disclosed pursuant to §164.504(f)(1)(ii) or (iii), or as authorized under §164.508, a group health plan must ensure that its plan documents provide that the plan sponsor will reasonably and appropriately safeguard electronic protected health information created, received, maintained, or transmitted to or by the plan sponsor on behalf of the group health plan....	(2) (REQUIRED) The plan documents of the group health plan must be amended to incorporate provisions to require the plan sponsor to- (i) Implement administrative, physical, and technical safeguards that reasonably and appropriately protect the confidentiality, integrity, and availability of the electronic protected health information that it creates, receives, maintains, or transmits on behalf of the group health plan; (ii) Ensure that the adequate separation required by §164.504(f)(2)(iii) is supported by reasonable and appropriate security measures; (iii) Ensure that any agent, including a subcontractor, to whom it provides this information agrees to implement reasonable and appropriate security measures to protect the information; and (iv) Report to the group health plan any security incident of which it becomes aware.

Policies and Procedures and Documentation Requirements

Sections	Standards	Implementation Specifications
164.316(a)	Policies and procedures. Implement reasonable and appropriate policies and procedures to comply with the standards, implementation specifications, or other requirements of this subpart, taking into account those factors specified in §164.306(b)(2)(i), (ii), (iii), and (iv). This standard is not to be construed to permit or excuse an action that violates any other standard, implementation specification, or other requirements of this subpart. A covered entity may change its policies and procedures at any time, provided that the changes are documented and are implemented in accordance with this subpart.	(REQUIRED)

Continued

504 Clinical Engineering Handbook

Table 104-1 HIPAA Final Security Rule—Standards and Implementation Specifications Matrix—Cont'd

Policies and Procedures and Documentation Requirements

Sections	Standards	Implementation Specifications
164.316(b)	(1) Documentation.	(i) Maintain the policies and procedures implemented to comply with this subpart in written (which may be electronic) form; and
	(ii) If an action, activity or assessment is required by this subpart to be documented, maintain a written (which may be electronic) record of the action, activity, or assessment...	(i) Time limit (REQUIRED). Retain the documentation required by paragraph (b)(1) of this section for 6 years from the date of its creation or the date when it last was in effect, whichever is later.
		(ii) Availability (REQUIRED). Make documentation available to those persons responsible for implementing the procedures to which the documentation pertains.
		(iii) Updates (Required). Review documentation periodically, and update as needed, in response to environmental or operational changes affecting the security of the electronic protected health information.

REQUIRED VS ADDRESSABLE
The entity must decide whether a given ADDRESSABLE implementation specification is a reasonable and appropriate security measure to apply within its particular security framework. This decision will depend on a variety of factors, such as, among others, the entity's risk analysis, risk mitigation strategy, what security measures are already in place, and the cost of implementation. Based upon this decision the following applies:
(a) If a given ADDRESSABLE implementation specification is determined to be reasonable and appropriate, the covered entity must implement it.
(b) If a given ADDRESSABLE implementation specification is determined to be an inappropriate and/or unreasonable security measure for the covered entity, but the standard cannot be met without implementation of an additional security safeguard, the covered entity may implement an alternate measure that accomplishes the same end as the addressable implementation specification. An entity that meets a given standard through alternative measures must document the decision not to implement the addressable implementation specification, the rationale behind that decision, and the alternative safeguard implemented to meet the standard.
A covered health care provider must comply with the applicable requirements of this subpart (e.g., Part 164 Security and Privacy, Subpart C – Security Standards for the Protection of Electronic Protected Information) no later than April 20, 2005
*Health Insurance Reform: Security Standards; Final Rule (68 FR 8334-8381 ~ Feb 20, 2003)
Part 164: Security and Privacy; Subpart C: Security Standards for the Protection of Electronic Protected Health Information

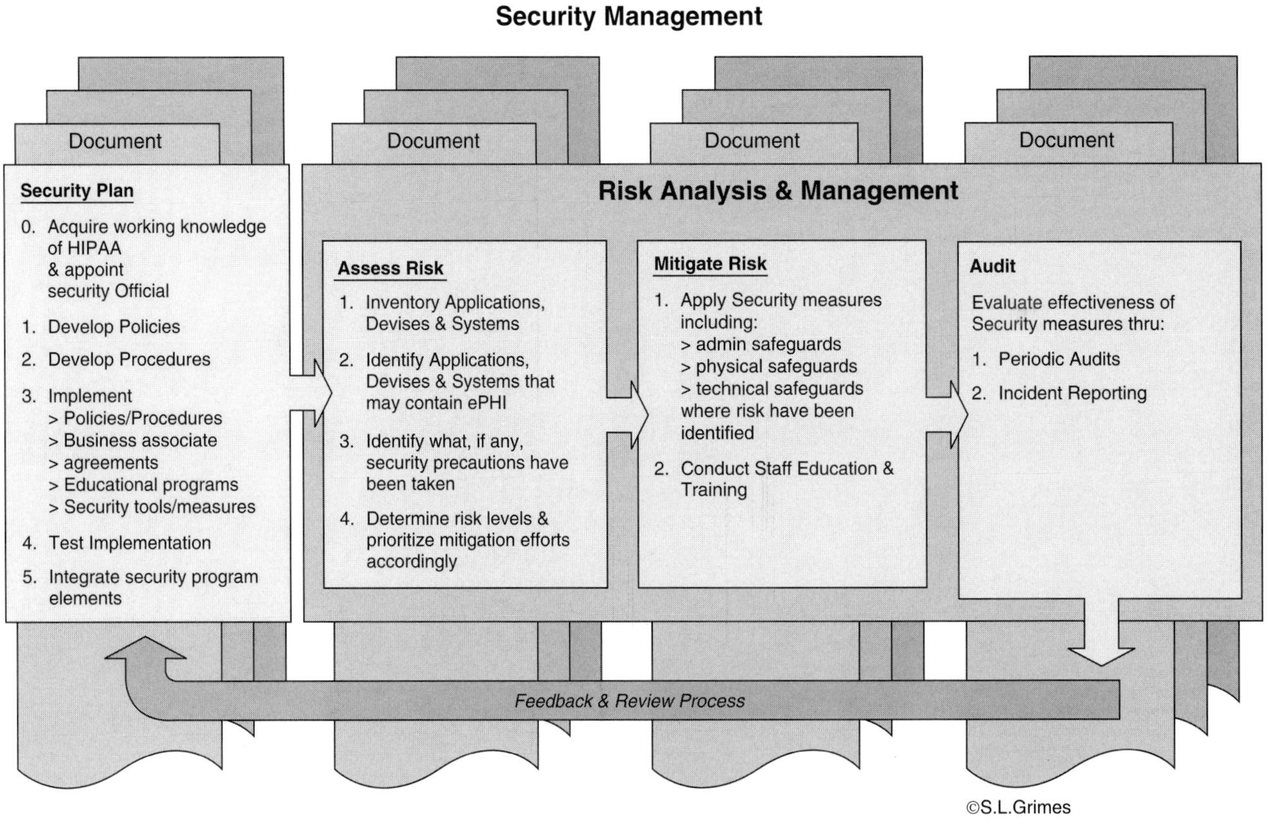

Figure 104-4 Security management and risk analysis and management process under HIPAA.

size, complexity, and capability) of the organization. These policies must address such issues as:

- Security management
- Sanctions
- Workforce security
- Information access management
- Security incidents
- Contingency planning
- Business associate agreements
- Periodic evaluations
- Facility access controls
- Device/system use (i.e., systems maintaining/transmitting ePHI)
- Device and media controls (i.e., devices and media maintaining/transmitting ePHI)
- Integrity (i.e., protecting ePHI from alternation or destruction)
- Documentation maintained for 6 years, reviewed and updated periodically

Level 2 Develop and adopt the necessary procedures, including physical and technical safeguards, per the security rule's implementation specifications.
After developing the policies that address the security rule's standards and that are appropriate for the organization, the provider must then develop complementary procedures, physical and technical safeguards that track the security rule's implementation specifications. Examples include the following:

- Risk analysis, risk management, and system activity review procedures to address the security-management policies
- Authorization/supervision, workforce clearance, and termination procedures to address the workforce security policies
- Access authorization, access establishment/modification procedures to address the access management policies
- Security reminders, log-in monitoring, and password management procedures to address security-awareness policies
- Security response and reporting procedures to address security incident policies
- Data backup, disaster recovery, emergency-mode operation, and testing procedures to address contingency plan policies
- Business associate procedures to address business associate policies
- Evaluation procedures to address evaluation policies
- Contingency operations, facility security plans, access control/validation, record maintenance, and physical safeguards to address facility-access policies
- Device/system use and security procedures and physical safeguards to address system-use and security policies
- Data disposal, media re-use, system/data tracking and data backup/storage procedures and physical safeguards to address device- and media-control policies
- User identification, emergency access, automatic logoff, and encryption/decryption procedures to address access control policies
- Audit control procedures and technical safeguards to address audit-control policies
- Data verification procedures and technical safeguards to address data integrity policies
- Integrity control and encryption procedures and technical safeguards to address transmission security policies
- Documentation procedures to address documentation policies

Level 3 Implement policies/procedures, business associate agreements, workforce education, and physical/technical safeguards
After developing an appropriate complement of policies, procedures, physical and technical safeguards to meet the security rules requirements, the provider must effectively implement these measures.

Level 4 Test implementation
After implementing policies, procedures, and physical and technical safeguards, the provider must conduct on-going testing to verify their effectiveness and to determine which changes are required to keep the measures effective.

Level 5 Integrate security measures into an organization-wide program
Finally, the provider must ensure that these measures are integrated into their overall operations and security programs. Only after achieving integration can the security management program be considered effective.

Risk Analysis and Management

Simultaneously with establishing a security plan, the security rule requires that the provider incorporate risk analysis and management elements into the process. For biomedical technology programs, these elements involve the following:

An Inventory

The provider must identify biomedical devices and systems that maintain and/or transmit ePHI. Typically, for each affected device/system the organization should answer the following questions:

- What type(s) of ePHI does the device/system contain?
- Who has access to the ePHI? Who needs access?
- What types of connections with other devices/systems exist?
- What types of security measures are currently employed with this device/system?

Assessing the Risk Associated with ePHI in Inventoried Items

The provider should categorize the levels of risk with respect to the potential for compromises to confidentiality, integrity and availability of ePHI. The level of risk is determined by looking at both the degree of criticality (should ePHI become compromised) and the probability of such a compromise occurring.

- *Criticality:* Level of risk/vulnerability (e.g., high, medium, low) with respect to confidentiality, integrity, and availability of ePHI for each affected device/system
- *Probability:* Likelihood of risk (e.g., frequent, occasional, or rare) with respect to confidentiality, integrity, and availability of ePHI for each affected device/system
- *Composite risk score:* Combining criticality and probability levels into a composite risk score for each device and system maintaining or transmitting ePHI

Establishing Priorities

The provider can use the scores to assign priorities after establishing a composite risk score for each device and system maintaining or transmitting ePHI. The provider would:

- Use criticality/probability composite scores to prioritize risk mitigation efforts
- Conduct a mitigation effort that gives priority to devices and systems with highest scores (i.e., devices/systems that represent the most significant risks)
- Determining gap

According to priorities determined for each device and system in the inventory:

- Review existing security measures identified during biomedical device and system inventory process
- Prepare gap analyses for devices and systems detailing additional security measures necessary to mitigate recognized risks (addressing devices and systems according to priority)

Formulating and Implementing a Mitigation Plan

After determining the gap between existing and needed security measures and prioritizing devices and systems based on their risk levels:

- Prepare a written plan incorporating
- Additional security measures as required
- Priority assessment
- Schedule for implementation
- Implement the plan and document the process
- Monitoring process

Ensure that the risk analysis and management process is continuously reviewed and improved as necessary by:

- Establishing on-going monitoring systems (including a security incident reporting system) to ensure mitigation efforts are effective
- Documenting results of regular audits of security processes

Summary

While the compliance process associated with HIPAA's security rule might seem onerous at first, consider that HIPAA has not so much placed an additional compliance burden on clinical engineers as it has legislated key elements of an effective clinical engineering or biomedical technology management program. Remember that HIPAA's administrative

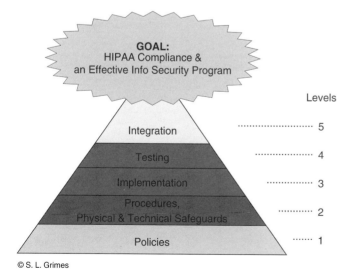

Figure 104-5 HIPAA security rule security management levels.

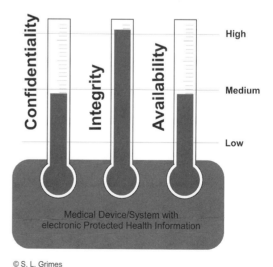

Figure 104-6 Levels of risk with respect to confidentiality, integrity, and availability of ePHI for an affected medical device or system.

simplification provisions are, first and foremost, about standards. The standards in this case are facilitating the rapid and accurate exchange of patient-related information among providers, insurers, and others involved in health care operations. HIPAA's security rule requires that providers insure the integrity, availability, and confidentiality of patient-related information, including diagnostic and therapeutic data. Consider that ensuring integrity and availability of data in diagnostic and therapeutic devices and systems is an intrinsic and principle objective of any effective clinical engineering program. The security rule gives that objective the force of law and adds confidentiality considerations.

HIPAA is a major force in the future development of the U.S. health care system, and as such it will be an area of focus for most health care administrators and planners for many years. Clinical engineers should recognize their opportunity to contribute to ways to address this issue.

References

Duncan M, Rishel W, Kleinberg K, et al. *A Common Sense Approach to HIPAA.* , GartnerGroup, March 2001.
45 CFR Parts 160, 162 and 164. *Federal Register* 68(34) , 2003. 45 CFR Parts 160, 162 and 164.
Heffler S et al. Trends: Health Spending Projections For 2002-2012. *Health Affairs* W3:54-65, 2003.
Lageman RC, Melick JR. *HIPAA: Wake-Up Call for Health Care Providers.* Fitch IBCA, Duff and Phelps, 2000.
Marietti C. Beyond Y2K. *Health Care Informatics* 11: , 1999.
Moynihan JJ. The Basics of HIPAA for Clinicians, Health Care Executives and Trustees, Compliance Officers, Privacy Officers, and Legal Counsel. *First National HIPAA Summit.* 2000,
Woolhandler S, Himmelstein DU. The Deteriorating Administrative Efficiency of the U.S. Health Care System. *New England Journal of Medicine* 324:1253-1258, 1991.

105

Y2K and Clinical Engineering

Stephen L. Grimes
Senior Consultant and Analyst, GENTECH
Saratoga Springs, NY

For want of a nail, a shoe was lost.
For want of a shoe, a horse was lost.
For want of a horse, a rider was lost.
For want of a rider, a message was lost.
For want of a message, a battle was lost.
For want of a battle, a kingdom was lost.
And all for want of a nail. –George Herbert

The British poet George Herbert used this verse to illustrate how a problem that at first might seem small (i.e., the lack of a nail) can sometimes have devastating consequences (i.e., the loss of a kingdom).

The Year 2000, or Y2K, technology issue had a simple genesis. Since the widespread use of computers in government, research, and business in the 1950s, computer programmers had used two digits in their software applications, rather than four, to represent the year. For example, the years 1952, 1977, and 1980 were stored and represented as "52," "77," and "80," respectively. At the time, this shortcut seemed to make sense. It saved keystrokes in data entry (e.g., it requires two fewer keystrokes to enter 01/01/52 than 01/01/1952), and required less computer memory and data storage capability, both of which were precious resources in early computer systems. When doing computations, it was safe to assume that only the last two digits of the year were significant (e.g., subtracting 52 from 80 yields the same result as subtracting 1952 from 1980). Whether calculating ages or interest rates, projecting dates, or scheduling routine activities such as planned maintenance, the use of two digits to represent the year was generally adequate.

Early programmers adopted the two-digit year convention, and subsequent generations of programmers and operators continued to use it without giving it further thought. Computers and their application programs proliferated throughout the latter half of the 20th century. They became faster, with higher capacities, while becoming smaller and more affordable. By the end of the 20th century, virtually every business in every industry was computerized. For most businesses, some computers and software applications had critical roles, i.e., the ability of the business to function depended on the ongoing, successful operation of the computer application. As businesses' needs changed over the years, they added staff programmers to modify existing software. As a consequence, core applications lasted beyond their projected life cycles, with supplements, modifications, and fixes added to meet business needs. Nearly all programs used the same two-digit year convention.

In the early 1980s, inexpensive personal computers appeared in the business environment and on the desks of key staff members, from the chief executive officer to the storeroom clerk. These personal computers had internal clocks and often ran spreadsheet or database software that had been customized using for the most part the two-digit year convention. Computers also began morphing into other forms, appearing in such diverse environments as automotive subsystems, household appliances, industry tools, and medical equipment. These hybrid (often referred to as "embedded") systems often were programmed with software on a silicon chip that kept track of time or dates using the two-digit convention.

The opportunities for the applications of computer technology seemed limitless. However, during the mid 1990s, reports of an inherent and perhaps fatal flaw in a broad spectrum of these systems began to surface. Some computer experts began speculating that many systems would produce unpredictable results and some might catastrophically fail as the world passed from the year 1999 to the year 2000. These failures would occur because many systems were not capable of recognizing that the year 2000 came after the year 1999. For example, when asked to determine the difference between the years 2000 and 1999, systems might calculate the difference between "00" from "99" yielding an erroneous result of "-99" (rather than the correct answer of "1"). The Y2K bug made its debut.

It Gets Complicated . . . "a kingdom was lost"

In the United States alone, it was estimated that 1.57 billion "function points" of software, roughly equivalent to 157 billion program lines of COBOL (COmmon Business Oriented Language–a popular business programming language of the 1960s and 1970s) existed by 1997 (Jones, 1997). This number did not include the millions of macro programs written by users for their spreadsheets or database applications. Affected software applications permeated all industries, including government, military, utilities, financial, manufacturing, and health care.

In the mid 1990s, as many industries began to assess their vulnerability to the Y2K bug, they realized the enormous breadth and depth of the potential impact. First, since adoption of the two-digit year convention by programmers had been nearly universal, virtually any function controlled by computer or containing an embedded computer might be affected. Problems could occur in missile defense systems, air traffic control, streetlights, telephones, water and electric power, and an endless list of other functions in other industries. Second, it was impossible to know which applications would be affected without testing. Applications were generally inadequately documented to identify which programs had the Y2K bug and where it existed within the program. Third, the problem was worldwide. Nearly every person on earth depended on services or businesses that were susceptible to the Y2K bug in some way. Most services and business were so interdependent that if some failed because they could in turn cause the failure of other business and services.

While many businesses and the public were unaware or unconcerned about Y2K in 1995, perspectives had changed by the end of 1998 and beginning of 1999. Governments and industries worldwide began to focus on solving the Y2K problem. A new industry of Y2K consultants and remediation services was spawned. As governments and industries assessed the scope of the Y2K problem, the following became clear:

1. The number of potentially affected computers and computer-based systems was so vast as to make comprehensive Y2K repair impossible. There were too few resources and too little time to effectively address the problem.
2. There was a wide disparity among governments, industries and the public in the perception of the seriousness of the Y2K problem and of the level of effort these groups were willing or able to spend on Y2K remediation.

In the United States, the executive branch of government took the problem seriously. Congress established special Y2K committees to oversee government and industry efforts to address the problem. Many state and local governments mirrored these activities. The United States Securities and Exchange Commission (SEC) required that publicly traded companies report their Y2K readiness status. Industry trade groups established guidelines for and provided educational services on Y2K remediation. Insurance companies amended their policies to limit their liability in not only in case of "acts of war" or "acts of God" but "acts of the Y2K bug." It is estimated that the United States spent between one and two trillion dollars to prepare for Y2K problems, of which $13 billion was spent by the federal government (Luening et al, 1999).

By the late 1990s, a significant number of supposed experts were claiming all efforts were "too little too late" and worldwide disruptions in key businesses and services were imminent (Bergeon et al, 1999; Yourdon and Yourdon, 1998; Yardeni, 1998). The general public was advised to prepare for the worst and hope for the best. Common recommendations included taking a sizeable amount of cash from the bank to keep on hand, stocking food and water for days or weeks, keeping extra clothing and bedding available, filling automobile gas tanks, purchasing batteries and battery-operated radios and lanterns, and obtaining extra supplies of any needed prescription medications. By the end of 1999, the manufacturers of gas-powered electric generators could not keep up with the demands from concerned homeowners.

Much of the concern over the Y2K problem occurred because the problem was unprecedented and of such a wide scope that it was impossible to accurately predict the consequences. There was some question whether a simple convention adopted by a few programmers few decades earlier might topple (or at least incapacitate) businesses, governments, and basic society functions.

Y2K's Effect on Health care

The health care industry was late in responding to the Y2K issue. Like most other industries, health care employed computers and computer-based systems extensively in its operations and would face interruptions of those operations if the Y2K bug caused applications to fail. Potential failures differed in severity as follows:

- Catastrophic failures, including systems that completely shut down and fail to function
- Operational failures, including critical calculation and other processing errors (e.g., miscalculations causing errors in diagnostic results and in medication and other therapy delivery systems)
- Non-critical failures, including systems that provide erroneous date information but whose operations are otherwise unaffected

Catastrophic and operational failures of some applications, particularly those affecting patient diagnosis or treatment, could seriously impact on a patient's health and safety. Health care vulnerabilities fell into the following categories:

- Public infrastructure: Power, water, police and fire protection, public transportation, telecommunications are subject to their own "Y2K" failures and interruptions.
- Business dependencies: Third parties on whom the provider relies may experience "Y2K" related-failures that would compromise the hospital's ability to conduct its operations (e.g., drug and food suppliers).
- Computer systems: Medical information management systems, medical record and billing systems, networks, personal computers, and computer-controlled devices.
- Embedded processors: Non-programmable microcircuits "hard wired" into other pieces of equipment that may be critical to patient services or hospital operations, many of which include date calculations in their programming logic.

The embedded processor devices were considered a special risk for three reasons. First, embedded processor devices far outnumbered traditional computing systems. Nearly $10 billion worth of microprocessors were manufactured and sold between 1991 and 1998, and of that number, only 10% were installed in traditional computers (Figure 105-1). The remaining 90% went into embedded processor devices. Second, since the processor and its application software were embedded, it was not obvious which devices had processors and might be affected by the Y2K bug. Finally, embedded processor devices that did have the Y2K bug could not simply be reprogrammed as traditional computers could. A device had to have its hardware changed or it had to be replaced. Typical embedded processor devices include the following:

- Medical devices and equipment, including infusion pumps, defibrillators, monitors, MRIs, CT scanners, dialysis machines, chemotherapy and radiation therapy machines, laboratory equipment, and other diagnostic and therapeutic systems
- Monitoring and control systems, including environmental and safety equipment
- Fire alarms, including detection and suppression units;
- Security systems, including badge readers and video and surveillance systems
- Telecommunications equipment, including telephones, call management and voice mail systems, pagers, cellular phones, and fax machines
- Building infrastructure, including HVAC, energy management and lighting controls, emergency generators and emergency lighting, backup power systems, elevators, and parking systems

Figure 105-1 Only 10% of processors manufactured between 1991 and 2000 were installed in computers. The remaining 90% were embedded in non-computers.

The interdependency of medical devices and systems added to the complexity of the Y2K problem. The weakest link would be the simple device in the chain that was not Y2K compliant and failure of that weak link threatened the entire system (see Figure 105-2).

The consequences of Y2K failures to health care providers fell into the three basic categories, illustrated by the following three figures:

Remedying Y2K in Medical Devices

The process of remedying Y2K in medical devices was daunting. Hospitals are enormous health care providers and own many medical devices. An average hospital might have 1000 to 5000 medical devices representing 500 to 1000 different device models and hundreds of manufacturers. Information technology (IT) staffs coordinated Y2K compliance efforts in most hospitals because they were responsible for the business computers and applications. However, IT professionals were often unfamiliar with medical technology and relied on the clinical engineering staff for assistance. Most clinical engineers broke the Y2K remediation process into five basic steps.

1. Limit further increases in risk exposure: All future medical device acquisitions had to be reviewed to ensure Y2K compliance. Y2K compliance terms were usually included on all purchase orders and manufacturers and required vendor certification.
2. Identify scope of existing Y2K risk: An inventory of all medical devices was created. That inventory was reviewed and a determination made as to whether devices were known to be compliant, non-compliant, or compliant status unknown. The United States Food and Drug Administration (FDA) had previously required all medical equipment manufacturers who sold their products in the U.S. to supply the compliance status of all their models. This FDA database was a major source of information about medical devices, followed by manufacturer web sites. In addition to compliance status, the inventory usually included an assessment of the critical nature of the medical device.

Typical critical assessment categories included the following:

- High risk and critical devices/systems are life support, resuscitation, or critical monitoring systems, or other devices that would seriously harm a patient if they fail.
- Moderate risk devices and systems would have significant impact on patient care if they failed, but failure does not pose immediate harm.
- Low risk devices and systems would have no serious impact on patient safety if they failed.
- No risk devices and systems, especially those not battery- or AC-powered, are unaffected by date changes and would have no impact on patient safety if they failed.

The inventory might include any known corrective measures (e.g., repairs, upgrades, or replacements that were necessary to secure compliance for non-compliant devices).

3. Develop/implement plan to address problem and insure compliance:

Following completion of the inventory, hospitals would prepare a plan to addressing all non-compliant devices and those with an unknown status. Priority was given to the most critical devices. If a device had an externally accessible clock/calendar and its Y2K compliance status could not be reliably assessed, it was tested. The testing typically involved verifying the devices ability to accomplish the following:

a. Accept a new date
b. Rollover between critical dates in both powered-up and powered-down states
c. Re-initialize on power-up after critical date change

Non-compliant devices were brought into compliance through upgrades, modifications, repairs, or replacement (Figure 105-2).

Contingency plans were made to back up critical systems, and support staff members were on hand on December 31, 1999, to ensure that all critical systems made a smooth transition into the new millennium.

4. Monitor implementation of plan through 2000:

The remediation plan required regular reviews into 2000 to ensure that it remained effective and reflected updated information. Devices would need testing to verify that they remained compliant after undergoing modifications, upgrades, or repairs. Procedures to bring devices and systems into compliance were rechecked to reflect current information.

5. Status reports:

Regular progress reports to organization's Y2K coordinator were made regarding the status of medical device compliance. These reports typically included the following:

a) Initial Y2K exposure and any subsequent changes in status
b) Corrective actions taken and additional actions pending
c) Preventive measures taken to limit additional exposure

The Outcome: Lessons Learned and Benefits Gained

As the clock ticked past 11:59 pm on December 31, 1999, and into the year 2000, all the fears of worldwide catastrophe were allayed. There were minor problems but nothing serious.

Was the Y2K problem over exaggerated? Perhaps it was. However, updates and upgrades on critical applications and systems did prevent some glitches from occurring. In addition, Y2K remediation provided many lessons and indirect benefits including the following:

- A better understanding of our reliance on technology and its vulnerability and perhaps a better appreciation of how small decisions can have unintended and significant consequences when constructing technical support systems.
- Contingency planning that incorporates means of addressing those vulnerabilities when elements of technology fail.
- An examination of existing technical systems and the overdue replacement of many that were obsolete or unsupportable.
- The development of more effective risk assessment and management methods to avoid problems and prioritize remediation efforts.

The American Hospital Association (AHA) estimated that hospitals spent approximately $8.5 billion in their efforts to remediate Y2K (Marietti, 1999). As a result, many hospitals have updated medical technologies and are more knowledgeable with respect to the risks associated with those technologies. For clinical engineers, their Y2K remediation efforts provided an opportunity to gain insight and to learn valuable lessons on what is required to effectively manage evolving medical technologies.

Figure 105-2 Testing, upgrading, modifying, repairing, or replacing devices and systems to bring them into compliance.

References

Bergeon RP, deJager P. *Countdown Y2K: Business Survival Planning for the Year 2000*. New York, Wiley, 1999.
Jones C. *The Year 2000 Software Problem: Quantifying the Costs and the Consequences*. Addison-Wesley, 1997.
Luening E, Ricciuti M, Yamamoto M. Everyone Pays a Price for Y2K Hype. *CNET News.com*, http://news.com.com/2009-1091-232056.html?legacy=cnet, November 4, 1999.
Marietti C. Beyond Y2K. *Health care Informatics* 11: , 1999.
Yardeni E. Prepared Testimony in Hearing on Disclosure of Year-2000 Readiness. In *Senate Banking, Housing and Urban Affairs Committee: Subcommittee on Financial Services and Technology*. Washington, DC, US Government Printing Office, http://www.senate.gov/~banking/98_06hrg/061098/witness/yardeni.htm, 1998.
Yourdon E, Yourdon J. *Time Bomb 2000: What the Year 2000 Crisis Means to You!* Prentice Hall, 1998.

106

The Integration and Convergence of Medical and Information Technologies

Ted Cohen
Manager, Clinical Engineering Department
Sacramento Medical Center, University of California
Sacramento, CA

Colleen Ward
Clinical Engineering Department,
Sacramento Medical Center, University of California
Sacramento, CA

Since the development of minicomputers in the early 1970s and the invention and implementation of the microprocessor in the late 1970s and early 1980s, medical products have become increasingly more dependent on computer technology. In fact, some clinical technologies (e.g., computerized tomography [CT] scanners) are entirely dependent on computers and could not function without them. More recently, microprocessors have become ubiquitous in medical technology and have been used in products including "smart" electric beds that weigh the patient and sense whether he or she is in or out of the bed; sophisticated implants such as implanted cardiac defibrillators; a large variety of medical systems that measure physiological parameters (*in vivo* and *in vitro*); and systems that image almost any portion of the anatomy. Today, many medical systems not only contain embedded microprocessors but also are capable of storing and communicating the clinical information they collect, over standard computer networks.

Information technology in health care has evolved from mainframe-based insurance and medical billing applications to a large variety of information systems, including laboratory, pharmacy, surgical, medical records, physician order entry, radiology, and picture archiving and communication systems. All of these information systems use acronyms that make them a little easier to remember (e.g., CPOE for Computerized Physician Order Entry, PACS for Picture Archiving and Communication Systems, and LIS for Laboratory Information Systems). Information technology in health care has adopted IT-standards-based data communication technologies (e.g., Ethernet, ATM, and category 5 cabling) that have allowed the relatively easy implementation of a standard data communications infrastructure throughout the modern health care facility.

Integration and Convergence

As clinical and information technologies have converged, two trends have emerged: (1) the widespread use of commercial off-the-shelf (COTS) hardware and software and (2) communication technologies that have interconnected the office, the enterprise, the community and the world (e.g., Transmission Control Protocol/Internet Protocol (TCP/IP), Internet. COTS technology significantly reduces manufacturers' costs and improves manufacturers' time-to-market for new products. These technologies allow many of the major medical systems that are sold today to operate as a computer system as well as a medical device. The modern hospital can interconnect these "medical devices" using standard data ports in patient care areas and standard wiring, hubs, switches and routers in data closets. These systems integrate information and clinical technology and support of these systems require an integrated approach by those who are trained and familiar with both the clinical and computer technology.

Combined together, the convergence and integration features result in systems like those shown in Figure 106-1. Such systems, with modified personal computers (PCs) as medical devices, are currently in use in a wide variety of inpatient (e.g., emergency room (ER), intensive care unit (ICU), acute care, operating room (OR), and outpatient settings (primary and specialty clinics) and include many different diagnostic and therapeutic devices and systems (e.g., lab, EEG, ECG, other vital signs, infusion pumps, and medical imaging).

With the use of standard PCs as the development platform, modern medical system design is now focused on system development with most of the effort going into software development and interfaces and, depending on the application, some additional development work being done on transducers (e.g., *in vivo* measurements of clinical laboratory parameters (e.g., blood gases, potassium, and sodium). For some systems, little hardware work, other than an occasional interface circuit, is required. For other devices (e.g., respiratory therapy ventilators and anesthesia machines), microcomputers are included within the medical product, but considerable additional hardware design is still required. This critical equipment also has additional design challenges in order to meet critical life support requirements such as assuring that internal processor and system reboot times are very short.

The use of COTS and modern data communication technologies allows many of these integrated medical and information systems to provide new and robust features including automatic data collection, transmission, analysis and reporting, dynamic reconfiguration for differing applications without buying new hardware (e.g., pediatric vs. adult physiological monitoring, and remote software-version upgrading).

Another trend is the use of computers, both general purpose and specialty, to access multiple information systems. With the large number of computer information systems in a health care facility, a PC or terminal cannot be separately deployed for each clinical location for each information system due to cost and infrastructure (e.g., data closet, wiring, requirements and lack of space). Therefore, multiple applications are integrated on one system in order to allow almost simultaneous access to multiple information systems.

Figure 106-2 graphically depicts a computerized fetal monitoring system that has been adapted to allow access to laboratory data and a hospital information system as well as continuing to collect and process real-time fetal monitor data. When medical devices are used for such multiple applications, they require integration testing either by the manufacturer or the end-user. End-user integration testing requires that the manufacturer provide the end-user with an integration test procedure or software suite. This procedure is then followed to test for the proper operation of the medical device while the "foreign" integrated medical applications are operating. For example, in the fetal-monitor application depicted in Figure 106-2, full compliance with the integration test procedure assures that the real-time collection, processing and archiving of the fetal monitor data, including all its important features such as alarming for abnormal fetal heart rates, can be assured while the "foreign" applications are running. The more critical and time dependent the original applications are the more restrictive and conservative an integration test procedure must be. See Figure 106-3 for a portion of a sample integration test procedure.

Reliability and Quality Control

In order to ensure that medical systems based on COTS operate reliably, the entire system (i.e., transducer, interface, COTS hardware, COTS software, and application software) must operate together and reliably. COTS hardware can be extremely reliable. Typically, the weakest points from a reliability standpoint are COTS software (operating systems) and the application software.

Operating system software reliability is often expressed in terms of "nines," with 99.0% up time equivalent to two nines. The reported reliability of Windows NT 4.0 is two

- *Convergence: ... **moving toward uniformity** ...*

- *Integration: ... **process of incorporating as equals** ...*

Figure 106-1 Definitions of convergence and integration.

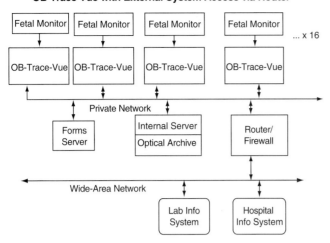

Figure 106-2 Fetal monitoring system with external access.

nines. For a continuously operating system, this equates to about 80 hours per year of down time. The reported reliability of Windows 2000 is 99.95%; i.e., three nines (five minutes per year of unscheduled downtime). Older versions of Windows were far less reliable. Further quantitative comparison of the reliability of various operating systems (e.g., UNIX vs. Windows 2000) is controversial and not yet well documented because there are no standards for software reliability measurement comparisons. Some companies have attempted to measure reboots per time period, but even that is suspect because different operating systems have differing scheduled needs for reboots, such as the reboot requirements that occur when new applications are installed in older versions of Windows. It is known and well documented that the newer versions of Windows (e.g., Windows 2000) are more reliable and require fewer reboots than the older versions (e.g., Windows 3.1, Windows 95, Windows 98). UNIX, and its many variants, is generally more reliable than the older versions of Windows. It remains to be seen whether the reliability of Windows 2000 can equal that of UNIX.

Integrated systems provide the medical device manufacturer with the challenge of ensuring quality at a level required for a medical device, while at the same time using COTS operating system software and COTS hardware that originally might not have been put through as rigorous a quality control protocol. In the United States, the FDA regulates medical device manufacturers and mandates that various quality control measures be in place. According to the FDA document entitled *Off-the Shelf Software Use in Medical Devices* (FDA, 1999), the medical device manufacturer who uses off-the-shelf software "still bears the responsibility for the continued safe and effective performance of the medical device." Like any medical device, the level of validation and verification required for software-based medical systems is based on the severity of the potential hazards to the patient, operators, and bystanders should there be a system failure, regardless of the failure cause, be it in hardware or software.

Software is very difficult to test exhaustively. Although it does not fatigue or catastrophically fail in the same way as a mechanical device or an electronic component, software problems occur regularly. These problems can range from applications that do not perform as designed and can restart with minimal problems, to operating a system stoppages that require reboots on a life support system, which could be catastrophic. Even when exhaustive testing has been performed, systems can experience software failures due to memory problems that develop over long periods of time (e.g., so-called memory leaks), user or operator error (e.g., inappropriate system recovery from erroneous keystroke sequences), lack of computer resources, and problems caused by foreign applications, viruses, or malicious intrusions. Medical device manufacturers must design systems to be as reliable as possible and so that failures are "soft" and do not harm the patient. As operating systems continue to evolve, their real-time functionality and reliability are also improving, and more critical applications and devices are using COTS-based systems.

Information System Security

A computer system or network can be considered secure only when its resources are available solely to authorized users and when use of those resources produces trusted results. A system that has been compromised by an intruder cannot be trusted. However, software bugs, user errors, or a malfunctioning sprinkler system are also threats to the security of a computer system. Clinical information system security is evolving. Security as related to patient data confidentiality in the United States is regulated by the Health Insurance Portability and Accountability Act (HIPAA) and is discussed elsewhere in this handbook (see Chapter 104). Designing security into medical information systems is an important feature and should include network-connectivity security issues, user name and password management, and update and version control, as well as physical security for the computer hardware, both servers and client workstations.

Computer security threats can be divided into errors of use and design and malicious attacks. Errors of use and design include authorized users making errors (e.g., accidental data deletion) and common software bugs (i.e., erroneous and/or incompletely tested software code). Malicious attacks include unauthorized users, authorized users maliciously viewing or altering data, authorized users knowingly or unknowingly giving away passwords, malicious code unknowingly placed on the computer (e.g., viruses, worms, trap doors or back doors, denial of service attacks, packet sniffing, or otherwise unauthorized electronic interception of data), and unauthorized physical access to data or systems.

Good system security design can preclude some of the malicious, as well as unintentional, security problems. For example, systems can force eight-digit passwords that include numbers, letters (upper and lower case), and special characters and therefore are far more difficult to crack than 3-digit numeric passwords.

OB TraceVue Archiving Functionality Test

Test Instructions   Follow these instructions, marking your results on the Worksheet

Archiving Functionality
Procedure 1: Partitioning the optical drive 1. Connect the optical drive to the SCSI interface 2. Insert an optical disk into the drive 3. Start the server 4. Start Disk Administrator from the Administrative Tools menu 5. Format the optical disk 6. Create a partition on the optical disk
Procedure 2: Checking the optical drive Archive 1. With the third party running, start OB Trace Vue on the server with the optical drive and optical disk 2. Create one user and one bed using default basic alert settings 3. Assign Bed 1 to serial port A 4. Turn on the fetal monitor's recorder 5. Admit a new patient into Bed 1 6. Obtain a 5 minute trace with a fetal heart rate of 190 and toco that switches between 60 and 10. 7. Close the episode in Bed 1 8. Wait for 15 minutes 9. Retrieve the archived episode into Bed 1 10. Make a printout of the retrieved trace

Figure 106-3 Sample portion of an integration test procedure (Philips, 2001).

In addition, health care systems and system administrators help to protect against security problems by implementing all of the following:

1. The system administrator should develop a security guideline that documents all physical security, access controls, backup policies and procedures, audit policies, and other key security-related use parameters.
2. The system administrator should implement access control with data access privilege-level control (e.g., appropriate view, add, modify and delete privileges by individual user, and mandate unique user names and a strong password policy, including required periodic password changes. Where additional security is required, biometrics such as retinal scans, fingerprints, or handprints should be considered. Systems should have automatic logouts implemented, and users should not leave systems logged in and unattended.
3. Physical security must be managed. Access to data closets and server rooms should be controlled.
4. Backups should be performed routinely, and backup media should be stored in a separate location from their computer system, and preferably in a fireproof safe. Backups are probably the single most important security task!
5. System administrators need to implement software update and version control in compliance with the guidance of the medical system manufacturer. When manufacturers and regulatory agencies approve, operating systems and application programs should be periodically updated with available security patches and updates (e.g., virus detection and correction updates). It is the administrator's responsibility to keep up to date regarding announcements of security issues affecting the operating system and application software. When update announcements are made, the administrator must assess the impact on the system and take action in an appropriately timed manner, depending on the nature of the problem and the update status. Interim precautions or actions can be required until a permanent correction is developed, tested, and released by the manufacturer. The gap in time between when a security defect is discovered by hackers and the time that action is taken by the system administrator must be kept as short as possible to lessen the chance of further intrusion or new attacks. However, timing is particularly a problem for systems cleared by the Food and Drug Administration Center for Devices and Radiological Health (FDA CDRH), where version control and significant manufacturer testing are required prior to update implementation.
6. In some cases, for network protection, a firewall or virtual private network (VPN) can be installed to control access into and out of specific locations via domain, Internet Protocol (IP), and other network access control methodologies. Firewalls can be programmed to control all access into and out of a local or wide area network. A VPN can be implemented to encapsulate and encrypt private data over a public network so that it cannot be intercepted easily.
7. Where additional security is required, it can be provided by various encryption techniques. For wireless systems, the Wired Encryption Privacy (WEP) is one common encryption standard.
8. Where available, the logging and auditing of all administrator accesses, possible intrusion attempts (e.g., failed log-in attempts), and other significant events should be implemented.

In general, what should be done if a major security breach is discovered? First, refer to the security policy and then back up the entire system to disk in a disk space that is not currently in use (system snapshot). If the system is networked, disconnect it from the network if that is the likely source of the security breach or if the source of the breach is unknown. Further evaluate the security breach to determine what occurred and what was compromised. This will determine what additional actions must be taken, to first correct any compromised data or programs (e.g., restore using latest data backups and original program disks, where needed, patch the security breach so that it is less likely to occur in the future, and then determine whether additional action needs to be taken, such as notification of appropriate administrators or law enforcement agencies.

Vendor access to a medical information system for troubleshooting and upgrades is an increasingly more common feature, but it presents security challenges. Common vendor-access methods include dial-up modems, network (e.g., WAN) access, and access via a VPN. Where staff are present, modems provide a simple connectivity method and allow the end-users to disconnect the modem when it is not in use. However, when the information system is in a secure or remote location, or a location that is not staffed, then it is not practical to turn the modem on and off; modems become security risks. WAN access is simple, but it can be quite insecure unless access is controlled by a firewall or other authorization methods. Installing VPN equipment provides a much more secure method as it uses public infrastructure but provides IP control, similar to a firewall, and also encapsulates and encrypts the data. Of course, with all of these external access methods, user name and password management are also important. Leaving a persistent Internet connection (non-VPN) constantly open with a generic user name and no password is an invitation to an unwanted intrusion.

In summary, the greatest security challenges are caused by the manufacturer or the end-user not complying with good computer security design and implementation practices as discussed above. Poor practices that can lead to corrupted systems include users and system administrators allowing "generic" user names; no passwords or passwords that are easily broken; no physical security, resulting in systems being available to unauthorized individuals for unsanctioned activities; network connections to an intranet that are really connections to the public Internet with no firewalls or other intervening security measures; networked systems without virus-scanning software; and nonexistent or inadequate audit trails that prevent adequate security breach follow-up action to take place. Of course, periodic backups can ease the pain of recovering from any data loss, whether malicious or unintentional.

Support of Computerized Medical Systems

Computerized medical systems offer several support advantages for both the manufacturer and the end-user. Built-in system self-tests allow devices to test themselves and to determine whether they are working properly or not. Therefore, some networked devices can self-test and, when they are not working properly, automatically "phone home" and tell their manufacturer that there is a problem. For software problems, remote access (automatic or manual) can then be used to fix the problem. Other advantages include online, fail-safe, "high-availability" systems that include a constantly running second computer that "mirrors" the operation of the primary system and takes over operation if a problem occurs.

Support Challenges

System integration provides a new set of challenges for the manufacturers and support staff for these complex systems. New methods must be developed to manage and support these systems. For example, the health care institutions and the manufacturers need to track and control software versions and upgrades more effectively, including manufacturer-provided COTS software and application programs. There is a conflict between the FDA requirements for version control and the ever-changing versions and service packs being added to COTS software. On network-connected information systems, and all their nodes, this conflict particularly impacts updates for security improvements for operating systems and virus-scanning software.

Other support challenges are further described in the following sections on hospital infrastructure, training and education, and documentation.

Infrastructure

IT standards-based medical systems allow communication via TCP/IP and other standards that hasten interconnectivity. Systems based on standards also allow common data infrastructure to be installed during construction and prior to knowing which specific clinical system will be purchased. Other advantages include installing computer hardware in the data closet and saving space in the clinical location. Challenges include building the data closets large enough to house more increasingly sensitive, equipment, color coding (or otherwise identifying) cables and other closet hardware, particularly for real-time medical systems, in order to separate them from office and other noncritical applications. Uninterruptible power supply (UPS) or emergency power must be provided to these systems in case of power failure and to ensure continuous operation during emergency generator tests. Access to the data closet must be controlled, but medical systems support staff must be allowed access.

Several wireless technologies have penetrated the health care market including 802.11b in clinical telemetry applications, micro-cellular internal-to-the-hospital cellular phone systems, and wireless personal digital assistants (PDAs) for medical staff. Both frequency management and access point (antenna) location management are required in order to avoid interference between all of the varieties of wireless technologies currently vying for the health care market. In order to reduce wireless infrastructure, standardization of the various wireless technologies is important, but it is currently difficult because of the large number of different wireless standards in use (e.g., IEEE 802.11a, IEEE 802.11b).

Training and Education

Information technology (IT), clinical engineering, and biomedical equipment technology professionals who support these converged medical and information systems have new training needs, with the IT staff needing more clinical knowledge and the biomedical/clinical engineering community requiring additional computer and IT training. Clinical engineer and BMET training must include fundamental computer technologies including operating systems e.g., Microsoft Windows, UNIX, databases, applications, communication technologies e.g., Ethernet, TCP/IP, Internet, Asynchronous Transfer Mode (ATM), and wireless, and new computer technologies e.g., storage area network (SAN), plus clinical information system education. Security, patient data confidentiality, and other public issues associated with data communication are issues that everyone must understand better.

Documentation

The clinical engineering literature contains many articles regarding required service documentation for medical instrumentation. However, little has been written about ways to document complex, computer-based medical instruments, particularly computerized information systems. One approach is to require the following: (1) as-built drawings; (2) operator manuals and specifications for each component; (3) service manuals and troubleshooting information for critical components; and (4) software tools to aid in troubleshooting.

As-built drawings provide a way to document the system after it is installed, showing all wiring, hubs, routers, servers, access points, and workstations. Computerized, as-built drawings based on Adobe Acrobat® and PDF files are one way to develop as-built drawings. These network drawings can include "hot" links to printer and other peripheral information and also include information regarding the equipment's physical location, model, data communication paths, TCP/IP addresses, modem phone numbers, and more.

Traditional user's manuals including setup and configuration information are required and typically supplied. Service manuals are difficult to obtain, but critical for all systems and components that are not "off-the-shelf" and therefore can be difficult and/or expensive to replace. Any software troubleshooting tools that the vendor will make available to the customer should also be obtained and appropriate documentation provided in order to operate these tools.

Conclusion

Information technology is changing rapidly, and medical technology, although more slowly changing than information technology, is evolving rapidly. In the near future, new and emerging medical technologies that are based on IT include new wireless networks (e.g., Bluetooth personal area networks, surgical robotics with tactile sensors, "smart" artificial limbs, advanced speech recognition, voice over IP telephones, reasonably priced digital broadcast quality video, and swallowable video endoscopes). Standards-based information and medical technology integration will easily allow workstations to communicate with multiple systems without special integration testing and concern over critical performance problems. Data-transmission rates will continue to increase, with costs continuing to decrease (e.g., gigabit Ethernet and faster digital subscriber line (DSL)). Data and voice infrastructure will merge, and data outlets will become as ubiquitous as electrical power outlets. Data closets and data infrastructure will continue to grow in size and complexity as the rate of equipment that moves from clinical spaces into the data closets increases faster than the size reduction of the equipment.

The clinical laboratory will move in two directions. For critical patients, point-of-care testing and indwelling sensors will become more commonplace. For the main lab, increasingly more tests will be performed via automated, robotics-based laboratories. Nursing unit central stations will become increasingly less important as physiological monitor alarms, "nurse call" requests, and other critical information are communicated directly to the assigned care givers. The acuity level of the inpatient will continue to increase, and technology will be moved to the inpatient's room, rather than the patient being moved to the technology.

Continuing education of all clinical and support staff is required in order to stay abreast of these changes. New paradigms in health care technology leadership are required for managing integrated clinical and information technology.

References

FDA. *Off-the-Shelf Software: Use in Medical Devices.* Bethesda, MD, FDA, 1999.

OB Trace-Vue Integration Test Protocol. Andover, MA, Philips Medical Systems North America, 2001.

Web sites

Adobe Acrobat: www.adobe.com

Bluetooth: www.bluetooth.com

Windows 2000 Server Family: Delivering the Level of Reliability You Need. Available at http://www.microsoft.com/windows2000/server/evaluation/business/overview/reliable/default.asp

Klaus CW, *Wireless LAN Security 802.11b and Corporate Networks,* Internet Security Systems. Available at www.iss.net/wireless

Section X

Engineering the Clinical Environment

Matthew F. Baretich
President, Baretich Engineering, Inc.
Fort Collins, CO

Clinical engineering is one among many engineering and technical professions that contributes to the design and operation of the clinical environment. At the most fundamental level, clinical engineers must coordinate their work with that of health care facilities engineers and environmental safety professionals. For this reason, clinical engineers should have at least a basic understanding of the principles and practices in these related areas.

However, there are good reasons to go beyond a basic level of knowledge. Clinical engineers often encounter opportunities to go from coordination to active cooperation, which means looking outside the typical definition of clinical engineering practice and finding ways to bring broader engineering principles to bear on health care facility design and operation. Only by fostering synergism among the various engineering and technical professionals within the health care delivery system can we maximize the contribution of engineering to patient care.

Clinical engineers are increasingly being asked to take responsibility for technical activities outside their usual areas of practice. Therefore, clinical engineers should develop an understanding of related professions so they can expand their roles within their organizations.

Beyond Clinical Engineering

The chapters in this section are not intended to make clinical engineers experts in the topics they address. They are, however, intended to provide an introduction to these topics and to point the reader towards additional sources of information. These chapters may be regarded as representative starting points for clinical engineers who wish to expand their knowledge of related fields, and perhapsto expand their range of professional practice.

The first group of topics introduces the basic utility systems found in health care facilities. Hyndman provides an overview of the variety of systems that are behind the scenes supporting the patient care environment. In the next chapter, Hyndman presents a detailed description of heating, ventilation, and air conditioning (HVAC) systems and their roles in patient comfort, infection control, and indoor air quality. Baretich addresses electrical power systems, including both normal and emergency systems, as well as the isolated power systems used in some health care facilities. Frank provides a practical overview of medical gas systems, including storage and distribution systems for oxygen, medical-grade air, vacuum, and other gas systems used in health care delivery.

The second group of topics describes a range of activities in health care facility design and operation. Cram discusses the wide range of support service programs that are vital to the operation of health care facilities. Baretich addresses the role of clinical engineering in facility design, construction, and renovation.

The third group of topics covers a representative set of environmental safety issues. Strzelczyk presents an extensive overview of radiation safety, including specific examples of radiation safety programs. Brito and Magagna discuss hospital sanitation activities, including infection control, sterilization, and waste-handling. Hernández addresses water distribution systems and their roles in disease prevention and infection control. Epstein and Harding conclude with an overview of disaster planning in health care facilities.

The topics addressed in this section represent the nuts and bolts that we use to build an environment that supports high-quality patient care. In a very real sense, "engineering the clinical environment" is what we do.

107

Physical Plant

Bruce Hyndman
Director of Engineering Services, Community Hospital of the Monterey Peninsula
Monterey, CA

The physical plant (i.e., the "plant" of a building or complex) is generally considered to comprise the infrastructure elements that provide the utilities systems to the building. This chapter describes some of the common utilities that are considered to be elements of the plant. The utilities described do not constitute an all inclusive list of elements of the plant in a hospital, but they are representative of major systems.

Major physical plant elements are frequently housed in a central location for efficiency of construction and operation. Those elements thus installed and constructed are commonly called the "central plant." A central plant may be housed in a separate but adjacent utility building or may exist inside the envelope of the hospital building. The central plant often houses hot water and steam boilers, chillers, and power generating equipment, or some subset of these elements. These large electromechanical devices have associated pumps, motors, pipes, conduits, wires, and flues. Other components of the plant are distributed and provide the pathways for distribution of the plant's commodities and energy. To familiarize the clinical engineer with the plant, the plant elements are described in this chapter.

Life Safety

Fire Detection

Modern fire detection systems consist of one or more microprocessor-equipped cabinets with individual field devices attached to the cabinet by wiring and communications protocol. The devices may include the following:

- Smoke detectors
- Heat detectors
- Control units for fans and doors
- Visual annunciating devices (strobes)
- Audio annunciating devices (speakers, horns)

Modern fire alarm systems, also known as fire alarm control panels, are associated with a personal computer (PC) that loads the site-specific information, including the assignment of devices, the naming of devices, the annunciating protocols and messages, and the sequence of operations for control devices and annunciators. Options for these systems operations can include the amount of time a detector remains in alarm condition before a general alarm is announced, the length of time audible signals are active with a general alarm, and the form of the messages displayed for any condition. Large buildings and building complexes are equipped with multiple cabinets containing supporting devices. The cabinets within zones are linked to a common network. These systems may be smart enough to monitor and announce such conditions as dirty smoke detectors, missing devices, and open or ground fault conditions. Installation and maintenance of a fire alarm system is governed by codes and standards, usually adopted from National Fire Protection Association (NFPA) standards.

Fire Protection/Suppression

Sprinklers

The automatic sprinkler system is the primary fire protection system for hospitals. An automatic sprinkler system is a piped water system, usually fed from a street water main into the building through a set of pipes and valves separate from the domestic water system, although at the street level the water likely stems from the same main piping system. The sprinkler system has valves for each branch that are accessible to the fire department and can be locked in the open position. It also has electronic monitoring systems to indicate a closed or partially closed condition. The valves outside the building may be of the type known as a post-indicating valve (PIV), which has a window showing the words "open" or "closed." Inside a building, the sprinkler system might have additional zone valves, but each valve is electronically monitored to create an alert condition if the valve is closed or partially closed. The sprinkler system piping is distributed throughout the building to heat-activated sprinkler heads. At a specified temperature and after a specified time, each sprinkler head opens and sprays water over the area it serves. Water continues to flow until a control valve is closed.

Some systems operate strictly on water main pressure from the street. Others may have booster pumps to increase the sprinkler pressure to a desired level or to provide water from alternate source, such as storage tanks or a pond. A fire department connection may be part of the outside piping system, allowing a pumping truck to attach to the automatic sprinkler system and increase the pressure in the system or to add additional water if the main supply is deficient.

In addition to wet automatic sprinkler systems, dry standpipe systems may be installed on the exterior of the building to allow a water source such as a fire engine pumper to be connected, to provide water for fire hose connections or sprinklers within the building. Other fire suppression systems are used for selected areas in hospitals. Computer rooms may be equipped with Halon systems that use compressed fire suppression chemicals. These systems are operated electrically based upon a specific detection pattern in the room. If activated, a valve opens that results in the rapid release of the compressed agent into a closed room. Cooking areas with grills and other grease producing devices are equipped with exhaust hoods with heat-initiated dry chemical extinguishing systems. These systems release a fire-retarding chemical on the cooking area if the temperature in the exhaust hood reaches a specified temperature.

Other extinguishing systems, including compressed carbon dioxide systems, are in less common in hospitals. However, all systems are linked to the fire alarm system in the hospital so that activation of any single suppression system results in a general fire alarm. In the case of the sprinkler system, flow detection devices in the sprinkler pipes initiate a fire alarm. Other systems provide alarm signals to the main fire alarm system through relay contacts or other means.

Barriers

Fire protection in hospitals is provided in part by the elements of the buildings' construction. Barriers to the spread of smoke and fire include walls, doors, and dampers.

Hospital construction requirements specify the materials and configurations to contain fire and smoke within rooms and compartments to allow time for the fire to be extinguished and to protect the paths of egress from the area or from the building.

Extinguishers

In addition to the automatic sprinklers and non-water suppression systems, hospitals have a large inventory of hand-held extinguishers of various sizes and specifications or ratings. These devices require regular inspection and testing.

Hoses

Some hospitals may have fire hoses that are attached to the automatic sprinkler piping so building occupants can suppress or extinguish a fire. Operation of these hoses results in a flow alarm from the fire alarm system.

Hydrants

Hospital properties have standard fire hydrants at specified intervals on the perimeter of the property for use by fire department trucks. The hydrants must be tested regularly for proper flow and static pressure.

Egress Signage and Lighting

The required "Exit" signs in hospital buildings are an important element of the life safety systems. The signs may be illuminated by LEDs or light bulbs, or may be self-illuminating with radioactive luminous material. Certain light fixtures along the path of egress from any building must be supported by an emergency power source to provide a lighted pathway during a fire or other emergency that interrupts normal power sources.

Emergency Electric Power

Specified electric loads in a hospital must be supported by a backup emergency electric supply. A common compliance method is the installation of diesel-fueled engine generator sets. These devices are connected through automatic transfer switches (ATS) that sense loss of normal power and automatically operate a mechanical switch to connect the required loads to the output of the engine generator set. In some cases, loads other than required loads may be connected. Gasoline, natural gas, or other fuels can power engine-generator sets. Steam or jet turbine-driven generators are also used to supply power to hospitals but are not commonly used as emergency backup generator systems. In one case, steam-driven generators were specified as normal power and the public utility system specified as the emergency backup. Code requirements generally include the ability of the engine generator set to start, come to full speed, operate the ATS, and connect to and support the required electrical loads within 10 seconds after failure of normal power sources. (See Chapter 109 for more information about emergency electrical power.)

Lighting

Lighting is a major energy user in a hospital and entails a variety of fixtures and incandescent and fluorescent lamps of various luminosities. Lighting controls may be sophisticated to improve energy efficiency, but generally lighting in hospitals remains at a constant level because of the nature of hospital operations.

Vertical Transport

Vertical transport may be accomplished by elevator, or by escalator in some cases. Elevators can operate independently or can have control systems that optimize the locations of elevators based upon some predetermined rules, such as time of day, floor priority, and user priority.

Normal Electric Power

A local utility company most often provides normal electric power to hospitals. Electric power can be purchased at a user voltage level or 208 or 480 volts, for example, and the transformed to other voltages for some hospital use. In this case, the utility company will own large transformers on or near the hospital property, because the utility company owns the electric power at the street at higher voltage levels, such as 21 kilovolts or 12 kilovolts. Hospitals may choose to purchase power at the higher voltage to obtain a discount, and then own a small, high-voltage electric system with their own transformers to reduce the voltage to a service level. Understanding utility rate structures is an important part of operating a hospital and minimizing utility bills. (See Chapter 109 for a further discussion of normal electrical power.)

Domestic Water

A local water company or utility sells water for bathing, drinking, cooking, and sanitary sewer systems. Much of the water used in a hospital on a daily basis may be for comfort control, and it evaporates into the atmosphere from cooling towers. A chilling system can use more than 10,000 gallons of water per day. Domestic water may require treatment before use in boilers and other equipment to reduce hardness or contaminants that damage to pipes or systems. (See Chapter 115 for further information on water.)

HVAC

Heating, ventilation, and air conditioning (HVAC) systems are comprised of the following major components:

- Heating hot water
- Chilling pumps, cooling towers, and chillers
- Water treatment
- Fans
- Controls
- Humidification

A full discussion of these elements appears in Chapter 108.

Sewers

Sanitary Sewer

Unlike home sewer systems, hospital sewer systems must accommodate caustic chemicals, grease from major kitchen operations, chemically treated water from boilers or cooling towers, silver-contaminated water from film development, and other effluents not found in household sewage. To accommodate the various effluents, sanitary sewer systems may include grease interceptors or traps, mixing or diluting systems, exotic pipe materials such as glass, and effluent treatment systems using enzymes or bacteria.

Storm Sewers

Storm sewers are different from the sanitary sewer systems. Storm sewers handle water runoff from rain, snow, or irrigation. Separation of sewer systems to comply with local codes may require special attention in some areas, including loading dock or trash compactor areas, where contaminants from washing or spills must be carefully separated from rainwater into different sewer systems. (See Chapter 114 for further information on sanitation.)

Medical Gases and Vacuum

Piped medical gas systems in hospitals include medical air and oxygen, as well as nitrous oxide and nitrogen in some areas. Piped medical vacuum systems are usually considered a part of these systems. Piped systems operation requires knowledge of pressure regulation, gas alarm systems, bulk liquefied gas storage systems, high-pressure gas storage cylinder handling, and manifold cylinder delivery systems. In addition, hospitals typically have a variety of non-piped gases supplied in cylinders, including carbon dioxide, carbon monoxide, and other gases used for testing in various applications. Careful maintenance and operation are required to preclude delivery of the wrong gas or any gas at the wrong pressure. Chapter 110 contains a further discussion of medical gas systems.

Steam and Natural Gas

Steam boilers in the central plant generally produce steam for use in heating, sterilization, cooking, and humidification and pipe it to the end user. Natural gas, fuel oil, propane, or other fuel may fire steam boilers. In some areas, steam can be provided as a commodity from a local utility and piped in from outside the hospital. Natural gas also can be used as a commodity to heat hot water, steam, domestic hot water, cooking, and other services.

Communications

Modern communications systems include a variety of hard-wired and wireless systems. The hospital might operate a telephone system that is provided as a service from a local telephone company, but it usually owns a telephone system termed a "Private Business Exchange (PBX)." PBX systems comprise a central "switch," i.e., the electronics that control the telephones, the calls coming in on wires or fiber from outside the hospital, and the local network inside the hospital. In addition, a hospital can operate a telephone system that is linked to the internal data network system using internet protocol devices. Other communications systems operated and maintained in a hospital can include nurse call systems, radiofrequency devices, such as two-way radios and repeaters, "beeper" systems, and cellular telephone systems for use inside or outside the hospital. Some hospitals use pneumatic tube systems in which carriers are moved through a system of tubes from location to location with vacuum or pressure. The pressure systems require a control system to sort and direct carriers to the correct destination and can have sophisticated routing computers to optimize their use.

108

Heating, Ventilation, and Air Conditioning

Bruce Hyndman
Director of Engineering Services, Community Hospital of the Monterey Peninsula
Monterey, CA

"Heating, ventilation, and air conditioning" (HVAC), is an umbrella term that encompasses many individual generic types of equipment and systems, usually with subsystems and subcomponents. The components of HVAC systems in general include fans, dampers, coils, filters, humidifiers, and controls. The engineering study of the design and operation of such systems falls largely within the realm of mechanical engineering. It is not the intent in this chapter to provide the engineering theory, design parameters, or specific operating guidelines for such systems. Such information is provided in large volumes of handbooks and texts and is too broad in scope to address here. The purpose of this chapter is to provide a simple overview of these systems that will allow clinical engineers to be familiar with the systems and equipment and their functions.

HVAC systems are generally integrated, although they can be separate. For example, a radiant heating system that does not employ heated air supplies as the primary method of heating may be used. Even in that example, it is likely that the minimal heating of supply air will use the same source of heat as the radiant heating system. In the simplest model, the central system to the HVAC operations is the air handler, shown in Figure 108-1 with its various elements. A building requires ventilation for fresh air and to exhaust the products of human ventilation and processes. Simple systems take fresh air from outside the building, using a fan, and they blow the fresh air through ducts to every room in the building, where it enters the rooms through vents. Air is similarly exhausted by being drawn out of the rooms through vents into exhaust ducts that blow it out of the building and into the atmosphere.

Associated with the fans and ducts that move air in or out of the building are systems that filter the air, humidify or dehumidify the air, heat the air, cool the air, and control the volume of air supplied to the duct work and, in some cases, to each room. Remembering the basic air supply system, the individual additional systems and elements are described below.

Supply and Exhaust Air Flow and Pressure

Codes and standards usually dictate the required amount of air to be supplied and exhausted into any room or area of a hospital. These will vary according to the use of the space. The design parameter is expressed in "room air changes per hour." Knowing the room volume allows the designer to specify a volume of air to be supplied (in cubic feet per minute (CFM)) for a room that might be divided among several duct outlets or vents in the room. The pressure relationship of each room, with respect to adjacent spaces, likewise probably will be dictated by codes or standards. Some medical circumstances require special ventilation. "Isolation" rooms might have a specific code definition and requirements. In general terms, there are needs to prevent air from entering a room or, in some cases, from leaving a room.

Patient rooms housing transplant patients with suppressed immune systems need to minimize the possibility of contamination from airborne pathogens that may be found in the hospital but at levels not harmful to a healthy individual. These rooms are typically maintained at positive pressure relative to the hallway and adjacent spaces and might include a two-door isolation system with an intervening space between the hall and the room. In this configuration, the room is positively pressurized with respect to the intervening space and the hallway so that air tends to flow out of the room, rather than into it through an open door. This pressure relationship is established by supplying more air to the room than is exhausted. Testing and monitoring such conditions can be accomplished by using differential pressure measuring systems with alarms or can be confirmed by using smoke to visualize the actual flow of air at the door when it is opened. More sophisticated testing can be performed using trace amounts of elements tagged to the ventilation air and can be monitored with specific detectors.

Rooms that house patients with infectious diseases (e.g., tuberculosis) will be designed to prevent airborne pathogens from leaving the room and entering the hospital's general air circulation. Such rooms are kept at a negative pressure relationship to the hall and adjacent areas. This is accomplished by exhausting more air than is supplied, it and can include high-efficiency filtration of exhaust air to reduce the possibility of contaminating outside air in the vicinity of other air intakes.

An office space requires the minimum amount of air exchanges per hour and is neutral in pressure with respect to adjacent spaces. An operating room requires greater air-exchange rates and might have an unbalanced supply-to-exhaust relationship. Special cases for isolation rooms, clean rooms, and other applications sometimes exist. Systems can be as simple as a fan running at a fixed speed for supply and exhaust with ducts equipped with internal dampers that are adjusted to "fine tune" the volumes being delivered from each branch of the duct. More complicated systems might have variable air volume (VAV) schemes that use variable speed motors to drive fans or variable aperture vents at the end of the ductwork supplying each area. Another variation of controls incorporates a constant volume vent at the end of the supply duct to compensate for any changes in pressure elsewhere in the ventilation system. One could model a sophisticated air supply and exhaust system against the human cardiovascular system, with flows and pressures changing in individual branches of the system, requiring compensatory changes in flow and pressure elsewhere in the system.

In many designs, return air systems are included, rather than simple exhaust systems. A forced air furnace for a home probably has a single fan that draws air from the house and blows it though a heating chamber and then back into the house. Return-air systems for hospitals function in a similar way but use separate fans to remove air from the building and return it to the intake of the supply fan, where some part of the returned air is exhausted to the atmosphere, and the difference is made up of fresh outside air. Again, codes and standards will dictate the minimum percentage of outside air that must be supplied to the building. Return air systems are used to conserve energy for either cooling or heating. Imagine, for example, that the outside air temperature is 30°F. This air is heated as it travels through the air handler and is then supplied to rooms in the building, where it may be heated even more to achieve a desired room temperature. In a system that uses 100% outside air for supply, the energy cost of heating the air would be higher than that of a return air system that used 20% outside air and 80% return air since the returned air already would be at or near the target room air temperature and would not need to be heated. Using a variable percentage of outside air (versus returned air) through control of dampers is known as an "economizer cycle". Economizer control routines might be simple and controlled at the air handler or might be part of a larger building automation and energy management system.

Exhaust systems for ordinary air handlers typically are not filtered, but exhaust systems for isolation rooms for tuberculosis patients, or for chemical or other fume hoods, might require specific filtering and special isolation from air intakes and staff or public exposure.

Filtration

All air supply to hospitals is filtered. Filtration, as other specifications, is dictated by codes and standards. General use areas have enough filtration to eliminate from the air particulate matter that could cause maintenance problems with components of the air supply system or general cleaning problems, and to eliminate the intake of animal life.

Air can be filtered at the intake and/or along the path to delivery to rooms in the building. An OR probably will be supplied with air that is filtered first at the outside air intake and then again as it enters the main duct in the air handler. The efficiency of the second stage of filtering will be higher than that of the first stage and thus will filter out smaller contaminants. For some rooms, final filtering at the duct entering the room might be required. These are special use rooms for clean room applications, for some types of surgical procedures, or to protect some kinds of patients who are extraordinarily susceptible to infection.

Filter testing and maintenance are typically based upon the measured differential pressure across the filter. With a known air flow rate, the resistance of the filter can be calculated based upon the area of the filter and the flow. Manufacturers of filters will specify a range of resistance or pressure differentials that constitute a normal range of operation. Higher resistances indicate occlusion of the filter, lower resistances, breaches in the filter, or leaks in or around the filter. Testing of "filter drop" using differential pressure gauges can be performed with handheld equipment, stationary monitoring meters, or remote monitoring and alarm systems. Handling of some exhaust filters requires special precautions, including filters from isolation rooms and from chemical and biological exhaust hoods.

Heating and Cooling

Heating of room spaces in a hospital can be direct, with local radiant heating, but it is more likely done by heating the air supplied to the room. One method of supplying heated air to spaces in a hospital is the "terminal reheat" method or design in which air taken into an air handler is heated to a target temperature in the main duct in the air handler. This heating is accomplished by passing the air through a heat exchange coil (as in a car radiator) that contains hot water supplied from a hot water system. The temperature of the water in the coil is controlled by valves and is based upon reaching a desired target for supply air temperature. For example, if the air supply system is designed for a constant air supply temperature of 55°F, the valve controlling the supply of hot water to the heating coil will close as the supply air temperature approaches 55° and will open wider if the supply air temperature begins to drop below 55°F. In the terminal reheat system, the 55°F air (or some other target temperature air) is blown down the ductwork until it nears a room (or group of rooms) the temperature of which is controlled by a thermostat (i.e., a reheat zone). There, the air passes through a second, smaller heat exchanger controlled by the zone thermostat. If the thermostat is set for 72°F, and the room air is 70°F, the thermostat will operate a valve at the local heating coil to allow hot water to flow and to "reheat" the air to some higher temperature until the room air (mixed) rises to 72°F, at which point the thermostat will control the heating coil valve to stop the flow of hot water.

The process for precooling is the same in such a system at the air handler with a separate cooling coil at the air handler for this purpose. Because building spaces usually produce some heat from the equipment and occupants, no secondary cooling is required. In the above example, 55°F air is blown into the rooms at the end of the ductwork. The designer would have calculated the worst-case outside air and heat generating characteristics of the room and would have supplied a sufficient number of air exchanges so that, in the worst case, 55°F supply air would keep the room at a desired temperature. If, on a hot day, the room were not warm enough, the air would be "reheated" by the local thermostat and coil combination (hence, the name for this type of system design). More sophisticated methods of controlling room temperatures can control the overall supply air temperature based upon room temperatures, or can vary the flow of air into the room to control the exchange of heat from outside to inside.

Heating of the water that flows through hot water coils in the air handlers and at the individual heat zones can be accomplished with hot water boilers, steam heat exchangers, or other methods. Cooling or chilling of the air may can be accomplished with chilled water or with direct heat exchange with refrigerant coils. A discussion of the physical plant in a hospital is addressed in Chapter 107.

Control Systems

The simplest notion of controlling an HVAC system is the control of room temperatures by controlling air temperatures, as in the case of adjusting a room thermostat to "turn up the heat" or "turn on the air conditioning". In the simple, forced air furnace system used in homes, a thermostat might or might not be equipped with a thermometer or temperature gauge but has some internal method of measuring temperature. The thermostat performs its function by completing an electrical circuit to turn on the furnace when the occupant moves the target or desired temperature indicator above the current temperature. The "sequence of operation" for such an event in a simple, gas heated, forced air furnace is as follows:

1. The gas supply valve to the furnace burner will open.
2. A timer will start.
3. An ignition system will start (unless there is a pilot light, in which case step 1 will include verification of a pilot flame).
4. Verification of ignition by temperature (or of timer limit reached, and gas valve closed).
5. Power will be supplied to the fan motor to supply warm air through ductwork.

This is a simple example of the most basic system of controls for a single function in the process of ventilating a building and providing comfort controls. Of course, a hospital, that has hundreds or thousands of zones or rooms to be controlled with heating, cooling, humidification, and other features requires a much more complicated control system.

The actual moving parts of a hospital HVAC system are usually mechanical and electromechanical. The most common methods of controlling the individual devices and the overall process are with pneumatic controls that use compressed air, or with electric or electronic signals that drive solenoid-operated devices or motors.

Pneumatic controls use differential pressures and flows to push flexible diaphragms connected to mechanical valves and similar devices to operate switches, open or close valves, or move dampers. An example of one element of such a system is shown schematically in Figure 108-1. In this example, an air heating device (a reheat coil) in the duct supplying air to a room is controlled by a pneumatic thermostat and a pneumatically actuated water valve.

Direct digital control (DDC) is a system of controls using software and firmware to control sequence of operation. In a DDC system, conditions in a system are measured as analog or digital inputs, and operating instructions to devices are created as digital output signals. In the case of controlling the same reheating device above, the room thermostat would provide a digital representation of the existing room air temperature, and the occupant operated temperature selection device could be an analog signal converted to a digital value to be compared to the digital value of the current room temperature. The consequence of a difference in the two values is whatever the system programmer chooses, but in a normal case it could be a digital output signal that operates an electronic relay supplying power to a solenoid-operated hot water valve. A more sophisticated option would be for the output to be a series of pulses to a stepper motor controlling the hot water valve, with the valve being proportionally controlled ad the differential between desired and actual temperature changes.

If one considers that there are hundreds of comfort control zones supplied by many fans with hot water boilers, electric water chillers, and steam or electric humidifiers, all with potential safety considerations (like the pilot light verification in a home furnace), then one can see that building system controls are complicated and dynamic, with constantly changing operating conditions throughout the building. Managing the control systems in the current environment of energy costs has become an important function. DDC systems are frequently called "energy management systems", although the control of sequences of operation is a necessity whether or not systems are operated economically.

Humidification

The psychrometric chart (Figure 108-2) is a tool for understanding the relationships between the various parameters of supply air and the relative humidity. This template allows a designer or operator to "work backwards" from a desired room relative humidity to the desired condition of the air as it enters the supply duct. Control of humidity is not required by most codes in all areas, but is for some in most geographic areas. The Uniform Mechanical Code (Uniform Mechanical Code, 2003) calls for OR relative humidities of 50%–60%. In a warm and humid climate, supply air can be drawn into a system and then exposed to cooling coils that drop the air temperature to 55°F. According to the psychrometric chart, if the relative humidity of the outside air is 80% and the outside air temperature is 85°F, when the air is cooled to 55°F, vapor will condense out of the air, and the relative humidity will be 100%. Then, if the air is heated to 70°F without the addition of water, the humidity will change to about 60%, according to the chart. In climates of low humidity, it is necessary to inject water vapor into the supply air to increase the humidity to the target level. This can be done with steam or other methods. Steam can be provided by steam boilers or electric steam generators that boil water directly.

Codes and Standards

Parameters for operation and maintenance of HVAC systems can be mandated by codes or can be designed on the basis of nonmandatory standards. Local jurisdictions probably will have adopted some national standards into their codes, such as the Uniform Mechanical Code and the Life Safety Code (NFPA, 2003). The JCAHO references some of the codes in the sections on the Management of the Environment of Care (JCAHO, 2000). For example, testing the control of supply air fans upon activation of some fire alarm devices refers the reader to the relevant standard, NFPA 90A, published by the National Fire Protection Association. (See Chapter 118.) Some of the building codes pertain primarily to the design of systems, rather than to their maintenance and operation. Hospital licensing laws might adopt and enforce some codes and standards or might simply incorporate some of the requirements of other documents. Because the applicable codes vary with location, no attempt is made here to provide a reference list for such documents. Starting with those codes referenced by the JCAHO is a good beginning. Design engineers in particular can direct one to the applicable locally adopted building codes for HVAC systems. Chapter 120 addresses hospital facility safety standards.

Indoor Air Quality

For many years, the press and media gave much attention to the "sick building" syndrome. Stories discussed the exposure of occupants to toxic materials from new carpeting, and the contamination of air supplies by biological contaminants such as legionella and molds. "Indoor air quality" is the general term for the examination and control of the things that might be contained in the air inside a building. Any hospital facility operator has occasion to answer complaints or questions about indoor air quality from staff, patients, visitors, and physicians.

The first priority in considering indoor air quality is to ensure that no contaminants are introduced into the building from outside, or created within the building and systems and then spread by the HVAC systems. Generally, outside air quality is a known quantity, and if sufficient outside air is supplied to the building, the indoor air quality should be about the same, except that filtering will reduce some contamination (primarily particulate). The internal sources of air contamination include toxic chemicals, dust from construction work, airborne infectious contaminants from patients, and opportunities for biological growth within the ventilation systems. Isolation rooms are used to prevent pathogens from spreading through the air circulation system.

Toxic chemicals that are necessary in the delivery of health care have specific procedural requirements for such provisions as exhaust hoods, to prevent them from mixing with the internal air supply. Improper use or storage of chemicals can lead to impermissibly high exposures. Special smoke evacuation systems for some surgical equipment remove airborne particulates and aerosols from the operating room air.

Air handlers and adjacent evaporative cooling towers have the potential to serve as incubators of some biological contaminants. This incubation typically occurs where

Heating, Ventilation, and Air Conditioning 519

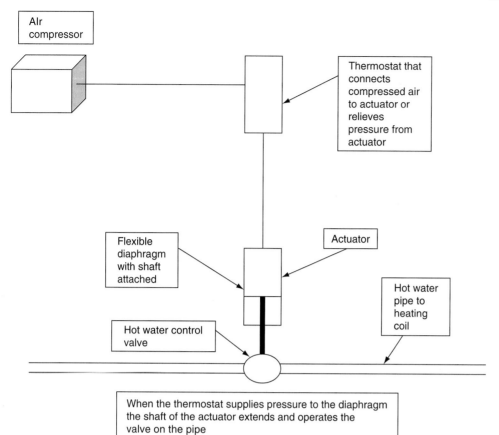

Figure 108-1 HVAC air handler.

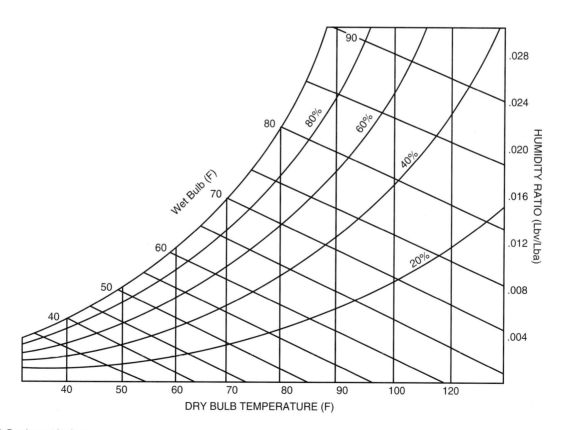

Figure 108-2 Psychrometric chart.

standing water stagnates and chlorine or other agents degrade. Condensation collection areas in air handlers and refrigeration devices must be maintained and monitored properly, and in some cases treated to avoid unwanted biologic growth.

Some molds that are ubiquitous in soil and dust and that are usually harmless to healthy individuals can cause problems for ill patients. Current practice requires that construction zones within hospitals be kept at a negative pressure with respect to surrounding hospital areas, to avoid contamination of patient areas.

References

JCAHO. *Comprehensive Accreditation Manual for Hospitals.* Chicago, JCAHO, 2000.
NFPA. *Life Safety Code 2003 Edition.* Quincy, MA, National Fire Protection Association, 2003.
International Conference of Building Officials, International Association of Plumbing and Mechanical Officials. *Uniform Mechanical Code, 2003.* Whittier, CA, and Walnut, CA, ICBO and IAPMO, 2003.

109

Electrical Power

Matthew F. Baretich
President, Baretich Engineering, Inc.
Fort Collins, CO

Electrical power systems have traditionally been regarded as falling within the province of the facilities engineering profession. However, there are several reasons for clinical engineers to be knowledgeable in this area. First, clinical engineers are increasingly involved in renovation and construction projects for health care facilities. The electrical power system is a key component of facility design, and a well-designed electrical power system is critical for the safe and effective performance of medical devices. Therefore, the clinical engineer must be cognizant of the major design issues and able to contribute to the design process. Second, in many health care facilities, the role of clinical engineering is expanding to encompass a wide variety of technological systems. Integrated management of technological systems in support of patient care is a role for which the clinical engineering professional is well prepared. Management of electrical power systems, including responsibility for system maintenance, can be a part of that expanded role. Third, clinical engineers are often called upon to investigate incidents in which a medical device or other technological system is suspected of having contributed to an adverse event. For example, investigation of incidents involving electrically powered devices may require consideration of the electrical power system. For this reason, it is important for clinical engineers to understand the design, function, maintenance, and clinical use of electrical power systems.

This chapter provides an overview of normal and emergency power systems, the two major categories of electrical power distribution in health care facilities. It also addresses isolated power systems that are used in some facilities. Basic principles are presented, along with references for further study.

Normal Electrical Power

Patient care increasingly relies on device-embodied medical technologies. The level of sophistication, number of products, and relative importance of medical devices have grown, placing greater demands on electrical power systems. These demands are not only for power capacity, but also for terms power quality and operational safety. The normal electrical power system in a health care facility distributes electricity provided by the community's electrical utility company. Although the utility company may give a limited degree of special consideration to health care facilities, the normal electrical power system is generally subject to the same quality and reliability issues affecting the larger community.

A qualified electrical engineer, in cooperation with the project architect, designs electrical power systems for construction and renovation projects. The primary role of the clinical engineer in system design is to provide information on the electrical requirements of any medical devices that will be used in the facility. When specifying the electrical requirements for medical devices, fundamental considerations include the voltage and current requirements of each device. This is a relatively simple issue for small, portable devices, but it is especially important to identify those medical devices that have more demanding requirements, such as imaging equipment systems.

Specification of the number and location of electrical outlets in a construction project might appear to be a minor issue. However, the number of receptacles is critical to accommodate not only the medical devices used today but also the number to be used in the future. The location of receptacles must be consistent with and supportive of clinical practice. Guidelines are available to help planners determine the number and location of electrical receptacles (AIA 2001).

Additional considerations include requirements for quality and reliability of electrical power. Power quality involves such factors as voltage stability and the limitation of anomalies such as spikes, surges, dropouts, noise, and waveform distortion. Power quality can be particularly critical for certain analytical and imaging devices, and extensive testing may be required before device installation to determine the steps necessary to ensure adequate quality. The reliability of electrical power is addressed in the following section of this chapter.

Maintenance of electrical power systems is addressed in the JCAHO (Joint Commission on Accreditation of Healthcare Organizations) Environment of Care standards for Utility Systems Management (JCAHO 2002). Publications by the National Fire Protection Association (NFPA 2002b) and, for clinical laboratories, the College of American Pathologists (CAP 2002) offer additional guidance about system maintenance.

Emergency Electrical Power

Emergency electrical power systems are designed to provide power to critical devices and systems when power from the community's utility company is not available. To accomplish this objective, an emergency power system consists of an alternative power source and a means for distributing electricity from that source to critical devices and systems.

Identification of critical devices and systems to receive emergency power is an important design function for clinical engineering. A comprehensive list of critical devices and systems ranges from life-support equipment, such as ventilators, and essential utility systems, such as medical air compressors, to safety and emergency systems, such as egress path lighting. The JCAHO Environment of Care standards (JCAHO 2002) provide a minimum list of emergency functions and areas to be included on such a list (Standard EC.1.7.1). However, development of a comprehensive list is a multidisciplinary process incorporating clinical and technical perspectives.

In some cases, a local battery-powered uninterruptible power supply (UPS) protects a critical device. However, NFPA standards (NFPA 2002a, NFPA 2002b) require certain health care facilities, including hospitals, to provide an essential power system that includes a centralized power source and a distribution system meeting stringent design criteria (NFPA 2002c).

Although some health care facilities are considering power sources of a more exotic nature, such as kinetic energy (flywheel) systems, fuel cell systems, and small gas turbine motor generator systems, the power source in most cases is a reciprocating diesel engine motor generator set. When power from the community utility company is lost, the motor-generator set starts automatically. Once the generator's voltage and frequency have stabilized at the appropriate level (generally within a few seconds) an automatic transfer switch (ATS) disconnects a portion of the facility's electrical power distribution system from the utility company lines to the facility generator.

Key design issues for the emergency power system include: The capacity of the motor-generator set and associated switchgear and the selection of circuits from the facility's distribution system that will be transferred to the generator. These issues depend, in turn, on the number and location of devices and systems that require emergency power and on the electrical requirements of those devices and systems. These are areas in which clinical engineering is key to the facility design process.

JCAHO Environment of Care standards (JCAHO 2002) include detailed requirements for testing and maintenance of emergency power systems (Standards EC.2.10.4 and

EC.2.10.4.1). Maintenance requirements are also addressed in NFPA publications (NFPA 2002c). Monthly testing of the motor generator set and associated switchgear remains at the heart of the maintenance program. When open transfer switchgear is used, a brief dropout of power occurs as emergency circuits are transferred from the community utility company to the generator and back (at the end of the test period). Closed transfer switchgear eliminates this power dropout and is preferred in new installations.

JCAHO Environment of Care standards (JCAHO 2002) also address emergency power systems as part of the facility's emergency management plan (Standard EC.1.4) and educational program (Standard EC.2.8). Clinical personnel need to know how to respond in case of an electrical system failure. In the most likely scenario, power is lost on normal circuits but continued (with a few seconds of delay) on emergency circuits. In this scenario, it is important for emergency circuits to be readily and consistently identifiable throughout the facility. These circuits are typically identified with red receptacles and/or cover plates. In a more serious scenario, the complete loss of electrical power occurs in all or part of the facility (when power is unavailable from either the community power utility or the facility's emergency power source). In this scenario, it is vital to have manual procedures and equipment in place to provide basic patient care. Planning for either scenario should be multidisciplinary and should include clinical and technical perspectives.

Isolated Power Systems

The isolated power system (IPS) is a three-wire power distribution system with one ground conductor and two power conductors that are isolated from ground. In an ideal IPS, no current would flow in a connection from either power conductor to the ground conductor. In a real-world IPS, the potential for current flow from either power conductor to the ground conductor is limited by design and is continuously monitored for deterioration.

Isolated power systems were developed as a means to reduce the risk of explosion in anesthetizing locations housing flammable anesthetics. In conjunction with conductive flooring to minimize build up of static charges and explosion-proof electrical receptacles to prevent explosions, isolated power systems reduced the likelihood of electrical sparks that could ignite explosive vapors. With the pervasive use of nonflammable anesthetics in current anesthetic practice, this rationale for isolated power is primarily of historical interest.

Isolated power systems are applicable in settings where the limitation of current provides an important measure of safety. NFPA standards (NFPA 2002b) require health care facilities to identify any "wet locations" in which the potential for ground currents must be controlled. The standards define a "wet location" as: A patient care area that is normally subject to wet conditions while patients are present, including standing fluids on the floor or fluids drenching the work area, either of which condition is intimate to the patient or staff (Section 3.3.179). Routine housekeeping procedures and incidental spillage of liquids do not define a wet location (Section A.3.3.179).

Many hospitals identify hydrotherapy rooms as wet locations. Although it is a matter of considerable controversy, some hospitals (notably, the Veterans Administration Health System hospitals) identify their operating rooms as wet locations.

In areas identified by the health care facility as wet locations, NFPA standards require the use of devices to limit the potential for current flow to ground. One option is to use ground fault circuit interrupter (GFCI) devices. However, since GFCI devices interrupt power when they detect excessive current to ground, they can be used only in areas where interruption of power is acceptable. For example, a hospital that identifies its hydrotherapy rooms as wet locations might install GFCI devices to protect patients and staff because it has determined that interruption of electrical power in this area is acceptable.

On the other hand, interruption of electrical power in an operating room is clearly unacceptable. If a health care facility identifies its operating rooms as wet locations, then GFCI devices are not appropriate and an isolated power system is required. Therefore, it is important to consider carefully whether or not the definition of wet location represents actual conditions within the operating rooms.

Whether or not to install an isolated power system is a critical design issue, and the initial decision in the construction of new operating rooms. The facility might wish to conduct a risk assessment to determine whether the installation costs, maintenance costs, and operational complexity for clinical and technical personnel associated with an isolated power system will produce a net benefit in terms of patient and staff safety. Maintenance of an isolated power system includes regular testing to confirm that the actual level of isolation meets appropriate standards and that line isolation monitors (LIMs) are functioning correctly. It is important for clinical personnel (such as operating room staff) and technical personnel, (such as the clinical engineering staff) to understand the function of isolated power systems. In particular, it is important to establish policies and procedures for staff response to LIM alarms.

Summary

Clinical engineers should take active roles in the designs of health care facilities, including contributions to the design of electrical power systems. They also should adopt a broad view of their roles and should apply their expertise to the entire spectrum of technological systems supporting patient care.

References

American Institute of Architects. *Guidelines for Design and Construction of Hospital and Health Care Facilities*. Washington, DC, American Institute of Architects, 2001.

American Society for Health care Engineering. *Electrical Standards Compendium*. Chicago, American Society for Health care Engineering, 1999.

College of American Pathologists. *Standards for Laboratory Accreditation*. Northfield, IL, College of American Pathologists, 2002.

Joint Commission on Accreditation of Healthcare Organizations. *Comprehensive Accreditation Manual for Hospitals*. Oakbrook Terrace, IL, Joint Commission on Accreditation of Healthcare Organizations, 2002.

National Fire Protection Association. *Emergency and Standby Power Systems (NFPA 110)*. Quincy, MA, National Fire Protection Association, 2002.

National Fire Protection Association. *Health Care Facilities (NFPA 99)*. Quincy, MA, National Fire Protection Association, 2002.

National Fire Protection Association. *National Electrical Code (NFPA 70)*. Quincy, MA, National Fire Protection Association, 2002.

110

Medical Gas Systems

William Frank
Medical Gas Services, Inc.
Webster, NH

Few data have been published about maintaining an existing medical gas system. As with most mechanical systems, common sense is a prerequisite, as are adequate manpower and tools. Unfortunately, the maintenance of health care facilities appears to be in decline. Medical gas systems pose a unique problem because the gases that leak from them cannot be seen or smelled. It is too late by the time leaks are audible, and repairs are significantly more costly by then. This chapter should encourage attentiveness to the financial savings and improved patient care that result from a medical gas system maintenance program. The following material is based primarily on first-hand experience.

History of Medical Gas Systems

In the late 1940s and early 1950s, health care workers recognized the hazards of moving heavy high pressure cylinders to various locations. Cylinders were placed in closets, equipped with a pressure-reducing regulator, and rubber hoses or soft copper tubing was run to several rooms. The terminal end was a valve from an oxy-acetylene torch, equipped with a flow meter. When oxygen was not needed, the torch valve was turned off, just like a water faucet. This arrangement was greatly improved when gas cylinders were connected by way of a common manifold, and use points were increased. Outlets and connectors were designed and manufactured by companies including NCG, Ohio Medical, Puritan Bennett, Oxequip, and Schrader. Some connectors were of the screw-on type, incorporating the diameter index safety system (DISS). Some were the "quick connect" type. Initially, any gas could be plugged into any outlet, but this meant that inappropriate gases could be connected.

In some instances, the volume of oxygen used justified the need for an oxygen generator in the hospital, and a large bank of high pressure cylinders was used as a reservoir. When the pressure neared depletion, the generator was started to refill the cylinders. Central piping systems using unclean copper pipes were installed, crude by today's standards. In most cases, the pipe was soft drawn and soft soldered at the connections. Blinking red light bulbs indicated pressure problems.

When the problems with portable suction pumps and cross contamination were recognized, central vacuum systems were installed, and the other gases, nitrous oxide, and medical air all followed suit in using these systems. National Fire Protection Association Standard NFPA 56F (now NFPA 99) became the standard for installation of medical gas systems.

Uses of Medical Gases in Patient Care

Some medical gases used in patient care include the following:

- Oxygen: Administered directly to patients via cannula, blenders, aerosol tents, ventilators, anesthesia machines, hyperbaric chambers, and other methods.
- Medical air: The oxygen used in patient care.
- Nitrous oxide: Administered via anesthesia machines, along with oxygen and various anesthetic agents.
- Carbon dioxide: Primarily used in open heart surgery and laparoscopy procedures.
- Nitrogen: A gas that drives various orthopedic tools and tourniquets.
- Vacuum: Not a gas, but this device provides the means for suctioning patients and for anesthesia waste gas evacuation.

"My suction doesn't work . . . do something!"

Medical vacuum systems are not complicated, yet they are often misunderstood and abused. The purpose of a vacuum system is to provide sufficient flow, at a negative pressure, to move fluids from the patient into a collection bottle. It generally consists of a pump system, a receiver for storage of vacuumed gases, piping, shut-off valves, and sometimes alarms and outlets or terminals. Secondary control equipment is attached to the terminals. Successful operation depends on flow rates under negative pressure. Flows diminish if any portion of the system becomes obstructed or is undersized, or if the pumps are worn or undersized. Frequently, the pump vacuum switch settings are adjusted because of flow problems, causing the pumps to operate at higher negative pressures, with minimal increase in flows. In addition, diminished flow occurs when terminals are added to the system with no consideration for existing pump or piping system capacities.

By far, the greatest flow problems that the author has encountered are associated with the terminals or secondary equipment. NFPA 99, paragraph 4-10.1.1.3, requires new vacuum terminals to have a minimal flow rate of 3 cubic feet per minute (cfm) with a pressure of at least 12 inches of mercury ("Hg) at the nearest adjacent terminal. In a properly designed system, new terminals easily attain this flow rate. Many older terminals cannot flow 3 cfm because of flow-restrictive components and passages. At best, the flow rates may be 1 to 2 cfm. The most important fact is knowing what terminals are capable of flowing at normal negative pressures and ensuring that all similar terminals yield those flows.

Causes of Low Flow Terminals and Corrective Actions

- Debris or improper parts in terminals. Disassemble, clean, and rebuild with proper parts and new seal rings. Leaking seal rings behind the faceplate draw in atmospheric air rebuild terminal with new seal rings.
- Body fluids that have been drawn through the terminal and into the riser. Flush with hot, soapy water. Repeated attempts at flushing will either remove the blockage or prove that the blockage has solidified. If flushing is not successful, the line probably will have to be replaced.

Tips on Flushing

There must be some flow to draw the solution into the terminal and riser. Try hot, soapy water. Good results have been obtained with a product called "sani treet," manufactured by Pascal Company. It comes in a concentrated form and is used in dentistry.

If there is a shut-off valve, draw some solution through the terminal into the riser. Close the valve and let the solution work for 10–15 minutes. If the blockage dissolves, one hears the increased flow when the shut-off valve is opened.

The author has seen old systems that are so full of solidified fluids that 3/4" laterals were reduced to the size of a pencil. Typically, such solidified fluids accumulated over many years ago when body fluid was drawn from the patient into the terminal without the use of a collection bottle.

Failure to shut off regulators when they are not used for patient care is a common abuse of vacuum systems. A suction regulator usually controls the amount of flow required to suction a patient. Contrary to most opinions, liquid should not flow through the pipes into the receiver. Liquid should remain in the canister. If liquid enters the terminals, the moisture evaporates leaving a powder-like residue that usually causes blockage in 3/8" and 1/4" risers. The author has found no obstructions in 1/2" and larger risers.

Various cylinder gas mixtures are used in pulmonary function testing and blood gas analysis. Quality standards for medical gases are found in the compressed gas data book published by the Compressed Gas Association, as well as in the United States Pharmacopoeia.

System Components

Sources

Sources are devices that produce or control the flow of medical gases through the piping network.

Bulk Systems

Bulk systems for oxygen, nitrous oxide, and carbon dioxide consist of special insulated vessels, vaporizers, and regulators. They have reserve systems and are wired to master alarms. These systems can be constructed with cryogenic vessels or a high pressure manifold, depending on usage. Typically, oxygen, nitrous oxide, and carbon dioxide are supplied to large facilities in cryogenic tanks.

Manifolds supply nitrogen, nitrous oxide, air, and carbon dioxide in high pressure cylinders may be found in smaller hospitals.

Cryogenic vessels are equipped with ambient air vaporizers that convert the liquids to a gaseous state before they enter the pipeline.

Manifold and bulk tanks are equipped with regulators to control the pressures, reserve systems, and pressure switches wired to master alarms.

Vacuum Pumps

Vacuum pumps are mechanized devices that create a negative pressure in the piping system. Use of a receiving tank allows for storage, which permits the pumps to cycle on and off rather than run continuously. Systems should be sized according to number of outlets and use factors. Larger facilities might have several pump systems or multiples of two or more pumps.

Medical Air Treatment Systems

Medical air treatment systems are usually two or more compressors equipped with a receiver, drivers, regulator, filters, dew point monitors, and carbon monoxide alarms. Unlike other medical systems, where usage varies only slightly, medical air usage is erratic in many hospitals. This often results in elevated dew points in low-use areas an occasional "stale" odor.

Piping Networks

Piping networks in most medical gas systems are a mixture of old and new pipes. Some facilities still have combinations of soft solder and silver solder joints, while the new sections have brazed joints. Although the replacement of existing pipes is usually an economic decision, it is wise to have all piping properly labeled with color-coded labels and flow arrows.

Valves

There are two categories of valves in piping systems. Zone valves are placed on corridor walls and should be labeled to indicate the rooms that they control. If in doubt, verify the areas controlled prior to labeling.

Service valves are valves in concealed areas such as pipe chases or in the ceiling, and have been a source of fatalities. NFPA 99 has specific labeling instructions for service valves, including that all valves should be verified to determine exactly what they control prior to labeling.

The following are several case histories of fatalities resulting from a lack of attention to medical gas system valves.

"Mystery of Turned-Off Hospital Oxygen Supply Solved by Denver Police" (Biomedical Safety & Standards, 1987)

"The holiday season at the children's hospital in Denver was marred by an unpleasant mystery. How did the hospital's main oxygen valve get turned off, and did the deaths of two babies in the neonatal intensive care unit (NICU) result from this act? The initial investigation was conducted by hospital-shared services of Colorado. Denver Police Chief Tom Coogan complained that the security service was slow in reporting details to the police.

Biomedical Safety & Standards® contacted the Denver police and the district attorney's office. Chief Deputy District Attorney Chuck Lepley explained the sequence of events. The main oxygen valve was located in a small room, which was kept locked. The room also was used to store laundry, and several employees had access to it. Police investigation revealed that one employee decided to eat his lunch there on November 29. Becoming warm, he turned off what he through was a heater valve. When he noticed that the temperature did not go down, and when he heard alarms and people rushing about, he left the room apparently in panic. The hospital's chief engineer turned the valve back on. The hospital employee eventually confessed his actions to the police. District Attorney Lepley told this newsletter that his office determined there was no criminal intent. Reportedly, the hospital has now installed a protective box around the valve.

Although the oxygen had been off for approximately 15 minutes, 19 babies in the NICU were deprived of supplemental oxygen for only about two minutes before backup oxygen was established. During this two-minute period, the babies were breathing room air. One infant, with serious congenital abnormalities and lung disorders, died three days later on December 1. Another infant, also inside an incubator at the time of the accident, died on December 2. Shortly thereafter, two more babies in the unit died, but neither had been in an incubator at the time of the accident. While there remains the possibility of litigation by the parents of the infants involved, District Attorney Lepley told this newsletter that his office currently had no evidence linking any baby death with the incident.

Another question in the case is why the initial investigations were not reported to the police earlier. Detective Marcus Chavez of the Denver police told this newsletter that hospital shared services has been cited for failure to report the incident, which is in violation of a city ordinance. Currently, there are no Colorado regulations that require health care facilities to report unexplained deaths to state health officials. However, some proposed regulations are being considered by the legislature. Index: Oxygen shut-off, hospital, accident."

"Hospital's Oxygen Cut; One in Coma" (Boston Globe, 1995)

"Melbourne, Florida—The wrong valve on a hospital's oxygen line was close on Monday, cutting off some patient's oxygen supply for 10 to 15 minutes, hospital officials said. One woman remained in critical condition yesterday.

The 55-year old patient, whose name was withheld at her family's request, was among 56 patients at Holmes regional medical center whose supplementary oxygen supply was interrupted early Monday.

'At this point, her prognosis is not good,' Valerie Davis, a Holmes spokeswoman, said yesterday. 'We don't necessarily anticipate a recovery, but it's too early to tell.'

Nurses alerted by an oxygen pressure alarm distributed portable oxygen units, but the woman went into cardiopulmonary arrest, said Mike Means, Holmes chief executive officer, said Tuesday.

The other patients suffered no adverse effects, Means said.

Staff members had planned to shut off oxygen only to the hospital's east wing for a routine construction project. Instead, a valve was closed that shut off oxygen to the east and west wings, hospital officials said.

'This was a purely human error,' Means said at a news conference Tuesday.

Though he did not turn the valve, the hospital's director of engineering, Glen E. Anderson, was reassigned and later submitted his resignation.

The hospital would not identify who turned the valve. 'This was a team failure,' means said. 'I don't want anyone to accept total responsibility.'

The shut off did not affect patients on life support and respirators, only those in the west wing needing supplementary oxygen, said Dr. Richard Baney, the hospitals medical director. Many of the patients were recovering from surgery or suffering conditions such as pneumonia.

The hospital said the woman, who was made comatose, had been recovering from surgery."

"Woman Dies after Hospital Error" (Concord Monitor, 1995)

"Melbourne, Florida – a woman died four days after a maintenance worker accidentally shut off her oxygen and put her in a coma.

The 55-year-old woman died Friday at Holmes regional medical center after her family requested that life support systems be disconnected, hospital officials said.

The woman, who was recovering from orthopedic surgery, went into cardiac arrest last Monday and never regained consciousness after the worker accidentally turned off a valve that cut off her supplemental oxygen supply.

The mistake also left 58 other patients without backup oxygen for up to 15 minutes. None of them suffered ill effects.

A hospital spokeswoman blamed an unidentified veteran maintenance worker who misread part of a construction project plan."

Older valves in these systems pose a problem during renovations; many no longer seal, and must be replaced.

Master Alarms

NFPA 99 specifies which signals must be in the panels and how they must be wired. These alarm points monitor the main gas lines and source conditions. Normally, the suppliers of cryogenic vessels test their alarm functions annually, but they do not test pressure switches inside the facility. It is the responsibility of every facility to test all master alarms.

Area Alarms

Area alarms, which monitor conditions in specific critical care areas, are found on alarm panels. NFPA 99 has specific requirements for locations and signals. Both area and master alarms should be reviewed to determine whether they are serviceable, whether parts are available, whether all signals alarm, and whether sensing switches exist.

Outlets and Inlets

Outlets are points at which connections can be made to the medical gas piping system to supply gases under pressure. Inlets supply vacuum. There are two styles of connections: Quick-connect and "twist-on." Quick connections are "plug-ins." The author has observed outlets sagging out of the wall from the weight of the equipment connected to them, which presents a safety hazard. Many of the old outlets do not provide the higher flows required by NFPA 99. Some can be updated, and some can not. Some manufacturers no longer exist. All outlets should be listed by make and model on an inventory list to ensure that parts are available.

Secondary Equipment

While hoses, flow meters, and vacuum regulators are not part of the pipeline system, they can contribute substantially to gas and vacuum consumption. These items should be checked as part of routine inspection procedures. Often, no single department seems to be responsible for inspecting them for either leakage or function. An emulsion of soap, water, and glycerin can be used to check for leaks in flow meters and fittings.

Anesthesia hoses are not considered part of the anesthesia machine's preventive maintenance, and are not usually tested during machine maintenance. It is not uncommon to find vacuum regulator seals completely missing, with the regulator pulling 3 cfm through the outlets.

System Maintenance

Periodically, one hears such phrases as "follow the manufacturer's instructions" and "follow NFPA 99 guidelines." One should not confuse initial certification with ongoing maintenance. Medical gas systems are dynamic, not static, and as with most things, the older the systems become, the more likely it will be to deteriorate and require increased attention, maintenance, and repair.

Medical Gas System Inspection Methodology

The purpose of an inspection is the following:

- To determine whether systems and components perform as designed, are leak free, and are functionally safe for patient care

- To provide unique inventory and location of all components
- To provide short-range and long-range summary conditions

Short-range conditions mean that problems should be addressed within 12 months, such as repair of leaks and verification and labeling of zone valves and alarms.

Long-range conditions mean items that require capitol budget money, such as the installation of medical air system monitors and alarms and the replacement of obsolete alarms and manifolds.

Testing Protocol

Bulk Systems

Document the main tank and reserve pressures, conduct a leak test of the piping on a pad, and recommend improvements.

Manifolds

Document a leak test on all components, record pressures, and recommend improvements.

Vacuum Pumps

Document time and pressure cycles, record pressures, and recommend improvements.

Air treatment systems

Document the location of the intake pipe document time and pressure cycles, record pressures, and recommend improvements.

Zone valves

Document an external leak test, and conduct an internal leak test with permission of department head.

Outlets

Document the model and manufacturer, check functions of latches and labels, check for external leaks and flow rate.

Master Alarms

Document the location of all master pressure switches and the location of all main service valves. Conduct testing only with hospital personnel present.

Area Alarms

Document a test actuation if permitted by hospital personnel.

Particulate Contamination and Remedies

Many systems have particulates in various forms, such as pink or black copper oxide, copper chips, and decayed "O" rings, which can usually be removed by high pressure, high-flow purging. When any existing lines are cut for expansion, they should be thoroughly purged prior to brazing in new sections.

There are as many descriptions as there are types of dirt in the lines. Terms such as "particulates," "contaminants in solid form," "dirt in the lines," and "foreign material" are frequently used. The author has encountered a wide variety of contaminants including lumps of soft solder, flux, mud, copper chips, copper oxide, and gray and black dust.

Detecting the Problem

The following tips can help detect particulate contamination in a piping system.

- Look for dirt in the anesthesia machine; in-line filters.
- Look for dirt in flow meters; look at the needle valve and seal rings.
- Look for dirt after purging outlets at maximum velocity into a white cloth.
- Check inside blenders for presence of dirt.
- Examine the 5 micron filters on volume ventilators.
- Disassemble the first outlet in any system and look for debris. Cover the body opening with white cloth—a face cloth will do. Then, open the zone valve for 5 seconds. Hold the cloth firmly or it will blow away. Look for stains. Reassemble the outlet. Repeat the procedure at the last outlet on that lateral. If debris is found, purge the pipe clean as previously described

Purging

The author has successfully purged systems with cryogenic source nitrogen at a pressure of 90 psi on all gases except air. However, in order to be done effectively, the outlets should be rebuilt. Significant amounts of dirt are often retained in the seal rings, springs, and pistons. The rebuilding process should be done sequentially from the first to last outlet to drive the dirt to the last outlet. If the dirt does not come out at 90 psi, and maximum flow, it will not break loose unless you pound on the pipes.

Water Sealed Pumps

Water sealed pumps pose a different problem. Minerals from untreated water leach out in the air stream after the drier. Unless the system has good filtration, the minerals migrate throughout the system, usually appearing first in clinical areas of heavy ventilator use, such as the intensive care unit (ICU) and neonatal intensive care unit (NICU).

A gray or white powder begins to occlude the 5-micron filter on the ventilator. What happens to blenders, aerosol tents and incubators? The author is aware of several systems with the distinct odor of chlorine in the air systems. They can be purged in the same way as other pressure systems, but filtration must be installed to prevent a reoccurrence.

Certified Systems

Problems may occur in newly installed and "certified" systems. If the testing company used industrial air sampling techniques (i.e., 10 lpm for 10 min = 100 lpm, instead of 100 lpm flows) the presence of undetected contaminants is likely.

Renovations and Modifications

When adding to an existing system, thoroughly purge the existing line before connecting to the new pipe.

Outlet Seals

Seal rings in outlets will dry out due to the dry gases. The only way to determine system leakage is with a pressure test, which is more than closing a zone valve. For a true pressure test, drop the pressure below the line pressure. Otherwise, a pressure loss will not be detected if the valve seal leaks. Be sure to remove all the secondary equipment when performing pressure tests, or else the leaks can not be identified. The common practice of having equipment attached to all outlets at all times, causes early failure of the front seal ring, but this failure is only evident when the equipment is disconnected.

Conclusion

Medical gas systems can be appreciated as life-support systems, drug dispensing systems, and utilities to be conserved. Patient deaths have resulted from failure to adequately control, maintain, and repair medical gas systems. Few facilities track the amounts that they pay for gas products versus the amounts that they actually use, and such inattention drains financial resources.

Common sense should dictate any gas maintenance program. JCAHO has always called for that "unique inventory." In 1987, JCAHO described some specific maintenance steps, such as pressure testing and flow rating. Clinical engineers should know what kinds of system components exist (e.g., outlets or alarms) and what type of maintenance they require, taking frequency of maintenance and parts inventory into consideration.

References

Biomedical Safety & Standards, February 1, 1987, p 19.
Boston Globe, June 15, 1995.
Concord Monitor, June 19, 1995.

Further Information

National Fire Protection Association (NFPA)
1 Battery March Street
Quincy, Massachusetts 02269-9101
(617) 770-3000

Compressed Gas Association
1235 Jefferson Davis Highway
Arlington, Virginia 22002

Pipeline Suppliers

Allied Health Care Products
1720 Sublet Avenue
St. Louis, Missouri 63110
(800) 444-3954
(NCG, Chemetron, Oxyquip)

Amico Corporation
21-121 Granton Drive
Richmond Hill, l4b 3n4
Ontario, Canada

Beacon Medical Products
13325a Carowinds Block
Box 7064
Charlotte, North Carolina 78241
(Puritan Bennett)(905) 764-0800

Hill Rom Architectural Products
Batesville, Indiana 47006-9167
(800) 445-3730
(Ohmeda, Medaes)

Tri-tech Medical Inc.
810 Center Road
Building EAvon, Ohio 44011
(800) 253-8692

111

Support Services

Nicholas Cram
Texas A&M University
College Station, TX

Jobs that fall under the heading of support services vary according to facility location, budget, size, and mission. Normally, support services are those services that are not clinically required, but are essential for efficient operations and improved clinical outcomes in a facility. Support services include both blue-collar and white-collar jobs. Examples blue collar support services include plant operations, maintenance and general engineering, laundry, housekeeping, dietary and food handling, and landscaping and volunteer services. White collar support services include human relations, social services, mail room organization, purchasing and receiving, employment, education, pharmacy, medical records, laboratory, administration, spiritual focus. In addition, technical support services include clinical engineering, telecommunications and private broadcast exchange, information services, and electronics and video support. Also, infection control services, quality management, and regulatory and legal affairs could be considered types of support services. If a facility were associated with a university, it would have an academic category with its own support services.

Health care costs could be reduced if support services were automated or run more efficiently. Essentially, support services are hospital employees not directly involved in patient care. The term "team nursing" was introduced by visionary nurses more than ten years ago (Alfaro-Lefevre, 1999). The concept suggested that everyone had multiple job functions, and instead of waiting for their specific task to appear, team nurses were continually occupied with tasks, with patient care their priority. The morale was high, and the work environment was jubilant. If it worked so well, why was it discontinued? Some nurses felt they were being overworked and failed to see the overall picture of better outcomes, less haphazard behaviors, and more empowerment, which resulted in lowered employee absenteeism, fewer needle sticks, and a pride in their work due to the team effort. The disadvantage of team nursing is that everyone becomes a generalist. American medicine has prided itself on becoming increasingly subspecialized, almost to a point where one physician cannot discuss an entire case with a non-subspecialized physician.

Undoubtedly, as a clinical engineer, the opportunities to augment patient care due to a specialized, separatized evolution are enormous. Clearly, support personnel must be downsized and/or educated for more challenging tasks. The clinical engineer is the natural educator in this environment because he has a clear knowledge of what must be done. Every automated task that places the nurse at the patient's bedside improves patient care. According to the United States Bureau of Census (2002), the United States will need approximately 500,000 nurses by 2020. Accordingly, a medical device or medicaltransmission system (such as telemedicine) that efficiently allows a nurse or other clinical specialist to care for additional patients without clinical risk will be one response to the rapidly worsening health care crisis.

Hospitals have become complex electronic facilities requiring specialized engineers to help them function without any down time (i.e., time when a machine cannot be used).

Common boards for similar transducers will lower the cost of repair, and auxiliary equipment will also become more complicated, serving larger numbers of patients and requiring more training to maintain and operate.

In addition to changes within the hospital environment, external changes will encourage the use of convenient home telemedical care. The health care specialists will not wait for the patients to walk into their clinics; specialized ambulances will be dispatched to care for the swarm of aging babyboomers who will have a significant impact on health care resources. For this affluent generation before them, any medical process, service or device that makes life more convenient will be a success.

The days of huge support staffs for hospitals may be a thing of the past along with the 500–1000 bed hospital. Health care must compete commercially to survive, and the current health care model must change as well in order to do so.

Director Plant Operations Maintenance & Engineering (POM&E)

The typical POM&E director has an educational background in construction science or mechanical engineering with a salary of $65,000–$100,000/year depending on location, expertise, responsibilities, and the number of team members supervised. Responsibilities include the overall building and grounds, HVAC (heating, ventilation, and air conditioning), electricity, water, hazardous waste disposal, fire safety, and the general maintenance of the building. The job is similar to that of a mayor operating a small city.

Laundry Supervisor

The supervisor or director of laundry usually learns on the job and has a high school diploma or graduate equivalent degree (GED). The laundry supervisor is usually promoted from within the pool of salaried blue-collar workers in the hospital system. Salary commensurate with experience ranges from $28,000–45,000 per year. Systems that contract out laundry services will place someone with a bachelor's degree in this position to gain managerial experience. In some hospitals this is a preparatory assignment as the employee moves up the management chain.

Housekeeping

The supervisor or director of housekeeping usually learns on the job and has a high school diploma or GED, similar to the laundry supervisor. The housekeeping supervisor is also usually promoted from within the pool of salaried blue-collar workers in the hospital system. Salary commensurate with experience ranges from $28,000–45,000/year. This is another training ground for contract services, and most hospitals place someone with at bachelor's degree in this position to provide exposure and managerial experience. This assignment, as with the laundry supervisor, is often a preparatory assignment as the employee moves up the management chain.

Dietary

This is sometimes considered a clinical job, since most dieticians now earn a master's degree or graduate degree in nutrition and physiology. A dietician is given parameters of calories, proteins, and carbohydrates that a patient is allowed to consume. In addition to planning a healthy diet with the appropriate amounts of minerals and vitamins, the dietician also attempts to make the meal visually and gastronomically pleasing. Salaries vary greatly depending on experience and education, as well as the priority the hospital administration places on the position.

Food Handlers

Cooks, servers, food stores, and management make up the bulk of positions in the food-handling department. Many hospitals take great pride in hiring the best chefs and in serving meals that are competitive with local delis and restaurants. The priority the administration places on this department by determines the salaries, but the hospitals generally pay better than the local fast-food chains, their major employee drain.

Landscaping and Groundskeeping

This area requires little skill or training unless the hospital is a high profile, large, or well-known facility associated with an academic institution. Landscaping and grounds-keeping duties are usually managed under POM&E or contracted out. This area presents an opportunity for a third-party outsourcing group employee to move up the management chain, but it is more difficult to move up than with laundry or housekeeping. This work may be seasonal, and hospitals often employ high school students for labor, and college students as managers.

Volunteer Services

Many hospitals could not function if community volunteers did not devote hours of service to the hospital. Many volunteers are former hospital employees who find gratification through giving back to their former employers. The baby boomers provide more volunteer services than any previous generation (US Bureau of Census, 2002). Volunteers usually answer phones; deliver books, newspapers, and videos; and work in the hospital library. Some volunteers walk with patients who require daily exercise, chatting with them and lifting their spirits. Volunteers also operate hospital gift shops and information desks, and those with expertise can enter patient information forms into the main server. With more baby boomers retiring, hospitals can anticipate even more help, which might somewhat check the loss of nursing, clinical, and technology positions.

Human Relations

The human relations, or human resources (HR), department provides stability in hospital employment and maintains policies that preserve the legal liability of the hospital from civil and criminal actions. The goal of all human relations departments should be to maintain a joyous workplace and ensure that employees are paid in an appropriate and timely manner. This goal is idealistic; however, HR departments are beginning to realize that a good work environment maintains employee longevity better than bonuses. Some hospitals still provide bonuses to attract employees, which causes a pattern of turnover and resentment among those who have remained loyal to their current hospital. As soon as the contract period ends, the bonus employee looks for the next bonus. An attraction bonus paid out as a 401(k) after 5 years, and again at 10 years, might better serve the hospital. The author has heard HR directors comment that this plan would attract fewer employees. Although this could be true, the commitment obtained would end the bonus cycle.

Social Services

Many customers who are treated at hospitals cannot pay for their treatment(s). This situation has created a two-tiered hospital system in the United States known as "for-profit" and "not-for-profit" hospitals. Not-for-profit hospitals are usually affiliated with a religious sect and have a commitment to turn no one away who needs healing, in keeping with the teachings of St. Francis. For-profit hospitals run their hospitals like a business and have shareholders to finance growth. Everyone who enters a hospital must be cared for and stabilized before they are released, according to the provisions of the Emergency Medical Treatment and Active Labor Act (EMTALA).

Whether for-profit or not-for-profit, certain help must be provided to all citizens. Finding care and housing for patients and families is the responsibility of the social services department. Social workers have at least a bachelor's degree, and their aids may hold an associate's degree. Salaries tend to be lower than those for other white-collar jobs.

Mail Room

A health care system is much like a small city, with all attendant services, including mail delivery. Most correspondence takes place by way of the Internet, telephone, voicemail, and pager, but typewritten, handwritten, and word processor generated communications still must be delivered within and outside the hospital. This process is sometimes a responsibility of volunteer services or long-time employee of the hospital. A misdelivered or undelivered piece of mail could be catastrophic. This position generally falls in the medium pay range.

Purchasing, Receiving, and Warehouse

Every item that is bought, received, and stored must have a transaction with the purchasing and receiving department and the warehouse department. In large facilities, these areas are comprise two distinct departments. However, they are often combined, because the ordering, handling, and storage of goods is usually done in one process, and reporting to a single director makes tracking easier. Department directors have discretion to purchase inexpensive items in order to keep normal hospital processes flowing. Expensive items must be reviewed by equipment and budget committees and approved before they are purchased. Managing this dynamic department is similar to managing a Wal-Mart or other large multi-department franchise. The primary difference is that all orders are processed on a computer, and no customers walking through the front door. Most transactions occur via fax, phone, or e-mail.

There are several levels of management in this department. The director of a large group of facilities is well compensated, and the position requires at least a bachelor's degree in business or advertising, and possibly a master's in business administration.

Employment Office

Most employment offices are a division of the human relations department. However, many facilities have split off the employment office as its own department because finding qualified personnel, reviewing references, and interviewing prospective candidates has become so demanding. . This separation allows the physical building to be located away from the congestion of the day-to-day hospital operations. . Usually, several associates recruit clinical personnel such as intensive care nurses, or other specialized categories of employees. This job may seem like that of a telemarketer, but the personal fulfillment can be very rewarding. The department director should have a bachelor's degree in marketing or business. Salary depends on the size of the organization and the demands to fill certain positions.

Education Department

The education department is responsible for educating the staff and the patients about the health care facility. All clinical team members must have annual training to keep their life-support skills current and accurate. All clinical and non-clinical employees receive testing and training about fire, waste management, infection control, local and external disaster, and security annually. The education department is responsible for documenting and coordinating all of these activities.

Pharmacy

The pharmacy maintains all medications, including oral (those that are swallowed, such as pills or syrups), external (those rubbed on the skin, creams and ointments), intravenous (those given with sterile tubing and/or a syringe), sprays and drops for the nose and eyes, and suppositories (those that are placed in the rectum). Certain medications are not pre-manufactured, and must be prepared in a sterile environment using specific calculations. Aside from dispensing medications for patients, pharmacists also check for drug interactions, when one drug given to a patient taking one or more other drugs will cause an unwanted effect or chemical interactions, when the chemistry of one drug may disable another drug. Pharmacists also check the dosages of drugs prescribed by physicians to ensure that they are within an appropriate range. Pharmacy technicians assist the pharmacist in filling physician orders and delivering medications to the patients. A pharmacy degree is a six-year degree consisting of five years of classroom training and one year spent working with a mentor. There is a shortage of pharmacists due to the rigorous demands. Pharmacists may earn as much as $75 per hour and command a $5000-10,000 sign-on bonus.

Medical Records

The medical records department is rarely seen or known unless except among those who work in health care. Every medication, nursing, or physician procedure and/or surgery is recorded in the medical record. Entire laboratory tests, diagnostic tests using medical devices, and physician-nursing, physician-physician consults appear in the medical records. Different states have different requirements as to length of time medical records must be maintained. Most states require at least 15 years retention. Due to the immense amount of information created in just one medical visit, the medical records department stores records on compact disks. Voice recognition and other storage and retrieval technology, now available and widespread in the commercial sector, will soon be affordable enough for all medical records departments to translate physician records and to store their contents on a server-like network. Many people working in the medical records department have clerical type jobs, while others have associated managerial tasks and report to the director of medical records.

Clinical Laboratory

Few departments have embraced new technology to the extent of the clinical laboratory. Better and more accurate tests, along with cheaper and faster results have resulted in an automated, assembly line-like lab environment. Each department in the lab has a specialist responsible for a distinct set of lab tests. The rapid pace and precise performance creates a high stress level. Most technologists have a four-year degree, while technicians have a two-year degree. Some members of the team are trained on the job, such as phlebotomists (technicians who draw blood).

Administration

Administration is the top level of health care management, including the president, chief executive officer, chief financial officer, chief technology officer, and the vice presidents who assist them. The administrators creates strategic plans and goals so that the health care facility will be an asset to the community while maintaining a financial environment that allows it to continue to provide services and jobs for many years into the future.

Spiritual Care

Health care facilities associated with religious organizations as well as some community-oriented facilities often have a chaplain, priest, or other clergy ministering to the sick within the health care facility. Spiritual care is often associated with improved clinical outcomes, and the author's observations from 28 years of experience in health care

Information Services/Information Technology Department

From its inception in the late 1960s until the early 1980s, the information services department was referred to as "computer services" or "data processing." The exponential growth in this sector has radically changed how health care processes and medical devices operate. Adaptations from advances in computer and network technology continue to permeate health care. Hospitals now have their own web pages and intranets, and departments are connected by both intranet and Internet e-mail and common servers. In reality, most medical devices are computers with external sensors that monitor functions or provide curative or palliative care. As a result of the combination of computers and networks in the same workspace, clinical engineering and information services departments find themselves sharing responsibilities. The division of responsibilities is normally the patient-device interface, which is the domain of the clinical engineer. Some health care facilities find it efficient to combine all technology-related services into one branch, the technology department, members of which report to the chief technology officer. Information technology experts maintain and expand the information network, servers, picture archiving, and communications services, while clinical engineers and biomedical equipment technicians maintain, acquire, and consult on medical equipment.

Telecommunications Department

When data processing centers were only closets within health care facilities, the telecommunications department was only a small but important segment of the hospital support group. Each health care facility maintained its own telecommunications system, sometimes rivaling small telephone companies. The result of these large systems was a feature known as the "private branch exchange" (PBX). The PBX was the hub of all incoming and outgoing calls within the hospital. In the late 1990s, innovations such as voiceover internet protocol (VOIP) and packet information exchange made it possible to combine both voice and data on the same transmission line. This concept is being incorporated into health care infrastructure, with tremendous savings. Most hospitals operate under the top-down management scheme and could be skeptical of new contraptions that benefit the public in general (Cram, 2001). The typical scenario is still a separate telecommunications and information services department, since those who are empowered to implement the new technology hesitate to change or implement technology that is not well understood.

References

Alfaro-Lefevre R. Critical *Thinking in Nursing: A Practical Approach, 2nd Edition*. Philadelphia, W.B. Saunders, 1999.
US Census. Year 2000 Census of the United States. www.census.gov, 2002.
Cram N. *BMEN 310 course manual: Technology Management & Health Care Concepts*. College Station, TX, Texas A&M University Biomedical Engineering Department, 2001.

Glossary

Housekeeper (environmental service): An employee who mops, dusts, waxes floors, and also changes sheets and towels for patients.
Dietary aid: An employee who delivers carts of food to the nursing units, picks up discarded eating utensils, changes dressings, and bathes patients.
Nursing assistant (NA). An employee who works with nurses but is not required to have a formal education beyond a high school diploma or GED. This person is taught to take a patient's blood pressure and temperature and to write these values in the patient's chart. This person may also change dressings or bathe patients.
Respiratory therapy assistant: This person must be able to lift large shifting loads, assist patients with exercise, change dressings, operate whirlpools, massage muscles, and bathe patients.
Groundskeeper: An employee who mows lawns, trims hedges, sweeps main areas, waters plants, replaces in-ground irrigation systems, and uses compressor to remove water from flooded areas.

112

Construction and Renovation

Matthew F. Baretich
President, Baretich Engineering, Inc.
Fort Collins, CO

Health care delivery organizations construct new facilities and renovate existing ones for a variety of reasons. Population growth and population aging call for more capacity. Consumers demand health care facilities with more amenities in convenient locations. New technologies require new infrastructure and facility support (Hall, 1998). As a result, clinical engineers are likely to be associated with organizations undergoing extensive construction and renovation underway. Because many construction and renovation projects have a medical—technology component, they provide opportunities for clinical engineers to participate in the selection, acquisition, acceptance, and installation of medical equipment. Time and effort invested in this process will pay dividends over many years as the new or renovated facility is used to provide high-quality patient care. The impact of construction and renovation on patient care is addressed in other articles in this clinical engineering handbook.

Additional opportunities are available for clinical engineers who are prepared to assume a larger role on the facility design team. Effective translation of clinical concerns into to technical requirements is an critical success factor for health care facility design. Clinical engineers apply this skill on a daily basis in medical—technology management. Knowledge of clinical practice and the clinical environment enables clinical engineers to educate the design team during the entire design process and to see that the resulting facility meets the needs of the clinicians providing patient care. However, making the most of this opportunity requires an understanding of the design and construction process.

Project Stages

A construction or renovation project typically proceeds through the following stages:

- Conceptual planning: Preliminary definition of the project concept, schedule, and budget, based on a business plan that defines the purpose of the project.
- Programming: Development of the project program that specifies the objectives and design criteria for the project.
- Design:
  Schematic Design (SD): Definition of basic design parameters and features
  Design Development (DD): Detailed elaboration of the schematic design
  Construction Documents (CD): Preparation of documents to be used in construction
- Construction: Actual construction of the building
- Occupancy: Acceptance and occupancy of the building

Leaders of successful projects put significant effort into the initial stages of this process to establish a clear consensus among all parties regarding the purpose and objectives of the project. Although it is possible to change course during the later stages, such changes are increasingly expensive and constrained as the project proceeds.

Clinical engineers should be part of the project from the beginning to be of the greatest assistance. Familiarization with basic concepts and current practices will maximize the contributions of a clinical engineer (JCAHO 1997, Miller and Swensson 1995, Miller and Swensson 2002). Additional information regarding the design and construction process is available on the American Institute of Architects web site (www.e-architect.com).

Design and Construction Standards

There are numerous codes and standards that affect the design and construction of health care facilities. Many of these are highly detailed and are used by architects and engineers who specialize in facility design and construction. Discussion of these codes and standards is beyond the scope of this handbook and generally is not relevant to clinical engineers who are involved in projects for their own health care organizations.

However, there are some references with which clinical engineers should be familiar. These references include standards and guidelines from the Joint Commission on Accreditation of Healthcare Organizations (JCAHO), the American Institute of Architects (AIA), and the Association for Professionals in Infection Control and Epidemiology (APIC). A working knowledge of the material in these guidelines will help a clinical engineer become an effective member of the health care facility design team.

JCAHO Standards

In the United States, most health care delivery organizations are guided by the Environment of Care standards of the Joint Commission on Accreditation of Healthcare Organizations. For example, the primary JCAHO standard regarding facility design and construction in hospitals, EC.3.2.1 (JCAHO 2002), states:

When planning for the size, configuration, and equipping of the space of renovated, altered, or new construction, the hospital uses one of the following:

a) Guidelines for Design and Construction of Hospitals and Health Care Facilities, 1996 ed, published by the American Institute of Architects
b) Applicable state rules and regulations
c) Similar standards or guidelines

In practice, most hospitals use the current edition of the American Institute of Architects guidelines (AIA, 2001) as a baseline for facility design. However, adjustments to meet the requirements of local authorities having jurisdiction (AHJs) may be necessary and appropriate in carefully considered cases (with appropriate documentation) to depart from the AIA guidelines to achieve particular organizational objectives.

JCAHO Standard EC.3.2.1 also requires hospitals to conduct a proactive risk assessment as part of demolition, construction, and renovation projects. A discussion of the risk assessment process follows.

JCAHO Environment of Care standards also include broad guidelines describing how the "built environment" should contribute to high-quality patient care. Standard EC.3.1 calls for an environment that "meets the needs of patients, encourages a positive self-image, and respects their human dignity." Standard EC.3.2 requires hospitals to provide "an environment with appropriate space and equipment." Standard EC.3.3 mandates "appropriate privacy to patients." Standard EC.3.4 addresses the role of the environment in supporting "the development and maintenance of the patient's interests, skills, and opportunities for personal growth."

Additional issues for health care facility design can be found throughout the Environment of Care standards. For example, Standard EC.1.6 requires the equipment management plan to include provisions for "selecting and acquiring equipment." Standard EC.1.4 requires the emergency management plan to identify "facilities for radioactive or chemical isolation and decontamination." Standard EC.1.7.1 specifies where emergency electrical power must be provided in the hospital.

The requirements of Standard EC.1.5 to implement "interim life safety measures" (ILSMs) are of particular relevance to the construction phase of a project in order to compensate for the temporary life safety deficiencies that often occur during construction and renovation activities. Construction and renovation projects may also require preparation (for new facilities) or updating (for renovated facilities) of the Statement of Conditions compliance documents required by Standard EC.1.5.1.

Many of these specific design and construction related issues are covered in more detail elsewhere in this handbook. The main point is that the design of a fully JCAHO-compliant health care facility requires familiarity with the entire set of JCAHO Environment of Care standards.

AIA Guidelines

The American Institute of Architects (AIA) publication, *Guidelines for Design and Construction of Hospital and Health Care Facilities* (AIA 2001), is a valuable reference that provides concise information in a format that is accessible to clinical engineers and other health care professionals who are not facility design specialists. It is also a primary resource for health care facilities, and is cited in JCAHO Environment of Care standards.

The document provides specific guidelines for a vast array of design issues, including the minimum size for a patient room, the number of oxygen outlets at an ICU bed location, and the number of air changes per hour for an operating room. It also highlights key aspects of the design and construction process and provides guidelines for handling critical issues.

Infection control is one critical issue addressed in the 2001 edition of the AIA guidelines. The guidelines now mandate an Infection Control Risk Assessment (ICRA) process for each construction and renovation project. The ICRA process is defined as "a determination of the potential risk of transmission of various [infectious] agents in the facility." During the design phase of the project, infection control requirements identified in the ICRA process become design parameters (e.g., the number of airborne infection isolation [AII] or "negative pressure" rooms required). During the construction phase, construction policies and procedures are implemented to mitigate the infection control risks identified in the ICRA process (e.g., the installation of airtight barriers between the construction area and adjacent patient care areas).

The Association for Professionals in Infection Control and Epidemiology (APIC) offers extensive information on infection control issues associated with construction and renovation, including practical procedures for implementing the ICRA process (Bartley 1999). The ICRA portion of the APIC material may be downloaded from the American Society for Health care Engineering web site (www.ashe.org).

Risk Assessment

As mentioned above, JCAHO Environment of Care standards call for a proactive risk assessment to "identify hazards that could potentially compromise patient care in occupied areas of the hospital's buildings" due to demolition, construction, or renovation. The items to consider include air quality requirements, infection control, utility requirements, noise, vibration, and emergency procedures. The ICRA process described in the AIA guidelines is only one component of the wide-ranging JCAHO risk assessment process.

In the health care delivery system, many risk assessment processes involve the translation of clinical concerns into technical requirements (e.g., a description of clinical procedures to be performed is translated into specifications for mechanical and electrical systems). The inputs to the process come from clinical professionals and the outputs go to technical professionals. Clinical engineers have the opportunity to bridge the communication gap between these two professional groups and to ensure that the output accurately represents the input.

Clinical Engineering in Construction and Renovation

At first glance, the role of clinical engineering in construction and renovation might appear to be limited to medical equipment selection. This is a critical role for clinical engineers and it is thoroughly addressed other articles in this handbook. (See Chapters 30-34.) However, the purpose of this article is to expand the perceived role of clinical engineering in two dimensions. First, clinical engineers should be involved at every phase of the project from conceptual planning (e.g., investigation of emerging technologies) through occupancy (e.g., planning for the safe transfer of patients to the new facility). Clinical engineers can make valuable contributions at every stage and should be brought onto the design team at the onset when the project direction is being set.

Second, clinical engineers should be involved in aspects of the project beyond medical technology. Any aspect of the project that involves translating clinical concerns into technical specifications is an opportunity for clinical engineering. The ICRA process previously outlined is one example; many other activities (e.g., design of mechanical and electrical systems, specification of department orientation and patient room configurations, definition of space and equipment requirements for support areas) can benefit from a clinical engineer's involvement.

Summary

Clinical engineers have many opportunities to become valued members of the design team for construction and renovation projects. To make the most of these opportunities, clinical engineers must expand their knowledge of the design and construction process and should see their roles as encompassing more than medical equipment. They also need to insist on active involvement at every stage of the design and construction process. Through broad-scale participation in construction and renovation activities, the clinical engineering field expands its contributions to safe and effective patient care.

References

American Institute of Architects. *Guidelines for Design and Construction of Hospital and Health Care Facilities.* Washington, DC, American Institute of Architects, 2001.

Bartley J. *Infection Control Tool Kit Series: Construction and Renovation.* Washington, DC, Association for Professionals in Infection Control and Epidemiology, 1999.

Hall RR. Technology and the Impact on the Design of Health Care Facilities. *Health care Facilities Management Series 055163.* Chicago, American Society for Health care Engineering, 1998.

Joint Commission on Accreditation of Healthcare Organizations. *Planning, Design, and Construction of Health Care Environments.* Oakbrook Terrace, IL, Joint Commission on Accreditation of Healthcare Organizations, 1997.

Joint Commission on Accreditation of Healthcare Organizations. *Comprehensive Accreditation Manual for Hospitals.* Oakbrook Terrace, IL, Joint Commission on Accreditation of Healthcare Organizations, 2002.

Miller RL, Swensson ES. *New Directions in Hospital and Health care Facility Design.* New York, McGraw-Hill, 1995.

Miller RL, Swensson ES. *Hospital and Health care Facility Design.* New York, Norton, 2002.

113

Radiation Safety

Jadwiga (Jodi) Strzelczyk
Assistant Professor of Radiology, Radiological Sciences Division, University of Colorado, Health Sciences Center, Denver, CO

Brief Historical Review of Major Radiation Discoveries

The radiation story began at the end of 19th century. In November 1895, a German/Dutch physicist, Wilhelm Conrad Roentgen discovered mysterious penetrating rays while investigating the passage of an electrical discharge from an induction coil through a partially evacuated glass tube,. He called them "X rays." Just a few months later, in March 1896, a French professor of physics, Antoine Henri Becquerel, discovered a peculiar property of certain fluorescent materials; they were capable of creating their own images on photographic plates wrapped in light-tight black paper. The newly discovered phenomenon was coined "radioactivity" by Mme Maria Sklodowska-Curie. This Polish scientist, while working with her husband, the French professor Pierre Curie, on purification of uranium pitchblende ore, separated the radioactive element radium in December 1898 (Strzelczyk, 1999). These discoveries found immediate applications, perhaps the most significant ones being in medicine.

Any progress comes at a price. As it frequently happens with new technologies, harmful effects of prolonged radiation exposures were not immediately known. While certain adverse effects became obvious just months after Roentgen's discovery, others came later due to a latency period. The latter was, no doubt, one of the major contributors to the delay in the acceptance of radiation protection procedures. Lack of radiation measurement units was another; for a number of years, erythema on patient's skin was considered an indicator of an appropriate dose. General radiation protection measures were eventually adopted following the establishment of the International Radiation Protection Committee (ICRP) in 1928. Its initial recommendations were directed at medical practice, but current ones are applicable to a wide range of uses of radiation. Shortly after the formation of the ICRP, the National Council on Radiation Protection and Measurements (NCRP) was established in the United States with a mission similar to that of the ICRP.

Then came World War II. Marking the fiftieth anniversary of Roentgen's discovery, the year 1945 saw the devastating detonations of three new weapons. The first test atomic bomb exploded over the Alamogordo military grounds, in the White Sands desert of New Mexico (Trinity site). It was fueled with the fissionable element plutonium, which was discovered by a team of scientists lead by Glen Seaborg, an American scientist of Swedish ancestry, in 1941. The next two bombs, containing uranium and plutonium, respectively, leveled two Japanese cities, Hiroshima and Nagasaki. In spite of the enormous human toll of these detonations, more than on100 test explosions above ground were held by the United States, the Soviet Union, Great Britain, China, and France following WWII. These events took place until the "Non-Proliferation Treaty" (NPT) was negotiated by the members of the United Nations (UN). The NPT came into effect in 1970 (Henderickx, 1998).

There is a common misconception that research during WWII was limited to the development of the atomic bomb. While it is true that this aspect was emphasized, research of the WWII and post-war era encompassed work on new peaceful applications of discoveries such as clean generation of electric power, the development of radiation measuring techniques and instrumentation, and in-depth studies of radiation effects. The world continues to benefit from these discoveries.

In the field of medicine, radiation and radioisotopes far exceeded the potential that early researchers had hoped for. Currently, more than half of the patients admitted to U.S. hospitals are diagnosed or treated using radioisotopes. Many of these isotopes are the "byproducts" that occur during the operation of nuclear reactors.

The use of radiation and radioactive materials requires special precautions, but many benefits are possible when such uses are carried out in a responsible fashion. With a growing number of applications of radiation and radioisotopes in medicine for diagnostic and therapeutic purposes, the role, scope, organization, and structure of radiation protection programs in medical facilities continues to evolve.

Radiation Safety Program Philosophy

Recent ICRP Publication 73 (ICRP, 1996) reviewed the recommended framework for radiological protection. It reinforced the view that the primary aim of radiological protection is to provide an appropriate standard of protection for man without unduly limiting the beneficial practices giving rise to radiation exposure.

The ICRP system of protection is based on the following three objectives:

1. To prohibit radiation exposure to individuals unless a benefit to the exposed individuals and to the society can be demonstrated, and be sufficient to balance any detriment
2. To provide adequate protection to maximize the net benefit, taking into consideration economic and social factors
3. To limit the dose (other than from medical exposures) received by individuals as a result of all uses of radiation

In the United States and in many countries throughout the world, regulations promulgated by regulatory authorities require medical facilities to maintain radiation safety programs. While these programs differ from one another, depending on the type of industry and scope of activities, each is based on a generic template and should contain the key components described in the ICRP system. In addition, such programs must satisfy numerous prescriptive requirements.

Licensing and Registration

Establishing and maintaining a well-organized medical radiation safety program requires operational and technical knowledge of an experienced medical health physicist. A model program for medical uses of radioactive materials can be found in the U.S. Nuclear Regulatory Commission (NRC) Guide 10.8 (USNRC, 1987).

Any individual or institution in the U.S. who intends to engage in such uses must obtain the services of a qualified physicist who will, on their behalf, prepare and file an application with an appropriate regulatory agency, the NRC or an Agreement State agency. At the time of writing, the majority of states have entered into an agreement with the NRC to develop their own state radiation control programs that oversee radioactive materials licensees with their respective states. Although preparation of radioactive materials license (RAM) applications is straightforward, it is quite a laborious process. In the completion of such applications for medical uses, one should refer to the NRC Guide 10.8 as well as appropriate parts of the U.S. NRC (Code of Federal Regulations) and/or state regulations pertinent to radiation control.

Support of an institution's management in this process cannot be overemphasized. Of note is the fact that any statement made in the application potentially will become part of the license conditions that are subject to regulatory compliance inspections. Thus, while the goal is to have a sound radiation safety program, one should not commit to policies and procedures for which the resources are not available or to which the institution's management has not explicitly agreed (Trueblood and Furr, 1993).

Slightly different requirements apply to ionizing radiation-producing equipment. The Food and Drug Administration (FDA) within the U.S. Department of Health and Human Services (HHS) requires that such equipment, intended for human use, must meet the criteria of the Federal Performance Standard (FDA). The owner or operator of radiation-producing equipment is obligated to register it with an appropriate state agency and to operate it in accordance with the regulatory provisions.

Regulations, Policies, and Procedures

When considering the issue of protection and prevention of disease, differences must be considered between the objectives and parameters of public health and "private health" (clinical medicine). In the former, the stakeholder is society or a community; in the latter, it is an individual. Public health sets as a goal prevention of diseases by promoting healthful conditions such as the environment and group behavior for the community at large, while the objective of clinical medicine is to cure a sick individual (Laska, 1997), which may include a behavioral change on an individual level.

Radiation protection is a public health issue for a number of reasons. Health effects of radiation are not unique. Except when acute radiation exposures are involved, radiation effects in a population can be detected only on the basis of statistics and epidemiology. Another reason for considering radiation protection to be a public health issue is that individuals have only a limited ability to structure or control their own environment (Cember, 1998).

Societal, i.e., governmental, regulation of radiation sources became a necessity to ensure that the health of the public would not be compromised by the new technology that rose from the ashes of World War II. The U.S. Atomic Energy Commission (AEC), created by the Atomic Energy Act adopted by the U.S. Congress in 1946, acquired broad authority to regulate nuclear reactors and the use of "byproducts," radioactive materials resulting from the operation of the reactors (Cember, 1998). At the international level, as applications of nuclear technology developed and became worldwide, the International Atomic Energy Agency (IAEA), a specialized agency within the United Nations (UN), was formed in 1956.

The U.S. NRC was established as an independent federal agency by the Energy Reorganization Act of 1974. A description of NRC's mission can be found in its 1997–2002 Strategic Plan. It reflects the ICRP philosophy:

NRC's mission is to regulate the Nation's civilian use of byproduct, source, and special nuclear materials to ensure adequate protection of the public health and safety, to promote the common defense and security, and to protect the environment (Jones and Raddatz, 1998).

NRC has an extensive web site containing full text of press releases, summaries of proposed and final rules, and other relevant information (www.nrc.org). NRC's mission is reflected in the Agreement States' radiation safety programs that control most uses of byproduct material. Nuclear reactors, federal facilities such as Veterans Administration hospitals and army bases, and facilities distributing exempt quantities of byproduct material are some examples of uses that remain under the NRC jurisdiction. The Conference of Radiation Control Program Directors (CRCPD) established in 1968 serves as a common forum for state radiation control programs.

Good radiation safety programs at the licensee level must be well balanced among engineered safety and personnel training, considering technical, scientific, economic, human and ethical aspects of radiation use. Written policies and procedures based on applicable regulations, endorsed by the management and major users are at the core of a successful program.

Integrated Radiation Safety Program: Organization and Structure

The medical radiation safety program must adequately protect patients, care givers, visitors, and the general public. Its development, implementation, and maintenance are grounded in cooperation on all administrative levels, efficient coordination of efforts, and excellent communications.

Management

In a medical facility, regardless of size, management is ultimately responsible for the success or failure of the radiation safety program. Management might consist of a senior-level administrator or a team of administrators in a large organization, or it might be limited to a radiologist or owner in a small independent practice (Vetter, 1998). In its excellent treatment of management of medical programs, the NRC has developed the concept of a management triangle to emphasize three responsible entities: executive management, the Radiation Safety Officer (RSO), and the Radiation Safety Committee (RSC) (USNRC, 1997). There is clear expectation that executive management will rely on the RSC and the RSO for day-to-day operation but that it might not delegate ultimate responsibility for the program.

The Radiation Safety Committee (RSC)

The U.S. NRC and Agreement States require medical institutions to establish a Radiation Safety Committee (RSC). In facilities that include inpatient care, the executive management appoints membership in the RSC. The committee should represent a cross section of medical use areas and expertise. The NCR requires that membership include the RSO, an administrator who has the authority to commit resources, a nurse, and representatives from each major area that utilizes radioactive materials, such as nuclear medicine or diagnostic radiology, radiation oncology, cardiology, and clinical research (USNRC, 1997). The RSC is charged with overseeing, with the assistance of the RSO, uses of all radioactive materials and machine produced radiation. In some cases, this entity oversees the sources and uses of non-ionizing radiation, as well. The RSC should have the expertise to carry out all responsibilities with which it is charged.

The RSC chair often has considerable influence over the tenor and effectiveness of the committee. The chair should be a leader as well as an expert in radiation (Vetter, 1998). The function should be rotated every few years to provide new energy and ideas. Similarly, some members of the committee should be rotated to allow for fresh support for radiation safety advocacy within the radiation user community.

While it works well in some organizations, management should be cautious about appointing the RSO to chair the RSC. The RSO is responsible for day-to-day operations and might be too busy to run the committee effectively, or might be too closely involved with licensed activities to be objective. Depending on the culture of the institution, an authorized user from a major user department (a physician), a medical physicist, or a management representative may be appointed. The selection of an authorized user has a potential of leading to a conflict of interest if a radiation safety problem develops in this person's department. Medical physicists with leadership skills usually are effective chairs because they have considerable knowledge of the applications of radiation and the principles of radiation safety. Whether management appoints a physician or a physicist to chair the RSC, they must delegate an appropriate level of authority to support and empower the radiation safety program (Vetter, 1998).

Regulations require the RSC to meet at least quarterly and to record and maintain the minutes (Code of Federal Regulations). A typical agenda of these meetings includes: RAM license amendment requests, modifications to the program, incidents/accidents, occupational radiation doses of an institution's personnel, radioisotopes' inventory, approval of authorized users, review of human-use protocols, and the Radioactive Drug Research Committee (RDRC) items if the facility has an approved RDRC (FDA). In most programs, the RSO must assemble this information for the RSC and should ensure that members receive the agenda and any additional information that will facilitate efficient discussion and consensus decisions. In large, broad scope medical programs, the RSO is generally assisted by an alternate (i.e., deputy, or assistant) RSO. Such a person oversees designated elements of the program, participates in the RSC activities, and might be assigned to prepare the RSC agenda items.

The RSC and the RSO have a joint responsibility for the development, implementation, and maintenance of the radiation safety program. The committee, with RSO's assistance, must review the entire program, including the "As Low As Reasonably Achievable" (ALARA) policy on an annual basis.

The Radiation Safety Officer (RSO)

A firm commitment to radiation safety on the part of management is a condition of licensing. The chain of command for the medical radiation safety program and designated RSO is of utmost importance. The RSO should be reporting directly to the upper level of administration as this prevents placing the RSO in a compromising position of being answerable to potential violators of good radiation safety practices (Miller, 1992). Responsibilities and the authority of the RSO should be carefully considered and explicitly expressed in writing. Access to all areas where radiation sources are used or stored must be provided. The RSO must be granted discretional authority to terminate activities that violate established safety practices immediately and without consulting management or the RSC. Such instances require a review by management, the RSC, or both, as soon as possible. The goals of these reviews are to determine the root cause(s) of the problem, to identify effective corrective actions, to examine the program for the potential existence of a similar problem in another area, and, if possible, to authorize resumption of activities (Vetter, 1998).

The principal responsibility of the RSO is to ensure that radiation is used safely, in accordance with regulatory and accreditation requirements and approved policies and procedures which reflect those requirements. The RSO should be given the authority and responsibility to audit activities performed under the RAM license to ensure compliance with established policies and procedures.

The RSO is responsible for the implementation of a wide variety of duties and tasks that are common to specific and broad scope licenses. These duties encompass administrative, technical, clinical, and teaching areas. Listed below are some of the most essential duties.

Regulatory Activities Related to RAM License

These activities include obtaining and becoming familiar pertinent regulations, keeping abreast of regulatory changes, preparing and filing of radioactive-materials-license applications, license renewals, and amendments. Following receipt of the license, it is the RSO's duty to ensure that license conditions and statements made in the application are adhered to. This is accomplished through the development and implementation of practical procedures, by maintaining appropriate documentation and/or development and completion of forms. These records are subject to regulatory inspection. The RSO and radiation safety representatives of areas where radioactive materials are utilized participate in internal audits and in regulatory inspections.

These activities consist of examination of procedures and representative records, measurements of radiation levels in selected areas, interviews with personnel, and/or observations. If the findings of an audit or inspection indicate existence of items of noncompliance, whether or not they are cited as violations or recommendations, these conditions must be addressed as soon as possible. Authorized users and the RSO have a joint responsibility to see that corrective actions are taken and that the possibility of recurrence is minimized.

If corrective items require unplanned expenditure or the need for other resources that are not available to the RSO, the RSO will present these items and will propose the course of action to management, the RSC, or both. Approval or disapproval (with explanation) must be documented in the RSC minutes.

Multilevel Radiation Safety Training

It is imperative that all individuals involved in the use of radiation sources receive appropriate area/modality specific training and education. Federal (5) and state regulations specify these requirements, and the NCRP Report No. 105 (NCRPM, 1989) provides general guidelines. The RSO is responsible for ensuring that authorized users have the required (Code of Federal Regulations) credentials and that other personnel receive instructions. Depending on the scope of operations, training can take on various forms, ranging from formal lecture series (live or computer-assisted), to topical in-services, to supervised viewing of videotapes followed by question-and-answer sessions. The training must be provided for all individuals who are likely to receive an occupational dose of 1 milliSievert (mSv) (100 millirem (mrem)) in a year.

The following general topics must be included: Information on the location of sources of radiation, the basics of radiation, health effects of radiation exposure, principles of radiation protection, area-specific policies and procedures based on regulations and on scientific grounds, information on posting, and radiation safety contacts. More detailed information is provided as part of an area-/modality-specific instruction (i.e., patient and visitor restrictions for nursing staff, and operational techniques that minimize patient and personnel dose for staff participating in fluoroscopic procedures). Personnel must be instructed to report to the RSO any condition that constitutes or may lead to an unsafe situation and/or is in violation of established procedures.

Many medical procedures require specialized patient instruction. Depending on the type of procedure, the instruction may be provided by the RSO, by the physician in charge of the case, or by other trained health care personnel.

Monitoring of Occupational Radiation Doses of Personnel

Each authorized user of radioactive materials and each registrant of equipment that generates ionizing radiation is required to provide personnel monitoring devices to certain categories of individuals who could be exposed to radiation during the course of their employment. Specifically, these categories include individuals who are likely to receive a dose of radiation in excess of 10% of the appropriate permissible occupational limit and individuals who enter "High Radiation" and "Very High Radiation" Areas (Code of Federal Regulations) (Special limits are set for minors and "declared pregnant" female employees.) Various types of dosimeters are available on the market. The RSO makes a determination regarding suitability and types of dosimeters issued to individuals working in different areas. For example, collar dosimeters are appropriate for personnel who work with fluoroscopy, and finger or wrist dosimeters are issued to personnel working with radiopharmaceuticals in nuclear medicine. A number of technologies applied for this purpose include film, thermoluminescence (TLD), and optically stimulated luminescence (OLS). Whatever the choice, the provider of dosimetry service is required to hold an accreditation from the National Voluntary Laboratory Accreditation Program (NVLAP) of the National Institute of Standards and Technology (NIST).

The RSO is charged with the responsibility of tracking radiation doses of personnel and investigating all cases when these doses exceed established ALARA levels. Note that while the existence of ALARA program in a medical setting and its annual review by the RSC are mandated, the levels of doses at which an investigation and/or other action is taken are based on the judgment call by the RSC.

Control of Radioactive Materials

The RSO is responsible for establishing and maintaining a system of control wherein radioactive materials will be ordered and received only by authorized users, and in accordance with license conditions. Generally, there are one or more designated areas where the actual receipt takes place. In addition to user verification, personnel who receive radioactive materials ensure that there is no external contamination on packages.

Authorized users have the responsibility of using materials in accordance with established area-specific procedures. When not in use, radioactive materials must be safely stored in a fashion that prevents unauthorized use. It is generally advisable to keep a "running" inventory of all materials. All areas where radioactive materials are used must be surveyed to ensure that there is no removable radioactive contamination. Types and frequencies of these surveys are area specific. Transfers of radioactive materials (internal and external) can be done only with the RSO's authorization. In all cases, such transfers only can be between specifically authorized or licensed parties.

Used (spent) radioactive materials must be properly disposed of. It is the RSO's responsibility to develop and oversee a radiological waste (radwaste) management program. Except for certain long-lived isotopes (e.g., Carbon-14) utilized in clinical research, the majority of radioisotopes used in routine clinical applications fall into two categories: Short-lived isotopes used in diagnostic (i.e., Technetium-99m) and therapeutic procedures (Iodine-131), and sealed sources used for therapeutic purposes (i.e., Iridium-192). Radwaste containing long-lived isotopes is generally disposed of through the use of outside waste contractors. Disposal of radwaste containing short-lived isotopes can be accomplished through on-site DIS (decay-in-storage) method. Sealed sources that are no longer needed are generally returned to their respective suppliers. In all approaches, all materials must be accounted for; stored waste must be properly labeled (Code of Federal Regulations); and vehicles that are used to transport waste and packages containing no-longer-needed sources must meet appropriate local and federal (USDOT) criteria. The facilities where radioisotopes are no longer used must be properly decontaminated, if needed, and decommissioned under the RSO's supervision.

Active Participation in Planning of New or Remodeled Facilities

Before any structural changes are made in areas housing radioactive materials and/or radiation producing equipment, the RSO must be notified to afford the assessment of radiation-protection adequacy of the proposed changes. To facilitate this assessment, management must provide all required details to the RSO. Assessment of adequacy and correction recommendations, if deemed needed, must include the aspect of general public protection (Code of Federal Regulations). Any radiation producing equipment that is no longer in use must be disabled.

Response to and Investigation of Incidents/Accidents

All incidents that involve radiation (e.g., misadministration and occupational doses exceeding the institution's ALARA levels) must be investigated immediately. Certain instances are required to be reported to regulatory agencies within 24 hours of their occurrence (Code of Federal Regulations). In addition to making efforts to identify the cause, and to taking and recommending corrective action, the RSO must develop a plan to prevent recurrence of such incidents.

Similarly, all accidents, e.g., loss of control over radioactive materials, a patient who has been radioactively contaminated off site, and presenting in the emergency room, will require the prompt response of the RSO and/or designated radiation safety staff. In most medical facilities, radiological response is part of the overall emergency (or disaster) plan. Appropriate training and drills are necessary to test the efficacy of such plans.

Quality Management Program (QMP)

Most medical radiation safety programs require individual user departments to develop and implement a quality management Program (QMP). This approach aims at prevention of unnecessary radiation exposures and extends continuity of protection activities from personnel and the general public to patients without infringing on physicians' rights to practice. To accomplish these tasks, management of the organization must ensure that the radiation safety program has appropriate resources.

Medical Radiation safety Program Resources and their Utilization

What is needed to run an efficient medical radiation safety program will depend upon the size and scope of the organization. However, certain generic resources are necessary for the program to function. These include the following:

1. Appropriate staffing
2. Radiation safety office
3. Health physics laboratory (or access to one in shared programs that exist at many university centers)
4. Storage and radwaste handling facility and equipment
5. Financial resources and logistical means to conduct the program

Appropriate Staffing

The designated RSO should be a medical-health physicist who has the pertinent technical and operational knowledge of the field. In small settings, a physician (i.e., a radiologist or radiation oncologist) who has experience in handling radioactive materials could be designated as an RSO on the license. This person must have appropriate credentials, specified by the federal and/or state regulations, to be accepted for this position (5). It is desirable for them to be well-organized, to have people skills, and to be a good communicator. In addition, the RSO must have (or develop) a good bedside manner, which is essential in a clinical setting.

In any size of radiation safety program, the RSO must have an appropriate backup. Such person(s) will need to be knowledgeable to fill in for the RSO in routine, as well as emergency, functions to ensure continuity of the program in all instances. Large programs have multiple assistants (i.e., deputies or alternates). In addition, the program needs technical and clerical personnel who perform clearly designated day-to-day duties. The number and level of personnel are determined by the scope of operation and the number and type of in-house services required in a particular setting.

The Radiation Safety Office

The office of the RSO is at the heart of the radiation safety program operation. It is the place where all appropriate documentation is assembled and stored. The most essential documentation includes the following:

1. Radioactive materials licenses, their amendments, renewals, and support materials
2. Records of communications with regulatory agencies regarding all aspects of the program (e.g., licensing, inspection)
3. Records of interactions with management, the RSC (including meeting minutes and support materials), authorized users, and other personnel, as they pertain to radiation safety
4. Records of internal audits and inspections
5. Personnel radiation monitoring records
6. Records of incidents (e.g., maladministrations) and accidents (e.g., loss of control of radioactive material) including date, location, names of involved/responsible/affected individuals, event description, results of investigation, action taken/pending, resolution, decision/action taken/pending to prevent recurrence
7. Records pertinent to control of radioactive materials, from receipt to proper disposal
8. Compliance records reflecting routine and periodic radiation safety procedures (e.g., area surveys, instrument calibration, results of testing sealed sources' integrity)
9. Radiation safety training materials and records of multilevel in-services and lectures
10. Drawings of facilities and shielding reports (when appropriate)

A number of these documents now can be generated and stored as computer files; however, certain records (e.g., original licenses and occupational dose requests) bear signatures must be stored as hard copy files.

Health Physics Laboratory

The laboratory is necessary for the radiation safety staff to implement monitoring and evaluations of work environment conditions. Some or all of the following equipment must be available (15):

1. Portable monitoring equipment such as G-M counters, ionization chambers, neutron detectors, air-sampling apparatus, and velometers
2. Sample collection and preparation equipment such as air samplers or monitoring systems, shielded containers, other specialized sampling equipment such as hot cells, and remote handling devices for radiopharmacy
3. Counting equipment such as liquid scintillation counters, gamma spectrometers and associated computerized analysis equipment and software, and radioactivity calibrators/analyzers
4. Specialized vehicles if transportation of materials of any kind, and sampling or monitoring outside environment are planned

In addition, a medical facility must have instrument calibration/repair capability. This may achieved by contracting an outside firm or by providing an in-house service. In the latter case, a close association of the radiation safety program with the clinical engineering program is essential.

Storage and Radwaste Handling Facilities

Again, depending on the scope and size of the operation, radioactive waste storage and processing require careful consideration and space/equipment allocation. Provisions for incineration, controlled space for short-lived radioisotopes, and on-site storage of volume-reduced (compaction, solidification) long-lived isotopes must now be considered, due to escalating costs of waste disposal and uncertainties of ground burial.

Financial Resources and Logistical Means to Conduct the Program

Financial resources are necessary to the purchase of appropriate equipment, services, and supplies that are needed in order to implement the program (e.g., personnel monitoring devices, equipment repair/calibration). Additionally, there are licensing and inspection fees. The RSO must be provided with logistical means to obtain necessary funds in an expedient fashion.

Continuing education is an essential component of radiation safety programs. It involves financial expenditures, which must be included in the overall budget. The RSO must be provided financial and time resources to attend professional conferences and courses, and to purchase and/or develop educational materials. Whenever a new procedure is introduced, the RSO must develop specific radiation safety procedures and provide introductory training and retraining for involved staff.

The Future of Radiation Safety

The ever-growing field of medical uses of radioactive materials and radiation-producing equipment for both diagnostic and therapeutic purposes requires continuous updates and improvements of radiation safety programs. New uses necessitate the RSO and radiation safety staff not only to stay abreast of pertinent regulations but also to be aware of the new developments in medical applications. For the most part, this is accomplished through participation in professional societies' conferences and courses, e.g., Health Physics Society (HPS), American Association of Physicists in Medicine (AAPM), and Nuclear Medicine Society (NMS). In some cases, manufacturers or distributors of specialized equipment such as High Dose Rate Brachytherapy systems provide extensive training on-site as part of the sales or service contract. It is essential for the RSO to participate in such training and to ensure the participation of all personnel who will be involved in clinical uses of such equipment.

Many new applications of radiation and radioactive materials involve uses other than diagnosis or cancer treatment. The most recent example is intravascular brachytherapy, which involves the insertion of radioactive sources into the coronary system to prevent vessel restenosis. These new applications require the RSO to be a part of a multidisciplinary team.

Another area to which the role of the RSO is extending is community involvement. Because of the versatility of experiences, individuals who hold these positions in medical institutions are perfectly suitable to serve on regulatory advisory boards and to share their knowledge of radiation with local, national, and international organizations (e.g., local fire departments and National Radiological Emergency Preparedness), with local high schools and teacher associations, and, most importantly, with the media. As we begin the 21st century and an era of energy resources shortage, it becomes crucial for radiation professionals to work diligently in efforts to stem the tide of ignorance, unreasonable fear, and misunderstanding of radiation.

References

Cember H. Evolution of Radiation Safety Standards and Regulations. In Roessler CE (ed). *Management and Administration of Radiation Safety Programs, Health Physics Society 1998 Summer School*. Madison, WI, Medical Physics Publishing, 1998.
Code of Federal Regulations, Energy, Title 10: NRC, Parts 1–50. Washington, DC, US NRC,
FDA. *Code of Federal Regulations, Federal Performance Standards, Title 21*. Rockville, MD, FDA,
FDA. *The Code of Federal Regulations, Food and Drug Administration, Title 21, Part 361*.
Henderickx H. (Roos E. transl) *Plutonium: Blessing or Curse?* Denver, The Cooper Beach, 1998.
International Commission on Radiological Protection. *Radiological Protection and Safety in Medicine*. ICRP Publication 73; Ann ICRP 26(2). Oxford, Pergamon Press, 1996.
Jones CG, Raddatz CT. The U.S. Nuclear Regulatory Commission's Regulatory Program. In Roessler CE (ed). *Management and Administration of Radiation Safety Programs, Health Physics Society 1998 Summer School*. Madison, WI, Medical Physics Publishing, 1998.
Laska RD. *Medicine and Public Health*. New York, New York Academy of Medicine, 1997.
Miller KL. *Handbook of Management of Radiation Protection Programs*, 2nd Edition. Boca Raton, FL, CRC Press, 1992.
National Council on Radiation Protection and Measurements. *Radiation Protection for Medical and Allied Health Personnel*, NCRP Report No. 105. Bethesda, MD, NCRP, 1989.
Rockville, MD, FDA,
Strzelczyk J. Radiogenic Health Effects: Communicating Risks to the General Public. In Kappas C, Del Guerra A, Kolitsi Z, et al (eds). *VI International Conference on Medical Physics, Patras, Greece, September 1999*. Bologna, Italy, Monduzzi Editore, 1999.
Trueblood JH, Furr WL. Hospital Health Physics Organization. In Eichholz GG, Shonka JJ (eds). *Hospital Health Physics: Proceedings of the 1993 Health Physics Summer School*. Richland, WA, Research Enterprises, 1993.
USDOT. *The Code of Federal Regulations, U.S. Department of Transportation, Title 49*. Washington, DC, USDOT,
USNRC. Management of Radioactive Material Safety Programs at Medical Facilities. Washington, DC, National Technical Information Service, 1997.
U.S. Nuclear Regulatory Commission. *Regulatory Guide 10.8: Guide for the Preparation of Applications for Medical Use Programs*. Washington, DC, USNRC Office of Nuclear Regulatory Research, 1987.
Vetter RJ. Radiation Safety in a Medical Facility. In Roessler CE (ed). *Management and Administration of Radiation Safety Programs, Health Physics Society 1998 Summer School*. Madison, WI, Medical Physics Publishing, 1998.

114

Sanitation

Lúcio Flávio de Magalhães Brito
Certified Clinical Engineer, Engenhária Clínica, Limitada
São Paulo, Brazil

Douglas Magagna
Certified Clinical Engineer, Engenhária Clínica, Limitada
São Paulo, Brazil

Nosocomial Infection Control

History

It is impossible to consider a history of hospital acquired (i.e., nosocomial) infections without discussing how hospitals came to be. Today's large hospital is a modern, complex institution that offers a wide variety of diagnostic and therapeutic services. To many people, hospitals are synonymous with high-quality, albeit expensive, health care. Such was not always the case. In earlier days, only the poor went to hospitals, whereas the upper classes were cared for at home. But by the end of the 18th century, hospitals were accepted as the place where all patients could receive treatment. Physicians soon became aware that certain diseases, like smallpox, could spread to hospitalized patients, and the practice of segregating certain patients was accepted in many hospitals. By 1850, the demand for hospital beds had increased dramatically in European cities.

Resources were stretched and were often insufficient to provide acceptable hygienic practices.

Ignaz P. Semmelweis, Florence Nightingale, and William Farr were early pioneers in nosocomial infection control (Wenzel, 1983). Semmelweis reviewed maternal deaths in the two divisions of Vienna Lying-in Hospital. Division I was a medical student teaching service. Division II was staffed by midwife trainees. He noted that almost 10% of women delivered by Division I physicians and students died, while only 3% off women bedded in Division II delivered by midwives died. If sick women in Division I were transferred to a general hospital and died at general hospital, their deaths were not included in the Division I Lying-in statistics. Semmelweis showed that contact spread with cadaveric or necrotic material (Division I) could account for virtually all cases of puerperal fever. Semmelweis was keenly aware that he had been responsible for many cases of puerperal fever, noting, "Because of my convictions, I must here confess that God only knows the number of patients that have gone prematurely to their graves by my fault. I have handled cadavers extensively, more than most accoucheurs. As painful and depressing, indeed, as such an acknowledgment is, still the remedy does not lie in concealment and this misfortune should not persist forever, for the truth must be made known to all concerned" (Semmelweis,1861).

Nightingale and Farr, against a hostile, entrenched military bureaucracy showed that safe food and water and a clean environment could result in a major decrease in death rates in a military hospitals. Analyzing mortality data from hospitals during the Crimean War showed that the higher incidence of mortality in military hospitals, as compared to civilian hospitals, resulted from contagious diseases and crowding. Their work led to improve hygienic practices and a standardized reporting system for army deaths. Health care professionals are now well aware of nosocomial infection and continue to develop methods to reduce, control, or eliminate it. (Fernandes, 2000)

Clinical Engineers and Infection Control Practices

Clinical engineers can play a principal role in infection control by involvement in water quality, waste management, sterilization, infection control standards and committee work, and health professional training in areas of sterilization and medical device maintenance. Areas of work include corrective and preventive maintenance and calibration of manometers in autoclaves, analysis of the cost of sterilization, sterilization productivity analysis, assessing the effect of medical device technology as it relates to the infection control program, construction and renovation of sterilization facilities, and the development of engineering controls to reduce nosocomial infection. The clinical engineers skills in technology management are an asset to the infection control program of the health care facility. For example, good equipment assessment and procurement processes ensure that all devices related to infection control are accompanied by the operator's manual, the service and maintenance manual, replacement parts price lists, and the drawings and schematics.

The Operator's Role in Maintenance

Users of medical devices and systems are responsible for performing day-to-day "operator's maintenance." Described in operators manuals, these tasks and necessary information include initial preparation; operational use; associated risks (including the biological ones); precautions and warnings; procedures and methods of cleaning; disinfection and sterilization; assembly and functional verification; resistance to, and compatibility with, chemical agents; heat resistance; failure modes and troubleshooting; specifications such as weight, dimensions, power supply necessities; and a list of replacement parts with prices. The technical documentation enables the user to perform superficial, but important, verification of the quality of the technology.

The Maintenance Team

Maintenance activity in health care facilities should be conducted in conformity with hospital infection control policies, procedures, and objectives. The maintenance team, composed of clinical engineers, biomedical engineering technicians, and other maintenance personnel, should strive to achieve and improve measurable quality levels. Documentation of all maintenance activity is required. The team should work in conformity with the objectives and goals of the institution and its infection control committee to ensure standardized, safe, and effective results.

Physical Area for Maintenance

The infrastructure of a hospital contains systems and equipment whose proper functioning is vital for effective infection control. These systems provide water; sewers, medical gases and vacuum; laundry; sterilization; heating; ventilation and air conditioning; and refrigeration. In maintaining and servicing these systems, it is often necessary to work at the site of system installation, rather than a remote workshop. Appropriate areas should be maintained in a centralized workshop and in satellite locations throughout the hospital. The complexity of the maintenance area should be compatible with the complexity of the technology installed and with the maintenance activities to be developed. Adequate space is required for work on large systems (such as steam sterilizers), and care must be taken to minimize the dangers such as working with high temperature and high pressure steam.

Necessary Equipment

Maintenance of sterilizers typically involves measurement of: Electrical parameters; water and air flow; static and dynamic pressures; position; force; speed and acceleration; distance; frequency; weight; and temperature. To maintain sterilizers and other equipment related to infection control, the following tools should be available:

- Pressure meters for steam; water; medical gases; compressed air and clinical vacuum; and heating ventilation and air conditioning (HVAC) systems
- Manometers to measure differential pressure across the filters of HVAC systems, water, steam and pneumatic systems, ventilators, and anesthesia equipment
- Flow meters for water; medical gases; medical compressed air and vacuum; and hydraulic circuits
- Particulate air meter for monitoring the quality of treated air atmospheres with ventilation systems or conditioned air and other isolated areas
- Air sample collectors
- Thermometers or thermocouples for temperature measurement of air, water, and steam at specific points inside equipment or systems
- Chronometers to measure cycle times
- Vacuum cleaner
- Chemical products for cleaning, disinfecting, and sterilizing surfaces of equipment
- Electrical safety analyzer

Sterilization Concepts for Clinical Engineers

A hospital must have sterilization capability. Treating patients using sterile material reduces the risks of nosocomial infection. A clinical engineer can develop a program to reduce the risks and costs of sterilization while maximizing the performance of the sterilization equipment. Absence of sterilization capability from inappropriate planning, maintenance, or use directly effects the health of patients and the financial position of the hospital, as treatments and procedures are necessarily delayed or postponed. This section describes the most common sterilization methods: Steam, ethylene oxide, gamma radiation, hydrogen peroxide, and formaldehyde.

Steam Sterilization

There are three basic methods of moist heat sterilization: Saturated steam, steam-air mixture, and super-heated water (Linvall, 1999), ll of them with the same basic necessities; i.e., to remove air from the chamber, to define temperature and time of the sterilization phase, and to set the parameters for the drying phase. The air should be removed at the beginning of the cycle to facilitate steam entering every space inside the chamber and the loads. Steam transfers its thermal energy to microorganisms, which results in their death. Temperature, time, and pressure are the critical parameters of the sterilization process, and they vary depending on the specific loads, or group of loads, to be sterilized.

Removing Air from the Internal Chamber

Air in the chamber impedes heat transfer to microorganisms. Several methods are available to remove air from the chamber. Among them, single air evacuation and pulsating-air evacuation are used to reach a specific cycle configuration. Single air evacuation is less effective and obtains residual values on the order of 1% of the initial amount of air inside the chamber. Pulsating air evacuation is applied in three steps. At the end of the first pulse, 15% of the amount of initial air remains in the chamber, and at the end of the second and third pulses, 2.25%–0.4%, respectively, of the initial air is retained.

The configuration of the sterilization cycle must account for the permeability of wrappers and other packaging material. Appropriate use of these will reduce the processing time.

Typical Cycle Curve

A typical steam sterilization cycle is shown in Figure 114-1. The data in Figure 114-1 were obtained with a data acquisition system during an inspection and document the performance of the equipment at a particular sterilization cycle configuration. An array of temperature sensors should be placed as follows:

- Establish three imaginary planes parallel to the frontal plane of the chamber.
- In the case of a square chamber, the imaginary plane
- will make intersection with the four corners inside the same, defining other 4 points.
- Place one sensor at each point of the corner, 12 (4 inches the front plane, 4 inches the plane of the middle, and 4 inches the plane of the bottom).
- One sensor is placed at the coldest point of the chamber, typically the drain.
- The last sensor is placed in the geometric center of the chamber.

The data thus obtained allow calculating the equivalent time; the F_o-value; and the differences between the maximum and center temperatures, between lower and center temperatures, and between lower and maxim temperatures. The analysis of these data enables the assessment of the performance of the cycle.

To create specific parameters for different types of loads, it is necessary to use the concept of F_o-value (i.e., the lethality of the cycle expressed in terms of the equivalent time), in minutes, at a temperature of 121°C with reference to microorganisms possessing a Z-value (see Glossary) of 10°C.

Sterilization Control Process

The safest and most economical (e.g., fastest) method is heat treatment; i.e. steam under pressure in a sterilizer, achieving sterilization within a minimum of 15 minutes at +121°C (250 °F) or 3 minutes at 134 °C (273 °F). Other times and temperatures can achieve the same killing effect (e.g., 121 °C for 20 minutes). The F_o-value is expressed in equation 114-1. F is the time in minutes at a temperature, T (°C). Z is the Z-value.

$$F_o = F \times 10^{(T - 121)/Z} \quad (114\text{-}1)$$

Example 1: How long should a product be heat-treated at 115 °C to achieve $F_o = 8$ if Z-value = 10 °C?

$8 = F \times 10^{(115 - 121)/10}$
$8 = F \times 10^{-0.6}$
$F = 32$ minutes

According to practice, it is assumed that a bacterial population has Z-value = 10 and D_{121}-value = 1 minute. It is also assumed that the population of microorganisms is 100 bacteria/container ($N_o = 100$) and that this container should be sterilized at 121°C to achieve a probability of survival $N = 10^{-6}$. Equation 114-2 applies:

$$\begin{aligned} F_o &= D_{121} (\log N_o - \log N) \\ F_o &= 1 (\log 100 - \log 10^{-6}) \\ F_o &= 1 \times (2 - (-6)) \\ F_o &= 8 \text{ minutes} \end{aligned} \quad (114\text{-}2)$$

Example 2: How long should a product be heat-treated at 114 °C to achieve $F_o = 3.65$ if the Z-value = 15°C?

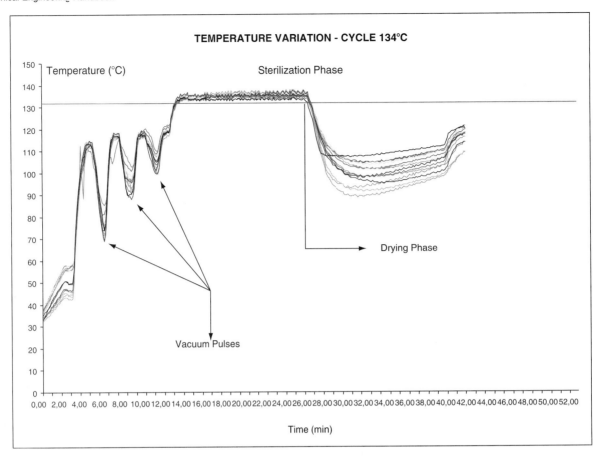

Figure 114-1 Typical steam strerilization cycle.

$8 = F \times 10^{(114-121)/15}$
$8 = F \times 10^{-0.47}$
$F = 10.7$ minutes

According to practice, it is assumed that a bacterial population has Z-value = 15 °C and D_{121}-value = 0.5 minute. It is also assumed that there is a population of 20 bacteria/container ($N_o = 20$) and this container shall be sterilized at 121°C to achieve a probability of survival $N = 10^{-6}$.

Using equation 114-2,

$F_o = D_{121}$ (Log N_o –log N)
$F_o = 0.5$ (Log 20 – log 10^{-6})
$F_o = 0.5 \times (1.3 - (-6))$
$F_o = 3.65$ minutes

Steam Sterilization Equipment

Figure 114-2 presents all components of an autoclave ad permit to understand the principle of operation. Today, all phases of the cycle are automated and computer controlled.

Ethylene Oxide

EtO is also known as "epoxitane" or "dimethylene oxide." It is a colorless gas and reacts with several chemicals, such as alcohols, amines, organic acids and amides. It is soluble in water to 10°C, and it forms polyglycols in the presence of bases. It is highly inflammable and explosive. While it is occasionally used in the pure form, EtO is usually used in diluted mixtures. Because ethylene oxide (EtO) can sterilize at low temperatures, the main advantage of this gas for hospital sterilization is that it will not damage heat sensitive materials, such as plastics.

Gas concentration, sterilization temperature, chamber humidity, and exposure time affect EtO sterilization. Adjusting these parameters according to the load can optimize the sterilization process. Basically, there are two types of sterilization cycles. The first uses positive pressure inside the chamber because EtO is diluted with other gases in order to diminish its explosivity. The second uses negative pressure inside the chamber, because they use EtO 100% pure. In this case, if there is a hole in the chamber, the gas will not escape easily.

The main disadvantage of EtO is its toxicity to patients and personnel if it is not properly utilized. EtO brings serious risks of chemical burns to body tissues if a device, such as a plastic or rubber face mask, sterilized in EtO is not aerated (aeration cycle) after the sterilization cycle to allow the EtO absorbed during the cycle to elute from the mask, is placed in contact with the patient (Dyro and Tai, 1976). The use of EtO in hospitals is decreasing because of such factors as risks to health, environmental concerns over waste sterilant disposal, and longer turnaround time. Hospitals that utilize contract sterilizer companies must be concerned with the quality of that company's sterilization process.

Gamma Radiation

Gamma radiation is pure energy generated by the spontaneous decay of radioisotopes. Exposure to gamma rays sterilizes the product by disrupting the DNA structure of microorganisms on or within the product, thereby eliminating the ability to reproduce life sustaining cells. The ionizing energy produced by gamma rays is deep penetrating, thus making it an ideal solution for products with various densities and product packaging types. No area of the product, its components, or packaging is left with uncertain sterility after treatment. Even high-density products, such as pre-filled syringes, can be readily processed and used with confidence. The gamma process is repeatable and easy to use, with a proven track record. Its deep penetrating capabilities make it a viable solution for a wide range of packaged products. Accompanying dosimetric-release procedures allow for the immediate shipment of products after sterilization processing.

Gamma radiation has been recognized as a safe, cost-competitive method for the sterilization of health care products, components, and packaging since the 1950s and has gained in popularity over the years, due to its simplicity and reliability. Gamma radiation effectively kills microorganisms throughout the product and its packaging, with little temperature effect. Benefits of gamma radiation are precision dosing, rapid processing, uniform-dose distribution, system flexibility, and the immediate availability of the product after processing.

Gamma radiation of single-use, disposable medical devices for the purpose of sterilization is being used by an ever-increasing percentage of the health care industry. Gamma radiation, which once accounted for only 5% of the sterilization market, has grown to nearly 50%. Although cost and reliability have been identified as contributing factors to the industry's conversion to gamma radiation, there are other key elements to be considered.

Packaging remains intact with gamma processing. As there is no requirement for pressure or vacuum, seals are not stressed. In addition, gamma radiation eliminates the need for permeable packaging materials. Packaging and raw materials suppliers have recognized the shift to radiation and are continuing to develop products that are specifically formulated for radiation stability. Tough, impermeable packaging materials are available to provide a strong, long-term, sterile barrier.

Gamma processing, is a highly reliable procedure, mostly due to its simplicity. Because time is the only variable to control, the possibility of deviation is reduced to a minimum. Gamma radiation allows for the immediate release of the product after processing, through a procedure known as "dosimetric release." This procedure is accepted by the

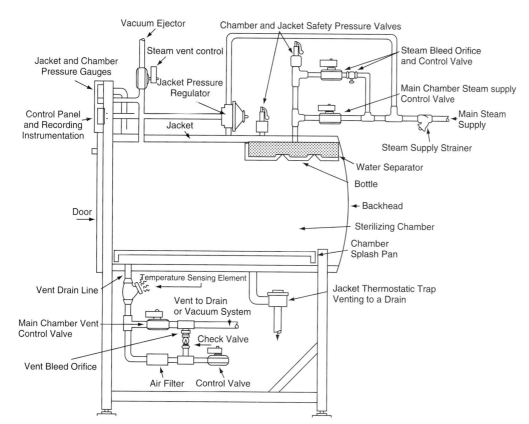

Figure 114-2 Schematic diagram of a typical steam sterilizer.

U.S. Food and Drug Administration (FDA) due to the inherent reliability of the radiation process and is outlined, in detail, in the standards document *ANSI/AAMI/ISO 11137–1994, Sterilization of Health Care Products—Requirements for Validation and Routine Control—Radiation Sterilization.*

Dosimetric release is a product release mechanism, based solely on the dosage of radiation delivered to the product. This measurement, usually identified in kiloGrays (kGy), is obtained using dosimeters, which are placed on product containers during processing. Upon completion of the gamma process, the dosimeters are removed from the product containers and are read, using a specialized instrument to verify the minimum and maximum dosages received by the product. Once the delivered dose is verified, the product is released for shipment.

Hydrogen Peroxide

A hydrogen peroxide gas cloud or low temperature plasma produced by a strong electrical field is suitable for sterilization of surfaces. Low temperature hydrogen peroxide gas plasma sterilization has been available in the United States since 1993. Hydrogen peroxide plasma is formed directly in the sterilization chamber, rather than in a separate reaction chamber (Figure 114-3). During low temperature hydrogen peroxide gas plasma sterilization, medical devices and surgical instruments are placed in the sterilization chamber, and a vacuum is applied to remove the air. A solution of 58% hydrogen peroxide in water is injected into an outer chamber from a self-contained cassette inserted by the operator at the beginning of every 10 sterilization cycles. The solution is then vaporized and allowed to diffuse throughout the sterilization chamber, surrounding the items to be sterilized. Radio frequency (RF) energy is then applied to create an electromagnetic field, which in turn initiates the generation of the plasma. At the conclusion of the sterilization cycle after approximately one hour, the electromagnetic field is turned off, and the reactive hydrogen peroxide species rapidly revert primarily to water vapor and oxygen. The sterilization chamber is repressurized, and the air is passed through an activated charcoal filter to remove any residual hydrogen peroxide.

Advantages of low temperature hydrogen peroxide gas plasma system include an excellent safety profile for employees. Unlike EtO, liquid chemical sterilants, and other alternative sterilization technologies, low temperature hydrogen peroxide gas plasma sterilizer poses little risk to operators and the environment. Exposure to hydrogen peroxide has been limited to 1 ppm over eight hours by Occupational Safety and Health Agency (OSHA) regulations. Monitoring by OSHA of the external environment of the sterilizer during operation has demonstrated that the average concentration of hydrogen peroxide in the atmosphere over eight hours is 0.018 ppm, while the personal sample exposure is 0.013 ppm.

Safety for patients has been established through laboratory tests of the low temperature hydrogen peroxide gas plasma technology prior to market clearance in 1993. These tests have demonstrated that this technology destroys a broad spectrum of micro organisms, including Gram-negative and Gram-positive vegetative bacteria, mycobacterium, yeast, fungi, and viruses, as well as highly resistant aerobic and anaerobic bacterial spores. In addition, as the hydrogen peroxide breaks down into water and oxygen, there are no concerns about toxic residues following the completion of a sterilization cycle.

Studies have demonstrated that items processed by this technology are nonirritating and nontoxic to cells and tissues.

Significant economic advantages have been associated with low temperature gas plasma sterilization as well. For example, because of its extremely rapid cycle time (about one hour), inventory of expensive surgical instruments such as rigid endoscopes can be reduced while still ensuring that every patient receives a sterilized, not just disinfected, device at any given time of the day. In addition, for a wide range of medical and surgical instruments processed by low temperature hydrogen peroxide gas plasma sterilization, significantly less deleterious effects on metal items have been observed, as compared with steam sterilization, reducing replacement costs of expensive surgical instruments. Because there are no installation requirements except a modified 208-volt electrical outlet, the system can be placed almost anywhere, including the surgical suite, to facilitate the distribution of sterile instruments. Finally, because preparation of items for sterilization is similar to EtO processing, consisting of instrument cleaning, reassembly, and packaging in commercially available, nonwoven polypropylene wraps, employee training costs can be kept to a minimum.

Environmental concerns are addressed in the fundamental manner in which the technology operates: Self-contained ampoules of hydrogen peroxide are used that virtually eliminate concerns about exposure to the chemical. In addition, following the completion of the sterilization cycle, the hydrogen peroxide degrades primarily into vaporized water and oxygen, free of toxic materials associated with other low temperature technologies.

Disadvantages of this technology include the inability to sterilize linens and other cellulose containing materials, as the hydrogen peroxide reacts with the organic material found in these items. In addition, because the technology relies upon diffusion, materials such as powders and liquids cannot be sterilized. Special attention should be given when considering applications for narrow lumen devices.

As the shift occurs away from traditional low temperature sterilization technologies (such as EtO and liquid chemical germicides) to the newer methods, the full spectrum of medical sterilization technology comes into focus. While no single technology can fulfill all of the sterilization needs of a hospital, a range of safe, economical choices is imperative. Dry heat and steam probably will always have an appropriate place in the provision of this service. Low temperature systems that are potentially dangerous and costly might not.

Hydrogen peroxide is not recommended for the following materials because they can become brittle, and they have absorption problems (the sterilization is effective, but the material degrades with the time):

- Biphenol and epoxy or components done of polysulfone or polyurethane
- Nylon and cellulose
- Polymethylmethacrylate, policarbonate, and vinylacetate
- Cellulose

It can be used for temperature sensitive materials and other materials such as the following:

- Catheters with an internal diameter minimum of 1 mm and up to 2 meters in length
- Metallic instruments

- Electric equipment
- Rigid endoscopes
- Pneumatic equipment
- Endoscopic equipment
- Organic matter

Formaldehyde

Formaldehyde is a colorless gas, but it is normally distributed as a solution (generally referred to as "formalin"), and it is known to most people within hospitals as an important disinfectant that has been in use since the late 1800s (Weber-Tschopp et al., 1977).

Typical equipment suitable for processing with low temperature steam formaldehyde sterilizers include the following:

- Most types of endoscopes—both rigid (straight) and flexible (e.g., arthroscopes, cystoscopes, laparoscopes, bronchoscopes, coloscopes, gastroscopes, duodenoscopes, choledochoscopes, laryngoscopes, and nephroscopes)
- All heat sensitive instruments for advanced eye surgery
- Most plastic materials (e.g., syringes, coils, tubing, and diathermy cables)

There are, however, some products made of, or including, parts with materials that cannot stand the heat of such sterilization processes, normally at +121°C or +134°C.

Whatever the process, it must result in an approved sterile product, free from hazardous levels of residuals. It must be reasonably easy to use; must be capable of physical monitoring; must have a short process time; must be possible for the normal packaging staff to operate and control; and must be inexpensive. Moreover, the process should allow the product to be packaged in normal wrapping material so as not to create additional costs. It must result in a product that is available for immediate use, must be safe to use with standard preprocessing, and must have good safety margins. Low temperature steam formaldehyde sterilization fulfils all these requirements.

Formaldehyde solutions are widely used in autopsy, surgical, and pathology departments and, to a lesser extent, in dermatology and surgical clinics, X-ray departments and other health care units. The principal use in hospitals is for fixation of tissues.

Formaldehyde solutions are known to be toxic, irritating, and allergenic (WHO, 1986; NRC, 1980). The solution is irritating to the skin and can cause allergic reaction on long or frequent contact exposure. However, even exposure to low, harmless doses causes irritation in the eyes, nose, and throat (Gamble, 1983). This irritation compels the person exposed to avoid further contact; i.e. formaldehyde has a built-in warning signal. Consequently, allergic reactions due to presence of formaldehyde in the air are unusual.

Research on possible carcinogenic effects, based on animal studies conducted by IARC in the United States, showed little to indicate such effects of formaldehyde. The tests were conducted by exposing mice and rats to extremely high concentrations of formaldehyde for two years. In mice, no changes were found, even from exposure to a concentration level of 2-ppm. (The level of painful nasal irritation in human being starts at 0.01 ppm.) (Gamble, 1983.) When subjecting mice to extremely high concentrations (5.6–14.3 ppm), it was possible to induce chronic changes in nasal tissues. This included two cases of nasal cancer. Because the same incidence of nasal cancer is found in unexposed mice, it was concluded that the study gave no clear evidence of any carcinogenic effects of formaldehyde (CIIT, 1981).

Extensive epidemiological studies on industrial workers, pathologists, and embalmers who are regularly exposed to concentrations higher than 10 ppm have shown no long-term adverse effects. Because the extremely irritating smell of formaldehyde at very low levels tends to prevent exposure to higher concentrations, long-term exposure to carcinogenic levels is extremely rare.

Sterilization by Low Temperature Steam and Formaldehyde (LTSF)

In the LTSF process, the pure heat energy of steam sterilization is replaced by a mixture of steam and formaldehyde gas at temperatures of 80°, 65°, 60°, 55°, or 50°C (176°, 149°, 140°, 131°, or 122°F). The presence of steam allows the formaldehyde to penetrate and kill any microorganisms. This sterilization process is intended for heat sensitive goods, especially plastic and hollow instruments (e.g., rigid and flexible endoscopes), which the high temperature of a conventional steam sterilizer could damage.

The Process of Sterilization by LTSF

The LTSF process occurs in four stages:

1. Pretreatment

Before the formaldehyde is admitted, the goods are subjected to pretreatment consisting of repeated evacuations and steam flushes. Pretreatment removes air from the goods and the chamber, while simultaneously humidifying the microorganisms to make them susceptible to formaldehyde. The effectiveness of the humidifying part of the pretreatment is essential to the rest of the process.

2. Formaldehyde Admission

A formalin solution is injected from a sealed bottle. The formalin is then evaporated and enters the chamber as a gas. A vacuum in the chamber assists the admission of the gas. Steam is then added to keep the temperature at the predetermined level. The admission is repeated several times to enhance the penetration into long, narrow lumens and cavities.

3. Sterilization

During the exposure time, the chamber is maintained at the specified temperature, sterilant concentration, pressure, and humidity.

4. Post-Treatment

After the pre-determined sterilizing exposure time, the formaldehyde is effectively removed from the goods by repeated vacuum and intermediate steam flushes. The post-treatment process ends with a deep vacuum, followed by a siginificant number of pulsating air flushes via the air admission filter. This part of the process removes residual formaldehyde in the goods and the chamber (Handlos, 1977).

The formaldehyde concentration in the air by the sterilizer door is generally a maximum of 0.5 ppm and drops to zero ppm within a few minutes. Formaldehyde in the sterilizer drain is diluted to 0.01% and is then decomposed quickly by microorganisms in the sewage.

Indicators

Medical authorities worldwide have recognized that it is not enough to check the sterilization process by physical means (e.g., charts and gauges). Additional controls—primarily chemical and biological—are also required. Various biological and chemical control methods, with varying degrees of accuracy and reliability, are available today. In order to have a parameter to control the quality of process in a daily frequency, operators must use complementary tests. They use indicators to evaluate the results of the sterilization process.

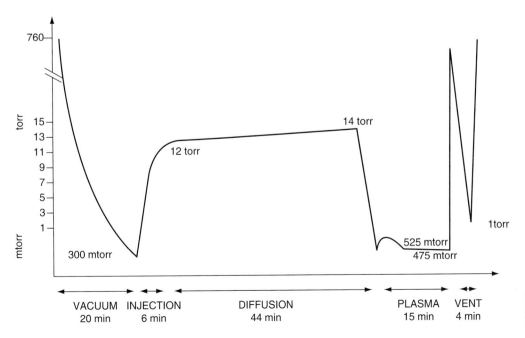

Figure 114-3 A typical cycle of the hydrogen peroxide sterilization process.

Chemical Indicators

This control method gives a clear, instant pass-or-fail indication by a change of color. Indicators are available as spots or strips. Spots are normally used as indicators on the outside of a pack, while the strip indicator should be placed inside the pack or inside a challenge device. It is used to show, via color change, that formaldehyde gas has penetrated and has been present in sufficient quantity, and for a sufficient period of time, to achieve sterilization.

Biological Indicators

As mentioned above, vital factors for good sterilization results are the removal of air and the access of humidity and sterilizing agent. This can be difficult to achieve if a sterilizer with a poorly designed vacuum system is used. Biological indicators, prepared from Bacillus stearothermophilus and resistance tested against formaldehyde, are placed inside a pack and provide reliable controls for steam sterilization.

The following two types of biological indicators are available:

1. Challenge devices (for laboratory cultivation): Because the low temperature steam formaldehyde sterilization process is mainly used for narrow and optical instruments, a special challenge test method has been developed to simulate a complex instrument. The challenge test unit consists of a narrow (2 mm inside diameter), one-way lumen, 1.5 m long, with an indicator at the closed end of the tube.
2. Self-contained biological indicators (for local testing): They include both the test organism and the cultivation broth.

Residual Tests

This challenge method of testing for formaldehyde residuals is based on filter paper. Evaluation is performed in a chemical laboratory, according to standard practice. In this test, residuals represent a worst case (absorbing materials) and are not representative of residuals on nonabsorbent instruments.

Waste Management

Hospital waste is considered to be hazardous residue. Its risk comes from the possibility of disease transmission such as AIDS, hepatitis B and C, and numerous bacteria resistant infections. Health workers and their families are most at risk. Next at risk are patients, visitors, municipal employees, recycling personnel, and neighboring inhabitants.

Not all trash dumped by health centers is dangerous. It is calculated that 80% could be considered as general or common, as it is composed of the same material and has the same quantity and type of bacteria as domestic trash generated city homes. Only 20% have health risks because of bacteria, viruses, fungi, parasites, toxic chemical products, medicine remains, radioactive material, and sharp objects (e.g., sharps). Sharps are the highest risk because viruses and bacteria can be easily transmitted through cuts and perforations. (See Chapter 61 for a discussion of the disposal of sharps.)

Management of Health Care Waste

The first aspect to consider in a program of residue administration in a hospital refers to legal demands of a country, state, or municipal district. Information from other qualified organizations can contribute to the effective implantation of this program (LWRA, 1994; WHO, 1997). In a general these organizations establish obligations that should be considered in the project, implantation, and maintenance of program. It is important to consider that legal aspects can be used to educate, and therefore that they should be used as learning instruments. A concentrated effort in assisting the orientations of these organizations allows the planner to obtain information and to conduct a series of considerations on the following:

- Processes or technologies that are in use, in disuse, and prohibited
- The details of a specific program as geographical aspects
- Recommendations from technical norms that must be considered in the elaboration of the a program
- The specification of advisable and obligatory technologies
- Classifications of residues
- Applicable definitions to this type activity
- The procedures to be adopted during hospital internal processing
- The collection procedure intern and it expresses to the hospital
- Obtaining licenses, technological viability, adaptation of equipments, fiscalization, treatment, and final destination of the residues
- Applicable simbology
- Other attributions and responsibilities of others related organs

The basis of management is to avoid the mixture of various types of waste because infectious material easily spreads into organic trash and multiplies within it at great speed, which makes it impossible to separate at the end. Handling stages comprise separation from its original place, gathering, separate transportation, dangerous material treatment, and final secure dump. If the technical norms of separation have not been executed, all trash generated at the hospital could be infectious, and the production of dangerous wastes, therefore, is five times greater than expected.

Categories of Health Care Waste

Infectious waste: Waste that is suspected to contain pathogens; e.g., laboratory cultures; waste from isolation wards; or tissues (swabs), materials, or equipment that has been in contact with infected patients

Pathological waste: Human tissues or fluids; e.g., body parts; blood and other bodily fluids; Sharps waste; e.g., needles; infusion sets; scalpels; knives; blades; and broken glass

Pharmaceutical waste: Waste containing pharmaceuticals; e.g., pharmaceuticals that are expired or no longer needed; items contaminated by, or containing, pharmaceuticals

Genotoxic waste: Waste that contains substances with genotoxic properties; e.g., waste containing cytostatic drugs (often used in cancer therapy)

Chemical waste: Waste containing chemical substances e.g., laboratory reagents; film developer; disinfectants that are expired or no longer needed; or solvents

Wastes with high content of heavy metals: Batteries, broken thermometers, and blood-pressure gauges.

Pressurized containers: Gas cylinders; gas cartridges; and aerosol cans

Radioactive waste: Waste containing radioactive substances e.g., unused liquids from radiotherapy or laboratory research; contaminated glassware, packages, or absorbent paper; urine and excreta from patients treated or tested with unsealed radionuclides; and sealed sources

Major sources of health care waste:

- Hospitals: (university hospitals, general hospitals, district hospitals, and other health care establishments)
- Emergency medical care services
- Health care centers and dispensaries
- Obstetric and maternity clinics
- Outpatient clinics
- Dialysis centers
- First-aid posts and sick bays
- Long-term health care establishments and hospices
- Transfusion centers
- Military medical services
- Medical and biomedical laboratories
- Biotechnology laboratories and institutions
- Medical research centers
- Mortuary and autopsy centers
- Animal research and testing
- Blood banks and blood collection services
- Nursing homes for the elderly

Laboratories: Mainly pathological (including some anatomical) produce highly infectious waste (small pieces of tissue, microbiological cultures, stocks of infectious agents, infected animal carcasses and blood), sharps and some radioactive and chemical waste.

Health care waste generation depends on numerous factors, such as established waste-management methods, type of health care establishment, hospital specialty, the proportion of reusable items employed in health care, and proportion of patients treated on a day-care basis. In middle- and low-income countries, health care waste generation is usually lower than in high-income countries. The amount of radioactive health care waste is generally extremely small, as compared with the radioactive waste produced by the nuclear industry.

Waste Segregation and Packaging

The keys to minimization and effective management of health care waste are segregation (separation) and identification of the waste. Appropriate handling, treatment, and disposal of waste by type reduce costs and do much to protect public health. Segregation always should be the responsibility of the waste producer, should take place as close as possible to where the waste is generated, and should be maintained in storage areas and during transport. The same system of segregation should be in force throughout the country.

The most appropriate way of identifying the categories of health care waste is by sorting the waste into color-coded plastic bags or containers. The recommended color-coding scheme is given in Table 114-1. In addition to the color coding of waste containers, the following practices are recommended:

- General health care waste should join the stream of domestic refuse for disposal.
- Sharps should be collected together, regardless of whether or not they are contaminated. Containers should be puncture-proof (usually made of metal or high-density plastic) and fitted with covers. They should be rigid and impermeable so that they safely retain not only the sharps but also any residual liquids from syringes. To discourage abuse, containers should be tamper-proof, and needles and syringes should be rendered unusable. Where plastic or metal containers are unavailable or too costly, containers made of dense cardboard are recommended.

Table 114-1 Recommended color-coding for health care waste

Type of waste	Color of container and markings	Type of container
Highly infectious waste	Yellow, marked "highly infectious"	Strong, leak-proof plastic bag, or container capable of being autoclaved
Other infectious waste, pathological and anatomical waste	Yellow	Leak-proof plastic bag or container
Sharps	Yellow, marked "sharps"	Puncture-proof container
Chemical and pharmaceutical waste	Brown	Plastic bag or container
Radioactive waste	-	Lead box, labelled with the radioactive symbol
General health care waste	Black	Plastic bag

- Bags and containers for infectious waste should be marked with the international infectious substance symbol, as described in Figure 114-4.
- Whenever possible, highly infectious waste should be sterilized immediately by autoclaving. It therefore must be packaged in bags that are compatible with the proposed treatment process: Red bags, suitable for autoclaving, are recommended.
- Cytotoxic waste, most of which is produced in major hospital or research facilities, should be collected in strong, leak-proof containers clearly labeled "Cytotoxic wastes."
- Small amounts of chemical or pharmaceutical waste can be collected together with infectious waste.
- Large quantities of obsolete or expired pharmaceuticals stored in hospital wards or departments should be returned to the pharmacy for disposal. Other pharmaceutical waste generated at this level, such as spilled or contaminated drugs or packaging containing drug residues, should not be returned because of the risk of contaminating the pharmacy; it should be deposited in the correct container at the point of production.
- Large quantities of chemical waste should be packed in chemical-resistant containers and sent to specialized treatment facilities (if available). The identity of the chemicals should be clearly marked on the containers: Hazardous chemical wastes of different types should never be mixed.
- Waste with a high content of heavy metals (e.g., cadmium or mercury) should be collected separately.
- Aerosol containers may be collected with general health care waste once they are completely empty, if the waste is not destined for incineration.
- Low-level radioactive infectious waste (e.g., swabs, syringes for diagnostic or therapeutic use) may be collected in yellow bags or containers for infectious waste if these are destined for incineration.

Because costs for safe treatment and disposal of hazardous health care waste typically are more than 10 times higher than those for general waste, all general (i.e., non-hazardous) waste should be handled in the same manner as domestic refuse and collected in black bags. No health care waste other than sharps should be deposited in sharp containers, as these containers are more expensive than the bags used for other infectious waste. Measures of this sort help to minimize the costs of health care waste collection and treatment. When a disposable syringe is used, for example, the packaging should be placed in the general waste bin, and the used syringe in the yellow sharps container. Under most circumstances, the needle should not be removed from the syringe because of the risk of injury. If removal of the needle is required, special care must be taken.

Appropriate containers or bag holders should be placed in all locations where particular categories of waste can be generated. Instructions on waste separation and identification should be posted at each waste collection point to remind staff of the procedures. Containers should be removed when they are three-quarters full. Ideally, they should be made of combustible, nonhalogenated plastics.

Staff should never attempt to correct errors of segregation by removing items from a bag or container after disposal or by placing one bag inside another bag of a different color. If general and hazardous wastes are accidentally mixed, the mixture should be treated as hazardous health care waste.

Cultural and religious constraints in certain countries make it unacceptable for anatomical waste to be collected in the usual yellow bags; such waste should be disposed of in accordance with local custom, which commonly specifies burial (MoH, 1995).

On-Site Collection

Nursing and other clinical staff should ensure that waste bags are tightly closed or sealed when they are about three-quarters full. Tying the neck can close light-gauge bags, but heavier-gauge bags probably require a plastic sealing tag of the self-locking type. Bags should not be stapled. Sealed sharps containers should be placed in a labeled, yellow, infectious health care waste bag before removal from the hospital ward or department. Wastes should not be allowed to accumulate at the point of production. A routine program for their collection should be established as part of the health care waste management plan.

- Waste should be collected daily (or as frequently as required) and transported to the designated central storage site.
- No bags should be removed unless they are labeled with their point of production (i.e., hospital and ward or department) and contents.
- The bags or containers should be replaced immediately with new ones of the same type.
- A supply of fresh collection bags or containers should be readily available at all locations where waste is produced.

On-Site Storage

A storage location for health care waste should be designated inside the health care establishment or research facility. The waste, in bags or containers, should be stored in a separate area, room, or building of a size appropriate to the quantities of waste produced and the frequency of collection. Recommendations for the storage area and its equipment are listed below.

- The storage area should have an impermeable, hard-standing floor with good drainage; it should be easy to clean and disinfect.
- There should be a water supply for cleaning purposes.
- The storage area should afford easy access for staff in charge of handling the waste.
- It should be possible to lock the store to prevent access by unauthorized persons.
- Easy access for waste-collection vehicles is essential.
- There should be protection from the sun.
- The storage area should be inaccessible for animals, insects, and birds.
- There should be good lighting and at least passive ventilation.
- The storage area should not be situated in the proximity of fresh food stores or food preparation areas.
- A supply of cleaning equipment, protective clothing, and waste bags or containers should be located conveniently close to the storage area.

- Unless a refrigerated storage room is available, storage times for health care waste (i.e., the delay between production and treatment) should not exceed the following:

Temperate climate: 72 hours in winter and 48 hours in summer
Warm climate: 48 hours during the cool season, and 24 hours during the hot season

- Cytotoxic waste should be stored separately from other health care waste in a designated secure location.
- Radioactive waste should be stored in containers that prevent dispersion, and behind lead shielding. Waste that is to be stored during radioactive decay should be labeled with the type of radionuclide, the date, and details of required storage conditions.

Transportation

Health care waste should be transported within the hospital or other facility by means of wheeled trolleys, containers, or carts that are not used for any other purpose and that meet the following specifications:

- Ease of loading and unloading
- No sharp edges that could damage waste bags or containers during loading and unloading
- Ease of cleaning

The vehicles should be cleaned and disinfected daily with an appropriate disinfectant. All waste bag seals should be in place and intact at the end of transportation.

The health care waste producer is responsible for safe packaging and adequate labeling of waste to be transported offsite and for authorization of its destination. Packaging and labeling should comply with national regulations governing the transport of hazardous wastes and with international agreements if wastes are shipped abroad for treatment.

The control strategy for health care waste should have the following components:

- A consignment note should accompany the waste from its place of production to the site of final disposal. On completion of the journey, the transporter should complete the part of the consignment note (especially reserved for him) and return it to the waste producer.
- The transporting organization should be registered with, or known to, the waste regulation authority.
- Handling and disposal facilities should hold a permit, issued by a waste regulation authority, allowing the facilities to handle and dispose of health care waste.

The consignment note should be designed to take into account the waste control system in operation within the country. The "Multimodal Dangerous Goods Form" recommended by the United Nations can be taken as an example.

If a waste regulation authority is sufficiently established, it might be possible to prenotify the agency about the planned system of transport and disposal of the health care waste and to obtain the agency's approval.

Anyone involved in the production, handling, or disposal of health care waste has a general "duty of care," (i.e. an obligation to ensure that waste handling and associated documentation comply with the national regulations).

Transportation Vehicles and Containers

Waste bags may be placed directly into the transportation vehicle, but it is safer to place them in further containers (e.g., cardboard boxes or wheeled, rigid, lidded plastic, or galvanized bins). This has the advantage of reducing the handling of filled waste bags but results in higher disposal costs. These secondary containers should be placed close to the waste source.

Any vehicle used to transport health care waste should fulfill the following design criteria:

- The body of the vehicle should be of a suitable size, commensurate with the design of the vehicle, with an internal body height of 2.2 meters.
- There should be a bulkhead between the driver's cabin and the vehicle body that is designed to retain the load if the vehicle is involved in a collision.
- There should be a suitable system for securing the load during transport.
- Empty plastic bags, suitable protective clothing, cleaning equipment, tools, and disinfectant, together with special kits for addressing liquid spills, should be carried in a separate compartment in the vehicle.
- The internal finish of the vehicle should allow it to be steam-cleaned, and the internal angles should be rounded.
- The vehicle should be marked with the name and address of the waste carrier.
- The international hazard sign should be displayed on the vehicle or container, as well as an emergency telephone number.

Vehicles or containers used for the transportation of health care waste should not be used for the transportation of any other material. They should be kept locked at all times, except when loading and unloading.

Articulated or demountable trailers (temperature-controlled if required) are particularly suitable as they can be easily left at the site of waste production. Other systems may be used, such as specially designed large containers or skips; however, open-topped skips or containers never should be used for transporting health care waste.

Where the use of a dedicated vehicle cannot be justified, a bulk container that can be lifted on to a vehicle chassis may be considered. The container may be used for storage at the health care establishment and replaced with an empty one when collected. Refrigerated containers may be used if the storage time exceeds the above recommendations for on-site storage or if transportation times are long. The finish of these bulk containers should be smooth and impervious and should permit easy cleansing or disinfection.

The same safety measures should apply to the collection of hazardous health care waste from scattered small sources. Health care establishments that practice minimal programs of health care waste management should either avoid off site transportation of hazardous waste or at least use closed vehicles to avoid spillage. The internal surfaces of any vehicle used for this purpose should be easy to clean.

Health care waste should be transported by the quickest possible route, which should be planned before the journey begins. After departure from the waste production point, every effort should be made to avoid further handling. If handling cannot be avoided, it should be pre-arranged and should take place in adequately designed and authorized premises. Handling requirements can be specified in the contract established between the waste producer and the carrier.

Special Packaging for Offsite Transport

In general, the waste should be packaged according to the recommendations provided above (see Table 114-1) in sealed bags or containers, to prevent spilling during handling and transportation. The bags or containers should be appropriately robust for their content (e.g., puncture-proof for sharps, for example, or resistant to aggressive chemicals) and for normal conditions of handling and transportation, such as vibration or changes in temperature, humidity, or atmospheric pressure.

In addition, radioactive material should be packed in containers whose surfaces can be easily decontaminated. The United Nations recommends further packing requirements for infectious substances (UN, 1997). For infectious health care wastes, it is recommended that packaging should be design-type tested and certified as approved for use. Health care wastes that are known or suspected to contain pathogens that are likely to cause human disease should be considered "infectious substances" and should comply with the packaging requirements indicated in Table 114-2.

The packaging should include the following essential elements:

- An inner packaging comprising:
  o Watertight primary receptacle of metal or plastics with leak-proof seal (e.g., a heat seal, a skirted stopper, or a metal crimp seal);
  o A watertight secondary packaging;
  o Absorbent material in sufficient quantity to absorb the entire contents placed between the primary receptacle and the secondary packaging.
- An outer packaging of adequate strength for its capacity, mass, and intended use, and with a minimum external dimension of 100mm.
- A list of contents should be enclosed between the secondary packaging and the outer packaging. The outer packaging should be appropriately labeled.

The packaging recommended for most health care wastes, with a relatively low probability that infectious substances are present and which are not likely to cause human disease.

There are two possibilities for packaging:

- Rigid and leak-proof packaging.
- Intermediate bulk containers—large, rigid or flexible, bulk containers made from a variety of materials such as wood, plastic, or textile.

Packaging or intermediate bulk containers tht are intended to contain sharp objects such as broken glass and needles shall be resistant to puncture and shall undergo additional performance tests.

Labeling

All waste bags or containers should be labeled with basic information on their content and on the waste producer. This information may be written directly on the bag or container or on preprinted labels, securely attached.

For health care waste, the following additional information should be marked on the label:

- Waste category
- Date of collection
- Place in hospital where produced (e.g., ward)
- Waste destination

In case of problems involving questions of liability, full and correct labeling allows the origin of the waste to be traced. Labeling also warns operative staff and the public of the hazardous nature of the waste. The hazards posed by container contents can be quickly identified in case of accident, enabling emergency services to take appropriate action.

Cytotoxic waste should be marked with the label "Cytotoxic waste."

Three labels have been designed by the UN/IAEA for radioactive material, providing information on the levels of activity of a given package (IAEA, 1996).

Unless the package is large (and it is assumed here that all packages containing radioactive waste do not exceed 1m^2 in cross-sectional area), radiation symbol the labels should be chosen according to Table 114-2. If the two types of conditions in Table 114-2 differ, the package should be assigned to the higher category.

Treatment and Disposal Technologies for Health Care Waste

Incineration was formerly the method of choice for most hazardous health care waste and is still widely used. However, recently developed alternative treatment methods are becoming increasingly popular. The final choice of treatment system should be made carefully on the basis of various factors, many of which depend on local conditions, listed below:

- Disinfection efficiency
- Health and environmental considerations
- Volume and mass reduction
- Occupational health and safety considerations
- Quantity of wastes for treatment and disposal/capacity of the system
- Types of waste for treatment and disposal
- Infrastructure requirements
- Locally available treatment options and technologies
- Options available for final disposal
- Training requirements for operation of the method
- Operation and maintenance considerations
- Available space
- Location and surroundings of the treatment site and disposal facility
- Investment and operating costs
- Public acceptability
- Regulatory requirements

Certain treatment options may effectively reduce the infectious hazards of health care waste but, simultaneously, give rise to other health and environmental hazards. Land disposal may result in groundwater pollution if the landfill site is inadequately designed. In choosing a treatment or disposal method for health care waste, particularly if there is a risk of toxic emissions or other hazardous consequences, the relative risks, as well as the integration into the overall framework of comprehensive waste strategy, therefore should be carefully evaluated in the light of local circumstances.

Incineration

Incineration is a high temperature, dry oxidation process that reduces organic and combustible waste to inorganic, incombustible matter and results in a significant reduction of waste volume and weight. This process is usually selected to treat wastes that cannot be recycled, reused, or disposed of in a landfill site.

The combustion of organic compounds produces mainly gaseous emissions, including steam, carbon dioxide, nitrogen oxides, and certain toxic substances (e.g., metals, halogen acids), and particulate matter, plus solid residues in the form of ashes. If the conditions of combustion are not properly controlled, toxic carbon monoxide also will be produced. The ash and wastewater produced by the process also contain toxic compounds, which have to be treated to avoid adverse effects on health and the environment.

Most large, modern incinerators include energy recovery facilities. In cold climates, steam or hot water from incinerators can be used to feed urban district heating systems, and in warmer climates the steam from incinerators is used to generate electricity. The heat recovered from small hospital incinerators is used for preheating of waste to be burnt.

Characteristics of Waste Suitable for Incineration

- Content of combustible matter above 60%
- Content of noncombustible solids below 5%
- Content of noncombustible fines below 20%
- Moisture content below 30%

Waste types not to be incinerated

- Pressurized gas containers
- Large amounts of reactive chemical waste
- Silver salts and photographic or radiographic wastes
- Halogenated plastics such as polyvinyl chloride (PVC)
- Waste with high mercury or cadmium content, such as broken thermometers, used batteries, and lead-lined wooden panels
- Sealed ampoules or ampoules containing heavy metals

Types of Incinerators

Incinerators can range from extremely sophisticated, high temperature operating plants to basic combustion units that operate at much lower temperatures. All types of incinerators, if operated properly, eliminate pathogens from waste and reduce the waste to ashes. However, certain types of health care waste, e.g., pharmaceutical or chemical waste, require higher temperatures for complete destruction. Higher operating temperatures and

Table 114-2 Categories of packages for radioactive waste (IAEA, 1996)

Categories of packages for radioactive waste		
Conditions		Category
Maximum radiation level at a distance of 1 m from the external surface of the package	Maximum radiation level at any point on the external surface	
Not more than 0,0005 mSv/h	Not more than 0,005 mSv/h	I-WHITE
More than 0,0005 mSv/h but not more than 0.01 mSv/h	More than 0,005 mSv/h but not more than 0.5 mSv/h	II-YELLOW
More than 0,01 mSv/h but not more than 0.1 mSv/h	More than 0,5 mSv/h but not more than 2 mSv/h	III-YELLOW

cleaning of exhaust gases limit the atmospheric pollution and odors produced by the incineration process.

Incinerators are designed especially for treatment of health care waste and operated at temperatures between 900°C and 1200°C. Low cost, high temperature incinerators of simple design are currently being developed, and a system designed specifically for health care and pharmaceutical waste in low-income countries is currently under test. Mobile incinerators for health care waste have been tested; those units permit on-site treatment in hospitals and clinics, thus avoiding the need to transport infectious waste through city streets. Test results for units with a capacity of 30 kg/hour were satisfactory in terms of function, performance, and air pollution.

High temperature incineration of chemical and pharmaceutical waste in industrial cement or steel kilns is practiced in many countries and is a valuable option; no additional investments are required, and industry benefits from a supply of free combustible matter.

Incineration equipment should be carefully chosen on the basis of the available resources and the local situation. Three basic kinds of incineration technology are of interest for treating health care waste:

- Double-chamber pyrolytic incinerators, which may be especially designed to burn infectious health care waste
- Single-chamber furnaces with static grate, which should be used only if pyrolytic incinerators are not affordable
- Rotary kilns operating at high temperature, capable of causing decomposition of genotoxic substances and heat-resistant chemicals

Pyrolytic incinerators—The most reliable and commonly used treatment process for health care waste is pyrolytic incineration, also called "controlled-air incineration" or "double-chamber incineration." The pyrolytic incinerators may be especially designed for hospitals. The pyrolytic incinerator comprises a pyrolytic chamber and a post-combustion chamber and functions as follows:

1. In the pyrolytic chamber, the waste is thermally decomposed through an oxygen-deficient, medium-temperature combustion process (800°C to 900°C), producing solid ashes and gases. The pyrolytic chamber includes a fuel burner, which is used to start the process. The waste is loaded in suitable waste bags or containers
2. A fuel burner in the post-combustion chamber using an excess of air to minimize smoke and odors burns the gases produced in this way at high temperatures (900°C to 1200°C)

Larger pyrolytic incinerators (capacity: 1 to 8 tons/day) are usually designed to function on a continuous basis. They also may be capable of fully automatic operation, including loading of waste, removal of ashes, and internal movement of burning waste.

Adequately maintained and operated pyrolytic incinerators of limited size, as commonly used in hospitals, do not require exhaust-gas-cleaning equipment. Their ashes will contain less than 1% unburnt material, which can be disposed of in landfills. However, to avoid dioxin production, no chlorinated plastic bags (and preferably no other chlorinated compounds) should be introduced into the incinerator and therefore none should be used for packaging waste before its incineration.

Capital costs for pyrolytic incinerators, suitable for treating health care waste, vary widely. The operating and maintenance costs for a small-scale hospital pyrolytic incinerator may reach about $19,000 per ton of waste incinerated.

Rotary kilns–A rotary kiln, which comprises a rotating oven and a post combustion chamber, may be used specifically to burn chemical wastes, and it is suited for use as a regional health care waste incinerator. The kiln rotates two to five times per minute and is charged with waste at the top. Ashes are evacuated at the bottom end of the kiln. The gases produced in the kiln are heated to high temperatures to burn off gaseous organic compounds in the post combustion chamber and typically have a residence time of two seconds. Rotary kilns may operate continuously and are adaptable to a wide range of loading devices. Those designed to treat toxic wastes should be operated by specialist waste disposal agencies and should be located in industrial areas.

Incineration in Municipal Incinerators

It is economically attractive to dispose of infectious health care waste in municipal incinerators if these are located reasonably close to hospitals. As the heating value of health care waste is significantly higher than that of domestic refuse, the introduction of relatively small quantities of health care waste will not affect the operation of a municipal incinerator. Municipal incinerators are usually of a double chamber design, with an operating temperature of 800°C in the first combustion chamber and gas combustion in the second chamber at temperatures of, typically, 1000°C–1200°C.

A number of rules and recommendations apply to the disposal of health care waste in municipal facilities:

1. When health care waste is delivered to the incineration plant, the packaging should be checked to ensure that it is undamaged.
2. Health care waste should not be packed in cylindrical containers because these could roll on the grids where they are placed for combustion.
3. Facilities should be available at the incineration site for the cleaning and disinfection of transportation equipment, including vehicles.
4. Deposit of health care waste in the normal reception bunker is not recommended: There is a risk of waste bags being damaged during transfer to the furnace by the overhead crane. Health care waste therefore should be loaded directly into the furnace.
5. Use of an automatic loading device for bags and containers of health care waste, rather than manual loading, would protect the safety of workers.
6. Health care waste should not be stored for more than 24 hours at an incineration plant; longer storage would require cooling facilities to prevent the growth of certain pathogens and the development of odors.7) The combustion efficiency should be checked. It should be at least 97% during incineration of health care waste.8) Health care waste should be introduced into the furnace only when the normal conditions of combustion have been established, never during startup or shut-down of the combustion process.
9. The process should be designed to prevent contamination of ashes or wastewater by the health care waste.

Incineration Options that Meet Minimum Requirements

If a pyrolytic incinerator cannot be afforded, health care waste may be incinerated in a static-grate, single-chamber incinerator. This type of incinerator treats waste in batches while loading and de-ashing operations are performed manually. The combustion is initiated by the addition of fuel and should then continues unaided.

Atmospheric emissions of that type of incinerators usually include acid gases such as sulfur dioxide, hydrogen chloride, and hydrogen fluoride, black smoke, fly ash (particulate), carbon monoxide, nitrogen oxide, heavy metals, and volatile organic chemicals. To limit these emissions, the incinerator should be properly operated and carefully maintained, and sources of pollution should be excluded from the waste to be incinerated whenever possible.

The drum incinerator is the simplest form of single-chamber incinerator. It should be used only as a last resort, as it is difficult to burn the waste completely without generating potentially harmful smoke. The option is appropriate only in emergencies during acute outbreaks of communicable diseases and should be used only for infectious waste.

The drum incinerator should be designed to allow the intake of sufficient, air and the addition of adequate quantities of fuel is essential, to keep the temperature as high as possible. A 210-liter steel drum should be used, with both ends removed; this will allow the burning of one bag of waste at a time. To operate the drum incinerator, a fire should first be established on the ground underneath it. One bag of waste should be lowered in to the drum. Tying the bag to a stick with string will help to avoid burns. Wood should be added to the fire until the waste is completely burned. After burning is complete, the ashes from both the fire and the waste itself should be collected and buried safely inside the premises of health care facilities.

The efficiency of this type of incinerator may reach 80%–90% and may thus result in destruction of 99% of microorganisms and a dramatic reduction in the volume and weight of waste. However, many chemical and pharmaceutical residues will persist if temperatures do not exceed 200°C. In addition, the process will cause massive emission of black smoke, fly ash, and potentially toxic gases.

Land Disposal

If a municipality or medical authority genuinely lacks the means to treat wastes before disposal, the use of a landfill must be regarded as an acceptable disposal route. Allowing health care waste to accumulate at hospitals or elsewhere constitutes a far higher risk of the transmission of infection than careful disposal in a municipal landfill, even if the site is not designed to the standard used in higher-income countries. The primary objections to landfill disposal of hazardous health care waste, especially untreated waste, may be cultural or religious or based on a perceived risk of the release of pathogens to air and water, or on the risk of access by scavengers.

Safe Burial in Hospital Premises

In health care establishments that use minimal programs for health care waste management, particularly in remote locations, in temporary refugee encampments, or in areas experiencing exceptional hardship, the safe burial of waste on hospital premises may be the only viable option available at the time. However, the hospital management should establish certain basic rules:

1. Access to the disposal site should be restricted to authorized personnel only.
2. The burial site should be lined with a material of low permeability, such as clay, if available, to prevent pollution of any shallow groundwater that may subsequently reach nearby wells.

Collection and Disposal of Wastewater

Characteristics and Hazards of Wastewater from Health Care Establishments
Wastewater from health care establishments is of a similar quality to urban wastewater, but may also contain various potentially hazardous components, discussed below.

Microbiological Pathogens

The principal area of concern is wastewater with a high content of enteric pathogens, including bacteria, viruses, and helminthes, which are easily transmitted through water. Contaminated wastewater is produced by wards that treat patients with enteric diseases and is a particular problem during outbreaks of diarrhea disease.

Hazardous Chemicals

Small amounts of chemicals from cleaning and disinfection operations are regularly discharged into sewers.

Pharmaceuticals

Small quantities of pharmaceuticals are usually discharged to the sewers from hospital pharmacies and from the various wards.

Radioactive Isotopes

Small amounts of radioactive isotopes will be discharged into sewers by oncology departments but should not pose any risk to health.

Related Hazards

In some developing and industrializing countries, outbreaks of cholera are reported periodically. Sewers of the health care establishments where cholera patients are treated are not always connected to efficient sewage treatment plants, and sometimes municipal sewer networks do not even exist. Although links between the spread of cholera and unsafe wastewater disposal have not been sufficiently studied or documented, they have been strongly suspected; for example, during recent African outbreaks (e.g., the Democratic Republic of the Congo and Rwanda), and during the 1991–92 cholera epidemic in southern America. In collection and disposal of wastewater, little information is available on the transmission of other diseases through the sewage of health care establishments.

In developed countries, water use is commonly high, and the sewage therefore greatly diluted; effluents are treated in municipal treatment plants, and no significant health risks should be expected, even without further specific treatment of these effluents. Only in the unlikely event of an outbreak of acute diarrhea diseases should excreta from patients be collected separately and disinfected. In developing countries, where there might be no connection to municipal sewage networks, discharge of untreated or inadequately treated sewage to the environment inevitably will pose major health risks.

The toxic effects of any chemical pollutants contained in wastewater on the active bacteria of the sewage purification process may give rise to additional hazards.

Wastewater Management

The basic principle underlying effective wastewater management is a strict limit on the discharge of hazardous liquids to sewers.

Connection to a Municipal Sewage Treatment Plant

In countries that do not experience epidemics of enteric disease and that are not endemic for intestinal helminthiasis, it is acceptable to discharge the sewage of health care establishments to municipal sewers without pretreatment, if the following requirements are met:

- The municipal sewers are connected to efficiently operated sewage treatment plants that ensure at least 95% removal of bacteria
- The sludge resulting from sewage treatment is subjected to anaerobic digestion, leaving no more than one helminthes egg per liter in the digested sludge.
- The waste management system of the health care establishment maintains high standards, ensuring the absence of significant quantities of toxic chemicals, pharmaceuticals, radionuclides, cytotoxic drugs, and antibiotics in the discharged sewage.
- Excreta from patients being treated with cytotoxic drugs may be collected separately and adequately treated (as for other cytotoxic waste).

If these requirements cannot be met, the wastewater should be managed and treated as recommended below in the section on on-site treatment or pretreatment of wastewater.

Under normal circumstances, the usual secondary bacteriological treatment of sewage, properly applied, complemented by anaerobic digestion of sludge, can be considered to be sufficient. During outbreaks of enteric disease, however, or during critical periods (usually during the summer because of warm weather, and during the autumn because of reduced river water flow), effluent disinfection by chlorine dioxide (ClO_2) or by any other efficient process is recommended. If the final effluent is discharged into coastal waters close to shellfish habitats, disinfection of the effluent will be required throughout the year.

When the final effluents or the sludge from sewage treatment plants are reused for agricultural or aquacultural purposes, the safety recommendations of the relevant WHO guidelines should be respected. (See following section.)

OnSite Treatment or Pretreatment of Wastewater

Many hospitals (in particular, those that are not connected to any municipal treatment plant) have their own sewage treatment plants.

Efficient on-site treatment of hospital sewage should include the following operations:

- Primary treatment
- Secondary biological purification. Most helminthes will settle in the sludge resulting from secondary purification, together with 90%–95% of bacteria and a significant percentage of viruses; the secondary effluent thus will be almost free of helminthes, but will include infective concentrations of bacteria and viruses.
- Tertiary treatment. The secondary effluent probably will contain at least 20 mg/liter of suspended organic matter, which is too high for efficient chlorine disinfection. It therefore should be subjected to a tertiary treatment, such as lagooning; if no space is available for creating a lagoon, rapid sand filtration may be substituted to produce a tertiary effluent with a much-reduced content of suspended organic matter (<10mg/litre).
- Chlorine disinfection. To achieve pathogen concentrations comparable to those found in natural waters, the tertiary effluent will be subjected to chlorine disinfection, to the breakpoint. This may be done with chlorine dioxide (which is the most efficient), sodium hypochlorite, or chlorine gas.
- Ultraviolet light disinfection

Disinfection of the effluents is particularly important if they are discharged into coastal waters close to shellfish habitats, especially if local people are in the habit of eating raw shellfish

The sludge from the sewage treatment plant requires anaerobic digestion to ensure thermal elimination of most pathogens. Alternatively, it may be dried in natural drying beds and then incinerated together with solid infectious health care waste. On-site treatment of hospital sewage will produce a sludge that contains high concentrations of helminthes and other pathogens.

According to the relevant WHO guidelines, the treated wastewater should contain no more than one helminthes egg per liter and no more than 1000 fecal coliforms per 100 ml if it is to be used for unrestricted irrigation. It is essential that the treated sludge contain no more than one helminthes egg per kilogram, and no more than 1000 fecal coliforms per 100g. The sludge should be applied to fields in trenches and then covered with soil (WHO, 1997).

Options for Establishments that Apply Minimal Waste Management Programs

Lagooning

In a region or an individual health care establishment that cannot afford sewage treatment plants, a lagooning system is the minimal requirement for treatment of wastewater. The system should comprise two successive lagoons to achieve an acceptable level of purification of hospital sewage.

Lagooning may be followed by infiltration of the effluent into the land, benefiting from the filtering capacity of the soil. There is no safe solution for disposal of sewage from a hospital that cannot afford a compact sewage treatment plant.

Minimal Safety Requirements

For health care establishments that apply minimal programs and that are unable to afford any sewage treatment, the following measures should be implemented to minimize health risks:

- Patients with enteric diseases should be isolated in wards where their excreta can be collected in buckets for chemical disinfection; this is of the utmost importance in case of cholera outbreaks, for example, and strong disinfectants will be needed.
- No chemicals or pharmaceuticals should be discharged into the sewer.
- Sludges from hospital cesspools should be dehydrated on natural drying beds and disinfected chemically (e.g., with sodium hypochlorite, chlorine gas, or, preferably, chlorine dioxide).
- Sewage from health care establishments never should be used for agricultural or aquacultural purposes.
- Hospital sewage should not be discharged into natural water bodies that are used to irrigate fruit or vegetable crops, to produce drinking water, or for recreational purposes.

Small-scale rural health care establishments that apply minimal waste management programs may discharge their wastewater into the environment.

One acceptable solution would be natural filtration of the sewage through porous soils, but this must take place outside of the catchment area of aquifers that are used to produce drinking water or to supply water to the health care establishment.

Sanitation

In many health care establishments in developing countries, patients have no access to sanitation facilities. Excreta are usually disposed of in the environment, creating a high direct or indirect risk of infection to other people. Human excreta are the principal vehicle for the transmission and spread of a wide range of communicable diseases, and excreta from hospital patients may be expected to contain far higher concentrations of pathogens, and therefore to be far more infectious than excreta from households. This underlines the prime importance of providing access to adequate sanitation in every health care establishment, and of handling this issue with special care. The fecal-oral transmission route and other routes such as penetration of the skin must be interrupted to prevent continuous infection and reinfection of the population.

Vector Control

Hygiene is fundamentally important in the hospital environment. Its importance should not be limited to the concepts of cleaning, disinfection, and sterilization. It should reach wider concepts, as the one of rationalization of hospital garbage and special cares. For instance, it should be guaranteed that eating in the hospital environment should only occur in hospital rooms, cafeterias, and snack bars. It is extremely important to applycontinuous and in-service education programs. These programs should reach employees, patients, and visitors. This part is designated to the health care professionals entrusted with the recognition, evaluation, and control of vectors of diseases in the hospital environment.

Rodents

The mouse and rat are the most harmful rodents. The most common species are the mouse *(Mus musculus)*, the lining mouse or domestic mouse *(Rattus rattus)*, and the sewer rat *(Rattus norvegicus)*. All live close to man, in homes, barns, docks, ships and garbage deposits. They are well-known bearers of diseases, intestinal parasites, and fleas. They are especially responsible for the transmission of bubonic plague, leptospirosis (Leptospira sp.), fever of Haverhill *(Streptobacillus moniliformes)*. It still transmits "SODOKU" *(Spirilum minus)*, characterized by a hardened ulcer around the point of the bite, recurrent fever, and exanthema cutaneous.

Knowing some characteristics of these rodents, as follows, will aid in their recognition:

- Their vision is weak and thus they move in contact with the walls, leaving marks from the dirt of their bodies.
- They have an excellent sense of smell, so they do not return to places where other mice have died.
- They defecate in the place where they have eaten. The species can be determined from the dimensions of the feces.

The evaluation of the amount of existent rodents in a certain place is made by considering the damages that they cause. Specialist professionals of pest control and sanitation can better quantify this. The evaluation is concluded with the location of the nest.

During the critical phase (infestation), control is accomplished through the use of rat venom in bait form. This venom normally acts by inhibiting blood coagulation, factors leading to the death of the mice from hemorrhages, days after the ingestion. Bait should be distributed close to walls. After the critical phase, control can be established with the maintenance of hygiene, eating in appropriate places in the hospital, and improvement in the quality of the sewer.

Insects

The insects are the dominant group of animals in the earth (Borror, 1964). They surpass in number all other terrestrial animals and can be seen everywhere. Many insects are extremely valuable to man, particularly in crop pollination. They also produce honey, beeswax, silk, and other products of commercial value. Insects also have been useful in medical and scientific research. However, some insects are noxious and human and animal annually, related to agricultural crops, stored products, and human and animal health. They have existed on earth for more than 300 million years. They adapt to several habitat types. Many possess social organization and high reproductive capacity; and each generation can last from days to years.

Cockroaches

Cockroaches belong to the order *Orthoptera* and the suborder Blattidae. They have an oval form and are flat. They are omnivorous, and they have domestic habits. The most common species are the American cockroach and Germanic Blatella. Cockroaches are not known as specific vectors of diseases. However, they feed off a great variety of products, contaminating victuals. They have an unpleasant scent and frequently become serious pests.

Fleas

Fleas belong to the order *Siphonaptera*. They are blood-feeding insects without wings. The bites of some species are quite irritating. Some serve as vectors of diseases, other as intermediate hosts of certain taenia. Some species penetrate into the skin of animals and of man.

The flea's body is flattened sidelong; some fleas have eyes, and some do not. They do not have specific hosts, and they can feed on several animals. They can live for up to one year and can survive several weeks without feeding. They lay their eggs in the ground or in the nest of the host.

Fleas transmit to man three plagues: The bubonic, the pneumonic, and the septicemia. The most important disease transmitted by fleas is bubonic plague, an infectious disease caused by the Bacillus pasteurella pests. It is a disease of rodents and is transmitted from one to another by way of fleas.

Three modes of transmission are known: Regurgitation of bacilli during a bite by an infested flea, scrubbing on the skin, and ingestion of infected fleas.

Lice

Lice belong to the order *Anophlura*. They are ectoparasites that feed on blood. Two species infest man: The head louse *(Pediculus humanus capitis)* and the corporal louse *(Pediculus humanus corporis)*. The adults' length varies from 2.5–3.5 millimeters; the head louse lays its eggs in the hair of the head and body and in the seams of clothes; its reproductive cycle lasts about one month; it feeds frequent, and the meal lasts for minutes.

Head lice are transmitted person to person by the casual use of combs, hairbrushes, and caps. Body lice are transmitted through personal clothes or beds; They can migrate during the night, from one place to another. It is an important vector of man's diseases, epidemic typhoid fever being the the most important disease. Often it attains serious epidemic proportions and can cause a mortality rate of up to 70%. Another important disease is recurrent fever, the transmission of which is made when the infected louse is squeezed against the skin. The feces and the bites are not infecting. Another disease, known as "fever of the trenches," was common during World War I.

Flies and Black Flies

Flies and black flies belong to the order *Diperous*. They constitute one of the largest orders of insects, and its representatives are plentiful in individuals. Species are founded almost everywhere. Most of the *Diperous* are relatively small insects with soft bodies; some have great economic importance. The black flies of the stables and others are hematophagous and they constitute severe plagues to the man and animals.

Many of the *Dipterous hematophagous* and *saprophagous*, as the domestic fly and the blowflies, are important vectors of diseases. The organisms that cause malaria, yellow fever, filariasis, sleeping sickness, typhoid fever, and dysentery are transported and disseminated by dipterous. Sleeping sickness is a deadly disease and currently affects as many as half a million people in sub-Saharan Africa, while an estimated 60 million people are at risk of contracting the disease in 36 countries.

Control

The control of insects can be made first through good hygiene practices in the hospital environment. Favorable results can be obtained by using screens in the windows of kitchens or places where they can obtain food. Pesticides are applied periodically in openings, dark places, and baseboards—the usual hiding places. An effective chemical product, piretróide, acts upon the neuron membrane, causing a chemical imbalance in the concentrations of sodium and potassium.

A specialist team should be established to control vectors in the hospital environment. Collaboration of all hospital departments promotes the success of the control program. This team should relate to the local health authorities. If outside companies are utilized, they should be specialized and registered in their areas of specialization.

The risks of pesticide application without appropriate criteria include low efficiency, human toxicity, and increased resistance of the insects. Effective control of vectors in the hospital environment requires a knowledge of the vector's habitat, alimentary habits, reproductive cycle, and other factors that could serve as vulnerability points.

Finally, the basic strategy of intervention should consist of treating the existent breeding places, expelling the insects, and impeding their return to these places for recolonization.

Clinical Engineering and Medical Equipment

Infection control related to medical equipment is important to nurses, physicians, and clinical engineers. Nurses and physicians are concerned with occupational health, and clinical engineers are concerned with evaluations of devices and sterilization processes.

Below, several medical devices and their relationship to infection control are described.

Ventilators

Ventilators breathe for patients who cannot breathe on their own because of such afflictions as disease, trauma, and congenital defects. Ventilators are connected to patients by means of such devices as breathing circuits and endotracheal tubes. Problems associated with ventilators include wounds caused during the patient's intubation, excessive pressure in the lungs, safety valves damaged or not adjusted, cross contamination, and infection. Usually, the breathing circuits are manufactured of plastic material and sterilized by ethylene oxide or a liquid sterilant. Most international manufacturers offer breathing circuits made of material that can be steam sterilized. When buying a breathing circuit, it is necessary to consider its material in order to minimize additional costs of sterilization.

Heated Humidifiers

Adequate humidity and temperature of the gas mixture to the lungs of a patient is accomplished through the use of a heated humidifier in the patient's breathing circuit. In normal respiration, the mouth, nostrils, and pharynx heat and humidify inhaled air and retain heat and vapors during exhalation. When the patient is under tracheostomy or using endotracheal tubes, this process of natural humidification does not occur. Heated humidifiers help to alleviate dehydration, provide comfort to membranes of respiratory system, and prevent loss of body heat.

The most common problems relating to this technology are hyperhydratation, dehydratation, hypothermia, hyperthermia, thermal lesions of the mucous membranes, melting of breathing circuits, and infection. Infection is a concern with all humidifiers and is usually the result of the inadequate or inappropriate handling, cleaning, and disinfecting of breathing circuits (see Figure 114-4).

Electrosurgical Units

During surgical procedures, surgeons use electrosurgical units (ESU) to cut and coagulate tissues. ESUs generate electric current at high frequency at the end of an active electrode. This current cuts and coagulates tissue. The advantages of this technology over the conventional scalpel are simultaneous cutting and coagulating and ease of use in several procedures (including surgical endoscopy procedures).

The most common problems are burns, fire and electric shock. This type of burn usually occurs under the electrode of ECG equipment, under the ESU grounding, also known as return or dispersive electrode), or on various parts of the body that may be in contact with a return path for the ESU current, e.g., arms, chest, and legs. Fires occur when flammable liquids come in contact with sparks from the ESU in the presence of an oxidant. Usually these accidents begin the development of an infectious process in the place of the burn. This can bring serious consequences to the patient and usually increase the patient's stay in the hospital.

Aspirators

Aspirators are used to aspirate substances such as blood, tissue, and fat. Aspirator maintenance procedures should consider biological hazards. Engineering and nursing must develop a plan to ensure that equipment is free from pathogenic microorganisms when delivered to maintenance. Mobile aspirators can pollute the environment because they can discharge particulate material.

Infant Incubators

An infant incubator is a closed atmosphere where temperature, humidity, and oxygen concentration are controlled. Usually, the values of relative humidity are not controlled and its value inside it will depend on the values of humidity found in the atmosphere outside the incubator. Humidification devices integral to incubators or used in conjunction with them may harbor pathogenic organisms. Clinical engineering should work with nursing

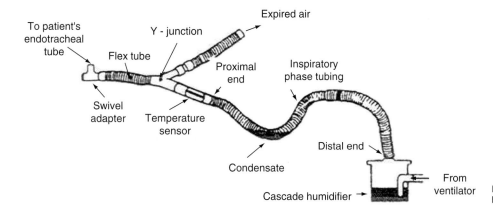

Figure 114-4 Representation of a ventilator breathing circuit and its components.

to ensure that infant incubators are routinely cleaned and disinfected and that air filters are changed at appropriate intervals.

Anesthesia Equipment

Anesthesia machines are discussed in detail in Chapter 90. The contamination of any part of the breathing circuit, including the wye connector, face mask, and reservoir bag can spread hospital infection. The related to this equipment include infections of respiratory superior tract and lungs. The Centers for Disease Control and Prevention (CDC) recommends the use of high level disinfection of reusable breathing circuits between each patients use to prevent cross contamination. Universal agreement has not been reached on the advisability of using bacterial filters to prevent cross contamination between patients.

Nebulizers

Nebulizers administer aerosol medication to patients. Nebulizers therapy is particularly effective in treatment of cystic fibrosis, emphysema, bronchitis, and severe asthma, in which viscous secretions occur. The aerosol disperses these secretions, stimulates coughing and expectoration, and alleviates respiratory tract congestion. Compressed air and ultrasonic nebulizers are most frequently used and both produce aerosols capable of effectively carrying medication to the lungs.

Ttuberculosis patients require special care. As aerosol treatment induces coughing it may also increase the potential for transmission of *M. tuberculosis*. Health care personnel should be particularly alert to prevent *M.tuberculosis* transmission where immunocompromised persons (e.g., HIV-infected persons) work or receive care especially if cough-inducing procedures, such as sputum induction and aerosolized pentamidine treatments are being performed.

M. tuberculosis is carried by airborne particles, or droplet nuclei, that can be generated when persons who have pulmonary or laryngeal TB sneeze, cough, speak, or sing. The particles are about 1 mm in diameter and normal air currents can keep them airborne for prolonged periods and spread them throughout a room or building. Infection occurs when a susceptible person inhales droplet nuclei containing *M. tuberculosis* and these droplet nuclei traverse the mouth or nasal passages, upper respiratory tract, and bronchi to reach the alveoli of the lungs. Once in the alveoli, the organisms are taken up by alveolar macrophages and spread throughout the body. Usually within 2–10 weeks after initial infection with *M. tuberculosis*, the immune response limits further multiplication and spread of the tubercle bacilli; however, some of the bacilli remain dormant and viable for many years.

Hemodialysis Units

The kidneys maintain body fluid and acid/base balance even as metabolic activity and external conditions change. Hemodialysis units are used to treat patients with compromised renal function and resultant retention of water and metabolites. Hemodialysis therapy removes ions and organic salts from the blood.

Infection is the most common complication and the largest causes of the morbid mortality associated with the use of hemodialysis units. The antigen of the hepatitis B (HBsAg) has been detected at several locations in hemodialysis centers including the control buttons of hemodialysis units. Consequently, application of strict measures and procedures that reduce infection risks should be used. Necessary measures to reduce this risk include disinfection and sterilization, hygiene and cleaning, maintenance, removal of residues, isolation precautions, and universal precautions. Water used in hemodialysis should be periodically tested to ensure that water treatment is adequate. In addition to performing their preventive maintenance routines, centers that use reverse osmosis equipment should tests at least once a year. See Chapter 115 for more information on water treatment in hemodialysis.

Hospital Engineering and Hospital Installations

Although some hospitals have hospital engineers, those that do not often employ clinical engineers to take care of infrastructure systems such as medical vacuum, compressed air, laundry and healting, ventilation, and air conditioning (HVAC).

Medical Vacuum

Medical vacuum is used to aspirate secretions and fluids from the human body during surgery or other medical and nursing procedures. Central of vacuum generates a negative pressure in a piping network sufficient to aspirate secretions from the human body. The aspiration bottles retain the heaviest particles aspirated. Smaller and lighter particles in aerosols form are aspirated into the piping network of the vacuum system may reach the outside atmosphere. Filters on equipment and systems that connect to the medical vacuum system reduce the possibility of air contamination. For additional information on medical vacuum, see Chapter 110.

Vacuum system should be inspected and serviced as follows:

- Air analysis: Tests to determine the contamination level of the air coming from the pipe network. Bacteriology laboratory tests are recommended.
- Analysis of pollutants in the reservoir: The vacuum reservoir should possess an opening for easy access. This permits the inspection, cleaning and disinfecting of the internal mechanical structure of the reservoir, as well as the collecting of samples for microbiological analyses.
- Whenever possible, redundancy should be incorporated in vacuum piping systems be without turning it off. Two reservoirs interlinked by specially located piping and valves can achieve this end.
- Analysis of seal water: Some pump models need a certain volume of water to work. This water comes in contact with human body fluids in form of aerosols coming from the piping network. This water should not be reused unless appropriate treatment is received.

Vacuum systems should never be located near HVAC systems or medical air compressors lest these systems become contaminated by infectious agents released by the vacuum system.

Medical Compressed Air

Medical compressed air is mainly used to supply pure air directly to patients or to pneumatic systems of medical devices such as ventilators, anesthesia machines, and inhalation devices. It must be free of pathogenic microorganisms and other contaminants (see Chapter 110).

The air supplied should be dry because water can damage pneumatic circuits. Dry air is required for proper functioning of patient humidifiers.

Compressors should be located where intake air is of a high-quality. This reduces the risks of contamination of the piping network and the cost of filtration of air generated in the compression process. Independent of the generating source of air, the following maximum levels of pollutants are recommended: 19%-23% for oxygen, 5 ppm for carbon monoxide, 500 ppm for carbon dioxide, 5 ppm for nitrous oxide, and 1 mg/m^3 for oil and other particles. Clinical engineering departments advise maintenance teams as to the special maintenance requirements of medical compressed air systems.

Laundry Equipment

To ensure that clothes and linens are cleaned and disinfected adequately, the laundry must have properly functioning steam and hot water generation systems. Temperature and steam pressure variances can reduce the effectiveness of certain products or chemical disinfectant agents. The clinical engineering department should contact the manufacturer of cleaning and disinfecting products in the laundry so that the optimal temperature and pressure of steam and hot water is achieved. The ventilation and exhaust systems of the laundry should be evaluated periodically. Exhaust should be directed to contaminated areas and intake air obtained from clean areas.

Heating, Ventilation, and Air Conditioning

Heating, ventilation, and air conditioning (HVAC) systems are often used in hospitals at least in the operating rooms (OR), intensive care unit (ICU), and other areas where patients are receiving care. (For a discussion of HVAC, see Chapter 108.)

Clinical engineering should be concerned with ensuring that all air conditioning variables are controlled, especially the purity of the air. In systems that make use of cooling towers and reservoir of water to air humidification, verification of the microbiologic quality of the water stored in these reservoirs is necessary because they can shelter pathogenic microorganisms. The purity of the air should be assessed by monitoring particulate material in the air every six months. Understanding that a filter's useful life is influenced by the quality of the outside air, the efficacy of air filters can be determined by measurement of differential pressure across the filters.

Conclusion

The activities of clinical engineering in infection control are varied. Clinical engineers should consider this chapter to be an introduction to the contribution that they can make in this field and should realize that much more can be accomplished. Clinical engineers are particularly well suited, by education and experience, to collaborate with other health care professionals in planning projects, training activities, construction projects, acquisition of equipment and systems, and working to control the spread of infection.

Glossary

Adherence: The behavior of patients when they follow all aspects of the treatment regimen as prescribed by the medical provider. Also refers to the behavior of health care workers and employers when they follow all guidelines pertaining to infection control.

Aeration: The method by which absorbed ethylene oxide gas is allowed to dissipate from sterilized items by the use of warm air circulation in a specially designed enclosed cabinet.

Aerosol: The droplet nuclei that are expelled by an infectious person (e.g., by coughing or sneezing); these droplet nuclei can remain suspended in the air and can transmit M. tuberculosis to other persons.

AIA: The American Institute of Architects, a professional body that develops standards for building ventilation.

Air changes: The ratio of the volume of air flowing through a space in a certain period of time (i.e., the airflow rate) to the volume of that space (i.e., the room volume); this ratio is usually expressed as the number of air changes per hour (ACH).

Air mixing: The degree to which the air supplied to a room mixes with the air already in the room, usually expressed as a mixing factor. This factor varies from 1 (for perfect mixing) to 10 (for poor mixing), and it is used as a multiplier to determine the actual airflow required (i.e., the recommended ACH multiplied by the mixing factor equals the actual ACH required).

Alveoli: The small air sacs in the lungs that lie at the end of the bronchial tree; the site where carbon dioxide in the blood is replaced by oxygen from the lungs and where TB infection usually begins.

Anteroom: A small room leading from a corridor into an isolation room. This room can act as an airlock, preventing the escape of contaminants from the isolation room into the corridor.

Area: A structural unit (e.g., a hospital ward or laboratory) or functional unit (e.g., an internal medicine service) in which health care workers provide services to, and share air with, a specific patient population or work with clinical specimens that may contain viable M. tuberculosis organisms. The risk for exposure to M. tuberculosis in a given area depends on the prevalence of TB in the population served and the characteristics of the environment.

Asepsis: Absence of infectious organisms.

ASHRAE: The American Society of Heating, Refrigerating and Air Conditioning Engineers, Inc., a professional body that develops standards for building ventilation.

Asymptomatic: Without symptoms, or producing no symptoms of disease.

Bioburden (N_o): The number of microorganisms that will be admitted prior to begin of the configuration process. It is the initial population of spores to be reduced.

Biological indicator: A sterilization process monitoring device that is commercially prepared with a known population of highly resistant spores to test the effectiveness of a sterilization method. The indicator is used to demonstrate that the conditions necessary to achieve sterilization were met during the cycle being monitored.

Bronchoscopy: A procedure for examining the respiratory tract that requires inserting an instrument (a bronchoscope) through the mouth or nose, and into the trachea. The procedure can be used to obtain diagnostic specimens.

Cavity: A hole in the lung, resulting from the destruction of pulmonary tissue by TB or other pulmonary infections or conditions. TB patients who have cavities in their lungs are referred to as having "cavitary disease," and they are often more infectious than TB patients without cavitary disease.

Chemotherapy: Treatment of an infection or disease by means of oral or injectable drugs.

Contact: A person who has shared the same air with a person who has infectious TB for a sufficient amount of time to allow possible transmission of mycobacterium tuberculosis (M. tuberculosis).

Culture: The process of growing bacteria in the laboratory so that organisms can be identified.

Chemical indicator: A sterilization monitoring device (e.g., chemically treated paper, pellets sealed in a glass tube, or pressure-sensitive tape) that is used to monitor certain parameters of a sterilization process by means of a characteristic color change.

Cleaning: This is the process (also known as "low level inactivation") in which a biocide inactivates vegetative bacteria, lipid-enveloped viruses, and fungi only.

Commissioning: Obtaining and documenting evidence that equipment has been provided and installed in accordance with its specifications and that it functions within pre-determined limits when operated in accordance with is operational instructions.

Concurrent disinfection: Done after the expulsion of infectious material from the body of the patient to an area or equipment.

Contamination: Transference of an infectious agent to an organism or substance. At this point, there is no infection process.

Control (risks): To adopt technical, administrative, preventive or corrective measurements of several natures, which tend to eliminate or to lessen the existent risks in the workplace.

Critical devices: Devices that carry a high risk of infection associated with their use as they are introduced directly into human tissue. Examples include needles, catheters, and implants. These goods should be sterilized, protecting the patient against contamination. Devices used in the care of an infectious patient should be sterile before use on another patient.

Disinfection: An intermediate-level of inactivation, in which a biocide process inactivates all organisms but more resistant spores. Steam and hot water are the most common and efficient agents used in the health care environment.

Droplet nuclei: Microscopic particles (i.e., 1–5 μm in diameter) produced when a person coughs, sneezes, shouts, or sings. The droplets produced by an infectious TB patient can carry tubercle bacilli and can remain suspended in the air for prolonged periods of time and be carried on normal air currents in the room.

Drug resistance, acquired: A resistance to one or more anti-TB drugs that develops while a patient is receiving therapy and that usually results from the patient's nonadherence to therapy or the prescription of an inadequate regimen by a health care provider.

Drug resistance, primary: A resistance to one or more anti-TB drugs that exists before a patient is treated with the drug(s). Primary resistance occurs in persons exposed to, and infected with, a drug resistant strain of M. tuberculosis.

D-Value: The D-value is defined as the time required to reduce the microbial population by one decimal or one log unit. It expresses the heat resistance of the microorganism (see Figure 114-5).

Endemy: Habitual incidence of an infectious agent or a disease in a determinate area. It can mean the usual prevalence of a determinate disease in this area.

Edotoxins: High molecular complexes from the outer cell membrane of Gram-negative bacteria.

Evaluation of risks: Quantification and verification, in agreement with certain techniques, the magnitude of the risk (e.g., calculating or measuring air change rate in an environment and comparing with standards).

Event-related sterility: Shelf life based on the quality of the packaging material, storage conditions during transportation, and amount of handling of the item.

Exposure: The condition of being subjected to something (e.g., infectious agents) that could have a harmful effect. A person exposed to M. tuberculosis does not necessarily become infected. (See definition of "transmission.")

F_o-Value: The lethality of the cycle expressed in terms of the equivalent time in minutes at a temperature of 121°C with reference to micro-organisms possessing a Z-Value of 10°C.

Fixed room air HEPA recirculation systems: Nonmobile devices or systems that remove airborne contaminants by recirculating air through a HEPA filter. These may be built into the room and permanently may be ducted, or they may be mounted to the wall or ceiling within the room. In either situation, they are fixed in place and are not easily movable.

Fomites: Linens, books, dishes, or other objects used or touched by a patient. These objects are not involved in the transmission of M. tuberculosis.

Gastric aspirate: A procedure sometimes used to obtain a specimen for culture when a patient cannot cough up adequate sputum. A tube is inserted through the mouth or nose and into the stomach to recover sputum that was coughed into the throat and then swallowed. This procedure is particularly useful for diagnosis in children, who are often unable to cough up sputum.

HEPA: Stands for "high efficiency particulate air filter." A specialized filter that is capable of removing 99.97% of particles 0.3 μm in diameter and that may assist in controlling the transmission of M. tuberculosis. Filters may be used in ventilation systems to remove particles from the air, or in personal respirators to filter air before it is inhaled by the person wearing the respirator. The use of HEPA filters in ventilation systems requires expertise in installation and maintenance.

Human immunodeficiency virus (HIV) infection: Infection with the virus that causes acquired immunodeficiency syndrome (AIDS). HIV infection is the most important risk factor for the progression of latent TB infection to active TB.

Immunosuppression: A condition in which the immune system is not functioning normally (e.g., severe cellular immunosuppression resulting from HIV infection or immunosuppressive therapy). Immunosuppressed persons are at greatly increased risk for developing active TB after they have been infected with M. tuberculosis. No data are available regarding whether these persons are also at increased risk for infection with M. tuberculosis after they have been exposed to the organism.

Infection: The condition in which organisms capable of causing disease (e.g., M. tuberculosis) enter the body and elicit a response from the host's immune defenses. TB infection may or may not lead to clinical disease. Penetration and development or multiplication of an infectious agent in the human or animal organism. An immunologic reaction should occur.

Infectious: Capable of transmitting infection. When persons who have clinically active pulmonary or laryngeal TB disease cough or sneeze, they can expel droplets containing M. tuberculosis into the air. Persons whose sputum smears are positive for AFB are probably infectious.

Infectious agent: Organism capable of causing an infection or infectious disease.

Lethality: Characterization of the performance of a sterilization cycle by a parameter known as "F_o-Value."

Master plan for validation: A plan comprising an introduction; definitions and terminology; organization and responsibility; time schedule; objects to be validated or not

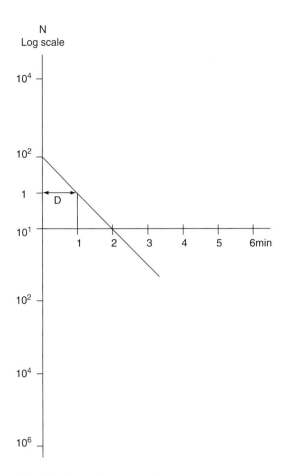

Figure 114-5 Microbial death rate curve: D-value.

validated; type of validation; document requirements; education requirements; and plan approval.

Multidrug resistant tuberculosis (MDR-TB): Active TB caused by M. tuberculosis organisms that are resistant to more than one anti-TB drug; in practice, it often refers to organisms that are resistant to both INH and rifampin, with or without resistance to other drugs. (See definitions of "drug resistance, acquired" and "drug resistance, primary.")

Negative pressure: The relative air pressure difference between two areas in a health care facility. A room that is at negative pressure has a lower pressure than adjacent areas, which keeps air from flowing out of the room and into adjacent rooms or areas.

Nosocomial: An occurrence, usually an infection, that is acquired in a hospital and/or as a result of medical care

Non-critical device: They are devices that come in contact with mucous membranes and nonintact skin and do not necessarily penetrate surfaces. Sterilization is not obligatory, although it is desirable because these devices (e.g., endotracheal tube, breathing equipment, and vaginal speculum) can become critical during procedures.

Operational qualification: The activities developed to control of the limits of the process, parameters of the software, specifications of the raw material, procedures for operation of the process, change of control of the process, and calibration (e.g., pressure, temperature, and time).

Pathogenesis: The pathologic, physiologic, or biochemical process by which a disease develops.

Pathogenicity: The quality of producing or the ability to produce pathologic changes or disease. Some nontuberculous mycobacteria are pathogenic (e.g., *Mycobacterium kansasii*), and others are not (e.g., *Mycobacterium phlei*).

Portable room air HEPA recirculation units: Free-standing, portable devices that remove airborne contaminants by recirculating air through a HEPA filter.

Performance qualification: Obtaining and documenting evidence that the equipment as commissioned will produce acceptable product ("goods") when operated in accordance with the operating instructions ("process specification").

Pyrogens: Foreign bodies or material in the body that cause an increase in body temperature.

Qualification: Establishing documented evidence through appropriate testing that the finished produced by a specified process (es) meet(s) all release requirements for functionality and safety.

Qualification of the installation: A control process that should include the characteristics of project of the equipment; conditions of the facilities (e.g., air, water, vapor, or electricity); calibration, preventive maintenance, cleaning; safety characteristics; software documentation; spare parts list; and environmental conditions.

Radiography: A method of viewing the respiratory system by using radiation to transmit an image of the respiratory system to film. A chest radiograph is taken to view the respiratory system of a person who is being evaluated for pulmonary TB. Abnormalities (e.g., lesions or cavities in the lungs, and enlarged lymph nodes) may indicate the presence of TB.

Recirculation: Ventilation in which most or all of the air that is exhausted from an area is returned to the same area or other areas of the facility.

Recognition (risks): To identify, characterize, and know to assign risk agents affecting patients and health care workers.

Resistance: The ability of some strains of bacteria, including *M. tuberculosis,* to grow and multiply in the presence of certain drugs that ordinarily kill them; such strains are referred to as "drug resistant strains."

Room air HEPA recirculation systems and units: Devices (either fixed or portable) that remove airborne contaminants by recirculating air through a HEPA filter.

Saturated steam: Vapor whose temperature is the same as the water from which it originated.

Specimen: Any body fluid, secretion, or tissue sent to a laboratory where smears and cultures for M. tuberculosis will be performed (e.g., sputum, urine, spinal fluid, and material obtained at biopsy).

Sputum: Phlegm coughed up from deep within the lungs. If a patient has pulmonary disease, an examination of the sputum by smear and culture can be helpful in evaluating the organism responsible for the infection. Sputum should not be confused with saliva or nasal secretions.

Sputum induction: A method for obtaining sputum from a patient who is unable to cough up a specimen spontaneously. The patient inhales a saline mist, which stimulates a cough from deep within the lungs.

Sterilization: The absence of viable microorganisms, or the absence of living organisms. It kills all viable microorganisms, including spores.

Superheated Steam: Vapor whose temperature is above the point at which it became steam.

Semi-critical device: Devices that do not enter the body of the patient, but do contact the skin. Low risk of hospital infection is associated with the use of these items such as thermometers, silverware, and X-ray tables. Consequently, depending on the article and degree of contamination, intermediate or low level disinfection can be achieved by washing with water and soap.

Sterilization: High level inactivation, in which bacterial spores, acid-fast bacteria, other nonsporulating bacteria, fungi, and lipid-enveloped viruses are killed.

Surgical site infection: Infection involving body layers that have been incised.

TB infection: A condition in which living tubercle bacilli are present in the body, but the disease is not clinically active. Infected persons usually have positive tuberculin reactions, but they have no symptoms related to the infection and are not infectious. However, infected persons remain at lifelong risk for developing disease unless preventive therapy is given.

Transmission: The spread of an infectious agent from one person to another. The likelihood of transmission is directly related to the duration and intensity of exposure to *M. tuberculosis.* (See definition of "exposure.")

Treatment failures: TB disease in patients who do not respond to chemotherapy and in patients whose disease worsens after having improved initially.

Tuberculosis (TB): A clinically active, symptomatic disease caused by an organism in the M. tuberculosis complex (usually M. tuberculosis or, rarely, M. bovis or M. africanum).

Terminal disinfection: Disinfection of clothes, personal use objects, medical devices, and immediate environment used by infected persons after death or suspension of isolation practices.

Ultraviolet germicidal irradiation (UVGI): The use of ultraviolet radiation to kill or inactivate microorganisms.

Ultraviolet germicidal irradiation (UVGI) lamps: Lamps that kill or inactivate microorganisms by emitting ultraviolet germicidal radiation, predominantly at a wavelength of 254 nm (intermediate light waves between visible light and X rays). UVGI lamps can be used in ceiling or wall fixtures or within air ducts of ventilation systems.

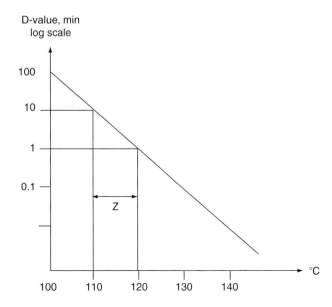

Figure 114-6 Z-value.

Validation: A documented procedure for obtaining, recording and interpreting data required to show that a process will consistently comply with predetermined specifications (configuration). For most heat sterilization, validation is considered as a total program, which consists of commissioning and performance qualification. It is a way of demonstrating, in accordance with the principles of good manufacturing practice (GMP), that any procedure, process, equipment, material, activity, or system actually leads to the expected result. It is used when sterilizer equipment is new; when there is a new product or load, and after changes or repair of the equipment.

Ventilation, dilution: An engineering control technique to dilute and remove airborne contaminants by the flow of air into and out of an area. Air that contains droplet nuclei is removed and replaced by contaminant free air. If the flow is sufficient, droplet nuclei become dispersed, dimishing their concentration in the air.

Ventilation, local exhaust: Ventilation that is used to capture and remove airborne contaminants by enclosing the contaminant source (i.e., the patient) or by placing an exhaust hood close to the contaminant source.

Virulence: The degree of pathogenicity of a microorganism as indicated by the severity of the disease produced and its ability to invade the tissues of a host. M. tuberculosis is a virulent organism.

Z-value: The Z-value (see Figure 114-6) is defined as the number of degrees temperature change to change the D-value by a factor of ten (10), or one log unit.

References

Borror JD. *An Introduction to the Study of Insects.* New York, Holt, Rinehart and Wiston, 1964.
CIIT (Chemical Industry Institute of Toxicology). Final Report on a Chronic Inhalation Toxicology Study in Rats and Mice Exposed to Formaldehyde, CIIT docket 10922. Columbus, OH, Battelle Columbus Laboratory, 1981.
Dyro JF, Tai S. Evaluation of Ethylene Oxide Sterilizers. In *Proceedings of the 29th Annual Conference on Engineering in Medicine and Biology.* Bethesda, MD, Alliance for Engineering in Medicine and Biology, 1976.
Fernandes, AT. Infecção Hospitalar e Suas Interfaces na Área da Saúde/Editor. São Paulo, Editora Atheneu, 2000.
Gamble J. Effects of Formaldehyde on the Respiratory System. In Gibson JE (ed). *Formaldehyde Toxicity.* Washington, DC, Hemisphere Publishing Corporation, 1983.
Handlos V. Determination of Formaldehyde Residuals in Autoclave Sterilized Materials. *Arch Pharm Chemi Sci Ed* 5:163, 1977.
IAEA. Regulations for the Safe Transport of Radioactive Material (IAEA Safety Standards Series, No. ST-1). Vienna, International Atomic Energy Agency,
Linvall L. Water Treatment and Steam Sterilization., GETING BRASIL/GETING AB, 1999.
London Waste Regulation Authority. *The London Waste Regulation Authority's Clinical Waste Guidelines, 2nd Ed.* London, London Waste Regulation Authority, 1994.
Thailand Ministry of Health. *Handbook of Hazardous Health Care Waste Management in 10-bed and 30-bed Community Hospitals.* Bangkok, Thailand Ministry of Health, 1995.
National Research Council. Formaldehyde: An Assessment of its Health Effects. Washington DC, National Academy of Sciences, 1980.
Semmelweis IP. The Etiology, the Concept and the Prophylaxis of Childbed Fever. Republished from *Pest, CA Hartleben's Verlag-Expedition,* 1861. Birmingham, England, Classics of Medicine Library, 1981.
United Nations. UN Recommendations on the Transport of Dangerous Goods—Model Regulations, 10th Revised Edition. ST/SG/AC.10/1/Rev. 10. New York, United Nations, 1997.
Weber-Tschopp A, Fischer T, Grandjean E. Reizwirkungen des Formaldehyds (HCHO) auf den Menschen. *Int Arch Occup Health* 39:207–218, 1977.
Wenzel RP. *Prevention and Control of Nosocomial Infection.* Philadelphia, Williams & Wilkins, 1983.
WHO. Product Information Sheets: Equipment for Expanded Programme on Immunization, Acute Respiratory Infections, Blood Safety, Emergency Campaigns, Primary Health Care. Geneva, World Health Organization, 1997.
WHO. Indoor Air Quality: Radon and Formaldehyde. *Environmental Health Series* 13:19–36, 1986.

115

Water Systems in Health Care Facilities

Diógenes Hernández
PAHO/WHO Panama City, Panama

Water and its supply system are vital for the operation of health care facilities. An appropriate system for capture, storage, treatment, conditioning, use, and disposal of water must be present in order to prevent hazards to patients, health professionals, and the public in general. It is also necessary to protect the installed medical and other equipment and to prevent the fast deterioration that could result from the use of inadequate or contaminated water. Water is one of the most complex resources to manage in a health care facility, given its implications to the health of users, the operation of the facility, and the impact on health services in general.

The health care facility water system described here applies to both developed and developing countries. Nevertheless, differences do exist between countries regarding the availability of water through public distribution systems, water characteristics, mechanisms of capture and distribution in the facility, different uses, treatment prior to discharge, norms and standards at national and local levels, and regulations. In developed countries, agencies regulate and control water quality, particularly water used by health care facilities, although even developed countries include localities and rural areas with deficiencies in water supply quality. In developing countries, the lack of regulation and enforcement is common, even in urban centers. This situation is of great significance in countries vulnerable to natural disasters and extreme climatic phenomena. Natural disasters usually affect the availability, supply, and quality of the water, hindering the normal operation of a health facility. The vulnerability to natural disasters should be handled as a critical factor in the process of planning a health care facility.

Water Use Cycle

The water system in a health care facility operates through several stages. It starts with the capture of the water from the municipal or public water supply system or from groundwater pumped from wells. Next, the storage, treatment, and conditioning occurs, depending on the expected use by clinical and support services. Water then passes through the distribution system or piping system in the facility. The demand for water by different services and the groups of users depends on many issues but must always consider potential risks and negative impacts of water use by people (e.g., infections) and equipment (e.g., fast deterioration and damage); the collection and treatment prior to its final disposal and discharge into the sewer system, rivers, and sea; and the possible environment impact upon the water use cycle (e.g., contamination).

The water use cycle, from capture from the public distribution system up to disposal, has implications for the health of the people and the community, Environmental and economic implications are inherent in the use of water. Unfortunately, water is one of the most wasted resources, and studies on the costs of water wastage are rare. Unjustified water loss costs can be high and can include water that has been treated and conditioned in order to serve specific needs of the different health care facility clinical and support services. The water-management plan should include responses to emergency situations relating to contaminated supply, pipe system failure, sewage system malfunction, and a water rationing program.

Planning

The components and management of the water system should be defined and established when the health facility is planned and constructed. The remodeling or expansion of the physical plant presents a good opportunity to update and optimize the water system. If problems with the supply and the water quality for the different services or in critical circumstances exist, establish an emergency project to improve the water system and its management.

The size of the health facility, its level of specialization, the clinical services it provides, the installed medical and industrial equipment, and the environmental conditions of the area where the facility is located determine the amount and characteristics of water needed. The patterns of consumption and the class of treatment that should be given to the water before and after its use and prior to its final disposal to the public sewerage system should be also considered.

When planning a facility, evaluate the availability and options for water supply. Water is obtained from a public or municipal supply system, a private supplier, or a facility owned well. The first two options are most common in urban centers. The water is usually available 24 hours per day and has adequate levels of pressure and potability, making it suitable for immediate distribution and consumption within the facility. The third option is more common in rural areas or in remote regions without access to a public or municipal water supply system. In these cases, the implemented water supply system must guarantee the continuous availability of potable water at a pressure required for the operation of the facility.

Once the water supply source is defined, determine the chemical, physical, and biological characteristics of the water at the facility entry point. Depending on the values obtained by laboratory tests, and considering the foreseen uses by the different clinical and support areas of the facility, the water must be treated and conditioned to fit specific needs and demands. This is especially important in countries that do not have regulatory agencies or the capacity to monitor and control the water system.

Design Parameters

The study of the water demand should be conducted for each individual health care facility, but some basic indicators of consumption patterns can influence the design of the facility. As a rule, a regional hospital with an average of 250 beds providing four basic specialties will have a daily water consumption ranging from 450 to 600 liters per bed (PAHO/CEPIS, 1996). The estimate is adjusted upward if the health care facility includes services or equipment with high water demand.

The storage capacity for a health care facility depends on the specific local conditions related to water availability and distribution. The facility should have a storage capacity of three or four days in the event of a breakage or disruption in the main water line connected to the public network. If the facility does not have an adequate water supply on a regular basis, or is located in a region subject to natural disasters, the storage capacity should be increased. The facilities should have additional storage capacities when located in regions that are subject to long periods of drought, earthquakes, hurricanes, and cyclones, or prone to fire.

In designing a building's water distribution system or piping system, areas and services that are vital for the operation of the facility will require access to reliable water even during emergency situations or main supply water breakage or disruptions. These areas should be clearly defined, and the internal supply system should be designed and constructed so that these areas always have access to water. Points of access to water should also be provided in the noncritical sectors.

To guarantee adequate operation of the different areas and services in the facility, the water system should provide water in quantity, with the flow and pressure that people and equipment require. The operation of some equipment requires specific levels of pressure and flow, and the equipment will not function if these levels are not achieved or maintained. For example, some sterilizers have valves operated by water pressure. If the valves do not have the required water pressure for activation, the equipment will function erratically, or not at all.

If the public water supply system does not provide adequate pressure, or if necessary pressure can not be produced by gravity at the facility, some areas or sectors will not have enough pressure and flow. In such cases, alternate systems must be used to provide pressure and flow to the facility distribution system. This situation is more critical in taller buildings than in shorter ones.

Various engineering solutions can compensate for a lack of water pressure and flow. In health care facilities, especially those more horizontal than vertical, the use of elevated tanks is the most cost-effective way to achieve adequate pressure and flow. Elevated tanks should have 25%–30% of the total water storage capacity of the facility (FNH, 1981).

If the characteristics, location, or safety limitations do not allow water storage in elevated tanks, other options include the use of pressure tanks or hydro-pneumatic systems that provide pressure and flow as required at different areas of the facility.

Leaks are among the most common causes of water wastage. They happen due to the lack of an adequate preventive maintenance program to detect water leaks, decreased flow of water at delivery points, or a sudden and unexplained increase in water consumption by equipment or areas in the facility.

Water Treatment and Conditioning

Water provided by a municipal or public supply system is treated to meet local, national, or international standards. In the health care facility, water needs additional treatment and conditioning to meet the specific needs of the different users or services. The treatment procedure allows modification of the chemical composition of the water to minimize or eliminate chemical and biological contamination. This is of utmost importance in the health care facility, since some of the water will be used in clinical procedures, in direct contact with patients.

Before reaching the facility, the water has been treated by the municipality or the public water supply agency. Water from a facility-owned well should follow the same process as that used at municipality treatment plants. The following steps are the most common in the treatment process:

- Use of screens to remove large particulates at the intake point of the water supply.
- Reduction of turbidity and suspended matter. This part of the process is known as "clarification" and includes the addition of chemicals and the use of filters to remove particles larger than 25 microns.
- Reduction of the level of calcium and magnesium in a process known as "lime softening." The purpose of lime softening is to minimize the hardness of the water and further clarify it.
- Use of chlorine gas to disinfect water with the objective of killing bacteria. The chlorine level must be constantly monitored to ensure that no harmful levels reach the population.
- Adjust the water pH. Adjusted pH to between 7.5 and 8.0 prevents corrosion of water pipes and prevents leaching of lead into the water supply.

Once the water is delivered to the facility, it requires further treatment and conditioning to meet specific needs. The processes needed for further treatment very with the end-user water requirements. Some treatment processes minimize risks and protect people, while others protect equipment. Some of the more common steps in on-site treatment and conditioning are as follows:

- Removal of particulates larger than 20 microns using sediment filtration, followed by removal of particulates larger than 10 microns using multimedia filtration.
- Adjustment of the water pH to prevent damage of pH sensitive materials. For example, adjustment of the pH to a level between 8.3 and 9.0 to prevent acid corrosion in boilers.
- Adding chemicals to soften the water in order to prevent hardness and the deposition of calcium, magnesium, ferrous, manganese and aluminum.
- Removal of chloramines, chlorines, and other low molecular weight organic chemicals using carbon beds.

Options to treat and condition water to suit specific needs are as follows:

- Use of filters. Sand filters remove turbidity but can not stop smaller impurities from passing through. Calcium carbonate calcite medium filters (also known as neutralizing fitters) neutralize low-pH water. Activated carbon filters absorb low-molecular-weight organic chemicals and reduce chlorine or other halogens from water, but they do not remove any salts. There are several types of disposable filters to trap fine particles in the range of 1 to 100 microns, and ultra-filtration solutions remove particles in the range of 0.005 to 0.15 microns.
- Reverse osmosis. Reverse osmosis (RO) is a crossflow membrane separation process that removes virtually all organic compounds (Osmonics, 1992). A large selection of RO membranes is available to remove most particles and microorganisms.

For specific clinical procedures such as hemodialysis, water must be further conditioned for patient use (ECRI, 1999). Some methods for additional conditioning are as follows:

- Removal of salts, bacteria, pyrogens, metal ions, and particles using RO filters.
- Removal of charged particles such as metals and ions using deionization.
- Removal of bacteria, pyrogens, and particles larger than 0.05 microns using ultra-filtration.
- Use of ultraviolet light to kill microorganisms.

The main centers of water consumption should be determined based on the volume of water to be treated and the different requirements of end users. For example, water treatment plants must be built for water to be used in boilers for steam production and water used in the laundry service. However, for use in laboratories and clinical areas, Water can be obtained by filtering, adding chemicals, or via distillation or the use of other special equipment.

In order to prevent outbreaks and nosocomial infections, protocols for water use should be part of a risk management program. Some common problems include infections from rinsing equipment with potable water, rinsing burn patients with tap water, and preparing baby formulas with tap water. A water quality control system should be implemented to avoid these problems.

End Users

Most health care facilities, have six major areas of water consumption: Sanitation; heating, ventilation, and air conditioning (HVAC); clinical and medical; laundry; food service; and miscellaneous uses. Sanitation and HVAC consume approximately 60% of the water in most facilities. Water use should not include the water required for the fire extinguishing system. This water is normally raw or untreated and it is stored in a separate tank.

The main end-users of water are patients and health care workers; clinical and support services; and the building itself. Most of the water used by patients and health care workers is for drinking and personal hygiene.

Water is used in many clinical services areas including the following:

- Clinical and pathology laboratories: Laboratories perform many operations with different types of equipment. Water is used and discharged by automated laboratory equipment, water baths, emergency showers, eyewash stations, and sinks. Some of the common pollutants in wastewater include solvents, mercury (Eppstein, 2000), zinc, and other heavy metals, strong acids and toxic chemicals, radionuclides, proteins, blood products, and body fluids.
- X-ray department: X-ray film processors need water. Ammonia and developer products are the most common chemical pollutants in film processor wastewater.

- Dental services: The hand-held tools in the dental care unit are the main users of water. The equipment water lines and cold wastewater from these tools become contaminated with microorganisms.
- Respiratory therapy department: Respiratory care equipment, nebulizers, and humidifiers are the devices that use the most water.
- Central supply and sterilization department: Water is used to clean and disinfect surgical instruments in conjunction with ethylene oxide sterilization. Some of the more common effluent pollutants are soap, detergent, disinfectants, and ethylene oxide.
- Operating theaters and emergency rooms: Water is used in surgical scrub sinks, where blood, body fluids, and glutaraldehyde can pollute the wastewater.
- Other areas with significant water usage include the pharmacy, for the preparation of formulas and parenteral solutions, and the physiotherapy unit, for hydrotherapy equipment. Both of these areas are susceptible to infection and contamination.

Water is used in the following support services areas:

- HVAC system: Cooling towers and humidifiers are the heavy consumers of non-potable water.
- Boiler House: Boilers and hot water tanks are the greatest water consumers in this area. Water treatment chemicals and oils are common pollutants.
- Laundry service: Washing machines are large consumers of water. An efficient operational policy supported by a preventive maintenance program can save water and the costs associated with its use and discharge in the laundry. Some of the more common pollutants are soaps, detergents, and bleaches are common pollutants.
- Food services and kitchen: Sinks, dishwashers, ice makers, and food preparation are users of water. Soaps, detergents, disinfectants, solvents, pesticides, and cleaning solutions are among the common pollutants.
- Medical waste incinerators: The most common pollutants are heavy metals and particulate mater.

Water used in the physical plant includes the following areas:

- Sanitary: Sinks, toilets, showers, and bathtubs are some of the appliances that consume water. Improper cleaning procedures will allow bacterial growth and contamination.
- Housekeeping: Cleaning and disinfecting will generate pollutants from soaps, detergents, cleaners, solvents, and disinfectants.
- Physical plant maintenance activities will generate pollutants from paint, adhesives, soaps, detergents, solvents, and oils.

Wastewater

Wastewater collected from different facility services and end-users carries a variety of chemical and biological pollutants, several of which are hazardous. Wastewater should be included in a facility's water management and water quality program and should be closely monitored and pretreated prior to its discharge into the public or municipal sewer system. Some developed countries have pretreatment standards and guidelines for health care facilities' wastewater; others apply pretreatment standards established for industrial waste disposal. Developing countries might need to strengthen their capacities for monitoring and controlling health care facilities' wastewater.

Hemodialysis

Hemodialysis is a clinical procedure for patients with chronic renal failure. The procedure partially replaces the kidney function in case of kidney failure by removing bodily fluids and toxins that result from normal metabolic processes. In dialysis, blood is pumped and circulated through an external device called a "hemodialysis unit" (ECRI, 1996) and is kept separated by a semipermeable membrane, from a solution that allows the removal of toxins and excess water from the patient's body fluids without damaging the blood.

The hemodialysis unit consists of three main components: The dialyzer, or semipermeable membrane, that allows the separation of the impurities from the blood; the dialystate system consisting of the solution (solution with electrolyte composition similar to the normal cellular fluid) and an associated circuit for monitoring and controlling the dialystate; and an extracorporeal blood circulation circuit that carries blood from the patient to the dialyzer and back to the patient (ECRI, 1991).

The dialystate, or electrolytic fluid, is customized to correct the body electrolyte disorders of the specific patient. Water is added to the concentrated dialysate to produce the appropriate formula. The preparation of the dialysate requires extremely pure water.

Studies conducted by ECRI (1996, 1999) indicate that patients in hemodialysis therapy are dialyzed an average of three times per week for three to four hours. During this time, the patient is exposed to approximately 360 liters of water per week—more than 25 times the amount of water a person normally ingests. The membrane or filter is the only barrier between the blood and the dialyzer. If the water used in the process has microorganisms or a high mineral content, there is a very high risk to patient's safety and well being.

The Association for the Advancement of Medical Instrumentation (AAMI) has prepared chemical and bacteriological standards and recommended practices for dialysis (AAMI, 1998). The standard includes the "water quality for dialysis" and provides maximum allowable levels of contaminants. In order to obtain the required levels of quality and purity, water must be treated as described in the water treatment and conditioning section.

For example, the maximum level of aluminum allowed in drinking water by the United States Environmental Protection Agency (EPA) is between 0.05 to 0.2 mg/L. By comparison, the maximum level of aluminum in water for hemodialysis, determined by the AAMI, is 0.01 mg/L. Exceeding the specified levels of aluminum in hemodialysis water could harm the patient. Possible adverse effects include anemia, bone disease, encephalopathy, and other neurological disorders. Hemodialysis departments should have a quality control program to closely monitor and control water contaminants. (see Table 115-1).

Table 115-1 AAMI and EPA maximum levels of contaminants in water (AAMI, 1998; Health Devices, 1999)

Contaminant	AAMI standard for dialysis water, mg/L	EPA standard for drinking water, mg/L
	AAMI and EPA Maximum Levels of Contaminants Water	
Aluminum	0.01	0.05 to 0.2
Arsenic	0.005	0.05
Barium	0.1	2.0
Cadmium	0.001	0.005
Calcium	2.0	not regulated
Chloramines	0.1	4.0
Chorine	0.5	4.0
Chromium	0.014	0.1
Cooper	0.1	1.3
Fluoride	0.2	4.0
Lead	0.005	0.015
Magnesium	4.0	not regulated
Mercury	0.0002	0.002
Nitrate	2.0	10.0
Potassium	8.0	
Selenium	0.09	0.05
Silver	0.005	0.10
Sodium	70.0	not regulated
Sulfate	100	400
Zinc	0.1	5.0

Conclusions

Water systems have profound implications for the performance of health care facilities and the expected outcomes of health care processes and procedures. Facilities should have an adequate management and quality control system for used and discharged water in order to prevent problems for people, equipment, and the environment. From the health perspective, water is one of the main sources of nosocomial infections (Rutala and Weber, 1997) and from the engineering perspective, one of the main causes for equipment deterioration and breakdown.

Water is a strategic energy resource that should be used and managed efficiently. Water waste or improper usage and disposal have a negative environmental impact and an adverse economic effect because of the resource expended in the costly process of treatment and conditioning.

Protection of the environment is an increasing concern worldwide. Governments and society are looking more closely into issues that negatively affect the environment. Usage and disposal of water in health care facilities is one of those issues, especially for developing countries that have but where health care is part of a small industrial sector but is one of the larger consumers of water.

Acknowledgement

The author thanks his friend and colleague, Antonio Hernandez, for the considerable time and effort he expended in assisting with the writing of this chapter. Antonio is Regional Advisor on Health Services, Pan American Health Organization/World Health Organization (PAHO/WHO), Washington, DC

References

AAMI. Water Quality for Dialysis. *AAMI Standards and Recommended Practices* 3, 1998.
ECRI. Water Quality for Hemodialysis Centers. *Health Devices Systems* 28(11), 1999.
ECRI. Hemodialysis Machines. *Health Devices Systems* 20(6), 1991.
ECRI. Dialysis: Risk Analyses. *HRC* 4. January 1996.
Eppstein D. *MASCO*. Boston, Mercury Work Group, 2000
Fondo Nacional Hospitalario/Ministerio de Salud. Aspectos Básicos para el Diseño de Areas Específicas en Instituciones Hospitalarias. Bogota, Colombia, *VI Seminario Nacional I Simposio Internacional de Arquitectura Hospitalaria*, 1981.
Pan American Health Organization. Curso de Saneaminto Ambiental Intrahospitalario. Lima, Peru, PAHO, 1996.
Rutala WA, Weber DJ. Water as a Reservoir of Nosocomial Pathogens. *Infection Control and Hospital Epidemiology J* 18:, 1997.
Osmonics, Inc. *Pure Water Osmonics Handbook.*, MN, Osmonics Inc., 1992.

Further Information

AAMI/ANSI, American National Standard for Hemodialysis Systems. 1992.
American Hospital Association/American Society for Health care Engineering. *Plant Services Management for Health care Facilities.*, AHA, 1990.
American Hospital Association/American Society for Health care Engineering. *Waste Management for Health care Facilities.*, AHA, 2003.
American Hospital Association/American Society for Health care Risk Management. *Risk Management Handbook for Health care Facilities.*, AHA, 1990.

American Institute of Architects. *Guidelines for Design and Construction of Hospitals and Health care Facilities.*, AIA, 2001.
Department of Human Services, Public Health Division. Melbourne, Victoria.
Hospital Infections, 4th Edition. Philadelphia, Lippincott-Raven, 1990.
FDA. Manual on Water Treatment for Hemodialysis. Rockville, MD, FDA, 1997.
Murphy JC. Materials Compatibility for Ozone. *Water Conditioning and Purification* 40(5): 1998.
National Fire Protection Association. *Health Care Facilities Handbook, 6th Edition.* Quincy, MA, NFPA, 1999.
University of Kentucky. *Standard Operation Procedure for Dialysis Water System Maintenance Program.* Louisville, KY, University of Kentucky, 2000.
LaBarge S. RO: How Does it Figure into a Dialysis Water System? *Contemporary Dialysis and Nephrology Journal*, 1990.
Wagner KD. Environmental Management in Health care Facilities. *Water Supply and Wastewater Discharge.*, WB Saunders, 1998.
Fresenius Medical Care. *What is Dialysis?* Fresenius Medical Care North America.

116

Disaster Planning

Gary H. Harding
Greener Pastures, Inc.
Durango, CO

Alice L. Epstein
CNA HealthPro
Durango, CO

In today's uncertain environment, every organization should have a disaster preparedness or emergency preparedness plan. While recent events such as the attacks on the World Trade Center surely have brought attention to the need for disaster plans in private and public facilities across the country, the health care industry might be slightly ahead of others in preparedness efforts because the involvement of health care providers as a resource and support for handling disasters preceded recent events by centuries.

The U.S. government has required disaster plans in health care facilities since 1967. They have since evolved into Emergency Management Plans, which encompass disaster preparedness, planning, mitigation, and recovery. While the safety officer might have had, and might continue to have, primary responsibility for the plan, certainly much greater emphasis on the part of administrative management and affected departments (e.g., epidemiology, infection control, radiation safety, and emergency medicine) has been mandated by need and regulation. Clinical engineers can, and should, assume primary responsibility for portions of the disaster plan that are directly related to their service and should be prepared to assume back-up responsibility for other related activities. Training and experience to become, and remain, an effective clinical engineer provides a strong base on which to provide support and to assume direct responsibility if the need arises. Lessons learned in facility management theory and operation, critical thinking, logistics, product acquisition, and product support can be applied to many other services if the disaster dictates. Yet, in time of need, we all should remember that clinical engineers might just need to be another pair of hands, a Good Samaritan, or a friendly body willing to listen or act.

The Engineer Professional Advisory Committee of the U.S. Public Health Service decided in 2000 that managers responsible for the organization of the Public Health Service response to disasters and emergencies needed a primer on the capabilities of engineering disciplines to facilitate the inclusion of engineers in emergency situations. The Handbook developed states, "They (engineers) possess strong problem solving and organizational skills. They have the analytical and technical skills to evaluate problems, develop solutions, and bring those solutions to reality . . . While engineers specialize, their capabilities tend to transcend their specialization." (USPHS, 2000).

This handbook identifies (for managers) and promotes the different types of engineering duties and tasks that might be best solved by an engineer. The hope of the authors of the handbook is that every engineer who participates in health care-related emergencies, regardless of specialization, will become familiar with (and preferably certified through) training in the following areas:

- Administrative services
- Information services
- Patient record management
- Communications systems, emergency communication frequencies
- FEMA, NDMS
- Disaster planning
- Medical product logistics
- Hazardous materials management
- Water and waste management treatment systems and testing
- Facility management and utility systems
- Structural integrity
- Emergency level clinical skills, including mandatory cardio-pulmonary resuscitation (CPR) and first aid certification, with emergency medical technician (EMT) certification desirable
- Field management of mass casualties (USPHS, 2000).

By definition, disaster plans can not be all encompassing. Unfortunately, no expert can forecast all types or special nuances of possible disasters, internal or external. The purposes of the overall plan are to identify potential disaster scenarios and to offer contingency responses, while maintaining flexibility to allow professionals to modify or improvise to address any nuances that actually occur. The objectives of the disaster plan should reflect the purpose and goals of the organization. If the organization delivers medical care, examples of objectives could include the following:

- Attending promptly and effectively to all individuals requiring medical attention
- Protect patients, visitors, and staff from injury
- Protect property, facility, and equipment
- Correlating with, and participating in, community disaster plans
- Outlining each department's responsibilities
- Preparing the staff and hospital resources for optimal performance in an emergency
- Satisfying all applicable regulatory requirements

Emergency situations and/or disasters that have a high degree of probability of one or more occurring within the community or the hospital are:

- Fire
- Chemical spill with toxic release—chemical accident
- Major traffic accident
- Multiple casualty incident
- Tornado
- Earthquake
- Blizzard (snow or inclement weather)
- Terrorist activity
- Airport disaster
- Bomb threat
- Industrial accident
- Building or structure collapse
- Riot
- Hostage situation
- Protestors
- Chemical hazards
- Biohazards

The intent of this chapter is to review the responsibilities of clinical engineering in the disaster planning, implementation, and recovery processes. It is not intended to provide a detailed disaster plan. The full disaster plan should be developed in conjunction with all internal and external participants (e.g., regional public entities, utility management services, police, fire, and emergency management). The following sections discuss issues surrounding planning, implementation, and recovery from known disaster types, with the objectives of the health care organization in mind, but specifically focusing on the role of the clinical engineer.

Framework—Emergency Preparedness Regulations and Accreditation Requirements

Governmental Bodies

The basic governmental framework under which disaster planning is expected to occur is being developed as a result of recent events. Creation of a new, cabinet-level agency, the U.S. Department of Homeland Security (DHS), has occurred and is expected to oversee federal involvement "in the event of terrorist attack, national disaster, or other large-scale emergency." The framework and method of implementation is currently under construction, although it appears that DHS intends to manage efforts centrally, but also to delegate responsibilities to a number of other federal agencies(e.g., the Centers for Disease Control and Prevention, the Office of Emergency Preparedness, and the Federal Emergency Management Agency) and, ultimately, to each state.

Actual federal laws and regulations mandating disaster planning by health care entities are slim. Typically, they are vague and not easily accessed, understood, or particularly enlightening. Title III of the Superfund Amendments and Reauthorization Act of 1976 mandated that every state form local emergency planning committees (LEPCs) and a state emergency response commission. This remains the basic tenet under which the federal government delegates responsibility for disasters to states.

An additional federal requirement for developing and implementing disaster planning in health care facilities is buried within the Medicare Conditions of Participation (at one time, under the auspices of the Health Care Financing Administration (HCFA), now termed the Centers for Medicare & Medicaid Services (CMS)), requiring health care facilities that receive federal Medicare/Medicaid funding to have an emergency preparedness program. CMS surveyors address staff preparation for emergencies and disasters, querying staff as to what they would do in the event of an emergency and evaluating staff responses against what surveyors consider the correct answers. Surveyors question the availability of emergency power and whether staff are aware of the electrical outlets that are powered by emergency power.

DHS offers disaster assistance under The Robert T. Stafford Disaster Relief and Emergency Assistance Act, which historically has been administered by the Federal Emergency Management Agency (FEMA). In the event of an emergency, the governor of a state determines that the features of the disaster are beyond the abilities of the state to handle; the then governor contacts the president to request relief and, if granted, financial, administrative, and other federal resources are made available. However, new changes include multiagency participation such as the CDC's National Pharmaceutical Stockpile, which ensures "the availability and rapid deployment of life-saving pharmaceuticals, antidotes, other medical supplies and equipment..." While the program suggests that it was developed in response to the September 11, 2001 terrorists attacks the agency states that the stockpile includes those items that are "necessary to counter the effects of nerve agents, biological pathogens, and chemical agents." A 12-hour "Push Package" is noted as being readily available, although tailored packages, noted as Vendor Managed Inventory (VMI), also can be requested. Unfortunately, the itemization of what is included in the package or able to be requested under the VMI is not provided in an easily accessible format. Accordingly, whether the CDC expects to offer packages for all types of disasters versus just nerve agent, biological , toxin or chemical disasters, and what exactly is available in the way of medical supplies and equipment remains unclear.

Also embedded within the federal government's agency function is the Department of Health and Human Services (DHHS) Office of Emergency Preparedness (OEP). It is charged with being the lead agency for health and medical response to terrorist activities, natural disasters, transportation disasters, and technological disasters. The OEP directs and manages the National Disaster Medical System (NDMS), a cooperative, asset sharing partnership among DHHS, the Department of Defense, the Department of Veterans Affairs, FEMA, state and local governments, private businesses, and civilian volunteers. The framework under which this agency works, how it is accessed, what it can provide, how it can provide it, and how it is differentiated from other agencies are under construction. Currently, very little useful information related to disaster planning is provided. Directions for participation require contacting their LEPC directly. While the NDMS offerings are somewhat nebulous, health care facilities voluntarily participate by providing bed availability information to a local coordinating center, agreeing to treat NDMS patients, and participating in NDMS exercises.

The Occupational Safety and Health Administration provides regulations for employers, including health care entities. 29 CFR § 1910 requires health care facilities to address certain man-made disasters for the protection of employees. An internal (employee-safety) disaster plan must be developed and must include emergency escape procedures, procedures for employees who must stay to perform critical tasks, and procedures to account for all employees after the emergency.

The Environmental Protection Agency (EPA) under the Emergency Planning and Community Right to Know Act (42 USC §11001) discusses the release of hazardous substances that can cause an emergency. It requires states to establish an emergency response commission and HAZWOPER plan. Hospitals that are designated to receive victims of hazardous substance emergencies must have an emergency response plan, decontamination capabilities, personal protective equipment, and trained employees.

Many state and local agencies that have jurisdiction, (e.g., fire departments) mandate compliance with National Fire Protection Association (NFPA) standards. NFPA 99 (Standard for Health Care Facilities) and NFPA 101 (Life Safety Code) include requirements for an emergency management plan. Within the Life Safety Code, all health care facilities are required to keep written copies of a plan for the protection of all persons in the event of a fire, evacuation to an area of refuge, and evacuation from the building when necessary. A copy of the plan must be readily available in the telephone operator's location or security center. The written plan must include the use of alarms, the transmission of alarms to the fire department, response to alarms, isolation of the fire, evacuation of immediate the area and smoke compartment, and extinguishment of the fire.

Requirements and authority to develop and implement disaster plans, conduct training and exercises, and establish mutual aid agreements can be found in most state department of health regulations, as well as in health care licensing organizations, accreditation standards, and organizational bylaws and policies. Engineers should review state-specific information because it is likely to be more detailed and, thereby, more useful. Following is a synopsis of the standards.

JCAHO

The Joint Commission for Accreditation of Healthcare Organizations (JCAHO) modified its standards in 2001 to include emergency management and community involvement concepts. Organizations were directed to assume an "all hazards" approach to disaster planning, with a responsibility to review, analyze, and address all hazards that are likely to occur and to be a serious threat to the community. While JCAHO has always required preparation for disasters, JCAHO mandated after September 11, 2001, that organizations communicate and coordinate with one another (effective since January 2003).

The JCAHO requirements apply to the full range of hospitals from small rural facilities to academic medical centers in urban locations. The preparedness standards are focused on the following four areas:

1. Emergency preparedness management plan
2. Security management plan
3. Hazardous materials and waste management plan
4. Emergency preparedness drills

JCAHO identifies an emergency as a natural or man-made event that suddenly or significantly disrupts the environment of care or actual care and treatment, as well as changes or increased demands for the organization's services. Four phases (mitigation, preparedness, response, and recovery) must be addressed. The organization also must participate in two drills annually, preferably community wide.

JCAHO recommends that the plan be general, but also that it provide the flexibility to allow responses specific to disaster type; separate plans for each disaster type are discouraged. Because JCAHO believes that disasters will occur at the local level, they emphasize that organizations might be on their own for 24–72 hours following the disaster. As a result, emergency medical services, fire departments, police, public health agencies, and local leaders must be prepared to respond.

JCAHO standards for the "Emergency Preparedness Management Plan" require hospitals to address the following:

- Specific procedures in response to a variety of disasters or emergencies
- Role in community wide plan
- Role of external authorities
- Space, supplies, security
- Radioactive or chemical isolation & decontamination
- Notifying and assigning personnel
- Evacuation and alternative care site
- Managing patients
- Backup for utilities and communication
- Orientation and education, including emergency procedures in the event of equipment failure
- Performance monitoring
- Annual evaluation

It is important for clinical engineers to address each element as it applies to their department and individual staff members. JCAHO surveyors schedule on-site assessments and will examine how the organization develops, implements, and improves the emergency management plan. They will examine whether a variety of possible events have been identified and addressed and whether staff members understand their roles and responsibilities. A new emphasis on managing the logistics of critical supplies and transportation of equipment critical for the clinical engineer.

Basics of Disaster Planning and Mitigation

Recognizing now that there is a responsibility to "be prepared," the following sections discuss methods that can be used in planning, and some of the specific hazards that are likely to be encountered. Remember that the discussion does not represent a detailed discussion of these matters and that it must be supplemented by local, state and other resources. There is no single hierarchy for emergency management structure (e.g., committees, terminology, and decision tree) that is applicable to all states, organizations, or situations. Expect the process, participants, methods, and most other aspects of emergency management to vary. However, one basic methodology (Hazard Vulnerability Analysis) does apply under most circumstances.

Hazard Vulnerability Analysis

JCAHO requires accredited health care facilities to assess proactively their vulnerabilities in order to improve disaster readiness. The analysis should be performed annually or amended if there are significant changes that could affect the analysis. The facility should be using the information gathered from the Hazard Vulnerability Analysis to refine its emergency preparedness plan, based upon logistics specific to the organization. A full vulnerability analysis requires the skills and knowledge of many different professionals. The clinical engineer can contribute to the assessment in collaboration with, among others, structural engineers, radiation specialists, plumbers, and electricians.

The Pan American Health Organization (PAHO), a regional office of the World Health Organization (WHO), offers an authoritative disaster mitigation book that includes algorithms to conduct hazard vulnerability analyses and to determine priorities (PAHO, 2000). Vulnerabilities are separated into the following categories: Structural, medical and support equipment and installations, and administrative and organizational. Vulnerability assessments can be used to improve disaster preparedness. WHO stresses the importance of disaster mitigation through the adoption of preventive measures.

Assessing the hazards and vulnerability of the organization first requires calculating the probability that a disaster will occur. Factors to consider in assessing probability include impact of geography, proximity to hazardous industries, disaster history, warning systems, and social or political unrest. Following an assessment of disaster probability, the team should estimate the potential damage, loss, and harm that could result. While there are a myriad of potential negative impacts, primary categories of damage include human, structural/property, loss of mechanical and electrical systems, financial, and operational.

Structural assessment begins with a visual inspection of the facilities. (While not a part of the analysis, mitigation could start immediately by repairing existing structural problems and deficiencies that are noted.) Assessment of the vulnerabilities of medical and support equipment and installations is an area in which clinical engineering contribute significantly. Are there sufficient medical devices on hand and adequate if evacuation becomes necessary? For example, will fixed equipment stand the stress of an earthquake? Can existing critical life support equipment function or are there medical alternatives readily available in the event of power loss or evacuation? Financial vulnerability includes the cost of lost business, the cost of providing emergency services, cash reserves on hand for the purchase of emergency supplies, and adequate insurance. Operational vulnerability includes the loss of essential personnel, management survivability, and crisis/media management.

Recognized areas of vulnerability of importance to the clinical engineer include the following.

Electricity

Specific to essential or critical life support equipment in areas such as nursing, pharmacy, blood bank, central suction, central sterile supply, surgical suite, labor/delivery, intensive care units, nursery, cardiac catheterization laboratory, emergency department, radiology, treatment and patient bed space

Emergency Generator

Emergency outlet identification and testing, length of backup electrical power duration, limitations on fuel availability and delivery, service to piped medical gas alarms, cardiology alarms, dedicated lines to emergency department (ED), operating room (OR), obstetrical suite (OB), and intensive care unit (ICU), blood storage, central vacuum systems, medical air compressors, and availability of extension cords

Water Supplies

Vacuum systems (if connected to water), sterilization, parental systems, radiology film development equipment, dialysis, portable suction options

Gas systems

Availability of portable cylinders, gas shutoff, alarms for medical gas supplies, alarms for accidental release of ethylene oxide and other hazardous gases, loss of piped oxygen due to bulk supply failure, automatic transfer to reserve oxygen supply system, low pressure alarm signal

Communications

Wire operated, radio wave operated (e.g. radios, pagers, televisions, telemetry, cellular phones, handheld PCs) availability of batteries for all systems, availability of PC not on hospital network, public address systems

Medical Devices

In addition to a current inventory of equipment in use, the clinical engineer also should know the levels of equipment that are stockpiled and ready for use in the event of an emergency. Are there alternative means to provide the necessary clinical equipment or support and/or how many days would it take to acquire same? Critical equipment includes but is not limited to:

- Anesthesia equipment
- Autoclave/sterilizers
- Blood bank freezers
- Cardiology/respiratory monitors
- CT scanners
- Defibrillators
- Dialysis equipment
- EEG monitors
- EtO sterilizers
- Infant incubators
- Infusion pumps and controllers
- Laboratory equipment
- Laparoscopy equipment
- Laryngoscopes with extra batteries
- Manual resuscitators
- MRI
- OR tables
- Orthopedic instruments, such as hand-powered drills and saws
- Oxygen concentrators
- Plate developers and processors
- Portable oxygen
- Pulmonary function analyzer
- Pulse oximeters
- Respirators
- Sterilizers
- Suction machine/pump
- Traction equipment
- Ultrasound equipment
- Ventilators (adult and pediatric)
- X-ray equipment

Algorithms are not the only process to assess vulnerability. Organizations, including the American Hospital Association, suggest questions to measure vulnerability. Following are some questions that could be asked:

What could go wrong that is under the domain of your department?

- Is there anything (e.g., equipment and service) under your domain that could trigger or contribute to a disaster?
- If utilities failed, how would the failure interrupt services for which you are responsible? Will utility service failure trigger backup power? Has the emergency power/utility crossover been tested for each piece of life-support equipment?
- If demand for equipment and service requests increases, is there a plan for acquiring increased equipment and parts that identifies sources and capacity?
- In the event of a large-scale disaster, is there a method and decision algorithm for allocating and/or re-allocating scarce critical equipment (e.g., ventilators)?
- If supplies and parts did not arrive when needed, how would clinical engineering cope?
- Where are backup and replacement equipment to come from? Who is responsible for arranging and triggering delivery?
- Are internal systems in place to notify suppliers of an emergency situation? Have the mechanics of delivery been worked out for the duration of the disaster?
- Are there backup supplies identified for critical equipment and components?

Hundreds of additional questions could be asked. Part of the plan development process certainly includes all involved parties examining their own departments/functions with respect to intra- and interdepartmental issues that might arise. Asking and answering these questions can improve the actual response in the event of disaster.

Departmental response is important to plan and understand. Chain of command and delineation of span of control are as important at the department level as they are in top management. It is important that all department members understand the role of clinical engineering during a disaster, as well as their specific responsibilities and that these may change as needs of the crisis vary. Management should keep a current roster of clinical engineering emergency contact telephone numbers. A contact plan should be developed and detailed in the disaster plan.

Clinical engineering should be responsible for development of their section of the disaster plan. Often departments develop their own disaster plans under the umbrella of the facility plan, with standard operating procedures specific to clinical engineering. It is imperative that the department-specific plans are not developed outside the umbrella and oversight of the complete plan. It is important, for example, that other departments (e.g., nursing and respiratory therapy) understand and have in writing the conditions under which they should contact clinical engineering. Specific examples can help individuals within those other departments to respond more appropriately and effectively to disasters. As a result, clinical engineering should contribute to the development of such departmental plans, policies, and procedures.

Similarly, other departments that have information or a possible impact on clinical engineering should participate in the development of the clinical engineering departmental plan. One of the most important relationships that clinical engineering should forge as part of the disaster plan is with facilities engineering. There are so many cross-over issues and combined knowledge bases between these two departments that a close relationship, understanding of function and interaction, and methods of cooperation are essential.

It is imperative that the clinical engineering contribution to each aspect of the plan be reviewed, understood, and approved by all parties affected by the plan. It should be understood that this contribution includes the disaster planning, implementation, and recovery processes.

Following are discussions relating to specific disasters, issues, and clinical engineering.

Fire

Clinical engineering does not typically bear primary responsibility for fire response. However, there are specific aspects of fires that could involve clinical engineering. For example in an internal disaster, equipment under the clinical engineer's control could have been the cause of the fire or could become involved as the fire propagates. Similarly, a fire could require the evacuation of parts or all of the facility and the ability to continue providing critical support during and subsequent to evacuation. Maintaining an awareness of equipment that poses a recognized risk of fire, steps that can minimize the likelihood, and procedures to follow in the event of an equipment-related fire should be developed. The role of clinical engineering in the event of a fire and evacuation should be determined, including: Whether evacuation of a critical care unit is required; the location, availability, and provision of battery-operated equipment, e.g., transport incubators, or alternative manual modes of support with related equipment, e.g., manual resuscitators, should be identified. It will not suffice for the clinical department's disaster plan to specify, for example, a switch to manual resuscitators if, in fact, those resuscitators are not readily available under a known protocol.

Natural Disasters–Earthquakes, Tornadoes, Hurricanes, Typhoons

Some natural disasters offer the opportunity for early initiation of the disaster plan. For example, hurricanes typically must develop over time, and authorities can provide warning levels accordingly. Some, like earthquakes, occur without warning. While our experience in the U.S. with natural disasters has been characterized by significant property damage, but limited casualties, mass casualties are a distinct possibility. In 2000, the American Hospital Association, with the support of the Office of Emergency Preparedness, issued *Hospital Preparedness for Mass Casualties*. Along with some review and background information, this work provides recommendations for community-wide preparedness, staffing, communications and public policy. It also provides an appendix: "Suggested Issues for Hospital Mass Casualty Preparedness" that includes over five pages of listed issues to be considered. These issues could be operational (e.g., coordination of volunteers); equipment-related (e.g., stockpiling of supplies); problematic (e.g., a weak relationship between hospital and health department); or financial (e.g., a hospital bearing costs of preparedness for events that originate outside its campus). A number of these issues (e.g., stockpiling of equipment,) could directly involve clinical engineering.

Man-Made Disaster of Terrorism

Man's inhumanity to man has affected the United States in the Oklahoma City, New York City, and Pentagon disasters. Much has been learned from the emergency management experience surrounding these disasters. From a health care perspective, most of the lessons learned revolve around mass casualties, but as the sophistication and resources of terrorism increase, it seems only a matter of time before disasters related to the release of chemical, biological, or other agents or weapons of mass destruction may follow. Much work is currently being undertaken in planning for these types of disasters, but in 1999 the Association for Professionals in Infection Control and Epidemiology (APIC) and the CDC completed *Bioterrorism Readiness Plan: A Template for Healthcare Facilities*. This document provides an overview of activities relating to:

Reporting requirements and contact information

- Potential agents
- Detection of outbreaks
- Infection control practices for patient management
- Post-exposure management
- Laboratory support and confirmation
- Patient, visitor, and public information

Some of these activities (e.g., transport of patients with communicable disease) certainly can impact on clinical engineering. Examples include the need for equipment decontamination or the use of equipment within cohorted patient populations who present with similar symptoms.

The document also singles out and discusses anthrax, botulism, plague and smallpox. Description of the agent/syndrome; preventive measures; infection control practices for patient management; post-exposure management; laboratory support and confirmation; and patient, visitor, and public information for each is provided. References and resources (e.g., state-specific telephone and other contact information for the FBI and public health directors) are included.

Certainly, there are a great number of other types of disasters (e.g., chemical spills with toxic releases, major traffic accidents, airport disasters, industrial accidents, building or structural collapses, and riots) that must be identified and addressed individually during plan development. References specific to each of them should be identified and considered at that time.

Loss of Communication

Disaster can result in loss of communication within a health care facility. In such an event, external paging systems and cellular telephones can provide an alternative means of communication inside and outside of the facility. While many facilities have strict criteria to eliminate the use of cellular telephones within the structure, it can be essential to set aside this restriction in the event of a disaster that results in loss of conventional communication. If cellular telephone use must be approved (and, in fact, encouraged), potential users also should be counseled to maintain a distance of at least 1 meter from medical equipment. Disaster plans could specify locations in which there is little or no concern for cellular telephone use and interference. Other stand-alone products (e.g., two-way walkie-talkies) typically require greater distances (6–8 meters, or 20–25 feet) from instrumentation for safe use. Specific examples of devices that have been adversely affected by cellular telephone use include cardiac pacemakers (e.g., reversion to default mode) and ventilators (e.g., device shut-off). However, operation of cellular telephones in close proximity (i.e., under 1 meter) to most medical devices (e.g., monitors and infusion control devices) should be avoided because there remains the possibility of such events as parameter changes and alarm adjustment.

Biological and Chemical Contamination of Equipment

If equipment requires decontamination, it is imperative that strict protocols be followed. Check manufacturers' requirements and work with your infection control practitioner and radiation safety officer to develop protocols specific to the equipment.

When possible, noncritical patient care equipment should be dedicated to a single patient (or cohort of patients exposed to the same agent). If shared use of items is unavoidable on noncohort patients, all potentially contaminated, reusable equipment should be appropriately cleaned and reprocessed before use. Policies should be in place and monitored for compliance. Equipment and supplies contaminated with blood, body fluids, secretions, or excretions should be handled in accordance with Universal Precautions, the blood-borne pathogens standard, and handling protocols specific to the agent, to prevent staff or bystander exposures to skin and mucous membranes, contamination of clothing, and transfer of agents to staff, other patients, and other persons.

What Should the Clinical Engineer Do in the Event of a Disaster?

The primary responsibility of the clinical engineer in the event of a disaster is to implement the features of the plan that is specific to the department and the disaster. With no plan, or a poorly conceived or poorly understood plan, one can expect chaos. If the plan has been carefully thought out, contributions to mitigate the disaster will be apparent. Notifying emergency contacts, acquiring an up-to-the-minute inventory and location list, and ensuring function of critical life-support equipment will fall into place naturally. Clinical engineers should expect greater demand on staff and resources. Troubleshooting and problem solving are likely to be the watchwords, with flexibility to respond to the situation paramount.

Being prepared, to the extent practicable, to anticipate need, the potential problems that may result, and alternative methods to address both, can help to make the process of disaster mitigation more successful. Anticipating the need for evacuation and challenges to equipment and medical support can help avert situations in which no reasonable alternative is available or in which the impact of the need is underestimated. Clinical engineers can be a powerful resource, especially if they choose to receive other training in disaster-related services (e.g., CPR, disaster volunteering, or biohazard and chemical hazard response).

Recovery

Recovery is a major part of addressing a disaster. It involves addressing issues remaining after the disaster has been mitigated. The cost for services and replacement of structures that might occur during recovery often involve FEMA and the identification of a disaster area by the president. As noted previously, the governor of the state(s) in which the disaster(s) occur must request the assistance on behalf of those who have suffered as a result of the disaster, according to the requirements of the Stafford Act,.

Private reimbursement for loss may be available through insurers. Insurance policies must have been in place prior to the disaster and must cover appropriate exposures (e.g., property, flood, business interruption, earthquake, terrorism/acts of war, equipment replacement, and equipment rental expenses) in order for many of the financial ramifications of disaster to be addressed. Assuming that the insurance purchasing authority for the organization has properly secured insurance, it is important to have up-to-date equipment inventories with replacement values assigned to each piece of equipment. Inventories including photographs of all fixed and possibly expensive) equipment should note the age of the equipment and the expected life cycle. A master log and tracking system for all equipment that leaves the building during a disaster with patients and staff should be maintained. Locating displaced equipment can be almost impossible without a tracking system. Accurately record all expenses. This information will be necessary in order to secure insurance, FEMA, and state reimbursement.

Recovery assessment should be performed as soon as possible after the event. The first assessment should be a safety assessment of the structural integrity of the buildings. Clinical engineering should complete an inventory of all medical equipment for damage assessments. Take pictures of all equipment prior to cleanup. Decisions will need to be made regarding the repair of equipment. Identify the condition of existing equipment and develop a strategy for replacement/repair of equipment, what can be repaired cost effectively, and who should be performing the repair. Evaluate the cost and requirements of in-house versus contracted repair services. Track the repairs, cost, and time expended. Evaluate the effects of sudden power loss when bringing equipment back online. All settings and alarms must be tested before patient use. Document decisions to repair or discard equipment.

The engineering department should be debriefed as a team on the status of the facilities equipment, the staff's contribution to the disaster efforts, and next steps. In order to capture information about how effective staff activities were during the disaster, it is important to document the disaster-related activities of the department staff. A secondary benefit from this process might be that equipment that has not been located might be captured in staff activity documents. Ask the team to identify efforts of disaster assistance that could benefit from changes or improvement. Identify what worked well.

Richter (2003) identifies a number of common postdisaster problems that can occur. Some of the following can impact on clinical engineers:

- Failure of water pressure
- Failure of backflow protection systems
- Failure of emergency generators, air conditioning, and public utility systems
- Difficulties with special needs patients (e.g., ventilator, dialysis)
- Detrimental effect on operating systems (e.g., technology availability) due to volume of patients, evacuees, and residents

- Failure of telecommunications systems/staffing systems
- Flooding of mechanical rooms, (e.g., patient floors, elevator shafts)
- Waste management problems
- Loss of equipment on, and damage to, hospital roofs
- Obstruction from debris
- Inability to secure electronic doors and alarm systems manually

Conclusion

Disasters are no longer events that occur to others. They are far more diversified from the standard flood, earthquake, and natural disasters of the past. In the recent past, development of disaster plans might have been viewed as necessary to comply with federal, local, and licensing authority regulations, but recent man-made disasters have shown all of us that these events can occur anywhere and at any time. Thoroughly and thoughtfully addressing planning, mitigation, and recovery issues on a comprehensive, detailed basis can save lives.

Clinical engineers can expect disasters to place great demands on staff and resources. Reliance on simpler, alternative means of therapy and support can be necessary in some cases. Activating federal, state, and local support channels as rapidly as possible (and consistent with the needs of the disaster) will be important. Special challenges might need to be faced, and perhaps hard decisions will need to be made. Emergency management plans can be a useful base from which to operate and make decisions as long as the plan has been properly thought out and staff understand and follow to the extent that practicable protocols are defined within the plan.

Fortunately, clinical engineers have a strong academic and experiential base on which to participate in disaster planning, mitigation and recovery, as well as direct portions of it. Success depends on the willingness of the clinical engineer to not only offer suggestions, but also to receive them–both offered and received in a constructive, open forum for the benefit of those who could suffer in a disaster. While preplanning and a strong sense and knowledge of responsibility are important during a disaster, it is important to remember that all of us just might need to be an extra set of hands, a willing listener, or prepared to assume primary responsibility for a task not typically within our purview, but necessitated by events.

References

PAHO. *Emergency Preparedness and Disaster Relief Coordination Program,* Disaster Mitigation Series. Washington, DC, Pan American Health Organization, 2000.

Richter PV. *Hospital Disaster Preparedness: Meeting a Requirement or Preparing for the Worst?* American Society for Health care Engineering, http://www.hospitalconnect.com/ashe/currentevent/disready_hospdisasterpreparednesstechdocgp.html, 2003.

USPHS. *Public Health Service Engineering Capabilities During Disaster Responses: Handbook for Deploying Appropriate Public Health Service Engineers.* Washington, DC, U.S. Public Health Service, www.usphsengineers.org, 2000.

Further Information

USDHHS. *Hospital Preparedness for Mass Casualties.* Washington, DC, U.S. Department of Health and Human Services Office of Emergency Preparedness, 2000.

Association for Professionals in Infection Control and Epidemiology, Inc. and Center for the Study of Bioterrorism and Emerging Infections. *Mass Casualty Disaster Plan Checklist: A Template for Health care Facilities.* (http://www.apic.org/bioterror/checklist.doc)

APIC. *Bioterrorism Readiness Plan: A Template for Health care Facilities.*, APIC Bioterrorism Task Force, 1999.

ECRI. *Emergency Preparedness Hazard Vulnerability Assessment.* Plymouth Meeting, PA, ECRI, 2002.

AHA. *American Hospital Association Chemical and Bioterrorism Preparedness Checklist.*, AHA, 2001.

PAHO. *Principles of Disaster Mitigation in Health Facilities*, Disaster Mitigation Series. Washington, DC, PAHO, 2000.

VHA. *Veterans Health Administration Emergency Management Strategic Health care Group and Emergency Management Program Guidebook.* Washington, DC, Veterans Health Administration, http://www.va.gov/emshg/ and http://www.va.gov/emshg/emp/emp.htm.

Auf Der Heide E. *Disaster Response: Principles of Preparation and Coordination.*, http://216.202.128.19/dr/flash.htm, 1989.

JCAHO. *Facts about the Emergency Management Standards.* Chicago, JCAHO, http://www.jcaho.org/accredited+organizations/hospitals/standards/ems+facts.htm.

JCAHO. *Health Care at the Crossroads: Strategies for Creating and Sustaining Community-Wide Emergency Preparedness Systems.* Chicago, JCAHO, http://www.jcaho.org/newsb+room/news+release+archives/emergency+preparedness.pdf.

Section XI

Medical Device Standards, Regulations, and the Law

David A. Simmons
Health Care Engineering, Inc.
Glen Allen, VA

The health care community in general and clinical engineering community in particular face a daunting challenge in the form of countless laws, statutes, regulations, standards, and guidelines. Our qualified and experienced contributing authors provide a robust examination of the many types of laws, regulations, standards, and guidelines, both national and international. It is important to begin by differentiating between the several categories of documents. Laws can be national (federal), state, or local, as in city or town. In these situations, laws are in effect and must be complied with throughout the governing area, e.g., nation, state, or local municipality. While there are no international laws regarding medical devices, there are international standards in place that pertain to the control and service of medical devices. One such standard is ISO 9001:2000, containing the first health care quality management system standard, IWA1. However, standards do not have the effects of laws, since they become compliance documents only when "an authority having jurisdiction" requires compliance. One example would be when a city requires compliance with one or more of the National Fire Protection Association (NFPA) standards.

The first chapter is a primer for understanding the differences between standards, regulations, and laws, as well as the need for standards and the standards development process. It addresses conformity assessment for national and international standards. The next chapter, "Regulation and Technology Assessment Agencies," provides insight into 43,000 mandatory and voluntary applicable standards, clinical guidelines, laws, and regulations promulgated by more than 1400 organizations and 300 state and federal agencies. It provides a comprehensive list of agencies and organizations that have the most significant clinical engineering and technology involvement. Also included is a list of worldwide technology assessment organizations. The chapter, "Hospital Facilities Safety Standards," discusses the roles of several organizations that promulgate standards for hospital facilities. These include the Centers for Medicare and Medicaid Services (CMS), formerly Health Care Finance Administration (HCFA), Occupational Safety and Health Administration (OSHA), National Fire Protection Association (NFPA), and Joint Commission on Accreditation of Healthcare Organizations (JCAHO). This is a comprehensive comparison of organizations and standards. JCAHO and its hospital-accreditation program, described in the next chapter, provide essential information for the clinical engineers to understand their roles with respect to accreditation. Following that, the chapter "Medical Equipment Management Program and ANSI/AAMI EQ56," details the development and the substance of a recommended practice document, ANSI/AAMI EQ56, entitled, "Recommended Practice for a Medical Equipment Management Program." This standard is the culmination of a collaborative effort of JCAHO, the American National Standards Institute (ANSI), and the Association for the Advancement of Medical Instrumentation (AAMI). The EQ56 recommends a program that addresses all aspects of a medical device's life cycle, from device selection to retirement within a health care organization. Both EQ56 and ISO 9000, discussed in a previous chapter, allow each organization to develop its own program. Both documents embody similar philosophies: EQ56's is "Say what you do, and then do what you say," and ISO 9000's is "Say what you do, do what you say, prove it, improve it."

The authors of the chapter "Clinical Engineering Standards for Canada," compare the Canadian Council on Health Services Accreditation (CCHSA) to its United States counterpart, JCAHO. They provide a detailed analysis of the standards development process in Canada, including a section on the elements of the Canadian standard. Peer review is also discussed at length and is worth considering in the United States and elsewhere. "Regulations and the Law" deals with risk management and the regulation of medical devices. Stages of regulatory controls are addressed. The chapter entitled "The EU Medical Device Directives and Vigilance System" describes three primary areas, or directives, for the European Community. These are active implantable medical devices, medical devices, and *in vitro* diagnostics. The directives focus on the essential requirements that constitute the fundamentals of compliance. The description of the two European standards bodies, CENELEC and CEN, leads into a discussion of the harmonization of standards and the European Vigilance System.

The chapter "United States Food and Drug Administration" is a comprehensive and detailed look at the FDA. This is required reading for any clinical engineer to understand the impact and importance of the FDA involvement with medical devices. It is one of the most comprehensive documents regarding FDA activities that this author has seen. Finally, the chapter "Tort Liability for Clinical Engineers and Device Manufacturers" is a must for clinical engineers and manufacturers. The chapter addresses preparation for JCAHO surveys and changes in an organization's culture. The material also addresses continuous quality-improvement programs and operating-cost reductions, as well as my personal favorite, eliminating the triennial ramp-up for JCAHO surveys.

117

Primer on Standards and Regulations

Michael Cheng
Health Technology Management
Ottawa, Ontario, Canada

The work of clinical engineers has a direct effect on health, safety and the environment. These public health domains are heavily guided by standards and regulations. This chapter provides a brief overview of standards and regulations with special reference to medical devices. Terms such as standards and regulations often are used loosely. More formal meanings of these terms or understandings from current usage among regulatory professionals are given here. One should keep in mind that the fields of standards, in general, and regulations of medical devices, in particular, are evolving rapidly. This chapter provides a frame of reference for other chapters in this section of the handbook on Standards, Regulations and the Law. Pointers will be made where a term or topic introduced is described in other chapters. In other chapters in this handbook, regulations and the law; regulating medical devices; the Global Harmonization Task Force (see Chapter 124), and post-market surveillance and vigilance on medical devices (see Chapter 139) will be discussed. A general understanding of the different aspects of standards is presented followed by reference to biomedical standards and the importance of clinical engineering participation in standards development.

Standards Defined

Various technical definitions of standards exist, but it would be difficult to find one that can satisfy all applications of the term. Following are three meanings for "standard" taken from the *Webster's International Dictionary*, and a fourth definition from the International Standardization Organization (ISO):

1. Something that is established by authority, custom, or general consent as a model or example to be followed;
2. A definite level or degree of quality that is proper and adequate for a specific purpose;
3. Something that is set up and established by authority as a rule for the measure of quantity, weight, extent, value, or quality; and
4. Documented agreements containing technical specifications or other precise criteria to be used consistently as rules, guidelines, or definitions of characteristics, to ensure that materials, products, processes, and services are fit for their purpose.

Standards can establish a wide range of specifications for products, processes, and services.

- Prescriptive specifications obligate product characteristics; for example, device dimensions, biomaterials, test or calibration procedures as well as definitions of terms and terminologies.
- Design specifications set out the specific design or technical characteristics of a product, for example, operating room facilities or medical gas systems.
- Performance specifications ensure that a product meets a prescribed test; for example, strength requirements, measurement accuracy, battery capacity, or maximum defibrillator energy.
- Management specifications set out requirements for the processes and procedures companies put in place, such as for quality systems for manufacturing or environmental management systems.

A standard might contain a combination of specifications. Performance, prescriptive, and design specifications have existed for many years. Management specifications, however, are rapidly gaining prominence.

Recent years have seen the development and application of what are known as "generic management system standards," where "generic" means that the standards' requirements can be applied to any organization, regardless of the product it makes (or whether the "product" is actually a service activity), and "management system" refers to what the organization does to manage its processes. Two of the most widely known series of international standards falling into this category are the ISO 9000 series (see Chapter 119) for managing quality systems, and the ISO 14000 series for environmental management systems. Wide-ranging information and assistance relating to these standards and their application is available on the web site www.ISO.org. The ISO quality systems standards for medical device manufacturing are the ISO13485 in conjunction with ISO9001 and ISO13488 in conjunction with ISO9002.

One might hear terms such as "outcome-oriented standards," "objectives standards," "function-focused standards," and "results-oriented standards." Essentially, these terms indicate that the standards specify the objectives, or "ends," to be achieved, while leaving the methods, or "means," to the implementers. This can minimize possible constrictive effects of standards (see Limitation of Standards below). One management standard for medical equipment is the ANSI/AAMI EQ56 described by Ethan Hertz (see Chapter 122), which also uses a results-oriented approach.

Need for Standards

Standards can serve the following different purposes:

- Providing specifications or other criteria that a product, process, or service must meet.
- Providing information that enhances safety, reliability, and performance of the products, processes, and services.
- Assuring consumers about reliability or other characteristics of goods or services provided in the marketplace.
- Giving consumers more choice by allowing one firm's products to be substituted for, or combined with, those of another.

Although one often takes for granted the advantage of being able to order shoes or clothes simply by referring to a size, this is only possible because manufacturers follow some industrial standards in making shoes and clothes. In contrast, incompatibility between electrical plugs and receptacles is a prime example of different countries not following the same standard. When North Americans want to use a portable computer or other electrical appliance in Europe or Asia, they find that the plug and voltage are not compatible, and this can be a frustrating experience. Clinical engineers are well aware of incompatible consumables or replacement parts in medical devices of similar function, made by different manufacturers. The lack of standard consumables and repair parts is an important cause of medical equipment problems in developing countries.

With the world becoming a global village, the need for, and benefits of, standardization are becoming more and more important internationally for manufacturing, trade, and communications. Quality systems and other management standards can provide common references to the kind of process, services, or management practice expected. Global communication will be difficult without international standardization. The Internet functions because of globally agreed-upon interconnection protocols, for example

Most medical devices, like drugs, are health care products made available globally through trade or international co-operation. The safety, effectiveness, and consistent quality of medical devices are now international public health issues. Therefore, global standardization of medical device standards and regulations has become increasingly important.

Standards Vs. Regulations

Most standards are voluntary. However, a standard may be mandated by a company, a professional society, an industry, or a government. A standard may be called a "regulation" when it is mandatory. The mandate might or might not be legally based.

When a standard is mandated by a government, it normally becomes legally obligatory based on regulations or a law established by the government. Governments in Canada, the European Union, and the United States have recently established policies to recognize hundreds of voluntary standards that are applicable to medical device regulations. These standards, however, are not mandatory. Manufacturers can choose between conformance with the recognized standards or other means to demonstrate product compliance with regulatory requirements.

Standards Development Process

A standard can be set and mandated by an authority. The current trend, however, is to adopt standards established by consensus from all interested parties (the stakeholders). In general, to be accepted voluntarily, standards should have the following attributes:

- Their development should be overseen by a recognized body to ensure that the process is transparent and not dominated by undue interests.
- The development process should be open to input from all interested parties and the resulting document should be based on consensus. Consensus, in a practical sense, means that substantial agreement is reached by interested parties involved in the preparation of a standard. This includes steps taken to resolve all objections and implies much more than the concept of a simple majority, but not necessarily unanimity.
- Good technical standards should be based on the consolidated results in science; and technology and experience and aimed at the promotion of optimum community benefits.
- Standards should not hinder innovations and should be reviewed periodically to remain updated with technological advances.

For many standards development organizations, the designation of the term "standard" to a document requires specific conditions. Figure 117-1 shows the twelve steps taken by creditable standards development organizations in developing and maintaining a standard.

This formal process, particularly the consensus steps 4, 5 and 6, is the norm for the resulting document to be recognized as a standard. There are many types of "standard-like" documents from standard organizations, but the documents may not have wide acceptance as standards because their development procedures did not satisfied the above norm. For example, technical specification (TS) and technical report (TR) from ISO and Technical Information Report (TIR) from the Association for the Advancement of Medical Instrumentation (AAMI) are standards-like documents. The reader should refer to the definitions for such types of documents given by the parent organizations.

Conformity Assessment with Standards

The four common industrial methods for assessing conformity to a standard are listed below:

1. A product's conformity to standards is commonly assessed by direct testing.
2. A process can be assessed by audit. Certification organizations or regulatory authorities attest that products or processes conform to a standard by authorizing the display of their certification mark.
3. The conformity to a management standard by an organization is known as "management systems registration," a relatively new term used primarily in North America. Formally established audit procedures are followed by certified auditors who are supported by technical experts of the domain under audit. Management system registration bodies (registrars) issue registration certificates to companies that meet a management standard such as the ISO9000 family, or to medical device manufacturers that meet the ISO13485/ISO9001 standards. Note that in North America, the term "registration" is used for an organization, while "certification" is reserved for products. Many other countries use "certification" for both a product and an organization.
4. Accreditation is used in Canada and the United States for affirming that a health care organization meets requirements set by a service standard. Usually, the health care facility is visited and reviewed by a team of appointed peers. Accreditation is also used by national authorities to recognize third-party assessment agencies and standards-development and auditor-training organizations.

National and International Standards Systems

A country might have many voluntary standards bodies. However, normally there is one official national organization that coordinates and accredits the standards-development bodies in that country. This official national organization would have the authority to endorse a document as a national standard in accordance with official criteria, and it also represents the country in the international standards organizations. In the United States, the American National Standards Institute (ANSI), a non-profit organization, is such an organization. In Canada, it is the Standards Council of Canada (SCC), a crown (i.e., government) organization. In Europe, it is a committee composed of Comité Européen de Normalisation (CEN), the European Committee for Electrotechnical Standardization (CENELEC) and the European Telecommunication Standards Institute (ETSI) that regroups the different previous European national standards bodies.

Three major international standardization organizations are the International Organization for Standardization (ISO), the International Electrotechnical Commission (IEC), and the International Telecommunication Union (ITU). Generally, ITU covers telecommunications, IEC covers electrical and electronic engineering, and ISO covers the rest. The work in information technology, risk management, quality systems, and many other areas carried out by a joint ISO/IEC technical committee.

There are other organizations that produce documents on international standardization. Their documents are usually adopted by ISO/IEC/ITU as international standards if those documents were developed in accordance with international consensus criteria. Any group of five member countries can propose a standard to the ISO for adoption as an international standard. Contact with these organizations can be made through the following websites: www.ISO.ch, www.IEC.ch, and www.itu.int/home/index.html. These web sites link to global standards.

Identification of Standards

Standards are generally designated by an alphabetical prefix and a number. The acronyms (e.g., ISO, IEC, ANSI, CAN, EN, and DIN) indicate the standard body that has approved them. The numbers identify the specific standard and the year in which it was finalized. The standards reference number quite frequently gives an indication of adoption where standards are equivalent. For example:

- CAN/CSA-Z386-94 refers to a standard developed in 1994 by the Canadian Standards Association (CSA), one of four accredited Canadian standards development organizations, and designated by the Standards Council of Canada (SCC) as a Canadian national standard.
- ANSI/AAMI/ISO 15223:2000 refers to the international standard ISO 15223 (established in 2000) adopted by AAMI, which in turn is designated by ANSI as an American national standard.
- UNI EN ISO 9001 indicates an Italian national standard (UNI) that is an adoption of a European standard (EN), which is an adoption of the International Standard, ISO9001.

Limitation of Standards

One major limitation is that standards are based mainly on retrospective experience, and this may inhibit innovation or limit progress. Furthermore, standards can cover only a limited scope at the time of development, and they could be inappropriate for unforeseen situations. Therefore, it is important to know the origin and background of the standards to be used. Table 117-1 presents a checklist in using a standard.

Another possible problem associated with standardization is that participation in the development of standards requires funding and time, which many individuals, small businesses, or organizations cannot afford. Therefore, standards development can fall under the strong influence of those who can afford the cost and time to participate (e.g., big businesses).

References for Biomedical Standards

Hundreds of biomedical standards exist, and the reader can identify most of them through relevant web sites. The richest source is ISO: www.ISO.ch. For American national standards, refer to www.ANSI.org, and for Canadian national standards, www.SCC.ca. For other national standards, non-national standards, or standard-like documents on medical devices, refer to the extensive sources provided by Bruley and Coates (Chapter 118). "The Guide to Biomedical Standards" by Aspen Publications, which has been published for two

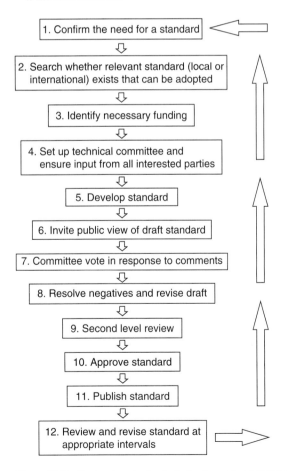

Figure 117-1 Twelve steps in standards development and maintenance.

decades, lists standards that concern medical devices and facilities throughout the world. While it is a handy hard copy reference, one must check the Web for the latest standards.

Increasingly, regulatory authorities recognize voluntary standards. Standards that are recognized by the United States Food and Drug Administration Center for Devices and Radiological Health (FDA CDRH) (see Chapter 126), www.fda.gov/cdrh/modact/fr0225ap.pdf, and those by Health Canada can be found at www.hc-sc.gc.ca/hpb-dgps/therapeut/htmleng/policy.html.

Conformity to such standards can serve as evidence in meeting certain regulatory requirements by the corresponding authorities.

Standards and the Clinical Engineer

As a professional user of standards, the clinical engineer should be aware of the limitations of standards (Table 117-1). A standard is a useful reference in decision making, but it should never replace professional judgment and responsible decision making.

Participation in Standards Development

Not so long ago, standardization was a field reserved for only a few specialists. Today, standardization has emerged as a major technical and commercial element. Clinical engineers must play an active role in standards development or be prepared to accept standardization that is established without them or without consideration of their interests. As a professional user of medical devices and manager of health care services, the clinical engineer makes important contributions in developing standards that govern medical devices and the health care environment.

Participation in standards development can be made directly by becoming involved with technical committees. Some standards development organizations will fund an expert's attendance at their technical committees. The clinical engineer also may seek funding from his or her employer or professional societies. Input can be made during the public consultation phase of the standards development process. (See step 6 in the standards development process (Figure 117-1).) The Internet has made such public access and input to documents convenient.

Standards development organizations normally are covered by insurance, which in turn should cover individual committee members. Those who are concerned about legal liability can make enquiries to the organization before participation.

Table 117-1 Checklist in Using a Standard

Who developed it?
How was it developed?
When was it developed?
When was it last revised?
When will it be revised?
Where has it been used?
Is it mandatory? According to whom?
What are possible limitations?
Why should it be used?

Further Information

Wald A. A Standards Primer for Clinical Engineers. In Bronzino JD (ed). *The Handbook of Biomedical Engineering*. Boca Raton, FL, CRC Press, 1995.
The Guide to Biomedical Standards. Gaithersburg, MD, Aspen Publishers,
International Standards Organization (ISO) www.ISO.ch.
American National Standards Institute (ANSI) www.ANSI.org.
Standards Council of Canada (SCC) www.SCC.ca.
The European Union (EU) http://europa.eu.int/comm/enterprise/medical_devices/index.htm.

118

Medical Device Regulatory and Technology Assessment Agencies

Mark E. Bruley
Vice President, Accident and Forensic Investigation ECRI,
Plymouth Meeting, PA

Vivian H. Coates
Vice President, Technology Assessement and Information Service, ECRI
Plymouth Meeting, PA

To be effective, clinical engineering department management and development requires a basic knowledge of relevant regulatory and technology assessment agencies. Regulatory agencies set standards of device performance, problem reporting, and recordkeeping for the clinical departments and technologies for which they are responsible. Furthermore, in the ever-expanding role of the clinical engineer in the technology decision making processes of hospital administration, the engineer must be aware of the technology assessment agencies that are valuable information resources in the rapidly developing, market-driven health care technology arena.

This chapter presents an overview of regulatory and technology assessment agencies in the United States, Canada, Europe, and Australia that are germane to clinical engineering. Because of the extremely large number of such agencies and information resources, this chapter focuses only on those of greatest relevance and/or informational value. The reader is directed to the references and sources of further information presented at the end of the chapter.

Regulatory Agencies

Within the health care field, there are over 43,000 applicable standards, clinical practice guidelines, laws, and regulations (ECRI, 2001). Voluntary standards are promulgated by more than 1,400 organizations; mandatory standards by more than 300 state and federal agencies. Many of these organizations and agencies issue guidelines that are relevant to the vast range of health care technologies within the responsibility of clinical engineering departments. Although many of these agencies also regulate the manufacture and clinical use of health care technology, such regulations are not directly germane to the management of a clinical department and are not presented.

For the clinical engineer, many agencies promulgate regulations and standards in areas including electrical safety, fire safety, technology management, occupational safety, radiology and nuclear medicine, clinical laboratories, infection control, anesthesia and respiratory equipment, power distribution, and medical-gas systems. In the U.S., medical device-

Copyright 2004, ECRI. All rights reserved.

problem reporting is also regulated by many state agencies and by the U.S. Food and Drug Administration (FDA) via its MEDWATCH program. It is important to note that, at present, the only direct regulatory authority that the FDA has over U.S. hospitals is in the reporting of medical device related accidents that result in serious injury or death. Other chapters in this section discuss in detail many of the specific agency citations. Presented below are the names and addresses of the primary agencies whose codes, standards, and regulations have the most direct bearing on clinical engineering and technology management.

American Hospital Association
One North Franklin
Chicago, IL 60606
(312) 422-3000
Web site: www.aha.org

American College of Clinical Engineering
5200 Butler Pike
Plymouth Meeting, PA 19462
(610) 825-6000
Web site: www.accenet.org

American College of Radiology
1891 Preston White Drive
Reston, VA 22091
(703) 648-8900
Web site: www.acr.org

American National Standards Institute
11 West 42nd Street
13th Floor
New York, NY 10036
(212) 642-4900
Web site: www.ansi.org

American Society for Hospital Engineering
840 North Lake Shore Drive
Chicago, IL 60611
(312) 280 5223
Web site: www.ashe.org

American Society for Testing and Materials
1916 Race Street
Philadelphia, PA 19103
(215) 299-5400
Web site: www.astm.org

Association for the Advancement of Medical Instrumentation
3330 Washington Boulevard
Suite 400
Arlington, VA 22201
(703) 525-4890
Web site: www.aami.org

Australian Institute of Health and Welfare
GPO Box 570
Canberra, ACT 2601
Australia
(61) 06-243-5092
Web site: www.aihw.gov.au

British Standards Institution
2 Park Street
London, W1A 2BS
United Kingdom
(44) 071-629-9000
Web site: www.bsi.org.uk

Canadian Healthcare Association
17 York Street
Ottawa, ON K1N 9J6
Canada
(613) 241-8005
Web site: www.canadian-healthcare.org

Canadian Standards Association
CSA International
178 Rexdale Boulevard
Etobicoke, ON M9W 1R3
Canada
(416) 747-4000
Web site: www.csa-international.org

Center for Devices and Radiological Health
U.S. Food and Drug Administration
9200 Corporate Boulevard
Rockville, MD 20850
(301) 443-4690
Web site: www.fda.gov/cdrh

Compressed Gas Association, Inc.
1725 Jefferson Davis Highway
Suite 1004
Arlington, VA 22202
(703) 412-0900
Web site: http://www.cganet.com

ECRI
5200 Butler Pike
Plymouth Meeting, PA 19462
(610) 825-6000
Fax: (610) 834-1275
Web sites: www.ecri.org
www.mdsr.ecri.org

Environmental Health Directorate
Health Protection Branch
Health Canada
Environmental Health Centre
19th Floor, Jeanne Mance Building
Tunney's Pasture
Ottawa, ON K1A 0L2
Canada
(613) 957-3143
Web site: www.hc-sc.gc.ca/hpb/index_e.html

Environmental Health Directorate
Therapeutic Products Programme
Health Canada
Holland Cross, Tower B
2nd Floor
1600 Scott Street
Address Locator #3102D1
Ottawa, ON K1A 1B6
(613) 954-0288
Web site: www.hc-sc.gc.ca/hpb-dgps/therapeut

European Committee for Standardization (CEN)
European Committee for Electrotechnical Standardization (CENELEC)
35 Rue de Stassart
B-1050 Brussels
+32(0)2.519.68.71
Web site: www.cenelec.org

Food and Drug Administration
MEDWATCH, FDA Medical Products
Reporting Program
5600 Fishers Lane
Rockville, MD 20857-9787
(800) 332-1088
Web site: www.fda.gov/cdrh/mdr.html

Institute of Electrical and Electronics Engineers
445 Hoes Lane
P.O. Box 1331
Piscataway, NJ 08850-1331
(732) 562-3800
Web site: www.standards.ieee.org

International Electrotechnical Commission
Box 131
3 rue de Varembe, CH 1211
Geneva 20
Switzerland
(41) 022-919-0211
Web site: www.iec.ch

International Organization for Standardization
1 rue de Varembe
Case postale 56, CH 1211
Geneva 20
Switzerland
(41) 022-749-0111
Web site: www.iso.ch

Joint Commission on the Accreditation of Health care Organizations
One Renaissance Boulevard
Oakbrook Terrace, IL 60181
(630) 792-5600
Web site: www.jcaho.org

Medical Devices Agency
Department of Health
Room 1209
Hannibal House
Elephant and Castle
London, SE1 6TQ

United Kingdom
(44) 171-972-8143
Web site: www.medical-devices.gov.uk

National Council on Radiation Protection and Measurements
7910 Woodmont Avenue, Suite 800
Bethesda, MD 20814
(310) 657-2652
Web site: www.ncrp.com

National Fire Protection Association
1 Batterymarch Park
PO Box 9101
Quincy, MA 02269-9101
(617) 770-3000
Web site: www.nfpa.org

Nuclear Regulatory Commission
11555 Rockville Pike
Rockville, MD 20852
(301) 492-7000
Web site: www.nrc.gov

Occupational Safety and Health Administration
U.S. Department of Labor
Office of Information and Consumer Affairs
200 Constitution Avenue, NW, Room N3647
Washington, DC 20210
(202) 219-8151
Web site: www.osha.gov

ORKI
National Institute for Hospital and Medical Engineering
Budapest dios arok 3, H-1125
Hungary
(33) 1-156-1522

Radiation Protection Branch
Environmental Health Directorate
Health Canada
775 Brookfield Road
Ottawa, ON K1A 1C1
Web site: www.hc-sc.gc.ca/ehp/ehd/rpb

Russian Scientific and Research Institute
Russian Public Health Ministry
EKRAN, 3 Kasatkina Street
Moscow
Russia 129301
(44) 071-405-3474

Society of Nuclear Medicine, Inc.
1850 Samuel Morse Drive
Reston, VA 20190-5316
(703) 708-9000
Web site: www.snm.org

Standards Association of Australia
PO Box 1055
Strathfield, NSW 2135
Australia
(61) 02-9746-4700
Web site: www.standards.org.au

Therapeutic Goods Administration
PO Box 100
Wooden, ACT 2606
Australia
(61) 2-6232-8610
Web site: www.health.gov.au/tga

Underwriters Laboratories, Inc.
333 Pfingsten Road
Northbrook, IL 60062-2096
(847) 272-8800
Web site: www.ul.com

VTT
Technical Research Center of Finland
Postbox 316
SF-33101 Tampere 10
Finland
(358) 31-163300
Web site: www.vti.fi

Technology Assessment Agencies

Technology assessment is the practical process of determining the value of a new or emerging technology in and of itself or against existing or competing technologies, using safety, efficacy, effectiveness, outcome, risk management, strategic, financial, and competitive criteria. Technology assessment also considers ethics and law as well as health priorities and cost-effectiveness compared to competing technologies. A "technology" is defined as devices, equipment, related software, drugs, biotechnologies, procedures, and therapies; and systems used to diagnose, treat, or monitor patients. The technology assessment process is discussed in Chapters 30, 32, and 33.

Technology assessment is not the same as technology acquisition/procurement or technology planning. The last two are processes for determining equipment vendors, soliciting bids, and systematically determining a hospital's technology-related needs based on strategic, financial, risk management, and clinical criteria. The informational needs differ greatly between technology assessment and the acquisition/procurement or planning processes. This section focuses on the resources that are applicable to technology assessment.

Worldwide, there are nearly 400 organizations (private, academic, and governmental) that provide technology assessment information, databases, or consulting services. Some are strictly information clearinghouses, some perform technology assessment, and some do both. For those that perform assessments, the quality of the information generated varies greatly from superficial studies to in-depth, well-referenced analytical reports. In 1997, the U.S. Department of Health and Human Services' Agency for Healthcare Research and Quality (AHRQ), formerly known as the Agency for Health Care Policy and Research, designated 12 "Evidence-based Practice Centers" (EPC) to undertake major technology assessment studies on a contract basis. Each of these EPCs is noted in the list below, and general descriptions of each center may be viewed on the AHRQ's web site: http://www.ahcpr.gov/clinic/epc.

Language limitations are a significant issue. In the final analysis, the ability to undertake technology assessment requires assimilating vast amounts of information, most of which exists only in the English language. Technology assessment studies published by the International Society for Technology Assessment in Health Care (ISTAHC), by the World Health Organization, and other umbrella organizations are generally written in English. The new International Health Technology Assessment database being developed by ECRI in conjunction with the U.S. National Library of Medicine contains more than 30,000 citations to technology assessments and related documents.

Below are the names, mailing addresses, and web addresses of some of the most prominent organizations that undertake technology assessment studies:

Agence Nationale pour le Developpement de l'Evaluation Medicale 159 Rue Nationale
Paris 75013
France
(33) 42-16-7272
Web site: www.upml.fr/andem/andem.htm

Agencia de Evaluacion de Technologias Sanitarias
Ministerio de Sanidad y Consumo
Instituto de Salud Carlos III, AETS
Sinesio Delgado 6, 28029 Madrid
Spain
(34) 1-323-4359
Web site: www.isciii.es/aets

Alberta Heritage Foundation for Medical Research
125 Manulife Place
10180-101 Street
Edmonton, AB T5J 345
(403) 423-5727
Web site: www.ahfmr.ab.ca

American Association of Preferred Provider Organizations
601 13th Street, NW
Suite 370 South
Washington, DC 20005
(202) 347-7600

American Academy of Neurology
1080 Montreal Avenue
St. Paul, MN 55116-2791
(612) 695-2716
Web site: www.aan.com

American College of Obstetricians and Gynecologists
409 12th Street, SW
Washington, DC 20024
(202) 863-2518
Web site: www.acog.org

Australian Institute of Health and Welfare
GPO Box 570
Canberra, ACT 2601
Australia
(61) 06-243-5092
Web site: www.aihw.gov.au

Battelle Medical Technology Assessment and Policy Research Center (MEDTAP)
901 D Street, SW
Washington, DC 20024
(202) 479-0500
Web site: www.battelle.org

Blue Cross and Blue Shield Association
Technology Evaluation Center
225 N Michigan Avenue
Chicago, IL 60601-7680
(312) 297-5530
(312) 297-6080 (publications)
Web site: www.bluecares.com/new/clinical
(An EPC of AHRQ)

British Columbia Office of Health Technology Assessment
Centre for Health Services & Policy Research
University of British Columbia
429-2194 Health Sciences Mall
Vancouver, BC V6T 1Z3
Canada
(604) 822-7049
Web site: www.chspr.ubc.ca

British Institute of Radiology
36 Portland Place
London, W1N 4AT
United Kingdom
(44) 171-580-4085
Web site: www.bir.org.uk

Canadian Coordinating Office for Health Technology Assessment
110-955 Green Valley Crescent
Ottawa ON K2C 3V4
Canada
(613) 226-2553
Web site: www.ccohta.ca

Canadian Health care Association
17 York Street
Ottawa, ON K1N 9J6
Canada
(613) 241-8005
Web site: www.canadian-health care.org

Catalan Agency for Health Technology Assessment
Travessera de les Corts 131-159
Pavello Avenue
Maria, 08028 Barcelona
Spain
(34) 93-227-29-00
Web site: www.aatm.es

Centre for Health Economics
University of York
York Y01 5DD
United Kingdom
(44) 01904-433718
Web site: www.york.ac.uk

Center for Medical Technology Assessment
Linköping University
5183 Linköping, Box 1026 (551-11)
Sweden
(46) 13-281-000

Center for Practice and Technology Assessment
Agency for Health care Research and Quality (AHRQ)
Suite 501
2101 East Jefferson Street
Rockville, MD 20852
(301) 594-4015
Web site: www.ahrq.gov
(301) 594-1364

Committee for Evaluation and Diffusion of Innovative Technologies
3 Avenue Victoria
Paris 75004
France
(33) 1-40-273-109

Conseil d'evaluation des technologies de la sante du Quebec
201 Cremazie Boulevard East
Bur 1.01
Montreal, PQ H2M 1L2
Canada
(514) 873-2563
Web site: www.msss.gouv.qc.ca

Danish Hospital Institute
Landermaerket 10
Copenhagen K
Denmark DK1119
(45) 33-11-5777

Danish Medical Research Council
Bredgade 43
1260 Copenhagen
Denmark
(45) 33-92-9700

Danish National Board of Health
Amaliegade 13
PO Box 2020
Copenhagen K
Denmark DK1012
(45) 35-26-5400

Duke Center for Clinical Health Policy Research
Duke University Medical Center
2200 West Main Street, Suite 320
Durham, NC 27705
(919) 286-3399
Web site: www.clinipol.mc.duke.edu
(An EPC of AHRQ)

ECRI
5200 Butler Pike
Plymouth Meeting, PA 19462
(610) 825-6000
(610) 834-1275 fax
Web sites: www.ecri.org
            www.mdsr.ecri.org
(An EPC of AHRQ)

Finnish Office for Health Care Technology Assessment
PO Box 220
FIN-00531 Helsinki
Finland
(35) 89-3967-2296
Web site: www.stakes.fi/finohta

Frost and Sullivan, Inc.
106 Fulton Street
New York, NY 10038-2786
(212) 233-1080
Web site: www.frost.com

Health Council of the Netherlands
PO Box 1236
2280 CE
Rijswijk
The Netherlands
(31) 70-340-7520

Health Services Directorate Strategies and Systems for Health
Health Promotion
Health Promotion and Programs Branch
Health Canada
1915B Tunney's Pasture
Ottawa, ON K1A 1B4
Canada
(613) 954-8629
Web site: www.hc-sc.gc.ca/hppb/hpol

Health Technology Advisory Committee
121 East 7th Place, Suite 400
PO Box 64975
St. Paul, MN 55164-6358
(612) 282-6358

Hong Kong Institute of Engineers
9/F Island Centre
No. 1 Great George Street
Causeway Bay
Hong Kong

Institute for Clinical PET
7100-A Manchester Boulevard
Suite 300
Alexandria, VA 22310
(703) 924-6650
Web site: www.icpet.org

Institute for Clinical Systems Integration
8009 34th Avenue South
Minneapolis, MN 55425

(612) 883-7999
Web site: www.icsi.org

Institute for Health Policy Analysis
8401 Colesville Road, Suite 500
Silver Spring, MD 20910
(301) 565-4216

Institute of Medicine (U.S.)
National Academy of Sciences
2101 Constitution Avenue, NW
Washington, DC 20418
(202) 334-2352
Web site: www.nas.edu/iom

International Network of Agencies for Health Technology Assessment
c/o SBU, Box 16158
S-103 24 Stockholm
Sweden
(46) 08-611-1913
Web site: http://www.inahta.org

Johns Hopkins Evidence-based Practice Center
The Johns Hopkins Medical Institutions
2020 E Monument Street, Suite 2-600
Baltimore, MD 21205-2223
(410) 955-6953
Web site: www.jhsph.edu/Departments/Epi/

(An EPC of AHRQ)
McMaster University Evidence-Based Practice Center
1200 Main Street West, Room 3H7
Hamilton, ON L8N 3Z5
Canada
(905) 525-9140 ext. 22520
Web site: http://hiru.mcmaster.ca.epc/
(An EPC of AHRQ)

Medical Alley
1550 Utica Avenue, South
Suite 725
Minneapolis, MN 55416
(612) 542-3077
Web site: www.medicalalley.org

Medical Devices Agency
Department of Health
Room 1209
Hannibal House
Elephant and Castle
London, SE1 6TQ
United Kingdom
(44) 171-972-8143
Web site: www.medical-devices.gov.uk

Medical Technology Practice Patterns Institute
4733 Bethesda Avenue, Suite 510
Bethesda, MD 20814
(301) 652-4005
Web site: www.mtppi.org

MEDTAP International
7101 Wisconsin Avenue, Suite 600
Bethesda MD 20814
(301) 654-9729
Web site: www.medtap.com

MetaWorks, Inc.
470 Atlantic Avenue
Boston, MA 02210
(617) 368-3573 ext. 206
Web site: www.metawork.com
(An EPC of AHRQ)

National Institute of Nursing Research, NIH
31 Center Drive, Room 5B10. MSC 2178
Bethesda, MD 20892-2178
(301) 496-0207
Web site: www.nih.gov/ninr

National Commission on Quality Assurance
2000 L Street NW, Suite 500
Washington. DC 20036
(202) 955-3500
Web site: www.ncqa.org

National Committee of Clinical Laboratory Standards (NCCLS)
940 West Valley Road, Suite 1400

Wayne, PA 19087-1898
(610) 688-0100
Web site: www.nccls.org

National Coordinating Center for Health Technology Assessment
Boldrewood (Mailpoint 728)
Univ of Southampton SO16 7PX
United Kingdom
(44) 170-359-5642
Web site: www.soton.ac.uk/~hta/address.htm

National Health and Medical Research Council
GPO Box 9848
Canberra, ACT
Australia
(61) 06-289-7019

New England Medical Center
Center for Clinical Evidence Synthesis
Division of Clinical Research
750 Washington Street, Box 63
Boston, MA 02111
(617) 636-5133
Web site: www.nemc.org/medicine/ccr/cces.htm
(An EPC of AHRQ)

New York State Department of Health
Tower Building
Empire State Plaza
Albany, NY 12237
(518) 474-7354
Web site: www.health.state.ny.us

NHS Centre for Reviews and Dissemination
University of York
York Y01 5DD
United Kingdom
(44) 01-904-433634
Web site: www.york.ac.uk

Office of Medical Applications of Research
NIH Consensus Program Information Service
PO Box 2577
Kensington, MD 20891
(301) 231-8083
Web site: http://odp.od.nih.gov/consensus

Ontario Ministry of Health
Hepburn Block
80 Grosvenor Street
10th Floor
Toronto, ON M7A 2C4
(416) 327-4377

Oregon Health Sciences University
Division of Medical Informatics and Outcomes Research
3181 SW Sam Jackson Park Road
Portland, OR 97201-3098
(503) 494-4277
Web site: www.ohsu.edu/epc
(An EPC of AHRQ)

Pan American Health Organization
525 23rd Street NW
Washington, DC 20037-2895
(202) 974-3222
Web site: www.paho.org

Physician Payment Review Commission (PPRC)
2120 L Street NW
Suite 510
Washington, DC 20037
(202) 653-7220

Prudential Insurance Company of America
Health Care Operations and Research Division
56 N Livingston Avenue
Roseland, NJ 07068
(201) 716-3870

Research Triangle Institute
3040 Cornwallis Road
PO Box 12194
Research Triangle Park, NC 27709-2194
(919) 541-6512
(919) 541-7480
Web site: www.rti.org/epc/
(An EPC of AHRQ)

San Antonio Evidence-based Practice Center
University of Texas Health Sciences Center
Department of Medicine
7703 Floyd Curl Drive
San Antonio, TX 78284-7879
(210) 617-5190
Web site: www.uthscsa.edu/
(An EPC of AHRQ)

Saskatchewan Health
Acute and Emergency Services Branch
3475 Albert Street
Regina, SK S4S 6X6
(306) 787-3656

Scottish Health Purchasing
Information Centre
Summerfield House
2 Eday Road
Aberdeen AB15 6RE
Scotland

(44) 0-1224-663-456, ext. 75246
Web site: www.nahat.net/shpic

Servicio de Evaluacion de Technologias Sanitarias
Duque de Wellington 2
E01010 Vitoria-Gasteiz
Spain
(94) 518-9250
E-mail: osteba-san@ej-gv.es

Society of Critical Care Medicine
8101 E Kaiser Boulevard
Suite 300
Anaheim, CA 92808-2259
(714) 282-6000
Web site: www.sccm.org

Swedish Council on Technology Assessment in Health Care
Box 16158
S-103 24 Stockholm
Sweden
(46) 08-611-1913
Web site: www.sbu.se

Southern California EPC-RAND
1700 Main Street
Santa Monica, CA 90401
(310) 393-0411 ext. 6669
Web site: www.rand.org/organization/health/epc/
(An EPC of AHRQ)

Swiss Institute for Public Health
Health Technology Programme
Pfrundweg 14
CH-5001 Aarau
Switzerland
(41) 064-247-161

TNO Prevention and Health
PO Box 2215
2301 CE Leiden
The Netherlands
(31) 71-518-1818
Web site: www.tno.nl/instit/pg/index.html

University HealthSystem Consortium
2001 Spring Road, Suite 700
Oak Brook, IL 60523
(630) 954-1700
Web site: www.uhc.edu

University of Leeds
School of Public Health
30 Hyde Terrace
Leeds L52 9LN

United Kingdom
Web site: www.leeds.ac.uk

USCF-Stanford University EPC
University of California, San Francisco
505 Parnassus Avenue, Room M-1490
Box 0132
San Francisco, CA 94143-0132
(415) 476-2564
Web site: www.stanford.edu/group/epc/
(An EPC of AHRQ)

U.S. Office of Technology Assessment
(former address) 600 Pennsylvania Avenue SE
Washington, DC 20003

Note: OTA closed on September 29, 1995. However, a complete set of OTA publications is available on CD-ROM; contact the U.S. Government Printing Office (www.gpo.gov) for more information.

Veterans Administration Technology Assessment Program
VA Medical Center (152M)
150 S Huntington Avenue
Building 4
Boston, MA 02130
(617) 278-4469
Web site: www.va.gov/resdev

Voluntary Hospitals of America, Inc.
220 East Boulevard
Irving, TX 75014
(214) 830-0000

Wessex Institute of Health Research and Development
Boldrewood Medical School
Bassett Crescent East
Highfield, Southampton SO16 7PX
United Kingdom
(44) 01-703-595-661
Web site: www.soton.ac.uk/~wi/index.html

World Health Organization
Distribution Sales
CH 1211
Geneva 27
Switzerland 2476
(41) 22-791-2111
Web site: www.who.ch
Note: Publications are also available
from the WHO Publications Center, USA,
at (518) 436-9686.

References

ECRI. *Health care Standards Official Directory*. Plymouth Meeting, PA, ECRI, 2001.
Eddy DM. *A Manual for Assessing Health Practices & Designing Practice Policies: The Explicit Approach*. Philadelphia, American College of Physicians, 1992.
Goodman C. *Medical Technology Assessment Directory*. Washington, DC, National Academy Press, 1988.
Marcaccio KY. *Gale Directory of Databases, Volume 1: Online Databases*. London, Gale Research International, 1993.
van Nimwegen C. *International List of Reports on Comparative Evaluations of Medical Devices*. Leiden, The Netherlands, TNO Centre for Medical Technology, 1993.

Further Information

A comprehensive listing of health care standards and the issuing organizations is presented in the *Health care Standards Directory* published by ECRI. The Directory is well organized by keywords; organizations and their standards; federal and state laws; and legislation and regulations; and contains a complete index of names and addresses.
The *International Health Technology Assessment* database is produced by ECRI. A portion of the database is also available in the U.S. National Library of Medicine's new database called *HealthSTAR*. Internet access to *HealthSTAR* is through http://igm.nlm.nih.gov. A description of the database is available at http://www.nlm.nih.gov/pubs/factsheets/healthstar.html.

119

Health Care Quality and ISO 9001:2000

David A. Simmons
Health Care Engineering, Inc.
Glen Allen, VA

"ISO" is recognized as the short name for the International Organization for Standardization (IOS), an internal agency located in Geneva, Switzerland, consisting of almost 100 member countries. Each country, no matter large or small, has one "equal" vote. The United States' representative to ISO is the American National Standards Institute (ANSI). From 1994 to 2000, the core of the ISO 9000 Quality Systems Standard was a series of five international standards that provide guidance in the development and implementation of an effective quality management system. Not specific to any particular product or service, these standards are applicable to manufacturing and service industries alike. In 2000, the International Standards Association in Geneva, Switzerland revised the international ISO Core Standards. Effective in 2000, there are four documents; ISO 900:2000, ISO 9001:2000, ISO 9004:2000 and ISO 19011. ISO 9002:2000 is a requirement standard. ISO 9000:2000, ISO 9004:2000 and ISO 19011 are guideline documents.

A quality management system refers to the activities one carries out within one's organization to satisfy the quality-related expectations of one's customers. To ensure that an organization has a quality management system in place, customers or regulatory agencies may insist that the organization demonstrate that its quality management system conforms to the ISO 9001:2000 quality system model. Then, the customer "second party" or an independent "third party" registrar goes to the organization to "audit", or verify, that such a system is in place. When a registrar determines that the organization fulfills the requirements of the ISO 9001:2000 standard, that organization becomes "registered" and receives a certificate that is accepted by many of the organization's customers. Organizations that are not concerned with becoming registered might, nevertheless, want to comply with ISO 9001:2000.

This chapter describes the ISO 9000 Core Standards, ISO 9001:2000, in particular, and the Industry Workshop Agreement, which aided in the expeditious development of the standards. The standard's application to health care and rationale for adoption will be explained. Health care quality management principles are illustrated in the light of ISO 9001:2000.

The ISO 9000 Core Standards

There are four Core Documents, as follows:

ISO 9000:2000 provides quality management principles and fundamentals. It describes what the series is about, and lists basic definitions of terms for use by any organization.

ISO 9001:2000 states requirements for quality management systems when it is necessary to demonstrate that an organization is capable of effectively meeting customer and regulatory requirements, such as those of the Joint Commission on the Accreditation of Health care Organizations (JCAHO), the National Committee for Quality Assurance (NCQA), and the URAC, or state and federal requirements.

ISO 9004:2000 provides guidance for establishing a quality management system that goes beyond ISO 9001 requirements to meet and exceed customer expectations efficiently.

ISO 19011 provides guidance on planning and conducting quality audits.

Clauses in the ISO 9001 Quality Systems Standard

ISO 9001:2000 consists of five primary clauses that contain 23 subclauses. These requirements spell out what an organization (and, possibly, an individual in that organization) must do to conform to the standard. All of the requirements must be documented and controlled. The clauses are as follows:
Sections 1 through 3–Administrative and Actual Quality Management System Standards
Section 4–Quality Management System
4.1 General Requirements
4.2 Documentation requirements
    4.2.1 General
    4.2.2 Quality Manual
    4.2.3 Control of Documents
    4.2.4 Control of Records

Section 5–Management Responsibility
5.1 Management Commitment
5.2 Customer Focus
5.3 Quality Planning
5.4 Planning
    5.4.1 Quality Objectives
    5.4.2 Quality Management System Planning
5.5 Responsibility, Authority and Communication
    5.5.1 Responsibility and Authority
    5.5.2 Management Representative
    5.5.3 Internal Communication
5.6 Management Review
    5.6.1 General
    5.6.2 Review Input
    5.6.3 Review Output
Section 6–Resource Management
6.1 Provision of Resources
6.2 Human Resources
    6.2.1 General
    6.2.2 Competence, Awareness and Training
6.3 Infrastructure
6.4 Work Environment
Section 7–Product Realization
7.1 Planning of Product Realization
7.2 Customer-related Processes
    7.2.1 Determination of Requirements related to the Product (or Service)
    7.2.2 Review of Requirements Related to the Product (or Service)
    7.2.3 Customer Communication
7.3 Design and Development
    7.3.1 Design and Development Planning
    7.3.2 Design and Development Inputs
    7.3.3 Design and Development Outputs
    7.3.4 Design and Development Review
    7.3.5 Design and Development Verification
    7.3.6 Design and Development Validation
    7.3.7 Control of Design and Development Changes
7.4 Purchasing
    7.4.1 Purchasing Process
    7.4.2 Purchasing Information
    7.4.3 Verification of Purchased Product
7.5 Production and Service Provision
    7.5.1 Control of Production and Service Provision
    7.5.2 Validation of Processes for Production and Service Provision
    7.5.3 Identification and Traceability
    7.5.4 Customer Property
    7.5.5 Preservation of Product
7.6 Control of Monitoring and Measurement Devices
Section 8–Measurement, Analysis and Improvement
8.1 General
8.2 Monitoring and Measurement
    8.2.1 Customer Satisfaction
    8.2.2 Internal Auditing
    8.2.3 Monitoring and Measurement of Process
    8.2.4 Monitoring and Measurement of Product
8.3 Control of Nonconforming Product
8.4 Analysis of Data
8.5 Improvement
    8.5.1 Continual Improvement
    8.5.2 Corrective Action
    8.5.3 Preventive Action

The model of a process-based quality management system shown in Figure 119-1 illustrates the process linkages presented in clauses 4 to 8. This illustration shows that interested parties play a significant role in defining requirements as inputs. Monitoring the satisfaction of interested parties requires the evaluation of information relating to the perception of interested parties as to whether the organization has met their requirements. The model shown in Figure 119-1 does not show processes at a detailed level.

Industry Workshop Agreement (IWA1)

In 2000, the IOS acknowledged that the standards-writing and -revision processes were time consuming, often taking years. ISO 9000 was, as are all ISO Systems Standards, quite generic in nature, by design. During the 1990s, some industries in the United States felt that ISO 9000 was too generic, and they proceeded to develop their own industry-specific standards without the direct approval of, or processing by, the IOS in Geneva. The automotive industry, for example, developed QS 9000 for itself. The aerospace industry created AS 9000. The telecommunications industry created TL 9000. The health care industry recognized a need to have some form of quality-system standard. In 1999, the author, as chair of the Health Care Division of the American Society, created a standards committee with the expressed intent of initiating activity of creating a health care industry standard, tentatively to be called HC 9000. Over the next two years, the Health Care Division of the American Society of Quality (ASQ), with participation and support from the Automotive Industry Action Group (AIAG), continued to develop a draft standard for process improvements in health care organizations. The guidelines that they developed were based upon the same quality management principles as ISO 9001 and encouraged the adaptation of a process approach to quality management.

During this same period, the IOS decided that it had to take action to attempt to reduce the proliferation of industry-specific standards. It developed and implemented a new "fast track" process that could produce industry specific "standards" in a much faster way without compromising the already-existing IOS system and procedures for generating standards. It came up with the Industry Workshop Agreement (IWA). The ASQ Health Care Division and the AIAG recognized that this could be a faster way to put a health care document in place, and it decided to accept the IWA. Furthermore, as there was no pre-existing health care quality system standard in place, they decided that the document needed to be comprehensive in scope and detail. The committee decided to base the health care quality management systems standard on ISO 9004:2000, the guidance document.

The IWA process calls for the creation of a draft document, or IWA draft, by the creating industry or special interest group. Next, the group is expected to call a public "workshop" to review, revise, and come to a consensus. When a consensus is achieved, the group is required to vote within a 60-day period. A three-day IWA Health Care Workshop was held in January, 2000 in Detroit, Michigan. During the 60-day voting period, the members overwhelmingly voted for approval of the IWA. The IWA was sent to the IOS in Geneva and has been issued as the first Industry Workshop Agreement, IWA-1. This IWA-1 document is the first quality management standard (QMS) developed for health care provider organizations. It is based on two ISO Standards; ISO 9001:2000, Quality Management Systems—Requirements, and ISO 9004:2000, Quality Management Systems—Guidelines for Performance Improvement. ISO 9001:2000 is the basic, or core, Standard for Quality Management Systems. ISO 9004:2000 is a more expansive standard that provides guidance for implementing ISO 9001:2000. While JCAHO imposes prescriptive standards on health care providers, the three ISO standards are voluntary and are used to implement or to modify a quality management standard for existing organizations, based on the uniqueness of their particular system and structure. ISO is a valuable system tool to assist in better, faster, and less costly way to achieve JCAHO accreditation.

With the JCAHO participating in the development process, the ASQ guidelines, in the form of IWA-1, meld JCAHO standards with the ISO standards requirements within JCAHO to provide guidance on the application and improvement of quality management systems in health care organizations. This is demonstrated by the adoption of specific ISO 9001:2000 requirements into forthcoming JCAHO medications standards.

Health Care Applications

The adoption and use of the ISO 9001:2000 series standards and guidelines by institutional, managed care, service providers, and physician practice health care providers can result in many benefits. This material contains several unique approaches useful in the implementation of ISO 9001:2000 in the health care setting.

As everyone in the health care community knows, medical errors are an ongoing problem for which there has been little or few major breakthroughs toward solution or reduction (IOM, 2000). It is well understood that one of the ways to reduce medical system failures is the proper collection, aggregation, and analysis of several categories of data. Such analysis can then lead to root cause analysis (RCA) and continuous quality improvement (CQI) through the use of Plan-Do-Check-Act (PDCA) cycles. The Institute of Medicine's March 2001 report, *Crossing the Quality Chasm: A New Health System for the 21st Century,* calls for "the elimination of handwritten clinical information by 2010 and refocusing the health care system on treating chronic illness" (IOM, 2001).

In an effort to improve the health care management system and to reduce medical errors, there have been a few small and localized attempts to use hand-held personal data assistants (PDAs) to collect, store, and analyze clinical and administrative data. The ultimate solution is to combine the use of hand-held devices that transmit data in real time with larger, fixed computer systems. These systems can be those of an institution or of an applications service provider (ASP). Health care professionals can access full and complete real-time data in order to treat patients more effectively and, in the process, to identify and reduce medication and medical errors.

Institutional health care providers are those hospitals, medical centers, and clinics that currently submit to, and comply with, JCAHO for the primary purpose of obtaining Medicare reimbursement for providing health care services to Medicare patients. The Medicare system is administered by the Federal Department of Human Services Department's Centers for Medicare and Medicaid Services (CMS); formerly the Health Care Finance Administration (HCFA).

Because CMS does not have sufficient staffing to perform site surveys of all Medicare provider facilities, CMS has given the JCAHO what is known as "deemed status," which means that if a hospital or dissimilar institutional facility or program has received JCAHO accreditation, CMS will accept that accreditation in lieu of its own survey. CMS does bear primary survey responsibility under the law, however can and has delegated survey and compliance monitoring to state or other authorized health care jurisdictional authorities. The benefit to the use of ISO 9001:2000 compliance and registration is that once ISO 9001:2000 is implemented, the very nature of the process assures continuous quality improvement (CQI), whereby problems and process variations are addressed quickly and permanently. With ISO 9000 registration, periodic internal and external audits are needed, to monitor continuing compliance. There is relatively little interruption in the delivery of health care as experienced with the three-year JCAHO accreditation cycle. Those who work in the hospital, long-term care facility, or clinical environment are well aware of the typical six- to nine-month preparation process that takes place before a JCAHO survey. Health care, allied health care professionals, and management leaders are diverted from providing health care and spend considerable time "gearing up" for surveys. The number of hours converted into dollars is considerable, as is the cost to the institution for the actual JCAHO survey.

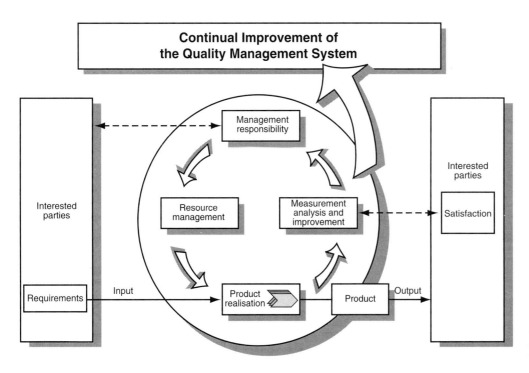

Figure 119-1 Model of a process-based quality management system showing key, value-added activities information flow.

Managed care and health maintenance organizations (HMO) experience similar kinds of problems in preparation for the NCQA Health Plan Employer Data and Information Set (HEDIS) Report Card System. NCQA performs both accreditation and performance measurement. While the standards are different, the "process" is essentially the same as far as the long lead time and costly preparation time and survey costs.

Physician group practices (e.g., institutional provider organizations (IPOs), preferred provider organizations (PPOs)) can benefit for two significant reasons. First, implementation of the ISO 9001:2000 process can create a more efficient practice that can identify problems early and can initiate corrective action to solve problems more quickly and to better monitor, control, and reduce costs in the process. The second major reason is that a practice that is ISO 9001:2000-registered can point to its registration as a benchmark of an efficient and controlled operation. Increasingly, this is what insurance companies, payers, and larger IPOs seek when assessing physician group practice providers.

Medical equipment service providers include the original equipment manufacturers (OEM), in-house clinical engineering programs, independent service organizations (ISOs), and equipment remanufacturers. The U.S. Food and Drug Administration requires medical equipment manufacturers to comply with a federal regulation known as "Good Manufacturing Practice" (GMP). The FDA is currently attempting to pass legislation to require all such service providers to comply with laws that will impose requirements upon providers in the area of tracking of devices after sale and during the life of the medical device until it is replaced. Given the large volume of medical devices in use today, as well as their portable nature, the ability to comply with such requirements is virtually impossible. New legislation would prove to be a cross between the Safe Medical Device Amendments of 1990 and the GMPs. After significant service-provider protests, the FDA has retreated for the time being, while urging the service providers to implement a quality-systems standard of their own development or to face GMPs or equivalent legislation. ISO 9001:2000 is a quality management system that can solve the problem with less effect on service costs than any other system currently available can.

Overall, it has been well demonstrated that if a health care provider is ISO 9001:2000 qualified or registered, any other survey process will be much simpler and less costly regarding preparation and compliance demonstration. Although ISO 9001:2000 registration is not intended to replace JCAHO or NCQA accreditation, it does make the compliance-demonstration process appreciably less difficult, less time-consuming, and less costly.

Rationale for Adoption of ISO 9001:2000 in Health Care

Health care organizations adopt ISO 9001:2000 standards for a variety of different reasons. an organization's decision to do so may include any of the following reasons:

- To improve a quality system
- To minimize repetitive auditing by accrediting organizations
- To improve subcontractors' and vendor's performance

Both the clinical engineer and the organization benefit because use of IS0 90001:2000 serves as a basis to:

- Achieve better understanding of all quality practices throughout the organization
- Ensure continued use of the required quality system year after year
- Improve documentation and records
- Improve quality awareness
- Strengthen both supplier and customer confidence and relationships
- Yield cost savings and improve profitability
- Form a foundation for improvement activities within total quality management (TQM).

Complying with ISO 9001:2000 requirements does not indicate that every product or service meets the customers' requirements. It merely means that the quality system in use is capable of meeting them. That is why an organization must continuously assess customer satisfaction and must constantly improve the processes that produce the products or services.

Health Care Quality Management Principles

The following principles have been identified within the ISO 9001:2000 and IWA-1 standards to facilitate achievement of quality objectives:

1. Customer Focus

- Understanding current customer needs
- Understanding future customer needs
- Meeting customer requirements
- Striving to exceed customer expectations

2. Leadership

- Establishing unity or purpose and direction for the organization
- Establishing the organization's internal environment

3. Involvement of People

- Developing abilities fully
- Using abilities toward maximum benefit

4. Process Approach

- Managing resources as a process
- Achieving desired results more efficiently

5. Systems Approach to Management

- Identifying
- Understanding and managing the interrelated processes of a system to reach objectives effectively and efficiently

6. Continual Improvement

- Making improvement a permanent objective

7. Factual Approach to Decision Making

- Analyzing data and information logically

8. Mutually Beneficial Supplier Relationships

- Creating value through mutually beneficial, interdependent relationships

Because every health care organization is different, the implementation of ISO 9001:2000 will be different in each organization. The following three basic and fundamental requirements exist:

1. Document processes that affect quality
2. Retain records and data that describe the product or service
3. Ensure that processes produce consistent quality

While JCAHO and NCQA have prescriptive standards, ISO does not have specific requirements in its quality management system. Each organization must develop its own system and then must comply with it. The imperatives that describe how ISO 9000 and, now, ISO 9001:2000 are to be implemented can be expressed as the following:

Say what you do.
Do what you say.
Prove it.
Improve it.

ISO 9001:2000 Documentation

Because documentation is so important to ISO 9001:2000 compliance, it is important to understand the levels of documentation. There are four levels, as follows:
Level I: Quality manual
The statement of the organization's quality, written by managers from top-level management and the quality department.
Level II: Procedures
Defines activities at the department level and written by department managers and supervisors.
Level III: Work/job instructions
Describes ways in which jobs are accomplished; are usually written by the operators and trainers.
Level IV: Other documentation (records)
Quality records include such items as forms (both hard copy and electronic), tags, labels, and other such forms of records that are usually created by quality and/or other middle managers. Level II procedures are the backbone of quality documentation because they are the documents that define and describe processes of the operation of hospital and health care provider operations. Typical process procedures should contain:

- Purpose/objective: The aim of the procedure
- Scope: What the procedure does and does not cover
- Responsibilities: Who, by job function, has responsibilities for specific tasks or actions
- References: To all documents covered under the procedure
- Definitions: Key terms or acronyms
- Procedures: Description of the actions or tasks to be carried out; by whom; and in what sequence
- Documentation: Documentation and records that are needed

Historically, most Level II process procedures are created in the form of flow charts rather than text documents.

Summary

The new ISO 9001:2000 standard stresses documentation control. Organizations that implement ISO 9001:2000 can do so by using by using the new ISO IWA 1, Health Care Guideline Document, which is an adaptation of ISO 9004:2000, specifically developed for health care applications and improve your document control system. This approach to implementing ISO 9001:2000 requires less consultation time and provides a comprehensive, integrated solution to address "what to do," "how to do it," and "how to be able to do it."

The goal of this document is to aid in the development or improvement of a fundamental quality management system for health service organizations, providing for continuous improvement, emphasizing error prevention and the reduction of variation and organizational waste (e.g., non-value-added activities).

This guide incorporates much of the text of ISO 9004:2000—"Quality Management Systems—Guidelines for Performance Improvements" and provides guidance on quality management systems, including the processes for continual improvement that contribute to the satisfaction of a health service organization's customers and other interested parties. The quality management system should provide for all customers of a health service organization, regardless of the product or service provided.

References

Institute of Medicine. *To Err Is Human: Building a Safer Health System.* Washington, DC, National Academy Press, 1999.
Institute of Medicine. *Crossing the Quality Chasm: A New Health System for the 21st Century;* Washington, DC, National Academy Press, 2001.

Further Information

ANSI/ISO/ASQ Q 9001-2000, Quality Management Systems—Requirements, Milwaukee, WI, American Society for Quality, 2000.
ANSI/ISO/ASQ Q 9004-2000, *Quality Management Systems—Guidelines for Quality Improvement,* Milwaukee, WI, American Society for Quality, 2000.
ANSI/ISO/ASQ Q 9000-2000, *Quality Management Systems-Fundamentals and Vocabulary,* Milwaukee, WI, American Society for Quality, 2000.
ANSI/ISO/ASQ Q ITA1, *Quality Management Systems-Guidelines for Process Improvement in Health Services Organizations,* Milwaukee, WI, American Society for Quality, 2000.
Simmons DA. Medical Instrumentation Maintenance Committee. *Guideline for Establishing and Administering Medical Instrumentation Maintenance Program,* AAMI Recommended Practice, AAMI, 1984.
Simmons DA. *Quality Management for Health Care Technology,* Oakton, VA, Health Care Engineering, Inc., 1995.
Simmons DA. *Quality Management for Clinical Engineering,* Oakton, VA, Health Care Engineering, Inc., 1995.
Simmons DA. *Health Care Quality Management Systems,* Oakton, VA, Health Care Engineering, Inc., 1995.
Simmons DA. *Quality Management: Auditing and Operational Sampling,* Oakton, VA, Health Care Engineering, Inc., 1995.
Simmons DA. *ISO 9000 and Health care Quality Analytical Tools, Techniques and Applications,* Richmond, VA, Virginia Management Systems Society, March 13, 1998.
Simmons DA. *Health Care Quality and ISO 9000,* Oakton, VA, Health Care Engineering, Inc., 1998.
Simmons DA. *ISO 9000 in Health Care,* Drexel University, Philadelphia, October 23, 1998.
Simmons DA. *Health Care Quality and ISO 9000,* Chief Information Officer' Council, University Health care Consortium Annual Meeting, April 27, 1999.
Simmons DA. *ISO 9002: Is It Really Applicable to Clinical Engineering Departments?,* HealthTech 99, Baltimore, MD, April 28, 1999.
Simmons DA. *Health Care Quality and ISO 9000,* Arlington, VA, Allied Association Information Resources Network, April 29, 1999.
Simmons D, Lloyd R. *Health Care Quality Tools, Techniques and Applications,* ASQ Pre-Conference Tutorial, ASQ 1999 Annual Quality Congress, Anaheim, CA, May 22, 1999.
Simmons DA, *ISO 9000: What We've Learned; ISO 9001:2000, Process, Progress and Profits,* Washington, DC, American Association of Blood Banks, November 4, 2000.

120 Hospital Facilities Safety Standards

Gerald Goodman
Assistant Professor, Health Care Administration
Texas Women's University
Houston, TX

The numerous hospital facilities safety standards in the United States include statutes promulgated the United States Congress and state legislatures. They also include those regulations developed by private agencies and designated as national standards because of their adoption by public agencies. This chapter addresses the major standard making organizations involved in hospital-facility safety. Because regulatory responsibility for device and facility safety and employee safety overlap, these regulatory programs are included for completeness.

Public facility safety standard making agencies include the Health Care Financing Administration (HCFA) and the Occupational Safety and Health Administration (OSHA). Private standard making agencies include the Joint Commission on Accreditation of Health Care Organizations (JCAHO) and the National Fire Protection Association (NFPA).

National Fire Protection Association

The mission of the international nonprofit National Fire Protection Association (NFPA) is to reduce the worldwide burden of fire and other hazards on the quality of life by providing and advocating scientifically based consensus codes and standards, research, training, and education. NFPA's focus on true consensus has helped the association's code-development process to earn accreditation from the American National Standards Institute (ANSI).

ANSI accreditation means that codes and standards developed by the NFPA have achieved the status of national consensus standards. Because of this standing, government agencies and many private organizations automatically adopt NFPA codes and standards as their own. HCFA and JCAHO use NFPA codes, as do the majority of state and local fire authorities.

The codes and standards of the NFPA apply to life safety and fire safety and, under the umbrella of fire safety, NFPA documents contain important design features for medical device safety. These include codes related to the use of medical gases, hyperbaric chambers, and high electrical-frequency devices.

Authority with Jurisdiction

The term "authority with jurisdiction" is not a facility safety organization *per se,* but rather is a term used by the NFPA to indicate the responsible local authority that takes regulatory responsibility for life safety compliance. Typically, this responsibility is held by a local fire marshal's office or a similar governmental entity. NFPA standards refer to the authority with jurisdiction in order to provide businesses with a specific local authority for interpreting compliance with life safety codes.

NFPA Code Revision Date

NFPA codes and standards are revised periodically. However, regulatory organizations including the JCAHO, HCFA, and the authority with jurisdiction adopt a specific revision of an NFPA code as the code they will enforce. The most recent NFPA code will not necessarily be the one enforced. In fact, government agencies such as HCFA cannot respond quickly to code changes. It is possible to find a government agency enforcing a code that is several revisions behind the present version.

Joint Commission on Accreditation of Healthcare Organizations

The Joint Commission on Accreditation of Healthcare Organizations (JCAHO) evaluates and accredits more than 17,000 health care organizations and programs in the United States (see Chapter 121). An independent, not-for-profit organization, JCAHO is the nation's predominant standards-setting and accrediting body in health care. Since 1951, JCAHO has developed state-of-the-art, professionally based standards and has evaluated the compliance of health care organizations against these benchmarks (JCAHO, 2000). JCAHO accreditation is recognized nationwide as a symbol of quality that reflects an organization's commitment to meeting certain performance standards. To earn and maintain accreditation, an organization must undergo an on-site survey by a JCAHO survey team at least every three years. Laboratories must be surveyed every two years.

JCAHO has separate accreditation criteria and manuals for each of its types of provider organizations. These include general, psychiatric, children's, and rehabilitation hospitals; health care networks, including health maintenance organizations (HMOs), integrated delivery networks (IDNs), preferred provider organizations (PPOs), and managed behavioral health care organizations; home care organizations, including those that provide home health services, personal care and support services, home-infusion and other pharmacy services, durable-medical equipment services and hospice services; and nursing homes and other long-term-care facilities.

The JCAHO standards for hospitals are typical of those for other provider types and will be used here for discussion purposes. The principal section of the JCAHO hospital accreditation manual that is relevant to facility safety is the chapter on "The Environment of Care" (JCAHO, 2000). The Environment of Care (EC) chapter includes those requirements related to device safety and life safety. Responsibilities of the clinical engineer may cross over to aspects of the EC chapter typically included in a facilities management program, such as the pneumatic tube system or fire alarm system.

The key facilities safety document under the JCAHO EC is the "Statement of Construction" (SOC). This document is intended to be an ongoing, historical summary of the facility's compliance with the NFPA Life Safety Code (NFPA 101). The SOC brings into focus varying codes for fire alarm and suppression systems, building construction standards, building maintenance related to fire safety, and building occupancy characteristics such as obstructed fire exit corridors.

Standards under the EC for device and facility safety are summarized in Tables 120-1 and 120-2. JCAHO addresses Table 120-1 to clinical engineering and Table 120-2 to facilities management.

Table 120-1 Clinical Engineering: JCAHO Environment of Care (EC) Standard

JCAHO section	Requirement	Source of data
EC.1.8	A management plan addresses medical equipment.	1. Medical equipment management plan 2. Minutes of a medical–equipment-safety subcommittee 3. Clinical engineering plan for providing patient care
EC.2.1	Personnel can describe their role in each of the EC plans: • Safety • Fire response • Hazardous materials • Emergency preparedness • Utility outage • Technical-training records • Process for reporting medical equipment problems • Whom to contact for an after-hours medical equipment emergency	1. Job descriptions 2. Required education matrix 3. Competency checklist 4. Education department training records and individual personnel files
EC.2.7	The medical equipment management plan is implemented.	Minutes of a medical equipment safety subcommittee
EC.2.13	Medical equipment is maintained, tested, and inspected.	1. Inventory of medical equipment 2. Records of medical equipment preventive maintenance program
EC.3.2	ICES	1. Clinical equipment user error log 2. Safety committee minutes

From 2000 JCAHO Manual. Sections can change with later revisions.

Table 120-2 Facilities Management: JCAHO Environment of Care (EC) Standard

JCAHO Section	Requirement	Source of data
EC.1.1	Newly constructed and existing facilities are designed and maintained to comply with the LSC.	1. SOC with improvement plan 2. Plans for building improvement
EC.1.2	When designing the EC, the organization considers design criteria referenced by the health care community.	1. Construction documents reference NFPA Life Safety Code (LSC) 1997 after January 1, 1998, or LSC 1994 prior to January 1998 2. All construction has been reviewed and approved by the authority with jurisdiction
EC.1.7	A management plan addresses life safety.	1. Life safety management plan. 2. Minutes of a facilities safety subcommittee 3. Facilities Management Department plan for providing patient care
EC.1.9	A management plan addresses utility systems.	1. Utility management plan 2. Minutes of a facilities safety subcommittee 3. Surveillance, Prevention, and Control of Infections Committee annual plan 4. Minutes of the Surveillance, Prevention, and Control of Infections Committee
EC.2.1	Personnel can describe roles in each of the EC plans: • Safety • Fire response • Hazardous materials • Emergency preparedness • Utility outage • Technical training records • Process for reporting utility problems • Location and use of emergency cutoffs • Who to contact in an emergency	1. Job descriptions 2. Required education matrix 3. Competency checklist 4. Education department training records and individual personnel files
EC.2.6	The life safety management plan is implemented.	1. Current record drawings 2. Policy on Interim Life Safety Measures (ILSM) 3. Documented use of ILSM 4. Inventory of life safety equipment and systems 5. Records of testing of automatic fire extinguishing systems 6. Records of inspection of smoke management systems—dampers, doors, wall penetrations 7. Wall penetration preventive maintenance summary 8. Records of portable fire extinguisher inspection 9. Records of fire alarm system testing
EC.2.8	The utility management plan is implemented.	1. Plans showing layout of utilities 2. Plans showing distribution of emergency power to all critical functions
EC.2.14	Utility systems are maintained, tested, and inspected.	1. Inventory of utility systems. 2. Records of utility preventive-maintenance program 3. Records of monthly emergency generator and transfer-switch testing 4. Record of annual load bank test. 5. Log of emergency generator testing
EC.3.2	ICES	1. Utility interruption/failure summary log 2. Safety committee minutes

From 2000 JCAHO Manual. Sections can change with later revisions.

Health Care Financing Administration

The Health Care Financing Administration (HCFA) is a federal agency within the U.S. Department of Health and Human Services. HCFA administers the Medicare and Medicaid programs. Recently, HCFA was renamed the Centers for Medicare & Medicaid Services. HCFA regulates all laboratory testing (except research) performed on humans in the United States. Approximately 158,000 laboratory entities fall within HCFA's regulatory responsibility.

HCFA has a number of quality assessment and performance improvement initiatives related to improving the quality of health care provided to Medicare, Medicaid and Child Health Insurance Program beneficiaries. These include the following:

- Developing and enforcing standards through surveillance
- Measuring and improving outcomes of care
- Educating health care providers about quality improvement opportunities
- Educating beneficiaries to make good health care choices

The surveillance program includes the hospital-safety standard compliance activities of HCFA, principally life safety compliance using standards developed by the National Fire Protection Association. HCFA surveillance is typically performed by state hospital-licensing personnel. HCFA's regulatory programs can be found in Title 42 of the Code of Federal Regulations (CFR).

Occupational Safety and Health Administration

The Occupational Safety and Health Administration (OSHA) is responsible for establishing and enforcing workplace-safety rules. In an era of increased focus on patient safety, it is important to remember that OSHA concerns itself with employee safety only. OSHA safety regulations do not address facility safety *per se*. Rather, OSHA regulations specify safety practices that are to be enforced while maintaining the facility. These regulations are typically identified for specific hazards, such as respiratory protection when using aerosolized paint. OSHA does have a "general duty clause," which is a catch-all regulation that requires employers to provide employee-safety programs regarding hazards of which the employers should have been aware, even if no specific safety program is identified.

Safety programs under OSHA that are particularly relevant to clinical engineering include those that concern employee protection from blood-borne pathogens and rules that relate to worker protection from energized equipment, ladder safety, and hearing and eye protection when using tools such as grinders and drills.

Rules regarding blood-borne pathogens include those that are applicable to all hospital personnel who have the potential for exposure to blood and body fluids from direct patient contact. They also include requirements that clinical engineering departments develop and enforce rules for the decontamination of patient equipment prior to repair. Rules for blood-borne pathogens can be found in 29 CFR 1910.

OSHA requires that worker protection be provided when repairs are being made to "energized" devices. These can be powered by an electrical, pneumatic, or other power source. The rules are classified as "lockout-tagout" rules. A typical example for the application of lockout-tagout rules is the repair of any permanently wired electrical equipment such as X-ray systems. The rules do not apply to those devices that can be disconnected from their power source prior to repair. In such cases, department policy must be clear that such devices must not be repaired while connected to their power source unless specifically approved on a case-by-case basis.

Reference

JCAHO. 2000 *Accreditation Manual for Hospitals*. Chicago, JCAHO Press, 2000.

121

JCAHO Accreditation

Britt Berek
JCAHO
Oakbrook Terrace, IL

The Joint Commission on Accreditation of Healthcare Organizations (JCAHO) is an independent, non-profit organization committed to improving the level of health care delivery in the United States and abroad. The overarching purpose of JCAHO is to continuously improve the safety and quality of care provided to the public in organized delivery settings. Setting standards and evaluating whether organizations seeking accreditation comply with those standards accomplishes this goal.

JCAHO History

Inspection of hospitals in the United States began as early as 1918 under the "Hospital Standardization Program" developed and administered by the American College of Surgeons (ACS). Although the first standards used by the ACS consisted of a single page of requirements, fewer than a third of the initial group of hospitals inspected were able to successfully meet them. As interest in hospital standardization and inspection became more prevalent, other professional medical societies became involved in hospital standards and compliance activities in addition to ACS, including the American Hospital Association (AHA), American Medical Association (AMA), and American College of Physicians (ACP).

The Joint Commission on Accreditation of Hospitals (JCAH) was founded in 1951, with the joint collaboration of the four leading provider organizations (i.e., the ACS, AMA, CMA, and ACP) as a non-profit organization that provided voluntary accreditation for hospitals.

In 1953, the JCAH published their first standards for hospital accreditation. By 1960, 2900 hospitals participated in the accreditation program. In 1965, Congress passed the Social Security Amendments of 1965 (commonly known as the Hill-Burton Act/Medicare Act) with a provision that hospitals accredited by JCAH are also in compliance with the Medicare Conditions of Participation for Hospitals and can participate in the Medicare and Medicaid programs. The JCAHO currently accredits the majority of hospitals in the United States, largely because of this accreditation arrangement. In addition, JCAHO has expanded the number and types of health care organizations it accredits, and in 1987 it changed its name to the Joint Commission on the Accreditation of Healthcare Organizations (JCAHO) to reflect this expanded scope of activities.

The JCAHO now accredits more than 18,000 health care organizations in the United States and many other countries. Accredited organizations include hospitals, long-term care facilities, ambulatory care facilities, home care organizations, clinical laboratories, and behavioral health care facilities.

Benefits of Accreditation

In addition to meeting the requirements for participation in the Medicare and Medicaid program for hospitals, JCAHO accreditation has accrued a number of other benefits for accredited organizations over the years. Health care organizations seek JCAHO accreditation for benefits including the following:

- Enhanced community confidence
- Provision of a progress report for the public
- Stimulation of the organization's quality improvement efforts
- Assistance with professional staff recruitment
- Expedition of third-party payment
- Fulfillment of state licensure requirements
- Possible improvement of liability insurance premiums
- Favorable influence on managed care contract decisions

By seeking accreditation, an organization agrees to be measured against national standards set by health care professionals. An organization that is "accredited" demonstrates that it substantially complies with JCAHO standards and continuously works to improve the care and services it provides.

The Accreditation Process

Trained surveyors evaluate each health care organization's compliance with JCAHO standards and identify the organization's strengths and weaknesses. This evaluation is performed at the organization. The surveyors' goal is not merely to find problems, but also to educate and advise; so health care organizations can improve. Summary results of the surveys are posted on the JCAHO's website for public disclosure.

Most on-site reviews occur every three years (except for clinical labs on two-year cycles). However, an organization may also undergo a number of unscheduled, intra-cycle events. These special surveys may be performed "for-cause" if the organization is identified for an on-site investigation through the JCAHO's Quality Monitoring process, or if the organization is selected for a random unannounced survey. These events may be announced (scheduled with the organization's knowledge) or unannounced (scheduled without the organization's knowledge as a validation and quality check, or as part of the investigation of a potentially serious accreditation issue).

Departments and Programs

The JCAHO currently employs approximately 500 field surveyors that conduct on-site surveys at organizations seeking accreditation or re-accreditation. In addition, there are approximately 500 other employees in the central office in Oak Brook, Illinois, that support the field surveyors, develop or improve accreditation programs and standards, and assist organizations in completing accreditation activities.

In the process of obtaining accreditation, organizations may interact with a number of departments and programs at the JCAHO's central office. These departments include the following:

Accreditation Operations

The Division of Accreditation Operations (AO) administrates accreditation services. Each organization seeking accreditation will be assigned to an account representative, who will guide the organization through the accreditation process from completing the application, scheduling the survey, assigning surveyors, completing the on-site event, and assisting with the survey report, as well as handle any subsequent follow-up. The JCAHO account representative will typically interact with a designated organizational liaison. The designated liaison is frequently involved in quality management or a similar administrative function

Office of Quality Monitoring

The Office of Quality Monitoring (OQM) receives and evaluates sentinel events and complaints from organizations, outside agencies, media, and the public. Organizations may need to formally respond to certain types of serious incidents. "Sentinel events" are rare, but serious, events that result in patient death or significant disruption of health care delivery. Analysis of these incidents is the source of "Sentinel Event Alerts" and the "National Patient Safety Goals" that are provided to health care organizations to assist them in reducing the odds of these events occurring or recurring.

Department of Education Programs

The Department of Education Programs (DEP) provides educational programs and teleconferences. DEP also maintains a speaker's bureau that coordinates the assignment of JCAHO speakers for external organizations.

Publications

To assist organizations in understanding and complying with JCAHO standards, the Publications department provides printed education materials and software. A catalog is available on the JCAHO's web site (www.jcaho.org or www.jcrinc.com) through "Infomart."

JCAHO Resources

On-site consulting services and international accreditation activities are provided by a for-profit subsidiary called JCAHO Resources (JCR). Organizations may request mock surveys and other consultative services to identify problem areas and to better prepare for their surveys.

Oryx

In 1997, the Oryx initiative was launched in an effort to integrate clinical outcomes and other performance data into the accreditation process. Oryx data are transmitted to JCAHO for integration and analysis. Currently, Oryx data focus on clinical measures.

Accreditation Decisions

The report that accompanies each accreditation decision details areas where an organization's performance must improve, and includes recommendations for how to meet the standards. If a health care organization has problems, it may receive accreditation contingent on those problems being fixed in a reasonable amount of time. The JCAHO closely monitors organizations with more substantial deficiencies to ensure that they are trying to solve their problems.

The JCAHO currently has eight accreditation-decision categories.

1. Accreditation without Type I Recommendations

Effective January 1, 2001, Accreditation without Type 1 Recommendations (previously known as Accreditation), is awarded to a health care organization that complies with applicable JCAHO standards in all performance areas.

2. Accreditation with Type I Recommendations

Effective January 1, 2001, Accreditation with Type 1 Recommendations (previously known as Accreditation with Recommendations for Improvement), is awarded to a health care organization that complies with applicable JCAHO standards in most performance areas, but has deficiencies in one or more areas or accreditation policy requirements that require resolution within a specified time.

3. Provisional Accreditation

An accreditation decision that results when a health care organization complies with a subset of standards during a preliminary on-site evaluation. This decision remains in effect until another official accreditation decision category is assigned, approximately six months later, based on a complete survey against all applicable standards.

4. Conditional Accreditation

An accreditation decision that results when a health care organization performs as follows:

- Fails to comply with applicable JCAHO standards in multiple performance areas, but is believed to be capable of achieving acceptable standards compliance within a stipulated time period
- Remains unable or unwilling to comply with one or more JCAHO standards
- Fails to comply with one or more specified accreditation policy requirements

5. Preliminary Denial of Accreditation

Effective January 1, 2001, Preliminary Denial of Accreditation (previously known as Preliminary Nonaccreditation), results when justification is found to deny accreditation to a health care organization because it has failed to comply with applicable JCAHO standards in multiple performance areas or failed to comply with accreditation policy requirements, or for other reasons. This accreditation decision is subject to subsequent review.

6. Accreditation Denied

Effective January 1, 2001, Accreditation Denied (previously known as Not Accredited), results when a health care organization has been denied accreditation. This accreditation decision becomes effective only when all available appeal procedures have been exhausted.

7. Accreditation with Commendation

Accreditation with Commendation, (eliminated effective January 1, 2000), was awarded to a health care organization that demonstrated more than satisfactory compliance with applicable JCAHO standards in all performance areas on a complete accreditation survey. Although this decision category has been discontinued as of January 1, 2000, organizations awarded this decision as a result of surveys conducted during 1997, 1998 and 1999 will retain this designation until their next complete surveys, unless it is lost based on an intracycle evaluation.

8. Accreditation Watch

Accreditation Watch, (though not a separate accreditation decision) is a publicly disclosable attribute of an organization's existing accreditation status. An organization is placed on Accreditation Watch when a sentinel event has occurred and a thorough and credible root cause analysis of the sentinel event, with an action plan, has not been completed within a specified time. Following JCAHO determination by that the organization has conducted an acceptable root cause analysis and developed an acceptable action plan, the Accreditation Watch designation is removed from the organization's accreditation status.

The reader is encouraged to see any of the JCAHO accreditation manuals for additional information on accreditation decisions, policies, and procedures (www.JCAHO.org).

JCAHO Standards

JCAHO standards address a comprehensive array of a health care organization's activities, from the medical staff to the care and treatment of patients, even the physical environment. For the hospital program alone, the accreditation manual standards that may each contain five to ten attributes. Chapters that relate to specific structures and functions within an organization organize this range of standards. Compliance with all the standards in the accreditation manual requires a planned and continuous effort at all levels and department within the organization.

In 1995 under an initiative called the "Agenda for Change," JCAHO standards were rewritten and reorganized to be less department-focused and more organization-focused. This shift required department managers to consider standards throughout the manual instead of a limited set for their departments. For example, clinical engineering personnel are now aware of requirements in the human resources, performance improvement, and infection control chapters, in addition to the standards in the environment of care chapter under Medical Equipment Management. Department managers should coordinate their compliance efforts with the organization's accreditation liaison to ensure that all applicable standards are covered. The standards are constantly revised in order to be consistent with changes in health care delivery and environmental conditions. For example, emergency management standards emphasize bio-terrorism preparedness as a result of the events surrounding September 11, 2001.

The Environment of Care Standards

The standards that apply most directly to clinical engineers appear in the *Environment of Care* chapter (EC) and are known as the medical equipment management standards. The medical equipment sub-group is one of seven major sub-groupings of standards in the EC chapter. The other sub-groupings include: safety, security, emergency management, hazardous materials and waste, life safety, and utilities management. In addition, supporting standards require each of these areas to be organized within a performance improvement framework.

The performance improvement element of the standards in the EC chapter follows a Design/Teach/Implement/Measure/Improve framework (see Table 121-1). Each of the seven major sub-groupings has a program design component, a teaching or educational component, an implementation and documentation component, a measurement and assessment component, and an improvement component. As a result, the seven areas will generate several of the same types of documents even though their specific content will differ.

For example, in the design step, the organization will need to develop a management plan for each of the seven EC areas. The management plan will describe how the organization will assign responsibility and carry out the necessary functions in each area as described by the JCAHO standard. The plan may reference organizational policies and procedures, describe reports, designate assignments, indicate reporting relationships, or reference relevant committees. Table 121-2 gives examples of how these EC documents may be organized in a moderately complex health care facility, such as a 200-bed hospital.

There are two ways to view the standards in the EC chapter that organizations may find useful. One view relates to the way in which all the standards within the chapter were designed to work together within the Design/Teach/Implement/Measure/Improve framework. The other way is to group the standards by the seven structural areas mentioned earlier, and consider the planning, teaching, implementation and documentation, and measurement, and strive to improve within a single functional area. The matrix of Table 121-1 shows how these two views can be presented simultaneously. The columns of Table 121-1 group the EC standards by functional area, while the rows group the standards by the Design/Teach/Implement/Measure/Improve cycle. Department managers should see the utility of dividing the EC standards by functional area, since responsibilities are typically divided this way within the organization.

Medical Equipment Management Standards

The EC's medical equipment management standards are of greatest interest to clinical engineers. Although there other standards apply to department operations, as previously mentioned, the standards found in the medical equipment group are the most relevant.

Table 121-1 Common Elements in the EC Cycle: Design/Teach/Implement/Measure/Improve

Design	Management plans
Teach	Orientation continuing education
Implement	Routine activities failure correction
Measure and assess	Data collection
improve	Organization-wide performance improvement

Table 121-2 Organization of EC documents in a 200-bed hospital

JC function	Documents referenced in EC standards	Commonly referenced organization documents
Design	Management plan	Policy and procedures committees / Reporting structures personnel assignments
Teach		Training logs competency assessments
Implement	EC documentation	Test logs
Measure and access	Monitors of performance	Data related to operational features of EC activities (e.g., effectiveness of staff training and timely completion of maintenance activities)
Improve	Annual evaluation	Data on program effectiveness Management reports

The medical equipment management standards require programs to address and document the following tasks:

- Selecting and acquiring medical equipment
- Evaluating and minimizing physical and clinical risks
- Monitoring and acting on adverse incidents
- Reporting under mandatory programs (e.g., Safe Medical Devices Act of 1990)
- Establishing and assessing user and maintainer competency
- Establishing emergency procedures for equipment failures
- Maintaining an accurate inventory of all equipment
- Developing maintenance strategies for equipment
- Documenting maintenance activities

The medical equipment management plan should address the elements referenced in the standards, as appropriate. In larger settings, this will include all the elements (as described above); for smaller organizations certain elements may not apply, and the plan should not any discrepancies. The plan is meant to serve as a reference document that demonstrates the organizations understanding of the JCAHO's requirements and shows how their operations will meet those requirements. The plan is a bridge between the JCAHO standards and the management of the organization. A well-constructed management plan should look like a business plan for department operations, and it will form the basis of subsequent on-going activities. Given the plan's importance in the accreditation process and to the organization, the JCAHO standards require on-going monitoring and an annual evaluation of the plan's scope, objectives, performance, and effectiveness. The continuous monitoring and re-evaluation is paramount to the JCAHO performance improvement model.

Key Program Attributes for Medical Equipment

The key program attributes for medical equipment are acquisition, inventory, maintenance strategies, and documentation of maintenance outcomes.

Acquisition

The organization must have a process for acquiring medical equipment. The process should address, at a minimum, elements of safety and clinical effectiveness. In order to minimize risk in subsequent use, the acquisition process is also an ideal time to address issues of patient safety and human factors.

Inventory

Organizations must decide which items to place into their structured medical equipment management plan (MEMP) inventory. The standards allow organizations to use risk-based criteria to decide which items may be included (or excluded) from this inventory.

If used, the organization must consider the following points:

- Equipment function (diagnosis, care, treatment, and monitoring)
- Physical risks associated with use
- Equipment incident history

Smaller organizations may choose to inventory all medical devices regardless of risk, but a larger organization will typically adopt inclusion criteria based on the attributes previously mentioned. Typically, a numerical algorithm is used to assign each type of equipment a "risk number," which will be used for inventory inclusion. In addition, this "risk number" may subsequently be used to identify the higher risk items in prioritizing activities or contingency planning. The idea is that items with a higher risk score are more important for patient safety and should receive preferential attention.

Maintenance Strategies

Equipment included in the inventory section of the MEMP must be appropriately maintained to minimize risk to patients. The JCAHO standards allow organizations to determine which maintenance strategies will be most effective for their organization in providing functional, safe, and available equipment.

These maintenance strategies may include ways to provide effective corrective maintenance, preventive maintenance, and inspections for damage or abuse, predictive maintenance, calibration, or metered maintenance based on hours of use. The organization must measure effectiveness and decide which maintenance strategy or combination of strategies will yield the optimal result.

While most maintenance activities have traditionally included some element of periodic, scheduled maintenance, this tactic is not always optimal for contemporary medical equipment. For example, a modern solid-state patient bedside monitor may not have moving parts that require replacement or periodic adjustment. However, it may be susceptible to unpredictable electronic failure. To address this failure mode, a strategy that includes spare monitors for immediate replacement may be more effective than a quarterly inspection of a seemingly functional monitor. On the other hand, if organizational data indicated that, based on their use, the Ni-Cad batteries in a particular portable device fall below required capacity after two years, a program for scheduled battery replacement every 20 months may be used to reduce failures, improve reliability, and minimize risk to patients.

Documentation of Maintenance Outcomes

Results of maintenance activities must be documented in order to improve the effectiveness of maintenance strategies. This activity is critically important for life-support equipment and other high-risk devices. In addition, tracking the results of operational efforts in other areas (even low-risk activities) will ensure that limited resources are used as effectively as possible. Summary data will be reported to the safety committee, quality improvement committee, or similar leadership oversight committees to ensure minimization of patient risk. Surveyors often ask for maintenance data as a first step in exploring department compliance.

Future Accreditation Initiatives

The JCAHO is constantly striving for ways to improve the accreditation process. In 2000, the accreditation process improvement initiative began to implement changes to the accreditation process, including the following:

- Revising the random unannounced survey process
- Addressing staffing effectiveness in the survey process
- Revising accreditation-decision categories
- Eliminating summary scores for organizations
- Reviewing and rewriting standards
- Enhancing use of the extranet
- Conducting periodic performance review (PPR)
- Priority focus process (PFP)
- Developing a new on-site survey agenda
- Enhancing surveyor development initiatives
- Developing new accreditation decision and reporting approaches

JCAHO looks forward to the implementation in 2004 of the new accreditation process: *Shared Visions–New Pathways* (SVNP). The JCAHO mission, "to continuously improve the safety and quality of care provided to the public through the provision of health care accreditation and related services," will not change with the introduction of SVNP, but the path JCAHO will take to achieve this mission will change.

The current survey process has allowed organizations to focus on survey preparation and organization-specific accreditation scores. Under the new accreditation process, less focus will be placed on ramp-up for survey, and instead, on continuous standards compliance. Joint Commission surveyors and staff will work with health care organizations under the new model to continually improve their systems and operations, eliminating the need for "ramp-up" or survey preparation. The components of the SVNP initiative fit together to guide us toward a better accreditation process and, as a result, better health care.

The components of the SVNP initiative were developed through extensive research and testing. They use technology and processes that are common in many industries. Most accrediting bodies use some form of self-assessment, which has proven effective not only in the health care industry, but in academia as well. The priority focus process (PFP) is a decision support tool that uses data to provide a focus on specific information. This type of tool is based on technology used in many industries to help consumers make decisions based on provided information. Similarly, surveyors will use PFP data to select areas of focus during a survey, based on presurvey data that was fed into the PFP tool. The tracer methodology is a further advancement of the individual-centered evaluation approach (ICE) that the JCAHO has been using effectively for several years.

The objects of SVNP are summarized as follows:

- Shift the paradigm from survey prep to systems improvement
- Focus away from the exam and score
- Focus toward using the standards to achieve and maintain excellent operational systems
- Enhance relevance of standards and accreditation process
- Focus on safety and quality of care
- Maintain rigorous but fair evaluation
- Enhance surveyor skill and consistency
- Maximize educational benefit
- Control or lower health care organization (HCO) costs
- Improve customer service

122

Medical Equipment Management Program and ANSI/AAMI EQ56

Ethan Hertz
Duke University Health System
Durham, NC

Most clinical engineers are involved in some way with the management of medical equipment. Modern medical care relies upon a wide variety of equipment to diagnose, monitor, and treat patients. Clinical engineers are often charged with keeping that equipment in good working condition.

In the U.S., the Joint Commission on Accreditation of Healthcare Organizations (JCAHO) accredits most hospitals. The JCAHO's Environment of Care standards require that a hospital design and implement an equipment management program to ensure that the medical equipment used by the hospital is safe, accessible, and effective.

In general, the JCAHO standards describe the broad elements the JCAHO expects to find when reviewing an equipment management program. For the most part, JCAHO standards focus on outcomes. They describe the types of expected results, but leave the methods for reaching those results up to the hospital. It is the hospital's responsibility to demonstrate that its actions actually meet the JCAHO's standards.

In the early 1990s, the JCAHO made the transition from department-focused, prescriptive requirements to function-focused, outcome-oriented standards. More recently, some individuals responsible for meeting the standards expressed frustration about the general nature of the JCAHO's requirements and asked for specific guidance. In response, the JCAHO has added explanatory materials to its accreditation manual, including sections describing the intent of the standards and examples of implementation actions taken by some hospitals to meet those standards. However, even with these improvements, some members of the clinical engineering community saw the need to develop a document to provide a more detailed set of guidelines that would help hospitals understand the elements of an equipment management program from the perspective of equipment management experts.

In 1994, the Association for the Advancement of Medical Instrumentation (AAMI) formed a committee to write such a document. The committee included hospital representatives, equipment manufacturers, independent service organizations, the JCAHO, the American Society for Hospital Engineering, the American College of Clinical Engineering, the Society of Biomedical Equipment Technicians, the United States Food and Drug Administration, and the United States military. After several years of work, the document was approved as an AAMI recommended practice, and accepted by the American National Standards Institute (ANSI) as an American national standard. In 1999, the committee's work was published as ANSI/AAMI EQ56: 1999, "Recommended Practice for a Medical Equipment Management Program."

The intended audience for EQ56 is anyone working in a health care setting who is responsible for any part of an equipment management program, including hospital administrators, clinical engineers (CEs), and biomedical equipment technicians (BMETs). The EQ56 tries to complement the JCAHO standards by fleshing out some of the details that define a truly effective equipment management plan.

EQ56 Basics

The objective of an equipment management program is to effectively utilize the limited resources available to a hospital to keep its equipment functioning safely and effectively. The EQ56 is designed to provide a framework to help the hospital examine its needs and resources and balance them effectively.

Except in a few areas, EQ56, like the JCAHO standards, does not prescribe specific actions. For example, to determine how often any particular medical device should be inspected and what specific activities should be included in that inspection is left up to the hospital. EQ56 details the considerations that should be taken into account when a hospital develops its equipment management program. The document attempts to lead each hospital to consider the most appropriate equipment management choices when designing and implementing a program for that hospital.

Every hospital is different. Hospitals support different patient populations, have different mixes of equipment, and have access to different levels of resources. Therefore, EQ56 does not mandate a single level of activity for each hospital. Instead, the philosophy of EQ56 is best summarized as: "Say what you do, and then do what you say."

Neutrality is a central feature of EQ56. The hospital as a whole is responsible for meeting EQ56's requirements, rather than just a particular department or service provider. Some hospitals meet their equipment management needs by hiring in-house personnel to monitor the condition of equipment and then perform the necessary inspections and maintenance. Other hospitals choose to contract with equipment manufacturers or independent service organizations for similar services. Still others meet their needs by mixing in-house, manufacturer and independent service organizations. Regardless of how the results are achieved, in keeping with the JCAHO focus on outcomes, EQ56 expects the hospital to take the steps necessary to ensure that its particular service arrangements meet EQ56's requirements for tracking and maintaining the hospital's equipment.

A hospital is expected to develop documents that address the elements of EQ56. Throughout EQ56's clauses are questions that a hospital should answer as it develops its equipment management plan.

EQ56 is divided into several sections, each addressing a different aspect of equipment management. Issues addressed in the document include program design, program implementation, including equipment inventory and procedure development, service agent training, acceptance testing, scheduled inspections, staffing, space, financial resources, tools, assessment and planning, equipment selection, removal from service, communication, incident investigation, and the Safe Medical Devices Act.

While this chapter describes many of the requirements of EQ56, it is not a substitute for the actual document. That publication describes the requirements of an effective equipment management program in detail and includes a discussion of the rationale for each requirement. In addition, EQ56 contains situations in which exceptions to its requirements may be appropriate. Anyone responsible for developing an equipment management program can obtain a copy of EQ56 from AAMI. The AAMI website (www.aami.org) provides specific information about ordering this publication.

Specific Requirements of EQ56

The specific requirements of EQ56 are summarized as follows, with a brief description of each of the major areas addressed.

Inclusion Criteria (4.1)

Any equipment management program must answer the fundamental question of what is and is not part of the program. Section 4.1 discusses how the hospital should determine which equipment to include in its program. Specifically, the hospital is expected to develop criteria that consider the equipment's function, physical risks, maintenance requirements, and known incident history. These criteria must be documented. The hospital is expected to examine each medical device it uses and determine whether it should be included in the inventory of managed devices based upon these criteria. The hospital is expected to document whether each new medical device has been evaluated against the program's criteria. Finally, as the hospital changes its inclusion criteria, it is expected to document these changes and the impact they will have on its program inventory.

Inventory (4.2)

The purpose of the inventory is to help the hospital schedule routine inspections for the equipment included in the maintenance program and retrieve the equipment quickly if it should be recalled. Section 4.2 discusses the actual listing of the inventory. The hospital is expected to maintain a list that shows all of the items it has chosen to include in its equipment management program. This list must be comprehensive; all equipment that meets the inclusion criteria developed in 4.1 must be listed, regardless of ownership. While most hospitals have little trouble maintaining records for the equipment owned by the hospital, documentation becomes more of a challenge when the equipment involved belongs to another party, as is the case with loaner, leased, rented, or research equipment. However, a patient should expect that any equipment used to provide hospital-related medical treatments is safe, regardless of its ownership. Most equipment must become part of the inventory, but equipment that is expected to be in the hospital for less than 15 days is an exception to this rule.

The minimum requirements for information to be included in the inventory database are specified in EQ56. These requirements include a unique identification number, the name of the equipment manufacturer, model and serial numbers, a description, the usual location, the department that owns the equipment, identification of the primary service provider, and the acceptance date. EQ56 does allow more information to be collected if the hospital finds it useful. For example, some hospitals include information such as software revision, acquisition cost, estimated useful life, and warranty expiration in their equipment inventory databases.

The JCAHO standards require that the equipment inventory be accurate. EQ56 expands on that requirement by identifying a methodology for auditing accuracy. EQ56 requires that an audit be conducted annually, based on statistical analysis. In this audit, a certain number of devices must be randomly selected, and the inventory database information about the selected devices must be compared with the actual information from the device. For example, if the equipment inventory includes between 501 and 1000 items, then 35 randomly selected pieces of equipment must be included in the audit in order to make it statistically valid. Before starting the audit, the hospital must develop a document that establishes its standards for inventory accuracy and outlines the steps needed to bring the inventory up standards if the audit identifies a deficiency.

Finally, regardless of how the inventory is maintained, the hospital must be able to produce a written copy of the entire inventory listing upon demand. This inventory must be separate from other inventories, such as the hospital's capital asset inventory. In addition, if different service providers maintain separate inventories, all of the inventories should be readily available to combined so that the hospital or surveyor can quickly determine the overall equipment inventory.

Inspection and Repair Procedures (4.2.2)

Section 4.2.2 discusses the requirements for the procedures used to inspect and repair equipment. The hospital is required to develop or adopt procedures that describe what a technician is expected to do when inspecting a device. In addition, each inspection procedure must include an estimate of the time required to perform that procedure. This time estimate determines the adequacy of the equipment inspection and maintenance staff.

In addition to inspection procedures, the hospital is expected to develop or adopt repair procedures. These differ from inspection procedures in that inspection procedures generally provide a significant level of detail. Repair procedures, on the other hand, should include general guidelines for completing a repair and descriptions of the documentation that must be kept about a repair. The hospital's repair procedures are expected to describe how an equipment user obtains repair services and the expectations for alternate care procedures to follow when a critical medical device is not available. Finally, this section asks the hospital to develop a records retention policy for the documentation of completed inspections and repairs.

Service Agents Training (4.3)

Section 4.3 describes the training of service agents. These are the employees who perform equipment maintenance activities. These may be in-house employees, or employees of an outside service group. Regardless, it is the hospital's responsibility to assure that anyone maintaining its equipment is qualified to do so.

Generally, documentation of each service agent's training must be maintained, along with an evaluation process that determines the training needs of each agent. EQ56 recognizes that different agents will have different training needs. Employees responsible for performing equipment inspections are expected to have sufficient training to ensure that they are capable of following maintenance procedures. Other factors to consider when developing a training program include the required safety training and the experience level of each employee. Other elements include the knowledge required to advance to other positions in the organization and training related to industry certifications, such as CBET certification.

There are two quantitative requirements in this section. First, each employee should receive at least 72 hours of training over a three-year period. In addition, a new employee is expected to receive at least 36 hours of training during the first year of employment. This requirement recognizes the need to orient and train a new employee in the practices of the hospital and service organization and to introduce the procedures that the employee is expected to use during maintenance activities.

Acceptance Testing (4.4)

As new equipment arrives in the hospital, it must be acceptance tested (also referred to as an incoming inspection) to show that is in proper working condition before it is used on patients. These requirements are discussed in section 4.4. Acceptance testing should follow procedures established by the hospital. The hospital must consider whether communication should take place between the department or service organization performing the acceptance test and other departments in the hospital. For example, the hospital may want equipment users to know that their equipment has arrived and is ready for installation. In some cases, a separate department is responsible for training equipment users, and that department must know that the equipment has arrived in order to schedule appropriate training. The hospital also might alert the accounting department, in order to release payment and/or initiate appropriate capital asset tracking.

Inspection Intervals (4.5)

Scheduled inspections should take place once the equipment has become part of the hospital's inventory. Section 4.5 addresses the requirements related to assigning and modifying inspection intervals. The hospital also should track failed inspections to justify changes in inspection intervals.

Staffing (4.6)

Staffing requirements are covered in section 4.6. Staff members must be available to perform the scheduled inspections discussed in section 4.2. In addition, the hospital should plan to have sufficient staff available to meet the workload associated with repairs. It is important to note that this requirement, like all the others in EQ56, applies to in-house departments and outside service organizations. For example, if a hospital decides to contract with an outside service provider, it must show that it has reviewed that provider's staffing plan, and determined that it will meet the hospital's anticipated needs for maintenance activities. Generally, the total time required to perform inspections based upon the number of items in the inventory and the estimated inspection procedure times determines staff adequacy. . In addition, time for anticipated repairs and for support activities such as documentation and training, must be factored in.

Space (4.7)

Space is always at a premium in a hospital. Section 4.7 addresses the requirements related to space. Generally, the hospital must show it has allocated sufficient space for work areas, storage of equipment awaiting repair, any replacement parts inventory, and documentation. In addition, the hospital must identify space in which to keep equipment that is not available for use, such as equipment awaiting acceptance tests or equipment under investigation as part of an incident. Finally, safety and sanitation issues must be considered when identifying equipment maintenance areas, meaning that hand-washing facilities are readily available, and adequate ventilation is present in the work area to quickly and safely dissipate any fumes from chemicals used during maintenance.

Financial Management (4.8)

Financial support is critical to the success of an equipment management program. In order to demonstrate adequate financial support for an equipment management program, the hospital must establish a methodology to determine the cost for services and show a budget that covers those costs.

Test Equipment (4.9)

The inspection and service of equipment requires many different types of tools. For example, many procedures require a technician to remove the equipment case in order to test internal components, a job that often requires screwdrivers and pliers. Section 4.9 discusses the availability of tools, ranging from basic hand tools to the sophisticated instruments needed to perform tests, such as checking defibrillator-output levels. In addition to the tools needed for routine work, personal protective equipment must be available. Finally, this section identifies the need for periodic recalibrations of test equipment.

The areas discussed come to mind as the day-to-day responsibilities of medical equipment management. The remainder of EQ56 concerns other components of an equipment management program that are also important but that are perhaps more global in their impact.

JCAHO (4.10)

Section 4.10 addresses the JCAHO standards for performance improvement. The hospital is expected to demonstrate its performance of scheduled equipment inspections, along with another indicator that tracks repairs. At least annually, the hospital should assess its equipment management program to determine improvements, and to plan for any upcoming environmental changes that might impact the program.

Equipment Selection (4.11)

Hospital purchases new equipment as equipment ages and new technologies become available. Section 4.11 addresses the equipment selection process, with the expectation that a 'hospital considers its experience with its current vendors when purchasing new equipment.

Obsolescence (4.12)

At some point, a device no longer functions. To ensure that obsolete or unsafe equipment is no longer used by the hospital, Section 4.12 discusses the development of a "removal from service policy."

Communication (4.13 and 4.14)

Effective communication is important to controlling the activities of an equipment management program. The next two sections of EQ56 deal with communication. Specifically, Section 4.13 addresses the communication between service agents and service providers; i.e., the employees who actually take care of the equipment maintenance activities and the service group or department to which they report. Service agents are expected to respond to changes within their own organizations, as well as to changes within the hospital. To help everyone keep aware of those changes, EQ56 recommends at least one staff meeting per month, at which management will inform employees of any changes. However, it is also the responsibility of employees to inform their managers of any changes they that they thing might affect the service organization's activities. The communication is expected to always be two-way.

Complementing 4.13, Section 4.14 addresses communication between the service organization and the hospital. Changes in the hospital's management, policies, or needs should be communicated to the management of the service organization so it can continue to meet the needs of the hospital. Two major communication responsibilities of the service organization are outlined in this section: to keep the hospital informed about the inspection status of its equipment, and to let the hospital know, if any deficiencies are found that indicate a need for user training.

Incident Investigation (4.15)

Section 4.15 addresses the hospital's approach to incident investigation. EQ56 requires that a qualified person participate in an incident investigation, and the hospital should have appropriate policies to allow the investigation to take place. For example, the policies should preserve evidence and allow for sequestration of the equipment involved. They should also address when and to whom the equipment can be released in order for the investigation to proceed.

Probably most important requirements in this section are those that a hospital takes steps to learn from any incidents that occur, and makes appropriate changes to prevent similar incidents in the future. To facilitate this process, Section 4.15 requires that information about an incident be shared with the people within the organization who can make changes, such as the safety committee and risk management personnel. In addition, the hospital should review the severity of the incident and determine whether the findings related to it warrant reporting information to any outside organization, thus allowing the entire health care industry to learn from the hospital's experiences.

Safe Medical Devices Act of 1990 (4.16)

The final section of EQ56 addresses the Safe Medical Devices Act of 1990. Essentially, Section 4.16 requires that the hospital comply with its legal obligations under this federal legislation.

Exclusions

While EQ56 is fairly comprehensive, it specifically excludes activities related to user training. Many equipment management programs are, quite properly, involved in user training, but the committee found that there was not enough consistency within hospitals to include user training in the final publication. Some hospitals rely on special departments, such as nursing education, to train users. In other cases, hospitals adopt a peer-training model, where one user trains other users. In a few cases, the hospital's in-house clinical engineering department is directly responsible for user training. While the EQ56 acknowledges the importance of effective user training, the variety of models found for user training and skills assessment led to the exclusion of user training requirements from the final document.

Summary

The "Recommended Practice for a Medical Equipment Management Program" is a document that is helpful to anyone involved with a medical equipment management program. It is an American national standard, developed by a committee of equipment experts representing hospitals, manufacturers, professional organizations, JCAHO, and the federal government.

The document identifies the key elements of an equipment management program, and establishes a community standard for how these elements should be incorporated in a hospital's program.

EQ56 discusses the documentation that should support the planning and implementation that are part of an equipment management program. It discusses the key aspects of a program, including program design, equipment inclusion criteria, equipment inventory, inspection and maintenance procedures, training, resources, communications, equipment selection, and incident investigation.

Reference

Association for the Advancement of Medical Instrumentation. *Recommended Practice for a Medical Equipment Management Program,* EQ56. Arlington, VA, The Association, 1999.

123
Clinical Engineering Standards of Practice for Canada

Tony Easty
Head, Medical Engineering Department
The Toronto Hospital
Toronto, Ontario, Canada

William M. Gentles
BT Medical Technology Consulting
Toronto, Ontario, Canada

This chapter outlines the development process of the *Clinical Engineering Standard of Practice for Canada* and the rationale behind what was included and what was omitted. It also outlines the development of a peer review process based on the standard practice, which is meant to act as a quality improvement tool.

The practice of clinical engineering has evolved over the years as the needs of the health care systems that it serves have changed. Unlike many professions with a support role in health care, such as physiotherapists, respiratory therapists, or pathologists, the roles and responsibilities of clinical engineers have been relatively ill defined. Clinical engineering services evolved in different directions depending on the skills and interests of the staff. As the profession has matured, there is a growing consensus that a core set of activities should be included in any clinical engineering service. This chapter outlines the development of a standard of practice for clinical engineering by clinical engineering professionals in Canada.

The Canadian clinical engineering community is relatively close-knit, despite the size of the country. Most engineers are employees of hospitals or groups of hospitals, rather than private industry. This situation has evolved due to the single-payer government-funded health care system in Canada. Consequently, the high degree of information-sharing and collaboration has made the development of a standard of practice easier than it would be in a for-profit health care environment.

Identification of the Need for a Standard of Practice

The Canadian equivalent of the United States' Joint Commission on Accreditation of Healthcare Organizations (JCAHO) is the Canadian Council on Health Services Accreditation (CCHSA). Prior to 1990, the CCHSA accreditation document had a specific and comprehensive section on medical devices. This section outlined what hospitals should have in place to manage technology. It had been written with input from the Canadian Medical and Biological Engineering Society (CMBES), with a committee led by Jim McEwen.

In 1990, the CCHSA and the JCAHO shifted the focus of accreditation to a patient-centered model. The emphasis on prescriptive structural elements was reduced and replaced with an emphasis on patient outcomes and quality of care. It was left to the individual organizations to organize support services and manage technology. It was also left up to individual professions to set their own standards. This shift led to a fear that the justification for well-organized clinical engineering services would diminish.

This shift in accreditation standards caused much discussion within the clinical engineering community in the U.S. and Canada. At the Canadian Medical and Biological Engineering Conference in Vancouver in 1994, the CMBES clinical engineering special interest group agreed that clinical engineering would benefit from a published standard of practice that would provide clinical engineering departments or services across the country with a common yardstick against which they could measure themselves. At the same time, the standard should provide the basis for a peer review mechanism, which would provide stronger incentives for services to meet it. In addition, a peer review process would provide a reference point for the CCHSA, which was interested in any external reviews of services within a health care organization. This process was an important step in the development of clinical engineering as a recognized profession, since self-regulation is a cornerstone of most professions.

Development and Approval of the Canadian Standard of Practice

A committee consisting of Tony Easty and Bill Gentles was appointed the task of setting up a process for developing a standard that could be adopted by the CMBES. They recruited the assistance of Ted McLeod, a clinical engineering master's student at the Institute of Biomaterials & Biomedical Engineering at the University of Toronto, to assist them in the project.

This group had a mandate to develop a draft standard of practice for the society. They reported to the chair of professional affairs of the society, who at that time was Gordon Campbell. They began in 1994 by investigating standards of practice in other health disciplines in Canada and around the world. They obtained and reviewed copies of standards related to social workers, psychologists, medical radiation technologists, occupational therapists, respiratory therapists, dieticians, nurses, physicians, and pathologists; two professions from which the group borrowed heavily in the development of the standard were physiotherapists and pathologists. The physiotherapists had developed standards of practice and a peer review mechanism that seemed to align well with the clinical engineers' stated mandate to define the scope and role of clinical engineering in Canadian health care organizations. The College of American Pathologists (CAP) has operated an accreditation program for laboratories for the past 35 years. JCAHO accepts the inspection results from CAP accreditation. Participating laboratories must agree to provide an inspection team equal in size and complexity to that required for its own inspection.

The next step in the development process was to write a draft in collaboration with a review committee of approximately 18 clinical engineers from across the country. This first draft was presented at the CMBES annual conference in Montreal in September 1995. A lively and extended discussion in a packed room indicated great interest within the profession, and the meeting led to numerous amendments of the draft standard.

Amended drafts were circulated to the working group in 1995 and early 1996 for fine-tuning. A polished draft was presented at the CMBES annual conference in Charlottetown, Prince Edward Island, in June 1996. After further revision, Draft Five was circulated to the full membership of the society in November 1996, requesting a vote of acceptance. An overwhelming majority of members voted in favor, and the society adopted and published the standard.

Underlying Philosophy of the Standard

Over the course of development of the standard, a fairly clear underlying philosophy evolved that dictated what should be included and what should be omitted. The first objective was to define the scope and role of clinical engineering in Canadian health care organizations. Of equal importance was the definition of standards suitable for peer review measurement and evaluation program similar to that of the College of American Pathologists.

After much debate, a consensus emerged that the standards should describe only what should be done, not who should do it or what their qualifications should be. The wide range of qualifications of people managing clinical engineering services across the country necessitated this approach. In smaller hospitals, the manager of the clinical engineering group was usually a technologist. In larger hospitals or regional services, the manager was usually an engineer. It was important to develop a standard that could be used by an organization of any size of organization and management structure.

Elements of the Standard

The standard has eight major sections, outlined as follows.

1. PREAMBLE

States the goals of the standard as follows:

- To define the scope and role of clinical engineering in Canadian health care organizations
- To define standards for measurement and evaluation

2. SERVICE MANAGEMENT

- Organization
- Personnel requirements
- Policies and procedures
- Facilities

3. Medical DEVICE TECHNOLOGY Management

- Device tracking and inventory
- Acquisition
- Unscheduled maintenance
- Scheduled maintenance.

4. TECHNOLOGY ASSESSMENTS and Planning

- Planning and pre-purchase evaluation
- Assessment of safety, efficacy and cost-effectiveness
- Long-range device planning
- Knowledge of emerging technologies.

5. RISK MANAGEMENT

- Ensure that devices conform to relevant safety standards.
- Re-use policy development
- Management of hazard alerts and recalls
- Organization of risk management program
- Incident investigation

6. QUALITY MANAGEMENT

- Identification of quality goals
- Consider customer input
- Establish a quality system structure
- Integration with the organization's quality programs
- Staff training on quality issues
- Failure analysis
- Annual internal review of quality program

7. EDUCATION

- Education of service staff
- Education of device users
- Education of clinical engineering interns and students

8. RESEARCH AND DEVELOPMENT and the MODIFICATION of medical devices

- Consider scope of involvement
- Devices are in alignment with organization mission and goals
- Participation of service staff
- Publications and presentations
- Ethics requirements of the organization
- Modification of devices
- User involvement in design and modification

Issues Left Out of the Standard

Numerous items were deliberately left out of the standard as a reflection of the realities of the profession. It was agreed that any mention of the necessary qualifications for the head of service would be left out, since it was clear that in many cases a highly skilled and experienced technologist managed a clinical engineering service. An attempt to mandate qualifications was deemed self-defeating because the standards did not have the authority of government legislation.

In Canada, engineering is a regulated profession. This means that by law, no one can work as an engineer without a license. Technicians and technologists are also regulated to a limited extent. The standards were designed to avoid any overlap with existing legislation. Since existing legislation already governed the ethical behaviour of engineers and technologists, a code of ethics was not incorporated or mentioned in the standard.

The standard mentions certification as a valid educational objective for service staff, but it does not require that any staff be certified. After much debate, certification was deemed outside the scope of a standard of practice that does not have the force of law.

Prescriptive specifications, such as number of technical staff, number of square feet, and frequency of preventive maintenance (PM) interventions, were omitted from the standard. These were seen as moving targets that would quickly make the document obsolete. Such practices would be assessed during a peer review to confirm adherence to accepted practice. The reviewers would be expected to have knowledge of the current specifications.

Development of a Peer Review Process Based on the Standard

From the outset, the standard assumed that a peer review process would be required. Peer review remains important for the following reasons:

1. It is an effective quality improvement tool.
2. It ensures that experienced professionals assess clinical engineering services.
3. It enhances the sharing of ideas in the clinical engineering community.

4. A standard that is not associated with an ongoing review process will be ignored.
5. Changes in accreditation standards mean that clinical engineering services will get less scrutiny in an accreditation survey.

Debates addressed the risks involved in subjecting an in-house service to peer review. Ill-informed administrators might use it as a way to cut service or hand clinical engineering over to an outside vendor. The counterargument accepted by the majority of CMBES members was that in times of restraint, any hospital support service might be subjected to an external review. If the profession does not have its own review process, administrators will find so-called experts outside the field to perform reviews, which will inevitably be ill informed.

Once the standard was adopted by the society, the same task force started to draft a workable peer review process that could be used by large and small services alike.

It was agreed that the development of the peer review process should go through the following stages to national implementation:

1. Definition of process
2. Create of supporting documents
3. Recruit test sites
4. Conduct pilot reviews
5. Refine of process
6. Implement nationally

In May 2001, the development had reached the implementation phase. Four pilot reviews were conducted in the spring of 2000, and all participants considered them highly successful and valuable. . The process was refined and approved at the CMBES annual meeting in Halifax in October 2000.

Process Definition Document

An overview of how the peer review process should function appears in the Peer Review Process Definition document. The sections of the document are as follows:

1. Policies

- Peer review process is managed by CMBES peer review committee
- Peer review committee reports to CMBES professional affairs committee
- Peer review committee members are a mix of engineers and technologists
- Peer review committee responsibilities
  - Oversees nomination of survey team and approves composition of team
  - Defines responsibilities of survey team
  - Responsible for training surveyors
  - Reviews pre-survey questionnaire to determine if site is ready to be surveyed
  - Reviews post-survey questionnaires
  - Maintains database of peer review activities
  - Revises and updates process support documents (see below) as required
- Standards. The process is based on *Clinical Engineering Standards of Practice for Canada*
- Survey team: The process of selecting surveyors, and determining size of the survey team.
- Peer review interval: Three years between reviews

2. Peer review process

- Application for survey by a clinical engineering service
- Preparation for survey by the service
  - First step is a self-appraisal using the pre-survey questionnaire
  - Completed self-appraisal is submitted to the peer review committee with a list of proposed surveyors.
- Survey process
  - Preparation activities of peer review committee and survey team
  - Timelines for the survey process
  - Activities during the survey team visit
  - Post-visit reporting requirements
- Peer review process is self-funding
  - No survey fees
  - Surveyors pay own expenses
- Final product
  - A written report outlining strengths, weaknesses, and priorities for improvement, is sent to the device that was surveyed.

Process Support Documents

The supporting documents that make the process function smoothly are as follows:

1. Pre-survey questionnaire

- Includes a checklist based on the standards of practice, to be filled out by those being surveyed
- Requests additional information about the service, such as organizational charts and policy and procedure manuals

2. Standards of practice (document on which the survey questionnaire is based)
3. Survey questionnaire

- Checklist based on standards of practice
- Similar to pre-survey questionnaire
- Completed by surveyors

4. Surveyors' post-survey questionnaire

- Asks surveyors for suggestions on how to improve the survey process

5. Surveyee post-survey questionnaire

- Asks site that was surveyed for suggestions on how to improve the survey process
  Surveyor guidelines
- Training tool for surveyors
- Promotes a collaborative, rather than judgmental, approach

Contentious Issues—Fees and Awards

Two elements of the above-mentioned process have been the subject of considerable debate. The first is the lack survey fees. This recommendation is based on the difficulty that smaller departments would have in securing funds. In addition, the benefit to the surveyors from conducting surveys is at least as great as the benefit to the site, given the exchange of ideas that takes place. The rule that no fees are charged will force participants to find the most economical way to conduct the surveys. For example, it prevents sites in Newfoundland, on the east coast, from seeking surveyors in British Columbia, on the west coast, due to travel costs.

The second controversial element is awards. After the four pilot surveys, the consensus of all participants was that the chief value of the process was as a quality improvement tool, with a checklist for areas of improvement from experienced outside reviewers. Hospital administrators saw the value of an award as questionable at this point in time, partly due to the lack of understanding of its significance. In addition, the absence of an award meant that the site being surveyed was not purposefully putting the best foot forward and glossing over weaknesses. Without the complication of an award, both the surveyors and surveyees are working to find ways to improve the service.

Summary

This section describes the development of a standard of practice and peer review process. The changing accreditation standards were among the forces that led to the development of this standard. Another stimulus was a consensus about the essential core activities of a clinical engineering service.

The importance of associating a peer review process with a standard of practice cannot be overemphasized. One of the principal activities of a profession is self-regulation, and peer review is one form of self-regulation. In addition, any standard that is not associated with a periodic review mechanism is often ignored.

The development of the documents and processes described in this chapter is an important step in the maturity of clinical engineering as a profession.

Further Information

Canadian Council on Health Services Accreditation (CCHSA) Steps to Accreditation. www.cchsa.ca.
Canadian Medical and Biological Engineering Society. www.cmbes.ca.
Joint Commission on Accreditation of Health care Organizations (JCAHO) Accreditation Information. www.jcaho.org.
McEwen J. *Clinical Engineering and Hospital Accreditation in Canada.* Saskatoon, CMBES, 1981.

124

Regulations and the Law

Michael Cheng
Health Technology Management
Ottawa, Ontario, Canada

Webster's International Dictionary defines "regulate" as "to govern or direct according to rule." Regulations are mandatory, but the mandate might or might not be legally based. When the mandate comes from the government, regulations are based on laws established by the government. This chapter describes, in the North American context, the relationship between regulations and the law.

Different countries and different levels of governments have different legislative systems and procedures. In the U.S. and Canada, medical devices are regulated by the federal governments, where the first-level laws, called "acts" or "statutes," are established by the Congress (in the U.S.) or the Federal Parliament (Canada). Acts and statues are usually brief statements covering a wide scope of applications. Medical device regulations derive from the legal power of relevant acts and statutes, and the regulations provide specific and detailed statements. For example, a statement such as "No person shall sell unsafe or ineffective medical devices" in an act or a statute can lead to a volume of medical device regulations specifying requirements for product approval, vendor registration, and after-sales obligations.

Guidance documents often are written after the regulations or standards are established. They normally use plain language to provide interpretations of the regulations and standards, usually with added specific details. See, for example, Chapter 122 as a guideline. Figure 124-1 shows the relationships and differences among acts/statutes, regulations, standards and guidelines relevant to medical devices.

Note that Governments normally consult all stakeholders, but that they do not necessarily need to seek consensus as in most international standards (Chapter 117). A voluntary standard can be recognized fully or partially by a government as part of regulatory requirements by a government.

Regulating Medical Devices

Compared with drug regulations, medical device regulations are relatively new. Canada and the United States started to regulate medical devices in the mid 1970s (see Chapter 126). The first harmonized European Union Directive on Active Implantable Medical Devices came into force in 1993. One can see that this field is evolving, particularly to keep pace with the rapid technological innovations. Because of the explosive growth in medical devices, estimated by the U.S. Food and Drug Administration Center for Devices and Radiological Health (USA FDA CDRH) in June of 2001 to be 10,000 new devices per year, and the concerns for safety, effectiveness, quality, and the impact on health care costs, many other countries are only beginning to set up the regulatory or monitoring programs for medical devices.

Different countries have been following different standards and regulatory systems. This chapter describes the general concepts and approach used in the U.S., the European Union, and Canada. Detailed descriptions of the U.S. system and the European Union system are found elsewhere in this handbook (see Chapters 125 and 126).

Glossary

In regulations, terms are legally binding and therefore have restricted meanings. For example, *manufacturer, distributor, vendors,* and *retailers* all have precise definitions in regulations; and their definitions vary in the regulations of different countries. A regulation normally has an accompanying list of definitions of terms used. This chapter provides an integrated description of three different medical device regulations. It is written to promote a *general understanding* of medical device regulations; therefore, the words used are nonbinding but carry general meanings. The meaning of several terms that could cause confusion are described below for the purpose of this chapter

Manufacturer   Any person who makes medical devices

Vendor   Any person who sells medical devices. This person could be a manufacturer, an importer, a distributor, a wholesaler, or a retailer.

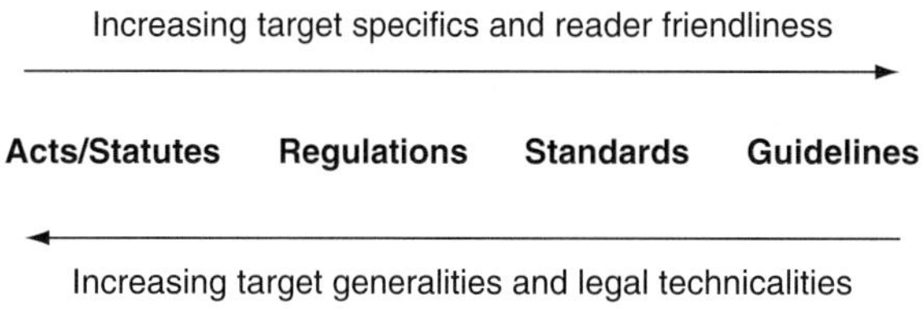

Increasing target specifics and reader friendliness →

Acts/Statutes   Regulations   Standards   Guidelines

← Increasing target generalities and legal technicalities

Figure 124-1 Relationships among acts/statutes, regulations, standards, and guidelines relevant to medical devices.

Person   Includes an establishment (in that case, person in charge or person responsible).

Effectiveness   A device is clinically effective when it produces the effect intended by the manufacturer relative to the medical conditions. For example, if a device is intended for pain relief, one would expect the device to actually relieve pain and would expect the manufacturer to possess objective evidence, such as clinical test results, that the device in fact does relieve pain. Effectiveness can be thought of as efficacy in the real-world clinical environment.

Efficacy   Not used in this guideline; generally means effectiveness under an ideal, controlled setting.

Performance   Means technical performance plus effectiveness.

Risk Management in Medical Device Regulations

Most of the new medical device regulations follow the risk-management philosophy. All devices carry a certain degree of risk that could cause problems under specific circumstances. Many medical device problems cannot be known until extensive market experience is gained. For example, an implantable device may fail in a manner that was not predictable at the time of implantation; the failure could be caused by conditions that are unique to certain patients. For nonimplant devices, component failure can occur in an unpredictable, random manner. One practical approach to device safety is to estimate the potential of a device becoming a hazard that could cause safety problems. This approach is often referred to as the *risk-assessment* of devices.

Hazard means a potential for an adverse event, a source of danger. *Risk* is a measure of the combination of (1) the hazard; (2) the likelihood of occurrence of the adverse event; and (3) the severity or overall impact. *Risk assessment* begins with *risk analysis* to identify all possible hazards, followed by *risk evaluation* to estimate the risk of each hazard. In general, risk assessment is based on experience, evidence, computation, or simply guesswork. Risk assessment is complex; it is influenced largely by personal perception and other factors, such as cultural background, economic conditions, and political climates. This field of knowledge is still developing.

In practice, the risk assessment of medical devices is based on the experience of health care professionals and safety-design engineering. In the United States, the government risk assessment of medical devices is based mainly on recommendations from members of 16 medical-specialty classification panels. Devices are categorized into three classes. In the European Union and Canada, the classification schemes for medical devices are predominantly rule based. These rules categorize medical devices according to their perceived potential hazard factors. Canada assigns four classes of devices. The European Union assigns three classes, with class II divided into IIa and IIb (effectively, thus, four classes). The Global Harmonization Task Force (GHTF) (see below) is proposing a harmonized scheme for Medical Devices Classification (see www.GHTF.org document SG1/N015R14).

In general, potential hazard areas that need to be considered for device classification include the degree of invasiveness, the duration of contact, the body system affected, and local-versus-systemic effects. Invasive devices are usually considered to have higher potential hazards than equivalent noninvasive devices (e.g., blood pressure monitors). Similarly, devices that have longer duration of contact, devices affecting vital organs (such the heart or the great arteries), and devices with systemic effects are assigned higher potentials of hazard or risk classes. The degree of regulation imposed on any device is proportional to its potential hazard. This approach is known as *risk management*.

Risk management should extend throughout the life span of medical devices and it requires the co-operations from all stakeholders: manufacturer, government, vendors, users and the public (Cheng, 2003).

Critical Elements for Regulatory Attention

The critical factors that affect safety and effectiveness of a medical device and ways that regulatory authorities control these elements through different stages are identified here. The common regulatory terms *pre-market* and *post-market* will be introduced and later illustrated with the device life span diagram (Figure 124-4). The term *on-market* is not an official regulatory term but is introduced here to provide a logical understanding of the regulatory mechanism.

The safety and effectiveness of medical devices depend on the two intrinsic critical elements of *product* and *use* as seen in Figure 124-2.

Governments use *pre-market review* to control the product and *post-market surveillance/vigilance* to ensure that medical devices in use continue to be safe and effective.

Representation of the product to the user is the important third element (Figure 124-3) and is controlled through labeling (during the pre-market stage) and advertisement (during the on-market stage). A powerful component of product representation, however, verbal presentation by the vendors, is difficult to document and control. Here, user and public education is essential to guard against misrepresentation. Here, the clinical engineer can play an important role.

Stages of Regulatory Control

During the life span of a medical device, these three critical elements are affected by regulatory control (Cheng, 2003) illustrated in Figure 124-4.

These phases of activities are simplified for the ease of understanding the regulatory system. For example, the development phase may include pro-type testing and clinical trials, and in practice, these phases may overlap and interact. The terms *pre-market* and *post-market* cover different phases and activities in different countries. Placing on-market is introduced in this guide for easy overall understanding.

- *Pre-market control* is performed on the device to ensure that the product complies with acceptance criteria before entering the market.
- *Establishment (company) control*, in effect, imposes on the vendor the obligations of post-market surveillance and provides the government with records for tracking purposes.
- *Advertising* control is maintained to prevent misleading or fraudulent claims by uninformed or unethical merchants.
- *Post-market surveillance/vigilance* includes the maintenance of official and systematic procedures and records for device safety and performance throughout its useful life.

Regulatory Tools and Requirements

The tools and requirements used at these stages in the medical device regulatory systems of Canada, the European Union, and the U.S. are described and summarized in Table 124-1.

Pre-Market Review

Although different authorities have different systems of pre-market review, they all use the risk-management philosophy. All medical devices must satisfy safety and performance, quality-system (some low-risk devices may be exempted), and labeling requirements. However, the degree of regulatory scrutiny increases with the potential risks of the medical devices. This approach is demonstrated by using the risk-based device-classification system (SG1-N015R14) proposed by the Global Harmonization Task Force (described below in this chapter).

In Canada, devices of classes III and IV are subject to in-depth regulatory scrutiny, while class II devices require only the manufacturer's declaration of device safety and effectiveness before sale. Class I devices are exempted from any pre-market submission, but they still must satisfy the safety, effectiveness, and labelling requirements.

In the European system, manufacturers of devices of classes II and III, as well as devices of class I with either measuring function or sterile requirements, must submit to the regulator (i.e., the competent authority): (1) a Declaration of Conformity to the appropriate EC Directives; and (2) the conformity-assessment procedure followed. In addition, for higher-risk class devices that require design examination or type examination, the

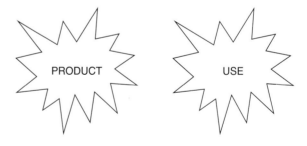

Figure 124-2   Product and use: Two critical elements.

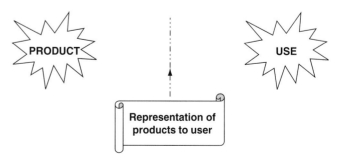

Figure 124-3   Product representation: Third critical element.

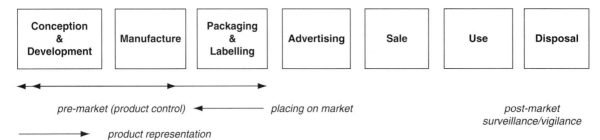

Figure 124-4 Common stages of government control.

corresponding EC-certificates issued by a Notified Body must be submitted to the competent authority. Other medical devices of class I are exempted from pre-market submissions, although they must follow labelling requirements and the essential principles of safety and performance in their design and construction.

In the United States, most Class III and new devices that cannot be found to be substantially equivalent to a legally marketed product that does not require an approved pre-market approval (PMA) application require clearance through the PMA or Product Development Protocol (PDP) process. Most class II, and some class I, devices require pre-market entry notification under paragraph 510(k) of the Medical Devices Act of 1976. A *510(k) submission* to the FDA comprises an information package that is subject to less-stringent review than the PMA process. Most class I, and some class II, (low-risk) devices are exempted from 510(k) submissions before sale but are still subject to general control requirements. The 510(k) submission is a presentation of the proposed medical device that is intended to show how it is substantially equivalent to a medical device that is already on the U.S. market (Winston, 2003).

Different authorities acknowledge product clearance for the market in different ways. In Canada, it is the Device License given by the Therapeutic Products Directorate (TPD). In the EU, after receiving the EC certificate for a notified body, the CE mark is placed by the manufacturer on or with the device. In the United States, the manufacturer of the device receives a 510(k) Marketing Clearance or Approval Letter (PMA) from the FDA.

Vendor Establishment Control

Vendor information allows the government to track vendors of medical devices in the country. In Canada, any individual or company wishing to sell medical devices must apply for permission to obtain an establishment license. The EU requires that a responsible person from the vendor establishment with a physical address in Europe be registered. In the United States, the establishment (i.e., manufacturer, initial importer, specifications developer, contract sterilizer, repackager and/or relabeller) must be registered with the FDA. With all three authorities, the licensing or registration process also imposes an obligation on the vendor for post-market surveillance/duties.

Advertising Control

Advertising control is an important tool to ensure that the public is protected from misleading and fraudulent claims put forth by unethical or uninformed merchants.

Post-Market Surveillance/Vigilance

The continued assessment of safety and effectiveness of medical devices in use is as important as pre-market scrutiny (see Chapter 139). The proof of a medical device is in the way it stands up to the conditions of use. No amount of rigor in any pre-marketing review process can predict all possible device failures or problems arising from device misuse. It is through actual use that unforeseen problems related to safety and effectiveness can occur.

For some devices, post-market surveillance serves to complement pre-market data. The FDA has issued the following statement:

"Pre-market testing cannot address all devices-related concerns. While post-market surveillance will not be used in lieu of adequate pre-market testing, post-market surveillance can serve to complement pre-market data. Certain issues that arise during pre-market evaluation of a device may be more appropriately addressed through data collection in the post-market period rather than prior to approval/clearance for marketing. FDA will consider the potential to collect post-market surveillance data to allow more rapid progress to market."

Post-market vigilance generally includes a system for registering and investigating adverse incidents relating to the use of a device and for requiring the manufacturer to recall or modify a defective device. All three authorities have mandatory requirements for vendors and manufacturers to report all device-related incidents that have resulted, or that could result, in serious injury or death. In the U.S. and some EU member states, mandatory problem-reporting is extended to users.

Quality System Requirements

A *quality system* is defined as the organizational structure, responsibilities, procedures, processes, and resources needed to implement quality management. A closely related term is *Good Manufacturing Practices* (see Chapter 126).

The international quality system standards for medical devices are *ISO13485:1996* and *ISO13488:1996*. *ISO13485:1996* includes all of the elements of *ISO9001:1994* plus a set of minimum supplementary requirements for medical devices (Simmons, 2003). *ISO13488:1996* is the same as *ISO13485:1994* without the design control requirements. A standard alone *ISO13485:2003* is being developed to become the new international quality standard for medical devices.

Regulatory requirements for quality systems may cover the methods, facilities, and controls used by the manufacturer in the design, manufacture, packaging, labelling, storage, installation, and servicing, and post-market of medical devices. Therefore, the influences of quality system requirements can extend throughout all phases in the life span of medical devices. Applicable requirements depend upon the risk class of the device and on the regulatory system of the country. Design control is normally not required for regulatory scrutiny in medium- to low-risk class devices.

When applied to the manufacturing process, quality-system requirements impose strict quality control assurance on every aspect of production. The result is a tightly controlled manufacturing system, commonly known as Good Manufacturing Practices" (GMPs), which reduces the chance of nonconforming products. This practice ensures consistency in the quality of the products and provides the basis for a greater reliability on device safety and performance. Elements of the quality system are periodically subject to audits, management review, and corrective or preventive actions that will maintain the quality of the product. The continuous monitoring and corrective action requirements are interrelated to post-market surveillance described above.

The key advantage regarding quality systems is that they represent a preventive approach to ensuring medical device quality versus the previous reactive approach by inspection and rejection at the end of the manufacturing line. Prevention has been proven to be more efficient and cost-effective in controlling processes and in maintaining medical device quality.

Because the majority of medical devices are in the medium- to low-risk classes, their compliance to regulations often depends upon the declarations of manufacturers, and therefore the question of quality assurance naturally arises. This is why it is important for manufacturers to conform with quality system standards. Conformity with quality-system standards is subject to periodic audit by government or third party agencies.

Table 124-1 Tools and General Requirements of Three Different Authorities

Country	Pre-market	Placing on-market		Post-market
	Product control Tools for acknowledging product cleared for the market	Medical device establishment control	Advertising control	Vendor after-sale obligations Examples of common requirements
Canada	Device license	Establishment license	Generally, prohibition of advertisement before a device is cleared to enter the market. Prohibition of any misleading or fraudulent advertisement	• Problem reporting • Implant registration • distribution records • recall procedure • complaint handling
European Union	Compliance label (the CE mark)	Responsible person registration		
United States	Approval Letter (PMA) or Marketing Clearance (510k)	Establishment registration		

Table 124-2 Standards used by Canada, the EU, the U.S

Country	Standards/regulations	Conformity assessment
Canada	ISO13485, ISO13488	Third party
European Union	EN46001* or ISO13485 EN46002* or ISO13488 EN46001, EN46002	Third party
United States	QS (21 CFR part 820)	Government

Table 124-2 shows the standards used by the three authorities. The applicable standard is determined by the risk class of the device.

The Global Harmonization Task Force (GHTF)

The purpose of the Global Harmonization Task Force, established in 1993, is to encourage convergence in requirements and regulatory practices related to ensuring the safety, effectiveness, performance and quality of medical devices, promoting technological innovation and facilitating international trade. The primary way in which this is accomplished is via the publication and dissemination of harmonized guidance documents on basic regulatory practices. These documents, developed by study groups formed by experts from different countries, can then be adopted and implemented by member national regulatory authorities.

The founding members of the GHTF are Australia, Canada, Japan, the European Union, and the United States. Technical committee members include representatives from national medical device regulatory authorities and the regulated industry. The tasks of the four study groups are as follows (Cheng, 2002):

Group 1 has been charged with comparing operational medical device regulatory systems around the world and from that comparison, isolating the elements and principles that are suitable for harmonization and those that may present obstacles to uniform regulations. In addition, the group is also responsible for developing a standardized format for pre-market submissions and harmonized product-labelling requirements.

Group 2 is charged with the task of reviewing current adverse-event reporting, post-market surveillance and other forms of vigilance for medical devices and performing an analysis of different requirements among countries with developed device regulatory systems with a view to harmonizing data collection and reporting systems.

Group 3 is responsible for the task of examining existing quality-system requirements in countries having developed device regulatory systems and identifying areas suitable for harmonization.

Group 4 has been charged with the task of examining quality system auditing practices (initially among the founding members of the GHTF) and developing guidance documents laying harmonized principles for the medical device auditing process.

Guidance documents developed by these groups can be found at www.ghtf.org

Post-Market Surveillance/Vigilance on Medical Devices

No amount of scrutiny in any pre-marketing review can predict all possible device failures. The continued assessment of medical devices in use is as important as product control. Post-market surveillance and vigilance has a vital role in ensuring the safety and effectiveness of medical devices. The clinical engineer, with technical knowledge and direct user experience, is in a unique position in this vital role (Chapter 139).

Clinical engineers are urged to play an active role in post-market surveillance/vigilance on medical devices and report problems or potential problems to manufacturers and regulatory authorities. However, problem reporting in medical devices has become a controversial issue.

There seems to be a fear that information in medical device problem reporting can become a personal legal liability if it turns out that the problem was actually caused by incorrect use of the device. Therefore, there is a general reluctance to report problems associated with the use of medical devices, even though such reporting is mandatory by law in the United States and some countries in Europe. Regulatory authorities in Europe and North America are experiencing a lack of co-operation from device users.

The clinical engineer is in a position to bring a solution to this dilemma. The job description of a clinical engineer normally would include responsibilities such as technology assessment, selection, acquisition, installation, maintenance, and consultation for hospital staff on safe and effective use of medical equipment. Thus, if a problem or accident associated with medical devices occurs, it would be the duty of the clinical engineer to participate in the investigation and to distinguish between a device problem or a use error. If it is a use error, it is the clinical engineer's duty to inform or instruct the user. If it is a device problem, there should be no hesitation to report to the manufacturer and the regulator. Frequently, use error can be prevented or minimized by design, using human factors engineering principles (Chapter 83) (see www.fda.gov/cdrh/humfac/1497.html). Therefore, unless it is clearly a user blunder, a use error should be followed up with an investigation for possible corrective actions by design.

References

Cheng M. *A Guide for the Development of Medical Device Regulations.* Geneva, World Health Organization, 2003.
Cheng M. *A Guide for the Development of Medical Device Regulations.* Washington, DC, Pan American Health Organization/World Health Organization, 2002.

125

European Union Medical Device Directives and Vigilance System

Nicolas Pallikarakis
Department of Medical Physics
University of Patras
Patras, Greece

Over the past two decades, an explosion of new technologies and applications has occurred in medicine, opening up amazing possibilities for diagnosis and therapy, but also raising questions of appropriateness, safety, and effectiveness (van Gruting, 1994). The activity and accelerated development in this field are often based on an uncontrolled spread and use of medical technology, as well as a noncritical acceptance of innovations. Additionally, the rising cost of health care, due in part to new technology, is a pressing issue for many governments. This situation is expected to continue, and by the year 2010, advances in biosensors, molecular and cellular engineering, and artificial organs will change the shape of health care, as will developments in more traditional areas such as signal processing, imaging modalities, and the application of telematics. The increasing cost and sophistication of biomedical technology inevitably leads to a greater need for regulation and risk monitoring. Patient and user protection should not be compromised, but neither should progress be inhibited.

The global market for medical technology products (excluding pharmaceuticals) was estimated at about $120 billion in 2000, with Europe sharing more than one-fourth of this world market (Bethune, 1995; Mattox, 1995; Jennett, 1992). However, fragmentation of the European market due to the inconsistencies in national regulations did not allow European manufacturers to take full advantage of this opportunity. In order to remove these internal barriers and to facilitate the free movement of goods, the European Union (EU) has adopted the "New Approach" on technical harmonization.

Under this "New Approach," the Commission of the European Communities is working on a harmonized, community-wide legislation, which is addressed in the medical device sector by three directives: the Active Implantable Medical Device Directive (AIMD), the Medical Device Directive (AIMD), the Medical Device Directive (MD), and the In Vitro Diagnostics Directive (IVDs). The EU medical device legislation covers a large range of products. Therefore, it must address a range of risks associated with medical devices. The legislation relates to such items as electromedical equipment, surgical instruments, implants, single-use products, reagents for in vitro diagnosis, and laboratory equipment for clinical laboratories, all of which are covered by the term *medical device*. The relevant EU directives govern the placement on the market of products not only within the 15 member states of the European Union but, in practice, the whole European economic area, including Norway, Iceland, and other associated countries.

The "New Approach" and the "Essential Requirements"

The ultimate aim of the harmonization process was to introduce a set of common rules throughout the EU under which medical devices would be sold and used, thus creating the right conditions for a single market. This policy influences the European medical device manufacturers or their authorized representatives by allowing the free movement of products across national barriers. Most of these barriers existed as a result of national laws and regulations concerning the safety of products. In order to lift these barriers, the European Community introduced the directives, a set of community legal acts enacted by the Council of Ministers that oblige member states to harmonize their national regulations and administrative measures in response.

Under the "New Approach" scheme for the implementation of these directives, each member state must nominate competent authorities (CAs), which in turn, designate notified bodies (NBs). The "New Approach is as follows:"

- Directives should contain essential requirements (ERs) with which products must comply.
- Detailed technical provisions will be contained in the standards adopted by the European Committee for Standardization (CEN), online at www.cenorm.be, and the European Committee for Electrotechnical Standardization (CENELEC), online at www.cenelec.org.
- Standards are voluntary, but a product that conforms to them is presumed to comply with the corresponding ERs.
- A product that complies with the requirements of a directive is in free circulation, i.e., authorities of a member state may not stop its entry into the country under normal circumstances.

All medical devices must bear the CE-mark to be placed on the EU market. This mark is also important in the context of mutual recognition. Mutual recognition means that all 15 member states of the European Union as well as those members of the European Free Trade Association (EFTA), Iceland, Norway, Liechtenstein, and Switzerland, must accept the products into their markets and will be required to allow CE-marked products to circulate freely throughout these markets.

The general rule is that medical devices may only be placed on the market and put into service if they do not compromise the safety and health of patients, users, and third persons. According to this rule, the manufacturer must adhere to the following:

- Eliminate or reduce the risks associated with the use of the device.
- Provide other appropriate protection mechanisms and measures when some risks cannot be eliminated.

Risks addressed in specific requirements relate to chemical, physical, mechanical, and microbiological properties, as well as biocompatibility or emission of radiation. Special emphasis is placed on the information and labeling provided with a device. If residual risks exist in spite of the previously mentioned measures, patients and users shall be informed through the label and instructions accompanying the device.

It is obvious that medical device regulation requires the manufacturer to perform a risk analysis. In general, a risk that may be associated with the use of the device should be acceptable when compared to the benefit to the patient. The manufacturer must document the risk analysis and the results.

In practice, the manufacturer must demonstrate that the product complies with the ERs. A product that complies with the ERs and has followed conformity-assessment procedures is marked with a CE mark, which is its passport for free circulation throughout the European Union.

Medical Devices Directives

One of the first fields of application of the "New Approach" is medical technology, including three separate directives: Active Implantable Medical Devices, Medical Devices, and In Vitro Diagnostics.

The first directive on Active Implantable Medical Devices (AIMD) was adopted in 1990 and appeared as national legislation in 1993 (CD, 1990). The AIMD directive applies to a small number of devices that are partially or totally implanted in the human body and rely on a power source (i.e., infusion pumps, pacemakers, implantable drug pumps neurostimulators, cochlea implants). The directive contains essential safety requirements including sterility, patient protection, or electrical hazards. The technical details are included in the harmonized European standards published by the CEN and CENELEC.

The European parliament approved the second directive on Medical Devices (MD) in 1993 (CD, 1993). The MD directive became effective in January 1995 and had a transition period until June 1998. At that time, the CE mark became mandatory. It is estimated that this directive covers approximately 80% of the medical devices currently available. The MD directive contains the essential requirements for safety and includes a set of rules for the division of the medical devices in four classes (I, IIa, IIb, and III) as along with the conformity-assessment procedures specific to each class.

The third directive relates to In Vitro Diagnostics (IVDs), intended for in vitro analysis of human samples. Implementation of this directive marks the first European regulation of reagents, as well as equipment, intended for in vitro diagnosis. The IVDs directive was adopted in 1998 and became applicable by the end of the year 2003.

The cornerstone of all three medical devices directives is the common definition of "medical device." A medical device, according to Article 1 § 2a of the medical device directive 93/42/EEC (CE, 1993) is defined as follows:

"A medical device means any instrument, apparatus, appliance, material, or other article, whether used alone or in combination, including the software necessary for its proper application intended by the manufacturer to be used by human beings for the purposes of:

- Diagnosis, prevention, monitoring, treatment, or alleviation of disease
- Diagnosis, monitoring, treatment, alleviation of or compensation for an injury or handicap
- Investigation, replacement, or modification of the anatomy or of a physiological process
- Control of conception

A medical device does not achieve its principal intended action in or on the human body by pharmacological, immunological, or metabolic means, but it may be assisted in its function by such means."

Medical devices are articles with medical purpose that is determined by the manufacturers. The manufacturer determines the specific medical purpose of a given device through the label, instructions for use, and promotional material related to it. The medical purpose generally relates to finished products regardless of whether they will be used alone or in combination with other products.

The protection ensured by the directives is valid for products having a stage of manufacture in which they are supplied to the final users. The medical purpose may be achieved either by using a product alone or by using a medical device in combination with other medical devices or other products (e.g., infusion pump used together with necessary tubing and needle). Therefore, the manufacturer may place a product on the market as a complete system, or it may follow the requirements of the directive for different subsets of a system, which can be placed on the market as medical devices in their own right.

Harmonized Standards and the Conformity-Assessment Procedures

European standards are adopted by the CEN and CENELEC and remain voluntary. Therefore, the manufacturer is free to apply relevant European standards or to fulfill appropriate legal requirements by other means. However, the adherence to relevant European standards brings an important advantage to the manufacturer: certification bodies and competent authorities are, in accordance with the directives, obliged to presume compliance with the directives, as far as the manufacturer has met the relevant requirements of the European standards (Article 5 AIMD MDD). Therefore, adherence to the relevant European standards provides a presumption of conformity with regard to legal requirements. Hundreds of European standards have been reviewed by CEN and CENELEC. The European Commission so far has published references of approximately 40 European standards in the official journal. Relevant standards that were published in the official journal have obtained a harmonized status and compliance with them creates the presumption of conformity.

The fulfillment of essential requirements and of requirements relating to the design and manufacture of medical devices must be established through the "conformity assessment procedures." The manufacturer that places the medical devices on the market must conduct the conformity assessment procedure in accordance with the directives. By affixing a CE mark to each medical device, manufacturers confirm that they have met all requirements of the directive and that they have correctly followed conformity-assessment procedures.

The details of conformity assessment procedures were adopted in 1990 and reconfirmed in 1993 by the Council of Ministers, who established eight typical procedural modules for conducting conformity assessment. These modules have been used in different EC directives governing machinery, personal protective equipment, telecommunication equipment, and pressure vessels. The use of conformity-assessment modules allows for a graduated approach to determine how much an independent certification body should intervene in a conformity assessment procedure. According to the MDD, medical devices are classified in relation to the potential hazards associated with their use. Depending on the class assigned to each device, the corresponding conformity-assessment module(s) vary.

Classification

The classification concept is based on a general risk-analysis approach that estimates potential hazards related to the use of the device under normal conditions or during failures. The classification criteria consider whether a device comes into contact with the human body, whether it is invasive, duration of use, whether it is in contact with vital organs, and whether it is activated by an energy source. According to this system, medical devices are placed in one of four classes as follows:

- Class I: Lowest level/responsibility for compliance under the manufacturer
- Class IIa: Mainly control of production
- Class IIb: Both design and production control
- Class III: Highest level/pre market approval

Since the adoption of council directive 93/42/EEC (CD, 1993), the responsibility of classification of medical devices rests with the manufacturers, who must group their products into the previously mentioned four classes in accordance with rules stated in annex IX of the MDD. The Commission of the European Union has issued guidelines for the classification of medical devices into one of the four risk classes using a decision tree.

The rules are based on criteria related to duration of contact between the device and the patient, the degree of invasiveness, and the anatomy affected by the use of the device. The duration of use is divided into three periods: transient use (less than 1 hour), short term use (from 1 hour to 30 days), and long-term use (more than 30 days). A medical device is classified as invasive if it penetrates inside the body, either through the surface of the body or through a body orifice. For the noninvasive devices, contact with blood is an important classification criterion. Modification of the biological or chemical composition of blood increases the risk associated with a device. Active medical devices are defined as those that are driven by electrical energy and differentiated in terms of the amount of energy that they exchange with the body.

According to the MDD and the guidelines, the classification rules apply accordance with the intended use of the device, not its technical characteristics. It is possible that two manufacturers would produce similar devices and assign them two different uses. Therefore, the intended, not accidental, use of a device factors into the classification of it. If a user uses a device in a manner not intended by the manufacturer, he or she does so under his or her own 'responsibility and does not change the class assigned for the conformity-assessment procedure.

Whenever there is a doubt as to classification based on the MDD, the manufacturer should ask an NB. If the manufacturer and the NB responsible for the conformity-assessment dispute the classification, the manufacturer should refer the matter to the competent authority, which decides the classification according to Article 9 of the directive. In addition, the directive states that the medical device committee is the final decision-maker in an extreme case.

Conformity assessment procedures, as previously mentioned, are the procedures that must be completed by a manufacturer before the CE mark can be placed on a product and a product placed on the market. It is not feasible to subject all medical devices to the most rigorous conformity-assessment procedures available. A graduated system of control is more appropriate, and a medical device classification system serves to channel medical devices into the proper conformity-assessment route.

For class I devices (simple dressings, corrective glasses, operating tables, and wheelchairs), the manufacturers are entirely responsible for conformity assessment. The manufacturers must ensure that devices comply with ERs using risk analysis to establish whether any risks present are acceptable. For class I products, the manufacturers must maintain technical documentation and create a declaration of conformity—available to CAs upon request—before placing the device on the market.

In the case of class IIa products (syringes, contact lenses, dental filling materials, and hearing aids), the manufacturer is responsible for the conformity assessment during the design of the devices, but an independent certification body must approve the products during the manufacturing stage. Class IIb products (e.g., implants, intraocular lenses, X-ray equipment, anesthesia machines, ventilators, and high-frequency surgical equipment) and class III (heart valves, resorbable implants, and medicated devices) are subjected to certification by an NB at the design and the manufacturing stages.

For class IIa, IIb and III devices when a third party intervention is required, the manufacturer may choose between at least two applicable procedures: the certification of products or the certification of the manufacturing system. In the latter case, the continuous maintenance of the manufacturer's quality system is subject to a third-party certification by an NB. In this context, compliance with European standards (series EN/ISO 29000 in conjunction with EN 46000) facilitates the establishment of conformity.

Nomenclature Issues

Differences in the nomenclature and codification systems for medical devices, as well as nonuniform terminology applied in different countries, are some of the problems that should be resolved in order to effectively exchange information about medical devices (CEC, 1994). An agreed-upon nomenclature system to identify and describe medical device terms would facilitate data exchange across Europe and encourage implementation of the MDDs. The two most well-established systems worldwide were developed by U.S. Food and Drug Administration, online at www.fda.gov and ECRI, online at www.ecri.org. In addition, many countries in Europe have developed their own systems, which are in use at national or regional levels.

Standardization of medical device nomenclature in Europe will resolve problems with the existing systems. In order to meet this goal, standardization procedures have been initiated, and a number of working groups are active in CEN and CENELEC. The initiative CEN/TC 257/SC1, known as "Coding Systems and Nomenclature for Medical Devices," was formed to produce a European nomenclature system. Following a request by the Commission of the European Union, the technical committee prepared European standards known as "Nomenclature: Recommendations for an Interim System and Rules for a Future System." The same working group prepared a report on "Identification, coding, nomenclature, and regulatory data sets for medical devices," which set the stage to elaborate on medical device standards (CEC, 1994). Device types will be identified by both the device name and manufacturer name. The Global Medical Device Nomenclature (GMDN) system follows a three-level hierarchical structure as shown in Table 125-1. The levels that can be used for administration, regulation, purchasing, aggregation, and quality assurance (QA) purposes are as follows:

- Device category (top level). This level contains 12 categories, with the potential to add more. The categories are divided on the basis of knowledge sets or professional disciplines and are as follows: active implantable devices, anesthetic/respiratory devices, dental devices, electro-medical/mechanical devices, hospital hardware, in vitro diagnostics, nonactive implantable devices, ophthalmic and optical devices, reusable surgical devices, single use devices, technical aids for disabled persons, and X-ray and imaging devices. These device categories facilitate the management of the nomenclature system.
- Generic device group (middle level). The generic device group contains sets of devices with the same or similar intended uses or common technologies, features, or functions. This level contains approximately 13,000 terms divided in three categories: preferred terms, templates, and synonyms.
- Device type (lower level). This level identifies the particular make and model of medical device and contains several hundred thousand items.

The commission will appoint a management agency in order to administer and maintain the GMDN system and promote international standardization. The Universal Medical

Table 125-1 General structure of the CEN medical device nomenclature system

Approx. Number of Terms	Structure
1020	Device category
<10,000	Generic device group
>500,000	Device type

Device Nomenclature System (UMDNS) developed and maintained by ECRI will continue to be used until the new system becomes an EN/ISO standard.

European Medical Devices Vigilance System

One of the most important tasks under the "New Approach" is the implementation of vigilance procedures concerning medical devices. The current regulations concentrate primarily on pre-marketing requirements but do not always ensure that the products will be safe or effective during use. From the perspective of the health care facility, a medical device should be subject to continuous monitoring and quality control of its functional status in the postmarketing life cycle whenever appropriate. In addition, the responsibility of the manufacturers should not stop when the device is put on the market. They should encourage the correct application of their products through appropriate guidance or user training, and monitor devices to ensure that they are performing according to their specifications.

Users can report unexpected adverse effects or research any deviation of the device's expected performance by going back to the manufacturers (post-market surveillance) or to the competent authorities by means of a user reporting system' (Figure 125-1). Post-market surveillance is extremely important to the manufacturer, and feedback from users helps to improve reliability (Donawa, 1994). The EU directives contain some provisions regarding post-market surveillance. The obligations rest especially with manufacturers, who should institute and update a post-market surveillance system including incident reporting and recalls. The AIMD directive contains requirements for reporting device incidents. Instead of the term post-market surveillance system, the MD directives call for "a systematic procedure to review experience gained from devices in the post-production phase and to implement appropriate corrective action," with which manufacturers must comply (see Chapter 139).

Even if the manufacturer is assigned with follow-up, maintenance remains the responsibility of health care facilities and professionals. The health care facility should make the necessary arrangements to ensure that devices are properly maintained and used. Member states are responsible for introducing appropriate measures in the framework of prevailing quality assurance that meet the user's safety requirements.

According to the MDD, every competent authority in the European Union must implement a user-reporting system to investigate adverse incidents involving medical devices and to take appropriate measures to prevent recurrence. In addition, manufacturers are legally obliged to report any serious incidents, deaths, or severe injuries involving devices they produce and sell. Based on these requirements, the commission has prepared an adverse-incident notification and evaluation system in the form of the medical devices vigilance system guidelines (CEC, 1993). The aims of the guidelines are as follows:

- To prevent the reoccurrence of incidents with the same type of medical devices, at another place, at another time
- To encourage manufacturers to perform investigation and take corrective actions if necessary
- To enable the superior authority to monitor the investigation procedures and intervene when necessary

Although, the guidelines are voluntary and are not laws, they provide guidance for the application of the directives, and help to turn the directives into laws. Figure 125-2 is a flowchart that shows procedures and responsibilities of the competent authorities and manufacturers in the medical device vigilance system. Under the vigilance system guidelines, manufacturers should submit reports to the CA in the country where the incident occurred. Manufacturers must notify the CA within 10 working days when death has occurred, or within 30 days for all other incidents, by completing an initial report form. Each initial report is evaluated by the CA, who follows up with the manufacturer when necessary (ECRI, 1991; ECRI, 1993; ECRI, 1995).

Manufacturers are encouraged to perform investigations and take corrective actions. The CA should monitor a manufacturer's investigation and intervene if necessary. At the end of the investigation, any information that is necessary for the prevention of further incidents (or the limitation of their consequences) should be shared in the form of a final report to all other CAs and the commission. Exchange of information between member states is essential, especially when measures must be taken as a consequence of an incident report (Pallikarakis et al., 1996). In cases of a recall, the responsible CA should ensure that information is given in a manner that avoids negative effects on those involved.

Confidentiality is important as well. The initial reports received by a CA are confidential, but the outcome of an investigation should be shared with other CAs in order to prevent similar incidents. On the other hand, dissemination of information that has not been verified can result in serious negative consequences for both manufacturers and patients. In order to meet these requirements, most member states have national medical device vigilance systems.

Although implanted devices receive regulatory approval before clinical use, many of them can create problems when applied to the patients, due to poor biocompatibility or inherent physiological human variability. Some devices have a finite performance time but, their use should be judged based on the benefits for the patient, as compared to the risks. It is often necessary to assess device performance after its application to the patient. Tracking systems for both patients and medical devices follow devices and assessing the clinical outcomes.

Tracking a device or implant means following the item from the manufacturer or importer to the user. Device traceability requires the gathering of data from manufacturers or importers, distributors, physicians, hospitals, and patients. Manufacturers and importers have a great responsibility, although they may have limited authority over the distribution network. The history of a device reflects product design, patient life expectancy, device cost, and other useful parameters. Maintaining a tracking system for implanted medical devices implies that it is possible to have information about the patient; i.e., to be able to have the patient under continuous observation, throughout the patient's or device's life, and to maintain the information in a confidential patient registry.

As previously mentioned, the approaches to medical device vigilance in the European Union countries prior to the implementation of the directives had no formal common communication structure between the involved authorities. The MDD has imposed requirements, and authorities recognize the need for a harmonized approach and cross-national exchange of information. In this era of rapid electronic communication, telematics will likely provide the appropriate means and tools for an effective, commonly accepted approach to information exchange. Such an approach has already been implemented in the field of pharmacovigilance, and a similar approach was explored with the European medical device information exchange system (EUROMEDIES) concerted action initiative (Pallikarakis et al., 1996). The information is available online at http://www.ehto.org/aim/volume2/euromedis.html. The project investigated requirements for a common telematics approach to information exchange in the field of medical device vigilance.

Conclusions

Although there is a strong global harmonization effort, the approaches followed in the field of medical device vigilance in different countries still vary, and no formal communication structure between authorities exists. The first step is the evaluation, acceptance, and use of a common medical device nomenclature system, which is a prerequisite for information exchange (Pallikarakis et al., 1996).

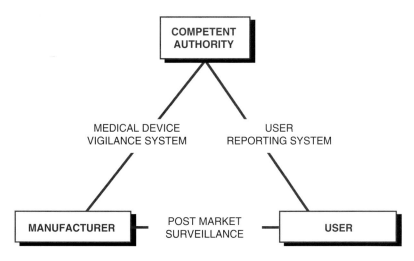

Figure 125-1 Schematic diagram of the parties involved in medical device vigilance procedures.

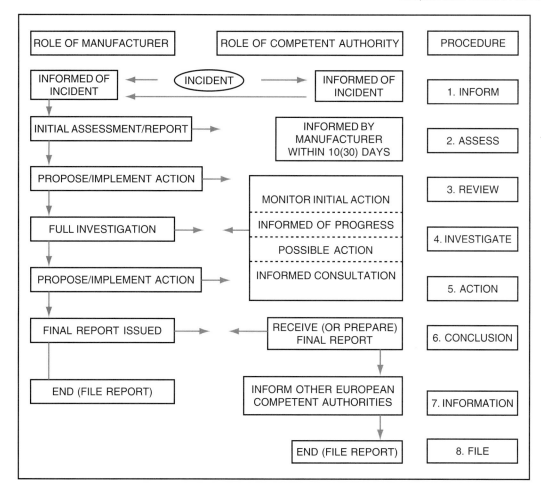

Figure 125-2 Flowchart showing procedures and responsibilities of the competent authorities and manufacturers in the medical device vigilance system.

References

Bethune J. Business outlook: US companies look abroad for growth, Medical Device and Diagnostic Industry 1995; pp 57-67.
CD. Council Directive 90/385/EEC, *Official Journal of the European Commission* L-189, 1990.
CD. Council Directive 93/42/EEC, *Official Journal of the European Commission* L-169, 1993.
CEC. Guidelines on a Medical Device Vigilance System, Commission of the European Communities, May 1993.
CED. *Guidelines to the Classification of medical devices, 4th Draft,* Commission of the European Communities, July 1994.
Donawa M. Handling Adverse Incident Reports and Complaints, Part I, European Requirements, *Medical Devices Technology*, 1994; Vol. 5, No. 9, pp 1216.
ECRI. ECRI Problem Reporting System, Health Devices 24: May/June 1995.
ECRI. Safe Medical Devices Act of 1990, FDA Issues SMDA and Voluntary Report Forms, ECRI Advisory. Plymouth Meeting, PA ECRI, 1993.
ECRI. *Safe Medical Devices Act requires User Reporting, Health Technology Management.* Plymouth Meeting, PA ECRI, 1991.
Jennett EL. The US medical device industry: Competitive with the EC and Japan? *ASTM Standardization News* Feb. 1992.
Mattox K. Growing opportunities for medical device markets outside the US, *Medical Device and Diagnostic Industry*, March 1995, pp 4854.
Pallikarakis N, Anselmann N, Pernice A (eds). *Information Exchange for Medical Devices.* Amsterdam, IOS Press, 1996.
van Gruting CWD. *Medical Devices—International Perspectives on Health and Safety,* Amsterdam, Elsevier, 1994.

Further Information

Webster JG (ed). *Encyclopedia of Medical Devices and Instrumentation,* New York, John Wiley & Sons, 1988.

United States Food & Drug Administration

F. Blix Winston
Mount Airy, MD

The U.S. Food and Drug Administration (FDA) regulates medical devices primarily through the law known as the Federal Food, Drug, and Cosmetic Act, as Amended ("the FD&C Act," or "the Act"). Regulations promulgated by FDA in Title 21 of the Code of Federal Regulations (CFR) spell out the broad provisions contained in the Act.

The FDA is made up of headquarters and field offices. The Offices of the Commissioner and functional centers, located in and around Washington D.C., include the centers for: Drug Evaluation & Research (CDER), Food Safety & Applied Nutrition (CFSAN), and Devices and Radiological Health (CDRH). FDA Regional and District Offices carry out the FDA's field activities in the U.S., including manufacturing inspections and import examination. Foreign inspections are carried out from the headquarters office.

CDRH, currently located in Rockville, Maryland, is organized into functional offices. Device Evaluation (ODE) has primary responsibility for review of clinical and marketing applications; Compliance (OC) is responsible for inspections and enforcement; Science and Technology (OST) performs laboratory research and provides scientific support for Center activities; Surveillance and Biometrics (OSB) manages the problem-reporting programs and provides statistical support to CDRH. The Office of Health and Industry Programs (OHIP) is responsible for outreach, education, manufacturer, and consumer assistance, and the mammography quality standards program. These offices administer the programs and regulations that impact medical devices, and the companies that make them, throughout their life cycle: starting with the concept and design stages; during research, development and testing; during the application process; during manufacture; and through the device's useful life.

Supplementing the regulations are hundreds of CDRH guidance documents, including manuals covering major processes like 510(k) and the Quality System requirements. All of the topics discussed in this chapter can be found on the CDRH web site (see References). Direct assistance is readily available through several mechanisms.

Laws and Regulations

Food, Drug and Cosmetic Act

The FDA's legal authority, the 1938 FD&C Act, is a flexible law that has been modified several times. It defines broad concepts (e.g., adulteration and misbranding), sketches out procedures (e.g., Pre-market Notification), and directs the agency to develop implementing regulations. Section 201(h) of the Act defines the term "device" as "an instrument, apparatus, implement, machine, contrivance, implant, *in vitro* reagent, or other similar or related article including any component, part, or accessory, which is:

- Intended for use in the diagnosis of disease or other conditions, or in the cure, mitigation, treatment, or prevention of disease, in man or other animals; or
- Intended to affect the structure or any function of the body of man or other animals, and, which does not achieve any of its primary intended purposes through chemical action within or on the body of man or other animals, and which is not dependent upon being metabolized for the achievement of any of its primary intended purposes."

The Act includes rules for classifying medical devices, provides for FDA review prior to marketing, governs device design and manufacturing, includes various rules for notifying FDA of certain actions, and contains provisions that apply both before and after marketing clearance is obtained. These requirements are enumerated and discussed throughout the chapter.

Major Amendments to the FD&C Act

1968 Radiation Control for Health and Safety Act

Before 1968, the FDA did not have the authority to regulate devices. Radiation-emitting consumer and medical devices, including X-ray machines, microwave products, lasers, cathode ray tubes, and sunlamps came became subject to FDA regulation with the 1968 RCHSA.

1976 Medical Device Amendments

The 1976 Medical Device Amendments gave FDA specific authority to regulate general medical devices. Along with other new provisions, Section 513 of the Act was changed to:

Describe a tiered classification scheme (class I, II and III), set forth requirements for each class of device "sufficient to provide reasonable assurance of the safety and effectiveness of the device," and provide extensive detail about the way the device classification process works.

Safe Medical Devices Act of 1990

Major revision of the 1976 device provisions did not occur until 1990 with passage of the Safe Medical Devices Act (SMDA). SMDA included classification changes, changes to the 510(k) process, the authority to order recalls, adverse incident reporting by user facilities such as hospitals and outpatient treatment centers, post-market requirements such as device tracking, and reports of corrections and removals. The SMDA was followed in 1992 by the Medical Device Improvements Act, which largely only corrected conflicting provisions of SMDA.

1997 FDA Modernization Act

While SMDA focused only on medical device regulation, the 1997 FDA Modernization Act (FDAMA) affected all of FDA and had a strong impact on CDRH. Its effect can be seen in significant modifications to the PMA process and opportunities for meetings between the FDA and manufacturers before and during the review process. It clearly restricted the FDA's role in regulating the "practice of medicine," and required the agency to take a "least burdensome" approach when requesting data to support marketing applications.

Medical Device User Fee and Modernization Act of 2002—MDUFMA

MDUFMA authorized FDA to collect user fees to perform pre-market reviews and gave FDA authority over reuse of devices lableled "for single use only."

Title 21, Code of Federal Regulations (CFR)

Regulations are "promulgated" by the FDA to flesh out the general provisions of the Act. The device volume, Parts 800–1299, contains nearly all of the regulations that affect medical devices. Other regulations that impact devices include Part 54 on Financial Disclosure Rules, Part 50 covering Patient Protection and Informed Consent, and Part 56 covering Institutional Review Boards.

Parts 862–892 in the device volume of the CFR cover the16 classification panels that describe and classify about 1700 device types into class I, II, or III. Separate parts also cover Labeling (801), Medical Device Reporting (803), Establishment Registration, Medical Device Listing, Pre-market Notification (807), Pre-market Approval (814), Investigational Device Exemptions (812), and Quality Systems (820).

Medical Device Regulation

Device Classification

The Act devised a tiered approach to classification based on risk and the amount of regulatory control needed to provide "reasonable assurance of the safety and effectiveness of the device." It places devices into one of three classes:

Class	Risk	Subject to	Approximate number in each class as of 1990*
I	Low	General controls	782
II	Medium	General controls and special controls	799
III	High	General controls, special controls, and pre-market approval	119
			1750

*These numbers are shifting as reclassification efforts proceed.

While most device types have already been classified, classification is a flexible mechanism that allows for reclassification when needed. Determining the classification for most devices can be done by using the Classification Database on the CDRH web site or searching through the classification panels in the CFR.

Each classification regulation in Parts 862–892 provides a brief description of the device and its intended use and gives its class and any particulars regarding marketing requirements. Proper classification is essential to understanding which regulatory requirements apply to a device. Center staff are always available to help determine classification.

General Controls

Class I devices, known as "General Controls," are subject to the least regulatory control. They present minimal potential for harm to the user and are often simpler in design than Class II or Class III devices. Examples include manual surgical instruments and bandages. Class I devices are subject to "General Controls," as are Class II and Class III devices. General Controls include:

- Prohibition against adulteration (e.g., contamination) and misbranding (e.g., the wrong label on a device)
- Establishment registration, which tells the FDA where a firm is located
- Medical device listing, to notify the FDA that the device is being sold in the U.S.
- Manufacturing devices in accordance with the Quality Systems regulation, also known as the current "Good Manufacturing Practices" (GMP)
- Labeling
- Submission of a Pre-market Notification, or 510(k) application, before marketing a device

Over half of the 1700 classified device types, including most class I devices and a few class II devices, are exempt from 510(k). These exemptions can be found on the CDRH web site.

Special Controls

In 1976, amendments to Section 514 of the Act directed the FDA to develop mandatory performance standards for all Class II devices (radiological device performance standards are promulgated under the RCHSA). The Safe Medical Devices Act of 1990 changed the designation of Class II from "Performance Standards" to "Special Controls." Special Controls apply to class II and III devices and may include one or more of the following: mandatory performance standards, guidelines, special labeling requirements, and post-market surveillance studies.

A Special Control becomes a requirement for a particular device, or a group of devices, when The FDA includes this designation in the CFR or requires a special control during the review process. Examples of current Special Controls include the classification for the Cranial Orthosis, 882.5970, and the performance standard for electrode lead wires and cables, 21 CFR Part 898.

Pre-Market Approval (PMA)

Class III devices are known as "PMA." Compared to 510(k), PMA is a more rigorous, data-intensive approval process. Once the reclassification efforts are completed (by 2005), all devices remaining in class III will require a PMA (or a Product Development Protocol). In addition to devices classified as class III by statute, the 510(k) mechanism, which is itself a classification process, may result in a device being classified into class III and thus subject to PMA.

Standards

The FDA currently has promulgated one medical device mandatory-performance standard in Part 898 for electrode lead wires and cables. Performance standards also exist for radiation-emitting devices such as X-ray equipment. Recently, the FDA "recognized" hundreds of U.S. industry and international performance standards that can be used to establish safety and effectiveness in 510(k)s and PMAs. The standards page on the CDRH web site describes the use of voluntary standards, lists the ones recognized by FDA, and provides source information.

The Regulations Process, Advisory Panels, and the *Federal Register*

New regulations, or those being modified, are "promulgated" through a process that begins with a "notice of proposed rulemaking," published in the *Federal Register* (FR). These FR notices usually ask for comments that are used by the FDA to modify proposed regulations. The modified regulation is usually published in the FR as a "final rule." Note: Final rules contain "preambles," which list the public comments received and the FDA's response to them. They often contain valuable insights into the way the FDA develops its regulations. Advisory panels, whose names conform to the classification panels (e.g., the Cardiovascular panel), are composed of experts in their respective fields of specialization. Panels also include consumer representatives. They participate in the PMA review process and are called upon to advise the FDA when new device types are classified.

Pre-Marketing Requirements

Investigational Device Exemption

Clinical (human) trials of unapproved medical devices are subject to CFR Part 812, the Investigational Device Exemption (IDE) requirements. Investigational devices are exempted from most of the general controls, with the exception of adulteration. They involve scrutiny by Institutional Review Boards (IRBs) and/or FDA, require signed informed consent from each subject who participates in the study, and are usually conducted to gather data on safety and effectiveness to support further development or marketing-clearance applications. Most PMAs involve clinical studies, whereas only about 15% of all 510(k) applications contain clinical data.

The two types of IDEs, significant risk (SR) and nonsignificant risk (NSR) studies, have slightly different approval routes. SR devices could be implants, could support or sustain life, or could present serious risks to health. SR devices require both IRB approval and an FDA-approved IDE application. Changes to an IDE application, made before the FDA has approved it, are referred to as "IDE amendments." After the IDE is approved, related submissions are called "IDE supplements." All types of IDE applications require a response.

NSR device studies are subject to the "abbreviated requirements." The sponsor presents the study protocol to a local Institutional Review Board (IRB) and explains the reasons why the study is NSR. If the IRB concurs in the risk assessment and grants permission, the study may begin. Note: The FDA is not aware of NSR studies. While it is always advisable to schedule pre-IDE meetings with the FDA before submitting an IDE for a significant risk study, it may be advisable, as well, to request FDA review of the study design and protocol before beginning NSR studies. Both NSR and SR studies require informed consent of all patients in the study.

Premarket Notification—510(k)

Most devices are cleared through the Pre-market Notification or 510(k) process. FDA does not test devices as part of the review process; it assesses safety and effectiveness by reviewing information submitted in applications.

A 510(k) is not a form but a package of information, averaging about 100 pages. It includes: classifying and descriptive information about the device; certain information elements; technical information such as specifications, engineering drawings, and schematic diagrams; data from bench (and perhaps animal or clinical) testing; labeling and advertising (including instructions and operating manuals); and information comparing the device to one, or more, similar "predicate" devices that have been legally marketed in the U.S. 510(k) applications often contain information relating to performance standards, as well. The point of a 510(k) application is to establish "substantial equivalence (SE)" between the subject device and one or more "predicate" devices. A device is SE if it:

- Has the same intended use and the same technological characteristics as the predicate device, and information in the 510(k) does not raise any new issues of safety and effectiveness or
- Has different technological characteristics, and the information in the 510(k) demonstrates that the device is at least as safe and effective as the predicate.

Equivalence (and thereby safety and effectiveness), is demonstrated by comparing specifications and labeling, with performance data, or by showing conformance to FDA-recognized performance standards. For example, electromagnetic compatibility can be shown by conformance to IEC-60601-1-2.

When the 510(k) is submitted, an acknowledgment letter with a 510(k) number will be sent to the applicant. If the application is incomplete, the submitter will be asked to send the required information. A 510(k)s is place on formal or informal "hold" if the reviewer needs more information.

A "substantial equivalence" determination by FDA serves to classify the new device into the same category as the predicate. A "not substantially equivalent" (NSE) determination results in automatic classification into class III. The review process, with flow diagrams, is discussed in the 510(k) Manual and is also described in Sec. 807.100.

The FDA Modernization Act introduced a new classification mechanism called "Reclassification of Automatic Class III Designation," known informally as the *de novo* process. It allows low-risk devices the opportunity to receive marketing clearance following an NSE decision in lieu of going through the expensive and cumbersome PMA process.

FDA accepts three types of 510(k) applications: Traditional 510(k)s (as described above), Special 510(k)s, and Abbreviated 510(k)s. The traditional 510(k) is discussed in the guidance memorandum K86-3 and in the *Pre-market Notification, 510(k) – Regulatory Requirements Manual*.

An Abbreviated 510(k) includes a voluntary declaration of conformity with an FDA-recognized performance standard. Test data collected to demonstrate conformance do not have to be included in the 510(k), hence the content is somewhat abbreviated. A Special 510(k) can be submitted for modifications to existing devices that require a new 510(k). It includes a declaration that the applicant has followed design-control procedures and has completed necessary verification and validation activities in making the change. The FDA occasionally inspects records to verify that standards-related test data exist, or that design controls cited in a Special 510(k) were followed. The guidance document *Deciding When to Submit a New 510(k) for Change to an Existing Device* provides convenient flow diagrams that tell whether a new 510(k) should be submitted for the change. Special 510(k)s are reviewed within 30 days. Two guidance documents describe these alternative applications: *A New 510(k) Paradigm*, and *Questions and Answers Based on the New Paradigm*.

Although FDA maintains a 90-day "clock" on 510(k) review, 510(k)s are not subject to mandatory review times, as are IDEs and PMAs. Total review time averages approximately 110 days for all 510(k)s. The current median review time is 70–80 days. In the 2000, FDA received 4,202 510(k)s. Pre-inspection is not part of the usual 510(k) review process unless it is for a class III device. A 510(k) tutorial program, *Device Advice*, is available on the CDRH website. The Standard fee for 510(k) is $3,490.00; the Small Business fee is $2,784.00.

Pre-Market Approval

A Pre-market Approval application may be submitted as an original PMA or in modular format. It must provide data demonstrating that the device is safe and effective for its intended use and that it will be manufactured in accordance with current GMP. By way of contrast, the FDA receives about 40 to 60 PMAs per year versus roughly 4,500 510(k)s. Rather than demonstrating safety and effectiveness (S&E) by establishing equivalence with another device, a PMA relies on performance data to demonstrate S&E. PMAs are larger and more complex than 510(k) applications. Six copies of the PMA must be submitted to the FDA. They are subject to a 180-day statutory review time. In addition to describing the PMA application process, FDA review, and post-approval requirements, Part 814 in the CFR also describes the "Humanitarian Use Device" process.

As part of the PMA review process, the FDA may seek an advisory review of portions of the PMA from an expert advisory panel. Following FDA approval, a "Summary of Safety and Effectiveness" will be published in the *Federal Register* and will appear in the PMA database online as well. Pre-PMA meetings are strongly encouraged. Numerous PMA-guidance documents exist, including the *PMA Manual*, device-specific guidance documents, and PMA administrative-review guidance. Average elapsed PMA review time, from log-in to final approval, averaged 364 days in FY-2002.

Changes affecting the safety or effectiveness of the device, such as design or labeling changes, require submission of a PMA supplement. Some PMA supplements can be as complex as the original application. Although the statutory FDA review is 180 days for PMAs and supplements, the FDA is committed to reviewing them in shorter timeframes, which it has reduced by accepting modular submissions and through the use of a real-time supplement process, 30-day notices, and expedited reviews. The Standard fee for PMA review is $206,811.00; Small Business is $78,508.00.

Product Development Protocol (PDP) and the Humanitarian Device Exemption (HDE)

Two alternatives to the PMA exist for class III devices: the "Product Development Protocol" (PDP) and the "Humanitarian Device Exemption" (HDE). The PDP process is based upon early consultation between the sponsor and the FDA, leading to a device development and testing plan that is acceptable to both parties. Once the PDP has been accepted, periodic reports are submitted to the FDA followed by a "Notice of Completion." If the FDA finds that all requirements have been met, the PDP will be declared complete, and the device may go to market.

An HDE application for a Humanitarian Use Device is essentially the same as a PMA in both form and content but is exempt from the effectiveness requirement of a PMA. Though the HDE does not require scientifically valid data demonstrating effectiveness the application must contain sufficient information for the FDA to determine that the device does not pose an unreasonable or significant risk of illness or injury to patients. It must also demonstrate that the probable benefit to health outweighs the risk of injury or illness from its use.

Labeling

The CFR contains a number of labeling regulations, beginning with Part 801, which covers the general labeling requirements that apply to all medical devices. Section 801.1 requires that the name and place of business of the manufacturer appear conspicuously on the device's label. Note: Section 801.4 defines "intended use" quite broadly, relating it to "...the objective intent of the persons legally responsible for the labeling..." and notes that labeling can include advertising, oral, and written statements made about the device. Usually, a device requires adequate directions for use by a layperson. "Restricted" devices that bear the prescription label (801.109), restricting their use to persons with professional training, are exempted from this requirement. The FDA does not have a list of restricted device types. In uncertain cases, CDRH reviewers will advise 510(k) or PMA applicants when the prescription label is required.

In addition to the general labeling requirements in Part 801, specific labeling requirements are found in the following CFR Parts: 812.5 for IDE devices, 809.10 for in vitro devices, 820.120 and 820.130 in the GMP regulation, and Parts 1010 through 1050 found in the performance standards for radiation-emitting products.

Registration and Listing

Initial Registration of Medical Device Establishment, form FDA-2891, which tells where a firm is located is required for U.S. and foreign manufacturers, specifications developers, contract manufacturers of finished devices, and importers (initial distributors). Registration is updated once a year on the FDA-2891a form, *Annual Registration of Device Establishment*, which the FDA mails out. Medical Device Listing on form FDA-2892, which tells FDA the firm has started selling the product, is required for any device marketed in the U.S., whether 510(k)-exempt or not. At the present time, original paper copies of the forms must be submitted to the FDA. Registration is not required prior to 510(k) submission; the forms must be submitted within 30 days of going to market.

Manufacturing Requirements

GMPs, Inspections, and Audits

The "Quality System" (QS) regulation in Part 820 (the current Good Manufacturing Practices (GMPs)) requires that most domestic or foreign manufacturers establish a system covering the design and manufacture of medical devices distributed in the United States. Some class I devices are exempt (by statute) from most of the Quality System requirement. GMPs address such topics as management organization and functions, design activities, control of materials and manufacturing processes, corrective and preventive actions, and label control. The Quality System regulation has been closely harmonized with European device requirements.

The CDRH web site contains extensive information on GMPs including the *Quality Systems Manual, Design Control Guidance for Device Manufacturers* and the *Quality System Inspection Technique (QSIT) Manual*. The Design Control portions of the GMPs (820.30) come into play well before the stage is reached when a 510(k) or PMA is submitted to the FDA. Engineering, regulatory, and R&D staff should carefully study and implement these requirements at the R&D stage. They not only are required by FDA but when properly applied also greatly reduce the probability that design flaws will be missed and that expensive rework will be needed once the device goes to manufacturing. PMAs contain manufacturing information, and a current GMP inspection is required as part of the PMA approval process. While 510(k)s do not contain manufacturing information, they do contain information on the design, including schematics and drawings. A Special 510(k) includes a declaration that the applicant followed Design Control procedures when making significant changes to their product.

The FDA conducts inspections of domestic and foreign manufacturers of class II and III devices on a regular basis. Class I device manufacturers are inspected on an as-needed basis. The GMPs require that manufacturers conduct periodic audits of their quality systems by qualified personnel. Audits help companies to identify GMP deficiencies and make corrections to their quality systems. While audit results are designed to help management to do its job and are not accessible to the FDA, audit procedures are subject to FDA review. Materials such as the QSIT guide can help companies to prepare for FDA inspections. Note: Synergy between 510(k)/PMA data requirements and design control requirements can be exploited. Design documents, such as engineering drawings and data produced by design Verification and Validation steps are the same information submitted in 510(k) or PMA applications.

Post-Marketing Requirements

Reports of Corrections and Removals

Part 806 in the CFR applies to certain actions taken to correct or to remove products from the market. Many corrections or removals are done for economic reasons or to rotate stock. However, when they are done to reduce a risk to health posed by the device or to remedy a violation of the Act that could pose a risk to health (e.g., if the device is misbranded), a report must be sent to the local FDA office. Section 806.10 outlines the information that must be included in the report.

Records of all corrections or removals must be kept, even when they do not require a report to the FDA.

Recalls

Before the 1990 SMDA, the FDA did not have the authority to order companies to recall hazardous products. Following SMDA, Part 810 was added to the CFR. When the FDA finds that there is reasonable probability that a device will cause serious adverse health consequences or death, it may issue an order requiring the company to cease marketing the product and to submit certain information to the agency. It may also order the company to recall any violative product in the distribution chain. Part 810 goes into rather extensive administrative procedures that both parties follow, and notes specifically that the recall information may appear in the weekly FDA Enforcement Reports on the FDA web site. While recalls and corrections do occur, a well-managed Quality System will minimize their occurrence.

Post-Market Surveillance

The FDA may require manufacturers to conduct post-market surveillance studies to gather safety and efficacy data for certain high-risk Class II and Class III devices. 510(k) or PMA applicants are notified at the time of submission as to whether their device will be subject to post-market surveillance. This requirement comes into play primarily with Class III devices. Section 814.82 discusses post-approval requirements.

Tracking

Tracking is another regulation that was added following the 1990 SMDA. Part 821 describes the procedures that manufacturers of certain Class II or Class III devices (such as heart valves or infusion pumps) must set up so that they can track their medical devices down to the patient level. Like post-market surveillance, applicants will be notified during 510(k) or PMA submission as to whether their devices are subject to the tracking requirements. Cards or other feedback instruments are issued with a tracked device and are returned to the manufacturer by distributors and patients so that the manufacturer can provide a recipient list to the FDA upon request.

Medical Device Reporting (MDR)

The MDR regulation in Part 803 requires manufacturers and importers to report to the FDA within 30 days, whenever the firm becomes aware of information that reasonably suggests that one of its marketed devices:

1. Has, or may have, caused or contributed to a death or serious injury, or
2. Has malfunctioned, and that the device or a similar device marketed by the manufacturer or importer would be likely to cause or contribute to a death or serious injury if the malfunction were to recur

User facilities are required to submit reports of deaths to the FDA within 10 days, and to the manufacturer within 10 days for serious injuries.

MedWatch

Reports are submitted to the FDA on MedWatch (3500) forms. Manufacturers who receive a MedWatch report from a user facility or health care practitioner are required to submit a

3500A report to the FDA. There are two versions of the MedWatch form for individual reports of adverse events. Both are available on the CDRH web site. FDA form 3500A is the mandatory reporting form used by manufacturers and user facilities. FDA form 3500 is used by health professionals and consumers for voluntary reports. MedWatch forms allow trends to be identified and corrections to made by manufacturers that otherwise might have been delayed. In addition to MedWatch forms, Part 803 requires manufacturers to submit baseline and five-day reports, if required. Information from MDR reports, Postmarket Surveillance reports, and other sources, such as the voluntary Product Reporting Program (PRP) and MeDSuN program, together with marketing-clearance data, contribute to the knowledge that the FDA analyzes in order to protect the health of the public.

International Harmonization, Service Providers and Reconditioners, and Enforcement Provisions

When the original GMP regulation was updated as the Quality System regulation, it was harmonized with the ISO 9001 series requirements. A new harmonization effort is the U.S./European Union Mutual Recognition Agreement, designed to allow the U.S. and participating countries to conduct quality-system inspections on behalf of each other. The FDA also has implemented the Third Party Review program, whereby 510(k)s for many devices may be reviewed by accredited bodies both in the U.S. and abroad. Information on both of these programs can be found under *A–Z Index* on the CDRH web site.

Service Providers and Reconditioners

In general, the FDA does not regulate service providers, such as firms that modify wheelchairs owned by patients. The FDA also does not actively regulate reconditioners and refurbishers who merely bring devices back to original specifications. One notable exception is entities reusing or reconditioning devices labeled "For Single Use Only." Under MDUFMA FDA considers such entities, including hospitals and "Third Party Reprocessors," as manufacturers in some cases. All regulatory requirements, including 510(k) and PMA apply to these new manufacturing entities.

Enforcement Provisions

The FDA can apply a variety of enforcement actions to U.S. and foreign manufacturers and the devices that they distribute within the U.S. when devices or manufacturing processes are discovered to be in violation of the Act.

When significant deviations from GMP requirements are found during an inspection, the FDA may issue a warning letter which lists the deviations and notifies the firm that further action may be taken if the deviations are not corrected. Further action could include citations, seizure of goods, prosecution of the firm, or the imposition of civil penalties (fines). The FDA also may utilize these enforcement provisions when it learns of imminent hazards through adverse incident reporting. Imported goods can be detained for testing or refused entry when found violatory by the FDA.

Guidance

The *Good Guidance Practices* (GGP) database can be used to find most guidance documents. Alternatively, *Topic Index* can be searched by keyword. Numerous "pages" have been added to the web site, such as the GMP or MDR pages, which provide a central location for guidance.

IDE and PMA premeetings can be scheduled with the appropriate review divisions in the Office of Device Evaluation (ODE). The Division of Small Manufacturers, International, and Consumer Assistance (DSMICA) also offers direct assistance via telephone (800) 638–2041 or (301) 443–6597, via e-mail (dsmica@cdrh.fda.gov), or via fax to: (301) 443–8818.

Further Information

All existing CDRH guidance and references can be ordered in hard copy from DSMICA via fax: (301) 443–8818, Attn: Publications.

FDA Homepage: http://www.fda.gov

The Act and related FDA laws can be found on the FDA homepage under "Laws FDA Enforces."

CDRH Homepage: http://www.fda.gov/cdrh

CDRH databases are cross-linked to provide a powerful tool when searching for information on device classification, 510(k), PMA, registration, and listing information.

Device-related organizations include: Advanced Medical Technology Association—AdvaMed—www.advamed.org American Society for Quality—ASQ—www.asq.org Association for the Advancement of Medical Instrumentation—AAMI—www.aami.org The Food & Drug Law Institute—FDLI—www.fdli.org The Regulatory Affairs Professional Society—RAPS—www.raps.org

Useful publications include: Medical Device Development—A Regulatory Overview, Jonathan S. Kahan, PAREXEL International Corporation, Waltham, MA 02154. The Medical Device Indstry, N.F. Estrin (ed), New York, Marcel Dekker.

127
Tort Liability for Clinical Engineers and Device Manufacturers

Edward P. Richards, III
Director, Program in Law, Science, and Public Health
Paul M. Hebert Law Center, Louisiana State University
Baton Rouge, LA

Charles Walter
College of Engineering, University of Houston
Houston, TX

This chapter discusses the legal liability of clinical engineers (CEs) whose negligence causes persons to be injured by the medical devices for which the engineer is responsible, and the liability of the manufacturers of medical devices that injure people. These injuries are usually to patients but they also can be to the medical care providers who use the devices, and even to bystanders. The magnitude of injuries can range from minor skin irritation due to an improper electrode jelly on an EKG electrode, to catastrophic injuries and property damage from the explosion of an operating suite from the untimely combination of an ignition source and explosive anesthesia agents. The chapter is divided into two sections, reflecting the different legal rules for the personal liability of individual engineers (malpractice) and the corporate liability (products liability) of medical device manufacturers. While the legal risks to manufacturers are much higher, both in terms of the number of claims filed and the dollars at issue in those claims, the personal impact of a claim against an individual engineer can be devastating. Understanding the legal basis for liability is a necessary part of effective clinical engineering risk management.

Personal Liability for Malpractice

There are relatively few malpractice claims against CEs for injuries caused by their actions, although there is reason to expect that such claims will increase. Traditionally, most CEs worked for corporate entities, usually hospitals and supply companies, who were the target defendants when a medical device related injury claim was filed. With changes in the industry, an increasing number of CEs work for contract service agencies and as independent contractors. This will encourage plaintiffs to sue the engineer personally because there is no longer a "deep-pocketed" corporate employer to pay the claims. Even if the plaintiff does not expect to collect damages from the engineer, the pressure of a lawsuit and personal liability can encourage the engineer to help the plaintiff's claim against the hospital or supply company as a condition of being allowed out of the lawsuit.

As indicated above, most personal claims against CEs will be for professional malpractice, which is based on negligence law. To establish negligence, a plaintiff must prove four things:

1. A duty
2. Breach of a standard of care
3. Causation
4. Damages

"Duty" is the legal relationship between the injured person and the CE. The hospital or other health care provider who supplies the device has a duty to the patient to ensure that the device is properly maintained and working correctly. This duty is shared by the CEs, who actually work on the device. CE supervisors have a duty to carry out their responsibilities correctly, which includes ensuring that those under their supervision carry out their work properly. For example, assume that a patient is injured by a negligently maintained medical device in the CCU at St. Elsewhere Hospital. The patient sues both the hospital and Jane, the CE who works for the hospital. If there had been negligence, the hospital would be liable regardless of which CE was responsible. Jane's liability depends on whether she worked on the device personally or was responsible for maintaining that device directly or through supervising other CEs. If Jane works in the outpatient department, did not work on the device, and has no responsibility for CCU devices, then she does not have a legal duty to the injured patient. If, however, she is the supervisor of the clinical engineering department, then she would have a legal duty to the patient to ensure that her employees properly maintained the device.

Duty can be more complicated for third-party service groups. CEs who perform third-party service always have a duty to others for the work that they do and for their supervision of others. Unlike hospital-based CEs, service contracts may limit the CE's access to the devices and the work that they are allowed to do. Thus, while the head of the hospital CE department might have an ongoing duty to monitor all of the hospital's devices, a contractor CE will only be responsible for the devices specified in the contract, unless it includes all of the hospital. Contractor CEs must ensure that the hospital gives them adequate access to the devices that they contract to maintain. Otherwise, they could be legally liable for devices to which they do not have physical access. It is likely that the courts will find that while hospitals may contract for outside service for devices, the hospital retains liability if that service is not rendered correctly. This is because maintaining safety devices is usually a "nondelegable duty," which means that the party that is originally charged with ensuring safety might not escape that obligation.

"Standard of care" refers to what a reasonable, competent CE would have done under the same circumstances. There are three ways for the plaintiff to establish the standard of care. The most common is to call another CE to testify as an expert witness as to what he or she believes the CE should have done. Standard of care also can be established through published standards, such as those promulgated by governmental agencies or organizations such as the Association for the Advancement of Medical Instrumentation (AAMI), or through manufacturers' maintenance manuals. CEs also must abide by any internally promulgated standards, such as hospital maintenance schedules or specific contracts for service. For example, if hospital guidelines are stricter than industry standards, courts have found that the internal guidelines are binding. The CE can present evidence that refutes the plaintiff's proffered standard, and the jury will decide which to apply.

Once the plaintiff has established the alleged standard of care, he or she must show that the defendant did not adhere to the standard of care. The defendant might argue that he or she did follow the standard of care, or that he or she had a valid reason for not following the standard of care in the particular situation. Such defenses are stronger if the valid reasons for deviating from the standard of care were documented at the time the work was done.

"Causation" is the evidence that links the defendant's actions to the plaintiff's injuries. For example, the defendant might have negligently maintained a ventilator, though the patient died of a pulmonary embolus that was unrelated to any potential problems with the ventilator. Causation is especially controversial when there is an alleged intermittent device failure that cannot be detected after the accident, and where the patient's injury was likely due to the underlying disease.

"Damages" are measured by the costs of compensating the plaintiff and the plaintiff's family for the injuries caused by the defendant's negligence. They can be very high when an otherwise healthy person is killed or permanently incapacitated by an injury and can include lost wages, past and future medical care, and other expenses related to the consequences of the injury. Conversely, if the patient is severely ill, with no chance of recovery, or elderly, the courts will award relatively little compensation because the patient's financial and physical condition is not worsened as much, even by death.

To avoid liability, CEs must ensure that their responsibilities are well defined and manageable. They must see that their work is properly documented and that those whom they supervise are competent and that their work is performed with the appropriate standard of care. CEs must avoid situations where they are not given adequate resources or control to ensure that the equipment for which they are responsible is properly maintained. For example, accepting a job that takes 12 hours but that must be done in 8 is likely to lead to injury and liability for CE malpractice.

Products Liability

Historically, liability for injuries caused by defectively designed or constructed products, or products for which there were inadequate warnings of dangers, was determined by the same negligence principles as CE malpractice discussed above. An injured person had to prove that the manufacturer did not follow the appropriate standard of care in designing or building the product and that the injury was caused by this failure to follow the appropriate standard of care. More difficult, the plaintiff would have to show that he was in "privity" with the manufacturer, which usually meant that he had purchased the product from the manufacturer. These rules dated from before the industrial revolution and anticipated that individuals would purchase products directly from the artisans who build them.

In an industrial society where products flow through many layers of distribution, the requirement of privity made it impossible to sue the manufacturer for most product-related injuries. Driven by the then-new phenomenon of automobile accidents due to poorly manufactured cars, courts abolished the privity rule in the early 20th century so that injured persons could sue the car manufacturer as well as the dealer where they purchased the car. Courts then extended the right to sue to nonpurchasers who were injured by defective products, such as passengers in cars, or cars driven by persons other than the purchaser.

Abolishing privity helped persons who had been injured by products such as cars that pass through the distribution chain with little change or risk of hidden damage by others before sale. Cheaper, more easily damaged consumer products still left the plaintiff with the problem of proving that the defect causing the injury was caused by the manufacturer and not by subsequent mishandling. The classic example is injuries caused by exploding Coke bottles. Old-style Coke bottles were made of heavy glass so that they could be recycled and refilled. (Older readers might remember checking bottles for place of origin and being amazed at how far bottles traveled in their lifetime; some might even recall finding roach carcasses lodged in the bottom of the recycled bottles.) While this practice was environmentally friendly, it increased the risk of an undetected defect in the glass, and the thickness of the glass made it particularly dangerous when it exploded under the pressure of the carbonated liquid. This combination led to many serious hand and eye injuries.

Plaintiffs then would sue the local Coca-Cola bottler and distributor, arguing that bottles broke in normal use because of either a defect in the glass or over-carbonation, another common problem. The manufacturer would deny responsibility, arguing that the defect was caused by mishandling by the delivery truck, the grocery stockers, the bar owners, or others who served as an intermediary between the customer and the manufacturer, or by the plaintiff himself. Because the manufacturer's duty was to produce a bottle that met the industry's standard of care, which it did, the plaintiff could not argue that bottles should be designed so that they were safe to use, even if mishandled.

The courts first attacked this problem with an old doctrine called *res ipsa loquitur*. Roughly translated, this means "the thing speaks for itself." It originated in a case where a barrel of flour rolled out of a second-floor warehouse loading door and fell on a bystander. The warehouse owner argued that the victim had to show why the barrel had rolled out of the window, to prove that the owner was negligent. The court ruled that plaintiff did not have to prove a specific negligent action if plaintiff could show that the thing causing the injury was in the defendants' exclusive control and that such accidents generally did not occur in the absence of negligence. Thus if the jury believed that the defendants must have been negligent if a barrel had rolled out of the warehouse, then the burden would shift to the defendants to provide an explanation for the accident, wherein they were not negligent.

As applied to the Coke bottle cases, the plaintiff did not have to prove a specific defect in the exploding bottle. Unfortunately, subsequent mishandling still was a defense, and the plaintiff would have to show that the defect had occurred while the bottle was under the control of the defendant. The more intermediaries between the injured customer and the bottler, the more difficult is was to rule out subsequent mishandling. This fact made these cases very difficult to prove and left the manufacturer free to keep using a dangerous design as long as it met the standards of the industry. Economists and legal scholars argued that this was inefficient in that it did not provide an appropriate incentive to reduce injuries by building safer products. This led courts and legislatures to turn to another old doctrine—strict liability.

Strict Liability

Historically, strict liability was applied to persons engaged in extremely hazardous activities such as keeping wild animals and constructing blasting. If these activities caused injuries to others through no fault of the injured person (e.g., that the plaintiff had entered the tiger's cage without permission would be a defense), there would be liability without any need to show negligence. More importantly, the defendant could not argue that it had complied with the standard of care and thus should not be liable. For strict liability to be found, the only question is whether the defendant's actions injured the plaintiff though no fault of the plaintiff's. The rationale underlying the adoption of strict liability is that there are some commercial activities with significant benefits that cannot be made safe. Rather than ban them, it makes better economic sense to allow them, but to require those who engage in them to pay for injuries that they cause, without regard to fault. This thinking influenced the California Supreme Court when it decided the case of *Greenman v. Yuba Power Products Inc.*, 377 P.2d 897 (Cal. 1963), which is seen as the first modern products liability case.

Greenman was a do-it-yourselfer who owned a multipurpose machine called a Shopsmith. The Shopsmith could be a saw, drill press, lathe, or other power tool, depending on how the user set it up. When Greenman was using it as a wood lathe, the chuck that held the wood came off the machine, and the turning piece struck him in the head, causing severe injuries. Investigation showed that the part had failed because it was attached with inadequately sized bolts. Greenman filed a lawsuit several months later, claiming that the machine was defective and violated the commercial warranty implied by the California Commercial Code. The manufacturer objected, pointing out that Greenman had not filed his claim in time to meet the statutory requirements. The court agreed that Greenman could not bring the warranty claim because the filing deadline had passed, but allowed Greenman to sue on the theory that the product was not fit for the intended use, irrespective of whether it met the standard of care or violated the statutory warranty. Thus the court launched the modern products liability claim that allows a jury to find a manufacturer liable if the product is defective and dangerous, even if it meets the standard of care. Applied to the Coke bottle cases, the standard of care would require the manufacturer to design bottles that could withstand mishandling, even though other manufacturers did not do so.

Shortly after *Greenman,* the American Law Institute, a private body that promulgates suggested legal standards, incorporated the rationale from *Greenman* into an influential document called the *Restatement of Torts*, 2nd edition, (1965).

402A. SPECIAL LIABILITY OF SELLER OF PRODUCT FOR PHYSICAL HARM TO USER OR CONSUMER

1. One who sells any product in a defective condition unreasonably dangerous to the user or consumer or to his property is subject to liability for physical harm thereby caused to the ultimate user or consumer, or to his property, if
   (a) the seller is engaged in the business of selling such a product, and
   (b) it is expected to and does reach the user or consumer without substantial change in the condition in which it is sold.
2. The rule stated in Subsection (1) applies although
   (a) the seller has exercised all possible care in the preparation and sale of his product, and
   (b) the user or consumer has not bought the product from or entered into any contractual relation with the seller.

Nearly all states have adopted some form of strict liability as defined in Section 402a. The policy behind Section 402a and the various modifications adopted by the states is to encourage manufacturers to engineer products so that they do not fail, or if they do fail, they fail in a way that does endanger users or others. The key provision is (2)(a)—strict liability applies even if the manufacturer "exercised all possible care in the preparation and sale of his product." It is no defense for a manufacturer to claim that it built the device to industry standards, or even that the product was built better than any other device on the market. Thus, if an EKG machine shocks a patient because the isolation transformer is defective, it will not be a defense for the manufacturer to claim that it bought the best transformers on the market.

Product Liability Defenses

Strict liability is not absolute liability. There are several defenses for products in general, and additional defenses for medical devices. The most fundamental is that the defective product did not cause the injury at all. This defense arises frequently in clinical engineering because the most critically ill patients tend to be the most instrumented. If a pacemaker fails and the patient dies of an arrhythmia that the pacemaker was implanted to control, then it is highly likely that the defective device caused the injury. The hospital itself also could be liable if the nursing staff failed to notice the arrhythmia because they were not monitoring the patient properly. If an electronic thermometer fails, masking a patient's fever, it is unlikely that it is what caused the patient's death from pulmonary embolism a few minutes later.

Misuse

Another common defense in clinical engineering is misuse. Thus, if a physician incorrectly sets up an anesthesia machine and injures a patient, the device manufacturer might not be liable. Misuse is a complicated defense because the courts have created the doctrine of foreseeable misuse; i.e., manufacturers have a duty to design products to prevent foreseeable misuse. Foreseeable misuse includes known common practices, such as turning off alarms, and practices that come to the attention of the manufacturer through incident reports or investigations. For example, when it became known that some physicians improperly assembled anesthesia machines and thus endangered patients, the manufacturer was found to have a duty to redesign the machine so that it could not be assembled improperly.

Learned Intermediary

Another facet of misuse is the expected expertise of the user. Historically, pharmaceutical manufacturers were allowed to assume that their drugs would be prescribed by physicians who were trained in pharmacology. This allowed the drug manufacturers to assume that warnings and other information would be read by an expert who could understand medical terminology and who would independently determine the suitability of the drug for the individual patient. This is called the "learned intermediary doctrine," and it insulates drug manufacturers from liability for physicians who incorrectly prescribe drugs, as long as the physician was provided the necessary information to prescribe the drug properly. In contrast, when drugs are sold over the counter (i.e., without a prescription), they must be labeled so that any consumer can use them safely. As drug manufacturers increase their direct marketing of prescription drugs to consumers, the courts are limiting their ability to claim the learned-intermediary defense. In a recent case, a state court found that the manufacturer of Norplant, a subcutaneous, long-term contraceptive, would be limited in the way it used the learned-intermediary defense because it advertised the product directly to consumers. The court allowed the patients to argue that the manufacturer should have provided safety information directly to the patients, rather than only to the physicians prescribing Norplant.

The courts have never clearly adopted the learned intermediary defense for medical devices. This is partially because, unlike pharmacology, physicians do not receive any training in clinical engineering. It also is due to the broader user base for medical devices. Unlike prescription drugs, which until recently were only prescribed by physicians, medical devices were frequently used by nurses and technicians who came from various backgrounds. Without a learned intermediary defense, medical device manufacturers must design and label devices so that they can be safely used by nonphysicians. This has become a particular problem as medical devices are routinely used in home health care, often without a medically trained user. Such uses demand well-designed training materials, device manuals, and labels on the device for safe use. The courts consider these explanatory materials as part of the device itself. If they are inadequate or do not provide appropriate information for the actual users of the device, they will be considered defective. Many medical device claims are based on defective labeling, rather than defects in the physical device itself.

Federal Preemption

In 1976, the FDA was given the authority to regulate medical devices through the Medical Device Amendments (MDA). Devices that were on the market at that time were "grandfathered" and could continue to be sold, pending future review by the FDA. New devices had to be submitted to the FDA before sale. If the manufacturer can show that the new device is substantially equivalent to a device that has been on the market since before 1976, and that the device is manufactured to the FDA's standards, then the FDA must permit the device to enter the stream of commerce with only a cursory review to determine whether it is the same as a pre-1976 device and passes basic safety standards (e.g., it will not shock anyone or come apart inside the patient. It is called 510(k) clearance, for the statutory section that establishes the procedure. Devices that are not equivalent to pre-1976 devices are subjected to an intensive new device approval process that is similar to the new drug approval process. This process can require extensive clinical trials and many redesigns of the device and its labels and manuals. While this intensive review of new medical device technologies is costly and time-consuming, it has an important legal advantage to the manufacturer: Design elements and labels specifically considered and approved by the FDA receive some protection from state product liability claims. For example, if a pacemaker lead fails, killing a patient, the plaintiff must show that the lead was defective, if he is to recover. If the lead's insulation failed because it became damaged during manufacture, then the plaintiff has a case. In many situations, however, the product was properly manufactured. Because pacemaker leads are delicate and can fail without being defective, the plaintiff will claim that the design is defective, perhaps because the manufacturer did not use a sufficiently strong alloy for the wire. If the FDA reviewed and approved the wire used as part of its review of the device, the courts have ruled that the plaintiff cannot second-guess the FDA by claiming that the wire should have been designed differently. If the FDA approved the label for the device, then the plaintiff cannot claim that the label was defective because it did not provide adequate instructions or warnings.

References

For general materials on clinical engineering liability and biotechnology law, see: http://biotech.law.umkc.edu

Abbey J, Shepherd MD. The Abbey-Shepherd Device Education Model. *Hosp Mater Manage Q* 13(4):69–81, 1992.

Benson D, Bell D, Kehner M, et al. Complying with the SMDA Medical Device Tracking Regulations: A Clinical Engineering Responsibility. *Biomed Instrum Technol* 28(5):376–80, 1994.

Forsell RD. The Clinical Engineer's Role in Incident Investigation. *Biomed Instrum Technol* 27(5):378–83, 1993.

Geddes LA, Salvendy G. Evaluation of Product Failure Rates for Litigation. *Biomed Instrum Technol* 33(5):417–22, 1999.

Grant LJ. Product Liability Aspects of Bioengineering. *J Biomed Eng* 12(3):262–66, 1990.

Gravenstein JS. How Does Human Error Affect Safety in Anesthesia? *Surg Oncol Clin N Am* 9(1):81–95, vii, 2000.

Hyman WA. Legal Liability in the Development and Use of Medical Expert Systems. *J Clin Eng* 14(2):157–63, 1989.

Hyman WA, Neigut JS. Device-Related Litigation & Clinical Engineering. *J Clin Eng* 19(6):441–45, 1994.

McCunney RJ. The Academic Occupational Physician as Consultant: A 10-Year Perspective. *J Occup Med* 36(4):438–42, 1994.

Morris RL. Consulting for Clinical Engineers. *Biomed Instrum Technol* 32(1):71–76, 1998.

Piehler HR. Innovation and Change in Medical Technology: Interactions Between Physicians and Engineers. *J Invest Surg* 5(3): 179–84, 1992.

Richards EP, Walter C. How Is an Anesthesia Machine Like a Lawn Mower?: The Problem of the Learned Intermediary. *IEEE Engineering in Medicine and Biology Magazine* 8;(2):55, 1989.

Richards EP. The Supreme Court Rules on Medical Device Liability—Or Does It? *IEEE Eng Med Biol Mag* 16(1):87–88, 90, 1997.

Richards EP, Walter C. Science in the Supreme Court: Round Two. *IEEE Eng Med Biol Mag* 17(2):124–25, 1998.

Richards EP, Walter C. The Supreme Court Sets Standards for Engineering Expert Testimony. *IEEE Eng Med Biol Mag* 18(6):83–84, 88, 1999.

Richards EP, Walter C. When Are Expert Witnesses Liable for Their Malpractice? *IEEE Eng Med Biol Mag* 19(2):107–09, 2000.

Saha S, Saha PS. Biomedical Ethics and the Biomedical Engineer: A Review. *Crit Rev Biomed Eng* 25(2):163–201, 1997.

Senders JW. Theory and Analysis of Typical Errors in a Medical Setting. *Hosp Pharm* 28(6):505–08, 1993.

Shepherd M. Reportability of Incidents. *Biomed Instrum Technol* 27(4):277–78, 1993.

Shepherd M. Clinical Engineers Have a Role to Play in Risk Management. *Biomed Instrum Technol* 32(6):60566, 1998.

Shepherd M. Eliminating the Culture of Blame: A New Challenge for Clinical Engineers and BMETs. *Biomed Instrum Technol* 34(5):37074, 2000.

Shepherd M, Brown R. Utilizing a Systems Approach to Categorize Device-Related Failures and Define User and Operator Errors. *Biomed Instrum Technol* 26(6):461–75, 1992.

Soller I. Legal Testimony: The Dos and Don'ts. *Biomed Instrum Technol* 35(1):61–63, 2001.

Stiefel RH. The Future of Clinical Engineering Is—"Appropriate." *Biomed Instrum Technol* 30(1):84–85, 1996.

Taylor K. Hospital Engineers and the Law. *Health Estate J* 46(3): 2–4, 1992.

Walter C, Richards EP. Employment Obligations Part I: Duties of an Employee to His Employer. *IEEE Eng Med Bio* 9(2):72, 1990.

Walter C, Richards EP. Employment Obligations Part II: Duties of an Ex-Employee. *IEEE Eng Med Bio* 9(3):72, 1990.

Walter C, Richards EP. Employment Obligations Part III: Who Is an Independent Contractor? *IEEE Eng Med Bio* 9(4):48, 1990.

Walter C, Richards EP. Employment Obligations Part IV: University Faculties and Expert Consultants. *IEEE Eng Med Bio* 10(1):85, 1991.

Walter C, Richards EP. Keeping Junk Science Out of the Courtroom. *IEEE Eng Med Bio* 17(4):78–81, 1998.

128

Professionalism

Gerald Goodman
Assistant Professor, Health Care Administration
Texas Women's University
Houston, TX

The word "profession" is defined in any standard dictionary as a calling requiring specialized knowledge, and often long and intensive academic preparation—a principle calling, vocation, or employment. "Professional" and "professionalism" are, likewise, defined by referring to some aspect of the root word, "profession." The dictionary definitions of "profession," "professional," and "professionalism" are not of real consequence in that most people, when defining these words, refer to a perception of what these words connote, to a mental picture or image of what a profession is, or how a professional looks (for example, abstract thinking, as opposed to manual labor; a suit and tie, as opposed to a uniform; a learned person such as a physician or a lawyer). This perception of a professional is cultural, having been nurtured carefully by those who most directly benefit from the mystique of the professions.

The background study of professions that is relevant to an analysis of the clinical engineering profession has four parts. The first part develops a baseline definition of profession. The second part traces the historical development of profession, looking for a developmental history of professions. The third part looks at the relationship of a professional to an employer. The fourth part looks at legal restraints such as licensure. This chapter concludes with a discussion of the role of professional societies in the professionalization process.

Definition of Profession

There is a dominant, contemporary mode of thinking about professionalism termed the "attribute approach" (Dingwall, 1976). The attribute approach begins from the basic assumption that it is possible to list fixed criteria for recognizing a profession on which there will be a general consensus. These criteria can be used to distinguish between professions and other occupations in a relatively clear and unproblematic fashion. Parsons (1954), Freidson (1971), and Illich (1973) have followed the attribute approach heavily. The following paragraphs discuss these three writers' definition of "profession."

Sociological Definition

"Profession" can be defined in sociological terminology as a cluster of occupational roles, or roles in which the incumbents perform certain functions valued in the society in general (Parsons, 1954). The profession is distinguished by the largely independent trusteeship exercised by the incumbents. Its roles are one important part of the major cultural tradition of the society. The profession's typical member is trained in that tradition, usually by a formally organized educational process, so that only those with the proper training are considered qualified to practice the profession. Furthermore, only the members of the profession are treated as qualified to interpret the tradition authoritatively and, if it admits of this, to develop and improve it. Finally, although there usually is considerable division of labor within such a group, a substantial portion of the members of the profession will be concerned largely with the "practical application" of the tradition to a variety of situations. Here, it can be more useful to others than to the members of the profession itself. The professional person is thus a "technical expert" of some order by virtue of his mastery of the tradition and the skills of its use.

The profession may be regarded sociologically as a mechanism of social control. The teaching profession helps to socialize the young, to bring them into accord with the expectations of full membership in the society, or to bring the young back into accord when they have deviated. The legal profession can be presumed to do this and two other things. First, the legal profession can forestall deviance by advising the client in the ways that will keep the client better in line and cool him off in many cases. Second, if it comes to a serious case, the lawyer can implement the procedure by which a socially sanctioned decision about the status of the client is arrived at. In the dramatic cases of criminal law, the determination is whether the client is innocent or guilty of a crime.

Another attribute of a profession is that of integrity (Parsons, 1954). The lawyer tends to be both permissive and supportive in her relationships with her client. However, there is more to the picture. As a member of this great profession, she accepts responsibility for its integrity, and her whole position in society focuses that responsibility upon her.

Classical Definition

Freidson (1971) notes that the discussions have been so fixed on a definition of "profession" that not much analysis has been made of the significance and consequences of some of the elements common to most definitions. The most critical elements are organizational in character and are related to the organization of practice and the division of labor. These elements are critical because they concern the facets of a professional occupation that are independent of individual motivation or intention. This may minimize the importance to behavior of the personal qualities of intelligence, ethicality, and trained skill that are imputed to professionals by most definitions.

The key institutional element is autonomy, which is the quality or state of being independent, free, and self-directing. For the professional, control over the content and the terms of work are self-directing his work. Self-direction is seen to be the key institutional element included in most definitions of "professional." The more self-directing an occupational group has become, the more likely a legal or political position of privilege has been obtained to protect it from encroachment by other occupations. This is stated to be the primary function of licensure.

A second institutional element related to self-direction is control of production and application of skill and knowledge. Control and knowledge is through the extended period of education controlled by the profession in an exclusively segregated professional school, rather than in a liberal arts school, and through curriculum that includes some special knowledge and skill necessary for the occupation. The professional school and its curriculum also constitute convenient institutional criteria for licensure, registration, or other exclusionary legal devices.

A third institutional element is the adoption of a code of ethics, which is a method of persuading society to grant the special status of autonomy. Thus, most of the commonly cited attributes that are used to define professions could be seen either as consequences of their autonomy or as conditions that are useful for persuading the public to grant such autonomy. The attributes that are used to define a profession contribute markedly to the mystique of that profession. It becomes difficult to say which came first—a definition or a mystique. In actuality, they have developed together (Freidson, 1971).

Political Definition

Profession is a concept that is used solely to achieve the successful mystification of a class interest (Dingwall, 1976). Illich (1973) defines the professional as someone who tells another what they need, and claims the power to prescribe. Professionals not only recommend what is good, but also actually ordain what is right. Neither income, nor long training, nor delicate tasks, nor social standing are the marks of the professional. Rather, it is his authority to define a person as a client, to determine that person's need, and to hand the person a prescription. This professional authority comprises three roles: The salient authority to advise, instruct and direct; the ethical authority that makes its acceptance not just useful but obligatory; and charismatic authority that allows the professional to appeal to some supreme interest of his client that outranks conscience. "By not taking the craftsman's counsel, a person is a fool. For not taking liberal counsel, someone is a masochist. Now the heavy arm of the law may reach out when they escape from the care that their surgeon or 'shrink' have decided upon for them" (Illich, 1973).

A Limitation of the Attribute Approach

Dingwall (1976) finds a central problem with the attribute approach to defining profession in that such attempts assume that the concept has a fixed meaning. Words do not have fixed and unequivocal uses according to some calculus of rules. Their sense is found by a process of filling in until someone can say that they understand. No one can define what profession is. All that anyone can do is elaborate what it appears to mean, to use the terms, and to list the occasions on which various elaborations are used.

Historical Development of Professions

The development of occupational groupings into professions can be traced from two aspects. The first traces the elements of the professionalization process. The second traces a natural developmental history of the professionalization process. Timperley and Osbaldeston (1975) and Wilensky (1964) wrote on the developmental and historical background of professions. A review of some of their work follows.

Elements of the Professionalization Process

Timperley and Osbaldeston (1975) identify five key elements in the process of professionalization. They are the level of activities to be performed, the achievement of status and prestige, the existence of a territory (and, hence, a boundary), the performance of service, and the prerequisite of prescribed educational training.

The educational background of members of an occupational association with professional aspirations is an important determinant of that association's status. Proper educational background is usually a prerequisite for entry (and, increasingly, for practice) in a profession. However, in the period of growth of an occupation in terms of numbers working within it and the range of functions performed, entry into that occupation will be loosely controlled, particularly given the need to build up the occupation and the range of tasks and functions.

The educational background of members of any formal occupational association is, to a large extent, governed by such factors as the speed of growth and development of an occupation in demand terms and the link with labor supply, the perception by people entering the occupational areas of the need for a formal identity, and particularly the amalgam of motives for setting up and perpetuating an occupational association.

Such motives might include the need for identity or status, the need for protection, and the need for improving standards in the area. All of these factors would create demands for controls once an occupational group has reached a point where it became necessary for competence to be defined and for functions to be ordered. In fact, it is important to distinguish between the control aspects of forming an occupational organization, which implies the mechanisms to be used to reach an end, and the achievement of a formal professional status.

Although the formal aims of professional associations might not emphasize collective bargaining as an activity, the individual member can, and does, use membership in, and qualifications from, a professional body as a bargaining instrument in his relationship with his employer or potential employer. The development of a strong identity enables professional associations to obtain considerable influence over the personnel activities of employers (Timperley and Osbaldeston, 1975).

A Natural Developmental History of Professions

Wilensky (1964) sought a natural development history of professionalization through the analysis of the developmental histories of 18 occupations. He hoped to document an invariant progression of events, a path along which the occupations had all traveled to professional recognition. Did the less established and marginal professions display a different pattern?

The progression of the occupation toward professionalization can be determined by noting first the present professional status of the occupation, and then the sequence in which six crucial factors occurred in the progression of the occupation toward professionalization. The six factors are the appearance of (1) the first training school; (2) the first university school; (3) the first local professional association; (4) the first national professional association; (5) the first state licensure law; and (6) the first formal code of ethics.

Schools and Associations

Before any of these six factors came into play, a job needed to be performed. The sick needed nursing, and hospitals needed managing. The original practitioners, by necessity, came from other occupations. The question of training soon arose. In four of the six established professions (in the study), university training schools appeared on the scene before national professional associations did. In the less-established professions, the reverse pattern is typical. The early appearance of the training school underscores the importance of the cultivation of a knowledge base and the strategic innovative role of universities and the early teachers in linking knowledge to practice and creating a rationale for exclusive jurisdiction (Wilensky, 1964).

Where professionalization has gone farthest, the occupational association does not typically set up a training school; the schools usually promote an effective professional association. Those pushing for prescribed training continue to form a professional association. The association defines the tasks of the profession; how to raise the quality of recruits; redefining of function to include the use of less-technical people to perform the more routine, less involved tasks; and the management of internal and external threats. This last item would include internal conflict between the old-guard committed to the local establishment, and newcomers committed to practicing the work wherever it takes them. External conflict is primarily competition with neighboring professions for claims to exclusive competence. This conflict with neighboring occupations seems to go with later stages of professionalization.

Licensing

A separate area of study is licensing and certification. Two lessons may be inferred from the study of patterns in licensure and certification. The turn toward legal regulation may be an expedient of an occupation "on the make" where internal debate persuades members that it will enhance status or protect jobs; or it may be forced on an occupation by some clear and present danger where public debate persuades lawmakers to protect the layman. Legal certification and licensure protection is apparently not an integral part of any "natural history" of professionalism (Wilensky, 1964).

Code of Ethics

The last area of professional development is the establishment of a formal code of ethics. The code of ethics usually includes rules to eliminate the unqualified and unscrupulous; rules to reduce internal competition; and rules to protect clients and to emphasize the service ideal. In 10 of 13 established professions or professions in process, a code of ethics comes at the end of the professionalization process (Wilensky, 1964).

Newer and more marginal professions often adopt new titles, announce elaborate codes of ethics, or set up paper organizations on a national level long before an institutional and technical base has been formed. Also, the tactical and strategic situation of an occupation, old or new, may demand early licensure or certification, whatever the actual level of development of the technique, training, or association (Wilensky, 1964).

Relationship of a Profession to an Employer

Understanding the behavior of a professional in relation to type of organization is a prerequisite to understanding motivations for joining a professional association, and the use of such an association. There is a relationship in terms of outcomes between organizational values and subsequent actions taken to control entry, progression, rewards, and other forms of behavior (Timperley and Osbaldeston, 1975). It is at this stage that the relationship between a professional organization and an employing organization becomes important, for it brings into an employer's control system a further dimension; through the creation of a dependency upon the professional organization; through the use an employer might make of a professional organization's role and status system in his own career structures; and through a possible limitation of an employer's control over employees as a result of controls exercised by a professional organization on its members.

The above relationship between professional organizations and employing organizations is an extremely complex one, for it brings into play a large number of institutional and personal interactions and considerations. This is especially so with the newer "professions" where the possible ideal of employment as a professional is often subordinated to the reality of employment as an organizational member who happens to be a member of a professional organization. However, one of the characteristics of an occupational grouping attempting to gain professional status appears to be its attempt to increase the dependency of an employer on the occupational norms operated and reinforced by the formal occupational association.

Parsons (1954) notes the following about the legal profession: Its members are trained and interested in a distinctive part of the cultural tradition, having a fiduciary responsibility for its maintenance, development, and implementation. They are expected to provide a service to a client, within limits, without regard to immediate self-interest. The lawyer has a position of immediate independent responsibility so that he is neither a servant only of the client (though he represents that client's interests) nor of any other group, in the lawyer's case, of public authority.

Above all, the lawyer stands between two major aspects of social structure: Between public authority and its norms, and the private individual or group whose conduct or intentions may or may not be in accord with the law. The lawyer is schooled in the great tradition of the law. As a member of a great profession, he accepts responsibility for its integrity, and his whole position in society focuses that responsibility upon him.

Reiser (1988) notes the following about the medical profession: Why is it that people visit a physician, often a stranger to them, and are willing to reveal to this individual all details of their past and their present, and then to permit this individual to physically exam them? These actions are based on a concept of trust. We can put ourselves in the hands of medical strangers because we trust them. But why should we? Because we have become convinced that those by whom we call physicians and nurses live by a set of ethical standards that binds them to a commitment to helping us and not harming us. It is the public's perception that medicine seeks to subject its theories and practices to unbiased evaluation, which separates fact from belief, which is at the heart of its support of the enterprise.

Noel and Jose Parry (1977) developed an historical analysis of the formation of the British Medical Association from its beginnings in 1832 to the present. Their analysis notes, in particular, how the presence of the Association furthered the interests of the medical profession in England. For nearly 200 years, professionalism has been a powerful ideology of a growing section of the middle class. Translated into practical activity, it involves a quest by an occupational association for self-governing autonomy in which control is exercised collectively by the occupation over its practitioners and over occupational recruitment.

The professional searches for legitimacy from the state. He hopes that through legislation, the occupation may be granted some degree of monopoly over the services it provides and a recognition, in legal terms, of self-governing autonomy. The establishment of a register of qualified persons has been a typical way in which insiders can be distinguished from unqualified outsiders. Moreover, the qualifying examination, conducted in educational institutions controlled by the profession, has been used as the mechanism by which closure of the occupation is achieved (Parry and Parry, 1977).

The autonomy of the physician is legendary among professions. As noted by Parry, their professional history reflects an early concern with the development of the profession as an entity, and the inherent self-direction and self-regulation of an autonomous profession (Parry and Parry, 1977). In contrast, engineering is a profession in name only. Engineers, as professionals, have little of the autonomy of the physician.

The contrast between the two professions is also reflected in the respective publics served by each profession. For the physician, the public is, in general, the individual patient. The level of concern is at the individual level, with an independent provider/independent client relationship. The patient/practitioner relationship is codified, as it were, in the Hippocratic Oath and subsequent codes of ethics.

Engineers are typically employed by large organizations, with a somewhat abstract provider/customer relationship. Except for those engineers in independent private practice, the customer is typically the general public.

The Role of Legal Restraints

Professions search for legitimacy from the state. They hope that through legislation, the occupation may be granted some degree of monopoly over the services it provides and recognition in legal terms of self-governing autonomy (Parry and Parry, 1977). Autonomy, the quality or state of being independent, free, and self-directing is a key element in the definition of profession. For the professional, control over the content and the terms of work are self-directing his work.

The more self-directing an occupational group has become, the more likely a legal or political position of privilege has been obtained to protect it from encroachment by other occupations. This is the primary function of licensure (Freidson 1971). One of the crucial factors in physician's achieving of self-governing professionalism in England was that the state machinery for regulating and controlling medical care in the mid-l9th century was weak. This lack of strength was still true in the early 1890's, at a time when the medical profession desired from the state a strengthening of protection from the competition of unqualified practitioners. Because the physician was well organized prior to the major entry of the state into the medical market place in 1911, their organized power as a profession was fundamental to securing an important measure of professional control over the state scheme. This control was guaranteed by state legislation and by administrative practices that grew under its umbrella (Parry and Parry, 1977).

Credentialing can occur through either licensing or certification. Licensure is a state function. It is defined as "the process by which an agency of government grants permission to an individual to engage in a given occupation upon finding that the applicant has attained the minimal degree of competence necessary to ensure that the public health, safety, and welfare will be reasonably protected" (Friedman, 1981). Licensure laws vary among professions and among states. Licensure laws pertaining to engineers exist in all 50 states, however licensure is not required, to practice as an engineer, except when an individual offers his services to the public as an "engineer" (Favoritti, 1980). This type of individualized engineering consultation is rare. Licensure in engineering is essentially "title protection".

Certification is a voluntary process. Nongovernmental agencies grant recognition of competence to an individual who has met certain predetermined qualifications (Freidson, 1971). Certification provides no legal protection for either the consumer or the practitioner (Favoritti, 1980). In 1977, the National Commission for Health Certifying Agencies (NCHCA) was chartered by the Department of Health and Human Resources as a voluntary body charged with developing standards for certifying groups. The standards ranged from the need for certification itself to the composition of organizations' governing boards. Issues that the NCHCA faced included the demand of third-party payers to reimburse only if a practitioner is credentialed, and the higher salaries asked by credentialed individuals.

Overshadowing all issues is a resolution to the question of whether credentialing in any form guaranteed quality and patient protection. Credentialing is heavily dependent on mechanisms for periodic recredentialing. Recredentialing, in turn, depends upon the continuing-education unit. William Turner of North Carolina State University, and several associates from both industry and academia proposed the continuing-education unit in 1968. The continuing-education unit is essentially a system of awarding credits for nonacademic credit courses. More than 1,200 colleges and 120 professional societies award continuing-education units. The continuing-education unit has been formalized through the National University Extension Association.

The continuing-education unit is defined as "ten contact hours" of participation in an organized continuing education activity under responsible sponsorship, capable direction, and qualified instructions (Patton, 1979). The continuing-education unit has been criticized as a public-relations gimmick designed to keep educators employed. Continuing-education unit programs may require examinations just as if they were degree courses. Other continuing-education programs award credit simply for chair-warming (Patton, 1979).

The Role of Professional Societies in the Professionalization Process

For nearly 200 years, occupational associations have sought self-governing autonomy in which control is exercised collectively over its practitioners and over occupational recruitment (Parry and Parry, 1977). Because the physician was well organized prior to the major entry of the state into the medical market place in England, the physician's organized power as a profession was fundamental to securing an important measure of professional control over the state scheme.

The role of the professional society in the professionalization process has been discussed in the work of Wilensky (1964). The professional society, through the traditional early influence of the academic community, establishes the knowledge base on which the profession is to develop.

In the Wilensky study, successful, established professions followed a developmental pattern where university training schools appeared before national professional associations. This pattern underscores the importance of the cultivation of a knowledge base. Teachers are the link between knowledge of practice and the creating of a rationale for exclusive jurisdiction. Where professionalization has gone farthest, the occupational association does not typically set up training schools; the schools usually promote an effective professional association.

Although the formal aims of professional associations may not emphasize collective bargaining as an activity, the individual member can and does use membership of, and qualifications obtained from, a professional body as a bargaining instrument in his relationship with his employer or potential employer. The development of a strong identity enables professional associations to obtain considerable influence over the personnel activities of the employer (Timperley and Osbaldeston, 1975).

As discussed earlier, university training schools appeared in the developmental history of the successful profession before professional associations did. Schools usually promote an effective professional association (Wilensky, 1964). Those individuals who push for prescribed training form a professional association. The association defines the tasks of the profession; how to raise the quality of recruits; redefining of function to include the use of less technical people to perform the more routine, less-involved tasks; and the management of internal and external threats.

Conclusion

In the Wilensky and the Timperley and Osbaldeston studies, status and education are identified. The educational system is the single most significant factor in determining the performance of the practicing professional. For example, the medical student is socialized to become a physician through the medical education system (Freidson, 1971). The educational background of members of an occupational association with professional aspirations is an important determinant of that association's status.

Education cultivates a knowledge base on which the knowledge to practice and exclusive jurisdiction of the profession is established. Knowledge and exclusive jurisdiction, in turn, enhance the status of the profession.

The early appearance of university training schools in the professionalization process underscores the importance of the cultivation of a knowledge base and the strategic innovative role of universities and the early teachers in linking knowledge to practice and creating a rationale for exclusive jurisdiction. Where the professionalization process has gone farthest, the occupational association typically does not set up a training school; the schools usually promote an effective professional association (Wilensky, 1964).

Legal certification and licensure protection is not a part of any natural developmental history of professionalism (Wilensky, 1964). However, much of a profession's strength is based on a legally supported monopoly over practice operating through a system of licensing. The mystique of a profession revolves around its ability to institutionalize legally its place in society (Illich, 1977).

References

Dingwall R. Accomplishing Profession. *The Sociological Review* 27:331–349, 1976.
Favoritti R. The Legal Significance of Clinical Engineering Certification. *Medical Instrumentation* 14:141–143, 1980.
Freidson E. *Profession of Medicine*. New York, Dodd, Meade, 1971.
Friedman E. The Dilemma of Allied Health Professions Credentialing. *Hospitals* 47–51, 1981.
Illich I. The Professions as a Form of Imperialism. *New Society* 13:20–27, 1973.
Illich I. *Disabling Professions*. London, Marion Boyers, 1977.
Parry N, Parry J. Professionalism and Unionism: Aspects of Class Conflict in the National Health Service. *The Sociological Review* 25:823–841, 1977.
Parsons T. *Essays in Sociological Theory*. Glencoe, IL, The Free Press, 1954.
Patton M. Continuing Education for Employed Clinical Engineers. *IEEE/EMBS Newsletter* 18:55–57, 1979.
Reiser J. A Perspective on Ethical Issues in Technology Assessment. *Health Policy* 1988.
Timperley SR, Osbaldeston MD. The Professionalization Process: A Study of an Aspiring Occupational Organization. *The Sociological Review* 607–627, 1975.
Wilensky HL. The Professionalization of Everyone. *American Journal of Sociology* 70:137–158, 1964.

129
Clinical Engineering Advocacy

Thomas J. O'Dea
Hemoxy, LLC
Shoreview, MN

This chapter concerns the advocacy of clinical engineering and the ethics associated with its practice. The bylaws of the American College of Clinical Engineering (ACCE) address both of these topics. For example, ACCE (Bauld, 1991) defines a clinical engineer as "A professional who supports and advances patient care by applying engineering and management skills to health care technology."

ACCE amplifies this definition in its seven-point code of ethics, as follows:

1. Prevention of injury in the clinical environment
2. Accurate representation of the clinical engineer's knowledge, level of responsibility, education, authority and experience
3. Revelation of conflict of interest
4. Protection of confidential information
5. Improvement of patient care delivery
6. Cost containment by technology utilization
7. Promotion of the profession of clinical engineering

This chapter will address the above points based on the knowledge of the author, a clinical engineer with 35 years of experience, and describes the relationship of these areas to ethics and advocacy.

Basis for Advocacy

The basis for advocacy is the perception that the clinical engineer, defined by the ACCE and the certification process, adds value to the equipment safety and technology functions frequently performed by various personnel in the health care environment. The justification of this perception forms the basis for advocacy, both within the medical technology community and to the public, using the experience of one average clinical engineer. The goal of improved patient care justifies efforts to preserve and to strengthen the profession. Whether the clinical engineering profession develops in influence and recognition depends, in part, on whether those in the profession are demonstrating and teaching other professionals in health care and the general public about the critical role that clinical engineering plays in health care.

The work of clinical engineers is often best appreciated in the long term, because many of their responsibilities entail strategic planning, systems development, training in the application of health care technologies, technology assessment and evaluation, health care technology management, and product development. Perhaps it is easier to explain the benefits of the work of biomedical equipment technicians (BMETs), whose main responsibility is to obtain immediate results in such areas as equipment repair and maintenance.

Clinical Engineering Expertise

Clinical engineers have unique expertise in medical device applications and are valuable to medical device technologies research and development efforts. Clinical engineers can start with an idea for a device and see that idea develop into prototypes, clinical trial models and finished products. An example of this is detailed in "A Novel Device for Measuring the Effect of Cholesterol on the Release of Oxygen from Red Blood Cells into Myocardial Tissue" (O'Dea et al., 2000). In this project, the author worked with a team to study the effects of cholesterol on the diffusion of oxygen through the red blood cell membrane. Clinical engineering expertise contributed to the creation of a device to measure the diffusion times for red blood cells by verifying the device's operation with laboratory results preparing a quality plan to satisfy existing regulations, and participating in clinical trials.

In addition to creating devices, existing devices must be introduced to patients by people knowledgeable in both engineering and the clinical environment. This requires adaptation to the physical environment, staff capabilities, and patient needs of an individual health care institution.

Patient Safety

Patient safety requires knowledgeable professionals in technical areas for the discovery, evaluation, and mitigation of patient risk. The author's work (O'Dea et al., 1998) on the interaction of infant warmers and bilirubin lights with the regulation of temperature of newborn infants is an example of an unexpected safety issue discovered during the ordinary duties of a clinical engineer. The nurses in the Neonatal Intensive Care Unit (NICU) experienced problems with using infant incubators in the servo-controlled temperature mode on some neonates. The manual mode triggered an audible alarm every few minutes, which added to the already high sound level in the NICU. In looking at the entire situation, interaction between the infant warmer and the bilirubin lights used simultaneously on low weight newborns prevented effective control of body temperature, since the uncontrolled energy delivered by the broad spectrum bilirubin lamps was significant compared to that delivered by the incubator. Once the cause of the problem was determined, operational changes in the simultaneous use of the two devices allowed effective care.

An example of combining the disciplines of clinical engineering and medical physics is quantification of patient skin radiation dose during neurointerventional procedures (O'Dea et al., 1999). This study determined the probability of patient skin injury due to radiation received during such procedures as cerebral angiograms and embolizations. Using the data from this study, a method of estimating the probability of patient injury was proposed, using fluoroscopy time and the number of digital angiographic images, both easily obtained from the radiographic equipment. Some neurointerventional procedures had a small likelihood of patient injury, and might be omitted from an elaborate monitoring program, while others had a higher likelihood of injury, and should be part of such a program. Another collaboration effort with a medical physicist yielded a quality control system for automated dosimetry and the estimation and mitigation of error (Li et al., 1999).

Engineering the Environment

Securing environmental conditions that contribute to a positive patient outcome through evaluation, specifications, design, and management is part of the practice of clinical engineering. The author demonstrated the way in which the clinical engineer can engineering the environment when he determined the isolation requirements of immunocompromised patients, assessed the ventilation system, and developed methods of measurement and practical solutions to balance isolation from spores against patient comfort and system efficiency (O'Dea, 1996).

The examples of how clinical engineering provides a safe patient environment have three characteristics in common:

1. They arose from conditions observed during the performance of ordinary clinical engineering duties.
2. They affected areas normally not mentioned in clinical engineering practice, i.e., facility engineering, radiology, and nursing. Collaboration with health care professionals in those areas is essential.
3. Resolution of the issues was not possible with standard repair or technology management methods. Solutions required in-depth study of the medical issues, engineering methods, and applications of practical hardware to safety.

These cases illustrate how clinical engineering has a unique and positive effect on safety in the health care setting.

Health Care Technology Management

The management of the health care institution is affected by applying clinical engineering principles to maintenance, performance verification, and the identification and adaptation of new technology. Demonstrating the role of clinical engineering in management, O'Dea and Marshall (1995) conducted a cost-saving study on the maintenance of sterilizers and operating room tables. They describe the steps taken to introduce in-house technology maintenance to an area previously served by outside contractors. The process resulted not only in cost savings, but also in the re-engineering of a washer-sterilizer to be safely used as a pressure vessel. It also allowed the facility greater flexibility in purchasing operating room tables that surgeons wanted but had not been able to acquire due to lack of manufacturer support.

Simply repairing a medical device and placing it back in service without looking into the reasons for failure could result in a missed opportunity to modify a hospital-wide system or practice that routinely contributes to device failure. O'Dea et al. (1997) developed a verification procedure for video-endoscope systems. This work was done after we had discovered, while performing a routine repair, that performance verification was not done, that educational programs for residents were deficient, and cleaning procedures were inadequate. These issues led to increased cost and decreased reliability of technology systems, but management changes significantly decreased operating costs.

Advocacy plans must continue to stress that clinical engineers can analyze problems and synthesize solutions. For example, clinical engineers recommended using frequency domain multiplexing and carrying clinical local networks, terminals, mainframe wiring, and security video over the wideband TV distribution system, rather than recabling an entire building for data. By postponing this recabling for ten years, the useful life of existing technology was extended at significant savings to the hospital. Clinical engineers often have solved system problems such as interfacing research-oriented data gathering systems and existing clinical systems. Clinical engineers have designed four-catheter, 16-channel research cardiac electro-stimulation systems for use in the cardiac catheterization laboratory, enabling the hospital to offer more services to the community.

These examples clearly demonstrate the benefits of clinical engineers. Advocates must relate these examples, and others like them, to those in the health care business and to the general public.

ACCE Advocacy Committee

ACCE has formed an advocacy committee to assist members in educating the health care profession and the general public in the value of clinical engineering to patient care (see Chapter 130 for a detailed description of ACCE). Among the committee's current activities is the awarding of the annual clinical engineering advocacy awards in technical areas, management areas, and safety. Other committee work involves developing tools for the celebration of National Engineering Week, assisting displaced clinical engineers, and responding to those who mischaracterize the role or importance of clinical engineers. For example, if a hospital seeking a clinical engineer advertised in a journal but listed inappropriate qualifications, such as a two-year associate's degree requirement, the committee would respond to the hospital and the journal running the ad with the correct educational requirements for a clinical engineer: A four-year engineering degree. The committee also would provide guidance on actions to take when individuals more accurately described as biomedical engineering technicians inappropriately list themselves as clinical engineers.

Conclusion

The care of the patient is paramount, and the clinical engineer can create a safer and more competent environment for that care. Clinical engineers have achieved recognition and shown value by applying engineering and management skills to health care as previously described. Advocacy efforts through the ACCE Advocacy Committee continue to dispense information regarding the role that clinical engineers play in health care.

References

Bauld TJ. The Definition of a Clinical Engineer. *J Clin Eng* l6:403–05, 1991.

Li S, O'Dea TJ, Geise RA. Establishing a Quality Control Program for an Automated Dosimetry System. *Medical Physics* 8:1732–1737, 1999.

O'Dea TJ, Koch S, Holte JE, et al. Improving Video-Endoscope Reliability at a Large University Hospital. *Biomed Instrum Technol* 31:579–598, 1997.

O'Dea TJ, Saly G, Holte JE. Safety Investigation: Interaction of Infant Radiant Warmers and Bilirubin Phototherapy Lights in the Regulation of Temperature of Newborn Infants. *Biomed Instrum Technol* 32:355–369, 1998.

O'Dea TJ, Menchaca HJ, Michalek VN, et al. A Novel Device for Measuring the Effect of Cholesterol on the Release of Oxygen from Red Blood Cells Into Myocardial Tissue. *Biomed Instrum Technol* 34:283–289, 2000.

O'Dea TJ, Geise RA, Ritenour ER. Estimation of Patient Skin Dose During Cerebral Angiograms and Embolizations from Fluoroscopy Time and Number of DSA Frames: A Study of 256 Cases. *Medical Physics* (submitted for publication).

O'Dea TJ, Marshall JW. Cost-Savings Case Study: Clinical Engineering Management; Maintenance of Sterilizers and Operating Room and Procedure Tables. *Biomed Instrum and Technol* 29:24–26, 1995.

O'Dea TJ. Protecting the Immunocompromised Patient: the Role of the Hospital Clinical Engineer. *J Clin Eng* 21:466–482, 1996.

O'Dea TJ, Geise RA, Ritenour ER. The Potential for Radiation-Induced Skin Damage in Interventional Neuroradiological Procedures: A Review of 522 Cases Using Automated Dosimetry. *Medical Physics* 9:2027–2033, 1999.

130 American College of Clinical Engineering

Jennifer C. Ott
Director, Clinical Engineering Department
St. Louis University Hospital
St. Louis, MO

Joseph F. Dyro
President, Biomedical Resource Group
Setauket, NY

Health Care

The American College of Clinical Engineering (ACCE), which is committed to enhancing the profession of clinical engineering, is the only internationally recognized professional society for clinical engineers with members in the United States and around the world. Founded in 1990, ACCE has developed and promoted clinical engineering leadership, guidelines, recommendations, and advocacy efforts in a short time. Extensive membership benefits include the *ACCE News*, the foremost newsletter of clinical engineering in the world, and educational programs, including annual symposia, meetings, teleconference series, and advanced clinical engineering workshops (ACEWs). ACCE relies upon the strength of its members, who ensure a vibrant voice for clinical engineering worldwide through participation on boards, committees, and special programs. The ACCE recognizes and rewards those who have excelled in the pursuit of clinical engineering excellence.

Establishment and History

The American College of Clinical Engineering was established in 1990 following two years of planning that included dialogue among colleagues in conjunction with other society meetings, phone calls, faxes, and planned group discussions, with the sole purpose of forming such a society. As with any organizational beginnings, there were growing pains that included some successes and some failures. Overall, the mission of the ACCE has been maintained, and the clinical engineering profession has matured more during ACCE's tenure than before it existed.

Rationale for Establishment

During the late 1980s, it became apparent through existing professional engineering organizations (such as the Institute of Electrical and Electronics Engineers (IEEE), Engineering in Medicine and Biology Society (EMBS), and the Association for the Advancement of Medical Instrumentation (AAMI)) that no single organization was dedicated exclusively to the promotion and support of clinical engineering. Many clinical engineers desired such an organization. In 1988, a session at the Annual Meeting and Exposition of AAMI in St. Louis, Missouri, entitled "Do We Need Another Professional Engineering Society?" convened members of IEEE, EMBS, AAMI, and others to discuss their current mission and how they would promote and develop the clinical engineering profession. Throughout the discussion, it became apparent that no organization had the resources or dedication to promote the profession of clinical engineering while maintaining their own mission and breadth of membership, and that the clinical engineering profession sorely needed sufficient representation. One astute session attendee pointed this out during the question-and-answer session and suggested that another society might be needed. The wheels were set into motion and the rest is history.

Over the course of the following year, many clinical engineers furthered the conversation on the need for such an organization, through ad hoc phone calls, letters, and meetings, both nationally and internationally. Many experts in the field discussed the topic and pursued the development of the organizational name, mission, bylaws, and application for formation. The important question to address was where this new society would fit in the current realm of professional and trade societies. Many of these groups crossed into the clinical engineering realm, but none was solely dedicated. There was much discussion as to the title of the organization, as many were sensitive to the international component and ways in which the formation of this society would play well into international support. Some suggested names included the North American Clinical Engineering Society, the Hospital Engineering Society, the Health Care Engineering Society, and the International College of Clinical Engineering.

After initial discussions that created the vision for ACCE, significant work remained to be done before the society could form. Other interested clinical engineers were invited to assist in crafting documents to present at the 1989 AAMI meeting. This presentation consisted of the core group of founders and other supporters. This group provided the interest and sustenance to continue with this endeavor. In February of 1990, paperwork was submitted to the state of Washington to form the American College of Clinical Engineering.

The First Clinical Engineering Organization

ACCE was the first organization to be dedicated solely to the field of clinical engineering, both nationally and internationally. At its inception, many requests for advocacy, support, and professional development were received, and the necessity to streamline and further develop the activities of the charter members became apparent. The newly approved board of directors and other charter members of the ACCE did not take the job lightly. The first step in ensuring the success was to develop a mission and definition of clinical engineering.

Exclusivity

The founding members of the ACCE were clear as to the purpose and future support. Within the clinical engineering domain there are many functions, all of which benefit the profession as a whole. The purpose of this group was to promote those individuals who fell within the definition of "clinical engineer" (Bauld, 1991). The ACCE definition, which most clinical engineering organizations worldwide now accept, is the following:

A clinical engineer is a professional who supports and advances patient care by applying engineering and managerial skills to health care technology.

While ACCE must remain focused on benefiting its members, it desires to support (to the extent possible) other professionals who work closely with clinical engineers. Such individuals include biomedical engineering and equipment technicians (BMETs). Over the years, there have been many discussions–some friendly, some not–regarding the role of ACCE and its support of the BMET. Currently, there is no national organization of BMETs in the United States. Some perceived the ACCE as an organization that neither wanted BMET involvement nor recognized the BMET's role with respect to the clinical engineering profession. Some ACCE members want an organization focused on clinical engineers only and others want to support BMETs as they work so closely with clinical engineers and share many of their goals. The ACCE Board of Directors has always striven for middle ground. While the need for support of the BMET group is keenly felt by most BMETs, ACCE must stay true to its mission. The ACCE is not a large enough organization, nor does it have the resources, to take on the complete support of the BMET in a national setting. Their requirements and needs are different. BMETs are encouraged to join ACCE and, depending upon their qualifications, can belong at one of many different levels. ACCE promotes the clinical engineering profession through educational opportunities in which BMETs can attend, and even lead. The ACCE Board of Directors would entertain the concept of assisting a group of BMETs in the development of its own professional society and determining how ACCE and BMETs can work more closely together to further the promotion of clinical engineering and the mission of ACCE.

Growing Pains

The early years of ACCE were indeed a struggle. Cash flow was tight, and the request for professional advancement was high, both nationally and internationally. There was much antagonism from other societies who treaded into the clinical engineering domain, such as AAMI, the American Society for Healthcare Engineering (ASHE), and IEEE EMBS. The ACCE board of directors and other members who crossed over into those societies worked hard to explain the mission of ACCE and that it was exclusive of the roles that these societies supported in clinical engineering. The other societies simply could not provide sufficient dedicated clinical engineering representation.

Charter Members

Many dedicated clinical engineering professionals deserve credit for the formation of the American College of Clinical Engineering (Table 130-1). Without their support of their profession and their contribution of time and resources, ACCE never would have been formed and would not have survived the early years. The list of charter members who founded the organization and showed support by joining during the first year of existence (1990) is shown below (ACCE News, 1991). All should be recognized for their aggressive support in seeing the establishment of the ACCE through to fruition.

Mission

The first board of directors and other founding members worked diligently to develop a mission of the ACCE that was the purpose for establishing ACCE in the first place. It was important that the mission kept the focus of a *clinical engineering only* organization while promoting the field of health care technology management, important components of which are medical device inspection, maintenance and repair, which constitute the bulk of BMET responsibilities.

The following four points were approved and have remained the sole mission of the ACCE since inception:

1. To establish a standard of competence and to promote excellence in clinical engineering practice
2. To promote safe and effective application of science and technology in patient care
3. To define the body of knowledge on which the profession is based
4. To represent the professional interests of clinical engineers

Examples of fulfilling the mission can be seen throughout the organization and will be discussed in detail within this chapter

Code of Ethics

During the formative years, the requirement for a professional code of ethics was determined to be necessary, to establish parameters of professional duties and to explain what ACCE was trying to achieve. This is different from the mission and provides a means for the members to determine the ways in which their work affects the promotion and application of the profession. Following is the Code of Ethics.

As a member of the American College of Clinical Engineering, I subscribe to the established Code of Ethics in that I will:

- Strive to prevent a person from being placed at risk of injury due to dangerous or defective devices or procedures
- Accurately represent my level of responsibility, authority, experience, knowledge, and education
- Reveal any conflicts of interest that may affect information provided or received
- Protect the confidentiality of information from any source
- Work towards improving the delivery of health care to all who need it
- Work towards the containment of costs by the better utilization of technology
- Promote the profession of clinical engineering

Membership Benefits

The available membership benefits have changed throughout the years to reflect the changing profession. In general, ACCE is building a strong, credible, dynamic, and flexible profession. ACCE membership provides advantages that will enhance a clinical engineering career now and for many years to come. Benefits include the following:

- Access to a network of experts and peers
- Representation of interests to legislators and regulatory agencies
- Instant access to valuable technical information via the ACCE web site at http://www.accenet.org
- Up-to-date information (via the ACCE newsletter) that is so critically needed in the rapidly changing health care environment

- Engaging in special projects from time to time, such as advanced clinical engineering workshops
- Providing opportunities to share expertise with other professionals

Marketing Relationship

In 1996, the ACCE Board of Directors entered into a marketing venture with Morse Medical. Wayne Morse, CCE, president of Morse Medical and founding member of the ACCE, assisted ACCE in marketing and promoting ACCE products. The list below indicates products and their description that were available throughout this arrangement, although some may no longer exist. For further information on products that are currently available, please refer to the ACCE website at http://www.accenet.org.

Products

- *Lapel Pin:* Attractive pin is designed for ACCE members and is suitable for use with a lab coat or suit jacket.
- *Logo:* The ACCE logo (Figure 130-1) can be added to apparel such as a hat or shirt.
- Clinical Engineer Definition Plaque: An 8″ x 12″ wooden plaque displaying the ACCE definition of a clinical engineer for an office or department to promote the profession.
- *Code of Ethics Plaque:* An 8″ x 12″ wooden plaque displaying the ACCE Code of Ethics for an office or department to promote the profession.
- *Membership Directory:* A current annual directory that is a very valuable resource and reference document. In these pages are the names, addresses, phone and fax numbers, and e-mail addresses of the finest group of clinical and biomedical engineers in the world.
- *Guidelines for Medical Equipment Donation:* In an attempt to address problems with donating equipment, the ACCE has developed guidelines to improve the effectiveness of donations of medical equipment.
- *CE Certification Study Guide:* Based upon the International Certification Commission (ICC) and AAMI program, this study guide offers a quick and easy self-evaluation for those professionals who plan to take the CE certification exam.
- Teleconference Audiocassette Tapes (including handouts): The audiocassette tape and handouts from some of the best in the ACCE teleconference series.
- *ACCE Brochures:* Brochures developed for the membership include the following: *ACCE: The American College of Clinical Engineering, What is a Clinical Engineer?,* and *Providing Clinical Engineering Support to the Countries Worldwide,* from the International Committee.
- *ACCE News*: A bimonthly publication and the foremost newsletter of clinical engineering in the world dedicated to promoting the activities of ACCE and its members.
- ACCE website: Access to a wealth of information regarding ACCE through the ACCE web site. Members-only access is under development.

Programs

- The educational teleconference series is held each year, with a variety of topics presented once per month. Members receive a discount and the ability to have multiple participants from one site.
- Advanced clinical engineering and health technology workshops are held at various times throughout the world. The goal is to improve the skills and knowledge of those who are involved in medical technology management and facilities planning.
- INFRATECH was created in 1999. It is a listserv sponsored by World Health Organization (WHO), hosted by the Pan American Health Organization (PAHO), and coordinated by ACCE—an Internet discussion group for the exchange of information on health care infrastructure and technology for health services.
- ACCE symposia are held once a year dedicated to a topic that is essential to the future of clinical engineering.
- ACCE works with other societies and conferences such as AAMI, ASHE, and HealthTech to develop programs that promote the profession of clinical engineering. Members are often provided discounts to attend these educational meetings. ACCE is always looking to broaden the interaction with other societies and conferences.

Development

ACCE has been successful since its inception in 1990 in fulfilling the mission set forth by the founding members. In order to see the progress of the organization, the complete assessment and overview of all the ACCE activities must be taken into account. Presented here, however, is a brief summary of all activities held since establishment, thus providing a key into the development of the profession of clinical engineering and the success of ACCE.

Highlights of 10 Years

- *Mission:* The first clinical engineering only organizational mission was mentioned earlier in full. It is an essential developmental step when trying to establish any organization and provides direction to the founding group and future members.
- *Definition of Clinical Engineer:* Complementing the mission (and also mentioned earlier), this helped to define the profession and the membership of ACCE. It has been applicable throughout the history of ACCE.
- *Code of Ethics:* Addressing the need for professional liability (and mentioned earlier), this code has helped to formulate ways in which ACCE members should carry out the mission of ACCE.
- *Vision 2000:* Begun in 1995, this was ACCE's process to review the current and future environment of health care and develop strategies, objectives, and specific initiatives that will prepare our membership for the changes that continue to occur at a rapid pace.
- *PACE Initiative:* This name stood for Preparing ACCE for Cost Effectiveness and took place during the Vision2000 project. It served as a guide for the college. There were seven initiatives under the PACE banner (Bauld and Judd, 1995).

Table 130-1 American College of Clinical Engineering charter members

Matthew F. Baretich	A. Pieter du Toit	Thomas M. Judd	David Natale
Thomas Bauld	Joseph F. Dyro	Philip Katz	Roger Neifert
David Bell	Larry Fennigkoh	Gary Kotter	Allen F. Pacela
Seymour Ben-Zvi	Steven Friedman	Alan Lipschultz	Frank Painter
William Betts	Gerald Goodman	Denver A. Lodge	Bryanne Patail
Jon Blasingame	Gailord Gordon	Salvador E. Longo	Al Plourde
Mark Brody	Dan Hare	Mahmoud A. Madani	Michael Shaffer
Joseph D. Bronzino	David Harrington	Dennis Minsent	David A Simmons
Michael Carver	Gary Haugen	Henry Montenegro	Ira Tackel
Richard Daken	Ethan Hertz	Mark Moody	F. Scott Varnum
Yadin David	Larry Hertzler	Robert Morris	Alvin Wald
David M. Dickey	Jan P. Ingebrigtsen	Wayne Morse	Binseng Wang
Frieda du Toit	James Jablonski	Joanne S. Munger	James Wear
			Steve Wixson

- *1996 FDA GMP Regulation Response:* Drafted by current ACCE president Thomas J. Bauld, Ph.D., it voiced the concern of the ACCE membership on the proposed regulations (Bauld, 1996).
- *1998 FDA ANPR Response:* Drafted by current ACCE president Frank Painter, CCE it discussed the expertise of ACCE and its members (Painter, 1998).
- *Teleconference Series:* An annual series of monthly, hour-long presentations covering topics of interest to ACCE members.
- *ACCE News:* A bimonthly publication and the foremost newsletter of clinical engineering in the world dedicated to promoting the activities of ACCE and its members
- *Advocacy Awards:* Allows ACCE to recognize members and nonmembers for professional achievement and professional development.
- *ACEW:* Advanced Clinical Engineering Workshops are international educational opportunities where ACCE provides faculty and shares expertise throughout the world. Workshops are often sponsored by World Health Organization (WHO) or the Pan American Health Organization (PAHO).
- *Membership Surveys:* Designed to take the pulse of the membership to ensure that the organization is meeting the needs of the membership and to plan further projects.
- *Web Site:* An Internet-based method to communicate to the membership. It allows members and nonmembers the opportunity to learn more about ACCE, the membership, and activities.
- *White Papers:* Topics of interest to the clinical engineering community and beyond are explored by an ad hoc task force developed by the board of Directors. The white paper developed in 2001 was on patient safety.
- *Certification:* In 1999, the AAMI board of directors approved the suspension of the current Certified Clinical Engineer (CCE) program. ACCE has worked diligently to put the program back into place in a much better format.
- *Joint Educational Activities:* ACCE has worked throughout the years with other groups such as ASHE, AAMI, HealthTech, and others to develop educational sessions geared towards clinical engineering. Expansion of the societies with which ACCE interacts is always encouraged.
- *BMET Task Force:* Formed originally in 1997 because AAMI dissolved the Society for Biomedical Equipment Technicians (SBET), the group explores ways in which ACCE can assist our fellow members of the health care technology community. These activities are ongoing.
- *AHA Telemetry Task Force:* Many ACCE members were asked to serve on the AHA telemetry task force to define the telemetry issues, to determine current and projected needs, and to develop educational materials. All of this was done in response to the digital-television technology shift and the subsequent allocation of broadcast frequencies by the U.S. government (Campbell, 1998).
- *Symposia:* Begun in 1998, the annual educational meeting discusses topics of interest to ACCE members in particular the future of clinical engineering and the profession.

Membership

Categories

There are five membership categories as of the approved bylaws in August 2002: Individual, Fellow, Emeritus, Associate, and Candidate. Each has different criteria for approval by the Membership Committee and the ACCE Board of Directors.

Individual

A person who has demonstrated evidence of professional practice of engineering in a clinical environment for at least three (3) years, and who meets one or more of the following three conditions:

1. Possession of a baccalaureate degree in an engineering discipline or engineering technology from an accredited college or university (or foreign equivalent)
2. Certification as a Clinical Engineer (CCE), by the International Certification Commission for Clinical Engineering and Biomedical Technology
3. By recommendation of the Membership Committee in recognition of exceptional contributions, consistent with criteria established by the board, to the profession of clinical engineering.

Fellow

An Individual member may be advanced to Fellow status in recognition of distinguished service to the profession or achievement in the field of clinical engineering. The member must have been active in the College business over a minimum period of three years.

Emeritus

An Individual or Fellow member may be advanced to Emeritus status:

- In recognition of distinguished service to the profession or achievement in the field of clinical engineering
- After at least five continuous years membership in good standing in the College, and upon reaching the age of 62 and retirement from full-time employment in the profession of clinical engineering.

Associate

An individual committed to the mission of this organization, who has demonstrated a contribution to the advancement of the clinical engineering profession, and meets other requirements established by the board, but does not meet the other conditions required for Individual membership.

Candidate

An individual who is interested in the purpose of this organization and who meets one of the following two conditions:

1. Current enrollment at least half-time in an accredited baccalaureate or graduate program in engineering, engineering technology, or related course of study; or
2. Completing the three-year clinical experience requirement for Individual membership after receiving a baccalaureate or graduate engineering degree.

Renewal

Membership renewal occurs on an annual basis in conjunction with the calendar and fiscal year of ACCE. Initial renewal notices are sent in November with reminders in January and February. March 1 is the deadline and the membership roster reflects the current membership at that time. Future mailings and notifications are dependent upon the current membership roster. Renewal costs are set by the board and have remained reasonable in comparison to other professional societies since inception. Renewal and dues costs for 2002 were $60 for Individual, Fellow, and Associate members, $30 for Candidate members, and $0 for Emeritus members.

Board

Responsibility for management of the College is vested in the board, which consists of 10 members: The president, the president elect, the vice president, the secretary, the treasurer, the immediate past president, and four members-at-large. Board members will be elected as specified in the bylaws. Each member of the board is entitled to one vote. Further explanation of the individual officer duties, election process, and terms of office can be found in the bylaws section in this chapter. Board meetings are held at least quarterly, usually bimonthly. They are conducted via conference call with items being distributed by e-mail. On an annual basis, there is one meeting conducted live, usually in conjunction with the annual membership meeting.

The executive board consists of the president, president elect, vice president, secretary, treasurer, and immediate past president. Meetings are held via conference call on a monthly basis to ensure progress continues on various ACCE activities and to develop the agenda for the bimonthly board meetings.

Committees

An organization with ACCE's focus and breadth could not survive without the help of committees. Many ACCE members have participated on committees, and their assistance is always greatly appreciated. Further specifics on the terms and conditions for committees can be found in the bylaws. There are two types of committees as specified in the bylaws: Standing and general.

Standing Committees

The current standing committees include the Membership and Nominating Committees. Standing committee members must be members of the College in good standing, at the membership level of Individual, Fellow, or Emeritus.

Membership Committee

The Membership Committee reviews all applications for new memberships and membership changes to the College and makes recommendations to the board regarding the qualifications of the applicants. It also establishes and maintains criteria and procedures for admission and resignation of members of the College, in addition to developing and recommending application processes and forms. The president with the approval of the board,, appoints a committee chair. The committee chair appoints four committee members with the approval of the president.

Nominating Committee

The Nominating Committee prepares a list of nominees for all offices that expire the following August 1. Only one nominee will be listed for each office, as others will be solicited at the annual membership meeting. The committee should list only nominees who are willing to serve if elected and to attend board meetings. The immediate past president serves as chair of the Nominating Committee. The president will appoint four committee members with the approval of the board.

General Committees

The board may create any other standing and special committees as may be necessary to carry out the business of the College. Committee chairs must be Individual, Fellow, or Emeritus members. Any member or nonmember may serve as a Committee member. The president will appoint committee chairs with the approval of the board. The committee chair will appoint members to the committee. All appointments expire at the end of the chair's term, but committee members may be re-appointed for an unlimited number of consecutive terms. The president will appoint a board member to serve as a nonvoting liaison member of each general committee. The president may appoint himself to serve in this role. Committees have included: Advocacy, Education, International, Government Relations, Inter-Society, Vision2000, and Clinical Engineering Certification.

Task Forces

In addition to formal, general committees, the board has requested members to assist with other projects that revolved around society relations or the development of white papers.

These have included intersociety relations, the BMET Task Force, and the Medical Error Task Force.

Demographics

ACCE membership has been diverse from the start. The strength of ACCE has been due to the support received throughout the United States and the world. This also provides a challenge to the board to develop programs and leadership that can reach the distance. As of August 2002 there are 189 current paid members: 158 Individual, 5 Fellow, 2 Emeritus, 18 Associate, and 6 Candidate. The breakdown of membership categories over the years is shown in Table 130-2.

Over the years the ACCE membership has represented 36 states plus the District of Columbia and Puerto Rico and 23 countries including: Albania, Bermuda, Brazil, Canada, Columbia, Denmark, Egypt, Germany, India, Italy, Kuwait, Malaysia, Mexico, Moldova, Nigeria, Pakistan, Poland, Saudi Arabia, Switzerland, Ukraine, the United Arab Emirates, the United Kingdom, and Venezuela. ACCE always continues to work in fostering national and international clinical engineering as reflected in the membership demographics.

Surveys

Throughout the years, the board has wanted to determine the best path for the organization and membership. This is often accomplished by taking a survey of the membership that allows member feedback and provides guidelines that ACCE leadership can concentrate resources. Questions asked include future focus as well as an evaluation of the current focus. The board reviews the results and dispenses activities to individual members or committee chairs to implement. Critical items are added to the board agenda for continual follow-up.

1995 Survey

There was a 50% response of the active membership. The scale was from 1–5, with 5 noting that the item was of "vital importance to the profession and of continued professional growth of the member." The following activities indicated strong support: *ACCE News* (average rating 3.9); Vision 2000 Project (average rating 3.8); advocacy of the profession (average rating 3.6); collaboration with other organizations (average rating 3.3); annual awards program (tie, average rating 3.1); and audio conference series (tie, average rating 3.1). The membership also suggested holding the annual membership meeting in conjunction with the AAMI meeting (Brody, 1996).

2001 Survey

Sixty-five members responded. A large proportion of them were Individual members. All ACCE benefits were rated as highly valued, except for clothing, which was also rated low for knowledge of existence. The board will work closely to promote this benefit. The preferred methods of communication were e-mail and web site access. Members felt that collaboration with other societies (including local BMET societies) was important. Comments included a concern for web site updating, improving advocacy efforts, and certification. Volunteers were solicited, and their names were forwarded to the appropriate committees (Zambuto, 2002).

Bylaws

The Bylaws for the American College of Clinical Engineering were established in 1990. They serve as an overall guide for the organization and membership on the way in which the society is managed carried forth. The membership has approved revisions. The version included within this chapter was approved the ACCE membership and board of directors in August 2002.

Table 130-2 ACCE members by membership categories

Year	Individual	Fellow	Emeritus	Associate	Candidate
1989	4				
1990	45	3*			
1991	26	2*			
1992	8				
1993	9				
1994	8			7	
1995	8			2	
1996	11				2
1997	25			5	2
1998	16			10	1
1999	15			6	1
2000	13			7	5
2001	16			2	2
2002†	7		2*	1	3
TOTAL	211	5	2	31	16

*Please note that this table shows a combination of the year joined and the membership category, which may have been upgraded throughout the years.
†2002 is a partial-year listing.

Bylaws

Article I: NAME

The name of this non-profit organization will be "American College of Clinical Engineering," hereinafter referred to as the "College." The College will also be known by the abbreviation "ACCE."

Article II: PURPOSE

1. To *establish* a standard of competence and to promote excellence in Clinical Engineering practice.
2. To *promote* safe and effective application of Science and Technology to patient care.
3. To *define* the body of knowledge on which the profession is based.
4. To *represent* the professional interests of Clinical Engineers.

Article III: MEMBERSHIP

There shall be five (5) classifications of membership, as follows:

Individual

A person demonstrating evidence of professional practice of engineering in a clinical environment for at least three (3) years, and meeting one or more of the following three conditions:

- Possession of a Baccalaureate degree in an Engineering discipline or Engineering Technology from an accredited College or University; (or Foreign equivalent); or
- Certification as a Clinical Engineer (CCE), by the International Certification Commission for Clinical Engineering and Biomedical Technology; or
- By recommendation of the Membership Committee in recognition of exceptional contributions, consistent with criteria established by the Board, to the profession of Clinical Engineering.

Fellow

An Individual member may be advanced to a Fellow status in recognition of distinguished service to the profession or achievement in the field of Clinical Engineering. The member must be active in the College business over a minimum period of three years.

Emeritus

An Individual or Fellow member may be advanced to Emeritus status

- In recognition of distinguished service to the profession or achievement in the field of Clinical Engineering; and
- After at least five continuous years membership in good standing in the College; and
- Upon member's reaching age of 62 and retirement from full time employment in the profession of Clinical Engineering.

Associate

An individual committed to the mission of this organization, who has demonstrated a contribution to the advancement of the clinical engineering profession, and meets other requirements established by the Board, but does not meet the other conditions required for Individual membership.

Candidate

An individual interested in the purpose of this organization and meeting one of the following two conditions:

- Current enrollment at least half-time in an accredited baccalaureate or graduate program in engineering, engineering technology, or related course of study; or
- Completing the three-year clinical experience requirement for Individual membership after receiving a baccalaureate or graduate engineering degree.

5. Only Individual, Fellow, and Emeritus members are entitled to vote on matters presented to the College membership.

Article IV: BOARD

Responsibility for management of the College is vested in the Board, which consists of ten (10) members: The President, the President Elect, the Vice President, the Secretary, the Treasurer, the Immediate Past President, and four Members-at-Large. Board Members will be elected as specified in these Bylaws. Each member of the Board is entitled to one (1) vote.

6. Officers:
    Officers are elected from the Individual and Fellow membership.
    President
        The President presides at all meetings of the College and the Board. The President is responsible for carrying out all orders and resolutions.
    President Elect
        The President Elect performs all duties of the President whenever the President is absent or otherwise unable to perform these duties. The President Elect will oversee the Member Services and Education functions including implementation of ongoing strategic objectives approved by the Board in these areas. The President Elect will perform other duties as requested by the President of the Board. After completing a term as President Elect, the President Elect shall become the

President, unless the President is elected for a second term. In such a case, the President Elect is eligible to serve an additional term as President Elect. In the event that the President Elect can not assume the role of President at the end of his or her term as President Elect, a new President shall be elected from current members of the Board. The Immediate Past President and officers, other that the President, are eligible for this election.

Vice President
The Vice President will oversee the Marketing/Advocacy and Strategic Alliances functions including implementation of ongoing strategic objectives approved by the Board in these areas. The Vice President will perform other duties as requested by the President or the Board.

Secretary
The Secretary is responsible for keeping minutes of all membership and Board meetings. The Secretary also coordinates the membership roster, and is responsible for all notices sent to the College's membership and Board.

Treasurer
The Treasurer of the College is responsible for the receipt, disbursement, and record keeping for all of the College's funds. The Treasurer will present an accounting of income, expenditures, and fund balances at all Board meetings and the Annual Membership Meeting(s). The Treasurer will disburse funds only in accordance with policies established by the Board. The Treasurer will present an annual written summary of the College's financial status to the membership.

Immediate Past President
The President will become the Immediate Past President upon completion of the President's term of office. The Immediate Past President chairs the Nominating Committee.

7. Members-at-Large
   Four (4) Members-at-Large are elected from the Individual and Fellow membership. Each Member-at-Large serves a two-year term. The Nominating Committee will try to ensure that in any one election only 2 Members-at-Large positions are being filled. Vacancies and special exceptions will be reviewed and approved by the Board.

8. Vacancies
   Any office declared vacant by the Board, whether through resignation, death, disability, or inactivity, will be filled by the Board through appointment. New appointments will fill the term being vacated.
   If the office of Immediate Past President becomes vacant, the Board will appoint a new Immediate Past President from among the former President's of the College. If none of the former Presidents are willing to serve, and appointment will be made from among the Fellows of the College.
   All Board Members are expected to attend meetings regularly. The Board will assume that a Board Member has vacated his office if the Board Member misses two consecutive meetings without cause (as determined by the Board).

9. Quorum for Business
   A. A majority of Board Members must be represented at a meeting to conduct business. Board Members may be represented either through actual presence at a meeting, through a telephone connection, through a proxy assigned to another Board Member, or through an absentee ballot must contain either an actual, a facsimile signature, or an e-mail from the Board Member. The validity of a proxy or absentee ballot will be determined by the officer presiding at the meeting.
   B. A simple majority of the Board Members represented at the meeting is required to approve motions.
   C. Emergency actions may be taken by the President whenever necessary to protect the assets or reputation of the College. If this becomes necessary, the President must consult with as many other Board members as practical before taking such emergency action. After such action is taken, the President must notify all Board members within five days. Transmission of a facsimile message or e-mail will be considered adequate notification under the provisions of this section.

10. Corporate Representative
    The Board will designate a resident of the State of Washington to be its Corporate Representative within that state, to fulfill all the legal obligations required by that state. The Corporate Representative may hold another office in the College. The Corporate Representative is not a member of the Board (except by virtue of any other office held by that person), and has no responsibility for the governance of the College.

Article V: ELIGIBILIITY AND TENURE OF OFFICERS

1. Officers and Members-at-Large must be members of the College in good standing at the membership level of Individual or Fellow.
2. No officer or Member-at-Large may hold more than one office at a time.
3. The President, President Elect, and Vice President will serve for a one-year term of office.
4. The Secretary and Treasurer will serve for a two-year term of office.
5. All Members-at-Large will serve for a two-year term of office.
6. The President, President Elect, and Vice President may serve two consecutive terms if elected for the office again.
7. The Secretary and Treasurer may serve two consecutive terms.
8. Member-at-Large may serve two consecutive terms.
9. A partial Term of Office assumed by appointment of the Board will not be considered a term under the provisions of this article.
10. The Term of Office of an elected officer or Member-at-Large begins on August 1 or 15 days after the closing date of the election, whichever is later. Closing date occurs by Board vote at the first Board Meeting after elections.

Article VI: VOTING

1. Each Individual, Fellow, and Emeritus member in good standing is entitled to one (1) vote on any business coming before meetings of the College, and in elections.
2. Votes may be cast in person, by written proxy, or by absentee ballot. The Board will establish policies to assure the validity of proxies or absentee ballots.
3. One-fifth (1/5) of the total number of Individual, Fellow, and Emeritus members in good standing must be represented at a meeting in order to conduct official business. Members may be represented whether through actual presence at a meeting, through a proxy assigned to another member, or through an absentee ballot presented to either the President or Secretary. A proxy or absentee ballot must contain a signature (either actual or facsimile) or an e-mail of the Member. The validity of a proxy or absentee ballot will be determined by the officer presiding at the meeting.
4. A simple majority of those voting is required to approve motions.
5. Elections are conducted as required to approve motions.

Article VII: FISCAL CONTROL

1. The Board controls all fiscal matters. Its actions are subject to review by the College membership.
2. The fiscal year will be the calendar year.
3. Dues are set annually by the Board, at a level consistent with an annual budget of expenditures and income, as well as other policies adopted by the Board.
4. The Board will establish a membership year consistent with the fiscal objectives of the College.

Article VIII: MEETINGS

1. An Annual Membership Meeting will be held during April, May, or June.
2. Board meetings will be held at least quarterly.
3. Special meetings of the Board can be called by the President, or upon written request of three members of the Board.
4. Notification of all Board meetings will be made by the Secretary at least 15 days prior to the meeting, either by mail, telephone, facsimile transmission, or e-mail.
5. Special meetings of the College may be called by the President or upon written request of at least one-fifth (1/5) of the voting members.
6. Written notification of the Annual Membership Meeting and any Special Meetings must be mailed to the membership at least thirty (30) days prior to the meeting date.
7. Parliamentary procedures to be followed in business meetings of the College, its Board, and its committee's will be those specified in "Robert's Rules of Order, Revised".

Article IX: COMMITTEES

1. Standing Committees
   The Membership and Nominating Committees will be standing committees. Standing committee members must be members of the College in good standing at the membership level of Individual, Fellow, or Emeritus.

   Membership Committee
   The Membership Committee will review all applications for membership to the College, and make recommendations to the Board regarding the qualifications of the applicants. The Membership Committee will also review all applications made by College members for a change in their membership category and will make recommendations to the Board regarding the applicant's qualification for the applied-for category. College members may apply for a change in member category from:

   - Candidate to Associate or Individual
   - Associate to Individual
   - Individual to Fellow or Emeritus
   - Fellow to Emeritus

   The committee chair will be appointed by the President, with the approval of the Board, for a two-year term. Four committee members will also be appointed by the committee chair, with the approval of the President, for staggered two-year terms. The Committee chair and the committee members may serve no more that two consecutive terms.

   The Membership Committee will establish and maintain criteria and procedures for admission and resignation of members of the College in accordance with these Bylaws.

   Upon recommendation by the Membership committee, applications for admission to the College will be forwarded to the Board for review and final approval. Upon acceptance by the Board, an applicant will become a member.

   The Membership Committee considers and recommends to the Board candidates for elevation to Fellow status from among those Individual members who apply and who have held active membership in the College and professionally contributed to the College function for at least three years. Upon approval by the Board, a candidate will be elevated to Fellow status.

   Nominating Committee
   The Immediate Past President serves as Chair of the Nominating Committee. Four committee members will also be appointed by the President, with the approval of the Board, for a one-year term. These appointments will be made no later than January 1 each year. No more that one (1) member of the Nominating Committee, in addition to the Nominating Committee Chair, may be a member of the incumbent Board.

   The Nominating Committee prepares a list of nominees for all offices which expire the following August 1. Only one nominee will be listed for each office.

The committee will only list nominees willing to serve if elected and to attend Board meetings.

The Nominating Committee will report on its recommendations to the Board at a Board meeting to be held in February or March. The Board will review this report. Upon acceptance of the report, the nominees will be listed on the ballot as candidates for office. This list of candidates will be sent to the College membership with the notice of the Annual Membership Meeting.

At the Annual Membership Meeting, the floor will be opened for additional nominations. The Nominating Committee will confirm that any person nominated from the floor is eligible for office, and is willing to serve if elected and attend Board meetings. Upon confirmation, these nominees will be added to the ballot.

The Secretary will send election ballots to all Individual, Fellow, and Emeritus members no later than twenty-one (21) days after the Annual Membership Meeting. Members will be allowed fifteen (15) days after the mailing of the ballots to return them. The Secretary will count the ballots and notify the Board and all candidates of the results within fifteen (15) days after the closing date of the election.

Elected candidates will assume office on August 1 or 15 days after the closing date of the election, whichever, is later. Closing date occurs by Board vote at the first Board Meeting held after elections.

2. General Committees
   A. The Board may create any other standing and special committees as may be necessary to carry on the business of the College. Committee chairs must be Individual, Fellow, or Emeritus members. Any member or non-member may serve as a Committee member.
   B. The committee chair will be appointed by the President, with the approval of the Board. A chair will serve for one (1) year and may serve no more than five consecutive terms.
   C. The committee chair will appoint members to the committee. All appointments expire at the end of the chair's term, but committee members may be re-appointed for an unlimited number of consecutive terms.
   D. The President will appoint a Board member to serve as a non-voting liaison member of each general committee. The President may appoint himself to serve in this role.

Article X: AMENDMENTS

1. Amendments may be proposed by any Board member, or by a written petition submitted by at least three Individual, Fellow, Emeritus members.
2. All proposed amendments to these Bylaws will be reviewed by the Board. Amendments proposed by Board members will be submitted to the membership only upon approval by the Board. Amendments submitted by written petition of the membership must be submitted to the membership, regardless of the Board's recommendation, unless the petitioners withdraw their proposed amendment.
3. All proposed amendments will be sent to the membership with the Board's recommendations.
4. A ballot containing the entire text of the proposed amendment will be sent to the membership by the Secretary no later than thirty (30) days after the Board meeting at which the amendment was considered. Members will have thirty (30) days after the mailing of the ballots to return them. An amendment will be adopted only if a majority of the total number of Individual, Fellow, or Emeritus members of the College in good standing approve.
5. The Secretary will promptly notify the Board and the membership of the voting results.

Article XI: INUREMENT AND DISSOLUTION

1. The assets of this organization will never inure to the benefit of any member of the College, nor will any assets or properties be used for the personal benefit of any member of any other person except in pursuit of the objectives stated in these Bylaws.
2. Should the College dissolve, all assets and property remaining after meeting the obligations of the College will be donated to a substantially similar non-profit corporation or organization having objectives compatible with those of the College.

Leadership

The true success of an organization depends upon its leadership. Within the clinical engineering community there has existed a strong group of cutting edge people. Many were influential in the inception of ACCE, and through their selfless works ACCE has excelled. Some of these important volunteers took on the leadership roles on the board of directors or various committees. Table 130-3 lists the presidents of ACCE from its inception to 2003.

Table 130-3 Past Presidents of the American College of Clinical Engineering

1990–1991	Yadin David	1998–1999	Robert L. Morris		
1991–1992	Matthew F. Baretich	1999–2001	Jennifer C. Ott		
1992–1994	Joseph F. Dyro	2001–2002	Elliot Sloane		
1994–1996	Thomas Bauld	2002–2003	Raymond Zambuto		
1996–1998	Frank Painter	2003–2004	Izabella Gieras		

Education

Among the major benefits of membership in the ACCE are the educational opportunities. They are directly related to the mission of the organization. Two general committees help with this task: The Education Committee and the International Committee. The committee chairpersons assist the board of directors and other ACCE committees in carrying out the educational mission of ACCE. They address requests from the constituency as well as support of clinical engineering educational activities throughout the world.

Teleconference Series

Annual meetings held by organizations are often highly beneficial to those who can afford the time away from work and the travel expenses. The Education Committee realized early on that an opportunity existed to provide an ongoing educational program in which attendees could participate in an hour-long teleconference. Thus, the annual ACCE Teleconference Series was born.

Each year, the Education Committee develops a list of eight to ten critical and timely topics and locates an expert who is willing to speak for 40–45 minute. This presentation is followed by a short question-and-answer session. ACCE members can join for a discounted rate that allows up to four participants per site. Additional persons may participate for $10 each. Each site receives dialing instructions, handouts, a session evaluation, and a sign-in sheet. Depending upon the topic, some sites will invite other members of their organization, such as clinical staff who will appreciate the topic being covered, or they may invite other members of a local biomedical society. The speaker receives an honorarium from ACCE for their hard work. Previous topics are shown in Table 130-4 (see Chapter 72 for a complete discussion of the role of the ACCE Teleconference Series in distance education).

Advanced Clinical Engineering Workshops (ACEWs)

The Education Committee, in conjunction with the International Committee and sponsoring international organizations such as WHO and PAHO, will provide a three-day (or up to a two-week) educational conference specifically targeting the education of clinical engineering and health care technology management. The purpose is to provide an educational forum whereby clinical engineers and related health professionals in other countries can be introduced to the clinical engineering management systems and working methods practiced in the United States, and to exchange information and ideas pertinent to clinical engineering here. These advanced clinical engineering workshops (ACEWs) also provide opportunities for engineers in the host country to learn about ACCE and to become members (ACCE web site, International Page, 2002). To date, over 15 workshops have occurred throughout the world. The workshop faculty comprise volunteers from the ACCE membership, and interested members are encouraged to volunteer. A fee paid to ACCE covers the travel expenses for the faculty of the workshop. Locations of previous workshops include: Brazil, China, Costa Rica, Cuba, the Dominican Republic, Ecuador, Italy, Lithuania, Mexico, Nepal, Panama, Peru, Russia, South Africa, the United States, and Venezuela (see Chapters 70 and 71 for complete descriptions of the ACEW programs).

Infratech

Infratech was created in January 1999 by Dr. Andrei Issakov of the World Health Organization and Antonio Hernandez of the Pan American Health Organization as an Internet discussion group for exchange of information on health care infrastructure and technology for health services. Information gathered through this exchange can be used in developing a knowledge base for use in future projects (Jakniunas, 2000). This is an English-language listserv that provides a forum for international and national members to share clinical engineering information. There are an estimated 150 members from 50 countries, with a 20% ACCE membership. ACCE is the coordinator of this PAHO- and WHO-sponsored activity (International Committee Report, 2002). Future considerations include a calendar of worldwide health care technology management related events, a

Table 130-4 ACCE audio-teleconference series past programs

Presenter	Title
Tom Judd	Clinical Engineering Involvement in Managed Care Quality Improvement (Wear, 1996)
Dave Simmons	Contract Management (Wear, 1996)
Frank Painter	Marketing Your Services Within and Outside Your Health Care Organization (Wear, 1997)
Tom O'Dea	Building Teamwork Between Clinical Engineering and Maintenance Staff (Wear, 1997)
Jennifer C. Ott	Non-Profit to Profit (Wear, 1998)
Steve Wexler	Year 2000 Issues (Wear, 1998)
Ode R. Keil	JCAHO Update (Wear, 1999)
Tim Adams	Development of a Customer-Focused Clinical Engineering Team (Wear, 1999)
Binseng Wang	Equipment Management Inclusion Criteria: An improved method for including equipment into the program and determining inspection frequency (Wear, 2000)
Joseph F. Dyro	Investigating an Equipment Incident (Wear, 2000)
Stephen L. Grimes	HIPAA's Impact on Clinical Engineering (Levenson, 2001)
Jeffrey L. Silverberg	Managing Electromagnetic Compatibility in the Hospital (Levenson, 2001)
David Harrington	Remote Diagnostic –Where Are We today? (Levenson, 2002)
Binseng Wang	Repair or Replace? The Hospital CFO's Point of View (Levenson, 2002)

listing of participants, a resource library, a message board, and utilization of participants to facilitate communications between the developer and users of the Essential Health Technology Package (EHTP) to share common experiences and to build common understanding (Jakniunas, 2000; Judd, 2002; Nunziata, 2003).

Publications

ACCE has been involved with many publications through the organization itself and its members. The main ACCE publication, the *ACCE News*, will be discussed later. Other ACCE publications were done at the request of the board of directors to benefit the membership or the clinical engineering profession. Some examples include: *Guidelines for Medical Device Donations* and the *Patient Safety White Paper*.

Conferences

ACCE has always been involved with other organizations that help to foster the profession of clinical engineering, such as AAMI, ECRI, HealthTech, ASHE, and IEEE EMBS. ACCE has worked closely with these organizations to develop content and to provide speakers for various meetings and annual expositions, paying particular attention to topics that are of interest to the ACCE membership. The success of many of these conferences has been due to ACCE involvement. ACCE has run individual sessions and entire educational tracks. Occasionally, ACCE members will request ACCE sponsorship in the form of name support when presenting at conferences. The board of directors, under strong negotiations with other organizations, has provided ACCE member discounts to the conference and a forum for members to gather and discuss ACCE business. It is important for ACCE to continue these collaborative efforts to promote the profession and to provide further opportunities for the ACCE membership. Future projects might include providing speakers and material for other non-clinical engineering-based conferences, to introduce them to field clinical engineering and the role they play in the health care setting.

Certification

ACCE and the clinical engineering community at large gave the United States Board of Examiners for Clinical Engineering Certification a mandate to establish and maintain a credible certification program. The program, established in 2002, serves to provide meaningful certification for clinical engineering professionals. Prior to ACCE's involvement with the certification of clinical engineers, certification was administered by the Association for the Advancement of Medical Instrumentation. For various reasons, the AAMI program did not thrive, and AAMI discontinued its role in clinical engineering certification.

The ACCE program (described below) has provisions to recognize individuals who are certified under the previous system. Up-to-date information with regard to the mechanics of the certification program, e.g., applications, fees, examination dates and locations, those interested may visit the web site certification@ACCEnet.org or contact ACCE directly.

Governance

The governance structure of the program is now in place and is depicted in Figure 130-1. The relationships, bylaws, and constitution of the organizations in the governance structure were developed to comprehensively address the requirements of a valid and defensible certification program, fitting together in a cohesive manner. The required development work ensures the appropriate authority of each entity related to the certification program. In this structure, the following responsibilities exist:

- The foundation is responsible for the financial management of the commission and the board.
- The commission is responsible for establishing performance criteria for the certification program, communication between the board and the foundation, including about financial and budgetary matters, and is the authority by which all certifications will be issued.
- The board is responsible for developing and implementing all policies and operating procedures that are necessary to maintain the certification program in compliance with the standards set forth by the commission.

In accordance with its responsibilities, the commission has worked with the foundation and the board to project the financial needs to establish and maintain the certification program over its first few years. Although a large amount of volunteer labor is invested in the program, there are considerable expenses to bear, including insurance costs, administrative-support labor, document storage costs, and services that lie outside the realm of the expertise of the program participants. One of the required services to support a defensible examination process is psychometric analysis. This analysis confirms that the examination truly assesses the competence of the clinical engineers.

National Accreditation of Certifying Organizations Standards

The Commission has adopted standards for the certification program, based on those established by the National Organization for Competency Assurance (NOCA), which is a membership association of certification organizations that provide technical and educational information concerning certification practices. These standards address the following:

- Purpose, governance, and resources
- Responsibilities to stakeholders
- Assessment instruments
- Recertification
- Maintaining accreditation

The certification program can be evaluated directly by NOCA for compliance with these standards only after the certification examination has been given twice at a national level. The commission has directed the board to develop a program based on these standards.

Body of Knowledge

In accordance with its defined responsibilities, the board has been drafting policies, procedures, and practices that address all of the standards as well as the pertinent legal issues. This includes developing justifiable eligibility criteria and an application form with supporting documents. One of the important early board decisions was to adopt the body of knowledge as defined by the American College of Clinical Engineering as the current performance domain against which competence will be measured. According to the certification program standards, it will be necessary for ACCE to update that body of knowledge periodically in order for the board to continue to use it as the defined performance domain. Another important activity of the board is to develop questions around the body of knowledge using recognized references as the basis for the questions. Using this approach, the questions (and answers) are defensible. As the questions are written, they are catalogued in such a way as to facilitate updating them in the future while remaining true to the breadth of the body of knowledge. The board is currently evaluating test services to psychometrically analyze the questions and the examinations as well as to help establish a valid cut score. Finally, the board is establishing appeals processes for any candidate who does not agree with a decision of the board. While the examination-development process described above consumes much of the board's time, other issues are being addressed in parallel, including:

- Development and maintenance of a registry of certified clinical engineers
- Processes related to recertification
- Administrative processes for storage and retrieval of appropriate documents
- Development of an orientation and education program for board members

While all of this work is in progress, the foundation, the commission, and the board are all committed to providing information to clinical engineers and the public. Because of the newness of the clinical engineering certification program, these entities are also sensitive to the need maintain credibility and hence they strive to release information only when the interactive development of specific pieces of the program is nearing completion. The foundation, commission, and board appreciate the patience of the ACCE members and their colleagues and looks forwards to disclosing the details about the certification program in the near future.

Validity and Value

Time spent reading about certification on the Internet reveals that the value of certification is debated in many diverse fields, including martial arts, human resources, and information technology. Within health care, the debate about the value of clinical engineering certification is several years old. In all industries, the debate must be separated into two issues: The validity of a certification program and the value of certification to others i.e., the public, the employer, and the individual. Certification can have value only if it is valid. If a certification is valid, market forces drive the value of the certification.

In the absence of an active, viable program, clinical engineering certification has been devalued. In order to restore the potential for value, an active, valid clinical engineering certification program must be restored. The validity of the certification program depends on the ability of the program to authenticate that an individual can perform a specific set of tasks at an established level. That validity is largely dependent on the program's

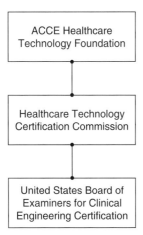

Figure 130-1 Governance structure of the Clinical Engineering Certification Program.

ability to define the specific skill set of interest and to use an assessment instrument that truly assesses that skill set in a consistent and nondiscriminatory manner. Following standards such as those established by the NOCA helps to ensure the validity of a certification program.

Once a valid certification program is established, the market will determine the value of clinical engineering certification. Because of technology convergence, perhaps a glimpse at the value of certification in information technology is pertinent to the clinical engineering certification discussion. The Computing Industry Technology Association has provided the results of its 2001 Global Training and Certification Study at www.comptia.org/cla.

The results of that study demonstrate that 83% of IT professionals found certification to be helpful in acquiring a new position, while 66% of certified IT professionals reported that their salaries increased after becoming certified. Against the background of a maligned health care industry and a declining economy, these results suggest that employers will increasingly value certification, thereby causing one's professional viability and marketability to be strengthened through certification. This value is only possible if a valid program exists.

The United States Board of Examiners for Clinical Engineering Certification (USBE-CEC) continues its efforts to develop a valid certification program. While this development process is laborious and time-consuming, the program will add value by promoting health care delivery improvement in the United States through the certification and continuing assessment of the competency of professionals who support and advance patient care by applying engineering and management skills to health care technology. In turn, the patience of the clinical engineering audience will be rewarded with increased value to clinical engineering certification.

Meetings

Symposia

In 1998, ACCE held its first of five annual symposiums. These are day-long educational endeavors for the ACCE membership, held in conjunction with the annual AAMI meeting and exposition dedicated, to hot topics and to providing critical in-depth information. The first symposium was titled *The Future of Clinical Engineering* and included panelists Malcolm Ridgway, Larry Hertzler, Tom Bauld, Greg Davis, and moderator Ira Tackel. Together, they provided informative talks and sparked lively discussion on where clinical engineering is headed (Dyro, 1998).

The second symposium was titled *The Future of Clinical Engineering–Clinical Engineering and Information Systems* and included moderator Brian Porras and panelists Tom Bauld, Dean Athanassiades, and Richard Schrenker, who discussed the future relationships between clinical engineering and hospital-information systems and methods to develop partnerships (Porras, 1999).

The third symposium was titled *Frequency Allocation Issues in Medical Telemetry* and included a panel of experts from government, the medical device industry, and the hospital community, such as moderator Brian Porras, Caroline Campbell, Steven Juett, David Paperman, Don Witters, Mary Beth Savary Taylor, Stan Wiley, and David Pettijohn. There was extensive discussion of the new regulatory changes to extend protections to electromagnetic frequencies used by medical telemetry (Porras, 2000).

The fourth symposium was titled *The Health Insurance Portability and Accountability Act (HIPAA)*. Its goal was to educate ACCE members on the HIPAA federal regulations that will require health care providers to protect the privacy and security of individually identifiable health care information. The final rules will affect virtually all high technology in health care and will impact many departments. Moderators Steve Grimes and Ray Zambuto lead panelists Madelyn Quattrone, M. Peter Adler, Ely Lezzer, Alan S. Goldberg, Charles Parisot, and Tom Walsh.

The fifth symposium was titled *Perspective for Successful Leadership in Clinical and Information Technology Services*. Fourteen speakers presented an outstanding program focusing on the information technology and clinical engineering perspective and included case studies (Cohen, 2002).

General Membership Meetings

According to the ACCE bylaws, at least one annual membership meeting is held in April, May, or June. Generally, it is held in conjunction with another society annual meeting, usually AAMI. There might be additional general membership meetings at other conferences held throughout the year if there will be an ACCE contingency attending. Meeting announcements are sent to members and are posted on the web site and in the *ACCE News*. The annual membership meeting begins with a social hour that includes a wine and cheese reception and then a summary of ACCE activities from the previous year is presented to the membership. In addition, the Advocacy Award winners are announced and celebrated, as are outgoing board members. The meeting concludes with a presentation of the slate of officers for the upcoming year and an open forum to address membership issues, questions, and concerns.

Board Meetings

The board of directors meets at least quarterly and usually bimonthly. They are conducted via conference, call with items being distributed by e-mail. On an annual basis, there is one live meeting prior to the annual membership meeting. The executive board meets monthly via conference call. (Please see the bylaws for other specific meeting guidelines.)

Committee Meetings

Committee chairmen organize and develop their own meeting schedules to address the tasks at hand. The bylaws provide some guidance on deadlines for standing committees, such as Membership and Nominations. Most committees use teleconferencing and e-mail as the primary methods of communication among committee members.

Cooperating Organization Meetings

ACCE works very closely with other organizations in developing educational materials and content for their meetings as well as honoring requests to participate in their organization-driven projects. Some of these co-operating organizations include: ASHE, AAMI, HealthTech, ECRI, the American Institute of Medical and Biological Engineering (AIMBE), IEEE, World Health Organization (WHO), Pan American Health Organization (PAHO), and the International Federation for Medical and Biological Engineering (IFMBE) Clinical Engineering Division (CED).

Advocacy Committee

The Advocacy Committee has one of the most important roles within ACCE and is integral to the mission of ACCE. Clinical engineering is a relatively small, but important, profession that is often misunderstood. A group of forward-thinking clinical engineers understood these limitations and formed ACCE. The Advocacy Committee recognizes those members who have fulfilled the ACCE mission, and it provides opportunities for ACCE members to further develop the profession by educating younger generations and colleagues on the field of clinical engineering and its importance to the health care team.

Advocacy Awards

The Advocacy Committee administers the Advocacy Awards program. Through this program, clinical engineering excellence and advocacy are recognized. If the nominee is a clinical engineer, he or she must be a member in good standing with ACCE. The categories of awards are described below.

Professional Achievement

The Tom O'Dea Professional Achievement Award recognizes work in defining the exclusive limits for the practice of clinical engineering; i.e., identifying unique functions, roles, activities, duties, and responsibilities of clinical engineers. The work should be either published in a journal of professional standing or at a conference of a professionally recognized, health-related organization. Publications and presentations must incorporate the words "clinical engineer" or "clinical engineering" in their titles. Terms used must be consistent with the ACCE definition of "clinical engineer" (Johnston, 1995). Past winners of the Professional Achievement Award are listed in Table 130-5. In 2002, the name of this award was changed to the Tom O'Dea Professional Achievement Award in recognition of all of Tom O'Dea's great work and contributions to the clinical engineering profession (Gieras, 2002).

Professional Development

The Professional Development Award recognizes accomplishments of professional advocacy which promote awareness and appreciation of clinical engineering within other health care professions. These actions are mainly through publications and presentations in a distinctly non-engineering health related publication or conference. Publications and presentations must incorporate the words "clinical engineer" or "clinical engineering" in their titles. Terms used must be consistent with the ACCE definition of "clinical engineer" (Johnston, 1995). Past recipients of this award and the title of their papers are listed in Table 130-6.

Managerial Excellence Award

Begun in 1999, the Managerial Excellence Award has been granted to those who have published articles promoting clinical engineering in the field of management development (see Table 130-7).

Table 130-5 Recipients of the Professional Achievement Award

	Professional Achievement Winners	
Year	Author	Title
1995	Wayne Morse	Career Opportunities in Clinical Engineering
1995	Monique Frize and Michael Shaffer	Clinical Engineering in Today's Hospital: Perspectives of the Administrator and the Clinical Engineer (Johnston, 1995)
1996	Lucio Flavio Brito	Seguranca No Ambiente Hospitalar (Johnston, 1996)
1997	Tom O'Dea	Protecting the Immunocompromised Patient (Johnson, 1997)
1998	Marvin Shepherd	Investigating Equipment-Related Incidents and an SMDA Update Risk Managers and Clinical Engineers Working as a Team (O'Dea, 1998)
2000	Manny Furst	Sixteen Years as Chairman of the AAMI's Clinical Engineering Management and Productivity Committee (Campbell, 2000)
2001	Eric Rosow and Joseph Adam	Virtual Instrumentation Tools for Real Time Performance Indicators. (Campbell, 2001)
2002	Tom O'Dea	For his committee leadership and the article he wrote in response to the confusing ad for a clinical engineer, in which it was clear the employer did not know what a clinical engineer was.

Table 130-6 Past recipients of the Professional Development Award

	Professional Development Winners	
Year	Author	Title
1995	Enrico Nunziata and Awa Diouf	Gestation Des Infrastructures et Des Equipment's Sanitaires (Johnston, 1995)
1996	Lucio Flavio Brito	Exposaúde '95 and Guia de Fornecedores Hospitalares (Johnston, 1996)
1997	Ira Tackel	Non-Traditional Support: Patient TV System and Biomedical Equipment Service–An International Incentive? (Johnston, 1997)
1998	Michael J. Shaffer	Clinical Engineers: A Vanishing Hospital Resource (O'Dea, 1998)
1999	Ira Soller	Contribution to the *Encyclopedia of Electrical and Electronics* edited by John Webster and published by Wiley (O'Dea, 1999)
2002	Joseph D. Bronzino	For his development of the Graduate Program in Clinical Engineering and his work with BEACON plus the numerous articles and books throughout his career (Gieras, 2002)

Table 130-7 Recipients of the Managerial Excellence Award

	Managerial Excellence Award Winners	
Year	Author	Title
1999	Binseng Wang	Letter to Consumer Reports taking to task certain Y2K issues raised concerning the compatibility of ECG Devices. (O'Dea, 1999)
1999	Dave Dickey and Larry Hertzler	Cover Your Assets. (O'Dea, 1999)
2001	Binseng Wang and Al Levenson	Equipment Inclusion Criteria–A New Interpretation of JCAHO's Medical Equipment Standard. (Campbell, 2001)
2002	Ted Cohen	For his several articles written over the years on Maintenance Management Systems and other related topics. (Gieras, 2002)

Devteq

In 1999, the Devteq Award was created, thanks to the support and encouragement of Marvin Shepherd, who wanted to reward those clinical engineering professionals who contributed to the rational balance between safety and resource management. Past winners of this award are listed in Table 130-8.

Devteq Challenge Award

In conjunction with the stipulation of the Devteq Award, ACCE Board of Directors awards at least one complimentary membership to a clinical engineering professional who is not currently a member, in hopes of enticing them to start and then continue their membership in future years. Past recipients of this award are shown in Table 130-9.

Guidelines and Recommendations

The Advocacy Committee has always taken on the role of monitoring clinical engineering activities in government and publications. Occasionally, a member will be asked to

Table 130-8 Recipients of the Devteq Award

	Devteq Award Winners	
Year	Author	Title
1999	Dave Francoeur	Impact upon the College of American Pathologists (CAP) to change their lab equipment electrical safety testing requirements from arbitrary annual inspection to (1) incoming inspection; (2) after major repair; (3) at a risk-based criteria frequency consistent with the equipment-management program of the institution (O'Dea, 1999).
2000	Steve Juett	Recognition of his identification of DTV interference at the Baylor University Medical Center in February 1998, his interaction with the press concerning that event, and his participation on the AHA Telemetry Task Force (Campbell, 2000)
2001	Jeffrey Cooper	His work in anesthesia safety in particular his response to the report on medical errors, which has significantly advanced the cause of patient safety (Campbell, 2001).
2002	Leslie A. Geddes	For his numerous patient-safety efforts through research, education, and publications (Gieras, 2002)

Table 130-9 Recipients of the Devteq Challenge Award

	Devteq Challenge Award Recipients
1999	Steve Wexler and Bob Stiefel
2000	Steve Juett
2001	Jeffrey Cooper
2002	Leonard Klebanov and John Czap

draft a response to an article, editorial, or position advertisement in order to clarify ACCE's position and to further educate the community at large on ACCE and its activities. Letters to the FDA have been written by the ACCE president, and other lobbying efforts occur as needed and as allowed under our current organizational structure.

Publicity

ACCE has developed educational and publicity material to assist members in the efforts to promote the profession and has set up booths at meetings. The first brochure developed, *The American College of Clinical Engineering*, was on ACCE itself and provides an introduction, mission statement, definitions, membership information, and contact information. The second brochure was titled *What's a Clinical Engineer?* It was provides information for clinical engineers to educate others who did not know or understand the profession. A third brochure, entitled *American College of Clinical Engineering Providing Clinical Engineering Support to Countries Worldwide*, describes the International Committee and the role of ACCE as a coordinating body for improvement of clinical engineering skills and for exchange of technical information with colleagues around the world. The generous assistance of ECRI made possible the publication of these brochures.

Health Care Technology Foundation

Formation

ACCE, at the request of the board of Directors, established the ACCE Healthcare Technology Foundation, an independent foundation to assist in the oversight of the certification process and other advocacy efforts. In August of 2001, the ACCE board of directors passed a motion to create an exploratory task force to investigate the establishment of a not-for-profit foundation. When ACCE was founded, it was known that in order to enable the organization to reach further professional goals and to continue to build the legacy on which the field of practice is based, the creation of a not-for-profit foundation (and its associated fundraising) potential was needed.

When the collective wisdom of senior members of ACCE, the exploratory task force, was put to the test, a report to the board resulted, showing that a foundation will significantly enhance the ACCE mission and that it should be created as soon as possible. Additional support for this recommendation was provided by the AAMI decision to discontinue its support of the clinical engineering certification process. With ACCE as the only organization that supports and recognizes the important role of clinical engineering in the safe and efficient development, integration, and use of modern health care technologies, a closely related foundation was considered to be like an oasis in the middle of a desert.

In May 2002, the ACCE board gave its approval for the creation of a foundation and provided the necessary seed money needed for filing and registering the foundation at the state and federal levels. A slate of directors and officers were selected, and work began on the foundation mission, objectives, and by-laws. This moment could not have been reached had it not been for the dedication of individuals like Matthew F. Baretich, Thoms J. Bauld, Yadin David, Joseph F. Dyro, Wayne Morse, Jennifer C. Ott, Frank Painter, Marvin Shepherd, Elliot B. Sloane, Ira Tackel, and Ray Zambuto. Assistance provided by Joe Welsh made this dream a reality. The effect of the work of these dedicated individuals probably will not be felt immediately, but future generations of clinical engineers and other practicing professionals in health care delivery system stand to benefit greatly. Some are known as doers, others as professionals, and yet others are dreamers. This above group comprised each of these. The first order of business was to organize a way to jump-start and sustain the certification process.

At the end of 2002, the ACCE Healthcare Technology Foundation was inaugurated and registered in the Commonwealth of Pennsylvania and proceeded to file with the federal authorities under Section 501(c)(3) of the tax code as a not-for-profit foundation. The foundation's purpose is as follows:

Improving of Health Care Delivery by Promoting the Development and Application of Safe and Effective Health Care Technologies Through the Global Advancement of Clinical Engineering Research, Education, Practice, and Other Related Activities.

Officers are Yadin David, president; Ira Tackel, treasurer; and Jennifer Ott, secretary. Other board members include Matt Baretich, Tom Bauld, Joe Dyro, Wayne Morse, Frank Painter, Elliot Sloane, Marvin Shepherd, Joe Welsh, and Ray Zambuto.

Robert M. Morris Humanitarian Award

In 2001, Elliot B. Sloane, in conjunction with other ACCE members and the ACCE board of directors, pursued through the AAMI foundation the development of an award to recognize a longtime ACCE leader, Robert L. Morris. During the next 12 months, funding was secured through the Manual and Beatrice Sloane Foundation and numerous contributions from ACCE members and others who have worked with Bob Morris through the years. The hope was to present the first award to Bob, but unfortunately he passed away in March 2001. He was granted the award posthumously. Table 130-10 lists award recipients and the reasons for their recognition.

Table 130-10 Robert M. Morris Humanitarian Award

Year	Recipient	Recognition
2001	Robert L. Morris	Bob Morris was posthumously honored as the first recipient of the annual humanitarian award that now bears his name at the AAMI Awards Luncheon in June of 2001 in Baltimore, Maryland. Ms. Julie Morris, Bob and Colleen Morris's younger daughter, received a check for $1,000 and the inaugural plaque with Bob's name on it. Julie expressed her family's appreciation for honoring her father by creating this award in his name and for recognizing Bob's humanitarian efforts and his passion for clinical engineering (Sloane, 2002)
2002	Herman R. Weed	A retired professor of electrical engineering and preventive medicine from Ohio State University; for over 30 years he has been involved with many humanitarian activities, a majority of them sponsored by Project Hope, a nonprofit foundation. Professor Weed served as director of Project Hope's Biomedical Engineering Programs. He was responsible for designing, supervising, and monitoring clinical engineering assignments of all sizes and scopes. In this role, he served as an inspiration to Bob and many other clinical engineers who participated in these projects throughout the world (Ott, 2002)

Newsletter

ACCE News is the official newsletter of ACCE. The *News* is published bimonthly and keeps the membership current with regard to important developments in the organization (e.g., election of officers), significant board decisions (e.g., establishment of Healthcare Technology Foundation), ACCE programs (e.g., Annual Symposium and Audio-Teleconference Series), and events in the clinical engineering world (e.g., National Patient Safety Movement, HIPAA, Telemetry Standards, and FDA Proposed Rules). The *News* has benefited from the editorial skills of four members who, as volunteers, solicited material; threatened, wheedled, and cajoled members to put their wisdom on the printed page; organized the material into an attractive format; and edited the final product. ACCE stands in deep appreciation of the efforts of Editors David Harrington (1990–1993), David Simmons (1993–1994), Mark Brody (1995–1996), and Joseph F. Dyro (1995–2003).

Content

Regular columns include the "President's Message," "Certification Update," "HIPAA Update," "Perspectives from ECRI," "The View from the Penalty Box," "Highlights of Board Meetings," "Calendar of Events," "People on the Move," "In the News," and "Advertisements." In addition to the regular columns, as news comes in, the following items are published in the *News*: "CE around the Globe," "Clinical Engineering Department Profiles," "ACCE Teleconference Series Announcements," "Reports of Meetings," "Reports of Advanced Clinical Engineering Workshops," "Conferences and Symposia," "Guest Editorials," and "Member Survey Results."

Highlights

Since its inception in 1990, much has been written in the *News*. Some of the highlights are mentioned here. A special issue of *ACCE News* chronicled the life and times of Robert M. Morris (Vol. 11, No. 2–March 2001). Below, the authors list articles that they find to be particularly interesting and insightful.

- Dyro JF. Advanced Clinical Engineering Workshop in Russia. *ACCE News* 9(5–6):12–17, 1999.
- Harrington D. The View from the Penalty Box. *ACCE News* 10(1):9, 2000.
- Judd T. EHTP in Kyrgyzstan. *ACCE News* 12(2):20–23, 2000.
- Kermit E. Medical Adhesive Tape Should be a Controlled Substance. *ACCE News* 10(2):15–16, 2000.
- Nunziata E. CE Around the Globe: EHTP in Mozambique. *ACCE News* 13(3):78, 2003.
- Shepherd M. Get on Board the Safety Train! *ACCE News* 10(2):3–4, 2000.
- Shepherd M. Health Care Systems and the Al-Qaida. *ACCE News* 12(2):6–7, 2002.
- Shepherd M. Mending the Way of our Errors. *ACCE News* 10(3): 3–4, 2000.
- Sloane E. Bad Medicine. *ACCE News* 12(1):3–4, 2002.

One may view the *ACCE News* over the Internet by going to the ACCE site at www.ACCEnet.org. The site should be accessed to obtain up-to-date information about ACCE and its many programs.

References

Editors of ACCE News. Membership List. *ACCE News* 1(1):4, 1991.
Bauld T, Judd T. P.A.C.E. Initiative. *ACCE News* 5:8, 1995.
Bauld T. New FDA GMP Regulations. *ACCE News* 6:5–10, 1996.
Bauld T. The Definition of a Clinical Engineer. *J Clin Eng* l6:403–405, 1991.
Brody M. Survey Results. *ACCE News* 6:5, 1996.
Campbell C. 10th annual membership meeting. *ACCE News* 10(4):6, 2000.
Campbell C. ACCE annual meeting. *ACCE News* 11(3,4,5):8–9, 2001.
Campbell C. American College of Clinical Engineering. *ACCE News* 8(6):1,11–12, 1998.
Cohen T. The ACCE Symposium. *ACCE News* 12(3):5, 2002.
David Y. ACCE Health Care Technology Foundation. *ACCE News* 13(1):10, 2003.
Dyro J. First ACCE Symposium. *ACCE News* 8(4):4–5, 1998.
Gieras I. ACCE Annual General Meeting. *ACCE News* 12(3):5–6, 2002.
Grimes S, Zambuto R. Fourth Annual Symposium Explores Impact of HIPAA on Clinical Engineering. *ACCE News* 11(3,4,5):6–7, 2001.
Jakniunas A. INFRATECH Meets in DC. *ACCE News* 10(2):5–6, 2000.
Jakniunas A. INFRATECH. *ACCE News* 10(5):5, 2000.
Jakniunas A. INFRATECH. *ACCE News* 10(6):5, 2000.
Johnston G. 1996 ACCE Advocacy Awards. *ACCE News* 6(4):13–14, 1996.
Johnston G. Advocacy Awards. *ACCE News* 7(4):11, 1997.
Johnston G. Advocacy Awards. *ACCE News* 5:14, 1995.
Judd T. EHTP in Kyrgyzstan. *ACCE News* 12(2):20–23, 2002.
Levenson A. 2002 ACCE Educational Teleconference Program. *ACCE News* 12(3):16, 2002.
Levenson A. ACCE 2001 Education Program. *ACCE News* 11(3,4,5):13–14, 2001.
Nunziata, E. CE Around the Globe: EHTP in Mozambique. *ACCE News* 13(3):7–8, 2003.
O'Dea T. Advocacy Award Update. *ACCE News* 8(5):9, 1998.
O'Dea T. Advocacy Awards. *ACCE News* 9(4):7, 1999.
Ott J. Robert Morris Humanitarian Award. *ACCE News* 12(4):6, 2002.
Ott J. American College of Clinical Engineering. *ACCE News* 13(2):12–24, 2003.
Painter F. ACCE Responds to FDA. *ACCE News* 8(4):11–14, 1998.
Porras B. Second ACCE Symposium. *ACCE News* 9(4):6, 1999.
Porras B. ACCE Symposium 2000. *ACCE News* 10(4):4–5, 2000.
Sloane E. Bob Morris Honored as the First Recipient of the AAMI Foundation/ACCE Robert L. Morris Humanitarian Award. *ACCE News* 12(2):10, 2002.
Wear J. ACCE Teleconference Series. *ACCE News* 6:15, 1996.
Wear J. Clinical Engineering for the Millennium: ACCE 1999 Educational Program. *ACCE News* 9(2):4–5, 1999.
Wear J. Clinical Engineering for the Millennium: ACCE 2000 Education Program. *ACCE News* 10(4):11–12, 2000.
Wear J. The Business of Clinical Engineering: ACCE 1998 Education Program. *ACCE News* 8(3):4–5, 1998.
Wear J. ACCE Teleconferences '97. *ACCE News* 7(2):11, 1997.
Zambuto R. Fall 2001 Membership Survey. *ACCE News* 12(3):10, 2002.

131 The New England Society of Clinical Engineering

Nicholas T. Noyes
Director, Department of Clinical Engineering
University of Connecticut Health Center
Farmington, CT

The New England Society of Clinical Engineering (NESCE) is an example of a local organization for clinical engineers, biomedical electronics technicians, and medical product sales and service representatives. NESCE organizes educational opportunities, such as guests, lecturers, and site visits to medical device manufacturing facilities, and conducts regular meetings in a social setting. Every several years NESCE is one of several New England clinical engineering and BMET societies that host the Northeast Biomedical Engineering Symposium.

Goals

As stated in the bylaws of the New England Society of Clinical Engineering, the goals of the Society are as follows:

- To promote better patient care through improved maintenance, operation, and application of medical instrumentation
- To promote the mutual exchange of information and technical assistance among clinical engineers (CEs) and biomedical electronics technicians (BMETs)
- To promote the development of the clinical engineering profession
- To organize and promote an educational program for CEs and BMETs
- To disseminate uniform interpretations of codes and standards pertaining to clinical engineering.

These goals were developed in the formative years of the Society and have remained to the present.

History

As reported by Mark English (Hartford Hospital), the New England Society of Clinical Engineering was launched on December 10, 1975. Several clinical engineering managers representing Connecticut and western Massachusetts hospitals held an informal meeting at Manchester Memorial Hospital. Among others, the original members included Allan Hoffman (Manchester Hospital), Norm Bertera (Baystate Medical Center), David Reese (Middlesex Memorial Hospital), Thomas Hayes (St. Francis Hospital and Medical Center), Robert Pisicane (Providence Hospital, Holyoke, MA), Mark English (Hartford Hospital), Dr. Ernie Guignon (U. of CT Health Center), and John Gilbert (New Britain Hospital). These first meetings were hosted by local hospitals on a monthly basis, typically with lunch provided. During these early years, the thrust of the meetings was to share problems and ideas and network at the management level.

The 1980s brought growth to the Society, with the membership roster expanding to over 200, including CEs, BMETs, service personnel from medical device manufacturers, medical-product sales representatives, and independent biomedical equipment service professionals. Formal bylaws for the New England Society of Clinical Engineering were established on November 20, 1987. Many would agree that the 1980s were golden years for the profession. In-house clinical engineering departments were valued and supported with sufficient resources and personnel to allow time, energy, and finances for clinical engineering extracurricular activities.

In addition to the original Society members listed above, some of the key personnel serving as officers and leaders of NESCE during the 1980s were as follows:

Frank Painter (Bridgeport Hospital), Nicholas Noyes, Anton Hebenstreit and Anna Marie Mercurio (U. of CT Health Center), Roger Debaise (Gaylord Hospital), Raymond Acosta and Tony Acavalla (Yale New Haven Hospital), Michael Romeo (Manchester Memorial Hospital), Dr. Joseph Bronzino (Trinity College), Robert Capetta (New Britain General), Robert Zbuska (St. Francis Hospital and Medical Center), Joseph Cane (St. Mary's Hospital), Mark Brody and Joanne Belden (Baystate Medical Center), Merrill Allen (Biotek, Inc.), and Martin Shapiro (Norwalk Hospital).

During the 1990s, NESCE's focus shifted to reflect the increasing financial pressures on health care institutions. As hospital directors changed their titles to "chief executive officers" and administrators became "vice presidents," the in-house clinical engineering programs were challenged to justify their role in supporting the hospital's bottom line. In response to these challenges, the Society's topics of interest became finance, benchmarking, and productivity. Asset management, equipment maintenance management, and maintenance insurance became the buzzwords of the profession and presented new vehicles for financing the hospital's medical technology programs. Society members were challenged to re-examine their niche in the health care landscape.

As a consequence, in the new millennium, several clinical engineering departments have been aligned with information technology services departments and also have assumed responsibility for telecommunications, nurse call systems, and patient television services. Traditional service areas are changing and with it. NESCE educational programs have been broadened to include such topics as "Networking, Hubs, Switches, and Servers" and "Wireless Networking Architectures."

To encourage the exchange of information and to open new lines of communication among the members, NESCE initiated its own web site, www.nesce.org with newsletters, meeting announcements, job listings, and other postings and links. Frank Painter (Technology Management Solutions, LLC) created and licensed the web site in the summer of 1999 and has been expanding it ever since. Our primary method of communication for the Society is now via the Web and e-mail distributions.

Past officers and active members since the 1990s include the following:

Anthony Guilietti and David Dittrich (Hospital of St. Raphael), Henry Montenegro (Waterbury Hospital), Shelley Steffy (Lawrence Memorial Hospital), Dave Crawfort and Dave Reihl (Novamed Biomedical Technology Services), John Elwood, Rebecca Thomson, and Vinnie DeFrancesco (Hartford Hospital), David Wilder and Robert Payne (Yale New Haven Hospital), Tom Chenail (Technology in Medicine), Dave Francoeur (Fischer Consulting), and Carolyn Mahoney (VA Medical Center).

Affiliation

Since the early 1980s, NESCE has been associated informally with three other northeastern clinical engineering societies: The Medical Device Society (MDS) in greater Boston, the Iroquois Biomedical Society (IBS) in the Albany area, and the Northern New England Society of Biomedical Equipment Technology (NNSBET), which encompasses Vermont, New Hampshire, and Maine. These alliances have been maintained over a period of 20 years. Each year, the four societies cosponsor the Northeastern Biomedical Symposium. The responsibility for planning and hosting the Symposium rotates among the four societies each year (Dyro, 1997). The hosting society manages the finances, hotel, vendor, and program development, while the other three provide mutual aid in the form of shared membership lists, presentations, friendly competition, and attendance.

The first Symposium was held in Danvers, MA, in 1981. The 20th was held in Sturbridge, MA, in 2001. NESCE hosted Symposia #1, 2, 5, 9, 13, 17 (Dyro, 1997), and 20 (Francoeur, 2002). The most recent Symposium, which was highlighted by 350 in attendance, 52 vendors, two keynote speakers, and the standard Certified Biomedical Electronics Technician (CBET) Review course and certification examination. The infamous Biobowl (a *Jeopardy*-like contest among the Societies, vendors and students) was replaced by a hayride and fireside dinner roast–a possible sign of a mellowing among members, who value fine food over academic and troubleshooting prowess. The author holds the distinction of initiating the Biobowl Competition at the 9th Symposium in 1989, borrowing the concept from the Respiratory Therapist's Sputum Bowl, which was patterned after the College Bowl format.

Since 2000, the Northeastern Biomedical Symposium alliance has suffered in that IBS and MDS are no longer active. The format and frequency of the Northeastern Biomedical Symposium undoubtedly will change in future years.

NEW ENGLAND SOCIETY OF CLINICAL ENGINEERING BYLAWS

ARTICLE I.
Name

The name of this Society shall be the New England Society of Clinical Engineering.

ARTICLE II
Objectives

The objectives of this Society shall be:

- To promote better patient care through improved maintenance, operation. and application of medical instrumentation;
- To promote the mutual exchange of information and technical assistance among Clinical Engineers and BMETs;
- To promote the development of the Clinical Engineering profession;
- To organize and promote an educational program for Clinical Engineers and BMETs; and
- To disseminate uniform interpretations of codes and standards pertaining to Clinical Engineering.

ARTICLE III
Membership

Section 1. This Society shall have three types of membership: Full member, associate member, and student member.

Full members shall be persons activity involved in the practice of Clinical Engineering who are employed by a health care institution.

Associate members shall be persons with an interest in Clinical Engineering.

Student members shall be full-time students with an interest in Clinical Engineering.

Associate and student members will have all of the privileges and rights of full members except holding office. Membership in the Society shall become effective upon approval by the Executive Committee of a competed application form and upon receipt of specified dues.

Section 2. The annual dues for all classes of membership shall be as set by the Executive Committee and approved by the membership. Dues shall be payable prior to or at the first meeting of the calendar year. All members must pay their dues in order to retain voting and attendance privileges.

ARTICLE IV
Officers

Section 1. The officers of this Society shall be a President, Vice President, Secretary, and Treasurer. These officers shall perform the duties prescribed by these bylaws.

Section 2. A Nominating Committee composed of at least two (2) full members shall be appointed by the Executive Committee sixty (60) days prior to the elections in April. It shall be the duty of this Committee to nominate candidates to the offices to be filled. The state of nominees selected shall be made know at the March meeting, at which time additional nominations from the floor can be made. Only full members can be nominated. All nominees will be required to submit a brief statement to the Secretary outlining their reasons for seeking office along with a description of their experience and qualifications. No member of the Executive Committee can be a member of the Nominating Committee.

Section 3. A Teller Committee composed of at least two (2) full members shall be appointed by the Executive Committee after the close of nominations. Members currently holding office (except the President) or candidates for office are not eligible for membership on this Committee. The Teller Committee is responsible for receiving and tabulating all ballots during the election. This Committee shall elect a Chairman from amongst its membership.

Section 4. Ballots will be mailed to all members listing all candidates prior to April 1. The ballot will include a brief statement describing each candidate's affiliation, reasons for seeking office, experience and qualifications. Returned ballots shall be sent to the Chairman of the Teller Committee so that they are received prior to or on the last day of April. Officers will be elected by simple majority. Election results will be announced at the next business meeting by the Chairman of the Teller Committee.

Section 5. Elections will be held every year for the position of Vice President.

The Vice President automatically becomes President at the end of his/her term of office. Elections will be held every even-numbered year for the position of Secretary and every odd-numbered year for the position of Treasurer.

Section 6. Term of Office shall begin at the first business meeting following the end of the elections and shall run until the meeting following the next election.

No member shall hold more than one office concurrently and no member shall be eligible to serve more than two (2) consecutive terms in the same office.

ARTICLE V
Executive Committee

Section 1. Officers, the immediate past President and Standing Committee Chairmen shall constitute the Executive Committee.

Section 2. The Executive Committee shall have general supervision of the affairs of the Society between its business meetings, fix the hour and place of meetings, make recommendations to the Society and shall perform such other duties as specified in these bylaws.

The Executive Committee shall appoint a full or associate member, other than the Treasurer, to perform an audit of the Society's financial records to be completed by the annual meeting.

The Executive Committee shall be subject to the orders of the Society, and none of its acts shall conflict with action taken by the Society.

ARTICLE VI
Committees

Section 1. There shall be such standing and ad hoc committees as are needed, to be established and appointed by the Executive Committee.

Section 2. The appointment and composition of the Nominating and Teller Committees will be as designated in Article IV, Sections 2 and 3.

ARTICLE VII
Parliamentary Authority

Roberts Rules of Order Newly Revised shall govern the meetings and business of the Society in all cases to which they are applicable which are not specifically covered by these bylaws or any special rules the Society may adopt.

ARTICLE VIII
Meetings

Section 1. Meetings of the Society shall be held at a time and place determined by the Executive Committee. One meeting (preferably the May meeting) shall be designated the Annual Meeting.

Section 2. Special meetings can be called by the President, Executive Committee or upon written petition by ten (10) members. At least ten (10) days notice shall be given and the purpose of the meeting stated in the meeting notice.

Section 3. Ten (10) full members of the Society shall constitute a quorum.

ARTICLE IX
Amendment of the Bylaws

Section 1. A motion to amend the by taws can be made at any business meeting and must be seconded.

Section 2. If at least two-thirds (2/3) of the total eligible voting membership is present, the amendment may be voted upon. Two-thirds (2/3) of the total eligible membership must vote in favor' of the amendment for passage.

Section 3. If the conditions stated in Article IX, Section 2 are not present, the Secretary shall mail a notice and ballot to all eligible members detailing the proposed amendment.

Section 4. Said ballots are to be retuned to the Secretary within thirty (30) days from the mailing date. A two-thirds (2/3) majority of the total eligible membership must vote in favor of the amendment for passage.

Structure

Executive Committee

The NESCE Bylaws provide the structure of the Society. An executive committee supervises the affairs of the Society and comprises the officers and past president. The term of each officer extends for two years, and each year the vice president automatically succeeds to the presidency. A new vice president is elected, and on alternating years a new secretary or treasurer is elected. This arrangement provides a training period for new officers and ensures continuity in NESCE leadership.

Officers

The duties of the officers are also defined in the Bylaws. The president presides over all general meetings and executive committee meetings, often providing the driving force in NESCE activities. The vice president serves as program chair and arranges the logistics of the general meetings, including meal arrangements, reservations, and coordinating an educational presentation for the membership. The secretary records minutes of the executive committee meetings; prepares and distributes mailings, announcements, and the *NESCE Newsletter*; and generally maintains communications with the membership. The treasurer bills and receives annual dues, maintains a checking account for NESCE, and reports to the membership on the Society's finances.

Meetings

The schedule, location, and program of the general meetings are determined by the executive committee and vary each year, although the meetings typically occur every other month between September and June. A dinner-meeting format has predominated through the years, meeting at restaurants in the urban centers of Connecticut as well as Springfield, Massachusetts, during mid-week evenings. The meeting format is as follows:

5:00 pm Cocktails
6:00 pm Dinner

7:00 pm Business meeting
7:30 pm Educational program
9:00 pm Adjourn

DUTIES OF THE OFFICERS

A. Duties of the President
1. Preside over all meetings of the Society.
2. Preside over at meetings of the Executive Committee.
3. Shall serve as an ex-officio member of all committees (except the Nominating Committee).
4. Prepares an Annual Report outlining the current status and activities of the Society for the year for presentation at the Annual Meeting.
5. Shall maintain a liaison with other regional Clinical Engineering societies.

B. Duties of the Vice President
1. Perform the duties and have the powers of the President in the absence of the President.
2. With the approval of the Executive Committee, the Vice President shall assume the duties of any other officer unable to perform his/her duties.
3. Serves as Program Committee Chairman, responsible for producing the educational programs held in conjunction with the monthly business meetings.

C. Duties of the Secretary
1. Records minutes of Executive Committee meetings and also general member-ship meetings.
2. Prepares and mails announcements for general membership meetings at least two (2) weeks before each meeting, including program information, time, place, form of acknowledgment, and minutes of previous meetings, both general meeting and Executive Committee meeting.
3. Updates membership lists and forwards a copy to each member periodically.
4. Maintains correspondence file.
5. Maintains supply of New England Society of Clinical Engineering supplies, e.g., stationery, except bill heads.
6. Submits bills to Treasurer for payment, e.g., printing, posting, and stationery.
7. Maintains a close liaison with Committee Chairmen and other members of the Executive Committee.
8. Arranges for printing and distribution of membership certificates.
9. Notifies general membership in writing, in advance of the election meeting, the slate of nominees.

D. Duties of the Treasurer
1. Bills and receives dues.
2. Deposits moneys and pays accounts.
3. Maintains on file, bank statements and records distribution of moneys.
4. Maintains an accurate financial statement in readiness for report at every business meeting.
5. Requires authorization of Executive Committee before incurring any indebtedness for the organization.
6. Prepares a statement of accounting for the fiscal year for-presentation to the membership at the final meeting.
7. Informs Secretary promptly in writing of paid members.

Executive committees have tried different approaches to meeting locations, to enhance attendance. Some years, the location was varied widely to reflect the geographic make-up of the membership. Other years, the same restaurant in the center of Connecticut was used for each meeting. NESCE members will forever remember the prime rib dinners at Angelico's in New Britain. With respect to general meeting locations, no consistent formula seemed to guarantee high attendance. The dinner expenses are subsidized by the Society each meeting, which limits the typical membership contribution to $15. Some years, new-member drives have been held in which anyone who brings a new member pays $5 for dinner. The usual attendance has been 20–40, with this number almost doubling on these special membership drives. Another enticement at each meeting is a drawing, with the winner receiving $25 to cover the cost of their travel and dinner.

The cocktail hour and dinner provide time for informal networking. During the business meeting, officer reports are presented; experiences with JCAHO, State Department of Health, and OSHA surveys are shared; job openings announced; and potential new hires are introduced to the membership.

As part of the general meetings, the educational program content has focused on current topics that are germane to the times and the profession, as well as programs expanding life skills. The technical programs are typically offered by manufacturers on new concepts in health care technology, or a review of the clinical aspects of medical instrumentation taught by a therapist, nurse, or physician. These topics have ranged from "Clinical Aspects of Mechanical Ventilation" to "Point-of-Care Laboratory Medicine" to "Medical Telemetry and Wireless Networks." The life skills programs have included "Time Management," "Dealing with Change," and "Humor in Medicine."

Occasionally, joint meetings have been held with the Connecticut Hospital Engineering Society, pooling our resources to sponsor a nationally known speaker on the JCAHO Environment of Care or environmental health and safety issues such as "Ethylene Oxide Use" or "Anesthetic Waste Gas Management." These meetings are exceptionally well attended and foster good interchange with the hospital engineering community. By extending the hand of friendship to a group of hospital engineers, a clinical engineering society can enlightening them to the group's perspectives of the important aspects of health care technology.

Membership

There are three types of memberships in NESCE: Full member, associate member, and student member. Full members are individuals who are actively involved in the field of clinical or biomedical engineering and are working in the health care technology arena. Associate members are individuals who have an interest in clinical and biomedical engineering and would include medical product sales representatives. Student members are full-time students who have an interest in the profession. Associate and student members have all privileges and rights of full members, other than holding office.

In 2002, NESCE membership stood at 352, which included all registrants from the last Symposium. The members who are actually located in Connecticut, Western Massachusetts, and Rhode Island total approximately 250 and are generally considered the active participants

Finances

There are no paid Officers in NESCE; all serve as volunteers. However, there are expenses associated with operating the Society. These include mailings, web site fees, subsidizing meetings, and sponsoring training programs. The finances to operate the Society come from annual membership dues and the shared revenues from the Northeastern Biomedical Symposia.

Annual membership dues have held constant at the following amounts:
$15: Full membership
$5: Student membership
$100: Corporate membership for up to 10 members from the same facility.

Many hospitals exercise the corporate-membership option, which ensures that all of their CEs and BMETs are members, receive meeting announcements, and have the opportunity to attend any of the society meetings or special training courses during the year. Some years, membership dues have been included with the annual Symposium registration.

The vendor and membership support from the Northeastern Biomedical Symposiums have lightened the financial burden of all four northeastern societies. NESCE's strong financial position allows the executive committee to waive membership dues occasionally, subsidize meetings, and sponsor service schools and training programs each year.

Ongoing Education

To support the biomedical education in Connecticut, NESCE member hospitals have provided practicum experience for biomedical electronics technicians from Gateway Community Technical College (formerly Greater New Haven Technical College) in North Haven, CT, and Springfield Technical Community College (STCC). NESCE members Thomas McGrath and Moshin Matar from Gateway, and Paul Barufaldi from STTC, have been instrumental in educating entry-level BMETs for the profession. Each year, NESCE has provided a $100 award to the best student biomedical instrumentation design project at Gateway's Annual Symposium. The selected student traditionally has been invited to present the project at a NESCE general meeting and to receive the award at that time.

NESCE also has coordinated educational tours of several local medical equipment manufacturers. General meetings have been scheduled at these facilities with catered meals followed by a tour of the manufacturing plant and presentations by applications, design, production, and quality assurance engineers. Tours have been held at Advanced Medical Systems, Ivy Biomedical, Novametrix, and Soma Technologies.

Summary

The New England Society of Clinical Engineering has served a vital role in the advancement of the clinical engineering presence in the northeastern United States, significantly raising the bar as a dynamic, high-quality, regional professional society. This is directly related to the contributions of all who have served as officers, helped with symposia, and participated as active members. NESCE recognizes and expresses its appreciation to all of these individuals for their time and talents expended to make NESCE what it has become over the past 27 years.

Unfortunately, NESCE's future is being challenged, as is the case for many regional biomedical professional societies today. As members' hospitals continue to experience increasing financial constraints, and as resources evaporate, the extracurricular activities of the clinical engineering community have suffered significant attrition. Although the actual membership numbers continue to remain strong, the general meeting attendance has diminished significantly, and it grows more difficult each year to find individuals who are willing to dedicate the time, energy, and drive to serve as officers and to maintain NESCE activities.

CEs and BMETs must appreciate that regional clinical engineering professional societies are worthy endeavors and that they must commit to supporting these organizations with their time and talents.

References

Dyro JF. 16th Northeastern Biomedical Symposium. *ACCE News* 7(1):8–9, 1997.
Dyro JF. Northeastern Biomedical Symposium. *ACCE News* 7(6):6–8, 1997.
Francoeur D. 20th Northeastern Biomedical Symposium. *ACCE News* 12(2):11–12, 2002.

132

New York City Metropolitan Area Clinical Engineering Directors Group

Ira Soller
Director, Scientific and Medical Instrumentation
SUNY Health Science Center at Brooklyn
Brooklyn, NY

Michael B. Mirsky
Clinical Engineering Solutions
Yorktown Heights, NY

This chapter provides an example of a local clinical engineering organization and describes its goals, history, affiliations, structure, meetings; membership, finances, ongoing education, speaker program, and networking.

Goals

The primary goal of the New York City Metropolitan Area Clinical Engineering Directors Group is to provide a local forum for clinical engineering directors in the greater metropolitan area. This organization provides a venue for the directors to meet periodically to informally address common concerns. Such networking among peers is vital in the ever-changing clinical engineering environment.

In addition to directly benefiting the clinical directors themselves, participation in the group is also indirectly beneficial to the health care institutions that the directors represent. It allows members facing similar business pressures to interact directly with peers, in the same (or very similar) health care provider markets.

A secondary goal of the organization is to provide technical/business presentations, in order to assist the membership in staying current in the clinical engineering field.

Throughout the years, these goals have been met.

History

Dr. Seymour Ben-Zvi (SUNY-Downstate Medical Center) and Walter Buchsbaum (Brookdale University Hospital Medical Center) started the New York City Metropolitan Area Clinical Engineering Directors Group in the early 1980s. By the 1990s, group literature sent to potential lecturers described the group as follows:

"The New York City Clinical Engineering Directors Group consists of 38 directors of clinical engineering departments, representing all of the major medical centers in the greater New York City area. We are directly responsible for managing the medical equipment stock in our respective medical centers. As a group, we are overseeing equipment used for a total of over 25,000 patient beds.

Our group is composed of engineers with various backgrounds and experience. Some have doctoral degrees; while others hold master's, bachelor's or associate's degrees. Some are heavily involved in medical research, while others manage departments that provide a broad range of repair and maintenance services to their facilities. In order for us to keep abreast of this dynamic field, we have group meetings every four to six weeks. A manufacturer or equipment vendor generally sponsors our meetings. A representative from the sponsor provides a technical presentation (definitely not a sales talk) on one of their latest products. The presentation usually lasts about an hour, which includes time for questions and answers."

AAMI Affiliation

In 1998, the group became affiliated with the Association for the Advancement of Medical Instrumentation (AAMI) when it was granted membership. The organization's group coordinator serves on the AAMI biomedical organizations committee. This AAMI committee is open to local, state, national, or international organizations (nonprofit and tax-exempt) representing biomedical equipment technicians, clinical engineers, or related technical service personnel in an educational or related context that have an interest in medical instrumentation.

The Clinical Engineering Directors Group contributed to the AAMI book *How to Establish and Maintain a Local Biomedical Organization* (AAMI, 1999). This contribution included a sample meeting announcement sent to the membership and the meeting sponsor (Figure 132-1); a confirmation letter sent to the meeting sponsor (Figure 132-2); group information and lecture guidelines sent to the meeting sponsor (Figure 132-3); and a press announcement (Figure 132-4).

Structure

This organization has no formal structure. There are no officers or dues. The functioning of the organization relies solely on shared responsibilities. The group coordinator arranges for the technical presentation, and informally coordinates the meetings. The coordinator also maintains the group membership mailing list and is responsible for group meeting notification mailings. Past group coordinators were Dr. Seymour Ben-Zvi (SUNY-Downstate Medical Center) and Paul Fried (Brookdale University Hospital Medical Center). The current group coordinator is Ira Soller (SUNY-Downstate Medical Center).

The director of clinical engineering at the hospital at which the meetings are held acts as meeting host, reserving a room, providing necessary audio-visual equipment, and assisting the sponsor with lunch arrangements. He or she also maintains a book that is used to maintain an attendance list for each meeting. This book serves as documentation should any of the members require proof of continuing education, which might be needed for clinical engineering certification renewal.

All participation in the organization is voluntary. The members contribute as their other commitments allow, often by recommending possible speakers and/or topics for future meetings and providing lecturer contacts based on their own experiences at other technical forums that they have participated in to the Group Coordinator.

Meetings

When the organization first started, the meeting location rotated among member hospitals. At least one meeting was held in conjunction with the post-graduate anesthesiology show in Manhattan. Ultimately, the group settled on St. Luke's-Roosevelt Hospital Center as the most convenient location for all attendees was due to its central Manhattan location. This location has proven quite successful, due in large part to Michael Mirsky's efforts both as host and in as substitute group coordinator as required.

The meetings, usually held every six weeks, are scheduled to last for three hours. Initially they were held in the evenings, but the members determined that a weekday afternoon from 2:00 PM to 5:00 PM encouraged the greatest attendance. The meetings are structured to allow for informal conversation, followed by a one-hour technical presentation by an invited vendor or guest lecturer. A formal meeting of group members only occurs after the technical presentation.

MEETING ANNOUNCEMENT

**SCIENTIFIC AND MEDICAL
INSTRUMENTATION CENTER**

January 19, 1999

Dear Colleagues:

The next New York City Metropolitan Area Clinical Engineering Director's Group meeting is scheduled for <u>Tuesday Feb. 9th</u> at 2:00 p.m. in the afternoon. Mr. xxxx and Mr. xxxxx of xxxxx Corp. will sponsor our meeting. A lecture will be given on "Principles of Operation for Vaporizers". Lunch will be served!

The meeting will take place at St. Luke's Roosevelt Hospital, Roosevelt Division, in the Winston's Conference Room. Enter the new building on 10th Avenue, between 58th and 59th Streets. Take the MAIN elevator to the 1st Floor and walk down the long twisting corridor (toward the Winston building) passing through three sets of doors. Immediately after the third door turn left and stop. Conference Room is on your left.

Parking will be available at a reduced rate in the Medical Center's lot on 59th Street, just West of 10th Avenue in the Concerto Garage. The new building is just a short walk from the lot. Make sure to have your parking ticket stamped by the Security Guard, in the main lobby of the hospital.

Please contact my Secretary, Ms. Emma Johnson via telephone at (718) 270-3192, or by filling in the bottom portion of this letter and faxing it back to us at (718)_270-3194 to confirm your attendance at the meeting. <u>Please respond by Thursday, Feb. 4, 1999.</u>

I wish to thank Mr. Mike Mirsky for acting as host for the meeting, and I am looking forward to seeing all of you there.

Sincerely,

Ira Soller
Director, Biomedical Engineering

NAME: _____ ☐ WILL ATTEND   ☐ WILL NOT ATTEND

Figure 132-1 A sample meeting announcement sent to the membership and the meeting sponsor.

Membership

Membership is open to all practicing clinical engineering directors working in hospital environments in the New York City metropolitan area. In reality, clinical engineering associate and assistant directors and other clinical engineering supervisors also attend. The limitation of membership encourages frank and open discussion. Members come from locations in New Jersey, the five boroughs of New York City, Long Island, Westchester, and other nearby areas. There is no distinction made between directors working directly for a hospital or for a third-party service provider. Manufacturers' representatives and equipment vendors are not permitted to join.

At present, the group includes approximately 80 members on the group mailing list. An updated copy is distributed at each meeting in order to solicit changes. This list is considered proprietary, and is not distributed outside the organization (never to vendors or manufacturers). Typically 20-25% of the membership attends the meetings at any given time.

Finances

The organization has no budget, and thus no need for membership fees, dues, or a treasurer. Group membership in AAMI includes three individual memberships, so the three director's group members selected for the individual memberships split the cost of the $100 AAMI group membership fee. The meeting room is provided free of charge by the host hospital. The technical lecturers typically underwrite the cost of a modest luncheon, such as sandwiches and sodas, but participants bring their own lunches when this is not feasible.

Ongoing Education

The technical lecture portion of the meetings provides the membership with ongoing technical and business training. "The environment in which clinical engineering functions changes daily as new technologies such as telemedicine, robotics, and wireless LANs are introduced into the clinical setting. In this dynamic field, continuing education is the rule" (Soller, 1999).

One method of obtaining further education is to attend technical expositions and professional organization meetings such as those of AAMI, the American College of Clinical Engineering (ACCE), the American Society of Health care Engineering (ASHE) and the Institute of Electrical and Electronics Engineers (IEEE) Engineering in Medicine and Biology Society (EMBS). However, as demonstrated by the New York City Metropolitan Area Clinical Engineering Directors Group, a local organization can fill a similar need, albeit on a smaller scale.

Many presentations directly address diverse areas of medical technology, while other topics such intellectual property do not. However, all topics are of importance to the group members. Some recent topics sponsored by vendors are as follows:

- Challenges of monitoring in a MRI environment
- "Controlled fresh gas delivery"

CONFORMATION LETTER
SENT TO MEETING SPONSOR

**SCIENTIFIC AND MEDICAL
INSTRUMENTATION CENTER**

Jan. 1, 1998

Mr. xxxxx xxxxxx
Clinical Sales Specialist
xxxxxxxxx
Fax: xxx xxx xxxx

Dear Mr. xxxxxx:

This note is to confirm that you have agreed to sponsor our next meeting, to be held on Tuesday, Feb. 3, 1998 at 6:00 p.m. The lecture will be a technical presentation on the "Biophysics of electro-surgery, stray energy and electra shield monitoring". The meeting will be hosted by Mr. Michael Mirsky, Director of Biomedical Engineering at St. Lukes Roosevelt Hospital 10th Avenue between 58th & 59th Sts., Phone #212-523-7020 or 212-523-7024. If you need any audio visual aid, Mr. Mirsky will be able to provide it.

A meeting notice specifying room location etc., is enclosed.

Also enclosed is information about the New York City Metropolitan Area Clinical Engineering Directors Group, and a general lecture outline that may be of some help to you.

I am looking forward to seeing you at the meeting.

Yours truly,

Ira Soller
Director Biomedical Engineering

cc: M. Mirsky

Figure 132-2 A confirmation letter sent to the meeting sponsor.

GROUP INFORMATION AND LECTURE GUIDELINES SENT TO MEETING SPONSOR

Brief History

The New York City Clinical Engineering Director's Group consists of 38 Directors of Clinical Engineering Departments, representing all of the major medical centers in the greater New York City area. We are directly responsible for managing the medical equipment stock in our respective medical centers. As a group, we are overseeing equipment used for a total of over 25,000 patient beds.

*Our group is composed of engineers with various backgrounds and experience. Some of us have PhD.'s; others Masters or Bachelor of Science Degrees. Some of us are heavily involved in medical research, while others perform more administrative type functions. In order to keep ourselves abreast of this dynamic field, we have group meetings every four to six weeks. Our meetings are generally sponsored by manufacturers who are responsible for making a __**TECHNICAL PRESENTATION**__ (definitely not a sales talk) by an engineer on one of their latest products. The presentation should be about one hour, with questions and answers.*

LECTURE OUTLINE

1. A brief history of the development of the type of equipment presented and significant milestones in it's development.
2. A theoretical presentation of the technology used, i.e. type of sensors, transducers necessary for the functioning of your product, one-of-a-kind design features, etc.
3. Physiological indication for and response to your product.
4. A general block diagram of the product with all major component functions.
5. Different technologies used to achieve the same end with pros and cons of each technological solution.
6. Future advances and theoretical limitations (what is on your drawing board now?).
7. Service information (self-diagnostics, board exchange program, service manuals and service support).

Figure 132-3 Group information and lecture guidelines sent to the meeting sponsor.

PRESS ANNOUNCEMENT

New York City Metropolitan Area
Clinical Engineering Directors Group
Ira Soller

The New York City Metropolitan Area Clinical Engineering Directors Group, consists of Directors of Biomedical/Clinical Engineering representing all of the major medical centers in the greater New York City area. The group holds biomed society membership status in AAMI.

At the Feb 9th meeting, hosted by Mr. Mike Mirsky of St. Luke's – Roosevelt, Mr. xxxxx and Mr. xxxxx of xxxx presented a lecture on Principals & Operation of Vaporizers. Subsequent member discussion included sharing of information relating to Y2K including SIIIM, and the FDA public health message dealing with use of "protected" patient cables.

The next meeting is scheduled for March 23rd at 2:00 PM. Mr. xxxxx and Mr. xxxxx of xxxxx will give a lecture on "Principles of Volume & Pressure Ventilation including Pathology".

For information, or manufacturers/vendors interested in making presentations contact: Ira Soller, Director of Biomedical Engineering, State University of New York, Downstate Medical Center, 450 Clarkson Ave, SMIC Box 26, Brooklyn, NY 11203. Phone: (718) 270-3192, Fax: (718) 270-3194.

Figure 132-4 Press announcement.

- IEEE 1073 medical information bus
- Infrared technology integrated locator systems
- Infrared thermometry
- Intellectual property
- New modes of ventilation in anesthesia delivery systems, and electronically-controlled fresh gas delivery
- Paperless systems
- Principles of volume and pressure ventilation including pathology
- Robotics and materials transport
- "Technical advances in monitoring"

At the members-only portion of the meeting, members discuss common concerns and diverse topics, a practice that helps them keep pace with changes in regulatory requirements and biomedical standards that impact their operations. The latest JCAHO requirements, FDA recommendations and recalls, and other regulatory issues are all relevant examples of discussion topics. In addition, general issues relating to departmental personnel and financial management, as well as equipment-specific issues, are addressed during this forum. Members might also share information gathered at conferences and lectures or gleaned from professional literature. This friendly and cooperative atmosphere among peers also allows mutually beneficial informal benchmarking.

In addition to the interaction that takes place at the meetings, members frequently contact each other outside of the meetings to discuss management technology or other issues related to clinical engineering. The group coordinator often acts as a clearinghouse of information, directing member queries to others in the organization who might be able to assist.

Conclusion

When first formed, the New York City Metropolitan Area Clinical Engineering Directors Group promised to provide a local forum where clinical engineering directors could network, and also help them keep current with developments in their field. The group has succeeded in filling both those needs, and continues to function as a viable organization, providing support at a local level.

References

Association for the Advancement of Medical Instrumentation. *How to Establish and Maintain a Local Biomedical Organization*. Arlington, VA, AAMI, 1999.

Soller I. Clinical Engineering. In *Wiley Encyclopedia of Electrical and Electronics Engineering, Vol 3*. New York, Wiley, 1999.

133

Clinical Engineering Certification in the United States

Thomas Nicoud
Design Planning Information
Tempe, AZ

Eben Kermit
Biomedical Engineer and Consultant
Palo Alto, CA

The first question one must ask is, "Why pursue Certification as a Clinical Engineer (CCE)? What value is a CCE?" The answers are complex, for if one were to ask a dozen clinical engineers why they became certified, one would get a dozen different responses. In the late 1990s, the number of individuals seeking certification declined, leading to the abandonment of the certifying program by the organization that served as the secretariat for the process, the Association for the Advancement of Medical Instrumentation (AAMI). AAMI continues to manage a renewal program for those previously certified clinical engineers who wish to remain affiliated with the former certification program.

The American College of Clinical Engineering (ACCE) has developed a replacement process for certification. (For a description of the ACCE initiative, see Chapter 130.) This chapter describes prior certification programs up until the time the process was discontinued by the AAMI. The history and rationale for establishment, as well as the application for, examination in, and renewal process of, clinical engineering certification is reviewed.

A Brief History

The 1960s saw the space race, the invention of the integrated circuit, and headline news about the number of patients who were being electrocuted in hospitals. Resistor-Transistor Logic (RTL) and Diode-Transistor Logic (DTL) devices arrived and were succeeded by Transistor-Transistor Logic (TTL). Operational amplifiers became solid state, and the Digital Equipment Corporation's PDP series of minicomputers were powerful new tools featuring 8K core memories and Teletype ASR33 units as I/O devices. A Computer of Average Transients (CAT) was a (nuclear channel analyzer—256 words— used for evoked response studies) then. Graduate-level programs in Biomedical Engineering began to appear. In 1971, Ralph Nader wrote an article that appeared in the *Ladies Home Journal*, citing unsubstantiated claims that 1200 people per year were electrocuted by medical devices.

Electronics for Medicine ("E for M") was a dominant supplier of catheterization laboratory physiological data recorders, and Statham was the prevalent provider of blood pressure transducers. Herb Mennen and Wilson Greatbatch teamed up to create Mennen Greatbatch. Sanborn Instruments was absorbed to become the core of Hewlett Packard's medical division. General Electric began producing implantable pacemakers.

In 1969, AAMI held its fourth annual meeting in Chicago as a joint event with the 8th International Conference on Medical and Biological Engineering and the 22nd Annual Conference on Engineering in Medicine and Biology. Nine years later, the Safe Current Limits for Electromedical Apparatus Standard (ANSI/AAMI SCL 12-78) was adopted.

Rationale for Certification

The 1970s was a time of rapid technological change, but health care was having difficulty coping. Perhaps the situation can be painted by a set of paraphrases and quotes from a paper titled "The Horror of Common Practices" by John MR Bruner in the *Electric Hazards in Hospitals Workshop* (April 4-5, 1968) (Bruner, JMR, 1970). The paradoxes of modern medicine form the basis of this horror.

- First, the electrical environment is in the hands of the least competent people.
- Virtually no attention is given to electrical safety education—even in teaching hospitals.
- In a sea of information, the real data (about incidents) are missing or hidden.
- Expediency in care overrides consideration of patient welfare.
- Buildings, which house medical practice, are antiquated.
- Innovation has been pushed, while the basics have been ignored.

A number of people from a variety of backgrounds were enlisted by health care organizations to address these basic issues, as well as to assist in the ongoing diffusion, development, and refinement of the technological tools. The term "clinical engineer" is believed to have been coined by physician Cesar Caceres in the early 1970s.

Certification was seen as a means of identifying and recognizing persons who have common skill sets, who have been demonstrative and successful in the applied technology (in direct patient care) arena. Under the auspices of AAMI, the Commission for Clinical Engineering and Biomedical Equipment Technology was established as the official body to oversee the application, examination, and renewal process and to confer the "certified" designation. Biomedical equipment technicians (BMETs) were the first to be examined and certified ("CBET") in 1972. Many had been trained formally and had acquired extensive experience in the U.S. Armed Services, (primarily the Army and Air Force), performing medical equipment repair.

Clinical engineers were a different matter. There were few formal training programs or academic degree programs in clinical engineering. Clinical engineering was still poorly defined, and no professional organization existed to define and promote clinical engineering and to educate and guide clinical engineers. Biomedical engineering organizations existed, but this broad discipline did not focus specifically on the application of engineering to medical device technologies in hospitals. In fact, biomedical engineering was then (and, to some extent, still is) seeking to promote an increased awareness among professionals and the general public of its identity and its role in society.

Rather than defining a specific skill set as a starting point, a group of practitioners who best fit the image were identified and deemed "certified" by the Certification Commission in 1975. These individuals became certified without examination in a process referred to as "grandfathering" (i.e., granting certification on the basis of extensive experience in the profession as demonstrated by work experience, education, publications, and general contribution to the field). Most references indicate that the number of grandfathered clinical engineers was 49. This core group became the model for skill set identification and testing.

At the time of the formation of the AAMI-based commission, another clinical engineering certifying body existed, known as the American Board of Clinical Engineering (ABCE). This entity was not sponsored (i.e., supported) by any other organization. It was sustained by the efforts of its volunteers and placed a heavy emphasis on formal engineering credentials. It differed in its lack of willingness to accept engineering credential equivalencies and in its willingness to accept funding from any single organization that might influence any aspect of its decision making.

By the end of 1977, 196 individuals were certified as clinical engineers by the Certification Commission. In 1983, the ABCE and the Certification Commission agreed to dissolve and to form the International Certification Commission for Clinical Engineering and Biomedical Technology (ICCCEBT, or ICC). Within the ICC, various country certification commissions were created; e.g., the United States Certification Commission (USCC) and the South African Certification Commission (SACC). By June, 1997 the meeting minutes of the U.S. Board of Examiners for Clinical Engineering noted that individuals 467 were certified in clinical engineering. Of those, 293 had been certified by the Certification Commission, and 50 had been certified by the ABCE prior to the merger into the ICC in 1983. Fifty had been certified by the Canadian Board of Examiners for Clinical engineering. By June 1999, the number had risen to 474.

Process

Based on the name "clinical engineering," it was not surprising that the process that was used as a model for certification in the United States was that used by the states in the registration of professional engineers. Engineering was a common element in the skill sets of the group first certified. It was a focus for credentialing by the ABCE. In addition, Canada required that any individual who was designated an "engineer" must be a registered "Professional Engineer" (PE). Thus, being certified as a PE became a prerequisite for Canadian certification. In an international context, it would have been difficult for the United States to be a leader without a credibly analogous focus on engineering (albeit lacking the PE registration requirement).

Three basic elements were involved in the ICC program: The application, written examination, and oral interview. Under the ICC, from 1983 through 1999, the process of becoming certified involved a lengthy application, including documentation of education (i.e., transcripts and diplomas, with a Bachelor of Science degree the minimum acceptable credential), and a listing of inventions, books, patents, journal articles, and meeting presentations. There was also a review of work history and evaluations by at least three supervisors or managers. A written statement of why certification was of value to the applicant, and ways in which certification would benefit the hospital or employer, was required. An application fee (in U.S. dollars) was necessary to offset the cost of administration. A history of at least five years in medical-related engineering work was another minimum requirement.

Information was conveyed to, and documented by, the certification applicant on the multipage application form. The preface to the actual data form contains sections addressing general information (including the definition of a clinical engineer), eligibility requirements, the certification process, certification renewal, available readings (reference books), and sample questions. Required data focused on the applicant's formal education and relevant professional experience. The experience section calls for references (from which written response is to be solicited by the Board of Examiners) and essay responses detailing experience and accomplishments.

Applicants who satisfy the eligibility requirements of the application, as determined by review of the submitted information by a panel of three members of the Board of Examiners, could request scheduling of the written examination. While most frequently these examinations had been administered in conjunction with the AAMI Annual Meeting, other arrangements were sometimes made to offset travel hardships or logistics problems. The test included 150 multiple-choice and true/false questions covering a number of subjects in the basic areas of medical science, clinical engineering, and engineering. Questions were drawn from a question bank on areas ranging from anatomy and physiology to regulatory issues, hospital practices, and engineering. The United States Board of Examiners (USBE) was responsible for the maintenance of this question bank, continuously updating it to keep pace with changes in technology, regulations, and practice. Three and a half hours were allowed for this portion of the examination. The last portion of this examination consisted of essay responses to five questions out of eleven. Two and a half hours are allowed for this portion. Correct multiple-choice questions carry a point value of one; the essay questions carry a maximum point value of ten. The minimum overall passing grade is 60% with a minimum of 50% in any one area.

Successful completion of the written examination qualified the applicant to take the oral interview. Like the written exam, this interview was most frequently administered in conjunction with the AAMI Annual Meeting. At least two members of the Board of Examiners conducted this interview, evaluating the candidates' experience, their thinking skills, and their ability to demonstrate good judgment in hypothetical scenarios. The interview usually lasted an hour and focused on the applicant's ability to make judgments on engineering issues. It also was partially devoted to clarifying abilities that appeared to be notably weak on the written exam. This last element was perhaps the most important, and the least objective to evaluate. Each examiner prepared an independent recommendation for certification, following the interview. If both examiners agreed, certification was recommended to the chair of the Board for subsequent submission to the ICC for granting of certification. If both concurred negatively, certification is not recommended. Lack of agreement between the two examiners required review by the full Board of Examiners. On rare occasions, re-interview by a different interview team conducted.

Recommendations regarding certification were conveyed to the members of the ICC at the annual meeting, which also was held in conjunction with the AAMI Annual Meeting. These recommendations were discussed and formally endorsed, as deemed appropriate by the members, for transmittal to the applicant through the Secretariat (i.e., AAMI).

For a variety of reasons, applications for certification declined precipitously in the late 1990s, the program became more difficult to support financially, and AAMI discontinued accepting new applications for certification in 1999.

Renewal of Certification

Renewal and/or recertification was considered for implementation (and actually communicated via letter) by the chairman of the Board of Examiners (Thomas Hargest) to all CCEs on December 20, 1977. The letter read, in part:

"At the March 1977 meeting of the United States Board of Examiners for Clinical engineering Certification, it was decided that those clinical engineers originally certified in March 1975 and March 1976 will be eligible for recertification in March 1979. After the latter date, recertification eligibility will occur at three year intervals."

A renew policy was adopted 1992. It states that applicants certified after 1992 who do not renew can be revoked but those who were certified before 1992 will become delisted. As of 2002, the AAMI web site "listed" fewer than 100 CCEs.

The Value of Certification

Given that demonstration of competency in the field of clinical engineering, and knowledge of medical devices, hospital practice, and regulatory affairs are important, why, then, has the certification process been foundering? In many other professions, the ability to demonstrate professional competence or proficiency is accomplished by a certification examination or a license, and is common practice. Lawyers, psychologists, FAA airframe mechanics, and certified public accountants (CPAs) are used here as examples of professions that require certification. All have an infrastructure for demonstration of competency requirements via examination prior to practice in their respective fields.

This was the thought process or impetus for the creation of the CCE title. The CCE examination was a peer review process to qualify individuals by a review of education, training, experience and critical thinking. Job placement, project assignments and professional activities were often determined (either included or excluded) based on CCE standing. Yet, the need for certification did not flourish.

Engineers continue to provide a much-needed and highly valued service to the health care field. However, it is possible, even typical, to work in the field of clinical engineering without a formally recognized degree or specialized training. This increases the likelihood of accidents, misuse of medical devices, and substandard maintenance and repair of equipment because of a lack of experience, training, and expertise. There is a reason why physicians attend medical school, why pilots take written and simulation exams, and why truck drivers must pass a proficiency test before they can haul freight. It makes sense for clinical engineers to follow the pattern set by other professions and to create a peer-peer approval process to ensure that clinical engineers continue to provide good advice, management, and safety to the health care profession.

The Future

The American College of Clinical Engineering (see Chapter 130) began a comprehensive approach to reinstating certification in 2000. The program also created a window of opportunity for recognition of persons certified under the AAMI program.

Reference

Bruner, JMR. *The Horror of Common Practices in Electric Hazards In Hospital Workshop Proceedings.* Washington, DC, National Academy of Sciences, 1970.

134 Clinical Engineering Certification in Germany

Vera Dammann
Department of Hospital and Medical Engineering,
Environmental, and BioTechnology
University of Applied Sciences Giessen-Friedberg
Giessen, Germany

Since entrance requirements, duration, curriculum, examination, and quality assurance of all educational institutions leading to a certified profession are government-regulated in Germany, an additional professional certification has not been necessary. Nevertheless the Deutsche Gesellschaft für Biomedizinische Technik e.V. founded a certification system for clinical engineers, and the Fachverband Biomedizinische Technik e.V.—the professional association of clinical engineers in Germany—is preparing a certification system for technicians. This chapter describes these programs.

General System of Professional Certification

To date, a system of professional certification by non-governmentcertification institutions such as professional associations is nearly unknown in Germany. A professional proves his or her knowledge, ability, and experience in the following ways:

- A certificate from educational institutes attesting to successful completion.
- A certificate/references from a former employer.
- Written confirmations of attendance at courses, seminars, and workshops, issued by the organizer
- A personal statement or report.

A certificate, issued by an institute of education, is acknowledged only if the institution is affiliated with or certified by the government. Primary and secondary schools, vocational training institutes, technician education programs, and and each course of study at every university require recognition by the respective ministry. Thus, subjects, quality, and duration are officially controlled. A university degree can not be obtained without passing all obligatory examination in the course, and no exceptions are possible.

Someone who lives in Germany and wants to use an academic degree obtained at a university abroad must request permission from the ministry of culture of the state where he or she lives. Fortunately, all degrees from recognized universities in Europe are acknowledged in every other member country of the EU.

Every employee has the right to obtain a written reference from his former employer upon leaving a job. This reference must contain the facts of the employment, i.e., the beginning and ending dates of employment, the position held, and the tasks performed. This is called an unqualified reference. A qualified reference includes a description of the tasks, plus statements concerning the quality of work, conduct, and managerial and supervisory skills. However, the legal prohibition on obviously negative statements is a problem. That is why all judgments are pronounced positively in a scale from satisfactory (meaning rather bad) to excellent. A written or oral reference from the former employer addressed to a future employer is very unusual.

An acknowledgement of the university certificate is necessary for certain jobs (e.g., physician, lawyer, or architect) or positions (e.g., teacher at a certified school or high level government officer). Employer and employee are both responsible for ensuring the necessary qualification of the employee.

Requirement of Certification in Clinical Engineering

In the clinical engineering field few tasks officially require special education, training, or experience. The following are examples:

- Radiation protection in radiotherapy: The so-called expert in medical physics must hold a degree in physics or engineering, prove knowledge in medical physics, obtained in certified courses, and possess experience on the job.
- Maintenance, safety checks, calibration of medical devices: The German directive for operation of medical devices (VEBAM, 1998) demands a suitable education plus practical experience. The owner of a medical device, who gives the order, and the person executing the order for device use are both responsible for the necessary qualification. However, a defined certificate is not required.

The start of calibration work on medical devices with measuring functions must be announced to the trade-supervisory board. This board has the right to check for personal qualifications and for adequate experience with equipment and procedures, but no formal certificate of qualification is required.

A private workshop for repair of electrically driven—not necessarily medical—devices must be registered at the local chamber of handicrafts. The responsible person must hold a certificate as "Meister" in electrical craftsmanship or prove an equivalent qualification. A degree in clinical engineering including proof of knowledge in electrical engineering is an example of equivalent qualification.

- A medical devices' consultant (i.e., the sales person who informs and instructs the owners or users of medical devices) must have a scientific, medical, or technical education, plus instruction by the manufacturer or one year of relevant professional experience (GM, 1994). A special certificate is not required.
- According to the German law on medical devices (GM, 1994) a manufacturer or importer of medical devices must appoint an authorized person to handle all safety information and act as liaison with the authorities. This person informs the government authorities about the hazards of medical devices. He or she must have a university degree in engineering, science, or medicine, or the equivalent plus a professional experience of at least two years. A special certificate is not required.

Certification by the Biomedical/Clinical Engineering Associations

Though neither the law nor employers ask for a certificate in clinical engineering other than a diploma from an educational institution, most of the biomedical engineering associations offer an additional certification or are planning to do so.

Certification of Clinical Engineers

Since the 1980s, the German Biomedical Engineering Society (Deutsche Gesellschaft für Biomedizinische Technik e.V., or DGBMT) has offered a certificate in accordance with the statutes of the International Federation of Medical and Biological Engineering (IFMBE). At the moment, the regulations of 1999 are in effect, requiring the following:

- A university degree in science or engineering
- Knowledge of German and English language
- Proof of professional practice in a biomedical environment
- Theoretical knowledge in at least six subjects from a catalogue, gained in at least 360 hours of lectures. Two of the six subjects must be medical, and one must be in the organizational field. Each subject must include at least 20 hours of lectures.
- Catalog of Subjects:
  Medical subjects:
   Physiology
   Anatomy
   Hygiene
   Emergency medicine
- Technical subjects:
   Biomaterials and artificial organs, clinical dosimetry and radiation therapy planning, radiation protection
   Clinical laboratory and analytics

Biosignal processing
Electromedicine and medical measurement
Imaging
Computer science in medicine
Statistical methods
Biomechanics
Organization and law
Technical safety and quality assurance in medical engineering and medicine,
Clinical trials
Approval of medical devices according to law

The candidate may propose additional subjects. Proof of knowledge requires certificates issued by certified institutions, based on appropriate oral or written exams.

Persons with at least eight years of practical experience in the clinical engineering field but without a degree may also apply for certification. They must pass an oral examination in four additional subjects from the catalogue, which takes at least two hours. A failed exam may be repeated once per year.

Certification of Biomedical Engineering Technicians

The German professional association of clinical engineers, Fachverband Biomedizinische Technik e. V. (FBMT), plans to introduce a certification system for skilled craftsmen and technicians working in the field of clinical engineering. Details on requirements and procedure will be published on the Internet: www.fbmt.de.

References

GM. Gesetz über Medizinprodukte (Medizinproduktegesetz) vom 2.8.1994 (Bundesgesetzblatt. I S. 1963) zuletzt geändert durch das Erste Gesetz zur Änderung des Medizinproduktegesetzes (1.MPG-ÄndG) vom 6.8.1998 (Bundesgesetzblatt. I S. 2005).

VEBAM. Verordnung über das Errichten, Betreiben und Anwenden von Medizinprodukten (Medizinprodukte-Betreiberverordnung-MPBetreibV) Bundesgesetzblatt. I vom 29.6.1998, S.1762 ff.

Further Information

Fachanerkennung als Klinik-Ingenieur (Clinical Engineering Certification)—Prüfungsordnung September 15, 1999. Deutsche Gesellschaft für Biomedizinische Technik e.V., published in internet: see www.dgbmt.de.

Richtlinie 97/43 Euratom des Rates vom 30.6.1997 über den Gesundheitsschutz von Personen gegen die Gefahren ionisierender Strahlung bei medizinischer Exposition und zur Aufhebung der Richtlinie 84/466 Euratom. Amtsblatt der Europäischen Gemeinschaften L 180/26 bom 9.7.1997.

Verordnung über den Schutz vor Schäden durch ionisierende Strahlen (Strahlenschutzverordnung – StrlSchV) in der Fassung der Bekanntmachung vom 30. Juni 1989 (Bundesgesetzblatt. I S. 1321, 1926), zuletzt geändert durch Verordnung vom 18. August 1997 (Bundesgesetzblatt. I S.2113).

Section XIII The Future

Jennifer C. Ott
Director, Clinical Engineering Department
St. Louis University Hospital
St. Louis, MO

At times it is overwhelming what can encompass the field of clinical engineering, a field that is as dynamic as the people in it. Those who understand the concept of clinical engineering get it, however, it is the transposition that is the constant struggle. Every clinical engineer has had to explain the profession at one time or another. The American College of Clinical Engineering (ACCE) definition provides a clear concise explanation that can then be tailored to the specific role: Clinical engineers support and advance patient care by applying engineering and management skills to health care technology.

In reviewing the chapters for this section I am in awe of where we have been, where we are, and where we can go. There are so many clinical engineers out there it is amazing. The goal is to have the non-traditional ones consider themselves part of the clinical engineering field. This handbook is a large step in accomplishing just that by linking all the specialties of clinical engineering under one cover.

This section focuses on the many avenues clinical engineers can explore to further their skill set or embark on a new career path that utilizes their ability of combining engineering and health care technology. The authors have done a wonderful job of exploring future opportunities while maintaining the core of the clinical engineering discipline. The initial chapter has the responsibility of reviewing the technical advances and how clinical engineers can apply themselves to their success. I found the concept of the *Inflection Point* to be very insightful and right on the mark for the development of future generations of clinical engineers. It is interesting that the avenues explored for the future roles really support the skills and continue them to the next level. In the following chapter, virtual instrumentation as applied to health care is illustrated by several practical examples (see Chapter 136). The next chapter covers historical and future non-traditional clinical engineering roles with a humorous entrance in which many traditional types may have to look in the mirror! The success of applying engineering skills for the improvement of health care is embodied in many careers that really are clinical engineers by definition but not typical or historical environment.

The following chapter (Chapter 138) on SWOT analysis takes inflection to the military-engineering level. SWOT stands for strengths, weaknesses, opportunities and threats. The results make you think how this applies in your particular career path. The author is very resilient and still on the cutting edge in the field with design in the health care setting.

Clinical engineers and biomedical engineering technicians have the capability and knowledge to assist with the user representation of the Global Harmonization Task Force and in particular their capacity to provide key information in post market surveillance and vigilance of medical devices. This role has always existed but behind closed doors, yet with the current trends in reporting and the required ending of the punitive aura of medical device mishaps, the clinical engineer is at a unique vantage point to fulfill this role quite well (Chapter 139).

Every clinical engineer can add a new tool to their managerial toolbox especially an outline to succeed and sell your future. Business plan development is a special skill and one often overlooked by the technical set. The author provides the fundamentals to assist anyone in breaking down the necessary parts to make your future successful (Chapter 140).

Reviewing an actual case study showing how clinical engineers affect the quality management of a primary health care concern (Chapter 141) expands the potential future of the profession. The technical skills of a clinical engineer can cross many paths including part of a health care delivery team in researching disease management. This is truly an avenue that is not explored enough within the resources of the health care community. The closing chapter (Chapter 142) blasts through the eighth dimension to 2050. The trends are certainly justified by looking at health care today. The role of the clinical engineer still has a place in 2050 as the meshing of engineering and health care technology continues plus the fact that the world is becoming a smaller place.

In addition to exploring the many potential futures of clinical engineering this section highlights the important role that clinical engineering will have for many years. Often it is difficult for technical people to see outside the box and assume a metamorphosis in their role. Staying within the traditional clinical engineering role will not slow the future just temporarily mask and one can only embrace the excitement of what is to come. The future is very bright for the field of clinical engineering so put on your shades and hold-on tight for what is sure to be an exciting ride. I look forward to the journey and hope to see you in the future.

135 The Future of Clinical Engineering: The Challenge of Change

Stephen L. Grimes
Senior Consultant and Analyst
GENTECH
Sarasota Springs, NY

"It's not the strongest of the species that survives, nor the most intelligent, but the one most responsive to change." —Charles Darwin

"It's not the progress I mind, it's the change I don't like." —Mark Twain

"Change is good. You go first...."—Dilbert

What will the future of clinical engineering look like? The question is an important one for clinical engineers to consider because the answer determines how they should be preparing now for that future. What level of education and what skill sets should be acquired? Anticipating the future enables us to effectively prepare to be proactive, rather than reactive, as changes take place in our industry.

To reasonably predict the future role of clinical engineering, we must consider the future nature of health care. Clinical engineering services evolve within the context of our health care delivery system. Consequently, significant changes in health care delivery are likely to result in a corresponding need for changes in the character of clinical engineering services. This chapter provides an overview of the changes in health care, the forces affecting those changes, and recommendations for clinical engineers to adapt and prosper in the environment to come.

Clinical Engineering's Future Linked to Developments in Health Care

We must recognize that there are revolutionary changes occurring within the health care delivery system. These changes are the results of a combination or confluence of technological, economic, cultural/demographic, and regulatory forces. Clinical engineering's ability to contribute, or perhaps even survive professionally will depend on how effectively it adapts to the changes brought about by these forces.

Technological Forces

Today we have the human genome mapped. As we continue our research in this area, we gain the ability to screen and identify individuals who possess genes that predispose them to certain diseases. Knowing who is predisposed to various diseases will enable us to focus our preventive efforts on those who are most at risk. Our understanding of the genome will enable us to refine our treatments. We will have the ability to develop some treatments that target affected genes while still other treatments can be optimized for an individual patient, based on what we know to be effective for someone of their genetic make-up.

Significant scientific developments have been occurring in the areas of micro- and nanotechnology (Fonash, 2002). Nanoparticle vectors are being developed to aid in drug delivery and DNA modification. Micro- and nano-scale devices are being designed to function as artificial organs and surgical instruments. Micro- and nano-sensors under development can serve as probes and detectors at an organ, tissue, cellular, or even molecular scale-level. These technologies are designed to be minimally invasive, and minimally disruptive, and to mimic the body's own natural systems closely.

Information technology promises to play a greater role in the business and the clinical aspects of health care. As a consequence of our ability to use technology to gather more data about a patient's condition, there has been a corresponding need to process that information in a way that is meaningful to the diagnostician and therapist.

The capability and availability of the integrated circuit have dramatically increased since personal computers appeared in the early 1980s. Over that period of time, the processing power of the computer has roughly doubled every 12 months. While the processing capability of each new generation of computer has grown exponentially over the past 20 years, the relative cost of those systems has dropped by half in the same period. In 1981, an IBM PC with a 4.77 MHz 8088 with 64Kb of RAM and 320Kb of storage sold for about $3,000. A typical system in 2001 with a 2 GHz Pentium 4, 512Mb of RAM, 120Gb of storage also sold for approximately $3,000, but that is equivalent to $1,454 in 1981 dollars.

Growing capabilities and the falling cost of integrated circuits have led both to the proliferation of computers in medicine and to the widespread adoption of those integrated circuits in medical devices and systems. Incorporating these technological advancements has greatly increased the amount of diagnostic and therapeutic data that these systems can collect, store, and process. Further, it is spurring the development of a broad array of experience and knowledge-based expert systems—systems that are designed to collect data and to suggest diagnoses and courses of treatment based on "pre-selected rules for decision making within specialized domains of knowledge (Frenster, 1989)." In medical devices and systems, these advancements have had the added benefit of improving system reliability and incorporating self-diagnostic capabilities.

Connectivity is another trend that is rapidly transforming the health care technology landscape. The number of diagnostic, therapeutic, and expert systems being linked is on the rise. Clinical information systems are also tying into the business-oriented hospital-information systems. The overall effect is synergistic, where the benefits gained from integrated systems far exceed the benefits available when the individual devices and systems are used in their stand-alone mode. Connectivity has also been advancing at another level. Networking and the Internet have the potential to bring health care resources to any near or remote location and to facilitate medical data and personal (i.e., voice & video) communications among a combination of patients, providers and payors.

The information technology industry is moving toward the development of what IBM calls "autonomic" systems (IBM, 2001; Gibbs, 2002). Like the involuntary nervous system that allows the human body to adjust to environmental changes, external attacks, and internal failures, future autonomic technical systems will:

- Be self-aware
- Adapt to environmental changes
- Continuously adjust to optimize performance
- Defend against attack
- Self-repair
- Exchange resources with unfamiliar systems
- Communicate through open standards
- Anticipate users' actions

The use of autonomic systems will enable us to realize the benefit of increasingly complex technologies that, without their autonomic abilities, would quickly overwhelm us with their needs for management and support.

We have just identified several key technological developments affecting health care. These have included the mapping of the human genome, advancements in micro- and nanotechnology, the exponential growth in information processing capacity and availability, the connectivity of technologies, and, finally, the introduction of autonomic systems. These are emerging medical technologies that have the potential to greatly improve the quality and availability of health care, and to so at a reasonable cost. (George, 1999)

As powerful forces for change, these emerging technologies will likely be critical in determining the future role of clinical engineering.

Economic Forces

The U.S. health care industry is experiencing a financial crisis. U.S. health care costs have been spiraling out of control for the past 10 years. In 1997, 13.5% of the U.S. gross domestic product (GDP) was spent on health care (considerably higher than Germany, Switzerland, and France, whose spending is the next highest as a percentage of their GDP, at 10%) (Health Forum, 2002). Total U.S. health care expenditures rose from $888 billion in 1993 to $1425 billion in 2001. Those expenditures are expected to grow to $3.1 trillion and 17.7% of GDP by 2012.(Heffler, 2003). Spending on health benefits for their employees cost American companies $177 billion in 1990, and by 1996 the amount had soared to more than $252 billion (rising at more than twice the rate of inflation) (PBS. 2000). Health insurance premiums increased by 11% in 2001—the fifth straight year of rising premiums, and the highest increase since 1993 (CSHSC, 2000). The U.S. now spends more money on health care than any other nation, but its care ranks 37th in quality, according to a recent World Health Organization survey of 191 countries (WHO, 2001).

One study, reported in the *New England Journal of Medicine* found that 19–24 percent of U.S. health care expenditures go for administrative costs (Woolhandler, 1991). This figure sharply contrasts with that of Canada, where it is estimated that administrative costs represent only 11% of total expenditures, and most European countries, where the average is 7%.

Adoption of new technologies has significantly contributed to industry cost increases. One recent study estimated that between 1998 and 2002, new medical technology contributed 19% of the increases in inpatient health care spending (Hay et al., 2001). Another recent study, conducted by the University of California at Berkeley researchers on the true total cost of ownership (TCO) for computer-based systems, calculated that TCO the first three years after acquisition now represents between 3.6 and 18.5 times the initial purchase cost of hardware and software (UC Berkeley, 2002). The same study suggests that "a third to half of TCO is recovering from or preparing against failures." Without innovative design (e.g., autonomic systems) and careful management, the disproportionate relationship between the initial purchase and ongoing support costs will only increase as new and more complex technologies are adopted.

Clearly, these trends cannot continue if we are to have a healthy economy and a first-rate health care industry. We must find a means of reducing costs for the level of health care that we are receiving. If we cannot reduce these costs, we must be prepared to adjust our expectations and to settle for the level of care that we can afford.

Demographic/Cultural Forces

The U.S. is undergoing significant cultural and demographic changes that will have an important impact on the health care industry. Within the next decade, the "baby-boomers" will begin reaching the age of retirement. Between 2011 and 2030, the number of Americans age 65 and older will jump from 13% to over 20% of the total US population (U.S. Department of Commerce, 2000). As this group reaches age 65, they can look forward to the prospect of living much longer lives than previous generations did, due to advancements in medical science and technology. Due to the aging population, there will be a growing shift from acute, episodic care to care for chronic conditions (Institute of Medicine, 2001). Currently,100 million Americans are treated for chronic conditions. Of them, 40% have multiple chronic conditions (Guralnick et al., 1989). Today, the medical care costs of people with chronic diseases account for more than 60% of the nation's medical care costs (Centers for Disease Control and Prevention, 2003). By the year 2020, 157 million Americans will have a chronic condition, and 80 percent of the country's total medical care spending will be associated with treatment of these individuals (Wu and Green, 2000). As a consequence of these trends, we need to be concentrating most of our efforts on the development of a health care system that provides long-term treatment programs for patients with multiple, chronic diseases.

Given their exposure to advances in technology and the Internet, today's consumers are better informed and have higher expectations regarding health care quality and availability than any previous generation. These consumers will demand that industry and government work together to ensure that quality health care is readily available at a reasonable cost. As the U.S. population looks forward to a longer lifespan and the prospect of long-term treatment for chronic diseases, the availability of effective and affordable health care becomes a critical quality-of-life issue.

Regulatory Forces

In recent years, issues related to health care quality and costs have led to government and industry initiatives directed at improving the quality and availability of health care, and at reducing its costs. In the early 1990s, a respected industry group reported to Congress that $73 billion per year could be saved if health care organizations would adopt the use of standardized data formats to exchange patient information (Duncan et al., 2001). Congress reacted by including the administrative simplification provisions in the Health Insurance Portability and Accountability Act (HIPAA) and passing this legislation in 1996. In the HIPAA legislation, the U.S. Department of Health and Human Services (HHS) was directed to develop regulations requiring health care organizations to adopt the use of standardized data, to ensure privacy of patient-related health information, and to implement safeguards to ensure the integrity, availability, and confidentiality of those data. The various provisions of HIPAA were scheduled for implementation between 2002 and 2005. Adoption of these provisions should help to facilitate the exchange of data between a rapidly growing array of information and biomedical technology systems.

In 2001, the Institute of Medicine (IOM) released its watershed report, *Crossing the Quality Chasm: A New Health System for the 21st Century*, which examined the state of the U.S. health care system (Institute of Medicine, 2001). That report details recommendations for the industry, including a number of major recommendations on "applying advances in information technology to improve clinical and administrative processes." In fact, many of the report's main recommendations can be accomplished only through the effective integration of information and clinical or biomedical technologies. A year earlier, the IOM had released another major report, *To Err Is Human: Building a Safer Health System*, which suggested that as many as 98,000 Americans die annually as the result of medical errors (Kohn et al., 2000). In that report, the use of increasingly sophisticated and complex technologies is cited as a contributory factor in many of these errors. The report goes on to assert that technology must be recognized as a member of the health care team and that among its roles are enhancing human performance and automating processes so as to remove opportunities for humans to make errors. However, system failures can occur, and where technology is employed, humans still must find ways to effectively monitor the processes that they automate. Both of these IOM reports have had a major impact on the health care industry and are very likely to influence federal legislation and industry initiatives for the foreseeable future.

In 1999, an industry-sponsored initiative called "Integrating the Healthcare Enterprise" (IHE) was launched. IHE brought together medical professionals and the health care information and imaging systems industry "to agree upon, document and demonstrate standards-based methods of sharing information in support of optimal patient care (RSNA, 2001)." The initiative was sponsored by the Radiological Society of North America (RSNA) and the Health care Information and Management Systems Society (HIMSS). After successful efforts in medical imaging, this program is now attempting to broaden its scope into clinical laboratory, cardiology, and other areas that would benefit from the effective integration of biomedical and information technology systems.

HIPAA, the IOM reports, and IHE represent perhaps the most significant regulatory and pseudo-regulatory initiatives impacting the adoption of technology in the health care industry in current times. As we look to technology to help address health care's growing number of problems and issues, they undoubtedly will turn out to be only the precursors of many future initiatives.

Net Impact of these Forces

Because of the forces described above, health care in the U.S. will undergo substantial changes within the next 5–20 years. The health care industry will focus increasingly on the long-term treatment of chronic conditions for an aging patient population. This population will expect high-quality care that is both readily available and reasonably priced. Technological advances will facilitate the industry's ability to meet these demands, and regulatory pressures will foster better integration.

Strategic Inflection Point for Clinical Engineering

Andrew Grove, Intel's chairman, defined "strategic inflection point" as a term that describes the time in which extreme forces forever alter the landscape of an industry, creating opportunities and challenges (Grove, 1999). In Grove's model, businesses and industries progress along at a steady, smooth fashion until hitting a subtle point where the business dynamics force a change in the curvature of that progression. At this "inflection point," the transition is so smooth and subtle that there are no obvious profound, major, or cataclysmic signs. However depending on the actions it takes, a business will progress through the inflection point along a path to potentially unprecedented heights, or it will find itself going down the path toward obscurity. If a business misses the opportunity and begins the descending branch of the curve, it is exceedingly difficult to reset the progression and to correct for the action not taken at the inflection point. It is, therefore, extremely important to anticipate and to act before reaching that inflection point.

Clinical engineering programs, and perhaps the profession, are arriving at a strategic inflection point, as shown in Figure 135-1. The long-term viability of clinical engineering as a distinct profession and service depends on the model that clinical engineering programs adopt for the future. Selecting the right model ensures clinical engineering's future role as a successfully productive and important element in health care. Selecting the wrong model will result in a declining role until other technical-service programs assimilate any clinical engineering function that remains.

Historical Perspective

Clinical engineering encountered another strategic inflection point in the past. Beginning in the mid to late 1960s, hospitals began to significantly increase their adoption and use of biomedical instrumentation. In 1970, the consumer activist Ralph Nader wrote an article for the *Ladies Home Journal*, claiming that at least 1200 Americans were electrocuted annually in hospitals during routine diagnostic and therapeutic procedures (Nader, 1971). The Emergency Care Research Institute (ECRI), a nonprofit organization evaluating the safety and effectiveness of medical devices, reported that "a disturbing proportion of ... medical devices is demonstrably ineffective, of inferior quality, or dangerous." Approximately half the new medical devices being purchased and delivered to hospitals were defective to some degree (Health Devices, 1971). ECRI's reports of medical equipment quality issues, along with Nader's comments on electrical safety, served to raise public interest, and a series of congressional hearings were subsequently held on medical device safety. Over the next five years, the U.S. Food and Drug Administration (FDA), the Joint Commission on Accreditation of Hospitals (JCAH), the National Fire Protection Association (NFPA), Underwriters Laboratories (UL), the Association for the Advancement of Medical Instrumentation (AAMI), and many states adopted standards or regulations pertaining to the design, manufacture, and/or testing of medical devices.

During this period, the combination of the rapid adoption of new technologies, rising public concerns about safety, and the promulgation of regulations resulted in the occurrence of an inflection point. This inflection point would have a significant effect on the

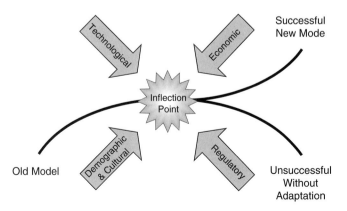

Figure 135-1 Strategic inflection point in clinical engineering programs.

growth of the relatively new field of clinical engineering. Few of the nearly 7000 U.S. hospitals had clinical engineering programs in the early 1970s. By 1975, JCAH (the largest hospital-accrediting organization in the U.S.) had established a requirement that hospitals conduct incoming inspections of all new medical equipment and perform routine electrical-safety testing of all medical equipment used in their facilities. When this inflection point occurred in the early 1970s, there were many opportunities for those individuals and businesses that were qualified and prepared to deliver biomedical equipment services. Larger hospitals employed clinical engineers and biomedical equipment technicians to develop and operate in-house medical equipment management programs. Smaller hospitals that could not afford to hire dedicated staff typically contracted with clinical engineering service organizations to obtain their medical equipment services.

The primary focus of clinical engineering services in the early years was on incoming and routine inspections (with an emphasis on electrical-safety testing) and on repairs of biomedical equipment. While clinical engineering's role over the past 25 years has broadened, biomedical equipment inspection, electrical safety testing, and repairs still represent a substantial portion of most program efforts. Today, clinical engineering program services may include the following:

Equipment management services

- Inventory management
- Risk analysis
- Evaluation of new devices and systems prior to acquisition
- Vendor management
- Compliance (e.g., government regulations, accrediting standards)
- Educational services (for equipment users and clinical engineering staff)
- Device tracking (hazards and recalls)
- Root-cause analysis of adverse outcomes, incident investigation, and reporting
- Quality assurance
- Evaluation of investigational devices, processes, and procedures
- Design, fabrication, and modification of new or existing devices

Technical services

- Inspection and testing (functional, safety, performance)
- Calibration
- Preventive maintenance
- Corrective maintenance (i.e., repair)

Other than a broadening of the services offered, clinical engineering has not significantly changed. Given the technological, economic, cultural, and regulatory dynamics at work in the health care industry now, clinical engineering will be transformed before 2010. The nature and success of that transformation depends on the action that we take now as clinical engineers.

Take Action Now!

"If you want to prosper on the other side of a strategic inflection point, you must take action before you get there." —Andrew S. Grove, Intel chairman and cofounder

While still being mindful of our current business, we must lay the foundation for the future now. We cannot wait for events to overtake us, or we will risk irrelevance.

Clinical engineering must . . .

Adopt a Systems and Process Approach

Clinical engineering services traditionally have been oriented toward the management of discrete devices (i.e., equipment management). Systems and processes (i.e., technology management) were managed only marginally, when they were managed at all. Consideration of systems and processes requires looking at the big picture; not focusing on discrete devices but understanding ways in which individual devices must interconnect to accomplish a technical process. Clinical engineering must become more systems- and process-oriented. Increasingly, biomedical devices are becoming part of integrated technology systems. Technology will significantly contribute to quality of care, patient safety, patient outcomes, health-data integrity, and availability issues. These issues involve processes and require a systems approach in their management.

Add Basic Information Technology and Telecommunications Skills

For years, the trend has been for biomedical devices and systems to process increasing amounts of data and for these systems to be networked together to share these data. Today, "microprocessor," "RAM," "firmware," "software," "I/O port," and "Ethernet" are terms equally applicable to biomedical and IT systems. This confluence of biomedical, IT, and telecommunications technologies will continue in health care. IT and telecommunication technologies will supply the backbone along which integrated biomedical systems will operate. Clinical engineering must develop basic proficiencies in IT and telecommunications. To ensure support and coverage of these merging technologies, clinical engineering also must be prepared to integrate its services at an appropriate level with those of IT and telecommunications.

Monitor Technological, Regulatory, Economic and Other Developments

Clinical engineering must monitor the developments that are likely to impact the value of its contribution to the organization. Clinical engineering services should be structured to take these developments into consideration and to provide the maximum value. Clinical engineering should become proficient at new key technologies and be prepared to address new regulatory pressures and to account for economic issues in its technology planning.

Become Conversant with the "Business" of Technology

Clinical engineering must acquire expertise in the economic nuances pertaining to the adoption and use of technology, including cost/benefit analyses, return on investment (ROI) analyses, and life-cycle-cost analyses. These considerations typically are the prima facie support for technology-related decisions made by health care executives. Clinical engineering must be prepared to supply these decision-makers with this information.

Plan for the Integration of Existing and New Medical Technologies

"Legacy" (i.e., existing and new) technologies will be integrated. Clinical engineering must anticipate the need for integration, understand the implications, and possess the necessary skills to manage the integration process successfully.

Develop Systems and Infrastructure to Support Technology in nontraditional Venues

Health care will be delivered increasingly outside of traditional venues (e.g., hospitals and clinics). Clinical engineering must be prepared to incorporate and support medical technology in non-traditional locations (e.g., patients' homes, assisted-living facilities, offices, schools, and public areas) by developing the necessary systems and infrastructure.

Closely Examine Existing Clinical Engineering Services and Practices

Clinical engineering must closely examine its existing services and practices to determine those that are necessary for the future, and those that should be discarded. Clinical engineering cannot afford to expend resources or to continue providing services for which there is little or no demonstrable benefit.

Incorporate Continuing Education

The pace of the health care technology revolution is quickening. Clinical engineering's only hope of making a contribution to the successful adoption of new technologies is to embrace a regimen of continuing education. Such education should include programs offered by universities, professional peer organizations, trade groups, and manufacturers. This education also should include regular literature research to identify current developments in technology and issues related to its adoption.

Build Relationships with Other Stakeholders

Teamwork is critical if the health care industry is to any hope of effectively addressing the technological, economic, regulatory, and social issues that it must address in the coming years. Clinical engineering should identify stakeholders in the technology-implementation process and should work to establish effective relationships with these key individuals and groups.

Develop a Plan to Transition from Existing to Future Services

Clinical engineering must develop and begin implementing a plan to transition from existing to future services. The plan must provide for acquisition of necessary skills and resources, education of clients as well as staff, and schedules to ensure a smooth transition.

Formulate a Vision for Clinical Engineering Within the Organization

Clinical engineering must develop and articulate a clear vision that is closely aligned with the vision and mission of the organization(s) served but also ensures that the vision promotes quality, service, and innovation.

Future Scope of Clinical Engineering Services

Education and experience typically endow the clinical engineer with a fundamental understanding not only of relevant technologies but also of the physiological systems on which that technology is applied, of the health care environment in which the technology is used, and of the regulatory framework in which the technology exists. Clinical engineers can be expected to develop organizational, project management, strategic planning, and investigative skills to ensure constant availability of safe and effective health care technology.

Given the forces for change and the clinical engineer's inherent and unique aptitudes, clinical engineering's future services are likely to include management consulting services, support services, and technical services.

Management and Consulting Services

Inventory and Asset Management

To remain viable, any business must effectively manage its assets. An accurate inventory is a fundamental component of any effective asset management program. Clinical engineering has assumed primary responsibility for maintaining the detailed inventory information on medical devices and systems, and it should continue to do so. The basic inventory information includes quantity, type, owner, acquisition/warranty dates, and monetary value of medical devices and systems. Additionally, clinical engineering must regularly assess and update a broad range of information on these devices and systems, including the following information:

- Current location
- Associated devices and systems (i.e., a record of interconnections to other devices and systems)
- Current physical and operating condition
- Degree of obsolescence

- Operating and service requirements and responsibilities
- Operating and service history
- Safety risk to patients and staff associated with device or system misapplication, failure, or lack of availability
- Financial risk to organization in case of device/system failure (i.e., is device or system mission critical?)
- History of recalls and reported hazards

Once collected, clinical engineering should use the above asset information to do the following:

- Provide a document trail regarding the condition, use, and servicing of devices and systems.
- Prepare capital budgets and schedule upgrades or replacements of worn or obsolete devices and systems.
- Encourage the standardization of devices and systems.
- Facilitate integration between other devices and systems.
- Plan the type, schedule and frequency, and source of service (i.e., calibration, inspection, and pm).
- Prepare service and operating budgets for medical devices and systems.
- Conduct risk analyses and facilitate risk mediation efforts (including disaster planning).
- Analyze performance, integrity, and reliability of devices and systems.
- Analyze quality and effectiveness of technical services provided to devices and systems.
- Establish the real financial value for the organization's medical technology assets.

Strategic Planning

The evolution of health care technology is rapidly accelerating on a number of fronts. As this increasingly wide array of technology becomes available, it has the potential to impact health care delivery positively by improving quality, safety, and availability, while reducing costs. However, these positive effects can only be realized through careful planning, adoption, and integration of appropriate technologies into the health care delivery process. Clinical engineering is uniquely suited to contribute to the strategic planning process that is due. Clinical engineers must be prepared to use their unique abilities to do the following:

- Continually work to sharpen their awareness of existing and newly available technologies
- Evaluate the technical strengths and limitations in the context of the intended applications
- Apply their knowledge of the environment where the devices or systems are to be used, to the appropriate selection and configuration of devices and systems
- Plan for installation, integration with other systems, training, and on-going service
- Contribute to cost-benefit and life cycle cost analyses

Quality & Safety

In recent years, quality and safety have become, and are likely to continue to be, a major concern of the U.S. health care industry. Technology has the potential to contribute to quality and safety, but it can be detrimental if not properly managed. To ensure that technology's impact is positive, clinical engineering must:

- Adopt a quality-management system (e.g., ISO 9000, Six Sigma, or Malcolm Baldrige) that will facilitate identifying:
  - Performance criteria for technical systems and processes
  - Target goals and objectives (benchmarks) associated with use of technology
  - Methods for achieving these goals and objectives
  - Techniques for measuring progress toward goals and objectives
  - A process for analyzing and improving the effectiveness of methods of achieving goals and objectives
- Implement a risk-management program including:
  - An assessment program that identifies risks by evaluating the implications of failure of technical systems and their related processes on patient health, patient/staff safety, and the financial well-being of the organization
  - A mitigation program that prioritizes identified risks and provides methods for reducing them to an acceptable level
- Provide root-cause analysis, investigation, and reporting support when technology and technical processes are involved in adverse outcomes or incidents
  - Identify contributing causes
  - Propose recommendations to prevent reoccurrence

Compliance

Government and industry produce an ever-changing array of regulations and standards that impact health care technology. A clinical engineer's role includes ensuring compliance with the relevant government, accreditation, and industry standards and initiatives (e.g., FDA, SMDA, HIPAA, JCAHO, IOM, and IHE). This requires them to understand existing laws, standards, codes, and practices, and to remain current on changes. Clinical engineering also must take steps to ensure that technical systems and processes are used and maintained in a manner that ensures compliance with all relevant regulations and standards.

Vendor Management

Vendors include medical-technology manufacturers, distributors, and independent service organizations. Because health care organizations necessarily work with vendors to obtain medical technology and technical services, clinical engineering needs to ensure that the working relationship is an effective one for all parties. Clinical engineering should serve as technical liaison, ensuring that vendors have all of the information and access that is necessary for them to deliver appropriate technology and services. Clinical engineering also should:

- Evaluate vendors and their products and services, recommending those that best meet the needs and standards of the organization.
- Ensure inclusion of necessary terms and conditions in agreements with the vendor and regularly ensure vendor compliance with those terms and conditions.
- Ensure that the vendor makes available any information, documentation, software, specialized tools, and education that is necessary for operating or servicing technology.
- Verify integrity of technology supplied.
- Monitor vendor quality and the integrity of services delivered.

Support Services

Education

Clinical engineering's role in technical education should continue to change. It is important to train users in the proper operation of technical systems. Their expertise in technology and understanding of applications make clinical engineers uniquely well-suited to complement application training with education of users on the effective use and basic troubleshooting of devices and systems.

Help Desk

Clinical engineering should adopt the "help desk" concept. The concept of a help desk comes from the information technology industry. The proliferation of business computers onto the desks of non-IT personnel resulted in the need for a readily identifiable and available source of technical expertise (i.e., the help desk). When computer users experienced inevitable computer problems, they could telephone a software or hardware consultant at a help desk who could talk the users through a solution, thereby solving the problem more quickly and less expensively then could be done by an on-site visit from IT personnel. The use of remote access and remote control software in recent years has further enhanced the ability of the help desk personnel to assist, troubleshoot, and solve problems remotely. If clinical engineering were to adopt the help desk approach, it could complement existing services by providing some operating and basic technical support for system and device users (e.g., clinicians and patients). As increasingly more medical systems are connected via telecommunications, networks and the Internet, the help desk can perform remote troubleshooting, diagnostics, and system upgrades. Senior technical personnel on such a help desk also would be available to support less-experienced technical staff at remote locations.

Technical Services

Installation and Integration

Installation of new technology today frequently requires software/hardware configuration and integration into existing systems. The level of integration of medical systems will increase substantially in the future. Clinical engineering's expertise in a wide variety of technical systems and applications makes it well suited to provide guidance and technical assistance in installation and integration services.

Upgrades

System upgrades will become more commonplace as advances occur in medical hardware and software. Technology owners will attempt to curb costs by staggering upgrades of components and software, rather than upgrading an entire system at once. Clinical engineering must be prepared to provide advice and assistance in this upgrade process.

Testing, Inspection and Preventive Maintenance

The roles of testing, inspection, and preventive maintenance (PM) will significantly decrease. Devices and systems will continue to require testing and inspection upon installation. However, routine testing and inspection on devices and systems where it has been demonstrated that the test/inspection results never vary is questionable, both in value and in the use of limited resources. Medical technology has become sufficiently reliable that visual inspections, operational checks, and monitoring of self-diagnostics by technical staff and users will largely replace traditional forms of inspection, testing, and PMs.

Repair

Clinical engineering will continue to offer repair services for medical technology, but the amount and nature of repair services certainly will change. Repair services will occupy fewer clinical engineering resources as the technology continues to become more reliable and to incorporate self-diagnostics. Repairs also will focus increasingly on troubleshooting and solving problems associated with systems of interconnected devices.

Conclusion

Clinical engineering is at a strategic inflection point. Technical, economic, regulatory and cultural dynamics are at work shaping the future of health care delivery. As the nature of health care delivery is transformed by these forces, the types and mix of technology management and support services needed by the industry are changing significantly. Clinical engineering has a relatively short opportunity to adopt a service model that will meet these changing needs. Delay or failure to adopt an effective service model as we pass through the inflection point will result in a diminished role for clinical engineering in health care technology management as other technical professionals move in to fill the need. The question is: Will clinical engineering rise to the challenge?

References

CDC. *Chronic Disease Prevention.* Atlanta, GA, Centers for Disease Control and Prevention (CDC), http://www.cdc.gov/nccdphp/about.htm, 2003.
CSHSC. *Tracking Health care Costs: Hospital Care Key Cost Driver in 2000.* Washington, DC, Center for Studying Health System Change, http://www.hschange.com/content/380/, 2000.
Duncan M, Rishel W, Kleinberg K, et al. *A Common Sense Approach to HIPAA.*, GartnerGroup, March 2001.
Fonash SJ. *Micro and Nano Technology: Impact on Biomedical Science and Practice.* Association for the Advancement of Medical Instrumentation (AAMI) Conference. Minneapolis, MN, AAMI, 2002.
Frenster JH. Expert Systems and Open Systems in Medical Artificial Intelligence. *Proc Am Assoc Med Syst Informatics*, 89(7):118–120, 1989.
George WW. *The Future of 21st Century Health: Getting the Right Care.* Harvard Business School Consumer-Driven Health Care Conference. Harvard, MA, Harvard Business School, 1999.
Gibbs W. Autonomic Computing. *Scientific American* May 6, 2002.
Grove AS. *Only the Paranoid Survive.* Random House, 1999.
Guralnick JM, et al. Aging in the Eighties: The Prevalence of Comorbidity and its Association with Disability. *Vital and Health Statistics* 170. Hyattsville, MD, National Center for Health Statistics, 1989.
Hay J, Forrest S, Goetghebeur M *Executive Summary: Hospital Costs in the U.S.,* Blue Cross Blue Shield Association, 2001.
Health Forum, Almanac of Policy Issues, Policy News Publishing http://www.policyalmanac.org/health/index.shtml, 2002.
Heffler S et al. Trends: Health Spending Projections for 2002–2012. *Health Affairs* W3:54–65, 2003.
Hospital Equipment Control Programs. *Health Devices* (1):75–93, 1971.
IBM. *Autonomic Computing: IBM Perspective on the State of Information Technology.* Armonk, NY, IBM, 2001.
Institute of Medicine. *Crossing the Quality Chasm: A New Health System for the 21st Century.*, Institute of Medicine, 2001.
Kohn LT, Corrigan JM, Donaldson MS. *To Err Is Human: Building a Safer Health System.* Washington, DC, National Academic Press, 2000.
Nader, R. Ralph Nader's Most Shocking Exposé. *Ladies Home Journal* 88:98ff, 176–179, 1971.
Public Broadcasting System. *Health care Crisis: Who's at Risk?* PBS, June 8, 2000. http://www.pbs.org/healthcarecrisis/healthinsurance.html.
Recovery Oriented Computing (ROC): Motivation, Definition, Techniques, and Case Studies. Berkeley, CA, UC-Berkeley, March 16, 2002.
Woolhandler S, Himmelstein DU. The Deteriorating Administrative Efficiency of the U.S. Health Care System. *N Engl J Med* 324:1253–1258, 1991.
Wu S, Green A. *Projection of Chronic Illness Prevalence and Cost Inflation.* CA, RAND Health, 2000.

Websites

RSNA. Radiological Society of North America. http://www.rsna.org/IHE/iheyr3_connectathon_2001.shtml
US Department of Commerce. Census Bureau Projections, U.S. Department of Commerce News, January 13, 2000. http://www.census.gov/Press-Release/www/2000/cb00-05.html
US Public Law. Health Insurance Accountability and Portability Act, US Public Law 104–191, August 21, 1996. http://aspe.hhs.gov/admnsimp/pl104191.htm
WHO. World Health Report 2000, World Health Organization, October 4, 2001. http://www.who.int/whr/2000/en/report.htm
This chapter was originally published in the March/April 2003 issue of IEEE Engineering in Biology in Medicine Magazine *and is reprinted here with permission.*

136

Virtual Instrumentation— Applications to Health Care

Eric Rosow
Director of Biomedical Engineering
Hartford Hospital
Hartford, CT

Successful organizations have the ability to measure and to act on key indicators and events in real time. By leveraging the power of virtual instrumentation and open-architecture standards, multidimensional executive dashboards can empower health care organizations to make better and faster data driven decisions. This chapter will highlight ways in which user-defined virtual instruments and dashboards can connect to hospital information systems (e.g., ADT systems and patient monitoring networks) and utilize statistical process control (SPC) to "visualize" information and to make timely, data driven decisions. The case studies described will illustrate enterprise-wide solutions for:

- Bed management and census control
- Operational management, data mining and business intelligence applications
- Clinical applications

Background

Virtual instrumentation (VI) allows organizations to effectively harness the power of the PC, to access, analyze, and share information throughout the organization. With vast amounts of data available from increasingly sophisticated enterprise-level data sources, potentially useful information is often left hidden, due to a lack of useful tools. Virtual instruments can employ a wide array of technologies, such as multidimensional analyses and statistical process control (SPC) tools, to detect patterns, trends, causalities, and discontinuities to derive knowledge and to make informed decisions.

Today's enterprises create vast amounts of raw data, and recent advances in storage technology; coupled with the desire to use these data competitively, has caused a data glut in many organizations. The health care industry, in particular, generates a tremendous amount of data. Tools such as databases and spreadsheets certainly help to manage and analyze these data, however databases, while ideal for extracting data, are generally not suited for graphing and analysis. Spreadsheets, on the other hand, are ideal for analyzing and graphing data, but this process can be cumbersome when working with multiple data files. Virtual instruments empower the user to leverage the best of both worlds by creating a suite of user-defined applications that allow the end-user to convert vast amounts of data into information that is ultimately transformed into knowledge to enable better, faster decision making.

Benefits of User-Defined Virtual Instruments

The benefits of virtual instrumentation are increased performance and reduced costs. Because the user controls the technology through software, the flexibility of virtual instrumentation is unmatched by traditional instrumentation. The modular, hierarchical programming environment of virtual instrumentation is inherently reusable and reconfigurable.

In effect, virtual instrumentation allows the user to "morph," (i.e., replicate and/or customize) the functionality of traditional "vendor-defined" instruments (such as the oscilloscope shown in Figure 136-1) into virtual "user-defined" instruments on a standard computer or personal digital assistant (PDA).

Virtual instrumentation applications have encompassed nearly every industry, including the telecommunications, automotive, semiconductor and biomedical industries. In the

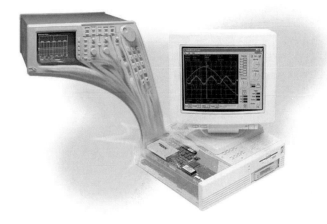

Figure 136-1 Virtual instrumentation: Replicating and enhancing traditional "vendor-defined" instruments with virtual "user-defined" instruments. (Courtesy of National Instruments.)

fields of health care and biomedical engineering, virtual instrumentation has empowered developers and end-users to conceive of, develop, and implement a wide variety of research-based biomedical applications and executive information tools. These applications fall into several categories, including process improvement, decision support, and clinical research. The following case studies are intended to illustrate the myriad of applications that are possible with this powerful technology.

The Lab VIEW Development Environment

For many of its VI solutions discussed in this chapter, an application development environment (ADE) called LabVIEW has been used. LabVIEW was created by National Instruments (Austin, TX). It is an off-the-shelf graphical development environment designed specifically for developing integrated measurement and automation systems. Developers assemble user interfaces and high-level functions for data acquisition and control, signal processing and analysis, and visualization, in the same way that flowcharts are constructed. With the modularity and hierarchical structure of LabVIEW, users can quickly and easily prototype, design, deploy, and modify systems. The LabVIEW ADE is compiled for maximum execution performance, contains hundreds of advanced analysis routines, and allows developers to quickly design and build advanced applications.

This chapter presents several "real-world" virtual instrument applications and tools that have been developed to meet the specific needs of health care organizations. Particular attention is placed on the use of quality control and "performance indicators," which provide the ability to trend and forecast various metrics and improve processes. The use of statistical process control (SPC) and modeling within virtual instruments is be demonstrated. Finally, several clinical research applications of virtual instrumentation are described.

Case Study #1: A Real-Time Bed-Management and Census-Control Dashboard

The Problem

Most health care institutions today manage patient flow via paper, whiteboards, and phone calls. As a result, hospitals often lack precise and timely information to match bed availability with patients' clinical needs. This causes inefficient use of beds, resources, and provider time, which leads ultimately to reduced throughput; emergency department (ED) overcrowding; lost admissions; operating room (OR) delays; unhappy physicians, staff, and patients; decreased revenue; and higher expenses.

Industry Trends

Concerns over patient access to hospital care are quickly becoming the largest, most urgent health care issue on the public's agenda. Many hospitals are now struggling to cope with surprising increases in the numbers of patients. Nationwide, hospitals report unprecedented inpatient and outpatient volumes, while average occupancy levels approaching 100% are not uncommon. These problems are compounded by two nationwide trends in hospitals: (1) increases in the numbers of patients; and (2) decreases in the numbers of beds.

Some hospitals are being forced to turn away ambulances and to cancel or delay elective surgeries because of a shortage of hospital beds. At other hospitals, ED patients wait hours, or even days, for rooms. Administration has no information to be able to match proactively the demand for patients with the appropriate level of nurse staffing. In addition, aging baby boomers are more likely to experience serious illness or injury, and medical advances are helping doctors to treat conditions that patients simply might have accepted in the past. There is an overall staff shortage of hospital workers, especially among nurses. With no sign of slowdown in these trends, these problems in hospital care appear set to escalate for the foreseeable future (AHA, 2001).

The Solution

The Bed Management Dashboard (BMD) is a real-time process improvement and decision support product used by hospital administrators, clinicians, and managers on a constant basis. It is an enterprise-wide system that directly and indirectly interacts with all departments throughout the continuum of care (see Figure 136-2).

Process Improvement

Process improvement is an outcome and benefit of a centralized Bed Management Process that maintains clinical input and incorporates technological support. The BMD streamlines the process of admitting, transferring, and discharging patients by:

- Optimizing patient placement processes
- Increasing staff efficiency (e.g., by reducing administrative expenses by eliminating paperwork and phone volume)
- Improving utilization of beds (e.g., by reducing inappropriate usage of inpatient beds without payer authorization, 23+ hour observation of outpatients by automatic alerts)
- Optimizing utilization of material resources (e.g., monitored beds and negative pressure rooms)
- Managing patient flow in the event of Admissions/Discharge/Transfer (ADT) system or network failure (disaster recovery)
- Enabling real-time and predictive capacity management
- Providing clinical attributes to supplement ADT data (e.g., telemetry and near nursing stations)

This results in:

- Reduced emergency room overcrowding
- Reduced operating room delays
- Increased physician and staff satisfaction
- Improved patient flow
- Time savings
- Better treatment of patients and improved customer satisfaction
- Reduced diversions and lost admissions
- Increased revenues and decreased expenses
- Reduced staff

Decision Support

The BMD allows administrators, clinicians, and managers to easily access, analyze, and display real-time patient and bed-availability information. It provides:

- On-demand historical, real-time, and predictive reports
- Alerts, warnings, and recommendations
- The ability to share information throughout the enterprise
- Provides point-in-time predicted bed availability
- The ability for decision-makers to move easily from "big-picture" analyses to transaction details

This results in:

- Timely data driven decisions
- Improved decisions regarding staffing levels
- Reduce reliance on third-party staffing companies to supplement in-house staff
- Improved crisis management during rapidly fluctuating census levels
- The improved ability to negotiate contracts with payers

The BMD has been running at Hartford Hospital since April 2001. It has over 900 trained users, and it can be accessed on the more than 2500 work stations throughout the organization. The system is interfaced to the Hospital's Admission/Discharge/Transfer (ADT) Information System (Siemens/SMS), pagers, phones, and e-mail.

An Air Traffic Control Tower for Beds

In many ways, the BMD is similar to an air traffic control tower. Like a real air traffic control tower, this application is real-time and mission-critical. It must handle scheduled and

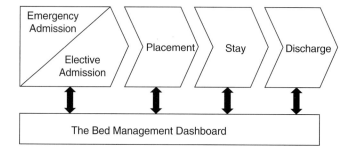

Figure 136-2 The Bed Management Dashboard across the Enterprise: The BMD interacts with all departments throughout the continuum of care (e.g., admitting, ED, OR, ICU, and administration).

emergency events. The system assists with the clinical and business decision processes that occur when a patient needs to be assigned to a specific bed location. Collectively, this system provides organizations with an array of enabling technologies to:

- Schedule/Reserve/Request patient bed assignments
- Assign and transfer patients from the emergency department and/or other clinical areas such as intensive care units, medical/surgical units, operating rooms, and post-anesthesia care units
- Reduce and eliminate dependency on phone calls to communicate patient and bed requirements
- Reduce and eliminate paper processes to manage varying census levels
- Apply statistical process control (SPC) and "Six Sigma" methodologies to manage occupancy and patient diversion
- Provide administrators, managers, and caregivers with accurate and on-demand reports and automatic alerts via pagers, e-mail, phone, and intelligent-software agents

How It Works

The Bed Management Dashboard is accessible via a web browser or via a client application. The supporting architecture of the BMD system is a standard N-tier server-based system, which, depending on the end users' needs, consists of one or more of the following: A web server, an application server, a database server, and an ADT interface server.

Key information from the hospital's ADT system is automatically fed to the BMD via a Health Level Seven (HL7) data stream. The system has been designed to accept inputs from patient monitoring and nurse-call networks. Figure 136-3 illustrates the flow of information from the ADT system to the user's desktop. The system typically receives up to 12,000 transaction messages per day, which are parsed into appropriate data elements by an HL7 parser and are stored on a database server.

The log-in authentication module accesses a hospital's central user log-in repository to validate the user's log-in information. This module enables single-user sign-on (SSO) capability by providing a user with the same username and password used by the other applications that authenticate to the central user log-in repository.

The integrated data-mining and report module consists of a number of standard reports that take advantage of the data warehouse nature of the BMD. These reports include, but are not limited to, a historical census report, a real-time census report, a bed manager report, a physician discharge report, and a discharge compliance report. In addition, access to the data by industry standard report writers, such as Crystal Reports and Access, give technical users the ability to create complicated or special reports as desired. For example, the BMD offers extensive ad hoc reporting capabilities ranging from length of stay (LOS) and "Care Day" metrics to asset management analyses (on beds and patient-monitor utilization), to biosurveillance reports that have been requested by state and federal organizations such as the Department of Public Health and the Centers for Disease Control and Prevention.

The utility and configuration module gives system operators the ability to administer BMD users (e.g., to add new users and to modify new or existing user security settings); to administer service, unit, rooms and beds (e.g., to add or modify clinical services, units, rooms, and beds, and their inter-relationships); to define automated alert thresholds; and to configure unit floor plan diagrams.

The embedded back-up utility module consists of a local version of the bed-management database that is constantly updated. Access to this database gives users a self-contained and mobile version of the system that can be used in the event of catastrophic failure of the system hardware or network hardware or in the event of a crisis that removes the users from direct access to the hospital network.

One of the most important features of the BMD is that it reformats information from the ADT system and presents it to the clinical user in a more "user-friendly" and process-oriented manner. Dynamic and interactive graphical presentations of data are used extensively. Figure 136-4 illustrates the way in which all of the patients from a given admitting source, such as the emergency department, can be displayed and selected from a dynamically sortable "smart table."

Hospital beds are classified as having predefined "attributes," such as being "monitored" or being assigned to the "surgery" service. The needs of patients are similarly described with attributes, such as "monitor required" or "scheduled for surgery." As illustrated in Figure 136-5, the BMD helps to find those available beds in the hospital that meet the specific needs of a patient by guiding the clinical staff through a set of process screens that perform the match.

Decisions for patient placement can be centralized or decentralized. The dashboard allows proper communication between the appropriate parties. Status of decisions is automatically tracked, and a monitoring process can detect and notify key stakeholders of any process delays. Admitting or emergency departments can be automatically notified of decisions, if appropriate. Reporting of information is provided by online screen views of data tailored to the needs of a particular class of system user. Unit personnel can view either detailed information or summary roll-ups about their patients. Figure 136-6 illustrates ways in which patient information can be viewed in a dynamic and interactive floor-plan mode.

Administrators and program directors can view data over a wider scope that encompasses multiple units, services, or physicians. An example of a summary report is shown in Figure 136-7.

A key feature of this system is its use of "Intelligent Agents." These online agents, as shown in Figure 136-8, are constantly monitoring and analyzing patient and census information, and they have the ability to detect key system situations, such as high census in a unit (i.e., no available beds), excessive ED placement time for a particular patient, or delays in responses to placement requests.

The BMD allows users to run real-time queries and reports on current and future hospital census (Figure 136-9). These reports are stratified by inpatients, outpatients, acute units, non-acute units, and/or services ("collaborative management teams"). In addition, budgeted discharge, length-of-stay, and care day statistics, along with average and peak occupancy indicators, are also profiled.

Patient Confidentiality

A full-security system is embedded within the dashboard to audit user access and to assign users to definable system roles. These roles are restricted to specified processes and are prohibited from viewing or changing certain data. The system is designed to be fully compliant with the evolving Health Insurance Portability and Accountability Act (HIPAA) regulations.

The BMD at Hartford Hospital on September 11, 2001

The dashboard played a critical role at Hartford Hospital on that terrible day of September 11, 2001. When the news came in from New York (just 100 miles away) Hartford Hospital was told to expect hundreds of injured patients. At that time, the hospital had only seven open beds, two of which were ICU beds. The BMD played an important role with respect to providing real-time occupancy information, predictive capacity management, and biosurveillance reporting.

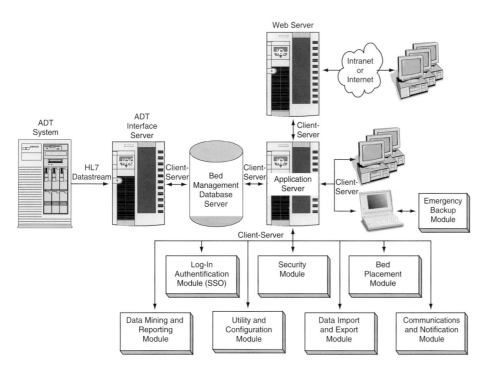

Figure 136-3 System diagram: The BMD consists of four major components: the ADT interface, the database server, an application server and a web server. The system has been designed to interface with any ADT system that provides an HL7 interface and utilizes TCP/IP as a communication protocol.

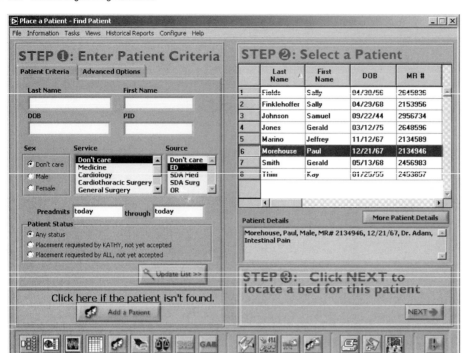

Figure 136-4 Find Patient: This screen is primarily used to request a bed for a patient. The user first enters various criteria to identify the patient. The system then displays all of the patients who meet the specified criteria. Finally, the user selects the patient in question and presses "NEXT" to move to another screen, where an available bed is located and requested.

Specifically, the dashboard helped in the following ways:

- It improved efficiency by eliminating the need for whiteboards and many phone calls
- It allowed for predictive occupancy reports to help free up 150 beds in a timely and orderly fashion
- It provided an efficient way to collect and report biosurveillance data that were required by the Connecticut Department of Public Health

The key factor in this success story was that the dashboard was directly linked to the hospital census database so that accurate data were constantly available to the command center. The dashboard displays were easily, rapidly, and repeatedly revised to facilitate iterative decision making.

Demonstrated Return on Investment (ROI)

Since its installation in April 2001, the system has more than paid for itself through achievement of the following indicators obtained from actual observations at Hartford Hospital.

- $200,000 annual expense avoidance by reducing the number of Bed Manager FTEs needed
- A 50% decrease in the number of phone calls per bed assignment/patient placement
- A 75%–90% decrease in the overall time needed for the patient-placement process
- A 50% decrease in the number of paper forms used
- A 25% increase in throughput by alerting bed managers to discharges sooner of those that are reported to bed managers

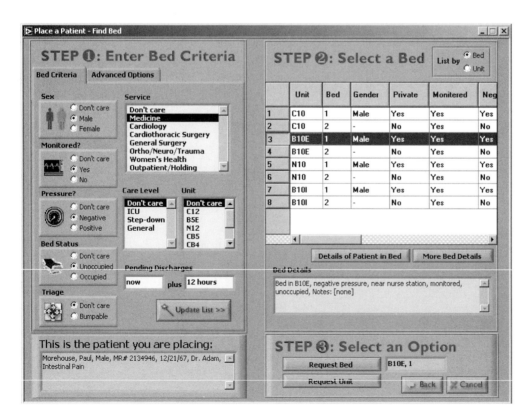

Figure 136-5 Find Bed: Once a patient has been selected, this "Bed Finder" screen is used to locate a specific bed or unit for the patient. The user first enters various criteria about the type of bed that is needed (e.g., patient gender, monitor required, and negative-pressure room required). The system then displays all of the available beds that meet the specified criteria. Finally, the user selects a particular bed or unit for the previously specified patient.

Virtual Instrumentation—Applications to Health Care 631

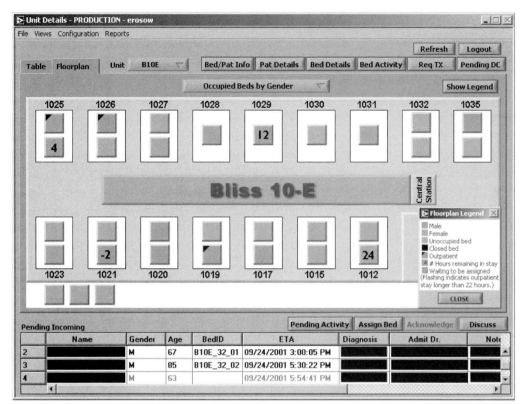

Figure 136-6 Unit Details (floor-plan view): This screen provides a graphical view of an intensive care unit. Each bed is presented as a square, using a simplified floor-plan view. Colors are used to indicate the selected attribute of the patient or bed. For example, the above screen indicates available beds in green, and occupied beds in red. The gray beds (which are flashing) are those with pending discharges, and the black bed is closed or inactive. Many other color-coded options are available via the pull-down selector. These include patient and bed attributes such as gender, monitored bed, negative-pressure room, and service (e.g., cardiology, surgery, and orthopedics).

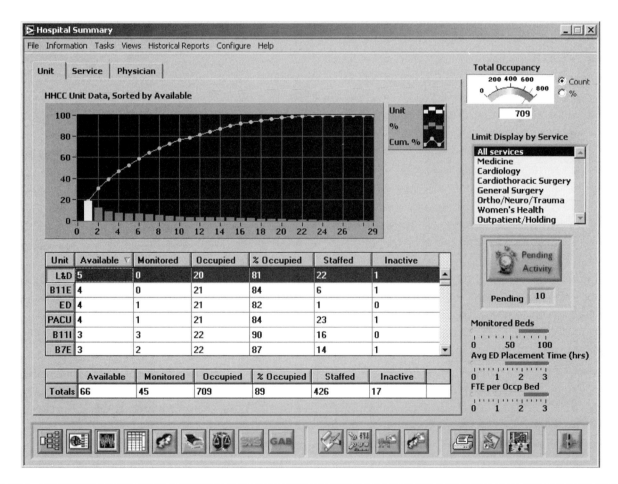

Figure 136-7 Hospital Summary: This screen is primarily used by administrators who need to see a global view of the hospital status. Table and Pareto charts profile the various units and the current status summaries of each unit. These patients also can be rolled up into services, or grouped by physician. In addition, patients can be aggregated in many other ways, such as time of admission, length of stay, and admitting diagnosis.

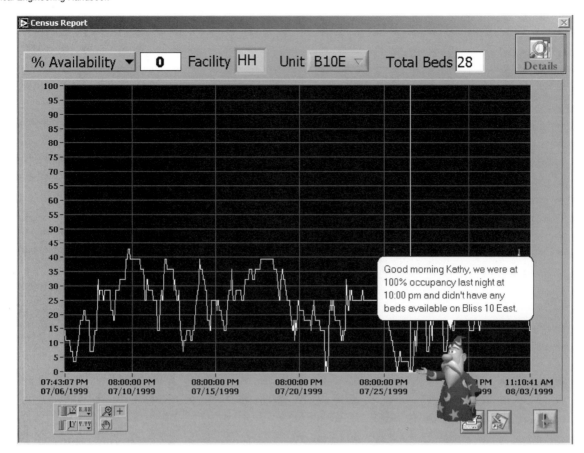

Figure 136-8 Intelligent Agents: The BMD can also employ online "intelligent agents" to provide assistance and to alert the user of important alarm conditions that otherwise might have gone unnoticed. Messages can be in the form of an on-screen agent (such as Merlin the Wizard) or via e-mail, pager, fax, and/or telephone.

	Inpatients					Care Days							Occupancy						
	Discharges			Census LOS		Inpatient		Outpatient		Total			Census				% of Avail		
Unit	Budget	Actual	Disch LOS	Budget	Actual	Budget	Actual	Budget	Actual	Budget	Actual	Prior	Average Midnight	Peak	Max	Avg Avail	Average Midnight	Peak	Peak Time
Acute Care																			
B10E	35	37	5.1	5.5	4.7	190	173	9	9	199	182		18.56	27	28	28	66.3	96.4	Jul 2 2002 6:00AM
B10I	4	1	16.6	0.0	93.0	82	93	0	0	82	93		10.00	12	12	12	83.3	100.0	Jul 1 2002 1:00PM
B11E	32	28	6.3	5.8	6.9	185	194	0	0	185	194		21.67	24	24	24	90.3	100.0	Jul 7 2002 11:00AM
B11I	5	3	58.8	0.0	34.7	84	104	0	0	84	104		11.33	12	12	12	94.4	100.0	Jul 1 2002 5:00AM
B11S	1	0	0.0	0.0	0.0	13	25	0	0	13	25		2.67	3	3	3	88.9	100.0	Jul 1 2002 3:00AM
B5E	20	23	10.1	7.6	8.3	150	191	1	3	151	194		21.22	23	42	22	96.5	104.5	Jul 1 2002 11:00AM
B7E	27	31	6.2	6.4	5.1	167	158	0	1	167	159		17.78	22	24	24	74.1	91.7	Jul 3 2002 9:00AM
B7I	3	1	49.2	0.0	96.0	92	96	0	0	92	96		10.67	12	12	12	88.9	100.0	Jul 1 2002 12:00AM
B7S	1	0	0.0	0.0	0.0	24	24	0	0	24	24		2.56	3	3	3	85.2	100.0	Jul 1 2002 5:00PM
B8	72	58	6.4	4.3	5.4	307	311	2	3	309	314		34.44	42	42	42	82.0	100.0	Jul 2 2002 12:00AM
B9E	30	31	9.2	5.9	4.8	172	150	0	3	172	153		17.22	21	24	24	71.8	87.5	Jul 1 2002 1:00PM
B9I	2	3	11.2	0.0	27.3	85	82	0	0	85	82		8.78	12	12	12	73.1	100.0	Jul 8 2002 11:00AM
B9S	0	1	26.1	0.0	21.0	0	21	0	0	0	21		2.44	3	3	3	81.5	100.0	Jul 1 2002 12:00AM
C10	27	26	5.5	6.7	6.4	181	167	2	4	183	171		19.11	25	26	26	73.5	96.2	Jul 2 2002 2:00AM
C12	30	34	5.5	6.6	6.1	197	209	1	0	198	209		23.22	26	26	26	89.3	100.0	Jul 2 2002 5:00AM
C9WI	3	3	6.3	0.0	30.0	70	90	1	0	71	90		10.11	11	10	10	101.1	110.0	Jul 7 2002 12:00AM
CB4	14	18	17.3	5.0	5.4	68	98	1	2	69	100		11.33	14	15	15	75.6	93.3	Jul 4 2002 2:00AM
CB5	26	20	11.5	6.6	10.3	171	205	12	6	183	211		23.00	27	26	26	88.5	103.8	Jul 1 2002 12:00PM
CB6	69	59	3.5	3.3	3.9	225	228	4	6	229	234		25.56	34	42	42	60.8	81.0	Jul 4 2002 8:00AM
N10	45	59	2.6	3.1	2.8	136	168	34	53	170	221		25.44	34	26	26	97.9	130.8	Jul 2 2002 8:00AM
N11	24	29	5.9	7.5	6.6	178	192	2	4	180	196		21.33	26	26	26	82.1	100.0	Jul 3 2002 4:00PM
N12	29	28	7.1	6.3	7.3	179	203	1	0	180	203		22.33	24	26	25	89.3	96.0	Jul 1 2002 9:00PM
N9	54	32	5.4	4.4	7.5	236	239	5	10	241	249		28.00	31	30	30	93.3	103.3	Jul 1 2002 12:00AM
Total Acute Care																			
	554	525	6.7	5.0	6.4	3,222	3,421	75	104	3,297	3,525		388.78	-	502	473	82.2	-	-
Non-Acute Care																			
B6M	100	86	3.0	2.2	2.7	219	235	1	0	220	235		25.33	37	38	38	66.7	97.4	Jul 6 2002 11:00AM
B6N	103	82	2.5	2.5	2.9	256	236	0	0	256	236		24.56	35	40	40	61.4	87.5	Jul 6 2002 11:00AM
CB5R	5	4	11.6	14.3	21.0	71	84	0	0	71	84		9.11	10	20	10	91.1	100.0	Jul 3 2002 7:00PM
D1N	13	13	13.1	14.0	16.5	176	214	0	0	176	214		24.00	27	30	30	80.0	90.0	Jul 1 2002 6:00PM
D1S	0	0	0.0	0.0	0.0	0	0	0	0	0	0		0.00	0	28	0	0.0	0.0	(no patients)
D2N	9	10	17.7	15.3	16.1	133	161	0	0	133	161		18.22	20	26	26	70.1	76.9	Jul 1 2002 12:00AM
D2S	21	17	15.6	10.5	14.2	220	242	0	0	220	242		27.11	29	30	30	90.4	96.7	Jul 1 2002 11:00AM
D3N	33	25	7.6	6.6	8.2	214	205	1	0	215	205		22.67	26	30	30	75.6	86.7	Jul 7 2002 10:00AM
D3S	20	26	5.2	7.1	7.5	141	194	1	0	142	194		21.44	24	32	0	0.0	0.0	Jul 1 2002 1:00PM
LDP6	3	0	0.0	0.0	0.0	47	0	5	0	52	0		0.00	0	5	0	0.0	0.0	(no patients)
LDR6	0	3	0.1	0.0	21.7	0	65	0	63	0	128		8.56	15	16	16	53.5	93.8	Jul 3 2002 11:00PM
N8	48	27	2.0	3.0	5.0	142	135	13	24	155	159		15.89	25	32	23	69.1	108.7	Jul 3 2002 6:00AM
N8N	1	0	0.0	1.0	0.0	1	0	0	0	1	0		0.00	0	12	0	0.0	0.0	(no patients)
Total Non-Acute Care																			
	356	293	5.1	4.5	6.0	1,620	1,771	21	87	1,641	1,858		196.89	-	339	243	81.0	-	-
Grand Total - Acute/Non-Acute Care																			

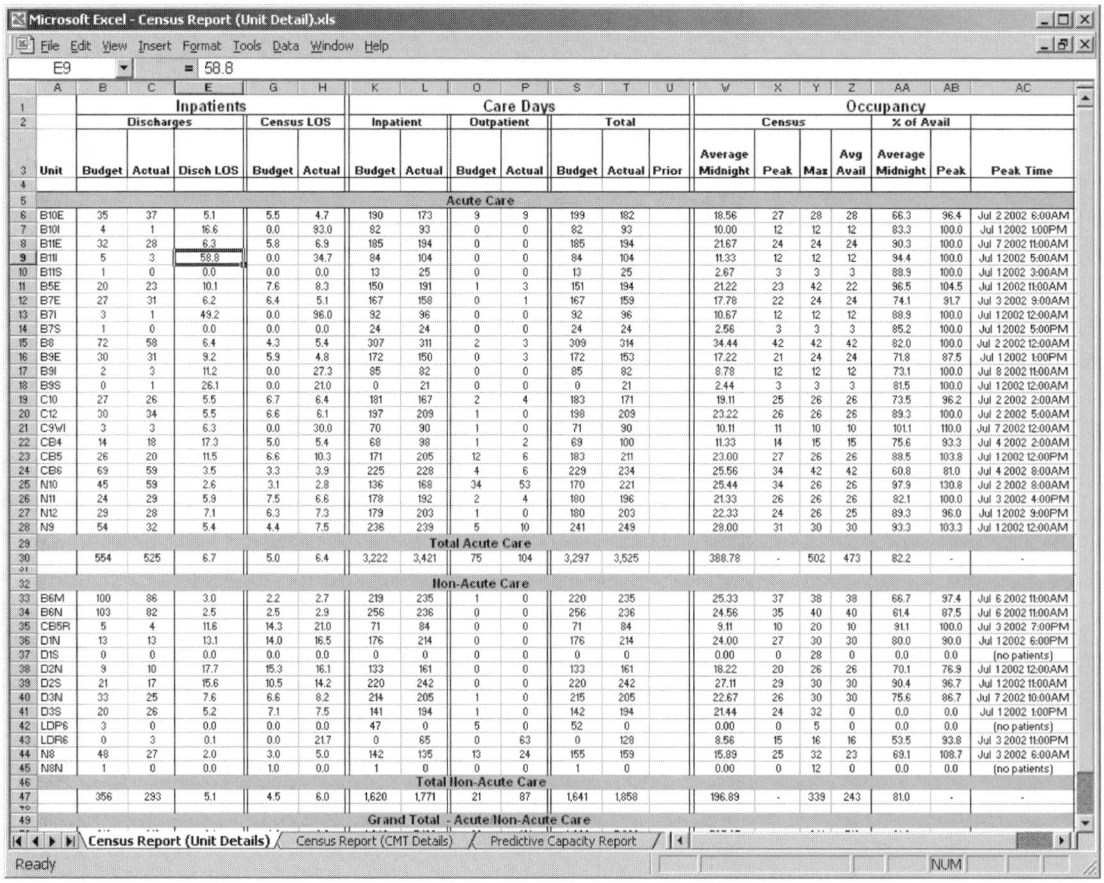

Figure 136-9 Detailed census and budget report.

- A 40-minute decrease in the length of the end-of-shift admission coordinator reports
- The hospital expects to save "thousands of dollars per day" by implementing the BMD outpatient alert system. (The specific dollars saved are still to be determined.)

Case Study #2: PIVIT™-Performance Indicator Virtual Instrument Toolkit

Most of the information management examples presented in the chapter are part of an application suite called PIVIT.™ The name is an acronym for "Performance Indicator Virtual Instrument Toolkit." PIVIT is an easy-to-use data acquisition and analysis product. PIVIT was developed specifically in response to the wide array of information and analysis needs throughout the health care setting. Figure 136-10 illustrates the main menu of PIVIT.

PIVIT applies virtual instrument technology to assess, analyze, and forecast clinical, operational, and financial performance indicators. Some examples include applications that profile institutional indicators (e.g., patient days, discharges, percentage of occupancy, ALOS, revenues, and expenses) and departmental indicators (e.g., salary, non-salary, total expenses, expense per equivalent discharge, and DRGs). Other applications of PIVIT include 360-degree peer review, customer satisfaction profiling, and medical equipment risk assessment.

PIVIT can access data from multiple data sources. Virtually any parameter can be easily accessed and displayed from standard spreadsheet and database applications (e.g., Microsoft Access, Excel, Sybase, and Oracle), using Microsoft's Open Database Connectivity (ODBC) technology. Furthermore, multiple parameters can be profiled and compared in real time to any other parameter via interactive polar plots and three-dimensional displays. In addition to real-time profiling, other analyses, such as statistical process control, can be employed to view large data sets in a graphical format. Statistical process control has been applied successfully for decades to help companies reduce variability in manufacturing processes. These SPC tools range from Pareto graphs to Run and Control charts. Although it will not be possible to describe all of these applications, several examples are provided below to illustrate the power of PIVIT.

Trending, Relationships, and Interactive Alarms

Figure 136-11 illustrates a virtual instrument that interactively accesses institutional and department-specific indicators and profiles them for comparison. Data sets can be acquired directly from standard spreadsheet and database applications (e.g., Microsoft Access, Excel, Sybase, and Oracle). This capability has proved to be quite valuable with respect to quickly accessing and viewing large sets of data. Typically, multiple data sets contained within a spreadsheet or database had to be selected, and then a new chart of these data had to be created. Using PIVIT, the user simply selects the desired parameter from any one of the pull-down menus, and this data set is instantly graphed and compared to any other data set.

Interactive "threshold cursors" dynamically highlight when a parameter is over and/or under a specific target. Displayed parameters can also be ratios of any measured value (e.g., "Expense per Equivalent Discharge" or "Revenue to Expense Ratio"). The indicator color will change, based on the degree to which the data value exceeds the threshold value (e.g., from green to yellow to red). If multiple thresholds are exceeded, then the entire background of the screen (normally gray) will change to red, to alert the user of an extreme condition.

Finally, multimedia has been employed by PIVIT to alert designated personnel an audio message from the personal computer or by sending an automated message via e-mail, fax, pager, or mobile phone.

PIVIT also is able to profile historical trends and to project future values. Forecasts can be based on user-defined history (e.g., "Months for Regression"), the type of regression (linear, exponential, or polynomial), the number of days, months, or years to forecast, and whether any offset should be applied to the forecast. These features allow the user to create an unlimited number of "what-if" scenarios and to allow only the desired range of data to be applied to a forecast. In addition to the graphical display of data values, historical and projected tables are also provided. These embedded tables look and function like a standard spreadsheet.

Data Modeling

Figure 136-12 illustrates another way in which virtual instrumentation can be applied to financial modeling and forecasting. This example graphically profiles the annual morbidity, mortality, and cost associated with falls within the state of Connecticut. Such an instrument has proved to be an extremely effective modeling tool due to its ability to highlight relationships and assumptions interactively, and to project the cost and/or savings of employing educational and other interventional programs.

Virtual instruments such as these are not only useful with respect to modeling and forecasting but, perhaps more importantly, they become a "knowledge base" in which interventions and the efficacy of these interventions can be statistically proved.

The example program in Figure 136-13 shows ways in which virtual instrumentation can employ standard Microsoft Windows® technology (in this case, Dynamic Data Exchange (DDE)) to transfer data to commonly used software applications such as Microsoft Access® or Microsoft Excel.® It is interesting to note that in this example, the virtual instrument can measure and graph multiple signals while sending these data to another application that could reside on the network or across the Internet.

In addition to utilizing DDE, virtual instrumentation can use other protocols for inter-application communication. These range from simple serial communication to TCP/IP and ActiveX.

Figure 136-14 illustrates the Communications Center. This application shows various methods by which the user can communicate information throughout an organization. The Communications Center can be used to simply create and print a report, or it can be used to send e-mail, faxes, or messages to a pager, or even to leave voicemail messages. This is a powerful feature in that information can be distributed easily and efficiently to individuals and groups in real time.

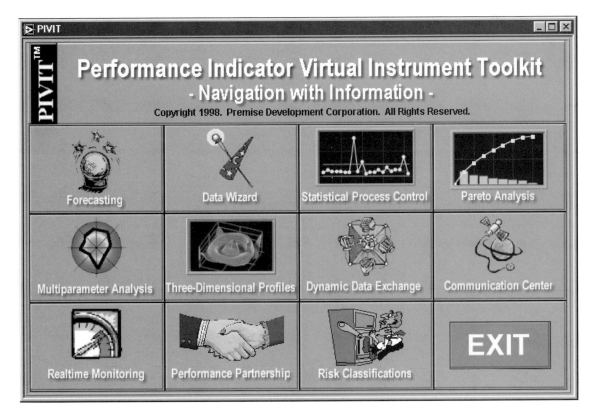

Figure 136-10 The Performance Indicator Virtual Instrument Toolkit (PIVIT™) Main Menu.

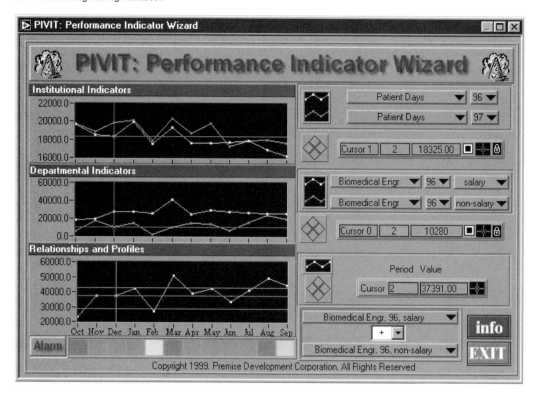

Figure 136-11 PIVIT—Performance Indicator Wizard Display Institutional and Departmental Indicators.

Additionally, Microsoft Agent technology can be used to pop up an animated help tool ("Merlin the Wizard"). Merlin then can be used to communicate a message or to indicate an alarm condition, or it can be used to help the user to solve a problem or to point out a discrepancy that otherwise might have gone unnoticed. Agents employ a "text-to-speech" algorithm to actually "speak" an analysis or alarm directly to the user or the recipient of the message. In this way, on-line help and user support can be provided in multiple languages.

In addition to real-time profiling of various parameters, more advanced analyses, such as statistical process control (SPC) can be employed to view large data sets in a graphical format.

SPC has enormous applications throughout health care. Such perceived applications were significant drivers to embed SPC tools into PIVIT's suite of virtual instruments. For example, Figure 136-15 is an example of a way in which Pareto analysis can be applied to a sample trauma database of over 12,000 records. The Pareto chart could show frequency or

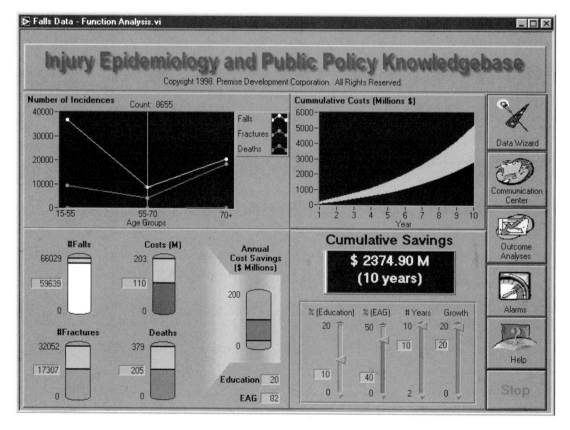

Figure 136-12 Injury Epidemiology and Public Policy Knowledgebase.

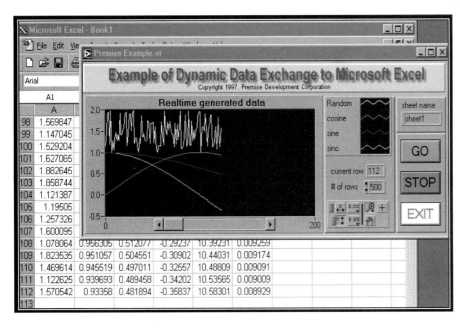

Figure 136-13 Example of Dynamic Data Exchange (DDE) for Interapplication Communication.

percentage, depending on front-panel selection, and the user can select from a variety of different parameters by clicking on the "pull-down" menu. This menu can be configured to automatically display each database field directly from the database. In this example, various database fields (e.g., DRG, Principal Diagnosis, Town, and Payer), can be selected for Pareto analysis. Other tools include run charts, control charts, and process capability distributions.

Medical Equipment Risk Criteria

Figure 136-17 illustrates a virtual instrument application that demonstrates the way that four "static" risk categories (and their corresponding values) are used to determine the inclusion of clinical equipment in the Medical Equipment Management Program at Hartford Hospital. Each risk category includes specific subcategories that are assigned points, which, when added together according to the formula listed below, yield a total score that will range from 4 to 25.

Considering these scores, the equipment is categorized into five priority levels (i.e., High, Medium, Low, Grey List, and Non-Inclusion into the Medical Equipment Management Program). The four static risk categories are Equipment Function (EF), Physical Risk (PR), Environmental Use Classification (EC), and Preventive Maintenance Requirements (MR).

Equipment Function (EF)

Stratifies the various functional categories (i.e., therapeutic, diagnostic, analytical and miscellaneous, of equipment). The specific rankings for this category are listed in Table 136-1.

Physical Risk (PR)

Lists the "worst-case scenario" of physical-risk potential to either the patient or the operator of the equipment.

Environmental Use Classification (EC)

Lists the primary equipment area in which the equipment is used.

Preventive Maintenance Requirements (MR)

Describes the level and frequency of required maintenance. The Aggregate Static Risk Score is calculated as follows:

Aggregate Risk Score = EF + PR + EC + MR

Using the criteria's system described above, clinical equipment is categorized according to the following priority of testing and degree of risk:

High Risk

Equipment that scores 18–25 points on the criteria's evaluation system. This equipment is assigned the highest risk for testing, calibration, and repair.

Figure 136-14 PIVIT's Communications Center.

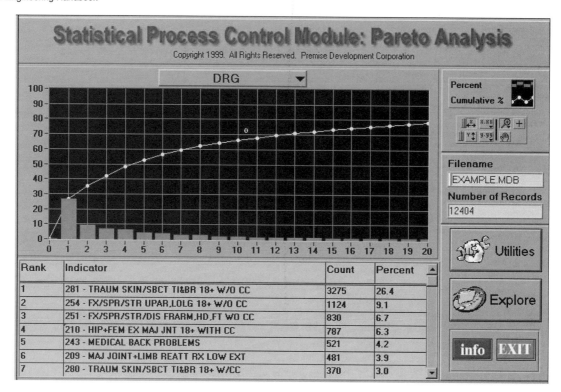

Figure 136-15 Statistical Process Control–Pareto Analysis of a sample trauma registry.

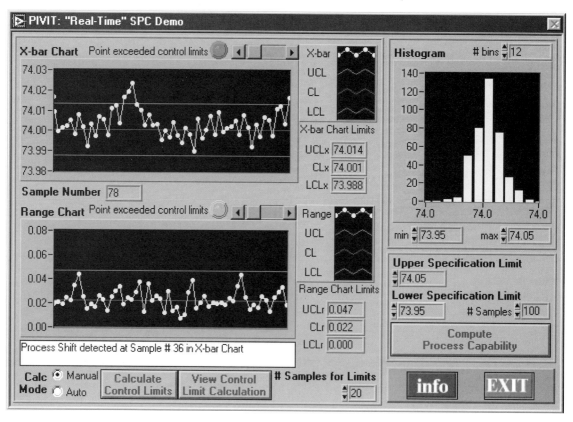

Figure 136-16 Statistical Process Control – "Real Time" SPC Application.

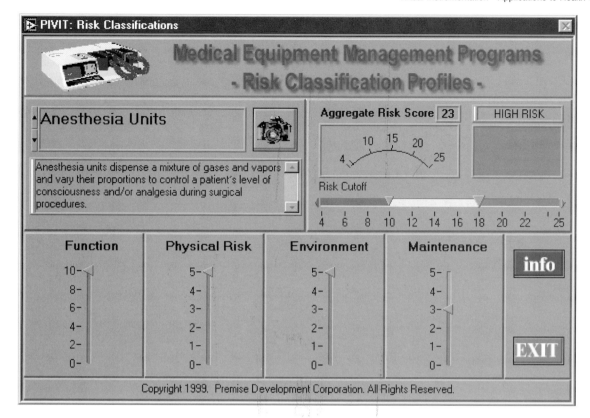

Figure 136-17 Medical Equipment Risk Classification Profiler.

Table 136-1 Equipment Function Ranking

	Risk Category I: Equipment Function (EF)
Point Score	Function Description
10	Therapeutic -Life Support
9	Therapeutic -Surgical or Intensive Care
8	Therapeutic -Physical Therapy or Treatment
7	Diagnostic -Surgical or Intensive Care Monitoring
6	Diagnostic -Other physiological monitoring
5	Analytical -Laboratory analytical
4	Analytical -Laboratory Accessories
3	Analytical -Computer and related
2	Miscellaneous -Patient-related
1	Miscellaneous -Non-patient related

Table 136-2 Physical Risk Ranking

	Risk Category II: Physical Risk (PR)
Point Score	Description of Use Risk
5	Potential patient death
4	Potential patient injury
3	Inappropriate therapy or mis-diagnosis
2	Equipment damage
1	No significant identified risk

Medium Risk

Equipment that scores 15–17 points on the criteria's evaluation system.

Low Risk

Equipment that scores 12–14 points on the criteria's evaluation system.

Hazard Surveillance (Gray)

Equipment that scores 6–11 points on the criteria's evaluation system are visually inspected on an annual basis during the hospital hazard-surveillance rounds.

Table 136-3 Environmental Use Classification Ranking

	Risk Category IV: Environmental Use Classification (EC)
Point Score	Primary Area of Equipment Use
5	Anesthetizing Locations
4	Critical Care Areas
3	Wet Locations/Labs/Exam Areas
2	General Patient Care Areas
1	Non-Patient Care Areas

Table 136-4 Preventive Maintenance Ranking

	Risk Category III: Preventive Maintenance (MR)
Point Score	PM Frequency
5	Monthly
4	Quarterly
3	Semi-annually
2	Annually
1	Not required

Medical Equipment Management Program Deletion

Medical equipment and devices that pose little risk and scores less than 6 points can be deleted from the management program as well as from the clinical equipment inventory.

Future versions of this application will also consider "dynamic" risk factors such as user errors, mean-time-between failure (MTBF), device failure within 30 days of a preventive maintenance or repair, and the number of years beyond the American Hospital Association's recommended useful life.

Peer Performance Reviews

The virtual instrument shown in Figure 136-18 has been designed to easily acquire and compile performance information with respect to institution-wide competencies. It has been created to allow every member of a team or department to participate in the evaluation of a co-worker (360-degree peer review).

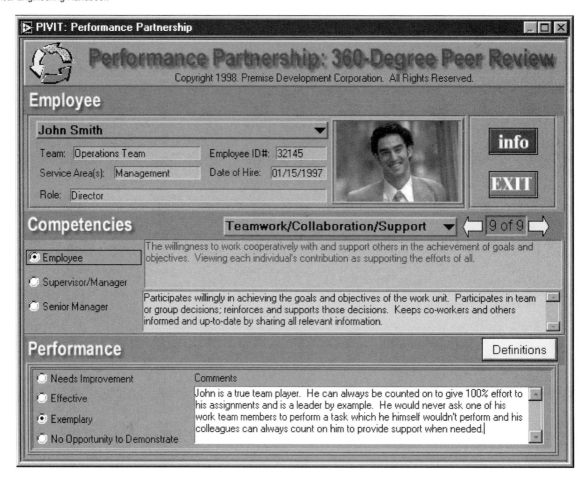

Figure 136-18 Performance reviews using virtual instrumentation.

Upon running the application, the user is presented with a "sign-in" screen where he or she enters a username and password. The application is divided into three components. The first (top section) profiles the employee and the relevant service information. The second (middle section) indicates each competency as defined for employees, managers, and senior managers. The last (bottom section) allows the reviewer to evaluate performance by selecting one of four "radio buttons" and also to provide specific comments related to each competency. This information is then compiled (with other reviewers) as real-time feedback.

Case Study #3: BioBench—A Virtual Instrument Application for Data Acquisition and Analysis of Physiological Signals

The biomedical industry relies heavily on the ability to acquire, analyze, and display large quantities of data. Whether researching disease mechanisms and treatments by monitoring and storing physiological signals, researching the effects of various drugs interactions, or teaching students in labs where they study physiological signs and symptoms, it was clear that there existed a strong demand for a flexible, easy-to-use, and cost-effective tool. In a collaborative approach, biomedical engineers, software engineers and clinicians and researcher created a suite of virtual instruments called BioBench.™

BioBench™ (National Instruments, Austin, TX) is a new software application designed for physiological data acquisition and analysis. It was built with LabVIEW™, the world's leading software development environment for data acquisition, analysis, and presentation.* Coupled with National Instruments data-acquisition (DAQ) boards, BioBench integrates the PC with data acquisition for the life sciences market.

Many biologists and physiologists have made major investments over time in data-acquisition hardware built before the advent of modern PCs. While these scientists cannot afford to throw out their investment in this equipment, they recognize that computers and the concept of virtual instrumentation yield tremendous benefits in terms of data analysis, storage, and presentation. In many cases, traditional medical instrumentation is too expensive to acquire and/or maintain. As a result, researchers and scientists are opting to create their own PC-based data-monitoring systems in the form of virtual instruments.

Other life scientists, who are just beginning to assemble laboratory equipment, face the daunting task of selecting hardware and software needed for their application. Many manufacturers for the life sciences field focus their efforts on the acquisition of raw signals and converting them into measurable linear voltages. They do not concentrate on digitizing signals or the analysis and display of data on the PC. BioBench™ is a low-cost, turnkey package that requires no programming. BioBench is compatible with any isolation amplifier or monitoring instrument that provides an analog output signal. The user can acquire and analyze data immediately because BioBench automatically recognizes and controls the National Instruments DAQ hardware, thus minimizing configuration headaches.

Some of the advantages of PC-Based Data Monitoring include the following:

- Easy-to-use software applications
- Large memory and the PCI bus
- Powerful processing capabilities
- Simplified customization and development
- More data storage and faster data transfer
- More efficient data analysis.

Figure 136-19 illustrates a typical setup of a data acquisition experiment using BioBench. BioBench also features pull-down menus through which the user can configure devices. Therefore, those who have made large capital investments can easily migrate their existing equipment into the Computer Age. Integrating a combination of old and new physiological instruments from a variety of manufacturers is an important and straightforward procedure. In fact, within the clinical and research settings, it is a common requirement to be able to acquire multiple physiological signals from a variety of medical devices and instruments that do not necessarily communicate with each other. Often, this situation is compounded by the fact that end-users would like to be able to view and analyze an entire waveform and not just an average value. In order to accomplish this, the end-user must acquire multiple channels of data at a relatively high sampling rate and have the ability to manage many large data files. BioBench can collect up to 16 channels simultaneously at a sampling rate of 1000 Hz per channel. Files are stored in an efficient binary format that significantly reduces the amount of hard disk and memory requirements of the PC. During data acquisition, a number of features are available to the end-user. These features include Data Logging, Event Logging, and Alarming.

Data Logging

Logging can be enabled prior to, or during, an acquisition. The application will either prompt the user for a descriptive filename or it can be configured to automatically

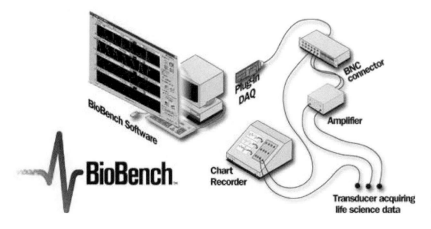

Figure 136-19 A typical biomedical application using BioBench.

assign a filename for each acquisition. Turning the data-logging option on and off creates a log data-event record that can be inspected in any of the analysis views of BioBench.

Event Logging

The capacity to associate and recognize user commands associated with a data file may be of significant value. BioBench has been designed to provide this capability by automatically logging user-defined events, stimulus events, and file-logging events. With user-defined events, the user can easily enter and associate date- and time-stamped notes with user actions or specific subsets of data. Stimulus events are also data- and time-stamped and provide the user information about whether a stimulus has been turned on or off. File-logging events note when data have been logged to a disk. All of these types of events are stored with the raw data when logging data to file, and they can be searched for when analyzing data.

Alarming

To alert the user about specific data values and thresholds, BioBench incorporates user-defined alarms for each signal displayed. Alarms appear on the user interface during data acquisition and notify the user if an alarm condition has occurred.

Figure 136-20 is an example of the Data Acquisition mode of BioBench. Once data have been acquired, BioBench can employ a wide array of easy-to-use analysis features. The user has the choice of importing recently acquired data or opening a data file that had been acquired for comparison or teaching purposes. Once a data set has been selected and opened, BioBench allows the user simply to select and highlight a region of interest and to choose the analysis options to perform a specific routine.

BioBench implements a wide array of scalar and array analyses. For example, scalar-analysis tools will determine the minimum, maximum, mean, integral, and slope of a selected data set, while the array analysis tools can employ Fast Fourier Transforms (FFTs), peak detection, histograms, and X-versus-Y plots.

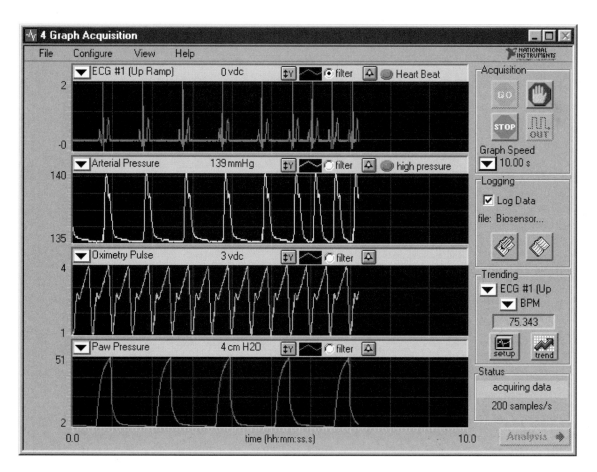

Figure 136-20 BioBench Acquisition mode with alarms enabled.

The ability to compare multiple data files is important in analysis, and BioBench allows the user to open an unlimited number of data files for simultaneous comparison and analysis. All data files can be scanned using BioBench's search tools in which the user can search for particular events that are associated with areas of interest. In addition, BioBench allows the user to employ filters and transformations to their data sets, and all logged data can be easily exported to a spreadsheet or database for further analysis. Finally, any signal acquired with BioBench can be played back, thus taking lab experience into the classroom. Figure 136-21 illustrates the analysis features of BioBench.

Case Study #4: A Cardiovascular Pressure—Dimension Analysis System

The intrinsic contractility of the heart muscle (myocardium) is the single most important determinant of prognosis in virtually all diseases affecting the heart (e.g., coronary artery disease, valvular heart disease, and cardiomyopathy). Furthermore, it is clinically important to be able to evaluate and track myocardial function in other situations, including chemotherapy, where cardiac dysfunction could be a side effect of treatment, and liver disease, where cardiac dysfunction could complicate the disease.

The most commonly used measure of cardiac performance is the ejection fraction. Although it does provide some measure of intrinsic myocardial performance, it is also heavily influenced by other factors, such as heart rate and loading conditions (i.e., the amount of blood returning to the heart and the pressure against which the heart ejects blood).

Better indices of myocardial function based on the relationship between pressure and volume throughout the cardiac cycle (pressure-volume loops) exist. However, these methods have been limited because they require the ability to track ventricular volume continuously during rapidly changing loading conditions. While there are many techniques to measure volume under steady-state situations, or at end-diastole and end-systole (the basis of ejection fraction determinations), few have the potential to record volume during changing loading conditions.

Echocardiography can provide online images of the heart with high temporal resolution (typically 30 frames per second). Because echocardiography is radiation-free and has no identifiable toxicity, it is ideally suited to pressure-volume analyses. Until recently however, its use for this purpose has been limited by the need for manual tracing of the endocardial borders, an extremely tedious and time-consuming endeavor.

The System

Biomedical and software engineers at Premise Development Corporation (Hartford, CT), in collaboration with physicians and researchers at Hartford Hospital, have developed a sophisticated research application called the "Cardiovascular Pressure-Dimension Analysis (CPDA) System". The CPDA system acquires echocardiographic volume and area information from the acoustic quantification (AQ) port, in conjunction with ventricular pressure(s) and ECG signals to perform pressure-volume and pressure-area analyses rapidly. The development and validation of this system have led to numerous abstracts and publications at national conferences, including the American Heart Association, the American College of Cardiology, the American Society of Echocardiography, and the Association for the Advancement of Medical Instrumentation. This fully automated system allows cardiologists and researchers to perform online pressure-dimension and stroke work analyses during routine cardiac catheterizations and open-heart surgery.

The system has been designed to work with standard computer hardware. Analog signals for ECG, pressure, and area/volume (AQ) are connected to a standard BNC terminal board. Automated calibration routines ensure that each signal is properly scaled and that it allows the user to immediately collect and analyze pressure-dimension relationships.

The development of an automated, online method of tracing endocardial borders (Hewlett-Packard's AQ Technology) (Hewlett-Packard Medical Products Group, Andover, MA) has provided a method for rapid online area and volume determinations. Figure 136-22 illustrates this AQ signal from a Hewlett-Packard Sonos Ultrasound Machine. This signal is available as an analog voltage (−1 to +1 volts) through the Sonos Dataport option (BNC connector).

Figure 136-23 illustrates the measured parameters and the specific hardware used for this application. Although this application was initially developed on a Macintosh platform, the system can run on multiple platforms, including Windows 9x//NT/2000/XP. The CPDA also takes advantage of the latest hardware developments, and form-factors and can be used on either a desktop or a laptop computer.

Data Acquisition and Analysis

Upon launching this application, the user is presented with a dialog box that reviews the license agreement and limited warranty. Next, the main menu is displayed, allowing the user to select from one of six options as shown in Figure 136-24.

When conducting a test, an automated calibration sequence for the pressure and ultrasound signals (acoustic quantification) is generally performed. If the user elects not to perform a calibration, the most recent calibration values are accessed from the integrated calibration log. Pressure-calibration data are retrieved from a "lookup table" containing specific gain and offset values for a variety of different manufacturers' pressure monitors. The AQ calibration procedure involves "scaling" and "mapping" the display image signal over the -1 to +1 volt output range. Figure 136-25 illustrates the front panel for the Hewlett-Packard Sonos calibration procedure. Sequential instructions in the form of a scrolling string indicator, as well as dialog boxes, are also available.

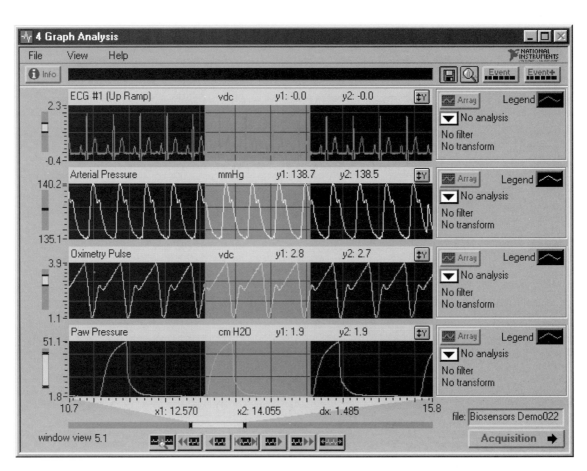

Figure 136-21 BioBench Analysis mode.

Virtual Instrumentation—Applications to Health Care

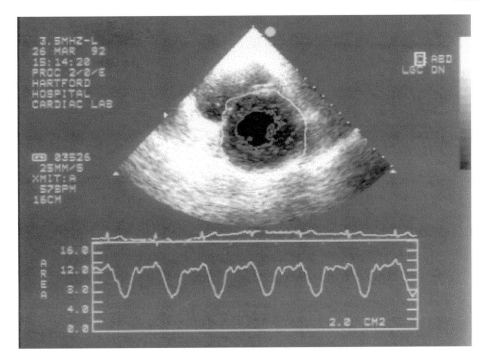

Figure 136-22 The Acoustic Quantification (AQ) Signal (Hewlett-Packard).

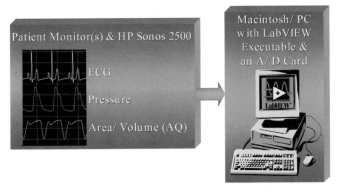

Figure 136-23 Schematic diagram of the cardiovascular pressure-dimension analysis system.

The default sampling frequency for each channel is 200 Hz. Data are typically collected for 20 to 60 seconds. The user is presented with a "Pre-Scan" panel to ensure that each signal is calibrated and tracking appropriately. When the user is ready to collect and store data, the "Cardiac DAQ" instrument is called (Figure 136-26). This sub-VI uses double-buffering to collect and display each channel for the predefined time. In order to maintain high temporal resolution, data are displayed in 10-second "sweeps." An indicator is provided to display the instantaneous and total collection time.

Once data are collected, the user is presented with the "Data Selection" sub-VI to define a particular range of data to save to a file. This option allows the user to store only the portion or subset of data that are useful among the entire collected data set (e.g., the last 25 seconds of a 60-second array). The default setting will store the entire data set. Interactive cursors are used to interactively set the initial and final indices of the data subset for analysis, as illustrated in Figure 136-27.

Clinical Significance

Several important relationships can be derived from these signals. Specifically, a parameter called the "End-Systolic Pressure-Volume Relationship" (ESPVR) describes the line of best fit through the peak-ratio (maximum pressure with respect to minimum volume) coordinates from a series of pressure-volume loops generated under varying loading conditions. The

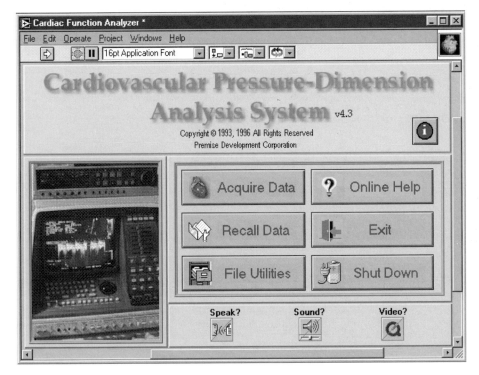

Figure 136-24 Cardiovascular Pressure-Dimension Analysis main menu.

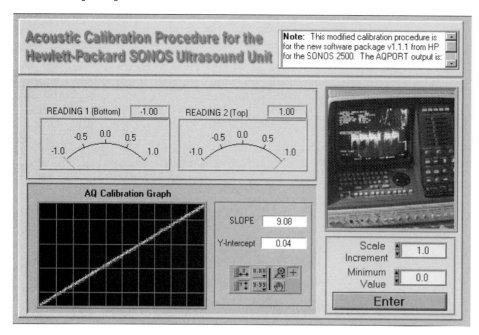

Figure 136-25 Hewlett-Packard's Sonos Ultrasound Machine Calibration front panel.

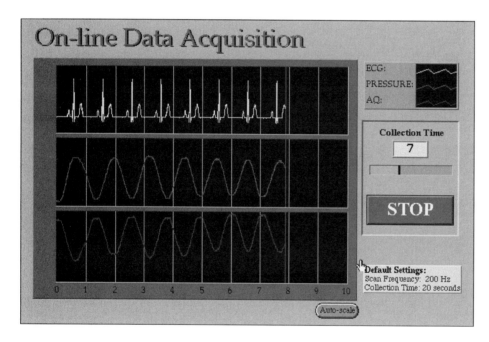

Figure 136-26 The Cardiac DAQ front panel.

slope of this line has been shown to be a sensitive index of myocardial contractility that is independent of loading conditions. In addition, several other analyses, including time-varying elastance (Emax) and stroke work, are calculated. Time-varying elastance is measured by determining the maximum slope of a regression line through a series of isochronic pressure-volume coordinates. Stroke work is calculated by quantifying the area of each pressure-volume loop. Statistical parameters are also calculated and displayed for each set of data. Figure 136-28 illustrates the pressure-dimension loops and each of the calculated parameters along with the various analysis options. Finally, the user has the ability to export data sets into spreadsheet and database files, and to export graphs and indicators into third-party-presentation software packages, such as Microsoft PowerPoint.®

Summary

Virtual instrumentation allows the development and implementation of innovative and cost-effective biomedical applications and information-management solutions. As the health care industry continues to respond to the growing trends of managed care and capitation, it is imperative for clinically useful, cost-effective technologies to be developed and utilized. As application needs surely will continue to change, virtual instrumentation systems will continue to offer users flexible and powerful solutions without requiring new equipment or traditional instruments.

Virtual instruments and executive dashboards allow organizations to effectively harness the power of the PC to access, analyze, and share information throughout the enterprise. The case studies discussed in this article illustrate ways in which various institutions have conceived and developed "user-defined" solutions to meet specific requirements within the health care and insurance industries. These dashboards support general operations, help hospitals manage fluctuating patient census and bed availability, and empower clinicians and researchers with tools to acquire, analyze, and display clinical information from disparate sources. Decision-makers can easily move from big-picture analyses to transaction-level details while at the same time safely sharing this information throughout the enterprise to derive knowledge and to make timely, data driven decisions. Collectively, these integrated applications directly benefit health care providers, payers, and, most importantly, patients.

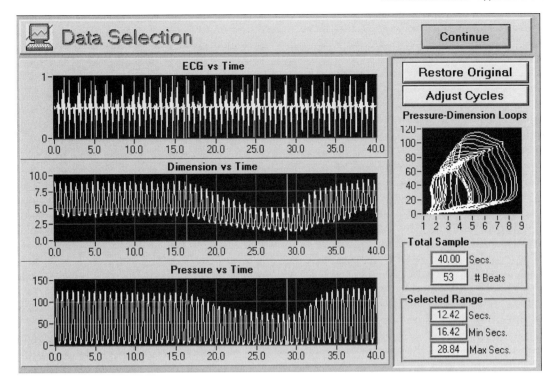

Figure 136-27 The Data Selection front panel.

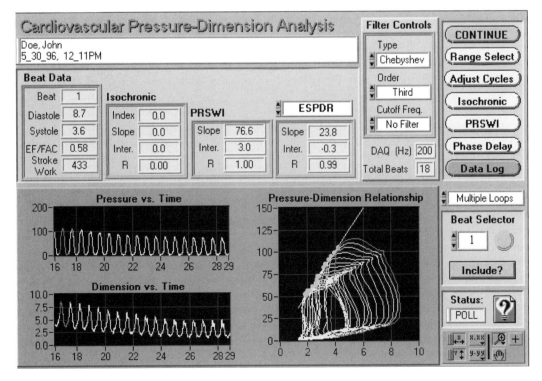

Figure 136-28 The Cardiac Cycle Analysis front panel.

Reference

AHA. *Statistics 2001: The Clinical Advisory Board, Capacity Command Center-Best Practices for Managing a Full House.* Chicago, American Hospital Association, 2001.

Further Information

American Society for Quality Control. *American National Standard: Definitions, Symbols, Formulas, and Tables for Control Charts, Publication number ANSI/ASQC A1-1987.* ANSI, 1987.
Breyfogle FW. *Statistical Methods for Testing, Development and Manufacturing.* New York, Wiley, 1982.
Carey RG, Lloyd RC. *Measuring Quality Improvement in Health care: A Guide to Statistical Process Control Applications.* 1995.
Frost, Sullivan. *Market Intelligence, File 765.* Mountain View, CA, The Dialog Corporation,
Fisher JP, Mikan JS, Rosow E, et al. Pressure-Dimension Analysis of Regional Left Ventricular Performance Using Echocardiographic Automatic Boundary Detection: Validation in an Animal Model of Inotropic Modulation. *Journal of the American College of Cardiology* 19(3):262A, 1992.
Fisher JP, McKay RG, Mikan JS, et al. *Human Left Ventricular Pressure-Area and Pressure-Volume Analysis Using Echocardiographic Automatic Boundary Detection.* Hartford, CT, American Heart Association, 1992.
Fisher JP, Mitchel JF, Rosow E, et al. *Evaluation of Left Ventricular Diastolic Pressure-Area Relations with Echocardiographic Automatic Boundary Detection.* Hartford, CT, American Heart Association, 1992.
Fisher JP, McKay RG, Mikan JS, et al. *A Comparison of Echocardiographic Methods of Evaluating Regional LV Systolic Function: Fractional Area Change Versus the End-Systolic Pressure-Area Relation.* Hartford, CT, American Heart Association, 1992.
Fisher JP, McKay RG, Rosow E, et al. On-Line Derivation of Human Left Ventricular Pressure-Volume Loops and Load-Independent Indices of Contractility Using Echocardiography with Automatic Boundary Detection: A Clinical Reality. *Circulation* 88:I–304, 1993.
Fisher JP, Chen C, Krupowies N, et al. Comparison of Mechanical and Pharmacologic Methods of Altering Loading Conditions to Determine End-Systolic Indices of Left Ventricle Function. *Circulation* 90(II):1–494, 1994.
Fisher JP, Martin J, Day FP, et al. Validation of a Less Invasive Method for Determining Preload Recruitable Stroke Work Derived with Echocardiographic Automatic Boundary Detection. *Circulation* 92:1–278, 1995.
Fontes ML, Adam J, Rosow E, Mathew J, DeGraff AC. Non-Invasive Cardiopulmonary Function Assessment System. *J Clin Monit* 13:413, 1997.
Johnson GW. *LabVIEW Graphical Programming: Practical Applications in Instrumentation and Control, 2nd Edition.*, McGraw-Hill, 1997.
Kutzner J, Hightower L, Pruitt C. Measurement and Testing of CCD Sensors and Cameras. *SMPTE J:* 325–327, 1992.
Mathew J, Adam J, Rosow E. Cardiovascular Pressure-Dimension Analysis System. *J Clin Monit* 13:423, 1997.
Montgomery DC. *Introduction to Statistical Quality Control, 2nd Edition.* New York, Wiley, 1992.
Rosow E, Adam J, Satlow M. *Real-time Executive Dashboards and Virtual Instrumentation: Solutions for Health care Systems.* 2002 Annual Health Information Systems Society (HIMSS) Conference and Exhibition, Georgia World Congress Center, Atlanta, GA, January 28, 2002.
Rosow E, Adam J. Virtual Instrumentation Tools for Real-Time Performance Indicators. *Biomed Instrum Technol* 34(2):99–104, 2000.
Rosow E, Olansen J. *Virtual Bioinstrumentation: Biomedical, Clinical, and Health care Applications in LabVIEW.*, Prentice-Hall, 2001.
Tufte ER. *Visual Explanations.*, Graphics Press, 1997.
Tufte ER. *Envisioning Information.*, Graphics Press, 1990.
Tufte ER. *The Visual Display of Quantitative Information.*, Graphics Press, 1983.
Walker B. *Optical Engineering Fundamentals.* New York, McGraw-Hill, 1995.
Wheeler DJ, Chambers DS. *Understanding Statistical Process Control, 2nd Edition.*, SPC Press, 1992.
*BioBench™ was developed for National Instruments (Austin, TX) by Premise Development Corporation (Hartford, CT).

137

Clinical Engineers in Non-Traditional Roles

Eben Kermit
Biomedical Engineer and Consultant
Palo Alto, CA

What is the first image that comes to mind with the term "clinical engineer"? For many, it is a middle-aged male with a college education and an advanced degree who manages a department of technicians and support staff. The primary function of a clinical engineer is to oversee the repair and maintenance of the many medical devices used in a modern hospital. Often, the repair facility is on the basement level, near the boiler room or the morgue. There is no apparent reason for this, but it is often the case. There is typically a line-up of equipment awaiting repair, often spilling into the hallway. Intravenous pumps, wheelchairs, and beds are often the first clue that one is approaching the clinical engineering department. Parts may have been ordered last week, but have not been received and installed. The office, or repair area, is likely to be littered with tools, cleaning supplies, assemblies, and sub-assemblies. Despite the clutter, there is a level of organization, distinguished by the individual technicians and the types of equipment under repair. The office contains reports and files, both paper and electronic, of repair orders, repair histories, safety inspections, and JCAHO audits. These documents are always present and always in a state of review and input.

There is also a steady stream of visitors. Some visit in person, others by telephone, and still more by computer. They are the "customers" to the clinical engineering office. They want to have a device repaired, ask how to use a new or forgotten item, or determine whether a screw driver or epoxy adhesive is available for a fix-up.

Welcome to the world of the hospital-based clinical engineer!

Of course this description is fictional, generalized to a great degree, and colored based on the author's experiences both working in and visiting many hospitals. Many situations are quite different. The definition of a clinical engineer is broad and inclusive and not limited to hospital-based job descriptions. The definition adopted by the American College of Clinical Engineers (ACCE) is, "A professional who supports and advances patient care by applying engineering and management skills to health care technology" (Bauld, 1991). The following is taken from the ACCE description of a clinical engineer:

"As clinical medicine has become increasingly dependent on wider use of highly sophisticated technology, and the ever more complex equipment associated with it, the clinical engineer's role in the health care system has continuously evolved as well. Serving initially as the health care provider's equipment manager, today's clinical engineer has become the technology officer and strategist, helping health providers and industry to plan for the manufacture or acquisition of technology, then ensuring its continued, safe and efficient utilization through a well-planned program of user training, maintenance and quality assurance.

In addition, a clinical engineer's education provides a solid understanding of the physical sciences necessary to apply those principles to the design and construction of instruments and devices for health care. With additional education, training, and experience in the life sciences and management, the clinical engineer is especially qualified for a variety of indispensable roles in the realm of health care, industry and academia."

Please note that there is not an expectation, nor is there a requirement, to work in a hospital, clinic, or other health care facility. The only requirement is the application of engineering skills for the improvement of health care. Given this interpretation, what other roles are filled by clinical engineers? What else can clinical engineers do besides work in a hospital or contract repair facility?

Ergonomics, the design of systems and controls that fit the human body, is one domain occupied by clinical engineers. The automotive, consumer electronics, and aerospace industries all use clinical engineers to design products that fit switches to fingers, levers or wheels to hands, and pedals to feet. All control interfaces must fit people for dimensions, strength, and reach, and must logically group controls together to avoid accidental selections. Human factors engineering, whether for seating, displays, or controls, is a subspecialty of clinical engineering. Most people do not realize that handles, doorknobs, gearshift grips, and bathtub faucets must all be designed to fit the human hand. Hands come in a range of sizes and designers must take this into account. Buttons on a car radio must be activated without impediment, even if the driver is wearing gloves, or can not see the controls due to darkness or concentration on traffic. Some engineers intuitively incorporate human factors into their designs; others are formally trained in specific college courses on this subject (see Chapter 83).

Another subspecialty of clinical engineering is that of expert witness. Attorneys in legal hearings or trials involving a medical device frequently retain clinical engineers as expert witnesses. Sometimes an accident or injury that results in a lawsuit; other times corporations claim of infringements of design elements and patent claims. The role of the expert is to teach or instruct the judge and jury. An expert may be asked to offer opinion on the typical use of a device or to project what might happen in hypothetical situations. The expert clinical engineer usually has sufficient educational background and experience (plus an advanced degree) and often a professional license in order to qualify as an expert for legal purposes (see Chapter 13).

Clinical engineers may also be called upon to investigate accidents involving medical devices and injury. In this role, they use analytical skills to understand the situation and how the attributes of a device design contributed to the injury. In the best tradition of Sherlock Holmes, gathering the facts, examining the clues, and arriving at the explanation is a form of clinical engineering. (See Chapter 64.)

A clinical engineer may be an "inventor" and "product designer" (see Chapter 82). Clinical engineers often use their logical and insightful reasoning to develop a new medical device or procedure, or improve on one that exists. For example, pulse oximetry uses a pair of light emitting diodes and a phototransistor to measure the absorbed light after it passes through tissue. The measurement is extremely useful as a measure of the amount of oxygen in the tissue. This technology was developed after an engineer recognized that a medical procedure for measuring cardiac output using a dye and photo detector was sensitive to changes in background absorption shifts due to changes in oxygen saturation. This was the foundation for the use of pulse oximetry, a transformation in health care.

Engineers are paid to create products and develop applications. The medical device industry is no different than other areas of product design. Engineers, typically working in small groups or teams, play a major role in the creation of ventilators, bandages, pulse oximeters, infusion pumps, artificial hearts, and CT scanners. Multiple elements must be addressed t for a product design to succeed. The fundamental application (requirement) must be identified to ensure there is demand (market) for the device. In addition, the product must meet regulatory standards. Prior to the introduction of any medical device, a validation/verification for both product design and process, including failure mode and effect analysis (FMEA), must be planned and executed to completion (Stamatis, 1995). Building prototypes differs from building large numbers of devices for distribution to the end user(s). Plans and designs for manufacturability become increasingly important as quantities increase. Fixtures, molds, assembly of fasteners, welds, and even packages must be produced in a way that is efficient and that reduces the likelihood of a manufacturing error during assembly.

Clinical engineers can also work in regulatory affairs. In the U.S., medical devices came under the purview of the U.S. Food and Drug Administration (FDA) in 1976 (see Chapter 126). In Europe, CE marking of products has been required for general sale and distribution of medical devices since 1996 (see Chapter 125). Clinical engineers often are key players in the review and approval processes of medical devices. The FDA requirement that medical devices be safe and effective is mandated in the US, and the regulatory process is intended to assure the public that medical devices work correctly and as designed.

Many clinical engineers seek careers in research and scientific development. They may be affiliated with a university, government center, or an industrial "think-tank." Familiarity with the scientific method—postulating a theory (hypothesis), designing an experiment with controls, collecting data, performing data analysis, and logically summarizing results—is the process expected for publication of findings at conferences or in journals. Clinical engineers are well suited to this process due to formal training in analysis in undergraduate or advanced degree programs.

Clinical engineers can and often do become entrepreneur. The skills and leadership abilities necessary for creating a business are complex. Finding money, researching intellectual property (patents), recruiting other engineers, scientists, programmers, or designers to develop a new product takes much commitment. Planning a marketing and sales strategy, finding a laboratory, building, or office space, writing contracts for sub-assembly, packaging, sterilization, etc., must also be considered (see Chapter 140). Also, managing investors, distributors, and sub-contractors, and, if the company grows large enough, an initial public offering (IPO) or acquisition by another firm are the responsibilities of the CEO/entrepreneur.

Rehabilitation is another area in which clinical engineers can use their skills and experience with a wide variety of devices ranging from simple to complex. A set of crutches is a simple but subtle device. The crutches must be able to support the patient's body weight, but also must be comfortable in the hands, accommodate a range of lengths, and must be durable, lightweight, inexpensive, cosmetically acceptable, possibly collapsible for storage, and capable of being manufactured in large quantities. All of these constraints have collapsed the number of design styles and materials to the few versions in use today.

Wheelchairs come in a wide variety of styles and shapes, and many more choices exist than for crutches. Some wheelchairs are motorized, others are designed for sport events, and still others include special features for climbing stairs or curbs. Some are recline. Some wheelchairs are essentially miniature electric carts, intended for travel over several miles on pavement or rough terrain. Patient support, controls, padding, footrests, brakes, weight, and cosmetics, e.g., plastic molded fairings, all must be designed. While the goal of mobility, wheelchairs are still relatively simple, and a single design is often acceptable for many users (see Chapter 95).

Artificial limb design involves both dedication to detail and craftsmanship. Each prosthetic (or orthotic) must be "custom fit" for every patient. As with wheelchairs, a wide range of choices and trade-offs are available. For example, an artificial below the elbow device may be a simple "hook" with a grasping claw, often selected by patients over more cosmetic motorized "hands". The reasons are subtle. While the "hook" does not look like a hand, it is quite functional. It is durable, easy to operate, works even (or despite) a dead battery, and can be used to pull muffins out of a hot oven without melting. There are also many adaptations, such as "hands" with hair brushes, ping-pong paddles, extension graspers, or *jai alai* baskets, which can be easily connected and disconnected with a simple retention coupling. Yet, despite these "advanced" features, the claw hook is the most common selection. The point is that design and patient acceptance must be in accord. Engineers continue to play a key role in filling the need for simple, well-designed devices, while providing patients with a range of prosthetic choices.

Implants or implantable devices bring a number of design issues into focus. Not only must this class of device be bio-compatible and sterile to go inside the body, it must also be durable in a highly corrosive environment. Consider vascular stents. These devices are metallic structures, with a complex and intricate pattern of cut-outs within a tubular shape. The stents are inserted inside the blood vessel in their collapsed state via a catheter, and advanced to the desired position. Often, a balloon catheter device is used to radially expand the tubular latticework. Once expanded, the stent keeps the vessel patent (open to flow). The design considerations are multiple. Not only must be bio-compatible, the latticework must be capable of radial expansion and yet not break, collapse, or deform. The device must be ductile enough to expand. Fibrin deposits, blood flow patterns, and clot formation are causes of device failures, and much research focuses on refining designs that are robust and resistant to occlusion. Also, it may be difficult or impossible to retrieve a device once it is deployed into the body, so the design engineer must conduct extensive testing, modeling, and validation prior to release of the device.

Teaching and education is a small but important niche sought by some clinical engineers (see Chapters 68 and 69). Often the teaching component is obvious, if clinical engineers are university professors or involved in academic research. However, many clinical engineers teach training seminars, give in-service lectures or classes, or travel abroad to developing countries to share knowledge and experience (see Chapters 70 and 71). Teachers can find great satisfaction in passing on methods and tools to others.

The military (directly or via sub-contractor firms) uses engineers extensively for research and the design of systems to protect soldiers from inhospitable environments (see Chapter 9). High-performance aircraft can pull G-forces that are greater than the limits of the human pilots. Undersea divers must have a respiratory gas mixture at high pressure that allows them to safely explore the ocean and its floor. In space flight, factors including artificial environment, energy management, waste disposal, exercise, "space sickness" or nausea must be considered and factored into the designs of space suits, submarines, tanks, space craft, and airplanes.

Space suit design is an apt example of clinical engineering. Space suits must meet many requirements to be effective. They must protect the human occupant from the hard vacuum of space, provide a breathable air mixture, keep the astronaut cool or warm, provide communications, and be lightweight and flexible. Engineers have been refining space suit design for decades, and materials and features continue to improve.

Telemedicine was once thought to be something from science fiction. Today, patients located in a rural clinic can "consult" with an expert in an urban medical center hundreds of miles away via a "telemedicine" link (see Chapter 101). Communications continue to improve and now can frequently support the transmission of clinically useful images and data . Soon, the ability to palpate the patient may be possibly via a robotic hand with tactile sensors and feedback servers. Someday, an astronaut in a spaceship may use telemedicine to receive medical care while traveling to the stars.

Engineers love robots because robots can perform tasks that either are repetitive or require superior precision. Try to make a screw thread with a steel rod and a hand file is much different from doing it with a lathe and motorized lead thread-cutting attachment. Orthopedic surgery for hip-socket replacement requires a precise match between the cavity in the femur and the spike of the orthotic. A robot creates the cavity with finer precision than the human surgeon. Stereotactic location used for neurosurgery can also be facilitated by the use of robots and space frame locators. Robots can also serve as couriers in hospitals; they never take a day off, call in sick, or demand a raise, even when called on at 2:00 a.m. to deliver materials.

The areas of interest for clinical engineers are limitless. The diverse fields described here merely reflect the complex interrelation of science, physics, materials, electronics, optics, computers, physiology, manufacturing, statistics, chemistry, anatomy, and mathematics that collectively form the practice of clinical engineering as it applies to the diverse set of problems of patients and health care.

If it were simple, anyone could be a clinical engineer, but it is not simple. This is both the challenge and the reward.

In conclusion, a wide variety of roles exist for clinical engineers. Some are based in the hospital, others in government or university settings, and many more in businesses. This field is truly international in scope, since technology knows no boundaries at borders. From the depths of the ocean to the vastness of space, clinical engineers have many diverse roles beyond working in hospitals.

References

Bauld TJ. The Definition of a Clinical Engineer. *J Clin Eng* 16:403–05, 1991.
Stamatis DH. *Failure Mode and Effect Analysis.* Milwaukee, WI, American Society for Quality, 1995.

138

Clinical Support: The Forgotten Function

Stan Scahill
Director of Biomedical Engineering
Concord Hospital
Concord, New South Wales, Australia

I recently read in the *Land* newspaper about the country town of Coolah, New South Wales. In 1995, the town was in serious decline following the closure of its main industry, saw milling. Town leaders called public meetings and undertook a strengths, weaknesses, opportunities, and threats (SWOT) analysis. The town residents realized that the tourist dollars were a way to keep people in the district. The town required beautification and an attraction, so the Coolah Spring Garden Festival was established. The town's main street and surrounding areas were refurbished, old heritage buildings were painted, and residents developed their gardens. The first spring festival raised $20,000. The second raised $30,000. This money was poured back into the town's development. This year the money raised will be going to the development of self-care units for the aged. The town has an interesting motto. "If you want a job well done, you have to do it yourself."

When I began working in hospital biomedical engineering 30 years ago (the term clinical engineering had not yet been coined), biomedical engineers and technicians were the technical wizards who were ushering in the new semiconductor technology. This advance enabled the medical profession to start measuring the electrical and physiological activity of the human body.

Alex Watson, David Jones, Don Melley, Richard Troughear, Rowley Hilder, Bruce Morrison, and Martin Dwyer were a few of those heavily involved in research in the fields of cardiology, neurology, psychology, and psychosurgery. We were mostly laboratory-based, but took our developing skills and knowledge into the clinical environment as well.

I will never forget the early Telectronics cardiac monitors, the HS1 and HS2, which were basically modified oscilloscopes. Rowley and I used to manufacture reusable ECG electrodes from small blue disposable electrodes. When the lead wire broke, a press stud was added and it was back in action. The electrode was attached with a double-sided adhesive disc and filled with ECG gel. Innovation and simplicity were the hallmarks of the day.

We designed, constructed, and tested our own equipment. Rowley would often have a device housed in a cardboard box or a cooking tray. Lawrie Knuckey from Melbourne would often tell of the manufacture of epoxy encased pacemakers and catching the tram to the Children's Hospital to implant them. We were unregulated and willing to try anything. We had the knowledge.

As electronics developed and science education for medical and nursing staff improved, engineers were muscled out of the clinical arena and soon became the repair team because we knew what happened inside the black box. The engineers and technicians in operating rooms and the clinics became centralized in biomedical engineering. Life had become simple. Our duties were confined to testing, safety assurance, purchasing, repairing, and fighting off independent service organizations. We became the service engineers with infrequent forays into the clinical environment.

However, service engineers may face extinction. With modern construction techniques you will not be able to repair boards. Board replacement will be the order of the day, and clinical engineers can not repair boards made with modern construction techniques. In addition, the equipment will probably be replaced. It is cheaper to replace a hand held oximeter than to repair it, and highly paid technical staff and engineers will no longer be needed to manage servicing.

What is our future? Let us do what the town of Coolah did and conduct an SWOT analysis on clinical engineering.

What are our strengths?

- Knowledge. Engineering, staff, equipment, clinical
- Multi skilled. Electronic, mechanical, computer literate
- Innovative. Can improvise, source alternate supplies
- Onsite. Generate quick response
- Support. Clinical support for new procedures

What are our weaknesses?

- Politically inept. Do not dance with the power brokers
- Inadequate clinically trained staff.
- Lack of organizational freedom.
- Poor self promotion.

What are our opportunities?

- Information technology
- Communications
- Clinical expertise
- Equipment development
- Research

What are our threats?

- External independent service organizations
- Predatory biomedical engineering services
- Limited budgets
- Internal self destruction due to the following:
  - Lack of knowledge
  - Lack of vision
  - Poor management skills

I will focus on the opportunity for knowledge. We are electronic and mechanical engineers and technicians in the life sciences. We should have an in-depth knowledge of ones engineering skills as applied to the medical environment. This gives us a uniqueness and advantage over other staff and external service groups.

Clinical engineering's actual clinical involvement is limited in many hospitals because engineers have failed to sell to our health care partners the fact that our knowledge base can be invaluable in the clinical environment. Clinical engineering has become a forgotten function.

However, I have observed that new opportunities are opening up for clinical engineers in the clinical environment. Modern medical equipment and devices have become more complex, and a modicum of technical knowledge is no longer adequate. Operator training has focused on the function rather than the system, leading to difficulties in fault resolution.

One complex device that comes to mind is the modern pacemaker. Outwardly, it is simple, but the technical knowledge of a clinical engineer or technician is invaluable when it is not performing as expected.

I will be employing a resident technician in the operating room (OR). What I have discovered is that the amount of equipment, the different types of equipment, and the systems built around equipment, are too complex for the nursing staff. They want to nurse, not find fault with equipment. Equipment malfunction may be due to a technical fault, an incorrect button push, or a reprogram of the equipment.

A senior surgeon at my establishment has calculated that a half-hour delay in operating room time costs a hospital more than $3000. Some of my colleagues suggest that there is insufficient work for a technician, but if the technician becomes involved in setting up rooms, testing the equipment before the surgery, rearranging the equipment to suit the surgeon, conducting safety tests, calibrating equipment, supplying printers with toner and paper, and managing repairs, I believe that there is adequate work. Many faults on anesthetic equipment are due to incorrect settings by the operator. Clinical engineers can provide onsite help.

When the anesthetist complains that the invasive mean pressure differs from the mean noninvasive blood pressure (NIBP) mean, onsite assistance may be available in the form of a clinical engineer. This person will become invaluable, and when he or she is not there, the main clinical engineering department will be expected to support the OR. This may become inconvenient, but it will happen when clinical engineers become indispensable, which many engineers want.

Quality improvement is another aspect of clinical engineering presence in the clinical area. How many times have you observed a video endoscopy procedure in the OR and seen the surgeon struggling with a poor image simply because the automatic gain control was turned off? A clinical engineering staff must be trained to observe procedures as well as equipment.

If any medical device is faulty, fix it or send it for repair. If a procedure is hampered by technical difficulties, either fix it or refer it. Near enough is never good enough.

I believe we must develop partnerships with clinical care providers. We can provide advice and technical support and we can be visible, but not threatening. A clinical partnership we have established with units is to observe what they do and provide engineering input when necessary to provide a safer, better working environment.

In my hospital, we are developing an "over-bed support" for the intensive care unit (ICU) so that ventilators and infusion devices can be attached to the bed when it is necessary to transport the patient to the OR, radiology, or nuclear medicine. Presently, several nurses are required to push the stretchers and intravenous (IV) poles. This over-bed support device can be attached and removed by anyone by using existing bed controls.

My department has two additional projects in development to aid patient care. The first is a trolley that attaches a large transport ventilator to the foot end of an ICU bed, using the bed's own up and down controls to lift the ventilator both on and off the trolley. Similar trolleys will be kept in OR and radiology so the patient can be transported with the ventilator when the patient is transferred to CT. The second item is an air bag lifter, which helps to position wheelchair patients in front of an image intensifier for swallowing studies. The medical air is used as the power source.

Clinical engineers have invaded neurology as well. My hospital is close to high-powered radio transmitters, and the clinical engineering department moved to a different location. It had not been shielded according to my instructions and consequently no procedures could be undertaken due to radio frequency (RF) breakthrough. Remedial repairs were out of the question as the budget had already been consumed. The solution was roof sarking.

I stapled sarking on the wall between the transmitter and the equipment and the RF disappeared. This process also gave me the opportunity to observe the engineers' techniques and suggest improvements.

I am investigating the effects of shielded (versus unshielded) recording leads. Neurology departments regularly measure microvolt signals with unshielded, excessively long leads, and continually battle interference. We would not attempt to do this with our high-fidelity audio systems, so why do it in the clinical environment? The fact that filters minimize the effect of interference is insufficient excuse for poor practice.

I have developed IS with nursing administration as well. Many of our patients are elderly and suffer from decubitus ulcers if they are bedridden for long periods of time. I have been commissioned by the director of nursing to develop a device that could resolve this problem and costs less than $150. I have two devices ready for testing. The case is a $6 fishing tackle box, and the control valve uses a surplus microwave oven motor and a fish tank pump. The total cost is less than $75.

Clinical engineers have developed continuous positive airway pressure (CPAP) machines using surplus oxygen-air blenders, with savings of $30,000 to the hospital. The management appreciates this effect the budget.

Research alliances include the following:

- Gait analysis
- Infection minimization using ozone
- Bed pans

Clinical engineering should cease to be a forgotten function and become an area of new opportunities, not necessarily with primary patient care, but through partnerships with hospital departments for either direct or indirect involvement with patients. For example, clinical engineering may train staff more thoroughly than the product sales representative can train them.

Clinical engineering should be seen as the department that exists to improve patient care and to make the working environment less arduous and hazardous.

We veteran engineers are responsible for passing on our knowledge to our replacements. There will be a considerable changing of the guard in the next 10 years, so it is important we share our 30-plus years' worth of knowledge. Clinical engineering staff must be trained, since many of the younger staff members have not had the clinical exposure or experience of an older engineer or technician. I firmly believe that if your department is an integral part of patient care, threats of extinction by financial bureaucrats no longer exist. The hospital's chief executive officer should be your strongest supporter.

If you are seen as knowledgeable, you become indispensable, and your future will be more secure against internal threats.

139 Postmarket Surveillance and Vigilance on Medical Devices

Michael Cheng
Health Technology Management
Ottawa, Ontario, Canada

The medical device regulations of the five founding members of the Global Harmonization Task Force (Australia, Canada, the European Union (EU), Japan, and the U.S.) are based on the philosophy of risk management (see Chapter 124). No amount of premarket regulatory scrutiny can predict all possible device failures, and it therefore is essential to have continued assessment of, and vigilance for, the safety and performance of medical devices in use. Thus, the regulations require post-market surveillance and vigilance (see Chapters 125 and 126). One important requirement is adverse event reporting to regulatory authorities for the possible dissemination of information to other users in order to reduce the likelihood of repetition of adverse events. Improvements in postmarket surveillance and vigilance are necessary. Such improvements can be effected through the expertise of clinical engineers and biomedical equipment technicians, particularly in the area of adverse event reporting.

The Global Harmonization Task Force (GHTF) primarily comprises representatives of manufacturers and government regulators. Noticeably absent are managers, users, and maintainers of medical devices. Within the scope of the GHTF, CEs and BMETs could make contributions of particular importance to developing countries where conditions are frequently quite different from those of the industrialized countries.

Concerns of Postmarket Surveillance and Vigilance

The current emphasis of the GHTF Study Group 2 seems to be mainly on high-risk issues such as those associated with cardiovascular devices or implants. However, statistics from the 2001 Adverse Incident Report from the Medical Devices Agency, United Kingdom (http://www.medical-devices.gov.uk/), have revealed that adverse events associated with low-risk devices such as those used for injection, drainage and suction, and daily living have significant adverse events (MDA, 2001). Adverse events from low-risk devices can be just as injurious and fatal as those from high-risk devices. For example, a suction device that malfunctions during surgery, or a patient hoist that fails, can result in serious injury to the patient. Moreover, low-risk devices are much larger in number, and they serve health care worldwide.

All founding members of the GHTF have mandatory requirements for manufacturers or their representatives to report to regulatory authorities all device-related incidents that have resulted, or could result, in serious injury or death. In the United States, some EU member states, and some Asian countries, mandatory-problem reporting is extended to users (see Chapter 126). However, problem reporting in the field of medical devices has become a controversial issue. There are concerns from the users and the manufacturers.

Some users are concerned that the information given in medical device problem-reporting might become an institutional or personal legal liability if it turns out that the problem was actually caused by the incorrect use of the device. Sometimes users are discouraged by their superiors to report problems when the cause of the problems is unclear. Institutions are also concerned about possible legal actions that can be taken by manufacturers when the latter feel they were wrongly implicated in device-problem reports. Some manufacturers are concerned that if the media reports news of an adverse event prematurely, they might cause unnecessary alarm to the public, and possible damage to the reputation of the manufacturer. Adverse events can require in-depth, thorough inquiries that take considerable time.

There is often a general reluctance to report problems associated with the use of medical devices to the regulators, even though such reporting is mandated by law. Some regulatory authorities in Europe, North America, and Asia are experiencing a lack of co-operations from device users and manufacturers. The CE and BMET, however, are in a position to provide a solution to this dilemma (Dyro, 2003).

The Role of the Clinical Engineer and Biomedical Equipment Technician

The assurance of safety and performance of medical devices is the responsibility of the clinical engineer and biomedical equipment technician within most health care facilities. The job descriptions of a clinical engineer and biomedical equipment technician normally include responsibilities such as technology assessment, selection, acquisition, installation, and maintenance, as well as consultation for hospital staff on safe and effective use of medical equipment. Thus, when an adverse event associated with medical devices occurs, the clinical engineer or technician (and, if available, a representative from the manufacturer) would normally conduct an investigation to determine whether there has been a device failure or use error (see Chapters 13 and 64). If it is a user error, it is the CE's or BMET's duty to inform or instruct the user. If it is a device problem, there should be no hesitation to report to the manufacturer and the regulator.

The following pertains to developing professionalism in handling this task.

- One major reason for a general apprehensiveness about problem-reporting is the traditional culture of "people blaming" rather than "problem solving." People are reluctant to report a problem because they fear that they may be held responsible for it. In reality, when an error occurs, blaming an individual does little to make the system safer or to prevent someone else from committing the same error. Therefore, one should use a systems or process approach to identify and solve problems (see Chapter 59). The focus must shift from blaming individuals for past errors to preventing future errors by designing safety not only into the device but also into the procedure of its use. This is consistent with the quality and continuous improvement principles. The report of The Institute of Medicine of the National Academy of Sciences of the USA provides valuable insight into the problem and suggests ways in which errors can be identified, reported, and analyzed in a constructive way. See "To Err Is Human—Building a Safer Health System" http://books.nap.edu/html/to_err_is_human. (Kohn et al. 2000)
- The fact of user interface in the use of medical devices brings out another aspect of medical devices with which the clinical engineer and biomedical equipment technician are familiar—the human factor in human-machine interface. Frequently, use errors occur because a machine was not designed with proper concerns for ergonomics, and use errors could have been prevented or minimized if the design had incorporated human factor engineering principles. Therefore, a use error should be followed up with an investigation that can determine whether possible corrective actions in the design are necessary. With their technical knowledge of medical devices and intimate user experience, the clinical engineer and biomedical equipment technician are again in positions to make positive contributions. An excellent reference on this subject is available on an FDA web site: http://www.fda.gov/cdrh/humfac/1497.html.
- The investigation of an adverse event is a complex task that often requires a multidisciplinary team approach (see Chapter 64). The clinician who uses the device, the device operator, and the manufacturer all should be involved in the investigation. Functioning as an effective team member are skills that can be developed. A good pocket guide for team members is "The Team Memory Jogger," a GOAL/QPC publication (http://www.goalqpc.com/) (GOAL/QPC, 2001).
- Discretion and confidentiality are, of course, essential. It is critical that the dissemination of information to the media be handled with discretion so that it does not damage the reputation of a health care professional, the health care institution, or the medical device company.

More information on adverse event reporting can be found in GHTF documents. To date, the Study Group 2 of the GHTF has provided documents on postmarket surveillance and vigilance issues. These documents named below are available from http://www.ghtf.org/. Please note that the documents are based on the regulatory requirements or practices existing in the participating countries, but that they are not identical to current regulatory requirements of individual countries. These documents represent a global model toward which existing systems should converge.

- SG2-N21R8: *Adverse Event Reporting Guidance for the Medical Device Manufacturer or Its Authorized Representative*
- SG2-N8R4: *Guidance on how to Handle Information Concerning Vigilance Reporting Related to Medical Devices*
- G2-N9R11: *Global Medical Devices Vigilance Report*
- SG2-N7R1: *Minimum Data Set for Manufacturer Reports to Competent Authority*
- SG2-N6R2: *Comparison of the Device Adverse Reporting Systems in the USA, Europe, Canada, Australia & Japan*

Please refer to: http://www.ghtf.org/ for updated information.

Postmarket Surveillance and Vigilance in Developing Countries

The importance of the role of CEs and BMETs in device vigilance is particularly important in developing countries. The formal postmarket regulatory requirements can be implemented only if a country has medical device regulations. Most developing countries, however, have not yet established effective medical device regulations and enforcement. Furthermore, additional problems are facing developing countries, such as the following:

- A large amount of donated, second-hand equipment is still being used, which is unlikely to have distribution records from vendors. (See Chapter 43.) Thus, any device alerts will have difficulty reaching the users.
- Local vendor representatives of medical devices are often not technically knowledgeable. Relying upon manufacturers to send technical personnel to investigate a medical device incident can result in a long wait.

Developing countries are home to approximately two-thirds of the world's population. The deployment of medical devices in developing countries is increasing at a rapid rate, and considerations on the safety and performance of medical devices with regard to global public health must include these countries. However, there is a general shortage of technical personnel who are familiar with medical device matters; the potential contributions to post-market surveillance and vigilance from CEs and BMETs in developing countries are important resources that should be cultivated and utilized.

Conclusions

At present, the role of CEs and BMETs in postmarket surveillance and vigilance has not been openly designated by regulatory authorities and the GHTF. CEs and BMETs should develop the professionalism discussed in this chapter because they are in unique positions to co-operate with governments and manufacturers in adverse event monitoring, investigation, and reporting. Professional associations such as the International Federation of Medical and Biological Engineering (IFMBE) and the American College of Clinical Engineering (see Chapter 130) could promote this role to the national bodies that are concerned with patient-safety and medical device regulations. The author believes that this role should receive official recognition in the health care community for maximum utilization.

Medical device regulation is still a relatively young field and is constantly evolving. The Global Harmonization Task Force is playing a critical role in the harmonization of global regulatory requirements and procedures. The excellent work by GHTF members has made impressive progress in arriving at harmonized recommendations. The next stage of advancement for the GHTF should be to include the user as a representative on the task force.

So far, GHTF documents have focused on issues between industry and governments from industrialized countries. This situation might be a result of differing legislative mandates of founding member countries. The emphasis might be on product control. Because the actual use is the most critical stage in the life span of a medical device, it is desirable to include the user in the process of deciding matters associated with medical devices. The CE and BMET are strong candidates to represent an important user input to the GHTF in developing further recommendations for the medical device community. One important consequence of including user input from developing countries will be to make the work of the GHTF more inclusive on a global scale.

References

Dyro JF. *JCAHO Patient Safety Goal #6: Improving the Effectiveness of Clinical Alarm Systems*. US Food and Drug Administration, Medical Product Surveillance Network (MedSun) Teleconference. April 10, 2003.

GOAL/QPC. *The Team Memory Jogger*. Salem, NH, GOAL/QPC, 2001.

Kohn LT, Corrigan JM, Donaldson, MS. *To Err Is Human: Building a Safer Health System*. Washington, DC, National Academy Press, 2000.

MDA. *Adverse Incident Report from the Medical Devices Agency*. United Kingdom, Medical Devices Agency, 2001.

Further Information

AHRQ. *Making Health Care Safer: A Critical Analysis of Patient Safety Practices*. Rockville, MD, Agency for Health care Research and Quality, 2001.

Cheng M. *Medical Device Regulation and Policy Development, International Forum for the Promotion of Safe and Affordable Medical Technologies in Developing Countries*, The World Bank, May 19–20, 2003.

Cheng M. *A Guide for the Development of Medical Device Regulations*. Washington DC, Pan American Health Organization/The World Health Organization (PAHO/WHO), 2002.

140

Small Business Development: Business Plan Development Fundamentals for the Entrepreneur

Peter W. Dyro
Principal, Re-Source Builders
Seattle, WA

Tristan Tzara once wrote, "Words are only postage stamps delivering the object for you to unwrap." Similarly, the written business plan is a construct of diligence, a derivative of hundreds of hours of blood, sweat, and tears. Research, phone calls, sleepless nights, early mornings, analysis and re-analysis—the business plan becomes your best friend and your worst enemy. It is essential to keep in mind that the written business plan is not the business. The business plan is a medium for communication, merely a reflection of your dedication and strategy. It is a large part of your representation when you finally stand in front of the potential investor to whom you will sell your idea.

The business plan is an interface between you and your investors; it is the evidence of your thorough research. Investors will know in an instant the difference between a plan stuffed with filler and a well-researched operation. As this chapter will stress, focus on the logistics of the business, the market, the economics, the product, the risks, and constant analysis of the industry. A successful business plan will follow.

Your business plan should be subject to constant revision. In today's Internet age, technologies and industries no longer evolve but see drastic turnovers in the way business is done. It has been said that the plan is obsolete at the printer; some would argue that it is obsolete before it even gets there! If one were to look at the business practices of new ventures today, none of them would match what was specified in the plan a year ago. Suffice it to say that your business plan is forever a work in progress.

To begin, the fundamental components of a business plan should include every aspect of your young venture. The document should be as lean as possible, containing only the essential information in the text. Necessary appendices may be added to the rear and referenced appropriately. The plan should be appealing and must not overwhelm the reader with technical lingo. Too much jargon could contribute to the ambiguity of the plan. One must be able to explain in common language the principal functions of the operation, the economics of the business, and the main idea without hiding behind fancy terms in order to remain vague and camouflage your lack of detail with rhetoric. Say what you want to say, and include genuine content. Do not make unsubstantiated claims such as estimating sales on the basis of what your company would like to produce. These claims will be sniffed out and will leave you high and dry as your would-be investor passes you over for a more truthful, realistic candidate.

Plan Format

The physical plan should follow the general format that is familiar to venture capitalists. Uniform formatting of the business plan will aide in establishing your credibility as a worthy candidate for capital. The following format was developed by two colleagues at the Venture Founders Corporation and is based on a over 20 years of observing entrepreneurs as well as actually preparing hundreds of plans (Timmons, 1999). This is not to say that the guidelines must be followed to the letter, because there are no absolutes in preparing a business plan. Every young venture is different, and some of the aspects of that venture could require a different take on the suggested outline.

1. Cover
2. Executive Summary
   a. Description of the Business Concept and the Business
   b. The opportunity and strategy
   c. The target market projections
   d. The competitive advantage
   e. The economics, profitability, and harvest potential
   f. The team
   g. The offering
3. The industry and the company and its product(s) and/or service(s)
   a. The industry
   b. The company and the concept
   c. The product(s) and/or service(s)
   d. Technology
   e. Entry and growth strategy
4. Market research and analysis
   a. Customers
   b. Market size and trends
   c. Competition and competitive advantage
   d. Estimated market share and sales
   e. Ongoing market evaluation
5. The economics of the business
   a. Gross and operation margins
   b. Profit and potential durability
   c. Fixed, variable, and semi-variable costs
   d. Months to break-even
   e. Months to reach positive cash flow
6. Marketing plan
   a. Overall marketing strategy
   b. Pricing
   c. Sales tactics
   d. Service and warranty policies
   e. Advertising and promotion
   f. Distribution
7. Design and development plans
   a. Development status and tasks
   b. Difficulties and risks
   c. Product improvement and new products
   d. Costs
   e. Proprietary issues
8. Manufacturing and operations plans
   a. Geographic location
   b. Facilities and improvements
   c. Operating cycle
   d. Manufacture process
   e. Regulatory and legal issues
9. Management team
   a. Organization
   b. Key management personnel
   c. Management compensation and ownership
   d. Other investors
   e. Employment and other agreements and stock options and bonus plans
   f. Board of directors
   g. Supporting professional advisors and services
10. Overall schedule
11. Critical risks, problems, and assumptions
12. The financial plan
    a. Actual income statements and balance sheets
    b. *Pro forma* income statements
    c. *Pro forma* balance sheets
    d. Pro forma cash flow analysis
    e. Break-even chart and calculation
    f. Cost control
    g. Highlights

13. Proposed company offering
    a. Desired financing
    b. Offering
    c. Capitalization
    d. Use of funds
    e. Investor's return
14. Appendixes
15. The next step

Cover

The cover of your business plan should include all pertinent information about your new business, including name, address, contact information, date, and, most importantly, a statement of confidentiality. You must note that the document has been submitted on a confidential basis and is to be returned to the specified address should an investor not agree to accept the offering.

Executive Summary

The executive summary is usually the first substantial bit of text your investors will read, thus it should not be boring or wordy. Rather, it should describe the product or service and why it will be better than its closest competitor. Investors use the executive summary to weed out the definite "nos," so it must be convincing. You will present your description of the business concept and how exactly your company will go about delivering on your promises. Include any information that will substantiate any claims to breakthrough technology, unique capabilities, or trade secrets that will provide you with a significant competitive advantage.

Summarize the opportunity and describe your strategy. Give a brief description of the market and the way you plan to approach it. Describe your competition, and list any of their weaknesses on which you can capitalize. Paint a picture of unfulfilled need by describing the shortcomings of the current industry. Be sure to note your plans for growth, and give concrete examples of the size of your market. Is your market growing? If not, what are you doing to create a market? From the size of your market and your positioning strategy, you should be able to project unit sales and expected revenues. Also include your pricing strategy and anticipated market share.

The investors interested in your company will be most interested in the economic details, so give your best estimate of timeframes until break-even and positive cash flow. Also give any key financial projections. Briefly describe your contribution analysis and the underlying operation and cash-conversion cycle.

The investor will also be interested in the composure of the team in which he is putting his money. Therefore, do not be modest. This is your opportunity to shine. Summarize any relevant experience you have that will contribute to this particular venture. Include key attributes from every member of the management team in order to portray a well-rounded experience pool. Note previous accomplishments, specific know-how, and people- and business-management skills, including sizes of divisions and projects.

You should conclude the executive summary with your offering. State the simple math: You invest this much, here's what you get. State the amount of capital you are looking for, and what portion of the company you are willing to surrender for that capital. You might want to include a simple breakdown of how the funding will be spent so as to clarify where all of the capital will be going.

The Industry, the Company, and its Products and/or Services

In this section of your plan, you will delve more deeply into the specifics of the industry and provide more detailed information on your product or service. This is your opportunity to tell investors why your company and product are great, and why they will do well in the market.

Present investors with an industry profile but make it concise as possible. Key points you will want to cover include industry structure and where in that structure the opportunity lies. Also discuss industry trends, overall growth of the industry, market size, and sustainability. Be sure to mention any new developments in the industry, such as new technologies that have revolutionized products or services, as well as manufacturing practices that enable more cost-effective production. What are the entry barriers? Why is everyone not starting a business in this industry?

Explain your business concept. Where in the industry are you positioning yourself as a company? You must have one (or more than one) target market. Describe your customers and what you are going to sell them. The investor also will want to know about the company's history, so include the date the business was created and the progress you have made to date. If your company has been in existence for several years, you will be liable for an explanation as to why you have not yet turned a profit, what you have learned, how your experiences have tempered your organization, and the steps you have taken to prevent future setbacks.

Next, describe the product in detail, provide a technical description, and emphasize any unique features that will carry the product to success in its market. Depending upon the product's stage of development, include either photographs or drawings and carefully describe the costs involved in development and testing.

It is important to list all of the uses of the product, including secondary uses. Is there any particular aspect of the product that equips it with a competitive advantage? If not, how are you going to beat the competition's price?

Coping with obsolescence is a major concern in today's market. It is important that you describe exactly what it is about your product that will keep you from being replaced. Will your product line be able to grow with technology or will technology grow around your product line? You must explain in depth what technologies you are incorporating in the design and production process. Give a very brief history of the technology if it is not common knowledge as well as any trends or new developments this technology has experienced. State the life expectancy of these technologies and your current plan for sustainability when the market catches up.

Define your entry strategy. What is your strategy for introducing the product or service to the market? There are four primary strategies to entering a market. One can enter with guns a-blazin', investing huge amounts of capital in a high-risk, potentially super-profitable venture. Such was the case with Hoffman La-Roche, a Swiss Pharmaceutical company that was marginal until it decided to invest all it had (and all it could borrow) to buy the patents for vitamins, hire the best discovers at unheard-of salaries, and manufacture and market over-the-counter vitamins for the first time, with astounding success. This all-or-nothing strategy capitalizes on being the first and the best in your market.

The second strategy capitalizes on leveraging on the ideas of others, taking a good idea and making it better, or finding a better use for it. Akio Morita of Sony flew to the United States after reading about the invention of the transistor by Bell Labs and bought a license to the transistor for $25,000. Five years later, the Japanese were the sole providers of radios to the world.

The third strategy seeks to gain a monopoly over a certain market niche. While this strategy offers limited growth dependent on its host industry, it offers shelter from competition for the same reasons. Alcon realized a niche when it invented an enzyme that quickly became an integral part of cataract surgery worldwide. As momentous as the discovery was, the market for this enzyme totaled $50 million per year worldwide, which was plenty for Alcon, but not enough to motivate competitors.

The fourth strategy creates a customer for itself, not by inventing a new product but by changing the utility, value, or economic characteristics of existing products. It capitalizes on creating a way to equip consumers with a good or service to which they would not otherwise have access. In the 1840s, Cyrus McCormick was one of many American who invented a harvesting machine. The American farmer, for whose benefit the machine was created, did not have the economic means to pay for it all at once, and banks were not willing to provide loans on such a new machine. All of the machines failed in the market, with the exception of Cyrus's machine, simply because Cyrus saw the need for installment buying and instituted it in his young company (Drucker, 1985).

Market Research and Analysis

This is the most crucial step in preparing the business plan. Your marketing research and analysis will lay the foundation for your entire plan, form your strategy, and determine resource requirements. Therefore, this step should be completed before all others. Research is essential for substantiating projected sales volumes and market size as well as finding out whether this product is something that your intended customer will actually buy.

Describe your customers. What is the economic and social reality of your target market? By accurately defining your customers, you begin honing in on your marketing strategy. Part of the research should focus on past industry trends. Examine what similar companies have done and what the outcome was. Use supporting facts and statistics to outline your marketing strategy. Consumer science can be a difficult and tedious process. You must examine socioeconomic trends, political developments, and demographic changes in order to determine what marketing strategy to use.

Compare your product to that of the competitors. What weaknesses leave them vulnerable? Are you capitalizing on those weaknesses? Also state the strengths of the competitors. Briefly demonstrate why companies have dropped out of the industry in past years, and how you will prevent it from happening to your company. Examine the four "Ps" of marketing: Product, price, place, and promotion. How are you doing in each arena? (Kotler, 1999) How are the competitors doing in each arena? Explain why customers will leave the competition for you.

As part of your market research, you should determine the value added by the product, who already has a similar product, and who will buy your product anyway. Identify every major customer group, and note the size of each. Based on the industry trends and your assessment of the advantages of your product over others, interpolate unit sales over the next three years. Be sure to show all assumptions.

A good marketing plan is forever a work in progress. Changing environments call for changing strategy. You must include how you plan to continually evaluate your marketing opportunities while improving your product, expanding your product line, adjusting your pricing, or entering new markets.

The Economics of the Business

Get out your finance textbook; this is your opportunity to impress the investor with some hard numbers. Show your gross margins and operating margins; your profit potential both before and after tax; and your fixed, variable, and semi-variable costs. Once you have laid out your entry plan, you should be able to estimate how much money you will spend, your costs, and your revenue from this. Give a timeline of your break-even analysis and explain any assumptions made. Also provide a timeline depicting when your company will run out of cash and when you will attain a positive cash flow.

Marketing Plan

Describe the marketing philosophy that you will use to accompany the entry strategy that you specified in the product section of the business plan. List the distribution channels that you will use, and support your decisions with sound reasons. Also specify geographic range; your range should not exceed that of you market research. Indicate whether you will sell nationally, internationally, or globally. Indicate the types of customers to whom your product will be sold and what that product will mean to them. What will be the main selling point of the product? Specify whether it will be the low price, the additional quality, the warranty the service, or something else.

In this section, you must include a comprehensive discussion of why you priced the product the way you did. You must discuss price in terms of cost to manufacture: How will it effect your break-even time, your gross margin, and your operating margin? Discuss price in terms of competition. If you plan to undersell your competition, how will you maintain profitability while providing value to the customer? Discuss price in relation to customer payback. Will the value provided to the customer save him or her money in the long run? Show how long it will take for the product to pay for itself. Discuss pricing in terms of market share. Decide whether you are better off making $100 dollars from each one and selling a million of them, or making $75 dollars from each one and selling 3 million of them.

Describe the sales tactics you will use. Explain the distribution process and how you will attain the proper channels. You must describe how you will select sales representatives and the way they will position the product for sale. Illustrate the breakdown of the selling costs and present a sales budget that includes all marketing promotion and service costs.

Service and warranty may be used as a major selling point for your product. Explain in depth the warranty or service program to be offered. Indicate how much it will cost the company to maintain the service and warranty policy. Compare your service or warranty to that of your competitors.

Describe the types of advertising that you will administer if your product is aimed at consumers and trade shows and professional organization participation if you are selling to industry. Include a schedule for the release of your ads or tradeshow appearances. Also give a detailed description of the costs of such events.

Research your distribution channels and devise a sensitivity analysis to show how sensitive your business will be to changes in distribution cost. If your product requires any special care in the shipping process, make a note of it and explain how you will accommodate that need.

Design and Development Plans

Explain exactly where the product lies in the development phase. Give a brief report on capital already invested in the product, and explain how much more you need in order to bring the product to market. Also include the time frame in which you plan to have certain steps completed. Give yourself plenty of time, as most new ventures underestimate the amount of time needed to complete the product.

List difficulties that you have faced in the past, and any difficulties you anticipate. Describe the risks facing your project. Are all of the technologies there to complete the product? Is your product dependent on a nonexistent technology? If so, how do you plan to cope with development issues?

Your business will not last without constant improvement and innovation. How will the product improvement coincide with company growth? Will you be able to keep up the R&D cycle while turning a profit? Explain how you will finance the cost of R&D. If you have not achieved a positive cash flow by the time new development costs begin to pile up, where will you get the money? Explain how your future plans for the product line relate to the sustainability of the company.

This is also the section of the business plan in which ownership of the patent should be discussed. Will you share a patent with anyone, or does your company have exclusive rights? How much time do you have until your patent expires? Will your company have enough of a head start to get their "foot in the door" before other companies catch on?

Manufacturing and Operations Plans

This is the section of the plan where you will list the requirements for the geography, facility type, operating cycle, manufacturing process, and legal issues concerning the production of the actual product. The logistics of operation begin with choosing a suitable location. Carefully consider proximity to resources as well as proximity to customers.

If you are operating from an existing facility, describe the facility in depth, including plant, storage, office space, accessibility, and equipment. If you are starting from scratch, describe exactly what equipment you will need for start-up and how much it will cost. Do not plan to buy everything at once! Rather, set a schedule for obtaining overhead as it is needed. Prepare a three-year time line for equipment and facility acquisition.

Your explanation of the operating cycle should include the lead and lag time that are specific to the type of manufacturing you will be doing. Discuss your inventory system and its role in coping with any seasonal fluctuations you may anticipate.

Your manufacture process should be explained in detail, including the numbers of workers and managers required to operate the plant, cost of labor, and necessary skills needed to operate the machinery. Discuss the possibility of outsourcing some of the manufacture or assembly process. Also discuss the implications of economies of scale, buying in bulk, and potential suppliers.

The legal implications of firing up a manufacturing plant can be extensive. Note any federal, state, or local (e.g., zoning) laws that pertain to you as a manufacturer. Note all costs of obtaining necessary licenses and permits as well as any political activity that could affect your organization in the near future.

Management Team

The organization of the management team must be arranged according to the need of the particular business. Your young venture most likely will find the basic functional structure to be the most efficient. Functional structures are most effective when one product line is being manufactured, which is often the case with fledgling corporations. This structure usually includes a president who oversees R&D, manufacturing, accounting, and marketing departments. Subsequent developments upon further growth of the business and its product line may warrant evolution of the organizational tree into a matrix or divisional structure. These structures allow for product-intensive management of the business (Daft, 2000).

List the key management personnel and their relevant skills and experiences as outlined in the executive summary. Include not only resumes of the principal management players, but also an in-depth analysis of their management performance. This analysis will help to determine their value to the company, which will translate directly to monetary form, as should be indicated in this section of the plan. List the salaries to be paid and the amount of equity to be given to each. Run the salaries of the management team against salaries earned at their last jobs, to show any contrasts and to emphasize dedication.

Investors will want to know with whom they are dealing, including other investors. List anyone who has invested in the company and what type of equity they own. List any employment agreements being negotiated and their corresponding compensation packages. Describe your plan for the board of directors. Indicate the ideal size and the representation you would like on the board. Also list any board members you have already chosen, along with a brief description of why.

Your young company may find it necessary to outsource much of the legal, consulting, advertising, marketing research, banking, or accounting. Be sure to include those whom you have chosen for these tasks, and why.

Overall Schedule

Prepare a realistic schedule for the completion of every task that is critical to the success of your venture. Do this by carefully listing all of the steps and then prioritizing them and assigning them to certain members of the team, along with due dates for each. Then lay the steps out on a time line and mark the most crucial points in the development of your venture. This will enable you to better visualize the crucial points and the necessary progress to reach those points. Discuss any weakness that you see in the time line. Point out any potential delays, and discuss how they will affect chain-reaction delays.

Critical Risks, Problems, and Assumptions

Starting your own small business is heavily laden with risk and uncertainty. One must think objectively and must address all sources of uncertainty, as well as any assumptions being made. This is your opportunity to identify all of the weak points in your business proposal and to explain what you will do to negotiate success. It also serves as an excellent opportunity to prove to the investor that you have thought of everything, carefully calculating your risks and not making any brash decisions. Explain your reasoning for making any decisions that may seem to compromise the success of the venture.

Investors must see that the management team is capable of identifying the risks and that it is honest in portraying them to the concerned parties. Do not try to hide risks or assumptions! A poor decision here will damage your credentials and will void any chance of funding.

Financial Plan

This section of the plan will include actual financial exhibits. If your plan involves an existing company, include actual income statements and balance sheets from the past two years to the present.

Construct *pro forma* income statements, balance sheets, and cash-flow analysis. Use your marketing research, your knowledge of the industry and seasonal trends to derive sales forecasts, production costs, and operating costs. Note all assumptions! Be sure to support your numbers with sound reasoning. Remember that every number you include in your plan is subject to the scrutiny and interrogation of the investor; make the numbers appealing, but be honest!

Building upon the information contained in your overall schedule, construct a break-even chart accompanied by a discussion of break-even sales volumes and price sensitivity (Scott et al., 1998).

Proposed Company Offering

This section summarizes the whole deal. Using your cash-flow projections, decide how much capital you are asking for, and devise a schedule demonstrating when and how it will be used.

Explain the stock offering, type of securities, and any nondistribution agreements by which they must abide, and how many shares of common stock will remain authorized, but unissued, after the initial offering.

Don't be vague! Indicate exactly how the investor will walk away from this venture with his compensation. Identify the exit mechanism, whether it is by IPO, sale, or merger.

Appendixes

Include any helpful material that is too lengthy or detailed to be placed in the body of the report. This may include technical drawings or descriptions, copies of permits or licenses, technical papers, research material, industry information, photographs, charts, and graphs.

The Next Step

Once you have all of the necessary parts and numbers, it is time to review. No matter how good you think you are, have another party look at it. Check for inconsistencies and feasibility. An outside perspective could save you thousands in costly mistakes. Finally, have the revised plan reviewed by an attorney in order that he or she may fine tune the language used and clarify any ambiguities. Now that your plan is complete, go forth and succeed!

References

Daft RL. *Organization Theory and Design, 17th Edition*, South-Western, 2001.
Drucker PF. *Innovation and Entrepreneurship*, Harper Business, 1985.
Kotler A. *Principles of Marketing, 8th Edition*, Prentice Hall.
Scott DF, Martin JD, Petty JW, et al. *Basic Financial Management, 8th Edition*, Prentice Hall.
Timmons J. *New Venture Creation, 5th Edition*, Irwin, 1999.

Further Information

Greenberg J, Baron RA. *Behavior in Organizations, 17th Edition*, Prentice Hall, 2000.
Larson KD. *Fundamental Accounting Principles, 14th Edition*, Irwin, 1996.

141

Engineering Primary Health Care: The Sickle Cell Business Case

Yancy Y. Phillips
The Southeast Permanente Medical Group
Atlanta, GA

Thomas M. Judd
Director, Quality Assessment, Improvement and Reporting
Kaiser Permanente Georgia Region
Atlanta, GA

Within Kaiser Permanente's (KP) Southern California operations, the West Los Angeles (LA) Medical Center's Sickle Cell Medical Program, a national model for comprehensive and culturally sensitive pediatric and adult sickle cell treatment, was the 2001 winner of KP's national James A. Vohs Award for Quality. The program resulted in a drop of 22% in West LA emergency room (ER) visits from 1998 to 1999, compared to a 1% decrease among non-West LA ER sickle cell patients. Similarly, West LA hospital-bed days were reduced by 26% while non-West LA hospital utilization increased by 37% for the same patient population.

This chapter describes how this award-winning, "best-practice" program can be replicated in KP Georgia. It is an example of clinical engineering at the large health care systems level in which methods and processes are engineered to improve health care delivery significantly.

Sickle cell disease (SCD) is significant, affecting historically perhaps an underserved portion of the KP membership, African-Americans, and a newly emerging portion, Hispanic-Americans. Proper SCD management meets several KP Georgia cross-cutting quality improvement themes, such as continuity and coordination of care, integration of primary and specialty care, and community interest in minority health issues. SCD presents an opportunity to improve the quality of life for members with SCD while incorporating service and affordability goals. Therefore, SCD management meets our key business goal of first in quality, first in service, and first in affordability.

SCD was designated by KP Georgia's Quality Forum a health management system (HMS) in December 2001. HMS components to be developed include: evidence-based practitioner clinical practice guidelines; key people resources and roles for physician champion, project manager, and care manager, as well as clear roles for primary care practitioners (PCPs), specialty care practitioners (SCPs), and community resources; education for practitioners, staff, patient, and family members, and patient self-management techniques; data systems, including registry, population risk stratification, and ability to monitor health outcomes; and process improvements led by an implementation team for streamlined care and systems' interventions.

Main expected results are reduced hospitalization, reduced ER visits, and improved quality of life (QOL) to be demonstrated by patient, nursing, and practitioner surveys.

The program will enhance continuity of care by clearly defining the PCP's role and closing the gap on patients who are not firmly anchored to their PCPs. It will be shown how SCD care should be coordinated with PCPs, SCPs, and community resources.

Sickle Cell Business Case

The objective of the sickle cell disease (SCD) health management system (HMS) program is to improve health care outcomes for KP Georgia members who have SCD, and to improve care options for their caregivers and their TSPMG (The Southeast Permanente Medical Group, Inc.) and affiliated practitioners through health education, enhanced self-efficacy, and the effective use of resources.

Market

The market for the SCD program are KP Georgia members with diagnosed SCD, TSPMG practitioners, and the KP Georgia health care delivery system. Program stakeholders are members; families and caregivers; pediatric, adult, and hematology practitioners; and the KP Georgia health care delivery system.

As of April 2002, KP Georgia provided care for 252 children and adults with SCD, or 0.04% of the membership. There were 138 members under 17 years of age and 114 members over 17 years of age. There were 149 females and 103 males. There was an approximately 25% growth in members with SCD from 1997–2001, while overall KP Georgia membership grew by about 10% during this time.

Financial Implications

SCD medical expenses per member per month (pmpm) are 2.7 times greater for adults, and 7 times greater for the pediatric population. Total pmpm of care for SCD members increased 35% from 1997–2001, an average of 8% annually. Total in-patient pmpm for SCD members increased 60% from 1997–2001, an average of 12% annually. The number of emergency room (ER) visits was 173 in 2000, and 231 in 2001, at approximately $500 per visit.

Current KP Georgia HMS programs for chronic diseases include asthma, diabetes, cardiovascular disease (CHF/CAD), and depression. All HMS programs have a primary care physician champion and are managed within the TSPMG PCP health care team (PCP-HCT) medical model.

While KP Georgia does not directly control the hospitals, current PCP-HCT patient-management practice does affect hospital admissions for chronic diseases. Current PCP-HCT patient management, within the context of active HMS disease processes, can reduce treatment delays, especially when an exacerbation of a disease process occurs, decreasing ER visits, hospital stays, and hospital costs, and increasing member satisfaction.

Goals

The program goals are to give members, families, and caregivers the following:

- Self-empowerment, self-efficacy, and shared decision making
- Increased PCP bonding, leading to improved clinical outcomes and quality of life
- Member and patient education classes and educational materials, including telephonic and web-based, designed for SCD and patterned after other HMS programs.
- Health promotion and prevention information and resources in both regular and after hours settings to provide "anticipatory guidance."
- Improved practitioner efficacy and member self-efficacy through use of care management resources

The goals for physicians and health care practitioners are increased TSPMG practitioners' knowledge of, and comfort in, treating patients with SCD, including:

- SCD chronic disease management processes
- Written, telephonic, or web-based educational material and resources
- National, KP Georgia, and community SCD and chronic disease resources

Outcomes

The anticipated outcomes of the SCD program are improved continuity and coordination of care (CCC) for members with SCD. This continuity and coordination of care will result in improved access to needed services as measured by the number of PCPs who attended to an SCD patient with greater than three primary care encounters in a given year. Research has shown that a PCP provides best overall CCC for chronic-disease patients with multiple annual visits. Another result will be appropriate and early treatment when acute and chronic complications arise, with improved integration between primary and specialty care, resulting in decreased ER visits and fewer hospital admissions, decreased average length of stay (ALOS), and overall decreased inpatient costs.

Other outcomes include no decrease (and probably an increase) in PCP and SCP visits and improved patient and practitioner satisfaction, increasing success of short-term and long-term SCD treatment and management.

Several KP Georgia Barriers

The SCD program must consider the following elements that could adversely affect the success of the program:

- The lack of PCP experience and efficacy to identify and treat SCD patients, because they refer to hematologists
- The need to identify SCD patients for purposes of population-based care management, to address preventive issues and systems issues, such as co-pay waivers for same-day visits to different practitioners for SCD
- The inability to readily control hospital admissions and costs, as is done by KP hospital-based regions
- Unique community resources that are not incorporated into care plan management
- The lack of practitioner knowledge about adult and pediatric SCD differences
- Inadequate specialty care relationships and feedback systems
- The lack of SCD-specific pain medication guidelines

SCD Program Description

Management and Key Roles

SCD Physician Champion

The program must have a physician champion. Such an individual is necessary to carry out the following responsibilities:

- Serving as the leader-facilitator for disease content authority
- Serving as the leader-facilitator for SCD awareness in primary care throughout KP Georgia
- Serving as the leader-facilitator for guideline development, review, renewal, and implementation
- Serving as the leader-facilitator in determining and implementing SCD measurement outcomes
- Serving as the national and local representative for TSPMG on SCD-related committees, task forces, and conventions

In the start-up and following year, this position would be 0.2 FTE, in keeping with other HMS programs. It will allow the SCD champion to give three disease-oriented, continuing medical education (CME) lectures to KP Georgia PCPs during the first year of the SCD program, then yearly to twice-yearly updates regarding SCD education, care, and management.

SCD RN Care Manager

Patterned after the West LA program, the SCD program will employ a 1.0 FTE RN care manager, who will provide a comprehensive "wrap around" service delivery to SCD members; an additional 0.2 FTE for case management and care-management assistance will be provided by Quantitative Risk Management (QRM) to ensure continuous coverage. The care manager will also work with the SCD Work Group, SCD champion, HMS Steering Committee, HMS Implementation Committee, and Quality Forum to identify ongoing delivery system modification needs and to help facilitate system changes.

Other responsibilities of the SCD RN care manager include the following:

- Ensuring that patients have regular appointments with a hematologist and their PCP to improve treatment plan compliance
- Providing educational counseling for the newly diagnosed patients on disease history, current treatments, primary care needs, and the care delivery system
- Educating patients and their families regarding the importance of following self care and home care plans to avoid complications
- Educating patients and their families regarding inappropriate ER use
- Providing patients and their families with information regarding health care resources and referrals within KP Georgia and the community at large; being knowledgeable about insurance coverage and the benefit structure for SCD patients
- Establishing and maintaining good rapports with patients and their families by being aware of their health care issues (e.g., making frequent follow-up calls, directing them to support groups, and making home visits)
- Serving as a contact person for subspecialist consultants, TSPMG practitioners, advice nurses and call center requests; include ability for Care Manager "book" same-day appointment with PCP for patient in crisis
- Collaborating with community agencies to promote information sharing, KP's community involvement, staying abreast of upcoming programs and events and developing resources
- Assisting in the facilitation of a smooth transition from a pediatric to an adult SCD patient to ensure continuity of care
- Participating in national and regional educational conferences; providing educational presentations to school and day-care nurses, staff, teachers, and administrators to help promote awareness of SCD
- Ensuring that pediatric SCD patients have a treatment plan on file at their school, which may include an individualized education plan, modified physical activity, and/or special consideration regarding classrooms' proximity
- Partnering and assisting SCD physician champion in arranging educational in-services for KP Georgia staff
- Being responsible for maintaining a disease registry and being aware of the progress of hospitalized SCD patients and their potential discharge dates, and acting as a referral source and point of contact for feedback with case managers in QRM Department
- Ensuring appropriate follow-up after acute care, hospitalization, or ER visits, using available next-day information about previous care
- Developing and implementing with Pharmacy an efficient prescription refill process (e.g., for antibiotics and pain medication).

SCD Program Manager

As in all KP Georgia health management system programs, there will be a 0.5 FTE program manager for SCD. The RN care manager will report to the SCD project manager, who will report to the director of population-based care (PBC). Responsibilities of the SCD program manager include:

- Facilitating the SCD Work Group, reporting to HMS Steering and Implementation Committees and the Quality Forum
- Overseeing development of all needed HMS components (i.e., guidelines, personnel, patient and practitioner and staff education, data systems, communications, and documentation of program interventions and outcomes) as appropriate

Other Resources Needed

Patient Registry, Tracking, and Follow-Up

At present, there is an informal registry of 252 members; a more formalized database will need to be developed. KP Georgia utilizes the pharmacy-based and web-based POINT Computerized Clinical Tracking System for several applications. POINT is used in KP Southern California and includes an automatic field for SCD data. Its web-based feature allows more flexibility in its access by practitioners and staff. Until it is implemented, KP Georgia will explore the use of the computerized clinical system that is currently used for dermatology and breast health tracking.

Tracking criteria include all patients who meet the designated SCD diagnostic codes. Clinical indicators to be tracked for each patient include at least childhood and influenza immunization history as well as other SCD items in POINT, such as pharmacy refills. Linkage of this tracking capability to the data warehouse will allow next-day follow-up related to TSPMG primary or specialty care or ambulatory health care (AHC) visits, as there is nightly download of this information from the KP CADENCE appointment system.

Sub-Specialty Care

Subspecialty care is essential for the care of SCD patients, due to a variety of factors. While related complications are relatively common in SCD patients, the management of complications frequently involves interventions that are uniquely targeted to SCD patients and that are gleamed only from a working knowledge of the SCD literature.

For pediatric patients, virtually 100% of the subspecialty care access will be directed through the Pediatric Hematologist at Children's Health care of Atlanta (CHOA) Scottish Rite Hospital. PCPs will rarely initiate subspecialty access; essentially, the pediatric hematologist is the co-gatekeeper.

For adult patients, the PCP will assume a much more central role in the subspecialty care access. All patients will see the adult hematologist two or four times per year. The adult PCPs will direct care to the subspecialty care physician based on practice guidelines or on the recommendation of the adult hematologist, other subspecialists, or the SCD physician champion.

Occasionally, adult SCD patients will be referred to the SCD program at Grady Hospital for development of a care plan. These referrals will be based on established criteria. Examples might include an SCD patient who has recalcitrant leg ulcers or chronic pain (not acute pain) that is unresponsive to a conventional outpatient protocol. Fewer than 10% of the adult patients are expected to need this type of referral.

Community Resources

The SCD Foundation of Georgia might be a resource for the families of children with SCD, for additional education as well as participation in Camp New Hope and other childhood SCD-related programs. The Sickle Cell Disease Association of America web site also provides key related information: http://sicklecelldisease.org/

The CHOA Pediatric Hematology SCD nurse practitioner and social worker will be resources, as well. At times, individualized education plans, homebound teaching, or modification of physical education or school classes will be needed, and they will be instituted though a joint effort of the KP Georgia SCD RN care manager and the CHOA pediatric hematology SCD nurse practitioner and social worker.

The schools need to be included in the SCD management team. The PCP and pediatric hematologist and the SCD RN care manager will develop treatment plans that all pediatric SCD patients will have on file at their schools.

The KP social worker will be available for adult patients who need help managing job-related issues that relate to having a chronic disease.

QRM Case Management

The care management of members with SCD will be a collaborative effort between the QRM and PBC departments. The role of the QRM case manager will be to partner with the SCD RN care manager to facilitate care coordination across the continuum, particularly to focus on hospitalized patients and to provide back-up care coordination when the care manager is not available.

Recommended Implementation Tasks and Timetable

At program initiation, the SCD champion will be named, and the SCD RN care manager and SCD project manager will be hired. The SCD champion will develop clinical practice guidelines in partnership with chiefs of medicine and pediatrics and KP Georgia population-based care department and will be in place and approved within six months after initiation. The CME schedule will be developed with the practitioners for months seven through nine. The SCD care manager will have an office at Gwinnett Medical Office, near the SCD physician champion, for the first six months, to facilitate rapid program development and a clear understanding of clinical issues involved.

The SCD RN care manager's next steps will be the following:

- Establishing contact with adult and pediatrics SCD subspecialists (e.g., hematology and pulmonology); ensuring that all patients on SCD registry have a PCP and hematologist; and meeting with relevant case managers, benefits representatives, pharmacy managers, social workers, and CHOA nurse practitioners
- Visiting Grady Hospital's SCD clinic and doing follow-up with Sickle Cell Foundation
- Creating SCD Program brochure; making educational and reference folders for newly diagnosed patients, school and childcare facilities, and new KP members; and preparing educational materials for teaching health care teams about SCD
- Compiling resource lists for the following services:

Genetic counseling	Support groups (parents, teens, and adults)	Stress management
Pain management	Proper nutrition-appropriate exercise	Tutorial services
Stroke prevention		Infection prevention
Smoking cessation	Depression screening	Forms: Self care, pain

- Developing a formal program development plan and overseeing its implementation, to meet established goals and the timetable set by Clinical Affairs and the SCD Work Group

Conclusions

The SCD HMS Program will increase health care options for SCD members, their families, and their primary and specialty care practitioners through health education, self-

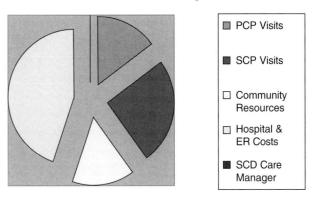

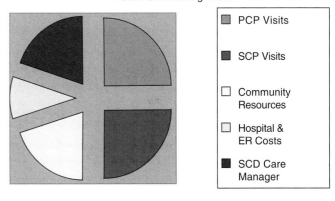

Figure 141-1 Expected differences in personnel, outcomes, and resource use between the current situation for SCD management in KP Georgia and the proposed SCD-management program.

efficacy, and the effective use of resources on the part of members and practitioners. By adopting this program, KP Georgia will realize a decrease in total health care costs and an increase in perceived quality of service and quality of care outcomes.

Projected hospital savings in the program's third year are $124,000 for pediatrics and $135,000 for adults; 30% and 40%, respectively. Projected ER savings in 2005 are $34,000 for pediatrics and $27,000 for adults, a 40% savings. There is a net $158,000 program savings in year three. Four-year net present value (NPV) for the program is $85,000.

Improved satisfaction and quality of life are to be demonstrated by surveys, using the West LA tools. Quality of care relating to continuity and coordination of care will be improved access to one's own PCP to direct appropriate services, resulting in decreased ER visits, hospital admissions, and ALOS. No decrease (or increase) in PCP and SCP visits is expected. Other SCD clinical management process and outcomes measures will be determined.

The KP Georgia health care delivery system will realize greater effectiveness and efficiency in service provision. This could impact not only SCD HMS members, but other HMS registry members, as well. This can occur as we overcome the barriers cited and provide solutions to our delivery system's continuity and coordination of care challenges.

In summary, Figure 141-1 shows the expected differences in personnel, outcomes, and resource use between the current situation for SCD management in KP Georgia and the proposed SCD-management program.

142

Global Hospital in 2050—A Vision

Yasushi Nagasawa
Department of Architecture
The University of Tokyo
Tokyo, Japan

Edward Sivak
Upstate Medical University
Syracuse, NY

Errki Vauramo
Care Facilities SOTERA
Research Institute for Health
Hut, Finland

As citizens of the world look ahead to the future, their expectations for health care should be viewed first from the standpoint of demands placed on health care systems.

The population of planet Earth grew to one billion by 1804. Then, 123 years later (1927), the population reached two billion; 33 years later (1960) three billion; 14 years later (1974) four billion; 13 years later (1987) five billion; and 12 years later (1999) six billion. Population is growing at an ever-increasing rate (Lewis, 2000). Recent projections show that the world population is estimated to be 8.9 billion in 2050, even though birth rates have fallen in many countries. Most of the gain will occur in the developing world, where countries are already overpopulated. Roughly 10 billion inhabitants in the latter half of the 21st century will have to cope with outstripping natural resources of the Earth.

Jonathan C. Lewis, Academy of International Health Studies, California USA, pointed out:

"[Six] billion people are now living on Village Earth. With me, imagine shrinking this multitude to just 100 people, all living together in a small village. One tiny village has 58 Asians, 13 Africans, 10 Latin Americans, 8 Europeans and 6 North Americans, 51 Villagers are female; 49 are male. Half the Village is under 25 years of age. The men live on average 61 years. The women live 3 years longer. Our village is a Tower of Babel, 20 villagers speak Chinese, 8 Speak English, 7 converse in Hindi, 6 in Spanish, 5 in Russian, 4 in Bengali, another 4 in Arabic, 3 in Portuguese, and 2 each communicate in Japanese, French and German. 30 villagers are Christians, 70 are not. 37 have jobs. 18 are illiterate and 1 has a college education. 80 live in substandard housing. 50 villagers are malnourished and do not have access to basic sanitary toilet. 96 people share 50% of the village's health care resources. The 20 poorest villagers consume just one and a half percent of all goods and services; the 20 richest villagers eat 45% of the meat and fish, use 59% of energy, 84% of all paper and own 87% of all vehicles. 5 people control 50% of the village's total wealth, and those 5 individuals live in a special neighbourhood called the United States. Village Earth is the world in which we live and the world in which health care must be delivered and financed (Lewis, 2000)."

Demands on health care systems will be partly influenced by diseases that arise from malnutrition. Populations in both India and China have already exceeded over one billion, and 53% of Indian children are malnourished. The observation of the Worldwatch Institute that half of the world population was starving in the year 2000 and 1.2 billion people had no access to clean water suggests that simple preventive measures can reduce strains on future health care systems. Health care expenditures of the world population, per capita, differ in large scale from U.S.$20 to $3700 in a year. The world health care system varies across populations. For example, about 70% of the world population does not benefit from the use of X-rays, although this technology is more than 100 years old. Medical education varies, and the delivery systems differ. One could easily believe that the situation will be the same after 50 more years. Conversely, equitable opportunities for economic development, peace and prosperity, inter-area, access to quality and affordable health care may not be critical issues. As the World Health Organization (WHO) recently redefined issues of human health to have physical, mental, social and spiritual aspects, the future may be associated with a societal commitment to improving the general well-being of all inhabitants of planet Earth. Appropriate provision of health care, medical care, and welfare services to all is important and must be sustained (Nagasawa, 1992). The major challenge is in knowing how to achieve this goal.

The challenge of making educated guesses as to what the situation will be in 2050 is demanding. On the other hand, those who will be in their 70s in 2050 are all living today. They have our genes, and they will suffer the current diseases: Infections in developing areas, and cancer and cardiovascular problems in the Western world. To develop a new medicine takes at least 10 years; to test it clinically takes 5–7 years; and to have it accepted by regulatory agencies and practitioners takes 5–7 years. The total process can take 10–20 years. The life cycle of hospital equipment can be between 10 and 25 years. On the other hand, the medical system is conservative and slow to change. Most of today's hospitals will work at the same site, but the building will be refurbished. Many treatments are changed, but slowly new devices may make procedures easier, and more effective pharmaceuticals may shorten treatment time dramatically. Thus, as the authors present their visions, the reader may have some disagreement. Nevertheless, the reader must place such directions in the abstract and must think imaginatively. To permit such thinking, detailed discussion of future diseases is left out because it would be beyond the scope of this handbook.

This chapter focuses on the global view in health care planning and architecture, both in developing and developed countries, on the basis of WHO documents and the authors' experiences as consultants to authorities in developing regions in Asia and in Baltic region. Additionally, experience is included from the current project of Global University Programs in Healthcare Architecture (GUPHA), an international organization linking and promoting educational and research programs of 11 technical universities in health care architecture (primarily aiming at projecting the scope of health care architecture in the year 2050) is included.

Factors influencing the prediction of development health care architecture can be grouped into three categories as follows: (1) country and culture; (2) provision of services that are critical to the delivery of care; and (3) environment of activities. Each of these factors is interrelated; if separated, they will almost certainly result in deficiencies of design or operation. Even if these deficiencies are overcome, social acceptance might be lacking.

Country and Culture

These criteria can include the scales of north/south, east/west, old/new, and developed/developing. Ten categories to consider are as follows:

1. Population (e.g., total number and age/sex structure)
2. Land area (e.g., natural border and man-made border)
3. Population density (e.g., rural/urban and major cities)
4. Climate (e.g., temperature, humidity, rainfall, constant wind, monsoon, savannah, and frigid areas)
5. Terrain (e.g., mountain, plane, forest, island, and earthquake-affected)
6. Demography (e.g., birth / death rate, population growth, infant mortality, life expectancy)
7. Epidemiological (e.g., infectious diseases and major cause of death)
8. Health financing (e.g., medical expenditure and budget, income/outcome)
9. Health provision system (e.g., insurance, network and health personnel)
10. Political/social situation (e.g., government, democratic, socialist, social classes, education, transportation and economy)

Global Population in 2050

At this time, there are several predictions available, all ending up with 9 billion people living on the globe in 2050. On the basis of analyzing world population growth in 2050, the distinction of four continents or geographic areas can be seen. Among the countries that will fully utilize high-tech, western medical technologies, the United States will increase its population in 50 years to about 400 million, while European countries and Japan will decrease their populations. The main increase of population will take place in the regions near the equator (e.g., Africa, India, and Brazil). The aging population, especially over 80 years old and mainly female, will grow all over the world, but particularly rapidly in Asia.

Healthy Environment

According to the Agenda 21 statement from the UNCED (1992) Rio de Janeiro environment summit, 60% of the Earth's population will live in cities in the year 2050. The critical issues facing these populations and their governing bodies are numerous and include the following: Water/air quality and pollution; nutrition and food supplies; congestion of communication and transportation systems; energy resources and consumption; noise, stress, and access to open spaces and recreation areas; responsive education, social, and criminal justice systems; sufficient employment and housing opportunities; and brutal commercial and residential development. Without proper intervention, the daily news will discuss the ecocatastrophe of the greenhouse effect, the rise of sea level, and perhaps unrelenting air pollution. Looking more carefully at the data, one finds that the sea level 20,000 year ago was 124 meters below the current level. The increase during the last century was 15 cm. The expectation is that sea level will increase 50 cm from the 1900 level by 2100. The increase could have an effect on societies located along the deltas of large rivers, such as the Ganges, Nile, or Mississippi. Some of the islands in the Caribbean and in Polynesia are having difficulty because of storms, and they may disappear. This may require fast mobile health care teams. The catastrophes, as difficult as they might be for an individual, might not affect the increased world population.

Beyond natural catastrophes are epidemic diseases, such as HIV, which produce dramatic effects south of the equator in Africa or in large cities in Eastern Europe. Of equal magnitude are new strains of tuberculosis that are resistant to modern drug regimens and that could turn into large, worldwide epidemics as larger numbers of the world's citizens travel the globe. Although the WHO and the United Nations are currently developing standards for treatment and consensus statements on such problems, a more uniform acknowledgement of their authorities should take place. Universal funding of worldwide problems would strengthen defense against the health care problems of the world.

Man-made air pollution will increase the incidence of lung diseases and will continue to cause serious problems to individuals. On the other hand, many people who live together with smokers will suffer, but with a relatively long life expectancy. There are many reports available from London, rural areas, and Tokyo demonstrating ways in which clean air has reduced health problems. We have to remember that the incidence of tuberculosis was reduced by improvement of urban sanitation systems well before effective treatment with curative drugs. These famous, well-documented success stories indicate the importance of a healthy environment also in the future. To plan healthy, urban environments in a future metropolis is perhaps the most effective tool in preventive medicine.

Funding of Services

The status of the economy in a geographical area will be dominant in the next 50 years. In most countries, resources in health care will be limited to under 10% of GNP. The need to use medical technology will grow perhaps 15% annually, but the GNP will grow less than 4%. Governments would like to reduce health care costs even below the present levels. Resources will be the same or less than today's levels. However, the societies are aging, so each elderly person will have fewer resources to use. The pressure on health care providers for more economical solutions is increasing, thus necessitating that priorities be set, and services optimized. Beyond developed health care systems, the economic problems of the Third World are critical. The gap between affluence and poverty is easily understood when one considers that health care in many African countries is based mainly on donations today that might not be available in the future. The focus in developing countries will be on preventive measures such as providing adequate nutrition, clean water and air, and immunization against diseases that have been eliminated in developed countries. Such measures are inexpensive when compared to measures that are intended to cure disease. On this optimistic note, we must be mindful that the diseases of developed countries could become health care nemeses as cultures advance in affluence. When health care funding is limited to U.S. $30–100 per person per year, clean air and water represent the most effective use of money for developing countries. Future research must focus on preventing diseases such as cancer and heart disease, which plague developed countries today.

Provision of Services

Variety of Health Care

These criteria may include the range of activities from a human's birth to death, health promotion/prevention/care/cure/rehabilitation, and if only 10 items were to be considered, the following would illustrate the point:

1. Health promotion (e.g., health education and physical fitness)
2. Family planning (e.g., birth control and pre-/post-natal care)
3. Disease prevention (e.g., vaccination and daily life)
4. Primary care (e.g., general practitioners, doctor's offices, community health, and first aid)
5. Secondary acute care (e.g., hospital services)
6. Tertiary medical care (e.g., teaching hospitals and specialized hospitals)
7. Rehabilitation (e.g., physical and stroke)
8. Accident/emergency treatment
9. Care for special-needs populations (e.g., handicapped and elderly, mentally deficient)
10. Terminal care (i.e., hospice care)

Trends in Health Care

Regardless of the technology, therapeutic intervention, or preventive measure utilized in improving the health of the world population, history demonstrates that the most sophisticated health care delivered by the sub-specialist in the most highly specialized institution becomes common place within 10–20 years. For example, the treatment of hypertension in the United States in the 1950s was provided by only the most knowledgeable researchers. By the 1970s, these researchers focused their efforts on the education of the primary care physician on the benefits of treatment of hypertension. By the dawn of the millennium, the worldwide treatment of hypertension had become the responsibility of the primary care physician. The dilution therapy of cardiac infarction today is often started in primary care units or in ambulances. Similar comments could be made about the diagnosis and treatment of asthma. We can expect that technology, therapeutic interventions, and preventive measures will diffuse in a similar way from the developed countries to those developing their economic profiles. Although one may take a pessimistic view, there is a hint of better things to come as developing countries benefit from the lessons of the well-to-do countries as they strive to improve the economics of health care. This concept also can be extended in the expectation that the technology in the tertiary health care center today will become the technology of the primary health care center of tomorrow. As one attempts to define the physical architecture of the hospital of the future, we must keep in mind that as the world grows smaller, communication improves and becomes more universal. In a sense, communications, architecture, diagnostic technologies, and therapeutic interventions begin to define the concept of global hospital 2050. To understand this statement, the following areas should be examined:

- Improved imaging technologies
- Improved diagnostic methods
- Improved surgical interventions
- Genetic and immunological therapies
- Improved preventive measures
- Reduced burdens of individual responsibility for the cost of care

Imaging Technology

At the present time, there is no doubt that imaging technologies have become more sensitive and specific. This trend can be seen easily when high-resolution CT images of the thorax are utilized in the diagnosis of inflammatory lung diseases or PET scans are utilized to image solitary pulmonary nodules. The former technique has made it possible to examine noninvasively the anatomy of the inflamed lung, and the latter to determine the presence of cancer. Magnetic resonance imaging (MRI) technology has advanced to the point where vascular imaging is now possible. Again, noninvasive methodology prevails. The operative theme is clearly a trend toward noninvasive diagnostic imaging. Over time, one could expect that these imaging technologies will become more portable and available in locations far beyond the walls of the traditional hospital. Access will become more universal. With respect to imaging modalities, the hospital of the future may become the command post from which images are ordered and analyzed. Improved communication networks will make it possible for the isolated practitioner to be connected to the most sophisticated medical centers. In the same sense that health care networks are developed today, the network of the future will de-emphasize physical facility and will emphasize communication and diffusion of technology to disparate locations.

Improved Diagnostic Methods

Improved imaging technology will make it possible for visualization of diseased organs and tissues by noninvasive techniques. The presence of cancer can be determined by such methods without tissue sampling. This trend will lend further credibility to minimally invasive surgery. For the future, one might project that laboratory studies done only in central laboratories can be performed in peripheral sites through the development of intelligent sensors that might relay abnormalities in blood detected by noninvasive methods to central processors. By the same token, one might foresee that tertiary health care facilities could become command posts from which medical care is directed. Technology will become more portable and smaller, to the point where diagnostic laboratories will be widespread and freely accessible. In one sense, this will reduce demands on central laboratories, but it could increase burdens on health care centers as requests for therapeutic intervention could increase. Improved imaging and diagnostic methods will be combined. Direct visualization of various organs and tissues by fiberoptic instruments will permit diagnosis of abnormal tissue without necessity for tissue biopsies. This will reduce demands for central laboratory services but may again increase demands for therapeutic services.

Improved Surgical Techniques

The future of surgical techniques is already being defined by the development of minimally invasive surgical centers. Within such centers, the disciplines of radiology, anesthesia, surgery, and medicine interact and use combined services. Imaging modalities appear in operating rooms. Radiologists may perform certain procedures that were performed only by surgeons in the past. Surgeons may utilize instruments that allow visualization of organs or cavities of the body and surgery on them through small incisions. Tissue damage during surgery is lessened, wound healing is hastened, and recovery time is shortened. In the future, utilization of acute care hospitals will be by patients with higher acuity. Recovery and rehabilitation will be accomplished in sites peripheral to the hospital and in the home.

Beyond the trend in minimally invasive surgery, there is expectation that some surgical procedures will be performed by robots directed by surgeons in remote locations. The possibility of a global hospital system based on imaging and real-time visualization will require mature communication systems. The architecture of the hospital of the future may be based on communication architecture. Again, demands for large centralized hospitals will lessen as technology becomes more portable and access to health care more widely distributed.

Current techniques in transplantation of solid organs such as lungs, hearts, and kidneys may change with the expectation that these organs will be obtained from laboratories rather than harvested from human donors. Stem cells could be programmed to generate new organs. Alternatively, these cells could be introduced into the body through infusion, with proper genetic programming, to replace diseased or damaged tissue or organs. Laboratories could be necessary to support organ development or maintain stem cells.

Genetic and Immunological Therapies

The Human Genome Project has made it possible for researchers to better understand the molecular basis of diseases such as asthma, inflammatory bowel disease, arthritis, and, to a certain extent, coronary artery disease. Therapeutic modalities are now directed at events occurring at the molecular and genetic levels. As the pathophysiology of disease is understood, the genetic predisposition to such disease can be determined. Thus, preventive medicine will be based on improving environment and eliminating some stresses that cause disease. Prevention will be based not only upon cleaning our environment but also upon identifying those individuals who should avoid additional environmental stresses.

Our current knowledge of the treatment of infectious diseases is based on prevention, immunization, and treatment with microbial agents if prevention and immunization fail. As medical science learns more about the human body's host defense mechanisms, treatment of infectious diseases can be based on the enhancement of host defense rather than with antimicrobial agents.

Improved Preventive Measures

The concept of a global hospital for 2050 implies that technology and the ability to apply medical knowledge will be universally distributed. The diseases that plague developed countries today will be better controlled and, perhaps, eliminated. At the beginning of the 21st century, such countries are already experiencing a reduction in deaths from coronary artery disease, but these same countries are plagued with infectious diseases (such as tuberculosis), not because treatment is not available, but because of reservoirs of infection created by immigrant populations and others infected with the human immunodeficiency virus (HIV). In the future, these same diseases will be controlled not because of the measures instituted in the developed countries, but rather by the efforts of these countries to help developing countries to treat and eliminate these diseases. In the natural evolution of countries and economies, developing counties will benefit from the lessons learned from developed countries that have conquered diseases that burden their health care systems. Economies of scale will be realized. Future generations will enjoy the benefits of better diets and maintenance of good physical condition. According to the Organization for Economic Cooperation and Development (OECD) health statistics, the life expectancy over the last 30 years has increased an average of 2–3 years for both males and females in most countries. This trend will continue, and healthy life will be longer, and the portion of inactive life before death will be a smaller percentage of the total life.

Reduced Burdens of Individual Responsibility for the Cost of Care

As the 21st century enters its infancy, the citizens of the world have developed a keen sense of responsibility in caring for themselves collectively. What is good for one also must be good for the majority, and vice versa. By the same token, each individual citizen should contribute to the good of the group. In the past, there has often been social upheaval because of discrepancy in the distribution of wealth and resources. As the world's health improves, so will its economy. Hopefully, this evolution will also create equitable distribution of health care facilities and reduction in the cost of care to the point where it no longer will be necessary for the individual to assume total responsibility for the cost of medical care. The concept of welfare will be replaced with community care.

Community care as a responsibility will develop as the number of people in the world over 80 years of age increases to about 400 million by 2050 (see Figure 142-1). When one considers that 50% of these individuals will live in Asia, it is not surprising that implementation of community care has started in Japan and that interest has intensified in China. Alternatives to acute-care-hospital delivery of medical care will be realized in the home, group homes, and nursing homes.

Medical Cultures

The Utopia for which citizens of Planet Earth have searched since the dawn of intelligent life may exist only as an ideal. Attempts to materialize the concept are influenced by the culture of those who attempt its expression, and, in a sense, can be extrapolated to the concept of "Global Hospital 2050." The moral and cultural issues that influence the delivery of health care can be identified in the United States, Central Europe, the Nordic countries, Russian, and Japan. Each has different priorities. As global communication advances, and as storehouses of human knowledge become standardized with methods of cataloging and retrieving information, the benefits and shortcomings of techniques and therapies will be understood more universally. Such understanding will reduce cultural differences in the delivery of health care. For example, what is considered to be Chinese medicine today will be integrated into western medicine. Anecdotal medicine based on valid experience will become increasingly more evidenced-based in the future.

Problems are different in the Third World, where resources are limited by economic and technological constraints. Beyond these obvious facts, one also must consider that technological development takes place in a stepwise fashion. Initial heavy monetary investment in the form of capital and manpower gives way to economic implementation and cost-effective utilization. Through all of these steps, medical personnel become educated as to methods that will affect the most cost-effective and satisfactory outcomes. What is known as state-of-the-art technology or medical practice today in developed countries could become the basis for best practices in 2050 in countries that are considered to be the Third World today. This seems to be a logical progression such that the time may come when worldwide communication and dissemination of knowledge will eliminate what we know as the Third World.

By 2050, we could see technology dividing itself into two cultures—one of developed countries, and one of developing countries. The former would be based on communication networks linking imaging technology, laboratory information, and tasks performed by robots in laboratories, operating and recovery rooms, and intensive care units. Wireless technology will reduce the expenses of hardwire connection of devices. For the latter, labor costs could be relatively inexpensive, and economic development in what we know today as the Third World could be possible by making health care in and of itself a component of each country's gross domestic product. In 2050, medical technology in these countries might be only at a 2001 level. However, as the communication networks of developed countries mature, they will reach worldwide acceptance, permitting linkages and dissemination of knowledge to the point where a universal health care system is available to all, regardless of the maturity of a country's economy.

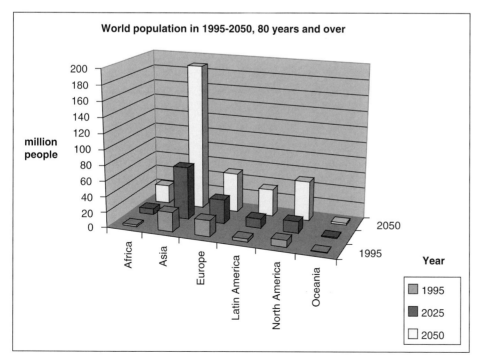

Figure 142-1 World population 80 years and older, 1995–2050. (From Graeme Hugo, *Over to the Next Century: Continuities and Discontinuities,* http://www.un.org/depts/escap/pop/apss141/chap8.htm).

As communication improves and medical knowledge becomes more universal, global efforts to improve the health status of entire populations will increase. Hospitals will form co-operative nets over state boundaries, allowing medical knowledge and skill to diffuse worldwide. This trend will accelerate the development of universal practice patterns, technology development, and, perhaps, hospital and health care facility architecture.

Environment of Health Care Activities

The Variety of Environments

A wide spectrum of facilities to be used for health care invites analysis. These areas include the following:

- Tools (e.g., installation and equipment)
- Rooms (e.g., bed rooms and component)
- Houses (e.g., private houses, group homes and collective houses)
- Community facilities (e.g., shops, schools, offices, health-promotion centers, sports centers, fitness centers and community centers)
- Ambulatory facilities (e.g., clinics, group practice offices, health centers, diagnostic centers, day care centers, sheltered workshops and day surgeries)
- Halfway facilities (e.g., halfway houses, community hospitals, and rehabilitation centers)
- Acute-care centers (e.g., emergency centers, general hospitals, women's centers, children's hospitals, cancer centers, tertiary medical centers, circulatory disease centers, and critical care centers)
- Assisted livings (e.g., nursing homes and homes for the elderly)
- Terminal care facilities (e.g., hospices and cemeteries)
- Mobile hospitals (e.g., hospital ships, flying hospitals, and space ships)

With the above areas in focus, we can expect the functionality and quality of buildings to improve to fit the medical cultures of the areas in which they exist. Over time, there should be agreement on standardized environments and medical practice that will define physical architecture. The end result will be a higher quality of care for all.

Trends in the Design of the Acute Care Hospital

Trends in the design and utilization of acute care hospitals are beginning to suggest that large centralized facilities can become inefficient and costly. As various medical technologies have become more widely distributed, networks of smaller facilities have become more efficient and have created the concept of regionalization. It is this regionalization that will add a human element and will permit systems to be adapted more to the culture of areas or regions served.

The modern hospital began to grow up during the 1950s. New functional units, such as radiology departments, clinical laboratory units, central sterile supply departments, and medical record centers were technology-intensive and required centralization for economical reasons. A compact hospital block for 800–1200 patients was developed. During the 1960s, British authorities and the WHO recommended building three 1000-bed acute-care hospitals for each one million inhabitants in order to improve economic efficiency and to use professional resources fully. The development in building technologies allowed the development of broad buildings with artificial illumination, clean-air ventilation, and electrical heating/cooling systems. At the same time, natural sunlight, fresh open-air circulation, and optimum room temperature was lost. Additionally, the centralization of hospital functions created the various complicated traffic systems for people and material inside the building. Hospital staff and patients may walk long distances through various departments during the day and may wait in each one for prolonged periods of time for testing or for service. With such obvious shortcomings, hospital designers began to explore more compact building shapes in order to reduce cost by reducing external wall-to-floor ratios. The aim was to shorten walking distances among relevant departments. The result is that the modern hospital building looks more like a factory than a human environment. Patients who present to such institutions often feel separated from their emotional and physical support systems.

Complicating the impersonality of large health care institutions, the development in medical technology continues, and no limit can be seen. New procedures will continuously change the space requirement. New methods implemented by newly trained medical teams will require a compact department with inpatient wards, outpatient departments, and diagnostics rooms with possibilities for procedures. The supporting services will receive more attention. The central sterilization systems, storages, and material supplies for kitchens and departments will work more on industrial principles. Logistics and process control will increase; for example, requirements for change, in the 1990s Nordic acute care hospitals, 3%–5% of floor area within these hospitals was annually under refurbishment. With the expectation of continuous change, buildings in the future must be able to accommodate periodic economic reconfiguration without major structural change. Changing technology might require installations to be changed, perhaps in 8–12-year intervals. Refurbishment is expensive and will more so be in the future. The hospital communication net will be complex and needs special knowledge for planning flexibility and for maintenance of the software. For detection of hospital-acquired infections, new automatic-control technology alarming through the hospital communication system will control epidemics and will identify potential epidemics before they start. The trend is toward an intelligent building where basic technical networks allow fast and easy reinstallation of inside walls and the reinstallation of equipment. Beyond versatility in modification lies the requirement for smaller buildings that are more widely distributed with extensive communication systems creating more of a global concept of health care. Beyond versatility in modification lies the requirement for smaller buildings where the function is centralized to departments that do not carry labels of medical specialties. Rather, patients will be grouped on the basis of need, such as critical care and orthopedics. The units will be more widely distributed in the form of a village with extensive communication systems to centralized departments, near or far, creating more of a global concept of health care.

The demands on health care facilities and environment vary. The hospitals to be used in 2050 in OECD countries already exist. New, effective, high-tech hospitals will be needed in the United States, where the population will increase. Similarly, in 2001, the British government announced a need for 100 new hospitals to meet the demand for medical service. Several new-model hospitals will be built in developing areas to be a model for local development. Conversely, in Europe and Japan, where populations are decreasing, few new hospitals will be built. By the same token, the absolute need of more health care services is characteristic of countries such as India, China, and Russia, where some 20,000 hospitals need refurbishment and enlargement to accommodate modern technology. At the lower end of the spectrum, provisions must be made for basic health care and environmental amenities such as food, potable water, and sanitary installations in the regions near the equator.

In this environment, hospital sites are stable, and managers have conceptual plans for future medical programs. Such programs call for modification and enlargement of existing buildings. Generally, strategic plans extend for 20–30 years, while a building's life expectancy is 30–50 years. Thus, architectural plans should allow for stability through two changes in strategic planning and two renovations and expansions.

Network of Smaller Scale Buildings

The compact hospital body with patient towers has remarkable difficulty in expanding its functions. Overall administrative costs in a large hospital complex are growing and difficult to control, but by the same token, the special care hospital functioning at a lower level works more efficiently but cannot implement all of the new methods. However, as knowledge bases become more widespread and communications networks improve, portable technology will make specialty centers more universally distributed. Such development will further strengthen the trend toward ambulatory care. The effect will be a global reduction in large, acute care facilities, a process now already well-known in Europe and the Unites States. In 2000, it started in China and it is expected to change the situation dramatically in Russian health care delivery.

The integration of medical and social services back to the society is foreseeable. Health care facilities will be placed where people will visit and stay with fewer physical and psychological (i.e., less institutional) barriers. The trend might well be toward a medical-mall concept. The basic elements are primary care units, hospitals, elderly care apartments, pharmacy, medical supplies, social service centers, club rooms, restaurants, shops, and music halls with cinemas. Individual clinics or service buildings also may form a hospital village where outside traffic lines are in parks or in Nordic areas with snow in a tunnel network. Some of the services may be far away but connected to the same communication net, as, for example, the term "X-ray images reading center" indicates. Administration of services provided to 3–5 million people through a network of 60–100 facilities would be managed by one organization. Such an organization might have an international profile joining hospitals in developing areas and permitting faster dissemination of technology and medical knowledge. This type of administration further strengthens the concept of Global Hospital 2050. The tendency in the modern service delivery system to bring services to the home may become more universally accepted as medical technology becomes more portable and information technology makes its application more feasible. Service network planning thus will be an important task for health care authorities.

More Human Design

Current health care buildings are not very conducive to human occupation. The environment is sterile. The accessibility to the institution is often problematic for a wheelchair user, and access to public transport is not universally available. The lighting is problematic for the visually impaired. The scale of the facility is too massive for human comfort. Patient rooms are based on programmatic needs. Surgical pavilions might be located at a great distance from emergency departments or intensive care units. Patient and personnel traffic patterns might not always be conducive to the best patient care. In the future, more human design is needed. The environment must be positive and healing. The natural and indirect lighting, human scale, landscaping, use of colors, and textiles are the tools of the designer. The trend is to analyze the functions in patient treatment and design space for processes. Individual rooms will decrease in numbers. Instead of individual rooms for procedures, the department will have the landscape space where large teams can work and equipment for several procedures can be stored.

Conclusions

Three aspects (country and culture, provision of services, and environment of activities) configure one cube (i.e., three-dimensional grid) representing our impressions of current health issues. The components of the cube will be developed according to the development of technology, improvement of education, and changes in social structure. For instance, the development of information technology will influence the availability of health services in less-developed areas. Societal consciousness of the global environment will increase awareness of what is necessary to make it more healthful. As this cube looks like the fortune-teller's crystal ball, we may call it the "crystal cube" through which the future is visualized.

Preventive methods, biomedical treatment methods, medical equipment technology and communication systems will continue to develop. No limits are in sight. Less-expensive technical solutions for developing countries are needed, and they are necessary to the implementation of new methods globally. This could open new possibilities for the health care technology industry. The growing functional and administrative networking of hospitals, the global tendency toward evidence-based medicine, and the implementation of quality systems will harmonize medical services globally. Perhaps in 2050 we can speak on Global Medicine in the Global Hospital. The facilities will be more human in design and will meet health care needs regardless of the culture of its occupants. Finally, global efforts to improve the environment of Planet Earth may very well reduce demands for health care in general.

Generally, it is very difficult to forecast our future. Arthur C. Clarke wrote his book *Profiles of the Future* in 1958 and showed us a chronological timetable of the future until the year of 2100, together with one of the past, back to 1800. He forecast what would happen 40 years after 1960. He anticipated human emigration to other planets, artificial intelligence (AI), global libraries, wireless energy, ocean mines, and an extended feeling of time passing. In addition, he predicted several achievements by 2050 (e.g. gravity control, memory recorder, and human hibernation). Clarke limited his work to the fields of natural science and technology, but it is amazing to find his ideas to be more or less accurate and in accord with our current views. This reveals that the forecast of our future is not impossible.

There is a distinct difference between what will be and what should be, 50 years hence. Our impressions are that health care is not the responsibility of individual nations but rather a global responsibility to be shared by all, with all contributing to the best of their abilities. Developed countries must facilitate advancement of those less developed through universal endorsement of efforts of socially acceptable organizations. The efforts of the World Health Organization and the United Nations must be universally endorsed as pathways through which the health of our global population will advance. With the recognition of organizations that have global profiles, monetary and human resources can be prioritized and equitably distributed. Advances in communication technology will facilitate dissemination of medical knowledge. Finally, we should recognize that even today's medical technology made available to developing countries in 2050 will represent a major advancement in the delivery of health care. What is known as the "community hospital" in the Western world today could tomorrow improve the quality of life of those living in areas in central Africa, India, and South America. Speculation is endless, but the concept of universal health care is not.

References

Lewis JC. Mapping International Challenges for Managed Care. *World Hospitals and Health Services* 36(1):28–37, 2000.
Nagasawa Y. *District Hospitals: Guideline for Development.* Geneva, World Health Organization, 1992.

Further Information

Nagasawa Y, Mann GJ. Opportunities, Problems, and Pitfalls in an International Architecture for Health Practice in the 21st Century. *Proceedings of the International Congress and Exhibition on Health Facility Planning, Design and Construction.* Tampa, Florida, 1995.
Skaggs, RJ, Sprague GJ. Designing for Health in the New Millennium. *Health Facilities Management* 10:, 1997.
Correspondence and Material on GUPHA courtesy of Prof. Y. Nagasawa, University of Tokyo, Department of Architecture, 7-3-1 Hongo, Bunkyoku, Tokyo, 113-8656 Japan.

Documents Available on the Internet

World Health Organization Annual Reports on World Health, Reports on World Population of United Nations, Department of Economic and Social Affairs; Population Division, Reports of United Nations Environment Programme (UNEP), Nairobi, Kenya, and similar.

Appendix 10-A: Sample Request for Proposal (rfp)

REQUEST FOR PROPOSAL

THIS IS A REQUEST FOR PROPOSAL TO PURCHASE AN 80-100 CHANNEL MEDICAL TELEMETRY SYSTEM

ISSUE DATE:

Preparer and Point of Contact: System Representative Clinical Engineer

Your Hospital, Bob's Lane. Anywhere, USA

TABLE OF CONTENTS

Introduction 1

Vendor Qualifications 2
   a) Vendor References
   b) Available Site Visits

Product Specifications

Vendor Presentation Schedule

Installation

Delivery Date & Payment Methods

Contract Submissions & Amendments
   A) Technical and Clinical Training
   B) Accessories
   C) Warranty
   D) Service After Warranty
   E) Miscellaneous

Final Contract and Closing Schedule

Today's Date

Your Hospital has one of the most dynamic cardiac programs in Central Texas. We require an expansion of our ECG telemetry capabilities. Your company is invited to bid on a proposed 80-100 channel, digital telemetry system. The product specifications are listed on page 4 and 5. *Bids will be accepted until close of business on August 30th, XXXX.* Delivery and installation must be completed by February 1st, XXXX.

Please direct all correspondence in this matter to:

    Director, Clinical Engineering & Telecommunications
    (Office) XXXXXXX
    (Fax) XXXXXXXX (e-mail) XXX@mail.yourhospital.org

PRODUCT SPECIFICATIONS

1) Product must operate in the UHF medical frequency band with digital transmission signal capabilities.
2) Open systems hardware – (nonproprietary)
3) Transmitters will be **water-resistant.**
4) Transmitters will operate on a standard AA or AAA battery.
5) Monitors will have color display
6) Product must conform to National Fire Protection Agency safety standards:
    A) Chapter 3 – Electrical systems – all sections
    B) Chapter 9-Manufacturer Requirements – all sections
        i) See Chapter 2 – Patient-Care-Related Electrical Appliance
7) Product must conform to American National Standard Institute and Association for the Advancement of Medical Instrumentation Standards:
    A) ES-1 1993 (Safe Current Limits for Electromedical Apparatus) – all sections
    B) ES11 – 1991 (Diagnostic Electrographic Devices) – all sections
    C) EC38 – 1994 (Ambulatory Electrocardiographs) – all sections
8) Product must operate in the UHF medical frequency band with digital transmission signal capabilities.
9) Open systems hardware – (nonproprietary)
10) Software license issued in accordance with hospital policy (see attachment B).
11) Transmitters will be **water-resistant.**
12) Transmitters will operate on a standard AA or AAA battery.
13) Monitors will have color display for 72 hour trending analysis and real-time alarms.
14) Analog signal ECG trace on monitors with corresponding digital values for heart rate.
15) ST segment analysis.

Our Company meets or exceeds the above required specifications
Signed _____

VENDOR QUALIFICATIONS

All bidding vendors must have product manufacturer experience in medical telemetry for at least five (5) years. All bidding vendors must have no unresolved FDA actions or pending FDA actions related to the Safe Medical Devices Act or the Good Manufacturing Practices act. Any company FDA incidents within the past five (5) years must be disclosed.

Our Company conforms to the above vendor qualifications

_____ (Signature) _____ (printed name) _____ (title)

List of any FDA related actions against your company in the past 5 years.

If none state: No FDA actions

Vendor References:
Provide a list of all sales of ECG telemetry, which your company has sold in the past 2 years that match the exact specifications required by SJRHC (e.g. same software revision, same hardware revision).

Vendor Site Visits:
Provide a list of available hospitals from the above vendor reference list located in Texas that would be available for a site visit.

INSTALLATION

The vendor must provide staff for the installation/application of the demonstrated product and the final installation of the purchased product. This will include antennae installation to allow coverage of the entire facility. Final installation will be completed by *February 1st, 2002.*

DELIVERY DATE & PAYMENT METHODS

The product will be delivered in coordination with the Clinical Engineering Department and the Purchasing/Materials Department in order to meet the February 1st, XXX installation date.

Contact for Final Negotiations:
Director, Purchasing/Materials Services Department

CONTRACT SUBMISSIONS & AMENDMENTS

The final contract will consider:

    A) Technical & Clinical Training
    B) Accessories
    C) Warranty
    D) Service After Warranty E) Miscellaneous

FINAL CONTRACT AND CLOSING SCHEDULE

The final contract must contain all information and inspections required on the checklist found on page X of this RFP.

Appendix 10-B
Position Description

CITY HOSPITAL
POSITION DESCRIPTION

POSITION TITLE: Clinical Engineer

DEPARTMENT: Clinical engineering   **CODE:**

Participates in the care of patients in the following age groups:

1 Neonate (0-1month)	1 Infant (1 month to 1 year)	1 Toddler (1yr-3 yrs)
1 Preschooler (3yr-6yr)	1 School Age (6yr-12yr)	1 Adolescent (13yr-18yr)
1 Early Adult (19yr-45yr)	1 Late Adult (45yr-64yr)	1 Geriatric (65yr +)

POSITION SUMMARY:

Performs a broad scope of duties under the general supervision of the Director Clinical Engineering. Provides sound leadership and management to assure services that include Clinical Engineering, Telecommunications, Engineering Internship and Cooperative Education Programs, technology assessment, and others are cost effective, efficient and highest quality so as to maintain the maximum standards for the customer. Performs a variety of leadership, technical and administrative tasks to support the Facility's Missions, Values, and Objectives. The employee in this position will be required to demonstrate competencies in providing service for the care of patients/customers between the age groups of neonate through geriatric.

EDUCATION AND EXPERIENCE:

Bachelor of Science in Engineering or Engineering Technology degree required. Preferential consideration for Certified Clinical Engineer (CCE) or obtain certification within one year after accepting position. At least three years of experience in Clinical Engineering or management activities in a Health Care environment. Fundamental understanding of schematics, repair diagrams, electronics, anatomy and physiology, telecommunications processing, medical equipment principles of operation, and related electronic test equipment. Technical knowledge and understanding of NFPA Codes, NEC Codes, Hospital Safety, Infection Control, and JCAHO Environment of Care standards. Has received training and/or is familiar with computer-based maintenance management systems, technology assessment matrixes, computer management spreadsheets, and computer graphics programs. Has the ability to prepare and coordinate lecture materials and present lectures and educational in-services relating to medical equipment and telecommunications. Has a general knowledge of Body Substance Isolation Techniques, using gloves, masks, and protective outerwear. Possesses the ability to establish and maintain effective working relationships with department staff, patients, visitors, and other hospital staff. Has the ability to effectively lead and manage a diversified staff providing equal opportunity for all to succeed with excellence.

Is willing to work additional hours other than those prescribed during a normal duty day in order to accomplish the tasks/jobs, assist in emergency situations, and or special events. Is willing to treat all people with dignity, courtesy, and fairness and promote the Hospital's values, missions, and philosophy, and objectives. Is able to pass a required physical examination and drug screening test.

PERFORMANCE STANDARDS:

Values

1. Exhibits willingness to understand, follow, and promote the hospital's mission, values, philosophy, and objectives.
2. Demonstrates ability to positively interact with persons of different ethnic, spiritual, moral and cultural beliefs.
3. Treats all people courteously, fairly, and with dignity.
4. Establishes, participates in, and fosters effective working relationships with our customers.
5. Ensures confidentiality of all customer information.
6. Uses, distributes, and conserves resources in consideration of the environment.
7. Willingness to support peers with their duties and responsibilities.

Safety

1. Shows willingness to read, understand, and follow all safety policies and procedures, according to organizational and departmental policies.
2. Performs essential work duties as identified in the attached Performance Evaluation Criteria.
3. Demonstrates proper body mechanics when performing tasks.
4. Utilizes standard precautions when performing tasks where contamination may be present.

Physical Requirements

1. Stands for long periods of time
2. Works in limited-space area
3. Carries, pushes/pulls, lifts up to 50 pounds

Mental Requirements

1. Understands and follows oral and written instructions/directions.
2. Organizes, prioritizes, and/or completes workload within designated time.
3. Communicates potential problems according to departmental and organizational policies.
4. Recognizes and responds to age specific needs of population served.
5. Participates in quality improvement activities-assesses, measures, improves.

_____   _____
Department Director   Human Resources   Date _____

CITY HOSPITAL
PERFORMANCE EVALUATION CRITERIA

POSITION: _____Clinical Engineer_____
Review Date: _____

TYPICAL DUTIES AND RESPONSIBILITIES

	STANDARD					
	Importance	Time	Weighted Value	Rating 1-5	Score	Comments
Policy development and enforcement Provides policy development and review related to medical equipment and the environment of care. Ensures new procedures are understood by all co-workers and enforced. Relates policies to mission and vision statements and practical applications of core values: Service, Reverence, and Stewardship	3	4	7			
Management of financial & personnel resources Supervises Clinical Engineering and Telecommunications Dept. personnel and facilitates department meetings. Reviews and coordinates schedules. Works as a team member and provides proactive solutions to scheduling conflicts. Maintains schedule with less than 3% overtime. Maintains operating budget with less than 5% budget variance overall in both departments. Completes personnel performance reviews within 10 working days of due date.	8	5	13			
Communication Provides nursing, physician and technical in-services on topics related to equipment operation and maintenance to improve the quality of care. Develops and reviews training objectives in conjunction with Director, Education Services to achieve Process Improvement where indicated. Coordinates and enhances relationships with Texas A&M University and other community organizations.	2	2	4			
Safety and legal oversight Demonstrates knowledge of hospital procedures for fire, severe weather, bomb threat, local disasters, and patient distress calls. Facilitates safety and values seminars as required in hospital policy, mission and vision statements. Identifies SMDA related incidents and completes proper documentation. Has no more than 1 deficiency per year in this area as a performance standard. Maintains a 90% injury-free work environment. Serves as a member of the Hospital Safety Committee and Medical Equipment Management and Standards Subcommitee. Attends 90% of safety as a performance standard.	5	7	12			
Special Projects Completes all assignments delegated by the VP Support Services in a timely manner.	4	6	10			
Construction and Renovation Attends weekly project meetings and reports concerns to VP Support Services. Ensures most cost-effective approach is taken and budget for projects is within projected cost. Reviews plans to insure federal and state health and safety codes are met.	6	8	14			
Personal and professional growth Continues education either through correspondence or traditional classroom courses related to clinical engineering with the goal of earning a master's degree. Participates in local and national organizations related to clinical engineering.	1	1	2			
Technology Assessment Provides system leadership and demonstrates technology assessment expertise in matters related to medical equipment acquisition and service contracts. Assists the VP Support Services and ORT in recommendations for equipment acquisition.	7	3	10			

Rating Scale: Excellent—5, Very Good—4, Good—3, Fair—2, Poor—1

Index

AABB (see American Association of Blood Banks)
AAMI (see Association for the Advancement of Medical Instrumentation)
AANA (see American Association of Nurse Anesthetists)
AAOS (see American Academy of Orthopaedic Surgeons)
AAP (see American Association of Physicists)
ABBA (see American Blood Bank Association)
ABCE (see American Board of Clinical Engineering)
ABIMO (see Brazilian Association of Industry of Medical Devices and Equipment)
Ablation, 40, 419
AC power, 431, 460, 496
Academic medical center, 3, 18-19, 21, 23, 25, 122, 484, 550
ACC (see American College of Cardiology)
ACCE (see American College of Clinical Engineering)
Acceptance testing, 19-25, 29, 55-56, 68, 104, 107, 120-121, 151, 202-203, 266-267, 323, 452, 574-575
Accident investigation (see also Incident investigation), 124, 225, 237, 269-271, 273-275, 277-279, 315, 323, 427
Accreditation (JCAHO), 8, 15, 19, 26, 37, 41, 64, 86, 107, 122-125, 127, 130, 176, 205, 212, 220, 225, 231, 237-241, 243, 245, 270, 280, 298, 310-311, 313, 315, 320, 322-323, 327, 349, 354-355, 383, 422, 436, 497, 520-521, 528, 550, 555, 560, 565-573, 576, 578, 626
Accreditation manual (JCAHO), 123, 130, 243, 245, 520-521, 528, 569-573
Accreditation process (JCAHO), 571-573
Accreditation standards (JCAHO), 245, 550, 576, 578
Accreditation survey (JCAHO), 571
ACEW (see Advanced Clinical Engineering Workshop)
ACEW Syllabus, 177, 302-305, 309
ACHE (see American College of Healthcare Executives)
ACP (see American College of Physicians)
Acquisition specifications (see also Purchase specifications), 99, 173
ACR (see American College of Radiology)
ACS (see American College of Surgeons)
Administrative safeguards (also see HIPAA), 500-502
Administrative simplification (also see HIPAA), 498-500, 505, 624
Advanced Clinical Engineering Workshop (ACEW), 15, 27, 42, 51, 71, 74, 77, 82, 97, 154, 172, 177, 224, 262, 285, 298, 301-306, 309, 593, 600-602, 605, 609
Advocacy committee (see also ACCE), 599, 607-608
AED (see Automatic External Defibrillator)
Africa, 9, 51, 53, 93, 155-156, 158-163, 167, 170, 172-173, 175, 224, 285, 304-309, 342, 542, 605, 656-657, 659
African Federation for Technology in Healthcare (AFTH), 156, 164, 170, 305, 307
AFTH (see African Federation for Technology in Healthcare)
Agency for Health Care Research and Quality (AHRQ), 114, 116, 118, 237, 240-241, 243, 435, 561-564, 648
AHA (see American Heart Association)
AHA (see American Hospital Association)
AHRQ (see Agency for Health Care Research and Quality)
AIA (see American Institute of Architects)
Air Force, 34-35, 207, 617
ALARA, 530-531
AMA (see American Medical Association)
American Academy of Orthopaedic Surgeons (AAOS), 240
American Association of Blood Banks (AABB), 19-21, 568
American Association of Nurse Anesthetists (AANA), 240

American Association of Physicists (AAP), 532
American Blood Bank Association (ABBA), 125
American Board of Clinical Engineering (ABCE), 617
American College of Cardiology (ACC), 640, 644
American College of Clinical Engineering (ACCE)
    ACCE News, 15, 18, 26, 51, 162, 280-281, 298, 305, 307, 309, 496-497, 600-603, 606-607, 609, 612
    Advocacy Committee, 599
    Board of Directors, 15, 600-602, 608
    Bylaws, 598, 600, 602-603, 605-607
    Code of Ethics, 15, 270, 601
    Definition of a clinical engineer, 92, 298, 346, 599, 601, 609, 644
    Healthcare Technology Foundation, 49, 608
    HIPAA Taskforce, 452
    Membership, 41, 51, 302, 601-603, 605-607
    Symposia, 298, 601, 609
    Teleconference, 271, 280-281, 601, 605, 609
    Website, 172, 601
American College of Clinical Engineering Healthcare Technology Foundation (AHTF), 49, 270, 280, 606, 608-609
American College of Physicians (ACP), 238, 564, 570
American College of Radiology (ACR), 173, 367, 492, 560
American College of Surgeons (ACS), 236-238, 240, 570
American Heart Association, (AHA), 3, 6, 15, 97, 122, 139, 153-154, 240, 345, 494-497, 508, 548, 553, 560, 570, 602, 608, 628, 640, 644
American Hospital Association (AHA), 3, 6, 9, 15, 97, 122, 139, 153-154, 229, 236, 238, 240, 282, 345, 349, 494-497, 508, 548, 551-553, 560, 570, 602, 608, 628, 637, 644
American Institute of Architects (AIA), 245, 378, 383, 413, 416, 520-521, 528, 544, 549
American Medical Association (AMA), 161, 165, 175, 237-238, 240, 282, 498, 570
American National Standard Institute (ANSI), 42, 108, 123, 156, 227, 255, 262, 267, 269, 353, 355, 358, 365, 430-431, 435, 495-498, 535, 548, 555, 557-560, 565, 568, 573, 575, 587, 617, 644
American Nurses Association (ANA), 240, 267, 280, 321-322, 327
American Society for Testing and Materials (ASTM), 14, 274, 279-281, 358, 386, 389, 391, 560, 585
American Society of Anesthesiologists (ASA), 240, 358, 386, 390-391, 442
American Society of Healthcare Engineering (ASHE), 15, 27, 42-43, 120, 122-123, 147, 152, 202, 205, 269, 298, 313, 493-494, 497, 528, 553, 560, 600-602, 606-607, 614
American Society of Mechanical Engineering (ASME), 42-43
American Society of Quality (ASQ), 566, 568
ANA (see American Nurses Association)
Analytic hierarchy process, 452, 455
Anesthesia machine, 9, 21, 24, 205, 220, 272, 279, 357, 364, 367, 371, 378-380, 383-389, 391, 415, 421, 442, 509, 522-524, 543, 583, 591-592
Anesthetizing locations, 282-283, 521, 637
Angiography, 8, 174, 381-382, 395, 400, 418, 420, 424
Angioplasty, 174, 345, 371, 400, 417, 419-420, 424
ANSI (see American National Standard Institute)
ANSI standards, 227
AORN (see Association of periOperative Registered Nurses)
Apnea monitor, 277, 321, 356, 371, 415-416
Argentina, 53, 85, 342
Armed Forces, 154, 218
Army, 8, 12, 34-36, 80-81, 153-154, 207, 275, 530, 533, 617
Arrhythmia monitoring, 373, 420

Artificial heart, 8-10, 322, 376, 645
Artificial intelligence, 3, 188, 449, 464, 467, 469, 471, 473, 475, 627, 659
Artificial organs, 8, 13, 485, 582, 619, 623
ASA (see American Society of Anesthesiologists)
ASHE (see American Society of Health care Engineering)
Asia, 53, 162, 333, 557, 648, 655-657
ASME (see American Society of Mechanical Engineering)
Aspirators, 25, 367-368, 382, 542
ASQ (see American Society of Quality)
Asset management, 10, 19, 94, 102-104, 107, 122-123, 136, 141, 143, 164, 172, 203, 223, 297, 318, 320, 477, 480, 610, 625, 629
Association for the Advancement of Medical Instrumentation (AAMI)
    Annual Meeting, 28, 39, 447, 600, 603, 607, 618
    Standards, 548
Association of periOperative Registered Nurses (AORN), 240, 327
ASTM (see American Society for Testing Materials)
Attorney, 42-43, 45-48, 124, 196, 215, 218, 277, 282, 358, 523, 593, 645, 652
Australia, 51, 53-58, 130, 218-219, 254, 285, 559-561, 563, 581, 646-648
Autoclave, 253, 282, 378, 534, 546, 551
Automatic external defibrillator (AED), 250
Automatic logoff, 503, 505
Autonomic systems, 623-624

Backup, 21, 34, 44, 120, 129, 133, 136, 154, 207, 234, 239, 254, 277-278, 325, 384-385, 387, 389, 397, 414-415, 425, 434, 451-452, 461, 477, 489-490, 500, 502, 505, 507, 511, 516, 523, 531, 549-551, 629, 654
Ballistocardiography, 8
Bar coding, 128-129, 241
Bar-code systems, 129
Batteries, 7-9, 126, 129, 131, 254, 276, 282, 300, 364, 389, 414-415, 420, 426, 430, 442, 452, 461, 485-486, 507, 524, 537, 539, 551, 557, 572, 645
Benchmarking, 22, 26, 60, 127, 137, 139, 143, 146, 189, 203, 205, 212, 311, 462, 610, 616
Best Practices, 14, 26, 108, 123, 162, 175, 202-203, 205, 212, 222, 225, 240-242, 657
BI&T (see Biomedical Instrumentation & Technology)
Bioengineering, 7, 10, 13, 22, 53, 58-59, 73-74, 88-89, 281, 298, 592
Biomaterials, 63, 297, 333-334, 557, 576, 619
Biomedical device
    Accidents, 323
    Acquisition, 323
    Design, 323, 625
    Hazards, 625
    Incidents, 323, 500, 505, 625
    Inventory, 103, 500, 505, 625
    Maintenance, 27, 30, 103, 323, 500, 505, 625
    Manufacturers, 625
    Purchasing, 27, 30, 103, 323
    Recalls, 625
    Regulations, 27, 625
    Repair, 27, 103, 323, 625
    Tracking, 500, 505, 625
    Users, 27, 30, 103, 323, 449, 500, 505, 625
Biomedical engineering
    Biomedical Engineering Department, 10, 18, 38, 62, 84, 86, 140, 188, 209, 213, 224, 264, 267-269, 297, 315, 323, 330, 431, 435, 527
    Definition, 3, 7, 61, 92, 179, 188, 211, 224, 298, 436, 527, 576, 599-600, 609, 621
    Program, 22, 39, 81, 90, 202, 297, 301, 323, 353

665

Index

Biomedical equipment technician (BMET), 1, 8, 18, 20, 27, 38, 41, 49, 59, 71, 82-83, 86, 91, 98, 125, 148, 153-154, 189, 207-208, 210, 212, 287, 289-292, 294, 299-301, 310-311, 313, 316, 322, 327, 331, 383, 385, 388, 449, 451-452, 511, 600-603, 610, 648

Biomedical equipment
  Accidents, 56, 59, 228, 315, 327, 331
  Acquisition, 1, 27, 29, 33-35, 38, 56, 74, 77, 92, 108, 125, 207, 306, 327, 527, 625, 648
  Design, 1, 4, 8, 16, 18, 35, 86, 97, 181, 205, 300, 307, 327, 331, 511, 527, 573, 598, 617, 625, 648
  Donations, 84, 125, 301, 306
  Hazards, 1, 56, 125, 307, 385, 617, 625, 647
  Incidents, 1, 125, 228, 307, 315, 322, 327, 617, 625, 647-648
  Inspection & PM, 125-126, 395, 626
  Inventory, 4, 8-9, 29, 35, 55-56, 59, 74, 108, 125-126, 207, 300, 306-307, 625
  Maintenance, 1, 4, 7-9, 16, 18, 27-31, 34-35, 38, 53, 55-56, 59, 64, 71, 74, 77, 84, 86, 91-92, 97-98, 108, 125-126, 136, 143, 179, 205-208, 212, 228, 287, 289-291, 299-301, 306-307, 315, 322, 598, 608, 610, 613, 625, 648
  Manufacturer, 7-9, 16, 28, 31, 35, 38, 56, 59, 84, 86, 91, 97, 126, 136, 143, 207, 212, 228, 291, 301, 306, 315, 511, 573, 610, 625, 647-648
  Manufacturing, 77, 108, 143, 610
  Operators, 84, 315, 511, 648
  Purchasing, 4, 9, 16, 27, 29-30, 35, 38, 56, 59, 71, 91, 97-98, 125-126, 206, 300, 306-307, 511
  Recalls, 125-126, 307, 316, 322, 625
  Regulations, 9, 27, 29, 84, 91, 108, 125, 181, 307, 598, 602, 625, 647-648
  Repair, 1, 4, 7-9, 27, 29, 34-35, 38, 56, 59, 74, 86, 91, 97, 125-126, 136, 205, 207-208, 212, 228, 287, 289, 291, 299-300, 306, 322, 598, 608, 613, 617, 625, 647
  Testing, 4, 9, 16, 18, 28-30, 34, 38, 55-57, 125-126, 136, 208, 228, 287, 289, 291, 299-300, 306, 316, 322, 331, 385, 511, 573, 608, 617, 625, 647
  Tracking, 9, 307, 327, 625
  Users, 1, 9, 27, 30, 33, 38, 56, 86, 97, 108, 125-126, 136, 143, 206-207, 306-307, 315-316, 322, 327, 385, 511, 625, 647-648

Biomedical Instrumentation & Technology (BI&T), 122, 130, 154, 287, 310, 315, 355, 370
Biometrics, 511, 586
Biophysics, 8, 10, 615
Biosensor, 3, 582
Biotechnology, 67, 294, 331, 537, 592, 619
Bluetooth, 128, 455, 497, 512
BMET (see Biomedical equipment technician)

Certification, 3, 6-7, 9, 15-16, 41-42, 46, 49, 51, 54, 59-60, 68, 71, 77, 80, 82, 86, 91-92, 116, 208, 212, 227, 236, 271, 279-280, 291, 301-302, 304, 307-308, 310, 319-320, 328, 332, 334, 349, 383, 394, 408, 498, 507, 523, 549, 558, 574, 577, 583, 586, 593, 596-598, 601-603, 606-610, 613, 617-620
Certification examination, 82, 92, 302, 310, 601, 606, 610, 618
Board of examiners, 3, 71, 80, 82, 334, 606-607, 617-618
Body of knowledge (see also ACCE), 1, 10, 15, 20, 46, 50, 245, 271, 280, 601, 603, 606
Brazil, 1, 51, 53, 69-72, 76-77, 85, 159-160, 162, 302, 304-305, 603, 605, 656
Brazilian Association of Industry of Medical Devices and Equipment (ABIMO), 69, 71-72
Break-even analysis, 197-198, 650
British Columbia, 63-64, 287, 292, 562, 578
Broadband, 128, 266-267, 484
Bronchoscope, 536, 544
Bulgaria, 53
Bureau of Labor Statistics, 181
Bureau of Medical Devices, 9, 235
Bureau of Radiological Health, 235
Business plan, 42-43, 68, 152, 173, 183, 199, 311, 527, 572, 621, 649-651
CAD (see Computer aided design)
Calibration equipment, 158
Calibration standards, 158
Calibration, 4, 26, 30, 35, 56, 67-68, 131, 135, 148, 155, 158, 189-190, 221, 250, 295, 317, 324, 330, 333-334, 364, 390, 401, 414-415, 434, 437, 442, 468, 470-471, 486, 489, 531-533, 545, 557, 572, 619, 625-626, 635, 640, 642

Call management, 143, 146, 464, 507
Canada Health Act, 62-63, 65
Canada, 8, 27, 51, 53, 62-65, 108, 179, 235, 240, 251, 254, 257, 260-262, 292, 358, 423, 436, 498, 524, 555, 557-563, 576-581, 603, 617, 624, 647-648
Canadian Board of Examiners, 617
Canadian Council on Health Services Accreditation (CCHSA), 64-65, 555, 576, 578
Canadian Medical and Biological Engineering Society (CMBES), 64-65, 134, 313, 576-578
CAP (see College of American Pathologists)
Cardiac catheterization laboratory, 371, 400, 417-419, 491, 551, 599
Cardiac output monitor, 371, 419
Cardiac pacemaker, 8-9, 13, 250, 254-255, 257-262, 281, 417, 420, 552
Cardiopulmonary bypass, 13, 383, 386, 388
Cardiovascular pressure-dimension analysis (CPDA), 640-641, 644
Career path, 1, 49, 318, 621
Caribbean, 76, 84, 86, 345, 656
C-arm, 319, 371, 396, 400, 418, 424
Cash flow analysis, 649
CBET (see Certified biomedical equipment technician)
CBI (see Computer based instruction)
CBR (see Cost benefit ratio)
CCE (see Certified clinical engineer)
CCHSA (see Canadian Council on Health Services Accreditation)
CDC (see Centers for Disease Control)
CDRH (see Center for Devices and Radiological Health)
CE (see Clinical engineer)
Certification Commission, 7, 15, 71, 82, 601-603, 617
Certification examination, 82, 92, 302, 310, 601, 606, 610, 618
Certification requirements, 16
Cellular telephone, 129, 249-250, 254-264, 268-269, 346, 357, 429, 461, 493, 507, 511, 516, 551-552
CE-mark, 67, 582
Center for Devices and Radiological Health (CDRH), 15, 45, 115, 118, 123, 156, 235, 237, 254, 268, 355, 357-358, 422, 429, 435-436, 492, 497, 511, 559-560, 578, 581, 586-589, 648
Centers for Disease Control (CDC), 115, 118, 323, 477, 543, 550, 552, 624, 627, 629
Centers for Medicare and Medicaid Services (CMS), 116, 188, 224, 498, 500, 550, 555, 566
Central America, 484
Central stations, 40, 234, 249, 300, 373-374, 414, 460-463, 512
Centrifuge, 333, 344, 383, 446
Certificate of Need, 9
Certified biomedical equipment technician (CBET), 155, 291, 574, 610, 617
Certified Clinical Engineer (CCE), 27, 92, 155, 205, 227, 304, 601-603, 606, 617-618
CFR (see Code of Federal Regulations)
CGA (see Compressed Gas Association)
Chartered engineer, 61
Chemical engineering, 3
Chief information officer, 38, 568
Chief technology officer, 2, 6, 10, 36, 38, 103, 110-112, 526-527
China, 12, 51, 172, 303, 305, 529, 605, 655, 657-658
Circulatory assist devices, 376
CIS (see Clinical information system)
Clinical alarms, 239, 280, 323-324, 648
Clinical data, 12, 31, 115, 125, 241, 367, 456-460, 463, 487, 587
Clinical Engineering (CE)
  Budgeting, 1, 5, 8, 10, 17, 21, 24, 27, 29, 31-32, 35-39, 49-50, 55-56, 59, 62, 65-68, 71-73, 75, 78, 85, 88, 93, 96, 102-104, 107-108, 111-112, 124, 136, 149-150, 152-153, 160, 170, 172-173, 177, 182, 188-195, 199, 201, 216, 228, 248, 298, 302-303, 306, 308-309, 318, 331, 343, 350, 360, 383, 451-452, 486, 524-527, 532, 575, 604, 606, 614, 626, 632, 646-647
  Certification, 3, 6-7, 9, 15-16, 41-42, 46, 49, 51, 54, 59-60, 68, 71, 77, 80, 82, 86, 91-92, 116, 212, 227, 236, 271, 279-280, 291, 301-302, 304, 307-308, 310, 320, 328, 332, 334, 349, 383, 394, 408, 498, 507, 549, 558, 574, 577, 593, 596-598, 601-603, 606-610, 613, 617-620
  Compensation, 43, 46, 198, 206-207, 210, 212, 216, 291, 302, 313, 356, 388, 390, 590

Consulting, 2, 6, 12-13, 19-20, 22, 25, 27, 29-33, 35-39, 41-44, 48-51, 53, 55-56, 62, 67-68, 71-75, 82, 87, 97, 114, 116, 119, 152, 160, 162, 170, 177, 188-190, 192-195, 199, 211, 214, 218, 225, 227, 252, 263-264, 266-267, 269-270, 274, 282, 297, 299, 309, 320, 323, 328, 341, 348, 356, 368-369, 374, 412, 426, 431, 433, 460, 466, 485, 501, 506, 526-527, 530, 559, 567, 576, 578, 588, 592, 597, 604, 610, 617, 619, 625-626, 644-645, 648
Definition, 3, 5-7, 17, 36, 56, 61, 92, 94, 102-103, 122, 170, 179, 184, 188, 196, 203, 211, 216, 224, 235, 244, 248, 298, 343, 346, 349, 356, 360, 364, 385, 395, 417, 422, 436, 456, 494, 497, 503, 513, 521, 527-528, 544, 549, 557, 576-577, 593, 595, 597-601, 607, 609, 618, 621, 627, 644-645
Department, 3-6, 16-19, 22, 26, 28, 36-41, 49, 53, 58-63, 80-82, 84, 86, 90-92, 97-98, 103-104, 107, 111, 119, 125-129, 143-144, 147-149, 151-153, 172, 179, 182, 188-199, 201-203, 205-207, 211-212, 227, 237, 239, 244, 249, 264, 267, 270, 277, 285, 294, 297-298, 306, 311, 313, 315-316, 324-325, 327, 343, 347, 350, 354, 370-371, 378, 383-384, 400, 421, 431, 433, 435, 441-443, 456, 460, 462-463, 497, 543, 551, 559, 568, 570, 575-576, 590, 609-610, 613, 615, 644, 646-647
Education, 1-3, 7, 16, 18-20, 22, 24-25, 27, 33, 39, 41-42, 46, 49, 51, 53, 55-56, 59, 61-62, 66-68, 72, 74, 77, 80-82, 86, 88-89, 91-92, 94, 98, 103, 124, 148, 160, 174, 202, 207, 214, 216, 222, 227, 238, 242-244, 248, 268-271, 278-279, 285, 290-291, 294-295, 297, 299, 301, 304, 306, 308-312, 314-318, 321-325, 327, 329, 332-335, 343, 346-349, 354, 371, 435, 463, 485, 496-497, 500, 504-505, 511-512, 525, 527, 530, 532, 544, 550, 568-569, 572, 575, 577, 592-593, 595,597-598, 601-603, 605-606, 608-609, 612-614, 617-619, 623, 625-626, 644-646, 652, 654, 656, 658
Function, 18, 24, 59, 92, 107, 199, 201, 385, 614, 624
International, 7, 12-13, 15, 18, 33, 42, 51, 53, 58, 61, 64, 66, 71-74, 77, 80, 82-83, 85, 89, 91-92, 97-98, 110, 116, 118, 124, 134, 155-156, 158, 164, 168, 170, 172-173, 175, 179, 182, 199, 205, 218, 222, 225, 235, 262, 280-281, 285, 295, 301-302, 304-307, 310, 313, 315, 329, 332, 334-335, 343, 345, 358, 426, 443, 454, 498, 520, 530, 532, 538, 542, 546, 555, 557-560, 564, 568, 578, 580, 593, 600-603, 605, 607-608, 613, 617, 619, 645, 648, 658
Management, 1-7, 10, 12-17, 19-24, 26-39, 41-42, 44-46, 48-51, 53, 55-56, 58-68, 71-72, 74-75, 77-80, 82, 84-94, 97-98, 102-104, 107-108, 110-114, 118-119, 122, 124-130, 132, 134-138, 140-143, 146-147, 149-153, 155-156, 158-160, 162, 164, 166, 168, 170-177, 179, 182-183, 188-189, 191-193, 195, 197-207, 210-214, 216, 218-220, 222, 224-225, 227, 232, 234-240, 242-245, 249-250, 252, 262, 267, 269-271, 280, 285, 287, 289-291, 294-295, 297-298, 300-302, 304-309, 311-312, 315, 317-318, 320, 322-335, 337, 341, 344, 346-347, 349, 352-355, 362, 369-371, 374, 378, 383-386, 400, 408, 411, 421, 431-433, 435-436, 449-452, 454, 456, 458, 460, 462-463, 474, 480, 485-486, 495-497, 500-501, 504-505, 507-508, 510-511, 518, 520-521, 525-528, 530-532, 538, 540, 546, 548-552, 555, 557-560, 566-569, 572-578, 580, 588, 590, 592-593, 596-599, 601-603, 605-608, 610, 612, 616, 618, 621, 623-628, 644-647, 650, 652, 654
Policies & procedures, 1, 5, 24, 45, 88, 124, 135, 170, 203, 236, 238, 240, 242-244, 264, 269-270, 325, 327, 343, 485, 500-505, 511, 521, 528, 530, 551, 572, 577, 606
Productivity, 200-201, 205
Profession, 1-2, 6-7, 10, 12-15, 19, 23-24, 35-39, 49, 51, 53, 58-59, 61-62, 64-65, 71, 82, 179, 199-200, 212, 237, 245, 282, 285, 290-291, 304, 307, 312-315, 318, 320, 322, 327, 341, 350, 370-371, 449, 456, 462, 520, 576-578, 593, 595-603, 606-608, 610-612, 617-619, 621, 624, 646
Service providers, 9, 20, 27, 56, 63, 67, 119, 124-126, 129, 136, 140, 143-144, 147-149, 151-152, 154, 191-192, 236, 269, 328, 346-347, 566-567, 574-575, 593, 614
Space allocation, 51, 55-57
Staff requirements, 66, 206, 574

Index 667

Staff productivity, 55, 96
Staffing levels (see also Staff requirements), 41, 55, 63, 173, 186, 206, 628
Clinical information system (CIS), 2, 241, 324, 327, 366, 368, 373-374, 449, 456-463, 487, 500, 510-511, 623
Clinical Laboratory Improvement Act, 586
Clinical laboratory, 3, 8-9, 27, 36, 56-57, 72-73, 75, 91, 114, 125, 148, 173, 189, 238, 291, 297, 299, 310, 328, 333, 342-345, 376, 449, 509, 512, 520, 526, 559, 563, 570, 582, 586, 619, 624, 658
Clinical Practice Guidelines (CPG), 159, 161-162, 174-175, 223, 520, 559, 652, 654
Clinical trial, 37, 104, 111, 238, 347, 349, 598
CM (see Corrective maintenance)
CMBES (see Canadian Medical & Biological Engineering Society)
CMMS (see Computerized Maintenance Management System)
CMS (see Centers for Medicare and Medicaid Services)
Code compliance, 102, 311
Code of ethics, 6, 15, 270, 577, 593, 595-596, 598, 601
Code of Federal Regulations (CFR), 110, 218, 224, 238, 269, 280, 318, 347, 349, 357, 363, 365, 422, 497, 506, 529-532, 550, 570, 581, 586-589
College of American Pathologists (CAP), 19-21, 36, 47, 125, 391, 429, 434, 520-521, 576, 608
Collimator, 17, 371, 392, 394-396, 399
Colombia, 51, 72-77, 177, 345, 548
Commercial off the shelf (COTS), 124, 435, 450-451, 487-491, 509-511
Communication protocol, 412, 453-454, 456, 458, 463, 477, 491, 629
Comparative evaluation, 337, 366-368, 564
Competency, 1, 10, 30, 34, 46, 49, 58, 60, 89, 101, 103, 107, 119, 161-162, 176, 189, 200, 206, 212-213, 221, 238, 242, 285, 306, 308, 315, 322-324, 326, 328, 347, 350, 366, 484-485, 565, 569, 572, 593, 596-597, 601, 603, 606-607, 618, 637-638
Component failure, 130, 132, 147, 246, 271, 323, 355-356, 371, 444-445, 579
Compressed Gas Association (CGA), 522, 524, 560
Computer aided design (CAD), 174, 294, 330, 348, 652
Computer based instruction (CBI), 310
Computer Telephony Integration (CTI), 128
Computerized Maintenance Management System (CMMS), 17, 26, 38, 110, 112-113, 122-130, 205, 212, 224, 315, 480
Computerized Patient Record (CPR), 6, 549, 552
Computerized Physician Order Entry (CPOE), 241, 458, 463, 509
Computerized tomography (CT), 3, 9, 32, 40, 71, 86, 133-134, 136, 140, 160, 172, 182, 189, 297, 303, 318-320, 330, 340, 367, 392, 399-401, 408, 410, 418, 469, 471, 476, 487, 489-490, 507, 509, 532, 551, 610, 612, 627, 640, 644-645, 647, 656
Connector, 271-272, 277, 357, 393-394, 412, 439, 441, 543, 640
Construction, 20, 22, 36, 73, 82, 85, 88-89, 107, 112, 129, 171, 245-246, 248, 251, 264, 274, 282, 310, 329, 331, 337, 339, 346, 367, 383, 386-387, 389, 397, 411, 413, 416, 426, 434-435, 490, 511, 513, 515, 518, 520-521, 523, 525, 527-528, 533, 544, 549-550, 569, 580, 644, 646, 659
Contingency plan, 85, 370, 502, 505
Contingency planning, 450, 504, 508, 572
Continuing education, 22, 24, 56, 86, 238, 285, 291, 294, 309-310, 312, 315-316, 318, 324, 332, 334-335, 463, 484, 497, 512, 532, 572, 597, 613-614, 625
Continuous Quality Improvement (CQI), 156, 202, 566
Corrective maintenance (CM), 16, 30-32, 36-37, 40, 55-57, 74-76, 82, 87-88, 94, 148, 164-165, 202, 255-261, 298, 386, 389, 392, 395, 414, 625, 656
Cost benefit ratio (CBR), 33, 101, 470, 475
Cost containment, 107, 598
Cost control, 6, 141, 147, 152, 649
Costa Rica, 51, 154, 304-305, 605
COTS (see Commercial off the shelf)
CPDA (see Cardiovascular pressure-dimension analysis)
CPG (see Clinical Practice Guidelines)
CPOE (see Computerized physician order entry)
CPR (see Computerized patient record)
CPT (see Current Procedure Terminology)
CQI (see Continuous quality improvement)
Critical care area, 21, 254, 261-262, 324, 333, 385, 412, 523, 637
Critical care units, 250, 412, 460, 551

CRM (see Customer relationship management)
CT scanner, 9, 32, 71, 86, 133, 140, 160, 172, 182, 189, 318, 367, 392, 469, 471, 507, 509, 551, 645
CT (see Computerized tomography)
CTI (see Computer Telephony Integration)
Cuba, 53-54, 167, 304-305, 605
Current Procedure Terminology (CPT), 161, 164-167, 175, 498
Customer Relationship Management (CRM), 475
Customer satisfaction, 90, 107, 136, 151, 200-201, 203, 213, 221, 224, 318, 323, 327, 346, 472, 480, 565, 567, 628, 633
Customer service, 181, 187, 201, 206, 221, 346, 465, 469, 573
Customized medical devices, 38
Cyprus, 53, 158, 305
Cystoscopy, 340, 400, 536

Dashboard, 127, 476-477, 480, 628-630
Data acquisition, 456-457, 459-461, 463-464, 480, 533, 628, 633, 638-640
Data analysis, 6, 86, 187, 223, 239, 369, 480, 638, 645
Data backup, 9, 452, 502, 505, 511
Data communication, 130, 371, 410, 412, 444, 451-452, 456, 509, 511
Data element, 107, 204, 477, 491-492, 629
Data entry, 16, 30, 125-126, 203, 221, 246, 456, 462, 490, 506
Data fields, 124, 127-128, 462
Data file, 627, 638-640
Data format, 126, 624
Data integrity, 124-126, 128, 501, 505
Data management, 125, 129, 300, 374, 408, 451, 456, 463
Data mining, 449, 476-477, 627, 629
Data network, 137, 146, 373, 412, 516
Data retrieval, 203
Data signal, 31-33, 453-454
Data storage, 416, 451-452, 457, 467, 506, 638
Decision support, 127, 129, 187, 222, 452, 458, 464, 469, 475-477, 573, 628
Defendant, 356, 590-591
Defibrillator, 8, 17, 25, 40, 45, 63, 92, 126, 250, 254-255, 257, 259-263, 279, 300, 331, 333, 340, 342-344, 353, 355, 371, 374-376, 379-380, 384, 400, 419-421, 442-443, 507, 509, 551, 557, 575
Denmark, 53, 65, 562, 603
Department of Veterans Affairs (VA), 3, 7, 26, 35-36, 123, 199, 202, 205-207, 211, 228-229, 231, 240, 249, 280-281, 297, 309-310, 315, 327, 345, 355, 358, 416, 422, 435-436, 447, 463, 495, 497, 550, 553, 555, 560-562, 564-565, 568, 575, 610, 616
Deposition, 46-48, 274, 277, 357, 547
Design considerations, 434, 645
Design criteria, 377, 424, 434, 490, 520, 527, 538, 569
Design defect, 46, 337, 356-357, 361
Design features, 248, 271, 421, 423-424, 426, 430, 568, 615
Design methodology, 337, 359-360
Design modifications, 4, 6, 83, 203, 209, 361
Design, health care facilities, 378, 513, 527-528
Destructive testing, 45, 274
Developing countries, 15, 26-27, 51, 84-86, 91, 93, 97-99, 108-110, 133-134, 155-156, 158-159, 162, 164, 170-173, 175, 177, 189, 199, 225, 251, 301-302, 304-307, 335, 345, 541, 546, 548, 557, 645, 647-648, 656-659
Developing nations, 98, 309, 332
Device classification, 20, 123, 343, 358, 579, 583, 586, 589
Diagnostic Related Groups (DRG), 9, 33, 67, 227, 480, 635
Dialysis, 8, 16-18, 21, 24, 27, 40, 59, 63-64, 91-92, 239, 272, 321, 333-335, 345, 371, 375-376, 421, 446, 507, 537, 548-549, 551-552
DICOM (see Digital Imaging and Communications In Medicine)
Digital image, 320, 329-331, 399, 407
Digital Imaging and Communications in Medicine (DICOM), 318-320, 367, 487, 490-492
Digital radiography, 368, 487
Digital radiology, 32, 490
Digital Subscriber Line (DSL), 128-129, 512
Disaster plan, 513, 531, 549-553, 626
Disease prevention, 11, 513, 627, 656-657
Documentation systems, 109

Dominican Republic, 304-305, 605
Doppler ultrasound, 418
Dosimeter, 443, 531, 535
Downtime, 6, 20, 96, 109, 153, 236, 434, 470, 489, 510
DRG (see Diagnostic Related Groups)
DSL (see Digital Subscriber Line)
Dye injector, 418

EC (see European Community)
ECG monitor, 40, 249-250, 257, 300, 391
Echocardiograph, 8, 384, 484, 491, 640, 644
Economic analysis, 31, 36, 77, 165
ECRI, 9, 14-15, 20, 26-27, 62, 85, 111, 113, 115-116, 118, 120, 122-125, 130, 155, 157-158, 190, 199, 221, 224, 231, 240, 248, 261-262, 264, 267, 269, 271-272, 280-282, 300, 302, 306, 313, 325, 327, 337, 342-345, 347-348, 358, 366-368, 370, 436, 490, 547-548, 553, 559-562, 564, 583-585, 606-609, 624
Ecuador, 51, 76, 78-79, 303-305, 605
EDI (see Electronic Data Interchange)
Education
Clinical engineering, 33, 294-295, 301, 304, 312, 314, 626
Staff, 39, 241-242, 504, 577, 653
User, 56, 124, 435, 459, 577
Educational opportunities, 216, 285, 318, 600, 602, 605, 610
EEG (see Electroencephalograph)
EEG monitor, 551
EHR (see Electronic health record)
EHTP (see Essential Health Care Technology Package)
EIA (see Electronic Industries Association)
EKG (see Electrocardiograph)
Electric bed, 126, 421-423, 436, 509
Electric field, 255-256, 300, 340
Electric generator, 507
Electric power, 413, 446, 506, 516, 529
Electric shock, 274, 282-283, 423, 431, 433, 446, 542
Electrical energy, 379, 431, 493, 583
Electrical engineering, 8, 20, 22-23, 49, 81, 87, 218, 227, 269, 287, 289-291, 294-296, 299-300, 417, 520, 609, 619
Electrical hazards, 280-281, 283, 357-358, 378-379, 383, 413, 446, 582
Electrical isolation, 282, 379
Electrical power, 25, 158, 323, 326, 343, 346, 378-379, 385, 389, 395, 410-411, 512-513, 516, 520-521, 528, 551
Electrical safety analyzers, 17, 190, 442-443, 533
Electrical safety testing, 6, 9, 21, 29-30, 129, 151, 282, 289, 291, 298, 300, 316, 608, 625
Electrical safety, 1, 3, 6, 9, 21, 26, 29-30, 68, 125, 129, 148, 151, 202, 204, 206, 225, 250, 261-262, 274, 281-282, 287, 289, 291, 298, 300, 315-317, 326, 379, 432, 434, 462, 559, 587, 608, 617, 624-625
Electrocardiograph (EKG), 250, 257, 279, 322, 326, 330, 333-334, 339, 345, 376, 380, 384, 416-418, 420, 457-458, 460, 462, 590-591
Electrocautery, 238, 282
Electrocution, 13, 281-282
Electroencephalograph (EEG), 7, 40, 330, 333-334, 339, 380, 509, 551
Electromagnetic compatibility, 254, 257, 261-262, 267-269, 280, 311, 357, 434-436, 492-493, 497, 587, 605
Electromagnetic energy, 18, 254, 263-264, 267, 392, 429, 493, 497
Electromagnetic field, 254, 262-263, 266-267, 343, 535
Electromagnetic interference (EMI), 1-2, 9, 23, 89, 147-148, 225, 239, 250, 254-269, 271, 273-274, 277, 355-358, 416, 429, 434, 436, 449, 492-497
Electromagnetic radiation, 263-264, 267, 429, 434
Electromyography (EMG), 40, 264, 266, 330
Electron microscope, 11, 300, 340
Electronic Data Interchange (EDI), 498
Electronic Industries Association (EIA), 358
Electronic medical record (EMR), 32, 371, 410, 416
Electronic patient record (EPR), 63, 449, 456-459, 462, 487
Electronic Protected Health Information (ePHI) (see also HIPAA), 500-505
Electronic stethoscope, 417, 485
Electronic whiteboard, 412, 414
Electronics engineer, 23, 26, 42, 51, 63, 128, 240, 279, 281, 298, 365, 560, 600, 614, 616
Electrosurgery, 358, 419
Electrosurgical generator, 380, 426

Electrosurgical unit (ESU), 17, 38, 40, 263, 279, 313, 322, 331, 342, 348-349, 366-367, 371, 379-381, 383, 421, 441-443, 542
Email, 27, 42, 53, 58, 64, 69, 78, 82, 128, 148, 214, 251, 285, 301-302, 307, 311-313, 358, 379, 453-456, 476-477, 480, 526-527, 564, 589, 601-604, 607, 610, 628-629, 632-633
Embedded software, 361, 364-365
EMBS (see Engineering in Medicine and Biology Society)
Emergency access, 500, 503, 505
Emergency management, 23, 243, 245, 521, 528, 549-550, 552-553, 572
Emergency power system, 41, 398, 520-521
Emergency preparedness, 532, 549-553, 569
EMG (see Electromyograph)
EMI (see Electromagnetic interference)
EMR (see Electronic medical record)
Encryption, 452, 454-455, 500, 503, 505, 511
Endoscope, 8, 35, 331, 333, 340, 343, 371, 382, 401-407, 512, 535-536
Endoscopic surgery, 10, 401
Energy management, 333-334, 507, 517-518, 557, 645
Engineering in Medicine and Biology Society (EMBS), 51, 98, 262, 298, 597, 600, 606, 614
England, 58, 61, 137, 159, 297-298, 318, 365, 498, 506, 546, 563, 593, 596-597, 610-612, 624
Entrapment, 115, 239, 273-274, 421-423, 426, 430, 435-436
Environment of Care (see also JCAHO), 14, 60, 123, 132, 240, 243-245, 310, 315, 323, 493, 497, 518, 520-521, 528, 550, 569, 572-573, 612
Environmental safety, 513
ePHI (see Electronic Protected Health Information)
EPR (see Electronic patient record)
Equipment control program, 107, 172, 627
Equipment donation, 155-158, 305-306, 308-309, 601
Equipment downtime, 20, 236
Equipment evaluation, 19, 25, 29, 86, 104, 202-203, 206, 281, 310, 346, 413
Equipment hazards, 237, 239, 330
Equipment history, 21, 127, 203, 437, 480, 483
Equipment management program, 5-6, 20, 24, 26, 39, 77, 82, 103-105, 107-108, 122-125, 135, 202-204, 220, 245, 298, 411, 555, 573-575, 625, 635, 637
Equipment planning, 19-20, 63, 68, 107, 147, 153, 171-172, 318
Equipment replacement, 20, 27, 78, 103, 150, 153-154, 161, 172, 198, 552
Error
 User, 15, 19, 21, 117, 122, 132, 147, 154, 173, 202-203, 220, 238, 244, 307, 315, 323-324, 326-327, 343, 348, 354-355, 391, 439, 480, 510, 569, 637, 648
 Human, 15, 225, 229, 246, 248-249, 269, 271, 280, 311, 348, 355, 361, 383, 385, 390, 523, 592
 Operator, 15, 109, 201, 249, 268, 281, 312, 315, 324, 355, 437, 510, 592
Essential Equipment List, 169
Essential Health Care Technology Package (EHTP), 159-170, 172, 175-177, 219, 222-224, 306-308, 606, 609
Estonia, 51, 65-66, 167
ESU (see Electrosurgical unit)
Ethernet, 32, 128-130, 318, 374, 453-455, 457, 460, 487, 489, 491, 509, 511-512, 625
Ethical conduct, 593
Ethylene oxide (EtO), 9, 251, 279, 378, 426, 442, 533-535, 542, 544, 546, 548, 551, 587, 612
EtO (see Ethylene oxide)
ETSI (see European Telecommunication Standards Institute)
EU (see European Union)
Europe, 53-56, 58, 172, 295, 418, 484, 498, 557-559, 580-583, 619, 645, 648, 656-658
European Alliance of Clinical Engineering, 68
European Community (EC), 122, 124, 240, 243-245, 355, 388, 493, 520-521, 528, 555, 569, 572, 579-580, 582-583, 585, 635, 637
European Economic Community (EEC), 67, 69, 582-583, 585
European Medical Devices Directives (MDD), 67, 69, 583-584
European Telecommunication Standards Institute (ETSI), 558
European Union (EU), 51, 65-66, 77, 173, 295, 305-306, 343, 555, 557, 559, 578-585, 589, 619, 647

Executive dashboards, 410, 449, 476-477, 481, 483, 627, 642, 644
Exhaust hood, 515, 518, 546
Expert systems, 222, 467-471, 474-475, 592, 623, 627
Expert witness, 19, 21, 23, 37, 41, 43, 45-48, 270, 280, 358, 590, 592, 645
Extracorporeal circulation, 8, 92
Extracorporeal counter-pulsation system, 420

Facilities engineering, 22, 27, 49, 117, 120, 240, 245, 513, 520, 551
Facilities management, 20, 26, 39, 50, 97, 114, 123, 138, 146, 236, 269, 349, 497, 528, 569, 659
Facilities planning, 99, 102, 141, 172, 187, 601
Factory training, 150
Failure mode and effect analysis (FMEA), 115, 118, 222, 224-225, 227, 229, 231, 234-235, 239, 271, 281, 370, 645
FCC (see Federal Communications Commission)
FDA (see Food and Drug Administration)
FDA Enforcement Report, 589
Federal Communications Commission (FCC), 15, 269, 492-497
Federal Emergency Management Agency (FEMA), 23, 549-550, 552
FEMA (see Federal Emergency and Management Agency)
Fetal monitor, 38, 371, 411, 413-414, 416, 442-443, 463, 509-510
Fiberoptic, 382, 401-403, 656
Field service, 18, 136, 145-146, 148, 150, 291, 400, 449, 464, 467-469, 471-475
Film processor, 342, 371, 398-400, 547
Filmless radiography, 320
Financial analysis, 6, 24, 311, 347
Financial management, 61, 173, 179, 188-189, 191-193, 195, 197-199, 318, 575, 606, 616, 652
Financial planning, 27, 43, 108, 181, 188, 192-198, 306, 649, 651
Finland, 51, 53, 65, 262, 561-562, 655
Fire safety, 13, 16, 525, 559, 568-569
Firewall, 128, 455, 510-511
Firmware, 459, 462, 518, 625
Fluoroscopy, 9, 318, 320, 331, 371, 382, 418, 485, 531, 598-599
FMEA (see Failure mode and effect analysis)
Food and Drug Administration (FDA), 5, 8 10, 13, 15, 19, 21, 28, 37-38, 42, 45, 47, 62, 71-72, 108, 111, 115, 118-119, 123-124, 127, 151, 156, 221, 227, 231, 235, 237-238, 240, 246, 248, 251, 254, 260-261, 268-271, 273-274, 277, 280-281, 311, 318, 321, 328, 346-348, 353-359, 362, 365-366, 369-370, 385, 391, 399, 416, 421-423, 429-430, 435-436, 451, 492-495, 497, 507, 510-512, 529-530, 532, 535, 549, 555, 559-560, 567, 573, 578, 580-581, 583, 585-589, 592, 602, 604-606, 616, 624, 626, 645, 648
France, 51, 410, 529, 561-562, 623
Frequency management, 511
FTE (see Full-time equivalent)
Full-disclosure, 373, 461-462
Full-time equivalent (FTE), 54-55, 62-64, 127, 135, 201, 653

Gantt chart, 182
Gas cylinder, 415, 446, 522, 537
Gas delivery, 380, 386, 614, 616
GDP (see Gross domestic product)
Generator, 17, 40, 74, 87, 95, 249-250, 300, 333, 361, 378, 380, 383, 391, 394-395, 398-400, 413, 420, 426, 485, 507, 511, 516, 518, 520-522, 551-552, 569
Genome, 341, 623, 657
German Biomedical Engineering Society, 619
Germany, 8, 51, 53, 65, 67-68, 77, 86, 88, 155-156, 158, 285, 294-295, 455, 593, 603, 619, 623
GFCI (see Ground fault circuit interrupters)
Global hospital, 655-659
Global Medical Device Nomenclature (GMDN), 583
Glucose monitor, 343-345, 367-368
GMDN (see Global Medical Device Nomenclature)
GMDTM (see Good Medical Device Technology Management Practices)
GMP (see Good manufacturing practices)
GMtP (see Good management practices)
Good management practices (GMtP), 108-110
Good manufacturing practices (GMP), 71, 108, 330, 359, 546, 567, 580, 587-589, 602, 609

Good Medical Device Technology Management Practices (GMDTM), 108-109
Greece, 11-12, 51, 53, 532, 582
Gross domestic product (GDP), 10, 12, 29, 62, 65-66, 77, 85, 94, 159, 171, 342, 449, 498, 623, 657
Ground fault circuit interrupters (GFCI), 379, 521
Guatemala, 302, 484

Hacker, 455, 511
Hazardous waste, 525, 539
HCFA (see Health Care Financing Administration)
Health and Human Services (HHS), 224, 248, 320, 422, 498-500, 529, 550, 553, 561, 570, 624, 627
Health Care Financing Administration (HCFA), 181, 188, 422, 436, 550, 555, 566, 568, 570
Health Care Quality Improvement Act, 237
Health care technology management, 27, 51, 65, 72, 86, 91-92, 99, 158, 163-164, 167, 169-171, 221, 223, 295, 304-305, 333, 598, 601, 605, 627
Health Devices Alerts, 358
Health Devices Sourcebook, 20, 115
Health Devices, 20, 45, 115, 122-123, 158, 190, 199, 224, 262, 269, 280, 299-300, 345, 358, 367-368, 370, 410, 492, 548, 585, 624, 627
Health Level Seven (HL7), 458, 462-463, 477, 498, 629
Health Maintenance Organizations (HMO), 1, 9-10, 34, 140, 181, 227, 346, 502, 567-568
Health technology management (HTM), 41, 51, 59, 65, 89-90, 97, 99, 108, 116, 118, 133, 137, 159-162, 164, 171-175, 177, 182, 218-221, 223, 239-240, 251-252, 285, 295, 301, 304-309, 408, 508, 553, 557, 559, 561, 563, 578, 585, 627, 647, 657
Healthcare Information and Management Systems Society (HIMSS), 487, 624, 654
Healthcare Information System (HIS), 7-8, 11-13, 18-19, 23, 25, 30, 40, 42, 46-48, 51, 72, 74, 94, 132, 179, 181, 187-189, 191-192, 198, 205, 213-217, 225, 227-228, 252, 258, 272, 274-276, 281-282, 285, 290-291, 294, 303-305, 308, 316, 321-322, 337, 339-340, 343, 346, 348-349, 357, 365, 370, 373-374, 392, 409-410, 419, 430-431, 449, 452, 456-458, 461-462, 465, 487, 492, 523, 548, 559, 583, 591-593, 595-597, 604, 607-609, 611-612, 619, 650-651, 659
Heating, ventilation and air conditioning (HVAC), 6, 156, 173, 249, 263, 307, 378, 383, 413, 449, 451, 469, 490, 507, 513, 516-519, 525, 533, 543, 547-548
Help desk, 142-143, 323, 452, 464, 466-467, 469-470, 626
Hemodialysis unit, 92, 345, 543, 548
HEPA (see High efficiency particulate air)
Hepatitis, 245, 251, 342, 537, 543
HHS (see Health and Human Services)
High efficiency particulate air (HEPA), 238, 378, 544-545
Hill-Burton Act, 8, 570
HIS (see Healthcare information system)
HL7 (see Health Level Seven)
HMO (see Health Maintenance Organizations)
Holter monitor, 371, 418
Home care, 15, 62, 238-239, 244, 280, 321, 344-345, 358, 416, 421-422, 431, 435, 447, 568, 570, 653
Hospital bed, 20, 28, 49, 54, 58, 62, 78, 84, 333, 346, 421-422, 431, 433, 435-436, 463, 477, 532, 628-629
Hospital engineering, 27-28, 42, 58-59, 68, 75, 80, 89, 122, 152, 199, 205, 305, 309-310, 328, 330, 333, 349, 543, 560, 573, 592, 600, 612
Hospital information system, 3, 6, 33, 368, 413, 449, 456, 458, 463, 476, 487, 509, 627
Hospital safety, 78-80, 225, 243, 245, 280, 310-311, 315
HTM (see Health technology management)
Human error, 15, 225, 229, 246, 248-249, 269, 271, 280, 311, 348, 355, 361, 383, 385, 390, 523, 592
Human factors engineering, 15, 181, 271, 281, 581, 644
Human factors, 1, 37, 117, 164, 171, 187, 225, 229, 234, 237, 248, 271-273, 280-281, 315, 323, 337, 348, 353-355, 367, 371, 391, 434, 451, 572, 645
Human resources, 60, 66, 72-75, 78-79, 84-86, 88-90, 93-94, 133-134, 156-157, 159-171, 173, 175, 179, 191, 207, 213-219, 222-223, 302, 304-306, 308, 332-334, 526, 565, 572, 597, 606, 659
Hungary, 51, 339, 561
Hygiene, 299, 333, 541-543, 547, 619
Hyper-hypothermia unit, 381

ICC (see International Certification Commission)
ICD (see Implantable cardioverter defibrillator)

Index 669

ICD (see International Classification of Diseases)
Iceland, 53-54, 582
ICPC (see International Classification of Primary Care)
IDN (see Integrated delivery network)
IEC (see International Electrotechnical Commission)
IEEE (see Institute of Electrical and Electronics Engineers)
IFMBE (see International Federation of Medical and Biological Engineering)
IHE (see Integrating the Healthcare Enterprise)
IIHI (see Individually Identifiable Health Information)
ILSM (see Interim Life Safety Measures)
Image intensifier, 8-9, 136, 318, 320, 330, 339, 395-396, 398, 400, 647
Imaging technology, 137, 142, 320, 334, 371, 381, 392, 407, 471, 656-657
Impact analysis, 172-173, 175, 177, 304
Implantable cardiac pacemaker, 9, 254-255, 257-258, 262, 420
Implantable cardioverter defibrillator (ICD), 164-167, 169, 254, 262, 498
Implantable device, 330, 579, 583, 645
Implementation specifications (see also HIPAA), 501-504
Incident investigation (see also Accident investigation), 1-2, 15, 20, 23, 62, 104, 198, 202, 220, 225, 227, 244, 246, 249, 267, 270-271, 273, 280-281, 298, 303, 307, 311, 321, 325, 327, 358, 383, 436, 574-575, 577, 592, 625
Incident report, 15, 44, 125, 202, 236-237, 254, 262, 268, 270, 274, 322-323, 357, 383, 504-505, 584-586, 589, 591, 647-648
Incoming Inspection, 19, 21, 36, 39, 56, 104, 107, 109, 116, 121, 125, 147-148, 151-152, 154, 202-203, 252, 271-272, 277, 298, 323, 430-432, 480, 574, 608, 625
Incubator, 257, 272-273, 277-281, 300, 331, 333, 371, 411-412, 414, 416, 421-424, 435-436, 439-440, 442, 518, 523-524, 542-543, 551-552, 598
Independent consultant, 37, 41, 114, 269-270
Independent service organization (ISO), 10, 28, 59, 63, 71, 136-137, 139-141, 143-144, 146, 148-149, 151, 156, 171, 173, 188-189, 222, 224, 235, 262, 298, 313, 318, 320, 359, 400, 431, 471, 491-492, 495, 497, 535, 555, 557-560, 565-566, 573-574, 583-584, 588-589, 626, 646
Independent Technology Consultant, 72
India, 11, 51, 155, 158, 304, 332-335, 603, 655-656, 658-659
Indicators, 7-8, 39, 53, 94, 96-97, 103, 105, 107, 112-113, 123, 127, 152, 161, 164, 169, 171-172, 175-177, 179, 199-200, 202-205, 214, 217, 219-221, 223, 236, 256, 271, 277, 304, 315, 318, 325, 327, 339, 347, 352, 357, 361, 388-389, 391, 395, 401, 426, 434, 438, 456, 468, 476, 480, 518, 529, 536-537, 544, 547, 575, 607, 627-630, 633-634, 640-642, 644, 653
Individually Identifiable Health Information (IIHI) (see also HIPAA), 499-500
Industrial engineering, 181, 187, 213
Industrial equipment, 74, 78-80, 547
Infant incubator, 257, 277, 279-280, 411, 421-423, 439-440, 542
Infant radiant warmer, 271, 273, 279, 415, 421, 599
Infant warmer, 22, 272, 280, 300, 411, 416, 598
Infection control, 1-2, 5, 14, 74, 177, 240, 245, 270, 289, 291, 307-308, 322, 383, 412, 414, 513, 525-526, 528, 532-533, 542, 544, 548-549, 552-553, 559, 572
Information access, 501, 504
Information management, 1, 12, 32, 60-61, 187, 205, 237, 328, 449, 456, 458, 462, 495, 507, 633
Information services (IS), 1-51, 53-69, 71-94, 96-99, 101-105, 107-112, 114-133, 135-144, 146-171, 173-176, 179, 181-225, 227-256, 259-264, 266-274, 276-282, 285, 287, 289-292, 294-295, 297-318, 320-325, 327-335, 337, 339-371, 373-403, 405-408, 410-433, 435-447, 449, 451-465, 467-472, 474, 476-480, 484-515, 515-518, 520-553, 555, 557-561, 563-584, 586-593, 595-619, 621, 623-631, 633-635, 638-642, 644-659
Information system, 2-3, 6, 13, 30-31, 33-34, 37, 74, 77-78, 80, 86, 88, 93-94, 99, 128, 137, 152, 160, 170, 172, 179, 181, 187-190, 229, 241, 291, 304, 306, 343, 366, 368, 373-374, 413, 449, 451, 453, 455-459, 461-463, 476, 487-489, 500-503, 509-511, 607, 623, 627-628, 644
Information technology (IT), 1, 3-51, 53-60, 62-63, 65-69, 71-75, 77-85, 87-94, 97, 99, 101-104, 107-112, 114-128, 130-144, 147-153, 155-164, 166-170, 174-176, 179, 181-207, 210-222, 225,
227-233, 235-241, 243-251, 255, 258, 260-264, 266-274, 277, 281-283, 285, 287, 289-291, 294-295, 297-304, 306, 308-313, 315-318, 320-322, 324-325, 328, 331-334, 337, 339-370, 373-379, 382-390, 392-401, 403, 405, 407-408, 410-415, 417-419, 421-422, 424, 426-427, 430-431, 433, 438-447, 449-465, 467-471, 476-477, 484-496, 498, 500-512, 515-518, 520-547, 549-553, 555, 557-560, 565-568, 570-584, 586-593, 595-602, 606-615, 617-618, 621, 623-630, 633-634, 637-638, 640, 642, 644-653, 655-659
Infusion controller, 249
Infusion device, 40, 300, 357, 368, 371, 375-376, 380, 647
Infusion pump, 22, 40, 62, 64, 111, 147, 203, 239, 250, 257, 279, 294, 300, 321, 323, 344, 350, 354, 357, 367, 375, 442-443, 453, 458, 460, 463, 507, 509, 551, 582-583, 589, 645
In-service education, 1-2, 19-20, 25, 209, 238, 315-316, 324, 435, 437, 541
In-service, 1-2, 19-22, 25, 37, 46, 86, 104, 107, 111, 206, 208-209, 211, 238, 285, 298, 315-317, 321-325, 327, 343, 347-348, 435, 437, 541, 645
Inspection frequency, 125, 311, 605
Inspection procedure, 3, 36, 523, 574
Inspection, 3-4, 15, 19, 21, 23-24, 26-27, 29, 35-36, 39, 44-48, 56, 67, 79, 86, 91-92, 104, 107, 109, 116, 119, 121-126, 129, 131, 147-148, 151-152, 154, 158, 171, 189, 192, 200-205, 208, 211-212, 218, 220-221, 224, 235, 237-239, 245, 247, 250, 252, 271-274, 276-277, 282, 291, 297-300, 311, 322-325, 332-334, 347, 356, 361, 371, 376, 383, 385, 395, 397, 400, 415, 423, 427, 430-432, 434-436, 447, 462, 480, 515, 523, 529-533, 543, 551, 569-570, 572-576, 580, 586, 588-589, 601, 605, 608, 625-626, 644
Installation requirements, 111, 120-121, 157-158, 535
Installation, 1, 6, 19-20, 23, 30, 68, 71, 74, 82, 90, 93, 103-105, 109, 111, 116, 118-122, 124, 126-129, 135, 147-148, 150-151, 156-158, 162, 166, 170-172, 183, 187, 202, 211, 262, 317-318, 340, 362, 368, 383, 389, 393, 397-400, 418, 431, 433, 435, 452-456, 461, 465-466, 484, 490, 496-497, 515-516, 520-522, 524, 527-528, 533, 543-545, 551, 574, 580-581, 626, 630, 648, 658
Institute of Electrical and Electronics Engineers (IEEE), 7, 42-43, 51, 61, 98, 128-129, 188, 202, 240, 262, 267, 274, 279, 281, 298, 315, 341, 359, 362-363, 365, 410, 412-413, 420, 435, 455, 463, 475, 497, 511, 560, 592-593, 597, 600, 606-607, 614, 616, 627
Institute of Medicine (IOM), 10, 13-15, 51, 99, 224, 233-235, 237, 240-241, 245, 269, 281, 353, 563, 566, 568, 624, 626-627, 648
Institutional review board, 20, 38, 42, 117, 238, 337, 348, 369, 586-587
Insufflator, 382
Integrated Delivery Network (IDN), 568
Integrated services digital network (ISDN), 128-129, 486-487
Integrating the Healthcare Enterprise (IHE), 624, 626-627
Intensive care unit, 3, 21, 23, 38, 40-41, 63, 68, 73, 86, 92, 183, 229, 236, 248-250, 254, 264, 271-272, 280, 299, 324, 340, 371, 373, 377, 385, 391, 410-411, 413, 459, 462, 476, 479, 489, 509, 523-524, 543, 551, 598, 629, 631, 647, 657-658
Intensive care, 1, 3, 9, 21, 23, 33, 38, 40-41, 62-63, 68, 73, 75, 86, 89, 92, 183, 207, 229, 236, 248-250, 254, 264, 271-272, 280, 299, 324-325, 330, 340, 353, 371, 373-375, 377, 385, 391, 410-411, 413, 416, 459, 462-463, 476, 479, 489, 509, 523-524, 526, 543, 551, 598, 629, 631, 637, 647, 657-658
Interconnectivity, 511
Interdisciplinary approach, 103, 321, 333
Interim Life Safety Measures (ILSM), 245, 528, 569
International Certification Commission (ICC), 7, 15, 71, 82, 208-209, 319, 601-603, 617-618
International Classification of Diseases (ICD), 161, 164-169, 175, 222, 254, 262, 498
International Classification of Primary Care (ICPC), 166-167, 169
International Electrotechnical Commission (IEC), 156, 235, 257, 261-262, 358-359, 558, 560
International Federation of Medical and Biological Engineering (IFMBE), 51, 53, 58, 61, 64, 83, 91-92, 302, 305, 332, 607, 619, 648
International Society of Technology Assessment, 306
International standard, 110, 491, 558

International Standards Organization (ISO), 28, 71, 136, 139-141, 143-144, 146, 148, 151, 156, 171, 173, 189, 222, 224, 235, 262, 318, 320, 359, 400, 491-492, 495, 497, 535, 555, 557-560, 565-568, 583-584, 588-589, 626
Internet, 9, 18, 27, 30, 32, 36-38, 42, 53, 78, 89, 112, 128, 132, 145, 161-162, 212, 217, 251, 285, 304, 309, 311-314, 317-318, 331, 333, 346-347, 379, 451-455, 470, 477, 484, 500, 509, 511-512, 516, 526-527, 557, 559, 564, 601, 605-606, 609, 620, 623-624, 626, 629, 633, 649, 659
Intra-aortic balloon pump, 92, 250, 300, 371, 374-375, 419-420
Intranet, 127, 129, 477, 488, 511, 527, 629
Inventory, 4-6, 9, 17, 19, 29, 36, 44-45, 55-56, 62, 67-68, 74, 79-81, 94-96, 103-105, 107, 109-111, 113-114, 122-129, 133-134, 145, 147, 151, 153-154, 164, 172, 182-183, 187-188, 190, 196-198, 207, 211, 223, 236, 239, 248, 252, 282, 300, 306-307, 321, 328, 347, 383, 400, 431-432, 464, 472, 496-497, 504-505, 507-508, 515, 523-524, 530-531, 535, 550-552, 569, 572, 574-575, 577, 625, 637, 651
IOM (see Institute of Medicine)
IS (see Information services)
ISDN (see Integrated services digital network)
ISO (see Independent service organization)
ISO (see International Standards Organization)
Isolated power system, 9, 283, 384, 513, 520-521
Isolated power, 1, 6, 9, 282-283, 384, 386, 513, 520-521
IT (see Information technology)
Italy, 28, 30, 32-33, 51, 53, 532, 603, 605

Japan, 8, 51, 77, 91-92, 267, 343, 392, 581, 585, 647-648, 655-659
JCAHO (see Joint Commission on Accreditation of Healthcare Organizations)
Job description, 1, 13, 16, 49, 206, 208, 210, 212-213, 300, 569, 581, 644, 648
Joint Commission on Accreditation of Healthcare Organizations (JCAHO), 8, 14-15, 19-21, 25-26, 37, 41, 43, 64, 86, 103, 107, 122-125, 127, 130, 132, 176, 202, 205, 212, 220-221, 225, 231, 237-241, 243-245, 269-270, 280-281, 298, 310-311, 313, 315, 318, 320, 322-323, 327, 347, 349, 354-355, 383, 422, 435-436, 493, 497, 518, 520 521, 524, 528, 550, 553, 555, 560, 565-576, 578, 605, 608, 612, 616, 626, 644, 648
Jordan, 51
Journal of Clinical Engineering, 3, 122, 207, 287, 300, 302, 345

Kellogg Foundation, 9, 18, 27-28
Kenya, 161, 170, 173, 306, 659
Kyrgyzstan, 51, 161-162, 175-177, 224, 304, 609

Latin America, 28, 53-56, 58, 69, 73, 78, 80, 83-84, 86, 89, 93, 159, 172, 177, 301, 345, 657
Latvia, 65
Lawyer, 13, 42, 146, 227-228, 281, 346, 358, 593, 595-596, 618-619
Leakage current, 21, 39, 129, 281, 316, 322, 355, 379, 434
Leapfrog Group, 240
Legal liability, 227, 235-236, 526, 581, 590, 592, 648
Liability insurance, 120, 307, 366, 571
Life-cycle cost, 127, 171-172
Life-cycle management, 34, 109-110, 159, 225, 251-252
Life safety code, 245, 518, 520, 550, 569
Life safety management, 243-245, 569
Life-support equipment, 63, 91, 238-239, 261-262, 385-386, 520, 551-552, 572
Life-support medical devices, 227, 413
Line isolation monitor (LIM), 379, 443, 521
Linear accelerators, 73, 333
List-serve, 64
Lithuania, 51, 304, 605
Litigation, 37, 43-48, 59, 116, 236, 238, 269-270, 280, 321, 356, 358, 457, 523, 592
Loaner, 20-21, 117, 119-120, 434, 574
Local area network, 6, 127-130, 254-255, 311, 413, 454, 456, 460, 464, 487
Local codes, 125, 516
Local regulations, 251
Luminance meters, 486

Index

Machine vision, 297, 371, 401-403, 405, 407, 409
Macroshock, 316
Magnetic field, 250, 254-256, 259, 261, 263, 300, 317, 340, 382, 389, 444-445, 493
Magnetic resonance imaging (MRI), 6, 9, 13, 18-20, 61, 68, 134, 136, 228, 250, 266, 317, 319-320, 330, 333, 340, 381-382, 385-386, 392, 399-400, 469-470, 487, 489-492, 551, 614, 656
Maintenance insurance, 9, 28, 62-63, 126, 136-137, 149-150, 152, 610
Maintenance management, 26, 34, 36, 38, 63-64, 78-80, 99, 110, 112, 122-125, 127, 129-130, 135, 147, 151-152, 172, 205, 212, 315, 457, 608, 610
Maintenance manuals, 29, 116, 229, 297, 434, 437, 533, 590
Maintenance requirements, 123, 125, 131, 148, 157-158, 165-166, 204, 220, 323, 521, 543, 574, 635
Malcolm Baldrige National Quality Award, 222
Malicious software, 501
Mammography systems, 135, 319
Mammography units, 399
Management engineering, 27, 179, 181, 183, 185, 187-188, 318
Management information system (MIS), 80, 137, 143, 472
Manuals
    Maintenance, 355-357
    Maintenance, 29, 116, 229, 297, 434, 437, 533, 590
    Operation, 147, 158, 416, 422
    Operator, 116, 128, 248, 511, 533
    Service, 16, 30, 97, 107, 109, 116-117, 120-121, 125-126, 128, 147, 156, 158, 208, 279, 347-348, 370, 395, 416, 432-433, 437, 511, 615
    User, 66, 115, 182, 236, 238, 326
Manufacturer's training, 238, 319, 325, 347, 432
Manufacturer's warranty, 356
Manufacturers and Users Device Experience (MAUDE), 45, 227, 231, 238, 274, 347-348, 356-357, 370, 470, 589
Manufacturers' field service, 291
Market approval, 583, 586
Market research, 143-145, 350, 649-651
Mass spectrometry, 373, 390
Materials management, 4, 53, 104, 107, 181, 187, 378, 549
MAUDE (see Manufacturers and Users Device Experience)
MDD (see European Medical Devices Directives)
MDR (see Medical Device Reporting)
Mean time between failure (MTBF), 111, 150, 183, 434, 472, 637
Mean-time-to-repair (MTTR), 471-472
Mechanical engineering, 1, 20, 42, 49, 61, 227, 294-296, 453, 517, 525
Mechanical failure, 447
Mechanical integrity, 251, 434
Mechanical safety, 116
Mechanical ventilator, 40, 374, 415
Media re-use, 502, 505
Medicaid, 9, 12, 110, 116, 224, 422, 498, 500, 550, 555, 566, 570
Medical care, 1, 3, 7, 11-13, 15, 21, 97, 179, 187, 228, 236, 335, 346, 410, 453, 486, 537, 545, 549, 573, 590, 597, 624, 645, 655-657
Medical Device Improvements Act, 357, 586
Medical Device Reporting (MDR), 15, 26, 45, 115, 118, 236, 277, 280-281, 354-355, 357, 423, 436, 497, 560, 586, 589
Medical device
    Accidents, 15, 59, 67-68, 107, 115, 124, 132, 225, 228, 235, 237, 246, 251, 269-271, 273-274, 277, 279-282, 315, 321, 323, 325, 327, 331, 337, 341, 345, 348, 353, 355-358, 416, 421, 426, 435, 442, 445, 447, 497, 510, 542, 551-552, 560, 581, 583, 590-591, 618, 644-645
    Acquisition, 9, 20, 29, 33, 36-38, 60, 62-63, 66, 74-75, 78-79, 82, 92, 99, 104, 107-112, 115, 118, 122, 127, 135, 141-142, 147-148, 150-153, 155, 160-161, 163-164, 169-170, 173, 189, 193, 207, 209, 221, 223, 237-238, 252, 298, 323-324, 327, 329-330, 337, 341, 344-345, 348, 366, 368, 370-371, 416, 431, 435, 451, 459-460, 462-463, 497, 507, 527, 533, 561, 572, 574, 577, 581, 624-625, 638, 644-645, 648
    Design, 3-5, 8, 12, 15, 17, 20-21, 36-37, 42, 45-46, 61, 66, 82-83, 107, 115, 123, 147, 149, 152, 168, 202, 209, 222, 225, 227, 235, 237-240, 245-247, 249-251, 260, 264, 269, 271-274, 277, 281, 294, 300, 302, 318, 323, 325, 327-331, 337, 346-349, 351, 353-357, 359, 361, 364-365, 371, 376, 378, 383-384, 410-413, 416, 420-421, 423-424, 426, 432-434, 442, 451, 453, 459, 462, 493, 495, 509-510, 520, 527, 557, 568, 572-575, 577, 579-581, 583-584, 586-588, 591-592, 598, 617, 621, 624-625, 644-645, 648
    Donations, 78, 80, 82, 155-158, 160, 301, 435, 601, 606
    Hazards, 13, 21, 26, 62, 99, 104, 107, 115, 156, 202, 221, 225, 227, 229, 237-239, 243, 245, 247, 250-251, 262, 272-274, 280-282, 325, 330, 337, 340-343, 348, 353-358, 364, 370, 378, 383-384, 413, 416, 421, 423-424, 426, 433-435, 437, 445-446, 495-496, 510, 525, 542, 551-552, 568, 572, 577, 579, 582-583, 589, 591, 617-619, 625-626, 647
    Incidents, 5, 14-15, 20-21, 23, 37, 44-45, 60, 62, 67-68, 104, 107, 122, 124, 202, 209, 225, 227-228, 235-240, 246, 249, 254, 260-264, 267-274, 277, 279-281, 298, 315, 321-325, 327, 347, 356-358, 383, 398, 421, 423, 426, 431, 436, 494-497, 505, 520, 572, 574-575, 577, 580, 584-586, 589, 591-592, 607, 617, 625-626, 647-648
    Inspection & PM, 126, 626
    Inventory, 4-5, 8-9, 17, 29, 36, 44-45, 51, 59, 62, 67-68, 74, 78-80, 104, 107-111, 122-124, 126-127, 134-135, 147, 151, 153, 164, 183, 190, 207, 211, 223, 235-236, 239, 252, 282, 300, 321, 328, 339-340, 347-348, 350-351, 383, 400, 431-432, 496-497, 505, 507-508, 535, 551-552, 572, 574-575, 577, 625, 645
    Maintenance, 4, 7-10, 12, 15, 17, 20-21, 26, 28-31, 36-38, 41, 44, 49, 59-63, 66-68, 71-72, 74-75, 78-80, 82-83, 91-92, 99, 104, 107-112, 122-124, 126-127, 130-132, 134-135, 137, 142, 147-153, 155-158, 160-162, 165-166, 169-170, 173, 179, 189-191, 202, 207, 209, 211, 221, 223, 228-229, 237-240, 245-247, 249, 252, 267-272, 274, 277, 282, 291, 294-295, 297-301, 304, 312, 315, 318, 322-323, 328, 333, 339, 342-344, 346-347, 353-354, 356, 359, 368, 378, 383, 412, 421, 423-424, 426, 431-437, 442, 451, 453, 459, 462, 495-497, 505, 520, 525, 533, 542-543, 545, 558, 567-568, 572, 574-575, 577, 579, 581, 583-584, 590, 598-599, 601, 606, 610, 618 619, 625 626, 644, 648
    Manufacturers, 3, 5, 7-10, 15, 17, 20-21, 23, 28, 31, 38, 41-47, 59-60, 66-68, 79, 91, 107, 109, 111, 115, 118, 123-124, 126-127, 130, 132, 135, 137, 147-152, 156, 158, 171, 173, 207, 221-222, 225, 227-228, 235-240, 247, 250-251, 257, 259-261, 263-264, 267, 269-274, 277, 280, 282, 291, 294-295, 301-302, 312-313, 315, 321, 324-325, 337, 342, 344, 346-349, 353-354, 356-357, 366-370, 400, 412-413, 415, 420, 426, 431-437, 443, 450, 453, 485, 491-495, 497, 507, 509-510, 542-543, 552, 555, 557-558, 567, 573-575, 578-586, 588-592, 599, 610, 619, 625-626, 638, 647-648
    Users, 3, 9, 15, 20-21, 26, 30, 33, 37-38, 45, 47, 49, 60, 62, 66-68, 75, 80, 82-83, 99, 104, 107-109, 111-112, 115, 118, 122-124, 126-127, 131-132, 135, 137, 141, 147-153, 155-156, 158, 161-164, 166, 170, 173, 183, 189, 191, 202, 207, 211, 223, 236, 238, 249, 252, 255, 260-261, 263, 272, 274, 277, 280-281, 298, 312-313, 315-316, 321-325, 327, 342-343, 346-349, 351, 353-357, 359, 364-371, 376, 378, 383, 410, 421, 426, 431, 434-435, 437, 439, 446, 449, 451, 453, 459-460, 462, 485, 491-497, 505, 510, 533, 552, 559, 572, 574-575, 577, 579-587, 589, 591-592, 606, 619, 621, 623, 625-626, 638, 644-645, 647-648
    Manufacturing, 10, 20, 37, 107-108, 147-148, 150, 160, 185, 222, 235, 272, 328, 330, 337, 346, 348-349, 356, 359, 434-435, 557, 567, 580, 583, 586-589, 610, 644-645
    Operators, 15, 40, 67, 78, 80, 109, 111, 124, 148, 155-156, 158, 225, 246-247, 249, 257, 268-269, 271-272, 277, 279, 281, 312, 315, 324, 355, 367, 370-371, 431-432, 434, 436-437, 439, 446, 491, 510, 533, 535, 567, 592, 648
    Purchasing, 4-5, 9-10, 20-21, 23, 26, 29-30, 36, 38, 42, 44, 59, 62-63, 65-68, 71, 80, 82-83, 91, 104, 107, 109-112, 115, 118, 124, 126-127, 135, 141, 147-153, 155, 158, 160, 164, 166, 171, 173, 189, 193, 221, 236-238, 261, 264, 272, 282, 300, 323, 342, 344, 346-347, 353, 356, 366-368, 371, 383, 413, 416, 421, 424, 426, 431-432, 435, 451, 491, 496, 507, 525-526, 551-552, 564, 575, 583, 591, 599, 624
    Recalls, 20-21, 23, 47, 62, 79, 107, 124, 126, 147, 155-156, 221, 223, 236-237, 239, 272, 274, 316, 322, 342-343, 348, 358, 426, 433, 435, 574, 577, 580, 584, 586, 589, 591, 625-626
    Regulations, 9, 12-13, 15, 26, 29, 37, 42, 46, 66-68, 72, 78-80, 82-83, 91, 108, 110, 123, 130, 134-135, 148, 151, 153, 160, 219, 221, 235, 237-240, 243, 245, 251, 269-270, 274, 277, 282, 294, 304, 313, 318, 328, 343-344, 348, 351, 354, 357, 359, 363, 369-370, 453, 497, 535, 555, 557, 559-560, 564, 567-568, 578-584, 586-589, 592, 598-599, 607, 618-619, 624-626, 647-648
    Tracking, 9, 17, 20, 45, 62, 107, 122, 124, 202, 221, 223, 327, 330, 363, 497, 505, 526, 552, 567, 572, 574, 577, 579, 584, 586, 589, 592, 625
Medical equipment management, 19-20, 26, 34, 39, 59, 77, 82, 103, 107-110, 122-124, 130, 132, 159-160, 162, 171, 173, 175, 183, 205, 239, 243-245, 309, 315, 324, 555, 569, 572-573, 575, 625, 635, 637
Medical Equipment Management Plan (JCAHO), 123, 239, 243-244, 315, 569, 572
Medical equipment
    Accidents, 15, 19, 59, 73, 81, 107, 124, 132, 237, 251, 282, 315, 325, 341, 355, 414, 497, 542, 552, 581
    Acquisition, 1, 20, 27, 29, 34-36, 38-39, 60, 73-75, 77, 79, 81-82, 92-93, 103-104, 107-112, 122, 125, 127, 135, 148, 151-152, 155, 159-160, 164, 169-170, 172-173, 175, 189, 193, 207, 209, 221, 223, 237-238, 252, 298, 324, 329, 341, 366, 368, 471, 480, 497, 507, 527, 572, 581, 624-625, 633, 648
    Design, 1-4, 8, 15, 18, 20-21, 25, 35-36, 42, 61, 73, 82, 86-87, 97, 107, 123, 152, 188, 202, 205, 209, 237-240, 245, 249, 251, 300, 307, 318, 325, 329, 335, 347, 349, 354-355, 379, 383, 411-412, 527-528, 557, 569, 572-573, 575, 581, 612, 615, 617, 624-625, 648, 658
    Donations, 80, 82, 84, 93, 125, 155-160, 305, 308-309, 601
    Hazards, 1, 19, 21, 24, 26, 39, 104, 107, 125, 156, 202, 221, 237-239, 243-245, 251, 262, 282, 307, 325, 341, 343, 354-355, 379, 383, 414, 446, 528, 542, 552, 569, 572, 617, 625, 637
    Incidents, 1-2, 15, 20-21, 23, 25, 60, 104, 107, 122, 124-125, 154, 202, 209, 220, 237-240, 244, 249, 261-262, 267, 298, 307, 311, 315, 324-325, 347, 383, 436, 497, 572, 575, 617, 625, 648
    Inspection & PM, 125-126
    Inventory, 4, 8, 19, 29, 35-36, 59, 74, 79-81, 94-95, 103-104, 107-111, 113, 122-127, 133-135, 151, 154, 164, 172, 183, 188, 199, 207, 223, 239, 252, 282, 300, 307, 339, 347, 383, 497, 507, 552, 569, 572, 575, 625, 637
    Maintenance, 1-2, 4, 8, 10, 15, 18-21, 26-31, 34-36, 38-39, 41, 49, 59-61, 71, 73-77, 79-82, 84, 86-87, 89, 91-95, 97, 103-104, 107-113, 122-127, 130-135, 137, 148, 151-152, 154-160, 162, 165-166, 169-170, 172-173, 189, 199, 202, 205, 207-209, 220-221, 223, 237-240, 244-245, 249, 252, 267, 282, 297-300, 305, 307-309, 311-312, 315, 318, 339, 343, 347, 354, 368, 383, 412, 414, 436, 480, 497, 506, 528, 542, 567, 569, 572, 575, 581, 601, 608, 613, 625, 635, 637, 648, 658
    Manufacturers, 3, 8, 10, 15, 19-21, 23-25, 28, 31, 35, 38, 41-42, 59-60, 79, 84, 86, 91, 97, 107, 109, 111, 123-124, 126-127, 130, 132, 135, 137, 148, 151-152, 154, 156, 158, 171, 173, 188, 207, 221, 237-240, 251, 261, 267, 282, 308, 312-313, 315, 324-325, 347, 349, 354, 368, 412, 414, 436, 480, 497, 507, 542, 552, 555, 557, 567, 573, 575, 581, 612, 615, 625, 648
    Manufacturing, 2, 10, 20, 77, 107-108, 148, 160, 199, 335, 349, 480, 506, 557, 567, 612, 633
    Operators, 15, 25, 80, 84, 109, 111, 124, 148, 155-156, 158, 249, 312, 315, 324, 355, 436, 446, 506, 567, 635, 646, 648
    Purchasing, 4, 10, 19-21, 23-24, 26-27, 29-30, 35-36, 38, 42, 59, 65, 71, 80-82, 91, 97, 103-104, 107, 109-112, 124-127, 135, 148, 151-152, 154-155, 158, 160, 164, 166, 171-173, 176, 189, 193, 221, 237-238, 261, 282, 300, 307-308, 347, 366, 368, 383, 414, 507, 552, 575, 624, 646

Recalls, 19-21, 23-24, 79, 107, 124-126, 154-156, 220-221, 223, 237, 239, 307, 316, 343, 625
Regulations, 15, 26-27, 29, 42, 79-80, 82, 84, 91, 93, 108, 110, 123, 125, 130, 134-135, 148, 151, 154, 159-160, 176, 219, 221, 237-240, 243, 245, 251, 282, 307, 313, 318, 343, 354, 480, 497, 528, 555, 557, 567, 581, 624-625, 648
Repair, 1, 4, 8, 15, 19-21, 24-25, 27, 29, 34-36, 38-39, 41, 49, 59-60, 73-74, 79-80, 86-87, 89, 91, 93, 97, 103-104, 107, 109, 113, 124-127, 130-135, 137, 148, 151, 155-156, 158-159, 170, 173, 183, 189, 193, 199, 202, 205, 207-209, 220-221, 223, 237-239, 297, 299-300, 305, 309, 312, 324-325, 349, 368, 383, 414, 436, 446, 471, 480, 507, 522, 557, 575, 601, 608, 613, 617, 625, 635, 637, 646
Testing, 4, 18-21, 23-26, 28-30, 34, 36, 38-39, 42, 49, 79, 104, 107, 109, 125-126, 130-132, 135, 148, 151, 158, 170, 175, 183, 189, 199, 202, 208-209, 221, 223, 237-240, 243, 245, 249, 261-262, 267, 282, 297-300, 316, 318, 324-325, 335, 347, 368, 412, 414, 436, 446, 497, 506, 552, 557, 569, 572-573, 575, 608, 617, 624-625, 635, 646, 658
Tracking, 9, 19-20, 24, 39, 107, 122, 124, 202, 221, 223, 307, 497, 552, 567, 572, 625
Users, 1, 3, 9, 15, 19-21, 24-27, 30, 38, 49, 60, 75, 80, 82, 86, 97, 103-104, 107-109, 111-112, 122-127, 131-132, 135, 137, 148, 151-152, 154-156, 158, 162, 164, 166, 170, 173, 183, 189, 202, 207, 220, 223, 238, 244, 249, 252, 261, 298, 307-308, 312-313, 315-316, 324-325, 343, 347, 349, 354-355, 366, 368, 383, 446, 471, 480, 497, 506, 552, 569, 572, 575, 581, 625, 633, 635, 637, 648, 658
Medical errors, 2, 10, 14, 43, 51, 227, 235, 237, 240-241, 245-246, 269, 353-354, 566, 603, 608, 624
Medical gas systems, 513, 516, 522-524, 557
Medical Informatics, 12-14, 563
Medical information system, 510-511
Medical physicist, 1, 16, 35, 53, 58-60, 68, 341, 492, 530, 598
Medical record, 5, 13, 32-33, 45-46, 236-237, 270, 274, 323, 357, 371, 373-374, 410, 416, 457-458, 462, 487, 491, 499, 507, 509, 525-526, 658
Medical technology management, 26, 92, 101, 103, 105, 107-108, 111, 113, 122-123, 132, 135, 152, 177, 306, 308-309, 346, 527, 601
Medical telemetry, 8, 15, 153, 255, 257, 262, 264, 449, 492-497, 607, 612
Medical vacuum, 411, 516, 522, 543
Medicare, 9-10, 12, 65, 110, 116, 224, 332, 334, 422, 498, 500, 550, 555, 566, 570
Medication error, 222, 237, 241-242, 353, 375, 458, 463
Mentoring, 37, 39, 212, 214-218, 248, 329, 484, 526
Mercury, 238, 277-278, 386, 413-414, 416, 522, 538-539, 547-548
Metal fatigue, 427
Mexico, 51, 53, 80-83, 159, 162, 287, 303-305, 529, 603, 605
Microscope, 11-12, 21, 133-134, 223, 300, 333, 340, 383
Microshock, 36, 225, 249, 281-283, 316
Microsoft Windows, 127, 511, 633
Microwave, 6, 8, 127, 429, 446, 454, 495, 497, 586, 647
Middle East, 51, 97, 172
Military, 8, 11, 16, 34-36, 90, 183, 207, 212, 240, 303, 310, 359, 426, 454-455, 484, 506, 529, 533, 537, 573, 645
MIS (see Management information system)
Mission critical, 107, 626
Modular design, 425, 459
Mongolia, 51, 110, 304
Monitor alarms, 229, 233, 247, 512
Monitoring, 4, 8, 13, 22-24, 29-30, 33, 35, 37-38, 40, 49, 79, 88, 91-94, 102, 107, 109-110, 118-119, 121-124, 131, 147-148, 151-152, 160, 166, 168, 171, 173-174, 181-183, 199, 203-204, 206-207, 217, 220-221, 233, 236-237, 239, 244, 247-250, 252, 255, 258, 261, 264, 269, 300, 306-307, 312-313, 322, 328-329, 333, 337, 340-341, 343, 345, 347, 353-354, 357, 363-364, 368, 371, 373-376, 380-381, 384-387, 389-391, 413-416, 418-420, 449, 456-464, 473, 476-477, 480, 484, 492-497, 501, 505, 507, 509-510, 515, 517-518, 531-533, 535-536, 544, 548, 550, 565-566, 571-572, 578, 580, 582-584, 591, 598, 608-609, 614-616, 626-627, 629, 637-638, 648
Monitoring equipment, 23, 119, 151, 173, 261, 300, 328, 353-354, 368, 373, 380, 387, 419, 493-495, 517, 531-532

Monitoring, bedside, 23, 33, 373, 449, 456, 458-459, 462-463
Monitoring, fetal, 8, 414, 416, 463, 509
Monitoring, patient, 8, 13, 23, 49, 91, 119, 233, 237, 269, 337, 340-341, 374, 380, 384-385, 390-391, 413, 415, 449, 458, 460-462, 476-477, 495, 497, 582, 591, 627, 629
Mozambique, 51, 93-96, 160, 162, 167, 305-306, 609
MRI (see Magnetic resonance imaging)
MTBF (see Mean time between failure)
MTTR (see Mean-time-to-repair)
Multidisciplinary analysis, 244
Multidisciplinary approach, 44, 170
Multimeters, 17, 45, 190, 276, 300, 442-443
Multiple service providers, 124

Namibia, 51, 167, 306
Nanotechnology, 128, 341, 623
NASA, 13
National Academy of Sciences, 12, 237, 281, 283, 546, 563, 618, 648
National Electrical Code (NFPA 70), 274, 281, 358, 384, 413, 521
National Electrical Manufacturers Association (NEMA), 367, 492
National Fire Protection Association (NFPA), 9, 37, 42, 125, 156, 245, 274, 281-283, 310, 353, 355, 358, 379, 383-384, 413, 416, 515, 518, 520-524, 549-550, 555, 561, 568-570, 624
National Institute of Science and Technology (NIST), 222, 531
National Library of Medicine (NLM), 12, 358, 561, 564
National Patient Safety Foundation, 237
National Quality Forum (NQF), 237, 240
Native American, 76-77
NATO, 65
Natural disaster, 248, 502, 546-547, 550, 552-553
Navy, 11, 34, 80-81, 182
NEMA (see National Electrical Manufacturers Association)
Neonatal intensive care, 38, 62, 236, 271-272, 371, 410, 416, 523-524, 598
Neonatal monitor, 371, 413, 416
Nepal, 51, 167, 303-306, 605
Netherlands, 53-54, 77, 562, 564
Network architecture, 457, 460, 462
Network diagram, 182
Network failure, 465, 628
Network infrastructure, 127-128, 462, 490-492
Network management system, 454
Network planning, 320, 658
Network protocol, 454, 491
Network security, 460
Networking, 27, 38, 43, 63, 90, 127-128, 132, 141, 206, 212, 263-264, 287, 291, 304, 318, 379, 453-454, 463, 486-487, 490, 610, 612-613, 623, 658
Newfoundland, 63, 578
NFPA (see National Fire Protection Association)
NFPA 70 (see National Electrical Code)
NFPA 99 (see Standard for Health Care Facilities)
NIST (see National Institute of Science and Technology)
Nitrogen, 379, 387, 390, 516, 522-524, 539-540
Nitrous oxide, 11, 17, 339, 379, 386-387, 390, 413, 442-443, 516, 522-523, 543
NLM (see National Library of Medicine)
Noise pollution, 354
Noninvasive blood-pressure monitoring, 300, 380
North America, 53-59, 108, 138, 291, 293, 512, 549, 558, 581, 624, 627, 648, 657
Norway, 53-54, 582
Nosocomial infection, 408, 532-533, 546-548
NQF (see National Quality Forum)
Nuclear medicine, 8-9, 32-33, 61, 73, 298, 300, 317, 319-320, 328, 330, 335, 339-340, 400, 418, 424, 471, 487-488, 490-491, 530-532, 559, 561, 647
Nurse call, 264, 477, 512, 516, 610
Nursery, 238, 277, 356, 411-413, 484, 551
Nursing care, 12, 65, 321
Nursing education, 148, 321-322, 324-325, 327, 575

Occupational Safety and Health Administration (OSHA), 9, 23, 37, 218, 221, 237-238, 240, 245, 270-271, 311, 313, 316, 383, 535, 550, 555, 561, 568, 570, 612
OEM (see Original equipment manufacturer)
Office of Emergency Preparedness, 550, 552-553

On-call, 19, 21, 23-24, 190, 206, 236
On-site service, 35, 146, 151, 464
On-the-job training, 22, 78, 207, 291, 294, 318, 322, 469
Open architecture, 449, 470, 476
Operating instructions, 158, 434, 518, 545
Operating room, 1, 3-4, 13, 19, 21, 23, 40-41, 46, 49, 63, 147, 183, 228, 250, 257, 279, 282-283, 298-299, 322, 324, 335, 339, 347, 357, 368, 371, 376-386, 391, 398-400, 406, 411, 417, 419-420, 424-426, 459, 476, 484, 490-491, 509, 517-518, 521, 528, 543, 551, 557, 590, 598-599, 628-629, 646, 656
Ophthalmic laser, 345
Ophthalmoscope, 11, 179, 223
OR table, 348, 380, 385, 424-426, 551
Original equipment manufacturer (OEM), 16, 21, 23, 35, 97, 111, 127, 132, 135-137, 139-141, 143-144, 146-149, 151-152, 184, 212, 221, 325, 567
Oscilloscope, 8, 17, 38, 190, 257, 300, 339, 417, 442-444, 627, 646
OSHA (see Occupational Safety and Health Administration)
Outsource, 10, 63, 135, 137, 139, 141, 143-144, 146, 489, 651
Outsourcing, 9-10, 12, 28, 32-33, 36, 63, 68, 135-137, 139-141, 143-146, 151, 187, 189, 307-308, 525, 651
Overhaul, 148, 237, 464
Oximeter, 8, 40, 250, 257, 263, 279, 331, 356-358, 390, 398, 416, 442, 551, 645-646
Oxygen concentrators, 551
Oxygen monitor, 17, 45, 300, 386, 390
Oxygen regulator, 411

Pacemaker, 8-9, 13, 30, 231, 250, 254-255, 257-262, 281, 300, 333, 343, 355, 371, 374-375, 383, 417, 419-420, 442-443, 453, 552, 582, 591-592, 617, 646
PACS (see Picture archiving and communication systems)
Paging, 129, 263-264, 268, 379, 461, 552
PAHO (see Pan American Health Organization)
Pakistan, 51, 603
Pan American Health Organization (PAHO), 27, 51, 71-72, 74, 78, 80, 84-87, 89, 155, 158-159, 162, 172, 177, 301-303, 307, 546-548, 551, 553, 563, 581, 601-602, 605, 607, 648
Panama, 76, 303-305, 605
Paraguay, 51, 84-86
Parallel processing, 467
Pareto analysis, 480-481, 634-636
Pareto chart, 480, 634
Parts inventory, 17, 35, 55, 129, 383, 472, 524, 575
Parts management, 126, 142, 172
Password management, 501, 505, 510-511
Password protection, 128
Password, 30, 128, 454-455, 457, 460, 477, 501, 505, 510-511, 629, 638
Patches, 182, 419, 452, 469, 511
Patient care areas, 262
Patenting, 20, 22, 37-38, 262, 318, 328, 337, 348, 350-352, 358, 392, 395, 412, 618, 645, 650-651
Pathology, 11-12, 101, 298, 486-487, 536, 547, 616
Patient areas, 19, 261, 520
Patient bed, 23, 41, 206, 249, 379, 411-412, 427, 463, 476, 551, 572, 613, 615, 629
Patient burns, 250, 280-281, 367, 436
Patient care area, 6, 16, 19, 49, 261, 272, 324, 450, 509, 521, 528, 637
Patient data management system, 300, 463
Patient education, 174, 241, 485, 653
Patient injury, 43-44, 123, 136, 203, 236-239, 241, 248, 271, 273, 323, 348, 354, 397, 421, 426, 430, 497, 598, 637
Patient lift, 421, 430-431, 435
Patient monitor, 8, 13, 40, 49, 91, 116, 119, 269, 300, 337, 340-341, 384, 412-413, 415, 421, 446, 449, 459-463, 476-477, 495, 497, 627, 629
Patient-owned equipment, 269
Patient safety goal, 239, 280, 648
Patient safety, 3, 6, 14-15, 61, 159, 176, 179, 219-220, 222, 224-225, 227-229, 231, 233-234, 236-242, 244-246, 254, 269-271, 280-281, 315, 322-324, 327, 349, 385, 387, 397, 400, 412, 421, 434-435, 493, 496, 507, 570-572, 598, 602, 606, 608-609, 648
Patient simulator, 17, 190, 257, 300, 443
Patient transport, 21, 377-378, 386
PBX (see Private Branch Exchange)
PC (see Personal computer)
PCA pump, 325

PDA (see Personal digital assistant)
Pediatric intensive care unit, 371, 410
Peer review, 43, 64, 236, 274, 480, 555, 576-578, 618, 633, 637
Performance specifications, 116-117, 156, 413, 557
Performance standard, 104, 216, 358, 426, 529, 532, 568, 587-589
Performance test, 4, 44-45, 107, 130-132, 148, 209, 237, 239, 300, 539
Performance testing, 4, 44, 107, 130-131, 148, 209, 237, 239
Performance verification, 415, 598-599
Perinatology, 26, 371, 410-411, 413, 415-416
Personal computer (PC), 6, 9, 37-38, 42, 62, 111, 124, 127-128, 311, 401, 410, 454-455, 458-459, 462, 464, 470-471, 475, 483, 490, 506-507, 509, 515, 551, 623, 627, 633, 638, 642
Personal digital assistant (PDA), 128-129, 143, 263, 285, 511, 627
Personal liability, 590
Personnel safety, 371
PERT (see Program evaluation review technique)
PERT chart, 182-183
Peru, 51, 76-77, 87-89, 304-305, 548, 605
PET (see Positron emission tomography)
Physical damage, 44, 107, 124, 220, 277, 324-325, 439, 454
Physical environment, 6, 271, 337, 353-354, 571, 598
Physical examination, 344, 384
Physical injury, 115, 125, 236, 246, 386
Physical plant, 6, 20-21, 49, 78-79, 84-87, 89, 128, 267, 270, 274, 310, 322, 515, 518, 547-548
Physical safeguards (see also HIPAA), 500-502, 504-505
Physical therapy, 5, 91, 266, 298, 408, 424, 431-432, 637
Physician order, 240-241, 377, 458, 462-463, 509, 526
Physician-owned equipment, 20, 125, 240
Physiologic monitor (see also Physiological monitor), 40, 367-368, 374, 456, 458-463, 493
Physiologic monitoring (see also Physiological monitoring), 40, 368, 374, 456, 458-459, 461-463, 493
Physiologic simulator (see also Physiological simulator), 442
Physiological monitor (see also Physiologic monitor), 21, 23, 25, 35, 147-148, 150-151, 173, 181, 229, 233, 264, 279, 300, 322, 342, 373-374, 380, 384-385, 390, 416, 418-419, 426, 449, 456, 459, 463, 509, 512, 637
Physiological monitoring (see also Physiologic monitoring), 23, 35, 147-148, 151, 173, 181, 300, 373-374, 380, 385, 390, 416, 418-419, 449, 456, 459, 463, 509, 637
Physiological simulator (see also Physiologic simulator), 276
Physiological waveforms, 495
Picture Archiving and Communication Systems (PACS), 1, 10, 137, 141, 212, 319-320, 343, 367-368, 373, 397-399, 449, 487-492, 509
Piezoelectric, 40, 340
Plaintiff, 47-48, 236, 238, 270, 356, 590-592
Planned maintenance (PM), 1, 18-21, 24-25, 29, 36, 51-117, 119-646, 648-660
Plant equipment, 6, 49
Plant operations, 49, 53, 525
Plethysmograph, 390
PLMRS (see Private Land Mobile Radio Service)
Positron emission tomography (PET), 10, 114, 328, 392, 400, 418, 420, 562, 656
Post-anesthesia care unit, 377, 386, 629
Power distribution system, 104, 379, 520-521
Power failure, 49, 248, 413, 511
Power fluctuation, 158, 323
Power interruption, 413
Power meter, 17, 443
Power plug, 414, 446
Power quality, 520
Power requirements, 158, 282, 346, 379, 411, 413
Power surge, 270, 469
Practice guideline, 103, 159, 161, 173-175, 223, 255, 346, 520, 559, 652, 654
Predictive maintenance, 88, 464-465, 469, 473, 572
Pre-market approval, 111, 115, 348, 357, 370, 580, 586-588
Premature failure, 123, 147, 254, 397, 442, 444
Prepurchase evaluation, 4, 16, 19-20, 62, 107, 111, 210, 323-324, 327, 347, 577
Pressure monitor, 24, 250, 380, 389, 391, 418, 439, 442, 579, 640

Pressure monitoring, 24, 250, 389, 391, 418
Pressure transducer, 183, 250, 389, 414, 439, 617
Pressure waveform, 250
Preventive maintenance, 1, 4, 6, 17-21, 24-25, 29-30, 32, 35-37, 51-117, 119-646, 648-660
Primary health care, 94, 97, 133-134, 160, 169-170, 546, 621, 652-653, 656
Privacy, 1, 13, 243, 452-455, 457, 463, 485, 498-500, 504, 506, 511, 528, 607, 624
Privacy Rule (see also HIPAA), 498-499
Private Branch Exchange (PBX), 516, 527
Private Land Mobile Radio Service (PLMRS), 493-494, 496-497
Private network, 128-129, 454, 460, 462, 501, 510-511
Process-modeling techniques, 241
Product design, 331, 347, 459, 584, 645
Product development, 187, 330, 337, 346, 350-352, 362-363, 365, 580, 587-588, 598
Product liability, 41, 235-236, 274, 337, 355-356, 591-592
Product quality, 140, 199, 252
Product safety, 239, 422
Product specification, 364
Productivity, 6, 9, 12, 14, 55, 61, 82-83, 86, 88, 90, 93-94, 96-98, 107, 127, 140, 142-143, 146, 173, 179, 181, 186-187, 191-193, 198-203, 205-206, 209-211, 213-218, 221, 306-308, 354, 360-361, 377-378, 464, 469, 472, 474-475, 490, 533, 607, 610
Professional association, 6, 46, 74, 189-190, 212, 596-597, 619-620, 648
Professional conferences, 112, 532
Professional development, 7, 35, 54, 60, 86, 173, 215, 285, 306, 318, 596, 600, 602, 607-608
Professional engineer, 3, 42, 46, 49, 59-61, 309, 593, 600, 617
Professional ethics, 1, 11, 13, 212, 593
Professional society, 6, 16, 42, 107, 116, 212, 216, 313, 358, 532, 557, 559, 595, 597, 600, 602, 612
Program Evaluation Review Technique (PERT), 154, 182-183
Programmable infusion pump, 357, 375, 442
Programming, 74, 80, 88, 111, 127, 183, 187, 297, 300, 330, 337, 359-360, 471, 476, 506-507, 527, 627, 638, 644, 657
Project Hope, 51, 97-98, 609
Project management, 5, 50, 55, 68, 182, 362, 365, 383, 470, 625
Pulmonary function analyzer, 551
Pulmonary function testing, 522
Pulse generator, 383, 391
Pulse oximeter, 40, 250, 257, 263, 279, 331, 356-358, 390, 398, 416, 442, 551, 645
Pulse waveform, 8, 12
Purchase agreement, 135, 147-149, 152, 236
Purchasing decision, 82, 103, 107, 353, 366

QA (see Quality assurance)
QC (see Quality control)
Quality assurance (QA), 5, 14, 19, 21, 30, 39, 55, 57, 61, 67-68, 79, 84, 97, 103, 107-108, 148, 152, 159, 171, 176, 205, 221, 224, 236-237, 252, 294-295, 307-308, 324-325, 328, 341, 357, 362-363, 365, 399, 456-457, 462, 563, 565, 580, 583-584, 612, 619-620, 625, 644
Quality control (QC), 16, 29, 32, 55-56, 66, 74, 122, 130, 135, 183, 187, 206, 238, 241, 331, 371, 399, 401, 435, 444, 453, 509-510, 547-548, 562, 580, 584, 598-599, 628, 644
Quality indicators, 94, 96, 175-176, 205, 219-220
Quality of care, 37, 58, 60, 71, 84, 101, 140, 160, 163-164, 171, 219, 227, 236, 240, 269, 307, 332-334, 354, 431, 485, 570, 573, 576, 625, 654, 658

Radiant warmer, 236, 271-273, 279, 371, 410, 413-416, 421, 423, 433, 440-441, 599
Radiation dose, 396, 530-531, 598
Radiation physics, 20, 58, 60-61, 371
Radiation safety officer, 307, 530, 552
Radiation safety, 16, 37, 41, 300, 307, 333, 400, 418, 513, 529-532, 549, 552
Radiation therapy equipment, 56
Radiation therapy, 57, 59, 68, 507, 619
Radio frequency, 8, 15, 40, 237, 250, 254-255, 257-259, 261-262, 264, 266, 380, 382, 385, 419, 429, 446, 455, 461, 492-493, 497, 516, 535, 647

Radio receiver, 441
Radio signals, 492-494
Radio transmitter, 261-262, 269, 357, 492-497, 647
Radioactive isotopes, 340, 541
Radioactive material, 529-532, 537, 539, 546
Radioactive substance, 537
Radioactivity, 418, 446, 529, 531
Radiographic equipment, 419, 598
Radiographic system, 368, 398, 418
Radiographic unit, 418
Radiological Society of North America (RSNA), 250, 624, 627
Radiology Information System (RIS), 449, 487-488, 490
Random component failure, 130, 323, 371
Rapid film changer, 396
RCA (see Root cause analysis)
Recall, 19-21, 23-24, 47, 107, 120, 125, 221, 223, 236-237, 272, 274, 316, 322, 326, 392, 435, 487, 580, 584, 589, 591
Receptacle tester, 275
Reconditioning, 221, 589
Recordkeeping, 24, 67, 124, 172, 344, 378, 400, 437, 559, 604
Recovery area, 411, 426
Registered professional engineer, 46, 617
Rehabilitation engineering, 58-59, 61, 81, 328, 332, 334
Remote access, 451, 511, 626
Remote control software, 626
Remote diagnostics, 112, 128, 142, 144-145, 311, 449, 464, 467, 469-475, 605
Renal dialysis, 63-64, 239
Repair, 1, 4, 6-9, 15, 17, 19-22, 24-25, 27, 29, 34-39, 41, 44-45, 49, 56, 59-60, 62-63, 68, 72-74, 79-80, 83, 86-89, 91, 93, 96-97, 103-105, 107, 109, 113-114, 120, 124-128, 130-137, 139, 141, 144-151, 153, 155-156, 158-159, 161, 170, 173, 183-184, 189, 191-195, 198-200, 202-203, 205, 207-212, 220-221, 223, 227-228, 231, 237-239, 248, 270-271, 274, 277, 287, 289, 291, 297, 299-300, 303, 305-306, 309-310, 312, 322-325, 332, 342, 346, 349-350, 353, 357, 368, 377, 383, 394-395, 398, 400-401, 406-407, 413-417, 419, 421, 424, 426-427, 430-437, 439-442, 444, 446-447, 451-453, 462, 464-474, 480, 502, 507-508, 522-525, 531-532, 546, 551-552, 557, 570, 574-575, 598-599, 601, 605, 608, 613, 617-619, 625-626, 635, 637, 644, 646-647
Request for proposal (RFP), 6, 37, 104, 120-122, 124, 150, 152, 347
Request for quote (RFQ), 20, 119-121
Res ipsa loquitur, 591
Respiration monitor, 277, 331
Respirator, 8, 17, 25, 80, 92, 161, 171, 175, 223, 231, 238, 274, 298, 310, 322, 324, 340, 342, 346, 357, 373-375, 377, 380, 386-389, 391, 415, 431, 442, 509, 523, 527, 542-546, 548, 551, 559, 570, 576, 583, 610, 645
Respiratory equipment, 340, 559
Retraining, 66, 109, 114, 285, 328-329, 331, 355, 435, 452, 532
RFP (see Request for proposal)
RFQ (see Request for quote)
RIS (see Radiology information system)
Risk analysis, 35, 131, 221, 235, 238, 337, 359-360, 362, 369-370, 495-496, 500-501, 504-505, 579, 582-583, 625
Risk assessment, 123, 203-204, 220, 229, 231, 233, 236, 238, 245, 412, 450, 495-496, 508, 521, 528, 579, 587, 593, 633
Risk management, 5, 15-16, 19, 21-22, 35, 37, 44-45, 48, 55-56, 58-62, 79, 101-103, 105, 107, 109-110, 127, 131, 170-171, 173, 202, 205, 215, 223, 225, 235-240, 255, 270-271, 291, 304, 307-308, 323-326, 354, 433, 457, 495-497, 500-501, 505, 547-548, 555, 558, 561, 575, 577, 579, 590, 592, 647, 653
Root cause analysis (RCA), 225, 228-231, 235, 239-240, 245, 269, 271, 277, 280-281, 303, 566, 571
RSNA (see Radiological Society of North America)

Safe Medical Devices Act (SMDA), 5, 15, 20, 26, 37, 43-46, 127, 151, 227, 237, 239, 270, 347, 349, 355, 357, 477, 572, 574-575, 585-587, 589, 592, 607, 626
Safety committee, 20, 242-245, 265, 495, 530, 569, 572, 575
Safety engineering, 187, 237
Safety testing, 6, 9, 21, 29, 36-37, 129, 148, 151, 208, 210, 239, 282, 289, 291, 298, 316, 608, 625

Satellite, 18, 22, 128, 285, 310-311, 411, 484, 486, 533
Scheduled maintenance (see also Planned maintenance, Preventive maintenance), 19, 62, 122-126, 201, 244, 347, 434, 572, 577
Schematic diagram, 257, 437, 535, 584, 587, 641
Security official (see also HIPAA), 501, 504
Security Rule (see also HIPAA), 449, 498, 500-506
Security, 1, 13, 44, 61, 72-73, 77, 80, 83, 88-89, 127-129, 217, 243-244, 251-252, 254-255, 257-262, 264, 268, 360, 371, 395, 398, 410, 412, 433, 449, 451-457, 460, 477, 480, 498-507, 510-512, 523, 526, 530, 550, 570, 572, 599, 607, 614, 629
Service agreement, 97, 135, 145, 147-152, 172, 193, 277
Service documentation, 16, 150, 511
Service manual, 16, 30, 97, 107, 109, 116-117, 120-121, 125-126, 128, 147, 156, 158, 208, 279, 347-348, 370, 395, 416, 432-433, 437, 511, 615
Service report, 126, 136, 147, 150, 152, 203, 221, 264, 277, 322, 400
Service vendor, 107, 135-136, 144, 146-147, 152, 203, 224, 237
Shared biomedical engineering service, 27
Shared clinical engineering service, 27
Shared service, 9, 26-28, 34, 116-117, 122, 143, 151, 301, 304, 523
Single use devices, 251, 344-345, 583
Six Sigma, 222, 235, 476, 626, 629
SMDA (see Safe Medical Devices Act)
Software Quality Assurance Plan (SQAP), 362-363
Sole source, 150
South Africa, 9, 51, 156, 159-163, 167, 170, 173, 175, 224, 285, 304-309, 342, 605
South America, 76, 177, 659
Spam, 455
Spare parts, 29, 55-57, 74, 86, 93, 97-98, 109, 117-118, 121, 132, 135, 144, 150, 154-155, 166, 173, 342, 399, 433-435, 444, 470, 545
Spectroscopy, 329-330, 390
Spectrum analyzer, 17, 255, 266-268, 340
Speech recognition, 128, 489, 512
Spot films, 396-398
Sprinkler, 379, 510, 515
Staff training, 97, 107, 130-132, 202, 204, 208, 220, 237, 322, 383, 572, 577
Staff turnover, 104, 148, 324
Staffing level, 41, 55, 63, 173, 186, 206, 461, 628
Staffing requirements, 169, 574
Standard for Health Care Facilities (NFPA 99), 274, 281, 310, 413, 521-523, 550
Standards development, 240, 274, 555, 557-559
Standards organizations (SO), 6-7, 12-13, 18, 25, 40-49, 54-55, 59, 73-74, 80, 91, 94, 101-102, 109, 111, 114, 116-117, 119-122, 135, 136, 140, 147-148, 151, 155-156, 158, 160, 163-164, 181-186, 189, 191, 197, 199-202, 204, 206-207, 210, 214-216, 218, 222, 225, 228-229, 232, 235-236, 243-244, 264, 266, 269-270, 272, 276, 280, 282, 285, 289, 291-292, 303, 306-309, 312-313, 315, 317, 320, 322, 324, 335, 341, 343-349, 353-354, 357, 360-361, 367-368, 370, 375-376, 378-379, 384, 386-387, 389-390, 392-395, 397, 399, 411-412, 414, 421, 426, 431, 438, 441-442, 444-446, 451-455, 458, 469, 488-489, 491, 495-498, 505-507, 510-511, 513, 515, 517-518, 522, 525-526, 536-537, 542-544, 547, 551, 558-559, 567, 571, 574-575, 583, 587-589, 591, 593, 595-596, 600-601, 611, 614, 619, 621, 623-625, 630, 645-648, 650-651, 656-658
Standby power systems, 521
Static electrical charge, 249, 379
Statistical analysis, 125, 131, 181, 465, 574
Steam sterilization, 253, 378, 533-537, 546
Sterilization methods, 378, 533, 544
Sterilizer, 8, 12, 133-134, 142, 239, 281, 333, 378, 533-537, 546-547, 551, 580, 598-599
Stethoscope, 11, 133-134, 250, 333, 342, 345, 366, 417, 485-486
Storage oscilloscope, 257
Strategic inflection point, 624-626
Strategic planning, 6, 29, 35, 37, 88, 93, 99, 102, 110-113, 153-154, 172, 182, 203, 222, 306, 308, 320, 598, 625-626, 658
Strengths, weaknesses, opportunities, threats (SWOT), 43, 198, 621, 646
Stretchers, 20, 371, 378-379, 386, 421, 423, 425-427, 429-431, 433, 435, 647
Structural integrity, 427, 430, 549, 552

Sub-Saharan Africa, 93, 155, 158, 170, 172, 224, 305, 542
Suction regulator, 413, 522
Surgical instrument, 8, 12, 62-63, 77, 162, 164, 282, 339-340, 371, 380, 382, 535, 548, 582, 587, 623
Surgical laser, 6, 114, 272, 300, 343
Surgical microscope, 383
Surgical table, 283, 379, 383, 424
Surgical telemetry, 348
Surgical Video System, 371, 401, 410
Swan-Ganz Catheter, 419
Sweden, 51, 53, 65, 262, 350, 562-564
Switzerland, 51, 99, 110, 163, 177, 305, 342, 358, 560, 564-565, 582, 603, 623
SWOT (see Strengths, weaknesses, opportunities, threats)
SWOT analysis, 152, 198, 621, 646
System analysis, 4, 202, 451, 470
System approach, 15, 46, 110, 219, 225, 228, 237, 245, 247, 249, 270-271, 279, 281, 303, 307, 321-322, 337, 343, 346, 354-355, 436, 567, 592, 625
System architecture, 330, 359, 456-457, 459, 475
Systems engineering, 10, 74, 82, 245

Tablet PCs, 455
Tachometer, 17, 443
TCO (see Total cost of ownership)
Technical safeguards (HIPAA), 181, 241-243, 311, 449, 452-455, 463, 480, 498-506, 510, 605, 607, 609, 624, 626-627, 629
Technical service, 6, 19, 26-28, 53, 68, 72, 78, 88, 107, 109, 116, 151, 153, 170, 172, 199, 235, 306, 317, 319, 613, 625-626
Technical support, 6, 17, 28, 38, 68, 98, 107, 131, 137, 143, 145, 147-150, 155-156, 163, 189, 297, 301, 375, 377-378, 410, 455, 457, 461, 463, 490, 508, 525, 626, 647
Technology assessment, 5-6, 10, 20, 34-38, 62, 68, 82, 85-86, 92, 99, 101-104, 109-113, 116, 118, 143, 150, 159, 164, 170-172, 202, 227, 237, 240, 281, 298, 304, 306, 308, 311, 313, 315, 337, 345, 351, 358, 371, 401, 435, 451, 555, 559, 561-564, 577, 581, 597-598, 648
Telecommunications, 6, 35-36, 38, 41, 87, 112, 125, 127-128, 137, 146, 187, 225, 248, 254-258, 260-262, 269, 289, 334, 453, 471, 476, 484-487, 495, 507, 525, 527, 553, 558, 566, 583, 610, 625-627
Teleconferencing, 1, 16, 150, 271, 278, 280-281, 285, 298, 309-311, 571, 600-602, 605, 607, 609, 648
Telehealth, 62, 64, 453, 484-487
Telemedicine, 1, 10, 24, 26, 32-33, 36-38, 110, 159, 212, 291, 449, 452-453, 484-487, 525, 614, 645
Telemetry monitoring, 460-461
Telemetry receiver, 461
Telemetry system, 111, 255, 257, 262, 266, 460-461, 463, 492, 495-497
Telemetry transmitter, 266, 414, 493-494, 496-497
Telemetry unit, 300, 459, 461
Telephony, 89, 128-129, 454
Teleradiology, 62, 319, 491
Test & measurement, 34, 189, 191, 194-195, 269
Test equipment, 5, 16-18, 25, 28, 39, 42, 45, 55-57, 107, 109, 129, 132, 136, 148, 158, 189-190, 208, 221, 264, 266-267, 297, 299, 320-321, 346, 359, 413-415, 442-444, 493, 575
Test instrument, 25, 217, 442
Testimony, 19, 21-22, 26, 46-47, 282, 331, 508, 592
Therapeutic devices, 56, 255, 267, 274, 337, 355-357, 371, 383, 506, 509
Thermometer, 12, 20, 45, 71, 132, 190, 238, 339, 341-342, 391, 414, 518, 533, 537, 539, 545, 591
Third Party maintenance (TPM), 137, 139-142, 144, 146, 471
Third-party service providers, 63, 614
Tilting table, 397
Time and materials (T&M), 9, 35, 136, 148-149, 151-152, 190-191
Tomography, 9, 26, 40, 73, 114, 136, 328, 330-331, 339-340, 343, 371, 381-382, 392, 399, 418, 420, 509
Topology
  Bus, 128-129, 454
  LAN, 128-129
  Network, 129
  Ring, 128-129
  Star, 129
Total cost of ownership (TCO), 109, 451-452, 624
Tourniquet, 371, 381, 522

TPM (see Third party maintenance)
Trace gas, 387
Transmission security, 500-501, 503, 505
Transport incubators, 435, 552
Transport monitor, 460-461
Transthoracic pacer, 40
Treadmill, 249-250, 418
Troubleshooting, 21-23, 136, 147-148, 150, 152, 156, 158, 173, 248, 291, 300, 304, 307, 312, 318, 320, 322-325, 371, 385, 390, 395, 398, 433, 436-447, 462-464, 467, 469-472, 475, 511, 533, 552, 610, 626

UL (see Underwriters Laboratories)
Ultrasonic disintegrator, 446
Ultrasonic scanner, 416
Ultrasound diagnostic unit, 148
Ultrasound equipment, 161, 343, 551
Ultrasound transducer, 413
UMDNS (see Universal Medical Device Nomenclature System)
Underwriters Laboratories (UL), 71, 116, 156, 221, 358, 383, 561, 624
Undesirable outcome, 269
Uninterruptible power, 398, 413, 511, 520
United Kingdom, 51, 58-59, 61, 135, 254, 560-564, 603, 647-648
United Nations, 87, 167, 334, 529-530, 538-539, 546, 656, 659
Universal Medical Device Nomenclature System (UMDNS), 165-166, 343, 584
UNIX, 291, 457, 510-511
Upgrades, 19, 90, 103-104, 109, 111, 120, 129, 135-136, 147-150, 154, 237, 318, 347, 368, 414, 457, 489-490, 507-508, 511, 626
Use error (see User error)
User education (see also User training), 56, 124, 435, 459
User error (see also Operator error), 15, 19, 21, 117, 122, 124, 132, 147, 154, 173, 202-203, 220, 237-238, 244, 307, 315, 323-324, 326-327, 343, 348, 354-355, 391, 439, 480, 510, 569, 581, 637, 648
User interface, 30, 129, 353, 456, 458-461, 628, 639, 648
User maintenance, 131, 148, 323
User support, 634
User training (see also User education), 37, 60, 83, 99, 104, 119, 135-136, 152, 155, 166, 170, 173, 182, 189, 192, 198, 203, 223, 306, 325-326, 368, 383, 575, 584, 644
User-friendliness, 32, 38, 66, 111, 125, 353, 408, 456, 458, 471, 477, 629
Utility systems, 44, 78, 173, 243, 245, 248, 304, 307-308, 513, 516, 520, 549, 552, 569

Vacuum regulator, 523
Vacuum system, 6, 173, 307, 379, 410, 413, 516, 522, 535, 537, 543, 551
Validation criteria, 104
Validation process, 362
Validation testing, 362-365, 393
Value-added service, 37, 136, 138, 143-145, 202, 354
Vaporizer, 9, 371, 380, 386-388, 522-523, 614, 616
Vendor management, 152, 625-626
Vendor service, 125-126, 148, 150, 152, 337
Ventilator analyzers, 442
Ventilator, 17, 21, 24-25, 39-40, 44, 86, 92, 126, 131, 229, 233, 238-239, 250, 269, 272-273, 277, 279, 321, 333, 365, 367, 371, 373-376, 384-390, 410, 412-416, 443, 456, 458, 509, 520, 522, 524, 533, 542-543, 551-552, 583, 590, 645, 647
Venture capital, 42, 188, 197, 649
Vibration analysis, 464
Video amplifier, 441
Video camera, 396-397, 401, 412
Video conferencing, 285, 312-313, 485
Video equipment, 386
Video monitor, 267, 339, 398, 401, 484
Video recorder, 382
Video system, 371, 395-396, 401, 410
Video teleconferencing, 1, 285, 309-310
Videoconferencing systems, 314, 485
Videoconferencing, 310, 313-315
Virtual instrumentation, 36, 38, 371, 401, 410, 416, 449, 476-477, 480-481, 483, 607, 621, 627-629, 631, 633-635, 637-639, 641-644
Virtual private network (VPN), 511
Virus detection, 511

Index

Virus, 455, 510-511
Visual inspection, 29, 44, 447, 551
Voice over Internet Protocol (VoIP), 454-455, 527
Voice recognition, 526
Voice-command controls, 425
VoIP (see Voice over Internet Protocol)
Voltmeter, 442
Voluntary standard, 261, 578
VPN (see Virtual private network)
Vulnerability assessments, 551, 553

WAN (see Wide area network)
Waste management, 173, 177, 243-244, 333-334, 526, 533, 537-541, 546, 548-550, 553
Web-enabled devices, 38, 287
WEDi (see Workgroup for Electronic Data Interchange)
WEP (see Wired Equivalent Privacy)
Western Europe, 53-56
Wet location, 386, 521, 637
White tape, 272-273, 279, 281, 322, 325, 433, 436, 440-441, 447
Whiteboard, 303, 361, 412, 414, 628, 630
WHO (see World Health Organization)
Wide area network (WAN), 127, 137, 454-455, 487, 511

Wired Equivalent Privacy (WEP), 511
Wireless access, 250, 455
Wireless communication, 17, 129, 143, 254, 261-262, 268, 455, 461, 474, 496-497
Wireless LAN, 225, 254-255, 257, 259-260, 262, 495-496, 512, 614
Wireless Medical Telemetry Service (WTMS), 492-495, 497
Wireless network, 128-129, 263, 452, 610, 612
Wireless patient monitoring, 495, 497
Wireless telecommunication, 225, 254-258, 260, 262
Witness, 9, 19, 21, 23, 37, 41, 43-48, 181, 270-271, 273-274, 276-277, 280, 325, 358, 508, 590, 592, 645
Work order, 17, 36, 38, 94, 113, 124-129, 152-154, 207, 325, 343, 480
Workforce security, 500-501, 504-505
Workgroup for Electronic Data Interchange (WEDi), 498
Workstation security, 502
World Bank, 71, 85-87, 97, 110, 134, 159, 162-163, 170, 334, 648
World Health Organization (WHO), 1-3, 6-13, 15-16, 18-19, 23-27, 29, 35, 37, 40-51, 58-61, 66-68, 71-74, 78, 80-81, 85-89, 91-92, 94, 97, 99, 101-102, 104, 108, 110, 114-121, 124-125, 127-128, 130-133, 135-136, 140-141, 146, 148, 150-152, 155-156, 158-167, 169-172, 174-177, 179, 181-183, 187-189, 192, 198-200, 203, 206-207, 210-219, 222, 225, 227-230, 233, 237-238, 242-244, 246-249, 251-252, 254, 259, 261, 264, 269-270, 272-274, 276-277, 282, 285, 287, 290-291, 294, 297, 301-316, 318, 320-322, 324-329, 331-335, 337, 339-354, 356-357, 360, 362-363, 366-368, 370-371, 373-374, 376-377, 384-385, 388-389, 391-392, 395, 408, 411, 415, 417-418, 420-422, 427, 430-431, 433, 446, 449-452, 455, 458, 460, 463, 469, 478-479, 484, 491, 494-495, 498-501, 505, 507, 509-511, 513, 517, 523, 525-532, 536-537, 541-546, 548-553, 558-559, 561, 564, 566-567, 569-571, 573-576, 578, 581, 583-584, 587, 589-593, 595-612, 615-619, 621, 623, 626-627, 630-631, 638, 644, 646, 648, 650-659
WORM, 221, 457
WTMS (see Wireless Medical Telemetry Service)

X-ray equipment, 206-207, 339, 341-344, 371, 378-379, 392, 395, 399, 551, 583, 587

Yellow safety box, 274-275, 442
Yoga, 333-335